Brad Courtaney CS-37

General data and fundamental constants

Quantity	Symbol	Value
Speed of light	c	$2.997\,925 \times 10^{8}$ m s^{-1}
Elementary charge	e	$1.602\,19 \times 10^{-19}$ C
Faraday constant	$F = eN_A$	$9.648\,46 \times 10^{4}$ C mol^{-1}
Boltzmann constant	k	$1.380\,66 \times 10^{-23}$ J K^{-1}
Gas constant	$R = kN_A$	8.314 41 J K^{-1} mol^{-1} $8.205\,75 \times 10^{-2}$ dm^{3} atm K^{-1} mol^{-1}
Planck constant	h $\hbar = h/2\pi$	$6.626\,18 \times 10^{-34}$ J s $1.054\,59 \times 10^{-34}$ J s
Avogadro constant	N_A	$6.022\,05 \times 10^{23}$ mol^{-1}
Atomic mass unit	m_u	$1.660\,56 \times 10^{-27}$ kg
Mass of electron	m_e	$9.109\,53 \times 10^{-31}$ kg
proton	m_p	$1.672\,65 \times 10^{-27}$ kg
neutron	m_n	$1.674\,95 \times 10^{-27}$ kg
nuclide	$M_r m_u$	
Vacuum permeability	μ_0	$4\pi \times 10^{-7}$ J s^{2} C^{-2} m^{-1} $4\pi \times 10^{-7}$ T^{2} J^{-1} m^{3}
Vacuum permittivity	$\varepsilon_0 = 1/c^2\mu_0$	$8.854\,19 \times 10^{-12}$ J^{-1} C^{2} m^{-1}
	$4\pi\varepsilon_0$	$1.112\,65 \times 10^{-10}$ J^{-1} C^{2} m^{-1}
Bohr magneton	$\mu_B = e\hbar/2m_e$	$9.274\,08 \times 10^{-24}$ J T^{-1}
Nuclear magneton	$\mu_N = e\hbar/2m_p$	$5.050\,82 \times 10^{-27}$ J T^{-1}
Bohr radius	$a_0 = 4\pi\varepsilon_0\hbar^2/m_e e^2$	$5.291\,77 \times 10^{-11}$ m
Rydberg constant	$R_\infty = m_e e^4/8h^3c\varepsilon_0^2$	$1.097\,37 \times 10^{5}$ cm^{-1}
Fine structure constant	$\alpha = \mu_0 e^2 c/2h$	$7.297\,35 \times 10^{-3}$
Gravitational constant	G	6.6720×10^{-11} N m^{2} kg^{-2}

Prefixes

f	p	n	μ	m	c	d	k	M	G
femto	pico	nano	micro	milli	centi	deci	kilo	mega	giga
10^{-15}	10^{-12}	10^{-9}	10^{-6}	10^{-3}	10^{-2}	10^{-1}	10^{3}	10^{6}	10^{9}

Physical Chemistry

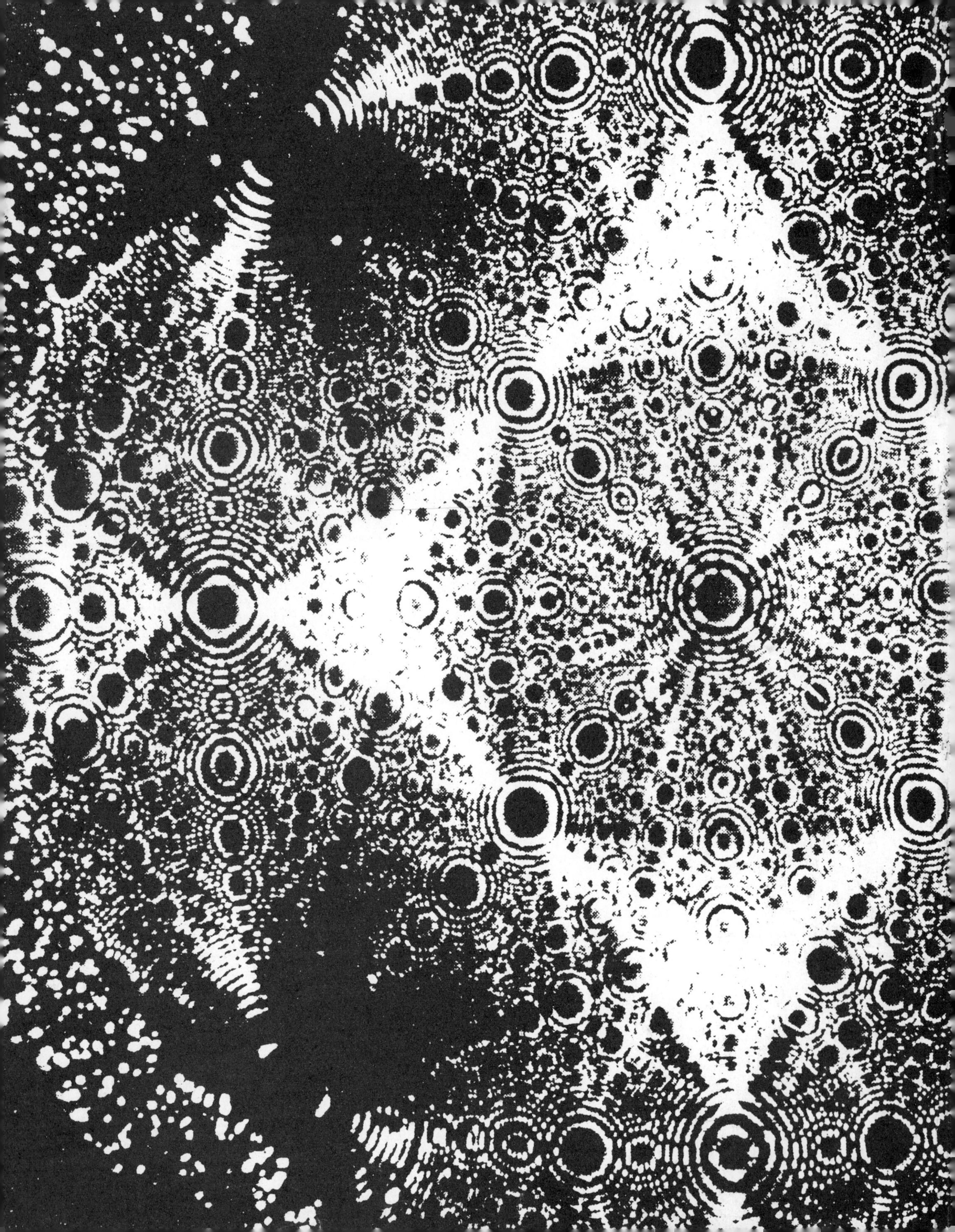

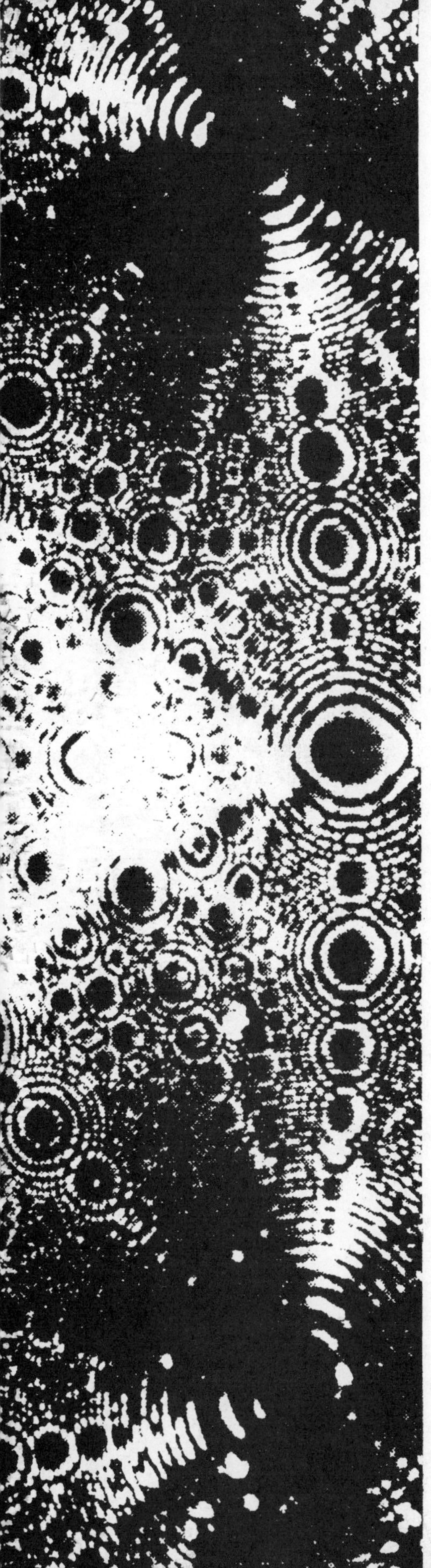

P. W. ATKINS

with introductory problems by J. C. MORROW

Physical Chemistry

Third Edition

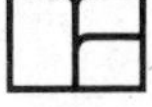

W. H. Freeman and Company
New York

Library of Congress Cataloging in Publication Data
Atkins, P. W. (Peter William), 1940–
Physical chemistry.
Includes bibliographies and index.
1. Chemistry, Physical and theoretical. I. Title.
QD453.2.A88 1985 541.3 85–7048
ISBN 0–7167–1749–2

Printed in the United States of America

2 3 4 5 6 7 8 9 0 VB 5 4 3 2 1 0 8 9 8 7

Preface to the Third Edition

Increasingly I find myself merely the chairman of an international team which is helping to shape the text and to keep it up to date both pedagogically and factually. The advice its many members have given has been acted on in many different ways, and the following is a summary of the principal changes in this edition. The aim throughout has been to modernize, clarify, and shorten the exposition, and no word or equation has gone unconsidered.

The structure of the book remains the same as in earlier editions, but I have acted on requests for more chemical kinetics and a higher level of quantum theory. The original chapter on kinetics has been divided into two, one treating simple reaction schemes, and the other complex reactions extended to include polymerization kinetics and oscillating reactions. The original chapter on quantum theory has also been divided into two, one treating the basic concepts and the other their applications. Applications of quantum theory are also treated in more detail in later chapters (for instance, I now outline the analytical solution of the hydrogen atom). However, I do recognize that not everyone needs that degree of detail, and so where I have included a mathematically difficult section I have summarized the main conclusions in the section immediately following, and so the details can be skipped.

There are several major changes of order (and many minor ones). For example, the chapter on the phase rule now precedes the discussion of chemical equilibria. The sequence (8) Solutions, (9) Phase rule, (10) Equilibria, (11) Electrochemistry is more logical, with the qualitative becoming the quantitative, and is often the way it is taught. Crystal symmetry has migrated from group theory to the chapter on diffraction, where it is more at home. The chapter on resonance methods now begins with NMR, the more important technique, and ESR has moved into second place.

I became increasingly unhappy with the electrochemistry chapters as I revised them, and ended up rewriting them completely (particularly Chapter 12 on cells). I now arrive at the Nernst equation as quickly as possible, and have dropped much of the arcane notation of earlier editions. All the electrochemistry should now be much easier to use.

I have constantly been aware of many people's wish that physical chemistry courses should have more points of contact with those taken by chemical engineers. To this end I have introduced elementary ideas related to chemical reactors and generalizations of the diffusion equation, including the material balance equation. The treatment is very introductory: I do a few examples, and relate it to topics such as spatially periodic reactions. I hope it will be seen as providing a touchstone for more detailed treatments in courses dealing with industrial and biochemical processes.

One trouble with a text of this size is that tables of data are scattered throughout, which is tedious when Problems draw on several; however, the reader needs tables to hand in mid-chapter. I have tried to resolve this conflict by providing a complete *Data Section* at the end of the book, with all the tables assembled in one place, and giving brief extracts in synoptic tables in the text, which should be sufficient to indicate the values of the properties under discussion. In compiling the thermodynamic data I have adopted the new recommended standard pressure of 1 bar in place of 1 atm: this is explained in the text.

The *Examples* (which are now numbered) have all been revised, partly with the aim of making them less arithmetical and more illustrative of the way problems should be tackled. I have tried to meet the criticism that the *Problems* are often a quantum leap in difficulty removed from the *Examples*. This has been done in two ways. First, each *Example* is followed by an *Exercise* along the lines of the *Example* itself; this adds about 190 new simple problems to the 1100 of the Problem sets themselves (the latter number reflecting the net result of deleting old problems and adding new ones). Secondly, Professor J. C. Morrow of the University of North Carolina, Chapel Hill, has provided an additional 300 'straightforward' problems (marked with an A), which should provide an easy transition to the more advanced set. There are now around 1600 problems throughout the text, and all the end-of-chapter *Problems* are solved in the *Solutions Manual*, which has been revised in collaboration with Professor Morrow and checked in detail by Professor M. Golde of the University of Pittsburgh and by Juvencio Robles of the University of North Carolina, Chapel Hill.

Throughout the text I have made remarks that reflect the increasing role of computers in modelling chemical processes and analysing data. A few of the *Examples* and *Problems* are specifically computer-related. A software package, compiled by Professor P. B. Ayscough of the University of Leeds, will also be available to accompany and extend the text.

As well as improving, I hope, the style of exposition, I have simplified the notation without lowering the level of mathematical rigour. As a result, the text should appear more straightforward and incisive. Various recommended notational devices (such as *Ŧ* to denote 298.15 K and $\Delta_r H$ for reaction enthalpy) have been adopted, and are explained in the text. All the illustrations have been reconsidered, and many have been redrawn. The *Objectives* are now much more explicit; the subsections are more numerous and are now numbered.

I hope very much that readers will enjoy reading the new edition, and will appreciate the steps that have been taken to improve it. I emphasize once again that much of the momentum for the revision has come from the army of people who have given time to comment on earlier editions and on my draft of this. Many people wrote with occasional points, and to most of them I have already expressed my thanks. More substantial comments were made by people whose names are listed separately: their contribution is deeply appreciated. So too is the advice and encouragement of the officers of Oxford University Press and W. H. Freeman & Company, who so willingly, and in so many ways, have done so much to help.

Oxford, 1985 P. W. A.

Extract from the Preface to the Second Edition

I owe a considerable debt to all those who wrote to me with comments either on the first edition or on the draft of the second. Extensive sections of the latter were commented on by W. J. Albery (London), A. D. Buckingham (Cambridge), A. J. B. Cruickshank (Bristol), A. A. Denio (Wisconsin), R. A. Dwek (Oxford), A. H. Francis and T. M. Dunn (Michigan), D. A. King (Liverpool), G. Lowe (Oxford), M. L. McGlashan (London), J. Murto (Helsinki), M. J. Pilling (Oxford), C. K. Prout (Oxford), H. Reiss (UCLA), H. S. Rossotti (Oxford), J. S. Rowlinson (Oxford), W. A. Wakeham (London), D. H. Whiffen (Newcastle), J. S. Winn (Berkeley), and M. Wolfsberg (Irvine) and I am grateful to them all. I should also like to thank the following, who made particularly helpful comments on the first edition, and whose remarks have been built into this: M. D. Archer (Cambridge), L. Brewer (Berkeley), D. H. Everett (Bristol), D. Husain (Cambridge), R. M. Lynden-Bell (Cambridge), I. M. Mills (Reading), and A. D. Pethybridge (Reading), as well as those others whom I have acknowledged privately. The correspondence with my translators, K. P. Butin (Moscow), G. Chambaud (Paris), H. Chihara (Osaka), M. Guardo (Bologna), and A. Höpfner (Heidelberg), has been a particularly fruitful source of advice.

Oxford, 1981

Extracts from the Preface to the First Edition

Authors should not preach to teachers. Textbooks should be flexible and adaptable, yet have a strong story-line. I have tried to conform to these demands by dividing the text into three parts, *Equilibrium*, *Structure*, and *Change*. Each part begins in an elementary way, drawing on the others only weakly. Of course they rapidly get tangled up with each other—as they should because the subject is a unity—but teachers will be able to match the text to their own needs without unduly burdening the student. The student, I hope, will be enticed to read his way into chemistry's web of interdependencies, and will find that he can master them without getting confused.

Physical chemistry possesses its mathematics for a purpose: there has to be enough mathematical spine in the subject to enable our ideas on the behaviour of molecules and systems to stand up to experimental verification. Ideas that cannot be tested do not belong to science. Nevertheless, in an introductory treatment the ideas must not be overborne by the mathematics. In this text I show how physical ideas can be developed mathematically, and I take care to interpret the mathematical statements I make. Only where the mathematics and the chemistry lose sight of each other is physical chemistry a difficult subject, so I try never to let that happen . . . .

A book such as this could not have appeared without the sustained help of a number of people. Chapters from an early draft were read and criticized by Professor G. Allen (University of London), Dr M. H. Freemantle (University of Jordan), Professor P. J. Gans (New York University), Dr R. J. Hunter (University of Sydney), Professor L. G.

Pedersen (University of North Carolina), Professor D. W. Pratt (University of Pittsburg), Dr D. J. Waddington (University of York), Dr S. M. Walker (University of Liverpool), and Professor R. W. Zuehlke (University of Bridgeport). From their remarks grew the second draft, which in turn owes a considerable debt to others. W. H. Freeman and Company of San Francisco played a key role in having the book extensively reviewed and in making sure that it would fit the requirements of courses in North America. They obtained comments and advice from Professor H. C. Andersen (Stanford University), Professor J. Simons (University of Utah), and Professor R. C. Stern (Lawrence Livermore Laboratory). I owe a particular debt of gratitude to Professor Stern who made penetrating remarks on almost every word, or so it seemed; to Miss A. J. MacDermott (University of Oxford) who read the whole and criticized acutely; and to Mr S. P. Keating (University of Oxford), who worked hard and thoughtfully on many of the Problems . . . .

Oxford, 1977

Acknowledgements

As well as the readers who helped with the earlier editions, and many who have written individually, the author and publisher thank the following for their advice in connection with this edition:

Professor L. D. Barron, University of Glasgow
Dr B. W. Bassett, Imperial College, London
Professor S. M. Blinder, University of Michigan, Ann Arbor
Professor R. D. Brown, Monash University
Professor A. D. Buckingham, University of Cambridge
Dr R. G. Egdell, Imperial College, London
Professor L. Epstein, University of Pittsburg
Dr C. J. Gilmore, University of Glasgow
Professor L. P. Gold, Pennsylvania State University, University Park
Dr A. Hamnett, University of Oxford
Professor F. Herring, University of British Columbia, Vancouver
Dr B. J. Howard, University of Oxford
Professor B. Huddle, Roanoke College, Salem
Dr B. P. Levitt, Imperial College, London
Professor I. M. Mills, University of Reading
Dr I. C. McNeill, University of Glasgow
Dr D. Nicholson, Imperial College, London
Professor J. Ogilvie, University of Bahrain
Professor L. G. Pedersen, University of North Carolina, Chapel Hill
Dr M. J. Pilling, University of Oxford
Professor C. Pritchie Jr., Tulane University, New Orleans
Professor J. S. Rowlinson, University of Oxford
Professor M. Schwartz, North Texas State University, Denton
Professor G. Shalhoub, La Salle University, Philadelphia
Dr M. Spiro, Imperial College, London
Professor B. Stevens, University of South Florida, Tampa
Professor J. M. Thomas, University of Cambridge
Professor D. H. Whiffen, University of Newcastle
Dr D. J. Williams, Imperial College, London
Professor R. Westrom, Michigan State University

Contents summary

Part 3: Change

Contents

Part 2: Structure

Part 3: Change

Units and notation

SI units are used throughout, but the *atmosphere* (atm) is retained where appropriate as a convenient unit of pressure and is augmented by the use of the *bar* (1 bar = 10^5 Pa); *standard pressure* (denoted $p^\ominus$) is taken as 1 bar in accord with international recommendations. The unit *Torr* has been used in place (almost everywhere) of mmHg; 760 Torr = 1 atm exactly. Concentrations have been expressed in mol dm^{-3}, abbreviated to M (so that 1 M ≡ 1 mol dm^{-3}).

The *conventional temperature* for reporting data is 298.15 K: it is denoted by the recommended symbol $\mathcal{T}$. *Avogadro's constant*, previously denoted L, is denoted N_A in this edition.

Some conversions, and some commonly encountered notational differences, are listed below. Others are given inside the front cover.

Units

1 L (litre) = 1 dm^3 = 10^3 cm^3 = 10^{-3} m^3

1 M (molar) = 1 mol dm^{-3}

1 Å (ångström) = 10^{-10} m = 10^{-8} cm = 100 pm

1 atm = 760 Torr (exactly) = 760 mmHg (almost exactly)

= 101.325 kPa = 1.013 25 × 10^5 Pa = 1.013 25 × 10^5 N m^{-2}

1 bar = 100 kPa (exactly)

Notation

U, internal energy (some texts use E)

G, Gibbs function (or Gibbs free energy, some texts use F)

A, Helmholtz function (or Helmholtz free energy, some texts use F)

N_A, Avogadro's constant (the recommended symbol is L)

μ, dipole moment (the recommended symbol is p)

$p^\ominus$ = 1 bar = 100 kPa = 10^5 N m^{-2}

$m^\ominus$ = 1 mol kg^{-1}

$\mathcal{T}$ = 298.15 K (exactly)

Numerical calculations

When showing *intermediate* arithmetical steps using a calculator, the figures beyond those justified by the data are shown as subscripts, as in 2.6/2.3 = 1.1_{304}.

Introduction

0

The nature of matter: orientation and background

Learning objectives

This introductory chapter presents in general outline a number of important concepts that are used throughout physical chemistry. Apart from some of the very elementary ideas, they are all treated in more depth later in the book, but they are so much a part of the language of the subject that it is sensible to meet them at the outset.

After studying this chapter you will:

(1) Be aware of the *relative sizes* of atoms and molecules, Fig. 0.3.

(2) Be able to describe some of the effects of the *quantization of energy*, Section 0.1(a).

(3) Be aware that the energy of translational, rotational, and vibrational motion is *quantized*, and be able to indicate the separation of the energy levels in each case, Section 0.1(a).

(4) Be aware that transitions between energy levels give rise to *spectral lines*, Section 0.1(b), and be able to use the *Bohr frequency condition*, eqn (0.1.1), to relate the energy separation to the *wavelength*, *frequency*, and *wavenumber* of the transition.

(5) Be able to define *mole* precisely (Box 0.1) and be able to distinguish and interrelate *molar mass*, *molecular mass*, and *relative molecular mass*, Box 0.1.

(6) Be able to state the formula for the *Boltzmann distribution*, eqn (0.1.2), and use it to calculate the relative populations of energy levels.

(7) Be aware of the distinction between *energy level* and *state*, Section 0.1(d), and know that several states may correspond to the same energy level.

(8) Be able to sketch the main features of the *Maxwell–Boltzmann distribution* of molecular speeds (Figs. 0.9 and 0.10).

(9) Be able to state the *equipartition theorem*, Section 0.1(f), and use it to estimate the mean energies of different modes of motion.

(10) Be aware of the three *states of matter* and the meaning of the expression *phase*, Section 0.2.

(11) Be aware of the classification of bonds as *ionic* and *covalent*, Section 0.1(g).

(12) Be aware that molecules may be either *polar* or *non-polar*, and be able to indicate the origin of *van der Waals' forces*, Section 0.1(g).

(13) Be aware of the *perfect gas law*, eqn (0.2.2), and be able to state the general features of a *perfect gas* and the *kinetic theory of gases*, Sections 0.2(d and e).

(14) Know the significance of the terms *mean free path* and *collision frequency*, and be able to indicate their magnitudes under normal conditions, Section 0.2(d).

(15) Be able to account for the general form of the *Arrhenius equation*, eqn (0.2.4), for the temperature-dependence of reaction rates.

(16) Be aware that the *laws of thermodynamics* concern the transformations of energy, and be able to indicate their general content, Section 0.2(f).

(17) Be able to specify and interrelate the SI and derived units for *force*, *pressure*, and *energy*, Section 0.3.

Introduction

We know that atoms and molecules exist because we can see them, Figs. 0.1 and 0.2. Their existence had been inferred long before the invention of the techniques used to obtain these pictures, and estimates of their sizes and shapes were made in the nineteenth century. Now we can measure the masses of molecules, atoms, and subatomic particles with great precision, and we can determine the shapes of molecules as complicated as those of enzymes and proteins. Whenever we want an explanation of some observation we shall build one in terms of atoms and molecules. If this is to be done successfully we must know something about their structures and properties, and how individual molecules contribute to the samples we normally encounter.

Fig. 0.1. An image of a platinum tip of about 150 nm radius. The technique employed (by Professor E. W. Müller) to obtain this photograph is called *field-ionization spectroscopy* (Chapter 31).

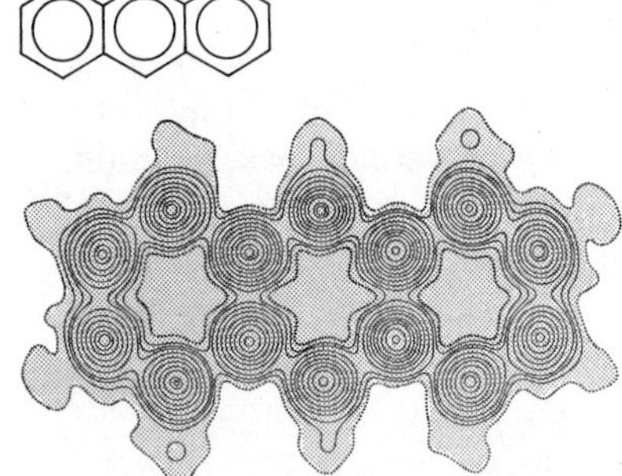

Fig. 0.2. Contours of constant electron density in a molecule of anthracene ($C_{14}H_{10}$); the conventional representation of the structure is shown as an inset. The picture was obtained (by V. L. Sinclair, J. M. Robertson, and A. McL. Mathieson, *Acta Crystallogr*. **3,** 254 (1950)) using *X-ray diffraction* (Chapter 23).

This chapter is an outline of the concepts we meet throughout chemistry. The material is a general background to the rest of the book, and all the ideas introduced here are developed with more explanation later.

0.1 The microscopic world

At the centre of an atom lies its nucleus. Almost the whole of the mass of the atom is concentrated in its nucleus even though it accounts for only a minute proportion of the atom's total volume. Around the nucleus cluster the electrons. They contribute very little to the total mass of the atom, but they occupy an appreciable volume and are responsible for its bulk.

How big are atoms? When we attempt to picture things happening on a molecular scale it is important to know *relative* sizes and masses. For example, a proton is about 2000 times heavier than an electron, an atom of carbon is about 12 times heavier than an atom of hydrogen, and in a molecule a chlorine atom has a diameter about 1.5 times that of an oxygen atom, Fig. 0.3. By dealing with relative sizes we develop an insight into the properties of atoms and molecules without being troubled by our inability to visualize objects as small as 10^{-10} m.

The *absolute* magnitudes of atoms and molecules, however, do play one very important role. One of the biggest surprises of 20th century physics was the discovery that *classical mechanics* (the mechanics of macroscopic particles) is an approximation: it is inapplicable to objects the size of atoms and has to be replaced by *quantum mechanics*.

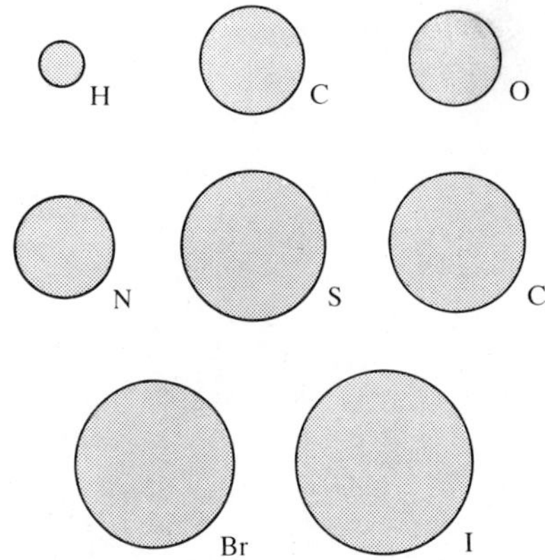

Fig. 0.3. The relative sizes of some atoms. The spheres represent the *covalent radii* of the atoms, the radii the atoms have when taking part in covalent bonding. Explicit values are listed in Table 23.3 at the end of the book.

0.1 (a) The quantum rules

According to classical physics, a pendulum can be made to swing with any energy. A small impulse sets it swinging with its natural frequency v, which is determined by its length, but the amplitude of the swing is small. A bigger impulse sets it swinging with the same frequency, but the amplitude is bigger because more energy has been transferred. An impulse of any reasonable magnitude can set the pendulum in motion, and by governing the impulse any amplitude of swing can be established. According to quantum mechanics, however, the pendulum can possess one of a precisely defined, discrete set of energies. The energy of the pendulum must have one of the values $\frac{1}{2}hv$, $\frac{3}{2}hv$, $\frac{5}{2}hv$, and so on, where v is its natural frequency and h, *Planck's constant*, is a fundamental constant with the value $h = 6.626 \times 10^{-34}$ J s. In a sense, only some amplitudes of swing of the pendulum are permitted, and it is impossible to set it swinging with an intermediate amplitude. Furthermore, since the minimum energy is $\frac{1}{2}hv$, according to quantum mechanics a pendulum can never be completely still.

Why did it take until the beginning of the twentieth century to discover quantum mechanics? The answer lies in the smallness of Planck's constant. For instance, a pendulum (or any other type of oscillator) of natural frequency 1 Hz can accept energy in steps of about 6.6×10^{-34} J, which is so small that for all practical purposes it seems that energy can be transferred continuously. Only for oscillators with very high natural frequencies does the quantum hv become large enough to give rise to observable differences from classical mechanics. A particle on a spring oscillates with simple harmonic motion, and its natural frequency increases as its mass decreases ($v \propto 1/m^{\frac{1}{2}}$). When the mass is extremely small, which is the case for atoms vibrating in molecules, the size of the quantum needed to excite the vibration to its next level is so great ($hv \propto 1/m^{\frac{1}{2}}$) that we must be prepared to encounter significant quantum effects.

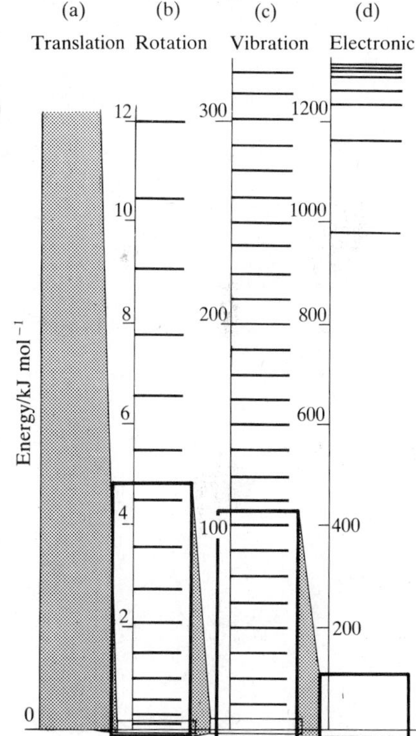

Fig. 0.4. An indication of the relative separations of the different types of energy levels of atoms and molecules. Notice that there is a scale change between each one. The translational energy levels are so close together that they may be taken as unquantized. The electronic levels in (d) are those of the hydrogen atom.

All kinds of motion are quantized, and Fig. 0.4 depicts the arrangement of energy *levels* (the permitted energy values) in typical systems. An object free to move in a container of ordinary size has translational kinetic energy, and can be accelerated to any of its translational energy levels. The separation between levels is so small, Fig. 0.4(a), however, that even for atoms it is usually permissible to disregard the quantization of translational energy. A molecule may rotate at different rates, but its energy of rotation must be one of the values indicated in Fig. 0.4(b). The separation between rotational levels is small but cannot be ignored. For instance, the separation between the two lowest rotational energy levels of CO is about 8×10^{-23} J (0.05 kJ mol^{-1}; energy units are discussed in Section 0.3). Molecules can also vibrate, and the permitted vibrational energy levels are determined by the masses of the atoms and the rigidity of the bonds between them. This is illustrated in Fig. 0.4(c). The effects of quantization are now very important, the separation of levels in CO being as much as 4×10^{-20} J, or 25 kJ mol^{-1}.

The energy of an electron in a hydrogen atom is restricted to the levels indicated in Fig. 0.4(d). (Other atoms show similar but more complex patterns.) Because the electron is so light we can expect the electronic energy levels of atoms (and molecules) to be much more widely spaced than the vibrational energy levels of molecules. This is the case, as the illustration shows. The separation between the lowest two energy levels of atomic hydrogen is 1.5×10^{-18} J, which corresponds to 1000 kJ mol^{-1}.

0.1 (b) Transitions

The results just quoted are theoretical conclusions, but there is plenty of experimental evidence for the quantization of energy. Compelling evidence comes from direct visual observation. Figure 0.5 shows the *spectrum* of light (its analysis into components of different frequencies) emitted by atoms that have been excited to high energy levels (for example, in a flame or a spark). The light is emitted at a series of definite frequencies, just as we might expect for atoms falling between different, discrete energy levels of the sort illustrated in Fig. 0.4(d), and emitting the excess energy as radiation. The frequency ν emitted (or absorbed) when a transition occurs and can be calculated from the *Bohr frequency condition*:

$$h\nu = E_{\text{upper}} - E_{\text{lower}}. \tag{0.1.1}$$

For example, when a small molecule changes from one rotational level to the next lower (and then rotates more slowly) the energy loss is about 0.1 kJ mol^{-1}, corresponding to an emission of radiation of frequency 2×10^{11} Hz or wavelength $\lambda \approx 1.5$ mm. (*Wavelength*, λ, is related to frequency by $\lambda = c/\nu$ for light in a vacuum, and by $\lambda = v/\nu$ in general, where v is the speed of propagation of the radiation.) This is the wavelength of microwave radiation, and so rotational transitions occur in the microwave region of the electromagnetic spectrum. Instead of frequency or wavelength, the *wavenumber* (symbol: $\tilde{\nu}$) of the radiation is often quoted: this is defined as $\tilde{\nu} = \nu/c$. In a vacuum (where the radiation has a speed c) it follows that $\tilde{\nu} = 1/\lambda$, and so the wavenumber may then be interpreted as *the number of waves per unit length*. In the present example, $\tilde{\nu} \approx 7\ \text{cm}^{-1}$ (7 wavelengths per centimetre).

Vibrational transitions give rise to, and can be caused by, infrared radiation. Electronic transitions, when electrons are redistributed in an atom or molecule, generally involve radiation in the visible and ultraviolet regions of the spectrum.

If the tightly bound inner electrons of heavy atoms can take part in a transition, we can expect radiation of very short wavelengths. This is the

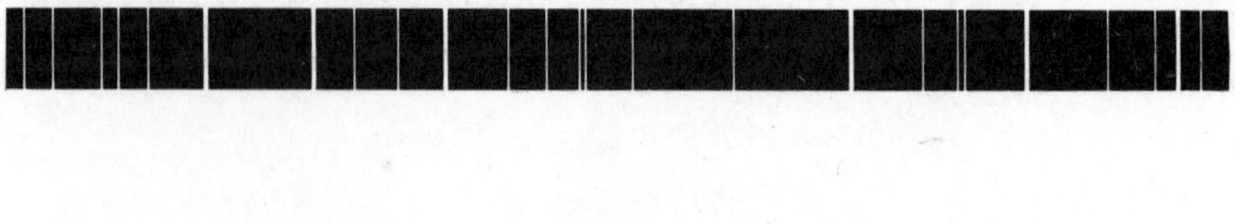

Fig. 0.5. The spectrum of light emitted by mercury atoms excited by an electric discharge. Each line arises from an atom that emits excess energy as it makes a transition between two of its quantized states.

the electrons. They contribute very little to the total mass of the atom, but they occupy an appreciable volume and are responsible for its bulk.

How big are atoms? When we attempt to picture things happening on a molecular scale it is important to know *relative* sizes and masses. For example, a proton is about 2000 times heavier than an electron, an atom of carbon is about 12 times heavier than an atom of hydrogen, and in a molecule a chlorine atom has a diameter about 1.5 times that of an oxygen atom, Fig. 0.3. By dealing with relative sizes we develop an insight into the properties of atoms and molecules without being troubled by our inability to visualize objects as small as 10^{-10} m.

The *absolute* magnitudes of atoms and molecules, however, do play one very important role. One of the biggest surprises of 20th century physics was the discovery that *classical mechanics* (the mechanics of macroscopic particles) is an approximation: it is inapplicable to objects the size of atoms and has to be replaced by *quantum mechanics*.

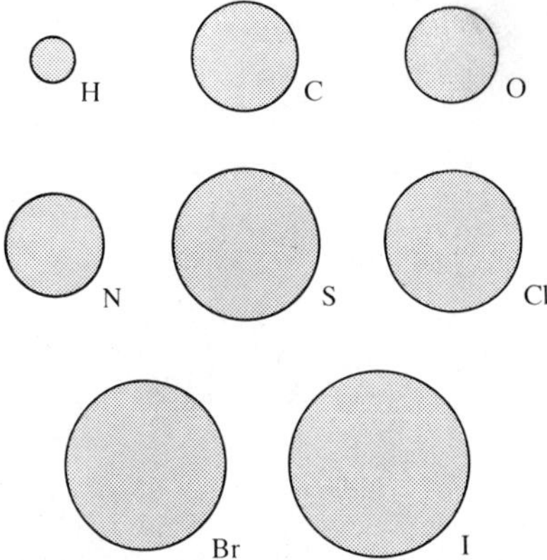

Fig. 0.3. The relative sizes of some atoms. The spheres represent the *covalent radii* of the atoms, the radii the atoms have when taking part in covalent bonding. Explicit values are listed in Table 23.3 at the end of the book.

0.1 (a) The quantum rules

According to classical physics, a pendulum can be made to swing with any energy. A small impulse sets it swinging with its natural frequency ν, which is determined by its length, but the amplitude of the swing is small. A bigger impulse sets it swinging with the same frequency, but the amplitude is bigger because more energy has been transferred. An impulse of any reasonable magnitude can set the pendulum in motion, and by governing the impulse any amplitude of swing can be established. According to quantum mechanics, however, the pendulum can possess one of a precisely defined, discrete set of energies. The energy of the pendulum must have one of the values $\frac{1}{2}h\nu$, $\frac{3}{2}h\nu$, $\frac{5}{2}h\nu$, and so on, where ν is its natural frequency and h, *Planck's constant*, is a fundamental constant with the value $h = 6.626 \times 10^{-34}$ J s. In a sense, only some amplitudes of swing of the pendulum are permitted, and it is impossible to set it swinging with an intermediate amplitude. Furthermore, since the minimum energy is $\frac{1}{2}h\nu$, according to quantum mechanics a pendulum can never be completely still.

Why did it take until the beginning of the twentieth century to discover quantum mechanics? The answer lies in the smallness of Planck's constant. For instance, a pendulum (or any other type of oscillator) of natural frequency 1 Hz can accept energy in steps of about 6.6×10^{-34} J, which is so small that for all practical purposes it seems that energy can be transferred continuously. Only for oscillators with very high natural frequencies does the quantum $h\nu$ become large enough to give rise to observable differences from classical mechanics. A particle on a spring oscillates with simple harmonic motion, and its natural frequency increases as its mass decreases ($\nu \propto 1/m^{\frac{1}{2}}$). When the mass is extremely small, which is the case for atoms vibrating in molecules, the size of the quantum needed to excite the vibration to its next level is so great ($h\nu \propto 1/m^{\frac{1}{2}}$) that we must be prepared to encounter significant quantum effects.

All kinds of motion are quantized, and Fig. 0.4 depicts the arrangement of energy *levels* (the permitted energy values) in typical systems. An object free to move in a container of ordinary size has translational kinetic energy, and can be accelerated to any of its translational energy levels. The separation between levels is so small, Fig. 0.4(a), however, that even for atoms it is usually permissible to disregard the quantization of translational energy. A molecule may rotate at different rates, but its energy of rotation

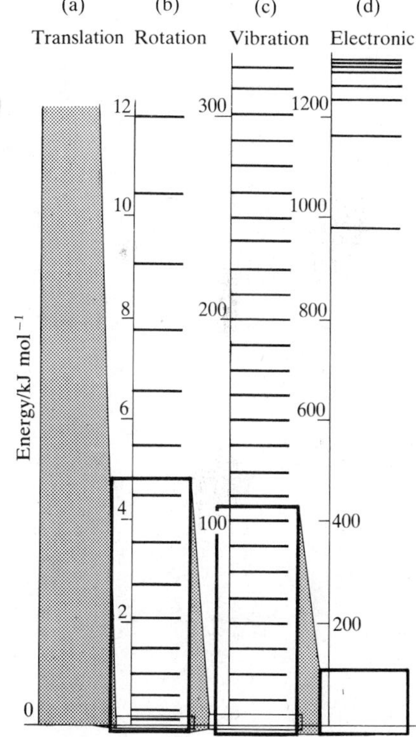

Fig. 0.4. An indication of the relative separations of the different types of energy levels of atoms and molecules. Notice that there is a scale change between each one. The translational energy levels are so close together that they may be taken as unquantized. The electronic levels in (d) are those of the hydrogen atom.

must be one of the values indicated in Fig. 0.4(b). The separation between rotational levels is small but cannot be ignored. For instance, the separation between the two lowest rotational energy levels of CO is about 8×10^{-23} J ($0.05\ \mathrm{kJ\ mol^{-1}}$; energy units are discussed in Section 0.3). Molecules can also vibrate, and the permitted vibrational energy levels are determined by the masses of the atoms and the rigidity of the bonds between them. This is illustrated in Fig. 0.4(c). The effects of quantization are now very important, the separation of levels in CO being as much as 4×10^{-20} J, or $25\ \mathrm{kJ\ mol^{-1}}$.

The energy of an electron in a hydrogen atom is restricted to the levels indicated in Fig. 0.4(d). (Other atoms show similar but more complex patterns.) Because the electron is so light we can expect the electronic energy levels of atoms (and molecules) to be much more widely spaced than the vibrational energy levels of molecules. This is the case, as the illustration shows. The separation between the lowest two energy levels of atomic hydrogen is 1.5×10^{-18} J, which corresponds to $1000\ \mathrm{kJ\ mol^{-1}}$.

0.1 (b) Transitions

The results just quoted are theoretical conclusions, but there is plenty of experimental evidence for the quantization of energy. Compelling evidence comes from direct visual observation. Figure 0.5 shows the *spectrum* of light (its analysis into components of different frequencies) emitted by atoms that have been excited to high energy levels (for example, in a flame or a spark). The light is emitted at a series of definite frequencies, just as we might expect for atoms falling between different, discrete energy levels of the sort illustrated in Fig. 0.4(d), and emitting the excess energy as radiation. The frequency ν emitted (or absorbed) when a transition occurs and can be calculated from the *Bohr frequency condition*:

$$h\nu = E_{\mathrm{upper}} - E_{\mathrm{lower}}. \tag{0.1.1}$$

For example, when a small molecule changes from one rotational level to the next lower (and then rotates more slowly) the energy loss is about $0.1\ \mathrm{kJ\ mol^{-1}}$, corresponding to an emission of radiation of frequency 2×10^{11} Hz or wavelength $\lambda \approx 1.5$ mm. (*Wavelength*, λ, is related to frequency by $\lambda = c/\nu$ for light in a vacuum, and by $\lambda = v/\nu$ in general, where v is the speed of propagation of the radiation.) This is the wavelength of microwave radiation, and so rotational transitions occur in the microwave region of the electromagnetic spectrum. Instead of frequency or wavelength, the *wavenumber* (symbol: $\tilde{\nu}$) of the radiation is often quoted: this is defined as $\tilde{\nu} = \nu/c$. In a vacuum (where the radiation has a speed c) it follows that $\tilde{\nu} = 1/\lambda$, and so the wavenumber may then be interpreted as *the number of waves per unit length*. In the present example, $\tilde{\nu} \approx 7\ \mathrm{cm^{-1}}$ (7 wavelengths per centimetre).

Vibrational transitions give rise to, and can be caused by, infrared radiation. Electronic transitions, when electrons are redistributed in an atom or molecule, generally involve radiation in the visible and ultraviolet regions of the spectrum.

If the tightly bound inner electrons of heavy atoms can take part in a transition, we can expect radiation of very short wavelengths. This is the

Fig. 0.5. The spectrum of light emitted by mercury atoms excited by an electric discharge. Each line arises from an atom that emits excess energy as it makes a transition between two of its quantized states.

origin of *X-rays*. Radiation of even shorter wavelengths constitute the *γ-rays*. For their origin we have to look beyond the atomic electrons and into the nucleus itself: transitions of the components of the nucleus between their extremely widely spaced quantum levels generate this high-energy, penetrating radiation.

The contributions made to the spectrum by transitions of different types are indicated in Fig. 0.6.

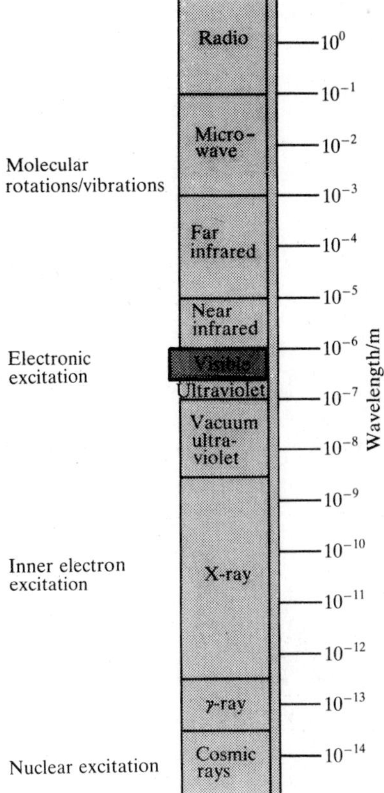

Fig. 0.6. The classification of different regions of the electromagnetic spectrum according to the wavelength of the radiation. Note that short wavelengths correspond to high-energy quanta.

0.1 (c) Assemblies of molecules

Most chemical systems are built from very large numbers of atoms or molecules. (From now on we shall refer to atoms and molecules in general as 'particles' and only specialize when necessary.) A very common measure of the amount of substance is the *mole* (symbol: mol). A mole of a substance contains 6.022×10^{23} of the specified particles: see Box 0.1 for a more precise statement.

Box 0.1 Moles and molecular masses

Moles

1 mole (1 mol) of substance contains as many objects (e.g. atoms, molecules, formula units, or other specified entities) as there are atoms in exactly 12 g of ^{12}C. This number is approximately 6.022×10^{23}. Therefore, if a sample contains N specified entities, the *amount of substance* (symbol: n) is $n = N/N_A$, where N_A is *Avogadro's constant*, $N_A = 6.022 \times 10^{23}\ \text{mol}^{-1}$. (Precise values are listed inside the front cover; note that N_A has dimensions: it is a constant, not a number.) Likewise, if the amount of substance is n (e.g. 2.0 mol O_2), the number of elementary objects is nN_A (e.g. 12×10^{23} O_2 molecules).

Mass

The *molecular mass* (symbol: m) is the mass of the molecule (and the atomic mass is the mass of the atom, and so on). The *relative molecular mass* (RMM, or *molecular weight*, symbol: M_r) is the mass of the molecule relative to ^{12}C taken as exactly 12:

$$M_r = 12 \times m/m(^{12}\text{C}).$$

The *relative atomic mass* (RAM, or *atomic weight*, symbol: A_r) is the analogous quantity for atoms. The *molar mass* (symbol: M_m) is the mass per unit amount of substance, i.e. the mass per mole of specified objects:

$$M_m = M/n.$$

In practice, the RMM is the numerical value of the molar mass expressed in g mol^{-1}:

$$M_r = M_m/\text{g mol}^{-1},$$

and the molecular mass can be calculated from the RMM by multiplication by the *atomic mass unit* (AMU; symbol m_u):

$$m = M_r m_u, \qquad m_u = 1.6606 \times 10^{-27}\ \text{kg}.$$

Avogadro's constant

Avogadro's constant has been determined in many ways. These include the measurement of the gas constant and its comparison with Boltzmann's constant ($N_A = R/k$), the measurement of Faraday's constant and its comparison with the electronic charge ($N_A = F/e$), and the determination, using X-ray diffraction, of the number of ions in a crystal of known density.

Typical quantities

The following examples illustrate how much substance constitutes 1 mol at 25 °C and 1 atm pressure:

(1) 1 mol of atoms of a perfect gas occupies 24.5 dm^3 (approximately 1 cubic foot).

(2) 1 mol H_2O (18 g of water) occupies 18 cm^3; 1 mol CH_3CH_2OH (46 g of ethanol) occupies 58 cm^3.

(3) 2 mol Fe (112 g of iron) is about 1 cubic inch.

(4) 1 mol NaCl (58 g of salt) is a conical pyramid 2 cm high and 4 cm in diameter. A pinch of salt is about 1 mmol (10^{-3} mol) NaCl.

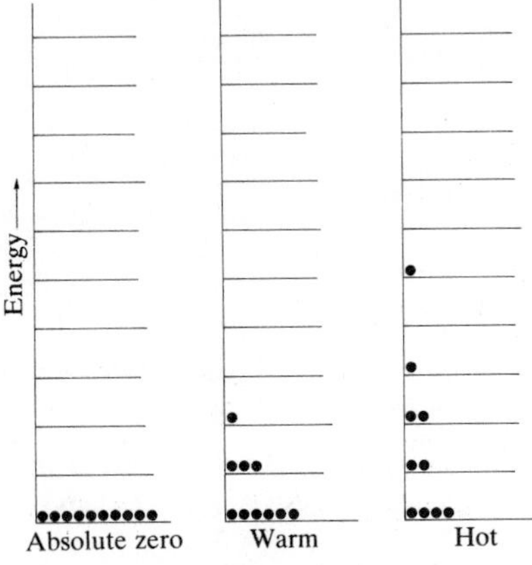

Fig. 0.7. The dependence on the temperature of the populations of the vibrational energy levels of a molecule. Near absolute zero only the lowest level is populated, but at high temperatures several levels are populated (as indicated by the number of blobs). In the case of typical molecules, even 'warm' would correspond to temperatures well above room temperature.

The continuous thermal agitation that the particles in a sample experience ensures that they are distributed among all the available energy levels. Some particles occupy the lowest energy level, others are in the excited levels. This is easy to discuss in the case of molecular vibrational motion, which we consider first.

At very low temperatures almost every molecule in a sample is in its lowest energy level, Fig. 0.7, which corresponds to the virtual absence of vibrational motion. Heating the sample stimulates some of the molecules to vibrate. Transitions up the ladder of energy levels correspond to acceleration into more vigorous motion (larger amplitude of vibration). As the sample becomes warmer more molecules are able to reach levels higher up the ladder. Not all occupy the higher levels because there is a predominance of molecules in the lower, less energetic levels, but the tail of the distribution gets longer and penetrates further into the high energy region as the temperature is raised, Fig. 0.7.

0.1 (d) The Boltzmann distribution

The formula for calculating the *populations* of the levels (i.e., the average number of particles in each one) is known as the *Boltzmann distribution*. This gives the ratio of the numbers of particles with energies E_i and E_j as

$$N_i/N_j = e^{-(E_i - E_j)/kT}. \qquad (0.1.2)$$

The fundamental constant k is Boltzmann's constant; $k = 1.381 \times 10^{-23}$ J K^{-1}. T is the *thermodynamic temperature*, the temperature on a fundamental scale, not on an arbitrary scale such as the Celsius scale; the two are related by $T/K = 273.15 + t/°C$ (T is sometimes called the *absolute temperature*; K denotes *kelvin*). The product of k and Avogadro's constant (N_A) is denoted R and called the *gas constant*; its value is $R = 8.314$ J K^{-1} mol^{-1}). When the Boltzmann distribution is expressed with the gas constant R in place of k, the energies in the exponential are molar energies, e.g. kJ mol^{-1}.

Consider first the distribution of atoms among their electronic levels. In a sample of atomic hydrogen at 25 °C, corresponding to 298 K, what proportion of the atoms are in the first excited electronic level? At 25 °C, which we shall refer to as 'room temperature', $RT = 2.48\ \text{kJ mol}^{-1}$.

The first electronically excited level of hydrogen lies about 1000 kJ mol^{-1} above the ground state. It follows that

$$\frac{N\ (\text{first excited electronic level})}{N\ (\text{lowest electronic level})} = \exp\{-1000\ \text{kJ mol}^{-1}/2.48\ \text{kJ mol}^{-1}\}$$
$$= e^{-403} \approx 10^{-175}.$$

This means that essentially every atom in the sample is in its lowest electronic level at 25 °C. The population of the excited level rises to about 1 per cent only when the temperature is raised to 10 000 °C. You should not, however, draw the conclusion that it is never necessary to consider the chemistry of electronically excited atoms. Very high temperatures are reached in rocket exhausts and explosions, and the absorption of light and the fracture of bonds can be used to generate temporary but significant concentrations of excited atoms. *The Boltzmann distribution applies only to systems in thermal equilibrium*.

The vibrational levels of molecules are closer together than the electronic levels, but even so the Boltzmann distribution shows that normal molecules are almost exclusively in their ground vibrational states at normal temperatures. In the case of CO the vibrational energy spacing is 25 kJ mol^{-1}, and so at room temperature the populations in the first excited and ground vibrational levels are in the ratio

$$\frac{N\ (\text{first excited vibrational level})}{N\ (\text{lowest vibrational level})} = \exp\{-25\ \text{kJ mol}^{-1}/2.48\ \text{kJ mol}^{-1}\}$$
$$= e^{-10} = 5 \times 10^{-5}.$$

This means that only 1 in 20 000 molecules are in their first excited vibrational level.

Since rotational levels are so closely spaced, even at room temperature a significant proportion of molecules are highly rotationally excited. The Boltzmann distribution, evaluated for CO, gives

$$\frac{N\ (\text{first excited rotational level})}{N\ (\text{lowest rotational level})} = \exp\{-0.05\ \text{kJ mol}^{-1}/2.48\ \text{kJ mol}^{-1}\}$$
$$= e^{-0.02} = 0.98.$$

However, this calculation is not quite right because the first rotationally excited level is really three *states* of the same energy. This complication comes about because quantum theory shows that in its first excited level a molecule can rotate in planes orientated in three different directions in space. All three orientations are equally likely, and so the ratio of populations of the upper level as a whole, not just the individual states, is $3 \times 0.98 = 2.9$. This means that there are *more* rotating molecules than non-rotating molecules. The next higher rotational level consists of five equal-energy states because quantum theory shows that five planes of rotation are permitted. As this level lies 0.14 kJ mol^{-1} above the ground level its relative population is

$$\frac{N\ (\text{second excited rotational level})}{N\ (\text{lowest rotational level})} = 5\ e^{-0.14/2.48} = 4.7.$$

This result, and the population distribution at room temperature, are illustrated in Fig. 0.8.

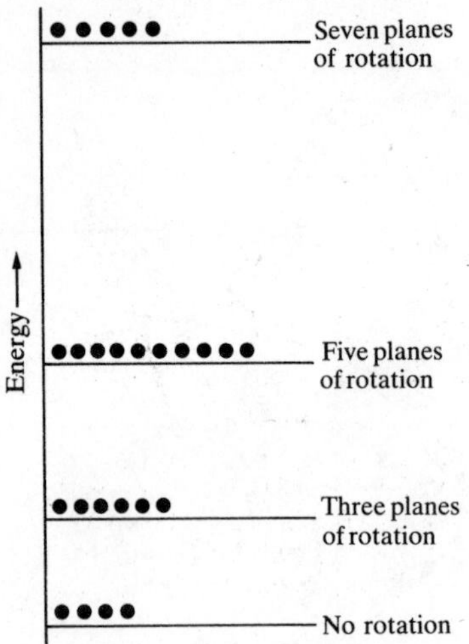

Fig. 0.8. The populations of the rotational energy levels of a linear molecule at a temperature well above absolute zero. The bulge in the distribution reflects the increasing number of *states* that belong to the higher energy levels (1, 3, 5, 7, . . .).

0.1(e) Molecular speeds

The Boltzmann distribution takes a special form when we consider the free translational motion of non-interacting gas particles. Different energies then correspond to different speeds, and so the Boltzmann formula can be used to predict the proportions of particles having a specified speed at any temperature. The distribution of speeds is called the *Maxwell–Boltzmann distribution*, and it has the features shown in Figs. 0.9 and 0.10. Figure 0.9 shows the distribution of speeds for identical particles at two widely different temperatures: notice how the tail towards high speeds is much longer at the higher temperature. The speed corresponding to the maximum of the curve is the *most probable speed*, the speed most likely to be observed for a particle selected at random. Figure 0.10 shows how the distribution depends on mass for some constituents of air at 25 °C. The lighter molecules move, on average, much faster than the heavy ones, but the most probable speed of even the heaviest molecule shown (CO_2) is about 330 m s^{-1}, or over 750 m.p.h.

An important deduction from the Maxwell–Boltzmann distribution concerns the total kinetic energy of a gas of particles. The kinetic energy of one particle of mass m is $\frac{1}{2}mv^2$. Different particles travel at different speeds, and to evaluate the total kinetic energy of a sample we must know the proportion of particles having each speed v; this can be calculated from the Maxwell–Boltzmann distribution. The result is:

$$\textit{total kinetic energy} = \tfrac{3}{2}nRT, \qquad (0.1.3)$$

n being the amount of substance (number of moles) of particles. Note that the result is independent of the mass of the particles, and so it applies to any gas.

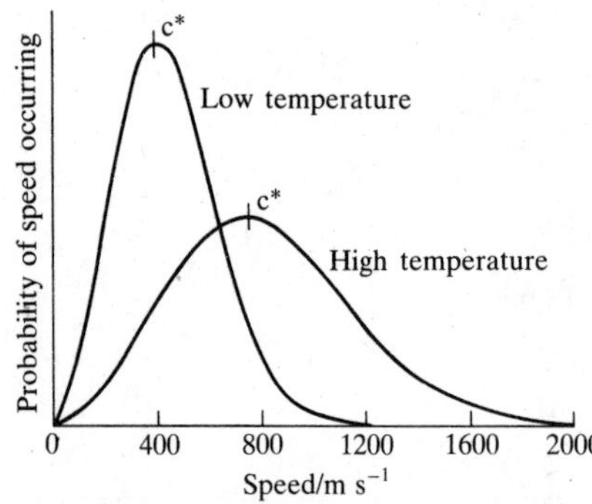

Fig. 0.9. The general features of the Maxwell–Boltzmann distribution of molecular speeds are illustrated by this and the next illustration. This shows the proportion of particles having various speeds at two different temperatures. Notice how the most probable speed (c^*) shifts to higher values, and the distribution becomes broader and extends more to high speeds, at high temperatures.

0.1 (f) Equipartition

The result in the last equation is a special case of the *equipartition theorem* of classical physics. This states, roughly, that *the average energy of each different mode of motion of a particle is* $\frac{1}{2}kT$. In the case of a gas of atoms there are three translational modes (corresponding to motion parallel to any three perpendicular axes), and so the average kinetic energy of each particle is expected to be $\frac{3}{2}kT$. Since the number of particles per mole is N_A, the total kinetic energy per mole of particles is $\frac{3}{2}N_AkT$, or $\frac{3}{2}RT$, in agreement with the expression above. The theorem also indicates that the average rotational energy of a molecule is the sum of $\frac{1}{2}kT$ for each of its axes of rotation. Therefore the mean energy of rotation of a methane molecule, which is free to rotate around three perpendicular axes, is expected to be $\frac{3}{2}kT$. This gives a further contribution to the total energy (on top of the translational kinetic energy) of $\frac{3}{2}RT$. The theorem also suggests that the mean vibrational energy of a bond ought to be kT ($\frac{1}{2}kT$ for the kinetic energy and $\frac{1}{2}kT$ for the potential energy), but this is hardly ever found in practice because the equipartition theorem, which is based on classical physics, fails when quantum effects are important. Yet, although the equipartition theorem does sometimes fail, it is a helpful yardstick for assessing the energy stored in various kinds of motion: at room temperature $kT = 4.1 \times 10^{-21}$ J, corresponding to $RT = 2.5$ kJ mol^{-1}.

The only tricky point about the application of the Boltzmann distribution is deciding how many states belong to each energy level. Rules can be constructed, and they are dealt with later (Part 2). The essential features of the Boltzmann distribution should not be obscured by this minor problem. It shows that the distribution of atoms and molecules among their states is an exponential function of energy and temperature, and that more states are significantly populated if they are close together in comparison with RT (like rotational and translational states) than if they are far apart (like vibrational and electronic states); moreover, more states are occupied at high temperatures than at low. Exponential expressions occur very widely in chemistry. For instance, many reactions proceed at rates that depend on temperature as $\exp(-E/RT)$, and the appearance of this factor can always be traced back to the Boltzmann distribution.

Figure 0.11 summarizes the form of the Boltzmann distribution for some typical sets of energy levels. This figure, and Fig. 0.9, carry all the information necessary at this stage.

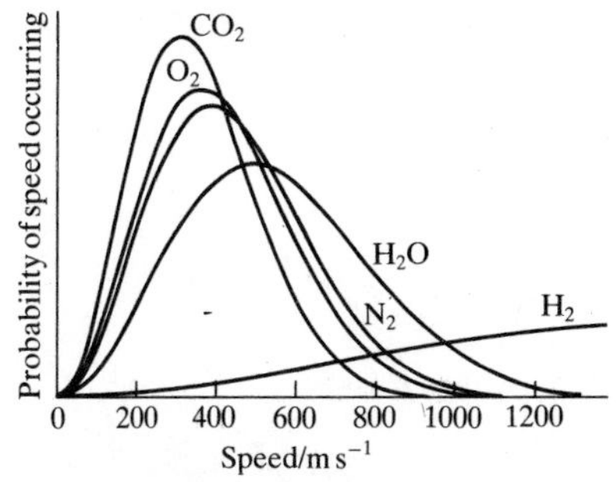

Fig. 0.10. The lighter the particle, the higher its average speed and the wider its spread of speeds. These are the distributions of speeds for H_2 and the major components of air at 25 °C.

0.1 (g) The association of atoms and molecules

The simplest type of bond is the *ionic bond*. This is formed when one or more electrons are transferred from one atom to another, and the resulting ions adhere by direct Coulombic attraction (with a force that depends on $1/r^2$ and a potential energy that depends on $1/r$). A typical example is the structure of sodium chloride, an aggregate of Na^+ cations and Cl^- anions. As in this example, ionic bonding often leads to extensive aggregates of ions and the formation of an ionic crystal.

The other extreme type of bond is the *covalent bond*. Its origin is the sharing of a pair of electrons, and the simplest example is the bond between hydrogen atoms in molecular hydrogen. When the electrons are shared equally the bond is *non-polar* (as in H_2 and Cl_2). When the electron pair is located nearer one atom than another the bond is *polar* (HCl, with the negative charge predominantly on the chlorine and the positive charge on the hydrogen, is an example). Whether or not a molecule as a whole is polar (and possesses a *dipole moment*) depends on the extent to which the polar bonds cancel each other when the symmetry of the molecule is taken into account. Thus benzene is a nonpolar molecule even though each C–H bond is polar, because the diametrically opposite C–H dipoles cancel. Covalently bonded molecules are usually discrete units, like N_2 or H_2O, but nevertheless may be very large. DNA, for instance, is a molecule several metres long.

That molecules congregate to form liquids and solids shows that there must be forces of attraction between them even though their normal valencies are satisfied. These *van der Waals forces* have an electromagnetic origin and arise in a variety of ways. In some cases molecules stick together because they are polar. Sometimes this is assisted by the formation of *hydrogen bonds* in which two atoms of different molecules are bridged by a hydrogen atom (as in water). In the case of non-polar molecules, like N_2 or benzene, the rapid fluctuations of the electron clouds give rise to instantaneous electric dipoles strong enough to result in stable solids.

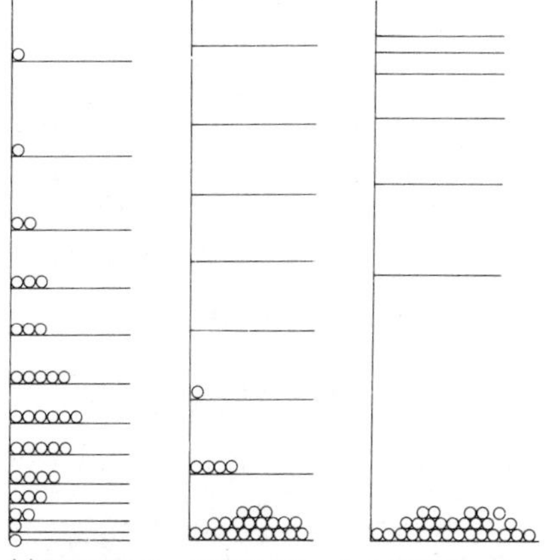

Fig. 0.11. The general features of the Boltzmann distributions of populations for rotational, vibrational, and electronic energy levels. Each diagram corresponds to a different vertical scale (as in Fig. 0.4).

0.2 The states of matter

Casual inspection of the familiar world reveals the existence of three *states* of matter: solids, liquids, and gases. Closer inspection shows that some materials can exist in different crystal forms (e.g. carbon can exist as diamond or graphite). These varieties of matter are called its *phases* and we speak of the solid, liquid, and gas phases of a substance, and also of its various solid phases. In rare instances, even the liquid state of a material may occur in different phases with sharply distinct properties.

0.2 (a) The solid state

Pure solids may exist with their particles arranged in an orderly way as *crystals* or arranged in a disorderly way as *glasses*, Fig. 0.12. Fine amorphous ('formless') dusts also occur, but in many cases these are nothing more than finely powdered crystals.

Crystals form under the influence of ionic and covalent bonding and van der Waals forces. In crystals composed of ions, the structure (and hence the crystal's external appearance, or *morphology*) is determined largely by the geometrical problem of packing together ions of different sizes into an almost infinite, symmetrical, and electrical neutral array. The rigidity of such solids can be traced to the difficulty of breaking large numbers of bonds simultaneously. In all real crystals there are imperfections in the uniformity of the array, and crystals fracture when these *defects* increase and spread under stress.

Aggregates of covalent molecules (which are called *molecular crystals* when the molecules form periodic, symmetrical arrays, but might also be amorphous conglomerates, like butter) can often be disrupted by the application of a little heat. Warming may cause enough molecular motion to overcome the weak intermolecular bonds, and these solids often have low melting and boiling points. In some cases van der Waals attractions are so weak that the crystals have no rigidity at all. For instance, *liquid crystals* are swarms of aligned, often rod-like, molecules that flow like liquids.

Metals form a special type of crystal, consisting of a collection of cations (e.g. Cu^{2+}) immersed in a sea of electrons. Their electric and thermal conductivities arise from the latter's mobility. Their malleability and ductility (the ability of the cations to move under stress, and then to remain in their new positions) can be traced to the ease with which the electron sea adjusts to permit different bonding arrangements.

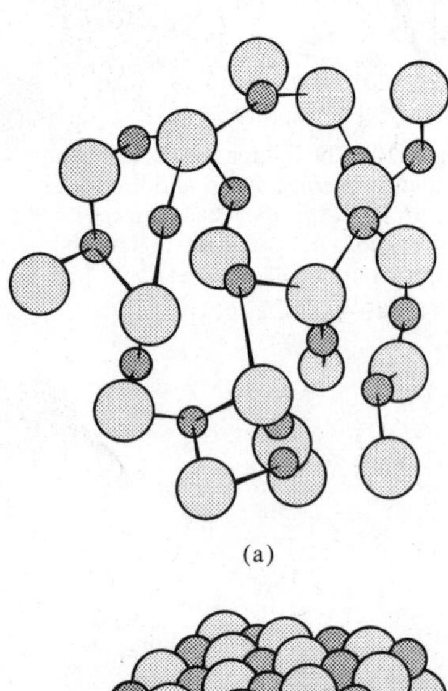

(a)

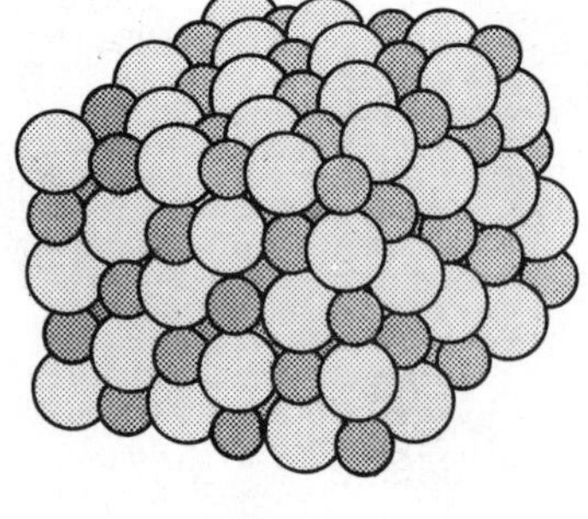

(b)

Fig. 0.12. In (a) glassy materials the particles (e.g. ions) are in a disorderly arrangement; in (b) crystals they are regularly ordered.

0.2 (b) The liquid state

When a solid is heated its particles vibrate with a greater amplitude, and at some temperature, the *melting point*, they are able to move away from their initial sites. Figure 0.13 shows a computer simulation of the melting process in the case of a two-dimensional crystal. Above the melting point the molecules move so much that the crystal lattice is no longer significant, and the whole sample becomes a mobile and almost structureless fluid.

The ordered structure of the solid might not be wholly lost. In the case of water, for example, the liquid can be pictured as a collection of ice-like and structureless regions, Fig. 0.14. The ice-like regions are continually forming and dispersing, and although at one moment a water molecule might be in one, at the next moment it might be in a structureless region.

The ability of a liquid to flow is measured by its *viscosity*: the greater the viscosity the less mobile the liquid. Glass and molten polymers are highly viscous because their large molecules get entangled. Water has a higher viscosity than benzene because its molecules bond together more strongly and this hinders the flow. Viscosities decrease with increasing temperature because the particles move more energetically and can escape from their neighbours more easily. Since moving a molecule from site to site involves breaking weak van der Waals bonds, the proportion of molecules with enough energy to move follows a Boltzmann distribution. This suggests that the ability of the liquid to flow ought to behave as

$$\textit{fluidity} \propto e^{-\Delta E/RT},$$

where ΔE is the energy to be overcome. The viscosity is the inverse of the fluidity, and so

$$\textit{viscosity} \propto e^{\Delta E/RT}. \qquad (0.2.1)$$

Viscosities are observed to follow this exponential form over a limited temperature range, and the value of ΔE is found to be similar to the intermolecular binding energy (a few kJ mol^{-1}; e.g. 11 kJ mol^{-1} for benzene and 3 kJ mol^{-1} for methane).

0.2 (c) Solutions

Ionic crystals often dissolve in solvents consisting of molecules that form an electrostatic association with the ions. This is called *solvation*. In the special case of water, Fig. 0.15, the solvation is called *hydration*. When an ionic crystal has dissolved, the solution consists of a collection of ions in a sea of solvent; this is an *electrolyte solution*. Such solutions conduct electricity because the positive and negative ions can migrate largely independently under the influence of an electric field. Non-ionic molecules often dissolve in non-polar or weakly polar solvents to form *solutions of non-electrolytes*. A typical example is methylbenzene dissolved in benzene. The dissolution proceeds well if the solvent and solute molecules have similar structures,

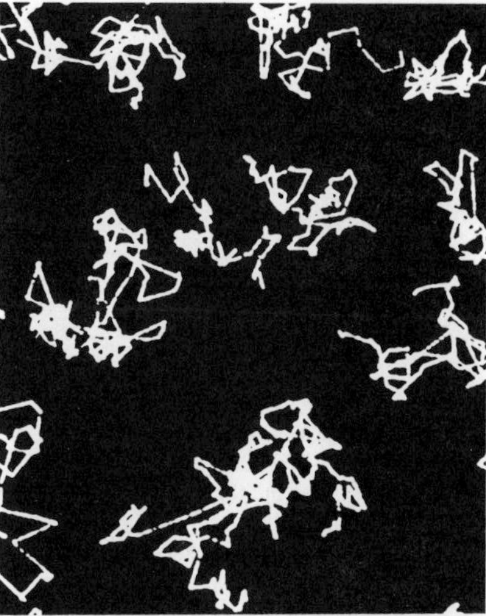
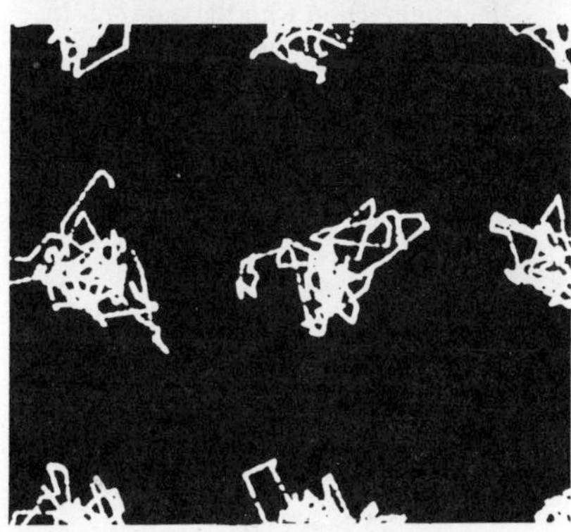

Fig. 0.13. The motions of particles in solids and liquids can be simulated by solving Newton's equations of motion on a computer. These diagrams (calculated by Professor B. J. Alder) show that in the solid (lower half) the particles jostle around their sites, but in the liquid (upper half) they move from place to place.

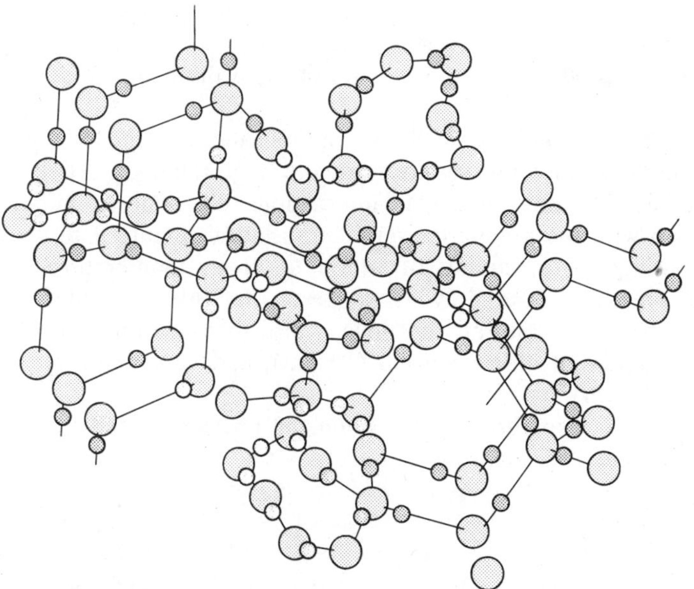

Fig. 0.14. A model of the structure of liquid water. In some regions the molecules are ordered much as in ice, but between them the molecules are disorganized. The 'ice-like' and 'liquid-like' regions are continually changing.

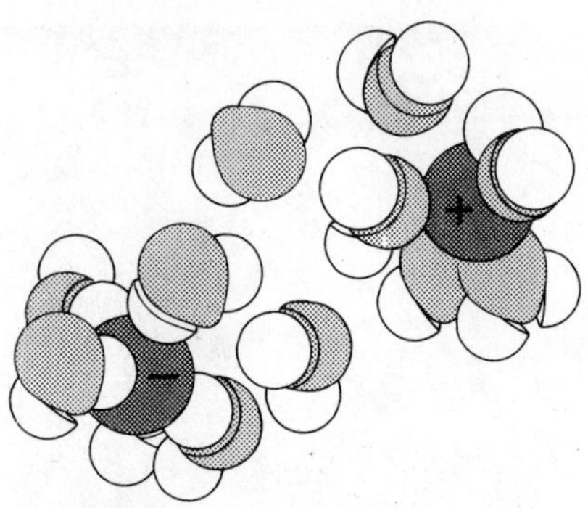

Fig. 0.15. Ions are *solvated* in solution. In this case, where the solvent is water, the cations (the positive ions) are *hydrated* as a result of their interaction with the oxygen of the water molecules, and the anions (the negative ions) are hydrated by their interaction with the more positive regions of the water molecules (the hydrogen atoms).

because the energy of the molecules in the pure liquid or solid solute is approximately the same as their energy when surrounded by solvent molecules.

The term *concentration* refers to the amount of substance per unit volume of the solution (e.g. so many moles per litre, or mol dm^{-3}; note that 1 litre is exactly 1 dm^3). *Molality* denotes the amount of substance divided by the mass of solvent. A 1 mol kg^{-1} solution (formerly and still often colloquially called a '1 molal' solution) is formed by dissolving 1 mol of substance in 1 kg of *solvent*. A 1 mol dm^{-3} solution (formerly and often colloquially called a '1 molar' solution) is formed by dissolving 1 mol of substance in enough solvent to produce 1 dm^3 of *solution*. Mixtures can be prepared more accurately by weighing, and for a given solution the volume, and hence the molarity, varies as the temperature changes, but the molality does not. Common (but officially deprecated) practice is to denote the units mol dm^{-3} by the single letter M (1 M $\equiv$ 1 mol dm^{-3}).

It is useful to be able to picture the structures of solutions of different concentrations. In the case of a 1 M aqueous NaCl solution the average distance between oppositely charged ions is about 1 nm, enough to accommodate about three water molecules. A 'dilute solution' often means a concentration of about 0.01 M or less, when the ions are separated by about 10 water molecules.

In an extremely dilute solution the cations and anions are so far apart that they have insignificant interactions, but as the concentration increases positive cations tend to congregate near the negative anions, and vice versa. This has the effect of reducing the conductivities of the ions and their ability to take part in reactions. Instead of the concentrations of the ions it then becomes more significant to talk in terms of their 'effective concentrations', or *activities*. We shall see how this concept can be given a precise and useful meaning.

0.2 (d) The gaseous state

The word 'gas' is derived from 'chaos'. We picture a gas as a swarm of molecules in constant, chaotic motion. Each particle travels in a straight line at high speed until it reaches another, when it is deflected, or until it collides with the wall of the vessel, when it might bounce back into the bulk or stick until dislodged by the vibration of the wall or the impact of another molecule.

In a gas the average distance between molecules and the average distance they travel between collisions are normally large relative to their diameter. This implies that intermolecular interactions play only a minor role in comparison with the kinetic energy of translation. Figure 0.16 shows three scale drawings of a sample of argon at room temperature and three different pressures. The drawing suggests that intermolecular forces are likely to be negligible at the lowest pressure (1 atm), significant for pressures somewhat above normal (10 atm), and very important at higher pressures (30 atm).

It is found by experiment that, when the pressure is sufficiently low, all gases satisfy the *perfect gas law*:

$$pV = nRT. \tag{0.2.2}$$

p is the pressure, V the volume, n the amount of substance of gas particles,

T the absolute temperature, and R the gas constant ($R = kN_A$). Deviations from this relation occur at high pressures, but it is a law to which every real gas conforms increasingly closely as its density is reduced. Intermolecular interactions decrease in significance as the density is reduced, and the limiting state, which is called a *perfect gas* (or an *ideal gas*), is one in which the molecules move freely, totally without interaction. This model is the basis of the *kinetic theory*, which regards a gas as a collection of particles of negligible size in constant chaotic independent motion.

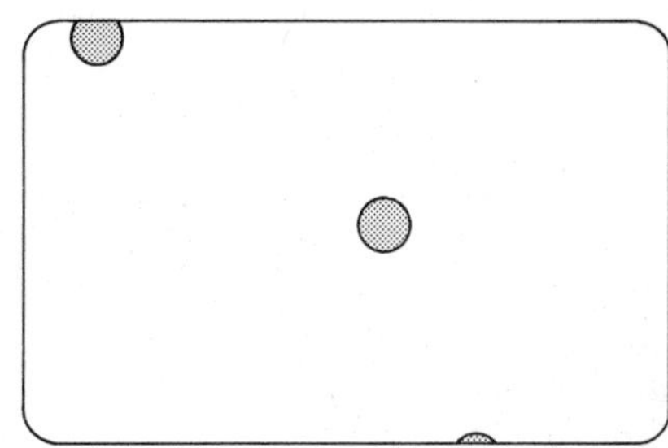

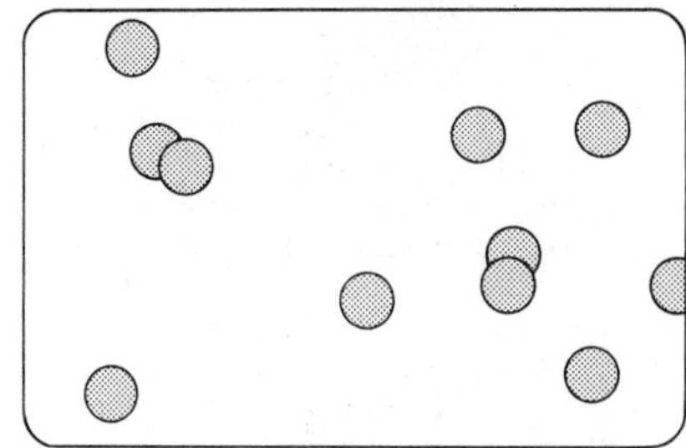

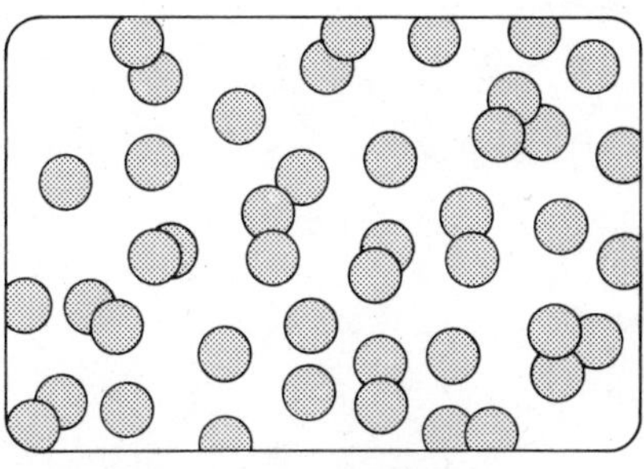

Fig. 0.16. Samples of argon gas at 1 atm (*top*), 10 atm (*middle*), and 30 atm (*bottom*) and 25 °C. The drawings, which are approximately to scale, represent the numbers of atoms in a slab 5 nm × 8 nm and 10 nm thick.

0.2 (e) Kinetic theory

The kinetic theory of gases leads to a number of useful conclusions. The key calculation is the relation of the mean speed to the temperature of the gas: this shows that the *root mean square speed* (RMS speed; symbol: c_{rms}) of the particles (the square root of the mean of the squares of their speeds) is

$$c_{\text{rms}} = \surd(3RT/M_{\text{m}}). \qquad (0.2.3)$$

That is, the RMS speed increases as the square root of the temperature and decreases as the square root of the molar mass. Typical RMS speeds calculated using this expression for a room temperature sample are $1360\ \text{m s}^{-1}$ for helium and $410\ \text{m s}^{-1}$ for carbon dioxide (10^{13} and 10^{12} molecular diameters per second respectively).

Several important results can be obtained once the connection between speed and temperature has been established. One of these is the average number of collisions per second made by a particle in the gas: this is its *collision frequency* (symbol: z). A rough estimate is that at room temperature and pressure an oxygen molecule makes about 6×10^9 collisions each second. The average distance a particle flies between collisions is called its *mean free path* (symbol: λ). Since a particle collides with a frequency z and travels with a speed c_{rms}, it follows that $\lambda \approx c_{\text{rms}}/z$. For oxygen the mean free path comes out as 70 nm (200 collision diameters) at room temperature and pressure.

0.2 (f) Transformations of matter

Two kinds of questions arise in connection with the physical and chemical transformation of matter. One is, 'Can it occur?' and the other is, 'How fast does it occur?'.

The second question is explored in *chemical kinetics* and, at a more fundamental level, in *molecular reaction dynamics*. It is possible at this stage, though, to construct a rudimentary model of a reaction as involving the collision of two particles with sufficient energy to break and form bonds. An expression for the rate of such a process is obtained as follows.

The frequency of collisions is z. The probability that when the collision occurs it does so with sufficient energy to react is determined by the Boltzmann distribution. Therefore, if E_{a} is the necessary energy (the *activation energy* of the reaction), the rate is given by

rate of reaction = (rate of collision) × (probability that collision carries enough energy);

so that

$$\textit{rate} = z\text{e}^{-E_{\text{a}}/RT}. \qquad (0.2.4)$$

The collision frequency depends on the temperature, but its variation is normally dominated by the exponential temperature dependence coming from the Boltzmann factor. The last expression, the *Arrhenius Law*, is often obeyed experimentally, and is a fair first approximation to the actual experimental result.

The answer to the first question, whether a change *can* occur, is the domain of *thermodynamics*. Thermodynamics deals with the possible transformations of energy, and embodies its conclusions in three laws.

The *First Law* can be summarized by the remark that 'energy can be neither created nor destroyed'. Therefore, any change must conserve the total energy in the universe. ('Universe' has a special meaning in thermodynamics, as we shall see.) The *Second Law* introduces another property, the *entropy*, and uses it to set up a criterion for assessing whether or not a transformation has a *natural* tendency to occur (like the flow of heat from hot to cold, or the rusting of iron): roughly, 'the entropy of the universe never decreases'. The entropy of a system is determined by the way its molecules are distributed over the available energy levels, and can be calculated once this is known. The important connection between the ideas of quantum theory and the laws of thermodynamics is the domain of *statistical thermodynamics*. The *Third Law* refers to processes at very low temperatures and (once again, very roughly) recognizes that 'absolute zero is unattainable'.

0.3 Force, pressure, and energy

The concepts of force, pressure, and energy occur throughout chemistry, and it is important to be familiar with their units.

If you can imagine the force necessary to move a mass of 1 kg so that it accelerates at the rate of $1\ \mathrm{m\,s^{-2}}$ (if the mass is at rest intially, that means it should have a velocity of $1\ \mathrm{m\,s^{-1}}$ after 1 s, $2\ \mathrm{m\,s^{-1}}$ after 2 s, and so on), then you have some appreciation of the basic unit of force, which is the *newton* ($1\ \mathrm{N} = 1\ \mathrm{kg\,m\,s^{-2}}$). For instance, this book, which has a mass of about 1.5 kg, if released would fall with an acceleration of $9.8\ \mathrm{m\,s^{-2}}$ (the *acceleration due to gravity*, g) and so would be subject to a force of 15 N (*force* is equal to *mass* × *acceleration*). Holding the book enables you to experience that force. A small apple on a tree experiences a force of about 1 newton.

Pressure is force per unit area. Imagining a force of 1 N applied to an area of $1\ \mathrm{m^2}$ gives an idea of the basic unit of pressure. When we come to do quantitative calculations we shall quote pressures in $\mathrm{N\,m^{-2}}$; this unit is also called the *pascal* ($1\ \mathrm{Pa} = 1\ \mathrm{N\,m^{-2}}$).

An important practical unit of pressure is the *atmosphere* (atm). There is international agreement that 1 atm is exactly $101\,325\ \mathrm{N\,m^{-2}}$ (101.325 kPa). Slightly different from 1 atm is the *bar*, 1 bar being exactly 10^5 Pa (so that 1 atm = 1.013 25 bar). The bar has the great advantage of being immediately converted to fundamental units without the nuisance of a lengthy numerical factor, and 1 bar is now the *standard pressure* (symbol: $p^{\ominus}$; $p^{\ominus} = 1$ bar exactly) for reporting thermodynamic data.

One way of becoming familiar with the magnitude of a pressure of 1 atm is to imagine the pressure exerted by a mercury column in a barometer: at 25 °C a column of mercury 760 mm high exerts a pressure of 1 atm. The use of mercury in barometers is responsible for two other practical units of pressure. The first is *millimetres of mercury*, written mmHg; this is the pressure exerted by a 1 mm high column of mercury of specified density in a gravitational field of specified strength. The conditions are chosen so that 1 atm corresponds closely to 760 mmHg. The other unit is the *torr* (symbol: Torr), which is named after Torricelli, who invented the barometer. The torr is defined as exactly 1/760 part of 1 atm, and so 760 Torr *is exactly equal to* 1 atm. 1 mmHg differs from 1 Torr by one part in 10^8 (see inside front cover), and so for most practical purposes Torr and mmHg can be used interchangeably.

The dimensions of *energy* are those of *force* times *distance* (but it has a much richer definition, as we shall see in Chapter 2). The unit for measuring energy is the *joule* (J). The formal definition of 1 J is that it is the energy needed to push against a force of 1 N for 1 m (1 J = 1 N m). There are several ways of envisaging the size of this unit. For instance, raising this book 1 m requires the expenditure of about 15 J of energy. Each pulse of the human heart consumes about 1 J of energy. Alternatively, when 1 kJ of energy is dissipated as heat in 50 cm^3 of water, its temperature rises by about 5 °C. This means that it needs about 15 kJ of energy to make a small cup of coffee.

Other units of energy are often encountered. One of the most common is the *thermochemical calorie* (symbol: cal or cal_{th}). This is defined as 4.184 J, and so it is easy to make the conversion. Much of the literature of chemistry quotes energies in kcal mol^{-1} (kilocalories per mole) even though the measurements were originally made in joules. Data in kilocalories per mole can easily be translated into kJ mol^{-1} by multiplication by 4.184. Another unit is the *electronvolt* (symbol: eV). This is the energy acquired by an electron when it is accelerated through a potential difference of 1 V. For chemical applications it is better to think in terms of 1 mol of electrons, then the conversion reads 1 eV $\triangleq$ 96.485 kJ mol^{-1}. (The sign $\triangleq$ means 'corresponds to'.)

We have remarked that spectral lines arise from transitions between energy levels. Since the energy of a transition can be expressed in terms of its frequency through the Bohr frequency condition, eqn (0.1.1), it is appropriate in some cases to express energies either as an equivalent frequency or as the corresponding wavenumber. The latter conversion reads 1 cm^{-1} $\triangleq$ 11.96 J mol^{-1}.

These units, and their interconversions, are collected inside the front cover.

Part 1 · Equilibrium

1

The properties of gases

Learning objectives

After careful study of this chapter you should be able to:

(1) Explain how *pressure* and *temperature* are measured, Sections 1.1(a and b).

(2) State the *Zeroth Law of thermodynamics* and explain its significance, Section 1.1(b).

(3) Explain the meaning of *limiting law*, Section 1.1(c).

(4) State and use *Boyle's Law*, Section 1.1(d), *Gay-Lussac's Law*, Section 1.1(e), and *Avogadro's hypothesis*, Section 1.1(f).

(5) Write the *perfect gas law*, eqn (1.1.1), and use it to calculate changes in pressure, volume, and temperature, Example 1.1.

(6) Describe how the *perfect gas temperature scale* is defined, Section 1.1(e).

(7) State and use *Dalton's Law*, Section 1.2(a) and Example 1.2, and define and calculate the *partial pressure* of a gas in a mixture, eqn (1.2.4).

(8) Explain why *real gases* differ from the perfect gas, Section 1.3(a), and distinguish between short-range repulsions and long-range attractions.

(9) Write the form of the *virial equation of state*, eqns (1.3.1) and (1.3.2), and define *virial coefficient* and *Boyle temperature*, Section 1.3(b).

(10) Define the terms *isotherm*, Section 1.1(d), *compression factor*, Section 1.3(a), and *vapour pressure*, Section 1.3(b).

(11) Define and explain the significance of the *critical constants* of a gas, Section 1.3(c), and use them to construct the *reduced variables*, eqn (1.4.6).

(12) Write down, justify, and use the *van der Waals equation of state*, eqn (1.4.2) and Example 1.4, and relate its parameters to the critical constants of a gas, eqn (1.4.3), and to the virial coefficients, eqn (1.4.4).

(13) State the *principle of corresponding states*, Section 1.4(c), and justify it in terms of the van der Waals equation.

(14) Use other equations of state to explore the properties of real gases, Box 1.1.

Introduction

The first purpose of this chapter is to see how the state of a gas is described, and how its properties depend on its condition. The second purpose is to show how to model molecular behaviour. This is important because understanding the way molecules behave makes it possible to suggest explanations for newly discovered phenomena and to judge the plausibility of others' explanations without getting trapped in lengthy calculations. Gases are so simple that they provide an excellent introduction to this technique.

1.1 Equations of state: the perfect gas

The basic quantities for the study of gases are *pressure* and *temperature*.

1.1 (a) Pressure

Pressure may be measured with a *manometer*, which in its simplest form is a U-tube filled with some liquid of low volatility (such as mercury). The pressure of the gas is given by the difference in heights of the liquid in the two arms (plus the external pressure if one tube is open to the atmosphere). This is the origin of the unit mmHg for the measurement of pressures. More sophisticated techniques are used at low pressures. The *McLeod gauge*, for instance, works by withdrawing a known volume of the sample gas, compressing it into a smaller volume, and measuring the pressure of the compressed sample. The compression magnifies the original pressure and makes it readily measurable; the pressure in the vessel is then calculated using the gas laws we develop later in the chapter. Methods that avoid the complication of having to account for the intrusion of vapour from the manometer fluid are also available. These include monitoring the deflection of a diaphragm, either mechanically or electrically, or the change in some pressure-sensitive electrical property.

1.1 (b) Temperature

The existence of the 'temperature' of a sample, and its measurement, depends on the validity of a generalization called the *Zeroth Law of thermodynamics*. This states that *if a system* A *is in thermal equilibrium with a system* B (in the sense that no change occurs when they are in thermal contact), *and if* B *is in thermal equilibrium with* C, *then* C *is also in equilibrium with* A whatever the composition of the systems. The Zeroth Law implies the existence of a property that is independent of the composition of a system and that signifies the existence of a condition of thermal equilibrium. We call this property the system's *temperature*. The Zeroth Law also guarantees that we can construct a device from any material and be confident that a particular property (such as the length of a column of mercury or the resistance of a length of wire) will give the same reading when it is in contact with any of the systems A, B, C, etc. that are in mutual thermal equilibrium (that 'have the same temperature').

The relation of the numerical value of the temperature to the property selected for monitoring it is arbitrary, but the common sense of practical workers in laboratories has resulted in the adoption of a scale that can be readily used and sharpened into precise significance. In the early days of thermometry, temperatures were related to the length of a column of liquid, and the difference in lengths shown when the thermometer was first in contact with melting ice and then with boiling water was divided into 100 steps, the lowest being labelled 0. That led to the *Celsius scale*. Different liquids, though, exhibit different expansion behaviour, and so thermometers constructed from different materials showed different numerical values of temperature. Thus, while the systems A, B, C, . . . may all have been ascribed the temperature 28.7 °C when a mercury-in-glass thermometer was used, they may have been ascribed 28.8 °C when an alcohol-in-glass thermometer was used. The variation of quoted temperatures is even greater when electrical measurements are included. One kind of material, however, gives rise to a temperature scale that is almost independent of the identity of the monitoring substance: the material is gas, the monitored property the pressure it exerts at constant volume. Furthermore, this approximate uniformity tends to exact uniformity as the density of the measuring gas is reduced, and is exact in the limit of zero density. This leads to the *perfect gas temperature scale*, which we shall define shortly.

The suspicion will arise that the measurement of temperature is based on a cyclic argument: temperature measurement depends on the properties of the idealization we refer to as a perfect gas; perfect gases show a simple temperature dependence. The argument *is* cyclic, but in due course we shall see that a temperature scale can be established independently of the properties of any class of substances, and of a perfect gas in particular. Happily, this *thermodynamic temperature scale* coincides with the perfect gas scale, and its introduction breaks the cycle. For the present, therefore, we regard the perfect gas scale as a common-sense refinement of practical temperature scales, and return to the rigorous discussion in Chapter 7.

1.1 (c) The gas laws

The experiments of Boyle, Gay-Lussac, and their successors showed that the pressure (p), volume (V), temperature (T), and amount of substance (n) of gases are related by the expression

$$pV = nRT, \qquad (1.1.1)^\circ$$

and that this expression is obeyed increasingly closely as the density is decreased. R, the *gas constant*, is a fundamental constant independent of the identity of the gas. A gas that obeys eqn (1.1.1) exactly is called a *perfect* or *ideal* gas. The equation is called a *limiting law* for the description of real gases. When an equation applies only to a perfect gas (including real gases behaving perfectly) a superscript ° will be attached to the equation number.

The perfect gas equation is one example of an *equation of state*. It is impossible to force a given amount of perfect gas into a state of pressure, volume, and temperature that does not satisfy this relation. We shall look briefly at its implications, and then interpret them in terms of molecules.

1.1 (d) Response to pressure: Boyle's Law

In 1661 Robert Boyle, acting on the suggestion of his assistant John Townley, verified that at constant temperature *the volume of a fixed amount of gas is inversely proportional to the pressure*:

$$\text{Boyle's Law: } V \propto 1/p, \text{ or } pV = \text{constant} \quad (\text{for constant } n, T). \qquad (1.1.2)^\circ$$

The p, V relation is illustrated in Fig. 1.1. Each curve corresponds to a single temperature and is called an *isotherm*. The early experiments were crude, and we now know that gases obey this law only in the limit of $p \to 0$.

Boyle's Law is used to predict the pressure of a gas when its volume is changed (or vice versa). If the initial values of pressure and volume are p_i and V_i, then because the product pV is a constant, the final values must satisfy

$$p_f V_f = p_i V_i \quad (\text{for constant } n, T). \qquad (1.1.3)^\circ$$

The molecular explanation of the law is based on the view that the pressure exerted by a gas arises from the impact of its particles on the walls of the vessel. If the volume is halved the density of particles is doubled, and so twice as many particles strike the walls in a given period of time. The

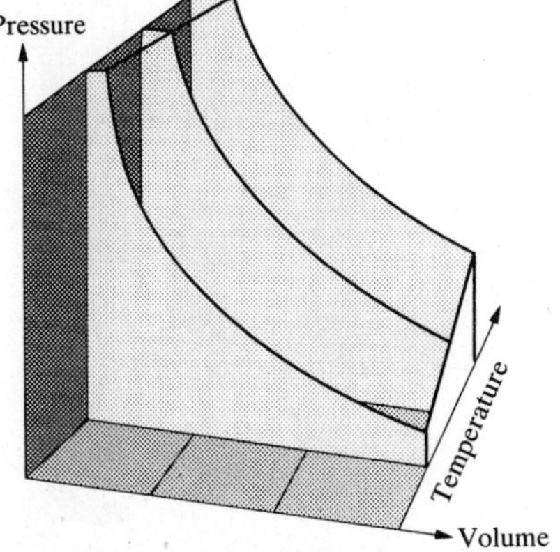

Fig. 1.1. The pressure–volume dependence of a perfect gas at three different temperatures (but same amount). Each curve is a hyperbola ($p \propto 1/V$) and is called an *isotherm*.

average force exerted by the gas is therefore doubled, and therefore so is the pressure it exerts, in accord with Boyle's Law. At very low densities the particles are so far apart that they exert negligible forces on each other on average, which accounts for the fact that the law is *universal* in the sense of applying to any gas without reference to its chemical composition.

1.1 (e) Response to temperature: The law of Gay-Lussac and Charles

The thermal expansion of gases was studied first by Jacques Charles (1787), the inventor of the hydrogen balloon (once known as the *Charlière*). He measured the temperature dependence of the volume of a fixed amount of gas, but did not publish his results. Joseph Gay-Lussac (1802) studied the dependence in greater detail. His observations led him to conclude that, at constant pressure, *the volume of a fixed amount of gas is proportional to its temperature*, and that, at constant volume, *the pressure of a fixed amount of gas is proportional to its temperature*, Fig. 1.2:

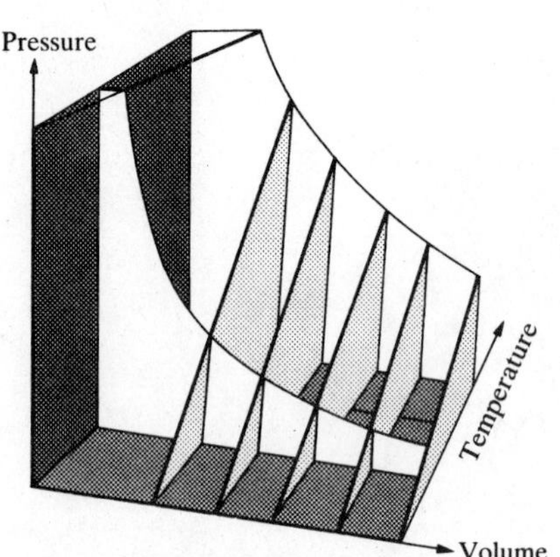

Fig. 1.2. The pressure–temperature dependence of a perfect gas at different volumes (but same amount). The straight lines ($p \propto T$) are called *isochores*.

$$\text{Gay-Lussac's Law: } V \propto T \quad (\text{for constant } n, p), \qquad p \propto T \quad (\text{for constant } n, V). \qquad (1.1.4)^\circ$$

Gay-Lussac's Law is used to predict the volume of a perfect gas as a fixed amount is heated (or cooled) at constant pressure. Equation (1.1.4) gives

$$V_f = (T_f/T_i)V_i \quad (\text{for constant } n, p). \qquad (1.1.5)^\circ$$

The alternative version is used to predict the pressure when a fixed amount is heated at constant volume:

$$p_f = (T_f/T_i)p_i \quad (\text{for constant } n, V). \qquad (1.1.6)^\circ$$

Example 1.1

In an industrial process, nitrogen is heated to 500 K in a vessel of constant volume. If it enters the vessel at a pressure of 100 atm and a temperature of 300 K, what pressure does it exert at the working temperature?

- *Method*. In the absence of more detailed instructions, assume that the gas behaves perfectly, and use eqn (1.1.6).
- *Answer*. $p_f = (500\ \text{K}/300\ \text{K}) \times (100\ \text{atm}) = 167\ \text{atm}$.
- *Comment*. Experiment shows that the pressure is actually 183 atm under these conditions, and so the assumption that the gas is perfect leads to a 10 per cent error.
- *Exercise*. What temperature would result in the same sample exerting a pressure of 300 atm? [900 K]

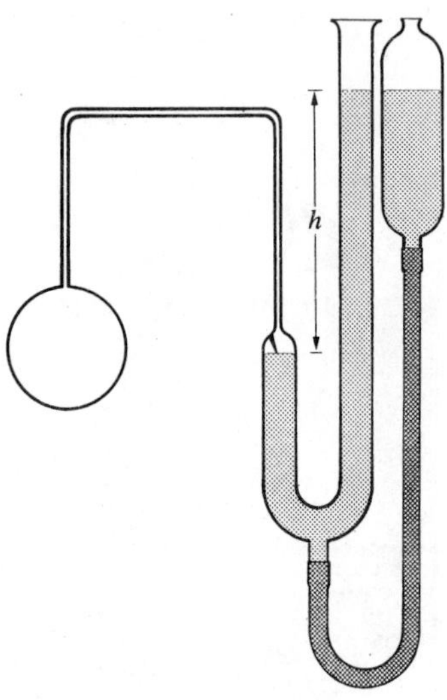

Fig. 1.3. A constant-volume gas thermometer. The bulb is put in contact with the sample and the pressure of the enclosed gas is measured in terms of the height *h* of the mercury column.

The temperature dependence of the pressure of a gas allows us to set up a temperature scale and to measure temperatures without having to rely on the vagaries of the properties of liquids in capillaries. Since a real gas behaves perfectly in the limit of zero pressure, a temperature can be measured with a *constant volume gas thermometer*, Fig. 1.3, by comparing the pressures when it is in thermal contact first with the sample of interest and then with a standard. The latter is taken as water at its *triple point* (the unique condition of temperature and pressure when ice, liquid water, and water vapour coexist at equilibrium), and its temperature is defined as

273.16 K exactly. (The freezing temperature of water under a pressure of 1 atm, the zero on the Celsius scale, is then observed experimentally to lie at 273.1500 ± 0.0003 K.) If the pressure measured when the gas thermometer is in contact with the sample is p, and the pressure when it is at the temperature of the triple point of water, T_3^*, is p_3, then the temperature of the sample is $T \approx (p/p_3)T_3^*$. This is exact only when the gas is behaving perfectly, and so readings are taken with decreasing amounts of gas in the thermometer, and the results extrapolated to zero pressure. The *perfect gas temperature*, which is identical to the *thermodynamic temperature*, Section 7.5, is then given by

$$T = \lim_{p\to 0} T(p), \quad \text{with} \quad T(p) = (p/p_3)T_3^* \quad \text{and} \quad T_3^* = 273.16\ \text{K}. \tag{1.1.7}$$

Ordinary, imperfect, but easier to use thermometers may then be calibrated against this measurement.

The molecular explanation of Gay-Lussac's Law lies in the fact that raising the temperature of a gas increases the average speed of its particles. If the volume is held constant, then at higher temperatures the particles collide with the walls more frequently, and do so with greater impact. Therefore they exert on them a greater average kinetic force, and hence a greater pressure. The reason why the pressure increases *linearly* with temperature lies in the dependence of the average speed of particles on the square root of the temperature, Section 0.2(e). The frequency of collisions and their impact *both* increase in proportion to the speed; consequently the pressure increases as $(\sqrt{T})^2$, or as T itself. This qualitative argument is made quantitative in the opening pages of Part 3.

1.1 (f) Avogadro's hypothesis

The only remaining part of the gas laws concerns the dependence on the amount of gas present. Amedeo Avogadro proposed that *equal volumes of gases at the same temperature and pressure contain the same number of particles*. Since the number of particles is proportional to n, the amount of substance, it follows that at fixed temperature and pressure the volume occupied by a gas is proportional to n:

$$V \propto n \quad \text{(for constant } p, T\text{)}. \tag{1.1.8}^\circ$$

1.1 (g) Collecting the fragments: the gas constant

The experimental observations that $V \propto 1/p$, $V \propto T$, and $V \propto n$ can be combined into $pV \propto nT$, or $pV_m \propto T$, where V_m is the *molar volume* (the volume occupied per unit amount of gas: $V_m = V/n$). We need the constant of proportionality, the *gas constant*, R, in the full equation, eqn (1.1.1). This may be obtained from the value of pV/nT for a real gas in the limit of zero pressure (so as to guarantee that it is behaving perfectly). Therefore, measurements of the pressure and volume of a known amount of gas are made at a series of pressures, and the products pV are extrapolated to zero pressure. The result is that $R = 0.082\,057\,5\ \text{dm}^3\ \text{atm}\ \text{K}^{-1}\ \text{mol}^{-1}$, or $8.314\,41\ \text{J}\ \text{K}^{-1}\ \text{mol}^{-1}$. The implication of this result is that the molar volume of a perfect gas at 25 °C (298.15 K) and 1 atm (101.325 kPa, $1.013\,25 \times$

$10^5\,\mathrm{N\,m^{-2}}$) is

$$V_m = RT/p = \frac{(8.314\,41\,\mathrm{J\,K^{-1}\,mol^{-1}}) \times (298.15\,\mathrm{K})}{1.013\,25 \times 10^5\,\mathrm{N\,m^{-2}}}$$

$$= 2.4465 \times 10^{-2}\,\mathrm{m^3\,mol^{-1}}, \quad \text{or} \quad 24.465\,\mathrm{dm^3\,mol^{-1}}.$$

Two sets of conditions are used as 'standard' values for reporting data. One is *standard temperature and pressure*, or STP, which corresponds to 0 °C and 1 atm. At STP the molar volume of a perfect gas is $22.414\,\mathrm{dm^3\,mol^{-1}}$. The most recent suggestion is that STP should be modernized to *standard ambient temperature and pressure*, or SATP, which corresponds to 298.15 K and 1 bar. At SATP the molar volume of a perfect gas is $24.789\,\mathrm{dm^3\,mol^{-1}}$.

1.2 Mixtures of gases: partial pressures

So far we have dealt only with pure gases. This is too restrictive for chemical considerations and so we now turn to mixtures.

1.2 (a) Dalton's Law

Suppose an amount of substance n_A of some gas A occupies a container of volume V; then according to the perfect gas law its pressure is $p_A = n_A(RT/V)$. If an amount n_B of another gas B occupies a container of the same volume and at the same temperature its pressure is $p_B = n_B(RT/V)$. But what would be the total pressure if gas B is injected into the container already containing gas A?

In the nineteenth century John Dalton‡ made observations which answer this question. *Dalton's law of partial pressures* states that the *pressure exerted by a mixture of gases behaving perfectly is the sum of the pressures exerted by the individual gases occupying the same volume alone*. This means that in the present example the total pressure is

$$p = p_A + p_B = (n_A + n_B)(RT/V). \qquad (1.2.1)^\circ$$

If the mixture consists of several gases A, B, C, . . . present in the amounts $n_A, n_B, n_C, \ldots$, then the total pressure is

$$p = p_A + p_B + p_C + \ldots, \qquad (1.2.2)^\circ$$

where the *partial pressure* p_J of each component J is

$$p_J = n_J(RT/V). \qquad (1.2.3)^\circ$$

Example 1.2

1.00 mol N_2 and 3.00 mol H_2 are in a container of volume $10.0\,\mathrm{dm^3}$ at 298 K. What are the partial pressures and the total pressure?

• *Method*. Assume the gases are perfect; calculate p_J using eqn (1.2.3) and the total pressure from eqn (1.2.2).

‡ Dalton is the Dalton of the atomic hypothesis, and also the Dalton of *daltonism*, or colour-blindness (from which he suffered and which he described). He was described as 'an indifferent experimenter, and singularly wanting in the language and power of illustration'. He paid another price: 'Into society he rarely went, and amusement he had none, with the exception of a game of bowls on Thursday afternoons'. Was that the source of the atomic hypothesis?

• *Answer*. $RT/V = 2.45$ atm mol^{-1} under the stated conditions; hence, $p(N_2) = (1.00\text{ mol}) \times (2.45\text{ atm mol}^{-1}) = 2.45$ atm. Likewise, $p(H_2) = (3.00\text{ mol}) \times (2.45\text{ atm mol}^{-1}) = 7.35$ atm. The total pressure is therefore $p = (2.45 + 7.35)$ atm $= 9.80$ atm.

• *Comment*. The precise value of RT at 298.15 K is 2.4789 kJ mol^{-1}, or 24.465 dm^3 atm mol^{-1}. This temperature, which we shall often refer to as '25 °C', is often chosen as a standard. Values of related quantities are listed inside the front cover.

• *Exercise* 1.00 mol N_2 and 2.00 mol O_2 were added to the same container (with the nitrogen and hydrogen still inside). Calculate the partial pressures and the total pressure at 298 K.

[N_2: 4.90 atm; H_2: 7.35 atm; O_2: 4.90 atm; $p = 17.2$ atm]

Another way of expressing Dalton's Law is in terms of the *mole fraction* of each component J. If the total amount is $n = n_A + n_B + \ldots$, then the mole fraction of J is defined as $x_J = n_J/n$, and can be interpreted as the number of J particles present expressed as a fraction of the total number of particles. When no J particles are present $x_J = 0$; when only J particles are present $x_J = 1$. It follows from the definition that whatever the composition of the mixture, the sum of the mole fractions of all the components is unity: $x_A + x_B + \ldots = 1$. Equation (1.2.3) may be expressed in terms of the mole fraction by writing $n_J = x_J n$, then

$$p_J = n_J(RT/V) = x_J nRT/V = x_J p. \qquad (1.2.4)°$$

That is, the *partial pressure of a component is proportional to its mole fraction*. Figure 1.4 shows how the partial pressures of a two-component mixture contribute to the total pressure as the mole fraction of one component increases from 0 to 1. In the case of real gases the relation $p_J = x_J p$ serves as the *definition* of the partial pressure even when the gas is not behaving perfectly. Then, however, we cannot interpret the partial pressure as the pressure the gas would exert if it were alone in the container.

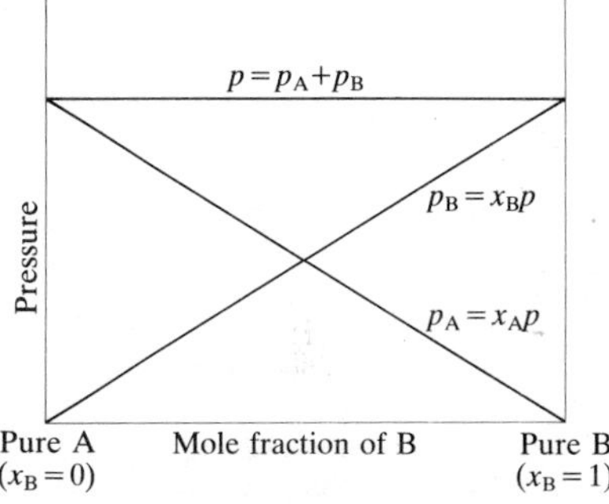

Fig. 1.4. The partial pressures p_A and p_B of a two-component mixture of (real or perfect) gases of total pressure p as the composition changes from pure A to pure B. The sum of the partial pressures is equal to the total pressure.

Example 1.3

The composition of dry air at sea level is approximately: N_2 75.5; O_2 23.2; Ar 1.3, in percentage by mass. What is the partial pressure of each component when the total pressure is 1.000 atm?

• *Method*. Use eqn (1.2.4) in the form $p_J = x_J p$. The mole fractions must first be calculated. Percentage composition by mass is converted to mole fraction as follows. The mass percentage of 75.5 for nitrogen means that in a sample of mass 100.0 g the mass of nitrogen is 75.5 g, and therefore that the amount of N_2 is $(75.5\text{ g})/M_m$, where M_m is the molar mass of N_2 (28.02 g mol^{-1}, data from the Periodic Table inside the back cover). Repeat this calculation for each component, and find the total amount of substance present (n). Form the mole fractions using $x_J = n_J/n$.

• *Answer*. The amounts present in 100 g of air are

$$n(N_2) = (75.5\text{ g})/(28.02\text{ g mol}^{-1}) = 2.69_4\text{ mol}$$
$$n(O_2) = (23.2\text{ g})/(32.00\text{ g mol}^{-1}) = 0.725\text{ mol}$$
$$n(\text{Ar}) = (1.3\text{ g})/(39.95\text{ g mol}^{-1}) = 0.032\text{ mol}.$$

The total amount is $n = 3.45_1$, and so the mole fractions and partial pressures are as follows:

	N_2	O_2	Ar
Mole fraction:	0.781	0.210	0.0093
Partial pressure/atm	0.781	0.210	0.0003

• *Comment*. We have not assumed that the gases are perfect: partial pressures are *defined* as $p_J = x_J p$ for any gas.

• *Exercise*. When carbon dioxide is taken into account the percentages are 75.52, 23.15, and 1.28 for N_2, O_2, and Ar and 0.046 for CO_2. What are the partial pressures when the total pressure is 0.900 atm? [0.703, 0.189, 0.0084, 0.0003 atm]

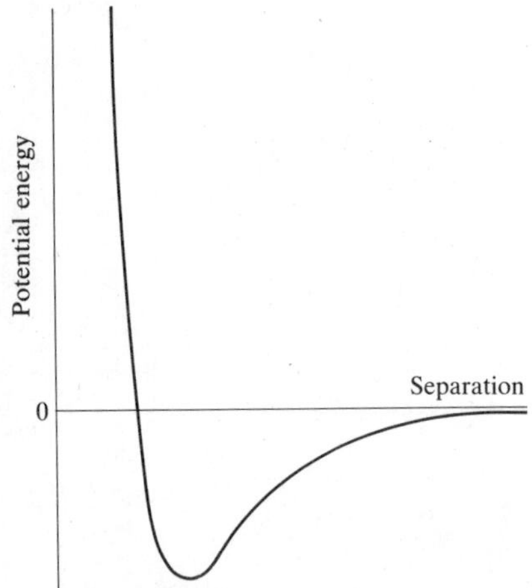

Fig. 1.5. The dependence of the potential energy of two atoms on their separation. High positive potential energy (at very small separations) indicates that the interactions between them are strongly repulsive. At intermediate distances, where the potential energy is negative, the interactions are attractive. At great distances the potential energy is zero and there are no interactions between them.

1.3 Incorporating imperfections

If a gas does not behave perfectly, its equation of state differs from the simple perfect gas law. We know that there are deviations from ideality, because a gas cannot be cooled to zero volume or squashed out of existence by the application of pressure. We also know that under some conditions gases turn into liquids and solids.

1.3 (a) Molecular interactions

The molecular origin of the deviations from ideality is the interaction between particles. Molecules and atoms are small but not infinitesimal, and they resist being squeezed together. We might therefore be tempted to draw the conclusion that it is more difficult to compress a real gas than a perfect gas. But we must remember that molecules also attract each other (that is the reason why liquids and solids form), and that attractions between molecules favour compression.

Which effect triumphs? Is a real gas more difficult to compress on account of the repulsions between molecules, or is it easier to compress on account of their attractions?

The key to the problem is that the repulsive forces come into operation only when the particles are almost in contact: *repulsive forces are short-range interactions*, even on a scale measured in molecular diameters, Fig. 1.5. They can be expected to be important only when the particles are close together on average, which means at high densities and pressures. *Attractive forces have a relatively long range*, and are effective over several molecular diameters. Therefore, they are important when on average the particles are fairly close but not necessarily touching (at intermediate separations in Fig. 1.5). Even they, however, are ineffective when the particles are far apart on average (well to the right in Fig. 1.5). It follows that at moderate densities the attractive forces dominate the repulsive, and that a real gas can be expected to be more readily compressed than a perfect gas. At high densities, however, the repulsive forces dominate, and the gas can be expected to be less readily compressible. At very low densities the particles are so far apart on average that the intermolecular forces play no significant role, and the gas behaves perfectly.

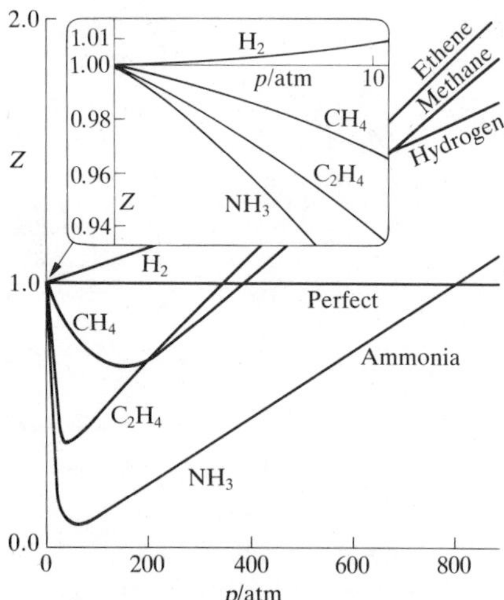

Fig. 1.6. The pressure dependence of the compression factor $Z = pV_m/RT$ for several gases at 0 °C. Note that a perfect gas has $Z = 1$ at all pressures.

That real gases behave in this way can be demonstrated by plotting the *compression factor*, $Z = pV_m/RT$, against pressure: for a perfect gas $Z = 1$ under all conditions, and so deviation from unity is a measure of imperfection. Some results are plotted in Fig. 1.6. At very low pressures all the gases shown have a compression factor close to unity, and so are behaving nearly perfectly. At high pressures all the gases have $Z > 1$, signifying that they are harder to compress than a perfect gas (the product pV_m is greater than RT): repulsive forces are now dominant. At intermediate pressures some of the gases have $Z < 1$, indicating that the attractive forces are dominating and favouring compression. For many gases marked deviations of Z from unity occur only at reasonably high pressures: this is pleasing because it shows how well the perfect gas law applies to many gases under normal conditions. But some common gases (ammonia, for instance) show large deviations even at low pressures. This emphasizes that we must be prepared to deal with gas imperfections if we want to make useful predictions.

1.3 (b) Virials and isotherms

The next step in the analysis of imperfections is to turn to the entire p, V, T-behaviour of real gases. Figure 1.7 shows some experimental isotherms for carbon dioxide. At low densities and high temperatures the real and perfect isotherms do not differ greatly. Nevertheless, there are differences, which suggests that the perfect gas law is only the first term of a more complicated expression of the form

$$pV_{\mathrm{m}} = RT\{1 + B'p + C'p^2 + \ldots\}. \tag{1.3.1}$$

In many applications a more convenient expansion is

$$pV_{\mathrm{m}} = RT\{1 + B/V_{\mathrm{m}} + C/V_{\mathrm{m}}^2 + \ldots\}. \tag{1.3.2}$$

These expressions are the *virial equations of state* (the name coming from the Latin word for force). The coefficients B, C, etc. are known as the *second*, *third*, etc. *virial coefficients*. The third virial coefficient is usually less important than the second.

Some values of the second virial coefficient are given in Table 1.1‡. The coefficients depend on the temperature, and depending on the gas there may be a temperature at which $B' = 0$. In that case $pV_{\mathrm{m}} \approx RT$ over an extended pressure range because $C'p^2$ and higher terms are negligibly small. This temperature, at which the gas behaves almost perfectly, is called the *Boyle temperature*, T_{B}. For helium $T_{\mathrm{B}} = 22.64$ K; for air $T_{\mathrm{B}} = 346.8$ K; some more values are given in Table 1.2.

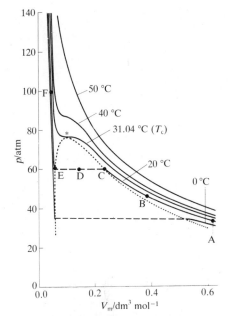

Fig. 1.7. The experimental isotherms for carbon dioxide at several temperatures. The *critical isotherm* is at 31.04 °C, and the *critical point* is marked by a star.

Table 1.1. Second virial coefficients, $B/\mathrm{cm^3\,mol^{-1}}$

	273 K	600 K
Ar	−21.7	11.9
Xe	−153.7	−19.6
N_2	−10.5	21.7
CO_2	−149.7	−12.4

Table 1.2. Critical constants

	p_c/atm	$V_{\mathrm{m,c}}/\mathrm{cm^3\,mol^{-1}}$	T_c/K	Z_c	T_{B}/K
He	2.26	57.8	5.2	0.305	22.6
Ar	48.0	75.3	150.7	0.292	411.5
N_2	33.5	90.1	126.2	0.292	327.2
CO_2	72.9	94.0	304.2	0.274	714.8

The virial equations can be used to make another important point. Consider the slope of the compression factor versus pressure curves. For a perfect gas $\mathrm{d}Z/\mathrm{d}p = 0$, but for a real gas

$$\mathrm{d}Z/\mathrm{d}p = B' + 2pC' + \ldots,$$

and in the limit of zero pressure this equals B', which is not necessarily zero (except at the Boyle temperature). Therefore, although the compression

‡ The tables in the text are abbreviated versions of the ones collected on the tinted pages at the end of the book.

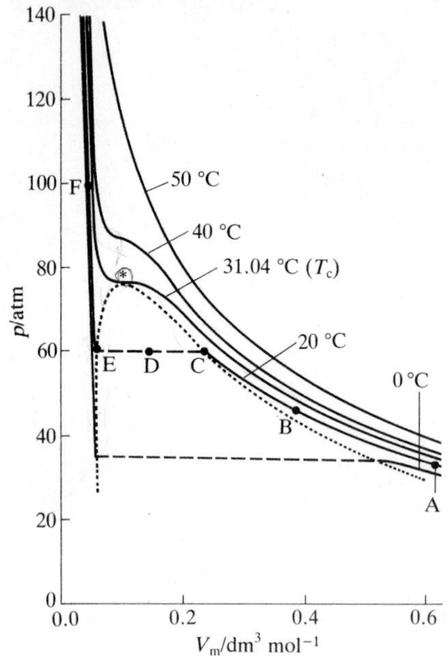

Fig. 1.7. The experimental isotherms for carbon dioxide at several temperatures. The *critical isotherm* is at 31.04 °C, and the *critical point* is marked by a star. (Repeated from p. 27.)

factor (and in general, the equation of state) of a real gas coincides with the perfect gas law as the pressure approaches zero, the slope of the graph does not. Other properties (as we shall see) also depend on derivatives like these, and so in general the *properties* of real gases do not always coincide with the perfect gas values at low pressures.

Although deviations from ideality may be small under many conditions, the complicated shape of the isotherms in Fig. 1.7 shows that considerable deviations do occur under other conditions. We shall now investigate the significance of the sharp deflections in the isotherms shown in the illustration.

Consider what happens when the volume of a sample of gas initially in the state marked A in Fig. 1.7 is decreased at constant temperature (e.g. by pushing in a piston). In the region close to A the pressure of the gas rises in approximate agreement with Boyle's Law. Serious deviations from that law begin to appear when the volume has been reduced to B. At C, which corresponds to about 60 atm in the case of carbon dioxide, all similarity to perfect behaviour is lost, for suddenly the piston slides in without any further rise in pressure: this is represented by the horizontal line CDE. Examination of the contents of the vessel shows that just to the left of C a liquid appears, and there are two phases separated by a sharply defined surface (which, in a narrow bore tube, is curved and called the *meniscus*). As the volume is decreased from C through D to E the amount of liquid increases. It is because the gas can respond by condensing that there is no additional resistance to the piston at this stage. The pressure corresponding to the line CDE is called the *vapour pressure* of the liquid at the temperature of the experiment: the liquid and the gas are in *equilibrium*.

At E all the gas has disappeared, the sample is entirely liquid, and the piston is resting on its surface. Any further reduction of volume requires the exertion of considerable pressure. This is reflected by the sharply rising line to the left of E. Even a small reduction of volume from E to F requires a great increase in pressure.

1.3 (c) Critical constants

The isotherm at the temperature T_c (31.04 °C for CO_2) plays a special role in the theory of the states of matter. An isotherm at a fraction of a kelvin below T_c behaves as we have already described: at some pressure a liquid condenses from the gas and is distinguishable from it by the presence of a visible surface. The phase separation occurs at a definite pressure, and even though the volume at which all the gas has been liquefied is only slightly less than the volume at which the liquid first appears, a separating surface is briefly visible. If, however, the compression takes place at T_c itself, a surface separating two phases does not appear and the volumes at each end of the horizontal part of the isotherm have merged to a single point, the *critical point* of the gas.

At and above T_c, the *critical temperature*, a liquid phase does not form. The pressure and molar volume at the critical point are called the *critical pressure*, p_c, and *critical molar volume*, $V_{m,c}$. Taken together, p_c, $V_{m,c}$, and T_c are referred to as the *critical constants* of the gas. Some values are given in Table 1.2. The value for nitrogen, for instance, signifies that it is impossible to produce liquid nitrogen by compression alone if its temperature is greater than 126.2 K: to liquefy it the temperature must first be lowered to below 126.2 K, and then the gas compressed.

Conclusions can be drawn from the virial equations of state only by inserting specific values of the coefficients. That is often too specialized, and tells us about only one gas at a time. We shall find it useful to have a broader, if less precise, view of all gases. Therefore, we introduce the approximate equation of state suggested by Johannes van der Waals. It is an excellent example of an equation that can be obtained by thinking scientifically about a mathematically complicated but physically simple problem. Van der Waals himself proposed it on the basis of experimental evidence available to him in conjunction with rigorous thermodynamic arguments.

Our aim is to find a simple expression to serve as an approximate equation of state of a real gas. The repulsive interactions between particles are taken into account by supposing that they cause the particles to behave as small but impenetrable spheres; the attractive forces are taken into account by supposing that they reduce the pressure exerted by the gas.

1.4 (a) Constructing the equation

The non-zero volume of the particles implies that instead of moving in a volume V they are restricted to a smaller volume $V-nb$, where nb is approximately the total volume taken up by the particles themselves. This suggests that the perfect gas law $pV=nRT$ should be replaced by

$$p(V-nb)=nRT, \quad \text{or} \quad p=nRT/(V-nb).$$

The attractive interactions hold the particles together and so reduce the pressure they exert. We have remarked that the pressure depends on both the frequency of collisions with the walls and the impulse of each collision. *Both* contributions are reduced by the attractive forces. Since the strength with which these forces act is roughly proportional to the density of particles, the pressure is reduced in proportion to the *square* of the density. If the reduction of pressure is written as $a(n/V)^2$, where a is a constant characteristic of each gas, then the combined effect of the repulsive and attractive forces is

$$p=nRT/(V-nb)-a(n/V)^2. \tag{1.4.1a}$$

This *van der Waals equation of state* is often reorganized into a form resembling $pV=nRT$:

$$(p+an^2/V^2)(V-nb)=nRT. \tag{1.4.1b}$$

For many purposes it is convenient to express it in terms of the molar volume $V_m=V/n$:

$$p=RT/(V_m-b)-a/V_m^2. \tag{1.4.2}$$

a/V_m^2 is called the *internal pressure* of the gas.

Example 1.4

Estimate the molar volume of nitrogen at 500 K and 100 atm by treating it as a van der Waals gas.

Table 1.3. Van der Waals constants

	a/dm^6 atm mol^{-2}	$b/10^{-2}$ dm^3 mol^{-1}
He	0.034	2.37
Ar	1.345	3.22
N_2	1.390	3.913
CO_2	3.592	4.267

• *Method*. Rearrange eqn (1.4.2) to an equation for V_m and solve it numerically (or graphically) for the stated conditions, using data from Table 1.3.

• *Answer*. The equation rearranges to the following cubic equation:

$$V_m^3 - (b + RT/p)V_m^2 + (a/p)V_m - (ab/p) = 0.$$

According to Table 1.3, $a = 1.390$ dm^6 atm mol^{-2} and $b = 3.913 \times 10^{-2}$ dm^3 mol^{-1}. Therefore, the coefficients in this equation are

$$b + RT/p = 0.449_4 \text{ dm}^3 \text{ mol}^{-1}$$
$$a/p = 1.39_0 \times 10^{-2} (\text{dm}^3 \text{ mol}^{-1})^2$$
$$ab/p = 5.439_1 \times 10^{-4} (\text{dm}^3 \text{ mol}^{-1})^3.$$

Then, on writing $x = V_m/\text{dm}^3 \text{ mol}^{-1}$, we must solve

$$x^3 - 0.449_4 x^2 + 1.39_0 \times 10^{-2} x - 5.439_1 \times 10^{-4} = 0.$$

The most elementary way of solving this cubic equation is by successive approximation starting from the perfect gas value $x = 0.410$: we find $x = 0.419_3$, which implies that $V_m = 0.419$ dm^3 mol^{-1}.

• *Comment*. Cubic equations can be solved analytically: the formula is given in Section 3.8.2 of *Handbook of mathematical functions*, M. Abramowitz and I. Stegun, Dover (1965), a rich source of this kind of information.

• *Exercise*. Calculate the molar volume of carbon dioxide at 100 °C and 100 atm pressure on the assumption that it is a van der Waals gas. [0.215 dm^3 mol^{-1}]

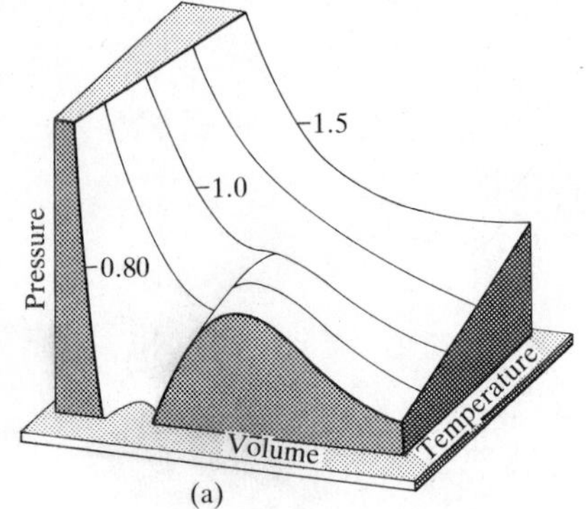

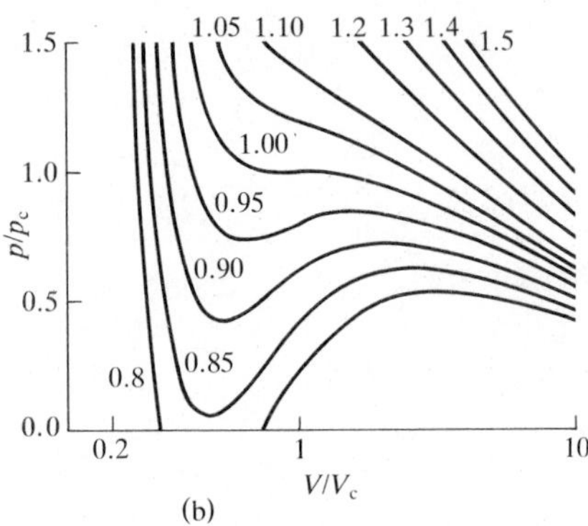

Fig. 1.8. The calculated van der Waals isotherms at several values of T/T_c. The solid diagram (a) shows the shape of the surface (compare it with the perfect gas surfaces starting to appear in Figs 1.1 and 1.2). The individual isotherms are shown in (b). The van der Waals' loops are normally replaced by horizontal straight lines. The critical isotherm is the one for $T/T_c = 1$.

We have built the van der Waals equation using vague arguments about particle volume and the effects of forces. It can be derived in other ways, but the present method has the advantage that it shows how to force the form of an equation out of a general idea. The derivation also has the advantage of keeping imprecise the significance of the coefficients a and b: they are much better regarded as adjustable parameters than as precisely defined molecular properties.

1.4 (b) The features of the equation

We now examine to what extent the van der Waals equation resembles the true equation of state, whatever that may be. We do so by comparing the isotherms it predicts with the experimental isotherms in Fig. 1.7. Some of the calculated isotherms are shown in Fig. 1.8, and apart from the peculiar oscillations (which are absent at high temperatures) they do resemble experimental isotherms. The oscillations, the *van der Waals loops*, are unrealistic because they suggest that under some conditions an increase of pressure results in an increase of volume. Therefore they are cut out and replaced by horizontal lines drawn so that the loops define equal areas above and below the lines (this is the *Maxwell construction*). The values of the coefficients (found by fitting the calculated curves to the experimental) for some gases are listed in Table 1.3.

The principal features of the van der Waals equation are as follows:

(1) *Perfect gas isotherms are obtained at high temperatures and low densities*. When the temperature is high RT may be so large that the first term in eqn (1.4.2) greatly exceeds the second. Furthermore, if the density is low, so that the molar volume is large, the denominator $V_m - b$ in the first term may be replaced by V_m itself. Hence, the equation reduces to $p = RT/V_m$, which is the perfect gas equation.

(2) *Liquids and gases coexist when cohesive and dispersing effects are in balance*. The van der Waals loops occur when both terms in eqn (1.4.2) have similar magnitudes. The first term arises from the kinetic energy of the particles (which is of order RT, Section 0.1(f)) and their repulsive interactions, and the second represents the effect of the attractive interactions.

(3) *The critical point can be located*. Below the critical temperature the isotherms oscillate and each one passes through a minimum followed by a maximum which get closer together as T approaches T_c. At T_c they coincide. Therefore, at the critical temperature the curve has a flat inflexion. From the properties of curves we know that an inflexion of this type occurs when both the slope and the curvature are zero, and so we can find the critical constants by setting the first and second derivatives of the equations equal to zero:

$$\left.\begin{aligned} &\textit{Slope:} && \mathrm{d}p/\mathrm{d}V_m = -RT/(V_m - b)^2 + 2a/V_m^3 = 0 \\ &\textit{Curvature:} && \mathrm{d}^2p/\mathrm{d}V_m^2 = 2RT/(V_m - b)^3 - 6a/V_m^4 = 0 \end{aligned}\right\} \text{ at } p_c, V_c, T_c.$$

Solving these two equations leads to

$$\left.\begin{aligned} V_{m,c} &= 3b, \\ p_c &= a/3V_{m,c}^2 = a/27b^2 \\ T_c &= 8p_cV_{m,c}/3R = 8a/27Rb \end{aligned}\right\} \quad Z_c = p_cV_{m,c}/RT_c = 3/8 = 0.375. \quad (1.4.3)$$

These relations can be tested by seeing whether the *critical compression factor*, Z_c, is equal to 3/8. From Table 1.2 we see that although Z_c is less than 0.375 it is approximately constant (at 0.3) and the discrepancy is gratifyingly small.

(4) *The Boyle temperature can be related to the critical temperature*. The van der Waals equation can be expanded into a virial equation. The first step is to express eqn (1.4.2) in the form

$$p = (RT/V_m)\{1/(1 - b/V_m) - a/RTV_m\}.$$

So long as $b/V_m < 1$ the first term inside the brackets can be expanded using $(1 - x)^{-1} = 1 + x + x^2 + \ldots$, which gives

$$p = (RT/V_m)\{1 + [b - (a/RT)](1/V_m) + \ldots\}.$$

We can immediately identify the second virial coefficient as

$$B = b - a/RT. \quad (1.4.4)$$

At the Boyle temperature $B = 0$, and so

$$T_B = a/bR = 27T_c/8. \quad (1.4.5)$$

1.4 (c) Comparing gases

When the properties of objects are compared an important technique is to choose a related fundamental property of the same kind and to set up a relative scale on that basis. This technique was used in a simple way when we expressed the distances between molecules as so many molecular diameters. We have seen that the critical constants are characteristic properties of gases, and so it may be that a scale can be set up using them as yardsticks. We therefore introduce the *reduced variables* by dividing the actual variable by the corresponding critical constant:

$$\begin{aligned} &\textit{Reduced pressure:} \quad && p_r = p/p_c \\ &\textit{Reduced volume:} \quad && V_r = V_m/V_{m,c} \\ &\textit{Reduced temperature:} \quad && T_r = T/T_c. \end{aligned} \tag{1.4.6}$$

Van der Waals, who first tried this, hoped that gases confined to the same reduced volume at the same reduced temperature would exert the same reduced pressure. The hope was largely fulfilled. Figure 1.9 shows the dependence of the compression factor Z on the reduced pressure for a variety of gases at various reduced temperatures. The success of the procedure is strikingly clear: compare this graph with Fig. 1.6, where similar data are plotted without using reduced variables.

The observation that real gases in the same state of reduced volume and temperature exert the same reduced pressure is called the *principle of corresponding states*. It is only an approximation, and works best for gases composed of spherical particles; it fails, sometimes badly, when the molecules are non-spherical or polar.

The van der Waals equation can shed some light on the principle. We can express it in terms of the reduced variables, which gives

$$p = p_r p_c = \frac{RT_r T_c}{V_r V_{m,c}} - \frac{a}{V_r^2 V_{m,c}^2}.$$

Then we express the critical constants in terms of the coefficients a and b using the relations in eqn (1.4.3):

$$ap_r/27b^2 = \frac{8aT_r}{27b(3bV_r - b)} - \frac{a}{9b^2 V_r^2}.$$

This reorganizes into

$$p_r = \frac{8T_r}{3V_r - 1} - \frac{3}{V_r^2}, \tag{1.4.7a}$$

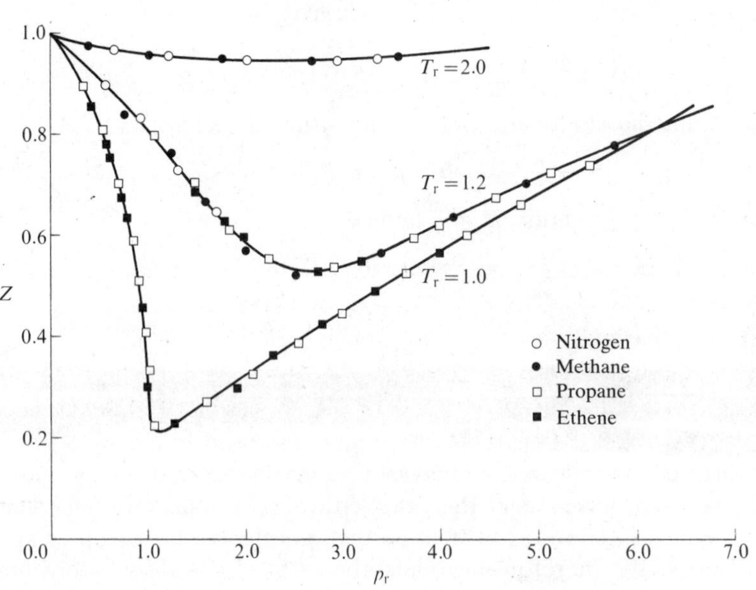

Fig. 1.9. The compression factors of four of the gases shown in Fig. 1.6 plotted in terms of the reduced pressure and temperature. The use of reduced variables organizes the data on to single curves.

or, alternatively,

$$\{p_r + (3/V_r^2)\}\{V_r - \tfrac{1}{3}\} = \tfrac{8}{3}T_r. \quad (1.4.7b)$$

These have the same form as the original equation, but the constants a and b, which differ from gas to gas, have disappeared. It follows that if the equation is plotted in terms of the reduced variables (as we did in fact in Fig. 1.7 without drawing attention to the fact), then the same isotherms should be obtained whatever the gas. This is precisely the content of the principle of corresponding states, and so the van der Waals equation is compatible with it.

Looking for too much significance in this apparent triumph is mistaken, because other equations of state also accommodate the principle. In fact, all we need are two parameters playing the roles of a and b, for then the equation can always be manipulated into reduced form. The observation that real gases obey the principle approximately amounts to saying that the effects of the attractive and repulsive interactions can each be approximated in terms of a single parameter. The importance of the principle is then not so much its theoretical interpretation but the way that it enables the properties of a range of gases to be coordinated on to a single diagram (e.g. Fig. 1.9 instead of Fig. 1.6).

1.4 (d) The status of the van der Waals equation

It is too optimistic to expect a single, simple expression to account for the p, V, T-behaviour of all systems. Accurate work on gases must resort to the virial equation, eqn (1.3.2), and rely on tabulated values of the coefficients at various temperatures. Such a procedure is clumsy and involves a good deal of numerical analysis. The advantage of the van der Waals equation is that it is analytical and allows us to draw some general conclusions about real gases. However, we have to be circumspect: we have to remember that it is an approximation and that under many quite common conditions (such as low temperatures and high densities) it is a poor approximation. When it fails we must use one of the other equations of state that have been proposed (some are listed in Box 1.1), invent a new one, or go back to the virial equation.

Box 1.1 Equations of state

Many people have proposed equations of state for real gases. The criteria of success are accurate representation of observed p, V, T-relations over moderate ranges of conditions, a simplicity of form (so that they can be differentiated and integrated reasonably easily), and the use of only a few adjustable parameters. The following small selection conforms to the first criterion with increasing success on passing down the list, but the number of parameters increases from zero in (i) to five in (v) and to an indefinite number in (vi).

(i) *Perfect gas equation:* $p = RT/V_m$

(ii) *Van der Waals equation:* $p = RT/(V_m - b) - a/V_m^2$
Critical constants: $p_c = a/27b^2$, $V_{m,c} = 3b$, $T_c = 8a/27Rb$, $Z_c = \tfrac{3}{8} = 0.375$.
Reduced form: $p_r = 8T_r/(3V_r - 1) - 3/V_r^2$.

(iii) *Berthelot equation:* $p = RT/(V_m - b) - a/TV_m^2$
Critical constants: $p_c = \frac{1}{12}(2aR/3b^3)^{\frac{1}{2}}$, $V_{m,c} = 3b$, $T_c = \frac{2}{3}(2a/3bR)^{\frac{1}{2}}$, $Z_c = \frac{3}{8} = 0.375$.
Reduced form: $p_r = 8T_r/(3V_r - 1) - 3/T_rV_r^2$.

(iv) *Dieterici equation:* $p = \{RT/(V_m - b)\} \exp(-a/RTV_m)$
Critical constants: $p_c = a/4e^2b^2$, $V_{m,c} = 2b$, $T_c = a/4bR$, $Z_c = 2/e^2 = 0.2706\ldots$
Reduced form: $p_r = \{e^2T_r/(2V_r - 1)\} \exp(-2/T_rV_r)$
(e is the exponential e, not a parameter).

(v) *Beattie–Bridgeman equation:* $p = (1 - \gamma)RT(V_m + \beta)/V_m^2 - \alpha/V_m^2$
$\alpha = a_0(1 + a/V_m)$, $\beta = b_0(1 - b/V_m)$, $\gamma = c_0/V_mT^3$.

(vi) *Virial equation (Kammerlingh Onnes):*
$p = (RT/V_m)\{1 + B(T)/V_m + C(T)/V_m^2 + \ldots\}$.

Further reading

Properties of matter. B. H. Flowers and E. Mendoza; Wiley, London, 1970.
Gases, liquids, and solids (2nd edn). D. Tabor; Cambridge University Press, 1979.
Three phases of matter (2nd edn). A. J. Walton; Clarendon Press, Oxford, 1983.
Determination of pressure and volume. G. W. Thomson and D. R. Douslin; in *Techniques of chemistry* (A. Weissberger and B. W. Rossiter, eds.) V, 23, Wiley-Interscience, New York, 1971.
The measurement of temperature. J. A. Hall; Barnes and Noble, New York, 1966.
Temperature. J. F. Swindells; NBS Special Publication 300, 1968.
Comparisons of equations of state. J. B. Ott, J. R. Coates, and H. T. Hall; *J. chem. Educ.* **48,** 515 (1971).
The virial coefficients of pure gases and mixtures. J. H. Dymond and E. B. Smith; Clarendon Press, Oxford, 1980.
International critical tables. *Vol. 3 (p, V, T data)*. McGraw-Hill, New York, 1928.

Introductory problems

A1.1. A perfect gas undergoes an isothermal compression which reduces its volume by 2.20 dm^3. The final pressure and volume of the gas are 3.78×10^3 Torr and 4.65 dm^3. Calculate the original pressure of the gas in (a) Torr (b) atm.

A1.2. A perfect gas at 340 K is heated at constant pressure until its volume has increased by 18%. What is the final temperature of the gas?

A1.3. A 255 mg sample of neon occupies a volume of 3.00 dm^3 at 122 K. Use the equation of state for perfect gases to calculate the pressure of the gas.

A1.4. A gas mixture consists of 320 mg of methane, 175 mg of argon, and 225 mg of neon. The partial pressure of neon at 300 K is 66.5 Torr. Calculate: (a) the volume of the mixture, (b) the partial pressure of argon, (c) the total pressure of the mixture.

A1.5. The density of a new gaseous compound is found to be 1.23 g dm^{-3} at 330 K and a pressure of 150 Torr. What is the RMM of the new compound?

A1.6. A gas at 250 K and 15 atm pressure has a molar volume 12% smaller than that calculated from the equation of state for perfect gases. Calculate: (a) the compression factor for this T and p, (b) the molar volume of the gas. At this T and p, which forces dominate, attractive or repulsive?

A1.7. At 300 K and 20 atm pressure, the compression factor for a gas is 0.86. Calculate: (a) the volume of 8.2 millimoles of the gas at this T and p, (b) an approximate value for the second virial coefficient, B, at 300 K.

A1.8. The critical volume and critical pressure of a gas are 160 cm^3 mol^{-1} and 40 atm, respectively. Estimate the critical temperature, assuming that the gas follows the Berthelot equation of state. Assume the gas molecules are spheres and calculate the radius of a gas molecule.

A1.9. Calculate the molar volume of Cl_2 at 350 K and 2.30 atm using (a) the perfect gas equation, (b) the van der Waals equation. Use the answer in part (a) to calculate a first approximation to the correction term for attraction and then use successive approximations to get a numerical answer for part (b).

A1.10. Use the van der Waals parameters of Cl_2 to calculate approximate values of: (a) the Boyle temperature of Cl_2, (b) the radius of a Cl_2 molecule.

Problems

1.1. A sample of air occupies 1 dm^3 at room temperature and pressure. What pressure is needed to compress it so that it occupies only 100 cm^3 at that temperature?

1.2. At sea level, where the pressure was 755 Torr, the gas in a balloon occupied 2 m^3. What volume will the balloon expand to when it has risen to an altitude where the pressure is (a) 100 Torr, (b) 10 Torr? Assume that the material of the balloon in infinitely extensible.

1.3. A diving bell has an air space of 3 m^3 when on the deck of a boat. What is the volume of the air space when it has been lowered to a depth of 50 m? Take the mean density of sea water as 1.025 g cm^{-3} and assume that the temperature is the same at 50 m as at the surface.

1.4. What pressure difference must be generated across the length of a vertical drinking straw of length 15 cm in order to drink a water-like liquid? Estimate the expansion of the lungs that is needed in order to create the appropriate partial vacuum at the upper end of the straw. Use $g = 9.81$ m s^{-2}.

1.5. To what temperature must a 1 dm^3 sample of a perfect gas be cooled from room temperature in order to reduce its volume to 100 cm^3?

1.6. A car tyre (i.e. an automobile tire) was inflated to a pressure of 24 lb in^{-2} (1 atm = 14.7 lb in^{-2}) on a winter's day when the temperature was −5 °C. What pressure will be found, assuming no leaks to have occurred, and that the volume is constant, on a subsequent summer's day when the temperature is 35 °C? What complications should be taken into account in practice?

1.7. A meteorological balloon had a radius of 1 m when released from sea level, and expanded to a radius of 3 m when it had risen to its maximum altitude, where the temperature was −20 °C. What is the pressure inside the balloon at that altitude?

1.8. The perfect gas law is a *limiting law*. First deduce the relation between *pressure* and *density* (ρ) of a perfect gas, and then confirm on the basis of the following data for dimethyl ether, CH_3OCH_3, at 25 °C,that perfect behaviour is approached at low pressures. Find the relative molar mass (RMM) of the gas.

p/Torr	91.74	188.98	277.3
$10^3\rho$/g cm^{-3}	0.225	0.456	0.664
p/Torr	452.8	639.3	760.0
$10^3\rho$/g cm^{-3}	1.062	1.468	1.734

1.9. Investigate some of the technicalities of ballooning using the perfect gas law. Suppose your balloon has a radius of 3 m and that it is a sphere when inflated. How much hydrogen is needed to inflate it to a pressure of 1 atm at an ambient temperature of 25 °C at sea level? What mass can the balloon lift at sea level, where the density of air is 1.22 kg m^{-3}? What would be the payload if helium were used instead of hydrogen? With you and a companion on board, the balloon ascends to 30 000 ft where the pressure is 0.28 atm, the temperature −43 °C, and the density of air 0.43 kg m^{-3}. Can you in fact attain that height with both hydrogen and helium? Do you need more hydrogen to rise further? (Leaks are encountered in Problem 26.22.)

1.10. In an experiment to determine the RMM of ammonia, 250 cm^3 of the gas was confined to a glass vessel. The pressure was 152 Torr at 25 °C and after correcting for buoyancy effects, the mass of the gas was 33.5 mg. What is (a) the molecular mass, (b) the RMM of the gas?

1.11. The RMM of a newly synthesized fluorocarbon gas was measured with a *gas microbalance*. This consists of a glass bulb forming one end of a beam, the whole surrounded by a closed container. The beam is pivoted, and the balance point is attained by raising the pressure of gas in the container, and so increasing the buoyancy of the enclosed bulb. In one experiment the balance point was reached when the fluorocarbon pressure was 327.10 Torr and for the same setting of the pivot a balance was achieved when CHF_3 was introduced at a pressure of 423.22 Torr. A repeat of the experiment with a different setting of the pivot required a pressure of 293.22 Torr of the fluorocarbon and 427.22 Torr of the CHF_3. What is the RMM of the fluorocarbon? Suggest a molecular formula. (Take M_r = 70.014 for trifluoromethane.)

1.12. A constant-volume perfect gas thermometer indicates a pressure of 50.2 Torr at the triple point of water (273.16 K). What change of pressure indicates a change of 1 K at this temperature? What pressure indicates a temperature of 100 °C (373.15 K)? What change of pressure indicates a change of temperature of 1 K at the latter temperature?

1.13. The synthesis of ammonia is an important process technologically, and it has features which make it useful for emphasizing and illustrating points made in the text. A simple problem is the following. A vessel of volume 22.4 dm^3 contains 2.0 mol of hydrogen and 1.0 mol of nitrogen at 273.15 K. What is the mole fraction of each component, their partial pressures, and the total pressure?

1.14. The question arose as to what the partial and total pressures would be if the whole of the hydrogen in the last Problem were converted to ammonia by reaction with the appropriate amount of nitrogen. What are these pressures?

1.15. Could 131 g of xenon in a vessel of capacity 1.0 dm^3 exert a pressure of 20 atm at 25 °C if it behaved as a perfect gas? If not, what pressure would it exert?

1.16. Now assume that the xenon in the last Problem behaves as a van der Waals gas (with the constants given in Table 1.3). What pressure will 131 g exert under the same conditions?

1.17. Calculate the pressure exerted by 1 mol of ethene behaving as (a) a perfect gas, (b) a van der Waals gas, when it is confined under the following conditions: (i) at 273.15 K in 22.414 dm^3, (ii) at 1000 K in 100 cm^3. Use the data in Table 1.3, even though those data refer to temperatures around 25 °C.

1.18. Use the data in Table 1.2 to suggest the pressure and temperature that 1 mol of (a) ammonia, (b) xenon, (c) helium will have in states corresponding to 1 mol of hydrogen at 25 °C at 1 atm.

1.19. Estimate the critical constants (p_c, $V_{m,c}$, T_c) of a gas with van der Waals parameters $a = 0.751$ atm dm^6 mol^{-2}, $b = 0.0226$ dm^3 mol^{-1}.

1.20. The critical constants of methane are $p_c = 45.6$ atm, $V_{m,c} = 98.7$ cm^3 mol^{-1}, and $T_c = 190.6$ K. Calculate the van der Waals parameters and estimate the size (volume and radius) of the gas molecules.

1.21. Estimate the radii of the rare-gas atoms on the basis of their critical volumes and the Dieterici equation of state, Box 1.1.

1.22. Estimate the coefficients a and b in the Dieterici equation of state for the critical constants of xenon. Calculate the pressure exerted by 1 mol of the gas when it is confined to 1 dm^3 at 25 °C. (Compare Problem 1.16.)

1.23. Express the van der Waals equation of state as a virial expansion in powers of $1/V_m$ and obtain expressions for B and C in terms of the parameters a and b. Use $1/(1-x) = 1 + x + x^2 + \ldots$.

1.24. Repeat the last Problem for a Dieterici gas.

1.25. The virial equation of state is often a convenient method of expressing both theoretical and experimental results on gases. For instance, measurements on the deviation of argon from ideality give the following virial expansion at 273 K:

$$pV_m/RT = 1 - (21.7\ \text{cm}^3\ \text{mol}^{-1}/V_m) + (1200\ \text{cm}^6\ \text{mol}^{-2}/V_m^2) + \ldots$$

Use the parameters in this expansion to predict the critical constants of argon using the results in the last two Problems and the relations set out in Box 1.1.

1.26. A scientist with a simple view of life proposes the following equation of state for a gas:

$$p = RT/V_m - B/V_m^2 + C/V_m^3.$$

Demonstrate that critical behaviour is accommodated by this equation. Express p_c, $V_{m,c}$, and T_c in terms of B and C, and find an expression for the critical compression factor Z_c.

1.27. The virial expansion can be expressed either as a series in powers of $1/V_m$ or in powers of p, see eqns (1.3.1) and (1.3.2). Express B' and C' in terms of B and C.

1.28. Find an expression for the Boyle temperature T_B in terms of the van der Waals parameters of a gas. Find the temperature at which 1 mol of xenon in a 5 dm^3 vessel has a compression factor of unity.

1.29. Express the Boyle temperature in terms of reduced variables of (a) a van der Waals gas, (b) a Dieterici gas.

1.30. The second virial coefficient B' can be obtained from measurements of the density of a gas at a series of pressures. Show that the graph of p/ρ versus p should be a straight line with slope proportional to B'. Use the data in Problem 1.8 to find B' and B for dimethyl ether. (The data relate to 25 °C.)

1.31. The *barometric formula* relates the pressure of a gas at some height h to its pressure p_0 at sea level (or any other base line). It can be derived in various ways, one involving the Boltzmann distribution, Section 0.1(d). Here we approach it from another viewpoint, and start from the change in pressure dp for an infinitesimal change in height dh, where the density is ρ: $dp = -\rho g dh$. Prove this relation, and then show that it integrates to $p(h) = p_0 \exp(-M_m gh/RT)$ for a perfect gas, where M_m is the molar mass of the gas molecules. The consequences of this expression are explored in the next few Problems.

1.32. One of the consequences of the barometric formula is the extra complication that no gas in any terrestrial laboratory has a uniform pressure. But how serious is the effect of the gravitational field? Find the pressure difference between top and bottom of (a) a laboratory vessel of height 15 cm and (b) the World Trade Center, 1350 ft; ignore temperature variations.

1.33. That (Problem 1.32) is not the only problem. If the gas is a mixture, its local composition depends on the altitude: the heavier molecules tend to sink to the bottom of a column because Mgh depends more strongly on height when M is big. As a first step in unravelling this effect, show that for a mixture of ideal gases the partial pressure of a component J follows $p_J = p_{J,0} \exp(-M_{J,m}gh/RT)$. Then assess the magnitude of the effect on the distortion of the composition of the atmosphere. Deduce the composition of the atmosphere at heights of (a) 1350 ft, (b) 29 000 ft, (c) 100 km in order to answer this sort of question. The sea-level composition is given in Example 1.3. Disregard the extra complication of temperature variation over this range of altitudes, and take it as 20 °C throughout.

1.34. The barometric formula also found application in the early determination of Avogadro's constant. Although the method has been superseded, there is some interest in seeing how simple macroscopic measurements can give values of atomic constants. As a first step, show that the

numbers of particles of mass m at two heights separated by h are in the ratio $\exp(-m'gh/kT)$, where m' is their effective mass in the solvent of density ρ. In an actual experiment, spheres of rubber latex of radius 2.12×10^{-4} mm and density $1.2049\ \text{g cm}^{-3}$ were distributed in water at 20 °C (when its density is $0.9982\ \text{g cm}^{-3}$). The average numbers of spherical particles at various heights were as follows:

h/mm	0	0.05	0.07	0.09	0.10	0.15	0.20
N	1000	400	280	190	160	60	25

Estimate Boltzmann's constant, and hence Avogadro's constant, from the known value of the gas constant R.

2

The First Law: the concepts

Learning objectives

After careful study of this chapter you should be able to:

(1) Define *system*, *surroundings*, and *universe*, and distinguish between a *closed* and an *isolated system*, Section 2.1.

(2) Define *work*, *energy*, and *heat*, Section 2.1(a).

(3) State the *First Law of thermodynamics* in words, Section 2.1(b), and in symbols, eqn (2.1.4).

(4) Explain why the First Law implies the existence of the *internal energy* of a system, Section 2.1(b).

(5) Write the expression for *mechanical work*, eqn (2.2.1), and deduce from it the expression for the work done when a gas expands, eqn (2.2.5).

(6) Obtain and use expressions for the work done when a gas expands against a constant pressure, eqn (2.2.7), and reversibly, eqn (2.2.9).

(7) Calculate the work of *reversible*, *isothermal expansion* of a perfect gas, eqn (2.2.11).

(8) Define thermodynamic *reversibility* and explain why it corresponds to *equilibrium* and the production of *maximum work*, Section 2.1(e).

(9) Define the *heat capacity* of a system, eqn (2.3.1), and state and justify the definitions of *heat capacity at constant volume*, eqn (2.3.3), and *heat capacity at constant pressure*, eqn (2.3.7).

(10) Demonstrate the relation between change of internal energy and *heat transferred at constant volume*, eqn (2.3.2).

(11) Define the *enthalpy* of a system, eqn (2.3.8).

(12) Demonstrate the relation between change of enthalpy and *heat transferred at constant pressure*, eqn (2.3.6).

(13) Explain and calculate the difference between internal energy and enthalpy, Section 2.3(c).

(14) Derive a relation between the heat capacities of a perfect gas, eqn (2.3.9).

(15) Explain the term *thermal motion* and distinguish between *heat* and *work* in terms of the chaotic and ordered motion of particles, Section 2.4.

Introduction

Energy is stored by molecules, and its release can be used to provide heat when a fuel burns, mechanical work when a fuel burns in an engine, and electrical work when a chemical reaction pumps electrons through a circuit. In chemistry we encounter reactions that can be harnessed to provide heat and work, reactions whose liberated energy might be squandered but which give products we require, and reactions that constitute the processes of life. *Thermodynamics*, the study of the transformations of energy, enables us to discuss all these matters rationally.

2.1 Heat, work, and the conservation of energy

The *system* is the part of the world in which we have a special interest, and may be a reaction vessel, an engine, an electric cell, etc. Around the system is its *surroundings*, in which we make our observations. The two parts may be in contact. When matter can be transferred between the system and its surroundings we call it an *open system*; otherwise it is a *closed system*. An *isolated system* is a closed system in neither mechanical nor thermal contact with its surroundings. The system and its surroundings together are called the *universe*. Normally a small part of the laboratory around the system of interest is sufficiently isolated from the rest of the actual universe to be regarded as the full extent of the surroundings.

2.1 (a) The basic concepts

The basic concepts of thermodynamics are work, heat, and energy.

Work is done if the process could be used to bring about a change in the height of a weight somewhere in the surroundings. Work has been done *by* the system if a weight has been raised in the surroundings; work has been done *on* the system if a weight has been lowered. When we need to measure the amount of work we shall use its definition as *force* × *distance*, which we develop below.

Energy is the capacity to do work. When we do work on an otherwise isolated system its capacity to do work is increased, and so its energy has been increased. When the system does work its energy is reduced because it is then capable of doing less work than before.

Experiments have shown that the energy of a system may be changed by means other than work itself. When there is a temperature difference between the system and its surroundings the energy will change if they are in thermal contact. *When the energy of a system changes as a result of a temperature difference we say that there has been a flow of heat*. Containers that permit energy transfer as heat are called *diathermic*. Containers that do not permit energy to be transferred as heat are called *adiabatic*: a Dewar flask is a good approximation. We can detect whether heat has been transferred by making an observation on the surroundings. For instance, in an *ice calorimeter* (an adiabatic container surrounding the diathermic system and containing a mixture of ice and water) the amount of ice will change as a result of a flow of heat.

2.1 (b) The First Law

Experiments have established that whereas *we* know how energy has been transferred in a particular case (because we can see whether a weight was raised or lowered in the surroundings, or whether ice melted), the *system* is blind to the mode employed. Heat and work are equivalent ways of changing a system's energy: energy is energy, however it is gained or lost. The system is like a bank: it accepts deposits in either currency, but stores its reserves as energy. In an adiabatic system, for instance, the same rise in temperature is brought about by the same quantity of any kind of work we choose to do on the system (stirring it with rotating paddles, electrical work, and so on).

The First Law of thermodynamics is a summary of the preceding remarks. It can be expressed in a number of ways. An involved but rich statement is:

When a system changes from one state to another along any adiabatic path, the quantity of work done is the same irrespective of the means employed.

The content of this statement can be unfolded as follows. Suppose a quantity w_{ad} of work is done on the system in some adiabatic process taking it from an initial state i to a final state f. The work may be of any kind and may take the system through different intermediate states. This would seem to require us to label w_{ad} with the type of path. However, the First Law implies that the value of w_{ad} is the same for all paths, and depends only on the initial and final states. This immediately suggests that *there is a property of the state of the system* such that w_{ad} may be expressed as the difference of its values for the initial and final states:

$$w_{ad} = U_f - U_i. \tag{2.1.1}$$

This is analogous to expressing the height through which someone climbs on a mountain as the difference of altitudes: the path-independence of the height implies the existence of the property 'altitude'. In this case the property U is called the *internal energy* of the system.

Next, suppose that the system changes between the same initial and final states as before, but now along a *non-adiabatic* path (the system being in thermal contact with its surroundings). The internal energy changes by the same amount because it depends on the state and not the path, but the work involved might not be the same as before. The difference between the work done and the change in internal energy, $\Delta U = U_f - U_i$, is defined as the heat absorbed in the process:

$$q = \Delta U - w. \tag{2.1.2}$$

The First Law therefore provides a *mechanical* and therefore truly fundamental definition of heat, because ΔU for the change $i \rightarrow f$ can be measured by measuring w_{ad} for the process of changing from i to f, the w can be measured for the non-adiabatic change between the same two states, and the heat is then the difference of the two mechanical quantities:

$$q = w_{ad} - w. \tag{2.1.3}$$

Finally, the First Law also implies that the internal energy of an *isolated* system cannot change. This is because there can be no transfers of energy as heat or as work, and when $w = 0$ and $q = 0$ it follows from $\Delta U = q + w$ that $\Delta U = 0$. An isolated system can undergo a change of state: for instance, a pendulum will come to rest in an isolated container, but the energy of the isolated system remains constant. This leads to a very succinct expression of the First Law:

First Law: *The energy of an isolated system is constant.*

In order to comprehend the full content of this statement, so that it is regarded as a *thermodynamic* statement and not merely a statement about mechanics, all that has gone before has to be borne in mind. The evidence for the law in this form is the impossibility of constructing perpetual motion machines, for if energy could arise spontaneously an engine could be constructed that would run without fuel. Experience has shown this to be unrealizable, and the First Law is based on the gloomy acceptance of its impossibility.

2.1 (c) Infinitesimal changes

The next step in the development is to express the First Law mathematically. By switching attention to *infinitesimal* quantities of heat and work we open up the way to powerful methods of calculation.

Let the work done on a system be $\mathrm{d}w$ and the heat supplied to it be $\mathrm{d}q$. Then in place of $\Delta U = q + w$ we have

$$\mathrm{d}U = \mathrm{d}q + \mathrm{d}w. \tag{2.1.4}$$

This is the mathematical expression of the First Law for a closed system. It expresses the observation that the internal energy of a closed system changes by an amount determined by the quantities of energy that pass through its walls as heat and work. If none passes through (when the system is isolated), then $\mathrm{d}U = 0$. In interpreting and using eqn (2.1.4) we have to choose a convention concerning signs: this is explained in Box 2.1.

Box 2.1 The sign convention

We use the convention that $\mathrm{d}w$ denotes work done *on* a system and $\mathrm{d}q$ denotes heat supplied *to* a system. Since it is sometimes more natural to discuss work done *by* a system, that is denoted $\mathrm{d}w'$. Therefore, $\mathrm{d}w' = -\mathrm{d}w$.

When $\mathrm{d}w$ or $\mathrm{d}q$ is *positive*, it signifies that energy has been supplied to the system as work or as heat, and has contributed to the internal energy ($\mathrm{d}U = \mathrm{d}w + \mathrm{d}q$).

When $\mathrm{d}w$ or $\mathrm{d}q$ is *negative*, it signifies that energy has left the system as work or as heat, and has contributed to a reduction of the internal energy.

The convention applies to measurable changes as well as to infinitesimal changes. Thus $w = +10\,\mathrm{kJ}$ signifies that 10 kJ of energy has been supplied to the system by doing work on it, and $q = -10\,\mathrm{kJ}$ signifies that 10 kJ of energy has left the system as heat. $w' = 10\,\mathrm{kJ}$ signifies that the system has done 10 kJ of work in its surroundings.

2.2 Work

In order to do anything useful with eqn (2.1.4) we must be able to relate $\mathrm{d}q$ and $\mathrm{d}w$ to events taking place in the surroundings. We begin by discussing *mechanical work*. This includes the work involved in compressing gases, and the work a gas does as it expands and drives back the atmosphere pressing on the walls of the system. Many chemical reactions involve the generation or the reaction of gases, and the characteristics of the reaction, such as the heat it generates, depend on the work the system does.

It is much more natural to think in terms of the work that a system does than in terms of the more abstract concept of the work done on it. Therefore, in all that follows we shall distinguish between the work that a system *does* (symbol: w') and the corresponding work done *on* it (symbol: w). The two are related, as explained in Box 2.1, by $w' = -w$.

2.2 (a) Mechanical work

The calculation of the quantity of work done on (w) or by (w') a system is based on the definition provided by elementary physics. In order to move

Fig. 2.1. The system, the contents of the box, does work when it raises a weight in the surroundings against the opposing force of gravity. 'Work' is always reducible to a process in which the height of a weight is changed.

an object through a distance dz against a force $F(z)$, Fig. 2.1, the system (the person in the box) has to do a quantity of work dw', where

$$\mathrm{d}w' = -F(z)\,\mathrm{d}z. \tag{2.2.1}$$

The force carries a sign which depends on the direction in which it acts: if it acts upwards, towards positive z, it is positive; if it acts downwards, towards negative z, it is negative. If the force is independent of position, the work a system does as it moves the object through dz is $\mathrm{d}w' = -F\,\mathrm{d}z$. Therefore, the work done by the system in moving the object through a measurable distance from z_i to z_f is the sum of the work required to move it through each dz segment of the path. Since the segments are infinitesimal, the sum is actually an integral:

$$w' = \int_{z_i}^{z_f} (-F\,\mathrm{d}z) = -(z_f - z_i)F. \tag{2.2.2}$$

In this case the work done by the system is proportional to the distance moved along the chosen path. If, like gravity, the force acts downwards, F is negative, and we can write $F = -|F|$. Consequently,

$$w' = (z_f - z_i)\,|F|. \tag{2.2.3a}$$

If the final position lies above the initial (so that $z_f > z_i$), w' is positive. Therefore, the work done *on* the system (according to the convention described in Box 2.1), is the negative of this, or

$$w = -(z_f - z_i)\,|F| = -h\,|F|, \tag{2.2.3b}$$

h being the height through which the object is raised. This w is negative: the internal energy of the system has decreased as a result of it doing work.

Example 2.1

Calculate the work a system (you, for instance) must do in order to lift this book (of mass approximately 1.5 kg) through a height of 10 cm.

• *Method*. At the surface of the Earth the force on a mass m is of magnitude mg and directed downwards (g is the acceleration due to gravity, $9.81\ \mathrm{m\,s^{-2}}$). Hence $F = -mg$, a constant (for small displacements). Use eqn (2.2.3a).

• *Answer*. $w' = mgh = (1.5\ \mathrm{kg}) \times (9.81\ \mathrm{m\,s^{-2}}) \times (0.10\ \mathrm{m}) = 1.5\ \mathrm{kg\,m^2\,s^{-2}} = 1.5\ \mathrm{J}$.

• *Comment*. The work done on the system is $w = -1.5\ \mathrm{J}$: the internal energy of the system (you, in this instance) falls by 1.5 J.

• *Exercise*. Calculate the work done on a mass of 1 g when it is raised through 1 cm on the surface of the Earth. [1×10^{-4} J]

If the force is not constant but varies along the path, the work done in each segment depends on the force acting there, and it must be calculated using eqn (2.2.1). The total work involved is the sum of all the infinitesimal contributions, and so in general

$$\textit{Work done by the system: } w' = -\int F(z)\,\mathrm{d}z, \qquad (2.2.4a)$$

$$\textit{Work done on the system: } w = \int F(z)\,\mathrm{d}z. \qquad (2.2.4b)$$

Once we know how the force depends on the location we can evaluate the integral and obtain the expression for the work.

Example 2.2

Calculate the work a system (e.g. an engine) must do in order to raise a rocket of mass 1.0×10^5 kg from the surface of the Earth to an altitude where the gravitational attraction is negligible.

• *Method*. We must take into account the decrease of gravitational force with distance (z) from the centre of the Earth. Newton's law of gravity is $F = -GmM/z^2$, where M is the Earth's mass (5.98×10^{24} kg), m the mass of the object, and G the gravitational constant ($G = 6.672 \times 10^{-11}\ \mathrm{kg^{-1}\,m^3\,s^{-2}}$). Use eqn (2.2.3a) setting $z_i = R$, the radius of the Earth (6.38×10^6 m) and $z_f = \infty$, a point infinitely far away.

• *Answer*. The work the system must do is

$$\begin{aligned} w' &= -\int_R^\infty (-GmM/z^2)\,\mathrm{d}z = GmM\int_R^\infty (1/z^2)\,\mathrm{d}z = GmM/R. \\ &= (6.672 \times 10^{-11}\ \mathrm{kg^{-1}\,m^3\,s^{-2}}) \times (1.0 \times 10^5\ \mathrm{kg}) \\ &\quad \times (5.98 \times 10^{24}\ \mathrm{kg})/(6.38 \times 10^6\ \mathrm{m}) = 6.3 \times 10^{12}\ \mathrm{J}. \end{aligned}$$

• *Comment*. The calculation neglects air resistance. In practice, the mass of the rocket decreases with altitude because fuel burns during the initial stage of ascent.

• *Exercise*. Calculate the work required to raise a commercial aircraft of mass 100 tonnes (1 tonne $= 10^3$ kg) to an altitude of 12 km. [1.17×10^{15} J]

2.2 (b) Work of compression and expansion

The next step is to apply eqn (2.2.4) to the work involved in compressing or expanding a gas. Consider the arrangement shown in Fig. 2.2. A gas is confined by a massless, frictionless, rigid, perfectly fitting piston. Choosing these properties simply means that we are focusing on the properties of the

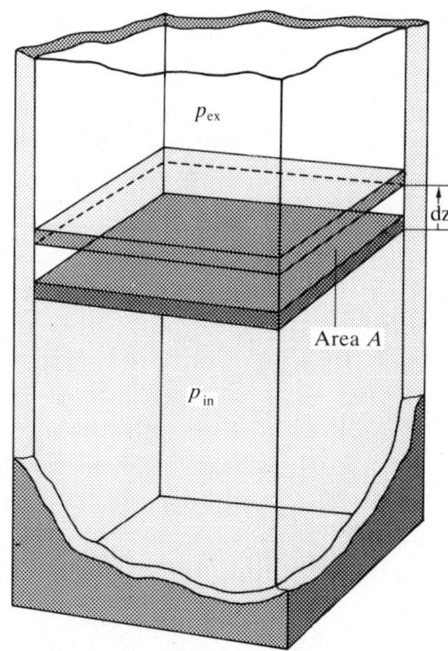

Fig. 2.2. When a piston of area A is pushed out through a distance $\mathrm{d}z$ by the enclosed gas it sweeps out a volume $\mathrm{d}V = A\,\mathrm{d}z$. The external pressure is equivalent to a weight pressing on the piston, and so work is done on expansion.

gas rather than any technological deficiencies: in this way we concentrate on the essentials of the problem. Inside the container is a gas. (It need not be a perfect gas. It need not even be a gas: any substance would do, but it is easier to think about gases.) Its pressure is p_{in}. The external pressure is p_{ex}. The area of the piston is A. Therefore the force exerted by the gas on its inner surface is $+p_{in}A$ (acting upwards). The force on the outer face of the piston is $-p_{ex}A$ (acting downwards).

When p_{ex} is less than p_{in} the force driving the piston out is greater than the force driving it in, and the tendency is towards expansion. If the gas expands through a distance dz it moves the piston against a constant force $-p_{ex}A$. The work done by the system (the gas) is given by eqn (2.2.1) as $dw' = -(-p_{ex}A)\,dz$. But Adz is the volume swept out in the course of the infinitesimal expansion, and we write it dV. Therefore, *the work involved when a gas expands through* dV *against a constant pressure* p_{ex} *is*

Work done by the system: $dw' = p_{ex}\,dV$, (2.2.5a)

Work done on the system: $dw = -p_{ex}\,dV$. (2.2.5b)

We have remarked that work is an effect detectable in the surroundings as a change in height of a weight. In the case of expansion work the effect of the external pressure can be reproduced by a weight resting on the piston. This weight is raised by the expanding gas, and so work is done by the system on its surroundings. This emphasis on the interpretation of work may seem pedantic, but its importance becomes clear when we turn to the work of compression. In the case of compression it is still the *external* pressure that determines the magnitude of the work even though it is the internal pressure that is opposing the insertion of the piston. This is because the act of compression corresponds to the lowering of a weight in the outside world, and hence p_{ex} occurs in the expression for dw because the weight still models the effect of the external pressure. In this case too, therefore, the work done on the system is given by eqn (2.2.5b), which therefore applies to compression as well as to expansion. Now, though, the sign of dV is negative (a reduction of volume), and so dw is positive: work is done on the system by compression and, so long as no other energy changes take place, its internal energy rises.

Other types of work (e.g. electrical) have analogous expressions, and some are collected in Box 2.2. For the present we continue with the work associated with changing the volume of systems, so-called *p,V-work*, and see what can be extracted from eqn (2.2.5).

2.2 (c) Free expansion

Free expansion occurs when $p_{ex} = 0$, so that there is no opposing force. Then no weight is raised, and eqn (2.2.5) gives $dw' = 0$ and $dw = 0$ for each stage of the expansion. Hence, overall:

Work done during free expansion: $w' = 0$ *and* $w = 0$. (2.2.6)

2.2 (d) Expansion against constant pressure

In this case the gas expands until it meets a mechanical stop or until the internal pressure falls to the external. Throughout the expansion p_{ex} is constant (e.g. the piston is pressed on by the atmosphere of the outside

Box 2.2 Varieties of work

In general, the work done on a system can be expressed in the form $dw = -F\,dz$, where F is a 'generalized force' and dz is a 'generalized displacement'. The table below gives some examples.

Type of work	dw	Comments	Units (for dw in J)
Change of volume	$-p_{ex}\,dV$	p_{ex} is the external pressure dV is the change in volume	Pa m^3
Change of surface area	$\gamma\,d\sigma$	γ is the surface tension $d\sigma$ is the change of area	$N\,m^{-1}$ m^2
Change of length	$f\,dl$	f is the tension dl is the change of length	N m
Electrical work	$Q\,d\phi$	Q is the electric charge $d\phi$ is the difference of potential	C V

The expression for electrical work is explained in detail in Chapter 11.

world) and so the work done by the system as it passes through each successive displacement dV is $dw' = p_{ex}\,dV$. The total work done by the system in the expansion from V_i to V_f is the sum (integral) of these successive equal contributions:

$$w' = \int_{V_i}^{V_f} p_{ex}\,dV = p_{ex}\int_{V_i}^{V_f} dV = p_{ex}(V_f - V_i).$$

Therefore, if we write the change of volume as $\Delta V = V_f - V_i$,

$$\text{Work done by system: } w' = p_{ex}\,\Delta V, \tag{2.2.7a}$$

$$\text{Work done on system: } w = -p_{ex}\,\Delta V. \tag{2.2.7b}$$

This result can be illustrated graphically, Fig. 2.3: w' is equal to the area beneath the horizontal line at $p = p_{ex}$ lying between the initial and final volumes. This type of graph is called an *indicator diagram* (James Watt first used one to indicate aspects of the operation of his steam engine). Note that the work of expansion depends on the *external* pressure: the pressure of the gas in the system might change in the course of the expansion (because its temperature might fall), but that does not affect the work. The only way that the internal pressure enters the calculation is in the requirement that it must exceed the external pressure in order that the system expand rather than contract.

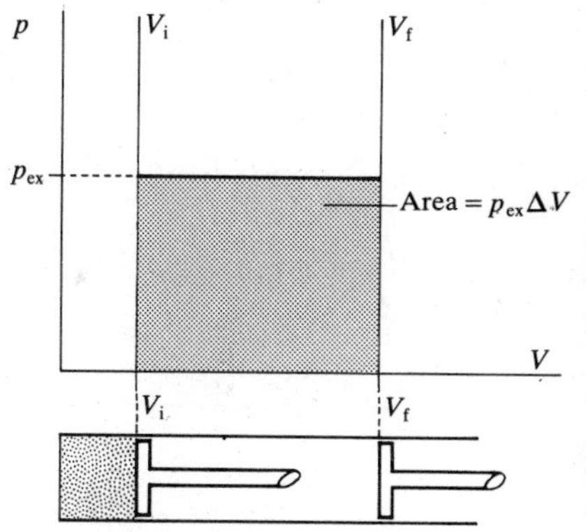

Fig. 2.3. The work done by a gas when it expands against a constant external pressure, $p_{ex}(V_f - V_i)$, is equal to the shaded area in the *indicator diagram*, the graph of opposing pressure versus volume.

Example 2.3

A greatly simplified model of an internal combustion engine is as follows. At the start of the power stroke the ignited gases exert a pressure of 20 atm, and drive the piston back against a constant force equivalent to 5.0 atm. In so doing, the piston sweeps out 250 cm^3. What is the power output of a six-cylinder engine working at 2000 RPM (with one power stroke from each cylinder every second revolution)?

• *Method*. The external equivalent pressure is constant, and so use eqn (2.2.7a) to calculate the work. Power (in watts, $1\,W = 1\,J\,s^{-1}$) is energy per unit time. Use $1\,atm = 1.013 \times 10^5\,N\,m^{-2}$.

• *Answer*. $w' = -(5.0 \times 1.013 \times 10^5 \,\mathrm{N\,m^{-2}}) \times (2.50 \times 10^{-4} \,\mathrm{m^3}) = -130 \,\mathrm{J}$. Six such cylinders produce $2000 \times 6 \times \frac{1}{2} \times w' = 7.8 \times 10^5 \,\mathrm{J}$ in one minute, and so the rate of energy production is $(7.8 \times 10^5 \,\mathrm{J})/(60 \,\mathrm{s}) = 13 \,\mathrm{kW}$.

• *Comment*. The calculation is extremely crude, but it gives a rough idea of the power output of an actual engine.

• *Exercise*. At how many RPM would an eight-cylinder engine, each cylinder being $200 \,\mathrm{cm^3}$, produce the same power (i.e. 13 kW), other conditions being equal? [1875 RPM]

2.2 (e) Reversible expansion

In thermodynamics a *reversible change is one that can be reversed by an infinitesimal modification of a variable*. The key word is 'infinitesimal' and it sharpens the everyday meaning of the word 'reversible'.

Suppose the gas is confined by a piston, and the external pressure is only infinitesimally less than the internal. The gas expands. However, if the external pressure is increased infinitesimally it rises above the internal pressure and the gas is compressed. The process is therefore reversible in the thermodynamic sense and at each stage the system is in *equilibrium* with its surroundings. In contrast, suppose the external pressure differs measurably from the internal, then changing p_{ex} by an infinitesimal amount will not decrease it below the internal pressure and so does not change the direction of the process. Therefore, expansion under these circumstances is *irreversible*.

If the expansion of a gas is to occur reversibly we must ensure that at each stage the external pressure is only infinitesimally less than the internal pressure. But the internal pressure might change as the gas expands. It follows that in the calculation of the work of reversible expansion we have to set p_{ex} virtually equal to the internal pressure at each stage of the expansion:

$$\mathrm{d}w = -p_{ex}\,\mathrm{d}V = -p_{in}\,\mathrm{d}V. \qquad (2.2.8)_r$$

(Equations valid only for reversible processes are labelled with a subscript r.) It is important to remember that although the pressure of the confined gas appears in this expression, it does so only because p_{ex} has been set equal to it to ensure reversibility. (p_{ex} is actually set equal to $p_{in} - \mathrm{d}p$, but $\mathrm{d}p\,\mathrm{d}V$ is doubly infinitesimal, and so can be ignored.)

The total work of reversible expansion is the sum of all the infinitesimal contributions:

$$w = -\int p_{in}\,\mathrm{d}V. \qquad (2.2.9)_r$$

p_{in} is not necessarily constant, and might fall as the piston is driven out: that is, p_{in} is in general a function of the volume of the system. This means that we can evaluate the integral only if we know how the pressure of the confined gas depends on its volume. This is the link with the material covered in Chapter 1: if we know the equation of state of the gas, we can express p_{in} in terms of V and evaluate the integral either analytically or numerically.

This can be illustrated in the case of the *isothermal expansion of a perfect gas*. The expansion is made isothermal by keeping the system in thermal contact with its surroundings (such as a constant-temperature bath). The equation of state is

$$pV = nRT, \qquad (2.2.10)^{\circ}$$

and so at each stage p_{in} is equal to nRT/V, V being the current volume. In an isothermal expansion T is constant, and so it is independent of V. It follows from eqn (2.2.9) that the work of expansion from V_i to V_f at a temperature T is

$$w = -\int_{V_i}^{V_f} (nRT/V)\,dV = -nRT\int_{V_i}^{V_f} (1/V)\,dV = -nRT\ln(V_f/V_i).$$

That is,

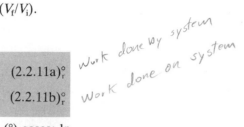

$$\textit{Work done by system: } w' = nRT\ln(V_f/V_i), \qquad (2.2.11a)^{\circ}_{r}$$

$$\textit{Work done on system: } w = -nRT\ln(V_f/V_i). \qquad (2.2.11b)^{\circ}_{r}$$

These results are limited to reversible ($_r$) processes and perfect (°) gases; ln denotes a natural logarithm.

The expressions conform to common sense. When the final volume is greater than the initial volume, as in expansion, V_f/V_i is greater than unity. This means that the logarithm is positive, and that w' is also positive: the system has done work on the outside world. Under the same circumstances, w is negative: the internal energy has decreased as a result of the system doing work. We also see that more work is done for a given change of volume when the temperature is increased. This is plausible, because it corresponds to the greater pressure of the confined gas needing a higher opposing pressure in order to ensure reversibility.

The result of the calculation can also be expressed in terms of an indicator diagram. In this case the work done by the system is given by the area under the isotherm $p = nRT/V$, Fig. 2.4. Superimposed on the diagram is the rectangular area obtained in the case of irreversible expansion against a constant external pressure fixed at the same final value as is reached in the reversible expansion (i.e. Fig. 2.3). We obtain *more* work when the expansion is reversible (the area is greater). This is because matching the external pressure to the internal pressure at each stage ensures that none of the system's pushing power is wasted. More work than the reversible amount cannot be obtained because increasing the external pressure even infinitesimally results in contraction.

We may infer from this discussion that, because some pushing power is wasted when $p_{\text{in}} > p_{\text{ex}}$, *the maximum work available from a system operating between specified initial and final states and passing along a specified path is obtained when it is operating reversibly*. Unfortunately, in practice this normally means that the path has to be traversed infinitely slowly. Consequently, reversible processes are called *quasi-static*. Nevertheless, this is not such a devastating restriction as it might seem. The principal point is that the use of reversible paths as *formal* (i.e. hypothetical, not necesarily practical) ways of taking a system from one state to another is, as we shall see, of great importance for calculating the changes that then occur in various thermodynamic properties. A secondary point is that even some practical processes may be regarded as occurring effectively reversibly. This is because what matters is not the absolute rate, but the rate relative to the rapidity with which the particles making up the system can adjust to the

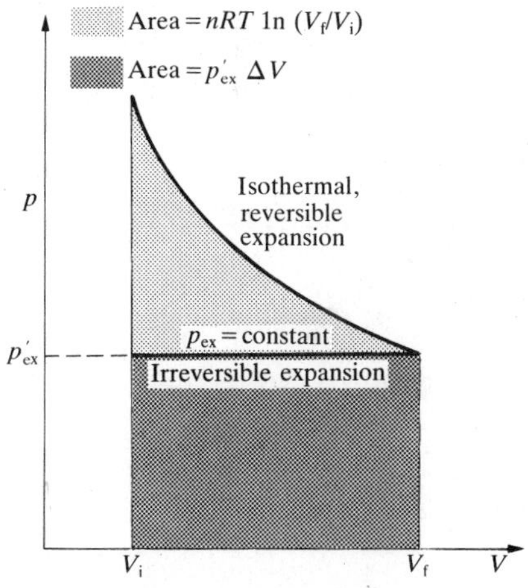

Fig. 2.4. The *reversible* isothermal work of expansion of a perfect gas is given by the area under the isotherm. When expansion occurs (irreversibly) against a constant external pressure equal to the same final value, the work is less, and is given by the area of the shaded rectangle.

changing conditions. For example, the propagation of sound through a gas is a sequence of very rapid compressions and expansions, but since the particles adjust to the changes so rapidly, the propagation can be treated as a sequence of reversible steps. On the other hand, if we do arrange for a process to occur infinitely slowly, then we can be sure that the particles do have time to respond (if the conditions are right, such as matching internal and external pressures). In other words, granted the correct balance of forces, infinite slowness is a sufficient condition for reversibility, but not a necessary condition for effective reversibility.

We have introduced the connection between reversibility and maximum work for the special case of a perfect gas undergoing expansion. However, it applies to all substances, and to all kinds of work, as we shall see when we turn to the consequences of the Second Law.

Example 2.4

These calculations have chemical applications. For instance, calculate the work done when 50 g of iron dissolves in hydrochloric acid in (a) a closed vessel, (b) an open beaker at 25 °C under 1 atm pressure.

• *Method*. The reaction is $Fe(s) + 2HCl(aq) \rightarrow FeCl_2(aq) + H_2(g)$. In the course of the reaction 1 mol $H_2(g)$ is generated when 1 mol Fe(s) is consumed. The gas drives back the atmosphere and therefore does the work $w' = p_{ex}\Delta V$, eqn (2.2.7a). Relate the change in volume to the amount of hydrogen generated by regarding it as a perfect gas and using $pV = nRT$. In (a) the volume cannot change. In (b) neglect the initial volume because the final volume (after the production of gas) is so much larger that $V_f - V_i \approx V_f = n_{H_2}RT$. The molar mass of Fe is 55.85 g mol^{-1} (inside back cover).

• *Answer*. In each case $w' = p_{ex}\Delta V$. In (a) $\Delta V = 0$, and so $w' = 0$. No work is done either by or on the system, as its volume is unchanged. In (b), $w' = p_{ex}\Delta V \approx p_{ex}V(H_2)$. Then, since $pV(H_2) = n_{H_2}RT$, and $n_{H_2} = n_{Fe}$, we have

$$w' = n_{Fe}RT = (50\,\mathrm{g}/55.85\,\mathrm{g\,mol^{-1}}) \times (8.314\,\mathrm{J\,K^{-1}\,mol^{-1}}) \times (298.15\,\mathrm{K})$$
$$= 2.2\,\mathrm{kJ}.$$

• *Comment*. The system, the dissolving iron, does 2.2 kJ of work driving back the atmosphere. When we apply thermodynamics to chemical reactions, especially those producing gases, these quantities must be taken into account because they affect the energy balance sheet of the process. This is explained in Chapter 4.

• *Exercise*. Calculate the expansion work done when 50 g of water is electrolysed under a constant pressure of 1 atm and 25 °C. [10 kJ]

2.3 Heat

When energy is transferred to a system as heat there is a change of state which may appear as a rise in temperature. When only an infinitesimal transfer takes place, the increase of temperature is proportional to the heat supplied, and so then

$$\mathrm{d}T \propto \mathrm{d}q, \quad \text{or} \quad \mathrm{d}T = \text{coefficient} \times \mathrm{d}q.$$

The magnitude of the coefficient depends on the size, composition, and state of the system. It proves more convenient to invert this relation and to write it as

$$\mathrm{d}q = C\,\mathrm{d}T. \tag{2.3.1}$$

The coefficient C is called the *heat capacity* (formerly the 'specific heat'). The *molar heat capacity* is $C_m = C/n$.

2.3 (a) Heat capacity

If we know the heat capacity we can measure the energy supplied as heat by monitoring the temperature rise that transfer produces. This is why the heat capacity of the local surroundings of a system (such as a water bath) is an essential piece of information in thermodynamics, because the heat absorbed or evolved by the system can be monitored by noting the temperature changes taking place there. The thermometer is consequently a most important thermodynamic instrument, and many thermodynamic properties are interpretations of temperature measurements.

When C is large a given amount of heating results in only a small temperature rise (the system has a large capacity for heat). When C is small the same amount of heating can result in a large temperature rise. Water has a large heat capacity: a lot of energy is needed to make it hot (i.e. to raise its temperature), and central heating systems take advantage of this property because a lot of energy can be transported by a slow flow of hot water. Similarly, ponds freeze slowly partly because a lot of energy has to be removed in order to lower the temperature of the water (and also because the freezing releases a large amount of energy).

The heat capacity depends on the conditions. Suppose the system is constrained to have constant volume and is unable to do any kind of work. The heat required to bring about a change of temperature $\mathrm{d}T$ is some quantity $C_V\,\mathrm{d}T$, where C_V is the *heat capacity at constant volume* (or the *isochoric heat capacity*). If instead the system is subjected to constant pressure and is allowed to expand or contract as it is heated, then the heat required to bring about the same change of temperature is $C_p\,\mathrm{d}T$, where C_p is the *heat capacity at constant pressure* (or the *isobaric heat capacity*). In the second case, but not in the first, the system may have changed its volume; therefore some of the energy supplied as heat may have been returned to the surroundings as work, and not used exclusively to raise the system's temperature. Hence in general C_V and C_p are different.

C_V has been expressed in terms of the heat supplied to a body under specified conditions (constant volume, no other forms of work). However, it can be related to the increase in internal energy that accompanies the heating. The argument runs as follows. The First Law can be written

$$\mathrm{d}U = \mathrm{d}q + \mathrm{d}w = \mathrm{d}q + \mathrm{d}w_e - p_{ex}\,\mathrm{d}V,$$

where $-p_{ex}\,\mathrm{d}V$ is the p,V-work and $\mathrm{d}w_e$ is any other work (e for extra, or electrical if that is the case). When no volume change is allowed $\mathrm{d}V = 0$ and so the last term in this equation is zero. When no other work is done $\mathrm{d}w_e = 0$ too. Therefore,

$$\mathrm{d}U = \mathrm{d}q \quad \text{at constant volume,} \quad \text{no other work.} \qquad (2.3.2)$$

We often write this as $\mathrm{d}U = (\mathrm{d}q)_V$. This result implies that at constant volume $C_V = \mathrm{d}U/\mathrm{d}T$. When one or more variables are held constant during the change of another, the derivatives are called *partial derivatives* with respect to the changing variable. The d is replaced by ∂ and the variables held constant are added as a suffix. In this case it follows that

$$\textit{Heat capacity at constant volume:}\ C_V = (\partial U/\partial T)_V \qquad (2.3.3)$$

for systems doing no work. This is now taken as the *definition of* C_V. Some experimental values are given in Table 2.1.

Table 2.1. Heat capacities at 25 °C and 1 atm.

	$C_{V,m}$/J K^{-1} mol^{-1}	$C_{p,m}$/J K^{-1} mol^{-1}
He	12.48	20.79
Ar	12.48	20.79
N_2	20.74	29.12
CO_2	28.46	37.11

Example 2.5

A function of several variables we have already encountered is the van der Waals equation of state, Box 1.1:

$$p = nRT/(V - nb) - an^2/V^2.$$

Find the partial derivatives of p with respect to V and T.

• *Method*. In order to find $(\partial p/\partial T)_V$, differentiate p with respect to T, regarding n and V as constants (like a, b, and R). To find $(\partial p/\partial V)_T$, differentiate with respect to V, holding n and T constant.

• *Answer*. $(\partial p/\partial T)_V = nR/(V - nb)$,

$$(\partial p/\partial V)_T = -nRT/(V - nb)^2 + 2an^2/V^3.$$

• *Comment*. The two partial derivatives are themselves functions of V and T, and so we could go on to find the four second derivatives. Later we shall find it useful to note that the order of differentiation is unimportant, and so $\partial^2 p/\partial V\,\partial T = \partial^2 p/\partial T\,\partial V$, as may be checked in this instance.

• *Exercise*. Calculate $(\partial p/\partial V)_T$ and $(\partial p/\partial T)_V$ for the Dieterici equation of state in Box 1.1.

[$\{nRT/(V - nb)\}\{(an/RTV^2) - 1/(V - nb)\}e^{-an/RTV}$; $\{nR/(V - nb)\}\{1 + an/RTV\}e^{-an/RTV}$]

2.3 (b) Enthalpy

But what of C_p? Is there a thermodynamic property of the system that can be identified with $(dq)_p$, the heat supplied at constant pressure? It turns out (we shall justify it in a moment) that it is easy to construct one simply by adding the product pV to the internal energy U. This gives a new quantity denoted H: $H = U + pV$.

In order to confirm that $dH = (dq)_p$, consider how H may change in general. When U changes to $U + dU$, p changes to $p + dp$, and V changes to $V + dV$, H changes to

$$\begin{aligned} H + dH &= (U + dU) + (p + dp)(V + dV) \\ &= U + pV + dU + V\,dp + p\,dV + dp\,dV. \end{aligned}$$

As we are considering infinitesimal changes, the last term, being doubly infinitesimal, can be ignored. Since $U + pV$ on the right is H itself, it follows that

$$dH = dU + p\,dV + V\,dp.$$

We can get this result more briskly. The differential of $H + pV$ is $dH = dU + d(pV)$; the second term is the differential of a product, and so can be written $d(pV) = p\,dV + V\,dp$, which reproduces the result above. We shall use this quick method from now on (the clumsy method is in fact its justification).

Now consider a system that is in mechanical equilibrium with its surroundings at a pressure p (so that $p_{ex} = p_{in} = p$). That being so we can substitute $dU = dq + dw$ in the form

$$dU = dq + dw_e - p\,dV \tag{2.3.4}$$

into the expression for dH, cancel the $p\,dV$ terms, and obtain

$$dH = dq + dw_e + V\,dp. \tag{2.3.5}$$

Now impose the conditions that (a) there is no non-p,V work, so that $dw_e = 0$, and (b) the heating occurs at constant pressure, so that $dp = 0$. Then

$$dH = dq \quad \text{at constant pressure, no non-}p,V\text{ work} \tag{2.3.6}$$

This is often written in the form $dH = (dq)_p$, the result we set out to confirm. That is, *when a system is heated at constant pressure, and no work other than p,V-work is allowed, H increases by an amount equal to the energy supplied as heat*. It follows from the introductory remarks about heat capacities, that

$$\textit{Heat capacity at constant pressure}: \; C_p = (\partial H/\partial T)_p. \tag{2.3.7}$$

This is now taken as the *definition* of C_p; some experimental values are given in Table 2.1.

All we seem to have done is to replace q by something called H. We have in fact done much more. We have replaced something measured in the surroundings (q), and which depends on the mode of energy transfer adopted (heating), by a change in a property of the system itself (H depends only on U, p, and V, each of them being properties of the state of the system). We shall soon see that H is a very useful quantity, and that inventing it has not been merely an exercise in generating equations. In fact, H plays a central role in chemistry because we are so often concerned with processes at constant pressure (reactions occurring in open vessels, including the body, are examples). Because H is so important it carries its own name, the *enthalpy* of the system:

$$\textit{Enthalpy}: \; H = U + pV. \tag{2.3.8}$$

We shall develop the full significance of the enthalpy in the following chapters. Note that p is to be taken as the pressure of the *system*, and that the form pV is a part of the general definition of H for any system, and does not imply a restriction to perfect gases.

Example 2.6

Water is brought to the boil under a pressure of 1 atm. When an electric current of 0.50 A from a 12 V supply is passed for 300 s through a resistance in thermal contact with it, it is found that 0.798 g of water is vaporized. Calculate the molar internal energy and enthalpy changes at the boiling point (373.15 K).

• *Method*. The change $H_2O(l) \rightarrow H_2O(g)$ is accompanied by a change of internal energy $\Delta U = q + w$. The heat absorbed during vaporization is equal to the electrical work (amps × volts × time) done on the resistance, and so $q = IVt$. The work done by the system on vaporization is the expansion work done when the liquid vaporizes, so that $w' = p\,\Delta V$, ΔV being the change of volume. If water vapour is assumed to be a perfect gas and the liquid volume is negligible, then $\Delta V = nRT/p$, and so $w = -w' = -nRT$. Therefore, calculate the change in internal energy of the system from

$$\Delta U = IVt - nRT,$$

and the molar internal energy change from $\Delta U/n$, n being calculated from the mass evaporated and $M_r = 18.02$. The vaporization occurs at constant pressure, and so the enthalpy change is equal to the heat supplied:

$$\Delta H = IVt.$$

• *Answer*. $IVt = (0.50\ \text{A}) \times (12\ \text{V}) \times (300\ \text{s}) = 1.8\ \text{kJ}$,

$$n = (0.798\ \text{g})/(18.02\ \text{g mol}^{-1}) = 0.0443\ \text{mol},$$

$$nRT = (0.0443\ \text{mol}) \times (8.314\ \text{J K}^{-1}\,\text{mol}^{-1}) \times (373.15\ \text{K}) = 137\ \text{J}.$$

Hence, $\Delta U = +1.7\ \text{kJ}$, $\Delta H = +1.8\ \text{kJ}$, and the molar quantities are

$$\Delta U_m = (1.7\ \text{kJ})/(0.0443\ \text{mol}) = +38\ \text{kJ mol}^{-1},\ \Delta H_m = +41\ \text{kJ mol}^{-1}.$$

• *Comment*. The plus sign is added to positive quantities to emphasize that they represent an increase in internal energy or enthalpy.

• *Exercise*. The molar enthalpy of vaporization of benzene at its boiling point (353.25 K) is 30.8 kJ mol^{-1}. What is the molar internal energy change? For how long would the same supply need to supply a 0.50 A current in order to vaporize a 10 g sample? [+27.9 kJ mol^{-1}, 660 s]

2.3 (c) Taking stock

We set out to express transfers of energy by heating in terms of changes of temperature. We found that the heat transferred at constant volume can be identified with the change of internal energy of the system, and the heat transferred at constant pressure with the change of its enthalpy, so long as in either case no extra work is being done. When a system is heated at constant pressure, the energy does not entirely remain inside the system: some passes back into the surroundings as work. It follows that *the changes in internal energy and the enthalpy for a given change of state differ by the amount of work involved in changing the volume of the system*. In many chemical applications we are interested in the heat evolved by reactions at constant pressure, and so it is important to know the enthalpy of a system rather than its internal energy. This is taken up in Chapter 4.

The isochoric and isobaric heat capacities differ on account of the work involved in changing the volume of the system. Since work is done in the case of C_p, it follows that for a given amount of heating, the temperature rise is less, and therefore that C_p is greater than C_V. The difference is greater for gases than for either liquids or solids because the latter change their volumes only slightly when heated, and therefore do little work. We shall see an important qualification of this remark in Chapter 3, but it remains true that C_p is greater than C_V for a given sample.

The difference between the two heat capacities can be calculated readily for a perfect gas because the difference between the enthalpy and the internal energy depends on only the temperature. Since $pV = nRT$,

$$H = U + pV = U + nRT.$$

When the temperature is increased by $\mathrm{d}T$ the enthalpy and internal energy changes are related by

$$dH = dU + nR\,dT.$$

Since $dH = C_p\,dT$ and $dU = C_V\,dT$, it follows that

$$C_p - C_V = nR. \qquad (2.3.9)°$$

(A more rigorous derivation is given in Section 3.2(d).) Molar heat capacities are of the order of R itself (see Table 2.1), and so the difference between the two heat capacities is important.

2.4 What is work, and what is heat?

In this section we introduce some simple and qualitative interpretations of thermodynamics, drawing on a knowledge of molecular behaviour no deeper than we needed for our discussion of gases. The description can be sharpened and made quantitative: that is the job of *statistical thermodynamics*, which we meet in Part 2.

We identify thermodynamic internal energy with the energy stored in molecular bonds, in molecular translation, rotation, and vibration, and in the energy of interaction of molecules.

Heating is the transfer of energy as a result of the vigorous *random* molecular motion in the surroundings. The chaotic motion of the atoms and molecules in the surroundings may stimulate the particles in the system to move more rapidly or to vibrate more energetically, and so the internal energy of the system is increased. We call this random motion (of either the system or the surroundings) *thermal motion*. When a system heats its surroundings, it stimulates their thermal motion at the expense of its internal energy.

Work involves *organized* motion. When a weight is raised or lowered, its particles move in an organized way, not chaotically. When a system does work it drives some of the atoms in its surroundings in an organized way. For example, when a spring is compressed by a system the atoms move together in a definite direction. Likewise, when we do work on a system, we transfer energy to it in an organized way.

In thermodynamics, the distinction between work and heat must be made in the *surroundings*. The fact that the falling of a weight gives rise to chaotic motion inside the system is irrelevant: work is identified as energy transfer making use of the organized motion of particles in the surroundings, and heat is energy transfer making use of their thermal motion. In the case of the adiabatic compression of a gas, for instance, work is done as the particles of the compressing weight are lowered, but the effect of the incoming piston is to accelerate the gas particles to higher average speeds. Since collisions between particles quickly randomize their directions, the organized motion of the weight is in effect stimulating thermal motion of the gas. Nevertheless, we observe the falling weight, the organized descent of its atoms, and report that work is being done even though it is stimulating thermal motion. Similarly, when we heat an iron rod clamped at one end, its atoms move in an organized way as it expands towards the unclamped end (it also gets hotter: thermal motion is also stimulated). Nevertheless, the *agent* we are using to transfer the energy is the thermal motion of the hot particles in the flame or electric heater. We observe in the surroundings, and report that we are heating.

Further reading

Elementary chemical thermodynamics. B. H. Mahan; Benjamin, New York, 1963.
Basic chemical thermodynamics (3rd edn). E. B. Smith; Clarendon Press, Oxford, 1982.
Engines, energy, and entropy. J. B. Fenn; W. H. Freeman & Co., New York, 1982.
Perpetual motion machines. S. W. Angrist; *Scientific American* **218** (1), 114 (1968).
Chemical thermodynamics (2nd edn). I. M. Klotz and R. M. Rosenberg; Benjamin, New York, 1972.
Chemical thermodynamics. P. A. Rock; University Science Books and Oxford University Press, 1983.
Chemical thermodynamics. M. L. McGlashan; Academic Press, London, 1979.
Heat and thermodynamics. M. W. Zemansky and R. H. Dittman; McGraw-Hill, New York, 1981.
Thermodynamics. G. N. Lewis and M. Randall, revised by K. S. Pitzer and L. Brewer; McGraw-Hill, New York, 1961.
Energy at the surface of the earth. D. H. Miller; Academic Press, New York, 1981.
Bibliography of thermodynamics. L. K. Nash; *J. chem. Educ.* **64,** 42 (1965).

Introductory problems

A2.1. In the isothermal reversible compression of 52.0 mmol of a perfect gas at 260 K, the volume of the gas is reduced to one-third of its initial volume. For this process, calculate w, w', and q.

A2.2. A sample of methane with a mass of 4.50 g has a volume of 12.7 dm^3 at 310 K. It expands isothermally against a constant external pressure of 200 Torr until its volume increases by 3.3 dm^3. Assume methane to be a perfect gas and calculate w, q, and ΔU for this process. Calculate these quantities for the same change in state carried out by a reversible process.

A2.3. Unlike gases, liquids are difficult to compress. A certain liquid sample has a volume of 0.450 dm^3 at 0 °C and 1 atm and shrinks by only 0.67% when subjected to isothermal compression under a constant external pressure of 95 atm. Calculate w.

A2.4. The temperature of a 3.0 mol $O_2(g)$ heated at a constant pressure of 3.25 atm, increases from 260 K to 285 K. Assume oxygen is a perfect gas, and calculate q, ΔH, and ΔU for this change in state.

A2.5. 5.0 mol $NH_3(g)$ at 375 K absorb 4.89 kJ of heat in a constant pressure process. What is the final temperature?

A2.6. The temperature of three moles of a perfect gas heated at constant pressure by absorption of 229 J of heat increases by 2.55 K. Calculate $C_{p,m}$ and $C_{V,m}$ for the gas.

A2.7. The second virial coefficient, B, of Kr at 373 K is $-28.7\ cm^3\ mol^{-1}$. A 70 mmol sample of Kr expands reversibly and isothermally at 373 K, and its volume increases from 5.25 cm^3 to 6.79 cm^3. The internal energy increases by 83.5 J with this change in state. Use the virial expansion as far as the second coefficient to calculate w, q, and ΔH for the change in state.

A2.8. Two moles of CO_2 occupying a volume held fixed at 15.0 dm^3 absorb 2.35 kJ as heat. As a consequence the temperature of the gas rises from 300 K to 341 K. Assume CO_2 obeys the van der Waals equation of state. Calculate w, ΔU, and ΔH.

A2.9. 25 g of a liquid is cooled from 290 K to 275 K in a constant pressure process by the extraction of 1200 J of heat. Calculate q, ΔH, and an approximate value of C_p.

A2.10. 0.5 mol of a liquid which has a molar heat of vaporization of 26.0 $kJ\ mol^{-1}$ is evaporated at 250 K and a constant pressure of 750 Torr. Assume the vapour follows the perfect gas law and calculate q, w, ΔH, and ΔU.

Problems

2.1. Calculate the work necessary to raise a mass of 1.0 kg through a height of 1.0 m on the surface of (a) the earth, where $g = 9.8\ m\ s^{-2}$, (b) the moon, where $g = 1.6\ m\ s^{-2}$.

2.2. How much work must a man do to climb a flight of stairs? Take his mass as 150 lb and suppose he climbs through 10 ft. Use 2.2 lb = 1 kg.

2.3. Calculate the work necessary to compress a spring by 1.0 cm if its force constant is $2.0 \times 10^5\ \mathrm{N\,m^{-1}}$. What work must be done to stretch the spring by the same amount?

2.4. In a spacecraft designed to land on the surface of Mars there was a sampling device that incorporated a spring of force constant $2.0 \times 10^6\ \mathrm{N\,m^{-1}}$ attached to a cam that alternately compresses and expands it through a displacement of 1.0 cm from equilibrium. Power supplies had to be planned very carefully, and the question arose as to how much power would be needed to drive the sampler. How much work would have to be done on the spring to compress and extend it 1000 times? What is the power (in watts) of an electric motor that can accomplish these 1000 oscillations in 1000 s? The spring is in contact with a thermally insulated mass of metal of heat capacity $4.2\ \mathrm{kJ\,K^{-1}}$: if its initial temperature is 20 °C what will be the metal's temperature after the 1000 oscillations?

2.5. In a machine of a particular design, the force acting on a mass of 2 g varies as $-F \sin(\pi x/a)$. Calculate the work required in order to move the mass (a) from $x=0$ to $x=a$, (b) from $x=0$ to $x=2a$.

2.6. A chemical reaction occurs in a vessel of cross-section $100\ \mathrm{cm^2}$ fitted with a loosely fitted seal. During the reaction the seal is pushed out through 10 cm against the external pressure of 1.0 atm. How much work does the reaction do on the outside world?

2.7. Using the same system as in the last Problem, calculate the work done when the pressure of the atmosphere is replaced by a mass of 5.0 kg acting downwards on the vertical piston. What is the work done when the same mass stands on a piston of cross-section $200\ \mathrm{cm^2}$ and moves through the same distance.

2.8. In another reaction occurring within the same apparatus employed in the last Problem, a contraction of 10 cm occurs. Discuss the work involved in the change.

2.9. A 5.0 g lump of solid carbon dioxide is allowed to evaporate in a $100\ \mathrm{cm^3}$ vessel maintained at room temperature. Calculate the work done when the system (treated as a perfect gas) is then allowed to expand at 1.0 atm pressure (a) isothermally against a pressure of 1.0 atm, (b) isothermally and reversibly.

2.10. 1.0 mol $CaCO_3$ was heated to 700 °C, when it decomposed. The operation was carried out in a container closed by a piston which was initially resting on the sample and was restrained throughout by the atmosphere. How much work was done during complete decomposition?

2.11. The same experiment as in the last Problem was repeated, the only difference being that the carbonate was heated in an open vessel. How much work was done during decomposition?

2.12. A strip of magnesium of mass 15 g is dropped into a beaker of dilute hydrochloric acid. What work is done on the surrounding atmosphere (1.0 atm pressure, 25 °C) by the subsequent reaction?

2.13. We have to be able to deal with the work involved in the expansion of real gases. In the case of reversible expansion or compression this involves knowing how the pressure varies with the volume of the enclosed system. Information of this kind is contained in the equation of state, and the next few problems explore the consequences of employing some of the approximate equations that are available. In the first place, calculate the work done in an isothermal, reversible expansion of a gas that satisfies the virial equation of state, $pV_m = RT(1 + B/V_m + \ldots)$.

2.14. The virial equation for argon at 273 K has $B = -21.7\ \mathrm{cm^3\,mol^{-1}}$ and $C = 1200\ \mathrm{cm^6\,mol^{-2}}$. Calculate (a) the work of reversible, isothermal expansion at this temperature, (b) the work of expansion against a constant pressure of 1 atm, (c) the work of expansion on the assumption that argon behaves perfectly. Take $V_i = 500\ \mathrm{cm^3}$, $V_f = 1000\ \mathrm{cm^3}$, and $n = 1.0$ mol.

2.15. The van der Waals gas is a useful model of a real gas, and we know the significance of the parameters a and b. It is instructive to see how these parameters affect the work of isothermal, reversible expansion. Calculate this work of expansion of 1 mol, and account physically for the way that a and b appear in the final expression.

2.16. Plot on the same graph the indicator diagrams for the isothermal expansion of (a) a perfect gas, (b) a van der Waals gas for which $b=0$ and $a = 4.2\ \mathrm{dm^6\,atm\,mol^{-2}}$, (c) the same but with $a=0$ and $b = 5.105 \times 10^{-2}\ \mathrm{dm^3\,mol^{-1}}$. This will show how the a and b parameters modify the area under the isotherm, and therefore the work. The values of a and b selected exaggerate imperfections for normal conditions, but the exaggeration is instructive because of the large distortions of the indicator diagram that result. Take $V_i = 1.0\ \mathrm{dm^3}$, $n = 1.0$ mol, and $T = 298$ K.

2.17. Show that the work of isothermal, reversible expansion of a van der Waals gas can be expressed in reduced variables, and that by defining the *reduced work* as $w_r = 3bw/a$ an expression can be obtained that is independent of the identity of the gas.

2.18. The technique of partial differentiation is straightforward once one realizes that only the variable stated explicitly, e.g. the x in $\partial f/\partial x$, is varied, all others being regarded as constants. In order to get further practice, take (a) the perfect gas equation, (b) the Dieterici equation, Box 1.1, and evaluate $(\partial p/\partial T)_V$ and $(\partial p/\partial V)_T$. Go on to confirm that $\partial^2 p/\partial V\,\partial T = \partial^2 p/\partial T\,\partial V$.

2.19. Partial derivatives can (and, in thermodynamics, do) have real physical meaning. For instance take the expressions for the two partial derivatives of p for a perfect gas obtained in the last Problem and answer the following. In an estimation of the pressure exerted by a gas, the temperature was uncertain by 1 percent and the volume was uncertain by 2 percent. What is the uncertainty in the prediction of the pressure arising from (a) the temperature uncertainty, (b) the volume uncertainty, (c) both.

2.20. The *equipartition principle*, Section 0.1(f), lets us calculate the internal energy of a perfect gas. In the case of a monatomic gas it gives $U=\frac{3}{2}nRT$. This result (and others like it for more complicated molecules) gives a very quick way of predicting the heat capacity of these materials. Deduce the value of the molar heat capacity of (a) a monatomic perfect gas, (b) a gas of rotating, translating, nonlinear, polyatomic molecules.

2.21. When 1 calorie of heat is transferred to 1.0 g of water at 14.5 °C under constant atmospheric pressure its temperature rises to 15.5 °C. What is the molar heat capacity of water at this temperature? Express your answer in cal K^{-1} mol^{-1} and in J K^{-1} mol^{-1}.

2.22. The constant-pressure heat capacity of helium is 20.79 J K^{-1}. How much heat is required to raise the temperature of a sample of 1.0 mol by 10 K at 25 °C (a) when it is in a constant volume vessel at a pressure of 10 atm, (b) when it is in a vessel fitted with a piston subjected to a constant external pressure of 10 atm? How much work is done in each case?

2.23. The heat capacity of air at room temperature and 1.0 atm pressure is approximately 21 J K^{-1} mol^{-1}. How much heat is required to heat an otherwise empty room through 10 K at room temperature? Let the room measure 5.0 m × 5.0 m × 3.0 m. Neglecting losses, how long will a 1.0 kW heater take to heat the room by that amount?

2.24. Assume that the heat capacity of water has the value calculated in Problem 2.21 over its whole liquid range. How much heat must be supplied to raise 1.0 kg of water from room temperature to its boiling point? For how long must a 1.0 kW heater be operated in order to supply this energy?

2.25. A kettle containing 1.0 kg of boiling water is heated until evaporation is complete. Calculate (a) w, (b) q, (c) ΔU, (d) ΔH for the process. ($\Delta H_{vap,m}=40.6$ kJ mol^{-1} at 373 K; treat $H_2O(g)$ as a perfect gas.)

2.26. The same amount of water as in the last Problem is evaporated in a large container where the pressure is only 0.10 atm. Calculate the same four quantities for the process. ($\Delta H_{vap,m}$ at the temperature of boiling water (46 °C) at that pressure is 44 kJ mol^{-1}.)

2.27. In each of the last two Problems the heat was supplied electrically. How much heat has to be supplied in each case, and what power of heater (in watts) is required to bring about total evaporation in 10 minutes in each case? From what height must a 10 kg mass drop to supply the required energy?

2.28. For water in the range 25–100 °C, $C_{p,m}=75.48$ J K^{-1} mol^{-1}. How much heat must be supplied to 1.0 kg of water initially at 25 °C in order to bring it to its boiling point at atmospheric pressure?

2.29. A piston exerting a pressure of 1.0 atm rests on the surface of water at 100 °C. The pressure is reduced infinitesimally, and as a result 10 g of water evaporate. This process absorbs 22.2 kJ of heat. What are the values of q, w, ΔU, ΔH, and ΔH_m for the vaporization?

2.30. One of the factors in the selection of refrigerant fluids is their enthalpy of vaporization. A new fluorocarbon of RMM 102 was synthesized and a small amount was placed in an electrically heated vessel. Under a pressure of 650 mmHg the liquid boiled at 78 °C. When it was boiling it was found that when a current of 0.232 A from a 12.0 V supply was passed for 650 s the quantity of distillate was 1.871 g. What are the values of the molar enthalpy and internal energy of vaporization of the fluorocarbon?

3

The First Law: the machinery

Learning objectives

After careful study of this chapter you should be able to:

(1) Define *extensive* and *intensive properties*, Section 3.3, *state function* and *path function*, Section 3.1(a), and *exact differential*, Section 3.1(b).

(2) Express changes of internal energy and enthalpy in terms of changes in pressure, volume, and temperature, eqns (3.1.3) and (3.2.5).

(3) Define *isobaric thermal expansivity*, eqn (3.2.9), and *isothermal compressibility*, eqn (3.2.11).

(4) Deduce an expression for the temperature dependence of the internal energy at constant pressure, eqn (3.2.10).

(5) Deduce an expression for the temperature dependence of the enthalpy at constant volume, eqn (3.2.16).

(6) Explain the thermodynamic significance of the *Joule experiment* to measure $(\partial U/\partial V)_T$, Section 3.2(a).

(7) Define the *Joule–Thomson coefficient*, eqn (3.2.14), and explain the thermodynamic significance of the *Joule–Thomson experiment*, Section 3.2(c).

(8) Explain the significance of the *inversion temperature* of a gas for the liquefaction of gases, Section 3.2(c).

(9) Derive and use the general relation between C_V and C_p, eqn (3.2.18).

(10) Deduce a general expression for the work done on a perfect gas during an adiabatic change, eqn (3.3.1).

(11) Derive and use an expression for the temperature change of a perfect gas as a result of an irreversible adiabatic expansion, eqn (3.3.2).

(12) Derive and use an expression for the work done during the adiabatic reversible expansion of a perfect gas, eqns (3.3.7) and (3.3.10).

(13) Derive and use an expression relating the pressure and volume of a perfect gas during a reversible adiabatic expansion, eqn (3.3.9).

Introduction

This chapter develops the elementary relations introduced in Chapter 2. The first section is important because it is essential to know the meaning of 'state function'. The next section develops the ways of manipulating thermodynamic expressions, and shows how they can relate apparently unrelated experimental quantities. We shall see that the measurement of a property can often be achieved indirectly by measuring and then combining others.

3.1 State functions and differentials

A common classification of properties is based on whether or not they depend on the size of the sample. For instance, doubling the size of the sample doubles its internal energy. This is an example of an *extensive property*. Other examples are mass, heat capacity, and volume. Properties that are independent of the size of the sample are called *intensive properties*: they include temperature, density, pressure, viscosity, and the molar properties (molar volume, molar heat capacity, etc.).

3.1 (a) State functions

Some properties depend only on the present state of the system but others are related to the way in which that state was prepared. The internal energy and the temperature are examples of the first type of property. The work done on a system is an example of the second kind: we do not speak of a system as possessing some quantity of 'work' or of its 'work' as having a particular value. Properties of the first kind, properties of the present state of the system, are called *state functions*. Properties of the system that depend on the path we call *path functions*.

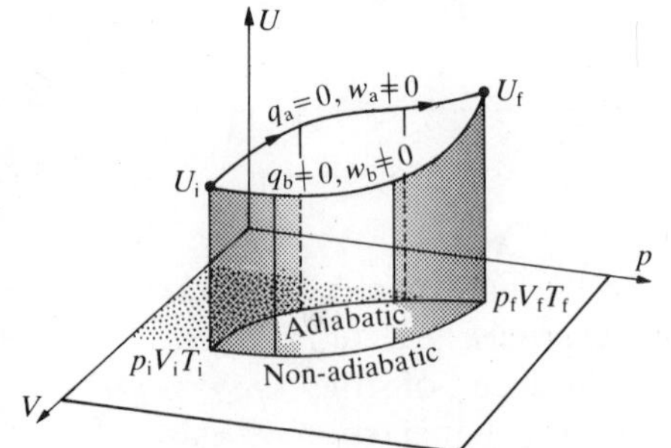

Fig. 3.1. As the pressure and volume of a system are changed (as indicated by the paths in the p,V plane) the internal energy changes (vertical axis). An adiabatic and a non-adiabatic path are shown: they correspond to different values of q and w but to the same value of U. (The paths on the plane can be regarded as indicator diagrams for the process, the work in each case being related to the area beneath them and the V-axis, as explained in Chapter 2.)

The classification of properties of the system as state or path functions is very important in thermodynamics. We can begin to see why by considering the implications of the classification for the First Law. The central point is that *a state function, being a property only of the present state of the system, is independent of how that state was prepared*. This has the following significance for a system undergoing the changes depicted in Fig. 3.1. The initial state of the system is p_i, V_i, T_i (together with whatever variables are needed to specify its composition), and in this state the internal energy is U_i. Work is done on the system so that it is compressed adiabatically to a state p_f, V_f, T_f. In this state it has an internal energy U_f and the work done on the system as it traversed the path is w_a. Notice our use of language: U is a property of the state, w_a is a property of the path.

Now consider another process in which the initial and final states are the same but in which the compression is not adiabatic. The internal energy of the initial state is U_i, the same as before (because it is a state function), and the internal energy of the final state is U_f, also the same as before (for the same reason). But a quantity of energy q_b enters the system as heat, and the work done on it, w_b, is different from w_a. Here we see the difference between state and path functions: both U_i and U_f are unchanged, hence U is a state function; w_a and w_b are different, and q_a and q_b are different; hence these are not state functions.

How do we know that U is a state function? It is implied by the First Law, for otherwise we could get work out of nothing: we could achieve perpetual motion. In order to see this implication, suppose that U is *not* a state function. It will then have a value that depends on the path the system is taken through. Suppose it is taken along Path 1 in Fig. 3.2: U changes from U_i to U_f. Now take it back to the initial state along Path 2: U changes from U_f to U_i', and if U is not a state function U_i' may differ from U_i. Therefore, simply by changing the state of the system from p_i, V_i, T_i to p_f, V_f, T_f and back to p_i, V_i, T_i again we have changed the internal energy from U_i to U_i'. If we choose paths so that U_i' is larger than U_i, the cycle of operations generates an internal energy $\delta U = U_i' - U_i$, which we could turn into useful work even though the system has returned to exactly the same state as before. But experience has shown that perpetual motion machines cannot be constructed. We are therefore forced to conclude that internal energy *is* a state function.

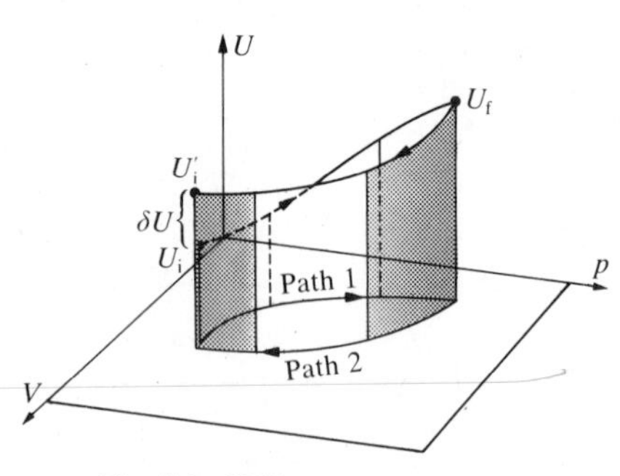

Fig. 3.2. If U were not a state function, an internal energy δU could be generated by going through the cycle of Path 1 followed by Path 2.

3.1 (b) Exact and inexact differentials

An infinitesimal change in U is denoted $\mathrm{d}U$. If a system is taken along a path (e.g. by compressing it isothermally) U changes from U_i to U_f, and the overall change is the sum of all the infinitesimal changes along the path:

$$\Delta U = \int_{i}^{f} \mathrm{d}U = U_f - U_i. \tag{3.1.1}$$

The value of U depends on the initial and final states but is independent of the path. We summarize this path independence of the integral of $\mathrm{d}U$ by saying that $\mathrm{d}U$ is an *exact differential*.

An infinitesimal transfer of energy as heat is denoted $\mathrm{d}q$. If a system is taken from one state to another along a path, the total heat transferred is the sum of all the individual contributions:

$$q = \int_{i,\text{path}}^{f} \mathrm{d}q. \tag{3.1.2}$$

Notice the difference between this and the preceding equation. First, we do not write Δq, because q is not a state function and the energy added as heat cannot be expressed in the form $q_f - q_i$. Secondly, we must specify the path of integration because q depends on the path selected. We summarize this by saying that $\mathrm{d}q$ is an *inexact differential* ('inexact' because to integrate it we must supply further information, the specification of the path). Often $\mathrm{d}q$ is written $\text{đ}q$ in order to emphasize that it is inexact.

An infinitesimal quantity of work done on the system is denoted $\mathrm{d}w$. We know that the work done depends on the path selected; therefore we know immediately that $\mathrm{d}w$ is an inexact differential. It is often written $\text{đ}w$.

Example 3.1

Consider a perfect gas inside a cylinder fitted with a piston. Let the initial state be T, V_i and the final state be T, V_f, so that the net change corresponds to isothermal expansion. The internal energy of a perfect gas is independent of its volume (a result we prove later). The change of state can be brought about in many ways, of which the two simplest are the following. *Path 1*, in which there is free, irreversible expansion against zero external pressure and an influx of whatever heat is necessary from the surroundings to maintain constant temperature. *Path 2*, in which there is reversible, isothermal expansion accompanied by the appropriate influx of heat. Find w, q, and ΔU for each process.

• *Method*. Since we are told that the internal energy of a perfect gas is independent of volume for an isothermal process, $\Delta U = 0$ for each path. We also have $\Delta U = q + w$; it follows that in each case $q = -w$. The work can be calculated from eqns (2.2.7) and (2.2.11) for each path.

• *Answer*. For Path 1, $\Delta U = 0$, $w = 0$ (as $p_{ex} = 0$); therefore $q = 0$. For Path 2, $\Delta U = 0$, $w = -nRT \ln (V_f/V_i)$, and $q = nRT \ln (V_f/V_i)$.

• *Comment*. The work and heat terms depend on the paths, but their sum, ΔU, is independent of the path. Hence w and q are path functions.

• *Exercise*. Calculate the values of q, w, and ΔU for an irreversible isothermal expansion against a constant non-zero external pressure. $[+p_{ex}\Delta V, -p_{ex}\Delta V, 0]$

3.1 (c) Changes in internal energy

The fact that $\mathrm{d}U$ is an exact differential is the basis of the manipulations mentioned in the introduction. We begin to unfold its consequences by noting that U is a function of volume and temperature and writing it $U(V, T)$. (Throughout this chapter we are dealing with a closed system of constant composition. U can be regarded as a function of V, T, and p; but since there is an equation of state it is possible to express p in terms of V and T, and so p is not an independent variable. We could choose p, T or p, V as independent variables, but V, T fit our purpose.)

When V changes to $V + \mathrm{d}V$ at constant temperature, U changes to

$$U(V + \mathrm{d}V, T) = U(V, T) + (\partial U/\partial V)_T\,\mathrm{d}V.$$

The coefficient $(\partial U/\partial V)_T$, the slope of U with respect to V at constant temperature, is the partial derivative of U with respect to V, see Fig. 3.3(a). If T changes to $T + \mathrm{d}T$ at constant volume, the internal energy changes to

$$U(V, T + \mathrm{d}T) = U(V, T) + (\partial U/\partial T)_V\,\mathrm{d}T,$$

and we encounter another partial derivative, Fig. 3.3(b). But now suppose that both V and T change by infinitesimal amounts, Fig. 3.3(c). The new internal energy, neglecting second-order infinitesimals, is

$$U(V + \mathrm{d}V, T + \mathrm{d}T) = U(V, T) + (\partial U/\partial V)_T\,\mathrm{d}V + (\partial U/\partial T)_V\,\mathrm{d}T.$$

The internal energy at $(V + \mathrm{d}V, T + \mathrm{d}T)$ differs from that at (V, T) by the infinitesimal amount $\mathrm{d}U$. Therefore, from the last equation we obtain the very important result that

$$\mathrm{d}U = (\partial U/\partial V)_T\,\mathrm{d}V + (\partial U/\partial T)_V\,\mathrm{d}T. \qquad (3.1.3)$$

The interpretation of this equation is straightforward: in a closed system of constant composition any infinitesimal change in the internal energy is proportional to the infinitesimal changes of volume and temperature, the coefficients of proportionality being the partial derivatives. Very often the

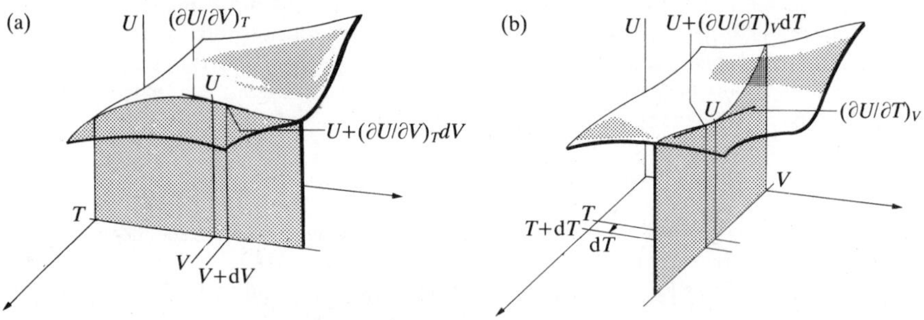

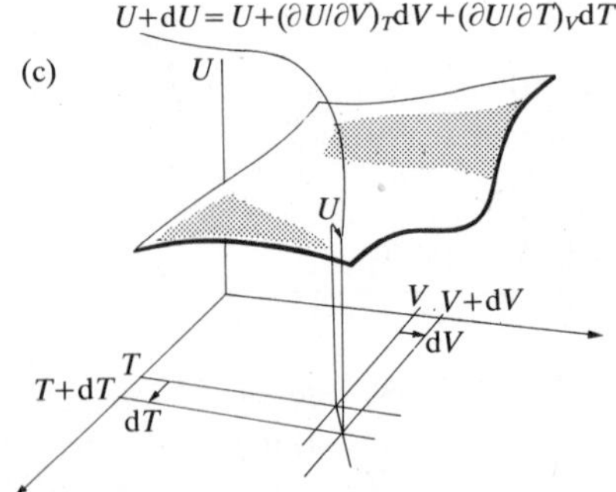

Fig. 3.3. Partial derivatives are slopes of a function with respect to one variable, the others being held constant. In (a) we see how U changes when V is changed, T being constant. In (b) we see the change of U with respect to T, with V constant. So long as we ignore second-order infinitesimals, these two individual changes can be used, as shown in (c), to find the total change of U, $\mathrm{d}U$, when both V and T change infinitesimally.

partial derivatives we meet have an easily discernible physical meaning, and thermodynamics gets shapeless and difficult only when that meaning is not kept in sight. In the present case $(\partial U/\partial T)_V$ has already been encountered in Chapter 2: it is the constant volume heat capacity C_V. We have not previously met the other coefficient, but it is the rate of change of the internal energy as the volume is changed isothermally. This quantity is of major importance in thermodynamics, and we consider it below.

Example 3.2

Measurements of $(\partial U_m/\partial V)_T$ for ammonia give the value 840 J m^{-3} mol^{-1} at 300 K (how this value can be estimated from the van der Waals' parameters is explained in *Example* 6.1). The value of $(\partial U_m/\partial T)_V$, which is $C_{V,m}$, is 27.32 J K^{-1} mol^{-1}. What is the change of molar internal energy of ammonia when it is heated through 2.0 K and compressed through 100 cm^3?

• *Method*. Infinitesimal changes in volume and temperature result in an infinitesimal change of internal energy as given by eqn (3.1.3). The changes in the present problem are small and may be regarded as virtually infinitesimal, and so the equation becomes

$$\Delta U_m \approx (\partial U_m/\partial V)_T \Delta V + (\partial U_m/\partial T)_V \Delta T.$$

• *Answer*. $\Delta U_m = (840\ \text{J m}^{-3}\ \text{mol}^{-1}) \times (-1.00 \times 10^{-4}\ \text{m}^3) + (27.32\ \text{J K}^{-1}\ \text{mol}^{-1}) \times (2.0\ \text{K})$

$$= -0.084\ \text{J mol}^{-1} + 55\ \text{J mol}^{-1} = 55\ \text{J mol}^{-1}.$$

• *Comment*. Note that the change is dominated by the effect of temperature. If ammonia behaved as a perfect gas, the coefficient $(\partial U_m/\partial V)_T = 0$ and the volume change would have no effect on U. In precise work, and for large changes of volume and temperature, dU must be integrated.

• *Exercise*. Confirm that $(\partial U_m/\partial V)_T$ has the same dimensions as pressure, and express the value for ammonia in atmospheres. [8.3×10^{-3} atm]

Partial derivatives have many useful properties, and in the development of thermodynamics we shall draw on them frequently. The kind of property we need to know is that partial derivatives such as $(\partial V/\partial T)_p$ and $(\partial T/\partial V)_p$ are related by expressions as straightforward as $(\partial V/\partial T)_p = 1/(\partial T/\partial V)_p$. There are a number of relations like this, and they are collected in Box 3.1. Adroit use of them can often turn some unfamiliar quantity into something we can recognize and interpret.

Box 3.1 Partial-derivative relations

If f is a function of x and y, then when x and y change by dx and dy, f changes by

$$df = (\partial f/\partial x)_y\, dx + (\partial f/\partial y)_x\, dy.$$

Partial derivatives may be taken in any order:

$$\partial^2 f/\partial x\, \partial y = \partial^2 f/\partial y\, \partial x.$$

In the following, z is a variable on which x and y depend (for example, x, y, z might correspond to p, V, T).

Relation 1. When x is changed at constant z:

$$(\partial f/\partial x)_z = (\partial f/\partial x)_y + (\partial f/\partial y)_x(\partial y/\partial x)_z.$$

Relation 2 (the Inverter).

$$(\partial x/\partial y)_z = 1/(\partial y/\partial x)_z.$$

Relation 3 (the Permuter).

$$(\partial x/\partial y)_z = -(\partial x/\partial z)_y(\partial z/\partial y)_x.$$

By combining this relation and Relation 2 we obtain *Euler's chain relation*:

$$(\partial x/\partial y)_z(\partial y/\partial z)_x(\partial z/\partial x)_y = -1.$$

Relation 4. This relation establishes whether df is an exact differential.

$$df = g\,dx + h\,dy \text{ is exact if } (\partial g/\partial y)_x = (\partial h/\partial x)_y.$$

If df is exact, then its integral between specified limits is independent of the path.

3.2 Manipulating the First Law

We begin by collecting together some of the information met so far. Everything we do will be based on the First Law written in the form:

$$dU = dq + dw. \qquad (3.2.1)$$

The First Law states essentially that U is a state function or, equivalently, that dU is an exact differential. It follows that, for a closed system of constant composition, any change of U can be expressed in terms of changes in V and T as

$$dU = (\partial U/\partial T)_V\,dT + (\partial U/\partial V)_T\,dV. \qquad (3.2.2)$$

Analogous expressions in terms of p and T or p and V are equally acceptable, but are often less convenient.

The work done on a system can be written

$$dw = -p_{ex}\,dV + dw_e, \qquad (3.2.3)$$

where the first term is the work accompanying a change of volume and the second is any other kind of work (e.g. electrical work).

The enthalpy H is defined as

$$H = U + pV. \qquad (3.2.4)$$

U, p, and V are all state functions; therefore H is also a state function and dH is an exact differential. It turns out to be convenient to regard H as a function of p and T (instead of V and T, the choice for U), and so, for a closed system of constant composition,

$$dH = (\partial H/\partial T)_p\,dT + (\partial H/\partial p)_T\,dp. \qquad (3.2.5)$$

The isochoric and isobaric heat capacities are defined as

$$C_V = (\partial U/\partial T)_V, \qquad C_p = (\partial H/\partial T)_p. \tag{3.2.6}$$

Insertion of these quanties into the expressions for dU and dH gives

$$dU = C_V\,dT + (\partial U/\partial V)_T\,dV, \tag{3.2.7}$$

$$dH = C_p\,dT + (\partial H/\partial p)_T\,dp. \tag{3.2.8}$$

3.2 (a) The temperature dependence of the internal energy

If we want to find out how the internal energy depends on the temperature when the *pressure*, rather than the volume, of the system is kept constant, we use the relations in Box 3.1 to extract an expression for $(\partial U/\partial T)_p$ from eqn (3.2.7).

Relation No. 1 takes an expression like eqn (3.2.7), divides through by dT (to give dU/dT on the left), and then imposes the condition of constant pressure. It yields

$$(\partial U/\partial T)_p = C_V + (\partial U/\partial V)_T(\partial V/\partial T)_p.$$

At this point, always inspect the output of a relation to see if it contains any recognizable physical quantity. The final differential coefficient in this expression is the rate of change of volume with increase of temperature (at constant pressure). This is a readily accessible property which is normally tabulated as the *isobaric thermal expansivity* or *isobaric coefficient of thermal expansion*, defined as

$$\alpha = (1/V)(\partial V/\partial T)_p, \tag{3.2.9}$$

(i.e. as the rate of change of volume with temperature per unit volume). Introduction of this definition gives

$$(\partial U/\partial T)_p = C_V + \alpha V(\partial U/\partial V)_T. \tag{3.2.10}$$

Example 3.3

At 300 K the isobaric coefficient of thermal expansion of neon is $3.3 \times 10^{-3}\,\mathrm{K}^{-1}$. What volume change occurs when a $50\,\mathrm{cm}^3$ sample is heated through 5.0 K?

- *Method*. For an infinitesimal change of temperature the volume changes by $dV = (\partial V/\partial T)_p\,dT = \alpha V\,dT$. For a very small change of temperature this can be approximated by $\Delta V = \alpha V\,\Delta T$.

- *Answer*. $\Delta V = (3.3 \times 10^{-3}\,\mathrm{K}^{-1}) \times (50\,\mathrm{cm}^3) \times (5.0\,\mathrm{K}) = 0.83\,\mathrm{cm}^3$.

- *Comment*. The approximation is valid so long as α is effectively constant over the temperature range considered. Note that the relation between dV and dT makes no assumption about whether the gas is perfect or not (and applies to liquids and solids too).

- *Exercise*. For copper $\alpha = 5.01 \times 10^{-5}\,\mathrm{K}^{-1}$. Calculate the change of volume when a $50\,\mathrm{cm}^3$ sample is heated through 5.0 K. [$12\,\mathrm{mm}^3$]

The last equation is entirely general (so long as the system is closed and its composition constant). It expresses the dependence of the internal energy on the temperature at constant pressure in terms of the constant volume heat capacity, which can be measured in one experiment, the coefficient of expansion α, which can be measured in another (Table 3.1), and the ubiquitous quantity $(\partial U/\partial V)_T$, which, if we knew what it was, could probably be measured in a third.

Table 3.1. Isobaric volume expansivities α and isothermal compressibilities κ

Substance	$10^4\ \alpha/\mathrm{K}^{-1}$	$10^6\ \kappa/\mathrm{atm}^{-1}$
Water	2.1	49.6
Benzene	12.4	90.5
Lead	0.861	2.21
Diamond	0.030	0.187

What, then, is this $(\partial U/\partial V)_T$? When a material is compressed isothermally, the interactions between its particles are increased. These interactions contribute to the internal energy of the system, and so the coefficient measures how the interactions change when the volume of the sample is changed. This interpretation is supported by noting that in the case of a van der Waals gas its value is a/V^2 (a result we obtain in Chapter 6). We might suspect that the coefficient is smaller for gases than for solids because we know that interactions are less important in gases. We might also suspect that it is zero for a perfect gas. James Joule thought that he could measure $(\partial U/\partial V)_T$ by observing the change in temperature of a gas when it is allowed to expand into a vacuum. He used two vessels immersed in a water bath, Fig. 3.4. One was filled with air at 22 atm, the other was evacuated. Using the same sensitive thermometer that he had used in his experiments on the 'mechanical equivalent of heat' he attempted to measure the change in temperature of the water of the bath when a stopcock was opened and the air expanded into a vacuum. He observed no change.

The thermodynamic implications of the experiment are as follows. No work was done in the expansion into a vacuum, and so $w=0$. No heat entered or left the system (the gas) because the temperature of the bath did not change, and so $q=0$. Consequently, within the accuracy of the experiment, $\Delta U=0$. It follows that U does not change much when a gas expands isothermally. Therefore, at constant temperature, U is (at least approximately) independent of volume. Expressed mathematically, $(\partial U/\partial V)_T=0$.

Joule's experiment was crude. In particular, the heat capacity of the water bath was so large that the temperature change that gases do in fact cause was too small to measure. His experiment was on a par with Boyle's: he observed a property characteristic of a perfect gas, without detecting the small deviations characteristic of real gases.

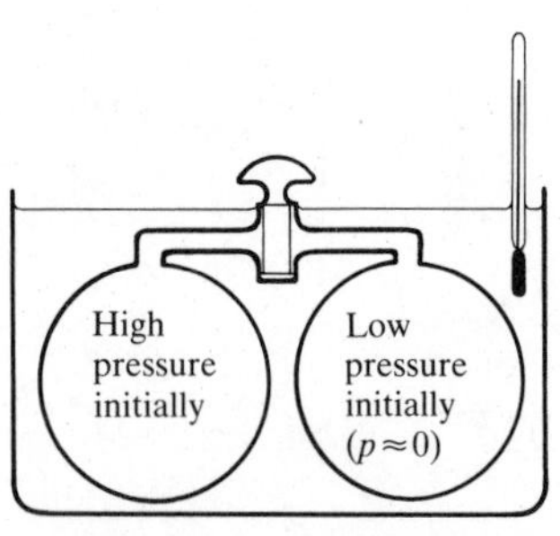

Fig. 3.4. A diagram of the apparatus used by Joule in an attempt to measure the change in internal energy when a gas expands isothermally. The heat absorbed by the gas can be related to the change in temperature of the bath.

3.2 (b) The temperature dependence of the enthalpy

We have seen how to express the temperature dependence of the internal energy under conditions of either constant volume or constant pressure. Now we turn to the temperature dependence of the enthalpy at constant volume. At the end of this piece of work, which also gives practice in manipulating thermodynamic relations, we shall have expressions for $(\partial U/\partial T)_V$, $(\partial U/\partial T)_p$, $(\partial H/\partial T)_V$, and $(\partial H/\partial T)_p$, and we shall be able to predict how U and H change when the temperature of any closed, constant composition system is changed.

We start as before, and act on eqn (3.2.8) with relation No. 1. This time we impose constant volume, and obtain

$$(\partial H/\partial T)_V=C_p+(\partial H/\partial p)_T(\partial p/\partial T)_V.$$

The final differential coefficient looks like something we ought to recognize, and is perhaps related to $(\partial V/\partial T)_p$, the thermal expansion coefficient. Relation No. 3 shuffles p, V, and T around inside partial differentials, and acting on $(\partial p/\partial T)_V$ it produces $-1/(\partial T/\partial V)_p(\partial V/\partial p)_T$. Unfortunately

$(\partial T/\partial V)_p$ occurs instead of $(\partial V/\partial T)_p$, but relation No. 2 inverts partial differentials, and so application of it leads to

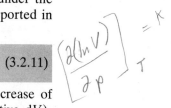

$$(\partial p/\partial T)_V = -(\partial V/\partial T)_p/(\partial V/\partial p)_T = -\alpha V/(\partial V/\partial p)_T.$$

The coefficient $(\partial V/\partial p)_T$ is a measure of the change of volume under the influence of pressure at constant temperature. This is normally reported in terms of the *isothermal compressibility*, defined as

$$\kappa = -(1/V)(\partial V/\partial p)_T. \qquad (3.2.11)$$

The negative sign here ensures that κ is positive, because an increase of pressure (positive dp) brings about a reduction of volume (negative dV), and $(\partial V/\partial p)_T$ itself is always negative. Some values of κ are listed in Table 3.1.

Example 3.4

The isothermal compressibility of water at 20 °C and 1 atm is $49.6 \times 10^{-6}\ \text{atm}^{-1}$. What change of volume occurs when a 50 cm^3 sample is subjected to an additional 1000 atm pressure?

• *Method*. For an infinitesimal change of pressure the volume changes by $dV = (\partial V/\partial p)_T\,dp = -\kappa V\,dp$. Therefore, for the state change, integrate both sides and obtain

$$\Delta V = -\int \kappa V\,dp.$$

If V and κ are effectively constant over the pressure range, they may be treated as constant, so that

$$\Delta V = -\kappa V\,\Delta p.$$

Check the assumption for self-consistency after calculating ΔV.

• *Answer*. $\Delta V = -(49.6 \times 10^{-6}\ \text{atm}^{-1}) \times (50\ \text{cm}^3) \times (1000\ \text{atm}) = -2.5\ \text{cm}^3$.

• *Comment*. The assumption of constant V and κ is probably marginally acceptable. Note that very high pressures are needed to bring about significant changes of volume.

• *Exercise*. A 50 cm^3 sample of copper is subjected to a additional pressure of 100 atm and a temperature increase of 5.0 K. Estimate the total change of volume. [8.8 mm^3]

Collecting all the fragments of the preceding calculation leads to

$$(\partial H/\partial T)_V = C_p + (\alpha/\kappa)(\partial H/\partial p)_T. \qquad (3.2.12)$$

The coefficient $(\partial H/\partial p)_T$ is the analogue of $(\partial U/\partial V)_T$ in eqn (3.2.10): V and $-p$ play analogous roles in U and H respectively. We might suspect that it can be measured in a similar way. The next few paragraphs explore this point.

Can the coefficient $(\partial H/\partial p)_T$ be changed into something recognizable? Shuffling H, p, and T leads to something in terms of $(\partial T/\partial p)$, which looks as though it is a directly measurable quantity. Using relation No. 3 we find

$$(\partial H/\partial p)_T = -1/(\partial p/\partial T)_H(\partial T/\partial H)_p.$$

A double use of the inverter, relation No. 2, then gives

$$(\partial H/\partial p)_T = -(\partial T/\partial p)_H(\partial H/\partial T)_p = -(\partial T/\partial p)_H C_p. \qquad (3.2.13)$$

We seem to be on the right track because the other term generated is simply the heat capacity C_p, and so we have not generated something more complicated than we started with (an occupational hazard in thermodynamics: when it happens be prepared to start again).

3.2 (c) The Joule–Thomson effect

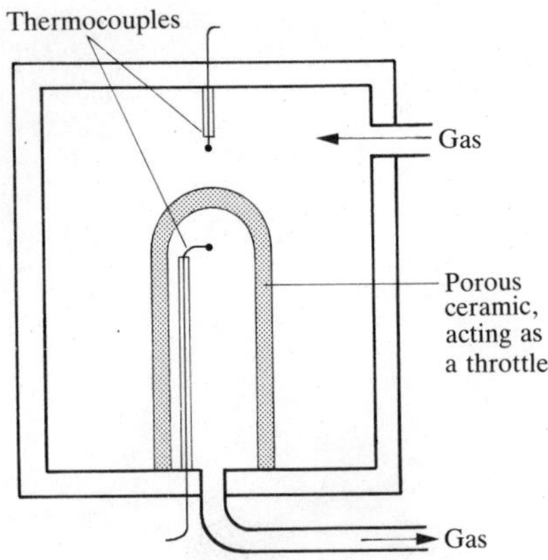

Fig. 3.5. A diagram of the apparatus used for measuring the Joule–Thomson effect. The gas expands through the throttle, and the whole apparatus is thermally insulated. As explained in the text, this arrangement corresponds to an *isenthalpic expansion* (i.e. an expansion at constant enthalpy). Whether the expansion results in a heating or a cooling of the gas depends on the conditions.

All would be well if we could measure $(\partial T/\partial p)_H$, the change in temperature accompanying a change of pressure at constant enthalpy: but how is the constraint imposed? The cunning required was provided by Joule and William Thomson (later Lord Kelvin), who not only solved this problem but found a way of making the experiment very sensitive. They recognized that the fault with the earlier Joule experiment was that the heat capacity of the water bath was too great, which resulted in the temperature changes being unmeasurably small. Joule and Thomson had the good idea to use the gas (with its low heat capacity) as its own heat bath and to establish a steady state flow so that the effect was easier to measure, Fig. 3.5. They let gas expand through a throttle from one constant pressure to another and monitored the difference of temperature. The whole apparatus was insulated so that the process was adiabatic. They observed a lower temperature on the low pressure side, the difference being proportional to the pressure difference they maintained. This is now called the *Joule–Thomson effect*.

The thermodynamic implication of the experiment is as follows. The system is adiabatic, and so $q = 0$. In order to calculate the work done as the gas passes through the throttle, consider the passage of a fixed amount from the high pressure side (where the pressure is p_i, the temperature T_i, and the gas occupies a volume V_i) to the low pressure side (where the same amount of gas is at a pressure p_f, a temperature T_f, and occupies a volume V_f. The gas on the left, Fig. 3.6, is compressed isothermally by the upstream gas acting as a piston. The relevant pressure is p_i and the volume changes from V_i to 0; therefore the work done on the gas is $-p_i(0 - V_i)$, or p_iV_i. The gas expands isothermally on the right of the throttle (but possibly at a different constant temperature) against the pressure p_f provided by the downstream gas acting as a piston to be driven out. The volume changes from 0 to V_f, and so the work done on the gas in this stage is $-p_f(V_f - 0)$, or $-p_fV_f$. The

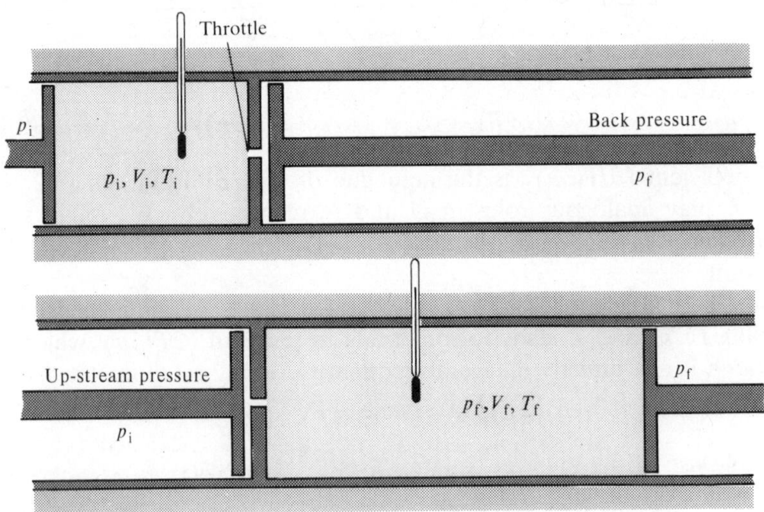

Fig. 3.6. A diagram representing the thermodynamic basis of Joule–Thomson expansion. The pistons represent the upstream and downstream gases, which maintain constant pressures either side of the throttle. The transition from the upper diagram to the lower, which represents the passage of a given amount of gas through the throttle, occurs without change of enthalpy.

total work done on the gas is the sum of these two quantities, or $p_iV_i - p_fV_f$. It follows that the change of internal energy of the gas as it moves from one side of the throttle to the other is $U_f - U_i = w = p_iV_i - p_fV_f$. Reorganizing this gives $U_f + p_fV_f = U_i + p_iV_i$, or $H_f = H_i$. Therefore, the expansion occurs without change of enthalpy: it is *isenthalpic*. The quantity observed is the temperature change per unit change of pressure, $\Delta T/\Delta p$. Adding the constraint of constant enthalpy and taking the limit of small Δp implies that the thermodynamic quantity measured is $(\partial T/\partial p)_H$. This is now called the *Joule–Thomson coefficient*:

$$\mu_{JT} = (\partial T/\partial p)_H. \quad (3.2.14)$$

The modern method of measuring μ_{JT} is indirect, and involves measuring the *isothermal Joule–Thomson coefficient*, the quantity $(\partial H/\partial p)_T$. According to eqn (3.2.13) the two coefficients are related by

$$(\partial H/\partial p)_T = -\mu_{JT}C_p. \quad (3.2.15)$$

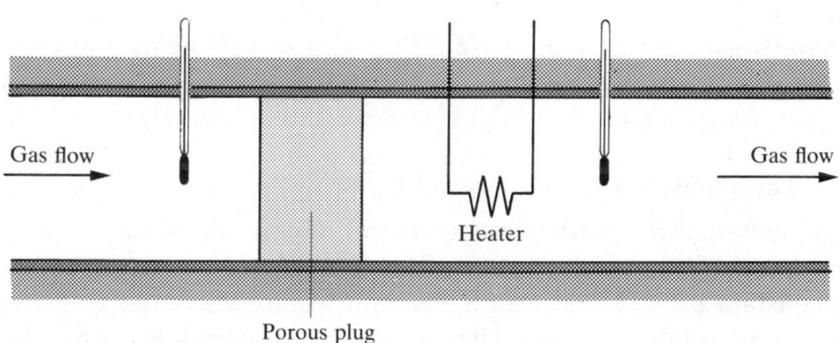

Fig. 3.7. A diagram of the apparatus used for measuring the isothermal Joule–Thomson coefficient. The electrical heating required to offset the cooling arising from expansion is interpreted as ΔH and used to calculate $(\partial H/\partial p)_T$, which is then converted to μ_{JT} as explained in the text.

The procedure is to pump the gas continuously at a steady pressure through a heat exchanger (which brings it to the required temperature), and then through a throttle inside a thermally insulated container. The steep pressure drop is measured, and the cooling effect is exactly offset by an electric heater placed immediately after the throttle, Fig. 3.7. The energy provided by the heater is monitored. Since the heat can be identified with the value of ΔH for the gas, and the pressure change Δp is known, the value of $(\partial H/\partial p)_T$ can be obtained from the limiting value of $\Delta H/\Delta p$ as $\Delta p \to 0$, and then converted to μ_{JT} using eqn (3.2.15). Some values of μ_{JT} obtained in this way are listed in Table 3.2.

Real gases have non-zero Joule–Thomson coefficients in general (even in the limit of zero pressure). The sign of the coefficient may be either positive or negative. A positive sign implies that $\mathrm{d}T$ is negative when $\mathrm{d}p$ is negative, in which case the gas cools on expansion. The sign and magnitude of μ_{JT} depend on the gas and the conditions. Gases that show a heating effect ($\mu_{JT} < 0$) show a cooling effect ($\mu_{JT} > 0$) when the temperature is lowered to below their *inversion temperature*, T_I, Table 3.2. In the case of a van der Waals gas $T_I = 2a/Rb$, and so we expect $T_I = 2T_B$, where T_B is the Boyle temperature, Section 1.3(b). This is in broad agreement with the data in the table.

Table 3.2. Inversion temperatures, normal freezing and boiling points, and Joule–Thomson coefficients at 1 atm and 298 K

	T_I/K	T_m/K	T_b/K	μ_{JT}/K atm^{-1}
He	40		4.6	−0.060
Ar	723	83.8	87.3	
N_2	621	63.2	77.4	0.25
CO_2	1500		194.6	1.11

The technological importance of the Joule–Thomson effect lies in its application to the cooling and liquefaction of gases. The *Linde refrigerator* works on the principle that below its inversion temperature a gas cools on expansion. It is then essential to know the conditions under which μ_{JT} is

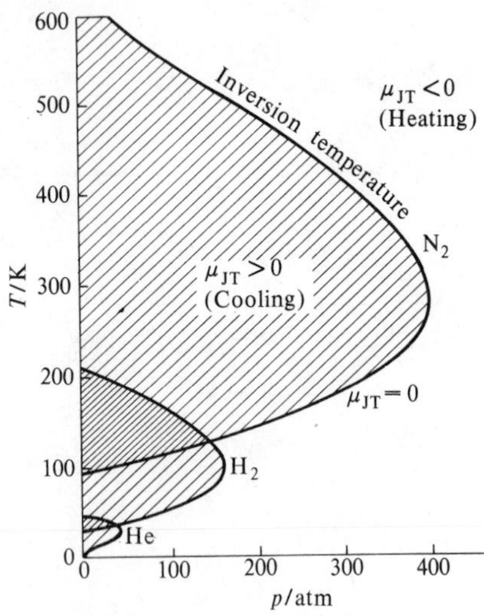

Fig. 3.8. The sign of the Joule–Thomson coefficient depends on the conditions. Inside the boundary, the shaded areas, it is positive and outside it is negative. The temperature corresponding to the boundary at a given pressure is called the *inversion temperature* of the gas at that pressure. Note that, for a given pressure, the temperature must be below a certain value if cooling is required, but if it becomes too low, the boundary is crossed again and heating occurs.

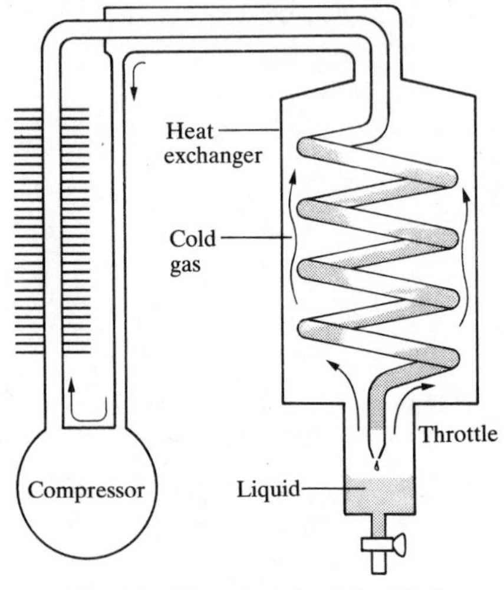

Fig. 3.9. The principle of the *Linde refrigerator* is shown in this diagram. The gas is recirculated, and so long as it is beneath its inversion temperature it cools on expansion through the throttle. The cooled gas cools the high-pressure gas, which cools still further as it expands. Eventually liquefied gas drips from the throttle.

positive: this is illustrated for three gases in Fig. 3.8. The principle is illustrated in Fig. 3.9: after recirculation of the cooling gas, and for a big enough pressure drop across the throttle, the temperature falls below the condensation temperature, and the liquid forms. Note the importance of working beneath the inversion temperature, and therefore the need to cool some gases by other means initially: using helium at room temperature would turn the refrigerator into an expensive oven.

For a perfect gas $\mu_{JT} = 0$ and its temperature is unchanged on Joule–Thomson expansion. This points clearly to the involvement of intermolecular forces in determining the size of the effect, and later we shall look at this in more detail. Note that the Joule–Thomson coefficient of a real gas does not necessarily approach zero as the pressure is reduced: μ_{JT} is an example of one of the properties mentioned in Section 1.3(b) that depends on derivatives and not on p, V, T themselves.

We now return to the development of eqn (3.2.12). On introducing eqn (3.2.15) we find

$$(\partial H/\partial T)_V = (1 - \alpha\mu_{JT}/\kappa)C_p. \tag{3.2.16}$$

This is the final equation for $(\partial H/\partial T)_V$, and it applies to any substance. Everything in it can be measured in suitable experiments, and so we now know how H varies with T.

3.2 (d) The relation between C_V and C_p

We have seen that properties under one set of conditions may be expressed in terms of properties under others. The heat capacities at constant volume and at constant pressure are examples, and we are now equipped to derive an expression relating them. The approach we adopt illustrates how to tackle similar problems.

C_p differs from C_V on account of the work needed to change the volume of the system when the pressure, not the volume, is held constant. This work arises in two ways. One is the work of driving back the atmosphere, the other is the work of stretching the bonds in the material, including the weak intermolecular (van der Waals) interactions. In the case of a perfect gas, the second makes no contribution. We shall now derive a general relation between the two heat capacities, and show that it reduces to the perfect gas result in the absence of intermolecular forces.

When tackling a problem in thermodynamics a useful rule is to *go back to first principles*. In the present problem we do this twice, first by expressing C_p and C_V in terms of their definitions:

$$C_p - C_V = (\partial H/\partial T)_p - (\partial U/\partial T)_V,$$

and then by inserting the definition $H = U + pV$:

$$C_p - C_V = (\partial U/\partial T)_p + (\partial pV/\partial T)_p - (\partial U/\partial T)_V.$$

The difference of the first and third quantities has already been calculated in eqn (3.2.10):

$$(\partial U/\partial T)_p - (\partial U/\partial T)_V = \alpha V(\partial U/\partial V)_T.$$

αV gives the change of volume when the temperature is raised, and $(\partial U/\partial V)_T$ converts this change of volume into a change of internal energy. The remaining term is simplified by noting that, since p is constant,

$$(\partial pV/\partial T)_p = p(\partial V/\partial T)_p = \alpha pV.$$

The middle term of this expression identifies it as the contribution to the work of pushing back the atmosphere: $(\partial V/\partial T)_p$ is the change of volume caused by a change of temperature, and multiplication by p converts this into work.

Collecting the two contributions gives

$$C_p - C_V = \alpha\{p + (\partial U/\partial V)_T\}V. \qquad (3.2.17)$$

This is an entirely general relation, applicable to all materials. It contains the ubiquitous quantity $(\partial U/\partial V)_T$. At this point we can go further using a result we prove in Chapter 6, that

$$(\partial U/\partial V)_T = T(\partial p/\partial T)_V - p.$$

When this is inserted in the last equation we obtain

$$C_p - C_V = \alpha TV(\partial p/\partial T)_V.$$

We can do even better than this. In the previous section we also encountered the coefficient $(\partial p/\partial T)_V$, and the use of relations Nos. 3 and 2 turned it into α/κ, eqn (3.2.12). Therefore,

$$C_p - C_V = (\alpha^2/\kappa)TV. \qquad (3.2.18)$$

This is a thermodynamic expression, which means that it applies to any substance (i.e. that it is 'universally true').

What does it tell us? Since thermal expansivities (α) of liquids and solids are small, it is tempting to say for them that C_p must be virtually equal to C_V. But this might be wrong, because the compressibility κ might also be small and so α^2/κ might be large. Put another way, we might look at eqn (3.2.17), and get as far as α, which might be small, but not look as far as $p + (\partial U/\partial V)_T$, which might be large. In some cases, in fact, the two heat capacities differ by as much as 30 per cent.

Example 3.5

The molar heat capacity of water at constant volume is 74.8 $\mathrm{K^{-1}\,mol^{-1}}$ at 25 °C. Estimate its constant pressure heat capacity.

• *Method*. For molar quantities eqn (3.2.18) reads

$$C_{p,\mathrm{m}} - C_{V,\mathrm{m}} = (\alpha^2/\kappa)TV_\mathrm{m}.$$

Use data from Table 3.1. The density of water at 25 °C is 0.997 04 $\mathrm{g\,cm^{-3}}$, and the molar mass of H_2O is 18.02 $\mathrm{g\,mol^{-1}}$. Hence the molar volume $V_\mathrm{m} = 18.07\ \mathrm{cm^3\,mol^{-1}}$. Express κ in coherent units (i.e. convert $\mathrm{atm^{-1}}$ to $\mathrm{N^{-1}\,m^2}$).

• *Answer*. Since 1 atm = $1.013\,25 \times 10^5\ \mathrm{N\,m^{-2}}$,

$$\kappa = (4.96 \times 10^{-5}\ \mathrm{atm^{-1}})/(1.013\,25 \times 10^5\ \mathrm{N\,m^{-2}\,atm^{-1}}) = 4.90 \times 10^{-10}\ \mathrm{N^{-1}\,m^2}.$$

Substitution of the data into the equation above then gives $C_{p,\mathrm{m}} = 74.8\ \mathrm{J\,K^{-1}\,mol^{-1}} + 0.48\ \mathrm{J\,K^{-1}\,mol^{-1}} = 75.3\ \mathrm{J\,K^{-1}\,mol^{-1}}$.

• *Comment*. We have used 1 N m = 1 J. The calculation shows that there is a small but significant difference between the two heat capacities at 25 °C.

• *Exercise*. The expansivity of benzene is much larger than for water (Table 3.1). Find the value of $C_{V,\mathrm{m}}$ at 25 °C given the following data: $C_{p,\mathrm{m}} = 134\ \mathrm{J\,K^{-1}\,mol^{-1}}$; density = 0.879 g $\mathrm{cm^{-3}}$; molar mass = 78.12 $\mathrm{g\,mol^{-1}}$. [89 $\mathrm{J\,K^{-1}\,mol^{-1}}$]

In the case of a perfect gas we can substitute $pV=nRT$ into the equations defining α and κ and deduce that

$$\alpha=(1/V)(\partial V/\partial T)_p=1/T \tag{3.2.19°}$$

$$\kappa=-(1/V)(\partial V/\partial p)_T=1/p. \tag{3.2.20°}$$

It follows immediately that

$$C_p-C_V=(1/T^2)pTV=nR. \tag{3.2.21°}$$

as we deduced by an elementary method in Section 2.3. If α and κ can be calculated from other equations of state, then C_p-C_V can be calculated for real gases too: this is explored in the Problems.

3.3 Work of adiabatic expansion

When a system expands reversibly the work done on it is

$$w=-\int_i^f p\,dV$$

(see eqn (2.2.9); p is the pressure of the gas, p_{in}). In the case of a perfect gas we can relate the pressure and volume through $p=nRT/V$. In the case of isothermal expansion T is constant, but during an adiabatic change the temperature changes. In other words, T depends on V. How, then, can we evaluate the integral and calculate w for an adiabatic change?

Another useful rule in thermodynamics is: *instead of calculating what you are asked for, find a state function related to it, and calculate the change in that function by the most convenient path*. In the present case, because the change is adiabatic, $dq=0$ at each stage of the expansion. Consequently, $dU=dw$. Therefore, instead of calculating the work done during the expansion, we can calculate the change in internal energy between the same initial and final states:

$$w=\int_i^f dU.$$

Equation (3.2.7) is a general expression for dU. In the case of a perfect gas $(\partial U/\partial V)_T=0$, and so

$$w=\int_i^f C_V\,dT.$$

For many gases C_V is almost independent of temperature (this remark is justified in Part 2), and so the integration is very simple:

$$w=C_V\int_{T_i}^{T_f} dT=C_V(T_f-T_i)=C_V\,\Delta T. \tag{3.3.1°}$$

Hence, the work done during an adiabatic expansion is proportional to the temperature difference between the initial and final states.

But what is that temperature difference? That is one of the items left out of the calculation. A few other things are missing too: we have not specified whether the change in the system is expansion or contraction, nor have we specified whether it has been done reversibly or irreversibly. The equation applies to *any* adiabatic expansion or contraction of a perfect gas, reversible or irreversible.

3.3 (a) Special cases

Before sorting out these points there is a general conclusion: if w is negative (so that the system has *done* work), then the equation shows that T_f is less than T_i irrespective of whether the change was reversible or irreversible. This should not be surprising: if the adiabatic system does work, then since no heat can enter, the internal energy must fall; but because the internal energy of a perfect gas is unaffected by changes of volume alone, a reduction of internal energy on expansion must mean that the temperature has fallen too.

3.3 (b) Irreversible adiabatic expansion

If expansion occurs against zero external pressure, then no work is done. Therefore, $w=0$, and consequently $\Delta T=0$. This a peculiar case: the expansion is simultaneously adiabatic and isothermal.

If the expansion occurs against fixed external pressure then the work done on the system is $w=-p_{ex}\Delta V$ (the work done *by* the system is $w'=p_{ex}\Delta V$). This is a general result, and applies equally to isothermal and adiabatic changes. Since this quantity of work must also equal the work calculated using eqn (3.3.1) for a perfect gas, we can find the change of temperature that accompanies this irreversible adiabatic expansion:

$$\Delta T=-p_{ex}\Delta V/C_V. \qquad (3.3.2)^\circ$$

Notice that on expansion ΔV is positive and so ΔT is negative: the temperature falls. The signs are looking after themselves, but it is important to learn their language. Also notice that if the external pressure is zero, the temperature does not change when the expansion occurs: this accords with the first case we considered. (It is always sensible to check an equation by seeing whether it reduces to a simpler, known result.)

Example 3.6

A sample of 2.0 mol Ar in a cylinder of 5.0 cm^2 cross-sectional area at a pressure of 5.0 atm is allowed to expand adiabatically against an external pressure of 1.0 atm. During the expansion it pushes a piston through 1.0 m. If the initial temperature is 300.0 K, what is the final temperature of the gas?

• *Method*. The expansion is (a) adiabatic, (b) irreversible; therefore use eqn (3.3.2). Take $C_{V,m}$ from Table 2.1.

• *Answer*. $\Delta T=-\dfrac{(1.013\times10^5\ \mathrm{N\ m^{-2}})\times(5.0\times10^{-4}\ \mathrm{m^2}\times1.0\ \mathrm{m})}{(2.0\ \mathrm{mol})\times(12.48\ \mathrm{J\ K^{-1}\ mol^{-1}})}=-2.0\ \mathrm{K}.$

Therefore, the temperature falls to 298.0 K.

• *Comment*. The temperature falls because energy has been extracted as work and none has returned as heat.

• *Exercise*. Calculate the final temperature and the internal energy when 2.0 mol NH_3 is used instead, the other conditions being unchanged. [299.1 K, −51 J]

3.3 (c) Reversible adiabatic expansion

If the expansion is reversible at every stage the external pressure is matched to the internal throughout the process. Consider some stage during which the pressure (inside and out) is p, then when the volume changes by $\mathrm{d}V$ the work is $\mathrm{d}w = -p\,\mathrm{d}V$. Since $\mathrm{d}q = 0$, it follows that $\mathrm{d}U = -p\,\mathrm{d}V$. However, by the argument above, for a perfect gas $\mathrm{d}U = C_V\,\mathrm{d}T$. These two terms must be equal ($\mathrm{d}U$ is an exact differential, independent of path), and so at each stage $C_V\,\mathrm{d}T = -p\,\mathrm{d}V$. At each stage the perfect gas obeys the equation of state $pV = nRT$, and so this equality becomes

$$C_V\,\mathrm{d}T/T = -nR\,\mathrm{d}V/V.$$

Since C_V may be regarded as independent of temperature (this is true of monatomic perfect gases, and approximately true of others), both sides may be integrated between the initial and final states:

$$C_V\int_{\mathrm{i}}^{\mathrm{f}} \mathrm{d}T/T = -nR\int_{\mathrm{i}}^{\mathrm{f}} \mathrm{d}V/V,$$

or

$$C_V \ln (T_{\mathrm{f}}/T_{\mathrm{i}}) = -nR \ln (V_{\mathrm{f}}/V_{\mathrm{i}}).$$

On writing $c = C_V/nR = C_{V,\mathrm{m}}/R$ and doing a little rearranging, we get

$$\ln (T_{\mathrm{f}}/T_{\mathrm{i}})^c = \ln (V_{\mathrm{i}}/V_{\mathrm{f}}),$$

which means that

$$V_{\mathrm{f}}T_{\mathrm{f}}^c = V_{\mathrm{i}}T_{\mathrm{i}}^c. \qquad (3.3.3)^{\circ}_{\mathrm{r}}$$

This is our initial goal. It means that we can now predict the temperature of a perfect gas that has expanded (or contracted) *adiabatically* and *reversibly* from a volume V_{i} to a volume V_{f}, its initial temperature being T_{i}:

$$T_{\mathrm{f}} = (V_{\mathrm{i}}/V_{\mathrm{f}})^{1/c}T_{\mathrm{i}}. \qquad (3.3.4)^{\circ}_{\mathrm{r}}$$

The work done as the volume changes is obtained by substituting this result into $w = C_V\,\Delta T$:

$$w = C_V(T_{\mathrm{f}} - T_{\mathrm{i}}) = C_V T_{\mathrm{i}}\{(V_{\mathrm{i}}/V_{\mathrm{f}})^{1/c} - 1\}. \qquad (3.3.5)^{\circ}_{\mathrm{r}}$$

This is normally expressed in terms of the *heat capacity ratio*

$$\gamma = C_p/C_V. \qquad (3.3.6)$$

Then, since for a perfect gas $C_p - C_V = nR$, it follows that $1/c = \gamma - 1$, and therefore that

$$w = C_V T_{\mathrm{i}}\{(V_{\mathrm{i}}/V_{\mathrm{f}})^{\gamma-1} - 1\}. \qquad (3.3.7)^{\circ}_{\mathrm{r}}$$

Example 3.7

A sample of argon at 1.0 atm pressure and 25 °C expands reversibly and adiabatically from 0.50 dm^3 to 1.00 dm^3. Calculate its final temperature, the work done during the expansion, and the change in internal energy.

• *Method*. The final temperature is calculated from eqn (3.3.4) and the work from the first equality in eqn (3.3.5). Take $C_{V,\mathrm{m}}$ from Table 2.1 and, in order to convert this to C_V, calculate n from $pV = nRT$. For ΔU use $\Delta U = q + w$ with $q = 0$ as the process is adiabatic.

• *Answer*. $c = (12.48\ \mathrm{J\,K^{-1}\,mol^{-1}})/(8.314\ \mathrm{J\,K^{-1}\,mol^{-1}}) = 1.501.$

$n = (1.0\ \mathrm{atm}) \times (0.50\ \mathrm{dm^3})/(0.0821\ \mathrm{dm^3\,atm\,K^{-1}\,mol^{-1}}) \times (298.15\ \mathrm{K}) = 0.020_4\ \mathrm{mol}.$

Then, from eqn (3.3.4):

$$T_f = (0.50\ \text{dm}^3/1.00\ \text{dm}^3)^{1/1.501} \times (298.15\ \text{K}) = 188\ \text{K},$$

and from eqn (3.3.5):

$$w = (0.020_4\ \text{mol}) \times (12.48\ \text{J K}^{-1}\ \text{mol}^{-1}) \times (188\ \text{K} - 298.15\ \text{K}) = -28\ \text{J}.$$

Since $q = 0$, $\Delta U = -28$ J.

• *Comment*. If the work had been done reversibly and isothermally, an amount $-nRT \ln 2 = -35$ J would have been done. That is, $w' = 35$ J in place of $w' = 28$: more is done on the outside world on account of the influx of energy as heat that helps to raise the pressure inside the system.

• *Exercise*. Calculate the final temperature, the work done, and the change of internal energy when ammonia is used in a reversible adiabatic expansion from 0.50 dm³ to 2.00 dm³, the other initial conditions being the same. [196 K, −57 J, −57 J]

It is now easy to find the relation between p and V when a perfect gas undergoes reversible adiabatic change. From the perfect gas equation and eqn (3.3.4) we have respectively

$$p_iV_i/p_fV_f = T_i/T_f \quad \text{and} \quad T_i/T_f = (V_f/V_i)^{1/c} = (V_f/V_i)^{\gamma-1}.$$

Combining the two gives

$$p_iV_i^\gamma = p_fV_f^\gamma, \qquad (3.3.8)^\circ_r$$

which is often expressed in the form

$$pV^\gamma = \text{constant}, \qquad (3.3.9)^\circ_r$$

where the constant is simply the initial value $p_iV_i^\gamma$.

For all gases γ exceeds unity. For a perfect monatomic gas $\gamma = 5/3$, a result found in Part 2 (but see Problem 3.27 for a straightforward proof). This means that an *adiabat*, the curve showing the dependence of the pressure on the volume, Fig. 3.10, falls more steeply ($p \propto 1/V^\gamma$) than the corresponding isotherm ($p \propto 1/V$). The physical reason is that in an isothermal expansion energy flows into the system as heat and maintains the temperature, and so the pressure does not fall as much as in an adiabatic expansion.

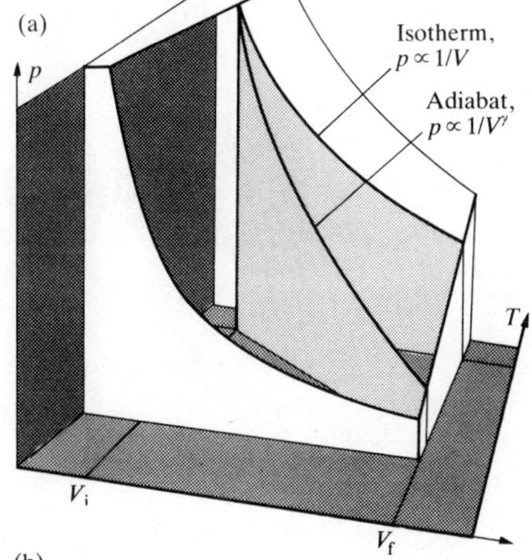

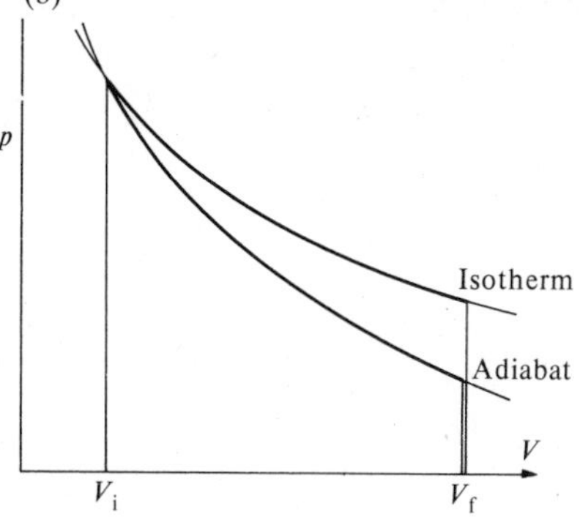

Fig. 3.10. The isotherms and adiabats corresponding to the expansion of a perfect gas are paths on the p, V, T surface, as shown in (a). The projections of these paths on to the p,V plane, (b), show that the pressure decreases more rapidly along an adiabat than along an isotherm, and therefore that the work corresponding to the process (the area under the curves) is less for adiabatic reversible expansion than for isothermal reversible expansion between the same initial and final volumes.

Example 3.8

Using the same data as in *Example* 3.7, calculate the final pressure of the gas and the change of enthalpy as a result of the expansion.

• *Method*. The expansion is adiabatic and reversible; therefore the final pressure may be calculated using eqn (3.3.8). Take $C_{V,m}$ from Table 2.1 and estimate $C_{p,m}$ (for calculating γ) from $C_{V,m} - C_{p,m} = R$. Since $\Delta H = \Delta U + \Delta(pV)$, and $pV = nRT$, calculate the change in enthalpy from $\Delta H = \Delta U + nR\,\Delta T$.

• *Answer*. Since $C_{V,m} = 12.48$ J K^{-1} mol^{-1}, $C_{p,m} = 20.79$ J K^{-1} mol^{-1} and $\gamma = 1.666$. From eqn (3.3.8):

$$p_f = (0.50\ \text{dm}^3/1.00\ \text{dm}^3)^{1.666} \times (1.0\ \text{atm}) = 0.32\ \text{atm}.$$

The enthalpy change is

$$\Delta H = \Delta U + nR\,\Delta T = nC_{V,m}\,\Delta T + nR\,\Delta T = nC_{p,m}\,\Delta T.$$

Since $\Delta T = -110\ \mathrm{K}$ (from *Example* 3.7),

$$\Delta H = (0.0204\ \mathrm{mol}) \times (20.79\ \mathrm{J\ K^{-1}\ mol^{-1}}) \times (-110\ \mathrm{K}) = -47\ \mathrm{J}.$$

• *Comment.* In an isothermal reversible change the final pressure would have been 0.50 atm and the enthalpy change zero.

• *Exercise.* Calculate the final pressure and the enthalpy change for the process described in the *Exercise* to *Example* 3.7. [0.16 atm, −74 J]

Finally, we can express the work of adiabatic change in terms of the *pressures* of the initial and final states. On combining eqns (3.3.8) and (3.3.7) it follows that

$$w = C_V T_i\{(p_f/p_i)^{(\gamma-1)/\gamma} - 1\}. \qquad (3.3.10)^{\circ}_{\mathrm{r}}$$

3.4 Comments on types of change

The algebraic results for the different types of change are collected in Box 3.2. It will be helpful to remember the following points.

Box 3.2 Work done on expansion

For any kind of expansion of any substance, the work done by the system as its volume changes from V_i to V_f is

$$w' = \int_i^f p_{ex}\, dV.$$

For reversible changes $p_{ex} = p_{in}$. The entries below marked ° relate only to a perfect gas; the rest apply to any substance. $c = C_{V,m}/R$. The work done *on* the gas is $w = -w'$ in each case. The symbol ΔX always means $X_{final} - X_{initial}$.

Type of work	w'	q	ΔU	ΔT
Expansion against vacuum				
Isothermal	0	0°	0°	0
Adiabatic	0	0	0	0
Expansion against constant pressure				
Isothermal	$p_{ex}\Delta V$	$p_{ex}\Delta V$°	0°	0
Adiabatic	$p_{ex}\Delta V$	0	$-p_{ex}\Delta V$	$-p_{ex}\Delta V/C_V$°
Reversible expansion or compression				
Isothermal	$nRT \ln (V_f/V_i)$°	$nRT \ln (V_f/V_i)$°	0°	0
Adiabatic	$-C_V\Delta T$°	0	$C_V\Delta T$°	$T_i\{(V_i/V_f)^{1/c} - 1\}$°

The work done on a system during an infinitesimal change is always given by $dw = -p_{ex}\, dV$, where p_{ex} is the pressure acting on the system. (The work done *by* the system is $dw' = p_{ex}\, dV$.)

In a reversible change the external pressure is always matched to the internal, and so $p_{ex} = p$; p depends on the temperature and volume of the gas (and the amount present).

In an irreversible expansion the external pressure is less than the internal, and the work done by the system is less than in the case of reversible expansion.

In an isothermal expansion the temperature of the gas is maintained by a flow of heat from the thermal bath representing the surroundings. For a *perfect gas* the internal energy is independent of the volume during an isothermal change, which implies that the same quantity of energy enters the system as heat as leaves it as work: $dq = -dw$.

In an adiabatic expansion no heat enters the system, but work is done. Therefore $dq = 0$, and the internal energy change is equal to the work done: $dU = dw$.

In the reversible expansion of a perfect gas through a given volume from the same initial state, the isothermal process produces more work than the adiabatic because the internal energy is continuously replenished by the flow of heat that maintains the initial temperature: the area on the indicator diagram is less beneath an adiabat than an isotherm, Fig. 3.10.

Further reading

Mathematical methods in elementary thermodynamics. S. M. Blinder; *J. chem. Educ.* **43,** 85 (1966).

The mathematics of physics and chemistry (*Vol 1*). H. Margenau and G. M. Murphy; Van Nostrand, New York, 1956.

Chemical thermodynamics. P. A. Rock; University Science Books and Oxford University Press, 1983.

Thermodynamics for chemical engineers. K. E. Bett, J. S. Rowlinson, and G. Saville; Athlone Press, London, 1975.

Heat and thermodynamics. M. W. Zemansky and R. H. Dittman; McGraw-Hill, New York, 1981.

Thermodynamics. G. N. Lewis and M. Randall, revised by K. S. Pitzer and L. Brewer; McGraw-Hill, New York, 1961.

Chemical thermodynamics. M. L. McGlashan; Academic Press, London, 1979.

Thermodynamics. J. G. Kirkwood and I. Oppenheim; McGraw-Hill, New York, 1961.

Introductory problems

A3.1. Show that the following functions have exact differentials: (a) $x^2y + 3y^2$, (b) $x \cos(xy)$, (c) $t(t + e^s) + 25s$.

A3.2. Express $(\partial C_V/\partial V)_T$ as a second derivative of U, and find its relation to $(\partial U/\partial V)_T$. From this relation, show that for perfect gases, $(\partial C_V/\partial V)_T$ vanishes.

A3.3. By direct differentiation of the definition of enthalpy obtain a relation between $(\partial H/\partial U)_p$ and $(\partial U/\partial V)_p$. Check the result by expressing $(\partial H/\partial U)_p$ as the ratio of derivatives with respect to volume and then using the definition of enthalpy.

A3.4. Write an expression for dV given that V is a function of p and T. Find an expression for $d \ln V$ in terms of the isobaric coefficient of thermal expansion and the isothermal compressibility.

A3.5. In the temperature range 300 K to 350 K, the volume of one gram of a certain liquid is given in terms of the volume at 300 K by the empirical function $V(T) = V_{300}\{0.75 + 3.9 \times 10^{-4}T + 1.48 \times 10^{-6}T^2\}$ with $T \equiv T/\mathrm{K}$. The density of the liquid is $0.875\ \mathrm{g\ cm^{-3}}$ at 300 K. Calculate the isobaric expansivity at 320 K.

A3.6. The isothermal compressibility of copper at 293 K is $7.35 \times 10^{-7}\ \mathrm{atm^{-1}}$. What pressure must be applied to copper at 293 K in order to increase its density by 0.08%?

A3.7. The Joule–Thomson coefficient μ_{JT} for N_2 is $0.25\ \mathrm{K\ atm^{-1}}$. Calculate the isothermal Joule–Thomson coefficient for N_2. Calculate the heat which must be supplied electrically to maintain constant temperature when 15 moles of N_2 flow through the throttle valve in an isothermal Joule–Thomson experiment. The observed pressure drop at the throttle is 75 atm.

A3.8. Four moles of O_2 originally confined to a volume of 20 dm^3 at 270 K undergo adiabatic expansion against a constant pressure of 600 Torr until the volume triples. Assume that O_2 obeys the perfect gas law and calculate: q, w, ΔT, ΔU, ΔH.

A3.9. Three moles of a perfect gas at 200 K and 2.00 atm pressure undergo reversible adiabatic compression until the temperature becomes 250 K. For the gas, $C_{V,\mathrm{m}}$ is 27.5 J K^{-1} mol^{-1} in this temperature range. Calculate: q, w, ΔU, ΔH, final V, and final p.

A3.10. A one mole sample of a monatomic perfect gas with a constant-pressure heat capacity of 20.8 J K^{-1} is initially at a pressure of 3.25 atm at 310 K. It undergoes adiabatic reversible expansion until its pressure becomes 2.50 atm. Calculate the final V, final T, and w.

Problems

3.1. Which of the following properties are *extensive*? Density, pressure, mass, temperature, enthalpy, refractive index, heat capacity, molar heat capacity.

3.2. The equipartition principle gives $\frac{3}{2}nRT$ for the internal energy of a perfect monatomic gas. What are the values of $(\partial U/\partial V)_T$ and $(\partial H/\partial V)_T$ for a perfect gas? These two quantities occur frequently in thermodynamic discussions, and this is one of several ways of arriving at their value.

3.3. The isothermal compressibility of lead is 2.3×10^{-6} atm^{-1}. Express this in N^{-1} m^2. A cube of lead of side 10 cm was to be incorporated into the keel of an underwater exploration TV camera, and the designers needed to know the stresses in the equipment. What change of volume will such a cube undergo on the floor of the sea at depths of (a) 100 ft, (b) 5000 fathoms? Take the mean density of sea water as $\rho \approx 1.03$ g cm^{-3}; $g = 9.81$ m s^{-2}; 1 fathom = 6 ft.

3.4. The isobaric coefficient of thermal expansion of lead is 8.61×10^{-5} K^{-1}. The temperature of the water where the camera will operate is -5 °C. What effect will this temperature reduction have on the volume of the block? What effect will both pressure and temperature have? Take $T = 298$ K initially.

3.5. In this chapter we encountered various properties of partial and complete differentials, and the next few Problems give some practice in dealing with them. In the first place, consider a case in which the volume and temperature of a perfect gas depend on the *time* t. Show that the rate of change of the pressure is related to the rate of change of the volume and the temperature by

$$\mathrm{d}\ln p/\mathrm{d}t = \mathrm{d}\ln T/\mathrm{d}t - \mathrm{d}\ln V/\mathrm{d}t.$$

Do this by finding an expression for $\mathrm{d}p$, with p regarded as a function of V and T.

3.6. As an illustration of the result in the last Problem, consider a system in which the temperature is cooling exponentially towards absolute zero with a time-constant τ_T: exponential cooling is termed *Newtonian cooling*. At the same time a piston is driven into the vessel containing the gas, and the design of the cam controlling the piston is based on a logarithmic spiral, so that the volume is also decreased exponentially towards zero with some time-constant τ_V. How rapidly does the pressure change? What is the time-dependence of the pressure when $\tau_T = \tau_V$?

3.7. The pressure of a given amount of a van der Waals gas is a function of V and T. Find an expression for $\mathrm{d}p$ for infinitesimal changes in V and T. In order to repeat the time-dependence calculation of the last Problem, what modification to the time-dependence of V is appropriate?

3.8. Some of the general relations between partial derivatives can be illustrated explicitly by taking the van der Waals equation of state. As a first step, express T as a function of p and V. This permits us to calculate $(\partial T/\partial p)_V$: confirm the relation $(\partial T/\partial p)_V = 1/(\partial p/\partial T)_V$, and then go on to confirm Euler's chain relation (Box 3.1).

3.9. Find expressions for the isothermal compressibility κ and the isobaric expansivity α of a van der Waals gas. Show from Euler's chain relation that κ and α should be related by $\kappa R = \alpha(V_\mathrm{m} - b)$ and confirm explicitly that this is so. In Chapter 1 we saw that reduced variables let us express relations in a way that is independent of the identity of the gas: show that the reduced compressibility and expansivity of a van der Waals gas are related by $8\kappa_\mathrm{r} = \alpha_\mathrm{r}(3V_\mathrm{r} - 1)$.

3.10. We shall see later that the Joule–Thomson coefficient μ_JT can be determined from the expression $\mu_\mathrm{JT}C_p = T(\partial V/\partial T)_p - V$. Derive an expression for μ_JT in terms of the van der Waals parameters and estimate its value for xenon at room temperature and pressure when $V_\mathrm{m} = 24.6$ dm^3 mol^{-1}.

3.11. The Joule–Thomson coefficient changes sign at the *inversion temperature* T_I. This is an important quantity, because it indicates the conditions at which a real gas will switch from heating to cooling on adiabatic expansion. Use the information in the last Problem to derive an expression for T_I of a van der Waals gas, and express your answer in both ordinary and reduced variables. Estimate the temperatures at which (a) hydrogen, (b) carbon dioxide switch to cooling from heating when their pressures are 50 atm.

3.12. In order to design a refrigerator it is necessary to know what temperature drop is brought about by adiabatic expansion of the refrigerant fluid. The value for one type of freon is about 1.2 K atm^{-1}: what pressure drop is needed to bring about a 5.0 K temperature decrease?

3.13. Another fluorocarbon was investigated with a view to using it in a commercial refrigerating system. The temperature drops for a series of initial pressures p_i expanding into 1 atm pressure were measured at 0 °C, with the following results:

p_i/atm	32	24	18	11	8	5
$-\Delta T$/K	22	18	15	10	7.4	4.6

What is the value of μ_{JT} for the gas at this temperature? Is it more suitable than the first freon (Problem 3.12) for the job? What other factors have to be taken into account?

3.14. One *thermodynamic equation of state* is given on p. 69. Find its partner, $(\partial H/\partial p)_T = -T(\partial V/\partial T)_p + V$, from it and the general relations between partial differentials.

3.15. Show that the value of μ_{JT} for a gas that is described by the virial equation of state $pV_m/RT = 1 + B/V_m$ is related to B when $4pB/RT \ll 1$ by

$$\mu_{JT} = (T^2/C_{p,m})\left\{\frac{\partial}{\partial T}\left(\frac{B}{T}\right)\right\}_p.$$

Estimate the value of μ_{JT} for argon at room temperature, using the information in Table 1.1.

3.16. Estimate the changes in volume that occur when $1\ cm^3$ blocks of (a) mercury, (b) diamond are heated through 5 K at room temperature. Use the data in Table 3.1.

3.17. Equation (3.2.18) is a general expression for the difference in heat capacities $C_p - C_V$ for any substance. What is the value of $C_{p,m} - C_{V,m}$ for a van der Waals gas? Show, using the result of Problem 3.9, that for such a gas, $C_{p,m} - C_{V,m} = \lambda R$, with $1/\lambda = 1 - (3V_r - 1)^2/4V_r^3T_r$, and estimate the value of the difference for xenon at 25 °C and 10 atm.

3.18. The heat capacities of (a) copper, (b) benzene at constant pressure and volume are related by eqn (3.2.18). Calculate this difference at room temperature and pressure using the data in Table 3.1 and estimate the difference in heat that is required to raise the temperature of a 500 g sample of each through 50 K at constant pressure and at constant volume.

3.19. Combine the results of the last Problem with eqn (3.2.17) to find values of $(\partial U/\partial V)_T$ for copper and benzene.

3.20. By how much does (a) the molar internal energy, (b) the molar enthalpy, of water change when it is heated through 10 K at room temperature and pressure? Data for the problem will be found in *Example* 3.5. Account for the difference between the two quantities.

3.21. Whereas the material of Chapter 2 let us deal with isothermal changes, this chapter has shown how to deal with another important class, the adiabatic changes. The next few Problems look at some examples of this kind of process. A straightforward application of the methods described in the text will orientate our ideas. Consider a system consisting of a 2.0 mol room-temperature (25 °C) sample of carbon dioxide (to be regarded as a perfect gas at this stage) confined to a cylinder of $10\ cm^2$ cross-section and at a pressure of 10 atm. It is allowed to expand adiabatically against a pressure of 1.0 atm. During the expansion the piston is pushed through 20 cm. Find the values of w, q, ΔU, ΔH, and ΔT for the expansion.

3.22. 65 g of xenon is held in a container at 2.0 atm. It is then allowed to expand adiabatically (a) reversibly to 1.0 atm, (b) against 1.0 atm pressure. If the initial temperature is 298 K, what will the final temperature be in each case?

3.23. A sample of a fluorocarbon was allowed to expand reversibly and adiabatically to twice its volume. In the expansion the temperature dropped from 298.15 K to 248.44 K. Assuming that the gas behaves perfectly, estimate the value of $C_{V,m}$.

3.24. We have just seen that the heat capacity of a gas C_V can be measured by observing the temperature drop on adiabatic expansion. If the pressure drop is also monitored we can use the same experiment to find γ, and hence C_p of the gas. In the experiment just described the initial pressure was 1522.2 Torr and after expansion the pressure was 613.85 Torr. What are the values of γ and $C_{p,m}$ for the fluorocarbon? Continue to assume perfect gas behaviour.

3.25. Find the change in the molar internal energy and the molar enthalpy of the fluorocarbon during the expansion described in Problem 3.23. Continue to assume that the gas behaves perfectly.

3.26. In *Example* 3.8 the value of ΔH for an adiabatic change was found by equating pV with nRT. However, $dH = V dp$ can be integrated directly so long as we know V as a function of p. Use this method to find ΔH for the reversible, adiabatic expansion of a perfect gas.

3.27. We have seen (Problem 2.20) that the equipartition theorem leads to $C_{V,m} = 3R/2$ for a perfect monatomic gas. Predict a value for γ of such a gas. What is the value of γ for a translating, rotating, non-linear, polyatomic molecule?

3.28. In Problem 3.17 we found an expression relating C_p and C_V of a van der Waals gas. Find an expression for γ of a monatomic gas and of γ for a non-ideal polyatomic gas, both being regarded as van der Waals gases.

3.29. Estimate γ for (a) xenon, (b) water vapour, both at 100 °C and 1 atm pressure, on the assumption that they are van der Waals gases.

3.30. The speed of sound in a gas is related to the ratio of heat capacities γ by $c_s = \sqrt{(RT\gamma/M_m)}$ where M_m is the molar mass. Show that this can be written $c_s = \sqrt{(\gamma p/\rho)}$ where p is the pressure and ρ the density. Calculate the speed of sound (a) in helium, (b) in air, both at 25 °C.

3.31. Measuring the speed of sound in a gas is a method of determining the heat capacity. The speed of sound in ethene at 0 °C was measured as 317 m s^{-1}. Calculate the values of γ and $C_{V,m}$ at this temperature.

3.32. An old man begins to blow an ocarina, and generates the note A (440 Hz). He goes on blowing, and as he blows the air he exhales turns more and more to pure carbon dioxide. His last gasp is pure carbon dioxide: what is his dying frequency? You need to know that the frequency of the ocarina is given by Kc_s, where K depends upon its dimensions, and c_s is the speed of sound.

4

The First Law in action: thermochemistry

Learning objectives

After careful study of this chapter you should be able to:

(1) Define *endothermic* and *exothermic*, Section 4.1(a).

(2) Define *reaction enthalpy* and *standard reaction enthalpy*, and specify the *standard state* of a pure substance, Section 4.1(b).

(3) Define the *standard enthalpy of formation* of a compound, Section 4.1(b), and expresses the standard reaction enthalpy in terms of the standard enthalpies of formation of the participants in a general reaction, eqn (4.1.4).

(4) State, derive, and use *Hess's Law*, Section 4.1(c) and Example 4.2.

(5) Derive and use *Kirchhoff's Law* for the temperature dependence of reaction enthalpy, eqn (4.1.6).

(6) State the relation between reaction enthalpy and reaction energy in general, eqn (4.1.9), and for a reaction involving perfect gases, eqn (4.1.11).

(7) Explain the operation of the *adiabatic bomb calorimeter* and the *adiabatic flame calorimeter*, Section 4.1(f).

(8) Define the enthalpies of *combustion*, *hydrogenation*, *atomization*, *sublimation*, *vaporization*, *melting*, *solution*, and *solvation*, Section 4.2.

(9) Define *bond dissociation enthalpy*, *mean bond dissociation enthalpy*, and *bond dissociation energy*, Section 4.2(a).

(10) Construct *Born–Haber cycles* to determine enthalpies from other data, Section 4.2(b).

(11) Define *ionization energy*, *electron affinity*, and *lattice enthalpy*, Section 4.2(b) and Example 4.7.

Introduction

Thermochemistry is the study of energy changes that occur during chemical reactions. The designer of a new chemical plant needs to know how much energy each reaction generates or absorbs as heat, and must make sure that the plant can cope. For a plant to work with maximum economy the designer must know how much heat to supply to one region and how much can be taken from another. Similar questions arise in biological systems: can a sequence of reactions release enough energy to drive a biological process and keep a cell alive? We shall see that questions like these are not solely a matter of the heat available from reactions, but the information that comes from the techniques we describe in this chapter is an indispensable part of the answer.

4.1 Heat in chemical reactions

A reaction vessel is a thermodynamic system in the sense described in Chapter 2. The energy transferred as heat during a reaction is q, which is a path function and therefore depends on how the reaction is carried out. It is much more convenient to discuss energy changes in terms of state functions, because they are independent of how a change occurs. In particular, they are independent of whether the reaction takes place reversibly or not. The next sections explore this shift of emphasis.

4.1 (a) The reaction enthalpy

If energy is transferred as heat at constant volume, and if no work is done, the change of internal energy produced is equal to the heat transferred, Section 2.3(a), and we can write $(\Delta U)_V = q_V$. For a specified change of state ΔU is independent of how the change occurs, and so the subscript V can be dropped from ΔU:

$$\Delta U = q_V. \tag{4.1.1}$$

This means that if we measure the heat transferred at constant volume, we can identify it with the change of a thermodynamic state function.

We also saw in Section 2.3(b) that if energy is transferred as heat at constant pressure, then the quantity transferred can be identified with the change of enthalpy, another state function. Therefore, so long as no work other than p,V-work is being done, by the same argument as above

$$\Delta H = q_p. \tag{4.1.2}$$

It follows that if a reaction occurs at constant pressure (as in many everyday processes), the heat transfer can be identified with the change of enthalpy.

Reactions for which $\Delta H > 0$ are called *endothermic*; those for which $\Delta H < 0$ are called *exothermic*. When the reaction takes place in an adiabatic container an endothermic reaction results in a lowering of temperature and an exothermic reaction results in a rise of temperature. If the reaction is maintained at constant temperature as a result of thermal contact with the surroundings, then an endothermic reaction results in a transfer of energy as heat into the system and an exothermic reaction produces heat in the surroundings. These points are summarized as follows:

	Endothermic $\Delta H > 0$	Exothermic $\Delta H < 0$
Isothermal conditions, p constant:	Heat enters $q_p > 0$	Heat exits $q_p < 0$
Adiabatic conditions:	T falls	T rises

One aspect of this discussion must be emphasized. At constant pressure, and when no work other than p,V-work is involved, the whole of the enthalpy change of a reaction is available as heat ($q_p = \Delta H$). However, this does not mean that a simple modification of the apparatus will allow the same quantity of *work* to be obtained, and in some cases the work a reaction system can do exceeds the enthalpy change. These points are taken up in Section 5.3(b).

4.1 (b) Standard enthalpy changes

The enthalpy change accompanying a reaction is called the *reaction enthalpy* (symbol: $\Delta_r H$). Tables of values generally refer to reactions in which the reactants and products are in their *standard states:*

> *Standard state: The standard state of a pure substance is the pure form at a pressure of 1 bar and the specified temperature.*

1 bar is 10^5 Pa (Section 0.3)‡. The *conventional temperature* for reporting thermodynamic data is 298.15 K (corresponding to 25 °C): since this temperature is frequently mentioned it is given the special symbol $\mathcal{T}$.

A further notational detail is that it is often appropriate to specify the physical state of the substance in an equation. We use s for solid, l for liquid, and g for gas, as in $H_2O(l)$, to signify liquid water. The precise definition of the standard state of a real gas involves a correction for non-ideality: this is discussed in Section 6.2.

The *standard enthalpy of reaction* (symbol: $\Delta_r H^\ominus$, the superscript $^\ominus$ will always denote a standard value, i.e. one relating to the standard states as defined above) is the change of enthalpy (per unit amount of reaction) when the reactants in their standard states change to products in their standard states. Since values depend on the temperature, the temperature must be specified. As an example, consider the reaction

$$C(s, \text{graphite}) + O_2(g) \rightarrow CO_2(g); \qquad \Delta_r H^\ominus(\mathcal{T}) = -393.5 \text{ kJ mol}^{-1}.$$

(The 'per mole' refers to the reaction as written; i.e., per mole of CO_2 produced.) The standard states at $\mathcal{T}$ (i.e. 298.15 K) are pure graphite and pure oxygen and carbon dioxide, each at 1 bar. In general, there are several reactants and products, and the standard enthalpy change then refers to the overall process (unmixed reactants) → (unmixed products) but, except in the case of ionic reactions in solution, the enthalpy changes accompanying mixing and unmixing are insignificant in comparison with the contribution from the reaction itself.

The reaction quoted above is an example of a *formation reaction*, in which a compound is formed from its elements. A *standard enthalpy of formation* (symbol $\Delta_f H^\ominus$) is the standard reaction enthalpy when the compound is formed from its elements in their *reference phases*. The reference phase of an element is *the thermodynamically most stable phase* under standard pressure (with one exception: for phosphorus, *white phosphorus* is taken as the reference phase). By definition, *the standard enthalpies of formation of elements are zero at all temperatures*. Some values are listed in Tables 4.1 and 4.2 (a consolidated version is given in the table of thermodynamic data at the end of the book).

Table 4.1. Standard enthalpies of formation, $\Delta_f H^\ominus(\mathcal{T})/\text{kJ mol}^{-1}$

$H_2O(l)$	−285.8	$H_2O_2(l)$	−187.8
$NH_3(g)$	−46.1	$N_2H_4(l)$	+50.6
$NO_2(g)$	+33.2	$N_2O_4(g)$	+9.2
$NaCl(s)$	−411.2	$KCl(s)$	−436.8

Table 4.2. Standard enthalpies of formation and combustion $\Delta H^\ominus(\mathcal{T})/\text{kJ mol}^{-1}$

	Formation	Combustion
$CH_4(g)$	−74.81	−890
$C_2H_6(g)$	−84.68	−1560
$CH_3OH(l)$	−238.66	−726
$C_6H_6(l)$	+49.0	−3268
Glucose(s)	−1268	−2808

‡ The recommendation to take 1 bar as the standard pressure for reporting thermodynamic data is recent and not universally accepted. One problem is that many published tabulations of data refer to the former and now declared obsolete standard pressure of 1 atm. At the level of accuracy of most data, especially those relating to liquids and solids, the differences in values are usually negligible except in work of the highest precision. In the rare cases where it is necessary to distinguish between the new standard and the old, we adopt the recommendation that the latter should be signified by a superscript ^{a} (for atm) in place of the $^\ominus$ we shall normally employ to signify the new convention. A meticulous analysis of the consequences of the change has been given by R. D. Freeman in *Bulletin of Chemical Thermodynamics* **25,** 523 (1982). Modern tables of data (e.g. the tables published by the National Bureau of Standards) now widely, but not universally, adopt the new standard. Note that the *standard state* should be distinguished from the *reference phase*, defined below.

4.1 (c) Hess's Law and reaction enthalpies

Standard reaction enthalpies can be calculated from tables of enthalpies of formation. This is because, since enthalpy is a state function, its value is independent of the path between the specified initial state (reactants) and the specified final state (products). Therefore, the reaction enthalpy for a process such as

$$2HN_3(l) + 2NO(g) \rightarrow H_2O_2(l) + 4N_2(g)$$

can be expressed as the appropriate sum and differences of the enthalpies of formation of all the components, because we can *formally* regard the reaction as proceeding by 'unforming' the reactants into their elements, and then forming those elements into the products. In this case, for example, at T:

$$\begin{aligned}\Delta_r H^{\ominus} &= \{(-187.8) + 4(0)\}\ \text{kJ mol}^{-1} - \{2(264.0) + 2(90.25)\}\ \text{kJ mol}^{-1}\\ &= -896.3\ \text{kJ mol}^{-1}.\end{aligned}$$

The formal way of generalizing this calculation is as follows. The trick is to express a reaction $aA + bB \rightarrow cC + dD$ as

$$0 = cC + dD - aA - bB, \quad \text{or} \quad 0 = \sum_J \nu_J J \text{ in general} \qquad (4.1.3)$$

This way of writing the reaction may look odd at first sight, but will turn out to be very useful. Note that the stoichiometric coefficients (ν_J) are **p**ositive for **p**roducts and negative for reactants. The standard reaction enthalpy is the sum of the standard enthalpies of formation of the products, $c\Delta_f H_C^{\ominus} + d\Delta_f H_D^{\ominus}$, less the sum of the standard enthalpies of formation of the reactants, $a\Delta_f H_A^{\ominus} + b\Delta_f H_B^{\ominus}$. That is,

$$\Delta_r H^{\ominus} = \{c\Delta_f H_C^{\ominus} + d\Delta_f H_D^{\ominus}\} - \{a\Delta_f H_A^{\ominus} + b\Delta_f H_B^{\ominus}\}.$$

This can be written neatly in terms of the coefficients ν_J as

$$\Delta_r H^{\ominus} = \sum_J \nu_J\, \Delta_f H_J^{\ominus}, \qquad (4.1.4)$$

$\Delta_f H_J^{\ominus}$ being the standard enthalpy of formation of substance J. This expression is easy to use, and enables the data listed in Tables 4.1 and 4.2 to provide thermochemical information on a wide variety of reactions.

Example 4.1

Express the standard reaction enthalpy of $2HN_3(l) + 2NO(g) \rightarrow H_2O_2(l) + 4N_2(g)$ in terms of the standard enthalpies of formation of the components.

- *Method*. Express the reaction in the form of eqn (4.1.3), identify the stoichiometric coefficients, and then use eqn (4.1.4) to express $\Delta_r H^{\ominus}$ in terms of the $\Delta_f H^{\ominus}$.

- *Answer*. The reaction is

$$0 = H_2O_2(l) + 4N_2(g) - 2HN_3(l) - 2NO(g).$$

Therefore, $\nu(H_2O_2) = +1$, $\nu(N_2) = +4$, $\nu(HN_3) = -2$, $\nu(NO) = -2$. The standard reaction enthalpy is therefore

$$\Delta_r H^{\ominus} = \Delta_f H^{\ominus}(H_2O_2) + 4\Delta_f H^{\ominus}(N_2) - 2\Delta_f H^{\ominus}(HN_3) - 2\Delta_f H^{\ominus}(NO).$$

• *Comment*. The expression for $\Delta_r H$ can normally be written down by inspection: remember that the stoichiometric coefficients of the products are positive.

• *Exercise*. Express the standard reaction enthalpy of $2C_3H_6 + 9O_2 \rightarrow 6CO_2 + 6H_2O$ in terms of enthalpies of formation.

$[6\Delta_f H^{\ominus}(CO_2) + 6\Delta_f H^{\ominus}(H_2O) - 2\Delta_f H^{\ominus}(C_3H_6) - 9\Delta_f H^{\ominus}(O_2)]$

Reaction enthalpies in general may also be combined. The generalization is *Hess's Law* (which is also sometimes called his law of 'constant heat summation'):

> *Hess's Law: The standard reaction enthalpy is the sum of the standard enthalpies of the reactions, at the same temperature, into which the overall reaction may formally be divided.*

The thermodynamic justification of the law is the path-independence of the value of ΔH. The way it is used is illustrated in the following example.

Example 4.2

The standard reaction enthalpy for the hydrogenation of propene, $CH_2{=}CHCH_2(g) + H_2(g) \rightarrow CH_3CH_2CH_3(g)$, has the value $-124\ kJ\ mol^{-1}$. The standard reaction enthalpy for the combustion of propane, $CH_3CH_2CH_3(g) + 5O_2(g) \rightarrow 3CO_2(g) + 4H_2O(l)$, is $-2220\ kJ\ mol^{-1}$. Calculate the standard reaction enthalpy for the combustion of propene.

• *Method*. Add and subtract the reactions given, together with any others needed, so as to reproduce the reaction required. Then add and subtract the reaction enthalpies in the same way. Additional data are in Tables 4.1 and 4.2.

• *Answer*. The combustion reaction we require is

$$C_3H_6(g) + \tfrac{9}{2}O_2(g) \rightarrow 3CO_2(g) + 3H_2O(l).$$

This can be recreated from the following sum:

$$\begin{array}{ll} C_3H_6(g) + H_2(g) \rightarrow C_3H_8(g), & \Delta_r H^{\ominus} = -124\ kJ\ mol^{-1} \\ C_3H_8(g) + 5O_2(g) \rightarrow 3CO_2(g) + 4H_2O(l), & \Delta_r H^{\ominus} = -2220\ kJ\ mol^{-1} \\ H_2O(l) \rightarrow H_2(g) + \tfrac{1}{2}O_2(g), & \Delta_r H^{\ominus} = +286\ kJ\ mol^{-1} \\ \hline C_3H_6(g) + \tfrac{9}{2}O_2(g) \rightarrow 3CO_2(g) + 3H_2O(l), & \Delta_r H^{\ominus} = -2058\ kJ\ mol^{-1} \end{array}$$

The reaction enthalpy for the water decomposition is the negative of its enthalpy of formation, Table 4.1.

• *Comment*. When equations are manipulated in this way they are sometimes called *thermochemical equations*. The skill to develop is the ability to assemble a given equation from others.

• *Exercise*. Calculate the enthalpy of hydrogenation of benzene from its enthalpy of combustion and the enthalpy of combustion of cyclohexane. $[-205\ kJ\ mol^{-1}]$

4.1 (d) The temperature dependence of reaction enthalpies

Many reactions do not take place at 298 K, and so we must examine how to adapt tabulated data to other conditions. The standard reaction enthalpies of many important reactions have been measured over a range of temperatures, and for serious work these accurate data must be used.

However, in the absence of this information, the enthalpies of reaction at different temperatures may be estimated as follows.

The isobaric heat capacity of a substance is the rate of change of enthalpy with respect to temperature at constant pressure: $C_p = (\partial H/\partial T)_p$. Therefore, when the temperature is changed from T_1 to T_2, the enthalpy of a substance changes from $H(T_1)$ to

$$H(T_2) = H(T_1) + (T_2 - T_1)(\partial H/\partial T)_p = H(T_1) + (T_2 - T_1)C_p.$$

(This assumes that C_p is independent of temperature in the range T_1 to T_2; we do a more precise calculation below.) The equation applies to each substance involved in the reaction, and so the reaction enthalpy changes from $\Delta_r H(T_1)$ to

$$\Delta_r H(T_2) = \Delta_r H(T_1) + (T_2 - T_1)\{C_{p,\text{products}} - C_{p,\text{reactants}}\}.$$

If the difference of the products' and the reactants' heat capacities is written $\Delta_r C_p$ and the temperature change as $\delta T = T_2 - T_1$, then this expression simplifies to

$$\Delta_r H(T_2) = \Delta_r H(T_1) + \delta T \Delta_r C_p. \quad (4.1.5)$$

Therefore, in order to estimate the temperature-dependence of the reaction enthalpy, all we need to know are the heat capacities of the substances involved.

Example 4.3

The enthalpy of formation of gaseous water at 25 °C is -241.82 kJ mol^{-1}. Estimate its value at 100 °C given the following values of the constant pressure molar heat capacities: H_2O(g): 33.58 J K^{-1} mol^{-1}; H_2(g): 28.82 J K^{-1} mol^{-1}; O_2(g): 29.36 J K^{-1} mol^{-1}.

- *Method*. Use eqn (4.1.5). The reaction is $H_2(g) + \frac{1}{2}O_2(g) \rightarrow H_2O(l)$.

- *Answer*. $\Delta_r C_p = C_{p,m}(H_2O, g) - C_{p,m}(H_2, g) - \frac{1}{2}C_{p,m}(O_2, g)$

$$= (33.58 - 28.82 - 14.68)\text{ J K}^{-1}\text{ mol}^{-1} = -9.96\text{ J K}^{-1}\text{ mol}^{-1}.$$

Then, since $\delta T = +75$ K,

$$\Delta_f H^\ominus(373\text{ K}) = \Delta_r H^\ominus(T) + (75\text{ K}) \times (-9.94\text{ J K}^{-1}\text{ mol}^{-1})$$
$$= -241.82\text{ kJ mol}^{-1} - 0.75\text{ kJ mol}^{-1} = -242.6\text{ kJ mol}^{-1}.$$

- *Comment*. The heat capacities have been assumed to be constant. Their temperature dependence is taken into account as described below.

- *Exercise*. Estimate the standard enthalpy of formation of ammonia at 400 K from the data in Tables 2.1 and 4.1. [-48.4 kJ mol^{-1}]

A more formal way of arriving at the result above, and one that avoids the assumption that the heat capacities are independent of temperature, runs as follows. When the temperature of a substance is changed by dT its enthalpy changes by dH, and at constant pressure $dH = C_p\, dT$. Integration between T_1 and T_2 gives

$$H(T_2) - H(T_1) = \int_{T_1}^{T_2} C_p(T)\, dT.$$

This expression applies to each substance J involved in the reaction, and so the reaction enthalpies are related by

$$\textit{Kirchhoff's Law: } \Delta_r H(T_2) = \Delta_r H(T_1) + \int_{T_1}^{T_2} \Delta_r C_p(T)\,dT, \quad (4.1.6)$$

where $\Delta_r C_p(T)$ is the difference in heat capacities of the individual substances at the temperature T:

$$\Delta_r C_p = \{cC_{p,\mathrm{C}} + dC_{p,\mathrm{D}}\} - \{aC_{p,\mathrm{A}} + bC_{p,\mathrm{B}}\}, \quad (4.1.7a)$$

and $C_{p,\mathrm{J}}$ is the molar heat capacity of substance J. More succinctly, this may be written in terms of the stoichiometric coefficients ν_J as

$$\Delta_r C_p = \sum_\mathrm{J} \nu_\mathrm{J} C_{p,\mathrm{J}}. \quad (4.1.7b)$$

Kirchhoff's Law can be applied when the temperature dependences of the heat capacities are known in the range of interest. Occasionally numerical tables are available. These are often expressed in the form

$$C_{p,\mathrm{J}}(T) = a_\mathrm{J} + b_\mathrm{J} T + c_\mathrm{J}/T^2. \quad (4.1.8)$$

and the temperature-independent coefficients a, b, c are listed. A selection of values is given in Table 4.3. When these values are inserted in eqn (4.1.6) in the appropriate combination the integral may be evaluated (see Problem 4.19).

Table 4.3. Molar heat capacities, $C_{p,\mathrm{m}} = a + bT + c/T^2$

	a/J K^{-1} mol^{-1}	b/10^{-3} J K^{-2} mol^{-1}	c/10^5 J K mol^{-1}
N_2(g)	28.58	3.77	−0.50
CO_2(g)	44.22	8.79	−8.62
H_2O(l)	75.48	0	0
C (graphite)	16.86	4.77	−8.54

4.1 (e) The relation between $\Delta_r H$ and $\Delta_r U$

The enthalpy of a substance differs from its internal energy by an amount pV. It follows that the reaction enthalpy and the *reaction energy* ($\Delta_r U$) are related by

$$\Delta_r H = \Delta_r U + (pV)_{\text{products}} - (pV)_{\text{reactants}}. \quad (4.1.9)$$

$\Delta_r H$ and $\Delta_r U$ differ because at constant pressure, but not at constant volume, the system does work on the surroundings in the course of the reaction. In reactions involving only solids or liquids the volumes of the products and reactants are approximately the same, and except under some geophysical conditions (when pressures are large), we have:

For reactions involving only solids and liquids: $\Delta_r H \approx \Delta_r U$. (4.1.10)

In the case of reactions involving gases $\Delta_r H$ and $\Delta_r U$ may be quite different because the volume changes markedly. In elementary applications it is adequate to assume that all the gases are perfect, and so the pV in $H = U + pV$ may be replaced by nRT for each gas present. If we denote the change in the number of molecules of gas in the reaction as written by $\Delta\nu_{gas}$ (e.g. $\Delta\nu_{gas} = +1$ in the 'reaction' $H_2O(l) \rightarrow H_2O(g)$), then

$$\Delta_r H = \Delta_r U + \Delta\nu_{gas} RT. \qquad (4.1.11)^\circ$$

This expression gives a simple way of relating the two quantities.

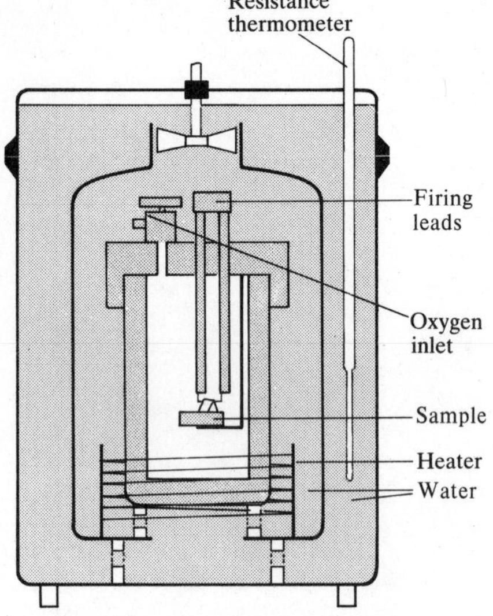

Fig. 4.1. A constant volume bomb calorimeter. The 'bomb' is the central vessel, which is massive enough to withstand high pressures. The calorimeter (for which the heat capacity must be known) is the entire assembly shown here. In order to ensure adiabaticity, the calorimeter may be immersed in a water bath with a temperature continuously readjusted to that of the calorimeter at each stage of the combustion.

Example 4.4

Estimate the value of $\Delta_f U^\ominus(\mathcal{T})$ for ammonia from its standard enthalpy of formation.

- *Method*. Treat the gases as perfect, and use eqn (4.1.11). The reaction is $\frac{3}{2}H_2(g) + \frac{1}{2}N_2(g) \rightarrow NH_3(g)$, and the enthalpy of formation is given in Table 4.1.

- *Answer*. The change in the number of gas molecules in the equation as written is $1 - \frac{1}{2} - \frac{3}{2} = -1$. At $\mathcal{T}$, $RT = 2.48\ \text{kJ mol}^{-1}$. Therefore,

$$\Delta_f U^\ominus(\mathcal{T}) = \Delta_f H^\ominus(\mathcal{T}) - \Delta\nu_{gas} R\mathcal{T}$$
$$= -46.1\ \text{kJ mol}^{-1} - (-1) \times (2.48\ \text{kJ mol}^{-1}) = -43.6\ \text{kJ mol}^{-1}.$$

- *Comment*. When a real gas is being considered, use eqn (4.1.9) directly, and evaluate the pV for the products and the reactants.

- *Exercise*. Calculate $\Delta_r U^\ominus(\mathcal{T})$ for the combustion of propene (using the information in *Example* 4.2). [$-2059\ \text{kJ mol}^{-1}$]

4.1 (f) Calorimetric measurements of $\Delta_r H$ and $\Delta_r U$

The measurements of $\Delta_r H$ and $\Delta_r U$ are based on the relations $\Delta H = q_p$ and $\Delta U = q_V$. The actual observation involved is the measurement of the change in temperature brought about by the heat produced or absorbed: if the reaction takes place in a thermally insulated system of heat capacity C and the temperature changes by δT, then the value of q is obtained from $C\,\delta T$. This q is then identified with $\Delta_r H$ or $\Delta_r U$, depending on the conditions of the reaction.

The most important device for measuring q_V is the *adiabatic bomb calorimeter*, Fig. 4.1. The reaction is initiated inside a constant volume container, the bomb, which, if the reaction is a combustion, contains oxygen at about 30 atm. The bomb is immersed in a stirred water bath, and the whole device is the calorimeter. The calorimeter is itself immersed in a water bath. The temperature of the water in the calorimeter is monitored. The temperature of the surrounding water bath is monitored simultaneously and adjusted to the same value: this ensures that there is no net loss of heat from the calorimeter to the surroundings (the bath); hence the name 'adiabatic' calorimeter. The temperature change of the calorimeter produced by the completed reaction is then converted to q_V using the known heat capacity of the calorimeter. The apparatus is often calibrated against a standard (e.g. the combustion of benzoic acid), or electrically. The value of $\Delta_r U$ is converted to $\Delta_r H$, if that is required, by using eqn (4.1.9). The *adiabatic flame calorimeter*, Fig. 4.2, works in a similar way, but the reaction proceeds at constant pressure, and so $\Delta_r H$ is obtained directly. Reaction enthalpies and energies may also be measured by non-calorimetric methods: these are described in Chapters 10 and 12.

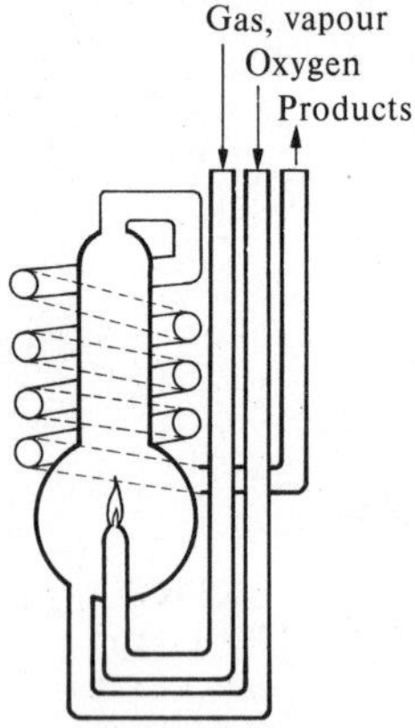

Fig. 4.2. A constant pressure flame calorimeter consists of this element immersed in a stirred water bath. Combustion occurs as a known amount of reactant is passed through to fuel the flame, and the rise of temperature is monitored.

4.2 Varieties of enthalpy

In chemistry we are often interested in the enthalpy changes in a variety of different types of reaction. The enthalpy changes then carry special names, and a few of these are reviewed in this section.

4.2 (a) The enthalpies of physical and chemical change

The *enthalpy of combustion* (symbol: $\Delta_c H$) is the reaction enthalpy for the oxidation of a substance to CO_2 and H_2O in the case of compounds of C, H, and O, and also to N_2 if N is present. Some standard values (with all substances in their standard states) are listed in Table 4.2. An important example is the enthalpy of combustion of glucose:

$$C_6H_{12}O_6(s) + 6O_2(g) \rightarrow 6CO_2(g) + 6H_2O(l); \qquad \Delta_c H^{\ominus} = -2808\ \text{kJ mol}^{-1}.$$

(We use the convention here and henceforth that *values refer to 298.15 K*, i.e. to $\mathcal{T}$, unless otherwise specified.) This large exothermic change of enthalpy is the basis of much cellular activity, and its value reveals something about the low efficiency of the cells of the less highly developed organisms. For instance, in *anaerobic fermentation* (the method of obtaining energy employed by some microorganisms, such as yeast) a central reaction is *glycolysis*, in which glucose is broken down into lactic acid:

$$C_6H_{12}O_6(s) \xrightarrow{\text{enzymes}} 2CH_3CH(OH)COOH(s).$$

The reaction enthalpy of

$$2CH_3.CH(OH).COOH(s) + 6O_2(g) \rightarrow 6CO_2(g) + 6H_2O(l); \qquad \Delta_r H^{\ominus} = -2688\ \text{kJ mol}^{-1}$$

is twice the enthalpy of combustion of lactic acid. Since the overall combustion of glucose corresponds to an enthalpy change of -2808 kJ mol^{-1}, according to Hess's Law the glycolysis reaction must correspond to a change of only -120 kJ mol^{-1}. Therefore the anaerobic cell succeeds in extracting only 120 kJ mol^{-1} from a fuel capable of providing 2808 kJ mol^{-1}. We shall see later that the reaction enthalpy is only one contribution we need consider when examining the efficient use of resources in the respiration process, but the result does suggest that aerobic respiration (such as occurs in our cells) is a more sophisticated process than the more primitive anaerobic fermentation.

Enthalpies of formation of compounds were introduced in Section 4.1(b). Since they refer to the overall enthalpy changes that occur when elements in their standard reference states are broken up into atoms and then reassembled into compounds, they do not give direct information about the strengths of bonds. That information is carried by the *bond dissociation enthalpy* (symbol: $\Delta H(\text{A–B})$), the reaction enthalpy for $AB(g) \rightarrow A(g) + B(g)$, where A and B may be *groups* of atoms. Some experimental values are listed in Table 4.4. Notice that the value of a dissociation enthalpy of a given bond depends on the structure of the rest of the molecule: in water, for instance, $\Delta H^{\ominus}(\text{HO—H}) = +492$ kJ mol^{-1} and $\Delta H^{\ominus}(\text{O—H}) = +428$ kJ mol^{-1}. This means that removing the first and second hydrogen atoms involves different enthalpy changes because the electronic structure of the molecule adjusts once the first atom is removed.

The bond dissociation enthalpy must be distinguished from two related

Table 4.4. Bond dissociation enthalpies $\Delta H(A-B)/\text{kJ mol}^{-1}$ at 298 K

H–H	436
H–O	428
H–OH	492
H–CH_3	435
H_3C–CH_3	368

Table 4.5. Bond enthalpies $E(\text{A}-\text{B})/\text{kJ mol}^{-1}$

	H	C	N	O
H	436			
C	413	348		
N	388	305	163	
O	463	360	157	146

Table 4.6. Standard enthalpies of atomization, $\Delta_a H^\ominus(T)/\text{kJ mol}^{-1}$

C(s)	716.68
Na(s)	108.4
K(s)	89.8
Cu(s)	339.3

quantities. One is the *bond dissociation energy* (symbol: $D(\text{A—B})$), the change in internal energy accompanying bond dissociation at $T=0$; this differs only slightly from the bond dissociation enthalpy at 298 K. The other is the *mean bond dissociation enthalpy*, or simply the *bond enthalpy* (symbol: $E(\text{A—B})$). This is the value of the bond dissociation enthalpy of the A—B bond averaged over a series of related compounds, Table 4.5. For instance, the O—H bond enthalpy is the mean of $\Delta H^\ominus(\text{HO—H})$ and $\Delta H^\ominus(\text{O—H})$, namely $+463\text{ kJ mol}^{-1}$, calculated using the data for water and similar compounds, such as $\Delta H^\ominus(\text{CH}_3\text{O—H})$ in methanol. Bond enthalpies are useful because they let us make estimates of enthalpy changes in reactions where data might not be available (see *Example* 4.5 below).

The enthalpy change on shattering a molecule completely into its component atoms is called the *enthalpy of atomization* (symbol: $\Delta_a H$). It follows from Hess's Law that the enthalpy of atomization of H_2O is the sum of the HO—H and H—O bond dissociation enthalpies, which from Table 4.4 is $+920\text{ kJ mol}^{-1}$. Values for other substances are given in Table 4.6. In the case of a metal, the enthalpy of atomization is the same as the *enthalpy of sublimation* (symbol: ΔH_{sub}), the enthalpy of the process $\text{M(s)} \rightarrow \text{M(g)}$, provided it vaporizes to a monatomic gas. The enthalpy of sublimation of carbon (the enthalpy of atomization of graphite) proved to be a notoriously difficult quantity to measure, but it is so important for the discussion of the energetics of organic molecules that a great deal of effort went into its determination. The currently accepted value is $+716.68\text{ kJ mol}^{-1}$.

Example 4.5

Use bond enthalpy and enthalpy of atomization data to estimate the standard enthalpy of formation of liquid methanol.

• *Method*. Atomize the reactants in the formation reaction (which is $\text{C(s)}+2\text{H}_2\text{(g)}+\frac{1}{2}\text{O}_2\text{(g)} \rightarrow \text{CH}_3\text{OH(l)}$) and then form the product, the reverse of its atomization. Note that the gaseous product must be condensed to form the stated product. Use data from Tables 4.4–4.7.

• *Answer*. The atomization involves the following enthalpy change:

$$\Delta_a H^\ominus = 2\Delta_a H^\ominus(\text{C, s}) + 2\Delta H(\text{H—H}) + \tfrac{1}{2}\Delta H(\text{O=O})$$
$$= \{(+717)+2(+436)+\tfrac{1}{2}(+497)\}\text{ kJ mol}^{-1} = 1838\text{ kJ mol}^{-1}.$$

The construction of $\text{CH}_3\text{OH(g)}$ involves the following decrease of enthalpy:

$$\Delta H_m^\ominus = -\{3E(\text{C—H})+E(\text{C—O})+E(\text{O—H})\} = -2062\text{ kJ mol}^{-1}.$$

The enthalpy of formation of $\text{CH}_3\text{OH(g)}$ is therefore the sum of the two enthalpy changes, -224 kJ mol^{-1}. The enthalpy of vaporization is $+37.99\text{ kJ mol}^{-1}$, and therefore the enthalpy of condensation is $-37.99\text{ kJ mol}^{-1}$. The overall enthalpy of formation of $\text{CH}_3\text{OH(l)}$ is therefore -262 kJ mol^{-1}.

• *Comment*. The measured value is $-239.0\text{ kJ mol}^{-1}$. The discrepancy is due to the use of mean bond enthalpies in the calculation of the enthalpy of construction of the CH_3OH.

• *Exercise*. Estimate the standard enthalpy of formation of ethanol.

[-276 kJ mol^{-1}]

The enthalpy of sublimation is also a special case of an *enthalpy of phase transition*. Phase transitions include vaporization (ΔH_{vap}, *enthalpy of vaporization*, also called 'latent heat' of vaporization), melting (ΔH_{melt}, the

Table 4.7. Enthalpies of fusion and vaporization at the transition temperature, $\Delta H_m^\ominus/\text{kJ mol}^{-1}$

	T_f/K	Fusion	T_b/K	Vaporization
He	3.5	0.021	4.22	0.084
Ar	83.81	1.188	87.29	6.51
H_2O	273.15	6.008	373.15	40.656 44.016 at $\mathcal{T}$
C_6H_6	278.6	10.59	353.2	30.8

enthalpy of melting, or the *enthalpy of fusion;* it is also called the 'latent heat' of melting), and changes of crystal form. For example, the vaporization of water corresponds to the process:

$$H_2O(l) \rightarrow H_2O(g); \qquad \Delta H^\ominus_{vap,m}(\mathcal{T}) = +44.02\ \text{kJ mol}^{-1}.$$

(At 100 °C the molar enthalpy of vaporization is $+40.66\ \text{kJ mol}^{-1}$.) The positive value indicates that vaporization is endothermic; further values are given in Table 4.7. There may be significant changes in volume during a phase transition, and so ΔU_{vap} may be significantly different from ΔH_{vap}. For example, at 100 °C and 1 bar, 1 mol $H_2O(l)$ occupies about 18 cm^3, but 1 mol $H_2O(g)$ occupies about 30 000 cm^3. It follows from eqn (4.1.11) with $\Delta \nu_{gas} = +1$ (and $RT = 3.10\ \text{kJ mol}^{-1}$ at 373 K), that

$$\Delta U^\ominus_{vap,m} = 40.66\ \text{kJ mol}^{-1} - 3.10\ \text{kJ mol}^{-1} = 37.56\ \text{kJ mol}^{-1},$$

and so the enthalpy and internal energy of vaporization differ by 8 per cent.

Example 4.6

Someone sits in a warm room and eats $\frac{1}{2}$ lb of cheese (an energy intake of about 4000 kJ). Suppose that none of the energy is stored in the body. Calculate the mass of water that must be perspired in order to maintain the original body temperature.

• *Method*. Perspiration cools the body because water requires energy in order to evaporate. Take the value of $\Delta H^\ominus_{vap,m}$ at 25 °C (Table 4.7) as sufficiently accurate for the present problem. The evaporation (i.e. vaporization) occurs at constant pressure, and so ΔH can be equated to q, the energy to be dissipated as heat. Therefore, find n in $n\,\Delta H^\ominus_{vap,m} = q$.

• *Answer*. $n = (4000\ \text{kJ})/(44.0\ \text{kJ mol}^{-1}) = 91\ \text{mol}$, corresponding to a mass of $(91\ \text{mol}) \times (18.02\ \text{g mol}^{-1}) = 1.6\ \text{kg}$.

• *Comment*. When calculating thermochemical aspects of food, it is important to remember that the 'Calories' of a diet are the kilocalories of science. Therefore, multiply by 4184 to convert to joules.

• *Exercise*. For how long must a 1000 W heater operate in order to vaporize 1.0 kg of water at 100 °C? [2.2×10^3 s]

The *enthalpy of hydrogenation* is the reaction enthalpy when an unsaturated organic compound is fully saturated. Two especially important cases are the hydrogenations of ethene and benzene:

$$CH_2{=}CH_2 + H_2 \rightarrow CH_3CH_3, \qquad \Delta_r H^\ominus = -137\ \text{kJ mol}^{-1},$$

$$C_6H_6\ (\text{benzene}) + 3H_2 \rightarrow C_6H_{12}\ (\text{cyclohexane}), \qquad \Delta_r H^\ominus = -205\ \text{kJ mol}^{-1}.$$

The interest in these two values lies in the observation that the second is not three times the first, as might be expected on the basis that benzene contains three double bonds. The value for benzene is less exothermic than $3 \times 137\ \text{kJ mol}^{-1}$ by $206\ \text{kJ mol}^{-1}$, which therefore represents a *thermochemical stabilization* of benzene (so that it lies closer in energy than expected to the fully hydrogenated, stable form). This stabilization is explained in Part 2.

4.2 (b) The enthalpies of ions in solution

The *enthalpy of solution* (symbol: ΔH_{soln}) of a substance is the enthalpy change when it dissolves in a specified amount of solvent. The *enthalpy of solution to infinite dilution* is the enthalpy change when the substance dissolves in an infinite amount of solvent and the interactions between the ions (or solute molecules) are negligible; in the case of water as solvent infinite dilution is denoted by the label 'aq' (but often 'aq' signifies an aqueous solution in general, not necessarily one that is infinitely dilute, and so the context is important). In the case of HCl:

$$\text{HCl(g)} \rightarrow \text{HCl(aq)}; \qquad \Delta H^{\ominus}_{\text{soln,m}}(\mathcal{T}) = -74.85\ \text{kJ mol}^{-1},$$

and so 75 kJ of heat is generated when 1 mol HCl(g) dissolves to produce an infinitely dilute solution; a different amount of heat is produced when HCl(g) dissolves in less water to give a more concentrated solution.

Once enthalpies of solution have been measured we can obtain values of *enthalpies of formation of species in solution* by combining them with the standard enthalpies of formation of the dissolving substances. These values are important because many reactions take place in solution. The standard enthalpy of formation of HCl in solution, for instance, is the sum of the enthalpies of the following two processes:

$$\tfrac{1}{2}\text{H}_2\text{(g)} + \tfrac{1}{2}\text{Cl}_2\text{(g)} \rightarrow \text{HCl(g)}; \qquad \Delta_r H^{\ominus} = -92.31\ \text{kJ mol}^{-1}$$
$$\text{HCl(g)} \rightarrow \text{HCl(aq)}; \qquad \Delta_r H^{\ominus} = -74.85\ \text{kJ mol}^{-1}$$
$$\tfrac{1}{2}\text{H}_2\text{(g)} + \tfrac{1}{2}\text{Cl}_2\text{(g)} \rightarrow \text{HCl(aq)}; \qquad \Delta_r H^{\ominus} = -167.16\ \text{kJ mol}^{-1}.$$

Similarly, the enthalpy of formation of NaCl(aq), for instance, can be obtained by combining the enthalpy of solution of NaCl(s) with its enthalpy of formation.

The enthalpy of formation of an ionic compound in solution can be regarded as the sum of the enthalpies of formation of the constituent ions. Thus, the enthalpy of formation of HCl(aq) can be thought of as being the sum of the enthalpies of formation of H^+(aq) and Cl^-(aq). The problem, though, is to decide how the value $-167.16\ \text{kJ mol}^{-1}$ is to be apportioned between the two kinds of ion. This is resolved by taking the *arbitrary* decision to set the standard enthalpy of formation of H^+(aq) equal to zero at all temperatures:

$$\tfrac{1}{2}\text{H}_2\text{(g)} \rightarrow \text{H}^+\text{(aq)}; \qquad \Delta_f H^{\ominus} = 0 \text{ by convention.}$$

On this basis it follows that the enthalpy of formation of Cl^-(aq) is $-167.16\ \text{kJ mol}^{-1}$. Combining that value with the enthalpy of formation of NaCl(aq) then leads to a value for Na^+(aq), and so on. This procedure leads to the values in Table 4.8.

Table 4.8. Standard enthalpies of formation of ions in solution at infinite dilution, $\Delta_f H^{\ominus}(\mathcal{T})/\text{kJ mol}^{-1}$

Cations		Anions	
H^+	0	OH^-	−230.0
Na^+	−240.1	Cl^-	−167.2
Cu^{2+}	+64.8	SO_4^{2-}	−909.3
Al^{3+}	−524.7	PO_4^{3-}	−1277.4

The enthalpy of formation of a substance in solution may be analysed into several contributions. In the case of NaCl(aq), for instance, the overall reaction $Na(s)+\frac{1}{2}Cl_2(g) \rightarrow NaCl(aq)$ can be regarded as the outcome of five steps:

1. $Na(s) \rightarrow Na(g)$, the sublimation of sodium metal. The standard molar enthalpy change is $\Delta H^{\ominus}_{\mathrm{sub,m}}$, Table 4.6.

2. $Na(g) \rightarrow Na^+(g) + e^-(g)$, the ionization of sodium atoms. The energy involved is the *ionization energy* (symbol: I), which is obtained from spectroscopy (Chapter 15); values for several atoms are given in Table 4.9. We should use the ionization enthalpy, $I+RT$, since $\Delta \nu_{\mathrm{gas}} = +1$; but not only is the difference small ($RT = 2.5\ \mathrm{kJ\ mol^{-1}}$ at $\mathcal{T}$), the RT is in fact cancelled by a similar term, as we see in a moment.

3. $\frac{1}{2}Cl_2(g) \rightarrow Cl(g)$, the dissociation of chlorine, the reaction enthalpy being half the bond dissociation enthalpy, Table 4.4. Its value is also obtained spectroscopically.

4. $Cl(g) + e^-(g) \rightarrow Cl^-(g)$, electron attachment to chlorine atoms. The energy is measured by the *electron affinity* (symbol: E_A), which is defined so that a *positive* electron affinity corresponds to the ion $X^-(g)$ having a *lower* energy than the atom X. The energy of the reaction $X(g) + e^-(g) \rightarrow X^-(g)$ is therefore $-E_A$. Some values are listed in Table 4.10: some are from spectroscopy, others from calculation. (The present type of analysis is sometimes used to determine the electron affinity if it is the only unknown.) In this step too we should deal with the enthalpy, which is $-E_A - RT$ because $\Delta \nu_{\mathrm{gas}} = -1$; this is the $-RT$ that cancels the RT in Step 2.

5a. $Na^+(g) + Cl^-(g) \rightarrow NaCl(s)$ is one possible step at this stage; it corresponds to the formation of the NaCl crystal, and its enthalpy change is the *lattice enthalpy*, $\Delta H^{\ominus}_{\mathrm{lattice}}$.

The sequence of steps is depicted in Fig. 4.3. When the step $Na + \frac{1}{2}Cl_2(g) \rightarrow NaCl(s)$ is included, a *cycle*, a closed sequence of processes, is completed: this is known as a *Born–Haber cycle*. The importance of a cycle is that, since ethalpy is a state function, *the sum of enthalpy changes round a cycle is zero*. It follows that if all but one of the enthalpy changes in a cycle are known, the unknown may be found.

Table 4.9. First and second ionization energies, $I/\mathrm{kJ\ mol^{-1}}$

H	1312
He	2372, 5251
Na	496, 4562
Mg	738, 1451

Table 4.10. Electron affinities, $E_A/\mathrm{kJ\ mol^{-1}}$

H	74		
O	141	O^-	−844
F	333		
Cl	349		

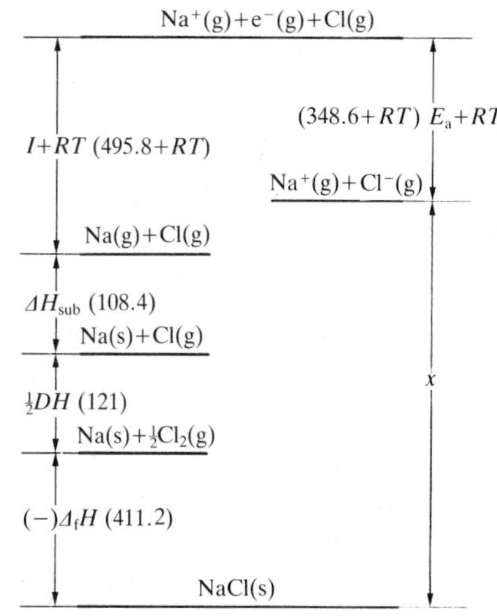

Fig. 4.3. A Born–Haber cycle for the determination of the lattice enthalpy. The sum of the enthalpy changes round the cycle is zero. In other words, the distance up on the left must be equal to the distance up on the right. From this equality, x may be determined.

Example 4.7

Calculate the lattice enthalpy of NaCl(s) at 298 K using the Born–Haber cycle in Fig. 4.3.

• *Method*. Find x in the illustration from the condition that the distance up on the left must equal the distance up on the right. The following data may be assembled from the tables:

$$\Delta H(\mathrm{Cl{-}Cl}) = +242\ \mathrm{kJ\ mol^{-1}}, \quad \Delta H^{\ominus}_{\mathrm{sub,m}}(\mathrm{Na, s}) = +108.4\ \mathrm{kJ\ mol^{-1}}$$
$$I(\mathrm{Na}) = +495.8\ \mathrm{kJ\ mol^{-1}}, \quad E_A(\mathrm{Cl}) = +348.6\ \mathrm{kJ\ mol^{-1}}$$
$$\Delta_f H^{\ominus}(\mathrm{NaCl, s}) = -411.2\ \mathrm{kJ\ mol^{-1}}.$$

• *Answer*. The distance up on the left, from NaCl(s) to the top of the cycle, is $\{411.2 + 121 + 108.4 + 495.8\} + R\mathcal{T} = 1136 + R\mathcal{T}$. The distance up on the right is $\{x + 348.6\}\ \mathrm{kJ\ mol^{-1}} + R\mathcal{T}$. The two distances are equal; therefore $x = 787$. Since the lattice enthalpy is positive (it corresponds to the up-going arrow in the cycle), $\Delta H^{\ominus}_{\mathrm{lattice,m}} = +787\ \mathrm{kJ\ mol^{-1}}$.

• *Comment*. The lattice enthalpy is the reaction enthalpy for $NaCl(s) \rightarrow Na^+(g) + Cl^-(g)$, and is endothermic. The internal energy is smaller by an amount $2R\mathcal{T}$; that is, $+782\ \mathrm{kJ\ mol^{-1}}$.

• *Exercise*. Calculate the lattice energy of $CaBr_2$. [2140 kJ mol⁻¹].

Another cycle, Fig. 4.4, is obtained when we use

5b. $Na^+(g) + Cl^-(g) \rightarrow Na^+(aq) + Cl^-(aq) = NaCl(aq)$.

The enthalpy change is the sum of the *enthalpy of solvation* (symbol: $\Delta H^\ominus_{solv}$), of each type of ion. Since all the other enthalpy changes are known, this quantity may be determined: the distance up on the left of Fig. 4.4 (1133 kJ mol^{-1}) is equal to the distance up on the right (348.6 kJ mol^{-1} $+x$). Hence $x = 784$ kJ mol^{-1}. The solvation process is the step *down* on the right, and so the enthalpy of solvation is $-x$, or -784 kJ mol^{-1}. Other solvation enthalpies may be obtained in this way, and some *enthalpies of hydration*, as they are called when the solvent is water, are listed in Table 4.11 (using slightly different data).

The enthalpy of solvation is the sum of the enthalpies of solvation of the individual $Na^+(g)$ and $Cl^-(g)$ ions. This time, however, we cannot adopt an arbitrary convention about the solvation enthalpy of some ion such as the proton because it would conflict with the choice already made about its enthalpy of formation: we have selected zero for the step $\frac{1}{2}H_2(g) \rightarrow H^+(aq)$, and we cannot simultaneously select zero for $H^+(g) \rightarrow H^+(aq)$. We are therefore forced to estimate the actual value of the enthalpy change in the step by calculating the energy of interaction of a proton and the surrounding water molecules, and there is some agreement that its value is about -1090 kJ mol^{-1}. Once that value is established the value for $Cl^-(g) \rightarrow Cl^-(aq)$ can be obtained from data on HCl and then combined with the value of $\Delta H^\ominus_{solv}$ for NaCl as obtained above, to arrive at the value of about -400 kJ mol^{-1} for $Na^+(g) \rightarrow Na^+(aq)$. The ion solvation enthalpies in Table 4.12 have been obtained in this way.

The data in the table conform with common sense (and with calculation, as we see in Chapter 12). Thus, small ions of high charge have the most exothermic hydration enthalpies because they attract the solvent so strongly. The hydration enthalpy of the electron, the enthalpy of the process $e^-(g) \rightarrow e^-(aq)$, is only about -170 kJ mol^{-1}. This suggests that when it is trapped in water (e.g. when water is irradiated with high energy radiation) it spreads over a volume about the size of two water molecules.

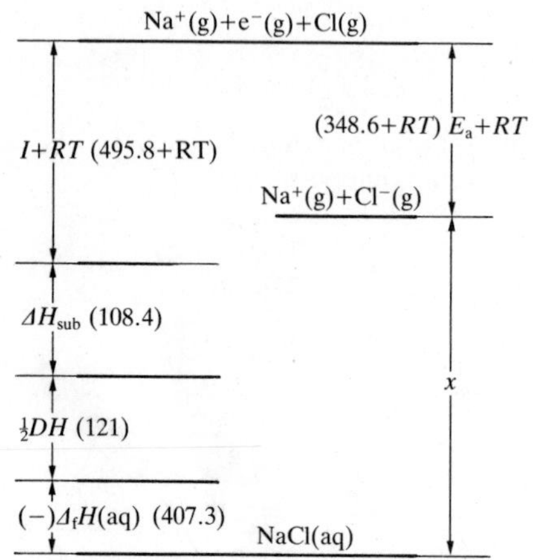

Fig. 4.4. A similar Born–Haber cycle for the determination of the enthalpy of solvation of Na^+ and Cl^- ions. The distance up on the left is equal to the distance up on the right.

Table 4.11. Standard molar enthalpies of hydration at infinite dilution, $-\Delta H^\ominus_{solv,m}(T)$/kJ mol^{-1}

	Li^+	Na^+	K^+
F^-	1026	911	828
Cl^-	884	769	685
Br^-	856	742	658

Table 4.12. Standard molar ion hydration enthalpies, $-\Delta H^\ominus_{solv,m}$ (T)/kJ mol^{-1}

Li^+	520	F^-	506
Na^+	405	Cl^-	364
K^+	321	Br^-	337

Further reading

Experimental techniques:

Calorimetry. J. M. Sturtevant; in *Techniques of chemistry* (A. Weissberger and B. W. Rossiter, eds.) V, 347, Wiley-Interscience, New York, 1971.

Experimental thermochemistry (*Vol. 2*). F. D. Rossini; Wiley-Interscience, New York, 1956.

Experimental thermochemistry (*Vol. 2*). H. A. Skinner (ed.); Wiley-Interscience, New York, 1962.

Experimental thermodynamics. J. D. McCullough and D. W. Scott; Butterworths, London, 1968.

Applications:

Energy changes in biochemical reactions. I. Klotz; Academic Press, New York, 1967.

Bioenergetics (2nd edn). A. Lehninger; Benjamin-Cummings, New York, 1971.

Inorganic energetics (2nd edn). W. E. Dasent; Cambridge University Press, 1982.

Some thermodynamic aspects of inorganic chemistry (2nd edn). D. A. Johnson; Cambridge University Press, 1982.

Data:

NBS tables of chemical thermodynamic properties. Supplement to Vol. 2, *J. phys. and chem. reference data*, 1982.

Physico-chemical constants of pure organic compounds. J. Timmermans; Elsevier, Amsterdam, 1956.

Selected values of chemical thermodynamic properties. NBS technical note 270, 1965–71 (six parts).

Bond energies, ionization potentials, and electron affinities. V. I. Vedeneyev, L. V. Gurvich, V. N. Kondrat'yev, V. I. Mendaradev, and Y. L. Frankevich; Edward Arnold, London, 1966.

Tables of physical and chemical constants. G. W. C. Kaye and T. H. Laby; Longmans, London, 1973.

Handbook of chemistry and physics (Vol. 65). R. C. Weast (ed.); CRC Press, Boca Raton, 1986.

American Institute of Physics handbook. D. E. Gray (ed.); McGraw-Hill, New York, 1972.

Conversion of standard (1 atm) thermodynamic data to the new standard-state pressure, 1 bar (10^5 Pa). R. D. Freeman; *Bull. chem. thermodynamics* **25,** 523 (1982).

Introductory problems

A4.1. Calculate the standard enthalpies of formation of $KClO_3(s)$, $NaHCO_3(s)$, and $NOCl(g)$ using Table 4.1, Table 4.2, and the following:

$$2KClO_3(s) \rightarrow 2KCl(s) + 3O_2(g) \qquad \Delta H^\ominus(T) = -89.4 \text{ kJ}$$
$$NaOH(s) + CO_2(g) \rightarrow NaHCO_3(s) \qquad \Delta H^\ominus(T) = -127.5 \text{ kJ}$$
$$2NOCl(g) \rightarrow 2NO(g) + Cl_2(g) \qquad \Delta H^\ominus(T) = +75.5 \text{ kJ}.$$

A4.2. The standard enthalpy of formation of ethylbenzene, $C_8H_{10}(l)$, at 298.15 K is -12.5 kJ mol^{-1}. Calculate its standard molar enthalpy of combustion at 298.15 K.

A4.3. The standard molar enthalpy of combustion of 1-hexene, $C_6H_{12}(l)$, at 298.15 K is -4003 kJ mol^{-1}. Calculate the standard molar enthalpy of hydrogenation of 1-hexene.

A4.4. The standard enthalpy of formation of methyl acetate, $CH_3COOCH_3(l)$, at 298.15 K is -442 kJ mol^{-1}. Calculate the standard internal energy of formation of methyl acetate.

A4.5. The molar heat capacity of ethane is represented in the temperature range 298 K to 400 K by the empirical expression $\{14.73 + 0.1272(T/K)\}$ J K^{-1} mol^{-1}. Calculate the standard enthalpy of formation of ethane at 350 K.

A4.6. When 120 mg of naphthalene(s), $C_{10}H_8(s)$, at 298 K were burned in a bomb calorimeter under conditions of constant volume, the temperature rose 3.05 K. Calculate the heat capacity of the calorimeter. When 100 mg of phenol, $C_6H_5OH(s)$, are burned in the calorimeter under the same conditions, how much will the temperature rise?

A4.7. Calculate the amount of heat required to melt 750 kg of sodium at 370.95 K.

A4.8. An object is cooled by the evaporation of $CH_4(l)$ at its normal boiling point (111.66 K). What volume of $CH_4(g)$ at 1.00 atm pressure must be formed from the liquid in order to remove 32.5 kJ of heat from the object?

A4.9. Calculate the standard reaction enthalpy at 298.15 K for

$$2Ag^+(aq) + Cu(s) \rightarrow Cu^{2+}(aq) + 2Ag(s).$$

A4.10. The standard enthalpy of formation at 298.15 K for AgCl(s) is -127.0 kJ mol^{-1}. Calculate the standard enthalpy of solution of AgCl(s) in water.

Problems

4.1. Which of the following reactions are *exothermic* and which *endothermic*?

(a) $CH_4(g) + 2O_2(g) \rightarrow CO_2(g) + 2H_2O(l)$ $\quad \Delta_r H^\ominus = -890$ kJ mol^{-1}

(b) $2C(s) + H_2(g) \rightarrow C_2H_2(g)$ $\quad \Delta_r H^\ominus = 227$ kJ mol^{-1}

(c) $NaCl(s) \rightarrow NaCl(aq)$ $\quad \Delta_r H^\ominus = 3.9$ kJ mol^{-1}.

4.2. Express the reactions in Problem 4.1 in the form $0 = \sum_J \nu_J J$ and identify the stoichiometric coefficients.

4.3. Use the tables of standard molar enthalpies, Table 4.1, to find the standard enthalpies of the following reactions at 298 K.

(a) $2NO_2(g) \rightarrow N_2O_4(g)$.

(b) $NH_3(g) + HCl(g) \rightarrow NH_4Cl(s)$.
(c) cyclopropane (g) → propene (g).
(d) $HCl(aq) + NaOH(aq) \rightarrow NaCl(aq) + H_2O(l)$.

4.4. A bomb calorimeter rose in temperature by 1.617 K when a current of 3.20 A was passed for 27.0 s from a 12.0 V source. What is its heat capacity?

4.5. The same calorimeter was used as in the last Problem. When 0.3212 g of glucose was oxidized at 298 K under conditions of constant volume the temperature rose 7.793 K. What is (a) the standard molar enthalpy of combustion of glucose, (b) ΔU for the combustion, and (c) the standard enthalpy of formation of glucose at $\mathcal{T}$?

4.6. The standard enthalpy of formation of the sandwich compound bis(benzene)chromium, where benzene is the bread and a chromium atom the meat, was measured in a microcalorimeter (J. A. Connor, H. A. Skinner, and Y. Virmani, *J. chem. Soc. Faraday Trans.* I, 1218 (1973)). The change in internal energy for the reaction $Cr(C_6H_6)_2(c) \rightarrow Cr(c) + 2C_6H_6(g)$ was found to be 8.0 kJ mol^{-1} at 583 K. Find $\Delta_r H^{\ominus}$, and estimate the standard enthalpy of formation of the sandwich compound at this temperature. Take the molar heat capacity of benzene as 140 J K^{-1} mol^{-1} in its liquid range and at 28 J K^{-1} mol^{-1} in its vapour phase.

4.7. A 0.727 g sample of the sugar D-ribose, $C_5H_{10}O_5$, was weighed into a calorimeter and then ignited in the presence of excess oxygen. The temperature of the calorimeter rose by 0.910 K from $\mathcal{T}$. In a separate experiment in the same apparatus, 0.825 g of benzoic acid, which has the accurately known value of −3251 kJ mol^{-1} for the internal energy of combustion, was ignited, and the rise in temperature was 1.940 K. Calculate the molar internal energy of combustion, the molar enthalpy of combustion, and the enthalpy of formation of D-ribose.

4.8. The standard enthalpy of combustion of naphthalene is −5157 kJ mol^{-1} at $\mathcal{T}$. What is its standard enthalpy of formation at that temperature?

4.9. The standard enthalpy of decomposition of the yellow solid adduct $NH_3.SO_2$ is 40 kJ mol^{-1} at $\mathcal{T}$. What is its standard enthalpy of formation at that temperature?

4.10. By how much do the standard molar internal energy and standard molar enthalpy of combustion of diphenyl differ at (a) room temperature, (b) 99 °C, (c) 101 °C?

4.11. Geophysical conditions are sometimes so extreme that quantities neglected in normal laboratory experiments assume overriding importance. For example, consider the formation of diamond under geophysically typical pressures. The densities of the two allotropes are 2.27 g cm^{-3} (graphite) and 3.52 g cm^{-3} (diamond) at the temperature of the environment. By how much does ΔU_m for the transition differ from ΔH_m in a region where the pressure is 500 kbar.

4.12. Several aspects of a hydrocarbon have to be considered when a fuel is being selected. Among them are the amount of heat evolved for every gram of fuel consumed: the advantage of a high molar enthalpy of combustion may be eliminated if a large mass of fuel has to be transported. Use Table 4.2 to find the following information on butane, pentane, and octane: (a) the amount of heat evolved per mole at 25 °C when each hydrocarbon is burnt at constant pressure, (b) the heat evolved per gram, (c) the cost per kilojoule of heat (for the last, use your initiative to find the current cost of the hydrocarbons).

4.13. What amount of heat is available per mole and per gram of the three hydrocarbons referred to in the last Problem when combustion takes place in a constant-volume vessel at the same temperature?

4.14. The enthalpy of combustion of glucose is −2808 kJ mol^{-1} at 25 °C. How many grams of glucose do you need to consume (a) to climb a flight of stairs rising through 3 m, (b) to climb a mountain of altitude 3000 m? Assume that 25% of the enthalpy can be converted to useful work.

4.15. Calculate the standard enthalpy of formation of N_2O_5 at $\mathcal{T}$ using the following data:

$$2NO(g) + O_2(g) \rightarrow 2NO_2(g) \quad \Delta_r H^{\ominus} = -114.1 \text{ kJ mol}^{-1}$$
$$4NO_2(g) + O_2(g) \rightarrow 2N_2O_5(g) \quad \Delta_r H^{\ominus} = -110.2 \text{ kJ mol}^{-1}$$
$$N_2(g) + O_2(g) \rightarrow 2NO(g) \quad \Delta_r H^{\ominus} = +180.5 \text{ kJ mol}^{-1}.$$

4.16. The enthalpy of combustion of graphite at 25 °C is −393.51 kJ mol^{-1} and that of diamond is −395.41 kJ mol^{-1}. What is the enthalpy of the graphite → diamond phase transition at this temperature?

4.17. Use the data in Table 4.1 to predict the standard reaction enthalpy of $2NO_2(g) \rightarrow N_2O_4(g)$ at 100 °C given its value at 25 °C, Problem 4.3.

4.18. The heat capacity of a substance can often be expressed with adequate accuracy by the expression $C(T) = a + bT + c/T^2$, and coefficients for some materials are reported in Table 4.3. Find an expression for the enthalpy of the reaction $0 = \sum_J \nu_J J$ at a temperature T_2 in terms of the enthalpy at T_1 and the coefficients a, b, and c for the heat capacities of the species involved.

4.19. Using the data in Tables 4.1 and 4.3 to predict the standard molar enthalpy of formation of water at (a) −0.1 °C, (b) 100.1 °C.

4.20. When the coefficients a, b, c, of the heat capacity formula are not available, or when only a rough answer is required, the assumption is made that the heat capacities of all the species are constant over the temperature range of interest. What is the error introduced by this assumption in the case of the enthalpy of formation of water at 99 °C? Calculate $\Delta_f H^{\ominus}(372 \text{ K})$ using the room-temperature heat capacities, and also using the a, b, c coefficients.

4.21. Devise an expression for the reaction internal energy at a temperature T_2 in terms of its known value at T_1.

4.22. Samples of D-arabinose and α-D-glucose were burnt to completion in separate experiments in the same constant pressure microcalorimeter at $\mathcal{T}$. 88 mg of D-arabinose ($M_r = 150.1$) led to a temperature rise of 0.761 K; 102 mg of the glucose ($M_r = 180.2$) led to a rise of 0.881 K. The standard enthalpy of formation of α-D-glucose is -1274 kJ mol^{-1}; what are the standard enthalpies of (a) formation, (b) combustion of D-arabinose at $\mathcal{T}$?

4.23. The standard enthalpy of combustion of sucrose at $\mathcal{T}$ is -5645 kJ mol^{-1}. What is the advantage of complete aerobic oxidation as compared with incomplete anaerobic hydrolysis to lactic acid at this temperature (and under standard conditions). Use the data in Table 4.2.

4.24. A typical sugar cube has a mass of 1.5 g. What is its standard enthalpy of combustion at 25 °C? To what height can you climb assuming 25 per cent of the enthalpy is available for such work?

4.25. Damp clothes can be fatal on a mountain. Suppose the clothing you were wearing had absorbed 1 kg of water, and a cold wind dried it. What heat loss does the body have to make good? How much glucose would have to be consumed to replace that loss? Suppose your body did not make good the heat loss, what would your temperature be at the end of the evaporation? (Assume your heat capacity is the same as that of water.)

4.26. The standard enthalpy of combustion of propane gas at 25 °C is -2220 kJ mol^{-1} and the standard molar enthalpy of vaporization of liquid propane at this temperature is 15 kJ mol^{-1}. What is the standard enthalpy of combustion of the liquid at this temperature? What is the value of $\Delta U^{\ominus}$ for the combustion?

4.27. The isobaric molar heat capacities of liquid propane and water are 39.0 J K^{-1} and 75.5 J K^{-1} mol^{-1} respectively, and gaseous O_2 and CO_2 have the values 29.3 J K^{-1} mol^{-1} and 37.1 J K^{-1} mol^{-1} respectively. Combine this information with that in the last Problem to find $\Delta H^{\ominus}(308\text{ K})$ and $\Delta U^{\ominus}(308\text{ K})$ for the combustion of liquid propane. For liquids, $C_p \approx C_V$; assume heat capacities to be constant over the temperature range.

4.28. The standard enthalpies of hydrogenation of ethene and benzene are -132 kJ mol^{-1} and -246 kJ mol^{-1} respectively at $\mathcal{T}$. Calculate the thermochemical stabilization energy (the 'resonance energy') of benzene at this temperature.

4.29. In an experiment to measure the enthalpy of solution of potassium fluoride in glacial acetic acid (J. Emsley, *J. chem. Soc.* 2702 (1971)), a known weight of the anhydrous salt was added to a known weight of acid in a Dewar flask fitted with a heating coil, stirrer, and thermometer. The heat capacity of the system was determined by supplying a known amount of electricity and monitoring the temperature rise. The experiment was repeated for the salt KF.AcOH, where AcOH stands for acetic acid (ethanoic acid). The following is a reconstruction of an experiment:

KF: Heat capacity 4.168 kJ K^{-1}				
molality/(mol KF/kg	0.194	0.590	0.821	1.208
AcOH) $\Delta T/\text{K}$	1.592	4.501	5.909	8.115

KF.AcOH: Heat capacity				
4.203 kJ K^{-1} molality/(mol	0.280	0.504	0.910	1.190
KF/kg AcOH) $\Delta T/\text{K}$	-0.277	-0.432	-0.866	-1.189

Calculate the enthalpies of solvation of the two species at these molalities and at infinite dilution, and find the best straight line of the form $\Delta H/\text{kJ mol}^{-1} = a + bm$, where m is the molality. Account for the difference between the enthalpies of the two salts.

4.30. The bond dissociation enthalpy of $H_2(g)$ is 436 kJ mol^{-1} and that of $N_2(g)$ is 945 kJ mol^{-1}. What is the atomization enthalpy of ammonia?

4.31. The table below gives the enthalpies of solvation of several alkali metal halides. Complete the table from the information it contains.

	KCl	KI	RbCl	RbI
Lattice energy/kJ mol^{-1}	701	629	675	609
$\Delta H^{\ominus}_{\text{soln}}$/kJ mol^{-1}	16	12	11	?

4.32. Set up the Born–Haber cycle for determining the enthalpy of solvation of Mg^{2+} ions by water given the following data. Enthalpy of sublimation of Mg(s), 167.2 kJ mol^{-1}; first ionization energy of Mg, 7.646 eV; second ionization energy 15.035 eV; dissociation enthalpy of $Cl_2(g)$, 241.6 kJ mol^{-1}; electron affinity of Cl(g), 3.78 eV; enthalpy of formation of $MgCl_2(s)$, -639.5 kJ mol^{-1}; enthalpy of solution of $MgCl_2(s)$, -150.5 kJ mol^{-1}; enthalpy of hydration of $Cl^-(g)$, -383.7 kJ mol^{-1}.

4.33. The hydrogen bond between F^- and acetic acid has been found to be exceptionally strong (J. Emsley, *J. chem.* 2702 (1971)) and an indication of this was encountered in Problem 4.29. In this Problem we set up the Born–Haber cycle which enables the strength of the bond to be ascertained. Use the following data. Lattice enthalpy of KF, 797 kJ mol^{-1}; lattice enthalpy of KF.AcOH, *ca.* 734 kJ mol^{-1}; enthalpy of vaporization of acetic acid, 20.8 kJ mol^{-1}; enthalpy of solution of KF, 35.2 kJ mol^{-1}; enthalpy of solution of KF.AcOH, -3.1 kJ mol^{-1}. Find the hydrogen bond energy between F^- and acetic acid in the gas phase.

5

The Second Law: the concepts

Learning objectives

After careful study of this chapter you should be able to:

(1) State the criterion for the direction of spontaneous change, Introduction.

(2) State the *Second Law of thermodynamics*, Section 5.1(a).

(3) Define the *entropy change* of a reference system, eqn (5.1.1), and use it to deduce an expression for the change of entropy of any system, eqn (5.1.2).

(4) Calculate the entropy change accompanying the isothermal expansion of a perfect gas, eqn (5.1.4).

(5) State and justify the *Clausius inequality*, eqn (5.1.5), and use it to show that spontaneous changes in isolated systems are accompanied by an increase of entropy, eqn (5.1.6).

(6) Use the Clausius inequality to show that entropy is a state function, Section 5.1(d).

(7) Deduce, state, and use an equation for the change in entropy when a system is heated, eqn (5.2.1).

(8) State and use an expression for the change of entropy of the surroundings, eqn (5.2.3).

(9) State and use an expression for the *entropy of phase transition*, eqn (5.2.4).

(10) Deduce and use an expression for the maximum thermodynamic *efficiency* of a heat engine, eqn (5.2.6).

(11) Specify the *Carnot cycle*, and use it to calculate the work produced by a heat engine, Section 5.2(e).

(12) Calculate the entropy change during *irreversible* processes, Section 5.2(d).

(13) Define the *Helmholtz function* and the *Gibbs function*, eqn (5.3.6), and use them to specify the criteria of spontaneous change, eqn (5.3.7).

(14) Relate the Helmholtz function to the *maximum work* available from a process, eqn (5.3.9), and prove that maximum work is done when a process occurs reversibly, eqn (5.3.8).

(15) Relate the Gibbs function to the *maximum non-p,V work* available from a process, eqn (5.3.11).

(16) Evaluate the entropy of a substance from thermochemical data, Section 5.4 and Example 5.9.

(17) State the *Third Law of thermodynamics* and the *Nernst heat theorem*, Section 5.4(a), and summarize the experimental evidence.

(18) Define *Third Law entropy* and *standard reaction entropy*, eqn (5.4.3) and Example 5.10.

(19) Define *standard Gibbs function of formation* and use tables of values to calculate the *standard reaction Gibbs function*, eqn (5.4.4) and Example 5.11.

Introduction

Some things happen spontaneously, some things don't. A gas expands to fill the available volume, it does not spontaneously contract into a smaller one. A hot body cools to the temperature of its surroundings, it does not spontaneously get hotter at their expense. A chemical reaction runs in one direction rather than another: burning diamonds gives hot carbon dioxide, hot carbon dioxide does not spontaneously form diamonds. Some aspect of the world determines the natural direction of change, the direction of *spontaneous change*. We can in fact compress a gas, refrigerators refrigerate, and we can make diamonds; but none of these things happens spontaneously: they have to be brought about by doing work.

What determines the direction of spontaneous change? It is not the total energy. The First Law states that energy is conserved in any process, and we cannot disregard it now and say that everything tends towards the state of lowest energy. The energy of the universe (in a thermodynamic sense) is constant.

Is it the energy of the system itself that tends towards a minimum? Two arguments show that this cannot be so. First, a perfect gas expands spontaneously into a vacuum, yet its internal energy remains constant (Box 3.2). Secondly, if the energy of a system does happen to decrease during a spontaneous change, the energy of the surroundings must increase by the same amount (by the First Law). The increase in energy of the surroundings is just as spontaneous a process as the decrease in energy of the system. Why should we favour one part of the universe over another?

When a change occurs the total energy remains constant, but it is parcelled out in different ways. Can it be, therefore, that the direction of change is related to the *distribution* of energy? We shall see that this is so, and that spontaneous changes are always accompanied by a reduction in the 'quality' of energy in the sense that it is degraded into a more dispersed, chaotic form. *Spontaneous changes are consequences of the natural tendency of the universe towards greater chaos*.

The role of the distribution of energy can be illustrated by thinking about a ball bouncing on a floor. We all know that after each bounce the ball does not rise as high. This is because there are frictional losses in the materials of the ball and floor. The direction of spontaneous change is towards a state in which the ball is at rest with all its energy degraded into the chaotic thermal motion of the particles of the virtually infinite floor, Fig. 5.1.

A ball resting quietly on a warm floor has never been observed to start bouncing. For bouncing to begin, something rather special would have to happen. In the first place, some of the thermal motion of the floor would have to accumulate in a single, small object, the ball. This requires a spontaneous localization of energy from the myriad of vibrations of the particles that constitute the floor into the much smaller number that constitute the ball. Furthermore, whereas the thermal motion is chaotic, in order for the ball to move upwards, its atoms must all move in the same direction. This localization of *ordered* motion is so unlikely that we can dismiss it as wholly improbable.

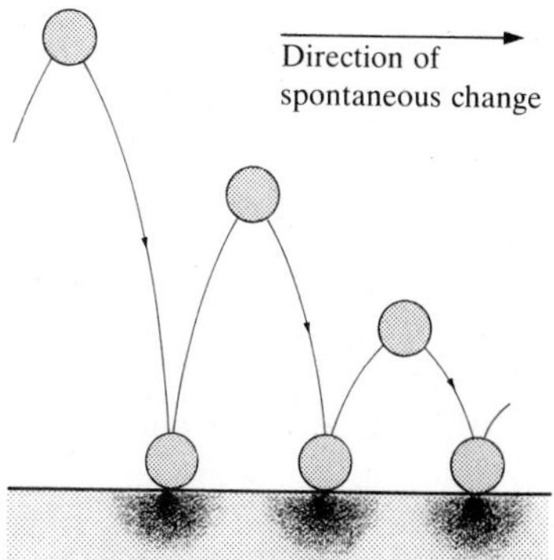

Fig. 5.1. The direction of spontaneous change for a ball bouncing on a floor. On each bounce some of its energy is degraded into the thermal motion of the floor, and that energy disperses. The reverse has never been observed.

We have found the signpost of spontaneous change: *we look for the direction of change that leads to a chaotic dispersal of the total energy*. This accounts for the direction of change of the bouncing ball, because its energy is dissipated into the enormous number of modes of thermal motion of the floor: this is a natural, spontaneous process. The reverse process is unnatural because the chaotic distribution of energy is extremely unlikely to coordinate itself by chance into a local, uniform motion.

Although we have identified the signpost on the basis of a single example, it can easily be developed to account for the changes mentioned in the opening paragraph. A gas does not spontaneously contract, because in order to do so the chaotic motion of its particles would have to take them all simultaneously into the same region of the container; the opposite change, however, is a natural consequence of chaos. A cool object does not spontaneously become warmer than its surroundings because it is extremely improbable that the jostling of randomly vibrating atoms will lead to the accumulation in it of excess thermal motion; the opposite change, however, the spreading of excess thermal motion from the object, is a natural consequence of chaos. Burnt diamonds turn into carbon dioxide because the energy locked in the highly localized small diamonds is carried away in the highly unlocalized gas molecules and dissipated into the rest of the universe as fast molecules collide with slower ones: the opposite change will not occur by chance alone.

Our everyday experience has led to the idea that chaotic dispersal determines direction. In order to develop the idea we must make it quantitative.

5.1 Measuring dispersal: the entropy

The First Law led to the introduction of the internal energy, U, a state function that lets us assess whether a change is feasible: only those changes may occur for which the total energy of the isolated system (the 'universe') remains constant. The Second Law also leads to a state function, the *entropy* (symbol: S), which lets us assess whether a state is accessible from another by a *spontaneous* change: the entropy of the universe is greater after the occurrence of a spontaneous change. The First Law uses the internal energy to identify *permissible* changes; the Second Law uses the entropy to identify the *spontaneous* changes among those permissible changes.

One way of introducing the entropy develops the view that the extent of the dispersal of energy can be calculated: this leads to the *statistical* definition of entropy. We shall deal with this approach in Part 2, by which time we shall know more about atomic and molecular energy levels. Another approach develops the view that the dispersal can be related to the heat involved in a process: this leads to the *thermodynamic* definition. This is the approach we develop here. In due course we shall see that the statistical and thermodynamic definitions coincide: that is an exciting moment in physical chemistry, because it establishes a link between the world of observations (the concern of thermodynamics) and the underworld of atoms.

5.1 (a) The Second Law

The most fundamental example of a spontaneous, irreversible process is the generation of heat in a reservoir as a result of dropping a weight. This, as we shall see, models all the spontaneous changes mentioned in the introduction. In the process, the 'quality' of energy is degraded in the sense of becoming less available for doing work. All spontaneous changes are accompanied by this kind of degradation, and so the definition of entropy to which it leads turns out to be universally applicable.

There are a few important preliminary remarks. Irreversible processes (like cooling and the free expansion of gases) are spontaneous processes: they cause a degradation of the quality of energy and hence an increase in the entropy of the universe. *Irreversible processes generate entropy*. In contrast, reversible processes are finely balanced changes, the system being in equilibrium with its surroundings at every stage. Each infinitesimal step along a quasi-static reversible path is reversible, and occurs without degrading the quality of energy, without dispersing energy chaotically, and without increasing the entropy of the universe. *Reversible processes do not generate entropy* (but they may transfer it from one part of the universe to another).

The second remark concerns an observation based on experience. No spontaneous change has ever been reversed without there being, somewhere in the universe, a degradation of energy. In other words, while it is certainly possible to restore a temperature difference between two objects, it can be done only at the expense of degrading energy elsewhere, for example by running a refrigerator on electricity generated by a falling weight. In particular, the reverse of the reference process, a weight rising at the expense of heat from a reservoir (e.g. a ball starting to bounce), with no other change elsewhere, has never been observed, Fig. 5.2. This remark is promoted into a general principle:

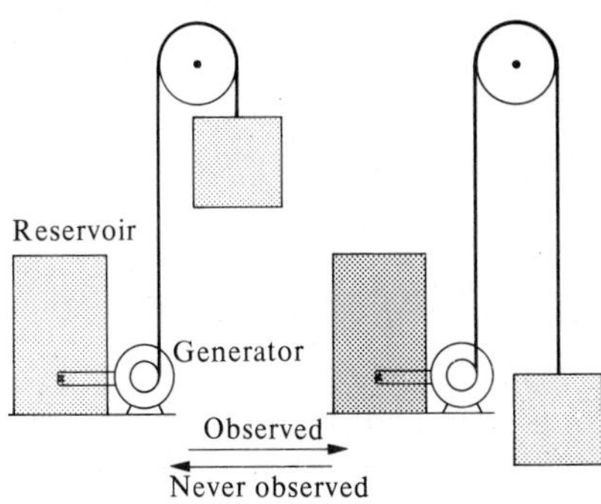

Fig. 5.2. The fundamental spontaneous process is represented by a falling weight and a thermal reservoir. The potential energy of the weight is transferred to the thermal motion of the reservoir, never vice-versa spontaneously. Note that a reservoir is a sink of infinite extent, and its temperature remains constant however much energy is transferred as heat.

Second Law: *No process is possible in which the sole result is the absorption of heat from a reservoir and its complete conversion into work.*

The importance of the words 'sole result' must be appreciated: It means that any system must have gone through a complete cycle, and be back in its starting condition, and any gas must have returned to its original volume, etc. Note that the law does permit natural processes in which the sole result is the conversion of work into heat. It is unsymmetrical: it cuts off one side of all possible processes allowed by the First Law.

5.1 (b) The definition of entropy

Now we move on to make all these matters quantitative. We do this by making explicit what is meant by the entropy change in the reference system. We use the sign † to denote the reference system; as in the case of any reservoir, its reservoir always remains at the same temperature however much energy is transferred to it (simply because it is so big).

Suppose the falling weight is coupled to the reservoir (e.g. by driving a generator connected to a heater, as in Fig. 5.2), and that when it falls a quantity of heat $q^{\dagger}$ is transferred to the reservoir. The degradation of the quality of energy is related to this quantity of heat, but we cannot simply equate it to the change of entropy. This is because it is necessary to take into account the temperature of the reservoir: the energy is degraded more completely if it is transferred directly to a cold reservoir than to a hot one. In the latter case we could go on to extract work from an engine driven by the flow of $q^{\dagger}$ from the hot to the cold reservoir, but this is not possible if the energy is transferred directly to a cold reservoir. It follows that less entropy is generated when a given quantity of heat is transferred to a hot reservoir than when it is transferred to a cold one.

The simplest way of taking the temperature into account is to define the entropy change of the reference system ($\Delta S^{\dagger}$) as

$$\Delta S^{\dagger} = q^{\dagger}/T^{\dagger}, \qquad (5.1.1)$$

where $T^\dagger$ is the reservoir's temperature. In the case of an infinitesimal change, the corresponding expression is $\mathrm{d}S^\dagger = \mathrm{d}q^\dagger/T^\dagger$. Of course, other forms such as $q^\dagger/T^{\dagger 2}$ or more complicated functions might also seem suitable because they also capture the preceding remarks. However, we shall see that the simplest form works; more importantly, only this form results in S being a state function (as we show in Section 5.1(d)).

Entropy often seems a mysterious and arbitrary quantity, and its definition hardly sheds much light. It is helpful to recall that it has been introduced as a measure of the extent to which a high quality form of energy (a weight at some height) has been degraded into a low quality form (the thermal motion in a reservoir) taking into account the temperature of the reservoir and hence its quality relative to some still lower temperature sink. Big changes of entropy occur when there is a lot of degradation, when a lot of thermal motion is generated at low temperature.

5.1 (c) The entropy change in the system

Now we turn to the problem of using the definition of the entropy change in the reference system to measure the entropy change accompanying a change in another system (which we shall call the *actual system*). The strategy is to use the reference system to restore the actual system to its initial state without further generation of entropy, and to inspect the reference system to see how much entropy has been transferred in the process. There are four steps.

1. Let the original change in the actual system (the change we want to measure) be $\mathrm{d}S^{\mathrm{sys}}$. Then, when that system is restored it undergoes a change of entropy $-\mathrm{d}S^{\mathrm{sys}}$ (because S is a state function and $\mathrm{d}S$ is independent of path). In the course of this restoration a quantity of heat $\mathrm{d}q^\dagger$ enters the reference system, and so its entropy changes by $\mathrm{d}q^\dagger/T^\dagger$ (by definition). The total entropy change during the restoration is zero (because it is carried out reversibly). Therefore:

$$(-\mathrm{d}S^{\mathrm{sys}}) + (\mathrm{d}q^\dagger_{\mathrm{rev}}/T^\dagger) = 0, \quad \text{or} \quad \mathrm{d}S^{\mathrm{sys}} = \mathrm{d}q^\dagger_{\mathrm{rev}}/T^\dagger.$$

We have attached the subscript 'rev' because the transfer of heat occurs during a reversible process.

2. The temperatures of the system (T) and the reference system are the same (because the restoration is reversible); consequently we can write $T^\dagger = T$ and hence $\mathrm{d}S^{\mathrm{sys}} = \mathrm{d}q^\dagger_{\mathrm{rev}}/T$.

3. $\mathrm{d}q^\dagger_{\mathrm{rev}}$ is the heat transferred to the reference system during the restoration; that heat comes reversibly from the system. Therefore $\mathrm{d}q^\dagger_{\mathrm{rev}} = -\mathrm{d}q^{\mathrm{sys}}_{\mathrm{rev,restoration}}$. At this stage we have $\mathrm{d}S^{\mathrm{sys}} = -\mathrm{d}q^{\mathrm{sys}}_{\mathrm{rev,restoration}}/T$.

4. $\mathrm{d}q^{\mathrm{sys}}_{\mathrm{rev,restoration}}$ is equal, on account of the reversibility of the path, to $-\mathrm{d}q^{\mathrm{sys}}_{\mathrm{rev}}$, the heat accompanying the change in the *original* direction. Therefore, finally,

$$\mathrm{d}S^{\mathrm{sys}} = \mathrm{d}q^{\mathrm{sys}}_{\mathrm{rev}}/T^{\mathrm{sys}}. \tag{5.1.2}$$

That is, we can determine the entropy change when a system changes from state i to state f by finding the heat necessary to take it along a *reversible* path between the same two states. For a measurable change, the entropy change is the sum (integral) of the infinitesimal changes:

$$\Delta S^{\mathrm{sys}} = S_{\mathrm{f}} - S_{\mathrm{i}} = \int_{\mathrm{i}}^{\mathrm{f}} \mathrm{d}q^{\mathrm{sys}}_{\mathrm{rev}}/T^{\mathrm{sys}}. \tag{5.1.3}$$

As a first application of this result, consider the change of entropy when a perfect gas changes its state from (V_i, T) to (V_f, T) in an isothermal expansion. According to eqn (5.1.3) we have to find a *reversible* path between the same two states. Since the temperature is constant,

$$\Delta S^{\text{sys}} = \int_i^f dq_{\text{rev}}^{\text{sys}}/T^{\text{sys}} = (1/T)\int_i^f dq_{\text{rev}}^{\text{sys}} = q_{\text{rev}}^{\text{sys}}/T.$$

The work in Chapter 3 led to the result (Box 3.2) that for the isothermal reversible expansion of a gas from V_i to V_f,

$$q_{\text{rev}}^{\text{sys}} = nRT \ln (V_f/V_i).$$

Therefore, the entropy change accompanying this change of state is

$$\Delta S^{\text{sys}} = nR \ln (V_f/V_i). \qquad (5.1.4)^\circ$$

Example 5.1

A sample of hydrogen is in a cylinder fitted with a piston of 50 cm² cross-section. The initial volume at 25 °C is 500 cm³, and the pressure is 2.0 atm. Calculate the change of entropy when the piston is withdrawn isothermally through 10 cm.

• *Method*. Assume that the gas is perfect, and use eqn (5.1.4); calculate n from $pV = nRT$.

• *Answer*. $n = pV/RT = \dfrac{(2.0\text{ atm})\times(0.50\text{ dm}^3)}{(0.0821\text{ J K}^{-1}\text{ mol}^{-1})\times(298.15\text{ K})} = 0.041\text{ mol}.$

Then, from eqn (5.1.4):

$$\Delta S = (0.041\text{ mol}) \times (8.314\text{ J K}^{-1}\text{ mol}^{-1}) \ln (1000\text{ cm}^3/500\text{ cm}^3)$$
$$= (0.34\text{ J K}^{-1}\text{ mol}^{-1}) \ln 2 = +0.24\text{ J K}^{-1}\text{ mol}^{-1}.$$

• *Comment*. Note the units for entropy. Sometimes entropies are quoted in 'entropy units' (symbol: e.u.), which means cal/(deg mol). These are not part of SI: 1 e.u. $\triangleq$ 4.184 J K^{-1} mol^{-1}. It is helpful to remember that the dimensions of molar entropy are the same as those of R and of molar heat capacity.

• *Exercise*. Calculate the entropy change when a 500 cm³ sample of air at 20 °C and 1 atm is compressed isothermally to 50 cm³. [-0.40 J K^{-1}]

The important *general* point to note is that *whatever* the change of interest, the entropy change of the system can be calculated from eqn (5.1.3) by finding some *reversible* path between the same two states, and then evaluating the integral along that path.

5.1(d) Natural events

Consider a system in thermal and mechanical contact with a reference system; let them be in thermal equilibrium (at the same temperature) but not necessarily at mechanical equilibrium (a gas might be at a pressure greater than the pressure exerted by the reference system's weight). Any change of state is accompanied by a change of entropy dS^{sys} in the actual system and $dS^\dagger$ in the reference system. The overall change of entropy is greater than zero in general, because the change of state might be irreversible:

$$dS^{\text{sys}} + dS^\dagger \geq 0, \quad \text{or} \quad dS^{\text{sys}} \geq -dS^\dagger.$$

(The equality applies if the systems happen to be in equilibrium.) Since $dS^\dagger = dq^\dagger/T^\dagger$ (by definition), and $dq^\dagger = -dq^{sys}$ (because the heat entering the reference system comes from the actual system), and $T^\dagger = T^{sys}$ (by arrangement), it follows that for *any* change

$$dS^{sys} \geqslant dq^{sys}/T^{sys}. \tag{5.1.5}$$

This is the *Clausius inequality*.

The Clausius inequality has the following important consequence for changes taking place in isolated systems. Since in an isolated system $dq^{sys} = 0$, eqn (5.1.5) implies that for any change $dS^{sys} \geqslant 0$. A measurable change is the sum of infinitesimal changes, and so it follows that in an isolated system $\Delta S^{sys} \geqslant 0$. A universe (in the thermodynamic sense) is an isolated system, and so for any change inside a universe:

$$\Delta S^{univ} \geqslant 0, \tag{5.1.6}$$

where S^{univ} is the total entropy of all parts of the universe. If there are changes going on inside an isolated system, they must be spontaneous (because our technology cannot penetrate inside to drive them). Hence, the inequality shows that *spontaneous processes lead to an increase in entropy of the universe*. Only when the universe is at equilibrium, so that every change is thermodynamically reversible, does the equality $\Delta S^{univ} = 0$ apply.

A further consequence of the Clausius inequality is that *the integral of dq/T for a system taken round any cycle cannot exceed zero*. This remarkable result is proved as follows. Consider a cycle in which a system changes from a state i to a state f along an irreversible path and is then brought back to i along a reversible path, Fig. 5.3. At each stage of the irreversible path a quantity of heat dq is transferred to the system at its temperature T (which may be different at different stages). At each stage of the reversible path the heat transferred is dq_{rev}. The integral of dq/T around the cycle, denoted $\oint$, is therefore

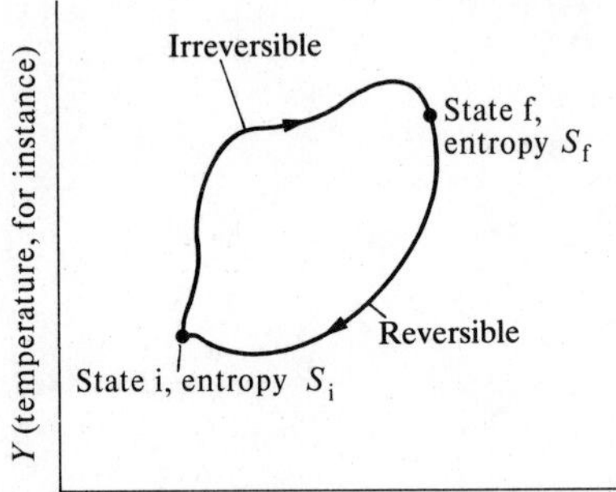

Fig. 5.3. The integration of dq/T around a cycle. The stage from i to f is traversed irreversibly, the return from f to i is reversible.

$$\oint dq/T = \int_i^f dq/T + \int_f^i dq_{rev}/T.$$

The limits on the second integral may be reversed, and so

$$\oint dq/T = \int_i^f dq/T - \int_i^f dq_{rev}/T = \int_i^f dq/T - \int_i^f dS = \int_i^f (dq/T - dS),$$

where dS is the entropy change dq_{rev}/T. According to the Clausius inequality $dS \geqslant dq/T$, or $dq/T - dS \leqslant 0$, implying that

$$\oint dq/T \leqslant 0, \tag{5.1.7}$$

as was to be proved. We shall make use of this result shortly. One immediate consequence is that since the integral *is* zero round a *reversible* path, then dq_{rev}/T (and therefore dS) is an exact differential; therefore, S is a state function. Note that dq_{rev} is an inexact differential, but when multiplied by $1/T$ it becomes exact. For this reason, $1/T$ is called an *integrating factor*.

It will be recalled that the First Law can be expressed in several ways.

One (Section 2.1(b)) was quite involved, and consisted of a report on direct experience. Another ('the energy of an isolated system is constant') is much more succinct, but has the same implications so long as the meaning of 'energy' is comprehended. The Second Law can be expressed similarly. The form we have seen in this chapter is of the 'report on experience' type. We are now in a position to express it succinctly, but with the same implications so long as we comprehend the meaning of 'entropy':

Second Law: *The entropy of an isolated system increases in the course of a spontaneous change.*

5.2 Entropy changes in the universe

By 'universe' is meant the system together with its surroundings. This emphasis is important because the entropy of the system may decrease in the course of a spontaneous change so long as the increase in the surroundings is greater. For example, living things are highly ordered, relatively low entropy structures, but they grow and are sustained because their metabolism generates excess entropy in their surroundings. Ultimately we shall wish to discuss entropy changes in complex systems, but as a first step we must see how to calculate the entropy production as a result of some simple, basic processes.

5.2 (a) The entropy change when a system is heated

The entropy of a system at a temperature T_f can be calculated if it is known at T_i by integrating eqn (5.1.2) between the two temperatures:

$$S(T_f) = S(T_i) + \int_i^f dq_{rev}/T.$$

(For the time being we drop the 'sys' superscript.) We shall be particularly interested in the entropy change when the system is subjected to a constant pressure (such as from the atmosphere) during the heating. Then from the definition of heat capacity,

$$(dq_{rev})_p = dH = C_p\, dT,$$

so long as the system is doing no non-p,V work. Consequently, at constant pressure:

$$S(T_f) = S(T_i) + \int_i^f (C_p/T)\, dT. \qquad (5.2.1a)$$

The corresponding expression for heating at constant volume is

$$S(T_f) = S(T_i) + \int_i^f (C_V/T)\, dT. \qquad (5.2.1b)$$

These expressions let us find the entropy of a substance at any temperature provided the heat capacity has been measured in the temperature range of interest. In the case of gases it is often the case that C_p is independent of temperature over moderate ranges, and so then, at constant pressure:

$$S(T_f) = S(T_i) + C_p \int_i^f (1/T)\, dT = S(T_i) + C_p \ln (T_f/T_i). \qquad (5.2.2)$$

Example 5.2

Calculate the entropy change when argon at 25 °C and 1.00 atm pressure in a container of volume 500 cm^3 is allowed to expand to 1000 cm^3 and is simultaneously heated to 100 °C.

• *Method*. Since S is a state function, choose the most convenient path from the initial state (n, 500 cm^3, 298 K) to the final state (n, 1000 cm^3, 373 K). One path is (a) isothermal expansion from 500 cm^3 to 1000 cm^3 at 298 K followed by (b) isochoric (i.e. constant volume) heating from 298 K to 373 K at the final volume, 1000 cm^3. For (a) use eqn (5.1.4) and for (b) use eqn (5.2.1b). Calculate n from $pV = nRT$ and take $C_{V,m}$ from Table 2.1.

• *Answer*. $n = \dfrac{(1.00\text{ atm}) \times (0.500\text{ dm}^3)}{(0.082\,06\text{ atm dm}^3\text{ K}^{-1}\text{ mol}^{-1}) \times (298\text{ K})} = 0.020_4\text{ mol}.$

Stage (a): $\Delta S = (0.020_4\text{ mol}) \times (8.314\text{ J K}^{-1}\text{ mol}^{-1}) \ln 2 = 0.118\text{ J K}^{-1}\text{ mol}^{-1}$.

Stage (b): $\Delta S = (0.020_4\text{ mol}) \times (12.48\text{ J K}^{-1}\text{ mol}^{-1}) \ln (373\text{ K}/298\text{ K}) = 0.057\text{ J K}^{-1}\text{ mol}^{-1}$.

Therefore, the overall entropy change is the sum of these two changes: $\Delta S = 0.175\text{ J K}^{-1}$.

• *Comment*. Always look for the simplest, reversible path between the specified initial and final states, and calculate the entropy change along it. Note that we have calculated the entropy change of the system, not of the entire universe; we have also assumed that argon behaves as a perfect gas.

• *Exercise*. Calculate the entropy change when the same initial sample is compressed to 50.0 cm^3 and cooled to −25 °C. [-0.44 J K^{-1}]

5.2 (b) Entropy changes in the surroundings

The surroundings of a system constitute a huge reservoir, and so they act as a reference system. It follows that we can use the definition in eqn (5.1.1) to write down the entropy change that occurs when a quantity of heat q^{surr} enters them when they are at a temperature T^{surr}:

$$\Delta S^{\text{surr}} = q^{\text{surr}}/T^{\text{surr}}. \tag{5.2.3}$$

Since this expression is a direct consequence of the definition, it is valid however the change comes about, reversibly or irreversibly, so long as the reservoir remains at internal equilibrium (i.e. no hot spots are formed which then cool irreversibly).

Equation (5.2.3) makes it very simple to calculate the changes in entropy of the surroundings in the course of any process. For instance, in the case of *any* adiabatic change, since $q^{\text{surr}} = 0$, then $\Delta S^{\text{surr}} = 0$ too. When a chemical reaction takes place at constant pressure with enthalpy change ΔH, the entropy change of the surroundings is $\Delta S^{\text{surr}} = -\Delta H/T^{\text{surr}}$, because then $(q^{\text{surr}})_p = -\Delta H$. This last conclusion will shortly be seen to play a vital role in determining the direction of spontaneous chemical change.

Example 5.3

Calculate the entropy change in the surroundings when 1 mol $H_2O(l)$ is formed from its elements under standard conditions at 298 K.

• *Method*. The standard enthalpy of the reaction $H_2(g) + \frac{1}{2}O_2(g) \rightarrow H_2O(l)$ is the standard enthalpy of formation of $H_2O(l)$; use the data in Table 4.1.

• *Answer*. Since $\Delta_f H^\ominus = -285.8\,\text{kJ mol}^{-1}$, for 1 mol $H_2O(l)$, $\Delta H = -285.8\,\text{kJ}$.

$$\Delta S^{\text{surr}} = -\Delta H/T^{\text{surr}} = -(-285.8\times10^3\,\text{J})/298.15\,\text{K}$$
$$= +958.6\,\text{J K}^{-1}.$$

• *Comment*. Note that the entropy of the surroundings *increases* in the course of an exothermic reaction. In the case of an endothermic reaction, their entropy decreases. Whether or not the reaction is spontaneous depends on the *total* entropy change, including the entropy change of the system (see below).

• *Exercise*. Calculate the entropy change in the surroundings when 1 mol $N_2O_4(g)$ is formed from 2 mol $NO_2(g)$ under standard conditions at $\mathcal{T}$. [$-192\,\text{J K}^{-1}$]

5.2 (c) The entropy of phase transition

Some change of order occurs when a substance freezes or boils, and so we expect there to be a change of entropy. The thermodynamic definition of entropy lets us arrive at a value very simply.

At constant pressure the 'latent heat' of a phase transition is its enthalpy of transition, $\Delta_t H$, Section 4.2(a). At the temperature of the transition, T_t, the system is in equilibrium with its surroundings (for instance, at the melting point solid and liquid are at equilibrium, and the temperatures of the system and the surroundings are the same). Heat is therefore transferred *reversibly* between the system and its surroundings. The change of entropy in the surroundings is $-\Delta_t H/T_t$; there is no entropy production overall; therefore the change of entropy of the system must be

$$\Delta S^{\text{sys}} = \Delta_t H/T_t. \tag{5.2.4}$$

Some experimental transition entropies are listed in Table 5.1. Melting and vaporization are endothermic processes ($\Delta_t H \geqslant 0$), and so both are accompanied by an increase of the system's entropy just as we would expect on the grounds that liquids are more disordered than solids, and gases more disordered than liquids.

Table 5.1. Entropies (and temperatures) of phase transitions at 1 atm pressure, $\Delta_t S_m/\text{J K}^{-1}\,\text{mol}^{-1}$

	Melting (at T_f)	Boiling (at T_b)
He	6.0 (at 3.5 K)	19.9 (at 4.22 K)
Ar	14.2 (at 83.8 K)	74.5 (at 87.3 K)
H_2O	22.0 (at 273.15 K)	109.0 (at 373.15 K)
C_6H_6	38.0 (at 279 K)	87.2 (at 353 K)

Example 5.4

The table below gives the molar enthalpies of vaporization and the boiling temperatures (at 1 atm) of several liquids. Calculate the entropy of vaporization of each one.

• *Method*. Use eqn (5.2.4) to calculate $\Delta S_{\text{vap,m}}$. Convert transition temperatures to kelvins.

• *Answer*. The standard molar entropy changes are as follows:

	$\Delta H_{\text{vap,m}}/\text{kJ mol}^{-1}$	$\theta_b/°\text{C}$	$\Delta S_{\text{vap,m}}/\text{J K}^{-1}\,\text{mol}^{-1}$
Methane	8.18	−161.5	+73.2
Carbon tetrachloride	30.00	76.7	+85.8
Cyclohexane	30.1	80.7	+85.1
Benzene	30.7	80.1	+87.2
Hydrogen sulphide	18.7	−60.4	+87.9
Water	40.7	100.0	+109.1

• *Comment*. A wide range of liquids give approximately the same molar enthalpy of vaporization (about $85\,\text{J K}^{-1}\,\text{mol}^{-1}$): this is *Trouton's rule*. This is because a comparable amount of disorder is generated when the same amount of liquid molecules evaporate. Some liquids, however, deviate sharply from the rule. This is often because they have structure, and so a greater change of disorder occurs when they evaporate. An example is water, where the large entropy change reflects the presence of hydrogen-bonding in the liquid. These bonds tend to organize the molecules in the liquid so that they are less random than, for example, the molecules in liquid hydrogen sulphide.

• *Exercise*. Predict the molar enthalpy of vaporization of bromine given that its boiling temperature is 59.2 °C. [28 kJ mol^{-1}]

5.2 (d) The entropy of irreversible change

We have already emphasized that as entropy is a state function, any change in the entropy of a system is independent of the path between the specified initial and final states. In contrast, the *overall* entropy change (of the system and its surroundings) depends on the path, for if the change is reversible the overall entropy change is zero, but if it is irreversible the overall change is greater than zero. We shall now see how to calculate the overall entropy change for various processes.

Consider the case of the isothermal expansion of a perfect gas from (V_i, T) to (V_f, T). The entropy change is $nR \ln (V_f/V_i)$ whether the change of state occurs reversibly or irreversibly. If it is reversible, the entropy change in the surroundings (which are in thermal and mechanical equilibrium with the system) must be such as to give an overall entropy change of zero. Therefore, in this case, the change of entropy of the surroundings is $-nR \ln (V_f/V_i)$. On the other hand, if the expansion occurs freely and irreversibly, since no work is done and the temperature remains constant, there is no heat transferred between the system and its surroundings. Consequently, the entropy of the surroundings does not change (because $q^{surr} = 0$), the entropy of the system changes by $nR \ln (V_f/V_i)$, the same as before (because entropy is a state function), and so the overall change of entropy of the universe is $nR \ln (V_f/V_i)$: entropy has been created in the course of an irreversible change.

A second example of entropy generation in an irreversible process is the flow of heat from a hot body to a cold: this is one of the primitive examples that opened the chapter and which the Second Law ought to be able to capture. In order to simplify the discussion, we take the bodies to be so big that they act as thermal reservoirs. Then when a quantity of heat q is removed from the hot source (of temperature T_h), its entropy changes by $-q/T_h$ (a decrease). When that quantity is added to a cold block (of temperature T_c), its entropy changes by $+q/T_c$ (an increase). The overall change of entropy is therefore $q\{(1/T_c) - (1/T_h)\}$, which is positive (because $T_h > T_c$). Hence, cooling (the transfer of heat from hot to cold) is spontaneous, as we know from experience. Notice that when the temperatures of the two blocks are the same, the entropy change is zero: they are then at thermal equilibrium.

5.2 (e) Carnot efficiency

The result just obtained can be extended in a way that lets us draw a very powerful conclusion about heat and work. Since the temperature of a cold reservoir is lower than that of a hot one, an overall increase in entropy can be produced even if q is withdrawn from the hot reservoir and *less* than q is transferred to the cold. Since the removal of q from the hot reservoir reduces its entropy by q/T_h, and a transfer of q' to the cold reservoir increases its entropy by q'/T_c, the overall entropy change

$$\Delta S_{\text{total}} = (q'/T_c) - (q/T_h) \tag{5.2.5}$$

is positive so long as q' is not less than $q(T_c/T_h)$, Fig. 5.4. Therefore, we are free to use some kind of device, an *engine*, to draw off the difference $q - q'$ as work, yet still have a spontaneous process. It then follows that the *maximum work* an engine may produce when it is working between a hot source of temperature T_h and a cold sink of temperature T_c is

$$w'_{\max} = q - q'_{\min} = q\{1 - (T_c/T_h)\}.$$

The *Carnot efficiency* (symbol: ε) is defined as (*maximum work generated*)/(*heat absorbed*), and so

$$\varepsilon = w'_{\max}/q = 1 - (T_c/T_h). \tag{5.2.6}_r$$

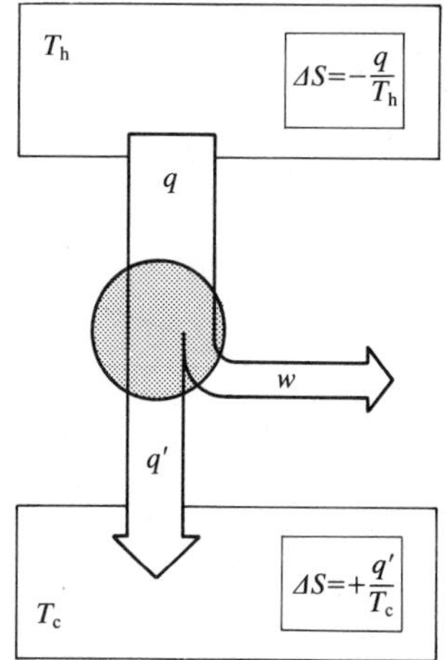

Fig. 5.4. The thermodynamic basis of engines. When a quantity of energy q is withdrawn from a hot reservoir the entropy decreases by q/T_h, and when q' is added to a cold reservoir its entropy is increased by q'/T_c. So long as $T_c < T_h$, the overall entropy change may still be positive even if less energy is returned to the cold reservoir than was removed from the hot. The difference $q - q'$ may be withdrawn as work, the overall process being spontaneous.

This Carnot efficiency formula is a result of major economic significance, for it shows that an engine cannot convert heat into work with 100% efficiency (except at $T_c = 0$). For instance, a typical generating station using superheated steam at around 550 °C and a cold sink at around 100 °C has a thermodynamic efficiency of only 55%, the remaining 45% of heat taken from the hot source having to be discarded into the surroundings in order to ensure that enough entropy is generated to make the overall process spontaneous. An internal combustion engine operates between about 3200 K (the high temperature being brought about by the combustion of fuel) and 1400 K (the temperature in the exhaust manifold): its thermodynamic efficiency is therefore only 56%, but other losses reduce this to around 25% in practice. The less than 100% conversion efficiency in this case is a consequence of needing to generate enough entropy to overcome the reduction in the entropy of the hot source: the vehicle is driven forward as a result of entropy being generated by discarding heat to the 'cold' sink. In other words, fuel must be 'wasted'.

The limit implied by the Carnot efficiency applies to all engines, whatever their construction, and whatever their working substance (its derivation made no mention of any specific substance). It is an upper limit to their conversion efficiency, and technological deficiencies (e.g. friction) reduce the actual efficiency of real engines. Its positive feature is that it indicates how high conversion efficiencies may be achieved: ε approaches unity as the temperature of the cold sink approaches zero or as the temperature of the hot source is raised. The former is not a realizable option, and so engines and turbines are designed to run at high temperatures.

An explicit calculation of the Carnot efficiency can be made in the case of an engine using a perfect gas as the working fluid. The *Carnot cycle* is illustrated in Fig. 5.5; each stage is reversible. The stage from A to B is an isothermal expansion; that from B to C is an adiabatic expansion; C to D is an isothermal compression; D to A is an adiabatic compression. The work done and the heat transferred in each stage can be taken from Box 3.2, and the resulting expressions are given in the illustration. The work done on going round the cycle is the sum of the work done in each stage:

$$w' = nRT_h \ln(V_B/V_A) - nRT_c \ln(V_D/V_C).$$

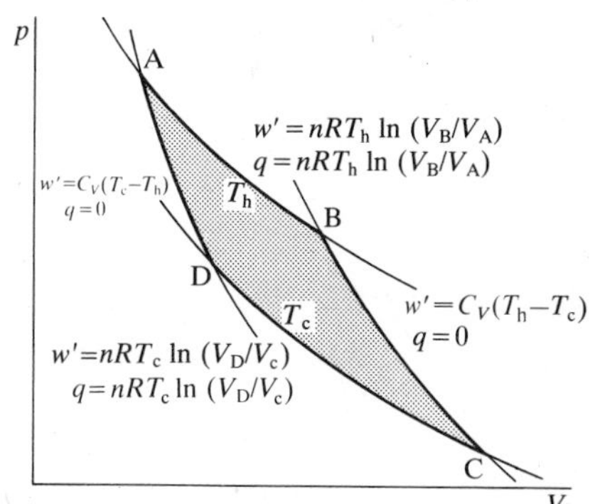

Fig. 5.5. The Carnot cycle for a perfect gas. All stages are reversible. A to B is an isothermal expansion at a temperature T_h, B to C an adiabatic expansion which lowers the temperature to T_c. C to D is an isothermal compression at T_c. D to A completes the cycle with an adiabatic compression that raises the temperature to T_h. The work done by the engine at each stage, and the heat absorbed, are shown.

(The adiabatic stages cancel.) The heat absorbed from the hot source is

$$q = nRT_h \ln (V_B/V_A).$$

The ratio w'/q can be simplified by using the relations between temperature and volume for reversible adiabatic processes (i.e. $V_B/V_C = (T_c/T_h)^c$, eqn (3.3.3)), and leads to

$$w'/q = 1 - (T_c/T_h),$$

which is exactly the Carnot efficiency, eqn (5.2.6), because since each stage is reversible, the cycle generates maximum work.

Example 5.5

Use the Carnot cycle to demonstrate explicitly that the integral of dq_{rev}/T around a cycle is zero (i.e. the special case of the Clausius inequality, eqn (5.1.7), for a reversible cycle).

• *Method*. The integral in this case is the sum of the heat transferred in the two isothermal stages (the adiabatic stages contribute zero). Use the information on reversible adiabatic changes derived in Section 3.3 to relate volumes and temperatures.

• *Answer*. From Fig. 5.5:
$$\int_{\text{cycle}} dq/T = nR \ln (V_B/V_A) - nR \ln (V_D/V_C)$$
$$= nR \ln (V_B V_C/V_A V_D).$$

From eqn (3.3.3), the ratio of volumes is

$$(V_B/V_D)(V_C/V_A) = (T_c/T_h)^c (T_h/T_c)^c = 1,$$

and so the integral is zero, in accord with eqn (5.1.7) for a cycle traversed entirely reversibly.

• *Comment*. This is a special case of a perfect gas as the working substance in the engine; the Clausius inequality (and the special case of an equality) applies to *any* working substance.

• *Exercise*. Evaluate the Carnot efficiency explicitly using the information in Fig. 5.5 and Section 3.3. [eqn (5.2.6)]

5.3 Concentrating on the system

Entropy is the basic concept for discussing the direction of natural change, but in order to use it we have to investigate changes in the system *and* its surroundings. We have seen that it is always very simple to calculate the entropy change in the latter, and we shall now see that it is possible to devise a simple method for taking that contribution into account automatically. This focuses attention on the system and simplifies discussions.

5.3 (a) The Helmholtz and Gibbs functions

Consider a system in thermal equilibrium with its surroundings (so that $T^{sys} = T^{surr} = T$); the Clausius inequality, eqn (5.1.5), then reads

$$dS^{sys} - dq^{sys}/T \geqslant 0. \tag{5.3.1}$$

The importance of the inequality in this form is that it expresses the criterion for spontaneous change solely in terms of the properties of the system. *From now on, the superscript denoting the system will be dropped, and everything will relate to the system* (unless explicitly stated otherwise). This convention greatly simplifies the notation.

The inequality can be developed in two ways. First, take the case of heat transferred at constant volume. Then, so long as there is no non-p,V work, $(dq)_V = dU$, and so $dS - dU/T \geq 0$, or

$$T\,dS \geq dU \quad \text{(constant } V\text{, no non-}p,V\text{ work).} \qquad (5.3.2)$$

At either constant internal energy ($dU = 0$) or constant entropy ($dS = 0$) this becomes

$$(dS)_{U,V} \geq 0, \qquad (dU)_{S,V} \leq 0. \qquad (5.3.3)$$

These are the criteria for natural changes in terms of properties relating to the *system*. Secondly, when heat is transferred at constant pressure, then so long as there is no non-p,V work, we can write $(dq)_p = dH$ in eqn (5.3.1) and obtain

$$T\,dS \geq dH \quad \text{(constant } p\text{, no non-}p,V\text{ work).} \qquad (5.3.4)$$

At either constant enthalpy or constant enthalpy or constant entropy this becomes

$$(dS)_{H,p} \geq 0, \qquad (dH)_{S,p} \leq 0. \qquad (5.3.5)$$

Since eqns (5.3.2) and (5.3.4) have the forms $dU - T\,dS \leq 0$ and $dH - T\,dS \leq 0$ respectively, they can be expressed more simply by introducing two more thermodynamic functions:

Helmholtz function: $A = U - TS$, *Gibbs function*: $G = H - TS$. (5.3.6)

All the symbols refer to the system. When the state of the system changes at constant temperature,

$$dA = dU - T\,dS, \qquad dG = dH - T\,dS.$$

Now introduce eqn (5.3.2) into the first of these, and eqn (5.3.4) into the second; then the criteria of spontaneous change are

$$(dA)_{T,V} \leq 0, \qquad (dG)_{T,p} \leq 0. \qquad (5.3.7)$$

(It must be understood that no non-p,V work is being done in either case.) *These inequalities are the most important conclusions from thermodynamics for chemistry*. They are developed in subsequent sections.

5.3 (b) Some remarks on the Helmholtz function

A change in a system at constant temperature and volume is spontaneous if $(dA)_{T,V} < 0$. That is, a change under these conditions is spontaneous if it corresponds to a *decrease* in the Helmholtz function. Such systems move spontaneously towards states of lower A, and the criterion of equilibrium is $(dA)_{T,V} = 0$.

The expression $dA = dU - T\,dS$ is sometimes interpreted as follows. A negative value of dA is favoured by a negative value of dU and a positive value of $T\,dS$. This suggests that the tendency of a system to move to lower A is due to its tendency to move towards states of lower internal energy and higher entropy. This interpretation is false (even though it is a good rule of thumb for remembering the expression for dA). This is because *the tendency to lower A is solely a tendency towards states of greater overall* *entropy*. Systems change spontaneously solely because that increases the entropy of the universe, not because they tend to lower internal energy. The form of dA gives the impression that systems favour lower energy, but that is misleading. dS is the system's entropy change, and $-dU/T$ is the entropy change of the surroundings: their total tends to a maximum.

5.3 (c) Maximum work

It turns out that A carries a greater significance than being simply a signpost. If we know the value of dA for a change (and of ΔA for a measurable change), then we can also state the maximum amount of work the system can do. This is why A is sometimes called the *maximum work function*, or the *work function* (*Arbeit* is the German for work; hence A). The proof runs in two stages.

First we prove that *a system does maximum work when it is working reversibly*. (This was demonstrated in Chapter 2 for the expansion of a perfect gas; now we prove its universal validity.) We combine the Clausius inequality $dS \geqslant dq/T$ in the form $T\,dS \geqslant dq$ with the First Law $dU = dq + dw$, obtaining $dU \leqslant T\,dS + dw$ (because we are replacing dq by something that in general is larger). This rearranges to $-dw \leqslant -dU + T\,dS$. The work done *by* the system is $dw' = -dw$, and so

$$dw' \leqslant -dU + T\,dS. \tag{5.3.8}$$

Then, since T, S, and U are all state functions, it follows that the work a system can do for a specified change of state cannot exceed some maximum value, $dw'_{max} = -dU + T\,dS$, and that this work is obtained only when the path is traversed reversibly (because then the equality applies). That proves the result we required.

Now for the final stage. When the path is traversed reversibly and the equality holds, since $-dU + T\,dS = -dA$ at constant temperature, we conclude that $dw'_{max} = -dA$. For a measurable change when the Helmholtz function changes by ΔA, it follows that

$$\textit{Maximum work at constant temperature:}\quad w'_{max} = -\Delta A. \tag{5.3.9$_r$}$$

In the case of an isothermal change between specified initial and final states,

$$w'_{max} = -\Delta A = -\Delta U + T\Delta S. \tag{5.3.10}$$

This shows that in some cases, depending on the sign of $T\Delta S$, not all the change in internal energy may be available for doing work. If the change occurs with a decrease in entropy (of the system), so that $T\Delta S$ is negative, the maximum work that can be done is *less* than the value of $-\Delta U$. This is because, in order for the change to be spontaneous, some of the energy must escape as heat in order to generate enough entropy in the surroundings to overcome the reduction in the system, Fig. 5.6(a). In this case Nature is demanding a tax on the internal energy as it is converted into work. On the other hand, if the change occurs with an increase in the entropy of the system (so that $T\Delta S$ is positive), then the maximum work that can be done is *greater* than the value of $-\Delta U$. The explanation of this apparent paradox is that the system is not isolated, and heat may flow in as work is done. Since the entropy of the system increases, we can afford a reduction of the entropy of the surroundings yet still have, overall, a

(a)

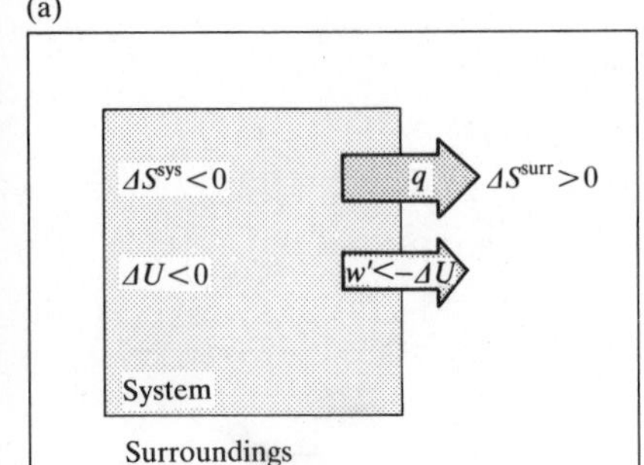

(b)

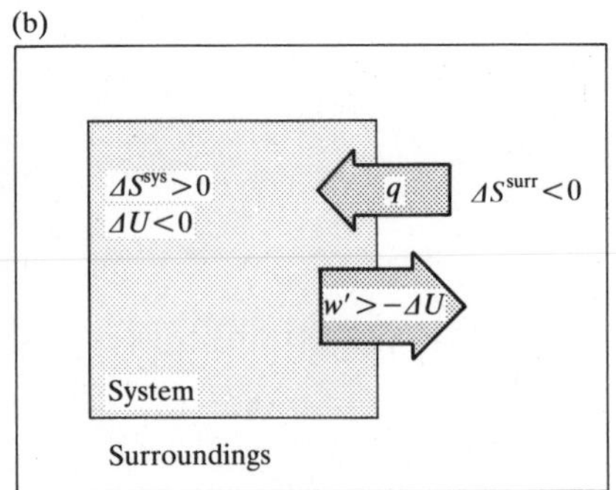

Fig. 5.6. In a system not isolated from its surroundings, the work done may be different from the change in internal energy. Moreover, the process is spontaneous if *overall* the entropy of the universe increases. In (a) the entropy of the system decreases, and so that of the surroundings must increase in order for the process to be spontaneous, which means that energy must pass from the system to the surroundings as heat. Therefore, less work than $-\Delta U$ can be obtained. In (b) the entropy of the system increases, hence we can afford to lose some entropy of the surroundings; that is, some of the energy of the surroundings may be lost as heat to the system. This energy can be returned as work. Hence the work done can exceed $-\Delta U$.

spontaneous process. Therefore, some heat (no more than the value of $T\Delta S$) may leave the surroundings and contribute to the work the change is generating, Fig. 5.6(b). Nature is now providing a tax refund.

Example 5.6

When glucose is oxidized to carbon dioxide and water according to the equation $C_6H_{12}O_6(s) + 6O_2(g) \rightarrow 6CO_2(g) + 6H_2O(l)$, calorimetric measurements give $\Delta_r U = -2810\ \text{kJ mol}^{-1}$ and $\Delta_r S = +182.4\ \text{J K}^{-1}\ \text{mol}^{-1}$ at $\mathcal{T}$. How much of this energy change can be extracted as (a) heat, (b) work at this temperature?

• *Method*. Use $\Delta U = q_V$, $\Delta H = q_p$, and $\Delta_r H = \Delta_r U$ (since the number of gas molecules is unchanged by the reaction). For the available work, use eqn (5.3.10).

• *Answer*. (a) Under conditions of constant volume or constant pressure, the quantity of energy available as heat is $2810\ \text{kJ mol}^{-1}$. (b) The value of ΔA is

$$\begin{aligned}\Delta A = \Delta U - T\Delta S &= (-2180\ \text{kJ mol}^{-1}) - (298.15\ \text{K}) \times (182.4\ \text{J K}^{-1}\ \text{mol}^{-1}) \\ &= -2864\ \text{kJ mol}^{-1}.\end{aligned}$$

Therefore, the combustion of glucose can be used to produce up to $2864\ \text{kJ mol}^{-1}$ of work.

• *Comment*. The maximum work available is greater than the change in internal energy on account of the positive entropy of reaction (which is partly due to the generation of a large number of small molecules from one big one). The system can therefore draw in energy from the surroundings (so reducing their entropy) and make it available for doing work.

• *Exercise*. Repeat the calculation for the combustion of methane under the same conditions using enthalpy data from Table 4.2 and $\Delta S^{\ominus} = -140.3\ \text{J K}^{-1}\ \text{mol}^{-1}$ for the reaction.

[$q_V = -887\ \text{kJ mol}^{-1}$, $q_p = -890\ \text{kJ mol}^{-1}$, $w'_{\text{max}} = 845\ \text{kJ mol}^{-1}$]

5.3 (d) Some remarks on the Gibbs function

The Gibbs function, which is also known as the *free energy*, is more common in chemistry than the Helmholtz function. This is because, at least in laboratory chemistry, we are usually more interested in changes occurring at constant pressure, not constant volume. The criterion $(dG)_{T,p} \leq 0$ carries over into chemistry as the remark that, at constant temperature and pressure, *chemical reactions are spontaneous in the direction of decreasing Gibbs function*. Therefore, if we want to know whether a reaction will run in some direction, the pressure and temperature being constant, we assess the change of Gibbs function (ΔG) for the reaction, where $\Delta G = G_{\text{products}} - G_{\text{reactants}}$. If this ΔG is negative, then the reaction has a spontaneous tendency to convert the reactants into products. If ΔG is positive, then the reverse reaction is spontaneous.

The interpretation of reactions as tending to fall to lower values of the Gibbs function until they attain equilibrium at some minimum value is the same as for the Helmholtz function, and the same mistake can be made. The *apparent* driving force is the tendency to move towards lower enthalpy and higher entropy. The actual driving force is the tendency for the universe to move towards greater entropy. This is achieved by maximizing the sum of the entropy of the system and the surroundings, and the latter is achieved if heat enters them as a result of an exothermic reaction (for which $\Delta H < 0$).

An illustration of the way that the Gibbs function plays its role is provided by the existence of spontaneous endothermic reactions. In such reactions $\Delta H > 0$, and so the system rises spontaneously to states of higher enthalpy. Since the reaction is spontaneous, ΔG must be negative. For this to be so even though ΔH is positive requires $T\Delta S$ to be positive and larger than ΔH. Endothermic reactions are therefore driven by the large increase of entropy of the *system*, and this entropy change overcomes the reduction of entropy brought about in the surroundings by the inflow of heat into the system ($\Delta S^{\text{surr}} = -\Delta H/T$).

Example 5.7

Calculate the change in Gibbs function when 1 mol $N_2O_4(g)$ forms 2 mol $NO_2(g)$ under standard conditions at $\mathcal{T}$. The change of entropy is +4.8 J K^{-1} for the reaction as specified.

- *Method*. The reaction is $N_2O_4(g) \rightarrow 2NO_2(g)$. Use the enthalpies of formation (Table 4.1) to calculate the standard reaction enthalpy. Then combine the values to give $\Delta G = \Delta H - T\Delta S$, which follows from eqn (5.3.6) at constant temperature.

- *Answer*. The standard reaction enthalpy is $\Delta_r H^\ominus = 2(33.2\ \text{kJ mol}^{-1}) - (9.2\ \text{kJ mol}^{-1}) = +57.2\ \text{kJ mol}^{-1}$. Therefore, at $\mathcal{T}$, for 1 mol of $N_2O_4(g)$ dissociated:

$$\Delta G^\ominus = \Delta H^\ominus - T\Delta S^\ominus = (57.2\ \text{kJ}) - (298.15\ \text{K}) \times (4.8\ \text{J K}^{-1}) = +55.8\ \text{kJ}.$$

- *Comment*. The reaction as specified is not spontaneous at $\mathcal{T}$, but we see that as the temperature increases, the negative entropy change of the system becomes more important and eventually may dominate the endothermic enthalpy change. We discuss this question in detail in Chapter 10 when we shall see that partial dissociation occurs even at $\mathcal{T}$.

- *Exercise*. Is the oxidation of iron to $Fe_2O_3(s)$ spontaneous given that the formation of 1 mol $Fe_2O_3(s)$ is accompanied by an entropy change of -272 J K^{-1}? [$\Delta G = -741$ kJ; yes]

The analogue of the maximum work interpretation of ΔA can be found for ΔG. In a general change $dH = dq + dw + d(pV)$; when the change is reversible $dq = T\,dS$ and $dw = dw_{max}$. Therefore,

$$dG = T\,dS + dw_{max} + d(pV) - T\,dS = dw_{max} + d(pV).$$

The work consists of p,V-work, which for a reversible change is given by $-p\,dV$, and possibly some other kind of work (e.g. the electrical work of pushing electrons through a circuit or the work of raising a column of liquid); this non-p,V work we denote dw_e. Therefore, with $d(pV) = p\,dV + V\,dp$,

$$dG = -p\,dV + dw_{e,\,max} + p\,dV + V\,dp = dw_{e,\,max} + V\,dp.$$

If the change occurs at constant pressure (as well as constant temperature), then $(dG)_{T,p} = dw_{e,max}$. This means that the change in the Gibbs function for a process is equal to the maximum non-p,V work done on the system. We are more interested in the work a system can do, and so we convert this expression using $dw'_{e,max} = -dw_{e,max}$. Then, for a measurable change,

$$\textit{Maximum non-}p,V\ \textit{work:}\quad w'_{e,max} = -\Delta G \quad (T, p\ \text{constant}). \qquad (5.3.11)_r$$

This expression is particularly useful for assessing the electrical work that may be produced by fuel cells and electrochemical cells, and we shall see many applications of it.

Example 5.8

Animals operate under conditions of constant pressure, and most of the processes that maintain life are electrical (in a broad sense). How much energy is available for sustaining this type of muscular and nervous activity from the combustion of 1 mol of glucose molecules under standard conditions at 37 °C (blood temperature)? The entropy change is $+182.4\ \mathrm{J\,K^{-1}}$ for the reaction as stated.

• *Method*. The non-p,V work available is obtained from the ΔG accompanying the combustion reaction. Ignore the temperature-dependence of the reaction enthalpy and obtain ΔH from Table 4.2. Then use $\Delta G = \Delta H - T\,\Delta S$.

• *Answer*. $\Delta G = (-2808\ \mathrm{kJ}) - (310\ \mathrm{K}) \times (182.4\ \mathrm{J\,K^{-1}}) = -2864\ \mathrm{kJ}$. Therefore, the combustion of 1 mol glucose molecules (180 g) can do up to 2864 kJ of non-p,V work.

• *Comment*. A 70 kg person would need to do 2.1 kJ of work to climb vertically through 3 m; therefore, at least 0.13 g of glucose is needed to complete the task (and in practice significantly more).

• *Exercise*. How much non-p,V work can be obtained from the combustion of 1 mol $CH_4(g)$ under standard conditions at T? Use $\Delta S = -140\ \mathrm{J\,K^{-1}}$. [849 kJ]

5.4 Evaluating the entropy and the Gibbs function

The entropy of a system at a temperature T can be related to its entropy at $T=0$ by measuring its heat capacity C_p as a function of temperature and evaluating the integral in eqn (5.2.2). At each phase transition encountered the entropy of transition $(\Delta_t H/T_t)$ must be added. For example, if a substance melts at T_f and boils at T_b, its entropy above its boiling temperature is given by

$$S(T) = S(0) + \int_0^{T_f} (C_p^{\text{solid}}/T)\,dT + \Delta H_{\text{melt}}/T_f + \int_{T_f}^{T_b} (C_p^{\text{liquid}}/T)\,dT + \Delta H_{\text{vap}}/T_b + \int_{T_b}^{T} (C_p^{\text{gas}}/T)\,dT. \quad (5.4.1)$$

All the quantities involved except $S(0)$ can be measured calorimetrically, and the integrals can be evaluated either graphically or, as is now more usual, by numerical integration on a computer. This is illustrated in Fig. 5.7: the area under the curve of C_p/T against T is the integral required. Since $dT/T = d\ln T$, an alternative procedure is to plot C_p against $\ln T$. Examples of this procedure are given in the Problems.

One problem is the difficulty of measuring heat capacities near $T=0$. There are good theoretical grounds for assuming that the heat capacity is proportional to T^3 when T is low, and this is the basis of the *Debye extrapolation* (which is justified in Part 2). In this method, C_p is measured to as low a temperature as possible, and a curve of the form aT^3 is fitted to it. That determines the value of a, and the expression $C_p = aT^3$ is assumed valid down to $T=0$. The entropy in that region is evaluated by integration:

$$S(T) - S(0) = \int_0^T (C_p/T)\,dT = \int_0^T aT^2\,dT = \tfrac{1}{3}C_p(T). \quad (5.4.2)$$

Since the heat capacity is very small at low temperatures, only small errors arise from this extrapolation. In the case of metals there is also a contribution to the heat capacity from the electrons, which is linearly proportional to the temperature when the temperature is low.

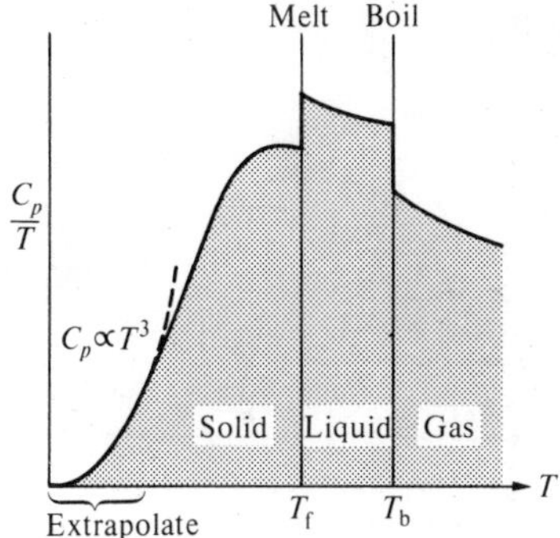

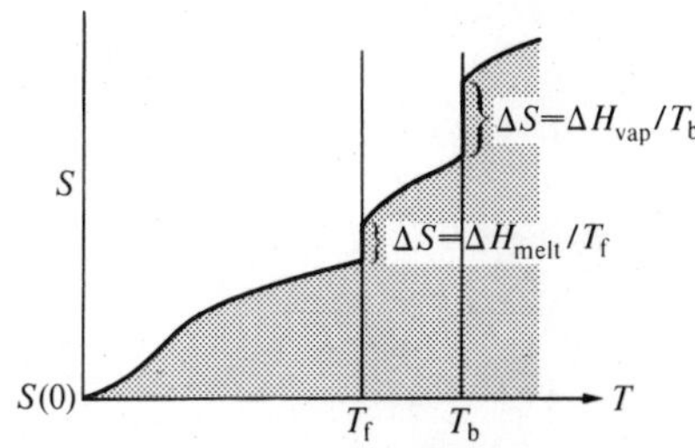

Fig. 5.7. The calculation of a Third Law entropy from heat capacity data. The upper diagram shows C_p/T as a function of temperature for a sample. The lower curve shows the entropy, which is equal to the area beneath the upper curve up to the corresponding temperature, plus the entropy of each phase transition passed.

Example 5.9

The standard entropy of nitrogen gas at $\mathcal{T}$ has been calculated from the following data:

	ΔS_m/J K^{-1} mol^{-1}
T^3 extrapolation (0–10 K)	1.92
Integration, eqn (5.2.1)	25.25
Phase transition (at 35.61 K)	6.43
Integration, eqn (5.2.1)	23.38
Phase transition (melting, 63.14 K)	11.42
Integration, eqn (5.2.1)	11.41
Phase transition (vaporization, 77.32 K)	72.13
Perfect gas expression, eqn (5.2.2) from 77.32 K to 298.15 K	39.20
Correction for gas imperfection	0.92
Total:	192.06

Hence, $S_m^{\ominus}(\mathcal{T}) = S_m(0) + 192.1$ J K^{-1} mol^{-1}. We deal with the value of $S(0)$ below.

5.4 (a) The Third Law of thermodynamics

The question remaining is the value of $S(0)$. At $T=0$ all quenchable energy has been quenched. In the case of a perfect crystal at $T=0$ all the particles are in a regular, uniform array, and the absence of disorder and thermal chaos suggests that such materials also have zero entropy. That, however, is a *molecular* view; can we draw a similar conclusion from generalizations of experience like those expressed by the First and Second Laws? As in the case of those two laws, the generalization, the *Third Law* of thermodynamics, can be expressed in an 'obvious', experience-related way, and in a way that demands more background.

The basic, generalized observation is as follows:

Third Law: *It is impossible to reach absolute zero in a finite number of steps.*

No one has ever succeeded in cooling any system to $T=0$, and the Third Law generalizes this experience into a principle. Once again in thermodynamics we have a law that recognizes an impossibility (First: energy cannot be created or destroyed; Second: universal entropy never decreases; Third: absolute zero is unattainable).

The Third Law is closely related to the following theorem:

Nernst heat theorem: *The entropy change accompanying a transformation between condensed phases in equilibrium approaches zero as the temperature approaches zero*‡.

In other words, if we consider the entropy change ΔS for the transformation between two solid states of a system (where 'transformation' includes

‡ Whereas the Third Law (in the form of a statement about the unattainability of absolute zero) is implied by the Nernst heat theorem (as we shall demonstrate with particular examples in Section 7.5), the opposite is not true except in restricted circumstances. This is a point of some subtlety, and is discussed by A. Münster in *Statistical thermodynamics*, Vol. 2, Springer, 1974, p. 79. Hence the unattainability statement and the Nernst theorem are not logically equivalent. We shall ignore this subtle point, and regard the 'unattainability' statement and the statement to be given in the next paragraph as 'equivalent' for our simple purposes. The question, though, is real and deep, and shows that thermodynamics is still alive.

chemical reaction) at internal thermodynamic equilibrium (e.g. the transformation of rhombic sulphur into monoclinic sulphur: one phase is thermodynamically more stable than the other but each phase alone is at *internal* thermodynamic equilibrium, each is a perfect crystal), then ΔS approaches zero as the temperature approaches zero. It follows that if the value zero is *arbitrarily* ascribed to the entropies of elements in their perfect crystalline form at absolute zero, then all perfect crystalline *compounds* also have zero entropy at $T = 0$. Hence, all perfect crystals may be taken to have zero entropy at absolute zero. Therefore, an alternative statement of the Third Law is as follows:

> Third Law: *If the entropy of every element in its stable state at $T = 0$ is taken as zero, every substance has a positive entropy which at $T = 0$ may become zero, and does become zero for all perfect crystalline substances, including compounds.*

Note that a non-crystalline perfect state, such as the superfluid state of helium, is included by the opening phrase. Note too, that the Third Law does not imply that entropies *are* zero at absolute zero: it merely implies that all perfect materials have the same entropy. Choosing this common value as zero is then a matter of convenience.

Evidence for the Third Law (in the second form) comes from a variety of observations. One is the measurement of heat capacities down to low temperatures. For example, the entropy of the transition *monoclinic sulphur* $\rightarrow$ *rhombic sulphur* can be measured by determining its enthalpy (-401.7 kJ mol^{-1}) at the transition temperature (368.5 K):

$$\Delta S_m = S_m(\text{rh}; 368.5\,\text{K}) - S_m(\text{monocl}; 368.5\,\text{K})$$
$$= (-401.7\,\text{kJ mol}^{-1})/(368.5\,\text{K}) = -1.09\,\text{J K}^{-1}\,\text{mol}^{-1}.$$

The two individual entropies can also be determined by measuring the heat capacities from $T = 0$ up to $T = 368.5$ K (the rate of transformation is so slow at low temperatures that the relative thermodynamic instability of the monoclinic form is unimportant; the forms are internally thermodynamically stable). It is found that

$$S_m(\text{rh}; 368.5\,\text{K}) = S_m(\text{rh}; 0) + (36.86 \pm 0.20)\,\text{J K}^{-1}\,\text{mol}^{-1},$$
$$S_m(\text{monocl}; 368.5\,\text{K}) = S_m(\text{monocl}; 0) + (37.82 \pm 0.40)\,\text{J K}^{-1}\,\text{mol}^{-1},$$

implying that at the transition temperature

$$\Delta S_m = S_m(\text{rh}; 0) - S_m(\text{monocl}; 0) - (0.96 \pm 0.60)\,\text{J K}^{-1}\,\text{mol}^{-1}.$$

On comparing this value with the one above, we conclude that, within experimental error, $S_m(\text{rh}; 0) - S_m(\text{monocl}; 0) = 0$. Therefore, the entropies of the two substances are identical at $T = 0$. We may then *ascribe* the value zero to both.

The choice $S(0) = 0$ will be made from now on, and entropies reported on that basis will be referred to as *Third Law entropies*. When the substance is in its standard state at the temperature T, the Third Law entropy is denoted $S^\ominus(T)$. A list of values at 298 K is given in Table 5.2.

Table 5.2. Standard Third Law entropies at T

	$S_m^\ominus$/J K^{-1} mol^{-1}
C (graphite)	5.7
C (diamond)	2.4
I_2(s)	116.1
Sucrose (s)	360.2
Hg(l)	76.0
H_2O(l)	69.9
C_6H_6(l)	173.3
He(g)	126.2
H_2(g)	130.7
CO_2(g)	213.7
NH_3(g)	192.5

5.4 (b) Third Law entropies

A glance at Table 5.2 is enough to give some familiarity with the magnitudes of entropies and to see how values vary between different types of substance. For example, the standard entropies of gases (except hydrogen and the monatomic noble gases) are all similar and generally bigger than for liquids and solids composed of particles of comparable complexity. This fits in well with the view that gases are more chaotic than either liquids or solids. The low values of the entropies of water and mercury are striking: the former accords with the view that water retains some ice-like structure on account of its hydrogen bonds; the value for mercury reflects its metal-like structure. Note the exceptionally low value of the entropy of diamond and the high value of a complex solid such as sucrose.

Once the Third Law entropies of substances are known, it is a simple matter to calculate the *standard reaction entropy* (symbol: $\Delta_r S^{\ominus}$). This is defined, like the standard reaction enthalpy, as the difference between the entropies of the pure, separated products and the pure, separated reactants, all substances being in their standard states at the specified temperature. Thus, in general:

For the reaction: $$0 = \sum_J \nu_J \mathrm{J},$$

Standard reaction entropy: $$\Delta_r S^{\ominus} = \sum_J \nu_J S_J^{\ominus}. \qquad (5.4.3)$$

Example 5.10

Calculate the standard reaction entropy of $H_2(g) + \frac{1}{2}O_2(g) \rightarrow H_2O(l)$ at $\mathcal{T}$.

- *Method*. Identify the stoichiometric coefficients in the reaction written in the form $0 = H_2O(l) - H_2(g) - \frac{1}{2}O_2(g)$. Then use eqn (5.4.3) with standard entropies taken from Table 5.2.

- *Answer*. The coefficients are $\nu(H_2O) = 1$, $\nu(H_2) = -1$, and $\nu(O_2) = -\frac{1}{2}$. Therefore:

$$\Delta_r S^{\ominus} = \nu(H_2O)S_m^{\ominus}(H_2O, l) + \nu(H_2)S_m^{\ominus}(H_2, g) + \nu(O_2)S_m^{\ominus}(O_2, g)$$
$$= \{69.9 - 130.7 - \tfrac{1}{2}(205.1)\}\ \mathrm{J\,K^{-1}\,mol^{-1}} = -163.4\ \mathrm{J\,K^{-1}\,mol^{-1}}.$$

- *Comment*. The large decrease in entropy is largely due to the formation of a low entropy liquid from two gases. Note that the formation of $H_2O(l)$ is opposed by this decrease of entropy of the system (but it is exothermic, and the entropy increase in the surroundings is dominatingly large).

- *Exercise*. Calculate the standard reaction entropy for the combustion of $CH_4(g)$ at $\mathcal{T}$. [$-140.3\ \mathrm{J\,K^{-1}\,mol^{-1}}$]

5.4 (c) Standard molar Gibbs functions

Once the reaction entropies have been determined they can be combined with reaction enthalpies to obtain the *reaction Gibbs function* (symbol: $\Delta_r G$) using $\Delta_r G = \Delta_r H - T\Delta_r S$. As in the case of enthalpies, it is convenient to define the *standard Gibbs function of formation* (symbol: $\Delta_f G^{\ominus}$):

The standard Gibbs function of formation is the change of Gibbs function per unit amount of substance when a compound is formed from its elements in their reference phases, all substances being in their standard states at the specified temperature.

The values for the elements are defined as zero. A selection of values for compounds is given in Table 5.3.

With $\Delta_f G^\ominus$ values established it is a simple matter to obtain the standard reaction Gibbs function by taking the appropriate combination

For the reaction: $$0 = \sum_J \nu_J J,$$

Standard Gibbs function of reaction: $$\Delta_r G^\ominus = \sum_J \nu_J \Delta_f G_J^\ominus. \qquad (5.4.4)$$

Table 5.3. Standard Gibbs functions of formation at $\mathcal{T}$

	$\Delta_f G^\ominus/\text{kJ mol}^{-1}$
C (diamond)	+2.9
NaCl(s)	−384.1
$H_2O(l)$	−237.1
$C_6H_6(l)$	+124.3
$CO_2(g)$	−394.4
$CH_4(g)$	−50.7
$NH_3(g)$	−16.5

Example 5.11

Calculate the standard reaction Gibbs function for $CO(g) + \frac{1}{2}O_2(g) \rightarrow CO_2(g)$ at $\mathcal{T}$.

• *Method*. Write the reaction in the form $0 = CO_2(g) - CO(g) - \frac{1}{2}O_2(g)$, identify the stoichiometric coefficients, and use eqn (5.4.4) and data from Table 5.3.

• *Answer*. The stoichiometric coefficients are $\nu(CO_2) = 1$, $\nu(CO) = -1$, and $\nu(O_2) = -\frac{1}{2}$. Therefore:

$$\begin{aligned}\Delta_r G^\ominus(\mathcal{T}) &= \nu(CO_2)\Delta_f G^\ominus(CO_2, g) + \nu(CO)\Delta_f G^\ominus(CO, g) + \nu(O_2)\Delta_f G^\ominus(O_2, g)\\ &= \{(-394.4) - (-137.2) - \tfrac{1}{2}(0)\}\ \text{kJ mol}^{-1} = -257.2\ \text{kJ mol}^{-1}.\end{aligned}$$

• *Comment*. This $\Delta_r G^\ominus$ is the change of Gibbs function per mole of reaction in which the pure, separated gases are converted completely into carbon dioxide. In due course we shall have to deal with changes of Gibbs function for systems in which there are mixtures of reactants and products.

• *Exercise*. Calculate the standard reaction Gibbs function for the combustion of $CH_4(g)$ at $\mathcal{T}$. [$-818\ \text{kJ mol}^{-1}$]

Calorimetry (for ΔH directly, and for ΔS via heat capacities) is only one of the ways of determining the values of Gibbs functions. They may also be obtained from equilibrium constants (Chapter 10) and electrochemical measurements (Chapter 12), and they may be calculated using data from spectroscopic observations (Chapter 22). The information in Table 5.3, however, together with the machinery we shall now construct, is all we need in order to draw far-reaching conclusions about reactions and other processes of interest in chemistry.

Further reading

The second law. P. W. Atkins; Scientific American Books, New York, 1984.
Chemical thermodynamics. P. A. Rock; University Science Books and Oxford University Press, 1983.
Entropy. J. D. Fast; McGraw-Hill, New York, 1963.
Engines, energy, and entropy. J. B. Fenn; W. H. Freeman & Co., New York. 1982.
The third law of thermodynamics. J. Wilks; Clarendon Press, Oxford, 1961.
Heat and thermodynamics. M. W. Zemansky and R. H. Dittman; McGraw-Hill, New York, 1981.
Thermodynamics. G. N. Lewis and M. Randall, revised by K. S. Pitzer and L. Brewer; McGraw-Hill, New York, 1961.
Bibliography of thermodynamics. L. K. Nash; *J. chem. Educ.* **42,** 71 (1965).

Introductory problems

A5.1. A sample of Al(s) with a mass of 1.75 kg is cooled at constant pressure from 300 K to 265 K. How much heat must be extracted from the sample? Calculate the entropy change for the process.

A5.2. A 25 g sample of $CH_4(g)$ at 250 K and 18.5 atm pressure expands isothermally until the pressure is 2.5 atm. Assume the gas obeys the perfect gas law and calculate the entropy change for the process.

A5.3. A sample of a perfect gas, initially occupying a volume of 15.0 dm^3 at 250 K and 1 atm, undergoes isothermal compression. To what volume must the gas be compressed in order to decrease its entropy by 5.0 $J\,K^{-1}$?

A5.4. The amount of heat required to vaporize a mole of $CHCl_3(l)$ at its normal boiling point 334.88 K is 29.4 $kJ\,mol^{-1}$. Calculate the molar entropy of vaporization at 334.88 K.

A5.5. In the temperature range 240 K to 330 K, $C_{p,m}$ for $CHCl_3(l)$ is given by $\{91.47+7.5\times10^{-2}(T/K)\}\,J\,K^{-1}\,mol^{-1}$. A mole of $CHCl_3(l)$ originally at 275 K is heated to 300 K by coming in contact with a very large solid sample at 300 K. Assume that the liquid and solid are isolated from the surroundings and that the heat capacity of the solid is large enough to keep its temperature essentially unchanged by the flow of heat to $CHCl_3(l)$. Calculate the change in entropy for $CHCl_3(l)$ and the change in entropy for the solid. Verify that $\Delta S>0$ for this process.

A5.6. Calculate the standard molar entropy at 298.15 K for the following:

(a) $2CH_3CHO(g)+O_2(g)\rightarrow 2CH_3COOH(l)$

(b) $2AgCl(s)+Br_2(l)\rightarrow 2AgBr(s)+Cl_2(g)$.

A5.7. The standard molar Gibbs function of formation at 298.15 K for $SO_2(g)$ is $-300.4\,kJ\,mol^{-1}$. Calculate the change in the standard Gibbs function at 298.15 K for

$$2H_2S(g)+3O_2(g)\rightarrow 2H_2O(l)+2SO_2(g).$$

A5.8. Assume that $H_2(g)$ and $O_2(g)$ obey the perfect gas law and calculate the standard Helmholtz function of formation for $CH_3OH(l)$ at 298.15 K.

A5.9. In the following reactions, all the materials are in their standard states at 298.15 K. Decide which reactions are spontaneous.

(a) $2NO_2(g)\rightarrow 2NO(g)+O_2(g)$

(b) $C_2H_4(g)+H_2(g)\rightarrow C_2H_6(g)$

(c) $N_2O_4(g)\rightarrow 2NO_2(g)$.

A5.10. The standard molar enthalpy of combustion of phenol, $C_6H_5OH(s)$, at 298.15 K is $-3054\,kJ\,mol^{-1}$, and its standard molar entropy at 298.15 K is 144.0 $J\,K^{-1}\,mol^{-1}$. Calculate $\Delta_f G^{\ominus}$ for $C_6H_5OH(s)$ at 298.15 K.

Problems

5.1. Calculate the change of entropy when 25 kJ of heat is transferred reversibly and isothermally to a system at (a) 0 °C, (b) 100 °C.

5.2. A block of copper of mass 500 g initially at 293 K is in thermal contact with an electric heater of resistance 1000 Ω and negligible mass. An electric current of 1.00 A is passed for 15.0 s. What is the change of entropy of the copper? Assume $C_{p,m}=24.4\,J\,K^{-1}\,mol^{-1}$ throughout.

5.3. The same block of copper as in the last Problem is immersed in a stream of water which maintains its temperature at 293 K. What is the change of entropy of (a) the metal, (b) the water when the same amount of electricity is passed through the resistance?

5.4. Find the molar entropy of neon at 500 K given that its standard molar entropy at 298 K is 146.22 $J\,K^{-1}\,mol^{-1}$. Take constant volume conditions, and C_V to be constant.

5.5. The standard molar entropy of ammonia at $\mathcal{T}$ is 192.4 $J\,K^{-1}\,mol^{-1}$. We have seen that its heat capacity follows $C_{p,m}=a+bT+c/T^2$, with the coefficients given in Table 4.3. What is the standard molar entropy of ammonia at (a) 100 °C, (b) 500 °C?

5.6. What is the change of entropy when 50 g of hot water at 80 °C is poured into 100 g of cold water at 10 °C in an insulated vessel? Take $C_{p,m}=75.5\,J\,K^{-1}\,mol^{-1}$.

5.7. Calculate the change of entropy when 200 g of water at 0 °C is added to 200 g of water at 90 °C in an insulated vessel. (Data in Table 4.3.)

5.8. When calculating entropy changes we have to be careful to take into account phase transitions. Calculate the change of entropy when 200 g of ice at 0 °C is added to 200 g of water at 90 °C in an insulated vessel.

5.9. Calculate the changes of entropy in the system, the surroundings, and the universe when a 14 g sample of (perfect) nitrogen gas at 298 K and 1 bar doubles its volume in (a) an isothermal reversible expansion against $p_{ex}=0$, (b) an isothermal irreversible expansion, (c) an adiabatic reversible expansion.

5.10. Calculate the differences in molar entropy (a) between water liquid at −5 °C and ice at −5 °C, (b) between water liquid and water vapour at 95 °C and 1 atm pressure. Distinguish between entropy differences of the system and entropy differences of the universe, and discuss the spontaneity of transitions between phases at these temperatures. (Difference of heat capacities on melting: $37.3\ \mathrm{J\,K^{-1}\,mol^{-1}}$; difference on vaporization: $-41.9\ \mathrm{J\,K^{-1}\,mol^{-1}}$.)

5.11. We saw that the Gibbs function concentrates attention on the system and automatically carries around the entropy changes in the surroundings. Calculate the difference in Gibbs function between water and ice at $-5\ °\mathrm{C}$ and between water and water vapour at $95\ °\mathrm{C}$, and see how ΔG incorporates the information of the last Problem.

5.12. Calculate the molar entropy change when (a) water, (b) benzene are evaporated at their boiling points under 1 atm pressure. What are the entropy changes in (i) the system, (ii) the surroundings, (iii) the universe?

5.13. Suppose now that the evaporated water and benzene of Problem 5.12 are compressed to half their volume and *simultaneously* heated to twice the absolute temperature at which they originally boiled. What change of entropy occurs? ($C_{V,\mathrm{m}} = 25.3\ \mathrm{J\,K^{-1}\,mol^{-1}}$ for water vapour, and $C_{V,\mathrm{m}} = 130\ \mathrm{J\,K^{-1}\,mol^{-1}}$ for benzene vapour.)

5.14. Calculate (a) the maximum work, (b) the maximum non-p,V work that can be obtained from the freezing of supercooled water at $-5\ °\mathrm{C}$ under atmospheric pressure. What are the corresponding values under 100 atm pressure? Use $\rho(\mathrm{water}) = 0.999\ \mathrm{g\,cm^{-3}}$ and $\rho(\mathrm{ice}) = 0.917\ \mathrm{g\,cm^{-3}}$ at $-5\ °\mathrm{C}$ to calculate ΔV.

5.15. In Part 3 we shall see that *fuel cells* are designed to extract electrical work from the direct reaction of readily available fuels. Even at this stage we can make a thermodynamic assessment of the maximum amount of electrical work available from various reactions. Methane and oxygen are readily available, and the standard reaction Gibbs function for $CH_4(g) + 2O_2(g) \rightarrow CO_2(g) + 2H_2O(l)$ is $-802.8\ \mathrm{kJ\,mol^{-1}}$ at $25\ °\mathrm{C}$. What is the maximum work and the maximum electrical work available from this reaction under these standard conditions?

5.16. In order to use entropies and Gibbs functions we need to be able to measure their values. Succeeding chapters will bring forward a number of methods for their measurement, but some were encountered in this chapter and can be illustrated here. The following table gives the heat capacity of lead over a range of temperatures. What is the standard molar Third Law entropy of lead at (a) $0\ °\mathrm{C}$, (b) $25\ °\mathrm{C}$?

T/K	10	15	20	25	30	50
$C_{p,\mathrm{m}}/\mathrm{J\,K^{-1}\,mol^{-1}}$	2.8	7.0	10.8	14.1	16.5	21.4
T/K	70	100	150	200	250	298
$C_{p,\mathrm{m}}/\mathrm{J\,K^{-1}\,mol^{-1}}$	23.3	24.5	25.3	25.8	26.2	26.6

5.17. In a thermochemical study of nitrogen the following heat capacity data were found

$$\int_0^{T_t} (C_{p,\mathrm{m}}/T)\,\mathrm{d}T = 27.2\ \mathrm{J\,K^{-1}\,mol^{-1}};$$

$$\int_{T_t}^{T_f} (C_{p,\mathrm{m}}/T)\,\mathrm{d}T = 23.4\ \mathrm{J\,K^{-1}\,mol^{-1}};$$

$$\int_{T_f}^{T_b} (C_{p,\mathrm{m}}/T)\,\mathrm{d}T = 11.4\ \mathrm{J\,K^{-1}\,mol^{-1}}; \qquad T_t = 35.61\ \mathrm{K},$$

the enthalpy of transition being $0.229\ \mathrm{kJ\,mol^{-1}}$; $T_f = 63.14\ \mathrm{K}$, $\Delta H^{\ominus}_{\mathrm{melt,m}} = 0.721\ \mathrm{kJ\,mol^{-1}}$; $T_b = 77.32\ \mathrm{K}$, $\Delta H_{\mathrm{vap,m}} = 5.58\ \mathrm{kJ\,mol^{-1}}$. What is the Third Law entropy of nitrogen gas at its boiling point?

5.18. Demonstrate that $\oint \mathrm{d}q/T < 0$ for a modification of the Carnot cycle in which the isothermal reversible expansion step replaced by an irreversible isothermal expansion against a pressure p_{ex}.

5.19. Develop Problem 5.18 by calculating the entropy difference between the states p_{ex}, V_{B}, T_{h} and p_{ex}, V_{A}, T_{h}, and show that the Clausius inequality is satisfied.

5.20. Find the thermodynamic coefficient of performance of a primitive steam engine operating on steam at $100\ °\mathrm{C}$ and discharging at $60\ °\mathrm{C}$. A modern steam turbine operates on steam at $300\ °\mathrm{C}$ and discharges at $80\ °\mathrm{C}$. What is its coefficient of performance?

5.21. Using the data in Problem 5.20, calculate the minimum amount of oil that must be burnt in order to raise a mass of 1000 kg through 50 m. Take the 'work content' of the fuel as $4.3 \times 10^4\ \mathrm{kJ\,kg^{-1}}$ and neglect all losses. Heat engines work at less than their thermodynamic efficiencies, but at least thermodynamics points to the ceiling of performance for a particular design.

5.22. The internal combustion engine is not noted for its efficiency, but some of its deficiencies are due to thermodynamic rather than technological constraints. Take the fuel to be an octane having an enthalpy of combustion of $-5512\ \mathrm{kJ\,mol^{-1}}$, and use 1 gallon ≈ 3.03 kg. Then examine the coefficient of performance from the point of view of an engine working with a cylinder temperature of $2000\ °\mathrm{C}$ and an exit temperature of $800\ °\mathrm{C}$. What is the maximum height, neglecting all forms of friction, to which a 2500 lb car can be driven on one gallon of fuel?

5.23. Represent the Carnot cycle on a temperature/entropy diagram, and show that the area enclosed by the cycle is equal to the work done.

5.24. Two equal blocks of the same metal, one at a temperature T_{h} and the other at T_{c}, are placed in contact and come to thermal equilibrium. Assuming the heat capacity to be constant over the temperature range at $24.4\ \mathrm{J\,K^{-1}\,mol^{-1}}$, calculate the change of entropy. Calculate the value for the case of two 500 g blocks of copper, with $T_{\mathrm{h}} = 500\ \mathrm{K}$ and $T_{\mathrm{c}} = 250\ \mathrm{K}$.

5.25. The enthalpy of the graphite $\rightarrow$ diamond phase transition, which under a pressure of 100 kbar occurs at 2000 K, is $+1.90\ \mathrm{kJ\,mol^{-1}}$. What is the entropy of the phase transition at the transition temperature? Does that fit your idea of the different crystal structures of graphite and diamond?

5.26. Solid hydrogen chloride undergoes a phase transition at 98.36 K for which $\Delta H_m^\ominus = +1.19$ kJ mol^{-1}. Calculate the molar entropy of transition. The sample is maintained in contact with a block of copper, the whole being thermally isolated. What is the change of entropy of the copper at the phase transition, and the change of entropy in the universe?

5.27. Use the Third Law entropies, Table 4.1, to deduce the standard reaction entropies for the following reactions at $\mathcal{T}$.

(a) $Hg(l) + Cl_2(g) \rightarrow HgCl_2(s)$
(b) $Zn(s) + CuSO_4(aq) \rightarrow Cu(s) + ZnSO_4(aq)$
(c) sucrose $+ 12O_2(g) \rightarrow 12CO_2 + 11H_2O(l)$.

5.28. Combine the results of the last Problem with the enthalpies given in Table 4.1 and find the standard reaction Gibbs functions at $\mathcal{T}$.

5.29. We shall encounter several methods for determining the Gibbs functions of reactions in later chapters, but even at this stage it is easy to use the information contained in Table 4.1 to draw conclusions about the spontaneity of chemical reactions. As a first example of this type of calculation, determine $\Delta_r S^\ominus$, $\Delta_r H^\ominus$, and $\Delta_r G^\ominus$ for the following reactions at 298 K, and state whether they are spontaneous in the forward or reverse directions.

(a) $H_2(g) + \frac{1}{2}O_2(g) \rightarrow H_2O(l)$
(b) $3H_2(g) + C_6H_6(l) \rightarrow C_6H_{12}(l)$
(c) $CH_3CHO(g) + \frac{1}{2}O_2(g) \rightarrow CH_3COOH(l)$.

5.30. The standard reaction Gibbs function

$$K_4Fe(CN)_6.3H_2O \rightarrow 4K^+(aq) + Fe(CN)_6^{4-}(aq) + 3H_2O(l)$$

is 26.120 kJ mol^{-1} (I. R. Malcolm, L. A. K. Staveley, and R. D. Worswick, *J. chem. Soc. Faraday Trans.* I, 1532 (1973)). The enthalpy of solution of the trihydrate is 55.000 kJ mol^{-1}. Find the standard molar entropy of solution at $\mathcal{T}$ and use the values in Table 4.1 to deduce a value of the standard molar entropy of the ferrocyanide ion in aqueous solution. ($S^\ominus$(ferrocy.) = 599.7 J K^{-1} mol^{-1}.)

5.31. The same authors referred to in the last Problem also measured the heat capacity (C_p) of the anhydrous potassium ferrocyanide (potassium hexacyanoferrate(II)) over a wide temperature range, and obtained the following results:

T/K	10	20	30	40
$C_{p,m}$/J K^{-1} mol^{-1}	2.09	14.43	36.44	62.55
T/K	80	90	100	110
$C_{p,m}$/J K^{-1} mol^{-1}	149.4	165.3	179.6	192.8
T/K	150	160	170	180
$C_{p,m}$/J K^{-1} mol^{-1}	237.6	247.3	256.5	265.1
T/K	50	60	70	
$C_{p,m}$/J K^{-1} mol^{-1}	87.03	111.0	131.4	
T/K	120	130	140	
$C_{p,m}$/J K^{-1} mol^{-1}	205.0	216.5	227.3	
T/K	190	200		
$C_{p,m}$/J K^{-1} mol^{-1}	273.0	280.3		

What is the entropy of the salt at these temperatures?

5.32. Integrate the data in the last Problem to find the enthalpy of the salt relative to its enthalpy at absolute zero; that is, find $H_m(T) - H_m(0)$. In Chapter 10 we shall meet the *Giauque function* $\Phi_0 = \{G_m(T) - H_m(0)\}/T$. This function varies more slowly than $G(T)$ itself, and so it is useful for recording the Gibbs function of substances at various temperatures. Calculate both $G_m(T) - G_m(0)$, which is the same as $G_m(T) - H_m(0)$, and the Giauque function for anhydrous potassium ferrocyanide using the data in the last Problem, and plot both as a function of temperature.

5.33. 1,3,5-trichloro-2,4,6-trifluorobenzene is an intermediate in the conversion of hexachlorobenzene to hexafluorobenzene. Its thermodynamic properties have been examined by measuring its heat capacity over a wide temperature range (R. L. Andon and J. F. Martin, *J. chem. Soc. Faraday Trans.* I, 871 (1973)).

T/K	14.14	16.33	20.03
$C_{p,m}$/J K^{-1} mol^{-1}	9.492	12.70	18.18
T/K	100.90	140.86	183.59
$C_{p,m}$/J K^{-1} mol^{-1}	95.05	121.3	144.4
T/K	31.15	44.08	64.81
$C_{p,m}$/J K^{-1} mol^{-1}	32.54	46.86	66.36
T/K	225.10	262.99	298.06
$C_{p,m}$/J K^{-1} mol^{-1}	163.7	180.2	196.4

Evaluate the Third Law entropy, the molar enthalpy relative to $H_m^\ominus(0)$, and the Giauque function at 298 K.

6

The Second Law: the machinery

Learning objectives

After careful study of this chapter you should be able to:

(1) State how the internal energy of a closed system changes when no non-p, V work is done, eqn (6.1.1).

(2) State how the internal energy changes when the volume and entropy change infinitesimally, eqn (6.1.2).

(3) Deduce and state the *Maxwell relations*, Section 6.1(b) and Box 6.1, and use them to deduce a *thermodynamic equation of state*, eqn (6.1.6).

(4) State how the Gibbs function of a closed system changes when no non-p, V work is done, eqn (6.2.1).

(5) Derive the *Gibbs–Helmholtz equation*, eqn (6.2.4), for the temperature dependence of the Gibbs function.

(6) State how the Gibbs function depends on the pressure for a general type of substance, eqn (6.2.6).

(7) State how the Gibbs functions of solids and liquids depend on the pressure, eqns (6.2.7) and (6.2.8).

(8) Derive an expression for the pressure-dependence of the Gibbs function of a perfect gas, eqn (6.2.9).

(9) Write an expression for the pressure-dependence of a real gas, eqn (6.2.11), and define its *fugacity* and *fugacity coefficient*, eqn (6.2.12).

(10) Explain how to measure the fugacity of a gas, eqn (6.2.14), and relate it to the interactions between particles, Example 6.3.

(11) Define the *standard state* of a real gas, Section 6.2(g).

(12) State how the Gibbs function of an open system changes when the composition changes, eqn (6.3.1).

(13) Define the *chemical potential* of a substance, eqn (6.3.3), and use it to express the composition-dependence of the Gibbs function, eqn (6.3.4).

(14) Explain the connection between non-p, V work and change of composition, eqn (6.3.5).

(15) Deduce the relations between the chemical potential and the composition-dependence of the state functions U, H, and A, eqns (6.3.7) and (6.3.8).

Introduction

In this chapter we discuss the dependence of the Gibbs function on temperature, pressure, and composition. We meet the *chemical potential*, the quantity on which almost all the most important applications of thermodynamics to chemistry are based. Recall that a *closed system*, the kind we consider first, is one in which there is no change of composition. In an *open system* there may be changes of composition.

Another important topic introduced here is the technique for treating real systems systematically. Thermodynamics results in exact expressions between physical properties, but in order to use them it is necessary to be able to relate these formal properties to observable properties, such as pressures and concentrations.

6.1 Combining the First and Second Laws

The First Law may be written $\mathrm{d}U = \mathrm{d}q + \mathrm{d}w$. For a reversible change in a closed system, and in the absence of any non-p, V work, $\mathrm{d}w$ may be replaced by $-p\,\mathrm{d}V$ and $\mathrm{d}q$ may be replaced by $T\,\mathrm{d}S$. Therefore, for a *reversible* change of this kind, $\mathrm{d}U = T\,\mathrm{d}S - p\,\mathrm{d}V$. However, since $\mathrm{d}U$ is an exact differential, it is independent of path, and so the same value of $\mathrm{d}U$ is obtained whether the change is brought about irreversibly or reversibly. Therefore, *for any reversible or irreversible change* of a closed system involving no non-p, V work,

$$\mathrm{d}U = T\,\mathrm{d}S - p\,\mathrm{d}V. \tag{6.1.1}$$

This combination of the First and Second Laws we call the *fundamental equation*.

The fact that the fundamental equation applies to both reversible and irreversible changes may appear puzzling at first sight. The reason is that only in the case of a reversible change may $T\,\mathrm{d}S$ be identified with $\mathrm{d}q$ and $-p\,\mathrm{d}V$ with $\mathrm{d}w$. When the change is irreversible $T\,\mathrm{d}S$ is bigger than $\mathrm{d}q$ (the Clausius inequality, eqn (5.1.5)) and $p\,\mathrm{d}V$ is bigger than $\mathrm{d}w$. The *sum* of $\mathrm{d}w$ and $\mathrm{d}q$ remains equal to the *sum* of $T\,\mathrm{d}S$ and $-p\,\mathrm{d}V$, provided the composition is constant: this must be the case, because U is a state function.

There are two ways of developing the fundamental equation. They make use of the fact that, because $\mathrm{d}U$ is an exact differential, it can be developed using the techniques described in Chapter 3 and collected in Box 3.1.

6.1 (a) One way of developing the fundamental equation

Equation (6.1.1) shows that the internal energy of a closed system changes in a simple way when S and V are changed ($\mathrm{d}U \propto \mathrm{d}S$ and $\mathrm{d}U \propto \mathrm{d}V$). This suggests that U should be regarded as a function of S and V. Therefore, we write it $U(S, V)$. We could regard it as a function of other variables, such as S and p or T and V, because they are all interrelated; but the simplicity of the fundamental equation suggests that $U(S, V)$ is the best choice when the system is closed and no non-p, V work is done.

The *mathematical* consequence of U being a function of S and V is that a change $\mathrm{d}U$ can be expressed in terms of the changes $\mathrm{d}S$ and $\mathrm{d}V$ by

$$\mathrm{d}U = (\partial U/\partial S)_V\,\mathrm{d}S + (\partial U/\partial V)_S\,\mathrm{d}V. \tag{6.1.2}$$

When this is compared to the *thermodynamic* relation in eqn (6.1.1), it follows that for systems of constant composition,

$$(\partial U/\partial S)_V = T \quad \text{and} \quad -(\partial U/\partial V)_S = p. \tag{6.1.3}$$

The first is a *purely thermodynamic definition of temperature* as the ratio of the changes in the internal energy and entropy of a constant-volume closed system. We are beginning to generate relations between the properties of a system and to discover the power of thermodynamics for establishing unexpected relations. That will be seen to be the theme of the entire chapter.

The coefficient $(\partial U/\partial V)_T$ played a central role in the manipulation of the First Law, and in Section 3.2 we used the relation $(\partial U/\partial V)_T = T(\partial p/\partial T)_V - p$, which was to be proved in this chapter. We are now ready to derive it from the relations we have just established. The coefficient required can be obtained from eqn (6.1.2) by relation No. 1 of Box 3.1, which gives

$$(\partial U/\partial V)_T = (\partial U/\partial S)_V(\partial S/\partial V)_T + (\partial U/\partial V)_S = T(\partial S/\partial V)_T - p. \quad (6.1.4)$$

This is beginning to look like the expression we want. The next few paragraphs will let us make further progress.

6.1(b) A second way of developing the fundamental equation

Since the fundamental equation, eqn (6.1.1), is an expression for an exact differential, the coefficients of dS and dV must pass the test of relation No. 4 in Box 3.1 (the test for exact differentials). With $g = T$ and $h = -p$ it follows that

$$(\partial T/\partial V)_S = -(\partial p/\partial S)_V. \quad (6.1.5)$$

We have generated another relation between quantities which, at first sight, would not seem to be related.

The equation just derived is an example of a *Maxwell relation*. Apart from being unexpected, it does not look particularly interesting. Nevertheless, it does suggest that there may be other similar relations and that the coefficient $(\partial S/\partial V)_T$ in eqn (6.1.4) might be related to something recognizable. It turns out that the fact that H, G, and A are state functions can be used to derive three more Maxwell relations. The argument to obtain them runs in the same way in each case: since H, G, and A are state functions, the expressions for dH, dG, and dA satisfy relation No. 4 in Box 3.1. All four relations are listed in Box 6.1. In the next section we derive one of them, but as no new principles are involved we shall not derive them all.

Box 6.1 The Maxwell relations

$$(\partial T/\partial V)_S = -(\partial p/\partial S)_V$$
$$(\partial T/\partial p)_S = (\partial V/\partial S)_p$$
$$(\partial p/\partial T)_V = (\partial S/\partial V)_T$$
$$(\partial V/\partial T)_p = -(\partial S/\partial p)_T$$

One of the Maxwell relations does the job of turning $(\partial S/\partial V)_T$ into something else: $(\partial S/\partial V)_T = (\partial p/\partial T)_V$. Using this relation in eqn (6.1.4) completes the proof of the relation

$$(\partial U/\partial V)_T = T(\partial p/\partial T)_V - p. \quad (6.1.6)$$

This is called a *thermodynamic equation of state* because it expresses a quantity in terms of the two variables T and p.

Example 6.1

Show thermodynamically that $(\partial U/\partial V)_T = 0$ for a perfect gas, and compute its value for a van der Waals gas.

• *Method*. Proving a result 'thermodynamically' means basing it entirely on general thermodynamic relations and equations of state, without drawing on molecular arguments (e.g., the existence of intermolecular forces). Use eqn (6.1.6) in conjunction with the two equations of state, Box 1.1.

• *Answer*. For a perfect gas, $p = nRT/V$. Therefore, from eqn (6.1.6) with $(\partial p/\partial T)_V = nR/V$,

$$(\partial U/\partial V)_T = T(nR/V) - p = 0.$$

For a van der Waals gas,

$$p = nRT/(V - nb) - n^2a/V^2.$$

Therefore, since $(\partial p/\partial T)_V = nR/(V - nb)$,

$$(\partial U/\partial V)_T = nRT/(V - nb) - \{nRT/(V - nb) - n^2a/V^2\} = n^2a/V^2 = a/V_m^2.$$

• *Comment*. This calculation shows that the internal energy of a van der Waals gas *increases* when it expands isothermally, and that the increase is related to the parameter that models the attractive interactions between the particles: a larger molar volume implies a weaker mean attraction.

• *Exercise*. Calculate the value of $(\partial U/\partial V)_T$ for a gas that obeys the virial equation of state, Box 1.1.

[$RT^2(\partial B/\partial T)_V/V_m + \ldots$]

6.2 Properties of the Gibbs function

The same simple arguments as we applied to the fundamental equation for the internal energy may be applied to the Gibbs function. This is a prelude to the discussion of chemical equilibria.

The Gibbs function was defined in eqn (5.3.6) as $G = H - TS$. When the system changes its state, G may change because H, T, and S change. For infinitesimal changes in each property,

$$dG = dH - T\,dS - S\,dT.$$

Since $H = U + pV$, we have

$$dH = dU + p\,dV + V\,dp.$$

For a closed system doing no non-p,V work dU can be replaced by the fundamental equation $dU = T\,dS - p\,dV$. The result of these steps is

$$dG = (T\,dS - p\,dV) + p\,dV + V\,dp - T\,dS - S\,dT.$$

That is, for a closed system doing no non-p,V work,

$$dG = V\,dp - S\,dT. \qquad (6.2.1)$$

This new equation suggests that G should be regarded as a function of p and T, and should be written $G(p, T)$. This confirms that it is an important quantity in chemistry because the pressure and temperature are usually the variables under our control. *G carries around the combined consequences of the First and Second Laws in a way that makes it particularly suitable for chemical applications*.

The same argument that led to eqn (6.1.3) when applied to dG now gives

$$(\partial G/\partial T)_p = -S \quad \text{and} \quad (\partial G/\partial p)_T = V. \qquad (6.2.2)$$

These relations show how the Gibbs function varies with temperature and pressure. Likewise, when we examine eqn (6.2.1) with the test for an exact differential using relation No. 4 of Box 3.1, and set $g = V$ and $h = -S$, then since we know that dG *is* exact we conclude that

$$(\partial V/\partial T)_p = -(\partial S/\partial p)_T.$$

This is another of the Maxwell relations in Box 6.1.

6.2 (a) The temperature dependence of the Gibbs function

Since $(\partial G/\partial T)_p = -S$, and S is positive, it follows that G *decreases when the temperature is raised* at constant pressure and composition. Moreover, it decreases most sharply when the entropy of the state is large. Therefore, the Gibbs functions of gases are likely to be more sensitive to temperature than those of liquids and solids.

The temperature dependence given by eqn (6.2.2) can be expressed in terms of the *enthalpy* rather than the entropy of the system. First, replace S by $S = (H - G)/T$, which follows from the definition of G; then

$$(\partial G/\partial T)_p = (G - H)/T, \quad \text{or} \quad (\partial G/\partial T)_p - G/T = -H/T. \qquad (6.2.3)$$

This simple rearrangement can be taken further. First, note that $(\partial G/\partial T)_p - G/T$ is proportional to $\partial(G/T)/\partial T$, which follows from the rules for differentiating a quotient:

$$\begin{aligned}\{\partial(G/T)/\partial T\}_p &= (1/T)(\partial G/\partial T)_p + G\{\partial(1/T)/\partial T\}_p \\ &= (1/T)(\partial G/\partial T)_p - G/T^2 = (1/T)\{(\partial G/\partial T)_p - G/T\}.\end{aligned}$$

This leads immediately to the

$$\textit{Gibbs–Helmholtz equation:}\ \{\partial(G/T)/\partial T\}_p = -H/T^2. \qquad (6.2.4)$$

(G–H is a helpful way of remembering what this equation relates.) It shows that if the enthalpy of the system is known, then the temperature dependence of G/T is also known.

The Gibbs–Helmholtz equation is most useful when it is applied to *changes*, including changes of physical state and chemical reactions. Then, with $\Delta G = G_f - G_i$ for the change of Gibbs function between the final and initial states, since the equation applies to both G_f and G_i, we can write

$$\{\partial(G_f/T)/\partial T\}_p - \{\partial(G_i/T)/\partial T\}_p = -\{(H_f/T^2) - (H_i/T^2)\}.$$

Therefore, with $\Delta H = H_f - H_i$,

$$\{\partial(\Delta G/T)/\partial T\}_p = -\Delta H/T^2. \qquad (6.2.5)$$

It may seem that $\Delta G/T$ is a clumsy quantity to deal with, but in Chapter 10 we shall see that it is exactly what is needed when discussing chemical equilibria.

6.2 (b) The pressure dependence of the Gibbs function

From eqn (6.2.2) we see that we know how the Gibbs function depends on the pressure if we know the system's volume. Since V is positive, G always increases when the pressure of the system is increased at constant temperature (and composition). The equation may be integrated to find the

Gibbs function at one pressure in terms of its value at another:

$$G(p) = G(p') + \int_{p'}^{p} V\,\mathrm{d}p. \tag{6.2.6}$$

In the case of a liquid or solid, the volume changes only slightly as the pressure changes, and so V may be regarded as a constant and taken outside the integral. Then, for molar quantities,

$$G_m(p) = G_m(p') + (p - p')V_m. \tag{6.2.7}$$

Under normal laboratory conditions $(p - p')V_m$ is very small and may be neglected. For instance, for water $V_m \approx 18\ \mathrm{cm^3\ mol^{-1}}$; therefore, on going from 1 atm to 10 atm the Gibbs function changes by $(10 - 1) \times (1.013 \times 10^5\ \mathrm{N\ m^{-2}}) \times (18 \times 10^{-6}\ \mathrm{m^3\ mol^{-1}})$, or only $0.02\ \mathrm{kJ\ mol^{-1}}$. This means that we may usually write

$$\textit{For liquids and solids: } G_m(p) = G_m(p'). \tag{6.2.8}$$

In other words, the Gibbs functions of solids and liquids are virtually independent of pressure. However, if we are interested in geophysical problems, then since pressures in the Earth's bowels are huge, their effect on the Gibbs function cannot be ignored and we have to use eqn (6.2.7) or even, if the pressures are so great that there are substantial volume changes, the complete expression, eqn (6.2.6).

Example 6.2

The pressure deep inside the earth is probably greater than 3×10^3 kbar, and the temperature there is around 4×10^3 °C. Estimate the change in the reaction Gibbs function on going from crust to core for a process in which $\Delta_r V = +1.0\ \mathrm{cm^3\ mol^{-1}}$ and $\Delta_r S = +2.1\ \mathrm{J\ K^{-1}\ mol^{-1}}$.

• *Method*. Use eqn (6.2.2) in the form $(\partial \Delta_r G/\partial p)_T = \Delta_r V$ and $(\partial \Delta_r G/\partial T)_p = -\Delta_r S$. Then make a rough estimate of the total change from

$$\Delta_r G_{\mathrm{core}} - \Delta_r G_{\mathrm{crust}} = (p_{\mathrm{core}} - p_{\mathrm{crust}})\Delta_r V - (T_{\mathrm{core}} - T_{\mathrm{crust}})\Delta_r S.$$

Note that $1\ \mathrm{kbar} = 10^3\ \mathrm{bar} = 10^8\ \mathrm{N\ m^{-2}}$.

• *Answer*. $(p_{\mathrm{core}} - p_{\mathrm{crust}})\Delta_r V = (3 \times 10^{11}\ \mathrm{N\ m^{-2}}) \times (1.0 \times 10^{-6}\ \mathrm{m^3\ mol^{-1}})$
$= 3 \times 10^5\ \mathrm{J\ mol^{-1}} = 3 \times 10^2\ \mathrm{kJ\ mol^{-1}}$.

$$(T_{\mathrm{core}} - T_{\mathrm{crust}})\Delta_r S = (4 \times 10^3\ \mathrm{K}) \times (2.1\ \mathrm{J\ K^{-1}\ mol^{-1}}) = 8\ \mathrm{kJ\ mol^{-1}}.$$

Therefore, the overall change is

$$\Delta_r G_{\mathrm{core}} - \Delta_r G_{\mathrm{crust}} = \{3 \times 10^2 - 8\}\ \mathrm{kJ\ mol^{-1}} = 3 \times 10^2\ \mathrm{kJ\ mol^{-1}}.$$

• *Comment*. The effect of pressure dominates, and results in a very large modification of the reaction Gibbs function. This is the thermodynamic reason why materials change their forms at great depths in the earth's interior.

• *Exercise*. Calculate the difference in molar Gibbs function between the top and bottom of a column of mercury in a barometer. The density of mercury is $13.6\ \mathrm{g\ cm^{-3}}$. [$1.5\ \mathrm{J\ mol^{-1}}$]

In the case of gases the molar volume is large, and so the correction term may be large even if the pressure difference is small. Furthermore, since the volume depends strongly on the pressure, we cannot treat it as a constant in the integral in eqn (6.2.6). In the case of a perfect gas we can substitute $V = nRT/p$ into the integral, and find

$$G(p) = G(p') + nRT\int_{p'}^{p} (1/p)\,dp.$$

Then, since $(1/p)\,dp = d\ln p$,

$$G(p) = G(p') + nRT\ln(p/p'). \qquad (6.2.9)^{\circ}$$

This shows that when the pressure is increased tenfold at room temperature, the Gibbs function increases by about 6 kJ mol^{-1}.

6.2 (c) The chemical potential of a perfect gas

The standard state of a perfect gas is established at a pressure of exactly 1 bar: this pressure we denote $p^{\ominus}$. The Gibbs function is then $G^{\ominus}$. According to the last equation, its value at any other pressure p is

$$G(p) = G^{\ominus} + nRT\ln(p/p^{\ominus}).$$

The molar Gibbs function, G/n, is therefore

$$G_m(p) = G_m^{\ominus} + RT\ln(p/p^{\ominus}).$$

Henceforth, for a pure substance, we write $\mu = G_m(p)$, so that

$$\mu = \mu^{\ominus} + RT\ln(p/p^{\ominus}). \qquad (6.2.10)^{\circ}$$

The quantity μ is called the *chemical potential*. The origin of the name is the analogy between mechanical systems, where particles tend to travel in the direction of decreasing potential, and the thermodynamic tendency of systems to shift in the direction of decreasing Gibbs function. The reason for introducing yet another name and symbol into thermodynamics is that the concept of chemical potential is exactly what we need in order to deal with mixtures of substances, as we see soon.

The chemical potential plays a central role in the remaining chapters of Part 1. For the time being it is only another name for the molar Gibbs function, but by the end of the chapter it will have been generalized to the point that it contains all the information we need about chemically reactive systems.

6.2 (d) Real gases: the fugacity

Equation (6.2.10) is a moderately good description of the pressure dependence of the Gibbs functions of real gases so long as they are behaving almost perfectly. But as departures from ideality occur under quite common conditions we need to be equipped to deal with them. It proves sensible to preserve the form of eqn (6.2.10) in the case of real gases, and to replace p by some effective pressure, the *fugacity* (symbol f) such that

$$\mu = \mu^{\ominus} + RT\ln(f/p^{\ominus}). \qquad (6.2.11)$$

(The name 'fugacity' comes from the Latin for 'fleetness'; f has the same dimensions as pressure.) This relation is exact, by definition. It has *two* new quantities (f and $\mu^{\ominus}$), and so we need a further definition before we can discuss either the fugacity or the standard value of the chemical potential.

First, we construct a formula relating the fugacity to the pressure. Equation (6.2.6) is true for all gases, whether real or perfect. Expressing it in terms of molar quantities and then using eqn (6.2.11) gives

$$\int_{p'}^{p} V_m \, dp = \mu(p) - \mu(p') = RT \ln (f/f').$$

(f is the fugacity when the pressure is p, f' the fugacity when it is p'.) If the gas were perfect we could write

$$\int_{p'}^{p} V_m^{\circ} \, dp = \mu^{\circ}(p) - \mu^{\circ}(p') = RT \ln (p/p'),$$

where the superscript $^{\circ}$ denotes quantities relating to a perfect gas. The difference of the two equations is

$$\int_{p'}^{p} \{V_m - V_m^{\circ}\} \, dp = RT \ln (f/f') - RT \ln (p/p'),$$

or

$$\ln \{(f/p)/(f'/p')\} = (1/RT) \int_{p'}^{p} \{V_m - V_m^{\circ}\} \, dp.$$

This untidy expression can be improved. In the first place one of the pressures, p' is allowed to approach zero. Then the gas behaves perfectly, and in this limit the fugacity f' becomes equal to the pressure p'. Therefore, f'/p' approaches 1 as p' approaches 0. In this limit the last equation becomes

$$\ln (f/p) = (1/RT) \int_{0}^{p} \{V_m - V_m^{\circ}\} \, dp. \tag{6.2.12}$$

This is a relation between f and p that is consistent with writing the chemical potential as in eqn (6.2.11). We shall henceforth use it as the *definition* of the fugacity.

Now that we have a definition of the fugacity, the standard value of the chemical potential can be found by rearranging eqn (6.2.11) to

$$\mu^{\ominus} = \mu - RT \ln (f/p) - RT \ln (p/p^{\ominus}). \tag{6.2.13}$$

All three terms on the right can now be measured, and so the standard value can be constructed.

6.2 (e) Measuring the fugacity

The expression for the fugacity corresponding to a pressure p, eqn (6.2.12), can be simplified as follows. For a perfect gas $V_m^{\circ} = RT/p$. For a real gas, since Z (the compression factor, Section 1.3(a)) is defined as $Z = pV_m/RT$, we have $V_m = RTZ/p$. (Remember that the compression factor depends on the pressure and temperature.) Therefore:

$$\ln (f/p) = \int_{0}^{p} \{(Z-1)/p\} \, dp,$$

which rearranges to

$$f = \gamma p, \qquad \gamma = \exp \int_0^p \{(Z-1)/p\}\, dp. \qquad (6.2.14)$$

The factor γ is called the *fugacity coefficient*. This is the recipe for determining the fugacity of a gas at any pressure. Experimental data are needed on the compression factor from very low pressures up to the pressure of interest. Sometimes this information is available in numerical tables, in which case the integral may be evaluated numerically. Sometimes an algebraic expression is available for Z (for instance, from one of the equations of state, Box 1.1) and it may be possible to evaluate the integral analytically. For example, if we have the virial coefficients in eqn (1.3.1), we can obtain the fugacity from

$$f = p e^{B'p + C'p^2/2 + \dots},$$

which is obtained by explicit evaluation of eqn (6.2.14).

Example 6.3

Supposing that the attractive interactions between gas particles can be neglected, find an expression for the fugacity of a van der Waals gas in terms of the pressure, and estimate its value for ammonia at 10.00 atm and 298.15 K.

• *Method*. Neglect a in the van der Waals equation of state, Box 1.1. Then $p = RT/(V_m - b)$. Form $Z = pV_m/RT$, and evaluate the integral in eqn (6.2.14). For the numerical part, use data from Table 1.3.

• *Answer*. From the equation of state,

$$V_m = (RT/p) + b, \qquad Z = 1 + pb/RT.$$

The integral required is therefore

$$\int_0^p \{(Z-1)/p\}\, dp = (b/RT)\int_0^p dp = bp/RT.$$

Consequently, from eqn (6.2.14) the fugacity at the pressure p is

$$f = p e^{pb/RT}.$$

From Table 1.3, $b = 3.707 \times 10^{-2}\, dm^3\, mol^{-1}$, and so

$$pb/RT = \frac{(10.00\, atm) \times (3.707 \times 10^{-2}\, dm^3\, mol^{-1})}{(0.082\,06\, dm^3\, atm\, K^{-1}\, mol) \times (298.15\, K)} = 0.015,$$

$$f = (10.00\, atm)e^{0.015} = 10.15\, atm.$$

• *Comment*. The effect of the repulsive term (as represented by the b term in the van der Waals equation) is to *increase* the fugacity above the pressure, and so the 'effective pressure' of the gas is greater than if it were perfect.

• *Exercise*. Find an expression for the fugacity when the attractive interaction is dominant in a van der Waals gas, and the pressure is low enough to make the approximation $4ap/R^2T^2 \ll 1$. Evaluate the fugacity for ammonia, as above. [$f = pe^{-ap/RT}$, 9.33 atm]

Compression factor less than unity at moderate pressures but greater than unity at higher pressures

It is clear from Fig. 1.6 that for most gases the compression factor is less than unity up to moderate pressure, but that it always becomes greater than unity at higher pressures. If Z is less than unity throughout the range of integration, then the exponent in eqn (6.2.14) is negative and the fugacity

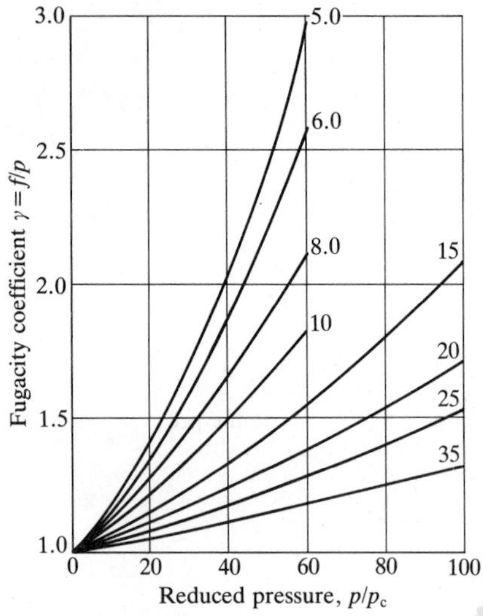

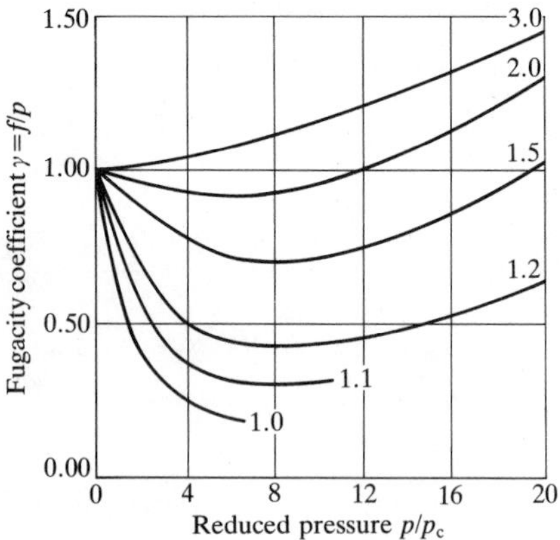

Fig. 6.1. The fugacity coefficient of a van der Waals gas as a function of its reduced variables. The fugacity itself is given by $f = \gamma p$. The curves are labelled with the reduced temperature of the gas. The two illustrations differ in the range of reduced pressures. (Reduced pressures are defined in eqn (1.4.6).)

Table 6.1. The fugacity of nitrogen at 273 K

p/atm	f/atm
1	0.999 55
10	9.956 0
100	97.03
1000	1839

coefficient is less than unity, which implies that the fugacity is less than the pressure (the particles tend to stick together). At higher pressures the region where Z exceeds unity dominates the region where it is less than unity. The integral is then positive, $\gamma > 1$, and the fugacity is greater than the pressure (the repulsive interactions are dominant and tend to drive the particles apart). Figure 6.1, which has been calculated using the full van der Waals equation of state, shows how the fugacity depends on the pressure in terms of the reduced variables. Since the critical constants are available in Table 1.1, the graphs can be used for quick estimates of the fugacities of a wide range of gases. Table 6.1 gives some explicit values for nitrogen.

6.2 (f) The importance of fugacity

We shall derive thermodynamically exact expressions for chemical properties such as equilibrium constants in terms of the chemical potentials of the species involved in the reaction. However, these equilibrium constants will be expressed in terms of fugacities, not the pressures. Therefore, only if the fugacities can be related to the pressures of the reacting species, as described in the last few paragraphs, will the equations be of practical use. Thermodynamic expressions in terms of the fugacity are exact; the relation of the fugacity to the pressure is sometimes approximate, but gives results of practical utility.

6.2 (g) Standard states of real gases

The standard state of a perfect gas is established at 1 bar (i.e. at $p = p^{\ominus}$). In the case of a real gas the chemical potential has its standard value when $f = p^{\ominus}$, eqn (6.2.11). At 298 K, $f = 1$ bar occurs at about $p = 1$ bar. However, we do not say that a real gas is *in* its standard state when $f = p^{\ominus}$; instead, we define the standard state as a *hypothetical* state. This can be understood as follows.

Introducing eqn (6.2.14) into eqn (6.2.11) gives

$$\mu = \mu^{\ominus} + RT \ln (p/p^{\ominus}) + RT \ln \gamma. \tag{6.2.15}$$

The first two terms on the right have the form of the chemical potential of a perfect gas, and so $RT \ln \gamma$ measures the deviations from ideality. If, however, the *whole* of the deviations from ideality are to be carried by γ, $\mu^{\ominus}$ itself must be characteristic of a perfect gas. Therefore we say:

> *The standard state of a real gas is a hypothetical state in which the gas is behaving perfectly and is at a pressure $p^{\ominus}$.*

The standard chemical potential (at the specified temperature) is the value of μ when the pressure is $p^{\ominus}$ (so that $\mu = \mu^{\ominus} + RT \ln \gamma$) *and* $\gamma = 1$ (so that $\ln \gamma = 0$ and $\mu = \mu^{\ominus}$), Fig. 6.2.

The advantage of this cumbersome definition is that the standard state of a *real* gas has the simple properties of a *perfect* gas: if we had defined the standard state as the one for which $f = p^{\ominus}$, the standard states of different gases would have had relatively complex properties. The choice of a hypothetical standard state literally standardizes the interactions between the particles (by setting them to zero and hence leaving us with a 'kinetic energy only' gas). Moreover, differences of standard chemical potential of different gases are now due solely to the *internal* structure and properties of the particles, not to the way they interact.

6.3 Open systems and changes of composition

So far we have treated the Gibbs function for the case in which it depends on only the pressure and temperature. In an *open system* (one in which the composition may vary) it also depends on the composition. It should therefore be written $G(p, T, n_1, n_2, \ldots)$, and a general change should be expressed as

$$\begin{aligned}dG = (\partial G/\partial p)_{T,n_1,n_2,\ldots}\,dp + (\partial G/\partial T)_{p,n_1,n_2,\ldots}\,dT \\ + (\partial G/\partial n_1)_{p,T,n_2,\ldots}\,dn_1 + (\partial G/\partial n_2)_{p,T,n_1,\ldots}\,dn_2 + \ldots \quad (6.3.1)\end{aligned}$$

This fearsome expression simply says that G may change because any of its variables may change. Our first job must be to simplify the notation.

As a first step, consider how G changes when the composition is constant but the pressure and temperature change infinitesimally:

$$dG = (\partial G/\partial p)_{T,n_1,n_2,\ldots}\,dp + (\partial G/\partial T)_{p,n_1,n_2,\ldots}\,dT.$$

(This is nothing more complicated than the analogous expression in eqn (6.1.2).) We already know that $dG = V\,dp - S\,dT$ under the same conditions, and since dG is an exact differential we may identify the coefficients:

$$(\partial G/\partial p)_{T,n_1,n_2,\ldots} = V, \qquad (\partial G/\partial T)_{p,n_1,n_2,\ldots} = -S. \quad (6.3.2)$$

These are the same as eqn (6.2.2), but are dressed more elaborately because the constraint of constant composition is specified explicitly.

Now suppose that only substance 1 is present. G can then be equated to $n_1 G_{1m}$, which is the same as $n_1\mu_1$ in the notation introduced in Section 6.2, μ_1 being the chemical potential of pure substance 1. It follows that at constant temperature and pressure, $dG = \mu_1\,dn_1$; consequently,

$$(\partial G/\partial n_1)_{p,T} = \mu_1.$$

Expressed in this way we see that the chemical potential carries information about how G changes as substance 1 is added to the system.

The last remark is now generalized. From now on we *define* the chemical potential of a substance in terms of the effect it has on the *total* Gibbs function of a system of *arbitrary* composition:

$$\textit{Chemical potential: } \mu_1 = (\partial G/\partial n_1)_{p,T,n_2\ldots}. \quad (6.3.3)$$

This reduces to the earlier definition in the case of a single component, because then $G = nG_m$. The definition means that the chemical potential of a substance tells us the change in the total Gibbs function of a mixture when an infinitesimal amount of a substance is added, the temperature and pressure and the amounts of all the other substances being constant: $dG = \mu_1\,dn_1$. Introduction of this definition (for each substance) into eqn (6.3.1) leads to

$$dG = V\,dp - S\,dT + \mu_1\,dn_1 + \mu_2\,dn_2 + \ldots \quad (6.3.4)$$

The chemical potential of a substance, like the overall volume and entropy, depends on the conditions, including the composition of the system. For instance, adding 10^{-3} mol CH_3OH (almost an infinitesimal amount of methanol) to 1 dm^3 of a 20 per cent methanol/water mixture leads to a change in the overall Gibbs function which is different from the

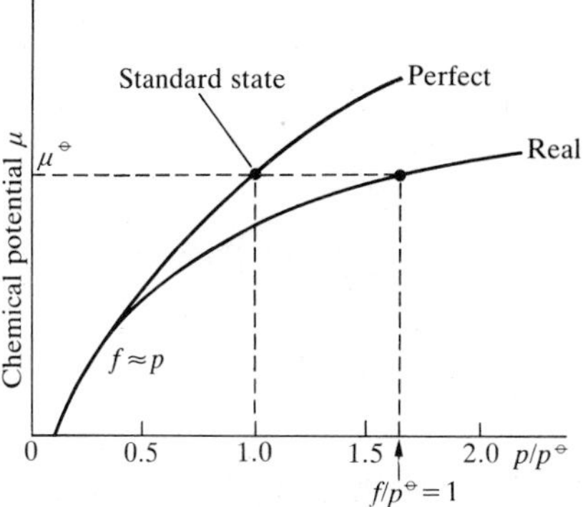

Fig. 6.2. The chemical potential of a perfect gas depends on the pressure as $\mu = \mu^\ominus + RT\ln(p/p^\ominus)$, the curve labelled 'Perfect'. The standard state is reached at $p = p^\ominus$. For a real gas, with a chemical potential given by eqn (6.2.15), the 'perfect' value is modified by the addition of $RT\ln\gamma$, which reduces the chemical potential below the 'perfect value' (so long as $\gamma < 1$). The real gas has its standard value at $f = p^\ominus$ (which might occur when the true pressure is 1.6 bar, as here), but it is not then in its standard state. That is a *hypothetical* state based on the assumption that all the interactions between particles have been extinguished, and is the same standard state as for a perfect gas.

change brought about by the addition of the same amount to an 80 per cent mixture. It should also be noted that a substance has a non-zero chemical potential even when it is absent! This is because the Gibbs function of a system changes when substance 1 is added even though it contains no substance 1 initially; therefore μ_1 is non-zero (negatively infinite, in fact) even at $n_1 = 0$.

Under conditions of constant pressure and temperature the last expression reduces to

$$dG = \mu_1\,dn_1 + \mu_2\,dn_2 + \ldots.$$

We saw in Section 5.3 that under the same conditions $dG = dw_{e,max}$. Therefore,

$$dw_{e,max} = \mu_1\,dn_1 + \mu_2\,dn_2 + \ldots \quad \text{(at constant } p, T). \qquad (6.3.5)$$

This means that non-p,V work arises from the change of composition of a system.

The generalized definition of the chemical potential has a wider significance. Since $G = U + pV - TS$, a general infinitesimal change in U for a system of variable composition can be written

$$\begin{aligned} dU &= -p\,dV - V\,dp + S\,dT + T\,dS + dG \\ &= -p\,dV - V\,dp + S\,dT + T\,dS + (V\,dp - S\,dT + \mu_1\,dn_1 + \mu_2\,dn_2 + \ldots) \\ &= -p\,dV + T\,dS + \mu_1\,dn_1 + \mu_2\,dn_2 + \ldots. \end{aligned} \qquad (6.3.6)$$

This is the generalization of the fundamental equation, eqn (6.1.1), to systems in which the composition may change. It follows that at constant volume and entropy, $dU = \mu_1\,dn_1 + \mu_2\,dn_2 + \ldots$, and so

$$(\partial U/\partial n_1)_{V,S,n_2,\ldots} = \mu_1. \qquad (6.3.7)$$

Therefore, not only does the chemical potential show how G changes when the composition changes, it shows how the internal energy changes too (but under a different set of conditions). In the same way it is easy to deduce that

$$(\partial H/\partial n_1)_{p,S,n_2,\ldots} = \mu_1, \qquad (\partial A/\partial n_1)_{V,T,n_2,\ldots} = \mu_1. \qquad (6.3.8)$$

Thus we see that the μ_J show how all the extensive thermodynamic properties U, H, A, and G depend on the composition. This is why they are so central to chemistry.

Equation (6.3.4) is the *fundamental equation of chemical thermodynamics*. Its implications and consequences are explored and developed in the next six chapters.

Further reading

Chemical thermodynamics (2nd edn). I. M. Klotz and R. M. Rosenberg; Benjamin, Menlo Park, 1972.

Chemical thermodynamics. P. A. Rock; University Science Books and Oxford University Press, 1983.

Chemical thermodynamics. M. L. McGlashan; Academic Press, London, 1979.

Heat and thermodynamics. M. W. Zemansky and R. H. Dittman; McGraw-Hill, New York, 1981.
Thermodynamics. G. N. Lewis and M. Randall, revised by K. S. Pitzer and L. Brewer; McGraw-Hill, New York, 1961.
Applications of thermodynamics (2nd edn). B. D. Wood; Addison-Wesley, New York, 1982.

Introductory problems

A6.1. Use Maxwell's relations to express the derivatives $(\partial S/\partial V)_T$ and $(\partial S/\partial p)_T$ in terms of α, the coefficient of thermal expansion, and κ, the isothermal compressibility.

A6.2. Use Euler's chain relation and Maxwell's relations to express $(\partial p/\partial S)_V$ and $(\partial V/\partial S)_p$ in terms of C_V, C_p, α, and κ.

A6.3. For a certain constant-pressure process, the change in the Gibbs function over the temperature range 300 K to 330 K is given by the empirical expression: $\Delta G = \{-85.40 + 36.5(T/\mathrm{K})\}$ J. By direct differentiation, obtain the value of ΔS for the process.

A6.4. Three millimoles of $N_2(g)$ occupy a volume of 36 cm^3 at 300 K. The gas expands isothermally until it has a volume of 60 cm^3. Assume nitrogen is a perfect gas, and calculate the change in the Gibbs function for this process.

A6.5. When the pressure on a 35 g liquid sample is increased isothermally from 1.0 atm to 3000 atm, its Gibbs function increases by 12 kJ. Calculate the density of the liquid.

A6.6. Use the tabulated values of the standard Gibbs function of formation (Chapter 5) and the tabulated values of the standard (molar) enthalpy of formation (Chapter 4) to calculate $\Delta_r G^\ominus$ at 375 K for $2CO(g) + O_2(g) \rightarrow 2CO_2(g)$.

A6.7. Two moles of a perfect gas at 330 K and 3.50 atm are subjected to isothermal compression, and the entropy of the gas decreases by 25.0 J K^{-1}. Calculate the final pressure of the gas, and the change in the Gibbs function for the compression.

A6.8. A gas at 200 K and 50 atm pressure has a fugacity coefficient of 0.72. Calculate the difference between the chemical potential and the standard chemical potential of the gas at 200 K and 50 atm.

A6.9. Calculate the change in the chemical potential of a perfect gas which undergoes compression from 1.8 atm to 29.5 atm pressure at a constant temperature of 40 °C.

A6.10. At 373 K, the second virial coefficient, B, for Xe is -81.7 cm^3 mol^{-1}. Calculate the virial coefficient B' and use it to estimate the fugacity coefficient of Xe at 50 atm and 373 K.

Problems

6.1. Show that $(\partial U/\partial S)_V = T$ and $(\partial U/\partial V)_S = -p$ in the case of a perfect gas.

6.2. Two of the four Maxwell relations were derived in the text, but two were not. Complete their derivation by showing that $(\partial S/\partial V)_T = (\partial p/\partial T)_V$ and $(\partial T/\partial p)_S = (\partial V/\partial S)_p$.

6.3. Use the first of the Maxwell equations derived in Problem 6.2 to show that, in the case of a perfect gas, the entropy depends on volume as $S = \mathrm{const} + nR \ln V$.

6.4. Derive the thermodynamic equation of state

$$(\partial H/\partial p)_T = V - T(\partial V/\partial T)_p.$$

6.5. Derive an expression for the dependence of the enthalpy on the pressure $(\partial H/\partial p)_T$, for (a) a perfect gas, (b) a van der Waals gas. Estimate its value for 1.0 mol of argon at 10 atm and $\mathcal{T}$. By how much does the enthalpy of the argon change when the pressure is changed to 11 atm without change of temperature?

6.6. Deviations from perfect behaviour are often listed in terms of the virial equation of state $pV_m/RT = 1 + B/V_m + \ldots$. Show that the important quantity $(\partial U/\partial V)_T$ can be obtained from tables of the coefficient B as a function of temperature, and if $\Delta B = B(T_2) - B(T_1)$ and $\Delta T = T_2 - T_1$, that $(\partial U/\partial V)_T \approx (RT^2/V_m^2)(\Delta B/\Delta T)$, where T is the mean of T_1 and T_2. In the case of argon, $B = -28.0$ cm^3 mol^{-1} at 250 K and -15.6 cm^3 mol^{-1} at 300 K. Estimate $(\partial U/\partial V)_T$ at 275 K and (a) 1 atm, (b) 10 atm.

6.7. Prove that the heat capacities of a perfect gas are independent of both volume and pressure. May they depend on the temperature?

6.8. Deduce an expression for the volume dependence of $C_{V,m}$ of a gas that is described by $pV_m/RT = 1 + B/V_m$, and calculate the change of molar heat capacity when a sample of argon at 10 atm is reduced isothermally to 1 atm. You will need the following values of the virial coefficients: -15.49 cm^3 mol^{-1} at 298 K, -11.06 cm^3 mol^{-1} at 323 K, and -7.14 cm^3 mol^{-1} at 348 K.

6.9. Confirm the relation used in Problem 3.10 that the Joule–Thomson coefficient for a gas is given by $\mu_{JT}C_p = T(\partial V/\partial T)_p - V$, and show that it may be expressed in terms of the isobaric expansivity α as $\mu_{JT}C_p = V(\alpha T - 1)$.

6.10. The *Joule coefficient* for the free expansion of a gas is defined by $\mu_J = (\partial T/\partial V)_U$. Show that it may be expressed in terms of C_V, α, and κ the isothermal compressibility.

6.11. Evaluate $(\partial U/\partial V)_T$ for a van der Waals gas, and justify the form of the expression obtained. Evaluate $(\partial U/\partial V)_T$ in the limit $V \to \infty$. Calculate the value of the coefficient for argon gas at (a) 1.0 atm, (b) 10.0 atm.

6.12. Derive an expression for $(\partial U/\partial V)_T$ for a Dieterici gas, and express the results in reduced variables by defining the reduced internal energy by $U = (a/e^2b)U_r$.

6.13. We have met the isothermal compressibility κ. If compression takes place reversibly and adiabatically the relevant quantity is the *adiabatic compressibility* κ_S. Why is the subscript S? Show that $p\gamma\kappa_S = 1$ for a perfect gas, where $\gamma = C_p/C_V$.

6.14. A very useful expression for $T\,dS$ can be deduced on the basis of the Maxwell relations and the properties of partial derivatives, and so this Problem has the double job of giving some experience in these manipulations and at the same time finding a valuable result. The work is spread over this and the next few Problems. First, show that $T\,dS = C_V\,dT + T(\partial p/\partial T)_V\,dV$ by starting at the statement 'S may be regarded as a function of T and V'. Hence calculate how much heat has to be transferred per mole of a van der Waals gas when it is expanded reversibly and isothermally from V_i to V_f.

6.15. An alternative expression for $T\,dS$ can be derived on the basis that S is a function of T and p. Show that $T\,dS$ may be written in terms of $C_p\,dT$ and dp. Hence find the heat transferred during the reversible, isothermal compression of a liquid or solid of constant isobaric expansivity α when the pressure is increased from p_i to p_f. Evaluate the heat transferred during the compression of 100 cm^3 of mercury at 0 °C when the pressure on the sample is increased by 1000 atm. ($\alpha = 1.82 \times 10^{-4}\,K^{-1}$.)

6.16. The properties of materials change markedly when they are subjected to high pressures, and there is now a great deal of interest in their behaviour. In one experiment 100 g of benzene at 298 K was subjected to a pressure that was increased reversibly and isothermally from zero to 4000 atm. Calculate (a) the heat transferred to the sample, (b) the work done, (c) the change in its internal energy. Take $\alpha = 1.24 \times 10^{-3}\,K^{-1}$, $\rho = 0.879\,g\,cm^{-3}$, $\kappa = 9.6 \times 10^{-5}\,atm^{-1}$. Assume that $V \approx V_i$ in the integrations.

6.17. Assuming that water is incompressible, estimate the change in Gibbs function of 100 cm^3 of water at 25 °C when the pressure is changed from 1 atm to 100 atm.

6.18. In fact the volume of a sample does change when it is subjected to pressure, and so we ought to see how to take the effect into account, and then judge whether it is significant. The volume varies in a way that can be determined by specifying the isothermal compressibility $\kappa = -(1/V)(\partial V/\partial p)_T$, and we may assume that κ is virtually constant over a pressure range of interest. (That is still an approximation, but it is a better approximation than assuming that the volume itself is constant.) Deduce an expression for the Gibbs function at p_f in terms of its value at p_i, the original volume of the sample, and the compressibility.

6.19. Now we can estimate the change of G with pressure, and judge the significance of assuming that the volume remains unchanged. The isothermal compressibility of copper is $0.8 \times 10^{-6}\,atm^{-1}$. What is the change in its molar Gibbs function when it is subjected to an extra pressure of (a) 100 atm, (b) 10 000 atm? What error is introduced by assuming that the block is incompressible? (The density of copper is 8.93 g cm^{-3}.)

6.20. Repeat the calculation in Problem 6.19, but for water. The isothermal compressibility is $4.94 \times 10^{-5}\,atm^{-1}$ at 25 °C and the density is 0.997 g cm^{-3}.

6.21. Much simpler expressions generally result when approximations are made, but it is always important to know whether they are plausible. In order to find the temperature dependence of the Gibbs function of a reaction it is sometimes supposed that ΔH is independent of temperature. In fact we know that ΔH is temperature dependent, and Kirchhoff's law is that $\Delta H(T) = \Delta H(T_i) + (T - T_i)\Delta C_p$. Find an approximate expression for $\Delta G(T_i)$ on the basis that ΔH is independent of temperature. Then take the temperature dependence into account by integrating eqn (6.2.5) with the Kirchhoff expression for $\Delta H(T)$. Assume that ΔC_p is independent of temperature.

6.22. The full expression just found is much more inconvenient than the simplification; but how important are the additional terms? The standard enthalpy of the reaction $H_2 + \frac{1}{2}O_2 \to H_2O$ is -285.8 kJ mol^{-1} at 298 K, and the heat capacities are H_2: 28.8 J K^{-1} mol^{-1}; O_2: 29.4 J K^{-1} mol^{-1}, and H_2O(l): 75.3 J K^{-1} mol^{-1}. The Gibbs function for the reaction is -237.2 kJ mol^{-1} at 298 K; what is its value at 330 K? What is the error arising from assuming that ΔH is independent of temperature?

6.23. The standard enthalpy of the reaction $N_2 + 3H_2 \to 2NH_3$ is -184.4 kJ mol^{-1} and the standard Gibbs function is -62.0 kJ mol^{-1}, both at 298 K. Estimate the standard Gibbs function at (a) 500 K, (b) 100 K. Is the reaction spontaneous at room temperature? Is the formation of ammonia favoured or disfavoured by a rise in temperature? Assume $\Delta_f H^\ominus$ to be independent of temperature.

6.24. At 25 °C the standard enthalpy of combustion of sucrose is -5645 kJ mol^{-1} and the Gibbs function -5797 kJ mol^{-1}. What is the extra amount of non-p,V work that can be obtained per mole by raising the temperature from 25 °C to 35 °C?

6.25. Calculate the change in Gibbs function of 1.0 mol of hydrogen regarded as a perfect gas, when it is compressed isothermally from 1.0 atm to 100 atm at 298 K.

6.26. The volume of a newly synthesized plastic was found to depend exponentially on the pressure, it being V_0 when there was no excess pressure, and decreased towards zero as $V(p) = V_0 \exp(-\bar{p}/p^*)$, where $\bar{p}$ is the excess pressure and p^* some constant. Deduce an expression for the Gibbs function of the plastic as a function of excess pressure. What is the natural direction of change of the compressed material?

6.27. The *fugacity f* of a gas at a pressure p is given by eqn (6.2.14). In order to evaluate it we have to know how the compression factor varies with pressure all the way up to the pressure of interest. This information can be given in several ways, and the next few Problems give some examples. One way of arriving at the fugacity is to measure the compression factor Z over a range of pressures, and then to do the integration numerically or graphically. Find the fugacity of oxygen at 100 atm and 200 K from the following data:

p/atm	1	4	7	10
Z	0.997 01	0.987 96	0.978 80	0.969 56
p/atm	40	70	100	
Z	0.8734	0.7764	0.6871	

6.28. At such low temperatures and high pressures as in the last Problem, it is not surprising that the fugacity differs markedly from the pressure. When the conditions are not so extreme some of the equations of state may be used and the integration performed explicitly. For instance, suppose that the gas obeys the virial equation of state, $pV_m/RT = 1 + B/V_m + C/V_m^2$. Find an expression for the fugacity and $\ln \gamma$ in terms of a power expansion in the pressure (Problem 1.27).

6.29. Use the expression derived in Problem 6.28 to estimate the fugacity of argon at 1.00 atm pressure at (a) 100 K, (b) 273 K from the following data. At 273 K: $B = -21.13\ \text{cm}^3\ \text{mol}^{-1}$, $C = 1054\ \text{cm}^6\ \text{mol}^{-2}$; at 373 K: $B = -3.89\ \text{cm}^3\ \text{mol}^{-1}$, $C = 918\ \text{cm}^6\ \text{mol}^{-2}$.

6.30. In order to see how the fugacity differs from the pressure over a typical range, plot the fugacity of argon at 273 K for the range 0–10^3 atm pressure, and on the same graph plot the fugacity of a perfect gas. At what pressure does argon have unit fugacity at 273 K?

6.31. The second virial coefficient of ammonia has the value $-261\ \text{cm}^3\ \text{mol}^{-1}$ at 298 K. What is its fugacity when its pressure is (a) 1 atm, (b) 100 atm? Use the condition $(C - B^2/2R^2T^2)p^2 \ll Bp/RT$.

6.32. A third way of arriving at the fugacity is to use one of the approximate equations of state introduced in Chapter 1. Derive an expression for the fugacity of a gas that obeys the equation of state $pV_m/RT = 1 + BT/V_m$ and plot γ against $4pB/R$.

7

Changes of state: physical transformations of pure materials

Learning objectives

After careful study of this chapter you should be able to:

(1) State the general condition for phase equilibrium in terms of chemical potentials, Section 7.1.

(2) State how the chemical potential depends on temperature, eqn (7.1.1) and on pressure, eqn (7.1.2).

(3) Define *normal* and *standard* melting and boiling points, Section 7.1.

(4) Define and explain the significance of *vapour pressure*, Section 7.2 and Example 7.1.

(5) Derive the *Clapeyron equation*, eqn (7.2.1), for the slope of a phase boundary, and use it to deduce an approximate expression for the equation of the solid–liquid boundary curve, eqn (7.2.2).

(6) Derive and use the *Clausius–Clapeyron equation*, eqn (7.2.4), for the temperature dependence of the vapour pressure.

(7) Draw and interpret the *phase diagrams* for water, carbon dioxide, carbon, and helium, Section 7.3.

(8) Define and state the characteristic properties of *first-* and *second-order phase transitions*, and state the characteristics of a *λ-transition*, Section 7.4.

(9) Define the *coefficient of performance* of a refrigerator, and derive an expression for its maximum value, eqn (7.5.1).

(10) Derive and use an expression for the power needed by a refrigerator to maintain low temperatures, eqn (7.5.2).

(11) Explain the operation of *a heat pump*, Section 7.5(a).

(12) Explain the basis of the *thermodynamic temperature scale*, Section 7.5(a).

(13) Describe the principles of the technique of *adiabatic demagnetization*, Section 7.5(b).

(14) Define *surface tension*, eqn (7.6.1), and relate changes in the Helmholtz function to changes in surface area, eqn (7.6.2).

(15) Derive the *Laplace equation*, eqn (7.6.3), for the pressure difference across a curved surface, and use it to derive the *Kelvin equation*, eqn (7.6.5), for the vapour pressure of droplets of liquid.

(16) Indicate the importance of *nucleation*, Section 7.6(b).

(17) Use the Laplace equation to derive an expression for the height to which a liquid can rise by capillary action, eqn (7.6.7).

Introduction

Boiling, freezing, and the onset of ferromagnetism are all examples of changes of phase without change of composition. This chapter describes such processes thermodynamically. The guiding principle is the tendency of systems at constant temperature and pressure to shift to lower Gibbs function. Since we are dealing with pure substances the molar Gibbs function of the system is the same as the chemical potential, and so the tendency to change is in the direction of decreasing chemical potential. It must never be forgotten, however, that these alternative ways of expressing the direction all stem from the tendency of the universe to greater disorder (i.e. for the universe to change in the direction of increasing entropy), and although a *system* may take on greater order, such as when it freezes, there is a compensating increase of entropy in the surroundings (as a result of the heat released into them) and the entropy of the universe is greater overall. The Gibbs function and the chemical potential are this total entropy in disguise.

If at a particular pressure and temperature a vapour has a lower chemical potential than the corresponding liquid, then the latter tends to evaporate. If a solid has a lower chemical potential than its corresponding liquid, then the solid is stable under those conditions and the liquid has a natural tendency to freeze. Note, however, that we are dealing with the thermodynamics of phase transitions, not their rates. This distinction is of considerable practical importance. For instance, at normal temperatures and pressures graphite has a lower chemical potential than diamond, and so there is a thermodynamic tendency for diamonds to crumble into graphite. But, in order for this to take place, the atoms have to change their locations and form a new crystal lattice, and this is an unmeasurably slow process except at high temperatures. The rate of attainment of equilibrium is a *kinetic* problem, and is outside the range of thermodynamics. In gases, liquids, and liquid solutions the mobilities of the molecules allow equilibrium to be reached rapidly, but in solids thermodynamic instability may be frozen in.

Initially we assume that the properties of the surface of the sample have no effect on the properties of the bulk phase, but we take the surface into account in Section 7.6.

7.1 The stabilities of phases

We base our discussion on the following principle:

> *The chemical potential of a sample is uniform when it is at equilibrium regardless of how many phases are present.*

The argument runs as follows. Consider a system in which the chemical potential is not uniform, and suppose that at one point its value is $\mu(1)$ and at another $\mu(2)$. When an amount of substance $\mathrm{d}n$ is transferred from one point to the other the Gibbs function changes by $-\mu(1)\,\mathrm{d}n$ in the first step and by $+\mu(2)\,\mathrm{d}n$ in the second. The overall change is therefore $\mathrm{d}G = \{\mu(2) - \mu(1)\}\,\mathrm{d}n$. If the chemical potential at 1 is higher than at 2 the transfer is accompanied by a decrease in the Gibbs function, and so it has a spontaneous tendency to occur. Only if $\mu(1) = \mu(2)$ is there no change in G, and so only then is the system at internal equilibrium. The argument applies when the transfer of matter occurs between different phases as well as between different locations within a single phase. Therefore, when a liquid and a solid are in equilibrium, the chemical potential is the same throughout the liquid and throughout the solid, and is the same in the solid as in the liquid.

The chemical potentials of the solid, liquid, and gas (vapour) phase of a substance are denoted $\mu(\mathrm{s})$, $\mu(\mathrm{l})$, and $\mu(\mathrm{g})$ respectively. In Chapter 6, eqn

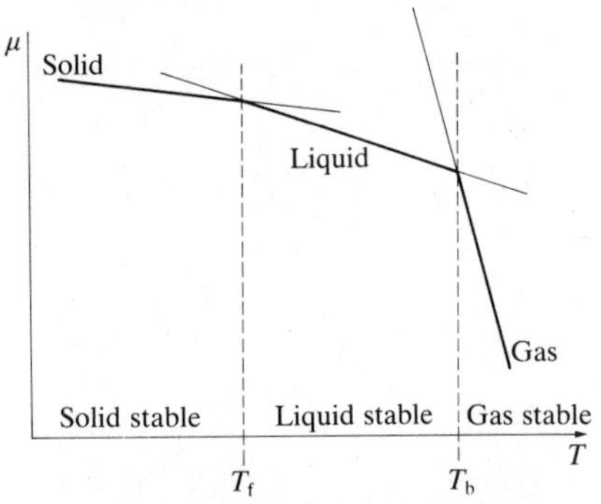

Fig. 7.1. The temperature dependence of the chemical potential of the solid, liquid, and gas phases of a substance. The slopes are determined by the different entropies of the phases. The phase with the lowest μ at a specified temperature is the most stable one at that temperature.

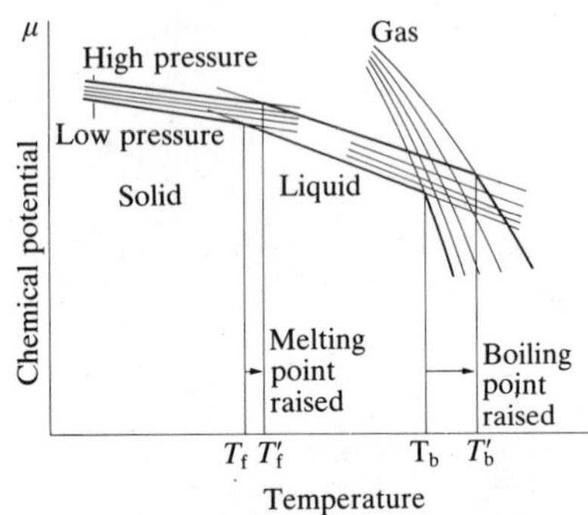

Fig. 7.2. The pressure dependence of the chemical potential of a substance depends on the molar volume of the phase. The curves show the effect of increasing pressure on the chemical potentials of the solid, liquid, and gas phases, and the corresponding effects on the melting and boiling temperatures.

(6.2.2), we saw that the temperature dependence of the chemical potential is

$$(\partial\mu/\partial T)_p = -S_m. \tag{7.1.1}$$

This implies that as the temperature is raised the chemical potential of a pure substance decreases (because S_m is always positive), and that the slope is steeper for gases than for liquids (because $S_m(g) > S_m(l)$), and steeper for a liquid than the corresponding solid (because $S_m(l) > S_m(s)$). These deductions let us draw the general features of the temperature-dependence of the chemical potentials of three phases of a pure system, Fig. 7.1.

A phase is thermodynamically stable over the range of temperature at which it has a lower chemical potential than any other phase. At low temperatures the solid phase is stable, but the greater negative slope of the liquid's potential results in it falling beneath the chemical potential of the solid, and so the liquid becomes the stable phase at a high enough temperature. The chemical potential of the vapour plunges downwards as the temperature is raised, and there comes a temperature at which it lies lowest. Then the vapour is the stable phase. The temperature at which, under a specified pressure, liquid and solid exist in equilibrium (i.e. have the same chemical potential) is called the *melting temperature* or *freezing temperature* (except in some very bizarre systems the two are identical). The melting temperature when the pressure is 1 atm is called the *normal melting point* (symbol: T_f). The temperature at which, under a specified pressure, the liquid and vapour are in equilibrium is called the *boiling temperature*. When the pressure is 1 atm it is called the *normal boiling point* (symbol: T_b). With the replacement of 1 atm by 1 bar as standard pressure, there is some advantage in modifying the definitions so that the transition temperatures refer to that pressure. We shall therefore use the terms *standard melting point* and *standard boiling point* when the phase equilibria relate to 1 bar. Fortunately, melting points change only slightly (e.g. the normal boiling point of water is 100.0 °C, the standard boiling point is 99.6 °C). As the chapter goes on, we shall see how to convert one to the other.

Figure 7.1 can be adapted to show the pressure dependence of the boiling and freezing temperatures. The chemical potential depends on pressure as

$$(\partial\mu/\partial p)_T = V_m. \tag{7.1.2}$$

(This is the second part of eqn (6.2.2).) An increase of pressure raises the chemical potential of a pure substance (because V is positive) and increases it much more for gases than for either solids or liquids (because typically $V_m(g) \approx 1000V_m(l)$). In most cases the volume of a solid increases on melting, and so at the melting temperature the molar volume of a liquid is usually greater than that of the solid. In these cases the equation predicts that an increase in pressure increases the chemical potential of the liquid slightly more than that of the solid. These responses to pressure are shown in Fig. 7.2. The effect is to raise the boiling temperature sharply, and also to raise the melting temperature, but only slightly.

7.2 Phase equilibria and phase diagrams

Figure 7.3 shows in a general way the regions of pressure and temperature where each phase is the most stable. The boundaries between regions lie at the values of p and T where two phases coexist. Since the phases are then in equilibrium, their chemical potentials are equal. Therefore, where the phases α and β are in equilibrium $\mu_\alpha(p, T) = \mu_\beta(p, T)$. This is an equation relating the values of p and T at which there is equilibrium between the two phases, and so by solving it for p as a function of T we shall have an equation for the boundary curve, the *phase boundary*. The pressure of a vapour in equilibrium with its condensed phase at a specified temperature is called the *vapour pressure* (VP) of the condensed phase, and so in the case of the liquid/vapour and solid/vapour phase boundaries the p versus T curves show the temperature-dependences of the vapour pressures of the liquid and the solid.

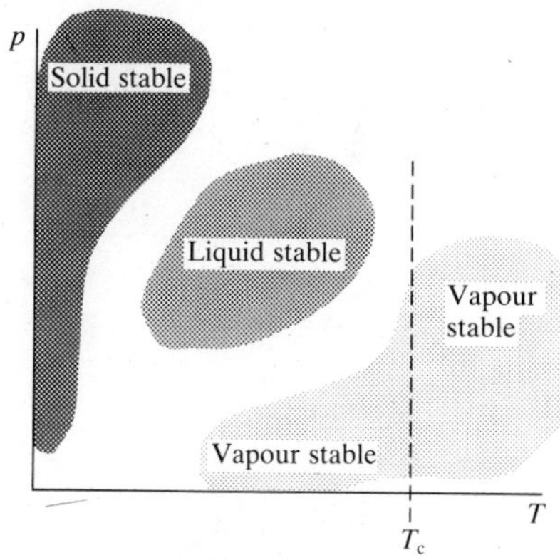

Fig. 7.3. The general regions of pressure and temperature where solid, liquid, or gas is stable (i.e. has minimum chemical potential). For example, the solid phase is the most stable phase at low temperatures and high pressures. In the following paragraphs we locate the precise boundaries between the regions.

Example 7.1

The practical significance of the VP can be appreciated by considering the system shown in Fig. 7.4(a). The vessel contains water at room temperature, and the pressure exerted by the piston can be controlled (e.g. by adding weights). Suppose the pressure exerted is 1 atm, and that we monitor the volume enclosed by the piston as the water is heated. The volume increases slightly until its temperature reaches 100 °C. At that point there is a large increase in volume, and although heating may continue the temperature does not increase until all the liquid has evaporated. While both liquid and vapour are present the temperature stays at 100 °C, and the proportions of the two phases can be modified arbitrarily either by cooling (the removal of energy as heat, which causes condensation, the temperature remaining at 100 °C) or by heating (which causes evaporation at 100 °C). At this temperature (100 °C) and pressure (1 atm) the liquid and its vapour are at equilibrium, and we report that the VP is 1 atm at 100 °C. When all the liquid has evaporated, further heating raises the temperature of the vapour, and it expands.

If weights are added to the piston sufficient to make it exert a pressure of 1.5 atm, the same kind of equilibrium is reached at 112 °C. Therefore, we report that the VP of water at 112 °C is 1.5 atm.

The same experiment can be repeated for any applied pressure, and the *VP curve* can be drawn as in Fig. 7.4(b). Note that we can use this curve to predict the boiling temperature of water at any pressure. For example, at an altitude of 10 000 ft, where the atmospheric pressure is 0.7 atm, water boils at only 90 °C.

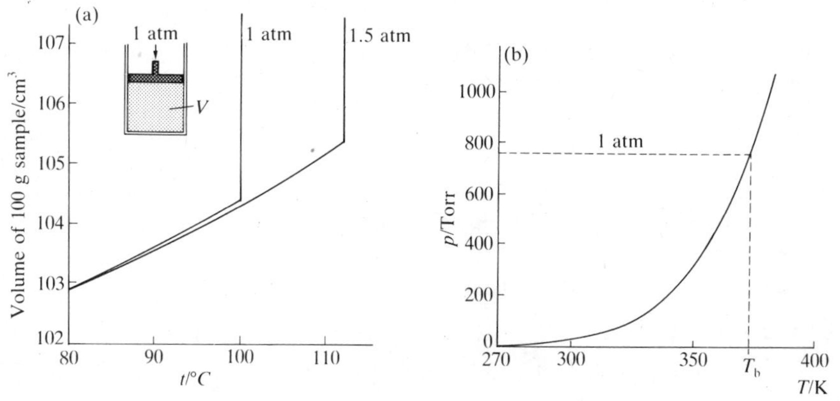

Fig. 7.4. A schematic illustration of the significance of the vapour pressure (VP) of water. (a) The volume of a 100 g sample under a pressure of 1.0 atm increases sharply at 100 °C, and under those conditions (so long as liquid and vapour are both present) can be controlled at will by the location of the piston because the liquid and vapour are then in equilibrium. When the pressure is 1.5 atm, the same equilibrium is reached at 112 °C. (b) The plot of the equilibrium pressure against the corresponding temperature (up to the critical point) gives the *VP curve* for water.

It turns out to be simplest to discuss the phase boundaries in terms of their slopes, and so we begin by finding an equation for $\mathrm{d}p/\mathrm{d}T$.

Let p and T be changed infinitesimally, but in such a way that the two phases, which we label α and β, remain in equilibrium. The chemical potentials of the phases were equal, and they remain equal; therefore the changes in them must be equal and we can write $\mathrm{d}\mu_\alpha = \mathrm{d}\mu_\beta$. Since from eqn

(6.2.1) we know that $d\mu = -S_m\,dT + V_m\,dp$ for each phase, it follows that

$$-S_{\alpha,m}\,dT + V_{\alpha,m}\,dp = -S_{\beta,m}\,dT + V_{\beta,m}\,dp,$$

where $S_{\alpha,m}$ and $S_{\beta,m}$ are the molar entropies of the phases and $V_{\alpha,m}$ and $V_{\beta,m}$ their molar volumes. Hence

$$\{V_{\beta,m} - V_{\alpha,m}\}\,dp = \{S_{\beta,m} - S_{\alpha,m}\}\,dT,$$

which rearranges into the

$$\textit{Clapeyron equation: } dp/dT = \Delta S_m/\Delta V_m, \qquad (7.2.1)$$

where $\Delta S_m = S_{\beta,m} - S_{\alpha,m}$ and $\Delta V_m = V_{\beta,m} - V_{\alpha,m}$ are the changes of molar entropy and molar volume when the transition occurs. This important result for the slope of the phase boundary at any point is *exact* and applies to any phase transition of a pure substance.

7.2 (a) The solid–liquid boundary

Melting involves a molar enthalpy $\Delta H_{melt,m}$ and occurs at a temperature T. The molar entropy of melting at T is therefore $\Delta H_{melt,m}/T$, and so the Clapeyron equation reads

$$dp/dT = \Delta H_{melt,m}/T_f\,\Delta V_{melt,m},$$

where $\Delta V_{melt,m}$ is the molar volume change on melting. The enthalpy of melting is positive, and the volume change is usually positive but always small. This means that the slope dp/dT is steep and usually positive. The curve itself can be obtained by integrating dp/dT on the basis of the assumption that ΔH_{melt} and ΔV_{melt} are insensitive to temperature and pressure. If the melting temperature is T^* when the pressure is p^*, and is T when the pressure is p, then the integration required is

$$\int_{p^*}^{p} dp = (\Delta H_{melt,m}/\Delta V_{melt,m})\int_{T^*}^{T}(1/T)\,dT.$$

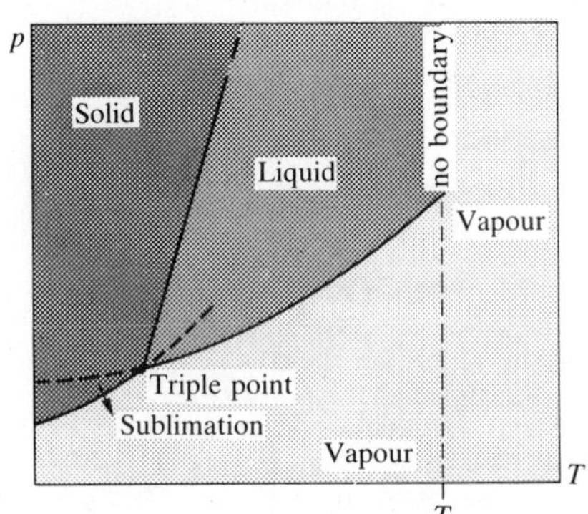

Fig. 7.5. The general location of the phase boundaries between solid, liquid, and vapour. On the boundaries the respective phases are in equilibrium. All three phases are in equilibrium at the *triple point*. There is no phase boundary between the liquid and the vapour above the critical temperature (as explained in Chapter 1).

Therefore, the *approximate* equation of the solid–liquid boundary curve is

$$p = p^* + (\Delta H_{melt,m}/\Delta V_{melt,m})\ln(T/T^*). \qquad (7.2.2)$$

This equation was originally obtained by yet another Thomson—this time James, the brother of Lord Kelvin. When T is close to T^* the logarithm can be approximated using $\ln(T/T^*) = \ln\{1 + (T - T^*)/T^*\} \approx (T - T^*)/T^*$, because $\ln(1 + x) \approx x$ when x is small; then

$$p = p^* + (\Delta H_{melt,m}/T^*\Delta V_{melt,m})(T - T^*), \qquad (7.2.3)$$

which is the equation of a steep straight line, Fig. 7.5.

Example 7.2

Construct the ice/liquid phase boundary for water at temperatures between −1 °C and +1 °C. What is the melting temperature of ice under a pressure of 1.5 kbar?

• *Method*. Use eqn (7.2.2) for both parts. For the fixed point (p^*, T^*) use the triple point of water (6.11 mbar, 273.16 K), and refer to Table 4.7 for the enthalpy of melting, assuming it to be constant over the pressure and temperature range of interest. Take $\Delta V_m = -1.7\,\text{cm}^3\,\text{mol}^{-1}$ from density measurements at 0 °C, and assume it to be constant. For the second part, take (p^*, T^*) as the normal melting point (1 atm, 273.15 K), and use eqn (7.2.2) to find T for $p = 1.5\,\text{kbar} = 1.5_2 \times 10^3\,\text{atm}$.

• *Answer*. Evaluation of the equation over the stated range gives the following values of p for each melting temperature:

t/°C	−1.0	−0.5	0.0	0.5	1.0
p/bar	131	66	1	−64	−128

This is plotted in Fig. 7.6 as the line EF. For the second part, rearrange eqn (7.2.2) to

$$T = T^* \exp\{(p - p^*)\,\Delta V_{\text{melt,m}}/\Delta H_{\text{melt,m}}\}$$

and obtain

$$T = (273.15\,\text{K}) \exp\left\{\frac{(1.5\times10^3 - 1)\times(1.013\times10^5\,\text{N m}^{-2})\times(-1.7\times10^{-6}\,\text{m}^3\,\text{mol}^{-1})}{6.01\times10^3\,\text{J mol}^{-1}}\right\}$$

$$= (273.15\,\text{K})\text{e}^{-0.043} = 262\,\text{K}, \quad \text{or } -11\,°\text{C}.$$

• *Comment*. Note the *decrease* in melting temperature with increasing pressure: water is denser than ice, and so ice responds to pressure by tending to melt. The negative pressures denote ice under tension.

• *Exercise*. Estimate the value of the difference between the standard melting point of ice and its normal melting point. [About +0.0001 K]

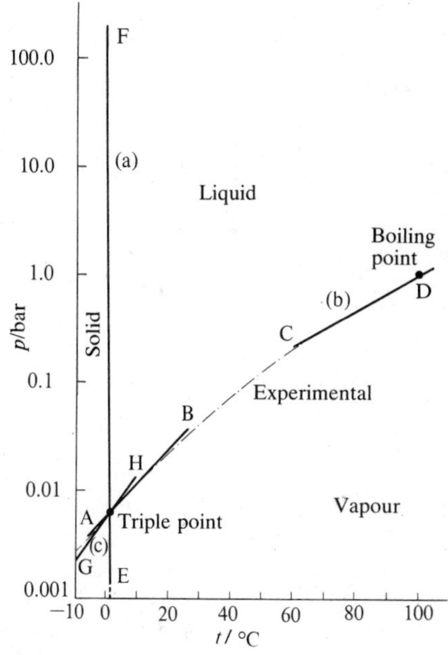

Fig. 7.6. The phase boundaries for water as calculated in *Examples* 7.2, 7.3, and 7.4. Note that the vertical axis is logarithmic, and so it has the effect of squashing down the upper parts of the diagram (compare the lower part of Fig. 7.7, which is on a linear scale).

7.2 (b) The liquid–vapour boundary

The molar entropy of vaporization at a temperature T is equal to $\Delta H_{\text{vap,m}}/T$; the Clapeyron equation for the liquid–vapour boundary is therefore

$$\text{d}p/\text{d}T = \Delta H_{\text{vap,m}}/T\,\Delta V_{\text{vap,m}}.$$

The enthalpy of vaporization is positive; $\Delta V_{\text{vap,m}}$ is large and positive. Therefore $\text{d}p/\text{d}T$ is always positive but it is much smaller than for the solid–liquid boundary. The slope of the boundary increases with pressure because the value of $\Delta V_{\text{vap,m}}$ decreases. This is why the line drawn in Fig. 7.5 curls upwards. The magnitude of the slope at the boiling point can be estimated by taking Trouton's constant ($85\,\text{J K}^{-1}\,\text{mol}^{-1}$, Example 5.4) for the entropy and approximating ΔV_m by the molar volume of a perfect gas (about $30\,\text{dm}^3\,\text{mol}^{-1}$ at 1 atm and near but above room temperature). This gives a slope of 0.04 atm/K, corresponding to a $\text{d}T/\text{d}p$ slope of 25 K/atm, which means that a change of pressure of +0.1 atm can be expected to change a boiling temperature by about +3 K.

Since the molar volume of a gas is so much greater than the molar volume of a liquid we can write $\Delta V_{\text{vap,m}} = V_m(\text{g}) - V_m(\text{l}) \approx V_m(\text{g})$. If we also assume that the gas behaves perfectly, then $V_m(\text{g}) = RT/p$. These approximations turn the exact Clapeyron equation into the

Clausius–Clapeyron equation: $\text{d}\ln p/\text{d}T = \Delta H_{\text{vap,m}}/RT^2$. (7.2.4)°

(We have used $\text{d}x/x = \text{d}\ln x$.) If we also assume that the enthalpy of vaporization is independent of temperature this equation integrates to

$$p = p^*\text{e}^{-C}, \qquad C = (\Delta H_{\text{vap,m}}/R)\{(1/T) - (1/T^*)\}, \qquad (7.2.5)°$$

where p^* is the VP when the temperature is T^* and p its value when the temperature is T. This is the curve plotted as the liquid–vapour boundary in Fig. 7.5. The line does not extend beyond the critical temperature T_c because above this temperature the liquid does not exist (Section 1.3).

Example 7.3

Construct the VP curve for water between −5 °C and 100 °C.

• *Method*. Use eqn (7.2.5) taking as the fixed point (p^*, T^*) the triple point (6.11 mbar, 273.16 K) for temperatures nearby and as the normal boiling point (1.01 bar, 373.15 K) for temperatures in its vicinity. Use $\Delta H_{vap,m}(273\text{ K}) = +45.05\text{ kJ mol}^{-1}$ and $\Delta H_{vap,m}(373\text{ K}) = +40.66\text{ kJ mol}^{-1}$ for the respective ranges.

• *Answer*. Evaluation of eqn (7.2.5) for several temperatures gives the following values:

t/°C	−5	0	5	10	20	30	...	70	80	90	100
p/atm	0.004	0.006	0.008	0.012	0.023	0.042	...	0.32	0.48	0.70	1.0

The curve is plotted as AB and CD in Fig. 7.6, and compared there with the experimental VP.

• *Comment*. The negative curvature of the curve as plotted comes from the use of a logarithmic scale. Note that we can use eqn (7.2.5) in conjunction with experimental VP data to determine enthalpies of vaporization.

• *Exercise*. Calculate the standard boiling point of water from its normal boiling point. [99.6 °C]

7.2 (c) The solid–vapour boundary

The only difference between this case and the last is the replacement of the enthalpy of vaporization by the enthalpy of sublimation, ΔH_{sub}. The approximations that led to the Clausius–Clapeyron equation give the following expressions for the temperature dependence of the *sublimation vapour pressure*:

$$\mathrm{d}\ln p/\mathrm{d}T = \Delta H_{sub,m}/RT^2, \qquad (7.2.6a)^\circ$$

$$p = p^* e^{-C'}, \qquad C' = (\Delta H_{sub,m}/R)\{(1/T) - (1/T^*)\}. \qquad (7.2.6b)^\circ$$

Since the enthalpy of sublimation is greater than the enthalpy of vaporization, the equation predicts a steeper slope for the sublimation curve than for the vaporization curve near where they meet, Fig. 7.5.

Example 7.4

Construct the ice/vapour phase boundary over the range −10 °C to +5 °C, using the information that at 273 K $\Delta H_{vap,m} = +45.05\text{ kJ mol}^{-1}$ and $\Delta H_{melt,m} = +6.01\text{ kJ mol}^{-1}$.

• *Method*. In order to use eqn (7.2.6b) we need to know the molar enthalpy of sublimation. From the First Law, H being a state function, we can write $\Delta H_{sub,m} = \Delta H_{melt,m} + \Delta H_{vap,m}$.

• *Answer*. The sublimation VP at several temperatures is as follows:

t/°C	−10	−5	0	5
p/bar	0.003	0.004	0.006	0.009

These points are plotted as GH in Fig. 7.6 and compared with the experimental curve.

• *Comment*. The small discrepancy between the experimental and calculated curves is a result of assuming that the enthalpy of sublimation is a constant.

• *Exercise*. Construct the phase diagram for carbon dioxide using the data given on the right. [Fig. 7.8]

Density of solid:	1.53 g cm^{-3}
Density of liquid:	0.78 g cm^{-3}
$\Delta H_{sub,m}$:	2.5 kJ mol^{-1}
$\Delta H_{melt,m}$:	8.3 kJ mol^{-1}
Triple point:	−57 °C, 5.2 bar
Critical constants:	31 °C, 74 bar

7.2 (d) The solid–liquid–vapour equilibrium

There is generally some set of conditions under which the solid, liquid, and vapour phases all simultaneously coexist in equilibrium. It is given by the values of p and T for which all three chemical potentials are equal. Geometrically it occurs at the intersection of the three boundary curves, Fig. 7.5: this is called the *triple point*. It is very important to note that the position of the triple point of a pure substance is completely outside our control. It occurs at a single definite pressure and temperature characteristic of the substance. In the case of water it lies at 273.16 K and 6.11 mbar (4.58 Torr). The three phases of water coexist in equilibrium at no other combination of pressure and temperature. This invariance is the basis of the use of the triple point in the definition of the Kelvin temperature scale (Section 1.1): experimentally it is found that the normal melting point of water lies 0.0100 K lower than the temperature of the triple point.

7.3 Four real systems

The experimental phase diagram for *water* is drawn in Fig. 7.7. The liquid–vapour line summarizes the temperature dependence of the vapour pressure; equally, it summarizes the pressure dependence of the boiling temperature. The solid–liquid curve shows how the melting temperature depends on the pressure, and indicates that enormous pressures are needed to bring about significant changes. Notice that the line slopes backwards (has a negative slope), which means that the melting temperature falls as the pressure is raised. The thermodynamic reason for this is that ice contracts on melting, and so ΔV is negative (as we saw in the discussion of the construction of Fig. 7.6). The molecular reason for the decrease of volume on melting is the very open structure of the ice crystal lattice: the water molecules are held apart (as well as together) by the hydrogen bonds, but the structure partially collapses on melting and the liquid is denser than the solid. A practical consequence is one mechanism for the advance of glaciers: where ice is pressed against the sharp edges of stones and rocks it melts, and the glacier inches forwards.

At high pressures other phases come into stability as the bonds between water molecules are modified. Some of these ice forms (which are called ice-II, III, V, VI, and VII; ice-IV was an illusion, like the once-fashionable alternative liquid phase called 'polywater') melt at high temperatures. For example, ice-VII melts at 100 °C, but exists only when the pressure exceeds 25 kbar.

The phase diagram for *carbon dioxide* is shown in Fig. 7.8. The features to notice include the 'orthodox' slope of the solid–liquid boundary, which indicates that the melting temperature of solid carbon dioxide rises as the pressure is increased. Notice also that at 1 atm pressure the liquid–vapour equilibrium cannot be established at any temperature. This means that the solid sublimes when left in the open. In order to obtain the liquid it is necessary to exert a pressure of at least 5.11 atm. Cylinders of carbon

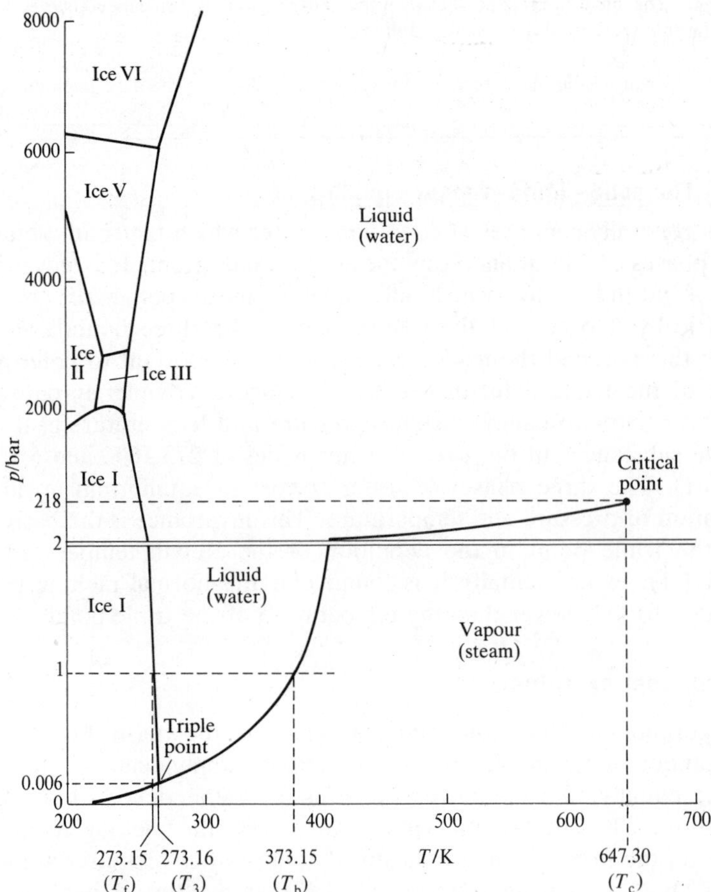

Fig. 7.7. The experimental phase diagram for water showing the different solid phases. Note the change of vertical scale at 2 bar.

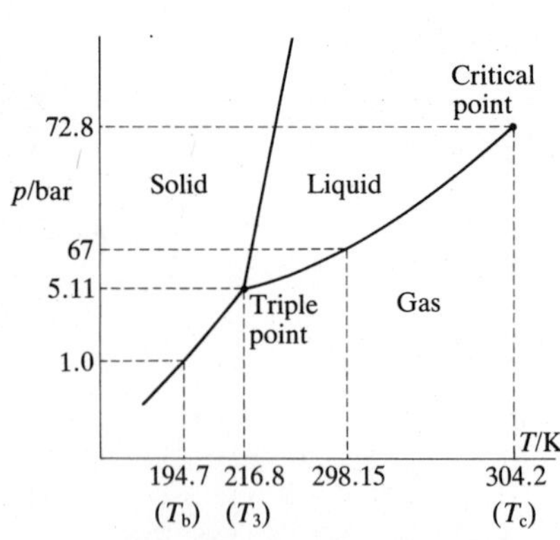

Fig. 7.8. The experimental phase diagram for carbon dioxide. Note that, as the triple point lies at pressures well above atmospheric, liquid carbon dioxide does not exist under normal conditions (a pressure of at least 5.1 bar must be applied).

dioxide generally contain the liquid; at room temperature that implies a vapour pressure of 67 atm. When the vapour squirts through the nozzle it cools by the Joule–Thomson effect, but because it emerges into a region where the pressure is only 1 atm, it condenses into a finely divided snowlike solid.

The phase diagram of *carbon* is shown in Fig. 7.9. It is ill-defined because the various phases come into stability at extremes of temperature and pressure, and gathering the data is very difficult. For instance, at atmospheric pressure carbon gas is the stable phase only at temperatures well over 4000 K. To get liquid carbon it is necessary to work at about 4500 K and 1 kbar (1000 atm), or at 2000 K and 1000 kbar. Making diamonds is a minor problem in comparison, because the diamond phase comes into stability at 10^4 atm and 1000 K. Small diamonds are synthesized and are widely used in industry, but the phase diagram does not reveal the full problem. The rate of conversion is an important factor, and pure graphite changes into diamond at a useful rate only when the temperature is around 4000 K and the pressure exceeds 200 kbar; but then the apparatus tends to disappear first. Therefore catalysts are added in commercial syntheses, and then the conversion proceeds at 70 kbar and 2300 K, which are attainable conditions. The contamination by the metal catalysts such as molten nickel (which also acts as a solvent for the carbon) enables commercial and natural diamonds to be distinguished.

The phase diagram of *helium* is shown in Fig. 7.10. Helium behaves unusually at low temperatures. For instance, the solid and gas phases are never in equilibrium however low the temperature. This is because the helium atoms are so light that they vibrate with a large-amplitude motion even at very low temperatures. and the solid simply shakes itself apart. Solid helium can be obtained, but only by holding the atoms together by applying pressure. A second peculiarity is that pure ^{4}He has a liquid–liquid phase transition at its λ-*line* (the reason for this name will be explained in the next Section). The liquid phase marked He-I behaves like a normal liquid. the other phase, He-II, is a *superfluid*, so called because it flows without viscosity. ^{3}He has a phase diagram quite similar to ^{4}He, but until recently it was thought that it lacked the λ-line and had only a normal liquid phase. Recent observations, however, have shown that there is a superfluid phase, and evidence is still being accumulated.

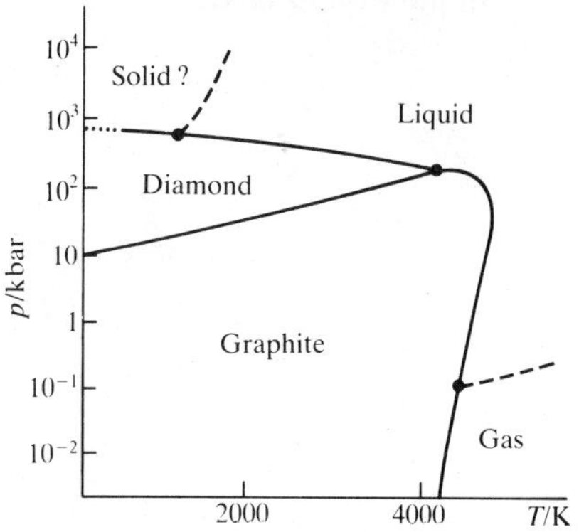

Fig. 7.9. The phase diagram for carbon. Note that the pressure axis is logarithmic and in kilobars (1 kbar = 10^3 bar, about 1000 atm). There are very large uncertainties about the precise form of this phase diagram because the data are so difficult to obtain.

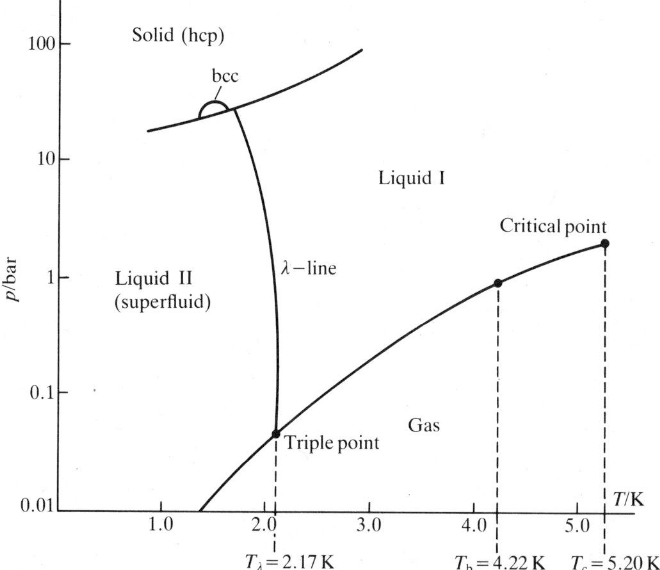

Fig. 7.10. The phase diagram for helium (^{4}He). The λ-line marks the conditions under which the two liquid phases are in equilibrium; He-II is the superfluid phase. Note that a pressure of over 20 bar must be exerted before solid helium can be obtained.

7.4 The classification of phase transitions

Familiar phase transitions, like melting and boiling, are accompanied by changes of enthalpy and volume. The slopes of the chemical potentials of the two phases are given by eqn (6.2.2), and so, at the transition,

$$(\partial\mu_\beta/\partial p)_T - (\partial\mu_\alpha/\partial p)_T = (\partial\,\Delta\mu/\partial p)_T = \Delta V_{\mathrm{m}} \neq 0,$$
$$(\partial\mu_\beta/\partial T)_p - (\partial\mu_\alpha/\partial T)_p = (\partial\,\Delta\mu/\partial T)_p = -\Delta S_{\mathrm{m}} = -\Delta H_{\mathrm{m}}/T \neq 0.$$

Since both ΔV_{m} and ΔS_{m} are non-zero, it follows that both slopes are different either side of the transition, Fig. 7.11(a). In other words, *the first derivatives of the chemical potentials are discontinuous at the transition*. This is the origin of the name *first-order phase transition*.

The heat capacity C_p of a substance is the slope of the enthalpy with respect to temperature. At a first-order phase transition H changes by a finite amount for an infinitesimal change of temperature, Fig. 7.11(a). Therefore, at the transition its slope, and therefore the heat capacity, is infinite. The physical reason is that heating drives the transition rather than raising the temperature (e.g. water boiling in a kettle stays at the same temperature even though heat is being supplied). It follows that a first-order phase transition is also characterized by an infinite heat capacity at the transition temperature, Fig. 7.11(a).

A first-order phase transition is characterized by a discontinuous first-derivative of the chemical potential. This suggests that there might exist a *second-order phase transition* in which the first-derivative is continuous but the second-derivative is discontinuous. If the slope of μ is continuous, then the entropy, enthalpy, and volume do not change at the transition, Fig. 7.11(b). Nevertheless, the heat capacity might still change discontinuously at the transition. This is shown in Fig. 7.11(b).

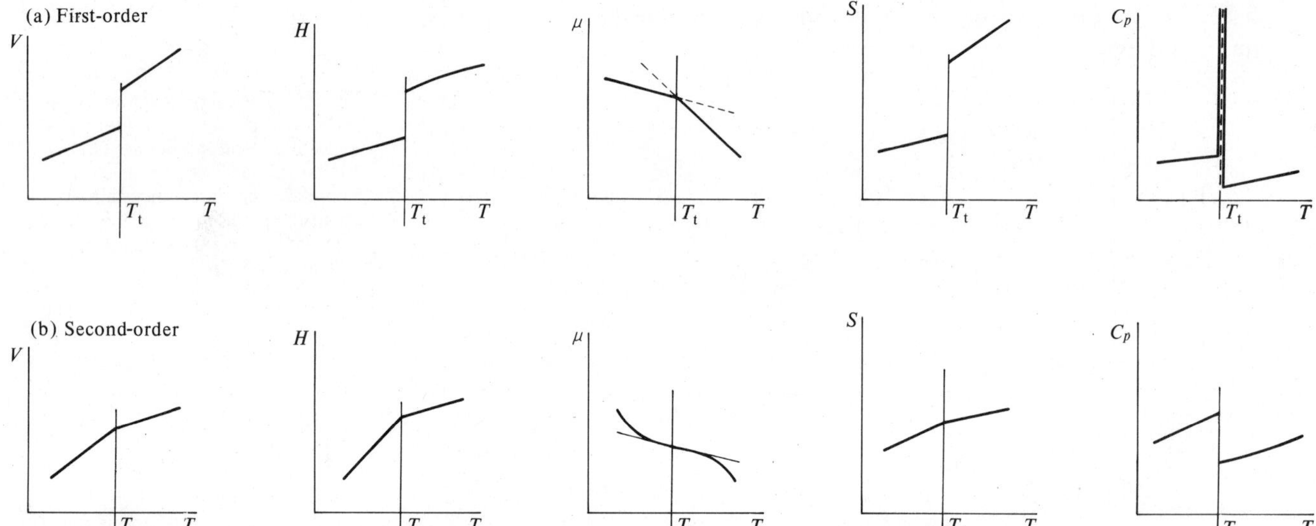

Fig. 7.11. The changes in thermodynamic properties accompanying (a) first-order and (b) second-order phase transitions.

It was once thought that there were many examples of second-order phase transitions, but the closer measurements were made to the transition temperature the greater the value of the heat capacity that was found. Only in the conducting–superconducting transition (when a metal like lead suddenly loses its electrical resistance) does the heat capacity appear to change through a finite step. The transitions that are not first-order yet approach infinite heat capacity include order–disorder transitions in alloys, the onset of ferromagnetism, and the fluid–superfluid transition of helium. Although the heat capacity approaches infinity ('has a singularity'), these transitions differ from first-order processes because the heat capacity does not change abruptly but begins to grow well before the transition: compare Figs 7.11 and 7.12. The shape of the heat capacity curve resembles the Greek letter λ, and so this type of phase change is called a *λ-transition*.

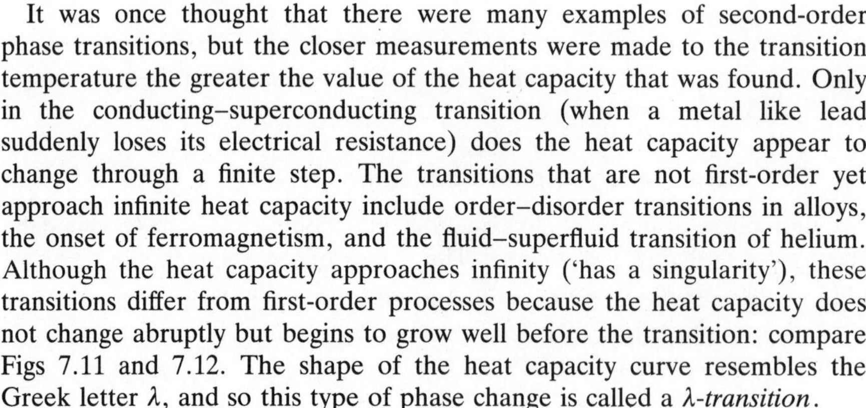

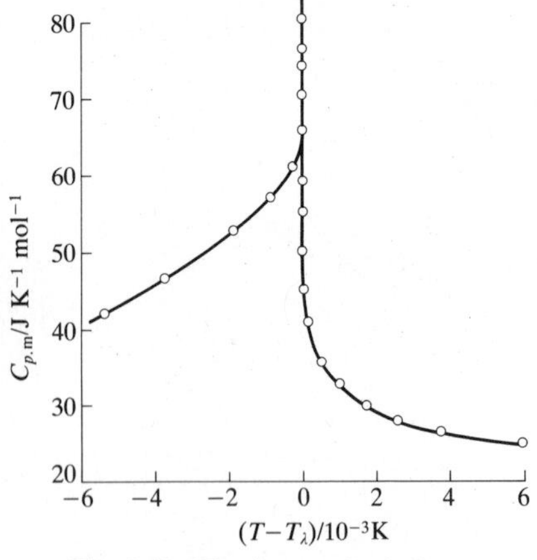

Fig. 7.12. The λ-curve for helium, where the heat capacity rises to infinity (and does not step through a finite range as it would in a second-order transition). The shape of this curve is the origin of the name λ-transition.

7.5 Refrigeration and heat pumps

In order to study some phase transitions, as well as to study the more general properties of matter, it is necessary to cool samples to low temperatures. Gases may be cooled by Joule–Thomson expansion below their inversion temperatures, and temperatures as low as about 4 K (the boiling point of helium) may be reached without great difficulty. Temperatures lower than 4 K can be reached by evaporating liquid helium by pumping rapidly through large diameter pipes: as the helium evaporates it withdraws energy from the object being cooled. Temperatures as low as about 1 K can be reached, but below this temperature the vapour pressure of helium is too low for the process to be effective, and the superfluid phase begins to interfere with the cooling by creeping around the apparatus.

In this section we consider two aspects of cooling. One is the general thermodynamic problem of the cost in terms of work of removing heat from a system using a *refrigerator*; that is, a cyclic engine operated in reverse. The other is the technique required to cool matter to the lowest temperatures yet reached.

7.5 (a) The energetics of refrigeration

Work must be done in order to transfer heat from a system to its cooler surroundings and to go against the natural tendency of change. The general problem, and its solution, are illustrated in Fig. 7.13. When heat is removed from a cool body its entropy is lowered. When the same quantity of heat is released into a hot reservoir that reservoir's entropy increases, but not by enough to cancel the loss, Fig. 7.13(a) (because the temperature is higher, and entropy change is inversely proportional to temperature). Therefore, in order to generate more entropy, the quantity of heat released must be greater than that absorbed. This can be achieved by augmenting the flow of energy by doing work, Fig. 7.13(b). Our task is to find the minimum quantity of work that has to be added to the stream in order for the process to take place. The outcome will be expressed as the *coefficient of performance* $c = q_c/w$, where q_c is the heat to be removed from the cold object and w the work required to bring that removal about. The less the work required, the more efficient the operation, and the greater the coefficient of performance.

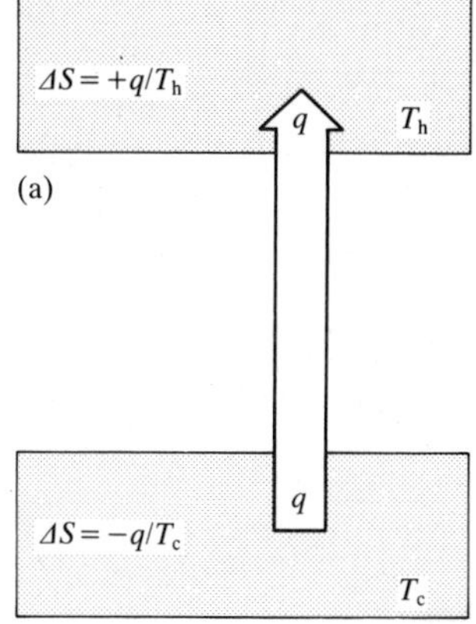

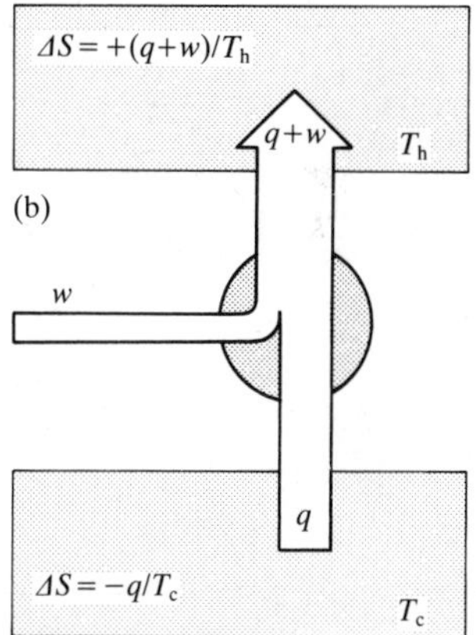

Fig. 7.13. (a) When heat leaves a cool body and enters a hotter one, there is an overall decrease in entropy, and so the process is not spontaneous. (b) If energy is supplied to the stream in the form of work and is dissipated into the hot sink as heat, then just enough extra entropy may be generated there to cancel the reduction in the cold source, and the overall process is spontaneous. The minimum work needed to augment the flow can be calculated as described in the text.

If a quantity of heat q_c is withdrawn from the cold object, and a quantity w of work is done, the quantity of heat to dissipate in the hot reservoir (such as the surrounding room) is $q_h = q_c + w$. Therefore,

$$1/c = w/q_c = (q_h - q_c)/q_c = (q_h/q_c) - 1.$$

The refrigerator is at its most efficient when it is working reversibly because then w is a minimum. The best coefficient of performance (c_0) is therefore given by

$$1/c_0 = (q_h/q_c)_{rev} - 1.$$

The ratio of q's is now expressed in terms of the temperatures of the hot and cold parts of the system. The entropy of the cold object changes by $-(q_c/T_c)_{rev}$; the entropy of the hot reservoir changes by $+(q_h/T_h)_{rev}$. Since the process is occurring reversibly there is no net entropy production (in other words, just enough work is being done to ensure that overall the process occurs without reduction of entropy). Therefore

$$\Delta S = (q_h/T_h)_{rev} - (q_c/T_c)_{rev} = 0 \quad \text{or} \quad (q_h/q_c)_{rev} = T_h/T_c.$$

It follows that *the coefficient of performance of a perfect refrigerator working reversibly between the temperatures T_c and T_h is*

$$c_0 = T_c/(T_h - T_c). \tag{7.5.1}_r$$

The ideal value makes no reference to the type of refrigerator or the working substance: it depends only on the temperatures involved. Practical refrigerators (which do not work reversibly) have coefficients of performance lower than c_0.

A related problem is the work required to *sustain* a low temperature once it has been reached. No thermal insulation is perfect, and so there is always a flow of heat from the warm surroundings back into the sample. The rate of this re-entry of energy is proportional to the temperature difference $T_h - T_c$. In order to sustain the low temperature, heat must be removed from the cold object as the same rate as it leaks in. If the rate at which heat leaks in is $dq_c/dt = A(T_h - T_c)$, where A is a constant that depends on the size of the sample and the insulation, then the minimum rate at which work must be done is

$$dw/dt = (dq_c/dt)/c_0 = A(T_h - T_c)/c_0.$$

The rate of doing work is the *power* (symbol: P); therefore, with the value of c_0 given in eqn (7.5.1), we find that the *minimum power needed to sustain an object at T_c when the surroundings are at T_h is*

$$P = A(T_h - T_c)^2/T_c. \qquad (7.5.2)_r$$

The important point is that the power increases as the *square* of the temperature difference that has to be sustained. For this reason air-conditioners are much more expensive to run on hot days than on days that are merely warm. Notice too that the power depends inversely on the temperature of the cold object: very high powers must be dissipated when the temperature is very low.

Example 7.5

Assess the relative powers that must be expended in order to maintain the same object at (a) 0 °C with surroundings at 20 °C and (b) 1.0 mK with surroundings at 1.0 K.

• *Method*. Calculate the ratio of powers for each case using eqn (7.5.2), remembering to convert the data in (a) to kelvins.

• *Answer*. For (a), $P = A\,(293\text{ K} - 273\text{ K})^2/(273\text{ K}) = 1.5A\text{ K}$. For (b), $P = A(1.0\text{ K} - 1.0 \times 10^{-3}\text{ K})^2/(1.0 \times 10^{-3}\text{ K}) = 1.0 \times 10^3 A\text{ K}$. Therefore, the ratio of powers required is

$$P(\text{b})/P(\text{a}) = 7 \times 10^2.$$

• *Comment*. This means that if 150 W is needed in (a), 100 kW is needed in (b). In practice the experimenter would use a much smaller sample in (b), and take great care with its insulation.

• *Exercise*. An air-conditioner required 50 W to maintain a room at 20 °C when the surroundings were 22 °C; calculate the power needed when the surroundings were at 32 °C.

[1.8 kW]

The thermodynamics of refrigeration also accounts for the operation of a *heat pump*, where work is done to pump heat from a cold reservoir (such as a river or the surrounding land) into a hot sink (such as a house). The device is a refrigerator, but our interest now focuses on the heat being dissipated rather than on the object being cooled. If indoors the temperature is $T_h = 300\text{ K}$ and outside $T_c = 290\text{ K}$, then $c_0 = 29$. This means that for each

1 kJ of energy used to drive the pump ($w = 1$ kJ), a quantity of heat $q_h = q_c + w = (1 + c_0)w = 30$ kJ will be supplied to the house. A 1 kW pump would therefore be a 30 kW heater if it worked reversibly. Commercial heat pumps have $c \approx 5$, which still amounts to a good yield in terms of the energy used to drive the equipment.

The coefficient of performance of refrigerators is extremely small at very low temperatures. For instance, if the hot sink is at 1 K and the sample is at 10^{-5} K, then $c_0 = 10^{-5}$, and to remove 1 J of heat involves doing 100 kJ of work. Whatever the temperature of the surroundings (other than zero) the coefficient of performance disappears as the temperature of the sample approaches zero, and so an infinite amount of work is needed to reach $T = 0$ (and an infinite amount of power would be needed to maintain that temperature even if it could be reached): *absolute zero is unattainable*.

A final aspect of this part of the discussion is that we have arrived at a *fundamental* basis of a temperature scale, one that is independent of the nature of any thermometric substance. This is because the coefficient of performance c_0 is *measurable* in terms of the observables q_h and q_c and *expressible* in terms of a property (T) which is independent of the nature of the working substance. The *thermodynamic temperature scale* is the one based on eqn (7.5.1), with the size of the divisions chosen so as to coincide with the perfect gas scale, as described in Section 1.1(a).

7.5 (b) Adiabatic demagnetization

Although $T = 0$ cannot be reached, the world record stands at about 2×10^{-8} K (20 nK). The method used to reach such a low temperature is *adiabatic demagnetization*.

In Part 2 we shall see that magnetic properties arise because electrons behave as tiny magnets. In normal circumstances in a *paramagnetic* material (one composed of molecules in which not all the electrons are paired) these magnets are orientated at random, but in a magnetic field more of them are aligned along the field than against it, Fig. 7.14. In thermodynamic terms, the application of a magnetic field lowers the entropy of the paramagnetic system on account of the ordering it induces.

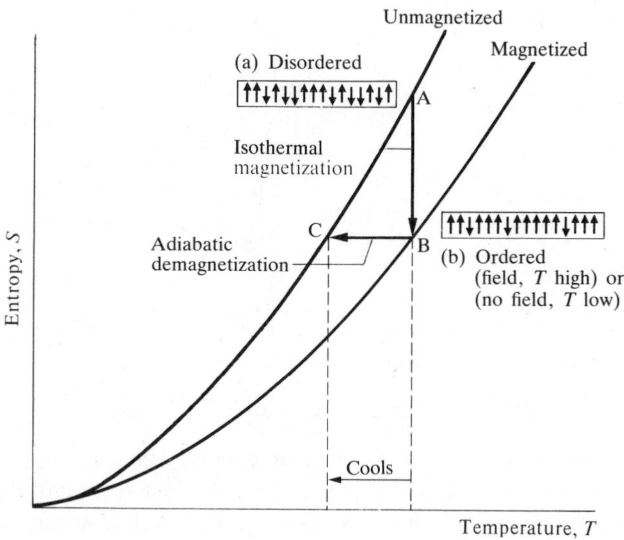

Fig. 7.14. The technique of adiabatic demagnetization is used to reach low temperatures. The upper curve shows the temperature dependence of the entropy of the system when no magnetic field is present. The lower shows the dependence when a field is present and has made the electron magnets more orderly. The isothermal magnetization step (at constant T) is from A to B, the adiabatic demagnetization step (at constant S) is from B to C.

Therefore, a sample has *two* entropy/temperature curves, Fig. 7.14, the lower one corresponding to the field being on, the upper one corresponding to it being off.

A sample of paramagnetic material, such as a transition metal salt, is cooled to about 1 K in the way already described. (Gadolinium sulphate is often used because each gadolinium ion carries several electrons, but is separated from its neighbours by a sheath of solvating water molecules.) Then the sample is magnetized by the application of a strong magnetic field. This magnetization occurs while the sample is surrounded by helium gas, which provides a thermal contact with a cold reservoir. The magnetization is therefore isothermal, and energy flows out of the sample (as heat) as the electron magnets align with the applied field. This stage is represented by the line AB in Fig. 7.14. The thermal contact between sample and surroundings is then broken by pumping away the gas, and then the magnetic field is reduced slowly to zero. Since there is no flow of heat in this last reversible, adiabatic step the entropy of the sample remains constant, and so its state changes from B to C in Fig. 7.14. It is now in a state corresponding to a lower temperature, and so the adiabatic magnetization step has cooled the sample.

Even lower temperatures can be reached if instead of electronic magnetic moments the magnetic moments of nuclei are used. This process of *adiabatic nuclear demagnetization* works on the same principle as the electronic method, and has been used to establish the world record (in copper).

From the Nernst heat theorem (Section 5.4) we know that entropy curves of substances coincide as T approaches zero. It follows that adiabatic demagnetization cannot be used to cool an object to absolute zero in a finite number of steps. For suppose the theorem were false, then the entropy might behave as in Fig. 7.15(a). The last step indicated could cool the system to $T=0$. However, we believe that the curves actually coincide at $T=0$ as shown in Fig. 7.15(b), and so no finite sequence of steps can cool the system to $T=0$. By this method (as for every method), absolute zero is unattainable.

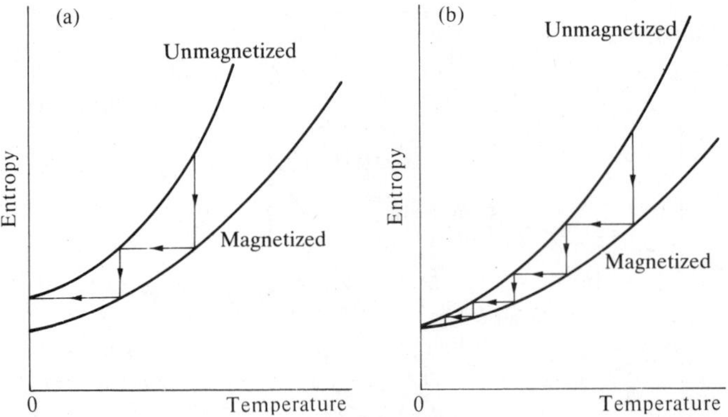

Fig. 7.15. The connection between the Nernst theorem and the unattainability of absolute zero is shown in this pair of illustrations. In (a) we assume that the Nernst theorem is false, and that the entropies do not coincide as $T \to 0$: absolute zero can be reached in a finite number of steps. In (b) we see that the steps down in temperature become progressively smaller if, as the Nernst theorem asserts, the entropies do coincide as $T \to 0$. A *finite* number of steps does not reach $T=0$.

7.6 The region between phases: surfaces

In this section we turn to the *interfacial region*, the region of a sample where one phase ends and another begins. We concentrate on the liquid–vapour interface, which is interesting because it is so mobile.

Liquids tend to adopt shapes that minimize their surface area because then the maximum number of molecules are in the bulk and the minimum are at the surface and surrounded by fewer neighbours. For this reason droplets are spherical, because a sphere is the geometrical object with the smallest surface/volume ratio. There may be other forces present that compete against the tendency to form this ideal shape, and in particular gravitational forces may flatten spheres into puddles or oceans. Nevertheless, the shape is always a consequence of the tendency of liquids to acquire minimum surface area.

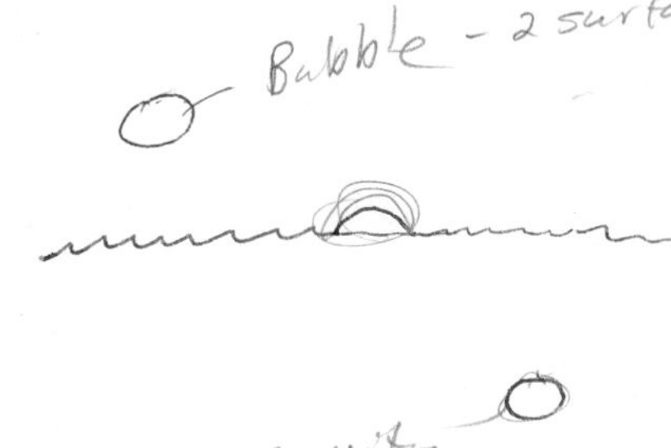

7.6 (a) Surface tension

Surface effects can be expressed in the language of Helmholtz and Gibbs functions and chemical potentials (although there are subtle problems arising if G is used). Changing the area of a surface by an infinitesimal amount $d\sigma$ involves doing an infinitesimal quantity of work dw, where

$$dw = \gamma\, d\sigma. \tag{7.6.1}$$

The coefficient γ is the *surface tension* of the material. Some values are given in Table 7.1. The work of surface formation is additional to p,V-work, and is therefore a contribution to the change of the Helmholtz function, Section 5.3(c). Therefore, for a system of variable surface area but constant composition,

$$dA = -S\,dT - p\,dV + \gamma\, d\sigma. \tag{7.6.2}$$

Table 7.1. Surface tensions of liquids at 293 K

	$\gamma/\mathrm{N\,m^{-1}}$
Water	7.275×10^{-2}
Methanol	2.26×10^{-2}
Benzene	2.888×10^{-2}
Mercury	47.2×10^{-2}

This is the link we require with thermodynamics, and the arguments about systems tending to lower Helmholtz function are applicable to this equation. For instance, at constant temperature and volume the Helmholtz function decreases ($dA < 0$) if the surface area decreases ($d\sigma < 0$): hence, surfaces have a natural tendency to contract. This is a more formal way of expressing what we have already described.

7.6 (b) Bubbles, cavities, and droplets

By *bubbles* we mean regions in which air and vapour are trapped by a thin film. *Cavities* are holes in a liquid filled with vapour. 'Bubbles' in liquids are therefore strictly cavities. True bubbles have two surfaces, cavities only one. The treatments of both are similar, but a factor of 2 is required for bubbles in order to take into account the doubled surface area. *Droplets* are spheres of liquid at equilibrium with their vapour.

Cavities in liquids are at equilibrium when the tendency for their surface area to decrease is balanced by the rise of internal pressure which would then result. If the pressure inside the cavity is p_{in} and its radius is r, then the outward force is $4\pi r^2 p_{\text{in}}$. The force inwards is due to the sum of the external pressure p_{out} and the surface tension. The *energy* of a surface of area σ is $\gamma\sigma$, or $4\pi r^2\gamma$ if the cavity is spherical; the work required to stretch this surface through dr is $d(\gamma\sigma)$, or $8\pi\gamma r\,dr$. But as (force) × (distance) is work, the force opposing stretching through a distance dr when the radius is r is $8\pi\gamma r$. Balancing the forces gives

$$4\pi r^2 p_{\text{in}} = 4\pi r^2 p_{\text{out}} + 8\pi\gamma r,$$

which rearranges into the

$$\textit{Laplace equation: } p_{in} = p_{out} + 2\gamma/r. \tag{7.6.3}$$

The Laplace equation shows that *the pressure inside a curved surface is always greater than the pressure outside*, but the difference decreases to zero as the radius of curvature becomes infinite (when the surface is flat). By 'inside' we mean the concave side of the interface. Small cavities have very small radii of curvature, and so the pressure difference across them is quite large. For instance, a 0.10 mm radius 'bubble' in champagne implies a pressure difference $p_{in} - p_{out} = 2 \times (7.4 \times 10^{-2}\,\mathrm{N\,m^{-1}})/(1.0 \times 10^{-4}\,\mathrm{m}) = 1.5\,\mathrm{kPa}$, which is enough to sustain a 15 cm column of water.

The thermodynamic properties of bubbles, cavities, and droplets can be established by exploring the consequences of the increased pressure inside the curved surface. As a first step we show that *when the pressure on a liquid is increased, its vapour pressure rises*. An increase in pressure is caused by curving the surface, as in a droplet; hence we expect droplets to have higher vapour pressures than bulk liquid. The pressure on a liquid can also be increased by introducing an inert gas into a sealed container, but this suffers from the complication that the pressurizing gas might dissolve and change the properties of the sample.

We can calculate the pressure dependence of the vapour pressure of a liquid by considering the effect of pressure on the chemical potential of the liquid and seeing how the equilibrium between vapour and liquid is affected. At equilibrium the chemical potentials $\mu(\mathrm{g})$ and $\mu(\mathrm{l})$ are equal, and for any change that preserves equilibrium the change in one is equal to the change in the other: $\mathrm{d}\mu(\mathrm{g}) = \mathrm{d}\mu(\mathrm{l})$. When the pressure on the liquid is increased by $\mathrm{d}p(\mathrm{l})$ its chemical potential changes by $V_m(\mathrm{l})\,\mathrm{d}p(\mathrm{l})$. The change in the pressure of the vapour, $\mathrm{d}p(\mathrm{g})$, corresponds to a change in its chemical potential of $V_m(\mathrm{g})\,\mathrm{d}p(\mathrm{g})$, and so, at constant temperature, the two are related by

$$V_m(\mathrm{g})\,\mathrm{d}p(\mathrm{g}) = V_m(\mathrm{g})\,\mathrm{d}p(\mathrm{g}).$$

If the vapour is assumed to be a perfect gas we can write $V_m(\mathrm{g}) = RT/p(\mathrm{g})$. Then

$$\mathrm{d}p(\mathrm{g})/p(\mathrm{g}) = V_m(\mathrm{l})\,\mathrm{d}p(\mathrm{l})/RT.$$

From now on we write the molar volume of the liquid as simply V_m. This expression can be integrated once we have the limits of integration. When there is no additional pressure acting on the liquid its pressure and that of the vapour are the same and equal to the normal vapour pressure p^*: that is, when $p(\mathrm{l}) = p^*$, $p(\mathrm{g}) = p^*$. When there is an additional pressure on the liquid so that $p(\mathrm{l}) = p^* + \Delta p$, the VP changes from p^* to p (the value we want to find). The integrations required are therefore as follows:

$$\int_{p^*}^{p} \mathrm{d}p(\mathrm{g})/p(\mathrm{g}) = (1/RT)\int_{p^*}^{p^*+\Delta p} V_m\,\mathrm{d}p(\mathrm{l}).$$

We now assume that the molar volume of the liquid is the same throughout the small range of pressures involved. Then

$$\ln(p/p^*) = \{V_m/RT\}\{(p^* + \Delta p) - p^*\} = V_m\,\Delta p/RT.$$

Therefore,

$$p = p^* \exp\{V_m \Delta p/RT\}. \qquad (7.6.4)°$$

This formula shows that the VP increases when the pressure acting on the liquid increases: particles are squeezed out into the vapour. (To be precise, Δp should be interpreted as the *sum* of the applied pressure and the increase in VP above p^* this increase brings about; the latter is generally small, and so can be neglected in Δp.)

We now link this result to the discussion of the effects of curved surfaces. The pressure difference across a curved surface is $2\gamma/r$, and so inserting this value for Δp in the last expression we obtain an equation for the VP of a liquid when it is dispersed as droplets of radius r: this is the

Kelvin equation: $$p(\text{mist}) = p(\text{bulk})e^{2\gamma V_m/rRT}, \qquad (7.6.5)°$$

where $p(\text{mist})$ is the VP of the mist and $p(\text{bulk})$ the VP of a bulk sample, with a plane surface.

The analogous expression for the VP inside a cavity can be written at once. The pressure of the liquid outside the cavity is less than the pressure inside, and so the only change is in the sign of the exponent in the last expression:

$$p(\text{cavity}) = p(\text{bulk})e^{-2\gamma V_m/rRT}. \qquad (7.6.6)°$$

In the case of droplets of water of radius 10^{-3} mm and 10^{-6} mm the ratios $p(\text{mist})/p(\text{bulk})$ at 25 °C are about 1.001 and 3.0 respectively. The second figure, although quite large, is unreliable because at that radius the droplet is only about 40 molecules in diameter and the calculation is suspect. The first figure shows that the effect is usually small; nevertheless it may have important consequences because it is responsible for the *kinetic* stabilization of a vapour against its condensation into a liquid. Consider, for example, the formation of a cloud. Warm, moist air rises into the cooler regions higher in the atmosphere. At some altitude the temperature is so low that the vapour becomes thermodynamically unstable with respect to the liquid and we expect it to condense into a cloud of liquid droplets. The initial step can be imagined as a swarm of water molecules sticking together into a microscopic droplet. Since the initial droplet is so small it has an enhanced vapour pressure. Therefore, instead of growing it evaporates. This effect stabilizes the vapour because an initial tendency to condense is overcome by a heightened tendency to evaporate. The vapour phase is then said to be *supersaturated* because it is thermodynamically unstable yet is prevented from condensing by kinetic effects.

Clouds do form, and so there must be a mechanism. Two processes are responsible. The first is that a sufficiently large number of molecules might stick together into a droplet so big that the enhanced evaporative effect is unimportant. The chance of one of these *spontaneous nucleation centres* forming is low, and in rain formation it is not a dominant mechanism. The more important process depends on the presence of minute dust particles or other kinds of foreign matter. These *nucleate* the condensation by providing surfaces to which the water molecules can attach. The *cloud chamber* works on the same principle. In a very clean environment a supersaturated mixture of water vapour and air does not condense, but when ionizing radiation in the form of a swiftly moving elementary particle passes through, the ions formed in its path nucleate the condensation and the trajectory is mapped as a streak of condensed water.

Liquids may be *superheated* above their boiling points and *supercooled* beneath their freezing points. In each case the thermodynamically stable phase is not achieved on account of the kinetic stabilization that occurs in the absence of nucleation centres. For example, superheating occurs because the vapour pressure inside a cavity is artificially low, and so any cavity that does form tends to collapse. This is encountered when an unstirred beaker of water is heated, for its temperature may be raised above its boiling point. Violent bumping often ensues as spontaneous nucleation leads to bubbles big enough to survive. In order to ensure smooth boiling at the true boiling temperature, nucleation centres, such as small pieces of sharp-edged glass or bubbles of air, should be introduced. The *bubble chamber* works on a similar principle, but depends on the nucleation of the evaporation of superheated liquid hydrogen by ionizing radiation

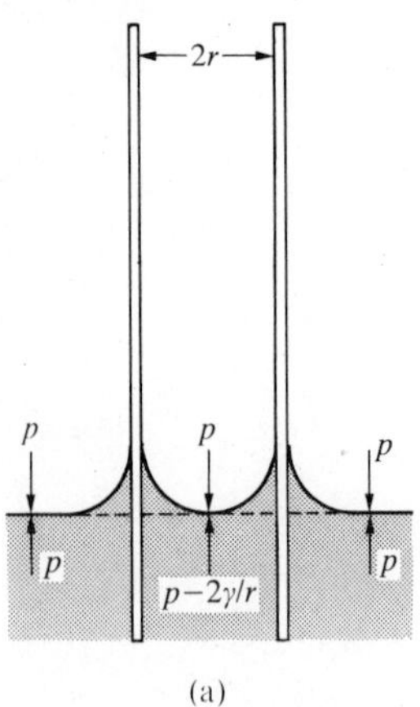

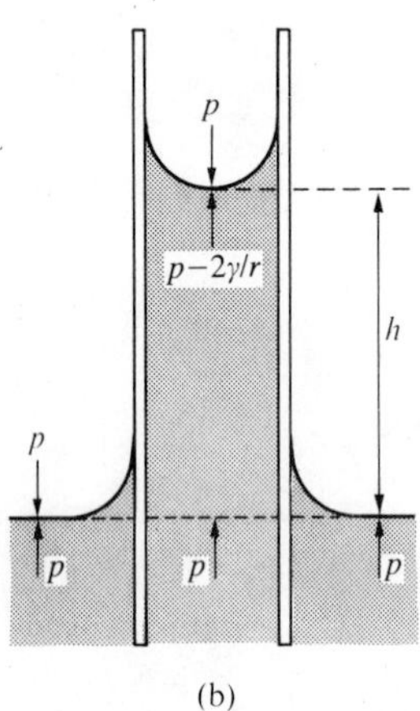

Fig. 7.16. (a) When a capillary tube is first stood in a liquid, the latter climbs up the walls, so curving the surface. The pressure just under the meniscus is then less than that due to the atmosphere by an amount $2\gamma/r$ (by the Laplace equation). (b) The equality of pressures at equal heights throughout the liquid is achieved if the liquid rises in the tube. The condition of equilibrium is when the hydrostatic pressure at the foot of the column (ρgh) cancels the pressure difference due to the curved surface.

7.6 (c) Capillary action

The tendency of liquids to rise up capillary tubes is a consequence of surface tension. Consider what happens when a glass capillary tube is first immersed in water or any liquid that has a tendency to stick to the walls. The energy is lowest when a thin film covers as much of the glass as possible. As this film creeps up the inside wall it has the effect of curving the surface of the liquid inside the tube, Fig. 7.16(a). This implies that the pressure just beneath the curving meniscus is less than the atmospheric pressure by approximately $2\gamma/r$, where r is the radius of the tube and we assume a hemispherical surface. The pressure immediately under the flat surface *outside* the tube is p, the atmospheric pressure; but *inside* the tube under the curved surface it is only $p-2\gamma/r$. The excess external pressure presses the liquid up the tube until hydrostatic equilibrium (equal pressures at equal depths, another consequence of the equality of chemical potential) has been reached, Fig. 7.16(b). This equilibrium is established when a column of liquid of density ρ has reached a height h such that

$$Pressure = \rho\pi r^2 hg/\pi r^2 = 2\gamma/r.$$

($\pi r^2 h$ is the volume of the liquid in the tube, $\rho\pi r^2 h$ its mass, $\rho\pi r^2 hg$ the downwards force, and πr^2 the area over which that force is exerted.) Therefore

$$h = 2\gamma/\rho gr. \tag{7.6.7}$$

This simple expression provides an accurate way of measuring the surface tension of liquids.

Example 7.6

In an experiment to measure the surface tension of water over a range of temperatures, a capillary tube of internal diameter 0.40 mm was supported vertically in the sample. The density of the liquid was measured in separate experiments. The following results were obtained:

$\theta/°C$	10	15	20	25	30
h/cm	7.56	7.46	7.43	7.36	7.29
$\rho/\mathrm{g\,cm^{-3}}$	0.9997	0.9991	0.9982	0.9971	0.9957

Find the temperature variation of the surface tension.

• *Method*. Use eqn (7.6.7) in the form $\gamma = \frac{1}{2}\rho hgr$, with $g = 9.81\ \mathrm{m\,s^{-2}}$ and $r = 0.20$ mm.

• *Answer*. The following values are obtained:

$\theta/°C$	10	15	20	25	30
$\gamma/\mathrm{N\,m^{-1}}$	7.4N2	7.3N2	7.3N2	7.2N2	7.1N2

• *Comment*. The answers have been expressed in the form 7.4N2, etc. which means 7.4×10^{-2}. A positive power of ten would be denoted by P (as in $2.5\mathrm{P}3 = 2.5 \times 10^3$). This notation is the recommended way of reporting data, and is becoming more widely used. Note that the surface tension generally decreases with increasing temperature.

• *Exercise*. Suppose that, over a limited range, the surface tension and the density decrease linearly with temperature. Find an expression for the temperature at which the height of a capillary column is a maximum (or a minimum).

[For $\gamma = \gamma_0(1 - \alpha T)$ and $\rho = \rho_0\,(1 - \beta T)$, $h = (\beta - \alpha)/\beta\alpha$.]

When the liquid and the material of the capillary tube attract each other less strongly than the liquid particles attract each other (mercury in glass is a common example) the liquid in the tube retracts from the walls. This curves the surface with the concave, high pressure side downwards. In order to equalize the pressure at the same depth throughout the liquid the surface must fall to compensate for the heightened pressure arising from its curvature. This gives rise to a capillary depression. In many cases there is a non-zero angle between the edge of the meniscus and the wall. If this *contact angle* is θ, eqn (7.6.7) is modified by multiplying the right hand side by $\cos\theta$. This modification is examined in the Problems.

Further reading

Experimental techniques:

Determination of melting and freezing temperatures. E. L. Skau and J. C. Arthur; *Technique of chemistry* (A. Weissberger and B. W. Rossiter, eds.) V, 105, Wiley-Interscience, New York, 1971.

Determination of boiling and condensation temperatures. J. R. Anderson; *Techniques of chemistry* (A. Weissberger and B. W. Rossiter, eds.) V, 199, Wiley-Interscience, New York, 1971.

Determination of surface and interfacial tension. A. E. Alexander and J. B. Hayter; *Techniques of chemistry* (A. Weissberger and B. W. Rossiter, eds.) V. 501, Wiley-Interscience, New York, 1971.

Low temperatures:

The quest for absolute zero. K. Mendelssohn; McGraw-Hill, New York, 1966.

Temperatures very low and very high. M. W. Zemansky; Dover, New York, 1964.

Properties:

Vapor pressure of organic compounds. T. E. Jordan; Interscience, New York, 1954.

Molecular theory of capillarity. J. S. Rowlinson and B. Widom; Clarendon Press, Oxford, 1982.

7.6 | Changes of state: physical transformations of pure materials

Introduction to phase transitions and critical phenomena (2nd edn). H. E. Stanley; Clarendon Press, Oxford, 1971.

Data:

Physico-chemical constants of pure organic compounds. J. Timmermans (ed.); Elsevier, Amsterdam, 1956.

Vapor pressure of organic compounds. T. E. Jordan; Interscience, New York, 1954.

Tables of physical and chemical contants. G. W. C. Kaye and T. H. Laby; Longman, London, 1973.

Handbook of chemistry and physics. R. C. Weast (ed.); CRC Press, Boca Raton, 1986.

Introductory problems

A7.1. The molar volume of a solid substance is $161.0\ \mathrm{cm^3\ mol^{-1}}$ at 1.00 atm and the melting point, 350.75 K. The molar volume of the liquid at this temperature and pressure is $163.3\ \mathrm{cm^3\ mol^{-1}}$. Under a pressure of 100 atm, the freezing point changes to 351.26 K. Calculate the molar entropy and molar enthalpy of fusion for the substance.

A7.2. Measured values of the VP of a certain liquid in the temperature range 200 K to 260 K can be represented by the empirical equation: $\ln(p/\mathrm{Torr}) = 16.255 - 2501.8(T/\mathrm{K})^{-1}$. Calculate the derivative $\mathrm{d}\ln p/\mathrm{d}T$ and obtain a numerical value of the molar enthalpy of vaporization.

A7.3. The VP of dichloromethane, $CH_2Cl_2(l)$, at 24.1 °C is 400 Torr, and the molar enthalpy of vaporization at 25 °C is $28.7\ \mathrm{kJ\ mol^{-1}}$. Estimate the temperature at which the VP is 500 Torr.

A7.4. The molar heat of vaporization of a liquid is found to be $14.4\ \mathrm{kJ\ mol^{-1}}$ at its normal boiling point, 180 K. The molar volume of the liquid at that temperature is $115\ \mathrm{cm^3\ mol^{-1}}$, and the molar volume of the gas is $14.5\ \mathrm{cm^3\ mol^{-1}}$. Use the Clapeyron equation to calculate $\mathrm{d}p/\mathrm{d}T$. Calculate the percent error in the value of $\mathrm{d}p/\mathrm{d}T$ obtained by using the approximations of the Clausius–Clapeyron equation.

A7.5. A refrigerator operating reversibly extracts 45 kJ of heat from a thermal reservoir and delivers 67 kJ as heat to a thermal reservoir at 300 K. Calculate the temperature of the reservoir from which heat was removed.

A7.6. A refrigerator operates reversibly between the temperatures 80 K and 200 K. Calculate the amount of work required to pump 2.10 kJ of heat from a thermal reservoir at 80 K using this refrigerator.

A7.7. Calculate the work done when the surface of liquid water at 25 °C is increased from $150\ \mathrm{cm^2}$ to $2500\ \mathrm{cm^2}$.

A7.8. From the observation that at 25 °C a certain liquid with density $0.871\ \mathrm{g\ cm^{-3}}$ rises 1.20 cm in a capillary tube with an internal diameter of 0.80 mm, calculate the surface tension of the liquid.

A7.9. At 20 °C the vapour pressure of $CCl_4(l)$ is 0.90 Torr higher in the form of droplets than in the form of bulk liquid. The VP and density of the bulk liquid are 87.05 Torr and $1.60\ \mathrm{g\ cm^{-3}}$ respectively at 20 °C. Calculate the radius of the droplets.

A7.10. Calculate the pressure inside a bubble exposed to an external pressure of 740 Torr. The diameter of the bubble is 0.25 mm. The bubble is made from an aqueous solution with a surface tension of $5.7\times10^{-2}\ \mathrm{N\ m^{-1}}$.

Problems

7.1. At 25 °C the enthalpy of the graphite → diamond phase transition is $1.8961\ \mathrm{kJ\ mol^{-1}}$, and the entropy is $-3.2552\ \mathrm{J\ K^{-1}\ mol^{-1}}$. What is the spontaneous direction at 25 °C. Which direction is favoured by a rise of temperature?

7.2. The densities of graphite and diamond are $2.27\ \mathrm{g\ cm^{-3}}$ and $3.52\ \mathrm{g\ cm^{-3}}$ respectively. What is the value of ΔA_m at 298 K for the transition when the sample is subjected to a pressure of (a) 1 atm, (b) 500 kbar?

7.3. The rate of change of the chemical potential with temperature, as expressed by the coefficient $(\partial\mu/\partial T)_p$, is an important quantity in the discussion of phase transitions and equilibria. What is the difference in slope on either side of (a) the freezing point of water, (b) the normal boiling point of water? Use the data in Table 4.7.

7.4. Another important quantity is the rate of change of chemical potential with pressure, $(\partial\mu/\partial p)_T$. Calculate the difference of this quantity on either side of the two transitions referred to in the last Problem. You need to know the densities of ice and water at 0 °C ($0.917\ \mathrm{g\ cm^{-3}}$ and $1.000\ \mathrm{g\ cm^{-3}}$) and of water and steam at 100 °C ($0.958\ \mathrm{g\ cm^{-3}}$ and $0.598\ \mathrm{kg\ m^{-3}}$).

7.5. By how much does the chemical potential of water supercooled to −5 °C exceed that of ice at that temperature?

7.6. A cloud chamber operates on the principle that vapour may be supercooled, and the passage of an ionizing particle can generate ions which induce condensation. By how much does the chemical potential of the water vapour exceed that of water at that temperature? Suppose, instead, that the pressure was increased to 1.2 atm at 100 °C. What then is the difference of chemical potentials?

7.7. We have just calculated the increase in chemical potential when the temperature of liquid water is lowered. When the pressure is increased the chemical potential of ice increases more rapidly than that of water, and so the chemical potentials of the two phases can be brought back into equality, and therefore the phases into equilibrium. Use this approach to estimate the freezing point of water under a pressure of 1000 atm; assume ΔV_m constant, and take data from Problem 7.4.

7.8. When benzene freezes at 5.5 °C its density changes from 0.879 g cm^{-3} to 0.891 g cm^{-3}. Its enthalpy of fusion is 10.59 kJ mol^{-1}. Use the same technique as in the last Problem to estimate the freezing point of benzene under a pressure of 1000 atm.

7.9. The Clapeyron equation, eqn (7.2.1), is sometimes more useful when it is upside down, for then dT/dp gives the dependence of the transition temperature on the pressure. What change in the boiling point of water at 1 atm is brought about by a 10 Torr increase of pressure?

7.10. The change in enthalpy is given by $dH = C_p\,dT + V\,dp$. The Clausius equation relates dp and dT for equilibria, and so this equation can be used to obtain an expression for the change in ΔH along the phase boundary when the temperature changes and the phases remain in equilibrium. Find this expression, and solve it for $\Delta H(T)$ in the case when C_p of both phases may be assumed independent of temperature. Use this temperature dependence of the transition enthalpy in the Clausius–Clapeyron equation to show that the vapour pressure depends on temperature as $p/p^* = (T/T^*)^a$, where $a = \Delta H_m(T^*)/RT^* + (\Delta C_{p,m}/2R)\ln(T/T^*)$, and plot the phase boundary for the vaporization of water using p^*, T^* for the normal boiling point.

7.11. The enthalpy of fusion of mercury is 2.292 kJ mol^{-1}, and it freezes at 234.3 K under 1 atm pressure, the change in volume being 0.517 cm^3 mol^{-1}. At what temperature will the bottom of a 10 m high column of mercury (of density 13.6 g cm^{-3}) freeze?

7.12. Several methods exist for the determination of vapour pressure. When it lies between 10^{-3}–10^3 Torr a pressure gauge can be used directly. When it is very low (as for an almost involatile solid) an *effusion method* can be used (p. 661). The *gas saturation method* is a straightforward technique for liquids, and we concentrate on it here. Let a volume V of some gas, measured at temperature T and pressure p be bubbled slowly through the sample maintained at the temperature T. The loss of mass by the sample is determined: let this be m, its RMM being M_r. Show that the VP of the liquid at that temperature, p', is related to the mass loss by $p' = Amp/(Am + 1)$, where $A = RT/M_m Vp$.

7.13. The vapour pressure of geraniol, which is a component of oil of roses and other perfumes, with $M_r = 154.2$, was measured at 110 °C. When 5.00 dm^3 of nitrogen at 760 Torr was passed through the heated liquid the loss of weight was 0.32 g. What is the vapour pressure of geraniol at this temperature? Assume perfect gas behaviour and that the vaporization is an equilibrium process.

7.14. The last experiment was repeated at 140 °C, the mass loss then being 243 mg for 1000 cm^3 of nitrogen passed. What is the molar enthalpy of vaporization of geraniol? Estimate its boiling point; make the same assumptions as in Problem 7.13.

7.15. 50.0 dm^3 of dry air were slowly bubbled through a thermally insulated beaker containing 250 g of water. If the initial temperature was 25 °C, what is the final temperature? The VP of water at 25 °C is 23.8 Torr and its molar heat capacity is 75.5 J K^{-1} mol^{-1}. Assume that the air leaving the beaker is saturated with moisture, that the VP is constant throughout, that the air is not heated or cooled, and that H_2O(g) is perfect.

7.16. In July in Los Angeles the incident sunlight at ground level has a power density of 1.2 kW m^{-2} at noon. A swimming pool of area 10 m × 5 m is directly exposed to the radiation. What is the maximum rate of loss of water?

7.17. An open vessel containing (a) water, (b) benzene, (c) mercury stands in a laboratory measuring 5 m × 5 m × 3 m. The temperature is 25 °C. What quantity (in g) of each material will be found in the air if there is no ventilation? (The vapour pressures are 24 Torr, 98 Torr, and 1.7×10^{-3} Torr respectively.)

7.18. The *relative humidity* of air is the ratio of the partial pressure of water vapour to its vapour pressure at that temperature. What is (a) the partial pressure, (b) the amount present in grams, when the relative humidity is 70% in the laboratory in Problem 7.17.

7.19. The VP of nitric acid depends on the temperature as follows:

θ/°C	0	20	40	50	70	80	90	100
p/Torr	14.4	47.9	133	208	467	670	937	1282

What is (a) the conventional boiling point, (b) the molar enthalpy of vaporization?

7.20. The ketone carvone is a component of oil spearmint. Its VP depends on the temperature as follows:

θ/°C	57.4	100.4	133.0	157.3	203.5	227.5
p/Torr	1.00	10.0	40.0	100	400	760

The RMM is 150.2. What is the molar enthalpy of vaporization and the conventional boiling point? (Make full use of the data by plotting a graph.)

7.21. Combine the barometric formula for the dependence of pressure on altitude, Problem 1.31, with the Clausius–Clapeyron equation, and predict how the boiling point of a liquid depends on the altitude and the ambient temperature (be careful to distinguish the various temperatures in the expressions). Taking the mean ambient temperature as 20 °C, predict the boiling point of water at 10 000 ft.

7.22. Vapour pressure is often reported in the form $\lg(p/\text{mmHg}) = b - 0.052\,23a/(T/\text{K})$ (see for example, *Handbook of chemistry and physics*, CRC Press). How is the molar enthalpy of vaporization related to the parameters a and b? The values for white phosphorus are $a = 63\,123$ and $b = 9.651\,1$ in the range 20–44 °C. What is the VP of white phosphorus at 25 °C, and what is the molar enthalpy of sublimation of the solid?

7.23. On a cold, dry morning after a frost the temperature was −5 °C. The partial pressure of water vapour in the atmosphere dropped to 2 mmHg. Will the frost sublime? What partial pressure of water would ensure that the frost remained? Use data from *Example* 7.3.

7.24. Construct the phase diagram for benzene in the vicinity of its triple point (36 Torr, 5.50 °C) using the following data:

$$\Delta H^{\ominus}_{\text{melt,m}} = 10.6\ \text{kJ mol}^{-1}, \qquad \Delta H^{\ominus}_{\text{vap,m}} = 30.8\ \text{kJ mol}^{-1},$$
$$\rho(\text{s}) = 0.91\ \text{g cm}^{-3}, \qquad \rho(\text{l}) = 0.899\ \text{g cm}^{-3}.$$

7.25. Refer to Fig. 7.7. What changes would be observed when water vapour at 1 atm and 400 K is cooled at constant pressure to 260 K? Suppose you observed the rate of cooling of the sample when it was in contact with a cool bath. What would you notice?

7.26. What modification to the observation would appear when the cooling was carried out at 0.006 atm?

7.27. Using the phase diagram in Fig. 7.8, state what you would observe when a sample of carbon dioxide, initially at 1 atm and 298 K, is subjected to the following cycle. (a) Isobaric heating to 320 K, (b) isothermal compression to 100 atm, (c) isobaric cooling to 298 K, (d) isothermal decompression to 80 atm, (e) isobaric cooling to 210 K, (f) isothermal decompression to 1 atm, (g) isobaric heating to 298 K.

7.28. What is the coefficient of performance of a refrigerator working with thermodynamic efficiency in a room of temperature 20 °C when the interior is at a temperature of (a) 0 °C, (b) −10 °C?

7.29. How much work is needed to freeze 250 g of water originally at 0 °C in a refrigerator standing in a room at 20 °C? What is the minimum time for bringing about total freezing when the power consumption of the refrigerator is 100 W?

7.30. Quite often we have to find out how much work has to be done in order to lower the temperature of a body; this is slightly more involved than the last Problem because the temperature of the object, and therefore the coefficient of performance, is changing. Nevertheless it is quite a simple job to find the work by writing $|dw| = |dq/c_0|$, and relating dq to dT through the heat capacity C_p. Then the total work is the integral of the resulting expression. First we assume that the heat capacity is independent of temperature in the range of interest. Find an expression for the amount of work needed to cool an object from T_i to T_f when the refrigerator is in a room with a temperature T_h.

7.31. Apply the calculation in the last Problem to find the work needed to freeze 250 g of water put into the refrigerator at 20 °C. How long will it take when the refrigerator operates at 100 W? Suppose the water was put in at 25 °C: what work is then required?

7.32. Calculate the minimum work needed to lower the temperature of a 1 g block of copper from 1.10 K to 0.10 K, the surroundings being at 1.20 K. In the first place, make a crude estimate by assuming that the heat capacity remains constant at $3.9 \times 10^{-5}\ \text{J K}^{-1}\ \text{mol}^{-1}$ and that the coefficient of performance can be evaluated at the mean temperature of the block.

7.33. Repeat the last calculation in a more realistic way by noting that the molar heat capacity varies as $AT^3 + BT$, where $A = 4.82 \times 10^{-5}\ \text{J K}^{-4}\ \text{mol}^{-1}$ and $B = 6.88 \times 10^{-4}\ \text{J K}^{-2}\ \text{mol}^{-1}$, and taking into account the variation of the coefficient of performance with temperature.

7.34. In a more ambitious experiment the aim was to cool the copper block to 10^{-6} K. How much work would be needed, and for how long, at least, must a 1 μW refrigerator operate?

7.35. How much work must be done in order to cool the air in an otherwise empty room of size 3.0 m × 5.0 m × 5.0 m from 30 °C to 22 °C when the ambient temperature is (a) 20 °C, (b) 30 °C? Assume c_0 is constant over the range, and take $C_{p,\text{m}} = 29\ \text{J K}^{-1}\ \text{mol}^{-1}$ for the air, which has a mean density of $1.2\ \text{mg cm}^{-3}$ and $M_r \approx 29$.

7.36. An electric heater rated at 1 kW is left on in the same room. An air conditioner is also left running. What is the minimum power rating of the air conditioner that is needed in order to maintain the room at 25 °C when the ambient temperature is 30 °C?

7.37. The assumption that the number of molecules in the surface of a sample is much less than the total number fails when the sample is dispersed as very small droplets. But how small do the droplets have to be before the effect is significant? Estimate the ratio of the number of water molecules ($r \approx 120$ pm) on the surface of a spherical droplet to the total number in the entire droplet, when the radius of the droplets is (a) 10^{-5} mm, (b) 10^{-2} mm, (c) 1 mm.

7.38. A sample of benzene of mass 100 g is dispersed as droplets of radius 1.0 μm. The surface tension of benzene is $2.8 \times 10^{-2}\ \text{N m}^{-1}$ and its density is $0.88\ \text{g cm}^{-3}$. What is the change of Helmholtz function? What is the minimum work necessary to bring about the dispersal?

7.39. By how much is the VP of benzene changed when it is dispersed in the form of small droplets of radius (a) 10 μm, (b) 0.10 μm, both at 25 °C.

7.40. A carburettor jet generates a fine spray of fuel in an air stream. Suppose the operating temperature is 60 °C and the fuel has a VP of 100 Torr (like heptane). First suppose that the jet is badly adjusted and simply provides puddles of fuel. If the air, at a pressure of 1 atm, passes through sufficiently slowly to become saturated, what mass of fuel will be carried into the engine for the passage of 10 dm^3 of air? Now adjust the jet so that it provides a spray of droplets of radius 10^{-4} mm. How much fuel is transported for the same flow of air? (In practice, the flow would transport the droplets themselves; here we assume that only their vapour is transported.) Take $\gamma = 2.8 \times 10^{-2}$ N m^{-1} and $\rho = 0.879$ g cm^{-3}.

7.41. The surface tension of water is 7.28×10^{-2} N m^{-1} at 20 °C and 5.80×10^{-2} N m^{-1} at 100 °C. The densities are respectively 0.998 g cm^{-3} and 0.958 g cm^{-3}. To what height will water rise in tubes of internal radius (a) 1.0 mm, (b) 0.10 mm at these two temperatures?

7.42. At 30 °C the surface tension of ethanol in contact with its vapour is 2.189×10^{-2} N m^{-1}, and its density is 0.780 g cm^{-3}. How far up a tube of internal diameter 0.20 mm will it rise? What pressure is needed to push the meniscus back level with the surrounding liquid?

7.43. A glass tube of internal diameter 1.00 cm surrounds a glass rod of diameter 0.98 cm. How high will water rise in the space between them at 25 °C?

7.44. The capillary rise when a fluid meets the material of the tube tangentially is given by eqn (7.6.7). Some liquids make a non-zero *contact angle* θ with the wall. Only the vertical component of the force due to surface tension draws the liquid up the tube; on this basis deduce an expression for capillary rise for a general contact angle.

7.45. In Section 6.1(b) we saw how to deduce the Maxwell relations. Now regard μ as a function of σ as well as T and p, and deduce the relation $(\partial V/\partial \sigma)_{p,T} = (\partial \gamma/\partial p)_{\sigma,T}$ on the basis that $d\mu$ is an exact differential. The derivative $\partial V/\partial \sigma$ can be evaluated very easily for spherical droplets of radius r. Derive this coefficient, and show that the new Maxwell type of relation leads at once to the Laplace equation, eqn (7.6.3).

7.46. Show that the velocity of formation of a hole in the skin of a bubble is of the order of $\surd(2\gamma/\rho\delta)$, where γ is the surface tension of the fluid, ρ its density, and δ the thickness of the film. Estimate the velocity in a film of (a) water, $\gamma = 7.2 \times 10^{-2}$ N m^{-1}, (b) soapy water, $\gamma = 2.6 \times 10^{-2}$ N m^{-1}.

8

Changes of state: physical transformations of simple mixtures

Learning objectives

After careful study of this chapter you should be able to:

(1) Define *partial molar quantity*, Section 8.1, and express a thermodynamic quantity in terms of its partial molar values, eqns (8.1.2) and (8.1.4).

(2) Explain the *method of intercepts* for measuring partial molar quantities, Appendix 8.1.

(3) Derive the *Gibbs–Duhem equation*, eqn (8.1.5), relating the changes in chemical potentials of the components of a system.

(4) Derive expressions for the thermodynamic functions of mixing of perfect gases, eqns (8.2.1–5).

(5) State *Raoult's Law* for the partial vapour pressure of a component of a mixture, eqn (8.2.7).

(6) Write down an expression for the chemical potential of a component in an ideal mixture, eqn (8.2.8).

(7) State *Henry's Law* for the partial pressure of a solute, eqn (8.2.9) and use it to deduce the solubilities of gases, Example 8.4.

(8) Define *ideal solution* and *ideal dilute solution*, Section 8.2(c).

(9) Deduce an expression for the thermodynamic functions of mixing of ideal solutions, eqn (8.2.10).

(10) Define the *excess functions* of mixing, eqn (8.2.11), and explain the term *regular solution*.

(11) Explain the term *colligative properties* and give examples, Section 8.3.

(12) Derive expressions for the *elevation of boiling point* and the *depression of freezing point* of an ideal solution, eqns (8.3.2) and (8.3.4), and explain the principles of *ebullioscopy* and *cryoscopy*, Sections 8.3(b and c).

(13) Derive and use an expression for estimating the *solubility* of an ideal solute, (8.3.5).

(14) Define *osmotic pressure*, Section 8.3(e), derive the *van't Hoff equation*, eqn (8.3.10), and explain the principles of *osmometry*, Example 8.7.

(15) Construct and interpret *vapour-pressure diagrams* for a mixture of two volatile liquids, Section 8.4(a).

(16) Relate the vapour pressure of a liquid to the vapour pressures of its components and the composition of the mixture, Section 8.4(a).

(17) Derive and use the *lever rule* to determine from a phase diagram the relative amounts of liquid and vapour present at equilibrium, eqn (8.4.5).

(18) Interpret a *temperature–composition diagram*, and use it to discuss the distillation of a mixture, Section 8.4(b).

(19) Explain the meaning of the term *azeotrope*, and distinguish between *high-boiling* and *low-boiling* azeotropes, Section 8.4(c).

(20) Write an expression for the chemical potential of a real solvent, eqn (8.5.3), define its *activity* and its *activity coefficient*, eqn (8.5.4) and eqn (8.5.5), and explain how these may be measured, Example 8.8.

(21) Write and justify an expression for the chemical potential of a solute, eqn (8.5.9), define its *activity* and its *activity coefficient*, eqns (8.5.10) and (8.5.11), and explain how these may be measured, Example 8.9.

(22) Write and justify an expression for the chemical potential of a solute in terms of its molality, eqn (8.5.16).

(23) Define the *standard states* of the components of ideal and real solutions, Section 8.5 and Box 8.1.

Introduction

We now leave pure materials and the limited but important changes they can undergo, and examine mixtures. In this chapter we consider only unreactive mixtures, and reactions are considered in the next. In each case the discussion is based on the chemical potential, and we shall see how this simple concept unifies much of chemistry. At this stage we deal only with *binary mixtures*, mixtures consisting of two components. This means that we shall often be able to simplify equations using $x_A + x_B = 1$ for the mole fractions. More complex mixtures are discussed in Chapter 9.

8.1 Partial molar quantities

Imagine a huge volume of pure water. When a further 1 mol H_2O is added the volume increases by 18 cm^3; the quantity 18 cm^3 mol^{-1} is the *molar volume* of pure water. When 1 mol H_2O is added to a huge volume of pure ethanol the volume increases by only 14 cm^3. The reason for the smaller increase is that the volume a given number of water molecules occupy depends on the molecules that surround them. Now there is so much ethanol present that each water molecule is surrounded by pure ethanol, and the packing of the molecules results in them occupying only 14 cm^3. The quantity 14 cm^3 mol^{-1} is the *partial molar volume* of water in pure ethanol. The partial molar volumes of the components of a mixture vary with composition because the environment of each type of molecule changes as the composition changes from pure A to pure B. The partial molar volumes of water and ethanol across the full composition range at 25 °C are shown in Fig. 8.1.

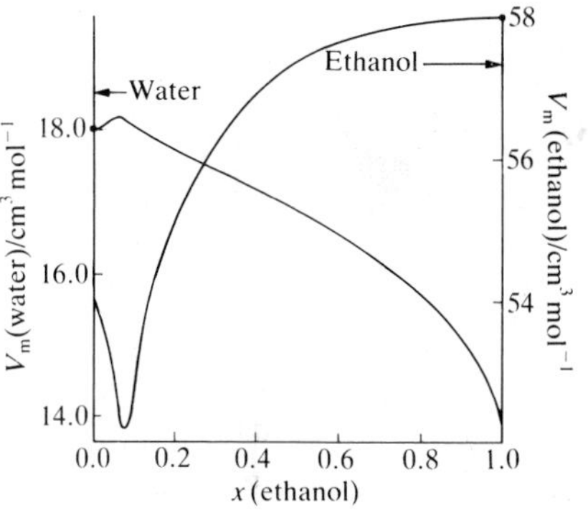

Fig. 8.1. The partial molar volumes of water and ethanol at 25 °C. Note the different scales (water on the left, ethanol on the right).

8.1 (a) Partial molar volume

The partial molar volume of a substance A at some general composition is defined formally as follows:

$$\textit{Partial molar volume: } V_{A,m} = (\partial V/\partial n_A)_{p,T,n_B\ldots}. \tag{8.1.1}$$

In words, it is the rate of change of volume when the amount of substance of A is increased, the pressure, temperature, and amount of substance of the other components being constant. The partial molar volume depends on the composition, and so in general we should write it $V_{A,m}(x_A, x_B)$, but the mole fractions will normally not be written explicitly. The definition implies that when the composition is changed by the addition of an amount dn_A of A and an amount dn_B of B, the total volume of the mixture changes by

$$dV = (\partial V/\partial n_A)_{p,T,n_B} dn_A + (\partial V/\partial n_B)_{p,T,n_A} dn_B = V_{A,m} dn_A + V_{B,m} dn_B.$$

Once we know the partial molar volumes of the two components of a mixture at the composition (and temperature) of interest we can state the *total* volume of the mixture. We shall now demonstrate that

$$V = n_A V_{A,m} + n_B V_{B,m}, \tag{8.1.2}$$

the partial volumes in this expression being the values relating to the composition $x_A = n_A/n$ and $x_B = n_B/n$, with $n = n_A + n_B$.

The reasoning behind this simple result is as follows. Consider a very large sample of the mixture of the specified composition. Then, when an amount n_A of A is added the composition remains virtually unchanged, and so the partial molar volume of A is the same throughout the addition: this means that the volume of the sample will change by $n_A V_{A,m}$. When n_B of B is added the volume changes by $n_B V_{B,m}$ for the same reason. The total change of volume is therefore $n_A V_{A,m} + n_B V_{B,m}$. The sample now occupies a larger volume, but the proportions of the components are still the same. Now scoop out of this enlarged volume a sample containing n_A of A and n_B of B. Its volume is simply $n_A V_{A,m} + n_B V_{B,m}$. Because V is a state function, the same sample could have been prepared simply by mixing the appropriate amounts of A and B. This justifies eqn (8.1.2).

Partial molar volumes (and partial molar quantities in general) can be measured in several ways. One method is to measure the dependence of the volume on the composition and to determine the slope dV/dn at the composition of interest. This is not very accurate. A better technique is the *method of intercepts*, which is outlined in Appendix 8.1.

Example 8.1

A corrupt barman attempts to prepare 100.00 cm^3 of some drink by mixing 30.0 cm^3 of ethanol with 70.0 cm^3 of water at 25 °C. Does he succeed? If not, what volumes should have been mixed in order to arrive at a mixture of the same strength but of the required volume?

• *Method*. The barman seems to have understood neither the concept nor the importance of partial molar quantities. In order to find the total volume use eqn (8.1.2) with partial molar volumes taken from Fig. 8.1 for the specified composition. The mole fraction composition of the mixture is determined from the densities (Table 0.1) and molar masses of the components. In order to find the true recipe, find the solution of $100.0\ cm^3 = n_A V_{A,m} + n_B V_{B,m}$ with n_A/n_B the same as in the corrupt attempt.

• *Answer*. $n(H_2O) = (70.0\ cm^3) \times (0.997\ g\ cm^{-3})/(18.02\ g\ mol^{-1}) = 3.87_3\ mol$.

Likewise, $n(CH_3CH_2OH) = 0.514$ mol and $n = 4.38_7$ mol. Therefore,

$$x(H_2O) = (3.87_3\ mol)/(4.38_7\ mol) = 0.883$$
$$x(CH_3CH_2OH) = 1.000 - 0.883 = 0.117.$$

From Fig. 8.1, the partial molar volumes at this composition are $V_m(H_2O) = 18.0\ cm^3\ mol^{-1}$ and $V_m(CH_3CH_2OH) = 53.6\ cm^3\ mol^{-1}$. Therefore, the total volume of the mixture is

$$V = (3.87_3\ mol) \times (18.0\ cm^3\ mol^{-1}) + (0.514\ mol) \times (53.6\ cm^3\ mol^{-1}) = 97.3\ cm^3.$$

A mixture of the same relative composition but of total volume 100.0 cm^3 will have the same mole fractions of the components, and therefore the same partial molar volumes, but a different overall amount n. Therefore,

$$100\ cm^3 = n\{0.883 \times (18.0\ cm^3\ mol^{-1}) + 0.117 \times (53.6\ cm^3\ mol^{-1})\} = 22.1_7 n\ cm^3\ mol^{-1},$$

so that $n = 4.51$ mol. The mixture should therefore contain $n(H_2O) = 0.883 \times 4.51\ mol = 3.98\ mol\ H_2O$, or 71.9 cm^3 of water, $n(CH_3CH_2OH) = 0.117 \times 4.51\ mol = 0.528\ mol\ CH_3CH_2OH$, or 30.8 cm^3 of ethanol.

• *Comment*. It would probably be unwise to attempt to explain this to the barman.

• *Exercise*. At 25 °C the density of a 50% by mass ethanol/water solution is 0.914 g cm^{-3}. Given that the partial molar volume of water in the solution is 17.4 $cm^3\ mol^{-1}$, what is the partial molar volume of the ethanol? [56.4 $cm^3\ mol^{-1}$]

8.1 (b) Partial molar Gibbs function

The concept of partial molar quantity can be extended to any of the extensive state functions. One already encountered, but under a different name, is the *partial molar Gibbs function*. If we refer back to eqn (6.3.3) we see that the chemical potential is defined in exactly the same way as the partial molar volume in eqn (8.1.1), and so *the chemical potential of a substance is its partial molar Gibbs function:*

Partial molar Gibbs function,

$$\text{or chemical potential: } \mu_A = (\partial G/\partial n_A)_{p,T,n_B\ldots}. \qquad (8.1.3)$$

By the same argument that led to eqn (8.1.2) it follows that the *total Gibbs function* of a mixture is

$$G = n_A\mu_A + n_B\mu_B, \qquad (8.1.4)$$

where μ_A and μ_B are the chemical potentials at the composition of the mixture.

We are now equipped to deal with physical and chemical changes of mixtures: the dependence of the Gibbs function of a mixture on its composition is given by eqn (8.1.4), and we know that systems tend towards lower Gibbs function. This is the link we need in order to apply thermodynamics to the discussion of spontaneous changes of composition.

One further property of partial molar quantities can now be introduced. Since the total Gibbs function of a mixture is given by eqn (8.1.4), and the chemical potentials depend on the composition, when the compositions are changed infinitesimally we might expect G to change by

$$dG = \mu_A\, dn_A + \mu_B\, dn_B + n_A\, d\mu_A + n_B\, d\mu_B.$$

However, according to eqn (6.3.4), at constant pressure and temperature the Gibbs function changes by

$$dG = \mu_A\, dn_A + \mu_B\, dn_B.$$

Since G is a state function, these two equations must be equal to each other. This implies that

$$n_A\, d\mu_A + n_B\, d\mu_B = 0,$$

which is a special case of the

$$\textit{Gibbs–Duhem equation: } \sum_J n_J\, d\mu_J = 0. \qquad (8.1.5)$$

The significance of this result is that the chemical potentials of a mixture cannot change independently: in a binary mixture, if one increases the other must decrease. Some of the consequences are explored in the Problems. The same line of reasoning applies to all partial molar quantities. For instance, in the case of partial molar volumes, in a binary mixture the analogue of eqn (8.1.5) is $n_A\, dV_{A,m} + n_B\, dV_{B,m} = 0$, or

$$n_A\, dV_{A,m} = -n_B\, dV_{B,m},$$

which means that where one partial molar volume increases the other must decrease. This can be seen in Fig. 8.1, where increases in the water partial molar volume are mirrored by decreases in the partial molar volume of ethanol, and vice versa.

One final word of warning. Molar volumes and molar entropies are always positive, but the corresponding *partial* molar quantities need not be. For example, the partial molar volume of $MgSO_4$ is $-1.4\,cm^3\,mol^{-1}$, which means that the addition of 1 mol $MgSO_4$ to a large volume of water results in a *decrease* of volume of $1.4\,cm^3$. The contraction occurs because the salt breaks up the open structure of water as the ions are hydrated, and so it collapses slightly.

8.2 The thermodynamics of mixing

If two gases, each at a pressure p, are in separate vessels, and then the vessels are connected, the gases mix. The mixing is spontaneous, and so it must correspond to a decrease of the system's Gibbs function. We now see how to make this idea precise.

8.2 (a) The Gibbs function of mixing

Let the amounts of gases in the two containers be n_A and n_B; both are at a temperature T and a pressure p. The Gibbs function of the total system is $n_A\mu_A + n_B\mu_B$, where μ_J is the chemical potential of the pure gas J at a pressure p and temperature T. We take each gas to be perfect, in which case the chemical potentials are given by eqn (6.2.10). Therefore, the initial Gibbs function is

$$G_i = n_A\{\mu_A^\ominus + RT\ln(p/p^\ominus)\} + n_B\{\mu_B^\ominus + RT\ln(p/p^\ominus)\}.$$

After mixing, each gas exerts a partial pressure p_J, with $p_A + p_B = p$. The total Gibbs function is then

$$G_f = n_A\{\mu_A^\ominus + RT\ln(p_A/p^\ominus)\} + n_B\{\mu_B^\ominus + RT\ln(p_B/p^\ominus)\}.$$

The change of Gibbs function on mixing, the *Gibbs function of mixing* (symbol: $\Delta_{mix}G$; $\Delta_{mix}G = G_f - G_i$), is therefore

$$\Delta_{mix}G = n_ART\ln(p_A/p) + n_BRT\ln(p_B/p).$$

Dalton's Law (Section 1.2(a)) lets us replace p_J/p by the mole fraction x_J for each component. We may also replace n_J by x_Jn because a mole fraction is defined as n_J/n. These substitutions lead to the following expression for the Gibbs function of mixing of two perfect gases:

$$\Delta_{mix}G = nRT\{x_A\ln x_A + x_B\ln x_B\}. \qquad (8.2.1)^\circ$$

Since mole fractions are always less than unity, the logarithms in this equation are negative, and so $\Delta_{mix}G$ is also negative. This confirms that perfect gases mix spontaneously in all proportions. However, the equation extends common sense by allowing us to discuss the process quantitatively. We see, for instance, that $\Delta_{mix}G$ is directly proportional to the temperature but is independent of the pressure.

Example 8.2

A container is divided into two compartments. One contains 3.0 mol H_2 at 1.0 atm and 25 °C, the other contains 1.0 mol N_2 at 3.0 atm and 25 °C. Calculate the Gibbs function of mixing when the partition is removed.

- *Method:* Calculate G_i and G_f from eqn (6.2.10). For the latter it is necessary to evaluate the final pressure from $p = nRT/V$, which requires knowledge of the volume $V = V(H_2) + V(N_2)$. These individual initial volumes are obtained from the individual $V = nRT/p$ expressions for each gas. The partial pressures are obtained from $p_J = x_J p$.

• *Answer*. The initial Gibbs function is

$$G_i = (3.0\,\text{mol})\{\mu^{\ominus}(H_2) + RT\ln(1.0\,\text{atm}/p^{\ominus})\} + (1.0\,\text{mol})\{\mu^{\ominus}(N_2) + RT\ln(3.0\,\text{atm}/p^{\ominus})\}.$$

The initial volumes occupied are

$$V(H_2) = (3.0\,\text{mol})RT/(1.0\,\text{atm}), \qquad V(N_2) = (1.0\,\text{mol})RT/(3.0\,\text{atm}),$$

and so the total volume of the container is

$$V = (1\,\text{mol}/1\,\text{atm})RT\{3.0 + 1.0/3.0\}$$

and the final pressure (with $n = 1.0\,\text{mol} + 3.0\,\text{mol} = 4.0\,\text{mol}$) is

$$p = nRT/V = (4.0\,\text{mol})RT/(1\,\text{mol}/1\text{atm})RT\{3.0 + 1.0/3.0\}$$
$$= (12.0/10.0)\,\text{atm} = 1.2_0\,\text{atm}.$$

The final partial pressures, using $x(H_2) = 0.75$ and $x(N_2) = 0.25$ are

$$p(H_2) = 0.75 \times 1.2_0\,\text{atm} = 0.90\,\text{atm}, \qquad p(N_2) = 0.30\,\text{atm}.$$

The final Gibbs function is

$$G_f = (3.0\,\text{mol})\{\mu^{\ominus}(H_2) + RT\ln 0.90\} + (1.0\,\text{mol})\{\mu^{\ominus}(N_2) + RT\ln 0.30\},$$

and so the Gibbs function of mixing is

$$\Delta_{mix}G = G_f - G_i$$
$$= (3.0\,\text{mol})RT\{\ln 0.90 - \ln 1.0\} + (1.0\,\text{mol})RT\{\ln 0.30 - \ln 3.0\}$$
$$= (1\,\text{mol})RT\ln\{0.90^{3.0} \times 0.30/3.0\} = -(2.6_2\,\text{mol})RT = -6.5\,\text{kJ}.$$

• *Comment*. The gases have been assumed to behave perfectly. This change of Gibbs function is the sum of two contributions: the mixing itself, and the changes in pressure of the two gases. When 3 mol H_2 mixes with 1 mol N_2 at the same pressure, the change of Gibbs function is given by eqn (8.2.1) as -5.6 kJ independent of the initial common pressure.

• *Exercise*. 2.0 mol H_2 at 2.0 atm and 25 °C and 4.0 mol N_2 at 3.0 atm and 25 °C were used instead. Calculate $\Delta_{mix}G$. What would be the value of $\Delta_{mix}G$ had the pressures been identical initially? [−9.8 kJ, −9.5 kJ]

8.2 (b) Other thermodynamic mixing functions

The quantitative expression for $\Delta_{mix}G$ lets us compute the *entropy of mixing* (symbol: $\Delta_{mix}S$). Since $(\partial G/\partial T)_{p,n} = -S$, it follows immediately from eqn (8.2.1) that, for a mixture of perfect gases,

$$\Delta_{mix}S = -(\partial\Delta_{mix}G/\partial T)_{p,n} = -nR\{x_A \ln x_A + x_B \ln x_B\}. \quad (8.2.2)^{\circ}$$

Since $\ln x$ is negative, $\Delta_{mix}S$ is positive. This is what we expect when one gas disperses into the other and the system becomes more chaotic. The entropy of mixing in the last *Example* is readily found to be $+22\,\text{J K}^{-1}$.

The *enthalpy of mixing* (symbol: $\Delta_{mix}H$) of two perfect gases may be found from $\Delta G = \Delta H - T\Delta S$ (because $\Delta T = 0$ for the isothermal process). From eqns (8.2.1) and (8.2.2) we find

$$\Delta_{mix}H = 0 \quad (\text{constant } p, T). \quad (8.2.3)^{\circ}$$

The enthalpy of mixing is zero, as we should expect for a system in which there are no interactions between particles. It follows that the whole of the driving force for mixing comes from the increase in entropy of the system (since $\Delta_{mix}H = 0$, the entropy of the surroundings is unchanged).

The *volume change of mixing* (symbol $\Delta_{mix}V$) can be found from $\Delta_{mix}G$ because $(\partial G/\partial p)_{T,n} = V$. But for perfect gases $\Delta_{mix}G$ is independent of pressure, and so $(\partial \Delta_{mix}G/\partial p)_{T,n} = 0$; this means that

$$\Delta_{mix}V = 0 \quad (\text{constant } p, T). \qquad (8.2.4)^{\circ}$$

This is also to be expected for a system without interactions. Since there is neither volume nor enthalpy change on mixing, and at constant pressure $\Delta H = \Delta U + p\Delta V$, it follows that the *internal energy of mixing* (symbol $\Delta_{mix}U$) is also zero:

$$\Delta_{mix}U = \Delta_{mix}H - p\Delta_{mix}V = 0 \quad (\text{constant } p, T). \qquad (8.2.5)^{\circ}$$

All these conclusions are changed when we turn to real gases because the temperature and composition dependence is more complicated. Nevertheless, $\Delta_{mix}G$ is normally negative, and so the mixing of real gases is normally spontaneous, but may be accompanied by changes of enthalpy, volume, and internal energy. Note that we say *normally*, because exceptions can be contrived. For instance, if two gases are in the same vessel but above their critical temperatures (and so are not liquids), then when the pressure is raised the intermolecular interactions may cause them to behave like immiscible liquids. Then the enthalpy of mixing is positive and may dominate $T\Delta_{mix}S$ so that $\Delta_{mix}G$ is positive; separation is then spontaneous.

8.2 (c) The chemical potentials of liquids

In order to discuss the equilibrium properties of liquid mixtures we need to know how the chemical potential of a liquid depends on its composition. We first note that, when the system is at equilibrium, the chemical potential of a substance present as a vapour must be equal to its chemical potential in the liquid. Then we use the expression for the chemical potential of a gas, $\mu = \mu^{\ominus} + RT\ln(p/p^{\ominus})$.

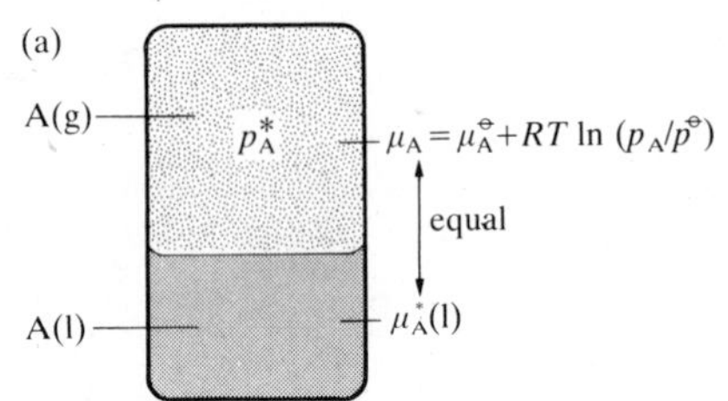

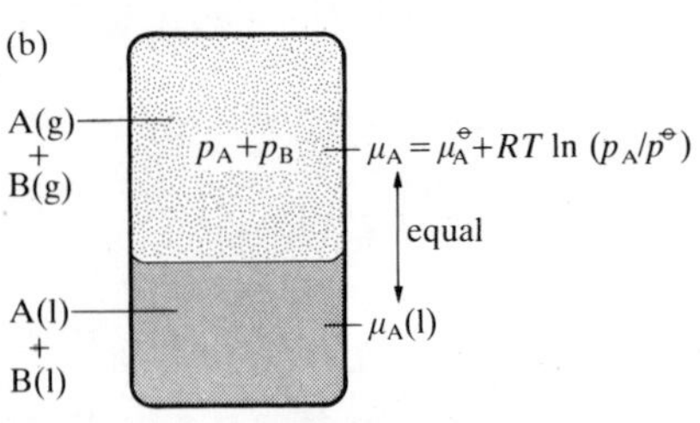

Fig. 8.2. (a) At equilibrium the chemical potential of a pure liquid is in equilibrium with its vapour at the vapour pressure p_A^* and the two chemical potentials are equal. (b) In a mixture A has the same chemical potential in the vapour and the liquid, but its partial VP (its contribution to the total VP) is now p_A.

Refer to Fig. 8.2. We denote quantities relating to pure substances by *. The chemical potential of pure liquid A is $\mu_A^*(l)$; since its VP is then p_A^* the chemical potential of A in the vapour is $\mu_A^{\ominus} + RT\ln(p_A^*/p^{\ominus})$, and we can write

$$\mu_A^*(l) = \mu_A^{\ominus} + RT\ln(p_A^*/p^{\ominus}).$$

If several components are present, the chemical potential of A in the liquid, $\mu_A(l)$, remains equal to its chemical potential in the vapour, but now its VP is p_A, and so $\mu_A(g) = \mu_A^{\ominus} + RT\ln(p_A/p^{\ominus})$. This means that we can write

$$\mu_A(l) = \mu_A^{\ominus} + RT\ln(p_A/p^{\ominus}).$$

The two equations can be combined to eliminate the standard chemical potential of the gas, giving

$$\mu_A(l) = \mu_A^*(l) + RT\ln(p_A/p_A^*). \qquad (8.2.6)^{\circ}$$

The final step depends on relating the ratio of vapour pressures to the composition of the liquid, and draws on experimental information. In a series of experiments on mixtures of closely related liquids (such as benzene and toluene), the French chemist François Raoult found that the *ratio* p_A/p_A^* *is proportional to the mole fraction of* A *in the liquid:*

$$\textit{Raoult's Law: } p_A = x_A p_A^*. \qquad (8.2.7)^\circ$$

This is illustrated in Fig. 8.3a.

Some mixtures obey Raoult's Law very well, especially when the components are chemically similar, Fig. 8.3(b). Mixtures that obey the law throughout the composition range from pure A to pure B are called *ideal solutions*. When we write equations relating to them we shall label them with the superscript °. In the case of ideal solutions it follows from eqns (8.2.6) and (8.2.7) that

$$\mu_A(\mathrm{l}) = \mu_A^*(\mathrm{l}) + RT \ln x_A. \qquad (8.2.8)^\circ$$

This important equation can also be turned round and used as the *definition* of an ideal solution (so that it implies Raoult's Law rather than stemming from it). In practice it lets us express the chemical potential of an ideal solution in terms of its composition. We shall use it frequently in the following sections.

Although some mixtures obey Raoult's Law very well, others depart significantly from it, Fig. 8.3(c). Nevertheless, even in the latter cases the law is obeyed increasingly closely for the component in excess (the *solvent*) as it approaches purity, Fig. 8.3(c). This means that the law is a good approximation for the solvent so long as the solution is dilute.

But what of the *solute*, the component present in low concentration? In ideal solutions the solute also obeys Raoult's Law, that its VP is proportional to its mole fraction; but in real solutions, although the VP of the solute is found to be proportional to its mole fraction at low concentrations, the slope is no longer equal to the VP of the pure substance, Fig. 8.4. This different linear dependence was discovered by the English chemist William Henry, and is now called *Henry's Law*. We can express it quantitatively as follows:

$$\textit{Henry's Law: } p_A = x_A K_A. \qquad (8.2.9)$$

x_A is the mole fraction of the solute and K_A is some constant (with the dimensions of pressure) chosen so that the plot of the VP of B against its mole fraction is tangential to the experimental curve at $x_A = 0$, Fig. 8.4. Mixtures obeying Henry's Law are less ideal than those obeying Raoult's Law throughout their composition range, and are called *dilute ideal solutions*.

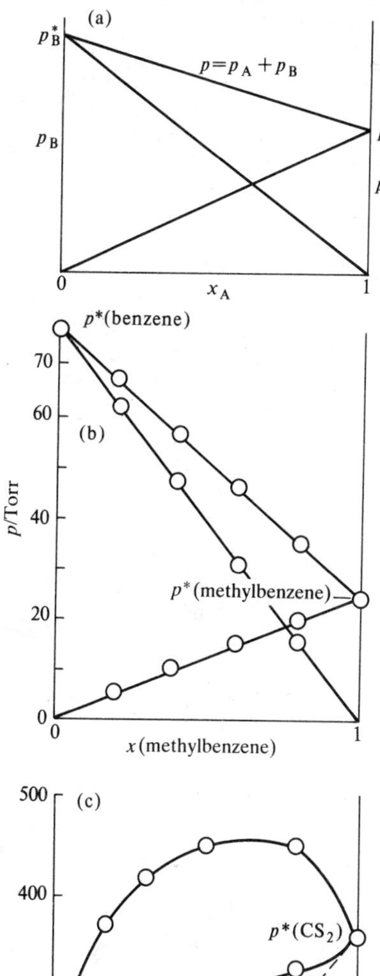

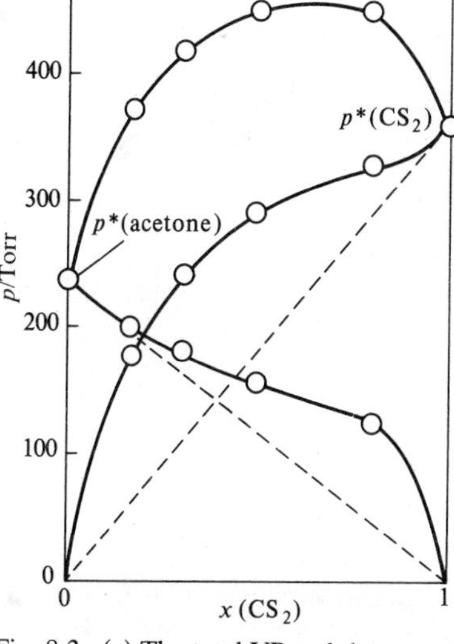

Fig. 8.3. (a) The total VP and the two partial VPs of an ideal binary mixture are proportional to the mole fractions of the components. (b) Two similar liquids, in this case benzene and methylbenzene, behave almost ideally, but (c) strong deviations from ideality are shown by dissimilar liquids (in this case carbon disulphide and acetone).

Example 8.3

The VP and partial VPs of a mixture of propanone (acetone, A) and chloroform (trichloromethane, C) were measured at 35 °C with the following results:

$x_{\text{chloroform}}$	0	0.20	0.40	0.60	0.80	1
p_C/Torr	0	35	82	142	219	293
p_A/Torr	347	270	185	102	37	0

Confirm that the mixture conforms to Raoult's Law for the component in large excess and to Henry's Law for the minor component. Find the Henry's Law constants.

• *Method*. Plot the partial VPs and the total VP against mole fraction. Raoult's Law is tested by comparing the data with the straight line $p = xp^*$ for each component. Henry's Law is tested by finding a straight line $p = xK$ that is tangential to each partial VP at low x.

• *Answer*. The data are plotted in Fig. 8.5 together with the Raoult's Law lines. Henry's Law requires $K_A = 175$ Torr and $K_C = 165$ Torr.

• *Comment*. Notice how the data deviate from both Raoult's and Henry's Laws even for quite small departures from $x = 1$ and $x = 0$ respectively. We deal with these deviations in Section 8.5.

• *Exercise*. The partial VP of methyl chloride at various mole fractions in a mixture at 25 °C was found to be as follows:

x	0.005	0.009	0.0019	0.0024
p/Torr	205	363	756	946

Estimate Henry's Law constant. [4×10^5 Torr]

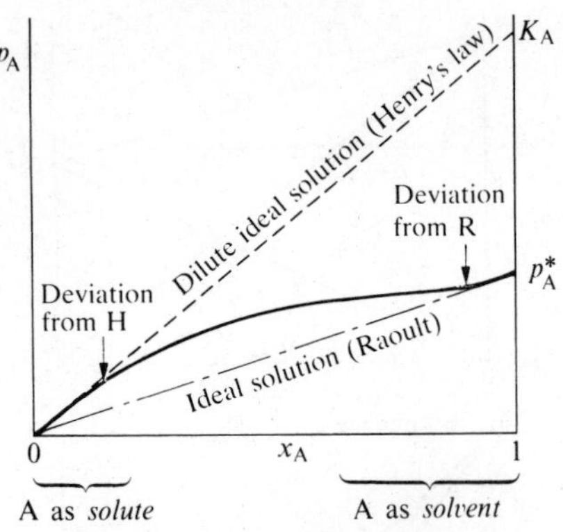

Fig. 8.4. When a component (the *solvent*) is nearly pure, it behaves according to Raoult's Law and has a VP that is proportional to mole fraction with a slope p_A^*. When it is the minor component (the *solute*) its VP is still proportional to the mole fraction, but the constant of proportionality is now K_A. This is Henry's Law.

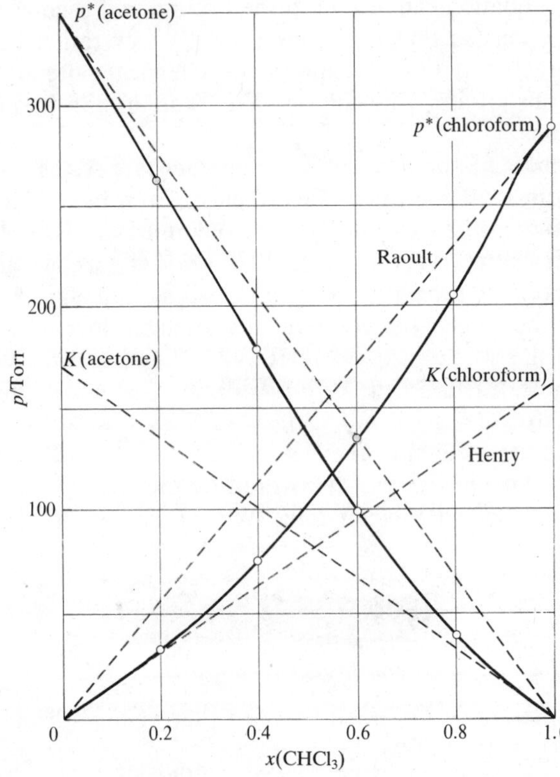

Fig. 8.5. The experimental partial VPs of a mixture of chloroform and acetone based on the data in *Example* 8.3. The values of K are obtained by extrapolating the dilute solution VPs as explained in the *Example*.

Some Henry's Law data are listed in Table 8.1. They are often used in calculations relating to gas solubilities (where the solute is a gas), as the following *Example* illustrates.

Table 8.1. Henry's Law constants for gases in water at 298 K

	K/Torr
H_2	5.34×10^7
N_2	6.51×10^7
O_2	3.30×10^7
CO_2	1.25×10^6

Example 8.4

Estimate the solubility of oxygen in water at 25 °C and a partial pressure of 190 Torr.

• *Answer*. The mole fraction of solute is given by Henry's Law as $x = p/K$, p being its partial pressure. Calculate the amount of O_2 dissolved in 1.00 kg water. Since the amount dissolved is small, approximate the mole fraction by writing

$$x(O_2) = n(O_2)/\{n(O_2) + n(H_2O)\} \approx n(O_2)/n(H_2O),$$

so that $n(O_2) = x(O_2)n(H_2O) = pn(H_2O)/K$.

• *Answer*. From the data in Table 8.1,

$$n(O_2) = (190\ \text{Torr}) \times (1.00\ \text{kg}/18.02\ \text{g mol}^{-1})/(3.30 \times 10^7\ \text{Torr})$$
$$= 3.2 \times 10^{-4}\ \text{mol}.$$

The molality is therefore 3.2×10^{-4} mol kg^{-1}, corresponding to a concentration of approximately 3.2×10^{-4} M.

• *Comment*. Knowledge of Henry's Law constants for gases in fats and lipids is important for the discussion of respiration, especially when the partial pressure of oxygen is abnormal, as in diving and mountaineering.

• *Exercise*. Calculate the concentration of nitrogen in water exposed to air at 25 °C; partial pressures were calculated in *Example* 1.2. [5.1×10^{-4} M]

8.2 (d) Liquid mixtures

The Gibbs function of mixing of two liquids to form an ideal solution is calculated as follows. When the liquids are separate the total Gibbs function is

$$G_i = n_A\mu_A^*(l) + n_B\mu_B^*(l).$$

When they are mixed the individual chemical potentials are given by eqn (8.2.8), and so the total Gibbs function is

$$G_f = n_A\{\mu_A^*(l) + RT \ln x_A\} + n_B\{\mu_B^*(l) + RT \ln x_B\}.$$

Consequently, the Gibbs function of mixing is

$$\Delta_{mix}G = G_f - G_i = nRT\{x_A \ln x_A + x_B \ln x_B\}, \qquad (8.2.10)°$$

where $n = n_A + n_B$. This expression is the same as that for two perfect gases, and so all the conclusions drawn there are valid here: the driving force for mixing is the increasing entropy of the system as the particles mingle, the enthalpy and internal energy of mixing are zero, and there is no change of volume. Note, however, that solution ideality means something different from gas perfection. In the latter there are no interactions between particles. In ideal solutions there *are* interactions, but the average A–B interactions in the mixture are the same as the average A–A and B–B interactions in the pure liquids. The dependences of the Gibbs function, entropy, and enthalpy of mixing on the composition of the mixture are shown in Fig. 8.6.

Real solutions are composed of particles for which A–A, A–B, and B–B interactions are all different. Not only may there be an enthalpy change when liquids mix, but there may also be an additional contribution to the entropy change arising from the way that the particles of one type might cluster together instead of mingling freely with the others. If the enthalpy change is large and positive (i.e. endothermic) or if the entropy change is adverse (because of a reorganization of the particles that results in an orderly mixture), the Gibbs function might be *positive* for mixing, in which case separation is spontaneous and the liquids may be *immiscible* or only

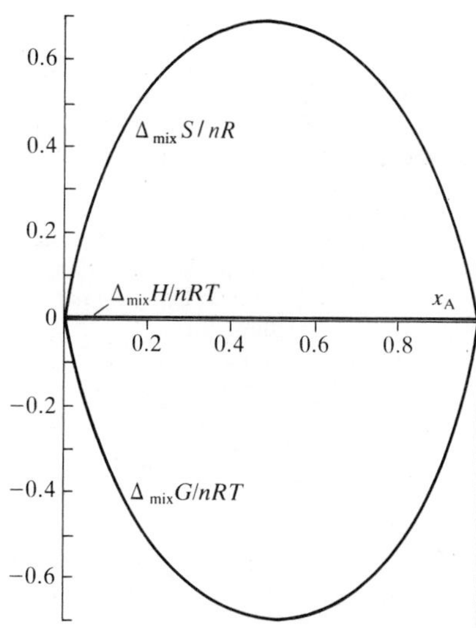

Fig. 8.6. The Gibbs function, enthalpy, and entropy of mixing of an ideal binary mixture, calculated using eqn (8.2.10). The entropy is expressed as a multiple of nR, the Gibbs function as a multiple of nRT.

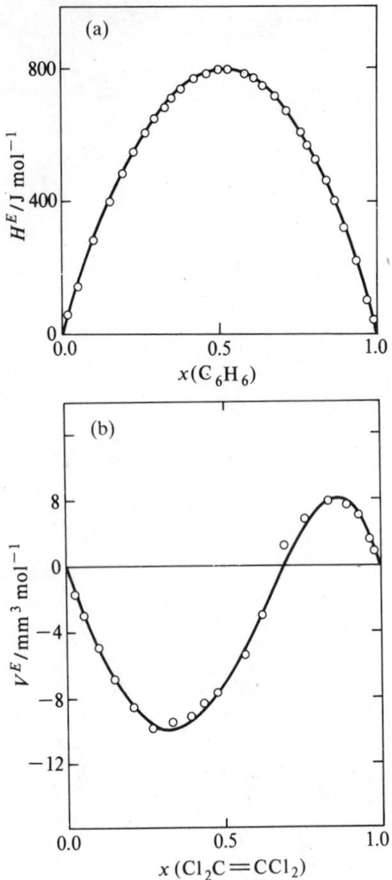

Fig. 8.7. Experimental excess functions at 25 °C. (a) H^E for benzene/cyclohexane; this shows that the mixing is endothermic (because $\Delta_{mix}H = 0$ for an ideal solution). (b) V^E for tetrachloroethane/cyclopentane; this shows that there is a contraction at low tetrachloroethane mole fractions, but an expansion at high mole fractions (because $\Delta_{mix}V = 0$ for a perfect mixture). [K. N. March and I. A. McLure, *Int. DATA Ser., Selec, Data Mixtures, Ser.* A, 1973, 2 and 9.1.]

partially miscible (when the Gibbs function is negative over a limited range of compositions). The thermodynamic properties of real solutions may be expressed in terms of the *excess functions*, G^E, S^E, etc. An excess function is defined as the *difference between the observed thermodynamic function of mixing and the function for an ideal solution*. In the case of the *excess entropy*, for example,

$$S^E = \Delta_{mix}S + nR\{x_A \ln x_A + x_B \ln x_B\}. \qquad (8.2.11)$$

Deviations of the excess functions from zero then indicate the extent to which the solutions are non-ideal. In this connection a useful model system is the *regular solution*, in which the excess enthalpy is non-zero, but the excess entropy is zero. A regular solution can be thought of as one in which the two kinds of particles are distributed randomly but have different interactions with each other. Two examples of the composition dependence of excess functions are shown in Fig. 8.7.

8.3 Colligative properties

In this section we see how to calculate the effect of a solute on the boiling and freezing points of mixtures, and how a solute gives rise to an osmotic pressure. All the properties we consider depend only on the number of solute particles present, not their identity, and for this reason they are called *colligative properties* (denoting 'depending on the collection').

We make two assumptions. The first is that *the solute is not volatile*, and so it does not contribute to the vapour. The second is that the *solute does not dissolve in the solid solvent*. The latter assumption is quite drastic, although it is true of many mixtures; it can be avoided at the expense of more algebra, but that introduces no new principles. Both assumptions are removed in the general but qualitative discussion of mixtures in Chapter 9.

8.3 (a) The common feature

The common feature of colligative properties is the reduction of the chemical potential of the liquid solvent as a result of the presence of solute. The reduction is from $\mu_A^*(l)$ for the pure solvent to $\mu_A^*(l) + RT \ln x_A$ when a solute is present (x_A being the mole fraction of the solvent, $\ln x_A$ is negative). The chemical potentials of the vapour and solid are unchanged by the presence of the non-volatile, insoluble solute. As can be seen from Fig. 8.8, this implies that the liquid–vapour equilibrium occurs at a higher temperature (the boiling point is raised) and the solid–liquid equilibrium occurs at a lower temperature (the freezing point is lowered).

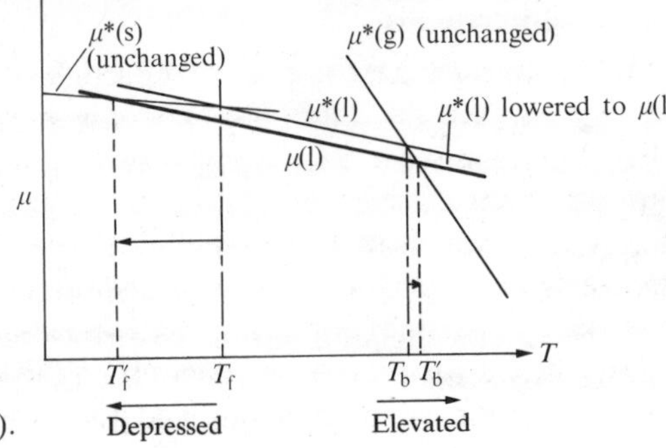

Fig. 8.8. The chemical potential of a solvent in the presence of a solute. Note that the lowering of the liquid's chemical potential has a greater effect on the freezing point than on the boiling point because of the angles at which the lines intersect (which are determined by entropies, recall Fig. 7.1).

The physical basis of the lowering of the chemical potential is not due to the energy of interaction of the solute and solvent particles, because it occurs even in ideal solutions (which have zero enthalpy of mixing). If it is not an enthalpy effect it must be an entropy effect. In the absence of a solute, the pure liquid solvent has an entropy that reflects its disorder. Its VP reflects the tendency of the universe towards greater entropy, which can be achieved if the liquid evaporates to form a more random gas. When a solute is present there is an additional contribution to the entropy of the liquid, even in an ideal solution, and so the tendency for the solvent to escape in order to acquire higher entropy is not so compelling. When there is already an additional randomness in the liquid, the universe reaches its maximum entropy when less liquid evaporates than when the solute, and the randomness it causes, is absent. This appears as a lowered VP, and hence a higher boiling point. Similarly, the enhanced randomness of the solution retards the tendency to freeze and a lower temperature must be reached before equilibrium between solid and solution is achieved. Hence, the freezing point is lowered.

The strategy for the quantitative discussion is to look for the temperature at which one phase (the pure vapour or the pure solid) has the same chemical potential as the solvent in the solution. This is the new equilibrium temperature.

8.3 (b) The elevation of boiling point

The equilibrium of interest is between the solvent vapour and the solvent in solution, Fig. 8.9. We denote the solvent by A and the solute by B. The equilibrium is established at a temperature given by

$$\mu_A(g) = \mu_A^*(l) + RT \ln x_A.$$

This equation rearranges into

$$\ln(1 - x_B) = \{\mu_A(g) - \mu_A^*(l)\}/RT = \Delta G_{vap,m}(T)/RT.$$

where $\Delta G_{vap,m}$ is the molar Gibbs function of vaporization of the pure solvent and B is the mole fraction of the solute (and we have used $x_A + x_B = 1$).

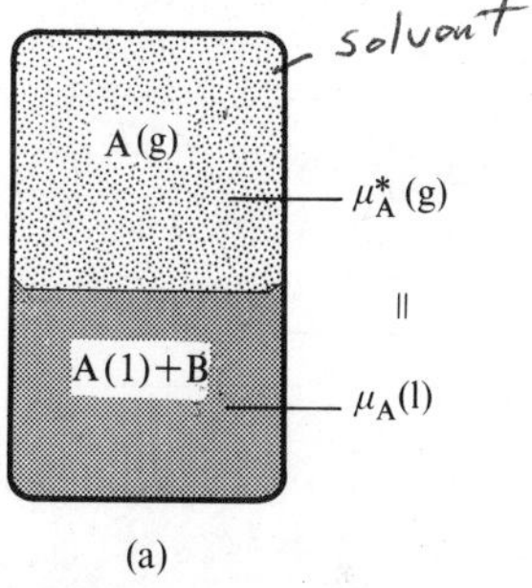

Fig. 8.9. The equilibrium involved in the calculation of the elevation of boiling point is between A in the pure vapour and A in the mixture, A being the solvent and B an involatile solute.

There are various ways of solving this equation for T in terms of x_B. The best uses the Gibbs–Helmholtz equation, eqn (6.2.4), and allows the temperature dependence of ΔG to be taken into account exactly (see Problem 8.25). A simpler procedure is as follows. When $x_B = 0$ the boiling point is T^*, and so in this case

$$\ln 1 = \Delta G_{vap,m}(T^*)/RT^*.$$

(In a moment we recognize that $\ln 1 = 0$.) The difference of this equation and the previous one, after making use of $\Delta G = \Delta H - T\Delta S$, is

$$\begin{aligned}\ln(1 - x_B) - \ln 1 &= \{\Delta G_{vap,m}(T)/RT\} - \{\Delta G_{vap,m}(T^*)/RT^*\} \\ &= \{\Delta H_{vap,m}(T)/RT - \Delta S_{vap,m}(T)/R\} \\ &\quad - \{\Delta H_{vap,m}(T^*)/RT^* - \Delta S_{vap,m}(T^*)/R\}.\end{aligned}$$

Notice that we have been careful to allow for the temperature dependence of the Gibbs function so far. We now suppose that the amount of solute present is so small that $x_B \ll 1$. This has two consequences. The first is that $\ln(1 - x_B) \approx -x_B$. The second is that the elevation of boiling point can be anticipated to be small. That being so, we can take the enthalpy and entropy of vaporization as independent of the temperature over the range of interest. Then the two entropy terms in the last expression cancel, and we are left with

$$x_B = -(\Delta H_{vap,m}/R)\{(1/T) - (1/T^*)\}.$$

This equation can be rearranged into an expression for the elevation of boiling point, $\delta T = T - T^*$. The two reciprocals combine into $(T^* - T)/TT^*$, but as T is so close to T^* this is almost the same as $(T^* - T)/T^{*2}$. It follows that

$$\delta T = (RT^{*2}/\Delta H_{vap,m})x_B. \qquad (8.3.1)°$$

The formula we have just derived makes no reference to the identity of the solute, only to its mole fraction. This distinguishes the elevation of boiling point as a colligative property. The value of δT does depend on the properties of the solvent, and the biggest changes occur for solvents with high boiling points but low molar enthalpies of vaporization. The elevation is normally expressed in terms of the molality, m_B, of the solute as

$$\textit{Elevation of boiling point: } \delta T = K_b m_B, \qquad (8.3.2)$$

where K_b is the *ebullioscopic constant* for the solvent. The derivation of this result is set out in the following *Example*, and some values of ebullioscopic constants are given in Table 8.2. Although (as eqn (8.3.1) shows) the constants are related to the properties of the solvent, they are best determined experimentally. Once measured they can be used to determine the molar masses of solutes, the technique being called *ebullioscopy*.

Table 8.2. Cryoscopic and ebullioscopic constants

	K_f/K kg mol^{-1}	K_b/K kg mol^{-1}
Benzene	3.90	3.07
Camphor	40	—
Phenol	7.27	3.04
Water	1.86	0.51

Example 8.5

Confirm that the elevation of boiling point can be written as in eqn (8.3.2) when the molality m_B is low. In an experiment, 10.0 g of a solid were dissolved in 100.0 g of benzene and the boiling point increased by 0.800 K. What is the RMM of the solute?

• *Method*. Begin at eqn (8.3.1) and use the smallness of the molality to write

$$x_B = n_B/n \approx n_B/n_A.$$

Consider 1 kg of a solution. Then $n_A = 1\text{ kg}/M_{A,m}$ and $n_B = (1\text{ kg}) \times m_B$. Take the value of K_b from Table 8.2.

• *Answer*. Since at low molalities

$$x_B = n_B/n_A = (1\text{ kg}) \times m_B/(1\text{ kg}/M_{A,m}) = m_B M_{A,m},$$

it follows from eqn (8.3.1) that

$$\delta T = (RT^{*2}/\Delta H_{vap,m})M_{A,m}m_B = K_b m_B,$$

with $K_b = RT^{*2}M_{A,m}/\Delta H_{vap,m}$.

For the numerical part, $\delta T = 0.800$ K and $K_b = 2.53$ K/(mol kg^{-1}). Therefore, the solute molality is

$$m_B = 0.800\text{ K}/2.53\text{ K mol}^{-1}\text{ kg} = 0.316\text{ mol kg}^{-1}.$$

If the solute's molar mass is $M_{B,m}$, the molality of a 10.0 g solution in 100.0 g of solvent (i.e.,

100 g solute in 1 kg) is $m_B = (100\ \text{g}/M_{B,m})\ \text{kg}^{-1}$, so that

$$M_{B,m} = 100\ \text{g}/(m_B\ \text{kg}) = 100\ \text{g}/(0.316\ \text{mol}) = 320\ \text{g mol}^{-1}.$$

Therefore, $M_{B,r} = 320$.

• *Comment*. Check that x_B is small (for consistency). In this case $x_B = m_B M_{B,m} = 0.02 \ll 1$. Note that the expression for K_b shows that it is large for solvents with high boiling points, high molar masses, and low enthalpies of vaporization.

• *Exercise*. 1.50 g of a solute raised the boiling point of 100 g of carbon tetrachloride by 0.652 K. Calculate its RMM. [114]

8.3 (c) The depression of freezing point

The equilibrium now of interest is between pure solid solvent and the solution with solute present at a mole fraction x_B, Fig. 8.10. At the freezing point the chemical potentials of A in the two phases are equal:

$$\mu_A^*(\text{s}) = \mu_A^*(\text{l}) + RT \ln x_A.$$

The only difference between this calculation and the last is the appearance of the solid's chemical potential in place of the vapour's. Therefore we can write the result directly from eqn (8.3.1)

$$\delta T = (RT^{*2}/\Delta H_{\text{melt,m}})x_B. \qquad (8.3.3)°$$

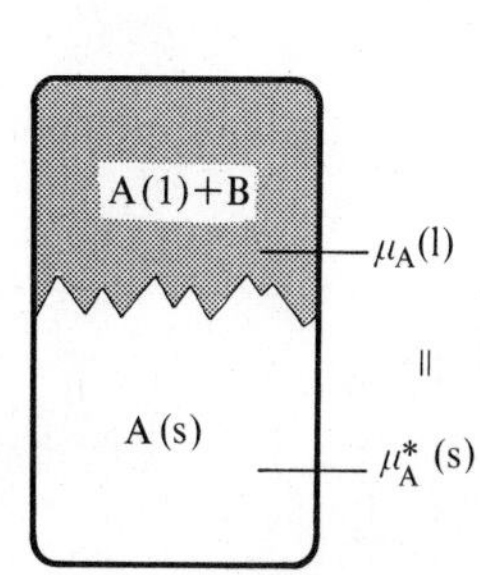

(b)

Fig. 8.10. The equilibrium involved in the calculation of the lowering of freezing point is between A in the pure solid and A in the mixture, A being the solvent and B a solute that is insoluble in solid A.

δT is now the freezing point depression, $T^* - T$, and $\Delta H_{\text{melt,m}}$ is the molar enthalpy of melting. When the solution is dilute the same calculation as set out in the last *Example* allows us to express the depression in terms of the molality:

$$\textit{Depression of freezing point: } \delta T = K_f m_B. \qquad (8.3.4)$$

K_f is the *cryoscopic constant*. It may be calculated from the enthalpy of melting, but it is normally determined experimentally; some values are given in Table 8.2. Big depressions are observed in solvents with low enthalpies of melting and high melting points (as may be deduced from eqn (8.3.3)). Once the cryoscopic constant of a solvent is known, the depression of freezing point may be used to measure the molar mass of a solute. Freezing point depressions are generally larger than boiling point elevations (as Fig. 8.8 suggests), and the measurement is less likely to damage and possibly degrade the solute; hence *cryoscopy* is more commonly used than ebullioscopy. The following *Example* shows how the method is used.

Example 8.6

Very precise measurements in the freezing point of a $1.00 \times 10^{-3}\ \text{mol kg}^{-1}\ CH_3CO_2H(aq)$ solution gave a freezing point depression of 2.1 mK. What can be inferred?

• *Method*. The expected depression can be calculated from the information in Table 8.2. A discrepancy indicates that a different number of particles is present in solution than the molality itself indicates. Assess the degree of ionization of the acid by considering the equilibrium

$$CH_3CO_2H(aq) \rightleftharpoons CH_3CO_2^-(aq) + H^+(aq).$$

• *Answer*. From $K_f = 1.86\ \text{K}/(\text{mol kg}^{-1})$ and $m_B = 1.00 \times 10^{-3}\ \text{mol kg}^{-1}$ we predict $\delta T = 1.86\ \text{mK}$. Since the observed value is greater, this suggests that more solute particles are present. For each CH_3CO_2H molecule that dissociates, one $CH_3CO_2^-$ and one H^+ are produced, and so the actual molality of the solution is $(1-\alpha)m_B + \alpha m_B + \alpha m_B = (1+\alpha)m_B$ if a fraction $1-\alpha$ remain non-ionized. Therefore, to account for a depression of 2.1 mK instead of 1.86 mK, we require $1+\alpha = 2.1/1.86 = 1.1$, suggesting that 10% of the acid molecules are ionized.

• *Comment*. We shall explore the consequences of the partial ionization of acid molecules in Chapter 12. Cryoscopy is one way of assessing the extent of ionization and its dependence on the molality of the acid. However, care has to be taken with the interpretation, because ions give strongly non-ideal solutions, as we shall shortly see.

• *Exercise*. 120 mg of benzoic acid ($C_6H_5CO_2H$) were dissolved in 100 g of water, and a depression of freezing point of 20 mK was observed. What proportion of the molecules are ionized? [8%]

8.3 (d) Solubility

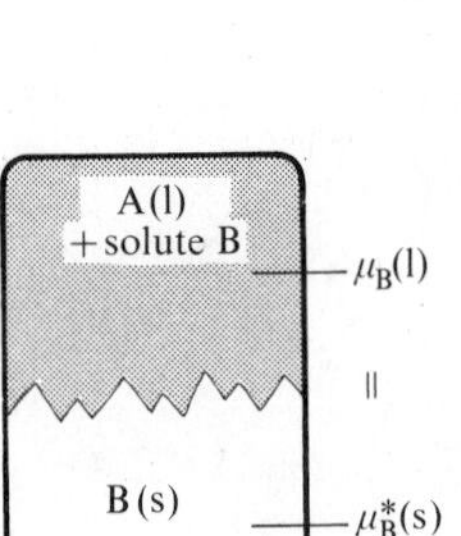

Fig. 8.11. The equilibrium involved in the calculation of the solubility is between pure solid B and B in the mixture.

Although it is not strictly a colligative property, the solubility of a solute may be estimated by the same techniques as we have been using. If a solid solute is left in contact with a solvent it dissolves until the solution is *saturated*. Saturation corresponds to equilibrium, the chemical potential of the pure solid solute, $\mu_B(s)$, being equal to its value in solution, $\mu_B^*(l) + RT \ln x_B$, Fig. 8.11:

$$\mu_B^*(s) = \mu_B^*(l) + RT \ln x_B.$$

This is the same as the starting equation of the last section, except that the quantities refer to the solute B, not the solvent A.

The starting point is the same but the aim is different. In the present case we want to find the mole fraction of B in solution at equilibrium when the temperature is T. At the melting point of the solute, T^*, we know that $\Delta G_{melt} = 0$, and so $\Delta G_{melt,m}(T^*)/RT^* = 0$ too. Since the last equation rearranges to

$$\ln x_B = -\{\mu_B^*(l) - \mu_B^*(s)\}/RT,$$

it is also true that

$$\ln x_B = -\{\Delta G_{melt,m}(T)/RT - \Delta G_{melt,m}(T^*)/RT^*\}.$$

If we now assume that ΔH and ΔS barely change over the temperature range of interest we may assume them to be constant, so that

$$\ln x_B = -(\Delta H_{melt,m}/R)\{(1/T) - (1/T^*)\}, \qquad (8.3.5)^\circ$$

where x_B is the mole fraction of B present in solution at saturation. This shows that the solubility of B decreases exponentially as the temperature is lowered from its melting point, and that solutes with high melting points and large enthalpies of melting have low solubilities at normal temperatures.

8.3 (e) Osmosis

The phenomenon of *osmosis* (from the Greek for 'push') is the tendency of a pure solvent to enter a solution separated from it by a *semipermeable membrane* (a membrane permeable to the solvent but not to the solute), Fig. 8.12. One of the most important examples of the process is transport of fluids through cell membranes, but it is also the basis of an important technique for the measurement of molar masses, especially of macromolecules. In the arrangement shown in Fig. 8.12(a) osmosis results in water flowing into the solution. The flow can be opposed by applying pressure to the solution, and the *osmotic pressure* (symbol: Π) is the pressure that just stops the flow. In the simple arrangement shown in Fig. 8.12(b) the opposing pressure arises from the head of solution that the osmosis itself produces, and equilibrium is reached when the hydrostatic pressure of the column of solution matches the osmotic pressure. The complication of this 'simpler' arrangement is that the entry of solvent into the solution results in its dilution, and so it is more difficult to treat than the arrangement in Fig. 8.12(a) where there is no flow and the concentrations remain unchanged.

The thermodynamic treatment of osmosis is straightforward. The strategy is to note that at equilibrium the chemical potential of the solvent must be the same on each side of the membrane. On the pure solvent side the chemical potential of the solvent is $\mu_A^*(1)$. On the solution side the chemical potential is lowered by the presence of the solute but it is raised on account of the greater pressure. The equilibrium may therefore be expressed as

$$\mu_A^*(1; p) = \mu_A(1; x_A; p + \Pi). \qquad (8.3.6)$$

The presence of solute is taken into account in the normal way:

$$\mu_A(1; x_A; p + \Pi) = \mu_A^*(1; p + \Pi) + RT \ln x_A. \qquad (8.3.7)^\circ$$

We saw in Section 6.2(b) how to take the effect of pressure into account (use eqn (6.2.6)):

$$\mu_A^*(1; p + \Pi) = \mu_A^*(1; p) + \int_p^{p+\Pi} V_{A,m}^* \, dp. \qquad (8.3.8)$$

When the last three equations are combined we have

$$-RT \ln x_A = \int_p^{p+\Pi} V_{A,m}^* \, dp. \qquad (8.3.9)^\circ$$

This expression relates the osmotic pressure to the mole fraction of solute ($x_B = 1 - x_A$), and so confirms that osmosis is another colligative property.

In the case of dilute solutions, $\ln x_A$ may be replaced by $\ln(1 - x_B) = -x_B$. We may also assume that the pressure range in the integration is so small that the molar volume of the solvent is a constant. That being so, $V_{A,m}^*$ may be taken outside the integral. This gives

$$RTx_B = \Pi V_{A,m}^*.$$

The mole fraction of the solute is $n_B/(n_A + n_B)$, which is almost the same as n_B/n_A when the solution is dilute. Since $n_A V_m^* = V$, the volume of the solvent, the equation simplifies to the

Van't Hoff equation: $\Pi V = n_B RT$. $\qquad (8.3.10)^\circ$

Since $n_B/V = [B]$, the *concentration* of the solute, a simpler form of this equation is $\Pi = [B]RT$. However, it applies only to dilute solutions where we can be confident that they are behaving ideally (and that eqn (8.3.7) is valid). One of the most common uses of *osmometry* (the measurement of molar masses on the basis of osmotic pressure) is to macromolecules, such

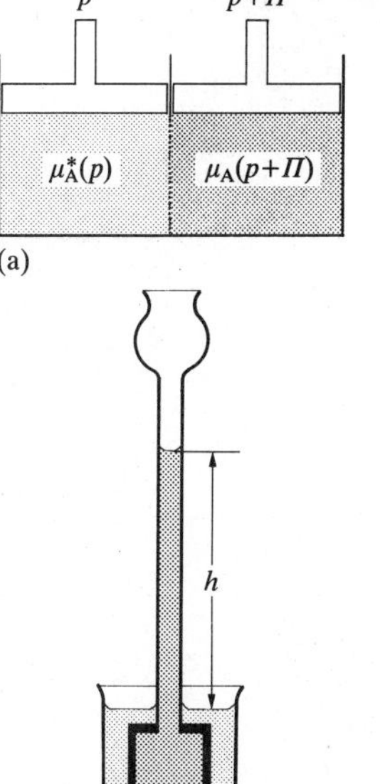

Fig. 8.12. (a) The equilibrium involved in the calculation of osmotic pressure is between pure solvent A at a pressure p on one side of the semipermeable membrane and A as a component of the mixture on the other side of the membrane, where the pressure is $p + \Pi$. (b) In a simple version, A is at equilibrium on each side of the membrane when enough has passed into the solution to cause a hydrostatic pressure difference. However, this leads to a dilution of the solution, and so the interpretation of Π is less straightforward.

as proteins and synthetic polymers. Their huge molecules dissolve to produce solutions that are far from ideal. In these cases it is assumed that the van't Hoff equation is only the first term of a virial-like expansion:

$$\Pi/[\mathrm{B}] = RT\{1 + B[\mathrm{B}] + \ldots\}, \qquad (8.3.11)$$

where the additional terms take the non-ideality into account. The osmotic pressure is measured at a series of concentrations, and a plot of $\Pi/[\mathrm{B}]$ against [B] is used to find the molar mass of B. The method is illustrated in the following *Example*, and will be returned to when we consider macromolecules in more detail in Part 2.

Example 8.7

The osmotic pressures of solutions of polyvinyl chloride (PVC) in cyclohexanone at 298 K are given below. The pressures are expressed in terms of the heights of solution (of density 0.980 g cm^{-3}) in balance with the osmotic pressure. Find the molar mass of the polymer.

$c/\mathrm{g\,dm^{-3}}$	1.00	2.00	4.00	7.00	9.00
h/cm	0.28	0.71	2.01	5.10	8.00

• *Method*. Use eqn (8.3.11) with $[\mathrm{B}] = c/M_{\mathrm{m}}$, where c is the mass concentration, and $\Pi = \rho gh$, $g = 9.81\ \mathrm{m\,s^{-2}}$. Since then

$$h/c = (RT/\rho g M_{\mathrm{m}})\{1 + (B/M_{\mathrm{m}})c + \ldots\},$$

plot h/c against c, and expect a straight line with intercept $RT/\rho g M_{\mathrm{m}}$ at $c = 0$.

• *Answer*. Draw up the following table:

$c/\mathrm{g\,dm^{-3}}$	1.00	2.00	4.00	7.00	9.00
$(h/c)/(\mathrm{cm/g\,dm^{-3}})$	0.28	0.36	0.503	0.739	0.889

The points are plotted in Fig. 8.13. The intercept is at 0.21. Therefore,

$$M_{\mathrm{m}} = (RT/\rho g)/(0.21\ \mathrm{cm\,g^{-1}\,dm^3})$$

$$= \frac{(8.314\ \mathrm{J\,K^{-1}\,mol^{-1}}) \times (298\ \mathrm{K})}{(0.980\ \mathrm{cm^{-3}}) \times (9.81\ \mathrm{m\,s^{-2}}) \times (0.21\ \mathrm{cm\,g^{-1}\,dm^3})}$$

$$= 120\ \mathrm{kg\,mol^{-1}} = 120\,000\ \mathrm{g\,mol^{-1}}.$$

Hence, $M_{\mathrm{r}} = 120\,000$.

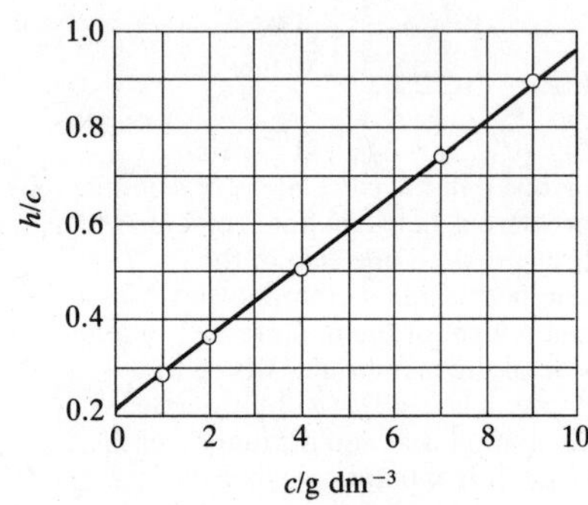

Fig. 8.13. The plot involved in the determination of molar mass by osmometry; see *Example* 8.7. The molar mass is calculated from the intercept at $c = 0$; in Chapter 25 we shall see that additional information comes from the slope.

• *Comment*. Osmometry is a very important technique for the measurement of molar masses of macromolecules, partly because the other colligative techniques give such small effects. We take up the discussion again in Chapter 25.

• *Exercise*. Estimate the depression of freezing point of the most concentrated of these solutions, taking K_{f} as about 10 K/(mol kg^{-1}). [0.8 mK]

8.4 Mixtures of volatile liquids

In this section we consider the relation of the boiling point of a binary liquid mixture to its composition. In the process we see how to break away from ideal solutions and describe the properties of real mixtures.

8.4 (a) Vapour pressure diagrams

In an ideal solution of two liquids the vapour pressures of the components are related to the composition by Raoult's Law:

$$p_{\mathrm{A}} = x_{\mathrm{A}} p_{\mathrm{A}}^*, \qquad p_{\mathrm{B}} = x_{\mathrm{B}} p_{\mathrm{B}}^*, \qquad (8.4.1)^\circ$$

where p_A^* is the VP of pure A and p_B^* that of pure B. The total VP of the mixture is therefore

$$p = p_A + p_B = x_A p_A^* + x_B p_B^* = p_B^* + (p_A^* - p_B^*)x_A, \qquad (8.4.2)°$$

which shows that the total VP (at some fixed temperature) changes linearly with the composition from p_A^* to p_B^*, Fig. 8.14.

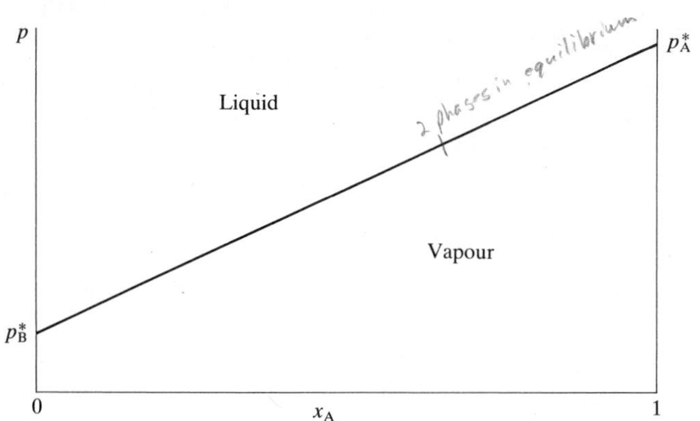

Fig. 8.14. The dependence of the total VP of a binary mixture on the mole fraction of A in the liquid when Raoult's Law is obeyed (recall Fig. 8.3(a) for a similar case).

Figure 8.14 is also a *phase diagram* because the straight line represents the pressures where the two phases are in equilibrium. All points above the line (when the pressure exceeds the VP) correspond to the liquid being the stable phase; all points below the line (when the pressure is less than the VP, so that the sample evaporates) correspond to the vapour being stable.

When liquid and vapour are in equilibrium their compositions are not necessarily the same. Common sense suggests that the vapour should be richer in the more volatile component. This can be confirmed as follows. The partial pressures of the components are given by eqn (8.4.1). It follows from Dalton's Law that the mole fractions *in the gas*, y_A and y_B, are

$$y_A = p_A/p, \qquad y_B = p_B/p.$$

The partial pressures and the total pressure may be expressed in terms of the mole fractions in the liquid using eqns (8.4.1) and (8.4.2), which give

$$y_A = \frac{x_A p_A^*}{p_B^* + (p_A^* - p_B^*)x_A} \qquad (8.4.3)°$$

and $y_B = 1 - y_A$. We now have to show that, if A is the more volatile component, its mole fraction in the vapour, y_A, is greater than its mole fraction in the liquid, x_A. Figure 8.15 shows the dependence of y_A on x_A for

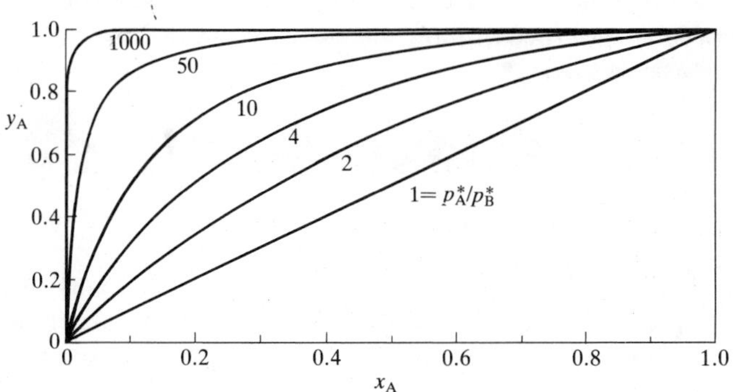

Fig. 8.15. The mole fraction of A in the vapour of a binary ideal solution expressed in terms of its mole fraction in the liquid, calculated using eqn (8.4.3), for various values of p_A^*/p_B^* with A more volatile than B. In all cases the vapour is richer than the liquid in A.

various values of p_A^*/p_B^* greater than unity: we see that in all cases y_A is greater than x_A, as we expect. Note that if B is non-volatile, so that $p_B^* = 0$ at the temperature of interest, then it makes no contribution to the vapour ($y_B = 0$).

Equation (8.4.2) shows how the total VP depends on the composition of the liquid. Since we can relate the composition of the liquid to the composition of the vapour through eqn (8.4.3) we can now relate the VP to the composition of the vapour itself. This leads to

$$p = \frac{p_A^* p_B^*}{p_A^* + (p_B^* - p_A^*)y_A}. \qquad (8.4.4)°$$

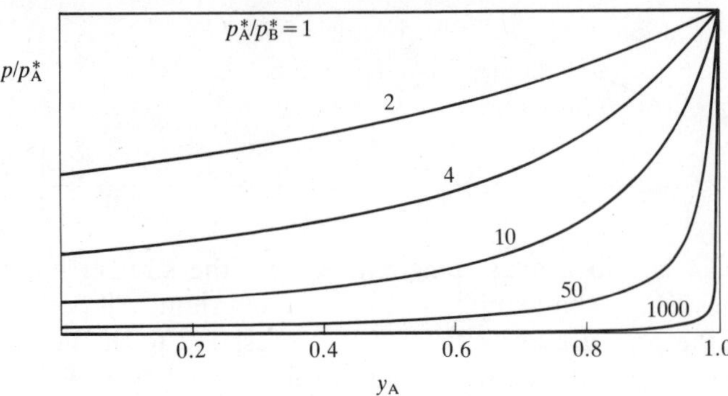

Fig. 8.16. The dependence of the VP of the same system as in Fig. 8.14, but expressed in terms of the mole fraction of A in the vapour using eqn (8.4.4).

This curve is plotted in Fig. 8.16. The illustration is another way of drawing the phase diagram because the line is the boundary of the regions of stability of the two phases.

We can use either diagram to discuss the phase equilibria of the mixture. However, if we are interested in distillation both the vapour and the liquid compositions are of equal interest, and it is then sensible to combine both diagrams into one. This is done in Fig. 8.17, where the composition axis is labelled z_A, the *overall* mole fraction of A in the system. On and above the upper line only liquid is present and then z_A is the same as x_A; on and below the lower line only vapour is present, and then z_A is the same as y_A. This leaves the region between the lines to be interpreted: a point in the unshaded region of the diagram indicates not only *qualitatively* that both liquid and vapour are present, but represents *quantitatively* the relative amounts of each.

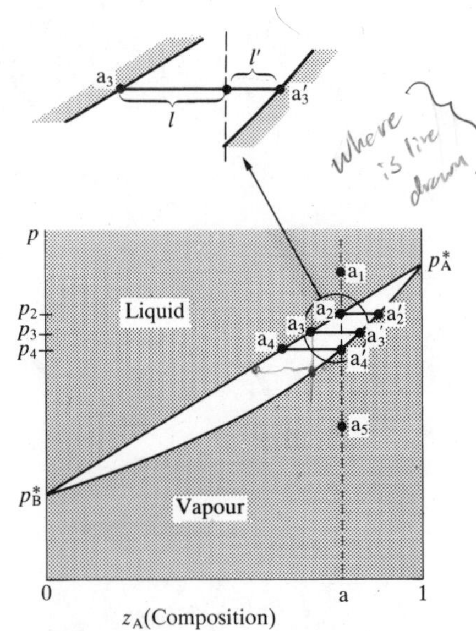

Fig. 8.17. The dependence of the total VP of an ideal solution on the mole fraction of A in the entire system. A point in the white sector corresponds to both liquid and vapor being present with proportions given by the lever rule. The enlargement shows the case discussed in the text, where the amount of liquid is about 0.7 times the amount of vapour (i.e. $l'/l = 0.7$).

Refer to Fig. 8.17. Let the overall composition be $z_A = \text{a}$ and the pressure be p_3. Since the point (a, p_3) lies in the unshaded sector, we know at once that both phases are present in equilibrium. In order to find the relative amounts, we measure the distances l and l' along the horizontal *tie-line*, and then use the

$$\textit{Lever rule: } n(\text{l})/n(\text{g}) = l'/l. \qquad (8.4.5)$$

In the present case, since $l' = 0.7l$, the amount of liquid, $n(\text{l})$ is about 0.7 the amount of vapour, $n(\text{g})$.

The proof of the lever rule is as follows. Let the amounts of liquid and vapour be $n(\text{l})$ and $n(\text{g})$ and their sum be n. The overall amount of A is nz_A. The overall amount of A is also the sum of its amounts in the two phases, $n(\mathrm{l})x_\mathrm{A}$ and $n(\mathrm{g})y_\mathrm{A}$:

$$nz_\mathrm{A} = n(\mathrm{l})x_\mathrm{A} + n(\mathrm{g})y_\mathrm{A}.$$

Furthermore, as $n = n(\mathrm{l}) + n(\mathrm{g})$, we can also write

$$nz_\mathrm{A} = n(\mathrm{l})z_\mathrm{A} + n(\mathrm{g})z_\mathrm{A}.$$

On equating these two expressions it follows that

$$n(\mathrm{l})(x_\mathrm{A} - z_\mathrm{A}) = n(\mathrm{g})(z_\mathrm{A} - y_\mathrm{A}),$$

or

$$n(\mathrm{l})/n(\mathrm{g}) = (y_\mathrm{A} - z_\mathrm{A})/(z_\mathrm{A} - x_\mathrm{A}) = l'/l,$$

as was to be proved.

In order to see in more detail how the rule is used, consider what happens when a mixture of composition a_1 in Fig. 8.17 is subjected to decreasing pressure. The sample remains entirely liquid until the pressure is reduced to p_2, when liquid and vapour can coexist. At p_2 the liquid has composition a_2 and the vapour has composition a_2'. Since $l'/l \approx \infty$ along the tie line at p_2, there is only a trace of vapour present. When the pressure is reduced to p_3 the composition of the liquid is a_3 and that of the vapour is a_3', but overall the composition is still a. The relative amounts of liquid and vapour are given by the value of l'/l measured along the tie-line at p_3. This means that the amount of liquid is about 0.7 times the amount of vapour. Furthermore, the vapour (of composition a_3') is richer in A, the more volatile component, than the liquid (which has composition a_3). When the pressure has fallen to p_4 the sample is almost completely gaseous: the vapour's composition is a_4' and the trace of liquid that remains (trace, because $l'/l \approx 0$) has the composition a_4 and so is rich in B. Only an infinitesimal reduction in pressure is now needed to eliminate the last trace of liquid, and from then on the sample is entirely gaseous and of composition a.

8.4 (b) The representation of distillation

Reducing the pressure at constant temperature is one way of doing distillation, but it is more common to distil at constant pressure by raising the temperature. In order to discuss distillation in this way we need a *temperature–composition diagram*. Figure 8.18 shows such a diagram for an ideal binary solution. The phase lines now mark off the *temperatures* and compositions at which the liquid and vapour are in equilibrium, the pressure being constant (and typically 1 atm). Note that the liquid phase region now lies in the lower part of the diagram.

Using the diagram involves the same kind of interpretation as we have already seen in connection with the pressure–composition diagram. For instance, consider what happens when a liquid of composition a is heated. Initially its state is a_1. It boils when the temperature reaches T_2. Then the liquid has composition a_2 and the vapour (which is present only as a trace) has composition a_2'. The vapour is richer in the more volatile component (A), as common sense leads us to expect. From the location of a_2' we can state the vapour's composition at the boiling point, and from the location of the tie line $\mathrm{a}_2, \mathrm{a}_2'$ we can read off the boiling temperature of the original liquid.

In a distillation experiment the vapour is withdrawn and condensed. If the vapour in this example is drawn off and completely condensed, then the

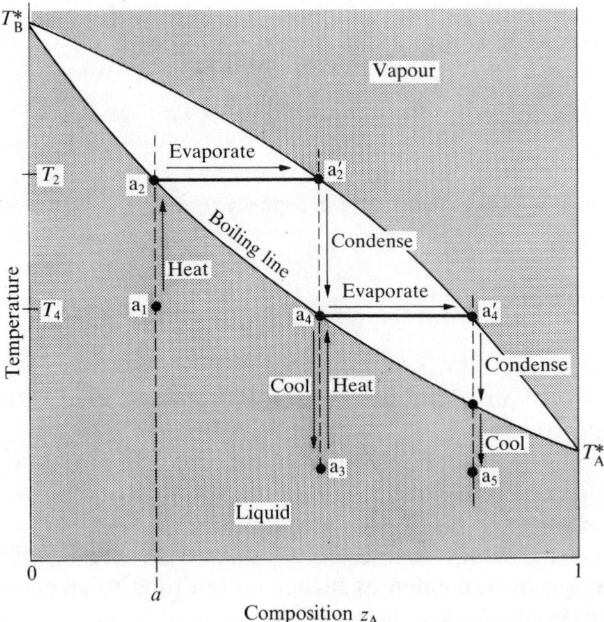

Fig. 8.18. The temperature–composition diagram corresponding to an ideal mixture with A more volatile than B. Successive boilings and condensations of a liquid originally of composition a_1 lead to a condensate that is pure A. This is the process of fractional distillation.

first drop gives a liquid of composition a_3, which is richer in the more volatile component than the original liquid. *Fractional distillation* repeats the boiling and condensation cycle several times. When the condensate of composition a_3 is reheated it boils at T_4 and yields a vapour of composition a_4' which is even richer in the more volatile component. That vapour is drawn off, and the first drop condenses to a liquid of composition a_5. The cycle can then be repeated until in due course almost pure A is obtained.

8.4 (c) Azeotropes

While many liquids have temperature–composition phase diagrams resembling the ideal version in Fig. 8.18, in a number of important cases there are marked deviations which may completely upset the distillation process. These deviations may lead to maxima or minima in the boiling point curve of the mixture. A maximum, Fig. 8.19, may occur when the interactions in the mixture reduce the VP below the ideal value, which suggests that the A–B interactions stabilize the liquid. In such cases the excess Gibbs function is negative (more favourable to mixing than ideal), and examples include chloroform/acetone and nitric acid/water mixtures. Boiling point curves showing a minimum, as in Fig. 8.20, indicate that the mixture is destabilized relative to the ideal solution, the A–B interactions then being unfavourable. For such mixtures the excess Gibbs function is positive (less favourable to mixing than ideal), and there may be contributions from both enthalpy and entropy effects. Examples include dioxane/water and ethanol/water.

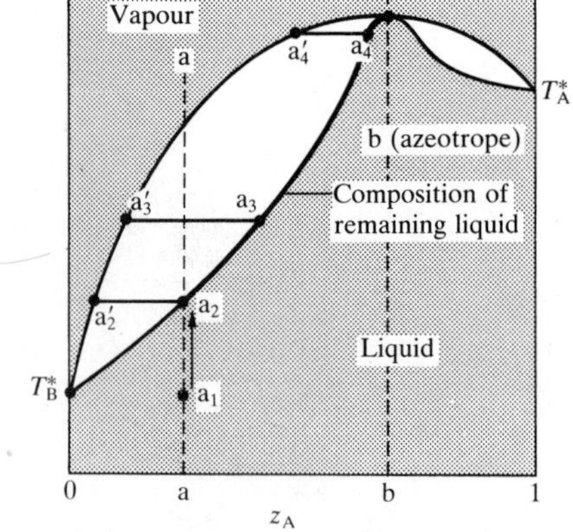

Fig. 8.19. A high-boiling azeotrope. When the mixture at a_1 is distilled, the composition of the remaining liquid changes towards b but no further.

Deviations from ideality are not always so strong as to lead to a hump or a trough in the boiling point curves, but when they do there are important consequences for distillation. Consider a liquid of composition a on the left of the maximum in Fig. 8.19. The vapour (at a_2') of the boiling mixture (at a_2) is richer in B. If that vapour is removed (and condensed elsewhere), the remaining liquid will be at a_3 and *its* vapour will be at a_3'. If that is removed, the composition of the boiling liquid shifts to a_4, and its vapour shifts to a_4'. This means that as evaporation proceeds, the composition of the remaining liquid shifts towards A as B is drawn off. The boiling point of the liquid rises, and the vapour becomes richer in A. When so much B has been evaporated that the liquid has reached the composition b, the vapour has the same composition as the liquid. Evaporation then occurs without change of composition. The mixture is said to form an *azeotrope* (which comes from the Greek for 'boiling without changing'). When the azeotropic composition has been reached distillation cannot separate the two liquids for the condensate retains the composition of the liquid. One example of azeotrope formation is hydrochloric acid/water, which is azeotropic at 80% water (by mass) and boils (unchanged) at 108.6 °C.

The type of system shown in Fig. 8.20 also forms an azeotrope, but one that shows itself in a different way. Consider starting with a liquid of composition a_1 and following the changes that occur in the *vapour* that rises through a fractionating column (essentially, a vertical glass tube packed with glass rings to give a large surface area). The mixture boils at a_2 to give a vapour of composition a_2'. This condenses in the column to a liquid of the same composition (now marked a_3). That liquid reaches equilibrium with its vapour at a_3', which condenses higher up the tube to give a liquid of the same composition, which we now call a_4. The fractionation therefore shifts the vapour towards the azeotropic composition, but not beyond, and the azeotropic vapour emerges from the top of the column. An example is ethanol/water, which boils unchanged when the water content is 4% and the temperature is 78 °C.

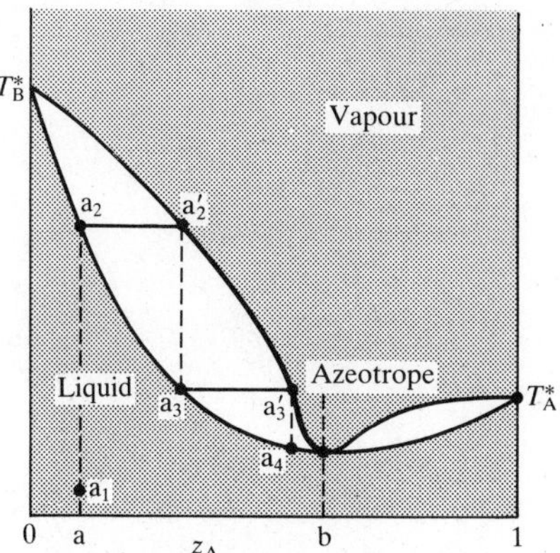

Fig. 8.20. A low-boiling azeotrope. When the mixture at a is fractionated, the vapour in equilibrium in the fractionating column moves towards b.

8.4 (d) Immiscible liquids

Finally we consider the distillation of two *immiscible liquids*, such as oil and water. As they are immiscible we can regard their 'mixture' as unscrambled with each component in a separate vessel, Fig. 8.21. If the VPs of the two pure components are p_A and p_B, then the total VP is $p = p_A + p_B$, and the mixture boils when $p = 1$ atm. The presence of the second component means that the 'mixture' boils at a lower temperature than either would alone because boiling begins when the *total* pressure reaches 1 atm, not when either VP reaches 1 atm. This is used in *steam distillation*, which enables some heat-sensitive organic compounds to be distilled at a lower temperature than their normal boiling point. The only snag is that the composition of the condensate is in proportion to the VPs of the components, and so oils of low volatility distil in low abundance. Steam distillation is also the basis of a messy way of determining molar masses.

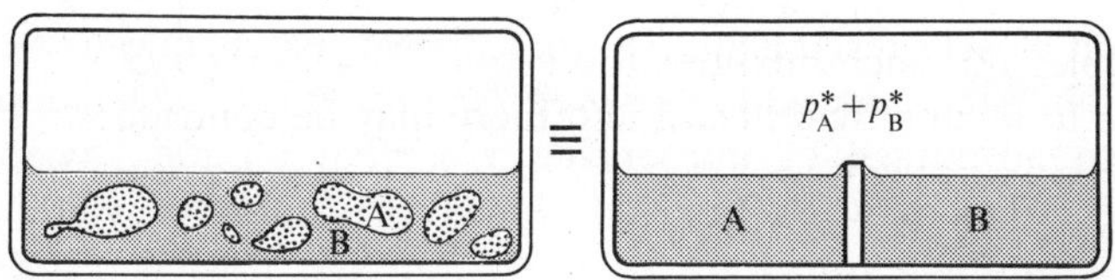

Fig. 8.21. The distillation of two immiscible liquids can be regarded as the joint distillation of the separated components, and boiling occurs when the *sum* of the partial pressures equals 1 atm.

8.5 Real solutions and activities

So far, our work on real solutions has been largely qualitative. Now we see how to make it quantitative so that calculations that we did for ideal solutions earlier in the chapter can be carried through for real solutions too.

We have dealt with solutions that obey Henry's and Raoult's Laws, Fig. 8.4. The general form of the chemical potential of a solute or a solvent is

$$\mu_A(l) = \mu_A^*(l) + RT \ln (p_A/p_A^*), \tag{8.5.1}$$

where p_A^* is the VP of pure A and p_A its VP when in a mixture. In the case of an ideal solution both solvent and solute obey Raoult's Law at all concentrations, and we can write

$$\mu_A(l) = \mu_A^*(l) + RT \ln x_A. \tag{8.5.2}°$$

The standard state is pure liquid A, and is obtained when $x_A = 1$.

When the solution departs from Raoult's Law the form of the last equation can be preserved by writing

$$\mu_A(l) = \mu_A^*(l) + RT \ln a_A. \tag{8.5.3}$$

a_A is the *activity* of A, a kind of 'effective' mole fraction. This is no more than a definition of activity; if the concept is to be useful we have to relate it to the actual composition of the solution. Since eqn (8.5.1) is true for both real and ideal solutions (the only approximation being the use of pressures rather than fugacities), we can conclude that

$$a_A = p_A/p_A^*. \tag{8.5.4}$$

This means that the activity of a component present at known concentration can be determined experimentally by measuring its VP when in a solution.

8.5 (a) The solvent activity

The *standard state of a component in a mixture* must now be defined more precisely. In the case of the component in excess (the solvent) there is no difficulty. All solvents obey Raoult's Law increasingly closely as they approach purity, and so the activity of the solvent approaches its mole fraction as $x_A \to 1$. As in the case of real gases, a convenient way of expressing this convergence is to introduce the *activity coefficient* (symbol: γ) by the definition

$$a_A = \gamma_A x_A. \tag{8.5.5}$$

Then as $x_A \to 1$ (pure solvent), $\gamma_A \to 1$. The chemical potential of the solvent is now written in terms of γ as

$$\mu_A = \mu_A^* + RT \ln x_A + RT \ln \gamma_A, \tag{8.5.6}$$

an equation closely resembling eqn (6.2.15) for the chemical potential of a real gas. The standard state, *the pure liquid solvent*, is established when $x_A = 1$. An advantage of introducing γ is that all the deviations from ideality are carried by $RT \ln \gamma$, and so investigations of non-ideality can concentrate on this term.

Example 8.8

Calculate the activity of water and its activity coefficient from the data below for the VP of water at different sucrose molalities (at 298 K)

m(sucrose)/mol kg^{-1}	0	0.200	0.500	1.000	2.000
p/Torr	23.75	23.66	23.52	23.28	22.75

• *Method*. The activity of the solvent is given by $a = p/p^*$, and the activity coefficient by $\gamma = a/x$. Begin by converting the molality of sucrose to a mole fraction of water by considering the amounts of each component in 1 kg of water.

• *Answer*. The mole fraction of water in a solution containing 1 kg of water (and therefore $1\,\text{kg} \times m_{\text{sucrose}}$) is

$$x = \frac{(1000\,\text{g}/18.02\,\text{g mol}^{-1})}{(1000\,\text{g}/18.02\,\text{g mol}^{-1}) + 1\,\text{kg} \times m_{\text{sucrose}}}$$
$$= 55.49/\{55.49 + (m_{\text{sucrose}}/\text{mol kg}^{-1})\}.$$

Construct the following table:

$m_{\text{sucrose}}/\text{mol kg}^{-1}$	0.000	0.200	0.500	1.000	2.000
x	1.000	0.996	0.991	0.982	0.965
a	1.000	0.996	0.990	0.980	0.958
γ	1.000	1.000	0.999	0.998	0.993

• *Comment*. Notice how the activity of the solvent remains close to unity. We shall often make the approximation that $a_{\text{solvent}} = 1$ in dilute solutions.

• *Exercise*. The following table gives the partial VPs of acetic acid (A) and benzene (B) at 50 °C. At that temperature the VPs of the pure components are 56.6 Torr and 270.6 Torr respectively. Find the activities and activity coefficients of both components.

x_A	0.016	0.044	0.084	0.114	0.171	0.298	0.370	0.583	0.660	0.844	0.993
p_A/Torr	3.63	7.25	11.51	14.2	18.4	24.8	28.7	36.3	40.2	50.7	54.7
p_B/Torr	263	257	250	245	232	211	153	135	75.3	3.5	

8.5 (b) The solute activity

The activity and standard state of the *solute* (the component in low abundance) need careful treatment. The problem is that the solution approaches ideal dilute (Henry's Law) behaviour at low concentrations of solute, concentrations far away from being pure liquid solute. The first step in setting up a definition is to produce one that works for a solute that obeys Henry's Law exactly, and the second is to allow for deviations.

A solute B that satisfies Henry's Law has a VP given by $p_B = K_B x_B$, where K_B is some empirical constant (which is equal to the VP of pure B only if the solute happens to obey Raoult's Law, Fig. 8.4). In this case eqn (8.5.1) for the chemical potential of B (which might not be a liquid in its pure state) is

$$\mu_B = \mu_B^* + RT \ln (p_B/p_B^*).$$

At low concentrations B obeys Henry's Law, and so we may replace p_B by $x_B K_B$ to obtain

$$\mu_B = \mu_B^* + RT \ln (K_B/p_B^*) + RT \ln x_B.$$

Both K_B and p_B^* are characteristics of the solute, and so the second term may be combined with the first to give a new standard chemical potential which we denote $\mu^\dagger$:

$$\mu_B^\dagger = \mu_B^* + RT \ln (K_B/p_B^*). \tag{8.5.7}$$

By comparing this with eqn (8.5.1) we see that the standard state (†) can be regarded as one in which the pure solute is giving rise to a VP of magnitude K_B. That is, the standard state is a *hypothetical state* in which the solute is pure, but behaving as though it still obeyed Henry's Law (i.e., it is an *extrapolation* of behaviour at lower concentrations). On inserting this standard value into the preceding equation we can write the following expression for the chemical potential at a general concentration, so long as the solute obeys Henry's Law, as

$$\mu_B = \mu_B^\dagger + RT \ln x_B. \qquad (8.5.8)°$$

We now take the important step of permitting deviations from ideal dilute, Henry's Law behaviour. Since the solute does obey Henry's Law at very low concentrations we write

$$\mu_B = \mu_B^\dagger + RT \ln a_B, \qquad (8.5.9)$$

and continue to take the same standard state as before, namely as *the hypothetical state in which the solute is pure and is behaving in accord with Henry's Law* (and therefore has the vapour pressure K_B). The last equation defines the activity of the solute. It is equivalent to writing

$$p_B = a_B K_B \qquad (8.5.10)$$

in place of Henry's Law in the derivation that led to eqn (8.5.8), and so we can measure the activity of the solute by measuring its partial VP. As in the case of the solvent it is sensible to introduce an activity coefficient through

$$a_B = \gamma_B x_B. \qquad (8.5.11)$$

Since the solute *does* behave ideally (in the Henry's Law sense) at infinite dilution, it follows that a_B coincides with x_B when the concentration is very low. This means that

$$a_B \to x_B \quad \text{and} \quad \gamma_B \to 1 \quad \text{as} \quad x_B \to 0. \qquad (8.5.12)$$

Notice that deviations from ideality disappear at infinite dilution in this definition, but for the solvent the deviations disappear as purity is approached.

Example 8.9

Use the information in *Example* 8.3 to calculate the activity and activity of chloroform in acetone at 25 °C on both the Raoult's Law and the Henry's Law basis.

• *Method*. For the Raoult's Law activity, form $a = p/p^*$ and $\gamma = a/x$, and for the Henry's Law activity form $a = p/K$ and $\gamma = a/x$.

• *Answer*. Since $p^* = 293$ Torr and $K = 165$ Torr, we can construct the following table:

x	0	0.20	0.40	0.60	0.80	1.00	
a	0	0.12	0.28	0.49	0.75	1.00	Raoult
γ	—	0.60	0.70	0.82	0.94	1.00	
a	0	0.21	0.50	0.86	1.33	1.78	Henry
γ	1	1.05	1.25	1.43	1.66	1.78	

• *Comment*. In the Henry's Law calculation $\gamma = 1$ is the *limiting* value, not 0/0. Notice how $\gamma \to 1$ as $x \to 1$ in the Raoult's Law case, but as $x \to 0$ in the Henry's Law case.

• *Exercise*. Calculate the activities and activity coefficients for acetone according to the two conventions.

The compositions of mixtures are often expressed as *molalities* (symbol: m, usually measured in $\mathrm{mol\,kg^{-1}}$) in place of mole fractions. It proves convenient to introduce yet another definition of activity, but one that follows naturally from what we have done so far. First, we note that in dilute solutions the amount of solute is much less than the amount of solvent, and so $n_B \ll n_A$. That is, a good approximation to the mole fraction of the solute is $x_B = n_B/n \approx n_B/n_A$. If the molality of the solute is m_B, the amount of B present in 1 kg of solvent is $n_B = (1\,\mathrm{kg}) \times m_B$ and the amount of A is $n_A = (1\,\mathrm{kg})/M_{A,m}$, where $M_{A,m}$ is the molar mass of A. Therefore, at low molalities, $x_B = m_B M_{A,m}$. In the region where Henry's Law is obeyed the chemical potential of B is given by eqn (8.5.9) as

$$\mu_B = \mu_B^{\dagger} + RT \ln m_B M_{A,m}.$$

The molality of B can be separated out from the second term. First, we introduce the useful symbol $m^{\ominus} = 1\,\mathrm{mol\,kg^{-1}}$ exactly. Then we write

$$\mu_B = \mu_B^{\dagger} + RT \ln (m^{\ominus} M_{A,m}) + RT \ln (m_B/m^{\ominus}).$$

(Introducing $m^{\ominus}$ keeps the terms inside the logarithms dimensionless.) The middle term on the right is independent of the molality of B, and so it is convenient to combine it with the first term and to define a new standard chemical potential as

$$\mu_B^{\ominus} = \mu_B^{\dagger} + RT \ln (m^{\ominus} M_{A,m}). \qquad (8.5.13)$$

Then the chemical potential at a general molality (in an ideal dilute solution) is

$$\mu_B = \mu_B^{\ominus} + RT \ln (m_B/m^{\ominus}). \qquad (8.5.14)^{\circ}$$

Note that the chemical potential of the ideal dilute solution has its standard value $\mu_B^{\ominus}$ when the molality of B is equal to $m^{\ominus}$ (i.e., at $m_B = 1\,\mathrm{mol\,kg^{-1}}$).

Now, as before, we incorporate deviations from ideality by introducing a (dimensionless) activity a and writing

$$a_B = \gamma_B m_B/m^{\ominus}, \qquad \gamma_B \to 1 \quad \text{as} \quad m_B \to 0. \qquad (8.5.15)$$

We then arrive at the following succinct expression for the chemical potential of a real solute at any molality:

$$\mu = \mu^{\ominus} + RT \ln a. \qquad (8.5.16)$$

The standard state remains unchanged in this last stage, and all the deviations from ideality are captured in the activity a. If we wish, we can picture the standard state as the *hypothetical* state in which the solute is at a molality $m^{\ominus}$ but the environment of each of its particles is the same as at infinite dilution. It is important to be aware of the different definitions of standard states and activities, and they are summarized in Box 8.1. We shall put them to work in the next few chapters, when we shall see that their applications are much less confusing than their definitions.

Box 8.1 Standard states

For the *solvent* (the major component), use Raoult's Law and take the standard state as the pure solvent. Then

$$\mu = \mu^* + RT \ln a,$$

with $a = p/p^*$ and $a = \gamma x$. Note that $\gamma \to 1$ as $x \to 1$ (i.e. pure solvent).

For the *solute* (the minor component), use Henry's Law, with the standard state as a hypothetical state of the pure solute. Then

$$\mu = \mu^\dagger + RT \ln a,$$

with $a = p/K$ and $a = \gamma x$. Note that $\gamma \to 1$ as $x \to 0$ (i.e. infinite dilution).

Alternatively, use

$$\mu = \mu^\ominus + RT \ln a,$$

with $a = \gamma m/m^\ominus$, $m^\ominus = 1\ \text{mol kg}^{-1}$. In this case $\gamma \to 1$ as $m \to 0$ (i.e. infinite dilution). The standard state is now a hypothetical state of the solution at a molality $m^\ominus$.

Appendix 8.1 The method of intercepts

Consider some extensive thermodynamic property X (such as the volume or the Gibbs function of a system). Let the binary system contain amounts n_A of A and n_B of B, the total amount being $n = n_A + n_B$ and the mole fractions $x_A = n_A/n$ and $x_B = n_B/n$. Define the *mean molar property* as $X_m = X/n$. The partial molar properties (which we seek to determine) are $X_{A,m} = (\partial X/\partial n_A)_{n_B}$ and $X_{B,m} = (\partial X/\partial n_B)_{n_A}$. Constant temperature and pressure are understood throughout.

The partial molar quantity $X_{A,m}$ can be written

$$X_{A,m} = (\partial X/\partial n_A)_{n_B} = (\partial nX_m/\partial n_A)_{n_B} = X_m + n(\partial X_m/\partial n_A)_{n_B}.$$

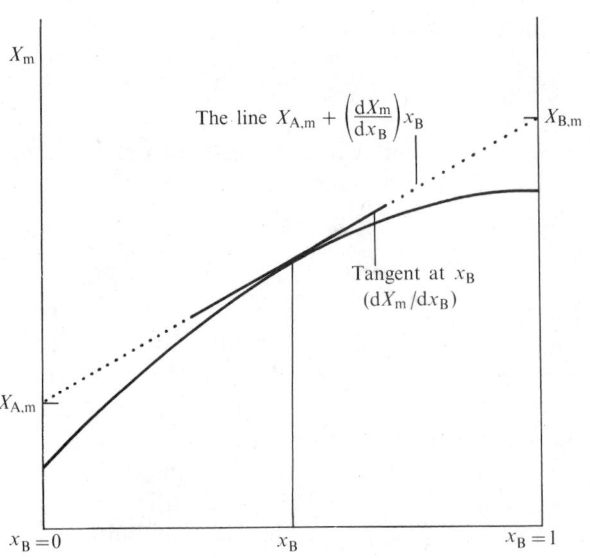

Fig. 8.22. The extrapolation needed to find the partial molar quantities $X_{A,m}$ and $X_{B,m}$ at the composition x_B.

Now, since $x_B = n_B/n$, it follows that

$$(\partial X_m/\partial n_A)_{nB} = (\partial x_B/\partial n_A)_{nB}(dX_m/dx_B) = (-n_B/n^2)(dX_m/dx_B),$$

and so

$$X_{A,m} = X_m + n(-n_B/n^2)(dX_m/dx_B), \quad \text{or} \quad X_m = X_{A,m} + (dX_m/dx_B)x_B.$$

The last equation is the key result. It is the equation of a straight line of slope (dX_m/dx_B) and intercept $X_{A,m}$ at $x_B = 0$. Therefore, plot X_m against x_B, and at the value of x_B of interest draw a tangent to the experimental curve, Fig. 8.22. This is the line of the equation, and so the partial molar quantity can be taken from the intercept at $x_B = 0$. The same argument, with A and B interchanged, shows that the intercept at $x_B = 1$ gives the value of $X_{B,m}$ at the selected composition.

Further reading

Experimental techniques:

Determination of osmotic pressure. J. R. Overton in *Techniques of chemistry* (A. Weissberger and B. W. Rossiter, eds.) V, 309, Wiley-Interscience, New York, 1971.

Determination of solubility. W. J. Mader and L. T. Brady in *Techniques of chemistry* (A. Weissberger and B. W. Rossiter, eds.) V, 257, Wiley-Interscience, New York, 1971.

Properties:

Liquids and liquid mixtures (3rd edn). J. S. Rowlinson and F. L. Swinton; Butterworths, London, 1982.

Solubilities of non-electrolytes. J. H. Hildebrand and R. L. Scott; Reinhold, New York, 1950.

Regular and related solutions. J. H. Hildebrand, J. M. Prausnitz, and R. L. Scott; Van Nostrand Reinhold, New York, 1970.

Properties of liquids and solutions. J. N. Murrell and E. A. Boucher; Wiley-Interscience, New York, 1982.

Chemical thermodynamics. M. L. McGlashan; Academic Press, London, 1979.

Solutions and solubilities. M. R. J. Dack (ed.) in *Techniques of chemistry* (A. Weissberger, ed.) VIII, Wiley-Interscience, New York, 1975.

Surfaces:

Physical chemistry of surfaces (3rd edn). A. W. Adamson: Wiley-Interscience, New York, 1976.

Physical surfaces. J. R. Bikerman; Academic Press, New York, 1970.

Interfacial phenomena. J. T. Davies and E. K. Rideal; Academic Press, New York, 1973.

Data:

International critical tables (*Vol. 4*). McGraw-Hill, New York, 1927.

Landolt–Börnstein tables (*Vol. 4*). McGraw-Hill, New York, 1927.

Physico-chemical constants of binary systems in concentrated solutions (*Vols. 1–4*). J. Timmermans; Interscience, New York, 1959.

Introductory problems

A8.1. At 300 K some equilibrium pressures (kPa) of HCl(g) over very dilute solutions of HCl in $GeCl_4$(l) are as follows: 32.0 for $x = 0.005$; 76.9 for $x = 0.012$; 121.8 for $x = 0.019$. Show that the system obeys Henry's Law in this range of concentrations and calculate the Henry's Law constant at 300 K.

A8.2. The temperature dependence of the Henry's Law constant, K, relating the mole fraction of HCl dissolved in $GeCl_4(l)$ to the equilibrium pressure of HCl(g) is given by: $K/\text{kPa} = \exp(12.137 - 1010(T/\text{K})^{-1})$. The numerical value of an enthalpy change can be derived from this information. What is its value? To what change in state does it correspond?

A8.3. The following table gives the mole fraction of methylbenzene, $C_6H_5CH_3$, denoted A, in liquid and vapour mixtures with methylethyl ketone, $CH_3COC_2H_5$, and the equilibrium pressure of the vapour at 303.15 K:

x_A	0	0.0898	0.2476	0.3577	0.5194
y_A	0	0.0410	0.1154	0.1762	0.2772
p/kPa	36.066	34.121	30.900	28.626	25.239
x_A	0.6036	0.7188	0.8019	0.9105	1
y_A	0.3393	0.4450	0.5435	0.7284	1
p/kPa	23.402	20.698	18.592	15.496	12.295

Consider the vapour perfect, calculate the partial pressures of the two components, and plot them as functions of their respective liquid mole fractions. Find the Henry's Law constants for the two components.

A8.4. On the basis of Raoult's Law, calculate the activity and activity coefficient for methylbenzene in the mixture described in the previous problem.

A8.5. Benzene freezes at 5.50 °C. Its enthalpy of fusion is 9.84 kJ mol^{-1}. Consider an ideal liquid solution which is saturated with benzene. If the mole fraction of benzene in the liquid is 0.905, what is the temperature of the system?

A8.6. The addition of 100 g of a compound to 750 g of CCl_4 lowers the freezing point of CCl_4 by 10.5 K. Calculate the molar mass of the solute.

A8.7. The osmotic pressure of an aqueous solution at 300 K is 120 kPa. Calculate the freezing point of the solution.

A8.8. The VP of pure liquid A at 300 K is 575 Torr and that of pure liquid B is 390 Torr. These compounds form ideal liquid and vapour mixtures. Consider equilibrium in a mixture with vapour composition $y_A = 0.350$. Calculate the total pressure of the vapour and calculate the composition of the liquid solution.

A8.9. The excess Gibbs function for some two-component liquid solutions has the form: $G^E = RTg_0x_1(1 - x_1)$, where g_0 is independent of composition. Differentiate the Gibbs function for the mixture and obtain an expression for the chemical potential of component 1 as a function of the composition x_1.

A8.10. The excess Gibbs function for solutions of methylcyclohexane, $CH_3C_6H_{11}$, and tetrahydrofuran, C_4H_8O, at 303.15 K are given by: $G^E/RT = x(1-x)[0.4857 - 0.1077(2x - 1) + 0.0191(2x - 1)^2]$, where x is the mole fraction of $CH_3C_6H_{11}$. Calculate the change in the Gibbs function on mixing 1 mole of $CH_3C_6H_{11}$ and 3 moles of C_4H_8O.

Problems

8.1. The volume of an aqueous solution of sodium chloride at 25 °C was measured at a series of molalities m, and it was found that the data could be fitted to the expression $V/\text{cm}^3 = 1003 + 16.62m + 1.77m^{\frac{3}{2}} + 0.12m^2$ (where $m \equiv m/$mol kg^{-1} and V refers to the volume of a solution formed from 1 kg of water). Find the partial molar volume of the components at $m = 0.10$ mol kg^{-1} by explicit differentiation.

8.2. Different behaviour is shown by an aqueous solution of magnesium sulphate. At 18 °C the total volume of a solution formed from 1 kg of water is given approximately by $V/\text{cm}^3 = 1001.21 + 34.69(m - 0.07)^2$, the expression applying up to about 0.1 mol kg^{-1}. What is the partial molar volume of (a) salt, (b) solvent at 0.05 mol kg^{-1}?

8.3. The *method of intercepts* is described in the Appendix. Use it to plot the partial molar volume of the HNO_3 in aqueous nitric acid at 20 °C using the following data, where w is the weight fraction of HNO_3.

$100w$	2.162	10.98	20.80	30.00	39.2	51.68
ρ/g cm^{-3}	1.01	1.06	1.12	1.18	1.24	1.32
$100w$	62.64	71.57	82.33	93.40	99.60	
ρ/g cm^{-3}	1.38	1.42	1.46	1.49	1.51	

8.4. The densities of aqueous solutions of copper sulphate at 20 °C were determined and are reported below. Determine and plot the partial molar volume of anhydrous copper sulphate in the range given.

%	5	10	15	20
ρ/g cm^{-3}	1.051	1.107	1.167	1.230

(% means the number of grams of the salt per 100 g of solution.)

8.5. The partial molar volumes of acetone and chloroform in a solution containing a mole fraction 0.4693 of chloroform are 74.166 cm^3 mol^{-1} and 80.235 cm^3 mol^{-1} respectively. What is the volume of a solution of mass 1.000 kg? What is the volume of the unmixed components? (For the second part you need to know that the molar volumes are 73.993 cm^3 mol^{-1} and 80.665 cm^3 mol^{-1} respectively.)

8.6. What proportions of ethanol and water should be mixed in order to form 100 cm^3 of mixture containing 50 percent by mass of ethanol? What change of volume is brought about by adding 1 cm^3 of ethanol to 100 cm^3 of a 50 percent (by weight) ethanol/water mixture? Base your answers on the information in Fig. 8.1.

8.7. When chloroform is added to acetone at 25 °C the

volume of the mixture varies with composition as follows:

x	0	0.194	0.385	0.559
$V_m/cm^3\,mol^{-1}$	73.99	75.29	76.50	77.55

x	0.788	0.889	1.000
$V_m/cm^3\,mol^{-1}$	79.08	79.82	80.67

x is the mole fraction of the chloroform. Determine the partial molar volumes of the two components at these compositions, and plot the result.

8.8. The *Gibbs–Duhem equation* for the chemical potentials of the components of a mixture is given in eqn (8.1.5). Show that for partial molar volumes the corresponding equation is $n_A\,dV_{A,m} + n_B\,dV_{B,m} = 0$. Equations of this type have important implications, and save a lot of effort. We illustrate this in the next few Problems. As a first step, show that the equation implies that the slopes of the graphs of partial molar volumes (or any such quantity) of the two components of a binary mixture must be equal and opposite when $x_A = x_B$. Check the consistency of your results for the last three Problems by seeing whether this is so.

8.9. The way in which the Gibbs–Duhem equation links the properties of the components of a mixture can be illustrated as follows. First, derive the *Gibbs–Duhem–Margules equation*, that $(\partial \ln f_A/\partial \ln x_A)_{p,T} = (\partial \ln f_B/\partial \ln x_B)_{p,T}$. This can be obtained as a straightforward development of the Gibbs–Duhem equation, and in the limit of gases that behave perfectly, reduces to a connection between vapour pressures and composition of the components of a mixture. Use it to show that if Raoult's Law applies to one component of a binary mixture, then it must also apply to the other.

8.10. As a third example of how the Gibbs–Duhem equation can be used, show that the partial molar volume (or any partial molar quantity) of a component B can be obtained if the partial molar volume (etc.) of component A is known for all compositions up to the one of interest. You should find an integral involving x_A, x_B, and $V_{A,m}$. Show that the partial molar volume of acetone in chloroform at $x = \frac{1}{2}$ can be obtained in this way from the data in Problem 8.7.

8.11. We now pass on to an exploration of how the thermodynamic properties of substances change when mixtures are prepared. In the first place, consider a container of volume 5.0 dm^3, which is divided into two compartments of equal size. In the left compartment there is nitrogen at 1.0 atm and 25 °C; in the right compartment there is hydrogen at the same pressure and temperature. What are the changes in the entropy and the Gibbs function of the system when the partition is removed and the gases are able to mix isothermally? Assume the gases to be perfect.

8.12. In a similar experiment, the nitrogen was initially at 3.0 atm and the hydrogen at 1.0 atm. Find the entropy and the Gibbs function changes in this case. Do gases always mix spontaneously? What would the changes of entropy and Gibbs function be if the hydrogen were replaced by nitrogen in this sample?

8.13. Air is a mixture with the composition listed in *Example* 1.3. What is the change of entropy when it is mixed from its constituents?

8.14. The same formulas for the thermodynamic functions of mixtures apply to ideal liquid mixtures as well as to perfect gas mixtures. Calculate the Gibbs function, entropy, and enthalpy of mixing when 500 g of hexane and 500 g of heptane are mixed at 298 K.

8.15. What proportions of hexane and heptane (a) by mole fraction, (b) by weight should be mixed in order to achieve the greatest change in entropy?

8.16. Henry's Law provides a simple way of estimating the solubilities of gases in liquids. For instance, what is the solubility of carbon dioxide in water at 25 °C when its partial pressure is (a) 0.1 atm, (b) 1.0 atm? Henry's Law data will be found in Table 8.1.

8.17. The mole fractions of nitrogen and oxygen in air at room temperature at sea level are approximately 0.782 and 0.209. What are the molalities in a vessel of water left open to the atmosphere at 25 °C?

8.18. A water carbonating plant is available for use in the home. It operates by providing carbon dioxide at 10 atm. Estimate the composition of the soda water it produces.

8.19. But how do we determine the value of Henry's Law constant? That can be found quite accurately by plotting VPs for various compositions and extrapolating the dilute solution data as in Fig. 8.5. The table below lists the VP of methyl chloride above its mixture with water at 25 °C. Find Henry's Law constant for methyl chloride.

$m/mol\,kg^{-1}$	0.029	0.051	0.106	0.131
$p/mmHg$	205.2	363.2	756.1	945.9

Don't forget to convert to mole fractions; the graph should be extrapolated analytically (i.e. by fitting a straight line to the data points), not on the graph paper itself.

8.20. At 90 °C the vapour pressure of toluene is 400 mmHg, and that of *o*-xylene is 150 mmHg. What is the composition of a liquid mixture that will boil at 90 °C when the pressure is 0.50 atm? What is the composition of the vapour produced?

8.21. In order to determine relative molar mass by the methods of depression of freezing point or elevation of boiling point it is necessary to know the cryoscopic or ebullioscopic constants K_f and K_b. In most cases these are available in the literature, or can be determined experimentally by dissolving a known amount of a material with known M_r. Sometimes, though, it is useful to be able to make a rapid assessment of their magnitudes. Calculate K_f and K_b for carbon tetrachloride using $\Delta H_{melt,m} = 2.5$ kJ mol^{-1}, $T_f = 250.3$ K, $\Delta H_{vap,m} = 30.0$ kJ mol^{-1}, $T_b = 350$ K, $M_r = 153.8$.

8.22. Colligative properties depend on the total number of particles dissolved in a solvent: sometimes this means keeping alert. Estimate the freezing point of 100 g of water containing 2 g of sodium chloride.

8.23. Instead of measuring boiling points we could measure the vapour pressure depression directly, and relate that to the RMM of the solute using Raoult's law. The vapour pressure of a sample of 500 g of benzene is 400 Torr at 60.6 °C, but it fell to 386 Torr when 19 g of an involatile organic compound was dissolved in it. What is the RMM of the compound?

8.24. The results obtained by depression of freezing point may be in error for a variety of reasons, such as the solution being too concentrated for the approximations involved to be valid, or because it behaves non-ideally. But what is the 'intrinsic' sensitivity of the method? A fluorocarbon is believed to be either $CF_3(CF_2)_3CF_3$ or $CF_3(CF_2)_4CF_3$; the cryoscopic constant of camphor is 40 K/mol kg^{-1}. To what precision must the temperature be measured if the freezing point depression of a solution of 1.0 g of the fluorocarbon in 100 g of camphor is to be used to distinguish the possibilities?

8.25. When we deduced the form of the expressions for the depression of freezing point and the elevation of boiling point we used a simple approximation of assuming that both ΔH and ΔS for the transitions are independent of temperature. This is quite a good approximation for small concentrations and some materials, but we ought to be able to deal with the more general case. We shall see, for example, that something rather odd happens in the case of water. As a first step return to $\ln x_A = \Delta G_{\text{melt,m}}(T)/RT$ on p. 171 and use the Gibbs–Helmholtz equation to find an expression for $d \ln x_A$ in terms of dT. Then integrate the left-hand side from $x_A = 1$ to the x_A of interest, and integrate the right-hand side from the transition temperature of interest for the pure material T_t^* to the transition point of the solution. In order to carry out the integration, assume that $\Delta H_{\text{t,m}}(T) \approx \Delta H_{\text{t,m}}(T_t^*)$. This leads in a more formal way to the equation already derived.

8.26. Differentiating and then integrating the differential might seem to be a waste of time. It is not really, because we need not have made the assumption that ΔH is independent of temperature. By putting in the appropriate dependence we can obtain a more accurate expression for the change of the transition temperatures. Assume now that the heat capacities of the solid and liquid pure solvent are independent of temperature. Derive an expression for the freezing point depression on this basis, and calculate the cryoscopic constant for water. Compare the result with that obtained by assuming ΔH to be independent of temperature. ($C_{p,\text{m}} = 75.3\ \text{J K}^{-1}\ \text{mol}^{-1}$ for water and 24.3 J K^{-1} mol^{-1} for ice.)

8.27. What is the freezing point of a 250 cm^3 glass of water sweetened with five cubes of sugar (7.5 g of sucrose, $C_{12}H_{22}O_{11}$)?

8.28. Common salt is spread on roads to prevent the formation of ice. The cost of the salt is £1.80/100 kg; $1.92/100 kg. A rich new source of $CaCl_2$ was discovered and mined at £1.40/100 kg; $1.49/100 kg. Which salt is more cost-effective? Assume all $\gamma \approx 1$.

8.29. Potassium fluoride is very soluble in glacial acetic acid, and the solutions have a number of peculiar properties. In an attempt to discover something about their structure, freezing point depression data were obtained by taking a solution of known molality and then diluting it several times. (J. Emsley, *J. chem. Soc.* A, 2702 (1971).) The following data were obtained:

$m(\text{KF})/\text{mol kg}^{-1}$	0.015	0.037	0.077	0.295	0.602
$\Delta T_f/\text{K}$	0.115	0.295	0.470	1.381	2.67

Find the apparent RMM of the solute, and suggest an interpretation. (This system was also used in Problem 4.29: the result there has a bearing on the data here.) Use $\Delta H_{\text{melt,m}} = 11.4\ \text{kJ mol}^{-1}$ and $T_f^* = 290\ \text{K}$.

8.30. In a study of the properties of aqueous solutions of thorium nitrate $Th(NO_3)_4$ (A. Apelblat, D. Azoulay, and A. Sahar, *J. chem. Soc. Faraday Trans.* I, 1618 (1973)) a freezing point depression of 0.0703 K was observed for a 0.0096 mol kg^{-1} aqueous solution. What is the apparent number of ions per solute molecule? Take $K_f = 1.86$ K/mol kg^{-1}.

8.31. Predict the ideal solubility of lead in bismuth at 280 °C, taking $T_f^* = 327$ °C and $\Delta H_{\text{melt,m}} = 5.2\ \text{kJ mol}^{-1}$.

8.32. Anthracene has an enthalpy of fusion of 28.8 kJ mol^{-1} and melts at 217 °C. What is its ideal solubility in benzene at 25 °C?

8.33. The osmotic pressures of solutions of polystyrene in toluene were measured in a determination of the mean RMM of the polymer. The pressure was expressed in terms of the height of the solvent h, its density being 1.004 g cm^{-3}. The following results were obtained at 25 °C:

$c/\text{g dm}^{-3}$	2.042	6.613	9.521	12.602
h/cm toluene	0.592	1.910	2.750	3.600

What is the mean RMM of the polymer? The density of toluene is 0.867 g cm^{-3}.

8.34. The RMM of a newly isolated enzyme was determined by dissolving it in water, measuring the osmotic pressure of various solutions, and then extrapolating the data to zero concentration. The following data were obtained at 20 °C.

$c/\text{mg cm}^{-3}$	3.221	4.618	5.112	6.722
h/cm water	5.746	8.238	9.119	11.990

What is the RMM of the enzyme? More elaborate Problems of this kind, and better ways of handling the data, will be found in Chapter 25.

8.35. An industrial process led to the production of a mixture of toluene (T) and octane (O), and the vapour pressure of the mixture was investigated with a view to designing a separation plant. The following temperature–composition data were obtained at 760 mmHg: x is the mole fraction in the liquid and y the mole fraction in the vapour at equilibrium.

$\theta/°C$	110.9	112.0	114.0	115.8	117.3
x_T	0.908	0.795	0.615	0.527	0.408
y_T	0.923	0.836	0.698	0.624	0.527
$\theta/°C$	119.0	120.0	123.0		
x_T	0.300	0.203	0.097		
y_T	0.410	0.297	0.164		

The boiling points are T: 110.6 °C, O: 125.6 °C. Plot the temperature–composition diagram for T and O. What is the composition of the vapour in equilibrium with liquid of composition (a) $x_T = 0.250$, (b) $x_O = 0.250$?

8.36. The table below lists the vapour pressures of mixtures of ethyl iodide (I) and ethyl acetate (A) at 50 °C. Find the activity coefficients of both components on (a) the Raoult's Law basis, (b) the Henry's Law basis with I regarded as solute.

x_A	0	0.0579	0.1095
p_I/mmHg	0	28.0	52.7
p_A/mmHg	280.4	266.1	252.3
x_A	0.1918	0.2353	0.3718
p_I/mmHg	87.7	105.4	155.4
p_A/mmHg	231.4	220.8	187.9
x_I	0.5478	0.6349	0.8253
p_I/mmHg	213.3	239.1	296.9
p_A/mmHg	144.2	122.9	66.6
x_I	0.9093	1.0000	
p_I/mmHg	322.5	353.4	
p_A/mmHg	38.2	0	

(The data are from *International critical tables*, Vol. 3, p. 288; McGraw-Hill, New York (1928).)

8.37. The industrial process referred to in Problem 8.35 produced a mixture containing a mole fraction 0.300 of toluene. At what temperature will it boil when the pressure is 760 mmHg? What is the composition of the first drop of distillate? What proportions of liquid and vapour are present at the boiling point? What is the composition of the liquid and the vapour at 1 °C above the boiling point? How much of each phase is present?

8.38. The next step in the analysis of the design problems for the separation plant was to set up a fractional distillation column. First we need some terminology. When the fractionation process was described in Section 8.4(b) we saw that it could be regarded as occurring in a sequence of zig-zag steps through the phase diagram: the horizontal lines in these steps are called *theoretical plates*. In a column set up in a pilot plant it was found that when a mixture of 0.300 toluene and 0.700 octane was heated, the distillate had a composition $x_T = 0.700$. How many theoretical plates had that column?

8.39. That degree of separation was not adequate for the subsequent syntheses involved in the plant. You needed a mixture with not less than $x_T = 0.900$. What is the minimum number of theoretical plates in the column that has to be designed?

8.40. Tabulated data can be analysed for their thermodynamic content. The next few Problems invite you to analyse some data taken directly from *International critical tables*, a principal source for this kind of material (see Vol. 3, p. 287). As a first step, take the vapour pressure data for benzene (B) in acetic acid (A) in the table below, and plot the vapour pressure composition curve for the mixture at 50 °C. Then confirm that Raoult's and Henry's Laws are obeyed in the appropriate regions.

x_A	0.0160	0.0439	0.0835
p_A/mmHg	3.63	7.25	11.51
p_B/mmHg	262.9	257.2	249.6
x_A	0.1138	0.1714	0.2973
p_A/mmHg	14.2	18.4	24.8
p_B/mmHg	244.8	231.8	211.2
x_A	0.3696	0.5834	0.6604
p_A/mmHg	28.7	36.3	40.2
p_B/mmHg	195.6	153.2	135.1
x_A	0.8437	0.9931	
p_A/mmHg	50.7	54.7	
p_B/mmHg	75.3	3.5	

8.41. Deduce the activities and activity coefficients of the components on the basis of Raoult's Law. Regarding B as solute, express its activity and activity coefficients on the basis of Henry's law.

8.42. Evaluate the excess Gibbs function of mixing of benzene and acetic acid over the composition range $x_A = 0$ to $x_A = 1$ at 25 °C.

9

Changes of state: the phase rule

Learning objectives

After careful study of this chapter you should be able to:

(1) State and derive the *phase rule*, eqn (9.1.1).

(2) Define the terms *phase*, *component*, and *variance*, Section 9.1(b and c).

(3) Count the number of components in a system, Example 9.1.

(4) Apply the phase rule to one-component systems, Section 9.2(a).

(5) Explain how *cooling curves* are used to construct phase diagrams, Sections 9.2(a) and 9.3(c).

(6) Interpret *liquid–liquid phase diagrams*, and explain the term *consolute temperature*, Section 9.3(a).

(7) Describe the distillation of partially miscible liquids in terms of phase diagrams, Section 9.3(b).

(8) Interpret *liquid–solid phase diagrams* and explain the terms *eutectic* and *eutectic halt*, Section 9.3(c).

(9) Interpret phase diagrams for systems in which reactions may occur, and explain the terms *peritectic reaction* and *incongruent melting*, Section 9.3(d).

(10) Explain the principles of *zone refining* and *zone levelling*, Section 9.3(e).

(11) Use *triangular coordinates* to show the composition of a three-component system, Section 9.4(a) and Example 9.4.

(12) Interpret three-component phase diagrams for partially miscible liquids and for solutions of two salts, Sections 9.4(b and c).

Introduction

In this chapter we set up a systematic way of discussing the changes mixtures undergo when they are heated and cooled and when their compositions are changed. We shall construct *phase diagrams*, which will let us judge whether two or three substances are mutually miscible, whether an equilibrium can exist over a range of conditions, or whether the system must be brought to a definite pressure, temperature, and composition before an equilibrium can be established.

9.1 The phase rule

We have seen that at equilibrium the chemical potential of a substance is the same in every phase (Section 7.1). In a system consisting of a single component (for which we write $C = 1$), when only one phase is present (for which we write $P = 1$) the pressure and temperature may be changed independently. We say that the *variance* of the system is 2, and write $F = 2$: the system is *bivariant*; there are two *degrees of freedom*. When two phases

α and β are present at equilibrium ($P=2$) the equality $\mu(\alpha;p,T)=\mu(\beta;p,T)$ can be regarded as an equation for p in terms of T. Since the pressure is now set by the choice of the temperature, the variance is 1: there is only one degree of freedom. When three phases α, β, and γ are present at equilibrium ($P=3$) there are three equations but one of them is redundant (because A = B with B = C implies A = C). The two remaining equations $\mu_A(\alpha;p,T)=\mu_A(\beta;p,T)$ and $\mu_A(\beta;p,T)=\mu_A(\gamma;p,T)$ have a solution only for *fixed* values of p and T (just like any pair of two simultaneous equations in two unknowns), and so the variance is $F=0$: there are no degrees of freedom. Overall, therefore, we see that the number of degrees of freedom reflects the number of equations that have to be satisfied, and the more equations there are, then the smaller the variance.

In one of the most elegant calculations of the whole of chemical thermodynamics, J. W. Gibbs‡ deduced the general relation between the variance (F), the number of components (C), and the number of phases at equilibrium (P) for a system of any composition:

The phase rule: $F=C-P+2$. (9.1.1)

The argument is given in the following section.

9.1 (a) The derivation of the phase rule

Consider a system in which there are C components and P phases. We begin by counting the number of intensive variables available and ignore at this stage the constraints (the equations that have to be satisfied as a result of phases being at equilibrium). The pressure and temperature count as two variables. In less conventional chemistry this number would increase to three or more if we were interested in the effects of magnetic or gravitational fields.

The compositions of the phases are specified if the relative amounts of each component are specified. The relative amounts in a given phase are specified by giving the mole fractions of the components in that phase. But since $x_1+x_2+\ldots+x_C=1$, all C mole fractions are known if all but one are specified. Therefore, specifying $C-1$ mole fractions completely specifies the composition of a phase. There are P phases, and so the total number of composition variables is $P(C-1)$.

At this stage the total number of variables (the variance) is $P(C-1)+2$. The equilibria between phases reduces our freedom to vary all these variables independently, and these constraints must now be counted.

At equilibrium the chemical potential of a component must be the same in every phase. If there are P phases, it follows that at equilibrium the chemical potentials of a component J are related by

$$\mu_J(\text{phase } 1)=\mu_J(\text{phase } 2)=\ldots=\mu_J(\text{phase } P).$$

‡ Josiah Willard Gibbs spent most of his working life at Yale, and may justly be regarded as the originator of chemical thermodynamics. He reflected for years before publishing his conclusions, and then did so in precisely expressed papers in an obscure journal (*The Transactions of the Connecticut Academy of Arts and Sciences*). He needed interpreters before the power of his work was recognized and before it could be applied to industrial processes. It may be judged from this that he was of retiring disposition and not of a practical turn; but that is certainly not so, for it is recorded that he prescribed and made his own spectacles, and that he held a number of patents.

That is, $P-1$ equations must be satisfied by component J. There are C components, and so the total number of equations is $C(P-1)$.

Each equation reduces our freedom to vary one of the $P(C-1)+2$ variables. It follows that the total number of degrees of freedom is

$$F = P(C-1)+2-C(P-1) = C-P+2,$$

which is the phase rule, eqn (9.1.1).

9.1 (b) The meaning of 'phase'

The formal definition of phase (which comes from the Greek word for 'appearance') is as follows:

> *Phase: a state of matter that is 'uniform throughout, not only in chemical composition but also in physical state'.*

(The words are Gibbs's.)

The definition is consistent with the common-sense meaning of the term. A gas, or a gaseous mixture, is a single phase, a crystal is a single phase, two totally miscible liquids form a single phase. Ice is a single phase ($P=1$) even though it might be chipped into small fragments. A slurry of ice and water is a two-phase ($P=2$) system even though it is difficult to map the boundaries between the phases.

An alloy of two metals is a two-phase system ($P=2$) if the metals are immiscible, but a single phase system ($P=1$) if they are miscible. This example shows that it is not always easy to decide whether a system consists of one phase or of two. A *solution* of solid A in solid B is uniform on a molecular scale, which means that particles of A are surrounded by particles of A and B, and any sample cut from it, however small, is representative of the composition of the whole. A *dispersion* is uniform only on a macroscopic scale, for close inspection shows that it consists of grains or droplets of one component in a matrix of the other. A tiny sample could come entirely from one of the minute grains of pure A, and so it would not be representative of the whole, Fig. 9.1. The problem is worse for very fine liquid dispersions because, as we saw in Section 7.6, vapour pressure depends on droplet size. What now constitutes a phase? What constitutes a phase in a gravitational field where the chemical potential depends on height? Fortunately, we do not need to deal with these tricky problems, but we should be aware that they do complicate the analysis of some systems.

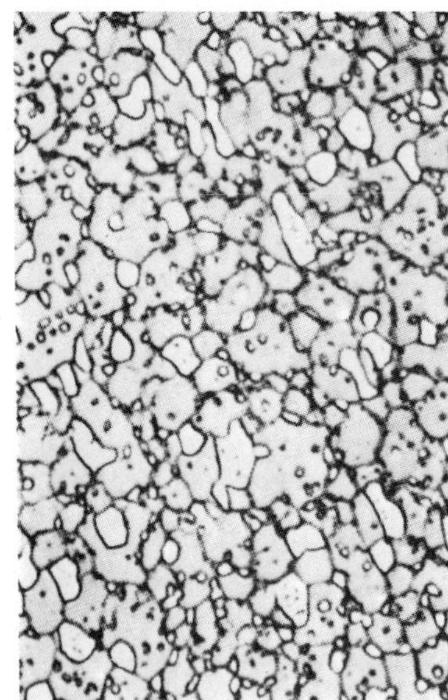

Fig. 9.1. A photomicrograph of a superplastic nickel–chromium–iron alloy. Its properties are discussed in the article by H. W. Hayden, R. C. Gibson, and J. H. Brophy (who kindly provided the illustration) in *Scientific American*, March 1969.

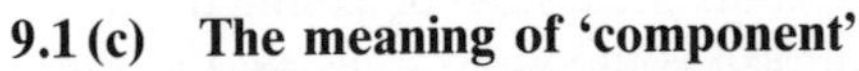

9.1 (c) The meaning of 'component'

The number of components (C) is defined as follows:

> *The minimum number of independent species necessary to define the composition of all the phases present in the system.*

The definition is easy to apply when the species do not react, for then we simply count their number. For instance, pure water is a one-component ($C=1$) system, and a mixture of ethanol and water is a two-component ($C=2$) system. If species do react, *and are at equilibrium*, we have to take into account the significance of the phrase 'all the phases' in the definition. Thus, in the case of ammonium chloride in equilibrium with its vapour (which consists of NH_3 and HCl molecules), since both phases have the formal composition 'NH_4Cl' it is a *one-component* system. However, if extra HCl(g) is added, the system has two components because now the relative amounts of HCl and NH_3 are arbitrary. In contrast, calcium carbonate in equilibrium with its vapour (CO_2 molecules) is a *two-component* system because we have to specify 'carbon dioxide' for the gas phase and 'calcium carbonate' for the solid. (Since the three species are related by an equilibrium, the concentration of calcium oxide is not independent.) In this case $C = 2$ whether we start from pure calcium carbonate, or equal amounts of calcium oxide and carbon dioxide, or arbitrary amounts of all three.

Example 9.1

How many components are present in the following systems: (a) sucrose in water; (b) sodium chloride in water; (c) aqueous phosphoric acid?

• *Method*. Identify the number of different kinds of *species* (e.g. ions) present in each single phase system; write this S. Identify the number of relations between the species (e.g. reactions at equilibrium, charge neutrality); write this R. Then the number of components is the number of kinds of species less the number of relations: $C = S - R$.

• *Answer*. (a) The species present are water molecules and sucrose molecules, so that $S = 2$. There are no relations between them, so that $R = 0$. Therefore, $C = 2$.

(b) The species present are water molecules, sodium ions, and chloride ions, and so $S = 3$. Since the solution is electrically neutral, the numbers of sodium and chloride ions are the same. Therefore, there is one relation, and so $R = 1$. Consequently, $C = 3 - 1 = 2$.

(c) The species present in aqueous phosphoric acid are H_2O, H_3PO_4, $H_2PO_4^-$, HPO_4^{2-}, PO_4^{3-}, H^+, and so $S = 6$. However, the acid ionizations are at equilibrium:

$$H_3PO_4(aq) \rightleftharpoons H_2PO_4^-(aq) + H^+(aq)$$
$$H_2PO_4^-(aq) \rightleftharpoons HPO_4^{2-}(aq) + H^+(aq)$$
$$HPO_4^{2-}(aq) \rightleftharpoons PO_4^{3-}(aq) + H^+(aq),$$

and so there are three relations between the species. Moreover, there is overall charge neutrality, and so the total number of cations must equal the total number of anions, whatever their kind. The total number of relations is therefore $R = 4$, and so the number of components is $C = 6 - 4 = 2$.

• *Comment*. Even if we were to include the vapour phase in each case, since that is 'water', and one of the components already specified is water, all three systems remain two-component systems.

• *Exercise*. Give the number of components in the following systems: (a) water, allowing for its ionization, (b) aqueous acetic acid, (c) magnesium carbonate in equilibrium with its vapour. [(a) 1, (b) 2, (c) 2]

9.2 One-component systems

One-component systems were discussed in detail in Chapter 7, and here we summarize their properties in terms of the phase rule. Since $C = 1$, the variance of a one-component system is $F = 3 - P$.

9.2 (a) Phase diagrams

When only *one phase* is present, $F = 2$. This means that p and T can be varied independently. In other words, a single phase is represented by an *area* on a *phase diagram*, a diagram of the kind first introduced in Chapter 7, showing the conditions at which each phase is thermodynamically the most stable. When *two phases* are in equilibrium, $F = 1$, which means that pressure is not freely variable if we have set the temperature. In other words, the equilibrium of two phases is represented by a *line* in the phase diagram. Instead of setting the temperature, we can set the pressure, but having done so the two phases come into equilibrium at a single definite temperature. Therefore, boiling (or any other transition) occurs at a definite temperature at a given pressure. When *three phases* are in equilibrium, $F = 0$, and so this special condition, the *triple point*, can be established only at a definite temperature and pressure, which is an intrinsic property of the system and not under our control. In other words, the equilibrium of three phases is represented by a *point* on the phase diagram. When *four phases* are present in equilibrium . . . but four phases cannot be in equilibrium in a one-component system because F cannot be negative.

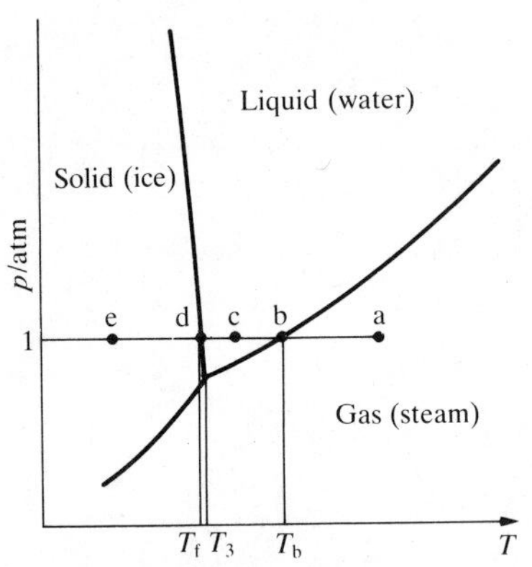

Fig. 9.2. The phase diagram for water, a simplified version of Fig. 7.7. T_3 marks the temperature of the triple point; T_b is the normal boiling point and T_f the normal freezing point.

This has been quite abstract. Now turn to the phase diagram for water shown in Fig. 9.2 and consider what happens as a system at a is cooled at constant pressure. It remains entirely gaseous until the temperature reaches b, when liquid appears. Two phases are now in equilibrium, and $F = 1$: we have set the pressure, which uses up the single degree of freedom, and so the temperature at which the equilibrium occurs is not under our control. Lowering the temperature takes the system to c in the one-phase, liquid region. The temperature can now be varied around the point c at will, and only when lumps of ice appear, at d, does the variance drop to unity again.

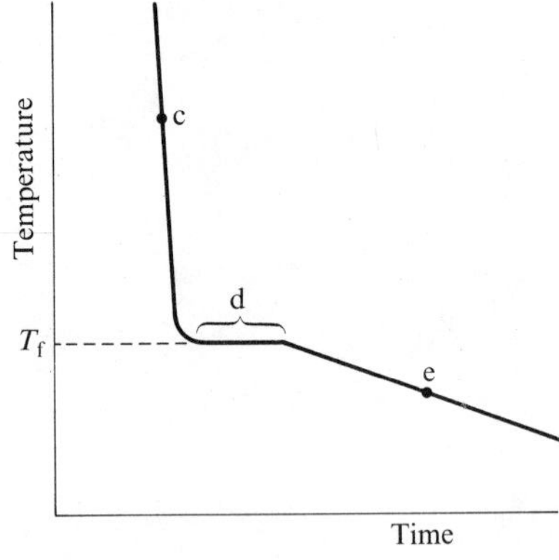

Fig. 9.3. The cooling curve for the isobar cde in Fig. 9.2. The halt marked d corresponds to the pause in the fall of temperature while the first-order exothermic transition (freezing) occurs. This enables T_f to be located even if the transition cannot be observed visually.

The determination of a phase diagram is straightforward in principle: the sample is subjected to constant pressure and the temperature of the phase change is detected. However, detecting a phase change is not always as simple as seeing a kettle boil, and so special techniques have been developed. One is *thermal analysis*, which takes advantage of the effect of the enthalpy change during a first-order transition. In this method a sample is allowed to cool, and its temperature is monitored. At a first-order transition, heat is evolved and the cooling stops until the transition is complete. (As remarked above, at fixed pressure the temperature can take only one value while two phases are present in equilibrium.) The cooling curve along the isobar cde in Fig. 9.2 therefore has the shape shown in Fig. 9.3. The transition temperature is obvious, and is used to mark point d on the phase diagram. This technique is useful for solid–solid transitions, where simple visual inspection of the sample is inadequate.

9.2 (b) Ultra-high pressures

Modern work on phase transitions deals with systems at very high pressures, and more sophisticated detection procedures must be adopted. For instance, some of the highest pressures currently attainable are produced in the *diamond-anvil cell* illustrated in Fig. 9.4. The sample is

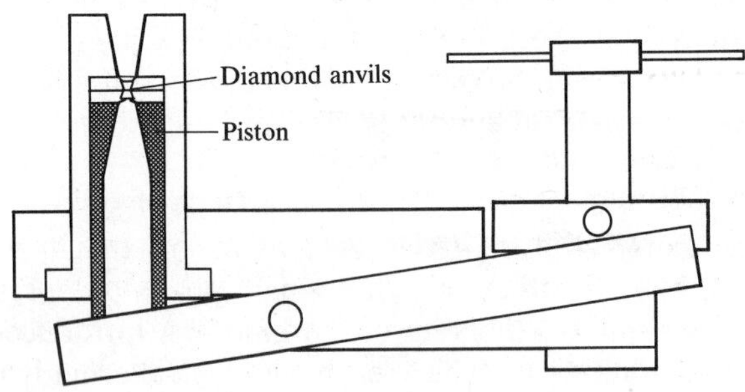

Fig. 9.4. Ultra-high pressures (up to about 1.7 Mbar) can be achieved using a diamond anvil. The sample, together with a ruby for pressure measurement and a drop of liquid for pressure transmission, are placed between two gem-quality diamonds. The principle of its action is like a nutcracker: the pressure is exerted by turning the screw by hand.

placed in a minute cavity between two gem-quality diamonds, and then pressure is exerted simply by turning the screw. The advance in design this represents is quite remarkable, for with a turn of the screw pressures (of up to about 1 million atmospheres) can be reached which a few years ago could not be reached with equipment weighing tons. The pressure is monitored spectroscopically by observing the shift of spectral lines in small pieces of ruby added to the sample, and the properties of the sample itself are observed optically through the diamond anvils. The technique is employed to examine the transition of covalent solids to metallic form: iodine, for instance, becomes metallic at around 200 kbar (200 000 atm), while remaining as I_2, but makes a transition to a monatomic metallic solid at around 210 kbar. Studies such as these are relevant to the structure of material deep inside the Earth (at the centre of the Earth the pressure is around 5 Mbar) and in the interiors of the giant planets, where even hydrogen is believed to be metallic.

9.3 Two-component systems

When two components are present, $C=2$ and so $F=4-P$. For simplicity we shall keep the pressure constant (at 1 atm, for instance), which discards one of the degrees of freedom. We write $F'=3-P$, the single prime denoting that one degree of freedom has been given up. The variance is greatest ($F'=2$) when $P=1$. One of these remaining degrees of freedom is the temperature, the other the composition (the mole fraction of a component). Note that all the points on any vertical line in a temperature–composition diagram have the same composition: any vertical line is therefore an *isopleth* (from the Greek for 'same abundance').

9.3 (a) Liquid–liquid phase diagrams

We deal in this section with binary systems ($C=2$, $F'=3-P$) at temperatures and pressures such that only liquids are present. We consider pairs of *partially miscible* liquids, liquids that do not mix in all proportions at all temperatures. In general we denote the liquids A and B: an explicit example is hexane and nitrobenzene.

To a sample of A at some temperature T' add a little B. It dissolves completely, and the binary system remains one phase. As more B is added a stage comes when no more dissolves. The sample now consists of two phases, the major one consisting of A saturated with B, the minor one consisting of a trace of B saturated with A. In Fig. 9.5 the appearance of two phases at the composition a′ is marked by the single horizontal line breaking into two.

When more B is added, A dissolves in it slightly. This means that the B-rich phase grows at the expense of the other. A stage is reached when so much B is present that it can dissolve all the A, and so the system reverts to a single phase. This is indicated by the fusion of the double line in Fig. 9.5 at the composition a″. The addition of more B simply dilutes the solution, and so from then on it remains one phase.

The temperature of the system affects the compositions at which phase separation and coalescence occur. In the case of hexane and nitrobenzene raising the temperature *increases* their miscibility. Phase separation does not occur until more B is present in A, and as less B is needed to soak up all the A the single phase is encountered sooner. The two-phase system is

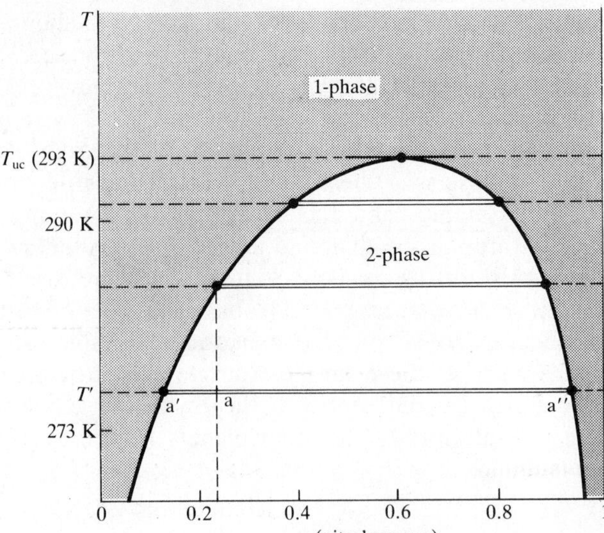

Fig. 9.5. The temperature–composition diagram for hexane and nitrobenzene at 1 atm. The upper consolute temperature (T_{uc}) is the temperature above which no phase separation occurs. For this system it lies at 293 K.

therefore less extensive, as shown by the lines corresponding to higher temperature in Fig. 9.5. The entire phase diagram can be constructed by repeating the observations at different temperatures and drawing the envelope of the two-phase region.

Example 9.2

A mixture of 50 g hexane and 50 g nitrobenzene was prepared at 290 K. What are the compositions of the phases, and in what proportions do they occur? To what temperature must the sample be heated in order to obtain a single phase?

• *Method*. Refer to Fig. 9.5 and use the lever rule, eqn (8.4.5). Begin by converting the masses to mole fractions using relative molecular masses of 86.2 for hexane and 123 for nitrobenzene.

• *Answer*. The amounts present are 0.58 mol and 0.41 mol for hexane and nitrobenzene respectively. Therefore the mole fractions are $x_{Hex} = 0.59$ and $x_{Nb} = 0.41$. The point $x_{Nb} = 0.41$, $T = 290$ K occurs in the two-phase region of the phase diagram in Fig. 9.5. The horizontal tie-line cuts the phase boundary at $x_{Nb} = 0.37$ and $x_{Nb} = 0.83$, and so those are the compositions of the two phases. The ratio of amounts is equal to the ratio of lengths (i.e. the distance of the intersections from the total composition point):

$$l'/l = (0.41 - 0.37)/(0.83 - 0.41) = 0.04/0.42,$$

or 1:10. Heating the sample to 292 K takes it into the single phase region.

• *Comment*. Since the phase diagram has been constructed experimentally, these conclusions are 'exact'. They would be modified if the system were subjected to a different pressure.

• *Example*. Repeat the problem for 50 g hexane and 100 g nitrobenzene at 273 K.
[$x_{Nb} = 0.09$ and 0.95 in ratio 1:1.3; 275 K]

Above the *consolute temperature* phase separation does not occur whatever the composition. Since it represents an upper limit to the temperatures at which two phases occur it is called the *upper* consolute temperature (symbol: T_{uc}), or *upper critical solution temperature*.

Although it may seem quite natural that there should be an upper consolute temperature (because the more violent thermal motion overcomes the tendency of molecules of the same type to stick together in swarms and therefore to form two phases), some systems show a *lower* consolute temperature (symbol: T_{lc}) *beneath* which they mix in all proportions and above which they form two phases. An example is water and triethylamine, Fig. 9.6. In this case, at low temperatures the two kinds of molecules form a weak complex, which increases their mutual solubility; at higher temperatures the complexes break up and the two types of molecules cluster together in swarms of their own kind.

Some systems show both upper and lower consolute temperatures. This is because after the weak complexes have been disrupted the thermal motion at higher temperatures homogenizes the mixture, just as in the case of ordinary partially miscible liquids. The most famous example is nicotine and water, which are partially miscible between 61 °C and 210 °C, Fig. 9.7.

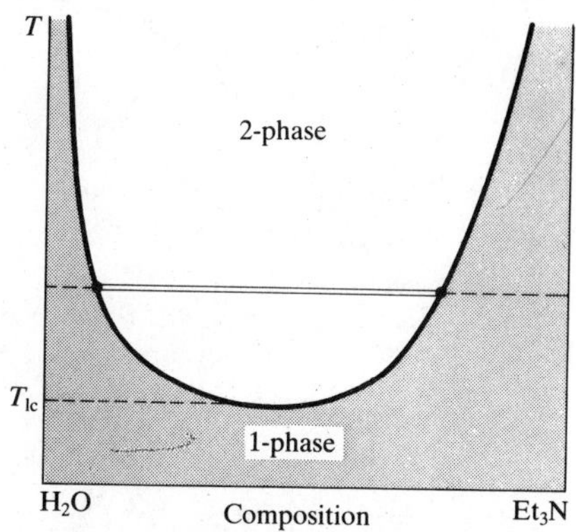

Fig. 9.6. The temperature–composition diagram for water and ethylamine. This shows a lower consolute temperature (the temperature *beneath* which phase separation does not occur) at $T_{lc} = 292$ K.

9.3 (b) The distillation of partially miscible liquids

We now consider what happens when the conditions are such that we have to take into account the presence of the vapour. We shall consider a pair of liquids that are partially miscible and form a low-boiling azeotrope (Section 8.4(c)). This combination of properties is quite common because both reflect the tendency of the two kinds of molecule to get out of each other's way.

The phase diagrams are constructed by combining the liquid–liquid diagram (Fig. 9.5) with the type of azeotrope liquid–vapour diagram we first encountered in Fig. 8.20. There are two possibilities, one representing the case when the liquids become fully miscible before they boil, the other when boiling occurs before mixing is complete.

Figure 9.8 shows the phase diagram for the case of liquids that become fully miscible before they boil. Distillation of a mixture of composition a_1 leads to a vapour of composition b_1, which condenses to the completely miscible single-phase solution at b_2. Phase separation occurs only when this distillate is cooled to b_3. (This applies only to the first drop of distillate. If distillation continues the composition of the remaining liquid changes. In the end, when the whole sample has evaporated and condensed, the composition is back to a_1.)

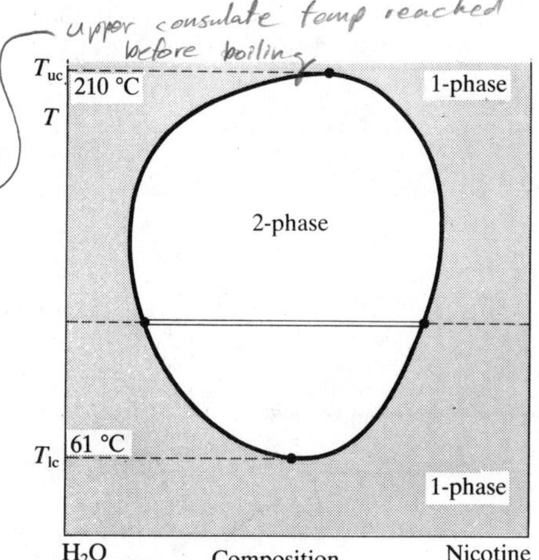

Fig. 9.7. The temperature–composition diagram for water and nicotine, which has both upper and lower consolute temperatures. Note the high temperatures for the liquid (especially the water): the diagram corresponds to a sample under pressure.

Figure 9.9 shows the second possibility, in which boiling occurs before the upper consolute temperature is reached. The distillate, of composition b_3, now drips out as a two-phase mixture, one phase of composition b_3' and the other of composition b_3''. The only odd thing about the diagram is the behaviour of a system of composition corresponding to the isopleth through e. A system at e_1 forms two phases, which persist, but in changing proportions, up to the boiling point at e. The vapour given off has the same composition as the liquid (the liquid is an azeotrope), and so condensing a vapour of composition e_2 gives a liquid of the same initial composition. At e, $P = 3$ (one vapour and two liquid phases), $C = 2$, and so $F' = 0$: e is an *invariant* point of the system (so long as the pressure is fixed).

The lesson to learn at this stage is that phase diagrams might look complicated, but they convey simple *experimentally established* information. In order to interpret them it is helpful to think *operationally*: this means that definite situations should be imagined, and the diagram should be progressed through bearing in mind how it was constructed originally: phases come and go, systems boil and freeze, and relative amounts change. Another useful rule is to concentrate on the *lines* rather than the

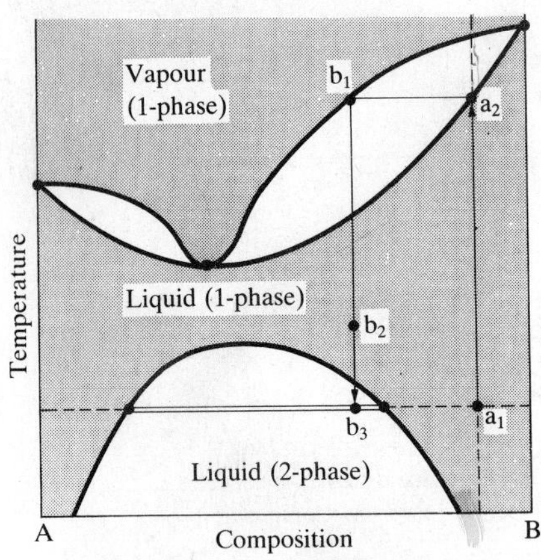

Fig. 9.8. The temperature–composition diagram for a binary system in which the upper consolute temperature is less than the boiling point at all compositions. The mixture forms a low-boiling azeotrope.

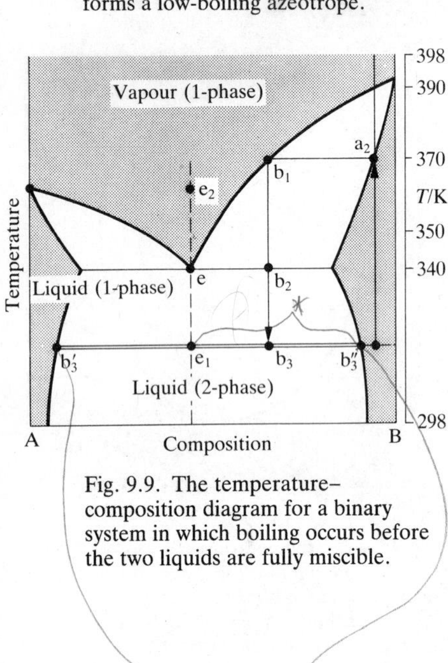

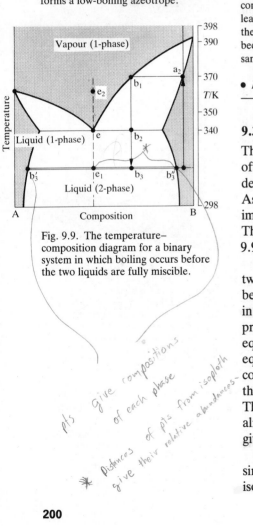

Fig. 9.9. The temperature–composition diagram for a binary system in which boiling occurs before the two liquids are fully miscible.

areas: *the points at the ends of the horizontal lines* passing through a point (e.g. the line through b_3 in Fig. 9.9) give the *compositions* of each phase, and the *distances* of the points from the isopleth give (through the lever rule, eqn (8.4.5) their *relative abundances*. These horizontal lines are called the *tie lines* of the phase diagram.

Example 9.3

State the changes that occur when a mixture of composition $x_B = 0.95$ (a_1) in Fig. 9.9 is boiled and the vapour is condensed.

- *Method*. The area occupied by the point gives the number of phases; the compositions of the phases are given by the points at the intersections of the horizontal tie-line with the phase boundaries; the relative abundances are given by the lever rule, eqn (8.4.5).

- *Answer*. The initial point is in the one-phase region. When heated it boils at 370 K (a_2) giving a vapour of composition $x_B = 0.66$ (b_1). If the vapour is not drawn off, further heating is required to maintain boiling. The liquid gets richer in B, and the last drop (of pure B) evaporates at 392 K. The boiling range of the liquid is therefore 370–392 K. If the initial vapour is drawn off, it has a composition $x_B = 0.66$. This would be maintained if the sample were very large, but for a finite sample it shifts to higher values and ultimately to $x_B = 0.95$. Cooling the distillate corresponds to moving down the $x_B = 0.66$ isopleth. At 350 K, for instance, the liquid phase has composition $x_B = 0.87$, the vapour $x_B = 0.49$, in relative proportions 1:1.3. At 340 K the sample is entirely liquid, and consists of two phases, one of composition $x_B = 0.44$, the other of composition $x_B = 0.84$ in the ratio 0.85:1. Further cooling leaves the system in the two-phase region, and at 298 K the compositions are 0.05 and 0.93 in the ratio 0.46:1. As further distillate boils over, the overall composition of the distillate becomes richer in B. When the last drop has been condensed the phase composition is the same as at the beginning.

- *Exercise*. Repeat the discussion, beginning at the point $x_B = 0.4$, $T = 298$ K.

9.3 (c) Liquid–solid phase diagrams

The phase diagrams for binary solid systems (e.g. alloys of two metals) often resemble the phase diagrams for liquids, but in place of regions denoting liquids and vapours there are regions denoting solids and liquids. As an example, consider a pair of metals that are almost completely immiscible right up to their melting points (e.g. antimony and bismuth). The phase diagram is shown in Fig. 9.10; note how closely it resembles Fig. 9.9.

Consider a liquid of composition a_1. When it is cooled to a_2 it enters the two-phase region 'liquid + A'. At this temperature almost pure solid A begins to come out of solution, and so the remaining liquid becomes richer in B. On cooling to a_3 more of the solid (and slightly impure) A precipitates, and the relative amounts of the solid and liquid (which are in equilibrium) are given by the lever rule: at this stage there are roughly equal amounts of each. The liquid phase is richer in B than before (its composition is given by b_3) because so much A has been deposited. At a_4 there is less liquid (lever rule) than at a_3, and its composition is given by e. This liquid now freezes to give a two-phase system of almost pure A and almost pure B. At a_5, for example, the compositions of the two phases are given by a_5' and a_5''.

The mixture of composition e plays a role in solid–liquid transitions similar to the one played by the azeotrope in liquid–vapour transitions. The isopleth at e corresponds to the *eutectic* composition, the word coming from the Greek for 'easily melted'. A liquid of eutectic composition solidifies, without change of composition, at the lowest temperature of any mixture. A solid of eutectic composition melts, without change of composition, at the lowest temperature of any mixture. Solutions of composition to the right of e precipitate A as they cool, and solutions to the left deposit B: only the eutectic mixture solidifies at a single definite temperature ($F' = 0$ when $C = 2$ and $P = 3$) without gradually unloading one or other of the components from the liquid.

The thermal analysis procedure outlined above is a very useful practical way of determining phase diagrams of this kind. Consider the rate of cooling down the isopleth a in Fig. 9.10. The liquid cools steadily, Fig. 9.11, until it reaches a_2, when A begins to be deposited. The rate of cooling is now less because the progressive solidification of A releases its enthalpy of melting, which retards the cooling. When the remaining liquid reaches the eutectic composition the temperature remains constant ($F' = 0$) until the whole sample has solidified: this is the *eutectic halt*. If the liquid has the eutectic composition e initially, then the liquid cools steadily down to the freezing temperature of the eutectic, when there is a long eutectic halt as the entire sample solidifies. Monitoring the cooling curves at different overall compositions therefore gives a clear indication of the structure of the phase diagram: the solid–liquid boundary is given by the points at which the rate of cooling changes, and the longest eutectic halt gives the location of the eutectic composition and its melting temperature.

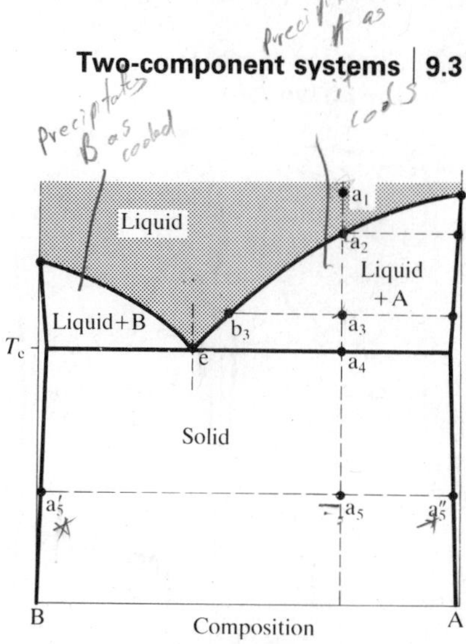

Fig. 9.10. The temperature–composition phase diagram for two almost immiscible solids and their completely miscible liquids. Note the similarity to Fig. 9.9. The isopleth through e corresponds to the eutectic composition, the mixture with lowest melting point.

Eutectics have some important practical applications. One technologically important eutectic is solder, which consists of 67% by mass of tin and 33% lead and melts at 183 °C. The eutectic formed by 23% sodium chloride and 77% water melts at −21.1 °C. This mixture has two important applications. When salt is added to ice under isothermal conditions (e.g., when spread on an icy road) the mixture melts if the temperature is above −21.1 °C (and the eutectic composition has been achieved). When salt is added to ice under adiabatic conditions (e.g., to ice in a vacuum flask) the ice melts but in doing so it absorbs heat from the rest of the mixture. The temperature falls, and if enough salt is added, cooling continues down to close to the eutectic temperature.

9.3(d) Phase diagrams for reactive systems

In some cases a mixture of A and B reacts to produce a compound C. An example is aniline and phenol. (The system is still binary, $C = 2$, because the compositions of the three species are related by an equilibrium constant.) We consider the case in which the equilibrium lies strongly towards C.

A system prepared by mixing an excess of A with B consists of C together with the unreacted A. This is a binary C, A system, the only change from Fig. 9.10 being that the whole of the diagram has to be squeezed into the range of compositions lying between pure A and equal amounts of A and B ($x_A = 0.5$), Fig. 9.12. On the left of that composition is another phase diagram, but in this region B is in excess. The interpretation of the information in the complete diagram is the same as for Fig. 9.10, but the solid deposited down isopleth a in Fig. 9.12 is the *compound* C slightly contaminated with A, and the two-phase solid beneath a_4 consists of C and A (each one slightly contaminated by the other).

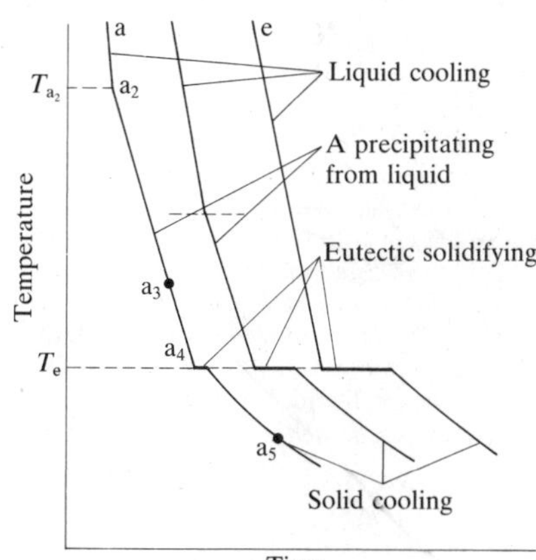

Fig. 9.11. The cooling curves for the system shown in Fig. 9.10. For isopleth a the rate of cooling slows at a_2 because solid A comes out of solution. There is a complete halt at e while the eutectic solidifies. This halt is longest for the eutectic isopleth, e. The cooling curves are used to construct the phase diagram.

An important modification arises when the compound is not stable as a

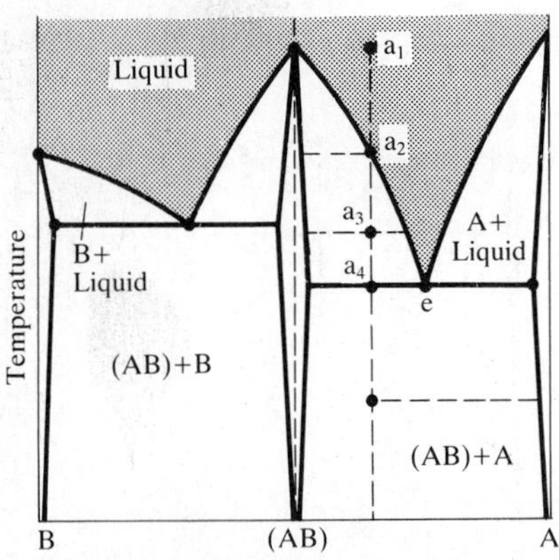

Fig. 9.12. The phase diagram for a system in which A and B react to form a compound AB. This is like two versions of Fig. 9.10 in each half of the diagram. Note that AB signifies a true compound, not just an equimolar mixture.

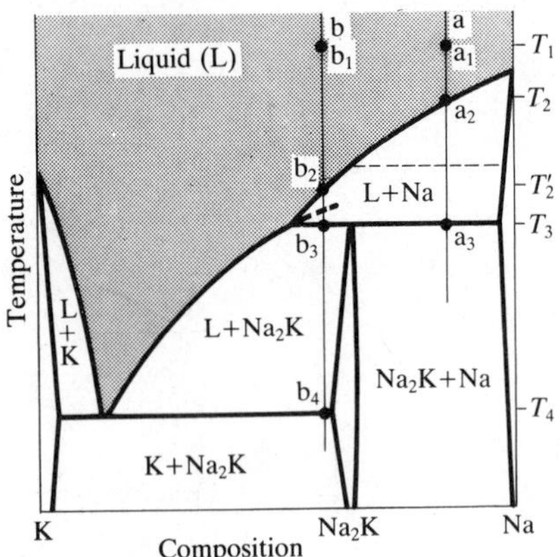

Fig. 9.13. The phase diagram for an actual system like that shown in Fig. 9.12, but with two differences. One is that the compound is Na_2K, corresponding to A_2B and not AB as in that illustration. The second is that the compound exists only as the solid, not as the liquid. It is an example of *incongruent melting*.

liquid. An example is the alloy Na_2K which survives only as a solid: the phase diagram is shown in Fig. 9.13. Consider what happens as a liquid at a_1 is cooled. At a_2 some solid sodium (slightly contaminated with potassium) is deposited, and the remaining liquid is richer in potassium. At the temperature corresponding to a_3 the sample is entirely solid, and consists of solid sodium and solid Na_2K (each slightly contaminated by the other). Now consider the isopleth b. The initial precipitation at b_2 is of dirty solid sodium. However, at b_3 a reaction forming Na_2K occurs. At this stage there is present liquid sodium and potassium and a little solid Na_2K, but there is still no liquid compound. As cooling continues, the amount of solid compound increases until at b_4 the liquid reaches its eutectic composition. It then solidifies to give a two-phase solid consisting of slightly dirty potassium and slightly dirty compound. No liquid Na_2K has occurred at any stage because it is too unstable to survive melting.

The behaviour just described is called *incongruent melting*, a *phase reaction*, or a *peritectic reaction*. The last name comes from the Greek for 'around' and 'melting' and reflects the appearance of the solid formed on cooling: it consists of a distribution of minute sodium crystals, each surrounded by a shell of compound.

9.3 (e) Ultrapurity and controlled impurity: zone refining

Advances in technology have called for materials of extreme purity. For example, semiconductor devices consist of almost perfectly pure silicon or germanium doped to a precisely controlled extent. If these materials are to operate successfully the impurity level must be kept down to less than 1 in 10^9 (which corresponds to about one grain of salt in 5 tons of sugar). The technique of *zone refining* was developed in response to this need, and the principle involved can be illustrated using the phase diagram in Fig. 9.14.

Consider a liquid of composition a: it is mainly B with some A impurity. On cooling to a_1 a solid of composition b_1 appears. Removing that solid gives a slightly purer material than the original, but not much of it (lever rule). That solid could be used as the starting substance for a second stage of this *fractional crystallization* process. In each stage the composition is shifted towards pure B, in the manner of fractional distillation; but the procedure is slow and wasteful. We should recognize, however, that Fig. 9.14 refers to an *ideal* process, and is based on the assumption that the freezing is so slow that the composition of the solid is always uniform and at any stage has the equilibrium composition expressed by the diagram. In a real system this is not the case, because A does not have time to disperse throughout the whole solid sample. Zone refining makes use of the *non-equilibrium* properties of the system. It relies on the impurities being more soluble in the molten sample than in the solid, and sweeps them up by passing a molten zone repeatedly from one end to the other along a sample.

Consider a liquid (this represents the molten zone) on the isopleth a, and let it cool without the entire sample coming to overall equilibrium. If the temperature falls to a_2 a solid of composition b_2 is deposited and the remaining liquid (the zone where the heater has moved on) is at a_2'. Cooling that liquid down an isopleth passing through a_2' deposits solid of composition b_3 and leaves liquid at a_3'. The process continues until the last drop of liquid to solidify is heavily contaminated with A. There is some everyday evidence that impure liquids freeze in this way: an ice cube is clear near the surface but misty in the core. This is because the water used to make ice normally contains dissolved air; freezing proceeds from the outside, and air is accumulated in the retreating liquid phase. It cannot escape from the interior of the cube, and so when that freezes it occludes the air in a mist of tiny bubbles.

In zone refining the sample is in the form of a narrow cylinder. This is heated in a thin disk-like zone which is swept from one end of the sample to the other. The advancing liquid zone accumulates the impurities as it passes. In practice a train of hot and cold zones are swept repeatedly from one end to the other. The zone at the end of the sample is the impurity dump: when the heater has gone by, it cools to a dirty solid which can be discarded.

A modification of zone refining is *zone levelling*. It is used to introduce controlled amounts of impurity (for example, indium into germanium). A sample rich in the required dopant is put at the head of the main sample, and made molten. The zone is then dragged repeatedly in alternate directions through the sample, where it deposits a uniform distribution of the impurity.

Zone refining is used in fields other than the semiconductor industry. Ultrapure materials often have markedly different properties from conventionally pure materials. For instance, bismuth is normally regarded as a hard, brittle metal; yet when it has been zone refined it forms rods that can be bent without fracture. A few per cent of carbon in iron produces cast iron, which is brittle; when all traces of carbon are removed from iron it retains its ductility down almost to absolute zero. Organic chemicals may also be zone refined, and substances thought to have a foul smell have been rendered odourless by purification.

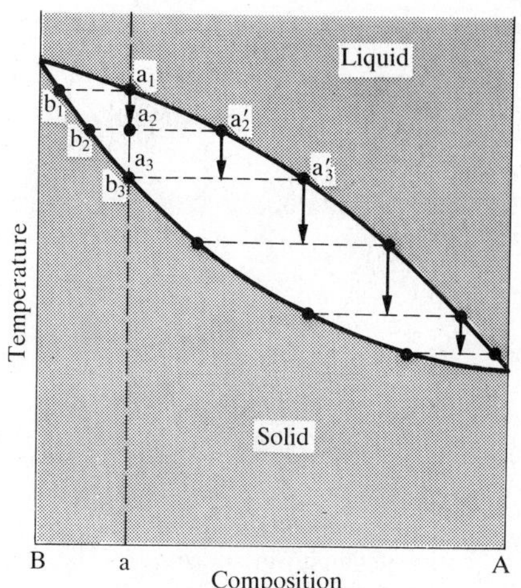

Fig. 9.14. A binary temperature–composition diagram can be used to discuss zone refining as explained in the text.

9.4 Three-component systems

In three-component systems the variance may reach 4. Imposition of the constraints of constant temperature and pressure still leaves two degrees of freedom, and the representation of the equilibria involved are on the verge of being absurdly complicated. We shall examine only the very simplest type of three-component systems in order to acquire some familiarity with triangular phase diagrams.

9.4 (a) Triangular phase diagrams

The mole fractions of the three components of a *ternary system* ($C=3$) sum to unity: $x_A + x_B + x_C = 1$. A *triangular phase diagram* based on the properties of equilateral triangles ensures that this property is satisfied automatically. This is because the sum of the distances to a point inside an equilateral triangle measured parallel to the edges is equal to the length of the side of the triangle, Fig. 9.15, which may be taken to have unit length. If the mole fractions of the three components are represented by these three distances, then a system of any composition may be represented by a single point.

Figure 9.15 shows how this works in practice. In the illustration the point P represents $x_A = 0.50$, $x_B = 0.10$, and $x_C = 0.40$. Note that the edge AB corresponds to $x_C = 0$, and likewise for the other two edges. This means that each of the three edges refer to one of the three binary systems (A, B), (B, C), and (C, A).

An important property of a triangular diagram relates to the straight line

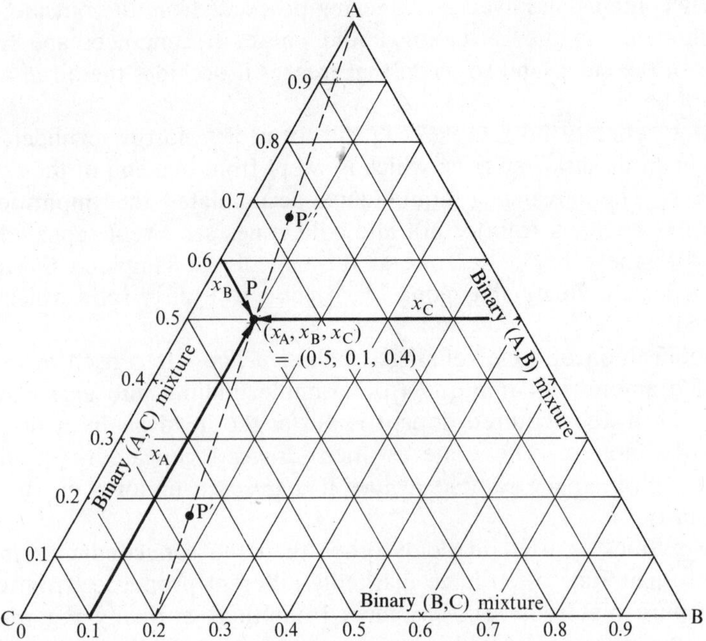

Fig. 9.15. The triangular coordinates used for the discussion of three-component systems. The edges correspond to binary systems. All points along the broken line correspond to mole fractions of C and B in the same ratio.

joining an apex to a point on the opposite edge (the broken line in Fig. 9.15). Any point on that line represents a composition which (a) is progressively richer in A the closer the point is to the A apex and (b) refers to the *same* proportions of B and C in the system. (This property depends on the properties of equilateral triangles, and on showing that x'_B/x'_C at P′ is equal to x''_B/x''_C at P″.) Therefore, if we wish to represent the changing composition of a system as A is added, all we need do is to draw a line from the A apex to the point on BC representing the initial binary system. Any ternary system formed by adding A then lies at some point on this line.

Example 9.4

Mark the following points on a triangular composition diagram:

(a) $x_A = 0.20, \quad x_B = 0.80, \quad x_C = 0$
(b) $x_A = 0.42, \quad x_B = 0.26, \quad x_C = 0.32$
(c) $x_A = 0.80, \quad x_B = 0.10, \quad x_C = 0.10$
(d) $x_A = 0.10, \quad x_B = 0.20, \quad x_C = 0.70$
(e) $x_A = 0.20, \quad x_B = 0.40, \quad x_C = 0.40$
(f) $x_A = 0.30, \quad x_B = 0.60, \quad x_C = 0.10$.

• *Method*. x_A is measured along either edge leading to apex A; likewise for x_B and B; x_C takes care of itself (but it is sensible to check).

• *Answer*. The points are plotted in Fig. 9.16.

• *Comment*. Note that the points (d), (e), and (f) have $x_A/x_B = 0.50$, and fall on a straight line, as stated in the text.

• *Exercise*. Plot the following points:

(a′) $x_A = 0.25, \quad x_B = 0.25, \quad x_C = 0.50$
(b′) $x_A = 0.50, \quad x_B = 0.25, \quad x_C = 0.25$

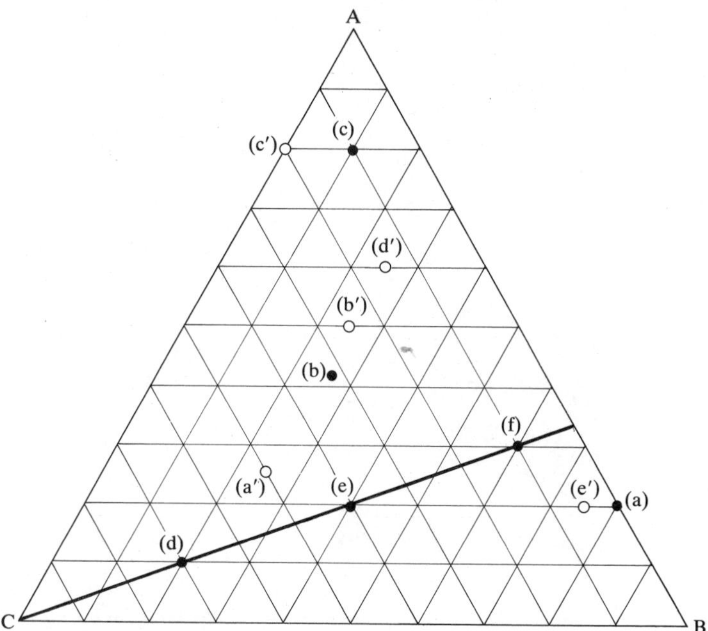

Fig. 9.16. The points referred to in *Example* 9.4 and its *Exercise*.

(c′) $x_A = 0.80$, $x_B = 0$, $x_C = 0.20$
(d′) $x_A = 0.60$, $x_B = 0.25$, $x_C = 0.15$
(e′) $x_A = 0.20$, $x_B = 0.75$, $x_C = 0.05$ [See Fig. 9.16.]

9.4 (b) Partially miscible liquids

A good example of a three-component system is water/chloroform/acetic acid. Water and acetic acid are fully miscible, and so are chloroform and acetic acid. Water and chloroform are only partially miscible. What happens when all three are present together?

The phase diagram at room temperature and pressure is shown in Fig. 9.17. This shows that while the two fully miscible pairs form single-phase regions, the water/chloroform system (along the base of the triangle) has a two-phase region at intermediate compositions. The base of the triangle corresponds to one of the horizontal lines in a two-component phase diagram, such as the one in Fig. 9.5.

A single-phase system is formed when acetic acid is added to the binary water/chloroform mixture. This is shown by following the line a_1, a_4 in Fig. 9.17. Starting at a_1 we have a binary, two-phase system, and the relative amounts of the two phases can be read off in the usual way (using the lever rule). Adding acetic acid takes the system along the line joining a_1 to the apex. At a_2 the solution still has two phases, but there is more water in the chloroform phase (a_2') and more chloroform in the water (a_2'') because the acid helps both to dissolve. Note that the phase diagram shows that there is more acetic acid in the water-rich phase than in the other (a_2'' is closer than a_2' to the acetic acid apex). Note also that the tie-lines must be drawn in on the basis of experiment (in a binary system they are horizontal lines). We need them in order to know which points on the boundary relate to the overall composition (e.g., a_2' and a_2'' relate to a_2). At a_3 two phases are

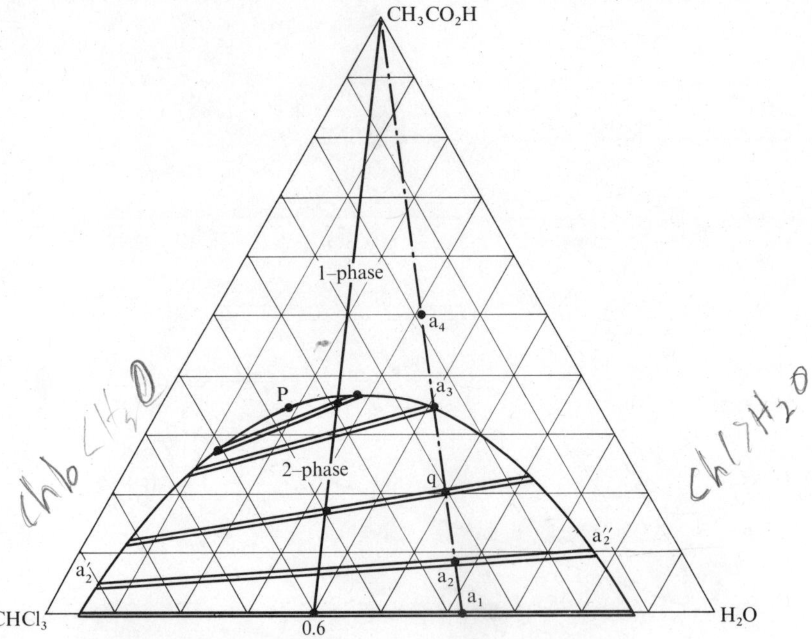

Fig. 9.17. The phase diagram, at fixed temperature and pressure, of the three-component system acetic acid, chloroform, and water. Only some of the tie-lines have been drawn in the two-phase region. All points along the line a correspond to chloroform and water present in the same ratio.

present, but the chloroform-rich layer is present only as a trace. Further addition of acid takes the system towards a_4, and only a single phase is present.

Example 9.5

A mixture is prepared consisting of chloroform ($x_C = 0.60$) and water ($x_W = 0.40$). Describe the changes that occur when acetic acid is added to the mixture.

• *Method*. Base the discussion on Fig. 9.17. The relative proportions of chloroform and water remain constant, and so the addition of acetic acid (A) corresponds to motion along the line from the point $x_C = 0.60$ on the base line opposite the A apex to the apex itself. The tie-lines give the compositions of the phases at their intersections with the boundaries; the lever rule gives their proportions. We shall denote compositions in the order (x_C, x_W, x_A).

• *Answer*. The initial composition is (0.60, 0.40, 0). This point lies in the two-phase region, the phase compositions being (0.95, 0.05, 0) and (0.12, 0.88, 0) with relative proportions 1.3 : 1. Addition of acetic acid takes the system along the straight line to A. When sufficient has been added to raise its mole fraction to 0.18 the overall composition is (0.49, 0.33, 0.18) and the system consists of two phases of compositions (0.82, 0.06, 0.12) and (0.17, 0.60, 0.23) in almost equal abundance. When enough acid has been added to raise its mole fraction to 0.37 the system consists of a trace of a phase of composition (0.64, 0.11, 0.25) and a dominating phase of composition (0.35, 0.28, 0.37). Further addition of acid takes the system into the single-phase region, where it remains right up to the point corresponding to pure acid.

• *Exercise*. Repeat the question, starting with a mixture of composition (0.70, 0, 0.30) to which water is added.

9.4 (c) The role of added salts

Ammonium chloride and ammonium sulphate are both individually soluble in water. What is the solubility of a mixture of the two salts?

Figure 9.18 shows the room-temperature phase diagram of the ternary system ammonium chloride/ammonium sulphate/water. The point b indi-

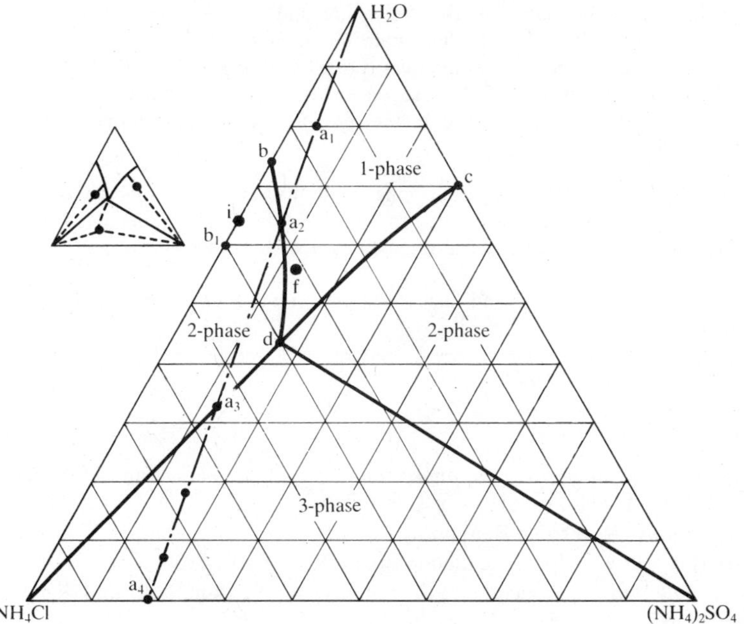

Fig. 9.18. The phase diagram, at constant temperature and pressure, for the ternary system H_2O, NH_4Cl, NH_4SO_4. The points i and f are the ones mentioned in *Example* 9.6. All tie-lines in the 2-phase regions terminate at an apex, and all tie lines in the 3-phase region terminate at the three corners of the 3-phase triangular area, as shown in the insert.

cates the solubility of the chloride: a mixture of composition b_1 consists of the undissolved chloride and a saturated solution of composition b. The point c similarly indicates the solubility of the sulphate.

Consider a ternary system of composition a_1. This is unsaturated, and forms a single phase. As water is evaporated the composition moves along the line a_1 to a_4. At a_2 it enters the two-phase region, and some solid chloride crystallizes (all the tie lines ending on that boundary also end at the pure chloride apex). The liquid becomes richer in sulphate, and its composition moves towards d. When enough water has been removed to bring the overall composition to a_3 the liquid composition is d. At this point (which is joined to the chloride apex and to the sulphate apex), the system consists of saturated solution in equilibrium with the two solids. Note that this point, which corresponds to the joint solubility of the two solids, corresponds to a smaller mole fraction of water than in either of the binary systems b and c. This means that the two salts form a more concentrated solution overall than either does alone.

If more water is removed after the system has arrived at d, all that happens is that the amount of solution decreases, but its composition remains constant (at d, the saturated solution). Both solids are precipitated, and the system has three phases: each point in the three-phase region is tied to d and the two solid apexes. When the composition arrives at a_4 we have a binary system, consisting of a mixture of the two solids, and all the water has gone.

Example 9.6

A solution of 50 g ammonium chloride in 30 g water is prepared at room temperature. 45 g ammonium sulphate is then added. Describe the initial and final states.

• *Method*. Use Fig. 9.18 after converting compositions to mole fractions. Relative molecular masses are as follows: H_2O, 18.02; NH_4Cl, 53.49; NH_4SO_4, 132.1. We shall write compositions in the order (x_W, x_C, x_S), for water (W), chloride (C), and sulphate (S). Use the lever rule for proportions of each phase.

• *Answer*. The initial *amounts* are $n_W = 1.66$ mol and $n_C = 0.93$ mol, and so the initial composition is (0.64, 0.36, 0). In the final state the amounts of W and S are the same but $n_S = 0.34$ mol. Therefore the final composition is (0.57, 0.21, 0.12). From Fig. 9.18 we see that (0.64, 0.36, 0) corresponds to a two-phase system consisting of solid C with a saturated solution of composition (0.64, 0.36, 0) in relative proportions 0.16:1. After addition of S there is only one phase.

• *Comment*. We see that the presence of one solute may affect the solubility of another. In the case of a gas (or other non-ionic substance) dissolving in water, its solubility is reduced if a salt is added to the water. This is called the *salting-out effect*.

• *Exercise*. Describe the initial and final states of a solution of compositions (0.80, 0, 0.20) and (0.40, 0.50, 0.10).

Further reading

Freezing points, triple points, and phase equilibria. R. C. Parker and D. S. Kristol; *J. chem. Educ*. **51,** 658 (1974).
Phase transitions. R. Brout; Benjamin, Reading, 1965.
Phase transitions. A. Alper; Academic Press, New York, 1970.
The phase rule and its applications. A. Findlay, A. N. Campbell, and N. O. Smith; Dover, New York, 1951.
The phase rule and heterogeneous equilibria. J. E. Ricci; Van Nostrand, New York, 1951.
Geochemistry. W. S. Fyfe; Clarendon Press, Oxford, 1974.
High pressure chemistry. R. S. Bradley and D. C. Munro; Pergamon Press, Oxford, 1965.
The diamond-anvil high-pressure cell. A. Jayaraman; *Scientific American* **250** (4), 42 (1984).

Introductory problems

A9.1. Sketch the phase diagram for the system ammonia–hydrazine (N_2H_4). These substances form no compounds with each other. Ammonia freezes at −78 °C, and N_2H_4 freezes at 2 °C. The eutectic composition is 7 mole % N_2H_4 and the eutectic temperature is −80 °C.

A9.2. The substances UF_4 and ZrF_4 melt at 1035 °C and 912 °C, respectively. They form a continuous series of solid solutions with a minimum melting temperature of 765 °C and a composition of 77 mole % ZrF_4. At 900 °C, the liquid solution which is 28 mole % ZrF_4 is in equilibrium with a solid solution which is 14 mole % ZrF_4. At 850 °C a liquid solution which is 87 mole % ZrF_4 is in equilibrium with a solid solution which is 90 mole % ZrF_4. Sketch the phase diagram for this system, and state what is observed if a liquid mixture which is 40 mole % ZrF_4 is cooled slowly from 900 °C to 500 °C.

A9.3. The compounds CH_4 and CF_4 do not form solid solutions with each other. As liquids, they are partially miscible. The melting points of CH_4 and CF_4 are 91 K and 89 K, respectively. The upper consolute temperature for the liquid system is 94 K at 43 mole % CF_4. The eutectic temperature is 84 K, and the eutectic composition is 88 mole % CF_4. At 86 K, the phase in equilibrium with CF_4-rich liquid solution changes from CH_4(s) to CH_4-rich liquid solution. At that temperature, the two liquid solutions in equilibrium are 10 mole % CF_4 and 80 mole % CF_4. Sketch the phase diagram.

A9.4. Describe the phase changes which occur when a liquid mixture of 4 moles of diborane, B_2H_6, and 1 mole of methylethyl ether, $CH_3OC_2H_5$, is cooled from 140 K to 90 K. These substances form a compound, CH_3O-$C_2H_5.B_2H_6$, which melts congruently at 133 K. The system exhibits a eutectic at 25 mole % B_2H_6 and 123 K and another eutectic at 90 mole % B_2H_6 and 104 K. The melting points of $CH_3OC_2H_5$ and B_2H_6 are 131 K and 110 K, respectively.

A9.5. Refer to the information in Problem A9.4 and sketch temperature–time cooling curves for liquid mixtures of $CH_3OC_2H_5$ and B_2H_6 in which the mole % of B_2H_6 is: (a) 10, (b) 30, (c) 50, (d) 80, (e) 95.

A9.6. At 25 °C, aqueous solutions containing Li_2SiF_6 and $(NH_4)_2SiF_6$ can be in equilibrium with Li_2SiF_6(s) or $(NH_4)_2SiF_6$ or the double salt $Li_2SiF_6.(NH_4)_2SiF_6$(s). Li_2SiF_6(s) can be in equilibrium with solutions ranging from 29.5 wt. % Li_2SiF_6 and 70.5 wt. % H_2O to 29.3 wt. %

Li_2SiF_6 and 67.6 wt. % H_2O. $(NH_4)_2SiF_6(s)$ can be in equilibrium with solutions ranging from 18.7 wt. % $(NH_4)_2SiF_6$ and 81.3 wt. % H_2O to 16.5 wt. % $(NH_4)_2SiF_6$ and 70.4 wt. % H_2O. Sketch the phase diagrams for this ternary system; be sure to include some tie lines. Describe the phase changes which occur when a solution, originally 3.0 wt. % Li_2SiF_6 and 83.0 wt. % H_2O, is gradually dehydrated.

A9.7. The substances hexane, C_6H_{14}, and perfluorohexane, C_6F_{14}, show limited liquid–liquid miscibility below the upper consolute temperature, 22.70 °C. The critical composition is 35.5 mole % C_6F_{14}. At 22 °C, the two solutions in equilibrium have C_6F_{14} concentrations of 24 mole % and 48 mole % respectively. At 21.5 °C, the concentrations are 22 and 51 mole % C_6F_{14}. Sketch the phase diagram. Describe the phase changes which occur when C_6F_{14} is gradually added to a fixed amount of C_6H_{14} at (a) 23 °C, (b) 22 °C.

A9.8. The substances $FeCl_2$ and KCl react at elevated temperature to form $KFeCl_3$, a solid which melts congruently at 399 °C, and K_2FeCl_4, a solid which melts incongruently at 380 °C. $FeCl_2$ melts at 677 °C, and KCl melts at 776 °C. The two eutectics are at 38 mole % $FeCl_2$ and 351 °C and at 54 mole % $FeCl_2$ and 393 °C. The KCl solubility curve intersects the K_2FeCl_4 curve at 34 mole % $FeCl_2$. Sketch the phase diagram for this system. State the phases which are in equilibrium when a mixture with composition 36 mole % $FeCl_2$ is cooled from 400 °C to 300 °C.

A9.9. At a certain low temperature, the solubility of iodine in liquid CO_2 is 3 mole % iodine. At the same temperature, its solubility in nitrobenzene is 4 mole % iodine. Liquid CO_2 and nitrobenzene are miscible in all proportions, and the solubility of iodine in their mixtures varies linearly with the concentration of nitrobenzene. Sketch a phase diagram for the ternary system iodine–nitrobenzene–liquid CO_2.

A9.10. The binary system nitroethane–decahydronaphthalene (DEC) shows limited liquid–liquid miscibility with the two-phase region between nitroethane concentrations of 8 and 84 mole %. The system liquid CO_2–DEC behaves similarly, and the two-phase region lies between 36 and 80 mole % DEC. The addition of liquid CO_2 to mixtures of nitroethane and DEC increases the range of miscibility, and it becomes complete with a critical point at 18 mole % liquid CO_2 and 53 mole % nitroethane. The addition of nitroethane to mixtures of DEC and liquid CO_2 similarly leads to a critical point at 8 mole % nitroethane and 52 mole % DEC. Nitroethane and liquid CO_2 are miscible in all proportions. (a) Sketch the phase diagram for the ternary system. (b) For some binary mixtures of nitroethane and liquid CO_2, the addition of arbitrary amounts of DEC will not cause phase separation. Find the range of concentrations for such binary mixtures.

Problems

9.1. The compound *p*-azoxyanisole forms a liquid crystal. 5 g of the solid was put into a tube which was then evacuated and sealed. Use the phase rule to prove that the solid will melt at a definite temperature, and that the liquid crystal phase will make a transition to a normal liquid phase at a definite temperature.

9.2. State how many *components* there are in the following systems. (a) NaH_2PO_4 in water in equilibrium with water vapour, but disregarding the possibility that the salt ionizes in solution; (b) the same, but taking into account the possibility of complete ionization into all possible ions; (c) $AlCl_3$ in water, noting that hydrolysis and precipitation of $Al(OH)_3$ occur.

9.3. Two phases are in equilibrium if the chemical potentials of the species present are the same in each phase. Show that two phases are in thermal equilibrium only if their temperatures are the same, and that they are in mechanical equilibrium only if they are at the same pressure.

9.4. Blue $CuSO_4.5H_2O$ crystals decompose and release their water of hydration when heated. How many phases and components are present in an otherwise empty heated vessel?

9.5. Ammonium chloride also dissociates when it is heated. How many phases and components are present when the salt is heated in an otherwise empty vessel? How many phases and components are present when ammonia is added before the salt is heated?

9.6. A saturated solution of sodium sulphate, with excess of the salt, is at equilibrium with its vapour in a closed vessel. How many phases and components are present? How many degrees of freedom are there, and what are they?

9.7. Now suppose that the solution referred to in the last Problem was not saturated. How many components, phases, and degrees of freedom are there at equilibrium, and what are the degrees of freedom?

9.8. Both MgO and NiO are highly refractory materials. Nevertheless, when the temperature is high enough they do melt, and the temperatures at which their mixtures melt are of considerable interest in the ceramics industry. Draw the temperature–composition diagram for the MgO/NiO system using the data below, where x is the composition of the solid, and y that of the liquid (as mole fraction of MgO).

θ/°C	1960	2200	2400	2600	2800
x	0	0.35	0.60	0.83	1.00
y	0	0.18	0.38	0.65	1.00

9.9. Use the MgO/NiO phase diagram just constructed, state (a) the melting point of composition $x = 0.30$; (b) the composition of the system, in terms of the nature, composition, and proportions of the phases, when a solid with $x = 0.30$ is heated to 2200 °C; (c) the temperature at which a liquid of composition $y = 0.70$ will begin to solidify.

9.10. The bismuth cadmium system is of interest in metallurgy, and we can use it to illustrate several points made in the chapter. The solid–liquid phase diagram calculated using the Clausius–Claperyron equation is very close to the experimental phase diagram throughout the composition range, and so we begin by constructing it. Use the information that $T_f(\text{Bi}) = 544.5\ \text{K}$, $T_f(\text{Cd}) = 594\ \text{K}$, $\Delta H_{f,m}(\text{Bi}) = 10.88\ \text{kJ mol}^{-1}$, $\Delta H_{f,m}(\text{Cd}) = 6.07\ \text{kJ mol}^{-1}$, and the knowledge that they are insoluble in each other in the solid state.

9.11. Use the phase diagram constructed in the last Problem, state what you would observe when a liquid containing $x(\text{Bi}) = 0.70$ is cooled slowly from 550 K. What relative amounts of solid and liquid are present at (a) 460 K, (b) 350 K? Describe the solid obtained by cooling the sample very rapidly.

9.12. Sketch the cooling curve (the temperature versus time, at constant rate of heat transfer) for the mixture treated in the last Problem.

9.13. Now we move on to more conventional liquid mixtures, and begin by constructing the phase diagram for *m*-toluidine and glycerol. In the experiments on which data below were obtained (at $p = 1$ atm), solutions of *m*-toluidine were made up in glycerol, and then warmed from room temperature. The mixture was observed to lose its turbidity at the temperature θ_1, and then on further heating to become turbid again at θ_2. Plot the phase diagram on the basis of the data, and find the upper and lower consolute temperatures.

$100w$	18	20	40	60	80	85
$\theta_1/°\text{C}$	48	18	8	10	19	25
$\theta_2/°\text{C}$	53	90	120	118	83	53

$100w$ is the percentage by weight of *m*-toluidine in the mixture.

9.14. Now use the phase diagram drawn in the last Problem to state what happens as *m*-toluidine is added dropwise to glycerol at 60 °C. State the number of phases present at each concentration, their composition, and their relative amounts.

9.15. In Fig. 9.5 is shown the phase diagram for two partially miscible liquids, and the water/isobutanol (2-methyl-1-propanol) system resembles it quite closely. State what happens when a mixture of water (A) and isobutanol (B) of composition b_3 is heated, at each stage stating the composition, number, and amounts of the phases in equilibrium.

9.16. The phase diagram for the silver/tin system is shown in Fig. 9.19. Label the regions and state what will be observed when liquids of compositions a and b are cooled to 200 K.

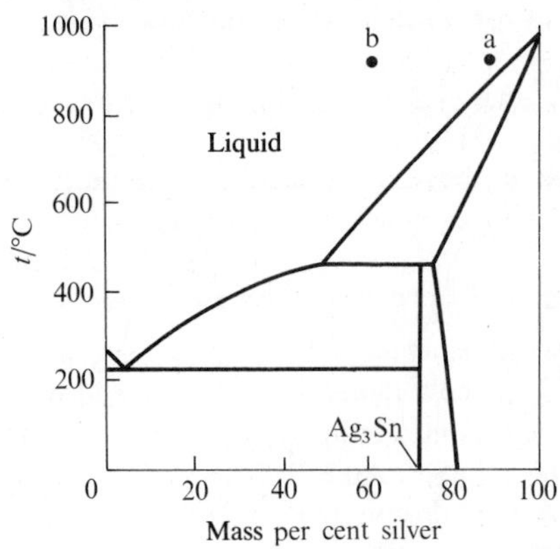

Fig. 9.19. The silver/tin phase diagram.

9.17. Indicate on the phase diagram the feature that denotes an incongruent melting point. What is the composition of the eutectic mixture? At what temperature does the eutectic melt? Sketch the cooling curves for the isopleths a and b.

9.18. Use the silver/tin phase diagram to state (a) the solubility of silver in tin at 800 °C, (b) the solubility of the compound Ag_3Sn in silver at 460 °C, (c) the solubility of the compound in silver at 300 °C.

9.19. Sketch the phase diagram for the magnesium/copper system using the following information $T_f(\text{Mg}) = 648\ °\text{C}$, $T_f(\text{Cu}) = 1085\ °\text{C}$; two intermetallic compounds are formed with melting points $T_f(\text{MgCu}_2) = 800\ °\text{C}$, $T_f(\text{Mg}_2\text{Cu}) = 580\ °\text{C}$; eutectics of composition and melting points 10%, 33%, 65% by weight magnesium, 690 °C, 560 °C, and 380 °C, respectively.

9.20. A sample of Mg/Cu alloy containing 25 percent magnesium by weight was prepared in a crucible heated to 800 °C in an inert atmosphere. Use the phase diagram drawn in the last Problem to state what will be observed if the melt is cooled slowly to room temperature. State what phases are in equilibrium at each temperature, their compositions, and their relative abundances.

9.21. Sketch the temperature–time cooling curve for the molten alloy referred to in the last Problem.

9.22. At first sight triangular coordinates may appear quite confusing, but the next few Problems will give practice in thinking about the information they contain. As a first step, mark the following features on triangular coordinates. (a) The point denoting the composition $x_A = 0.2$, $x_B = 0.2$, $x_C = 0.6$; (b) the same for $x_A = 0$, $x_B = 0.2$, $x_C = 0.8$; (c) the same for $x_A = x_B = x_C$; (d) the point denoting a mixture formed with 25 percent NaCl, 25 percent $Na_2SO_4.10H_2O$, the rest water; (e) the line denoting the composition where NaCl and $Na_2SO_4.10H_2O$ retain the same relative abundances as water is added to the solution.

9.23. One of the properties of triangular phase diagrams is that a straight line from an apex (A) to an opposite edge (BC) represents the mixtures where x_B and x_C are in the same relative proportions however much A is present. Prove this property on the basis of the properties of similar triangles.

9.24. Now we turn to the actual construction of a triangular phase diagram (you will find similar data in Vol. 3 of *International critical tables*). Methanol (M), diethyl ether (E), and water (W) form a partially miscible ternary system. The phase diagram at 20 °C was determined by adding the methanol to various binary ether–water mixtures, and noting the mole fractions of methanol x_M^* at which complete miscibility occurred. Plot the phase diagram using the following data:

$x_E(2)$	0.10	0.20	0.30	0.40	0.50
$x_M(3)$	0.20	0.27	0.30	0.28	0.26

$x_E(2)$	0.60	0.70	0.80	0.90
$x_M(3)$	0.22	0.17	0.12	0.07

Ether and water are completely immiscible at $x_E \leqslant 0.05$ and $x_E \geqslant 0.95$. $x_E(2)$ is the mole fraction of ether in the original binary mixture, and $x_M(3)$ is the mole fraction of methanol in the ternary mixture. How many phases will a mixture of 5 g methanol, 30 g diethyl ether, and 50 g water show at 20 °C? How many grams of water would have to be removed or added to change the number of phases?

9.25. Refer to the ternary mixture with the phase diagram shown in Fig. 9.17. How many phases are there, what are their compositions, and in what relative proportions do they occur, in a mixture formed from 2.3 g water, 9.2 g chloroform, and 3.1 g acetic acid?

9.26. State what would be observed if (a) water, (b) acetic acid is added to the mixture referred to in the last Problem.

9.27. The phase diagram for the ternary system NH_4Cl(A)/$(NH_4)_2SO_4$(B)/water (C) at 25 °C is illustrated in Fig. 9.18. What is the nature of the system that contains (a) $x_A = 0.2$, $x_B = 0.4$, $x_C = 0.4$; (b) $x_A = 0.4$, $x_B = 0.4$, $x_C = 0.2$; (c) $x_A = 0.2$, $x_B = 0.1$, $x_C = 0.7$; (d) $x_A = 0.40$, $x_B = 0.16$, $x_C = 0.44$?

9.28. What is the solubility of (a) NH_4Cl in water at 25 °C, (b) $(NH_4)_2SO_4$ in water at 25 °C?

9.29. Explain what happens as (a) $(NH_4)_2SO_4$ is added to a saturated solution of NH_4Cl in water; (b) $(NH_4)_2SO_4$ is added to a saturated solution of NH_4Cl in water in the presence of excess NH_4Cl; (c) water is added to a mixture of 25 g NH_4Cl, 75 g $(NH_4)_2SO_4$.

10

Changes of state: chemical reactions

Learning objectives

After careful study of this chapter you should be able to:

(1) Deduce the condition for chemical equilibrium in terms of the reaction Gibbs function, eqn (10.1.2).

(2) Define the terms *exergonic* and *endergonic*, and explain their significance, Section 10.1(b).

(3) Define the *reaction quotient* and write an expression for the reaction Gibbs function at an arbitrary stage in a reaction, eqn (10.1.16).

(4) Define the *equilibrium constant* of a reaction, and derive its relation to the standard reaction Gibbs function, eqn (10.1.17).

(5) Explain why a catalyst has no effect on the equilibrium composition of a reaction, Section 10.2(a).

(6) State *Le Chatelier's principle*, Section 10.2(b), and justify it for changes of pressure and temperature.

(7) Derive a relation between the mole-fraction equilibrium constant and the overall pressure of a gas phase reaction, eqn (10.2.3).

(8) Derive and use the *van't Hoff isochore* for the temperature dependence of the equilibrium constant, eqn (10.2.4).

(9) Derive and use an equation for the equilibrium constant at one temperature when it is known at another, eqn (10.2.6).

(10) Define and use the *Giauque function* to calculate the equilibrium constants of reactions at different temperatures, Example 10.5.

(11) Describe the construction and interpretation of an *Ellingham diagram*, Section 10.3(a).

(12) Define the *biological standard state*, Section 10.3(b), and relate biological standard values to normal standard values, Example 10.6.

Introduction

Chemical reactions move towards a *dynamic equilibrium* in which both reactants and products are present but have no tendency to undergo net change. Sometimes the concentration of products is so much greater than the concentration of unchanged reactants in the equilibrium mixture that for all practical purposes the reaction is 'complete'. However, in many important cases the equilibrium mixture has significant concentrations of both reactants and products. In this chapter we see how thermodynamics is used to predict the equilibrium composition under any reaction conditions.

All this is of vital importance. In industry it is obviously worse than useless to build a sophisticated plant if the overall reaction has a tendency to run in the wrong direction. If a plant is to be run economically we must know how to maximize yields. Does that mean raising or lowering the temperature or the pressure? Thermodynamics gives a very simple recipe for deciding what to do. Alternatively, we might be interested in the way that food is used in the complicated series of biochemical reactions involved in warming the body, powering muscular contraction, and energizing the nervous system. Some reactions (such as the oxidation of carbohydrates) have a natural tendency to occur and may be coupled to others to drive them in unnatural but necessary directions (as in the biosynthesis of proteins). With thermodynamics we can sort out the reactions that need to be driven, and calculate the spare driving force of reactions that occur spontaneously.

We have seen that the direction of natural change at constant temperature and pressure is towards lower values of the Gibbs function. The idea is entirely general, and in this chapter we apply it to the discussion of reactions. In the case of the reaction $A + B \rightarrow C + D$, for instance, we can decide whether there is a natural tendency for A and B to form C and D by calculating the Gibbs function of the mixture at various compositions, Fig. 10.1, and identifying the location of the minimum. We can then predict whether A and B will react to form C and D virtually completely (as in a), partially (as in b), or virtually not at all (as in c). In order to proceed, we have to relate the Gibbs function to the composition of the reaction mixture.

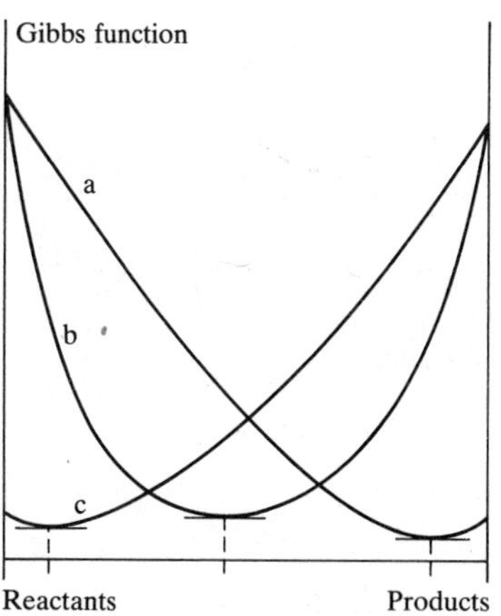

Fig. 10.1. The direction of spontaneous chemical change is towards the minimum Gibbs function. This illustration shows three possibilities: (a) equilibrium lies close to pure products (the reaction 'goes to completion'), (b) equilibrium corresponds to reactants and products present in similar proportions, and (c) equilibrium lies close to pure reactants (the reaction 'does not go').

10.1 Which way is downhill?

We begin with the simplest possible reaction equilibrium: $A \rightleftharpoons B$. Even though this looks trivial, there are many examples of it, such as the isomerization of pentane to 2-methylbutane (*n*-pentane to *iso*-pentane) and the conversion (racemization) of L-alanine to D-alanine.

10.1 (a) The Gibbs function minimum.

Suppose an infinitesimal amount $d\xi$ of A turns into B, then we can write

$$\textit{Change in amount of } \mathrm{A} \textit{ present} = -d\xi,$$

$$\textit{Change in amount of } \mathrm{B} \textit{ present} = +d\xi.$$

ξ is a measure of the progress of the reaction, and is called the *extent of reaction*. We shall choose it so that (for the reaction $A \rightarrow B$) pure A corresponds to $\xi = 0$; the value $\xi = 1$ mol signifies that 1 mol A has been destroyed and 1 mol B has been formed. At constant pressure and temperature the change this causes in the Gibbs function G of the system can be expressed in terms of the chemical potentials (the partial molar Gibbs functions) of the substances in the mixture:

$$dG = \mu_A\, dn_A + \mu_B\, dn_B = -\mu_A\, d\xi + \mu_B\, d\xi.$$

This equation can be reorganized into

$$(\partial G/\partial \xi)_{p,T} = \mu_B - \mu_A. \qquad (10.1.1)$$

We have to remember that the chemical potentials depend on the composition. Therefore, their values change, and so does the slope of G, as the reaction proceeds. If at some stage of the reaction μ_A is greater than μ_B, then the slope of G is negative, but it changes sign when μ_B exceeds μ_A. The chemical reaction proceeds in the direction of decreasing G; thus:

If $\mu_A > \mu_B$, *the reaction proceeds* $A \rightarrow B$;

If $\mu_A < \mu_B$, *the reaction proceeds* $A \leftarrow B$.

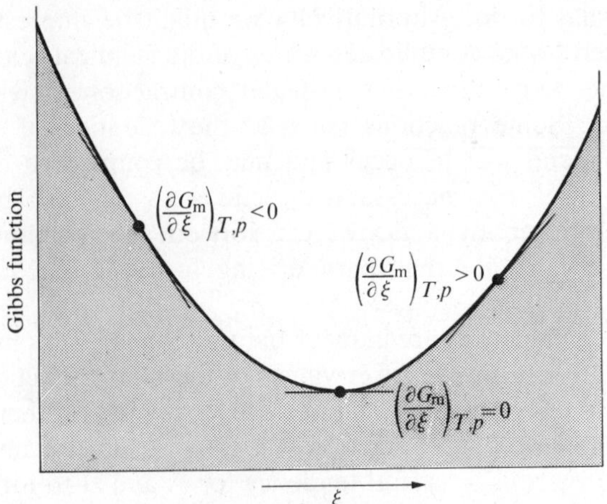

Fig. 10.2. As the reaction advances (represented by motion from left to right along the horizontal axis) the slope of the Gibbs function changes. Equilibrium corresponds to zero slope, at the foot of the valley.

This is illustrated in Fig. 10.2. We have switched attention from the Gibbs function to the chemical potential; now μ is fulfilling the role its name suggests. When $\mu_A = \mu_B$ the slope of G with respect to ξ is zero. This occurs at the minimum of the curve in Fig. 10.2, and corresponds to the position of chemical equilibrium:

If $\mu_A = \mu_B$, *the reaction is at equilibrium.*

Since $(\partial G/\partial \xi)_{p,T} = \mu_B - \mu_A$, and the chemical potentials are the partial molar Gibbs functions at a general stage of the reaction, $(\partial G/\partial \xi)_{p,T}$ is equal to the change in Gibbs function per mole of reaction at a *fixed composition* of the reaction mixture. That is, it is the *reaction Gibbs function at that composition* (symbol: $\Delta_r G$). This quantity must be distinguished from the *standard* reaction Gibbs function (symbol: $\Delta_r G^{\ominus}$) which is the change per mole of reaction when the reactants and products are all in their standard states (Section 5.4). Therefore, by writing $\Delta_r G = \mu_B - \mu_A$ the condition for equilibrium can be expressed as follows:

$$\textit{At equilibrium:}\ \Delta_r G = 0. \qquad (10.1.2)$$

In words, eqn (10.1.2) says that *the change in the Gibbs function per unit amount of substance is zero when products are formed from reactants at the equilibrium composition of the reaction mixture.*

10.1 (b) Exergonic and endergonic reactions

The general reaction Gibbs function can also be used to re-express the direction of spontaneous change at any stage in the reaction:

If $\Delta_r G < 0$, *the reaction proceeds* $A \rightarrow B$,
If $\Delta_r G > 0$, *the reaction proceeds* $A \leftarrow B$.

Reactions under conditions (e.g. composition) for which $\Delta_r G < 0$ are called *exergonic*, which signifies that, since they are spontaneous, they can be used to drive other processes (such as other reactions, or used to do non-p,V work). Reactions for which $\Delta_r G > 0$ under the prevailing conditions are called *endergonic*. They are spontaneous in the reverse direction.

Reactions at equilibrium are spontaneous in neither direction: they are neither exergonic nor endergonic.

10.1 (c) Perfect gas equilibria

The condition of equilibrium for the reaction $A \rightarrow B$ can be developed further by supposing that A and B are perfect gases and expressing the chemical potentials in terms of their partial pressures:

$$\Delta_r G = \mu_B - \mu_A = \{\mu_B^\ominus + RT\ln(p_B/p^\ominus)\} - \{\mu_A^\ominus + RT\ln(p_A/p^\ominus)\}$$
$$= \Delta_r G^\ominus + RT\ln(p_B/p_A).$$

The ratio of partial pressures is called the *reaction quotient* (symbol: Q_p), and so we arrive at

$$\Delta_r G_m = \Delta_r G^\ominus + RT\ln Q_p, \qquad Q_p = p_B/p_A. \qquad (10.1.3)^\circ$$

The standard reaction Gibbs function, which is the difference of the standard chemical potentials of A and B, can always be obtained from tables of standard Gibbs functions of formation, as explained in Section 5.4(c). In this case,

$$\Delta_r G^\ominus = \Delta_f G_B^\ominus - \Delta_f G_A^\ominus,$$

the values of the formation functions being those for the temperature of the reaction.

Now we can take the final step. At equilibrium $\Delta_r G = 0$. The value of the reaction quotient at equilibrium is called the *equilibrium constant* of the reaction (symbol: K_p):

$$K_p = Q_{p,\text{equilibrium}} = \{p_B/p_A\}_{\text{eq}}. \qquad (10.1.4)^\circ$$

Then, with $\Delta_r G = 0$ when $Q_p = K_p$, we conclude from eqn (10.1.3) that

$$RT\ln K_p = -\Delta_r G^\ominus. \qquad (10.1.5)^\circ$$

This is a special case of one of the most important equations in chemical thermodynamics: it is the link between tables of thermodynamic data, such as Table 5.3, and the chemically important equilibrium constant. We see that when $\Delta_r G^\ominus$ is *positive*, K_p is less than unity, and so at equilibrium the partial pressure of A exceeds that of B, which means that the reactant is favoured in the equilibrium. When $\Delta_r G^\ominus$ is *negative*, K_p is greater than unity, and so at equilibrium the partial pressure of B exceeds that of A, which means that the product is favoured in the equilibrium.

The derivation of the relation between the Gibbs function and the equilibrium constant of a general perfect gas reaction runs in a similar way. Consider the reaction $2A + 3B \rightarrow C + 2D$; this is a special case of the general reaction

$$0 = \sum_J \nu_J J, \qquad (10.1.6)$$

written using the notation introduced in Section 4.1(c), eqn (4.1.3), with $\nu_A = -2$, $\nu_B = -3$, $\nu_C = +1$, $\nu_D = +2$. In the following derivation we shall develop the special and general cases in parallel.

When the reaction advances by $d\xi$ the amounts of reactants and products change as follows:

$$\left.\begin{aligned}&\textit{Change in amount of } \mathrm{A} = -2\mathrm{d}\xi\\&\textit{Change in amount of } \mathrm{B} = -3\mathrm{d}\xi\\&\textit{Change in amount of } \mathrm{C} = +\mathrm{d}\xi\\&\textit{Change in amount of } \mathrm{D} = +2\mathrm{d}\xi\end{aligned}\right\} \nu_{\mathrm{J}}\,\mathrm{d}\xi \textit{ in general.}$$

At constant temperature and pressure the change in Gibbs function is

$$\begin{aligned}\mathrm{d}G &= \mu_{\mathrm{A}}\,\mathrm{d}n_{\mathrm{A}} + \mu_{\mathrm{B}}\,\mathrm{d}n_{\mathrm{B}} + \mu_{\mathrm{C}}\,\mathrm{d}n_{\mathrm{C}} + \mu_{\mathrm{D}}\,\mathrm{d}n_{\mathrm{D}}\\&= \{-2\mu_{\mathrm{A}} - 3\mu_{\mathrm{B}} + \mu_{\mathrm{C}} + 2\mu_{\mathrm{D}}\}\,\mathrm{d}\xi, \quad \text{and} \quad \mathrm{d}G = \left\{\sum_{\mathrm{J}} \nu_{\mathrm{J}}\mu_{\mathrm{J}}\right\}\mathrm{d}\xi \text{ in general.}\end{aligned}$$

Therefore,

$$\Delta_{\mathrm{r}}G = (\partial G/\partial\xi)_{p,T} = -2\mu_{\mathrm{A}} - 3\mu_{\mathrm{B}} + \mu_{\mathrm{C}} + 2\mu_{\mathrm{D}},$$

and

$$\Delta_{\mathrm{r}}G = \sum_{\mathrm{J}} \nu_{\mathrm{J}}\mu_{\mathrm{J}} \text{ in general.} \tag{10.1.7}$$

The chemical potentials may be expressed in terms of the partial pressures, and it is easy to deduce that

$$\Delta_{\mathrm{r}}G = \Delta_{\mathrm{r}}G^{\ominus} + RT\ln Q_p, \tag{10.1.8}^{\circ}$$

where

$$\Delta_{\mathrm{r}}G^{\ominus} = -2\mu_{\mathrm{A}}^{\ominus} - 3\mu_{\mathrm{B}}^{\ominus} + \mu_{\mathrm{C}}^{\ominus} + 2\mu_{\mathrm{D}}^{\ominus}, \quad \text{and} \quad \Delta_{\mathrm{r}}G^{\ominus} = \sum_{\mathrm{J}} \nu_{\mathrm{J}}\mu_{\mathrm{J}}^{\ominus} \text{ in general,} \tag{10.1.9}$$

and

$$Q_p = \frac{(p_{\mathrm{C}}/p^{\ominus})(p_{\mathrm{D}}/p^{\ominus})^2}{(p_{\mathrm{A}}/p^{\ominus})^2(p_{\mathrm{B}}/p^{\ominus})^3}, \quad \text{and} \quad Q_p = \prod_{\mathrm{J}} (p_{\mathrm{J}}/p^{\ominus})^{\nu_{\mathrm{J}}} \text{ in general.} \tag{10.1.10}^{\circ}$$

(Π is the sign for multiplication, as Σ is the sign for summation: since the reactants have negative stoichiometric coefficients, they automatically appear as the denominator when the product is written out explicitly, as in the line above.)‡ The standard reaction Gibbs function can be evaluated from tables of standard Gibbs functions of formation using eqn (5.4.4); that is, in general,

$$\Delta_{\mathrm{r}}G = \sum_{\mathrm{J}} \nu_{\mathrm{J}}\Delta_{\mathrm{f}}G^{\ominus} \tag{10.1.11}$$

at the temperature of the reaction.

At equilibrium $\Delta_{\mathrm{r}}G = 0$. The partial pressures then have their equilibrium values, and we can write

$$\begin{aligned}K_p = Q_{p,\text{equilibrium}} &= \left\{\frac{(p_{\mathrm{C}}/p^{\ominus})(p_{\mathrm{D}}/p^{\ominus})^2}{(p_{\mathrm{A}}/p^{\ominus})^2(p_{\mathrm{B}}/p^{\ominus})^3}\right\}_{\mathrm{eq}}\\&= \prod_{\mathrm{J}} \{(p_{\mathrm{J}}/p^{\ominus})^{\nu_{\mathrm{J}}}\}_{\mathrm{eq}} \text{ in general.}\end{aligned} \tag{10.1.12}^{\circ}$$

‡ The value of the reaction quotient depends on the choice of standard pressure (e.g. $p^{\ominus} = 1$ bar, as here, or $p^{\ominus} = 1$ atm, as previously). This is why it is sometimes written as $Q^{\ominus}$, the superscript being there to signify that a standard pressure (and in general, a standard state for systems other than gases) has been selected. Similarly, K is written $K^{\ominus}$ because its value also depends on the choice of standard in general. However, adding superscripts to equations already heavily burdened with them makes them look even more complicated, and so we shall not adopt this recommended practice even though it is wise. The reason underlying the recommendation, that the numerical values of Q and K depend on the choice of standard, must be borne in mind.

Setting $\Delta_r G = 0$ and replacing Q by K in eqn (10.1.8) leads at once to

$$RT \ln K_p = -\Delta_r G^{\ominus}, \qquad (10.1.13)°$$

an explicit recipe for calculating the equilibrium constant for reactions involving perfect gases.

Example 10.1

Calculate the equilibrium constant of the reaction $N_2(g) + 3H_2(g) \rightarrow 2NH_3(g)$ at 298 K.

• *Method*. Write the reaction in the form of eqn (10.1.6) and identify the stoichiometric coefficients. Then implement eqns (10.1.9–13) using data from Table 5.3.

• *Answer*. The reaction is

$$0 = 2NH_3(g) - N_2(g) - 3H_2(g),$$

so that $\nu(NH_3) = +2$, $\nu(N_2) = -1$, and $\nu(H_2) = -3$. Therefore:

$$K_p = \{(p_{NH_3}/p^{\ominus})^2(p_{N_2}/p^{\ominus})^{-1}(p_{H_2}/p^{\ominus})^{-3}\}_{eq}$$
$$= \{p_{NH_3}^2 p^{\ominus 2}/p_{N_2}p_{H_2}^3\}_{eq}.$$
$$\Delta_r G^{\ominus} = 2\Delta_f G^{\ominus}_{NH_3} - \Delta_f G^{\ominus}_{N_2} - 3\Delta_f G^{\ominus}_{H_2} = 2\Delta_f G^{\ominus}_{NH_3} = -2 \times 16.45 \text{ kJ mol}^{-1}.$$

Therefore, since $RT = 2.48 \text{ kJ mol}^{-1}$,

$$\ln K_p = -(-32.90 \text{ kJ mol}^{-1}/2.48 \text{ kJ mol}^{-1}) = 13.3.$$

Hence, $K_p = 6.0 \times 10^5$.

• *Comment*. Note that K_p and Q_p are dimensionless. This will be the case with all the equilibrium constants in the following pages.

• *Exercise*. Evaluate (1) the equilibrium constant for the reaction $\frac{1}{2}N_2(g) + \frac{3}{2}H_2(g) \rightarrow NH_3(g)$ and (2) the equilibrium constant for $N_2O_4(g) \rightarrow 2NO_2(g)$, both at 298 K. [770, 0.14]

10.1 (d) A recipe for equilibrium constants: real gases

We can readily convert the preceding equations so that they apply to real gases: the partial pressures p_J are replaced by the fugacities f_J. No other change need be made, and so we arrive at

$$\Delta_r G = \Delta_r G^{\ominus} + RT \ln Q, \qquad Q = \prod_J (f_J/p^{\ominus})^{\nu_J}, \qquad (10.1.14)$$

the thermodynamically exact version of eqn (10.1.8), and hence at

$$\Delta_r G^{\ominus} = -RT \ln K, \qquad K = \prod_J \{(f_J/p^{\ominus})^{\nu_J}\}_{eq}, \qquad (10.1.15)$$

the thermodynamically exact version of eqn (10.1.13). K is called the *thermodynamic equilibrium constant*. In order to use these expressions we have to be able to interpret the fugacities in terms of the partial pressures of the gases: that involves knowing their fugacity coefficients, which are normally available from tables of data, such as Table 6.1.

10.1 (e) Another recipe: reactions in general

Only a short step is now needed in order to obtain an expression for the equilibrium constant of any type of reaction. The general expression for the chemical potential of a substance J is

$$\mu_J = \mu_J^\ominus + RT \ln a_J,$$

where a_J is its activity when it is present at the concentration of interest. (Activities were discussed in Section 8.5; they are defined as dimensionless quantities.) All the previous analysis now follows through, with the result that at a general stage of the reaction

$$\Delta_r G = \Delta_r G^\ominus + RT \ln Q, \qquad Q = \prod_J a_J^{\nu_J}, \qquad (10.1.16)$$

and at equilibrium, when $\Delta_r G = 0$,

$$\Delta_r G^\ominus = -RT \ln K, \qquad K = \left\{\prod_J a_J^{\nu_J}\right\}_{eq}. \qquad (10.1.17)$$

In order to use these relations we use the conventions established in Chapter 8, see Box 8.1, that the activities of pure solids and liquids are unity.

The only remaining problem is to express the thermodynamic equilibrium constant in terms of the mole fractions or concentrations of the substances. This requires us to know the activity coefficients, and then to use $a_J = \gamma_J x_J$ (if we are using mole fractions, eqn (8.5.11)) or $a_J = \gamma_J m_J/m^\ominus$ (if we are using molalities, eqn (8.5.15)). For example, in the latter case, for an equilibrium of the form $A + B \rightleftharpoons C + D$, where all four species are solutes, the thermodynamic equilibrium constant can be expressed in terms of molalities as follows:

$$K = \left\{\frac{a_C a_D}{a_A a_B}\right\}_{eq} = \left\{\frac{\gamma_C \gamma_D}{\gamma_A \gamma_B}\right\}\left\{\frac{m_C m_D}{m_A m_B}\right\}_{eq} = K_\gamma K_m. \qquad (10.1.18)$$

The activity coefficients must be evaluated at the equilibrium composition of the mixture, which may involve a complicated calculation, because the latter is known only if the equilibrium constant K_m is already known. In elementary applications (and to begin the iterative calculation of the concentrations in a real example) the assumption is often made that the activity coefficients are all so close to unity that $K_\gamma = 1$. Then we obtain the result widely used in elementary chemistry that $K \approx K_m$, and equilibria are discussed in terms of molalities (or concentrations) themselves. In Chapter 11 we shall see a way of making better estimates of activity coefficients for equilibria involving ions.

10.2 The response of equilibria to the conditions

In this section we investigate how equilibrium constants (and therefore the equilibrium compositions) depend on the conditions. We shall be able to state whether a rise of temperature will shift the equilibrium towards products, or drag it back towards the reactants. The technological importance of this kind of calculation is obvious: it is important to know whether raising the temperature of a reaction in order to make it go faster will result in the equilibrium shifting away from the desired products.

10.2 (a) How equilibrium responds to a catalyst

It doesn't. A claim that an equilibrium can be shifted by a catalyst (a substance that changes the rate of a reaction without undergoing net chemical change itself) or an enzyme (a biological catalyst) can be dismissed by appealing to the laws of thermodynamics.

Consider an exothermic reaction, and a hypothetical catalyst which, it is claimed, shifts the reaction towards the products. Take an uncatalysed reaction at equilibrium, and add the catalyst. If the claim were true, the reaction would resume, more products would be formed, and the system would evolve heat until the new equilibrium was reached. The heat evolved could be converted at least partially into work. Now remove the catalyst. The equilibrium shifts back and heat is withdrawn from the surroundings. In the end, everything is in its original state, except that some heat of the surroundings has been converted completely into work. This is contrary to the Second Law, and so the claim is fraudulent. *Nevertheless* it would be wise not to base a court case on this argument: in industry reactions rarely reach perfect equilibrium, partly on account of the rates at which reactants mix, and under these *non*-equilibrium conditions catalysts can have some odd effects and may change the composition of the reaction mixture. For example, in the commercially important Fischer–Tropsch synthesis of hydrocarbon fuels from carbon monoxide and hydrogen, the choice of catalyst influences the distribution of chain lengths in the product.

The advantage of catalysts is that they increase the speed at which reactions take place without affecting the equilibrium. Therefore, if an equilibrium lies strongly towards products at low temperatures, but towards reactants at high temperatures (as is the case with the synthesis of ammonia), a catalyst may be found which allows the reaction to take place at a commercially acceptable rate at the lower temperature and therefore with the more favourable yield. The mechanism of catalysis is described in Part 3.

10.2 (b) How equilibria respond to pressure

The equilibrium constant depends on $\Delta_r G^\ominus$, which is defined at *a single, standard pressure*. It is a *constant*, independent of the pressure at which the equilibrium is established. The same is therefore true of K, and so *K is independent of pressure*. Formally we may express this as

$$(\partial K/\partial p)_T = 0. \tag{10.2.1}$$

This conclusion does not mean that the *amounts* of the species at equilibrium do not depend on the pressure. In the case of the perfect gas equilibrium $A \rightleftharpoons 2B$, for example, increasing the pressure decreases the number of B particles and increases the number of A particles, but it does so in such a way that the equilibrium constant remains unchanged. This is a special case of *Le Chatelier's principle:*

A system at equilibrium, when subjected to a perturbation, responds in a way that tends to minimize its effect.

When pressure is applied to a system at equilibrium the principle implies that it will adjust so as to minimize the increase: this it can do by reducing the number of particles in the gas phase, which implies a shift $A \leftarrow 2B$.

The response to the pressure can be expressed quantitatively (and Le Chatelier's principle can be justified). First consider the special case of the $A \rightleftharpoons 2B$ equilibrium. Suppose that there is an amount n of A present

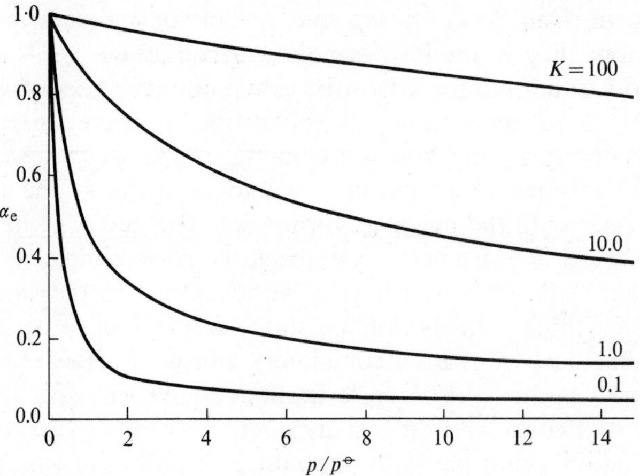

Fig. 10.3. The pressure dependence of the extent of an $A(g) \rightarrow 2B(g)$ reaction at equilibrium calculated using eqn (10.2.2) for different values of the equilibrium constant. $\alpha = 0$ corresponds to pure A. $\alpha = 1$ corresponds to pure B.

initially (and no B). At equilibrium the amount of A is $(1-\alpha_e)n$ and the amount of B is $2\alpha_e n$. It follows that the mole fractions present at equilibrium are

$$x_A = n(1-\alpha_e)/\{n(1-\alpha_e)+2n\alpha_e\} = (1-\alpha_e)/(1+\alpha_e), x_B = 2\alpha_e/(1+\alpha_e).$$

The equilibrium constant is

$$K_p = \left\{\frac{(p_B/p^{\ominus})^2}{(p_A/p^{\ominus})}\right\}_{eq} = \left\{\frac{(x_B p/p^{\ominus})^2}{(x_A p/p^{\ominus})}\right\}_{eq}$$
$$= \{x_B^2/x_A\}_{eq}(p/p^{\ominus}) = \{4\alpha_e^2/(1-\alpha_e^2)\}(p/p^{\ominus}).$$

(Dalton's Law was used to relate partial pressures and mole fractions.) This expression rearranges to

$$\alpha_e = \{K_p/(K_p + 4p/p^{\ominus})\}^{\frac{1}{2}}, \tag{10.2.2}$$

which shows that even though K_p is independent of pressure, the amounts of A and B do depend on pressure. This is shown in Fig. 10.3 for various values of K_p.

The effect of pressure can be expressed in a more general way by considering the general perfect gas equilibrium

$$0 = \sum_J \nu_J J, \qquad K_p = \left\{\prod_J (p_J/p^{\ominus})^{\nu_J}\right\}_{eq}.$$

On replacing each partial pressure by the mole fraction through $p_J = x_J p$ (where $p = \sum_J p_J$) we find

$$K_p = \left\{\prod_J (x_J p/p^{\ominus})^{\nu_J}\right\}_{eq} = \{x_A^{\nu_A} x_B^{\nu_B} \ldots\}_{eq}(p/p^{\ominus})^{\nu_A+\nu_B+\ldots}$$
$$= (p/p^{\ominus})^{\Delta\nu} K_x, \tag{10.2.3°}$$

where K_x is the equilibrium constant in terms of the mole fractions, and $\Delta\nu = \nu_A + \nu_B + \ldots$, which (since the reactants have negative coefficients) is the *difference* between the numbers of gas particles in the products and the reactants. Now, since K_p is independent of pressure, it follows that K_x must be proportional to $1/p^{\Delta\nu}$. This means that if $\Delta\nu$ is positive (an increase of gas phase particles) then K_x decreases with increasing pressure, implying a shift back towards reactants. If $\Delta\nu$ is negative, then the opposite is true. Only if $\Delta\nu = 0$ (the same numbers of gas phase particles on each side of the equation) is the equilibrium composition independent of pressure.

Example 10.2

Predict the effect of a tenfold increase of pressure on the composition of the ammonia synthesis reaction.

• *Method*. Use Le Chatelier's principle to predict the qualitative effect, and eqn (10.2.3) to predict the quantitative effect on K_x.

• *Answer*. Since the number of gas molecules decreases on formation of product (from 4 to 2), Le Chatelier's principle predicts that an increase of pressure will favour the product. The stoichiometric coefficients were written down for the reaction $N_2(g) + 3H_2(g) \rightarrow 2NH_3(g)$ in *Example* 10.1, and from there $\Delta\nu = 2 - 1 - 3 = -2$. Therefore, $K_p = (p/p^\ominus)^{-2}K_x$, so that $K_x = (p/p^\ominus)^2 K_p$. Since K_p is independent of p, K_x will increase 100-fold when the pressure is increased 10-fold.

• *Comment*. The Haber synthesis of ammonia is run at high pressure in order to make use of this result. In precise work it is necessary to base the argument on the pressure independence of the *thermodynamic* equilibrium constant, and therefore to take note of the pressure-dependence of the fugacity coefficients when discussing K_x.

• *Exercise*. Predict the effect of a tenfold pressure increase on the equilibrium composition of the reaction $\frac{3}{2}N_2(g) + \frac{1}{2}H_2(g) \rightarrow N_3H(g)$. [Tenfold decrease in K_x.]

10.2 (c) How equilibria respond to temperature

We have seen how to relate the equilibrium constant to the standard molar Gibbs function of reaction. A slight rearrangement of eqn (10.1.17) gives

$$\ln K = -\Delta_r G^\ominus / RT.$$

The variation of $\ln K$ with temperature is therefore given by

$$\mathrm{d}\ln K/\mathrm{d}T = -(1/R)\{\mathrm{d}(\Delta_r G^\ominus/T)/\mathrm{d}T\}.$$

(The differentials are complete because K and $\Delta_r G^\ominus$ depend only on temperature, not on pressure.) In Chapter 6 we derived the Gibbs–Helmholtz equation relating the temperature dependence of G to the enthalpy. The form of the equation given in eqn (6.2.5) is precisely what we need to develop this equation. In the present notation (dropping the constraint of constant pressure on the grounds that $\Delta_r G^\ominus$ is independent of pressure) it reads

$$\mathrm{d}(\Delta_r G^\ominus/T)/\mathrm{d}T = -\Delta_r H^\ominus/T^2,$$

where $\Delta_r H^\ominus$ is the standard reaction enthalpy at the temperature T. Then, on combining these two equations, we arrive at the

$$\textit{van't Hoff isochore: } \mathrm{d}\ln K/\mathrm{d}T = \Delta_r H^\ominus/RT^2. \qquad (10.2.4)$$

This is exact. Another form of the equation is obtained by recognizing that as $\mathrm{d}(1/T)/\mathrm{d}T = -1/T^2$, or $\mathrm{d}T/T^2 = -\mathrm{d}(1/T)$, we can write

$$\mathrm{d}\ln K/\mathrm{d}(1/T) = -\Delta_r H^\ominus/R. \qquad (10.2.5)$$

The first conclusion we can draw from the van't Hoff isochore concerns the direction of change of K when the temperature is raised. Equation (10.2.4) shows that the slope $\mathrm{d}\ln K/\mathrm{d}T$ (and therefore the slope $\mathrm{d}K/\mathrm{d}T$) is negative for a reaction that is exothermic under standard conditions ($\Delta_r H^\ominus < 0$) and positive for one that is endothermic ($\Delta_r H^\ominus > 0$). A negative slope (downhill from left to right) means that $\ln K$, and therefore K itself, decreases as the temperature rises. Therefore, in the case of an exothermic reaction the equilibrium shifts away from products. The opposite occurs in the case of endothermic reactions. This important conclusion can be summarized as follows:

Exothermic reactions: increased temperature favours the reactants.
Endothermic reactions: increased temperature favours the products.

Some insight into this behaviour can be found in the expression $\Delta G = \Delta H - T\Delta S$ but written in the form $-\Delta G/T = -\Delta H/T + \Delta S$, which emphasizes that we are really dealing with an *entropy*. When the reaction is exothermic $-\Delta H/T$ corresponds to a positive change of entropy of the surroundings, and is a driving force for the formation of products. When the temperature is raised $-\Delta H/T$ gets smaller, and the increasing entropy of the surroundings is a less potent driving force; as a result, the equilibrium lies less to the right. When the reaction is endothermic, the principal driving force is the increasing entropy of the system. The importance of the unfavourable change of entropy of the surroundings is reduced if the temperature is raised (because then $\Delta H/T$ is smaller), and so the reaction is able to shift towards products. The conclusions emphasized above are also special cases of Le Chatelier's principle. In this case it predicts that an equilibrium will tend to shift in the endothermic direction if the temperature is raised, and in the exothermic direction if it is lowered, for in each case the change of temperature is opposed.

A second application of the van't Hoff isochore is to the measurement of standard reaction enthalpies. Equilibrium compositions are measured over a range of temperatures and $\ln K$ is plotted against $1/T$. It follows from eqn (10.2.5) that the slope is $-\Delta_r H^\ominus/R$, Fig. 10.4. One drawback is that the reaction enthalpy is temperature dependent, and so the plot is not expected to be perfectly straight; however, the temperature dependence is weak in many cases, and so the plot is reasonably straight. This is a *noncalorimetric* method of determining $\Delta_r H^\ominus$; in practice it turns out not to be very accurate, but it is often the only method available.

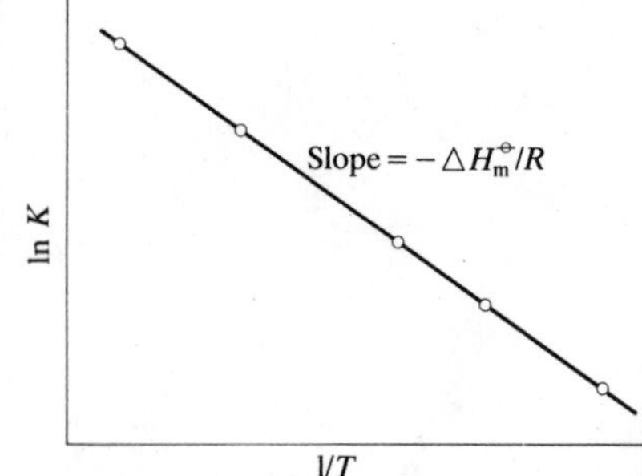

Fig. 10.4. When $\ln K$ is plotted against $1/T$, a straight line is expected with slope equal to $-\Delta_r H^\ominus/R$. This is a non-calorimetric method for the measurement of reaction enthalpies.

The weak temperature dependence of the reaction enthalpy is the basis of a third application of the van't Hoff isochore. If we wish to find the value of the equilibrium constant at a temperature T' in terms of its value K at another temperature T we begin by integrating eqn (10.2.5):

$$\int \mathrm{d}\ln K = -\int (\Delta_r H^\ominus/R)\,\mathrm{d}(1/T).$$

The integration limits on the right are $1/T$ and $1/T'$, and on the left they are $\ln K$ and $\ln K'$ respectively. The left-hand integral integrates easily to $\ln K' - \ln K$. We can integrate the right-hand expression only if we know how $\Delta_r H^\ominus$ depends on the temperature. Since it depends only weakly, if the temperature range is not too great we may make the approximation that

$\Delta_r H^\ominus$ is a constant, and take it outside the integral. Then

$$\ln K' - \ln K = -(\Delta_r H^\ominus/R)\int_{1/T}^{1/T'} d(1/T) = -(\Delta_r H^\ominus/R)\{(1/T') - (1/T)\}.$$

Therefore,

$$\ln K' = \ln K - (\Delta_r H^\ominus/R)\{(1/T') - (1/T)\}, \qquad (10.2.6)$$

an explicit (but approximate) expression for predicting the value of an equilibrium constant at one temperature when it is known at another.

Example 10.3

The equilibrium constant for the synthesis of ammonia at 298 K was calculated in *Example* 10.1. Estimate its value at 500 K.

• *Method*. This involves straightforward numerical substitution in eqn (10.2.6). The stoichiometric coefficients for the reaction as written in *Example* 10.1 (i.e. $N_2(g) + 3H_2(g) \rightarrow 2NH_3(g)$) are -1, -3, $+2$, and so the standard reaction enthalpy is $2\Delta_f H^\ominus_{NH_3}$, which can be taken from Table 4.1 and assumed to be constant over the temperature range in question.

• *Answer*. Since at 298 K $K_p = 6.0 \times 10^5$, and $\Delta_f H^\ominus = -46.1\ \mathrm{kJ\ mol^{-1}}$,

$$\ln K'_p = \ln(6.0 \times 10^5) - (-2 \times 46.1\ \mathrm{kJ\ mol^{-1}}/8.314\ \mathrm{J\ K^{-1}\ mol^{-1}})\{(1/500\ \mathrm{K}) - (1/298\ \mathrm{K})\}$$
$$= 13.3_0 - 15.0_3 = -1.7_3,$$

so that $K'_p = 0.18$.

• *Comment*. Note the considerable decrease in the value of the equilibrium constant for this exothermic reaction. This is in accord with Le Chatelier's principle, but now expressed quantitatively.

• *Exercise*. The equilibrium constant for the $N_2O_4(g) \rightarrow 2NO_2(g)$ reaction was calculated in the *Exercise* of *Example* 10.1. Estimate its value at 100 °C. [15]

In accurate work it is important to take into account the temperature dependence of the reaction enthalpy. One way is to use Kirchhoff's relation between reaction enthalpy and heat capacity, eqn (4.1.6). This is illustrated by the following *Example*.

Example 10.4

Deduce an expression for the temperature dependence of an equilibrium constant when the heat capacity of each substance in the reaction may be written in the form $C_{p,\mathrm{m}} = a + bT$.

• *Method*. Use eqn (4.1.6) to find an expression for the temperature dependence of $\Delta_r H^\ominus$, and then insert this expression into eqn (10.2.4) and integrate it from T to T'.

• *Answer*. If the standard reaction enthalpy is $\Delta_r H^\ominus$ at T then at a general temperature T'' it is

$$\Delta_r H''^\ominus = \Delta_r H^\ominus + \int_T^{T''} \Delta_r C_p\, dT = \Delta_r H^\ominus + \int_T^{T''} \{\Delta_r a + \Delta_r bT\}\, dT$$
$$= \Delta_r H^\ominus + \Delta_r a(T'' - T) + \tfrac{1}{2}\Delta_r b(T''^2 - T^2).$$

This is now substituted into eqn (10.2.4) and integrated (T'' is now the variable of integration):

$$\ln K' - \ln K = \int_T^{T'} (\Delta_r H''^{\ominus}/RT''^2)\,dT''$$

$$= \{\Delta_r H^{\ominus} - \Delta_r aT - \tfrac{1}{2}\Delta bT^2\}\int_T^{T'} (1/RT''^2)\,dT'' + (\Delta_r a/R)\int_T^{T'} (1/T'')\,dT'' + (\Delta_r b/2R)\int_T^{T'} dT''$$

$$= \{\Delta_r H^{\ominus} - \Delta_r aT - \tfrac{1}{2}\Delta_r bT^2\}\{(1/T) - (1/T')\}/R + (\Delta_r a/R)\ln(T'/T) + (\Delta_r b/2R)(T' - T).$$

• *Comment*. The resulting expression is obviously very much more complicated. As we saw in Section 4.1, it is normal to express the heat capacity with an additional c/T^2 term, and then to use a, b, and c from tables, such as Table 4.3.

• *Exercise*. First complete the calculation by including the additional c/T^2 term in the heat capacity. Then evaluate the equilibrium constant for the ammonia synthesis reaction at 500 K. [Additional term: $(\Delta_r c/2RT^2T'^2)(T' - T)^2$; $K_p = 0.16$]

10.2 (d) Giauque functions

A more accurate procedure depends on finding a way of tabulating values of $\Delta_r G^{\ominus}$ in such a way that they span a wide range of temperatures, and so that accurate interpolations may be made between the listed values. Then the value of K may be calculated directly from eqn (10.1.17) at any temperature in the range listed without having to rely on dubious approximations. It is therefore common to tabulate a quantity which is related to $\Delta_r G^{\ominus}$ but which varies more slowly with temperature (so that it can be interpolated more accurately): it is sometimes called the *Giauque function* (symbol: Φ). There are actually two such functions:

$$\Phi_0 = \{G_m^{\ominus}(T) - H_m^{\ominus}(0)\}/T; \quad \Phi = \{G_m^{\ominus}(T) - H_m^{\ominus}(\mathcal{T})\}/T, \qquad (10.2.7)$$

where, as usual, $\mathcal{T} = 298.15$ K. They are related as follows:

$$\Phi = \Phi_0 - \{H_m^{\ominus}(\mathcal{T}) - H_m^{\ominus}(0)\}/T. \qquad (10.2.8)$$

Table 10.1. Giauque functions, $-\Phi_0$/J K^{-1} mol^{-1}

	500 K	1000 K	$\{H_m^{\ominus}(\mathcal{T})$ $H_m^{\ominus}(0)\}$/ kJ mol^{-1}
H_2(g)	116.9	137.0	8.468
N_2(g)	177.5	197.9	8.669
NH_3(g)	176.9	203.5	9.92
H_2O(g)	172.8	196.7	9.908

The difference in enthalpies is normally listed in tables (such as in Table 10.1), and so the two functions may readily be interconverted. Note that G and H in the definition of Φ refer to a single substance: they are the *absolute* values of the molar Gibbs function and the enthalpy of a substance, not the Gibbs function and enthalpy of formation. However, whenever the functions are used they occur as differences, and so not knowing their individual absolute values is not troublesome. The values of Φ may be compiled either by calculation (using the techniques of statistical thermodynamics described in Part 2) or by measurement based on determining heat capacities down to very low temperatures, as explained in Section 5.4.

The form of the Φ functions makes them quite easy to use to calculate the equilibrium constant of a reaction at any temperature. The quantity controlling the magnitude of K is $\Delta_r G^{\ominus}/T$. In order to find this from the tables of Φ we form

$$G^{\ominus}/T = \Phi + H_m^{\ominus}(\mathcal{T})/T, \qquad (10.2.9)$$

which applies to each substance involved. For the reaction we then find

$$\Delta_r G^{\ominus}/T = \Delta_r \Phi + \Delta_r H^{\ominus}(\mathcal{T})/T. \qquad (10.2.10)$$

Everything in this apparently cumbersome expression is normally available.

Φ can be found from Table 10.1 (via eqn (10.2.8)). The second term can be calculated in the usual way by combining the standard enthalpies of formation (at 298 K), which are listed in Tables 4.1 and 4.2.

Example 10.5

Use the tables of the Φ functions to calculate the equilibrium constant for the ammonia synthesis reaction at 500 K.

• *Method*. Find the value of $\Delta_r G^\ominus/T$ at 500 K from eqn (10.2.10) and then use $\ln K = -\Delta_r G^\ominus/RT$. Use the reaction as written in the preceding *Examples*. Take Φ_0 from Table 10.1 and $\Delta_r H^\ominus(\mathcal{T})$ from Table 4.1.

• *Answer*. For the ammonia synthesis as written, $\Delta_r H^\ominus = 2\times(-46.1\ \text{kJ mol}^{-1}) = -92.2\ \text{kJ mol}^{-1}$. From Table 10.1,

$$\Delta_r\Phi_0 = \{2\times(-176.9) - 3\times(-116.9) - (-177.5)\}\ \text{J K}^{-1}\,\text{mol}^{-1}$$
$$= +174.4\ \text{J K}^{-1}\,\text{mol}^{-1}.$$

Then, from eqn (10.2.8) and Table 10.1,

$$\Delta_r\Phi = 174.4\ \text{J K}^{-1}\,\text{mol}^{-1} - \{[2\times(9.92) - 3\times(8.468) - (8.669)]\ \text{kJ mol}^{-1}\}/500\ \text{K}$$
$$= 202.9\ \text{J K}^{-1}\,\text{mol}^{-1}.$$

Consequently, from eqn (10.2.10),

$$\Delta_r G^\ominus/T = 202.9\ \text{J K}^{-1}\,\text{mol}^{-1} + (-92.2\ \text{kJ mol}^{-1})/500\ \text{K}$$
$$= 18.5\ \text{J K}^{-1}\,\text{mol}^{-1}.$$

Therefore,

$$\ln K = -(18.5\ \text{J K}^{-1}\,\text{mol}^{-1})/R = 2.21,\ \text{so that } K = 0.11.$$

• *Comment*. Note that Φ is in a useful form for equilibrium calculations because $\ln K$ depends on $\Delta_r G^\ominus/T$.

• *Exercise*. Evaluate the equilibrium constant at 1000 K. $[3\times10^{-7}]$

10.3 Applications to selected systems

In this section we look at some of the conclusions that can be drawn from the equilibrium equations, $\Delta_r G^\ominus = -RT\ln K$. For convenience, Fig. 10.5 shows how K depends on $\Delta_r G^\ominus$ for several different temperatures: if $\Delta_r G^\ominus$ is known at the temperature of interest, then the value of the equilibrium constant can be estimated from the diagram. Another general point is that since thermodynamic data are often listed for $\mathcal{T}$, equilibrium constants are often calculated at that temperature. Since $R\mathcal{T} = 2.48\ \text{kJ mol}^{-1}$, if we write $\Delta_r G^\ominus = g\ \text{kJ mol}^{-1}$, then $\ln K = -g/2.48$. Furthermore, since $\lg K = (\ln K)/2.303$:

$$\textit{At 25 °C}:\ K = 10^{-g/5.71}. \qquad (10.3.1)$$

This gives a quick estimate of K near room temperature. Note that when $\Delta_r G^\ominus$ (and therefore g) is negative, $K > 1$ and products dominate reactants.

10.3 (a) The extraction of metals from their oxides

Metals can be obtained from their oxides by reduction with carbon if either of the equilibria

$$MO(s) + C(s) \rightleftharpoons M(s) + CO(g), \qquad MO(s) + \tfrac{1}{2}C(s) \rightleftharpoons M(s) + \tfrac{1}{2}CO_2(g)$$

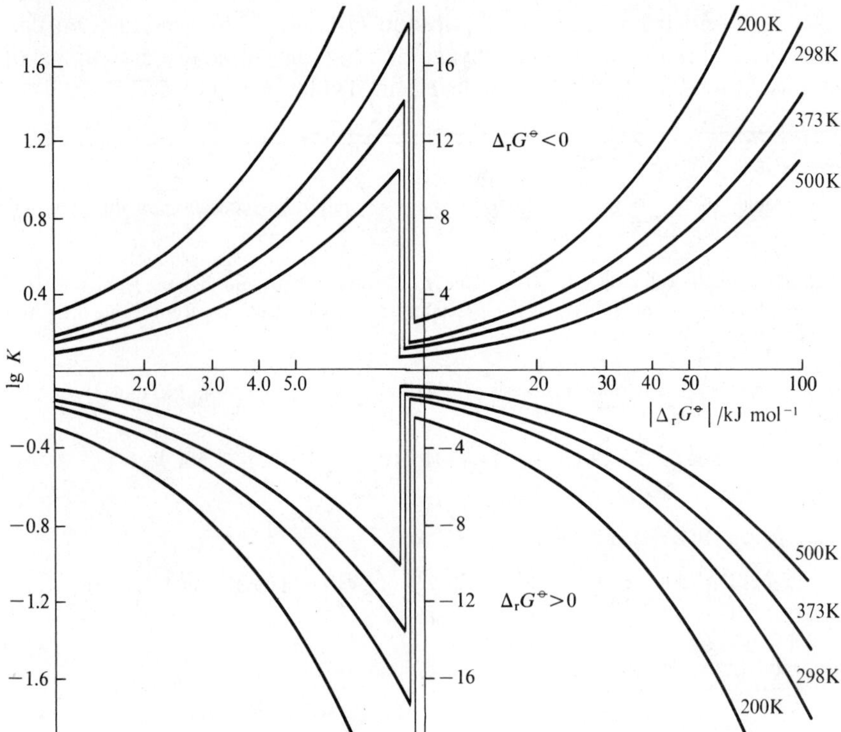

Fig. 10.5. The dependence of the equilibrium constant on the standard reaction Gibbs function at different temperatures. Note the change of scale. The curves have been calculated using eqn (10.1.17): the horizontal scale is logarithmic.

lies to the right. These equilibria can be discussed in terms of the thermodynamic functions for the reactions

$$\text{(i) } M(s)+\tfrac{1}{2}O_2(g)\rightarrow MO(s) \qquad \text{(ii) } \tfrac{1}{2}C(s)+\tfrac{1}{2}O_2(g)\rightarrow \tfrac{1}{2}CO_2(g)$$

$$\text{(iii) } C(s)+\tfrac{1}{2}O_2(g)\rightarrow CO(g) \qquad \text{(iv) } CO(g)+\tfrac{1}{2}O_2(g)\rightarrow CO_2(g).$$

The temperature dependences of the standard Gibbs functions of these reactions depend on the reaction entropy through $(\partial\Delta_r G^\ominus/\partial T)_p=-\Delta_r S^\ominus$. Since in reaction (iii) there is a net increase in the amount of gas, the standard reaction entropy is large and positive; therefore, $\Delta_r G^\ominus$ *decreases* sharply with increasing temperature. In reaction (iv) there is a similar net decrease in the amount of gas, and so $\Delta_r G^\ominus$ *increases* sharply with increasing temperature. In reaction (ii) the amount of gas is constant, and so the entropy change is small and $\Delta_r G^\ominus$ changes only slightly with temperature. These remarks are summarized in Fig. 10.6, which is called an *Ellingham diagram* (Note that $\Delta_r G^\ominus$ decreases upwards!).

The standard Gibbs function of reaction (i) indicates the metal's affinity for oxygen. At room temperature the contribution of the reaction entropy to $\Delta_r G^\ominus$ is dominated by the reaction enthalpy, and so the order of increasing $\Delta_r G^\ominus$ is the same as the order of increasing $\Delta_r H^\ominus$. This gives the order of values on the left of the diagram (Al_2O_3 is most exothermic, Ag_2O is least). The standard reaction entropy is similar for all metals because in each case gaseous oxygen is eliminated and a compact, solid oxide is formed. This implies that the temperature dependence of the standard Gibbs function of oxidation should be similar for all metals, as is shown by the similar slopes of the lines in the diagram. The kinks at high temperatures correspond to the evaporation of the metals; less pronounced kinks occur at the melting temperatures of the metals and the oxides.

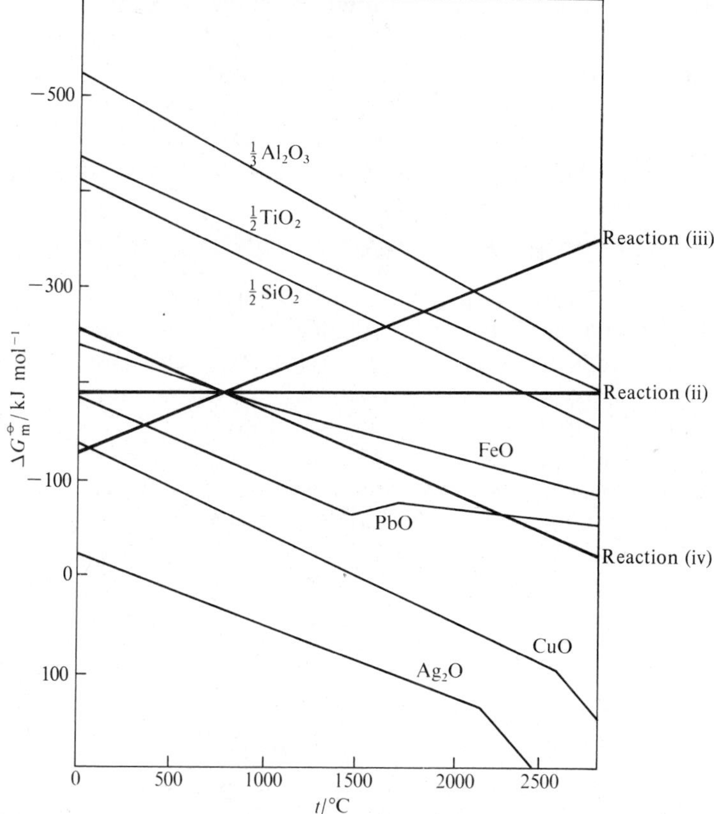

Fig. 10.6. The standard reaction Gibbs functions for the reactions involved in the discussion of metal ore reduction. This is called an *Ellingham diagram*. Note that $\Delta_r G^{\ominus}$ is most negative at the top of the diagram.

Reduction of the oxide depends on the competition of the carbon for the oxygen bound to the metal. The standard Gibbs functions for the reductions can be expressed in terms of the standard Gibbs functions for the reactions above:

$$MO(s) + C(s) \rightarrow M(s) + CO(g), \qquad \Delta_r G^{\ominus} = \Delta_r G^{\ominus}(iii) - \Delta_r G^{\ominus}(i)$$
$$MO(s) + \tfrac{1}{2}C(s) \rightarrow M(s) + \tfrac{1}{2}CO_2(g), \qquad \Delta_r G^{\ominus} = \Delta_r G^{\ominus}(ii) - \Delta_r G^{\ominus}(i)$$
$$MO(s) + CO(g) \rightarrow M(s) + CO_2(g), \qquad \Delta_r G^{\ominus} = \Delta_r G^{\ominus}(iv) - \Delta_r G^{\ominus}(i).$$

The equilibrium lies to the right if $\Delta_r G^{\ominus} < 0$. This is the case when the line for reaction (i) lies *below* (is more positive than) the line for one of the carbon reactions (ii)–(iv). We can predict the success of a reduction at any temperature simply by looking at the diagram: *a metal oxide is reduced by any carbon reaction lying above it*. For example, CuO can be reduced to Cu at any temperature above room temperature. Even in the absence of carbon, Ag_2O decomposes when heated above 200 °C because then the standard Gibbs function for reaction (i) becomes positive (so that the reverse reaction is then spontaneous). On the other hand, Al_2O_3 is not reduced by carbon until the temperature has been raised to above 2000 °C, and so it is a stable, refractory material up to that temperature. The equilibrium constants for the reductions can be obtained from the vertical separation of the appropriate lines on the diagram. Similar diagrams may be constructed for sulphides, nitrides, phosphides, etc., and their interpretation is the same as we have described.

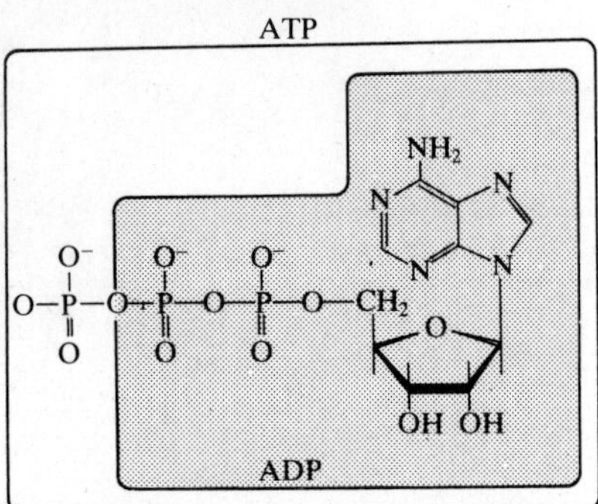

Fig. 10.7. The molecules adenosine triphosphate (ATP) and adenosine diphosphate (ADP). The exergonic hydrolysis of ATP to ADP drives many biochemical reactions. ATP is formed from ADP by reactions that are more exergonic.

10.3 (b) Biological activity: the thermodynamics of ATP

An important biochemical is adenosine triphosphate, ATP (Fig. 10.7). Its function is to store the energy made available when food is consumed and then to supply it on demand to a wide variety of processes, including muscular contraction, reproduction, and vision. ATP's action is due to its ability to lose its terminal phosphate group by hydrolysis and to form adenosine diphosphate (ADP):

$$\mathrm{ATP(aq)} + \mathrm{H_2O(l)} \rightarrow \mathrm{ADP(aq)} + \mathrm{P_i^-(aq)} + \mathrm{H^+(aq)}.$$

(P_i^- denotes an inorganic phosphate group, such as $H_2PO_4^-$.) This reaction is exergonic and can drive an endergonic reaction if suitable enzymes are available. If we were to continue with our earlier conventions, the standard states of the substances taking part in this reaction would include hydrogen ions at unit activity (around 1 M, pH = 0). However, this is not appropriate to normal biological conditions, and the *biological standard state* specifies that the hydrogen ion activity should be such that pH = 7 (an activity of 10^{-7}, corresponding to neutral solution). We shall adopt this convention in this section, and label the corresponding standard functions as $G^{\oplus}$, $H^{\oplus}$, and $S^{\oplus}$ (some texts use $X^{\circ\prime}$).

Example 10.6

Calculate the value of $\Delta_r G^{\oplus}(\mathcal{T})$ for the reaction

$$\mathrm{NADH(aq)} + \mathrm{H^+(aq)} \rightarrow \mathrm{NAD^+(aq)} + \mathrm{H_2(g)}$$

given that $\Delta_r G^{\ominus}(\mathcal{T}) = -21.8\ \mathrm{kJ\,mol^{-1}}$. NADH is the reduced form of nicotinamide adenine dinucleotide and NAD^+ is its oxidized form; the molecules play an important role in the later stages of the respiratory process.

• *Method*. Calculate the chemical potential of the hydrogen ion for pH = 7 (i.e. for an activity 10^{-7}), and then adjust the standard reaction Gibbs function by this amount.

• *Answer*. The chemical potential of H^+ (aq) is

$$\mu = \mu^{\ominus} + RT \ln a_{\mathrm{H^+}}$$

and so at pH = 7 (i.e. $\lg a_{\mathrm{H^+}} = -7$)

$$\mu = \mu^{\ominus} + 2.303RT \lg a_{\mathrm{H^+}} = \mu^{\ominus} - 2.303 \times 7 \times RT = \mu^{\ominus} - 16.1RT.$$

For the reaction as written, it follows that $\Delta_r G^{\oplus} = \Delta_r G^{\ominus} + 16.1R\mathcal{T} = (-21.8 + 16.1 \times 2.48)\ \mathrm{kJ\,mol^{-1}} = +18.2\ \mathrm{kJ\,mol^{-1}}$.

• *Answer*. Note the significantly different values of the two standard reaction Gibbs functions: it is obviously very important to make sure that a consistent set of data is being used. The biological standard value is the same as the usual standard value when hydrogen ions are not involved in the reaction.

• *Exercise*. Calculate $\Delta_r G^{\oplus}$ for the hydrolysis of ATP (reaction above) given that at 37 °C (blood temperature) $\Delta_r G^{\ominus} = -30\ \mathrm{kJ\,mol^{-1}}$. [$-27\ \mathrm{kJ\,mol^{-1}}$]

At 37 °C (310 K, blood temperature) the standard values for the ATP hydrolysis are $\Delta_r G^{\oplus} = -30\ \mathrm{kJ\,mol^{-1}}$, $\Delta_r H^{\oplus} = -20\ \mathrm{kJ\,mol^{-1}}$, and $\Delta_r S^{\oplus} = +34\ \mathrm{J\,K^{-1}\,mol^{-1}}$. The hydrolysis is therefore exergonic ($\Delta_r G^{\oplus} < 0$) under these standard conditions, and $30\ \mathrm{kJ\,mol^{-1}}$ is available for driving other reactions; because the entropy of reaction is large the value of the Gibbs function is sensitive to temperature. On account of its exergonicity the

ADP-phosphate bond has been called a *high-energy phosphate bond*. The name is intended to signify a high tendency to break, and so it should not be confused with 'strong' bond. In fact, even in the biological sense it is not of very 'high energy'. The action of ATP depends on it being intermediate in activity. Thus it acts as a phosphate donor to a number of acceptors (e.g. glucose), but is recharged by more powerful phosphate donors in the respiration cycle.

The efficiency of some biological processes can be gauged in terms of the value of $\Delta_r G^{\oplus}$ given above. The energy source of anaerobic cells is glycolysis, Section 4.2, and at blood temperature $\Delta_r G^{\oplus} = -218\ \mathrm{kJ\ mol^{-1}}$ (the standard reaction enthalpy is $-120\ \mathrm{kJ\ mol^{-1}}$, the exergonicity exceeding the exothermicity on account of the large increase of entropy accompanying the fracture of the glucose molecule). The glycolysis is coupled to a reaction in which two ADP molecules are converted into two ATP molecules:

$$\text{Glucose} + 2\mathrm{P_i^-} + 2\text{ADP} \rightarrow 2\text{lactate}^- + 2\text{ATP} + 2\mathrm{H_2O}.$$

The standard reaction Gibbs function is $(-218\ \mathrm{kJ\ mol^{-1}}) - 2(-30\ \mathrm{kJ\ mol^{-1}}) = -158\ \mathrm{kJ\ mol^{-1}}$. This is exergonic, and therefore spontaneous: the digestion of the food has been used to 'recharge' the ATP.

Aerobic respiration is much more efficient. The standard Gibbs function of combustion of glucose is $-2880\ \mathrm{kJ\ mol^{-1}}$, and so terminating its oxidation at the stage of lactic acid is a poor use of resources. In aerobic respiration the oxidation is carried out to completion, and an extremely complex set of reactions preserves as much of the energy released as possible. In the overall reaction 38 ATP molecules are generated for each glucose molecule consumed. Each mole of ATP extracts 30 kJ from the 2880 kJ supplied by 1 mol $C_6H_{12}O_6$ (180 g of glucose), and so 1140 kJ has been stored for later use.

Each ATP molecule can be used to drive an endergonic reaction for which $\Delta_r G^{\oplus}$ does not exceed $30\ \mathrm{kJ\ mol^{-1}}$. For example, the biosynthesis of sucrose from glucose and fructose can be driven (if a suitable enzyme system is available) because the reaction is endergonic to the extent $\Delta_r G^{\oplus} = +23\ \mathrm{kJ\ mol^{-1}}$. The biosynthesis of proteins is strongly endergonic not only on account of the enthalpy change but also on account of the large decrease in entropy that occurs when many amino acids are assembled into a precisely determined sequence. For instance, the formation of a peptide link is endergonic, with $\Delta_r G^{\oplus} = +17\ \mathrm{kJ\ mol^{-1}}$, but the biosynthesis occurs indirectly and is equivalent to the consumption of three ATP molecules for each link. In a moderately small protein like myoglobin, with about 150 peptide links, the construction alone requires 450 ATP molecules, and therefore about 12 mol glucose molecules for 1 mol protein molecules.

Further reading

Principles:

Chemical thermodynamics (2nd edn). I. M. Klotz and R. M. Rosenberg; Benjamin, Menlo Park, 1972.

Chemical thermodynamics. P. A. Rock; University Science Books and Oxford University Press, 1983.

Thermodynamics. G. N. Lewis and M. Randall, revised by K. S. Pitzer and L. Brewer; McGraw-Hill, New York, 1961.

Changes of state: chemical reactions

The principles of chemical equilibrium (4th edn). K. G. Denbigh; Cambridge University Press, 1981.
A thermodynamic bypass: *GOTO log K*. P. A. H. Wyatt; Royal Society of Chemistry, London, 1982.

Applications:
Some thermodynamic aspects of inorganic chemistry (2nd edn). D. A. Johnson; Cambridge University Press, 1982.
Energy changes in biochemical reactions. I. Klotz; Academic Press, New York, 1967.
Bioenergetics (2nd edn). A. Lehninger; Benjamin-Cummings, Menlo Park, 1971.

Data: *See Chapter* 4.

Introductory problems

A10.1. The equilibrium constant for the reaction in which *cis*-2-butene(g) changes into *trans*-2-butene(g) has the value 2.07 at 400 K. Calculate the standard Gibbs function at 400 K for the above reaction.

A10.2. The standard Gibbs function at 400 K for the isomerization of *cis*-2-pentene to *trans*-2-pentene is $-3.67\ \text{kJ mol}^{-1}$. Calculate the equilibrium constant for the reaction.

A10.3. For the reaction $Zn(g) + H_2O(g) \rightarrow ZnO(s) + H_2(g)$ the standard molar enthalpy change is $224\ \text{kJ mol}^{-1}$ over the temperature range 921 K to 1280 K. The change in the standard molar Gibbs function at 1280 K is $33\ \text{kJ mol}^{-1}$. Assume that the standard molar enthalpy change remains $224\ \text{kJ mol}^{-1}$ as the temperature is increased, and estimate the temperature at which the reaction begins to be spontaneous with all materials in their standard states.

A10.4. The equilibrium constant for the reaction:

$$2C_3H_6(g) \rightleftharpoons C_2H_4(g) + trans\text{-}C_4H_8(g)$$

is given by the empirical expression: $\ln K = -1.04 - 1088\ T^{-1} + 1.51 \times 10^5\ T^{-2}$ in the range 300 K to 600 K (with $T \equiv T/\text{K}$). Calculate the standard enthalpy change and the standard entropy change for the reaction at 400 K.

A10.5. The change in the standard molar Gibbs function for the isomerization of borneol(g), $C_{10}H_{17}OH(g)$, to isoborneol(g) at 503 K is $9.4\ \text{kJ mol}^{-1}$. Calculate the difference between the Gibbs function for isoborneol(g) and that for borneol(g) in a mixture of 0.15 mol of borneol and 0.30 mol of isoborneol at 503 K and a total pressure of 600 mmHg.

A10.6. At 500 K the equilibrium pressure of $H_2(g)$ over α-U(s) and β-$UH_3(s)$ is 1.04 Torr. Calculate the standard Gibbs function of formation for β-$UH_3(s)$ at this temperature.

A10.7. In the temperature range 450 K to 715 K, the experimental values of the equilibrium pressure of $H_2(g)$ over α-U(s) and β-$UH_3(s)$ are represented by the equation: $\ln p = 69.32 - 14.64 \times 10^3\ T^{-1} - 5.65 \ln T$ where p is in Pa and $T = T/\text{K}$. Find an expression which gives the standard enthalpy of formation of β-$UH_3(s)$ as a function of T and from it calculate $\Delta C^{\ominus}_{p,\text{m}}$.

A10.8. In the temperature range 350 K to 470 K, the standard enthalpy change for the dissociation of $CaCl_2.NH_3(s)$ into $NH_3(g)$ and $CaCl_2(s)$ is nearly constant at $78.15\ \text{kJ mol}^{-1}$. The equilibrium dissociation pressure of $NH_3(g)$ is 12.8 Torr at 400 K. Find an expression for the standard molar Gibbs function for the dissociation as a function of T in the above temperature range.

A10.9. Calculate the percentage change in the equilibrium constant K_x for each of the following systems: (a) $H_2CO(g) \rightleftharpoons CO(g) + H_2(g)$ and (b) $CH_3OH(g) + NOCl(g) \rightleftharpoons HCl(g) + CH_3NO_2(g)$ when the total pressure of the system is increased from 1 atm to 2 atm and the temperature is kept fixed.

A10.10. The equilibrium constant for the isomerization of borneol(g), $C_{10}H_{17}OH(g)$, to isoborneol(g) at 503 K is 0.106. A mixture of 7.5 g of borneol and 14.0 g of isoborneol in a 5 dm^3 container is heated to 503 K, and the system is allowed to come to equilibrium. Calculate the mole fraction of borneol at equilibrium.

Problems

10.1. Use the thermodynamic data in Table 4.1 to decide which of the following reactions are spontaneous at T in the direction written, the species being present in their standard states:
(a) $HCl(g) + NH_3(g) \rightarrow NH_4Cl(s)$
(b) $2Al_2O_3(s) + 3Si(s) \rightarrow 3SiO_2(s) + 4Al(s)$
(c) $Fe(s) + H_2S(g) \rightarrow FeS(s) + H_2(g)$
(d) $FeS_2(s) + 2H_2(g) \rightarrow Fe(s) + 2H_2S(g)$
(e) $2H_2O_2(l) + H_2S(aq) \rightarrow H_2SO_4(aq) + 2H_2(g)$.

10.2. Which of the reactions in Problem 10.1 are favoured by a rise in temperature at constant pressure?

10.3. What is the standard reaction enthalpy of a reaction for which the equilibrium constant (a) is doubled, (b) is halved when the temperature is increased by 10 K from $\mathcal{T}$?

10.4. Suppose you make an error of 10 per cent in the determination of an equilibrium constant at $\mathcal{T}$. Which error does that imply for the value of $\Delta_r G^\ominus$? Conversely, suppose you make an error of 10 per cent in the determination of $\Delta_r G^\ominus$: what error does that imply for the equilibrium constant at that temperature?

10.5. Given that for the reaction $\frac{1}{2}N_2 + \frac{3}{2}H_2 \rightarrow NH_3$ $\Delta_r G^\ominus = -16.5\ \mathrm{kJ\ mol^{-1}}$, find the equilibrium constant at $\mathcal{T}$ for (a) the reaction as written, (b) the reaction $N_2 + 3H_2 \rightarrow 2NH_3$, (c) the reaction $NH_3 \rightarrow \frac{1}{2}N_2 + \frac{3}{2}H_2$.

10.6. The standard Gibbs function for the synthesis of ammonia was given in the last Problem. What is the value of $\Delta_r G(\mathcal{T})$ when the pressures of the (perfect gas) components are $p(N_2) = 3$ atm, $p(H_2) = 1$ atm, $p(NH_3) = 4$ atm?

10.7. Calculate the equilibrium constant $\mathcal{T}$ for the reaction $CO(g) + H_2(g) \rightarrow H_2CO(g)$ given that $\Delta_r G^\ominus$ for the production of liquid is $28.95\ \mathrm{kJ\ mol^{-1}}$ at $\mathcal{T}$ and that the vapour pressure of formaldehyde is 1500 Torr.

10.8. When ammonium chloride is heated the vapour pressure at 427 °C is 608 kPa. At 459 °C the vapour pressure has risen to 1115 kPa. What is (a) the equilibrium constant for the dissociation at 427 °C, (b) the Gibbs function at 427 °C, (c) the enthalpy at 427 °C, (d) the entropy of dissociation at 427 °C? Assume the vapour behaves as a perfect gas and that $\Delta H^\ominus$ and $\Delta S^\ominus$ are independent of temperature in the range given.

10.9. In the *Dumas vapour density method* a liquid is allowed to evaporate in a glass bulb heated to some selected temperature; then the bulb is sealed and cooled, the amount of sample being determined by weighing or by titration. In this way a known volume is filled with vapour at a known temperature and pressure, and by a known amount of material. In one such experiment, acetic acid was evaporated at 437 K and the amount of acid in the bulb of volume 21.45 cm^3 was 0.0519 g when the external pressure was 764.3 Torr. In a second experiment in the same vessel heated to 471 K the amount present was found to be 0.038 g when the external pressure was the same. What is the equilibrium constant for the dimerization of the acid in the vapour? What proportion of the vapour is dimeric at each temperature? What is the enthalpy of dimerization?

10.10. The techniques described in the chapter can be applied to all kinds of reactions, including ionic reactions in solution. The only difficulty is that ionic solutions are strongly non-ideal, and thermodynamic equilibrium constants may have to be strongly corrected in order to get concentration equilibrium constants. We shall see how to do this in Chapter 11, but even now we can calculate solubility products and the solubilities of salts. For instance, use the information in Table 4.1 to evaluate the *solubility product* $K_{sp} = \{a(Ag^+)a(Cl^-)\}_{eq}$ and the solubility (as a molality) of silver chloride in water at 298 K. Ignore all effects of non-ideality in the latter part.

10.11. The *ionization constant* of water, $K_w = \{a(H^+) \times a(OH^-)\}_{eq}$ is another special type of equilibrium constant that plays an important role in governing the behaviour of acids and bases. At 20 °C $K_w = 0.67 \times 10^{-14}$, at 25 °C $K_w = 1.00 \times 10^{-14}$, and at 30 °C $K_w = 1.45 \times 10^{-14}$. Calculate the standard enthalpy of the ionization, $H_2O(l) \rightarrow H^+(aq) + OH^-(aq)$, at 25 °C. What would the error bounds be if K_w were known to within ± 0.01?

10.12. Hydrogen and carbon monoxide have been investigated with a view to using them in high temperature fuel cells, and so their solubility in various molten nitrates is of some technological interest. The solubility in a $NaNO_3/KNO_3$ mixture was examined (E. Desimoni, F. Paniccia, and P. G. Zambonin, *J. chem. Soc. Faraday Trans*. **1,** 2014 (1973)) with the following results:

$$\lg s(H_2) = -5.39 - 768(T/K)^{-1},$$

$$\lg s(CO) = -5.98 - 980(T/K)^{-1},$$

where s is the solubility expressed in units of $\mathrm{mol\ cm^{-3}\ bar^{-1}}$. Find the standard molar enthalpies of solution of the two gases at 570 K.

10.13. Everyone knows that when blue crystals of copper sulphate are heated they crumble and lose their colour as a result of the dehydration reaction $CuSO_4.5H_2O \rightarrow CuSO_4 + 5H_2O$. Given the data for 298 K in Table 4.1, discuss the process from the viewpoint of thermodynamics, and predict the temperature at which the vapour pressure of water reaches (a) 10 Torr, (b) 1 atm.

10.14. Calculations on the positions of chemical equilibrium are much more involved when the components are mixed in arbitrary proportions initially, and the equilibrium composition is required. A good way of working through problems of this kind is to draw up a table with columns headed by the names of all the species involved, and then successive rows labelled (1) the initial amounts, (2) the change in one of the components on reaching equilibrium; sometimes this is the unknown in the problem, and so it is just left as x, (3) the changes in the amounts of all components that are then implied by the reaction stoichiometry, (4) the final composition (perhaps still in terms of the unknown x), (5) the mole fractions. Then the equilibrium constant can be expressed in terms of this last line, and then solved for the unknown x. See how this works in practice with the following problems. First consider a flask which was filled with 0.30 mol $H_2(g)$, 0.40 mol $I_2(g)$, and 0.20 mol HI(g), the total pressure being 1.0 atm. What is the composition when the mixture has reached thermal equilibrium, the equilibrium constant at the temperature of the experiment being $K = 870$.

10.15. You will have noticed that the set of rules given in the last Problem is virtually in the form of a computer algorithm. If you have access to a computer, write a program that will predict the final composition of the reaction mixture for arbitrary initial compositions. You may find it instructive to do the same for some of the other examples that follow.

10.16. Now consider the general reaction $2A + B \rightarrow 3C + 2D$, all components gases. It was found that when 1.0 mol of A, 2.0 mol of B, and 1.0 mol of D were mixed and the whole allowed to come to equilibrium, the mixture contained 0.9 mol of C. Is this enough information to find the equilibrium constant?

10.17. The equilibrium constant for the reaction $N_2 + 3H_2 \rightarrow 2NH_3$ was treated in Problem 10.5. Express K in terms of an extent of the reaction ξ defined so that $\xi = 0$ corresponds to the initial state and $\xi = 1$ to pure $NH_3(g)$, given that the nitrogen and hydrogen were initially present in stoichiometric proportions when the catalyst was added and behave as perfect gases.

10.18. The thermodynamic equilibrium constant is independent of pressure, but the extent of reaction at equilibrium does depend on pressure (Section 10.2(b)). Find ξ_e at a pressure p. In fact, there are two ways of solving this type of problem. One is to attempt to solve the equilibrium constant expression for ξ_e in terms of p; but in the present case that means solving a quartic equation, which is tiresome. The other method involves noticing that if $K \propto 1/p^2$, then the constant of proportionality must be proportional to p^2 for K to be independent of pressure. Go on from there.

10.19. Plot ξ_e for the ammonia synthesis as a function of total pressure from 0.1 bar to 1000 bar; use a logarithmic scale. What is the composition of the equilibrium mixture when the pressure is 500 bar?

10.20. Triethylamine (TEA) and 2,4-dinitrophenol (DNP) form a molecular complex in chlorobenzene, the binding depending on proton transfer and electrostatic attraction. The equilibrium constant for the formation of the complex has been measured at various temperatures by monitoring the optical absorption of the solution (K. J. Ivin, J. J. McGarvey, E. L. Simmons, and R. Small, *J. chem. Soc. Faraday Trans*. **1,** 1016 (1973)) with the following results:

$\theta/°C$	17.5	25.2	30.0
K	29 670 ± 1230	14 450 ± 560	9270 ± 70
$\theta/°C$	35.5	39.5	45.0
K	870 ± 120	3580 ± 30	2670 ± 70

where $K = c(\text{complex})/c(\text{TEA})c(\text{DNP})$ and $c(\text{A}) \equiv c(\text{A})/\text{mol dm}^{-3}$. Calculate the standard enthalpy and entropy of formation of the complex from its components at 20 °C.

10.21. At high temperatures gaseous iodine dissociates and the vapour phase contains I_2 molecules and I atoms. The extent of dissociation can be measured by monitoring the pressure change, and three sets of results are as follows:

T/K	973	1073	1173
p/atm	0.06244	0.07500	0.09181
$10^4 n(I)/\text{mol } I_2$	2.4709	2.4555	2.4366

Find the equilibrium constants for the reaction and the molar enthalpy of dissociation as the mean temperature. Use $V = 342.68\ \text{cm}^3$.

10.22. Nitrogen(IV) oxide is in equilibrium with its dimer at room temperature: $2NO_2 \rightleftharpoons N_2O_4$. Since NO_2 is a brown gas and N_2O_4 is colourless, the amounts present at equilibrium may be studied either spectroscopically or by monitoring the pressure. The following data were obtained at two temperatures:

	$p(NO_2)$/mmHg	$p(N_2O_4)$/mmHg
298 K	46	23
305 K	68	30

Calculate the equilibrium constant for the reaction, the standard molar Gibbs function, enthalpy, and entropy of dimerization at 298 K.

10.23. When light passes through a cell of length l containing an absorbing gas at a pressure p the amount of absorption is proportional to pl. In the case of a cell in which there is the equilibrium $2NO_2 \rightleftharpoons N_2O_4$, with NO_2 the absorbing species, show that when two cells of lengths l_1 and l_2 are used, and the pressures needed to obtain equal amounts of absorption are p_1 and p_2 respectively, that the equilibrium constant is given by

$$K = (r^2p_1 - p_2)^2/r(r-1)(p_2 - rp_1), \qquad r = l_1/l_2.$$

The following data were obtained (R. J. Nordstrum and W. H. Chan, *J. phys. Chem.* **80,** 847 (1976)):

Absorbance	p_1/Torr	p_2/Torr
0.05	1.00	5.47
0.10	2.10	12.00
0.15	3.15	18.65

$l_1 = 395$ mm, $l_2 = 75$ mm

Find the equilibrium constant for the reaction.

10.24. At normal temperatures BF_3 acts as a catalyst for the equilibrium between acetaldehyde (ethanal, CH_3CHO) and paraldehyde (a trimer of CH_3CHO). The partial pressures of the components in the reaction $3CH_3CHO(g) \rightleftharpoons (CH_3CHO)_3(g)$ are too low for accurate determination of the equilibrium constant by direct measurement, but this can be overcome by ensuring that liquid forms of the two components are always present (W. K. Busfield, R. M. Lee, and D. Merigold, *J. chem. Soc. Faraday Trans*. I, 936 (1973);. On the basis of perfect behaviour, show that the equilibrium constant for the trimerization can be written $K = p_P^0(p_A^0 - p)(p_A^0 - p_P^0)^2/p_A^{03}(p - p_P^0)^3$, where p_A^0 and p_P^0 are the saturated vapour pressures of acetaldehyde and paraldehyde, and p is the total pressure. Use the enthalpies of vaporization $\Delta H_{\text{vap,m}} \approx 25.6\ \text{kJ mol}^{-1}$ (acetaldehyde) and $41.5\ \text{kJ mol}^{-1}$ (paraldehyde) and the following data to find the enthalpy of trimerization in the gas phase.

$\theta/°C$	20.0	22.0	26.0	28.0	30.0
p/kPa	23.9	27.3	36.5	42.6	49.9
$\theta/°C$	32.0	34.0	36.0	38.0	40.0
p/kPa	56.9	65.1	74.3	85.0	96.2

Use $\ln(p/\text{kPa}) = 15.1 - \Delta H_{\text{vap,m}}/RT$ for the saturated VP of acetaldehyde and $\ln(p/\text{kPa}) = 17.2 - \Delta H_{\text{vap,m}}/RT$ for paraldehyde.

10.25. The enthalpy and entropy of trimerization of acetaldehyde in the gas phase are $-133.5\,\text{kJ mol}^{-1}$ and $-457.5\,\text{J K}^{-1}\,\text{mol}^{-1}$ respectively (see the last Problem). Given the enthalpies of vaporization quoted there, and their boiling points of 294 K (acetaldehyde) and 398 K (paraldehyde), find the values of the enthalpy and entropy of the trimerization in the liquid phase. What is the ratio of the equilibrium constants for the trimerization in the gas and liquid phases at 25 °C?

10.26. In Chapter 6 we saw that if the attractive part of the van der Waals interaction dominates, the fugacity of a gas is related to the pressure by $f = p\exp(-ap/R^2T^2)$. First show that $(\partial \ln K_p/\partial p)_T = -(\partial \ln K_\gamma/\partial p)_T$. Then find an expression for the pressure dependence of K_p, the first of these two coefficients, in the case of the reaction $I_2 + H_2 \rightarrow 2HI$, and estimate its value at 25 °C and 500 atm. By how much will K_p change when the pressure is increased by 50 atm? Use $a(\text{HI}) = 6.2\,\text{dm}^3\,\text{atm mol}^{-2}$, $a(I_2) = 7.3\,\text{dm}^3\,\text{atm mol}^{-2}$, and $a(H_2)$ from Table 1.3.

10.27. We have also seen that ΔG tells us the maximum non-p,V work that may be available from a system. In an *electrochemical cell* this work may be tapped electrically, and we shall explore this in Chapter 12. At this point, though, we have enough information to be able to assess how much electrical work can be obtained from cells that depend on any type of reaction. Suppose that a cell is designed with copper and zinc electrodes and is driven by the reaction $Zn(s) + CuSO_4(aq) \rightarrow ZnSO_4(aq) + Cu(s)$. What is the maximum electrical work available from the consumption of 1 mol Zn(s) when all the materials are in their standard states? Use the data in Table 4.1 for the ions. What is the maximum amount of heat? Why do the two figures differ?

10.28. Suppose an electrical cell could be designed that converted coal directly into electrical work at 298 K; what work is available from the consumption of 100 kg of coal (regarded as graphite)? What is the maximum work available if the coal is burnt and the thermal output is converted into work in a plant operating between 150 °C (hot source) and 30 °C (cold sink)?

10.29. The next few Problems take a closer look at the way of dealing with the temperature dependence of the Gibbs function of a reaction, and hence with the temperature dependence of the equilibrium constants. First we derive explicit expressions for G at different temperatures. The entropy and enthalpy at a temperature T_2 can be related to their values at T_1 if the temperature dependence of the heat capacity is known. Find $\Delta_r G(T_2)$ for a reaction in terms of $\Delta_r G(T_1)$ and the parameters a, b, c in the expression $C_{p,\text{m}} = a + bT + c/T^2$.

10.30. Use the heat capacity data in Table 4.1 to find $\Delta_f G^\ominus(372\,\text{K})$ for water, given $\Delta_f G^\ominus(\mathcal{T}) = -237.2\,\text{kJ mol}^{-1}$ and $\Delta_f H^\ominus(\mathcal{T}) = -285.8\,\text{kJ mol}^{-1}$.

10.31. The Giauque functions $\Phi_0 = \{G_{\text{m}}^\ominus(T) - H_{\text{m}}^\ominus(0)\}/T$ or $\Phi = \{G_{\text{m}}^\ominus(T) - H_{\text{m}}^\ominus(\mathcal{T})\}/T$ are useful for discussing equilibria at different temperatures because they vary only slowly with temperature. One method of finding their values was described in Problem 5.32. Here we shall look at ways of using them. As a first step, show how to relate the entropy of a substance, $S^\ominus(T)$, to Φ_0 and the enthalpy difference $H_{\text{m}}^\ominus(T) - H_{\text{m}}^\ominus(0)$.

10.32. The Giauque functions Φ_0 and Φ are tabulated in Table 10.1. Use the information given there to calculate $\Delta_r G^\ominus$ for the following reactions at the stated temperatures, all substances gaseous:

(a) $N_2 + 3H_2 \rightarrow 2NH_3$ 1000 K
(b) $H_2O + CO \rightarrow H_2 + CO_2$ 500 K, 2000 K.

10.33. Find the equilibrium constants for the reactions listed in the preceding Problem.

10.34. Finally, link the two methods of arriving at the temperature dependence of equilibrium constants. Express Φ in terms of its value at $\mathcal{T}$ and the parameters a, b, c in the heat capacity expression.

11

Equilibrium electrochemistry: ions and electrodes

Learning objectives

After careful study of this chapter you should be able to:

(1) Define the *activity*, eqn (11.1.2), the *activity coefficient*, eqn (11.1.3), and the *mean activity coefficient*, eqn (11.1.5), of ions in solution, and write down an expression for their chemical potentials, eqn (11.1.6).

(2) Describe the physical basis of the *Debye–Hückel theory* of ionic solutions and the formation and role of the *ionic atmosphere* in determining the activity coefficients of ions, Section 11.2(a).

(3) State the form of the *shielded Coulomb potential*, eqn (11.2.4).

(4) Define and calculate the *ionic strength* of a solution, eqn (11.2.6).

(5) Derive an expression for the *Debye length*, eqn (11.2.8), in terms of the ionic strength of a solution, and account for its properties, Section 11.2(d).

(6) Derive and use the *Debye–Hückel Limiting Law* for the mean activity coefficient, eqn (11.2.11).

(7) Criticize the assumptions of the Debye–Hückel model, Section 11.2(f), and justify and write the form of the *extended Debye–Hückel Law*, eqn (11.2.14).

(8) Explain the terms *galvanic cell* and *electrolytic cell*, and define *cathode* and *anode*, Section 11.3(a) and Box 11.2.

(9) Write down an expression for the *electrochemical potential* of an ion, eqn (11.3.1).

(10) Derive an expression for the potential difference across an interface in terms of the *standard potential difference* and the activity of the ions, eqn (11.3.6).

(11) Describe the construction of, and derive expressions for the potential difference at a *gas | inert metal electrode*, eqn (11.4.2), and at an *ion | insoluble salt | metal electrode*, eqn (11.4.4), and use the general expression for electrode potential differences, eqn (11.4.3).

(12) Derive and use an expression for the potential difference at a *redox electrode*, eqn (11.4.5).

(13) Describe the formation of a *liquid junction potential*, Section 11.4(d).

(14) Derive and use an expression for the potential difference across a membrane, eqn (11.4.6).

Introduction

The study of the reactions of ions and molecules in environments where they can take part in the transfer of electrons has important applications to a wide variety of processes, including power production in cells and fuel cells, corrosion, catalysis, and biology. Though none of these is strictly an equilibrium process, a knowledge of equilibrium electrochemistry is a foundation for understanding them and has many useful applications in its own right.

Only two changes of detail are necessary in order to accommodate the presence of charged particles in a solution. In the first place, ions interact Coulombically (their potential energy depending on 1/*separation*) over long distances, and so their solutions are markedly non-ideal even at very low concentrations. In the second place, when several phases are present (e.g. a metal electrode immersed in an electrolyte solution) each phase may be at a different electric potential. We explore these two related points in this chapter: we see how ions respond to potentials arising from other ions (which leads to an expression for activity coefficients), and how they respond to the potential differences between different phases (which leads to an expression for electrode potentials).

11.1 The activities of ions in solution

X-ray diffraction studies of solids show that many compounds exist as ionic solids. Before that evidence was available, it had already been established that many compounds (the *electrolytes*) exist as ions in aqueous solution. For instance, the conduction of electricity by electrolyte solutions is explained very readily if ions are present. Evidence also came from colligative properties (Chapter 8): greater depressions of vapour pressure and greater osmotic pressures are observed than would be expected if molecules dissolved but remained associated. For example, when sodium chloride dissolves in water the osmotic pressure is about twice the value expected on the basis of the number of NaCl units present: this is readily explained if Na^+ and Cl^- ions are present, for then there are twice as many dissolved particles.

11.1 (a) Activities and standard states

The thermodynamic properties of ions in solution depend on their chemical potentials μ_i. If the ions formed an ideal dilute solution (one conforming to Henry's Law) their chemical potential would be related to their molarity m_i by (Box 8.1)

$$\mu_i = \mu_i^\ominus + RT \ln (m_i/m^\ominus), \qquad (11.1.1)^\circ$$

and the standard state would be established at unit molality (at $m_i = m^\ominus$; we use the notation that $m^\ominus = 1\ \text{mol kg}^{-1}$ exactly). The chemical potential of a *real* solution is written in a similar form by introducing the dimensionless *activity* (symbol: a_i):

$$\mu_i = \mu_i^\ominus + RT \ln a_i. \qquad (11.1.2)$$

This relation is useful only if we can relate the activity to the molality. As a first step we introduce the *activity coefficient* (symbol: γ_i) through

$$a_i = \gamma_i m_i/m^\ominus, \qquad (11.1.3)$$

where γ_i depends on the composition and molality of the solution. The aim of the first part of this chapter is to find the dependence of γ_i on m_i.

The standard state of a real solution of an electrolyte is defined in the same way as for non-electrolytes (Box 8.1), and can be summarized as follows:

> *The standard state is a hypothetical solution of molality $m^\ominus$ in which all the interactions leading to deviations from ideality have been extinguished.*

The advantage of this definition, as explained in Chapter 8, is that all the deviations from ideality are carried in the activity, and hence in the activity coefficient. Moreover, as in the case of non-electrolytes, although the chemical potential takes its standard value $\mu_i^{\ominus}$ when $a_i = 1$, the solution is not then in its standard state because that state is hypothetical. At a molality $m_i = m^{\ominus}$ of the actual solution, the activity coefficient might be far from unity. The solution approaches ideality (in the sense of Henry's Law) only at low molalities, and the activity approaches $m_i/m^{\ominus}$ as $m_i \to 0$. This means that the activity coefficient approaches unity in the same limit: $\gamma_i \to 1$ and $a_i \to m_i/m^{\ominus}$ as $m_i \to 0$.

11.1 (b) Mean activity coefficients

If the chemical potential of a univalent cation is denoted μ_+ and that of a univalent anion μ_-, then the total chemical potential of the ions in the electrically neutral solution is

$$\begin{aligned}\mu_+ + \mu_- &= \mu_+^{\ominus} + \mu_-^{\ominus} + RT\ln a_+ + RT\ln a_- \\ &= \mu_+^{\circ} + \mu_-^{\circ} + RT\ln\gamma_+ + RT\ln\gamma_-.\end{aligned}$$

The μ° are the 'ideal' values, and the effects of the interactions are contained in the last term. Although, as we shall see, individual ion activity coefficients can be calculated, there is no way of disentangling the product $\gamma_+\gamma_-$ experimentally and of assigning one part to the cations and another part to the anions. The best an experimenter can do is to assign the deviations from ideality equally to both kinds of ion. Therefore, for practical applications we introduce the *mean ionic activity coefficient* (symbol: $\gamma_\pm$):

$$\gamma_\pm = (\gamma_+\gamma_-)^{\frac{1}{2}}, \tag{11.1.4}$$

and write the total chemical potential as

$$\mu = \mu_+^{\circ} + \mu_-^{\circ} + 2RT\ln\gamma_\pm$$

so that the non-ideality term is shared equally.

The last result can be generalized to the case of a salt M_pX_q which dissolves to give a solution of the ions $M^{+|z_+|}$ and $X^{-|z_-|}$ in the ratio $p:q$. The total chemical potential is

$$\begin{aligned}\mu = p\mu_+ + q\mu_- &= p\mu_+^{\ominus} + q\mu_-^{\ominus} + pRT\ln a_+ + qRT\ln a_- \\ &= p\mu_+^{\circ} + q\mu_-^{\circ} + pRT\ln\gamma_+ + qRT\ln\gamma_-.\end{aligned}$$

Introducing the mean activity coefficient defined in general as

$$\gamma_\pm = (\gamma_+^p\gamma_-^q)^{1/s}, \qquad s = p + q \tag{11.1.5}$$

allows us to express this chemical potential as

$$\mu = p\mu_+^{\circ} + q\mu_-^{\circ} + sRT\ln\gamma_\pm, \tag{11.1.6}$$

and both types of ion now share equal responsibility for the non-ideality.

The principal problem remaining is the calculation of the mean activity coefficient. Once we know its dependence on the concentration and other properties of the solution, we can investigate its consequences for thermodynamic properties such as equilibrium constants and solubilities.

11.2 The Debye–Hückel theory

Electrolyte solutions are strongly non-ideal on account of the long range of the electrostatic interactions between ions. The energy of interaction between two neutral molecules decreases as approximately $1/R^6$, where R is their separation, but the Coulomb interaction energy decreases only as $1/R$. A simplifying feature, therefore, is that the Coulombic interaction is likely to be primarily responsible for the departures from ideality. This simplification is the basis of the Debye–Hückel theory of ionic solutions. Their theory is an excellent example of the way in which the essential physical features of a problem are identified and then expressed quantitatively: the model is intricate, but each step is guided by a clear appreciation of the underlying physical ideas.

11.2 (a) The model

Oppositely charged ions attract each other. As a result, cations and anions are not uniformly distributed in solutions, but anions are likely to be found near cations, and vice versa, Fig. 11.1. Overall the solution is electrically neutral, but near any given ion there is an excess of *counter-ions* (the ions of opposite charge). On the average, more counter-ions than like-ions pass by any given ion, and they come and go in all directions. This time-averaged, spherical haze of opposite charge around a given ion is called its *ionic atmosphere*.

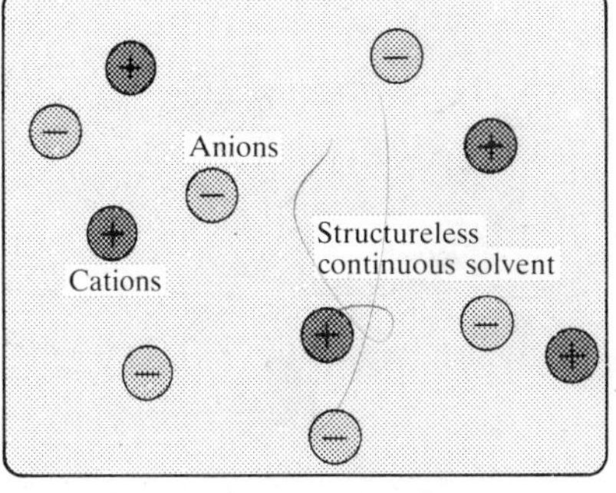

Fig. 11.1. The picture underlying the Debye–Hückel theory is of a tendency for anions to be found around cations, and of cations to be found near anions. The average non-zero charge distribution around an ion its called the *ionic atmosphere*.

The energy, and therefore the chemical potential, of any given central ion is lowered as a result of its electrostatic interaction with its ionic atmosphere. Our task is to formulate the effect quantitatively.

The whole of the departure from ideality is assumed to be due to the electrostatic interaction of an ion with its atmosphere. We therefore imagine a solution in which all the ions have their actual positions, but in which all the ion-atmosphere interactions have been turned off, Fig. 11.2. The calculation of the activity coefficient then reduces to the evaluation of the change in chemical potential when the charges of the ions are returned to their true values while their average distribution is held constant. This change in chemical potential is the difference in molar Gibbs function between the hypothetical, uncharged and true, charged states. Under conditions of constant temperature and pressure, this change of Gibbs function is equal to w_e, the electrical work of charging the system. This suggests the following strategy:

(1) Denote the chemical potential of ions in the ideal solution as $\mu°$. Then it follows from eqn (11.1.6) that the activity coefficient for the ions in the true solution is given by

$$\mu = \mu° + sRT \ln \gamma_{\pm}, \quad \text{or} \quad \ln \gamma_{\pm} = (\mu - \mu°)/sRT.$$

(2) The difference $\mu - \mu°$ is identified with the change in the molar Gibbs function for the process

hypothetical uncharged, ideal state $\rightarrow$ *real state*, $\Delta G_m = \mu - \mu°$.

(3) ΔG_m is identified with the electrical work, w_e, of charging the ions while they are maintained in their final average distribution.

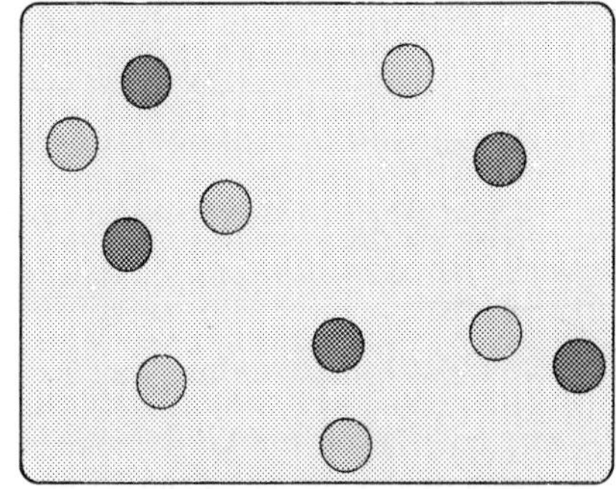

Fig. 11.2. In order to calculate the electrical interactions between the ions, they are assumed to be in their final average positions (as in Fig. 11.1) but with their charges extinguished. The work of charging the ions is then calculated.

These remarks combine to give

$$\ln \gamma_{\pm} = w_e/sRT. \tag{11.2.1}$$

This means that the problem has been reduced to finding the final average distribution of the ions and the work of charging them in that distribution.

Example 11.1

Measurements of the kind to be described later show that the mean activity coefficient of 0.010 mol kg^{-1} KCl(aq) at 298 K is 0.901. Estimate the work involved in charging the hypothetical uncharged ideal solution.

- *Method*. Use eqn (11.2.1) with $s = 2$ and $R\mathcal{T} = 2.48$ kJ mol^{-1}.

- *Answer*. $w_e = sRT \ln \gamma_\pm$
$= 2 \times (2.48 \text{ kJ mol}^{-1}) \ln 0.901 = -0.52 \text{ kJ mol}^{-1}$.

- *Comment*. Note that we are dealing with substantial energies. The work is negative because the ions attract each other on the average more than they repel each other.

- *Exercise*. The mean activity coefficient of 0.010 mol kg^{-1} $CaCl_2$(aq) is 0.732 at 298 K. Calculate the work of charging the hypothetical uncharged ideal solution. [-2.3 kJ mol^{-1}]

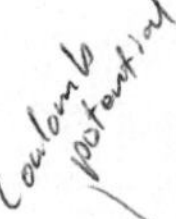

11.2 (b) The ionic atmosphere

The Coulomb potential at a distance r from an ion of charge $z_i e$ is

$$\phi_i = (z_i e/4\pi\varepsilon_0)(1/r). \quad (11.2.2)$$

(A brief review of electrostatics is given in Appendix 11.1.) This is the potential due to an isolated ion in a vacuum. In solution two modifications are needed. In the first place, the solvent decreases the strength of the potential, Fig. 11.3, and if the electric permittivity is ε the potential at r is

$$\phi_i = (z_i e/4\pi\varepsilon)(1/r). \quad (11.2.3)$$

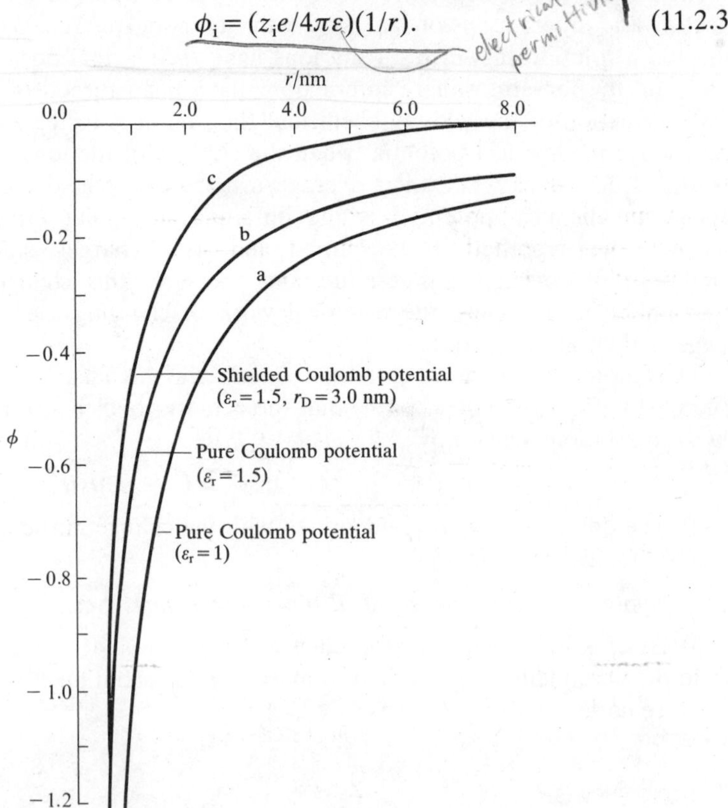

Fig. 11.3. The distance dependence of (a) the pure Coulomb potential in a vacuum, (b) the pure Coulomb potential in a medium of relative permittivity 1.5 (the value in (a) is reduced by a factor of 1/1.5 uniformly), and (c) a shielded Coulomb potential corresponding to $r_D = 3.0$ nm.

The factor $z_i e/4\pi\varepsilon$ will occur frequently in the following; we denote it Z_i and write

$$\phi_i = Z_i/r. \tag{11.2.3a}$$

The permittivity is usually expressed in terms of the *relative permittivity* (or *dielectric constant*; symbol: ε_r), through $\varepsilon = \varepsilon_r\varepsilon_0$. Some values of relative permittivities are given in Table 11.1. Since $\varepsilon_r > 1$ the potential is reduced from its vacuum value. This reduction is very important in many solvents. For example, $\varepsilon_r = 78.5$ for water at 25 °C, and so at a given distance the Coulomb potential is reduced from the vacuum value by nearly two orders of magnitude. This is one reason why water is such a successful solvent: the Coulombic interactions are so strongly reduced by the solvent that the ions interact only weakly with each other and do not aggregate into a crystal.

Table 11.1. Relative permittivities at 298 K

	ε_r
Water	78.54
Ammonia	16.9
	22.4 (at −33 °C)
Ethanol	24.30
Benzene	2.274

The second modification of the potential arises from the ionic atmosphere. An imaginary probe measuring the potential near an ion would enter the weak, oppositely charged ionic atmosphere as it moves away from the central ion, and so the potential decreases more rapidly than is predicted by eqn (11.2.3). The central ion is said to be *shielded* by the atmosphere. When shielding is present, and the ions can be regarded as point-like, the appropriate potential to use is the *shielded Coulomb potential*, in which $1/r$ is replaced by $(1/r)\exp(-r/r_D)$. The parameter r_D, which is called the *shielding length* or the *Debye length*, determines how strongly the potential is damped from its pure Coulomb value, Fig. 11.3. When r_D is very large, $r/r_D = 0$, and since $e^0 = 1$ the shielded potential is virtually the same as the unshielded potential. When r_D is small, the shielded potential is much smaller than the unshielded potential, even for short distances.

The shielded Coulomb potential in a medium of permittivity ε is

$$\phi_i = (Z_i/r)e^{-r/r_D}. \tag{11.2.4}$$

The unknown quantity is the Debye length r_D. It can be found by solving some equation for ϕ. In electrostatics the potential arising from a charge distribution is given by *Poisson's equation* (Appendix 11.1). In the case of a spherically symmetrical charge distribution the *charge density* (the charge per unit volume; symbol: ρ_i) at a distance r from a central ion i is the same in any direction, and the potential also depends only on distance. It is then permissible to use the following simplified form of Poisson's equation:

$$(1/r^2)(d/dr)\{r^2\, d\phi_i/dr\} = -\rho_i/\varepsilon.$$

Substituting eqn (11.2.4) for the potential then gives

$$\phi_i/r_D^2 = -\rho_i/\varepsilon, \quad \text{so that} \quad r_D^2 = -\varepsilon\phi_i/\rho_i. \tag{11.2.5}$$

This is an equation for the unknown parameter r_D; but in order to solve it we must first know the charge density ρ_i. We therefore need another equation for ρ_i.

11.2 (c) The charge density

Peter Debye and Erich Hückel took the view that the average charge density at any point arises from the competition between the electrostatic attraction of the central ion for its counter-ions and the disruptive effects of thermal motion. The difference in energy of an ion j of charge $z_j e$ at a position where the central ion i is giving rise to a potential ϕ_i relative to its energy at infinity (where the potential is zero) is $\Delta E = z_j e\phi_i$. The Boltzmann distribution (Section 0.1(d)) then gives the *proportion* of ions at this location relative to the proportion in the bulk solution (effectively at infinity):

$$\frac{\mathcal{N}_j}{\mathcal{N}_j^\circ} = \frac{\textit{number of j ions per unit volume where the potential is } \phi_i}{\textit{number of j ions per unit volume where the potential is zero}}.$$

$$= e^{-\Delta E/kT}, \quad \text{with} \quad \Delta E = z_j e\phi_i.$$

This means that

$$\mathcal{N}_j/\mathcal{N}_j^\circ = e^{-z_j e\phi_i/kT}.$$

In this expression $\mathcal{N}_j$ is the number density of ions of type j, the number of j ions per unit volume of solution. $\mathcal{N}_j^\circ$ is the number density in the bulk of the solution.

The charge density at a distance r from the ion i is the number density of each type of ion multiplied by the charge each type carries:

$$\rho_i = \mathcal{N}_+ z_+ e + \mathcal{N}_- z_- e$$
$$= \mathcal{N}_+^\circ e z_+ e^{-z_+ e\phi_i/kT} + \mathcal{N}_-^\circ e z_- e^{-z_- e\phi_i/kT}.$$

($z_j e$ is the charge of the ion j, and so z_+ is positive and z_- is negative.)

At this point a major simplification is achieved if we suppose that the average electrostatic interaction energy is small compared with kT. If it were large, then the ionic interaction would overcome the thermal motion and the ions would aggregate into a solid. On the assumption that it *is* small we may use the expansion $e^x = 1 + x + \ldots$ to write the last equation as

$$\rho_i = (\mathcal{N}_+^\circ z_+ + \mathcal{N}_-^\circ z_-)e - (\mathcal{N}_+^\circ z_+^2 + \mathcal{N}_-^\circ z_-^2)(e^2\phi_i/kT) + \ldots$$

The first term in the expansion is zero because it is the charge density in a uniform solution, and the solution is electrically neutral. The unwritten terms (represented by the dots) are assumed to be too small to be significant. This is the *linearization* step in the argument (because we keep only the term linear in ϕ and discard the rest).

11.2 (d) The ionic strength

The factor $\mathcal{N}_+^\circ z_+^2 + \mathcal{N}_-^\circ z_-^2$ can be simplified as follows. First, the number densities are expressed in terms of molalities through

$$\mathcal{N}_j^\circ = m_j \rho N_A,$$

where ρ is the solution density (which is virtually the same as the solvent density at the low molalities we consider), and N_A is Avogadro's constant. (m_j is the amount per unit mass of solvent, $m_j\rho$ is the amount per unit volume, and multiplication by N_A converts amounts into numbers.) Then we introduce the dimensionless *ionic strength* (symbol: I):

$$I = \tfrac{1}{2}\sum_j (m_j/m^\ominus)z_j^2. \qquad (11.2.6)$$

The ionic strength occurs widely wherever ionic solutions are being discussed, as we shall see (and the sum is not limited in general to one type of cation and one type of anion but extends over all the types present in the solution). In the case of two types of ion at molalities m_+ and m_- this general expression simplifies to

$$I = \tfrac{1}{2}(m_+ z_+^2 + m_- z_-^2)/m^{\ominus}. \qquad (11.2.6a)$$

Notice how I emphasizes the charges of the ions because the charge numbers occur as their squares. For instance, in the case of a completely dissociated (1,1)-electrolyte (an electrolyte in which $z_+ = 1$ and $|z_-| = 1$), the ionic strength is $I = \tfrac{1}{2}(m_+ + m_-)/m^{\ominus} = m/m^{\ominus}$, where m is the solute molality (and $m_+ = m_- = m$). On the other hand, for a completely dissociated (1,2)-electrolyte giving the ions $2M^+ + X^{2-}$ the ionic strength of a solution of molality m is $I = \tfrac{1}{2}(m_+ + 4m_-)/m^{\ominus} = 3m/m^{\ominus}$, because $m_- = m$ and $m_+ = 2m$, and so the ionic strength is greater even though the concentration is the same. Box 11.1 summarizes the relation of ionic strength and molality in an easily usable form.

Box 11.1 Ionic strength and molality

The ionic strength of a solution of a single salt may be written $I = km/m^{\ominus}$, where k is a number that depends on the valence types of the ions. The definition, eqn (11.2.6), then allows us to build the following table of values of k.

k	X^-	X^{2-}	X^{3-}	X^{4-}
M^+	1	3	6	10
M^{2+}	3	4	15	12
M^{3+}	6	15	9	42
M^{4+}	10	12	42	16

For example, the ionic strength of the salt M_2X_3, which is understood to be $M_2^{3+}X_3^{2-}$ of molality m is $15\,m/m^{\ominus}$ (and dimensionless).

In terms of ionic strength, the expression for the charge density around i is

$$\rho_i = -(2\rho e^2 I N_A m^{\ominus}/kT)\phi_i, \qquad (11.2.7)$$

which means that the charge density and potential are directly proportional to each other (but opposite in sign, because counter-ions predominate in the atmosphere). Equation (11.2.5) can now be solved for r_D without difficulty:

$$r_D^2 = \varepsilon RT/2\rho F^2 I m^{\ominus}, \qquad (11.2.8)$$

where $F = eN_A$ is *Faraday's constant*, the magnitude of the charge per mole of electrons ($1F = 9.648 \times 10^4\ \mathrm{C\,mol^{-1}}$). This value of r_D determines the potential through eqn (11.2.4) and then the charge density through eqn (11.2.7).

Example 11.2

Calculate the shielding length for a (1,1)-electrolyte in aqueous solution at 298 K when the molality is 0.01 mol kg^{-1}.

• *Method*. Use eqn (11.2.8). Take $\rho = 0.997\ \mathrm{g\,cm^{-3}}$ ($= 0.997 \times 10^3\ \mathrm{kg\,m^{-3}}$) from Table 0.1 and use $\varepsilon_r = 78.5$ from Table 11.1. The ionic strength of a (1,1)-electrolyte is related to the molality by $I = m/m^{\ominus}$, Box 11.1.

• *Answer*.

$$r_D^2 = \frac{(78.5\times 8.854\times 10^{-12}\,\mathrm{J^{-1}\,C^2\,m^{-1}})\times(8.314\times 298.15\,\mathrm{J\,mol^{-1}})}{2\times(0.997\times 10^3\,\mathrm{kg\,m^{-3}})\times(9.648\times 10^4\,\mathrm{C\,mol^{-1}})^2\times(1\,\mathrm{kg\,mol^{-1}})\times I}$$
$$= (9.28\times 10^{-20}\,\mathrm{m^2})/I,$$

so that $r_D = 0.305\,\mathrm{nm}/I^{\frac{1}{2}}$. In the present case, since $I = 0.01$, $r_D = 3.0\,\mathrm{nm}$.

• *Comment*. We demonstrate later that r_D can be interpreted as the 'thickness' of the ionic atmosphere' and the present result means that the atmosphere can be thought of as a spherical shell of charge at a distance 3.0 nm (about 10 ionic radii) from the central ion.

• *Exercise*. Calculate r_D for an 0.01 mol kg^{-1} $CaCl_2$(aq) at 298 K. [1.7 nm]

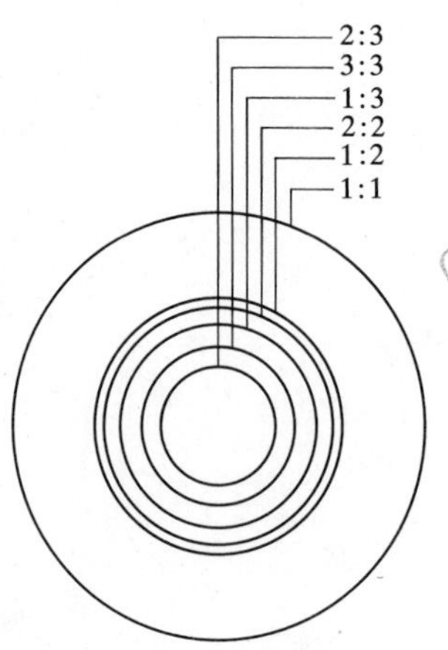

Fig. 11.4. The ionic atmosphere is equivalent to a spherical shell of charge of radius r_D surrounding the central ion. This diagram shows the size of the shell for different charge types and $m = 0.01$ mol kg^{-1}.

Before using these results we should check that the expression for the Debye length is physically plausible. It predicts that *if we ignore the temperature dependence of the solution density and permittivity, then r_D increases with temperature*. This is reasonable because thermal motion disperses the ionic atmosphere and weakens its shielding effect (corresponding to increasing r_D). In practice, though, the quantity $\varepsilon_r T/\rho$, which gives the *overall* temperature dependence and not just that arising from the Boltzmann distribution, actually decreases for water as the temperature is raised. Secondly, *the Debye length decreases with increasing ionic strength*: the higher the ionic concentration the more effective the shielding. Furthermore, since the ionic strength emphasizes the charge of the ions, even a low concentration of highly charged ions may form an effective shield. This is illustrated in Fig. 11.4. Finally, *the Debye length increases with increasing permittivity*: when ε_r is large, the marshalling effect of the central ion is weak and the ionic atmosphere is diffuse.

11.2 (e) The activity coefficient

The strategy for calculating the activity coefficient involves determining the electrical work of charging the central ion when it is already surrounded by its atmosphere. It follows that we need to know the potential at the ion due to its atmosphere, ϕ_{atmos}. This is the difference between the total potential (as given by eqn (11.2.4)) and the potential due to the central ion itself:

$$\phi_{\text{atmos}} = \phi_i - \phi_{\text{central ion}} = Z_i\{(e^{-r/r_D}/r) - (1/r)\}.$$

The potential at the ion (at $r = 0$) is then obtained by taking the limit of this expression as $r \to 0$:

$$\phi_{\text{atmos}}(0) = Z_i\{[1 - r/r_D + \tfrac{1}{2}(r/r_D)^2 + \ldots]/r - (1/r)\}_{r\to 0}$$
$$= -Z_i/r_D. \qquad (11.2.9)$$

This result suggests another interpretation of the Debye length. It shows that the potential at the central ion of charge $z_i e$ due to its own atmosphere is equivalent to the potential of a single charge of opposite charge, $-z_i e$, at a distance r_D. Thus the ionic atmosphere acts as if it were all concentrated on a spherical shell at a distance r_D from the central ion.

If the charge of the central ion were q and not $z_i e$ the potential due to its atmosphere would be

$$\phi_{\text{atmos}}(0) = -(q/4\pi\varepsilon)(1/r_D).$$

The work of adding a charge dq to a region where the electrical potential is

$\phi_{atmos}(0)$ is $dw_e = \phi_{atmos}(0)\,dq$ (Appendix 11.1). Therefore, the total work of fully charging unit amount of ions is

$$w_e = N_A\int_0^{z_i e}\phi_{atmos}(0)\,dq = -(N_A/4\pi\varepsilon r_D)\int_0^{z_i e} q\,dq$$
$$= -z_i^2 e^2 N_A/8\pi\varepsilon r_D = -z_i^2 F^2/8\pi\varepsilon N_A r_D. \qquad (11.2.10)$$

Example 11.3

In *Example* 11.1 the work of charging an 0.01 mol kg^{-1} (1,1)-electrolyte solution was calculated as -0.52 kJ mol^{-1}. Check that the Debye–Hückel theory predicts the right order of magnitude. Confirm that the linearization step is arithmetically valid in this case.

• *Method*. The first part of the calculation involves substitution in eqn (11.2.10) using the numerical expression for r_D obtained in *Example* 11.2. The total work is the sum of the work of charging the cations and the anions. For the second part, decide whether $z_+e\phi/kT$ is much smaller than unity at an average distance (i.e. r_D) from the central ion.

• *Answer*. From *Example* 11.2 with $I = 0.01$, $r_D = 3.0$ nm. Then, from eqn (11.2.10):

$$w_e = \frac{-(9.648\times10^4\,\mathrm{C\,mol^{-1}})^2}{8\pi\times(78.5\times8.854\times10^{-12}\,\mathrm{J^{-1}\,C^2\,m^{-1}})\times(6.022\times10^{23}\,\mathrm{mol^{-1}})\times(3.0\times10^{-9}\,\mathrm{m})}$$
$$= -0.29\ \mathrm{kJ\,mol^{-1}}.$$

Therefore, the total work required is -0.58 kJ mol^{-1}, in good agreement with the experimental value.

The value of $e\phi/kT$ when $r_D = 3.0$ nm is, from eqn (11.2.4),

$$e\phi/kT = e^2 e^{-1}/4\pi\varepsilon r_D kT = 0.088,$$

which is substantially less than 1, as the approximation requires.

• *Comment*. The experimental and theoretical values are in good agreement, suggesting that the Debye–Hückel theory is reliable at very low molalities. The value of r_D is also consistent with the linearization approximation.

• *Exercise*. Compare the values for 0.01 mol kg^{-1} $CaCl_2$(aq), for which $\gamma_\pm = 0.732$ at 298 K. Is the linearization valid? [-3.1 kJ mol^{-1}; barely]

It now follows from eqn (11.2.1) that the mean activity coefficient of the ions is

$$\ln\gamma_\pm = (pw_{e,+} + qw_{e,-})/sRT = -(F^2/8\pi\varepsilon sN_A RTr_D)(pz_+^2 + qz_-^2),$$

where $s = p + q$. However, for neutrality $pz_+ + qz_- = 0$, and so

$$\ln\gamma_\pm = -|z_+z_-|\,(F^2/8\pi\varepsilon r_D N_A RT).$$

The expression for r_D given in eqn (11.2.8) may now be used, and if logarithms are converted to base 10 this gives

$$\lg\gamma_\pm = -1.825\times10^6\,|z_+z_-|\,\{I(\rho/\mathrm{g\,cm^{-3}})/\varepsilon_r^3(T/\mathrm{K})^3\}^{\frac{1}{2}}$$
$$= -A\,|z_+z_-|\,I^{\frac{1}{2}}.$$

For water at 298 K, when the density is 0.997 g cm^{-3} and the relative permittivity is 78.54, $A = 0.509$. This gives the *Debye–Hückel Limiting Law* for water at 298 K as

$$\lg\gamma_\pm = -0.509\,|z_+z_-|\,\sqrt{I}. \qquad (11.2.11)$$

The name 'limiting law' is applied for the same reason as for gases. Ionic solutions of moderate molalities may have activity coefficients that differ from the values given by this expression, yet all solutions are expected to conform in the limit of arbitrarily low molalities. At least, they should conform if the model is correct and the approximations are tenable. In order to test it we turn to experiment.

11.2 (f) Experimental tests, improvements, and extensions

Table 11.2. Mean activity coefficients in water at 298 K

$m/m^{\ominus}$	KCl	$CaCl_2$
0.001	0.966	0.888
0.01	0.902	0.732
0.1	0.770	0.524
1.0	0.607	0.725

In this section we examine the results of measuring activity coefficients. The measurement techniques will be described in the next chapter, and some will be found in the Problems at the end of this. Some experimental values of activity coefficients for salts of various valence types are listed in Table 11.2. Figure 11.5 shows some of these values plotted against the square-root of the ionic strength, and compares them with the theoretical straight lines based on eqn (11.2.11). The agreement at low molalities (less than 0.01 to 0.001 mol kg^{-1}, depending on valence type) is impressive, and convincing evidence in support of the model. Nevertheless, the departures from the theoretical curves above these molalities are large, and show that the approximations are valid only at very low concentrations.

What are the approximations, and how may they be removed? Four approximations were made: (1) the bulk permittivity was used, (2) the sizes of the ions were ignored and they were treated as points, (3) the spherically symmetrical Poisson equation was combined with the Boltzmann distribution, and the resulting expression was linearized, and (4) the deviations from ideality were ascribed solely to Coulombic interactions. The simplest of these to remove is the assumption of negligible ion size, and we shall consider that initially.

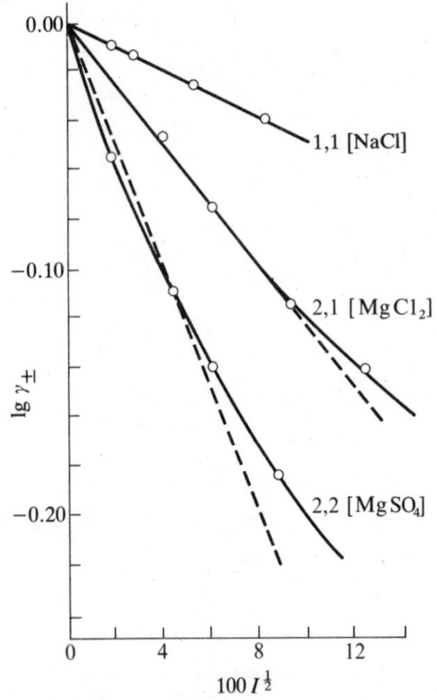

Fig. 11.5. An experimental test of the Debye–Hückel Limiting Law. Although there are marked deviations for moderate ionic strengths, the limiting slopes as $I \to 0$ are in good agreement with the theory, and so it can be used for extrapolating data to low molalities.

In the case of a (1,1)-electrolyte, $r_D \approx 100$ ionic radii when the molality is around 10^{-4} mol kg^{-1}, but $r_D \approx 10$ radii at 10^{-1} mol kg^{-1}. This warns us that the assumption of point-like ions is untenable even at moderate molalities, above about 10^{-2} mol kg^{-1}. If the ions have radius r_I we can expect the shielded Coulomb potential to take the form

$$\phi_i = (A'/r)e^{-(r-r_I)/r_D} \quad \text{for} \quad r > r_I,$$

where A' is independent of r (in fact: $A' = Z_i/(1 + r_I/r_D)$). This form allows the potential its full, Coulombic strength at the surface of the ion (when $r = r_I$), and for the shielding to operate only beyond this distance. The expression for the charging work is modified when this form of the potential is used because now the charge is applied to the surface of the ion where the potential is $\phi_{atmos}(r_I)$. The same manipulations that led to eqn (11.2.11) now lead to

$$\lg \gamma_\pm = -\{A/(1 + r_I/r_D)\}\,|z_+z_-|\,\sqrt{I}. \tag{11.2.12}$$

The Debye length in the denominator of the first factor depends on the square root of the ionic strength, eqn (11.2.8), and so this expression has the form

$$\lg \gamma_\pm = -\{A/(1 + A^*\sqrt{I})\}\,|z_+z_-|\,\sqrt{I}, \tag{11.2.13}$$

where A^* is another constant.

When the solution is very dilute the denominator in eqn (11.2.12) is almost unity, and we recover the limiting law. The criterion for this is $r_I/r_D \ll 1$, which is equivalent to assuming that the ions have negligible size.

When the concentration rises to the point where $r_{\rm I}$ is small but no longer negligible in comparison with $r_{\rm D}$ (so that $r_{\rm I} \approx r_{\rm D}/10$), the approximation $(1+x)^{-1} \approx 1-x$ leads to the *extended Debye–Hückel Law*:

$$\lg \gamma_{\pm} = -A\,|z_+z_-|\,\sqrt{I} + AA^*\,|z_+z_-|\,I. \qquad (11.2.14)$$

This predicts that the deviations from the limiting law should correspond to an increase in the activity coefficient (the second, new term is positive), and it accounts for the positive departures from the limiting law shown by the experimental curves in Fig. 11.5 (but not for the negative deviations for the (2,2)-electrolytes). The equation is fitted to the observed curves by choosing an appropriate value of the constant A^*, which is best regarded as an adjustable parameter. A curve drawn in this way is shown in Fig. 11.6. It is clear that eqn (11.2.14) accounts for some activity coefficients over a moderate range of dilute solutions; nevertheless it remains very poor for molalities in the vicinity of 1 mol kg^{-1}. At such high molalities the relative permittivity of the solution between the ions deviates strongly from its bulk values because the most of the water molecules between the ions are waters of hydration, and these molecules are not free to rotate in response to the fields of other ions; it is very difficult to take this effect into account quantitatively.

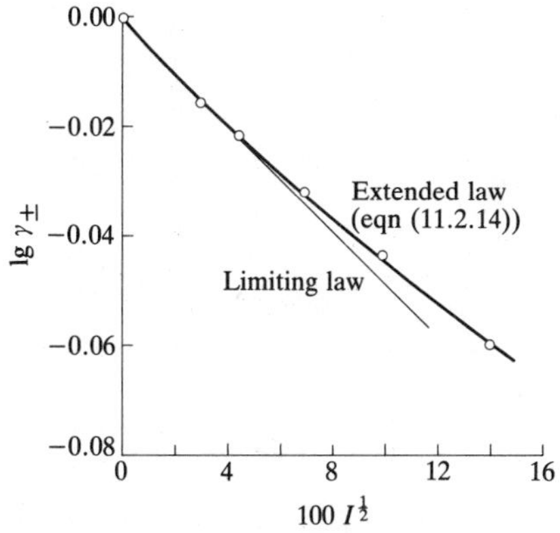

Fig. 11.6. The extended Debye–Hückel Law gives agreement with experiment over a wider range of molalities (as shown here for a (1,1)-electrolyte), but it still fails at higher molalities.

Several attempts have been made to improve on the limiting law, but the main difficulty lies deep in the structure of the model. The fundamental flaw is the combination of the Boltzmann distribution with the Poisson equation because this fails to allow for the interactions between ions in the same atmosphere: each ion is attempting to construct its own atmosphere at the same time as it is being collected around the ion of interest. The linearization step effectively ignores this competition, and so is valid only in the limit of infinite dilution, when the atmosphere is infinitely diffuse. In fact, a self-consistent theory of ionic solutions is still awaited. The theory will probably be based on a *dynamical* view of ions in solution, and it may draw on the techniques used to account for the properties of plasma (ionized gases) which are central to an understanding of controlled nuclear fusion and rocket exhausts. Until that theory is available we have to use the Debye–Hückel Limiting Law, eqn (11.2.11), or the extended theory, eqn (11.2.14), and accept that the predictions are valid only for solutions of low ionic strength. Nevertheless, although the theory is flawed for all but the lowest molalities, it remains useful in that region, especially for extrapolating data to low ionic strengths.

11.3 The role of electrodes

A large number of reactions can be regarded as the outcome of a reduction and an oxidation:

$$\textit{Reduction}: \ \mathrm{A} + \mathrm{e}^- \rightarrow \mathrm{A}^-; \qquad \textit{Oxidation}: \ \mathrm{B} \rightarrow \mathrm{B}^+ + \mathrm{e}^-.$$

One way of bringing about this *redox reaction* is to mix A and B, for then the electron released by B in the oxidation step is transferred to a nearby A, which undergoes reduction. In the bulk solution the spatial directions of the electron transfers are random, Fig. 11.7(a), and they occur without doing useful work (apart from p,V-work if the system expands or contracts). Another way of bringing about the reaction is to have solutions of A

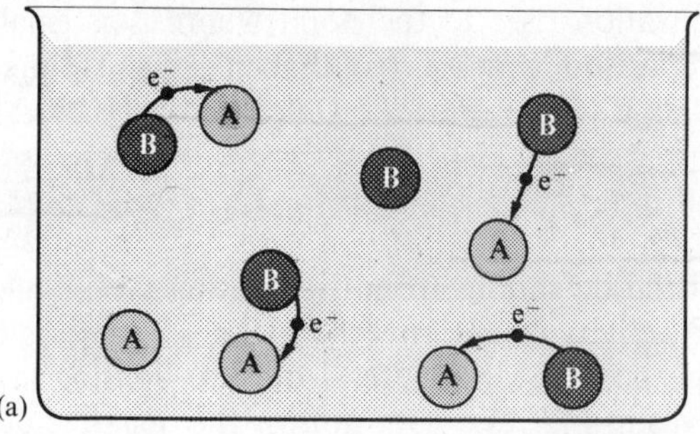

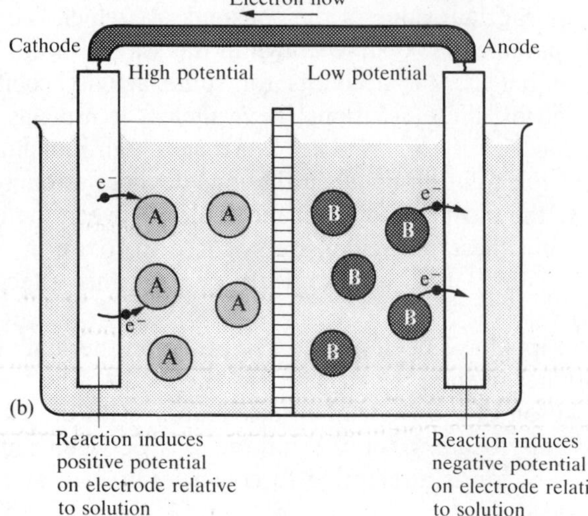

Fig. 11.7. When a reaction takes place in a bulk medium (a) the electron transfers are in random spatial directions and there is no net electric current. When the reaction takes place in a galvanic cell (b) electrons are deposited in one electrode and collected from another, and so there is a net flow of current which can be used to do work.

and B in separate compartments, and to connect them by an electric circuit making contact with the solutions via two electrodes, one of which acts as a source of electrons and the other as a sink, Fig. 11.7(b). Then B loses its electron to one electrode, and A collects its from the other. The two electrodes of this *electrochemical cell* need to be connected by a wire so that the electrons dumped by B can be carried to A. Then, as the reaction takes place, an electric current runs through the external circuit, and may be used to do work.

11.3 (a) Anodes and cathodes

The electrode where oxidation occurs is called the *anode*; the electrode where reduction occurs is called the *cathode*. In the case of an electrochemical cell acting as a source of electricity (this is called a *galvanic cell*) the cathode is at a higher potential than the anode. This is because the species undergoing reduction (A) withdraws electrons from its electrode (the cathode), so leaving a net positive charge on it (corresponding to a high potential). At the anode, oxidation corresponds to B transferring electrons to the electrode, so giving it a negative charge (corresponding to a low potential). In the case of electrolysis, where the reaction in the cell is driven by an externally generated electric current, the anode is still the location of oxidation (by definition), but now electrons must be withdrawn from the species, and at the cathode there must be a supply of electrons available for reduction. Therefore, the anode must now be made relatively positive with respect to the cathode. These remarks are summarized in Box 11.2.

Box 11.2 Electrode processes

Location:	Anode	Cathode	Device
Process:	Oxidation	Reduction	
Potential:	Low (−)	High (+)	Discharging galvanic cell
	High (+)	Low (−)	Electrolytic cell

Now consider what happens when a single metal electrode M is dipped into a solution containing the corresponding metal ions M^{z+}, Fig. 11.8. Some of the ions in the solid break away and go into solution as M^{z+}. Each ion that breaks away leaves z electrons behind, and so the process of dissolving builds up a negative charge on the electrode and a positive charge in the solution. Ions already in the solution attract electrons from the electrode, and the resulting neutral M atoms stick to its surface. This is equivalent to the condensation of metal cations on to the metal surface, and so it tends to acquire a positive charge. Very quickly (when only a tiny charge imbalance has occurred), the rates of escape and return become equal and the *redox equilibrium* $M^{z+}(aq) + ze^- \rightleftharpoons M(s)$ is set up; depending on where the equilibrium lies there will now be a net charge on the electrode. If that charge is positive it will appear to an observer as a positive electric potential, because it is found that more work is needed to bring up a positive test charge from infinity than to an uncharged electrode. If the charge is negative at equilibrium, the observer reports that the electrode has a negative potential, because less work is needed to bring up the positive test charge. Our task is to calculate these electric potentials of electrodes in contact with solutions of ions.

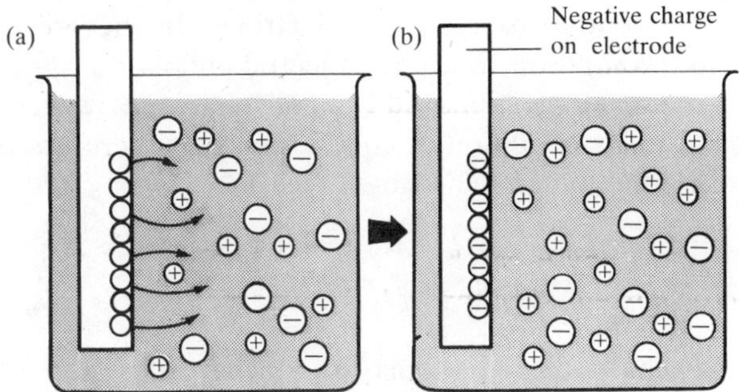

Fig. 11.8. The charge changes at the surface of electrodes. In (a), cations leave the metal electrode, and so it becomes relatively negatively charged while the solution near the electrode becomes slightly positively charged (b). When a positive test charge is brought up to the solution, more work has to be done than when it is brought up to the electrode, and so we report that the potential of the solution is higher than that of the electrode.

11.3 (b) The electrochemical potential

The work done when a charged species is added to a system in which the electric potential is ϕ is different from when the potential is zero. If the species is an ion of charge z_ie, the work per unit amount of ions brought up to the system from infinity is $z_ie\phi$. Therefore, the partial molar Gibbs function of the ions in the charged phase, the *electrochemical potential* (symbol: $\bar{\mu}_i$), is related to the chemical potential in the same uncharged phase (μ_i) by

$$\bar{\mu}_i = \mu_i + N_A z_i e\phi = \mu_i + z_i F\phi, \qquad (11.3.1)$$

where F is Faraday's constant.

When a positive ion (z_i positive) is in a region of positive potential, the electrochemical potential is greater than the chemical potential: this corresponds to the ion having a greater tendency to escape from the region or to undergo chemical change. The opposite is true for a negative ion in the same region: it is less reactive. This modification of the chemical potentials of ions by electric potentials underlies all electrochemistry. Everything that we have already discussed in relation to equilibrium properties applies when regions of different electric potentials are present, the only modification being that *at equilibrium, the electrochemical potential of each species must be the same in every phase*.

11.3 (c) The interfacial potential difference

In order to see how the electrochemical potential is used in a calculation, consider the simplest type of electrode. This is the metal ion | metal electrode, Fig. 11.9, denoted $M^{z+} | M$. There are numerous examples, including $Ag^+ | Ag$, a silver electrode in contact with a solution of silver ions (such as silver nitrate solution), and $Cu^{2+} | Cu$, a copper electrode in contact with copper ions (such as aqueous copper sulphate). The redox equilibrium of interest is between the M^{z+} ions in solution and the metal atoms and electrons of the solid electrode:

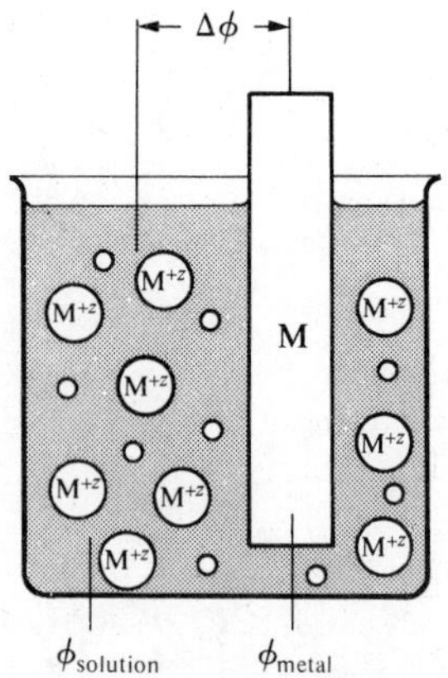

Fig. 11.9. The *ion | metal electrode* and the potentials at equilibrium. $\Delta\phi$ is the difference $\phi_{\text{metal}} - \phi_{\text{solution}}$.

$$M^{z+}(aq) + ze^- \rightleftharpoons M(s). \qquad (11.3.2)$$

At equilibrium, the appropriately weighted sums of the electrochemical potentials of the 'reactants' and 'products' are equal, and so by the same argument as in Chapter 10:

$$\textit{At equilibrium}: \ \bar{\mu}_M - \{\bar{\mu}_{M^{z+},\,\text{solution}} + z\bar{\mu}_{e^-,\,\text{metal}}\} = 0, \qquad (11.3.3)$$

where $\bar{\mu}_{\text{solution}}$ denotes an electrochemical potential in solution and $\bar{\mu}_{\text{metal}}$ an electrochemical potential in the metal electrode. In the present case, the M^{z+} ions are in the solution where the electric potential is ϕ_{solution}, and the electrons and metal atoms are in the metal electrode, where the electric potential is ϕ_{metal}. The condition of equilibrium may therefore be written in terms of these electric potentials using

$$\bar{\mu}_{M^{z+},\,\text{solution}} = \mu_{M^{z+}} + zF\phi_{\text{solution}},$$
$$\bar{\mu}_{e^-,\,\text{metal}} = \mu_{e^-} - F\phi_{\text{metal}},$$

and $\bar{\mu}_M = \mu_M$ because the M atoms are uncharged. Therefore, at equilibrium, the *electric potential difference* (symbol: $\Delta\phi$) across the metal/solution interface is

$$\Delta\phi = (1/zF)\{\mu_{M^{z+}} + z\mu_{e^-} - \mu_M\}. \qquad (11.3.4)$$

The chemical potential of the ions depends on their concentration in solution through $\mu_{M^{z+}} = \mu^{\ominus}_{M^{z+}} + RT \ln a_{M^{z+}}$, where $a_{M^{z+}}$ is their activity. The potential difference therefore depends on the activity as follows:

$$\Delta\phi = (1/zF)\{\mu_{M^{z+}} + z\mu_{e^-} - \mu_M\} + (RT/zF) \ln a_{M^{z+}}. \qquad (11.3.5)$$

This expression can be simplified. When the ions are at unit activity the *standard potential difference* (symbol: $\Delta\phi^{\ominus}$) is given by the first term on the right of the equation. Therefore, for a general value of the activity,

$$\Delta\phi = \Delta\phi^{\ominus} + (RT/zF) \ln a_{M^{z+}}. \qquad (11.3.6)$$

We can check that this equation gives physically plausible predictions. It states that as the activity of the ions in solution increases (e.g. when their molality is increased), the potential difference increases (with the metal becoming more positive relative to the solution). This is in accord with the view that as the activity of the ions in solution increases, they have a greater tendency to leave the solution for the electrode, and therefore tend to increase the electrode's positive charge. Since $R\mathcal{T}/F = 25.7\,\text{mV}$, increasing the activity by a factor of 10 at 25 °C increases the potential difference across an $M^+ \mid M$ interface by $(25.7\,\text{mV}) \ln 10 = 59.2\,\text{mV}$.

11.4 The electric potential at interfaces

The electric potential difference across an interface can be related to the electrochemical potentials of the species involved in the redox equilibrium. We saw in the last section how to do this in one case; here we generalize to other types of electrode.

11.4 (a) The gas | inert metal electrode

The construction of the gas | inert metal electrode is illustrated in Fig. 11.10. The inert metal acts as a source or sink of electrons, but takes no other part in the redox reaction. The gas is bubbled over the surface of the metal while it is bathed in a solution of ions related to the gas. For example, if (a) the gas is hydrogen the solution should contain H^+ ions, and if (b) it is chlorine it should contain Cl^- ions. The electrode is denoted either (a) $G^+ \mid G_2 \mid M$, or (b) $G^- \mid G_2 \mid M$. We shall deal explicitly with case (a), and then generalize the result to include case (b).

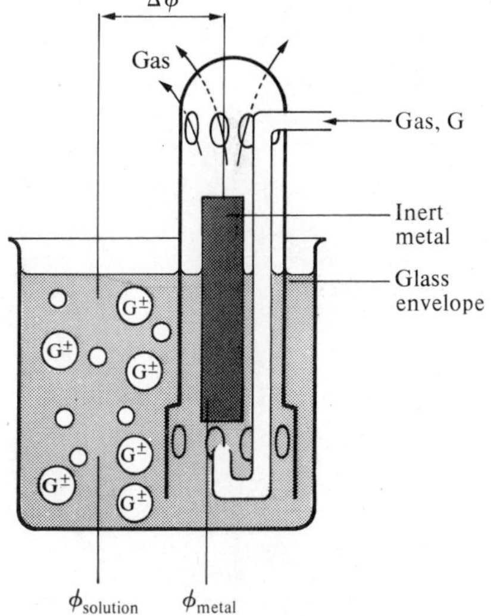

Fig. 11.10. The *gas | inert metal electrode* and the potentials at equilibrium. The gas is bubbled over the inert metal surface, and the equilibrium is between it and its ions (e.g. between H_2 and H^+ or between Cl_2 and Cl^-).

The redox equilibrium is

$$G^+(\text{solution}) + e^-(\text{metal}) \rightleftharpoons \tfrac{1}{2}G_2(\text{g}),$$

$$\bar{\mu}_{G^+,\,\text{solution}} + \bar{\mu}_{e^-,\,\text{metal}} = \tfrac{1}{2}\bar{\mu}_{G_2,\,\text{gas}}.$$

The electrochemical potential of the gas, which is uncharged, is equal to its chemical potential. The ions are in solution, where the electric potential is ϕ_{solution}. The electrons are in the metal electrode, where the potential is ϕ_{metal}. The relations between the electrochemical potentials and the electric potentials are therefore

$$\bar{\mu}_{G^+,\,\text{solution}} = \mu_{G^+} + F\phi_{\text{solution}},$$

$$\bar{\mu}_{e^-,\,\text{metal}} = \mu_{e^-} - F\phi_{\text{metal}},$$

and $\bar{\mu}_{G_2,\,\text{gas}} = \mu_{G_2}(\text{g})$. The equilibrium condition is therefore

$$\{\mu_{G^+} + F\phi_{\text{solution}}\} + \{\mu_{e^-} - F\phi_{\text{metal}}\} = \tfrac{1}{2}\mu_{G_2}(\text{g}). \qquad (11.4.1)$$

The chemical potentials are now expressed in terms of the activities of the ions and the fugacity of the gas:

$$\mu_{G^+} = \mu^{\ominus}_{G^+} + RT \ln a_{G^+}, \qquad \mu_{G_2}(\text{g}) = \mu^{\ominus}_{G_2} + RT \ln (f/p^{\ominus}).$$

Then rearranging eqn (11.4.1) for the potential difference $\Delta\phi = \phi_{\text{metal}} - \phi_{\text{solution}}$ gives

$$\Delta\phi = (1/F)\{\mu^{\ominus}_{G^+} + \mu_{e^-} - \tfrac{1}{2}\mu^{\ominus}_{G_2}\} + (RT/F) \ln (a_{G^+}/f^{\frac{1}{2}}),$$

where f, for simplicity, now means $f/p^{\ominus}$. When the gas is at unit fugacity and the ions are at unit activity, the metal/solution interfacial potential difference has its standard value $\Delta\phi^{\ominus}$. Then the potential difference at any fugacity and activity may be calculated from

$$\Delta\phi = \Delta\phi^{\ominus} + (RT/F)\ln\left(a_{G^+}/f^{\frac{1}{2}}\right). \tag{11.4.2}$$

This equation makes sense physically. Increasing the activity increases the tendency of the positive ions to discharge at the electrode, and so we should expect its potential to become more positive. Since a_{G^+} occurs in the numerator of the logarithmic term, eqn (11.4.2) does predict that $\Delta\phi$ increases as a_{G^+} is increased.

Example 11.4

Chlorine gas is bubbled over a platinum electrode dipping into aqueous sodium chloride at 298 K. Calculate the change in potential difference when the chlorine pressure is increased from 1.0 atm to 2.0 atm.

• *Method*. Find the expression corresponding to eqn (11.4.2) for a gas in equilibrium with its univalent negative ions. Assume that the gas behaves perfectly at the pressures specified, and so replace fugacity by pressure. Use $R\mathcal{T}/F = 25.69$ mV.

• *Answer*. The condition for equilibrium of the redox reaction $\frac{1}{2}G_2(g) + e^- \rightleftharpoons G^-(aq)$ is $\frac{1}{2}\bar{\mu}_{G_2}(g) + \bar{\mu}_{e^-,\,\text{metal}} = \bar{\mu}_{G^-,\,\text{solution}}$. By the same argument as above, this turns into

$$\Delta\phi = \Delta\phi^{\ominus} - (RT/F)\ln\left(a_{G^-}/f^{\frac{1}{2}}\right). \tag{11.4.2a}$$

Therefore, the change in potential difference at 298 K is

$$\Delta\phi(2\,\text{atm}) - \Delta\phi(1\,\text{atm}) = (R\mathcal{T}/F)\ln\sqrt{2} = 8.9\,\text{mV}.$$

• *Comment*. The potential difference increases when the gas pressure is increased because the equilibrium shift away from $Cl_2(g)$ towards $Cl^-(aq)$, and the resulting withdrawal of electrons from the electrode makes it more positive.

• *Exercise*. Find an expression for the equilibrium potential difference for the redox reaction $G_2(g) + 4e^- \rightleftharpoons 2G^{2-}(aq)$. $[\Delta\phi = \Delta\phi^{\ominus} - (RT/2F)\ln\left(a_{G^{2-}}/f^{\frac{1}{2}}\right)]$

The easiest way of remembering the expressions derived so far is as

$$\Delta\phi = \Delta\phi^{\ominus} + (RT/\nu F)\ln(\text{Ox}/\text{Red}), \tag{11.4.3}$$

where Red denotes the activity (or fugacity) of the reduced form raised to the appropriate power, Ox denotes the oxidized form, and ν is the number of electrons transferred in the redox reaction ($\nu = 1$ in the cases we have considered in the section so far). For example, for $\frac{1}{2}G_2(g) + e^- \rightleftharpoons G^-(aq)$, the reduced species is $G^-(aq)$, and so Red denotes a_{G^-}, and the oxidized species is $\frac{1}{2}G_2(g)$, so that Ox denotes $f^{\frac{1}{2}}$, and we recover eqn (11.4.2a) by noting that

$$\ln(\text{Ox}/\text{Red}) = \ln\left(f^{\frac{1}{2}}/a_{G^-}\right) = -\ln\left(a_{G^-}/f^{\frac{1}{2}}\right).$$

We shall see that eqn (11.4.3) applies to all forms of electrodes. (You should check that it includes the metal ion | metal electrode potential given by eqn (11.3.6).)

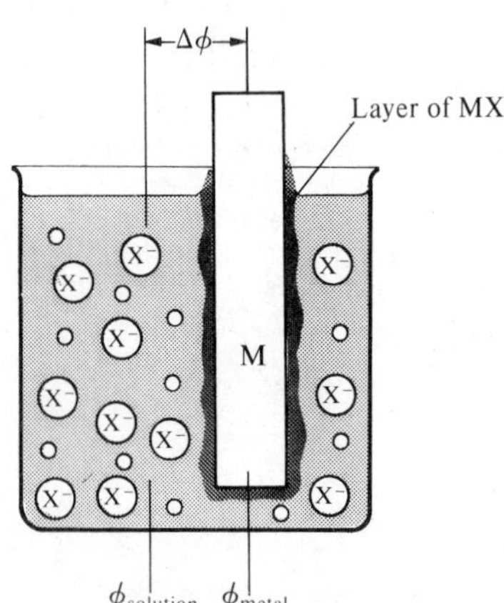

Fig. 11.11. The *ion | insoluble salt | metal electrode* consists of the metal (e.g. silver) coated with the insoluble salt (e.g. silver chloride) dipping into a solution of the salt's anions (e.g. chloride ions). The potential depends on the activity of the anions in solution.

11.4. (b) The ion | insoluble salt | metal electrode

The construction of this electrode is shown in Fig. 11.11. It consists of a metal M covered by a layer of insoluble salt MX, the whole immersed in a solution containing X^- ions. The electrode is denoted $X^-(aq)|MX(s)|M(s)$. A common example is the $Cl^-|AgCl(s)|Ag(s)$ electrode, which we look at in more detail below. The importance of this type of electrode is the dependence of its potential difference on the activity of the anion X^- in the solution.

In order to calculate the potential difference, we may picture the electrode as a system of two interfaces, one between the metal electrode and the metal ions in the insoluble salt, $MX(s)+e^- \rightleftharpoons M(s)+X^-(s)$, and the other between the $X^-(s)$ anions in the coating of insoluble salt and the $X^-(aq)$ anions in the solution: $X^-(s) \rightleftharpoons X^-(aq)$. The overall equilibrium is

$$MX(s)+e^-(\text{metal}) \rightleftharpoons M(s)+X^-(aq).$$

Now we identify Ox with $a_{MX(s)}$ and Red with $a_{M(s)}a_{X^-(aq)}$, so that from eqn (11.4.3) we obtain

$$\Delta\phi = \Delta\phi^{\ominus} + (RT/F)\ln(1/a_{X^-})$$

because the activities of the pure solids are both unity. This rearranges to

$$\Delta\phi = \Delta\phi^{\ominus} - (RT/F)\ln a_{X^-}. \qquad (11.4.4)$$

As we anticipated, the potential difference depends on the activity of the anions in the solution. For instance, in both the $Cl^-(aq)|AgCl(s)|Ag(s)$ electrode, where the equilibrium is $AgCl(s)+e^- \rightleftharpoons Ag(s)+Cl^-(aq)$ and the *calomel electrode*, $Cl^-(aq)|Hg_2Cl_2(s)|Hg(l)$, where the equilibrium is $\frac{1}{2}Hg_2Cl_2(s)+e^- \rightleftharpoons Hg(l)+Cl^-(aq)$, the potential difference depends on the chloride ion activity.

Example 11.5

Calculate the change that takes place in the potential difference across a $Cl^-(aq)|AgCl(s)|Ag(s)$ electrode when an excess of $0.010\ \text{mol kg}^{-1}$ KCl(aq) is added at 298 K. The mean activity is 1.3×10^{-5} in a saturated silver chloride solution at this temperature.

• *Method*. The potential difference is given by eqn (11.4.4), and depends on the chloride activity. When silver chloride is present alone, the saturated solution has a very low Cl^- activity. When potassium chloride is added the $Cl^-(aq)$ molality rises sharply to $0.010\ \text{mol kg}^{-1}$. Use the data in Table 11.2 to convert this to an activity. Use $R\mathcal{T}/F = 25.69$ mV.

• *Answer*. The mean activity coefficient is 0.906 for $0.010\ \text{mol kg}^{-1}$ KCl(aq), and so $a_{Cl^-} = 0.00906$. The change of potential difference is therefore

$$\Delta\phi_f - \Delta\phi_i = -(R\mathcal{T}/F)\{\ln 9.06\times10^{-3} - \ln 1.3\times10^{-5}\} = -0.17\ \text{V}.$$

• *Comment*. The solubility of silver chloride can be measured by an adaptation of this result, as can the activity of Cl^- used in the calculation. We see how in the next chapter.

• *Exercise*. Calculate the change in potential difference when an $0.05\ \text{mol kg}^{-1}$ KCl(aq) solution is added to a calomel electrode solution at 298 K (for which, initially, $a_{Cl^-} = 8.7\times10^{-7}$). [−0.28 V]

11.4 (c) Oxidation–reduction electrodes

All electrodes involve oxidation and reduction, but the term *oxidation–reduction electrode*, or *redox electrode*, is normally reserved for the case in which a species exists *in solution* in two oxidation states. The equilibrium is

$$Ox + \nu e^- = Red^{\nu-}.$$

where Ox is the oxidized form (e.g. $Fe^{3+}(aq)$) and Red is the reduced form (e.g., Fe^{2+}, in which case $\nu = 1$). The redox electrode is denoted Red, Ox | M, where M is an inert metal making electrical contact with the solution. Examples include Fe^{2+}, Fe^{3+} | Pt in which the equilibrium is $Fe^{3+} + e^- = Fe^{2+}$, and the *hydroquinone electrode* in which the redox equilibrium is **1**$\rightleftharpoons$**2**. The latter models some important biological reactions. The potential difference across the solution | metal interface can be written down immediately from eqn (11.4.3):

1 O=C1C=CC(=O)C=C1 2 HO-C6H4-OH

$$\Delta\phi = \Delta\phi^{\ominus} + (RT/\nu F)\ln(a_{Ox}/a_{Red}), \qquad (11.4.5)$$

where a_{Ox} and a_{Red} are the activities of the species in solution.

The wider importance of the last equation is due to the following alternative interpretation. Instead of treating the potential difference as arising from the redox equilibrium we can imagine controlling the position of equilibrium by modifying the potential difference using an external source of potential. If we can ensure that the interfacial potential difference is $\Delta\phi$, then the concentrations of reduced and oxidized species will adjust so that the equilibrium, as expressed by eqn (11.4.5), is satisfied. This gives us electrical control over equilibria.

11.4 (d) Liquid junctions and membrane potentials

Electric potential differences occur where two ionic solutions are in contact: this gives rise to the *junction potential*. A simple example is the junction between different concentrations of hydrochloric acid. At the junction the mobile hydrogen ions diffuse into the more dilute solution; the bulkier chloride ions follow, but initially do so more slowly. The initially different rates of diffusion result in a potential difference at the junction, the potential settling down to a value such that, after that brief initial period, the ions diffuse at the same rates. The interface is quite difficult to study because it blurs, and we shall see that in practice it is usually better to eliminate the junction potential than to try to assess its magnitude. The ways in which this can be done will be described when we deal with cells in Chapter 12.

The junction between two solutions can be stabilized if there is a membrane dividing them, and then the potential that arises can be calculated quite simply. Consider two solutions containing different concentrations of a salt MX, and let the membrane dividing them be permeable only to M^+. This ion tends to diffuse into the more dilute solution, but the anion X^- cannot follow. As a result, a potential difference is established, Fig. 11.12. The diffusion does not continue indefinitely because the charge separation that it builds retards the migration of the cations, and in due course, when the electrochemical potentials of the cations are the same on each side of the barrier, the system reaches equilibrium. If the solutions are denoted α and β, the equilibrium condition is $\bar{\mu}_\alpha = \bar{\mu}_\beta$, where $\bar{\mu}$ is the electrochemical potential of M^+. In terms of the electric potentials this condition is

$$\mu_\alpha + F\phi_\alpha = \mu_\beta + F\phi_\beta.$$

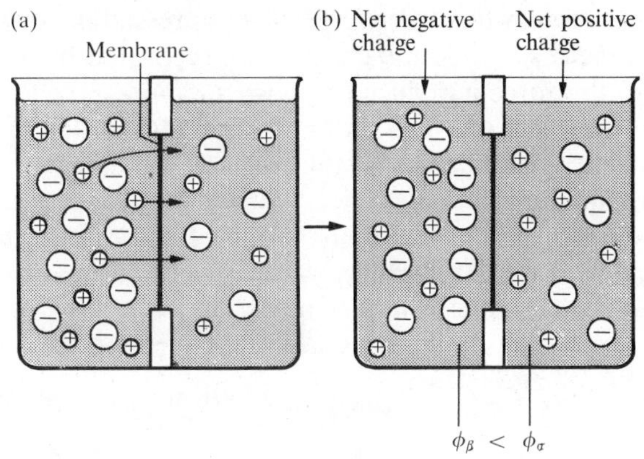

Fig. 11.12. The formation of a membrane potential. (a) There is initially a net migration of cations (in this case) from left to right through the membrane, and so a small charge imbalance is set up. (b) At equilibrium, when a potential difference between the solutions has been established, the rates of passage of the cations are the same in both directions.

Expressing this in terms of the activities of M^+ gives the equilibrium potential difference $\Delta\phi = \phi_\alpha - \phi_\beta$ as

$$\Delta\phi = (RT/F)\ln(a_\beta/a_\alpha), \tag{11.4.6}$$

where a_α is the activity of M^+ in solution α and a_β that in solution β. If β is the more concentrated solution this potential difference is positive: $\phi_\alpha > \phi_\beta$ because positive ions have passed into the solution and have discarded anions in the solution.

One important example of a membrane system that resembles this description is the biological cell wall, which is more permeable to K^+ than to either Na^+ or Cl^-. The concentration of K^+ inside the cell is about 20–30 times that on the outside, and is maintained at that level by a specific pumping operation fuelled by ATP and governed by enzymes. If the system is approximately at equilibrium the potential difference $\phi_{\text{inside}} - \phi_{\text{outside}}$ across the cell wall is predicted to be $\Delta\phi \approx (25.7\ \text{mV})\ln(1/20) = -77\ \text{mV}$, which accords quite well with the measured value. This potential difference plays a particularly interesting role in the transmission of nerve impulses. K^+ and Na^+ pumps occur throughout the nervous system, and when the nerve is inactive there is a high K^+ concentration inside the cells and a high Na^+ concentration outside. The potential difference across the cell wall is around -70 mV. When the cell wall is subjected to a pulse of about 20 mV the structure of the membrane adjusts and it becomes permeable to Na^+. This causes a drop in the membrane potential as the Na^+ ions flood into the interior of the cell. The change in potential difference triggers the neighbouring part of the cell wall, and the pulse of collapsing potential passes along the nerve. Behind the pulse the sodium and potassium pumps restore the concentration difference ready for the next pulse.

Appendix 11.1 Electrostatics

Electrostatic force (units: newton, N). The fundamental equation of electrostatics is

$$F = (q_1 q_2/4\pi\varepsilon_0)(1/r^2),$$

which gives the magnitude of the Coulomb force acting between two charges q_1 and q_2 separated by a distance r in a vacuum. ε_0 is a fundamental constant, the *vacuum permittivity* ($\varepsilon_0 = 8.854 \times 10^{-12}\ \text{J}^{-1}\ \text{C}^2\ \text{m}^{-1}$). The force is obtained in newtons when the charges are expressed in coulombs (C) and the distance in metres. When the charges are separated by a medium other than a vacuum, the force is reduced to $F=(q_1q_2/4\pi\varepsilon)(1/r^2)$, where ε is the *permittivity* of the medium. The latter is normally written $\varepsilon=\varepsilon_r\varepsilon_0$, where the dimensionless quantity ε_r is the *relative permittivity*, or *dielectric constant*, of the medium.

Electrical work (symbol: w_e; units: joule, J). Work is the integral of *force* × *distance*. It follows that the work involved in bringing up a charge q_1 from infinity to a distance r from a charge q_2 is

$$w_e=-\int_\infty^r F\,dr=-(q_1q_2/4\pi\varepsilon_0)\int_\infty^r(1/r^2)\,dr=(q_1q_2/4\pi\varepsilon_0)(1/r).$$

Electric potential (symbol: ϕ; units: volt, V, with $1\,V=1\,J\,C^{-1}$). Doing work of the amount just calculated increases the potential energy (symbol: V) of the system by the same amount. This potential energy is written

$$V=q_1\phi,\quad \text{with}\quad \phi=(q_2/4\pi\varepsilon_0)(1/r).$$

ϕ is called the *potential* arising from the charge q_2. (It is important to distinguish between potential and potential energy.) The particular form quoted is the *Coulomb potential*; in a medium of uniform permittivity it is reduced by a factor ε_r. When the charge distribution is more complex than a single point, it is expressed in terms of a *charge density* (symbol: ρ; units: $C\,m^{-3}$). The electric potential arising from a given charge distribution is the solution of *Poisson's equation*:

$$\nabla^2\phi=-\rho/\varepsilon_0,$$

where $\nabla^2=(\partial^2/\partial x^2)+(\partial^2/\partial y^2)+(\partial^2/\partial z^2)$. If the charge density is spherically symmetrical, so too is the potential it generates, and the equation then simplifies to the form used in the text, preceding eqn (11.2.5).

Electric field strength (symbol: $\boldsymbol{E}$; units: $V\,m^{-1}$). Just as the energy of a charge q_1 can be written $V=q_1\phi$, so the magnitude of the force on q_1 can be written $F=q_1E$, where E is the magnitude of the electric field strength arising from q_2. The electric field strength (which, like the force, is a vector quantity) is the negative gradient of the electric potential:

$$\boldsymbol{E}=-\nabla\phi.$$

∇ is the *gradient*, $\nabla=\hat{\mathbf{i}}(\partial/\partial x)+\hat{\mathbf{j}}(\partial/\partial y)+\hat{\mathbf{k}}(\partial/\partial z)$, with $\hat{\mathbf{i}},\hat{\mathbf{j}},\hat{\mathbf{k}}$ unit vectors along x, y, z.

Further reading

Electrolyte solutions (2nd edn). R. A. Robinson and R. H. Stokes; Academic Press, New York and Butterworth, London, 1959.

Modern electrochemistry. J. O'M. Bockris and A. K. N. Reddy; Plenum, New York, 1970.

The principles of electrochemistry. D. A. MacInnes; Dover, New York. 1961.

The physical chemistry of electrolyte solutions. H. S. Harned and B. B. Owen; Reinhold, New York, 1958.

Ionic solution theory. H. L. Friedman; Wiley-Interscience, New York. 1962.

Treatise on electrochemistry. G. Kortum; Elsevier, Amsterdam, 1952.

Ions, electrodes, and membranes. J. Koryta; Wiley-Interscience, New York, 1982.

Ion-selective electrodes. R. A. Durst; US Government Special Publication 314, 1974.

Introductory problems

A11.1. Calculate the ionic strength of a solution which is 0.04 mol kg^{-1} in $K_3Fe(CN)_6$, 0.03 mol kg^{-1} in KCl, and 0.05 mol kg^{-1} in NaBr.

A11.2. Calculate the ionic strength of a solution formed by mixing 800 g of 1.50 mol kg^{-1} aqueous NaCl solution with 300 g of 1.25 mol kg^{-1} aqueous Na_2SO_4 solution.

A11.3. Calculate the masses of the following salts which must be added to 500 g of 0.15 mol kg^{-1} aqueous KNO_3 solution in order to raise its ionic strength to 0.25 mol kg^{-1}: (a) $Ca(NO_3)_2$, (b) NaCl.

A11.4. The mean activity coefficient of $CaCl_2$ in aqueous solution at a concentration of 2.00 M is 1.554 at 25 °C. Estimate the activity of the chloride ion in this solution.

A11.5. To create 5.00 dm^3 of a 0.065 M aqueous solution of a 1:2 electrolyte by charging the hypothetical solute in the uncharged ideal state at 25 °C would require the expenditure of 1.04 kJ of electrical work. What is the mean activity coefficient of the electrolyte?

A11.6. The proportions of non-electrolytes in a liquid mixture were selected to provide a solvent in which the Debye length at 25 °C for a 2:1 electrolyte with a molality of 0.100 mol kg^{-1} would be 0.400 nm. The density of the mixture used was 0.850 g cm^{-3}. What was the relative permittivity of the mixture?

A11.7. For a 0.50 mol kg^{-1} aqueous solution, the mean activity coefficient of $LaCl_3$ at 25 °C is 0.303. Calculate the % error in the value given by the Debye–Hückel Limiting Law.

A11.8. The mean activity coefficients for HBr in three dilute aqueous solutions at 25 °C are: 0.930 at 0.005 mol kg^{-1}, 0.907 at 0.010 mol kg^{-1}, and 0.879 at 0.020 mol kg^{-1}. Use these data to estimate the value of A^* in the extended Debye–Hückel Law.

A11.9. The solubility product of CaF_2 is 3.9×10^{-11} at 25 °C. The standard molar Gibbs free energy of formation of CaF_2 is -1.162×10^3 kJ mol^{-1}. Calculate the standard molar Gibbs free energy of formation for aqueous CaF_2.

A11.10. Consider a hydrogen gas electrode in an aqueous HBr solution at 25 °C. The presure of hydrogen is maintained at 1.15 atm. Calculate the change in the electrode–solution potential difference when the HBr concentration is changed from 0.005 mol kg^{-1} to 0.020 mol kg^{-1}. Use the data in Problem A11.8 above.

Problems

11.1. The *ionic strength* was defined in eqn (11.2.6). It is a crucial quantity in the discussion of ionic solutions, and the first few Problems give practice in manipulating it. As a first step, find an expression for the ionic strength in terms of the concentration c expressed in mol dm^{-3}.

11.2. Show that the ionic strengths of solutions of KCl, $MgCl_2$, $FeCl_3$, $Al_2(SO_4)_3$, $CuSO_4$, are related to their molalities as follows: $I(\text{KCl})=m/m^{\ominus}$, $I(\text{MgCl}_2)=3m/m^{\ominus}$, $I(\text{FeCl}_3)=6m/m^{\ominus}$, $I[\text{Al}_2(\text{SO}_4)_3]=15m/m^{\ominus}$, $I(\text{CuSO}_4)=4m/m^{\ominus}$.

11.3. Calculate the ionic strength of a solution which is 0.10 mol kg^{-1} in KCl and 0.20 mol kg^{-1} in $CuSO_4$.

11.4. A solution was prepared by adding 5.0 g of KCl and 5.0 g of $FeCl_3$ to 100 g water. What is the ionic strength of the solution?

11.5. What molality of $CuSO_4$ has the same ionic strength as a 1.0 mol kg^{-1} solution of KCl?

11.6. The other set of quantities that appeared through the chapter were the mean activities and mean activity coefficients. Demonstrate, starting from $\mu(\text{M}_p\text{X}_q)=p\mu(\text{M})+q\mu(\text{X})$, that the activity of a solution of a salt M_pX_q can be written $a(\text{M}_p\text{X}_q)=p^pq^q\gamma_\pm^{p+q}m^{p+q}$, where m is its molality (actually $m/m^{\ominus}$) and $\gamma_\pm$ is the mean activity coefficient $(\gamma_+^p\gamma_-^q)^{1/(p+q)}$.

11.7. Express the activities of the salts KCl, $MgCl_2$, $FeCl_3$, $CuSO_4$, and $Al_2(SO_4)_3$ in terms of the molality m and the mean activity coefficients.

11.8. Estimate the electrical work of charging 2.00 dm^3 of a hypothetical uncharged 0.500 M solution of (a) NaCl, with $\gamma_\pm=0.679$ at 25 °C, (b) $In_2(SO_4)_3$, with $\gamma_\pm=0.014$.

11.9. The Coulomb potential has a very long range. Consider a spherical shell of electrons around a single proton, there being one electron in every square centimetre of the surface of the shell. What force does the proton exert on the *entire* shell of electrons, when the latter's radius is (a) 10 cm, (b) 1 m, (c) 10^6 km?

11.10. The shielded Coulomb potential has a much shorter range. Repeat the calculation in Problem 11.9 using the potential in eqn (11.2.24) with $\varepsilon_r=1$ but $r_D=10$ cm.

11.11. The Debye–Hückel theory lets us calculate the shielding length in dilute ionic solutions. Find its value in 0.0010 mol kg^{-1} solution of magnesium iodide at (a) 25 °C, (b) 0 °C.

11.12. The Debye–Hückel theory is not limited to water as solvent, but the latter's high relative permittivity and consequent suppression of the Coulombic forces between ions means that it is applicable at higher concentrations. What is the shielding length in the case of a 0.0010 mol kg^{-1} solution of magnesium iodide in liquid ammonia at −33 °C. The solvent density is 0.69 g cm^{-3}, and $\varepsilon_r=22$.

11.13. Calculate the value of the Debye–Hückel A-coefficient for solutions in liquid ammonia at $-33\,°C$, $\rho(NH_3) = 0.69\ g\ cm^{-3}$ and $\varepsilon_r = 22$ at this temperature.

11.14. Calculate the mean activity coefficients for aqueous solutions of NaCl at 25 °C at the molalities 0.001, 0.002, 0.005, 0.010, 0.020 mol kg^{-1}. The experimental values are 0.9649, 0.9519, 0.9275, 0.9024, and 0.8712. By plotting $\lg \gamma_\pm$ against $I^{1/2}$ confirm that the Debye–Hückel law gives the correct limiting behaviour.

11.15. But how are the activity coefficients actually measured? One of the most important methods makes use of the electrochemical cell, and we shall meet that in the next chapter. In the next few Problems, however, we shall see some alternative methods. Several methods hinge on the Gibbs–Duhem equation (eqn (8.1.5)) and so we begin by doing some manipulations involving it. The Gibbs–Duhem equation links the chemical potentials, and therefore the activities, of solvent and solute. It follows that the vapour pressure (and hence the colligative properties) of the solvent should depend on the activities of the ions it contains. As a first step in finding the connection, repeat the freezing point depression calculation in terms of the Gibbs–Helmholtz equation, Problem 8.25, but do not assume that the solution is ideal. Derive the expression $\mathrm{d}\ln a_A = (M_{A,m}/K_f)\,\mathrm{d}\delta T$, where δT is the depression, $M_{A,r}$ the RMM of the solvent, and a_A its activity. Use the Gibbs–Duhem equation to deduce that $\mathrm{d}\ln a_B = (-1/m_B K_f)\,\mathrm{d}\delta T$ where a_B is the activity of the solute and m_B its molality.

11.16. The *osmotic coefficient* for a 1:1 electrolyte is defined as $\phi = \delta T/2mK_f$, where m is the solute molality. It is important to remember in what follows that δT depends on m, and that $\phi \to 1$ in infinitely dilute solutions. There are two ways of dealing with the equation $\mathrm{d}\ln a = (-1/mK_f)\,\mathrm{d}\delta T$ derived above. The first assumes that the activity coefficient for the solute is given by the Debye–Hückel Limiting Law. Assume that this is so, and hence show that the osmotic coefficient at some molality m is given by $\phi = 1 - \frac{1}{3}A'm^{\frac{1}{2}}$, where $A' = 2.303A$.

11.17. The cryoscopic constant of water is 1.858 K/(mol kg^{-1}). When sodium chloride was dissolved in water the following feezing point depressions were observed. Confirm that the correct limiting law behaviour is obeyed at low concentrations (by plotting ϕ against $m^{\frac{1}{2}}$).

m/mol kg^{-1}	0.001	0.002	0.005	0.010	0.020
$10^3\,\delta T$/K	3.696	7.376	18.36	36.43	72.5

11.18. The other way of using the result of Problem 11.15 involves setting it up in such a way that it can be used to *measure* the activity coefficients, and not merely to test some theory about them. Begin by finding $\mathrm{d}\phi/\mathrm{d}m$, and then deducing that

$$-\ln \gamma_\pm = 1 - \phi(m) + \int_0^m \left(\frac{1-\phi(m)}{m}\right)\mathrm{d}m.$$

11.19. The following freezing point depressions were observed when KCl was dissolved in water. What is the mean activity coefficient for KCl in 0.050 mol kg^{-1} solution? Use the expression just derived, but take the limiting law value of ϕ (Problem 11.16) for the integration $m = 0$ to 0.010 mol kg^{-1}.

m/mol kg^{-1}	0.010	0.020	0.030	0.040	0.050
δT/K	0.0355	0.0697	0.1031	0.137	0.172

11.20. The *ionization constant* of an acid electrolyte MA is defined as $K_a = \{a(M^+)a(A^-)/a(MA)\}_{eq}$. The degree of ionization α of acetic acid was measured at 25 °C over a range of concentrations, the number of ions present being determined by measuring the conductivity of the solution. Use the data reported below to confirm that the Debye–Hückel Limiting Law correctly predicts the limiting behaviour of γ at low concentrations by demonstrating that $\lg K_c$ plotted against $\surd(\alpha c)$, where K_c is the ionization constant in terms of concentrations and c is the concentration of MA, should be a straight line.

$10^3\,c$/M	0.0280	0.1114	0.2184	1.0283	2.414	5.9115
α	0.5393	0.3277	0.2477	0.1238	0.0829	0.0540

11.21. The pK_a value of an acid is defined as $-\lg K_a$, where K_a is the ionization constant (in terms of activities). Use the Debye–Hückel Limiting Law to estimate the value of pK_a', where the prime indicates the ionization constant in terms of the concentration, for 0.10 M acetic acid at 25 °C given that its value at zero ionic strength is 4.756.

11.22. One of the properties of a salt that is modified by ionic interactions of various kinds is its solubility. The next few Problems explore some of these effects. The *solubility product*, K_{sp}, of a salt M_pX_q that dissolves to give the ions $M^{+|z_+|}$ and $X^{-|z_-|}$ is defined as $K_{sp} = \{a_+^p a_-^q\}_{eq}$, the activities being those of the ions in the saturated solution. Express K_{sp} in terms of the molality m and the mean activity coefficient of the ions in the saturated solution. In very dilute solutions (as of sparingly soluble salts) it may be possible to set the activity coefficients equal to unity. Assume this is the case for AgCl and $BaSO_4$ in water, and calculate their solubility products on the basis that their saturated solutions are found to have concentrations 1.34×10^{-5} M and 9.51×10^{-4} M, respectively.

11.23. The solubility product is an equilibrium constant, and so it can be calculated from tables of thermodynamic data. Calculate the solubility of AgBr in water at $\mathcal{T}$ using data from Table 4.1 and $\gamma_\pm \approx 1$.

11.24. The solubility of a salt is modified if the solution contains another salt with one ion in common: this is the *common-ion effect*, and was described in Section 9.4(c). Now we can deal with it quantitatively. Suppose that a 1:1 salt MX has a solubility product K_{sp} and that the saturated solution is so dilute that $\gamma_\pm \approx 1$; let the saturated concentration be c_0. Now consider the saturated solubility in a solution containing a concentration c of some freely soluble and fully dissociated 1:1 salt NX. Show that the solubility of MX shifts to $c_0' = \frac{1}{2}(c^2 + 4K_{sp})^{\frac{1}{2}} - \frac{1}{2}c$.

11.25. Investigate the effect of taking into account the deviations from ideality. Suppose that the ionic strength is dominated by the added freely soluble salt, but that it is still possible to use the Debye–Hückel Limiting Law. Show that $c_0' = K_{sp}10^{2A\sqrt{c}}/c$.

11.26. Calculate the solubility of AgBr in water containing of 0.01 M KBr; take account of activity coefficients.

11.27. The solubility of a salt is modified by the presence of another salt even though there are no ions in common because the addition of ions changes the ionic strength of the solution, and therefore the activity coefficients are changed too. The solubility product itself remains unchanged, but its relation to concentrations is modified. Derive an expression for the solubility of a sparingly soluble 1:1 salt MX.

11.28. What is the solubility of AgCl in the following solutions at 25 °C: (a) 0.1 M KCl, (b) 0.01 M KCl, (c) 0.01 M KNO_3? Take account of activity coefficients (Problem 11.25). Evaluate K_{sp} from Table 4.1.

11.29. Another way of determining activity coefficients is to write $K_{sp} = K_{sp}'K_\gamma$, where K_{sp}' is the solubility product expressed in molalities and K_γ the appropriate combination of activity coefficients, to measure the solubility at some concentration, and then, knowing K_{sp}, to find K_γ. The solubility of AgCl in aqueous magnesium sulphate was measured at 25 °C with the following results:

$m/(\text{mol MgSO}_4/\text{kg})$	0.001	0.002	0.003
$10^5 m'/(\text{mol AgCl/kg})$	1.437	1.482	1.547
$m/(\text{mol MgSO}_4/\text{kg}$	0.004	0.006	0.010
$10^5 m'/(\text{mol AgCl/kg})$	1.575	1.598	1.650

where m' is the saturation molality. Find the thermodynamic solubility product (by extrapolation to zero ionic strength) and the mean ionic activity coefficient when $m' = 0.004\ \text{mol kg}^{-1}$.

11.30. What is the difference in electrochemical potential when a Cu^{+2} ion moves from one region of a solution to another, the electrical potential difference being 2.0 V? Show that negative ions tend to move to regions of relatively positive potential.

11.31. The pH of a solution is defined as $\text{pH} = -\lg a(\text{H}^+)$, and we explore this quantity in more detail in the next chapter. Show that the potential difference across a Pt | H_2(g) electrode is proportional to the pH of the solution. By how much does the potential difference change when the electrode is transferred from a strongly acid solution (pH ≈ 0) to a strongly alkaline solution (pH ≈ 14), both at 298 K.

11.32. Devise an expression for the potential difference across a gas | metal electrode at which the reaction is $\frac{1}{2}G_2(g) + ze^- \rightleftharpoons G^{z-}(aq)$. How does the potential difference depend on the pressure of G_2? Does the potential difference increase or decrease when the pressure is raised? Account physically for your answer.

11.33. The antimony|antimony oxide electrode $Sb|Sb_2O_3|OH^-$ is reversible with respect to hydroxide ions and so it has played an important role in electrochemical measurements. Set up the expression for the dependence of the potential difference across the electrode solution interface, in terms of the OH^- activity in the solution. By how much does the interfacial potential difference change when the NaOH molality is changed from $0.01\ \text{mol kg}^{-1}$ to $0.05\ \text{mol kg}^{-1}$? Use the Debye–Hückel Limiting Law to estimate the activity coefficients you require.

11.34. Consider an electrode that responds to the equilibrium between chromium (III) ions and chromium (VI) oxide ions, $Cr_2O_7^{2-}$, according to the reaction

$$Cr_2O_7^{2-}(aq) + 14H^+(aq) + 6e^- \rightleftharpoons 2Cr^{3+}(aq) + 7H_2O.$$

Derive an expression for the potential difference across the electrode interface.

12

Equilibrium electrochemistry: electrochemical cells

Learning objectives

After careful study of this chapter you should be able to:

(1) Describe the construction of *electrochemical cells*, and distinguish between cells with and without a liquid junction, Section 12.1(a).

(2) Indicate how a *salt bridge* minimizes a junction potential, Section 12.1(a).

(3) Define the *electromotive force* (e.m.f.) of a cell and explain how it is measured, Section 12.1(c).

(4) Write the *cell reaction* for a given cell, and devise a cell that corresponds to a given cell reaction, Section 12.1(d) and Box 12.1.

(5) Explain the *sign convention*, and relate the e.m.f. of a cell to the spontaneous direction of the cell reaction, Section 12.1(e) and Box 12.1.

(6) Derive and use the *Nernst equation*, eqn (12.1.6), for the concentration dependence of the e.m.f. of a cell.

(7) Define the term *standard e.m.f.* of a cell, Section 12.2, and relate it to the equilibrium constant of the cell reaction, eqn (12.1.9) and Box 12.2.

(8) Define *standard electrode potential*, and explain how values may be measured, Section 12.2(b).

(9) Explain the significance of the value of the standard potential of a couple, and use the *electrochemical series*, Example 12.4.

(10) Explain how the mean activity coefficients of ions may be measured, Section 12.2(c).

(11) Relate the e.m.f. of a cell to the Gibbs function of the cell reaction, eqn (12.3.1).

(12) Explain the convention used for reporting the *standard Gibbs function of formation* of ions in solution, Section 12.3(c), and state the *Born equation*, eqn (12.3.3).

(13) Relate the *temperature-dependence of e.m.f.* of a cell to the cell reaction entropy, eqn (12.3.4), and explain the convention for reporting *standard partial molar ion entropies*, Section 12.3(c).

(14) Define the *solubility-product constant*, and calculate its value from electrode potentials, Section 12.4(a).

(15) Describe the electrochemical basis of *potentiometric titrations*, Section 12.4(b).

(16) State the *Brønsted–Lowry classification* of acids and bases, and define the pK_a of an acid or base and the pH of a solution, Section 12.4(c).

(17) Describe the measurement of pH, Section 12.4(d).

(18) Derive expressions for the pH in the course of weak acid/strong base and strong acid/weak base titrations, eqns (12.4.11–15).

(19) Explain the electrochemical basis of *indicator detection* of the end-points of titrations, Section 12.4(f), and predict the pH at the end-point, eqns (12.4.13–14).

(20) Explain the principles of *buffer action*, Section 12.4(f), and calculate the pH of a buffer solution using the *Henderson–Hasselbalch equation*, eqn (12.4.12).

Introduction

In this chapter we examine the information that can be obtained by studying the potential difference between two electrodes combined to form an *electrochemical cell*. We see that it is possible to discuss the equilibria of reactions in terms of the electrode potentials of the redox processes into which they may be analysed. We also see how to use electrochemical measurements to obtain values of the Gibbs functions, enthalpies, and entropies of reactions. Finally, we consider applications of electrochemistry to typical chemical problems, such as the determination of equilibrium constants and the strengths of acids and bases, and analysis.

12.1 Electrochemical cells

An electrochemical cell consists of two electrodes dipping into an electrolyte. When it produces electricity it is called a *galvanic cell*; when the reaction inside is driven by an external source of current it is called an *electrolytic cell*. In this chapter we consider the cell as a producer of electricity.

12.1 (a) Kinds of cell

The simplest type of cell has a single electrolyte common to both electrodes, Fig. 12.1. In this case the potential difference across the terminals is the algebraic sum of the potential differences across the two electrode/electrolyte interfaces (but we shall be more precise shortly). In some cases it is necessary to immerse the electrodes in different electrolytes, as in the *Daniell cell*, Fig. 12.2(a): the copper electrode is reversible with respect to Cu^{2+} ions, and the zinc electrode is reversible with respect to Zn^{2+}. In this case there is an additional source of potential difference at the junction of the two liquids: the Daniell cell is an example of a *cell with a liquid junction*.

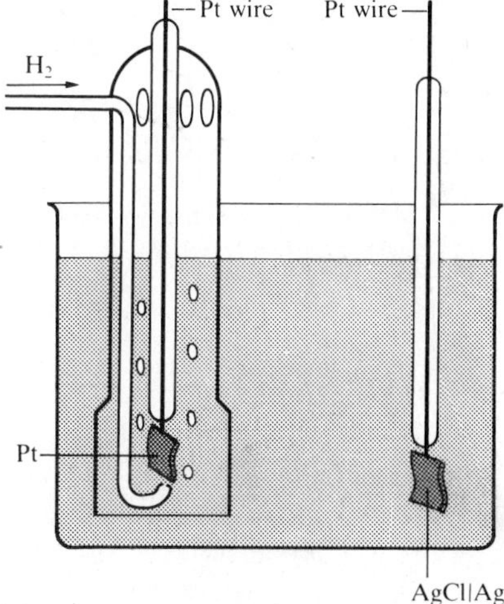

Fig. 12.1. A simple electrochemical cell without a liquid junction. Hydrogen is bubbled over the platinum electrode, which shares a common electrolyte (hydrochloric acid) with the other electrode (a silver/silver chloride electrode).

The contribution of the liquid junction to the potential can be reduced (to about 1–2 mV) by joining the electrolyte compartments through a *salt bridge*, a saturated solution (usually of potassium chloride) in agar jelly, Fig. 12.2(b). The reason for the success of the salt bridge is that (as we saw in Section 11.4), the diffusion of ions across a liquid junction gives rise to a potential difference, the *liquid junction potential* (symbol: E_{lj}). Since the salt is so concentrated (about 4 M), E_{lj} is determined largely by the KCl diffusing outwards and is barely affected by the ions of the less concentrated electrolyte solutions diffusing inwards. Moreover, since K^+ and Cl^- diffuse at similar rates (being almost the same size), the liquid junction potential they give rise to is in fact quite small. Finally, since the bridge introduces two similar, small, junction potentials in opposite directions, these partially cancel each other.

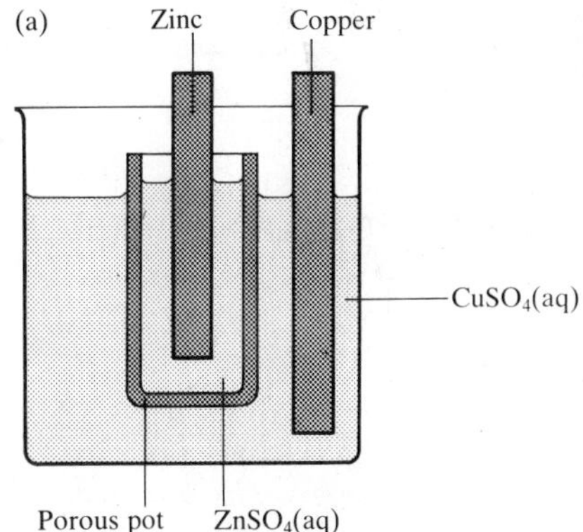

Fig. 12.2. A form of the *Daniell cell*. (a) In this form there is a liquid junction potential at the $ZnSO_4(aq)\,|\,CuSO_4(aq)$ phase boundary. (b, overleaf) The salt bridge, essentially an inverted U-tube full of concentrated salt solution in a jelly, minimizes the liquid junction potential.

The last kind of galvanic cell we shall consider is a *concentration cell*, Fig. 12.3, in which the halves are identical apart from the electrolytes having different concentrations. Concentration cells may or may not have a liquid junction.

12.1 (b) Notation

The notation for cells is based on that for the individual electrodes. A phase boundary is denoted by a vertical bar. For example, the cell in Fig. 12.1 is

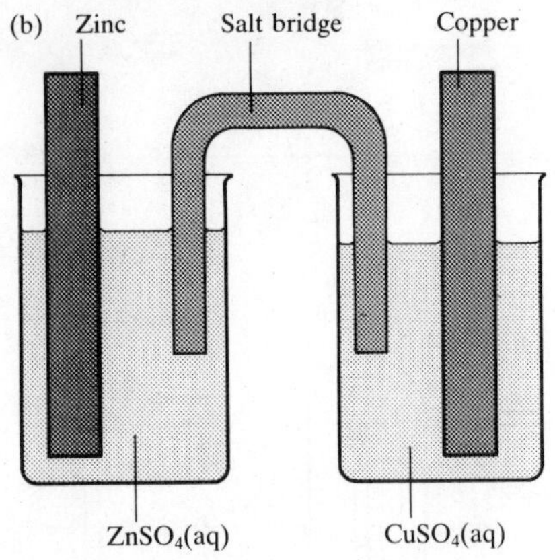

Fig. 12.2—(Continued)

denoted

$$\mathrm{Pt} \mid \mathrm{H_2}(\mathrm{g}, p) \mid \mathrm{HCl}(\mathrm{aq}, m) \mid \mathrm{AgCl}(\mathrm{s}) \mid \mathrm{Ag},$$

where p is the pressure of the hydrogen and m the molality of the solution. A cell with a liquid junction, such as the one in Fig. 12.2(a), would be denoted

$$\mathrm{Zn}(\mathrm{s}) \mid \mathrm{ZnSO_4}(\mathrm{aq}, m_1) \mid \mathrm{CuSO_4}(\mathrm{aq}, m_2) \mid \mathrm{Cu}(\mathrm{s}).$$

The elimination of a liquid junction is denoted $\|$, and so the cell without liquid junction in Fig. 12.2(b) is denoted

$$\mathrm{Zn}(\mathrm{s}) \mid \mathrm{ZnSO_4}(\mathrm{aq}, m_1) \,\|\, \mathrm{CuSO_4}(\mathrm{aq}, m_2) \mid \mathrm{Cu}(\mathrm{s}).$$

A concentration cell without a liquid junction, such as that in Fig. 12.3, is denoted

$$\mathrm{Pt} \mid \mathrm{H_2}(\mathrm{g}, p) \mid \mathrm{HCl}(\mathrm{aq}, m_1) \,\|\, \mathrm{HCl}(\mathrm{aq}, m_2) \mid \mathrm{H_2}(\mathrm{g}, p) \mid \mathrm{Pt}.$$

In many cases, the representations of cells can be abbreviated without ambiguity.

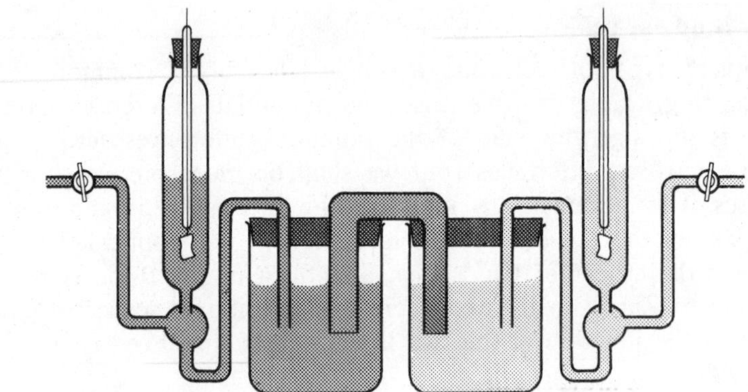

Fig. 12.3. In a concentration cell, the electrolyte (e.g. hydrochloric acid) is present at two different concentrations, but the electrodes are otherwise the same. This is a concentration cell without a liquid junction.

12.1 (c) Measurements

The fundamental observation on a cell is the measurement of the potential difference (PD) between the two electrodes, but this remark has to be interpreted with care. In the first place, we are dealing with *equilibrium* electrochemistry. If a current is allowed to flow, the concentrations of the components change, and the current ceases when the reaction reaches equilibrium (the cell is then 'exhausted'). The description of a changing system is outside the scope of equilibrium thermodynamics, and so we have to measure the PD when the cell is held at a constant composition. Therefore, it is necessary to make the measurement without allowing current to flow. This used to be done using a potentiometer, a device for comparing the PD of the cell with the PD of a standard cell such as the *Weston standard cell*:

$$\mathrm{Cd(Hg)} \mid \mathrm{CdSO_4}(\mathrm{aq, satd.}) \mid \mathrm{Hg_2SO_4}(\mathrm{s}) \mid \mathrm{Hg}.$$

The Weston cell's PD, 1.018 07 V at 298 K, is reproducible and depends only weakly on temperature. However, potentiometers have been superseded, and now an electronic digital voltmeter, which draws negligible current, is used to measure the potential difference directly.

The next point to establish is that the measured PD relates to a *reversible* process in the cell. Sources of irreversibility include the slowness with which the redox reactions take place at the electrodes, and therefore the slowness of the response of these reactions to changes of potential. This suggests one way of testing for reversibility using an external d.c. supply to balance the cell's PD, and an electronic microammeter or a multimeter in its current mode. If the electrode reactions are slow, then little current flows when the cell is slightly off balance, and so a sign of irreversibility is that very little current is produced near the balance point. Reversible cells, where the electrode reactions are rapid and responsive to changes of potential, often give appreciable currents when just off balance.

These remarks can be brought together in the following definition. By *electromotive force* (e.m.f.; symbol: E) we shall mean *the potential difference of a cell measured when there is no flow of current and the cell is operating reversibly*.

12.1 (d) E.m.f. and electrode potentials

Figure 12.4 illustrates the contributions to the e.m.f. of a cell without a liquid junction which we can designate in general

$$(\mathrm{Pt})\,|\,\mathrm{Red_L},\mathrm{Ox_L}\,||\,\mathrm{Red_R},\mathrm{Ox_R}\,|\,(\mathrm{Pt}).$$

the platinum electrodes being required, for instance, in a gas | inert metal electrode. For example, in the Daniell cell, the reduced and oxidized species are identified by comparing this general form of the cell with the specification

$$\mathrm{Zn(s)}\,|\,\mathrm{ZnSO_4}(\mathrm{aq},m_1)\,||\,\mathrm{CuSO_4}(\mathrm{aq},m_2)\,|\,\mathrm{Cu(s)}.$$

so that:

Red_L	Ox_L	Red_R	Ox_R
Zn(s)	Zn^{2+}(aq)	Cu(s)	Cu^{2+}(aq)

From the illustration, $E = \phi_{\mathrm{metal,R}} - \phi_{\mathrm{metal,L}}$. The overall PD can be expressed in terms of the PD at each interface:

$$E = \phi_{\mathrm{R,metal}} - \phi_{\mathrm{L,metal}} = \phi_{\mathrm{R,metal}} - \phi_{\mathrm{solution}} + \phi_{\mathrm{solution}} - \phi_{\mathrm{L,metal}}$$
$$= E_{\mathrm{R}} - E_{\mathrm{L}},$$

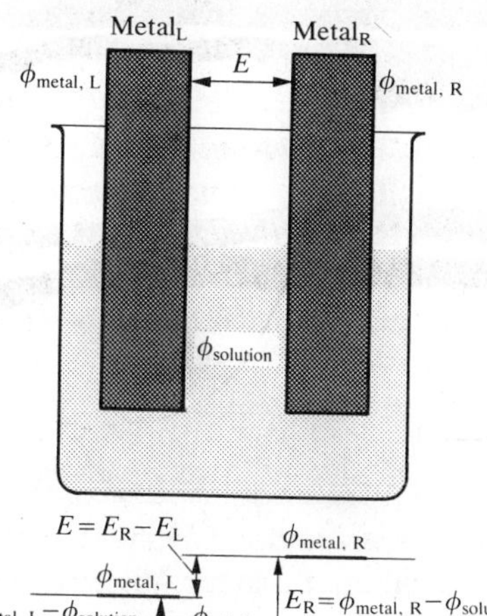

Fig. 12.4. The electric potentials and the potential differences that contribute to the e.m.f. of a cell. The latter is the potential of the right-hand electrode minus the potential of the left-hand electrode.

where $E_R = \phi_{metal,R} - \phi_{solution,R}$ (and similarly E_L) is an *electrode potential*. The labels L (left) and R (right) refer to the electrodes as they are written in the cell description. This means that reporting the e.m.f. of the cell Pt | H_2 | HCl(aq) | AgCl(s) | Ag | Pt, or more briefly H_2 | HCl(aq) | AgCl | Ag, as +0.2 V indicates that the Cl^- | AgCl | Ag electrode has a potential that is positive relative to the other electrode. The same cell written Ag | AgCl | HCl(aq) | H_2 would be found to have an e.m.f. equal to −0.2 V, indicating that the H^+ | H_2 | Pt electrode is negative relative to the other electrode. The point to appreciate is that *the sign of the e.m.f. always refers to the potential difference RIGHT–LEFT between the electrodes of the cell as written*.

12.1 (e) The sign convention

In a cell with E positive, the right-hand electrode has a higher electrical potential than the left ($E_R > E_L$). This indicates that it is positively charged relative to the left. This, in turn, indicates that if a circuit were completed between the terminals, electrons would flow from left to right, Fig. 12.5. (Electric current is conventionally regarded as the flow of a positively charged substance, and so the current flow is opposite to the electron flow. This convention is confusing, and so we shall always express processes in terms of the flow of electrons.)

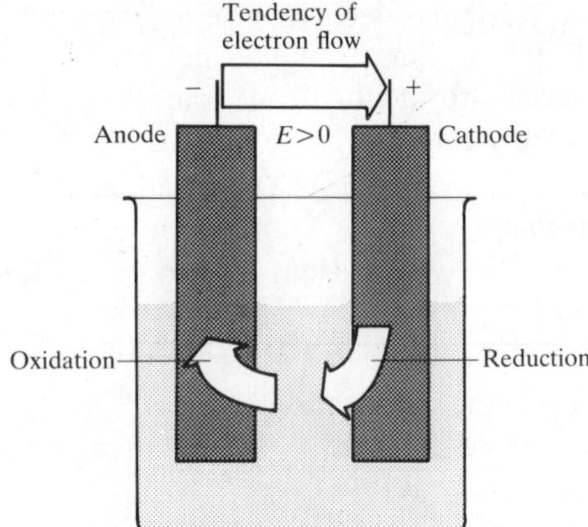

Fig. 12.5. The implication of the sign convention. When $E > 0$ there is a *tendency* for electrons to flow from left to right in the external circuit, and to bring about reduction at the right-hand electrode. In an e.m.f. measurement the current is not allowed actually to flow (except to a negligible extent).

A positive e.m.f. signifies a deficiency of electrons in the right-hand electrode. This deficiency is caused by the reaction taking place there tending to extract electrons from the electrode. Therefore, *$E > 0$ indicates a tendency for reduction*, $Ox_R + e^- \rightarrow Red_R$, to occur at the right-hand electrode, and for that electrode (by definition, Box 11.2) to be the *cathode*. The relative excess of electrons at the left-hand electrode (when $E > 0$) indicates that the tendency of the reaction there is oxidation, $Red_L \rightarrow Ox_L + e^-$. Overall, therefore, a positive e.m.f. arises from the tendency of the reaction to occur as follows:

$$E > 0: \quad Red_L + Ox_R \rightarrow Red_R + Ox_L. \qquad (12.1.1)$$

The positive e.m.f. indicates that the reaction has a *tendency* to run in the direction indicated; but only if the external circuit is completed, and the electrons that are released at the left electrode are delivered to the other, will the reaction proceed in the direction indicated. Conversely, if the cell has a negative e.m.f., then we know that the tendency of the reaction is in the opposite direction:

$$E<0: \quad \mathrm{Red_L} + \mathrm{Ox_R} \leftarrow \mathrm{Red_R} + \mathrm{Ox_L}. \tag{12.1.2}$$

This analysis suggests a simple scheme (Box 12.1) for relating the sign of a cell's e.m.f. to the tendency of the overall reaction.

Box 12.1 Cells and cell reactions

The *cell reaction* is always written in the 'standard form'

$$\mathrm{Red_L} + \mathrm{Ox_R} \rightarrow \mathrm{Red_R} + \mathrm{Ox_L},$$

where at this stage L and R have no particular significance. The Red and Ox species of the reaction of interest should be identified by comparing the reaction with this 'standard form'.

The corresponding cell is then constructed from these identified components:

$$(\mathrm{Pt})\,|\,\mathrm{Red_L}, \mathrm{Ox_L}\,\|\,\mathrm{Red_R}, \mathrm{Ox_R}\,|\,(\mathrm{Pt}).$$

The labels L and R now signify the left and right electrodes of the cell as written.

In order to write the cell reaction for given cell, reverse this procedure.

The e.m.f. of the cell is $E = E_R - E_L$.

The reaction is spontaneous as written (i.e., to the right) if $E > 0$.

The *half-cell reactions* are written as reductions:

$$\mathrm{Ox_L} + \nu e^- \rightarrow \mathrm{Red_L}, \qquad \mathrm{Ox_R} + \nu e^- \rightarrow \mathrm{Red_R},$$

which identifies the value of ν. The overall cell reaction is R–L.

The e.m.f. at a general composition is given by the Nernst equation, eqn (12.1.6).

Example 12.1

One of the reactions important in corrosion in an acid environment is $\mathrm{Fe(s)} + 2\mathrm{HCl(aq)} + \frac{1}{2}\mathrm{O_2(g)} \rightarrow \mathrm{FeCl_2(aq)} + \mathrm{H_2O(l)}$. Which is the spontaneous direction of this reaction when the activities of Fe^{2+} and H^+ are unity?

- *Method*. The general procedure for calculations of this kind is as follows:

(1) Identify the Red and Ox species on the left and right of the reaction, written in the 'standard form'

$$\mathrm{Red_L} + \mathrm{Ox_R} \rightarrow \mathrm{Red_R} + \mathrm{Ox_L}.$$

(2) Consider the cell written in its corresponding 'standard form'

$$(\mathrm{Pt})\,|\,\mathrm{Red_L}, \mathrm{Ox_L}\,\|\,\mathrm{Red_R}, \mathrm{Ox_R}\,|\,(\mathrm{Pt}).$$

(3) Find the electrode potentials from Table 12.1 (below). (Since the activities are unity, in the present case, use the standard values given in the table without further modification.)

(4) Form $E = E_R - E_L$. If $E > 0$, then the reaction is spontaneous L→R as written.

- *Answer*. (1) The Red/Ox identification is as follows:

$\mathrm{Red_L}$	$\mathrm{Ox_R}$	$\mathrm{Red_R}$	$\mathrm{Ox_L}$
Fe(s)	$2\mathrm{H^+(aq)} + \frac{1}{2}\mathrm{O_2(g)}$	$\mathrm{H_2O(l)}$	$\mathrm{Fe^{2+}(aq)}$.

(2) The cell is therefore

$$Fe(s)\,|\,Fe^{2+}(aq)\,||\,H^{+}(aq)\,|\,O_2(g)\,|\,Pt.$$

(3) The electrode potentials at unit activity are

$$E_{Fe^{3+},Fe^{2+}} = -0.44\ V; \qquad E_{H^{+},O_2,H_2O} = 1.23\ V.$$

(4) $E = (1.23\ V) - (-0.44\ V) = +1.67\ V$. Therefore, the reaction is spontaneous as written.

• *Comment*. The conclusion relates to unit activities of the ions. When the reaction has progressed there comes a stage when the activities, and therefore the electrode potentials, have changed so much that $E = 0$ and there is no longer a thermodynamic tendency of the reaction to the right. Then net reaction ceases. Read on.

• *Exercise*. Can zinc displace copper from solution when the ions are at unit activity?
[$Zn(s) + Cu^{2+}(aq) \rightarrow Zn^{2+}(aq) + Cu(s)$]

The e.m.f. of a cell depends on the composition of the electrolytes, and in particular on the molalities of the reactants. Suppose the molalities are adjusted so that $E = 0$, then the cell reaction has no tendency to run to either left or right. This means that the reaction is at equilibrium. This is a very important conclusion, because if we can predict the molalities that correspond to $E = 0$, then we can predict the equilibrium constant of the cell reaction. The next section explores this point.

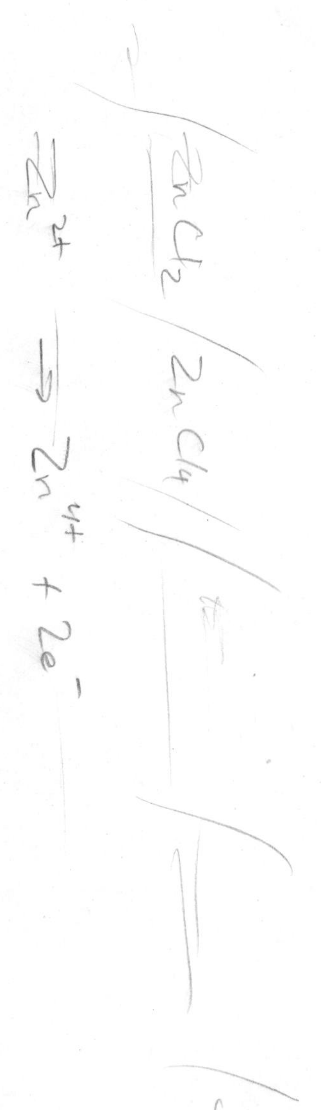

12.1 (f) The concentration dependence of the e.m.f.

Consider the 'standard form' of cell

$$(Pt)\,|\,Red_L, Ox_L\,||\,Red_R, Ox_R\,|\,(Pt).$$

and the corresponding 'standard form' of the cell reaction:

$$Red_L + Ox_R \rightarrow Red_R + Ox_L. \tag{12.1.3}$$

The overall reaction may be expressed in terms of the *difference* R–L of the *half-cell reactions* with both written as reductions:

$$L:\ Ox_L + \nu e^- \rightarrow Red_L, \qquad R:\ Ox_R + \nu e^- \rightarrow Red_R. \tag{12.1.4}$$

ν is the number of electrons transferred in the overall reaction. An example is the reaction of the Daniell cell:

$$Zn(s)\,|\,ZnSO_4(aq, m_1)\,||\,CuSO_4(aq, m_2)\,|\,Cu(s).$$

Red_L	Ox_L	Red_R	Ox_R
$Zn(s)$	$Zn^{2+}(aq)$	$Cu(s)$	$Cu^{2+}(aq)$

$$L:\ Zn^{2+}(aq) + 2e^- \rightarrow Zn(s), \qquad R:\ Cu^{2+}(aq) + 2e^- \rightarrow Cu(s);$$
$$R\text{–}L:\ Zn(s) + Cu^{2+}(aq) \rightarrow Zn^{2+}(aq) + Cu(s), \qquad \nu = 2.$$

The formalism readily includes more complicated reactions, such as the one taking place in the lead–acid battery;

$$Pb(s)\,|\,PbSO_4(s)\,|\,H_2SO_4(aq)\,|\,PbSO_4(s)\,|\,PbO_2(s)\,|\,Pb(s)$$

Red_L	Ox_L	Red_R	Ox_R
$Pb(s)$	$Pb^{II}(s)$	$Pb^{II}(s)$	$Pb^{IV}(s)$

$$\text{R: } Pb^{4+}(s) + 2e^- \rightarrow Pb^{2+}(s),$$
$$\text{or: } PbO_2(s) + SO_4^{2-}(aq) + 4H^+(aq) \rightarrow PbSO_4(s) + 2H_2O(l),$$
$$\text{L: } Pb^{2+}(s) + 2e^- \rightarrow Pb(s),$$
$$\text{or: } PbSO_4(s) + 2e^- \rightarrow Pb(s) + SO_4^{2-}(aq),$$
$$\text{R} - \text{L: } Pb(s) + PbO_2(s) + 2H_2SO_4(aq) \rightarrow 2PbSO_4(s) + 2H_2O(l),$$

and $\nu = 2$. More examples of the general interpretation of the redox reaction will be given shortly.

The PD across each electrode/electrolyte interface is given by eqn (11.4.3):

$$\Delta\phi = \Delta\phi^{\ominus} + (RT/\nu F)\ln(\text{Ox}/\text{Red}), \tag{12.1.5}$$

where Ox and Red denote the appropriate combination of activities for each half-cell reaction. Therefore, taking the difference of two expressions like this,

$$\Delta\phi_R - \Delta\phi_L = \{\Delta\phi_R^{\ominus} + (RT/\nu F)\ln(\text{Ox}_R/\text{Red}_R)\} - \{\Delta\phi_L^{\ominus} + (RT/\nu F)\ln(\text{Ox}_L/\text{Red}_L)\},$$

we obtain the *Nernst equation* for the cell e.m.f.:

$$E = E^{\ominus} - (RT/\nu F)\ln\left\{\frac{\text{Ox}_L\text{Red}_R}{\text{Red}_L\text{Ox}_R}\right\} \tag{12.1.6}$$

$E^{\ominus}$ is the *standard e.m.f.* of the cell, its e.m.f. when all the species are in their standard states at the specified temperature. The simplest way of remembering the Nernst equation is in the form:

$$E = E^{\ominus} - (RT/\nu F)\ln Q, \tag{12.1.7a}$$

with Q the reaction quotient (Section 10.1) for the cell reaction written in its 'standard form':

$$Q = \text{Ox}_L\text{Red}_R/\text{Red}_L\text{Ox}_R. \tag{12.1.7b}$$

Example 12.2

At what activity of $Fe^{2+}(aq)$ does iron stop dissolving in acid saturated with oxygen at 1 atm pressure when the hydrogen ion activity is 0.1 and the temperature 298 K?

• *Method*. This develops *Example* 12.1. Write down the Nernst equation for the cell specified there, then look for the value of $a_{Fe^{2+}}$ for which $E = 0$ (denoting the stage when the reaction is no longer spontaneous). Identify the value of ν by writing the half-cell reactions.

• *Answer*. The cell components are

Red_L	Ox_R	Red_R	Ox_L
Fe(s)	$H^+(aq) + \frac{1}{2}O_2(g)$	$H_2O(l)$	$Fe^{2+}(aq)$

and so the half-reactions are

$$\text{R: } 2H^+(aq) + \tfrac{1}{2}O_2(g) + 2e^- \rightarrow H_2O(l) \qquad E^{\ominus} = +1.23\text{ V}$$
$$\text{L: } Fe^{2+}(aq) + 2e^- \rightarrow Fe(s), \qquad E^{\ominus} = -0.44\text{ V}.$$

The overall reaction and its reaction quotient are:

$$\text{R–L: } Fe(s) + 2H^+(aq) + \tfrac{1}{2}O_2(g) \rightarrow Fe^{2+}(aq) + H_2O(l),$$
$$Q = a_{Fe^{2+}}a_{H_2O}/a_{Fe}a_{H^+}^2 f^{\frac{1}{2}},$$

where f is the fugacity of oxygen (specifically, $f/p^\ominus$). Since the metal iron is pure, $a_{Fe} = 1$. The activity of the water is close to unity in dilute solutions (pure water has unit activity). The fugacity of oxygen is very close to 1 atm when $p = 1$ atm, and so $f/p^\ominus$ is very close to unity (because 1 atm/1 bar = 1.01). Therefore, since $\nu = 2$, and the temperature is 298 K (i.e. $T = \mathcal{T}$),

$$E = E^\ominus - (R\mathcal{T}/2F)\ln\{a_{Fe^{2+}}/a_{H^+}^2\}.$$

With $a_{H^+} = 0.1$, $R\mathcal{T}/F = 0.025\,69$ V, and $E^\ominus = +1.67$ V, the equation to solve is

$$0 = 1.67\text{ V} - (0.012\,85\text{ V})\ln(100a_{Fe^{2+}}).$$

This solves to $a_{Fe^{2+}} = 3 \times 10^{54}$.

• *Comment*. The result shows that iron continues to dissolve virtually indefinitely, even in dilute, oxygen saturated acid.

• *Exercise*. What is the e.m.f. of a lead–acid battery at 25 °C when the hydrogen ion activity is (a) 1.0, (b) 0.01. Assume complete ionization of H_2SO_4 when calculating Q.

[+1.60 V, +1.23 V]

12.1 (g) Cells at equilibrium

If the reactants are all at their equilibrium concentrations, the e.m.f. of the cell is zero. Then Q is equal to the equilibrium constant K, and so the Nernst equation reads

$$0 = E^\ominus - (RT/\nu F)\ln K, \qquad K = \{Ox_L Red_R/Red_L Ox_R\}_{eq}. \quad (12.1.8)$$

Therefore,

$$E^\ominus = (RT/\nu F)\ln K. \quad (12.1.9)$$

That is, if we know the standard e.m.f. of the cell corresponding to the reaction of interest, we can easily find the equilibrium constant. The procedure is summarized in Box 12.2. Clearly, a great deal depends on a knowledge of cell e.m.f.s, and so we shall consider them next.

Box 12.2 Calculating the equilibrium constant

1. Write the reaction and the corresponding expression for the equilibrium constant K.
2. Regard the reaction as a cell reaction in its 'standard form' (Box 12.1), and identify the Red and Ox species.
3. Set up the cell in its 'standard form', Box 12.1, and identify the electrode couples. Evaluate $E^\ominus = E_R^\ominus - E_L^\ominus$.
4. Write the half-cell reactions as reductions, and identify the value of ν, which should be chosen so that the cell reaction is the difference R − L of the half-cell reactions.
5. Evaluate $\ln K = \nu F E^\ominus/RT$.

(Note that $R\mathcal{T}/F = 25.693$ mV.)

Example 12.3

Calculate the equilibrium constant for the disproportionation reaction $2Cu^+(aq) \rightarrow Cu(s) + Cu^{2+}(aq)$ at 298 K.

• *Method*. Work through the procedure set out in Box 12.2. The standard electrode potentials can be taken from Table 12.1; use $R\mathcal{T}/F = 25.69$ mV.

• *Answer*. (1) Since the reaction is $2Cu^+(aq) \rightarrow Cu(s) + Cu^{2+}(aq)$, the equilibrium constant is

$$K = \{a_{Cu^{2+}}/a^2_{Cu^+}\}_{eq}.$$

(2) The Red and Ox species are as follows:

Red_L	Ox_R	Red_R	Ox_L
$Cu^+(aq)$	$Cu^+(aq)$	$Cu(s)$	$Cu^{2+}(aq)$

(3) The cell is therefore

$$(Pt)\,|\,Cu^+(aq), Cu^{2+}(aq)\,\|\,Cu^+(aq)\,|\,Cu(s),$$

and so the standard electrode potentials we require are

$$\text{R: } Cu^+(aq)\,|\,Cu(s), \qquad E^\ominus = +0.52\ \text{V},$$
$$\text{L: } Cu^{2+}(aq), Cu^+(aq)\,|\,(Pt), \qquad E^\ominus = +0.15\ \text{V}.$$

The standard e.m.f. of the cell is therefore

$$E^\ominus = (+0.52\ \text{V}) - (+0.15\ \text{V}) = +0.37\ \text{V}.$$

(4) The cell half-reactions that reproduce the cell reaction are

$$\text{L: } Cu^{2+}(aq) + e^- \rightarrow Cu^+(s), \qquad \text{R: } Cu^+(aq) + e^- \rightarrow Cu(s),$$

and so $\nu = 1$.

(5) It follows that $\ln K = (+0.37\ \text{V})/(0.025\,69\ \text{V}) = 14.4$. Hence, $K = 1.8 \times 10^6$.

• *Comment*. The equilibrium lies strongly towards the right of the reaction as written, and so Cu^+ disproportionates almost totally in solution.

• *Exercise*. Calculate the equilibrium constant for the reaction $Sn^{2+}(aq) + Pb(s) \rightarrow Sn(s) + Pb^{2+}(aq)$ at 298 K. [0.46]

12.2 Standard electrode potentials

The *standard e.m.f.* of a cell (symbol: $E^\ominus$) is the difference $E^\ominus_R - E^\ominus_L$ of two *standard electrode potentials*. Each standard electrode potential is denoted $E^\ominus_{Ox,Red}$. For example, in the case of the Daniell cell $Zn\,|\,ZnSO_4(aq)\,\|\,CuSO_4(aq)\,|\,Cu$ without a liquid junction, $E^\ominus = E^\ominus_{Cu^{2+},Cu} - E^\ominus_{Zn^{2+},Zn}$. As in this example, we deal only with *differences* of standard potentials, and so the standard potential of one electrode may arbitrarily be assigned the value zero and all others tabulated on that basis. The one chosen to have zero potential is the *standard hydrogen electrode* (SHE). This electrode is constructed like the one shown in Fig. 11.10, the inert metal being platinum, the hydrogen ion and the hydrogen gas being in their standard states at the specified temperature (normally $T^\ominus$).

$$\text{SHE: } H^+(aq)\,|\,H_2(g)\,|\,Pt; \quad E^\ominus_{H^+,H_2} = 0.$$

The standard electrode potential of any other redox system (or *couple*) is then defined as *the e.m.f. of a cell without a liquid junction in which the couple forms the right-hand electrode and the SHE forms the left, all species being in their standard states*:

$$E^\ominus_{Cell} = E^\ominus_R - E^\ominus_{SHE} = E^\ominus_R.$$

Once the standard potential of an electrode has been measured by making it a part of a cell in conjunction with a SHE (in a manner we explain

Table 12.1. Standard electrode potentials at $\mathcal{T}$

Couple	$E^{\ominus}$/V
$Na^+ + e^- \rightarrow Na$	−2.71
$Mg^{2+} + 2e^- \rightarrow Mg$	−2.36
$Zn^{2+} + 2e^- \rightarrow Zn$	−0.76
$2H^+ + 2e^- \rightarrow H_2$	0
$AgCl + e^- \rightarrow Ag + Cl^-$	0.2223
$Cu^{2+} + 2e^- \rightarrow Cu$	0.34
$Ce^{4+} + e^- \rightarrow Ce^{3+}$	1.61

below), it can be used in combination with other electrodes to obtain their standard potentials. A list of values obtained in this way (at $\mathcal{T}$) is given in Table 12.1.‡

12.2 (a) The electrochemical series

The significance of the size and sign of the standard electrode potential is as follows. The standard e.m.f. is obtained from Table 12.1 by forming the difference $E^{\ominus} = E^{\ominus}_{R} - E^{\ominus}_{L}$, then the equilibrium constant for the cell reaction is obtained from eqn (12.1.9) in the form

$$\text{Red}_L + \text{Ox}_R = \text{Red}_R + \text{Ox}_L, \qquad \ln K = \nu F E^{\ominus}/RT. \qquad (12.2.1)$$

K is large (products dominant, Red_R formed, right-hand couple undergoes reduction) if $E^{\ominus}$ is positive. $E^{\ominus}$ is positive when $E^{\ominus}_{R} > E^{\ominus}_{L}$. Therefore, *couples with high standard electrode potentials are reduced by couples with low standard electrode potentials*. In other words:

> *Low reduces high.*

For example, $E^{\ominus}_{Zn^{2+},Zn} = -0.764$ V is lower than $E^{\ominus}_{Cu^{2+},Cu} = +0.34$ V, and so in the reaction $Zn(s) + CuSO_4(aq) \rightarrow ZnSO_4(aq) + Cu(s)$, the equilibrium lies to the right ($K = 10^{37}$ at 298 K): zinc reduces copper, and metallic copper precipitates from the solution as the zinc dissolves.

The layout of Table 12.1, which is the *electrochemical series* of the elements, is designed to show the order of reducing power: couples higher up the Table (with lower, more negative, standard electrode potentials) reduce couples lower down:

> *Couples higher in the Table reduce couples lower down.*

This is a qualitative conclusion. The quantitative value of K is then obtained by applying the set of rules summarized in Box 12.1.

Example 12.4

Can zinc displace (a) copper (b) magnesium from aqueous solutions at 298 K?

• *Method*. Displacement corresponds to a reduction of $M^{2+}(aq)$ by Zn(s). This may occur if the $E^{\ominus}_{Zn^{2+},Zn}$ is more negative (higher in the electrochemical series) than $E^{\ominus}_{M^{2+},M}$. Refer to Table 12.1.

• *Answer*. From Table 12.1 we see that the Zn^{2+} | Zn couple lies above Cu^{2+} | Cu but below Mg^{2+} | Mg. Therefore, zinc can displace copper but not magnesium.

• *Comment*. The quantitative values of the equilibrium constants of the displacement reactions can be obtained from $\ln K = \nu F E^{\ominus}/RT$. Note that these are *thermodynamic*

‡ Tables of standard electrode potentials are sometimes encountered, especially in the older literature, having signs opposite to those in Table 12.1. This indicates the adoption of a different convention. We are listing *standard reduction potentials*; the alternative is to list *standard oxidation potentials*. It is easy (and important) to check which convention is being used by noting the sign of the Cu^{2+}|Cu couple: in our convention it is positive. If the other source lists oxidation potentials, change all the signs and proceed as here. A further point is that with the recent redefinition of standard state in terms of 1 bar instead of 1 atm, all the tabulated standard electrode potentials have slightly modified values. However, the change is of the order of 0.1 mV, which is within the accuracy with which most can be measured, and so for all except the most precise calculations, the change can be ignored.

conclusions. There may be kinetic factors (e.g. passivation of the surface) which result in immeasurably slow rates of reaction.

• *Exercise*. Can lead displace (a) iron, (b) copper from solution at 298 K? [(a) No, (b) Yes]

12.2 (b) The measurement of standard electrode potentials

The procedure for measuring a standard electrode potential can be illustrated by considering a specific case, and we choose the important silver/silver chloride electrode, $Cl^-(aq) \mid AgCl(s) \mid Ag(s)$. The measurement is made on the *Harned cell*:

$$Pt \mid H_2(g) \mid HCl(aq, m) \mid AgCl(s) \mid Ag(s);$$
$$\tfrac{1}{2}H_2(g) + AgCl(s) \rightarrow HCl(aq) + Ag(s).$$

The observed e.m.f. is related to the standard electrode potential by

$$E = E^{\ominus}_{Ag,AgCl,Cl^-} - (RT/F)\ln(a_{H^+}a_{Cl^-}/f^{\frac{1}{2}}).$$

We shall set $f = 1$ from now on (corresponding to approximately $p = 1$ atm). The activities can be expressed in terms of the molality m and the mean activity coefficient $\gamma_{\pm}$ through eqn (11.1.6):

$$E = E^{\ominus}_{Ag,AgCl,Cl^-} - (RT/F)\ln(m/m^{\ominus})^2 - (RT/F)\ln\gamma_{\pm}^2.$$

This rearranges to

$$E + 2(RT/F)\ln(m/m^{\ominus}) = E^{\ominus}_{Ag,AgCl,Cl^-} - 2(RT/F)\ln\gamma_{\pm}. \quad (12.2.2)$$

From the Debye–Hückel Limiting Law for a (1,1)-electrolyte, $\lg\gamma_{\pm} = -0.509\,(m/m^{\ominus})^{\frac{1}{2}}$. Therefore,

$$E + 2(RT/F)\ln(m/m^{\ominus}) = E^{\ominus}_{Ag,AgCl,Cl^-} + 2.34(RT/F)(m/m^{\ominus})^{\frac{1}{2}}. \quad (12.2.3)$$

(In precise work, the $m^{\frac{1}{2}}$ term is brought to the left, and a higher order correction term from the extended Debye–Hückel theory is used on the right.) The expression on the left is evaluated at a range of cell molalities, plotted against $m^{\frac{1}{2}}$, and extrapolated to $m = 0$. The intercept is the value of $E^{\ominus}_{Ag,AgCl,Cl^-}$, as shown in Fig. 12.6.

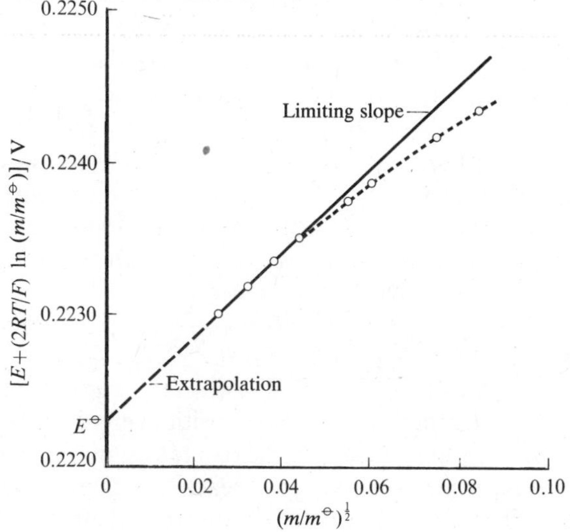

Fig. 12.6. The plot, and the $\sqrt{m}$ extrapolation, used for the experimental measurement of a standard e.m.f. of a cell. The intercept at $\sqrt{m} = 0$ is $E^{\ominus}$.

12.2 (c) The measurement of activity coefficients

Once the standard potential of an electrode in a cell is known, the activities of the ions can be determined simply by measuring the cell e.m.f. with the ions at the concentration of interest. For example, the mean activity coefficient of the ions in hydrochloric acid of molality m is obtained from eqn (12.2.2) in the form

$$\ln \gamma_{\pm} = (F/2RT)\{E^{\ominus}_{\mathrm{Ag,AgCl,Cl^-}} - E\} - \ln (m/m^{\ominus}), \qquad (12.2.4)$$

once E has been measured. Knowledge of activity coefficients is essential, as explained in Chapter 11, if equilibrium constants are to be interpreted precisely.

12.3 Thermodynamic data from cell e.m.f.s

The measurement of e.m.f. is a convenient source of data on the Gibbs functions, enthalpies, and entropies of reactions. In practice the standard values of these quantities are the ones normally determined.

12.3 (a) The measurement of $\Delta_r G^{\ominus}$

The quickest way to establish the connection between $\Delta_r G^{\ominus}$ and $E^{\ominus}$ is to combine $E^{\ominus} = (RT/\nu F) \ln K$ with the expression $\Delta_r G^{\ominus} = -RT \ln K$, eqn (10.1.17). Then

$$\Delta_r G^{\ominus} = -\nu F E^{\ominus}. \qquad (12.3.1)$$

Hence, knowing $E^{\ominus}$ for a cell lets us determine $\Delta_r G^{\ominus}$ for the cell reaction.

Another way of deriving the same result proves to be more general. We established in Chapter 5 that the change of Gibbs function accompanying a process is equal to the maximum electrical work that process can do: $\Delta G = -w'_e$, eqn (5.3.11). As explained in Section 10.1, if we consider a vast reaction mixture (a swimming-pool full) of some specified composition, then the change of Gibbs function per mole of reaction *at that fixed composition* is the reaction Gibbs function, $\Delta_r G$. When this change comes about, work is capable of being done, and if the reactants are separated so that the reaction takes place in an electrochemical cell working reversibly, then maximum *electrical* work can be done. Since *charge* × *potential difference* is work (Box 2.2), the maximum electrical work per mole of reaction is the product of the potential difference between the electrodes when the cell is operating reversibly (that is, the e.m.f., E) and the charge per mole of electrons (F, Faraday's constant) when one electron is involved in each redox event, and ν times that quantity when ν are involved. Therefore, the maximum electrical work done per mole of reaction is $w' = \nu FE$. Equating this with $-\Delta_r G$ for the reaction at the specified, effectively fixed composition, gives

$$\Delta_r G = -\nu FE. \qquad (12.3.2)$$

Hence, knowing the cell e.m.f. at *any* composition gives us the reaction Gibbs function at that composition. If the composition happens to correspond to the equilibrium of the cell reaction, then $\Delta_r G = 0$, and consequently $E = 0$: the cell is exhausted. If the reactants are all in their standard states, then $\Delta_r G$ has its standard value $\Delta_r G^\ominus$, and the e.m.f. has its: this recovers eqn (12.3.1).

Example 12.5

Given that the standard electrode potentials of the $Cu^{2} | Cu$ and $Cu^+ | Cu$ couples are

$$E^\ominus_{Cu^{2+},Cu} = +0.34\text{ V} \quad \text{and} \quad E^\ominus_{Cu^+,Cu} = +0.52\text{ V},$$

evaluate the standard electrode potential of the $Cu^{2+} | Cu^+$ couple at T.

• *Method*. The reaction Gibbs functions may be added (because G is a state function). Therefore, convert the $E^\ominus$ values to $\Delta_r G^\ominus$ values, using eqn (12.3.1), add them appropriately, and then convert to the required $E^\ominus$ using eqn (12.3.1) again.

• *Answer*. The electrode reactions are as follows:

(a) $Cu^{2+}(aq) + 2e^- \rightarrow Cu(s)$, $E^\ominus_{Cu^{2+},Cu} = +0.34\text{ V}$; $\Delta_r G^\ominus = -2F \times (0.34\text{ V}) = (-0.68\text{ V})F$.
(b) $Cu^+(aq) + e^- \rightarrow Cu(s)$, $E^\ominus_{Cu^+,Cu} = +0.52\text{ V}$; $\Delta_r G^\ominus = -F \times (0.52\text{ V}) = (-0.52\text{ V})F$.

The required reaction is

(c) $Cu^{2+}(aq) + e^- \rightarrow Cu^+(aq)$; $E^\ominus_{Cu^{2+},Cu^+}$.

Since (c) = (a) − (b), the standard Gibbs function of reaction (c) is

$$\Delta_r G^\ominus = \Delta_r G^\ominus(a) - \Delta_r G^\ominus(b) = \{(-0.68\text{ V}) - (-0.52\text{ V})\}F = (-0.16\text{ V})F.$$

Therefore, since now $\nu = 1$,

$$E^\ominus = -\Delta_r G^\ominus / F = +0.16\text{ V}.$$

• *Comment*. Note that we cannot combine the $E^\ominus$ values directly, as they are not extensive properties, and we must always work *via* $\Delta_r G^\ominus$.

• *Exercise*. Calculate the standard potential of the $Fe^{3+} | Fe$ couple from the values for the Fe^{3+}, Fe^{2+} and $Fe^{2+} | Fe$ couples. [−0.04 V]

12.3 (b) Gibbs functions of formation of ions

Values of the standard Gibbs functions of reactions involving ions in solution are normally reported in terms of the Gibbs functions of formation of the ions concerned. These tables are set up by defining the standard Gibbs function of formation of one type of ion to be zero, and then quoting all others on that basis. The convention adopted is as follows:

The standard Gibbs function of formation of the proton in water is zero:

$$\tfrac{1}{2}H_2(g) \rightarrow H^+(aq) + e^-, \qquad \Delta_f G^\ominus_{H^+(aq)} = 0.$$

The Gibbs functions of formation of other ions are then found from values of $E^\ominus$. For example, $\Delta_f G^\ominus_{Ag^+(aq)}$ is found from the standard Gibbs function of $Ag^+(aq) + \tfrac{1}{2}H_2(g) \rightarrow H^+(aq) + Ag(s)$, since

$$\Delta_r G^\ominus = \Delta_f G^\ominus_{H^+(aq)} - \Delta_f G^\ominus_{Ag^+(aq)} = -\Delta_f G^\ominus_{Ag^+(aq)}.$$

The reaction occurs in the cell $H_2 | H^+(aq) \| Ag^+(aq) | Ag$, the standard e.m.f. of which is $E^\ominus_{Ag^+,Ag} - E^\ominus_{SHE} = E^\ominus_{Ag^+,Ag} = 0.7996\text{ V}$. Therefore, from eqn (12.3.1) and $\nu = 1$, we find

$$\Delta_f G^\ominus_{Ag^+(aq)} = -\Delta_r G^\ominus = -(-FE^\ominus_{Ag^+,Ag}) = +77.10\ \mathrm{kJ\ mol^{-1}},$$

as in Table 12.2.

The standard Gibbs function of formation of an ion can be estimated theoretically using a result due to Max Born. The *Born equation* identifies the *Gibbs function of solvation* of an ion (the change of Gibbs function for the process $M^+(g) \rightarrow M^+(aq)$ in the case of water) with the electrical work of transferring an ion from the vacuum into the solvent, the latter being treated as a continuous dielectric of relative permittivity ε_r:

$$\Delta G^\ominus_{m,solvation} = -\tfrac{1}{2}(z_i^2 e^2 N_A / 4\pi\varepsilon_0 r_i)\{1 - (1/\varepsilon_r)\}, \qquad (12.3.3)$$

where z_i is the charge number of the ion and r_i its radius. Note that $\Delta G^\ominus_{m,solvation}$ is negative, and that it is strongly negative for small, highly charged ions in media of high relative permittivity. The standard Gibbs function of formation of an ion in solution is the sum of this solvation term and the standard Gibbs function of formation of the gaseous ions.

12.3 (c) The temperature dependence of the e.m.f.

The e.m.f. of a cell of given composition is related to the reaction Gibbs function through eqn (12.3.2). The variation of any ΔG with temperature is given by eqn (6.2.2): $(\partial\, \Delta G/\partial T)_p = -\Delta S$. Therefore, the variation of the cell e.m.f. with temperature is

$$(\partial E/\partial T)_p = \Delta_r S/\nu F. \qquad (12.3.4)$$

A similar equation gives the temperature dependence of the standard e.m.f. in terms of the standard entropy of reaction. Hence we have an electrochemical technique for obtaining reaction entropies. Furthermore, since $\Delta G = \Delta H - T\,\Delta S$, the electrochemical measurements can be combined to give the reaction enthalpy:

$$\Delta_r H = \Delta_r G + T\Delta_r S = -\nu F\{E - T(\partial E/\partial T)_p\}. \qquad (12.3.5)$$

This is a purely non-calorimetric method of measuring a reaction enthalpy. The enthalpies of formation of ions in solution were discussed in Section 4.2; some values are listed in Table 12.2 on the basis (as described in Section 4.2) of the choice $\Delta_f H^\ominus_{H^+(aq)} = 0$.

Table 12.2. Standard thermodynamic functions of ions in solution at T

Ion	$\Delta_f H^\ominus$/kJ mol^{-1}	$S^\ominus$/J K^{-1} mol^{-1}	$\Delta_f G^\ominus$/kJ mol^{-1}
H^+	0	0	0
Na^+	−240.1	59.0	−261.9
K^+	−252.4	102.5	−283.3
Cu^{2+}	+64.8	−99.6	+65.5
Cl^-	−167.2	+56.5	−131.2
PO_4^{3-}	−1019	−221.8	−1277.4

Example 12.6

The standard e.m.f. of the cell

$$\mathrm{Pt}\,|\,H_2(g)\,|\,HCl(aq)\,|\,Hg_2Cl_2(s)\,|\,Hg(l)$$

was found to be +0.2699 V at 293 K and +0.2669 V at 303 K. Evaluate the standard Gibbs function, enthalpy, and entropy of reaction at 298 K.

• *Method*. Write the cell reaction, and identify ν. Find $\Delta_r G^\ominus$ from eqn (12.3.1) at 298 K by linear interpolation between the two temperatures (i.e. in this case take the mean $E^\ominus$ because 298 K lies midway between 293 K and 303 K). Use eqn (12.3.4) for $\Delta_r S^\ominus$ and eqn (12.3.5) for $\Delta_r H^\ominus$. Use 1 C V = 1 J.

• *Answer*. The cell reaction is

$$Hg_2Cl_2(s) + H_2(g) \rightarrow 2Hg(l) + 2HCl(aq); \; \nu = 2.$$

The mean standard e.m.f. is +0.2684 V, and so

$$\Delta_r G^\ominus = -2FE^\ominus = -2 \times (9.6485 \times 10^4 \,\mathrm{C\,mol^{-1}}) \times (0.2684\,\mathrm{V}) = -51.8\,\mathrm{kJ\,mol^{-1}}.$$

The temperature coefficient of e.m.f. is

$$(\partial E/\partial T)_p = \{(+0.2669\,\mathrm{V}) - (+0.2699\,\mathrm{V})\}/10\,\mathrm{K} = -3.0 \times 10^{-4}\,\mathrm{V\,K^{-1}}.$$

Therefore, from eqn (12.3.4),

$$\Delta_r S^\ominus = 2 \times (9.6485 \times 10^4\,\mathrm{C\,mol^{-1}}) \times (-3.0 \times 10^{-4}\,\mathrm{V\,K^{-1}}) = -58\,\mathrm{J\,K^{-1}\,mol^{-1}}.$$

Now from $\Delta_r H^\ominus = \Delta_r G^\ominus + T\Delta_r S^\ominus$:

$$\Delta_r H^\ominus = -51.8\,\mathrm{kJ\,mol^{-1}} + (298\,\mathrm{K}) \times (-58\,\mathrm{J\,K^{-1}\,mol^{-1}}) = -69\,\mathrm{kJ\,mol^{-1}}$$

• *Comment*. One difficulty with this procedure lies in the accurate measurement of small temperature coefficients of e.m.f. Nevertheless, it is another example of the striking ability of thermodynamics to relate the apparently unrelated, in this case to relate electrical measurements to thermal properties.

• *Exercise*. Predict the standard e.m.f. of the Harned cell at 303 K from tables of thermodynamic data. [0.2191 V]

Electrochemical measurements always relate to electrically neutral collections of ions. Therefore, in tabulating their entropies in solution we are free to ascribe zero entropy to one arbitrarily selected kind of ion. The normal convention is as follows:

The standard partial molar entropy of the aquated proton is set at zero:

$$S^\ominus(\mathrm{H^+, aq}) = 0.$$

Some standard values based on this choice are listed in Table 12.2. They vary as expected if the entropy is related to the way in which the ions order the water molecules in their vicinity in the solution. (This is why it is important to appreciate that they are *partial* molar entropies, in the sense described in Section 8.1, because only then can the 'entropy of a solute' depend on the properties of the solvent.) Thus, small, highly charged ions induce local structure in the surrounding water, and the entropy of the solution is decreased more than in the case of large, singly charged ions.

The remarks about structure-forming refer to the effect relative to the structure induced by protons, because all entropies are relative to $S^\ominus_{\mathrm{m,H^+(aq)}} = 0$. The actual Third Law partial molar entropy of the proton in water can be estimated by proposing a model of the structure it induces, and there is some agreement on the value $-21\,\mathrm{J\,K^{-1}\,mol^{-1}}$, the negative value representing the order induced in the solvent.

12.4 Simple applications of e.m.f. measurements

Electrochemical techniques are fundamental to a wide variety of calculations and measurements in chemistry, and the next few paragraphs describe some of them.

12.4 (a) Solubility products

The *solubility* (the molality of the saturated solution, symbol: s) of a sparingly soluble salt MX may be discussed in terms of the equilibrium

$$MX(s) \rightleftharpoons M^+(aq) + X^-(aq), \qquad K_s = \{a_{M^+}a_{X^-}\}_{eq}. \tag{12.4.1}$$

The equilibrium constant K_s is called the *solubility-product constant*, but more usually just the *solubility product*, of the salt (the a_{MX} that might be expected in the denominator is unity for a pure solid). When the solubility is so low that the mean activity coefficient is unity, $a_{M^+} = m_{M^+}/m^{\ominus}$ and $a_{X^-} = m_{X^-}/m^{\ominus}$, and both molalities are equal to s in the saturated solution. Therefore,

$$K_s = (s/m^{\ominus})^2, \text{ so that } s = (\sqrt{K_s})m^{\ominus}. \tag{12.4.2}$$

The value of K_s can be found from the standard e.m.f. of a cell with a reaction corresponding to the solubility equilibrium.

Example 12.7

Evaluate the solubility-product constant of AgCl(s) from cell e.m.f. data at 298 K.

• *Method*. Find an electrode combination that reproduces the solubility equilibrium, and then identify the solubility-product constant with the equilibrium constant of the cell reaction. Use eqn (12.1.9) to relate K to $E^{\ominus}$.

• *Answer*. The equilibrium reaction is

$$AgCl(s) \rightleftharpoons Ag^+(aq) + Cl^-(aq).$$

By adding Ag(s) to both sides this can be written in the 'standard form' of a cell redox reaction (Box 12.1):

$$Ag(s) + AgCl(s) \rightleftharpoons Ag(s) + Ag^+(aq) + Cl^-(aq).$$

We can now make the following identifications:

Red_L	Ox_R	Red_R	Ox_L
Ag(s)	AgCl(s)	Ag(s) + Cl^-(aq)	Ag^+

and hence construct the corresponding 'standard form' of the cell:

$$Ag(s) \mid Ag^+(aq) \parallel Cl^-(aq) \mid AgCl(s) \mid Ag(s).$$

The standard e.m.f. of this cell is

$$E^{\ominus} = E^{\ominus}_{Ag,AgCl,Cl^-} - E^{\ominus}_{Ag^+,Ag} = 0.22\text{ V} - 0.80\text{ V} = -0.58\text{ V}.$$

The equilibrium constant of the cell reaction is

$$K = \{a_{Ag^+}a_{Cl^-}\}_{eq} = K_{sp}.$$

Since $v = 1$,

$$\ln K = (-0.58\text{ V})/(0.025\,69\text{ V}) = -23.$$

Therefore, $K_{sp} = 1.0 \times 10^{-10}$.

- *Comment*. It follows that the solubility of silver chloride is 1.0×10^{-5} mol kg^{-1}.

- *Exercise*. Calculate the solubility-product constant of mercury(I) chloride at 298 K.
[1.3×10^{-18}]

12.4 (b) Potentiometric titrations

In a *redox titration* the reduced form of some ion (e.g. Fe^{2+}) is oxidized by the addition of some oxidizer (e.g. Ce^{4+}) and the end-point is detected by monitoring the e.m.f. of the cell formed by a platinum electrode in the mixture and another electrode in electrical contact with the mixture through a salt bridge. As we shall see, there is a sharp change of e.m.f. at the end point, when exactly enough oxidant has been added to oxidize all the reduced ion.

We illustrate the technique with the iron(II)/cerium(IV) redox reaction, $Fe^{2+} + Ce^{4+} \rightarrow Fe^{3+} + Ce^{3+}$, all in aqueous solution, the two half-reactions being

$$Ce^{4+} + e^- \rightarrow Ce^{3+}, \qquad E^{\ominus}_{Ce^{4+},Ce^{3+}} = +1.61\text{ V}$$
$$Fe^{3+} + e^- \rightarrow Fe^{2+}, \qquad E^{\ominus}_{Fe^{3+},Fe^{2+}} = +0.77\text{ V}.$$

The equilibrium constant for the reaction (which has $\nu = 1$) is therefore 1.52×10^{14}, and so cerium is a good choice because it ensures that the equilibrium lies strongly towards products (Fe^{3+}). During the course of the titration the presence of the iron couple gives rise to an electrode potential $E_{Fe^{3+},Fe^{2+}}$, and the cerium couple gives rise to a potential $E_{Ce^{4+},Ce^{3+}}$. However, these couples share the same electrode, and therefore the potentials must be equal (a single electrode cannot be at two different potentials simultaneously). Consequently, at any stage of the titration the electrode potential may be written in either of the following forms:

$$E = E^{\ominus}_{Ce^{4+},Ce^{3+}} + (RT/F)\ln(a_{Ce^{4+}}/a_{Ce^{3+}}),$$
$$E = E^{\ominus}_{Fe^{3+},Fe^{2+}} + (RT/F)\ln(a_{Fe^{3+}}/a_{Fe^{2+}}),$$

and whichever is more convenient may be used. We shall simplify the following discussion by supposing that activities may be replaced by molalities, and we shall also ignore the dilution of the sample as the cerium solution is added: these are minor technical complications which do not affect the main conclusions.

Suppose that initially the amount of Fe(II) present in the sample is f. As Ce(IV) is added this is reduced to $(1-x)f$; simultaneously, the amount of Fe(III) rises to xf, where x depends on the amount of Ce(IV) added (if K were infinite, xf would be equal to the amount of oxidant added). It follows that at an intermediate stage of the titration, the electrode's potential is

$$E = E^{\ominus} + (RT/F)\ln\{x/(1-x)\}. \qquad (12.4.3)$$

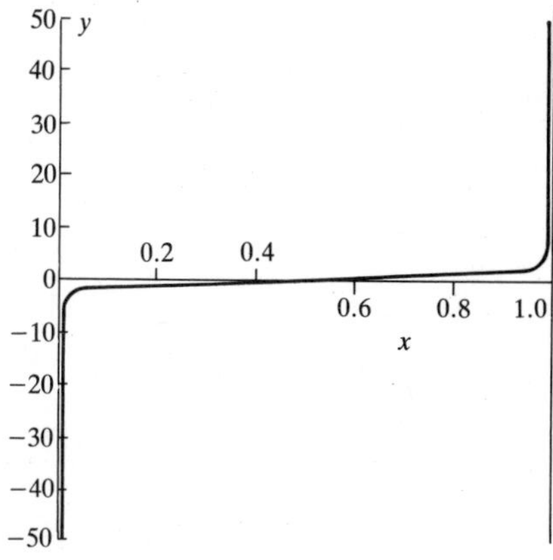

Fig. 12.7. The function $y = \ln\{x/(1-x)\}$. This quick–slow–quick variation occurs widely in electrochemistry.

The function $\ln\{x/(1-x)\}$ occurs widely in electrochemistry, and is plotted in Fig. 12.7: it has a large negative value when x is small ($x \ll 1$), is zero when $x = \frac{1}{2}$, and rises to a large positive value as x approaches unity. An initial rapid rise, a slow variation around $x = \frac{1}{2}$, and a further rapid rise, is characteristic of many electrochemical measurements, and it can usually be traced back to the properties of the $\ln\{x/(1-x)\}$ function.

In the context of a potentiometric titration, the shape of the $\ln\{x/(1-x)\}$ function indicates that the electrode potential is initially strongly negative, rises rapidly to the vicinity of $E^{\ominus}_{Fe^{3+},Fe^{2+}}$ as Ce(IV) is added, and is exactly equal to that value when enough has been added to equalize the molalities (actually the activities) of Fe(II) and Fe(III). As more cerium is added the electrode potential rises slowly at first, but then the rapid second rise of the $\ln\{x/(1-x)\}$ function takes over. The line in Fig. 12.7 barely leaves the zero axis (on the scale used) until x has reached about 0.98, and so the end-point of the titration is signalled by the abrupt rise in potential away from $E^{\ominus}_{Fe^{3+},Fe^{2+}}$.

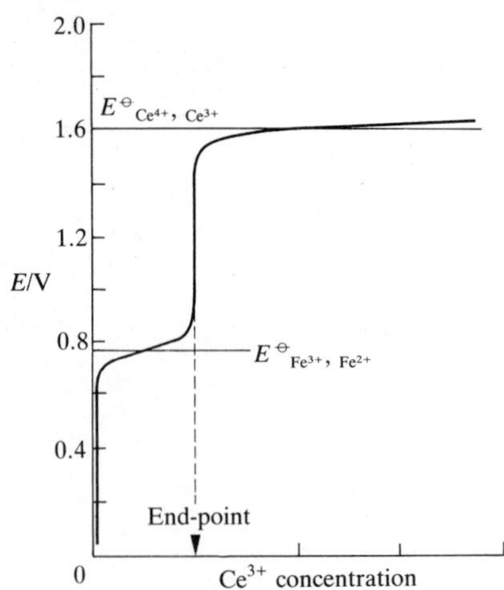

Fig. 12.8. The variation of an electrode potential during the titration of Fe(II) and Ce(IV). The end-point is recognized by the sudden change from the standard potential of the Fe^{3+}, Fe^{2+} couple to that of the Ce^{4+}, Ce^{3+} couple.

How high does the potential rise? When enough Ce(IV) has been added to eliminate all but a trace of the Fe(II), it is more convenient to discuss the potential in terms of the cerium molality. If an amount c of Ce(IV) has been added, the amount present in solution after the end-point is about $c-f$, because virtually all the Fe(II) has been oxidized (since the equilibrium constant is so large). The amount of Ce(III) present is about f however much Ce(IV) is added after the end-point. Therefore the potential of the electrode is

$$E = E^{\ominus}_{Ce^{4+},Ce^{3+}} + (RT/F)\ln\{(c-f)/f\}. \qquad (12.4.4)$$

This function rises to $E^{\ominus}_{Ce^{4+},Ce^{3+}}$ by the time $c=2f$ (about twice as much cerium as was needed to reach the end-point), and then climbs only very slowly as more cerium is added because when $c \gg f$, $\ln\{(c-f)/f\} \approx \ln c/f$, which increases very slowly as c increases.

The complete curve is sketched in Fig. 12.8: it is clear that the end-point can be detected by noting where there is a rapid change of potential from the vicinity of $E^{\ominus}_{Fe^{3+},Fe^{2+}}$ to the vicinity of $E^{\ominus}_{Ce^{4+},Ce^{3+}}$, or in general, for other redox systems, from the standard electrode potential of the couple in the sample to the standard electrode potential of the couple in the solution with which it is being titrated.

12.4 (c) p*K* and pH

The Brønsted–Lowry classification defines an acid as a *proton donor* and a base as a *proton acceptor*. These definitions can be generalized, but they are adequate for the present discussion. An acid HA takes part in the following equilibrium in water:

$$HA(aq) + H_2O(l) \rightleftharpoons A^-(aq) + H_3O^+(aq), \qquad K_a = \{a_{H^+}a_{A^-}/a_{HA}\}_{eq}.$$

A^- is a proton acceptor, and is therefore the *conjugate base* of the acid HA. Although a base is a proton acceptor, the resulting ion BH^+, its *conjugate acid*, is a proton donor, and so a characteristic equilibrium of bases in water is

$$BH^+(aq) + H_2O(l) \rightleftharpoons B(aq) + H_3O^+(aq).$$

This means that we can also treat bases in the same way as we treat acids, so long as we focus on the properties of the conjugate acid, BH^+ (e.g. on NH_4^+ when dealing with aqueous ammonia). The general form of the equilibrium in either case is therefore

$$\text{Acid(aq)} + H_2O(l) \rightleftharpoons \text{Base(aq)} + H_3O^+(aq). \qquad (12.4.5)$$

The equilibrium can be expressed in terms of an equilibrium constant, the *ionization constant* (symbol: K_a) in which the activity of the water is assumed to be unity (because the solutions are dilute):

$$K_a = \{a_{H_3O^+} a_{Base}/a_{Acid}\}_{eq}. \tag{12.4.6}$$

In the case of *polyprotic acids* (acids with more than one ionizable proton) it is necessary to distinguish between the first, second, etc. ionization constants. In the case of an acid with two ionizable protons (such as H_2SO_4), the equilibria are

$$H_2A(aq) + H_2O(l) \rightleftharpoons HA^-(aq) + H_3O^+(aq), \qquad K_{a1} = \{a_{H^+} a_{HA^-}/a_{H_2A}\}_{eq},$$
$$HA^-(aq) + H_2O(l) \rightleftharpoons A^{2-}(aq) + H_3O^+(aq), \qquad K_{a2} = \{a_{H^+} a_{A^-}/a_{HA^-}\}_{eq}.$$

Generally $K_{a2} < K_{a1}$ (typically by three orders of magnitude) because the second proton is more difficult to remove on account of the negative charge on HA^-.

A *strong acid* is a strong proton donor, and its ionization constant is effectively infinite in water (an example is the first ionization of H_2SO_4). A *strong base* is a strong proton acceptor, which is equivalent to saying that its conjugate acid is a very weak proton donor; hence its (acid) ionization constant is very small indeed (an example is OH^-, its conjugate acid being H_2O). A *weak acid* is incompletely ionized in solution (K_a much less than infinity; e.g. 1.7×10^{-5} for acetic acid). A *weak base* is only partly ionized, but its conjugate acid is stronger than that of a strong base. In the case of ammonia in water, for example, $K_a = 5.6 \times 10^{-10}$. (Note that CH_3CO_2H is a stronger proton donor than NH_4^+.)

Values of ionization constants fall over a wide range (e.g. $K_a = 5.6 \times 10^{-10}$ for NH_4^+ and 0.16 for HIO_3 at 298 K). It is therefore convenient to list them as their logarithms, and so we introduce

$$pK_a = -\lg K_a, \tag{12.4.7}$$

the logarithm being to the base 10. Hence p denotes the negative power of 10. In this notation $pK_a = 9.25$ for NH_4^+ and 0.80 for HIO_3. A list of values is given in Table 12.3.

Table 12.3. Ionization constants in water at $\mathcal{T}$

	pK_{a1}	pK_{a2}
H_2CO_3	6.37	10.25
H_2SO_4		1.92
CH_3COOH	4.75	
$NH_3(NH_4^+)$	9.25	
$C_6H_5NH_2$	4.63	

Water is *amphoteric*: it can act as both an acid and a base. In pure water there is the *autoionization equilibrium*

$$H_2O(l) + H_2O(l) \rightleftharpoons H_3O^+(aq) + OH^-(aq), \quad K_w = \{a_{H_3O^+} a_{OH^-}\}_{eq}, \tag{12.4.8}$$

K_w being the *ionization constant of water* (as before, the ion concentrations are so small that the water activity is almost exactly unity and does not appear in the denominator). At $T = \mathcal{T}$ (i.e. 25 °C), $K_w = 1.008 \times 10^{-14}$ ($pK_w = 14.00$), showing that only a very small proportion of the water molecules are ionized. The concentration of ions in pure water can then be calculated by approximating activities by molalities, writing $K_w = \{(m_{H^+}/m^\ominus)(m_{OH^-}/m^\ominus)\}_{eq}$, and noting that the two ion molalities must be equal. Consequently, the hydrogen ion molality is approximately

$$m_{H^+} = (K_w)^{\frac{1}{2}} m^\ominus = 1.004 \times 10^{-7}\ \text{mol kg}^{-1} \text{ at } 298\ \text{K}.$$

The hydrogen ion molality (more precisely, its activity) plays a central role in many processes, and its magnitude can vary over a wide range. For instance, in 1 M strong acid, m_{H^+} is approximately 1 mol kg^{-1}, in pure water it is about $10^{-7}\,\mathrm{mol\,kg^{-1}}$, and in 1 M strong base it is about $10^{-14}\,\mathrm{mol\,kg^{-1}}$. The wide span of values is compressed if we use the *pH scale*. The pH was originally defined as $-\lg(m_{\mathrm{H^+}}/m^{\ominus})$, but it is now defined in terms of the activity as

$$\mathrm{pH} = -\lg a_{\mathrm{H^+}}. \qquad (12.4.9)$$

Since $a_{\mathrm{H^+}} = \sqrt{K_w} = 1.004 \times 10^{-7}$ at 298 K, the pH of pure water at that temperature is 7.00. Hence, pH = 7.00 corresponds to *neutrality* at 298 K. (At blood temperature, 37 °C, when $K_w = 2.1 \times 10^{-14}$, neutrality corresponds to pH = 6.84.) In an acid aqueous solution $a_{\mathrm{H^+}}$ is greater than in pure water, and so then pH < 7; in basic solutions pH > 7.

The water autoprotolysis equilibrium is maintained, and the value of K_w is unchanged in the presence of acids and bases. This means that $a_{\mathrm{H^+}}$ or $a_{\mathrm{OH^-}}$ must adjust so that their product remains equal to K_w. Hence, in a basic solution (where we might know the OH^- molality) the pH can be obtained from $a_{\mathrm{H^+}} = K_w/a_{\mathrm{OH^-}}$. Note that it is legitimate to replace $a_{\mathrm{H^+}}$ and $a_{\mathrm{OH^-}}$ by molalities only if *all* the ions are present at low concentration, not merely H_3O^+ or OH^-: all ions contribute to the ionic strength and therefore to the activity coefficients. When concentrations are not low enough for molalities to be used, activity coefficients can be found from tables (or estimated from the Debye–Hückel Limiting Law, eqn (11.2.11), or its extension). For instance, in $1.0\,\mathrm{mol\,kg^{-1}}$ HCl(aq), $m_{\mathrm{H^+}} = 1.0\,\mathrm{mol\,kg^{-1}}$ (the acid is fully ionized) and $\gamma_{\pm} = 0.811$ (Table 11.2); therefore, $a_{\mathrm{H^+}} = 0.811 \times (1.0\,\mathrm{mol\,kg^{-1}}/m^{\ominus}) = 0.81$, implying pH = 0.092 in place of the value pH = 0 which would have been obtained from the use of molality alone. Note that there is nothing mysterious about a negative pH, for it merely corresponds to an activity greater than unity. For example, in $2.00\,\mathrm{mol\,kg^{-1}}$ HCl(aq), where the mean activity coefficient is 1.011 (Table 11.2), the hydrogen ion activity is 2.02_2, implying pH = −0.31.

12.4 (d) The measurement of pH and p*K*

The measurement of pH is simple in principle, for its is based on the measurement of the potential of a hydrogen electrode immersed in the solution. The left-hand electrode of the cell is typically a calomel electrode with potential E_{cal}; the right-hand electrode is the hydrogen electrode with potential $E_{\mathrm{SHE}} + (RT/F)\ln a_{\mathrm{H^+}}$. The e.m.f. of the cell is therefore

$$E = \{E_{\mathrm{SHE}} + (RT/F)\ln a_{\mathrm{H^+}}\} - E_{\mathrm{cal}} = (-59.15\,\mathrm{mV})\,\mathrm{pH} - E_{\mathrm{cal}},$$

the numerical value relating to 25 °C. It follows that the pH is given by $(E - E_{\mathrm{cal}})/(59.15\,\mathrm{mV})$ once the e.m.f. has been measured.‡

In practice, indirect methods are much more convenient, and the hydrogen electrode is replaced by the *glass electrode*. This electrode, Fig. 12.9, is sensitive to hydrogen ion activity, and has a potential that depends linearly

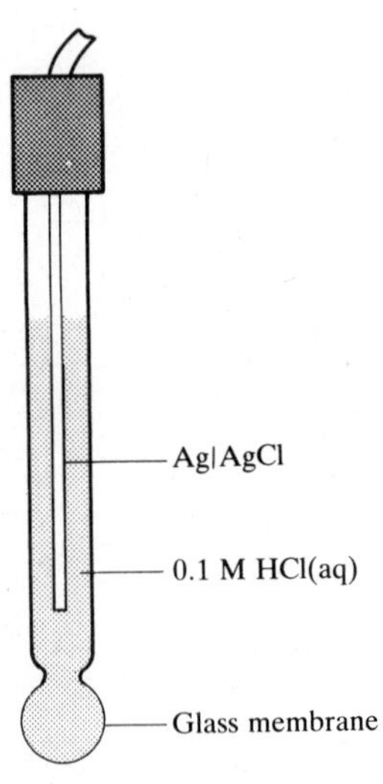

Fig. 12.9. The construction of the glass electrode. It is usually used in conjunction with a calomel electrode which makes contact with the test solution through a salt bridge.

‡ The *operational definition* of the pH of a solution X is that it is given by $\mathrm{pH(X)} = \mathrm{pH(S)} + E/2.3026(RT/F)$, where E is the e.m.f. of the cell

$$\mathrm{Pt}\,|\,\mathrm{H_2}\,|\,\mathrm{X(aq)}\,|\,3.5\,\mathrm{M\,KCl(aq)}\,\|\,\mathrm{S(aq)}\,|\,\mathrm{H_2}\,|\,\mathrm{Pt},$$

the solution S being a solution of standard pH. The primary standard is a 0.05 M aqueous solution of pure potassium hydrogen phthalate, of which the pH is *defined* as being exactly 4 at 15 °C and at other temperatures t/°C as $\mathrm{pH(S)} = 4 + (t - 15)^2/2 \times 10^4$ if t lies between 0 and 55 (e.g. 4.005 at 25 °C). The values of pH given by this definition differ very slightly from the *formal definition*, eqn (12.4.9).

on pH. It is much more convenient to handle than the gas electrode itself, and can be calibrated using solutions of standard pH values.

12.4 (e) Acid–base titrations

In the case of a strong acid/strong base (SA/SB) titration, the *end-point* (or *equivalence point*), when a stoichiometrically equivalent amount of acid has been added to the base, lies at pH = 7 (at 25 °C). In other cases, such as weak acid/strong base (WA/SB) titrations the end-point lies at different pH on account of *hydrolysis* (in general: *solvolysis*) by the salt. The next few paragraphs establish how to take hydrolysis into account, and show how to predict the pH at any stage of a WA/SB or SA/WB titration, and especially at the end-point. We shall deal with molalities, and ignore activity coefficients.

Consider a WA/SB titration in which the weak acid HA is being titrated with the strong base MOH. At any stage of the titration the acid molality is nominally $Am^{\ominus}$ and the salt molality is $Sm^{\ominus}$. Initially, before base has been added, $S=0$. At the end point, just enough base has been added to convert all the acid to salt. The problem is to find the pH at any stage bearing in mind

(1) the existence of the equilibria:

$$HA(aq) + H_2O(l) \rightleftharpoons A^-(aq) + H_3O^+(aq), \qquad K_a = \{m_{H^+}m_{A^-}/m_{HA}m^{\ominus}\}_{eq},$$
$$2H_2O(l) \rightleftharpoons OH^-(aq) + H_3O^+(aq), \qquad K_w = \{m_{H^+}m_{OH^-}/m^{\ominus 2}\}_{eq},$$

(2) the electrical neutrality of the solution:

$$m_{H^+} + m_{M^+} = m_{A^-} + m_{OH^-},$$

(3) the constancy of the number of A groups:

$$m_{HA} + m_{A^-} = (A+S)m^{\ominus}.$$

These relations can be combined to give an equation for $m_{H^+} = Hm^{\ominus}$ by eliminating the molalities of the other species. The final result is

$$H = A/\{1+(H/K_a)\} + K_w/H - S/\{1+(K_a/H)\}. \qquad (12.4.10)$$

This equation can be arranged into a quartic equation for H, and explicit solutions obtained. However, the forms of the solutions are very obscure, and it is more instructive to examine approximations based on the weakness of the acid and the complete dissociation of the salt.

At the start of the titration no salt is present, the hydrogen ions from the acid greatly outnumber those from the water autoionization (so that we can write $m_{H^+} = m_{A^-}$), and the weak acid is only slightly ionized (so that in the denominator of K_a we can write $m_{HA} = Am^{\ominus}$). This means that

$$K_a \approx H^2/A, \text{ so that } H = (AK_a)^{\frac{1}{2}}.$$

Therefore, *the pH of a solution of a weak acid is*

$$pH = \tfrac{1}{2}pK_a - \tfrac{1}{2}\lg A. \qquad (12.4.11)$$

In the case of an 0.10 mol kg^{-1} solution of acetic acid (so that $A = 0.10$), pH $= (\frac{1}{2})(4.75) - (\frac{1}{2})\lg(0.10) = 2.9$.

Near the half-way point the hydrogen ions from the acid still greatly outnumber those from the water autoionization (and so m_{H^+} can be calculated from K_a alone). The A^- molality arises almost entirely from the salt present (so that $m_{A^-} = Sm^{\ominus}$), and the acid is present mainly as non-ionized HA (so that $m_{HA} = Am^{\ominus}$). This means that

$$K_a = HS/A, \qquad \text{or } H = AK_a/S,$$

so that *the pH of a solution containing similar molalities of acid and salt is*

$$\mathrm{pH} = \mathrm{p}K_a - \lg(A/S). \qquad (12.4.12)$$

This is the *Henderson–Hasselbalch equation*. Note that when the molalities of acid and salt are the same, $\mathrm{pH} = \mathrm{p}K_a$, hence the $\mathrm{p}K_a$ of the acid can be measured directly from the pH of the mixture.

At the end point of the WA/SB titration all the acid has been replaced by salt, and so $A = 0$. The A^- molality is almost exactly that of the salt (and so we can write $m_{A^-} = Sm^{\ominus}$), the small discrepancy being due to the fact that A^- is the conjugate base of a weak acid, and so the equilibrium

$$A^-(aq) + H_2O(l) \rightleftharpoons HA(aq) + OH^-(aq);$$
$$K = \{m_{HA}m_{OH^-}/m_{A^-}m^{\ominus}\} = K_w/K_a$$

must be considered. The equilibrium increases the supply of HA, and their molality is equal to the molality of the OH^- ions produced by the same reaction. Therefore, $m_{HA} = m_{OH^-}$, the only error being the negligibly small number of OH^- ions involved in the water autoionization. The OH^- molality is related to the hydrogen ion molality through $m_{OH^-} = K_w m^{\ominus 2}/m_{H^+}$. Collecting these results leads to

$$K = \{m^2_{OH^-}/Sm^{\ominus}\} = K_w^2/SH^2.$$

On combining this with $K = K_w/K_a$ we obtain

$$H = \sqrt{(K_w K_a/S)},$$

so that *the pH of a solution of a salt MA formed from a weak acid HA and a strong base MOH is*

$$\mathrm{pH} = \tfrac{1}{2}\mathrm{p}K_a + \tfrac{1}{2}\mathrm{p}K_w + \tfrac{1}{2}\lg S. \qquad (12.4.13)$$

This is also the pH at the *end-point* of the WA/SB titration.

The corresponding expression for the pH of a solution of a salt $(BH)^+X^-$ formed from a strong acid HX and a weak base B, and therefore also for the pH of the end-point of a SA/WB titration, is

$$\mathrm{pH} = \tfrac{1}{2}\mathrm{p}K_a - \tfrac{1}{2}\lg S. \qquad (12.4.14)$$

Example 12.8

Calculate the pH at the end-point of a titration of 0.01 mol kg^{-1} butyric acid ($\mathrm{p}K_a = 4.82$) with a concentrated solution of strong base at 25 °C.

• *Method*. The end-point of the titration occurs when the amount of base added is equal to the amount of acid present initially (if, as here, the acid is monoprotic). All the acid is then present formally as the salt. Therefore, the pH of the solution is the same as the pH of a solution of the salt itself. Use eqn (12.4.13).

• *Answer*. Since $pK_a = 4.82$, $pK_w = 14.00$, and $S = 0.01$,

$$pH = \tfrac{1}{2}(14.00) + \tfrac{1}{2}(4.82) + \tfrac{1}{2}\lg 0.01 = 8.4.$$

• *Comment*. Note that in an actual titration, the volumes change as titrant is added. The S should be taken as 0.01 modified to account for the expansion of volume arising from the addition of base. If the base and acid solutions are of similar molality, the volume doubles, and so $S = 0.005$. However, this modifies the answer above by only a small amount, to $pH = 8.3$.

• *Exercise*. Calculate the pH of the end point of the titration of $0.10\ \text{mol kg}^{-1}$ $NH_3(aq)$ and $0.10\ \text{mol kg}^{-1}$ HCl(aq). [5.3]

When so much strong base has been added that the titration has been carried well past the end-point, the pH is determined by the excess base present. Since $H = K_w m^{\ominus}/m_{OH^-}$, if we write the excess base molality as $Bm^{\ominus}$, then

$$pH = pK_w + \lg B. \quad (12.4.15)$$

In this expression, as in all the preceding ones, the molalities must take into account the change of volume that occurs as the base is added to the acid sample.

The general form of the pH curve throughout a titration is illustrated in Fig. 12.10. The pH rises slowly from the value given by the 'weak acid alone' formula, eqn (12.4.11), until the end-point is approached, when it changes rapidly to the value given by the 'salt alone' formula, eqn (12.4.13). It then climbs less rapidly towards the value given by the 'base in excess' formula, eqn (12.4.15). The end-point can be detected easily by observing where the pH changes rapidly through $pH = pK_a$. This slow–quick–slow variation of the pH is another example of the form of the logarithmic dependence like that drawn in Fig. 12.7.

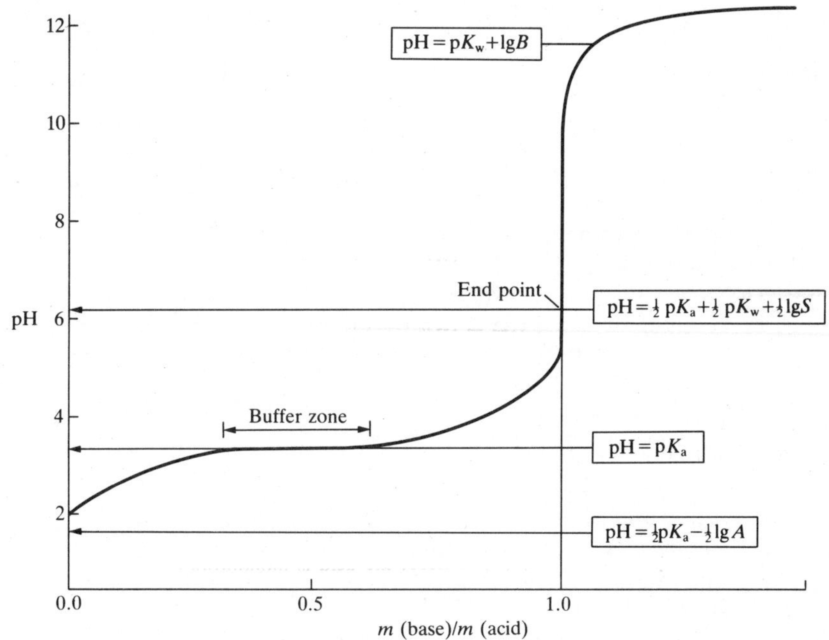

Fig. 12.10. Changes in the pH of a solution during the titration of a weak acid with a strong base. The expressions are taken from equations developed in the text.

12.4 (f) Buffers and indicators

The slow variation of the pH in the vicinity of $S = A$, when the salt and acid molalities are equal, is the explanation of *buffer action*. Buffered solutions maintain approximately the same pH when small amounts of acids or bases are added. The *mathematical* basis of buffer action is the logarithmic dependence given by the Henderson–Hasselbalch equation, eqn (12.4.12), which is quite flat in the vicinity of $\mathrm{pH} \approx \mathrm{p}K_a$. The *physical* basis is that the existence of an abundant supply of A^- ions (as a result of the presence of the salt MA) can remove any hydrogen ions brought by additional acid, and the existence of an abundant supply of non-ionized acid (as a result of the presence of the acid HA) can supply hydrogen ions to any base that is added.

The rapid change of pH near the end-point is used for *indicator detection*. An acid–base indicator is normally some large, water-soluble, organic molecule which can exist as acid (HIn) or conjugate base (In^-) with different colours. The two forms are in equilibrium in solution:

$$\mathrm{HIn(aq)} + \mathrm{H_2O(l)} \rightleftharpoons \mathrm{In^-(aq)} + \mathrm{H_3O^+(aq)}, \qquad K_{\mathrm{In}} = \{a_{\mathrm{In}^-} a_{\mathrm{H}^+}/a_{\mathrm{HIn}}\}_{\mathrm{eq}}.$$

At the end-point, the pH changes sharply through several pH units, so that a_{H^+} changes through several orders of magnitude. Therefore the indicator equilibrium changes so as to accommodate the change of hydrogen ion activity, with HIn the dominant species on the acid side of the end-point and In^- dominant on the basic side. The accompanying colour change signals the end-point of the titration. Care must be taken to choose an indicator that changes colour at the pH appropriate to the type of titration. Thus, in a WA/SB titration, the end point lies at the pH given by eqn (12.4.13), and so an indicator that changes at that pH must be selected; similarly, in a SA/WB titration, an indicator changing near the pH given by eqn (12.4.14) should be used.

Further reading

Principles:

Electrolyte solutions (2nd edn). R. A. Robinson and R. H. Stokes; Academic Press, New York and Butterworth, London, 1959.

Experimental approach to electrochemistry. N. J. Selley; Edward Arnold, London 1977.

Principles and applications of electrochemistry (2nd edn). D. R. Crow; Chapman and Hall, London, 1979.

Modern electrochemistry. J. O'M. Bockris and A. K. N. Reddy; Plenum, New York, 1970.

The principles of electrochemistry. D. A. MacInnes; Dover, New York, 1961.

Treatise on electrochemistry. G. Kortum; Elsevier, Amsterdam, 1952.

Comprehensive treatise of electrochemistry (*Vols 1–7*). J. O'M. Bockris, B. E. Conway, *et al*. (eds.); Plenum, New York, 1980.

Applications:

The study of ionic equilibria. H. S. Rossotti; Longman, London, 1978.

Chemical applications of e.m.f. H. S. Rossotti; Longman, London, 1978.

Acid–base equilibria. E. J. King; Pergamon Press, Oxford, 1965.

The proton in chemistry. R. P. Bell; Cornell University Press, 1959.

Determination of pH: theory and practice. R. G. Bates; Wiley, New York, 1973.

Data:

Electrochemical data. B. E. Conway; Elsevier, Amsterdam, 1952.
The oxidation states of the elements and their potentials in aqueous solutions. W. M. Latimer; Prentice-Hall, Englewood Cliffs, 1952.
Encyclopedia of electrochemistry of the elements (*Vols. 1–6*). A. J. Bard; Marcel Dekker, New York, 1973.

Introductory problems

A12.1. Devise a cell which has the cell reaction $Mn(s) + Cl_2(g) \rightarrow MnCl_2(aq)$. Write the left and right half-cell reactions. From the standard e.m.f. of the cell at 25 °C, namely, 2.54 V, obtain the standard electrode potential for the manganese–manganese(II) ion electrode.

A12.2. For each of the following cells write the cell reactions and the left and right half-cell reactions:
(a) $Zn|ZnSO_4||AgNO_3|Ag$
(b) $Cd|CdCl_2||HNO_3|H_2(g)|Pt$
(c) $Pt|K_3Fe(CN)_6, K_4Fe(CN)_6||CrCl_3|Cr$
(d) $Pt|Cl_2(g)|HCl(ag)||K_2CrO_4(ag)|Ag_2CrO_4|Ag$.

A12.3. Calculate the standard e.m.f. of the cell $Hg(l)|HgCl_2(aq)||TlNO_3(aq)|Tl$ at 25 °C. Calculate the e.m.f. at 25 °C when the activity of the mercuric ion is 0.15 and the activity of the thallous ion is 0.93. The standard potential of the Tl^+, Tl couple is −0.34 V at 25 °C.

A12.4. For the reaction $K_2CrO_4(aq) + 2Ag(s) + FeCl_3(aq) \rightarrow Ag_2CrO_4(s) + 2FeCl_2(aq) + 2KCl(aq)$, $\Delta G^\ominus$ (298 K) is −62.5 kJ. Calculate: (a) the standard e.m.f. at 25 °C for the cell for which the above reaction is the cell reaction, (b) the standard electrode potential at 25 °C for the silver–silver chromate electrode.

A12.5. For the cell $Ag|AgBr|KBr(aq)||Cd(NO_3)_2(aq)|Cd$, the molalities of the bromide ion and the cadmium ion are 0.050 and 0.010 respectively. Use the Debye Limiting Law and the Nernst Equation to calculate the e.m.f. of the cell at 25 °C.

A12.6. Use the information in Tables 12.1 and 12.2 to calculate for the cell $Ag|AgNO_3||Fe(NO_3)_2|Fe$ the standard e.m.f., $\Delta_r G^\ominus$, and $\Delta_r H^\ominus$ at 25 °C. Calculate the standard e.m.f. and $\Delta_r G^\ominus$ at 35 °C.

A12.7. The solubility product constant for $Cu_3(PO_4)_2$ is 1.3×10^{-37} at 25 °C. Calculate: (a) the concentration of the copper(II) ion in a saturated aqueous solution of $Cu_3(PO_4)_2$, (b) the e.m.f. of the cell $Pt|H_2(1\ atm)|HCl(a^+ = 1)||Cu_3(PO_4)_2(aq,\ m = s)|Cu$ where s is the solubility of $Cu_3(PO_4)_2$.

A12.8. Calculate the equilibrium constants for the following reactions at 25 °C: (a) $Sn(s) + Sn^{4+}(aq) \rightleftharpoons 2Sn^{2+}(aq)$, (b) $Sn(s) + 2AgCl(s) \rightleftharpoons SnCl_2(aq) + 2Ag(s)$, (c) $2Ag(s) + Cu(NO_3)_2(aq) \rightleftharpoons Cu(s) + 2AgNO_3(aq)$.

A12.9. For HCN in aqueous solution, $pK_a = 9.31$ at 25 °C. (a) Calculate K_a for HCN. (b) Calculate the pH of an aqueous solution of HCN with a concentration of 0.25 M. (c) Calculate the pH of an aqueous solution which is 0.25 M in HCN and 0.15 M in KCN.

A12.10. At the half-way point in the titration of a weak monobasic acid with sodium hydroxide solution, the pH determined from e.m.f. measurements is 5.40. What is the value of K_a for the acid? Calculate the pH of a solution which is made by dissolving 0.60 mol of the weak acid in enough water to give a solution volume of 2.50 dm³.

Problems

12.1. Write the cell reactions and the half-cell reactions for the following cells:
(a) $Pt|H_2(g)|HCl(aq)|AgCl|Ag$;
(b) $Pt|FeCl_2(aq),FeCl_3(aq)||SnCl_4(aq),SnCl_2(aq)|Pt$;
(c) $Cu|CuCl_2||MnCl_2(aq), HCl(aq)|MnO_2|Pt$;
(d) $Ag|AgCl|HCl(aq)||HBr(aq)|AgBr|Ag$.

12.2. Devise cells in which the following are the reactions:
(a) $Zn(s) + CuSO_4(aq) \rightarrow ZnSO_4(aq) + Cu(s)$;
(b) $AgCl(s) + \frac{1}{2}H_2(g) \rightarrow HCl(aq) + Ag(s)$;
(c) $H_2(g) + \frac{1}{2}O_2(g) \rightarrow H_2O(l)$;
(d) $Na(s) + H_2O(l) \rightarrow NaOH(aq) + \frac{1}{2}H_2(g)$;
(e) $H_2(g) + I_2(s) \rightarrow 2HI(aq)$.

12.3. Evaluate the standard e.m.f.s at $\mathcal{T}$ of the cells in the preceding two problems. Which electrode is the positive electrode?

12.4. Calculate $\Delta_r G^\ominus(\mathcal{T})$ for the following reactions from the electrode potential data in Table 12.1:
(a) $2Na(s) + 2H_2O(l) \rightarrow 2NaOH(aq) + H_2(g)$;
(b) $K(s) + H_2O(l) \rightarrow KOH(aq) + \frac{1}{2}H_2(g)$;
(c) $K_2S_2O_8(aq) + 2KI(aq) \rightarrow I_2(s) + 2K_2SO_4(aq)$;
(d) $Pb(s) + ZnCO_3(aq) \rightarrow PbCO_3(aq) + Zn(s)$.

12.5. Calculate the equilibrium constants for the following reactions at $\mathcal{T}$:
(a) $Sn(s) + CuSO_4(aq) \rightleftharpoons Cu(s) + SnSO_4(aq)$;
(b) $2H_2(g) + O_2(g) \rightleftharpoons 2H_2O(l)$;
(c) $Cu^{2+}(aq) + Cu(s) \rightleftharpoons 2Cu^+(aq)$.

12.6. The cell $Zn|ZnSO_4(aq, a = 1)||CuSO_4(aq, a = 1)|Cu$ was set up in a laboratory experiment. Use the information in Table 12.1 to state (a) its e.m.f. at $\mathcal{T}$, (b) the value of $\Delta_r G^\ominus$ for the cell reaction, (c) the equilibrium constant for the cell reaction. What is the ratio of the activities of the two electrolytes when the cell is 'exhausted'?

12.7. Write the cell reaction for $Al|Al^{3+}(aq)||Sn^{2+}(aq), Sn^{4+}(aq)|Pt$. State (a) the cell e.m.f. when the activities are all 0.10, 1.00, (b) the value of $\Delta_r G^\ominus$ for the reaction, (c) the equilibrium constant for the reaction. Which is the positive electrode? Which way do electrons tend to flow?

12.8. The same cell as in the last Problem was set up, the only difference being that the molalities of Al^{3+}, Sn^{2+}, and Sn^{4+} were each $0.010\ mol\ kg^{-1}$. What is the cell e.m.f.?

12.9. Devise a cell in which the overall reaction is $Pb(s) + Hg_2SO_4(s) \rightarrow PbSO_4(s) + 2Hg(l)$. What is its e.m.f. at $\mathcal{T}$ when the electrolyte it contains is saturated with the two salts? [$K_{sp}(PbSO_4) = 2.43 \times 10^{-8}$, $K_{sp}(Hg_2SO_4) = 1.46 \times 10^{-6}$.]

12.10. What is the e.m.f. at 25 °C of the cell $Pt|H_2(1\ atm)|HCl(m_1)||HCl(m_2)|H_2(1atm)|Pt$ when $m_1 = 0.10\ mol\ kg^{-1}$ and $m_2 = 0.20\ mol\ kg^{-1}$? What would be the e.m.f. of the cell if the pressure of gas at the right-hand electrode were increased to 10 atm? The mean activity coefficients of HCl(aq) are $\gamma_1 = 0.798$ and $\gamma_2 = 0.790$.

12.11. Hydrogen behaves imperfectly at high pressure, and its equation of state may be summarized by the virial expansion $pV_m/RT = 1 + 5.37 \times 10^{-4}(p/atm) + 3.5 \times 10^{-8}(p/atm)^2$. The e.m.f. of the cell $Pt|H_2|HCl\ (0.1\ mol\ kg^{-1})|Hg_2Cl_2|Hg$ has been measured up to pressures of 1000 atm (W. R. Hainsworth, H. J. Rowley, and D. A. McInnes, *J. Am. chem. Soc.* **46**, 1437 (1924)), and good agreement between experiment and a theoretical expression was obtained up to 600 atm on the assumption that the only significant volume change was that associated with the hydrogen. Above that pressure the volume change of the other components became significant. What is the e.m.f. of the cell when the pressure is 500 atm? Take $\gamma_\pm(HCl) = 0.798$, Table 11.2.

12.12. The fugacity of a gas can be determined electrochemically from the pressure dependence of the electrode potential. In the cell $Pt|H_2(1\ bar)|HCl(0.01\ mol\ kg^{-1})|Cl_2|Pt$ the following values of the e.m.f. were obtained at $\mathcal{T}$ for various pressures of chlorine. What is the fugacity and the fugacity coefficient of chlorine at the pressures quoted? Table 11.2 gives $\gamma_\pm$.

p/bar	1.000	50.00	100.0
E/V	1.5962	1.6419	1.6451

12.13. A leak developed in a cell which had been designed to operate as $Pt|H_2(1\ atm)|HCl(m)|AgCl|Ag$. As a result water was able to dribble in and dilute the hydrochloric acid so that its molality changed with time according to $m = m^\ominus/(1 + \kappa t)$. What is the time-dependence of the cell potential? Neglect activity coefficients. Sketch the result of the calculations, and calculate dE/dt.

12.14. A fuel cell converts chemical energy directly into electrical energy without involving the wasteful process of combustion, heating of steam for use in a turbine, and the conversion into electrical energy. What is the maximum e.m.f. obtainable from a cell in which hydrogen and oxygen, both at 1 atm pressure and 298 K, combine on a catalyst surface? What is the maximum electrical work available per mole of $H_2(g)$?

12.15. Much modern research goes into investigating the employment of hydrocarbon gases in fuel cells. Calculate the maximum amount of heat, the maximum work, and the maximum electrical work available from the oxidation of butane at $\mathcal{T}$ and 1 bar.

12.16. Electrochemical cells are very important for finding the values of activities and activity coefficients of electrolytes in solutions (not just aqueous solutions). For a 1:1 electrolyte in aqueous solution at 25 °C the extended Debye–Hückel Law is that $\lg \gamma_\pm = -0.509(m/m^\ominus)^{\frac{1}{2}} + B(m/m^\ominus)$. As a first step in seeing how activity coefficients are measured, show that a plot of $E + 0.1183 \lg(m/m^\ominus) - 0.0602(m/m^\ominus)^{\frac{1}{2}}$ against m, where E is the e.m.f. of the cell $Pt|H_2(1\ atm)|HCl(m)|AgCl|Ag$, should give a straight line with intercept $E^\ominus$ and slope $-0.1183B$. Once $E^\ominus$ has been determined it is a simple matter to measure E and relate it to the activity coefficient.

12.17. The e.m.f. of the cell referred to in the last Problem has been measured at a series of molalities of HCl. The data for 25 °C are below. What is (a) $E^\ominus$ of the cell, (b) $E^\ominus$ of the AgCl,Ag electrode?

m/mol kg^{-1}	0.1238	0.025 63	0.009 138
E/mV	341.99	418.24	468.60

m/mol kg^{-1}	0.005 619	0.003 215
E/mV	492.57	520.53

12.18. The e.m.f. of the same cell as in the last Problem is 0.3524 V at 25 °C when the HCl molality was $0.100\ mol\ kg^{-1}$. What is (a) the hydrogen ion activity, (b) the mean activity coefficient of the HCl at this concentration, (c) the pH of the solution.

12.19. The $Cl^-|AgCl|Ag$ electrode potential has been measured very carefully at a series of temperatures (R. G. Bates and V. E. Bower, *J. Res. Nat. Bur. Stand.* **53**, 283 (1954)), and the results were found to fit the expression

$$E^\ominus/V = 0.236\,59 - 4.8564 \times 10^{-4}(\theta/°C) - 3.4205 \times 10^{-6}(\theta/°C)^2 + 5.869 \times 10^{-9}(\theta/°C)^3,$$

$E^\ominus$ being the e.m.f. of the cell with an SHE as the second electrode. Calculate the Gibbs function, enthalpy, and entropy of the cell reaction, and those of the Cl^- ion in water at 298 K.

12.20. The cell $Pt|H_2(1\ atm)|HCl(aq)|AgCl|Ag$ has an e.m.f. of 0.332 V at 25 °C: what is the pH of the solution?

12.21. The solubility of AgBr is $2.1 \times 10^{-6}\ mol\ kg^{-1}$ at 25 °C. What is the e.m.f. of the cell $Ag|AgBr(aq)|AgBr|Ag$ at that temperature?

12.22. The standard e.m.f. of the cell $Ag|AgI(aq)|AgI|Ag$

is −0.9509 V at $\mathcal{T}$. What is (a) the solubility of AgI in water at this temperature, (b) the solubility product.

12.23. Measurements of the e.m.f. of cells of the type $Ag|AgX|MX(m_1)|M_xHg|\,|MX(m_2)|AgX|Ag$, where M_xHg denotes an amalgam and the electrolyte is an alkali metal halide dissolved in ethylene glycol, have been reported (U. Sen, *J. chem. Soc. Faraday Trans.* I **69,** 2006 (1973)) and a selection for LiCl at 25 °C is given below. Estimate the activity coefficient at the concentration marked * and then use this value to calculate the activity coefficients from the measured e.m.f. at the other concentrations.

$m/\text{mol kg}^{-1}$	0.0555	0.0914*	0.1652
E/V	−0.0220	0.0000	0.0263
$m/\text{mol kg}^{-1}$	0.2171	1.040	1.350
E/V	0.0379	0.1156	0.1336

Base your answer on the use of the extended Debye–Hückel expression $\lg \gamma_\pm = -Am^{\frac{1}{2}}/(1 + A^*m^{\frac{1}{2}}) + Bm$, $m \equiv m/m^\ominus$, using $A = 1.461$, $A^* = 1.70$, $B = 0.20$.

12.24. The e.m.f. of the cell $Pt|H_2(1\,\text{atm})|HCl(m)$ $|Hg_2Cl_2|Hg$ has been measured with high precision (G. J. Hills and D. J. G. Ives, *J. chem. Soc.*, 311 (1951)) with the following results at

$100m/m^\ominus$	0.160 77	0.307 69	0.504 03
E/mV	600.80	568.25	543.66
$100m/m^\ominus$	0.769 38	1.094 74	
E/mV	522.67	505.32	

Find the standard e.m.f. of the cell and the experimental values of the HCl activity coefficients at these molalities. (Make a least-squares fit of the data to the best straight line, and quote the coefficient of determination.)

12.25. In view of the importance of knowing the enthalpy of ionization of water, and because there were discrepancies between the published values, careful measurements of the e.m.f. of the cell $Pt|H_2(1\,\text{atm})|NaOH,$ $NaCl|AgCl|Ag$ have been reported (C. P. Bezboruah, M. F. G. F. C. Camoes, A. K. Covington, and J. V. Dobson, *J. chem. Soc. Faraday Trans.* I **69,** 949 (1973)). Among the data is the following information: $m(\text{NaOH}) = 0.0100$ mol kg^{-1}; $m(\text{NaCl}) = 0.011\,25$ mol kg^{-1};

$\theta/°\text{C}$	20.0	25.0	30.0
E/V	1.047 74	1.048 64	1.049 42

Calculate the values of pK_w at these temperatures, and the enthalpy and entropy of ionization of water at 25 °C.

12.26. There is some interest in the structure of water/urea mixtures on account of the protein denaturation that occurs in them. The e.m.f. of the cell $Pt|H_2(1\,\text{atm})|$ $HCl(m)$, water, urea$|AgCl|Ag$ has been measured for a variety of different mixtures and temperatures (K. K. Kundu and K. Mazumdar, *J. chem. Soc. Faraday Trans.* I **69,** 806 (1973)). The following data were obtained at 25 °C for a 29.64 weight per cent urea/water mixture (relative permittivity 91.76, density 1.0790 g cm^{-3}). Find the standard e.m.f. of the AgCl|Ag electrode in this medium.

$m(\text{HCl})/\text{mol kg}^{-1}$	0.005 58	0.013 10	0.019 2
E/V	0.561 6	0.518 7	0.499 9
$m(\text{HCl})/\text{mol kg}^{-1}$	0.024 6	0.034 9	0.041 1
E/V	0.487 8	0.470 8	0.462 9

12.27. Calculate the A-coefficient from the Debye–Hückel Limiting Law for the urea/water mixture referred to in the last Problem, and compare its value with experiment.

12.28. Calculate the pH of (a) a 0.10 M solution of ammonium chloride, (b) a 0.10 M solution of sodium acetate, (c) a 1.0 M solution of sodium acetate, (d) a 0.10 M solution of acetic acid. Data in Table 12.3.

12.29. An 0.10 M solution of sodium lactate was prepared at 25 °C. What is the pH of the solution? What is the pH of the end point of the titration of lactic acid with NaOH?

12.30. Plot the pH of a solution containing 0.10 M of sodium acetate and a variable amount of acetic acid. Notice that the pH changes least quickly in the region of pH ≈ pK_a, the buffer region, and that the stability is greatest for equimolar mixtures of acid and salt. In what region of pH will a mixture of boric acid, $pK_a = 9.14$, and sodium borate (Na_2HBO_3), molalities 0.1 mol kg^{-1} for both, be an effective buffer?

12.31. From the data in Table 12.3, select combinations of acids and salts that have their maximum buffering ability at (a) pH ≈ 2.2, (b) pH ≈ 7.0.

12.32. Quartic equations can be solved in a closed form, just like quadratic equations, but the expression for the four roots is very clumsy: see, for instance, M. Abramowitz and I. A. Stegun, *Handbook of mathematical functions*, Dover (1965), Section 3.8.3. If you have access to a computer or a programmable calculator, solve eqn (12.4.10) for H and hence pH, for a wide range of concentrations of added weak acid. Take $K_a = 2.69 \times 10^{-5}$ for the whole range of acid concentrations.

12.33. In order to deal with corrosion processes we have to treat them as dynamical processes. Nevertheless, the *tendency* of a system to corrode may be assessed thermodynamically, and we look into that here (the full subject is taken up in Part 3). Metals corrode by being oxidized, the driving reactions being the reductions

$$2H_3O^+ + 2e^- \rightarrow 2H_2O + H_2 \qquad E^\ominus = 0$$

$$O_2 + 4H^+ + 4e^- \rightarrow 2H_2O \qquad E^\ominus = 1.23\,\text{V}$$

(acid conditions)

$$O_2 + 2H_2O + 4e^- \rightarrow 4OH^- \qquad E^\ominus = 0.40\,\text{V, alkaline conditions}$$

Which of the following metals will corrode in water made slightly acid (pH ≈ 6.0)? Which will corrode in water made slightly alkaline (pH $\approx$ 8.0)? Which corrode only in strong acid (pH $\approx$ 1) and which corrode only in strong alkali (pH $\approx$ 14)? Take the criterion of corrosion to be the thermodynamic tendency to form *at least* a 10^{-6} M solution of ions in the vicinity of the metal surface. The metals to consider are (a) Al, (b) Cu, (c) Fe, (d) Ag, (e) Pb, (f) Au.

12.34. An industrial process produces a 1 M solution of sodium acetate. Can it be collected and stored in mild steel vessels?

12.35. Superheavy elements are now of considerable interest. Shortly before it was believed that the first had been discovered an attempt was made to predict the chemistry of element 115 (O. L. Keller, C. W. Nestor, and B. Fricke, *J. phys. Chem.* **78,** 1945 (1974)). In one part of the paper the enthalpy and entropy of the reaction $(115)^{+}(aq) + \frac{1}{2}H_2(g) \rightarrow (115)(s) + H^{+}(aq)$ were estimated from the following data: $\Delta H_{sub,m} = 1.5\ \text{eV mol}^{-1}$, $I = 5.2\ \text{eV}$, $\Delta H(115^{+}, aq)_{solv,m} = -3.22\ \text{eV}$, $S^{\ominus}(115^{+}, aq) = 1.34 \times 10^{-3}\ \text{eV K}^{-1}\ \text{mol}^{-1}$, $S^{\ominus}(115, s) = 0.69 \times 10^{-3}\ \text{eV K}^{-1}\ \text{mol}^{-1}$. Estimate the expected standard electrode potential of the 115^{+},115 couple.

Part 2 · Structure

13

Quantum theory: introduction and principles

Learning objectives

After careful study of this chapter you should be able to:

(1) List the characteristics of *black-body radiation* and explain how *Planck's hypothesis* of the quantization of energy explains them, Section 13.2(a).

(2) Explain how the characteristics of the *heat capacities* of solids at low temperatures conflict with classical physics but are explained by the quantum theory, Section 13.2(b).

(3) Describe the experimental features of the *photoelectric effect*, Section 13.2(c), and show how the quantum theory accounts for them in terms of *photons*, eqn (13.2.6).

(4) Describe the features of the *Compton effect* and explain how quantum theory accounts for them, Section 13.2(d).

(5) Summarize the evidence for the *wave-nature of matter*, Section 13.2(e), and state the *de Broglie relation*, eqn (13.2.8).

(6) Explain why *atomic and molecular spectra* provide evidence for quantization, Section 13.2(f).

(7) Write the *Schrödinger equation* for a general potential energy, eqn (13.3.1).

(8) Show that the solutions of the Schrödinger equation for a free particle lead to the de Broglie relation, eqns (13.3.2–4).

(9) Describe the relation between the *curvature* of the wavefunction and the *kinetic energy* of the particle it describes, Section 13.3(a).

(10) State the *Born interpretation* of the wavefunction, Section 13.3(b).

(11) Justify and write the *normalization condition*, eqn (13.3.7), and *normalize* a wavefunction, Example 13.3.

(12) List the conditions that must be satisfied by a wavefunction for it to be acceptable, and explain how these lead to quantization, Section 13.3(c).

(13) Explain what is meant by an *operator*, an *eigenvalue equation*, an *eigenvalue*, and an *eigenfunction*, eqn (13.4.3).

(14) Explain the relation between *operators* and *observables*, eqn (13.4.4).

(15) Write the operator for the *linear momentum*, eqn (13.4.5), and the *position* of a particle.

(16) Write the wavefunctions for a particle in a definite state of linear momentum, Section 13.4(a).

(17) Explain the term *superposition* and its interpretation, Section 13.4(b).

(18) Explain the significance of an *expectation value*, Section 13.4(b), and calculate its value for a given wavefunction, Example 13.6.

(19) State the *uncertainty principle*, eqn (13.4.8), and justify it in terms of a superposition of wavefunctions, Section 13.4(c).

(20) Explain the meaning of the expression *complementary observables*, Section 13.4(c).

Introduction

In Part 1 the properties of bulk matter were examined from the viewpoint of thermodynamics. In Part 2 we turn to the study of individual atoms and molecules from the viewpoint of quantum mechanics. The two viewpoints merge in Chapter 21.

13.1 | Quantum theory: introduction and principles

We are all familiar with the way everyday objects move along trajectories in response to forces, how they can be brought to rest and examined, and how they can be accelerated into definite states of motion. These properties can be expressed quantitatively in terms of *classical mechanics* by solving the laws of motion introduced by Newton. Until the present century it was assumed that these classical concepts and laws applied to objects as small as atoms. Experimental evidence was accumulated, however, which showed that classical mechanics failed when it was applied to very small particles, and it was not until 1926 that the appropriate concepts and equations for describing them were discovered: this new mechanics is called *quantum mechanics*.

13.1 Classical mechanics: some central ideas

The way in which classical mechanics describes systems can be illustrated by two equations. One equation expresses the total energy of a particle in terms of its kinetic energy $\frac{1}{2}mv^2$, where v is its speed at the time of interest and m is its mass, and the potential energy V at the location of the particle:

$$E = \tfrac{1}{2}mv^2 + V; \quad x \text{ and } v \text{ functions of } t.$$

In terms of the *linear momentum*, $p = mv$, this is

$$E = p^2/2m + V. \qquad (13.1.1)$$

This equation can be used in a number of ways. For example, since $p = m(\mathrm{d}x/\mathrm{d}t)$, it is a differential equation for x as a function of t, and its solution gives the position (and the momentum) of the particle as functions of time. A statement of both $x(t)$ and $p(t)$ is called the *trajectory* of the particle. The implication is that the future trajectory of the particle can be predicted exactly if its present position and momentum are known.

The simplest example of this procedure is the case of a uniform, constant potential, so that V is independent of x and t. Then, with V set equal to zero for simplicity, the equation is

$$E = p^2/2m, \quad \text{or} \quad (2E/m)^{\frac{1}{2}} = \mathrm{d}x/\mathrm{d}t,$$

the solution being

$$x(t) = x(0) + (2E/m)^{\frac{1}{2}}t.$$

The constant energy E can be expressed in terms of the initial momentum $p(0)$, and so the trajectory is

$$x(t) = x(0) + p(0)t/m, \qquad p(t) = p(0). \qquad (13.1.2)$$

Hence, knowing the initial position and momentum, all later positions and momenta can be predicted.

The second basic equation of classical mechanics is Newton's Second Law of motion:

$$\dot{p} = F, \qquad (13.1.3)$$

where $\dot{p} = \mathrm{d}p/\mathrm{d}t$, the rate of change of momentum (which is proportional to the acceleration, $\dot{p} = m(\mathrm{d}^2x/\mathrm{d}t^2)$) and F is the force acting on the particle. It follows that if we know the force acting everywhere and at all times, then solving this equation will also give the trajectory. This calculation is equivalent to the one based on E, but is more suitable in some applications. For example, consider the case of a particle that is subjected to a constant force F for a time τ, and is then allowed to travel freely. Newton's equation becomes

$$\mathrm{d}p/\mathrm{d}t = F, \text{ a constant, for times between } t=0 \text{ and } t=\tau,$$
$$\mathrm{d}p/\mathrm{d}t = 0 \text{ for times later than } t=\tau.$$

The first equation has the solution

$$p(t) = p(0) + Ft \qquad (0 \leqslant t \leqslant \tau),$$

and at the end of the period the particle's momentum is

$$p(\tau) = p(0) + F\tau.$$

The second equation has the solution $p = \text{constant}$, and so at all times later than $t = \tau$ the particle has a momentum $p(\tau)$ as given above. For simplicity, we shall suppose that the particle is initially at rest, and set $p(0) = 0$. The kinetic energy is $p^2/2m$, and so in this example its value is $F^2\tau^2/2m$ at all times after the force ceases to act. Therefore, the total energy of the accelerated particle has been increased to $F^2\tau^2/2m$ by the force, and since F and τ may take any values, *the energy of the particle may take any value*.

The same type of calculation may be applied to more complex cases. For example, we can calculate how much energy may be given to a rotating body. The *angular momentum* (symbol: J) is related to the *angular velocity* (symbol: ω) by $J = I\omega$, where I is the *moment of inertia*. (The analogous roles of m and I, of v and ω, and of p and J in the translational and rotational cases should be remembered, because they provide a ready way of constructing and recalling equations.) In order to accelerate a rotation it is necessary to apply a *torque* (a twisting force, symbol: T), and Newton's equation is $\dot{J} = T$. If the torque is applied for a time τ, then the rotational energy of the body is increased by $T^2\tau^2/2I$. This means that an appropriate torque and period for which it is applied can excite the rotation to an arbitrary energy.

The final example is the *harmonic oscillator*. Harmonic oscillation occurs when a particle experiences a restoring force with a strength linearly proportional to the displacement, so that $F = -kx$, k being the *force constant*: a strong spring has a large force constant. The negative sign in F signifies that the force is directed opposite to the displacement: when x is positive (displacement to the right), the force is negative (pushing towards the left), and vice versa. Newton's equation is now

$$m(\mathrm{d}^2x/\mathrm{d}t^2) = -kx,$$

and a solution is

$$x(t) = A \sin \omega t, \quad \text{with} \quad \omega = (k/m)^{\frac{1}{2}}. \tag{13.1.4}$$

The momentum is $m\dot{x}$, and so

$$p(t) = m\omega A \cos \omega t. \tag{13.1.5}$$

The properties of this motion are probably familiar: the position of the particle varies *harmonically* (as $\sin \omega t$) with a frequency $\nu = \omega/2\pi$; the momentum is least when the displacement has its maximum value (when

$x = A$, A being the *amplitude* of the motion), and is greatest when the displacement is least (when $x = 0$); the total energy is $\frac{1}{2}kA^2$.‡ Hence, the energy of the oscillating particle can be raised to any value by a suitably controlled impulse that knocks it to an amplitude A. It is important to note that the *frequency* of the motion depends only on the structure of the oscillator (as represented by k and m) and is independent of the energy; the *amplitude* governs the energy, through $E = \frac{1}{2}kA^2$, and is independent of the frequency.

The lessons of these examples are that classical physics (1) predicts a precise trajectory and (2) allows the translational, rotational, and vibrational modes of motion to be excited to any energy simply by controlling the forces, torques, and impulses we apply. These conclusions agree with everyday experience.

Everyday experience, however, does not extend to individual atoms, and careful experiments have shown that the laws of classical mechanics fail when we deal with the transfers of very small quantities of energy. Both conclusions mentioned in the last paragraph are overturned by quantum mechanics. Equation (13.1.2) fails because it turns out that it is impossible to know x and p simultaneously and that the concept of a definite trajectory is untenable. Equation (13.1.3) fails, because it turns out that it is impossible to transfer arbitrary amounts of energy. The results of classical mechanics are in fact only approximations to the true behaviour of particles, and the approximations fail when small masses, small moments of inertia, and small transfers of energy are involved.

13.2 The failures of classical physics

A number of experiments done late in the nineteenth century and early in this gave results totally at variance with the predictions of classical physics. All, however, could be explained on the basis that *classical physics is wrong in allowing systems to possess arbitrary amounts of energy*. When this key idea was pursued, quantum mechanics was discovered.

13.2 (a) Black-body radiation

A hot object emits electromagnetic radiation. At high temperatures an appreciable proportion of the radiation is in the visible region of the spectrum, and a higher proportion of short-wavelength, blue, light is generated as the temperature is raised. This is seen when an object glowing red hot becomes white hot when heated further. The precise dependence is illustrated in Fig. 13.1, which shows how the energy output depends on the wavelength at several temperatures. The curves are those of an ideal emitter, a *black-body*, one that is capable of emitting and absorbing all frequencies of radiation uniformly. A good approximation to a black body is a pin-hole in a heated container, because the radiation leaking out of the hole has been absorbed and reemitted inside so many times that it has come to thermal equilibrium with the walls.

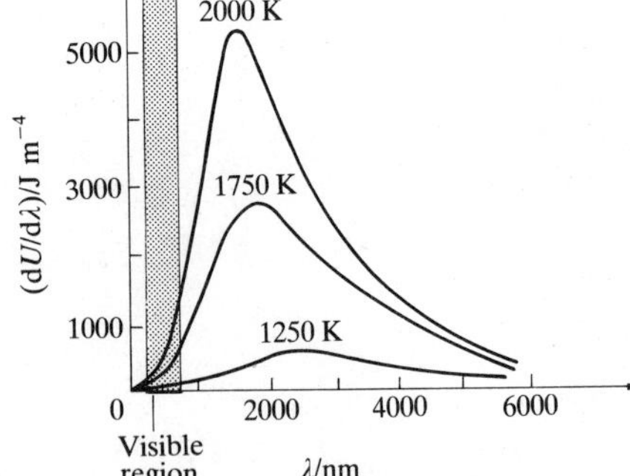

Fig. 13.1. The energy per unit volume per unit wavelength range in a black-body cavity at several temperatures. Note how the energy density increases in the visible region as the temperature is raised, how the peak shifts to shorter wavelengths, and how the total energy density (the area under the curve) increases.

‡ The total energy is the sum of the kinetic and potential energies. The potential energy is related to the force by $F = -\mathrm{d}V/\mathrm{d}x$, and since in this case $F = -kx$ it follows that $V = \frac{1}{2}kx^2$ if $V = 0$ at $x = 0$. The total energy is therefore

$$E = p^2/2m + V = p^2/2m + \tfrac{1}{2}kx^2 = \tfrac{1}{2}kA^2,$$

by substitution of eqns (13.1.4) and (13.1.5) and use of $\sin^2\theta + \cos^2\theta = 1$.

The illustration shows two main features. The first is that the peak shifts to shorter wavelengths as the temperature is raised, and the short-wavelength tail spreads through the whole of the visible region. This corresponds to the perceived colour shifting towards the blue, as already mentioned. Analysis of the data led Wilhelm Wien to the conclusion that the wavelength corresponding to the emission maximum is related to the temperature by

$$\textit{Wien's Displacement Law: } T\lambda_{\max} = \text{constant.} \qquad (13.2.1)$$

The experimental value of the constant is 2.9×10^{-3} m K, and so at 1000 K $\lambda_{\max} \approx 2900$ nm. The second feature was noticed by Josef Stefan. He considered the *total energy density* (symbol: $\mathcal{U}$), the energy per unit volume carried by the electromagnetic field summed over all wavelengths, and concluded that it follows what we now call

$$\textit{Stefan's Law: } \mathcal{U} = aT^4. \qquad (13.2.2)$$

An alternative form of this law is in terms of the *excitance* (symbol: M), the power emitted per unit area. The power emitted is proportional to the energy density of the emitter, and so M is also proportional to T^4, and we write $M = \sigma T^4$. The constant σ is called the *Stefan–Boltzmann constant*. Its experimental value is 5.67×10^{-8} W m^{-2} K^{-4}, so that 1 cm^2 of the surface of a black body at 1000 K radiates about 5.7 W when all wavelengths are taken into account. (The strong temperature-dependence of the excitance is said to be a factor in the adaptation of Merino sheep to the high temperatures they encounter in Australia: their dark coats function as good black-body radiators.)

Lord Rayleigh, with minor help from James Jeans, took a classical view, and thought of the electromagnetic field as a collection of harmonic oscillators, one for each possible frequency of light. Then the occurrence of light of some frequency ν (and therefore of wavelength $\lambda = c/\nu$) was interpreted as the exitation of the oscillator of that frequency. According to the equipartition theorem (Section 0.1(f)), the average energy of an oscillator at a temperature T is kT. Therefore, the energy density in the range of wavelengths λ to $\lambda + d\lambda$ is the number of oscillators per unit volume in that range, $d\mathcal{N}(\lambda)$, multiplied by their average energy kT; hence $d\mathcal{U}(\lambda) = kT\, d\mathcal{N}(\lambda)$. An explicit calculation of $\mathcal{N}$ then gives the

$$\textit{Rayleigh–Jeans Law: } d\mathcal{U}(\lambda) = \rho(\lambda)\, d\lambda, \quad \rho(\lambda) = 8\pi kT/\lambda^4. \qquad (13.2.3)$$

The factor ρ is the energy per unit volume per unit wavelength, and is called the *density of states*. This means that when ρ is multiplied by the wavelength range we get the energy density, $d\mathcal{U}(\lambda)$, in the range of wavelengths from λ to $\lambda + d\lambda$, and when that energy density is multiplied by the volume of the region, we get the total energy in the region due to radiation in that wavelength range. Unfortunately (for Rayleigh and Jeans), as λ decreases their $\rho(\lambda)$ increases without going through a maximum, Fig. 13.2. This implies that oscillators of extremely short wavelength (high frequency, corresponding to ultraviolet light, X-rays, and even γ-rays) are strongly excited even at room temperature. According to classical physics, objects should glow in the dark: there would in fact be no darkness. This

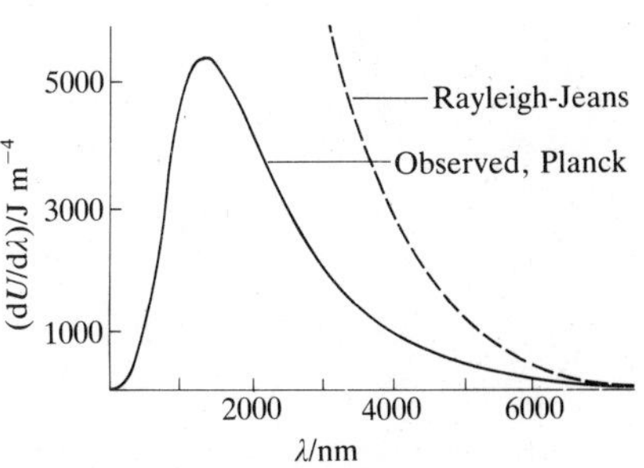

Fig. 13.2. Theoretical attempts to account for black-body radiation. The Rayleigh–Jeans Law, eqn (13.2.3), leads to an infinite energy density at short wavelengths: this is the *ultraviolet catastrophe*. The Planck distribution, eqn (13.2.4), is in good agreement with experiment.

absurd result is the *ultraviolet catastrophe;* but the conclusion is unavoidable if classical mechanics is used.

Max Planck studied this problem from the viewpoint of thermodynamics, in which he was an expert. He found that he could account for the experimental observations by proposing the hypothesis that *energy is quantized*. That is, he needed to suppose that *the energy of a radiation oscillator of given frequency is limited to discrete values and cannot be varied arbitrarily*. In particular, he supposed that in the case of an oscillator of frequency ν, its permitted energies are all integral multiples of $h\nu$, where h is a constant now known as *Planck's constant*. The modern value for h is 6.626×10^{-34} J s. Since an oscillator of frequency ν can possess only the energies 0, $h\nu$, $2h\nu$, and so on, a ray of light of that frequency can be thought of as consisting of a stream of particles, each one having an energy $h\nu$. These particles are called *photons*. This means that if the ray carries an energy E into some region, then the number of photons arriving is $E/h\nu$. For example, in 1 s a 100 W yellow lamp ($100\ \text{W} = 100\ \text{J s}^{-1}$, yellow light has $\lambda = 560$ nm, or $\nu = 5.4 \times 10^{14}$ Hz) generates $(1\ \text{s}) \times (100\ \text{J s}^{-1})/(6.626 \times 10^{-34}\ \text{J s}) \times (5.4 \times 10^{14}\ \text{Hz}) = 2.8 \times 10^{20}$ photons.

The implications of Planck's hypothesis for black-body radiation are as follows. The particles in the walls of the black body are in thermal motion, and this motion excites the oscillators of the electromagnetic field. At equilibrium there is no net flow of energy between the walls and the field. According to classical theory, all the oscillators of the field share equally in the energy supplied by the walls, and so even the highest frequencies are excited. In quantum theory, however, the oscillators are excited only if they can acquire an energy of at least $h\nu$. This is too large for the walls to supply in the case of the high-frequency oscillators, and so they remain unexcited. Therefore, *the effect of quantization is to eliminate the contribution from the high-frequency oscillators*, for they cannot be excited with the energy available. Detailed calculation (of the kind we do in Chapter 21) then shows that the energy density in the range λ to $\lambda + \mathrm{d}\lambda$ is given by the

$$\textit{Planck distribution:}\ \mathrm{d}\mathcal{U} = \rho(\lambda)\,\mathrm{d}\lambda, \tag{13.2.4a}$$

$$\rho(\lambda) = (8\pi hc/\lambda^5)\left\{\frac{\mathrm{e}^{-hc/\lambda kT}}{1 - \mathrm{e}^{-hc/\lambda kT}}\right\}. \tag{13.2.4b}$$

The Planck density of states is similar to the Rayleigh–Jeans expression apart from the all-important factor involving exponentials. When the wavelength is short the term $hc/\lambda kT$ is large and $\exp(-hc/\lambda kT) \approx 0$; then the energy density is zero, in agreement with observation. On the other hand, when the wavelength of interest is long, $hc/\lambda kT$ is small, and the exponentials can be approximated by $1 - (hc/\lambda kT)$. In this case, the Planck expression reduces to the Rayleigh–Jeans expression. Notice that the classical expression is also obtained if h is erroneously set equal to zero (take the limit of the Planck distribution as $h \to 0$ in the numerator and the denominator). The complete expression fits the experimental curve very well at all wavelengths, Fig. 13.2, and h may be obtained by adjusting its value for the best fit.

Example 13.1

A spherical cavity of volume $1.0\,\text{cm}^3$ (e.g. the interior of a small lamp bulb) is heated to 1500 K. Calculate the energy inside the cavity due to radiation in the wavelength range 550–575 nm.

• *Method*. Use eqn (13.2.4) for the density of states in the middle of the wavelength range, then multiply by $\delta\lambda = 25$ nm for the energy density. Finally, multiply by the volume of the cavity.

• *Answer*. Take $\lambda = 562.5$ nm, then

$$8\pi hc/\lambda^5 = 8\pi \times (6.6262 \times 10^{-34}\,\mathrm{J\,s}) \times (2.9979 \times 10^8\,\mathrm{m\,s^{-1}})/(5.625 \times 10^{-7}\,\mathrm{m})^5$$
$$= 8.866 \times 10^7\,\mathrm{J\,m^{-4}}.$$

$$e^{-hc/\lambda kT} = \exp\left\{\frac{-(6.6262\times 10^{-34}\,\mathrm{J\,s})\times(2.9979\times 10^9\,\mathrm{m\,s^{-1}})}{(5.625\times 10^{-7}\,\mathrm{m})\times(1.3807\times 10^{-23}\,\mathrm{J\,K^{-1}})\times(1500\,\mathrm{K})}\right\}$$
$$= e^{-17.05} = 3.94\times 10^{-8}.$$

Therefore,

$$\rho(562.5\,\mathrm{nm}) = 8.866\times 10^7\,\mathrm{J\,m^{-4}}\times\left\{\frac{3.94\times 10^{-8}}{1-3.94\times 10^{-8}}\right\}$$
$$= 3.49\,\mathrm{J\,m^{-4}}.$$

The energy density is therefore

$$\mathcal{U}(562.5\,\mathrm{nm}) = (3.49\,\mathrm{J\,m^{-4}}) \times (25 \times 10^{-9}\,\mathrm{m}) = 8.7 \times 10^{-8}\,\mathrm{J\,m^{-3}}.$$

The total energy due to the radiation in the range 550–575 nm in the 1.0 cm^3 cavity is therefore 8.7×10^{-14} J.

• *Comment*. Since each photon carries an energy hc/λ, it follows that there are about 2.5×10^5 yellow photons in the cavity. For more precise calculations, and when the wavelength range is wide, the energy density must be evaluated by integration of eqn (13.2.4) over the specified range (e.g. numerically).

• *Exercise*. Calculate the energy due to infrared radiation in the wavelength range 1000–1025 nm, and the number of infrared photons, in the same cavity.

[$9.0 \times 10^{-12}\,\mathrm{J\,m^{-3}}$, 5×10^7]

The Planck distribution accounts for the Stefan and Wien Laws. The Stefan Law is obtained by integrating the energy density from all wavelengths from $\lambda = 0$ to $\lambda = \infty$ for a given temperature. This gives $\mathcal{U} = aT^4$, with $a = 4\sigma/c$ and $\sigma = \pi^2k^4/60c^2h^3$. Substitution of the values of the fundamental constants then gives $\sigma = 5.670 \times 10^{-8}\,\mathrm{W\,m^{-2}\,K^{-4}}$, in accord with the experimental value. The Wien Law is obtained by looking for the wavelength at which $d\rho/d\lambda$ is zero: for high temperatures this gives the condition $\lambda_{max}T = hc/5k$. The value of the constant works out as 2.878×10^{-3} m K, which also agrees with experiment.

The important implication of the agreement that Planck found between his theory and experiment is that oscillators cannot possess arbitrary amounts of energy. This *quantization* (from the Latin *quantum*, amount) is wholly alien to classical theory. Its discovery meant that the whole structure of physics had to be revised.

13.2 (b) Heat capacities

Accounting for black-body radiation involves examining how energy is taken up by the electromagnetic field. Accounting for the heat capacities of solids involves examining how energy is taken up by the vibrations of particles. Therefore, a study of heat capacities can also be expected to show evidence of quantization.

If classical physics were valid, the mean energy of vibration of an atom oscillating in one dimension in a solid should be kT (this is the equipartition principle again, Section 0.1(f)). Each of the N atoms in a block is free to vibrate in three dimensions, and so the total vibrational energy of a block is expected to be $3NkT$. The molar vibrational energy is therefore $3N_AkT = 3RT$, N_A being Avogadro's constant and $R = N_Ak$ the gas constant. The molar heat capacity at constant volume (Section 2.3(a)) is $C_{V,m} = (\partial U_m/\partial T)_V$, and so classical physics predicts that $C_{V,m} = 3R$, independent of the temperature. This result is known as *Dulong and Petit's Law*, for it had been proposed by them on the basis of some experimental evidence.

Einstein turned his attention to heat capacities when technological advances made it possible to test Dulong and Petit's Law at low temperatures. Significant deviations were observed, and all metals were found to have molar heat capacities lower than $3R$ at low temperatures, and the values appeared to approach zero as $T \to 0$. Einstein assumed that each atom could vibrate about its equilibrium position with a single frequency ν, and then used Planck's hypothesis to assert that the energy of any oscillation is $nh\nu$, n being an integer. First he calculated the molar vibrational energy of the metal (by a method we describe in Section 2.23(a)) and obtained

$$U_m = 3N_Ah\nu\left\{\frac{e^{-h\nu/kT}}{1 - e^{-h\nu/kT}}\right\}.$$

Notice the similarity of this result to the Planck distribution. Then he found the heat capacity by differentiating with respect to T. This gave the

$$\textit{Einstein formula: } C_{V,m} = 3R(h\nu/kT)^2\left\{\frac{e^{-h\nu/kT}}{(1 - e^{-h\nu/kT})^2}\right\}. \quad (13.2.5)$$

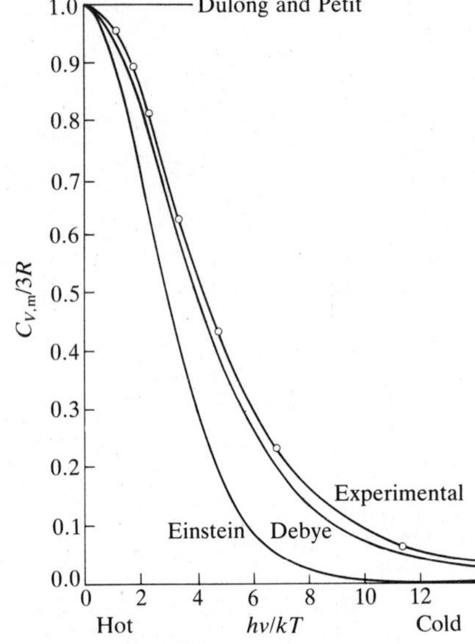

Fig. 13.3. Experimental low-temperature heat capacities and theoretical predictions. The Dulong and Petit Law predicts no falling off at low temperatures; the Einstein equation predicts the general temperature dependence fairly well, but is everywhere too low. Debye's modification of Einstein's calculation gives very good agreement with experiment. For copper, '2' corresponds to about 170 K, and so the detection of deviations from Dulong and Petit's Law had to await advances in low-temperature physics.

At high temperatures ($kT \gg h\nu$) the exponentials can be expanded as $1 - (h\nu/kT) + \ldots$ and higher terms ignored. The result is

$$C_{V,m} = 3R(h\nu/kT)^2\left\{\frac{1 - h\nu/kT + \ldots}{(1 - 1 + h\nu/kT - \ldots)^2}\right\} = 3R,$$

in agreement with the classical result. At low temperatures, $e^{-h\nu/kT} \to 0$, and so Einstein's formula accounts for the decrease of heat capacity at low temperatures. The physical reason for this success is that, as in the Planck calculation, at low temperatures only a few oscillators possess enough energy to begin oscillating. At higher temperatures there is enough energy available for all the oscillators to become active: all $3N$ oscillators contribute, and the heat capacity approaches its classical value.

The overall temperature dependence predicted by the Einstein formula is plotted in Fig. 13.3: the general shape of the curve is satisfactory, but the numerical agreement is poor. The poor numerical agreement arises from Einstein's assumption that all the atoms oscillate with the same frequency. In fact, they oscillate with a range of frequencies. This complication can be taken into account by averaging over all the frequencies present, the final result being the *Debye formula*. The details of this modification, which, as Fig. 13.3 shows, gives improved agreement with experiment, need not deflect us at this stage from the main conclusion that quantization must be introduced in order to explain the thermal properties of solids.

13.2 (c) The photoelectric effect

Evidence for quantization also comes from the measurement of the energies of electrons produced by the *photoelectric effect*, the ejection of electrons from metals when they are irradiated with ultraviolet light. The characteristics of the photoelectric effect are as follows:

(1) No electrons are ejected, regardless of the intensity of the light, unless its frequency exceeds a threshold value characteristic of the metal.

(2) The kinetic energy of the ejected electrons is linearly proportional to the frequency of the incident light.

(3) Even at low light intensities, electrons are ejected immediately if the frequency is above threshold.

These observations strongly suggest that the photoelectric effect depends on an electron being ejected if it is involved in a collision with a particle-like projectile carrying enough energy to knock it out of the metal. If we suppose that the projectile is a photon of energy $h\nu$, ν being the frequency of the light, then the conservation of energy requires that the kinetic energy of the electron should obey

$$\tfrac{1}{2}mv^2 = h\nu - \Phi, \qquad (13.2.6)$$

Φ being the *work-function* of the metal, the energy required to remove an electron. If $h\nu$ is less than Φ, photoejection cannot occur because the photon brings insufficient energy: this accounts for observation (1). The expression predicts that the kinetic energy of an ejected electron should be proportional to the frequency, in agreement with observation (2). When a photon collides with an electron, it gives up all its energy, and so we should expect electrons to appear as soon as the collisions begin, provided they have sufficient energy: this agrees with observation (3).

A *corpuscular theory* of light (i.e., one that regards light as consisting of particles) seems to be at variance with its well-established wave properties, such as diffraction. If the wave-theory of light is to be discarded, or at least modified, we need further evidence. The next topic provides it.

13.2 (d) The Compton effect

It is observed that when X-rays are scattered from electrons their wavelength is slightly changed: this is the *Compton effect*. According to classical physics, the electron is expected to be accelerated by the electric field of the light, and many different wavelengths are to be expected in the scattered ray. What in fact is observed is that the wavelength is increased by a single, definite amount which depends only on the angle through which the light is scattered; moreover, the shift is independent of the wavelength of the incident light.

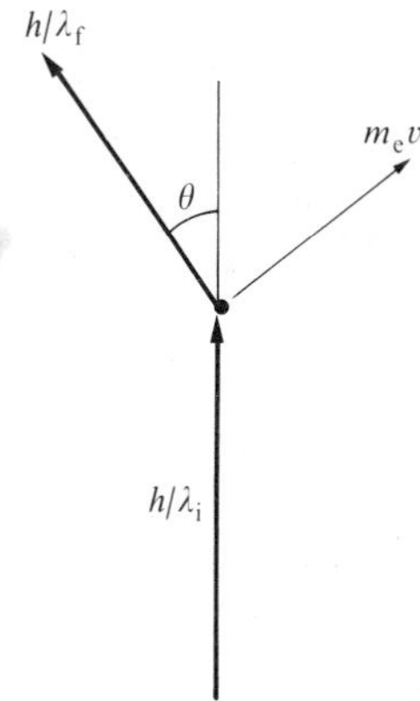

Fig. 13.4. The directions and momenta involved in the Compton effect. λ_i is the wavelength of the incident radiation, λ_f that of the scattered radiation. The electron acquires a momentum $m_e v$.

The photon theory of light explains the observations neatly if we suppose that, as well as an energy, a photon of light of frequency ν also has a *linear momentum* given by $p = h\nu/c$ (or $p = h/\lambda$). Then the scattering can be treated as a collision between a particle of momentum h/λ and another of mass m_e, Fig. 13.4, and by requiring both energy and linear momentum to be conserved in the collision the following expression is obtained for the shift of wavelength accompanying scattering through an angle θ:

$$\textit{Compton scattering:}\ \delta\lambda = (h/m_e c)(1 - \cos\theta). \qquad (13.2.7)$$

This expression is verified experimentally. The quantity $h/m_e c$ is called the

Compton wavelength of the electron; its magnitude is 2.43 pm, and so the maximum wavelength shift, which occurs for $\theta = 180°$ (back towards the light source) is only 4.86 pm, whatever the incident wavelength.

13.2 (e) The diffraction of electrons

Both the photoelectric effect and the Compton effect show that light has the attributes of particles. Although contrary to the long-established wave theory of light, a similar view had been held before, but dismissed. No significant scientist, however, had taken the view that matter is wavelike. Nevertheless, experiments in 1925 forced people to even that conclusion. The crucial experiment was performed by Clinton Davisson and Lester Germer, who observed the diffraction of electrons by a crystal, Fig. 13.5. Their experiment was a lucky accident, because a chance rise of temperature caused their polycrystalline sample to anneal, and the ordered planes of atoms then acted as a diffraction grating.

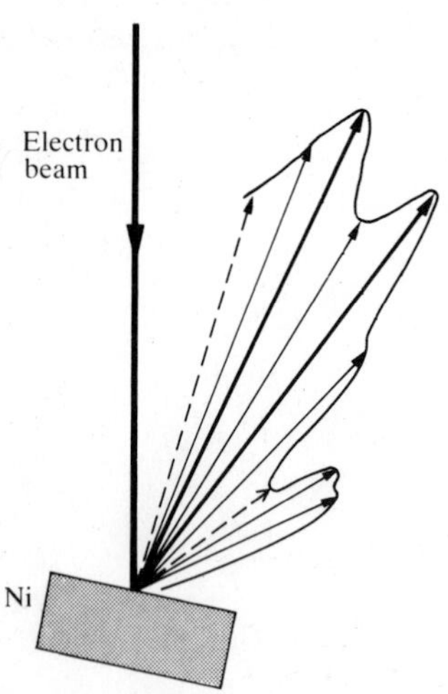

Fig. 13.5. The scattering of an electron beam from a nickel crystal shows a variation of intensity characteristic of a diffraction experiment in which waves interfere constructively and destructively in different directions.

This diffraction experiment, which has since been repeated with other particles (including molecular hydrogen), shows clearly that particles have wavelike properties. We have also seen that waves have particle properties. Thus we are brought to the heart of modern physics: *when examined on an atomic scale, the concepts of particle and wave melt together, particles taking on the characteristics of waves, and waves the characteristics of particles.*

Some progress towards coordinating these properties was made by Louis de Broglie when, in 1924, he suggested that *any* particle, not only photons, travelling with a momentum p, should have in some sense a wavelength given by the

$$\textit{de Broglie relation: } \lambda = h/p. \tag{13.2.8}$$

This expression was confirmed for particles by the Davisson–Germer experiment and for photons by the Compton effect. We shall build on it, and understand it more, in the next section.

13.2 (f) Atomic and molecular spectra

The most directly compelling evidence for the quantization of energy comes from the observation of the frequencies of light absorbed and emitted by atoms and molecules. A typical atomic spectrum is shown in Fig. 13.6, and

Fig. 13.6. The spectrum of light emitted by excited mercury atoms consists of radiation at a series of discrete frequencies (called 'spectral lines').

a typical molecular spectrum in Fig. 13.7. The obvious feature of both is that radiation of discrete rather than continuous frequencies is emitted or absorbed. This can be understood if the energy of the atoms or molecules is confined to discrete values, so that energy can be discarded or absorbed in discrete amounts, Fig. 13.8. Then if the energy of an atom decreases by ΔE, the energy is carried away as a photon of frequency $\nu = \Delta E/h$, and a definite line appears in the spectrum.

Classical mechanics utterly failed in its attempts to account for the appearance of spectra, just as it failed to account for the other experiments

Fig. 13.7. When a molecule changes its state, it does so by absorbing light at definite frequencies. This suggests that it can possess only discrete energies, not an arbitrary energy. This is a part of the absorption spectrum of ScF (provided by Dr R. F. Barrow).

described above. Such total failure showed that the basic concepts of classical mechanics were false. A new mechanics, *quantum mechanics*, had to be devised to take its place.

13.3 The dynamics of microscopic systems

We take the de Broglie relation $p = h/\lambda$ as the starting point, and abandon the classical concept of a localized particle. A particle from now on has a position distributed like the amplitude of a wave. The concept of *wavefunction* (symbol: ψ) is introduced to replace the classical concept of trajectory, and the new mechanics involves setting up a scheme for calculating and interpreting ψ.

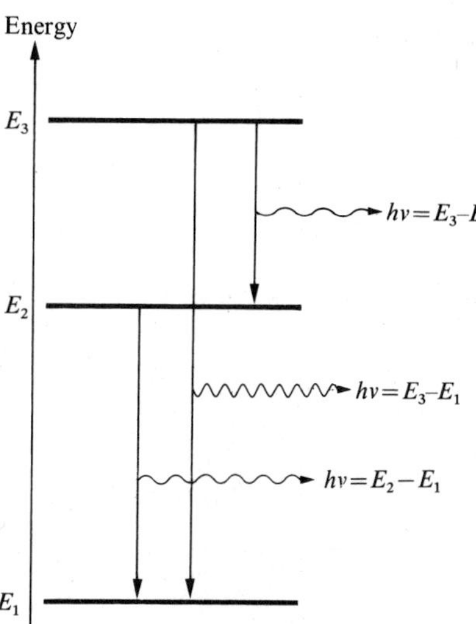

Fig. 13.8. Spectral lines can be accounted for if we assume that a molecule emits a photon as it changes between discrete energy levels. Note that high-frequency radiation is emitted when the energy change is large.

13.3 (a) The Schrödinger equation

In 1926 Erwin Schrödinger proposed an equation which, when solved, gives the wavefunction for any system. Its position is as central to quantum mechanics as Newton's equations are to classical mechanics. Just as Newton's equations were an inspired postulate which, when solved, give the trajectories of particles, so Schrödinger's equation can be regarded as an inspired postulate which, when solved, gives the wavefunction. For a particle of mass m moving in one dimension with energy E the equation is

$$(-\hbar/2m)(\mathrm{d}^2\psi/\mathrm{d}x^2) + V\psi = E\psi. \tag{13.3.1}$$

V, which may depend on the position, is the potential energy of the particle; $\hbar$ (which is read h-cross or h-bar) is a convenient modification of Planck's constant: $\hbar = h/2\pi$. Various ways of expressing this equation, of incorporating the time-dependence of the wavefunction, and of extending it to more dimensions, are collected in Box 13.1 (p. 300).

The form of the Schrödinger equation can be justified to a certain extent by the following remarks. Consider first the case of motion in a region where the potential energy is zero. Then

$$-(\hbar^2/2m)\,\mathrm{d}^2\psi/\mathrm{d}x^2 = E\psi, \tag{13.3.2}$$

and a solution is

$$\psi = \mathrm{e}^{ikx} = \cos kx + \mathrm{i}\sin kx, \qquad k = \surd(2mE/\hbar^2). \tag{13.3.3}$$

Cos kx (or sin kx) is a wave of wavelength $\lambda = 2\pi/k$: this can be seen by comparing cos kx with the standard form of a harmonic wave, cos $(2\pi x/\lambda)$. The energy of the particle is entirely kinetic (because $V = 0$ everywhere), and so $E = p^2/2m$; but since the energy is related to k by $E = k^2\hbar^2/2m$, it follows that $p = k\hbar$. Therefore, the linear momentum is related to the wavelength of the wavefunction by

$$p = k\hbar = (2\pi/\lambda)(h/2\pi) = h/\lambda,$$

which is de Broglie's relation. In the case of freely moving particles, the Schrödinger equation has led to an experimentally verified conclusion.

If the particle is in a region where its potential energy is uniform but non-zero, the Schrödinger equation is

$$(-\hbar^2/2m)\,\mathrm{d}^2\psi/\mathrm{d}x^2 = (E - V)\psi.$$

Box 13.1 The Schrödinger equation

For *one-dimensional* systems

$$(-\hbar^2/2m)\, \mathrm{d}^2\psi/\mathrm{d}x^2 + V\psi = E\psi, \qquad V = V(x), \qquad \psi = \psi(x),$$

or

$$\mathrm{d}^2\psi/\mathrm{d}x^2 + (2m/\hbar^2)(E-V)\psi = 0.$$

V is the potential energy of the particle. For example, for a free particle $V = 0$ (or some constant) and for a harmonic oscillator $V = \frac{1}{2}kx^2$.

For *three-dimensional* systems

$$(-\hbar^2/2m)\, \nabla^2\psi + V\psi = E\psi, \qquad V = V(x, y, z), \qquad \psi = \psi(x, y, z),$$

where ∇^2 ('del-squared') is

$$\nabla^2 = (\partial^2/\partial x^2) + (\partial^2/\partial y^2) + (\partial^2/\partial z^2).$$

In systems with spherical symmetry it is more appropriate to take ψ as a function of the spherical polar coordinates (Fig. 13.11). Then

$$\nabla^2 = (\partial^2/\partial r^2) + (2/r)(\partial/\partial r) + (1/r^2)\Lambda^2,$$

where

$$\Lambda^2 = (1/\sin^2\theta)(\partial^2/\partial\phi^2) + (1/\sin\theta)(\partial/\partial\theta)\sin\theta(\partial/\partial\theta).$$

In the *general case*, the Schrödinger equation is written

$$H\psi = E\psi,$$

where H is the *hamiltonian operator* for the system:

$$H = (-\hbar^2/2m)\, \nabla^2 + V.$$

When the system is time-dependent, use the *time-dependent Schrödinger equation*

$$H\psi = \mathrm{i}\hbar(\partial\psi/\partial t).$$

The solutions are the same as in eqn (13.3.3) but now $E - V = \hbar^2k^2/2m$. Now the relation $\lambda = 2\pi/k$ leads to

$$\lambda = h/\{2m(E-V)\}^{\frac{1}{2}}. \qquad (13.3.4)$$

This equation shows that the greater the difference between the total energy and the potential energy, the shorter the wavelength of the wavefunction. In other words, *the greater the kinetic energy, the shorter the wavelength*. A stationary particle, one with zero kinetic energy, has infinite wavelength, which means that its wavefunction has the same value everywhere: for a particle at rest, ψ = constant.

A more general aspect of the wavefunction, one that remains well-defined even when it is not possible to speak in terms of wavelength, is its *curvature*, which for our purposes we interpret as the second derivative $\mathrm{d}^2\psi/\mathrm{d}x^2$. When the wavefunction is sharply curved (when it has a short wavelength) the kinetic energy is large; when the wavefunction is not sharply curved (when its wavelength is long) the kinetic energy is low. The association of sharp curvature with high kinetic energy will turn out to be a valuable clue to the interpretation of wavefunctions. It is also a guide to guessing their shapes. For example, suppose we need to know the wavefunction for a particle with a potential energy that decreases with increasing x, as in the lower half of Fig. 13.9. If its total (constant) energy is E, since the difference $E - V$ increases from left to right, the wavefunction must become more sharply curved as x increases: its 'wavelength' is decreasing as its kinetic energy increases. We can therefore guess that the wavefunction will look like the function drawn in the upper half of the illustration, and more detailed calculation confirms this to be so.

Schrödinger's equation is a second-order differential equation, and therefore it has an infinite number of solutions. For example, in the case of a free particle, e^{ikx} is a solution, so is Ae^{ikx}, with A having any value, and k (and hence E) is arbitrary. The next step in the argument involves putting an interpretation on ψ. We shall see that the interpretation implies that many mathematical solutions are physically inadmissible. Discarding these solutions means rejecting some values of E, which brings us to the quantization of energy.

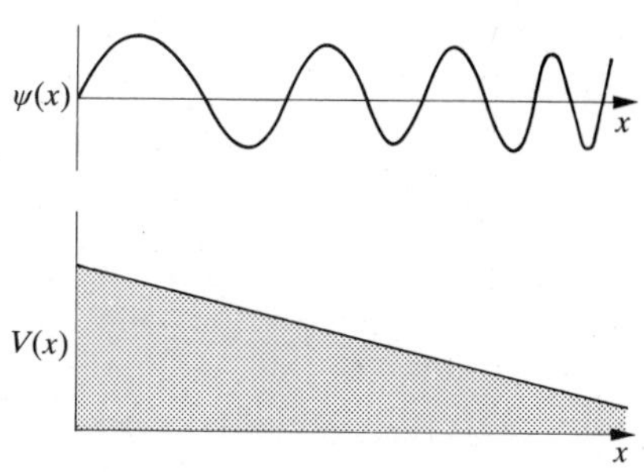

Fig. 13.9. The wavefunction of a particle in a constant gradient (i.e. a particle subjected to a constant force to the right). Only the real part of the wavefunction is shown, the imaginary part is similar, but displaced to the right.

13.3 (b) The interpretation of the wavefunction

The interpretation of ψ is based on a suggestion made by Max Born and based on an analogy with the wave theory of light in which the square of the amplitude of an electromagnetic wave is interpreted as the intensity and therefore, in quantum terms, as the number of photons present. The *Born interpretation* is that the square of the wavefunction (or $\psi^*\psi$ if ψ is complex) is proportional to the probability of finding the particle at each point in space. Specifically, for a one-dimensional system:

> *If the amplitude of the wavefunction of a particle is ψ at some point x, then the probability of finding the particle between x and $x + dx$ is proportional to $\psi^*\psi\, dx$.*

This means that $\psi^*\psi$ is a *probability density* (since it must be multiplied by the length of the infinitesimal region dx in order to arrive at a probability); ψ itself is called a *probability amplitude*. For a particle able to move in three dimensions (e.g. an electron near a nucleus in an atom) the wavefunction depends on the point $\boldsymbol{r}$ with coordinates x, y, z, and the interpretation of $\psi(\boldsymbol{r})$, Fig. 13.10, is then as follows:

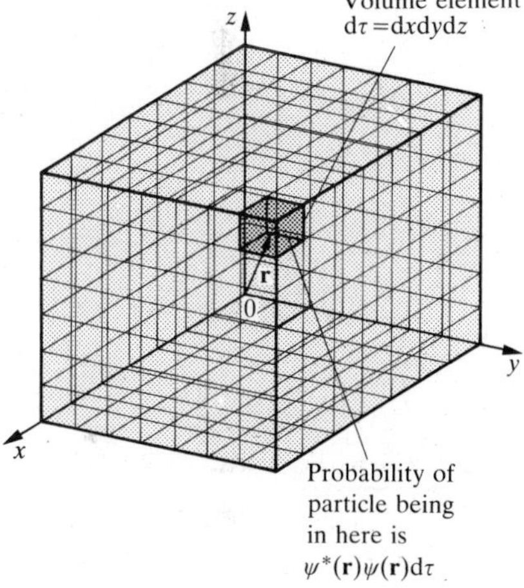

Fig. 13.10. The *Born interpretation* of the wavefunction in three-dimensional space implies that the probability of finding the particle in the volume element $d\tau = dx\, dy\, dz$ at some location $\mathbf{r}$ is proportional to the product of $d\tau$ and the value of $\psi^*\psi$ at that location.

> *If the amplitude of the wavefunction of a particle is ψ at some point $\boldsymbol{r}$, then the probability of finding the particle in an infinitesimal volume $d\tau = dx\, dy\, dz$ at the point $\boldsymbol{r}$ is proportional to $\psi^*\psi\, d\tau$.*

Example 13.2

The wavefunction of an electron in the lowest energy state of the hydrogen atom is the function $\psi(\boldsymbol{r}) \propto e^{-r/a_0}$, with $a_0 = 52.9$ pm and r the distance from the nucleus. Notice that the wavefunction depends only on this distance, not the angular position. Calculate the relative probabilities of finding the electron inside a small volume of magnitude $1.0\,\text{pm}^3$ located at (a) the nucleus, (b) a distance a_0 from the nucleus.

- *Method*. The probability is proportional to $\psi^2\, d\tau$ evaluated at the location in question. The volume is so small (even on the scale of the atom) that we can ignore the variation of ψ within it and write $d\tau = \delta\tau = 1.0\,\text{pm}^3$.
- *Answer*. (a) At the nucleus $r = 0$, and so there $\psi^2 \propto 1$ and $\psi^2\, \delta\tau \propto 1.0$.

(b) At a distance $r = a_0$ in an arbitrary (but definite) direction, $\psi^2\,\delta\tau \propto e^{-2} \times 1.0 = 0.14$. Therefore, the ratio of probabilities is $1.0/0.14 = 7.1$.

• *Comment*. Note that it is more probable (by a factor of 7) that the electron will be found at the nucleus than in the same volume element located at a distance a_0 from the nucleus. The probability of finding the electron in the same volume at a distance of 1 mm is strictly non-zero, but it is so small that it is completely negligible.

• *Exercise*. The wavefunction for the lowest energy orbital in the ion He^+ is $\psi \propto e^{-2r/a_0}$. Repeat the calculation for this ion. Any comment? [55; more compact wavefunction]

If ψ is a solution of the Schrödinger equation, then so is $N\psi$, where N is any constant. This means that we can always find a factor such that the 'proportionality' of the Born interpretation becomes an equality. This greatly simplifies discussions. The factor N, which is called the *normalization constant*, is chosen as follows. If we include N with the wavefunction, then the Born interpretation states that the probability that a particle is in the region dx is *equal* to $(N\psi^*)(N\psi)\,dx$. Furthermore, the sum over all space of these individual probabilities must be unity (the probability of the particle being somewhere in the system is unity). Expressed mathematically:

$$N^2\int_{-\infty}^{\infty}\psi^*\psi\,dx = 1, \quad \text{or} \quad N^2 = 1\Big/\int_{-\infty}^{\infty}\psi^*\psi\,dx. \qquad (13.3.5)$$

Therefore, by evaluating the integral, we can find the normalization constant N. This procedure is called *normalizing the wavefunction*. From now on, unless we state otherwise, *we always use normalized wavefunctions;* that is, from now on we assume that ψ already includes a factor such that

$$\int_{-\infty}^{\infty}\psi^*\psi\,dx = 1. \qquad (13.3.6)$$

In the case of three dimensions, the wavefunction is normalized if

$$\int_{-\infty}^{\infty}\psi^*\psi\,dx\,dy\,dz = 1, \quad \text{or, succinctly, if} \quad \int\psi^*\psi\,d\tau = 1. \qquad (13.3.7)$$

Example 13.3

The wavefunction used for the hydrogen atom in *Example* 13.2 is not normalized. Normalize it.

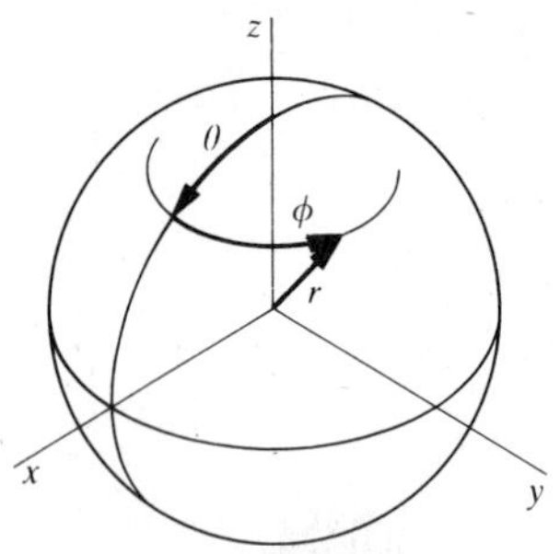

Fig. 13.11. Spherical polar coordinates. The *radius* r ranges from 0 to infinity; the *colatitude* θ ranges from 0 (North pole) to π (South pole), and the *azimuth* ϕ ranges from 0 to 2π.

• *Method*. Evaluate N from eqn (13.3.7). The volume element in three-dimensional space is $d\tau = dx\,dy\,dz$, but in problems with spherical symmetry (as here) it is far more convenient to work in *spherical polar coordinates*, Fig. 13.11:

$$x = r\sin\theta\cos\phi, \qquad y = r\sin\theta\sin\phi, \qquad z = r\cos\theta.$$

Then the volume element is

$$d\tau = r^2\,dr\,\sin\theta\,d\theta\,d\phi$$

with the *radius* r ranging from 0 to ∞, the *colatitude* θ from 0 to π, and the *azimuth* from 0 to 2π.

• *Answer*. From eqn (13.3.7) with $\psi = Ne^{-r/a_0}$,

$$\int \psi^* \psi \, d\tau = N^2 \int_0^\infty r^2 e^{-2r/a_0} \, dr \int_0^\pi \sin\theta \, d\theta \int_0^{2\pi} d\phi$$
$$= N^2\{(a_0^3/4) \times 2 \times (2\pi)\} = \pi a_0^3 N^2.$$

Therefore, in order for this to equal unity, $N = (1/\pi a_0^3)^{\frac{1}{2}}$, and the normalized wavefunction is

$$\psi = (1/\pi a_0^3)^{\frac{1}{2}} e^{-r/a_0}.$$

• *Comment*. If *Example* 13.2 is now repeated, we can obtain the actual probabilities of finding the electron in the volume element at each location, not just their relative values. This gives (a) 2.2×10^{-6} and (b) 3.1×10^{-7}.

• *Exercise*. Normalize the He^+ wavefunction given in *Example* 13.2. [$N = (8/\pi a_0^3)^{\frac{1}{2}}$]

13.3 (c) Quantization

Accepting the Born interpretation requires us to put severe restrictions on wavefunctions, the principal constraint being that *the wavefunction must not be infinite anywhere*.‡ For if it were, the integral in eqn (13.3.5) would be infinite, and the normalization constant would be zero. This would mean that the normalized function was zero everywhere, except where it is infinite, which would be absurd. This requirement rules out many possible solutions of the Schrödinger equation because many mathematically acceptable solutions curl up to infinity. We shall see examples shortly.

The requirement that ψ be finite everywhere is not the only restriction implied by the Born interpretation. We could imagine (and will shortly meet) a strange function that gave rise to more than one value of $\psi^*\psi$ at a single point. The Born interpretation implies that such functions are unacceptable, because it would be absurd to have more than one probability that a particle is at some point. This restriction is expressed by saying that the wavefunction must be *single-valued*.

The Schrödinger equation itself also implies some restrictions on the type of functions that will occur. Since it is a second-order differential equation, the second derivative of ψ must be well-defined if the equation is to be applicable everywhere. We can take the second derivative of a function only if it is continuous (so that there are no sharp steps in it, Fig. 13.12) and if its first derivative, its slope, is continuous (so that there are no kinks, Fig. 13.12).§ Therefore, the wavefunctions we shall meet must be (a) continuous, (b) have continuous first-derivatives.

At this stage we have assembled the following restrictions:

> *A wavefunction is acceptable only if it (a) is continuous, (b) has a continuous slope, (c) is single-valued, and (d) is finite.*

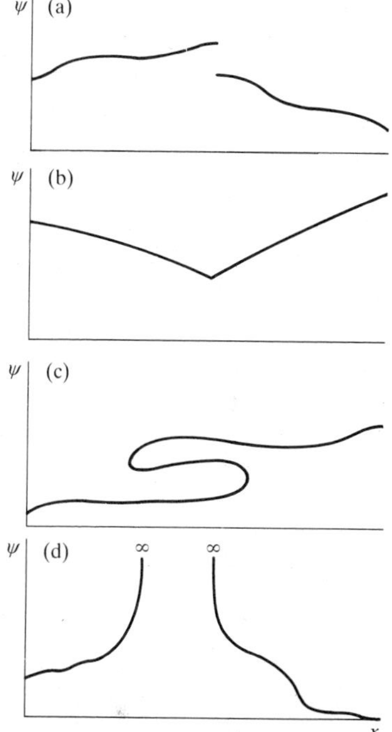

Fig. 13.12. The wavefunction must satisfy stringent conditions for it to be acceptable. (a) Unacceptable because it is not continuous; (b) unacceptable because its slope is discontinuous; (c) unacceptable because it is not single-valued; (d) unacceptable because it is infinite over a finite region.

‡ Infinitely sharp spikes are acceptable so long as they have zero width; the true constraint is that the wavefunction must not be infinite over any finite region. In elementary quantum mechanics the simpler restriction, to finite ψ, is sufficient.

§ Nothing, or almost nothing, is ever quite true. There are cases, and we shall meet them, where acceptable wavefunctions have kinks. These arise when the potential energy has peculiar properties, such as rising abruptly to infinity. When the potential energy is smoothly well-behaved and finite, the slope of the wavefunction must be continuous; if the potential energy becomes infinite, then the slope of the wavefunction need not be continuous. There are only two cases of this behaviour in elementary quantum mechanics, and the peculiarity will be mentioned when we meet them.

These are severe restrictions. They are so severe that acceptable solutions of the Schrödinger equation do not in general exist for arbitrary values of the energy E. In other words, a particle may possess only certain energies, for otherwise its wavefunction would be unacceptable. This is nothing other than quantization. What the acceptable energies are we shall discover by solving the Schrödinger equation for motion of various kinds, and selecting the solutions that conform to the restrictions listed above.

13.4 Quantum mechanical principles

A wavefunction contains *all there is to know* about the outcome of experiments that can be done on a system. Our job now is to see how to extract that information. This will reveal the strangeness of the quantum world.

Much of quantum theory can be illustrated in a simple way by considering its application to a free particle in one dimension. We have already written down a solution of the Schrödinger equation for free translational motion, eqn (13.3.3). That solution is a special case of the general solution of eqn (13.3.2):

$$\psi = A\mathrm{e}^{\mathrm{i}kx} + B\mathrm{e}^{-\mathrm{i}kx}, \qquad E = k^2\hbar^2/2m. \tag{13.4.1}$$

We have seen that the wavefunction having a given value of k corresponds to a particle with linear momentum $p = k\hbar$. One of the questions we must now address is the significance of the coefficients A and B. We shall make progress by setting up quantum mechanics in a more general and powerful way than we have presented it so far.

13.4 (a) Operators and observables

Consider the Schrödinger equation, eqn (13.3.1), rewritten in the succinct form

$$H\psi = E\psi, \quad \text{with} \quad H = -(\hbar^2/2m)(\mathrm{d}^2/\mathrm{d}x^2) + V. \tag{13.4.2}$$

H is an *operator*, something that operates on the function ψ, in this case by taking its second derivative. The operator H plays a special role in quantum mechanics, and is called the *hamiltonian* after the nineteenth-century mathematician William Hamilton, who developed a form of classical mechanics that could readily be converted into quantum mechanics.

When written as $H\psi = E\psi$, the Schrödinger equation has the form of an *eigenvalue equation*; that is, an equation of the form:

$$\text{(operator)(function)} = \text{(numerical factor)(same function)}. \tag{13.4.3}$$

In the case of $H\psi = E\psi$, the numerical factor on the right, which is called the *eigenvalue* of the operator, is the energy, and the function (which must be the same on each side in an eigenvalue equation) is the *eigenfunction* corresponding to that eigenvalue. In our case, the eigenfunction is the wavefunction corresponding to the energy E.

Example 13.4

Show that e^{ax} is an eigenfunction of the operator $\mathrm{d}/\mathrm{d}x$, and find the corresponding eigenvalue. Show that e^{ax^2} is not an eigenfunction of $\mathrm{d}/\mathrm{d}x$.

• *Method*. Operate on the function (f) with the operator ($\hat{O}$) and check whether the result is a numerical factor (O) times the *original* function: that is, see if f satisfies the eigenvalue equation $\hat{O}f = Of$. If it does, the O is the eigenvalue corresponding to the eigenfunction f.

• *Answer*. For $\hat{O} = \mathrm{d}/\mathrm{d}x$ and $f = \mathrm{e}^{ax}$:

$$\hat{O}f = (\mathrm{d}/\mathrm{d}x)\mathrm{e}^{ax} = a\mathrm{e}^{ax} = af.$$

Therefore, e^{ax} is an eigenfunction of $\mathrm{d}/\mathrm{d}x$, and its eigenvalue is a.
In the case of $f = \mathrm{e}^{ax^2}$,

$$\hat{O}f = (\mathrm{d}/\mathrm{d}x)\mathrm{e}^{ax^2} = 2ax\mathrm{e}^{ax^2} = 2a\{x\mathrm{e}^{ax^2}\},$$

which is not an eigenfunction equation (unless $a = 0$).

• *Comment*. Much of quantum mechanics boils down to looking for functions that are eigenfunctions of a given operator, especially of the hamiltonian operator for the energy.

• *Exercise*. Is $\cos ax$ an eigenfunction of (a) $\mathrm{d}/\mathrm{d}x$, (b) $\mathrm{d}^2/\mathrm{d}x^2$? [(a) No, (b) yes]

The importance of eigenvalue equations is that the pattern

(energy operator)(wavefunction) = (energy)(wavefunction)

exemplified by the Schrödinger equation itself is repeated for other properties (which are called *observables*) of the system. In general we can write

$$\text{(operator)(wavefunction)} = \text{(observable)(wavefunction)} \quad (13.4.4a)$$

Symbolically we write this

$$\hat{O}\psi = O\psi, \quad (13.4.4b)$$

where $\hat{O}$ is the operator (e.g. the hamiltonian H) corresponding to the observable O (e.g. the energy E). Therefore, if we know the wavefunction, ψ, and if we also know the operator corresponding to the observable of interest, we can predict the outcome of an observation of that property (e.g. an atom's energy) by picking out the factor O in the corresponding eigenvalue equation.

The first step is to find the operator corresponding to a given observable. The form of the operator for *linear momentum* is one of the *postulates* of quantum mechanics:

$$\hat{p} = (\hbar/\mathrm{i})\,\mathrm{d}/\mathrm{d}x. \quad (13.4.5)$$

That is, to find the value of the linear momentum possessed by a particle, we *differentiate* the wavefunction with respect to x, and then pick out the momentum p from the eigenvalue equation

$$(\hbar/\mathrm{i})(\mathrm{d}\psi/\mathrm{d}x) = p\psi. \quad (13.4.6)$$

The form of the operator for *position* is also one of the basic postulates of quantum mechanics. It is simply *multiplication* by the coordinate x.

For example, suppose we select $B = 0$ for one of the free-particle wavefunctions given in eqn (13.4.1), then $\psi = A\mathrm{e}^{\mathrm{i}kx}$ and

$$(\hbar/\mathrm{i})(\mathrm{d}\psi/\mathrm{d}x) = A(\hbar/\mathrm{i})(\mathrm{d}\mathrm{e}^{\mathrm{i}kx}/\mathrm{d}x) = A(\hbar/\mathrm{i})(\mathrm{i}k)\mathrm{e}^{\mathrm{i}kx}$$
$$= k\hbar A\mathrm{e}^{\mathrm{i}kx} = k\hbar\psi,$$

so that $p = k\hbar$, as we already knew. But suppose we chose the wavefunction with $A = 0$. By the same reasoning,

$$(\hbar/\mathrm{i})(\mathrm{d}\psi/\mathrm{d}x) = B(\hbar/\mathrm{i})(\mathrm{d}\mathrm{e}^{-\mathrm{i}kx}/\mathrm{d}x) = -k\hbar\psi,$$

which shows that a particle described by the wavefunction $\mathrm{e}^{-\mathrm{i}kx}$ has the same magnitude of momentum (and the same kinetic energy), but now its momentum is towards $-x$ (the momentum is a vector quantity, and the sign gives the direction).

13.4 (b) Superpositions and expectation values

Suppose now that the wavefunction of the free particle has $A = B$. What is the linear momentum of the particle? If we use the operator technique we quickly run into a problem. In the first place, the wavefunction is

$$\psi = A(\mathrm{e}^{\mathrm{i}kx} + \mathrm{e}^{-\mathrm{i}kx}) = 2A\cos kx,$$

which is a perfectly respectable wavefunction. However, when we operate with $\hat{p}$ we find

$$(\hbar/\mathrm{i})\,\mathrm{d}\psi/\mathrm{d}x = 2A(\hbar/\mathrm{i})\,\mathrm{d}(\cos kx)/\mathrm{d}x = -2kA(\hbar/\mathrm{i})\sin kx,$$

which is not an eigenvalue equation because the function on the right is different from the original one. When this occurs quantum mechanics instructs us to interpret it as meaning that the momentum of the particle is indefinite. However, the momentum is not completely indefinite, because the cosine wavefunction is a sum (the technical term is a *linear superposition*) of $\mathrm{e}^{\mathrm{i}kx}$ and $\mathrm{e}^{-\mathrm{i}kx}$ which, as we have seen, individually correspond to definite momentum states. Symbolically we can write the superposition as

$$\psi = \psi(\bigcirc\!\!\longrightarrow) + \psi(\longleftarrow\!\!\bigcirc),$$

and interpret it as follows: if the momentum of the particle is measured, its *magnitude* will be found to be $k\hbar$, but half the measurements will show that it is moving to the right, and half the measurements will show that it is moving to the left.

The same interpretation is to be made of any wavefunction written in the form of a superposition. For example, if the wavefunction is known to be a sum of many different linear momentum eigenfunctions and written in the form

$$\psi = c_1\psi_{\text{momentum 1}} + c_2\psi_{\text{momentum 2}} + \ldots$$

where the cs are coefficients, quantum mechanics tells us the following:

(1) When the momentum is measured, *one* of the values 'momentum 1', 'momentum 2', . . . will be found, but only if the corresponding eigenfunction occurs in the superposition.

(2) Which of the possible values will be found is unpredictable.

(3) The probability of measuring 'momentum 1' in a series of observations is proportional to the *square* of the coefficient c_1 (specifically, proportional to $c_1^*c_1$ if c_1 is complex), and likewise for the other values.

(4) The *average* value of a large number of observations is given by the *expectation value* of the observable.

The *expectation value* (symbol: $\langle O\rangle$) is calculated from the wavefunction as follows. If the system is described by a normalized wavefunction ψ, then the expectation value of some observable O is defined as the following integral:

$$\text{Expectation value of } O\text{: } \langle O\rangle = \int \psi^*\hat{O}\psi \, d\tau, \qquad (13.4.7)$$

where $\hat{O}$ is the operator corresponding to O.

Example 13.5

Calculate the average value of the distance of an electron from the nucleus in the hydrogen atom.

• *Method*. The normalized wavefunction was obtained in *Example* 13.3. The operator corresponding to its distance from the nucleus is *multiplication by r* (see the remark following eqn (13.4.6)). Therefore, the average value is given by the expectation value

$$\langle r\rangle = \int \psi^* r\psi \, d\tau.$$

Evaluate the integral using spherical polar coordinates. A very useful integral in calculations of this kind is

$$\int_0^\infty x^n e^{-ax} \, dx = n!/a^{n+1},$$

where $n! = n(n-1)(n-2)\ldots 1$.

• *Answer*. Using the normalized wavefunction from *Example* 13.3 gives

$$\begin{aligned}\langle r\rangle &= (1/\pi a_0^3)\int_0^\infty r^3 e^{-2r/a_0}\int_0^\pi \sin\theta \, d\theta \int_0^{2\pi} d\phi \\ &= (1/\pi a_0^3)(3!\, a_0^4/2^4)\times 2\times 2\pi = \tfrac{3}{2}a_0.\end{aligned}$$

Since $a_0 = 52.9$ pm (*Example* 3.1), $\langle r\rangle = 79$ pm.

• *Comment*. This result means that if a very large number of measurements of the distance of the electron from the nucleus are made, then the *mean* value will be 79 pm. However, each different observation will give a different individual result, because the wavefunction is not an eigenfunction of the operator corresponding to r.

• *Exercise*. Evaluate the root mean square distance (i.e., $r_{\text{rms}} = \surd\langle r^2\rangle$) of the electron from the nucleus in the hydrogen atom. $[a_0\surd 3]$

13.4 (c) The uncertainty principle

We have seen that if the particle's wavefunction is Ae^{ikx}, then it corresponds to a definite state of linear momentum, namely travelling to the right with momentum $k\hbar$. But we might also ask for the *position* of the particle when it is in this state. The Born interpretation instructs us to answer this question by forming the probability density $\psi^*\psi$. In this case:

$$\psi^*\psi = (Ae^{ikx})^*(Ae^{ikx}) = A^2(e^{-ikx})(e^{ikx}) = A^2.$$

This probability density is a constant, A^2, independent of x. Therefore, the particle has an equal probability of being found anywhere. In other words, *if the momentum is specified precisely, it is impossible to predict the location of the particle*. This is one half of *Heisenberg's uncertainty principle*, one of the most celebrated results of quantum mechanics, which states that *it is impossible to specify simultaneously, with arbitrary precision, both the momentum and the position of a particle*.

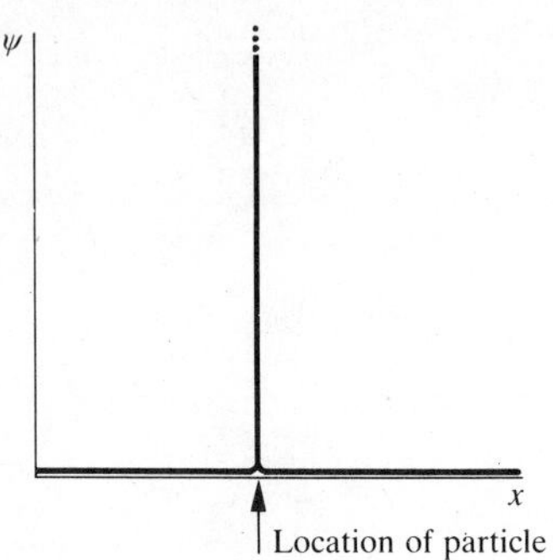

Fig. 13.13. The wavefunction for a particle at a well-defined location is a sharply spiked function which has zero amplitude everywhere except at the particle's position.

Before discussing this principle further, we must establish the other half: that if we know the position of a particle exactly, we can say nothing about its momentum. The argument draws on the idea of expressing a wavefunction as a superposition of eigenfunctions, and runs as follows. If we know that the particle is at a definite location, then its wavefunction must be large there and zero everywhere else, Fig. 13.13. Such a wavefunction can be recreated by adding together a number of harmonic (sine and cosine) functions, or, what is equivalent, a number of e^{ikx} functions. In other words, *we can recreate a sharply localized wavefunction by superimposing wavefunctions corresponding to different linear momenta.* The superposition of a few harmonic functions gives a broad, ill-defined wavefunction, Fig. 13.14(a), but as the number used increases, the wavefunction becomes much sharper because of the interference between the positive and negative regions of the components, Fig. 13.14(b). When an infinite number of components is used, the wavefunction is a sharp, infinitely narrow spike, Fig. 13.14(c), which corresponds to perfect localization of the particle. The particle is perfectly localized, but at the expense of

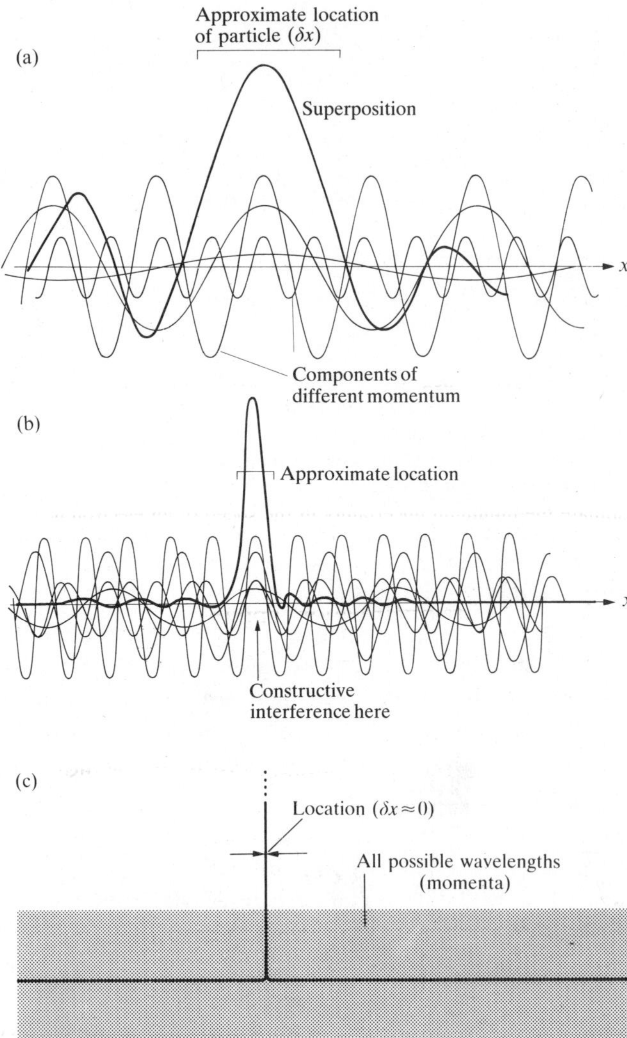

Fig. 13.14. (a) The wavefunction for a particle with an ill-defined location can be regarded as the sum (superposition) of several wavefunctions of definite wavelength which interfere constructively in one place but destructively elsewhere. (b) As more waves are used in the superposition, the location becomes more precise at the expense of uncertainty in the particle's momentum. (c) An infinite number of waves are needed to construct the wavefunction of a perfectly localized particle.

discarding all information about its momentum. This is because, as we saw above, a measurement of the momentum will give a result corresponding to any one of the infinite number of waves in the superposition, and which one it will give is unpredictable. Hence, if we know the location of the particle, its momentum is unpredictable.

Werner Heisenberg arrived at a quantitative version of this result by considering the expectation values of position and momentum. For our purposes, only his result is important. The quantitative version of the *position–momentum uncertainty relation* is

$$\delta p\,\delta q \geqslant \tfrac{1}{2}\hbar. \tag{13.4.8}$$

δp is the 'uncertainty' in the linear momentum (strictly, it is the root-mean-square, RMS, deviation of the momentum from its mean value) and δq is the 'uncertainty' in position (the RMS deviation from the mean position, essentially the half-width of the superposition in Fig. 13.14). p and q refer to the same direction in space, and so whereas position on the x axis and momentum parallel to it are restricted by the uncertainty relation, position on x and motion along y or z are not restricted.

Example 13.6

The speed of a 1.0 g projectile is known to within $10^{-6}\,\mathrm{m\,s^{-1}}$. Calculate the minimum uncertainty in its position.

- *Method*. Estimate δp from $m\,\delta v$, where δv is the uncertainty in the speed; then use eqn (13.4.8) to estimate the minimum uncertainty in position, δq.

- *Answer*. The minimum uncertainty in position is

$$\delta q = \hbar/2m\delta v = (1.055\times10^{-34}\,\mathrm{J\,s})/2\times(1.0\times10^{-3}\,\mathrm{kg})\times(1\times10^{-6}\,\mathrm{m\,s^{-1}})$$
$$= 5\times10^{-26}\,\mathrm{m}.$$

- *Comment*. This is completely negligible for all practical purposes. However, when the mass is that of an electron, then the same uncertainty in speed implies an uncertainty in position far larger than the diameter of an atom, and so the concept of a trajectory, the simultaneous possession of a precise position and momentum, is untenable.

- *Exercise*. Estimate the minimum uncertainty in the speed of an electron in a hydrogen atom (taking its diameter as $2a_0$). [$500\,\mathrm{km\,s^{-1}}$]

The Heisenberg uncertainty relation applies to a number of pairs of observables called *complementary observables*. A common feature of complementary observables is that their product has the same dimensions as Planck's constant (energy × time). Other than position and momentum, they include energy and lifetime and properties related to angular momentum (which we meet in the next chapter). With the discovery that some observables are complementary we are at the heart of the difference between classical and quantum mechanics. Classical mechanics supposed, falsely as we now know, that the position and momentum of a particle could be specified simultaneously with arbitrary precision: that, after all, is what is meant by a trajectory, as we saw in Section 13.1.

Quantum mechanics is built on the acceptance that we have to make choices: we can specify position at the expense of momentum, or momentum at the expense of position. The realization that some observables are complementary allows us to make considerable progress with the calculation of atomic and molecular properties; but it does away with some of classical physics' most cherished concepts.

Further reading

Historical background:

The strange story of the quantum. B. Hoffman; Dover, New York, 1959.
Black-body theory and the quantum discontinuity, 1894–1912. T. S. Kuhn; Clarendon Press, Oxford, 1978.
The conceptual development of quantum mechanics. M. Jammer; McGraw-Hill, New York, 1966.

Principles and concepts:

Quantization (lecture cassette and workbook). P. W. Atkins; Royal Society of Chemistry, London, 1981.
Quanta: a handbook of concepts. P. W. Atkins; Clarendon Press, Oxford, 1974.
Molecular quantum mechanics (2nd edn). P. W. Atkins; Oxford University Press, 1983.
Lectures in physics. R. P. Feynman, R. B. Leighton, and M. Sands; W. H. Freeman & Co., San Francisco, 1963.
Atoms and molecules. M. Karplus and R. N. Porter; Benjamin, Menlo Park, 1970.

Introductory problems

A13.1. Calculate the power radiated by a 2.0×3.0 m section of the surface of a black body heated to a temperature of 1500 K.

A13.2. The power delivered to a photodetector which collects 8.0×10^7 photons in 3.8 ms from monochromatic light is 0.72 microwatt. What is the frequency of the light?

A13.3. A diffraction experiment requires electrons with a wavelength of 0.45 nm. Calculate the velocity of such electrons.

A13.4. Calculate the momenta of photons with wavelength: (a) 750 nm, (b) 70 pm, (c) 19 m.

A13.5. The energy required for removal of an electron from a certain atom is 3.44×10^{-18} J. Absorption of a photon of unknown wavelength ionizes the atom and produces an electron with velocity 1.03×10^6 m s^{-1}. Calculate the wavelength of the radiation which is absorbed.

A13.6. In an experiment on Compton scattering of X-rays by electrons, the incident beam has a wavelength 7.078×10^{-2} nm. Calculate the wavelength of the X-rays scattered through an angle of 70°.

A13.7. The speed of a proton is 4.5×10^5 m s^{-1}. If the indeterminacy of the momentum of the proton is to be reduced to 0.0100%, what indeterminacy in the location of the proton must be tolerated?

A13.8. A particle with mass 6.65×10^{-27} kg is confined to an infinite square well of width L. The energy of the third level is 2.00×10^{-24} J. Calculate the value of L.

A13.9. For a particle of mass m in the lowest energy state in a one-dimensional box of length L, calculate the location for which the wavefunction has a value one-fourth of its maximum value.

A13.10. Calculate the spacing between the fourth and fifth energy levels for a mass of 3.3×10^{-27} kg in a one-dimensional box with length 5.0 nm.

Problems

13.1. Calculate the energy per photon, and the energy per mole of photons, when their wavelength is (a) 600 nm (red), (b) 550 nm (yellow), (c) 400 nm (blue), (d) 200 nm (ultraviolet), (e) 150 pm (X-ray), (f) 1 cm (microwave).

13.2. What are the momenta of the photons in the last Problem? What speed would a stationary hydrogen atom attain if the photon collided with it and was absorbed?

13.3. A glow-worm of mass 5.0 g emits red light (650 nm) with a power of 0.10 W entirely in the backward direction. What velocity will it have accelerated to after 10 years if released in free space (and assumed to live)?

13.4. A sodium light emits yellow light (550 nm). How many photons does it emit each second if its power is (a) 1.0 W, (b) 100 W?

13.5. The Planck distribution, eqn (13.2.4), gives the energy in a wavelength range $d\lambda$ at the wavelength λ. Estimate the energy density lying between the wavelengths 650 nm (regarding the range $\Delta\lambda$ as virtually infinitesimal) at equilibrium inside a cavity of volume 100 cm^3 when its temperature is (a) 25 °C, (b) 3000 °C.

13.6. One way of deducing the value of Planck's constant h is to fit his distribution law to the measured radiation energy emitted by a heated black body. Less accurate, but simpler, is to derive an expression for the wavelength corresponding to the emission maximum, and to fit that to experiment. Derive the expression $\lambda_{max}T = constant$ in the region of short wavelengths, and find an expression for the constant in terms of h, c, and k.

13.7. The wavelength of the emission maximum from a small pinhole in an electrically heated container was determined at a series of temperatures. From the data below, deduce a value of Planck's constant.

θ/°C	1000	1500	2000	2500	3000	3500
λ_{max}/nm	2181	1600	1240	1035	878	763

13.8. The peak in the sun's emitted energy occurs at about 480 nm. Assuming it to behave as a black-body emitter, what is the temperature of the surface?

13.9. Write a computer program to evaluate the Planck distribution at any temperature and wavelength (or frequency), and add to it a routine for evaluating the integral (e.g. using Simpson's rule) of the energy density between any two wavelengths. Use it to calculate the total energy density in the visible region (600 nm to 350 nm) for a black body at (a) 100 °C, (b) 500 °C, (c) 1000 °C.

13.10. The Einstein theory of heat capacities leads to the expression quoted in eqn (13.2.5). The Einstein frequency for copper is 7.1×10^{12} Hz. What is its molar heat capacity at (a) 200 K, (b) 298 K, (c) 700 K? What are the classical values at these temperatures?

13.11. The Einstein frequency is often expressed in terms of an equivalent temperature θ_E. When the actual temperature is much greater than θ_E the heat capacity is close to the classical value. To what temperature does 7.1×10^{12} Hz correspond?

13.12. The work function for caesium is 2.14 eV. What is the kinetic energy and the speed of the electrons emitted when the metal is irradiated with light of wavelength (a) 700 nm, (b) 300 nm?

13.13. The photoelectric effect is the basis of the spectroscopic technique known as *photoelectron spectroscopy* (Chapter 19). An X-ray photon of wavelength 150 pm ejects an electron from the inner shell of an atom. The speed of the latter was measured as 2.14×10^7 m s^{-1}. Calculate the electron's binding energy.

13.14. By how much does the wavelength of radiation change when it scatters from (a) a free electron, (b) a free proton, and is detected at 90° to the initial line of flight.

13.15. Derive the Compton formula, [illegible] follows. When the electron is at rest it p[illegible] m_ec^2. When it is in motion with a mom[illegible] p its energy is $\sqrt{(p^2c^2 + m_e^2c^4)}$. Let the [illegible] λ_i, strike the electron and be scattere[illegible] length λ_f through an angle θ, and the ele[illegible] stationary, move off with a momentum of magn[illegible] an angle θ' to the incoming photon. Set up the conservation expressions (energy, momentum along line o[illegible] approach, momentum perpendicular to the line of approach), eliminate θ', then eliminate p, and hence arrive at an expression for $\delta\lambda$.

13.16. Calculate the size of the quantum involved in the excitation of (a) an electronic motion of period 10^{-15} s, (b) a molecular vibration of period 10^{-14} s, (c) a pendulum of period 1 s. Express your results in kJ mol^{-1}.

13.17. What is the de Broglie wavelength of (a) a mass of 1 g travelling at 1 cm s^{-1}, (b) the same, travelling at 100 km s^{-1}, (c) a helium atom travelling at its r.m.s. speed at 25 °C, (d) an electron accelerated from rest through a potential difference of 100 V, 1 kV, 100 kV?

13.18. Calculate the minimum uncertainty in the speed of a ball of mass 500 g that is known to be within 10^{-6} m of a point on a bat. What is the minimum uncertainty in the position of a bullet of mass 5 g that is known to have a speed somewhere between 350.00 001 m s^{-1} and 350.00 000 m s^{-1}?

13.19. An electron is confined to a linear region with a length of the order of the diameter of an atom (≈0.1 nm). What are the minimum uncertainties in its linear momentum and speed?

13.20. In order to use the Born interpretation directly it is necessary that the wavefunction is normalized to unity. Normalize to unity the following wavefunctions: (a) $\sin(n\pi x/L)$ for the range $0 \leq x \leq L$, (b) c, a constant in the range $-L \leq x \leq L$, (c) $\exp(-r/a_0)$ in these dimensions, (d) $x \exp(-r/2a_0)$ in three-dimensional space. In order to integrate over the three dimensions you need to know that the volume element is $d\tau = r^2 dr \sin\theta \, d\theta \, d\phi$, with

$$0 \leq r \leq \infty, \quad 0 \leq \theta \leq \pi, \quad 0 \leq \phi \leq 2\pi.$$

Use

$$\int_0^\infty x^n e^{-ax} dx = n!/a^{n+1}.$$

13.21. A wavefunction for a particle confined to a one-dimensional box of length L is $\psi = (2/L)^{\frac{1}{2}} \sin(\pi x/L)$. Let the box be 10 nm long. What is the probability of finding the particle (a) between $x = 4.95$ nm and 5.05 nm, (b) between $x = 1.95$ nm and 2.05 nm, (c) between $x = 9.90$ and 10.00 nm, (d) in the right half of the box, (e) in the central third of the box?

13.22. The wavefunction for the electron in the ground state of the hydrogen atom is $\psi(r, \theta, \phi) = (1/\pi a_0^3)^{\frac{1}{2}} \exp(-r/a_0)$, where $a_0 = 53$ pm. What is the probability of electron somewhere inside a small sphere of ..0 pm centred on the nucleus? Now suppose the ..ny sphere is moved to surround a point at a distance ..m from the nucleus: what is the probability that the ..ectron is inside it?

13.23. Two excited state wavefunctions for the hydrogen atom are $\psi = (2 - r/a_0)e^{-r/2a_0}$ and $\psi' = r \sin\theta \cos\phi e^{-r/2a_0}$. Normalize both functions.

13.24. Which of the following functions are eigenfunctions of the operator d/dx: (a) $\exp(ikx)$, (b) $\cos kx$, (c) k, (d) kx, (e) $\exp(-\alpha x^2)$? Give the eigenvalue where appropriate.

13.25. Which of the functions in the last Problem are also eigenfunctions of d^2/dx^2, and which are eigenfunctions only of d^2/dx^2? Give the eigenvalues where appropriate.

13.26. A particle is in a state defined by the wavefunction $\psi = \cos\chi e^{ikx} + \sin\chi e^{-ikx}$ where χ is a parameter. What is the probability that the particle will be found with a linear momentum (a) $+k\hbar$, (b) $-k\hbar$. What form would the wavefunction take if it were 90% certain that the particle had a linear momentum $+k\hbar$?

13.27. Evaluate the kinetic energy of the particle described by the wavefunction in Problem 13.26.

13.28. The expectation value of momentum is evaluated by using eqn (13.4.7). What is the average momentum of a particle described by the following wavefunctions: (a) $\exp(ikx)$, (b) $\cos kx$, (c) $\exp(-\alpha x^2)$, each one in the range $-\infty \leq x \leq \infty$?

13.29. Evaluate the expectation values of r and r^2 for the excited state hydrogen wavefunctions given in Problem 13.23. What is the average potential energy of an electron in the ground state of a hydrogen atom?

13.30. The *commutator* of two operators A_{op} and B_{op} is written $[A_{op}, B_{op}]$ and is defined as the difference $A_{op}B_{op} - B_{op}A_{op}$. It can be evaluated by taking some convenient function ψ (which can be left unspecified) and evaluating both $A_{op}B_{op}\psi$ and $B_{op}A_{op}\psi$, and finding the difference in the form $C_{op}\psi$. Then C_{op} is identified as the commutator $[A_{op}, B_{op}]$. Quite often C_{op} turns out to be a simple numerical factor. An extremely important commutator is that of the operator for the components of momentum and the components of position. Find $[x_{op}, y_{op}]$, $[x_{op}, x_{op}]$, $[p_{x,op}, p_{y,op}]$, $[x_{op}, p_{x,op}]$, $[x_{op}, p_{y,op}]$. Are x and y complementary observables? Are x and p_x? Are x and p_y?

13.31. One of the reasons why the commutator is so important is that it lets us identify at a glance the observables that are restricted by the uncertainty relation. Thus, if A_{op} and B_{op} have a non-zero commutator, the observables A and B cannot in general be determined simultaneously. Can p_x and x be determined simultaneously? Can p_x and y? Can the three components of position be specified simultaneously?

13.32. Another important commutator is that for the components of angular momentum. From classical theory $l_x = yp_z - zp_y$, $l_y = zp_x - xp_z$, $l_z = xp_y - yp_x$; hence write the corresponding operators. Show that $[l_{x,op}, l_{y,op}] = i\hbar l_{z,op}$. Can l_x and l_y be determined in general simultaneously?

13.33. Write a computer program for constructing superpositions of cosine functions in order to generate wavepackets like those drawn in Fig. 13.14 and explore how the superposition becomes sharper as more components are added. Include routines that determine the probability that a given momentum will be observed. If you plot the superposition (which you should), set $x = 0$ at the centre of the screen, and build the superposition there. Include a routine that evaluates $\langle x^2\rangle^{\frac{1}{2}}$, e.g. by numerical integration.

14

Quantum theory: techniques and applications

Learning objectives

After careful study of this chapter you should be able to:

(1) Write down the Schrödinger equation for a *particle in a box*, eqn (14.1.3), justify and specify the boundary conditions, and find the allowed energies and wavefunctions, eqn (14.1.9).

(2) Describe the principal features of the solutions of a particle in a box, Section 14.1(b).

(3) Write down the Schrödinger equation for a particle confined to a rectangular surface, eqn (14.1.11), and use the *separation of variables* technique to find its solutions.

(4) State the meaning of *degeneracy*, Section 14.1(c).

(5) Describe the procedure for calculating the wavefunction of a particle in a system with *potential barriers*, Section 14.1(d), and find the wavefunctions in the case of a rectangular barrier, Example 14.2.

(6) Explain how *tunnelling* occurs, and state how the probability of tunnelling depends on the mass of the particle and the width of the barrier, Section 14.1(d).

(7) Write down the Schrödinger equation for a *harmonic oscillator*, eqn (14.2.1), obtain the form of its solutions at large displacements, and describe how the full equation is solved, Section 14.2(a).

(8) Write down an expression for the energy levels of a harmonic oscillator, eqn (14.2.4), and calculate their separation.

(9) Describe the wavefunctions of the harmonic oscillator, and use the properties of *Hermite polynomials* to calculate its properties, Example 14.4.

(10) State and apply the *virial theorem*, eqn (14.2.11).

(11) Calculate the probability that an oscillator will be found at classically forbidden displacements, eqn (14.2.12).

(12) Write down the Schrödinger equation for the motion of a *particle on a ring*, eqn (14.3.2), justify and specify the *cyclic boundary conditions*, and find the wavefunctions, eqn (14.3.4).

(13) Explain the origin of the *quantization of angular motion*, and write expressions for the permitted values of its z-component, eqn (14.3.6).

(14) Write down the operator for the z-component of angular momentum, eqn (14.3.9).

(15) Write down the Schrödinger equation for *rotation in three dimensions*, eqn (14.3.17).

(16) Show that the Schrödinger equation for rotation is separable, eqn (14.3.18).

(17) Write down the expression for the energy of a particle on a sphere in terms of its angular momentum quantum number, eqn (14.3.22).

(18) Write down expressions for the magnitude of the angular momentum, eqn (14.3.24), and the permitted values of its z-component, eqn (14.3.25). See also Box 14.3.

(19) State the meaning of *space quantization* and explain the significance of the *Stern–Gerlach experiment*, Section 14.3(g).

(20) Explain the meaning of *spin*, and state the spin properties of an electron, Section 14.3(h).

(21) Specify the *vector model* of angular momentum, Section 14.3(i).

Introduction

There are three basic types of motion: translation, vibration, and rotation. All three play an important role in chemistry because they are ways in which molecules can store energy. For example, molecules undergo translational motion in containers, and their kinetic energy is a contribution to the total internal energy of the sample. Molecules can rotate, and transitions between the allowed rotational energy levels are responsible for their rotational spectra. Their bonds can also vibrate: this is a store of energy, and transitions give rise to vibrational spectra. Electrons in atoms and molecules store energy in their translational kinetic energy and in the potential energy of their interactions with the nuclei and with each other. In this chapter we see how the concepts of quantum mechanics introduced in Chapter 13 can be developed into a powerful set of techniques for dealing with all these types of motion.

14.1 Translational motion

The quantum mechanical description of *free motion* was outlined in Section 13.3(a). The Schrödinger equation is

$$-(\hbar^2/2m)\,\mathrm{d}^2\psi/\mathrm{d}x^2 = E\psi, \tag{14.1.1}$$

and the general solutions are

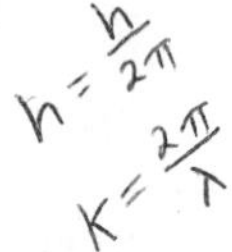

$$\psi = A\mathrm{e}^{\mathrm{i}kx} + B\mathrm{e}^{-\mathrm{i}kx}, \qquad E = k^2\hbar^2/2m. \tag{14.1.2}$$

When we set $B = 0$ we get a wavefunction that corresponds to the particle having a linear momentum $p = k\hbar$ towards positive x (the 'right') and when we set $A = 0$ we get a wavefunction that corresponds to the same momentum, but directed to the left. In either state the position of the particle is totally unpredictable. Note, however, that *all* values of k, and therefore all values of the energy E, are permitted. *The energy of a free particle is not quantized*.

14.1 (a) The particle in a box

Quantized energy is obtained as soon as the range of freedom of the particle is limited. In order to illustrate this feature, we consider the problem of a *particle in a box* in which a particle of mass m is confined between two walls at $x = 0$ and at $x = L$. In the *infinite square well* the potential energy of the particle is zero inside the box but rises abruptly to infinity at the walls, Fig. 14.1.

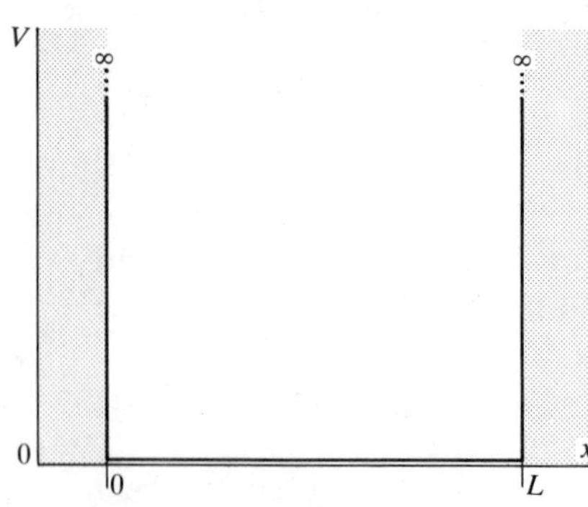

Fig. 14.1. A particle in a one-dimensional region with impenetrable walls. Its potential energy is zero between $x = 0$ and $x = L$, and rises abruptly to infinity as soon as it touches the walls.

The Schrödinger equation for the region between the walls (where $V = 0$) is

$$-(\hbar^2/2m)(\mathrm{d}^2\psi/\mathrm{d}x^2) = E\psi, \tag{14.1.3}$$

which is the same as for the free particle, and so the general solutions are the same. It is convenient‡ to write them as

$$\psi = A\sin kx + B\cos kx, \qquad E = k^2\hbar^2/2m. \tag{14.1.4}$$

‡ Equations (14.1.2) and (14.1.4) are equivalent. We have used $\mathrm{e}^{\mathrm{i}\theta} = \cos\theta + \mathrm{i}\sin\theta$ and $\mathrm{e}^{-\mathrm{i}\theta} = \cos\theta - \mathrm{i}\sin\theta$, and have absorbed all numerical factors into the coefficients A and B.

Now consider the Schrödinger equation for $x<0$ and $x>L$ where the potential energy is infinite. The easiest way of dealing with this region is to suppose that the potential energy is not infinite but very large, and then to allow V to become infinite later. The Schrödinger equation is

$$-(\hbar^2/2m)\,\mathrm{d}^2\psi/\mathrm{d}x^2 + V\psi = E\psi,$$

which rearranges to

$$\mathrm{d}^2\psi/\mathrm{d}x^2 = (2m/\hbar^2)(V-E)\psi. \tag{14.1.5}$$

The qualitative significance of this equation is as follows. If the second derivative of ψ is positive it means that the curvature of the wavefunction is ◡; if it is negative, then the curvature is ⌒. Suppose that ψ happens to be positive just inside the material of the walls, then since the right-hand side of eqn (14.1.5) is positive (because V is so large that it certainly exceeds E, and ψ is positive), the curvature of ψ is positive, and so ψ curls off rapidly towards infinite values as x increases. This makes it an unacceptable function according to the criteria set out in Section 13.3. If ψ has a negative value just inside the walls, the second derivative is negative, and so the wavefunction droops very rapidly down to negatively infinite values, which also makes it unacceptable. Since the wavefunction can be neither positive nor negative just inside the walls, it must be zero there, this requirement being increasingly stringent as the potential energy V approaches infinity.

The quantitative version of this discussion is as follows. The general solution of eqn (14.1.5) is

$$\psi = A\mathrm{e}^{kx} + B\mathrm{e}^{-kx}, \qquad k = \{2m(V-E)/\hbar^2\}^{\frac{1}{2}}. \tag{14.1.6}$$

(Note that the exponentials are real.) Since the first term increases without limit as x increases, the only way we can ensure that the wavefunction does not become infinite is to set $A=0$. Therefore, for $x>L$, the wavefunction is

$$\psi = B\mathrm{e}^{-kx}, \tag{14.1.7}$$

which decays exponentially towards zero as x increases. Note that the rate at which it approaches zero depends on two things: the mass of the particle, and the difference $V-E$. We shall return to the effect of the mass later on. As the potential energy approaches infinity, the decay of the wavefunction towards zero becomes infinitely rapid. When V is infinite, the decay is infinitely fast. Therefore, $\psi=0$ at $x=L$. A similar argument applies at the wall on the left: for an infinite potential energy, the wavefunction decays so rapidly inside the wall that $\psi=0$ at $x=0$. This means that we have established two *boundary conditions* (conditions a function must satisfy at stated locations):

For a particle in an infinite square well potential:
$\psi=0$ at $x=0$ and at $x=L$.

At this stage we know that the wavefunction has the general form given in eqn (14.1.4), but with the additional requirement that it must satisfy the boundary conditions set out above. Consider the wall at $x=0$. According to eqn (14.1.4), $\psi(0)=B$ (because $\sin 0=0$ and $\cos 0=1$). But the boundary condition there is that $\psi(0)=0$. Therefore, $B=0$, which implies that the wavefunction must be of the form

$$\psi = A \sin kx.$$

The amplitude at the other wall is $\psi(L) = A \sin kL$. This too must be zero. Taking $A = 0$ would give $\psi = 0$ for all x, which would conflict with the Born interpretation (the particle must be somewhere). Therefore, kL must be chosen so that $\sin kL = 0$. This requires kL to be an integral multiple of π (because $\sin \theta = 0$ for $\theta = 0, \pi, 2\pi, \ldots$). Consequently, *the only permitted values of k are those for which $kL = n\pi$, with $n = 1, 2, \ldots$* ($n = 0$ cannot occur, because it implies $k = 0$ and $\psi = 0$ everywhere, which is unacceptable, and negative values of n merely change the sign of $\sin(n\pi x/L)$). Since k and E are related as in eqn (14.1.4), it follows that *the energy of the particle is limited to the values*

$$E = n^2\hbar^2\pi^2/2mL^2 = n^2h^2/8mL^2, \qquad n = 1, 2, \ldots. \qquad (14.1.8)$$

The energy of the particle is *quantized*, the quantization arising from the boundary conditions the wavefunction must satisfy in order to be acceptable according to the Born interpretation.

Before discussing this result in more detail, we shall complete the derivation of the wavefunctions. We have to find the normalization constant (here written A). This is done by ensuring that the integral of ψ^2 over all x is equal to unity:

$$1 = \int_{-\infty}^{\infty} \psi^*\psi \, dx = A^2 \int_0^L \sin^2 kx \, dx = A^2L/2, \quad \text{or} \quad A = (2/L)^{\frac{1}{2}}.$$

Therefore, the complete solution to the problem is

$$\textit{Energies:} \qquad E_n = n^2h^2/8mL^2, \qquad n = 1, 2, \ldots \qquad (14.1.9a)$$

$$\textit{Wavefunctions:} \quad \psi_n = (2/L)^{\frac{1}{2}} \sin(n\pi x/L). \qquad (14.1.9b)$$

The energies and wavefunctions have been labelled with the *quantum number n*. A quantum number is an integer (in some cases, as we shall see, a half-integer) that labels the state of the system: knowing the quantum number we can calculate the energy corresponding to that state (through the expression for E_n) and specify the wavefunction.

Example 14.1

An electron is confined to a molecule of length 1.0 nm (about five atoms long). What is (a) its minimum energy and (b) the minimum excitation energy from that state? What is the probability of locating the electron between $x = 0$ and $x = 0.2$ nm?

• *Method*. Use eqn (14.1.9a) with $m = m_e$ and (a) $n = 1$. The minimum excitation energy is $E_2 - E_1$. The distribution of the electron is given by eqn (14.1.9b) with $n = 1$. The total probability of finding the electron in the specified region is the integral of $\psi_1^2\,dx$ over that region.

• *Answer*. For $L = 1.0$ nm,

$$h^2/8m_eL^2 = 6.0_2 \times 10^{-20}\ \text{J}.$$

Therefore, $E_1 = 6.0 \times 10^{-20}$ J (corresponding to 36 kJ mol^{-1}). The minimum excitation energy is

$$E_2 - E_1 = (2^2 - 1)h^2/8m_eL^2 = 18 \times 10^{-20}\ \text{J},$$

which corresponds to 108 kJ mol^{-1} (1.1 eV).

For the probability, we evaluate

$$P = (2/L)\int_0^l \sin^2(n\pi x/L)\,dx = (l/L) - (1/2n\pi)\sin(2n\pi l/L)$$

with $n = 1$ and $l = 0.2$ nm. This gives $P = 0.05$, or a chance of 1 in 20 of finding the electron in that region.

• *Comment*. The electron-in-a-box is a crude model of molecular structure; it can be used for estimating rough values of transition energies.

• *Exercise*. Calculate the first excitation energy of a proton confined to a region roughly equal to the diameter of a nucleus (10^{-15} m). Calculate the probability that it in its ground state it will be found between $x = 0.25L$ and $x = 0.75L$. [600 MeV, 0.18]

14.1 (b) The properties of the solutions

The shapes of some of the wavefunctions of a particle in a box are shown in Fig. 14.2. It is easy to see the origin of the quantization in pictorial terms: each wavefunction is a standing wave and, in order to fit into the cavity, successive functions must possess one more half-wavelength. Shortening the wavelength implies sharpening the curvature of the wavefunction, and therefore increasing the kinetic energy of the particle it describes. The linear momentum of a particle in a box is not well-defined because the wavefunction $\sin kx$ is a standing wave and not an eigenfunction of the linear momentum operator, as explained in Section 13.4. However, each wavefunction is a superposition of momentum eigenfunctions (because $\sin kx = (e^{ikx} - e^{-ikx})/2i$), and so measurement of the linear momentum would give the value $k\hbar = nh/2L$ half the time and $-k\hbar = -nh/2L$ the other half. This is the quantum mechanical version of the classical picture that a particle in a box rattles from wall to wall, and travels to the right half the time and to the left for the other half.

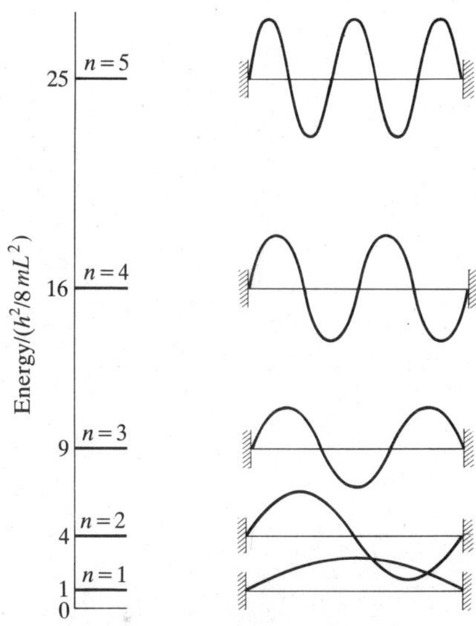

Fig. 14.2. The allowed energy levels and the corresponding (sine wave) wavefunctions for a particle in a box. Note that the energy levels increase as n^2, and so their separation increases as the quantum number increases. Each wavefunction is a standing wave, and successive functions possess one more half wave.

Because n cannot be zero, the lowest energy that a particle may possess when it is in a box is not zero (as would be allowed by classical mechanics) but is $E_1 = h^2/8mL^2$. This lowest, irremovable energy is called the *zero-point energy*. Its physical origin can be explained in two ways. First, the uncertainty principle requires a particle to possess kinetic energy if it is confined to a finite region. This is because the particle's location is not completely indefinite, and so its momentum cannot be precisely zero. Alternatively, if the wavefunction is to be zero at the walls, but smooth, continuous, and not zero everywhere, then it must be curved, and curvature in a function implies the possession of kinetic energy.

The separation between neighbouring energy levels is

$$\Delta E = E_{n+1} - E_n = (2n+1)h^2/8mL^2, \qquad (14.1.10)$$

which decreases as the length of the container increases. The separation ΔE is extremely small when the container is large, and becomes zero when the walls are infinitely far apart. Atoms and molecules free to move in laboratory-sized vessels may therefore be treated as though their translational energy were not quantized.

The distribution of the particle in a box is not uniform: the probability density at x is $(2/L)\sin^2(n\pi x/L)$. The effect is pronounced when n is small, and Fig. 14.2 shows that there is an apparent repulsion from the vicinity of the walls. At high quantum numbers the distribution becomes more uniform. This reflects the classical result that a particle bouncing between the walls spends, on the average, equal times at all points. That the quantum result corresponds to the classical prediction at high quantum numbers is an aspect of the *correspondence principle*, which states that classical mechanics emerges from quantum mechanics as high quantum numbers are reached.

14.1 (c) Motion in two dimensions

When a particle is confined to a rectangular surface of length L_1 in the x-direction and L_2 in the y-direction, the potential energy being zero everywhere except at the walls, where it is infinite, the Schrödinger equation is

$$-(\hbar^2/2m)\{(\partial^2\psi/\partial x^2)+(\partial^2\psi/\partial y^2)\}=E\psi, \qquad (14.1.11)$$

and ψ is a function of both x and y: $\psi=\psi(x,y)$.

In some cases, partial differential equations (differential equations in more than one variable) can be solved very simply by the method called the *separation of variables*, which has the effect of cutting the complete equation into two or more ordinary differential equations, one for each variable. The method works in this case, as we can see by writing the wavefunction as a product of functions, one depending only on x and the other only on y: that is, $\psi(x,y)=X(x)Y(y)$. Then, since $\partial^2\psi/\partial x^2=Y\,\mathrm{d}^2X/\mathrm{d}x^2$ (because X depends only on x), and similarly for $\partial^2\psi/\partial y^2$, the equation becomes

$$-(\hbar^2/2m)\{Y(\mathrm{d}^2X/\mathrm{d}x^2)+X(\mathrm{d}^2Y/\mathrm{d}y^2)\}=EXY.$$

On dividing by XY we obtain

$$-(\hbar^2/2m)\{(X''/X)+(Y''/Y)\}=E,$$

where $X''=\mathrm{d}^2X/\mathrm{d}x^2$ and $Y''=\mathrm{d}^2Y/\mathrm{d}y^2$. Now for the crucial step. X''/X is independent of y, and so if y is varied only the other term, Y''/Y, might vary. But the sum of these two terms is a constant; therefore, even the Y''/Y term cannot vary. In other words, Y''/Y is a constant. Similarly, X''/X is also a constant. Therefore, we can write

$$-(\hbar^2/2m)X''/X=E^X, \quad \text{or} \quad -(\hbar^2/2m)\,\mathrm{d}^2X/\mathrm{d}x^2=E^XX$$
$$-(\hbar^2/2m)Y''/Y=E^Y, \quad \text{or} \quad -(\hbar^2/2m)\,\mathrm{d}^2Y/\mathrm{d}y^2=E^YY,$$

with $E^X+E^Y=E$. Each of these equations is the same as the one-dimensional particle-in-a-box Schrödinger equation; hence we can adapt the results in eqn (14.1.9) without further calculation:

$$X_{n_1}=(2/L_1)^{\frac{1}{2}}\sin(n_1\pi x/L_1),$$
$$Y_{n_2}=(2/L_2)^{\frac{1}{2}}\sin(n_2\pi y/L_2).$$

Since $\psi=XY$ we obtain

$$\psi_{n_1,n_2}=(4/L_1L_2)^{\frac{1}{2}}\sin(n_1\pi x/L_1)\sin(n_2\pi y/L_2), \qquad (14.1.12a)$$
$$E_{n_1,n_2}=E^X_{n_1}+E^Y_{n_2}=\{(n_1/L_1)^2+(n_2/L_2)^2\}(h^2/8m), \qquad (14.1.12b)$$

with the quantum numbers permitted the values $n_1=1,2,\ldots$ and $n_2=1,2,\ldots$ independently. Some of these functions are plotted in Fig. 14.3; they are the two-dimensional versions of the wavefunctions shown in Fig. 14.2.

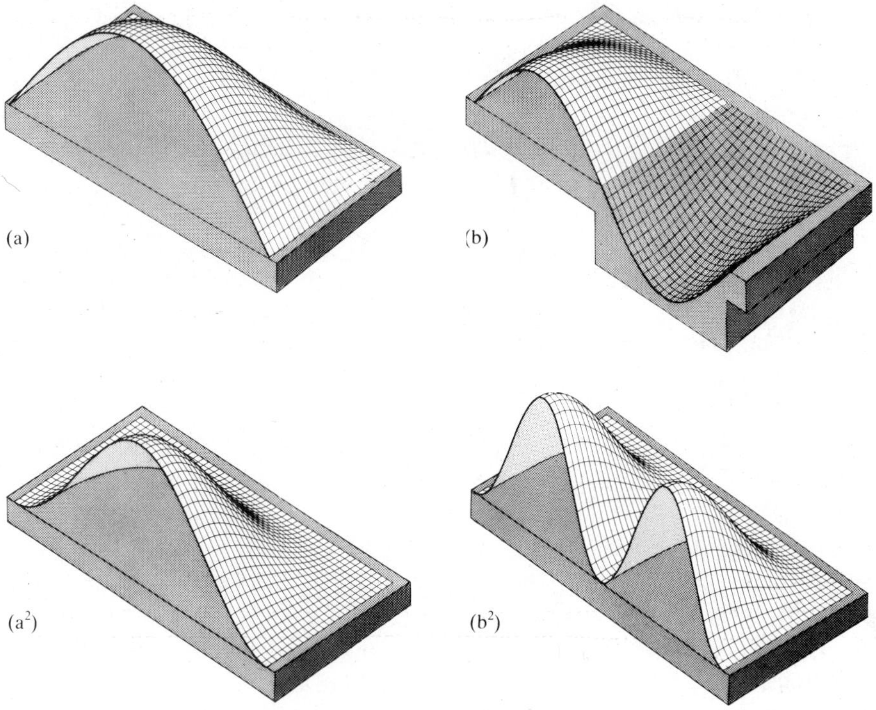

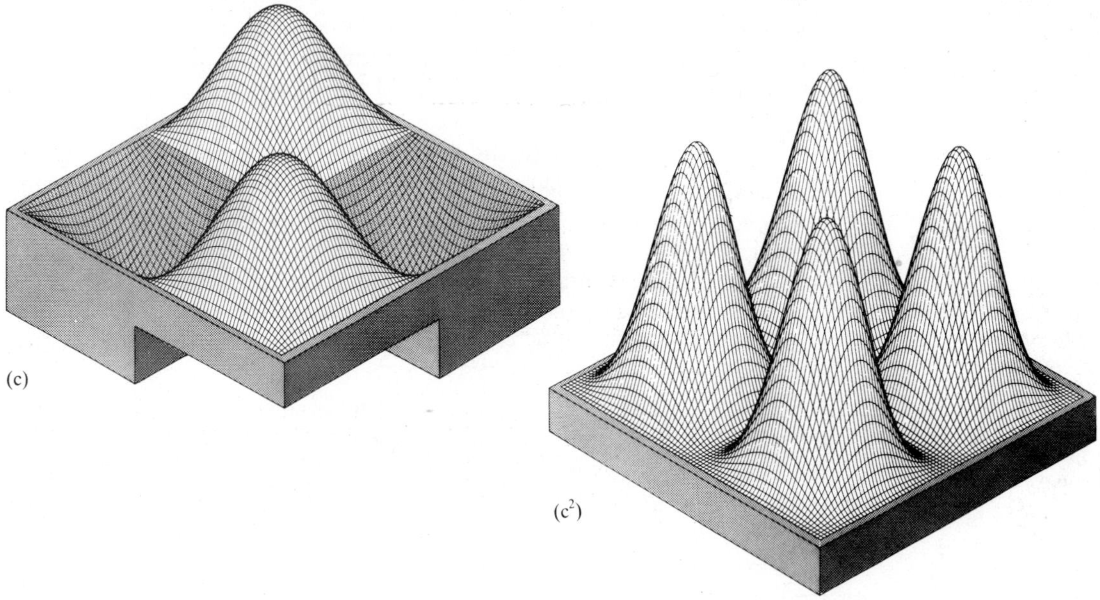

Fig. 14.3. The wavefunctions and probability densities for a particle confined to a rectangular surface. (a) $\psi_{1,1}$ section; (b) $\psi_{2,1}$ section (rotate by 90° for $\psi_{1,2}$). (c) $\psi_{2,2}$. The corresponding probability densities are labelled (a^2), (b^2), and (c^2). Each section is half the total function.

The three-dimensional case, of a particle in an actual box, can be treated in the same way, and the wavefunctions have another factor (for the z-dependence), and the energy has an additional term.

An interesting feature of the solutions is obtained when the plane surface is square, when $L_1 = L$ and $L_2 = L$. Then

$$\psi_{n_1,n_2} = (2/L)\sin(n_1\pi x/L)\sin(n_2\pi y/L),$$
$$E_{n_1,n_2} = \{n_1^2 + n_2^2\}(h^2/8mL^2).$$

(a)

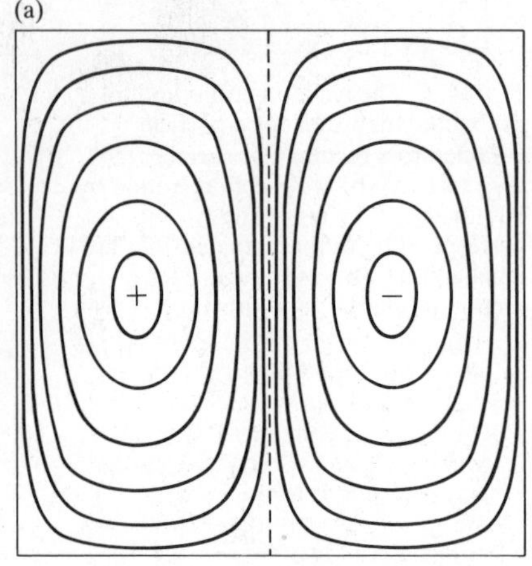

(b)

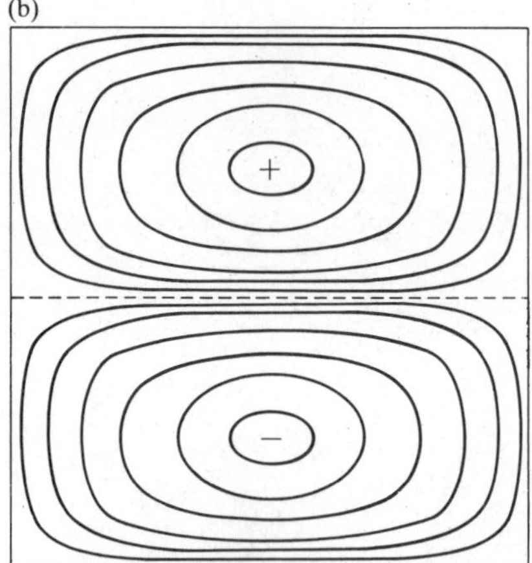

Fig. 14.4. It is easier to represent the functions in terms of contour diagrams. Here we show contour diagrams for (a) $\psi_{2,1}$ and (b) $\psi_{1,2}$ in a square well. Note that one can be converted into the other by a 90° rotation: we say that they are related by a *symmetry transformation*. These two functions are also *degenerate* (i.e. have the same energy).

Consider the cases $n_1 = 1, n_2 = 2$ and $n_1 = 2, n_2 = 1$:

$$\psi_{1,2} = (2/L)\sin(\pi x/L)\sin(2\pi y/L), \qquad E_{1,2} = 5(h^2/8mL^2),$$
$$\psi_{2,1} = (2/L)\sin(2\pi x/L)\sin(\pi y/L), \qquad E_{2,1} = 5(h^2/8mL^2).$$

The point to note is that *more than one wavefunction* (two in this case) *correspond to the same energy*. This is the condition of *degeneracy*. In this case, we say that the level with energy $5(h^2/8mL^2)$ is *doubly degenerate*.

The occurrence of degeneracy is related to the symmetry of the system. The two degenerate functions $\psi_{1,2}$ and $\psi_{2,1}$ are shown in Fig. 14.4: because the plane is square, we see that we can convert one into the other simply by rotating the plane by 90°. This is not possible when the plane is not square, and then $\psi_{1,2}$ and $\psi_{2,1}$ are non-degenerate. We shall see many examples of degeneracy in the pages that follow (e.g. in the hydrogen atom), and all of them can be traced to the symmetry properties of the system.

14.1 (d) Quantum leaks

If the potential energy of the particle does not rise to infinity when it is in the walls of the container, then the argument that led to eqn (14.1.6) allows the wavefunction to remain non-zero. If the walls are thin (so that the potential energy falls to zero again after a finite distance), the exponential decay of the wavefunction stops, and it begins to oscillate like the wavefunctions for the inside of the box, Fig. 14.5. This means that the particle might be found on the outside of a container even though according to classical mechanics it has insufficient energy to escape. This leaking through classically forbidden zones is called *tunnelling*.

The Schrödinger equation lets us calculate the extent of tunnelling and how it depends on the mass of the particle. In fact, from the result in eqn (14.1.6) we can see that, since the wavefunction decreases exponentially inside the wall, and does so with a rate that depends on $\sqrt{m}$, *light particles are more able to tunnel through barriers than heavy ones*. Tunnelling is very important for electrons, and moderately important for protons; for heavier particles it is less important. A number of effects in chemistry (e.g. some reaction rates) depend on the ability of the proton to tunnel more readily than the deuteron.

The kind of problem we can solve with the material developed so far is illustrated by the case of a projectile (such as an electron or a proton) incident from the left on a region where its potential energy increases sharply from zero to a finite, constant value V, remains there for a distance L, and then falls to zero again, Fig. 14.6. This is a model of what happens when particles are fired at an idealized metal foil or sheet of paper. We can ask for the proportion of incident particles that penetrate the barrier when their kinetic energy is less than V so that classically none can penetrate.

The strategy of the calculation (and of others like it) is as follows:

(1) Write down the Schrödinger equation for each zone of constant potential.

(2) Write down the general solutions for each zone using eqn (14.1.2) for the regions where $V < E$ and eqn (14.1.6) for regions where $V > E$.

(3) Find the coefficients by ensuring that (a) the wavefunction is continuous at each zone boundary, and (b) the first derivatives of the wavefunctions are also continuous at the zone boundaries.

The way this is carried through for the arrangement shown in Fig. 14.6 is described in the following *Example*.

Example 14.2

Deduce an expression for the probability that a particle of mass m and energy E will penetrate a potential energy barrier of height V (with $V > E$) and width L when it is incident from the left.

• *Method*. Follow the procedure set out above; the three zones are shown in Fig. 14.6. There are no further barriers to the right to reflect back particles, and so no particles have momentum to the left in that zone.

• *Answer*. The three general solutions are as follows:

$$\text{Zone A: } \psi_A = Ae^{ikx} + A'e^{-ikx}, \qquad k = \{2mE/\hbar^2\}^{\frac{1}{2}},$$
$$\text{Zone B: } \psi_B = Be^{\kappa x} + B'e^{-\kappa x}, \qquad \kappa = \{2m(V-E)/\hbar^2\}^{\frac{1}{2}},$$
$$\text{Zone C: } \psi_C = Ce^{ikx} + C'e^{-ikx}, \qquad k = \{2mE/\hbar^2\}^{\frac{1}{2}}.$$

Since there are no particles with negative momentum in Zone C, we can set $C' = 0$. The probability of penetration is proportional to $|C|^2$, and the probability of penetration relative to the incident probability, which is proportional to $|A|^2$, is $P = |C|^2/|A|^2$.

The boundary conditions are as follows:

$$\psi_A(0) = \psi_B(0), \qquad \psi'_A(0) = \psi'_B(0), \qquad \psi_B(L) = \psi_C(L), \qquad \psi'_B(L) = \psi'_C(L)$$

with $\psi' = d\psi/dx$. These imply

$$A + A' = B + B', \qquad ikA - ikA' = \kappa B - \kappa B',$$
$$Be^{\kappa L} + B'e^{-\kappa L} = Ce^{ikL}, \qquad \kappa Be^{\kappa L} - \kappa B'e^{-\kappa L} = ikCe^{ikL}$$

These four simultaneous equations are now solved for P, and the result is eqn (14.1.13).

• *Comment*. A *trap*, a region of negative V, has interesting properties as it becomes transparent for some values of E (see Problem 14.8): the classical analogy is the coating on a lens, chosen to have a refractive index and thickness to appear transparent to the incident light.

• *Exercise*. Find an expression for P in the case when the potential energy in Zone C is $V \neq 0$.

[Problem 14.6]

The *Example* shows that the probability of penetration of the barrier is

$$P = 1/(1+G), \quad \text{with} \quad G = \frac{\{e^{L/D} - e^{-L/D}\}^2}{4\varepsilon(1-\varepsilon)} \qquad (14.1.13)$$

where $D = \hbar/\{2m(V-E)\}^{\frac{1}{2}}$ and $\varepsilon = E/V$. When the barrier is high and long, so that $L/D \gg 1$, the first exponential dominates the second, and since then $G \gg 1$, $P \approx 1/G$. In this case

$$P = 4\varepsilon(1-\varepsilon)e^{-2L/D}, \qquad (14.1.14)$$

and so P depends exponentially on the square root of the mass of the particle (as we anticipated) and on the length of the barrier. Figure 14.7 shows how a proton's and a deuteron's tunnelling probabilities depend on the incident energy for a barrier of height 5 eV and width 0.1 nm.

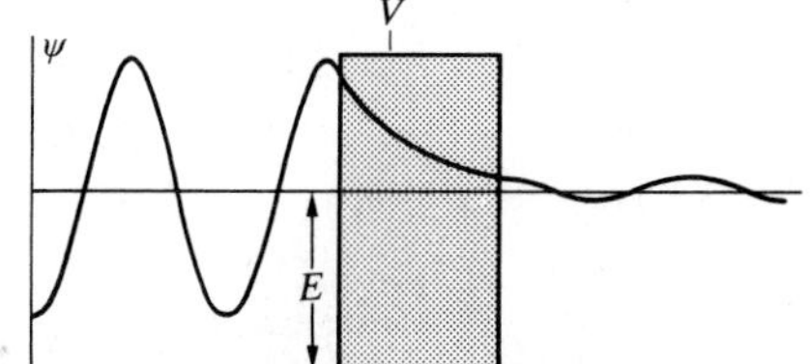

Fig. 14.5. A particle incident on a barrier from the left has an oscillating wavefunction, but inside the barrier it decays exponentially (for $E < V$). If the barrier is not too thick, the wavefunction is non-zero at its opposite face, and so oscillates there, which corresponds to the particle penetrating the barrier. (Only the real component is shown.)

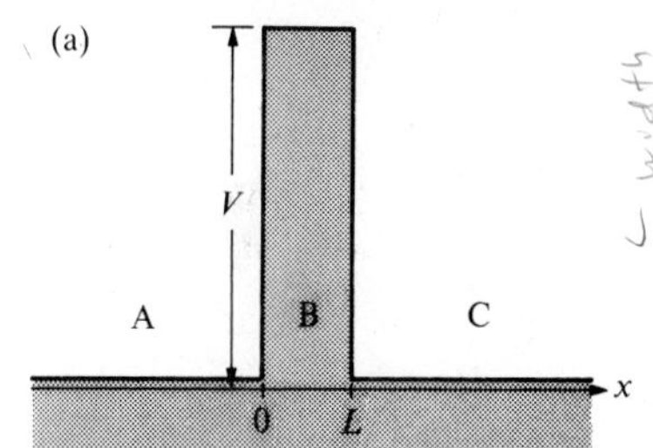

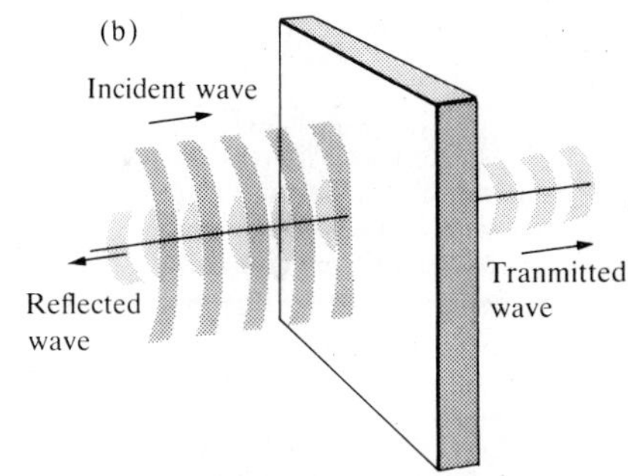

Fig. 14.6. (a) Shows the zones of constant potential energy used in the calculation of a tunnelling probability, and (b) represents the various components of the wavefunction: there is a strong incident wave (e^{-ikx}, momentum to the right), a weaker reflected wave (e^{-ikx}), and a possibly very weak transmitted wave (e^{ikx} in Zone C).

14.2 Vibrational motion

A particle undergoes *harmonic motion* if it experiences a restoring force proportional to its displacement from equilibrium (at $x = 0$). The force is

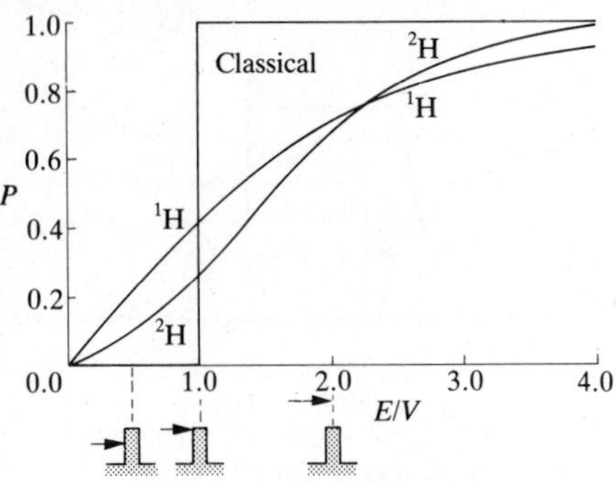

Fig. 14.7. The probability of transmission of a proton and a deuteron through a potential barrier for $E < V$, calculated using eqn (14.1.13). The diagrams underneath the horizontal axis indicate the relative values of E and V.

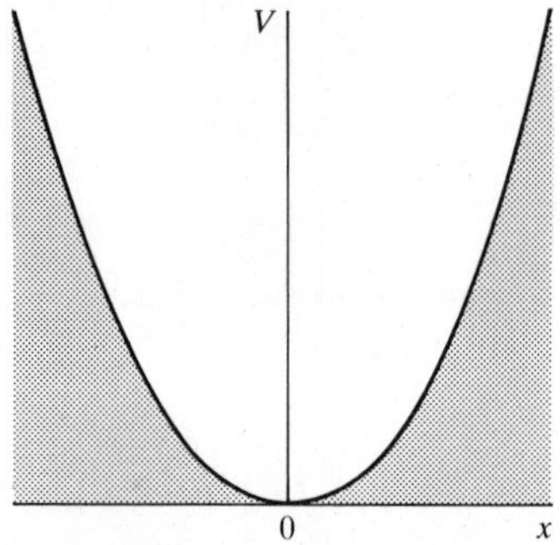

Fig. 14.8. The potential energy $V = \frac{1}{2}kx^2$ of a harmonic oscillator. The narrowness of the curve depends on the *force constant* k. x is the displacement from equilibrium. This is a *parabolic potential energy*.

written $F = -kx$, where k is the *force constant*; this means that the particle's potential energy at a displacement x is $V = \frac{1}{2}kx^2$ (as explained in Section 13.1). The Schrödinger equation for the particle is therefore

$$-(\hbar^2/2m)\,\mathrm{d}^2\psi/\mathrm{d}x^2 + \tfrac{1}{2}kx^2\psi = E\psi. \qquad (14.2.1)$$

It is helpful to identify the similarities between this problem and the particle in a box. As in that case, the particle is trapped in a symmetrical well in which the potential energy rises to large values (and ultimately to infinity) for sufficiently large displacements, Fig. 14.8. As boundary conditions have to be satisfied, we should expect the energy of the particle to be quantized. The wavefunctions should resemble the square-well wavefunctions, but with two minor differences. First, their amplitudes can fall towards zero more slowly at large displacements, because the potential energy climbs towards zero only as x^2 and not abruptly. Second, the kinetic energy of the particle depends on the displacement in a more complex way on account of the variation of the potential energy, and therefore the curvature of the wavefunction also varies in a more complex way.

We now sketch the way in which eqn (14.2.1) is solved. This is moderately technical, and so the important conclusions, which can be used without needing the details in the following section, are summarized in Section 14.2(b).

14.2 (a) The formal solution

The first step is to simplify the appearance of the Schrödinger equation. This is achieved by introducing the following dimensionless quantities:

$$y = x/\alpha, \qquad \varepsilon = E/(\tfrac{1}{2}\hbar\omega), \qquad (14.2.2)$$

where $\alpha^2 = \hbar/(mk)^{\frac{1}{2}}$ and $\omega = (k/m)^{\frac{1}{2}}$. These transform the equation into

$$\mathrm{d}^2\psi/\mathrm{d}y^2 + (\varepsilon - y^2)\psi = 0. \qquad (14.2.3)$$

The next step involves exploring the solutions at large displacements. When x (and therefore y) is very large, the equation simplifies to

$$\mathrm{d}^2\psi/\mathrm{d}y^2 - y^2\psi = 0.$$

One approximate‡ solution is $\mathrm{e}^{y^2/2}$, but it is obviously unacceptable because it increases rapidly with the displacement, and approaches infinity. The other solution is $\mathrm{e}^{-y^2/2}$, which is well-behaved for both positive and negative displacements, however large, and so is acceptable. Note that this *asymptotic form* of the wavefunction decays more rapidly with increasing displacement than an ordinary exponential function (e^{-kx}) because of the x^2 dependence. This is plausible for the following reason. In Section 14.1 we saw that in regions of constant potential energy with $V > E$, wavefunctions fall exponentially to zero. In the harmonic oscillator, however, the potential energy is increasing with displacement. Therefore, we expect the wavefunction to decrease more rapidly because it is responding to the increasing, rather than merely constant, potential energy.

‡ Substitution of this solution into the simplified equation gives $\mathrm{d}^2\psi/\mathrm{d}y^2 = \mathrm{e}^{y^2/2} + y^2\mathrm{e}^{y^2/2}$, but the first of these two terms is much smaller than the second (because we are interested only in the region where y is very large), and so may be ignored. This type of approximate solution, where only the largest terms are retained, is called an *asymptotic solution*.

The next stage is to suppose that the exact wavefunction has the form $\psi = f\mathrm{e}^{-y^2/2}$ where f is a function that does not increase more rapidly than $\mathrm{e}^{-y^2/2}$ decays: then $f\mathrm{e}^{-y^2/2}$ remains acceptable at large displacements. We substitute this trial solution into the exact wave equation, eqn (14.2.3), and find

$$f'' - 2yf' + (\varepsilon - 1)f = 0,$$

where $f' = \mathrm{d}f/\mathrm{d}y$ and $f'' = \mathrm{d}^2f/\mathrm{d}y^2$. This differential equation, which is called *Hermite's equation*, is one that has been thoroughly studied by mathematicians, and its solutions have been listed.

Acceptable solutions of Hermite's equation (those not going to infinity more quickly than the factor $\mathrm{e}^{-y^2/2}$ falls to zero, so that the product is never infinite) exist only for ε equal to a positive, odd integer. Therefore we write $\varepsilon = 2v + 1$, with $v = 0, 1, 2, \ldots$. This immediately leads to an expression for the energy levels of the oscillator: since $\varepsilon = E/(\frac{1}{2}\hbar\omega)$ and $\varepsilon = 2v + 1$, we have

$$E_v = (2v + 1)(\tfrac{1}{2}\hbar\omega).$$

14.2 (b) The energy levels

The energy levels are most tidily expressed in the form

$$E_v = (v + \tfrac{1}{2})\hbar\omega, \quad \text{with} \quad v = 0, 1, 2, \ldots \quad \text{and} \quad \omega = (k/m)^{\frac{1}{2}}. \qquad (14.2.4)$$

We conclude that the energy of the oscillator is quantized, as we anticipated, and that the separation between adjacent levels is

$$\Delta E = E_{v+1} - E_v = \hbar\omega, \qquad (14.2.5)$$

which is the same for all v. Therefore, the energy levels form a uniform ladder of spacing $\hbar\omega$, Fig. 14.9. Furthermore, there is a zero-point energy of magnitude $E_0 = \frac{1}{2}\hbar\omega$. The *mathematical* reason for the zero-point energy is that v cannot take negative values, for if it did the wavefunction would be ill-behaved. The *physical* reason is the same as for the particle in a square well: the particle is confined, its position is not completely uncertain, and therefore its momentum, and hence its kinetic energy, cannot be exactly zero. We can picture this zero-point energy state as the particle fluctuating incessantly around its equilibrium position; classical mechanics would allow it to be perfectly still.

The energy separation $\hbar\omega$ is negligibly small for macroscopic objects, but is of great importance for objects the mass of atoms. The force constant of a typical chemical bond is around $500\,\mathrm{N\,m^{-1}}$, and since the mass of a proton is about $1.7 \times 10^{-27}\,\mathrm{kg}$, $\omega \approx 5 \times 10^{14}\,\mathrm{s^{-1}}$ and $\hbar\omega \approx 6 \times 10^{-20}\,\mathrm{J}$, corresponding to $30\,\mathrm{kJ\,mol^{-1}}$, which is a chemically very significant quantity. The transition between two adjacent vibrational levels of such a bond releases a photon of energy $\Delta E = 6 \times 10^{-20}\,\mathrm{J}$, corresponding to a frequency $\nu = \Delta E/h = \omega/2\pi = 9 \times 10^{13}\,\mathrm{Hz}$, and wavelength $\lambda = c/\nu = 3000\,\mathrm{nm}$. Therefore, transitions between the vibrational energy levels of molecules give rise to infrared radiation, and this is the basis of infrared spectroscopy, as we describe in Chapter 18.

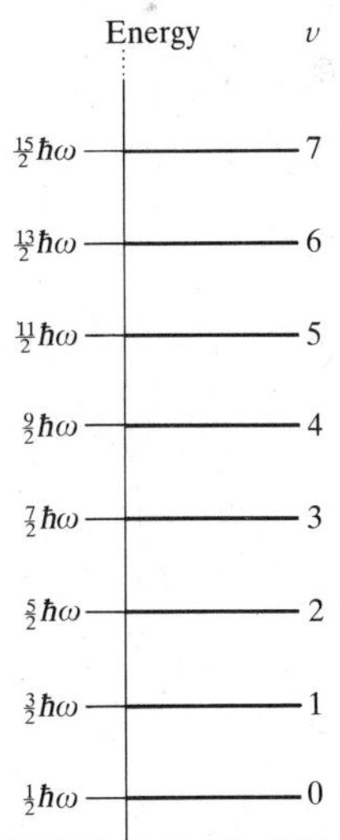

Fig. 14.9. The first few energy levels of a harmonic oscillator. Notice that the levels have constant separation, and that there is a minimum energy of $\frac{1}{2}\hbar\omega$, the *zero-point energy*.

14.2(c) The wavefunctions

The shapes of the wavefunctions can be obtained once we know the shapes of the solutions of Hermite's equation. The solutions for v a positive integer are the *Hermite polynomials* (symbol H_v). These are polynomials (i.e. functions of the form $a_0 + a_1 y + a_2 y^2 + \ldots + a_v y^v$ running to a *finite* number of terms, unlike $\cos y$, for example, which runs to an infinite number of terms when expressed in the same way) which can be generated by differentiating e^{-y^2} the appropriate number of times:

$$\textit{Hermite polynomials: } H_v(y) = (-1)^v e^{y^2}(d^v/dy^v)e^{-y^2}. \qquad (14.2.6)$$

The explicit forms of the first few polynomials are given in Box 14.1 together with some of their most useful properties.

Box 14.1 Harmonic oscillator wavefunctions

Write $x = \alpha y$, where x is the displacement from equilibrium and

$$\alpha^2 = \hbar/(mk)^{\frac{1}{2}}, \qquad \omega^2 = k/m;$$

then the normalized wavefunctions are

$$\psi_v = N_v H_v(y) e^{-y^2/2}, \qquad N_v^2 = 1/\alpha\pi^{\frac{1}{2}} 2^v v!$$

with $v = 0, 1, 2, \ldots$ and the $H_v(y)$ the following Hermite polynomials:

v	$H_v(y)$
0	1
1	$2y$
2	$4y^2 - 2$
3	$8y^3 - 12y$
4	$16y^4 - 48y^2 + 12$
5	$32y^5 - 160y^3 + 120y$
6	$64y^6 - 480y^4 + 720y^2 - 120$

The Hermite polynomials (which continue up to infinite v) satisfy the equation

$$H_v'' - 2yH_v' + 2vH_v = 0$$

and the recursion relation

$$H_{v+1} = 2yH_v - 2vH_{v-1}.$$

An important integral is

$$\int_{-\infty}^{\infty} e^{-y^2} H_v H_{v'}\, dy \begin{cases} = 0 & \text{if } v' \neq v \\ = \pi^{\frac{1}{2}} 2^v v! & \text{if } v' = v. \end{cases}$$

The wavefunction for the level with label v is the product of the Hermite polynomial $H_v(y)$ and $e^{-y^2/2}$. We need to normalize it to unity, but this is easily done using the properties of the Hermite polynomials, for they have simple, standard integrals, as the following *Example* shows.

Example 14.3

Find the normalization constant for the harmonic oscillator wavefunctions.

- *Method*. Write the wavefunctions as

$$\psi_v(x) = N_v H_v(y) e^{-y^2/2}$$

and choose N_v so that

$$\int_{-\infty}^{\infty} \psi_v^2\, dx = 1.$$

Begin by expressing the integral as an integration over the dimensionless variable $y = x/\alpha$. Use the integration properties given in Box 14.1.

• *Answer*. Since $dx = \alpha\, dy$, we must solve

$$\alpha\int_{-\infty}^{\infty}\psi_v^2\,dy = \alpha N_v^2\int_{-\infty}^{\infty}H_v^2 e^{-y^2}\,dy = \alpha N_v^2\pi^{\frac{1}{2}}2^v v! = 1.$$

Therefore, $N_v = 1/\{\alpha\pi^{\frac{1}{2}}2^v v!\}^{\frac{1}{2}}$.

• *Comment*. The Hermite polynomials are one of a class of functions called *orthogonal polynomials*. These have a wide range of important properties that allow a number of quantum mechanical calculations to be done with relative ease. See *Further reading* for a reference to their properties.

• *Exercise*. Confirm, by explicit evaluation of the integral, that ψ_0 is normalized to unity.

[Use $\int_0^{\infty} e^{-z^2}\,dz = \pi^{\frac{1}{2}}/2$]

It follows that the normalized wavefunctions of the harmonic oscillator are the functions

$$\psi_v(x) = N_v H_v(y)e^{-y^2/2}, \qquad N_v = \{1/\alpha\pi^{\frac{1}{2}}2^v v!\}^{\frac{1}{2}}, \tag{14.2.7}$$

where $y = x/\alpha$ and $\alpha^2 = \hbar/(mk)^{\frac{1}{2}}$, as specified in the substitutions that led to eqn (14.2.3). Some of these functions and the corresponding probability densities are illustrated in Fig. 14.10.

Since the Hermite polynomial for $v = 0$ is simply 1, the wavefunction for the *ground state* (the lowest energy state) of the harmonic oscillator is $N_0 e^{-y^2/2}$ and the probability density is the bell-shaped *gaussian function* $N_0^2 e^{-y^2}$. This has its maximum amplitude at zero displacement, and so it captures the classical picture of the zero-point motion as arising from the fluctuation of the particle about its equilibrium position. At high quantum numbers, the wavefunctions have their largest amplitudes near the limits of the range of displacements. This also captures the classical motion of the particle, because it is most likely to be found at the *turning points* (where $V = E$) where it is travelling most slowly, and is least likely to be found at zero displacement where it travels with maximum velocity. Once again we see classical properties emerging in the correspondence limit of high quantum numbers.

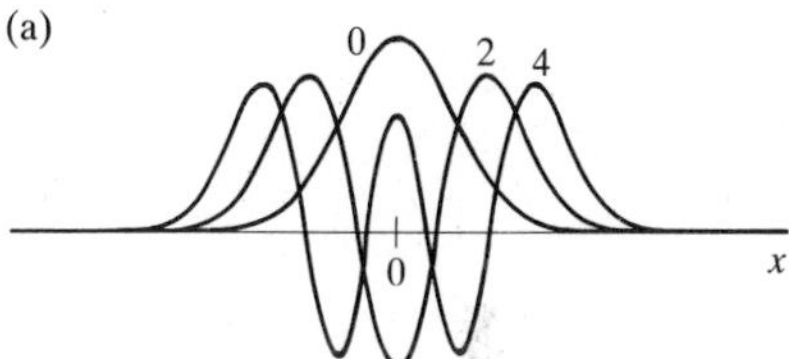

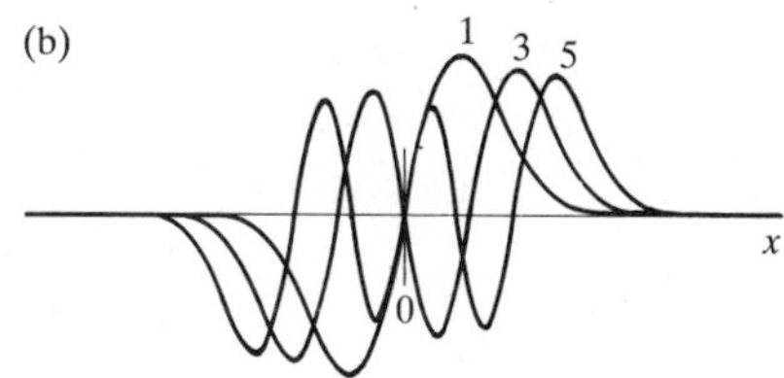

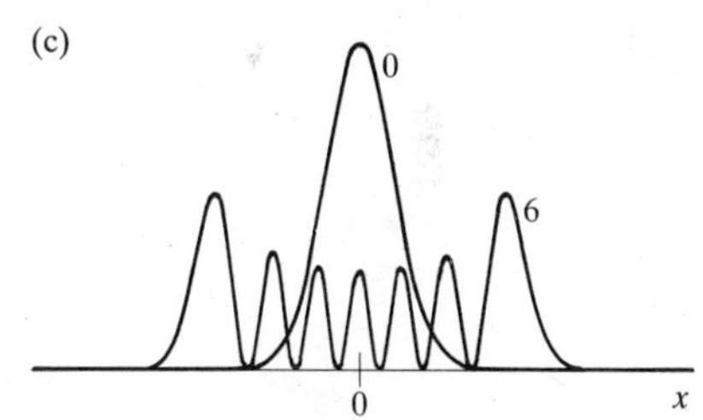

Fig. 14.10. Probability densities and wavefunctions for some states of a harmonic oscillator. (a) ψ for even v; (b) ψ for odd v; (c) ψ^2 for $v = 0,6$. Notice that the most probable location of the oscillator lies closer to the turning points of the classical motion as the quantum number v increases.

14.2 (d) Properties of the solutions

Once the wavefunctions are available we can start calculating the properties of the harmonic oscillator. For instance, we can calculate the expectation values of an observable O by evaluating integrals of the type

$$\langle O\rangle = \int_{-\infty}^{\infty}\psi^*\hat{O}\psi\,dx = N_v^2\int_{-\infty}^{\infty}(H_v e^{-y^2/2})\hat{O}(H_v e^{-y^2/2})\,dx.$$

(The wavefunctions are real, and so the * can be omitted.) These integrals might look fearsome, but the Hermite polynomials have many simplifying features. For instance, we show in the following *Example* that the *mean displacement* $\langle x\rangle$ and the *mean square displacement* of the oscillator from equilibrium when it is in the state with quantum number v are

$$\langle x \rangle = 0, \qquad \langle x^2 \rangle = (v + \tfrac{1}{2})\hbar(mk)^{-\frac{1}{2}}. \tag{14.2.8}$$

Example 14.4

Calculate the *mean displacement* of the oscillator from equilibrium when it is in a quantum level v.

• *Method*. Calculate the expectation value of x using the normalized wavefunctions. The operator for position along x is multiplication by the value of x (Section 13.4). Use the relations in Box 14.1 to manipulate the integrals.

• *Answer*.

$$\langle x \rangle = \int_{-\infty}^{\infty} \psi_v x \psi_v \, dx = N_v^2 \int_{-\infty}^{\infty} (H_v e^{-y^2/2}) x (H_v e^{-y^2/2}) \, dx$$

$$= \alpha^2 N_v^2 \int_{-\infty}^{\infty} H_v y H_v e^{-y^2} \, dy$$

(we have used $x = \alpha y$, so that $dx = \alpha \, dy$). Now use the *recursion relation* given in Box 14.1 to write

$$yH_v = vH_{v-1} + \tfrac{1}{2}H_{v+1},$$

which turns the integral into

$$\int_{-\infty}^{\infty} H_v y H_v e^{-y^2} \, dy = v \int_{-\infty}^{\infty} H_v H_{v-1} e^{-y^2} \, dy + \tfrac{1}{2} \int_{-\infty}^{\infty} H_v H_{v+1} e^{-y^2} \, dy.$$

But both these integrals are zero (Box 14.1), and so $\langle x \rangle = 0$.

• *Comment*. The result is in fact obvious from the probability densities drawn in Fig. 14.10: they are all symmetrical about $x = 0$, and so displacements to the right are as probable as displacements to the left. The reason for going through the calculation in detail even though the result is obvious is that the same technique is applicable to other observables for which the result is not obvious.

• *Exercise*. Calculate the *mean square displacement* of the particle from its equilibrium position. (Use the recursion relation twice.) [Eqn (14.2.8)]

The result summarized in eqn (14.2.8) shows that the mean square displacement increases with v. This is apparent from the probability densities in Fig. 14.10, and corresponds to the classical amplitude of swing increasing as the oscillator becomes more highly excited.

The *mean potential energy* of the oscillator can now be calculated very simply. Since $V = \tfrac{1}{2}kx^2$,

$$\langle V \rangle = \tfrac{1}{2}k\langle x^2 \rangle = \tfrac{1}{2}(v + \tfrac{1}{2})\hbar(k/m)^{\frac{1}{2}} = \tfrac{1}{2}(v + \tfrac{1}{2})\hbar\omega.$$

However, since the total energy in the state v is $(v + \tfrac{1}{2})\hbar\omega$, we can express this as

$$\langle V \rangle = \tfrac{1}{2}E_v. \tag{14.2.9}$$

With this result established it is easy to deduce the *mean kinetic energy* of the oscillator (symbol: $\langle T \rangle$). Since the sum of the mean potential and kinetic energies is the total energy, E_v, it follows at once that

$$\langle T \rangle = \tfrac{1}{2}E_v. \tag{14.2.10}$$

The result that the mean potential and kinetic energies are equal (and therefore both equal to half the total energy) is a special case of the *virial theorem:*

If the potential energy of a particle has the form $V = ax^b$, then its mean potential and kinetic energies are related by

$$2\langle T\rangle = b\langle V\rangle. \qquad (14.2.11)$$

In the case of a harmonic oscillator $b = 2$, and so $\langle T\rangle = \langle V\rangle$, as we have found. The virial theorem is a short cut to the result, and we shall use it again.

Finally, we can calculate *the probability that the particle occurs in a classically forbidden region*. According to classical mechanics, the turning point (x_{tp}) of an oscillating particle occurs when its kinetic energy is zero, which is when its potential energy $\frac{1}{2}kx^2$ is equal to its total energy E. This occurs when $x_{tp}^2 = 2E/k$, or $x_{tp} = \pm(2E/k)^{\frac{1}{2}}$. In quantum mechanics the wavefunction does not stop abruptly at these values, but glides towards zero as the displacement increases towards infinity. It follows that there is a non-zero probability of finding the particle where it is classically forbidden. The total probability of finding it beyond a displacement x_{tp} is the integral

$$P(x > x_{tp}) = 2\int_{x_{tp}}^{\infty} \psi^2\, dx.$$

The factor 2 arises because the classically forbidden region on the left (to negative displacements) contributes the same value as the region on the right. The variable of integration is best expressed in terms of $y = x/\alpha$, and then the turning point lies at $y_{tp} = (2E_v/k)^{\frac{1}{2}}/\alpha = (2v+1)^{\frac{1}{2}}$. In the case of the ground state, $y_{tp} = 1$ and the total probability is

$$P = 2\int_{x_{tp}}^{\infty} \psi_0^2\, dx = 2\alpha N_0^2\int_1^{\infty} \{H_0 e^{-y^2/2}\}^2\, dy = (2/\pi^{\frac{1}{2}})\int_1^{\infty} e^{-y^2}\, dy.$$

The integral is a special case of one that we shall encounter several times and which is called the *error function* (symbol: erf). It is defined as follows:

$$\textit{Error function:}\ \operatorname{erf}(z) = 1 - (2/\pi^{\frac{1}{2}})\int_z^{\infty} e^{-y^2}\, dy. \qquad (14.2.12)$$

The integral cannot be obtained in closed form, but it can be evaluated numerically, and extensive tables are available. A brief extract is given in Table 14.1. In the present case

$$P = 1 - \operatorname{erf}(1) = 1 - 0.843 = 0.157.$$

This means that in 15.7% of a large number of observations, an oscillator in the state $v = 0$ (whatever its mass and the value of the force constant) will be found at classically forbidden displacements. The probability decreases quickly with increasing v, and vanishes entirely as v approaches infinity: this is what we would expect from the correspondence principle. Since macroscopic oscillators (such as pendulums) are in states with very high quantum numbers, the probability that they will be found at a classically forbidden displacement is wholly negligible. Molecules, however, are normally in their vibrational ground states, and for them the probability is very significant.

Table 14.1. The error function

z	erf (z)
0	0
0.01	0.0113
0.05	0.0564
0.10	0.1125
0.50	0.5205
1.00	0.8427
1.50	0.9661
2.0	0.9953

14.3 Rotational motion

The treatment of rotational motion can be broken down into two problems. The first deals with motion of a particle on a ring, and the second with rotation in three dimensions.

14.3 (a) Rotation in two dimensions

We consider a particle of mass m moving in a circular path of radius r. The total energy is equal to the kinetic energy, because the potential energy is zero everywhere. We can therefore write $E = p^2/2m$. According to classical mechanics, the *magnitude of the angular momentum* (symbol: l) is $l = pr$, and so the energy can be expressed as $l^2/2mr^2$. The quantity mr^2 is called the *moment of inertia* (symbol: I) of the system. It follows that

$$E = l^2/2I. \tag{14.3.1}$$

We now show that not all the values of the angular momentum are permitted in quantum mechanics; consequently, the energy is quantized.

The argument runs as follows. Since $l = pr$, and $p = h/\lambda$, the angular momentum is related to the wavelength of the particle's wavefunction, and the shorter the wavelength the greater the angular momentum. Suppose for the moment that λ can take an aribitrary value, then the wavefunction will vary round the ring as the angle ϕ increases, as shown in Fig. 14.11(a). When ϕ increases beyond 2π the wavefunction continues to change, but for an arbitrary wavelength it gives rise to a different amplitude at each point. Different values are given on successive circuits, and so the wavefunction is not single-valued. This is unacceptable, because the Born interpretation fails. A satisfactory solution is obtained if the wavelength is such that the wavefunction reproduces itself on successive circuits, as in Fig. 14.11(b). Since only some wavefunctions are acceptable, only some angular momenta are acceptable, and therefore only some energies exist. Hence, the energy of the particle is quantized.

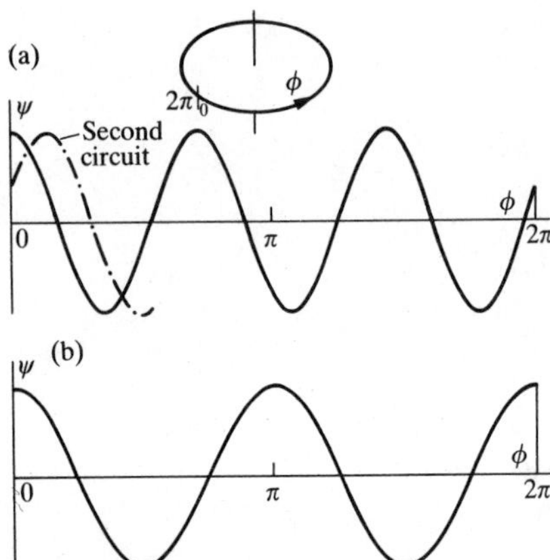

Fig. 14.11. Two solutions of the Schrödinger equation for a particle on a ring. The circumference has been opened out into a straight line, the points at $\phi = 0$ and 2 being identical. The solution in (a) is unacceptable because it is not single-valued. Moreover, on successive circuits it interferes destructively with itself, and does not survive. The solution in (b) is acceptable: it is single-valued, and on successive circuits it reproduces itself.

A quick estimate of the permitted energies can be obtained as follows. The wavelength must be a whole-number fraction of the circumference if its ends are to match after each circuit. That is, $\lambda = 2\pi r/n$, with $n = 0, 1, 2, \ldots$ ($n = 0$, which gives an infinite wavelength, corresponds to a uniform amplitude.) The angular momentum is therefore limited to the values $l = hr/\lambda = nhr/2\pi r$, or $l = n\hbar$ (because $\hbar = h/2\pi$). The energy is therefore limited to the values $E = l^2/2I = n^2(\hbar^2/2I)$, with $n = 0, 1, 2, \ldots$.

A moment's thought shows that this solution cannot be complete. The angular momentum can arise from motion in either direction, and so l, like p, ought to carry a sign, $l = +n\hbar$ indicating one sense of rotation and $l = -n\hbar$ indicating the other. We should therefore expect the quantum number for rotational motion to take both positive and negative values. How this comes about, and how it affects the expression for the energy, we can see by solving the Schrödinger equation explicitly.

14.3 (b) The formal solution

The two-dimensional Schrödinger equation for a particle in a plane (with $V = 0$) is

$$-(\hbar^2/2m)\{(\partial^2\psi/\partial x^2) + (\partial^2\psi/\partial y^2)\} = E\psi. \tag{14.3.2}$$

Instead of trying to solve this equation as it stands, it is better to transform it to polar coordinates, writing $x = r\cos\phi$ and $y = r\sin\phi$, because then the condition r = constant is much easier to impose. (In general, it is always a good idea to use coordinates that reflect the full symmetry of the system.) Since r is constant, $\partial^2/\partial x^2 + \partial^2/\partial y^2$ transforms to $(1/r^2)\,\mathrm{d}^2/\mathrm{d}\phi^2$ by standard manipulations of the differentials, and so

$$-(\hbar^2/2mr^2)\,\mathrm{d}^2\psi/\mathrm{d}\phi^2 = E\psi.$$

Notice that the moment of inertia $I = mr^2$ appears automatically. Therefore, the equation may be written

$$\mathrm{d}^2\psi/\mathrm{d}\phi^2 = -(2IE/\hbar^2)\psi. \qquad (14.3.3)$$

Its normalized solutions are

$$\psi_{m_l} = (1/2\pi)^{\frac{1}{2}}\mathrm{e}^{\mathrm{i}m_l\phi}, \qquad m_l = \pm(2IE/\hbar^2)^{\frac{1}{2}}. \qquad (14.3.4)$$

(m_l is some dimensionless number at this stage; it will soon be promoted to the status of a quantum number. The reason for the notation 'm_l' will be made clear later in the section.) This is the general form of the solution: the *permitted* solutions are found by imposing the condition that the wavefunction should be single-valued. That is, the wavefunction must match at points separated by a complete revolution. This is called the

Cyclic boundary condition: $\psi(\phi + 2\pi) = \psi(\phi)$.

On substituting the general form into this condition, we find

$$\psi_{m_l}(\phi + 2\pi) = (1/2\pi)^{\frac{1}{2}}\mathrm{e}^{\mathrm{i}m_l(\phi+2\pi)} = (1/2\pi)^{\frac{1}{2}}\mathrm{e}^{\mathrm{i}m_l\phi}\mathrm{e}^{2\pi\mathrm{i}m_l} = \psi_{m_l}(\phi)\mathrm{e}^{2\pi\mathrm{i}m_l}.$$

As $\mathrm{e}^{\mathrm{i}\pi} = -1$, it follows that $2m_l$ must be a positive or negative even integer, and therefore that $m_l = 0, \pm1, \pm2, \ldots$.

14.3 (c) Quantization of rotation

We can collect these conclusions together as follows. In the first place, since the energy is related to the value of m_l by eqn (14.3.4), and the cyclic boundary conditions confine m_l to integral values, we conclude that the *energy is quantized*, and the allowed values are given by

$$E_{m_l} = m_l^2(\hbar^2/2I), \qquad m_l = 0, \pm1, \pm2, \ldots. \qquad (14.3.5)$$

The occurrence of m_l as its square means that the energy of rotation is independent of the sense of rotation (the sign of m_l), as we expect physically. Furthermore, although the result has been derived for the rotation of a single mass point, it also applies to any body of moment of inertia I constrained to rotate about one axis. We can also conclude that *the angular momentum is quantized*. This is our first example of an observable other than the energy that is confined to discrete values. In the present case, the relation is established by comparing the last equation and eqn (14.3.1), which shows that the angular momentum is limited to the values

$$l_z = m_l\hbar, \qquad m_l = 0, \pm1, \pm2, \ldots. \qquad (14.3.6)$$

(The subscript z is there to remind us that the angular momentum corresponds to motion about the z-axis.) The increasing angular momentum is associated with the increasing number of nodes in the wavefunction: the wavelength decreases stepwise as $|m_l|$ increases, and so the momentum with which the particle travels round the ring increases.

In our discussion of free motion in one dimension we saw that the opposite signs in the wavefunctions e^{ikx} and e^{-ikx} corresponded to opposite directions of travel, and that the *linear* momentum was given by the eigenvalue of the linear momentum operator. The same conclusions can be drawn here, but now we need the eigenvalues of the *angular* momentum operator. In classical mechanics the angular momentum about the z-axis is defined as

$$l_z = xp_y - yp_x. \qquad (14.3.7)$$

p_x is the component of linear motion parallel to the x-axis and p_y the component parallel to the y-axis. The operators for the two linear momentum components are differentiation with respect to x and y respectively, and so in quantum mechanics the *operator for angular momentum about the z-axis* is

$$\hat{l}_z = (\hbar/i)\{x(\partial/\partial y) - y(\partial/\partial x)\}. \qquad (14.3.8)$$

This simplifies considerably when it is expressed in terms of polar coordinates, and becomes

$$\hat{l}_z = (\hbar/i)\,\partial/\partial\phi. \qquad (14.3.9)$$

With the angular momentum operator available, we can test the wavefunction in eqn (14.3.4):

$$\hat{l}_z\psi_{m_l} = (\hbar/i)(1/2\pi)^{\frac{1}{2}}(\mathrm{d}e^{im_l\phi}/\mathrm{d}\phi) = (\hbar/i)(im_l)(1/2\pi)^{\frac{1}{2}}e^{im_l\phi}$$
$$= m_l\hbar\psi_{m_l}. \qquad (14.3.10)$$

$m_l = +3$

$m_l = +2$

$m_l = +1$

$m_l = 0$

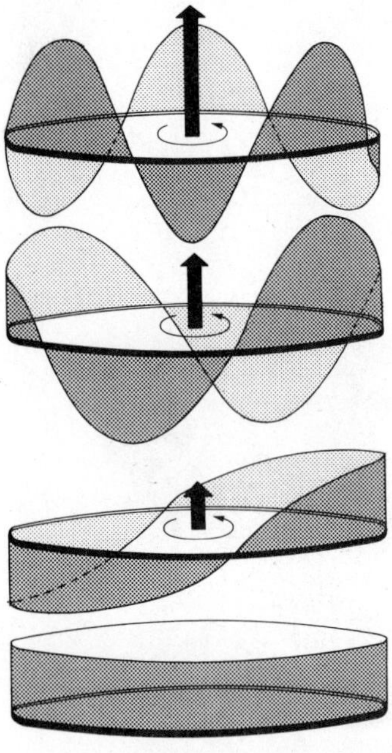

Fig. 14.12. The real parts of the wavefunctions of a particle on a ring. As shorter wavelengths are achieved, the angular momentum grows in steps of $\hbar$. This momentum can be represented by a vector of length m_l units along the z-axis. Note the sign convention (the *right-hand screw rule*) for the direction of motion.

That is, ψ_{m_l} is an eigenfunction of the angular momentum and corresponds to an angular momentum $m_l\hbar$. Furthermore, when m_l is positive the angular momentum is positive (clockwise when seen from below), and when m_l is negative the angular momentum is negative (counter-clockwise when seen from below). This is the basis of the *vector representation* of the angular momentum, when the magnitude is represented by the length of a vector and the direction of motion by its orientation, as shown in Fig. 14.12.

Example 14.5

Calculate the minimum rotational energy (above zero) and the minimum angular momentum of a disc the size of a benzene molecule ($I = 2.93 \times 10^{-45}$ kg m^2) when it is rotating in a plane.

- *Method*. Use eqn (14.3.5) for E_{m_l} and eqn (14.3.6) for l_z with $m_l = 1$.

- *Answer*. $E_1 = \hbar^2/2I$
$= (1.0546 \times 10^{-34}\ \mathrm{J\,s})^2/2 \times (2.93 \times 10^{-45}\ \mathrm{kg\,m^2}) = 1.9 \times 10^{-24}\ \mathrm{J}$.

The minimum angular momentum is $\pm\hbar = \pm 1.05$ J s.

- *Comment*. The actual minimum is zero in each case. Note that the energy corresponds to 1.2 J mol^{-1}, and so it is just large enough to be chemically significant. Benzene might rotate like this when adsorbed on to a surface.

• *Exercise*. Given that a solid disc of mass M and radius R has a moment of inertia $I = \frac{1}{2}MR^2$ about its axis, calculate the approximate quantum number of a gramophone record ($M = 150$ g, $R = 10$ cm) when it is rotating at $33\frac{1}{3}$ r.p.m. [-2.5×10^{31}]

Finally, we can explore the question of the position of the particle when it is in a state of definite angular momentum. As usual, we form the probability density:

$$\psi_{m_l}^*\psi_{m_l} = \{(1/2\pi)^{\frac{1}{2}}e^{im_l\phi}\}^*\{(1/2\pi)^{\frac{1}{2}}e^{im_l\phi}\} = 1/2\pi. \qquad (14.3.11)$$

That is, *the probability density is uniform*, and the probability of locating the particle is independent of the angle ϕ. This means that the location of the particle is indefinite, and knowing the angular momentum precisely involves discarding the possibility of specifying the particle's location. *Angular momentum and angle* are a pair of complementary observables (in the sense defined in Section 13.4) and the inability to specify them simultaneously to arbitrary precision is another example of the uncertainty principle.

14.3 (d) Rotation in three dimensions

We now consider a mass point free to move over the surface of a sphere of radius r. The requirement that the wavefunction should match as a path is traced over the poles as well as round the equator, Fig. 14.13, introduces another cyclic boundary condition and therefore another quantum number. This apparently artificial model underlies the description of electrons in atoms (Chapter 15) and rotating molecules (Chapter 18). The latter application arises from the fact that the rotation of a solid body of moment of inertia I can be represented by a single mass point M rotating at a radius R, the *radius of gyration* of the body, which is defined so that $I = MR^2$.

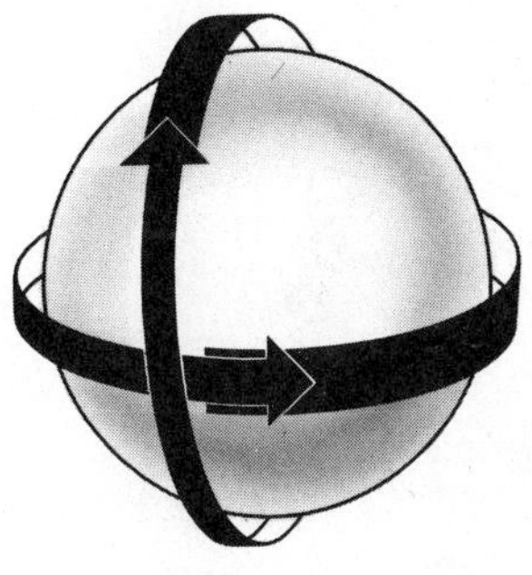

Fig. 14.13. A particle on the surface of a sphere must satisfy two cyclic boundary conditions, and this leads to two quantum numbers for its state of angular momentum.

The Schrödinger equation in three dimensions (Box 13.1) is

$$-(\hbar^2/2m)\nabla^2\psi + V\psi = E\psi, \qquad (14.3.12)$$

where ∇^2 is the *laplacian operator:*

$$\nabla^2 = \partial^2/\partial x^2 + \partial^2/\partial y^2 + \partial^2/\partial z^2. \qquad (14.3.13)$$

Since the radius of the particle's motion is constant, it is sensible to express this equation in terms of spherical polar coordinates, Fig. 13.11. Then standard manipulations of the differentials transform the laplacian into the form quoted in Box 13.1:

$$\nabla^2 = (1/r)(\partial^2/\partial r^2)r + (1/r^2)\Lambda^2, \qquad (14.3.14)$$

where Λ^2 is the *legendrian:*

$$\Lambda^2 = (1/\sin^2\theta)(\partial^2/\partial\phi^2) + (1/\sin\theta)(\partial/\partial\theta)\sin\theta(\partial/\partial\theta). \qquad (14.3.15)$$

The legendrian is the angular part of the laplacian. Note that when using it the operations have to be performed in order from right to left. The second term, for example, instructs you to differentiate with respect to θ, then to multiply by $\sin\theta$, then to differentiate that product with respect to θ, and then finally to multiply by $1/\sin\theta$.

Since the particle is moving on the surface of a sphere, the radius is not a variable, and so the differentiations in ∇^2 with respect to r can be ignored. On the surface of the sphere V is a constant and may be set equal to zero.

Therefore, eqn (14.3.12) simplifies to

$$-(\hbar^2/2mr^2)\Lambda^2\psi = E\psi. \qquad (14.3.16)$$

Notice that the moment of inertia $I = mr^2$ has appeared automatically. The equation may therefore be written

$$\Lambda^2\psi = -(2IE/\hbar^2)\psi, \qquad \psi = \psi(\theta, \phi). \qquad (14.3.17)$$

We now present a sketch of the way this equation is solved. Once again, this is moderately technical, and so the important conclusions, which can be used without the following details, are summarized in Section 14.3(f).

14.3 (e) The formal solution

As it stands, the Schrödinger equation looks formidable. However, we can draw on physical intuition to simplify it. It is likely that the rotation of the particle about the z-axis will resemble the two-dimensional motion we have already discussed. In order to test this suggestion, we try writing the wavefunction as the product $\psi = \Theta(\theta)\Phi(\phi)$, and using the separation of variables technique described in Section 14.1. On substituting the product into eqn (14.2.17) and using the full form of the legendrian, we find

$$\begin{aligned}\Lambda^2\psi &= \{(1/\sin^2\theta)(\partial^2/\partial\phi^2) + (1/\sin\theta)(\partial/\partial\theta)\sin\theta(\partial/\partial\theta)\}\Theta\Phi \\ &= (1/\sin^2\theta)\Theta(\mathrm{d}^2\Phi/\mathrm{d}\phi^2) + \Phi(1/\sin\theta)(\mathrm{d}/\mathrm{d}\theta)\sin\theta(\mathrm{d}\Theta/\mathrm{d}\theta) \\ &= -(2IE/\hbar^2)\Theta\Phi.\end{aligned}$$

As described in Section 14.1, the next step is to divide by $\Theta\Phi$, giving

$$(1/\sin^2\theta)(\Phi''/\Phi) + (1/\Theta)(1/\sin\theta)(\mathrm{d}/\mathrm{d}\theta)\{\sin\theta\Theta'\} = -2IE/\hbar^2,$$

where $\Theta' = \mathrm{d}\Theta/\mathrm{d}\theta$ and $\Phi' = \mathrm{d}\Phi/\mathrm{d}\phi$. The idea behind the separation of variables technique is to ensure that each term depends on only one variable. This is achieved if we multiply through by $\sin^2\theta$, for then we obtain

$$(\Phi''/\Phi) + (1/\Theta)\sin\theta(\mathrm{d}/\mathrm{d}\theta)\{\sin\theta\Theta'\} = -(2IE/\hbar^2)\sin^2\theta.$$

The first term depends only on ϕ, and the rest depend only on θ. Therefore, by the same argument as before, the first term must be equal to a constant, which we shall write $-m_l^2$. This means that the partial differential equation splits into two ordinary differential equations, which, after some rearrangement, are:

$$\Phi'' = -m_l^2\Phi \qquad (14.3.18a)$$

$$\sin\theta(\mathrm{d}/\mathrm{d}\theta)\{\sin\theta\Theta'\} + \{(2IE/\hbar^2)\sin^2\theta - m_l^2\}\Theta = 0. \qquad (14.3.18b)$$

The first of this pair of equations is familiar: it is the same as we encountered for a particle rotating in a plane. The cyclic boundary conditions are also the same, and so the acceptable solutions are the same, and we can write

$$\Phi_{m_l} = (1/2\pi)^{\frac{1}{2}}\mathrm{e}^{im_l\phi}, \qquad m_l = 0, \pm1, \pm2, \ldots. \qquad (14.3.19)$$

However, since m_l also occurs in eqn (14.3.18b), it may be the case that the boundary conditions on Θ also limit the values that m_l may take, and so the specification in eqn (14.3.19) is only temporary.

The equation for Θ is one that has been studied extensively by mathematicians. Its appearance can be simplified and turned into a standard form by making the substitution $\zeta = \cos\theta$ and writing $2IE/\hbar^2 = l(l+1)$, where l is a dimensionless number (shortly to be promoted to a quantum number). Since $\mathrm{d}/\mathrm{d}\theta = (\mathrm{d}\zeta/\mathrm{d}\theta)(\mathrm{d}/\mathrm{d}\zeta) = -\sin\theta(\mathrm{d}/\mathrm{d}\zeta)$, and $\sin\theta = \sqrt{(1-\cos^2\theta)} = \sqrt{(1-\zeta^2)}$, the equation transforms into

$$(1-\zeta^2)\Theta'' - 2\zeta\Theta' + \{l(l+1) - [m_l^2/(1-\zeta^2)]\}\Theta = 0, \qquad (14.3.20)$$

where now $\Theta' = \mathrm{d}\Theta/\mathrm{d}\zeta$. This is known as the *associated Legendre equation* (because it is associated with another equation known by Legendre's name). It has acceptable solutions (i.e. solutions that are single-valued and do not become infinite anywhere) so long as two conditions are fulfilled:

$$\text{(a) } l = 0, 1, 2, \ldots; \qquad \text{(b) } |m_l| < l. \qquad (14.3.21)$$

That is, acceptable solutions exist only for *positive* integral l, and the absolute value of the number m_l must not exceed l (this is the second constraint on m_l mentioned above). When these conditions are satisfied, the functions Θ, which now must be denoted $\Theta_{l,|m_l|}$, are the *associated Legendre functions*. They can be written in terms of sine and cosine functions, and the first few are listed in Box 14.2.

Box 14.2 Wavefunctions for a particle on a sphere (spherical harmonics)

The wavefunctions for a particle on a spherical surface are

$$Y_{l,m_l} = N\Theta_{l,|m_l|}\Phi_{m_l},$$

with $l = 1, 2, 3, \ldots$ and $m_l = 0, \pm1, \pm2, \ldots \pm l$. The functions Y_{l,m_l} are the *spherical harmonics*, and have the following form:

l	m_l	Y_{l,m_l}
0	0	$(1/4\pi)^{\frac{1}{2}}$
1	0	$(3/4\pi)^{\frac{1}{2}}\cos\theta$
	±1	$\pm(3/8\pi)^{\frac{1}{2}}\sin\theta\,\mathrm{e}^{\pm i\phi}$
2	0	$(5/16\pi)^{\frac{1}{2}}(3\cos^2\theta - 1)$
	±1	$\pm(15/16\pi)^{\frac{1}{2}}\cos\theta\sin\theta\,\mathrm{e}^{\pm i\phi}$
	±2	$(15/32\pi)^{\frac{1}{2}}\sin^2\theta\,\mathrm{e}^{\pm 2i\phi}$

14.3 (f) The properties of the solutions

The energy of the particle is related to l through $2EI/\hbar^2 = l(l+1)$, which rearranges to

$$E = l(l+1)\hbar^2/2I, \qquad l = 0, 1, 2, \ldots. \qquad (14.3.22)$$

This shows that the energy is quantized, and that l is the relevant quantum number, m_l playing no role in determining the energy. Since for a given l, m_l may take the values

$$m_l = 0, \pm1, \pm2, \ldots \pm l \qquad (14.3.23)$$

(which is the combination of the list in eqn (14.3.19) with the constraint in eqn (14.3.21)), or $2l+1$ values in all for a given value of l, it follows that there are $2l+1$ different wavefunctions (one for each value of m_l) that correspond to the same energy. That is, *a level with quantum number l is $(2l+1)$-fold degenerate*.

The energy of a rotating particle is related classically to the magnitude of its angular momentum by the expression in eqn (14.3.1). Therefore, on comparing that equation with eqn (14.3.22) we can deduce that *the magnitude of the angular momentum is quantized*, and confined to the values

$$\textit{Magnitude of angular momentum} = \{l(l+1)\}^{\frac{1}{2}}\hbar, \qquad l=0,1,2,\ldots. \tag{14.3.24}$$

We have already seen that the component about the z-axis is quantized, and that it has the values

$$z\textit{-component of angular momentum} = m_l\hbar, \qquad m_l=0,\pm1,\pm2,\ldots,l. \tag{14.3.25}$$

The wavefunctions are the normalized products of the Θ and Φ functions. They are normally denoted $Y_{l,m_l}(\theta,\phi)$ and called the *spherical harmonics:*

$$\textit{Wavefunctions:}\ \psi_{l,m_l} = Y_{l,m_l}(\theta,\phi), \qquad Y_{l,m_l} = N\Theta_{l,|m_l|}\Phi_{m_l}. \tag{14.3.26}$$

N is a normalization constant; it depends on the quantum numbers l and m_l. Some of the spherical harmonics are listed in Box 14.2, and their amplitudes at different points on the spherical surface are illustrated in Fig. 14.14. The dominant feature is the increase in the number of *nodal lines* (the positions at which $\psi=0$) as l increases. This reflects the fact that higher angular momentum implies higher kinetic energy, and therefore a more sharply buckled wavefunction. You should also note that the states corresponding to high angular momentum around the z-axis are those in which most nodes cut the equator: this indicates a high kinetic energy arising from motion parallel to the equator, because the curvature is greatest in that direction.

Example 14.6

Calculate the energies of the first five rotational levels of H_2 ($I=4.603\times10^{-48}\,\mathrm{kg\,m^2}$), the corresponding magnitudes of the angular momentum, and the number of different values of m_l in each case.

• *Method*. Use eqn (14.3.22) with $l=0,1,2,3,4$. For the magnitude, use eqn (14.3.24), and express the result as multiples of $\hbar$. For a state of given l there are $2l+1$ values of m_l ($l, l-1, \ldots, l$). In the case of rotating molecules the angular momentum quantum number is normally denoted J.

• *Answer*. We need $\hbar^2/2I = 1.208\times10^{-21}\,\mathrm{J}$ (corresponding to $0.727\,\mathrm{kJ\,mol^{-1}}$). Draw up the following table; $|\boldsymbol{J}|$ denotes the magnitude of the angular momentum.

J	0	1	2	3	4
$J(J+1)$	0	2	6	12	20
$E_J/\mathrm{kJ\,mol^{-1}}$	0	1.46	4.37	8.73	14.6
$\lvert\boldsymbol{J}\rvert/\hbar$	0	1.41	2.45	3.46	4.47
Components	1	3	5	7	9

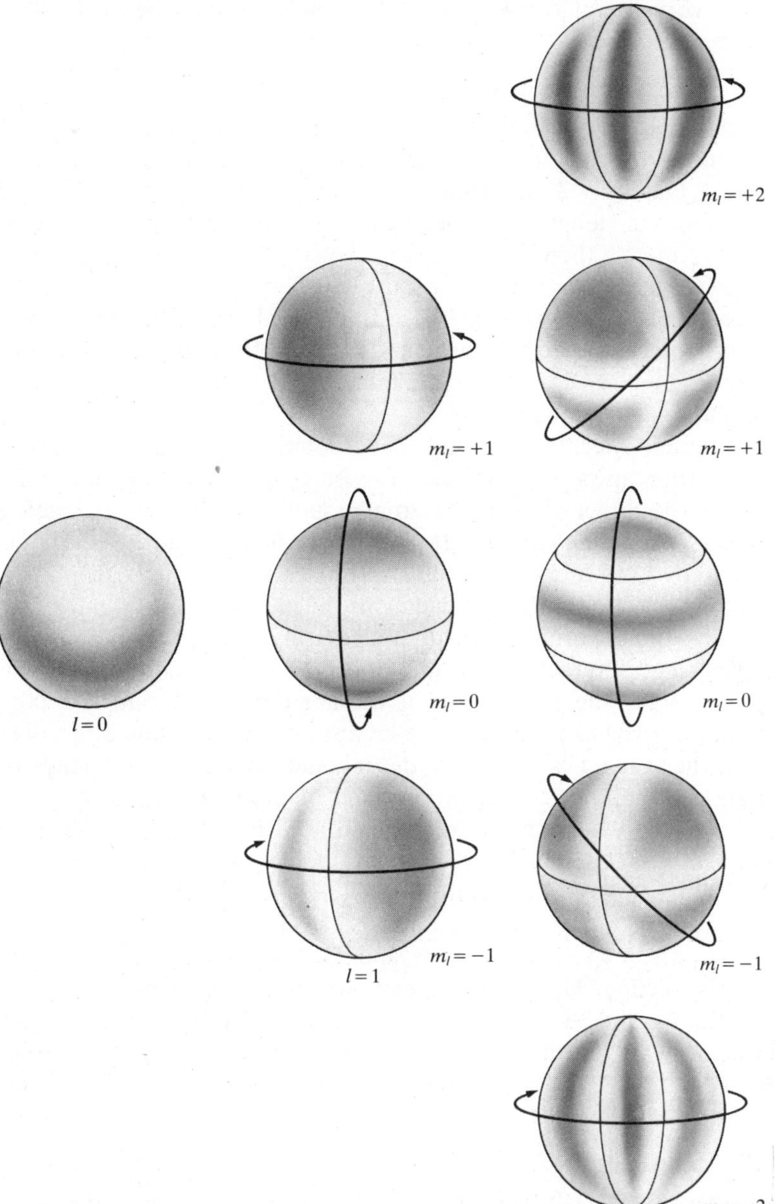

Fig. 14.14. The patterns of wavefunctions (represented by shading indicating the corresponding probability densities) for a particle on the surface of a sphere. Note that the number of nodes increases as the value of l increases, and that their orientation is determined by the value of m_l. The corresponding classical motion is indicated in each case.

• *Comment*. Note that the smaller the moment of inertia of the molecule, the greater the separation of the rotational energy levels. Since all components of a given J correspond to the same energy, the rotational levels are $(2J + 1)$-fold degenerate.

• *Example*. Repeat the calculation for a deuterium molecule (same bond length, approximately twice the mass).

Finally, notice that eqn (14.3.17) may be written in terms of l, so giving the differential equation satisfied by the spherical harmonics:

$$\Lambda^2 Y_{l,m_l} = -l(l+1) Y_{l,m_l}. \qquad (14.3.27)$$

This is a very useful result, which we shall draw on when discussing the hydrogen atom.

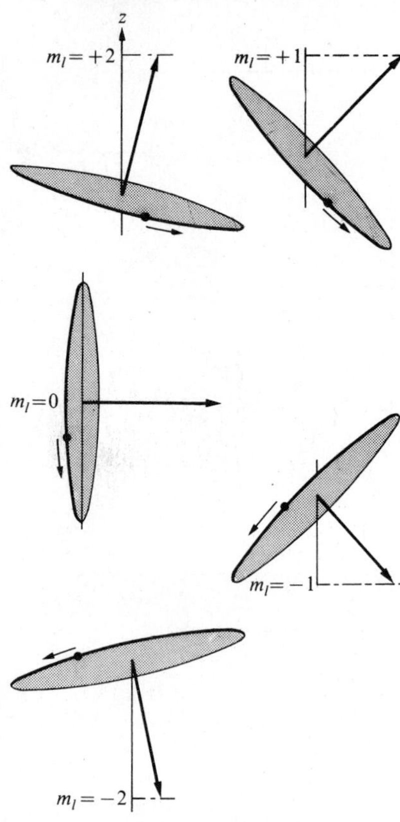

Fig. 14.15. The permitted orientations of angular momentum when $l = 2$. We shall see soon that this is too specific because the azimuthal orientation of the vector (its angle around z) is indeterminate.

14.3 (g) Space quantization

The result that m_l is confined to the discrete values $l, l-1, \ldots -l$ for a given value of l means that the component of angular momentum about the z-axis may take only $2l+1$ values. If the angular momentum is represented by a vector of length proportional to its magnitude (i.e. of length $\sqrt{\{l(l+1)\}}$ units), then to represent correctly the value of the component of angular momentum, the vector must be oriented so that its projection on the z-axis is of length m_l, Fig. 14.15. In classical terms, this means that the plane of rotation of the particle can take only a discrete range of orientations, Fig. 14.15. The remarkable implication is that the *orientation* of a rotating body is quantized.

The quantum mechanical result that a rotating body may not take up an arbitrary orientation with respect to some specified axis (e.g., an axis defined by the direction of an externally applied electric or magnetic field) is called *space quantization*. It was confirmed by an experiment first performed by Otto Stern and Walther Gerlach in 1921, who shot a beam of silver atoms through an inhomogeneous magnetic field, Fig. 14.16(a). The idea behind the experiment was that a rotating, charged body behaves like a magnet and interacts with the applied field. According to classical mechanics, since the orientation of the angular momentum can take any value, the associated magnet can take any orientation. Since the direction in which the magnet is driven by the inhomogeneous field depends on the orientation, it follows that a broad band of atoms is expected to emerge from the region where the magnetic field acts, Fig. 14.16(b). According to quantum mechanics, since the angular momentum is quantized, the associated magnet lies in a number of discrete orientations, and so several sharp bands of atoms are expected, Fig. 14.16(c).

In their first experiment, Stern and Gerlach appeared to confirm the classical prediction. However, the experiment is difficult because the atoms in the beam collide with each other and this blurs the bands. When the experiment was repeated with a beam of very low intensity (so that collisions were less frequent) they observed discrete bands, and so confirmed the quantum prediction.

14.3 (h) Spin

Using silver atoms as projectiles, Stern and Gerlach observed *two* bands. This seems to conflict with one of the predictions of quantum mechanics,

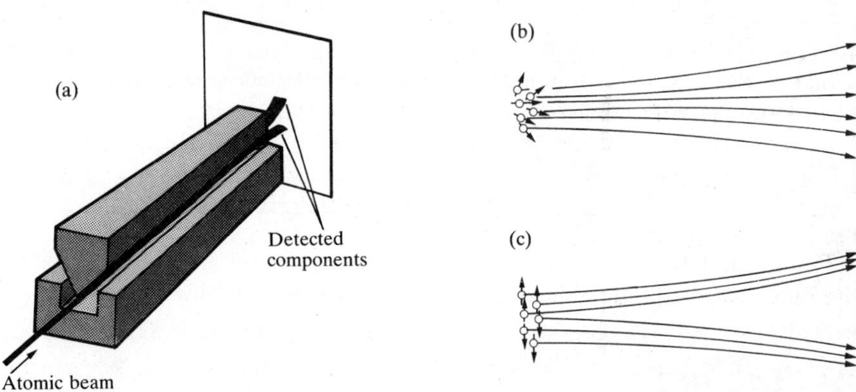

Fig. 14.16. (a) The experimental arrangement for the Stern–Gerlach experiment: the magnet provides an inhomogeneous field. (b) The classically expected result. (c) The observed outcome using silver atoms.

because an angular momentum l gives rise to $2l+1$ orientations, which is equal to 2 only if $l=\frac{1}{2}$, contrary to the conclusion that l must be an integer. The conflict was resolved by the suggestion that the angular momentum they were observing was not due to the atom's *orbital angular momentum* (which arises from the motion of an electron around the atomic nucleus) but arose from the motion of the electron about its own axis. This internal angular momentum of the electron is called its *spin*. The wavefunction of an electron spinning at a single point in space does not have to satisfy the same boundary conditions as those for a particle moving over the surface of a sphere, and so the quantum number l is not subject to the same restrictions. In order to distinguish this spin angular momentum from orbital angular momentum we use the quantum number s (in place of l) and m_s for the projection on the z-axis. The magnitude of the spin angular momentum is $\{s(s+1)\}^{\frac{1}{2}}\hbar$ and the component $m_s\hbar$ is restricted to the $2s+1$ values given by $m_s = s, s-1, s-2, \ldots, -s$.

The detailed analysis of the spin of a particle is quite sophisticated, and shows that the property should not be taken to be an actual spinning motion (but that picture can be very useful when used with care). For an electron it turns out that only one value of s is allowed, and that $s=\frac{1}{2}$, corresponding to an angular momentum of magnitude $(\frac{1}{2}\sqrt{3})\hbar = 0.866\hbar$. This spin angular momentum is an intrinsic property of the electron, like its rest mass and its charge, and every electron has exactly the same value. The spin may lie in any of $2s+1=2$ different orientations. One orientation corresponds to $m_s=+\frac{1}{2}$ (this is often called an α-electron); the other corresponds to $m_s=-\frac{1}{2}$ (and is called a β-electron), Fig. 14.17.

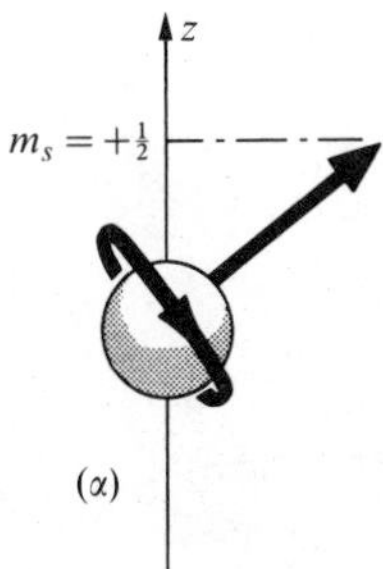

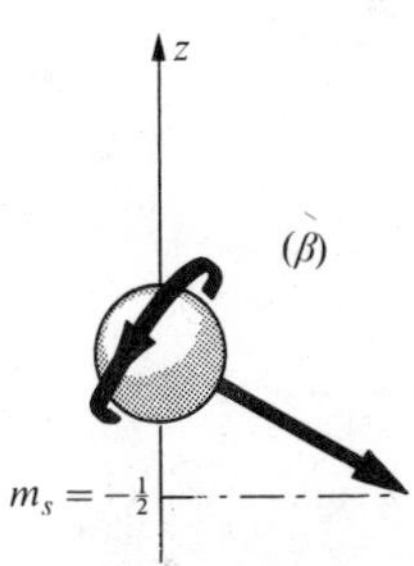

Fig. 14.17. An electron spin ($s=\frac{1}{2}$) can take only two orientations with respect to a specified axis. An α-electron is one with $m_s=+\frac{1}{2}$; a β-electron is one with $m_s=-\frac{1}{2}$.

The outcome of the Stern–Gerlach experiment can now be explained if we suppose that each silver atom possesses an angular momentum due to the spin of a single electron, because the two bands of atoms then correspond to the two spin orientations. The reason why the atoms behave like this will be explained in Chapter 15.

Like the electron, other elementary particles have characteristic spins. For example, protons and neutrons are spin-$\frac{1}{2}$ particles (i.e. $s=\frac{1}{2}$) and so invariably spin with angular momentum $(\frac{1}{2}\sqrt{3})\hbar = 0.866\hbar$. Since the masses of a proton and a neutron are so much greater than the mass of an electron, yet they all have the same spin angular momentum, the classical picture would be of particles spinning much more slowly than an electron. Some elementary particles have $s=1$, and so carry an irremovable angular momentum of magnitude $(\sqrt{2})\hbar$, or $1.414\hbar$. Some mesons are spin-1 particles (as are some atomic nuclei), but for our purposes the most important spin-1 particle is the photon. Why that value is so important we shall see in the next chapter.

The properties of angular momentum as far as we have developed them at this stage are set out in Box 14.3. As mentioned there, when we use the quantum numbers l and m_l we shall mean *orbital* angular momentum; when we use s and m_s we shall mean *spin*; and when we use j and m_j we shall mean either (or, in some contexts to be described in Chapter 15, a combination of orbital and spin momenta).

14.3 (i) The vector model

Throughout the preceding discussion we have referred to the z-component of angular momentum, and have made no reference to the x- and y-components. The reason for this omission is that *the uncertainty principle*

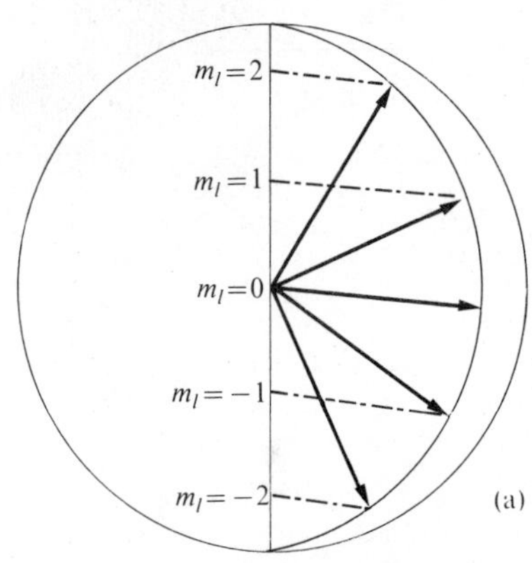

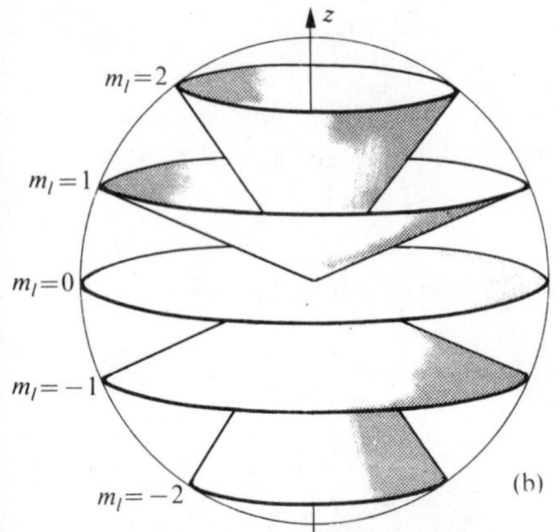

Fig. 14.18. (a) A summary of Fig. 14.15. However, since the azimuthal angle of the vector is indeterminate a better representation is as in (b), where each vector lies at an unspecified azimuth on its cone.

Box 14.3 Angular momentum

The quantum numbers:
Orbital angular momentum quantum number: l ($l = 0, 1, 2, \ldots$)

z-component quantum number: m_l ($m_l = 0, \pm 1, \pm 2, \ldots \pm l$).

Spin angular momentum quantum number: s ($s = \frac{1}{2}$ for an electron)

z-component quantum number: m_s ($m_s = \pm\frac{1}{2}$ for an electron)

m_l is also known as the *magnetic quantum number*.
In general, an angular momentum is denoted by the quantum numbers j and m_j.
The *magnitude of the angular momentum* is equal to $\{j(j+1)\}^{\frac{1}{2}}\hbar$.
The *z-component of angular momentum* is equal to $m_j\hbar$.

forbids the simultaneous, exact specification of more than one component. Therefore, if j_z is known, it is impossible to ascribe values to the other two components. It follows that the illustration in Fig. 14.15, which is summarized in Fig. 14.18(a), gives a false impression of the state of the system, because it suggests definite values for the x- and y-components. A better picture must reflect the indeterminance of j_x and j_y.

The *vector model of angular momentum* uses pictures like that in Fig. 14.18(b). The cones are drawn with side $\{j(j+1)\}^{\frac{1}{2}}$ units, and represent the magnitude of the angular momentum. Each cone has a definite projection (of m_j units) on the z-axis, representing the system's precise value of j_z. The j_x and j_y projections, however, are indefinite. The vector representing the state of angular momentum can be thought of as lying with its tip on any point on the mouth of the cone. (At this stage it should not be thought of as sweeping round the cone, that aspect of the model will be added later when we allow the picture to convey more information.) This vector model of angular momentum, although only a pictorial representation of aspects of the quantum mechanical properties, turns out to be surprisingly useful when we turn to the structure and spectra of atoms.

Further reading

Molecular quantum mechanics (2nd edn). P. W. Atkins; Oxford University Press, 1983.

Quantum mechanics. A. I. M. Rae; McGraw-Hill, London, 1981.

Quantum chemistry. D. A. McQuarrie; University Science Books and Oxford University Press, 1983.

Quantum chemistry (2nd edn). I. N. Levine; Allyn and Bacon, Boston, 1974.

Introduction to quantum mechanics. L. Pauling and E. B. Wilson; McGraw-Hill, New York, 1935.

Quantum mechanics (2nd edn). A. S. Davydov; Pergamon Press, Oxford, 1976.

Quantum mechanics. L. I. Schiff; McGraw-Hill, New York, 1968.

Principles of quantum mechanics (4th edn). P. A. M. Dirac; Clarendon Press, Oxford, 1958.

The tunnel effect in chemistry. R. P. Bell; Chapman and Hall, New York, 1980.

Introductory problems

A14.1. Calculate the probability that an electron will penetrate a potential barrier of height 2.00 eV and width 0.25 nm, when the total energy is 0.90 eV.

A14.2. Consider a particle of mass m in a cubic box with edge L and find the degeneracy of the level with energy which is three times that of the lowest level.

A14.3. Calculate the percent change in a given energy level for a particle in a cubic box when the edge of the cube is decreased by ten per cent.

A14.4. Calculate the zero point energy of a harmonic oscillator consisting of a particle with mass 2.33×10^{-26} kg and a force constant 155 N m^{-1}.

A14.5. For a harmonic oscillator consisting of a particle with mass 1.33×10^{-25} kg, the difference in the energies of levels with vibrational quantum numbers 5 and 4 is 4.82×10^{-21} J. Calculate the force constant k.

A14.6. Calculate the wavelength of light one quantum of which has the energy corresponding to the spacing between levels of a particle of mass 1.67×10^{-27} kg oscillating harmonically with a force constant of 855 N m^{-1}.

A14.7. Refer to the previous problem and find the change in wavelength which would result from doubling the mass of the particle while keeping the force constant fixed.

A14.8. Use the independent variable y, defined as $(m\omega/\hbar)^{\frac{1}{2}}x$, and write the wavefunction for a harmonic oscillator with vibrational quantum number 1. Find the value of y (other than zero) for which the wavefunction vanishes. Differentiate the function and find the value of y for which the wavefunction is a maximum.

A14.9. A point mass of 6.35×10^{-26} kg rotates about a centre. Its angular momentum is described by $l = 2$. The energy of rotation of the mass is 2.47×10^{-23} J. Calculate the distance of the mass from the centre of rotation.

A14.10. A point mass rotates about a centre and has an angular momentum described by $l = 1$. Calculate the value of the angular momentum and the values of the angular momentum about the z-axis.

Problems

14.1. The energy levels of a particle of mass m in a box of length L are given by eqn (14.1.9a). Suppose that the particle is an electron, and that the box represents a long conjugated molecule. What are the energy separations in J, kJ mol^{-1}, eV, and cm^{-1} between the levels (a) $n = 2$ and $n = 1$, (b) $n = 6$ and $n = 5$, in both cases taking $L = 1$ nm (10 Å).

14.2. A gas molecule in a flask has quantized translational energy levels, but how important are the effects of quantization? Calculate the separation between the lowest two energy levels for an oxygen molecule in a one-dimensional container of length 5 cm. At what value of the quantum number n does the energy of the molecule equal $\frac{1}{2}kT$, when $T = 300$ K? What is the separation of this level from the one below?

14.3. Set up the Schrödinger equation for a particle of mass m in a three-dimensional square well with sides L_X, L_Y, and L_Z (and volume $V = L_X L_Y L_Z$). Show that the wavefunction requires three quantum numbers for its specification, and that $\psi(x, y, z)$ can be written as the product of three wavefunctions for one-dimensional square wells. Deduce an expression for the energy levels, and specialize it to the case of a cubic box of side L.

14.4. The wavefunction inside a barrier of height V extending from $x = 0$ to $x = \infty$ is $\psi = Ae^{-\kappa x}$, where κ (and A) depends on the total energy. Calculate the probability of finding the particle inside the barrier. What is the average penetration depth of the particle?

14.5. Repeat the calculation in *Example* 14.2, but for a potential that falls to a non-zero value V' on the right of the barrier, and explore the dependence of P on V' as well as on E.

14.6. Repeat the calculation in *Example* 14.2 but for $E > V$.

14.7. Calculate the probability that a particle will be reflected back from a dip in an otherwise uniform potential. Take the depth of the well as V and its width as L. Plot the probability as a function of E/V with $V = 5$ eV, $L = 0.1$ nm, for a proton and a deuteron.

14.8. Particles can penetrate into regions where classical mechanics forbids them to go. This is because the wavefunction is continuous at the boundaries of containers, and so it seeps into regions of high potential before dropping to zero. Take the wavefunction for the ground state of a harmonic oscillator formed from a mass m on a spring of force-constant k. Plot the probability density on graph paper and estimate by graphical integration the probability that the mass will be found in regions where for the same total energy it is forbidden by classical mechanics. Take the mass to be that of (a) a proton, (b) a deuteron, and the force constant to be that of a typical bond, 500 N m^{-1}. An alternative procedure would be to evaluate the integrals numerically: devise a program for doing so.

14.9. The wavefunction for the lowest state of a harmonic oscillator has the form of a *Gaussian function*, e^{-gx^2}, where x is the displacement from equilibrium. Show that this function satisfies the Schrödinger equation for a harmonic oscillator, and find g in terms of the mass m and the force-constant k. What is the (zero-point) energy of the oscillator with this wavefunction? What is its minimum excitation energy?

14.10. Several types of motion are simple harmonic, and their energy levels are all given by eqn (14.2.4). Find the minimum excitation energies of the following oscillators: (a) a pendulum of length 1 m in a gravitational field; (b) the balance wheel of a watch ($\nu = 5$ Hz); (c) the 33 kHz quartz crystal of a watch; (d) the bond between two oxygen atoms ($k = 1177\ \mathrm{N\,m^{-1}}$).

14.11. The following molecules were found to absorb radiation at the wavenumbers quoted (G. Herzberg, *Spectra of diatomic molecules*, van Nostrand (1950)), the energy going into the excitation of the vibration of the bond. Since both atoms of the diatomic molecules move when the bond vibrates, the mass to use in the expression $\omega = (k/\mu)^{\frac{1}{2}}$ is $\mu = m_1m_2/(m_1 + m_2)$. Find the force constants of the bonds in the molecules, and arrange them in order of increasing stiffness, $H^{35}Cl$, 2989.74 $\mathrm{cm^{-1}}$; $H^{81}Br$, 2649.72 $\mathrm{cm^{-1}}$; HI, 2309.5_3 $\mathrm{cm^{-1}}$; CO, 2170.21 $\mathrm{cm^{-1}}$; NO, 1904.03 $\mathrm{cm^{-1}}$.

14.12. Calculate the mean kinetiç energy of a harmonic oscillator in the state v using the relations in Box 14.1.

14.13. Calculate the values of $\langle x^3 \rangle$ and $\langle x^4 \rangle$ for a harmonic oscillator using the recursion relations in Box 14.1.

14.14. In Chapter 18 we shall see that the intensity of spectroscopic transitions between the vibrational states of a molecule are proportional to the square of the integral $\int_{-\infty}^{\infty} \psi_{v'} x \psi_v \, dx$. Use the recursion relations (a) to show that the only non-zero intensity transitions are those for which $v' = v \pm 1$, and (b) to evaluate the integral in these cases.

14.15. Use the virial theorem to obtain an expression for the relation between the mean kinetic and potential energies of an electron in a hydrogen atom.

14.16. The rotation of an HI molecule can be visualized as the orbiting of the hydrogen atom at a distance of 160 pm from a stationary iodine atom (this is quite a good approximation, but to be precise we would have to take into account the motion of both atoms around their joint centre of mass). Suppose that the molecule rotates only in a plane. How much energy (in $\mathrm{kJ\,mol^{-1}}$ and $\mathrm{cm^{-1}}$) is needed to excite the stationary molecule into rotation? What, apart from zero, is the minimum angular momentum of the molecule?

14.17. Take the same model of the HI molecule as in the last Problem and suppose that it is known to possess (a) zero angular momentum, (b) one unit of angular momentum. Use the form of the wavefunction for the two states to decide on the location of the hydrogen atom.

14.18. Evaluate the z-component of angular momentum of a particle on a ring with (non-normalized) wavefunctions (a) $e^{+i\phi}$, (b) $e^{-2i\phi}$, (c) $\cos\phi$, and (d) $\cos\chi e^{i\phi} + \sin\chi e^{-i\phi}$ (where χ is a parameter). What is the particle's kinetic energy in each case?

14.19. A particle on a ring is in a state with wavefunction $ae^{i\phi} + be^{2i\phi} + ce^{3i\phi}$. Evaluate its mean angular momentum and kinetic energy. Evaluate the root mean square angular momentum and $\delta l_z = \{\langle l_z^2 \rangle - \langle l_z \rangle^2\}^{\frac{1}{2}}$.

14.20. What are the magnitudes of the angular momentum in the lowest four energy levels of a particle rotating on the surface of a sphere? How many states (as distinguished by the z-component of the angular momentum) are there in each case?

14.21. Calculate the energies of the first four rotational levels of HI, $R = 160$ pm, allowing it to rotate in three dimensions about its centre of mass. Express your answer in $\mathrm{kJ\,mol^{-1}}$ and $\mathrm{cm^{-1}}$. Use $I = \mu R^2$, where μ is the reduced mass.

14.22. Confirm that the following wavefunctions ψ_{l,m_l} for a particle on a spherical surface satisfy the Schrödinger equation, and that in each case the energy and angular momentum of the particle are given by eqns (14.3.22) and (14.3.24).

(a) $\psi_{0,0} = 1/2\pi^{\frac{1}{2}}$,
(b) $\psi_{1,0} = \frac{1}{2}(3/\pi)^{\frac{1}{2}} \cos\theta$,
(c) $\psi_{2,-1} = \frac{1}{2}(15/2\pi)^{\frac{1}{2}} \cos\theta \sin\theta \exp(-i\phi)$,
(d) $\psi_{3,3} = -\frac{1}{8}(35/\pi)^{\frac{1}{2}} \sin^3\theta \exp(3i\phi)$.

14.23. Confirm that $\psi_{3,3}$ is normalized to unity. Take the integration over the surface of a sphere.

14.24. In the vector model of angular momentum a state with quantum numbers l, m_l (or s, m_s) is represented by a vector length $\sqrt{\{l(l+1)\}}$ units and of z-component m_l units. Draw scale diagrams of the state of an electron with (a) $s = \frac{1}{2}$, $m_s = \frac{1}{2}$, (b) $l = 1$, $m_l = +1$, (c) $l = 2$, $m_l = 0$.

14.25. Derive an expression for the half-angle of the apex of the cone of precession in terms of the quantum numbers l, m_l (or s, m_s). What is its value for the α-state of an electron spin? Show that the minimum possible angle approaches zero as l approaches infinity.

14.26. Draw the cones of precession of an electron with $l = 6$. In what region of the spherical surface is an electron most likely to be found when $l = 6$ and (a) $m_l = 0$, (b) $m_l = +6$, (c) $m_l = -6$?

15

Atomic structure and atomic spectra

Learning objectives

After careful study of this chapter you should be able to:

(1) Describe the main features of the spectrum of atomic hydrogen and state the *Ritz combination principle*, Section 15.1(a).

(2) Write down the Schrödinger equation for the hydrogen atom in centre of mass coordinates, eqn (15.1.7), and show that it can be separated into radial and angular equations, eqn (15.1.8).

(3) State how the *energy levels* of hydrogen depend on the principal quantum number, eqn (15.1.22).

(4) Explain the significance of an *atomic orbital* and describe the shapes of s-, p-, and d-orbitals, Section 15.1(d).

(5) Define the *radial distribution function*, eqn (15.1.25), and explain its significance.

(6) State the *Bohr frequency condition*, eqn (15.1.26), and the *selection rules* for atomic transitions, Section 15.1(f).

(7) State the *Pauli principle* and the *Pauli exclusion principle*, Section 15.2(b).

(8) State and use the *Clebsch–Gordan series* for the coupling of angular momenta, eqn (15.2.1).

(9) State the rules of the *building-up principle and Hund's rule*, and use them to account for the electronic configurations of atoms, Section 15.2(d).

(10) Account for the variation of first ionization energies through the Periodic Table, Section 15.2(e).

(11) Describe the strategy for *self-consistent field* calculations of the electronic structures of many-electron atoms, Section 15.2(f).

(12) Describe the origin of *spin–orbit coupling*, and show how it is responsible for the *fine structure* of spectra, Section 15.3(a).

(13) Calculate the spin–orbit coupling constant from spectral information, Example 15.5.

(14) Describe the construction and significance of *term symbols*, Section 15.3(b).

(15) Describe *Russell–Saunders coupling*, and state the *selection rules* for many-electron atom spectra, Section 15.3(b).

(16) Write expressions relating the *magnetic moment* of an electron to its orbital and spin angular momenta, eqn (15.3.5), and for the energy of interaction with a magnetic field, eqns (15.3.6) and (15.3.7).

(17) Describe the meaning of *precession*, Fig. 15.22.

(18) Describe and explain the *Zeeman effect*, Section 15.3(c).

Introduction

The rudiments of the quantum mechanical description of the fundamental types of motion were described in Chapter 14. Now we examine the quantum mechanics of electrons in atoms. By the end of the chapter we shall see how to interpret information provided by spectroscopy in terms of atomic structure. Information about atomic energy levels is of central importance for a rational discussion of the structures and reactions of molecules, and the material we are about to cover has extensive chemical applications.

15.1 The structure and spectrum of atomic hydrogen

When an electric discharge is passed through hydrogen, the molecules are dissociated and the excited atoms emit light at a number of frequencies, Fig. 15.1. The first important contribution to the explanation was made by Johann Balmer, who pointed out that the wavelengths of the light in the visible region (the lines now called the *Balmer series*) fitted an expression which written in modern notation has the form

$$1/\lambda = R_H\{(1/2^2) - (1/n^2)\}, \qquad n = 3, 4, \ldots \qquad (15.1.1)$$

where R_H is a constant now called the *Rydberg constant* and having the value 109 677 cm^{-1}. As further lines were discovered in the ultraviolet (the *Lyman series*) and the infrared (the *Paschen series*), it was noted that all could be fitted to

$$1/\lambda = R_H\{(1/n_1^2) - (1/n_2^2)\}, \qquad (15.1.2)$$

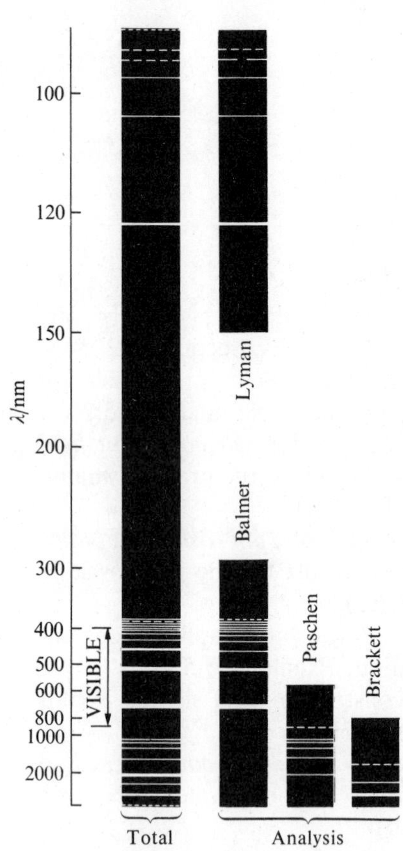

Fig. 15.1. The spectrum of atomic hydrogen. The spectrum is shown on the left, and is analysed into its overlapping series on the right. Note that the Balmer series lies in the visible region.

with $n_1 = 1$ (Lyman), 2 (Balmer), and 3 (Paschen), and that in each case $n_2 = n_1 + 1, n_1 + 2, \ldots$ The form of this expression strongly suggests that the inverse wavelength (the *wavenumber*, symbol: $\tilde{\nu}$; $\tilde{\nu} = 1/\lambda$) and, through $\nu = c/\lambda = c\tilde{\nu}$, the frequency, of each spectral line can be written as the difference of two terms, each of the form R_H/n^2. This statement amounts to the *Ritz combination principle*:

The wavenumber (or frequency) of any line can be expressed as the difference of two terms.

Two terms T_1 and T_2 are said to *combine* to produce a spectral line of wavenumber

$$\tilde{\nu} = T_1 - T_2. \qquad (15.1.3)$$

Three points arise. The first is to see why the terms in hydrogen have the form R_H/n^2. The second is to account for the value of R_H. The third is to explain why, as we shall see later, not all possible combinations of terms are observed.

15.1 (a) The structure of the hydrogen atom

The calculation of the electronic structure of the hydrogen atom depends on being able to solve the appropriate Schrödinger equation. At first sight this is a problem of considerable difficulty, because we have to deal with two particles, and therefore six coordinates (three for each particle). The equation is therefore a partial differential equation in six variables, which is taxing. However, physical intuition suggests that the full equation might separate into two equations, one for the motion of the atom as a whole through space, and the other for the electron around the nucleus. The first of these is of no direct interest (and we have solved it anyway, because it corresponds to the free motion of a particle in space, Section 14.1). The second is of considerable interest, but it still involves a partial differential equation in three variables (the coordinates of the electron relative to the nucleus). However, we can suspect that even that equation is separable, because the motion of a particle around a fixed point has already been solved (Section 14.3), and it is likely that the hydrogen atom's wavefunctions will contain those angular wavefunctions as a factor, leaving only the radial part of the total wavefunction to be found. This turns out to be the case. The strategy, then, is first to separate the relative motion of the electron and the nucleus from the motion of the atom as a whole, and then to separate their relative motion into angular and radial parts.

The separation of the internal motion from the motion of the atom as a whole is achieved by a series of straightforward steps that are given in detail in Appendix 15.1. In outline, we begin by writing the Schrödinger equation in the form

$$H\psi = E\psi \qquad (15.1.4a)$$

with

$$H = -(\hbar^2/2m_e)\nabla_e^2 - (\hbar^2/2m_N)\nabla_N^2 + V(r), \qquad (15.1.4b)$$

where m_e is the mass of the electron, m_N that of the nucleus (a proton in normal hydrogen), and $V(r)$ the Coulomb potential energy when an electron (of charge $-e$) is at a distance r from a nucleus of charge Ze (with $Z = 1$ for hydrogen):

$$V(r) = -(Ze^2/4\pi\varepsilon_0)(1/r). \qquad (15.1.5)$$

(ε_0 is the vacuum permittivity, $8.854 \times 10^{-12}\,J^{-1}\,C^2\,m^{-1}$.) The laplacians ∇^2 differentiate with respect to the electron and nucleus coordinates (so that ∇_e^2 has terms like $\partial^2/\partial x_e^2$, etc., where x_e is the location of the electron with respect to some x-axis). By the method described in Appendix 15.1, H is transformed into

$$H = -(\hbar^2/2m)\nabla_{c.m.}^2 - (\hbar^2/2\mu)\nabla^2 + V(r),$$

with m the total mass of the atom $(m_e + m_N)$ and μ the *reduced mass* defined through

$$\textit{Reduced mass:}\ 1/\mu = 1/m_e + 1/m_N. \qquad (15.1.6)$$

The laplacian $\nabla_{c.m.}^2$ relates to the coordinates of the centre of mass of the atom; the laplacian ∇^2 relates to the coordinates of the electron relative to the nucleus, Fig. 15.2. The former corresponds to the kinetic energy of the atom as a whole, while the latter corresponds to the internal structure of the atom. Therefore, we discard the former (as shown in Appendix 15.1, what we really do is go through a separation of variables procedure and isolate the two types of motion). Finally, we arrive at the following simplified form for the Schrödinger equation for the internal structure of the atom:

$$-(\hbar^2/2\mu)\nabla^2\psi + V(r)\psi = E\psi, \qquad (15.1.7)$$

which is the equation we have to solve.

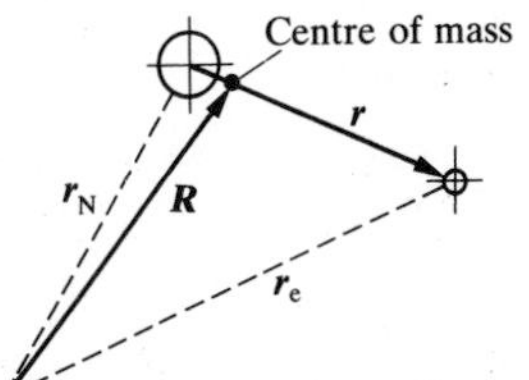

Fig. 15.2. Centre of mass coordinates for the hydrogen atom. Because the proton is so much more massive than the electron, the centre of mass of the atom lies very close to the proton (much closer than this diagram suggests).

The wavefunctions of the hydrogen atom underlie a great deal of chemistry and are therefore very important. However, they are quite tricky to find. We now sketch in outline the formal solution of eqn (15.1.7). The important conclusions, which can be used without the following details, are summarized in Section 15.1(c).

15.1 (b) The formal solution

The Schrödinger equation for the internal structure is still a partial differential equation in three variables, but it is separable into an equation for the radial and angular dependences of the wavefunction. We can prove this by writing the wavefunction as the product $\psi(r, \theta, \phi) = R(r)Y(\theta, \phi)$, and going through the separation of variables procedure.

First, rewrite the equation as

$$\nabla^2\psi + (Z\mu e^2/2\pi\varepsilon_0\hbar^2)(1/r)\psi = -(2\mu E/\hbar^2)\psi. \tag{15.1.7a}$$

The appearance of this equation can be simplified by introducing $\varepsilon = -2\mu E/\hbar^2$ and $\gamma = Z\mu e^2/2\pi\varepsilon_0\hbar^2$, for it then becomes

$$\nabla^2\psi + (\gamma/r)\psi = \varepsilon\psi,$$

which at least looks simple. Next, we use the polar coordinate form of the laplacian given in eqn (14.3.14) and obtain

$$(1/r)(\partial^2/\partial r^2)r\psi + (1/r^2)\Lambda^2\psi + (\gamma/r)\psi = \varepsilon\psi.$$

Now substitute $\psi = RY$, noting that $\partial^2/\partial r^2$ acts only on R while Λ^2, which is a collection of derivatives with respect to angle, acts only on Y:

$$(1/r)(\mathrm{d}^2rR/\mathrm{d}r^2)Y + (1/r^2)R\Lambda^2Y + (\gamma/r)RY = \varepsilon RY.$$

The next step in the separation of variables procedure is to divide through by RY so as to isolate each variable. This gives

$$(1/rR)(\mathrm{d}^2rR/\mathrm{d}r^2) + (1/r^2)(1/Y)\Lambda^2Y + (\gamma/r) = \varepsilon.$$

The second term still depends on r, but that can be eliminated by multiplying through by r^2, giving

$$r^2(1/rR)(\mathrm{d}^2rR/\mathrm{d}r^2) + (1/Y)\Lambda^2Y + \gamma r = \varepsilon r^2.$$

Now the second term depends only on the angles, and the remaining terms depend on the radius. By the usual argument (Section 14.1(c)), each set of terms is equal to a constant, and we can write

$$(1/Y)\Lambda^2Y = -constant, \tag{15.1.8a}$$

$$r^2(1/rR)(\mathrm{d}^2rR/\mathrm{d}r^2) + \gamma r - \varepsilon r^2 = constant. \tag{15.1.8b}$$

The first of these equations is the same as the one we solved for the angular momentum of a particle in three dimensions, eqn (14.3.27):

$$\Lambda^2Y = -l(l+1)Y, \tag{15.1.9}$$

and so the angular factor in the overall wavefunction is a spherical harmonic of the kind we met in Section 14.3. All the properties of a particle rotating in three dimensions carry over from that discussion, and so we know that the quantum numbers m_l and l are needed, and that $m_l = 0, \pm1, \pm2, \ldots, \pm l$, with $l = 0, 1, 2, \ldots$ However, just as m_l came under a new constraint when it appeared in an equation that involved l and its accompanying boundary condition, so l comes under a new constraint because it also appears in the radial equation and its associated boundary conditions.

By comparing eqn (15.1.8a) and eqn (15.1.9) we see that the constant in eqn (15.1.8) is equal to $l(l+1)$. Therefore, the radial equation is

$$r^2(1/rR)(\mathrm{d}^2rR/\mathrm{d}r^2) + \gamma r - \varepsilon r^2 = l(l+1).$$

This rearranges to

$$(\mathrm{d}^2rR/\mathrm{d}r^2)+\{(\gamma/r)-l(l+1)/r^2\}rR=\varepsilon rR. \qquad (15.1.10)$$

This is the *radial wave equation*.

Some idea of the form of the solutions can be obtained by recognizing that this equation is like a Schrödinger equation for a particle with a potential energy

$$V=-(Ze^2/4\pi\varepsilon_0 r)+l(l+1)\hbar^2/2\mu r^2, \qquad (15.1.11)$$

the first term being the Coulomb potential energy, and the second representing the effect of the centrifugal force arising from the angular momentum of the electron around the nucleus. When $l=0$ the electron has zero angular momentum, and the potential energy is pure Coulombic and attractive, Fig. 15.3. On the other hand, when l is non-zero, the centrifugal term gives a positive contribution to the effective potential energy, and when the electron is close enough to the nucleus (r small enough), this repulsive term dominates the attractive Coulomb component, Fig. 15.3. The two potential energies are qualitatively very different close to the nucleus, but similar at large distances. Therefore, we can expect the solutions to be quite different near the nucleus but similar far away.

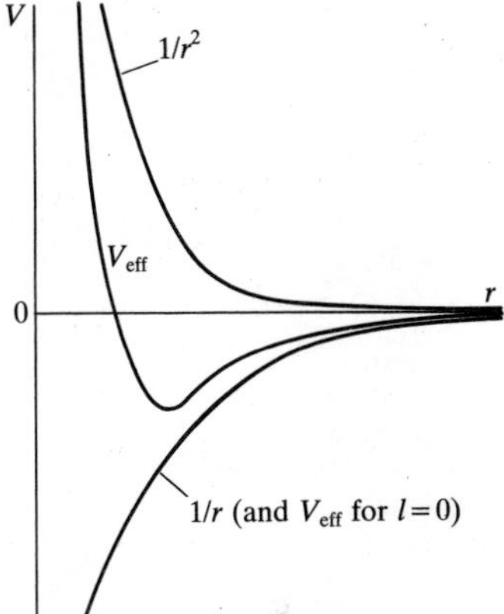

Fig. 15.3. The *effective potential energy* of an electron in the hydrogen atom. When it has zero angular momentum the V_{eff} is the pure Coulomb potential energy. When the electron has angular momentum, the centrifugal effect gives rise to a positive contribution which is very large close to the nucleus. We can expect the $l=0$ and $l\neq 0$ wavefunctions to be very different near the nucleus.

When r is large the radial equation becomes approximately

$$\mathrm{d}^2rR/\mathrm{d}r^2=\varepsilon rR.$$

The term on the left is equal to $r(\mathrm{d}^2R/\mathrm{d}r^2)+2(\mathrm{d}R/\mathrm{d}r)$, which in turn is approximately $r(\mathrm{d}^2R/\mathrm{d}r^2)$ when r is very large. Therefore, the asymptotic form of the radial equation is

$$\mathrm{d}^2R/\mathrm{d}r^2=\varepsilon R. \qquad (15.1.12)$$

This has the solutions $\mathrm{e}^{-r\sqrt{\varepsilon}}$ and $\mathrm{e}^{r\sqrt{\varepsilon}}$, but only the former is acceptable because the latter increases to infinity as r increases. Therefore, the radial wavefunction at large distances is of the form $\mathrm{e}^{-r\sqrt{\varepsilon}}$. We can immediately conclude that *electron distributions in atoms do not have sharp edges, but decay exponentially towards zero*.

As in the case of the harmonic oscillator, we build on the asymptotic solution by writing the exact solution as

$$R=f\mathrm{e}^{-r\sqrt{\varepsilon}} \qquad (15.1.13)$$

and looking for a solution f that does not increase to infinity more quickly than the exponential function falls to zero (so that the product remains acceptable everywhere). On substituting this equation into eqn (15.1.10) we obtain

$$f''+\{(2/r)-\sqrt{\varepsilon}\}f'+\{(\gamma/r)-[l(l+1)/r^2]-(2\sqrt{\varepsilon}/r)\}f=0,$$

with $f'=\mathrm{d}f/\mathrm{d}r$ and $f''=\mathrm{d}^2f/\mathrm{d}r^2$. This is badly in need of simplification. It is achieved first by introducing the substitution $r=\rho/2\sqrt{\varepsilon}$, which transforms it into

$$f''+\{(2/\rho)-1\}f'+\{(n-1)/\rho-l(l+1)/\rho^2\}f=0,$$

where now $f'=\mathrm{d}f/\mathrm{d}\rho$, etc., and $n=\gamma/2\sqrt{\varepsilon}$, and then by substituting $R=\rho^l L$, obtaining

$$L''+\{2(l+1)-\rho\}L'+\{n-(l+1)\}L=0.$$

Mathematicians recognize this as the *associated Laguerre equation*, and its solutions have been extensively studied. They are well-behaved (in the sense of leading to acceptable wavefunctions) only if n is a positive integer and if $n-(l+1)$ is not negative. This means that solutions are acceptable for

$$n=1,2,\ldots \quad \text{and} \quad l<n-1. \tag{15.1.14}$$

Since we know that l is an integer, this means that its values are limited to $l=0,1,2,\ldots,n-1$ for a given value of n. This is the second restriction on l we anticipated.

The restriction of n to positive integral values means that the energy is quantized. Since $n=\gamma/2\sqrt{\varepsilon}$, and γ and ε are given following eqn (15.1.7a), we see that

$$n=(\mu Ze^2/2\pi\varepsilon_0\hbar^2)/2(-2\mu E/\hbar^2)^{\frac{1}{2}}$$

which rearranges to

$$E=-(Z^2e^4/32\pi^2\varepsilon_0^2\hbar^2)(1/n^2). \tag{15.1.15}$$

This is of the form

$$E=-constant/n^2,$$

which is in agreement with the spectroscopic evidence.

The solutions of the associated Laguerre equation are the *associated Laguerre polynomials*, and are certain polynomials in ρ (for example, $6-6\rho+\rho^2$ is the one corresponding to $n=3$ and $l=0$). They depend on n and l (but not on m_l because that does not appear in the equation), and so may be denoted $L_{n,l}$. The radial wavefunction is related to them through eqn (15.1.13) and $f=\rho^l L$, which means that

$$R_{n,l}(\rho)=\rho^l L_{n,l}(\rho)\mathrm{e}^{-\rho/2}. \tag{15.1.16}$$

These functions are called the *associated Laguerre functions* (they are not finite polynomials on account of the exponential factor). As they stand here they are not normalized, but explicit, normalized versions are given in Box 15.1 and their dependence on the radius is illustrated in Fig. 15.4. Note that they all decay exponentially with distance at large distances from the nucleus. In interpreting these diagrams recall that since $\rho=2r\sqrt{\varepsilon}$, and $\varepsilon=-2\mu E/\hbar^2$, with E given in eqn (15.1.15), the dimensionless parameter ρ and the radius are related by

$$\rho=(\mu Ze^2/2\pi\varepsilon_0\hbar^2)(r/n).$$

The collection of fundamental constants multiplying r/n has dimensions of 1/length, and they are normally expressed in terms of the

$$\textit{Bohr radius:}\ a_0=4\pi\varepsilon_0\hbar^2/m_e e^2. \tag{15.1.17}$$

Its value is 52.9 pm; we shall see its significance later. Then, noting that the reduced mass appears in ρ rather than the electron's mass, we can write

$$r=\{n(m_e/\mu)/2Z\}a_0\rho \tag{15.1.18}$$

when we need to relate r and ρ. For a hydrogen atom, making the

Box 15.1 Hydrogenic atomic orbitals

Hydrogenic orbitals (i.e. orbitals for 1-electron atoms of atomic number Z) are written $\psi_{n,l,m_l} = R_{n,l}(r)Y_{l,m_l}(\theta, \phi)$, with the Y the spherical harmonics given in Box 14.2. The radial wavefunctions are as follows:

1s: $R_{1,0} = 2(Z/a_0)^{\frac{3}{2}}e^{-\rho/2}$
2s: $R_{2,0} = (1/2\sqrt{2})(Z/a_0)^{\frac{3}{2}}(2-\rho)e^{-\rho/2}$
2p: $R_{2,1} = (1/2\sqrt{6})(Z/a_0)^{\frac{3}{2}}\rho e^{-\rho/2}$
3s: $R_{3,0} = (1/9\sqrt{3})(Z/a_0)^{\frac{3}{2}}(6-6\rho+\rho^2)e^{-\rho/2}$
3p: $R_{3,1} = (1/9\sqrt{6})(Z/a_0)^{\frac{3}{2}}(4-\rho)\rho e^{-\rho/2}$
3d: $R_{3,2} = (1/9\sqrt{30})(Z/a_0)^{\frac{3}{2}}\rho^2 e^{-\rho/2}$.

In each case, $\rho = 2Zr/na_0$.
Further values can be found in L. Pauling and E. B. Wilson, *Introduction to quantum mechanics*, McGraw-Hill.

approximation that $m_e/\mu = 1$ (the actual value is 1.000 54), the relation is simply $r = na_0\rho/2$.

We now have the complete solution of the Schrödinger equation: the energies are given in eqn (15.1.15), and the wavefunctions are the products

$$\psi_{n,l,m_l}(r) = R_{n,l}Y_{l,m_l}, \tag{15.1.19}$$

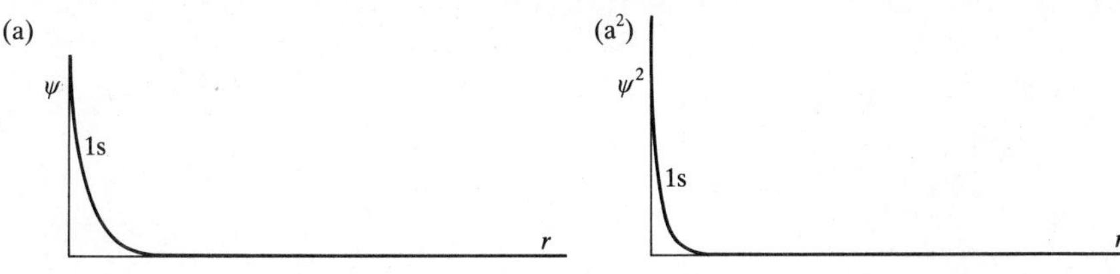

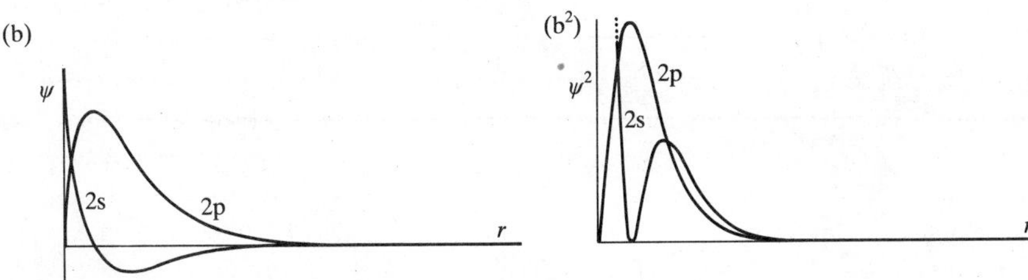

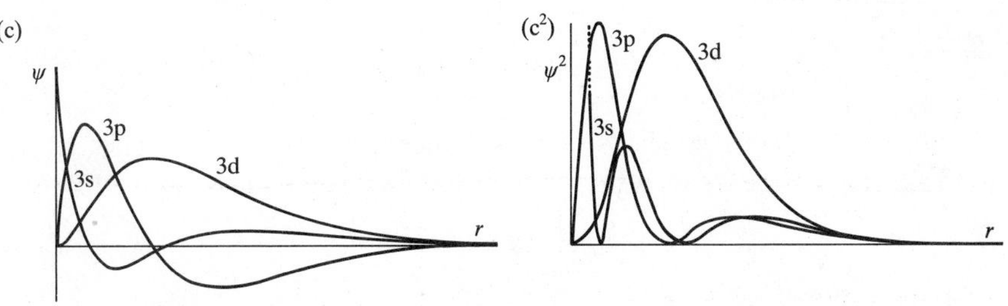

Fig. 15.4. (a, b, c) The radial wavefunctions of the first few states of the hydrogen atom and (a^2, b^2, c^2) the probability densities. Note that the s-orbitals have a non-zero and finite value at the nucleus. The vertical scales are different in each case.

R being the functions in Box 15.1 and Y the spherical harmonics listed in Box 14.2.

15.1 (c) The energy levels of atomic hydrogen

We can summarize the conclusions drawn above as follows. Solving the Schrödinger equation for the hydrogen atom and imposing the boundary conditions that ensure the wavefunctions are acceptable leads to *three* quantum numbers, Box 15.2. Two, the *angular momentum quantum number* l and the *magnetic quantum number* m_l, come from the angular solutions, and relate to the angular momentum of the electron around the nucleus (why m_l is called 'magnetic' we shall explain later). An electron with quantum number l has an angular momentum of magnitude $\{l(l+1)\}^{\frac{1}{2}}\hbar$ with l any of the values $0, 1, 2, \ldots, n-1$. The value of m_l which can be $0, \pm 1, \ldots, \pm l)$ denotes the orientation of this momentum, the z-component being $m_l\hbar$. The third quantum number, n, is called the *principal quantum number*. It is confined to the values $1, 2, 3, \ldots$, and determines the energy through

$$E_n = -(\mu Z^2 e^4/32\pi^2\varepsilon_0^2\hbar^2)(1/n^2). \qquad (15.1.20)$$

Box 15.2 The hydrogen atom

The wavefunctions of the hydrogen atom depend on three quantum numbers:

Principal quantum number:

$$n = 1, 2, 3, \ldots$$

Angular momentum quantum number (or azimuthal quantum number):

$$l = 0, 1, 2, \ldots, n-1.$$

Magnetic quantum number:

$$m_l = l, l-1, l-2, \ldots, -l \quad \text{(i.e. } 2l+1 \text{ values)}.$$

The energy is determined only by the principal quantum number and is related to it by

$$E_n = -hcR_{\mathrm{H}}/n^2, \qquad hcR_{\mathrm{H}} = \mu e^4/32\pi^2\varepsilon_0^2\hbar^2.$$

The quantum number l determines the *magnitude* of the angular momentum, through $\{l(l+1)\}^{\frac{1}{2}}\hbar$, and the quantum number m_l determines the projection of this angular momentum on some axis as $m_l\hbar$. There are n^2 different values of l and m_l for a given n, and so each energy level is n^2-fold degenerate.

The wavefunctions are products of radial and angular components. The angular wavefunctions are the spherical harmonics, Box 14.2, and the radial wavefunctions are the associated Laguerre functions, Box 15.1.

The selection rules for spectroscopic transitions are

$$\Delta n \text{ unrestricted}; \quad \Delta l = \pm 1.$$

It is a peculiarity of *hydrogen-like atoms* (atoms of general atomic number Z but with only one electron), that *the energy is independent of l and m_l, and depends only on n*. Note that all the energies are negative: this means that we are dealing with the *bound states* of the atom, in which the energy of the electron is lower than when it is infinitely far away and at rest (which corresponds to the zero of energy). There are also solutions of the Schrödinger equation with positive energies. These correspond to unbound states of the electron, the states to which an electron is raised when it is knocked out of the atom by a high-energy collision. The energies of the unbound electron are not quantized, and form the *continuum states* of the atom.

In the case of hydrogen itself $Z=1$, and the bound state energies (the only ones we consider from now on) are given by

$$E_n = -hcR_{\mathrm{H}}(1/n^2), \qquad hcR_{\mathrm{H}} = \mu e^4/32\pi^2\varepsilon_0^2\hbar^2. \qquad (15.1.21)$$

The levels are drawn in Fig. 15.5. The spectrum of atomic hydrogen can now be explained by supposing that it undergoes a transition from a state with principal quantum number n_2 (and energy $-hcR_{\mathrm{H}}/n_2^2$) to one with principal quantum number n_1 (and energy $-hcR_{\mathrm{H}}/n_1^2$), and discards the energy difference as a photon of energy $h\nu$ and frequency ν. Since energy is conserved overall, the frequency is given by

$$h\nu = hcR_{\mathrm{H}}/n_1^2 - hcR_{\mathrm{H}}/n_2^2, \quad \text{or} \quad \tilde{\nu} = \nu/c = R_{\mathrm{H}}/n_1^2 - R_{\mathrm{H}}/n_2^2, \qquad (15.1.22)$$

which is the form required by experiment. Accounting for the form of the terms was one of our objectives. The second was to account for the value of the Rydberg constant: inserting the values of the fundamental constants into the expression for R_{H} gives almost exact agreement with experiment.

A related experimental test is given by the *ionization energy* (symbol: I). This is the minimum energy required to ionize the atom in its lowest energy state (its *ground state*). The ground state of hydrogen is the state with the smallest value of n (because then E has its greatest negative value, representing greatest lowering of energy). This lowest energy is $E_1 = -hcR_{\mathrm{H}}$. The atom is ionized when the electron has been excited to the level $E=0$, corresponding to $n=\infty$, Fig. 15.5. Therefore the energy that must be supplied is hcR_{H}, or 2.179×10^{-18} J (corresponding to 13.60 eV).

The experimental determination of ionization energies of atoms depends on observing the convergence of the spectral lines, the detection of the *series limit*, the wavenumber at which the series terminates and becomes a continuum. Suppose that the upper state lies at an energy E_{upper}, then when the atom makes a transition to E_{lower} a photon is emitted of wavenumber

$$\tilde{\nu} = (E_{\mathrm{upper}} - E_{\mathrm{lower}})/hc.$$

Since the upper energy levels are given by an expression of the form $E_{\mathrm{upper}} = -hcR/n^2$, the emission lines will occur at

$$\tilde{\nu} = -R/n^2 - E_{\mathrm{lower}}/hc. \qquad (15.1.23)$$

The separation between successive lines decreases as n increases, Fig. 15.5, and so the lines converge. Moreover, plotting the wavenumbers against $1/n^2$ should give a straight line of slope $-R$ and intercept $-E_{\mathrm{lower}}/hc$. The value of $-E_{\mathrm{lower}}$ is the ionization energy of the atom in that state (and is the ground state ionization energy when the lower state is the ground state). This is illustrated by the following *Example*.

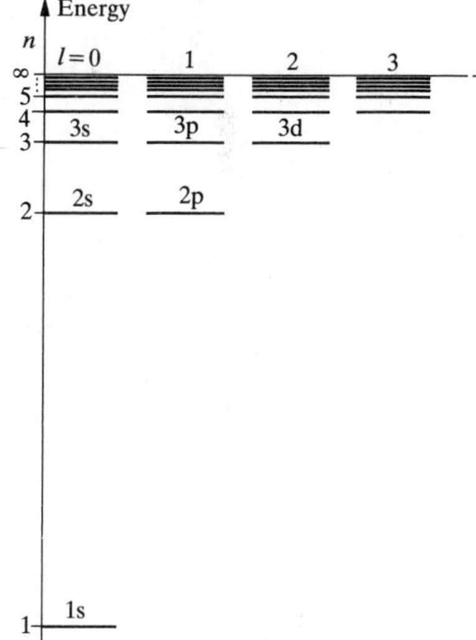

Fig. 15.5. The energy level diagram of the hydrogen atom. The values are relative to an infinitely separated, stationary electron and a proton. The degeneracy of the orbitals with the same n but different l is a special characteristic of a pure Coulomb potential.

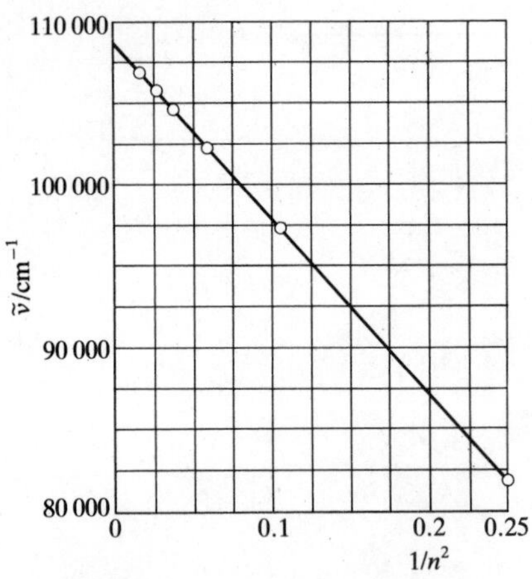

Fig. 15.6. The plot of the data in *Example* 15.1 used to determine the ionization energy of an atom (in this case, of H).

Example 15.1

The spectrum of atomic hydrogen shows lines at the following wavenumbers: 82 259, 97 492, 102 824, 105 292, 106 632, 107 440 cm^{-1}. Find (a) the ionization energy of the lower state, (b) the value of the Rydberg constant.

• *Method*. Plot wavenumber against $1/n^2$ with $n = 2, 3, \ldots$ (this assumes that the lower state is $n = 1$; if a straight line is not obtained, try $n = 2, 3, \ldots$, etc.). According to eqn (15.1.23), the intercept at $1/n^2 = 0$ is $-E_{\text{lower}}/hc$, or I/hc, and the slope is $-R_{\text{H}}$. Use a computer to make a least-squares fit of the data to get a result that reflects the precision of the data.

• *Answer*. The wavenumbers are plotted against $1/n^2$ in Fig. 15.6. The (least-squares) intercept lies at 109 679 cm^{-1}, and so the ionization energy is hc times this quantity, or 2.1788×10^{-18} J (1312.1 kJ mol^{-1}). The slope is, in this instance, numerically the same, and so $R_{\text{H}} = 109\,679\ \text{cm}^{-1}$.

• *Comment*. The Rydberg constant depends on the reduced mass of the atom, and so in principle it can be used to determine the mass of the nucleus.

• *Exercise*. The spectrum of atomic deuterium shows lines at the following wavenumbers: 15 238, 20 571, 23 039, 24 380 cm^{-1}. Determine (a) the ionization energy of the lower state, (b) the ionization energy of the ground state, (c) the mass of the deuteron.

[(a) 328.1 kJ mol^{-1}, (b) 1312.4 kJ mol^{-1}, (c) 3.4×10^{-27} kg]

15.1 (d) Atomic orbitals

The wavefunctions of hydrogen-like atoms are important because they are used to describe the structures of all atoms and molecules. One-electron wavefunctions for atoms and molecules are called *orbitals*, and we shall use that name from now on.

The atomic orbitals of hydrogen-like atoms depend on three quantum numbers and so are denoted ψ_{n,l,m_l}. They are the product of a radial and an angular component, $\psi_{n,l,m_l}(\boldsymbol{r}) = R(r)Y(\theta, \phi)$, where R is related to the associated Laguerre functions, as described above, and Y is a spherical harmonic, as described in Chapter 14. By combining the information in Boxes 14.2 and 15.1 we obtain the explicit forms of the first few hydrogen-like atomic orbitals, Box 15.2.

The probability of the electron being in an infinitesimal region $d\tau$ at $\boldsymbol{r}$ when it is in a state with quantum numbers n, l, m_l is given by $|\psi_{n,l,m_l}(\boldsymbol{r})|^2\, d\tau$. The probability density $|\psi_{n,l,m_l}(\boldsymbol{r})|^2$ can be represented by the density of shading in a diagram. In order to simplify their representation, atomic orbitals are normally shown as diagrams of a surface, the *boundary surface*, that captures about 90 per cent of the electron probability and shows the location of the angular nodes. This results in diagrams of the type shown on the right of Fig. 15.7.

The *ground state orbital*, the orbital corresponding to the ground state and for which the electron is most tightly bound to the nucleus, is the one corresponding to $n = 1$ (and therefore to $l = 0$ and $m_l = 0$). From Box 15.2 we can write (for $Z = 1$):

$$\psi = (1/\pi a_0^3)^{\frac{1}{2}} e^{-r/a_0} \tag{15.1.24}$$

Max value

This wavefunction is independent of angle, and so the orbital is spherically symmetrical. Its amplitude decays exponentially from a maximum value of $(1/\pi a_0^3)^{\frac{1}{2}}$ at the nucleus, and so the most probable point at which the electron will be found is at the nucleus itself. This is consistent with the

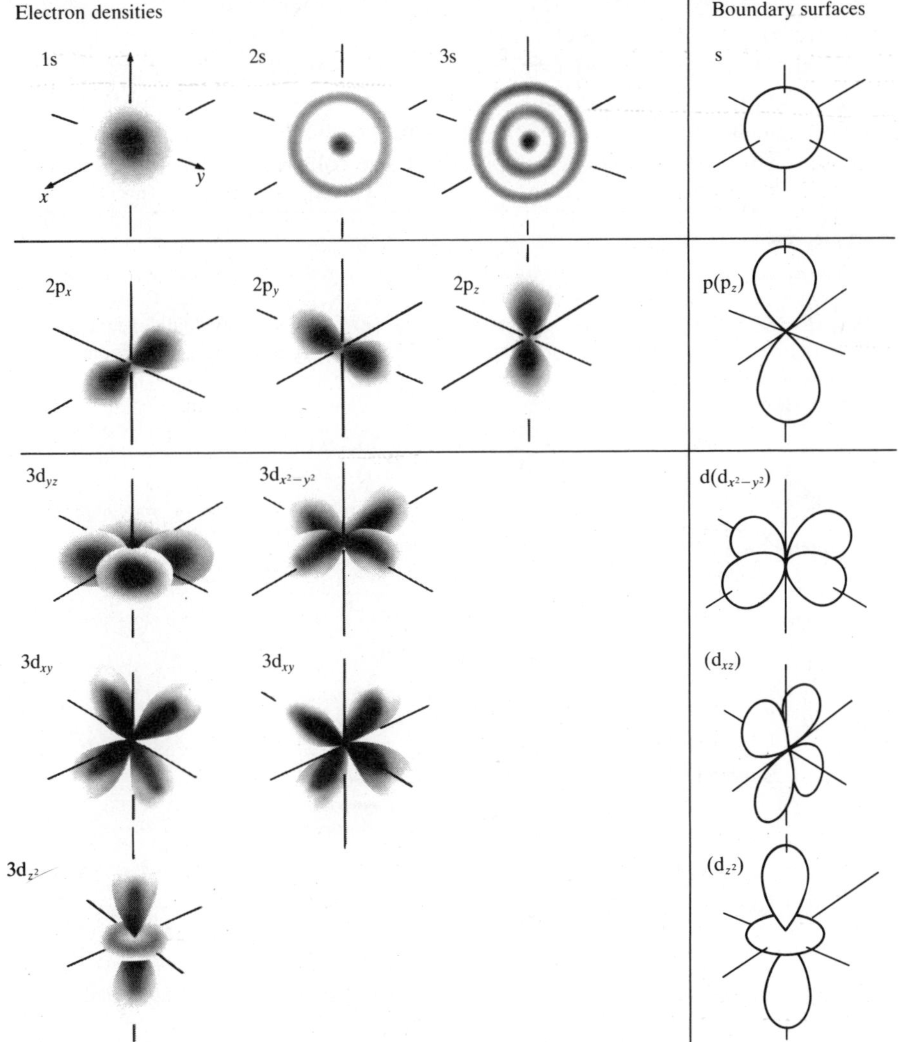

Fig. 15.7. Representations of the first few hydrogenic atomic orbitals in terms of the electron densities (as represented by the density of shading) and the *boundary surfaces*, within which there is a 90% probability of finding the electron.

view that the electron should be close to the nucleus in order to achieve its lowest potential energy. However, the minimum total energy must take into account the kinetic energy too, and this increases as the electron clusters more closely around the nucleus. This can be seen as follows. In order for the electron to be close to the nucleus for much of its time, its wavefunction must be sharply peaked there and small elsewhere, Fig. 15.8(a). However, this implies a high kinetic energy (because high curvature corresponds to high kinetic energy, Section 13.3). On the other hand, if we minimize the kinetic energy by allowing the wavefunction to have only little curvature, Fig. 15.8(b), then it will be diffuse, and the electron will often be far from the nucleus, which implies a high potential energy. The actual ground state is a compromise, Fig. 15.8(c). Note that the kinetic energy we are referring to is that associated with the electron's radial motion (classically, to a motion that swings it through the nucleus), and not to its angular motion: since $l = 0$, it has no orbital angular momentum.

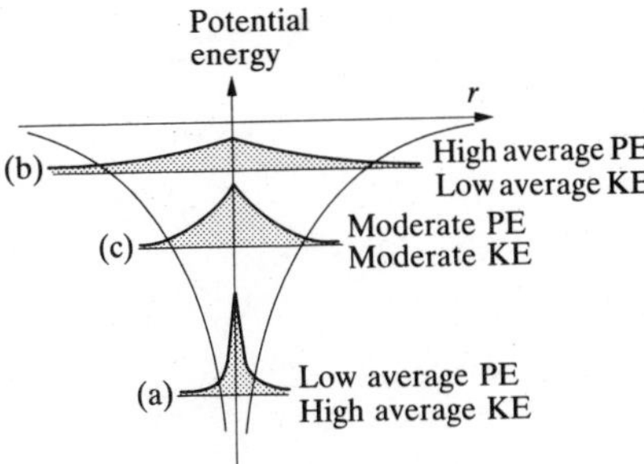

Fig. 15.8. The balance of kinetic and potential energies that accounts for the structure of the ground state of hydrogen (and similar atoms). (a) The sharply curved but localized orbital has high mean kinetic energy, but low potential energy; (b) the mean kinetic energy is low, but the potential energy is not very favourable; (c) the compromise of moderate kinetic energy and moderately favourable potential energy.

All spherically symmetrical ($l = 0$, and therefore zero angular momentum) orbitals are called *s-orbitals*. The lowest (with $n = 1$) is called a 1s-orbital. When $n = 2$ and $l = 0$ we have a 2s-orbital, and so on. Their energies increase (i.e. the electron becomes less tightly bound) as n increases because the average distance of the electron from the nucleus increases. This is demonstrated explicitly in the following *Example*.

Example 15.2

Use the hydrogen atom wavefunctions to calculate the mean radius of a 1s- and a 2s-orbital.

• *Method*. The mean value is the expectation value $\langle r \rangle$; therefore, evaluate the integrals $\int \psi^* r \psi \, d\tau$ using the wavefunctions given in Box 15.1. The integration over the angular part gives unity in each case (because the spherical harmonics are normalized to unity), and so the only work relates to the integration over r, which can be done explicitly using

$$\int_0^\infty x^n e^{-ax} \, dx = n!/a^{n+1}.$$

• *Answer*. $\langle r \rangle = \int r\psi^2 \, d\tau = \int r(RY)^2 r^2 \sin\theta \, d\theta \, d\phi \, dr$

$$= \int_0^\infty r^3 R^2 \, dr, \quad \text{as} \quad \int Y^2 \sin\theta \, d\theta \, d\phi = 1.$$

(a) For a 1s-orbital, $R = 2(1/a_0^3)^{\frac{1}{2}} e^{-r/a_0}$, and so

$$\langle r \rangle = (4/a_0^3) \int_0^\infty r^3 e^{-2r/a_0} \, dr = \tfrac{3}{2} a_0.$$

(b) For a 2s-orbital, $R = (1/8a_0^3)^{\frac{1}{2}}(2 - r/a_0) e^{-r/2a_0}$, and so

$$\langle r \rangle = (1/8a_0^3) \int_0^\infty r^3 (2 - r/a_0)^2 e^{-r/a_0} \, dr = 6a_0.$$

• *Comment*. The 2s-orbital has a mean radius four times larger than that of a 1s-orbital. The general expression for the mean radius is

$$\langle r \rangle_{n,l,m_l} = n^2\{l + \tfrac{1}{2}[1 - l(l+1)/n^2]\}(a_0/Z).$$

• *Exercise*. Evaluate the mean radius of a 3s-orbital by integration, and of a 3p-orbital using the general formula.

[$27a_0/2$; $25a_0/2$]

When $n = 1$ the only value allowed l is 0, but when $n = 2$, either $l = 0$ (giving the $2s$-orbital) or $l = 1$. Orbitals with $l = 1$ are called *p-orbitals*, and so $n = 2$, $l = 1$ is the 2p-orbital. An electron in a p-orbital has an angular momentum of magnitude $(\sqrt{2})\hbar$ on account of the presence of the angular node, which introduces angular curvature into the wavefunction. The possession of orbital angular momentum has a profound effect on the shape of the wavefunction close to the nucleus: whereas s-orbitals are non-zero at the nucleus, p-orbitals have zero amplitude there. This can be understood in terms of the centrifugal effect of the angular momentum flinging the electron away from the nucleus: the Coulombic force on the electron is proportional to $1/r^2$, but the centrifugal force is proportional to (angular momentum)$^2/r^3$, and so for small enough distances it always overcomes the attraction so long as the angular momentum is not identically zero (as in the case of s-orbitals, with $l = 0$). The centrifugal effect also appears in the *d-orbitals* ($l = 2$), the *f-orbitals* ($l = 3$), and the higher orbitals (*g*, *h*, *i*, . . .-*orbitals*). All orbitals with $l > 0$ have zero amplitude at the nucleus, and consequently zero probability of finding the electron there.

Since for $l = 1$ the magnetic quantum number m_l can take the values $+1, 0, -1$, there are *three* p-orbitals for a given n. Different values of m_l correspond to different components of angular momentum with respect to the z-axis. The orbital with $m_l = 0$, which has the form $f(r)\cos\theta$, is independent of ϕ and so corresponds to zero component of angular momentum on the z-axis, and is called a *p_z-orbital*, Fig. 15.9. Since the orbital depends on $\cos\theta$, the electron density depends on $\cos^2\theta$, which has its maximum value when $\theta = 0$ or 180° and is zero when $\theta = 90°$. This means that the maximum electron density occurs in regions on the z-axis, and that the xy-plane is a *nodal plane* on which there is zero probability of finding the electron. The orbitals with $m_l = \pm 1$ depend on ϕ and have the form $f(r)\sin\theta e^{\pm i\phi}$. They have zero amplitude where $\theta = 0$ and 90° (along the z-axis) and the maximum amplitude where $\theta = 90°$, which is in the xy-plane. They depend on ϕ, and the function with the factor $e^{+i\phi}$ corresponds to rotation in one direction and that with the factor $e^{-i\phi}$ corresponds to rotation in the opposite direction. In order to draw the functions it is usual to take the linear combination

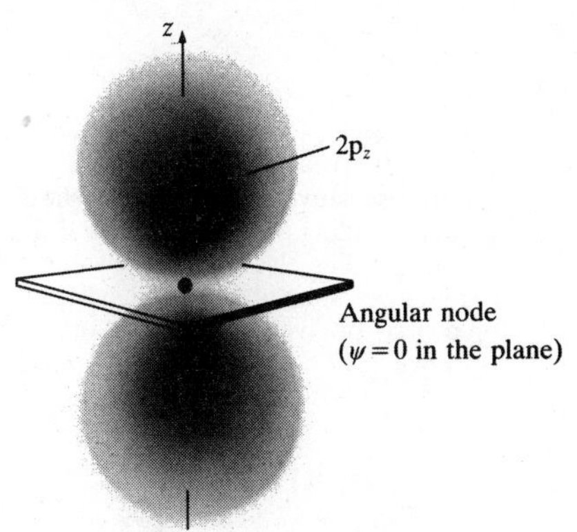

Fig. 15.9. The electron density in a 2p-orbital; the region of zero density (in the xy-plane) is called a *node*.

$$f(r)\sin\theta(e^{i\phi} + e^{-i\phi}) \propto f(r)\sin\theta\cos\phi \propto xf(r)$$
$$f(r)\sin\theta(e^{i\phi} - e^{-i\phi}) \propto f(r)\sin\theta\sin\phi \propto yf(r)$$

which (when normalized) are called the *p_x- and p_y-orbitals* respectively. They are standing waves (composed of equal and opposite angular momenta): p_x is the same shape as a p_z-orbital, but directed along the x-axis, Fig. 15.7, and p_y is similarly directed along the y-axis.

When $n = 3$, then $l = 0$, 1, or 2. This gives one 3s-orbital, three 3p-orbitals, and five 3d-orbitals (the last with $m_l = 2, 1, 0, -1, -2$) and corresponding to five different orientations of the angular momentum (of magnitude $\sqrt{6}\hbar$) with respect to the z-axis. The states with opposite values of m_l may be combined in pairs to give standing waves, and the resulting shapes are shown in Fig. 15.7.

All nine of the $n = 3$ orbitals have the same energy (because, as shown above, the energy of a hydrogen atom depends only on the principal quantum number n and is independent of l). This peculiar degeneracy is unique to hydrogen-like atoms and is due to the peculiarity of the pure Coulomb potential. Note that the pattern of degeneracy has now emerged: when $n = 1$ there is only one orbital, when $n = 2$ there are four orbitals (one 2s- and three 2p-orbitals), and with $n = 3$ there are nine. In general, *the number of degenerate orbitals is n^2*. For example, there are 16 orbitals with principal quantum number $n = 4$: one 4s, three 4p, five 4d, and seven 4f).

15.1 (e) The radial distribution function

The wavefunction tells us the probability of finding the electron in an infinitesimal volume at each point of space. We can imagine a probe, with a volume $d\tau$ and sensitive to electrons, moved around near the nucleus. Since the probability density in the ground state is

$$\psi^*\psi = (1/\pi a_0^3)e^{-2r/a_0},$$

the reading from the detector decreases exponentially as the probe is moved outwards along any radius, but it gives a constant reading if it is moved at constant radius, Fig. 15.10: the 1s-orbital (like any s-orbital) is

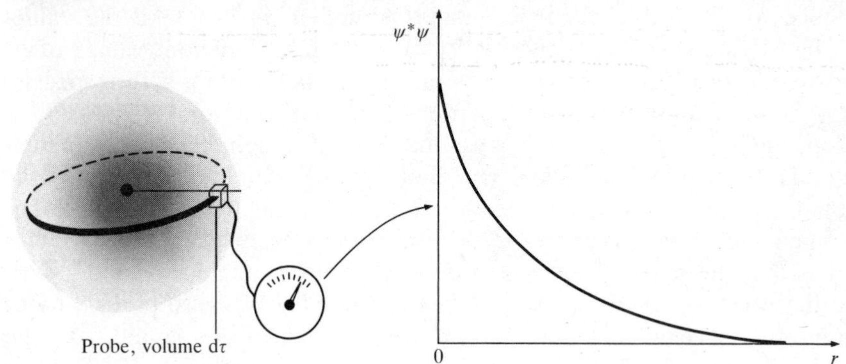

Fig. 15.10. A constant-volume electron-sensitive detector would show greatest reading at the nucleus, and a smaller reading elsewhere. The same reading would be obtained anywhere on a circle of given radius: the s-orbital is *isotropic*.

spherically symmetrical. The wavefunction can also be used to find the probability that the electron will be found anywhere in a spherical shell of thickness dr at a radius r. The sensitive volume of the probe is now the volume of the shell, Fig. 15.11, which is $4\pi r^2\,dr$. The probability that the electron will be found in this shell is the probability density multiplied by the probe volume:

$$\psi^*\psi(4\pi r^2\,dr) = P(r)\,dr, \quad \text{with} \quad P(r) = 4\pi r^2(1/\pi a_0^3)e^{-2r/a_0}. \quad (15.1.25)$$

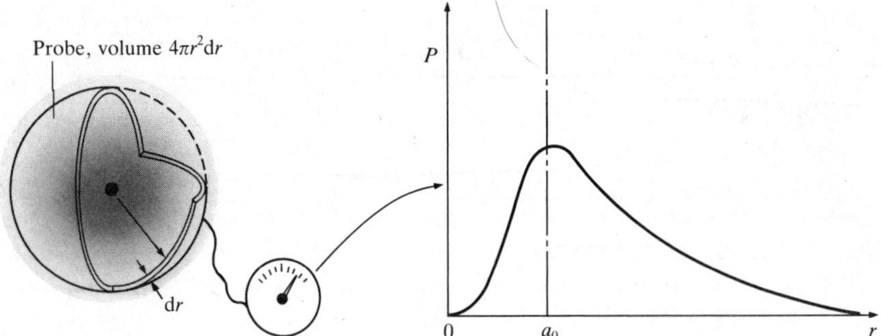

Fig. 15.11. The *radial distribution function P* gives the probability that the electron will be found *anywhere* in a shell of radius r. It is a maximum when r is equal to the Bohr radius a_0.

P is another kind of probability density: when multiplied by dr it gives the probability of finding the electron anywhere in a shell of thickness dr at the radius r. It is therefore called the *radial distribution function* (RDF). Since r^2 increases with radius and $\psi^*\psi$ decreases, the RDF has the form shown in Fig. 15.11. The *most probable radius* at which the electron will be found is the radius at which the RDF is a maximum, which in this case corresponds to $r = a_0$. In other words, the Bohr radius (53 pm) is the most probable radius at which the electron will be found when the hydrogen atom is in its ground state. When the RDF for the 2s-orbital is calculated (simply by forming $4\pi r^2\psi^*\psi$ with $\psi = \psi_{2s}$), the most probable radius is found to be $5.2a_0 = 276$ pm, which reflects the expansion of the atom as its energy increases.

Example 15.3

Calculate the *most probable radius* at which an electron will be found when it occupies a 1s-orbital of a hydrogen-like atom of atomic number Z, and tabulate the values for the atoms from H to Ne.

• *Method*. Find the radius at which the radial distribution function has a maximum value (by solving $dP/dr=0$). A hydrogenic 1s-orbital (i.e. a hydrogen-like orbital, but for nuclear charge Ze) is $(Z^3/\pi a_0^3)^{\frac{1}{2}}e^{-Zr/a_0}$, and the RDF is $P=4\pi r^2\psi^*\psi$.

• *Answer*. $P=4\pi r^2(Z^3/\pi a_0^3)e^{-2Zr/a_0}$,

$$dP/dr=(4Z^3/a_0^3)\{2r-(2Z/a_0)\}e^{-2Zr/a_0}=0 \quad \text{at} \quad r=r^*.$$

Therefore, $r^*=a_0/Z$. Then, with $a_0=52.9$ pm,

	H	He	Li	Be	B	C	N	O	F	Ne
r^*/pm	52.9	26.5	17.6	13.2	10.6	8.82	7.56	6.61	5.88	5.29

• *Comment*. Notice how the 1s-orbital is drawn towards the nucleus as the nuclear charge increases. At uranium the most probable radius is only 0.58 pm, almost 100 times closer than for hydrogen. (On a scale where $r^*=10$ cm for H, $r^*=1$ mm for U.) The electron then experiences strong accelerations, and relativistic effects are important. The significance of the Bohr radius is that it is the most probable radius in hydrogen.

• *Exercise*. Find the most probable distance of a 2s-electron from the nucleus in a hydrogen-like atom. $[(3+\sqrt{5})a_0/Z]$

15.1 (f) The spectral selection rules

We now know the energies of all the states of the hydrogen atom, and it is tempting to ascribe the atomic spectrum to transitions between all possible pairs of them. Then one could say that a change of energy ΔE in the atom gives rise to a photon of energy ΔE, and therefore of frequency ν such that $\Delta E=h\nu$. This is the

Bohr frequency condition: $h\nu=\Delta E$. (15.1.26)

This prescription, however, cannot be complete. We have remarked that the photon possesses an intrinsic spin angular momentum corresponding to $s=1$. If a photon is generated by an atom, the atom's angular momentum must change to compensate for the amount carried away by the photon. This means that an electron in a d-orbital ($l=2$, a 'd-electron') cannot drop into an s-orbital ($l=0$) because the photon cannot carry away enough angular momentum, and nor can an s-electron make a transition to another s-orbital, because then there is no change in the atom's angular momentum to make up for the one quantum carried away by the photon. It follows that some spectral transitions are *allowed* while others are *forbidden*. The statements of which transitions are allowed are the *selection rules*. In the present case they are

One-electron atom selection rules: $\Delta n=$ any integer, $\Delta l=\pm1$.

(The principal quantum number can change by any amount consistent with the electron's angular momentum quantum number because it does not relate directly to the angular momentum.)

These remarks enable us to construct the *Grotrian diagram* in Fig. 15.12, which summarizes neatly the energies and the allowed transitions (the densities of the lines denoting their relative intensities) that contribute to the spectrum.

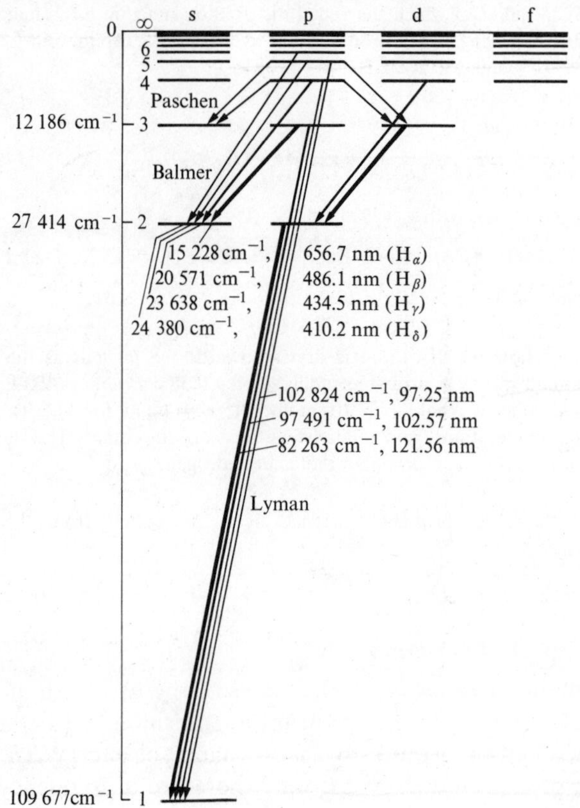

Fig. 15.12. A *Grotrian diagram* which summarizes the appearance and analysis of the spectrum of atomic hydrogen. The thicker the line, the more intense the transition.

15.2 The structures of many-electron atoms

The Schrödinger equations for atoms with more than one electron are extremely complicated because all the electrons interact with each other and there are three variables for each electron. Even in the case of helium, no analytical expression for the orbitals and energies can be given, but extremely accurate *numerical* solutions are available from calculations using large computers. Later in this section we shall see the kind of calculations involved, but for the purpose of understanding in a general way the structures of these atoms we can adopt a simpler approach based on what we already know about the structure of hydrogen.

15.2 (a) Building atoms

The ground state of hydrogen consists of one electron with a distribution described by the 1s-orbital. We say the *electronic configuration* is $1s^1$, and that the electron *occupies* the 1s-orbital. The helium atom has two electrons. We can imagine building the atom by adding first one electron to the bare nucleus (of charge $2e$), and then the second. The first electron occupies a 1s hydrogen-like orbital, but since $Z = 2$ it is more compact than in hydrogen itself. The second electron joins the first in the 1s-orbital, and so the electron configuration of the ground state is $1s^2$. Of course, this description is only approximate, because the two electrons repel each other and the orbitals are distorted (but the total electron distribution of the atom remains spherical). Nevertheless, the *orbital description* is a useful model for discussing the chemical properties of atoms, and is the starting point for more sophisticated versions of atomic theory.

The next atom to build is lithium, $Z = 3$. The first two electrons occupy a 1s-orbital drawn even more closely around the more highly charged nucleus. The third electron, however, does not join the first two in the 1s-orbital. The *exclusion principle* was introduced by Wolfgang Pauli in order to account for the absence of some lines in the spectrum of helium, but he was later able to derive a very general form of the principle from theoretical considerations. In the form we need it here it states:

No more than two electrons may occupy any given orbital, and if two do occupy one orbital, their spins must be antiparallel.

Electrons with antiparallel spins (that is, one α-electron and one β-electron, with cancelling total angular momentum) are said to be *paired*. This remarkable feature of nature is the key to the structure of complex atoms, to chemical periodicity, and to molecular structure.

15.2 (b) The Pauli principle

The *exclusion principle* is a special case of a general principle called the *Pauli principle* which, when applied to electrons, states the following:

Pauli principle: When the labels of any two electrons are exchanged, the wavefunction changes sign.

In fact the principle applies to any pair of identical particles with half-integral spins: such particles are called *fermions*, and include protons, neutrons, and some atomic nuclei. The 'opposite' principle (i.e. that the wavefunction does not change sign) applies to particles with integral spin: these are the *bosons*, and include photons. For the time being we need consider only the form as stated above.

The mathematical form of the principle is obtained by considering the wavefunction for two electrons, $\psi(1,2)$: this is a very complicated function in six variables, three being the coordinates of electron 1 and three the coordinates of electron 2. Then the Pauli principle states that it is a fact of nature (which has its roots in the theory of relativity) that the total wavefunction including spin must change sign if we interchange 1 and 2 wherever they occur: that is, $\psi(2,1) = -\psi(1,2)$.

The connection of this general form of the principle with the exclusion principle can be illustrated by the following argument, which has three stages.

Stage 1. A single electron has an energy represented by the hamiltonian H_1, and its wavefunction is a solution of the Schrödinger equation $H_1\psi(\boldsymbol{r}_1) = E_1\psi(\boldsymbol{r}_1)$. A second electron has an energy described by a hamiltonian H_2, and its wavefunction is a solution of $H_2\psi(\boldsymbol{r}_2) = E_2\psi(\boldsymbol{r}_2)$. When both electrons are present in the same atom the hamiltonian is

$$H = H_1 + H_2 + e^2/4\pi\varepsilon_0 r_{12},$$

the additional term being the potential energy arising from their mutual repulsion. The wavefunction for the pair of electrons is the solution of $H\psi = E\psi$, with $\psi = \psi(\boldsymbol{r}_1, \boldsymbol{r}_2)$, a complicated function of the coordinates of both electrons. Suppose, though, that as a first approximation we can ignore their repulsion, then the total hamiltonian is $H = H_1 + H_2$, and the solution of the corresponding Schrödinger equation is simply the *product*, $\psi(\boldsymbol{r}_1)\psi(\boldsymbol{r}_2)$, of the individual wavefunctions. This is proved as follows (where, for simplicity, we write $\psi(1)$ for $\psi(\boldsymbol{r}_1)$ and $\psi(2)$ for $\psi(\boldsymbol{r}_2)$):

$$\begin{aligned} H\psi &= (H_1 + H_2)\psi(1)\psi(2) = \{H_1\psi(1)\}\psi(2) + \psi(1)\{H_2\psi(2)\} \\ &= \{E_1\psi(1)\}\psi(2) + \psi(1)\{E_2\psi(2)\} = (E_1 + E_2)\psi(1)\psi(2) \\ &= E\psi, \quad \text{with} \quad E = E_1 + E_2. \end{aligned}$$

That is, the product $\psi(1)\psi(2)$ *is* an eigenfunction of the (simplified) hamiltonian H. If both electrons occupy the same orbital, ψ_a, for instance, then the overall wavefunction would be $\psi_a(1)\psi_a(2)$, which from now on we shall denote more simply as a(1)a(2).

Stage 2. In order to use the Pauli principle we have to deal with the *total* wavefunction, the wavefunction including spin. In this case, it might seem that there are several possibilities for two spins: they can be both α-spins (denoted $\alpha(1)\alpha(2)$), both β (denoted $\beta(1)\beta(2)$), or one can be α and the other β (denoted $\alpha(1)\beta(2)$ or $\alpha(2)\beta(1)$). Since we cannot tell which electron is α and which is β, in the last case it is appropriate to express the spin states as the linear combinations

$$\sigma_+(1,2) = \alpha(1)\beta(2) + \alpha(2)\beta(1), \qquad \sigma_-(1,2) = \alpha(1)\beta(2) - \alpha(2)\beta(1)$$

because these allow one spin to be α and the other β, but do not specify which is which. The *total* wavefunction of the system is therefore the product of the orbital part a(1)a(2) and one of the four spin states:

$$\begin{gathered} \mathrm{a}(1)\mathrm{a}(2)\alpha(1)\alpha(2), \\ \mathrm{a}(1)\mathrm{a}(2)\beta(1)\beta(2), \\ \mathrm{a}(1)\mathrm{a}(2)\sigma_+(1,2), \\ \mathrm{a}(1)\mathrm{a}(2)\sigma_-(1,2). \end{gathered}$$

Stage 3. Now we take the Pauli principle into account. It says that if a wavefunction is to be acceptable (for electrons), then it must change sign when the electrons are exchanged. In each case, exchanging the labels 1 and 2 converts the factor a(1)a(2) into a(2)a(1), which is the same, because the order of multiplying the functions does not change the value of the product. The same is true of $\alpha(1)\alpha(2)$ and $\beta(1)\beta(2)$. Therefore, the first two overall products are not allowed, because they do not change sign. The combination $\sigma_+(1,2)$ changes to $\alpha(2)\beta(1) + \alpha(1)\beta(2)$, which is simply the orginal σ_+ written in a different order, and so we also have $\sigma_+(2,1) = +\sigma_+(1,2)$, and the third overall product is also disallowed. Finally, consider $\sigma_-(1,2)$:

$$\sigma_-(2,1) = \alpha(2)\beta(1) - \alpha(1)\beta(2) = -\{\alpha(1)\beta(2) - \alpha(2)\beta(1)\} = -\sigma_-(1,2),$$

and so it does change sign (is 'antisymmetric'). Therefore the product $\mathrm{a}(1)\mathrm{a}(2)\sigma_-(1,2)$ also changes sign under particle exchange, and therefore it is acceptable.

Now we see that *only one* of the *four possible states is allowed by the Pauli principle*, and the one that survives has one α and one β spin. This is the content of the Pauli exclusion principle. The exclusion principle is irrelevant when the orbitals occupied by the electrons are different, and both electrons may then have (but need not have) the same spin. Nevertheless, even then the overall wavefunction must still be antisymmetric overall, and must still satisfy the Pauli principle itself: that is one reason why the Pauli principle is more general than the exclusion principle.

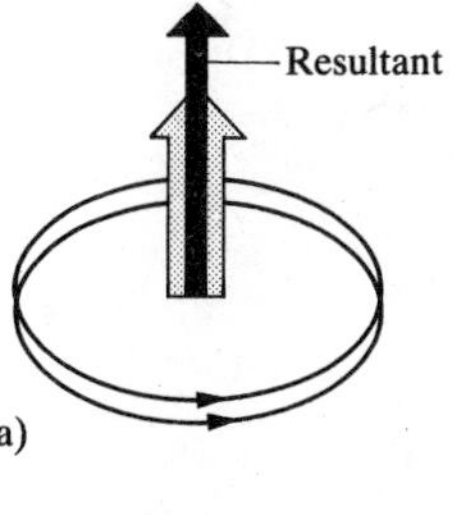

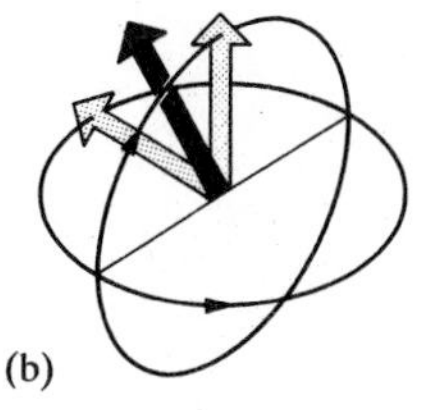

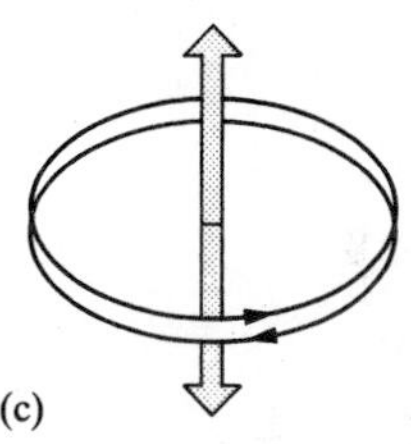

Fig. 15.13. Two classically spinning objects can have a combined angular momentum of any value between the sum (a) and difference (c) of the two contributions. According to quantum mechanics, only some total angular momenta are possible. In the case of two electrons (which have $s=\frac{1}{2}$), only two combined states are permitted ($S=0, 1$); e.g. (b) cannot occur.

15.2 (c) Total spin states

One feature brought out by this discussion is the nature of the four spin states $\alpha(1)\alpha(2)$, $\beta(1)\beta(2)$, $\sigma_+(1,2)$, and $\sigma_-(1,2)$. The first three are all symmetric under particle interchange, and are called a spin *triplet*. The last one, which is antisymmetric, and the only one available when the two electrons occupy the same orbital, is called a spin *singlet*. However, there is more to them than their symmetry, for they represent states of different *total* spin angular momentum.

In classical physics two angular momenta combine to give a total angular momentum which, depending on the orientation of the components, can take any intermediate value between the minimum, when the two angular momenta are opposed, and the maximum, when they are parallel, Fig. 15.13. We have already seen that angular momenta are space-quantized, and so it should not be surprising that in quantum mechanics, two angular momenta can take up only discrete orientations with respect to each other. This means that only some *total* angular momenta are permitted for given component momenta, Fig. 15.13(b). The quantum mechanical rule for calculating total angular momentum (as expressed by the angular momentum quantum number J, so that the magnitude is $\{J(J+1)\}^{\frac{1}{2}}\hbar$, in the usual way) given two contributing momenta described by the quantum numbers j_1 and j_2 is called the

Clebsch–Gordan series: $J=j_1+j_2, j_1+j_2-1, \ldots, |j_1-j_2|$. (15.2.1)

In the present case of two electron spins $s_1=\frac{1}{2}$ and $s_2=\frac{1}{2}$, the total spin quantum number (S) may take the values

$$S=\tfrac{1}{2}+\tfrac{1}{2}, \tfrac{1}{2}+\tfrac{1}{2}-1, \ldots, \tfrac{1}{2}-\tfrac{1}{2},$$

or simply $S=1,0$. The first of these has three values of M_S, namely $+1, 0, -1$ (because total angular momenta have the same quantum properties as ordinary angular momenta), which clearly correspond to the three states of the triplet: $\alpha\alpha$ (both electrons having a projection $m_s=\frac{1}{2}$) corresponds to $M_S=+1$, $\beta\beta$ (both having $m_s=-\frac{1}{2}$) to $M_S=-1$, and σ_+ (with one spin up and the other down) to $M_S=0$. On the other hand, the state with $S=0$, which has only a single value of M_S, namely 0, must correspond to the singlet combination σ_-.

Some idea of the reason why σ_+ and σ_- belong to states of different total spin angular momentum can be seen by reference to the vector diagrams in Fig. 15.14: in σ_+ the individual momenta lie in the same vertical plane, and have a non-zero resultant; in σ_- they lie in the same plane, but point in opposite directions and have zero resultant. The + and − are simply telling us that the vectors are in phase or out of phase respectively, and whether the individual momenta add or cancel. Note, incidentally, that in the singlet state, when the momenta cancel, the spins are perfectly antiparallel and are truly 'paired'. In σ_+, although one is α and the other β, the angle between

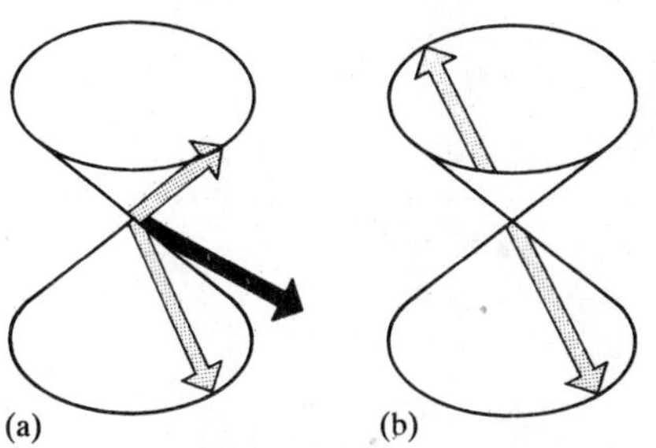

Fig. 15.14. (a) The two spins described by the σ_+ wavefunction are aligned so as to result in a net resultant angular momentum. (b) The two spins described by σ_- are aligned so that their angular momenta cancel. They are perfectly antiparallel.

them is acute, and they are not antiparallel. This is why we refer to σ_- as the 'spin-paired state', not σ_+, even though both are built from α and β spins.

We shall see that other angular momenta also add like spins, and in each case the allowed total states can be predicted using the Clebsch–Gordan series. This is the basis of the classification of the states of many-electron atoms, and we shall now move towards that application.

15.2 (d) The building-up principle

In the case of lithium ($Z = 3$) the third electron cannot enter the 1s-orbital, because that is already full: we say the *$n = 1$ shell is complete*. (The $n = 1$ shell of an atom is also called the *K-shell*.) Therefore, it enters the next available orbital, which is one with $n = 2$. The 2s- and 2p-orbitals are not degenerate in atoms with more than one electron because the Coulombic repulsion between the electrons interferes with the special characteristics of the centrosymmetric, Coulombic electron-nucleus attraction. It turns out that in general *s-electrons lie lower in energy than p-electrons, and p-electrons lie lower than d-electrons*.

The reason for the removal of the degeneracy is the combined effect of *penetration* and *shielding*. An electron in a many-electron atom is *shielded* from the full charge of the nucleus by the other electrons, especially those of inner shells, such as the K-shell in lithium. This means that any given electron experiences the charge of a nucleus with an *effective atomic number* (symbol: Z_{eff}), which is less than the true atomic number Z. Moreover, the effective atomic number depends on the type of electron involved, and an s-electron would normally experience a larger Z_{eff}, and be more tightly bound, than a p-electron of the same principal quantum number. This is because an s-electron has a non-zero probability of being found at the nucleus, and therefore it *penetrates* through the inner electron shell and is more likely to be found close to the nucleus than a p-electron of the same shell. Similarly, a d-electron penetrates less than a p-electron of the same shell, and therefore experiences more shielding and an even smaller Z_{eff}. Table 15.1 shows some of the effective atomic numbers that have been calculated for electrons in atoms. The different energies of orbitals in a many-electron atom means that the *shell* (all the orbitals of a given n) is split into s-, p-, d-, etc. *subshells*.

Table 15.1 Effective atomic numbers

	Z_{eff}
He 1s	1.69
Na 1s	10.6
2s	6.85
2p	6.85
3s	2.20

This all boils down to the result that we can predict the electronic configuration of lithium to be $1s^2 2s^1$, with the central nucleus surrounded by a complete K-shell of two electrons, and around that a more diffuse 2s-electron. The generalization of this result is called the *building-up principle* (or the *Aufbau principle, Aufbau* being German for building up). In order to construct the lowest energy electron configuration of an atom of atomic number Z (and therefore possessing Z electrons if it is neutral), we proceed as follows.

1. *Imagine the orbitals as already existing and as having the energy sequence* $1s < 2s < 2p < 3s < 3p < (4s, 3d)$.

(4s, 3d) indicates that the orbital approximation is misleading when the 3d-orbitals begin to be occupied. The lowest total energies of the ions of the d-block elements, taking into account the effects of electron–electron repulsions, are achieved with configurations best written in the order $3d^n 4s^2$.

2. *Feed the appropriate number of electrons into the lowest available*

orbital subject to the demand of the exclusion principle that no more than two can occupy any single orbital.
Since the p-subshell consists of three orbitals, it can accommodate up to six electrons, and the d-subshell, which consists of five orbitals, can accommodate up to ten.

There are two further points of detail. The carbon atom has $Z = 6$, and so its ground configuration is $1s^2 2s^2 2p^2$. On electrostatic grounds, we can expect the two 2p-electrons to occupy different p-orbitals (e.g. one can be thought of as occupying $2p_x$ and the other $2p_y$), because in this way they are further apart and repel each other less. This is a general principle:

3. *If a set of degenerate orbitals is available, electrons occupy different members of the set before doubly occupying any one.*

Consequently, a more precise statement of the electron configuration of the ground state of neutral carbon is $1s^2 2s^2 2p_x^1 2p_y^1$. Since the p-electrons occupy different orbitals there is no need for their spins to be paired. The question of which is the ground state, a singlet (with the spins paired) or a triplet (with the spins parallel) is resolved by a final rule:

4. *Hund's rule: An atom adopts a configuration with the greatest number of unpaired electrons.*

The rule applies at *this* stage of the argument, not earlier; the reason behind it is complicated, but it reflects the quantum mechanical property that electrons with parallel spins have a tendency to stay well apart: this is called *spin correlation*. We can now conclude that the carbon atom is in a triplet state, with its two outermost electrons having parallel spins. This is known to be the case.

Neon, with $Z = 10$, has the ground configuration $1s^2 2s^2 2p^6$, which completes the *L-shell* (the shell with $n = 2$). The next electron must enter the 3s-orbital and begin the *M-shell*. Sodium, with $Z = 11$, therefore has the configuration $1s^2 2s^2 2p^6 3s^1$ and, like lithium with $1s^2 2s^1$, consists of a single s-electron outside a complete core.

This analysis has brought us to the origin of chemical periodicity. The L-shell is completed by eight electrons, and so the element with $Z = 3$ (Li) should have similar properties to the element with $Z = 11$ (Na). Likewise, Be ($Z = 4$) should be similar to Mg ($Z = 12$), and so on up to the noble gases He ($Z = 2$), Ne ($Z = 10$), and Ar ($Z = 18$), which have complete shells. Above $Z = 18$ the *N-shell* ($n = 4$) begins to fill and the 3d-orbitals (formally of the M-shell) are energetically accessible. As a result, up to eighteen outer electrons can be accommodated (2 in 4s, 6 in 4p, and 10 in 3d), accounting for the first long period of the periodic table. The existence of the d-block elements (the 'transition elements') reflects the stepwise occupation of the 3d-orbitals, and the subtle shades of energy differences along this series gives rise to the rich complexity of inorganic transition metal chemistry.

15.2 (e) Ionization energies

The minimum energy necessary to remove an electron from an atom is its *first ionization energy* (symbol: I_1). The *second ionization energy* is the minimum energy needed to remove a second electron (from the singly charged cation). The variation of the first ionization energy through the periodic table is shown in Fig. 15.15. Numerical values are given in Table 4.9.

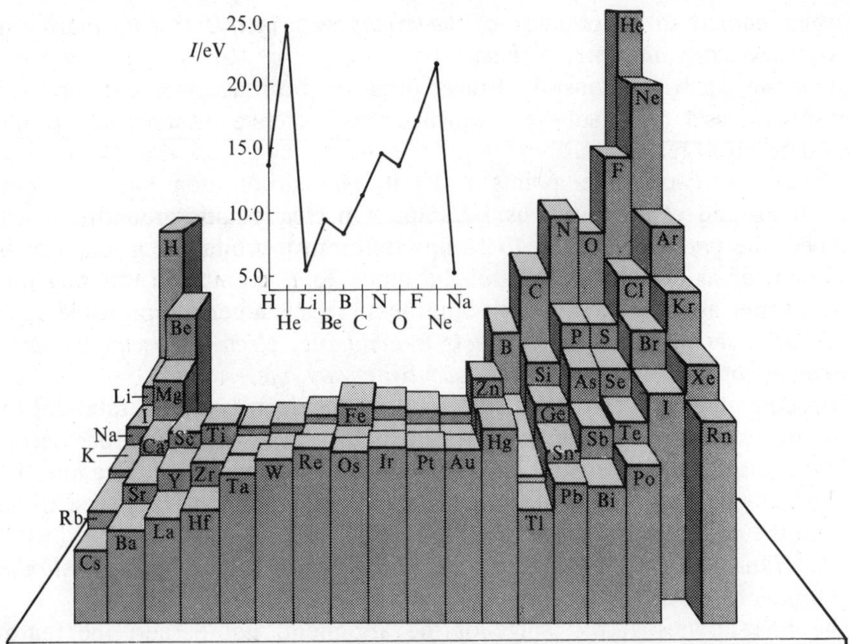

Fig. 15.15. The first ionization energies of the elements are represented by the heights of the bars. The inset gives more detail for the first and second row elements.

Lithium has a low first ionization energy: its outermost electron is well-shielded ($Z_{\text{eff}} = 1.3$ even though $Z = 3$) and easily removed. Beryllium has a higher nuclear charge than lithium, and its outermost electron (one of the two 2s-electrons) is more difficult to remove: its ionization energy is larger. The ionization energy drops from Be to B because in the latter the outermost electron occupies a 2p-orbital and is less strongly bound. The ionization energy increases to C because its outermost electron is also 2p, and the nuclear charge has increased and so the electron is more firmly bound. Nitrogen has a still higher ionization energy because of the further increase in nuclear charge.

There is now a kink in the curve which reduces the ionization energy of O below what would be expected by simple extrapolation. This is because at O for the first time a 2p-orbital must become doubly occupied, and the electron–electron repulsions are increased above what would be expected by simple extrapolation along the row. (The kink is less pronounced in the next row, between P and S, Fig. 15.15, because their orbitals are more diffuse.) The values for O, F, and Ne fall roughly on the same line, the increase of their ionization energies reflecting the increasing attraction of the nucleus for the outermost electrons.

At sodium the outermost electron is 3s. It is far from the nucleus, and the latter's charge is shielded by the compact, complete core. This is why the ionization energy of sodium is substantially lower than that of neon. The periodic cycle starts again along this row, and the variation of the ionization energy can be traced to similar reasons.

15.2 (f) Self-consistent field orbitals

The essential difficulty of the Schrödinger equation lies in the electron–electron interaction terms: the potential energy depends on $1/r_{ij}$, r_{ij} being the separation of electrons i and j, and there are many such terms in a many-electron atom. It is hopeless to expect to find analytical solutions, but very powerful computational techniques are available which give very detailed and reliable numerical solutions for the wavefunctions and energies. The scheme was originally introduced by Douglas Hartree and then modified by Vladimir Fock in order to take into account the indistinguishability of individual electrons. In broad outline, the *Hartree–Fock procedure* is as follows.

Imagine that we have a rough idea of the structure of the atom. In the sodium atom, for instance, the orbital approximation suggests the configuration $1s^22s^22p^63s^1$ with the orbitals approximated by hydrogen-like atomic orbitals corresponding to various effective nuclear charges (the inner electrons experiencing a stronger nuclear charge than the outer because they are less shielded). Now consider the 3s-electron. A Schrödinger equation can be written for this electron by ascribing to it a potential energy due both to the nuclear attraction and to the electronic repulsions from the other electrons distributed in their approximate orbitals. This equation has the form

$$-(\hbar^2/2m_e)\nabla^2\psi_{3s}-(Ze^2/4\pi\varepsilon_0 r)\psi_{3s}+V_{ee}\psi_{3s}=E\psi_{3s}, \qquad (15.2.2)$$

where V_{ee}, which depends on the distribution of all the other electrons, is the repulsion term. This equation may be solved for ψ_{3s} (using techniques of numerical integration), and the solution obtained will be different from the solution guessed initially. The procedure is then repeated for another orbital, such as ψ_{2p}: first the Schrödinger equation is written in a form like eqn (15.2.2) but with the improved ψ_{3s} orbital used in setting up the electron–electron repulsion term. The equation is then solved numerically, giving an improved version of ψ_{2p}. This procedure is repeated for the 2s- and 1s-orbitals, each time using the improved orbitals found at the earlier stage. Then the whole procedure is repeated using the improved orbitals, and a second improved set of orbitals is obtained. The cycling continues until the orbitals and energies obtained are insignificantly different from those used at the start of the latest cycle. The solutions are then *self-consistent* and accepted as solutions of the problem.

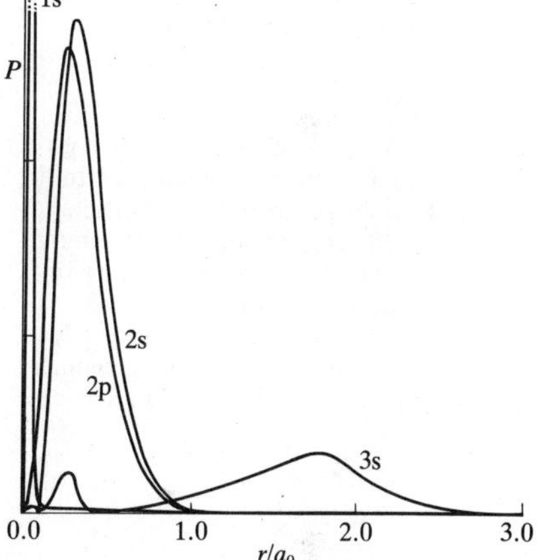

Fig. 15.16. The radial distribution functions for the orbitals of Na based on SCF calculations. Note the shell-like structure, with the 3s-orbital outside the inner K and L shells.

Some of the self-consistent field (SCF) Hartree–Fock (HF) atomic orbitals (AO) for sodium are illustrated in Fig. 15.16. They show the grouping of electron density into shells, as was anticipated by the early chemists, and the differences of penetration as discussed above. These SCF calculations therefore support (but also considerably extend, by providing detailed wavefunctions and precise energies) the qualitative discussions that are used to explain chemical periodicity.

15.3 The spectra of complex atoms

The spectra of atoms rapidly become very complicated as the number of electrons increases, but there are some important and moderately simple features. The general idea is very straightforward: lines in the spectrum (in either emission or absorption) occur when the atom undergoes a change of state involving a change of energy ΔE, and emits or absorbs a photon of frequency $\nu=\Delta E/h$ (or wavenumber $\tilde{\nu}=\Delta E/hc$). It follows that an interpretation of the spectrum should give information about the energies of electrons in atoms. That, however, is only a primitive approach, and several details help (and sometimes hinder) the analysis.

15.3 (a) Spin–orbit interaction

An electron possesses spin, and on account of this spin it possesses a magnetic moment, Fig. 15.17. An electron with orbital angular momentum is in effect a circulating current, and so there is also a magnetic moment arising from the orbital angular momentum. These two magnetic moments interact, and the energy of this *spin–orbit interaction* depends on their relative orientations, and therefore on the relative orientations of the spin and orbital angular momenta.

The strength of the spin–orbit interaction, which is a *magnetic interaction*, depends on the nuclear charge. This can be understood by imagining an observer riding on the orbiting electron. The observer sees a nucleus orbiting around, like the Sun rising and setting, and is therefore at the centre of a ring of current. The greater the nuclear charge the greater this current, and therefore the greater the magnetic field the observer detects. The spin magnetic moment interacts with this orbital magnetic field, and so the greater the nuclear charge, the stronger the spin–orbit interaction. The interaction energy increases sharply with atomic number (as Z^4), and while only small in hydrogen (giving rise to shifts of energy levels of no more than about $0.4\,\text{cm}^{-1}$) in heavy atoms like lead it is very large (giving shifts of the order of thousands of cm^{-1}).

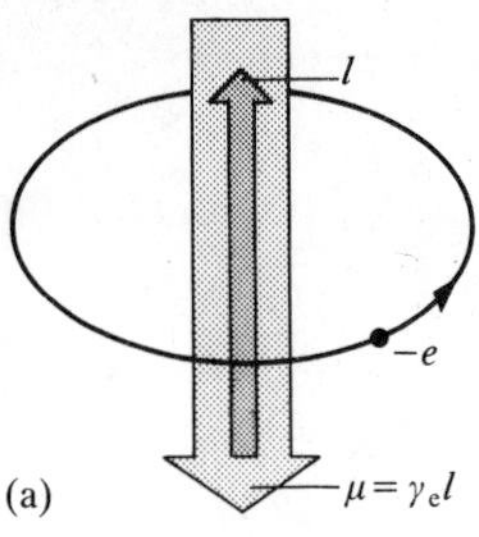

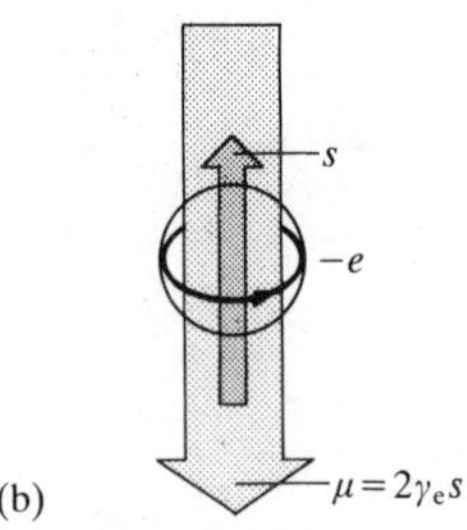

Fig. 15.17. Angular momentum gives rise to a magnetic moment (μ). In the case of an electron, (a) the magnetic moment is antiparallel to the orbital angular momentum, but proportional to it. In the case of spin angular momentum (b), there is a factor 2, which increases the magnetic moment. This anomaly is discussed in Section 15.3(c).

The effect of spin–orbit coupling on the states of the atom and the appearance of the spectrum can be illustrated by considering the alkali metals, which consist of a single electron outside a closed core. As a good first approximation we can ignore the core electrons (which have no net orbital angular momentum) and concentrate on the outermost, or *valence*, electron.

In the ground state the valence electron is an s-electron: it has zero orbital angular momentum, and so there is no spin–orbit interaction. If the electron is excited into a p-orbital, it acquires orbital angular momentum (from the incoming photon), and is in a state with $l=1$. In Chapter 14 we saw that an electron spin may take only two orientations with respect to any chosen axis, and so in this case the spin may be aligned parallel, Fig. 15.18(a), or antiparallel, Fig. 15.18(b), to the orbital momentum. The former gives the greater total angular momentum, because the two momenta are in the same direction.

As in the case of combining the angular momenta of two spins, Section 15.2, the total angular momentum of a spinning, orbiting electron is quantized. It is described by the quantum numbers j and m_j, and, as in that case too, the permitted values of j are given by the Clebsch–Gordan series, which now reads

$$j=l+s,l+s-1,\ldots|l-s|. \tag{15.3.1}$$

Example 15.4

Calculate the values of the total angular momentum quantum number that may arise from (a) a d-electron with spin, (b) an s-electron with spin.

- *Method*. Use the Clebsch–Gordan series, eqn (15.3.1), with $s=\frac{1}{2}$ and (a) $l=2$, (b) $l=0$. Decide on the minimum value $|l-s|$ first.

- *Answer*. (a) $|l-s|=|2-\frac{1}{2}|=\frac{3}{2}$; hence $j=2+\frac{1}{2},2+\frac{1}{2}-1\ldots\frac{3}{2}$; or $\frac{5}{2},\frac{3}{2}$; Fig. 15.19.
(b) $|l-s|=|0-\frac{1}{2}|=\frac{1}{2}$; hence $j=0+\frac{1}{2},\ldots,\frac{1}{2}$, or simply $\frac{1}{2}$.

• *Comment*. Note that j is always a *positive* number. There is only one value of j when $l = 0$ because there is only spin momentum present and so no question arises as to the various relative alignments of orbital and spin momenta.

• *Exercise*. Calculate the values of j for a p-electron and an f-electron. [p: $\frac{3}{2}, \frac{1}{2}$; f: $\frac{7}{2}, \frac{5}{2}$]

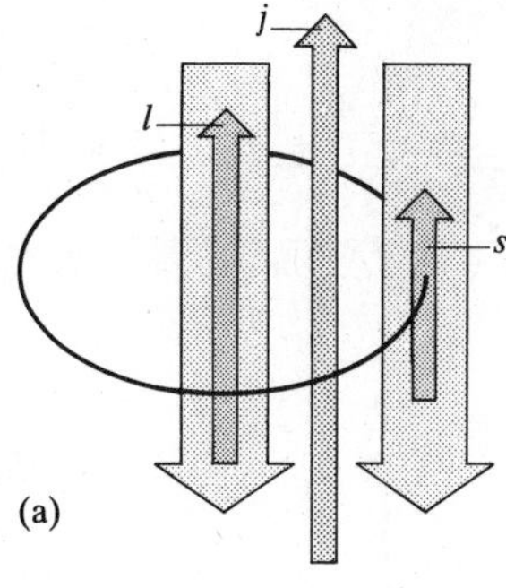

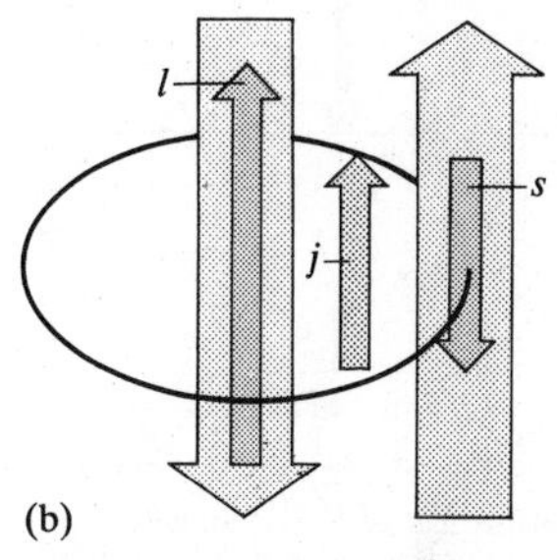

Fig. 15.18. The spin–orbit coupling is a magnetic interaction between spin and orbital magnetic moments. When the angular momenta are parallel, as in (a), the magnetic moments are aligned unfavourably; when they are opposed, as in (b), the interaction is favourable. This is the cause of the splitting of a configuration into *levels*.

In the case of a single p-electron, $l = 1$, $s = \frac{1}{2}$, and so the permitted values of j are $\frac{3}{2}$ and $\frac{1}{2}$. These specify the *levels* of a configuration. The former corresponds to the state in which the angular momenta are parallel, and the latter to the case when they are opposed. In the $j = \frac{3}{2}$ level the spin magnetic moment is in a higher energy orientation with respect to the orbital field than for $j = \frac{1}{2}$. Therefore, the $\ldots np^1$ configuration splits into two energy levels, Fig. 15.20. When the excited atom undergoes a transition and the np-electron falls into a lower s-orbital, *two* spectral lines are observed, depending on which of the two levels of the $\ldots np^1$ configuration was occupied initially. This splitting is called the *fine structure* of the spectrum. One place where the fine structure can be clearly seen is in the emission spectrum from sodium vapour excited by an electric discharge (for example, in one kind of street lighting). The yellow line at 589 nm (17 000 cm^{-1}) is seen, on careful analysis, to be composed of two closely spaced lines, one at 589.76 nm (16 956 cm^{-1}) and other at 589.16 nm (16 973 cm^{-1}). The transitions, Fig. 15.20, are from the $j = \frac{3}{2}$ and $\frac{1}{2}$ levels of the $1s^22s^22p^63p^1$ configuration to the ground configuration $1s^22s^22p^63s^1$. Therefore, in sodium, the spin–orbit interaction affects the energies by about 17 cm^{-1}.

The magnitude of the spin–orbit interaction is normally expressed in terms of the *spin–orbit coupling constant* (symbol: λ), and a quantum-mechanical calculation leads to the result that the energies of the levels with quantum numbers s, l, and j are given by

$$E_{l,s,j} = \tfrac{1}{2}hc\lambda\{j(j+1) - l(l+1) - s(s+1)\}. \qquad (15.3.2)$$

This expression makes it easy to relate observed fine structure splittings to the spin–orbit coupling constant, as the following *Example* shows.

Example 15.5

The sodium D-lines are analysed in Fig. 15.20. They lie at 16 956.2 cm^{-1} and 16 973.4 cm^{-1}. Calculate the spin–orbit coupling constant for the upper state of the sodium atom.

• *Method*. The splitting of the lines is equal to the splitting of the $j = \frac{3}{2}$ and $\frac{1}{2}$ levels of the excited configuration. Express this in terms of λ using eqn (15.3.2). Since there is a single p-electron, use $l = 1$ and $s = \frac{1}{2}$ (but the l and s contributions cancel when the energy difference is calculated).

• *Answer*. The two levels are split by

$$\{E_{1,\frac{1}{2},\frac{3}{2}} - E_{1,\frac{1}{2},\frac{1}{2}}\}/hc = \tfrac{1}{2}\lambda\{\tfrac{3}{2}(\tfrac{3}{2}+1) - \tfrac{1}{2}(\tfrac{1}{2}+1)\} = \tfrac{3}{2}\lambda.$$

The experimental value is

$$\{E_{1,\frac{1}{2},\frac{3}{2}} - E_{1,\frac{1}{2},\frac{1}{2}}\}/hc = 17.2\ \text{cm}^{-1}.$$

Hence, $\lambda = \frac{2}{3} \times 17.2\ \text{cm}^{-1} = 11.4\ \text{cm}^{-1}$.

• *Comment*. The same calculation repeated for the other alkali metal atoms gives Li: 0.23 cm^{-1}, K: 38.5 cm^{-1}, Rb: 158 cm^{-1}, Cs: 370 cm^{-1}.

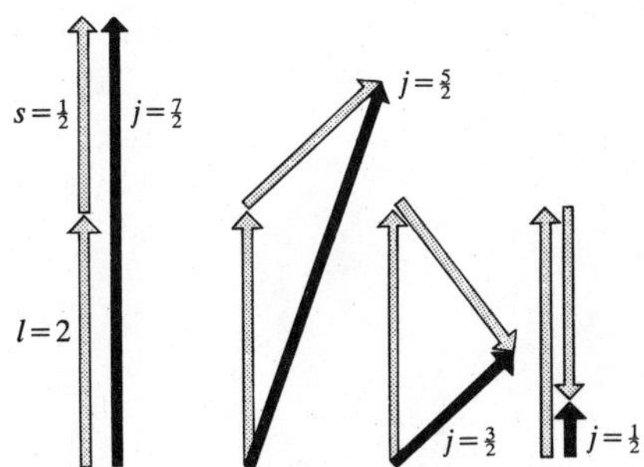

Fig. 15.19. The Clebsch–Gordan series for the coupling of the spin and orbital angular momenta of a d-electron ($l = 2$). Two values of j are permitted.

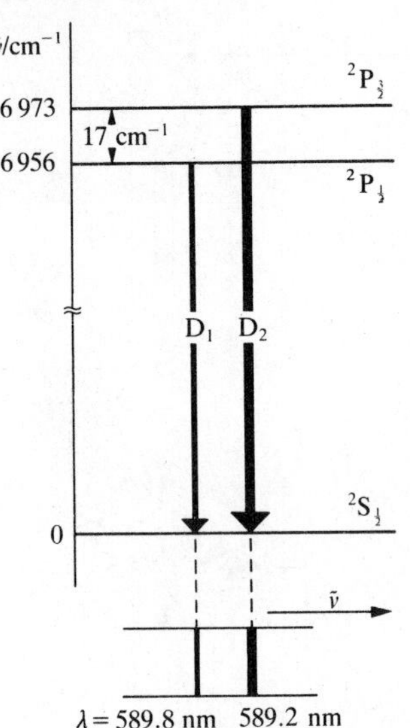

Fig. 15.20. The energy-level diagram for the formation of the D-lines of sodium. The splitting of the spectral lines (by 17 cm^{-1}) reflects the splitting of the levels.

• *Exercise*. The configuration . . . $4p^6 5d^1$ of rubidium has two levels at 25 700.56 cm^{-1} and 25 703.52 cm^{-1} above the ground configuration. What is the spin–orbit coupling constant in this excited state? [1.18 cm^{-1}]

15.3 (b) Term symbols and selection rules

We have used expressions such as the $j = \frac{3}{2}$ level of the 'configuration'. A *term symbol*, which is a symbol looking like $^2P_{\frac{3}{2}}$ or $^3D_2$, conveys this information much more succinctly. It gives three pieces of information:

(1) The *letter* (e.g. P or D in the examples) indicates the *total orbital angular momentum*. The total orbital angular momentum (symbol: L) is obtained by coupling the individual orbital angular momenta using the Clebsch–Gordan series. In the case of two electrons the possible L values are

$$L = l_1 + l_2, l_1 + l_2 - 1, \ldots, |l_1 - l_2|, \qquad (15.3.3)$$

so that for two p-electrons ($l_1 = l_2 = 1$) $L = 2, 1, 0$. The code for converting the value of L into a letter is the same as for the s, p, d, f, . . . designation of orbitals:

L:	0	1	2	3	4	. . .
	S	P	D	F	G	. . .

This means that a p^2 configuration can give rise to D, P, and S terms; they differ in energy on account of the different electrostatic interactions between the electrons. A closed shell has zero orbital angular momentum, and so in working out term symbols, only the electrons of the unfilled shell need be considered. In the case of a single electron outside a closed shell, the value of L is the same as the value of l, and so the configuration $1s^2 2s^2 2p^6 3s^1$ (or, more briefly KL$3s^1$, where KL denotes the complete K- and L-shells) can give only an S term.

Example 15.6

Find the values of the total orbital angular momentum quantum number that can arise from the configurations (a) d^2, (b) p^3. Write the term letters in each case.

• *Method*. Use the Clebsch–Gordan series, eqn (15.3.3), in each case; begin by finding the minimum value of L. In (b), couple two electrons, and then couple the third to each combined state.

• *Answer*. (a) Minimum value: $|l_1 - l_2| = |2 - 2| = 0$. Therefore, $L = 2 + 2, 2 + 2 - 1, \ldots, 0 = 4, 3, 2, 1, 0$ (corresponding to G, F, D, P, S terms respectively).

(b) First coupling: Minimum value: $|1 - 1| = 0$. Therefore,

$$L' = 1 + 1, 1 + 1 - 1, \ldots, 0 = 2, 1, 0.$$

Now couple l_3 with 2, to give $L = 3, 2, 1$; with 1, to give $L = 2, 1, 0$; and with 0, to give 1. The overall result is therefore

$$L = 3, 2, 2, 1, 1, 1, 0 \quad (\text{F, 2D, 3P, S}).$$

• *Comment*. In the case of several electrons, the overall coupling can be broken down into a chain of pairwise couplings.

• *Exercise*. Repeat the question for the configurations (a) f^1d^1 and (b) d^3.
[(a) H, G, F, D, P; (b) I, 2H, 3G, 4F, 5D, 3P, 2S]

(2) The *left superscript* in the term symbol (e.g. 2 in $^2P_{\frac{3}{2}}$) gives the *multiplicity* of the term. The multiplicity is the value of $2S+1$, where S is the *total spin* of the electrons. When $S=0$ (as for a closed shell) the electrons are all paired and there is no net spin: this gives a *singlet* term, such as ^1S. (Take care not to confuse the S of the spin and the S of the term symbol.) A single electron has $S=s=\frac{1}{2}$, and so a configuration such as $KL3s^1$ can give rise to a *doublet* term, ^2S. The configuration $KL3p^1$ likewise is a doublet, ^2P. When there are two unpaired electrons, $S=1$ and so $2S+1=3$, giving a *triplet* term, such as ^3D. The relative orientations of the spins in a triplet term were discussed in Section 15.2, see Fig. 15.14, and their energies differ on account of the different electrostatic interactions arising from the quantum-mechanical effect of spin-correlation, Fig. 15.21.

(3) The right subscript on the term symbol (e.g. the $\frac{3}{2}$ in $^2P_{\frac{3}{2}}$) is the value of the *total angular momentum quantum number* (symbol: J). It therefore also indicates the relative orientation of the total spin and orbital angular momenta (J is large when they are parallel, small otherwise). If there is a single electron outside a closed shell, then $J=j$, with j one of the values given by the Clebsch–Gordan series in eqn (15.3.1). The $KL3s^1$ configuration has $j=\frac{1}{2}$ (because $l=0$ and $s=\frac{1}{2}$), and so the ^2S term has a single *level*, $^2S_{\frac{1}{2}}$. The $KL3p^1$ configuration has $l=1, s=\frac{1}{2}$, and therefore $j=\frac{3}{2}$ and $\frac{1}{2}$; the ^2P term therefore has two levels, $^2P_{\frac{3}{2}}$ and $^2P_{\frac{1}{2}}$, and these lie at different energies on account of the magnetic spin–orbit interaction, Fig. 15.21.

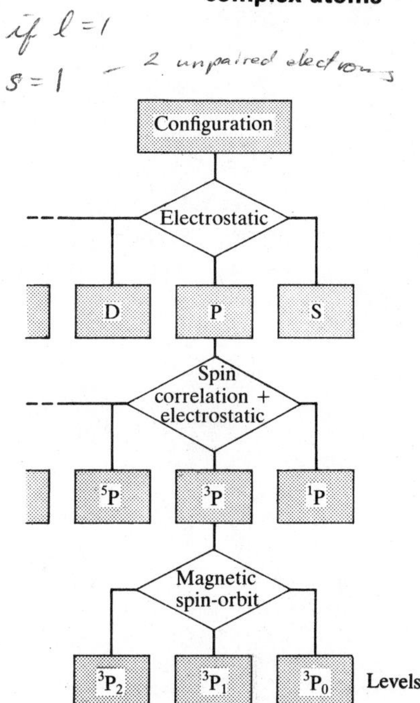

Fig. 15.21. The types of interaction that are responsible for the various kinds of splitting of energy levels in atoms. For light atoms, magnetic interactions are small, but in heavy atoms may dominate the electrostatic interactions.

If there are several electrons outside a closed shell, things are more complicated, but when the spin–orbit interaction is weak (for atoms of low atomic number) the *Russell–Saunders coupling scheme* can be used. This is based on the view that if the spin–orbit coupling is weak, then it is effective only when all the orbital momenta are operating cooperatively. The orbital angular momenta of all the electrons couple to give some total L (as in eqn (15.3.3)), all the spins are similarly coupled to give some total S, and only then do these momenta couple through the spin–orbit interaction to give a total J, the permitted values being given by the Clebsch–Gordan series

$$J=L+S, L+S-1, \ldots, |L-S|. \tag{15.3.4}$$

For example, in the case of the ^3D term of the configuration $KL2p^13p^1$, the permitted values of J are 3, 2, 1 (because ^3D has $L=2$ and $S=1$), and so the term has three levels, $^3D_3$, $^3D_2$, and $^3D_1$.

Note that *when $L \geqslant S$, the multiplicity gives the number of levels*, as in ^2P having the two levels $^2P_{\frac{3}{2}}$ and $^2P_{\frac{1}{2}}$ and ^3D the three levels $^3D_3$, $^3D_2$, and $^3D_1$. However, this is not the case when $L<S$: ^2S, for example, has only the single level $^2S_{\frac{1}{2}}$.

Example 15.7

Write the term symbols for the ground configurations of Na and F, and the excited configuration $1s^22s^22p^13p^1$ of C.

• *Method*. Write the configurations; ignore inner closed shells. Couple the orbital momenta to find L and the spins to find S. Couple L and S to find J. Express the term as $^{2S+1}\{L\}_J$, where $\{L\}$ is the appropriate letter. For F, treat the single gap in $2p^6$ as a single particle.

• *Answer*. For Na the configuration is $KL3s^1$, and we consider the single 3s-electron. Since $L=l=0$ and $S=s=\frac{1}{2}$, it is possible for $J=j=s=\frac{1}{2}$ only. Hence the terms symbol is $^2S_{\frac{1}{2}}$.

For F the configuration is $K2s^22p^5$, which we can treat as $KL2p^{-1}$. Hence $L = l = 1$, and $S = s = \frac{1}{2}$. Two values of $J = j$ are allowed: $J = \frac{3}{2}, \frac{1}{2}$. Hence the term symbols for the two levels are ${}^2P_{\frac{3}{2}}, {}^2P_{\frac{1}{2}}$.

For C the configuration is effectively $2p^13p^1$. This is a two-electron problem, and $l_1 = l_2 = 1$, $s_1 = s_2 = \frac{1}{2}$. It follows that $L = 2, 1, 0$ and $S = 1, 0$. The terms are therefore ${}^3D$ and ${}^1D$, ${}^3P$ and ${}^1P$, and ${}^3S$ and ${}^1S$. For ${}^3D$, $L = 2$ and $S = 1$; hence $J = 3, 2, 1$ and the levels are ${}^3D_3$, ${}^3D_2$, and ${}^3D_1$. For ${}^1D$, $L = 2$ and $S = 0$, so that the single level is ${}^1D_2$. The triplet of levels of ${}^3P$ is ${}^3P_2$, ${}^3P_1$, and ${}^3P_0$, and the singlet is ${}^1P_1$. For ${}^3S$ there is only a single level, ${}^3S_1$ (because $J = 1$ only), and the singlet term is ${}^1S_0$.

• *Comment*. The reason why we have treated an excited configuration of carbon is that in the ground configuration, $2p^2$, the Pauli principle forbids some terms, and deciding which survive (${}^1D$, ${}^3P$, ${}^1S$) is quite complicated.

• *Exercise*. Write down the terms arising from the configurations (a) $2s^12p^1$, (b) $2p^13d^1$.
[(a) ${}^3P_2$, ${}^3P_1$, ${}^3P_0$, ${}^1P_1$; (b) ${}^3F_4$, ${}^3F_3$, ${}^3F_2$, ${}^1F_3$, ${}^3D_3$, ${}^3D_2$, ${}^3D_1$, ${}^1D_2$, ${}^3P_2$, ${}^3P_1$, ${}^3P_0$, ${}^1P_1$]

Any state of the atom, and any spectral transition, can be specified in using term symbols. For example, the transitions giving rise to the yellow sodium doublet, Fig. 15.20, are $KL3p^1\ {}^2P_{\frac{3}{2}} \rightarrow KL3s^1\ {}^2S_{\frac{1}{2}}$ and ${}^2P_{\frac{1}{2}} \rightarrow {}^2S_{\frac{1}{2}}$. Note that the configuration is not always specified, and that *the upper term precedes the lower*. The corresponding absorptions would therefore be denoted ${}^2P_{\frac{3}{2}} \leftarrow {}^2S_{\frac{1}{2}}$ and ${}^2P_{\frac{1}{2}} \leftarrow {}^2S_{\frac{1}{2}}$.

The selection rules can be expressed in terms of the term symbols because they carry information about angular momentum, and we have seen that selection rules arise from the conservation of angular momentum and the unit spin of the photon. A detailed analysis leads to the following rules:

Selection rules for atomic spectra:
$\Delta S = 0$ (because the light does not affect the spin directly).
$\Delta L = 0, \pm 1$ with $\Delta l = \pm 1$ (the orbital angular momentum of an individual electron must change, but whether or not this results in an overall change of orbital momentum depends on the coupling).
$\Delta J = \pm 1, 0$ but $J = 0$ cannot combine with $J = 0$.

Russell–Saunders coupling fails when the spin–orbit interaction is large, and then individual spin and orbital momenta are coupled first into individual j values; then these momenta are combined into a grand total J. This is called *jj-coupling*. If we insist on labelling the terms with symbols like ${}^3D$, then we shall find that the selection rules progressively fail as the atomic number increases because the quantum numbers S and L are ill-defined. For this reason, transitions between singlet and triplet states ($S = 0 \leftrightarrow S = 1$), while forbidden in light atoms, are allowed in heavy atoms.

15.3 (c) The effect of magnetic fields

Since orbital and spin angular momenta give rise to magnetic moments, it can be expected that the application of a magnetic field should modify an atom's spectrum. This is called the *Zeeman effect*.

The component of orbital angular momentum on the z-axis (the direction of the applied field) is $m_l\hbar$. The magnetic moment is proportional to the angular momentum, and so the component on the z-axis may be written $\gamma_e m_l \hbar$, where γ_e is a constant called the *magnetogyric ratio* of the electron. Treating the magnetic moment as arising from the circulation of an electron of charge $-e$ gives $\gamma_e = -e/2m_e$, the negative sign (reflecting the sign of the charge) indicating that the magnetic moment is antiparallel to the orbital angular momentum. Denoting the z-component of magnetic moment as μ_z, the possible values are

$$\mu_z = \gamma_e m_l \hbar = -(e\hbar/2m_e)m_l = -\mu_B m_l. \quad (15.3.5)$$

The positive quantity $\mu_B = e\hbar/2m_e$ is called the *Bohr magneton*, and is often regarded as the fundamental quantum of magnetic moment.

If the magnetic field applied parallel to the z-axis is B, the energy of the magnetic moment is $-\mu_z B$ (this is a result from standard magnetic theory; B is actually the *magnetic induction*, and is measured in tesla, T, but gauss, G, are still widely used: $1\,\mathrm{T} = 10^4\,\mathrm{G}$). Therefore, an electron in a state with quantum number m_l has an additional contribution to its energy given by

$$E = -\mu_z B = \mu_B m_l B. \quad (15.3.6)$$

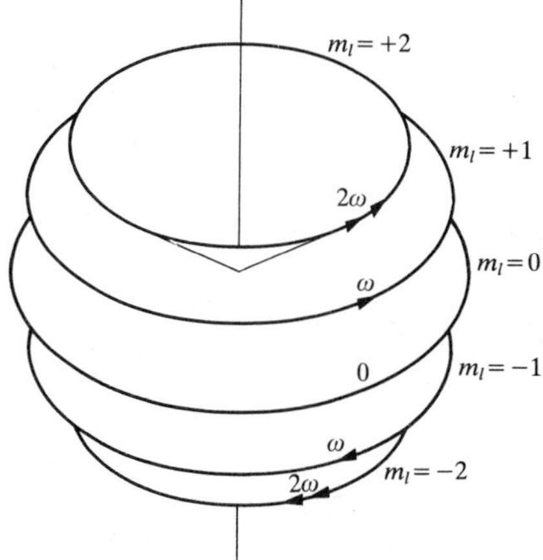

Fig. 15.22. The different energies of the m_l states in a magnetic field are represented by different rates of *precession* of the vectors representing the angular momentum.

The same expression, but with m_l replaced by M_L, applies when the orbital magnetic moment arises from several electrons.

A p-electron has $l = 1$ and $m_l = 0, \pm 1$. In the absence of a magnetic field these three states are degenerate, but when a field is applied the degeneracy is removed: the state $m_l = +1$ moves up in energy by an amount $\mu_B B$, the state $m_l = 0$ is unchanged, and the state $m_l = -1$ moves down by an amount $\mu_B B$. These different energies are sometimes represented on the vector model by picturing the vectors as sweeping round their cones, as shown in Fig. 15.22, and the higher the energy the greater the rate of the motion, which is called *precession*.

The *normal Zeeman effect* is the observation of three lines in the spectrum where, in the absence of the field, there was only one, Fig. 15.23. The splitting is in fact extremely small. Since $\mu_B = 9.273 \times 10^{-24}\,\mathrm{J\,T^{-1}}$, which corresponds to $0.467\,\mathrm{cm^{-1}\,T^{-1}}$ when expressed as a wavenumber, a field of 2 T (20 kG) is needed to produce a splitting of about $1\,\mathrm{cm^{-1}}$, which is very small in comparison with typical optical transition wavenumbers of $20\,000\,\mathrm{cm^{-1}}$ and more. The three lines observed in the normal Zeeman effect have different polarizations: the lines with $\Delta m_l = \pm 1$ (the *σ-lines*) correspond to changes of angular momentum in opposite directions, and the light emitted is circularly polarized in opposite directions, Fig. 15.23. The light from the $\Delta m_l = 0$ transition (the *π-line*) is linearly polarized, and observable only when the sample is viewed perpendicular to the field, Fig. 15.23.

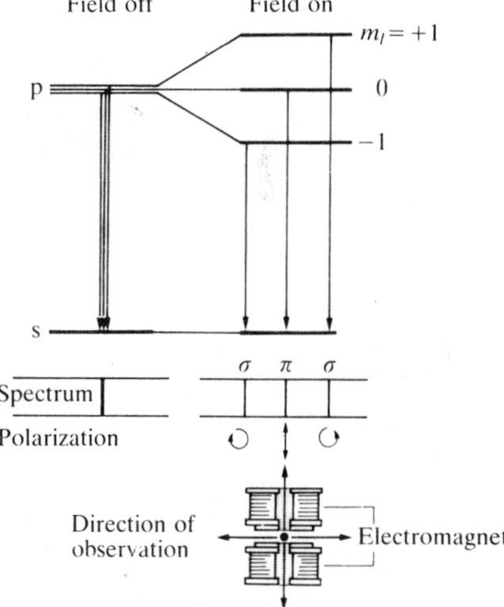

Fig. 15.23. The normal Zeeman effect. On the left, when the field is off, a single spectral line is observed. When the field is on, the line splits into three, with different polarizations. The circularly-polarized lines are called the σ-lines; the plane-polarized lines are called π-lines. Which line is observed depends on the orientation of the observer.

Much more common than the normal Zeeman effect is the *anomalous Zeeman effect* in which the original line splits into more than three components. The origin of the complexity is the spin of the electrons and the fact that its magnetic moment is not given by $\gamma_e m_s \hbar$ but by twice this value. The extra factor, the *g-factor* of the electron (symbol: g_e), is derived from the *Dirac equation*, the relativistic version of the Schrödinger equation. It is not exactly 2, but 2.0023 (but we shall often use $g_e = 2$ for simplicity). The magnetic moment is therefore $g_e \gamma_e m_s \hbar$, and the energy of an electron in a state m_s in a magnetic field B is

$$E = -g_e \gamma_e m_s \hbar B = g_e \mu_B m_s B. \quad (15.3.7)$$

The same expression, but with m_s replaced by M_S, applies to the magnetic moment arising from the spin of several electrons.

The anomalous value of the spin magnetic moment accounts for the anomalous Zeeman effect because the spin and orbital magnetic moments interact with the field in a more complicated way than their angular momenta would lead us to expect. Hence the pattern is more complex. When the field is very strong, the orbital and spin magnetic moments are uncoupled and align separately with respect to its direction. Only the orbital angular momentum is affected when a transition occurs, and the anomalous splitting reverts to the normal Zeeman effect with the anomalous magnetic moment of the spin playing no role. This reversion is called the *Paschen–Back effect*.

Appendix 15.1 Centre of mass coordinates

We want to show that the hydrogen atom Schrödinger equation, eqn (15.1.4b), can be expressed in terms of the separation of the particles, $\boldsymbol{r}$, and the location of the centre of mass, $\boldsymbol{R}$. We shall do the transformation of the derivatives with respect to x-coordinate: the others follow in the same way.

The location of the centre of mass is at

$$X=(m_e/m)x_e+(m_N/m)x_N,$$

with m the total mass. The separation of the particles is

$$x=x_e-x_N.$$

Since we can now write

$$\partial/\partial x_e=(\partial X/\partial x_e)(\partial/\partial X)+(\partial x/\partial x_e)(\partial/\partial x)=(m_e/m)(\partial/\partial X)+(\partial/\partial x),$$
$$\partial/\partial x_N=(m_N/m)(\partial/\partial X)-(\partial/\partial x),$$

the x-part of the sum of the two laplacians becomes

$$(1/m_e)(\partial^2/\partial x_e^2)+(1/m_N)(\partial^2/\partial x_N^2)$$
$$=(1/m)(\partial^2/\partial X^2)+\{(1/m_e)+(1/m_N)\}(\partial^2/\partial x^2).$$

The y- and z-components can be dealt with similarly, and so

$$(1/m_e)\nabla_e^2+(1/m_N)\nabla_N^2=(1/m)\nabla_{\text{c.m.}}^2+(1/\mu)\nabla^2,$$

with $1/\mu=1/m_e+1/m_N$, as required in the text.

Now write the wavefunction as the product $\psi(\boldsymbol{R})\psi(\boldsymbol{r})$, and carry through the separation of variables procedure. The total equation separates into an equation for the free motion of a particle of mass m and eqn (15.1.7).

Further reading

Structure and spectra of atoms. W. G. Richards and P. R. Scott; Wiley, London, 1976.

Atomic spectra and atomic structure. G. Herzberg; Dover, New York, 1944.

Molecular quantum mechanics (2nd edn). P. W. Atkins; Oxford University Press, 1983.

Introduction to quantum mechanics. L. Pauling and E. B. Wilson; McGraw-Hill, New York, 1935.

Atoms and molecules. M. Karplus and R. N. Porter; Benjamin, New York, 1970.

Atomic spectra and the vector model. C. Candler; Hilger and Watts, London, 1964.
Atomic spectra (2nd edn). H. G. Kuhn; Longman, London, 1969.
Atomic structure. E. U. Condon and H. Odabasi; Cambridge University Press, 1980.

Data:
Atomic energy levels. C. E. Moore; NBS-Circ. 467, Washington, 1949, 1952, and 1958.
Tables of spectral lines of neutral and ionized atoms. I. R. Striganov and N. S. Sventitskii; Plenum, New York, 1968.
Atomic energy levels and Grotrian diagrams. S. Bashkin and J. O. Stonor Jr.; North-Holland, Amsterdam, 1975 *et seq*.

Introductory problems

A15.1. Calculate the wavelength of the Balmer line for H for which $n_2 = 4$.

A15.2. The frequency of one of the lines in the Paschen series for H is 2.7415×10^{14} Hz. Calculate the quantum number n_2 for the transition which produces this line.

A15.3. One of the terms in the H atom has a value 27 414 cm^{-1}. What is the value of the term with which it combines to produce light of wavelength 486.1 nm?

A15.4. By differentiation, show that the radial part of the 2s hydrogen atomic orbital has a minimum and determine the value of r for this minimum.

A15.5. At what values of r does the radial part of the 3s hydrogen orbital vanish?

A15.6. An electron is known to have only the following values of total angular momentum quantum number: $\frac{3}{2}$, 1, $\frac{1}{2}$. What is the orbital angular momentum of the electron?

A15.7. What information does the term symbol $^1D_2$ provide about the angular momentum of an atom?

A15.8. Consider an electron in the ground state in the H atom, and calculate the value of r for which the probability density is 50% of its maximum value.

A15.9. Consider an electron in the ground state in the H atom, and find the numerical values of r for which the radial distribution function is (a) 50%, (b) 75% of its maximum value.

A15.10. In the normal Zeeman effect, the energy of one state of the electron rises by 2.23×10^{-22} J when a magnetic field of 12.0 T is applied. What is the magnetic quantum number m_l for this electron?

Problems

15.1. The *Humphreys series* is another of the series in the spectrum of atomic hydrogen. It begins at 12 368 nm and has been traced to 3281.4 nm. What are the transitions involved? What are the wavelengths of the intermediate transitions?

15.2. A series of lines in the spectrum of atomic hydrogen lies at the wavelengths 656.46 nm, 486.27 nm, 434.17 nm, and 410.29 nm. What is the wavelength of the next line in the series? What energy is required to ionize the hydrogen atom when it is in the lower state involved in these transitions?

15.3. The doubly ionized lithium ion, Li^{2+}, has only one electron, and its spectrum is expected to resemble hydrogen's From the data below (which are taken from C. E. Moore, *Atomic energy levels*, Natl. Bur. Stds., Circ. 467 (1949), a rich source of information of this kind) show that the energy levels do have the form K/n^2, and find the value of the constant K. Data: Lyman series at 740 747 cm^{-1}, 877 924 cm^{-1}, 925 933 cm^{-1}, ...

15.4. There is enough information in the last Problem for you to predict the wavenumbers of the Balmer series of Li^{2+}. What are the wavenumbers of the two lowest energy transitions of the series?

15.5. Calculate the mass of the deuteron on the basis that the first line of the Lyman series lies at 82 259.098 cm^{-1} for H, and at 82 281.476 cm^{-1} for D.

15.6. Positronium consists of an electron and a positron (same mass, opposite charge) orbiting around their common centre of mass. The broad features of the spectrum are therefore expected to resemble those of hydrogen, the differences arising largely from the mass relations. Where will the first three lines of the Balmer series of positronium lie? What is the binding energy in the ground state?

15.7. One of the most famous of the obsolete theories of the hydrogen atom was proposed by Bohr. It has been displaced by quantum mechanics, but by a remarkable coincidence (not the only one where the Coulomb potential is concerned) the energies it predicts agree exactly with those obtained from the Schrödinger equation. The *Bohr atom* is imagined as an electron circulating about a central nucleus. The Coulombic force of attraction, $Ze^2/4\pi\varepsilon_0 r^2$, is balanced by the centrifugal effect of the circular orbiting motion of the electron. Bohr proposed that the angular momentum was limited to some integral multiple of $\hbar = h/2\pi$. When the two forces are balanced, the atom remains in a 'stationary state' until it makes a spectral transition. Find the energies of the hydrogen-like atom on the basis of this model.

15.8. What features of the Bohr model are untenable in the light of quantum mechanics? How does the Bohr ground state differ from the actual ground state? If numerical agreement is exact, is there no experimental way of eliminating the Bohr model in favour of the quantum mechanical model?

15.9. A hydrogen-like 1s-orbital in an atom of atomic number Z is the exponential function $\psi_{1s} = (Z^3/\pi a_0^3)^{\frac{1}{2}} \exp(-Zr/a_0)$. Form the radial distribution function and derive an expression for the most probable distance of the electron from the nucleus. What is its value in the case of (a) helium, (b) fluorine?

15.10. In 1976 it was mistakenly believed that the first of the 'superheavy' atoms had been discovered in some mica: the atomic number was believed to be 126. What is the most probable distance of the innermost electrons from the nucleus in an atom of this element? (In such elements the Coulombic forces are so strong that relativistic corrections are very significant: ignore them here.)

15.11. What is the magnitude of the angular momentum of an electron that occupies the following orbitals: (a) 1s, (b) 3s, (c) 3d, (d) 2p, (e) 3p? Give the number of radial and angular nodes in each case, and infer a rule.

15.12. Is an electron on average further away from the nucleus when it occupies a 2p-orbital in hydrogen than when it occupies a 2s-orbital? What is the most probable distance of an electron in a 3s-orbital from the nucleus? Orbitals are given in Box 15.1.

15.13. Take the exponential 1s-orbital for the ground state of hydrogen and confirm that it satisfies the Schrödinger equation for the atom and that its energy is $-R_H$. Now modify the nuclear charge from e to Ze. What is the binding energy of the electron in the ion F^{8+}?

15.14. What is the most probable *point* (not radius) at which an electron will be found if it occupies a $2p_z$-orbital on hydrogen?

15.15. What is the *orbital degeneracy* of the level of the hydrogen atom that has the energy (a) $-R_H$, (b) $-R_H/9$, (c) $-R_H/25$?

15.16. Which of the following transitions are allowed in the normal electronic spectrum of an atom: (a) 2s → 1s, (b) 2p → 1s, (c) 3d → 2p, (d) 5d → 3s, (e) 5p → 3s?

15.17. How many electrons can enter the following sets of atomic orbitals: (a) 1s, (b) 3p, (c) 3d, (d) 6g?

15.18. Write the configurations of the first 18 elements of the periodic table ($Z = 1$ to 18).

15.19. Ionization energies I may be determined in a variety of ways. Spectroscopy may be used by looking for the energy of excitation at which the line structure is replaced by a continuum. In favourable cases, such as the alkali metals and some ions, I can be determined by extrapolating the gradually converging series of lines. Use the data in Problem 15.3 to find the ionization energy of Li^{2+}.

15.20. A series of lines in the spectrum of neutral lithium (the spectrum known as Li(I)) arise from combinations of the state $1s^2 2p\ ^2P$ with $1s^2 nd\ ^2D$, and occur at 610.36 nm, 460.29 nm, and 413.23 nm. The d-orbitals involved are hydrogen-like. It is known that the $1s^2 2p\ ^2P$ term lies at 670.78 nm above the ground term $1s^2 2s\ ^2S$. What is the ionization energy of the neutral atom in its ground state?

15.21. An alternative method for measuring the ionization energy is to expose the atom to high energy monochromatic radiation, and to measure the kinetic energy or the speed of the electrons it ejects (recall Problem 13.13). When 58.4 nm light from a helium discharge lamp is directed into a sample of krypton, electrons are ejected with a velocity of $1.59 \times 10^6\ \mathrm{m\,s^{-1}}$. The same radiation releases electrons from rubidium vapour with a speed of $2.45 \times 10^6\ \mathrm{m\,s^{-1}}$. What are the ionization energies of the two species?

15.22. By how much does the ionization energy of deuterium differ from that of ordinary hydrogen atoms?

15.23. What values of the total angular momentum quantum number j, and magnitude of the total angular momentum, may a single electron with $l = 3$ possess?

15.24. Suppose that an s-electron is part of a molecule which is rotating with an angular momentum corresponding to a quantum number $J_{mol} = 20$. What are the permitted angular momenta of the entire system?

15.25. An object rotates in space with an angular momentum corresponding to the quantum number j_1. Inside the rotating object there is another object rotating with an angular momentum corresponding to the quantum number j_2. The objects may be macroscopic, in which case j_1 and j_2 could be extremely large (as for a tornado on earth) but we shall confine attention to microscopic objects and small momenta. Suppose (a) $j_1 = 5$ and $j_2 = 3$, (b) $j_1 = 3$, $j_2 = 5$: what values may the total angular momentum quantum number take, and what values are permitted for the total angular momentum of the composite system?

15.26. The characteristic emission from potassium atoms when heated is purple and lies at 770 nm. On close inspection the line is seen to be composed of two closely spaced components, one at 766.70 nm and the other at 770.11 nm. Account for this observation, and deduce what quantitative information you can.

15.27. Suppose an atom has (a) two, (b) three, (c) four electrons in different orbitals. What values of the total spin quantum number S may the atom possess? What would be the multiplicity in each case?

15.28. What values of J may arise in the following terms: ^1S, ^2P, ^3P, ^3D, ^2D, ^1D, ^4D? How many states (distinguished by different values of M_J) occur in each level?

15.29. What (electric dipole) transitions are allowed between the terms encountered in the last Problem?

15.30. Write the possible term symbols for the following atomic configurations: (a) Li $(1s^2)2s^1$, (b) Na $(1s^22s^22p^6)3p^1$, (c) Sc $(1s^2\ldots)3d^1$, (d) Br $(1s^2\ldots)4p^5$. In each case the configuration in brackets denotes closed inner shells and subshells.

15.31. Calculate the magnetic induction B that is required in the Zeeman experiment in order to produce a splitting of $1.00\ \mathrm{cm}^{-1}$ between the states of a ^1P term.

16

Molecular structure

Learning objectives

After careful study of this chapter you should be able to:

(1) State and justify the *Born–Oppenheimer approximation*, Section 16.1(a).

(2) State what is meant by a *molecular orbital* and justify the *linear combination of atomic orbitals* approximation, Section 16.1(b).

(3) Write down expressions for *bonding* and *antibonding* molecular orbitals for the hydrogen molecule-ion, eqn (16.1.4), and analyse their electron density distributions, Section 16.1(c).

(4) Account for the shape of a *molecular potential energy curve*, Fig. 16.2.

(5) Define an *overlap integral*, eqn (16.2.1), explain the importance of the overlap of orbitals, and state what is meant by *orthogonality*, Section 16.2(c).

(6) Account for the importance of the *electron pair* in chemical bonding, Section 16.2(a).

(7) Construct *molecular orbital energy level diagrams*, Fig. 16.9, and use them to account for the electronic structures of diatomic molecules, Section 16.2(d).

(8) Explain the *g, u classification* of molecular orbitals, and construct *term symbols* for diatomic molecules, Section 16.2(e).

(9) State the *variation principle*, Section 16.2(g), and explain how it is used to obtain optimum molecular orbitals and energies.

(10) Explain the terms *secular equation*, eqn (16.2.9), and *secular determinant*, eqn (16.2.10), and obtain them for a trial function.

(11) Explain the term *hybridization*, and describe why it is a useful concept, Section 16.2(h).

(12) Relate hybridization to bond angle, eqn (16.3.4).

(13) State the *Hückel approximation* for conjugated double bonds, and obtain the molecular orbital energy level diagrams in the Hückel approximation, Section 16.4(a).

(14) Explain the origin of *aromatic stability*, Section 16.4(b), and calculate the *delocalization energy*, Example 16.5.

(15) Describe the *band theory* of the structure of metals, and account for electrical conductivity, Section 16.5(a).

(16) Outline the *valence-bond theory* of molecular structure, and write the valence-bond wavefunctions for simple molecules, Section 16.6(a).

(17) Explain the difference between the molecular orbital and valence-bond description of a bond, Section 16.6(a).

(18) Explain the significance of *ionic–covalent resonance* and of *resonance* between structures, Section 16.6(b).

Introduction

Now we turn to *valence theory*, the theory of the origin of the strengths, numbers, and three-dimensional arrangement of chemical bonds between atoms. The theory of valence has become so highly developed through the use of computers that it is now possible to account for, and to predict, the structures even of complex molecules.

In Chapter 15 we took the hydrogen atom as the primitive species, and based our discussion of complex atoms on what we learnt from it. Now we use the simplest molecule of all, the *hydrogen molecule-ion*, H_2^+, to discover the essential features of bonding, and then use it as a guide to more complex molecules.

16.1 The hydrogen molecule-ion

The hydrogen molecule-ion is a three-particle species: the two protons repel each other, but they attract the single electron, and the balance between the kinetic energy and these electrostatic interactions must account for its stability. The exact Schrödinger equation must consider all three particles simultaneously, but the nature of the molecule suggests a simplification, known as the *Born–Oppenheimer approximation*.

16.1 (a) The Born–Oppenheimer approximation

The Born–Oppenheimer approximation supposes that the nuclei, being so massive, move much more slowly than the electrons, and may be regarded as stationary. This being the case, we can choose the nuclei to have a definite separation (i.e. we can choose a definite *bond length*) and solve the Schrödinger equation for the electrons alone; then we can choose a different bond length, and repeat the calculation. In this way we can calculate how the energy of the molecule varies with bond length (and, in more complex molecules with angles too), and identify the equilibrium geometry of the molecule with the lowest point on this curve. This is far easier than trying to solve the complete Schrödinger equation by treating all three particles on an equal footing. In fact the approximation is quite good, for calculation suggests that the nuclei in H_2^+ move through only 1 pm while the electron speeds through about a thousand times that distance, and so the error of assuming that the nuclei are stationary is small. In molecules other than H_2^+ the nuclei are even heavier, and the approximation is generally (but not always) better.

The Born–Oppenheimer approximation reduces the full problem to a single-particle Schrödinger equation for an electron in the electrostatic field of two stationary protons at a separation R. The potential energy of the electron is

$$V = -(e^2/4\pi\varepsilon_0)\{(1/r_A) + (1/r_B)\}, \qquad (16.1.1)$$

where r_A and r_B are its distances from nuclei A and B. This expression is used in the one-particle Schrödinger equation

$$-(\hbar^2/2m_e)\nabla^2\psi + V\psi = E\psi, \qquad (16.1.2)$$

and exact solutions obtained. The total energy of the molecule at the selected separation R is the sum of the eigenvalue E and the nucleus–nucleus electrostatic repulsion

$$V_{\text{nuc–nuc}} = +e^2/4\pi\varepsilon_0 R. \qquad (16.1.3)$$

(There is no nuclear kinetic energy because they are stationary with respect to the centre of mass.) The wavefunctions obtained by solving eqn (16.1.2) are called *molecular orbitals*. Their squares give the probability density of the electron in the field of the nuclei. While it is interesting that there are exact, analytical solutions, they are extremely complicated, and we do not get much insight into the form of the orbitals and the contributions to the energy. Therefore, we shall adopt a simpler procedure that, while more approximate, gives more insight.

16.1 (b) The approximation of molecular orbitals

The exact molecular orbital (MO) gives, through its square, the exact distribution of the electron in the molecule (within the Born–Oppenheimer approximation). The MO is like an atomic orbital (an AO), but spreads throughout the molecule. When the electron is very close to nucleus A, the term $1/r_A$ in V is very much bigger than $1/r_B$, and so the potential energy in eqn (16.1.1) reduces to $-(e^2/4\pi\varepsilon_0 r_A)$. The Schrödinger equation then becomes the same as the one for a single hydrogen atom, and so its lowest energy solution is a 1s-orbital on A: we denote this $1s_A$. Therefore, close to A the MO resembles a 1s-orbital. Likewise, close to B the MO resembles a 1s-orbital on B, which we denote $1s_B$. This suggests that we can approximate the overall wavefunction as a *linear combination of atomic orbitals* (LCAO):

$$\psi = 1s_A + 1s_B. \qquad (16.1.4)$$

(ψ is not normalized, but that is a detail we take care of later.) This fits the preceding discussion, because when the electron is close to A its distance from B is large, the amplitude of the function $1s_B$ is small, and therefore the wavefunction is almost pure $1s_A$. Similarly, ψ is almost pure $1s_B$ close to B. Note the sequence of approximations: the Born–Oppenheimer approximation lets us separate the electronic and nuclear motions and therefore to talk in terms of molecular orbitals; the LCAO approximation goes one step further, and approximates an MO as a superposition of atomic orbitals.

The LCAO-MO in eqn (16.1.4) determines the distribution of the electron in the molecule. According to the Born interpretation, the probability density is proportional to its square:

$$\psi^*\psi = (1s_A + 1s_B)^2 = (1s_A)^2 + (1s_B)^2 + 2(1s_A)(1s_B). \qquad (16.1.5)$$

Since $1s_A$ is an abbreviation for the function

$$\psi_{1s}(r_A) = (1/\pi a_0^3)^{\frac{1}{2}} e^{-r/a_0}$$

with r_A the distance of the electron from A, and similarly for $1s_B$, it is easy to evaluate the probability density (and the wavefunction itself) at any point, and the result is shown in Fig. 16.1.

A very important feature of the MO appears when we examine the probability density in the internuclear region where both atomic orbitals have similar amplitudes. According to eqn (16.1.5), the total probability density is proportional to the sum of (a) the probability density it would have there if it were on A (this is the value of $(1s_A)^2$ evaluated in that region), (b) the probability density that it would have if it were on B (the value of $(1s_B)^2$), and (c) an *extra* contribution to the density, the term $2(1s_A)(1s_B)$. Hence, because the electron is free to move from one nucleus to the other, the electron probability density in the internuclear region is enhanced. The

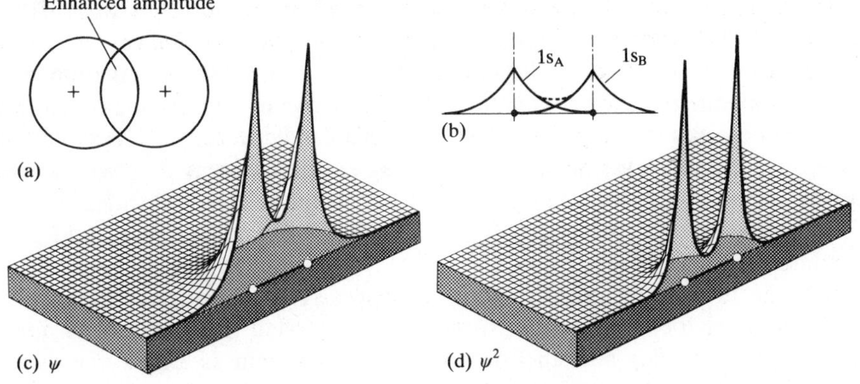

Fig. 16.1. (a) The orbital overlap responsible for bonding in the hydrogen molecule-ion and (b) the constructive interference in the internuclear region. (c) The orbital amplitude in a plane containing the two nuclei, and (d) the corresponding electron density. (b) is a side view of (c).

reason can be traced to the *constructive interference* of the two atomic orbitals: each has a positive amplitude in the internuclear region, and so the total amplitude is greater there than if the electron were described by a single atomic orbital. We shall constantly use the result that *electrons accumulate in regions where atomic orbitals overlap and interfere constructively*.

The following explanation of the strength of bonds is often given. The accumulation of electron density between the nuclei puts the electron predominantly in a position where it interacts strongly with both nuclei; hence the energy of the molecule is lower than that of the separate atoms, where each electron can interact strongly with only one nucleus. Unfortunately, this neat explanation is probably incorrect in the case of H_2^+ (at least). This is because shifting an electron *away* from a nucleus into the internuclear region *raises* its potential energy. The modern explanation is more subtle, and does not emerge from the simple LCAO treatment given here. The broad feature of the explanation is that at the same time as the electron shifts into the internuclear region, the atomic orbitals shrink. This *orbital shrinkage* improves the electron–nucleus attraction more than it is damaged by the migration to the internuclear region, and so there is a net lowering of potential energy. The kinetic energy of the electron is also modified, but it is dominated by the potential energy.

It is important to get the explanations in perspective, because the precise analysis has been done only for a very few molecules. Even though the details are obscure for more complex molecules, it is generally found that bonding occurs when electrons accumulate between nuclei even though the actual cause of the bonding may be an accompanying shrinkage of the orbitals. Therefore, throughout the following discussion *we ascribe the strength of chemical bonds to the accumulation of electron density in the internuclear region* and leave open the question whether in molecules more complicated than H_2^+ the true source of energy lowering is that accumulation itself or some indirect but related effect.

16.1 (c) Bonding and antibonding orbitals

From now on we denote the LCAO-MO in eqn (16.1.4) as σ. (In general, σ denotes an MO with axial symmetry; it is the analogue of s in atoms.) This σ-orbital is formed by the *constructive* interference of two atomic orbitals.

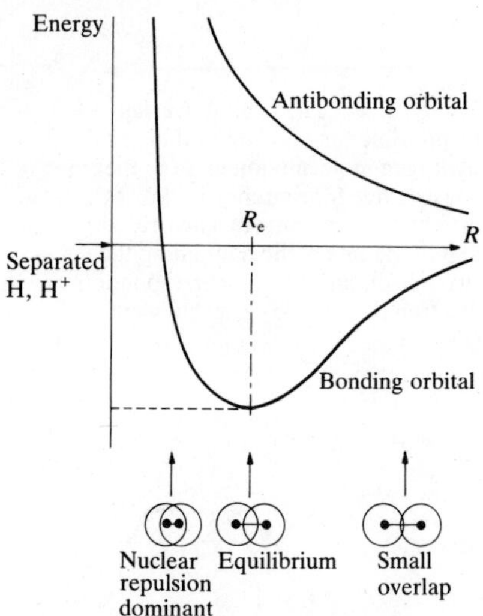

Fig. 16.2. The molecular potential energy curve for the hydrogen molecule-ion. (The upper curve is discussed later.) The equilibrium bond length corresponds to the energy minimum.

The energy of a molecule with a single σ-electron can be calculated for different internuclear distances R, and we obtain a *molecular potential energy curve*, Fig. 16.2. ('Potential energy' because we are within the Born–Oppenheimer approximation, and the nuclei are stationary.) The energy decreases initially as R decreases. This is because the accumulation of electron density in the internuclear region increases as the two atomic orbitals increasingly overlap. However, at small separations there is too little space between the nuclei for significant accumulation there, and also the nucleus–nucleus repulsion, eqn (16.1.3), becomes extremely large. As a result, the total energy rises at short distances. The internuclear separation corresponding to the minimum of the curve is called the *equilibrium bond length* (symbol: R_e) and the depth of the minimum is called the *bond dissociation energy* (symbol: D_e).‡ Calculations give $R_e = 130$ pm and $D_e = 0.9$ eV (87 kJ mol^{-1}); the experimental values are 106 pm and 2.6 eV, and so this simple LCAO-MO description of the molecule, while inaccurate, is not absurdly wrong.

The σ-orbital is called a *bonding orbital*. In general, a bonding MO is one which, if occupied, contributes to a lowering of the energy of a molecule. The MO of H_2^+ next higher in energy can be modelled by combining the two 1s-orbitals with opposite sign. This gives rise to the *antibonding orbital* $\sigma^* = 1s_A - 1s_B$ (this is not normalized; the asterisk means 'antibonding', not complex conjugate). It is so called because when it is occupied it raises the energy of the molecule relative to the separated atoms. It is easy to see why. Since the atomic orbitals are combined with opposite signs, they interfere *destructively* where they overlap, Fig. 16.3. The probability density of an electron in σ^* is proportional to

$$(\sigma^*)^2 = (1s_A - 1s_B)^2 = (1s_A)^2 + (1s_B)^2 - 2(1s_A)(1s_B), \qquad (16.1.6)$$

the important difference from eqn (16.1.5) being the *subtraction* of the third term. This means that the probability density is reduced between the nuclei. In fact, on a plane perpendicular to the internuclear axis and passing through the mid-point of the internuclear distance, $1s_A$ and $1s_B$ have equal amplitudes and so σ^* is zero: there is zero probability of finding the

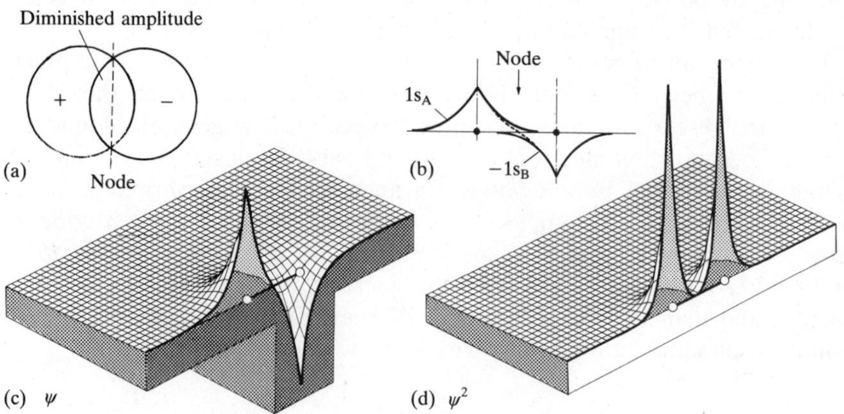

Fig. 16.3. (a) The orbital overlap responsible for antibonding in the hydrogen molecule-ion and (b) the destructive interference in the internuclear region. (c) The orbital amplitude in a plane containing the two nuclei, and (d) the corresponding electron density. (b) is a side view of (c).

‡ In Chapter 18 we shall see that the measured dissociation energy, the *true dissociation energy* (symbol: D_0) is always smaller than D_e because the vibration of the molecule must be taken into account. This is explained in Section 18.3; in H_2^+ the two values differ by 0.3 eV, 30 kJ mol^{-1}.

electron on this *nodal plane*, and so it is largely excluded from the important internuclear region.

An antibonding σ^*-electron destabilizes the molecule relative to the separated atoms, Fig. 16.2. This is partly because, since it is excluded from the internuclear region but has to be somewhere, it is distributed largely on the internuclear axis, but *outside* the bonding region. In effect, it pulls the nuclei apart from outside. There is no compensation for the nucleus–nucleus repulsion, and so the molecular potential energy curve simply rises as R is decreased. If we could excite an H_2^+ molecule from a σ^1 electronic configuration to a σ^{*1} it would immediately dissociate into H and H^+.

16.2 The structures of diatomic molecules

In Chapter 15 we used the hydrogen-like atomic orbitals in a building-up process for arriving at the lowest energy electronic configurations of atoms. The same procedure may be used for molecules, but using the hydrogen molecule-ion molecular orbitals as a framework. The procedure is as follows:

(1) Construct the MOs of the molecule by linear combination of the available AOs.

(2) Add the electrons to the lowest available MO subject to the constraint of the Pauli exclusion principle that no more than two electrons may occupy a single MO (and then must be paired).

(3) If degenerate MOs are available, add the electrons to each individual MO before doubly occupying any.

(4) If electrons occupy different MOs, then select the arrangement with most parallel spins (Hund's rule).

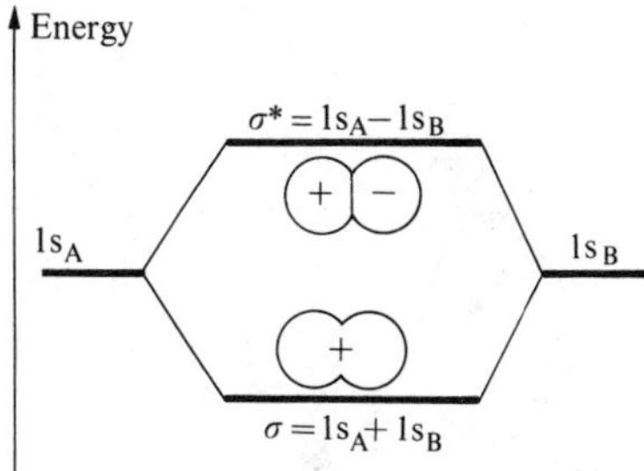

Fig. 16.4. A molecular orbital energy level diagram for orbitals constructed from (1s, 1s)-overlap, the separation of the levels corresponding to the equilibrium bond length.

16.2 (a) The hydrogen molecule

Each hydrogen atom of H_2 contributes a 1s-orbital (as in H_2^+). Therefore, form the σ and σ^* linear combinations. These are denoteu $1s\sigma$ and $1s\sigma^*$ respectively (we need to show their parentage), and at some internuclear separation will have the energies shown in Fig. 16.4. This is called a *molecular orbital energy level diagram*. There are two electrons to accommodate, and both can enter the bonding MO by pairing their spins. The ground state configuration is therefore $1s\sigma^2$, Fig. 16.5, and the atoms are joined by a bond consisting of an electron pair in a σ-orbital. *Spin-pairing is therefore not an end in itself, but the means by which the electrons can enter the lowest energy molecular orbitals*. This is the reason for the importance of electron pairs in chemical bonding.

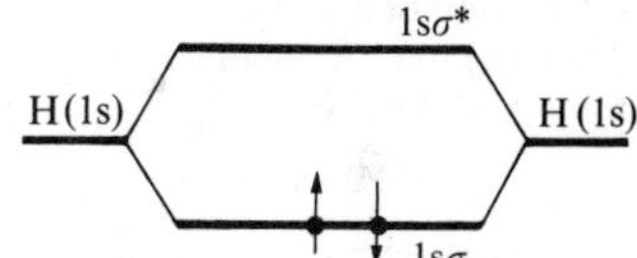

Fig. 16.5. The ground electronic configuration of H_2 is obtained by accommodating the two electrons in the lowest available orbital (the bonding orbital).

The same argument shows why helium does not form diatomic molecules. Each He atom contributes a 1s-orbital, and so the same two MOs can be constructed. (They differ in detail from the hydrogen MO because the He 1s-orbitals are more compact, but the general shape is the same, and we can use the same energy level diagram, Fig. 16.4, in the discussion.) There are four electrons to accommodate. Two can enter the $1s\sigma$-orbital, but then it is full, and the next two must enter the $1s\sigma^*$-orbital. The ground electronic configuration of He_2 would therefore be $1s\sigma^2 1s\sigma^{*2}$, Fig. 16.6, and so there is one bond and one antibond: we say that its net *bond order* is $1 + (-1) = 0$. In fact, an antibond is slightly more antibonding than a bond is bonding. Therefore a molecule with bond order zero has a higher energy than the separated atoms, and so He_2 does not form.

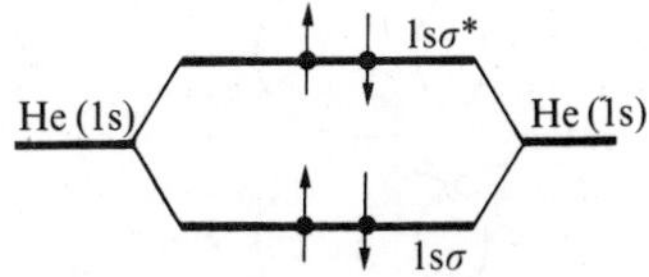

Fig. 16.6. The ground electronic configuration of the four-electron molecule He_2 has two bonding electrons and two antibonding electrons. It has a higher energy than the separated atoms, and so He_2 is unstable.

However, when one of the He atoms is electronically excited (for example, when one 1s-electron is promoted to 2s), the argument fails because the configuration of the diatomic molecule is $1s\sigma^2 1s\sigma^{*1} 2s\sigma^1$, where $2s\sigma$ is an MO formed from the overlap of 2s-orbitals. Now the antibonding effect of $1s\sigma^{*1}$ does not overcome the bonding effect of $1s\sigma^2 2s\sigma^1$, and so the HeHe* molecule survives until it discards its energy by radiation or collision. These weakly bonded *excimers* (excited dimers) of the noble gases have been detected.

16.2 (b) σ- and π-molecular orbitals

We now turn to the *homonuclear diatomic* molecules, molecules of the form A_2, of the second-row elements. Similar remarks apply to the homonuclear diatomics of the later rows.

In line with the building-up procedure, we first consider the MOs that may be formed without troubling about how many electrons are available. The atomic orbitals available are those of the inner, closed shells (the *core orbitals*), those of the valence shell (the *valence orbitals*), and those of the atom that are unoccupied in its ground state (these are the *virtual orbitals*). In elementary treatments (but not in the modern sophisticated treatments on main-frame computers), the core orbitals are ignored as being too compact to have significant overlap with orbitals on other atoms, and the virtual orbitals are ignored on the grounds that they are too high in energy. Therefore, the molecular orbitals are formed using only the valence orbitals. In second-row elements the valence orbitals are 2s and 2p. The 2s-orbitals on the two atoms overlap to give a bonding $2s\sigma$-orbital and an antibonding $2s\sigma^*$-orbital, in exactly the same way as we have already seen for 1s-orbitals. The new feature of the L-shell orbitals, however, is that p-orbitals are also available for bonding.

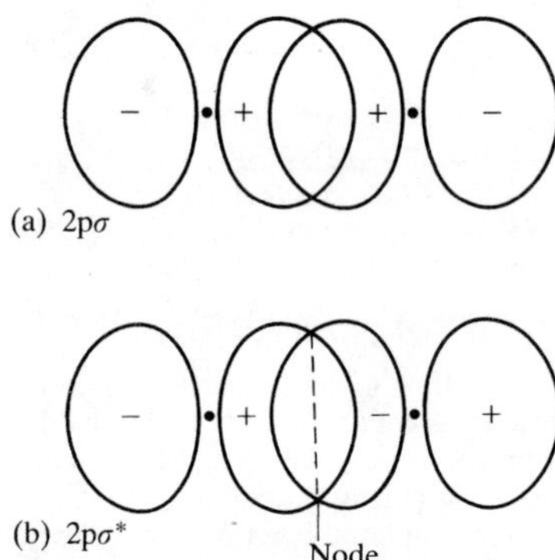

Fig. 16.7. (a) The interference leading to the formation of a $2p\sigma$-bonding orbital and (b) the corresponding antibonding orbital.

Consider first the overlap of two $2p_z$-orbitals directed along the internuclear axis. These overlap strongly, and may do so either constructively, to give an axially symmetrical $2p\sigma$-orbital, or destructively, to give an axially symmetrical $2p\sigma^*$-orbital, Fig. 16.7.

Now consider the $2p_x$- and $2p_y$-orbitals of each atom. Even though they are not directed towards each other, they can still overlap broadside-on and may do so constructively, to give a $2p\pi$-orbital, or destructively, to give a $2p\pi^*$-orbital, Fig. 16.8. The notation *π-orbital* is the analogue of p-orbital in atoms (when viewed along the axis of the molecule, a π-orbital looks like a p-orbital). The two $2p_x$-orbitals overlap to give a π_x-orbital (and its antibonding partner $2p\pi_x^*$), and the two $2p_y$-orbitals overlap to give a $2p\pi_y$ (and its partner $2p\pi_y^*$). The two $2p\pi$-orbitals are degenerate, and so are the two $2p\pi^*$-orbitals.

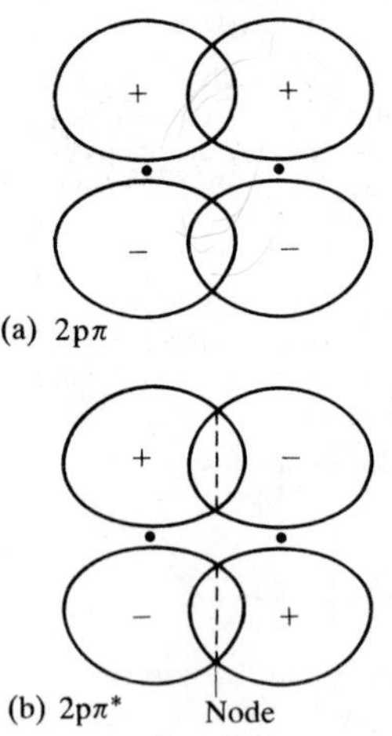

Fig. 16.8. (a) The interference leading to the formation of a $2p\pi$-bonding orbital and (b) the corresponding antibonding orbital.

The π-orbitals can be expected to be less strongly bonding than the σ-orbitals because the overlap occurs off-axis, and away from the optimum bonding region. This suggests that the MO energy level diagram ought to be as shown in Fig. 16.9(a). However, it is found (spectroscopically, and by more detailed calculation) that electron–electron interactions upset this order, and that it varies from molecule to molecule. It turns out that the order shown in Fig. 16.9(b) is often more appropriate, and we shall use it in the following discussion.

16.2 (c) s,p-overlap

The discussion so far has considered s,s-overlap and p,p-overlap as occurring distinctly, and has ignored the possibility of s,p-overlap. One reason for this is that *strong bonds arise from the overlap of atomic orbitals having similar energies*. In homonuclear diatomic molecules the energies of the 2s-orbitals of the two atoms are identical, and so are the 2p-orbitals, and so we have isolated the major contributions to the molecular orbitals by treating them separately. However, there is nothing in principle against the formation of bonds by s,p-overlap, and its effects must be taken into account. Indeed, when we come to molecules like H_2O we shall see that s,p-overlap is the major contribution to bonding.

Whether or not s,p-overlap plays a role depends on the symmetry of the arrangement. Thus, a 2s-orbital may overlap with a $2p_z$-orbital, Fig. 16.10(a), because they both have axial symmetry around the internuclear axis. However, $(2s, 2p_x)$-overlap makes no contribution to bonding. This is because the effect of the constructive overlap in one region, Fig. 16.10(b), is exactly cancelled by the effect of the destructive overlap in another, and so there is no net overlap and no net contribution to bonding.

A measure of the extent to which two orbitals overlap is the *overlap integral* (symbol: S):

$$\textit{Overlap integral: } S = \int \psi_A^* \psi_B \, d\tau. \qquad (16.2.1)$$

If the amplitude of the atomic orbital ψ_A on A is small wherever the orbital ψ_B on B is large, or vice versa, the product of their amplitudes is everywhere small and the integral—the sum of these products—is small. If ψ_A and ψ_B are simultaneously large in some region of space, then S may be large. If the two atomic orbitals are identical (e.g. 1s-orbitals on the same nucleus), then $S = 1$. This is illustrated in Fig. 16.11(a) and (b). In some cases, simple analytical expressions can be given for overlap integrals. For

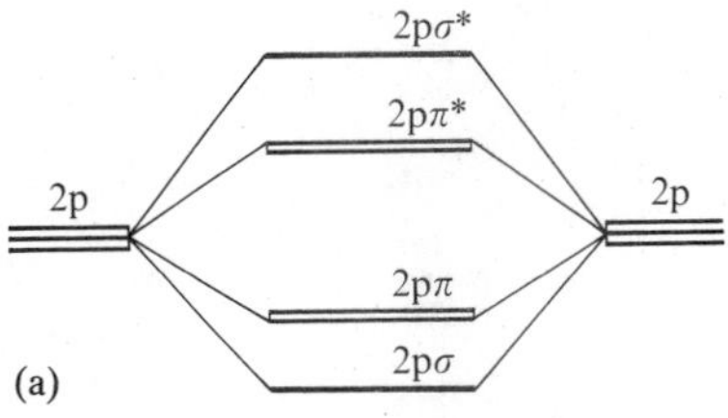

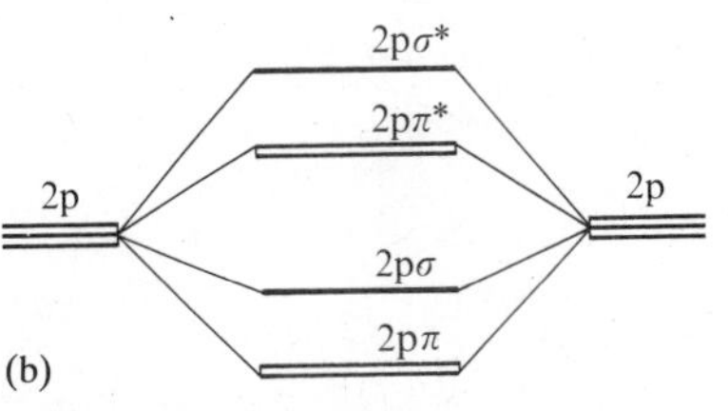

Fig. 16.9. The molecular orbital energy level diagram for (2p, 2p)-overlap. While simple overlap considerations suggest the order in (a), the order often found in practice is closer to that shown in (b). We use the latter in subsequent discussions.

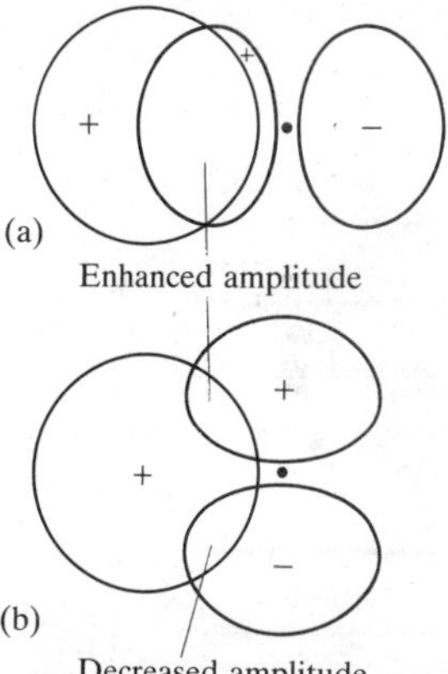

Fig. 16.10. Overlapping s- and p-orbitals. (a) End-on overlap leads to non-zero overlap and to the formation of an axially symmetric σ-bond. (b) Broadside overlap leads to no net accumulation of electron density.

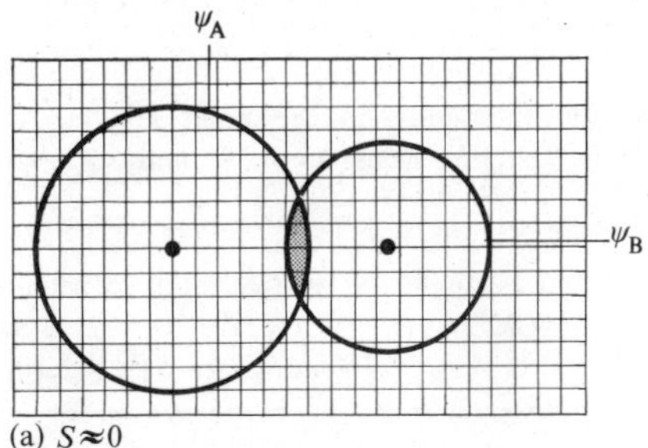

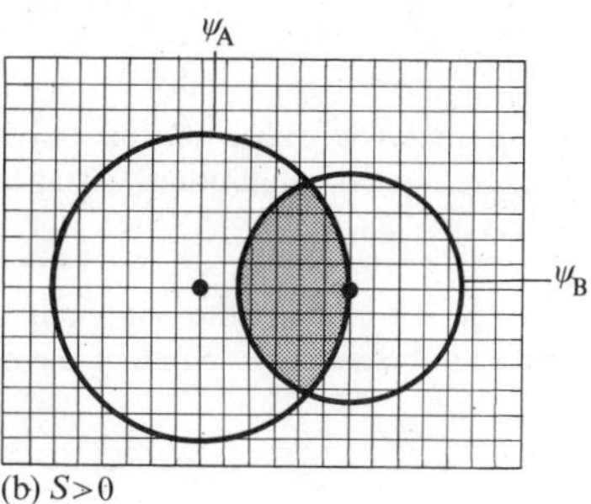

$\psi_A > 0$ $\quad \psi_A\psi_B = c$

$\psi_B > 0$

$\psi_B < 0$

(c) $S = 0$ $\quad \psi_A\psi_B = -c$

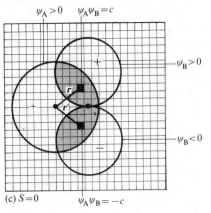

Fig. 16.11. A schematic representation of the contributions to the overlap integral. (a) $S \approx 0$ because the orbitals are far apart and their product is always small. (b) S is large (but less than 1) because $\psi_A\psi_B$ is large over a substantial region. (c) $S = 0$ because the positive region of overlap is exactly cancelled by the negative region.

example, in the case of two hydrogenic 1s-orbitals on nuclei separated by a distance R,

$$S = \{1 + (ZR/a_0) + \tfrac{1}{3}(ZR/a_0)^2\}e^{-ZR/a_0}. \qquad (16.2.2)$$

This evaluates to 0.59 for two 1s-orbitals at the equilibrium bond length in H_2^+, but this is an unusually large value, and typical values for L-shell orbitals are around 0.2–0.3.

Now consider the arrangement in Fig. 16.11(c) in which an s-orbital overlaps a p-orbital broadside on. At some point $\boldsymbol{r}_1$ the product $\psi_{1s}(\boldsymbol{r}_1)\psi_{2p}(\boldsymbol{r}_1)$ may be large. However, there is a point $\boldsymbol{r}_1'$ where the product has exactly the same magnitude but an opposite sign. When the integral is evaluated, these two contributions are summed, and cancel. For every point in the upper half of the diagram, there is a point in the lower that cancels, and so $S = 0$. Therefore, there is no net overlap between the s- and p-orbitals in this arrangement. Orbitals for which $S = 0$ are called *orthogonal*; two different AOs on the same atom are always orthogonal.

Example 16.1

Normalize the 1$s\sigma$-orbital for H_2^+.

• *Method*. Write the orbital as $N\{(1s_A) + (1s_B)\}$, where the 1s-orbitals are normalized. Then form $\int \psi^*\psi\, d\tau$, and find N such that the integral is equal to unity. The orbitals are real.

• *Answer*.
$$\int \psi^*\psi\, d\tau = N^2\int \{(1s_A) + (1s_B)\}\{(1s_A) + (1s_B)\}\, d\tau$$
$$= N^2\left\{\int (1s_A)^2\, d\tau + \int (1s_B)^2\, d\tau + 2\int (1s_A)(1s_B)\, d\tau\right\}$$
$$= N^2\{1 + 1 + 2S\} = 2(1+S)N^2 = 1.$$

Therefore, $N = 1/\{2(1+S)\}^{\frac{1}{2}}$, and the normalized wavefunction is $\psi = \{(1s_A) + (1s_B)\}/\{2(1 + S)\}^{\frac{1}{2}}$.

• *Comment*. The value of the normalization constant depends on the bond length. At the experimental equilibrium distance (106 pm) eqn (16.2.2) gives $S = 0.59$, and so $\psi = 0.56\{(1s_A) + (1s_B)\}$.

• *Exercise*. Normalize the antibonding 1$s\sigma^*$-orbital and evaluate the normalization constant at the equilibrium bond length of the molecule (using $S = 0.59$).

[$N = 1/\{2(1-S)\}^{\frac{1}{2}} = 1.10$]

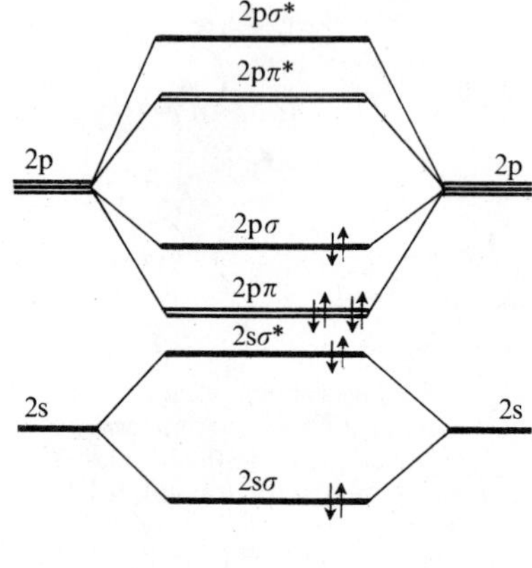

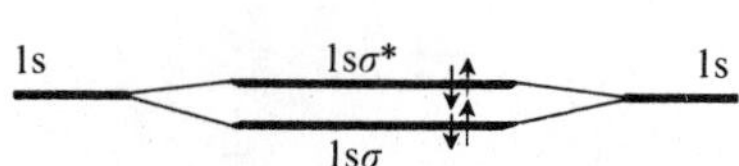

Fig. 16.12. The molecular orbital energy level diagram used for the discussion of the homonuclear diatomics of the second row. The electron configuration shown is that of N_2.

16.2 (d) The structures of homonuclear diatomic molecules

The atomic orbitals contributed by the second-row atoms are shown on the left and right of the MO energy level diagram in Fig. 16.12. The lines in the middle are a rough guide to the energies of the MOs that can be formed by overlap of corresponding orbitals. The ground configurations of the molecules are now obtained by adding the appropriate number of electrons and following the building-up rules. For a neutral homonuclear diatomic formed from atoms of atomic number Z, $2Z$ electrons have to be accommodated (and charged species need either more or less).

Consider N_2, and its 14 electrons. Two electrons pair and enter 1$s\sigma$; that is now full, and so the next two enter and fill 1$s\sigma^*$. Likewise four electrons complete 2$s\sigma$ and 2$s\sigma^*$. Six electrons remain. There are *two* 2$p\pi$-orbitals, and so four electrons can be accommodated. The last two enter 2$p\sigma$. The ground state configuration of N_2 is therefore $1s\sigma^2 1s\sigma^{*2} 2s\sigma^2 2s\sigma^{*2} 2p\pi^4 2p\sigma^2$. The bond order can now be calculated by counting the number of bonds and subtracting the number of antibonds: there is a $1s\sigma$ bond and antibond (net: 0), a $2s\sigma$ bond and antibond (net: 0) two $2p\pi$ bonds and a $2p\sigma$ bond (net: 3), giving a bond order of 3 overall. This accords very well with the chemical view of the bonding in N_2, that it is a triply-bonded molecule and can be denoted N≡N. Note that a *triple bond* consists of a σ-bond and two π-bonds.

The O_2 molecule ($2Z = 16$) has two more electrons than N_2. They must enter the $2p\pi^*$-orbital because the lower orbitals are full. Its configuration is therefore $\ldots 2p\pi^4 2p\sigma^2 2p\pi^{*2}$. The presence of the π-antibond cancels the effect of one π-bond, and so its bond order is 2. This accords with the classical view that oxygen has a double bond and may be denoted O=O. Note that a *double bond* consists of a single σ-bond and a single π-bond.

The MO description of O_2 has a further success. According to the building-up principle, the two $2p\pi^*$-electrons will occupy different orbitals: one will enter $2p\pi_x^*$ and the other $2p\pi_y^*$. Then, since they are in different orbitals, they will have parallel spins. Therefore, the MO theory predicts that the oxygen molecule will have a net spin angular momentum (in the language introduced in Section 15.2, that it will be in a *triplet state*). Furthermore, since electron spin is the source of a magnetic moment, we can go on to predict that *oxygen should be paramagnetic*‡. This is in fact the case, and liquid oxygen sticks to a magnet. This is excellent confirmation of the general correctness of the MO description of molecular bonding.

Molecular fluorine, $2Z = 18$, has the configuration $\ldots 2p\pi^{*4}$, and the extra pair of electrons, which completes the $2p\pi^*$-orbitals, reduces the bond order to 1. Therefore, F_2 is a singly bonded molecule, in agreement with classical views, and its dissociation energy is low ($154\ \mathrm{kJ\ mol^{-1}}$, compared with $942\ \mathrm{kJ\ mol^{-1}}$ for triply bonded N_2). Molecular neon, $2Z = 20$, has two further electrons: these enter the $2p\sigma^*$-orbital and reduce the bond order to 0. This agrees with the monatomic nature of neon.

Another use of the MO energy level diagram is to judge whether the ions (e.g. N_2^+ or N_2^-) are more or less strongly bonded than the neutral parent.

Example 16.2

Judge whether N_2^+ is likely to have a larger or smaller dissociation energy than N_2.

• *Method*. Compare their electronic configurations, and assess their bond orders. The molecule with the larger bond order is likely to have the larger dissociation energy.

• *Answer*. From Fig. 16.12 we can write the configurations:

$$N_2:\ 1s\sigma^2 1s\sigma^{*2} 2s\sigma^2 2s\sigma^{*2} 2p\pi^4 2p\sigma^2$$
$$N_2^+:\ 1s\sigma^2 1s\sigma^{*2} 2s\sigma^2 2s\sigma^{*2} 2p\pi^4 2p\sigma^1.$$

The bond orders are

$$N_2:\ (1-1)+(1-1)+(2+1)=3; \qquad N_2^+:\ (1-1)+(1-1)+(2+\tfrac{1}{2})=2\tfrac{1}{2}.$$

Since the cation has the smaller bond order, we expect it to have the smaller dissociation energy.

‡ A *paramagnetic* substance tends to move into a magnetic field; a *diamagnetic* substance tends to move out of one. Paramagnetism, the rarer property, arises when the molecules have unpaired electron spins. Both properties are discussed in more detail in Section 24.6.

• *Comment*. The experimental dissociation energies are 945 kJ mol^{-1} for N_2 and 842 kJ mol^{-1} for N_2^+.

• *Exercise*. Which can be expected to have the higher dissociation energy, F_2 or F_2^+? [F_2^+]

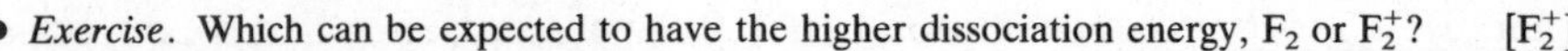

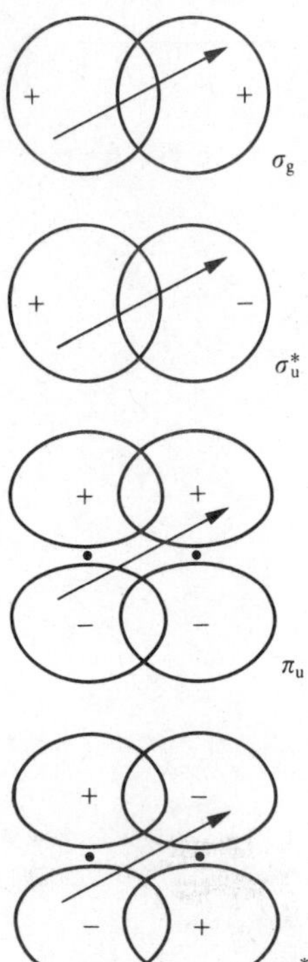

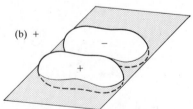

Fig. 16.13. The parity of an orbital: even (g) if its amplitude is unchanged under inversion in the centre of symmetry of the molecule, but odd (u) if the amplitude changes sign. Heteronuclear diatomic molecules do not have a centre of inversion, and so the g, u classification is irrelevant.

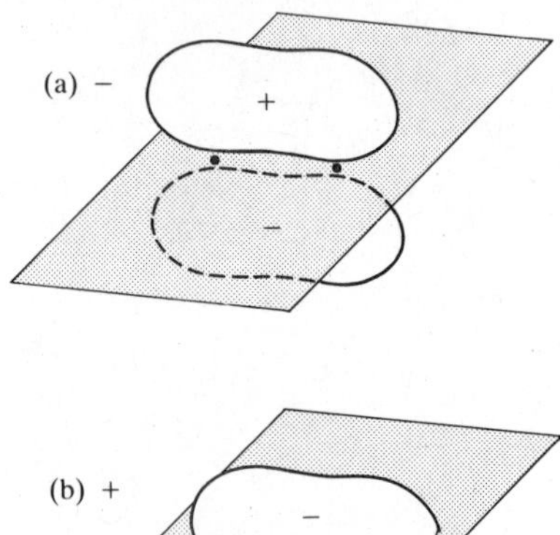

Fig. 16.14. The ± symmetry refers to the symmetry of an orbital when it is reflected in a plane containing the two nuclei.

16.2 (e) More about notation

In the description of homonuclear diatomic molecules it is common to see a subscript g or u attached to σ or π. This specifies the *parity*, the behaviour of the orbital under inversion. To see what this means, consider any point in a homonuclear diatomic, and note the sign of the amplitude of the wavefunction. Now travel through the centre of the molecule (this is the *centre of inversion*) and go to the corresponding point on the other side (this process is the operation of *inversion*). If the amplitude has the same sign, the orbital has *even parity* and is denoted g (from *gerade*, the German for even). If the amplitude has the opposite sign the orbital has *odd parity* and is denoted u (from *ungerade*, uneven). Inspection of Fig. 16.13 shows that bonding σ-orbital is g (and so is written σ_g) but an antibonding σ-orbital is u (and written σ_u^*). A π-orbital is u if it is bonding (π_u), but g if it is antibonding (π_g^*). Later we shall see the importance of the parity designation for the statement of spectroscopic selection rules.

The *term symbols* for molecules (the analogues of the symbols ^2P, etc. for atoms) are constructed in a similar way, but now we have to pay attention to the orbital angular momentum about the internuclear axis. A single electron in a σ-orbital (or a σ^*-orbital) has zero orbital angular momentum: the orbital is axially symmetrical and has no angular nodes. The term symbol for H_2^+ is therefore $^2\Sigma$, the superscript 2 being the multiplicity, the value of $2S+1$ (with $S=\frac{1}{2}$ in this case). The parity is added as a right subscript, and so in full dress it is $^2\Sigma_g$. The term symbol is $^1\Sigma_g$ for any closed-shell homonuclear diatomic because the spin is zero (all electrons paired), there is no orbital angular momentum from a closed shell, and the overall parity is g. The last remark follows from the rule that in a many-electron molecule the overall parity is obtained from the product of the parities of each electron, using $g\times g=g$, $u\times u=g$, and $g\times u=u$. Thus any closed-shell molecule (such as H_2 and N_2) has overall g parity.

A π-electron has one unit of orbital angular momentum about the internuclear axis (a π-orbital has one nodal plane) and gives rise to a Π term (the analogue of P). If there are two π-electrons (as in O_2) the term symbol may be either Σ (if the electrons are orbiting in opposite directions) or Δ (if they are orbiting in the same sense). In the case of O_2 it is known that Σ lies lower than Δ, and so its ground term is Σ. We have already seen that it is a triplet, and the overall parity is (closed shell) $\times g^2=g$. The term symbol is therefore $^3\Sigma_g$.

A superscript is added to the term symbol to denote the behaviour of orbitals under *reflection* in a plane containing the nuclei, Fig. 16.14. In the case of O_2, for example, if one electron is π_x^* (which changes sign under reflection) and the other is π_y^* (which does not change sign under reflection in the same plane), the overall reflection symmetry is (closed shell) $\times(+)\times(-)=(-)$, and so the full term symbol is $^3\Sigma_g^-$. The need for all this dressing of a basic symbol will become apparent when we deal with the spectroscopic selection rules in Chapter 19.

16.2 (f) The structures of heteronuclear diatomic molecules

A *heteronuclear diatomic molecule* is a molecule of the form AB; chemically interesting species include CO and HCl. The principal consequence of the presence of different atoms is that the electron distribution in the bond is no longer symmetrical between the atoms because it is energetically favourable for the charge to drift towards one of the atoms. When this happens, the *nonpolar* bond of homonuclear diatomic molecules is replaced by the *polar* bond of heteronuclear diatomic molecules. Thus HCl has a polar bond, in which the electron pair is closer to the chlorine, which therefore has a partial negative charge, leaving a partial positive charge on the hydrogen. In polar and nonpolar bonds the two participating atoms still share the electron pair, and neither loses complete control of it. That is, both are examples of *covalent bonds*, in which electron pairs are shared and occupy definite locations between the atoms. The extreme case of a polar bond is an *ionic bond*, in which one atom loses one or more electrons to another atom, the two ions then sticking together by the electrostatic attraction of opposite charges. Now one atom has lost control of its electron, and there is no region where the electron pair is located and shared.

The range of bond types, from nonpolar through polar to ionic, is captured in MO theory by writing the LCAO as

$$\psi = c_A\psi_A + c_B\psi_B, \tag{16.2.3}$$

where ψ_A and ψ_B are the atomic orbitals contributing to the MO and c_A and c_B are coefficients. The proportion of ψ_A in the bond is c_A^2, and the proportion of ψ_B is c_B^2. A nonpolar bond has $c_A^2 = c_B^2$, and a pure ionic bond has one coefficient zero (so that A^+B^- would have $c_A = 0$ and $c_B = 1$). A polar bond would have unequal, nonzero coefficients.

16.2 (g) The variation principle

The question, of course, is how to find the values of the coefficients, and relate them to the energy of the orbital. The key is the *variation principle:*

> *If an arbitrary wavefunction is used to calculate the energy, then the value obtained is never less than the true energy.*

The arbitrary wavefunction is called the *trial function*. The principle implies that if we guess the form of an MO (e.g. guess values of the coefficients in eqn (16.2.3)) and calculate the energy of an electron that occupies it, then we never get an energy lower than the true energy. Moreover, the trial MO with the lowest energy is the best of that kind. Of course, we might get a lower energy if we use a more complicated wavefunction (for example, by taking a linear combination of several atomic orbitals on each atom), but *of the kind* we shall have the optimum MO.

We need a systematic way of finding the optimum coefficients in an LCAO. The method can be illustrated by using the *minimal basis set* for an MO; this is a basis set that is the simplest possible for constructing an LCAO. In the case of HCl it would consist of a hydrogen 1s-orbital (H1s) and a chlorine 3p-orbital (Cl3p). In general, the minimal basis set for an AB molecule is (ψ_A, ψ_B), and the trial function is the LCAO given in eqn (16.2.3). This trial function is not normalized as it stands, because the coefficients can take arbitrary values, and so in the following we cannot assume that $\int \psi^2\,d\tau = 1$.

The energy of an electron described by a trial wavefunction is the expectation value of the energy operator (the hamiltonian, H): this was explained in Section 13.4(b). For a non-normalized wavefunction

$$E = \int \psi^* H\psi\,d\tau \bigg/ \int \psi^*\psi\,d\tau. \qquad (16.2.4)$$

From now on we consider only real wavefunctions, and so we can drop the star from ψ^*. We now search for values of the coefficients in the trial function that minimize the value of E. This is a standard problem in calculus: we seek the coefficients such that

$$\partial E/\partial c_A = 0 \quad \text{and} \quad \partial E/\partial c_B = 0. \qquad (16.2.5)$$

The first step is to express the two integrals in terms of the coefficients. The denominator is

$$\begin{aligned}\int \psi^2\,d\tau &= \int (c_A\psi_A + c_B\psi_B)(c_A\psi_A + c_B\psi_B)\,d\tau \\ &= \int (c_A^2\psi_A^2 + c_B^2\psi_B^2 + 2c_Ac_B\psi_A\psi_B)\,d\tau \\ &= c_A^2\int \psi_A^2\,d\tau + c_B^2\int \psi_B^2\,d\tau + 2c_Ac_B\int \psi_A\psi_B\,d\tau.\end{aligned}$$

The individual atomic orbitals are normalized, and so the first two integrals are 1. The third integral is the overlap integral S, eqn (16.2.1). Therefore, the denominator is

$$\int \psi^2\,d\tau = c_A^2 + c_B^2 + 2c_Ac_BS. \qquad (16.2.6)$$

The numerator is

$$\begin{aligned}\int \psi H\psi\,d\tau &= \int (c_A\psi_A + c_B\psi_B)H(c_A\psi_A + c_B\psi_B)\,d\tau \\ &= \int (c_A^2\psi_AH\psi_A + c_B^2\psi_BH\psi_B + 2c_Ac_B\psi_AH\psi_B)\,d\tau.\end{aligned}$$

There are some complicated integrals in this expression, but we can denote them by the constants

$$\alpha_A = \int \psi_AH\psi_A\,d\tau, \qquad \alpha_B = \int \psi_BH\psi_B\,d\tau, \qquad \beta = \int \psi_AH\psi_B\,d\tau. \qquad (16.2.7)$$

Then

$$\int \psi H\psi\,d\tau = \alpha_Ac_A^2 + \alpha_Bc_B^2 + 2\beta c_Ac_B. \qquad (16.2.8)$$

α_A, which is negative, is called a *Coulomb integral* and can be interpreted as the energy of the electron when it is in the orbital α_A on A (and likewise for α_B). In a homonuclear diatomic molecule $\alpha_A = \alpha_B$. β is called a *resonance integral* (for classical reasons), and vanishes when the orbitals do not overlap; at equilibrium bond lengths it is normally negative. We can therefore suspect that the strength of the bond will depend on β.

The complete expression for E is

$$E=\frac{\alpha_A c_A^2+\alpha_B c_B^2+2\beta c_A c_B}{c_A^2+c_B^2+2c_A c_B S}.$$

Its minimum is found by differentiation with respect to the two coefficients and using eqn (16.2.5). This involves elementary but slightly tedious work, the end result being

$$\begin{aligned}(\alpha_A-E)c_A+(\beta-ES)c_B&=0\\(\beta-ES)c_A+(\alpha_B-E)c_B&=0.\end{aligned}\qquad(16.2.9)$$

These are called the *secular equations* (the term arising originally in astronomy, where the same equations appear in connection with modifications of planetary orbits). They have a solution if the determinant of the coefficients, the *secular determinant* vanishes; that is, if

$$\begin{vmatrix}\alpha_A-E & \beta-ES\\ \beta-ES & \alpha_B-E\end{vmatrix}=0.\qquad(16.2.10)$$

This determinant expands to a quadratic equation in E, which may be solved. Its two roots give the energies of the bonding and antibonding MOs formed from the basis set and, according to the variation principle, these are the best energies for the given basis set. The corresponding values of the coefficients are then obtained by solving the secular equations using the two energies: the lower energy gives the coefficients for the bonding MO, the upper energy the coefficients for the antibonding MO. The secular equations give expressions for the *ratio* of the coefficients in each case, and so we need a further equation in order to find their individual values. This is obtained by demanding that the best wavefunction should be normalized, which means that we must also ensure (from eqn (16.2.6)) that

$$\int\psi^2\,d\tau=c_A^2+c_B^2+2c_A c_B S=1.\qquad(16.2.11)$$

The complete solutions are very cumbersome.‡ There are two cases where the roots can be written down very simply. First, when the two atoms are the same, and we can write $\alpha_A=\alpha_B=\alpha$, the solutions are

$$\begin{aligned}E_{\text{bonding}}&=(\alpha+\beta)/(1+S),\ c_A=1/\{2(1+S)\}^{\frac{1}{2}},\ c_B=c_A\\E_{\text{antibonding}}&=(\alpha-\beta)/(1-S),\ c_A=1/\{2(1-S)\}^{\frac{1}{2}},\ c_B=-c_A.\end{aligned}\qquad(16.2.12)$$

In this case the best bonding function has the form A + B and the corresponding antibonding function is A − B, in agreement with the discussion of homonuclear diatomics we have already given. However, we can now explore how the energies of the orbitals change with the parameter β and the overlap S. The easiest way of doing so is to suppose that β, which depends on the overlap, is directly proportional to the overlap, and to write

‡
$$\begin{aligned}E&=(A\pm B)/(1-S^2),\\A&=\tfrac{1}{2}(\alpha_A+\alpha_B)-\beta S,\\B&=\{\tfrac{1}{4}(\alpha_A-\alpha_B)^2-(\alpha_A+\alpha_B)\beta S+\alpha_A\alpha_B S^2+\beta^2\}^{\frac{1}{2}},\\c_A&=(\beta-SE)/C,\\C&=\{(\beta-SE)^2+(\alpha_A-E)^2-2S(\alpha_A-E)(\beta-SE)\}^{\frac{1}{2}}.\end{aligned}$$

$\beta = \kappa S$, κ being a constant; this is illustrated in Fig. 16.15. We see how the bonding and antibonding effects fall sharply with decreasing overlap. This justifies the neglect of 1s,1s-bonding in the second-row elements, because the overlap is so slight.

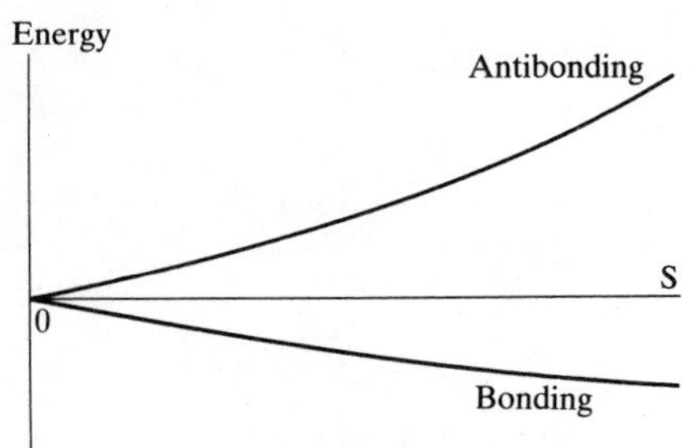

Fig. 16.15. The dependence of the bonding and antibonding orbital energies on the overlap, calculated using the expressions given in the footnote (p. 387) and making the assumption that the resonance integral is proportional to the overlap integral, $\beta = \kappa S$.

The other extreme case, when the energies of the atomic orbitals are widely different, supposes that $|\alpha_A - \alpha_B|$ is large in comparison with the resonance integral β. Then the solutions (with ψ_A taken as the orbital with lower energy) are

$$\left.\begin{aligned} E_{\text{bonding}} &= \{\alpha_A - X\}/(1-S^2) \\ c_A &= 1 - \{S\beta^2/(\alpha_A S - \beta)(\alpha_A - \alpha_B)\}, \\ c_B &= \beta^2/(\alpha_A S - \beta)(\alpha_A - \alpha_B). \end{aligned}\right\} \quad (16.2.13a)$$

$$\left.\begin{aligned} E_{\text{antibonding}} &= \{\alpha_B + X - 2\beta S\}/(1-S^2) \\ c_B &= 1 + \{S\beta^2/(\alpha_B S - \beta)(\alpha_A - \alpha_B)\}, \\ c_A &= -\beta^2/(\alpha_B S - \beta)(\alpha_A - \alpha_B). \end{aligned}\right\} \quad (16.2.13b)$$

$$X = \{\beta^2 - 2\alpha_B S + \alpha_A\alpha_B S^2\}/(\alpha_B - \alpha_A). \quad (16.2.13c)$$

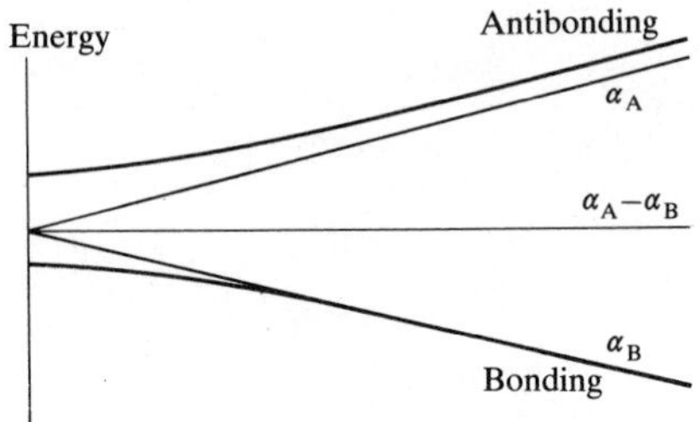

Fig. 16.16. The dependence of the bonding and antibonding orbital energies on the energy separation of the contributing atomic orbitals calculated using $\beta = \kappa S$.

These expressions, which have been used to construct the curves in Fig. 16.16, show that when the energy separation is very large (as in 1s, 2s-overlap of homonuclear diatomics), then the bonding orbital has $c_A = 1$, corresponding to pure ψ_A, and the antibonding orbital has $c_B = 1$, so that it is pure ψ_B. This justifies the neglect of $(1s_A, 2s_B)$-overlap in homonuclear diatomics, because the two orbitals remain effectively unchanged in the molecule. A classical analogy is the coupling between two pendulums sharing a common support: when they have very different frequencies they barely affect each other, and swing with their natural frequencies; when the frequencies are similar they interact more strongly and exchange energy freely.

We can now see how the LCAO coefficients for heteronuclear diatomics are found: the secular equations are solved for the energies, and those energies are used to obtain the optimum coefficients. There is still the problem of knowing the values of the Coulomb and resonance integrals. One approach has been to estimate them from spectroscopic information. This combination of empirical data and quantum mechanical calculations has given rise to the *semi-empirical methods* of molecular structure calculation. The modern tendency, however, particularly for small molecules but increasingly for bigger ones too, is to calculate the integrals in eqn (16.2.7) from first principles. These give rise to the *ab initio methods*, which are now widely used. They involve extensive numerical computation, and for this reason theoretical chemists (along with meteorologists and cryptanalysts) are among the heaviest users of computers.

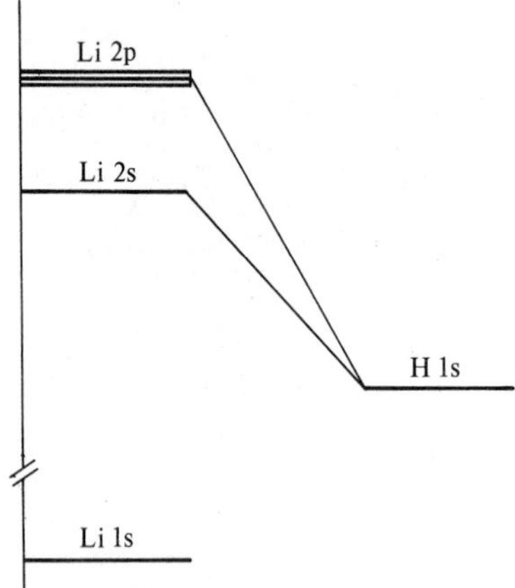

Fig. 16.17. Hydrogen and lithium *atomic* energy levels: H1s overlaps with both Li2s and Li2p, and the resulting orbital can be regarded as arising from the overlap of H1s with a (Li2s, Li2p)-hybrid orbital. Li1s is a core orbital, and plays only a minor role in the bonding.

16.2 (h) Hybridization

When dealing with heteronuclear diatomics there is often no clear-cut distinction between the energies of the atomic orbitals, and therefore no precise criterion about which ones should be combined. This is illustrated in Fig. 16.17 in connection with LiH. Although Li2s lies closer to H1s, the Li2p-orbitals are not far away and cannot be ignored. While the *minimal* basis set is (H1s, Li2s), a better description will be obtained if it is enlarged to (H1s, Li2s, Li2p) and the MOs are expressed as linear combinations of all three. A variational calculation then gives $\psi = 0.323(\text{Li2s}) + 0.231(\text{Li2p}) + 0.685(\text{H1s})$ as the optimum wavefunction of this form, showing that the 2p-orbitals make a substantial contribution.

At this stage, however, we seem to have lost an attractive feature of LCAO-MO theory, for it now seems impossible to think of an Li—H bond as being formed from the overlap of *two* orbitals, one on Li and the other on H. This is in fact not so, because we can still regard the MO as arising from the overlap of H1s and a *hybrid orbital* on Li. The MO can be expressed alternatively as $\psi = 0.397(\text{Li hybrid}) + 0.685(\text{H1s})$, with $(\text{Li hybrid}) = 0.813(\text{Li2s}) + 0.582(\text{Li2p})$. (We shall discuss the actual figures below.) Through *hybridization* we retain the simplicity of the basic LCAO-MO picture, but at the expense of a more complicated atomic contribution.

In a sense, hybridization is unnecessary, for it arises from a wish to find a formally simple description of a bond: there is no mathematical reason why it should be introduced, and no compelling physical reason. Nevertheless, hybridization does help to disentangle the contributions to the energies of bonds, as we can see by considering LiH more closely.

The shape of the Li hybrid is shown in Fig. 16.18: the bulk of the amplitude lies in the internuclear region. This 'distortion' arises from the interference between the Li2s and Li2p atomic orbitals, because they add on one side of the nucleus (where their amplitudes have the same sign) but partially cancel on the side (where their signs are opposite). The distortion gives rise to a stronger bond because the overlap with H1s is enhanced. Exactly the same enhancement is present in the 'three orbital' description of the bond, but hybridization focuses our attention on it.

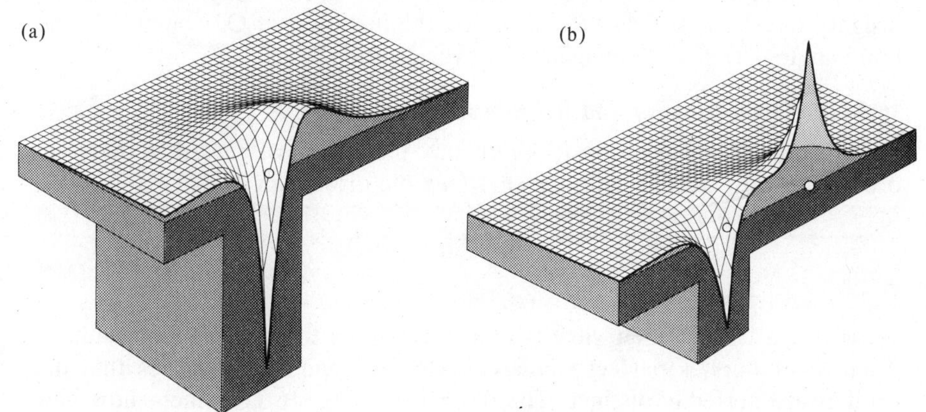

Fig. 16.18. (a) A cross-section through the (Li2s, Li2p)-hybrid showing the accumulation of amplitude on one side of the nucleus. (b) The H1s-orbital overlaps the hybrid strongly, and a stronger bond is formed than with Li2s alone.

Why, though, does the Li atom hybridize as far as $0.813(\text{Li2s}) + 0.582(\text{Li2p})$ and no further, such as to $0.71(\text{Li2s}) + 0.71(\text{Li2p})$, which is even more strongly distorted, and would give a better overlap? The reason lies in an opposing energy contribution. The actual hybrid is 66% Li2s and 34% Li2p (take squares of the coefficients), and so an electron in it can be thought of as spending 34% of its time in an excited state. In the alternative hybrid the proportion is higher, 50%. Therefore, in order to achieve a better overlap, it is necessary to invest more energy in the *promotion* of the atom to a higher energy configuration. The two effects are in opposition, and the outcome, the observed hybrid, is the resulting compromise. Its precise composition, and the optimum energy of the molecule, may be calculated by numerical evaluation of the integrals that occur in the expressions obtained from the variation principle.

16.3 The structures of polyatomic molecules

Polyatomic molecules are molecules consisting of more than two atoms, and include most of the interesting molecules of chemistry. Their bonds are built in the same way as in diatomic molecules: atomic orbitals overlap and give rise to molecular orbitals spreading over all the atoms. But why do they adopt their characteristic shapes? Why is H_2O triangular, NH_3 pyramidal, CH_4 tetrahedral, and so on?

The shape of H_2O illustrates the approach. The ground configuration of the oxygen atom is $K2s^22p_z^22p_y^12p_x^1$, Section 15.2. This suggests a minimal basis set ($O2p_y$, $O2p_x$, $H1s_A$, $H1s_B$), with four electrons to accommodate in the bonds (two from O, one from each H). The MOs are formed as each H1s overlaps an O2p, forming two σ-bonds, Fig. 16.19. Each bonding MO accommodates two electrons, and so all four are accounted for by the two σ-bonds. The configuration of H_2O is therefore $(K2s^22p_z^2)\sigma_A^2\sigma_B^2$. Since the O2p-orbitals are perpendicular, this predicts a 90° H—O—H bond angle. This simple picture is quite good: the H_2O molecule is bent; but its bond angle is 104.45°, and the discrepancy must be explained.

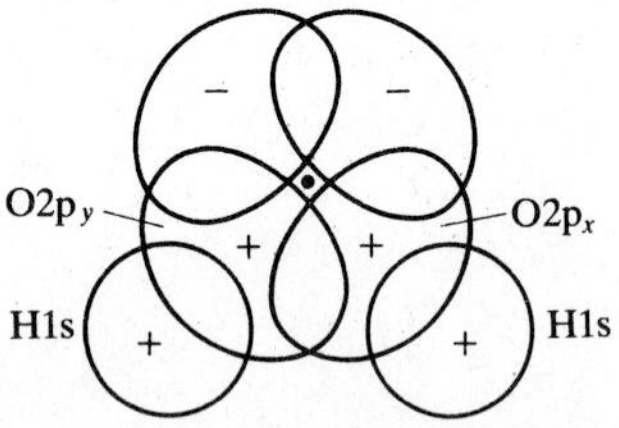

Fig. 16.19. A primitive description of the electronic structure of water. The H1s-orbitals overlap the O2p-orbitals, and result in a 90° molecule. When O2s is included, the bond angle increases to its observed value of 104°.

The bond angle changes when the basis set is enlarged to include the O2s-orbital, which is nearby in energy and cannot be ignored. The molecular energy can be computed using this extended basis set and the variation principle, and values found for various bond angles. The minimum energy of the molecule turns out to occur when the bond angle is 104.45°, in agreement with observation. However, this does not give much insight, and the best way of seeing how the inclusion of O2s opens out the bond angle is to use the language of hybridization.

16.3 (a) Orthogonality and hybridization

In quantum chemistry *orthogonality* has the technical meaning that one orbital has zero overlap with another (see the discussion of eqn (16.2.1)):

$$\textit{Orthogonality:} \quad \int \psi_A^* \psi_B \, d\tau = 0. \qquad (16.3.1)$$

However, a less abstract view is that S measures the *similarity* of orbitals. Then $S = 1$ implies perfect similarity (identity) and $S = 0$ implies that the orbitals are perfectly distinct. The diagrams in Fig. 16.11, which show how S changes as orbitals come together, also show their changing 'similarity'. The three 2p-orbitals on a single atom are also perfectly distinct from each other ($S = 0$), because even though they have the same shapes they are pointing in mutually perpendicular directions, like three mutually perpendicular (orthogonal) vectors. In what follows we use the interpretation that *orbitals that are orthogonal are distinct*.

Example 16.3

Confirm that a $2p_x$-orbital is orthogonal to a $2p_y$-orbital of the same atom.

• *Method*. Evaluate the integral $\int (2p_x)^*(2p_y)\, d\tau$, taking the form of the orbitals as xf and yf respectively, where f is a function only of r. Use polar coordinates for x and y, writing them as

$r \sin\theta \cos\phi$ and $r \sin\theta \sin\phi$ respectively. It is sufficient to show that the integral over either angle vanishes. Begin by integrating over ϕ.

• *Answer*. $\int(2p_x)^*(2p_y)\,d\tau = \int xyf^2\,d\tau = \int r^2 \sin^2\theta \cos\phi \sin\phi f^2\,d\tau;$

$$\int_0^{2\pi} \cos\phi \sin\phi \,d\phi = 0.$$

• *Comment*. All the spherical harmonics are mutually orthogonal, and so all hydrogenic AOs are mutually orthogonal so long as they belong to the same atom.

• *Exercise*. Show that $3d_{xy}$ is orthogonal to $2p_x$.

In the MO description of H_2O we seek to construct two O—H bonds that are chemically equivalent but spatially distinct. However, it is important to appreciate that this is not a *necessary* requirement, because we could forge ahead by setting up MOs that spread over all three atoms, and then compute the optimum coefficients and energy. This leads to *delocalized molecular orbitals*, which are a perfectly valid description of the molecule, and are the wavefunctions normally obtained in modern computational studies of molecular structure. The only reason we are pursuing this line of argument is to open up the discussion of the molecule to the kind of arguments chemists often employ, which relate to identifiable bonds. We obtain equivalent but distinct bonds if we construct molecular orbitals starting from hybrid orbitals on oxygen that are equivalent (in the sense of having the same s,p composition) but mutually orthogonal.

When only O2p-orbitals are used, the two MOs are equivalent, because each orbital is an (O2p, H1s)-overlap, and distinct, because the O2p_x- and O2p_y-orbitals are mutually orthogonal. When O2s-character is included, both MOs acquire some and therefore are no longer orthogonal (they are similar to the extent of both having O2s-character). Only if they bend further away from each other, so reducing their spatial resemblance, can they become distinct. Therefore, we can predict that hybridization not only increases the bond strength (because of the better O,H overlap) but also increases the bond angle.

The argument is made quantitative as follows. Suppose we want to construct a hybrid orbital pointing along a line making an angle $\frac{1}{2}\Phi$ to the x-axis, Fig. 16.20. The p_x, p_y combination required‡ is

$$p(\tfrac{1}{2}\Phi) = p_x \cos\tfrac{1}{2}\Phi + p_y \sin\tfrac{1}{2}\Phi. \quad (16.3.2a)$$

The equivalent orbital pointing along $-\frac{1}{2}\Phi$ is

$$p(-\tfrac{1}{2}\Phi) = p_x \cos\tfrac{1}{2}\Phi - p_y \sin\tfrac{1}{2}\Phi. \quad (16.3.2b)$$

Now we add s-character. The same proportion must be added to both (so that the hybrids remain equivalent). We therefore write

$$h = as + bp(\tfrac{1}{2}\Phi),\ h' = as + bp(-\tfrac{1}{2}\Phi). \quad (16.3.3)$$

The coefficients a and b are found by requiring the hybrids to satisfy two

‡ A p_x-orbital is proportional to $x = r\cos\phi$ and a p_y-orbital is proportional to $y = r\sin\phi$. A p-orbital directed along the line at $\frac{1}{2}\Phi$ is proportional to $\cos(\phi - \frac{1}{2}\Phi)$, because it is like a p_x-orbital but rotated through an angle $\frac{1}{2}\Phi$. Now note that $\cos(\phi - \frac{1}{2}\Phi) = \cos\phi\cos\frac{1}{2}\Phi + \sin\phi\sin\frac{1}{2}\Phi$, which is proportional to $p_x\cos\frac{1}{2}\Phi + p_y\sin\frac{1}{2}\Phi$.

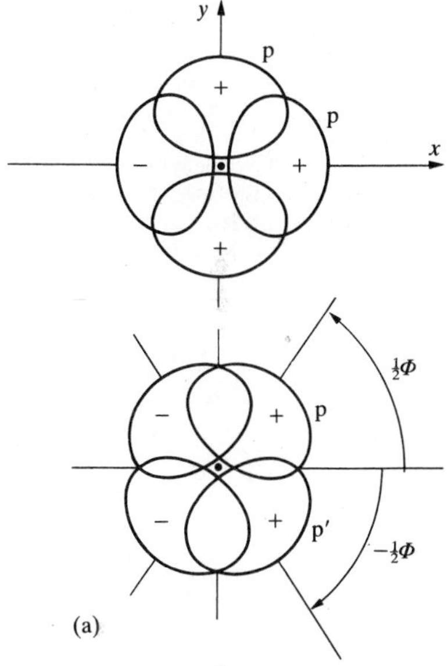

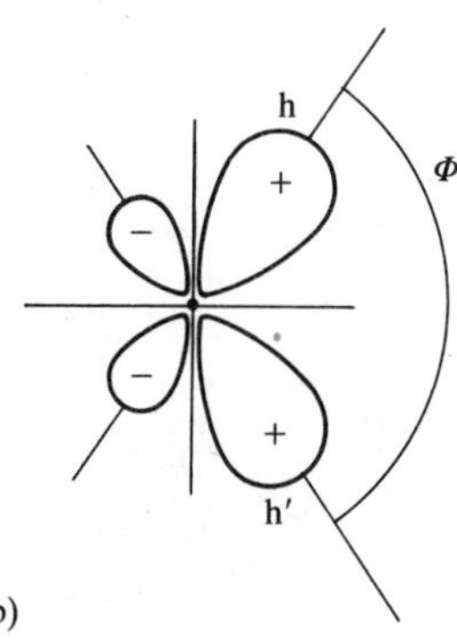

Fig. 16.20. (a) p and p′ can be expressed as linear combinations of p_x and p_y using eqn (16.3.2), but the combinations are not orthogonal. (b) The orthogonal hybrids, obtained by mixing s-character into p and p′.

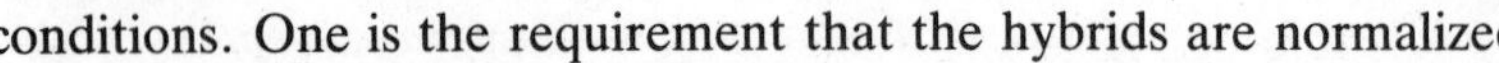

conditions. One is the requirement that the hybrids are normalized:

$$\int h^2\,d\tau = a^2\int s^2\,d\tau + b^2\int p^2\,d\tau + 2ab\int sp\,d\tau = a^2 + b^2 = 1.$$

(The s- and p-orbitals are individually normalized and mutually orthogonal; p stands for p($\frac{1}{2}\Phi$) or p($-\frac{1}{2}\Phi$).) The second requirement is for the hybrids to be distinct, i.e. orthogonal:

$$\int hh'\,d\tau = \int (as + bp_x\cos\tfrac{1}{2}\Phi + bp_y\sin\tfrac{1}{2}\Phi)(as + bp_x\cos\tfrac{1}{2}\Phi - bp_y\sin\tfrac{1}{2}\Phi)\,d\tau$$
$$= a^2 + b^2\cos^2\tfrac{1}{2}\Phi - b^2\sin^2\tfrac{1}{2}\Phi = a^2 + b^2\cos\Phi = 0.$$

(The sp- and p_xp_y-integrals are zero by orthogonality, and the s^2- and p^2-integrals are unity.) On combining these two results we get

$$\cos\Phi = -a^2/(1-a^2), \text{ or } a^2 = \cos\Phi/(\cos\Phi - 1). \qquad (16.3.4)$$

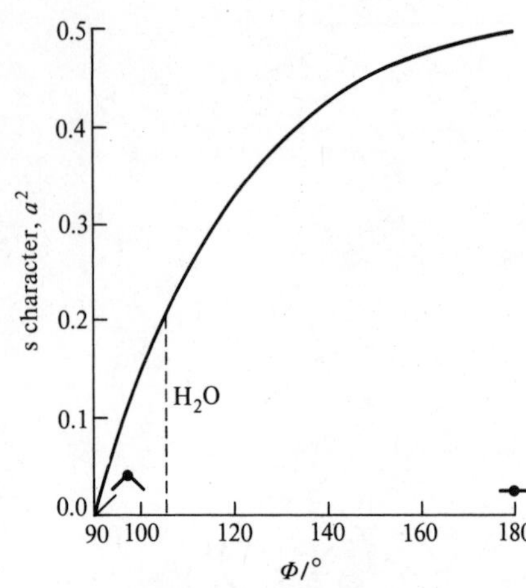

Fig. 16.21. The dependence of s-character on the angle between two equivalent hybrids calculated using eqn (16.3.4). The p-character is $1-a^2$.

This expression is plotted in Fig. 16.21. It shows that as the s-character of the hybrids increases (as a^2 increases) the angle between the hybrids, Φ, increases and changes from 90° ($a^2=0$, pure p) to 180° when $a^2=0.5$, a 50:50 mixture of s and p.

Example 16.4

Calculate the hybridization of the O–H bonds in H_2O, which has a bond angle of 104.45°.

- *Method*. Calculate the s-character of a hybrid from eqn (16.3.4), and its p-character from $1-a^2$. Hybrids are often denoted $s^{a^2}p^{b^2}$.

- *Answer*. When $\Phi = 104.45°$, $\cos\Phi = -0.25$; hence $a^2 = 0.20$. The p-character of the hybrid is therefore 0.80, and it can be denoted $s^{0.20}p^{0.80}$.

- *Comment*. Note that hybrids need not be of integral composition (such as 'sp^3'). The coefficients are the square roots of 0.20 and 0.80, and so the hybrid wavefunctions are of the form 0.45s + 0.89p, where p is a p-orbital directed from O to H.

- *Exercise*. What is the hybridization for a 120° bond angle? [$s^{\frac{1}{3}}p^{\frac{2}{3}}$, or sp^2 in terms of ratios]

The linear combination of three atomic orbitals leads to three orthogonal hybrids (and in general N orbitals give N hybrids). The third hybrid is $a's + b'p$, where p is an orbital directed along $-x$, Fig. 16.22. Normalization requires that $a'^2 + b'^2 = 1$, and the orthogonality of this hybrid to the other two requires $aa' + bb'\cos\Phi = 0$. After a little rearrangement these conditions lead to

$$a'^2 = (1+\cos\Phi)/(1-\cos\Phi) \qquad (16.3.5)$$

for the s-character of this non-bonding, lone pair orbital. In the case of H_2O, $a'^2 = 0.60$, implying 60% s-character and a hybrid orbital of the form 0.77s + 0.63p, Fig. 16.22.

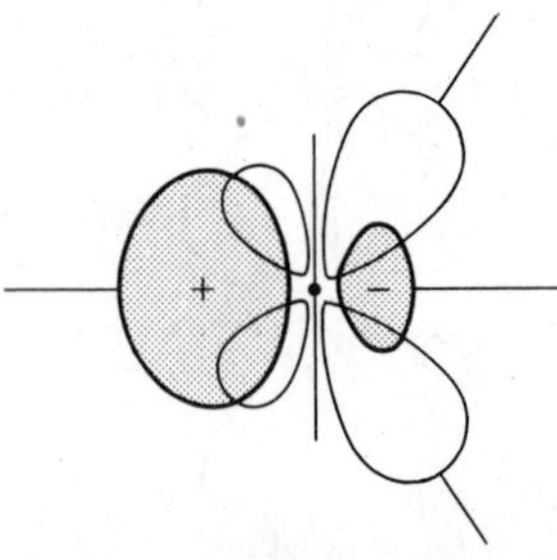

Fig. 16.22. Three orthogonal AOs hybridize to give three orthogonal hybrid orbitals. This is the third; its composition is given by eqn (16.3.5).

Now consider how these remarks apply to H_2O. If we ignore the O2s-contribution, there is no promotion energy, and the two bonds have moderately good (s,p)-overlap. When O2s-hybridization is allowed, the bond strength increases because the overlap improves, but a promotion energy is required because the 2s-electrons now take part in the bonding. The actual shape of the molecule, which is found by minimizing the total energy, is a compromise between strong bonding and promotion energy, and corresponds to a bond angle (104.45°) between these extremes.

In a number of important molecules (e.g. NH_3 and CH_4), there are more than two equivalent bonds. For instance, three equivalent hybrids may be formed from the basis $(2s, 2p_x, 2p_y)$. This is just a special case of the H_2O calculation with $\Phi = 120°$, and we find three hybrids of the form $(s+\sqrt{2}p)/\sqrt{3}$ directed towards the corners of an equilateral triangle, Fig. 16.23. The s- and p-orbitals are present in the proportion 1:2, and so the orbitals are called *sp^2-hybrids*. The basis $(2s, 2p_x, 2p_y, 2p_z)$ leads to four equivalent hybrids of composition $(s+\sqrt{3}p)/2$. These *sp^3-hybrids* point towards the corners of a regular tetrahedron, Fig. 16.23.

Numerous other forms of hybridization may occur, and those involving d-orbitals are particularly important. Some possibilities giving symmetrical arrangements are listed in Box 16.1.

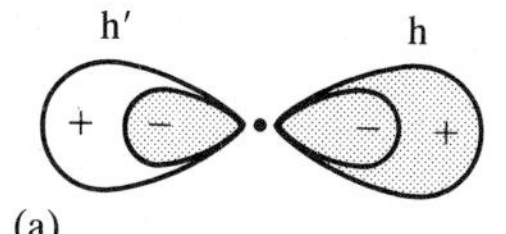

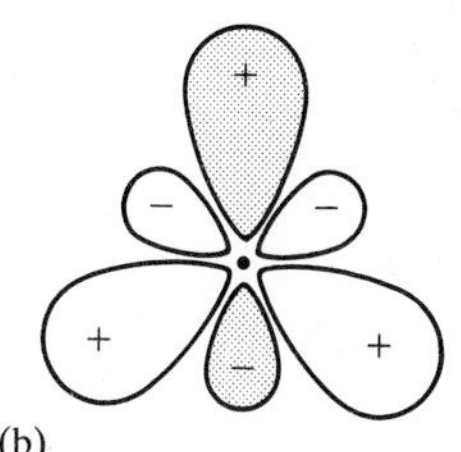

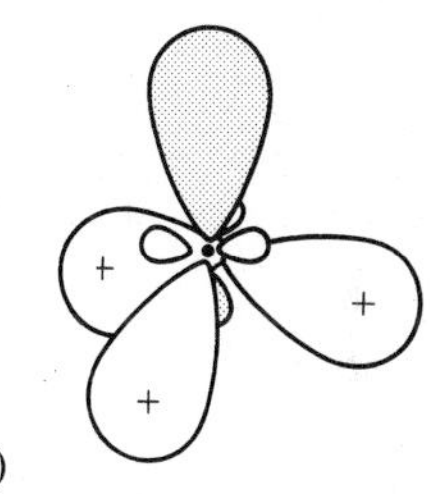

Fig. 16.23. Three important types of symmetrical equivalent hybrid orbitals: (a) linear (180°) sp-hybrids, (b) plane triangular (120°) sp^2-hybrids, and (c) tetrahedral (109°) sp^3-hybrids.

Box 16.1. Hybrid orbitals

Coordination number	Shape	Hybridization‡
2	Linear	**sp**, dp
	Bent	$\mathbf{p^2}$, ds, d^2
3	Trigonal planar	$\mathbf{sp^2}$, dp^2, ds^2, d^3
	Unsymmetrical planar	dsp
	Trigonal pyramidal	$\mathbf{p^3}$, d^2p
4	Tetrahedral	$\mathbf{sp^2}$, sd^3
	Irregular tetrahedral	spd^2, dp^3, pd^3
	Tetragonal pyramidal	d^4
5	Bipyramidal	$\mathbf{sp^3d}$, spd^3
	Tetragonal pyramidal	$\mathbf{sp^2d^2}$, sd^4, d^2p^3, pd^4
	Pentagonal planar	p^2d^3
	Pentagonal pyramidal	d^5
6	Octahedral	$\mathbf{sp^3d^2}$
	Trigonal prismatic	spd^4, pd^5
	Trigonal antiprismatic	p^3d^3

‡ Important forms are bold. More information may be found in H. Eyring, J. Walter, and G. E. Kimball, *Quantum chemistry*, John Wiley, New York.

16.4 Unsaturated and aromatic hydrocarbons

The structure of ethene (ethylene, $CH_2{=}CH_2$) is easy to understand in the light of what has already been described. The two carbon atoms are both approximately sp^2-hybridized, giving three almost equivalent orbitals in a plane, Fig. 16.24(a), and a single p-orbital perpendicular to it. Two hydrogen atoms form $(Csp^2, H1s)$-overlap σ-bonds to each atom, and the two CH_2 groups link through a (Csp^2, Csp^2)-overlap σ-bond. The latter draws the 2p-orbitals together: they overlap and form a π-bond. The *double bond* of unsaturated compounds is this $\sigma^2\pi^2$ configuration. The *torsional rigidity* of a double bond (its resistance to twisting) arises from the

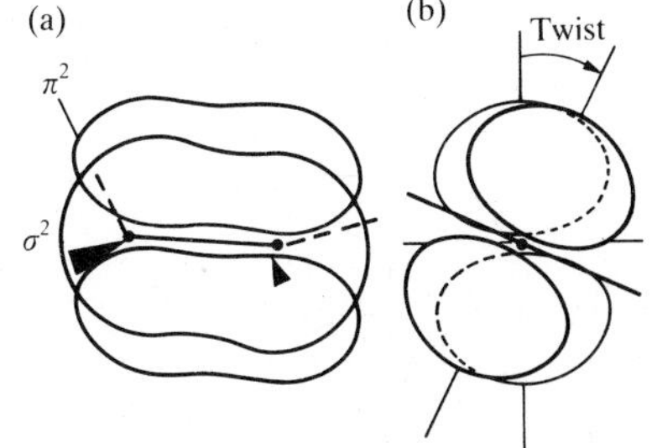

Fig. 16.24. The structure of a double bond consists of a σ-bond plus a π-bond. The bond is torsionally rigid because the (p, p)-overlap is reduced as the groups are twisted out of the plane.

reduction of the (C2p, C2p)-overlap and the rise of energy (the loss of bonding stabilization) that occurs as one CH_2 group is rotated relative to the other. The *reactivity* of a double bond, and unsaturated compounds in general, reflects the fact that the energy is lowered if the single π-bond is replaced by *two* σ-bonds, one on each atom.

16.4 (a) The Hückel approximation

The MO energy level diagrams of π-electron molecules, especially where there are *conjugated double bonds* (that is, an alternation of single and double bonds along a chain of carbon atoms) can be constructed using a set of approximations suggested by Erich Hückel. The MOs they lead to are examples of the delocalized orbitals mentioned above, orbitals that spread over all the atoms in the molecule and which cannot be identified with a bond between a particular pair of atoms.

The first two approximations are as follows:

(1) Separate the π-orbitals from the σ-orbitals, and treat the latter as a rigid framework. In other words, forget the σ-bonds except for the way that they determine the geometry of the molecule.

(2) Treat all the carbon atoms as identical. This means that all the Coulomb integrals (α in the notation used in Section 16.2(g)) are set equal.

For example, in ethene, we take the σ-bonds as given, and concentrate on finding the energies of the single π-bond and its companion π-antibond. In butadiene ($CH_2{=}CH{-}CH{=}CH_2$) the σ-framework $H_2C{-}CH{-}CH{-}CH_2$ is taken as given and we concentrate on finding the π-orbitals spreading across the four carbon atoms.

The next stage is to express the π-orbitals as LCAOs of the C2p-orbitals. In ethene we would write

$$\psi = c_A(\mathrm{C2p_A}) + c_B(\mathrm{C2p_B}), \tag{16.4.1}$$

and in butadiene

$$\psi = c_A(\mathrm{C2p_A}) + c_B(\mathrm{C2p_B}) + c_C(\mathrm{C2p_C}) + c_D(\mathrm{C2p_D}). \tag{16.4.2}$$

Next, the optimum coefficients and energies are found by the variation principle as explained in Section 16.2. This means that we have to solve the secular determinant, which in the case of ethene is eqn (16.2.10) with $\alpha_A = \alpha_B$. The determinant for butadiene is similar, but more atoms contribute and, being at various distances from each other, they have different overlap and resonance integrals:

$$\textit{Ethene:}\quad \begin{vmatrix} \alpha - E & \beta - ES \\ \beta - ES & \alpha - E \end{vmatrix} = 0$$

$$\textit{Butadiene:}\quad \begin{vmatrix} \alpha - E & \beta_{AB} - ES_{AB} & \beta_{AC} - ES_{AC} & \beta_{AD} - ES_{AD} \\ \beta_{BA} - ES_{BA} & \alpha - E & \beta_{BC} - ES_{BC} & \beta_{BD} - ES_{BD} \\ \beta_{CA} - ES_{CA} & \beta_{CB} - ES_{CB} & \alpha - E & \beta_{CD} - ES_{CD} \\ \beta_{DA} - ES_{DA} & \beta_{DB} - ES_{DB} & \beta_{DC} - ES_{DC} & \alpha - E \end{vmatrix} = 0.$$

The roots of the ethene determinant can be found very easily (they are the same as those in eqn (16.2.12)), but the roots of the butadiene determinant are obviously going to prove difficult to find. In a modern computation all the resonance integrals and overlap integrals would be computed, but a rough idea of the MO energy level diagram can be obtained very readily if we make the following additional Hückel approximations:

(3) All overlap integrals are set equal to zero.

(4) All resonance integrals between non-neighbours are set equal to zero.

(5) All remaining resonance integrals are set equal (to β).

These are obviously very severe approximations, but they let us calculate at least a general picture of the MO energy levels with very little work.

In the case of ethene, the Hückel approximations lead to

$$\begin{vmatrix} \alpha - E & \beta \\ \beta & \alpha - E \end{vmatrix} = 0 \qquad (16.4.3)$$

and its roots are $\alpha \pm \beta$ (compare this with eqn (16.2.12)). $\alpha + \beta$ corresponds to the bonding combination (β is negative) and $\alpha - \beta$ corresponds to the antibonding combination, Fig. 16.25. The building-up principle then leads to the configuration π^2, because each carbon atom supplies one electron to the π-system. This agrees with the previous qualitative discussion. However, it goes beyond, because we have an estimate of the π-bond energy (2β), and can see that an excited state of the molecule, when an electron is excited into the π^*-orbital, lies about 2β above the ground state. (β is often left as a parameter; a rough value for (C2p, C2p)-overlap π-bonds is about $-75\ \text{kJ mol}^{-1}$.)

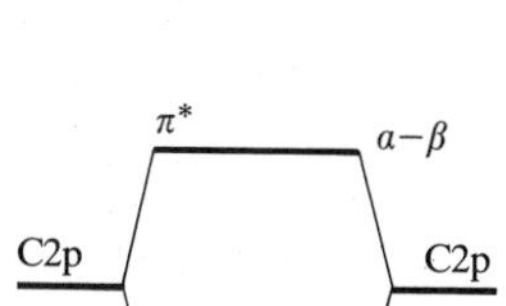

Fig. 16.25. The Hückel molecular orbital energy levels of ethene. Two electrons occupy the lower π-orbital.

In the case of butadiene, the approximations result in the determinant

$$\begin{vmatrix} \alpha - E & \beta & 0 & 0 \\ \beta & \alpha - E & \beta & 0 \\ 0 & \beta & \alpha - E & \beta \\ 0 & 0 & \beta & \alpha - E \end{vmatrix} = 0. \qquad (16.4.4)$$

This expands into the quadratic equation

$$x^2 - 3x + 1 = 0, \qquad x = (\alpha - E)^2/\beta^2$$

with the roots $x = 2.62,\ 0.38$. Therefore, the energies of the four LCAO-MOs are

$$E = \alpha \pm 1.62\beta,\ \alpha \pm 0.62\beta, \qquad (16.4.5)$$

as shown in Fig. 16.26. There are four electrons to accommodate, and so the ground state configuration is $1\pi^2 2\pi^2$.

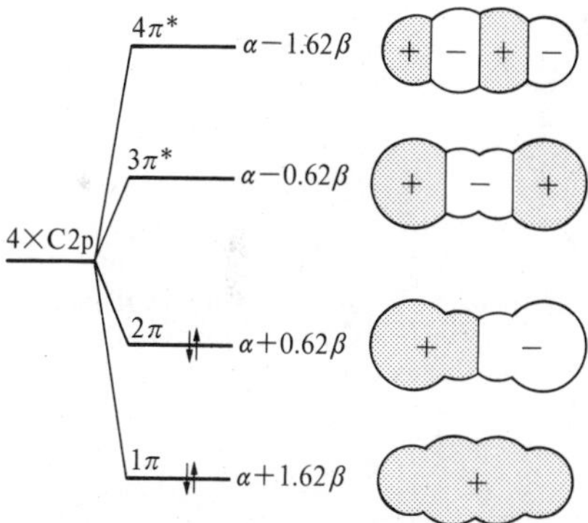

Fig. 16.26. The Hückel molecular orbital energy levels of butadiene and the top view of the corresponding π-orbitals. The four p-electrons (one supplied by each C) occupy the two lower π-orbitals. Note that the orbitals are delocalized.

There is an important point that emerges when we calculate the total π-electron binding energy in butadiene and compare it with what we find in ethene. In ethene the total energy is $2(\alpha + \beta) = 2\alpha + 2\beta$; in butadiene it is $2(\alpha + 1.62\beta) + 2(\alpha + 0.62\beta) = 4\alpha + 4.48\beta$. Therefore, the energy of the molecule lies lower by $(4\alpha + 4.48\beta) - 2(2\alpha + 2\beta) = 0.48\beta$ (roughly $-36\ \text{kJ mol}^{-1}$) than the sum of two π-bonds. This extra stabilization of a conjugated system is called the *delocalization energy*.

Example 16.5

Use the Hückel approximation to find the energies of the π-orbitals of cyclobutadiene, and estimate the delocalization energy.

• *Method*. Set up the secular determinant using the same LCAO as in eqn (16.4.2), but noting that A and D are now neighbours. Find the roots. Use the building-up principle to estimate the total π-bond energy. Subtract from this the energy of two localized π-bonds.

• *Answer*. The secular determinant, setting $z=(\alpha-E)/\beta$, is

$$\begin{vmatrix} z & 1 & 0 & 1 \\ 1 & z & 1 & 0 \\ 0 & 1 & z & 1 \\ 1 & 0 & 1 & z \end{vmatrix} = z\begin{vmatrix} z & 1 & 0 \\ 1 & z & 1 \\ 0 & 1 & z \end{vmatrix} - \begin{vmatrix} 1 & 1 & 0 \\ 0 & z & 1 \\ 1 & 1 & z \end{vmatrix} - \begin{vmatrix} 1 & z & 1 \\ 0 & 1 & z \\ 1 & 0 & 1 \end{vmatrix}$$

$$= z\left\{z\begin{vmatrix} z & 1 \\ 1 & z \end{vmatrix} - \begin{vmatrix} 1 & 1 \\ 0 & z \end{vmatrix}\right\} - \left\{\begin{vmatrix} z & 1 \\ 1 & z \end{vmatrix} - \begin{vmatrix} 0 & 1 \\ 1 & z \end{vmatrix}\right\} - \left\{\begin{vmatrix} 1 & z \\ 0 & 1 \end{vmatrix} - z\begin{vmatrix} 0 & z \\ 1 & 1 \end{vmatrix} + \begin{vmatrix} 0 & 1 \\ 1 & 0 \end{vmatrix}\right\}$$

$$= z(z^3-2z)-2z^2 = z^2(z^2-4)=0.$$

The solutions are $z^2=0$ and $z^2=4$, so that $z=0, 0, -2, +2$. The energies of the orbitals are therefore

$$E=\alpha+2\beta,\ \alpha,\ \alpha,\ \alpha-2\beta.$$

Four electrons must be accommodated. Two occupy the lowest orbital $(\alpha+2\beta)$, two occupy the doubly-degenerate orbitals at $E=\alpha$. The total energy, disregarding electron–electron repulsions, is therefore $4\alpha+4\beta$. Two isolated π-bonds would have an energy $4\alpha+4\beta$; therefore, in this case, the delocalization energy is zero.

• *Exercise*. Repeat the calculation for benzene. [Next subsection]

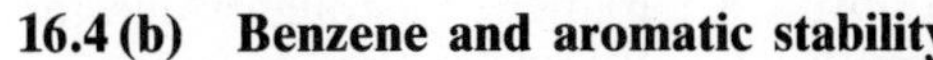

16.4 (b) Benzene and aromatic stability

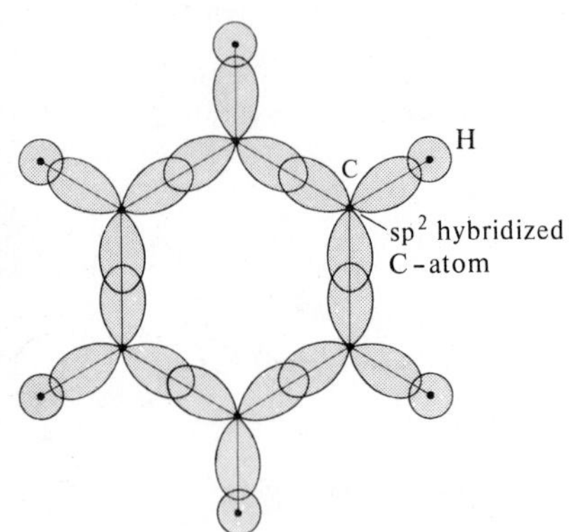

Fig. 16.27. The σ-framework of benzene is formed by the overlap of Csp^2-hybrids, which fit without strain into a hexagonal arrangement.

The most notable example of delocalization conferring extra stability is benzene and the aromatic molecules based on its structure. Its six carbon atoms are sp^2-hybridized and each has a single perpendicular 2p-orbital. One hydrogen is (Csp^2, H1s)-overlap σ-bonded to each carbon, and the remaining hybrids overlap to give a regular hexagon of atoms, Fig. 16.27. The internal angle of a regular hexagon is 120°, and so the sp^2-hybridization is ideally suited to forming σ-bonds. This is the first contribution to the stability of the aromatic ring: hexagonal geometry permits strain-free σ-bonding.

The six C2p-orbitals overlap to give six π-orbitals that spread all round the ring. Their energies can be calculated within the Hückel approximation by solving the secular determinant

$$\begin{vmatrix} \alpha-E & \beta & 0 & 0 & 0 & \beta \\ \beta & \alpha-E & \beta & 0 & 0 & 0 \\ 0 & \beta & \alpha-E & \beta & 0 & 0 \\ 0 & 0 & \beta & \alpha-E & \beta & 0 \\ 0 & 0 & 0 & \beta & \alpha-E & \beta \\ \beta & 0 & 0 & 0 & \beta & \alpha-E. \end{vmatrix} = 0. \quad (16.4.6)$$

This produces the energies

$$E=\alpha\pm2\beta,\quad \alpha\pm\beta,\quad \alpha\pm\beta, \quad (16.4.7)$$

as shown in Fig. 16.28. The orbitals there have been given special labels, which we shall explain in Chapter 17.

The building-up principle is now applied. There are six electrons to accommodate (one from each carbon), and so the three lowest orbitals (a_{2u} and the doubly-degenerate pair e_{1g}) are fully occupied, giving the ground state configuration $a_{2u}^2e_{1g}^4$. This raises the second feature contributing to aromatic stability: the only MOs occupied are those with net bonding character.

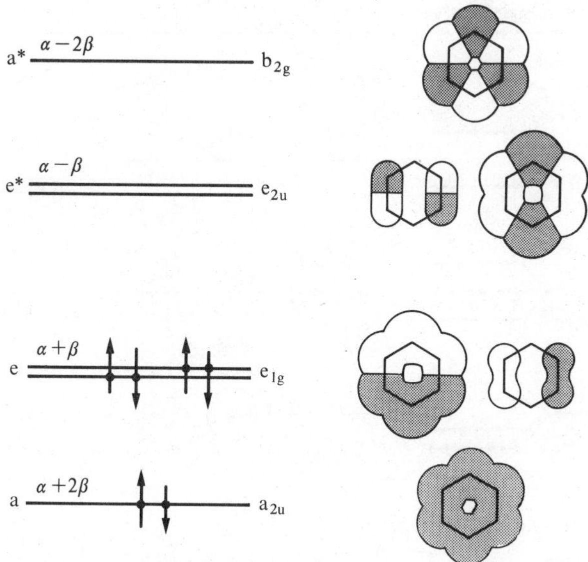

Fig. 16.28. The Hückel orbitals of benzene and the corresponding energy levels. The labels are explained in Chapter 17. The bonding and antibonding character of the delocalized orbitals reflects the numbers of nodes between the atoms. In the ground state, only the bonding orbitals are occupied.

The π-bond energy of benzene is $2(\alpha+2\beta)+4(\alpha+\beta)=6\alpha+8\beta$. If we ignored delocalization and thought of the molecule as having three isolated π-bonds, then it would be ascribed a π-bond energy $3(2\alpha+2\beta)=6\alpha+6\beta$. The delocalization energy is therefore 2β, around $-150\,\text{kJ mol}^{-1}$, considerably more than for butadiene.

Aromatic stability can therefore be traced to two main contributions. First, the geometry of the regular hexagon is ideal for the formation of strong σ-bonds: the σ-framework is relaxed and without strain. Second, the π-orbitals are such as to be able to accommodate all the electrons in bonding orbitals, and the delocalization energy is large.

16.5 Metals

The extreme case of conjugation is a metal, in which atom after atom lies in a three-dimensional array and takes part in bonding spreading throughout the sample. We shall consider a single, infinitely long line of atoms (a *one-dimensional metal*), each one carrying a single valence electron in an s-orbital. We can construct the LCAO-MOs of the metal by adding atoms to a line. Then we find the electronic structure using the building-up principle.

16.5 (a) Band theory

One atom contributes one s-orbital at some energy, Fig. 16.29(a). When a second atom is brought up it overlaps the first, and forms a bonding and antibonding orbital, Fig. 16.29(b). The third atom overlaps its nearest neighbour (and only slightly the next-nearest), and from these three atomic orbitals, three MOs are formed, Fig. 16.29(c). The fourth atom leads to the formation of a fourth MO, Fig. 16.29(d). At this stage we can begin to see that the general effect of bringing up successive atoms is to spread slightly the range of energies covered by the MOs, and also to fill in the range of energies with more and more orbitals (one more for each atom). When N atoms have been added to the line, there are N molecular orbitals covering

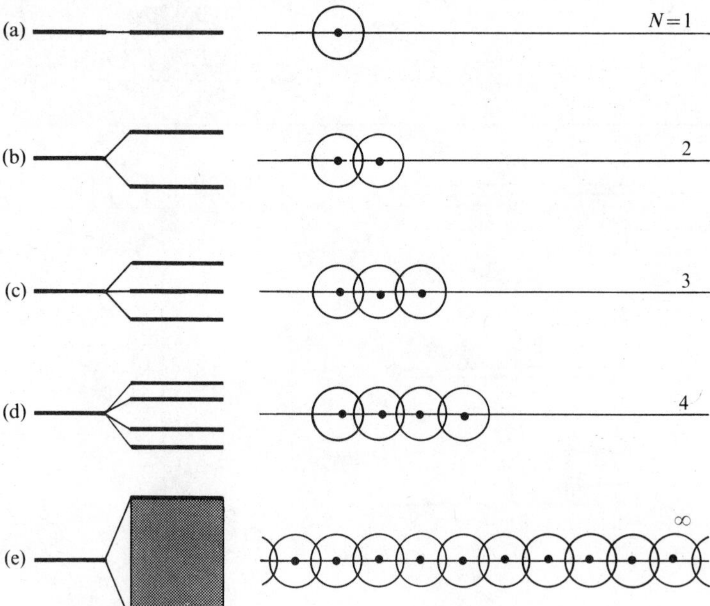

Fig. 16.29. The formation of a band of N orbitals by successive addition of atoms to a line. Note that the band remains of finite width, and although it looks continuous, it consists of N different orbitals.

a band of finite width and the Hückel secular determinant is

$$\begin{vmatrix} \alpha - E & \beta & 0 & 0 & 0 & 0 & \dots & 0 \\ \beta & \alpha - E & \beta & 0 & 0 & 0 & \dots & 0 \\ 0 & \beta & \alpha - E & \beta & 0 & 0 & \dots & 0 \\ \vdots & \vdots & \vdots & \vdots & \vdots & \vdots & & \vdots \\ 0 & 0 & 0 & 0 & 0 & 0 & \dots & \alpha - E \end{vmatrix} = 0 \qquad (16.5.1)$$

where β is now the (s, s)-resonance integral. The theory of determinants applied to such a symmetrical example as this (technically a *tridiagonal determinant*) leads to the following expression for the roots:

$$E = \alpha + 2\beta \cos\{k\pi/(N+1)\}, \qquad k = 1, 2, \dots, N. \qquad (16.5.2)$$

When N is infinitely large, the difference between neighbouring energy levels (the energies corresponding to k and $k+1$) is infinitely small, but the *band* still has finite width (the difference between the energies for $k = N$ and $k = 0$ is finite, and equal to -4β for $N \to \infty$). We can think of this band as consisting of N different MOs, the lowest-energy orbital ($k = 0$) being fully bonding, and the highest-energy orbital ($k = N$) being fully antibonding between neighbouring atoms, Fig. 16.30.

The band formed from (s, s)-overlap is called the *s-band*. If the atoms carry p-orbitals, the same procedure leads to a *p-band*, Fig. 16.30. If the atomic p-orbitals lie higher in energy than the s-orbitals, then the p-band lies higher than the s-band, and there may be a *band gap*.

Now consider the electronic structure of a metal formed from atoms each able to contribute one electron (e.g. the alkali metals). There are N atomic orbitals and therefore N molecular orbitals squashed into an apparently continuous band. There are N electrons to accommodate. Therefore, only the lowest $\frac{1}{2}N$ MOs are occupied, Fig. 16.31(a). The important feature is

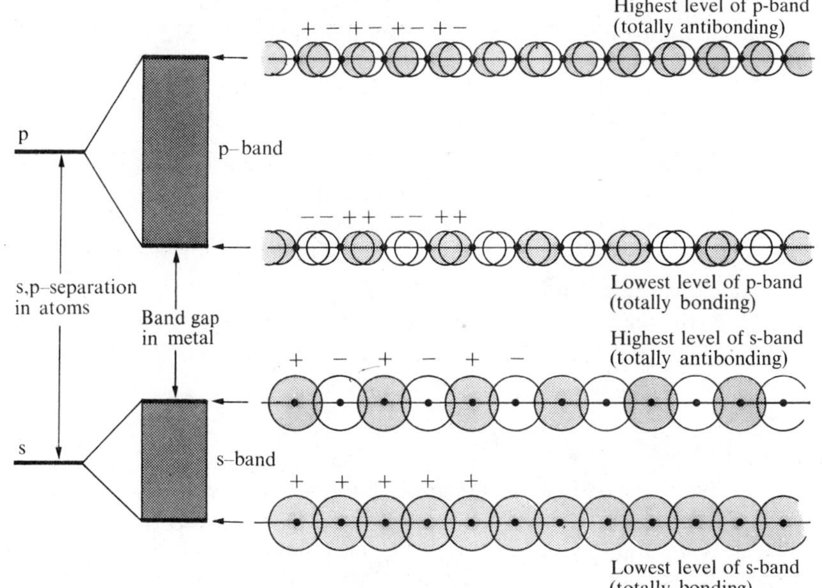

Fig. 16.30. The overlap of s-orbitals gives rise to an s-band, and the overlap of p-orbitals gives rise to a p-band. In this case the s- and p-orbitals of the atoms are so widely spaced that there is a band gap. In many cases the separation is less, and the bands overlap.

that there are empty levels lying very close to the uppermost filled level, the *Fermi level*, and so it requires hardly any energy to excite the uppermost electrons. Some of the electrons are therefore very mobile, and give rise to electrical conductivity. *Electrical conductivity is a characteristic of partially filled bands of orbitals*.

16.5 (b) Insulators and semiconductors

When each atom provides two electrons, the $2N$ electrons fill the N orbitals of the s-band. The Fermi level now lies at the top of the band, and there is a gap before the next band begins, Fig. 16.31(b). This gives rise to an *insulator*, because the electrons are no longer mobile. An example would be a chain of helium atoms in solid helium. In the elements of Group 2, the p-band overlaps the s-band, and so although beryllium provides two electrons per atom and apparently fills its 2s-band, the empty orbitals of the 2p-band are available, and the solid is a conductor.

In some materials there is only a small band gap, Fig. 16.31(c). If some of the electrons can be excited into the upper band the solid will be a conductor. This is called *semiconductivity*. It can be brought about in several ways. For example, thermal excitation (if the band gap is no greater than about kT) can generate enough of these *carriers*, and raising the temperature increases the conductivity (in normal metals the conductivity decreases as the temperature is raised). Another method is to introduce impurities into otherwise ultrapure materials. If these impurities can trap electrons, they withdraw electrons from the filled band, leaving holes which allow the remaining electrons to move. This gives rise to *p-type semiconductivity* (the *p* indicating that the holes are *p*ositive relative to the electrons in the band). Alternatively, an impurity (a *dopant*, since it is present by design, not accidentally) might carry excess electrons (e.g. P atoms introduced into germanium), and these additional electrons occupy otherwise empty bands, giving *n-type semiconductivity* (where *n* denotes

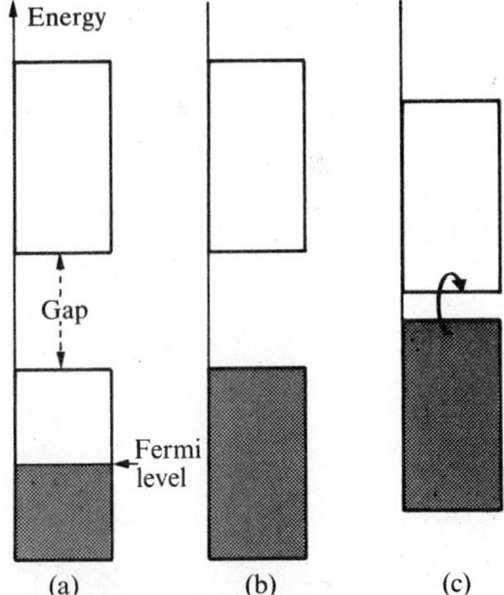

Fig. 16.31. (a) When N electrons occupy a band of N orbitals, it is only half full and the electrons near the Fermi level (the top of the filled levels) are mobile. (b) When $2N$ electrons are present, the band is full and the material is an insulator. (c) When the band gap is small, semiconduction may occur. In this case it is due to the thermal excitation of electrons across the gap, but it may be induced chemically with a dopant.

the *negative* charge of the carriers). The preparation of doped but otherwise ultrapure materials was described in Section 9.3(e).

16.6 Valence-bond theory

Molecular orbital theory is a natural extension of atomic orbital theory, and is the most widely used description of molecular structure. However, it was preceded by *valence-bond theory*, and some of the language and ideas this theory introduced are still widely encountered. Valence-bond (VB) theory developed from the insight of the early chemists, and concentrates on the role of electron pairs from the outset instead of seeing them as a natural consequence of the building-up principle. Thus, it considers the two-electron bond in H_2 as the primitive structure.

16.6 (a) The electron pair

When two hydrogen atoms are far apart, their wavefunctions are $1s_A$ or $1s_B$, and if electron 1 is on A and electron 2 is on B, then the two atoms have the overall wavefunction $\psi = \psi_{1s_A}(\boldsymbol{r}_1)\psi_{1s_B}(\boldsymbol{r}_2)$, or A(1)B(2) for short. (That wavefunctions of independent systems can be multiplied was demonstrated in Section 15.2.) The product wavefunction is accurate at large nuclear separations; VB theory assumes that it can be retained when the atoms are close together, so long as the possibility of electron 1 then escaping to atom B (and electron 2 to atom A) is taken into account. This is done by allowing the wavefunction to consist of equal mixtures of A(1)B(2) and A(2)B(1) and writing it as

$$\psi_{VB} = A(1)B(2) + A(2)B(1). \tag{16.6.1}$$

This wavefunction (which is not normalized) is symmetrical under particle interchange. Therefore, in order to fulfil the Pauli principle, it must be multiplied by the antisymmetrical spin wavefunction $\alpha(1)\beta(2) - \alpha(2)\beta(1)$. As we have seen (Section 15.2), this corresponds to the two electron spins being paired, and hence, as in MO theory, the bond is formed by an electron pair.

The probability density of the electrons in the bond is proportional to the square of the VB wavefunction:

$$\psi_{VB}^2 = A(1)^2B(2)^2 + A(2)^2B(1)^2 + 2A(1)B(1)A(2)B(2). \tag{16.6.2}$$

This may be compared with the (non-normalized) probability density given by the MO for the hydrogen molecule:

$$\begin{aligned}\psi_{MO}^2 &= \{\sigma(1)\sigma(2)\}^2 = \{[A(1) + B(1)][A(2) + B(2)]\}^2 \\ &= A(1)^2B(2)^2 + A(2)^2B(1)^2 + 4A(1)A(2)B(1)B(2) \\ &\quad + A(1)^2A(2)^2 + B(1)^2B(2)^2 + \ldots \end{aligned} \tag{16.6.3}$$

Terms involving A(1)B(1) and A(2)B(2) are common to both probability densities. As we saw in Section 16.1, these represent the enhancement of electron density in the overlap region. Therefore, as in MO theory, the strength of the bond in VB theory can be traced in large measure to the effects arising from the accumulation of electron density in the internuclear region. The differences between the descriptions are due to the extra terms in the MO density, such as $A(1)^2A(2)^2$. These terms represent a distribution in which electrons 1 and 2 both occupy the same atomic orbital. $A(1)^2A(2)^2$, for instance, represents the *ionic contribution* H^-H^+. The corresponding terms do not appear in the VB wavefunction, and so the latter treats the bond as purely covalent.

The exclusion of ionic terms in the description of the bond is too stringent (because there must be some chance that both electrons will be near the same nucleus), and so we can anticipate that the VB description of H_2 (and any diatomic) can be improved by adding additional ionic terms of the form

$$\psi^{\text{ionic}} = A(1)A(2) + B(1)B(2) \tag{16.6.4}$$

to the original wavefunction and obtaining the better description

$$\psi = \psi_{\text{VB}} + c\psi^{\text{ionic}}. \tag{16.6.5}$$

The amount of ionic character is proportional to c^2, and the mixing is called *ionic–covalent resonance*. If $c = 1$, the improved wavefunction is identical to the MO wavefunction. This wavefunction therefore gives equal weight to the ionic and covalent structures, which is also unreasonable. Therefore, the MO description *over-emphasizes* the contribution of ionic structures.

16.6 (b) Resonance and aromaticity

The VB description of the π-orbitals of benzene examines a single Kekulé structure and identifies three bonds, Fig. 16.32. If the atoms are labelled A, B, . . . , F and the electrons 1, 2, . . . , 6, then one bond is $A(1)B(2) + A(2)B(1)$ because formally it is like an isolated hydrogen molecule (but A and B are now C2pπ-orbitals). Another bond is $C(3)D(4) + C(4)D(3)$ and the third is $E(5)F(6) + E(6)F(5)$. Therefore, the overall wavefunction is

$$\psi_{\text{VB}}^{\text{Kek1}} = \{A(1)B(2) + A(2)B(1)\}\{C(3)D(4) + C(4)D(3)\}\{E(5)F(6) + E(6)F(5)\} \tag{16.6.6}$$

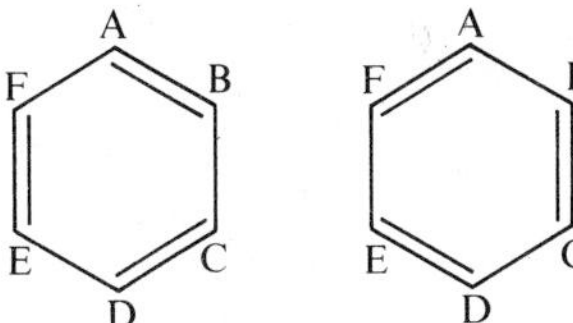

Fig. 16.32. The two Kekulé structures of benzene.

because the independent bond wavefunctions multiply together. This may look complicated, but it can be analysed in the same way as for the hydrogen molecule. As it stands it forbids bonding electron density between carbon atoms B and C, D and E, and F and A. This can be allowed by considering the alternative Kekulé structure, and writing

$$\psi = \psi_{\text{VB}}^{\text{Kek1}} + \psi_{\text{VB}}^{\text{Kek2}}. \tag{16.6.7}$$

The mixing of structures of equal (or almost equal) energy is called *resonance*. It results in a lower energy (it is a better description) because the electrons are distributed over all the bonding regions.

Bonding may also occur between non-neighbouring atoms, and a better description is obtained by allowing the *Dewar structures*, Fig. 16.33(a). The variation principle leads to the result that each Dewar structure contributes about 6%, and each Kekulé structure contributes about 40%. However, resonance among these covalent structures is not the complete story, because we should also allow ionic–covalent mixing, taking into account structures of the kind shown in Fig. 16.33(b). There are many such ionic terms, especially for molecules more complex than benzene, and the difficulty of taking them all into account is one of the reasons why VB theory has undergone less development than MO theory.

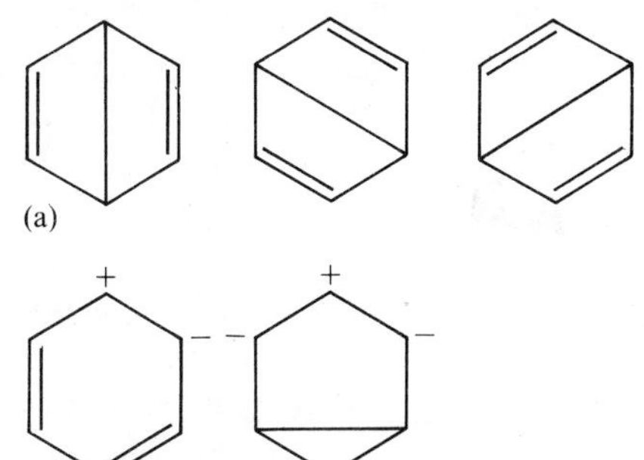

Fig. 16.33. Many structures resonate in the VB description of benzene. They include (a) the Dewar structures and (b) singly and doubly excited ionic structures (and many more).

Further reading

Principles:

The shape and structure of molecules (2nd edn). C. A. Coulson (revised by R. McWeeny); Oxford University Press, 1982.
Coulson's Valence. R. McWeeny; Oxford University Press, 1979.
The nature of the chemical bond. L. Pauling; Cornell University Press, 1960.
Valence theory. J. N. Murrell, S. F. A. Kettle, and J. M. Tedder; Wiley, New York, 1965.
Molecular quantum mechanics (2nd edn). P. W. Atkins; Oxford University Press, 1983.
Quantum chemistry. D. A. McQuarrie; University Science Books and Oxford University Press, 1983.
Chemical structure and bonding. R. L. De Kock and H. B. Gray; Benjamin Cummings, Menlo Park, 1980.

Results of calculations:

Atoms and molecules. M. Karplus and R. N. Porter; Van Nostrand, New York, 1970.
The organic chemist's book of orbitals. W. L. Jorgensen and L. Salem; Academic Press, New York, 1973.
Molecular wavefunctions. E. Steiner; Cambridge University Press, 1976.
Quantum chemistry (the development of ab initio methods in molecular electronic structure theory). H. Shaefer III; Clarendon Press, Oxford, 1984.

Solids:

Band theory of metals. S. L. Altmann; Pergamon Press, Oxford, 1970.
Electronic structure and the properties of solids. W. A. Harrison; W. H. Freeman & Co., San Francisco, 1980.

Introductory problems

A16.1. Give the ground state configuration of the following species: Li_2, Be_2, C_2. Give the bond order in each case.

A16.2. Use the ground state configurations of B_2 and C_2 to predict which should have the greater bond dissociation energy. Explain.

A16.3. The term symbol for the ground state of N_2^+ is $^2\Sigma_g$. Find the spin and total angular momentum for the molecule. Give the electron configuration of the molecule.

A16.4. One of the upper states of the C_2 molecule has the electron configuration $(1s\sigma_g)^2(1s\sigma_u^*)^2(2s\sigma_g)^2(2s\sigma_u^*)^2(2p\pi_u)^3$ $(2p\pi_g)^1$. Give the multiplicity and the parity of this state.

A16.5. Use the electron configurations of NO and O_2 to predict which should have the shorter internuclear distance.

A16.6. One of the excited states of the H_2 molecule has the term symbol $^3\Pi_u$. In this state, the molecule is stable with respect to a hydrogen atom in the ground state and a hydrogen atom in an excited state. Give the electron configuration of the molecule.

A16.7. One of the sp^2-hybrid orbitals has the form $3^{-\frac{1}{2}}(s + 2^{\frac{1}{2}}p_x)$. Show that it is normalized.

A16.8. The sp^2-hybrid orbital that extends in a direction lying in the xy-plane and making an angle of 120° with the x-axis has the form $3^{-\frac{1}{2}}(s - 2^{-\frac{1}{2}}p_x + 3^{\frac{1}{2}}2^{-\frac{1}{2}}p_y)$. Using hydrogen 2s, $2p_x$, and $2p_y$ orbitals, write the wave function explicitly. Show it is symmetric about the above direction by identifying the angular dependence of the hybrid orbital with that of the projection of the vector **r** on a line in the above direction.

A16.9. The bond angle in H_2S is 92.2°. Estimate the per cent s-character in the H_2S bonding hybrid orbital constructed from s, p_x, and p_y orbitals. Estimate the per cent s-character in the non-bonding hybrid orbital obtained in the above hybridization.

A16.10. Verify the fact that the hydrogen $2p_x$ and $2p_y$ orbitals are orthogonal by showing that the ϕ portion of their overlap integral vanishes.

Problems

16.1. Wavefunctions are just like any other mathematical functions, and they can be manipulated in the same way. One of their aspects that was encountered throughout this chapter was their interference (just like waves in any kind of medium). The first few Problems in this chapter give some practice with calculations of this kind. First, show that a wave $\cos k_1x$ centred on A (so that, for this wave, x is measured from A) interferes with a similar wave $\cos k_2x$ centred on B (so that, for this wave, x is measured from B) to give an enhanced amplitude half-way between A and B when $k_1 = k_2 = \pi/2R$, but that the waves interfere destructively if $k_1R = \frac{1}{2}\pi$, $k_2R = \frac{3}{2}\pi$, (R is the separation of A and B).

16.2. In molecules the atomic wavefunctions are not extended waves (like $\cos kx$) but are localized around the nuclei. Their amplitudes spread into common regions of space, and interference occurs that is essentially of the same kind as in the last Problem. The expression in eqn (16.1.4) gives the form of the superposition involved in the H_2 molecule. Take $1s_A = \exp(-r/a_0)$ with r measured from A (and questions of normalization ignored at this stage), and $1s_B = \exp(-r/a_0)$ with r measured from B. Plot the amplitude of the bonding molecular orbital along the internuclear axis for an internuclear distance of 106 pm.

16.3. Repeat the calculation for the antibonding combination $1s_A - 1s_B$, and notice how the node appears in the internuclear region.

16.4. Sketch the forms of the interference patterns that arise for $\psi_A + \lambda\psi_B$ with (i) $\lambda = 0.5$, (ii) $\lambda = -0.5$ using (a) $\psi_{A,B} = \cos kx$, $k = \pi/2R$, (b) $\psi_{A,B} = e^{-r/a_0}$ centred on each atom.

16.5. Normalize the orbital $(1s_A) + \lambda(1s_B)$ in terms of the parameter λ and the overlap integral.

16.6. One orbital is *orthogonal* to another if the integral $\int \psi'^*\psi \, d\tau = 0$. Confirm that the $1s\sigma$- and $1s\sigma^*$-bonding and antibonding orbitals are mutually orthogonal, and find the orbital orthogonal to $(1s_A) + \lambda(1s_B)$, with λ arbitrary.

16.7. Plot the electron density along the internuclear axis (both between and beyond the nuclei) for the H_2^+ molecule using the orbitals constructed above. Make sure that the orbitals are normalized to unity: this is easily done by dividing the value of $1s_A + 1s_B$ by 1218 $pm^{\frac{3}{2}}$ and that of $1s_A - 1s_B$ by 622 $pm^{\frac{3}{2}}$. Superimpose on this diagram the simple sum of electron densities $(1s_A^2 + 1s_B^2)/9.35 \times 10^5\ pm^3$ that would be obtained if we disregarded interference effects (the divisor ensures that the normalization is correct).

16.8. The diagram just constructed shows the densities along the line joining the nuclei. In order to clarify the shifts of electron density that take place on molecule formation it is usual to plot the *difference density*. Plot the change in electron density that occurs on the formation of a bond and an antibond, by subtracting the appropriate curves.

16.9. Imagine a small electron-sensitive probe of volume 1 pm^3 being inserted into a H_2^+ molecule in its ground state. What is the probability that it will register the presence of an electron when it is inserted at the following positions: (a) at nucleus A, (b) at nucleus B, (c) half-way between A and B, (d) at a point arrived at by moving 20 pm from A along the axis towards B and then 10 pm perpendicularly?

16.10. What probabilities would occur if the same probe were inserted into a H_2^+ molecule in which the electron had just been excited into the antibonding orbital (and the molecule had not yet dissociated)?

16.11. The energy of an H_2^+ molecule with an internuclear distance R is given by the expression $E = E_H - [V_1(R) + V_2(R)]/[1 + S(R)] + e^2/4\pi\varepsilon_0 R$, where E_H is the energy of an isolated hydrogen atom, V_1 is the attraction between the electron in the orbital centred on one nucleus and the other nucleus, V_2 is the attraction between the overlap density and a nucleus, and S is the overlap integral between the two atomic orbitals. All the terms depend on the internuclear distance (except E_H) and their values are given below. Plot the potential energy curve for the molecule and find (a) the bond dissociation energy (in eV), (b) the equilibrium bond length.

R/a_0	0	1	2	3	4
$V_1/\bar{R}_H$	1.000	0.729	0.473	0.330	0.250
$V_2/\bar{R}_H$	1.000	0.736	0.406	0.199	0.092
S	1.000	0.858	0.587	0.349	0.189

You will need to evaluate the nuclear repulsion term: $E_H = -\frac{1}{2}\bar{R}_H$, $\bar{R}_H = 27.3$ eV and $a_0 = 53$ pm. Why does V_2 decrease so rapidly?

16.12. From the same data, plot the energy of an H_2^+ molecule when the electron occupies the antibonding orbital. (The energy expression changes by $V_2 \to -V_2$ and $S \to -S$.)

16.13. Derive the expression used in Problem 16.11. First, normalize the LCAO wavefunctions. Then evaluate the expectation value of the hamiltonian. Write the Schrödinger equation for H_2^+. We know, however, that $1s_A$ and $1s_B$ both individually satisfy the hydrogen atom Schrödinger equation. You should recognize the occurrence of the atom equation inside the H_2^+ equation, and therefore you will be able to replace parts of the left-hand side by E_H. Proceed from that point, and deduce the bonding and antibonding energy expressions. The nuclear repulsion term is then tacked on at the end.

16.14. The *overlap integral* can be calculated by graphical, numerical, or analytical integration. The simplest case is when we are interested in the overlap of two hydrogen 1s-orbitals on nuclei separated by a distance R, for then $S = [1 + (R/a_0) + (R^2/3a_0^2)] \exp(-R/a_0)$. Plot this function for $0 \leq R \leq \infty$, and confirm the entries in the table of data for Problem 16.11.

16.15. When an s-orbital approaches a p-orbital along the latter's axis, the overlap increases and then drops to zero when the centres of the orbitals coincide. Why? The analytical expression for 1s, 2p overlap of this kind is

$(R/2a_0)\ [1+(R/a_0)+(R^2/3a_0^2)]\exp(-R/a_0)$. Plot this function, confirm the above remark, and find the separation for which the overlap is a maximum.

16.16. Use Fig. 16.12 to give the configurations of the following species: H_2^-, N_2, O_2, CO, NO, CN.

16.17. Which of the species N_2, NO, O_2, C_2, F_2, CN would you expect to be stabilized (a) by the addition of an electron to form AB^-, (b) by ionization to AB^+?

16.18. A transition metal was vaporized in a furnace, and spectroscopic analysis showed the presence of diatomic molecules and ions. What types of molecular orbitals may be formed when atoms stick together by using their d-orbitals? Sketch the orbitals. Give the configurations of the diatomics formed when each atom brings up (a) 1, (b) 3, (c) 4 d-electrons.

16.19. Draw the molecular orbital diagram for (a) CO, (b) XeF, and use the *aufbau* principle to put in the appropriate number of electrons. Is XeF^+ likely to be more stable than XeF?

16.20. Give the g and u character of the following types of molecular orbital: (a) π^* in F_2, (b) σ^* in NO, (c) δ in Tl_2, (d) δ^* in Fe_2, (e) the six orbitals of the benzene molecule (Fig. 16.28).

16.21. What is the energy required to remove a potassium ion from its equilibrium distance of 294 pm in the K^+Br^- ion pair? Assume negligible repulsive forces.

16.22. The ionization energies of some alkali metal atoms and the electron affinities of some halogen atoms are listed below. Draw up a table showing the separation of the metal and halogen atoms at which it becomes energetically favourable for M^+X^- to form.

$I/\mathrm{kJ\ mol^{-1}}$	520.3 (Li)	495.8 (Na)	418.9 (K)
$E_a/\mathrm{kJ\ mol^{-1}}$	332.7 (F)	348.6 (Cl)	328 (Br)

16.23. Derive an expression for the extent of promotion of a $1s^22p^3$ atom needed in order to achieve three equivalent and coplanar bonds making an angle 120° to each other.

16.24. Calculate the extent of promotion of a $2s^22p^3$ atom in order to achieve pyramidal hybridization with three equivalent bonds making angles Θ to each other and Φ to a lone pair. Express your answer in terms of Θ alone, and evaluate the expression for NH_3 ($\Theta = 106.7°$).

16.25. Which of the following species do you expect to be linear: CO_2, NO_2, NO_2^+, NO_2^-, SO_2, H_2O, H_2O^{2+}? Give reasons.

16.26. Which of the following species do you expect to be planar: NH_3, NH_3^{2+}, CH_3, NO_3^-, CO_3^{2-}? Give reasons.

16.27. Construct the molecular orbital diagram of (a) ethene, (b) ethyne (acetylene) on the basis that the molecules are formed from the appropriately hybridized CH_2 or CH fragments.

16.28. A simple model of conjugated polyenes allows their electrons to roam freely along the chain of atoms. The molecule is then regarded as a collection of independent particles confined to a box, and the molecular orbitals are taken to be the square-well wave functions. This is the *free-electron molecular orbital* (FEMO) approximation. As a first example of employing it, take the butadiene molecule with its four π-electrons and show that for its FEMO description we require the lowest two particle-in-a-box wavefunctions. Sketch the form of the orbitals required, and compare them with the orbitals in Fig. 16.26.

16.29. An advantage of the FEMO approach is that it lets us draw some quantitative conclusions with very little effort. What is the minimum excitation energy of butadiene?

16.30. Consider the FEMO description of the molecule $CH_2{=}CHCH{=}CHCH{=}CHCH{=}CH_2$ and regard the electrons as being in a box of length $8R_{CC}$ (as in this case, an extra $\frac{1}{2}R_{CC}$ is often added at each end of the molecule). What is the minimum excitation energy of this molecule? What colour does the molecule absorb from white light? What colour does it then appear? Sketch the form of the uppermost filled orbital. Take $R_{CC} = 140$ pm.

16.31. As an example of the application of the variation principle, take a trial function for the hydrogen atom of the form (a) e^{-kr}, (b) e^{-kr^2} and find the optimum value of k in each case. Observe that e^{-kr} has the form of the true ground state wavefunction, and that the energy it corresponds to is lower than that of the optimum form of e^{-kr^2}. In the calculation, use the following form for the kinetic energy operator: $T = (-\hbar^2/2\mu)(1/r)(\mathrm{d}^2/\mathrm{d}r^2)r$, which is the radial part of the full operator.

16.32. Write down the secular determinant for (a) linear H_3, (b) cyclic H_3 using the Hückel approximation. Solving even this simple system involves dealing with a cubic equation unless the symmetry is taken into account (using the techniques described in Chapter 17). The analytical solutions for cubic equations are given in M. Abramowitz and I. A. Stegun, *Handbook of mathematical functions*, Dover, 1965 (Section 3.8.2), but their roots can also be found numerically.

16.33. Predict the electronic configuration of (a) the benzene negative ion, (b) the benzene cation. Estimate the π-bond energy in each case.

16.34. Write the valence-bond wavefunction for the HF molecule (regarding it as being formed from an H1s-orbital and an F2p_z-orbital) (a) supposing it to be purely covalent, (b) supposing it to be purely ionic, (c) supposing it to be 80 per cent covalent and 20 per cent ionic.

16.35. Draw the covalent and ionic structures that should be included in a complete description of the π-electron structure of cyclobutadiene.

16.36. Draw the covalent and singly polar (i.e. one positive and one negative charge) valence-bond structures of the π-electrons in naphthalene. (Go as far as covalent structures with one long bond.)

17

Symmetry: its description and consequences

Learning objectives

After careful study of this chapter you should be able to:

(1) Define *symmetry element* and *symmetry operation*, Section 17.1(a).

(2) Identify an *axis of symmetry*, a *plane of symmetry*, a *centre of symmetry*, and an *axis of improper rotation* in a molecule, Section 17.1(a).

(3) Classify a molecule according to its symmetry group, Section 17.1(b) and Box 17.1.

(4) Decide, from the point group of the molecule, whether it can be polar or optically active, Section 17.1(c).

(5) Explain what is meant by a *group multiplication table* and construct one for any molecule, Section 17.2(a).

(6) Explain what is meant by a *basis*, a *representative*, and a *matrix representation*, and find the matrix representatives for a given basis, Section 17.2(a) and Example 17.2.

(7) Define the *character* of an operation, Section 17.2(b), and explain the content of a *character table*, Section 17.2(c).

(8) State the meaning of an *irreducible representation* and its *symmetry species*, Section 17.2(c).

(9) Deduce the transformation properties of functions under the symmetry operations of a group, Section 17.2(d) and Example 17.4.

(10) State and use the criterion for an integral being necessarily zero, Section 17.3(a) and Example 17.5.

(11) Use character tables to select orbitals having non-zero overlap, Section 17.3(b) and Example 17.6.

(12) Explain what is meant by a *selection rule*, and use character tables to deduce the rules for spectral transitions, Section 17.3(c) and Example 17.7.

(13) Explain what is meant by a *symmetry-adapted orbital*, and use character tables to construct them from a given basis, Section 17.3(d).

(14) Explain the origin of the symmetry labels of molecular orbitals, Section 17.3(d) and Example 17.6.

Introduction

We have seen the quantum-mechanical reasons for the shapes adopted by molecules, and in the following chapters we shall see how their structures are determined experimentally. In this chapter we sharpen the concept of 'shape' and show that the symmetries of molecules may be discussed systematically. We shall see how to classify any molecule according to its symmetry, and to use this classification to discuss molecular properties without the need for detailed calculation.

The detailed discussion of symmetry is called *group theory*. Much of group theory is a systematic summary of common sense about the symmetries of objects, and this common-sense background should never be forgotten. However, because group theory is systematic, its rules can be applied in a straightforward, mechanical way, and in some cases unexpected results are obtained. In most cases, the theory gives a simple, direct method for arriving at useful conclusions with the minimum of calculation, and this is the aspect we stress here.

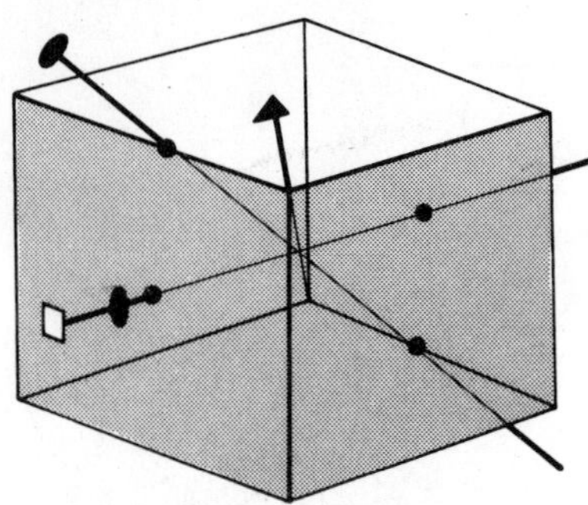

Fig. 17.1. Some of the symmetry elements of a cube. The twofold, threefold, and fourfold axes are labelled with the conventional symbols.

17.1 The symmetry elements of objects

Some objects are 'more symmetrical' than others. For example, a sphere is more symmetrical than a cube because it looks the same after it has been rotated through any angle about any diameter, whereas a cube looks the same only if it is rotated through 90°, 180°, or 270° about axes passing through the centres of its faces, Fig. 17.1, or by 120° or 240° about axes passing through opposite corners. Similarly the NH_3 molecule is 'more symmetrical' than H_2O because it looks the same after rotations of 120° or 240° about the axis shown in Fig. 17.2, whereas H_2O looks the same only after a rotation of 180°. Operations that leave objects looking the same are called *symmetry operations*. There is a corresponding *symmetry element* for each operation: these are the points, lines, and planes with respect to which the symmetry operation is performed. Objects can be classified into symmetry *groups* by identifying all their symmetry elements. This puts a sphere into a different group from a cube, and NH_3 into a different group from H_2O.

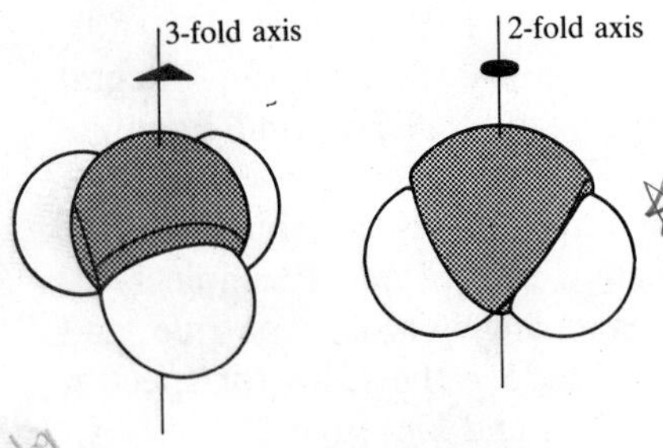

Fig. 17.2. (a) The NH_3 molecule has a threefold (C_3) axis and (b) the H_2O molecule has a twofold (C_2) axis.

17.1 (a) Operations and elements

There are five kinds of symmetry operation (and five kinds of symmetry element) that leave at least a single point unchanged. These give rise to the *point groups*. When considering crystals (Chapter 23), the symmetries arising from the translation of objects through space give rise to the *space groups*.

(1) The *identity* (symbol: E). This operation consists of doing nothing; the corresponding element is the entire object. Since every molecule is indistinguishable from itself if nothing is done to it, every object possesses at least this symmetry. One reason for including it is that some molecules (e.g. CHClBrF) have only this symmetry element; another is technical and connected with the formulation of group theory.

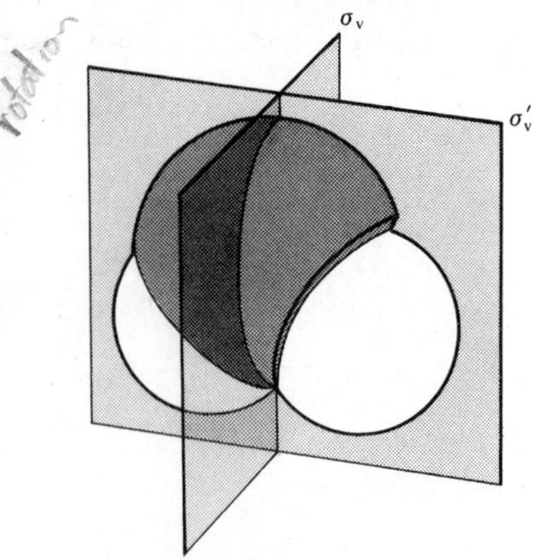

Fig. 17.3. The H_2O molecule has two mirror planes. They are both vertical (i.e. contain the principal axis), and so are denoted σ_v and $\sigma_{v'}$.

(2) A *rotation* (the operation) about an *axis of symmetry* (the corresponding element). If a rotation through $360°/n$ leaves the molecule apparently unchanged, it is said to have an *n-fold axis of symmetry* (symbol: C_n). This means that H_2O has a twofold axis (C_2). NH_3 has one threefold axis (C_3) but *two* corresponding threefold rotations (also symbolized C_3) because we must distinguish a clockwise from a counter-clockwise rotation. (There is only one twofold rotation corresponding to a twofold axis because clockwise and counter-clockwise rotations are indistinguishable.) A cube has three C_4 axes (and six fourfold rotations), four C_3 axes, and six C_2 axes, but even this high symmetry is capped by a sphere, which possesses an infinite number of symmetry axes (any diameter) of all possible orders of n. An object may possess several rotation axes; then the one (or more) with the greatest value of n is called the *principal axis*. From now on we shall denote the operation and the element by the same symbol (C_n), and the context will always make the meaning clear.

(3) A *reflection* in a *plane of symmetry* or a *mirror plane* (symbol: σ). If reflection in a plane passing through the molecule leaves it apparently unchanged, then it has a *plane of symmetry*. If the plane contains the principal axis, then it is called *vertical* and denoted σ_v. H_2O has two vertical planes of symmetry, Fig. 17.3; NH_3 has three. When the plane of symmetry is perpendicular to the principal axis it is called *horizontal* and denoted σ_h. The benzene molecule has a C_6 principal axis and a horizontal mirror plane (as well as several other elements). When a mirror plane is vertical (i.e. contains the principal axis) and bisects the angle between two C_2 axes that are themselves perpendicular to the principal axis, Fig. 17.4, it is called a *dihedral* plane and denoted σ_d.

(4) An *inversion* through a *centre of symmetry* (symbol: i). Imagine taking each point in an object, moving it to its centre, and then moving it out the same distance on the other side. If the object looks unchanged, it has a centre of inversion. Neither H_2O nor NH_3 possesses the element i, but both the sphere and the cube do; so do benzene and a regular octahedron, Fig. 17.5, but a regular tetrahedron and CH_4 do not.

(5) An *improper rotation* (or a *rotary-reflection*) about an *axis of improper rotation* (or a *rotary-reflection axis*). An n-fold improper rotation (symbol: S_n) consists of an n-fold rotation followed by a horizontal reflection. CH_4, Fig. 17.6, has three S_4 axes (and six corresponding operations, three clockwise and three counter-clockwise).

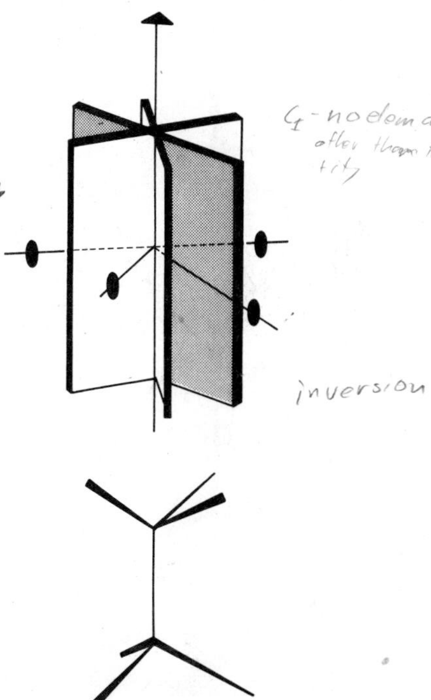

Fig. 17.4. Dihedral mirror planes (σ_d) bisect the C_2 axes perpendicular to the principal axis. The staggered conformation of ethane provides an example.

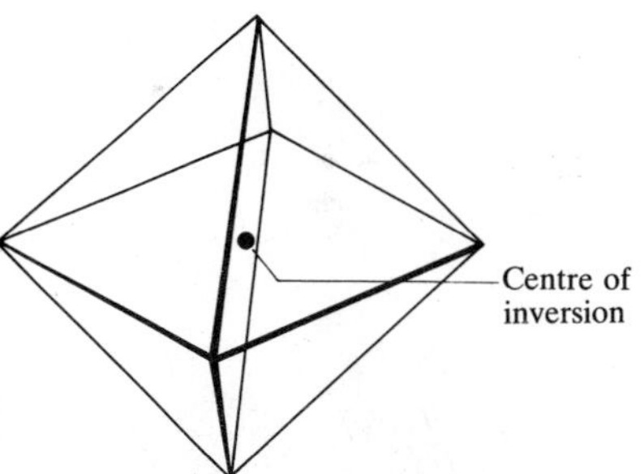

Fig. 17.5. A regular octahedron has a centre of *inversion* (i); a regular tetrahedron does not.

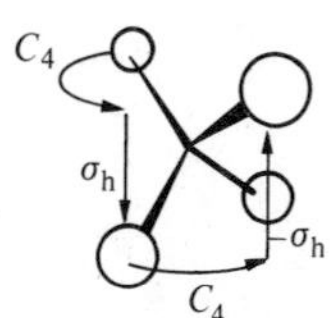

Fig. 17.6. A fourfold rotary reflection axis (S_4) is possessed by CH_4: the molecule is indistinguishable after a 90° rotation followed by a reflection across the horizontal plane.

17.1 (b) The symmetry classification of molecules

In order to classify molecules according to their symmetries, we list their symmetry elements and group together those with the same list. This puts CH_4 and CCl_4, which both have the same symmetry as a regular tetrahedron, into the same group, and H_2O into another.

The name of the group to which a molecule belongs is determined by its list of symmetry elements. There are two systems of notation. The *Schoenflies system* is more common for the discussion of individual molecules, and the *Hermann–Mauguin system* (or *International system*) is used almost exclusively in the discussion of crystal symmetry. We shall describe the Schoenflies system in detail; the translation to the Hermann–Mauguin system is described in Box 17.1.

(1) *The groups C_1, C_i, C_s.* A molecule belongs to C_1 if it has no element other than the identity (e.g. CHFClBr, Fig. 17.7(a)), to C_i if it has the identity and the inversion (e.g. *meso*-tartaric acid, Fig. 17.7(b)), and to C_s if it has the identity and a plane of reflection (e.g. the quinoline molecule, Fig. 17.7(c)).

(2) *The groups C_n.* A molecule (or any object) belongs to C_n if it possesses the identity and an n-fold axis. (Note that C_n is now playing a triple role: as the label of a symmetry element, a symmetry operation, and a group.) The H_2O_2 molecule, Fig. 17.8, has the elements (E, C_2) and so belongs to the group C_2.

(3) *The groups C_{nv}.* Objects in these groups possess the identity, a C_n axis, and n vertical mirror planes σ_v. H_2O, for example, has the symmetry elements $(E, C_2, 2\sigma_v)$, and so belongs to the group C_{2v}. NH_3 has the elements $(E, C_3, 3\sigma_v)$, and so belongs to the group C_{3v}.

(4) *The groups C_{nh}.* Objects in these groups possess the identity, an n-fold principal axis, and a horizontal mirror plane. An example is

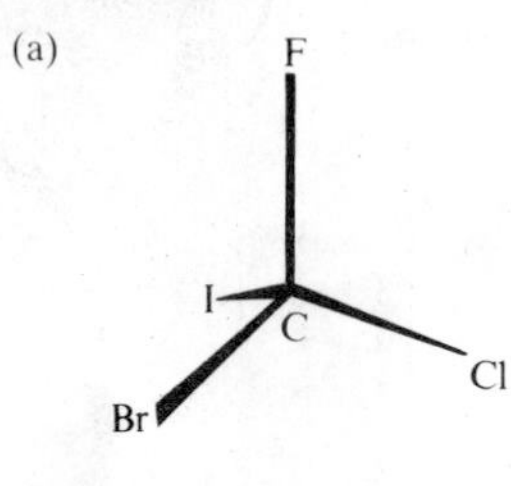

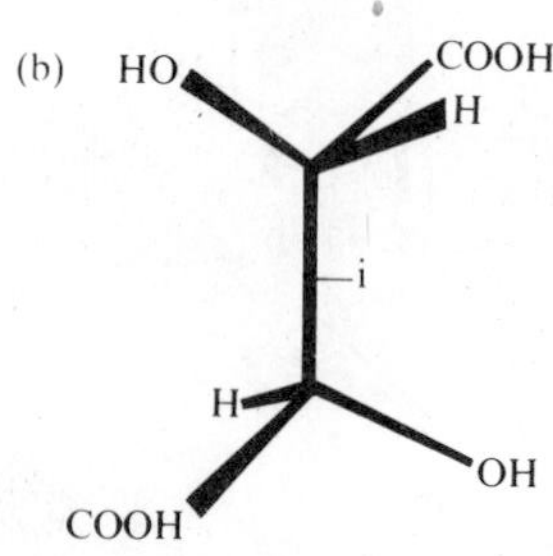

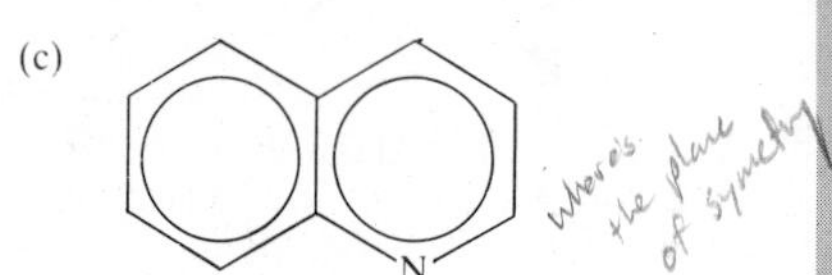

Fig. 17.7 Examples of molecules belonging to the groups (a) C_1, (b) C_i, and (c) C_s.

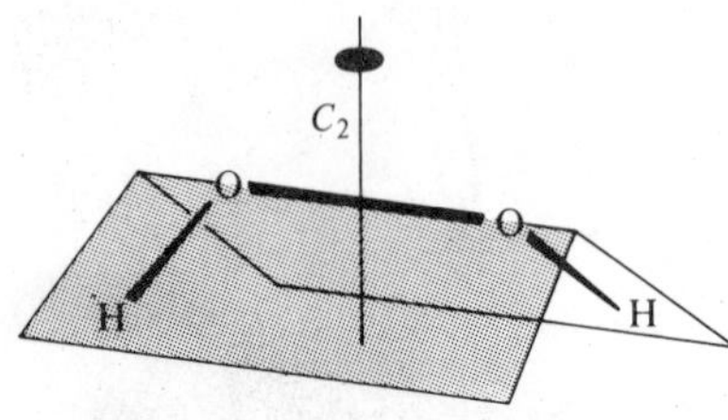

Fig. 17.8. The H_2O_2 molecule belongs to the group C_2 when it is in the conformation shown.

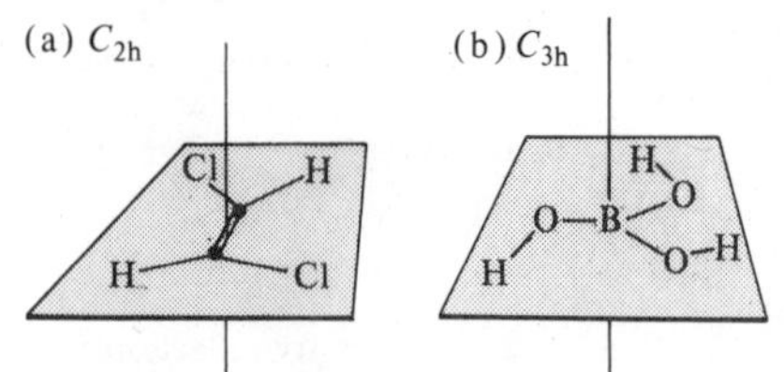

Fig. 17.9. (a) *trans*-dichlorethene belongs to C_{2h} and (b) $B(OH)_3$ belongs to C_{3h}.

Box 17.1 The International System for point groups

In the *International System*, which is also called the *Hermann–Mauguin System*, a number n denotes the presence of an n-fold axis and a letter m denotes a mirror plane. A diagonal line (/) indicates that the mirror plane is perpendicular to the symmetry axis (so that $2/m$ corresponds to C_{2h}). It is important to distinguish symmetry elements of the same type but in different classes (so that D_{4h}, with its three classes of mirror plane, σ_h, σ_v, and σ_d, is denoted $4/mmm$). A bar over a symbol indicates that that element is combined with an inversion (so that $\bar{6}$ translates to C_{3h}).

The following table translates the Schoenflies System into the International System. The only groups listed are the 32 crystallographic point groups (Section 23.1(b)).

C_i $\bar{1}$				
C_s m				
C_1 1	C_2 2	C_3 3	C_4 4	C_6 6
	C_{2v} $2mm$	C_{3v} $3m$	C_{4v} $4mm$	C_{6v} $6mm$
	C_{2h} $2/m$	C_{3h} $\bar{6}$	C_{4h} $4/m$	C_{6h} $6/m$
	D_2 222	D_3 32	D_4 422	D_6 622
	D_{2h} mmm	D_{3h} $\bar{6}2m$	D_{4h} $4/mmm$	D_{6h} $6/mmm$
	D_{2d} $\bar{4}2m$	D_{3d} $\bar{3}m$	S_4 $\bar{4}$	S_6 $\bar{3}$
T 23	T_d $\bar{4}3m$	T_h $m3$	O 43	O_h $m3m$

The group D_2 is sometimes denoted V and called the *Vierer group* (the group of four).

*trans*CHCl=CHCl, Fig. 17.9, which possesses the elements (E, C_2, σ_h, i), and so belongs to the group C_{2h}. Note that sometimes the presence of a symmetry element is implied by others: in this case C_2 and σ_h imply the presence of the inversion.

(5) *The groups D_n.* Objects in these groups possess an n-fold principal axis and n twofold axes perpendicular to C_n, Fig. 17.10.

(6) *The groups D_{nh}.* Objects belong to D_{nh} if they belong to D_n and possess a horizontal mirror plane, Fig. 17.11. The planar, triangular BF_3 molecule has the elements (E, C_3, $3C_2$, σ_h), with the C_2 axes along each B—F bond, and so belongs to D_{3h}. The benzene molecule possesses the elements (E, C_6, $6C_2$, σ_h), together with some others that these imply, and so it belongs to D_{6h}. A uniform cylinder belongs to $D_{\infty h}$, but a cone belongs to $C_{\infty v}$. This implies that homonuclear diatomic molecules all belong to $D_{\infty h}$, and that heteronuclear diatomics belong to $C_{\infty v}$.

(7) *The groups D_{nd}.* An object belongs to these groups if it belongs to D_n and has n dihedral mirror planes. The staggered form of ethane, Fig. 17.12(b), belongs to D_{3d}.

(8) *The groups S_n.* Objects possessing an S_n axis belong to the group S_n. An example is shown in Fig. 17.13. Molecules belonging to S_n with $n > 4$ are rare. Note that the group S_2 is the same as C_i.

(9) *The cubic groups.* A number of very important molecules (e.g. CH_4)

possess more than one principal axis and belong to the *cubic groups*, and in particular to the *tetrahedral groups* T, T_d, T_h and the *octahedral groups* O, O_h, Fig. 17.14. The group T_d is the group of the regular tetrahedron (e.g. CH_4), and the group O_h is the group of the regular octahedron. If the object possesses the *rotational* symmetry of the tetrahedron or the octahedron, but none of their planes of reflection, then it belongs to the simpler groups T or O. The group T_h is based on T but also contains a centre of inversion.

(10) *The full rotation group*, R_3. A sphere and an atom belong to R_3, but no molecule does. Exploring the consequences of R_3 is a very important way of applying symmetry arguments to atoms.

The identification of a molecule's group is simplified by referring to the flow diagram in Box 17.2 and the shapes shown in Fig. 17.15.

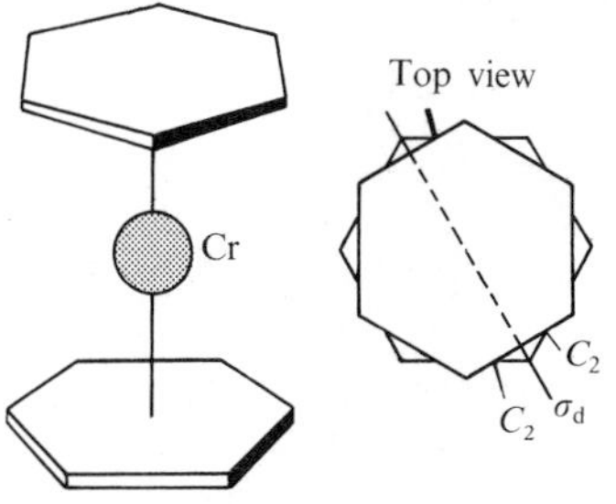

Fig. 17.10. The sandwich compound $Cr(C_6H_6)_2$ has a staggered conformation and belongs to the group D_{6d}. The top view shows two of the C_2 axes and a dihedral plane.

Example 17.1

Identify the point group to which the sandwich molecule ruthenocene (two eclipsed cyclopentadiene rings) belongs.

- *Method*. Trace a path through the flow diagram in Box 17.2.

- *Answer*. The path is shown by a dotted line; it ends at D_{nh}. Since the molecule has a fivefold axis, it belongs to the group D_{5h}.

- *Comment*. If the rings were staggered, as in ferrocene, the horizontal reflection plane would be absent, but dihedral planes would be present.

- *Exercise*. Classify (a) ferrocene and (b) benzene. [(a) D_{5d}, (b) D_{6h}]

17.1 (c) Some immediate consequences of symmetry

As soon as the point group of a molecule has been identified, it is possible to make some statements about its properties. For instance, only molecules belonging to the groups C_n, C_{nv}, and C_s may have an *electric dipole moment*, and in the case of C_n and C_{nv} that dipole must lie along the rotation axis. This can be understood as follows. If the molecule is C_n it cannot possess a charge distribution corresponding to a dipole moment perpendicular to the axis. However, since the group makes no reference to operations relating the 'top' and 'bottom' of the molecule, a charge distribution resulting in a dipole along the axis may exist. The same remarks apply to C_{nv}. In all the other groups, such as C_{3h}, D_n, etc., there are symmetry operations that correspond to turning the molecule upside down. Therefore, as well as having no dipole perpendicular to the axis, such a molecule can have none along the axis, for otherwise 'turning it upside down' (by reflection or rotation) would not be a symmetry operation.

An *optically active* molecule is one that rotates the plane of polarized light (a property discussed in more detail in Section 24.1(e)). A molecule is optically active only if it cannot be superimposed on its mirror image. This implies that it must not possess an axis of improper rotation, S_n. It is important to check whether a group contains an *implied* S_n axis. For example, the groups C_{nh} include S_n implicitly because they include C_n and

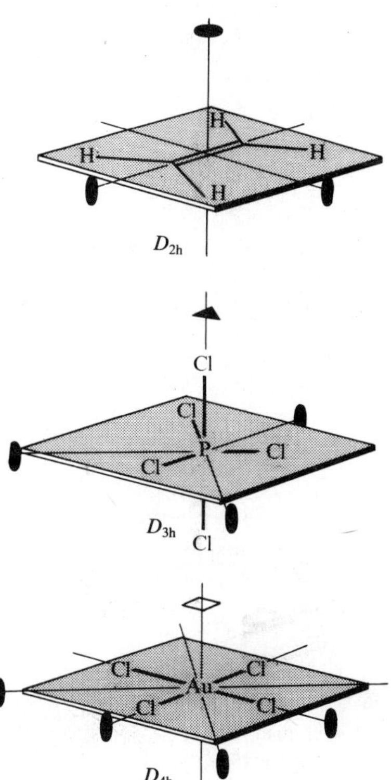

Fig. 17.11. Three examples of molecules belonging to D_{nh}. (a) C_2H_4 belongs to D_{2h}, (b) PCl_5, which has different axial and equatorial bond lengths, belongs to D_{3h}, and (c) the square-planar complex $[AuCl_4]^-$ belongs to D_{4h}.

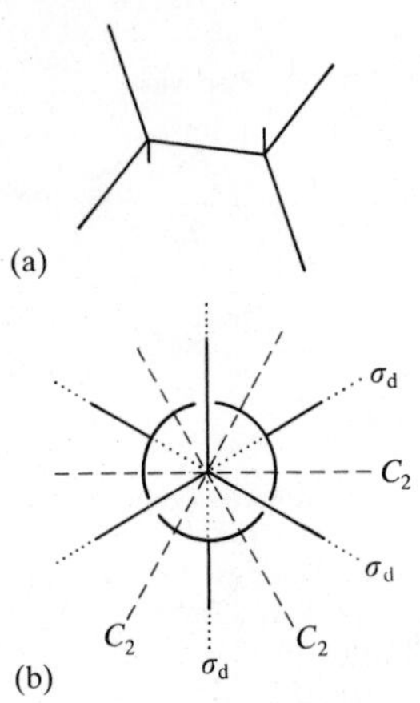

Fig. 17.12. (a) The staggered conformation of ethane belongs to D_{3d}: the three C_2 axes and dihedral planes are shown in (b).

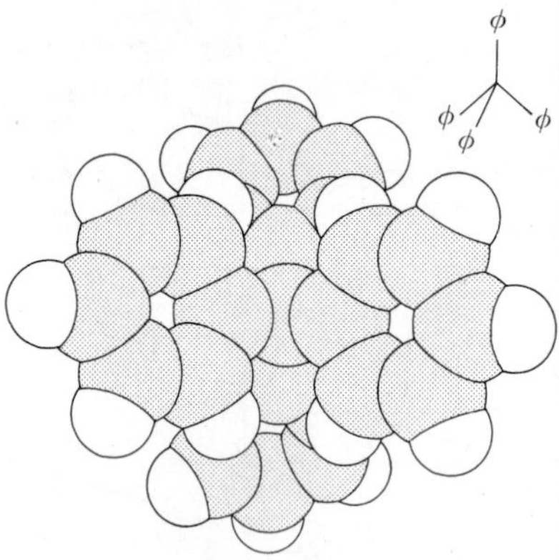

Fig. 17.13. Tetraphenylmethane is an example of a molecule that belongs to the group S_4.

Box 17.2 The determination of the point group

In order to arrive at the point group of a given molecule, work through the following table. (The dotted line shows the path used in *Exercise* 17.1.) Use of a *subgroup* (i.e. not travelling to the end of the table) is permissible, but it gives less complete information.

- Molecule → Linear?
  - Y → i?
    - Y → $D_{\infty h}$
    - N → $C_{\infty v}$
  - N → Two or more C_n $n>2$?
    - Y → i?
      - Y → C_5?
        - Y → I_h
        - N → O_h
      - N → T_d
    - N → C_n?
      - Y → ‡
        - Y → σ_h?
          - Y → D_{nh}
          - N → $n\sigma_d$?
            - Y → D_{nd}
            - N → D_n
        - N → σ_h?
          - Y → C_{nh}
          - N → $n\sigma_v$?
            - Y → C_{nv}
            - N → S_{2n}?
              - Y → S_{2n}
              - N → C_n
      - N → σ?
        - Y → C_s
        - N → i?
          - Y → C_i
          - N → C_1

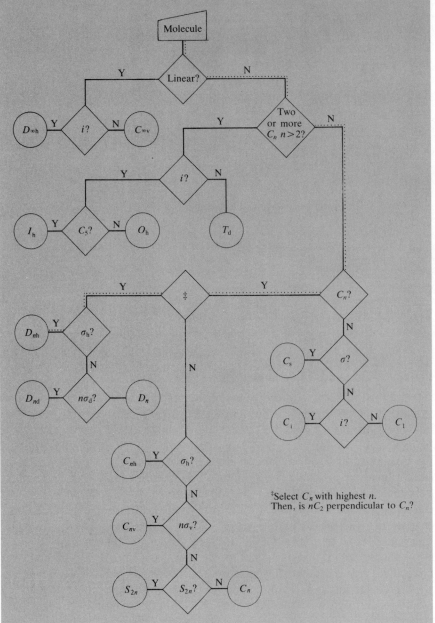

‡Select C_n with highest n. Then, is nC_2 perpendicular to C_n?

σ_h. Any group containing the inversion also possesses S_2 because i is equivalent to C_2 followed by σ_h, which is S_2, Fig. 17.17. It follows that all molecules with centres of inversion are optically inactive. A molecule may be inactive even though it has no centre of inversion. For example, an S_4 molecule, Fig. 17.18, is inactive, because though it lacks an i it possesses an S_4 axis.

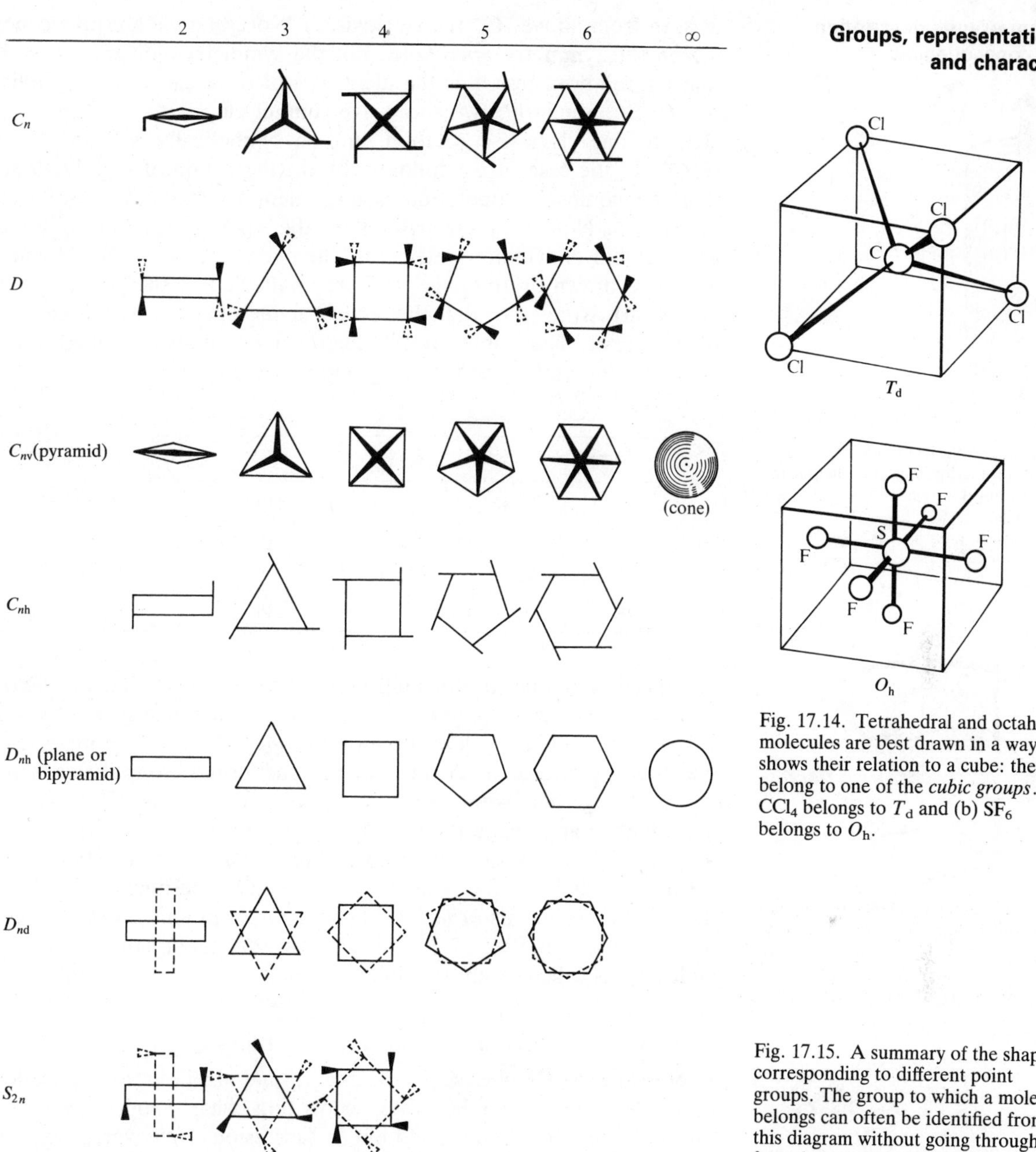

Fig. 17.14. Tetrahedral and octahedral molecules are best drawn in a way that shows their relation to a cube: they belong to one of the *cubic groups*. (a) CCl_4 belongs to T_d and (b) SF_6 belongs to O_h.

Fig. 17.15. A summary of the shapes corresponding to different point groups. The group to which a molecule belongs can often be identified from this diagram without going through the formal procedure in Box 17.2.

Other molecular properties may be analysed once the point group is known, but the information is often buried more deeply. In order to extract it we have to turn to the numerical aspects of group theory, but even so we shall do no more than skim the surface of this very subtle and powerful subject.

17.2 Groups, representations, and characters

Consider NH_3, which belongs to C_{3v} and has the symmetry operations (E, C_3^+, C_3^-, σ_v, σ_v', σ_v''), Fig. 17.19 (C_3^+ is the counter-clockwise 120° rotation

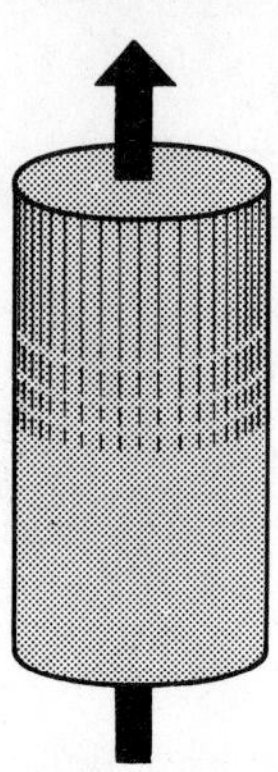

Fig. 17.16. A molecule with a C_n axis cannot have a dipole perpendicular to the axis, but it may have one parallel to the axis.

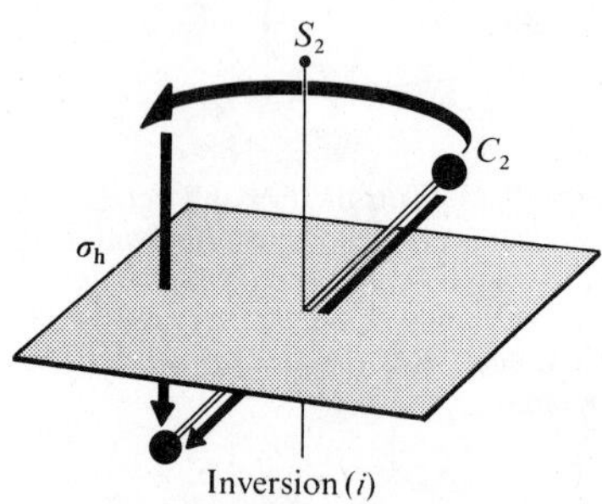

Fig. 17.17. Some elements are implied by the other elements in a group. Any group containing an inversion, also possesses at least an S_2 element because i and S_2 are equivalent.

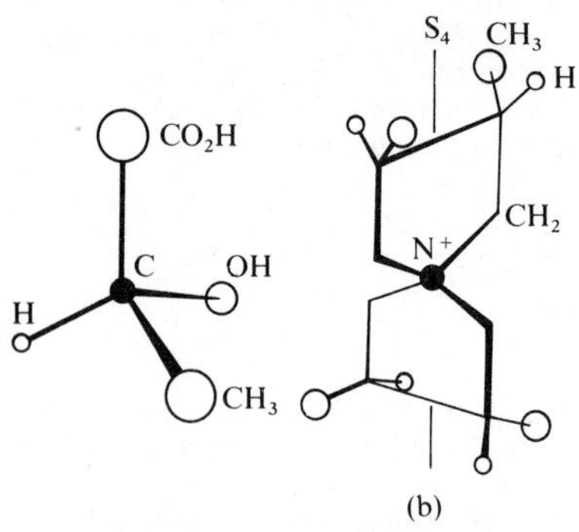

Fig. 17.18. (a) An optically active molecule without a centre of inversion. (b) Although this molecule has no centre of inversion (i.e. no S_2 axis) it is not optically active (it has an S_4 axis).

as seen from above, C_3^- the clockwise‡). Note that we are now concentrating on the symmetry *operations*, not the symmetry elements. It is obvious that C_3^+ followed by C_3^- is the identity, which we can write symbolically as $E = C_3^- C_3^+$. Similarly, two successive counter-clockwise rotations are equivalent to one clockwise rotation, and so symbolically we can write $C_3^- = C_3^+ C_3^+$. In the case of C_3^+ followed by σ_v, inspection of Fig. 17.19 confirms that the composite operation is equivalent to σ_v'', and so we can write $\sigma_v'' = \sigma_v C_3^+$. Note that throughout all the operations refer to some fixed arrangement of symmetry elements (the planes and axes remain where they were first drawn on the page, and are unaffected by the performance of an operation). Note, too, that the *second* acting operator is written to the *left* of the first. The table of all such combinations is called the *group multiplication table*, and for this group it is as follows:

Second	First	E	C_3^+	C_3^-	σ_v	σ_v'	σ_v''
E		E	C_3^+	C_3^-	σ_v	σ_v'	σ_v''
C_3^+		C_3^+	C_3^-	E	σ_v'	σ_v''	σ_v
C_3^-		C_3^-	E	C_3^+	σ_v''	σ_v	σ_v'
σ_v		σ_v	σ_v''	σ_v'	E	C_3^-	C_3^+
σ_v'		σ_v'	σ_v	σ_v''	C_3^+	E	C_3^+
σ_v''		σ_v''	σ_v'	σ_v	C_3^-	C_3^+	E

Inspection of the group multiplication table shows that *the outcome of successive symmetry operations is equivalent to a single symmetry operation of the group*. This is called the *group property*, and is the main feature of the structure of groups. A set of operations form a group if they satisfy this property together with some other mild conditions (Box 17.3). It is important to appreciate the slightly confusing point that the *group elements* are the symmetry *operations* (the reflections and rotations) not the symmetry elements (the planes and axes, etc). The number of elements in the group is called its *order*. All symmetry operations on molecules satisfy the conditions in Box 17.3, which is why the theory of the symmetry of molecules is called group theory.

17.2 (a) The representation of transformations

Expressions like $E = C_3^+ C_3^-$ look like normal algebraic multiplications, but they are really symbolic ways of writing what happens when various physical operations are carried out in succession. However, it is possible to give them an actual algebraic significance. Doing that means that we can deal with numbers instead of abstract symbols for operations, and by dealing with numbers we arrive at precise conclusions.

The following sections make use of the properties of matrices, which are reviewed in Appendix 17.1. The remainder of the section establishes the language and sets out the background of group theory: the rules needed are specified and used in Section 17.3.

Consider a C_{3v} molecule (such as NH_3) with s-orbitals on each atom, Fig. 17.20. Consider what happens to these functions under a symmetry

‡ This sign convention may seem odd, but it matches the convention used for angular momentum, when a clockwise rotation as seen from below (i.e., counter-clockwise as seen from above) is associated with positive values of m_l.

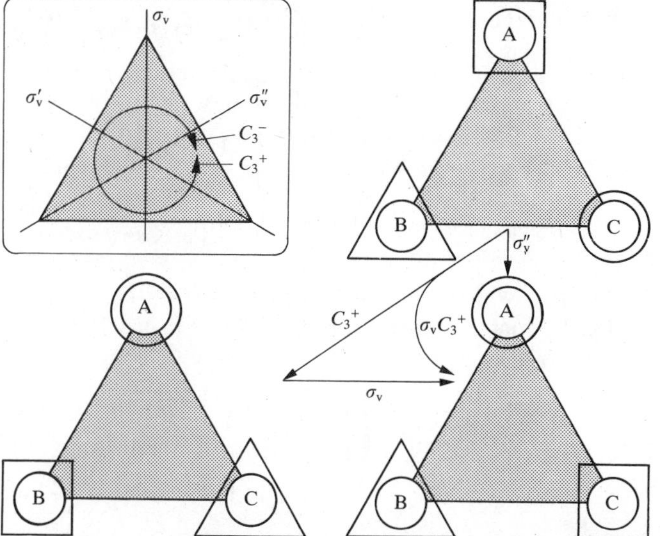

Fig. 17.19. The symmetry operations of the group C_{3v} are shown in the box, and the equivalence $\sigma_v C_3^+ = \sigma''_v$ is constructed by considering the effect of successive operations.

Box 17.3 The definition of a group

A group of *order h* is a set of *h* objects, its *elements* (such as the symmetry operations of a molecule) $G = \{g_1, g_2, \ldots, g_h\}$, together with a rule of combination, so that the symbol $g_i g_j$ has a well-defined meaning (such as symmetry operation g_j followed by symmetry operation g_i), and which satisfy the following criteria:

(1) The set includes the *identity* element. This is an element normally denoted E such that $Eg_i = g_iE = g_i$ for all the elements in the set.

(2) The set includes the *inverse* of each element in the set. The inverse of g_i (denoted g_i^{-1}) is the element for which $g_i g_i^{-1} = g_i^{-1} g_i = E$.

(3) The rule of combination is *associative*, so that $(g_i g_j)g_k$ is the same as $g_i(g_j g_k)$.

(4) The combination of any two elements of the set is itself a member of the set. That is, $g_i g_j = g_k$, where g_k is a member of G. This is called the *group property*.

Note that the definition does not require $g_i g_j = g_j g_i$, except in the special cases defining E and g_i^{-1}. Groups for which $g_i g_j = g_j g_i$ are called *commutative groups* or *Abelian groups*.

operation. Under σ_v the change $(s_N, s_A, s_C, s_B) \leftarrow (s_N, s_A, s_B, s_C)$ takes place. This transformation can be expressed using matrix multiplication (Appendix 17.1). We can find some matrix, $\mathbf{D}(\sigma_v)$ such that the transformation is reproduced:

$$(s_N, s_A, s_C, s_B) = (s_N, s_A, s_B, s_C)\begin{pmatrix} 1 & 0 & 0 & 0 \\ 0 & 1 & 0 & 0 \\ 0 & 0 & 0 & 1 \\ 0 & 0 & 1 & 0 \end{pmatrix} = (s_N, s_A, s_B, s_C)\mathbf{D}(\sigma_v). \tag{17.2.1}$$

The matrix $\mathbf{D}(\sigma_v)$ is called a *representative* of the operation σ_v. The same technique may be used to find matrices that reproduce the other symmetry

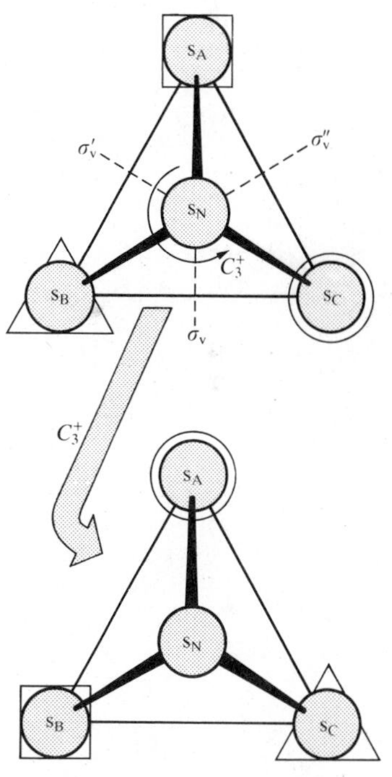

Fig. 17.20. The *basis* used for the discussion of the transformation properties of a C_{3v} molecule. We use the convention that the basis is *always* written in the order {□, △, ○} and that the operations move the shapes *on the diagram* without affecting the N, A, B, C labels. The interpretation of the effect of C_3^+ is shown.

operations. For instance, C_3^+ has the effect $(s_N, s_B, s_C, s_A) \leftarrow (s_N, s_A, s_B, s_C)$. This can be expressed as

$$(s_N, s_B, s_C, s_A) = (s_N, s_A, s_B, s_C)\begin{pmatrix} 1 & 0 & 0 & 0 \\ 0 & 0 & 0 & 1 \\ 0 & 1 & 0 & 0 \\ 0 & 0 & 1 & 0 \end{pmatrix} = (s_N, s_A, s_B, s_C)\mathbf{D}(C_3^+). \tag{17.2.2}$$

The operation σ_v'', which causes $(s_N, s_C, s_B, s_A) \leftarrow (s_N, s_A, s_B, s_C)$ can be represented by the matrix multiplication

$$(s_N, s_C, s_B, s_A) = (s_N, s_A, s_B, s_C)\begin{pmatrix} 1 & 0 & 0 & 0 \\ 0 & 0 & 0 & 1 \\ 0 & 0 & 1 & 0 \\ 0 & 1 & 0 & 0 \end{pmatrix} = (s_N, s_A, s_B, s_C)\mathbf{D}(\sigma_v''). \tag{17.2.3}$$

The matrix representative of the identity leaves (s_N, s_A, s_B, s_C) unchanged, and so it is

$$\mathbf{D}(E) = \begin{pmatrix} 1 & 0 & 0 & 0 \\ 0 & 1 & 0 & 0 \\ 0 & 0 & 1 & 0 \\ 0 & 0 & 0 & 1 \end{pmatrix}. \tag{17.2.4}$$

A very important property of these matrices may now be identified. Using the rules of matrix multiplication gives

(a)

$$\mathbf{D}(\sigma_v)\mathbf{D}(C_3^+) = \begin{pmatrix} 1 & 0 & 0 & 0 \\ 0 & 1 & 0 & 0 \\ 0 & 0 & 0 & 1 \\ 0 & 0 & 1 & 0 \end{pmatrix}\begin{pmatrix} 1 & 0 & 0 & 0 \\ 0 & 0 & 0 & 1 \\ 0 & 1 & 0 & 0 \\ 0 & 0 & 1 & 0 \end{pmatrix} = \begin{pmatrix} 1 & 0 & 0 & 0 \\ 0 & 0 & 0 & 1 \\ 0 & 0 & 1 & 0 \\ 0 & 1 & 0 & 0 \end{pmatrix} = \mathbf{D}(\sigma_v'').$$

The importance of this result is that its structure, $\mathbf{D}(\sigma_v)\mathbf{D}(C_3^+) = \mathbf{D}(\sigma_v'')$, is exactly the same as the group multiplication rule $\sigma_v C_3^+ = \sigma_v''$. Whichever group elements are chosen, the matrix representatives multiply in an analogous way (we say the multiplications are *homomorphous*). Therefore, *the whole of the group multiplication table is reproduced by the matrix multiplication of the representatives*. The set of six matrices is called a *matrix representation* of the C_{3v} group for the *basis* (s_N, s_A, s_B, s_C), and its discovery means that a link has been established between the symbolic manipulations of the group and algebraic manipulations involving numbers.

Example 17.2

Consider the four H1s-orbitals of CH_4. Find matrix representatives for the operations C_3^+ and S_4^+, and confirm that they satisfy the group multiplication property.

• *Method*. CH_4 belongs to the group T_d. The axis corresponding to C_3^+ runs along a C—H bond (e.g. C–H_A), and so it rotates the other three hydrogen atoms into each other in a counter-clockwise sense seen from above. S_4^+ rotates counter-clockwise by 90° about a bisector of a CH_2 angle (e.g. H_ACH_C) and then reflects across the perpendicular plane. Find the 4×4

matrices that reproduce these changes. Identify $C_3^+S_4^+$ by assessing the effect on the basis of successive symmetry operations, and confirm the homomorphism by multiplying the representatives. A good plan is to put different shaped receptacles on each atom location (the objects in Fig. 17.21), and to allow the operations to move these (the letters remaining stationary): the row vector representing the basis is then determined by writing the appropriate letter in the receptacle always written in the same order (□, ◇, △, ▯).

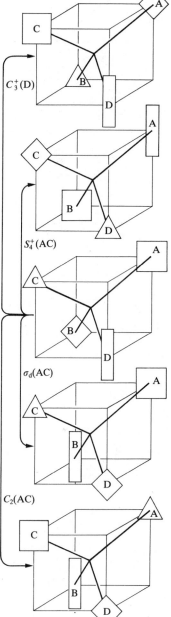

• *Answer*. Refer to Fig. 17.21. The effects can be represented as follows:

$$C_3^+(\mathrm{H_A},\mathrm{H_B},\mathrm{H_C},\mathrm{H_D})=(\mathrm{H_A},\mathrm{H_C},\mathrm{H_D},\mathrm{H_B})$$

$$=(\mathrm{H_A},\mathrm{H_B},\mathrm{H_C},\mathrm{H_D})\begin{pmatrix}1&0&0&0\\0&0&0&1\\0&1&0&0\\0&0&1&0\end{pmatrix}=(\mathrm{H_A},\mathrm{H_B},\mathrm{H_C},\mathrm{H_D})\mathbf{D}(C_3^+),$$

$$S_4^+(\mathrm{H_A},\mathrm{H_B},\mathrm{H_C},\mathrm{H_D})=(\mathrm{H_B},\mathrm{H_C},\mathrm{H_D},\mathrm{H_A})$$

$$=(\mathrm{H_A},\mathrm{H_B},\mathrm{H_C},\mathrm{H_D})\begin{pmatrix}0&0&0&1\\1&0&0&0\\0&1&0&0\\0&0&1&0\end{pmatrix}=(\mathrm{H_A},\mathrm{H_B},\mathrm{H_C},\mathrm{H_D})\mathbf{D}(S_4^+).$$

The effect of the joint operation $S_4^+C_3^+$ is

$$S_4^+C_3^+(\mathrm{H_A},\mathrm{H_B},\mathrm{H_C},\mathrm{H_D})=(\mathrm{H_B},\mathrm{H_D},\mathrm{H_A},\mathrm{H_C})=S_4^-(\mathrm{H_A},\mathrm{H_B},\mathrm{H_C},\mathrm{H_D}),$$

where the axis of the operation S_4^- is $\mathrm{H_ACH_D}$. The product of the representatives is

$$\begin{pmatrix}0&0&0&1\\1&0&0&0\\0&1&0&0\\0&0&1&0\end{pmatrix}\begin{pmatrix}1&0&0&0\\0&0&0&1\\0&1&0&0\\0&0&1&0\end{pmatrix}=\begin{pmatrix}0&0&1&0\\1&0&0&0\\0&0&0&1\\0&1&0&0\end{pmatrix}$$

which is the representative of S_4^-.

• *Comment*. The representation depends on the basis selected: in this case it is four-dimensional because the basis has four members.

• *Exercise*. Find the representatives for $S_4^-(\mathrm{CD})$ and $C_3^+(\mathrm{B})$, and confirm that they multiply homomorphously with the elements of the group.

$$\left[\mathbf{D}(S_4^-)=\begin{pmatrix}0&0&1&0\\0&0&0&1\\0&1&0&0\\1&0&0&0\end{pmatrix},\quad \mathbf{D}(C_3^+)=\begin{pmatrix}0&0&1&0\\0&1&0&0\\0&0&0&1\\1&0&0&0\end{pmatrix},\quad S_4^-C_3^-=S_4^+(\mathrm{AC})\right]$$

17.2 (b) The character of symmetry operations

In common parlance, the rotations C_3^+ and C_3^- of the group C_{3v} have the same 'character', but differ in direction. Likewise, the three reflections have the same character, but are different from the rotations. This notion can be given precise, numerical significance.

Inspection of the matrix representation of C_{3v} for the s-orbital basis shows a remarkable fact. If we sum the diagonal elements of each representative (i.e. take the *trace* of each matrix) we get the following numbers:

	$\mathbf{D}(E)$	$\mathbf{D}(C_3^+)$	$\mathbf{D}(C_3^-)$	$\mathbf{D}(\sigma_v)$	$\mathbf{D}(\sigma_v')$	$\mathbf{D}(\sigma'')$
$\chi=$	4	1	1	2	2	2

Fig. 17.21. The symmetry transformations used in *Example* 17.2. Note that we are using the convention defined in the caption of Fig. 17.20.

Matrices representing operations of the same type are seen to have identical diagonal sums. We call the sum of the diagonal elements of a representative the *character* (symbol: χ) of the operation. *Symmetry operations in the same class have the same character*. The two rotations and the three reflections therefore belong to different classes of operation, as intuition suggests.

Example 17.3

Calculate the characters of the operations C_3, S_4^+, and S_4^- in the basis used in *Example* 17.2.

• *Method*. Refer to the **D**-matrices calculated in the *Example* and sum their diagonal elements.

• *Answer*. $\chi(C_3^+) = 1+0+0+0 = 1,$
$\chi(S_4^+) = 0+0+0+0 = 0,$
$\chi(S_4^-) = 0+0+0+0 = 0.$

• *Comment*. A quick rule for determining the character is to count 1 each time a basis function is left unchanged by the symmetry operation, because only these functions give a non-zero entry on the diagonal. In some cases there is a sign change, $(\ldots f \ldots) \to (\ldots -f \ldots)$; then -1 occurs on the diagonal, and so count -1. The character of the identity is always the dimension of the basis.

• *Exercise*. Calculate the character of (a) C_2 and (b) σ_d in the same basis. [(a) 0, (b) 2]

The character of an operation depends on the basis of the representation. For example, if instead of considering the four s-orbitals, we used only s_N as the basis, then since each operation results in $s_N \leftarrow s_N$, which may be written $s_N = s_N 1$, with 1 regarded as a matrix, the characters of the operations would be

$$\begin{array}{ccccccc} & \mathbf{D}(E) & \mathbf{D}(C_3^+) & \mathbf{D}(C_3^-) & \mathbf{D}(\sigma_v) & \mathbf{D}(\sigma_v') & \mathbf{D}(\sigma_v'') \\ \chi = & 1 & 1 & 1 & 1 & 1 & 1 \end{array}.$$

It is still true that the characters of the operations of the same class are equal, but the example emphasizes that the characters of different classes may be the same. Furthermore, it is obvious that because $1 \times 1 = 1$, the matrices do reproduce the group multiplication table, but they do so in a trivial and uninformative way. For this reason, the representation with 1 representing each element is called the *unfaithful representation* of the group.

17.2 (c) Irreducible representations

The unfaithful representation of the group, although apparently trivial, *is* a representation, and should not be discarded as being of no importance. In the next few sections, in fact, we shall see that it is the most important representation for many chemical applications.

The representatives in the basis (s_N, s_A, s_B, s_C) are *four-dimensional* (i.e., they are 4×4 matrices), but inspection shows that they are all of the form

$$\begin{pmatrix} 1 & 0 & 0 & 0 \\ 0 & \blacksquare & & \\ 0 & & & \\ 0 & & & \end{pmatrix}$$

and that the symmetry operations never mix s_N with the other three basis functions. This suggests that the basis can be cut into two parts, one consisting of s_N alone and the other of (s_A, s_B, s_C). The s_N is a basis for the unfaithful representation, as we have seen, and the other three are a basis for a three-dimensional representation consisting of the following matrices:

$$\begin{array}{cccccc} \mathbf{D}(E) & \mathbf{D}(C_3^+) & \mathbf{D}(C_3^-) & \mathbf{D}(\sigma_v) & \mathbf{D}(\sigma_v') & \mathbf{D}(\sigma_v'') \\ \begin{pmatrix}1&0&0\\0&1&0\\0&0&1\end{pmatrix} & \begin{pmatrix}0&0&1\\1&0&0\\0&1&0\end{pmatrix} & \begin{pmatrix}0&1&0\\0&0&1\\1&0&0\end{pmatrix} & \begin{pmatrix}1&0&0\\0&0&1\\0&1&0\end{pmatrix} & \begin{pmatrix}0&1&0\\1&0&0\\0&0&1\end{pmatrix} & \begin{pmatrix}0&0&1\\0&1&0\\1&0&0\end{pmatrix} \\ \chi = \quad 3 & 0 & 0 & 1 & 1 & 1 \end{array}$$

Notice that the characters still satisfy the rule about symmetry operations of the same class. The matrices are the same as those of the four-dimensional representation, except for the elimination of the first row and column. We say that the original four-dimensional representation has been *reduced* to the sum (more precisely, the *direct sum*) of a one-dimensional representation *spanned* by s_N and a three-dimensional representation spanned by (s_A, s_B, s_C). This fits the common-sense view that the central orbital plays a role different from the other three. The reduction is denoted symbolically by

$$D^{(4)} = D^{(1)} + D^{(3)},$$

the $D^{(l)}$ standing for the l-dimensional representations.

The one-dimensional representation $D^{(1)}$, the six trivial matrices 1, 1, 1, 1, 1, 1, obviously cannot be reduced any further, and is called an *irreducible representation* (or *irrep*) of the group. The question we now face is whether $D^{(3)}$ is reducible to representations of lower dimension.

We can demonstrate that $D^{(3)}$ is reducible by switching attention from s_A, s_B, s_C to the linear combinations

$$s_1 = s_A + s_B + s_C, \qquad s_2 = 2s_A - s_B - s_C, \qquad s_3 = s_B - s_C. \quad (17.2.5)$$

These are sketched in Fig. 17.22 (and their form will be justified later). Even at this stage it is clear that, because of the presence of the node in s_2 and s_3, these two have different symmetry from s_1. The decomposition $D^{(3)} = D^{(1)} + D^{(2)}$ is beginning to emerge.

The representatives in the new basis can be constructed from the old. For example, since under σ_v, $(s_A, s_C, s_B) \leftarrow (s_A, s_B, s_C)$, it follows that $(s_1, s_2, -s_3) \leftarrow (s_1, s_2, s_3)$, which is achieved by

$$(s_1, s_2, -s_3) = (s_1, s_2, s_3)\begin{pmatrix}1&0&0\\0&1&0\\0&0&-1\end{pmatrix},$$

which gives the representative $\mathbf{D}(\sigma_v)$ in the new basis. The representative of C_3^+ takes a little more calculation, but depends on the transformation $(s_B, s_C, s_A) \leftarrow (s_A, s_B, s_C)$. We know how the individual s_Q transform, and so by substitution in the expressions defining the new basis gives the transformation of the s_n:

$$(s_1, -\tfrac{1}{2}s_2 + \tfrac{3}{2}s_3, -\tfrac{1}{2}s_2 - \tfrac{1}{2}s_3) = (s_1, s_2, s_3)\begin{pmatrix}1&0&0\\0&-\tfrac{1}{2}&-\tfrac{1}{2}\\0&\tfrac{3}{2}&-\tfrac{1}{2}\end{pmatrix},$$

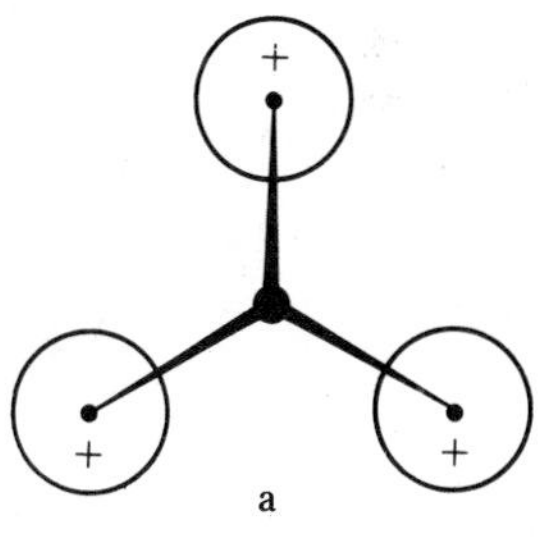

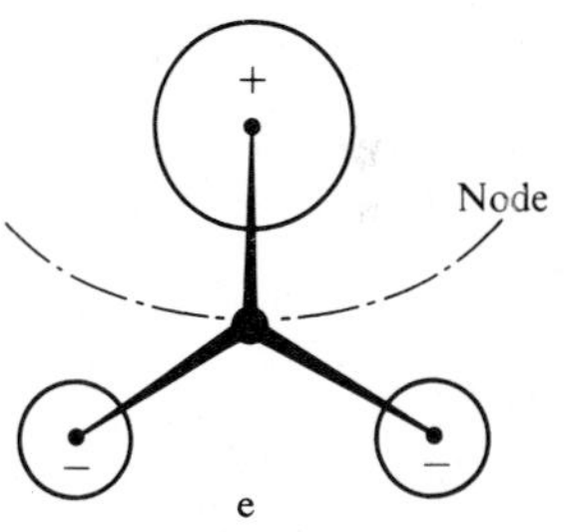

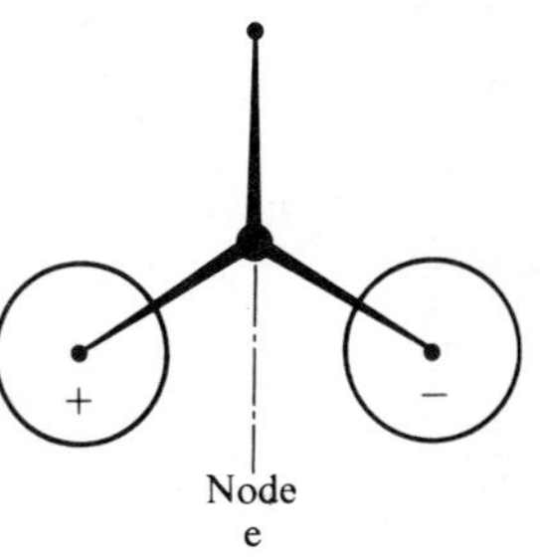

Fig. 17.22. The three linear combinations defined in eqn (17.2.5). The symmetry species they span (see later) have been added.

so giving $\mathbf{D}(C_3^+)$. The complete representation and its characters may be found similarly, and we obtain

$$\begin{array}{cccccc} \mathbf{D}(E) & \mathbf{D}(C_3^+) & \mathbf{D}(C_3^-) & \mathbf{D}(\sigma_v) & \mathbf{D}(\sigma_v') & \mathbf{D}(\sigma_v'') \\ \begin{pmatrix}1&0&0\\0&1&0\\0&0&1\end{pmatrix} & \begin{pmatrix}1&0&0\\0&-\frac{1}{2}&-\frac{1}{2}\\0&\frac{3}{2}&-\frac{1}{2}\end{pmatrix} & \begin{pmatrix}1&0&0\\0&-\frac{1}{2}&\frac{1}{2}\\0&-\frac{3}{2}&-\frac{1}{2}\end{pmatrix} & \begin{pmatrix}1&0&0\\0&1&0\\0&0&-1\end{pmatrix} & \begin{pmatrix}1&0&0\\0&-\frac{1}{2}&\frac{1}{2}\\0&\frac{3}{2}&\frac{1}{2}\end{pmatrix} & \begin{pmatrix}1&0&0\\0&-\frac{1}{2}&-\frac{1}{2}\\0&-\frac{3}{2}&\frac{1}{2}\end{pmatrix} \\ \chi=3 & 0 & 0 & 1 & 1 & 1 \end{array}$$

A number of important features emerge. First, the characters conform to the principle about operations of the same class. Second, the characters are the same as for the original three-dimensional basis: *taking linear combinations of a basis set leaves the characters unchanged*. Third, and most important of all, the representatives are all in *block diagonal form*

$$\begin{pmatrix} \boxed{1} & 0 & 0 \\ 0 & \blacksquare & \\ 0 & & \end{pmatrix}$$

and the s_1 combination is not mixed with the other two by any symmetry operation of the group. This means that the transformation has brought about the reduction $D^{(3)} = D^{(1)} + D^{(2)}$, with s_1 a basis for the same one-dimensional irrep $(1, 1, 1, 1, 1, 1)$ as before and with (s_2, s_3) a basis for a two-dimensional representation formed from $D^{(3)}$ by slicing off the first row and column of each representative:

$$\begin{array}{cccccc} \mathbf{D}(E) & \mathbf{D}(C_3^+) & \mathbf{D}(C_3^-) & \mathbf{D}(\sigma_v) & \mathbf{D}(\sigma_v') & \mathbf{D}(\sigma_v'') \\ \begin{pmatrix}1&0\\0&1\end{pmatrix} & \begin{pmatrix}-\frac{1}{2}&-\frac{1}{2}\\\frac{3}{2}&-\frac{1}{2}\end{pmatrix} & \begin{pmatrix}-\frac{1}{2}&\frac{1}{2}\\-\frac{3}{2}&-\frac{1}{2}\end{pmatrix} & \begin{pmatrix}1&0\\0&-1\end{pmatrix} & \begin{pmatrix}-\frac{1}{2}&\frac{1}{2}\\\frac{3}{2}&\frac{1}{2}\end{pmatrix} & \begin{pmatrix}-\frac{1}{2}&-\frac{1}{2}\\-\frac{3}{2}&\frac{1}{2}\end{pmatrix} \\ \chi=2 & -1 & -1 & 0 & 0 & 0 \end{array}$$

It is easy to check that these matrices are a representation by multiplying pairs together and seeing that they reproduce the original group multiplication table.

Is the representation $D^{(2)}$ reducible? In fact *no* linear combination of s_2 and s_3 exists that reduces $D^{(2)}$ to two one-dimensional representations, and so it is irreducible. We can conclude that while s_N and s_1 'have the same symmetry' (i.e. are the bases for the *same* irrep of the group), the pair s_2, s_3 'have different symmetry' (i.e. span a different irrep) and must be treated as a pair (because the irrep is two-dimensional). These features agree with what common sense tells us by inspection of the diagrams in Fig. 17.22.

But how do we know that $D^{(2)}$ is irreducible? The list of the characters of *all possible* irreps of a group is called its *character table*. The C_{3v} character table is shown on the left. The columns are labelled by the operations of the group: it is not necessary to show the character for every individual operation because those in the same class have the same value. The number of operations in the class is given (e.g. the 2 in $2C_3$). The column on the left labels the *symmetry species* of the irrep: the convention is to use A (or, when the character is -1 under a principal rotation, B) for one-dimensional irreps, E for two-dimensional irreps, and T for three-dimensional irreps (there are none in C_{3v}). There are two species of one-dimensional irrep in C_{3v}, and both have $+1$ for the character of the principal rotation; they are therefore distinguished as A_1 and A_2. Now we can see that the characters of

C_{3v}	E	$2C_3$	$3\sigma_v$
A_1	1	1	1
A_2	1	1	-1
E	2	-1	0

the $D^{(2)}$ representation spanned by (s_2, s_3) are those of E: this is an irreducible representation, and so $D^{(2)}$ is irreducible too.

Perhaps the most surprising feature of the character table is that there are so few symmetry species: the three given exhaust all possibilities. This is confirmed by an elegant theorem in group theory which states that

Number of species of irrep = Number of classes.

In C_{3v} there are three classes (three columns in the character table), and so there are only three species of irrep.

Although we have introduced these points through the group C_{3v}, they are entirely general, and the characters of all possible species of irrep of any group may be listed. A selection of these very important tables is given at the end of the data section.

17.2 (d) Transformations of other bases

The development so far has concentrated on a basis formed from orbitals on each atom. We now show that the functions (x, y, z), which can be thought of as p-orbitals on a central atom, are a basis for another three-dimensional representation.

Under the reflection σ_v in C_{3v} the functions (x, y, z) are carried into $(-x, y, z)$, Fig. 17.23. (Note that the coordinate system is an unchanging background on which these transformations are played out.) This can be expressed as a matrix equation:

$$(-x, y, z) = (x, y, z)\begin{pmatrix} -1 & 0 & 0 \\ 0 & 1 & 0 \\ 0 & 0 & 1 \end{pmatrix} = (x, y, z)\mathbf{D}(\sigma_v).$$

Similarly, under C_3^+, $(-\tfrac{1}{2}x + \tfrac{1}{2}\sqrt{3}y, -\tfrac{1}{2}\sqrt{3}x - \tfrac{1}{2}y, z) \leftarrow (x, y, z)$, which can be expressed as

$$(-\tfrac{1}{2}x + \tfrac{1}{2}\sqrt{3}y, -\tfrac{1}{2}\sqrt{3}x - \tfrac{1}{2}y, z) = (x, y, z)\begin{pmatrix} -\tfrac{1}{2} & -\tfrac{1}{2}\sqrt{3} & 0 \\ \tfrac{1}{2}\sqrt{3} & -\tfrac{1}{2} & 0 \\ 0 & 0 & 1 \end{pmatrix} = (x, y, z)\mathbf{D}(C_3^+).$$

All the representatives can be compiled in this way, and so we obtain the complete representation

$$\begin{array}{ccc} \mathbf{D}(E) & \mathbf{D}(C_3^+) & \mathbf{D}(C_3^-) \\ \begin{pmatrix} 1 & 0 & 0 \\ 0 & 1 & 0 \\ 0 & 0 & 1 \end{pmatrix} & \begin{pmatrix} -\tfrac{1}{2} & -\tfrac{1}{2}\sqrt{3} & 0 \\ \tfrac{1}{2}\sqrt{3} & -\tfrac{1}{2} & 0 \\ 0 & 0 & 1 \end{pmatrix} & \begin{pmatrix} -\tfrac{1}{2} & \tfrac{1}{2}\sqrt{3} & 0 \\ -\tfrac{1}{2}\sqrt{3} & -\tfrac{1}{2} & 0 \\ 0 & 0 & 1 \end{pmatrix} \\ \chi = 3 & 0 & 0 \end{array}$$

$$\begin{array}{ccc} \mathbf{D}(\sigma_v) & \mathbf{D}(\sigma_v') & \mathbf{D}(\sigma_v'') \\ \begin{pmatrix} -1 & 0 & 0 \\ 0 & 1 & 0 \\ 0 & 0 & 1 \end{pmatrix} & \begin{pmatrix} \tfrac{1}{2} & -\tfrac{1}{2}\sqrt{3} & 0 \\ -\tfrac{1}{2}\sqrt{3} & -\tfrac{1}{2} & 0 \\ 0 & 0 & 1 \end{pmatrix} & \begin{pmatrix} \tfrac{1}{2} & \tfrac{1}{2}\sqrt{3} & 0 \\ \tfrac{1}{2}\sqrt{3} & -\tfrac{1}{2} & 0 \\ 0 & 0 & 1 \end{pmatrix} \\ 1 & 1 & 1 \end{array}$$

This shows that the three-dimensional representation is reducible because

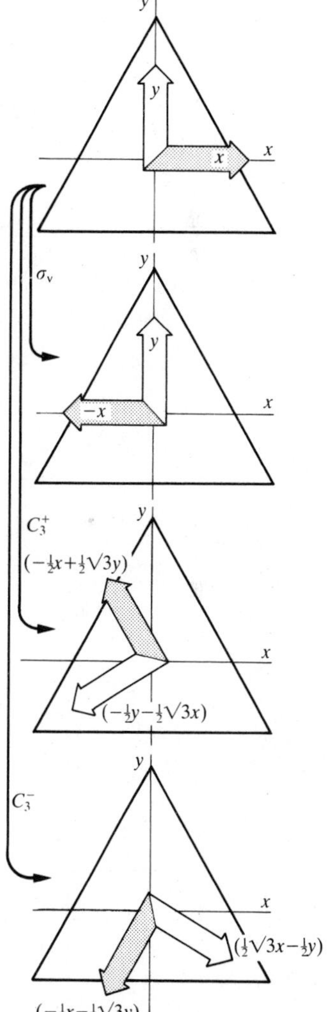

Fig. 17.23. The transformations of the functions x and y under the operations of the group C_{3v}. Notice that they take place against a fixed coordinate system (just as the N, A, B, C are fixed in Fig. 17.20).

all the matrices have block-diagonal form and the parts relating to z may be sliced off. The characters of the matrices remaining are 2, −1, −1, 0, 0, 0, and by comparing this with the C_{3v} character table it is clear that (x, y) spans an irrep of symmetry species E.

Example 17.4

Establish the symmetry species of the irreps spanned by (x, y, z) in the group C_{2v}.

• *Method*. Establish the effect of the operations E, C_2, σ_v, and σ_v' on the three functions. Write the matrix representation in the basis, and identify the symmetry species from the characters.

• *Answer*. The four transformations are

$$E(x,y,z)=(x,y,z)=(x,y,z)\mathbf{D}(E)$$
$$C_2(x,y,z)=(-x,-y,z)=(x,y,z)\mathbf{D}(C_2)$$
$$\sigma_v(x,y,z)=(-x,y,z)=(x,y,z)\mathbf{D}(\sigma_v)$$
$$\sigma_v'(x,y,z)=(x,-y,z)=(x,y,z)\mathbf{D}(\sigma_v').$$

The matrix representation is therefore

$$\overset{E}{\begin{pmatrix}1&0&0\\0&1&0\\0&0&1\end{pmatrix}}\overset{C_2}{\begin{pmatrix}-1&0&0\\0&-1&0\\0&0&1\end{pmatrix}}\overset{\sigma_v}{\begin{pmatrix}-1&0&0\\0&1&0\\0&0&1\end{pmatrix}}\overset{\sigma_v'}{\begin{pmatrix}1&0&0\\0&-1&0\\0&0&1\end{pmatrix}}$$

This is in block-diagonal form, and may be decomposed into the following one-dimensional irreps:

$$\begin{array}{lrrrr} x: & 1 & -1 & -1 & 1\\ y: & 1 & -1 & 1 & -1\\ z: & 1 & 1 & 1 & 1\end{array}$$

The characters of the representatives are the numbers themselves (because they are 1×1 matrices), and so the irreps spanned by x, y, and z are B_1, B_2, and A_1 respectively.

• *Comment*. This approach can always be used. Fortunately, for most groups the symmetry species of the irreps spanned by (x, y, z) are tabulated.

• *Exercise*. What symmetry species does xy span in C_{2v}? [A_2]

The conclusion of the analysis given above is that in C_{3v} z spans A_1 and (x, y) spans E. This will turn out to be so important that the symmetry species of the irreps spanned by x, y, and z are normally specified in the character table. The same technique may be applied to the quadratic forms $x^2, xy, xz, \ldots, z^2$ (and to other simple functions), and the symmetry species of the irreps they span are also usually listed. A complete character table therefore looks something like the following:

C_{3v}	E	$2C_3$	$3\sigma_v$				
A_1	1	1	1	z	$x^2+y^2+z^2$	$2z^2-x^2-y^2$	
A_2	1	1	−1				R_z
E	2	−1	0	(x, y)	(xz, yz)	(xy, x^2-y^2)	(R_x, R_y)

The R_x, etc. denote *rotations*, and their listing shows how they transform under the operations of the group. They are deduced from the transformation properties of the angular momentum; for instance, R_z transforms as

$l_z = xp_y - yp_x$, eqn (14.3.7), with the linear momentum components transforming like x, y, z.

17.3 Using character tables

Although the characters do not carry all the information contained in the representatives (they are only the diagonal sums), they do contain enough to make them of central importance in chemistry. One of the reasons for this importance is that they let us say, almost at a glance, whether an integral is zero.

17.3 (a) Vanishing integrals

Suppose we had to evaluate the following integral:

$$I = \int f_1 f_2 \, d\tau, \tag{17.3.1}$$

where f_1 and f_2 are wavefunctions. For example, f_1 might be an atomic orbital on one atom and f_2 one on another, and then I would be their overlap integral; then if we knew that the integral was zero we could say at once that a molecular orbital does not result from (f_1, f_2)-overlap in that molecule. The key point in dealing with eqn (17.3.1) is that *the value of an integral is independent of the orientation of the molecule*, Fig. 17.24. In group theoretical language we say that I is unchanged by any symmetry transformation of the molecule, so that each symmetry operation brings about the trivial transformation $I \leftarrow I$. Since the volume element $d\tau$ is unchanged by a symmetry transformation, it follows that the integrand $f_1 f_2$ must also be unchanged by the symmetry operations if I is not to disappear. Since under any operation $f_1 f_2 \leftarrow f_1 f_2$, all the representatives are 1, and the characters are also all equal to 1. Therefore, *the integrand $f_1 f_2$, must be a basis for the totally symmetric* (A_1) *irrep of the molecular point group*. If $f_1 f_2$ does *not* transform as A_1, then the integral must vanish. It follows that we need a rule to be able to deduce which symmetry species are spanned by the product $f_1 f_2$. The following procedure is used:

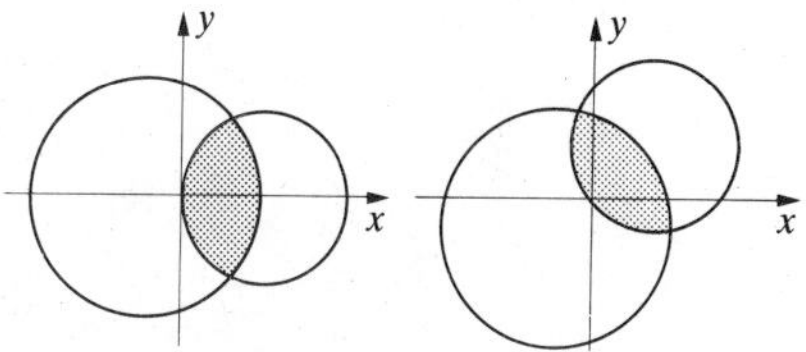

Fig. 17.24. The value of an integral I (e.g. an area) is independent of the coordinate system used to evaluate it. (That is, I is a basis for A_1.)

> (1) *Decide on the symmetry species of the functions by reference to the character table, and write the characters in two rows in the same order as in the table.*

For example, if f_1 is the s_N-orbital in NH_3 and f_2 is the linear combination s_3, Fig. 17.22, then since in C_{3v} s_N spans A_1 and s_3 is a member of the basis spanning E, we refer to the C_{3v} character table and write

$$\begin{array}{lrrr} f_1\colon & 1 & 1 & 1 \\ f_2\colon & 2 & -1 & 0. \end{array}$$

> (2) *Multiply the numbers in each column, writing the results in the same order.*

For example, for the NH_3 calculation,

$$f_1 f_2\colon \quad 2 \quad -1 \quad 0.$$

(3) Inspect the row so produced, and see if it can be expressed as a sum of characters of the irreps of the group. If this sum does not contain A_1, then the integral must be zero.

For example, in C_{3v} the numbers produced can always be expressed as $c_1\chi(A_1)+c_2\chi(A_2)+c_3\chi(E)$, and the integral must be zero if $c_1=0$. In the present example, the characters 2, −1, 0 are those of E alone, and so the integral must be zero. Inspection of the form of the functions shows why this is so: s_3 has a node running through s_N, Fig. 17.22. Had we taken $f_1=s_N$ and $f_2=s_1$, then since each spans A_1 with characters 1, 1, 1, the product would span 1, 1, 1, which is A_1 itself. Therefore, s_1 and s_N may have non-zero overlap.

It is important to note that group theory is specific about when an integral *must* be zero; integrals that it allows to be non-zero *may* be zero for other (non-symmetry) reasons. For example, the N–H distance may be so great that the s_1, s_N-overlap integral is zero simply because the orbitals are so far apart.

The same technique may be used to decide whether integrals of the form

$$I=\int f_1f_2f_3\,d\tau \qquad (17.3.2)$$

necessarily disappear. In this case the triple product $f_1f_2f_3$ must contain a component that spans A_1, and to test whether this is so, the characters of all three functions are multiplied together in the same way as in the rules set out above.

Example 17.5

Does the integral $\int(3d_{z^2})x(3d_{xy})\,d\tau$ vanish in a tetrahedral molecule?

• *Method*. Refer to the T_d character table (end of data section). Find the characters of the irreps spanned by $3z^2-r^2$ (the form of the d_{z^2}-orbital), x, and xy; then use the procedure set out above (with one more row of multiplication). Note that $3z^2-r^2=2z^2-x^2-y^2$.

• *Answer*. Draw up the following table:

	E	$8C_3$	$3C_2$	$6S_4$	$6\sigma_d$
$f_3=d_{xy}(T_2)$	3	0	−1	−1	1
$f_2=x(T_2)$	3	0	−1	−1	1
$f_1=d_{z^2}(E)$	2	−1	2	0	0
$f_1f_2f_3$	18	0	2	0	0

The characters are the sum of $A_1+A_2+2E+2T_1+2T_2$. Therefore, the integral is not necessarily zero.

• *Comment*. Closer inspection of the integral (e.g. using the representation itself) shows that it vanishes: *be warned that arguments based on the character tables, since they carry only incomplete information in general, show only when an integral is necessarily zero*. A second point is that the decomposition of the result to discover if A_1 is included is sometimes a lengthy job. Here is a simple procedure:

(1) Multiply the characters (18, 0, . . .) by the number of elements in each class (1, 8, . . .).
(2) Add together the numbers this produces (18 + 0 + . . . = 24).
(3) Divide the sum by the order of the group (the number of elements, 24).

This gives the number of times A_1 appears in the reducible representation (1).

• *Exercise*. Does the integral $\int(2p_x)(2p_y)(2p_z)\,d\tau$ necessarily vanish in an octahedral environment? [No; spans A_1]

17.3 (b) Orbitals with non-zero overlap

The rules just given let us decide which atomic orbitals may have non-zero overlap in a molecule. We have seen that s_N (specifically a N2s-orbital) may have non-zero overlap with s_1 (specifically $H1s_A + H1s_B + H1s_C$), and so (s_N, s_1)-overlap bonding and antibonding MOs can form, Fig. 17.25(a). The general rule is that *only orbitals of the same symmetry species may have non-zero overlap*, and so *only orbitals of the same symmetry species form bonding and antibonding combinations*. When such MOs form, we label them with the lower-case letter corresponding to the symmetry species. Thus, the (s_N, s_1)-overlap orbitals are called a_1-orbitals (and a_1^* if they are antibonding).

The s_2 and s_3 linear combinations span an irrep of symmetry species E. Does the nitrogen atom have orbitals that have non-zero overlap with them (and give rise to an e-orbital)? Intuition, Fig. 17.25(b) and (c), suggest that $N2p_x$ and $N2p_y$ should be suitable. This can be confirmed as follows. In the hydrogen atom, the $2p_x$-orbital has the algebraic form $xe^{-r/2a_0}$, and although the $N2p_x$-orbital is different in detail, it is still of the form xf, where f is a function of radius alone. Likewise, the $N2p_y$-orbital has the form yf. Since the distance of a point from the central nitrogen nucleus is unaffected by any symmetry transformation of C_{3v}, the function f is also unchanged. Therefore, the transformation properties of the N2p-orbitals are the transformation properties of x and y. The character table shows that in C_{3v} x and y jointly span an irrep of symmetry species E. Therefore $N2p_x$ and $N2p_y$ also span E, and may have non-zero overlap with s_2 and s_3 (check this by multiplying the characters: $E \times E = A_1 + A_2 + E$ in C_{3v}). The two e-orbitals that result are shown in Fig. 17.25 (there are also two antibonding e*-orbitals).

The power of the method can be illustrated by asking whether any d-orbitals on the central atom can take part in bonding. The d-orbitals have the forms

$$d_{z^2} = (3z^2 - r^2)f, \quad d_{x^2-y^2} = (x^2 - y^2)f, \quad d_{xy} = xyf, \quad d_{yz} = yzf, \quad d_{xz} = xzf.$$

Their symmetries can be taken from the character tables by noting how the quadratic forms xy, xz, etc. transform. Reference to the C_{3v} table show that d_{z^2} has A_1 symmetry, and the pairs $(d_{x^2-y^2}, d_{xy})$ and (d_{xz}, d_{yz}) each transform as E. It follows that MOs may be formed by (s_1, d_z)-overlap and by overlap of the s_2, s_3 combinations with the E d-orbitals. Whether or not the d-orbitals are in fact important is a question group theory cannot answer because that depends on energy considerations, not symmetry.

Although we have illustrated the technique with the group C_{3v}, it is entirely general, and the importance of knowing which s-, p-, and d-orbitals overlap is one of the reasons why the transformation properties of x, xz, etc. are listed in the character tables.

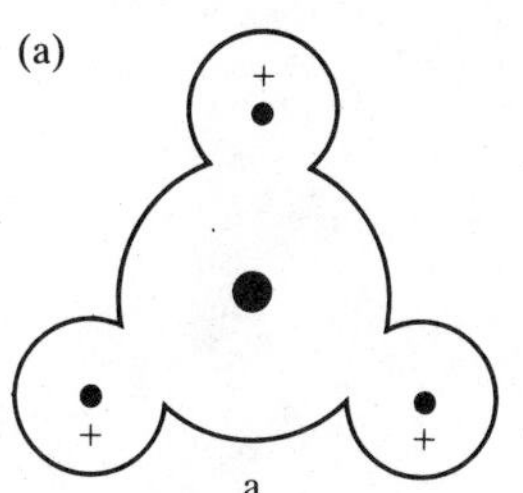

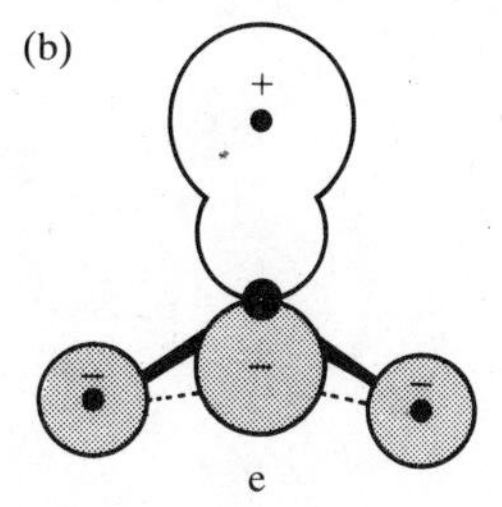

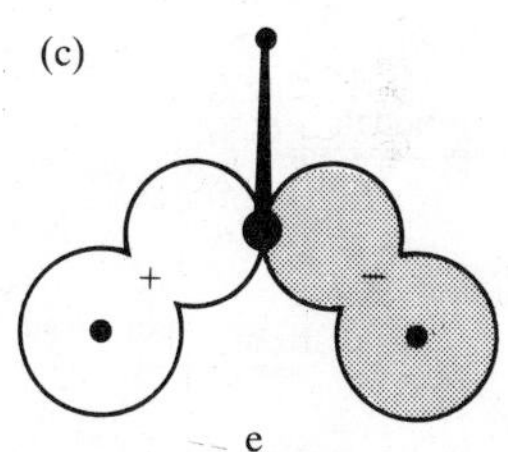

Fig. 17.25. Orbitals of the same symmetry species may have non-vanishing overlap. This diagram illustrates the three bonding orbitals that may be constructed from (N2s, H1s)- and (N2p, H1s)-overlap in a C_{3v} molecule, and their symmetry labels. (There are also three antibonding orbitals of the same species.)

Example 17.6

The four H1s-orbitals of methane span $A_1 + T_2$. With which of the carbon orbitals can they overlap? What if the carbon had d-orbitals available?

- *Method*. Refer to the T_d character table (end of data section). Look for s-, p-, and d-orbitals spanning A_1 or T_2.

- *Answer*. An s-orbital spans A_1, and so it may have non-vanishing overlap with the A_1

combination of H1s-orbitals. The C2p-orbitals span T_2, and so they may have non-vanishing overlap with the T_2 combination. The d_{xy}, d_{yz}, and d_{zx} orbitals span T_2, and so may overlap the same combination. Neither of the other two d-orbitals span A_1 (they span E), and so they remain non-bonding orbitals.

• *Comment*. It follows that in a delocalized picture of the bonding in methane, there are (C2s, H1s)-overlap a_1-orbitals and (C2p, H1s)-overlap t_2-orbitals. The C3d-orbitals might contribute to the latter. The lowest energy configuration is probably $a_1^2t_2^6$, with all bonding orbitals occupied.

• *Exercise*. Consider the octahedral SF_6 molecule, with the bonding arising from overlap of S orbitals and a 2p-orbital on each F directed towards the central S. The latter span $A_{1g}+E_g+T_{1u}$. What S orbitals have non-zero overlap? Suggest what the ground configuration is likely to be. [3s(A_{1g}), 3p(T_{1u}), 3d(E); $a_{1g}^2t_{1u}^6e_g^4$]

17.3 (c) Selection rules

In Chapters 18 and 19 we shall see that the intensity of a spectral line arising from a molecular transition between some initial state with wavefunction ψ_i and a final state with wavefunction ψ_f depends on the (*electric*) *transition dipole moment* (symbol: $\boldsymbol{\mu}$). The z-component of this vector is defined through

$$\mu_z = -e\int \psi_f^* z \psi_i \, d\tau, \qquad (17.3.3)$$

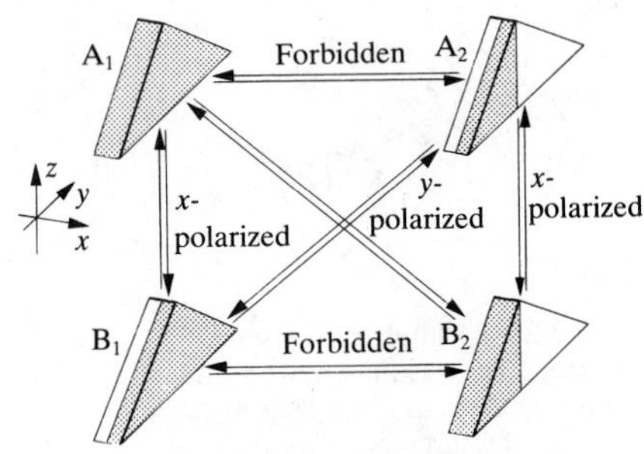

Fig. 17.26. The polarizations of the allowed transitions in a C_{2v} molecule. The shading indicates the structure of the orbitals of the specified symmetry species.

$-e$ being the charge of the electron. Stating the conditions for this quantity (and the x- and y-components) to be zero amounts to specifying the *selection rules* for the transition. The transition moment has the form of the integral in eqn (17.3.2), and so, once we know the symmetries of the states, we can use group theory to decide which transitions have zero transition dipole moment and therefore cannot occur.

As an example, we investigate whether an a_1-electron in H_2O (which belongs to C_{2v}) can make an electric dipole transition to a b_1-orbital (Fig. 17.26). We have to examine all three components of the transition dipole, and take f_2 in eqn (17.3.2) as x, y, and z in turn. Reference to the C_{2v} character table shows that these transform as B_1, B_2, and A_1 respectively. The three calculations run as follows:

	x-component				y-component				z-component			
$f_1(A_1)$	1	1	1	1	1	1	1	1	1	1	1	1
$f_2(x, y, \text{or } z)$	1	−1	1	−1	1	−1	−1	1	1	1	1	1
$f_3(B_1)$	1	−1	1	−1	1	−1	1	−1	1	−1	1	−1
$f_1f_2f_3$	1	1	1	1	1	1	−1	−1	1	−1	1	−1
	A_1				A_2				B_1			

Only the first product (with $f_2 = x$) spans A_1, and so only the x-component of the transition dipole may be non-zero. Therefore, we conclude that the electric dipole transition $b_1 \rightarrow a_1$ is allowed. We can go on to state that the radiation emitted (or absorbed) has its electric field vector in the x-direction (i.e. that it is x-*polarized*), because that form of radiation couples with the x-component of a transition dipole.

Example 17.7

Is $p_x \rightarrow p_y$ an allowed transition in a tetrahedral molecule?

• *Method*. Determine whether $\int p_y q p_x \, d\tau$, with $q = x, y, z$, spans A_1. Assess whether the integrand spans A_1 using the technique set out in *Example* 17.5. Use the T_d character table at the end of the data section.

• *Answer*. The procedure works out as follows:

$f_3(p_x; T_2)$	3	0	−1	−1	1
$f_2(q; T_2)$	3	0	−1	−1	1
$f_1(p_y; T_2)$	3	0	−1	−1	1
$f_1 f_2 f_3$	27	0	−1	−1	1

A_1 occurs (once) in this set of characters, and so $p_x \rightarrow p_y$ is allowed.

• *Comment*. Closer analysis (using the representatives) shows that only $q = z$ gives a non-zero integral, and so the transition is z-polarized.

• *Exercise*. What are the allowed transitions, and their polarizations, of a b_1-electron in a C_{4v} molecule? [$b_1 \rightarrow b_1(z)$; $b_1 \rightarrow e(x, y)$]

17.3 (d) Symmetry-adapted orbitals

So far we have only asserted the forms of the linear combinations (such as s_1, etc.) that act as bases for irreducible representations. Group theory also provides machinery which takes an arbitrary basis (s_A, etc.) as input and generates combinations of the specified symmetry. Because these combinations are adapted to the symmetry of the molecule, they are called *symmetry-adapted orbitals*.

The rules for building symmetry-adapted orbitals are as follows.

(1) *Construct a table showing the effect of each operation on each orbital of the original basis.*

For example, from the (s_N, s_A, s_B, s_C) basis in NH_3 we form the following table:

Original basis:		s_N	s_A	s_B	s_C
Under	E	s_N	s_A	s_B	s_C
	C_3^+	s_N	s_B	s_C	s_A
	C_3^-	s_N	s_C	s_A	s_B
	σ_v	s_N	s_A	s_C	s_B
	σ_v'	s_N	s_B	s_A	s_C
	σ_v''	s_N	s_C	s_B	s_A

(2) *To generate the orbital of a specified symmetry species, take each column in turn and:*

(i) *Multiply each member of the column by the character of the corresponding operation.*

(ii) *Add together all the orbitals in each column with the factors as determined in* (i).

(iii) *Divide the sum by the order of the group.*

In our example, in order to generate the a_1-orbital we take the characters for A_1 (1, 1, 1, 1, 1, 1), and so rules (i) and (ii) lead to $s_N + s_N + \ldots = 6s_N$. The order of the group (the number of elements) is 6, and so the orbital of A_1-symmetry that can be generated from s_N is s_N itself. Applying the same technique to the column under s_A gives $\frac{1}{6}(s_A + s_B + s_C + s_A + s_B + s_C) = \frac{1}{3}(s_A + s_B + s_C)$. The same orbital is built from the other two columns, and so they give no further information. This is the combination s_1 used before (apart from the numerical factor), and so the a_1-orbital will be some combination $c_N s_N + c_1 s_1$, the coefficients having to be found by solving the Schrödinger equation because they do not come directly from the symmetry of the system.

Suppose now we try to generate a symmetry-adapted orbital of species A_2. The previous work has shown that there is no such combination. The characters for A_2 are 1, 1, 1, −1, −1, −1. The column under s_N generates zero, and so do the other three. Therefore, we find that we generate no orbital of A_2 symmetry.

When we try to generate the E basis orbitals, we run into a problem. This is because, for representations of dimension two or more, since the characters are the sums of the diagonal elements, the rules generate sums of the symmetry-adapted orbitals. (A more detailed rule based on the representatives themselves gives the individual orbitals.) This can be seen as follows. The E characters are 2, −1, −1, 0, 0, 0, and so the column under s_N gives $2s_N - s_N - s_N + 0 + 0 + 0 = 0$. The other columns give $\frac{1}{6}(2s_A - s_B - s_C)$, $\frac{1}{6}(2s_B - s_A - s_C)$, and $\frac{1}{6}(2s_C - s_B - s_A)$. However, any one can be expressed as a sum of the other two (they are not *linearly independent*). The difference of the second and third gives $\frac{1}{2}(s_B - s_C)$, and this and the first, $\frac{1}{6}(2s_A - s_B - s_C)$, are the two (now linearly independent) orbitals we have used in the discussion.

The following chapters will show many more examples of the way that the systematic use of symmetry using the techniques of group theory can greatly simplify the analysis of molecular structure and spectra.

Appendix 17.1: matrices

Matrices are arrays of numbers with special rules for combining them together. We shall consider $n \times n$ square matrices (symbol: $\mathbf{M}$) with n^2 *elements* (symbol: M_{rc}):

$$\mathbf{M} = \begin{pmatrix} M_{11} & M_{12} & \ldots & M_{1n} \\ M_{21} & M_{22} & \ldots & M_{2n} \\ \ldots & \ldots & \ldots & \ldots \\ M_{n1} & M_{n2} & \ldots & M_{nn} \end{pmatrix}.$$

M_{rc} is the element in row r, column c (r, c is a map reference). Two matrices $\mathbf{M}$ and $\mathbf{N}$ are equal (written $\mathbf{M} = \mathbf{N}$) only if *all* corresponding elements are equal: $M_{rc} = N_{rc}$ for all r, c.

The *addition* of two matrices, $\mathbf{M} + \mathbf{N} = \mathbf{P}$, is defined through $P_{rc} = M_{rc} + N_{rc}$ (i.e. add corresponding elements). For example,

$$\mathbf{M} = \begin{pmatrix} 1 & 2 \\ 3 & 4 \end{pmatrix} \quad \mathbf{N} = \begin{pmatrix} 5 & 6 \\ 7 & 8 \end{pmatrix}, \quad \mathbf{M} + \mathbf{N} = \begin{pmatrix} 1 & 2 \\ 3 & 4 \end{pmatrix} + \begin{pmatrix} 5 & 6 \\ 7 & 8 \end{pmatrix} = \begin{pmatrix} 6 & 8 \\ 10 & 12 \end{pmatrix}.$$

The *multiplication of two matrices is written* $\mathbf{MN} = \mathbf{P}$ and defined through

$P_{rc} = \Sigma_q M_{rq}N_{qc}$. This rule can be remembered in terms of the following diagram:

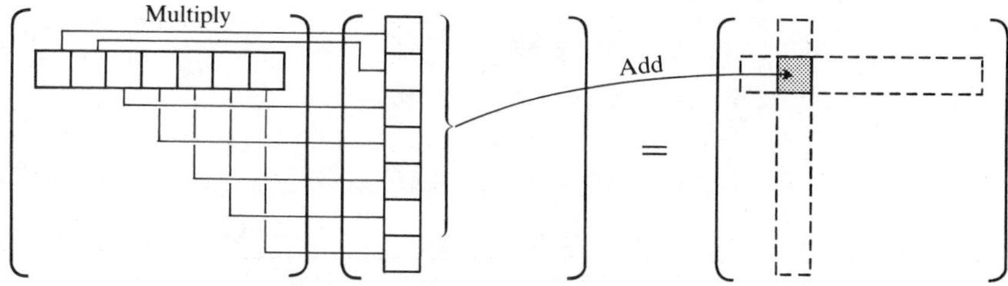

For example, using the same matrices as above,

$$\mathbf{MN} = \begin{pmatrix} 1 & 2 \\ 3 & 4 \end{pmatrix}\begin{pmatrix} 5 & 6 \\ 7 & 8 \end{pmatrix} = \begin{pmatrix} 1\times5+2\times7 & 1\times6+2\times8 \\ 3\times5+4\times7 & 3\times6+4\times8 \end{pmatrix} = \begin{pmatrix} 19 & 22 \\ 43 & 50 \end{pmatrix}.$$

Note that in this case $\mathbf{NM} \neq \mathbf{MN}$, and so matrix multiplication is *non-commutative* (i.e. depends on the order of multiplication) in general.

There are several types of matrix having special names of properties. Among them are the following:

A *diagonal matrix* is one in which all $M_{rc} = 0$ unless $r = c$. For example, $\begin{pmatrix} 1 & 0 \\ 0 & 2 \end{pmatrix}$ is diagonal but $\begin{pmatrix} 0 & 1 \\ 2 & 0 \end{pmatrix}$ is not. The *unit matrix* ($\mathbf{1}$) is diagonal with all diagonal elements equal to 1. Thus $\mathbf{1} = \begin{pmatrix} 1 & 0 \\ 0 & 1 \end{pmatrix}$ is the two-dimensional unit matrix.

The *transpose* of a matrix ($\tilde{\mathbf{M}}$) is related to $\mathbf{M}$ by interchange of rows and columns ($\tilde{M}_{rc} = M_{cr}$). Thus, if $\mathbf{M} = \begin{pmatrix} 1 & 2 \\ 3 & 4 \end{pmatrix}$, $\tilde{\mathbf{M}} = \begin{pmatrix} 1 & 3 \\ 2 & 4 \end{pmatrix}$.

The *inverse* of a matrix ($\mathbf{M}^{-1}$) satisfies $\mathbf{MM}^{-1} = \mathbf{M}^{-1}\mathbf{M} = \mathbf{1}$. It can be constructed as follows:

(1) Calculate the *determinant* ($\det \mathbf{M}$) of the matrix $\mathbf{M}$.

$$\text{If } \mathbf{M} = \begin{pmatrix} 1 & 2 \\ 3 & 4 \end{pmatrix}, \qquad \det \mathbf{M} = \begin{vmatrix} 1 & 2 \\ 3 & 4 \end{vmatrix} = 1\times4 - 2\times3 = -2.$$

If $\det \mathbf{M} = 0$ the matrix is *singular* and $\mathbf{M}^{-1}$ does not exist (just as 0^{-1} is not defined in ordinary arithmetic).

(2) Form the transpose of $\mathbf{M}$.

$$\text{If } \mathbf{M} = \begin{pmatrix} 1 & 2 \\ 3 & 4 \end{pmatrix}, \qquad \tilde{\mathbf{M}} = \begin{pmatrix} 1 & 3 \\ 2 & 4 \end{pmatrix}.$$

(3) Form the matrix $\tilde{\mathbf{M}}'$, where $\tilde{M}'_{rc}$ is the *cofactor* of M_{rc} (the signed minor, the determinant formed from the matrix with row r, column c struck out and multiplied by $(-1)^{r+c}$).

$$\text{If } \tilde{\mathbf{M}} = \begin{pmatrix} 1 & 3 \\ 2 & 4 \end{pmatrix}, \qquad \tilde{\mathbf{M}}' = \begin{pmatrix} 4 & -2 \\ -3 & 1 \end{pmatrix}.$$

(4) Form the inverse as $\mathbf{M}^{-1} = \mathbf{M}'/\det \mathbf{M}$.

$$\text{If } \mathbf{M} = \begin{pmatrix} 1 & 2 \\ 3 & 4 \end{pmatrix}, \qquad \mathbf{M}^{-1} = \begin{pmatrix} 4 & -2 \\ -3 & 1 \end{pmatrix}\Big/(-2) = \begin{pmatrix} -2 & 1 \\ \frac{3}{2} & -\frac{1}{2} \end{pmatrix}.$$

As a check, form $\mathbf{MM}^{-1} = \mathbf{1}$.

An important application of matrices (apart from their role as representatives of symmetry operations) is in the solution of simultaneous equations. A set of equations

$$\begin{aligned} M_{11}x_1 + M_{12}x_2 + \ldots + M_{1n}x_n &= c_1 \\ M_{21}x_1 + M_{22}x_2 + \ldots + M_{2n}x_n &= c_2 \\ \cdots\cdots\cdots\cdots\cdots\cdots\cdots\cdots \\ M_{n1}x_1 + M_{n2}x_2 + \ldots + M_{nn}x_n &= c_n \end{aligned}$$

can be expressed in the compact form $\mathbf{Mx} = \mathbf{c}$, where $\mathbf{M}$ is the matrix of coefficients and $\mathbf{x}$ and $\mathbf{c}$ are the $1 \times n$ matrices (or *column vectors*)

$$\mathbf{x} = \begin{pmatrix} x_1 \\ x_2 \\ \vdots \\ x_n \end{pmatrix} \qquad \mathbf{c} = \begin{pmatrix} c_1 \\ c_2 \\ \vdots \\ c_n \end{pmatrix}.$$

Then, since $\mathbf{M}^{-1}\mathbf{M} = \mathbf{1}$, on multiplying both sides of $\mathbf{Mx} = \mathbf{c}$ by $\mathbf{M}^{-1}$ we obtain $\mathbf{x} = \mathbf{M}^{-1}\mathbf{c}$. Hence, solving the set of equations amounts to finding the inverse of the matrix of coefficients.

Further reading

Symmetry. H. Weyl; Princeton University Press, 1952.

Symmetry in chemistry. H. H. Jaffe and M. Orchin; Wiley, New York, 1965.

Chemical applications of group theory. F. A. Cotton; Wiley, New York, 1971.

Symmetry and spectroscopy. D. C. Harris and M. D. Bertolucci; Oxford University Press, 1978.

Molecular quantum mechanics (2nd edn). P. W. Atkins; Oxford University Press, 1983.

Group theory and chemistry. D. M. Bishop; Clarendon Press, Oxford, 1973.

Group theory and quantum mechanics. M. Tinkham; McGraw-Hill, New York, 1964.

Tables for group theory. P. W. Atkins, M. S. Child, and C. S. G. Phillips; Clarendon Press, Oxford, 1970.

Introductory problems

A17.1. The point group D_5 has four symmetry species. How many classes does this group have?

A17.2. The molecule CH_3Cl has symmetry C_{3v}. List the symmetry elements possessed by this group. Locate these elements in CH_3Cl.

A17.3. From the symmetry of the following molecules, determine which may have electric dipole moments: pyridine (C_{2v}), $C_2H_5NO_2$ (C_s), $HgBr_2$ ($D_{\infty h}$), $B_3N_3H_6$ (D_{3h}), CH_3Cl (C_{3v}), $HW_2(CO)_{10}$ (D_{4h}), $SnCl_4$ (T_d).

A17.4. Use symmetry properties to determine whether or not the integral $\int p_x z p_z \, d\tau$ is necessarily zero in a molecule with symmetry C_{4v}.

A17.5. For a molecule with symmetry C_{3v}, show that an electric dipole transition between a state of symmetry A_1 and one of symmetry A_2 is forbidden.

A17.6. Show that a set of five hybrid orbitals in a molecule with symmetry C_{4v} and a representation:

E	$2C_4$	C_2	$2\sigma_v$	$2\sigma_d$
5	1	1	3	1

may be made of one p-orbital and four d-orbitals.

A17.7. Refer to the hybrid orbitals in the group C_{4v} in the previous problem. Do they have non-vanishing net overlap with s-orbitals or with p-orbitals?

A17.8. Show that the function xy transforms as B_2 in the group C_{4v}.

A17.9. Molecules with symmetry D_{2h}, C_{3h}, T_h, and T_d are among those which cannot show optical activity. For each of these four point groups, list symmetry elements which are sufficient to preclude optical activity.

A17.10. The group D_2 consists of the following elements: E, C_2, C_2', and C_2''. The three C_2 axes are mutually perpendicular. Construct the group multiplication table.

Problems

17.1. Name the point groups to which the following objects belong: a sphere, an isosceles triangle, an equilateral triangle, an unsharpened pencil, a sharpened pencil, a three-bladed propellor, a snow flake, a table, yourself.

17.2. List the symmetry elements of the following molecules and name the point groups to which they belong: NO_2(bent), CH_3Cl, CCl_3H, $CH_2{=}CH_2$, *cis* $CHCl{=}CHCl$, *trans* $CHCl{=}CHCl$, naphthalene, anthracene, chlorobenzene.

17.3. Do the same for the following molecules: CH_3CH_3 (staggered), cyclohexane (chair), B_2H_6, CO_2, $Co(en)_3^{3+}$ ('en' means ethylenediamine: disregard its detailed structure), S_8 (crown).

17.4. The group C_{2h} consists of the elements E, C_2, σ_h, i. Construct the group multiplication table. Find an example of a molecule possessing the symmetry of this group.

17.5. The group D_{2h} has a C_2 axis perpendicular to the principal twofold axis, and it also has a horizontal mirror plane. Show that the group must also have a centre of inversion.

17.6. Which of the molecules in Problems 17.2 and 17.3 can have a permanent electric dipole moment?

17.7. Which of the molecules in Problems 17.2 and 17.3 may be optically active?

17.8. Consider a water molecule (which has C_{2v} symmetry), and take as a basis for constructing molecular orbitals the two hydrogen 1s-orbitals, and the oxygen 2s- and 2p-orbitals. Set up the 6×6 matrices that reproduce the effects of the symmetry operations of the group in this basis.

17.9. Use the matrices derived in the last Problem to confirm that they represent the group multiplications (a) $C_2\sigma_v = \sigma_v'$, (b) $\sigma_v\sigma_v' = C_2$ correctly.

17.10. The (one-dimensional) matrices $\mathbf{D}(C_3) = 1$, $\mathbf{D}(C_2) = 1$ and $\mathbf{D}(C_3) = 1$, $\mathbf{D}(C_2) = -1$ are different matrix representatives for the group C_{6v} in the sense that they reproduce correctly the multiplication $C_3C_2 = C_6$, with $\mathbf{D}(C_6) = 1$ for the first case and -1 for the second. Use the character table to confirm these remarks. What are the representatives of σ_v and σ_d in each case?

17.11. One of the advantages of character tables is that one can use them to arrive at conclusions very quickly with the minimum of work. As the first of a set of exercises in their use, consider the NO_2 molecule (C_{2v}). The combination $p_{x1} - p_{x2}$, where p_{x1} and p_{x2} are the orbitals on the two oxygens (x perpendicular to the plane), spans A_2. Are there any orbitals on the central nitrogen atom with which that combination can have a net overlap? What if the central atom were sulphur?

17.12. The ground state of NO_2 is of symmetry A_1. To what excited states may it be excited when it absorbs light (by electric dipole transition), and what is the polarization of the light that it is necessary to use?

17.13. The ClO_2 molecule (which has C_{2v} symmetry) was trapped in a solid. Its ground state is known to be of B_1 symmetry (it has a single electron in a p_x-orbital outside a closed core, and p_x has B_1 symmetry). Light polarized parallel to the y-axis excited the molecule to an upper electronic state. What is the symmetry of that state?

17.14. What states of (a) benzene, (b) naphthalene may be reached by the absorption of light from their A_{1g} ground states, and what are the polarizations of the transitions?

17.15. What irreducible representations of the group T_d do the hydrogen 1s-orbitals span in methane? Are there both s- and p-orbitals on the central atom that may form bonds with them? Could d-orbitals, even if they were available on carbon, play a role in bonding in methane?

17.16. Suppose methane were distorted to (a) C_{3v} symmetry, by lengthening one bond, (b) to C_{2v} symmetry, by opening it out with a scissors action: would more d-orbitals become available for bonding?

17.17. Find the appropriate symmetry adapted combinations of the basis used in the discussion of H_2O in Problem 17.8. Deal with the oxygen and the hydrogen pair separately, and state which oxygen orbitals may have net overlap with which hydrogen orbital combinations.

17.18. Return to Problem 17.8 and confirm, by determining the trace of all the matrices, (a) that symmetry elements of the same class have the same character, (b) that the representation is reducible, (c) that the basis spans $3A_1 + B_1 + 2B_2$.

17.19. The f-orbitals are less familiar than the s-, p-, and d-orbitals, but they play an important role in lanthanide chemistry. Even though their shapes are unfamiliar, we can still draw conclusions about their properties from the viewpoint of their symmetry. But what are their symmetries? Their algebraic forms are $z(5z^2-3r^2)f$, $y(5y^2-3r^2)f$, $x(5x^2-3r^2)f$, $z(x^2-y^2)f$, $y(x^2-z^2)f$, $x(z^2-y^2)f$, and $xyzf$, where f is a function of r. What irreducible representations do these orbitals span in molecules with (a) C_{2v}, (b) C_{3v}, (c) T_d, (d) O_h symmetry?

17.20. Consider a lanthanide ion at the centre of (a) a tetrahedral, (b) an octahedral complex. What sets of orbitals will the seven f-orbitals break up into?

17.21. Which of the following transitions is allowed in (a) a tetrahedral complex, (b) an octahedral complex: (i) $d_{xy} \rightarrow d_{z^2}$, (ii) $d_{xy} \rightarrow f_{xyz}$?

17.22. In the text we arrived at the symmetry properties of rotations by a simple pictorial argument. Now we do it slightly more formally. This approach recognizes that the angular momentum about the z-direction is defined as $xp_y - yp_x$, where the p_x and p_y are components of linear momentum. Confirm that this z-component of angular momentum is a basis for A_2 in C_{3v}.

17.23. As what irreducible representations do translations

and rotations transform in the following groups: (a) C_{2v}, (b) C_{3v}, (c) C_{6v}, (d) O_h?

17.24. The response of a molecule to an electric field is governed by the latter's ability to stretch the electronic structure (this is a translational distortion). The result of this stretching is a tendency to mix excited states into the ground state of the molecule (thus p-orbitals may be mixed into s-orbitals, and the molecule bulges on the side where the two orbitals interfere constructively). Is it necessary to consider p-, d-, and f-orbitals being mixed into the s-orbitals of the central atom of (a) octahedral complexes, (b) tetrahedral complexes?

17.25. A magnetic field tends to distort a molecule by twisting it around the direction along which it is applied. A twisting distortion has the symmetry of a rotation. Which states does the magnetic field tend to mix into the ground state in (a) NO_2, (b) an octahedral Ti^{3+} complex, (c) benzene, the field perpendicular to the plane?

17.26. Group theory gives a rapid way of detecting when an integral is zero. Here are two examples. The integral of $\sin\theta\cos\theta$ over a range that is symmetrical about $\theta=0$ is zero. Prove this by a symmetry argument, beginning at $f_1=\sin\theta$, $f_2=\cos\theta$, showing that these span different irreducible representations of the group C_s, and using the argument described in Section 17.3. Why might the integral not disappear when the range is not symmetrical about the origin?

17.27. Does the product xyz vanish when integrated over (a) a cube, (b) a tetrahedron, (c) a hexagon prism, each centred on the origin?

17.28. Regard the naphthalene molecule as belonging to the group C_{2v}, with the C_2 axis perpendicular to the plane. Classify the irreps spanned by the $C2p_z$-orbitals, and find the symmetry-adapted linear combinations.

17.29. The NO_2 molecule belongs to the group C_{2v}, the C_2 axis bisecting the O–N–O angle. Taking a basis of N2s-, N2p-, and O2p-orbitals, identify the irreps they span, and construct the symmetry adapted linear combinations.

17.30. Confirm that the symmetry adapted orbitals for the benzene molecule constructed in the text lead to Hückel determinants that solve to the energy levels given in Chapter 16. This problem illustrates one of the great powers of group theory: large secular determinants are factorized into smaller ones, which can be solved much more readily.

18

Determination of molecular structure: rotational and vibrational spectra

Learning objectives

After careful study of this chapter you should be able to

(1) List the basic features of *absorption*, *emission*, and *Raman spectroscopy*, Section 18.1.

(2) Relate the intensity of a spectral line to the population of the initial state, eqn (18.1.3).

(3) State the *gross selection rules* for rotational and vibrational spectroscopy, Section 18.1(b).

(4) Account for the *widths* of spectral lines in terms of the Doppler effect and lifetime broadening, eqn (18.1.6), and state what is meant by *natural linewidth*, Section 18.1(c).

(5) Derive the *rotational energy levels* of spherical rotors, eqn (18.2.3), symmetric rotors, eqn (18.2.5), and linear rotors, eqn (18.2.6).

(6) Explain the significance of the quantum numbers J and K, and state the degeneracies of rotational levels, Section 18.2(c).

(7) Explain how *centrifugal distortion* affects rotational energy levels, and indicate the information it provides, eqn (18.2.8).

(8) Outline how a *microwave spectrum* is obtained, state the selection rules, eqn (18.2.10), and deduce expressions for the wavenumbers of the transitions, eqn (18.2.12).

(9) State the selection rules for rotational Raman transitions, Section 18.2(h), and derive expressions for the *Stokes* and *anti-Stokes* lines in the spectrum, eqns (18.2.15) and (18.2.16).

(10) Justify the *parabolic approximation* for the molecular potential energy curve, eqn (18.3.1), and write expressions for the vibrational energy levels in the harmonic approximation (18.3.4).

(11) Indicate how *anharmonicity* affects the energy levels of an oscillator, eqn (18.3.7).

(12) Derive expressions for the wavenumbers of vibrational transitions in the harmonic approximation, eqn (18.3.9), and allowing for anharmonicity, eqn (18.3.11).

(13) Explain how anharmonicities affect a vibrational spectrum, Section 18.3(b), and account for the occurrence of *harmonics*, eqn (18.3.12).

(14) Explain and use the *Birge–Sponer method* to find the dissociation energy of a molecule, Example 18.6.

(15) Account for the appearance of *vibration–rotation spectra* and derive expressions for the *P*-, *Q*-, and *R-branches* of the spectrum, eqns (18.3.16–18).

(16) State the selection rules for the vibrational Raman spectra of diatomic molecules, Section 18.3(a).

(17) Derive the number of *normal modes* of vibration, and explain why they are significant, Section 18.4(a).

(18) Determine the symmetry species of the normal modes of a molecule, Example 18.7.

(19) State the selection rules for infrared spectroscopy, Section 18.4(b), and use the symmetry species of vibrations to deduce which modes are infrared active, Example 18.8.

(20) State the selection rules for vibrational Raman spectroscopy, Section 18.4(c), and use the symmetry species of vibrations to deduce which modes are Raman active, Example 18.9.

(21) State the *exclusion rule*, Section 18.4(c).

18.1 | Determination of molecular structure: rotational and vibrational spectra

Introduction

The origin of spectral lines in molecular spectroscopy is the emission or absorption of a photon when the energy of a molecule changes. The difference from atomic spectroscopy is that a molecule's energy can change not only as a result of electronic transitions but also because its rotational and vibrational states may change. This means that molecular spectra are more complex than atomic spectra; but they also contain information relating to more properties, such as bond strengths and molecular geometry.

18.1 General features of spectroscopy

In *emission spectroscopy* a molecule undergoes a transition from a higher to a lower energy state and emits the excess energy as a photon of frequency ν given by

$$h\nu = E_1 - E_2. \tag{18.1.1}$$

This relation is often expressed in terms of the *vacuum wavelength* $\lambda = c/\nu$ or the *wavenumber* $\tilde{\nu} = \nu/c$. The units of the latter are almost always chosen as cm^{-1}. The relation of frequency, wavelength, and wavenumber to the various regions of the electromagnetic spectrum are summarized in Fig. 18.1. In *absorption spectroscopy* the absorption of incident radiation is monitored as it is swept over a range of frequencies, the presence of an absorption at a frequency ν signifying the presence of two energy levels separated by $h\nu$ as expressed by the equation above. Emission and absorption spectroscopy give the same information about energy level separations, but practical considerations generally determine which technique is employed.

Raman spectroscopy explores energy levels by examining the frequencies present in the light scattered by molecules. A monochromatic (single frequency) incident beam consists of a stream of photons of the same energy. When the beam passes through the sample some of the photons collide with the molecules, give up some of their energy, and emerge (in a different direction) with a lower energy and therefore a lower frequency.

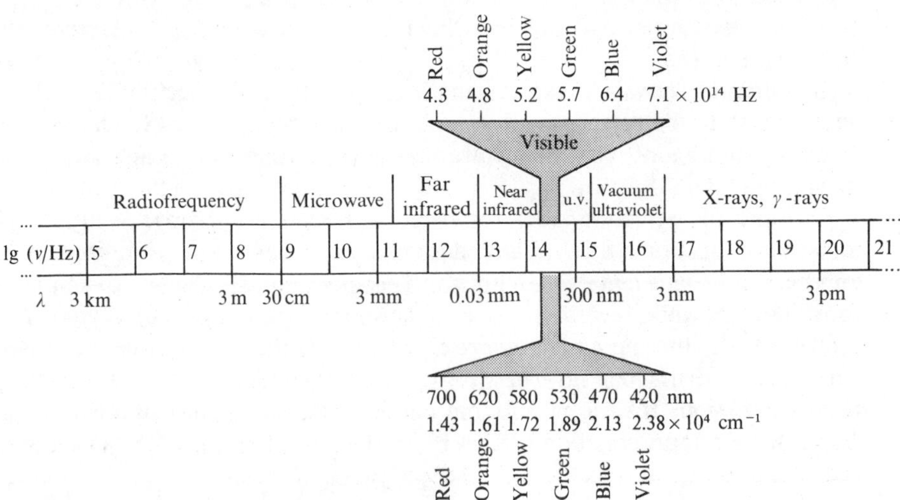

Fig. 18.1. The electromagnetic spectrum and the classification of the regions.

These *scattered* photons give rise to the *Stokes radiation* from the sample. Other photons may collect energy from the molecules (if they are already excited), and emerge as higher frequency *anti-Stokes radiation*. By analysing the frequencies present in the Stokes and anti-Stokes scattered radiation, the energy levels of the molecules can be deduced. The shifts in frequency are quite small, and the incident radiation must be very monochromatic if they are to be observed; moreover, the intensity of scattered radiation is low, and so very intense incident beams are needed. Lasers are ideal in both respects, and have entirely displaced the mercury arcs used originally.

A glance at the spectra (whether emission, absorption, or Raman) illustrated in this chapter and the next shows that their lines occur with a variety of intensities, and we shall also see that some lines which might be expected to occur do not appear at all. In order to account for these features we have to see how the intensities of spectral lines depend on the numbers of molecules in various states and how strongly individual molecules are able to interact with the electromagnetic field and generate or absorb photons.

18.1 (a) Population and intensity

The intensity of a line resulting from the transition of a molecule from a level of energy E depends on the *population* of the level (the number of molecules possessing that energy). When the sample of N molecules is at a temperature T the number with energy E is given by the Boltzmann distribution (Section 0.1(d)):

$$N(E) \propto Ng\mathrm{e}^{-E/kT}, \tag{18.1.2}$$

where g is the *degeneracy* of the level (i.e. the number of states corresponding to that energy). Another line in the spectrum might correspond to a transition from a level of energy E' and degeneracy g'. It follows that the intensities of the two lines should be in the ratio

$$I/I' \propto N(E)/N(E') = (g/g')\mathrm{e}^{-(E-E')/kT}. \tag{18.1.3}$$

(We write the intensity ratio as proportional, not equal, to the population ratio because there are other factors that affect the intensities.)

If $E - E' \gg kT$, then the population of the level with energy E is very much less than that of the level with energy E', and so the intensity of the transition starting from E is correspondingly less than that starting from E'. For example, at room temperature kT corresponds to about $200\,\mathrm{cm}^{-1}$, but as the first electronically excited state of a molecule is usually of the order of $10^4\,\mathrm{cm}^{-1}$ above the ground state, the electronic absorption spectrum is normally due entirely to transitions originating from the ground electronic state because the upper states are not significantly populated. Likewise, vibrational energy levels are separated by around 10^2–$10^3\,\mathrm{cm}^{-1}$, and so the principal transitions are those from the ground vibrational state. In contrast, rotational energy levels are separated by 1–$10^2\,\mathrm{cm}^{-1}$, and many states are occupied even at room temperature; consequently, rotational transitions occur from a wide range of levels, not only the lowest.

Molecules are often prepared in excited states as a result of chemical reaction, electric discharge, or photolysis. In these cases the populations may be quite different from those at thermal equilibrium, and the spectra then arise from transitions from all the populated levels.

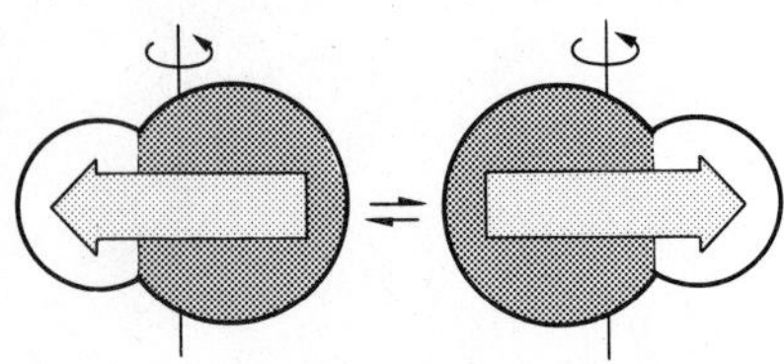

Fig. 18.2. To a stationary observer, a rotating polar molecule looks like an oscillating dipole which can stir the electromagnetic field into oscillation. This is the classical basis of the gross selection rule for rotational transitions.

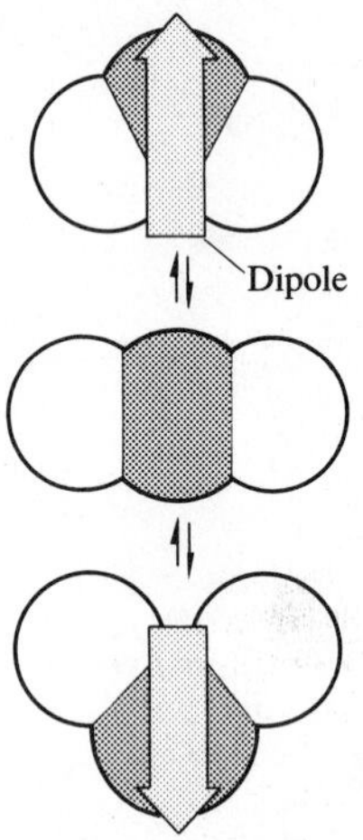

Fig. 18.3. The oscillation of a molecule, even if it is non-polar, may result in an oscillating dipole which can interact with the electromagnetic field.

18.1 (b) Selection rules and intensity

In Section 15.1(f) we met the concept of the *selection rule*, which determines whether a transition is *forbidden* or *allowed*. Selection rules also apply to molecular spectra, and the form they take depends on the type of transition. The underlying classical idea is that, for the molecule to be able to interact with the electromagnetic field and absorb or create a photon of frequency ν, it must possess, at least transiently, a dipole oscillating at that frequency. In the case of emission and absorption spectra (we treat Raman spectra later) this transient dipole is expressed quantum mechanically in terms of the *transition dipole moment*, which for a transition between states ψ_i and ψ_f is defined as

$$\boldsymbol{\mu}_{fi} = -e\int \psi_f^* \boldsymbol{r} \psi_i \, d\tau. \tag{18.1.4}$$

The *intensity* of the transition is proportional to the square of the transition dipole moment ($I \propto |\boldsymbol{\mu}_{fi}|^2$), and so only if the transition moment is non-zero does that transition contribute to the spectrum. Therefore, in order to determine the selection rules, we have to establish the conditions for which $\boldsymbol{\mu}_{fi} \neq 0$.

The *gross selection rules* state the general features a molecule must have in order to give rise to a spectrum of a given kind.

In the case of rotational transitions, the transition moment is zero unless the molecule has a permanent electric dipole (i.e. unless it is polar). The classical basis of this rule is that a polar molecule appears to possess a fluctuating dipole when rotating, Fig. 18.2. The permanent dipole can be regarded as a handle with which the molecule stirs the electromagnetic field into oscillation (and vice versa for absorption).

In the case of vibrational transitions, the transition moment is zero unless the electric dipole moment of the molecule changes during the vibration. The classical basis is that the vibrating molecule shakes the electromagnetic field into oscillation if its dipole changes as it vibrates, Fig. 18.3 (note that the molecule need not have a *permanent* dipole: the rule requires only a change, possibly from zero). Some vibrations do not affect the molecule's dipole moment (e.g. the stretching motion of a homonuclear diatomic molecule), and so they neither absorb nor generate radiation: such vibrations are *inactive*.

Example 18.1

State which of the following molecules have (a) a rotational absorption spectrum, (b) a vibrational absorption spectrum: N_2, CO_2, OCS, H_2O, $CH_2{=}CH_2$, C_6H_6.

• *Method*. Molecules that give rise to rotational spectra have a permanent dipole moment; therefore, select the polar molecules. Molecules that give rise to vibrational spectra have dipole moments that change during the course of a vibration. Therefore, judge whether a distortion of the molecule can change its dipole moment (including changing it from zero).

• *Answer*. (a) Only OCS and H_2O are polar, and so only these two give rise to a rotational absorption spectrum. (b) All the molecules except N_2 possess at least one vibrational mode that results in a change of dipole moment, and so all except N_2 can show a vibrational absorption spectrum.

• *Comment*. Not all the modes of complex molecules are vibrationally active. For example, the *symmetric stretch* of CO_2, in which the O—C—O bonds stretch and contract symmetrically, is inactive because it leaves the dipole moment unchanged (at zero),

- *Exercise*. Repeat the question for H_2, NO, N_2O, CH_4.

[(a) NO, N_2O; (b) NO, N_2O, CH_4]

A more detailed study of the transition moment leads to the *specific selection rules*, which express the allowed transitions in terms of the changes in quantum numbers (such as the rule $\Delta l = \pm 1$ for atoms). These can often be interpreted in terms of the changes of angular momentum involved when a photon (with its intrinsic spin angular momentum $s = 1$) enters or leaves a molecule, and we shall discuss them when we have set up the quantum numbers needed to describe molecular rotation and vibration. Electronic transitions are described in Chapter 19.

18.1 (c) Linewidths

Spectral lines are not infinitely narrow, and various processes contribute to their widths. One important broadening process in gaseous samples is the *Doppler shift*, the effect that makes radiation appear of different wavelength when the source is moving towards or away from the observer. When a source emitting radiation of wavelength λ recedes from an observer with a speed v, the observer detects radiation of wavelengths $(1 + v/c)\lambda$, where c is the speed of the radiation (the speed of light for electromagnetic radiation, the speed of sound for sound waves). A source approaching the observer appears to be emitting radiation of wavelength $(1 - v/c)\lambda$.

Molecules reach high speeds in all directions in a gas, and a static observer detects the corresponding Doppler shifted range of wavelengths. Some molecules approach the observer, some move away; some move quickly, others slowly. The detected spectral 'line' is the absorption or emission profile arising from all the resulting Doppler shifts. This profile reflects the Maxwell distribution of molecular speeds parallel to the line of sight (see Sections 0.1(e) and 26.1), which is a bell-shaped gaussian curve (of the form e^{-x^2}). The *Doppler line shape* is therefore also a gaussian curve, Fig. 18.4, and calculation shows that when the temperature is T and the mass of the molecule is m, the width of the line at half-height is

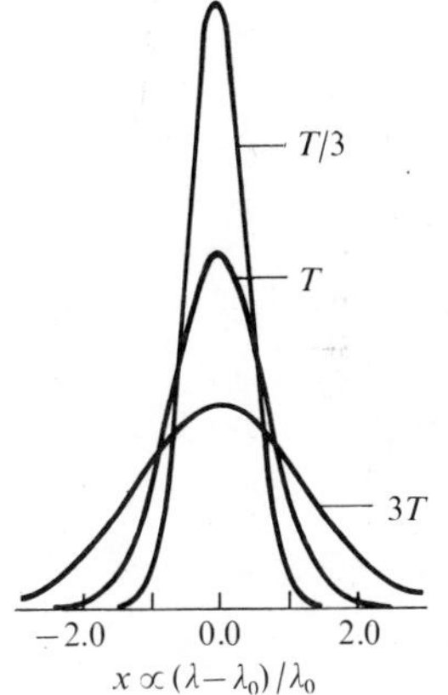

Fig. 18.4. The shape of a Doppler-broadened spectral line reflects the Maxwell distribution of speeds in the sample at the temperature of the experiment. Notice that the line broadens as the temperature is increased. The width at half height, $\delta\lambda$, is given by eqn (18.1.5).

$$\delta\lambda = (2\lambda/c)\{(2kT/m)\ln 2\}^{\frac{1}{2}}, \quad \text{or} \quad \delta v = (2v/c)\{(2kT/m)\ln 2\}^{\frac{1}{2}}. \quad (18.1.5)$$

One application of this expression is to the measurement of the surface temperatures of stars.

Example 18.2

The sun emits a spectral line at 677.4 nm which has been identified as arising from a transition in highly ionized ^{57}Fe. Its width at half-height is 5.3 pm. What is the temperature of the sun's surface?

- *Method*. Rearrange eqn (18.1.5) to

$$T = \{mc^2/8k \ln 2\}(\delta\lambda/\lambda)^2 = 1.949 \times 10^{12} M_r(\delta\lambda/\lambda)^2 \text{ K},$$

where M_r is the RMM of the emitter.

- *Answer*. Since $M_r = 57$, and $\delta\lambda/\lambda = 7.8_2 \times 10^{-6}$,

$$T = (1.949 \times 10^{12} \text{ K}) \times 57 \times (7.8_2 \times 10^{-6})^2 = 6.8 \times 10^3 \text{ K}.$$

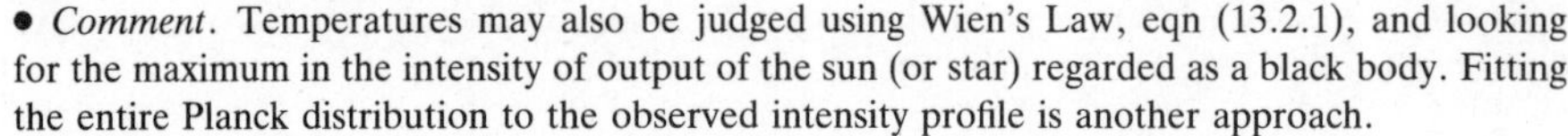

• *Comment*. Temperatures may also be judged using Wien's Law, eqn (13.2.1), and looking for the maximum in the intensity of output of the sun (or star) regarded as a black body. Fitting the entire Planck distribution to the observed intensity profile is another approach.

• *Exercise*. What is the Doppler linewidth of the spectrum of an interstellar NH_3 molecule in the 240 GHz region, the temperature being 10 K? [50 kHz]

The Doppler width increases with temperature because the molecules then possess a wider range of speeds. This means that in order to obtain spectra of maximum sharpness, it is best to work with cold samples. However, with the advent of lasers with their extremely high monochromaticity and of radiofrequency techniques with precise frequency control, a novel approach to the elimination of Doppler broadening has become available. When an intense, monochromatic beam with a frequency slightly higher than the absorption maximum passes through a gaseous sample, only the molecules that happen to be moving away from the source at some precise speed absorb some radiation. If the beam is then reflected back through the sample, Fig. 18.5, more is absorbed, but this time by the molecules that happen to be moving at the same precise speed but away from the mirror. The detector therefore observes a double dose of absorption. However, when the incident radiation is at the absorption peak, only those molecules moving perpendicular to the line of the beam (and therefore having no Doppler shift) absorb on the first passage, and the same ones absorb on the reflected path. Because some of those molecules were excited on the first passage, fewer are available to absorb the light on its second passage, and so a less intense absorption is observed. This appears as a dip, the *Lamb dip* (named after its discoverer), in the absorption curve, and its position gives a very precise location of the transition frequency. The precise location of absorption frequencies in this way is called *Lamb-dip spectroscopy*.

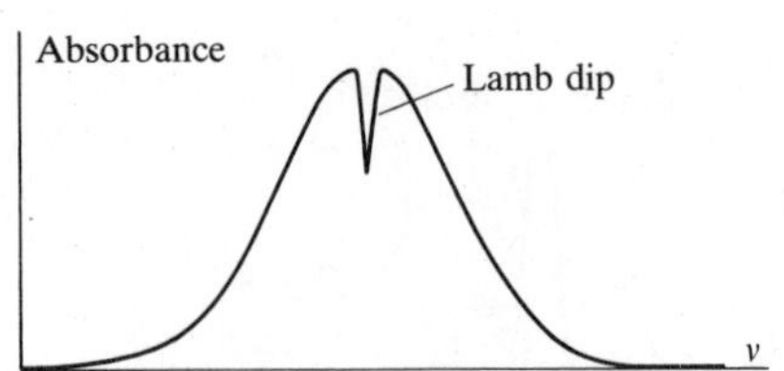

Fig. 18.5. A Lamb dip. The origin is explained in the text: Lamb-dip spectroscopy enables the positions of the centres of absorption lines to be pinpointed very precisely even if there is Doppler broadening.

Even when the Doppler broadening has been largely eliminated, either by working at low temperatures or by Lamb-dip spectroscopy, it is found that spectral lines are still not infinitely sharp. This is due to *lifetime broadening*. When the Schrödinger equation is solved for a system that is changing with time, it is found that it is impossible to specify its energy levels exactly. If on average a system survives in a state for a time τ (this is called the *lifetime* of the state), then its energy levels are blurred to an extent of order δE, where

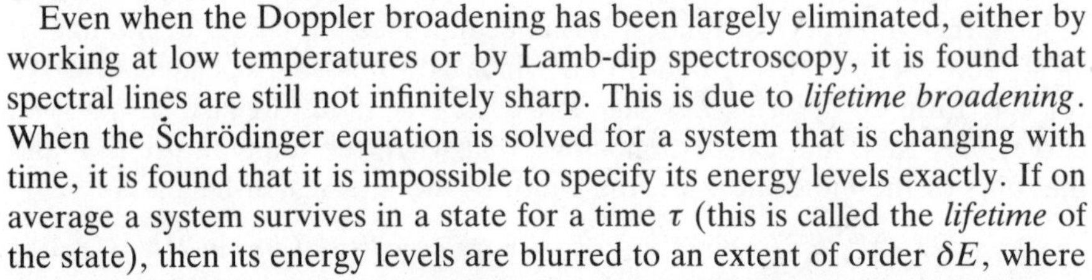

$$\textit{Lifetime broadening: } \delta E \approx \hbar/\tau. \tag{18.1.6}$$

This relation is reminiscent of the Heisenberg uncertainty principle, eqn (13.4.8), and although the connection is tenous lifetime broadening is often called *uncertainty broadening*.

No excited state has an infinite lifetime; therefore, all states are subject to some lifetime broadening, and *the shorter the lifetimes of the states involved in a transition, the broader the spectral lines*. Expressing the energy spread in wavenumbers through $\delta E = hc\,\delta\tilde{\nu}$ and using the values of the fundamental constants gives the practical form of eqn (18.1.6) as

$$\delta\tilde{\nu} = (5.31\ \text{cm}^{-1})/(\tau/\text{ps}). \tag{18.1.7}$$

Two processes are principally responsible for the finite lifetimes of excited states. The dominant one is *collisional deactivation*, which arises from the collisions of the molecules with other molecules or the walls of the container. If the collisional lifetime is $\tau_{\text{collision}}$, the resulting collisional linewidth is $\delta E_{\text{collision}} = \hbar/\tau_{\text{collision}}$. The collisional lifetime can be lengthened by working at low pressures. *Spontaneous emission*, the process by which an excited state discards a photon, is responsible for a natural limit on the lifetime, τ_{natural}, and gives rise to the *natural linewidth* of the transition. This is an intrinsic property of the transition, and cannot be changed by modifying the conditions. Natural linewidths depend strongly on the transition frequency (they increase as ν^3), and so low-frequency transitions (e.g. the microwave transitions of rotational spectroscopy) have very small natural linewidths, and collisional and Doppler line-broadening processes are dominant. The natural lifetimes of electronic transitions are very much shorter than for vibrational and rotational transitions, and so the natural linewidths of electronic transitions are much greater than those of vibrational and rotational transitions. For example, a typical electronic excited state natural lifetime is about 10^{-8} s (10^4 ps), corresponding to a natural width of about $5 \times 10^{-4}\ \text{cm}^{-1}$ (15 MHz); a typical rotational natural lifetime is about 10^3 s, corresponding to a natural linewidth of only $5 \times 10^{-15}\ \text{cm}^{-1}$ (10^{-4} Hz).

When several processes contribute to the line broadening the total linewidth is the sum of the individual linewidths:

$$\delta E = \delta E_{\text{collision}} + \delta E_{\text{natural}} + \ldots$$

Therefore, in terms of the lifetime associated with each process,

$$1/\tau_{\text{total}} = 1/\tau_{\text{collisional}} + 1/\tau_{\text{natural}} + \ldots \qquad (18.1.8)$$

18.2 Pure rotation spectra

First we find expressions for the rotational energy levels of molecules, then we calculate the spectral transition frequencies by applying the selection rules. Finally, we predict the appearance of the spectrum by taking into account the rotational level populations.

18.2 (a) The rotational energy levels

Rotational energy levels may be obtained by solving the Schrödinger equation for the molecule. Fortunately there is a short-cut. This depends on noting that the classical expression for the energy of a body rotating about some axis x is $E = \frac{1}{2}I_{xx}\omega_x^2$, where ω_x is the angular velocity (radians/s) about that axis and I_{xx} is the moment of inertia, Box 18.1. A body free to rotate about three axes has an energy

$$E = \tfrac{1}{2}I_{xx}\omega_x^2 + \tfrac{1}{2}I_{yy}\omega_y^2 + \tfrac{1}{2}I_{zz}\omega_z^2.$$

The classical angular momentum about x is $J_x = I_{xx}\omega_x$; therefore

$$E = J_x^2/2I_{xx} + J_y^2/2I_{yy} + J_z^2/2I_{zz}. \qquad (18.2.1)$$

This is the key equation. The quantum-mechanical properties of angular momentum were given in Section 14.3, Box 14.1, and can be used in conjunction with this equation to obtain the rotational energy levels.

By disregarding the possibility that they distort under the stress of rotation, all the molecules we consider will be treated initially as *rigid*

Box 18.1 Moments of inertia

The moment of inertia around the centre of mass is defined as follows:

$$I_{xx} = \sum_{\mathrm{J}} m_{\mathrm{J}} x_{\mathrm{J}}^2,$$

and likewise for I_{yy} and I_{zz}, where x_{J} is the x-coordinate of the particle of mass m_{J} from the centre of mass. The expressions given below are the moments of inertia of molecules treated as a collection of mass points. in each case m is the total mass of the molecule.

1. *Diatomics*

$$I = (m_1 m_2/m)R^2.$$

2. *Linear rotors*

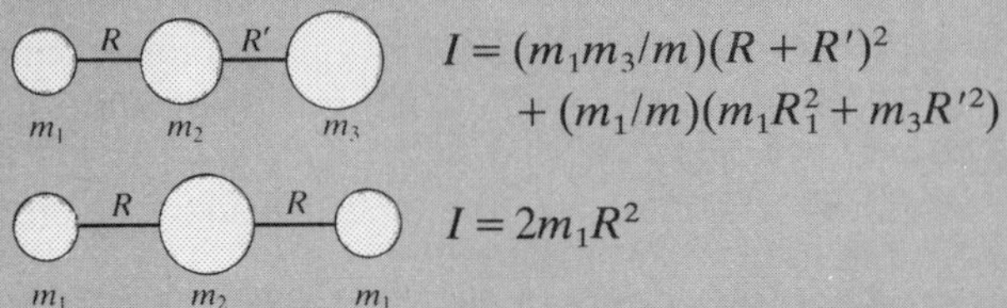

$$I = (m_1 m_3/m)(R+R')^2 + (m_1/m)(m_1 R_1^2 + m_3 R'^2)$$

$$I = 2m_1 R^2$$

3. *Symmetric rotors*

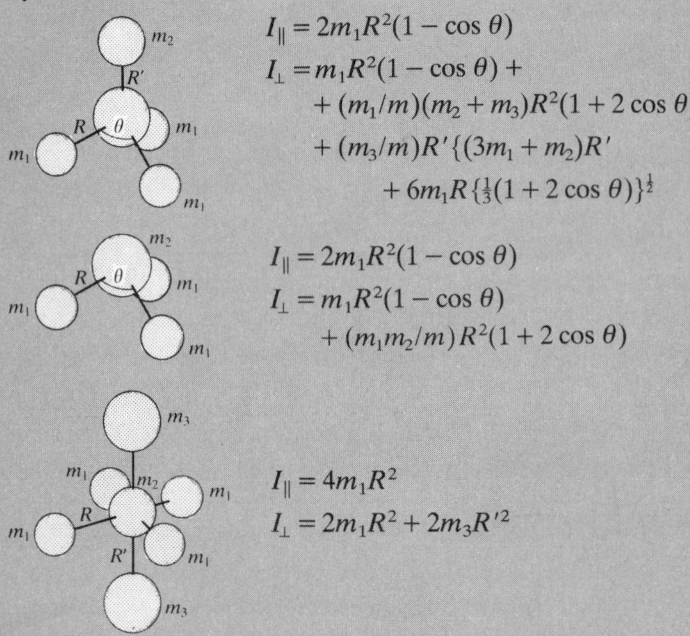

$$I_{\parallel} = 2m_1 R^2(1-\cos\theta)$$
$$I_{\perp} = m_1 R^2(1-\cos\theta) + (m_1/m)(m_2+m_3)R^2(1+2\cos\theta) + (m_3/m)R'\{(3m_1+m_2)R' + 6m_1 R\{\tfrac{1}{3}(1+2\cos\theta)\}^{\frac{1}{2}}$$

$$I_{\parallel} = 2m_1 R^2(1-\cos\theta)$$
$$I_{\perp} = m_1 R^2(1-\cos\theta) + (m_1 m_2/m)R^2(1+2\cos\theta)$$

$$I_{\parallel} = 4m_1 R^2$$
$$I_{\perp} = 2m_1 R^2 + 2m_3 R'^2$$

4. *Spherical rotors*

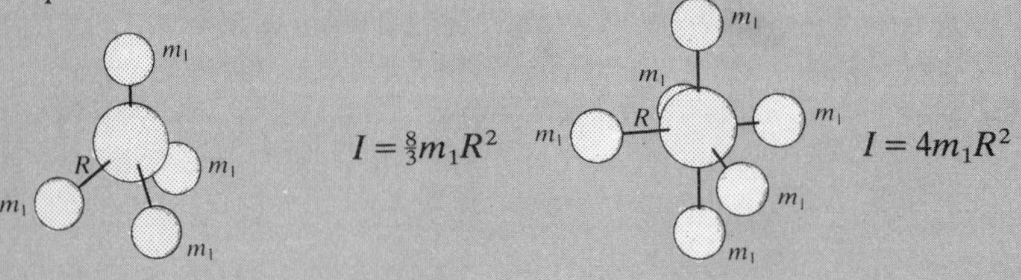

$$I = \tfrac{8}{3}m_1 R^2$$

$$I = 4m_1 R^2$$

rotors. *Spherical rotors* are molecules with all three moments of inertia equal (such as CH_4); *symmetric rotors* have two equal moments of inertia (e.g. NH_3), and *linear rotors* have one moment of inertia (the one about the axis) equal to zero (e.g. CO_2 and HCl). All diatomic molecules are linear rotors. The energy levels of *asymmetric rotors*, which have three different moments of inertia (e.g. H_2O), are complicated and we shall not consider them further. The relation of the moments of inertia to the bond lengths and bond angles are set out in Box 18.1. The convention is to label the moments of inertia I_A, I_B, I_C, with $I_C \geqslant I_B \geqslant I_A$. In group theoretical language, a spherical rotor is a molecule belonging to a cubic point group and a symmetric rotor is one with at least a three-fold axis of symmetry.

18.2 (b) Spherical rotors

When $I_{xx} = I_{yy} = I_{zz} = I$ (as in CH_4 and SF_6) the expression for the energy becomes

$$E = (1/2I)(J_x^2 + J_y^2 + J_z^2) = J^2/2I.$$

This is the classical expression, J being the magnitude of the angular momentum. It can be turned into the quantum expression by replacing J^2 by $J(J+1)\hbar^2$, with $J = 0, 1, 2, \ldots$ Therefore, the energy of a spherical rotor is confined to the values

$$E_J = J(J+1)\hbar^2/2I, \qquad J = 0, 1, 2, \ldots \qquad (18.2.2)$$

The factor $\hbar^2/2I$ is usually written hcB and B, the *rotational constant* of the molecule, is expressed as a wavenumber (cm^{-1}). Then we have the simple expression

$$E_J = hcBJ(J+1); \qquad B = \hbar/4\pi cI; \qquad J = 0, 1, 2, \ldots \qquad (18.2.3)$$

The separation between neighbouring rotational levels is

$$E_J - E_{J-1} = 2hcBJ, \quad \text{or} \quad \tilde{\nu}_J - \tilde{\nu}_{J-1} = 2BJ. \qquad (18.2.4)$$

This separation decreases as I increases: large molecules have closely spaced rotational energy levels. The magnitude of the separation can be estimated by considering CCl_4: from the bond lengths and masses of the atoms we find $I = 4.85 \times 10^{-45}$ kg m^2, and so $hcB = 1.04 \times 10^{-22}$ J (J is joule, J is the quantum number!), or $B = 5.24$ cm^{-1}.

18.2 (c) Symmetric rotors

In symmetric rotors $I_{xx} = I_{yy} \neq I_{zz}$ (as in CH_3Cl, NH_3, and C_6H_6); we shall write $I_{zz} = I_{\parallel}$ (z is called the *figure axis*), and $I_{xx} = I_{yy} = I_{\perp}$. If $I_{\parallel} > I_{\perp}$ the rotor is *oblate* (like a pancake, such as C_6H_6); if $I_{\parallel} < I_{\perp}$ it is *prolate* (like a cigar, such as PCl_5). The classical expression for the energy is

$$E = (1/2I_{\perp})(J_x^2 + J_y^2) + (1/2I_{\parallel})J_z^2.$$

This can be expressed in terms of $J^2 = J_x^2 + J_y^2 + J_z^2$ by adding and subtracting $J_z^2/2I_{\perp}$:

$$\begin{aligned} E &= (1/2I_{\perp})(J_x^2 + J_y^2 + J_z^2) - (1/2I_{\perp})J_z^2 + (1/2I_{\parallel})J_z^2 \\ &= (1/2I_{\perp})J^2 + \{(1/2I_{\parallel}) - (1/2I_{\perp})\}J_z^2. \end{aligned}$$

In order to generate the quantum expression, J^2 is replaced by $J(J+1)\hbar^2$. The quantum theory of angular momentum (Box 14.3) shows that the component about any axis is restricted to the values $K\hbar$, $K=0, \pm 1, \ldots, \pm J$. ($K$ is the quantum number used to signify a component on the molecular figure axis; M signifies a component on a laboratory axis.) Therefore we also replace J_z^2 by $K^2\hbar^2$. Consequently, the energy levels are

$$E_{JK}/hc = BJ(J+1) + (A-B)K^2; \tag{18.2.5}$$

$$J = 0, 1, 2, \ldots; K = 0, \pm 1, \ldots, \pm J; A = \hbar/4\pi c I_{\parallel};\ B = \hbar/4\pi c I_{\perp}.$$

When $K=0$ there is no component of angular momentum about the figure axis, Fig. 18.6, and the energy levels depend only on $I_{\perp}$. When $K=J$, almost all the angular momentum is due to rotation around the figure axis, and the energy levels are determined largely by $I_{\parallel}$. Note that the *sign* of K does not affect the energy: this is because opposite values of K correspond to opposite senses of rotation, and the sense of rotation does not affect the energy.

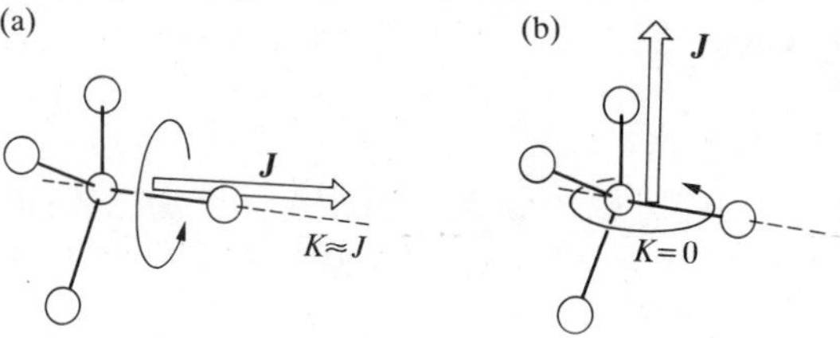

Fig. 18.6. The significance of K. (a) When $|K|$ is close to its maximum value, J, most of the molecular rotation is around the figure axis. (b) When $K=0$ the molecule has no angular momentum about its figure axis: it is undergoing end-over-end rotation.

Example 18.3

The NH_3 molecule is a symmetric rotor with bond length 101.2 pm and HNH bond angle 106.7°. Calculate its rotational energy levels.

• *Method*. Calculate the rotational constants A and B using the expressions for moments of inertia given in Box 18.1. Use data for $^{14}N^{1}H_3$ (inside back cover). The energy levels are given by eqn (18.2.5).

• *Answer*. Substitution of $m_1 = 1.0078 m_u$, $m_2 = 14.0031 m_u$, $R = 101.2$ pm, and $\theta = 106.7°$ into the second of the symmetric rotor expressions in Box 18.1 gives

$$I_{\parallel} = 4.412_8 \times 10^{-47}\ \text{kg m}^2, \qquad I_{\perp} = 2.805_9 \times 10^{-47}\ \text{kg m}^2.$$

Hence, with $A = \hbar/4\pi c I_{\parallel}$ and $B = \hbar/4\pi c I_{\perp}$,

$$A = 6.344\ \text{cm}^{-1}, \qquad B = 9.977\ \text{cm}^{-1}.$$

It follows from eqn (18.2.5) that

$$(E_{JK}/hc)/\text{cm}^{-1} = 9.997 J(J+1) - 3.633 K^2,$$

with $J = 0, 1, 2, \ldots$ and $K = 0, \pm 1, \ldots, \pm J$.

• *Comment*. In the case of $J=1$, the energy needed for the molecule to rotate mainly about its figure axis ($K=J$) is equivalent to $16.32\ \text{cm}^{-1}$, but end-over-end rotation ($K=0$) corresponds to $19.95\ \text{cm}^{-1}$.

• *Exercise*. The $CClH_3$ molecule has a C—Cl bond length of 178 pm, a C—H bond length of 111 pm, and an HCH angle of 110.5°. Calculate its rotational energy levels.

[$(E_{JK}/hc)/\text{cm}^{-1} = 0.444J(J+1) + 4.58K^2$]

18.2 (d) Linear rotors

In a linear rotor (such as CO_2, HCl, and C_2H_2) in which the atoms are regarded as mass points, the rotation occurs only about an axis perpendicular to the line of atoms: there is zero angular momentum around the line. Therefore we can use eqn (18.2.5) but set $K=0$. The energy levels are therefore

$$E_J/hc = BJ(J+1), \qquad J = 0, 1, \ldots \qquad (18.2.6)$$

18.2 (e) A note on degeneracies

The symmetric rotor has an energy that depends on J and K, and each level except those with $K=0$ is doubly-degenerate (the K, $-K$ degeneracy). However, we must not forget that an angular momentum has a component on a laboratory axis, and that this component is quantized, its permitted values being $M_J\hbar$, with $M_J = 0, \pm1, \ldots, \pm J$ (i.e. $2J+1$ values in all). This quantum number does not appear in the expression for the energy, but it is still necessary for a complete specification of the state of the rotor. Therefore, a symmetric rotor level should be regarded as $2(2J+1)$-fold degenerate (for $K \neq 0$).

A linear rotor has K fixed at 0, but the angular momentum may still have $2J+1$ components on the laboratory axis; hence each level of a linear rotor is $(2J+1)$-fold degenerate. A spherical rotor can be thought of as the limit of a symmetric rotor with A vanishingly different from B: K may still take any one of $2J+1$ values, but the energy in eqn (18.2.3) is independent of which value it takes. Therefore, as well as having a $(2J+1)$-fold degeneracy arising from the orientation in space, it also has a $(2J+1)$-fold degeneracy arising from the orientation with respect to an axis selected in the molecule. The overall degeneracy of a symmetric rotor is therefore $(2J+1)^2$ for a level with quantum number J. This degeneracy increases very rapidly: when $J=10$, for instance, there are 441 states of the same energy.

The M_J-degeneracy is removed when an electric field is applied to a molecule with a permanent electric dipole moment (e.g. HCl or NH_3) because now the energy of the molecule depends on its orientation in space. This is called the *Stark effect*. In the case of a linear rotor in an electric field of strength $\mathscr{E}$ the energy is given by

$$E_{J,M_J} = hcBJ(J+1) + \frac{\mu^2\mathscr{E}^2\{J(J+1) - 3M_J^2\}}{2hcBJ(J+1)(2J-1)(2J+3)}. \qquad (18.2.7)$$

Note that it depends on the square of the permanent electric dipole moment (μ), and so the measurement of these energies is a way of measuring this property. Furthermore, because the energy now depends on M_J as its square, the states $\pm|M_J|$ remain degenerate.

18.2 (f) A note on centrifugal distortion

When molecules rotate, their atoms are subject to centrifugal forces that tend to distort the molecular geometry and therefore to change the moments of inertia. In the case of a diatomic molecule, the centrifugal distortion stretches the bond, and therefore increases the moment of inertia; as a result, the energy levels are less far apart than predicted by eqn (18.2.6). The effect is usually taken into account by subtracting a term from the energy by writing

$$E_J/hc = BJ(J+1) - DJ^2(J+1)^2, \qquad (18.2.8)$$

where D is the *centrifugal distortion constant:* it is large when the bond is easily stretched. In the case of a diatomic molecule, the centrifugal distortion constant is related to the vibrational wavenumber of the bond (which, as we shall see later, is a measure of its stiffness):

$$D = 4B^3/\tilde{\nu}^2, \qquad (18.2.9)$$

and so the observation of the convergence of the rotational levels as J increases can be interpreted in terms of the rigidity (specifically, the force constant) of the bond.

The expression corresponding to eqn (18.2.8) for a symmetric rotor is more complicated because it must take into account the possibility that rotation about the figure axis can distort the molecule to a different extent from rotation about a perpendicular axis.

18.2 (g) Rotational transitions

Typical values of B for small molecules are in the region of 1–10 cm^{-1} (e.g. 0.356 cm^{-1}, 10.7 GHz for NF_3 and 10.59 cm^{-1}, 317 GHz for HCl),and so the transitions lie in the *microwave region* of the spectrum. The transitions are observed by monitoring the absorption of microwave radiation generated either by a *klystron* or, in modern instruments, by a *backward wave oscillator*, a device that has the advantage of being tunable over a wide range of frequencies. For technical reasons related to the detection system it is desirable to modulate the energy levels so that the absorption intensity, and therefore the detected signal, oscillates. This is achieved by applying an oscillating electric field (of strength of order 10^2 V cm^{-1} and frequency between 50 and 100 kHz) to the sample, the energy levels responding through the Stark effect; the technique is known as *Stark modulation*.

If a constant electric field is applied, the levels shift to an extent determined by the magnitude of the molecular dipole moment (eqn (18.2.7)), and so the modification in the spectrum brought about by a known field can be used to measure the dipole moment. Extremely high precision is available in the measurement of frequencies, and microwave spectroscopy is one of the most precise spectroscopic techniques.

Which transitions are observed depends on the selection rules. We have already seen (Section 18.1) that the gross selection rule is that *if a molecule is to have a pure rotation spectrum, then it must possess a permanent electric dipole moment*. This means that homonuclear diatomic molecules and symmetrical ($D_{\infty h}$) linear molecules are rotationally inactive. Spherical rotors cannot have electric dipole moments unless they are distorted by rotation, and so they are also inactive except in special cases. The exceptions include SiH_4 which, as a result of its rotation, is slightly distorted from a regular tetrahedron and has a dipole moment of magnitude 8.3×10^{-6} D (by contrast, HCl has a dipole moment of 1.1 D). The pure rotational spectrum of SiH_4 has been detected, but only by using long path lengths (10 m) through high pressure (4 atm) samples.

The specific selection rules can be found by investigating the values of the transition dipole moment between the two states. In the case of a linear molecule the transition moment vanishes unless the following conditions are fulfilled:

$$\Delta J = \pm 1; \; \Delta M_J = 0, \pm 1. \tag{18.2.10}$$

The change in J matches what we already know about the role of the photon angular momentum and the conservation of angular momentum when one is emitted or absorbed. When these conditions are fulfilled the total $J+1 \leftarrow J$ transition intensity (i.e. the intensity summed over all the values of M_J that contribute to the line) is proportional to

$$|\mu_{J+1,J}|^2 = \mu^2(J+1)/(2J+1), \tag{18.2.11}$$

μ being the permanent electric dipole moment of the molecule. The fact that the transition dipole, and therefore the intensity of the transition, depends on J will be referred to again shortly.

The only extension needed for symmetric rotors is a selection rule on K. If a symmetric rotor has a dipole it must lie parallel to the figure axis, as in NF_3. (Note that a symmetric rotor need not have a dipole moment; e.g. the D_{3h} BF_3 molecule is a nonpolar symmetric rotor.) Therefore it cannot be accelerated into different states of rotation around the figure axis by the absorption of radiation. Quantum mechanically this implies that $\Delta K = 0$ in addition to the rules set out above.

When these selection rules are applied to the expressions for the energy levels, it follows that the wavenumbers of the allowed $J+1 \leftarrow J$ absorptions are

$$\tilde{\nu} = 2B(J+1), \qquad J = 0, 1, 2, \ldots \tag{18.2.12}$$

More complicated expressions arise when centrifugal distortions are taken into account, for then we must use expressions like eqn (18.2.8) in place of the rigid rotor expressions. The transition energies in the presence of a Stark field are also more complicated: then we must use expressions like eqn (18.2.7) with the appropriate selection rule on M_J (for a common arrangement of the direction of the electric field, the selection rule is $\Delta M_J = 0$). We shall use the simple form in eqn (18.2.12) in what follows, because the more accurate expressions are unwieldy, and can easily be explored in the same way.

Example 18.4

Predict the form of the rotational spectrum of NH_3.

- *Method*. Energy levels were calculated in *Example* 18.3. The NH_3 molecule is a polar symmetric rotor, and so the selection rules $\Delta J = \pm 1$, $\Delta K = 0$ apply. For absorption, $\Delta J = +1$, and eqn (18.2.12) can be used. Intrinsic intensities are proportional to $|\mu_{J+1,J}|^2$, eqn (18.2.11).

- *Answer*. Since $B = 9.977\ \text{cm}^{-1}$, we can draw up the following table for the $J+1 \leftarrow J$

transitions.

$J =$	0	1	2	3
$\tilde{\nu}/\text{cm}^{-1}$	19.95	39.91	59.86	79.82
$\lvert\mu_{J+1,J}\rvert^2/\mu^2$	1	0.67	0.60	0.57

The line spacing is 19.95 cm^{-1}.

• *Comment*. The intensities are the intrinsic intensities of the lines; that is, they do not take into account the different populations of the initial rotational levels. A complication that we have ignored is the modification of the numbers of possible rotational states by the Pauli principle: this allows only some of the states to be occupied.

• *Exercise*. Repeat the problem for $^{35}\text{CClH}_3$ (see *Example* 18.3 for details).

[Lines of separation 0.888 cm^{-1}]

The form of the spectrum predicted by eqn (18.2.12) is shown in Fig. 18.7. The most significant feature is that it consists of a series of lines with wavenumbers $2B, 4B, 6B, \ldots,$ the separation being $2B$. The intensities

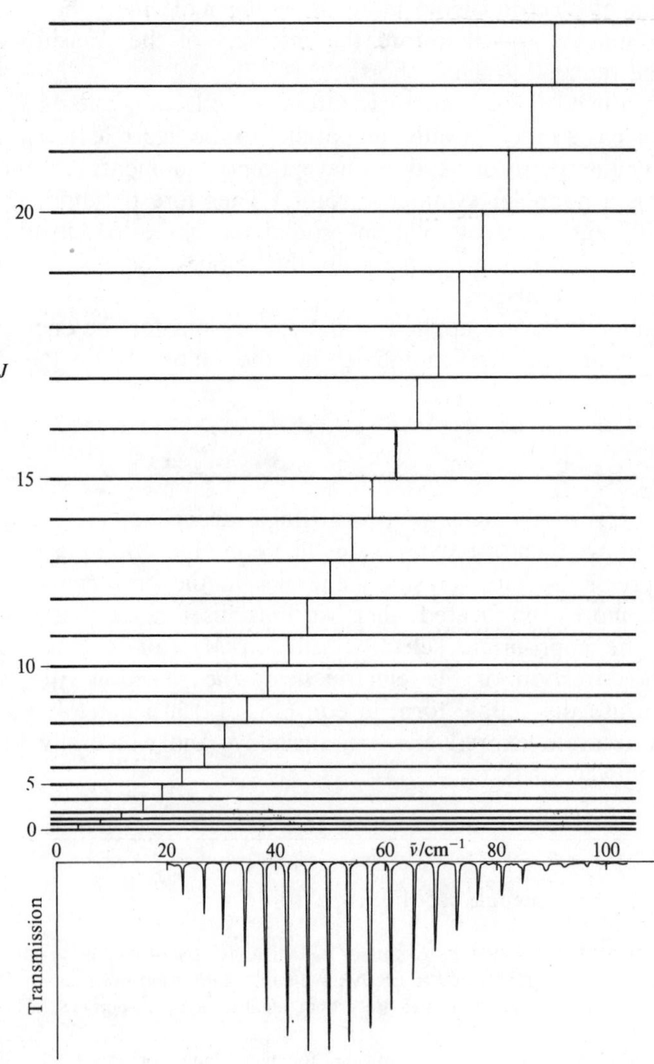

Fig. 18.7. The rotational energy levels of a linear rotor, the transitions allowed by the selection rule $\Delta J = \pm 1$, and a typical pure rotational absorption spectrum. The intensities reflect the populations of the initial level in each case and the strengths of the transition dipole moments.

increase with increasing J and pass through a maximum before tailing off as J becomes large. This reflects the Boltzmann populations of the initial state involved in each transition. For example, in the case of a linear molecule, since the degeneracy is $g_J = 2J + 1$, eqn (18.1.2) gives the population of a level of energy $E_J = hcBJ(J+1)$ as

$$N(E_J)/N \propto (2J+1)\mathrm{e}^{-hcBJ(J+1)/kT}, \tag{18.2.13}$$

which increases to a maximum at

$$J_{\max} \approx \{kT/hcB\}^{\frac{1}{2}} - \tfrac{1}{2} \tag{18.2.14}$$

and then decreases towards zero. Note, however, that eqn (18.2.13) gives the *populations*, but the intensity also depends on the square of the transition moment which, as we saw in eqn (18.2.11), also depends on J. It follows that the intensity depends on J in a way that, at low J at least (because for $J \gg 1$ eqn (18.2.11) settles down to a constant value of $\frac{1}{2}$), reflects only approximately the populations of the initial level, and care has to be taken when interpreting the details of the intensities; in particular, eqn (18.2.14) is only a rough guide to the location of the most intense transition.

The measurement of the line spacing gives B, and hence the moment of inertia perpendicular to the figure axis. Since the masses of the atoms are known, it is a simple matter to deduce the bond length of a diatomic molecule. However, in the case of a polyatomic molecule such as OCS or NH_3, since the analysis gives only a single quantity ($I_\perp$), there is insufficient information to obtain both the bond lengths (in OCS) or the bond length and bond angle (in NH_3). This difficulty can be overcome by using isotopically substituted molecules, such as ABC and A′BC; then, by assuming that $R(\text{A—B}) = R(\text{A}'\text{—B})$, both A—B and B—C bond lengths can be extracted from the two moments of inertia. A famous example of this procedure is the study of OCS, and the actual calculation is worked through in Problems 18.23–4.

18.2 (h) Rotational Raman spectra

The gross selection rule for rotational Raman spectra is that the molecule must possess an *anisotropic polarizability*. The meaning of this is as follows. When an atom or molecule is in an electric field, it is distorted, and the extent of distortion is determined by its *polarizability* (symbol: α; polarizabilities are dealt with in detail in Chapter 24). A xenon atom, for example, has a greater polarizability than a helium atom because its outer electrons are less tightly under the control of the central nucleus. In the case of an atom the polarizability is *isotropic*: the same distortion is induced whatever the direction of the applied field. The polarizability of a spherical rotor (i.e. a molecule belonging to one of the cubic point groups) is also isotropic. However, in general, molecules have polarizabilities that depend on the direction of the field: these are the molecules with *anisotropic polarizabilities*. The electron distribution in H_2, for example, is more distorted when the field is applied parallel to the bond than when it is applied perpendicular to it. All linear molecules and diatomics (whether homonuclear or heteronuclear) are rotationally Raman active. This is the major reason for the importance of rotational Raman spectroscopy: it enables us to examine many of the molecules that are inaccessible to pure rotational microwave spectroscopy. CH_4 and SF_6, however, being spherical rotors, are rotationally Raman inactive as well as microwave inactive.

The specific rotational Raman selection rule is $\Delta J = \pm 2$.‡ The rule $\Delta J = 0$ is also allowed for scattering, but in pure rotational spectroscopy it does not lead to a shift of the scattered photon's frequency: it contributes to the unshifted *Rayleigh scattered* light.

The form of the Raman spectrum, Fig. 18.9, can be predicted by applying the selection rule $\Delta J = \pm 2$ to the rotational energy levels. When the molecule makes a transition with $\Delta J = +2$, the scattered radiation leaves it in a higher rotational state, and so the wavenumber of the radiation (initially $\tilde{\nu}_0 = \nu_0/c$) is decreased. These transitions account for the *Stokes lines* in the spectrum:

$$\text{Stokes lines } (J+2 \leftarrow J)\text{: } \tilde{\nu} = \tilde{\nu}_0 - 2B(2J+3). \qquad (18.2.15)$$

The lines appear to low-frequency of the incident light and at displacements $6B, 10B, 14B, \ldots$ from $\tilde{\nu}_0$ for $J = 0, 1, 2, \ldots$ When the molecule makes a transition with $\Delta J = -2$ the scattered photon emerges with increased energy. These transitions account for the *anti-Stokes lines* of the spectrum:

$$\text{Anti-Stokes lines } (J \rightarrow J-2)\text{: } \tilde{\nu} = \tilde{\nu}_0 + 2B(2J-1). \qquad (18.2.16)$$

The anti-Stokes lines occur at displacements of $6B, 10B, 14B, \ldots$ (for $J = 2, 3, \ldots$; $J = 2$ is the lowest state that can contribute under the selection rule $\Delta J = -2$) to the high-frequency side of the incident radiation. The separation of the lines in both the Stokes and the anti-Stokes regions is $4B$, and so from its measurement $I_\perp$ can be determined and then used to find the bond lengths and angles exactly as in the case of microwave spectroscopy.

Example 18.5

Predict the form of the rotational Raman spectrum of NH_3 when it is irradiated with monochromatic 336.732 nm laser light.

• *Method*. The molecule is rotationally Raman active because end-over-end rotation modulates its polarizability as viewed by a stationary observer. The value of B was calculated in *Example* 18.2. The Stokes and anti-Stokes lines are given by the expressions above. Work in wavenumbers, but finally convert to vacuum wavelengths using $\lambda = 1/\tilde{\nu}$.

• *Answer*. Since $B = 9.977\ \text{cm}^{-1}$ and $\lambda_0 = 336.732$ corresponds to $\tilde{\nu}_0 = 29\,697.2\ \text{cm}^{-1}$, eqns

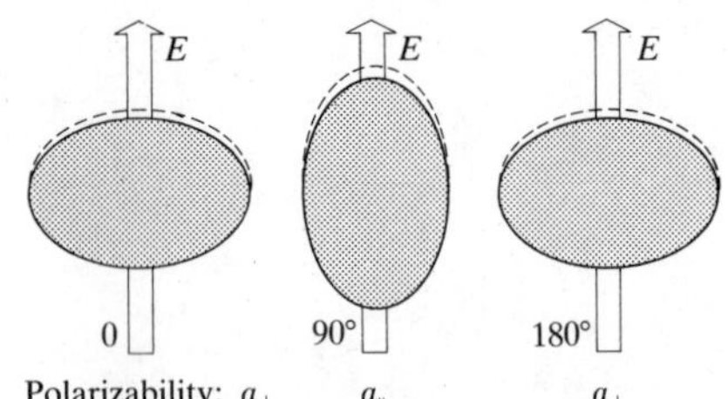

Fig. 18.8. The distortion induced in a molecule by an applied electric field returns to its initial value after a rotation of only 180° (i.e. twice a revolution). This is the origin of the $\Delta J = \pm 2$ selection rule in rotational Raman spectroscopy.

‡ The classical origin of the 2 in the selection rule is as follows. In an electric field $\mathscr{E}$ a molecule acquires a dipole moment of magnitude $\alpha\mathscr{E}$, where α is the plarizability. If the electric field is that of a light wave of frequency ω_0, then the induced dipole moment is time dependent and has the form $\mu = \alpha\mathscr{E}(t) = \alpha\mathscr{E}_0 \cos \omega_0 t$. If the molecule is rotating its polarizability in the direction of the field is also time dependent (if it is anisotropic), and we can write $\alpha = \alpha_0 + \Delta\alpha \cos 2\omega_R t$, the 2 appearing because the polarizability returns to its initial value twice per revolution, Fig. 18.8. Substituting this expression into the expression for the induced dipole moment gives

$$\begin{aligned}\mu &= \{\alpha_0 + \Delta\alpha \cos 2\omega_R t\}\{\mathscr{E}_0 \cos \omega_0 t\}\\ &= \alpha_0\mathscr{E}_0 \cos \omega_0 t + \mathscr{E}_0\,\Delta\alpha \cos 2\omega_R t \cos \omega_0 t\\ &= \alpha_0\mathscr{E}_0 \cos \omega_0 t + \tfrac{1}{2}\,\Delta\alpha\mathscr{E}_0\{\cos(\omega_0 + 2\omega_R)t + \cos(\omega_0 - 2\omega_R t)\}.\end{aligned}$$

This shows that the induced dipole has a component oscillating at the incident light frequency (so that it radiates Rayleigh radiation), and that it also has two components at $\omega_0 \pm 2\omega_R$, which gives rise to to the shifted Raman lines. Note that these lines appear only if $\Delta\alpha \neq 0$; hence the need for anisotropy in the polarizability.

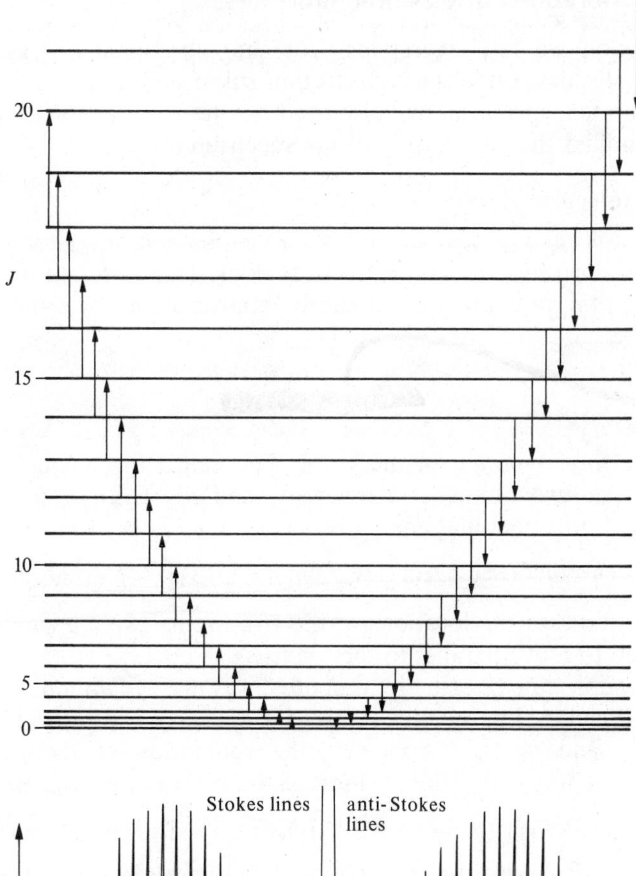

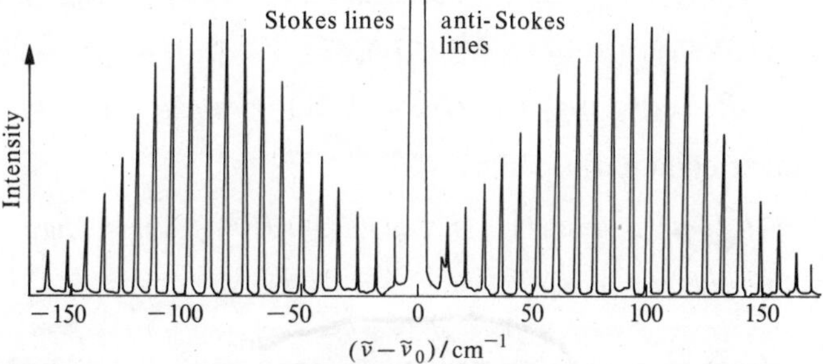

Fig. 18.9. The rotational energy levels of a linear rotor and the transitions allowed by the $\Delta J = \pm 2$ Raman selection rules. The form of a typical rotational Raman spectrum is also shown.

(18.2.15) and (18.2.16) give the following line positions:

$J =$	0	1	2	3
Stokes: $\tilde{\nu}/\mathrm{cm}^{-1}$	29 637.4	29 597.5	29 557.6	29 517.7
λ/nm	337.412	337.867	338.323	338.780
Anti-Stokes: $\tilde{\nu}/\mathrm{cm}^{-1}$			29 757.1	29 797.0
λ/nm			336.055	335.605

• *Comment*. The appearance of the spectrum will be as follows. There will be a strong central line at 336.732 nm accompanied on either side by lines of increasing and then decreasing intensity (as a result of transition moment and population effects). The spread of the entire spectrum is very small (about 500 cm^{-1} at room temperature), and so the incident light must be highly monochromatic.

• *Exercise*. Repeat the calculation for the rotational Raman spectrum of $CClH_3$ (data in *Example* 18.3).

18.3 The vibrations of diatomic molecules

In this section we find expressions for the vibrational energy levels of diatomic molecules, establish the selection rules, and then discuss the form of the vibrational spectrum. We also see how the simultaneous excitation of rotation modifies the appearance of the spectrum.

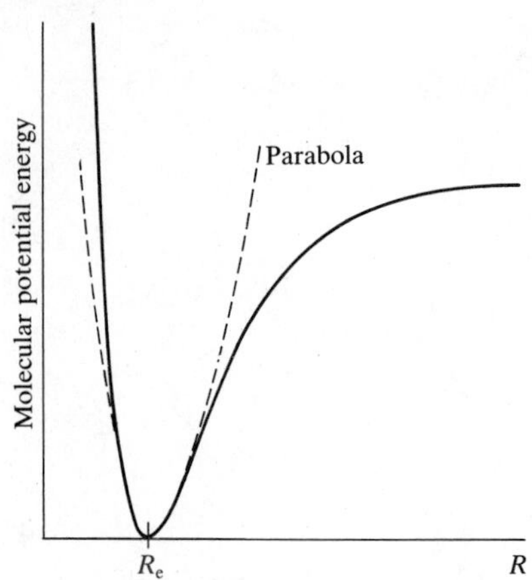

Fig. 18.10. The molecular potential energy curve can be approximated by a parabola near the bottom of the well. The parabolic potential leads to harmonic oscillations. At high excitation energies the parabolic approximation is poor, and is totally wrong near the dissociation limit.

18.3 (a) Molecular vibrations

A typical molecular potential energy curve (Section 16.1) for a diatomic molecule is illustrated in Fig. 18.10. In regions close to R_e, at the minimum of the curve, the potential energy can be approximated by a parabola, and we can write

$$V = \tfrac{1}{2}k(R - R_e)^2, \tag{18.3.1}$$

k being the *force constant* of the bond. The Schrödinger equation for the motion of the two atoms of masses m_1 and m_2 (with total mass $m = m_1 + m_2$) with this potential energy is

$$-(\hbar^2/2m_1)(\partial^2\psi/\partial x_1^2) - (\hbar^2/2m_2)(\partial^2\psi/\partial x_2^2) + V\psi = E\psi,$$

where x_1 and x_2 are the locations of the two atoms (we are supposing that they are free to move parallel to the x-axis).

This equation can be greatly simplified by the separation of variables technique (Section 14.1(c) and Appendix 15.1). We transform from the variables x_1 and x_2 to $R = x_1 - x_2$ (the separation of the atoms) and $x = (m_1/m)x_1 + (m_2/m)x_2$ (the location of the centre of mass). Then, since

$$\partial/\partial x_1 = (\partial R/\partial x_1)(\mathrm{d}/\mathrm{d}R) + (\partial x/\partial x_1)(\mathrm{d}/\mathrm{d}x) = \mathrm{d}/\mathrm{d}R + (m_1/m)\,\mathrm{d}/\mathrm{d}x$$
$$\partial/\partial x_2 = (\partial R/\partial x_2)(\mathrm{d}/\mathrm{d}R) + (\partial x/\partial x_2)(\mathrm{d}/\mathrm{d}x) = -\mathrm{d}/\mathrm{d}R + (m_2/m)\,\mathrm{d}/\mathrm{d}x,$$

the Schrödinger equation becomes

$$-(\hbar^2/2m)(\partial^2\psi/\partial x^2) - (\hbar^2/2\mu)(\partial^2\psi/\partial R^2) + V\psi = E\psi,$$

where μ is the *reduced mass*:

$$1/\mu = 1/m_1 + 1/m_2. \tag{18.3.2}$$

This Schrödinger equation can be separated. We write the complete wavefunction as the product of two functions, one, ψ_{trans}, depending on x and the other, ψ_{vib}, depending on R, and note that V depends only on R. The equation separates as follows:

$$-(\hbar^2/2m)(\partial^2\psi_{\text{trans}}/\partial x^2) = E_{\text{trans}}\psi_{\text{trans}}$$
$$-(\hbar^2/2\mu)(\partial^2\psi_{\text{vib}}/\partial R^2) + V\psi_{\text{vib}} = E_{\text{vib}}\psi_{\text{vib}}.$$

The first is an equation for the free motion of a particle of mass m, and so we need not consider it further. The second is an equation for the *internal motion* of the molecule. From now on we omit the 'vib' subscripts, so that the equation we must solve is

$$-(\hbar^2/2\mu)(\partial^2\psi/\partial R^2) + V\psi = E\psi \tag{18.3.3}$$

with V given by eqn (18.3.1).

The Schrödinger equation we have obtained is that for a particle of mass μ undergoing harmonic motion. Therefore, we can use the results of

Section 14.2(b) directly, and immediately write down the permitted vibrational energy levels:

$$E_v = (v + \tfrac{1}{2})\hbar\omega; \quad \omega = (k/\mu)^{\frac{1}{2}}; \; v = 0, 1, 2, \ldots \tag{18.3.4}$$

The vibrational wavefunctions are the same as in eqn (14.2.7) with y proportional to the displacement $R - R_e$, and consist of Hermite polynomials centred on R_e and multiplied by the bell-shaped Gaussian function.

It is important to note that the energies depend on the *reduced* mass μ of the molecule, not on its *total* mass. This is physically reasonable, because if atom 1 were as heavy as a brick wall, then $1/\mu \approx 1/m_2$, so that $\mu \approx m_2$, the mass of the lighter atom, and the vibration would be that of a light atom relative to that of a stationary brick wall (this is approximately the case in HI, for example, where the I atom barely moves and $\mu \approx m_H$). In the case of a homonuclear diatomic molecule, for which $m_1 = m_2$, the reduced mass is half the total mass: $\mu = m/2$.

18.3 (b) Anharmonicity

The energy levels given in eqn (18.3.4) are only approximate because they are based on the parabolic approximation to the actual potential energy curve (and, deeper in the background, there is the Born–Oppenheimer approximation which allows us to separate the nuclear motion from the electronic, and to talk in terms of potential energy curves). At high vibrational excitations the swing of the atoms (i.e. the spread of the vibrational wavefunction) allows the molecule to explore regions of the curve where the parabolic approximation is poor. The motion then becomes *anharmonic*. In particular, because the actual curve is less confining than a parabola, Fig. 18.10, we can anticipate that the energy levels become less widely spaced at high excitations.

One approach to calculating the energy levels over a wider range is to use a function that resembles the true potential energy more closely. The *Morse potential energy* is

$$V = D_e\{1 - e^{-a(R-R_e)}\}^2, \tag{18.3.5}$$

D_e being the depth of the potential minimum and $a = (\mu/2D_e)^{\frac{1}{2}}\omega$; the shape of the Morse curve is shown in Fig. 18.11. Near the well minimum it resembles a parabola (as can be checked by expanding the exponential as far as the first term), but unlike a parabola it allows for dissociation at high energies. The Schrödinger equation can be solved with this form of V, and the permitted energy levels are

$$E_v = (v + \tfrac{1}{2})\hbar\omega - (v + \tfrac{1}{2})^2 x_e\hbar\omega; \quad x_e = a^2\hbar/2\mu\omega. \tag{18.3.6}$$

x_e is called the *anharmonicity constant*. The number of vibrational levels of a Morse oscillator is finite, and $v = 0, 1, 2, \ldots, v_{max}$, as shown in Fig. 18.11. The second term in eqn (18.3.6) subtracts from the first, and gives rise to the convergence of the levels at high quantum numbers.

While the Morse oscillator is quite useful theoretically, in practice the more general expression for the energy levels,

$$E_v = (v + \tfrac{1}{2})\hbar\omega - (v + \tfrac{1}{2})^2 x_e\hbar\omega + (v + \tfrac{1}{2})^3 y_e\hbar\omega + \ldots \tag{18.3.7}$$

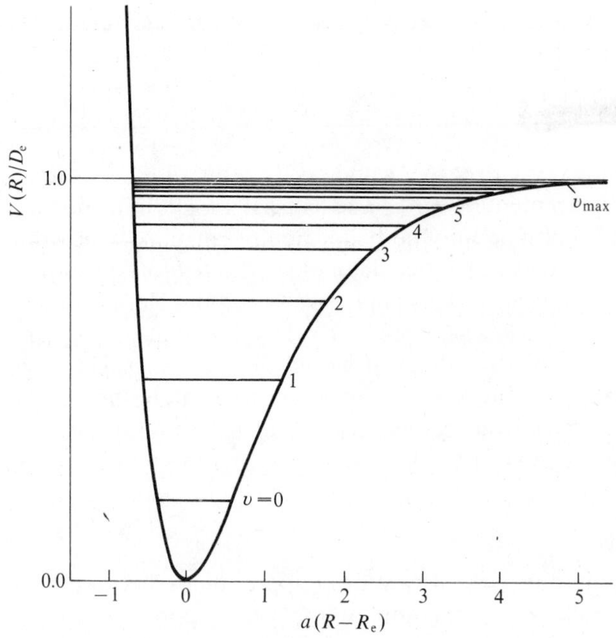

Fig. 18.11. The Morse potential energy curve reproduces the general shape of a molecular potential energy. The corresponding Schrödinger equation can be solved, and the values of the energies obtained. Note that the number of bound levels is finite.

is used to fit the experimental data and to find the dissociation energy of the molecule.

18.3 (c) The vibrational spectra of diatomic molecules

We have seen that the gross selection rule for vibrational transitions is that the electric dipole moment of the molecule must change in the course of a vibration. Homonuclear diatomic molecules are therefore inactive, because their dipole moment remains zero however long the bond. Heteronuclear diatomic molecules are active. The specific selection rule is obtained from an analysis of the expression for the transition moment (and the properties of integrals over Hermite polynomials), and is

$$\textit{Vibrational transitions: } \Delta v = \pm 1. \tag{18.3.8}$$

It follows that the allowed transitions correspond to changes of energy given by

$$\Delta E = E_{v+1} - E_v = \hbar\omega \quad \text{for} \quad v+1 \leftarrow v, \tag{18.3.9}$$

in the harmonic approximation, and by

$$\Delta E = \hbar\omega - 2(v+1)x_e\hbar\omega + \dots \tag{18.3.10}$$

if anharmonicities are taken into account using eqn (18.3.7). The energy is supplied as a photon of energy $h\nu$ (or $hc\tilde{\nu}$), and so the radiation absorbed has wavenumber

$$\tilde{\nu}(v+1 \leftarrow v) = \tilde{\nu}_0 - 2(v+1)x_e\tilde{\nu}_0 + \dots, \tag{18.3.11}$$

with $\tilde{\nu}_0 = \omega/2\pi c$. This shows that the lines converge as v increases. In the harmonic approximation all lines lie at the same wavenumber $\tilde{\nu} = \tilde{\nu}_0$.

HCl has a force constant of $516\,\mathrm{N\,m^{-1}}$, a reasonably typical value. The reduced mass of $^1H^{35}Cl$ is 1.63×10^{-27} kg (note that this is very close to the mass of the hydrogen atom, 1.67×10^{-27} kg, and so Cl is behaving like a brick wall). These values imply that $\omega=5.63\times10^{14}\,\mathrm{s^{-1}}$, corresponding to $\nu=8.95\times10^{13}$ Hz, $\tilde{\nu}=2990\,\mathrm{cm^{-1}}$, and $\lambda=3.35\,\mu\mathrm{m}$. This radiation lies in the *infrared region* of the spectrum, and so vibrational spectroscopy is an infrared technique.

At room temperature $kT/hc=200\,\mathrm{cm^{-1}}$, and so the Boltzmann distribution implies that almost all the molecules will be in their vibrational ground states initially, and therefore that the dominant spectral transition will be from $v=0$ to $v=1$. As a result, the spectrum is expected to consist of a single absorption line. From its position the force constant may be calculated using $\omega=(k/\mu)^{\frac{1}{2}}$. The line is denoted $1\leftarrow0$, the convention being to write the upper state first. The corresponding emission is $1\rightarrow0$.

There are two reasons why more lines are sometimes observed in infrared spectra. One is the modification of the populations from their thermal equilibrium values coupled with the effects of anharmonicity on the energy levels; another is the failure of the harmonic oscillator selection rules.

If the sample molecules are formed in a vibrationally excited state, such as when vibrationally 'hot' HF is formed in the reaction $H_2+F_2\rightarrow2HF^*$, the transitions $5\rightarrow4$, $4\rightarrow3$, etc. may also appear (in emission). In the harmonic approximation, all these lines lie at the same frequency, and the spectrum is a single line. However, the presence of anharmonicity causes the transitions to lie at slightly different frequencies, and so several lines are observed.

Anharmonicity also accounts for the appearance of additional weak absorption lines corresponding to the transitions $2\leftarrow0$, $3\leftarrow0$, etc., even though these second, third, ... *harmonics* are forbidden by the selection rule $\Delta v=\pm1$. The reason is that the selection rule is derived using harmonic oscillator wavefunctions, and these are only approximately valid when anharmonicity is present. Therefore, the selection rule is also only an approximation. For an anharmonic oscillator all values of Δv are allowed, but $\Delta v>1$ is allowed only weakly if the anharmonicity is slight. The second harmonic, for example, gives rise to an absorption at

$$\tilde{\nu}(v+2\leftarrow v)=2\tilde{\nu}_0-2(2v+3)x_e\tilde{\nu}_0+\dots \qquad (18.3.12)$$

When several vibrational transitions are detectable, a graphical technique, the *Birge–Sponer extrapolation*, may be used to determine the dissociation energy (D_0) of the bond. The basis of this method is that the sum of successive energy separations $\Delta E(v+1\leftarrow v)$, the separation of the rungs of a ladder, from the zero-point level to the dissociation limit is the dissociation energy, the height of the ladder:

$$D_0=\Delta E(1\leftarrow0)+\Delta E(2\leftarrow1)+\dots \qquad (18.3.13)$$

When only the x_e anharmonicity constant is taken into account, the successive terms decrease linearly, Fig. 18.12, and the construction shows that the *area* under the plot of $\Delta E(v+1\leftarrow v)$ aginst v is equal to the sum, and therefore to D_0.

In practice, often only the first few energy separations can be determined, and the measured values must be extrapolated to the point where the curve cuts the axis, Fig. 18.12. The area, the sum of the energy

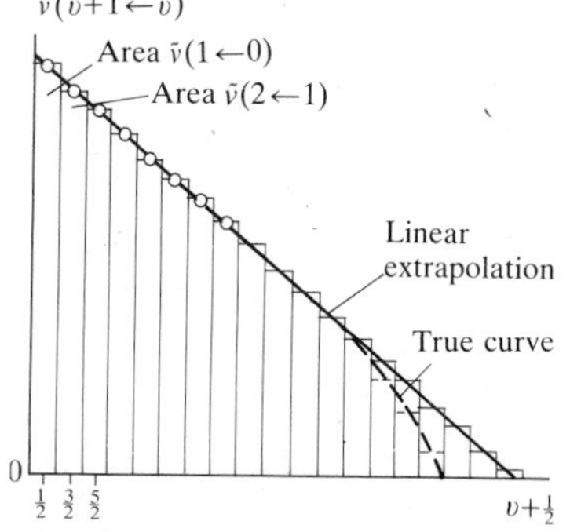

Fig. 18.12. The area under a plot of energy difference against vibrational quantum number is equal to the dissociation energy of the molecule. The assumption that the differences approach zero linearly is the basis of the Birge–Sponer extrapolation.

steps, still gives the dissociation energy, but the assumption that the linear extrapolation is valid often the leads to significant error. The only way to obtain reliable values is to measure energy separations up to as close to the dissociation limit as possible.

The molar bond dissociation enthalpy (ΔH_m) is related to the bond dissociation energy by $\Delta H_m = N_A D_0 + RT$ (because two particles are formed when the molecule dissociates). The depth of the potential well, D_e, differs from D_0 by an amount equal to the zero-point energy:

$$D_e = D_0 + \tfrac{1}{2}(1 - \tfrac{1}{2}x_e)\hbar\omega_0. \qquad (18.3.14)$$

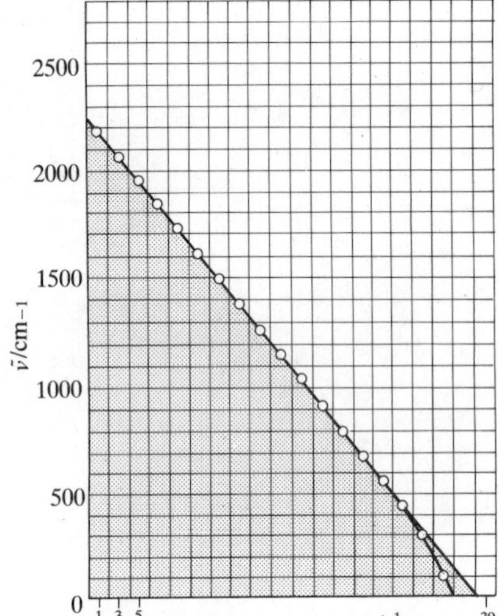

Fig. 18.13. The Birge–Sponer plot used in *Example* 18.4. The area is obtained simply by counting the squares beneath the line.

Example 18.6

The observed vibrational energy level separations of H_2^+ lie at the following values for $1 \leftarrow 0, 2 \leftarrow 1, \ldots$ respectively (in cm^{-1}):

2191, 2064, 1941, 1821, 1705, 1591, 1479, 1368, 1257, 1145,
1033, 918, 800, 677, 548, 411, 265, 117.

Use a Birge–Sponer plot to determine the dissociation energy of the molecule.

- *Method*. Plot the separations against v, and extrapolate linearly to the point cutting the v axis. Determine the area under the curve.

- *Answer*. The points are plotted in Fig. 18.13; the straight line is the Birge–Sponer linear extrapolation. The area under the curve (count the squares) is 214. Each square corresponds to 100 cm^{-1} (refer to the scale of the vertical axis); hence the dissociation energy is 21 400 cm^{-1} (corresponding to 256 kJ mol^{-1}).

- *Comment*. When only the first few level separations are known, the Birge–Sponer extrapolation leads only to *approximate* values of the dissociation energy.

- *Exercise*. The vibrational levels of HgH converge rapidly, and successive separations are 1203.7, 965.6, 632.4, and 172 cm^{-1}. Estimate the dissociation energy. [40 kJ mol^{-1}]

18.3 (d) Vibration–rotation spectra

When the vibrational spectrum of a gas-phase heteronuclear diatomic molecule is analysed under high resolution, each line is found to consist of a large number of closely spaced components, Fig. 18.14. For this reason, molecular spectra are often called *band spectra*. The separation between the components is of the order of 1 cm^{-1}, which suggests that the structure is due to a rotational transition accompanying the vibrational transition. A rotational change should be expected because classically we can think of

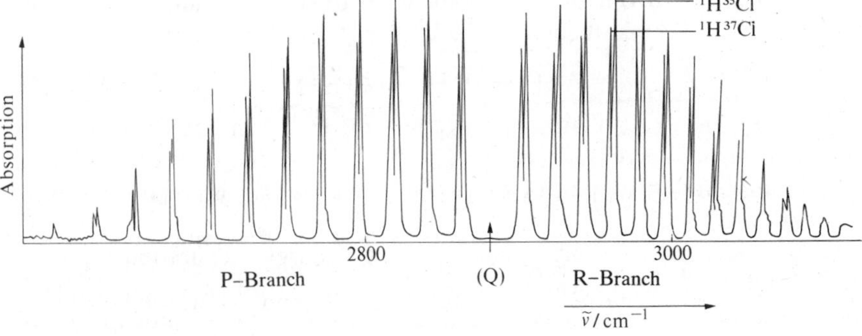

Fig. 18.14. A high-resolution vibration–rotation spectrum of HCl. The lines appear in pairs because $H^{35}Cl$ and $H^{37}Cl$ both contribute (their abundance ratio is 3:1). There is no Q-branch, because that is forbidden.

the transition as leading to a sudden increase or decrease in the instantaneous bond length. Just as in ice-skating you rotate more rapidly when your arms are brought in, and more slowly when they are thrown out, so the molecular rotation is either accelerated or retarded. A detailed analysis of the quantum mechanics of the process shows that the rotational quantum number J changes by ± 1 when a vibrational transition occurs. In exceptional circumstances (when the molecule possesses angular momentum about its axis, as in the case of the electronic orbital angular momentum of the Π molecule NO), the selection rules also allow $\Delta J = 0$.

The appearance of the *vibration-rotation spectrum* of a diatomic molecule can be discussed in terms of the energy levels

$$E_{v,J} = (v + \tfrac{1}{2})\hbar\omega + hcBJ(J+1). \qquad (18.3.15)$$

In a more detailed treatment the anharmonicity corrections are included, and the rotational constant B is allowed to depend on the vibrational state because as v increases the molecule swells slightly and the moment of inertia changes. Centrifugal distortion, and its effect on the rotational energy levels, is also taken into account. However, these features lengthen the algebra without introducing anything new, and so we shall continue with the simple expression above.

When the vibrational transition $v + 1 \leftarrow v$ occurs, J changes by ± 1 (and sometimes 0). The absorptions then fall into *branches*, called the *P*-, *Q*-, and *R-branches* of the spectrum.

$$P\text{-branch } (\Delta J = -1): \; \tilde{\nu}(v+1 \leftarrow v; J-1 \leftarrow J) = \tilde{\nu}_0 - 2BJ. \qquad (18.3.16)$$

This branch, Figs. 18.14 and 18.15, consists of lines at $\tilde{\nu}_0 - 2B$, $\tilde{\nu}_0 - 4B, \ldots$ with an intensity distribution reflecting both the populations of the initial rotational levels and the $J - 1 \leftarrow J$ transition moment.

$$Q\text{-branch } (\Delta J = 0): \; \tilde{\nu}(v+1 \leftarrow v; J \leftarrow J) = \tilde{\nu}_0. \qquad (18.3.17)$$

This branch, when it is allowed, looks like a single line at the vibrational transition wavenumber. In practice, since the rotational constants of the two vibrational levels are slightly different, the Q-branch appears as a cluster of closely spaced lines. In Fig. 18.14 there is a gap at the expected location of the Q-branch because it is forbidden in HCl.

$$R\text{-branch } (\Delta J = +1): \; \tilde{\nu}(v+1 \leftarrow v; J+1 \leftarrow J) = \tilde{\nu}_0 + 2B(J+1). \qquad (18.3.18)$$

This branch consists of lines displaced from $\tilde{\nu}_0$ to high wavenumber by $2B, 4B, \ldots$, Fig. 18.15.

The separation between the lines in the P- and R-branches of a vibrational transition gives the value of B, and so the bond length can be deduced without needing to take a pure rotation microwave spectrum (although the latter is more accurate).

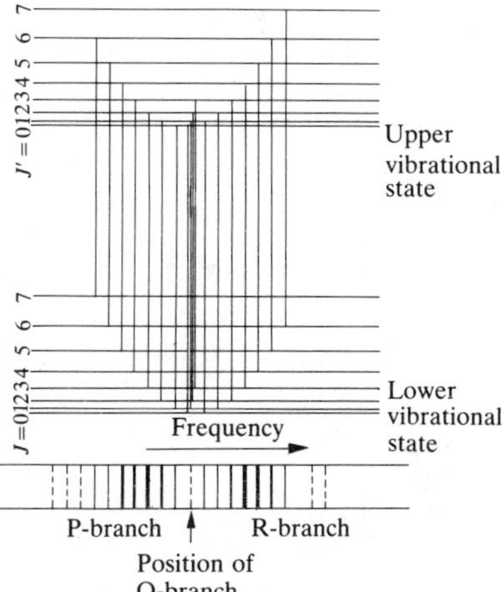

Fig. 18.15. The formation of P-, Q-, and R-branches in a vibration–rotation spectrum. The intensities reflect the populations of the initial rotational levels.

18.3 (e) Vibrational Raman spectra of diatomic molecules

The gross selection rules for vibrational Raman transitions is that *the polarizability should change as the molecule vibrates*. Both homonuclear and heteronuclear diatomic molecules swell and contract during a

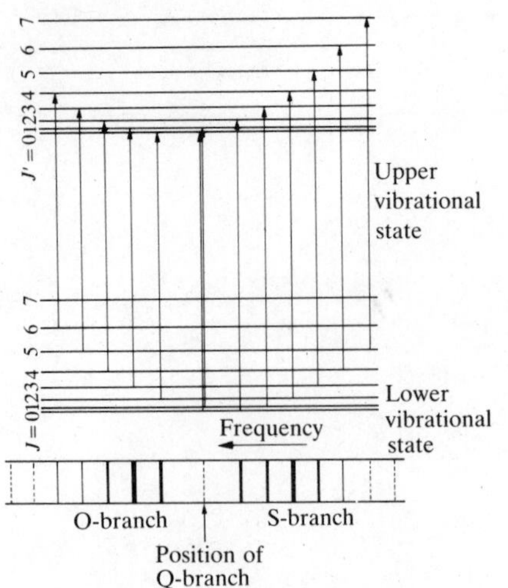

Fig. 18.16. The formation of O-, Q-, and S-branches in a vibration–rotation Raman spectrum. Note that the frequency scale runs in the opposite direction to that in Fig. 18.15, because the higher-energy transitions (on the right) extract more energy from the incident beam and leave it at lower frequency.

Table 18.1. Properties of diatomic molecules

	$\tilde{\nu}_0/\mathrm{cm}^{-1}$	B/cm^{-1}	$k/\mathrm{N\,m}^{-1}$
$^1H_2$	4400	60.86	575
$^1H^{35}Cl$	2991	10.59	516
$^1H^{127}I$	2308	6.51	314
$^{35}Cl_2$	560	0.244	323

vibration, and the control of the nuclei over the electrons, and hence the molecular polarizability, changes too. Both types of diatomic molecule are therefore vibrationally Raman active.

The specific selection rule is $\Delta v = \pm 1$. The lines to high frequency of the incident light, the anti-Stokes lines, correspond to $\Delta v = -1$. They are usually weak because very few molecules are in an excited vibrational state initially. The lines to low frequency, the Stokes lines, correspond to $\Delta v = +1$. Superimposed on these is a branch structure arising from the simultaneous rotational transitions that accompany the vibrational excitation. The selection rules are $\Delta J = 0, \pm 2$ (as in pure rotational Raman spectroscopy), and give rise to the *O-branch* ($\Delta J = -2$), the *Q-branch* ($\Delta J = 0$), and the *S-branch* ($\Delta J = +2$), Fig. 18.16.

The information available from vibrational Raman spectra augments that from infrared spectroscopy because homonuclear diatomics can be studied. The spectra can be interpreted in terms of the force constants, dissociation energies, and bond lengths, and some of the information obtained is included in Table 18.1.

18.4 The vibrations of polyatomic molecules

In a diatomic molecule there is only one mode of vibration: the bond *stretch*. In polyatomic molecules there are several modes because bonds may stretch and angles may bend.

18.4 (a) Normal modes

We begin by calculating the number of vibrational modes of an N-atomic molecule. The total number of coordinates needed to specify the locations of all N atoms is $3N$. Each atom may change its location by varying one of its coordinates, and so the total number of degrees of freedom is $3N$. It is possible to group these degrees of freedom together in a physically sensible way. For example, three coordinates are needed to specify the location of the centre of mass of the molecule, and so three of these degrees of freedom correspond to the translational motion of the molecule as a whole. The remaining $3N-3$ are non-translational modes of the molecule.

In the case of a non-linear molecule, three coordinates are needed to specify its orientation in space (only two are needed for a linear molecule): this is illustrated in Fig. 18.17. Therefore three (or two) of these $3N-3$ non-translational degrees of freedom are rotational. This leaves $3N-6$ (or $3N-5$) degrees of freedom for displacements of the atoms relative to each other: these are the vibrational degrees of freedom. It follows that a non-linear N-atomic molecule has $3N-6$ modes of vibration, while an N-atomic linear molecule has $3N-5$. For example, H_2O is a triatomic non-linear molecule, and has three modes of vibration (and three modes of rotation); CO_2 is a triatomic linear molecule, and has four modes of vibration (and only two modes of rotation). Even a middle-sized molecule such as naphthalene ($C_{10}H_8$) has 48 distinct modes of vibration.

The next step is to find the best description of the modes. In the case of CO_2, for example, one choice might be the ones shown in Fig. 18.18(a). This shows the stretching of one bond (the mode ν_L), the stretching of the other (ν_R), and the two perpendicular bending modes (ν_2). This choice, while permissible, has a serious disadvantage: when one C–O vibration is excited, the motion of the C atom sets the other C–O in motion, and so energy flows backwards and forwards between ν_L and ν_R. The description is much simpler if linear combinations of ν_L and ν_R are taken. For example, one combination is ν_1 in Fig. 18.18(b): this is the *symmetric stretch*, and in it the C atom is buffetted simultaneously from each side, and the motion continues indefinitely. Another mode is ν_3, the *antisymmetric stretch*, in which the two C—O bonds vibrate out-of-phase (in opposite senses). Both modes are independent in that if one is excited, then it does not excite the other. They are two of the *normal modes* of the molecule. The two other normal modes are the bending modes ν_2. In general, a normal mode is an independent, synchronous motion of atoms or groups of atoms that may be excited without leading to the excitation of any other normal mode.

The four normal modes of CO_2 (and the $3N-6$ or $3N-5$ normal modes of polyatomics in general) are the key to the description of molecular vibrations. Each normal mode behaves like an independent harmonic oscillator (if anharmonicities are neglected), and so each may be excited to an energy $(v_Q+\frac{1}{2})\hbar\omega_Q$, where ω_Q is the frequency of mode Q. This frequency depends on the force constant k_Q for the mode (which in general is made up from a combination of stretches and bends, the bends usually contributing a smaller resistance to distortion than a stretch), and on the reduced mass μ_Q of the mode, with $\omega_Q=(k_Q/\mu_Q)^{\frac{1}{2}}$. The reduced mass of the mode depends on how much mass is being swung about by the vibration. For example, in the symmetric stretch of CO_2 the carbon atom is static, and the reduced mass depends on the masses of only the O atoms; in the antisymmetric stretch and in the bends, all three atoms move, and so all contribute to the reduced mass. The three normal modes of H_2O are shown in Fig. 18.19: note that the predominantly bending mode (ν_2) has a lower frequency than the others, which are predominantly stretching modes.

One of the most powerful ways of dealing with normal modes, especially of complex molecules, is to classify them according to their symmetries. This is illustrated by the following *Example*.

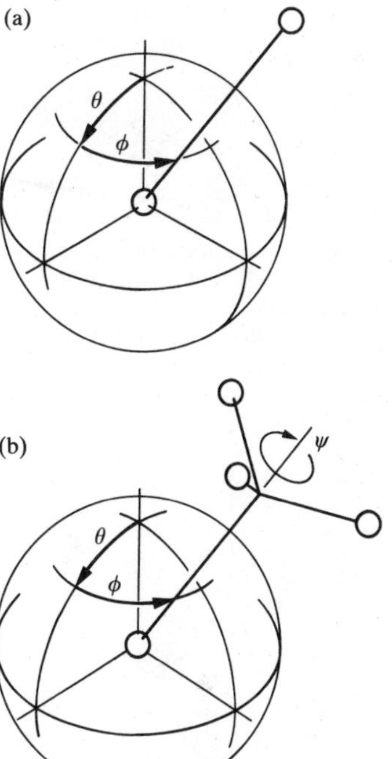

Fig. 18.17. The specification of the centre of mass of a molecule uses up three degrees of freedom. The orientation of a linear molecule (a) requires the specification of two angles; the orientation of a non-linear molecule (b) requires the specification of three angles.

Example 18.7

Establish the symmetry species of the normal mode vibrations of CH_4.

• *Example*. There are $3\times5=15$ degrees of freedom, of which $3\times5-6=9$ are vibrations. Classify all the displacements of the atoms, by regarding each atom as having an x-, a y-, and a z-component of displacement, and establishing the characters of the reducible representation they span in the molecular point group (T_d). Do this (as explained in *Example* 17.3) by counting 1 if the displacement is unchanged under a symmetry operation, -1 if it changes sign, and 0 if it is left unchanged. Identify the symmetry species of the irreps this representation spans, and subtract the symmetry species of the translations (the symmetry species spanned by x, y, and z), and those of the rotations. (These are established by reference to the last column of the character table.)

• *Answer*. Refer to Fig. 18.20. Under E no displacement coordinates are changed, and so the character is 15. Under C_3 no displacements are left unchanged, and so the character is 0. Under a C_2 rotation the z-displacement of the central atom is left unchanged, while its x- and y-components both change sign. Therefore $\chi(C_2)=1-1-1+0+0+\ldots=-1$. Under S_4 the z-displacement of the central atom is reversed, and so $\chi(S_4)=-1$. Under σ_d the z-displacements of C, H_3, and H_4 are left unchanged, three of the H displacements are left unchanged and three are reversed; hence $\chi(\sigma_d)=3+3-3=3$. The characters are therefore 15 0 −1 −1 3, corresponding to $A_1+E+T_1+3T_2$. The translations span T_2; the rotations span T_1. Hence the vibrations span A_1+E+2T_2.

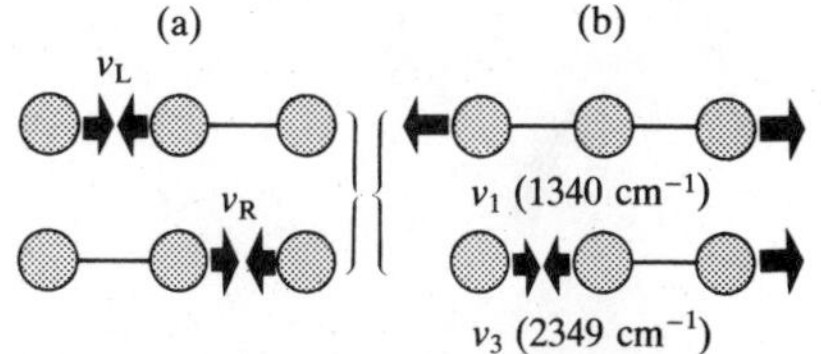

Fig. 18.18. Alternative descriptions of the vibrations of CO_2. (a) The stretching modes are not independent, and if one C–O is excited the other begins to vibrate. (b) The symmetric and antisymmetric stretches are independent, and one can be excited without affecting the other: they are *normal modes*.

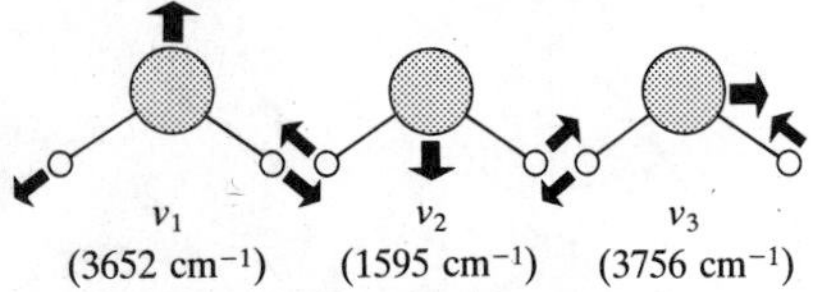

Fig. 18.19. The normal modes of H_2O. The mode ν_2 is predominantly bending, and occurs at lower wavenumber than the other two.

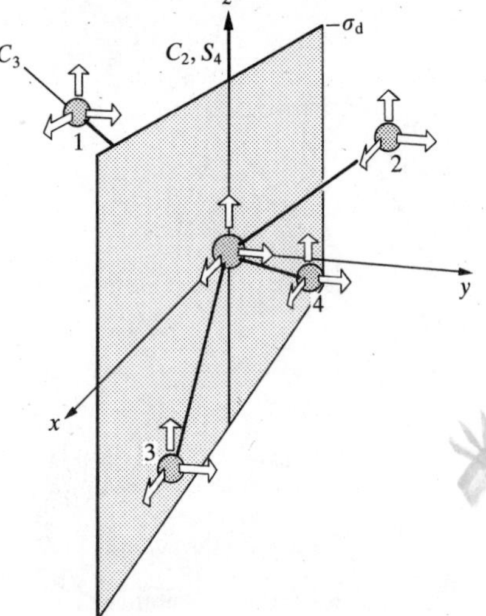

Fig. 18.20. The atomic displacements of CH_4 and the symmetry elements used to calculate the characters. Since all operations of the same class have the same character, there is no need to consider more than one element of each class.

• *Comment*. We shall soon see that this sort of symmetry analysis gives a quick way of deciding which modes are active.

• *Exercise*. Establish the symmetry species of the normal modes of H_2O. [$2A_1 + B_2$]

18.4 (b) The vibrational spectra of polyatomic molecules

The gross selection rule for infrared activity is that *the motion corresponding to a normal mode should be accompanied by a change of dipole moment*. Deciding whether this is so can sometimes be done by inspection. For example, the symmetric stretch of CO_2 leaves the dipole moment unchanged (at zero), and so this mode is infrared inactive. The antisymmetric stretch, however, changes the dipole moment because the molecule becomes unsymmetrical as it vibrates, and so this mode is infrared active. Since in this case the dipole moment changes parallel to the figure axis, the transitions arising from this mode give rise to a *parallel band* in the spectrum. Both bending modes are infrared active: they are accompanied by a changing dipole perpendicular to the figure axis, and so transitions involving them lead to a *perpendicular band* in the spectrum.

When the modes are more complex it is best to use group theory to judge their activities. This is easily done by checking the character table of the molecular point group for the symmetry species of the irreps spanned by x, y, and z, for these are also the symmetry species of the components of the electric dipole moment. Then the rule to apply is as follows:

If the symmetry species of a normal mode is the same as any of the symmetry species of x, y, or z, then the mode is infrared active.

Example 18.8

Which modes of CH_4 are infrared active?

• *Method*. Refer to the T_d character table to establish the symmetry species of x, y, and z. The symmetry species of the normal modes were established in *Example* 18.7. Then use the rule above.

• *Answer*. The symmetry species of x, y, and z is T_2. The normal modes have symmetry species $A_1 + E + 2T_2$. Therefore, only the T_2 modes are infrared active.

• *Comment*. The distortions accompanying the T_2 modes lead to a changing dipole moment. The A_1 mode, which inactive, is the symmetrical *breathing mode* of the molecule.

• *Exercise*. Which of the normal modes of H_2O are infrared active? [All three]

The active modes are subject to the specific selection rule $\Delta v_Q = \pm 1$, and so the energy of the *fundamental transition* (the first harmonic) of each active mode is $\hbar\omega_Q$, and an absorption is observed at $\tilde{\nu}_Q = \omega_Q/2\pi c$. From the analysis of the spectrum, a picture may be constructed of the stiffness of various parts of the molecule: that is, we can establish its *force field*.

Superimposed on this simple scheme are the complications arising from anharmonicities and the effects of molecular rotation. Very often the sample is a liquid or a solid, and the molecules are unable to rotate freely. In a liquid, for example, a molecule may be able to rotate for only a very short time (through a few degrees) before it is struck by another, and so it changes its rotational state frequently. Since the lifetimes of rotational states in liquids are very short, the rotational energies are ill-defined. Collisions occur at a rate of about $10^{13}\,s^{-1}$, and even allowing for only a 10% success rate in knocking the molecule into another rotational state, a lifetime broadening (Section 18.1(c)) of more than $1\,cm^{-1}$ can easily result. The rotational structure of the vibrational spectrum is blurred by this effect, and so the infrared spectra of molecules in condensed phases usually consist of broad lines spanning the entire range of the resolved gas-phase spectrum, and showing no branch structure.

One very important application of infrared spectroscopy to condensed phase samples, and for which blurring of the rotational structure by random collisions is a welcome simplification, is to chemical analysis. The vibrational spectra of different groups in a molecule give rise to absorptions at characteristic frequencies. Their intensities are also transferable between molecules. Consequently, the molecules in a sample can often be identified by examining its infrared spectrum and accounting for all the bands by referring to a table of characteristic frequencies and intensities, Table 18.2. This is illustrated in Fig. 18.21.

Table 18.2. Typical vibrational wavenumbers, $\tilde{\nu}/cm^{-1}$

C—H stretch	2850–2960
C—H bend	1340–1465
C—C stretch	700–1250
C═C stretch	1620–1680

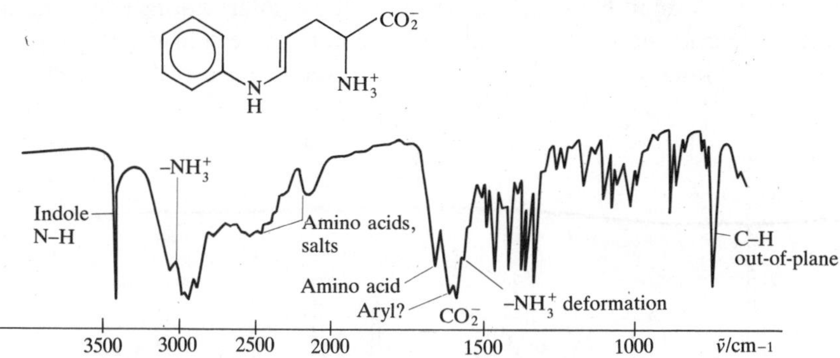

Fig. 18.21. The infrared absorption spectrum of an amino acid and a partial assignment.

18.4 (c) Vibrational Raman spectra of polyatomic molecules

The normal modes of vibration of molecules are Raman active if they are accompanied by a changing polarizability. It is quite difficult to judge by inspection when this is so. The symmetric stretch of CO_2, for example, alternately swells and contracts the molecule: this changes its polarizability, and so it is Raman active. The other modes of CO_2 leave the polarizability unchanged, and so they are Raman inactive.

Group theory provides an explicit recipe for judging the Raman activity of a normal mode. In this case, the symmetry species of the quadratic forms (x^2, xy, etc.) listed in the character table are noted (they transform in the same way as the polarizability), and then we use the following rule:

If the symmetry species of a normal mode is the same as the symmetry species of a quadratic form, then the mode is Raman active.

Example 18.9

Which of the vibrations of CH_4 are Raman active?

• *Method*. Refer to the T_d character table. The symmetry species of the normal modes were established in *Example* 18.7. Use the rule above.

• *Answer*. The quadratic forms span $E + T_2$. The normal modes have symmetry species $A_1 + E + 2T_2$. Therefore the E and T_2 normal modes are Raman active.

• *Comment*. Notice that the A_1 breathing mode is neither infrared nor Raman active.

• *Exercise*. Which of the vibrational modes of H_2O are raman active? [All three]

An important rule concerning normal modes is the following:

> *The exclusion rule: If the molecule has a centre of symmetry, then no modes can be both infrared and Raman active.*

(A mode may be inactive in both.) The rule applies to CO_2 but to neither H_2O nor CH_4 because they have no centre of symmetry.

One application of Raman spectroscopy other than to the determination of the structures of symmetrical molecules such as XeF_4 and SF_6, is to the identification of organic and inorganic species in solution. In order for this to be successful, it is important to use not only the positions of the vibrational Raman lines, but also to note their polarizations (the orientation of the electric vector in the scattered light) because that is also determined by the symmetry of the normal mode responsible for the scattering. An example of the technique is shown in Fig. 18.22, which shows the

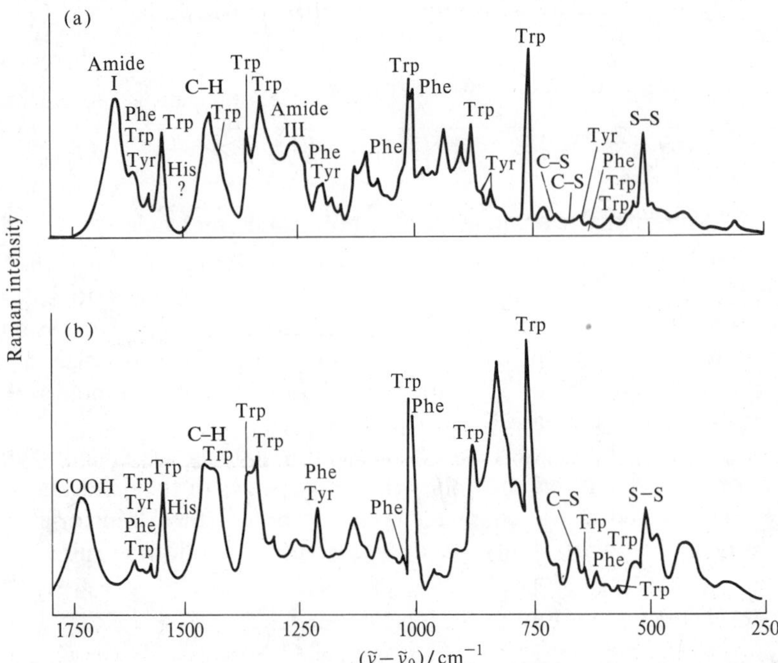

Fig. 18.22. The vibrational Raman spectrum of lysozyme in water and the superposition of the Raman spectra of the constituent amino acids. (From *Raman spectroscopy* by D. A. Long. Copyright © 1977, McGraw-Hill Inc. Used with the permission of the McGraw-Hill Book Company.)

vibrational Raman spectrum of an aqueous solution of lysozyme and, for comparison, a superposition of the Raman spectra of the constituent amino acids. The differences are indications of the effects of conformation, environment, and specific interactions (such as S—S linking) in the enzyme molecule.

Further reading

General:

Fundamentals of molecular spectroscopy. C. N. Banwell; McGraw-Hill, New York, 1972.

The determination of molecular structure (2nd edn). P. J. Wheatley; Clarendon Press, Oxford, 1968.

Spectroscopy. D. H. Whiffen; Longman, 1972.

Molecular quantum mechanics (2nd edn). P. W. Atkins; Oxford University Press, 1983.

Spectroscopy and molecular structure. G. W. King; Holt, Rinehart, and Winston, New York, 1964.

High resolution spectroscopy. J. M. Hollas; Butterworth, London, 1982.

Molecular structure and dynamics. W. H. Flygare; Prentice-Hall, Englewood Cliffs, 1978.

Microwave:

Microwave spectroscopy of gases. T. M. Sugden and C. N. Kenney; Van Nostrand, London, 1965.

Microwave spectroscopy. W. H. Flygare; *Techniques in chemistry* (A. Weissberger and B. W. Rossiter, eds.) IIIa, 439, Wiley-Interscience, New York, 1972.

Microwave spectroscopy. C. H. Townes and A. L. Schawlow; McGraw-Hill, New York, 1955.

Infrared:

Infrared spectroscopy. D. H. Anderson and N. B. Woodall; *Techniques of chemistry* (A. Weissberger and B. W. Rossiter, eds.) IIIB, 1, Wiley-Interscience, New York, 1972.

Vibrating molecules. P. Gans; Chapman and Hall, London, 1971.

The infrared spectra of complex molecules. L. J. Bellamy; Chapman and Hall, London, 1975.

Infrared and Raman spectra of polyatomic molecules. G. Herzberg; Van Nostrand, New York, 1945.

Molecular vibrations. E. B. Wilson, J. C. Decius, and P. C. Cross; McGraw-Hill, New York, 1955.

Raman:

Raman spectroscopy. J. R. Durig and W. C. Harris; *Techniques of chemistry* (A. Weissberger and B. W. Rossiter, eds.) IIIB, 85, Wiley-Interscience, New York, 1972.

Raman spectroscopy. D. A. Long; McGraw-Hill, New York, 1977.

Laser Raman spectroscopy. T. R. Gilson and P. J. Hendra; Wiley, New York, 1970.

Applications:

Spectroscopic methods in organic chemistry (3rd edn). D. H. Williams and I. Fleming; McGraw-Hill, London, 1980.

Physical methods in chemistry. R. Drago; Saunders, Philadelphia, 1977.

Chemical applications of infrared spectroscopy. C. N. R. Rao; Academic Press, New York, 1963.

Introductory problems

A18.1. Calculate the reduced mass of the following molecules and in each instance identify the atom which makes the greater contribution: $^{1}H^{35}Cl$, $^{2}H^{35}Cl$, $^{133}Cs^{35}Cl$.

A18.2. The bond length in the molecule $^{79}Br^{81}Br$ is 228 pm. Calculate the moment of inertia ($I_{\perp}$) of the molecule.

A18.3. The rotational constant, B, for $^{127}I^{35}Cl$ is $0.1142\ cm^{-1}$. Calculate the moment of inertia of this molecule.

A18.4. The exciting light used in observing the pure rotational spectrum of $^{14}N^{14}N$ in a Raman scattering experiment is $20\,487\ cm^{-1}$. Find the wavenumber of the scattered light corresponding to the Stokes line involving the lowest rotational state.

A18.5. Infrared absorption by $^{1}H^{81}Br$ gives rise to an R-branch line involving a lower state with $v = 0$ and $J = 2$. Find the wavenumber of the spectral line.

A18.6. What is the percentage difference in the fundamental vibrational frequencies of the molecules $^{23}Na^{35}Cl$ and $^{23}Na^{37}Cl$? Assume that the two molecules have identical force constants.

A18.7. The fundamental vibrational wavenumber $\tilde{\nu}_0$ for $^{35}Cl_2$ is $564.9\ cm^{-1}$. Find the force constant of the bond.

A18.8. For $^{127}I^{35}Cl$, $\tilde{\nu} = 384.3\ cm^{-1}$ and $\tilde{\nu}x = 1.5\ cm^{-1}$. Which fundamental pure vibrational transition has the greatest wavenumber? Calculate its wavenumber and that of the next highest.

A18.9. The bond dissociation energy of $^{127}I^{35}Cl$ is 2.153 eV. Use the information in the previous problem and find the depth of the molecular potential energy curve for this molecule.

A18.10. The molecule CH_2Cl_2 has symmetry C_{2v}. Its vibrational displacements correspond to the representation $5A_1 + 2A_2 + 4B_1 + 4B_2$. Find the symmetries of the normal modes. Indicate which are infrared active and which are Raman active.

Problems

18.1. Which of the following molecules may show a pure rotational microwave spectrum: H_2, HCl, CH_4, CH_3Cl, CH_2Cl_2, H_2O, H_2O_2, NH_3, NH_4Cl?

18.2. Which of the following molecules may show absorption in the infrared: H_2, HCl, CO_2, H_2O, CH_3CH_3, CH_4, CH_3Cl, N_2, N_3^-?

18.3. Which of the following molecules may show a pure rotational Raman spectrum: H_2, HCl, CH_4, CH_3Cl, CH_2Cl_2, CH_3CH_3, H_2O, SF_6?

18.4. Which of the following molecules can show a vibrational Raman spectrum: H_2, HCl, CH_4, H_2O, CO_2, CH_3CH_3, SF_6?

18.5. What is the Doppler-shifted wavelength of a red ($\lambda = 660$ nm) traffic light when approaching it at 50 m.p.h.? At what speed of approach would it appear to be green ($\lambda = 520$ nm)?

18.6. A spectral line of $^{48}Ti^{8+}$ in a distant star was found to be shifted from 654.2 nm to 706.5 nm and to be broadened to 61.8 pm. What is the speed of recession and the surface temperature of the star?

18.7. Microwave spectroscopy is a very precise technique, but the breadth of lines may obscure details. What is the Doppler width (expressed as a fraction of the wavelength of the transition) for any kind of transition in (a) HCl, (b) ICl at 298 K? What would be the widths of rotational (in MHz) and vibrational (in cm^{-1}) transitions of these species? ($B(\text{ICl}) = 0.114\ cm^{-1}$, $\tilde{\nu}_0(\text{ICl}) = 384\ cm^{-1}$.)

18.8. From the relation $\tau\delta E \approx \hbar$ deduce the lifetime of a state that gives rise to a line of width (a) $0.1\ cm^{-1}$, (b) $1\ cm^{-1}$, (c) 100 MHz.

18.9. A molecule in a liquid undergoes about 10^{13} collisions each second. Suppose (a) that every collision is effective in deactivating the molecule vibrationally, (b) that one in a hundred collisions is effective. What is the width (in cm^{-1}) of vibrational transitions to this excited state?

18.10. The number of collisions a molecule undergoes in unit time in a gas where the pressure is p is $4\sigma(kT/\pi m)^{\frac{1}{2}}p/kT$. What is the collision-limited lifetime of an excited molecular state at 298 K, assuming that every collision quenches? What is the width of the rotational transitions in HCl ($\sigma = 0.30\ nm^2$) when the pressure of the gas is 1 atm? What must be the pressure in order that collision broadening is less important than Doppler broadening;

18.11. What relative proportions of chlorine molecules ($\tilde{\nu} = 559.7\ cm^{-1}$) are in the ground and first excited vibrational states at (a) 273 K, (b) 298 K, (c) 500 K?

18.12. What is the value of J in the most highly populated rotational level of ICl at room temperature? Do not forget that there are $2J + 1$ states of the same energy for each value of J. ($B = 0.114\ cm^{-1}$.)

18.13. What is the value of J in the most highly populated rotational level of CH_4 at room temperature? There is a mild trick here: although the energy levels are given by the same expression as for a linear molecule we have to remember that there are $(2J + 1)^2$ states of the same energy for every value of J. Why? Use $B = 5.24\ cm^{-1}$.

18.14. The *reduced mass* was a quantity that appeared at several points in the chapter, and it is important to know how to calculate it. Calculate (in kg) the reduced masses and the moments of inertia of (a) $^{1}H^{35}Cl$, (b) $^{2}H^{35}Cl$, (c) $^{1}H^{37}Cl$. Take $R = 127.45$ pm.

18.15. The rotational constant of NH_3 has the value $cB = 298$ GHz. Compute the separation of the lines in GHz, cm^{-1}, and mm, and show that the value of B is consistent with an N–H bond length of 101.4 pm and a bond angle of 106° 47′. (Refer to Box 18.1 for information about moments of inertia.)

18.16. A space probe was designed to look for signs of CO in the atmosphere of Saturn. It was decided to employ a microwave technique from an orbiting satellite. Given the bond length of the molecule as 112.82 pm, at what wavenumbers do the first four transitions of $^{12}C^{16}O$ lie? What resolution is needed if it was desired to distinguish the 1–0 line in the $^{12}C^{16}O$ spectrum from that of $^{13}C^{16}O$ in order to examine the relative abundances of the two carbon isotopes?

18.17. Rotational absorption lines from HCl gas were found at the following wavenumbers (R. L. Hausler and R. A. Oetjen, *J. chem. Phys.* **21,** 1340 (1953)): 83.32, 104.13, 124.73, 145.37, 165.89, 186.23, 206.60, 226.86 cm^{-1}. Find the moment of inertia and bond length of the molecule. (The species used was $^{1}H^{35}Cl$.)

18.18. Predict the positions of the microwave absorption lines in DCl ($^{2}H^{35}Cl$) from the data in the preceding Problem.

18.19. Is HCl the same size as DCl? The data from the rotational structure of the infrared spectra of the species are as follows (I. M. Mills, H. W. Thompson, and R. L. Williams, *Proc. R. Soc.* **A218,** 29 (1953); J. Pickworth and H. W. Thompson, *Proc. R. Soc.* **A218,** 37 (1953)).

	J:	0	1	2	3
$^{1}H^{35}Cl$	R-branch	2906.25	2925.92	2944.99	2963.35
	P-branch	—	2865.14	2843.63	2821.59
$^{2}H^{35}Cl$	R-branch	2101.60	2111.94	2122.05	2131.91
	P-branch	—	2080.26	2069.24	2058.02

	J:	4	5	6
$^{1}H^{35}Cl$	R-branch	2981.05	2998.05	3014.50 cm^{-1}
	P-branch	2799.00	2775.77	2752.01
$^{2}H^{35}Cl$	R-branch	2141.53	2150.93	2160.06
	P-branch	2046.58	2034.95	2023.12

18.20. Show that the moment of inertia of a diatomic molecule composed of two atoms of masses m_A and m_B and of bond length R is given by $I=[m_A m_B/(m_A+m_B)]R^2$. What is the moment of inertia of (a) $^{1}H_2$, (b) $^{127}I_2$?

18.21. The pure rotation spectrum of HI consists of a series of lines separated by 13.10 cm^{-1}. What is the bond length of the molecule?

18.22. From a thermodynamic point of view the copper monohalides CuX are expected to exist mainly as polymers in the gas phase, and so it proved difficult to get enough intensity to measure the spectra of the monomers. This was overcome by flowing the halogen gas over chips of copper heated to 1000–1100 K (E. L. Manson, F. C. de Lucia, and W. Gordy, *J. chem. Phys.* **63,** 2724 (1975)). For CuBr the $J=13$–14, 14–15, 15–16 transitions occurred at 84 421.34, 90 449.25, 96 476.72 MHz. Find the rotational constant and the bond length of CuBr.

18.23. The microwave spectrum of $^{16}O^{12}C^{32}S$ shows absorption lines at 24.325 92, 36.488 82, 48.651 64, 60.814 08 GHz. What is the moment of inertia of the molecule? Note that the individual bond lengths cannot be obtained from these data; but read on.

18.24. In a study of the microwave spectra of various isotopically substituted OCS molecules (C. H. Townes, A. N. Holden, and F. R. Merritt, *Phys. Rev.* **74,** 1113 (1948)) the following transitions were observed:

$J\to J+1$	$1\to 2$	$2\to 3$	$3\to 4$	$4\to 5$
$^{16}O^{12}C^{32}S$	24.325 92	36.488 82	48.651 64	60.814 08 GHz
$^{16}O^{12}C^{34}S$	23.732 33		47.462 40	

Using the expression for the moment of inertia given in Box 18.1, and assuming that the bond lengths are unchanged in isotopic substitution, find the lengths of the C-O and C-S bonds in OCS.

18.25. The set of equations known as *Kraitchman's equations* relate the change of rotational constant (or moments of inertia) to the positions at which isotopic substitution is made. Show that when isotopic substitution is made at a distance z along the axis from the original centre of mass of a symmetric top molecule, z and ΔB, the change of the perpendicular rotational constant, are related by $(z/\text{pm})^2 = 1.685\,90\times 10^5\ (\Delta B/\text{cm}^{-1})/[(B/\text{cm}^{-1})(B'/\text{cm}^{-1})(\Delta M_r)]$, where B is the rotational constant before substitution, B' that after, and $\Delta M_r = M_r(M_r'-M_r)/M_r'$.

18.26. The microwave spectra of various isotopic species of $ClTeF_5$ show rigid symmetric top behaviour (A. C. Legon, *J. chem. Soc. Faraday Trans.* II, 29 (1973)). Four fluorines lie in a square, the tellurium atom lies just above the plane they form, and the other fluorine lies beneath this plane and the chlorine above. Use Kraitchman's equation derived in the last Problem to deduce the Te–Cl bond length on the assumption that all four TeF bond lengths have the same length. Data: 11–10 transition in $^{35}Cl^{126}TeF_5$: 30 711.18 MHz; the same in $^{35}Cl^{125}TeF_5$: 30 713.24 MHz; the same in $^{37}Cl^{126}TeF_5$: 29 990.54 MHz.

18.27. The moments of inertia of the linear mercury(II) halides are very large, and as a consequence the O- and S-branches of the vibrational Raman spectra show little structure. Nevertheless the peaks of both branches can be distinguished and have been used to measure the rotational constants (R. J. H. Clark and D. M. Rippon, *J. chem. Soc. Faraday Trans.* II **69,** 1496 (1973)). Show, from a knowledge of the value of J corresponding to the population maximum, that the separation of the peaks of the O- and S-branches is given by the *Placzek–Teller relation* $\delta\tilde{\nu} = \{32BkT/hc\}^{\frac{1}{2}}$. The following widths were obtained at the temperatures stated: $HgCl_2$(282 °C): 23.8 cm^{-1}; $HgBr_2$(292 °C): 15.2 cm^{-1}; HgI_2(292 °C): 11.4 cm^{-1}. Calculate the rotational constants of these molecules.

18.28. The hydrogen halides have the following fundamental vibration wavenumbers: HF (4141.3 cm^{-1}), $H^{35}Cl$ (2988.9 cm^{-1}), $H^{81}Br$ (2649.7 cm^{-1}), $H^{127}I$ (2309.5 cm^{-1}). Find the force constants of the hydrogen–halogen bonds.

18.29. From the data in the last Problem, predict the fundamental vibration wavenumbers of the deuterium halides.

18.30. The Morse curve, eqn (18.3.5), is very useful as a

simple representation of the molecular potential energy curve. When RbH was studied it was found that $\tilde{\nu} = 936.8\ \text{cm}^{-1}$ and $x\tilde{\nu} = 14.15\ \text{cm}^{-1}$. Plot the potential energy curve from 50 pm to 800 pm around $R_e = 236.7$ pm.

18.31. The rotation of a molecule may weaken its bonds. This can be illustrated by plotting the potential energy curve allowing for the kinetic energy of rotation of the molecule. Thus $V^*(R) = V(R) + B(R)J(J+1)$ is plotted, where V is the Morse potential and $B = \hbar/4\pi cmR^2$, m being the reduced mass. Plot these curves for RbH on the same graph as in the last Problem, taking $J = 40$, 80, and 100. Notice how the dissociation energy is affected by the rotation. (Taking $hcB(R_e) = 3.020\ \text{cm}^{-1}$ will simplify the calculation of $B(R)$.)

18.32. The first five vibrational energy levels of HCl were found to lie at 1481.86, 4367.50, 7149.04, 9826.48, 12 399.8 cm^{-1}. What is the dissociation energy of the molecule?

18.33. The vibrational energy levels of NaI lie at the following wavenumbers: 142.81, 427.31, 710.31, 991.81 cm^{-1}. Show that they fit an expression $(v+\frac{1}{2})\tilde{\nu} - (v+\frac{1}{2})^2 x\tilde{\nu}$, and deduce the force-constant, zero-point energy, and dissociation energy of the bond.

18.34. The HCl molecule has a potential energy that is quite well described by a Morse curve with $D_e = 5.33$ eV and $x\tilde{\nu} = 52.05\ \text{cm}^{-1}$, $\tilde{\nu} = 2989.7\ \text{cm}^{-1}$. Assuming that the potential remains unchanged on deuteration, predict the dissociation energies D_0 of (a) HCl, (b) DCl.

18.35. The expressions for the P-, Q-, and R-branches, Section 18.3(d), assume that the moment of inertia of the molecule is the same in the excited vibrational states as in the ground state. Derive expressions for the positions of the lines without making these assumptions.

18.36. Lines in the P-branch of $^1H^{35}Cl$ were observed at 2865.1, 2843.6, and 2821.6 cm^{-1} for $J = 1$, 2, and 3, and in the R-branch at 2906.2, 2925.9, 2945.0, 2963.3 cm^{-1} for $J = 0$, 1, 2, 3 (I. M. Mills, H. W. Thompson, and R. L. Williams, *Proc. R. Soc.* **A218,** 29 (1953)). Find the force-constant of the bond and the bond lengths of the upper and lower vibrational states.

18.37. The vibrational Raman spectrum of $^{35}Cl_2$ shows a series of Stokes lines separated by 0.9752 cm^{-1} and a similar series of anti-Stokes lines. What is the bond length of Cl_2?

18.38. Which of the three vibrations of a general AB_2, molecule are infrared or Raman active when the molecule is (a) bent, (b) linear?

18.39. Consider the vibrational mode that corresponds to the uniform expansion of the benzene ring. Is it (a) Raman active, (b) infrared active?

19

Determination of molecular structure: electronic spectroscopy

Learning objectives

After careful study of this chapter you should be able to:

(1) Define *molar absorption coefficient*, *absorbance*, and *transmittance*, Section 19.1.

(2) State and use the *Beer–Lambert Law* to relate the transmittance of a sample to the concentration of the sample, eqn (19.1.3) and Example 19.1.

(3) Define the *integrated absorption coefficient*, eqn (19.1.5), and relate it to the *oscillator strength* of a transition, eqn (19.1.6) and Example 19.2.

(4) Account for the occurrence of d–d *vibronic transitions* in centrosymmetric molecules, Section 19.2(a).

(5) Describe the transitions responsible for absorption by the C=C double bond and the carbonyl group, Section 19.2(b).

(6) State the *Franck–Condon principle* and use it to account for the appearance of the vibrational structure of electronic transitions, Section 19.3(a).

(7) Indicate how *Franck–Condon factors* are calculated, Example 19.4.

(8) Describe the mechanisms of *fluorescence*, Section 19.4(a), and *phosphorescence*, Section 19.4(b).

(9) Describe *laser action* and give examples of different types of lasers, Section 19.4(c) and Figs. 19.13–16.

(10) Explain the terms *super-radiant*, *Q-switching*, and *mode-locking*, and describe how they affect the laser output, Section 19.4(c).

(11) Explain the terms *dissociation*, *dissociation limit*, and *predissociation*, and describe how dissociation and predissociation affect the appearance of a spectrum, Section 19.4(d).

(12) Describe the technique of *photoelectron spectroscopy*, Section 19.5(a), and outline the information that can be obtained from *UPS*, Section 19.5(b), and *XPS* (*ESCA*), Section 19.5(c).

Introduction

The energies involved in changing the electron distributions of molecules are of the order of several electron volts (1 eV corresponds to about 8000 cm^{-1}), and so the photons emitted or absorbed lie in the visible and ultraviolet regions of the spectrum, which spreads from about 14 000 cm^{-1} for red light to 21 000 cm^{-1} for blue, and on to 50 000 cm^{-1} for the ultraviolet, Table 19.1 Studying the electronic spectra of molecules not only lets us account for the colours of substances, but reveals important information about the excited states of molecules. The green of vegetation, for example, is due to chlorophyll, a substance that can absorb in both the red and blue regions of the spectrum, Fig. 19.1, so leaving the green component of the incident white sunlight to be reflected; the examination of the details of the absorption is an important step in understanding, and then

Table 19.1. Colour, frequency, and energy of light

Colour	λ/nm	$\nu/10^{14}$ Hz	E/kJ mol^{-1}
Infrared	>1000	<3.0	<120
Red	700	4.3	170
Yellow	580	5.2	210
Blue	470	6.4	250
Ultra-violet	<300	>10	>400

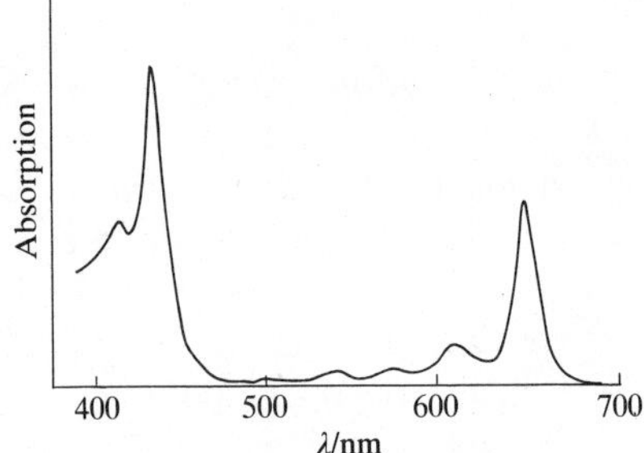

Fig. 19.1. The absorption spectrum of chlorophyll in the visible region. Note that it absorbs in the red and blue regions, and that green light is not absorbed.

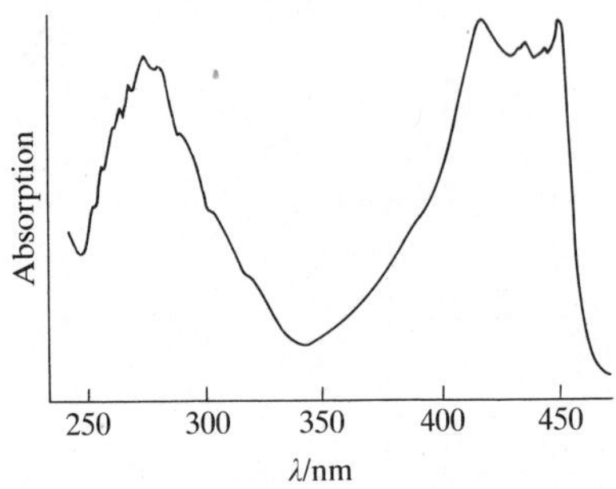

Fig. 19.2. The electronic absorption spectrum of biacetyl, $CH_3COCOCH_3$, in the ultraviolet and blue regions. Note the blurred vibrational structure of the absorption bands.

perhaps reproducing synthetically, the process of photosynthesis. The electronic energy levels of molecules can also be studied by measuring the energies needed to expel electrons from them: this branch of electronic spectroscopy, which is called *photoelectron spectroscopy*, is dealt with in Section 19.5.

There are two principal complicating features in electronic spectroscopy. One is that because the electron distribution is changed, the nuclei are subjected to different forces after the transition, and so the molecule responds by bursting into vibration. The vibrational structure of electronic transitions can be resolved in a gas, but in a liquid or solid the lines usually merge together and result in a broad, almost featureless band, Fig. 19.2. The vibrational transitions that accompany the electronic transition are themselves accompanied by rotational transitions; the electronic spectra of gaseous samples are therefore very complicated, but rich in information. The second complicating feature is that the molecules might absorb enough energy to fall apart. This is important because *photodissociation* is one of the fundamental processes of photochemistry.

19.1 Measures of intensity

The reduction of intensity (symbol: I for the intensity, $\mathrm{d}I$ for the change) that occurs when light passes through a sample of thickness $\mathrm{d}x$ containing absorbing species J at a concentration [J] is proportional to the thickness, the concentration, and the incident intensity, and we can write

$$\mathrm{d}I = -\alpha[\mathrm{J}]I\,\mathrm{d}x, \quad \text{or} \quad \mathrm{d}\ln I = -\alpha[\mathrm{J}]\,\mathrm{d}x. \qquad (19.1.1)$$

where α is the proportionality coefficient. This equation applies to each successive layer into which the sample can be regarded as being divided, and to obtain the intensity (I_f) that emerges from a sample of thickness l when the incident intensity is I_i we sum all the successive changes:

$$\int_i^f \mathrm{d}\ln I = -\int_0^l \alpha[\mathrm{J}]\,\mathrm{d}x. \qquad (19.1.2)$$

If the concentration is uniform, [J] is independent of x, and the expression integrates to the

$$\textit{Beer–Lambert Law: } I_f = I_i e^{-\alpha[\mathrm{J}]l}. \qquad (19.1.3)$$

We see that the intensity decreases exponentially with the sample thickness, and that it decreases exponentially with the concentration. The Beer–Lambert Law is often expressed as

$$I_f = I_i 10^{-\varepsilon[\mathrm{J}]l}, \quad \text{or} \quad \lg(I_f/I_i) = -\varepsilon[\mathrm{J}]l. \qquad (19.1.4)$$

ε (which is related to α by $\varepsilon = \alpha/\ln 10 = \alpha/2.303$) is called the *molar absorption coefficient* (formerly the *extinction coefficient*) of the species at the stated frequency. ε depends on the molecule and the frequency of the light. Its units are those of 1/(concentration × length), and it is normally convenient to express them as $\mathrm{M^{-1}\,cm^{-1}}$. However, alternative units are $\mathrm{cm^2\,mol^{-1}}$ (or $\mathrm{cm^2\,mmol^{-1}}$, with $1\,\mathrm{M^{-1}\,cm^{-1}} = 1\,\mathrm{cm^2\,mmol^{-1}}$) because this brings out the point that the molar absorption coefficient is a molar *cross-section* for absorption, and the greater the cross-section of the molecule for absorption, the greater the attenuation of the intensity of the beam. The dimensionless product $A = \varepsilon[\mathrm{J}]l$ is called the *absorbance* (formerly the *optical density*) of the sample, and I_f/I_i is called the *transmittance* (symbol: T).

Example 19.1

Light of wavelength 256 nm passes through a 1.0 mm thickness cell containing an 0.050 M C_6H_6 solution. The light intensity is reduced to 16% of its initial value. Calculate the absorbance and the molar absorption coefficient of the sample. What would be the transmittance through a 2.0 mm cell?

• *Method*. Use eqn (19.1.4) for the molar absorption coefficient (ε) and then convert to absorbance using $A = \varepsilon[\mathrm{J}]l$. For the final part, use eqn (19.1.4) again.

• *Answer*. Since $I_f = 0.16I_i$,

$$\begin{aligned}\varepsilon &= -\{1/[\mathrm{C_6H_6}]l\}\,\lg(I_f/I_i)\\ &= -\{1/(0.050\ \mathrm{M})\times(0.10\ \mathrm{cm})\}\,\lg 0.16 = 160\ \mathrm{M\ cm^{-1}}.\end{aligned}$$

The absorbance is

$$A = \varepsilon[\mathrm{C_6H_6}]l = (160\ \mathrm{M\ cm^{-1}})\times(0.050\ \mathrm{M})\times(0.10\ \mathrm{cm}) = 0.80.$$

The transmittance through a 2.0 mm sample is given by

$$T = I_f/I_i = 10^{-(160\ \mathrm{M\ cm^{-1}})\times(0.050\ \mathrm{M})\times(0.20\ \mathrm{cm})} = 0.025.$$

That is, the emergent light is reduced to 2.5% of its incident intensity.

• *Comment*. The molar extinction coefficient may be expressed as $160\ \mathrm{cm^2\ mmol^{-1}}$, and so one benzene molecule effectively acts as a screen of cross-sectional area $0.27\ \mathrm{pm^2}$ to light of this wavelength.

• *Exercise*. The transmittance of a 0.10 M $CuSO_4$(aq) solution at 600 nm was measured as 0.30 in a 5.0 mm cell. Calculate the molar absorption coefficient of Cu^{2+}(aq) at that wavelength and the absorbance of the solution. What would be the transmittance through an 1.0 mm cell?

[$10\ \mathrm{M^{-1}\ cm^{-1}}, A = 0.50, T = 0.79$]

Absorption bands generally spread over a range of frequencies, and so quoting the absorption coefficient at a single frequency might not give a true indication of the intensity of a transition. The *integrated absorption coefficient* (symbol: $\mathscr{A}$) is the sum of the absorption coefficients for all the frequencies covered by the band:

$$\mathscr{A} = \int \varepsilon(\nu)\,\mathrm{d}\nu, \qquad (19.1.5)$$

and does measure the total strength of a transition. Its dimensions are 1/(concentration × length × time), the 1/time coming from the frequency integration. It is normally expressed as either $\mathrm{M^{-1}\ cm^{-1}\ s^{-1}}$ or $\mathrm{cm^2\ mmol^{-1}\ s^{-1}}$. Sometimes the integrated absorption coefficient is expressed in terms of the dimensionless *oscillator strength* (symbol: f) defined as

$$f = (4m_e c\varepsilon_0 \ln 10/N_A e^2)\mathscr{A} \qquad (19.1.6)$$

where ε_0 is the vacuum permittivity (Appendix 11.1). The practical form of this relation is

$$f = 1.44\times10^{-19}(\mathscr{A}/\mathrm{cm^2\ mmol^{-1}\ s^{-1}}). \qquad (19.1.6a)$$

Table 19.2. Absorption characteristics of some groups and molecules

Group	ν_{max}/cm^{-1}	λ_{max}/nm	$\varepsilon_{max}/M^{-1}\,cm^{-1}$
C=C(π^*, π)	62 000	160	10 000
	57 300	174	1 700
C=O(π^*, n)	34 000	295	10
$H_2O(\pi^*, n)$	60 000	167	7 000

It turns out that intense transitions have $f \approx 1$. Another indication of the intensity of a transition is the maximum value of the molar absorption coefficient (ε_{max}). Typical values for strong transitions are of the order 10^4–$10^5\,M^{-1}\,cm^{-1}$ (10^4–$10^5\,cm^2\,mmol^{-1}$), indicating that in a 1 M solution the intensity of light (of the appropriate frequency) falls to 10 per cent of its initial value after passing through about 0.1 mm of solution. Some values are listed in Table 19.2.

Example 19.2

A transition in benzene has a molar absorption coefficient of 160 $cm^2\,mmol^{-1}$ (*Example* 19.1) and spreads over about 4000 cm^{-1}. Estimate its oscillator strength.

• *Method*. Approximate the area of the absorption band as (height) × (width) after expressing the width as a frequency using $\nu = c\tilde{\nu}$; then use eqn (19.1.5) to estimate the absorbance and from that use eqn (19.1.6a) to estimate the oscillator strength.

• *Answer*. The integrated absorption coefficient is approximately

$$\mathscr{A} = (160\ cm^2\,mmol^{-1}) \times (4000\ cm^{-1}) \times (2.997 \times 10^{10}\ cm\,s^{-1})$$
$$= 1.9 \times 10^{16}\ cm^2\,mmol^{-1}\,s^{-1}.$$

Therefore, the oscillator strength is

$$f = (1.44 \times 10^{-19}) \times (1.9 \times 10^{16}) = 2.8 \times 10^{-3}.$$

• *Comment*. Note that f is a dimensionless quantity. In the present example $f \ll 1$, which indicates that the transition is 'forbidden'. In a more accurate treatment, the area of the absorption curve would be measured as precisely as possible, not estimated in this crude manner.

• *Exercise*. The absorption band of Cu^{2+}(aq), *Example* 19.1, spreads over about 3000 cm^{-1}; estimate its oscillator strength. [1.3×10^{-4}]

The advantage of introducing the oscillator strength is that it has a direct connection with the wavefunctions of the initial and final states involved in the transition. The transition dipole between the initial and final states is defined in the usual way as

$$\boldsymbol{\mu}_{fi} = \int \psi_f^* \boldsymbol{\mu} \psi_i \, d\tau, \qquad \boldsymbol{\mu} = -e\boldsymbol{r}, \tag{19.1.7}$$

and is a measure of the dipole moment associated with the shift of charge that occurs when the electron redistribution takes place, and the oscillator strength is related to this moment by

$$f = (8\pi^2/3)(m_e \nu/he^2)|\boldsymbol{\mu}_{fi}|^2. \tag{19.1.8}$$

Therefore we may estimate the intensity of a transition if we know the wavefunctions. This is an important link between theory and experiment.

Example 19.3

Find the value of the transition dipole moment for the benzene transition discussed in *Examples* 19.1 and 19.2.

• *Comment*. The oscillator strength is estimated in *Example* 19.2. Use eqn (19.1.8), taking for ν the frequency corresponding to the wavelength of the band.

• *Answer*. We need the factor

$$3he^2/8\pi^2 m_e = 7.094\,58 \times 10^{-43}\,\mathrm{m^2\,s^{-1}\,C^2}.$$

The frequency of the transition at 256 nm is $\nu = c/\lambda = 1.17 \times 10^{15}$ Hz. Therefore, from eqn (19.1.8):

$$|\mu_{fi}|^2 = (7.095 \times 10^{-43}\,\mathrm{m^2\,s^{-1}\,C^2}) \times (2.8 \times 10^{-3})/(1.17 \times 10^{15}\,\mathrm{Hz})$$
$$= 1.7 \times 10^{-60}\,\mathrm{C^2\,m^2},$$

so that $|\mu_{fi}| = 1.3 \times 10^{-30}$ C m.

• *Comment*. Express the transition dipole moment in debye (D) using 1 D = 3.34×10^{-30} C m; this gives 0.39 D. The transition can be visualized as corresponding to the movement of about 40% of an electron through a distance about equal to the radius of an atom.

• *Exercise*. Estimate the value of the transition dipole moment for the Cu^{2+}(aq) transition treated in the *Examples* 19.1 and 19.2. [0.13 D]

The expression for the transition dipole moment is also the origin of the selection rules for electronic transitions. We saw in Section 17.3 that the transition dipole moment must be zero unless the integral is totally symmetric under the operations of the molecular point group. An example of how this condition leads to the classification of transitions as forbidden or allowed according to the symmetries of the initial and final states was given in Section 17.3(c). There we also saw that the non-vanishing component of the transition dipole corresponds to the *polarization* (the orientation of the electric field) of the light emitted or absorbed in the transition.

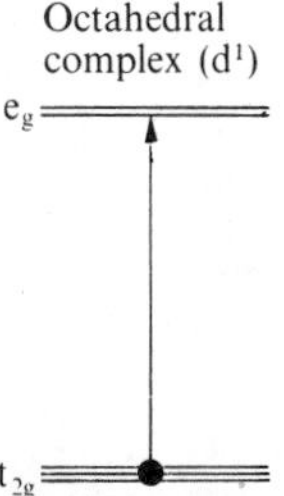

Fig. 19.3. In an octahedral complex the d-orbitals are split into two groups by their interactions with the ligands. In a d–d transition of a d[1] complex, a t_{2g}-electron is excited ino an e_g-orbital.

19.2 Chromophores

In a number of molecules the absorption of a photon can be traced to the excitation of the electrons of a small group of atoms. For example, when a carbonyl group is present, an absorption at about 290 nm is normally observed, although its precise location depends on the nature of the rest of the molecule. Groups of this kind are called *chromophores* (from the Greek for 'colour bringer'), and their presence often accounts for the colours of substances.

19.2 (a) d–d transitions

The colours of *transition metal complexes* can normally be ascribed to the presence of d-electrons, which therefore act as chromophores. The presence of ligands splits the d-orbitals of the central atom into two sets, Fig. 19.3. The separation is not large, and the excitation $e_g \leftarrow t_{2g}$ typically occurs in the visible region of the spectrum; an example, the spectrum of $[Ti(H_2O)_6]^{3+}$ near 20 000 cm^{-1}, 500 nm, is shown in Fig. 19.4.

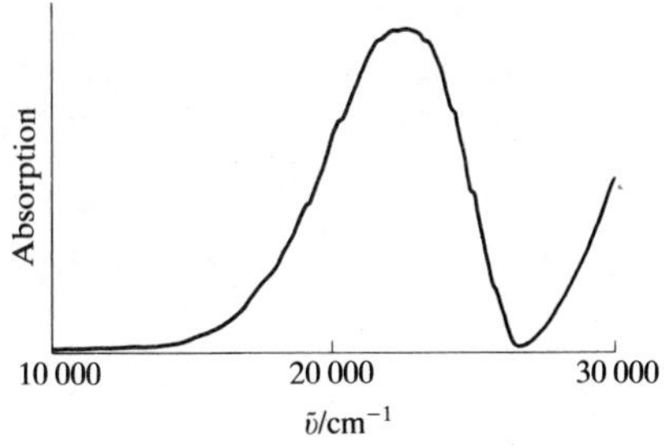

Fig. 19.4. The electronic absorption spectrum of $[Ti(H_2O)_6]^{3+}(Cl^-)_3$ in aqueous solution. The transition responsible is the $e_g \leftarrow t_{2g}$ transition illustrated in Fig. 19.3.

The only problem is that d–d transitions are forbidden in octahedral complexes. The *Laporte selection rule* for centrosymmetric complexes (and atoms) is that *the only allowed transitions are those accompanied by a change of parity*. That is, u↔g and g↔u transitions are allowed, but g↔g and u↔u transitions are forbidden‡.

An $e_g \leftarrow t_{2g}$ transition becomes allowed if the symmetry of the molecule is reduced by a vibration. For example, the vibration shown in Fig. 19.5 removes the centre of symmetry of the complex, and so eliminates the g, u classification of the states. In a non-centrosymmetric molecule d–d transitions are not parity-forbidden, and so the $e_g \leftarrow t_{2g}$ transition becomes weakly allowed (with $f \approx 10^{-4}$). When it depends on the vibration of the complex it is called a *vibronic transition*.

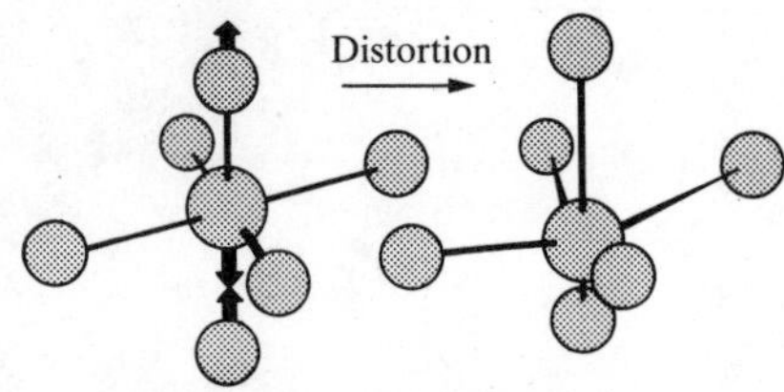

Fig. 19.5. A d–d transition is parity-forbidden because it corresponds to a g–g transition. However, a vibration of the molecule can destroy the inversion symmetry of the molecule so that the g, u classification no longer applies. This gives rise to a *vibronically allowed transition*.

A transition metal complex may also absorb light as a result of the transfer of electron density from the ligands into the d-orbitals of the central atom, or vice versa. In such *charge-transfer transitions* the electron moves through a considerable distance, which means that the transition dipole moment may be large and the absorption correspondingly intense. This mode of chromophore activity is shown by the manganate(VII) ion, MnO_4^-, and accounts for its intense violet ($\lambda = 420$–700 nm, $f \approx 0.03$).

19.2 (b) (π^*, π)- and (π^*, n)-transitions

In the case of the C═C *double bond*, absorption lifts a π-electron into an antibonding π^*-orbital, Fig. 19.6. The chromophore activity is therefore due to a (π^*, π)-*transition* (which is normally read 'π to π-star transition'). Its energy is around 7 eV for an unconjugated double bond, which corresponds to an absorption at 180 nm, in the ultraviolet. When the double bond is part of a conjugated chain, the energies of the molecular orbitals lie closer together and the (π^*, π)-transition shifts into the visible. We shall see an important example of this in a moment.

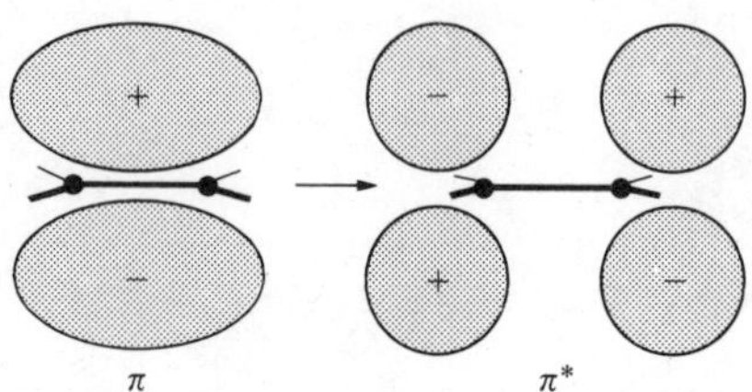

Fig. 19.6. The C═C double bond acts as a chromophore. One of its important transitions is the $\pi^* \leftarrow \pi$ transition illustrated here, in which an electron is promoted from a π-orbital to the corresponding antibonding orbital.

The transition responsible for absorption in *carbonyl compounds* can be traced to the oxygen lone-pairs. One of these electrons may be excited into an empty π^*-orbital of the carbonyl, Fig. 19.7, which gives rise to a (π^*, n)-*transition* (an 'n to π^* transition'). Typical absorption energies are about 4 eV (290 nm), and because (π^*, n)-transitions in carbonyls are symmetry forbidden, the absorptions are weak (f around 2×10^{-4} to 6×10^{-4}).

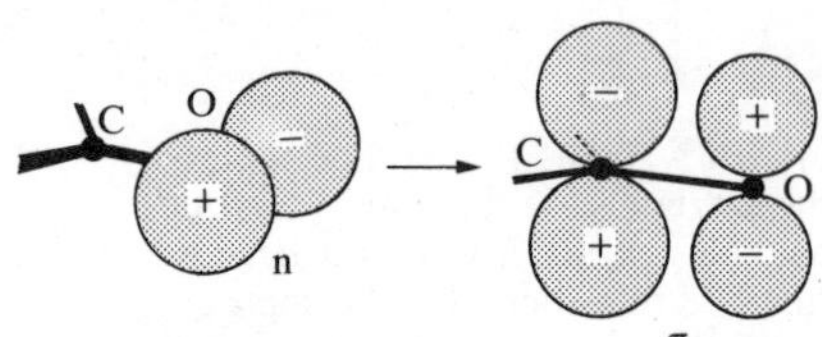

Fig. 19.7. The >C═O group acts as a chromophore primarily on account of the excitation of a non-bonding O lone-pair electron to an antibonding C—O π-orbital.

An important example of (π^*, π)- and (π^*, n)-transitions is provided by the photochemical mechanism of vision. The retina of the eye contains 'visual purple', which is a protein in combination with 11-*cis*-retinal (**1**). The 11-*cis*-retinal acts as a chromophore, and is the primary receptor for photons entering the eye. A solution of 11-*cis*-retinal absorbs at about 380 nm, but in combination with the protein (a link which might involve the elimination of the terminal carbonyl) the absorption maximum shifts to about 500 nm and tails into the blue. The conjugated double bonds are responsible for the ability of the molecule to absorb over the entire visible

O

1

‡ The group theoretical argument is as follows. The transition dipole moment in eqn (19.1.7) vanishes unless the integral is a basis for the totally symmetric irrep of the group. This means that in a centrosymmetric complex it must have g parity (in O_h it needs to be A_{1g}, but the g symmetry is important for this argument). The three components of the dipole moment operator transform like x, y, and z and are all u. Therefore, for a d ← d transition (a g ← g transition), the overall parity of the transition dipole is g × u × g = u, and so it must be zero. Likewise, for a u ← u transition the overall parity is u × u × u = u, and it must also vanish. Hence, transitions without a change of parity are forbidden.

region, but they also play another important role. In its electronically excited state the conjugated chain can isomerize, one half of the chain being able to twist about an excited C=C bond and forming all-*trans*-retinal (**2**). On account of its different shape, the new isomer cannot fit into the protein. The primary step in vision therefore appears to be photon absorption followed by isomerization: the uncoiling of the molecule then triggers a nerve impulse to the brain.

19.3 Vibrational structure

The width of electronic absorption bands in liquid samples is due to their unresolved vibrational structure. This structure, which can be resolved in gases and weakly interacting solvents, is due to the simultaneous excitation of a vibrational transition in the course of an electronic excitation.

19.3 (a) The Franck–Condon principle

The appearance of the vibrational structure of a band can be explained in terms of the *Franck–Condon principle*. This states that, because the nuclei are so much more massive than the electrons, *an electronic transition takes place faster than the nuclei can respond*. As a result of the transition, electron density is rapidly built up in new regions of the molecule, and the initially stationary nuclei suddenly experience a new force field. The classical picture of the process is that the nuclei respond to the new force by beginning to vibrate, and swing backwards and forwards from their original separation (which was maintained during the rapid electronic excitation). The stationary equilibrium separation of the nuclei in the initial electronic state therefore becomes the *turning point* (the point of a vibration when the nuclei are stationary, at the end point of their swing) in the final electronic state.

The quantum mechanical version of the Franck–Condon principle refines this picture to the point of letting us calculate the intensities of the transitions to different vibrational levels of the electronically excited molecule, and therefore to account for the shape of the absorption band.

Consider the molecular potential energy curves shown in Fig. 19.8. Before the absorption the molecule is in the ground vibrational state of its ground electronic state. The form of the vibrational wavefunction shows that the most probable location of the nuclei is at their equilibrium separation R_e. This means that the electronic transition is most likely to take place when the nuclei have this separation. When the transition occurs, the molecule is excited to the state represented by the upper curve. According to the Franck–Condon principle, the nuclear framework remains constant during this excitation, and so we may imagine the transition as being represented by the vertical line in Fig. 19.8. This is the origin of the expression *vertical transition*, which is used to denote an electronic transition that occurs without change of nuclear geometry.

The vertical transition cuts through several vibrational levels of the upper electronic state. The level marked * is the one in which the nuclei are most probably at the initial separation R_e (because the vibrational wavefunction has maximum amplitude there), and so this is the most probable level for the termination of the transition. However, it is not the only vibrational level for the termination of the transition, because several nearby levels have an appreciable probability of the nuclei being at the separation R_e.

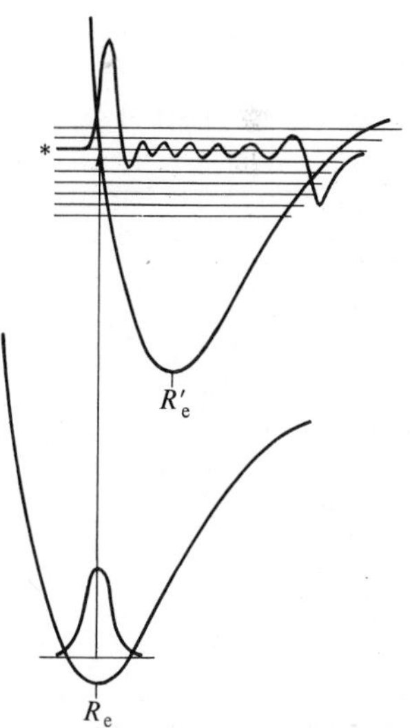

Fig. 19.8. According to the *Franck–Condon principle*, the most intense vibronic transition is from the ground vibrational state to the vibrational state lying vertically above it. Transitions to other vibrational levels also occur, but with lower intensity.

Therefore, transitions occur to all the vibrational levels in this region, but most intensely to the level with a wavefunction that peaks most strongly in the vicinity of R_e.

The vibrational structure of the spectrum depends on the relative displacement of the two potential energy curves, and a long *progression* of vibrations (a lot of vibrational structure) is stimulated if the two states are appreciably displaced. The upper curve is usually displaced to longer equilibrium bond lengths because excited states usually have more anti-bonding character than ground states. It should be noticed that the splitting between the vibrational lines of an electronic *absorption* spectrum depends on the vibrational energies of the *upper* electronic state: hence electronic absorption spectra may be used to assess the force fields and dissociation energies (e.g. via a Birge–Sponer plot, Section 18.3(c)) of electronically excited molecules.

19.3 (b) Franck–Condon factors

The quantitative basis of the Franck–Condon principle is the transition dipole moment. In order to deal with simultaneous electronic and vibrational transitions (i.e. with *vibronic transitions*, a marginally different usage of the term from the one in Section 19.2), we have to consider the total electronic and vibrational wavefunctions of the initial and final states. In terms of the Born–Oppenheimer approximation, these can be approximated as a product of electronic and vibrational wavefunctions, and so the total wavefunction for an electronic state ε and vibrational state v is approximately $\psi_v(\boldsymbol{R})\psi_\varepsilon(\boldsymbol{r})$, where $\boldsymbol{r}$ stands for the electronic coordinates and $\boldsymbol{R}$ for the nuclear coordinates. The transition dipole moment for the excitation $\varepsilon', v' \leftarrow \varepsilon, v$ is therefore approximately

$$\boldsymbol{\mu} = -e\int \{\psi_{\varepsilon'}(\boldsymbol{r})\psi_{v'}(\boldsymbol{R})\}^*\boldsymbol{r}\{\psi_\varepsilon(\boldsymbol{r})\psi_v(\boldsymbol{R})\}\,\mathrm{d}\tau_{\text{elec}}\,\mathrm{d}\tau_{\text{nuc}}$$

$$= -e\int \psi_{\varepsilon'}^*(\boldsymbol{r})\boldsymbol{r}\psi_\varepsilon(\boldsymbol{r})\,\mathrm{d}\tau_{\text{elec}}\int \psi_{v'}^*(\boldsymbol{R})\psi_v(\boldsymbol{R})\,\mathrm{d}\tau_{\text{nuc}}.$$

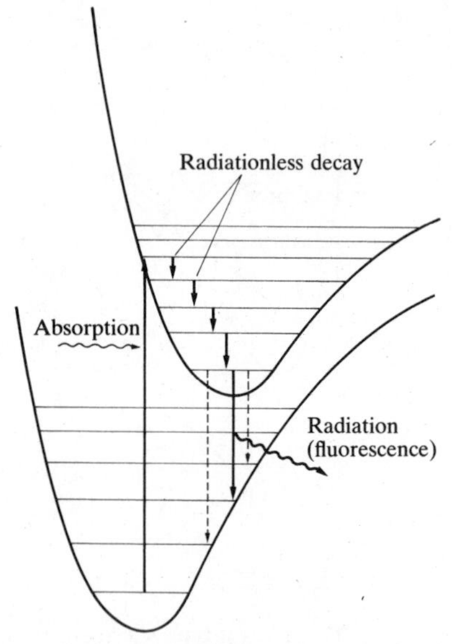

Fig. 19.9. The sequence of steps leading to *fluorescence*. After the initial absorption the upper vibrational states undergo radiationless decay by giving up energy to the surroundings. A radiative transition then occurs from the ground state of the upper electronic state.

The second integral is the one to concentrate on. It is the overlap integral (symbol: $S_{v',v}$) between the initial and final vibrational state wavefunctions. Since the intensity of a transition depends on the square of the transition dipole moment, the vibronic transition intensity depends on $S_{v',v}^2$, which is known as the *Franck–Condon factor* for the $\varepsilon', v' \leftarrow \varepsilon, v$ transition. It follows that the greater the vibrational overlap the greater the absorption intensity. Furthermore, by calculating the overlap we can predict the intensity.

Example 19.4

Consider the transition from one electronic state to another, their bond lengths being R_e and R_e' and their force constants equal. Calculate the Franck–Condon factor for the 0–0 transition and show that the vibronic transition is most intense when the bond lengths are equal.

- *Method*. Calculate $S_{0,0}$, the overlap integral of the two ground state harmonic oscillator wavefunctions, Box 14.1.

- *Answer*. $\psi_0 = (1/\alpha\pi^{\frac{1}{2}})^{\frac{1}{2}}\mathrm{e}^{-y^2/2}$,
  $\psi_0' = (1/\alpha\pi^{\frac{1}{2}})^{\frac{1}{2}}\mathrm{e}^{-y'^2/2}$,

where $y = (R - R_e)/\alpha$ and $y' = (R - R_e')/\alpha$, with $\alpha^2 = \hbar/(mk)^{\frac{1}{2}}$, as explained in Section 14.2. The overlap integral is

$$S_{0,0} = \int_{-\infty}^{\infty} \psi_0'(R)\psi_0(R)\,dR = (1/\pi\alpha^{\frac{1}{2}})\int_{-\infty}^{\infty} e^{-(y^2+y'^2)/2}\,dR.$$

It is not difficult to manipulate this into an expression proportional to an integral over e^{-R^2}, the value of which is $\pi^{\frac{1}{2}}$. Therefore, the overlap integral is

$$S_{0,0} = e^{-(R_e-R_e')^2/4\alpha^2},$$

and the intensity is proportional to $S_{0,0}^2$. This is unity when $R_e = R_e'$ and decreases as the equilibrium bond lengths diverge.

• *Comment*. In the case of Br_2, $R_e = 228$ pm and there is an upper state with $R_e' = 266$ pm. Taking the vibrational wavenumber as 250 cm^{-1}, gives $S_{0,0}^2 = 5\times10^{-10}$, and so the intensity of the 0–0 transition is only 5×10^{-10} of what it would have been if the potential curves had been directly above each other.

• *Exercise*. Suppose the vibrational wavefunctions can be approximated by rectangular functions of width W and W', centred on the equilibrium bond lengths. Find the corresponding Franck–Condon factors when the centres are coincident and $W' < W$. [W'/W]

19.4 The fate of electronically excited states

The energy of an electronically excited state may be lost in a variety of ways. A common fate is for it to be transferred into the vibration, rotation, and translation of the surrounding molecules. This *thermal degradation* converts the excitation energy into thermal motion of the environment (i.e. to 'heat'). Much more interesting, however, is the possibility that the excited molecule can take part in a chemical reaction: the general subject of *photochemistry* is taken up again in Part 3. In this section we shall be concerned mainly with *radiative decay*, which occurs when a molecule discards its excitation energy as a photon.

Two mechanisms of radiative decay have been identified. In *fluorescence* the emitted radiation ceases immediately the exciting radiation is extinguished. In *phosphorescence* the emission may persist for long periods (even hours, but characteristically seconds or fractions of seconds). The difference suggests that fluorescence is an immediate conversion of absorbed light into re-emitted energy, but that in phosphorescence the energy is stored in a reservoir from which it slowly leaks.

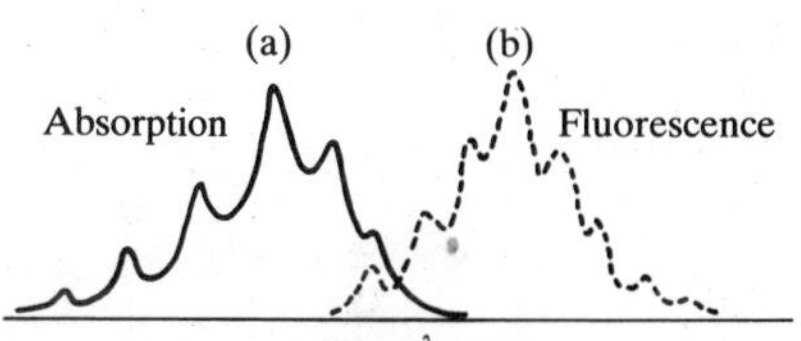

Fig. 19.10. The absorption spectrum (a) shows a vibrational structure characteristic of the upper state. The fluorescence spectrum (b) shows a structure characteristic of the lower state; it is also displaced to lower frequencies (but the 0–0 transitions are coincident) and resembles a mirror image of the absorption.

19.4 (a) Fluorescence

The sequence of steps involved in fluorescence is depicted in Fig. 19.9. The initial absorption takes the molecule to an excited electronic state, and if the absorption spectrum were monitored it would look like Fig. 19.10(a). In the excited state the molecule is subjected to collisions with the surrounding molecules (e.g. of the solvent), and as it gives up energy it steps down the ladder of vibrational levels. The surrounding molecules, however, might be unable to accept the larger energy difference needed to lower the molecule to the ground electronic state, and it may therefore survive long enough to undergo spontaneous emission, emitting the remaining excess energy as radiation. The downward electronic transition is vertical (in accord with the Franck–Condon principle) and the *fluorescence spectrum*, Fig. 19.10(b), possesses a vibrational structure characteristic of the *lower* electronic state (which may therefore be studied in this way).

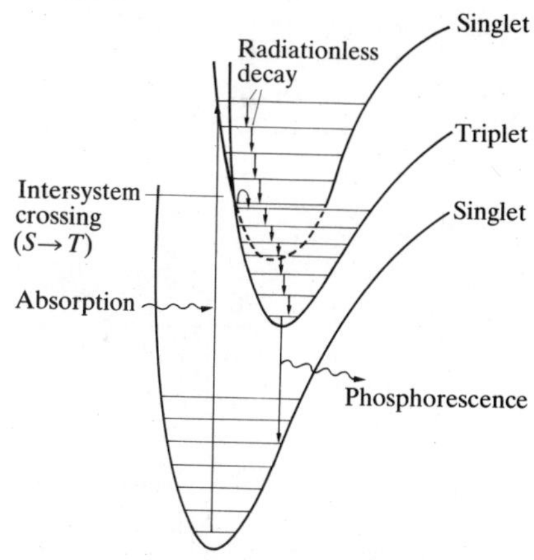

Fig. 19.11. The sequence of steps leading to *phosphorescence*. The important step is the *intersystem crossing*, the switch from singlet to triplet state brought about by spin–orbit coupling. The triplet state acts as a slowly radiating reservoir because the return to the ground state is spin-forbidden.

The mechanism accounts for the observation that fluorescence occurs at a lower frequency than the incident light: the radiation occurs after some vibrational energy has been discarded into the surroundings. The vivid oranges and greens of fluorescent dyes are an everyday manifestation of this effect: they absorb in the ultraviolet and blue, and fluoresce in the visible. The mechanism also suggests that the intensity of the fluorescence ought to depend on the ability of the solvent molecules to accept the electronic and vibrational quanta. It is indeed found that a solvent composed of molecules with widely spaced vibrational levels (such as water) may be able to accept the large quantum of electronic energy and extinguish the fluorescence.

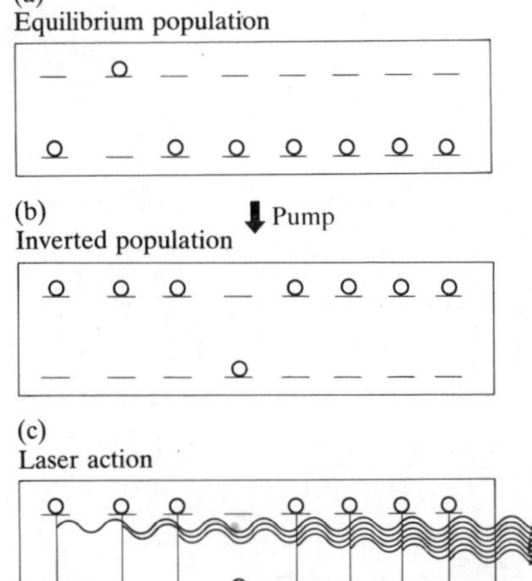

Fig. 19.12. A schematic illustration of the steps leading to laser action. (a) The Boltzmann population of states, with more atoms in the ground state. (b) When the initial state absorbs, the populations are inverted (the atoms are *pumped* to the excited state). (c) A cascade of radiation then occurs, as one emitted photon stimulates another atom to emit, and so on. The radiation is *coherent* (phases in step).

19.4 (b) Phosphorescence

The sequence of events leading to phosphorescence is depicted in Fig. 19.11 (p. 471). The first steps are the same as in fluorescence, but the presence of a *triplet* excited state plays a decisive role. (Triplet states were first mentioned in Section 15.2: they are states in which two electrons have parallel spins.) At the point where the potential energy curves intersect, the two states share a common geometry, and if there is a mechanism for unpairing two electron spins, then the molecule may cross into the triplet state. This is called *intersystem crossing*. We saw in the discussion of atomic spectra (Section 15.3) that singlet–triplet transitions may occur in the presence of spin–orbit coupling, and the same is true in molecules. Therefore, we expect intersystem crossing to occur when a molecule contains a heavy atom (such as S), because then the spin–orbit coupling is large and electrons can become unpaired.

If an excited molecule crosses into a triplet state, it continues to deposit energy into the surroundings and to step down the vibrational ladder. However, it is now stepping down the triplet's ladder, and at the lowest vibrational energy level it is trapped. The solvent cannot extract the final, large quantum of electronic excitation energy, and the molecule cannot radiate its energy because return to the ground state involves a forbidden triplet→singlet transition. This radiative transition, however, is not totally forbidden because the spin–orbit coupling that was responsible for the intersystem crossing also breaks the selection rule. The molecules are therefore able to emit weakly, and the emission may continue long after the original excited state is formed.

This mechanism accords with the experimental observation that the excitation energy seems to get trapped in a slowly leaking reservoir. It also suggests (as is confirmed experimentally) that the phosphorescence should be most intense from solid samples, because then energy transfer is less efficient and the intersystem crossing has time to occur as the singlet excited state steps slowly past the intersection point. The mechanism also suggests that the phosphorescence efficiency should depend on the presence of a heavy atom (with strong spin–orbit coupling), and this is also in agreement with observation. Finally, it also predicts that, on account of the presence of unpaired spins in the triplet, the reservoir state should be magnetic. This has been confirmed experimentally by observing magnetism in phosphorescent, excited molecules using the sensitive resonance techniques described in the next chapter.

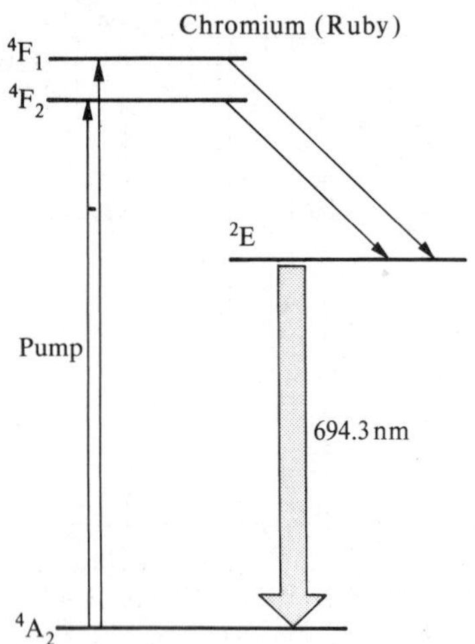

Fig. 19.13. The transitions involved in the *ruby laser*. The laser medium, ruby, consists of Al_2O_3 doped with Cr^{3+} ions.

19.4 (c) Laser action

Both fluorescence and phosphorescence are radiative decays by spontaneous emission. Laser action, as the acronym **l**ight **a**mplification by **s**timulated **e**mission of **r**adiation suggests, depends on emission by a *stimulated* process.

Laser action is shown schematically in Fig. 19.12. By some means a majority of molecules are excited (*pumped*) into an upper state. The sample is contained in a cavity between two mirrors, and when a molecule emits spontaneously the photon it generates is reflected backwards and forwards between them. Its presence stimulates other molecules to emit, and they add more photons of the same frequency to the cavity, and these stimulate more molecules to emit. The cascade of energy builds up rapidly, and if one of the mirrors is half-reflecting, the radiation may be tapped. Its characteristics reflect how it is generated: it is monochromatic (because photons stimulate emission of more photons of the same frequency), coherent (because the phases of the electric fields of the radiation are in step), and non-divergent (because photons travelling at an angle to the cavity axis are not trapped and do not stimulate others).

The first successful laser was based on transitions in ruby (Al_2O_3 doped with Cr^{3+}). The sequence of steps is shown in Fig. 19.13. The *population inversion* results from pumping a majority of Cr^{3+} ions into an excited state (^4F) using an intense flash from another source (e.g. a xenon lamp or another laser) followed by a radiationless transition to another excited state (^2E). The transition from this to the ground state ($^2E \rightarrow {}^4A_2$) is the laser transition, and gives rise to red 694 nm radiation. The neodymium laser works on a slightly different principle, for the laser transition ($^4F \rightarrow {}^4I$ in Fig. 19.14) is between two excited states. The advantage is that, as the population inversion is between two excited states, it is much easier to achieve than between an excited state and a heavily populated ground state.

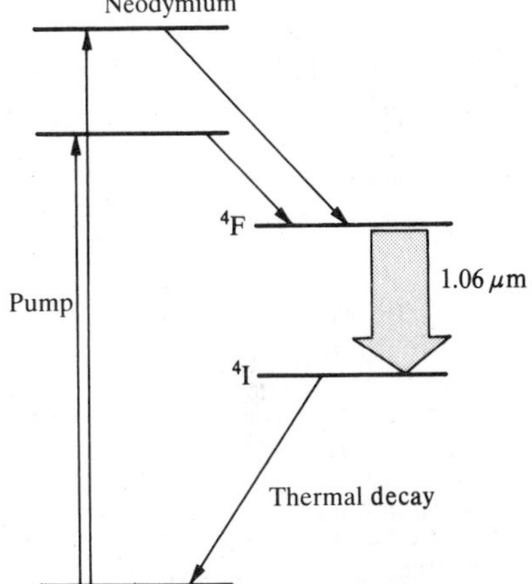

Fig. 19.14. The transitions involved in the *neodymium laser*. The laser action takes place between two excited states, and the population inversion is easier to achieve.

Laser action occurs in all kinds of media. *Dye lasers* use solutions of intensely absorbing organic dyes (such as Rhodamine 6 G in methanol) as the active medium. They have the advantage of operating over a wide range of wavelengths, and by modifying the design of the cavity (e.g. by incorporating a diffraction grating) the radiation can be tuned to a selected wavelength. Gas lasers are widely used, and since they can be cooled by a rapid flow of the gas through the cavity, they can be used to generate high powers. The pumping is normally achieved with a species other than the laser material. For example, in the *helium–neon laser*, Fig. 19.15, the initial step is the excitation of a helium atom to 1s2s ^1S using an electric discharge (the rapid collisions of electrons and ions cause transitions that are not restricted by electric dipole selection rules, and so 1s2s $^1S \leftarrow 1s^2\,{}^1S$ occurs). By chance, the excitation energy happens to match an excitation energy of neon, and during an He–Ne collision an efficient transfer of energy may occur, leading to the production of highly excited Ne atoms with unpopulated intermediate states. Laser action generating 633 nm radiation (among about 100 other lines) then occurs. The *carbon dioxide laser*, Fig. 19.16 (p. 474), works on a similar principle. Most of the working gas is nitrogen, which becomes vibrationally excited by electronic and ionic collisions in an electric discharge. The vibrational levels happen to coincide with the ladder of antisymmetric stretching (ν_3) vibrational energy levels of CO_2, which pick up the energy during a collision. Laser action then occurs from the lowest excited level of ν_3 to the lowest excited level of the symmetric stretch (ν_1), which has remained unpopulated during the collisions. This generates 10.6 μm radiation (in the infrared). In some cases (the N_2 gas laser at 337 nm) the efficiency of the stimulated transition is so great that a single passage of a pulse of radiation is enough to generate laser radiation, and mirrors are unnecessary: such lasers are said to be in a *super-radiant mode*.

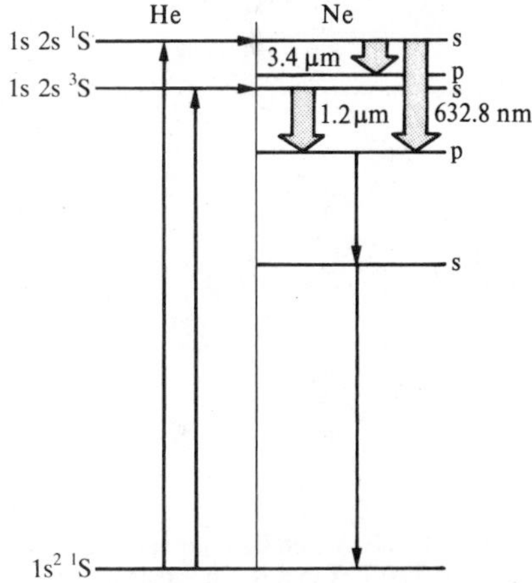

Fig. 19.15. The transitions involved in the *helium–neon laser*. The pumping (of the neon) depends on a coincidental matching of the helium and neon energy separations, so that excited He atoms can transfer their excess energy to Ne atoms during a collision.

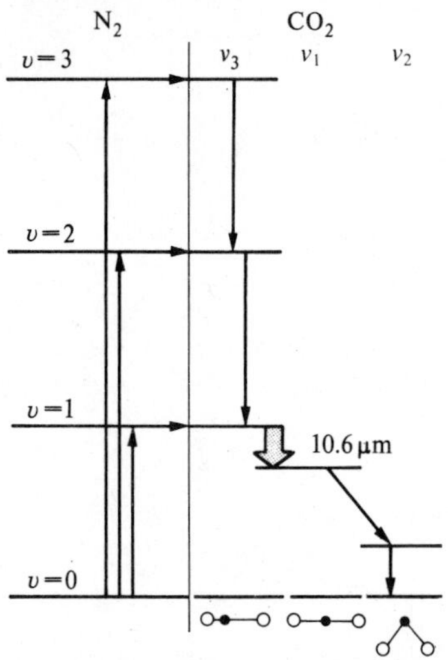

Fig. 19.16. The transitions involved in the *carbon dioxide laser*. The pumping also depends on the coincidental matching of energy separations; in this case the vibrationally excited N_2 molecules have excess energies that correspond to a vibrational excitation of the ν_3 (antisymmetric stretch) of CO_2. The laser transition is from $\nu_3 = 1$ to $\nu_1 = 1$.

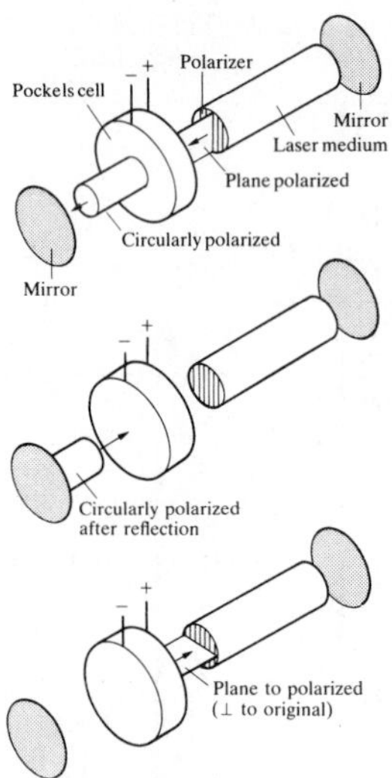

Fig. 19.17. When light passes through a *Pockels cell* that is 'on', its plane of polarization is rotated and so the laser cavity is non-resonant (its *Q-factor*, a measure of its resonant quality, is reduced). When the cell is turned off, no change of polarization occurs, and the cavity is resonant. The laser action then takes place in a single pulse. This is one method of *Q-switching*.

Chemical reactions may also be used to generate molecules with non-equilibrium, inverted populations. For example, the photolysis of Cl_2 leads to the formation of Cl atoms that attack H_2 molecules in the mixture. This produces HCl and H, the latter attacking Cl_2 and producing vibrationally excited ('hot') HCl molecules. Since these HCl molecules have non-equilibrium vibrational populations, laser action can result as they return to lower states. Such processes are remarkable examples of the direct conversion of chemical energy into coherent electromagnetic radiation.

Lasers may work continuously (these are the *continuous wave*, CW, lasers) if the pumping process continues during the lasing action and the population inversion can be sustained. They may also be operated in pulses, and techniques are now available that give rise to pulses so short that they are measured in picoseconds ($1\text{ ps} = 10^{-12}\text{ s}$) and even femtoseconds ($1\text{ fs} = 10^{-15}\text{ s}$). Pulses on a 10 ns time scale are produced by *Q-switching*, in which the pumping takes place while the resonance characteristics of the cavity are impaired. A *Pockels cell* is an electro-optical device based on the ability of crystals of ammonium dihydrogen phosphate to convert plane-polarized light to circular polarization when a potential difference is applied. If one is made part of a laser cavity, then as a result of the change of polarization that occurs when light is reflected, the active cell converts light polarized in one plane into reflected light polarized in the perpendicular plane (Fig. 19.17). As a result, when the potential difference is applied the reflected light does not stimulate more emission. However, if the potential difference is suddenly turned off, the polarization effect is extinguished and all the energy stored in the cavity can emerge as a giant pulse of high intensity stimulated radiation. An alternative technique is to use a *saturable dye*. A saturable dye loses its power to absorb when many of its molecules have been excited by intense radiation. It then suddenly becomes transparent, and the cavity becomes resonant.

The refinement that produces pulses lasting picoseconds and less is *mode locking*. The feature of laser action lying behind this technique is that a laser radiates at a number of different frequencies, depending on the precise details of the resonance characteristics of the cavity and in particular on the number of half-wavelengths of radiation that can be trapped between the mirrors (these are the cavity *modes*). Normally the modes, which differ in frequency by multiples of $c/2d$, d being the length of the cavity, have random phases relative to each other, but by modulating the active medium at the frequency $c/2d$ (e.g. with ultrasonic waves generated electrically) the modes can be brought into a definite phase relation to each other. Then, as a result of their mutual interference, the laser radiation emerges packed into a series of extremely sharp pulses, Fig. 19.18. Techniques based on mode locking are opening up new insights into the properties of molecules, especially the way that energy is used in chemical reactions (Part 3).

19.4 (d) Dissociation and predissociation

Another fate for an electronically excited molecule is *dissociation*, Fig. 19.19. It is detected in the spectrum by the termination of the vibrational structure of a band at the *dissociation limit:* the provision of more energy than is needed to dissociate the bond simply results in the appearance of a continuous band of absorption because the excess is taken up by the translational motion of the fragments. Locating the dissociation limit is a valuable way of determining the bond dissociation energy.

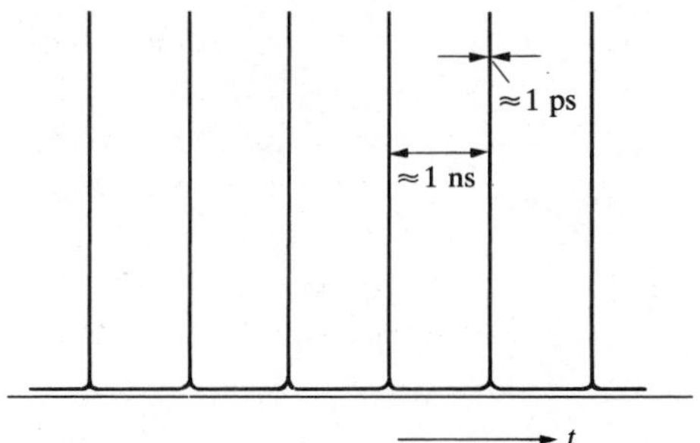

Fig. 19.18. The output from a mode-locked laser consists of a train of pulses of very short duration, typically picoseconds, but under special circumstances femtoseconds.

In some cases the vibrational structure disappears but returns at higher photon energies. This *predissociation* can be interpreted in terms of the potential energy curves shown in Fig. 19.20 (p. 476). When a molecule is excited to a vibrational level near the intersection of the two upper curves its electrons may undergo a reorganization which results in it undergoing an *internal conversion* to the other state. (The nuclear geometries of the two states are the same at the intersection.) The molecule may emerge into a dissociative state, even though its energy is apparently less than is needed for dissociation (hence the name *pre*dissociation), and on account of the reduction of its lifetime its energy is imprecise. As a result, the absorption spectrum is blurred in the vicinity of the intersection. When the incoming photon brings enough energy to excite the molecule to a vibrational level high above the intersection, the internal conversion does not occur (the nuclei are unlikely to have the same geometry). Consequently, the levels resume their well-defined, vibrational character with correspondingly well-defined energies. This means that the line structure resumes on the high-frequency side of the blurred region.

19.5 Photoelectron spectroscopy

Photoelectron spectroscopy measures the ionization energies of electrons in molecules, and uses the information to build up a picture of the orbital

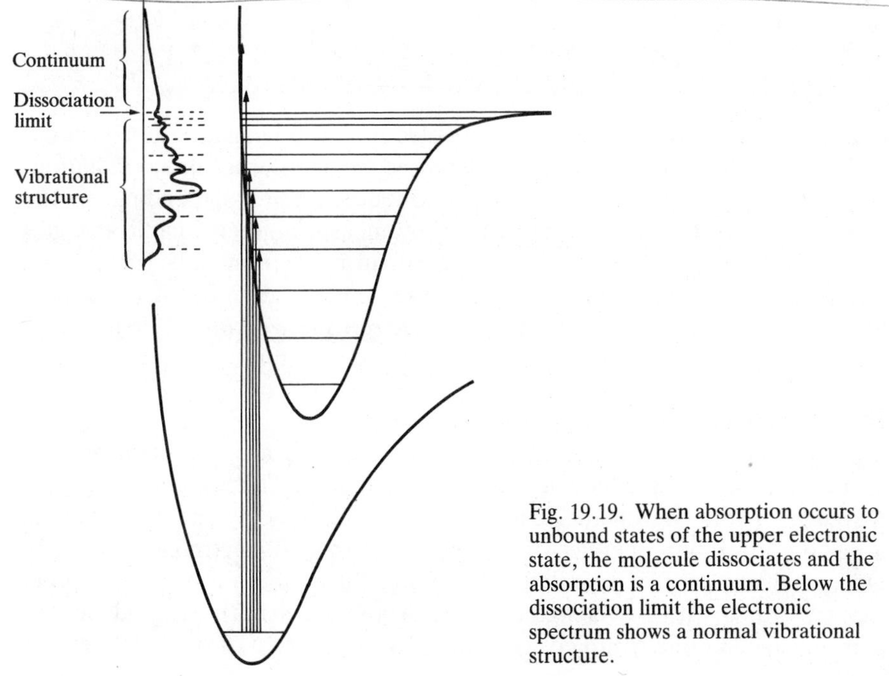

Fig. 19.19. When absorption occurs to unbound states of the upper electronic state, the molecule dissociates and the absorption is a continuum. Below the dissociation limit the electronic spectrum shows a normal vibrational structure.

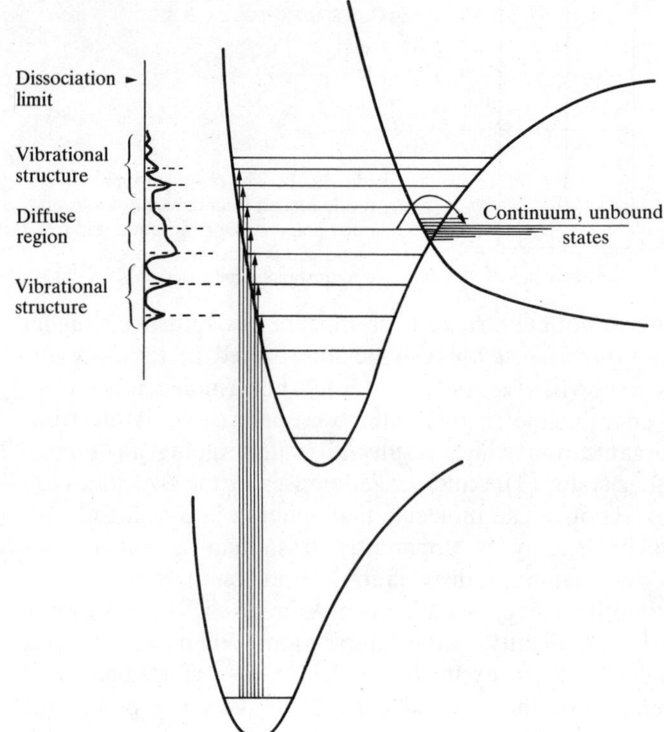

Fig. 19.20. When a dissociative state crosses a bound state, as in the upper part of the illustration, molecules excited to levels near the crossing may dissociate. This leads to *predissociation*, and is detected in the spectrum as a loss of vibrational structure which resumes at higher frequencies.

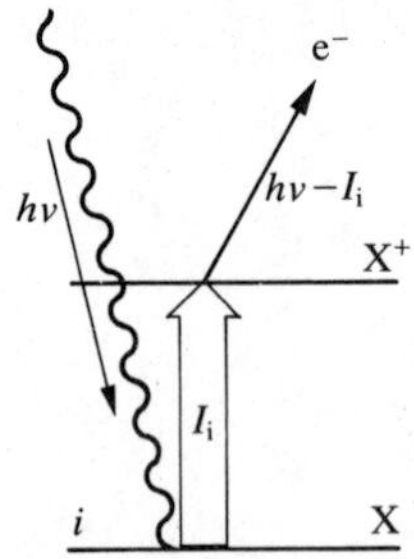

Fig. 19.21. An incoming photon carries an energy $h\nu$; an energy I_i is needed to remove an electron from an orbital i, and the difference appears as the kinetic energy of the electron.

energies. The technique is based on an expression analogous to that used for the photoelectric effect (eqn (13.2.6)), where the energy brought by the photon ($h\nu$) is equated to the sum of the work function and the excess kinetic energy of the ejected electron. In the present case, Fig. 19.21, we write

$$h\nu = \tfrac{1}{2}m_e v^2 + I_i, \qquad (19.5.1)$$

where I_i is the ionization energy for an electron occupying an orbital i in the molecule. When the molecule is exposed to monochromatic radiation, electrons will be ejected with various kinetic energies, depending on the orbital they initially occupy. Therefore, by measuring their kinetic energies and knowing ν, these ionization energies can be determined. The ejected electrons are called *photoelectrons*, and the determination of their spectrum of energies gives rise to the name *photoelectron spectroscopy* (PES).

19.5 (a) The technique

The experimental technique is straightforward in principle but difficult to refine to the point of achieving very high resolution. The first requirement is a source of monochromatic, energetic (high frequency), intense radiation. Since the ionization energies of molecules are several electronvolts even for valence electrons, it is essential to work in at least the ultraviolet region of the spectrum (with wavelengths less than about 200 nm). A great deal of work has been done with light generated by a discharge through helium: the

He(I) line (1s2p → 1s^2) lies at 58.43 nm, corresponding to a photon energy of 21.22 eV. Its use gives rise to the technique of *ultraviolet photoelectron spectroscopy* (UPS). If the core electrons are being studied, then photons of even higher energy are needed to expel them: X-ray frequencies are used, and the technique is denoted *XPS*. A modern version of PES involves using *synchrotron radiation*, the radiation emitted by packets of electrons circulating in a magnetic field after injection from a particle accelerator: one advantage is that synchrotron radiation can be tuned over a wide range of frequencies; the disadvantage is that the experiments have to be done at a national facility.

Example 19.5

Photoelectrons ejected from N_2 with He(I) radiation had kinetic energies of 5.63 eV. What is their ionization energy?

• *Method*. Use eqn (19.5.1). A useful conversion (inside front cover) is $1\,\text{eV} \triangleq 8065.5\,\text{cm}^{-1}$. He(I) radiation has wavelength 58.43 nm.

• *Answer*. He(I) radiation at 58.43 nm has wavenumber

$$\tilde{\nu} = 1/(58.43\,\text{nm}) = 1.711 \times 10^5\,\text{cm}^{-1}.$$

It therefore corresponds to an energy of 21.22 eV. Then, from eqn (19.5.1):

$$21.22\,\text{eV} = 5.63\,\text{eV} + I_i \quad \text{so that} \quad I_i = 15.59\,\text{eV}.$$

• *Comment*. This ionization energy (corresponding to $1504\,\text{kJ mol}^{-1}$) is the energy needed to remove an electron from the uppermost filled MO in the N_2 molecule, the $2s\sigma_g$ bonding orbitals (Fig. 16.12).

• *Exercise*. Under the same circumstances 4.53 eV electrons are also detected. To what ionization energy does that correspond? Suggest an origin. [16.7 eV, $2p\pi_u$]

As well as an ionizing source, it is necessary to have an *analyser* to measure the energies of the ejected electrons. This normally takes the form of an electrostatic deflector, Fig. 19.22, the extent of deflection of an electron as it passes between charged plates being determined by its speed. As the field strength is increased, electrons of different speeds, and therefore kinetic energies, reach the detector, and so the electron flux can be recorded and plotted against kinetic energy to obtain the photoelectron spectrum. The detector itself is an electron multiplier, which gives a cascade of electrons when struck by the incoming electrons. Present techniques give energy resolutions of about 5 meV, corresponding to about $30\,\text{cm}^{-1}$.

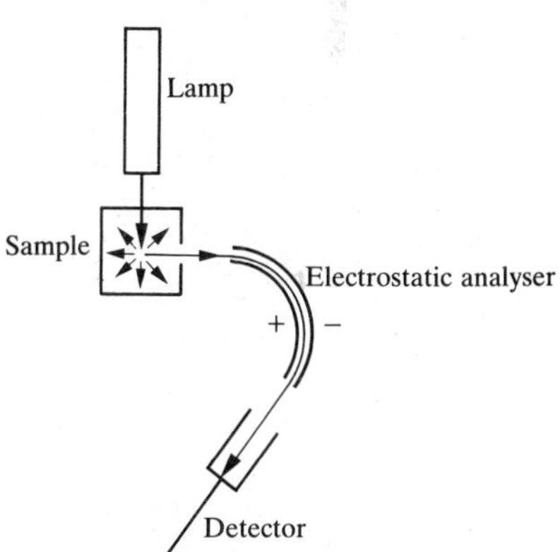

Fig. 19.22. A photoelectron spectrometer consists of a source of ionizing radiation (such as a helium discharge lamp for UPS and an X-ray source for XPS), an electrostatic analyser, and an electron detector. The deflection of the electron path caused by the analyser depends on their speed.

19.5 (b) UPS

A typical photoelectron spectrum (of HBr) is shown in Fig. 19.23. Its main features can be interpreted in terms of an approximation called *Koopmāns's theorem* which states that the ionization energy I_i is equal to the orbital energy of the ejected electron (formally: $I_i = -\varepsilon_i$). However, this is only an approximation because it ignores the fact that the remaining electrons rearrange their distributions when ionization occurs.

If we disregard the fine structure, we see that the HBr lines fall into two main groups. The least tightly bound electrons (those with the lowest ionization energies and highest kinetic energies when ejected) are those in

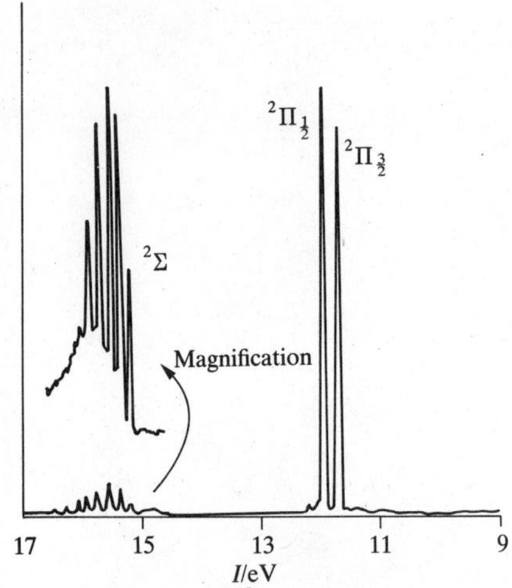

Fig. 19.23. The photoelectron spectrum of HBr. The lowest ionization energy band corresponds to the ionization of a bromine lone-pair electron. The higher ionization energy band corresponds to the ionization of a bonding σ-electron. The structure on the latter is due to the vibrational excitation of HBr^+ that results from the ionization.

the non-bonding lone-pairs of Br. The next ionization energy lies at 15.2 eV, and corresponds to the removal of an electron from the H–Br σ-bond.

The fine structure in the spectrum can be interpreted in terms of the excitation of the vibration of the ion caused by the ionization. If electron ejection vibrationally excites the ion, then some of the photon's energy does not appear in the ejected electron's kinetic energy, and the energy conservation equation becomes

$$h\nu = \tfrac{1}{2}m_e v^2 + I_i + \Delta E^+_{\text{vib}}, \qquad (19.5.2)$$

ΔE^+_{vib} being the energy used to excite the ion into vibration.

The HBr spectrum in Fig. 19.23 shows that ejection of a σ-electron is accompanied by a long vibrational progression. The Franck–Condon principle would account for this if ejection were accompanied by an appreciable change of equilibrium bond length between HBr and HBr^+ because the ion is formed in a compressed state. This is consistent with the important bonding effect of the σ-electrons. On the other hand, the lack of much vibrational structure in the other band is consistent with the non-bonding role of the Br2p-orbital lone-pair electrons, for the equilibrium bond length is little changed when one is removed.

19.5 (c) XPS and ESCA

In XPS the energy of the incident photon is so great that electrons are ejected from cores. As a first approximation we would not expect core ionization energies to be sensitive to the bonds between atoms because they are too tightly bound to be greatly affected by the changes in the distributions of valence electrons that accompany bond formation. This turns out to be largely true, and since the inner-shell ionization energies are characteristic of the individual atom rather than the overall molecule, XPS gives lines characteristic of the elements present in a compound or alloy. For instance, the K-shell ionization energies of the second row elements are:

Li	Be	B	C	N	O	F	
50	110	190	280	400	530	690	eV

and detection of one of these values (and values corresponding to ejection from other inner shells) indicates the presence of the corresponding element. Figure 19.24, for example, shows the result of an XPS analysis of lunar rock brought back on an Apollo mission. This chemical analysis application is responsible for the alternative name *ESCA* (electron spectroscopy for chemical analysis). The technique is mainly limited to the study of surface layers (as we shall explore in Chapter 31) because, even though X-rays may penetrate into the bulk sample, the ejected electrons cannot escape except from within a few nanometres of the surface.

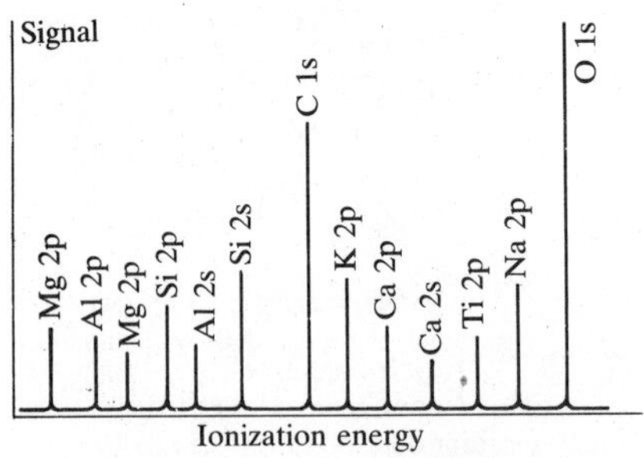

Fig. 19.24. The X-ray photoelectron spectrum of a sample of moon dust, with the assignment. (Adapted from *Physical methods and molecular structure*, The Open University Press, 1977.)

While it is largely true that core ionization energies are unaffected by bond formation, it is not entirely true, and small but observable shifts can be detected and interpreted in terms of the environments of the atoms. For example, the azide ion (N_3^-) gives the spectrum shown in Fig. 19.25, and although it is in the region of 400 eV (and hence typical of N1s-electrons), it has a doublet structure with splitting 6 eV. The structure of the ion is N=N=N, with charge distribution (−, +, −), and the presence of the negative charges on the terminal atoms lowers the core ionization energies, while the positive charge on the central atom raises it: hence, the two lines in the spectrum with intensities in the ratio 2:1. This kind of observation can be used to obtain valuable information about the presence of chemically inequivalent but otherwise identical atoms.

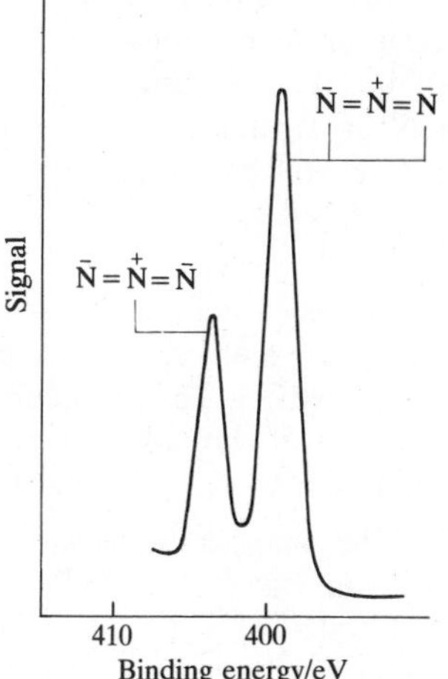

Fig. 19.25. The photoelectron spectrum of solid NaN_3 excited by Al K-radiation showing the region of N core ionization and the assignment. (K. Siegbahn *et al.*, *Science*, **176,** 245 (1972).)

Further reading

For references to books on general aspects of spectroscopy, *see* Chapter 18.

Ultraviolet and visible spectra:

Theory and applications of ultraviolet spectroscopy. H. H. Jaffe and M. Orchin; Wiley, New York, 1962.

Ultraviolet and visible spectroscopy. C. N. R. Rao; Butterworth, London, 1967.

Visible and ultraviolet spectroscopy. F. Grum; *Techniques of chemistry* (A. Weissberger and B. W. Rossiter, eds) IIIB, 207, Wiley-Interscience, New York, 1972.

Spectra of diatomic molecules. G. Herzberg; Van Nostrand, 1950.

Electronic spectra and electronic structure of polyatomic molecules. G. Herzberg; Van Nostrand, New York, 1966.

Excited state processes:

Determination of fluorescence and phosphorescence. N. Wotherspoon, G. K. Oster, and G. Oster; *Techniques in chemistry* (A. Weissberger and B. W. Rossiter, eds) IIIB, 429, Wiley-Interscience, New York, 1972.

Dissociation energies. A. G. Gaydon; Chapman and Hall, London, 1952.

Photochemistry. J. G. Calvert and J. N. Pitts; Wiley, New York, 1966.

Photophysics of aromatic molecules. J. B. Birks; Wiley, New York, 1970.

Photochemistry: past, present, and future. R. P. Wayne (ed); *J. photochem*. **25** (1), (1984).

Lasers:

Lasers. B. A. Lengyel; Wiley-Interscience, New York, 1971.

Handbook of laser science and technology. M. J. Weber; CRC Press, Boca Raton, 1982.

High resolution spectroscopy. M. Hollas; Butterworth, London, 1982.

Applications of lasers to chemical problems. T. R. Evans (ed); *Techniques of chemistry Vol. XVII*, Wiley-Interscience, New York, 1982.

Photoelectron spectra:

Photoelectron spectroscopy. J. H. Eland; Open University Press, Milton Keynes, 1977.

Photoelectron spectroscopy. A. D. Baker and D. Betteridge; Pergamon Press, Oxford, 1977.

Molecular photoelectron spectroscopy. D. W. Turner, C. Baker, A. D. Baker, and C. R. Brundle; Wiley-Interscience, New York, 1970.

Introductory problems

A19.1. The molar absorption coefficient for a substance dissolved in hexane is known to be $855\,M^{-1}\,cm^{-1}$ for a wavelength of 270 nm. Calculate the percentage reduction in the intensity when the light of that wavelength passes through a solution with concentration 3.25×10^{-3} M in a cell with a path length 2.5 mm.

A19.2. When a beam of 400 nm light passes through 3.5 mm of a solution with an absorbing solute at a concentration of 6.75×10^{-4} M, the per cent transmission is 65.5. Calculate the molar absorption coefficient of the solute at this wavelength, and express the answer in $cm^2\,mol^{-1}$.

A19.3. The molar absorption coefficient of a solute absorbing at a wavelength of 540 nm is known to be $286\,M^{-1}\,cm^{-1}$. When 540 nm light passes through a 6.5 mm cell containing the solute at an unknown concentration,

46.5% of the light is absorbed. What is the concentration of the solute?

A19.4. In a certain laboratory, the precision of equipment for measuring light intensity in absorption experiments is limited to 2%. The only available sample cells have a thickness of 1.5 mm. What is the limit of precision in determining concentrations with this equipment, if the absorbing species has a molar absorption coefficient of $275\ M^{-1}\ cm^{-1}$.

A19.5. The absorption associated with a certain transition begins approximately at 230 nm, peaks sharply at 260 nm, and ends at 290 nm. The maximum value of the absorption coefficient is $1.21 \times 10^4\ M^{-1}\ cm^{-1}$. Assume a triangular line shape and estimate the oscillator strength.

A19.6. The magnitude of a transition dipole moment obtained from wave functions thought to describe the initial and final states of a transition was found to be 2.65×10^{-30} C m. The wavenumber of the transition is $35\,000\ cm^{-1}$. What is the oscillator strength of the transition?

A19.7. Theoretical calculations of oscillator strengths for a group of transitions produced the following values: 2.9×10^{-2}, 0.75, 6.2×10^{-5}, 3.2×10^{-9}, 0.91. Classify these predictions using the terms: strong, weak, forbidden.

A19.8. The two compounds, 2,3-dimethylbutene-2 and 2,5-dimethylhexadiene-2,4, are to be distinguished by observation of their ultraviolet spectra. The maximum in one spectrum is at 192 nm: in the other, the peak is at 243 nm. Match the maxima with the compounds and briefly state the basis for identification.

A19.9. In the ultraviolet region, the compound $CH_3CH{=}CHCHO$ has a strong absorption at $46\,950\ cm^{-1}$ and a weak absorption at $30\,000\ cm^{-1}$. Show that these spectral features are consistent with the structure of the compound.

A19.10. The photoionization of H_2 by 21 eV photons produces H_2^+. Which of these species has the larger internuclear distance? Use the difference in the bond lengths to explain why the intensity associated with the $v = 0$ to $v = 2$ transition observed in the photoelectron spectrum is greater than that involving the two ground vibrational states.

Problems

19.1. The following data were obtained for the absorption of light by a sample of bromine dissolved in carbon tetrachloride using a 2.0 mm path-length cell. What is the molar absorption coefficient (ε) of bromine at the wavelength employed?

$[Br_2]$/M	0.001	0.005	0.010	0.050
Transmission/%	81.4	35.6	12.7	3.0×10^{-3}

19.2. In another experiment the same cell was filled with 0.010 M benzene, and the wavelength of the spectrometer changed to 256 nm, where there is a maximum in the absorption. At this wavelength the transmitted intensity was 48% of the incident intensity. What is the molar absorption coefficient of benzene at this wavelength?

19.3. What are the absorbances (optical densities) of the two samples in the last two Problems?

19.4. What would be the percentage of transmission by (a) 0.010 M benzene, (b) a 0.0010 M bromine solution in carbon tetrachloride when the sample cell is (i) 0.10 cm, (ii) 10 cm thick, the wavelengths being the same as above.

19.5. A swimmer enters a gloomier world (in one sense) on diving to greater depths. Given that the mean molar absorption coefficient for sea water in the visible region is $6.2 \times 10^{-5}\ M^{-1}\ cm^{-1}$, calculate the depth at which a diver will experience (a) half the surface intensity of light, (b) one tenth that intensity.

19.6. The absorption bands of many molecules in solution have the widths at half-height of about $5000\ cm^{-1}$. In such cases, estimate (by assuming a triangular line shape) the integrated absorption coefficient for a band in which (a) $\varepsilon_{max} \approx 1 \times 10^4\ M^{-1}\ cm^{-1}$, (b) $\varepsilon_{max} \approx 5 \times 10^2\ M^{-1}\ cm^{-1}$. (The width at half-height means the width of the band where $\varepsilon = \frac{1}{2}\varepsilon_{max}$.)

19.7. What are the oscillator strengths of the transitions in the last Problem?

19.8. A glance at any optical spectrum of a molecule in solution shows that the line shape is only poorly approximated by a triangle. In many cases it is better to assume that it is a Gaussian (proportional to e^{-x^2}) centred on the wave number or frequency corresponding to ε_{max}. Assume such a line shape, and show how to express the integrated absorption coefficient in terms of the values of ε_{max} and $\Delta\tilde{\nu}_{\frac{1}{2}}$, the width at half-height; deduce the relation $\mathscr{A} \approx 1.0645c\varepsilon_{max}\Delta\tilde{\nu}_{\frac{1}{2}}$. Both these quantities may be measured quite readily, and so we have a quick method of determining both $\mathscr{A}$ and the oscillator strength of the band.

19.9. Quick it may be, but how reliable is the assumption of a Gaussian line shape? The absorption spectrum of azoethane ($CH_3CH_2N_2$) between $24\,000\ cm^{-1}$ and $34\,000\ cm^{-1}$ is shown in Fig. 19.26. First estimate $\mathscr{A}$ and f for the band by assuming that its shape is Gaussian. Then integrate the absorption band graphically. This can be done either by counting squares on graph paper (tedious) or by tracing on to paper, cutting out the shape, and weighing (a general procedure for integrating tiresome shapes). Is the transition forbidden or allowed?

19.10. A more complex absorption spectrum, that of biacetyl in the range 250–470 nm, is shown in Fig. 19.2. Estimate the oscillator strengths of both principal transitions, and decide whether they are forbidden or allowed.

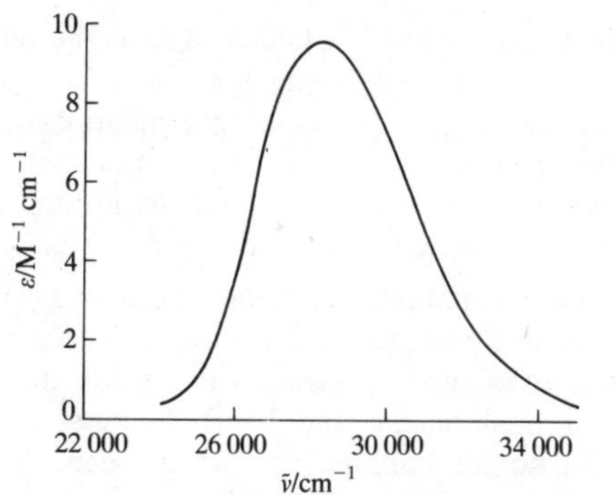

Fig. 19.26. The absorption spectrum of azoethane ($CH_3CH_2N_2$).

You will notice that the spectrum is plotted on a wavelength scale: this means that extra work has to be done before arriving at the f-values of the transitions. There are three approaches. One is to estimate $\Delta\tilde{\nu}_{\frac{1}{2}}$ from the wavelength data and to use the Gaussian line shape approximation (Gaussian in wavenumber, that is). The second is to transfer the spectrum to wavenumber scale, and to proceed by graphical integration. The third is to find an expression for f in terms of wavelength integration, and to use the spectrum directly. Use all three methods. ε_{max} (280 nm peak) $= 11\ M^{-1}\ cm^{-1}$, ε_{max} (430 nm) $= 18\ M^{-1}\ cm^{-1}$.

19.11. A lot of information about the energy levels and wavefunctions of small inorganic molecules can be obtained from their ultraviolet spectra. An example with considerable vibrational structure is shown in Fig. 19.27: it is the spectrum of gaseous SO_2 at 25 °C. Find the oscillator strength of the transition. Is it allowed or forbidden? What electronic states are accessible from the A_1 ground state of this C_{2v} molecule?

19.12. The oscillator strength can be measured from experimental spectra. It can also be calculated from the molecular orbitals of the ground and excited states involved in the transition. We can get some practice in this kind of calculation by dealing with some simple models of molecules. First, consider an electron in a one-dimensional square well of length L. Calculate the oscillator strength for the transitions $n+1 \leftarrow n$ and $n+2 \leftarrow n$. Begin by calculating the transition dipole moment, eqn (19.1.7), in the form $\mu_x = -e\int_0^L \psi_{n'} x \psi_n\, dx$, and then using eqn (19.1.8) to find f. (Only the x-component of the transition moment is nonzero.)

19.13. The FEMO model (Problem 16.28) gives a rough idea of the energies of spectroscopic transitions; now we shall see that it can be used to estimate their intensities (but not very accurately). The molecule **1** (β-carotene) is

1

responsible for the orange colour of carrots. Consider the polyene chain up to the mark as the chromophore. Estimate the excitation energy using $R = 140$ pm, suggest what colour that implies the carrots will appear in white light, and then use the result obtained in the last Problem to estimate the extinction coefficient of the molecule. Estimate the thickness of carrot soup (supposed clear and molar in the above molecule) that gives 50% absorption of incident light.

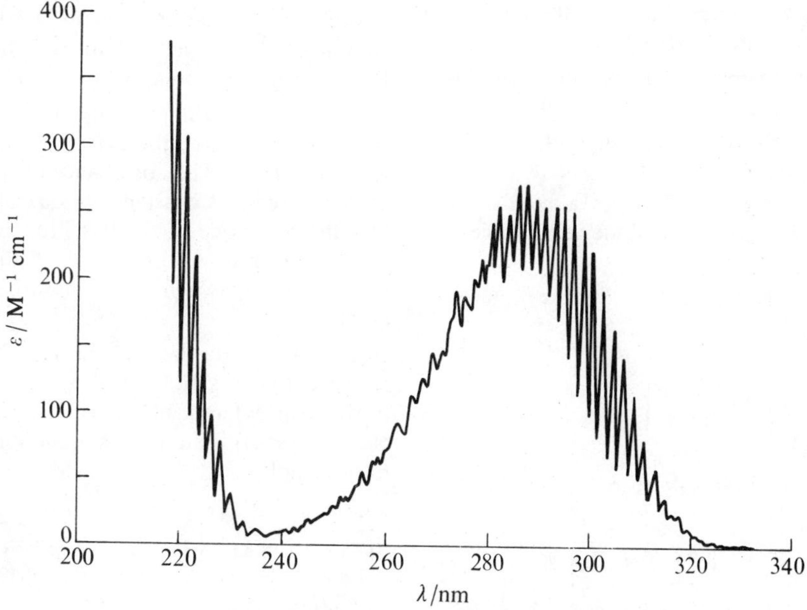

Fig 19.27. The absorption spectrum of SO_2.

19.14. Suppose you were a colour chemist and had been asked to intensify the colour of a dye without changing its nature, and the colour of the dye of interest depended largely on a transition involving a long polyene chain. Would you choose to lengthen or to shorten the chain? Would that shift the apparent colour of the dye towards the red or the blue?

19.15. Consider an electron in an atom to be oscillating with a simple harmonic motion (that was an early model of atomic structure). The wavefunctions for the electron are given in Box 14.1. Show that the oscillator strength for the transition of this electron from its ground vibrational state to its first vibrational state is exactly $\frac{1}{3}$. If the electron were able to vibrate in three dimensions the oscillator strength for all possible transitions out of the ground state would be unity: this is the source of the name *oscillator strength* for the quantity f.

19.16. In the text we saw that the class of transitions that give rise to *charge-transfer spectra* involves the migration of an electron from one region to another in the molecule. An example is the transfer of electrons from ligand orbitals to central metal ion orbitals in some transition metal complexes (like the purple MnO_4^- ion). A rough idea of the oscillator strength, and hence ε_{max}, can be obtained by a simple calculation based on the following model. Consider two hydrogen 1s-orbitals separated by a distance R, but close enough for there to be some overlap. Let the initial state of the molecule be with the electron entirely in one of these orbitals, and the final, charge-transferred, state be with the electron entirely in the other. The transition dipole is $\mu = -e\int \psi_f r \psi_i \, d\tau$, and this can be approximated by $-eRS$, where S is the overlap integral of the two orbitals. Plot the oscillator strength for the transition as a function of R, using the value of $S(R)$ in eqn (16.2.2). At what separation is the intensity greater? Why does the intensity drop to zero both as $R \to 0$ and as $R \to \infty$?

19.17. Use simple group-theoretical arguments to decide which of the following transitions are allowed: (a) the (π^*, π)-transition in ethene, (b) the (π^*, n)-transition in a (C_{2v}) carbonyl group.

19.18. The spectrum in Fig. 19.28 shows a number of features that will be examined in this and the next Problem.

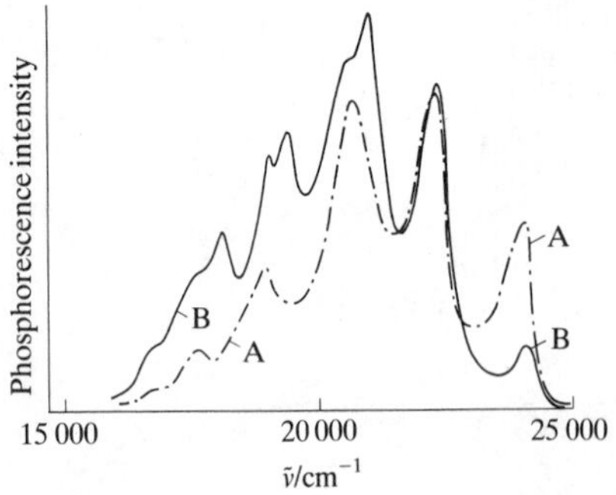

Fig. 19.28. Phosphorescence of naphthalene and benzophenone.

Concentrate on the line marked A. This is the phosphorescence spectrum of benzophenone in solid solution in ethanol at low temperatures during irradiation with 360 nm light. What can be said about the vibrational energy levels of the carbonyl chromophore (a) in its ground electronic state, (b) in its excited electronic state?

19.19. When naphthalene is irradiated with 360 nm light it does not absorb; but the line marked B in the diagram is the phosphorescence spectrum of a solid solution of a mixture of naphthalene and benzophenone, and a component due to the naphthalene can be seen. Account for this observation.

19.20. The fluorescence spectrum of anthracene vapour shows a series of peaks of increasing intensity with individual maxima at 440 nm, 410 nm, 390 nm, and 370 nm, and followed by a sharp cut-off at shorter wavelengths. The absorption spectrum rises sharply from zero to a maximum at 360 nm with a trail of peaks of lessening intensity at 345 nm, 330 nm, and 305 nm. Account for these observations.

19.21. Consider an indicator, such as bromophenol blue, which follows the equilibrium $InH \rightleftharpoons H^+ + In^-$ (HIn might itself be charged) and where the InH and In^- forms absorb in different parts of the spectrum (so that there is a colour change in response to the pH). Suppose that $A(InH)$ and $A(In^-)$ are the absorbances of a sample of the indicator when it is in the acid (InH) and base (In^-) forms. When a mixture of the forms is present let the absorbance at the same wavelength be $A(mix)$. Show that the degree of dissociation α of the indicator is given by $[A(mix) - A(InH)]/[A(In^-) - A(InH)]$. Hence show that the pH of the solution can be expressed in terms of the dissociation constant K_{In} of the indicator and the absorbance $A(mix)$. Sketch the form of the pH dependence of $A(mix)/A(InH)$ in the case of a colourless In^- form in order to appreciate the rapidity with which the colour change occurs. Take $pK_{In} \approx 4$.

19.22. When pyridine is added to a solution of iodine in carbon tetrachloride the 520 nm band of absorption shifts towards 450 nm. The absorbance of the solution at 490 nm, however, remains constant: this is called an *isobestic point*. Use the technique of the last Problem, and especially the relation of α to $A(mix)$, to show that an isobestic point should occur when two absorbing species are in equilibrium.

19.23. The data below are for the molar absorption coefficient ($10^{-3}\,\varepsilon/M^{-1}\,cm^{-1}$) of a solution of p-nitrophenol in water at different pH (A. I. Biggs, *Trans. Faraday Soc.* **50,** 800 (1954)). Find the pK value for the ionization of the nitrophenol.

pH	4	5	6	7	8	9	10
317 nm	8.33	8.33	7.64	4.16	0.42	—	—
407 nm	—	—	1.66	9.16	17.50	18.33	18.33

19.24. A transition of particular importance in the oxygen

molecule gives rise to the *Schumann–Runge band* in the ultraviolet spectrum. The wavenumbers (cm^{-1}) of transitions from the ground state to the vibrational levels of the upper electronic state (which is formally $^3\Sigma_u^-$) are as follows: 50 062.6, 50 725.4, 51 369.0, 51 988.6, 52 579.0, 53 143.4, 53 679.6, 54 177.0, 54 641.8, 55 078.2, 55 460.0, 55 803.1, 56 107.3, 56 360.3, 56 570.6. What is the dissociation energy of the upper electronic state? You should answer this by Birge–Sponer extrapolation.

19.25. The excited electronic state of the O_2 molecule referred to in the last Problem is known to dissociate into one ground state O atom and one excited state atom with an energy 190 kJ mol^{-1} above the ground state; this excited atom is responsible for a great deal of photochemical mischief in the atmosphere. Ground state oxygen dissociates into two ground state atoms. Combine this information with that in the last Problem to find the dissociation energy of ground state oxygen.

19.26. Spin angular momentum is conserved when a molecule breaks up into atoms. What atomic multiplicities are permitted when (a) an oxygen molecule, (b) a nitrogen molecule breaks up into two atoms?

19.27. The photoelectron spectra of N_2 and CO are shown in Fig. 19.29. Ascribe the lines to the ionization processes involved, and classify the orbitals into bonding (or anti-bonding) and non-bonding in the light of the amount of vibrational structure. Analyse the bands near 4 eV in terms of the vibrational spacing of the ions.

19.28. The photoelectron spectrum of NO can be described as follows (D. W. Turner, *Physical Methods in Advanced Inorganic Chemistry* (eds. H. A. O. Hill and P. Day), Wiley–Interscience (1968)). Using He 584 pm (21.21 eV) radiation a single strong peak is observed at 4.69 eV and there is a long series of 24 lines starting at 5.56 eV and ending at 2.2 eV. A shorter series of 6 lines begins at 12.0 eV and ends at 10.7 eV. Account for this spectrum. Data refer to the kinetic energies of the photoelectrons.

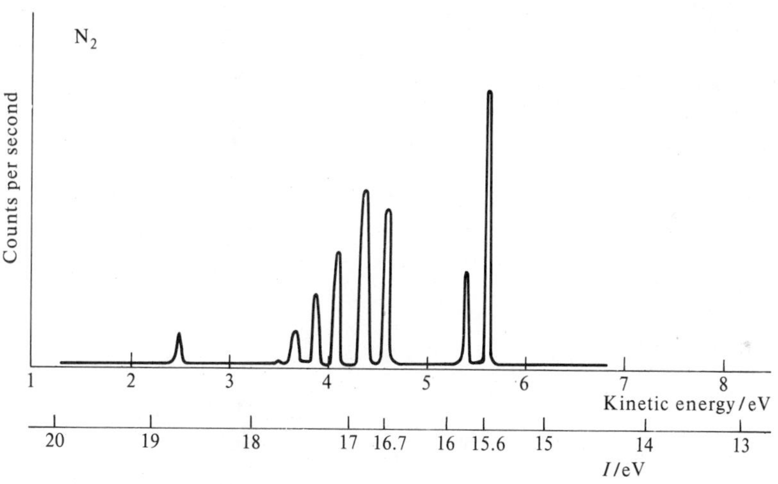

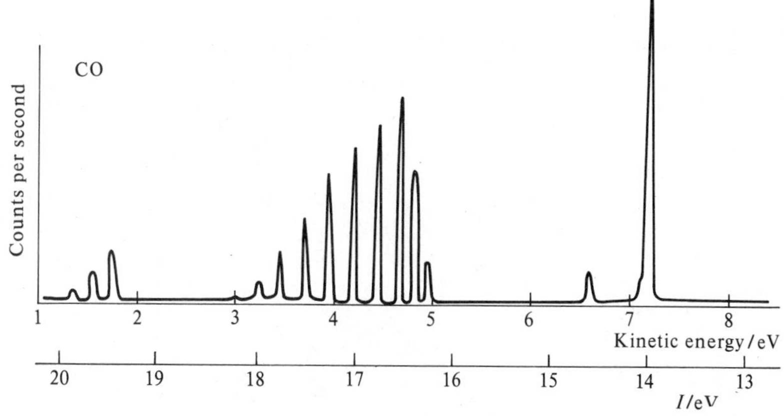

Fig. 19.29. Photoelectron spectra of N_2 and CO.

19.29. The highest energy electrons in the photoelectron spectrum of water (using 21.21 eV He radiation) are at about 9 eV and show a large vibrational spacing of 0.41 eV. The symmetric stretching mode of non-ionized water lies at 3652 cm^{-1}. What conclusions can you draw about the nature of the orbital from which the electron is ejected?

19.30. In the same spectrum of water the band at about 7 eV shows a long vibrational series with an interval of 0.125 eV. The bending mode of water lies at 1596 cm^{-1}. What conclusions can you draw about the bonding characteristics of the orbital occupied by the ejected electron?

20

Determination of molecular structure: resonance techniques

Learning objectives

After careful study of this chapter you should be able to:

(1) Explain the meaning of *resonance*, and write the *resonance condition*, eqn (20.1.3).

(2) Describe the *nuclear magnetic resonance* technique, Section 20.1(a).

(3) Explain the significance of the *shielding constant* and the *chemical shift*, Section 20.1(b).

(4) Explain the origin and effect of *spin–spin coupling*, Section 20.1(c).

(5) Outline the features of *Fourier transform NMR* and explain why a *free induction decay* signal carries spectral information, Section 20.1(d).

(6) State and use the condition for *line broadening*, eqn (20.1.7) and Example 20.2.

(7) Distinguish between the *longitudinal* and the *transverse* relaxation times, Section 20.1(e).

(8) State the resonance condition in *electron spin resonance*, eqn (20.2.4), and explain the significance of the *g-value*, Section 20.2(b).

(9) Explain the origin of *hyperfine structure* and distinguish between dipole–dipole interactions and the *Fermi contact interaction*, Section 20.2(c).

(10) State and use the *McConnell relation* for mapping the spin density, eqn (20.2.7) and Example 20.5.

(11) Describe the *Mössbauer effect* and explain the technique of *Mössbauer spectroscopy*, Sections 20.3 (a and b).

(12) Explain the terms *isomer shift* and *quadrupole splitting*, and indicate the kind of information they provide, Section 20.3(c).

Introduction

When two pendulums are joined by the same slightly flexible support and one is set in motion, the other is forced into oscillation by the motion of the common axle and energy flows backwards and forwards between the two. This process occurs most efficiently when the frequencies of the two oscillators are identical. The condition of strong effective coupling when the frequencies are identical is called *resonance*, and the excitation energy is said to *resonate* between the coupled oscillators.

Resonance is the basis of a number of everyday phenomena, including tuning radios to the weak oscillations of the electromagnetic field caused by a distant transmitter. In this chapter we explore some spectroscopic applications. These have the common characteristic of depending on the matching of a set of energy levels to a source of monochromatic radiation, and observing the strong absorption when resonance occurs.

20.1 Nuclear magnetic resonance

Nuclei with spin possess magnetic moments. A nucleus with spin I may take $2I+1$ different orientations in a magnetic field, the orientations being distinguished by the quantum number m_I, which may take the values $I, I-1, \ldots, -I$. (These conclusions follow from the general discussion of the quantum theory of angular momentum, Chapter 14, especially Box 14.3.) Just as the energy of an *electron* spin in a magnetic field B is confined to the values $g_e\mu_B m_s B$ (this is eqn (15.3.7)), where g_e is the electron g-value and μ_B is the Bohr magneton ($\mu_B = e\hbar/2m_e$), so a nuclear spin in a field B is confined to the energies

$$E_{m_I} = -g_I\mu_N m_I B, \qquad m_I = I, I-1, \ldots, -I, \qquad (20.1.1)$$

where g_I is the *nuclear g-factor* and μ_N is the *nuclear magneton*, $e\hbar/2m_p$. The change in sign of the expression for the energy compared with that for the electron spin reflects the opposite charges of nuclei and electrons. Some experimental values of g_I are given in Table 20.1. Note that the nuclear magneton is about 2000 times smaller than the Bohr magneton, and so nuclear magnetic moments are about that much weaker than the electron spin magnetic moment. This can be understood on the grounds that the angular moments of nuclei and electrons are about the same (when $I=\frac{1}{2}$ they are identical), but the masses of nuclei are much greater; therefore, on a classical picture of spin, the nuclei could be regarded as rotating much more slowly than electrons and therefore as giving rise to a weaker magnetic moment.

Table 20.1. Nuclear spin properties

Nuclide	Natural abundance	Spin I	g-value g_I
^{1}n	—	$\frac{1}{2}$	−3.826
^{1}H	99.98	$\frac{1}{2}$	5.586
^{2}H	0.02	1	0.857
^{13}C	1.11	$\frac{1}{2}$	1.405
^{14}N	99.64	1	0.404

In the case of protons $I=\frac{1}{2}$, and so $m_I=+\frac{1}{2}$ (the α-spin state) or $m_I=-\frac{1}{2}$ (the β-spin state). The energy separation between the two states, Fig. 20.1, is

$$\Delta E = E_\beta - E_\alpha = \{-(-\tfrac{1}{2})g_I\mu_N B\} - \{-(\tfrac{1}{2})g_I\mu_N B\} = g_I\mu_N B. \quad (20.1.2)$$

If the sample is bathed in radiation of frequency ν, the energy separations come into resonance with the radiation when the magnetic field is adjusted to satisfy the *resonance condition*:

$$h\nu = g_I\mu_N B. \qquad (20.1.3)$$

When this condition is satisfied there is strong coupling between the nuclear spins and the radiation, and strong absorption occurs as the nuclear spins make the transition $\beta \leftarrow \alpha$. In its simplest form, *nuclear magnetic resonance* (NMR) is the study of the properties of molecules containing magnetic nuclei by observing the magnetic fields at which they come into resonance with an applied field of definite frequency. When applied to proton spins the technique is sometimes called *proton magnetic resonance* (PMR): in the early days of the technique only protons could be studied, but now a wide variety of nuclei (especially ^{13}C and ^{31}P) are studied routinely.

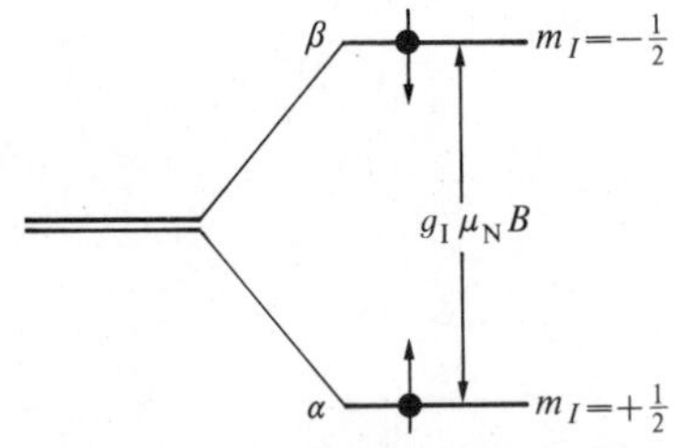

Fig. 20.1. The nuclear spin energy levels of a spin-$\frac{1}{2}$ nucleus (e.g. ^{1}H or ^{13}C) in a magnetic field. Resonance occurs when the energy separation of the levels matches the energy of the photons in the electromagnetic field.

20.1(a) The technique

In a field of 1 T (10 kG) the resonance condition is fulfilled for protons with 42.578 MHz radiation, and so NMR is a *radiofrequency technique*. Early NMR spectrometers used fields of about 1.5 T (*ca.* 60 MHz), but much higher fields are available using superconducting magnets, and modern spectrometers operate at up to about 10 T (100 kG), which corresponds to resonance at over 400 MHz. There are several advantages of working at high fields. One relates to the population differences of the spin states. According to the Boltzmann distribution (Section 0.1(d)), since the β-spin state lies at an energy $g_I\mu_N B$ above the α-spin state, at a temperature T the populations are related by

$$N_\beta/N_\alpha = e^{-g_I\mu_N B/kT},$$

and therefore the difference of populations relative to the total number of nuclei present is given by

$$\frac{N_\alpha - N_\beta}{N_\alpha + N_\beta} = \frac{1 - e^{-g_I\mu_N B/kT}}{1 + e^{-g_I\mu_N B/kT}} = \frac{1 - \{1 - g_I\mu_N B/kT + \ldots\}}{1 + \{1 - g_I\mu_N B/kT + \ldots\}} \approx g_I\mu_N B/2kT. \tag{20.1.4}$$

At 1.5 T and 300 K this evaluates to only 2.6×10^{-6} for protons (and less for other common nuclei as their g-values are smaller), and so there is only a very tiny excess of α-spins over β-spins. Since the radiofrequency field stimulates emission as well as absorption, the *net* absorption is very weak even at resonance, because almost the same numbers of spins are available to emit and to absorb. When a field of 10 T is available the population difference is nearly seven times greater, and the net absorption is correspondingly more intense. (In the following sections we shall see that there are more reasons for working at high fields.)

The simplest NMR spectrometers consist of a magnet and a radiofrequency source and detector. Because the spectrum has components split by very small energies (as described below), the magnetic field must be exceptionally well controlled and homogeneous over the bulk of the sample. One way of ensuring homogeneity is to rotate the sample rapidly so that inhomogeneities are averaged out. In this simple version, the experiment is performed by sweeping the magnetic field over a small range and monitoring the absorption of the monochromatic radiofrequency radiation.

The new generations of spectrometers use very sensitive *Fourier transform techniques*: how these work is best explained after we have seen some NMR spectra. One typical example (for ethanol) is shown in Fig. 20.2. Two features need explanation: the splitting of the spectrum into several blocks, and the detailed structure of the individual blocks.

20.1 (b) The chemical shift

The nuclear magnetic moments interact with the *local* magnetic field. The local field differs from the applied field because the latter induces electronic orbital angular momentum, which gives rise to a small additional magnetic field δB. The magnitude of δB is proportional to the applied field, and it is conventional to express it as $\delta B = -\sigma B$, where σ is the *shielding constant*. The ability of the applied field to induce currents depends on the details of the electronic structure near the proton of interest, and protons in different chemical groups have different shielding constants. Since the total local field is $B + \delta B = (1 - \sigma)B$, the resonance condition for the nucleus,

$$h\nu = g_I\mu_N B_{\text{local}} = g_I\mu_N(1 - \sigma)B, \tag{20.1.5}$$

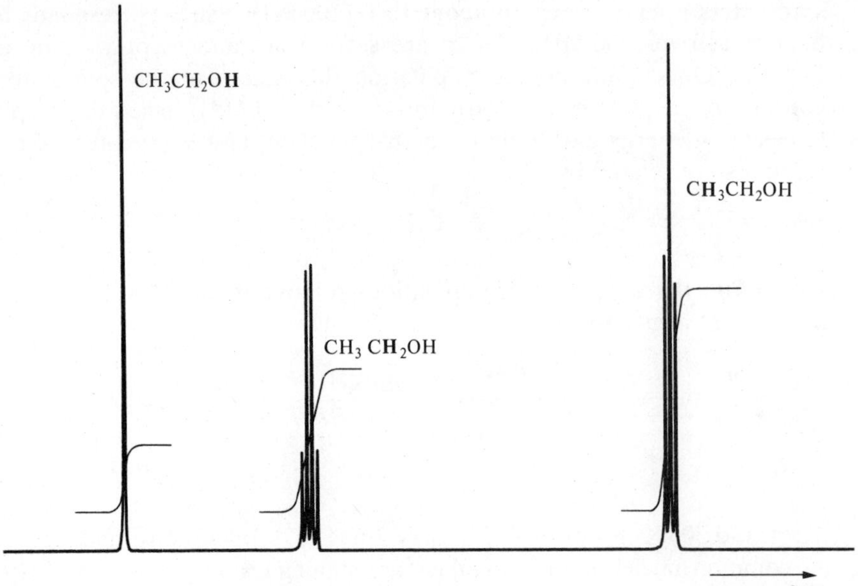

Fig. 20.2. An NMR spectrum of ethanol. The bold letters denote the protons giving rise to the resonance peak, and the step-like curve is the integrated signal.

is satisfied at different values of B for nuclei in different environments. The quantity σB is called the *chemical shift* of the nuclei.

The chemical shift explains the general features of the spectrum shown in Fig. 20.2. The methyl protons form one group of magnetically equivalent nuclei, and come into resonance at a field determined by their shielding constant. The two methylene protons are in a different part of the molecule, have a different shielding constant, and therefore come into resonance at a different applied field. Finally, the hydroxyl proton is in another environment, and has its characteristic shielding constant, and therefore resonates at yet another applied magnetic field. Note that we can use the relative intensities of the absorptions (the areas under the absorption lines) to distinguish which group of lines corresponds to which chemical group: the group intensities are in the ratio 3:2:1 because there are three methyl protons, two methylene protons, and one hydroxyl proton per molecule. Modern spectrometers are equipped to *integrate* the absorption (evaluate the areas under the absorption curve) automatically, Fig. 20.2.

The shielding constant is independent of the applied field but the chemical shift increases with applied field. This is another reason for using high magnetic fields in ^{1}H-NMR: groups of lines with different shielding constants are more widely separated (have bigger chemical shifts) when the applied field is strong. Chemical shifts are normally expressed on the *δ-scale*. This is defined as follows. First, a standard is selected: tetramethylsilane ($Si(CH_3)_4$, commonly referred to as TMS), which bristles with magnetically equivalent protons and which dissolves without reaction in many samples, is often used. If the chemical shift (expressed as a frequency) of a resonance line from the resonance line of TMS is $\delta\nu$, then $\delta = (\delta\nu/\nu_0) \times 10^6$, where ν_0 is the operating frequency of the spectrometer. $\delta\nu$ (and hence δ) is taken as *positive* when the resonance occurs at lower magnetic fields than TMS. Sometimes the *τ-scale* is used: this is defined as $\tau = 10 - \delta$, and most values are positive. Typical chemical shifts on the δ-scale are shown in Fig. 20.3.

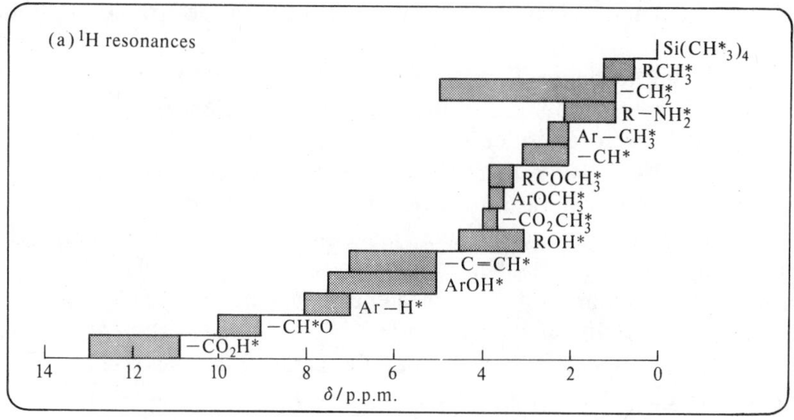

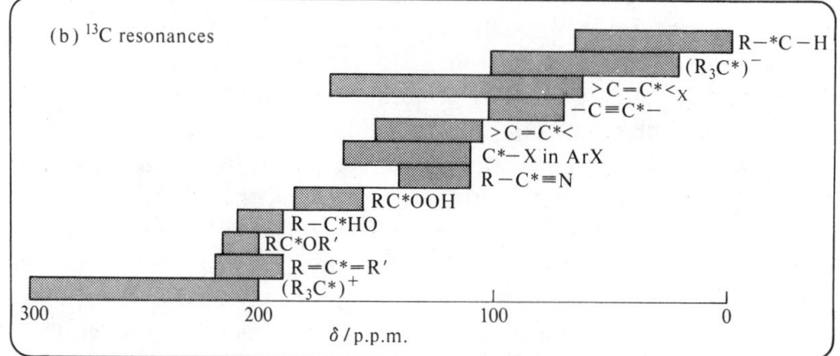

Fig. 20.3. The range of typical chemical shifts for (a) ^{1}H resonances and (b) ^{13}C resonances.

20.1 (c) The fine structure

The splitting of groups into individual lines, as shown in Fig. 20.2, is called the *fine structure* of the spectrum. It arises from the magnetic interactions between the nuclei in the molecule, because each magnetic nucleus (i.e. each nucleus with non-zero spin) gives rise to a magnetic field which contributes to the local field experienced by the resonating nucleus.

A nucleus with spin projection m_I gives rise to a magnetic field B_{nuclear} at a distance R, where

$$B_{\text{nuclear}} = -g_I \mu_N (\mu_0/4\pi)(1/R^3)(1-3\cos^2\theta)m_I. \qquad (20.1.6)$$

The angle θ is defined in Fig. 20.4; μ_0 is the *vacuum permeability*, a fundamental constant. Its value is given inside the front cover; in this connection the more convenient form is $\mu_0 = 4\pi \times 10^{-7}\,\mathrm{T^2\,J^{-1}\,m^3}$ because μ_N is expressed in $\mathrm{J\,T^{-1}}$ and so the product $\mu_N\mu_0/R^3$ has the units $(\mathrm{T^2\,J^{-1}\,m^3}) \times (\mathrm{J\,T^{-1}})/\mathrm{m^3} = \mathrm{T}$. The magnitude of this nuclear field is about 0.1 mT when $R = 0.3$ nm, and is of the order of magnitude of the splitting observed in *solid* samples. However, in a liquid the angle θ sweeps over all values as the molecule tumbles, and $1-3\cos^2\theta$ averages to zero.‡ This means that, although the *direct dipole–dipole interaction* makes an important contribution to the spectra of solid samples and to the spectra of molecules that tumble only slowly in solution, such as biological and synthetic macromolecules, and can be used to determine internuclear

Molecule
Proton B
Proton A
Proton B rotates over this spherical surface

Fig. 20.4. The dipole–dipole interaction between two nuclei depends on the angle θ, and when the molecule rotates, θ sweeps through all values. As a result, $1-3\cos^2\theta$ averages to zero, and the dipole–dipole interaction averages to zero.

‡ The volume element in polar coordinates is proportional to $\sin\theta\,d\theta$, and θ ranges from 0 to π. Therefore the average value of B_{nuclear} for a tumbling molecule is proportional to $\int_0^\pi (1-3\cos^2\theta)\sin\theta\,d\theta = 0$.

distances, it cannot account for the fine structure of rapidly tumbling molecules.

The principal mechanism for *spin–spin coupling* in molecules in solution is a magnetic interaction transmitted through the electrons of the bonds. The simplest case to consider is that of the two spin-$\frac{1}{2}$ nuclei joined by a bond, such as $^{13}C—^{1}H$. If the ^{13}C nucleus is α, then the electron spin that has a favourable magnetic interaction with it will tend to be found in its vicinity. The other electron, which must have the opposite spin, will be found mainly near the ^{1}H. The latter electron's spin interacts magnetically with the proton's spin, and the lower energy arrangement is for the proton to be β, Fig. 20.5. In other words, the antiparallel arrangement of nuclear spins lies lower in energy than the parallel arrangement as a result of their magnetic coupling with the bond electrons. The energy separation of the two arrangements is normally denoted J, which is called the *spin–spin coupling constant*, and it is usually expressed as a frequency (in Hz).

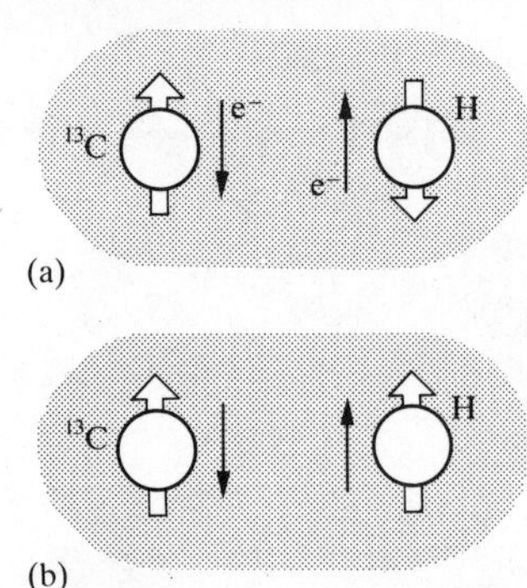

Fig. 20.5. The *polarization mechanism* for spin–spin coupling. The arrangement in (a) has a slightly lower energy than that in (b), and so there is an effective coupling between the spins of magnitude J. In this case J is positive.

The coupling of two protons joined to the same carbon atom (as in the CH_2 group of ethanol) occurs by a similar mechanism, but the nucleus of the carbon atom plays no direct role. In this case, Fig. 20.6, one proton polarizes the electrons in its bond, and the electrons in the neighbouring bond respond to the polarization as shown in the illustration. This is because the more favourable arrangement of two electrons on the same atom is with their spins parallel (Hund's rule, Section 15.2). The second proton responds to the polarized electron spins by its magnetic interaction with the nearer one, and the lower energy arrangement is now with its spin parallel to the first proton. This is opposite to the case when the two spins were joined directly, and so J is now negative. A similar mechanism accounts for the coupling of magnetic nuclei separated by several bonds.

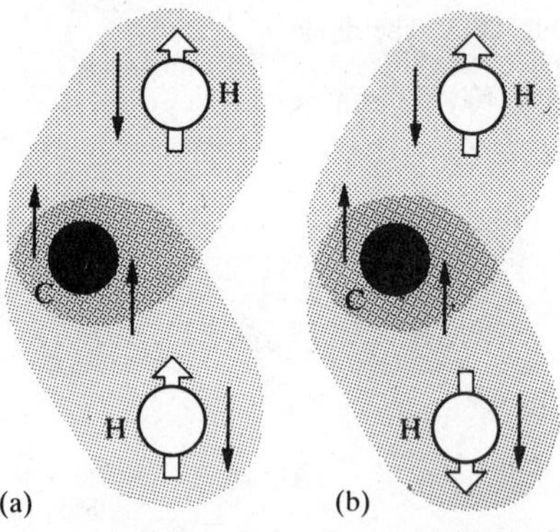

Fig. 20.6. The arrangement in (a) has slightly lower energy than that in (b) and so there is an effective coupling between the proton spins; in this case the spin–spin coupling constant is negative.

Although the difference of energy arising from the magnetic spin–spin interactions accounts for the fine structure of NMR spectra, there are various complications. One is that *magnetically equivalent nuclei do not show fine structure even though their spins are coupled*. This is because the radiofrequency field reorientates *groups* of equivalent nuclear spins collectively, and does not change the relative orientations of the spins within the group. Hence, an isolated CH_3 group gives a single, unsplit line because all the allowed transitions of the group of three protons occur without change of their relative orientations.

In ethanol the methyl group is not isolated, because there is a CH_2 group next to it. During a transition the spins of the methyl protons are reorientated as a group relative to the CH_2 protons, and so the transitions of the methyl group protons show a splitting due to their spin–spin interactions with the CH_2 protons. The single methyl resonance is split into two by one of the CH_2 protons, and each of those resonances is split into two again by the second CH_2 proton. Since the two spin–spin interactions are the same, the four lines appear as a 1:2:1 triplet (because the two central lines coincide), Fig. 20.7(a). This accounts for the form of the methyl resonance shown in Fig. 20.2.

Now consider the methylene resonance. If the two methylene protons were isolated they would resonate at a single position. The fine structure of the CH_2 absorption, Fig. 20.2, is due to the spin–spin interaction of the protons with the three neighbouring methyl protons. Three equivalent protons give rise to four lines in the intensity ratio 1:3:3:1, Fig. 20.8(b), and this accounts for the observed fine structure of the methylene resonance in ethanol. In general, N equivalent protons split the resonance of a nearby group into $N+1$ lines with an intensity distribution given by Pascal's triangle, as shown in the margin (the subsequent rows are formed by adding together the two neighbouring numbers in the line above). The 1:3:3:1 intensity ratio caused by a CH_3 group, for example, arises from coupling to its $(\alpha\alpha\alpha)$; $(\alpha\alpha\beta)$, $(\alpha\beta\alpha)$, $(\beta\alpha\alpha)$; $(\alpha\beta\beta)$, $(\beta\alpha\beta)$, $(\beta\beta\alpha)$; $(\beta\beta\beta)$ spin states.

The hydroxyl proton in ethanol splits the resonance of the other groups into doublets, and is itself split into a 1:2:1 triplet by the CH_2 and each line of that triplet is split into a 1:3:3:1 quartet by the CH_3. In many cases the splitting due to a hydroxyl proton cannot be detected, and is not shown in Fig. 20.2, for reasons we explain in the following section.

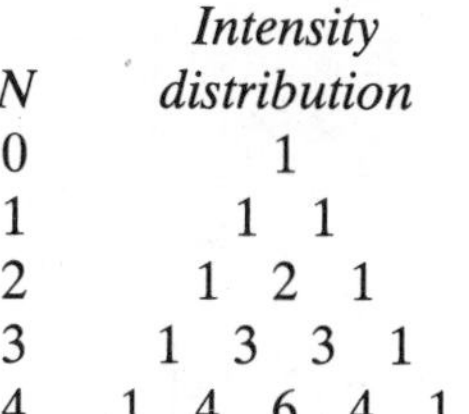

N	*Intensity distribution*
0	1
1	1 1
2	1 2 1
3	1 3 3 1
4	1 4 6 4 1

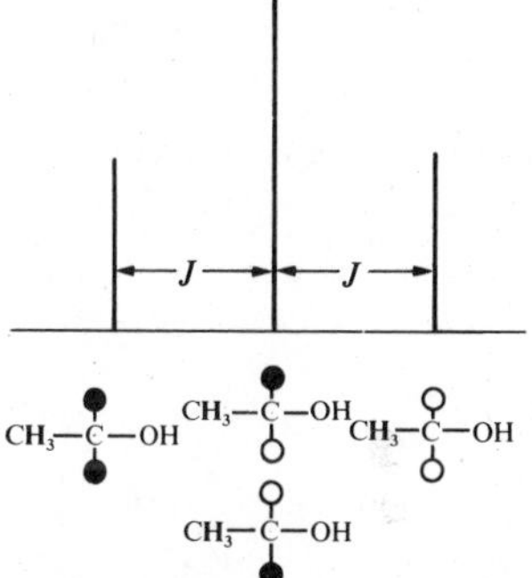

Fig. 20.7. The origin of the 1:2:1 triplet in the CH_3 resonance is its spin–spin coupling to the two CH_2 protons. The latter may have spin arrangements $(\alpha\alpha)$; $(\alpha\beta)$, $(\beta\alpha)$; $(\beta\beta)$, the middle two being responsible for coincident resonances. Full circles indicate α-spins, open circles indicate β-spins. The resonant protons are bold.

Example 20.1

Suggest an interpretation of the NMR spectrum shown in Fig. 20.9.

• *Method*. Look for groups with characteristic chemical shifts; refer to Fig. 20.3. Account for the fine structure as was done for ethanol. Report δ-values and J-values, and identify the sample.

• *Answer*. The resonance at $\delta = 3.4$ corresponds to $>CH_2$ in an ether; that at $\delta = 1.2$ corresponds to CH_3 in CH_3CH_2. The fine structure of the $>CH_2$ resonance (a 1:3:3:1 quartet) is characteristic of splitting caused by CH_3; the fine structure of the CH_3 resonance is characteristic of splitting caused by CH_2. The spin–spin splitting constant is $J = \pm 60$ Hz (the same for each group). The compound is probably $(CH_3CH_2)_2O$.

• *Comment*. When the chemical shifts and the spin–spin coupling constants are similar, the spectra are very much more complicated and cannot be analysed so simply.

• *Exercise*. What changes in the spectrum would be observed on recording it at 300 MHz?

[Groups of lines five times further apart; no change in spin–spin splitting]

NMR spectra are usually much more complex than this simple analysis suggests. We have described the extreme case in which the chemical shifts are much greater than the spin–spin couplings, and in which it is simple to say which nuclei are equivalent and which are not, and to think of complete groups of nuclei reorientating relative to each other. When the chemical shifts are comparable to the spin–spin couplings these simple pictures fail, and the absorption lines no longer fall into definite groups. Then it is essential to use a computer to evaluate all the energy levels and transition probabilities and to calculate the *second-order spectra*. This introduces another advantage of working at high fields: the chemical shifts increase with field, but the spin–spin couplings are independent of it. Therefore, at high fields a spectrum may be simplified because groups of nuclei are separated more completely.

Another technique for simplifying the appearance of a spectrum is *proton decoupling*. For example, if the methyl protons of ethanol are irradiated with a second, strong resonant radiofrequency source they undergo rapid spin reorientations and the nearby methylene protons sense an average orientation: as a result the methylene quartet collapses to a single line.

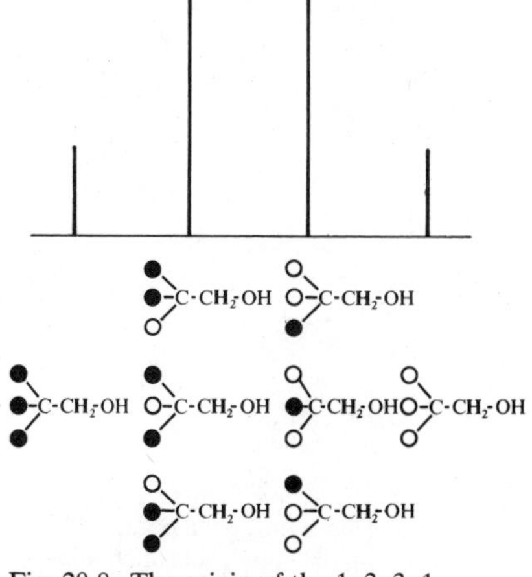

Fig. 20.8. The origin of the 1:3:3:1 CH_2 quartet is due to its spin–spin coupling with the eight possible spin arrangements of the CH_3 protons. The same code is used as in Fig. 20.7.

20.1 (d) Fourier transform spectroscopy

One of the main problems with NMR is the low intensity of the transitions on account of the extremely small population differences between the

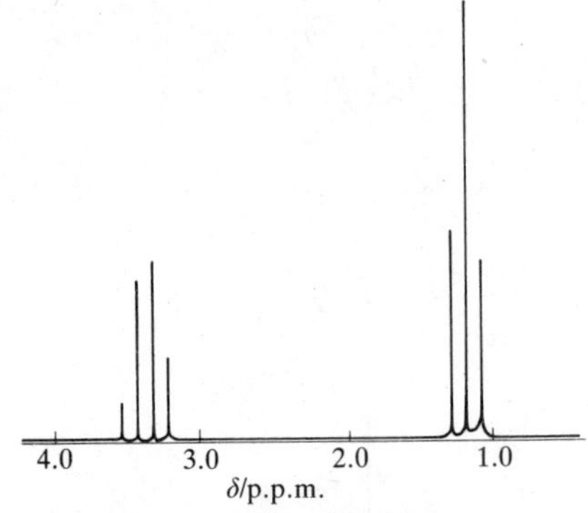

Fig. 20.9. The NMR spectrum referred to in *Example* 20.1. The field increases to the right.

nuclear spin states. This is particularly severe in the case of ^{13}C, a nucleus of prime interest in organic and biochemical studies. The sensitivity problem is worsened by the low natural abundance of the isotope (1.1 per cent) and its small nuclear magnetic moment ($0.702\mu_N$, only one-fourth that of the proton) which results not only in a population difference smaller than protons by a factor of four but also in a reduction of the strength of the coupling between the nuclei and the radiofrequency field by a factor of about 4^2.

The modern version of NMR relies on computer-based acquisition and transformation techniques. The essential feature of *Fourier transform NMR* (FT-NMR) is the observation of the time evolution of the nuclear spin states of the sample from a specially prepared initial state, and then the transformation of this evolution into information about the energy levels present.

The initial state is prepared by exposing the sample to a brief (1 μs) burst of high power (1 kW) radiofrequency radiation. This disturbs the equilibrium populations of the spin states. After the irradiation stops, the sample emits its excess energy as the populations return towards equilibrium. The resulting radiation consists of components at all the allowed transition frequencies, and is detected as a decaying, oscillating signal called the *free induction decay*, the oscillations arising from the beats between all the contributing frequencies, Fig. 20.10. All the spectral information is contained in the signal (just as dropping a piano generates a signal that contains complete information about its frequency spectrum), and it can be analysed by finding the superposition of frequencies that gives the observed shape. This analysis step is the Fourier transformation that gives the technique its name, and it is a standard mathematical procedure that can be performed on a computer (which is normally an integral part of the spectrometer). The transform of Fig. 20.10 is in fact the proton spectrum of ethanol given in Fig. 20.2.

When ^{13}C is the nucleus being studied it is not normally necessary to take into account ^{13}C–^{13}C spin–spin coupling because the low natural abundance of the isotope makes it unlikely that two such nuclei will occur in the same molecule (unless it has been enriched or the molecule is large). If the molecule contains many protons the spectrum (and the time evolution) may be excessively complex. In order to avoid this difficulty ^{13}C FT-NMR spectra

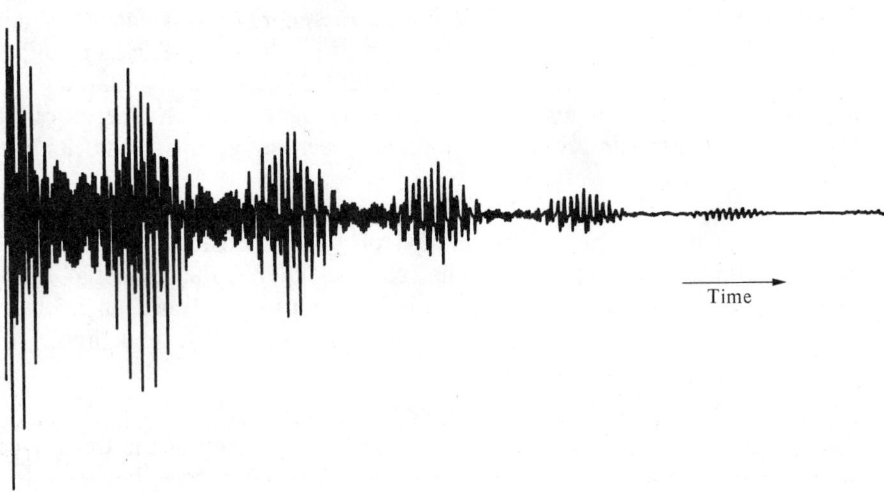

Fig. 20.10. A *free induction decay* signal. This is the signal obtained by saturating the spin levels of ethanol, and then monitoring their return to equilibrium. Its Fourier transform, that is, its analysis in terms of harmonic waves, is the spectrum in Fig. 20.2.

are normally taken with the protons decoupled either totally or selectively in groups. The proton decoupling has the additional advantage of enhancing sensitivity, because the intensity is concentrated into a single transition instead of being spread over several. A disadvantage, however, is that the signal strength is no longer proportional to the number of nuclei, and so signal integration is unhelpful.

20.1 (e) Linewidths and rate processes

The linewidths of NMR spectra, in common with other spectroscopic techniques, provide information about the rates of processes relating to the molecules in the sample. As an example, consider a molecule that can jump between conformations, such as the boat–chair inversion of a substituted cyclohexane, Fig. 20.11. In one conformation the substituent is axial, in another it is equatorial. The chemical shifts of the axial and equatorial ring protons are different, and so they are in effect jumping between two magnetic environments. When the inversion rate is slow the spectrum shows two sets of lines, one from molecules with an axial substituent, and one from molecules with an equatorial substituent. When the inversion is fast, the spectrum shows a single line at the mean of the two chemical shifts. At intermediate inversion rates the line is very broad. This broadening occurs when the lifetime of a conformation, τ_J, is so short that the lifetime broadening, eqn (18.1.6), is comparable to the difference of chemical shifts, $\delta\nu$; that is, *the collapse of structure occurs when*

$$\tau_J < 1/2\pi\delta\nu. \qquad (20.1.7)$$

For example, if the chemical shifts differ by 100 Hz, the spectrum collapses into a single line when the conformation lifetime is less than about 2 ms.

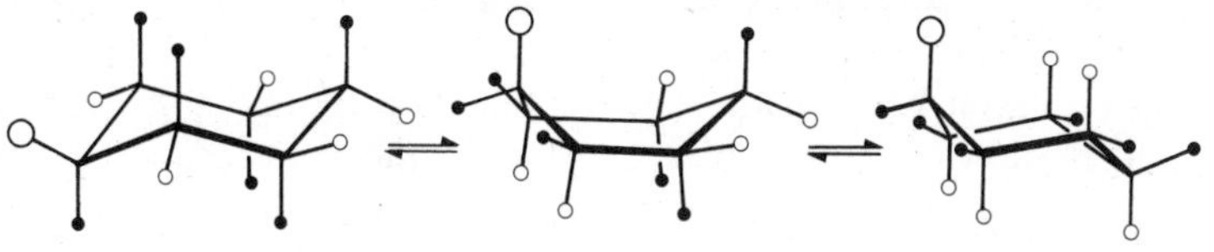

Fig. 20.11. When a substituted cyclohexane molecule undergoes inversion, the axial and equatorial protons are interchanged. This means that they are effectively jumping between magnetically distinct environments.

Example 20.2

The NO group in *N,N*-dimethylnitrosamine, $(CH_3)_2N—NO$, rotates and, as a result, the magnetic environments of the two CH_3 groups are interchanged. In a 60 MHz spectrometer the two CH_3 resonances are separated by 39 Hz. At what rate of interconversion will the resonance collapse to a single line?

• *Method*. Use eqn (20.1.7) for the average lifetimes of the tautomers. The rate of interconversion is the inverse of the average lifetime.

• *Answer*. With $\nu = 39$ Hz, eqn (20.1.7) gives

$$\tau_J < 1/2\pi(29 \text{ Hz}) = 4.1 \text{ ms}.$$

Therefore, the signal will collapse to a single line when the interconversion rate exceeds 250 s^{-1}.

• *Comment*. The dependence of the rate of collapse on the temperature is used to determine the energy barrier to the interconversion.

• *Exercise*. What would the minimum lifetime need to be in a 300 MHz spectrometer for a single line to be observed? [0.8 ms]

A similar explanation accounts for the loss of structure in solvents able to exchange protons with the sample. For example, alcohol hydroxyl protons are able to exchange with water protons. When this *chemical exchange* occurs a molecule ROH with an α-spin proton (we write this $ROH^{(\alpha)}$) rapidly converts to $ROH^{(\beta)}$ and then to $ROH^{(\alpha)}$ because the protons provided by the solvent molecules in successive exchanges have random spin orientations. Therefore, instead of seeing a spectrum composed of contributions from both $ROH^{(\alpha)}$ and $ROH^{(\beta)}$ molecules (that is, a spectrum showing the doublet structure due to the OH proton) all that is seen is a single, unsplit line at the mean position, as in Fig. 20.2. The effect is observed when the lifetime of a molecule due to chemical exchange, τ_{exchange}, is so short that the lifetime broadening is greater than the doublet splitting ($\tau_{\text{exchange}} < 1/2\pi\delta\nu$). Since this splitting is often very small (about 1 Hz), a proton must remain attached to the same molecule for longer than about $\tau_{\text{exchange}} \approx 1/2\pi \text{ Hz} \approx 0.1 \text{ s}$ if the splitting is to be observable. In water the exchange rate is much faster than that, and so alcohols show no splitting from the hydroxyl protons. In very dry alcohol the exchange rate may be slow enough for the splitting to be detected.

The importance of chemical exchange effects is that they may be used to study rapid chemical processes, and to see how their rates vary with the conditions.

20.1 (f) Spin relaxation

When nuclei absorb radiation they make transitions to their upper spin states. This tends to equalize the populations and therefore, if it continues, to reduce the net absorption intensity towards zero. The fact that resonance spectra do not vanish as they are observed implies that there must be a mechanism for returning the nuclei non-radiatively to their lower spin states. These mechanisms are called *relaxation processes*. If they are fast enough, the original population is maintained even though the sample is absorbing radiation because the energy absorbed by the spins is dissipated as thermal motion of the sample. Only when the radiofrequency power is increased to the point that the relaxation processes are unable to cope with the influx of energy does the population difference disappear: this condition is called *saturation* of the transition.

Relaxation increases the widths of spectral lines because the lifetimes of the spin states are reduced. This is the lifetime broadening effect discussed in Section 18.1. The broadening of the lines can be analysed into two contributions, one being *longitudinal relaxation* (or *spin-lattice relaxation*) and the other *transverse relaxation*.

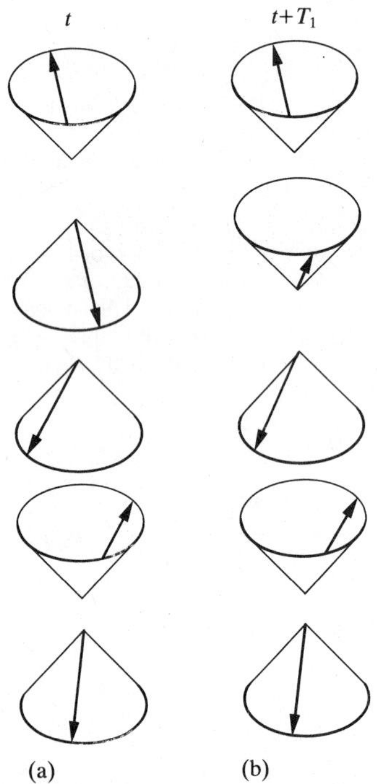

Fig. 20.12. In *longitudinal relaxation* the spins relax back towards their thermal equilibrium populations. In (a) we see the precessional cones representing spin-$\frac{1}{2}$ angular momenta, and they do not have their thermal equilibrium populations (there are more β-spins than α-spins). The time constant for return to the Boltzmann distribution in (b) is T_1.

Longitudinal relaxation is a true energy relaxation process. Nuclei in their upper spin states (e.g. β-protons) relax back into their lower spin states, and transfer energy to the surroundings, Fig. 20.12. The total component of spin along the direction of the magnetic field decreases towards its very small equilibrium value as β-spins revert to α-spins, which is the source of the name *longitudinal* (i.e. along the field). The reversion occurs exponentially with a time constant T_1, which is called the *spin-lattice* (or *longitudinal*) *relaxation time*. It occurs because the local magnetic fields arising from the motion of the molecule and its neighbours fluctuate, and the fluctuating magnetic field induces the transition $\beta \rightarrow \alpha$.

Transverse relaxation can be understood in terms of the relaxation of the relative *phases* of the spins in the sample. This is illustrated in Fig. 20.13, which represents a sample in which at some instant all the spins have the same orientation in the plane perpendicular (i.e. transverse) to the field direction. If the spins are in exactly the same magnetic environment, they precess at the same rate, and so retain their relative orientations. However, if they are all in slightly different local magnetic fields, the magnetic moments precess at slightly different rates, and in due course the orientations become random. Random relative phases is the equilibrium condition, and the phases relax exponentially towards it with a time constant T_2'. The fact that $\beta \rightarrow \alpha$ transitions are also occurring is an additional mechanism of transverse relaxation, and so we can anticipate that the *total* rate of transverse relaxation should depend on T_1 as well as T_2'. The overall time constant for transverse relaxation is T_2, which is called the *transverse relaxation time*. Detailed calculation shows that for a single spin-$\frac{1}{2}$ nucleus

$$1/T_2 = 1/T_2' + 1/2T_1, \qquad (20.1.8)$$

but more complicated relations hold in general.

In the absence of saturation the line-width is determined solely by T_2; it is therefore also called the *linewidth parameter*. From the general account of lifetime broadening (Section 18.1) we can estimate the width of the lines from the energy spread $\delta E = \hbar/T_2$ as being $\delta\nu = 1/2\pi T_2$. Typical values of T_2 in proton NMR are of the order of seconds, and so line widths of around 0.1 Hz can be anticipated, in broad agreement with observation.

The measurement of relaxation times (in principle by measuring the widths of individual spectral lines, but also by more sophisticated *spin-echo* spectroscopic techniques), gives valuable information about the motion, such as the rotation, of molecules in solution, and is used to determine the shapes and size of molecules, such as proteins. This application is described in Section 25.1(k).

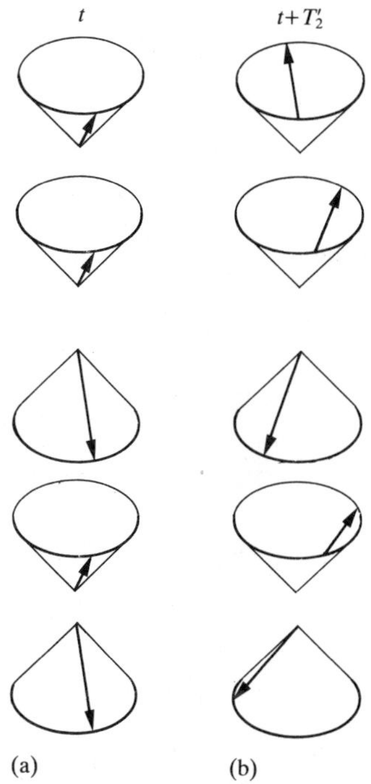

Fig. 20.13. The time it takes for the *phases* of the spins to become randomized (another condition for equilibrium) and to change from the orderly arrangement shown in (a) to the disorderly arrangement in (b), is the *transverse relaxation time*, T_2.

20.2 Electron spin resonance

The energy levels of an electron spin in a magnetic field B, Fig. 20.14, are given by

$$E_{m_s} = g_e\mu_B m_s B; \qquad m_s = \pm\tfrac{1}{2}. \qquad (20.2.1)$$

This shows that the energy of an α-electron ($m_s = +\frac{1}{2}$) increases and the energy of a β-electron ($m_s = -\frac{1}{2}$) decreases as the field is increased, and that the separation of the levels is

$$\Delta E = E_\alpha - E_\beta = g_e\mu_B B. \qquad (20.2.2)$$

The resonance condition is therefore

$$h\nu = g_e\mu_B B, \qquad (20.2.3)$$

and when this is fulfilled strong absorption of the radiation occurs. *Electron spin resonance* (ESR), or *electron paramagnetic resonance* (EPR), is the study of molecules containing unpaired electrons by observing the magnetic fields at which they come into resonance with monochromatic radiation.

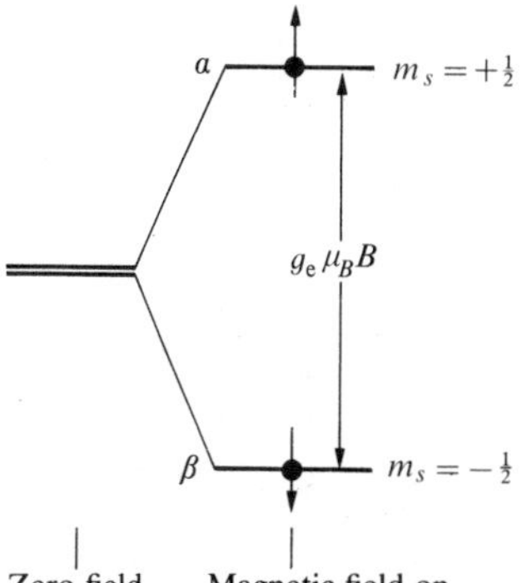

Fig. 20.14. Electron spin levels in a magnetic field. Note that the β-state is the lower. Resonance is achieved when the frequency of the incident radiation matches the frequency corresponding to the energy separation.

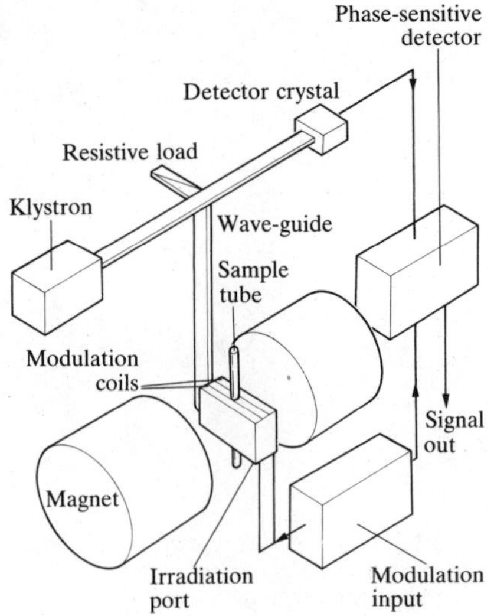

Fig. 20.15. The layout of an ESR spectrometer. A typical magnetic field is 0.3 T, which requires 9 GHz (3 cm) microwaves for resonance.

20.2 (a) The technique

Magnetic fields of about 0.3 T correspond to resonance with an electromagnetic field of frequency 10 GHz (10^{10} Hz) and wavelength 3 cm. This is the magnetic field used in most commercial ESR spectrometers; 3 cm radiation falls in the *X-band* microwave region.

The layout of the spectrometer is shown in Fig. 20.15. It consists of a microwave source (a klystron), a cavity in which the sample is inserted in a glass or quartz container, a microwave detector, and an electromagnet with a field that can be varied in the region of 0.3 T. The ESR spectrum is obtained by monitoring the microwave absorption as the field is changed, and a typical spectrum (of the benzene radical anion, $C_6H_6^-$) is shown in Fig. 20.16. The peculiar appearance of the spectrum is due to the detection technique: the *first derivative* of the absorption signal is usually displayed, Fig. 20.17.

The sample molecules must possess unpaired spins, and so ESR is less widely applicable than NMR. It is used to study free radicals formed during chemical reactions or by radiation, many transition metal complexes, and molecules in triplet states. It is insensitive to normal, spin-paired molecules. The sample may be a gas, a liquid, or a solid, but the free rotation of molecules in the gas phase gives rise to complications.

20.2 (b) The *g*-value

Just as in NMR, the spin magnetic moment interacts with the *local* magnetic field, and so the resonance condition should be written

$$h\nu = g_e\mu_B B_{\text{local}} = g_e\mu_B(1-\sigma)B.$$

However, in ESR it is conventional to write $g = (1-\sigma)g_e$, where g is the *g-factor* of the radical or complex. Then the resonance condition is

$$h\nu = g\mu_B B. \tag{20.2.4}$$

Example 20.3

The centre of an ESR spectrum of the methyl radical occurred at 329.4 mT in a spectrometer operating at 9.233 GHz. What is the g-value?

• *Method*. Use eqn (20.2.4) in the form $g = h\nu/\mu_B B$.

• *Answer*. A convenient conversion factor is

$$h/\mu_B = 7.144\,48\times10^{-11}\ \text{T s} = 71.4448\ \text{mT/GHz}.$$

Therefore, from the data,

$$g = (71.44\ \text{mT/GHz})\times(9.233\ \text{GHz}/329.4\ \text{mT}) = 2.0026.$$

• *Comment*. Many organic radicals have g-values close to 2.0026; inorganic radicals have g-values typically in the range 1.9–2.1; transition metal complexes have g-values in a wider range (e.g. 0–4).

• *Exercise*. At what field would the methyl radical come into resonance in a spectrometer operating at 9.468 GHz? [337.8 mT]

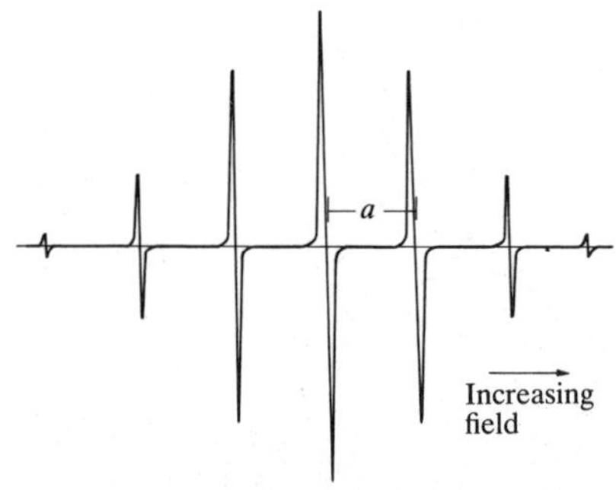

Fig. 20.16. The ESR spectrum of the benzene radical anion, $C_6H_6^-$, in fluid solution. a is the *hyperfine splitting* of the spectrum; the centre of the spectrum is determined by the *g-value* of the radical.

The deviation of g from $g_e = 2.0023$ depends on the ability of the applied field to induce local electron currents in the radical, and therefore its value gives some information about electronic structure. However, since g-values differ very little from g_e in many radicals (e.g. 2.002 26 for H, 1.999 for NO_2, 2.01 for ClO_2), its main use in chemical applications is to aid the identification of the species present in a sample.

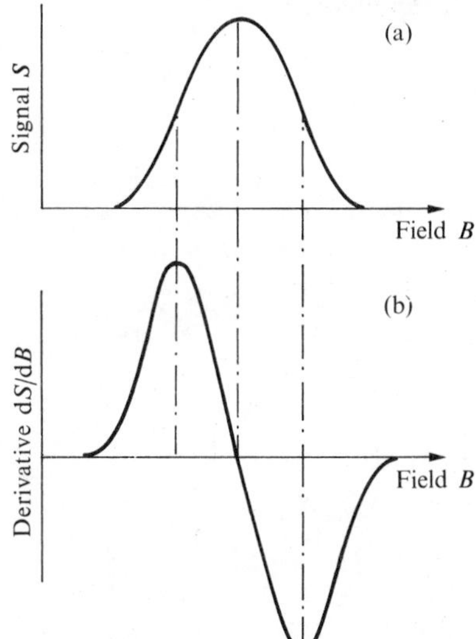

Fig. 20.17. When *phase-sensitive detection* is used, the signal is the first-derivative of the absorption. (a) The absorption, (b) the signal, the slope of the absorption signal at each point.

20.2 (c) Hyperfine structure

The most important feature of ESR spectra is their *hyperfine structure*, the splitting of individual resonance lines into components. The source of the splitting is the magnetic field arising from nuclei with spin.

Consider the effect on the ESR spectrum of a single hydrogen atom in a radical. The proton spin is a source of magnetic field, and depending on its orientation it adds to or subtracts from the applied field. The total local field is therefore

$$B_{\text{local}} = B + am_I; \qquad m_I = \pm\tfrac{1}{2}, \tag{20.2.5}$$

where a is the *hyperfine coupling constant*. Half the radicals in a sample have $m_I = +\frac{1}{2}$, and so half resonate when the applied field satisfies the condition

$$h\nu = g\mu_B(B + \tfrac{1}{2}a), \quad \text{or} \quad B = (h\nu/g\mu_B) - \tfrac{1}{2}a, \tag{20.2.6a}$$

and the other half (which have $m_I = -\frac{1}{2}$) resonate when

$$h\nu = g\mu_B(B - \tfrac{1}{2}a), \quad \text{or} \quad B = (h\nu/g\mu_B) + \tfrac{1}{2}a. \tag{20.2.6b}$$

Therefore, instead of a single line, the spectrum shows two lines of half the original intensity separated by a and centred on the field determined by g, Fig. 20.18.

If the radical contains a nitrogen atom ($I = 1$) the spectrum is split into three lines of equal intensity. This is because the ^{14}N nucleus has three possible spin orientations, and each spin orientation is possessed by one-third of all the radicals in the sample. In general, a spin-I nucleus splits the spectrum into $2I + 1$ hyperfine lines of equal intensity.

When there are several magnetic nuclei present in the radical, each one contributes to the hyperfine structure. In the case of *equivalent protons* (for example, the two methylene protons in the radical $CH_3CH_2\cdot$) some of the hyperfine lines are coincident. It is not hard to show that if the radical

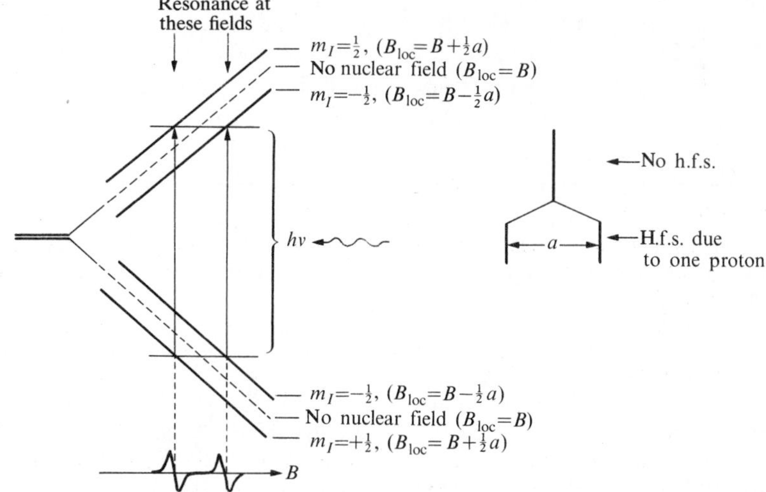

Fig. 20.18. The hyperfine interaction between an electron and a spin-$\frac{1}{2}$ nucleus results in four energy levels in place of the original two. As a result, the spectrum consists of two lines (of equal intensity) instead of one. This can be summarized by a simple stick diagram.

contains N equivalent protons, there are $N+1$ hyperfine lines with a *binomial intensity distribution* (i.e. the intensity distribution given by Pascal's triangle, Section 20.1(c)). The spectrum of the benzene radical anion in Fig. 20.16, which has seven lines of intensities 1:6:15:20:15:6:1, is consistent with a radical containing six equivalent protons.

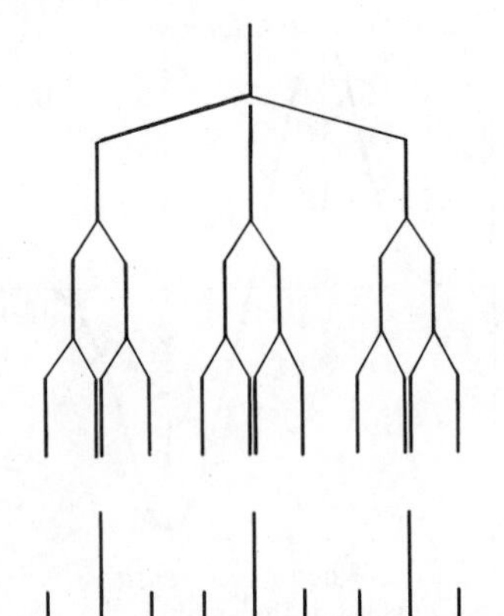

Example 20.4

A radical contains one ^{14}N nucleus ($I=1$) with hyperfine constant 1.03 mT and two equivalent protons ($I=\frac{1}{2}$) with hyperfine constant 0.35 mT. Predict the form of the ESR spectrum.

- *Method*. Deduce the hyperfine structure due to ^{14}N; split that with one of the protons; then split that with the other proton.

- *Answer*. ^{14}N gives three lines of equal intensity separated by 1.03 mT. Each line is split into doublets of spacing 0.35 mT by the first proton, and each line of these doublets is split into doublets with the same 0.35 mT splitting (see illustration on left). The central lines of each split doublet coincide, and so the proton splitting gives 1:2:1 triplets of internal splitting 0.35 mT. Therefore, the spectrum consists of three equivalent 1:2:1 triplets.

- *Comment*. Often it is quicker to realize that a group of equivalent protons give characteristic patterns (two giving a 1:2:1 triplet, in this case), and so superimpose the patterns directly.

- *Exercise*. Predict the form of the ESR spectrum of a radical containing three equivalent ^{14}N.
[1:3:6:7:6:3:1 septet]

The hyperfine structure of an ESR spectrum is a kind of fingerprint, and the radicals present in a sample can be identified. Moreover, since the magnitude of the splitting depends on the distribution of the unpaired electron into the vicinities of the magnetic nuclei present, the spectrum can be used to map the molecular orbital occupied by the unpaired electron. For example, since the hyperfine splitting in $C_6H_6^-$ is -0.375 mT, and one proton is close to a carbon with $\frac{1}{6}$ the unpaired electron *spin density* (because the electron is spread uniformly around the ring), the hyperfine splitting of an electron spin entirely confined to a single carbon should be $6\times(-0.375\text{ mT}) = -2.25$ mT. If in another aromatic radical we find a hyperfine splitting constant a, then the spin density there can be calculated from the *McConnell equation*:

$$a = Q\rho; \qquad Q = -2.25\text{ mT}, \tag{20.2.7}$$

where ρ is the unpaired electron spin density on a carbon atom and a is the hyperfine splitting observed for the hydrogen atom to which it is attached.

Example 20.5

The hyperfine structure of the ESR spectrum of (naphthalene)$^-$ can be interpreted as arising from two groups of four equivalent protons. Those at the α-positions in the ring have $a = -0.490$ mT and for those in the β-positions $a = -0.183$ mT. Map the unpaired spin density round the ring.

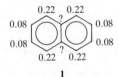

- *Method*. Use the McConnell equation in eqn (20.2.7).

- *Answer*. For the α-positions, $\rho = 0.22$, and for the β-positions $\rho = 0.08$, (**1**).

- *Comment*. Notice how the unpaired electron density accumulates in the α-positions. The

0.22 0.22 / 0.08 0.08 / 0.08 0.08 / 0.22 0.22 / ? ?

1

same value of Q may be used for the *approximate* mapping of spin density in heterocyclic radicals.

• *Exercise*. The spin density in (anthracene)$^-$ is shown in **2**. Predict the form of its ESR spectrum.
[A 1:2:1 triplet of splitting 0.43 mT split into a 1:5:6:5:1 quintet of splitting 0.22 mT, split into a 1:5:6:5:1 quintet of splitting 0.11 mT; 375 lines in all]

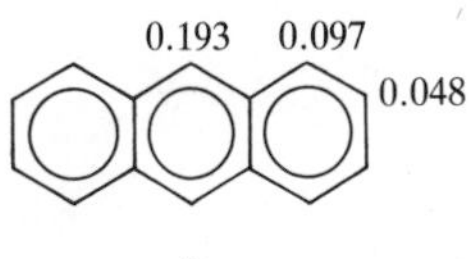

2

The hyperfine interaction is an interaction between the magnetic moments of the unpaired electron and the nuclei. There are two contributions to the interaction.

An electron in a p-orbital does not approach the nucleus very closely, and so it experiences a field that appears to arise from a point magnetic dipole. This is the *dipole–dipole interaction*, and its magnitude is given by an expression like that in eqn (20.1.6). A characteristic of this type of interaction is that it is *anisotropic*: that is, its magnitude (and sign) depends on the orientation of the radical with respect to the applied field. Furthermore, just as in the case of NMR, the dipole–dipole interaction averages to zero when the radical is free to tumble. Therefore, it is observed only for radicals trapped in solids.

An s-electron is spherically distributed around a nucleus and so has zero average dipole–dipole interaction with it even in a solid sample. However, since an s-electron has a non-zero probability of being at the nucleus, it is incorrect to treat the interaction as one between two *point* dipoles. An s-electron has a *Fermi contact interaction* with the nucleus: this is simply a magnetic interaction that occurs because the point dipole approximation fails. The contact interaction is *isotropic* (that is, independent of the radical's orientation), and consequently is shown even by rapidly tumbling molecules in fluids (so long as the spin density has some s-character).

The dipole–dipole interactions of p-electrons and the Fermi contact interactions of s-electrons can be quite large. For example, an N2p-electron experiences an average field of about 3.4 mT from the ^{14}N nucleus. An H1s-electron experiences a field of about 50 mT as a result of its Fermi contact interaction with the proton. More values are listed in Table 20.2. The magnitudes of the contact interactions in radicals can be interpreted in terms of the s-orbital character of the MO occupied by the unpaired electron, and the dipole–dipole interaction can be interpreted in terms of the p-character. This gives information about the composition of the MO, and especially the hybridization of the atomic orbitals (see Problem 20.33).

Table 20.2. Hyperfine coupling constants for atoms, a/mT

Nuclide	Isotropic coupling	Anisotropic coupling
^1H	50.8 (1s)	
^2H	7.8 (1s)	
^{14}N	55.2 (2s)	4.8 (2p)
^{19}F	1720 (2s)	108.4 (2p)

This still leaves the source of the hyperfine structure of the benzene negative anion and other aromatic radicals unexplained: the sample is fluid (and so it cannot be due to the dipole–dipole interaction), and the protons lie in the nodal plane of the π-orbital occupied by the unpaired electron (and so the structure cannot be due to a Fermi contact interaction). The explanation lies in a *polarization mechanism* similar to the one responsible for spin–spin coupling in NMR. There is a magnetic interaction between a proton and the σ-electrons that results in one of the electrons tending to be found with a greater probability in its vicinity, Fig. 20.19. The other electron is therefore more likely to be close to the carbon at the other end of the bond. The unpaired electron on the carbon has a lower energy if it is parallel to that electron (Hund's rule favours parallel electrons on atoms), and so the unpaired electron can detect the spin of the proton indirectly.

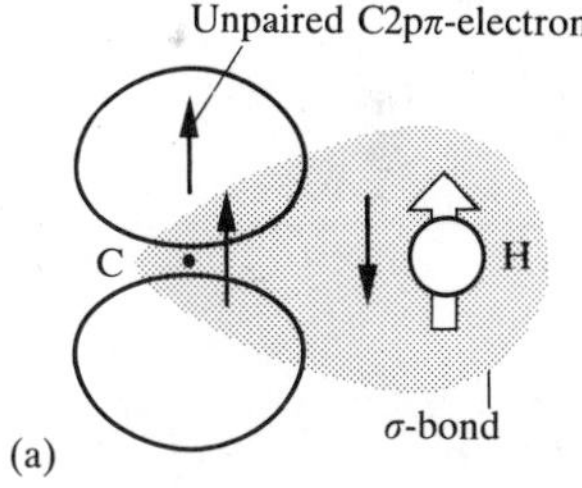

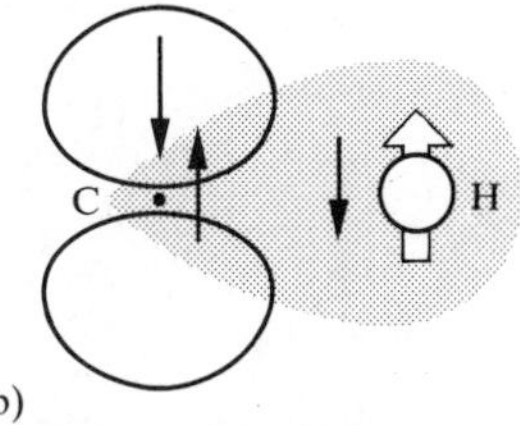

Fig. 20.19. The *polarization mechanism* for the hyperfine interaction in π-electron radicals. The arrangement in (a) is lower in energy than that in (b), and so there is an effective coupling between the unpaired electron and the proton.

Calculation using this model leads to a hyperfine interaction in agreement with the observed value of 2.3 mT.

20.3 Mössbauer spectroscopy

NMR depends on the absorption of photons of about the lowest energy used in chemistry. Mössbauer spectroscopy lies at the other extreme: it involves γ-ray photons with frequencies of the order of 10^{19} Hz and wavelengths of around 100 pm. But although the photon energies are so different, Mössbauer spectra are so sharply resolved that they show splittings comparable to those in NMR.

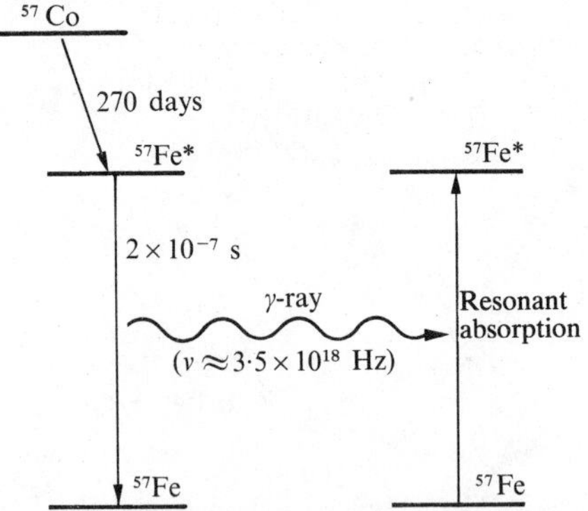

Fig. 20.20. The nuclear transitions involved in the Mössbauer effect using ^{57}Fe. The times are half-lives. The levels on the left are those of the source; the levels on the right are those of the sample.

20.3 (a) The emission and absorption of γ-rays

Mössbauer spectroscopy depends on the resonant absorption of a γ-ray photon by a nucleus. A typical experiment uses a sample of ^{57}Co as the primary source of an isotope of iron, ^{57}Fe. The ^{57}Co decays slowly ($t_{\frac{1}{2}} = 270$ days) to form an excited state of ^{57}Fe denoted ^{57}Fe*. This decays very rapidly to its ground state ($t_{\frac{1}{2}} = 0.2$ μs) and emits a γ-ray, Fig. 20.20. When this photon strikes a ground-state ^{57}Fe nucleus in another part of the spectrometer it is absorbed resonantly if its energy matches the ^{56}Fe*–^{57}Fe energy separation, but not otherwise.

The reason why the resonance is so sharp can be traced to a number of factors. In the first place, the ^{57}Fe* lifetime is long enough for its energy to be well-defined (a 0.2 μs half-life corresponds to an uncertainty of only 2 MHz). The γ-ray photon emitted by ^{57}Fe* has a frequency of 3.5×10^{18} Hz (corresponding to an energy of 14.4 keV), and so the linewidth is only $1/10^{12}$ the photon frequency. However, there is another factor. A photon of such high energy possesses a significant linear momentum (of magnitude $h\nu/c$). When ^{57}Fe* decays and emits a photon there is a recoil which gives the newly formed ^{57}Fe a velocity of about 100 m s^{-1}. This recoil has a damaging effect on the purity of the photon's frequency on account of the Doppler shift: if the source of a photon moves with a speed v, an observer reports a shift of frequency of magnitude $\nu v/c$ (recall Section 18.1). Even for $v = 1$ cm s^{-1} the photon from ^{57}Fe* would suffer a frequency shift of about 100 MHz, which is enough to obliterate the natural linewidth. One crucial aspect of the Mössbauer experiment overcomes the effect of recoil. If the ^{57}Fe* of the source and the ^{57}Fe of the target are held rigidly in a crystal lattice, the momentum is taken up by the whole crystal. This implies an extremely small recoil velocity, and therefore an extremely small Doppler broadening. The *Mössbauer effect* is this recoil-free emission and resonant absorption of γ-rays.

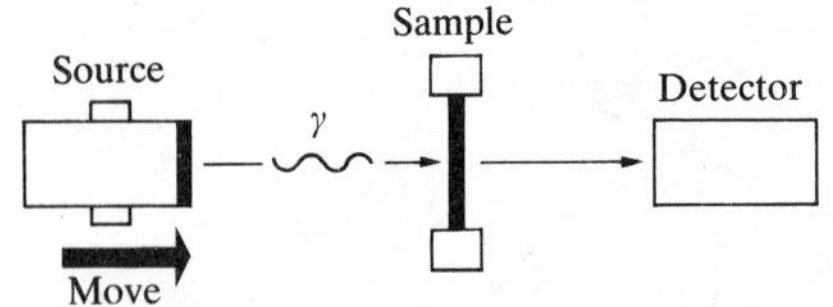

Fig. 20.21. The layout of a Mössbauer spectrometer. The source is fixed to a diaphragm which can be moved electromagnetically to and fro at a known speed, and the Doppler shift is used to achieve resonant matching of the γ-ray frequency with the nuclear energy levels of the sample.

20.3 (b) The technique

The γ-ray source is a thin foil of steel alloy containing ^{57}Co, Fig. 20.21. The sample is also in the form of a foil, and is placed in front of a counter. The resonant absorption occurs when the nuclear energy levels of the source and absorber match. The energy of the emitted photon can be modified by using the Doppler effect in a controlled way: this is achieved by mounting the source on a support and moving it at a known velocity. A range of a few mm s^{-1} is often enough to match the emitter and absorber levels. A screw drive can be used, but quite often the foil is driven electromagnetically, like the diaphragm of a loudspeaker.

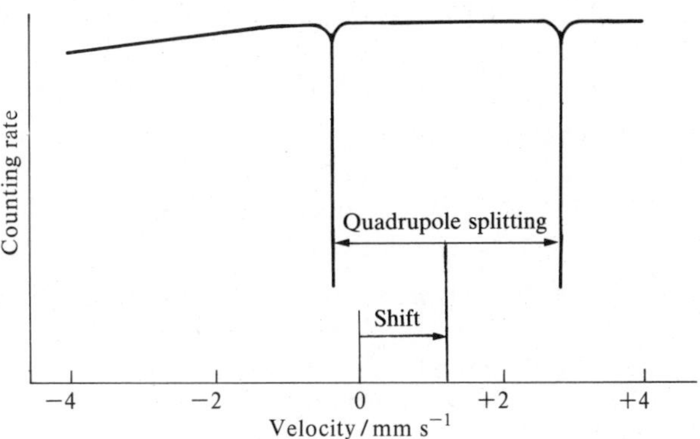

Fig. 20.22. The Mössbauer spectrum of $FeSO_4(s)$. The splitting of the resonance is due to the quadrupole moment of the iron nucleus.

Resonant absorptions are observed at several velocities, and a typical spectrum is shown in Fig. 20.22: positive velocities indicate motion of the source towards the absorber, negative away. Absorption at positive velocities indicates that the $^{57}Fe^{*}$–^{57}Fe separation is greater in the absorber than in the source.

Example 20.6

In a $^{57}Fe^{*}$ Mössbauer experiment it was found that the source had to be moved towards the sample at $2.2\,\mathrm{mm\,s^{-1}}$ in order to attain resonant absorption of the γ-ray. What is the shift (in MHz) between sample and source?

• *Method*. The magnitude of the Doppler shift is $\nu v/c$. The $^{57}Fe^{*}$ γ-ray has an energy of 14.4 keV. Convert this to a frequency using $1\,\mathrm{eV} \triangleq 8065.5\,\mathrm{cm^{-1}}$ followed by multiplication by c (but the c in the Doppler shift cancels this c, and so do not carry out the c multiplication explicitly).

• *Answer*. $\nu = (14.4\,\mathrm{keV}) \times (8065.5\,\mathrm{cm^{-1}\,eV})c$.
Therefore, $\delta\nu = (14.4 \times 8065.5\,\mathrm{cm^{-1}}) \times (2.2\,\mathrm{mm\,s^{-1}}) = 26\,\mathrm{MHz}$.

• *Comment*. This frequency corresponds to the wavenumber $8.5 \times 10^{-4}\,\mathrm{cm^{-1}}$. A remarkable feature of the Mössbauer experiment is that such small shifts are measured using photons of such enormous energy.

• *Exercise*. Another sample required a velocity of $1.8\,\mathrm{mm\,s^{-1}}$ away from the sample. What is the shift? [−21 MHz]

20.3 (c) Information from the spectra

We need to understand two features of the spectra. The first is why the resonance is shifted relative to the source frequency; the second is why the spectra consist of several lines.

The change in position of the resonance is called the *isomer shift*. The excited and ground state nuclei differ in radius to a small but significant extent (^{57}Fe shrinks by about 0.2% when it is excited to $^{57}Fe^{*}$), and so its electrostatic interaction with the surrounding electrons changes on excitation. The effect is important only for s-electrons (because they come close to the nucleus) and so the energy of the nucleus changes by an amount proportional to the change in nuclear radius (δR) and the s-electron density at the nucleus ($\psi_s^2(0)$), and we can write $\delta E \propto \psi_s^2(0)\delta R$. Although δR is the same in source and absorber, the s-electron density might be different because the ^{57}Fe nuclei are in different chemical environments. The isomer

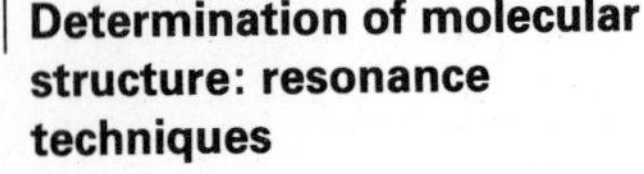

shift is given by

$$\delta E \propto \{[\psi_s^2(0)]_{\text{absorber}} - [\psi_s^2(0)]_{\text{source}}\}\,\delta R. \qquad (20.3.1)$$

Therefore, by measuring the isomer shift, the contribution of s-orbitals to the electron density at the nucleus can be assessed.

As an example, the Mössbauer spectra of ^{119}Sn compounds show an isomer shift of about +20 MHz relative to ^{119}Sn as white tin in the case of Sn(II) covalent molecules (such as $SnCl_2$), but of about −50 MHz for Sn(IV) covalent molecules (such as $SnCl_4$). The presence of the two extra s-electrons in Sn(II) leads to a more positive shift than in Sn(IV). This technique provides a very direct way of determining the valence of an element in a compound.

The *structure* of a Mössbauer spectrum arises from two principal interactions. In a number of cases (in ^{57}Fe in particular) the nuclear spins of the excited and ground state isotopes are different. For example, $I(^{57}\text{Fe}) = \frac{1}{2}$ but $I(^{57}\text{Fe}^*) = \frac{3}{2}$. This change of spin is accompanied by a change in the charge distribution in the nucleus, and whereas ^{57}Fe has a spherical distribution of charge, the charge of $^{57}Fe^*$ is concentrated at its poles. That is, $^{57}Fe^*$ has an *electric quadrupole moment*. In an octahedral environment (such as in $Fe(CN)_6^{4-}$) the energy of a quadrupolar nucleus is independent of its orientation, Fig. 20.23(a), but if the environment has only axial symmetry (such as in $Fe(CN)_5NO^{2-}$) the energy depends on the orientation, Fig. 20.23(b). The analysis of the *electric quadrupole splitting* of the Mössbauer spectrum gives a picture of the symmetry of the electron distribution close to the active nucleus.

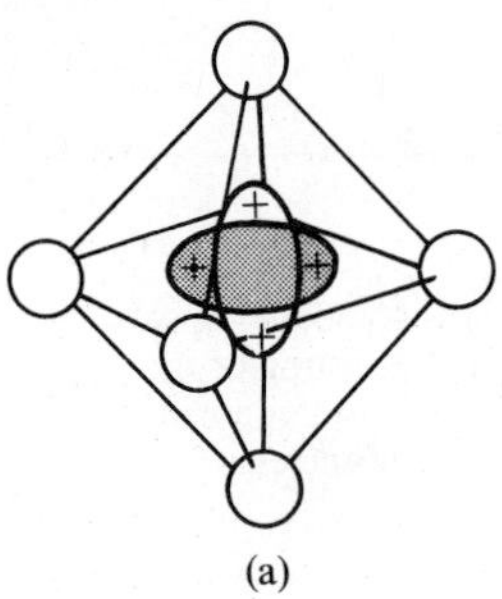

(a)

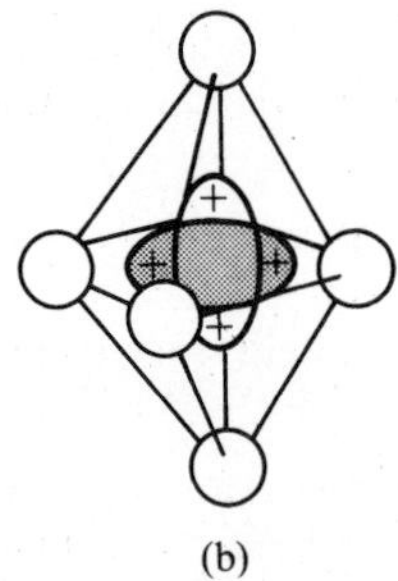

(b)

Fig. 20.23. (a) Two orientations of a quadrupolar nucleus in a regular octahedral environment: they have the same energy. (b) Two orientations of the same quadrupolar nucleus in a distorted octahedral environment: they now have different energies.

The importance of ^{57}Fe Mössbauer spectra reflects more than the fact that it is a convenient material to work with. Iron occurs in a variety of biologically and technologically important substances (such as blood and rust) and Mössbauer spectroscopy has been used to analyse them and to explore the mechanisms of corrosion and catalysis.

Further reading

NMR:

Nuclear magnetic resonance spectroscopy. R. K. Harris; Pitman, London, 1983.

High resolution NMR. E. D. Becker; Academic Press, New York, 1969.

Nuclear magnetic resonance. N. Muller; *Techniques of chemistry* (A. Weissberger and B. W. Rossiter, eds.) IIIA. 599. Wiley–Interscience, New York, 1972.

Spectroscopic methods in organic chemistry. D. H. Williams and I. Fleming; McGraw-Hill, New York, 1980.

Fourier transform NMR spectroscopy (2nd edn). D. Shaw; Elsevier, Amsterdam, 1984.

Fourier transform NMR techniques: a practical approach. K. Mullen and P. S. Pregosin; Academic Press, New York, 1976.

NMR and its applications to living systems. D. Gadian; Clarendon Press, Oxford, 1982.

ESR:

Theory and applications of electron spin resonance. W. Gordy (ed.). *Techniques of chemistry*, XV, Wiley-Interscience, New York, 1980.

Electron spin resonance: elementary theory and practical applications. J. E. Wertz and J. R. Bolton; McGraw-Hill, New York, 1972.

Chemical and biochemical aspects of electron spin resonance spectroscopy. M. C. R. Symons; Van Nostrand Reinhold, New York, 1978.

Mössbauer spectroscopy:

Experimental aspects of Mössbauer spectroscopy. R. H. Herber and Y. Hazony; *Techniques of chemistry* (A. Weissberger and B. W. Rossiter, eds.) IIID, 215, Wiley-Interscience, New York, 1972.

Physical methods in advanced inorganic chemistry. H. A. O. Hill and P. Day (eds.); Interscience, London, 1968.

Introductory problems

A20.1. The isotope ^{33}S has a nuclear spin of 3/2 and a nuclear g-factor of 0.4289. Calculate the allowed nuclear spin energies for ^{33}S in a magnetic field of 7.500 T.

A20.2. For the ^{14}N nucleus, calculate the difference between the lowest and highest energies of its spin states in a magnetic field of 15.00 T. This isotope has a nuclear spin of 1 and a nuclear g-factor of 0.4036.

A20.3. Calculate the magnetic field which satisfies the nuclear resonance condition for protons in a radio frequency field of 150.0 MHz.

A20.4. The tautomeric interconversion lifetime of a molecule which jumps between two conformations is 0.200 ms. At 100 MHz, the difference between the two forms is reflected in a separation of two resonances by 90.0 Hz. Determine whether or not the two lines will be merged or resolved in the observed spectrum.

A20.5. Find the magnetic field which produces the resonance condition for a sample with a g-value of 3.10 in an ESR spectrometer which uses 3.00 cm radiation.

A20.6. The ESR spectrum of a radical with a single magnetic nucleus is split into four lines of equal intensity. What is the nuclear spin of the magnetic nucleus?

A20.7. Compare the number of lines in the ESR spectra of the radicals X^1H_2 and X^2H_2, where the atom X has a nuclear spin of 5/2.

A20.8. At what temperature would the population of electron spins in the β-level be 0.500 % greater than in the α-level for a magnetic field of 0.800 T?

A20.9. The radiation of ^{129}I used in a Mössbauer experiment is 26.8 keV. The Mössbauer shift is 30.5 MHz. What sample velocity must be used to obtain resonance absorption?

A20.10. In an experiment with tin atoms in the compound $CdCr_2(SnS)_4$, the Mössbauer isomer shift for ^{119}Sn, which has radiation of energy 23.8 keV, corresponds to a speed of 1.15 mm s^{-1}, when the separation of the source and sample is increased. Determine the sign and magnitude of the shift in MHz. What information does this result provide about the valence state of tin in the sample?

Problems

20.1. Use the table of nuclear properties, Table 20.1, to predict the magnetic fields at which ^1H, ^2H, ^{13}C, ^{14}N, ^{19}F, and ^{31}P nuclei come into resonance with a radio-frequency field of (a) 60 MHz, (b) 300 MHz.

20.2. Calculate the relative population difference $\delta N/N$ for a proton in fields of 0.3 T, 1.5 T, and 7.0 T at (a) 4 K, (b) 300 K.

20.3. What magnetic field would be required in order to use an ESR X-band (9 GHz) spectrometer to observe proton NMR spectra, and a 60 MHz spectrometer to observe ESR?

20.4. A scientist investigates the possibility of *neutron* spin resonance, and has available a commercial NMR spectrometer operating at 60 MHz. Use the information in Table 20.1 to calculate the field required for resonance, and estimate the relative population difference at room temperature. Which is the lower spin state?

20.5. The chemical shift of the methyl protons in acetaldehyde (ethanal) is $\delta = 2.20$ p.p.m.; that of the aldehydic proton is $\delta = 9.80$ p.p.m. What is the difference in local magnetic field between the two regions of the molecule when the applied field is (a) 1.5 T, (b) 7.0 T?

20.6. What is the splitting (in Hz) between the methyl and aldehyde proton resonances when the spectrometer is operating at (a) 60 MHz, (b) 300 MHz?

20.7. Sketch the form of the ethanal NMR spectrum using the δ-values quoted above and a spin–spin coupling constant of 2.90 Hz. Indicate the relative intensities of the lines. How does the appearance of the spectrum change when it is recorded at 300 MHz instead of 60 MHz?

20.8. The solid-state spectrum of gypsum shows the presence of a direct dipole–dipole magnetic interaction which varies with angle in accord with eqn (20.1.6). The spectrum can be interpreted in terms of an extra magnetic field of 0.715 mT generated by one proton and experienced by the other. What is the distance separating the protons in the solid?

20.9. In a normal liquid, molecular rotation averages the direct dipolar interaction to zero. A molecule dissolved in a solution of a liquid crystal might not rotate freely in all directions, and so the dipolar interaction might not average to zero. Suppose a molecule is trapped so that, although the vector separating two protons may rotate freely around the direction of an applied field, the colatitude varies only between 0 and θ_{max}. Average the expression for the dipolar interaction over this restricted range of orientations, and confirm that the average disappears when θ_{max} is allowed to be π (free rotation over a sphere). What is the average value of the local dipolar field for the molecule in the last Problem if it were dissolved in a liquid crystal that enabled it to rotate up to $\theta_{max} = 30°$?

20.10. Two groups of protons are made equivalent by the isomerization of a molecule. At low temperatures, where the interconversion is slow, one group resonates with $\delta = 4.0$ p.p.m. and the other with $\delta = 5.2$ p.p.m. At what rate of interconversion will the two signals merge into a single line when the spectrometer is operating at 60 MHz?

20.11. The same compound as in the last Problem was investigated at different temperatures and in different spectrometers. It was found that the 60 MHz spectrum collapsed into a single line at 280 K, but at 300 MHz the temperature had to be raised to 300 K for the same effect. Estimate the activation energy for the interconversion.

20.12. The next three Problems give some practice with Fourier transform techniques. The shape of a spectral line $J(\omega)$ is related to its Fourier transform (essentially the time-dependence of the oscillations responsible for the absorption) by the integral $J(\omega) \propto \mathrm{re} \int_0^\infty G(t)\mathrm{e}^{\mathrm{i}\omega t}\,\mathrm{d}t$, where re is the operation 'take the real part'. Calculate the line shape corresponding to an oscillating, decaying function $G(t) = \cos \omega_0 t \mathrm{e}^{-t/\tau}$, where τ is the time-constant for the decay. Plot the shape for different values of τ: this shows that as τ increases, the line becomes sharper.

20.13. Show that if $G(t) = A \cos \omega_1 t \mathrm{e}^{-t/\tau} + B \cos \omega_2 t \mathrm{e}^{-t/\tau}$, with τ long, that the spectrum consists of two lines of intensities proportional to A and B and located at $\omega = \omega_1$ and $\omega = \omega_2$ respectively.

20.14. Devise a computer program to construct $G(t)$ for spectra with absorption lines at arbitrary frequencies, and explore the different free induction decays associated with different spectra.

20.15. Some commercial ESR spectrometers use 8 mm microwave radiation. What magnetic field is then needed to bring an electron spin into resonance?

20.16. Commercial ESR spectrometers can detect resonance from a sample containing as few as 10^{10} spins per cm^3. What is the molar concentration corresponding to this detection limit? Suppose a sample is known to contain 2.5×10^{14} spins: calculate the difference in populations. Take $B = 0.3$ T and $T = 298$ K.

20.17. Calculate the relative population difference, $\delta N/N = (N_\beta - N_\alpha)/(N_\beta + N_\alpha)$, for electron spins in a magnetic field of 0.3 T (3 kG) at (a) 4 K, (b) 300 K.

20.18. The centre of the ESR spectrum of atomic hydrogen lies at 329.12 mT in a spectrometer operating at 9.2231 GHz (1 GHz $= 10^9$ Hz). What is the g-value of the electron in a hydrogen atom?

20.19. The benzene radical anion, $C_6H_6^-$, has a g-value of 2.0025. At what field would you search for resonance in a spectrometer operating at (a) 9.302 GHz, (b) 33.67 GHz?

20.20. The mean g-value of the stable ClO_2 radical is 2.0102. What is the value of the additional local field that this implies when the applied field is (a) 3400 G, (b) 12 300 G?

20.21. The triangular molecule NO_2 has a single unpaired electron and it can be trapped in a solid matrix or prepared inside a crystal by radiation damage of nitrite ions. With the radical held rigidly in a definite orientation the magnetic field may be applied at selected orientations. When it is parallel to the O–O direction, the centre of the spectrum lies at 333.64 mT in a spectrometer operating at 9.302 GHz, but when the field lies along the bisector of the O–N–O angle the resonance shifts to 331.94 mT. What are the g-values in the two directions?

20.22. The ESR spectrum of atomic hydrogen consists of two lines. In a spectrometer working at 9.302 GHz one line appeared at 357.3 mT and the other at 306.6 mT. What is the hyperfine coupling constant for the atom?

20.23. A radical containing two equivalent protons shows a three-line ESR spectrum with an intensity distribution in the ratio 1:2:1. The lines occurred at 330.2 mT, 332.5 mT, and 334.8 mT. What is the hyperfine coupling constant for each proton?

20.24. The spectrometer used to obtain the spectrum described in the last Problem used a klystron generating microwaves of frequency 9.319 GHz. What is the g-value of the radical, and what are the coupling constants in MHz? (You should derive the useful rule that 0.10 mT$\triangleq$2.8 MHz.)

20.25. A radical containing two inequivalent protons with splitting constants 2.0 mT and 2.6 mT gives a spectrum centred on 332.5 mT. At what fields do the hyperfine lines lie, and what are their relative intensities?

20.26. The radical referred to in the last Problem isomerizes, and the motion interchanges the two protons. At low temperatures the spectrum is as described above. At room temperature the central pair of lines merge into a single line of double the intensity, and the spectrum is then as described in Problem 20.33. What is the rate of isomerization at which the spectrum switches from one form to the other?

20.27. Predict the intensity distribution in the hyperfine lines of the ESR spectrum of the radicals $\cdot CH_3$ and $\cdot CD_3$. ($I = 1$ for ^2H.)

20.28. The hyperfine coupling constant in the methyl radical is 2.3 mT. Use the information in Table 20.2 to predict the splitting between the hyperfine lines of the $\cdot CD_3$ radical's spectrum. What is the overall width of the spectra in each case?

20.29. Predict the appearance of the ESR spectrum of the ethyl radical, CH_3CH_2, using $a(CH_2) = 2.24\,mT$ and $a(CH_3) = 2.68\,mT$ for the respective protons. What changes will occur in the spectrum when deuterons are substituted for methylene protons?

20.30. The *p*-dinitrobenzene radical anion can be prepared by reduction of the nitrobenzene. The radical anion has two equivalent nitrogen nuclei ($I = 1$) and four equivalent protons. Predict the form of the ESR spectrum using $a(N) = 0.148\,mT$, $a(H) = 0.112\,mT$.

20.31. When an electron occupies a 2s-orbital on nitrogen it has a hyperfine interaction of 55.2 mT with the nucleus (this is the magnetic field generated by the nucleus and averaged over the region of the s-orbital where its presence is experienced). The spectrum of NO_2, on the other hand, shows an isotropic contact hyperfine interaction of 5.7 mT. For what proportion of its time can the unpaired electron of NO_2 be regarded as occupying the nitrogen 2s-orbital?

20.32. The nuclear field experienced by an electron when it occupies a 2p-orbital in the nitrogen atom has a magnitude of 3.4 mT. In NO_2 the anisotropic part of the hyperfine coupling has a magnitude of 1.3 mT. What proportion of its time does the unpaired electron spend in the nitrogen 2p-orbital in NO_2?

20.33. The information in the last two Problems is sufficient to map the unpaired electron distribution over the molecule. What is the total probability that the unpaired electron will be found on the nitrogen? What is the probability that it will be found on either oxygen atom? What is the hybridization of the central nitrogen atom? Does this hybridization support the suggestion that NO_2 is a bent molecule? Use the discussion of hybridization ratio, Section 16.3(a), to interpret p/s in terms of the bond angle by showing that $\cos\Theta = -\lambda/(\lambda+2)$, where $\lambda = b'^2/a'^2$. Estimate Θ; the experimental value is 134°.

20.34. The hyperfine coupling constants observed in various radical anions **1, 2, 3** are listed (in mT). Use the benzene value, and symmetry, to map the probability of finding the unpaired electron in the 2pπ-orbital on each carbon atom.

1: NO_2, NO_2; 0.011, 0.172, 0.172, 0.011

2: NO_2, NO_2; 0.450, 0.272, 0.108, 0.450

3: NO_2, NO_2; 0.112, 0.112, 0.112, 0.112

20.35. The pyridyl radical **4** has the proton hyperfine coupling constants shown (in mT), and the carboxy derivative **5** has the coupling constants also shown (H. Zemel and R. W. Fessenden, *J. phys. Chem.* **79,** 1419 (1975)):

4: N; 0.856, 0.412, 0.499, 0.128

5: N, CO_2; 0.968, 0.370, 0.156

Calculate the unpaired electron spin density around the rings.

20.36. The motion of nitroxide radicals trapped in a solid clathrate cage was examined at low temperatures and the hyperfine interaction of the electron with the nitrogen nucleus was found to depend on the orientation of the radical in the applied field (A. A. McConnell, D. D. MacNicol, and A. L. Porte, *J. chem. Soc.* A, 3516 (1971)). When the field was perpendicular to the bond the coupling constant varied between 113.1 MHz and 11.2 MHz, and when it was parallel to the bond the coupling constant was 14.1 MHz. When the temperature was raised to 115 K motion about the parallel axis began to influence the appearance of the spectrum. How rapidly does the molecule rotate in the clathrate cage at that temperature? (This Problem is taken further in Problem 30.24.)

20.37. The Mössbauer effect depends upon the recoilless emission of a γ-ray. How important is it to embed the emitter nucleus in a bulky, rigid, crystal lattice? Calculate the velocity of recoil of (a) a free ^{57}Fe atom, (b) a ^{57}Fe atom that is part of a rigid 100 mg crystal. Find the Doppler shift of the 14.4 keV γ-ray in each case.

20.38. The Mössbauer spectrum of $Na_4Fe(CN)_6$ consists of a single line, while that of $Na_2Fe(CN)_5NO$ shows a pair of lines, each of about half the intensity and approximately symmetrically displaced about the original resonance. Account for these observations.

20.39. The magnitudes of the isomer shifts relative to Au in Pt in the Mössbauer spectra of AuI, AuBr, and AuCl were found to be $1.25\,mm\,s^{-1}$, $1.43\,mm\,s^{-1}$, and $1.61\,mm\,s^{-1}$ respectively (V. G. Bhide, G. K. Shencry, and M. S. Multani, *Solid State Commun.* **2,** 221 (1964)). What does this imply about the ionic nature of these species?

21

Statistical thermodynamics: the concepts

Learning objectives

After careful study of this chapter you should be able to

(1) Define the terms *population*, *configuration*, and *weight*, and justify confining attention to the most probable configuration, Section 21.1(a).

(2) Derive the *Boltzmann distribution*, eqn (21.1.8).

(3) Define the *molecular partition function*, eqn (21.1.9), and interpret the significance of its value, Section 21.1(d).

(4) Express the *internal energy* of a system of independent particles in terms of the molecular partition function, eqn (21.1.12).

(5) Derive an expression for the *translational partition function* of independent particles, eqn (21.1.15).

(6) Define the *canonical ensemble* and state the *principle of equal a priori probabilities*, Section 21.2.

(7) Derive an expression for the *canonical partition function*, eqn (21.2.2), and relate it to the *internal energy* of the system, eqn (21.2.4).

(8) Write expressions relating the canonical partition function to the molecular partition functions for systems composed of distinguishable or indistinguishable particles, eqn (21.2.5).

(9) Define the *statistical thermodynamic entropy*, eqn (21.3.6), and relate it to the most probable populations of the states of the system, eqn (21.3.7).

(10) Derive and use an expression relating the statistical thermodynamic entropy to the canonical partition function, eqn (21.3.8) and Example 21.5.

(11) Derive and use the *Sackur–Tetrode equation* for the entropy of a monatomic gas, eqn (21.3.11) and Example 21.6.

(12) Justify the method of *undetermined multipliers*, Appendix 21.1.

Introduction

The last few chapters have shown how the energy levels of molecules can be calculated, measured spectroscopically, and related to their sizes and shapes. The next major step is to see how a knowledge of these individual molecular energies can be used to account for the properties of matter in bulk. In this chapter we set up the link between quantum theory and thermodynamics.

The crucial step in going from the quantum mechanics of individual molecules to the thermodynamics of bulk samples is to recognize that the latter deals with *average* behaviour. For example, the pressure of a gas is interpreted as the *average* force per unit area exerted by its particles, and there is no need to specify which particles are striking the wall at any instant. Nor is it necessary to follow the fluctuations in the pressure as different numbers of particles happen to collide with the wall at different moments, because the amplitude of these fluctuations is very small: it is highly improbable that there will be a sudden lull in the number of collisions, or a sudden storm. *Statistical thermodynamics* is based on the principle that *thermodynamic observables are averages of molecular properties*, and it sets about calculating these averages.

This chapter introduces statistical thermodynamics in three stages. The first, the derivation of the Boltzmann distribution for individual particles, is of restricted applicability, but it has the advantage of taking us directly to a result of central importance in a straightforward and elementary way: statistical thermodynamics can be *used* once the Boltzmann distribution has been deduced. Then, in Section 21.2, the arguments are elaborated a little (but not much) in order to deal with systems composed of interacting particles. Finally, in Section 21.3, we extend the techniques to the calculation of entropies and, through them, to all the functions of thermodynamics.

21.1 Molecular energy levels and the Boltzmann distribution

Consider a system composed of N particles. Although the total energy can be specified, it is not possible to be definite about how that energy is distributed. Collisions take place, and result in the ceaseless redistribution of energy not only between the particles but also among their different modes of motion. The closest we can come to a description of the distribution of energy is the statement of the average numbers of particles having each allowed energy. Thus on average there may be n_i particles in a state of energy ε_i, and in an equilibrium system this average *population* remains constant even though there may be small momentary fluctuations around it and the identities of the particles in that state may change at every collision. The problem we address in this section is the calculation of these populations for any type of particle in any mode of motion at any temperature: the only restriction is that the particles should be *independent*. That is, the total energy is a sum of the individual particle energies. We are discounting (at this stage) the possibility that in a real system a contribution to the total energy may arise from interactions between particles.

21.1 (a) Configurations and weights

Any individual particle may exist in states with energies $\varepsilon_0, \varepsilon_1, \ldots$ We shall almost always take ε_0, the lowest state, as the zero of energy ($\varepsilon_0 \equiv 0$), and measure all other energies relative to it. This means that we shall have to add some constant to the calculated energy of the system in order to obtain the internal energy (symbol: U) based on some other energy origin. For example, we may have to add the total zero-point energy of all the oscillators in the sample. In the sample as a whole, there will be n_0 particles in the state with energy ε_0, n_1 with ε_1, and so on. The specification of the set of populations $n_0, n_1, \ldots$ amounts to the statement of the *configuration* of the system.

The instantaneous configuration of the system fluctuates with time because the instantaneous populations change. We can picture a large number of different instantaneous configurations. One, for example, might be $n_0 = N$, $n_1 = 0$, $n_2 = 0, \ldots$, or $\{N, 0, 0, \ldots\}$, corresponding to every molecule being in its ground state. Another might be $\{N-2, 2, 0, \ldots\}$, in which two particles are in a higher state. This configuration is intrinsically more likely because it can be achieved in more ways: $\{N, 0, 0, \ldots\}$ can be achieved in only one way, $\{N-2, 2, \ldots\}$ can be achieved in $\frac{1}{2}N(N-1)$ different ways.‡ If the system were free to fluctuate between the configurations (as a result of collisions), it would almost always be found in the second (especially if N were large). In other words, a system free to switch between the two configurations would show properties characteristic almost exclusively of the second.

A general configuration $\{n_0, n_1, \ldots\}$ can be achieved in W different ways. W is called the *weight* of the configuration, and is given by

$$W = N!/n_0!\, n_1!\, n_2! \ldots \tag{21.1.1}$$

(Bear in mind that $0! = 1$.) This expression is a generalization of $W = \frac{1}{2}N(N-1)$, and reduces to it in the case of $\{N-2, 2, 0, \ldots\}$. It comes from elementary probability theory, being the number of distinguishable ways in which N objects can be sorted into bins with n_i in bin i (the objects are now particles, and the bins are their available states).

We have seen that the configuration $\{N-2, 2, 0, \ldots\}$ dominates $\{N, 0, 0, \ldots\}$, and it should be easy to believe that there may be others that greatly dominate both. Is there, in fact, a configuration with so great a weight that it overwhelms all the rest in importance to such an extent that supposing that the system *always* will be found in it introduces negligible error? If that configuration exists, then the properties of the system will be characteristic of it.

We can look for the dominating configuration by looking for the values of n_i that lead to a maximum value of W. Since W is a function of all the n_i we can do this by varying the n_i and looking for the values that correspond to $\mathrm{d}W = 0$ (just as in the search for the maximum of any function). However, there are two difficulties with this procedure. The first is that *we must confine attention to configurations that correspond to the specified, constant, total energy of the system*. This rules out many configurations, because $\{N, 0, 0, \ldots\}$ and $\{N-2, 2, 0, \ldots\}$ correspond to the different energies $N\varepsilon_0$ and $(N-2)\varepsilon_0 + 2\varepsilon_1$ respectively, and so these two cannot be achieved by the same system. It follows that in looking for the configuration with the greatest weight, we have to ensure that it also satisfies the *total energy criterion*:

$$\sum_i n_i\varepsilon_i = E, \tag{21.1.2}$$

where E is the total energy of the system. The second constraint is that we cannot arbitrarily vary *all* the populations simultaneously because the total number of particles present is also fixed (at N). Therefore, our search for the maximum value of W is subject to the *total number criterion*:

$$\sum_i n_i = N. \tag{21.1.3}$$

‡ The number of ways of arriving at the configurations is calculated as follows. One candidate for entering state 2 can be selected in N ways. There are then $N-1$ candidates for the second choice, corresponding to $N(N-1)$ choices overall. However, we should not distinguish the choice (Jack, Jill) from the choice (Jill, Jack), and so only half the choices lead to distinguishable configurations. It should also be noted that at this stage of the argument, we are ignoring the requirement that the total energy of the system should be constant: the energy of the second configuration is greater than that of the first. We impose the constraint of constant total energy later.

21.1 (b) The most probable configuration

We are now ready to begin. We are looking for the set of numbers $n_0^*, n_1^*, \ldots$ for which W has its maximum value W^*. It is equivalent, and it turns out to be simpler, to find the criterion for which $\ln W$ is a maximum. Since $\ln W$ depends on all the n_i, when a configuration changes so that all the n_i change to $n_i + \mathrm{d}n_i$, $\ln W$ changes to $\ln W + \mathrm{d}(\ln W)$, where

$$\mathrm{d}(\ln W) = \sum_i (\partial \ln W/\partial n_i)\, \mathrm{d}n_i. \qquad (21.1.4)$$

At a maximum, the change $\mathrm{d}(\ln W)$ vanishes, but the presence of the two equations involving E and N means that when the n_i change they do so subject to the *constraints*

$$\sum_i \varepsilon_i\, \mathrm{d}n_i = 0 \quad \text{and} \quad \sum_i \mathrm{d}n_i = 0. \qquad (21.1.5)$$

These two equations prevent us from solving the equation for $\ln W$ simply by setting all $(\partial \ln W/\partial n_i) = 0$ because the $\mathrm{d}n_i$ in eqn (21.1.4) are not all independent. The way to take constraints into account was devised by Lagrange and is called the *method of undetermined multipliers*. The technique is described in Appendix 21.1: all we need here is the rule that *a constraint should be multiplied by a constant and then added to the main variation equation*, eqn (21.1.4). The variables are then treated as though they are all independent, and the constants are evaluated at the end of the calculation.

The technique is used as follows. The two constraints are multiplied by the constants $-\beta$ and α respectively and then added to eqn (21.1.4):

$$\begin{aligned}\mathrm{d}(\ln W) &= \sum_i (\partial \ln W/\partial n_i)\, \mathrm{d}n_i + \alpha \sum_i \mathrm{d}n_i - \beta \sum_i \varepsilon_i\, \mathrm{d}n_i \\ &= \sum_i \{(\partial \ln W/\partial n_i) + \alpha - \beta\varepsilon_i\}\, \mathrm{d}n_i.\end{aligned}$$

All the $\mathrm{d}n_i$ are now treated as independent. This means that the only way of satisfying $\mathrm{d}(\ln W) = 0$ is to require that for each i

$$(\partial \ln W/\partial n_i) + \alpha - \beta\varepsilon_i = 0 \qquad (21.1.6)$$

when the n_i have their most probable values n_i^*.

In order to solve the last equation we use *Stirling's approximation* in the form‡

$$\ln x! \approx x \ln x - x. \qquad (21.1.7)$$

‡ A simple justification of Stirling's approximation is to write

$$\ln x! = \ln 1 + \ln 2 + \ldots + \ln x = \sum_{y=1}^{x} \ln y$$

$$\approx \int_1^x \ln y\, \mathrm{d}y = \{y \ln y - y\}\Big|_1^x \approx x \ln x - x,$$

where the approximations depend on x being large. The precise form of Stirling's approximation is

$$x! \approx (2\pi)^{\frac{1}{2}} x^{x+\frac{1}{2}} \mathrm{e}^{-x}$$

and is good when x is large (greater than about 10). We are dealing with far larger values of x, and the simplified version in eqn (21.1.7) is normally adequate.

Introduction of this approximation into the expression for $\ln W$, eqn (21.1.1), gives

$$\begin{aligned}\ln W &= \ln\{N!/n_0!\,n_1!\ldots\} = \ln N! - \ln\{n_0!\,n_1!\ldots\}\\ &= \ln N! - \sum_j \ln n_j! \approx \{N\ln N - N\} - \sum_j \{n_j \ln n_j - n_j\}\\ &\approx N\ln N - \sum_j n_j \ln n_j,\end{aligned}$$

because the sum of the n_j is equal to N. Since N is a constant, differentiation with respect to n_i gives

$$\begin{aligned}(\partial \ln W/\partial n_i) &\approx -\sum_j \{\partial(n_j \ln n_j/\partial n_i)\}\\ &\approx -\sum_j \{(\partial n_j/\partial n_i)\ln n_j + n_j(\partial \ln n_j/\partial n_i)\} = -\{\ln n_i + 1\}\end{aligned}$$

because, for $j \neq i$, n_j is independent of n_i so that $(\partial n_j/\partial n_i) = 0$; for $j = i$, $(\partial n_j/\partial n_i) = (\partial n_j/\partial n_j) = 1$; and $(\partial \ln n_j/\partial n_i) = (1/n_j)(\partial n_j/\partial n_i)$. The 1 in $\ln n_i + 1$ may be neglected (because $\ln n_i$ is so large), and so eqn (21.1.6) becomes

$$-\ln n_i + \alpha - \beta\varepsilon_i = 0 \quad \text{when} \quad n_i = n_i^*.$$

The most probable population of the state of energy ε_i is therefore

$$n_i^* = \mathrm{e}^{(\alpha - \beta\varepsilon_i)}.$$

The final step is to evaluate the constants α and β. Since

$$N = \sum_j n_j^* = \sum_j \mathrm{e}^{\alpha}\mathrm{e}^{-\beta\varepsilon_j},$$

it follows that $\mathrm{e}^{\alpha} = N/\sum_j \mathrm{e}^{-\beta\varepsilon_j}$, which leads to the

$$\textit{Boltzmann distribution: } n_j^* = N\mathrm{e}^{-\beta\varepsilon_i}/\sum_j \mathrm{e}^{-\beta\varepsilon_j}. \qquad (21.1.8)$$

The parameter β is equal to $1/kT$, where T is the thermodynamic temperature; this is justified later (Section 21.3(a)).

21.1 (c) The molecular partition function

In eqn (21.1.8) we have a result central to statistical thermodynamics. The sum in the denominator is called the *molecular partition function* (symbol: q):

$$\textit{Molecular partition function: } q = \sum_j \mathrm{e}^{-\beta\varepsilon_j}, \qquad \beta = 1/kT. \qquad (21.1.9)$$

The sum is sometimes expressed slightly differently. Note that it is a sum over the *states* of an individual molecule. It may happen that several states have the same energy, and so give the same contribution to the sum. If, for example, g_j states have the same energy ε_j (i.e. that energy level is g_j-fold degenerate), then we could write

$$q = \sum_j g_j \mathrm{e}^{-\beta\varepsilon_j}, \qquad (21.1.10)$$

where the sum is now over *energy levels*, not individual states. We shall normally employ the sum-over-states form.

Example 21.1

Write an expression for the rotational partition function of a linear rotor such as HCl.

• *Method*. The energy levels of a linear rotor are $\varepsilon_{JM_J} = hcBJ(J+1)$. Each J level consists of $2J+1$ degenerate states (distinguished by the $2J+1$ values of M_J). Therefore, either use eqn (21.1.9) and sum over all *states*, or use eqn (21.1.10), with $g_J = 2J + 1$.

• *Answer*. The second approach is much easier in this case, and we can write at once that

$$q = \sum_J (2J+1)\mathrm{e}^{-hc\beta BJ(J+1)}.$$

• *Comment*. The sum can be evaluated numerically by supplying the value of B (from spectroscopy or calculation) and the temperature. The expression is confined to unsymmetrical linear rotors (e.g. HCl, not CO_2) for reasons explained in Section 22.1(c).

• *Exercise*. Write down the partition function for a two-level system, the lower state being non-degenerate, and the upper state (at an energy ε) doubly degenerate. [$1 + 2\mathrm{e}^{-\beta\varepsilon}$]

The importance of the molecular partition function is that it contains all the thermodynamic information about a system of independent particles at thermal equilibrium. We have already seen that it appears in the Boltzmann expression for the most probable populations, which in terms of q is

$$n_i^*/N = \mathrm{e}^{-\beta\varepsilon_i}/q. \qquad (21.1.11)$$

With the populations established, it is a simple matter to obtain the total energy of the system, $E = \sum_i n_i\varepsilon_i$, because we can identify the populations with the most probable populations and write

$$E = \sum_i n_i^*\varepsilon_i = (N/q)\sum_i \varepsilon_i \mathrm{e}^{-\beta\varepsilon_i}.$$

The sum can be manipulated into a form involving only q. Since

$$(\mathrm{d}/\mathrm{d}\beta)\mathrm{e}^{-\beta\varepsilon_i} = -\varepsilon_i\mathrm{e}^{-\beta\varepsilon_i},$$

$$E = (N/q)\sum_i (-\mathrm{d}/\mathrm{d}\beta)\mathrm{e}^{-\beta\varepsilon_i} = -(N/q)(\mathrm{d}/\mathrm{d}\beta)\sum_i \mathrm{e}^{-\beta\varepsilon_i} = -(N/q)(\mathrm{d}q/\mathrm{d}\beta).$$

There are several points in relation to this expression. Since $\varepsilon_0 \equiv 0$, the value of E should be interpreted as the value of the internal energy *relative* to its value at $T=0$: that is, to obtain the conventional internal energy (U) we should add an amount corresponding to the internal energy at $T=0$ and write $U = U(0) + E$. Furthermore, the partition function may depend on variables other than the temperature (e.g. the volume), and so the derivative with respect to β is really a *partial* derivative with these other variables held constant. The final expression relating the molecular partition function to the thermodynamic internal energy of a system of independent particles is therefore

$$U - U(0) = -(N/q)(\partial q/\partial\beta)_V = -N(\partial \ln q/\partial\beta)_V. \qquad (21.1.12)$$

The last equation confirms that we need know only the partition function in order to calculate the internal energy. Section 21.3 will expand this by showing that *all* thermodynamic functions can be calculated once we know q (and its generalization for interacting molecules). In this respect q plays a role in thermodynamics very similar to that played by the wavefunction in quantum mechanics: it is a kind of thermal wavefunction.‡

21.1 (d) An interpretation of the partition function

Some insight into the partition function can be obtained by considering how it depends on the temperature. When the temperature is close to zero the parameter $\beta = 1/kT$ is close to infinity. Then every term except one in the sum $q = \sum_j e^{-\beta\varepsilon_j}$ is zero because each one has the form e^{-x} with $x \to \infty$. The exception is the term with $\varepsilon_0 \equiv 0$ (or the g_0 terms if the ground state is g_0-fold degenerate), because then $\varepsilon_0/kT \equiv 0$ whatever the temperature. As there is only one surving term when $T = 0$, and its value is g_0, it follows that

$$\text{As} \quad T \to 0, \qquad q \to g_0.$$

At the other extreme, consider the case when T is so large that for each term in the sum $\varepsilon_j/kT \to 0$. Since $e^{-x} = 1$ when $x = 0$, each term in the sum contributes 1. It follows that the sum is equal to the number of molecular states, which in general is infinite:

$$\text{As} \quad T \to \infty, \qquad q \to \infty.$$

(In some idealized cases the molecule may have only a finite number of states; then the upper limit of the partition function is equal to the number of states.) We see that the molecular partition function gives an indication of *the average number of states that are thermally accessible to a molecule at the temperature of the system*. At $T = 0$, only the ground level is accessible and $q = g_0$. At very high temperatures, virtually all states are accessible, and q is correspondingly large.

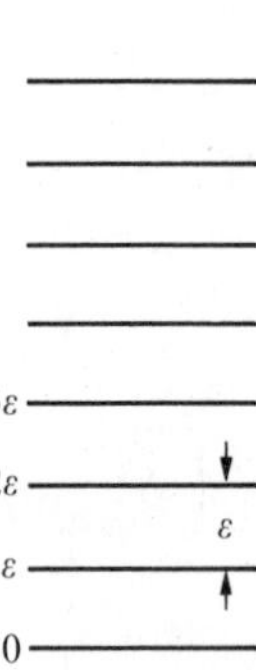

Fig. 21.1. The equally spaced infinite array of energy levels used in the calculation of the partition function, eqn (21.1.13). A harmonic oscillator has the same spectrum of levels.

As an explicit example, consider a molecule with a set of equally spaced non-degenerate energy levels, Fig. 21.1. (These can be thought of as the vibrational energy levels of a diatomic molecule.) If the separation between levels is ε, the partition function is

$$q = \sum_j e^{-\beta\varepsilon_j} = 1 + e^{-\beta\varepsilon} + e^{-2\beta\varepsilon} + \ldots = 1 + (e^{-\beta\varepsilon}) + (e^{-\beta\varepsilon})^2 + \ldots$$

This geometrical progression can be summed immediately§, giving

$$q = 1/(1 - e^{-\beta\varepsilon}). \qquad (21.1.13)$$

‡ The analogy can be seen very clearly by writing the expression for the mean energy per molecule, $\varepsilon = E/N$, in the form $\varepsilon q = -\partial q/\partial\beta$ and noticing the striking resemblance to the time-dependent Schrödinger equation $H\psi = -(\hbar/i)\,\partial\psi/\partial t$, Box 13.1. In the sense that we can go from one to the other by replacing q by ψ and β by $it/\hbar$, there is pleasure to be had from noticing that inverse temperature is imaginary time.

§ The sum of a geometric series $a + ar + ar^2 + \ldots + ar^{n-1}$ is $a(1 - r^n)/(1 - r)$ for $r < 1$. When the series is infinite the sum is $a/(1 - r)$. Two versions we need frequently are

$$1 - x + x^2 - \ldots = 1/(1 + x); \qquad 1 + x + x^2 + \ldots = 1/(1 - x).$$

The second of these special cases is used here.

It follows that the proportion, P_i, of molecules in the state with energy ε_i is almost certain to be

$$P_i = n_i^*/N = (1 - e^{-\beta\varepsilon})e^{-\beta\varepsilon_i}, \qquad (21.1.14)$$

Numerical values can be obtained for various temperatures, and are depicted in Fig. 21.2. We see that at very low temperatures q is close to unity, and only the lowest state is significantly populated. As the temperature is raised, the population breaks out of the lowest state, and the upper states become progressively more highly populated. At the same time the partition function rises from unity, and its value gives a broad indication of the range of states populated.

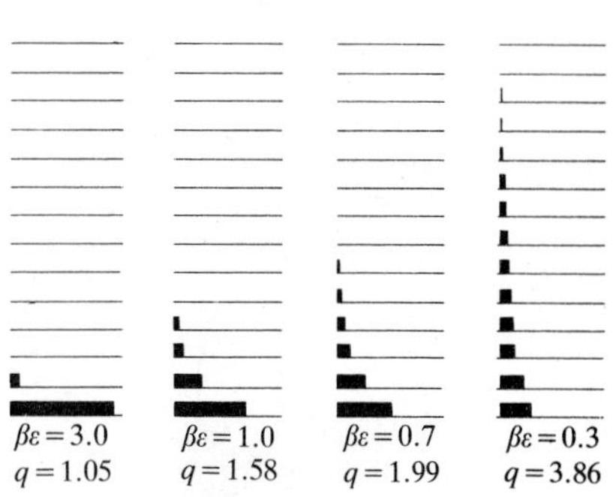

Fig. 21.2. The populations of the energy levels of the system shown in Fig. 21.1 at different temperatures, and the corresponding values of the partition function calculated using eqn (21.1.13). Note that $\beta = 1/kT$.

Example 21.2

Calculate the proportion of I_2 molecules in their ground, first excited, and second excited vibrational states at 25 °C. The vibrational wavenumber is $\tilde{\nu} = 214.6\ \text{cm}^{-1}$.

• *Method*. Vibrational energy levels have a constant separation (in the harmonic approximation, Section 18.3), and so the partition function is given by eqn (21.1.13) and the populations by eqn (21.1.14). Identify j with the quantum number v, and calculate P_v for $v = 0, 1, 2$. The following quantity is useful in this and similar calculations:

$$k\mathcal{T}/hc = 207.22_3\ \text{cm}^{-1}.$$

(We have used the notation $\mathcal{T} = 298.15$ K introduced in Part 1.)

• *Answer*. Since $\varepsilon\beta = 214.6\ \text{cm}^{-1}/207.22_3\ \text{cm}^{-1} = 1.036$, the relative populations are

$$P_v = (1 - e^{-\beta\varepsilon})e^{-v\beta\varepsilon} = 0.645e^{-1.036v}.$$

Therefore: $P_0 = 0.645$, $P_1 = 0.229$, $P_2 = 0.081$.

• *Comment*. The iodine bond is not stiff and the atoms are heavy: as a result the vibrational energy separations are small and even at room temperature several vibrational levels are significantly populated. The value of the partition function, $q = 1/(1 - e^{-1.036}) = 1.550$, reflects this small but significant spread.

• *Exercise*. At what temperature would the $v = 1$ level have (a) half the population of the ground state, (b) the same population as the ground state? [(a) 445 K, (b) infinite]

21.1 (e) The translational partition function

As an example of the calculation of q, we consider a perfect monatomic gas in a container of volume V. The length of the container in the x-direction is X (and similarly Y, Z for the other two directions, with $XYZ = V$). The molecular partition function is $q = \sum_j e^{-\beta\varepsilon_j}$ with ε_j a translational energy of the particle in the container, and j the label of its state. Since ε_j is the sum of the energies of translation in all three directions, $\varepsilon_j = \varepsilon_{j(X)} + \varepsilon_{j(Y)} + \varepsilon_{j(Z)}$, the partition function factorizes:

$$\begin{aligned} q &= \sum_{\text{All } j} \exp\{-\beta\varepsilon_{j(X)} - \beta\varepsilon_{j(Y)} - \beta\varepsilon_{j(Z)}\} \\ &= \Big\{\sum_{j(X)} \exp(-\beta\varepsilon_{j(X)})\Big\}\Big\{\sum_{j(Y)} \exp(-\beta\varepsilon_{j(Y)})\Big\}\Big\{\sum_{j(Z)} \exp(-\beta\varepsilon_{j(Z)})\Big\} \\ &= q_X q_Y q_Z. \end{aligned}$$

The energy levels of a particle of mass m in a container of length X were calculated in Section 14.1, eqn (14.1.9):

$$\varepsilon_n = n^2h^2/8mX^2, \qquad n = 1, 2, 3, \ldots$$

The lowest level has energy $h^2/8mX^2$, and so the energies relative to that are $(n^2 - 1)h^2/8mX^2$. The sum we have to evaluate is therefore

$$q_X = \sum_{n=1}^{\infty} \exp\{-(n^2-1)h^2\beta/8mX^2\}.$$

In a container the size of a typical laboratory vessel, the translational energy levels are very close together, and so the sum can be replaced by an integral:

$$q_X = \int_1^{\infty} \exp\{-(n^2-1)h^2\beta/8mX^2\}\, dn.$$

Extending the lower limit to $n = 0$ and replacing $n^2 - 1$ by n^2 introduces negligible error but turns the integral into standard form. We make the substitution $x^2 = n^2h^2\beta/8mX^2$, implying $dn = (8mX^2/h^2\beta)^{\frac{1}{2}}\, dx$:

$$q_X = (8mX^2/h^2\beta)^{\frac{1}{2}} \int_0^{\infty} e^{-x^2}\, dx = (8mX^2/h^2\beta)^{\frac{1}{2}}(\pi^{\frac{1}{2}}/2).$$

This is the molecular partition function for translational motion in the x-direction. The only change for the other two directions is to replace X by Y or Z, and so the partition function for motion in three dimensions is

$$q = (2\pi mX^2/h^2\beta)^{\frac{1}{2}}(2\pi mY^2/h^2\beta)^{\frac{1}{2}}(2\pi mZ^2/h^2\beta)^{\frac{1}{2}} = (2\pi m/h^2\beta)^{\frac{3}{2}}V.$$

This can be written more briefly as

$$q = V/\Lambda^3, \qquad \Lambda = h(\beta/2\pi m)^{\frac{1}{2}}. \qquad (21.1.15)^{\circ}$$

Example 21.3

Calculate the translational partition function for a hydrogen molecule ($^1H_2$) confined to a 100 cm^3 vessel at 25 °C.

• *Method*. Use eqn (21.1.15). Begin by calculating Λ (which has the dimensions of length, which serves as a check on the calculation). Use $\beta = 1/kT$ (which is yet to be established). Write $m(H_2) = 2M_r(H)m_u$, where m_u is the atomic mass unit (inside front cover).

• *Answer*. For this molecule

$$2\pi mkT = 2\pi \times (2 \times 1.0078 \times 1.6606 \times 10^{-27}\,\text{kg}) \times (1.3807 \times 10^{-23}\,\text{J K}^{-1}) \times (298.15\,\text{K})$$
$$= 8.657 \times 10^{-47}\,\text{J kg}$$

so that $\Lambda = h/(8.657 \times 10^{-47}\,\text{kg}^2\,\text{m}^2\,\text{s}^{-2})^{\frac{1}{2}} = 7.122 \times 10^{-11}$ m. Therefore,

$$q = (100 \times 10^{-6}\,\text{m}^3)/(7.122 \times 10^{-11}\,\text{m})^3 = 2.8 \times 10^{26}.$$

• *Comment*. This result shows that about 10^{26} quantum states are thermally accessible, even at room temperature and for this light molecule. It is always sensible to break these calculations down into small steps and to keep track of units.

• *Exercise*. What is the corresponding value for $^2H_2$? $[q_D = 2^{\frac{3}{2}}q_H]$

Several applications of this simple result will be described in the next chapter. For the present we simply show how to calculate the internal energy of the gas using eqn (21.1.12). First, calculate $(\partial q/\partial\beta)_V$:

$$(\partial q/\partial\beta)_V = (\mathrm{d}\Lambda^{-3}/\mathrm{d}\beta)V = -(3V/\Lambda^4)\,\mathrm{d}\Lambda/\mathrm{d}\beta$$
$$= -3V/2\beta\Lambda^3, \quad \text{since} \quad \mathrm{d}\Lambda/\mathrm{d}\beta = \Lambda/2\beta.$$

Thus, by eqn (21.1.12),

$$U - U(0) = -N(-3V/2\beta\Lambda^3)/(V/\Lambda^3) = 3N/2\beta. \qquad (21.1.16)°$$

Therefore, since we are anticipating that $\beta = 1/kT$, the internal energy is $3NkT/2$, or $3nRT/2$, the equipartition value quoted in Section 0.1(f).

Example 21.4

Calculate the constant-volume heat capacity of a monatomic gas.

- *Method*. By definition, $C_V = (\partial U/\partial T)_V$. Use the expression just derived for U.

- *Answer*. $C_V = (\partial\frac{3}{2}nRT/\partial T)_V = \frac{3}{2}nR$.
The molar heat capacity is therefore $C_{V,\mathrm{m}} = 12.5\ \mathrm{J\,K^{-1}\,mol^{-1}}$.

- *Comment*. This agrees exactly with experimental data on monatomic gases at normal pressures. In more complex molecules other modes of motion contribute: this is treated in Section 22.3(b). Note that differentiation with respect to T can often more conveniently be treated in terms of differentiation with respect to β using

$$\mathrm{d}/\mathrm{d}T = (\mathrm{d}\beta/\mathrm{d}T)\,\mathrm{d}/\mathrm{d}\beta = (-1/kT^2)\,\mathrm{d}/\mathrm{d}\beta.$$

- *Exercise*. Calculate the heat capacity of N two-level systems of the kind described in the *Exercise* of *Example* 21.1.

$[C = 2Nkf^2,\ f = \mathrm{e}^{-\beta\varepsilon/2}/(1+\mathrm{e}^{-\beta\varepsilon})]$

21.2 The canonical ensemble

In this section we see how the conclusions in Section 21.1 can be generalized to include systems composed of interacting particles. The crucial new concept is the *ensemble*. Like so many scientific terms, this has basically its normal meaning, but sharpened and refined into a precise significance.

In order to set up an ensemble we take a system of specified volume, composition, and temperature, and think of it as being reproduced $\mathbb{N}$ times. We can think of this ensemble as consisting of identical *closed* systems in thermal contact with each other. This thermal contact ensures that they are all at the same temperature, but allows them to exchange energy with each other; the *total* energy of the ensemble ($\mathbb{E}$), however, is constant. This imaginary collection of replications is called the *canonical ensemble*‡.

The important point about an ensemble is that it is a collection of *imaginary* replications of the system, and so we are free to let the number of

‡ There are two other important ensembles. In the *microcanonical ensemble* the condition of constant temperature is replaced by the requirement that all the systems should have exactly the same energy: each system is individually *isolated*. In the *grand canonical ensemble* the volume and temperature of each system is the same, but they are *open*. This means that matter can be imagined as able to pass between the systems; the composition of each one may fluctuate, but now the chemical potential is the same in each system. We can summarize these ensembles as follows:

Microcanonical: N, V, E common
Canonical: N, V, T common
Grand canonical: μ, V, T common.

members be as large as we like; when appropriate, we can let $\mathbb{N}$ become infinite. (Note that $\mathbb{N}$ is unrelated to N: N is the number of particles in the actual system; $\mathbb{N}$ is the number of imaginary replications of that system.) The reason for introducing an ensemble is as follows. In the first place, as a result of the exchange of energy of the actual system with its actual surroundings, its detailed state is constantly changing. This suggests that we should identify the *time average* of the properties of the system as the observable thermodynamic properties (and in particular that we identify the time average energy of the system as its internal energy U). But time averages are difficult to calculate because the changes are so complicated. Therefore, we make the *hypothesis* that the same answer is obtained if, instead of evaluating the time average of a single real system, we evaluate the average of a property over all the members of an infinitely large ($\mathbb{N}=\infty$) ensemble at a single instant. This converts a tricky time-dependent problem into a much easier time-independent problem; the validity of the procedure, however, is still the subject of a great deal of erudite discussion.

Another concept lying behind this development is the *principle of equal a priori probabilities*: we are assuming that *all possibilities for the distribution of energy are equally probable*. In other words, we assume that states corresponding to a given vibrational energy, for instance, are as likely to be populated as states corresponding to the same rotational energy, and so on. *How* the energy is stored is regarded as being wholly irrelevant to the discussion.

21.2 (a) Dominating configurations

Once the ensemble has been conceived, the argument runs along the same lines as in Section 21.1, but instead of having to think of some overall energy E of the system as being distributed over molecules, we have to think of some overall ensemble energy $\mathbb{E}$ as being distributed over the members of the ensemble (i.e. over the replications of the system). If the total energy is $\mathbb{E}$, and there are $\mathbb{N}$ members, the average energy of any member is $\mathbb{E}/\mathbb{N}$, and it is this quantity that is to be identified with the internal energy of the system in the limit of $\mathbb{N}$ (and $\mathbb{E}$) approaching infinity.

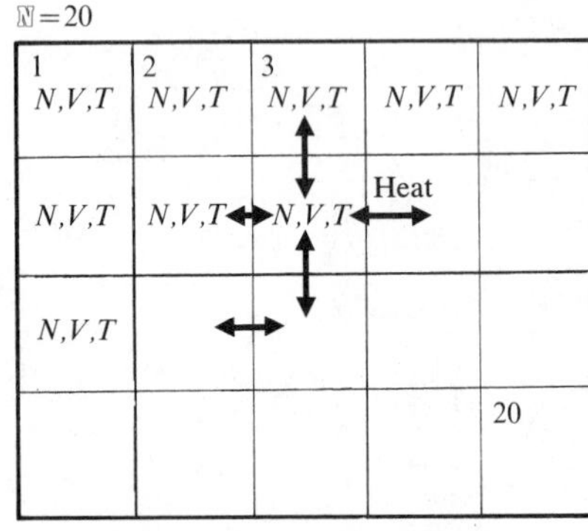

Fig. 21.3. A representation of the canonical ensemble, in this case for $\mathbb{N}=20$. The individual replications of the actual system all have the same composition and volume. They are all in mutual thermal contact, and so all have the same temperature. Energy may be transferred between them as heat, and so they do not all have the same energy. The *total* energy ($\mathbb{E}$) of all 20 replications is a constant because the ensemble is isolated overall.

All the members are in thermal contact, Fig. 21.3, and energy may migrate between them. If the number of members of the ensemble being in a state with energy E_i is denoted $\mathbb{n}_i$, we can speak of the configuration of the ensemble (by analogy with the configuration of the system used in Section 21.1) and its weight $\mathbb{W}$. Just as in Section 21.1, some of the configurations will be very much more probable than others. For instance, it is very unlikely that the whole of the energy $\mathbb{E}$ will accumulate in one system (giving the configuration $\{\mathbb{N}, 0, 0, \ldots\}$). By analogy with the earlier discussion, we can anticipate that there will be a dominating configuration (of composition $\{\mathbb{n}_0^*, \mathbb{n}_1^*, \ldots\}$ and weight $\mathbb{W}^*$), and that we can evaluate the thermodynamic properties by taking the average over the ensemble using that single, most probable, composition. In the thermodynamic limit of $\mathbb{N}\rightarrow\infty$, this dominating configuration is overwhelmingly the most probable, and it dominates the properties of the system virtually completely.

The quantitative discussion follows the argument in Section 21.1(b) with the modification that N, n_i is replaced by $\mathbb{N}$, $\mathbb{n}_i$. The weight of a configuration $\{\mathbb{n}_0, \mathbb{n}_1, \ldots\}$ is

$$\mathbb{W}=\mathbb{N}!/\mathbb{n}_0!\,\mathbb{n}_1!\ldots \qquad (21.2.1)$$

and the configuration of greatest weight subject to the constraints that the total energy of the ensemble is constant and that the total number of members is fixed at $\mathbb{N}$ is given by the *canonical distribution*

$$\mathbb{n}_i^*/\mathbb{N} = e^{-\beta E_i}/Q, \qquad Q = \sum_j e^{-\beta E_j}. \tag{21.2.2}$$

Q is called the *canonical partition function*.

The shape of the canonical distribution is only *apparently* an exponentially decreasing function of the energy of the system. It must be noted that eqn (21.2.2) gives the probability of occurrence of members in a single *state* i (of the entire system) of energy E_i. There may in fact be numerous states with almost identical energies. For example, in a gas the identities of the particles moving slowly or quickly can change without necessarily affecting the total energy. The *density of states*, the number of states per unit energy range, is a very sharply increasing function of energy. It follows that the probability of a member of an ensemble having a specified *energy* (as distinct from a specified state) is given by eqn (21.2.2), a sharply decreasing function, multiplied by a sharply increasing function, Fig. 21.4. Therefore, the overall distribution is a sharply peaked function near some energy: *most members of the ensemble have an energy very close to the mean value*.

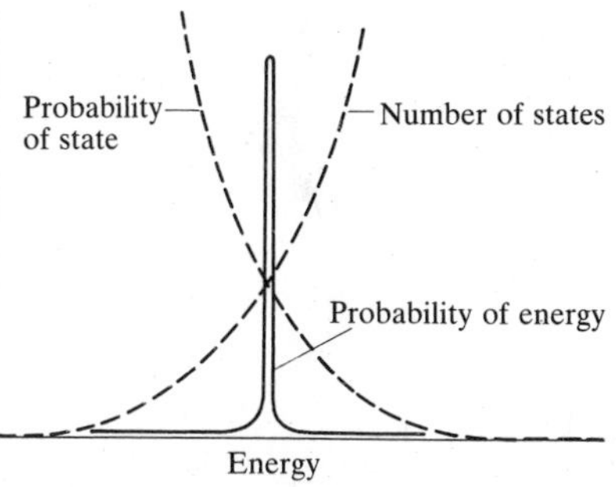

Fig. 21.4. In order to construct the form of the distribution of members of the canonical ensemble in terms of their energies, we multiply the probability that any one is in a state of given energy, eqn (21.2.2), by the number of states corresponding to that energy (a steeply rising function). The product is a sharply peaked function at the mean energy, which shows that almost all the members of the ensemble have that energy.

Like the molecular partition function, the canonical partition function carries all the thermodynamic information about a system, but it is more general because it is not based on the assumption that the particles are independent. The proportion of members of the ensemble in a state i with energy E_i is given by eqn (21.2.2) as

$$\mathbb{P}_i = \mathbb{n}_i^*/\mathbb{N} = e^{-\beta E_i}/Q, \tag{21.2.3}$$

and the internal energy is given by

$$U - U(0) = \mathbb{E}/\mathbb{N} = (1/\mathbb{N})\sum_i \mathbb{n}_i E_i = (1/Q)\sum_i E_i e^{-\beta E_i}.$$

This sum can be expressed in terms of the derivative of Q using the same argument as led to eqn (21.1.12), and in this case

$$U - U(0) = -(1/Q)(\partial Q/\partial\beta)_V = -(\partial \ln Q/\partial\beta)_V. \tag{21.2.4}$$

21.2 (b) The canonical partition function for independent particles

Now we see how to how to recover the molecular partition function from the more general canonical partition function when the particles are independent. The energy of a system is then the sum of the energies of its component particles:

$$E_j = \varepsilon_j^{(1)} + \varepsilon_j^{(2)} + \ldots + \varepsilon_j^{(N)},$$

where $\varepsilon_j^{(1)}$ is the energy of particle 1 in the member of the ensemble in the state j with energy E_j. The canonical partition function is then

$$Q = \sum_j \exp\{-\beta(\varepsilon_j^{(1)} + \varepsilon_j^{(2)} + \ldots + \varepsilon_j^{(N)})\},$$

where the sum is over all the states of the system. All these states can be covered by letting each particle enter all its own individual states (although we meet an important proviso shortly). Therefore, instead of summing over the collective states j of the system, we can sum over the *individual* states j of molecule 1, j of molecule 2, and so on. This leads to

$$Q = \left\{\sum_j e^{-\beta\varepsilon_j^{(1)}}\right\}\left\{\sum_j e^{-\beta\varepsilon_j^{(2)}}\right\}\dots\left\{\sum_j e^{-\beta\varepsilon_j^{(N)}}\right\}.$$

If all the particles are the same it is unnecessary to label them, and so

$$Q = \left\{\sum_j e^{-\beta\varepsilon_j}\right\}\left\{\sum_j e^{-\beta\varepsilon_j}\right\}\dots\left\{\sum_j e^{-\beta\varepsilon_j}\right\} = \left\{\sum_j e^{-\beta\varepsilon_j}\right\}^N.$$

We immediately recognize the molecular partition function on the right, and so it is tempting to write $Q = q^N$.

Unfortunately, the derivation fails in a very important case. This can be seen as follows. If all the particles are identical and free to move, we cannot distinguish them. Suppose that molecule 1 is in some state a, molecule 2 in b, and molecule 3 in c, then one member of the ensemble has an energy $E = \varepsilon_a + \varepsilon_b + \varepsilon_c$. This member, however, is indistinguishable from one formed by putting molecule 1 in state b, molecule 2 in state c, and molecule 3 in state a, or some other permutation (there are six in all, and $N!$ in general). In the case of *indistinguishable molecules*, it follows that we have counted too many states in going from the sum over system states to the sum over molecular states, and so writing $Q = q^N$ overestimates Q. The detailed argument is quite involved, but at all except very low temperatures it turns out that the correction factor is $1/N!$. Therefore:

For distinguishable particles: $Q = q^N$; (21.2.5a)

For indistinguishable particles; $Q = q^N/N!$. (21.2.5b)

When are the particles (a) distinguishable or (b) indistinguishable? To be indistinguishable, they must be of the same kind: an argon atom is never indistinguishable from a neon atom. Their identity, however, is not the only criterion. If identical particles are in a crystal lattice, each one can be 'named' with a set of coordinates. Identical particles in a lattice are therefore distinguishable, and we use eqn (21.2.5a). On the other hand, identical particles in a gas are free to move, and there is no way of keeping track of the identity of a given particle; we therefore use eqn (21.2.5b).

21.3 The statistical entropy

We have concentrated on the connection between the partition function and the First Law concept of internal energy. If it is true that the partition function contains *all* thermodynamic information, it must be possible to use it to calculate the Second Law concept of entropy.

21.3 (a) Heat, work, and entropy

In order to establish the relation between entropy and the partition function, we analyse the contributions to the total energy more closely; in particular, we consider how the internal energy changes when heat is supplied or work is done.

In terms of the picture we have been building, and in particular the expression

$$U - U(0) = (1/\mathbb{N}) \sum_i \mathbb{n}_i^* E_i, \qquad \mathbb{N} \to \infty$$

for the internal energy, a change in U may arise from either a modification of the energy levels of a system (so that E_i changes to $E_i + dE_i$) or from a modification of the populations (so that $\mathbb{n}_i^*$ changes to $\mathbb{n}_i^* + d\mathbb{n}_i$). The most general change is therefore

$$dU = dU(0) + (1/\mathbb{N}) \sum_i \mathbb{n}_i^* \, dE_i + (1/\mathbb{N}) \sum_i E_i \, d\mathbb{n}_i. \qquad (21.3.1)$$

From thermodynamics, we know that for a reversible change

$$dU = dq_{\text{rev}} + dw_{\text{rev}} = T \, dS + dw_{\text{rev}}, \qquad (21.3.2)$$

and a connection is beginning to emerge. The connection is established when we realize that *the energy levels of a system change when its size is changed by doing work*; they do not change when it is heated at constant volume, Fig. 21.5. On the other hand, *the populations of the states change when a system is heated*, the levels themselves remaining unchanged. This points to the identifications

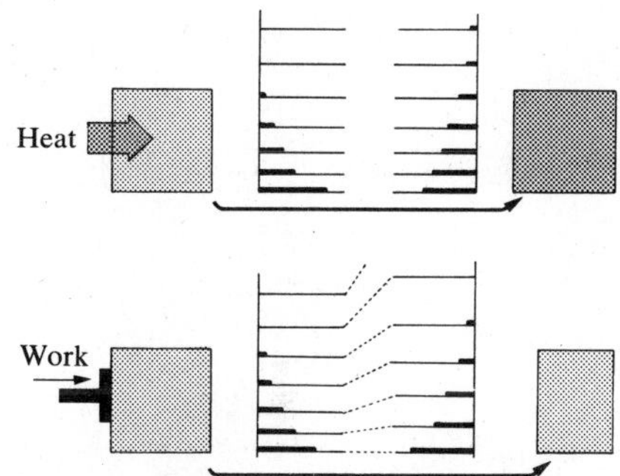

Fig. 21.5. (a) When a system is *heated* the energy levels are unchanged but their populations are changed. (b) When *work* is done on a system, the energy levels themselves are changed. The levels in this case can be regarded as the particle-in-a-box energy levels of Chapter 14, which depend on the size of the container and move apart as its volume is decreased.

$$dq_{\text{rev}} = T \, dS = (1/\mathbb{N}) \sum_i E_i \, d\mathbb{n}_i, \qquad (21.3.3)$$

so that *reversible heating corresponds to the redistribution of populations among fixed energy levels*, and

$$dw_{\text{rev}} = dU(0) + (1/\mathbb{N}) \sum_i \mathbb{n}_i^* \, dE_i, \qquad (21.3.4)$$

so that reversible work of compression or expansion corresponds to the modification of the energy levels, the populations remaining fixed. The first of these allows us to identify the change in entropy as

$$dS = (1/\mathbb{N}T) \sum_i E_i \, d\mathbb{n}_i. \qquad (21.3.5)$$

Thermodynamics deals with *changes* in properties, not their absolute values. Statistical thermodynamics, however, goes beyond classical thermodynamics by allowing us to calculate the absolute values of some properties. We now show that if an entropy is *defined* by the expression

$$S = (k/\mathbb{N}) \ln \mathbb{W}^*, \qquad \mathbb{N} \to \infty, \qquad (21.3.6)$$

then not only can we derive eqn (21.3.5), but we can also calculate the absolute values of the entropy.

$\mathbb{W}^*$ is the weight of the most probable configuration of the ensemble. When a system is heated reversibly a change in S arises from a change in $\ln \mathbb{W}^*$, and that in turn arises from a change in the configuration of the ensemble. This means that we can write

$$dS = (k/\mathbb{N}) \, d(\ln \mathbb{W}^*) = (k/\mathbb{N}) \sum_i (\partial \ln \mathbb{W}^*/\partial \mathbb{n}_i) \, d\mathbb{n}_i.$$

Since $\mathbb{W}^*$ is the weight of the most probable configuration, it satisfies eqn (21.1.6), which we now write as $(\partial \ln \mathbb{W}^*/\partial \mathbb{n}_i) + \alpha - \beta E_i = 0$. The equation for dS is therefore

$$dS = (k/\mathbb{N}) \sum_i (-\alpha + \beta E_i)\, d\mathbb{n}_i = (k\beta/\mathbb{N}) \sum_i E_i\, d\mathbb{n}_i.$$

We have used $\sum_i d\mathbb{n}_i = 0$ because the number of members remains constant; the sum $\sum_i E_i\, d\mathbb{n}_i$ is still a constraint, but it is no longer equal to zero because the system is being heated. The last equation is the same as eqn (21.3.5), confirming that eqn (21.3.6) is an appropriate definition of entropy so long as we identify β as $1/kT$. This is the formal demonstration of the relation $\beta = 1/kT$ that we had anticipated.

As the temperature is lowered, the value of $\mathbb{W}^*$, and hence of S, decreases because fewer configurations are compatible with the total energy. In the limit $T = 0$, $\mathbb{W}^* = 1$ so that $\ln \mathbb{W}^* = 0$, because only *one* configuration (every member of the ensemble in the lowest level) is compatible with $\mathbb{E} = 0$. It follows that $S \to 0$ as $T \to 0$. This is consistent with the Third Law of thermodynamics, that the entropies of all perfect crystals approach the same value as $T \to 0$ (Section 5.4(a)).

21.3 (b) The entropy and the partition function

The expression for the entropy can be converted into a useful form, and related to the partition function, by the following manipulations:

$$\begin{aligned}
S &= (k/\mathbb{N}) \ln \{\mathbb{N}!/\mathbb{n}_0^*!\mathbb{n}_1^* \ldots\} = (k/\mathbb{N})\left\{\ln \mathbb{N}! - \sum_i \ln \mathbb{n}_i^*!\right\} \\
&= (k/\mathbb{N})\left\{\mathbb{N} \ln \mathbb{N} - \mathbb{N} - \sum_i \mathbb{n}_i^* \ln \mathbb{n}_i^* + \sum_i \mathbb{n}_i^*\right\} \\
&= (k/\mathbb{N})\left\{\mathbb{N} \ln \mathbb{N} - \sum_i \mathbb{n}_i^* \ln \mathbb{n}_i^*\right\} \\
&= (k/\mathbb{N})\left\{\left(\sum_i \mathbb{n}_i^*\right) \ln \mathbb{N} - \sum_i \mathbb{n}_i^* \ln \mathbb{n}_i^*\right\} \\
&= k \sum_i \{(\mathbb{n}_i^*/\mathbb{N}) \ln \mathbb{N} - (\mathbb{n}_i^*/\mathbb{N}) \ln \mathbb{n}_i^*\} = -k \sum_i (\mathbb{n}_i^*/\mathbb{N}) \ln (\mathbb{n}_i^*/\mathbb{N}).
\end{aligned}$$

The ratio $\mathbb{n}_i^*/\mathbb{N} = \mathbb{P}_i$, the proportion of members of the ensemble having the energy E_i (eqn (21.2.3)) in the most probable configuration, and so

$$S = -k \sum_i \mathbb{P}_i \ln \mathbb{P}_i. \tag{21.3.7}$$

This succinct expression (which should remind you of the expression for the entropy of mixing, eqn (8.2.2), of two perfect fluids) may be developed further. It follows from eqn (21.2.3) that $\ln \mathbb{P}_i = -\beta E_i - \ln Q$, and therefore that

$$S = -k\left\{-\beta \sum_i \mathbb{P}_i E_i - \sum_i \mathbb{P}_i \ln Q\right\} = k\beta\{U - U(0)\} + k \ln Q,$$

because the sum over the $\mathbb{P}_i$ is unity, and the sum over $\mathbb{P}_i E_i$ is equal to $U - U(0)$ in the thermodynamic limit. We have already established that $\beta = 1/kT$, and so we find the following very important expression for the statistical entropy:

$$S = \{U - U(0)\}/T + k \ln Q. \quad (21.3.8)$$

Example 21.5

Calculate the entropy of a collection of N independent harmonic oscillators, and evaluate it using vibrational data for I_2 at 25 °C.

• *Method*. Stitch together several of the results already obtained. The canonical partition function is related to the molecular partition function using eqn (21.2.5a). The molecular partition function is given in eqn (21.1.13) for a particle with evenly spaced energy levels (as in a harmonic oscillator). The internal energy is obtained using eqn (21.1.12). Data for I_2 are given in *Example* 21.2.

• *Answer*. $Q = q^N$ in this case, and so

$$S = k\beta\{U - U(0)\} + Nk \ln q.$$

Since $q = 1/(1 - e^{-\beta\varepsilon})$,

$$U - U(0) = -(N/q)(\partial q/\partial\beta)_V = N\varepsilon e^{-\beta\varepsilon}/(1 - e^{-\beta\varepsilon}).$$

The entropy is then obtained by substituting this expression in the preceding one. For I_2 at 25 °C, $\beta\varepsilon = 1.036$ and $q = 1.440$ (*Example* 21.2). Therefore, since 214.6 cm^{-1} corresponds to 2.567 kJ mol^{-1}, and setting $N = nN_A$,

$$U - U(0) = (2.567\ \text{kJ mol}^{-1})ne^{-1.036}/1.550 = (1.412\ \text{kJ mol}^{-1})n$$
$$Nk \ln q = nR \ln 1.550 = (3.644\ \text{J K}^{-1}\ \text{mol}^{-1})n.$$

Hence,

$$S_m = (1.412\ \text{kJ mol}^{-1}/298.15\ \text{K}) + 3.644\ \text{J K}^{-1}\ \text{mol}^{-1}$$
$$= 8.38\ \text{J K}^{-1}\ \text{mol}^{-1}.$$

• *Comment*. We shall see in Section 22.3(a) that the total entropy of a molecule may be ascribed to its different modes of motion, and this value will turn out to be the contribution arising from the vibrational mode of the I_2 molecule.

• *Exercise*. Evaluate for $\beta\varepsilon = 1$ the molar entropy of the two-level system introduced in *Example* 21.1 and treated in subsequent *Exercises*. [10.7 J K^{-1} mol^{-1}]

21.3 (c) The entropy of a monatomic gas

In the case of a gas of independent particles, the canonical partition function may be replaced by $q^N/N!$, with the result that eqn (21.3.8) becomes

$$S = \{U - U(0)\}/T + kN \ln q - k \ln N!.$$

Since the number of particles ($N = nN_A$) in a typical sample is large, we can use Stirling's approximation (and $R = kN_A$):

$$S = \{U - U(0)\}/T + nR \ln q - k\{N \ln N - N\}$$
$$= \{U - U(0)\}/T + nR\{\ln q - \ln N + 1\}. \quad (21.3.9)^\circ$$

We have already seen how to calculate the molecular partition function, and so the last expression is an equation for the entropy in terms of molecular properties.

In the case of a gas of monatomic particles, the only motion is translation, and the molecular partition function is $q = V/\Lambda^3$, eqn (21.1.15). The internal energy is given by eqn (21.1.16) as $U - U(0) = 3N\beta/2 = 3nRT/2$, and so the entropy is

$$S = \{\tfrac{3}{2}RT\}/T + nR\{\ln(V/\Lambda^3) - \ln nN_A + 1\}$$
$$= nR\{\ln e^{\frac{3}{2}} + \ln(V/\Lambda^3) - \ln nN_A + \ln e\}$$
$$= nR \ln\{e^{\frac{5}{2}}V/nN_A\Lambda^3\}, \qquad (21.3.10)°$$

with $\Lambda = h/(2\pi mkT)^{\frac{1}{2}}$. This is the *Sackur–Tetrode equation* for the entropy of a monatomic perfect gas. Since the gas is perfect, $V = nRT/p$, and so an alternative form is

$$S = nR \ln\{e^{\frac{5}{2}}kT/p\Lambda^3\}. \qquad (21.3.11)°$$

A connection with the material in Part 1 can now be made: the Sackur-Tetrode equation implies that when a perfect gas expands isothermally from V_i to V_f, its entropy changes by

$$\Delta S = nR \ln aV_f - nR \ln aV_i = nR \ln(V_f/V_i),$$

where aV is the collection of quantities inside the logarithm of eqn (21.3.10). This is exactly the expression we obtain in classical thermodynamics.

Example 21.6

Evaluate the translational entropy of gaseous argon at standard ambient temperature and pressure ($\mathcal{T}$ and 1 bar).

- *Method*. Use eqn (21.3.11) with $m = 39.95m_u$.

- *Answer*. $2\pi mk\mathcal{T} = 39.95 \times (2\pi m_u k\mathcal{T}) = 39.95 \times (4.2949 \times 10^{-47}\ \text{J kg})$

$$= 1.716 \times 10^{-45}\ \text{J kg},$$

so that $\Lambda = 1.600 \times 10^{-11}$ m. Then, since

$$e^{\frac{5}{2}}k\mathcal{T}/p^{\ominus} = 5.0148 \times 10^{-25}\ \text{m}^3,$$
$$S_m = R \ln(5.0148 \times 10^{-25}\ \text{m}^3/4.096 \times 10^{-33}\ \text{m}^3)$$
$$= 18.62R = 155\ \text{J K}^{-1}\ \text{mol}^{-1}.$$

- *Comment*. It is often very convenient to express molar entropies (and molar heat capacities) as multiples of R.

- *Exercise*. Anticipate, on the basis of the number of accessible states for a lighter particle, whether the standard molar entropy of neon is likely to be larger or smaller than for Ar, and then calculate its value. [$17.60R$]

Appendix 21.1 Undetermined multipliers

Suppose we need to find the maximum (or minimum) value of some function f that depends on several variables $x_1, x_2, \ldots, x_n$. When the variables undergo a small change $x_i \to x_i + \delta x_i$ the function changes from f to $f + \delta f$, where

$$\delta f = \sum_i (\partial f/\partial x_i)\,\delta x_i.$$

At an extremum $\delta f = 0$, and so then

$$\sum_i (\partial f/\partial x_i)\,\delta x_i = 0. \qquad (21.A1.1)$$

If the x_i are all independent, all the δx_i are arbitrary, and so this equation can be solved by setting each $(\partial f/\partial x_i) = 0$ individually. When the x_i are not all independent (in this chapter one condition the n_i had to satisfy was $n_0 + n_1 + \ldots = N$) the δx_i are not all independent, and so the simple solution is no longer valid. We proceed as follows.

Let the *constraint* connecting the variables be an equation of the form $g = 0$. For example, in this chapter, one constraint was $n_0 + n_1 + \ldots = N$ which can be written $(n_0 + n_1 + \ldots) - N = 0$, so that in this case $g = (n_0 + n_1 + \ldots) - N$. This constraint is always valid, and so g remains unchanged when the x_i are varied:

$$\delta g = \sum_i (\partial g/\partial x_i)\,\delta x_i = 0. \qquad (21.A1.2)$$

Multiply this equation by some parameter λ and add it to eqn (21.A1.1):

$$\sum_{i=1}^{n} \{(\partial f/\partial x_i) + \lambda(\partial g/\partial x_i)\}\,\delta x_i = 0. \qquad (21.A1.3)$$

This can be regarded as an equation for one of the δx, δx_n for instance, in terms of all the other δx_i. All those other δx_i $(i = 1, 2, \ldots n-1)$ are independent, because there is only one constraint on the system. But here is the trick: λ is arbitrary; therefore we can choose it so that the coefficient of δx_n in eqn (21.A1.3) vanishes. That is, we choose it so that

$$(\partial f/\partial x_n) + \lambda(\partial g/\partial x_n) = 0. \qquad (21.A1.4)$$

Then eqn (21.A1.3) becomes

$$\sum_{i=1}^{n-1} \{(\partial f/\partial x_i) + \lambda(\partial g/\partial x_i)\}\,\delta x_i = 0.$$

Now the $n-1$ variations δx_i are independent, and so the solution of this equation is

$$(\partial f/\partial x_i) + \lambda(\partial g/\partial x_i) = 0, \qquad i = 1, 2, \ldots n-1.$$

But eqn (21.A1.4) has exactly the same form, and so the extremum can be found by solving

$$(\partial f/\partial x_i) + \lambda(\partial g/\partial x_i) = 0, \qquad \text{all } i.$$

This was illustrated in the text in the case of two constraints and therefore two undetermined multipliers λ_1 and λ_2 (α and $-\beta$).

The multipliers cannot always remain undetermined. One approach is to solve eqn (21.A1.4) instead of incorporating it into the minimization scheme. In this chapter we used the alternative procedure of keeping λ undetermined until a property was calculated for which the value was already known. Thus α was found by ensuring that $n_0 + n_1 + \ldots = N$, and β was found by relating $\mathrm{d}S$ to a known result.

Further reading

Elementary statistical thermodynamics. L. K. Nash; Addison–Wesley, Reading, 1968.

The second law. P. W. Atkins; Scientific American Books, New York, 1984.
The second law. H. A. Bent; Oxford University Press, 1965.
Statistical thermodynamics. B. J. McClelland; Wiley, New York, 1973.
Statistical thermodynamics. D. A. McQuarrie; Harper & Row, New York, 1976.
Statistical thermodynamics. N. Davidson; McGraw-Hill, New York, 1962.
An introduction to statistical mechanics. T. L. Hill; Addison–Wesley, Reading, 1960.
Statistical thermodynamics. A. Münster; Springer, Berlin, 1974.

Introductory problems

A21.1. Calculate the translational partition function at 300 K and at 400 K for a molecule with RMM = 120, when the volume of the gas is 2.00 cm^3.

A21.2. Find the ratio of the translational partition function of the deuterium molecule to that of the hydrogen molecule at the same temperature and volume.

A21.3. An atom has a threefold degenerate ground state, a singly degenerate electronic state at 3500 cm^{-1}, and a threefold degenerate electronic state at 4700 cm^{-1}. Calculate the electronic partition function for the atom at 1900 K.

A21.4. Calculate the electronic part of $U - U(0)$ at 1900 K for a 1 mol sample of the atoms in Problem A21.3.

A21.5. A molecule has a nondegenerate excited state lying at 540 cm^{-1} above the nondegenerate ground state. At what temperature will ten per cent of the molecules be in the excited state?

A21.6. Test the approximation $\ln N! = N \ln N - N$, widely used when N is very large, for $N = 13$. Compare the value obtained from the approximation with the value obtained by direct multiplication of integers. For $N = 13$, verify that Stirling's approximation in the form $\ln N! = (N + \frac{1}{2}) \ln N - N + \frac{1}{2} \ln (2\pi)$ is a much better approximation.

A21.7. What value of the entropy does the Sackur–Tetrode equation give for the entropy of a mole of neon atoms at 200 K and 1 atm pressure?

A21.8. Find the vibrational contribution to the entropy of a mole of a diatomic gas which has a fundamental vibrational frequency of 732 cm^{-1}. The temperature of the gas is 300 K.

A21.9. Calculate the translational entropy of a mole of Ne at 25 °C and 1 atm pressure.

A21.10. An atom has an energy level system consisting of two levels which are 450 cm^{-1} apart. The degeneracy of the lower level is 2 and that of the upper level is 4. In a certain collection of these atoms at 300 K, the fraction of atoms with the higher energy is 0.30. Show that this collection is not at equilibrium.

Problems

21.1. A sample of 5 molecules has a total energy $U = 5(\varepsilon_0 + \varepsilon)$. Each molecule is able to occupy states of energy $\varepsilon_0 + j\varepsilon$, j an integer $(0, 1, \ldots)$. What is the weight of the configuration corresponding to the energy being equally shared among the molecules?

21.2. In the first Problem we calculated the weight of one configuration, but there are several others, all of greater importance. Draw up a table with columns headed by the energy of a molecule (from ε_0 up to $\varepsilon_0 + 5\varepsilon$) and write beneath them all the configurations of the system compatible with the total energy $5(\varepsilon_0 + \varepsilon)$. Start, for example, with 4, 0, 0, 0, 0, 1 (only one member can have the energy $\varepsilon_0 + 5\varepsilon$ and the other four must then all have ε_0). Find the weight of each configuration (use eqn (21.1.1)). What is the most probable configuration?

21.3. In the case of a sample composed of nine molecules we are approaching the region of numbers where the averages are just beginning to have thermodynamic significance, but where averages can still be dealt with exactly and numerically. Draw up a table of configurations for $N = 9$, total energy $9(\varepsilon + \varepsilon_0)$, and therefore with each molecule, on the average, having the same energy as in the last problem.

21.4. Before working out the weights of the configuration in the $N = 9$ system, look at the ones you have drawn up in the table and guess (by looking for 'exponential' form) which of the configurations will turn out to be the most probable (with the greatest weight). Now calculate the weights of all the configurations (not nearly such a long task as it might seem at first sight) and identify the most probable one.

21.5. The most probable configuration is characterized by the *temperature* of the system. But what is the temperature of the system we have been considering? It must be such as to give a mean value $\varepsilon_0 + \varepsilon$ for the energy of each molecule and a total energy $N(\varepsilon + \varepsilon_0)$, the molecules being allowed only the energies $\varepsilon_0 + j\varepsilon$. Show that the temperature of a system can be obtained by plotting $\ln (n_j/n_0)$ against j, the n_j being the populations for the most probable configuration (in the thermodynamic limit). Apply this method to the nine molecule system in order to see how closely it corresponds to the thermodynamic limit (by assessing the quality of the straight line). Let the spacing of the molecular energy levels be $\varepsilon \triangleq 50\ \mathrm{cm}^{-1}$. What is the temperature of this system?

21.6. When N is large, show that if the mean energy per molecule is $\varepsilon_0 + a\varepsilon$ when molecules can occupy the ladder of energy levels $\varepsilon_0 + j\varepsilon$, then the temperature of the sample is given by $\beta = (1/\varepsilon)\ln(1 + 1/a)$. Find the temperature for the case treated in the last set of problems ($a = 1$, $\varepsilon \triangleq 50\ \mathrm{cm}^{-1}$). What is the value of q at the temperature for which the mean molecular energy is $\varepsilon_0 + a\varepsilon$?

21.7. The 'temperature' is a parameter that has significance only for the most probable configuration, and we should not expect to get a good straight line for others. Choose two other configurations for the nine molecule system, one close to most probable, one far from it, and plot $\ln(n_j/n_0)$ against j. The straight lines should be very poor.

21.8. Calculate the molecular partition function for the nine-molecule system at 104 K (the result of Problem 21.6), 100 K, and 108 K. Then find the gradient $\mathrm{d}q/\mathrm{d}\beta$ at 104 K and so confirm explicitly that the total energy (which we know to be $9(\varepsilon_0 + \varepsilon)$ at that temperature) is given by $-N(\mathrm{d}\ln q/\mathrm{d}\beta)$. Any discrepancy comes from the estimation of the temperature of the ensemble.

21.9. We now begin to explore the calculation and manipulation of the molecular partition function. As a first step, state for which systems it is essential to include a factor of $1/N!$ in going from Q to q: (a) a sample of helium gas; (b) a sample of carbon monoxide gas; (c) the same sample of carbon monoxide, but now frozen to a solid; (d) water vapour; (e) ice; (f) an electron gas; (g) an electron gas in a metal.

21.10. An argon atom is trapped in a cubical box of volume V. What is its translational partition function at (a) 100 K, (b) 298 K, (c) 10 000 K, (d) 0, when the box is of side 1 cm?

21.11. The form of the translational partition function, as normally used, is valid when a huge number of energy levels are accessible. When does the normal expression become invalid and when do we have to resort to the explicit summation, eqn (21.1.10)? Estimate what the temperature of the argon in the last Problem would have to be in order for the partition function to drop to about 10. What is the exact value of the partition function at that temperature?

21.12. There are several types of partition function that can be calculated by direct summation of exponentials, using spectroscopic data for the molecular energy levels involved. First we deal with the tellurium atom, which has several low-lying excited states. Find the electronic partition function for atoms at (a) 298 K, (b) 5000 K using the following data from atomic spectroscopy: ground state (fivefold degenerate), $4751\ \mathrm{cm}^{-1}$ (threefold degenerate), $4707\ \mathrm{cm}^{-1}$ (singly degenerate), $10\,559\ \mathrm{cm}^{-1}$ (fivefold degenerate).

21.13. What proportion of the tellurium atoms are in (a) their ground level, (b) the level at $4751\ \mathrm{cm}^{-1}$ at the two temperatures of the last Problem?

21.14. Most of the molecules we shall encounter have electronic states that are so high in energy above the ground state that only the latter need be considered for its thermodynamic properties. There are several exceptions, one interesting case being NO, for this has an electronically excited state at only $121.1\ \mathrm{cm}^{-1}$ above the ground state. Both this and the ground state are doubly degenerate. Calculate and plot the electronic partition function of NO from zero to 1000 K. What is the distribution of populations at room temperature?

21.15. What is the molar electronic internal energy of the NO molecule at 298 K? Answer this first by calculating an expression for U at a general temperature (use the partition function found above), and then substitute the value $T = 298$ K.

21.16. *Example* 21.5 assumed that the vibration of the I_2 molecules were harmonic. In fact it has vibrational energy levels at the following wavenumbers above the zero-point level: $213.30\ \mathrm{cm}^{-1}$, $425.39\ \mathrm{cm}^{-1}$, $636.27\ \mathrm{cm}^{-1}$, $845.93\ \mathrm{cm}^{-1}$, $1054.38\ \mathrm{cm}^{-1}$. Calculate the vibrational partition function by explicit summation of the expression for q at (a) 100 K, (b) 298 K.

21.17. What proportion of the iodine molecules are in the ground and first two excited vibrational states at the two temperatures in the last Problem?

21.18. What is the molar vibrational energy of molecular iodine at those two temperatures? Evaluate the sum for U, eqn (21.1.12), explicitly using the data in Problem 21.15.

21.19. An electron spin can adopt two orientations in a magnetic field, the energies being $\pm\mu_B B$. Find an expression for the electron spin partition function and mean energy, and plot these as a function of applied field. What are the relative populations of the spin levels at (a) 4 K, (b) 298 K and 300 mT?

21.20. A three-level system, such as a nitrogen nucleus in a magnetic field, can also be expressed in a simple closed form. Derive an expression for the partition function and mean energy of a nitrogen nucleus ($I = 1$) in a magnetic field at 4 K. Take the $M_I = 0$ state as the zero of energy.

21.21. A peculiarity of two-level systems can now be explored. Suppose by some artificial means we contrive to invert the populations of the spin levels (methods do exist for this and lasers make use of them). More electrons will then be in the upper energy state than in the lower. This is a non-equilibrium condition, and the distribution of populations is far from being the most probable. Nevertheless, we can still *formally* express the ratio of populations in terms of a single parameter, which we are free to call 'temperature'. Demonstrate that this temperature must be *less* than absolute zero. What 'temperature' corresponds to (a) an inversion of the equilibrium population at 298 K, (b) an inversion of the equilibrium populations at 10 K, (c) total inversion, with all electrons in the upper spin state?

21.22. Under what circumstances is it permissible or meaningful to speak of negative temperatures in connection with three-level systems?

21.23. What is the average entropy per molecule when N is large enough for eqn (21.3.8) to be applied to the system treated in Problem 21.6; that is, when the average energy per molecule is $\varepsilon_0 + a\varepsilon$? Show that $S/Nk = (1+a)\ln(1+a) - a\ln a$, and hence that $S/Nk = 2\ln 2$ when the mean molecular energy is $\varepsilon_0 + \varepsilon$.

21.24. Calculate the electronic contribution to the molar entropy of the tellurium atom at (a) 298 K, (b) 5000 K; use the data in Problem 21.12.

21.25. Calculate the electronic contribution to the molar entropy of the NO molecule at (a) 298 K, (b) 500 K; use the data in Problems 21.14–15.

21.26. Calculate the vibrational contribution to the molar entropy of the iodine molecule at (a) 100 K, (b) 298 K; use the data in Problems 21.16–18.

21.27. Calculate the dependence of the molar entropy of a collection of independent, localized electron spins on the strength of an applied magnetic field. What do you *expect* the entropy of the spins to be at (a) $B = 0$, (b) $B = \infty$, and what do you *calculate* it to be?

21.28. Suppose 1 mol of argon atoms are first held rigidly in position, and then allowed to move freely under 1 atm pressure: what is the change of entropy?

21.29. What is the entropy of 1 mol of a dilute electron gas at (a) 298 K, (b) 5000 K? Take $V = 10\ \mathrm{dm}^3$.

21.30. Confirm that the statistical thermodynamic expression for the entropy of a monatomic ideal gas accounts correctly for the dependence of the entropy on (a) its pressure, (b) its temperature.

22

Statistical thermodynamics: the machinery

Learning objectives

After careful study of this chapter you should be able to:

(1) Show that the partition function may be factorized into contributions from each mode of motion, eqn (22.1.2).

(2) Calculate the *translational partition function*, eqn (22.1.3) and Box 22.1.

(3) Explain the origin of the *symmetry number* in the case of molecular hydrogen, and calculate its value in general, Section 22.1(c).

(4) Calculate the *rotational partition function*, eqn (22.1.4), and derive and use high temperature approximate forms, Box 22.1.

(5) Derive and use an expression for the *vibrational partition function*, eqn (22.1.8) and Box 22.1.

(6) State the *electronic partition function*, eqns (22.1.9) and (22.1.10), in simple cases.

(7) Relate the thermodynamic functions U, S, H, A, p, and G to the partition function, Section 22.2 and Box 22.2.

(8) Use spectroscopic and structural data to calculate thermodynamic functions, Box 22.2 and Example 22.4.

(9) Justify, state, and use the *equipartition principle*, and state its limitations, Section 22.3(a).

(10) Calculate the *mean energies* of modes of motion, Section 22.3(a).

(11) Use the partition function to calculate the *heat capacities* of samples, and explain the term *characteristic temperature* of a mode, Section 22.3(b).

(12) Use the equipartition principle to estimate the high-temperature value of the heat capacity, eqn (22.3.14) and Example 22.5.

(13) Define *residual entropy* and calculate its value for simple disordered systems, Section 22.3(c).

(14) Relate the *equilibrium constant* of a reaction to the partition functions of the participants, eqn (22.3.21), and calculate equilibrium constants from spectroscopic and structural data, Example 22.6.

(15) Relate the equilibrium composition to the densities of states and the relative energies of the reactants and products, Section 22.3(d) and Figs. 22.8–9.

Introduction

The partition function is the bridge between thermodynamics, spectroscopy, and quantum mechanics because, once it is known, the thermodynamic functions may be calculated. In this chapter we explore the ways in which statistical thermodynamics is used to discuss problems of chemical significance, and we see how to calculate the molecular partition function q from spectroscopic data or structural information. We treat heat capacities, entropies, and equilibrium constants, and in the process discover new ways of thinking about their significance.

22.1 How to calculate the partition function

Consider a perfect gas of particles with internal structure. Since the molecules are indistinguishable and independent, $Q = q^N/N!$, and so we can concentrate on evaluating the molecular partition function:

$$q = \sum_j e^{-\beta\varepsilon_j}, \qquad \beta = 1/kT. \tag{22.1.1}$$

The first step is to note that the energy of a molecule is the sum of contributions from its different modes of motion:

$$\varepsilon_j = \varepsilon^{\text{translation}} + \varepsilon^{\text{rotation}} + \varepsilon^{\text{vibration}} + \varepsilon^{\text{electronic}}.$$

This separation is only approximate (except in the case of translation) because the modes are not perfectly independent, but in most cases it is satisfactory (the separation of the electronic and nuclear motions, for example, is justified by the Born–Oppenheimer approximation). The advantage of the separation is that the partition function factorizes into contributions from each mode:

$$\begin{aligned} q &= \sum_j \exp(-\beta\varepsilon^{\text{T}} - \beta\varepsilon^{\text{R}} - \beta\varepsilon^{\text{V}} - \beta\varepsilon^{\text{E}}) \\ &= \left\{\sum_j e^{-\beta\varepsilon^{\text{T}}}\right\}\left\{\sum_j e^{-\beta\varepsilon^{\text{R}}}\right\}\left\{\sum_j e^{-\beta\varepsilon^{\text{V}}}\right\}\left\{\sum_j e^{-\beta\varepsilon^{\text{E}}}\right\} = q^{\text{T}}q^{\text{R}}q^{\text{V}}q^{\text{E}}. \end{aligned} \tag{22.1.2}$$

22.1 (a) The translational contribution

The translational partition function of a particle of mass m in a container of volume V was derived in Section 21.1, eqn (21.1.15):

$$q^{\text{T}} = V/\Lambda^3, \qquad \Lambda = h(\beta/2\pi m)^{\frac{1}{2}}. \tag{22.1.3}^\circ$$

Notice that $q^{\text{T}} \to \infty$ as $T \to \infty$ because an infinite number of states become accessible. Even at room temperature $q^{\text{T}} \approx 2 \times 10^{28}$ for a molecule of oxygen in a 100 cm^3 vessel.

22.1 (b) The rotational contribution

The direct method of calculating q^{R} is to substitute the experimental values of the rotational energy levels into the expression for the partition function, and to sum the exponentials numerically. The degeneracy of the levels must be taken into account: in the case of a linear molecule, for example, each J level, of energy $E_J = hcBJ(J+1)$, is $(2J+1)$-fold degenerate (Section 18.2). The partition function is therefore

$$q^{\text{R}} = \sum_J (2J+1)e^{-\beta E_J}, \qquad E_J = hcBJ(J+1). \tag{22.1.4}$$

Example 22.1

Evaluate the rotational partition function of $^1H^{35}Cl$ at 25 °C. ($B = 10.591$ cm^{-1}.)

- *Method*. Use eqn (22.1.4), and evaluate it term by term. Use $k\mathcal{T}/hc = 207.22$ cm^{-1}.

• *Answer*. With $hcB/kT = 0.051\,11$ we draw up the following table:

J	0	1	2	3	4	...	10
$J(J+1)$	0	2	6	12	20	...	110
$x = 2J+1$	1	3	5	7	9	...	21
$y = e^{-0.05111J(J+1)}$	1	0.903	0.736	0.542	0.360	...	0.004
xy	1	2.71	3.68	3.79	3.32	...	0.08

Then form the sum required by eqn (22.1.4). This is equal to 19.9; hence $q^R = 19.9$ at this temperature. Taking J up to 16 gives 19.902.

• *Comment*. Notice that about 10 J-levels are significantly populated (but the number of populated *states* is larger on account of the $(2J+1)$-fold degeneracy of each level). We shall shortly encounter the approximation that $q^R \approx kT/hcB$. In the present case this gives $q^R = 19.6$, in good agreement with the exact value, and with much less work. Calculations based on eqn (22.1.4) are easily programmed for a microcomputer.

• *Exercise*. Evaluate the rotational partition function for HCl at 0 °C. [18.26]

22.1 (c) Symmetry numbers

Care must be taken not to include too many rotational states in the sum. This is a subtle problem, and can be traced to the fact that the Pauli principle forbids the appearance of some states. The difficulty arises only with a molecule with identical atoms that are interchanged as it rotates. Therefore it does not affect HCl, but it does affect homonuclear diatomics and other symmetrical molecules. Fortunately, this complication can be taken into account very simply when the temperature is so high that many rotational levels are populated. We shall begin by showing the nature of the problem, and then, in eqn (22.1.5), show how it may easily be taken into account at high temperatures.

Consider $^1H_2$. The hydrogen nuclei are *fermions*; that is, they are particles with half-integral spin (like electrons). Therefore they must obey the form of the Pauli principle set out in Section 15.2(b), and the total wavefunction of the molecule must change sign when the two nuclei are interchanged. Since a rotation of the molecule by 180° does interchange them, the total wavefunction of H_2 must change sign when this half-rotation occurs. The total wavefunction is the product of several factors. One is the electronic wavefunction. Since this is symmetrical when the molecule is rotated by 180°, we can ignore its effect from now on. The rotational wavefunctions are the spherical harmonics, Chapter 14. We can picture them as being like the s-, p-, . . . etc. orbitals of the hydrogen atom, but in place of the quantum number l we have the rotational quantum number J. When $J = 0$ (a spherical harmonic like an s-orbital), rotation by 180° results in no change of sign; when $J = 1$ (as for a p-orbital) there is a change of sign; when $J = 2$ (as for a d-orbital), there is no change of sign, and so on. The pattern emerging is that *the rotational wavefunction is multiplied by* $(-1)^J$ *when a rotation by 180° takes place*.

Now consider the wavefunctions representing the states of the two spin-$\frac{1}{2}$ nuclei. They are the same as for two spin-$\frac{1}{2}$ electrons, and are the ones specified in Section 15.2(c):

$$\alpha(1)\alpha(2), \qquad \sigma_+(1,2), \qquad \beta(1)\beta(2), \qquad \sigma_-(1,2).$$

As remarked there, the first three are *symmetric* under interchange of the particles (i.e. they do not change sign when 1 and 2 are exchanged), while σ_- is *antisymmetric* and does change sign.

We now have two factors in the overall H_2 wavefunction that may change when the molecule turns through 180°. However, the Pauli principle instructs us that *only totally antisymmetric wavefunctions are allowed* (for spin-$\frac{1}{2}$ particles and other fermions). Therefore we must take the symmetric (even J) rotational wavefunctions with an antisymmetric spin function (σ_-), and the antisymmetric rotational functions (odd J) with any of the three symmetric spin functions. This means that there are two kinds of molecular hydrogen:

para-hydrogen: Paired nuclear spins (⇅), $J = 0, 2, 4, \ldots$
ortho-hydrogen: Parallel nuclear spins (⇈), $J = 1, 3, 5, \ldots$

and there are three states of o-H_2 to each value of J (because there are three parallel spin states of the nuclei).

Now we are ready to set up the rotational partition function. In order to do so, we have to take note of the fact that 'ordinary' molecular hydrogen is a mixture of one part p-H_2 (with only its even-J rotational states occupied) and three parts o-H_2 (with only its odd-J rotational states occupied). Therefore, the *average* partition function per molecule is

$$q^{\mathrm{R}} = \frac{1}{4}\left\{\sum_{J\ \mathrm{even}} (2J+1)\mathrm{e}^{-hc\beta BJ(J+1)} + 3\sum_{J\ \mathrm{odd}} (2J+1)\mathrm{e}^{-hc\beta BJ(J+1)}\right\},$$

so that the odd-J states are more heavily weighted than the even-J states. This is illustrated in Fig. 22.1. From that illustration we see that we would obtain *approximately* the same answer for the partition function (the sum of all the populations) if each J term contributed *half* its normal value to the sum. That is, the last equation can be approximated as

$$q^{\mathrm{R}} = \tfrac{1}{2}\sum_{J} (2J+1)\mathrm{e}^{-hcB\beta J(J+1)},$$

and this approximation is very good when many terms contribute (at high temperatures).

At room temperature $kT/hc \approx 200\ \mathrm{cm}^{-1}$. The rotational constants of small molecules are around $1\ \mathrm{cm}^{-1}$ (HCl: $10.6\ \mathrm{cm}^{-1}$; CH_4: $5.2\ \mathrm{cm}^{-1}$; CO_2: $0.39\ \mathrm{cm}^{-1}$; I_2: $0.04\ \mathrm{cm}^{-1}$); hence many levels are populated, and so for

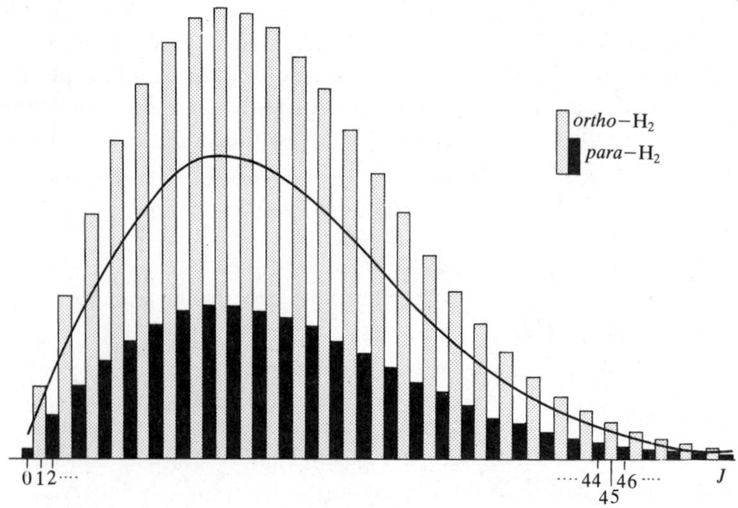

Fig. 22.1. The values of the individual terms $(2J+1)\mathrm{e}^{hcB\beta J(J+1)}$ contributing to the mean partition function of a 3:1 mixture of *ortho*- and *para*-H_2. The partition function is the sum of all these terms. At high temperatures the sum is approximately equal to the sum of the terms over *all* values of J, each with a weight of $\frac{1}{2}$. This is the sum of the contributions connected with a dotted line.

homonuclear diatomics the approximation of dealing with the Pauli principle by dividing by two is likely to be valid. (We have demonstrated it explicitly for H_2, but similar arguments apply to nuclei with other spins, including *bosons*, particles with integral spin.) Moreover, because many rotational states are occupied, the sum in the last equation may be approximated by an integral:

For a homonuclear diatomic:

$$q^{R} \approx \tfrac{1}{2}\int_0^{\infty} (2J+1)e^{-hcB\beta J(J+1)}\,dJ.$$

In the case of a heteronuclear diatomic, rotation by 180° does not interchange indistinguishable particles, the Pauli principle has nothing to say, and the factor $\frac{1}{2}$ does not appear. These remarks can be combined by writing

$$q^{R} \approx (1/\sigma)\int_0^{\infty} (2J+1)e^{-hcB\beta J(J+1)}\,dJ \qquad (22.1.5)$$

where σ, the *symmetry number*, is 1 for heteronuclear diatomics and 2 for homonuclear diatomics.

Although the integral in eqn (22.1.5) looks complicated, it can be evaluated without much difficulty by noticing that it can also be written as

$$q^{R} \approx (1/\sigma)\int_0^{\infty} (-1/\beta hcB)\{(d/dJ)e^{-hcB\beta J(J+1)}\}\,dJ.$$

The integral of a derivative of a function is the function itself, and so

$$q^{R} \approx (1/\sigma)\{(-1/\beta hcB)e^{-hcB\beta J(J+1)}\}\big|_0^{\infty} = 1/\sigma\beta hcB.$$

The approximate form of the rotational partition function of a linear molecule is therefore

$$q^{R} \approx kT/\sigma hcB = 2IkT/\sigma\hbar^2. \qquad (22.1.6)$$

The same problem arises for other types of symmetrical molecule, and a symmetry number can be introduced to avoid overcounting (so long as the temperature is high enough for the integral approximation to be valid). In the case of CO_2, for example, $\sigma = 2$ because it is a symmetrical linear molecule, but for OCS $\sigma = 1$. For H_2O $\sigma = 2$ and for NH_3 $\sigma = 3$. Notice that these numbers are the order of the *rotational subgroup* of the molecules. (The rotational subgroup is the point group of the molecule with all but the identity and the rotations shaved off, and its order is the number of symmetry operations remaining.) Thus, the rotational subgroup of H_2O is $\{E, C_2\}$, and so $\sigma = 2$; the rotational subgroup of NH_3 is $\{E, 2C_3\}$, and so $\sigma = 3$. This recipe makes it easy to find the symmetry numbers for more complicated molecules. The rotational subgroup of CH_4 is obtained from the T character table at the end of the data section as $\{E, 8C_3, 3C_2\}$, and so $\sigma = 12$ (physically that corresponds to three indistinguishable CH_3 orientations about each of four C—H axes). For benzene, the rotational subgroup of D_{6h} is $\{E, 2C_6, 2C_3, C_2, 3C_2', 3C_2''\}$, and so $\sigma = 12$.

The approximate form of the rotational partition function for these more complex molecules may be found by integration, as in eqn (22.1.5), and for a molecule with rotational constants A, B, and C,

$$q^{R} = (1/\sigma)(kT/hc)^{\frac{3}{2}}(\pi/ABC)^{\frac{1}{2}}. \tag{22.1.7}$$

Example 22.2

Estimate the rotational partition function for ethene at 25 °C. ($A = 4.828\ \text{cm}^{-1}$, $B = 1.0012\ \text{cm}^{-1}$, $C = 0.8282\ \text{cm}^{-1}$.)

• *Method*. Use eqn (22.1.7). Refer to the D_{2h} character table for the rotational subgroup. Use $k\mathcal{T}/hc = 207.223\ \text{cm}^{-1}$.

• *Answer*. The rotational subgroup $\{E, C_{2x}, C_{2y}, C_{2z}\}$ is of order 4 and so $\sigma = 4$. Therefore, since $ABC = 4.003_3\ \text{cm}^{-3}$, from eqn (22.1.7):

$$q^{R} = \tfrac{1}{4} \times (207.223\ \text{cm}^{-1})^{\frac{3}{2}} \times (\pi/4.003_3\ \text{cm}^{-3})^{\frac{1}{2}} = 661.$$

• *Comment*. Ethene is quite a big molecule, the energy levels are close together (compared with $k\mathcal{T}$), and so many are populated at room temperature.

• *Exercise*. Evaluate the rotational partition function of pyridine, C_5H_5N, at room temperature ($A = 0.2014\ \text{cm}^{-1}$, $B = 0.1936\ \text{cm}^{-1}$, $C = 0.0987\ \text{cm}^{-1}$). [$43 \times 10^3$]

The general conclusion at this stage is that *molecules with large moments of inertia* (small rotational contants) *have large rotational partition functions*. This reflects the closeness (compared with kT) of the rotational levels in large, heavy molecules, and the large number of them that are populated at normal temperatures.

22.1 (d) The vibrational contribution

In the case of a diatomic molecule, the vibrational partition function can be evaluated by substituting the measured vibrational energy levels into the exponentials appearing in the definition, and summing them numerically. In a polyatomic molecule each normal mode has its own partition function (so long as the anharmonicities are so small that the modes are independent), and the overall vibrational partition function is the product of these individual ones: $q^{V} = q^{V}(1)q^{V}(2)\ldots$, where $q^{V}(K)$ is the partition function for the Kth normal mode and is calculated by direct summation of the observed spectroscopic levels.

So long as the vibrational excitation is not too great, the harmonic approximation may be made, and the vibrational energy levels written as $E_v = (v + \frac{1}{2})hc\tilde{v}$, $v = 0, 1, 2, \ldots$ (Section 18.3). If we measure energies from the zero-point level (our general rule), the permitted values are $\varepsilon_v = vh\tilde{v}$. Then the partition function is

$$q^{V} = \sum_{v=0}^{\infty} e^{-vhc\tilde{v}\beta} = \sum_{v=0}^{\infty} \{e^{-hc\tilde{v}\beta}\}^{v}.$$

This series was encountered in Section 21.1(d) (which is no accident: the ladder-like array of levels in Fig. 21.1 is exactly the same as that of a harmonic oscillator). The series can be summed in the same way (see footnote on p. 512), and gives

$$q^{V} = \frac{1}{1 - e^{-hc\tilde{v}/kT}} \tag{22.1.8}$$

where $\tilde{\nu}$ is the wavenumber of the mode. In a polyatomic molecule, each normal mode gives rise to a partition function of this form.

Example 22.3

The wavenumbers of the three normal modes of H_2O are $3656.7\,cm^{-1}$, $1594.8\,cm^{-1}$, and $3755.8\,cm^{-1}$. Evaluate the vibrational partition function at 1500 K.

• *Method*. Use eqn (22.1.8) for each mode, and then form the product. At 1500 K, $kT/hc = 1042.5\,cm^{-1}$.

• *Answer*. Draw up the following table:

Mode:	1	2	3
$\tilde{\nu}/cm^{-1}$	3656.7	1594.8	3755.8
$hc\tilde{\nu}/kT$	3.508	1.530	3.603
q^V	1.031	1.276	1.028

The overall vibrational partition function is therefore

$$q = 1.031 \times 1.276 \times 1.028 = 1.353.$$

• *Comment*. The vibrations of H_2O are at such high wavenumber that even at 1500 K most of the molecules are in their vibrational ground state.

• *Exercise*. Repeat the calculation for CO_2, where the vibrational wavenumbers are 1388 cm^{-1}, $667.4\,cm^{-1}$, and $2349\,cm^{-1}$, ν_2 being the doubly-degenerate bending mode. [6.71]

In many molecules the vibrational wavenumbers are so great that $hc\tilde{\nu}\beta \gg 1$. For example, the lowest vibrational wavenumber of CH_4 is $1306\,cm^{-1}$, and so $hc\tilde{\nu}\beta = 6$ at room temperature. C–H stretches normally lie in the range $2850–2960\,cm^{-1}$, and so for them $hc\tilde{\nu}\beta = 14$. In these cases $e^{-hc\tilde{\nu}\beta}$ may be neglected in the denominator of q^V (for example, $e^{-6} = 0.002$), and the vibrational partition function for a single mode is simply $q^V \approx 1$ (implying that only the zero-point level is occupied).

22.1 (e) The electronic contribution

Electronic energy separations from the ground state are usually very large, and so for most cases $q^E = 1$. An important exception arises in the case of atoms and molecules having degenerate ground states, in which case

$$q^E = g^E, \tag{22.1.9}$$

where g^E is the dengeracy. The alkali metal atoms, for example, have doubly degenerate ground states (corresponding to the two orientations of their electron spin), and so $q^E = 2$.

Some atoms and molecules have low-lying electronically excited states (at high enough temperatures, all atoms and molecules have thermally accessible excited states). One case is nitrogen(II) oxide, NO, which has the configuration . . . π^1 (refer to Fig. 16.12, and add 15 electrons). The orbital angular momentum may take two orientations with respect to the molecular axis, and the spin angular momentum may also take two. This gives four states, Fig. 22.2. The energy of the two states in which the orbital and spin momenta are parallel (giving the $^2\Pi_{\frac{3}{2}}$ term) is slightly greater than that of

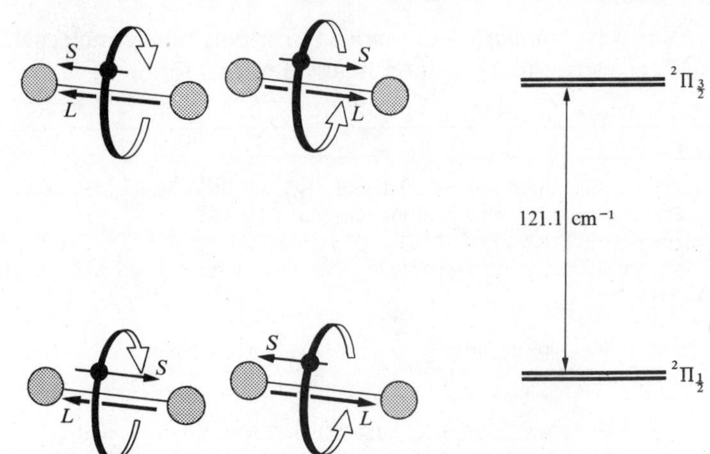

Fig. 22.2. The doubly degenerate ground electronic level of NO (with the spin and orbital angular momentum around the axis in opposite directions) and the doubly degenerate first excited level (with the spin and orbital momenta parallel). The upper level is thermally accessible at room temperature.

the two other states in which they are antiparallel (giving the $^2\Pi_{\frac{1}{2}}$ term) on account of spin-orbit coupling, Section 15.3. The separation between the two terms is only 121 cm^{-1}, and so at normal temperatures all four states are thermally accessible, but the doubly degenerate lower level is more heavily populated.

The electronic partition function of NO may be obtained as follows. Denoting the energies of two levels as $\varepsilon_{\frac{1}{2}}$ and $\varepsilon_{\frac{3}{2}}$ and writing $\varepsilon_{\frac{1}{2}} = 0$ and $\varepsilon_{\frac{3}{2}} = \varepsilon$, leads to

$$q^{\mathrm{E}} = \sum_j g_j \mathrm{e}^{-\beta\varepsilon_j} = 2 + 2\mathrm{e}^{-\beta\varepsilon}. \qquad (22.1.10)$$

At $T=0$, $q^{\mathrm{E}}=2$, because only the doubly degenerate ground state is accessible. At very high temperatures, q^{E} approaches 4 because all four states are accessible. At 25 °C, $q^{\mathrm{E}}=2.8$. Some of the consequences are explored in Problem 22.31.

22.1 (f) The overall partition function

The various contributions to the overall partition function of a molecule are collected in Box 22.1. The overall partition function is the product of each contribution. For a diatomic molecule this is

$$q = g^{\mathrm{E}}(V/\Lambda^3)(1/\sigma hcB\beta)\left\{\frac{1}{1-\mathrm{e}^{-hc\tilde{\nu}\beta}}\right\}. \qquad (22.1.11)$$

The overall partition functions obtained in this way are approximate because they assume that the rotational levels are very close and that the vibrational levels are harmonic. These approximations are avoided by evaluating the sums explicitly using spectroscopic data.

22.2 How to calculate the thermodynamic functions

In eqns (21.2.4) and (21.3.8) we have formulas for calculating the two principal thermodynamic functions:

$$U - U(0) = -(1/Q)(\partial Q/\partial\beta)_V = -(\partial \ln Q/\partial\beta)_V, \qquad (22.2.1)$$

$$S = \{U - U(0)\}/T + k \ln Q. \qquad (22.2.2)$$

Box 22.1 Contributions to the molecular partition function

In all the following expressions, $\beta = 1/kT$. It is often useful to note that

$$hc/k = 1.438\,79 \text{ cm K}.$$

See also inside front cover for further information.

1. *Translation*

$$q^{\mathrm{T}} = V/\Lambda^3, \qquad \Lambda = h(\beta/2\pi m)^{\frac{1}{2}}.$$
$$\Lambda = (1.749 \times 10^{-9}\,\mathrm{m})/(T/\mathrm{K})^{\frac{1}{2}} M_{\mathrm{r}}^{\frac{1}{2}}.$$
$$(q_{\mathrm{m}}^{\ominus})^{\mathrm{T}}/N_{\mathrm{A}} = kT/p^{\ominus}\Lambda^3 = 2.561 \times 10^{-2} (T/\mathrm{K})^{\frac{5}{2}} M_{\mathrm{r}}^{\frac{3}{2}}.$$

2. *Rotation*

(a) Linear molecules:

$$q^{\mathrm{R}} = 1/\sigma hc\beta B = (0.6950/\sigma)(T/\mathrm{K})/(B/\mathrm{cm}^{-1}).$$

(b) Non-linear molecules;

$$q^{\mathrm{R}} = (\pi^{\frac{1}{2}}/\sigma)(1/hc\beta)^{\frac{3}{2}}(1/ABC)^{\frac{1}{2}}$$
$$= (1.0270/\sigma)(T/\mathrm{K})^{\frac{3}{2}}/(ABC/\mathrm{cm}^{-3})^{\frac{1}{2}}.$$

(3) *Vibration*

$$q^{\mathrm{V}} = 1/(1 - \mathrm{e}^{-hc\tilde{\nu}\beta}) = 1/(1 - \mathrm{e}^{-a}), \quad a = 1.4388(\tilde{\nu}/\mathrm{cm}^{-1})/(T/\mathrm{K}).$$

(4) *Electronic*

$$q^{\mathrm{E}} = g,$$

where g is the degeneracy of the electronic ground state (when that is the only accessible level); at high temperatures, evaluate explicitly.

In the case of independent molecules these are simplified by making the substitutions $Q = q^N$ (distinguishable particles) or $Q = q^N/N!$ (indistinguishable particles). All the thermodynamic functions are based on U and S, and so we have a route to their calculation. The precise relations can be built by drawing on some of the results established in Part 1.

Since the *Helmholtz function* is given by $A = U - TS$, and $A(0) = U(0)$, substitution for U and S leads to

$$A - A(0) = -kT \ln Q. \qquad (22.2.3)$$

The *pressure* can now be related to Q using the relation $(\partial A/\partial V)_T = -p$ (which is derived in the same way as eqn (6.2.2)). This leads to

$$p = kT(\partial \ln Q/\partial V)_T. \qquad (22.2.4)$$

This relation is entirely general, and may be used for any type of substance, including real gases and liquids. Since it relates the pressure to the volume and temperature (because these occur in Q), it is a very important route to the equations of state of real gases and their relation to intermolecular forces (which have to be built into Q, see Section 24.4(a)). That it gives the correct equation of state for a perfect gas may be checked by substituting $Q = q^N/N!$ and noting that only $q^T = V/\Lambda^3$ depends on V:

$$p = kT(1/Q)(\partial Q/\partial V)_T = NkT(1/q)(\partial q/\partial V)_T$$
$$= nN_A kT(1/q^T q^R q^V q^E)(\partial q^T q^R q^V q^E/\partial V)_T$$
$$= nRT(1/q^T)(\partial q^T/\partial V)_T = nRT(\Lambda^3/V)(1/\Lambda^3) = nRT/V,$$

as required. (This can be regarded as yet another way of deducing that $\beta = 1/kT$.)

Since U and p have now both been related to Q, the *enthalpy*, H, can also be related to it through $H = U + pV$, noting that $H(0) = U(0)$:

$$H - H(0) = -(\partial \ln Q/\partial \beta)_V + kTV(\partial \ln Q/\partial V)_T. \qquad (22.2.5)$$

Then, since the *Gibbs function*, G, is related to H by $G = H - TS$, we find

$$G - G(0) = -kT \ln Q + kTV(\partial \ln Q/\partial V)_T. \qquad (22.2.6)$$

Since $G = A + pV$, and in the case of a perfect gas $pV = nRT$, use of eqn (22.2.3) leads to

$$G - G(0) = -kT \ln Q + nRT. \qquad (22.2.7)^\circ$$

Furthermore, since for non-interacting particles $Q = q^N/N!$,

$$G - G(0) = -NkT \ln q + kT \ln N! + nRT$$
$$= -nRT \ln q + kT\{N \ln N - N\} + nRT.$$

Therefore, for a perfect gas,

$$G - G(0) = -nRT \ln (q/N), \qquad (22.2.8)^\circ$$

with $N = nN_A$. It turns out to be convenient to define the *molar partition function* (symbol: q_m; units: 1/mol) as $q_m = q/n$. Then

$$G - G(0) = -nRT \ln (q_m/N_A). \qquad (22.2.9)^\circ$$

G is the central function of chemical thermodynamics: this equation gives a way of calculating it from spectroscopic data.

These important relations are collected in Box 22.2.

Box 22.2 Statistical thermodynamic relations

In terms of the canonical partition function Q:

$$U - U(0) = -(\partial \ln Q/\partial \beta)_V$$
$$S = \{U - U(0)\}/T + k \ln Q$$
$$p = kT(\partial \ln Q/\partial V)_T$$
$$H - H(0) = -(\partial \ln Q/\partial \beta)_V + kTV(\partial \ln Q/\partial V)_T$$
$$A - A(0) = -kT \ln Q$$
$$G - G(0) = -kT \ln Q + kTV(\partial \ln Q/\partial V)_T.$$

In the case $Q = q^N/N!$ (indistinguishable particles):

$$U - U(0) = -N(\partial \ln q/\partial\beta)_V$$
$$S = \{U - U(0)\}/T + nR\{\ln q - \ln N + 1\}$$
$$G - G(0) = -nRT \ln(q/N).$$

In the case $Q = q^N$ (distinguishable particles):

$$U - U(0) = -N(\partial \ln q/\partial\beta)_V$$
$$S = \{U - U(0)\}/T + nR \ln q$$
$$G - G(0) = -nRT \ln q.$$

In general, for non-interacting indistinguishable particles,

$$Q = (q_{\text{external}}q_{\text{internal}})^N/N! = \{(q_{\text{external}})^N/N!\}(q_{\text{internal}})^N.$$

The thermodynamic functions are then the sums of internal and external (translational) contributions.

Example 22.4

Calculate the value of the function $\Phi_0 = \{G_m^\ominus - H_m^\ominus(0)\}/T$ for H_2O(g) at 1500 K.

- *Method*. Use eqn (22.2.9) with $G(0) = H(0)$, so that

$$\Phi_0 = -R \ln(q_m^\ominus/N_A).$$

Use the expressions in Box 22.1 for $q_m^\ominus/N_A$. The vibrational partition function was calculated in *Example* 22.3. Use the expressions in Box 22.1 for the other contributions. Use $A = 27.8778\ \text{cm}^{-1}$, $B = 14.5092\ \text{cm}^{-1}$, $C = 9.2869\ \text{cm}^{-1}$. For this C_{2v} molecule, $\sigma = 2$. $M_r = 18.015$.

- *Answer*. For the translational contribution,

$$(q_m^\ominus)^T/N_A = 0.02561 \times (18.015)^{\frac{3}{2}} \times (1500)^{\frac{5}{2}} = 1.706 \times 10^8.$$

For the vibrational contribution we have already found that $q^V = 1.353$. For the rotational contribution,

$$q^R = \tfrac{1}{2} \times 1.0270 \times (1500)^{\frac{3}{2}} \times (1/27.8778 \times 14.5092 \times 9.2869)^{\frac{1}{2}} = 486.7.$$

The complete molar partition function is therefore

$$q_m^\ominus/N_A = (1.706 \times 10^8) \times (1.353) \times (486.7) = 1.123 \times 10^{11},$$

and so $\Phi_0 = -R \ln(1.123 \times 10^{11}) = -211.5\ \text{J K}^{-1}\ \text{mol}^{-1}$.

- *Comment*. This is in agreement with the value quoted in Table 10.1 ($-211.7\ \text{J K}^{-1}\ \text{mol}^{-1}$), the small discrepancy arising from the use of slightly different data. The other values in that table were calculated similarly.

- *Exercise*. Repeat the calculation for CO_2. The vibrational data are given in the *Exercise* of *Example* 22.3; $B = 0.3902\ \text{cm}^{-1}$; $M_r = 44.01$. [$-244.4\ \text{J K}^{-1}\ \text{mol}^{-1}$]

22.3 Using statistical thermodynamics

Any thermodynamic quantity can now be calculated from a knowledge of the energy levels of molecules: thermodynamics and spectroscopy have been combined. In this section we indicate how to do the calculation in four important applications: more will be described later (especially in Section 30.3).

22.3 (a) Mean energies and the equipartition principle

It is often useful to know the mean energy stored in various modes of motion. In the case of a gas of independent particles, and when the molecular partition function can be factorized into contributions from each mode, we can write

$$U-U(0)=-N(1/q)(\partial q/\partial\beta)_V=-N(1/q^{\mathrm{T}}q^{\mathrm{R}}q^{\mathrm{V}}q^{\mathrm{E}})(\partial q^{\mathrm{T}}q^{\mathrm{R}}q^{\mathrm{V}}q^{\mathrm{E}}/\partial\beta)_V$$
$$=N\{-(1/q^{\mathrm{T}})(\partial q^{\mathrm{T}}/\partial\beta)_V-(1/q^{\mathrm{R}})(\partial q^{\mathrm{R}}/\partial\beta)_V+\ldots\}.$$

Since $\{U-U(0)\}/N$ is the sum of the mean energies of all the modes, and

$$\{U-U(0)\}/N=\langle\varepsilon^{\mathrm{T}}\rangle+\langle\varepsilon^{\mathrm{R}}\rangle+\ldots, \qquad (22.3.1)^\circ$$

it follows that each mean energy can be expressed in terms of the corresponding partition function as

$$\langle\varepsilon^{\mathrm{mode}}\rangle=-(1/q^{\mathrm{mode}})(\partial q^{\mathrm{mode}}/\partial\beta)_V. \qquad (22.3.2)^\circ$$

The *mean translational energy* of a molecule can now be found from the partition function. In order to see a pattern emerging, consider first a one-dimensional system of length X, for which $q^{\mathrm{T}}=X/\Lambda$, with $\Lambda=h(\beta/2\pi m)^{\frac{1}{2}}$ (as in the lines deriving eqn (21.1.15)). Then

$$\langle\varepsilon^{\mathrm{T}}\rangle=-(\Lambda/X)(\mathrm{d}\Lambda^{-1}/\mathrm{d}\beta)X=1/2\beta=\tfrac{1}{2}kT.$$

For three dimensions, the same calculation leads to

$$\langle\varepsilon^{\mathrm{T}}\rangle=\tfrac{3}{2}kT. \qquad (22.3.3)^\circ$$

In both cases the mean energy is independent of the mass of the particle and of the size of the container. (This is consistent with the thermodynamic result that the internal energy of a perfect gas is independent of its volume: $(\partial U/\partial V)_T=0$, Section 6.1.) In classical mechanics the kinetic energy of a particle of mass m is related to the components of its velocity by

$$E_{\mathrm{kinetic}}=\tfrac{1}{2}mv_x^2+\tfrac{1}{2}mv_y^2+\tfrac{1}{2}mv_z^2.$$

This suggests that we can arrive at the mean energy by ascribing the value $\frac{1}{2}kT$ to each quadratic term. This is a general result, as we shall now see by considering the other modes of motion.

The *mean rotational energy* of a linear molecule is

$$\langle\varepsilon^{\mathrm{R}}\rangle=-(1/q^{\mathrm{R}})(\partial q^{\mathrm{R}}/\partial\beta)_V,\qquad q^{\mathrm{R}}=\sum_J(2J+1)\mathrm{e}^{-hc\beta BJ(J+1)}. \qquad (22.3.4)$$

When the temperature is low we have to deal with q term by term. In the case of a heteronuclear diatomic molecule, all J values contribute and (with $b=hcB\beta$) $q^{\mathrm{R}}=1+3\mathrm{e}^{-2b}+5\mathrm{e}^{-6b}+\ldots$. Hence

$$\langle\varepsilon^{\mathrm{R}}\rangle=\frac{hcB\{6\mathrm{e}^{-2b}+30\mathrm{e}^{-6b}+\ldots\}}{1+3\mathrm{e}^{-2b}+5\mathrm{e}^{-6b}+\ldots}, \qquad (22.3.5)$$

and the mean energy decreases to zero as $T\to 0$ (corresponding to $b\to\infty$). When the temperature is so high that many rotational energy levels are occupied, we can use the approximate form of the partition function given in Box 22.1 and write

$$\langle\varepsilon^{\mathrm{R}}\rangle=-(\sigma hcB\beta)(1/\sigma hcB)(-1/\beta^2)=1/\beta=kT. \qquad (22.3.6)$$

The classical expression for the energy of a linear rotor is $\frac{1}{2}I\omega_x^2+\frac{1}{2}I\omega_y^2$ (there is no z-component because there is no rotation around the line of atoms), and so once again we see that *in the classical limit* the mean energy kT can be obtained by ascribing $\frac{1}{2}kT$ to each quadratic term. Likewise, the mean rotational energy of a non-linear molecule can be obtained as $\frac{3}{2}kT$.

The *mean vibrational energy* is obtained using the vibrational partition function in Box 22.1. Since

$$(\partial q^{\mathrm{V}}/\partial\beta)_V = (\mathrm{d}/\mathrm{d}\beta)\{1/(1-\mathrm{e}^{-hc\tilde{\nu}\beta})\} = -hc\tilde{\nu}\mathrm{e}^{-hc\tilde{\nu}\beta}/(1-\mathrm{e}^{-hc\tilde{\nu}\beta})^2,$$

we obtain

$$\langle\varepsilon^{\mathrm{V}}\rangle = hc\tilde{\nu}\left\{\frac{\mathrm{e}^{-hc\tilde{\nu}\beta}}{1-\mathrm{e}^{-hc\tilde{\nu}\beta}}\right\}. \qquad (22.3.7)$$

This is exact, apart from the zero-point energy of $\frac{1}{2}hc\tilde{\nu}$ (which can simply be added). When the temperature is so high that $hc\tilde{\nu}\beta \ll 1$ it simplifies as follows:

$$\langle\varepsilon^{\mathrm{V}}\rangle = hc\tilde{\nu}\left\{\frac{1-hc\tilde{\nu}\beta+\ldots}{1-(1-hc\tilde{\nu}\beta+\ldots)}\right\} \approx 1/\beta = kT. \qquad (22.3.8)$$

The classical energy of a one-dimensional oscillator is $\frac{1}{2}mv_x^2 + \frac{1}{2}kx^2$, and so the high-temperature mean value can be obtained once again by ascribing $\frac{1}{2}kT$ to each quadratic term, even though this time one relates to the potential energy and the other to the kinetic.

These conclusions are summarized by the *equipartition theorem*:

When quantum effects can be ignored, the mean energy of each quadratic term in the energy expression is the same, and has the value $\frac{1}{2}kT$.

This is a very simple rule for arriving at the internal energy of a system, so long as the temperature is high enough for quantum effects to be negligible. We shall see an example below.

22.3 (b) Heat capacities

The heat capacity at constant volume is $C_V = (\partial U/\partial T)_V$. Since $\mathrm{d}\beta/\mathrm{d}T = -1/kT^2 = -k\beta^2$, an alternative form is

$$C_V = -k\beta^2(\partial U/\partial\beta)_V. \qquad (22.3.9)$$

Since the internal energy of a perfect gas is a sum of contributions, eqn (22.3.1), the heat capacity is also a sum:

$$C_V = C_V^{\mathrm{T}} + C_V^{\mathrm{V}} + \ldots;\ C_V^{\mathrm{mode}} = -Nk\beta^2(\partial\langle\varepsilon^{\mathrm{mode}}\rangle/\partial\beta)_V. \qquad (22.3.10)^\circ$$

The temperature is always high enough (so long as the gas is above its condensation temperature) for the mean translational energy to be given by the equipartition value. Therefore,

$$C_V^{\mathrm{T}} = -Nk^2(\mathrm{d}/\mathrm{d}\beta)(3/2\beta) = \tfrac{3}{2}Nk = \tfrac{3}{2}nR. \qquad (22.3.11)^\circ$$

In the case of a monatomic gas, this is the only contribution, and so molar constant volume heat capacities equal to $\frac{3}{2}R = 12.47\ \mathrm{J\,K^{-1}\,mol^{-1}}$ are expected (and observed). Helium, for example, has this value over a range of 2000 K.

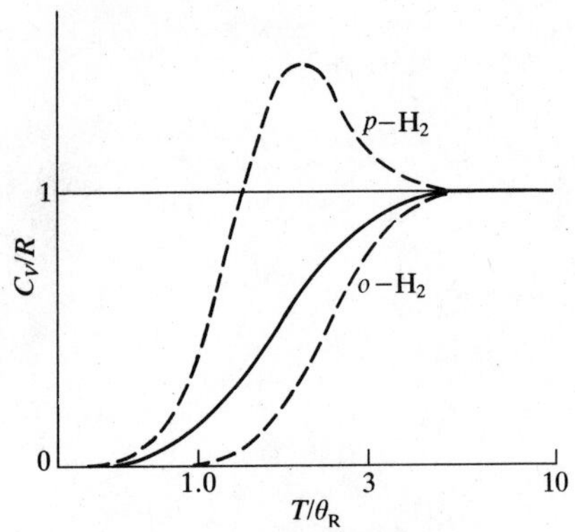

Fig. 22.3. The rotational heat capacity of hydrogen (a) pure o-H_2, (b) pure p-H_2, and (c) a 3:1 mixture. The curves have been calculated without making the high-temperature approximation. Note that the exact value is close to the high-temperature value when $T > \theta_R$, which for H_2 is 85 K.

When the temperature is high enough for the rotations of the molecules to be highly excited, the equipartition value $\langle \varepsilon^R \rangle = kT$ (for a linear rotor) can be used to obtain

$$C_V^R = nR. \tag{22.3.12}$$

When the temperature is very low, only the lowest rotational state is occupied, and the rotations do not contribute to the heat capacity ($C_V^R = 0$). At intermediate temperatures the rotational heat capacity can be calculated by differentiating eqn (22.3.5). The resulting (untidy) expression, which is plotted in Fig. 22.3, shows that the contribution rises from zero (when $T = 0$) to the equipartition value (when $kT \gg hcB$). Since the translational contribution is always present, we can expect the molar heat capacity of a gas of diatomic molecules ($C_V^T + C_V^R$) to rise from 12.5 J K^{-1} mol^{-1} to 20.8 J K^{-1} mol^{-1} as the temperature is increased above the *characteristic rotational temperature* $\theta_R = hcB/k$. For H_2, $\theta_R = 87.5$ K, an abnormally high value; for I_2, $\theta_R = 0.054$ K. The experimental heat capacity of hydrogen (over a part of the range) is shown in Fig. 22.4, confirming this prediction (the small hump in the curve is due to quantum effects relating to p- and o-H_2). In the case of non-linear molecules, the mean rotational energy rises to $\frac{3}{2}kT$, and so the molar rotational heat capacity rises to $\frac{3}{2}R$.

Molecular vibrations contribute to the heat capacity, but only when the temperature is high enough for them to be significantly excited. The equipartition mean energy is kT per mode, and so the maximum contribution to the heat capacity is R. However, it is very unusual for the vibrations to be so highly excited that equipartition is valid, and it is more appropriate to use the full expression for the vibrational heat capacity, which is obtained by differentiating eqn (22.3.7). This gives

$$C_V^V = nR\left\{\frac{(\theta_V/T)e^{-(\theta_V/2T)}}{1-e^{-\theta_V/T}}\right\}^2, \tag{22.3.13}$$

where $\theta_V = hc\tilde{\nu}/k$ is the *characteristic vibrational temperature* (in Chapter 13 we met it as the Einstein temperature); for H_2 θ_V is over 6323 K, another abnormally high value; for I_2 $\theta_V = 309$ K. The curve in Fig. 22.5 shows how C_V^V depends on temperature: note that it is close to the equipartition value when $T > \theta_V$.

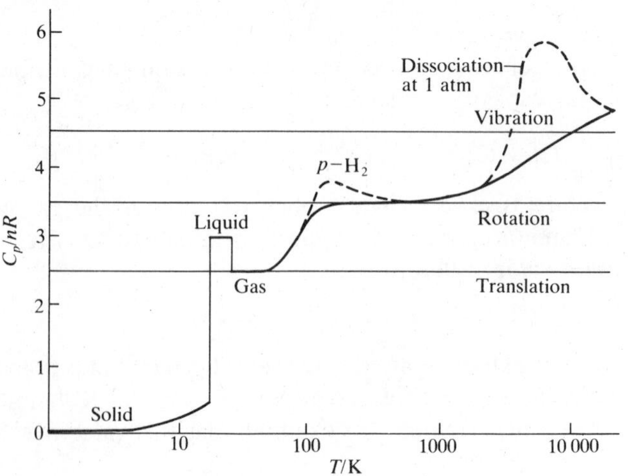

Fig. 22.4. The full experimental heat capacity curve for hydrogen. This molecule shows quantum effects up to abnormally high temperatures, but for most molecules at all normal temperatures rotational heat capacities are classical. Translational heat capacities are always classical.

The total heat capacity of a system is the sum of these contributions, Fig. 22.6. When equipartition is valid (when the temperature is well above the characteristic temperature of the mode, $T > \theta_{\text{mode}}$) the heat capacity can be estimated by counting the numbers of modes that are active. In gases, all three translational modes are always active, and contribute $\frac{3}{2}R$ to the molar heat capacity. If we denote the number of *active* rotational modes by ν_R^* (so that for most molecules at normal temperatures $\nu_R^* = 2$ for linear molecules, and $\nu_R^* = 3$ for non-linear molecules) then the rotational contribution is $\frac{1}{2}\nu_R^* R$. If the temperature is high enough for ν_V^* vibrational modes to be active (which is rarely likely), the vibrational contribution to the molar heat capacity is $\nu_V^* R$. Therefore, the total molar heat capacity is

$$C_{V,\text{m}} = \tfrac{1}{2}(3 + \nu_R^* + 2\nu_V^*)R. \tag{22.3.14}$$

An interesting consequence is that if a vibrational mode should convert to a rotational mode (such as when a hindered rotation of a CH_3 group in a molecule acquires enough energy to undergo free rotation, Fig. 22.7), the contribution of that mode to the heat capacity should *decrease* from about R (for vibration) to $\frac{1}{2}R$ (for rotation about a single axis).

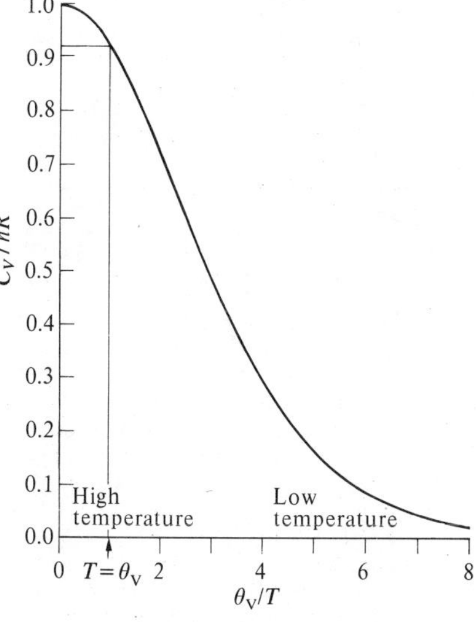

Fig. 22.5. The temperature dependence of the vibrational heat capacity calculated using eqn (22.3.13). Note that temperature decreases from left to right.

Example 22.5

Estimate the molar heat capacity of steam at 100 °C.

• *Method*. Use eqn (22.3.14). Assess whether the rotational and vibrational modes are active by computing their characteristic temperatures from the data in *Examples* 22.3 and 22.4. Use $hc/k = 1.439$ cm K.

• *Answer*. The characteristic temperatures (in round numbers) of the vibrations are 5300 K, 2300 K, and 5400 K and are therefore inactive at 373 K. The three rotational modes have characteristic temperatures 40 K, 21 K, and 13 K, and so are fully active, like the three translational modes. The translational contribution is $\frac{3}{2}R = 12.5\ \text{J K}^{-1}\,\text{mol}^{-1}$. Fully active rotations contribute a further $12.5\ \text{J K}^{-1}\,\text{mol}^{-1}$. Therefore a value close to $25\ \text{J K}^{-1}\,\text{mol}^{-1}$ is predicted.

• *Comment*. The experimental value is $26.1\ \text{J K}^{-1}\,\text{mol}^{-1}$. The discrepancy is probably due to deviations from ideality.

• *Exercise*. Estimate the molar heat capacity of I_2(g) at room temperature ($B = 0.037\ \text{cm}^{-1}$).

[$21\ \text{J K}^{-1}\,\text{mol}^{-1}$]

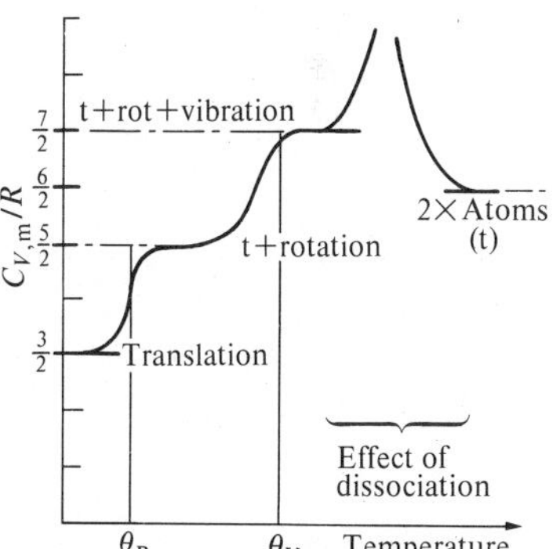

Fig. 22.6. The general temperature dependence of the heat capacity of diatomic molecules is as shown here. Each mode becomes active when its characteristic temperature is exceeded. The heat capacity becomes very large when the molecule dissociates because the energy is used to cause dissociation and not to raise the temperature. Then it falls back to the translation-only value of the atoms.

22.3 (c) Residual entropies

Entropies may be calculated from spectroscopic data; they may also be measured experimentally (Chapter 5). In many cases there is good agreement, but in some the experimental entropy is less than the calculated value. One possibility is that the experimental determination failed to take a phase transition into account. Another is that some disorder is present in the solid even at $T = 0$. The entropy at $T = 0$ is then greater than zero, and is called the *residual entropy*.

The origin and magnitude of the residual entropy can be understood by considering a crystal composed of AB molecules, where A and B are similar atoms (e.g. CO). There may be so little energy difference between . . .AB AB AB AB. . . and . . .AB BA BA AB. . . and other random arrangements that the molecules adopt either orientation at random in the solid.

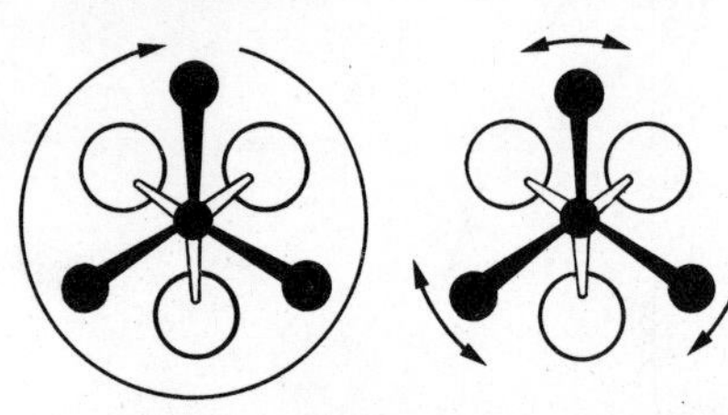

Fig. 22.7. At high temperatures a CH_3 group has effective rotational freedom (a) and the equipartition heat capacity value of $\frac{1}{2}R$ can be expected. At low temperatures (b) its motion is that of an oscillator, and the equipartition value of R is *larger* than in (a). However, it it unlikely that the full equipartition value will be reached.

The entropy arising from this residual disorder can be calculated using the fundamental equation $S = (k/\mathbb{N}) \ln \mathbb{W}$. Suppose that two orientations are equally probable, and that the sample consists of N molecules. In one member of the ensemble the same energy can be achieved in 2^N different ways (because each molecule can take either of two orientations). There are $\mathbb{N}$ members of the ensemble, and each one can be formed in 2^N ways. Therefore, the weight of the configuration, the total number of ways of achieving the same energy, is $(2^N)^{\mathbb{N}}$. It follows that the entropy is

$$S = (k/\mathbb{N}) \ln (2^N)^{\mathbb{N}} = k \ln 2^N = Nk \ln 2 = nR \ln 2. \qquad (22.3.15)$$

A residual molar entropy of magnitude $R \ln 2 = 5.8\ \mathrm{J\,K^{-1}\,mol^{-1}}$ is therefore predicted for solids composed of molecules that can adopt either of two orientations at $T = 0$. If p orientations are possible, then the residual molar entropy will be $R \ln p$.

The $FClO_3$ molecule can adopt four orientations with about the same energy, and the calculated residual entropy of $R \ln 4 = 11.5\ \mathrm{J\,K^{-1}\,mol^{-1}}$ is in good agreement with the experimental value ($10.1\ \mathrm{J\,K^{-1}\,mol^{-1}}$). In the case of CO the measured residual entropy is $5\ \mathrm{J\,K^{-1}\,mol^{-1}}$, which is close to $R \ln 2$, the value expected for a random structure of the form . . .CO CO OC CO OC OC. . . . Subsequent work, however, has questioned whether the calorimetric experiments missed a phase change at low temperatures. Although this is only a speculation, it does emphasize the care that must be taken when obtaining and interpreting experimental data.

The final example is the residual entropy of ice, which experimentally is found to be $3.4\ \mathrm{J\,K^{-1}\,mol^{-1}}$. This value can be explained in terms of the hydrogen-bonded structure of the solid. Each O atom is surrounded tetrahedrally by four H atoms, two of which are attached by short σ-bonds, the other two being attached by long hydrogen bonds. The randomness lies in which two of the four bonds are short, and an approximate analysis leads to a residual entropy of magnitude $R \ln \frac{3}{2} = 3.4\ \mathrm{J\,K^{-1}\,mol^{-1}}$, in good agreement with the experimental value.

22.3 (d) Equilibrium constants

The Gibbs function of a system of independent particles is given in Box 22.2:

$$G - G(0) = -nRT \ln (q_\mathrm{m}/N_\mathrm{A}), \qquad (22.3.16)^\circ$$

where q_m is the molar partition function (q/n). The equilibrium constant K_p is related to the *standard* Gibbs function of reaction by eqn (10.1.13):

$$\Delta_\mathrm{r} G^\ominus = -RT \ln K_p. \qquad (22.3.17)^\circ$$

In order to calculate the equilibrium constant, we must combine these two equations.

In the first place, we need the molar Gibbs function of each component: this is G/n. Next, we need its standard value. This means we need the value of the molar partition function when $p = p^\ominus = 1$ bar: we denote this $q_\mathrm{m}^\ominus$. Since only the translational component depends on the pressure (through $V_\mathrm{m} = V/n$) we evaluate q_m with $V_\mathrm{m} = RT/p^\ominus$. For a component J it follows that

$$G_{\mathrm{J,m}}^\ominus - G_{\mathrm{J,m}}^\ominus(0) = -RT \ln (q_{\mathrm{J,m}}^\ominus/N_\mathrm{A}). \qquad (22.3.18)^\circ$$

The standard Gibbs function of the reaction

$$0 = \sum_J \nu_J J$$

(in the notation used in Section 10.1(e)) is

$$\Delta_r G^\ominus = \sum_J \nu_J G^\ominus_{J,m} = \sum_J \nu_J G^\ominus_{J,m}(0) - RT \sum_J \nu_J \ln (q^\ominus_{J,m}/N_A). \quad (22.3.19)°$$

It follows that, since $G(0) = U(0)$, the first term on the right can be written

$$\sum_J \nu_J G^\ominus_{J,m}(0) = \Delta_r G^\ominus(0) = \Delta_r U^\ominus(0). \quad (22.3.20)$$

The reaction internal energy evaluated at $T = 0$ is the difference between the (molar) zero-point energy levels of the components involved in the reaction (weighted by the stoichiometric coefficients), and is what we normally write ΔE_0:

$$\Delta E_0 = \Delta_r U(0).$$

Next, the appearance of eqn (22.3.19) can be improved:

$$\Delta_r G^\ominus = \Delta E_0 - RT \sum_J \ln (q^\ominus_{J,m}/N_A)^{\nu_J}$$
$$= -RT\left\{-\Delta E_0/RT + \ln \prod_J (q^\ominus_{J,m}/N_A)^{\nu_J}\right\}.$$

This lets us pick out an expression for K_p by comparison with eqn (22.3.17):

$$\ln K_p = -\Delta E_0/RT + \ln \prod_J (q^\ominus_{J,m}/N_A)^{\nu_J}.$$

Hence

$$K_p = \left\{\prod_J (q^\ominus_{J,m}/N_A)^{\nu_J}\right\} e^{-\Delta E_0/RT}. \quad (22.3.21)°$$

We illustrate the application of this expression to the case of a reaction in which a diatomic molecule X_2 dissociates into its atoms: $X_2(g) \rightleftharpoons 2X(g)$. The reaction is written in the form

$$0 = 2X(g) - X_2(g), \qquad \nu_X = 2, \nu_{X_2} = -1.$$

The equilibrium constant is therefore

$$K_p = \left\{\prod_J (p/p^\ominus_J)^{\nu_J}\right\}_{eq} = \{(p_X/p^\ominus)^2(p_{X_2}/p^\ominus)^{-1}\}_{eq} = \{(p_X^2/p_{X_2}p^\ominus\}_{eq}.$$

In terms of the partition functions

$$K_p = (q^\ominus_{m,X}/N_A)^2(q^\ominus_{m,X_2}/N_A)^{-1}e^{-\Delta E_0/RT}$$
$$= \{(q^\ominus_{m,X})^2/q^\ominus_{m,X_2}N_A\}e^{-\Delta E_0/RT}. \quad (22.3.22)°$$

ΔE_0 is the (molar) dissociation energy D_0 of the molecule (measured from the zero-point vibrational level of X_2). The atomic partition functions relate only to their translational motion and any electronic degeneracy:

$$q^\ominus_{X,m} = g_X(V^\ominus_m/\Lambda^3_X), \qquad V^\ominus_m = RT/p^\ominus.$$

On the other hand, the diatomic molecule also has rotational and vibrational degrees of freedom:

$$q^{\ominus}_{X_2,m} = g_{X_2}(V^{\ominus}_m/\Lambda^3_{X_2})q^R_{X_2}q^V_{X_2}.$$

It follows that the equilibrium constant is given by

$$K_p = (g^2_X/g_{X_2})(\Lambda^3_{X_2}V^{\ominus}_m/\Lambda^6_X N_A)(1/q^R_{X_2}q^V_{X_2})e^{-D_0/RT}. \qquad (22.3.23)^{\circ}$$

In the case of $Na_2(g) \rightleftharpoons 2Na(g)$, the values of all the parameters are known, and in the following *Example* we deduce that $K_p = 2.4$.

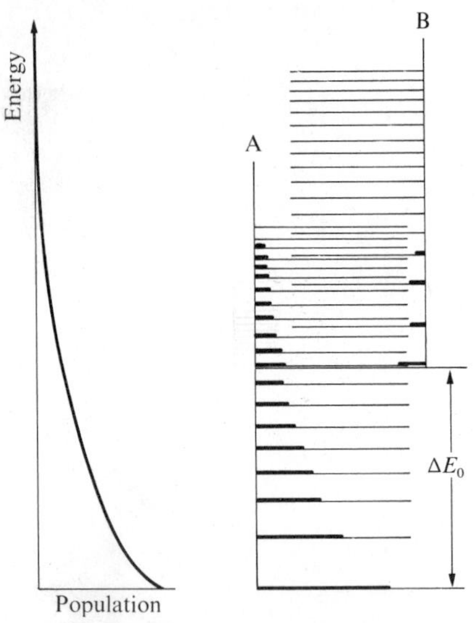

Fig. 22.8. The array of A and B energy levels. At equilibrium all are accessible, and the equilibrium composition of the system reflects the *overall* Boltzmann distribution of populations. As ΔE_0 increases, A becomes dominant.

Example 22.6

Evaluate the equilibrium constant K_p for the $Na_2(g) \rightleftharpoons 2Na(g)$ equilibrium at 1000 K. Use the following spectroscopic data: $B = 0.1547\ \mathrm{cm}^{-1}$, $\tilde{\nu} = 159.2\ \mathrm{cm}^{-1}$, $D_0 = 70.4\ \mathrm{kJ\ mol}^{-1}$ (0.73 eV). Use $M_r = 22.99$ for Na. The atoms have doublet ground terms.

• *Method*. Evaluate the partition functions using the expressions in Box 22.1, and then evaluate eqn (22.3.23).

• *Answer*. Substitution in the expressions in Box 22.1 gives:

$$\Lambda^3_{Na} = 5.40 \times 10^{-34}\ \mathrm{m}^3; \qquad \Lambda^3_{Na_2} = 1.53 \times 10^{-33}\ \mathrm{m}^3$$

$$q^R_{Na_2} = 2246; \qquad q^V_{Na_2} = 4.885; \qquad q_{Na} = 2; \qquad q_{Na_2} = 1.$$

Then, with $V^{\ominus}_m = 8.206 \times 10^{-2}\ \mathrm{m^3\ mol^{-1}}$, eqn (22.3.23) evaluates to $K_p = 2.42$.

• *Comment*. Note that the procedure leads to K_p. Convert to concentrations using $[J] = p_J/RT$.

• *Exercise*. Evaluate K_p at 1500 K. [52]

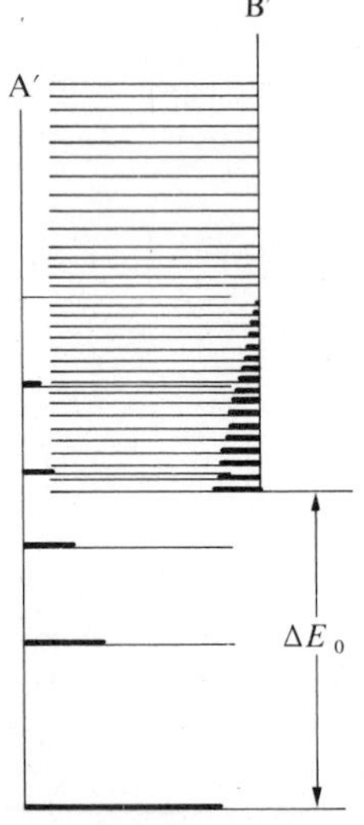

Fig. 22.9. It is important to take into account the densities of states of the molecules. Even though B might lie well above A in energy (i.e. ΔE_0 is large and positive), B might have so many states that its total population dominates in the mixture. In classical thermodynamic terms, we have to take entropies into account as well as enthalpies when considering equilibria.

We are now in a position to appreciate the physical basis of equilibrium constants. Consider a simple $A \rightleftharpoons B$ equilibrium. In Fig. 22.8 one set of levels correspond to A, and the other to B. The populations are given by the Boltzmann distribution, and are independent of whether any given level happens to belong to A or to B. We can therefore imagine a single Boltzmann distribution spreading, without distinction, over the two sets of energy levels. If the spacings of A and B are similar (as in Fig. 22.8), and B lies above A, then the diagram indicates that A will dominate in the equilibrium mixture. If, however, B has a high density of states (as in Fig. 22.9), then even though its zero-point energy lies above A's, it might still dominate at equilibrium.

We can in fact show that an analysis of diagrams like those in Figs. 22.8 and 22.9 leads to an expression for the equilibrium constant in terms of partition functions. The population in some state i of the composite (A + B) system is

$$n_i = N(e^{-\beta\varepsilon_i}/q),$$

where N is the total number of molecules. The total number of type A molecules is the sum of these populations taken over the states ε_a belonging to A, and the total number of B is the sum over ε'_b:

$$N_A = \sum_{i\ \mathrm{of\ A}} n_i = (N/q)\sum_a e^{-\beta\varepsilon_a}; \qquad N_B = \sum_{i\ \mathrm{of\ B}} n_i = (N/q)\sum_b e^{-\beta\varepsilon'_b}.$$

The sum over the states of A is nothing other than the partition function for A, and so $N_A = Nq_A/q$. The sum over the states of B is also a partition function, but the energies are measured from the ground state of the total system, which happens to be the ground state of A. However, since $\varepsilon_b' = \varepsilon_b + \Delta\varepsilon_0$, where $\Delta\varepsilon_0$ is the separation of zero-point energies (Fig. 22.8),

$$N_B = (N/q)\sum_b e^{-(\varepsilon_b'+\Delta\varepsilon_0)\beta} = (Nq_B/q)e^{-\Delta E_0/RT}.$$

The equilibrium constant of the $A \rightleftharpoons B$ reaction is proportional to the ratio of the numbers of the two types of molecule, and so

$$K_p = N_B/N_A = \{q_B/q_A\}e^{-\Delta E_0/RT},$$

which is exactly the same as the expression given by eqn (22.3.22).‡

The content of the last equation can be seen most clearly in the case when A has only a single accessible level (so that $q_A = 1$) and B has a large number of evenly, closely spaced levels, Fig. 22.10, so that its partition function is

$$q_B = 1/\{1 - e^{-\beta\varepsilon}\} = 1/\{1 - (1 - \beta\varepsilon + \ldots)\} \approx 1/\beta\varepsilon = kT/\varepsilon.$$

Then the $A \rightleftharpoons B$ equilibrium constant is

$$K_p = (kT/\varepsilon)e^{-\Delta E_0/RT}.$$

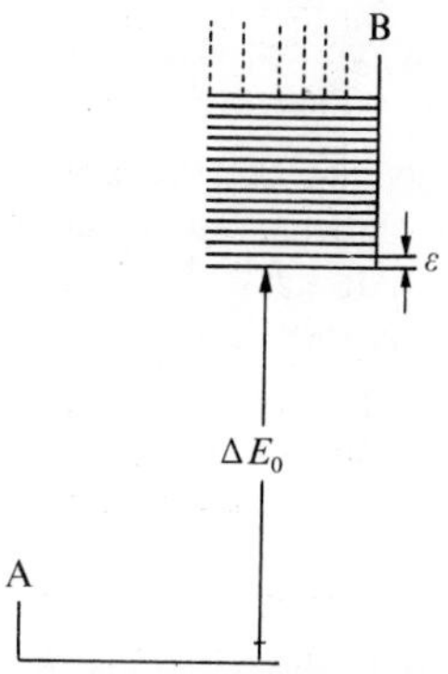

Fig. 22.10. The model used in the text for exploring the effects of energy separations and densities of states on equilibria. B can dominate so long as ΔE_0 is not too large.

When ΔE_0 is very large, the exponential term dominates and $K_p \approx 0$, which implies that very little B is present at equilibrium. When ΔE_0 is small but still positive, K_p can exceed unity, reflecting the predominance of B at equilibrium on account of its high density of states. At low temperatures $K_p \approx 0$, and the system consists entirely of A. At high temperatures the exponential approaches unity and the pre-exponential factor is large. Hence B becomes predominant. We see that in this endothermic reaction (endothermic because B lies above A), a rise in temperature favours B, because its states become accessible: this is what we saw, from the outside, in Chapter 10.

Further reading

Elementary statistical thermodynamics. L. K. Nash; Addison-Wesley, Reading, 1968.
Statistical thermodynamics. B. J. McClelland; Wiley, New York, 1973.
Statistical thermodynamics. D. A. McQuarrie; Harper & Row, New York, 1976.
Statistical thermodynamics. N. Davidson; McGraw-Hill, New York, 1962.
An introduction to statistical mechanics. T. L. Hill; Addison-Wesley, Reading, 1960.
Statistical thermodynamics. A. Münster; Springer, Berlin, 1974.

Introductory problems

A22.1. Give the symmetry number for each of the following molecules: (a) CO (b) O_2 (c) H_2S (d) SiH_4.

A22.2. The bond length in O_2 is 120.75 pm. Use the high temperature approximation to calculate the rotational partition function for O_2 at 300 K.

A22.3. The NOF molecule is an asymmetric rotor with

‡ For an $A \rightleftharpoons B$ equilibrium, the V factors in the partition functions cancel, and so the appearance of q in place of $q^\ominus$ has no effect. In the case of a more general reaction, the conversion from q to $q^\ominus$ comes about at the stage of converting the pressures that occur in K_p into numbers.

rotational constants $3.1752\ cm^{-1}$, $0.3951\ cm^{-1}$, and $0.3505\ cm^{-1}$. Calculate the rotational partition function at 100 °C.

A22.4. Find the range of vibrational wavenumbers for which the contribution to the vibrational partition function at 500 K may be considered unity within a limit of 0.10 % error.

A22.5. Carbon dioxide is a linear symmetrical molecule with fundamental vibrational wavenumbers at $1388.2\ cm^{-1}$, $667.4\ cm^{-1}$, and $2349.2\ cm^{-1}$. The lowest of these is doubly degenerate. The rotational constant of CO_2 is $0.3902\ cm^{-1}$. Calculate the contributions of vibration and rotation to $G - G(0)$ at 300 K.

A22.6. Calculate the electronic contribution to the heat capacity of chlorine atoms at 500 K and at 900 K. The ground level of Cl is $^2P_{\frac{3}{2}}$. The only other nearby level, which is $^2P_{\frac{1}{2}}$, lies $881\ cm^{-1}$ higher.

A22.7. Find the electronic contribution to $G - G(0)$ for molecular oxygen at 400 K. Its first electronically excited level, which is a doubly degenerate $^1\Delta_g$ level, lies $7918.1\ cm^{-1}$ above the ground state, which is $^3\Sigma_g^-$.

A22.8. In $CoSO_4.7H_2O$ the ground state of Co(II) is $^4F_{\frac{9}{2}}$. At temperatures below 1 K, the entropy of the solid is derived almost entirely from electron spin. Estimate the entropy in this region of temperature.

A22.9. Use equipartition theory to calculate the molar heat capacity at constant pressure for NH_3(g) and compare the calculated value with the experimental value $35.63\ J\ K^{-1}\ mol^{-1}$ at 298 K. Explain where the equipartition theory fails to give a good account of the heat capacity of NH_3.

A22.10. Calculate the equilibrium constant K_p at 300 K for the isotope exchange reaction: $2\ {}^{79}Br^{81}Br \rightleftharpoons {}^{79}Br^{79}Br + {}^{81}Br^{81}Br$. These molecules have $^1\Sigma$ ground electronic states with no upper electronic states nearby. Assume that the internuclear distances in the molecules are identical. For $^{79}Br^{81}Br$, the fundamental vibrational wavenumber is $323.33\ cm^{-1}$.

Problems

22.1. Estimate the rotational partition function of HCl at (a) 100 K, (b) 298 K, (c) 500 K, on the basis of the high-temperature approximation.

22.2. The pure rotational, microwave spectrum of HCl has absorption lines at the following wavenumbers (in cm^{-1}): 21.19, 42.37, 63.56, 84.75, 105.93, 127.12, 148.31, 169.49, 190.68, 211.87, 233.06, 254.24, 275.43, 296.62, 317.80, 338.99, 360.18, 381.36, 402.55, 423.74, 444.92, 466.11, 487.30, 508.48. Calculate the rotational partition function at (a) 100 K, (b) 298 K by direct summation.

22.3. On the basis of the high-temperature approximation, calculate the rotational partition function of the water molecule at 298 K using the following rotational constants: $A = 27.878\ cm^{-1}$, $B = 14.509\ cm^{-1}$, $C = 9.287\ cm^{-1}$. Above what temperature is the high temperature approximation valid?

22.4. The methane molecule is a spherical top with bond length of 109 pm. Calculate its rotational partition function at (a) 298 K, (b) 500 K, using the high-temperature approximation in each case.

22.5. Calculate the rotational partition function of methane by direct summation of the rotational energy levels at (a) 298 K, (b) 500 K.

22.6. The Sackur–Tetrode equation, eqn (21.3.10) gives the theoretical entropy of a monatomic gas. Derive the corresponding expression for a gas confined to move in two dimensions. Hence find an expression for the molar entropy of condensation of a gas to form a mobile surface film.

22.7. Calculate the rotational entropy of benzene that is free to rotate in three dimensions at 362 K. Its moments of inertia are $I_A = 2.93 \times 10^{-38}\ g\ cm^2$, $I_B = I_C = 1.46 \times 10^{-38}\ g\ cm^2$. What would be the change of rotational entropy if the molecule were adsorbed on to a surface and could rotate only about its six-fold axis?

22.8. In an experimental study of the thermodynamics of adsorption of organic molecules on graphite at 362 K (D. Dollimore, G. R. Heal, and D. R. Martin, *J. chem. Soc. Faraday Trans*. I 1784 (1973)) a change in entropy of $-111\ J\ K^{-1}\ mol^{-1}$ was observed when there was only little surface coverage, but it dropped to $-52\ J\ K^{-1}\ mol^{-1}$ for complete coverage. Use the results of Problems 22.6 and 22.7 to propose a model of the motion of the benzene molecules on the surface where each molecule occupies $4.08 \times 10^{-2}\ nm^2$.

22.9. The *symmetry number* σ can be calculated simply by counting the number of indistinguishable orientations of the molecule that can be reached by rotational symmetry operations. What is the value of σ in the case of (a) N_2, (b) NO, (c) benzene, (d) methane, (e) chloroform?

22.10. Calculate the room temperature (25 °C) entropy of ClO_2 (a bent molecule with an unpaired electron): OClO angle 118.5°, ClO bond length 149 pm, C_{2v} symmetry, $^2B_1$ electronic state.

22.11. Calculate and plot the equilibrium constant for the reaction $CD_4 + HCl \rightleftharpoons CHD_3 + DCl$ in the gas phase from the following data: $\tilde{\nu}(CHD_3)$: 2993(1), 2142(1), 1003(3), 1291(2), 1036(2) cm^{-1}; $\tilde{\nu}(CD_4)$: 2109(1), 1092(2), 2259(3), 996(3) cm^{-1}; $\tilde{\nu}(HCl)$: 2991 cm^{-1}; $\tilde{\nu}(DCl)$: 2145 cm^{-1}. (Numbers in brackets are degeneracies.) $B(HCl) = 10.59\ cm^{-1}$, $B(DCl) = 5.445\ cm^{-1}$, $B(CHD_3) = 3.28\ cm^{-1}$, $A(CHD_3) = 2.63\ cm^{-1}$. Take $300\ K \leqslant T \leqslant 1000\ K$.

22.12. The exchange of deuterium between acid and water is an important type of equilibrium, and we can deal with it using spectroscopic data on two types of molecule. Calculate the equilibrium constant at (a) 298 K, (b) 800 K for the gas-phase exchange reaction $H_2O + DCl \rightleftharpoons HDO + HCl$ using the following vibrational and rotational data: $\tilde{\nu}(H_2O)$: 3656.7, 1594.8, 3755.8 cm^{-1}; $\tilde{\nu}$(HDO): 2726.7, 1402.2, 3707.5 cm^{-1}; rotational constants: H_2O 27.88, 14.51, 9.29 cm^{-1}; HDO 23.38, 9.102, 6.417 cm^{-1}; HCl 10.59 cm^{-1}; DCl 5.449 cm^{-1}.

22.13. The equilibrium constants for $I_2 \rightleftharpoons 2I$ at 1000 K was treated from an experimental point of view in Problem 10.21. The spectroscopic data for I_2 is as follows: $B = 0.0373$ cm^{-1}, $\tilde{\nu} = 214.36$ cm^{-1}, $D_e = 1.5422$ eV. The iodine atoms have $^2P_{\frac{3}{2}}$ ground states, implying fourfold degeneracy. Calculate a statistical thermodynamic value of the equilibrium constant at 1000 K.

22.14. Iodine atoms have quite low-lying excited states at 7603 cm^{-1} (twofold degenerate). At 2000 K these are significantly populated. What effect does their inclusion have on the value of the equilibrium constant calculated in the last Problem?

22.15. Now return to the simpler version of the calculation in Problem 22.13. Can a magnetic field have an appreciable effect on the equilibrium constant of the dissociation equilibrium? Investigate this possibility by calculating the effect of a magnetic field on the partition functions of the atoms. What magnetic field would be needed to affect the equilibrium constant by 1 percent? The electronic g-value of $^2P_{\frac{3}{2}}$ iodine is 4/3.

22.16. The harmonic oscillator plays a specially important role in statistical thermodynamics (as well as in other subjects) because closed forms of the partition function and thermodynamic properties may be obtained; furthermore, molecular vibrations are normally well approximated by harmonic motion, and so the vibrational contribution to various thermodynamic properties may be calculated very easily. As a first step, deduce expressions for the internal energy, enthalpy, entropy, Helmholtz function, and Gibbs function of a harmonic oscillator, and plot the results as a function of $x = \hbar\omega/kT$.

22.17. The Giauque function Φ_0 is $\{G_m(T) - H_m(0)\}/T$, and its applications were explored in Chapter 10. Find an expression for Φ_0 of a harmonic oscillator.

22.18. Now put the harmonic oscillator calculation to use. Calculate the vibrational contribution at 1000 K to the Giauque function of (a) ammonia, which has vibrational modes of frequencies 3336.7(1), 950.4(1), 3443.8(2), 1626.8(2) cm^{-1}, (b) methane, with vibrational modes at 2916.7(1), 1533.6(2), 3018.9(3), 1306.2(3) cm^{-1}. (Numbers in brackets are degeneracies.) The easiest procedure is to use the graphs constructed in the earlier Problems, and to find x for each mode.

22.19. The Giauque function is well suited for equilibrium calculations under conditions where data have not been tabulated. Calculate the total values of Φ_0(1000 K) for (a) H_2, (b) Cl_2, (c) NH_3, (d) N_2, (e) NO using data assembled from Table 18.1, *Example* 18.3, Problem 22.18, and $B(\text{NO}) = 1.7406$ cm^{-1}, $\tilde{\nu}(\text{NO}) = 1904$ cm^{-1}.

22.20. Find the equilibrium constant for the $N_2 + 3H_2 \rightleftharpoons 2NH_3$ equilibrium at 1000 K using data from the last Problem.

22.21. Although expressions like $d \ln q/d\beta$ are useful for formal manipulations in statistical thermodynamics, and for arriving at neat expressions for thermodynamic quantities, they are sometimes more trouble than they are worth in practical applications. If you are presented with a table of energy levels it is often much more convenient to evaluate the following sums directly: $q = \sum_j \exp(-\beta\varepsilon_j)$, $\dot{q} = \sum_j (\varepsilon_j/kT)\exp(-\beta\varepsilon_j)$, $\ddot{q} = \sum_j (\varepsilon_j/kT)^2 \exp(-\beta\varepsilon_j)$. As a first step in seeing how these sums are employed, find expressions for U, S, and C_V in terms of q, $\dot{q}$, and $\ddot{q}$.

22.22. Practical applications can be simplified still further if we separate the internal modes of the molecule from its translation. Show that $U = U_{ex} + U_{int}$, and $S = S_{ex} + S_{int}$, and deduce expressions for U_{int}, S_{int}, and $C_{V,int}$ in terms of q, $\dot{q}$, and $\ddot{q}$ for the internal molecular modes.

22.23. The thermodynamic properties of monatomic gases may be calculated from a knowledge of their electronic energy levels obtained from spectroscopy. Calculate the electronic contribution to (a) the enthalpy $H_m(T) - H_m(0)$, (b) the Giauque function Φ_0, (c) the electronic contribution to C_V of magnesium vapour at 5000 K from the data given below. Use the direct summation procedures outlined in the last two Problems.

Term:	^1S	^3P	$^3P_1$
Degeneracy:	1	1	3
Wavenumber/cm^{-1}:	0	21 850	21 870
Term:	$^3P_2$	$^1P_1$	^3S
Degeneracy:	5	3	3
Wavenumber/cm^{-1}:	21 911	35 051	41 197

22.24. A rich source of data on atomic energy levels is *Atomic Energy Levels*, C. E. Moore, N.B.S. Circ. No. 476 (1949). Calculations of this kind are ideal for programming on to a computer or programmable calculator, especially when many electronic states are accessible at the temperature of interest. If you have access to such a calculator, calculate the value of Φ_0 from room temperature to 5000 K using the data below, which has been reproduced from Moore's collection. If you have no such calculator, evaluate the electronic Φ_0 and C_V at 3000 K.

Na I

Config.	Desig.	J	Level/cm^{-1}	Interval
3s	3s ^2S	$\frac{1}{2}$	0.000	
3p	3p $^2P°$	$\frac{1}{2}$	*16 956.183*	17.1963
		$1\frac{1}{2}$	*16 973.379*	
4s	4s ^2S	$\frac{1}{2}$	25 739.86	
3d	3d ^2D	$2\frac{1}{2}$	29 172.855	−0.0494
		$1\frac{1}{2}$	29 172.904	
4p	4p $^2P°$	$\frac{1}{2}$	*30 266.88*	5.63
		$1\frac{1}{2}$	*30 272.51*	

(continued)

(*continued*)

Config.	Desig.	J	Level/cm^{-1}	Interval
5s	5s ^{2}S	$\frac{1}{2}$	33 200.696	
4d	4d ^{2}D	$2\frac{1}{2}$	34 548.754	−0.0346
		$1\frac{1}{2}$	34 548.789	
4f	4f ^{2}F°	$\begin{Bmatrix} 2\frac{1}{2} \\ 3\frac{1}{2} \end{Bmatrix}$	*34 588.6*	
5p	5p ^{2}P°	$\frac{1}{2}$	*35 040.27*	2.52
		$1\frac{1}{2}$	*35 042.79*	
6s	6s ^{2}S	$\frac{1}{2}$	36.372.647	
5d	5d ^{2}D	$2\frac{1}{2}$	37 036.781	−0.0230
		$1\frac{1}{2}$	37 036.805	
5f	5f ^{2}F°	$\begin{Bmatrix} 2\frac{1}{2} \\ 3\frac{1}{2} \end{Bmatrix}$	*37 057.6*	
5g	5g ^{2}G	$\begin{Bmatrix} 3\frac{1}{2} \\ 4\frac{1}{2} \end{Bmatrix}$	37 060.2	
6p	6p ^{2}P°	$\frac{1}{2}$	*37 296.51*	1.25
		$1\frac{1}{2}$	*37 297.76*	
7s	7s ^{2}S	$\frac{1}{2}$	38 012.074	
6d	6d ^{2}D	$2\frac{1}{2}$	38 387.287	−0.0124
		$1\frac{1}{2}$	38 387.300	

22.25. Compute the electronic contribution to the heat capacity of monatomic sodium vapour from room temperature to 5000 K.

22.26. Sodium boils at 1163 K, and the vapour consists of both monomers and dimers. Calculate the equilibrium constants for the dimerization at this temperature and the proportion of dimers in the vapour at the boiling point using the Giauque function $\Phi_0(1163\ \text{K})$ for the atoms and molecules. (The molecule has a singlet electronic state, $B = 0.1547\ \text{cm}^{-1}$, and $\tilde{\nu} = 159\ \text{cm}^{-1}$.) What experiment could be done to confirm the prediction of this calculation?

22.27. Use equipartition theory to predict the likely molar heat capacities of the following molecules at room temperature: (a) I_2, (b) H_2, (c) CH_4, (d) benzene vapour, (e) water vapour, (f) carbon dioxide.

22.28. Plot $C_{V,\text{m}}$ as a function of $x = \hbar\omega/kT$ for a harmonic oscillator, and predict the heat capacities of ammonia and methane at (a) 298 K, (b) 500 K. (Vibrational data in Problem 22.18.)

22.29. The fundamental frequencies of the seven normal modes of acetylene (ethyne) are 612, 612, 729, 729, 1974, 3287, 3374 cm^{-1}. What is the molar heat capacity of the gas at (a) 298 K, (b) 500 K?

22.30. Obtain an expression for the heat capacity of a system in which there are only two levels separated by Δ. Draw the temperature dependence as a function of Δ/kT.

22.31. The NO molecule has a doubly degenerate ground state and a doubly degenerate electronically excited state 121.1 cm^{-1} above. Calculate the electronic contribution to the molar heat capacity of this molecule at (a) 50 K, (b) 298 K, (c) 500 K.

22.32. Can an applied magnetic field modify the heat capacity of a molecule? Investigate the question by deducing an expression for the heat capacity of NO_2 (which has a single unpaired electron) in an applied magnetic field. What change of heat capacity is brought about by a 5.0 T field at (a) 50 K, (b) 298 K?

22.33. The energy levels of a methyl group attached to a larger molecule are given by the expression for a particle on a ring so long as it is rotating freely. What is the high-temperature contribution to the heat capacity and the entropy of such a freely rotating group? (Its moment of inertia is $5.341 \times 10^{-47}\ \text{kg m}^2$.)

22.34. Calculate the temperature dependence of the heat capacity of *para*-hydrogen at low temperatures on the basis that its rotational levels constitute, in effect, a two-level system (but note the degeneracy of the $J = 2$ state). Use $B = 60.864\ \text{cm}^{-1}$ and sketch the heat capacity curve. The experimental heat capacity does in fact show a hump at low temperatures.

22.35. The heat capacity of a gas determines the speed of sound. Since we can calculate heat capacities from molecular data it follows that we also have a route to the calculation of this speed. We need to know that $c_s = (\gamma RT/M_\text{m})^{\frac{1}{2}}$, where $\gamma = C_p/C_V$ and M_m is the molar mass. Deduce an expression for c_s for an ideal gas of diatomic molecules (a) at high temperatures (translation, rotation, but no vibration), (b) for *para*-hydrogen at low temperatures.

22.36. Use the result of the last Problem to estimate the speed of sound in air at room temperature.

22.37. What is the residual entropy of a crystal in which the molecules can adopt (a) three, (b) five, (c) six orientations of equal energy at absolute zero?

22.38. The hexagonal molecule $C_6H_nF_{6-n}$ might have a residual entropy on account of the similarity of the hydrogen and fluorine atoms. What might be the residual value of the entropy for each value of n and for each isomer?

22.39. The thermochemical entropy of nitrogen gas at 298 K was discussed in *Example* 5.9 and the value 192.1 J K^{-1} mol^{-1} was obtained. On the basis that the rotational constant $B = 1.9987\ \text{cm}^{-1}$, and the vibrational wavenumber is 2358 cm^{-1}, calculate the statistical thermodynamic value of the entropy at this temperature. What does the value suggest about the crystal at absolute zero?

23

Determination of molecular structure: diffraction techniques

Learning objectives

After careful study of this chapter you should be able to

(1) List and describe the seven *crystal systems*, Section 23.1(a) and Fig. 23.1.

(2) Explain the meaning of the *crystal classes*, Section 23.1(a).

(3) Define *asymmetric unit*, *space lattice*, *crystal structure*, and *unit cell*, Section 23.1(b).

(4) Explain the significance of and draw the fourteen *Bravais lattices*, Section 23.1(b) and Fig. 23.8.

(5) Specify lattice planes by their *Miller indices*, Section 23.1(c), and calculate the separation of specified planes, eqn (23.1.2).

(6) Explain how X-rays are generated and why they are suitable for crystal structure determination, Section 23.2.

(7) Derive and use the *Bragg condition* for diffraction from lattice planes, eqn (23.2.1).

(8) Describe the *Laue method*, the *Bragg method*, and the *Debye–Scherrer method* of X-ray structural analysis, Section 23.2.

(9) *Index* reflections and identify the unit cell from the powder diffraction pattern, Section 23.2(a) and Example 23.3.

(10) Relate the structures of simple lattices to the intensities of the X-ray reflections, eqn (23.2.6).

(11) Define and calculate the *structure factor*, eqn (23.2.7) and Example 23.4, and relate it to the reflection intensities, Section 23.2(b).

(12) State how the structure factor is used to determine the electron density distribution, eqn (23.2.8) and Example 23.5.

(13) Describe the general procedure for an X-ray structural analysis, Section 23.2(c).

(14) State the *phase problem*, and outline the techniques for overcoming it, Section 23.2(d).

(15) Describe the *hexagonal* and *cubic close-packed* arrangements of identical spheres, Section 23.3(a).

(16) Describe the *diamond*, *ice*, *CsCl* and *NaCl* (*rock salt*) *structures*, Figs. 23.24–31, and state the *radius ratio rule*, Section 23.3(b).

(17) Describe how anomalous scattering may be used to determine the *absolute configurations* of molecules, Section 23.3(c) and Fig. 23.33.

(18) Indicate what information is available from a *neutron diffraction* experiment, Section 23.4.

(19) Describe the *electron diffraction* experiment and indicate how the *Wierl equation* is used to deduce molecular geometry, Section 23.5.

Introduction

The diffraction of waves is the basis of several powerful techniques for the determination of the structures of molecules and solids. X-rays have wavelengths comparable to the spacing of atoms in crystals (about 100 pm), and so they are diffracted by crystal lattices. By analysing the intensities of the diffraction pattern, it is possible to draw up a detailed picture of the location of atoms even of such complex molecules as proteins. Electrons moving at about 20 000 km s^{-1} (after acceleration through about 4 kV) have wavelengths of 40 pm, and may also be diffracted by molecules. Neutrons generated in a nuclear reactor, and then slowed to thermal energies, have similar wavelengths and may also be used.

In this chapter we examine the information that comes from these techniques, particularly X-ray diffraction. In order to do so, we begin by examining the orderly ways in which atoms, ions, and molecules can stack together to form crystals and see how these structures, and the diffraction patterns they give rise to, can be discussed systematically.

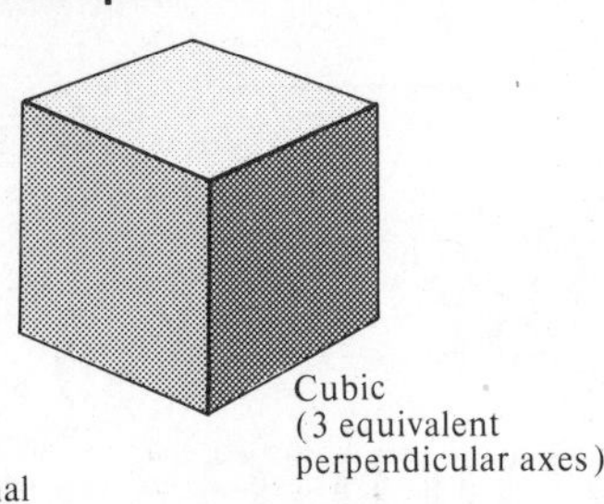

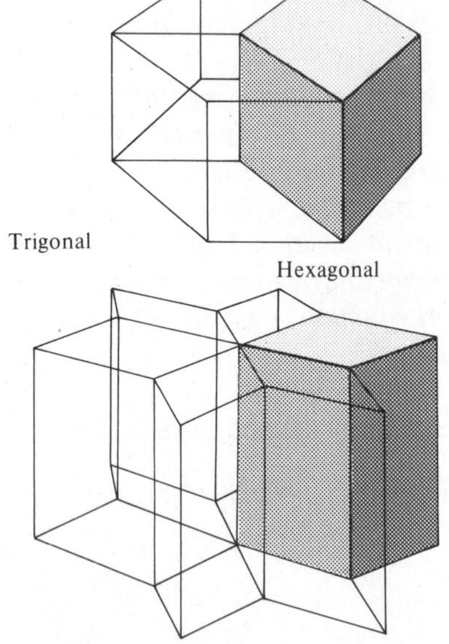

Fig. 23.1. The seven *crystal systems* and the essential symmetry elements that distinguish them.

23.1 Crystal lattices

Early in the history of modern science it was suggested that the regular external form of crystals (their *morphology*) implied an internal regularity. That crystals and their internal constitution are ordered and symmetrical suggests that the techniques described in Chapter 17 can be used both for their classification and for the discussion of their physical properties.

23.1 (a) Crystal morphology

Inspection of many different crystals led to the realization that they all correspond to one of only seven regular shapes called the *seven crystal systems*, Fig. 23.1 and Box 23.1. Which system a crystal belongs to is determined by measuring the angles between its faces, and deciding how many axes are needed to define the principal features of its shape. For example, if three equivalent, mutually perpendicular axes are required, Fig. 23.2(a), then the crystal belongs to the *cubic system*. If one axis (b in Fig. 23.2(b)) perpendicular to two that make an obtuse angle are required, then the crystal belongs to the *monoclinic system*.

Box 23.1 The 7 crystal systems and the 32 crystal classes

The systems	The classes	
	Schoenflies notation	International notation
Triclinic	C_1, C_i	1, $\bar{1}$
Monoclinic	C_s, C_2, C_{2h}	m, 2, $2/m$
Orthorhombic	C_{2v}, D_2, D_{2h}	$2mm$, 222, mmm
Rhombohedral	C_3, C_{3v}, D_3, D_{3h}, S_6	3, $3m$, 32, $\bar{6}2m$, $\bar{3}$
Tetragonal	C_4, C_{4v}, C_{4h}, D_{2d}, D_4, D_{4h}, S_4	4, $4mm$, $4/m$, 422, $\bar{4}2m$, $4/mmm$, $\bar{4}$
Hexagonal	C_6, C_{6v}, C_{6h}, D_{3d}, D_6, D_{6h}	6, $6mm$, $6/m$, $\bar{6}$, $\bar{3}m$, 622, $6/mmm$
Cubic	T, T_d, T_h, O, O_h	23, $\bar{4}3m$, $m3$, 43, $m3m$

Rhombohedral is also called *trigonal*, and *cubic* is also called *regular*. A detailed discussion can be found in *Group theory*, M. Hamermesh; Addison Wesley, 1962.

The crystal systems can be discussed in terms of the symmetry elements they possess. The *cubic system*, for example, has four threefold axes and three fourfold axes, Fig. 23.3(a), while the *monoclinic system* has one twofold axis. The *essential* symmetries are listed in Fig. 23.1. They are essential in the sense that the specified elements must be present for the crystal to belong to the system, but other elements may also be present, in which case they belong to a particular *crystal class* of that *crystal system*. For example, the crystals shown in Fig. 23.3 both belong to the cubic system, but one has a fourfold axis where the other has a twofold axis, and has the symmetry of a tetrahedron. When these additional symmetry elements are taken into account, it turns out that there are 32 crystal classes, and that they are distributed over the crystal systems as shown in Box 23.1. Crystallographers prefer the International (Hermann–Mauguin) nomenclature over the Schoenflies, and so the table contains a translation dictionary (see also Box 17.1).

The difficulty with the classification procedure is that crystal faces grow at different rates, and so the appearance of the crystal may be distorted. All the crystals in Fig. 23.4, for example, belong to the hexagonal system. Moreover, in some cases there may be accidental equivalences of axes. Therefore, in order to classify crystals unambiguously, we must do so on the basis of the *internal* symmetry elements they possess.

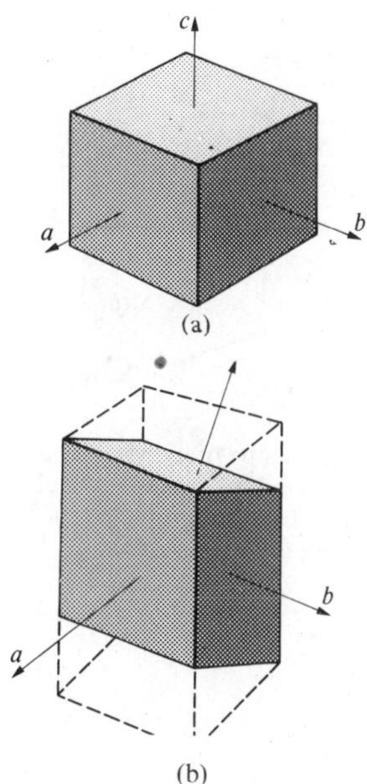

Fig. 23.2. (a) A crystal belonging to the *cubic* system has three equivalent perpendicular axes. (b) A crystal belonging to the *monoclinic* system has two axes (a and c) making an obtuse angle, and another axis (b) perpendicular to the ac-plane.

23.1 (b) Lattices and unit cells

A crystal is an orderly array of particles, and they are arranged systematically and, in some sense, 'symmetrically'. In order to be precise about the internal organization of the crystal we need to distinguish the individual units from which it is built from the patterns in which these units are present. Therefore, we introduce the following definitions:

(1) The *asymmetric unit* is the particle (e.g. ion or molecule) from which the crystal is built.

(2) The *space lattice* is a three-dimensional, infinite array of points, each of which is surrounded in an identical way by its neighbours.

That is, the space lattice defines the basic structure of the crystal. In some cases there may be an asymmetric unit at each lattice point, but that is not necessary: for instance, each lattice point might be at the centre of a cluster of three asymmetric units. The space lattice is, in effect, the abstract scaffolding for the crystal structure.

(3) The *crystal structure* is obtained by associating with each lattice point an assembly of asymmetric units in a symmetrical arrangement which is identical for each lattice point.

(4) The *unit cell* is the fundamental unit from which the entire crystal may be constructed by purely translational displacements (like bricks in a wall).

A unit cell must possess the overall symmetry of the crystal. This may be seen by considering the two-dimensional lattices shown in Fig. 23.5. The cubic unit cell, Fig. 23.5(a), has a C_4 axis, as has the crystal to which it gives rise. It follows that we should expect to be able to account for the external morphology of a crystal in terms of the symmetry of its unit cell. Some of the ways of stacking together unit cells to produce crystal faces are illustrated in Fig. 23.6: the morphology depends on the relative rates of growth of the different types of face, but the underlying unit cell structure is

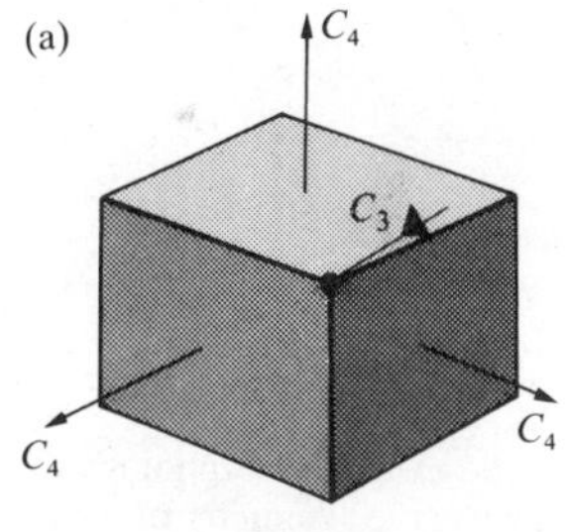

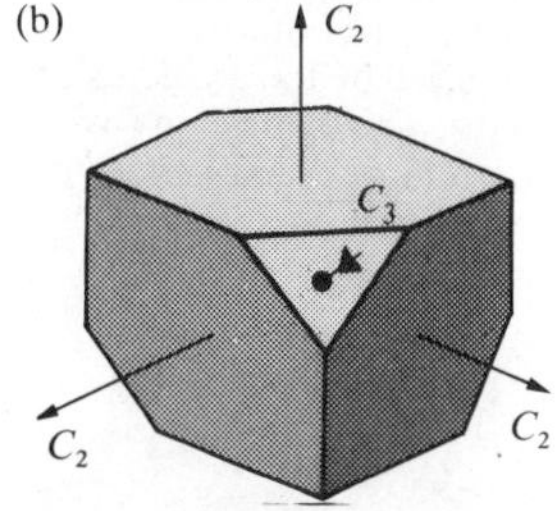

Fig. 23.3. Both these crystals belong to the cubic system, but they have different rotational symmetry elements: (a) has fourfold axes but (b) has twofold axes. They belong to different crystal *classes*.

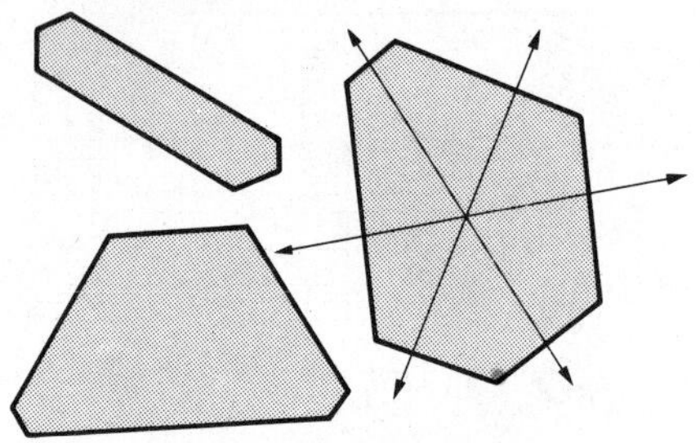

Fig. 23.4. The shape of a crystal depends on the relative rates at which its faces grow. In this example all the crystals belong to the hexagonal system.

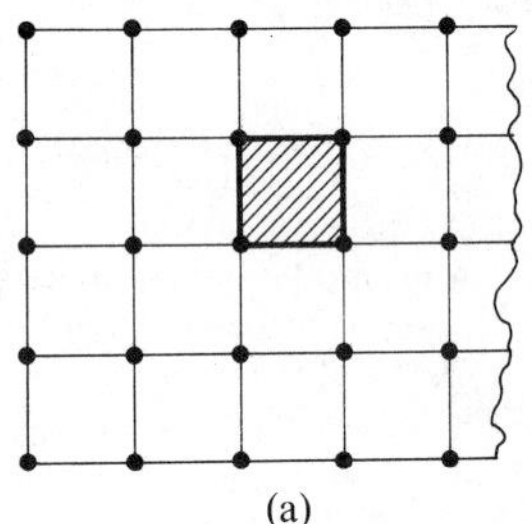

(a)

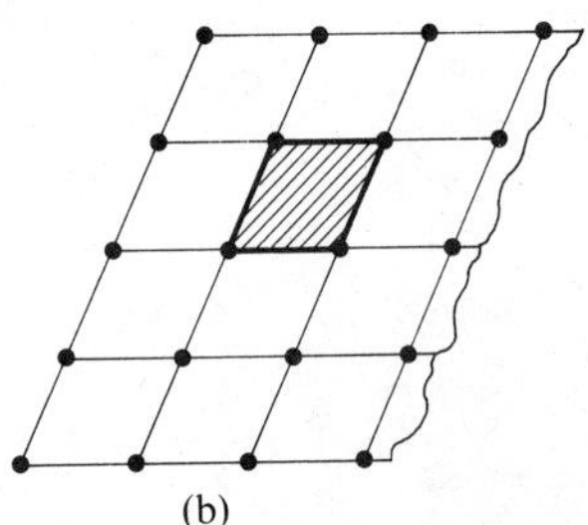

(b)

Fig. 23.5. The external morphology of a crystal reflects the symmetry of its unit cells. The unit cell in (a) has a fourfold axis, which is also possessed by the overall crystal (although it might be disguised by the accidents of crystal growth). The unit cell in (b) lacks the fourfold axis, and so does the crystal it constitutes.

uniform. Therefore, if we can identify the symmetry elements of the unit cell, we shall have an unambiguous classification of the crystal.

A unit cell cannot have the symmetry of an arbitrarily chosen point group. This is because the cells must have shapes that allow them to be stacked together to give indefinitely large structures. In the case of a two-dimensional lattice, for example, there are only five shapes capable of filling a plane, and they have the rotational symmetries C_n, with $n = 1, 2, 3, 4, 6$. No other rotational symmetries are possible: this corresponds to the impossibility of covering a floor with regular pentagons ($n = 5$) or regular n-gons with $n > 6$, Fig. 23.7.

In three dimensions there are *fourteen* types of unit cell that can stack together and give rise to a space lattice that fills all space: they are the *Bravais lattices*, Fig. 23.8. Those with lattice points only at the corners are called *primitive* (denoted P); a *body-centred* (I) unit cell has a point at the centre, and *face-centred* (F) unit cells have lattice points on their faces. Note that the Bravais lattices fall into seven groups (as represented by each primitive cell), and that these seven regular figures correspond precisely to the seven crystal systems. In other words, we see that *the seven systems correspond to the existence of the seven regular shapes that may be packed together to fill all space*. Furthermore, *the occurrence of the crystal classes reflects the presence of the corresponding symmetry elements in the unit cells*. Therefore, a tetrahedral crystal morphology indicates the presence of a unit cell with tetrahedral symmetry, and so on.

Once we know the symmetry of the unit cell we can draw immediate conclusions about the properties of the solid, just as we did for individual molecules in Chapter 17. For example, in order to assess whether a crystal is optically active, we see if its unit cell lacks an improper rotation axis. The unit cell of quartz, for example, has symmetry 32 (T in the Schoenflies system), and is optically active. On the other hand, the unit cell of calcite has symmetry $3m$ (D_{3d}), and so we know at once that it is not optically active.

A full analysis of the symmetry properties of crystals depends on taking into account their symmetries under translation, as well as the local symmetry of a single unit cell. This involves the study of *space groups* as distinct from the point groups studied in Chapter 17 (see *Further reading*). For our purposes, though, we have enough qualitative information about space lattices to be able to move on to the next stage: the labelling of the lattice planes and the quantitative discussion of their spacing.

23.1 (c) The Miller indices

Even in a rectangular lattice a large number of planes can be identified, Fig. 23.9, and so we need to be able to label them.

Consider the two-dimensional rectangular lattice formed from a unit cell of sides a, b, Fig. 23.9. Four sets of planes have been drawn, and it is clear that they can be distinguished by the distances along the axes where one representative member of each set intersects them. One scheme is to denote each set by the least intersection distances. For example, the four sets in the illustration can be denoted respectively $(1a, 1b)$, $(3a, 2b)$, $(-1a, 1b)$, and $(\infty a, 1b)$. If we agree to quote distances along the axes in terms of the lengths of the unit cell, these planes can be specified more simply as $(1, 1)$, $(3, 2)$, $(-1, 1)$, and $(\infty, 1)$. If the lattice in Fig. 23.9 is the top view of a three-dimensional rectangular lattice in which the unit cell has a length c in the z-direction, all four sets of planes intersect the z-axis at infinity, and so the full labels are $(1, 1, \infty)$, $(3, 2, \infty)$, $(-1, 1, \infty)$, and $(\infty, 1, \infty)$. The appearance of ∞ is inconvenient, and a way of eliminating it is to deal with the *reciprocals* of the indices (this turns out to have further advantages, as we shall see). The *Miller indices* are the reciprocals of the numbers in the brackets, with fractions cleared. For example, the plane $(1a, 1b, \infty c)$, the $(1, 1, \infty)$ plane, becomes (110) in the Miller system. This label, (110), is then used to refer to the complete set of equally spaced planes parallel to this one. Similarly, the $(3, 2, \infty)$ plane becomes $(\frac{1}{3}, \frac{1}{2}, 0)$ on taking reciprocals and $(2, 3, 0)$ when fractions are cleared, and so it is referred to as the (230) plane in the Miller system. The (230) planes are all the equally spaced planes parallel to this one. The planes in Fig. 23.9(d) are the (010) planes. Negative indices are written with a bar over the number: Fig. 23.9(c) shows the $(\bar{1}10)$ planes.

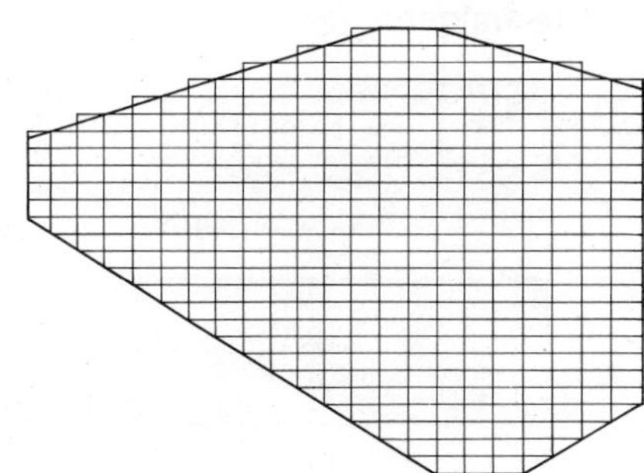

Fig. 23.6. When unit cells stack together they can give rise to crystals seemingly very different from the underlying symmetry because the faces might grow at different rates. The classification of the crystal is based on the symmetry of the unit cell, not the external appearance.

Two points should be remembered about the Miller indices in order to judge quickly which planes are under discussion. Refer to Fig. 23.10. First, the unit cell need not be rectangular: the procedure also works when the axes are not perpendicular. For that reason the axes are called a, b, c rather than x, y, z. The (111) planes of a triclinic lattice are shown in Fig. 23.10(d). Second, the smaller the value of h in (hkl), the more nearly parallel the plane to the a-axis. (The same is true of k and the b-axis and l and the c-axis.) When $h = 0$, the planes intersect the a-axis at infinity, and so $(0kl)$ are parallel to the a-axis. (Similarly, the $(h0l)$ planes are parallel to b and $(hk0)$ are parallel to c.)

The Miller indices are very useful for expressing the separation of planes. In the case of the square lattice shown in Fig. 23.11, it is easy to show that the separation of the $(hk0)$ planes is given by

$$1/d_{hk0}^2 = (1/a)^2\{h^2 + k^2\}, \quad \text{or} \quad d_{hk0} = a/(h^2 + k^2)^{\frac{1}{2}},$$

and, by extension to three dimensions, that the separation of the (hkl) planes of a cubic lattice is given by

$$1/d_{hkl}^2 = (1/a)^2\{h^2 + k^2 + l^2\}, \quad \text{or} \quad d_{hkl} = a/(h^2 + k^2 + l^2)^{\frac{1}{2}}. \quad (23.1.1)$$

The corresponding expression for a general orthorhombic lattice (a rectangular lattice based on a unit cell with different sides) is the generalization of the last equation:

$$1/d_{hkl}^2 = (h/a)^2 + (k/b)^2 + (l/c)^2. \quad (23.1.2)$$

Example 23.1

An orthorhombic cell has the following parameters:

$$a = 0.82 \text{ nm}, \qquad b = 0.94 \text{ nm}, \qquad c = 0.75 \text{ nm}.$$

What are the spacings of the (123) and the (246) planes?

- *Method*. Simply substitute in eqn (23.1.2).

- *Answer*. $1/d_{123}^2 = (1/0.82 \text{ nm})^2 + (2/0.94 \text{ nm})^2 + (3/0.75 \text{ nm})^2 = 22.04 \text{ nm}^{-2}$.

Hence, $d_{123} = 0.21$ nm. Similarly, $d_{246} = 0.11$ nm.

- *Comment*. Note that $d_{246} = \frac{1}{2}d_{123}$. In general, increasing the indexes uniformly by a factor n *decreases* the separation by the same factor.

• *Exercise*. What is the volume of the space enclosed by neighbouring (200), (030), and (001) planes of the same orthorhombic lattice?

[0.10 nm^3]

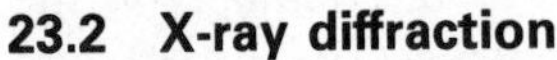

23.2 X-ray diffraction

X-rays are produced by bombarding a metal with high-energy electrons. When the electrons plunge into the metal they are decelerated and radiate a range of wavelengths. This continuum radiation is called *Bremsstrahlung* (*Bremsse* is German for brake, *Strahlung* for ray). Superimposed on the continuum are a few high-intensity, sharp peaks. These peaks arise from the

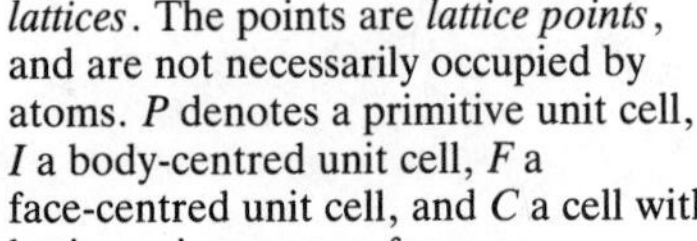

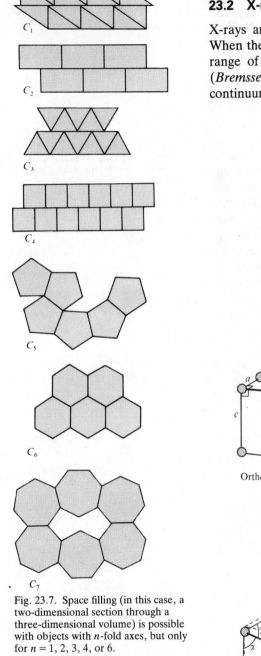

Fig. 23.7. Space filling (in this case, a two-dimensional section through a three-dimensional volume) is possible with objects with n-fold axes, but only for $n = 1, 2, 3, 4$, or 6.

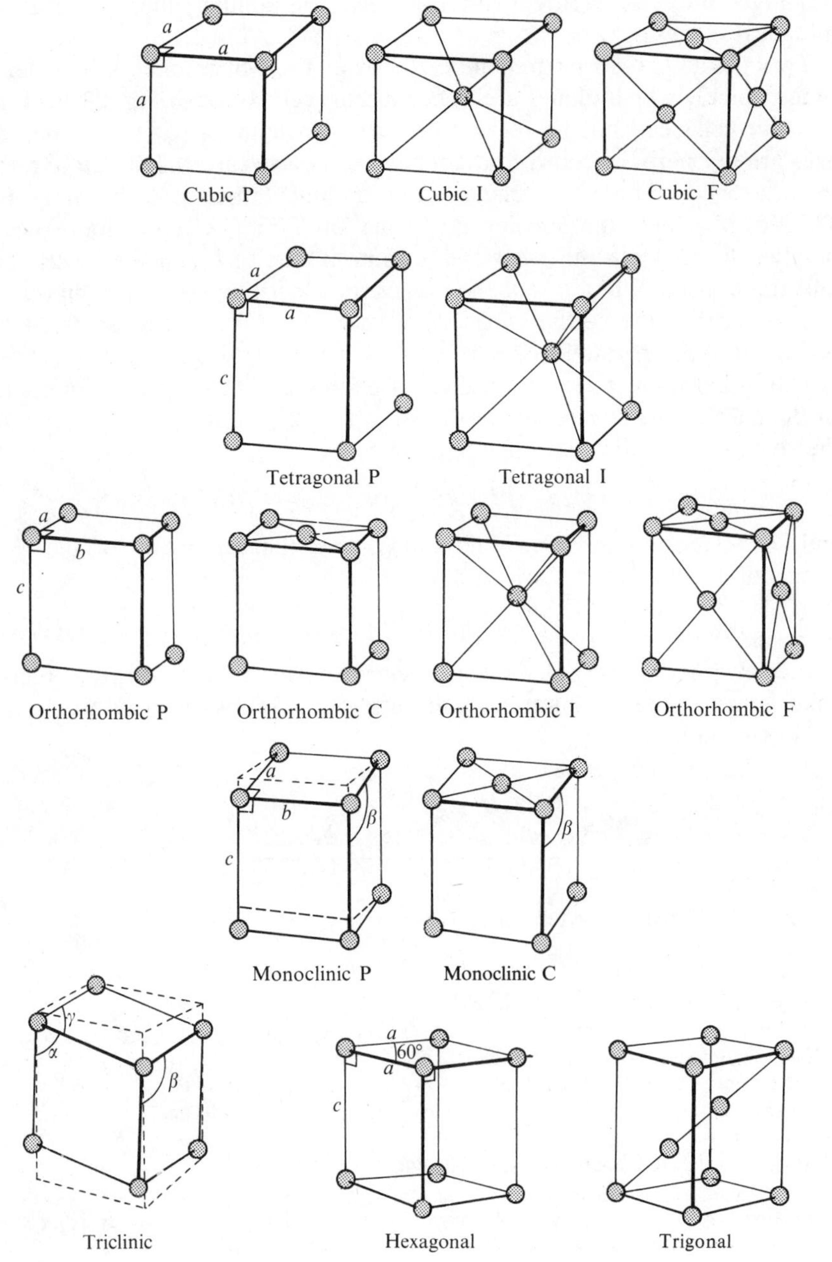

Fig. 23.8. The fourteen *Bravais lattices*. The points are *lattice points*, and are not necessarily occupied by atoms. *P* denotes a primitive unit cell, *I* a body-centred unit cell, *F* a face-centred unit cell, and *C* a cell with lattice points on two faces.

interaction of the incoming electrons with the electrons in the inner shells of the atoms: a collision expels an electron and an electron of higher energy drops into the vacancy, emitting the excess energy as an X-ray photon.

On the basis of fragments of evidence that the then recently discovered X-rays might have wavelengths comparable to the atomic spacing in crystals, Max von Laue suggested in 1912 that they might be diffracted when passed through a crystal. This was confirmed almost immediately by Walter Friedrich and Paul Knipping.

Diffraction arises as a result of the interference between waves. Where their amplitudes are in-phase the waves augment each other and the intensity is enhanced; where their amplitudes are out-of-phase they cancel, and the intensity is decreased. If the waves start at a common source, their relative phases depend on their path lengths. For instance, the Young's slit experiment, Fig. 23.12, can be explained in terms of the path difference of the two rays, different points of the screen corresponding to places where the waves from the two slits are successively in phase (bright) and out of phase (dark).

Now consider a stack of reflecting lattice planes, Fig. 23.13. The path-length difference of the two rays shown is AB + BC, which is equal to $2d \sin \theta$, where d is the layer spacing and θ is the *glancing angle*. For many glancing angles the path-length difference is not an integral number of wavelengths, and there is destructive interference. However, when the path-length difference is an integral number of wavelengths ($AB + BC = n\lambda$), the reflected waves are in phase and interfere constructively. It follows that a bright reflection should be observed when the glancing angle fulfils the

$$\textit{Bragg condition: } n\lambda = 2d \sin \theta. \qquad (23.2.1)$$

The Bragg condition (William and his son Laurence Bragg were pioneers of X-ray diffraction—they shared a Nobel prize for their work) is the basic equation of *X-ray crystallography*, the study of the internal structure of crystals with X-rays. Its primary use is in the determination of the spacing between the layers in the lattice, because once the angle θ corresponding to maximum intensity has been determined, d may readily be calculated.

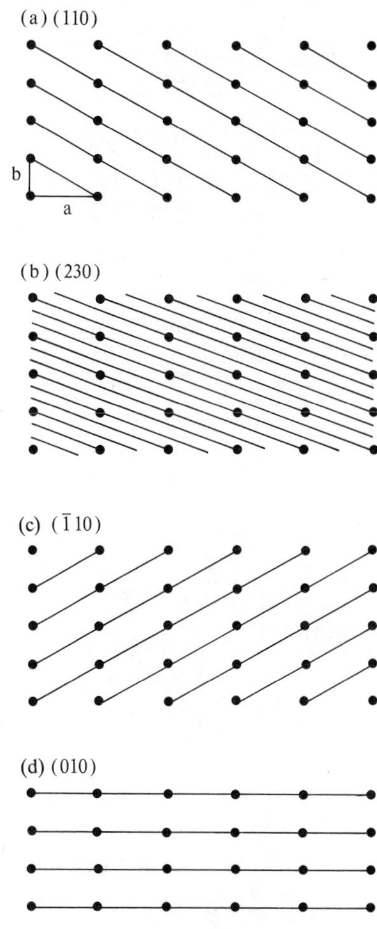

Fig. 23.9. Some of the planes that can be drawn through the points of the space lattice and their corresponding Miller indices (hkl). a and b denote the lengths of the sides of the unit cell.

Example 23.2

The separation of the lattice layers in a crystal is 404 pm. At what glancing angle will a reflection occur with CuK_{α} X-rays (wavelength 154 pm)?

- *Method*. Use eqn (23.2.1). Use $n = 1$.

- *Answer*. $\theta = \arcsin(\lambda/2d) = \arcsin 0.191 = 10° 59'$.

- *Comment*. Reflections with $n = 2, 3, \ldots$ are called *second-order*, *third-order*, and so on. In modern work it is normal to absorb the n into d, to write the Bragg Law as $\lambda = 2d \sin \theta$, and to regard the nth order reflection as arising from the (nh, nk, nl) planes (see *Comment* on *Example* 23.1).

- *Exercise*. Calculate the glancing angle for (a) the first and second order reflections from the (123) planes in the crystal treated in *Example* 23.1, and (b) the first-order reflections from the (246) planes. [(a) 21.2°, 46.0°, (b) 46.0°]

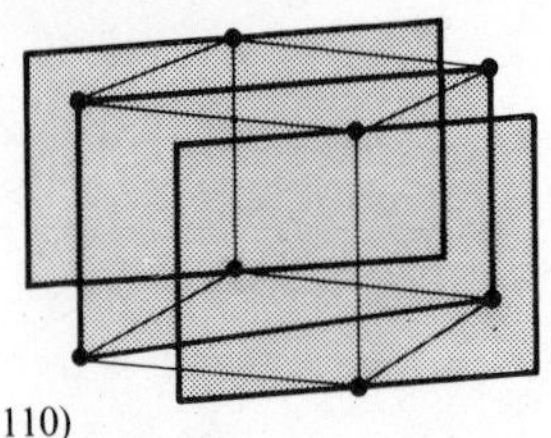
(110)

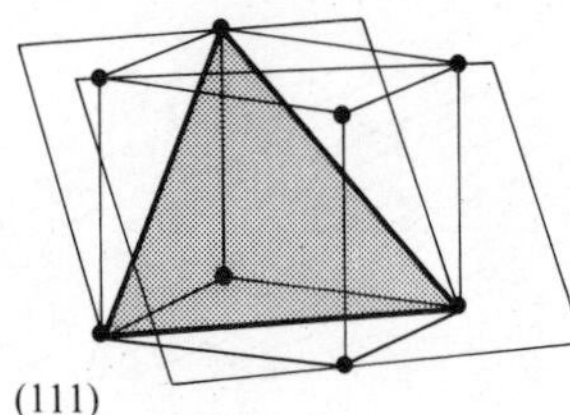
(111)

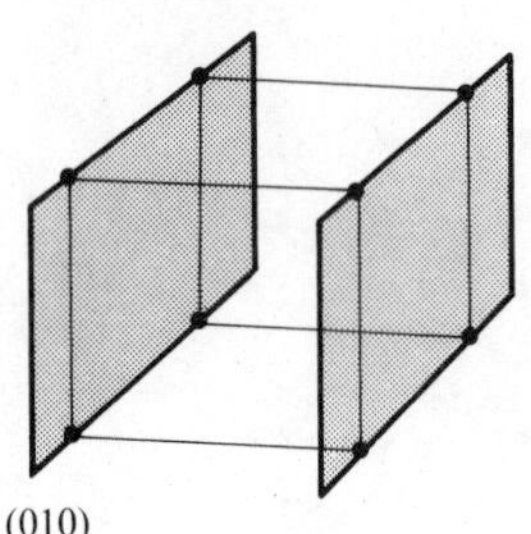
(010)

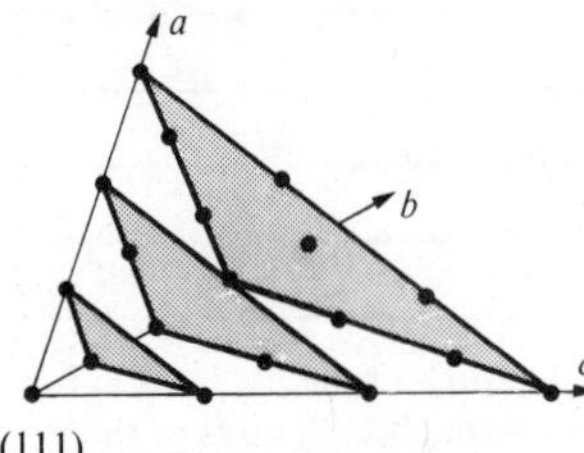

(111)

Fig. 23.10. Some representative planes and their Miller indices. Note that a 0 indicates that a plane is parallel to the corresponding axis, and that the indexing may also be used for unit cells with non-orthogonal axes.

Laue's original method for doing X-ray diffraction consisted of passing a beam of X-rays covering a broad band of wavelengths into a single crystal, and recording the diffraction pattern photographically. The idea behind the approach was that a single crystal might not be suitably oriented to act as a diffraction grating for a single wavelength, but whatever its orientation the Bragg diffraction condition would be satisfied for at least one of the wavelengths if a range was used. The Bragg's modification was to simplify the diffraction pattern by using a monochromatic beam and providing a means to rotate the crystal into different orientations. An alternative technique was developed by Peter Debye and Paul Scherrer and independently by Albert Hull. They used monochromatic radiation and a *powdered* sample in place of single crystal. Since a powder is a random heap of tiny crystals, a proportion of them will satisfy the Bragg relation even for monochromatic radiation. The powder technique is useful for a qualitative analysis of the sample, and for an initial determination of the dimensions and symmetry of the unit cell, but it cannot provide the detailed information about the electron density distribution that is available from the Bragg single-crystal, single-wavelength method.

23.2 (a) The powder method

When the sample is a powder, a proportion of the microcrystallites will satisfy the Bragg condition $\lambda = 2d \sin \theta$. (As explained in the *Comment* in *Example* 23.2, we consider all reflections as being first order, and so set $n = 1$ in the Bragg equation.) For example, some of them will be oriented so that their (111) planes, of spacing d_{111}, give rise to diffracted intensity at the angle 2θ to the incident beam, Fig. 23.14. The (111) planes of other crystallites may be at the angle θ to the beam, but at an arbitrary angle about the line of its approach. It follows that the diffracted beams lie on a cone of half-angle 2θ. Other crystallites will be oriented so that, for example, their (211) planes, with spacing d_{211}, satisfy the diffraction condition and give rise to another cone of diffracted intensity with half-angle $2\theta'$, Fig. 23.14. In principle, each set of (hkl) planes gives rise to a diffraction cone, because some of the randomly orientated crystallites can diffract the incident beam.

The original *Debye–Scherrer method* is illustrated in Fig. 23.15. The sample is in a capillary tube, which is rotated to ensure that the crystallites are randomly orientated. The diffraction cones are photographed as arcs of circles where they cut the strip of film, and some typical patterns are shown in Fig. 23.16. In modern diffractometers the sample is spread on a flat plate and the diffraction pattern is monitored electronically. The major application is now to qualitative analysis because the diffraction pattern is a kind of fingerprint, and many have been recorded.

The angle θ can be measured from the location of the diffraction intensity (the 'reflection'), and if the value of (h, k, l) is known—this is called *indexing the reflection*—the value of d_{hkl} can be deduced from the Bragg relation. The crux of the technique is its indexing. Fortunately, some types of unit cell give characteristic and easily recognizable patterns of lines. For example, in a cubic lattice of unit cell dimension a the spacing is given by eqn (23.1.1), and so the angles at which the (hkl) planes diffract are given by

$$\sin \theta_{hkl} = (\lambda/2a)(h^2 + k^2 + l^2)^{\frac{1}{2}}. \qquad (23.2.2)$$

The reflections are then predicted by substituting the values of h, k, l. However, not all integer values of $h^2 + k^2 + l^2$ are obtained:

(hkl):	(100)	(110)	(111)	(200)	(210)	(211)	(220)	(300)	(221)	(310)
$h^2 + k^2 + l^2$	1	2	3	4	5	6	8	9	9	10

etc. Notice that 7 (and 15, . . .) is missing because the sum of the squares of three integers cannot equal 7 (or 15, . . .). Therefore the pattern has omissions which are characteristic of the primitive cubic structure.

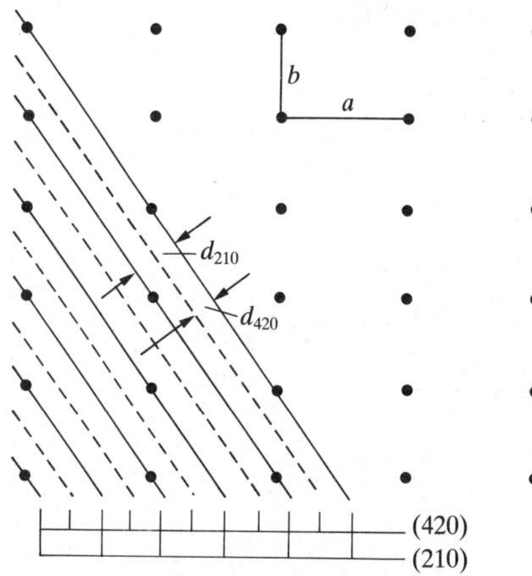

Fig. 23.11. The separation between the planes (hkl) is written d_{hkl}. This shows the values of d_{210} and d_{420} for a rectangular lattice. Note that the second separation is half the first.

Example 23.3

A powder diffraction photograph of KCl gave lines at the following distances (in mm) from the centre spot when 70.8 pm Mo X-rays were used in a camera of radius 5.74 cm:

13.2, 18.4, 22.8, 26.2, 29.4, 32.2, 37.2, 39.6, 41.8.

Identify the unit cell and determine its dimensions.

• *Method*. From eqn (23.2.2) we write

$$\sin^2\theta = A(h^2 + k^2 + l^2), \qquad A = \lambda/2a.$$

Therefore, convert distances to θ, then to $\sin^2\theta$, then find the common factor A, and then find $h^2 + k^2 + l^2$. Express this as (hkl). For the cell dimensions, solve A for a. Note that the angle θ (in radians) is related to the distances by $\theta = D/2R$. Convert to degrees by multiplication by $360/2\pi$. Since $R = 57.4$ mm, we have

$$\theta/\text{degrees} = \tfrac{1}{2}(D/57.4\text{ mm})(360/2\pi) = \tfrac{1}{2}D/\text{mm},$$

and so the conversion is very simple.

• *Answer*. Draw up the following table:

D/mm	13.2	18.4	22.8	26.2	29.4	32.2	37.2	39.6	41.8	43.8	46.0
θ/deg	6.60	9.20	11.4	13.1	14.7	16.1	18.6	19.8	20.9	21.9	23.0
$100\sin^2\theta$	1.32	2.56	3.91	5.24	6.44	7.69	10.2	11.5	12.7	13.9	15.3

The common divisor is 1.32/100. Divide through to identify $h^2 + k^2 + l^2$:

$h^2 + k^2 + l^2$	1	2	3	4	5	6	8	9	10	11	12

The corresponding indexes are

(100) (110) (111) (200) (210) (211) (220) (300) (310) (311) (222)

This has indexed the lines. Note the absence between 6 and 8: this indicates a primitive cubic cell. From $\lambda/2a = 0.132$, we find $a = 308$ pm.

• *Comment*. Later we shall see that additional information comes from the intensities of the lines. Note that a cunning choice of R saves a lot of work.

• *Exercise*. In the same camera, another cubic crystal gave reflections at the following distances (in mm): 2.3, 3.2, 4.0, 4.6, 5.1, 5.6. Identify the cell (refer to Fig. 23.17) and its dimensions. [Cubic I; 550 pm]

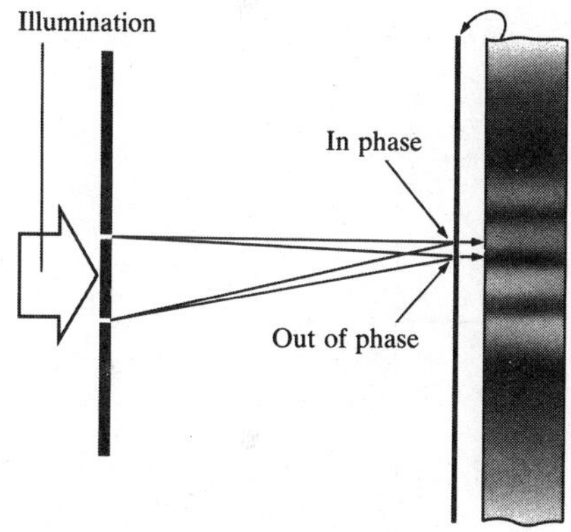

Fig. 23.12. A 'classic' diffraction experiment is the *Young's slit experiment* in which a beam of light passes through two slits in one screen and illuminates another. Depending on the relative path-lengths of the two rays, their interference is constructive (bright band) or destructive (dark band) where they reach the second screen.

23.2 (b) Systematic absences

Example 23.3 confirms the features of the analysis above, and enables us to determine the size of the KCl unit cell. When the same method is used on the very similar NaCl crystal, the diffraction pattern is surprisingly different. The reason is that the K^+ and Cl^- ions, having the same electronic configurations, act as almost identical scatterers of X-rays, and so although the crystal contains two kinds of ions, the X-rays scatter as though the crystal were a cubic lattice of identical ions. Because it has fewer electrons, the scattering power of Na^+ is less than that of Cl^-, and the analysis must be modified to take into account the presence of two kinds of ion in the crystal.

The problem can be discussed qualitatively by regarding the NaCl crystal as consisting of two interpenetrating cubic lattices, one of Na^+ ions, the other of Cl^- ions. Some reflections from the Na^+ lattice are out of phase with some Cl^- reflections, and the two tend to cancel. For other orientations the two sets of reflections are in phase, and intense lines appear. Cancellation is complete only if the ions have identical scattering power (as is nearly true for KCl, the precise extent of the similarity depending on the scattering angle), but in NaCl the cancellation is incomplete and an alternation of lines, some intense and others faint, occurs. This accounts for the general appearance of the NaCl diffraction pattern shown in Fig. 23.16.

The details of the diffraction pattern can be discussed by considering the two-dimensional lattice shown in Fig. 23.17. This is composed of A and B atoms with scattering powers measured by their *scattering factors* (f_A and f_B). These are related to the electron density distribution in the atom, $\rho(r)$, by

$$f = 4\pi \int_0^\infty r^2 \rho(r)\{\sin kr/kr\}\,\mathrm{d}r, \qquad k = (4\pi/\lambda)\sin\theta. \qquad (23.2.3)$$

The electrons are responsible for the scattering, and for forward scattering ($\theta = 0$, so that, by taking the limit as $k \to 0$, $\sin kr/kr = 1$) the scattering factor is proportional to the total electron density; that is, f is then proportional to the atomic number. A complication apparent from eqn (23.2.3) is that f also depends on the angle through which the scattering occurs, being less for sidewards than for forward scattering, and an analysis of intensities must always take this dependence on direction into account.

Consider a single unit cell in which there is an A atom at the origin and a B atom at the location $(xa, yb, 0)$. When the incident beam is at a glancing angle θ with respect to the $(hk0)$ planes of A atoms, it is also at that same angle with respect to the $(hk0)$ planes of B atoms. The A and B planes are separated by d_{hk0}, and so reflections from the B planes contribute waves with phases shifted relative to the reflections from the A planes. Reference

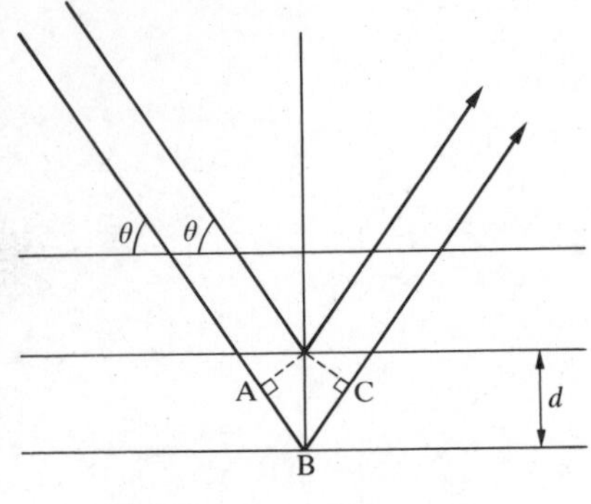

Fig. 23.13. The conventional calculation of the *Bragg Law* treats each lattice plane as reflecting the incident radiation. The path lengths differ by AB + BC, which depends on the angle θ. Constructive interference (a 'reflection') occurs when AB + BC is an integral number of wavelengths.

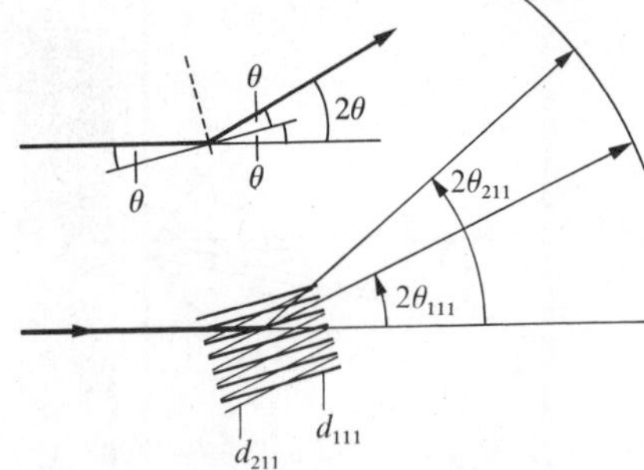

Fig. 23.14. Diffraction from lattice planes occurs when the Bragg Law is satisfied. This shows how two sets of planes (211) and (111) each gives rise to a reflection for a single setting of the crystal orientation. The full diffraction pattern consists of reflections from all the sets of (hkl) planes that satisfy the Bragg Law. (A reflection at a glancing angle θ gives rise to a reflection at an angle 2θ to the direction of the incident beam; see inset.)

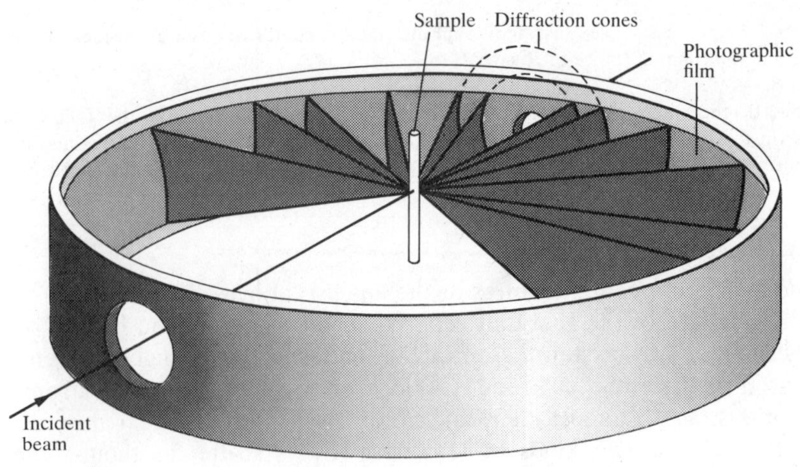

Fig. 23.15. In the *Debye–Scherrer method* a monochromatic X-ray beam is diffracted by a powder sample. The crystallites give rise to cones of intensity, which are detected by a photographic film wrapped round the circumference of the camera.

to Fig. 23.17 shows that the path-length difference is $2d'_{hk0}\sin\theta$. However, since θ is a diffraction angle, it is given by $\sin\theta = \lambda/2d_{hk0}$. Consequently, the A, B path-length difference is $d'_{hk0}\lambda/d_{hk0}$. Simple trigonometry gives $d'_{hk0} = (hx + ky)d_{hk0}$, and so the path difference is simply $(hx + ky)\lambda$. It follows that the A, B phase difference is $\{(hx + ky)\lambda\}(2\pi/\lambda)$, or $2\pi(hx + ky)$. In three dimensions:

$$\text{A, B phase difference} = 2\pi(hx + ky + lz). \qquad (23.2.4)$$

When the phase difference is π (180°) the amplitudes of the two reflections cancel and if the atoms have the same scattering power the intensity is completely annihilated. For example, if the unit cells are body centred cubic (b.c.c.), Fig. 23.8, each one containing a B atom at $x = y = z = \frac{1}{2}$, the A, B phase difference is $\pi(h + k + l)$, and so all reflections corresponding to odd values of the sum $h + k + l$ vanish because the phases are 180° displaced in phase. This means that *the diffraction pattern for a b.c.c. lattice can be constructed from that for the primitive cubic lattice by striking out all the reflections with $h + k + l$ odd*. Recognition of these *systematic absences* in a powder spectrum, Fig. 23.18, immediately indicates a b.c.c. lattice. A similar calculation can be done for a f.c.c. lattice (cubic F), Fig. 23.18.

If the amplitude of the waves scattered from A is f_A at the detector, the amplitude of the waves scattered from B is $f_B e^{2i\pi(hx+ky+lz)}$ because of the extra phase difference. The total amplitude at the detector is therefore

$$F = f_A + f_B e^{2\pi i(hx+ky+lz)}, \qquad (23.2.5)$$

and the intensity there is

$$I_{hkl} \propto F^*F = |f_A + f_B e^{2\pi i T}|^2, \qquad T = hx + ky + lz,$$

because the intensity is proportional to the square modulus of the amplitude of the wave. This expression expands to

$$I_{hkl} \propto f_A^2 + f_B^2 + f_A f_B\{e^{2\pi i T} + e^{-2\pi i T}\} = f_A^2 + f_B^2 + 2f_A f_B \cos 2\pi T. \qquad (23.2.6)$$

The cosine term either adds to or subtracts from $f_A^2 + f_B^2$, depending on the value of T, and so there is a variation in the intensities of the lines with different hkl indexes (as well as the smooth variation of intensity that arises

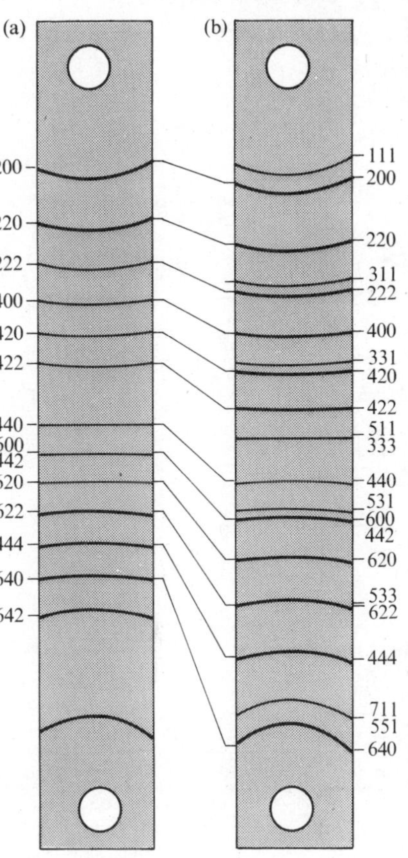

Fig. 23.16. X-ray powder photographs of (a) KCl, (b) NaCl and the indexed reflections. The smaller number of lines in (a) is a consequence of the similarity of the K^+ and Cl^- scattering lengths, as discussed in the text. Note that the film covers one-half of the circumference of the powder camera, the lines at the top corresponding to the largest diffraction angles.

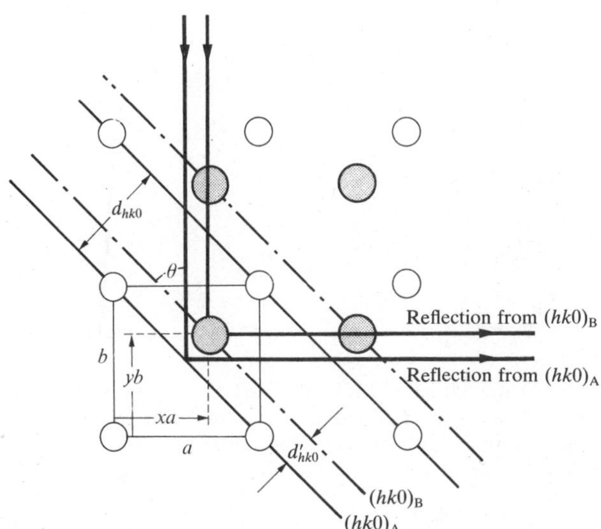

Fig. 23.17. Diffraction from a crystal containing two kinds of atom. The interference between the waves scattered from the B atom (of scattering length f_B) and of the A atoms (of scattering length f_A) depends on the relative locations of B in the unit cell.

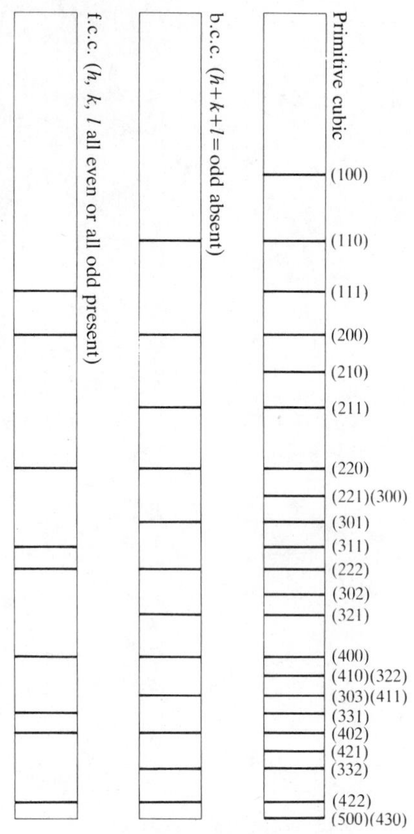

Fig. 23.18. The powder photographs and the systematic absences of the b.c.c. and f.c.c. unit cells. Comparison of the observed photograph with patterns like these enable the unit cell to be identified. The locations of the lines give the cell dimensions.

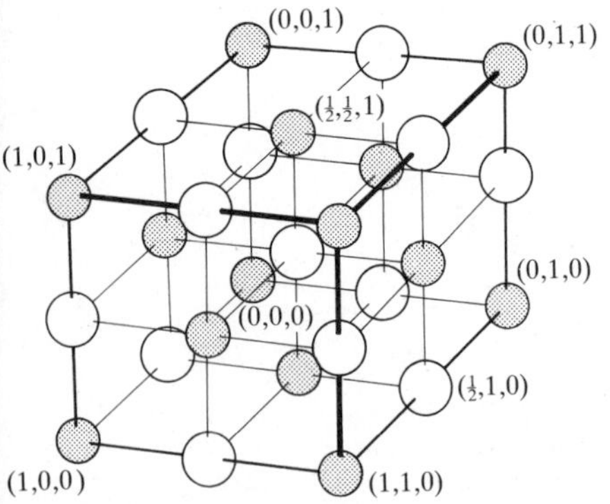

Fig. 23.19. The location of the atoms for the structure factor calculation in *Example* 23.4. The tinted circles are Na^+, the open circles are Cl^-.

from the angle dependence of the scattering factors). This is exactly what is observed for NaCl.

23.2 (c) Fourier synthesis

If the unit cell contains several atoms with scattering factors f_i and coordinates $(x_i a, y_i b, z_i c)$, the overall amplitude of a wave diffracted by the (hkl) planes is a generalization of eqn (23.2.5):

$$F_{hkl} = \sum_i f_i e^{2\pi i(hx_i + ky_i + lz_i)}, \tag{23.2.7}$$

the sum being over all the atoms in the unit cell. F_{hkl} is called the *structure factor*. The intensity of the *(hkl)-reflection* (the intensity of the reflection from the set of (hkl) planes) is proportional to $|F_{hkl}|^2$.

Example 23.4

Calculate the structure factors for a rock-salt NaCl unit cell (two interpenetrating f.c.c. lattices).

• *Method*. First, consider the Na^+ lattice. The unit cell has atoms at the coordinates $(0,0,0)$, $(0,1,0)$, $(0,1,1)$, etc., Fig. 23.19. The Cl^- lattice consists of ions at $(0,\frac{1}{2},0)$, $(\frac{1}{2},1,0)$, $(\frac{1}{2},\frac{1}{2},\frac{1}{2})$, etc. Write f^+ for the Na^+ scattering length and f^- for the Cl^- scattering length. Note that ions on faces are shared between two cells (use $\frac{1}{2}f$), those on edges by four (use $\frac{1}{4}f$), and those at corners by eight (use $\frac{1}{8}f$). Use eqn (23.2.7), summing over the coordinates of all 27 atoms in the illustration.

• *Answer*. We shall write only the first few terms of each line:

$$F_{hkl} = f^+\{\tfrac{1}{8} + \tfrac{1}{8}e^{2\pi il} + \ldots + \tfrac{1}{8}e^{2\pi i(h+k+l)} + \ldots + \tfrac{1}{2}e^{2\pi i(h/2+k/2+l)}\}$$
$$+ f^-\{e^{2\pi i(h/2+k/2+l/2)} + \tfrac{1}{4}e^{2\pi i(h/2)} + \ldots + \tfrac{1}{4}e^{2\pi i(h/2+l)}\}.$$

Now use $e^{2\pi ih} = e^{2\pi ik} = e^{2\pi il} = 1$ because h, k, l are all integers and $e^{2\pi i} = 1$. The expression then simplifies considerably, and becomes

$$F_{hkl} = f^+\{1 + \cos(h+k)\pi + \cos(h+l)\pi + \cos(k+l)\pi\}$$
$$+ f^-\{(-1)^{h+k+l} + \cos k\pi + \cos l\pi + \cos h\pi\}.$$

This may also be simplified, because $\cos h\pi = (-1)^h$, and so

$$F_{hkl} = f^+\{1 + (-1)^{h+k} + (-1)^{h+l} + (-1)^{l+k}\}$$
$$+ f^-\{(-1)^{h+k+l} + (-1)^h + (-1)^k + (-1)^l\}.$$

Now note that:

if h, k, l are all even,

$$F_{hkl} = f^+\{1+1+1+1\} + f^-\{1+1+1+1\} = 4(f^+ + f^-);$$

if h, k, l are all odd,

$$F_{hkl} = 4(f^+ - f^-);$$

if one is odd, two even, or vice versa,

$$F_{hkl} = 0.$$

• *Comment*. Note that the *hkl* all odd reflections are less intense than the *hkl* all even. For $f^+ = f^-$, which is the case for identical atoms in a simple cubic arrangement, the *hkl* all odd have zero intensity, corresponding to the systematic absences of simple cubic unit cells.

• *Exercise*. Deduce the rule for the systematic absences of a b.c.c. unit cell.

[$h+k+l$ odd]

The general idea behind an X-ray diffraction examination of a crystal should now be clear. The structure of the unit cell is contained in the structure factor because it depends on the atoms present (through f_i) and on their locations (through $hx_i + ky_i + lz_i$ and the angle dependence of f). The precise relation we need is the following. Given that we know the structure factors for all the reflections, then they may be used to construct the *electron density*, $\rho(\boldsymbol{r})$, in the unit cell using the relation

$$\rho(\boldsymbol{r}) = \sum_{hkl} F_{hkl}\mathrm{e}^{-2\pi i(hx+ky+lz)}. \qquad (23.2.8)$$

This is called a *Fourier synthesis* of the electron density. The sum may be evaluated in order to find the electron density at the point $\boldsymbol{r} = (xa, yb, zc)$, and therefore throughout the crystal, if we know the values of the structure factors for all the reflections.

Example 23.5

Consider the (h00) planes of a crystal extending indefinitely in the x-direction. In an X-ray analysis the structure factors were found as follows:

h:	0	1	2	3	4	5	6	7	8	9	10	11	12	13	14	15
F_h	16	−10	2	−1	7	−10	8	−3	2	−3	6	−5	3	−2	2	−3

(and $F_{-h} = F_h$). Construct a plot of the electron density along the x-axis of the unit cell.

• *Method*. Use eqn (23.2.8). Since $F_{-h} = F_h$,

$$\rho(x) = F_0 + 2\sum_{h>0} F_h \cos(2\pi hx).$$

Evaluate the sum for $x = 0, 0.1, 0.2, \ldots, 1.0$ (or more finely if a computer is available).

• *Answer*. The results are plotted in Fig. 23.20 (the full line).

• *Comment*. The positions of two or three atoms can be discerned very readily. The more terms there are included, the more accurate the density plot. Terms corresponding to high values of h (short wavelength cosine terms in the sum) account for the finer details of the electron density; low values of h account for the broad features.

• *Exercise*. Write a short computer program to evaluate these Fourier sums, and then experiment with different structure factors (including changing signs as well as amplitudes).

The problem at this stage is to obtain the intensities of all the diffraction spots, and to index them. This is a considerable problem, and it is essential to use single-crystal techniques.

The preliminary investigation of a crystal, to identify the symmetry of its unit cell and its dimensions, makes use of a photographic technique. One method is to use the *oscillation camera*, in which the crystal (which typically might be of side 0.1 mm) is set on a mount called a *goniometer head* and oscillated through about 10°. The diffraction pattern is recorded on a cylindrical film surrounding the sample. A difficulty is that many of the spots overlap, but the pattern can be simplified by placing a screen in front of the film so that only one *layer line* of spots is exposed, and gearing the film to the oscillating camera head so that it moves parallel to the axis of oscillation of the crystal. This is called the *Weissenberg technique*. Although the

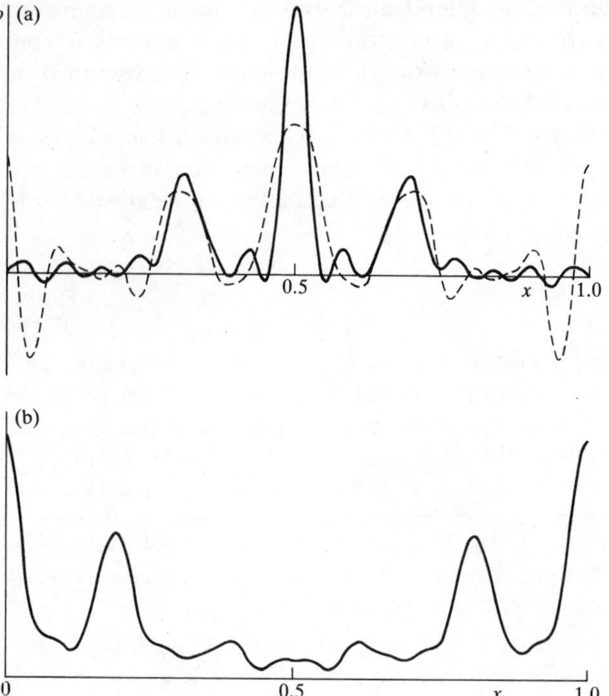

Fig. 23.20. (a) The plot of the electron density calculated in *Examples* 23.5 (full line) and 23.6 (broken line). Note how a different choice of phases for the structure factors leads to a markedly different structure (and, in the second case, for an unacceptable negative electron density in two regions). (b) The Patterson synthesis for the same data. The peaks correspond to the separations of the atoms in the actual structure.

indexing is greatly simplified, the photographs are severely distorted. A refinement that overcomes this difficulty is to have a different coupling between the motions of the crystal and the screened film; this *precession technique* gives an undistorted pattern that can be indexed quite readily.

Computing techniques are now available that lead not only to automatic indexing but also to the automated determination of the shape, symmetry, and size of the unit cell. The most sophisticated of the current techniques uses the *four-circle diffractometer*, Fig. 23.21. The crystal is set in an

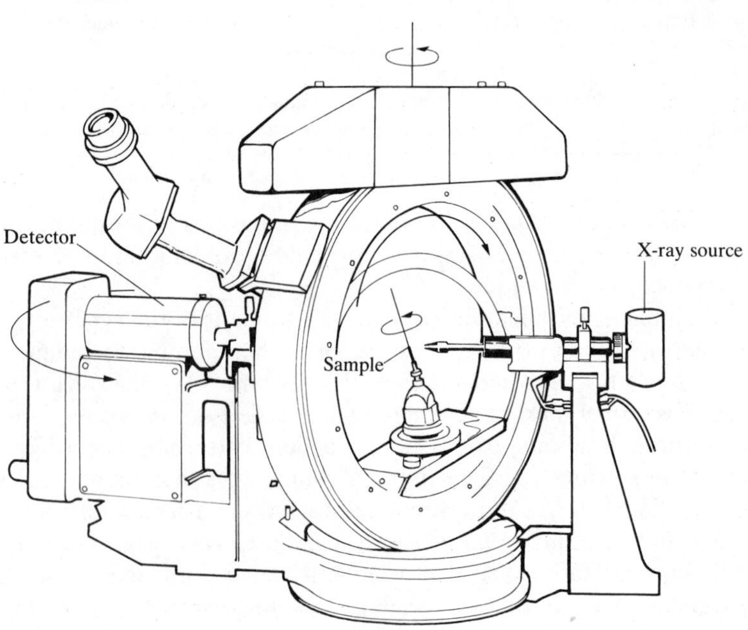

Fig. 23.21. A *four-circle diffractometer*. The settings of the orientations of the components are controlled by computer; each (hkl) reflection is monitored in turn, and their intensities are recorded.

arbitrary orientation in the goniometer head. Since its unit cell dimensions will already have been determined (e.g. using a precession camera) the settings of the diffractometer's four angles that are needed to observe any particular (hkl)-reflection can be computed. This requires knowledge of the orientation of the crystal, but that can be obtained by finding reflections that can be identified from the diffraction photographs taken earlier. The computer controls the settings, and moves the diffractometer to each one in turn. There the diffraction intensity is measured (using some kind of crystal detector or photomultiplier) and background intensities are assessed by making measurements at slightly different settings. The accumulated data for all the (hkl)-reflections monitored are then converted into the structure factors F_{hkl}.

23.2 (d) The phase problem

From the measured I_{hkl} we get the F_{hkl}, and then do the sum in eqn (23.2.8) to find ρ. Unfortunately, I_{hkl} is proportional to the *square modulus* $|F_{hkl}|^2$, and so we cannot say whether we should use $+|F_{hkl}|$ or $-|F_{hkl}|$ in the sum. In fact, the situation is worse, because if we write F_{hkl} as the complex number $|F_{hkl}|e^{i\phi}$, where ϕ is the phase of F_{hkl} and $|F_{hkl}|$ its magnitude, then the intensity lets us determine $|F_{hkl}|$ but tells us nothing of its phase. This indeterminance is called the *phase problem*. Some way must be found to assign phases to the structure factors, for otherwise the sum for ρ could not be evaluated and the method would be useless.

Example 23.6

Repeat *Example* 23.5, but suppose that all the structure factors after $h=6$ have the same positive phase.

• *Method*. Repeat the evaluation of the 16-term sum, but with positive values of F_h after $h=6$.

• *Answer*. The resulting electron density is plotted in Fig. 23.20 with a broken line.

• *Comment*. This illustrates the importance of the phase problem: different phase choices give quite different electron densities.

• *Exercise*. Explore the consequence of summing the cosine terms with coefficients equal to F_h^2. This is an example of the *Patterson synthesis* discussed below.

[Fig. 23.20(b)]

The phase problem can be overcome to some extent by a variety of methods, one of the most important of which has been the *Patterson synthesis* in which the intensities I_{hkl} (the square modulus of the structure factors, $|F_{hkl}|^2$) are used in place of the structure factors themselves in eqn (23.2.8). When this is done, the function produced is a map of *separations* of atoms in the unit cell. For example, if the unit cell has the structure shown in Fig. 23.22(a), the Patterson synthesis would be the map shown in Fig. 23.22(b), where the location of each spot relative to the origin gives the orientation and separation of each pair of atoms in the original structure. Although the number of pairs is very large, so that the pattern is

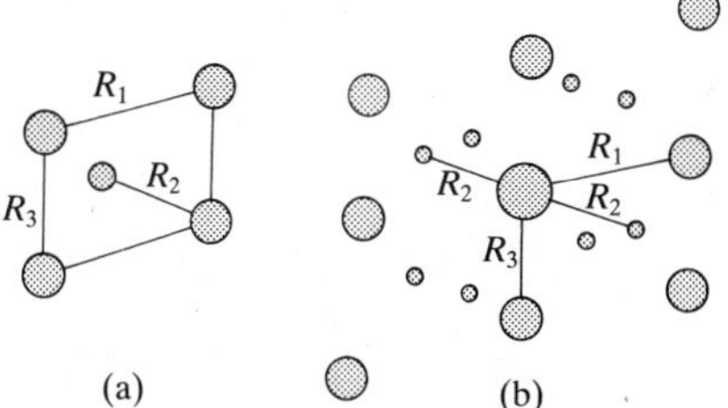

Fig. 23.22. The Patterson synthesis corresponding to the pattern in (a) is the pattern in (b). The distance and orientation of each spot from the origin gives the orientation and separation of each atom–atom separation in (a).

complex, if some atoms are heavy they dominate the scattering and their locations may be deduced quite readily. Such atoms may be introduced artificially into a complex molecule without affecting its structure significantly (the technique known as *isomorphous replacement*), and the principal features of its structure can be established before the locations of the lighter atoms are explored. About 60% or more of current structure determinations make use of this approach.

Modern structural analyses also make extensive use of the *direct method*, which is based on statistical procedures and depends on the possibility of treating the atoms in a unit cell as being virtually randomly distributed (from the radiation's point of view), and then to use the techniques of mathematical statistics to arrive at probabilities for phases having a particular value.‡ It is possible to deduce relations between some structure factors and sums (and sums of squares) of others, which have the effect of constraining the phases to have particular values, at least with high (over 99%) probability. Since the relations can be built into computer programs, the determination of a crystal structure from a collection of observed intensities becomes almost completely automated. Unfortunately, the reliability decreases as $1/N^{\frac{1}{2}}$, where N is the number of atoms per unit cell, and so it is unsuitable when the molecules are large ($N > 100$).

23.2 (e) The overall procedure

The X-ray technique depends on the availability of a single crystal of the sample, and this has led to a great deal of effort to obtain crystals of large, biologically important molecules, such as proteins. Once that has been achieved, the indexing and determination of the unit cell characteristics are carried out with a precession camera, and then the intensities are measured on a four-circle diffractometer and interpreted as structure factors with signs assigned by the direct method. The electron density can then be calculated, and drawn out as a contour diagram, such as in Fig. 23.23. Various *refinement* techniques may be employed. For example, the atoms vibrate about their mean positions, which blurs their location. The magnitude of this effect can be calculated, and it can be eliminated from the electron density map.

The remarkable and striking results obtained from X-ray diffraction suggest that it is an ideal technique. This would be true were it not for a number of limitations. The first is its restriction to the solid state. It is important to take into account that in their natural environment biological molecules might unwind significantly, and that in the solid the packing of molecules might impose constraints on their stereochemistry. In this respect NMR is often more suitable, because it can be used to study fluid samples, and the development of special techniques (Section 25.1(k)) has enabled enzymes to be studied in their natural environment. Another limitation (which makes NMR complementary) is the poor response of X-ray diffraction to the presence of hydrogen atoms (because they have so few scattering electrons). However, the sensitivity of modern instruments is such that the detection of hydrogen atoms has now become largely routine.

‡ Some of the techniques appear to have originated from the cryptanalysis (code-breaking) techniques developed during World War II. Letters in words occur in statistically predictable sequences that can be expressed more definitely as other parts of a message become deciphered, and so the structure factors in X-ray analysis become increasingly well-established.

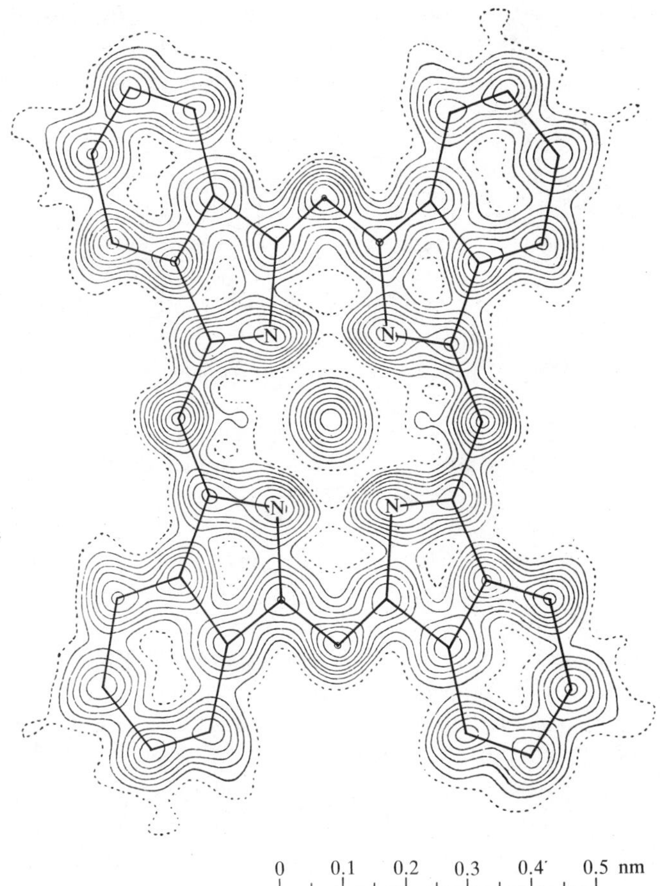

Fig. 23.23. The electron density contours of nickel phthalocyanine computed by Fourier synthesis from the observed structure factors (J. M. Robertson, *Organic crystals and molecules*, Cornell University Press, 1953). The conventional structure has been superimposed. Notice that in this (fairly early) example, the hydrogen atoms do not show.

23.3 Information from X-ray analysis

The bonds between the particles of a crystal may be of various kinds. Simplest of all (in principle) are *metals*, where a sea of electrons floods between arrays of identical cations and binds the whole together into a rigid but malleable structure. Their crystal structures are determined largely by the way spherical metal cations can pack together into an orderly array.

In *ionic crystals* the crystal structure is largely governed by the geometrical problem of packing the ions into the lowest energy arrangement taking account of their electrostatic Coulombic interactions and their different sizes.

In *covalent crystals*, covalent bonds in a definite spatial orientation link the atoms together in a network extending through the crystal. The stereochemical demands of valence override the simple geometrical problem of packing spheres together, and elaborate and extensive structures may be formed. A famous example of this is diamond, Fig. 23.24, in which each sp^3-hybridized carbon is bonded tetrahedrally to its four neighbours. Materials formed in this way are often hard and unreactive. The silicates, where there is a mixture of ionic and covalent bonding (both between the silicate groups and the cations, and within the SiO_4 groups on account of the difference of electronegativities of the silicon and oxygen atoms) are intermediate between the two extremes of ionic and covalent crystals.

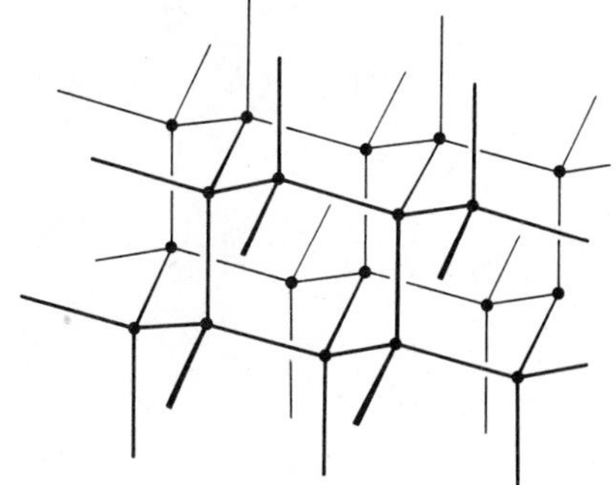

Fig. 23.24. A fragment of the structure of diamond. Each C atom is tetrahedrally sp^3-bonded to four neighbours. This results in a rigid crystal.

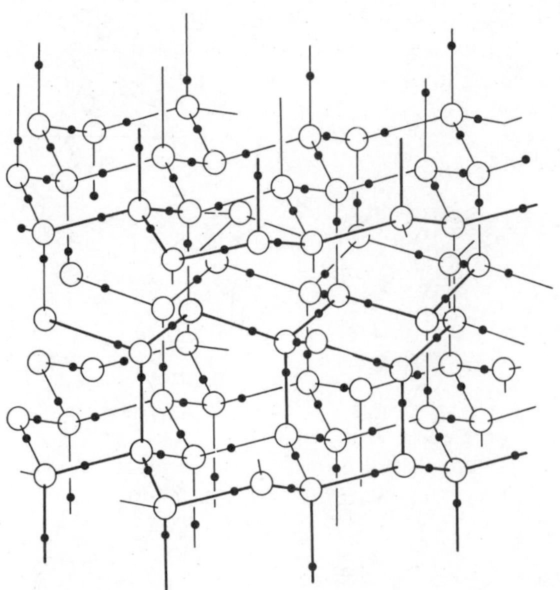

Fig. 23.25. A fragment of the crystal structure of ice (ice-I). Each O atom is at the centre of a tetrahedron of four O atoms at a distance of 276 pm. The central O atom is attached by two short O—H bonds to two hydrogens, and by two long H-bonds to the H atoms of two of the neighbouring molecules. Overall, it consists of planes of hexagonal puckered rings of H_2O molecules (like the chair form of cyclohexane).

Molecular crystals, which are the subject of the overwhelming majority of modern structural determinations, are formed when discrete molecules stick together on account of their van der Waals interactions (which we treat in detail in the next chapter). The crystal structure is then the solution of the problem of condensing objects of various shapes into an aggregate of minimum energy (actually, minimum Gibbs function), and the prediction of the structure is very difficult and rarely possible. The problem is made more complicated by the role of *hydrogen bonds*, where a proton acts as a link between two electronegative atoms. In some cases, the hydrogen bonds dominate the crystal structure, as in ice, Fig. 23.25, but in others (e.g. phenol) they distort a structure which is determined largely by the van der Waals interactions. Crystals intermediate between molecular and covalent in character are those formed by polymers: the long chains are formed from covalently bonded monomer units, but the chains lie together and interact through van der Waals interactions: the more uniform the lengths of the polymer molecules, the more crystalline (and denser) the solid.

(a)

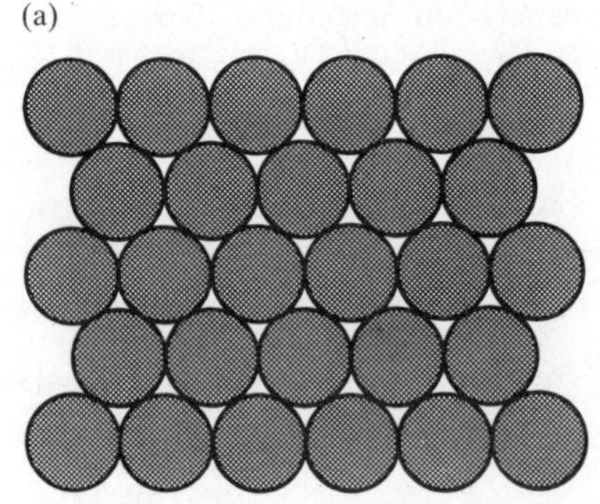

(b)

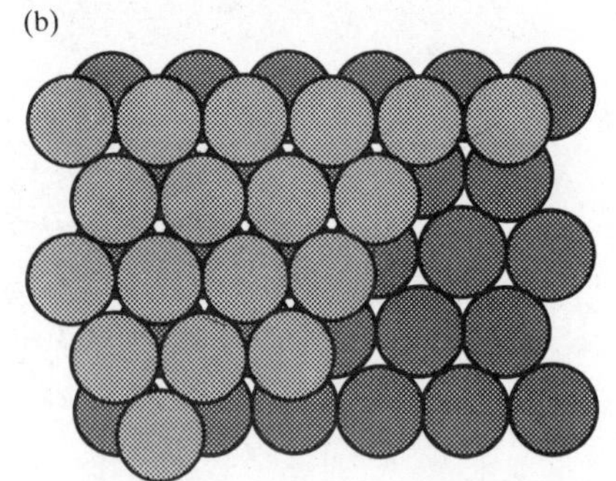

Fig. 23.26. The close-packing of identical spheres. (a) The first layer of close-packed spheres. (b) The second layer of close-packed spheres occupies the dips of the first layer. This is, so far, the AB component of the structure.

23.3 (a) The packing of identical spheres: metal crystals

It is found that most pure metals crystallize in one of three simple forms, two of which can be explained in terms of organizing spheres into the closest possible packing.

A close-packed layer of identical spheres can be formed as shown in Fig. 23.26(a). A similar layer can be formed by placing spheres in the depressions of this layer, Fig. 23.26(b). A third layer can be laid on top of this layer in two different ways. If it is placed so that it reproduces the first layer, Fig. 23.27(a) (so that the layer structure is ABABAB . . . as the process continues), the spheres are *hexagonally close-packed* (h.c.p.). Alternatively, the third layer can be placed so that it does not reproduce the first, Fig. 23.28(a) (so that the layer structure is then ABCABC . . .): this gives the *cubic close-packed* (c.c.p.) arrangement. Note that the c.c.p. structure gives rise to face-centred unit cells, and so it may also be denoted f.c.c.

The h.c.p. and c.c.p. are closely packed, dense structures. An indication of this is the *coordination number*, the number of atoms immediately surrounding any selected atom: in both cases it is 12. Metals possessing these structures are listed in Figs. 23.27 and 23.28.

Example 23.7

Calculate the proportion of hexagonally close-packed unit cell that is empty space.

• *Method*. Assume the ions to be perfect spheres of radius R. Refer to Fig. 23.29 and use some trigonometry.

• *Answer*. Area of the base of the unit cell (tinted) is $3^{\frac{1}{2}}R \times 2R = (2\sqrt{3})R^2$. The height is obtained as follows. The next layer of atoms lies above the point marked B. This is at a distance $R/\cos 30° = 2R/3^{\frac{1}{2}}$ from the centre of the neighbouring ion. The overlaying ion has a centre a distance $2R$ from the ions just referred to, and so it is at a height $\{(2R)^2 - (2R/3^{\frac{1}{2}})^2\}^{\frac{1}{2}} = 2(\frac{2}{3})^{\frac{1}{2}}R$. The height of the unit cell is twice this value, or $4(\frac{2}{3})^{\frac{1}{2}}R$. The volume of the unit cell is (area) × (height) $= (2\sqrt{3})R^2 \times 4(\frac{2}{3})^{\frac{1}{2}}R = (8\sqrt{2})R^3$. There are the equivalent of two complete ions per unit cell, and so the volume occupied per ion is $(4\sqrt{2})R^3$. The volume of a spherical ion is $(4\pi/3)R^3$. Therefore, the fraction of 'full' space is $(4\pi/3)R^3/(4\sqrt{2})R^3 = \pi/3\sqrt{2} = 0.74$: therefore, 26% is free space.

• *Comment*. The same type of argument may be applied to the other unit cells. In an f.c.c. unit cell the free volume is also 26% (equally close-packed); in a b.c.c. cell the free volume is 32% and in a simple cubic lattice the free volume is 48%. The *packing fractions* are 0.74, 0.68, and 0.52 respectively, and the structures are increasingly less close-packed.

• *Exercise*. Confirm the value of the packing fraction for the b.c.c. unit cell. [0.68]

In the third arrangement shown by a number of common metals (Ba, Cs, Cr, Fe, K, W), the atoms of the first layer are less than close-packed, and an ABABAB . . . structure forms as the successive layers of spheres sit in the dips of their predecessors. Inspection of the structure shows that the coordination number is only 8 (showing that is less than close-packed) and that the unit cell is body-centred cubic (b.c.c.).

23.3 (b) Ionic crystals

Not only do the spheres now have different radii, they also have different charges. Even if, by chance, the ions have the same size, it is still impossible to achieve dense 12-coordination. The best that can be done is to achieve the 8-coordination of the *CsCl structure*, Fig. 23.30.

When the sizes of the ions differ more than in CsCl, the packing reverts to 6-coordination of the type exhibited by the *rock-salt (NaCl) structure*, Fig. 23.31). This can be pictured as the interpenetration of two slightly expanded f.c.c. lattices. The switch from the CsCl to the NaCl structure occurs (sometimes) in accord with the *radius ratio rule*, which states that the CsCl structure should be expected when $r_{\text{Ion}}/r'_{\text{Ion}} > 0.732$ (i.e. $\sqrt{3} - 1$), and that the NaCl structure should be expected when this ratio lies between 0.732 and 0.414 (i.e. $\sqrt{2} - 1$). For ratios of less than 0.414 the most efficient packing leads to 4-coordination of the type exhibited by the *sphalerite* or *zincblende* forms of ZnS. The radius-ratio rules are fairly well supported by observation, even though they are based on simple geometrical considerations. In fact, deviation of a structure from the prediction can be taken to indicate a shift from ionic towards covalent bonding.

(a)

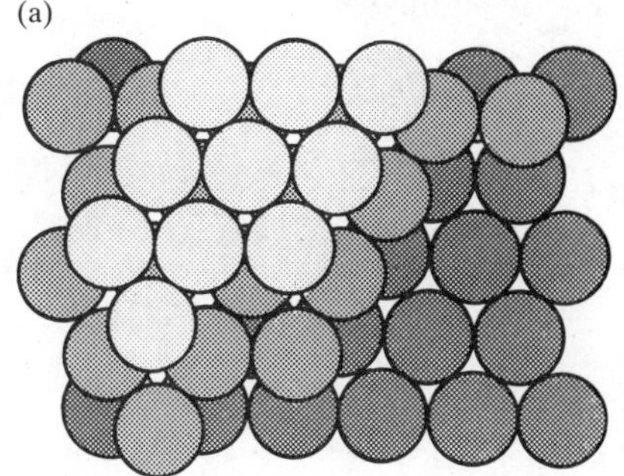

(b)

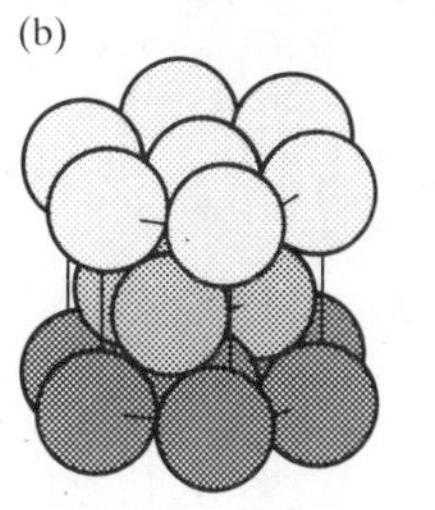

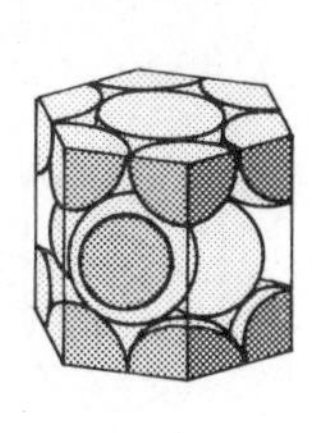

Fig. 23.27. The third layer of close-packed spheres might occupy the dips lying directly above the spheres in the first layer. This results in an ABA structure (a) which corresponds to hexagonal close-packing (b). This h.c.p. structure is shown by Be, Cd, Co, He, Mg, Ti, Zn.

(a)

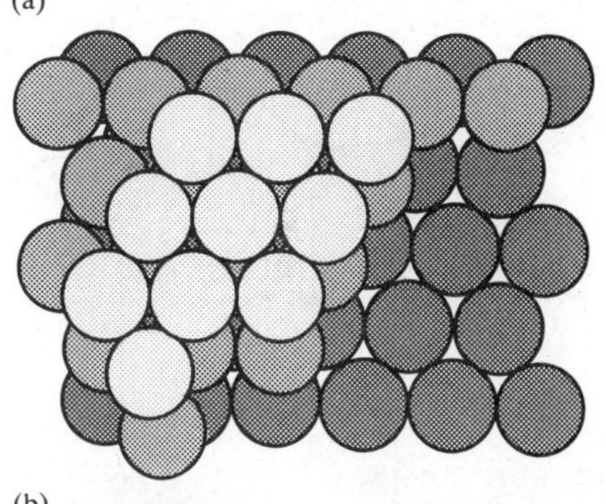

(b)

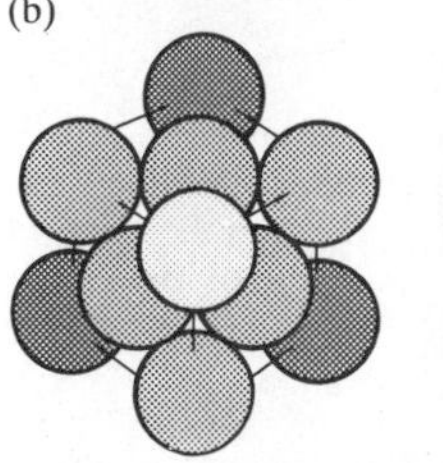

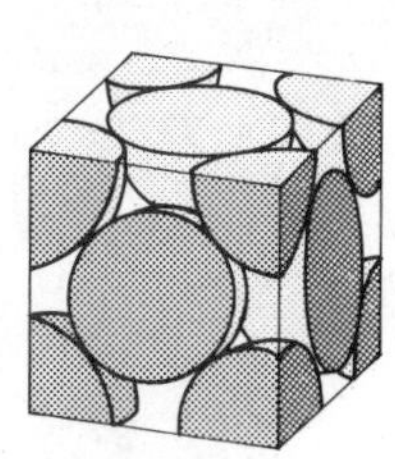

Fig. 23.28. Alternatively, the third layer might lie in the dips that are not above the spheres in the first layer. This results in an ABC structure (a) which corresponds to cubic close-packing, (b). This c.c.p. (or f.c.c.) structure is shown by Ag, Al, Ar, Au, Ca, Cu, Ne, Ni, Pb, Pt, and Xe.

The *ionic radii* to use in estimating the radius ratio, and wherever else it is important to know the sizes of ions, can be established by collecting a large number of X-ray measurements of crystal structures and noting that the distances between ions (i.e. between their nuclei) can be expressed as the sum of values that can be transferred between different compounds, Table 23.1. Thus Na^+ contributes to Na^+–X^- separations an almost constant amount independent of the identity of X, so long as the coordination number is the same. The same analysis can be applied to molecular crystals, and a list of *van der Waals radii* drawn up, Table 23.2.

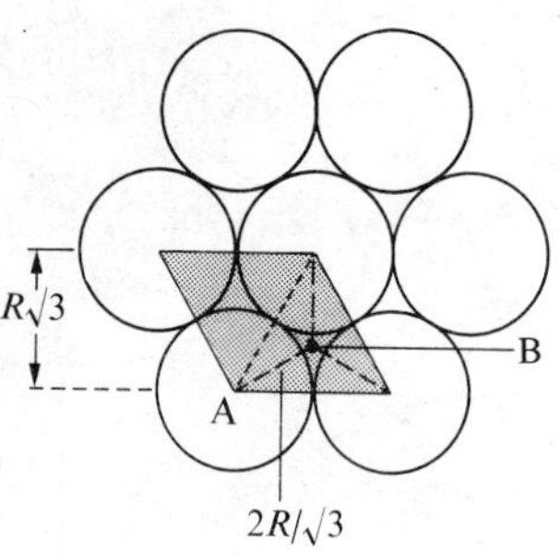

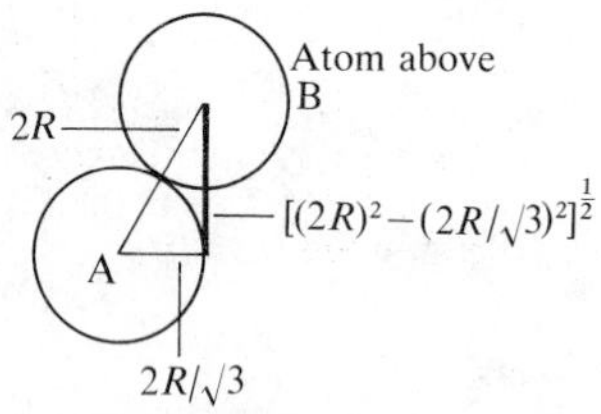

Fig. 23.29. The calculation of the *packing fraction*, the fraction of occupied space in a unit cell, for an h.c.p. unit cell (see *Example* 23.7).

23.3 (c) Absolute configurations

Optically active molecules cannot be superimposed on their mirror images (Section 17.1). Although it has long been possible to separate enantiomers (mirror image isomers), it was not until X-ray crystallography was developed that the *absolute* stereochemical configuration of an isomer could be determined. It is now possible to state, for example, that the L-tartaric acid shown in Fig. 23.32 is the isomer responsible for rotating light clockwise (i.e. it is the (+) isomer). The X-ray method is not trivial, because a crystal of one isomer gives an almost identical intensity pattern to its enantiomer. The information about the absolute configuration is contained in the phase of the diffracted radiation, and its extraction is based on a technique introduced by J. M. Bijvoet.

Consider first the diagrams in Fig. 23.33(a), which represent an idealized crystal and its mirror image. Each plane of atoms gives rise to a scattered wave, and the superpositions are shown. Note that the two superpositions have the same amplitude, but differ in phase. The diffraction pattern therefore has the same intensity for each enantiomer and, at this stage, cannot be used to distinguish them. The essence of the original Bijvoet method was to incorporate a heavy atom (such as Rb) into the molecule. Such an atom causes an extra phase shift in the scattered X-ray because (as a simple way of picturing it) the X-rays tend to excite it, and get delayed in the process. This is called *anomalous scattering*. If the layer marked A in the crystal contains the anomalous scatterers, the scattered waves are as shown in Fig. 23.33(b). The essential point is that the superpositions now differ slightly in amplitude, not just phase, and so the diffracted intensities are slightly different in each case. This means that the enantiomers can in fact be distinguished because the scattering intensities differ. Modern diffractometers are so sensitive that the incorporation of a heavy atom is no longer strictly necessary: it is now possible to detect the small intensity variations arising from the light atoms normally present, but the procedure is much easier and more reliable if some moderately heavy atoms (such as S or Cl) are present.

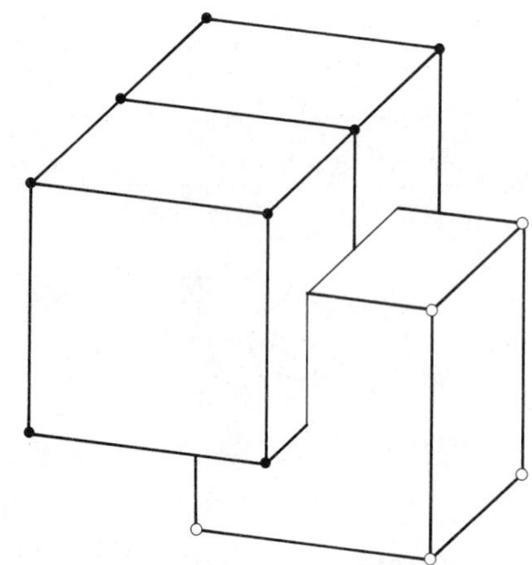

Fig. 23.30. The *CsCl lattice* consists of two interpenetrating simple cubic lattices, one of cations and the other of anions, so that each cube of ions of one kind has a counter-ion at its centre.

23.4 Neutron diffraction

A neutron generated in a reactor and slowed to thermal energies by repeated collisions with a *moderator* (such as graphite) until it is travelling at about $4\,km\,s^{-1}$ has a wavelength of about 100 pm. This is comparable to X-ray wavelengths, and so similar diffraction phenomena can be expected.

The scattering of X-rays is caused by the oscillations and incoming wave generates in the electrons of atoms. The scattering of neutrons is a *nuclear* phenomenon: neutrons pass through the electronic structures of atoms and interact with the nuclei through the forces that are responsible for binding nucleons together. As a result, the intensity with which neutrons are scattered is independent of the number of electrons and is not dominated by the heavy atoms present in a molecule (in contrast to X-rays, which are scattered more by electron-rich heavy atoms). Neutron diffraction therefore shows up the positions of hydrogen nuclei much more clearly than do X-rays and can be used to distinguish atoms with similar atomic numbers. This difference in sensitivity can have a pronounced effect on the measurement of C—H bond lengths: because X-rays respond to accumulations of electrons, the weak peaks in an X-ray diffraction map represent the locations of the bulk of the electron density in the bonds, and these may be shifted towards the C atom. For example, X-ray measurements on sucrose give $r(\text{C—H}) = 96\,\text{pm}$; neutron measurements, which respond to the location of the nuclei, give $r(\text{C—H}) = 109.5\,\text{pm}$. The O—H bond lengths in sucrose show similar differences: 79 pm by X-rays but 97 pm by neutrons.

Another attribute of neutrons that distinguishes them from X-ray photons is their possession of a magnetic moment due to their spin. This magnetic moment can couple to the magnetic fields of ions in a crystal (if they have unpaired electrons) and modify the diffraction pattern. A simple example of *magnetic scattering* is provided by metallic chromium. The lattice is b.c.c., and so systematic absences for $h + k + l$ odd are expected (as for X-rays, Section 23.2 and Fig. 23.16). In fact these absences are not observed because the structure is such that atoms at the body-centre location have magnetic moments opposite to those at the corners, and the structure is better regarded as consisting of two interpenetrating lattices, Fig. 23.34, of magnetically different atoms. Therefore, although the atoms are identical as far as X-rays are concerned, they are different for neutrons, and diffraction intensity is observed at the predicted systematic X-ray absences. Neutron diffraction is especially important for investigating these *magnetically ordered* lattices, and this is another reason why neutron diffraction experiments are done, even though the sources are expensive and (because a monochromatic, single-velocity ray has to be detected) the beam is weak.

23.5 Electron diffraction

Electrons can be accelerated to precisely controlled energies by a known potential difference. When accelerated through 10 keV they acquire a wavelength of 12 pm, which makes them suitable for molecular diffraction studies. Because they are charged, electrons are scattered strongly by their interaction with the charges of electrons and nuclei, and so cannot be used to study the interiors of solid samples. However, they can be used to study molecules in the gas phase, on surfaces, and in thin films.

A typical electron diffraction apparatus is illustrated in Fig. 23.35. Electrons are emitted from the hot filament on the left and accelerated through the potential difference. They then pass through the stream of gas, and on to a fluorescent screen. The modifications used when studifying surfaces are described in Section 31.1(c).

The sample, being a gas, presents all possible orientations of the scattering atom–atom separations to the electron beam, and the diffraction pattern is like an X-ray powder photograph. The pattern consists of a series of concentric undulations on a decreasing background, Fig. 23.36(a). The undulations are due to the sharply defined scattering from the nuclear

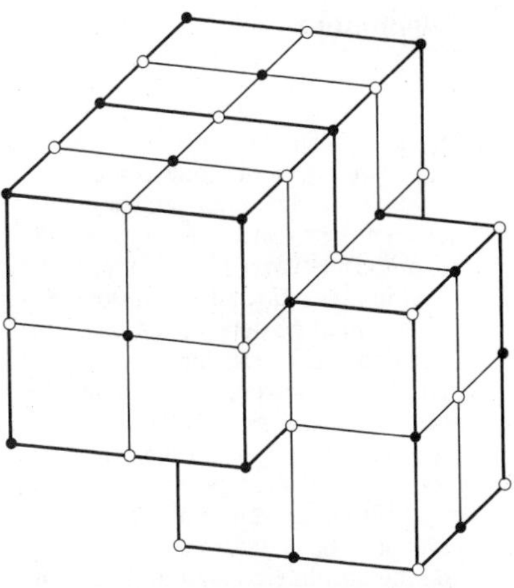

Fig. 23.31. The *rock salt (NaCl) lattice* consists of two interpenetrating slightly expanded face-centred lattices.

Table 23.1. Ionic radii, R/pm

Na^+	102 (6‡), 116 (8)
K^+	138 (6), 151 (8)
F^-	128 (2), 131 (4)
Cl^-	181

‡ Coordination number.

Table 23.2. Van der Waals radii, R/pm

H	120
O	140
Cl	180
Ar	286 (gas viscosity)
	383 (close packing)

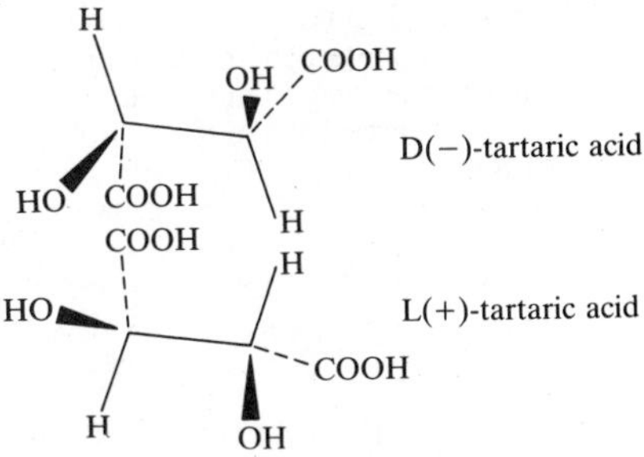

Fig. 23.32. An enantiomeric pair of optical isomers of tartaric acid with the correct absolute configurations as determined by anomalous X-ray scattering. (+) signifies that the enantiomer rotates the plane of polarization clockwise (from the viewpoint of the detector).

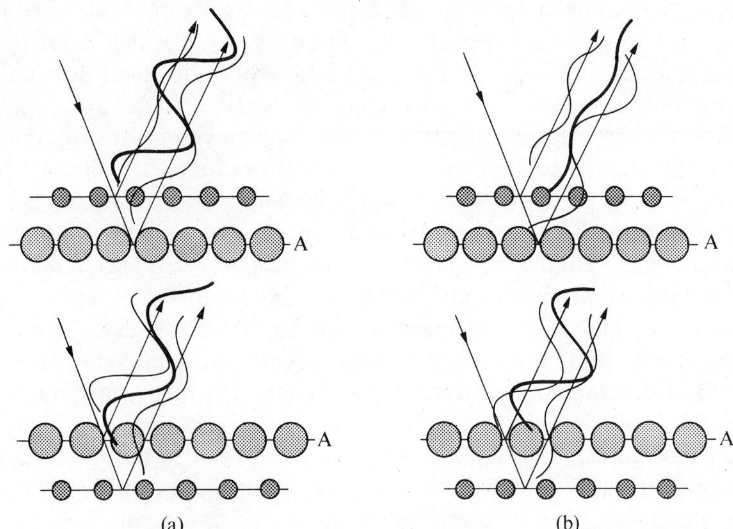

Fig. 23.33. The two versions (D and L) of the two layers of atoms represent enantiomers. The interference between their scattered waves results in composite waves (heavy line) that differ in phase (a), but the absolute phase cannot be determined, and the intensities of the reflections are identical. If, however, A modifies the phase of the waves it scatters, as represented by the light lines in (b), then the resultant superpositions differ in amplitude as well as phase, the reflections have different intensities, and the absolute configuration can be established.

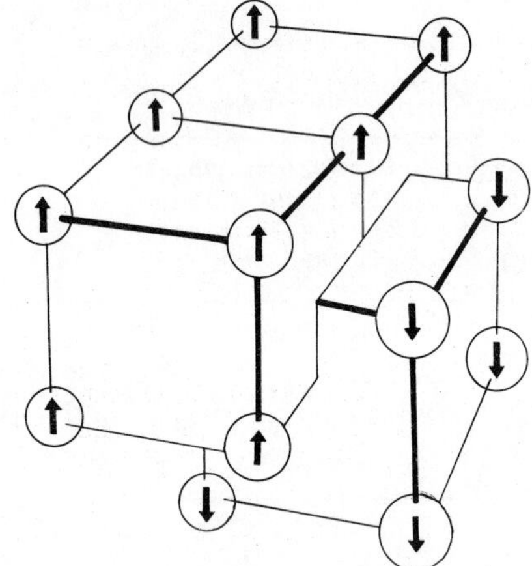

Fig. 23.34. If the magnetic moments of atoms at lattice points are orderly, as in this *antiferromagnetic* material, where the spins of one set of atoms are aligned antiparallel to those of the other set, then neutron diffraction detects two interpenetrating simple cubic lattices on account of the magnetic interaction of the neutron with the atoms, but X-ray diffraction would see only a single b.c.c. lattice.

positions, and the background is due to scattering from the less well-defined continuous electron density distribution in the molecule. One way of eliminating this unwanted background is to insert a rotating heart-shaped disk in front of the screen: this exposes the outer parts more than the inner, and helps to emphasize the undulations, Fig. 23.36(b).

The scattering from a pair of nuclei separated by a distance R_{ij} and orientated at a definite angle to the incident beam can be calculated. The overall diffraction pattern is then calculated by allowing for all possible orientations of this pair of atoms. In other words, we integrate over orientations. This leads to the expression

$$I_{ij}(\theta) = 2f_i f_j\{1 + \sin(sR_{ij})/sR_{ij}\}; \qquad s = (4\pi/\lambda)\sin\tfrac{1}{2}\theta, \qquad (23.5.1)$$

where λ is the wavelength of the electrons in the beam and θ is the scattering angle. The electronic scattering factors f are a measure of the scattering power of the atoms. When the molecule consists of a number of atoms, the total intensity has an angular variation given by the *Wierl*

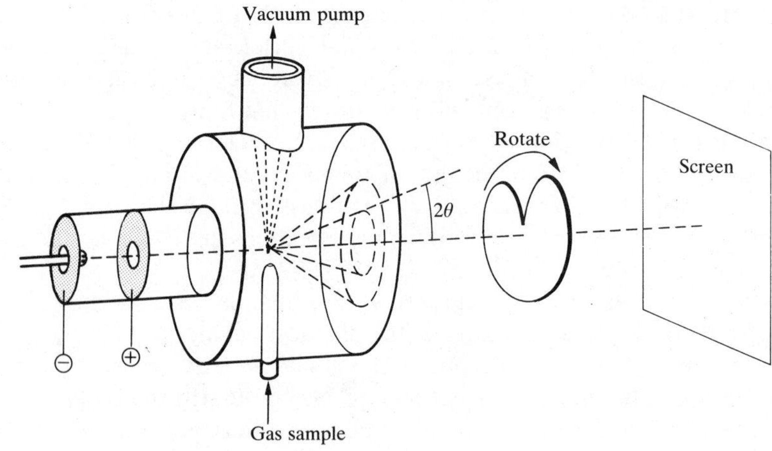

Fig. 23.35. The layout of an electron diffraction apparatus. The diffraction pattern is photographed from the fluorescent screen. A rotating heart-shaped sector emphasizes the scattering from the nuclear positions and suppresses the smoothly varying background due to scattering from the continuous electron distribution in the molecules.

equation:

$$I(\theta) = \sum_{i,j} f_i f_j \sin(sR_{ij})/sR_{ij}. \qquad (23.5.2)$$

The electron diffraction pattern can be interpreted in terms of the distances between all possible pairs of atoms in the molecule (not just those bonded together). The analysis may therefore be very complicated. When there are only a few atoms, the peaks can be analysed reasonably quickly, and the analysis proceeds by assuming a geometry and calculating the intensity pattern using the Wierl equation. The best fit is then taken as the appropriate geometry.

Electron diffraction has provided precise bond lengths and angles for a wide variety of molecules that can be obtained in the gas phase. In conjunction with the other diffraction techniques, we can use the data to ascribe *covalent radii* to atoms, Table 23.3, and hence predict the sizes and shapes of molecules.

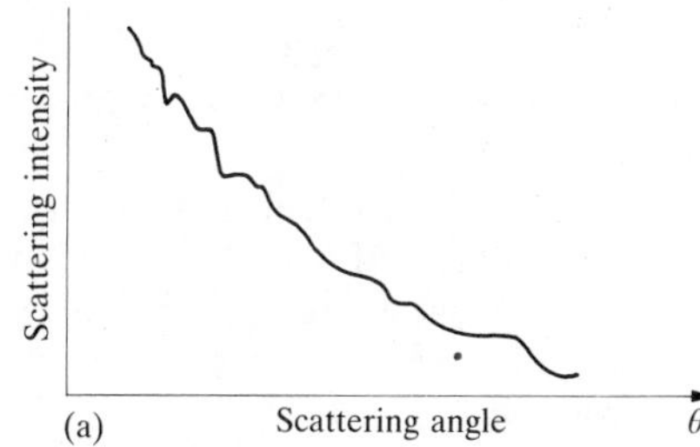

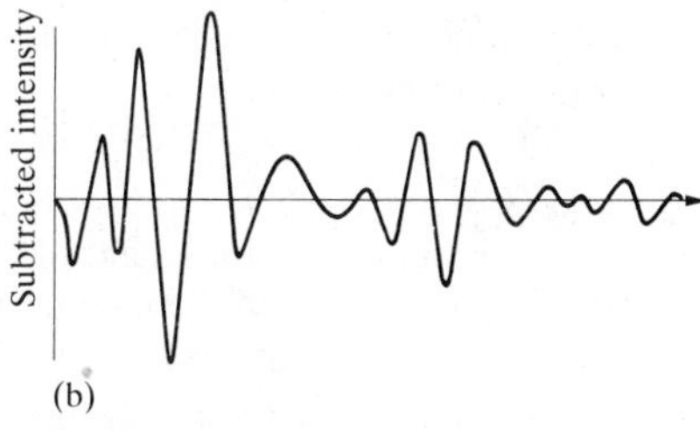

Fig. 23.36. (a) The scattering intensity consists of a smoothly varying background with undulations superimposed. (b) The undulations are emphasized if a sector is rotated in front of the screen, and then the densitometer trace taken from the photograph plotted against $s = (4\pi/\lambda) \sin \frac{1}{2}\theta$.

Table 23.3. Covalent radii, R/pm

H—	37	O—	74
C—	77	O=	57
C=	67	Cl—	99
C=	69.5‡	N—	74

‡ Benzene.

Further reading

Crystals:

The third dimension in chemistry. A. F. Wells: Clarendon Press, Oxford, 1956 (reissued 1968).

An introduction to crystallography. F. C. Phillips; Longman, London, 1979.

Elementary crystallography. M. J. Buerger; Wiley, New York, 1956.

Introduction to crystal geometry. M. J. Buerger; McGraw-Hill, New York, 1971.

X-ray diffraction:

The determination of molecular structure (2nd edn). P. J. Wheatley; Clarendon Press, Oxford, 1968.

Crystal structure analysis. J. P. Glusker and K. N. Trueblood; Oxford University Press, 1972.

X-ray crystal structure analysis. W. N. Lipscomb and R. A. Jacobson; *Techniques of chemistry* (A. Weissberger and B. W. Rossiter, eds) IIID, 1, Wiley-Interscience, New York, 1972.

An introduction to X-ray crystallography. M. M. Woolfson; Cambridge University Press, 1970.

Electron diffraction:

Electron diffraction by gases. L. S. Bartell: *Techniques of chemistry* (A. Weissberger and B. W. Rossiter, eds) IIID, 125, Wiley-Interscience, New York, 1972.

Electron diffraction. T. B. Rymer; Chapman and Hall, London, 1970.

Neutron diffraction:

Neutron diffraction. G. E. Bacon; Oxford University Press, 1975.

Data:

Tables of interatomic distances and configurations of molecules. L. E. Sutton (ed); Chem. Soc. Special Publication 11, 1958 (Supplement, Special Publication 18, 1965).

Structural inorganic chemistry (5th edn). A. F. Wells; Clarendon Press, Oxford, 1984.

Crystal structure (5 sections and supplements). R. W. G. Wycoff; Wiley-Interscience, New York, 1959.

Introductory problems

A23.1. The Bragg angle of a first-order X-ray reflection from a set of crystal planes with an interplanar spacing of 99.3 pm is 20.85°. What is the wavelength of the X-radiation?

A23.2. Cu Kα X-radiation consists of two lines, one at 154.051 pm and one at 154.433 pm. Find the separation of the two lines which these two wavelengths produce in a powder pattern recorded in a camera with radius 5.74 cm, when the interplanar spacing is 77.8 pm.

A23.3. The substance Rb_3TlF_6 has a tetragonal unit cell with dimensions: $a = 651$ pm, $c = 934$ pm. Calculate the volume of the unit cell.

A23.4. The unit cell of $NiSO_4$ is orthorhombic with dimensions: $a = 633.8$ pm, $b = 784.2$ pm, and $c = 515.5$ pm. An approximate measurement of the density of the crystals gives 3.9 g cm^{-3}. Find the number of formula units per unit cell and calculate the density.

A23.5. Crystals of $SbCl_3$ are orthorhombic with unit cell parameters: $a = 812$ pm, $b = 947$ pm, $c = 637$ pm. Calculate the spacing of the (411) planes.

A23.6. The orthorhombic unit cell of Li_3AlAs_2 has cell dimensions: $a = 1198$ pm, $b = 1211$ pm, $c = 1186$ pm. Find the distance from the origin to each of the intercepts of the (523) plane on the three coordinate axes.

A23.7. A substance known to have a cubic unit cell gives reflections with Cu X-radiation (154 pm) at the following Bragg angles: 19.4°, 22.5°, 32.6°, 39.4°. The reflection at 32.6° is known to be (220). Index the other reflections. Calculate the average of the values of unit cell edge derived from these data.

A23.8. The coordinates of the four I atoms in the unit cell of KIO_4 are: $(0, 0, 0)$, $(0, \frac{1}{2}, \frac{1}{4})$, $(\frac{1}{2}, \frac{1}{2}, \frac{1}{2})$, $(\frac{1}{2}, 0, \frac{3}{4})$. By calculating the exponential part of the I term in the structure factor, show that the I atoms contribute nothing to the intensity of the (114) X-ray reflection.

A23.9. A diffraction experiment requires a neutron beam with wavelength 70 pm. What is the average kinetic energy of neutrons in such a beam?

A23.10. What accelerating voltage must be applied to electrons which are to have a wavelength of 18 pm in an electron diffraction experiment?

Problems

23.1. Three crystals of pyrite, or fool's gold, FeS_2, are shown in Fig. 23.37. They have been quite carefully drawn from the accepted viewpoint, which K. Lonsdale (*Crystals and X-rays*, Bell (1948)) describes 'as if the eye were looking down from a little above and slightly to the right, like a benevolent Conservative government'. Check that they all have all the features that make them members of the cubic system. But of what class are they?

23.2. A crystal of sphalerite (ZnS) is shown in Fig. 23.38. To what crystal system and class does it belong?

23.3. The only element of symmetry possessed by blue crystals of copper sulphate pentahydrate is a centre of inversion. To what system and class do they belong?

23.4. Crystals of sucrose are found to possess one twofold axis as their only symmetry element. To what system do they belong?

23.5. When a crystal of rock salt is examined externally it is found to possess three fourfold axes, four threefold axes, six twofold axes, three mirror planes, and a centre of inversion. To what system and what class do such crystals belong?

23.6. The morphology of crystals tells us a great deal about the inner symmetry, but even simple unit cells may give rise to a wide variety of external forms (the drawings of pyrite illustrate this). In order to experiment with the variety of forms that may stem from even a cubic unit cell, procure 1 lb or so of cubical sugar lumps and see what

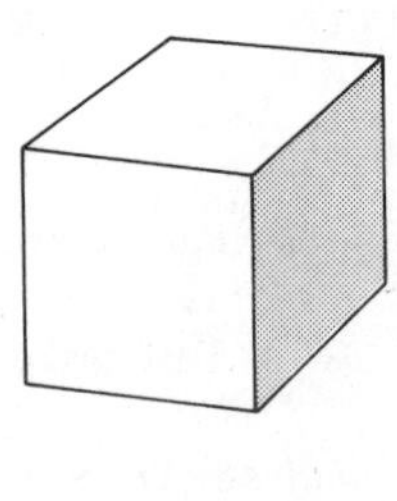

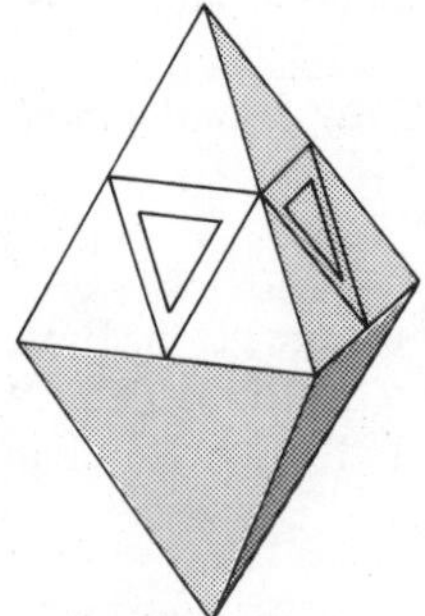

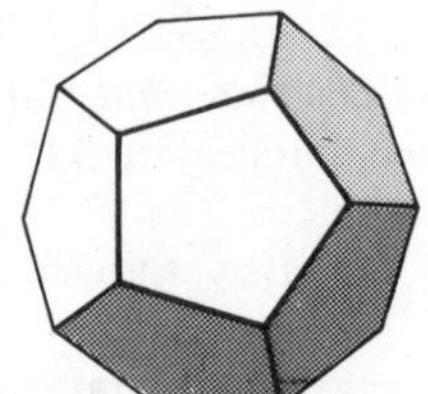

Fig. 23.37. Crystals of pyrite (FeS_2).

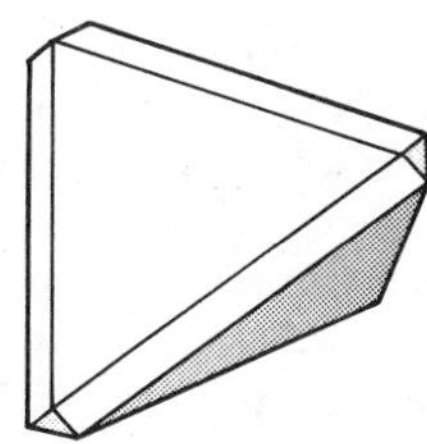

Fig. 23.38. A crystal of sphalerite (ZnS).

external forms may be constructed by stacking them together. Show, using sugar cubes, that an equilateral triangular face may be formed at a corner of an otherwise perfect cube, and calculate the angle that it makes to the uppermost flat surface.

23.7. Draw an array of points as a two-dimensional rectangular lattice based on a unit cell of sides a, b. Mark planes with Miller indices (10), (01), (11), (12), (23), (41).

23.8. Redraw the lattice in the last Problem so that the b axis makes an angle of 60° to the a axis. Mark the same planes as before.

23.9. Calculate the distances between the (11) planes for the lattices in the preceding Problems.

23.10. Crystal planes cut through the crystal axes at ($2a$, $3b$, c), (a, b, c), ($6a$, $3b$, $3c$), ($2a$, $-3b$, $-3c$). What are their Miller indices?

23.11. Draw an orthorhombic unit cell and mark the (100), (010), (001), (011), (101), and (111) planes. Draw a triclinic unit cell and mark on it the same planes.

23.12. What are the separations of the planes with indices (111), (211), and (100) in a crystal in which the cubic unit cell is of side 432 pm?

23.13. In the early days of X-ray crystallography there was an urgent need to know X-ray wavelengths. One technique was to measure the diffraction angle from a mechanically ruled grating, with the X-rays approaching it at a glancing angle. Another method was to estimate the separation of lattice planes from the measured density of a crystal. The density of NaCl is 2.17 g cm^{-3} and the (100) reflection using palladium K_α radiation occurred at 6° 0′. What is the wavelength of the X-rays?

23.14. In their book of *X-rays and crystal structures* (which begins 'It is now two years since Dr. Laue conceived the idea . . .') the Braggs give a number of simple examples of X-ray analysis. For instance, they report that the first-order reflection from (100) planes of KCl occurs at 5° 23′, but for NaCl it occurs at 6° 0′ for the rays of the same wavelength. If the side of the NaCl unit cell is 564 pm, what is the size of the KCl cell? The densities of KCl and NaCl are 1.99 g cm^{-3} and 2.17 g cm^{-3} respectively: do these values support the X-ray analysis?

23.15. Potassium nitrate single crystals have orthorhombic unit cells of dimensions $a = 542$ pm, $b = 917$ pm, $c = 645$ pm. Calculate the diffraction angle for first-order reflections from the (100), (010), and (111) planes using Cu K_α radiation (154.1 pm).

23.16. Copper(I) chloride, forms cubic crystals with four molecules per unit cell. The only reflections present in a powder photograph are those either with all even indices or with all odd. What is the nature of the unit cell? Note that this type of question about the form of the unit cell can be answered without knowing anything about the wavelength of the radiation: it is determined by symmetry, not size.

23.17. A powder diffraction photograph from tungsten shows lines which index as (110), (200), (211), (220), (310), (222), (321), (400), ... What type of unit cell is present?

23.18. Rock salt (NaCl) forms crystals with a simple cubic lattice of side 564 pm. What are the distances between the (100) planes, the (111) planes, and the (012) planes?

23.19. Show that the separation of the (hkl) planes in a crystal with an orthorhombic unit cell of sides a, b, and c is given by $1/d_{hkl}^2 = (h/a)^2 + (k/b)^2 + (l/c)^2$.

23.20. We often need to know the volumes of unit cells. Show that the volume of a triclinic unit cell of sides a, b, and c and angles α, β, and γ is $V = abc[1 - \cos^2\alpha - \cos^2\beta - \cos^2\gamma + 2\cos\alpha\cos\beta\cos\gamma]^{\frac{1}{2}}$. Note that by making the appropriate choice of angles and lengths this expression includes monoclinic and orthorhombic cells. Use the result from vector analysis that $V = \mathbf{a}\cdot\mathbf{b}\wedge\mathbf{c}$, and calculate V^2 initially.

23.21. Naphthalene forms a monoclinic crystal with two molecules in each unit cell. The sides are in the ratio 1.377:1:1.436, and $\beta = 122°\,49'$. The specific gravity is 1.152. What are the dimensions of the cell?

23.22. The type of calculation used in the last Problem can be turned round, and X-ray analysis combined with density measurements to find a value of Avogadro's constant. The density of LiF is 2.601 g cm^{-3}. The (111) reflection occurs at 8° 44′, when 70.8 pm X-rays from molybdenum are used. Find a value of Avogadro's constant given that there are four LiF in a unit cell.

23.23. How many lattice points are there in (a) a primitive cubic unit cell, (b) a b.c.c. unit cell, (c) an f.c.c. cell, (d) a unit cell for the diamond lattice?

23.24. Show that the maximum proportion of the available volume that may be filled by hard spheres is (a) 0.52 for a simple cubic lattice, (b) 0.68 for a b.c.c. lattice, (c) 0.74 for an f.c.c. lattice. These numbers are the *packing fractions* of the lattices. What method of packing is most economical for shipping (hard) oranges?

23.25. Calculate the maximum possible packing fraction for long, straight cylindrical molecules (or pipes).

23.26. The only example of a metal that crystallizes into a primitive cubic lattice is polonium. The unit cell is of side 334.5 pm. Predict the form of the powder diffraction pattern using 154 pm X-rays. What do you predict for the density of polonium? Suppose it packed into an f.c.c.

lattice, but with the same atomic radius: what would be its density?

23.27. The separation of the (100) planes of lithium metal is 350 pm. Its density is $0.53\ \mathrm{g\ cm^{-3}}$. Is the lattice f.c.c. or b.c.c.?

23.28. Copper crystallizes in an f.c.c. lattice of side 361 pm. Predict the form of the powder diffraction pattern using 154 pm X-rays. What is the density of copper?

23.29. Calculate the coefficient of thermal expansion of diamond given that the (111) reflection shifts from 22° 2′ 25″ to 21° 57′ 59″ on heating the crystal from 100 K to 300 K and 154.0562 pm X-rays are used.

23.30. A powder diffraction pattern of KCl in a camera of radius 28.7 mm gave lines at the following distances from the centre spot of the film 14.5, 20.6, 25.4, 29.6, 33.4, 37.1, 44.0, 47.5, 50.9, 54.4, 58.2, 62.1, 66.4, 78.1 mm. Index the lines. The radiation was Cu K_α (154 pm). What is the size of the unit cell?

23.31. In the same apparatus with the same wavelength radiation, LiF gave diffraction lines at the following distances from the central spot: 18.9, 22.1, 31.9, 38.3, 40.4, 48.9, 55.0, 58.0, 68.8, 83.3 mm. Deduce what you can about the unit cell.

23.32 The powder diffraction patterns of (a) tungsten, (b) copper are shown in Fig. 23.39. Both were obtained with 154 pm X-rays and the scale of the plate is indicated. Identify the cubic cell in each case, and find the lattice spacing. What are the crystal radii of the tungsten and copper atoms?

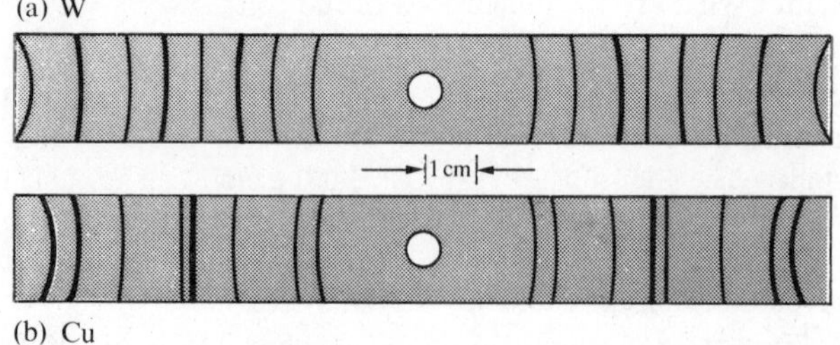

Fig. 23.39. Powder diffraction patterns of (a) tungsten, (b) copper.

23.33. Mercury(II) chloride is orthorhombic and with Cu K_α radiation the (100), (010), and (001) reflections occur at 7° 25′, 3° 28′, and 10° 13′ respectively. The density of the crystal is $5.42\ \mathrm{g\ cm^{-3}}$. What are the dimensions of the unit cell, and how many $HgCl_2$ units does it contain?

23.34 Genuine pearls consist of concentric layers of calcite crystals in which the trigonal axes are oriented along the radii. The nucleus of a cultured pearl is a piece of mother-of-pearl that has been worked to a sphere on a lathe. The oyster then deposits concentric layers of calcite on this central seed. Devise an X-ray method for distinguishing between real and cultured pearls.

23.35. The coordinates, in units of a, of the atoms in a simple cubic lattice are (0, 0, 0), (0, 1, 0), (0, 0, 1), (0, 1, 1), (1, 0, 0), (1, 1, 0), (1, 0, 1), and (1, 1, 1). Calculate the structure factors F_{hkl} when all the atoms are the same.

23.36. The coordinates of the atoms in a cubic unit cell are the same as in the last Problem for species A with scattering length f_A, but an atom B, with scattering length f_B, lies at $(\frac{1}{2}, \frac{1}{2}, \frac{1}{2})$. Calculate the scattering factors F_{hkl} and predict the form of the powder diffraction pattern in the cases (a) $f_A = f, f_B = 0$, (b) $f_B = \frac{1}{2}f_A$, (c) $f_A = f_B = f$.

23.37. Refer to Fig. 23.16. Why is the (311) reflection much less intense than the (222) reflection? You will need to think about structure factors.

23.38. The *Patterson synthesis* results in spots corresponding to the length and direction of the vectors joining scattering centres in the unit cell. Sketch the Patterson pattern that corresponds to (a) a planar, triangular BF_3 molecule, (b) the carbon atoms in benzene.

23.39. Calculate the wavelength of electrons that have been accelerated through (a) 1 kV, (b) 10 kV, (c) 40 kV.

23.40. What velocity should neutrons have if they are to have a wavelength of 50 pm? What is the wavelength of neutrons that have been *thermalized* by collision with a moderator (heavy water) at 300 K?

23.41. As a first step in doing an actual electron diffraction analysis, take the Wierl equation, eqn (23.5.2), and calculate the expected electron diffraction intensity distribution for a CCl_4 molecule of (as yet) undetermined bond length but of known tetrahedral symmetry. Take $f_{Cl} = 17f$ and $f_C = 6f$ and note that $R(\mathrm{Cl,Cl}) = (8/3)^{\frac{1}{2}}R(\mathrm{C,Cl})$. Plot I/f^2 against $x = sR(\mathrm{C,Cl})$.

23.42. In an actual electron diffraction experiment on CCl_4 electrons were used that had been accelerated through 10 kV. The positions of the maximal and minima were observed at the following scattering angles. Maxima: 3° 10′, 5° 22′, 7° 54′; Minima: 1° 46′, 4° 6′, 6° 40′, 9° 10′. Check that this pattern of maxima and minima confirms that the molecule is tetrahedral (compare with Problem 23.42) and then calculate the bond length of CCl_4.

24

The electric and magnetic properties of molecules

Learning objectives

After careful study of this chapter you should be able to:

(1) Define the electric *dipole moment* of a molecule, Fig. 24.1, and the *polarization* of a dielectric, Sections 24.1(a and b).

(2) Define the *polarizability* and the *polarizability volume*, eqn (24.1.10).

(3) Derive the *Debye equation* relating the relative permittivity, the molar polarizability, the dipole moment, and the polarizability, eqn (24.1.13).

(4) Deduce polarizabilities and dipole moments from experimental data, Example 24.2.

(5) Relate the *refractive index* of a substance to its polarizability, and define and use the *molar refractivity*, eqn (24.1.14) and Example 24.4.

(6) Explain the origin of *optical activity*, eqn (24.1.17) and *optical rotatory dispersion*.

(7) Explain the source of, and assess the magnitudes of, the *dipole/dipole interaction*, eqn (24.2.1), the *dipole/induced-dipole interaction*, eqn (24.2.2), and the *induced-dipole/induced-dipole dispersion interaction*, eqn (24.2.3).

(8) State and criticise the *Lennard–Jones potential*, eqn (24.2.8).

(9) Describe how *molecular beams* are formed, and indicate how they are used to explore intermolecular interactions, Section 24.3(a).

(10) Define *impact parameter* and *differential scattering cross-section*, Figs. 24.18 and 24.19.

(11) Explain the terms *glory scattering* and *rainbow angle*, Figs. 24.22 and 24.23.

(12) Define the *configuration integral*, eqn (24.4.4), and the *Mayer f-function*, eqn (24.4.7), and indicate how they are used to relate intermolecular potentials to gas imperfections, Section 24.4(a).

(13) Calculate a *virial coefficient* for a given intermolecular potential, eqn (24.4.8) and Example 24.5.

(14) Define the *radial distribution function* and the *pair distribution function* of a liquid, Section 24.4(b), and indicate their significance.

(15) Explain the *Monte Carlo* and *molecular dynamics* approaches to the simulation of molecular structures, Section 24.4(c).

(16) Describe the different phases of *liquid crystals* that may be obtained, Section 24.4(d).

(17) Account for the temperature dependence of the viscosity of a liquid, eqn (24.4.12).

(18) Define the *Madelung constant*, eqn (24.5.3), explain its significance, and use tabulated values to calculate lattice enthalpies of ionic solids, Section 24.5.

(19) Define *volume magnetic susceptibility*, eqn (24.6.1), and *mass magnetic susceptibility*, eqn (24.6.2), and relate them to the permanent magnetic dipole moment and *magnetizability* of a substance, eqn (26.6.4).

(20) Explain the distinction between *diamagnetism*, *paramagnetism*, *ferromagnetism*, and *antiferromagnetism*, Section 24.6(a).

(21) Calculate the *spin-only paramagnetic susceptibility* of a substance, eqn (24.6.7) and Example 24.6.

Introduction

In this chapter we examine some of the electric and magnetic properties of molecules and interpret them in terms of electronic structure. The properties we consider include the electric dipole moment and polarizability, and we see that they are related to refractive index, optical activity, and intermolecular forces. We also consider the magnetizability, and its relation to the magnetic susceptibilities of molecules.

24.1 Electric properties

We consider both the *permanent electric dipole moment* and the dipole moment induced by an electric field. An applied field distorts the electronic structure and changes the equilibrium positions of the nuclei. Hence it gives rise to an additional dipole moment with a magnitude proportional to its strength and the responsiveness, the *polarizability*, of the molecule.

24.1 (a) The permanent electric dipole moment

R

$-q$ $+q$ $\mu = qR$

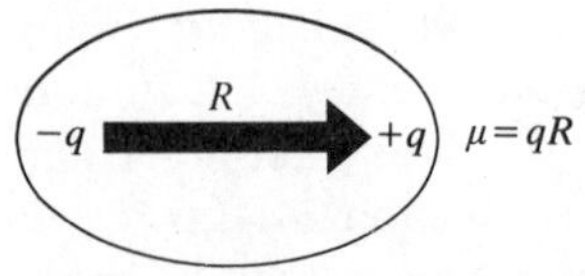

Fig. 24.1. The convention for the electric dipole moment. Two equal and opposite charges q and $-q$ separated by a distance R have a dipole moment $\mu = qR$ directed towards the positive charge.

When two charges q and $-q$ are separated by a distance R, they constitute an electric *dipole*. The dipole can be represented by a vector called the *dipole moment* (symbol: $\boldsymbol{\mu}$): its magnitude (symbol: μ) is qR and it is directed from the negative charge towards the positive, Fig. 24.1. Magnitudes of dipole moments are generally quoted in *debye* (D), where $1\,\mathrm{D} = 3.336 \times 10^{-30}\,\mathrm{C\,m}$, a unit named after Peter Debye, a pioneer in the study of polar molecules. When the charge q has the magnitude of the electronic charge, and the separation is of atomic dimensions (about 100 pm), the magnitude of the dipole moment is about $1.6 \times 10^{-29}\,\mathrm{C\,m}$, corresponding to 4.8 D. Dipole moments of small molecules are of roughly this order, but generally somewhat smaller. Molecules with non-zero permanent electric dipole moments are called *polar molecules*.

Dipole moments reveal information about molecular structure, and are used to test the reliability of wavefunctions. In practice, however, they are important through their role in intermolecular forces (see later) and in the way they determine the suitability of a solvent for a solute. Polar molecules play a double role in dissolving ionic solids. One is that one end of a dipole may be electrostatically attracted so strongly to an ion of opposite charge that the reduction of energy contributes to the solubility of the solid, Fig. 24.2. Their other role is to reduce the strengths of the Coulombic interactions between the ions in the solution. This comes about as follows. When two ions are separated by a distance r their potential energy of interaction is proportional to $1/4\pi\varepsilon_0 r$ when the medium is a vacuum, but $1/4\pi\varepsilon r$ when it is a solvent of relative permittivity (dielectric constant) $\varepsilon_r = \varepsilon/\varepsilon_0$. The value of ε_r is determined in part by the dipole moment of the solvent molecules, and it can have a very significant effect on the strength of the ionic interactions. For instance, in water $\varepsilon_r = 78$ at 25 °C, which means that the long-range interionic Coulomb interaction energy is reduced by nearly two orders of magnitude from its vacuum value. The consequences of this reduction were examined in Chapter 11.

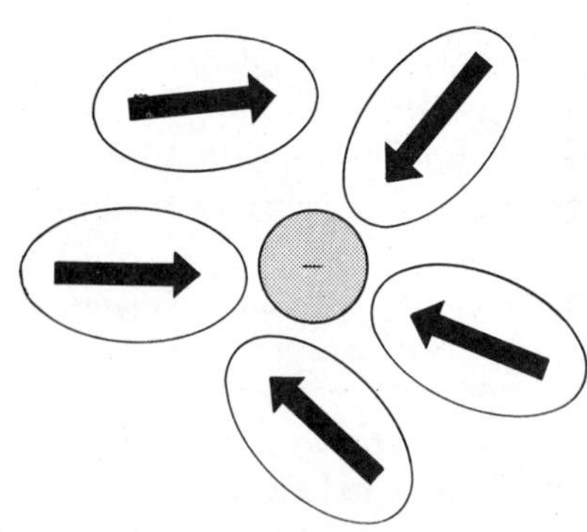

Fig. 24.2. An ion in a polar solvent is *solvated* as a result of its interaction with the dipoles of the solvent molecules.

24.1 (b) The measurement of dipole moments

Dipole moments may be measured in several ways. When a rotational spectrum can be observed, the Stark effect may be used, as described in

Section 18.2. When this technique cannot be used (because the sample is involatile or decomposes, or consists of too complex molecules), the dipole moment may be obtained from measurements of the relative permittivity of the bulk sample, as the following paragraphs explain.

We begin by defining the *capacitance* (symbol: C) of a capacitor. Consider the two plane, parallel plates depicted in Fig. 24.3. Let the charges on the plates be q and $-q$. If their areas are A we can write $q=\sigma A$, where σ is the *surface charge density* (the charge per unit area). On account of the charges, there is a potential difference $\Delta\phi$ between the plates. The capacitance is then defined as

$$\text{Capacitance: } C=q/\Delta\phi. \tag{24.1.1}$$

Therefore, in the present case, $C=\sigma A/\Delta\phi$. The potential difference can be written in terms of the charge density and the separation, d, of the plates. First, we accept the standard result of elementary electrostatics that the electric field in the region between the plates is σ/ε_0 if the medium is a vacuum and σ/ε if it is a dielectric. The presence of an electric field of strength E_0 (typically in $\mathrm{V\,m^{-1}}$; we shall use a 0 subscript to denote a vacuum value) indicates the existence of a potential difference of magnitude E_0d between the plates. Therefore, in the absence of a dielectric the capacitance is $C_0=\sigma A/E_0d=\sigma A/(\sigma d/\varepsilon_0)=\varepsilon_0A/d$. In the presence of a dielectric it is $C=\varepsilon A/d$. Therefore, since

$$C/C_0=(\varepsilon A/d)/(\varepsilon_0A/d)=\varepsilon/\varepsilon_0=\varepsilon_r, \tag{24.1.2}$$

the relative permittivity can be measured by measuring the capacitance of a capacitor with and without a dielectric present. Commercial instruments using an AC bridge are available.

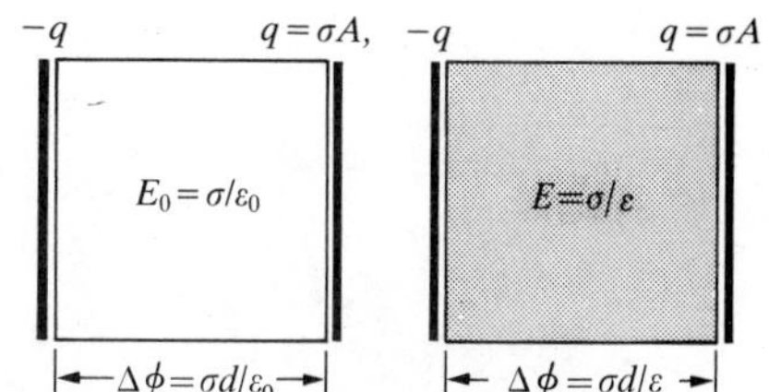

Fig. 24.3. (a) Two planes, each of surface area A and carrying charges $\pm\sigma A$ give rise to a uniform electric field of magnitude $E_0=\sigma/\varepsilon_0$. (b) When a dielectric is present the field is reduced to $E=\sigma/\varepsilon$, with $\varepsilon=\varepsilon_r\varepsilon_0$.

Example 24.1

The capacitance of an empty cell is 5.01 pF; when it was filled with a sample of camphor at 25 °C its capacitance was 57.1 pF. Calculate the relative permittivity (dielectric constant) of camphor at 25 °C.

- *Method*. Use eqn (24.1.2). The relative permittivity of dry air (1.000 536) may be taken as unity.

- *Answer*. $\varepsilon_r=C/C_0=57.1\text{ pF}/5.01\text{ pF}=11.4$.

- *Comment*. This trivial example introduces material we shall consider in more detail later. It is also an opportunity to emphasize that relative permittivities are temperature-dependent and dimensionless; F denotes a *farad*: $1\,\mathrm{F}=1\,\mathrm{C\,V^{-1}}$. The value for air will also prove useful shortly.

- *Exercise*. The same cell when filled with ethanol at 25 °C had a capacitance of 121.7 pF. Calculate the relative permittivity of ethanol at this temperature. [24.3]

The next step is to relate ε_r to a molecular property. This is done by introducing the *polarization* (symbol: P) of the dielectric: the polarization is the *average dipole moment per unit volume*. The illustration in Fig. 24.4 allows us to go on to show that the polarization may also be interpreted as the *charge per unit area* on the faces of the *sample* (i.e. that P is equal to σ', σ' being the surface charge density of the sample, not that on the charged

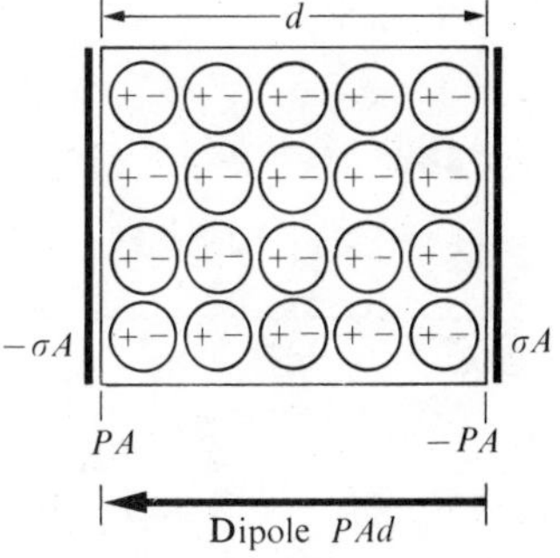

Fig. 24.4. The charges on the planes polarize the medium. On account of the cancellation of the induced charges in the bulk, the sample can be regarded as having charges $\pm PA$ on its opposite surfaces, and therefore as having a dipole of moment PAd.

plates). In Fig. 24.4 the dipoles are shown aligned in the applied field, and it can be seen that their charges cancel in the bulk but give rise to a charge at the two opposite surfaces. If this charge is $\sigma' A$ on one face and $-\sigma' A$ on the other, then the sample possesses a dipole moment of magnitude $\sigma' Ad$, and since its volume is Ad, the dipole moment per unit volume is $\sigma' Ad/Ad = \sigma'$. But dipole moment per unit volume is polarization, and so $\sigma' = P$, as we set out to prove.

Now we relate P and ε_r. In the absence of dielectric the electric field between the plates is $E_0 = \sigma/\varepsilon_0$. In the presence of the dielectric the field is reduced to E. There are two ways of expressing E. One is to write it in terms of the relative permittivity and the full surface charge of the plates: $E = \sigma/\varepsilon = \sigma/\varepsilon_0\varepsilon_r$. The other is to write it in terms of the net charge $\sigma - \sigma'$, but to treat the dielectric as a vacuum: $E = (\sigma - \sigma')/\varepsilon_0$, or $E = (\sigma - P)/\varepsilon_0$, since $\sigma' = P$. These two expressions must be equal, and so by eliminating σ we find

$$P = (\varepsilon_r - 1)\varepsilon_0 E, \qquad (24.1.3)$$

which relates the polarization to the field inside the dielectric. The quantity $\varepsilon_r - 1$ is called the *electric susceptibility* (symbol: χ_e) of the medium, and so we can write

$$P = \chi_e \varepsilon_0 E. \qquad (24.1.4)$$

We now concentrate on a single molecule somewhere inside the sample. This molecule experiences the field E together with a field that arises from the charges on the surface of the cavity that surrounds it, Fig. 24.5. Calculation shows that the extra contribution to the local field is $P/3\varepsilon_0$ if the cavity is assumed to be spherical and the medium continuous. This is the *Lorentz local field correction*. Therefore, the total field at the molecule is

$$E^* = E + P/3\varepsilon_0 = \frac{P(\varepsilon_r + 2)}{3\varepsilon_0(\varepsilon_r - 1)}. \qquad (24.1.5)$$

Even if the molecules are polar, the mean dipole moment is zero in a fluid sample when no external field is present because the rotational motion averages it to zero. When a field is present some orientations correspond to lower energies than others, and the mean dipole moment (and therefore the polarization) depends on a competition between the aligning influence of the field and the randomizing influence of thermal motion. The net moment can be evaluated using the Boltzmann distribution (Section 21.1) for a sample at a temperature T using the fact that the energy of a molecule with dipole moment of magnitude μ making an angle θ to an electric field of strength E^* is $-\mu E^* \cos\theta$. This gives

$$\mu_{\text{mean}} = \mu\mathscr{L}, \qquad \mathscr{L} = \frac{e^x + e^{-x}}{e^x - e^{-x}} - \frac{1}{x}, \quad \text{with} \quad x = \mu E^*/kT. \qquad (24.1.6)$$

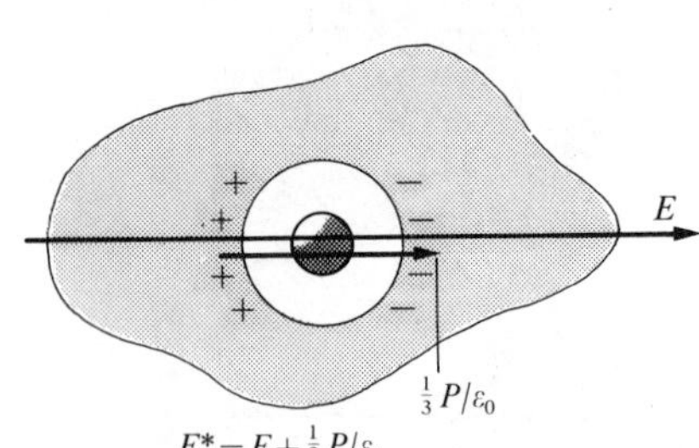

Fig. 24.5. A molecule surrounded by a dielectric experiences the sum of the applied field and the local field arising from the induced charges on the surrounding cavity. An estimate of its magnitude is the *Lorentz local field*, which is equal to $P/3\varepsilon_0$.

$\mathscr{L}$ is called the *Langevin function*. Under most circumstances x is very small (e.g. if $\mu = 1$ D and $T = 300$ K, x exceeds 0.01 only if the field strength exceeds 100 kV cm^{-1}, and most measurements are done at much lower strengths). When $x \ll 1$ the exponentials in the Langevin function can be expanded, and the largest terms that survive give $\mathscr{L} = \frac{1}{3}x + \ldots$. Therefore, the average molecular dipole moment is

$$\mu_{\text{mean}} = \mu^2 E^*/3kT. \tag{24.1.7}$$

If the *number density* of molecules (the number per unit volume) is $\mathcal{N}$, the mean dipole moment per unit volume in a field E^* is

$$P = \mathcal{N}\mu_{\text{mean}} = \mathcal{N}\mu^2 E^*/3kT, \tag{24.1.8}$$

By substituting this expression into eqn (24.1.5) and rearranging after cancelling the E^* we find

$$\mu^2 = (3\varepsilon_0 kT/\mathcal{N})\left\{\frac{\varepsilon_r - 1}{\varepsilon_r + 2}\right\}, \tag{24.1.9}$$

from which μ can be obtained by measuring ε_r and the density (for $\mathcal{N}$). However, read on before using this expression.

24.1 (c) Polarizability

Non-polar molecules may acquire a dipole moment in an electric field as a result of the distortion of their electronic distributions and nuclear positions. So long as the field is not too strong the magnitude of the induced moment is proportional to the local field, and we write

$$\mu_{\text{induced}} = \alpha E^*, \tag{24.1.10}$$

where the constant α is called the *polarizability* of the molecule. In general, the induced moment depends on the orientation of the molecule with respect to the field, and when the field is strong (such as for the electric field in laser beams) the induced moment also depends on E^{*2}, and the coefficient in βE^{*2} is called the *hyperpolarizability*.

First, a note on units. The polarizability α has the units $\text{J}^{-1}\,\text{C}^2\,\text{m}^2$ when the field strength is expressed in V m^{-1} and μ is in C m. That is awkward. Polarizability is therefore usually converted to *polarizability volume* (symbol: α'), using the relation $\alpha' = \alpha/4\pi\varepsilon_0$. Since the units of $4\pi\varepsilon_0$ are $\text{J}^{-1}\,\text{C}^2\,\text{m}^{-1}$, α' turns out to have the dimensions of volume (hence its name), and values are usually quoted in cm^3 (and Å^3: $1\,\text{Å}^3 = 10^{-24}\,\text{cm}^3$), Table 24.1.‡

Table 24.1. Dipole moments (μ) and polarizability volumes (α')

	μ/D	$\alpha'/10^{-24}\,\text{cm}^3$
H_2	0	0.819
HCl	1.08	2.63
H_2O	1.85	1.48
CCl_4	0	10.5

Now we consider the contribution of the polarizability to the polarization. Since each molecule contributes $\mu_{\text{induced}} = \alpha E^*$ to the total average dipole moment of the sample, when the number density of molecules is $\mathcal{N}$ the polarization is $P = \mathcal{N}\mu_{\text{induced}} = \alpha\mathcal{N}E^*$. The total polarization of a sample is therefore the sum of this contribution and that arising from the permanent moments, eqn (24.1.8):

$$P = \mathcal{N}(\alpha + \mu^2/3kT)E^*. \tag{24.1.11}$$

When this expression is combined with eqn (24.1.5), in place of eqn (24.1.9) we obtain the *Debye equation*:

$$(\varepsilon_r - 1)/(\varepsilon_r + 2) = (\mathcal{N}/3\varepsilon_0)\{\alpha + \mu^2/3kT\}. \tag{24.1.12}$$

‡ When using older compilations of data, it is useful to note that polarizability volumes have the same numerical values as the 'polarizabilities' reported using c.g.s. electrical units, and so the tabulated values previously called 'polarizabilities' can be used directly.

The same expression, but without the permanent dipole moment contribution, is called the *Clausius–Mossotti equation*. The Debye equation is normally written in terms of the density ρ of the sample using $\mathcal{N} = \rho N_A/M_m$, M_m being the molar mass of the molecules, and the *molar polarizability* (symbol: P_m):

$$P_m = (N_A/3\varepsilon_0)\{\alpha + \mu^2/3kT\}, \tag{24.1.13a}$$

$$\frac{\varepsilon_r - 1}{\varepsilon_r + 2} = \rho P_m/M_m. \tag{24.1.13b}$$

It follows from eqn (24.1.13) that both the polarizability and permanent dipole moment can be determined by measuring the temperature dependence of the molar polarizability and plotting it against $1/T$: the slope of the graph is $N_A\mu^2/9\varepsilon_0 k$ and the intercept at $1/T = 0$ is $N_A\alpha/3\varepsilon_0$. The physical reason for this result is that at high temperatures the disordering effect of the thermal motion eliminates the contribution of the permanent dipole, leaving the induced contribution alone. This induced dipole moment lies in the direction of the field inducing it, and it remains in that direction even though the molecule itself might be tumbling. Therefore, it is not averaged to zero by the thermal motion and survives to contribute to the polarization even at the highest temperatures.

Example 24.2

A series of experiments on camphor were made in the same sample cell as in *Example* 24.1 but at different temperatures. Use the following data to find the dipole moment and the polarizability volume of the molecule.

$\theta/°C$	0	20	40	60	80	100	120	140	160	200
$\rho/g\,cm^{-3}$	0.99	0.99	0.99	0.99	0.99	0.99	0.97	0.96	0.95	0.91
C/pF	62.6	57.1	54.1	50.1	47.6	44.6	40.6	38.1	35.6	31.1

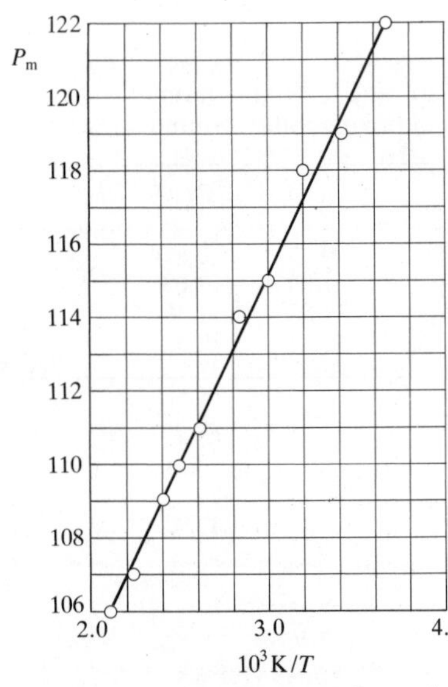

Fig. 24.6. The plot of $P_m/cm^3\,mol^{-1}$ against $10^3/(T/K)$ used in *Example* 24.2 for the determination of the polarizability and dipole moment of camphor.

• *Method*. Calculate ε_r at each temperature and form $(\varepsilon_r - 1)/(\varepsilon_r + 2)$. Multiply by M_m/ρ to form P_m (via eqn (24.1.13b)). For camphor, $M_m = 152.23\ g\ mol^{-1}$. Plot P_m against $1/T$. The intercept at $1/T = 0$ is $N_A\alpha/3\varepsilon_0 = (4\pi N_A/3)\alpha'$; the slope is $N_A\mu^2/9\varepsilon_0 k$.

• *Answer*. Draw up the following table:

$\theta/°C$	0	20	40	60	80	100	120	140	160	200
$10^3/(T/K)$	3.66	3.41	3.19	3.00	2.83	2.68	2.54	2.42	2.31	2.11
ε_r	12.5	11.4	10.8	10.0	9.50	8.90	8.10	7.60	7.11	6.21
$(\varepsilon_r - 1)/(\varepsilon_r + 2)$	0.793	0.776	0.766	0.750	0.739	0.725	0.703	0.688	0.670	0.634
$P_m/cm^3\,mol^{-1}$	122	119	118	115	114	111	110	109	107	106

The points are plotted in Fig. 24.6. The intercept lies at 82.7, and so

$$\alpha' = (3/4\pi N_A) \times (82.7\ cm^3\ mol^{-1}) = 3.3 \times 10^{-23}\ cm^3.$$

The slope is 10.9, so that

$$N_A\mu^2/9\varepsilon_0 k = 10.9\ cm^3\ mol^{-1}/(K^{-1}/1000) = 1.09 \times 10^4\ cm^3\ K\ mol^{-1}.$$

This solves to $\mu = 4.46 \times 10^{-30}\ C\,m$, corresponding to 1.34 D.

• *Comment*. The odd thing about the data is that they show that camphor, which does not melt until 175 °C, is rotating even in the solid. It is an approximately spherical molecule.

• *Exercise*. The relative permittivity of chlorobenzene is 5.71 at 20 °C and 5.62 at 25 °C. Assuming a constant density ($1.11\ g\ cm^{-3}$), estimate its polarizability volume and dipole moment. [$1.4 \times 10^{-23}\ cm^3\ mol^{-1}$; 1.2 D]

Some experimental polarizabilities of molecules are given in Table 24.1. The magnitudes reflect the strengths with which the nuclear charges control the electron distributions and prevent their distortion by the applied field. If the molecule has few electrons, their distribution is tightly controlled by the nuclear charges and the polarizability is low. If the molecule contains large atoms with electrons some distance from the nucleus, the nuclear control is less, the electron distribution is flabbier, and the polarizability is greater. Note that, in general (i.e. for all molecules other than those belonging to one of the cubic groups), the polarizability depends on the orientation of the molecule with respect to the field, and the values reported in Table 24.1 are mean values.

24.1 (d) Polarizabilities at high frequencies: the refractive index

When the applied field oscillates, the permanent dipole moments may be unable to reorientate themselves sufficiently rapidly to align with its ever-changing direction. At such high frequencies the permanent dipole makes no contribution to the molar polarization, and so the Debye equation reverts to the Clausius–Mossotti equation. Since a molecule takes about 10^{-12} s to rotate in a fluid, the change occurs when measurements are made at frequencies greater than about 10^{11} Hz (in the microwave region). Inverting this argument suggests that it might be possible to study molecular tumbling in liquids by observing the variation of the polarization with frequency. This can be done, and the technique called *dielectric relaxation* is based on the idea.

The polarizability contribution to the molar polarization survives at microwave frequencies, but it is modified as the frequency is taken higher. The polarizability arises in part from the *distortion polarization* of the molecule, the modification of the positions of the nuclei: the molecule is bent and stretched by the applied field. The time it takes for a molecule to bend is approximately the time it takes for it to vibrate, and so this distortion polarization disappears when the frequency of the radiation is increased through the infrared. The disappearance occurs in a number of steps, each one corresponding to a vibrational absorption band. At even higher frequencies, in the visible region, only the electrons are light enough to respond to the rapidly changing direction of the applied field, and then only the *electronic polarizability* contributes.

The *Maxwell equations*, which describe the properties of electromagnetic radiation, lead to the conclusion that the *refractive index* (symbol: n_r) of a non-magnetic medium (the ratio of the speed of light in a vacuum to its speed in the medium: $n_r = c/v$) at a frequency ν is related to the relative permittivity at that frequency by $n_r = \varepsilon_r^{\frac{1}{2}}$. The molar polarizability, and hence the molecular polarizability, can therefore be measured at optical frequencies simply by measuring the refractive index of the sample using a refractometer (Table 24.2) and using the Clausius–Mossotti equation. This is normally expressed in terms of the *molar refractivity* (symbol: R_m):

$$R_m = (M_m/\rho)\{(n_r^2 - 1)/(n_r^2 + 2)\}, \qquad (24.1.14)$$

so that, from eqn (24.1.12) with the μ^2 term omitted:

$$\alpha = (3\varepsilon_0/N_A)R_m, \quad \text{or} \quad \alpha' = (3/4\pi N_A)R_m. \qquad (24.1.15)$$

Table 24.2. Refractive indices relative to air at 20 °C

	434 nm	589 nm	656 nm
H_2O(l)	1.340	1.333	1.331
C_6H_6(l)	1.524	1.501	1.497
CS_2(l)	1.675	1.628	1.618
KI(s)	1.704	1.666	1.658

Example 24.3

The refractive index of water at 20 °C is 1.3330 for light of wavelength 589 nm. Calculate the polarizability volume of the molecule at this frequency.

• *Method*. Use eqn (24.1.14) to calculate R_m and then convert the value to α' using eqn (24.1.15). The density of water is 0.9982 g cm^{-3} at this temperature, and $M_m = 18.015$ g mol^{-1}.

• *Answer*. Since $(n_r^2 - 1)/(n_r^2 + 2) = 0.2057$,

$$R_m = (18.015\text{ g mol}^{-1}/0.9982\text{ g cm}^{-3}) \times 0.2057 = 3.699\text{ cm}^3\text{ mol}^{-1}.$$

Therefore, from eqn (24.1.15),

$$\alpha' = (3/4\pi N_A)R_m = 1.467 \times 10^{-24}\text{ cm}^3.$$

• *Comment*. The same calculation repeated for 434 nm light, for which the refractive index is 1.3404, gives 1.501×10^{-24} cm^3. The H_2O molecule is more polarizable at the higher frequency because the photons of incoming light carry more energy.

• *Exercise*. Calculate the polarizability volume of ethanol at the frequency corresponding to the sodium D lines, given that its refractive index is 1.361 at 20 °C and its density is 0.789 g cm^{-3}. [5.12×10^{-24} cm^3]

It is found that, to a fair approximation, molar refractivities can be analysed into contributions from the individual groups making up a molecule or solid. Thus, *bond refractivities* are contributions that can be ascribed to the bonds in covalent molecules, and *ion refractivities* are contributions that can be ascribed to ions in ionic solids. Some values are given in Table 24.3. When these values are added they may be used in eqn (24.1.14) modified to give n_r in terms of R_m:

Table 24.3. Molar refractivities at 589 nm, R_m/cm^3 mol^{-1}

C—H	1.65	Na^+	0.46
C—C	1.20	K^+	2.12
C═C	2.79	F^-	2.65
C═O	3.34	Cl^-	9.30

$$n_r^2 = \frac{V_m + 2R_m}{V_m - R_m}, \qquad V_m = M_m/\rho. \tag{24.1.16}$$

Example 24.4

Estimate the refractive index of ethanoic acid for sodium D light. Its density is 1.046 g cm^{-3}.

• *Method*. Use the bond refractivity information in Table 24.3, taking the CH_3COOH molecule to consist of 3(C—H) + (C—C) + (C═O) + (C—O) + (O—H). Then use eqn (24.1.16).

• *Answer*. $R_m/\text{cm}^3\text{ mol}^{-1} = 3(1.65) + 1.20 + 3.34 + 1.41 + 1.85 = 12.75.$

$$V_m = M_m/\rho = (60.05\text{ g mol}^{-1})/(1.046\text{ g cm}^{-3}) = 57.41\text{ cm}^3\text{ mol}^{-1}.$$

Therefore, by substitution in eqn (24.1.16), $n_r = 1.36$.

• *Comment*. The experimental value is 1.37, and so in this case the agreement is very good.

• *Exercise*. Estimate the refractive index for ethanol for sodium D light ($\rho = 0.789$ g cm^{-3}). [1.34]

The refractive index determines the speed of light in the medium. The reason why it is related to the molecular polarizability is that the propagation of light can be imagined as proceeding by the incident light inducing an oscillating dipole moment, which then re-radiates. The radiation has the same frequency as the incident light, but (so long as it is at

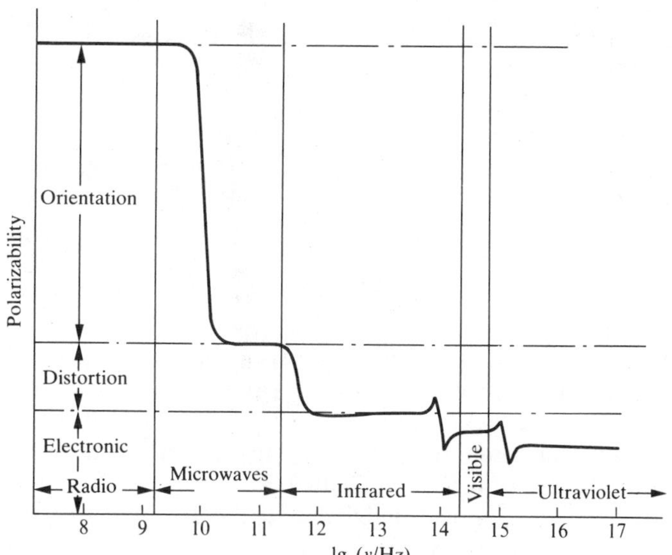

Fig. 24.7. The general form of the dependence of the polarizability on the frequency of the applied field. Note the considerable reduction in polarizability when the field is reversing direction so rapidly that the (polar) molecules cannot reorientate quickly enough to follow it.

lower frequency than an absorption band) its phase is delayed by the interaction. This delay of phase, which increases as the molecules respond more strongly (i.e. as their polarizabilities increase) delays the propagation of the light, and therefore slows its passage through the medium. This slowing corresponds to an increase of refractive index. Photons of high-frequency light carry more energy than low-frequency photons, and so they can distort the electronic distributions of the molecules in their path more effectively. This means that at optical frequencies, and at frequencies lower than an absorption band, we should expect the polarizabilities of molecules to increase with increasing frequency, and the refractive index to increase too. This is the origin of the dispersion of white light by a prism: the refractive index is greater for blue light than for red, and therefore the blue rays are bent more than the red. *Dispersion* is a name carried over from this phenomenon to mean the variation of the refractive index, or of any property, with frequency. Figure 24.7 shows the dispersion of the polarizability of a sample.

24.1 (e) Optical activity

An optically active substance rotates the plane of polarization of plane polarized light. The effect arises from the difference of the refractive indices for right and left circularly polarized light, n_R and n_L respectively. (We use the normal convention that in right-handed circularly polarized light the electric vector rotates clockwise from the viewpoint of the observer facing the oncoming beam, Fig. 24.8.)

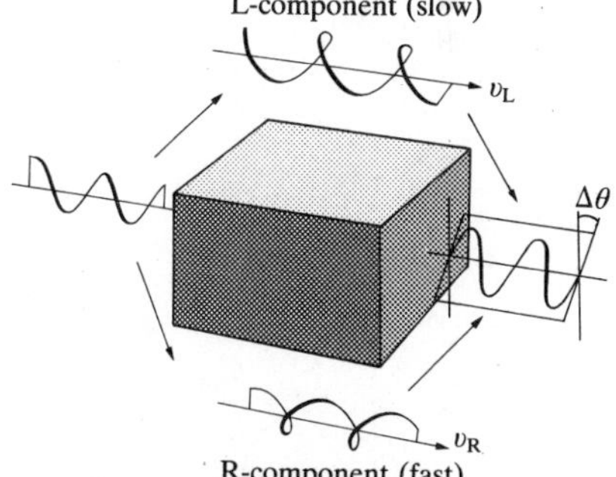

Fig. 24.8. Linearly polarized light entering a sample (from the left) can be regarded as the superposition of two counter-rotating circularly polarized components with a definite phase relation. If one component propagates more rapidly than the other in the medium, when they emerge the phase relation is changed, and the resultant is plane-polarized light rotated through an angle $\Delta\theta$ to its original orientation.

That optical rotation depends on $n_R \neq n_L$ may be seen by reference to Fig. 24.8. Before entering the medium the beam is plane polarized at an angle $\theta = 0$ to some axis. This plane-polarized light may be resolved into a superposition of two oppositely rotating circularly polarized components, Fig. 24.8. On entering the medium one component propagates faster than the other if their refractive indices are different, and if the sample is of length l the difference in the times of passage is $l/v_R - l/v_L$, where v_R and v_L are the speeds of the components. In terms of the refractive indices this difference is $(n_R - n_L)l/c$. The electric vectors are in phase at a different angle when they leave the sample, and so their superposition gives rise to a plane-polarized beam rotated through an angle $\Delta\theta$ relative to the plane of the incoming beam, where

$$\Delta\theta = 2\pi(n_L - n_R)l/\lambda, \qquad (24.1.17)$$

with λ the wavelength of the light.

In order to explain why the refractive indices depend on the handedness of the light, we must examine why the polarizabilities depend on the handedness. The full theory of optical rotation is complicated, but one interpretation is that if a molecule has a helical structure then its polarizability depends on whether or not the electric field of the incident radiation rotates in the same sense as the helix. Molecules having a helical structure are not superimposable on their mirror image, the criterion for optical activity discussed in Section 17.1(c).

The variation of the optical activity with the frequency of the light is called the *optical rotatory dispersion* (ORD). It arises from the individual dispersions of the polarizabilities (and refractive indices) for left- and right-circularly polarized light, and can be used to investigate the stereochemistry of molecules. Associated with the differences in the two refractive indices (the *optical birefringence*) is a difference of absorption intensities: this *circular dichroism* is also used to obtain structural information.

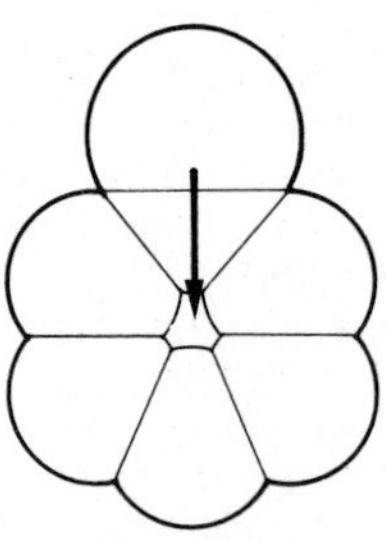

(a) $\mu_{obs} = 1.57$ D

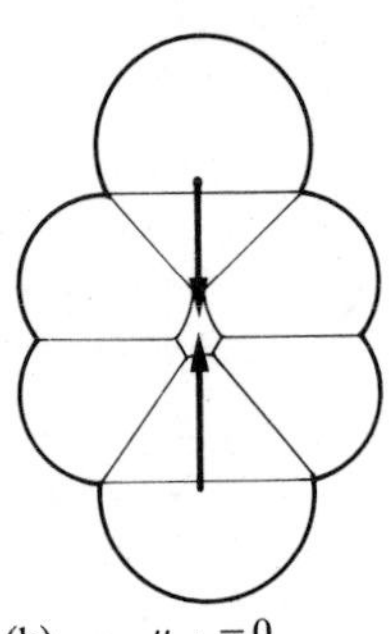

(b) $\mu_{calc} = 0$
$\mu_{obs} = 0$

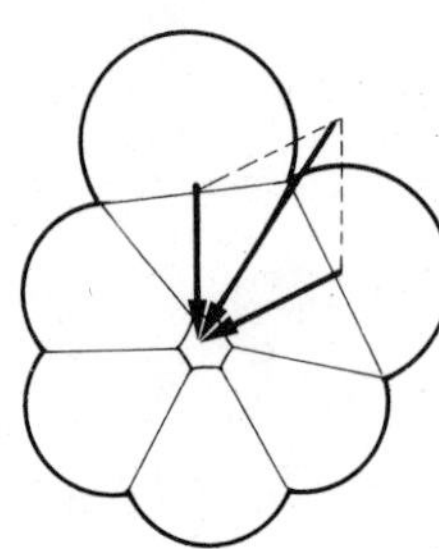

(c) $\mu_{calc} \approx 2.7$ D
$\mu_{obs} \approx 2.25$ D

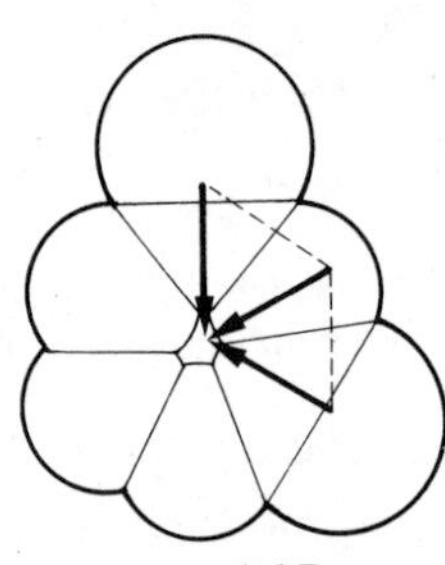

(d) $\mu_{calc} \approx 1.6$ D
$\mu_{obs} = 1.48$ D

Fig. 24.9. The dipole moments of the dichlorobenzene isomers can be obtained *approximately* by vectorial addition of two chlorobenzene dipole moments (1.57 D).

24.1 (f) Additive properties

To some extent it is possible to resolve the dipole moment of a molecule into contributions of various components. This is illustrated by the chlorobenzenes, Fig. 24.9. The experimental dipole moment of chlorobenzene is 1.57 D. *p*-Dichlorobenzene is non-polar because of the cancellation of the two equal but opposing moments. *o*-Dichlorobenzene has a dipole moment with a direction and magnitude that is approximately the resultant of two monochlorobenzene dipole moments arranged at 60°. The dipole moment of *m*-dichlorobenzene can be accounted for similarly, but with 120° between the contributing moments. This technique of vector addition can be applied with fair success to other series of related molecules. As already illustrated, molar polarizations and refractivities are also approximately additive.

24.2 Intermolecular forces

In this section we consider the attractive van der Waals forces between molecules, and see how they are related to the electrical properties treated in Section 24.1.

24.2 (a) Dipole/dipole interactions

When two polar molecules are close to each other, their dipoles interact and affect their potential energy. If the molecules took up all relative orientations with equal probability, then the average interaction energy would be zero because the attractive interaction (when the dipoles are head to tail, Fig. 24.10(a)) is cancelled by the repulsive interaction (when the dipoles are head to head, Fig. 24.10(b)). However, in a fluid the attractive orientations, because they correspond to lower energy, are more likely to occur than the repulsive, and so on average the attractions slightly dominate the repulsions. The Boltzmann distribution can be used to make this idea quantitative, and the repetition of a calculation like that leading to eqn (24.1.6) gives the following expression for the average interaction energy between two polar molecules with dipole moments μ_1 and μ_2 separated by a distance R:

$$V = -C_{\text{dip/dip}}/R^6, \qquad C_{\text{dip/dip}} = (2/3kT)(\mu_1\mu_2/4\pi\varepsilon_0)^2. \qquad (24.2.1)$$

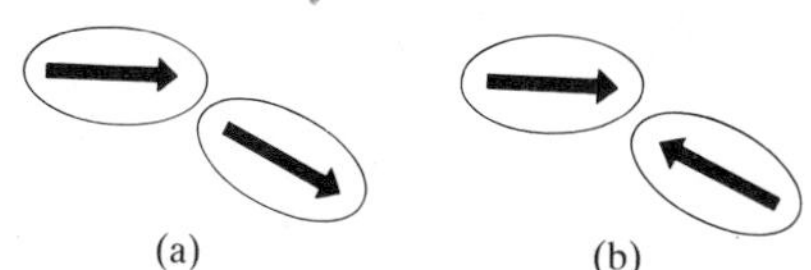

Fig. 24.10. When a pair of molecules can adopt all relative orientations with equal probability, the favourable orientations (a) and the unfavourable ones (b) cancel, and the average interaction is zero. In an actual fluid, the interactions in (a) slightly predominate.

The important feature is the dependence of this energy on the inverse *sixth* power of the separation, and its inverse dependence on the temperature. The latter reflects the way that the greater thermal motion overcomes the mutual orientating effects of the dipoles at higher temperatures.

At 25 °C the average interaction energy for pairs of molecules with $\mu = 1$ D is about -1.4 kJ mol^{-1} when the separation is 0.3 nm. This should be compared with the average molar kinetic energy of $\frac{3}{2}RT = 3.7$ kJ mol^{-1} at the same temperature: the two are not very dissimilar but they are both much less than the energies involved in the formation and breaking of chemical bonds.

24.2 (b) Dipole/induced-dipole interactions

A polar molecule near a polarizable molecule (which may itself be either polar or non-polar) induces a dipole in the latter. This induced dipole interacts with the permanent dipole of the first molecule, and the two are attracted together. The strength of the interaction depends on the dipole moment of the polar molecule and the polarizability of the second molecule. Since the induced dipole follows the direction of the inducing dipole, Fig. 24.11, we do not need to take account of the effects of thermal motion: both dipoles are aligned, however fast the tumbling. The average interaction energy when the separation of the molecules is R is

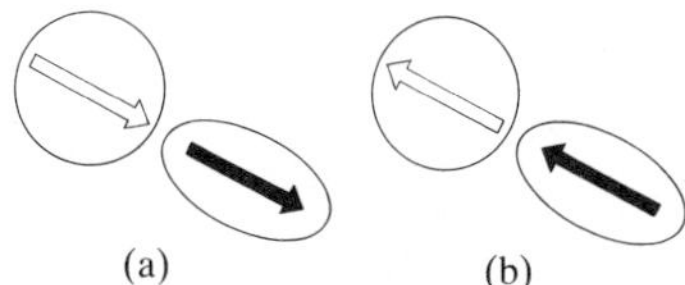

Fig. 24.11. A polar molecule (dark arrow) can induce a dipole (light arrow) in a non-polar molecule, and the latter's orientation follows the former's, so that the interaction does not average to zero.

$$V = -C_{\text{dip/induced}}/R^6, \qquad C_{\text{dip/induced}} = \mu_1^2\alpha_2'/4\pi\varepsilon_0, \qquad (24.2.2)$$

where α_2' is the polarizability volume of molecule 2 and μ_1 the permanent dipole moment of molecule 1. This interaction energy is independent of the temperature (because thermal motion has no effect) and, like the dipole/dipole interaction, depends on $1/R^6$. For a molecule with $\mu \approx 1$ D (such as HCl) near a molecule of polarizability volume $\alpha' \approx 10 \times 10^{-24}$ cm^3 (such as benzene, Table 24.1) the average interaction energy is about -0.8 kJ mol^{-1} when the separation is 0.3 nm.

24.2 (c) Induced-dipole/induced-dipole interactions

Consider two non-polar molecules separated by R. Although they have no permanent moments their electron locations are changing (with a time average corresponding to the electron distribution given by the molecular orbitals). As a result, we may think of them as having instantaneous dipoles which are constantly changing in magnitude and direction, Fig. 24.12. Suppose that one molecule flickers into an electronic arrangement that gives it an instantaneous dipole moment μ_1^*. This dipole polarizes the other molecule, and induces in it an instantaneous dipole moment μ_2^*. The two dipoles stick together, and so the two molecules have an attractive interaction. Although the first molecule will go on to change the direction of its dipole, the second will follow it, and because of this correlation the

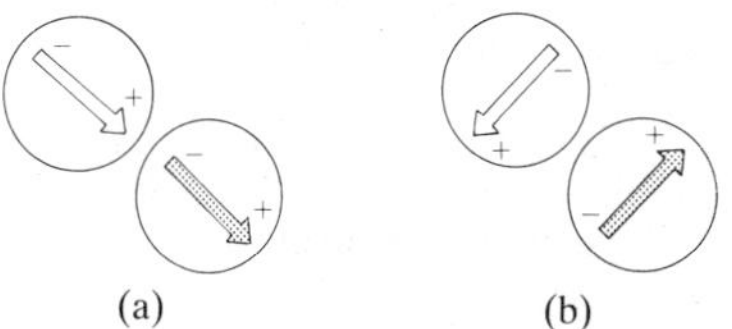

Fig. 24.12. In the *dispersion interaction*, an instantaneous dipole (shaded arrow) on one molecule induces a dipole (light arrow) on another molecule, and the two dipoles then interact to lower the energy. The two instantaneous dipoles are *correlated*, and although they occur in different orientations at different instants, the interaction does not average to zero.

attraction does not average to zero. This induced-dipole/induced-dipole interaction is also called the *dispersion interaction*.

The strength of the dispersion interaction depends on the polarizability of the first molecule because the instantaneous dipole moment μ_1^* depends on the looseness of the nuclear charge's control over the outer electrons. It also depends on the polarizability of the second molecule, because the size of μ_2^* depends on the extent to which it can be induced. The actual calculation of the dispersion interaction is quite involved, but a reasonable approximation to the interaction energy is given by the *London formula*:

$$V = -C_{\text{disp}}/R^6, \qquad C_{\text{disp}} = \tfrac{3}{2}\alpha_1'\alpha_2'\{I_1 I_2/(I_1 + I_2)\}, \qquad (24.2.3)$$

where I_1 and I_2 are the ionization energies of the two molecules (Table 4.9).

In the case of two methane molecules we can substitute $\alpha' = 2.6 \times 10^{-24}\,\text{cm}^3$ and $I = 7\,\text{eV}$ to obtain $V = -5\,\text{kJ mol}^{-1}$ for $R = 0.3\,\text{nm}$. A very rough check on this figure is the enthalpy of vaporization of methane, which is $8.2\,\text{kJ mol}^{-1}$. (The comparison is very rough because the enthalpy of vaporization is a many-body quantity.)

The London formula also applies to polar molecules: these also have dispersion interactions because they possess instantaneous dipoles, the only difference being that the time average of each fluctuating dipole does not vanish, but corresponds to the permanent dipole. The total attractive interaction energy between molecules is therefore the sum of the three contributions discussed above. All three vary as $1/R^6$, and so we may write

$$V = -C_6/R^6, \qquad (24.2.4)$$

C_6 being a coefficient that depends on the identity of the molecules. This is the way that attractive intermolecular interactions are normally expressed, but we must remember two limitations. The first is that we have taken into account only *dipolar* interactions: in a complete treatment we should also consider *quadrupolar* and higher-multipolar interactions reflecting the interactions between charge distributions more complicated than dipoles. For example, the symmetrical, linear molecule CO_2 has no permanent dipole but it does have a permanent electric quadrupole (like two dipoles back to back, Fig. 24.13): the quadrupole–quadrupole interaction is an important contribution to intermolecular energies, especially in non-dipolar molecules. The second limitation is that the expressions given above relate to the interactions of *pairs* of molecules, and there is no reason to suppose that the energy of interaction of three (or more) molecules is the sum of the pairwise interaction energies. In the case of three atoms the total dispersion energy is given by the *Axilrod–Teller formula:*

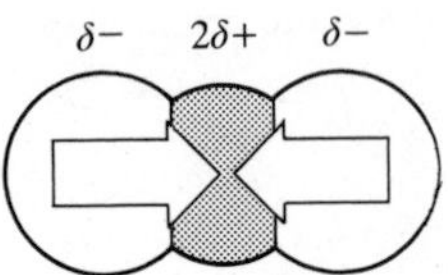

Fig. 24.13. Although CO_2 has no permanent dipole moment, it does have a permanent *quadrupole moment*, which can be regarded as two dipole moments back-to-back. These make an important contribution to the intermolecular forces, but are shorter range than dipole–dipole interactions.

$$V = -C_6/R_{AB}^6 - C_6/R_{BC}^6 - C_6/R_{CA}^6 + C'/(R_{AB}R_{BC}R_{CA})^3, \qquad (24.2.5)$$

where $C' = a(3\cos\theta_A\cos\theta_B\cos\theta_C + 1)$, a being a constant (which is approximately equal to $\tfrac{3}{4}\alpha' C_6$) and the angles θ the internal angles of the triangle formed by the three atoms. The final term (which represents the non-additivity of the pairwise interactions) is negative for a linear arrangement of atoms (and so that arrangement is stabilized) but positive for an equilateral triangular cluster. It is found that the three-body term contributes about 10% of the total interaction energy in liquid argon.

24.2 (d) Repulsive and total interactions

When molecules are squeezed together, the nuclear and electronic repulsions and the rising electronic kinetic energy begin to dominate the attractive forces. The repulsions increase steeply with decreasing separation, Fig. 24.14, and do so in a way that can be deduced only by very extensive, complicated molecular structure calculations of the kind described in Chapter 16. In many cases progress can be made by using a greatly simplified representation of the potential energy, where the details are ignored and the general features expressed by a few adjustable parameters.

One approximation is the *hard-sphere potential*, where it is assumed that the potential energy rises abruptly to infinity as soon as the particles come within some separation σ called the *collision diameter*:

$$V=0 \quad \text{for} \quad R>\sigma; \qquad V=\infty \quad \text{for} \quad R\leqslant\sigma. \tag{24.2.6}$$

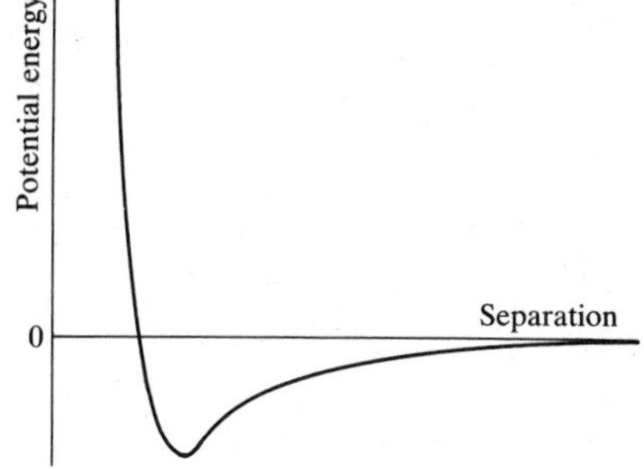

Fig. 24.14. The general form of an intermolecular potential energy curve. At long range the interaction is attractive, but at close range the repulsions dominate.

This very simple potential is surprisingly useful for assessing a number of properties, as we shall see. Another widely used approximation is to express the short-range repulsive potential energy as a function of a high power of R:

$$V=C_n/R^n. \tag{24.2.7}$$

Typically $n=12$, in which case the repulsion dominates the $1/R^6$ attractions strongly at short separations because C_{12}/R^{12} is then much larger than C_6/R^6. The sum of the repulsive interactions and the attractive interaction given by eqn (23.2.4) is called the *Lennard–Jones potential*:

$$V=C_n/R^n-C_6/R^6. \tag{24.2.8}$$

This is the two-parameter Lennard–Jones $(n,6)$-potential. Quite often the (12, 6)-potential is written in the form

$$V=4\varepsilon\{(\sigma/R)^{12}-(\sigma/R)^6\}. \tag{24.2.9}$$

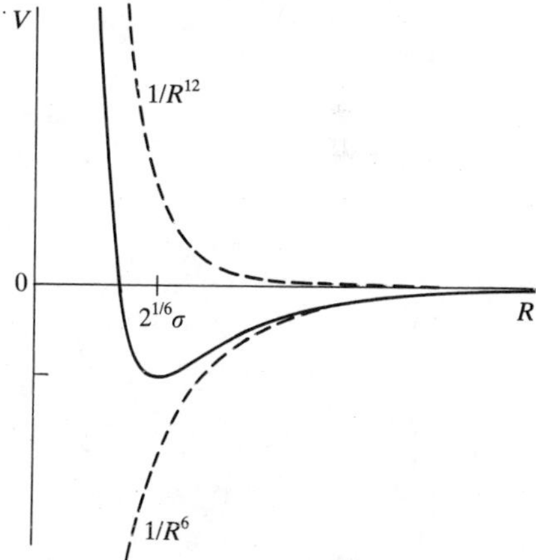

Fig. 24.15. The *Lennard–Jones* (12,6)-*potential*, and the relation of the parameters to the features of the curve. The dotted lines are the two contributions.

This is drawn in Fig. 24.15. The two parameters are now ε, which corresponds to the depth of the well, and σ, which is the separation at which $V=0$. The well minimum occurs at $R_e=2^{\frac{1}{6}}\sigma$. Some typical values are listed in Table 24.4. How they are measured and how they are used are described below. Although the (12, 6)-potential has been used in many calculations, there is plenty of evidence to show that $1/R^{12}$ is a very poor representation of the repulsive potential, and that an exponential form $(e^{-R/\sigma})$ is greatly superior.

Table 24.4. Lennard–Jones (12,6) parameters

	$(\varepsilon/k)/\text{K}$	σ/pm
He	10.22	258
Ar	124	342
N_2	91.5	368
CCl_4	327	588

24.3 Molecular interactions in beams

In recent years a notable advance in the experimental study of intermolecular forces has come from the development of *molecular beams*. A molecular beam is a narrow stream of molecules in an otherwise empty vessel. The beam is directed towards other molecules, and the scattering that occurs on impact is related to the intermolecular interactions.

24.3 (a) The basic principles

The basic arrangement is shown in Fig. 24.16. The slotted disks make up the *velocity selector*: they rotate in the path of the beam and allow only those molecules having a certain speed to pass. There are also more sophisticated devices for generating molecules with a desired velocity. Among the most important are *supersonic nozzles*, in which the molecules stream out of the source with a very narrow velocity distribution and are then stripped down to a beam by passing through a skimmer shaped like an inverted cone, Fig. 24.17. Other types of selection are also possible: for example, electric fields may be used to deflect polar molecules and to obtain a beam of aligned molecules. The target gas may be either a bulk sample or another molecular beam. The latter *crossed beam* technique gives a lot of information because the states of both the target and projectile particles may be controlled.

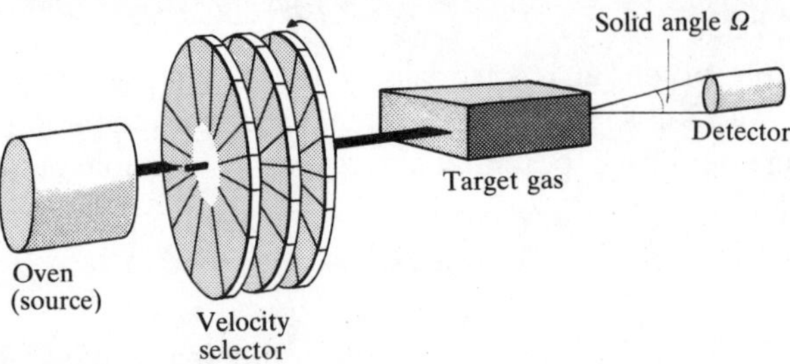

Fig. 24.16. The basic arrangement of a molecular beam apparatus. The atoms or molecules emerge from a heated source, and pass through the velocity selector, a train of rotating disks. The scattering occurs from the target gas (which might take the form of another beam), and the flux of particles entering the detector set at some angle is recorded.

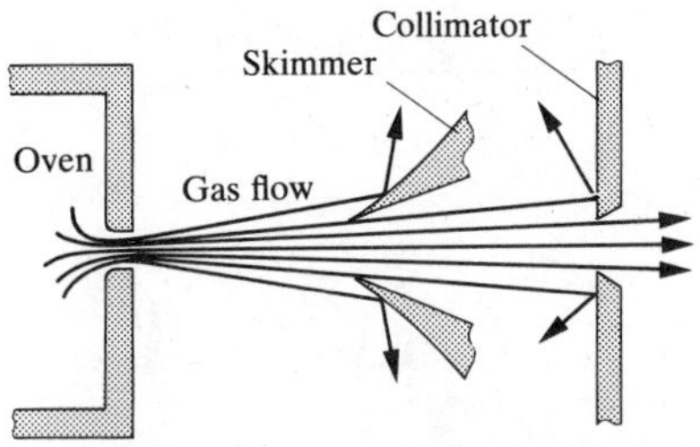

Fig. 24.17. A *supersonic nozzle* skims off some of the molecules of the beam and leads to a beam with well-defined velocity.

Collisions with the target particles scatter the beam molecules away from their initial direction, and the angle through which they are scattered is a major piece of experimental information. The detectors may consist of a chamber fitted with a sensitive pressure gauge or an ionization detector, in which the incoming molecule is first ionized and then detected electronically. The state of excitation of the scattered molecules may also be determined (e.g. spectroscopically), and is of interest when the collisions change their vibrational or rotational states.

24.3 (b) The experimental observations

The two principal quantities used in the discussion of molecular beam results are the *impact parameter* (symbol: b) and the *differential scattering cross-section* (symbol: $d\sigma$). The impact parameter is the initial perpendicular separation of the paths of the colliding particles, Fig. 24.18. The differential scattering cross-section measures the extent of scattering through different angles and is defined as

$$d\sigma = \frac{\textit{Number of particles scattered into } d\Omega \textit{ at } \Omega \textit{ per unit time}}{\textit{Incident beam flux}}.$$

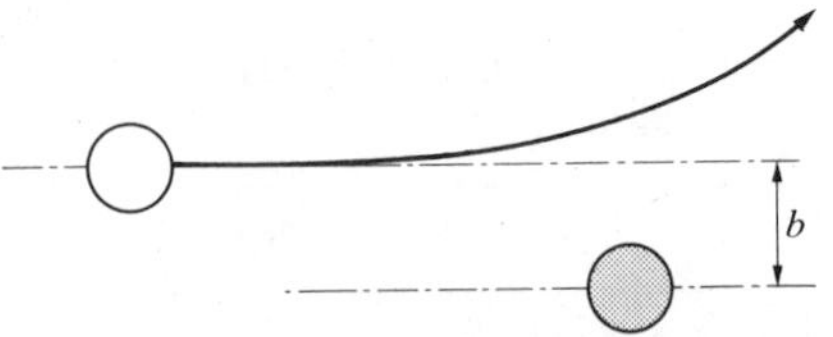

Fig. 24.18. The definition of the *impact parameter b* as the perpendicular separation of the initial paths of the particles.

$d\Omega$, which is equal to $\sin\theta\, d\theta\, d\phi$ in the polar coordinates defined in Fig. 24.19, is the infinitesimal solid angle at the angles θ, ϕ (collectively denoted Ω), Fig. 24.19, and so the differential cross-section measures the rate at which particles arrive in this region relative to the number arriving at the target in the incident beam. Since the incident flux is the number of particles per unit area per unit time, and $d\Omega$ is dimensionless, $d\sigma$ has the dimensions of area.

The value of $d\sigma$ depends on the impact parameter and the details of the intermolecular potential. This is most easily seen by considering the impact of two hard spheres, Fig. 24.20. If the impact parameter is zero, the lighter projectile is on a trajectory that leads to a head-on collision, and so all the scattering intensity is in the solid angle $d\Omega$ at $\theta = 180°$. If the impact parameter is so great that the spheres do not make contact ($b > R_A + R_B$), then there is no scattering and the scattering cross-section is zero at all angles. Glancing blows, with $0 < b < R_A + R_B$, lead to scattering intensity in cones around the forward direction, Fig. 24.20(c).

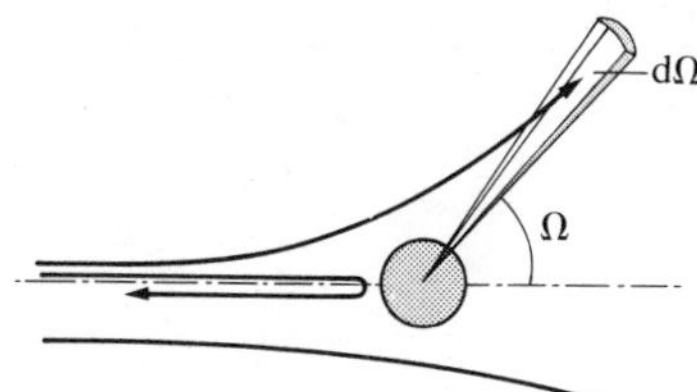

Fig. 24.19. The definition of the *differential scattering cross-section* $d\sigma$. The angles θ, ϕ are collectively denoted Ω, and the solid angle $d\Omega$ is equal to $\sin\theta\, d\theta\, d\phi$.

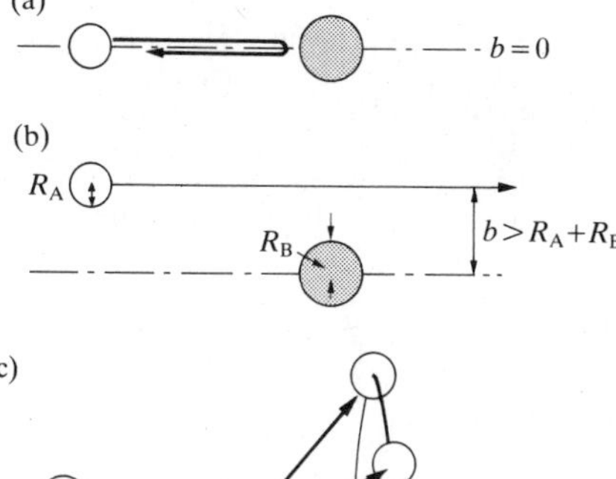

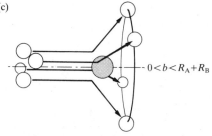

Fig. 24.20. The collisions of two hard spheres as the impact parameter changes. (a) $b = 0$, giving backward scattering; (b) $b > R_A + R_B$, giving forward scattering; (c) $0 < b < R_A + R_B$, leading to scattering into one direction on a ring of possibilities. (The shaded sphere is taken to be so heavy that it remains virtually stationary.)

24.3 (c) Scattering effects

This elementary discussion must be modified in order to treat the collisions of real molecules, which are not hard spheres. The scattering pattern then depends on the details of the intermolecular potential, including the anisotropy that is present when the molecules are non-spherical. Furthermore, the scattering depends on the relative speed of approach of the two particles, for a very fast particle might pass through the interaction region without much deflection, while a slower one on the same path might be temporarily captured and undergo considerable deflection, Fig. 24.21. This means that the dependence of the scattering cross-section on the relative speed of approach should give information about the strength and range of the intermolecular potential: that is why it is important to be able to control the speeds of molecular beams.

A further point is that the outcome of collisions is determined by quantum, not classical, mechanics. We saw a little of the strange things that can happen when we treated one-dimensional barriers in Chapter 14; now we must see what happens when the colliding particles can move in three dimensions. Just as in the Young's slit interference experiment mentioned in Section 23.1, the wave nature of the particles can be taken into account, at least to some extent, by drawing all classical trajectories that take the projectile particle from source to detector, and then considering the effects of interference between them.

Two quantum mechanical effects are of great importance. A particle with some impact parameter might approach the attractive region of the potential in such a way that it is deflected towards the repulsive core, Fig. 24.22, which then repels it out through the attractive region so that it continues its flight in the forward direction, just like particles with impact parameters so large that they are undeflected. The paths interfere, and the particle intensity in the forward direction is modified. This is called *glory scattering* because it is the same phenomenon that accounts for the optical glory effects where a bright halo can sometimes be seen surrounding an illuminated object (the coloured rings around the shadow of an aircraft cast on clouds by the sun, and often seen in flight, is an example of a glory). The other quantum effect is the observation of a strongly enhanced scattering in some non-forward direction: this is called *rainbow scattering* because the same mechanism accounts for the appearance of an optical rainbow. The phenomenon is illustrated in Fig. 24.23: as the impact parameter decreases, there comes a point where the scattering angle passes through a maximum, and the interference between the paths then results in a strongly scattered beam. The *rainbow angle* is the angle for which $d\theta/db = 0$ and the scattering is strong.

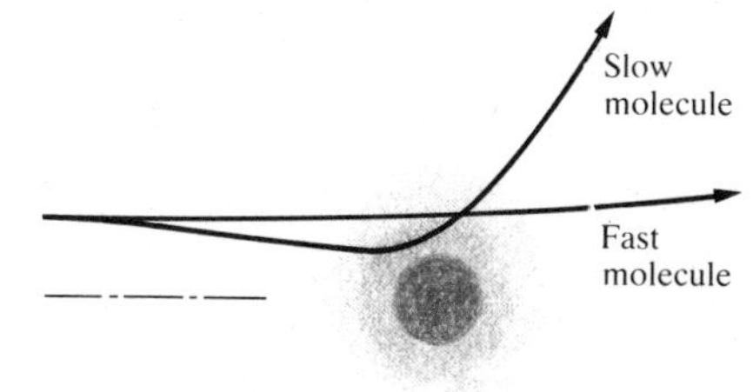

Fig. 24.21. The extent of scattering may depend on the relative speed of approach as well as the impact parameter. The dark zone represents the repulsive core; the fuzzy outer zone represents the long-range attractive potential.

The detailed analysis of scattering data can be very complicated, and we shall not go into it, but the main outcome should be clear: the intensity distribution of scattered particles can be related to the intermolecular potential, and a detailed picture can be built up of its radial and angular variation. One outcome is the value of C_6 in the van der Waals interaction and the testing of the Lennard–Jones potential (or some more elaborate version). It is found, for example, that although $1/R^6$ is quite a good representation of the attractive component of the potential, as remarked before, the repulsive component is not at all well described by $1/R^{12}$.

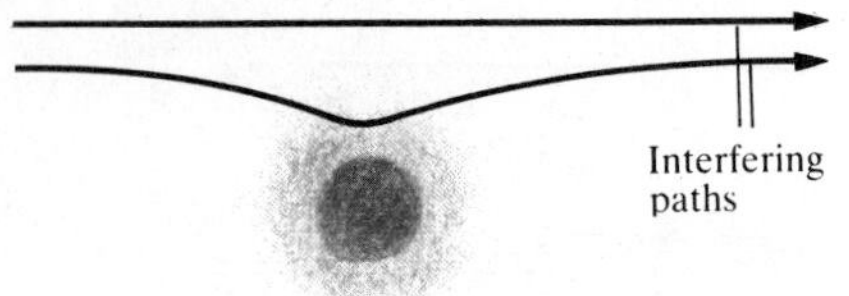

Fig. 24.22. Two paths leading to the same destination will interfere quantum-mechanically; in this case they give rise to *glory scattering* in the forward direction.

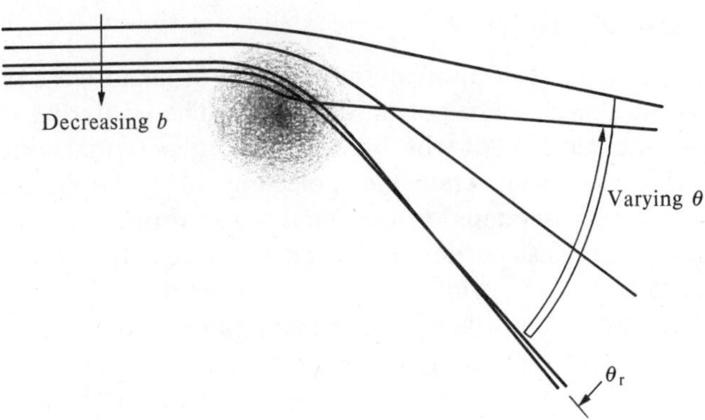

Fig. 24.23. The interference of paths leading to *rainbow scattering*. θ_r is the maximum scattering angle as b is decreased, and the interference between the numerous paths at that angle modifies the scattering intensity markedly.

24.4 Gases and liquids

Intermolecular forces are responsible for the deviations of real gases from perfect gas behaviour and for the cohesion of molecules into liquids. In this section we examine to what extent the properties of these two fluid phases may be related to the interactions between their particles.

24.4 (a) Gas imperfections

Real gases differ from perfect gases in their equations of state (Section 1.1) and in various transport properties (which we leave until Chapter 26). The meeting place of theory and experiment is the virial equation of state, eqn (1.3.2):

$$pV_m/RT = 1 + B/V_m + C/V_m^2 + \ldots, \tag{24.4.1}$$

where B is the *second virial coefficient* and C the third; both are temperature-dependent. The virial coefficients are consequences of the intermolecular forces, and so we have a route to the latter's measurement.

The connection between the virial coefficients and the intermolecular potential is quite involved, but the principle is straightforward. In Section 22.2 we saw that pressure is related to the canonical partition function Q by

$$p = kT(\partial \ln Q/\partial V)_T. \tag{24.4.2}$$

(This is eqn (22.2.4).) As we saw there, in the absence of interactions the canonical partition function is related to the molecular partition function $q = V/\Lambda^3$ by $Q = q^N/N!$, so that

$$Q = (V^N/N!)/\Lambda^{3N}, \qquad \Lambda = h(\beta/2\pi m)^{\frac{1}{2}}, \qquad \beta = 1/kT.$$

This led to $p = RT/V_m$, corresponding to the virial coefficients $B, C, \ldots$ all being zero. In the presence of interactions the canonical partition function cannot be expressed so simply because the particles are not independent.

However, it remains true that the total kinetic energy of the particles is the sum of the individual kinetic energies of the particles, and therefore that the partition function factorizes into a part arising from the kinetic energy, which is the same as for the perfect gas, and a new factor that depends on the intermolecular potentials:

$$Q = Z/\Lambda^{3N}. \tag{24.4.3}$$

Z is called the *configuration integral*, and is an integral of $e^{-\beta V_N}$ over all the positions of all the particles:

$$Z = (1/N!)\int e^{-\beta V_N}\, d\boldsymbol{r}_1\, d\boldsymbol{r}_2 \ldots d\boldsymbol{r}_N, \tag{24.4.4}$$

V_N being the total potential energy of interaction of all the particles.

When there is no interaction $V_N = 0$ and $e^{-\beta V_N} = 1$. The configuration integral is then

$$Z_1 = (1/N!)\int d\boldsymbol{r}_1\, d\boldsymbol{r}_2 \ldots d\boldsymbol{r}_N = (1/N!)V^N, \qquad (24.4.5)^\circ$$

because $\int d\boldsymbol{r} = V$, the volume; hence $Q = (V/\Lambda^3)^N/N!$, the form it has for a perfect gas.

When we consider only pairwise interactions (e.g. dispersion interactions between pairs of particles) the configuration integral becomes

$$Z_2 = \tfrac{1}{2}\int e^{-\beta V_2}\, d\boldsymbol{r}_1\, d\boldsymbol{r}_2, \tag{24.4.6}$$

and the second virial coefficient, which is a consequence of these pairwise interactions, turns out to be

$$B = -(N_A/2V)\int f\, d\boldsymbol{r}_1\, d\boldsymbol{r}_2, \qquad f = e^{-\beta V_2} - 1. \tag{24.4.7}$$

f is called the *Mayer f-function*: it goes to zero when the two particles are so far apart that $V_2 = 0$. When the intermolecular potential depends only on the separation R of the particles, as in the dispersion interaction of spheres, the last expression simplifies to

$$B = -2\pi N_A \int_0^\infty f R^2\, dR, \tag{24.4.8}$$

which can be evaluated (usually numerically) by substituting one of the intermolecular potential expressions discussed earlier. One reason why the Lennard–Jones potential is so widely used is that this integral may be evaluated analytically (but the result is quite complicated).

As an example, we consider the simplest potential of all: the hard sphere potential defined in eqn (24.2.6). Since $e^{-\beta V_2} = 0$ (and $f = -1$) when $R \leq \sigma$, and $e^{-\beta V_2} = 1$ (so that $f = 0$) when $R > \sigma$, the second virial coefficient is

$$B = 2\pi N_A \int_0^\sigma R^2\, dR = \tfrac{2}{3}\pi N_A \sigma^3. \tag{24.4.9}$$

Long ago, eqn (1.4.4), we saw that the part of the van der Waals equation of state that represented the hard-sphere repulsive part of the intermolecular potential (i.e. b) contributed to B in exactly this way. This raises the question as to whether there is potential which, when the virial coefficients are evaluated, gives the full van der Waals equation of state. Such a potential can be found: it consists of a hard-sphere repulsive core and a long-range, shallow attractive region.

Example 24.5

Take the Lennard–Jones potential for Ar in the form of eqn (24.2.9), with $\varepsilon/k = 120$ K and $\sigma = 340$ pm, and evaluate the second virial coefficient at 298 K.

• *Method*. Evaluate the integral in eqn (24.4.8). Do this numerically (the technical term is *numerical quadrature*) using Simpson's rule. This rule is as follows:

(1) Consider the range of integration as being covered by an even number of points (y_m) of separation h.

(2) Form the sum

$$(h/3)\{y_0 + 4(y_1 + y_3 + \ldots + y_{m-1}) + 2(y_2 + y_4 + \ldots + y_{m-2}) + y_m\}.$$

This is the value of the integral required.

Since the integrand rises from zero up to a maximum, then crosses through zero at 340 pm, and has a long tail out to several 1000 pm, it is sensible to do the integration in two ranges, one from $R = 0$ to 340 pm, the second from 340 pm out to about 5000 pm. The integration is readily programmed for a microcomputer.

• *Answer*. Integration in the first range gives $B/2\pi N_A = +1.130 \times 10^7\ \text{pm}^3$, and in the second range gives $B/2\pi N_A = -1.54 \times 10^7\ \text{pm}^3$. The complete integral is therefore

$$B/2\pi N_A = -4.19 \times 10^6\ \text{pm}^3, \qquad B = -15.9\ \text{cm}^3\ \text{mol}^{-1}.$$

• *Comment*. The two domains of integration, and the opposite signs of their contributions to B, reflect regions where the repulsions and attractions are dominant respectively.

• *Example*. Repeat the determination for 273 K. [$-21.5\ \text{cm}^3\ \text{mol}^{-1}$]

24.4 (b) The structures of liquids

The starting point for the discussion of solids is the well-ordered structure of perfect crystals. The starting point for the discussion of gases is the totally chaotic distribution of the particles of a perfect gas. With liquids we are between these two extremes: there is some structure and some chaos.

The particles of a liquid are held together by intermolecular forces, but their kinetic energies are comparable to their potential energies. As a result, the whole structure is very mobile. The best description of the average locations of the particles come from the *radial distribution functions*. The *pair distribution function* (PDF, symbol: g) is the most important, and we shall concentrate on it.

Knowing the PDF allows us, by forming $g\,dR$, to state the probability that a particle will be found in the range dR at a distance R from another. In a crystal, g is a periodic array of sharp spikes, representing the certainty (in the absence of defects and thermal motion) that particles lie at definite locations. This regularity continues out to large distances, and so we say that crystals have *long-range order*. When the crystal melts the long-range order is lost, and wherever we look at long distances from a given particle, there is equal probability of finding a second particle. Close to the first particle, though, there may be a remnant of order. This is because its nearest

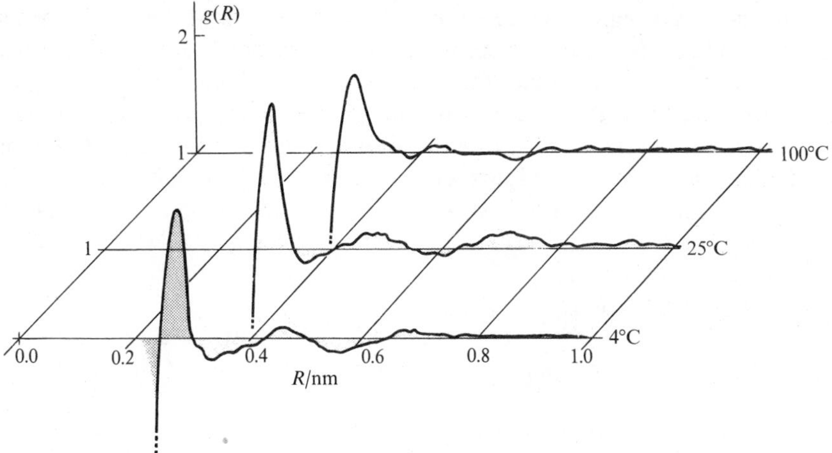

Fig. 24.24. The radial distribution function of the oxygen atoms in liquid water at three temperatures. Note the expansion as the temperature is raised. (A. H. Narten, M. D. Danford, and H. A. Levy, *Discuss. Faraday. Soc.* **43,** 97, 1967.)

neighbours might still adopt approximately their original positions, and even if they are displaced by newcomers, the new particles might adopt their vacated positions. It may still be possible to detect a sphere of nearest neighbours at a distance R_1, and perhaps beyond them a sphere of next-nearest neighbours at R_2. The existence of this *short-range order* means that the PDF can be expected to oscillate close to $R = 0$, with a peak at R_1, a smaller peak at R_2, and perhaps some more structure beyond that.

The PDF can be measured experimentally by X-ray diffraction. Crystals give sharp diffraction patterns (so long as there is little thermal motion). Liquids give a diffraction pattern consisting of blurred rings. The PDF can be extracted from the pattern in much the same way as a crystal structure is obtained, and the ones obtained for water are shown in Fig. 24.24. The shells of local structure are unmistakable. Closer analysis shows that any given H_2O molecule is surrounded by other molecules at the corners of a tetrahedron, similar to the arrangement in ice (Section 23.3), and the 100 °C result shows that the intermolecular forces (in this case, largely H-bonds) are strong enough to affect the local structure right up to the boiling point.

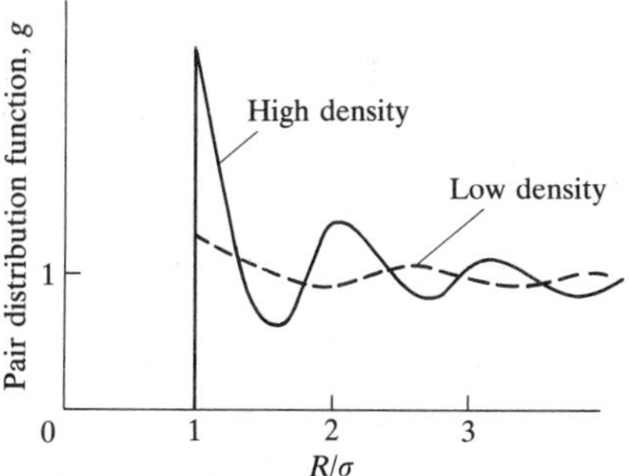

Fig. 24.25. The radial distribution function for a simulation of a liquid using impenetrable hard spheres (ball bearings).

The PDF can be calculated by making assumptions about the intermolecular forces, and so it can be used to test theories of liquid structure. However, even a fluid of hard spheres without attractive interactions (a box of ball-bearings) gives a PDF which oscillates near the origin, Fig. 24.25. This means that one of the factors influencing, and sometimes dominating, the structure of a liquid is simply the geometrical problem of stacking together a lot of hard spheres. The attractive part of the potential modifies this basic structure by gathering and trapping particles into each other's vicinity. One of the reasons behind the difficulty of describing liquids theoretically is the importance of both the attractive and repulsive, hard-core, parts of the potential.

24.4 (c) Simulations of structure

There are several ways of building the intermolecular potential into the calculation of g. Numerical methods take a box of between 100 and 1000 particles (the number increases as computers grow more powerful), and the rest of the liquid is simulated by surrounding the box with replications. This is illustrated for a two-dimensional system in Fig. 24.26. Then, whenever a

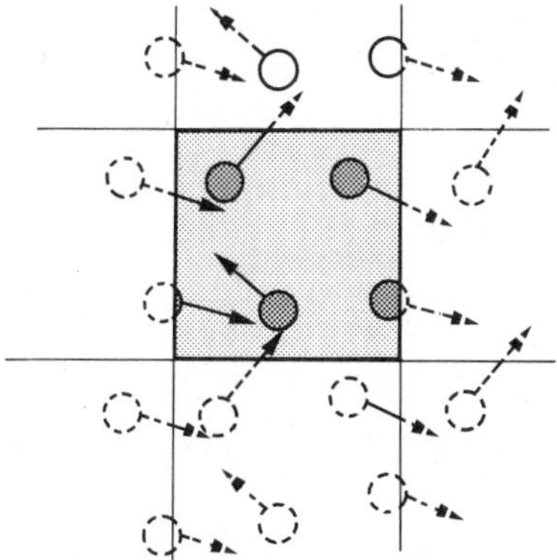

Fig. 24.26. In a two-dimensional molecular dynamics simulation, when one particle leaves the cell its mirror image enters through the opposite face.

particle leaves through a face, its image arrives through the opposite face so that the total number remains constant.

In the *Monte Carlo method* the particles in the box are moved by a random amount, and the total potential energy is calculated using one of the intermolecular potentials discussed earlier (including, as they have been found important, the non-additive three-body terms). Whether or not this new configuration is accepted is then judged on the basis of the following criteria:

(1) if the potential energy is not greater than before the change, the configuration is accepted;

(2) if it is greater, then it is accepted in proportion to the value of $e^{-\Delta V_N/kT}$, ΔV_N being the change in total potential energy of the N particles in the box.

This procedure ensures that the probability of occurrence of any configuration is proportional to the Boltzmann factor $e^{-V_N/kT}$. The configurations generated in this way can then be analysed for the value of g simply by counting the number of particles at a distance R from a selected one, and averaging the result over the whole collection.

In the *molecular dynamics* approach, the history of an initial arrangement is followed by calculating the trajectories of all the particles under the influence of the intermolecular potentials. Newton's laws are used to predict where each particle will be after a short time interval (about 10^{-15} s, which is shorter than the time between collisions), and then the calculation is repeated for tens of thousands of such steps. This procedure gives a series of snapshots of the liquid, and the PDF can be calculated as before.

Once the PDF is known it can be used to calculate the thermodynamic properties of liquids. For example, the contribution of the pairwise additive intermolecular potential to the internal energy is given by the integral

$$U_{\text{intermolec}} = 2\pi(N^2/V)\int_0^\infty gV_2R^2\,dR. \tag{24.4.10}$$

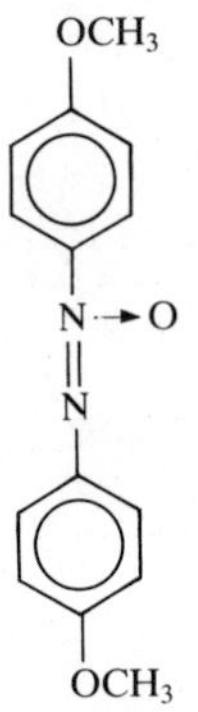

That is, it is the average two-particle potential energy weighted by g, the probability that the pair of particles have a separation R. Likewise, the pressure is given by

$$pV/nRT = 1 - (2\pi N/VkT)\int_0^\infty g\{R(dV_2/dR)\}R^2\,dR. \tag{24.4.11}$$

The quantity $R(dV/dR)$ is called the *virial* (hence the term *virial equation of state*).

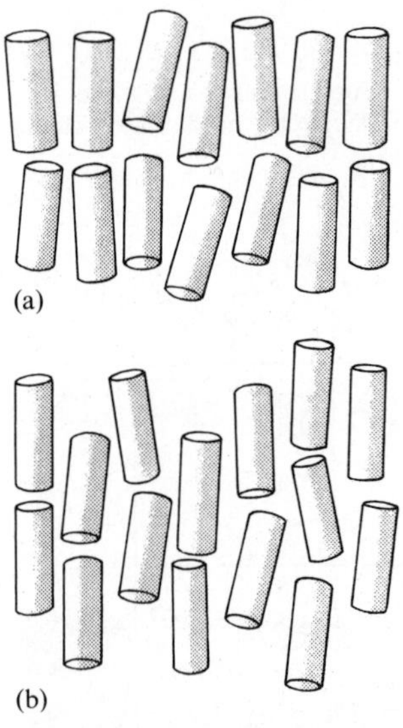

OCH$_3$ – C$_6$H$_4$ – N(→O)=N – C$_6$H$_4$ – OCH$_3$

1

(a)

(b)

Fig. 24.27. The arrangement of molecules in (a) the *smectic phase* and (b) the *nematic phase* of a liquid crystal. They both have elements of order, but it is different in different directions.

24.4 (d) Liquid crystals

A feature that makes calculations even more difficult is the possibility that molecules have strongly anisotropic interactions. This is the case when they are polar, but it is also the case when the molecules are long and thin, such as for p-azoxyanisole (**1**). When the solid melts, some aspects of the long-range order are maintained. This gives rise to *liquid crystals*, substances having liquid-like imperfect long-range order but some crystal-like aspects of short-range order. One type of retained long-range order gives rise to a *smectic* phase (from the Greek for soapy), in which the molecules align themselves in layers, Fig. 24.27(a). Other materials, and some smectic liquid crystals at higher temperatures, lack the layered structure but retain a parallel alignment, Fig. 24.27(b): this is the *nematic phase* (from the Greek for thread). In the *cholesteric phase* (from, unhelpfully, the Greek for bile solid) the molecules lie in sheets at angles that change slightly between each sheet. As a result, they form helical structures with a pitch that depends on the temperature. These cholesteric liquid crystals diffract light, and have colours that depend on the temperature. The strongly anisotropic optical properties of nematic liquid crystals, and their response to electric fields, is the basis of their use as data displays in calculators and watches.

24.4 (e) Movement in liquids

The PDF is a static property. Since a major property of liquids is their ability to flow, we need a method for studying their time-dependent properties. This can be done theoretically using the molecular dynamics simulation, because the detailed information it gives includes the velocities of the particles. Experimentally, liquids may be studied by a variety of methods. Relaxation time measurements in NMR and ESR (Section 20.1(e)) can be interpreted in terms of the mobilities of the particles, and have been used to show that some types of molecule rotate in a series of small (about 5°) steps, but that others jump through about 1 radian (57°) in each step. Another important technique is *inelastic neutron scattering*, in which the energy neutrons collect or discard as they pass through a sample is interpreted in terms of the motion of its particles, and very detailed information can be obtained.

Table 24.5. Viscosities of liquids at 298 K, $\eta/10^{-3}\ \mathrm{kg\ m^{-1}\ s^{-1}}$

Water	0.891
Benzene	0.601
Mercury	1.53
Pentane	0.224

Much more mundane than these experiments are viscosity measurements, Table 24.5. In order to move, a particle in a liquid must escape from its neighbours, and so it needs a minimum energy. The probability that it can acquire at least an energy E_a is proportional to $e^{-E_a/RT}$, and so the *mobility* of the liquid of the particles should follow this Boltzmann temperature dependence. Since the viscosity (η) is inversely proportional to the mobility of the particles, we should expect that

$$\eta \propto e^{+E_a/RT}, \tag{24.4.12}$$

implying that the viscosity should decrease exponentially with temperature (in contrast, as we shall see in Chapter 26, to the viscosity of gases, which increases). This is roughly true, Fig. 24.28. The intermolecular potentials govern the magnitude of E_a, but the problem of calculating it is immensely difficult and still largely unsolved.

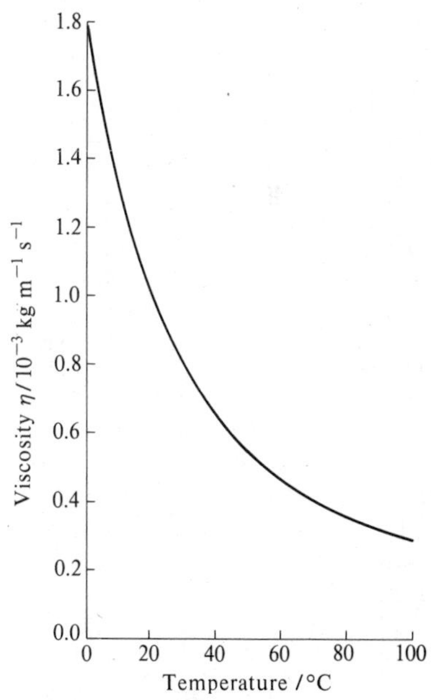

Fig. 24.28. The experimental temperature dependence of the viscosity of water. As the temperature is increased, more molecules are able to escape from the potential wells provided by their neighbours, and so the liquid becomes more fluid.

24.5 Ionic solids

At the opposite extreme from gases, where the average intermolecular potential energy is small compared with the kinetic energy of the particles, lie the ionic solids, where the Coulombic interaction between the ions is dominant and results in a rigid crystal. The lattice enthalpies of ionic crystals (Section 4.2(b)) can be calculated from the ion separations obtained from X-ray crystallography.

We suppose that the attractive contribution to the cohesive energy arises principally from the Coulombic interaction between ions of opposite charge; the repulsive interactions that oppose them arise from the electrostatic forces between ions of the same charge and the interactions described in Section 24.2. The equilibrium lattice spacing is the result of the balance of all these effects.

The potential energy V of a pair of ions of charges z_ie and z_je separated by a distance R_{ij} is

$$V_{ij} = z_i z_j e^2/4\pi\varepsilon_0 R_{ij}. \tag{24.5.1}$$

The potential energy of a single cation (of charge z_+e) as a result of its interactions with all the anions and other cations is

$$V^{(+)} = (z_+e^2/4\pi\varepsilon_0)\sum_i z_i/R_{+,i},$$

where $R_{+,i}$ is the distance from that cation to the ion i. If the shortest cation–anion distance in the crystal is R_0 we may use it as a unit of length, and write $R_{+,i} = \rho_{+,i}R_0$. Then

$$V^{(+)} = (z_+e^2/4\pi\varepsilon_0 R_0)\sum_i z_i/\rho_{+,i}.$$

Likewise, a single anion may be selected, and its potential energy written as

$$V^{(-)} = (z_-e^2/4\pi\varepsilon_0 R_0)\sum_i z_i/\rho_{-,i},$$

where $\rho_{-,i}R_0$ is the distance of i from that anion. The total electrostatic energy of the crystal of $N = nN_A$ M^+A^- units is therefore

$$V = \tfrac{1}{2}N\{V^{(+)} + V^{(-)}\} = na\mathcal{M}/R_0, \tag{24.5.2}$$

$$a = N_A z_+ z_- e^2/4\pi\varepsilon_0, \qquad \mathcal{M} = \frac{1}{2}\sum_i\left\{\frac{(z_i/z_-)}{\rho_{+,i}} + \frac{(z_i/z_+)}{\rho_{-,i}}\right\}. \tag{24.5.3}$$

The factor $\frac{1}{2}$ arises because we must not add each ion–ion interaction twice (which would happen if we formed $V^{(+)} + V^{(-)}$ and left it at that). $\mathcal{M}$ is called the *Madelung constant*. Its usefulness lies in the fact that it depends on the lattice *type*, and is independent of its dimensions, Table 24.6. We can therefore select the lattice type from the table, and state its Coulombic energy in terms of the single parameter R_0.

Table 24.6. Madelung constants

Lattice	$\mathcal{M}$
Rock salt	1.748
CsCl	1.773
Fluorite	2.519
Rutile	2.408

The total molar lattice energy is the sum of the expression just obtained and the repulsive potential energy. The latter is often taken to depend exponentially on the lattice parameter, and is written

$$V = nK\mathrm{e}^{-R_0/R^*}. \tag{24.5.4}$$

K and R^* are parameters, of which more later. The total molar energy of the solid is therefore

$$V/n = K\mathrm{e}^{-R_0/R^*} - a\mathcal{M}/R_0. \tag{24.5.5}$$

The minimum energy occurs at $\mathrm{d}V/\mathrm{d}R_0 = 0$, which enables K to be expressed as

$$K = (a\mathcal{M}R^*/R_0)\mathrm{e}^{R_0/R^*}. \tag{24.5.6}$$

because we can take R_0 as known experimentally. The molar potential energy is therefore

$$V/n = -(a\mathcal{M}/R_0)\{1 - (R^*/R_0)\}, \qquad a = N_A z_+ z_- e^2/4\pi\varepsilon_0. \tag{24.5.7}$$

The only remaining parameter is R^*: it may be related to the compressibility (κ) and molar volume (V_m) of the crystal:

$$R^* = \tfrac{2}{5}R_0 + bR_0^2/\kappa, \qquad b = (18\pi/5)\varepsilon_0 V_m/N_A e^2, \qquad (24.5.8)$$

a result obtained in Problem 24.31.

So long as we ignore both kinetic energy contributions (which is strictly valid only at $T=0$) and the zero-point energy of the lattice, V/n can be identified as the change of molar internal energy, ΔU_m, when the gaseous ions form the crystal. It follows that the lattice enthalpy (the standard enthalpy change for $M^+(g) + A^-(g) \rightarrow MA(s)$), which is related to the internal energy change by $\Delta H_m^\ominus = \Delta U_m^\ominus - 2RT$, may be calculated from eqn (24.5.7) using experimental values of the lattice parameter R_0 and the compressibility κ. Some values are given in Table 24.7; they were used in the Born–Haber cycles described in Chapter 4.

Table 24.7. Lattice enthalpies, $-\Delta H_m^\ominus(\mathcal{T})/\mathrm{kJ\,mol^{-1}}$

LiF	1022
NaCl	771
KCl	701
RbI	609

24.6 Magnetic properties

The magnetic and electric properties of molecules are analogous. In the first place, some molecules possess permanent magnetic dipole moments. Furthermore, an applied magnetic field can induce a magnetic moment. The analogue of the polarization is the *magnetization* (symbol: M).

24.6 (a) Magnetic susceptibility

The magnetization induced by a field of strength H is called the *volume magnetic susceptibility* (symbol: κ):

$$M = \kappa H. \qquad (24.6.1)$$

A closely related quantity is the *mass magnetic susceptibility* (symbol: χ):

$$\chi = \kappa\rho^\ominus/\rho, \qquad \rho^\ominus = 1\,\mathrm{kg\,m^{-3}}, \qquad (24.6.2)$$

where ρ is the density. The practical form of this expression is $\chi = 1000\kappa/(\rho/\mathrm{g\,cm^{-3}})$. The distinction between κ and χ is developed below. The *magnetic flux density* (symbol: B) is related to the applied field strength and the magnetization by

$$B = \mu_0(H + M) = \mu_0(1+\kappa)H. \qquad (24.6.3)$$

μ_0 is the *vacuum permeability*, and is defined to have the value $4\pi \times 10^{-7}$ $\mathrm{J\,C^{-2}\,m^{-1}\,s^2}$ (see inside the front cover for alternatives). B can be thought of as the density of magnetic lines of force permeating the medium. This density is increased if M adds to H (when $\kappa > 0$), but is decreased if M opposes H (when $\kappa < 0$). Materials for which κ (or χ) is positive are called *paramagnetic*; those for which κ (or χ) is negative are called *diamagnetic*.

Just as electric polarization is electric dipole moment per unit volume, the magnetization is the magnetic dipole moment per unit volume. The molecules might possess permanent magnetic dipole moments of magnitude m, which contribute to the magnetization an amount proportional to $m^2/3kT$, as in the electric case, eqn (24.1.13). An applied field can also induce a magnetic moment to an extent determined by the molecular *magnetizability* (symbol: ξ); a similar effect gave rise to the chemical shift of

NMR (Section 20.1(b)). The magnetic analogue of eqn (24.1.13) is therefore

$$\kappa = \mathcal{N}\mu_0(\xi + m^2/3kT). \tag{24.6.4}$$

The reason why χ is introduced can now be explained: the density in its definition cancels the density in this expression ($\mathcal{N} = \rho N_A/M_m$), and the cancellation eliminates the density dependence of the susceptibility.

The magnetic susceptibility may be measured with a *Gouy balance*, which consists of a sensitive balance from which the sample hangs in the form of a narrow cylinder, Fig. 24.29, and lies between the poles of a magnet. If the sample is paramagnetic, it is drawn into the field, and its apparent weight is increased: a diamagnetic sample tends to be expelled from the field. The balance is normally calibrated against a sample of known susceptibility. A typical paramagnetic volume susceptibility is about 10^{-3}, and a typical diamagnetic volume susceptibility is about $(-)10^{-5}$, Table 24.8.

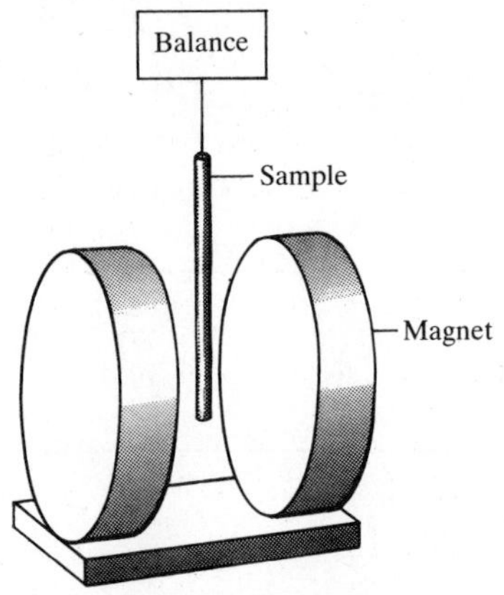

Fig. 24.29. The arrangement of the *Gouy balance* for measuring magnetic susceptibilities. A paramagnetic sample appears to weigh more and a diamagnetic sample appears to weigh less, when the magnetic field is on.

Table 24.8. Magnetic susceptibilities at 298 K

	$\kappa/10^{-6}$	$\chi/10^{-8}$
$H_2O(l)$	−90	−0.90
NaCl(s)	−13.9	−0.64
Cu(s)	−9.6	−0.107
$CuSO_4.5H_2O$	+176	+7.7

24.6 (b) The permanent magnetic moment

The permanent magnetic moment can be extracted from susceptibility measurements by plotting κ against $1/T$. The *Curie Law*,

$$\kappa = A + B/T, \tag{24.6.5}$$

is often obeyed, and by comparison with eqn (24.6.4) B can be identified with $\mathcal{N}\mu_0 m^2/3k$.

The permanent magnetic moment of a molecule is due to any unpaired electron spins in the molecule. We saw in Section 15.3 that the magnetic moment of an electron is proportional to its spin, and the square of the magnitude of the magnetic moment is proportional to the square of the magnitude of the angular momentum, $s(s+1)\hbar^2$:

$$m^2 = g_e^2\gamma_e^2\{s(s+1)\hbar^2\},$$

where g_e is the electron's g-factor and $\gamma_e = -e/2m_e$. In terms of the Bohr magneton, $\mu_B = e\hbar/2m_e$, this becomes

$$m^2 = g_e^2 s(s+1)\mu_B^2. \tag{24.6.6}$$

If there are several electron spins in each molecule they combine to a total spin S, and so then $s(s+1)$ in the last expression should be replaced by $S(S+1)$. It follows that the spin contribution to the magnetic susceptibility is

$$\kappa = \mathcal{N}g_e^2 S(S+1)\mu_0\mu_B^2/3kT. \tag{24.6.7}$$

This expression shows that κ and χ are positive, and so the spin magnetic moments contribute to the paramagnetic susceptibilities of materials. Note that the contribution decreases with increasing temperature because the thermal motion randomizes the spin orientations. In practice a contribution to the paramagnetism also arises from the *orbital* angular momenta of electrons: we have discussed the *spin-only* contribution.

Example 24.6

Calculate the volume and mass paramagnetic susceptibilities of a sample of a complex salt with three unpaired electrons at 298 K, given that its density is 3.24 g cm^{-3} and its molar mass is 200 g mol^{-1}.

• *Method*. Use eqn (24.6.7), and convert to mass susceptibility using eqn (24.6.2) and $\mathcal{N} = \rho N_A/M_m$, which gives

$$\chi = (N_A\rho^{\ominus}/3M_m kT)g_e^2\mu_0\mu_B^2 S(S+1).$$

A convenient quantity is

$$\rho^{\ominus}N_A g_e^2\mu_0\mu_B^2/3k = 6.3001\times10^{-3}\ \mathrm{K\,g\,mol^{-1}}.$$

Consequently,

$$\chi = 6.3001\times10^{-3}S(S+1)/M_r(T/\mathrm{K}),$$

and the numerical values can be obtained by substitution using $S=\frac{3}{2}$.

• *Answer*. Substituting the data gives

$$\chi = 6.3001\times10^{-3}\times(15/4)/(200\times298) = 3.96\times10^{-7}.$$

Then, from eqn (24.6.2),

$$\kappa = \rho\chi/\rho^{\ominus} = (3.96\times10^{-7})\times(3.24\times10^3\ \mathrm{kg\,m^{-3}}/1\ \mathrm{kg\,m^{-3}}) = 1.28\times10^{-3}.$$

• *Comment*. Note that χ and κ are dimensionless and (in this case) positive, indicating paramagnetism. Note also that the density is not needed for the calculation of χ.

• *Exercise*. Repeat the calculation for a complex with five unpaired spins, a density of $2.87\ \mathrm{g\,cm^{-3}}$, and a molar mass of $324.4\ \mathrm{g\,mol^{-1}}$ at 273 K. [$\chi = 6.2\times10^{-7}$, $\kappa = 1.79\times10^{-3}$]

At low temperatures some paramagnetic solids make a phase transition to a state in which large domains of spins align with parallel orientations. This cooperative alignment gives rise to a very strong magnetization, and is called *ferromagnetism*, Fig. 24.30(a). In other cases the cooperative effect leads to alternating spin orientations, and so they are locked into a low-magnetization arrangement, Fig. 24.30(b). This is called *antiferromagnetism* (and was encountered in connection with neutron diffraction, Section 23.6). The ferromagnetic transition occurs at the *Curie temperature*, and the antiferromagnetic transition occurs at the *Néel temperature*.

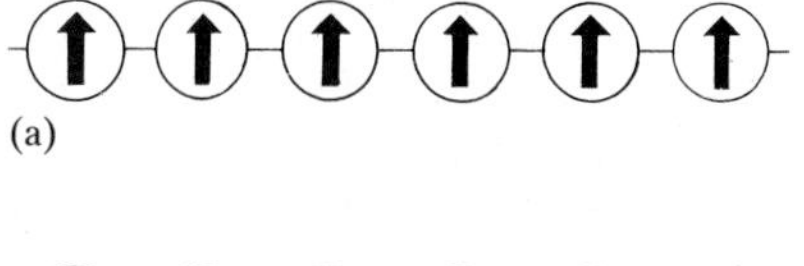

Fig. 24.30. In (a) a *ferromagnetic material* electron spins are locked into a parallel alignment over large domains; in (b) an *antiferromagnetic material* the electron spins are locked into an antiparallel arrangement (see also Fig. 23.34).

24.6 (c) Induced magnetic moments

Whereas an electric field polarizes a molecule by stretching it, a magnetic field magnetizes a molecule by inducing the circulation of electronic currents. These currents give rise to a magnetic field which usually opposes the applied field, so that the susceptibility is diamagnetic. In a few cases the induced field augments the applied field, and the susceptibility is then paramagnetic.

The great majority of molecules with no unpaired electron spins are diamagnetic. This is because the diamagnetic flow occurs within the ground state orbitals of the molecule, whereas the induction of a paramagnetic susceptibility depends on currents being stimulated by forcing the electrons to pass through excited state orbitals. When induced paramagnetism does occur it can be distinguished from spin paramagnetism by the fact that it is temperature independent: this is why it is called *temperature-independent paramagnetism* (TIP).

We can summarize these remarks as follows. All molecules have a diamagnetic component to their susceptibility, but this is dominated by spin paramagnetism if the molecules have unpaired electrons. In a few cases (where there are low-lying excited states) TIP is strong enough to make the molecules paramagnetic even though all their electrons are paired.

Further reading

Dipole moments and polarizabilities:

Determination of dipole moments. C. P. Smyth; *Techniques of chemistry* (A. Weissberger and B. W. Rossiter, eds.) IV, 397, Wiley-Interscience, New York, 1972.

Determination of dielectric constant and loss. W. E. Vaughan, C. P. Smyth, and J. C. Powles; *Techniques of chemistry* (A. Weissberger and B. W. Rossiter, eds.) IV, 431, Wiley-Interscience, New York, 1972.

Tables of experimental dipole moments. A. L. McClellan; W. H. Freeman & Co., San Francisco, 1963.

Intermolecular forces:

Molecular forces. B. Chu; Wiley-Interscience, New York, 1967.

Intermolecular forces: their origin and determination. G. C. Maitland, M. Rigby, E. B. Smith, and W. A. Wakeham; Clarendon Press, Oxford, 1981.

Molecular beams in chemistry. M. A. D. Fluendy and K. P. Lawley; Chapman and Hall, London, 1974.

Molecular beams. J. Ross (ed.); *Adv. chem. Phys*. **10** (1966).

Elastic scattering. U. Buck; *Adv. chem. Phys*. **30,** 313 (1975).

Liquids:

The structure of liquids. J. S. Rowlinson; *Essays in chemistry* (J. N. Bradley, R. D. Gillard, and R. F. Hudson, eds.), **1,** 1 (1970).

Properties of liquids and solutions. J. N. Murrell and E. A. Boucher; Wiley-Interscience, New York, 1982.

Computer simulation of liquids. D. Tildesley and M. P. Allen; Clarendon Press, Oxford, 1986.

Magnetism:

Introduction to magnetochemistry. A. Earnshaw; Academic Press, New York, 1968.

Instrumentation and techniques for measuring magnetic susceptibility. L. N. Mulay; *Techniques of chemistry* (A. Weissberger and B. W. Rossiter, eds.) IV, 431, Wiley-Interscience, New York, 1972.

Introductory problems

A24.1. Find the capacitance of a condenser when the space between the plates is filled with a substance of relative permittivity 35.5. When the space is evacuated, the capacitance is 6.2 pF.

A24.2. The molar polarizability of fluorobenzene vapour is a linear function of T^{-1}. It is 70.62 $cm^3\,mol^{-1}$ at 351.0 K and 62.47 at 423.2 K. Calculate the dipole moment and the polarizability of the molecule.

A24.3. At 0 °C the molar polarizability of ClF_3(l) is 27.18 $cm^3\,mol^{-1}$ and its density is 1.89 $g\,cm^{-3}$. Find its relative permittivity at 0 °C.

A24.4. A molecule like ClF_3 has five pairs of electrons around the central atom. If the electron pairs have a trigonal bipyramidal arrangement, two important possible structures are: (1) three equatorial F atoms, and (2) two axial and one equatorial F. Since ClF_3 has a non-zero dipole moment, which structure is the more likely.

A24.5. Assume that the linear dimerization of a molecule with dipole moment μ gives a molecule with dipole moment 2μ. What change does this make in the temperature dependent part of the polarization? What is the apparent moment of the dimer?

A24.6. Gaseous acetic acid forms planar hydrogen bonded dimers. The relative permittivity of pure liquid acetic acid is 7.14 at 290 K and increases with increasing temperature. Provide an explanation for this temperature dependence. What effect should isothermal dilution have on the relative permittivity of benzene solutions of acetic acid?

A24.7. Use the tabulated molar refractivities for 589 nm light to estimate the refractive index of diethylether. Its density is 0.715 $g\,cm^{-3}$ at 20 °C.

A24.8. The index of refraction of CH_2I_2 is 1.7320 for 656 nm light. Its density at 20 °C is 3.318 $g\,cm^{-3}$. Calculate its polarizability at this wavelength.

A24.9. The observed magnetic moment of $CrCl_3$ is 3.81

Bohr magnetons. How many unpaired electrons does chromium have in this compound?

A24.10. The ligand field in the complex compound $K_3Co(CN)_6$ is strong, and it is weak in the compound K_3CoF_6. Calculate the magnetic moments of these compounds in Bohr magnetons.

Problems

24.1. The dipole moment of toluene is 0.4 D; estimate the dipole moments of the three xylenes. Which answer can you be sure about?

24.2. The dipole moment of water is 1.85 D. Regard this as being the resultant of two bond dipoles at an angle of 104.5°, and predict the dipole moment of hydrogen peroxide as a function of the azimuthal angle between two OH groups (assume the OOH angle is 90°). The experimental dipole moment is 2.13 D: to what angle does this correspond?

24.3. What are the magnitudes of the fields that atoms and molecules are subjected to? Allow a water molecule ($\mu = 1.85$ D) to approach an ion. What is the favourable orientation of the molecule when the ion is an anion? Calculate the electric field experienced by the ion when the centre of the water dipole is at (a) 1.0 nm, (b) 0.3 nm, (c) 30 nm from its centre. Express your answer in V m^{-1}.

24.4. A water molecule is aligned by an external electric field of strength 1.0 kV m^{-1} and an argon atom ($\alpha' = 1.66 \times 10^{-24}$ cm^3) is brought up slowly from one side. At what separation is it favourable for the water molecule to flip over and point towards the approaching argon atom?

24.5. The relative permittivity of gaseous hydrogen halides is given by the expression $\varepsilon_r = 1 + \Delta/v$ in the range $0 \leqslant \theta/°\text{C} \leqslant 300$, the values of Δ being given below and v being a relative specific volume equal to unity at 273.15 K and 1 atm, so that $v = T/273.15$ K. What are their dipole moments and static polarizability volumes?

θ/°C	0	100	200	300
$10^3\Delta$(HCl)	4.3	3.5	3.0	2.6
$10^3\Delta$(HBr)	3.1	2.6	2.3	2.1
$10^3\Delta$(HI)	2.3	2.2	2.1	2.1

24.6. The polarizability volume of a water molecule at optical frequencies is 1.5×10^{-24} cm^3. Estimate the refractive index of water. The experimental value is 1.33: what may be the explanation of the discrepancy?

24.7. The relative permittivity of chloroform was measured over a range of temperatures with the following results:

θ/°C	−80	−70	−60	−40	−20	0	20
ε_r	3.1	3.1	7.0	6.5	6.0	5.5	5.0
ρ/g cm^{-3}	1.65	1.64	1.64	1.61	1.57	1.53	1.50

The melting point is −64 °C. Account for these results, and find the polarizability volume and dipole moment of the molecule.

24.8. The relative permittivities of methanol (m.p. −95 °C) corrected for density variation are given below. What molecular information can be deduced from these values? Take $\rho = 0.791$ g cm^{-3} at 20 °C.

θ/°C	−185	−170	−150	−140	−110
ε_r	3.2	3.6	4.0	5.1	67
θ/°C	−80	−50	−20	0	20
ε_r	57	49	42	38	34

24.9. Show that in a gas (where the refractive index is close to unity) the refractive index depends on the pressure as $1 + (2\pi\alpha'/kT)p$. Show how to deduce the polarizability volume from measurements of the refractive index of a gas.

24.10. The refractive index of benzene is constant (at 1.51) from 0.4 GHz up to about 0.55 GHz (note that these are microwave frequencies), but then shows a series of oscillations between 1.47 and 1.54. Throughout the same frequency range toluene shows a greater refractive index (of about 1.55), but it also shows oscillations in the same place as benzene, and additional oscillations between 1.52 and 1.56 in the vicinity of 0.4 GHz. Account for these observations.

24.11. The dipole moment of chlorobenzene is 1.57 D and its polarizability volume is 1.23×10^{-23} cm^3. Estimate its relative permittivity at 25 °C, taking its density as 1.1732 g cm^{-3}.

24.12. Use eqn (24.1.16) to calculate and plot the refractive index of water between 0 °C and 100 °C using the following density data:

θ/°C	0	20	40
ρ/g cm^{-3}	0.999 87	0.998 23	0.992 24
θ/°C	60	80	100
ρ/g cm^{-3}	0.983 24	0.971 83	0.958 41

Take $\alpha' = 1.50 \times 10^{-24}$ cm^3.

24.13. What is the refractive index of steam at 100 °C, 1 atm pressure? Estimate the refractive index of crystals of $CaCl_2$, NaCl, and solid Ar from the data in Table 24.3. The densities are 2.15, 2.163, and 1.42 g cm^{-3} respectively: $\alpha'(H_2O) \approx 1.5 \times 10^{-24}$ cm^3.

24.14. In his classic book on *Polar molecules*, Debye reports some early measurements on the molar polarizability of ammonia. From the selection below, find the dipole moment and polarizability volume of the molecule.

T/K	292.2	309.0	333.0
P_m/cm^3 mol^{-1}	57.57	55.01	51.22
T/K	387.0	413.0	446.0
P_m/cm^3 mol^{-1}	44.99	42.51	39.59

24.15. The refractive index of ammonia at 273 K and 1 atm (STP) is 1.000 379 (this is for yellow sodium light). What is the molar polarizability of the gas at (a) this temperature, (b) 292.2 K? Combine this information with the static molar polarizability at this temperature and deduce a value for the molecular dipole moment from these two measurements alone.

24.16. Show that the mean interaction energy of N atoms of diameter d interacting with a potential energy of the form $-C_6/R^6$ is given by $U=-2\pi N^2C_6/3Vd^3$, where V is the volume to which they are confined, and all effects of clustering are ignored. Hence find a connection between van der Waals parameter a and C_6 from $n^2a/V^2=(\partial U/\partial V)_T$. Investigate the consistency of this calculation in the case of argon showing that $C_6\approx\frac{3}{4}I\alpha'^2$.

24.17. The expression for the second virial coefficient B in terms of the intermolecular potential $V(R)$ is given by eqn (24.4.8). Suppose that the atoms have a distance of closest approach d, and outside that range they are attracted together by a $-C_6/R^6$ potential. Suppose also that when the atoms are not in contact, V/kT is so small that the exponential e^{-x} can be approximated by $1-x$. Find an expression for B in terms of C_6 and d.

24.18. Calculate the second virial coefficient for an intermolecular potential that has the following form:

$$V=\infty \quad \text{for} \quad R<\sigma_1, \qquad V=-\varepsilon \quad \text{for} \quad \sigma_1\leqslant R\leqslant\sigma_2,$$
$$V=0 \quad \text{for} \quad R>\sigma_2.$$

Explore the relation of this potential to the van der Waals a and b parameters.

24.19. Write a computer program for the evaluation of eqn (24.4.8) using the method outlined in *Example* 24.4. Use the program to evaluate the second virial coefficient of argon at various temperatures. Plot the result, and locate the Boyle temperature.

24.20. The Lennard–Jones $(n,6)$ potential gives the intermolecular potential energy. What is the *force* between the molecules? At what separation is the force zero?

24.21. Relate B to the polarizability volume of the gas atoms by connecting C_6 with α. Then estimate the value of B at 298 K for argon using data in Tables 4.9, 23.2, and 24.2.

24.22. Throughout the chapter we ignored totally the effect of gravitational forces between molecules. Is that justifiable? Estimate their relative importance for two argon atoms almost in contact.

24.23. The *cohesive energy density* is defined as $-U/V$, where U is the mean potential energy of attraction within the sample and V is its volume. Show that this quantity may be written as $-\frac{1}{2}\mathcal{N}^2\int V(R)\,d\tau$ where $\mathcal{N}$ is the number density of molecules and $V(R)$ their attractive potential energy, the integral ranging from d to infinity. Go on to show that the cohesive energy density of a uniform distribution of molecules that are held together by a van der Waals attraction of the form $-C_6/R^6$ is given by $(2\pi/3)(N_A^2/d^3M_m^2)\rho^2C_6$ where ρ is the density of the solid and M_m the molar mass of the molecules.

24.24. The cohesive energy density is approximately equal to the enthalpy of vaporization per unit volume. Estimate the molar enthalpy of vaporization of carbon tetrachloride given that $\rho\approx1.594\ \text{g cm}^{-3}$, $I\approx10\ \text{eV}$, and $\alpha'=10.5\times10^{-24}\ \text{cm}^3$.

24.25. Long ago, in Chapter 6, we saw that $(\partial U/\partial V)_T=n^2a/V^2$ for a van der Waals gas. If we identify U with the average cohesive energy we can use this relation to find a relation between a and C_6. Find this relation, and then express the critical constants in terms of d and C_6.

24.26. Molecular beams are an important way of obtaining information about intermolecular forces, and this and the next Problem give a very simple introduction to the complicated task of unravelling the richness of information they provide and extracting the piece required. Consider the collision between a hard-sphere molecule of radius R_1 and mass m_1 and an infinitely heavy spherical atom of radius R_2 which is also an impenetrable sphere. (Imagine them as H and I, for example.) Plot the scattering angle θ as a function of the impact parameter b. Do the calculation on the basis of simple geometrical considerations.

24.27. The scattering characteristics of atoms depend on the energy of the collision. We can model the situation as follows. Let both atoms always behave as impenetrable hard spheres, but let the effective radius of the heavy atom vary with the relative speed of approach of the light atom. Suppose its effective radius depends on the speed v as $R_2\exp(-v/v^*)$, where v^* is some constant. With this form the effective radius of the heavy atom is R_2 at slow speeds of approach of the other atom, and less when the speed of approach is great. Take $R_1=\frac{1}{2}R_2$ for simplicity, and an impact parameter $b=\frac{1}{2}R_2$ and plot the scattering angle as a function of (a) speed, (b) kinetic energy of approach.

24.28. The Madelung constant is defined in eqn (24.5.3). One simple case where it can be calculated without much effort is in the case of a line of alternating singly charged positive and negative ions separated by R. What is its value for this array?

24.29. The internal energy of an ionic crystal, as given by eqn (24.5.7), requires some knowledge of the parameter R^*. This can be obtained from a knowledge of the isothermal compressibility of the crystal, and this Problem explores this relation. First, from the thermodynamic equation of state $(\partial U/\partial V)_T=T(\partial p/\partial T)_V-p$, evaluated at absolute zero, show that $1/\kappa$ is proportional to $\partial^2U/\partial V^2$, and then deduce R^* in terms of V, κ, and R_0. (At some stage you may find it helpful to write $V=cR_0^3$, where c is a constant that can be eliminated at a later stage.)

24.30. Set up a Born–Haber cycle for determining the cohesive energy of KCl and evaluate it from the data in Chapter 4 and $\Delta H_{sub}(K)=82.6\ \text{kJ mol}^{-1}$. Calculate the cohesive energy of KCl from eqn (24.5.7), using $\kappa = 1.1 \times 10^{-5}\ \text{atm}^{-1}$ for the isothermal compressibility and $\rho = 1.984\ \text{g cm}^{-3}$.

24.31. The ionic radii of Na^+ and F^- ions are 102 pm and 133 pm respectively. Calculate the molar internal lattice energy of NaF? Use $R^* \approx 29$ pm.

24.32. The ionic radii of Cs^+ and Cl^- are respectively 182 pm and 181 pm. Calculate the molar internal lattice energy of CsCl. Use $R^* \approx 40$ pm.

24.33. The magnetizability ξ and then the mass and volume magnetic susceptibilities (χ and κ) can be calculated from the wavefunctions of molecules. For instance, for a one-electron atom $\xi \approx -(e^2/6m_e)\langle r^2 \rangle$. Calculate ξ and χ for hydrogen atoms.

24.34. Calculate the paramagnetic and total susceptibilities (χ) for a hydrogen atom at 25 °C.

24.35. As we saw in Chapters 21 and 22, the NO molecule is peculiar in so far as it has thermally accessible electronically excited states. It also has an unpaired electron, and so may be expected to be paramagnetic. In fact, its ground state is not paramagnetic. This is because the magnetic moment due to the orbital motion of the electron around the molecular axis cancels the spin magnetic moment. The first excited electronic state (at $121.1\ \text{cm}^{-1}$) is paramagnetic because the orbital moment augments, rather than cancels, the spin moment, and its magnetic moment is 2 Bohr magnetons. Since this upper state is thermally accessible, the paramagnetic susceptibility of the molecule will show a peculiar temperature dependence even at room temperature. Calculate the mass paramagnetic susceptibility of NO and plot it as a function of temperature. What is its value at 298 K?

25

The structures and properties of macromolecules

Learning objectives

After careful study of this chapter you should be able to:

(1) List the techniques available for the determination of the shapes and sizes of macromolecules, Section 25.1.

(2) Use *osmotic pressure* measurements to determine the molar masses of macromolecules, Section 25.1(a) and Example 25.1.

(3) Distinguish between *number average RMM*, eqn (25.1.7), and *mass average RMM*, eqn (25.1.23) and Example 25.6.

(4) Explain the terms *θ-temperature* and *θ-solution*, Section 25.1(a).

(5) Explain the terms *monodisperse* and *polydisperse* and indicate how they relate to the properties of synthetic polymers, Section 25.1(b).

(6) Account for the *Donnan effect* and deduce and use expressions for the effect of ionic charges on osmotic equilibria, eqn (25.1.12) and Example 25.2.

(7) Describe the operation of the *ultracentrifuge* and relate the rate of sedimentation to the shapes and molar masses of macromolecules, eqn (25.1.17) and Examples 25.3 and 25.4.

(8) Describe how to obtain molar masses from observations on *sedimentation equilibria*, eqn (25.1.18).

(9) Explain the basis and applications of the techniques of *electrophoresis* and *gel filtration*, Sections 25.1(g and h).

(10) Define *intrinsic viscosity*, eqn (25.1.20), describe how it is measured, and explain how to use the information to determine molar mass, Example 25.5.

(11) Describe the applications of *light scattering* measurements to the determination of the molar masses and the shapes of macromolecules, Section 25.1(j).

(12) Describe the information about macromolecules available from magnetic resonance, and the use of *shift reagents*, *broadening reagents*, and *spin labels*, Section 25.1(k).

(13) Explain the terms *primary*, *secondary*, *tertiary*, and *quaternary structure*, and the meaning of *denaturation*, Section 25.2.

(14) Describe a *random coil* in terms of a random walk, and use the distribution function, eqn (25.2.1), to calculate its properties, Example 25.7.

(15) Describe the structure of the *peptide link* and the *α-helix*, and describe how *conformational energy analysis* is used to account for the conformations of macromolecules, Section 25.2(c).

(16) Explain the term *colloid* and distinguish between *sols*, *aerosols*, and *emulsions*, and between *lyophilic* and *lyophobic* colloids, Section 25.3(a).

(17) Explain the term *hydrophobic interaction*, Section 25.3(c).

(18) Account for the stability of colloids in terms of the *electric double-layer*, Section 25.3(d).

(19) Define *surface excess*, eqn (25.4.3), and derive and use the *Gibbs surface tension equation*, eqn (25.4.6).

(20) Account for the elasticity of rubber in terms of the *conformational entropy* of a random coil, Appendix 25.1.

Introduction

There are macromolecules everywhere, inside us and outside us. Some are natural: these include the proteins, cellulose, DNA, and so on. Others are synthetic, and include the polymers such as nylon and polystyrene formed by stringing together and cross-linking smaller units, the *monomers*. Now that we are succeeding in synthesizing proteins, however, the distinction between natural and synthetic is becoming old-fashioned. Molecules and atoms sometimes swarm together under the influence of intermolecular forces, and the large conglomerates, the *colloids*, behave like macromolecules. Varieties of colloids, the sols, emulsions, and foams, are intimately involved in the processes of the world. Life in all its forms, from its intrinsic nature to its technological interaction with its environment, is the chemistry of macromolecules.

Although many concepts of physical chemistry apply to macromolecules as well as small molecules, macromolecules do give rise to special questions and problems. These include the determination of their sizes, the shapes and the lengths of polymer chains, and the large deviations from ideality of their solutions. These special characteristics are the ones we concentrate on here. The synthesis of macromolecules, and the relation of the rate of their formation to their physical properties, are treated in Section 29.1(b).

25.1 Size and shape

X-ray diffraction, Chapter 23, can reveal the position of almost every atom even in highly complex molecules. However, there are several reasons why other techniques must also be used. In the first place, the sample might be a mixture of polymers of different chain lengths and extents of cross-linking, in which case sharp X-ray images are unobtainable. Even if all the molecules in the sample are identical, it might prove impossible to obtain a single crystal. Furthermore, although the work on enzymes, proteins, and DNA has shown how immensely stimulating the data can be, the information is incomplete. For instance, what can be said about the shape of the molecule in its natural environment, the cell? What can be said about the way its shape changes in response to its environment? Shape and function go hand in hand, and it is essential to know how biological macromolecules, which often carry both acidic and basic groups, respond to the pH of the medium. It is often important to follow the *denaturation* of a macromolecule to a less orderly form: denaturation is often accompanied by loss of function, but it is sometimes an essential step in the fulfilment of function, as in the replication of DNA.

A direct method for the observation of the sizes and shapes of macromolecules is the *electron microscope*. Resolution of about 500 pm may now be achieved, and so in principle the fine details of the shape of a molecule may be discerned, Fig. 25.1. However, there are severe limitations: the sample is in a highly unnatural shape, for in order to obtain an image a replica of the molecule must be made. The replica is often a cast taken by spraying the sample with metal atoms, either directly (the method used to obtain Fig. 25.1) or after coating it with small protein or detergent molecules. The *scanning electron microscope* differs from the transmission microscope in that it obtains an image by scanning a metal cast with a spot-focused electron beam and monitoring the intensity of the electrons

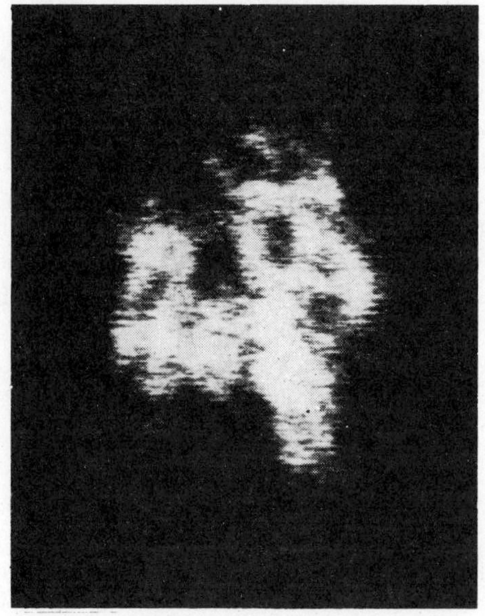

Fig. 25.1. A scanning electron microscope photograph of (deformed) haemoglobin at a magnification of 2.75×10^6 diameters. (Provided by Professor A. V. Crewe.)

ejected by and scattered from the beam. All these methods, however dramatic the pictures and useful the information they reveal, suffer from the limitation that the sample is modified.

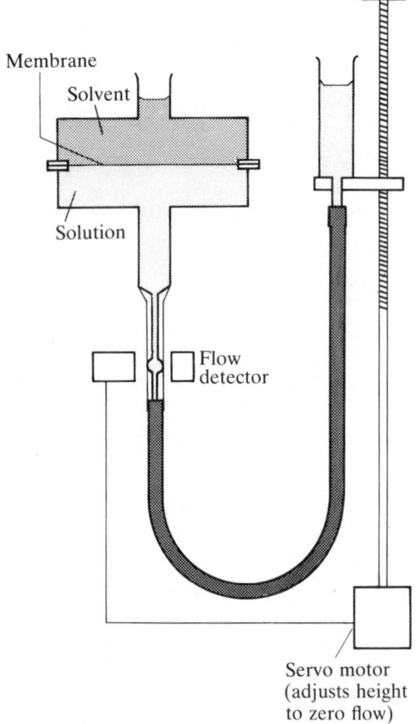

Fig. 25.2. One version of an osmometer used to determine the molar masses of macromolecules. The pressure on the solution is adjusted until there is no flow through the semipermeable membrane; its value is the osmotic pressure.

25.1 (a) Osmosis

The classical methods of determining the RMM of molecules employ the colligative properties (Chapter 8). In the case of macromolecules, where the *number* of molecules in solution may be very small even though their mass may be appreciable, only osmometry, Fig. 25.2, is sufficiently sensitive.

The fundamental relation we need is the extension of the van't Hoff equation to non-ideal solutions, eqn (8.3.11):

$$\Pi/[\mathrm{P}] = RT\{1 + B[\mathrm{P}] + \ldots\}, \qquad (25.1.1)$$

[P] is the concentration ($\mathrm{mol\,dm^{-3}}$) of the macromolecule P, and Π is the osmotic pressure. Since [P] is related to the mass concentration c_P by $[\mathrm{P}] = c_\mathrm{P}/M_\mathrm{m}$, where M_m is the molar mass of P, another version of this equation is

$$\Pi/c_\mathrm{P} = (RT/M_\mathrm{m})\{1 + (B/M_\mathrm{m})c_\mathrm{P} + \ldots\}. \qquad (25.1.2)$$

Therefore, by plotting Π/c_P against c_P and extrapolating to $c_\mathrm{P} = 0$, M_m can be obtained from the intercept and B, the *osmotic virial coefficient*, can be obtained from the slope (Fig. 25.3). The procedure was illustrated in *Example* 8.7.

Macromolecules give strongly non-ideal solutions. This is partly because, being so large, they displace a large quantity of solvent instead of dissolving simply by replacing individual solvent molecules with negligible disturbance. Furthermore, their great bulk means that there is a large excluded volume: a molecule is unable to swim freely through the solution since it is excluded from the regions occupied by others. In thermodynamic terms, this implies that the entropy change is important when a macromolecule dissolves.

Another complication is that there may also be a significant enthalpy of solution on account of the interaction of the solvent with a large number of component monomer units in the polymer.

The most straightforward thermodynamic justification of the osmotic virial expansion, and an interpretation of B, can be obtained by returning to the basic expressions for the osmotic pressure in terms of the chemical potential. Refer back to eqn (8.3.9): if $RT\ln x_A$ is replaced by $RT\ln a_A$, where a_A is the solvent activity, then that equation becomes

$$-RT\ln a_A = \int_p^{p+\Pi} V_m^* \, dp. \qquad (25.1.3)$$

The integral is equal to ΠV_m^* if the solvent is incompressible, and so the osmotic pressure is given by

$$\Pi V_m^* = -RT\ln a_A. \qquad (25.1.4)$$

If the solution is ideal, $a_A = x_A = 1 - x_P$, where x_P is the mole fraction of the solute macromolecule. When $x_P \ll 1$ we can replace $\ln(1-x_P)$ by $-x_P$; hence, $\ln a_A = -x_P$. This leads to the van't Hoff equation. When the solution is non-ideal, we suppose that $\ln a_A = -x_P$ is the start of a series:

$$\ln a_A = -x_P\{1 + B'x_P + \ldots\}, \qquad (25.1.5)$$

Then substitution of this expansion into eqn (25.1.4) leads to the osmotic virial expansion. The virial expansion for $\ln a_A$ is not a trivial result (the Debye–Hückel theory, Chapter 11, leads to the result that for solutions of *electrolytes* the activity expansion depends on the *square root* of the concentration of the ions), but it is confirmed by the *MacMillan–Meyer theory* of solutions of non-electrolytes.

The coefficient B arises largely from the effect of excluded volume. This is reminiscent of a van der Waals gas, for which $B = b - a/RT \approx b$ when the excluded volume effect is dominant. If we imagine a solution of a macromolecule as being built by the successive addition of macromolecules, each one being excluded by the ones that preceded it, the value of B turns out to be

$$B = \tfrac{1}{2}N_A v_P, \qquad (25.1.6)$$

v_P being the excluded volume due to a single molecule.

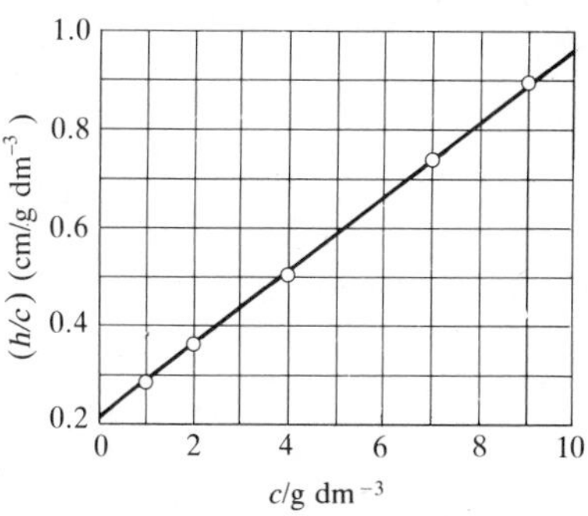

Fig. 25.3. The plot of h/c_P against c_P used for the determination of molar mass (from the intercept) and the osmotic virial coefficient (from the slope). Note that all axes are unitless numbers, and that the intercept and slope are also unitless. The manner of converting the numbers is described in *Example* 25.1.

Example 25.1

Use the information in *Example* 8.7 to estimate the volume of the polymer molecules regarded as impenetrable spheres.

• *Method*. The excluded volume of spherical molecules of volume v_{mol} is $v_P = 8v_{mol}$ because the minimum separation of two spheres is the sum of their radii. Estimate v_P from the osmotic virial coefficient B using eqn (25.1.6). Identify B from the slope of the graph plotted in Fig. 8.13. The practical form of eqn (25.1.2), as used in *Example* 8.7, is

$$h/c_P = (RT/\rho g M_m)\{1 + (B/M_m)c_P\},$$

and if we write $a = (RT/\rho g M_m) \times (\text{cm/g dm}^{-3})$, $b = (B/M_m) \times (\text{g dm}^{-3})$, this has the form

$$y = a + abx; \quad y = (h/c_P)/(\text{cm/g dm}^{-3}), \quad x = c_P/(\text{g dm}^{-3}).$$

Therefore, the slope is ab; the intercept a was found in *Example* 8.7 to be 0.21.

• *Answer*. The slope of the straight line in Fig. 8.13 is 0.073; therefore $b = 0.073/0.21 = 0.35$. Since $B/M_m = b/(\text{g dm}^{-3})$, it follows that

$$B/M_m = 0.35/(\text{g dm}^{-3}),$$

so that $B = 0.35 \times (123 \times 10^3 \text{ g mol}^{-1})/\text{g dm}^{-3} = 43 \times 10^3 \text{ dm}^3 \text{ mol}^{-1}$. Therefore, from eqn (25.1.6),

$$v_P = 2B/N_A = 1.4 \times 10^{-22} \text{ m}^3.$$

The molecular volume is therefore approximately $1.8 \times 10^4 \text{ nm}^3$.

• *Comment*. The radius of the molecule is therefore approximately 16 nm. The following material shows how the shapes of the molecules can be determined by different kinds of experiments.

• *Exercise*. Another sample in the same solvent resulted in the following heights of solution at the same temperature: 0.22, 0.53, 1.39, 3.32, 5.02 cm. Calculate the RMM and molecular volume of the solute. [144 000, 9×10^{-23} m^3]

There are both entropic (excluded volume) and enthalpic (attractions and repulsions) contributions to non-ideality, and for most solute/solvent systems there is a unique temperature (which is not always experimentally attainable) where their effects cancel. This temperature (the analogue of the Boyle temperature for real gases) is called the *θ-temperature*, and at it the osmotic virial coefficient B is zero. A solution at its θ-temperature is called a *θ-solution*. The cancellation of effects at the θ-temperature means that the solution then behaves ideally even though the concentration is not low, and its thermodynamic and structural properties are easier to describe. As an example, the θ-temperature for polystyrene in cyclohexane is around 306 K, the exact value depending on the molar mass of the polymer.

25.1 (b) Number average RMM

A further complication affecting the interpretation of the results obtained from osmometry is that the sample may have molecules covering a range of molar masses. A pure protein is a well-defined species with a single, definite molar mass: it is *monodisperse*. (But even then there may be small variations, such as one amino acid replacing another, depending on the source of the sample.) On the other hand, a synthetic polymer is a mixture of various chain lengths, and the sample contains a range of molar masses: it is *polydisperse* (see Section 29.1(b)). In the latter case osmometry gives the *number average RMM* (symbol $\langle M_r \rangle_N$) which is defined as follows. Suppose there are N_i molecules with RMM $M_{r,i}$, and N molecules in all. Then the number average RMM is the RMM of each one weighted by the numerical proportion, N_i/N, having that RMM:

$$\langle M_r \rangle_N = \sum_i (N_i/N) M_{r,i} = (1/N) \sum_i N_i M_{r,i}. \qquad (25.1.7)$$

This is the same type of average taken when calculating the mean size of a family or score in a test, etc. We shall meet more complicated averages shortly, but the reason why osmometry leads to the number average RMM is that osmosis is a colligative property: it depends on the number, not the nature, of the solute molecules.

25.1 (c) Polyelectrolytes and dialysis

A third complication is the presence of charge on some types of macromolecule. Some polymers are strings of acid groups (as in polyacrylic acid, $\text{+CH}_2\text{CHCOOH+}_n$), or strings of bases (as in nylon, $\text{+NH(CH}_2)_6\text{-NHCO(CH}_2)_4\text{CO+}_n$), and proteins have both acid and base groups. Macromolecules may therefore be *polyelectrolytes*, and, depending on their state of ionization, *polyanions*, *polycations*, or *polyampholytes* (of mixed cation and anion character). One consequence of dealing with polyelectrolytes is that it is necessary to know the extent of ionization before osmotic data can be interpreted. For example, suppose the sodium salt of a polyelectrolyte dissociates into ν sodium ions and a single polyanion $P^{\nu-}$, then it gives rise to $\nu+1$ particles for each salt molecule that dissolves, and if we guess that $\nu = 1$ when in fact $\nu = 10$, the estimate of the RMM will be seriously wrong (by about an order of magnitude). The way out of this difficulty can be found by considering another consequence of dealing with charged macromolecules.

Suppose that the solution of $Na_\nu P$ also contains added NaCl, and that it is in contact through a semipermeable membrane (such as cellophane or a cell wall) with another salt solution. Furthermore, suppose that the membrane is permeable to the solvent *and* to the salt ions, but not to the polyanion. This arrangement, the basis of *dialysis*, is one that actually occurs in living systems, where osmosis is an important feature of cell operation. We can investigate the effect of the salt on the osmotic pressure, which is called the *Donnan effect*, and at the same time solve the problem of coping with the unknown extent of ionization of the polyelectrolyte. An effect of the salt can be anticipated because the anions and cations cannot migrate through the membrane to an arbitrary extent, since electrical neutrality must be preserved on both sides: if an anion migrates, a cation must accompany it.

Suppose that $Na_\nu P$ is at a concentration [P] on one side of the membrane, and that sodium chloride is added to each side. On the left (L) there are $P^{\nu-}$, Na^+, and Cl^- ions; on the right (R) there are Na^+ and Cl^- ions. The condition for equilibrium is that the chemical potential of NaCl should be the same on both sides of the membrane, and so a net flow of Na^+ and Cl^- ions occurs until $\mu_{NaCl}(L) = \mu_{NaCl}(R)$. This requires

$$\mu^\ominus_{NaCl} + RT\ln\{a_{Na^+}a_{Cl^-}\}_L = \mu^\ominus_{NaCl} + RT\ln\{a_{Na^+}a_{Cl^-}\}_R. \quad (25.1.8)$$

If activity coefficients are disregarded this is equivalent to the condition

$$\{[Na^+][Cl^-]\}_L = \{[Na^+][Cl^-]\}_R. \quad (25.1.9)^\circ$$

The sodium ions are supplied by the polyelectrolyte as well as the added salt. The conditions for electrical neutrality are

$$[Na^+]_L = [Cl^-]_L + \nu[P]; \qquad [Na^+]_R = [Cl^-]_R. \quad (25.1.10)$$

These conditions can now be combined to obtain expressions for the differences in ion concentrations across the membrane:

$$[Na^+]_L - [Na^+]_R = \frac{\nu[P][Na^+]_L}{[Na^+]_L + [Na^+]_R} = \frac{\nu[P][Na^+]_L}{2[Cl^-] + \nu[P]} \quad (25.1.11a)^\circ$$

$$[Cl^-]_L - [Cl^-]_R = \frac{\nu[P][Cl^-]_L}{[Cl^-]_L + [Cl^-]_R} = \frac{-\nu[P][Cl^-]_L}{2[Cl^-]}. \quad (25.1.11b)^\circ$$

In obtaining these expressions we have used the definition

$$[\mathrm{Cl^-}] = \tfrac{1}{2}\{[\mathrm{Cl^-}]_L + [\mathrm{Cl^-}]_R\}.$$

$[\mathrm{Cl^-}]$ can be measured by analysing the two solutions.

The final step is to note that the osmotic pressure depends on the difference of the numbers of particles on each side of the membrane, and so the van't Hoff equation, $\Pi = RT[\text{solute}]$, becomes

$$\begin{aligned}\Pi &= RT\{([\mathrm{P}] + [\mathrm{Na^+}]_L + [\mathrm{Cl^-}]_L) - ([\mathrm{Na^+}]_R + [\mathrm{Cl^-}]_R)\}\\ &= RT[\mathrm{P}]\{1 + B[\mathrm{P}]\}, \qquad B = \nu^2/(4[\mathrm{Cl^-}] + \nu[\mathrm{P}]).\end{aligned} \qquad (25.1.12)^\circ$$

When the amount of added salt is so great that $B[\mathrm{P}] \ll 1$, this equation reduces to $\Pi = RT[\mathrm{P}]$, a result *independent* of the value of ν. This means that if the osmotic pressure is measured in the presence of high concentrations of salt, then the RMM may be obtained unambiguously.

A second point arises from eqn (25.1.11). There is often interest in the extent to which ions are bound to macromolecules, especially when a membrane (such as a cell wall) separates two regions. The equations show, however, that cations will dominate the anions in the compartment containing the polyanion (because the concentration difference is positive for Na^+ and negative for Cl^-) simply as a result of the equilibrium and electroneutrality conditions. Therefore, it would be wrong to conclude without further evidence that an increased Na^+ concentration in the macromolecule solutions implies that Na^+ ions bind to it.

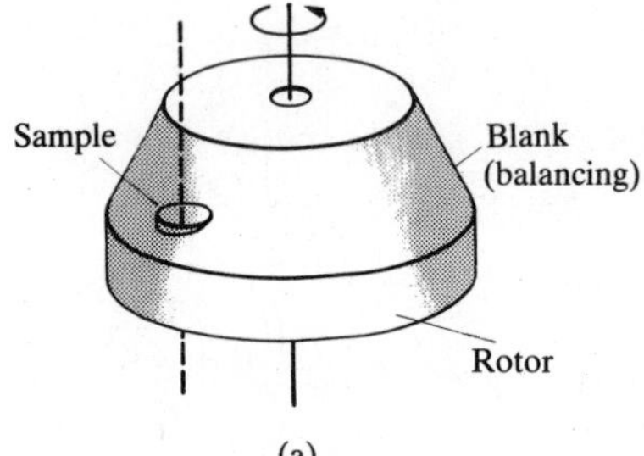

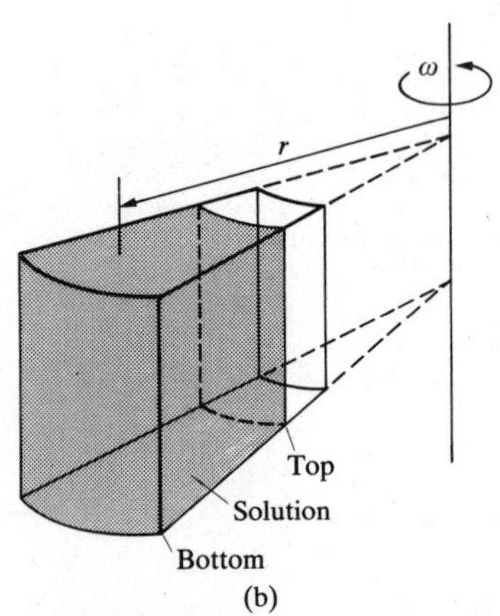

Fig. 25.4. (a) An ultracentrifuge head. The sample on one side is balanced by a blank diametrically opposite. (b) Detail of the sample cavity: the 'top' surface is the inner surface, and the centrifugal force causes sedimentation towards the outer surface; a particle at a radius r experiences a force of magnitude $mr\omega^2$.

Example 25.2

Two equal volumes of 0.200 M NaCl(aq) are separated by a membrane. A macromolecule with $M_r = 55\,000$, which cannot pass through the membrane, is added as its sodium salt Na_6P to a concentration of 50 g dm^{-3} to the left-hand compartment. What are the equilibrium concentrations of Na^+ and Cl^- in each compartment?

• *Method*. Use eqn (25.1.11) to calculate the concentration differences and eqn (25.1.10) to calculate their sum as

$$[\mathrm{Na^+}]_L + [\mathrm{Na^+}]_R = [\mathrm{Cl^-}]_L + [\mathrm{Cl^-}]_R + \nu[\mathrm{P}] = 2[\mathrm{Cl^-}] + \nu[\mathrm{P}].$$

Then use $[\mathrm{Cl^-}] = 0.200$ M.

• *Answer*. Since $[\mathrm{P}] = (50\ \mathrm{g\,dm^{-3}})/(55\,000\ \mathrm{g\,mol^{-1}}) = 9.1\times10^{-4}$ M, eqn (25.1.11) gives

$$\begin{aligned}[\mathrm{Na^+}]_L - [\mathrm{Na^+}]_R &= 6\times(9.1\times10^{-4}\ \mathrm{M})[\mathrm{Na^+}]_L/\{0.400\ \mathrm{M} + 6\times(9.1\times10^{-4}\ \mathrm{M})\}\\ &= 1.35\times10^{-2}[\mathrm{Na^+}]_L\end{aligned}$$

and the sum above gives

$$[\mathrm{Na^+}]_L + [\mathrm{Na^+}]_R = \{0.400 + 6\times(9.1\times10^{-4})\}\ \mathrm{M} = 0.405\ \mathrm{M}.$$

Solving these equations give $[\mathrm{Na^+}]_L = 0.204$ M and $[\mathrm{Na^+}]_R = 0.201$ M. Then $[\mathrm{Cl^-}]_R = [\mathrm{Na^+}]_R = 0.201$ M and $[\mathrm{Cl^-}]_L = [\mathrm{Na^+}]_L - 6[\mathrm{P}] = 0.199$ M.

• *Comment*. Note how the Na^+ accumulates slightly in the compartment containing the macromolecule.

• *Exercise*. Repeat the calculation for 0.300 M NaCl(aq), and a polyelectrolyte $Na_{10}P$ of RMM 33 000 at a concentration 50 g dm^{-3}. [Na: 0.311 M, 0.304 M]

25.1 (d) Sedimentation

In a gravitational field heavy particles settle towards the foot of a column of solution. The rate of this *sedimentation* depends on the strength of the field, the masses of the particles, and the shapes of the molecules, spherical (and compact molecules in general) sedimenting faster than rod-like or extended ones. For example, DNA helices sediment much faster when they are denatured to a random coil, and so sedimentation rates can be used to study denaturation. When the sample is at equilibrium the particles are dispersed over a range of heights in accord with the Boltzmann distribution (because the gravitational field competes with the stirring effect of thermal motion). The spread of heights depends on the masses of the molecules, and so the equilibrium distribution is another way of determining molecular mass.

Sedimentation is normally very slow, but it can be accelerated by replacing the gravitational field by a centrifugal field. This can be achieved in an *ultracentrifuge*, which is essentially a cylinder that can be rotated at high speed about its axis, the sample being held in a cell near its periphery, Fig. 25.4. Modern ultracentrifuges can produce accelerations equivalent to about 10^5 times that of gravity ('10^5 g'). Initially the sample is uniform, but the 'top' (innermost) boundary of the solute moves outwards as sedimentation proceeds, and its rate of movement can be monitored by making use of the effect of the concentration on the refractive index of the sample. Since there is a sharp change of refractive index between the solution and the solvent left behind by the sinking solute, the sample behaves like a prism and bends the light passing through it. The *Schlieren optical system* (which is also used to study air flow in wind tunnels and shock tubes) turns a refractive index gradient into a detectable image, Fig. 25.5. Alternatively, in the *interference technique* the concentration profile is monitored through the effect of refractive index on the interference on two beams of light, one coming through the sample and the other through a blank, Fig. 25.6.

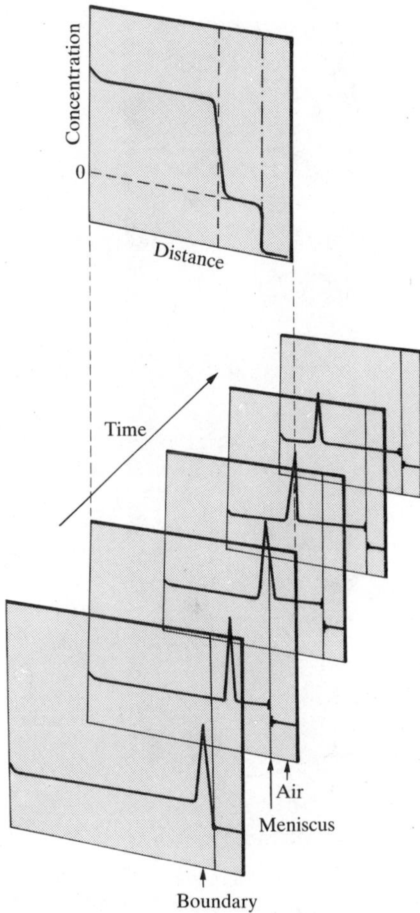

Fig. 25.5. A Schlieren photograph indicates regions where the refractive index is changing. This set, corresponding to a series of times, shows the sedimentation of the solute; one has been interpreted in terms of the concentration profile through the cell.

25.1 (e) The rate of sedimentation

A solute particle of mass m has an *effective mass* $m_{\text{eff}} = bm$ on account of the buoyancy of the medium, with $b = 1 - \rho v_s$. ρ is the solution density and v_s is the solute's specific volume (actually its *partial specific volume*, in the sense described in Section 8.1), its volume per unit mass (so that $\rho m v_s$ is the mass of solvent displaced by the solute). The solute particles at a distance r from the axis of a rotor spinning at an angular velocity ω experience a centrifugal force $m_{\text{eff}} r\omega^2$. The acceleration outwards this causes is countered by a frictional force proportional to the speed, s, of the particles through the medium. This force is written fs, f being the *friction coefficient*. The particles adopt a *drift speed* which is found by equating the two forces: $m_{\text{eff}} r\omega^2 = fs$ implies

$$s = m_{\text{eff}} r\omega^2/f = mbr\omega^2/f. \qquad (25.1.13)$$

The drift speed depends on the angular velocity and the radius, and it is convenient to focus on the ratio $S = s/r\omega^2$, which is called the *sedimentation constant*. Then, since the mass of an individual molecule is related to the molar mass through $m = M_m/N_A$,

$$S = s/r\omega^2 = M_m b/fN_A. \qquad (25.1.14)$$

Example 25.3

The sedimentation of bovine serum albumin (BSA) was monitored at 25 °C. The initial radius of the solute surface was 5.50 cm and during centrifugation at 56 850 r.p.m. it receded as

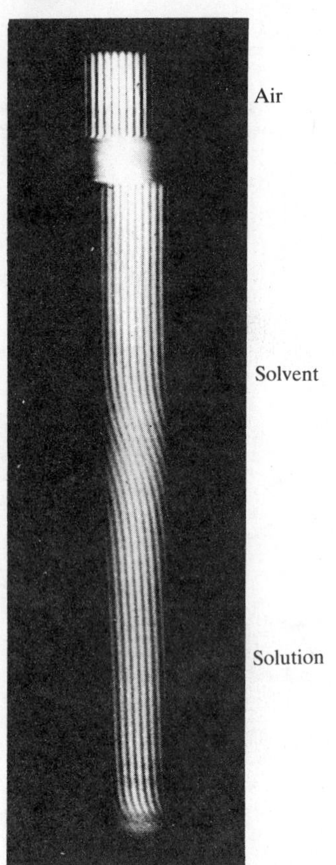

Fig. 25.6. An interference photograph also shows the concentration profile, but in a different manner. The regions of air, solvent, and solution have been marked. (Provided by Professor D. Freifelder, from his *Physical biochemistry*, W. H. Freeman & Co., 1976.)

follows:

t/s	0	500	1000	2000	3000	4000	5000
r/cm	5.50	5.55	5.60	5.70	5.80	5.91	6.01

Calculate the sedimentation coefficient.

• *Method*. From eqn (25.1.14), with $s = dr/dt$,

$$dr/dt = r\omega^2 S,$$

which integrates to

$$\ln(r/r_0) = \omega^2 St.$$

Therefore, a plot of $\ln(r/r_0)$ against t should be a straight line of slope $\omega^2 S$. Use $\omega = 2\pi\nu$, where ν is cycles/second.

• *Answer*. Draw up the following table:

t/s	0	500	1000	2000	3000	4000	5000
$100 \ln\{r/r_0\}$	0	0.900	1.80	3.57	5.31	7.19	8.87

The plot has slope $1.7_9 \times 10^{-5}$; and so $\omega^2 S = 1.7_9 \times 10^{-5}\,\mathrm{s}^{-1}$. Since $\omega = 2\pi \times (56\,850/60)\,\mathrm{s}^{-1} = 5.95 \times 10^3\,\mathrm{s}^{-1}$, it follows that $S = 5.1 \times 10^{-13}$ s.

• *Comment*. We develop this result below. The unit 10^{-13} s is sometimes called the *svedberg* and denoted S; in this case $S = 5.1$ S. Accurate results are obtained by extrapolating to zero concentration.

• *Exercise*. Calculate the sedimentation constant given the following data (the other conditions being the same as above):

t/s	0	500	1000	2000	3000	4000	5000
r/cm	5.65	5.68	5.71	5.77	5.84	5.90	5.97

[3.08 S]

In order to make progress we need to know f. For a spherical particle of radius a in a solvent of viscosity η it is given by *Stokes' relation*, $f = 6\pi a\eta$. Therefore, for spherical molecules,

$$S = bM_m/6\pi\eta a N_A, \qquad (25.1.15)$$

and S may be used to determine either M_m or a. If the molecules are not spherical, different values of f must be used, Box 25.1. As always when dealing with macromolecules, the measurements must be extrapolated to zero concentration in order to avoid the complications that arise from the interference between the bulky molecules.

In order to use sedimentation rates to determine RMM it would appear necessary to know the molecular radius a (and in general the frictional coefficient f). Fortunately, the problem can be avoided by drawing on the *Stokes–Einstein relation* (Section 27.2(a)) between f and the *diffusion coefficient*, D:

$$f = kT/D. \qquad (25.1.16)$$

The diffusion coefficient measures the rate at which molecules spread down a concentration gradient, and can be measured by observing the rate at which a boundary spreads, or the rate at which a more concentrated solution diffuses into a less concentrated one. Some typical values are given

Box 25.1 Frictional coefficients and molecular geometry

Sphere; radius a, $c = a$	f_0
Prolate ellipsoid; major axis $2a$, minor axis $2b$, $c = (ab^2)^{1/3}$	$\left\{\dfrac{(1-b^2/a^2)^{1/2}}{(b/a)^{2/3}\ln\{[1+(1-b^2/a^2)^{1/2}]/(b/a)\}}\right\}f_0$
Oblate ellipsoid; major axis $2a$, minor axis $2b$, $c = (a^2b)^{1/3}$	$\left\{\dfrac{(a^2/b^2-1)^{1/2}}{(a/b)^{2/3}\arctan[(a^2/b^2-1)^{1/2}]}\right\}f_0$
Long rod; length l, radius a, $c = (3a^2l/4)^{1/3}$	$\left\{\dfrac{(1/2a)^{2/3}}{(3/2)^{1/3}\{2\ln(l/a)-0.11\}}\right\}f_0$

In each case $f_0 = 6\pi\eta c$ with the appropriate value of c.

For prolate ellipsoid:

a/b	2	3	4	5	6	7	8	9	10	50	100
f/f_0	1.04	1.11	1.18	1.25	1.31	1.38	1.43	1.49	1.54	2.95	4.07

For oblate ellipsoid:

a/b	2	3	4	5	6	7	8	9	10	50	100
f/f_0	1.04	1.10	1.17	1.22	1.28	1.33	1.37	1.42	1.46	2.38	2.97

Source: K. E. Van Holde, *Physical biochemistry*, Prentice Hall.

in Table 25.1. It may also be measured using light-scattering, as we shall see later. Then, from eqn (25.1.14), it follows that

$$M_m = fSN_A/b = SN_AkT/bD = SRT/bD, \qquad (25.1.17)$$

a result independent of the shape of the solute molecules. Therefore, we can find the molar mass by combining measurements of sedimentation and diffusion rates (for S and D respectively).

Table 25.1. Diffusion coefficients of macromolecules in water at 293 K

	M_r	$D/\text{cm}^2\,\text{s}^{-1}$
Sucrose	342	4.586×10^{-6}
Lysozyme	14 100	1.04×10^{-6}
Haemoglobin	68 000	6.9×10^{-7}
Collagen	345 000	6.9×10^{-8}

Example 25.4

Use the result from *Example* 25.3 in combination with the data below to calculate the RMM of BSA. Estimate its axial ratio supposing that it is a prolate ellipsoid. Take $D = 6.97 \times 10^{-7}$ $\text{cm}^2\,\text{s}^{-1}$, $\rho = 1.0024\ \text{g cm}^{-3}$, $v_s = 0.734\ \text{cm}^3\,\text{g}^{-1}$, $\eta = 0.890\ \text{kg m}^{-1}\,\text{s}^{-1}$, and the temperature as 25 °C.

• *Method*. Use eqn (25.1.17) for M_m with $S = 5.1$ S. For the axial ratio refer to Box 25.1. First, find f from eqn (25.1.13) and $f_0 = 6\pi\eta c$ from the assumption that the molecule is a sphere of radius c. Obtain c from v_s using $v_{mol} = (4\pi/3)c^3$. Finally, select the value of b/a from Box 25.1 that gives the observed value of f/f_0 for a prolate ellipsoid.

• *Answer*. Substitution into eqn (25.1.17) with $b = 1 - \rho v_s$ gives $M_m = 69 \times 10^3\ \text{g mol}^{-1}$, so that $M_r = 69\,000$. Then, using $f = kT/D = 5.91 \times 10^{-11}\ \text{kg s}^{-1}$ and

$$v_{mol} = (0.734\ \text{cm}^3\,\text{g}^{-1}) \times (69\,000\ \text{g mol}^{-1})/N_A = 8.4 \times 10^{-26}\ \text{m}^3,$$
$$c = (3v_{mol}/4\pi)^{\frac{1}{3}} = 2.7 \times 10^{-9}\ \text{m};$$
$$f_0 = 6\pi\eta c = 4.5 \times 10^{-11}\ \text{kg s}^{-1}.$$

Therefore, $f/f_0 = 1.3$. Reference to Box 25.1 shows that this corresponds to an axial ratio of about 6.

• *Comment*. The ellipsoid is like a cigar, six times longer than it is broad. More accurate analysis (using extrapolation to zero concentration) gives an axial ratio of 4.4.

• *Exercise*. Use the result in the *Exercise* of *Example* 25.3 to find the RMM and the axial ratio of that macromolecule. Use the following additional data: $D = 5.89 \times 10^{-7}\ \mathrm{cm^2\,s^{-1}}$, $\rho = 1.0024\ \mathrm{g\,cm^{-3}}$, $v_s = 0.728\ \mathrm{cm^3\,g^{-1}}$, $\eta = 0.890 \times 10^{-3}\ \mathrm{kg\,m^{-1}\,s^{-1}}$, and the same temperature. [48 000, $f/f_0 = 1.7$]

25.1 (f) Sedimentation equilibria

The difficulty with using sedimentation rates to find RMMs lies in the inaccuracies inherent in the determination of diffusion coefficients, such as the blurring of the boundary by convection currents. The problem of needing to know D can be avoided by allowing the system to reach equilibrium. Since the number of solute molecules with any given potential energy E is proportional to $e^{-E/kT}$, the ratio of the concentrations at different heights (or radii in a centrifuge) can be used to determine their masses. This is because the potential energy of a molecule of mass m_{eff} is $\frac{1}{2}m_{\text{eff}}r^2\omega^2$ when it is travelling in a circle of radius r with angular velocity ω. Therefore, the ratio of the concentrations at radii r_1 and r_2 is

$$\begin{aligned} c_P(1)/c_P(2) &= N(1)/N(2) = e^{-E(1)/kT}/e^{-E(2)/kT} \\ &= \exp\{-mb\omega^2(r_1^2 - r_2^2)/2kT\}, \end{aligned}$$

so that

$$M_m = \frac{2RT \ln\{c_P(2)/c_P(1)\}}{(r_1^2 - r_2^2)b\omega^2}. \qquad (25.1.18)$$

In this technique, the centrifuge is run more slowly than in the sedimentation rate method, because it is no use having all the solute pressed in a thin film against the bottom of the cell.

25.1 (g) Electrophoresis

Many macromolecules are charged, and so they move in an electric field: this is called *electrophoresis*. The solution is supported on paper, but in *gel electrophoresis* the migration takes place through a cross-linked polyacrylamide gel. The mobility of the macromolecules depends on their masses and their shapes, and a constant drift speed is attained when the driving force (ezE, where z is the charge number and E the field strength) is matched by the viscous retarding force, fs. One way of avoiding the problem of knowing neither the hydrodynamic shape of the molecules nor their charge is to denature them in a controlled way. The detergent sodium dodecylsulphate has been found to be very useful in this respect. In the first place, it denatures proteins into rod-like shapes by forming a complex with them, and so all proteins, whatever their initial shapes, are made rod-like. Moreover, most proteins have been found to bind a constant amount of the detergent per unit mass, and so the charge per protein molecule is well regulated. The RMM of the protein is determined by comparing its mobility in its rod-like complexed form with standard samples.

25.1 (h) Gel filtration

Beads of porous polymeric material about 0.1 mm in diameter can capture molecules selectively, according to their size. Thus, if a solution is filtered through a column, the small molecules require a long *elution time*, while the larger ones, which are not captured, pass through rapidly. The RMM of a macromolecule may therefore be determined by observing its elution time in a column calibrated against standard samples. The range of RMMs that can be determined can be altered by selecting columns made from polymers with different degrees of cross-linking. The elution time depends on shape in a complicated way, and the technique works best if the macromolecules are globular (spherical).

25.1 (i) Viscosity

Macromolecules affect the viscosity of a solvent. The effect is large even at low concentrations, because big molecules affect the fluid's flow over a long range. At low concentrations the viscosity of the solution, η, is related to the viscosity of the pure solvent, η^*, by

$$\eta = \eta^*\{1 + [\eta]c_P + \ldots\}, \tag{25.1.19}$$

where $[\eta]$, the *intrinsic viscosity*, is a kind of virial coefficient (note that it has the dimensions of 1/concentration). It follows that the intrinsic viscosity can be measured by taking the following limit:

$$[\eta] = \lim_{c_P \to 0} \{[(\eta/\eta^*) - 1]/c_P\}. \tag{25.1.20}$$

Viscosities are measured in several ways. In the *Ostwald viscometer*, Fig. 25.7(a), the time taken for the solution to flow through the capillary is noted, and compared with a standard sample. Care must be taken to ensure that the temperature is both constant and uniform. The method is very suitable for determining $[\eta]$ because the ratio of the viscosities of the solution and the pure solvent is proportional to the drainage times t_{drain} (so long as a correction for different densities ρ and ρ^* is made):

$$\eta/\eta^* = (t_{\text{drain}}/t^*_{\text{drain}})(\rho/\rho^*). \tag{25.1.21}$$

This ratio can then be used directly in eqn (25.1.20). Viscometers in the form of rotating concentric cylinders are also used, Fig. 25.7(b), the torque on the inner cylinder being monitored while the outer one is rotated. Rotating drum viscometers have the advantage over the Ostwald type that the shear gradient between the cylinders is simpler than in the capillary and so non-Newtonian flow (of the kind shortly to be described) can be studied more easily.

There are many complications in interpreting viscosity measurements and much of (but not all) the work is based on empirical observations. The measurement of RMMs is usually based on comparisons with standard samples. Some regularities are observed. For example, it is found that θ-solutions of random coil, spherical polymers obey $[\eta] \propto M_r^{\frac{1}{2}}$, and that macromolecules of other shapes obey the *Mark–Houwink equation*, $[\eta] \approx KM_r^a$, where K and a are constants that depend on the solvent and type of macromolecule, Table 25.2. For solid spheres $a = \frac{1}{2}$, while for rigid rods $a = 2$. Therefore, by determining the value of a, the shape of the macromolecule in solution can be inferred. As an example, solutions of poly(γ-benzyl-L-glutamate) in its rigid, rod-like form has an intrinsic viscosity which is four times greater than when it is denatured and the rods collapse into random coils. Conversely, solutions of natural ribonuclease are less viscous than the denatured form: this suggests that the natural protein is more compact than when it is denatured.

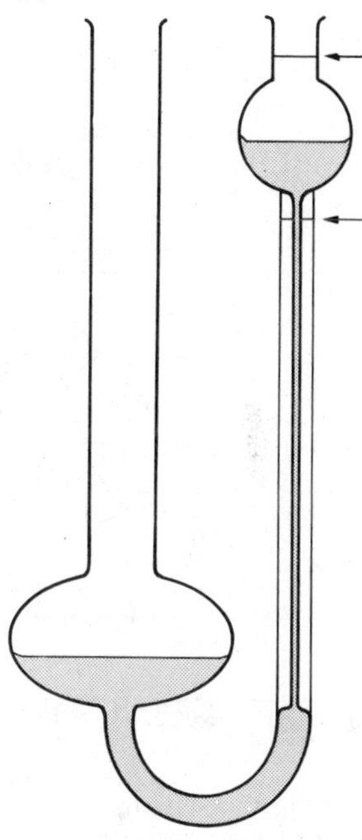

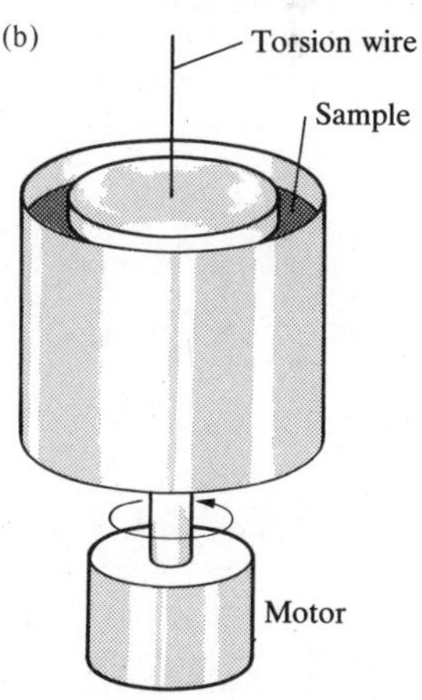

Fig. 25.7. Two types of viscometer. (a) In the *Ostwald viscometer* the time for the liquid to drain between the two marks is recorded. (b) In the *rotating drum viscometer* the torque on the inner drum is observed when the outer container is rotated.

Table 25.2. Intrinsic viscosity

Macromolecule	Solvent	θ/°C	K/cm^3 g^{-1}	a
Polystyrene	Benzene	25	9.5×10^{-3}	0.74
Polyisobutylene	Benzene	23†	8.3×10^{-2}	0.50
Various proteins‡	Guanidine hydrochloride + β-mercaptoethanol		7.16×10^{-3}	0.66

† θ-temperature.
‡ Use $[\eta] = KN^a$, N the number of amino acid residues.

Example 25.5

The viscosities of a series of solutions of polystyrene in toluene were measured at 25 °C with the following results:

c_P/(g dm^{-3})	0	2.0	4.0	6.0	8.0	10.0
$\eta/10^{-4}$ kg m^{-1} s^{-1}	5.58	6.15	6.74	7.35	7.98	8.64

Calculate the intrinsic viscosity and estimate the RMM of the polymer using the Mark–Houwink equation with $K = 3.80 \times 10^{-5}$ dm^3 g^{-1} and $a = 0.63$.

• *Method*. The intrinsic viscosity is the limit of $[(\eta/\eta^*) - 1]/c_P$ as $c_P \rightarrow 0$; therefore, form this ratio and extrapolate to $c_P = 0$.

• *Answer*. Draw up the following table:

c_P/(g dm^{-3})	0	2.0	4.0	6.0	8.0	10.0
η/η^*	1	1.102	1.208	1.317	1.430	1.549
$100[(\eta/\eta^*) - 1]/(c_P$/g dm$^{-3})$	–	5.11	5.20	5.28	5.38	5.49

The points are plotted in Fig. 25.8. The extrapolated intercept at $c_P = 0$ is 0.0504, and so $[\eta] = 0.0504$ dm^3 g^{-1}. Therefore, $M_r = \{[\eta]/K\}^{1/a} = (0.0504/3.80 \times 10^{-5})^{1/0.63} = 90 \times 10^3$.

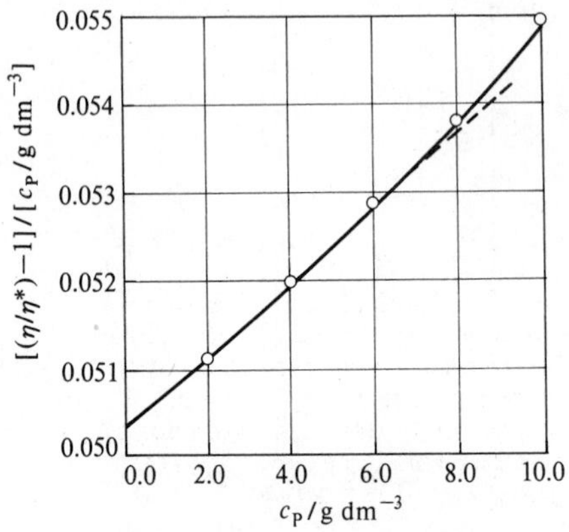

Fig. 25.8. The plot used for the determination of intrinsic viscosity, which is taken from the intercept at $c_P = 0$; see *Example* 25.5.

• *Comment*. This is an average RMM. Note that when $\eta \approx \eta^*$

$$\ln(\eta/\eta^*) = \ln[1 + (\eta - \eta^*)/\eta^*] \approx (\eta - \eta^*)/\eta^* = (\eta/\eta)^*) - 1.$$

Therefore, $[\eta]$ can also be defined as the limit of $(1/c_P)\ln(\eta/\eta^*)$ as $c_P \rightarrow 0$. The intercept can be identified more precisely by plotting both functions.

• *Exercise*. Evaluate the RMM using the second plotting technique. [90×10^3]

One complication in viscosity measurements is that in some cases it is found that the viscosity decreases as the rate of flow increases. This is an example of *non-Newtonian flow*, and it indicates the presence of long rod-like molecules that are orientated by the flow so that they slide past each other more freely. In some cases the stresses set up by the flow are so great that long molecules are broken up, with further consequences for the viscosity.

25.1 (j) Light scattering

When light falls on an object it drives the electrons into oscillation, and they radiate. If the medium is perfectly homogeneous (e.g. a perfect crystal) all the secondary waves interfere destructively except in the original propagation direction. Therefore an observer sees the beam only when looking towards the source along the initial direction. If the medium is inhomogeneous (e.g. an imperfect crystal or a solution containing foreign bodies, such as macromolecules in a solvent or smoke in air) radiation is scattered into other directions. A familiar example is light scattered by specks of dust in a sunbeam (and in advertisers' photographs of laser beams).

Light scattering by particles much smaller than the wavelength of the light is called *Rayleigh scattering*. The intensity of scattered radiation depends on $1/\lambda^4$, and so short wavelengths are scattered more intensely than long. This accounts for the blue of the sky, which arises from the predominant scattering of the blue component of white sunlight by the atoms and molecules of the atmosphere. The intensity also depends on the scattering angle θ, and is proportional to $1+\cos^2\theta$ when the light is unpolarized and to $\sin^2\theta$ when it is polarized, Fig. 25.9. In practice it turns out to be easier to make observations in a non-forward direction. The intensity also depends on the strength of the interaction of the light with the molecules, the interaction being large when their polarizability is large. This is the reason why light-scattering is so useful for studying macromolecules: they are large and more polarizable than the surrounding medium; as a result, they dominate the scattering.

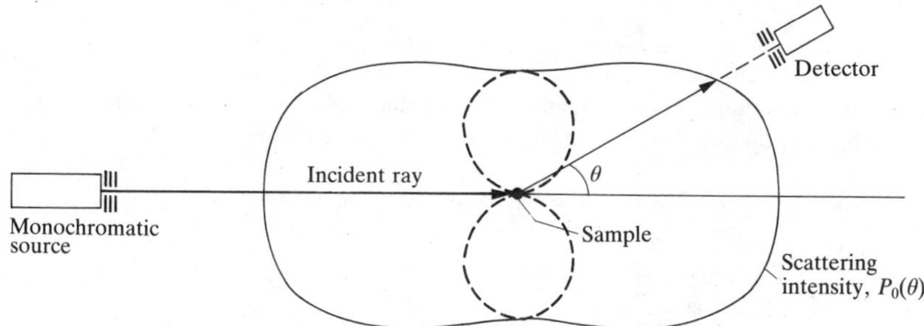

Fig. 25.9. Rayleigh scattering from a sample of point-like particles follows a $1+\cos^2\theta$ dependence (full line) when unpolarized light is used, but a $\sin^2\theta$ dependence (broken line) when plane polarized light is used (e.g. when the source is a laser).

When all these remarks are combined into a quantitative theory, it turns out that the scattering intensity, I, at the angle θ is

$$I = AI_0c_PM_rg(\theta) \qquad (25.1.22)^\circ$$

with $g=1+\cos^2\theta$ for unpolarized light (the type formerly used) and $g=\sin^2\theta$ for polarized light (from a laser, the type now generally used), and where I_0 is the incident intensity, c_P is the solute concentration, M_r its RMM, and A a constant that depends in a known way on the refractive index of the solution, the wavelength, and the distance of the detector from the sample. This is an 'ideal' result in the sense that it ignores the complications that arise from the interactions between solute particles, and in an actual experiment it is important to extrapolate to zero concentration.

The scattered intensity I is measured at several concentrations, and M_r is obtained from the limiting value of I/c_P. In the case of a polydisperse sample this procedure leads to the *mass average RMM* (symbol $\langle M_r\rangle_M$). This is defined in terms of the RMM of the individual molecules weighted according to their mass fractions, M_i/M, in the sample, M_i being the total mass of molecules of RMM $M_{r,i}$, and M the total mass of the sample:

$$\langle M_r\rangle_M = \sum_i (M_i/M)M_{r,i} = (1/M)\sum_i M_iM_{r,i}. \qquad (25.1.23)$$

The reason why light scattering gives the mass average RMM is that the intensity of scattering is greater for larger particles, and so their contribution is emphasized.

Example 25.6

A sample of polymer consists of two components present in equal masses, one having $M_r = 30\,000$ and the other $M_r = 12\,000$. What are the values of the mass average and number average RMMs?

• *Method*. Use eqn (25.1.23) for the mass average and eqn (25.1.7) for the number average. Use $M = M_1 + M_2$ with $M_1 = M_2$ so that the proportions by mass are each $\frac{1}{2}$. The proportions by number are obtained using the RMMs.

$$N_J/N = (M_J/M_{r,J})/\{(M_1/M_{r,1}) + (M_2/M_{r,2})\},$$

so that $N_1/N = M_{r,2}/(M_{r,1} + M_{r,2})$, $N_2/N = M_{r,1}/(M_{r,1} + M_{r,2})$, (because $M_1 = M_2$).

• *Answer*. From eqn (25.1.23):

$$\langle M_r \rangle_M = \sum_J (M_J/M)M_{r,J} = \tfrac{1}{2}(M_{r,1} + M_{r,2}) = 21\,000.$$

From eqn (25.1.7):

$$\langle M_r \rangle_N = \sum_J (N_J/N)M_{r,J} = 2M_{r,1}M_{r,2}/(M_{r,1} + M_{r,2}) = 17\,143.$$

That is, the two averages are respectively 21 000 and 17 000.

• *Comment*. Note that the two averages are widely different, their ratio being about 1.2. Since $M_J = N_J M_{r,J}/N_A$, the mass average may also be expressed as follows:

$$\langle M_r \rangle_M = \left(1\Big/\sum_J N_J M_J\right)\sum_J N_J M_J^2,$$

and so it is also a (number) *mean square RMM*. Sedimentation experiments give yet another average RMM: the *Z-average RMM*,

$$\langle M_r \rangle_Z = \left(1\Big/\sum_J N_J M_J^2\right)\sum_J N_J M_J^3,$$

a mean cubic RMM.

• *Exercise*. Evaluate the Z-average RMM of the sample. [25 000]

The *Example* shows that while at first sight it might appear troublesome to have two types of average, the observation that they have different values gives additional information about the range of RMMs in the sample. In the determination of protein RMMs we expect the two averages to be the same because the sample is monodisperse (unless there has been degradation). In synthetic polymers there is normally a range of RMMs, and the two averages are expected to be different. Typical synthetic materials have $\langle M_r \rangle_M / \langle M_r \rangle_N$ close to 3; the term 'monodisperse' is conventionally applied to synthetic polymers in which the ratio is smaller than about 1.1. One consequence of a narrow RMM distribution is a higher crystallinity, and therefore density and melting point, of synthetic polymers, the spread of values being controlled by the choice of catalyst and reaction conditions (Section 29.1(b)).

Since the solute scatters light away from the forward direction, the transmitted intensity is reduced. The intensity in the forward direction is given by a type of Beer–Lambert law (Section 19.1): if the incident intensity is I_0, then the intensity that survives after passing through a solution of length l is

$$I_l = I_0 e^{-\tau l}, \qquad (25.1.24)$$

where τ is the *turbidity*. For macromolecules of moderate size the turbidity is related to the concentration by

$$Hc_P/\tau = (1/M_r)\{1 + 2Bc_P + \ldots\}, \quad H = (32\pi^3/3\lambda^4 N_A)n_0^2(dn/dc_P), \qquad (25.1.25)$$

n_0 and n being the solvent and solution refractive indices respectively. Therefore, by plotting Hc_P/τ against c_P, the intercept gives the RMM. Typical values of τ are $10^{-5}\,cm^{-1}$ for pure transparent liquids (so that the sample would need to be 1 km long before the intensity drops to 1/e of its initial value as a result of scattering alone: absorption in fact then dominates), $10^{-3}\,cm^{-1}$ for polymers at 1 per cent concentration, and $10\,cm^{-1}$ for milk (which, loosely interpreted, means that you can see about 1 mm into a glass of milk).

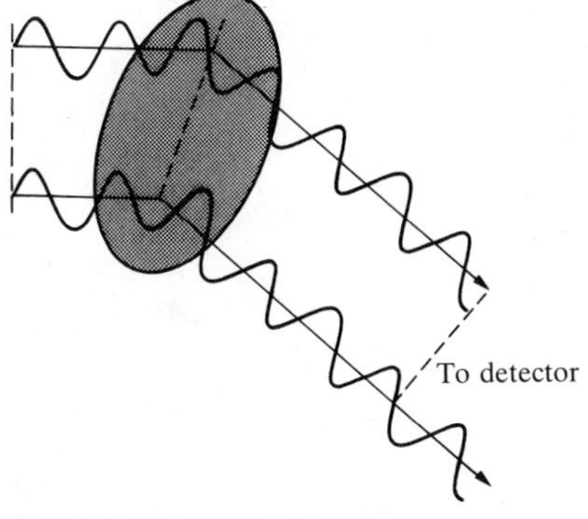

Fig. 25.10. When light scatters from different regions of the same molecule the emergent waves interfere and modify the Rayleigh scattering intensity distribution. The modification can be interpreted in terms of the shape of the macromolecules.

When the wavelength of light is comparable to the size of the scattering particles, scattering occurs at different sites of the same molecule, Fig. 25.10, and the interference between different rays is important‡. As a result, the scattering intensity is distorted from the $\sin^2\theta$ form characteristic of small-particle, Rayleigh scattering. One measure of the distortion is the *dissymmetry factor* (symbol: Z), the ratio of the scattered intensities at $\theta = 45°$ and 135°. More detailed information is contained in the ratio $P = I_{obs}/I_{Rayleigh}$ measured at several angles, I_{obs} being the observed intensity and $I_{Rayleigh}$ that predicted for Rayleigh scattering. If we treat the molecule as being composed of a collection of atoms i at distances R_i from some convenient origin, interference occurs between the light scattered by each pair, Fig. 25.10. The scattering from all the particles is then calculated by allowing for contributions from all possible orientations of each pair of atoms in each molecule. This description is very much like that used when discussing electron diffraction, Section 23.5, and so we can expect the intensity pattern to be described by a kind of Wierl equation. This turns out to be so, and if there are N atoms in the macromolecule, and if all are assumed to have the same scattering power, then

$$P = (1/N^2)\sum_{i=1}^{N}\sum_{j=1}^{N} \sin(sR_{ij})/sR_{ij}, \qquad s = (4\pi/\lambda)\sin\tfrac{1}{2}\theta. \qquad (25.1.26)$$

(Compare eqn (23.5.2).) R_{ij} is the separation of atoms i and j, and λ is the wavelength of light. The observed intensity is equal to $I_{Rayleigh}P$, with $I_{Rayleigh}$ given by eqn (25.1.22).

When the molecule is much smaller than the wavelength of light $sR_{ij} \ll 1$ (e.g. if $R = 5$ nm, and $\lambda = 500$ nm, all the sR_{ij} are about 0.1), and we can

‡ The effect accounts for the appearance of clouds, which, although we see them by scattered light, look white, not blue like the sky. The reason is that the water molecules group together into droplets of a size comparable to the wavelength of light, and scatter cooperatively. Although blue light scatters more strongly, more molecules can contribute cooperatively when the wavelength is longer (as for red light), and so the net result is uniform scattering for all wavelengths: white light scatters as white light. This paper looks white for the same reason. Cigarette smoke is blue before it is inhaled, but brownish after it is exhaled because the particles aggregate in the lungs.

use the expansion $\sin sR_{ij} = sR_{ij} - \frac{1}{6}(sR_{ij})^3 + \ldots$. Then

$$P = (1/N^2)\sum_i \sum_j \{1 - \tfrac{1}{6}(sR_{ij})^2 + \ldots\} \approx 1 - \tfrac{1}{6}(s/N)^2 \sum_{i,j} R_{ij}^2. \quad (25.1.27)$$

The sum over the squares of the separations is the square of the *radius of gyration* (symbol: R_g) of the molecule‡:

$$R_g^2 = (1/2N^2)\sum_{i,j} R_{ij}^2. \quad (25.1.28)$$

Then,

$$P \approx 1 - \tfrac{1}{3}s^2R_g^2 = 1 - (16\pi^2/3\lambda^2)R_g^2 \sin^2 \tfrac{1}{2}\theta. \quad (25.1.29)$$

The last equation shows that an analysis of the deviation of the scattering from the Rayleigh form leads to the value of R_g for the molecule in solution. This in turn can be interpreted in terms of the size of the molecule. For example, a solid sphere of radius R has $R_g = (\frac{3}{5})^{\frac{1}{2}}R$, and a long thin rod of length L has $R_g = L/2\sqrt{3}$. Once again, it must be emphasized that the analysis must be performed on data obtained by extrapolation to zero concentration. The appearance of both θ and c_P as variables leads to a special extrapolation procedure known as a *Zimm plot*. Some experimental values are listed in Table 25.3.

Table 25.3. Radius of gyration of some macromolecules

	M_r	R_g/nm
Serum albumin	66 000	2.98
Polystyrene	3.2×10^6	49.4 (in poor solvent)
DNA	4×10^6	117.0

The use of laser light has led to further refinements in the application and interpretation of light scattering. There has been a shift of emphasis towards the investigation of the time-dependence of the positions of atoms and the orientation of macromolecules in solution. These can be studied by measuring the shift of frequency that occurs when monochromatic light is scattered by a moving target, the general technique being called *dynamic light scattering*. In particular, laser light scattering can be used for the direct determination of the diffusional characteristics of macromolecules, and provides a fast, direct, and reliable method for the measurement of diffusion coefficients, even of macromolecules of low stability.

25.1 (k) Magnetic resonance

Both NMR and ESR (Chapter 20) have been applied to macromolecules in solution, and it is now possible to determine the locations of their magnetic nuclei and to study details of their motion.

The data from magnetic resonance spectra are principally line positions and line shapes (although modern data acquisition procedures give more extensive information, and relaxation times are often obtained directly). The position of NMR lines depend on the local magnetic fields at the protons or ^{13}C nuclei (or other magnetic nuclei) being examined. These local fields are modified if there are other magnetic dipoles present, such as transition metal ions deliberately incorporated into the molecule. The resonant nuclei at a distance R from the dipole experience an additional magnetic field with a strength that depends on distance as $1/R^3$, and so the distances of the nuclei from the ion may be determined. One complication is that in a tumbling molecule the shift gives distance and not direction, but this difficulty can be resolved by substituting an ion into different places and

‡ In Problem 25.24 this definition is shown to be equivalent to another and more easily visualized one in the case of a chain of identical atoms: the radius of gyration is the *root mean square distance* of the atoms from the centre of mass.

constructing a three-dimensional array of positions from separate experiments.

The *shift reagents* used may be of several kinds. The technique was developed using lanthanide ions, but it is not necessary to use paramagnetic species because the applied field can induce magnetic moments into sufficiently magnetizable groups (e.g. aromatic rings) already a part of the molecule, and these then act as sources of local field.

Line shapes are related to relaxation times (Section 20.1(f)), and modern techniques measure the latter directly. As explained in Chapter 20, the relaxation time depends on the strength of the local field causing the relaxation and the rate at which it is fluctuating. The relaxing efficiency of a magnetic dipole at a distance R from a resonant nucleus varies as the square of the field strength, and therefore it depends on $1/R^6$. Consequently, by studying relaxation times of nuclei in the presence of a paramagnetic *broadening reagent*, their locations can be deduced. As in the case of shift reagents, if this is done for a variety of substitutional sites the network of separations can be used to build a three-dimensional map of positions. If two closely related lanthanide ions are used, one being a shift reagent (e.g. Eu(III)) and the other a broadening reagent (e.g. Gd(III)), which can reasonably be supposed to bind to the same sites, then since the former explores $1/R^3$ and the latter explores $1/R^6$, a detailed map of the locations of the magnetic nuclei can be constructed.

The paramagnetic centre itself can be examined by ESR. The electron spin relaxation times depend on the tumbling rate of the molecule. If it is supposed that the molecule tumbles like a sphere of radius a in a medium of viscosity η, then the time it takes to rotate through about 1 rad (57°) in a large number of small steps, the *rotational correlation time* (symbol: τ_{rot}), is given by the *Debye expression:*

$$\tau_{rot} = 4\pi a^3\eta/3kT. \tag{25.1.30}$$

For example, a sphere of radius 2 nm in water at room temperature has $\tau_{rot} \approx 8$ ns, and the ESR technique is sensitive to processes on this time scale.

The use of *spin labels*, a paramagnetic nitroxide group, —R(NO)R′, attached to a molecule, has been particularly fruitful. They have been used to show, for example, that some molecules rotate at different rates around different axes. This *anisotropic rotational diffusion* can be detected from the shapes of ESR spectral lines and interpreted in terms of the shapes of the molecules. The same technique can be used to study the mobilities of groups within a macromolecule and to show that some groups can wave around loosely but others are sterically trapped into rigid conformations. In this way we can explore not only the size and shape of a molecule, but also the flexibility of its structure.

25.2 Conformation and configuration

The *primary structure* of a macromolecule is the sequence of chemical segments making up the chain (or network if there is cross-linking). In the case of a synthetic polymer virtually all the segments are identical, and it is sufficient to name the monomer used in the synthesis. Thus, the primary structure of polyethylene (poly(ethene)) is the —CH_2CH_2— unit, and the chemical constitution of the chain is specified by denoting it as $\text{-(}CH_2CH_2\text{)-}_N$. The occasional end groups, where the chains end, can often be treated as impurities.

The concept of primary structure ceases to be trivial in the case of biological macromolecules, for these are often chains of different molecules. Proteins, for example, are *polypeptides*, the name signifying chains formed from numbers of different amino acids (about twenty occur naturally) strung together by the *peptide link*, —CO—NH—. The determination of the primary structure is then a highly complex problem of chemical analysis called *sequencing*.

The *secondary structure* of macromolecules refers to the (often local) spatial, well-characterized arrangement of the basic structural units. The secondary structure of an isolated molecule of polyethylene is a random coil, while that of a protein is a highly organized arrangement determined largely by hydrogen bonds, and taking the form of helices or sheets in various segments of the molecule. When the hydrogen bonds in a protein are destroyed (for instance, by heat, as when cooking an egg) the structure collapses (*denatures*) into a random coil.

The difference between primary and secondary structure is closely related to the difference between the configuration and the conformation of a chain. The *configuration* refers to the structural features that can be changed only by breaking chemical bonds and forming new ones. Thus, the chains —A—B—C— and —A—C—B— have different configurations. The *conformation* of a chain refers to the spatial arrangement of its different parts, and one conformation can be changed into another by rotating one part of a chain round the bond joining it to another. The denaturation of a protein is a conformational, not a configurational change (unless it also *degrades*).

The *tertiary structure* of a protein refers to the overall three-dimensional structure of the molecule. For instance, many proteins have a helical secondary structure, but in many the helix is bent and distorted and the molecule has a globular tertiary structure.

The *quaternary structure* is the manner in which some molecules are formed by the aggregation of others. Haemoglobin is a famous example: it consists of four subunits of two types (the α- and β-chains).

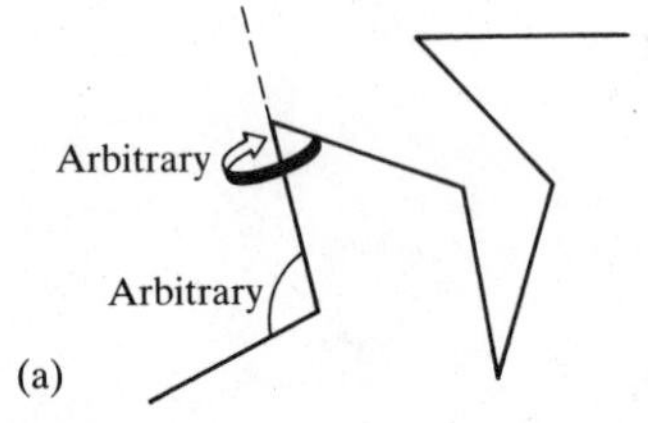

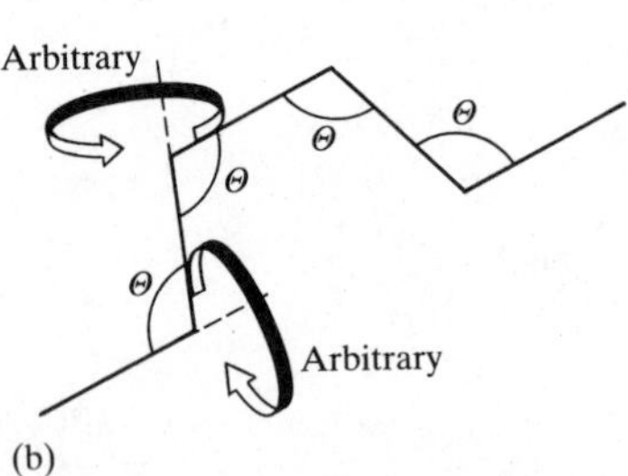

Fig. 25.11. (a) A freely jointed chain is like a three-dimensional random walk, each step being in an arbitrary direction but of the same length. (b) A better description is obtained by fixing the bond angle (e.g. at the tetrahedral angle) and allowing free rotation about a bond direction.

25.2 (a) Random coils

As the first step in unravelling the various aspects of structure, we consider the most likely conformation of a chain of identical units that are incapable of forming hydrogen bonds or any other type of specific bond. Polyethylene is a basic example, but the general idea applies to a denatured protein. The simplest model is the *freely jointed chain*, in which any bond is free to make any angle with respect to the preceding one, Fig. 25.11(a). This is obviously a great simplification, because a bond is actually constrained to a cone of angles around a direction defined by its neighbour, Fig. 25.11(b).

The main point to notice is that the model is the same as the *three-dimensional random walk*, each bond representing a step taken in a random direction. The mathematics of the random walk will be dealt with in Appendix 27.1; for our purposes all we need is the central result that the probability of the ends of the chain lying in the range R to $R + dR$ (i.e. the probability that the random walker has travelled a distance between R and

$R+\mathrm{d}R$, whatever the direction) is $f\,\mathrm{d}R$, where

$$f=4\pi(a/\pi^{\frac{1}{2}})^3R^2\mathrm{e}^{-a^2R^2}, \qquad a^2=3/2Nl^2, \qquad (25.2.1)$$

with N the number of bonds (the number of paces) and l the bond length (the length of each pace)‡. This expression is very much like the Maxwell distribution of molecular speeds in a gas (Section 0.1(e)), but with distance replacing speed: the Maxwell distribution can therefore be thought of as the result of a random walk in speed, with each collision accelerating the particle randomly. The last equation shows that in some coils (the proportion being given by the value of f with R large), the ends may be far apart, while in others their separation is small. An alternative interpretation is to regard each coil as writhing continually from one conformation to another; then $f\,\mathrm{d}R$ is the probability that at any instant it will be found with its ends lying at separations between R and $R+\mathrm{d}R$.

The *root mean square separation* (symbol: R_{rms}) is a measure of the *average* separation of the ends of a random coil. It is the square root of the mean value of R^2, calculated by weighting each possible value of R^2 with the probability that R occurs:

$$R_{\mathrm{rms}}=\sqrt{\left\{\int_0^\infty R^2f\,\mathrm{d}R\right\}}=N^{\frac{1}{2}}l. \qquad (25.2.2)$$

The essential feature of this result is that as the number of monomer units increases, the size of the random coil increases as $N^{\frac{1}{2}}$ (and its volume increases as $N^{\frac{3}{2}}$). This is a general feature of the scaling of random coils. For example, the radius of gyration (defined in eqn (25.1.28)) can be calculated using f:

$$R_{\mathrm{g}}=N^{\frac{1}{2}}l/6^{\frac{1}{2}}. \qquad (25.2.3)$$

Example 25.7

Calculate the mean separation of the ends of a freely-jointed polymer chain of N bonds of length l.

- *Method*. Use eqn (25.2.1) in the expression

$$\langle R^n\rangle=\int_0^\infty R^nf\,\mathrm{d}R$$

with $n=1$. The resulting integral is standard:

$$\int_0^\infty x^3\mathrm{e}^{-a^2x^2}\,\mathrm{d}x=1/2a^4.$$

- *Answer*.

$$\langle R\rangle=4\pi(a/\pi^{\frac{1}{2}})^3\int_0^\infty R^3\mathrm{e}^{-a^2R^2}\,\mathrm{d}R=2/a\pi^{\frac{1}{2}}=(8/3\pi)^{\frac{1}{2}}N^{\frac{1}{2}}l.$$

- *Comment*. When the chain is not freely jointed the result must be multiplied by a factor: see below.

- *Exercise*. Evaluate the root mean square separation of the ends of the chain. [$N^{\frac{1}{2}}l$]

---

‡ Here and elsewhere we are ignoring the fact that the chain cannot be longer than Nl. Although eqn (25.2.1) gives a non-zero probability for $R>Nl$, the values are so small that the errors in pretending that R can range up to infinity are negligible.

Before making use of these conclusions we must remove the obvious absurdity of allowing bonds to make all angles to each other. This is simple in the case of long chains, for it is possible to take groups of neighbouring bonds, and to think about the direction of their resultant. Although individual bonds are constrained to a single cone of angle Θ, the resultant of several lies in a random direction. By concentrating on groups rather than individuals, it turns out that the average values given above should be multiplied by $\{(1-\cos\Theta)/(1+\cos\Theta)\}^{\frac{1}{2}}$. In the case of tetrahedral bonds, for which $\cos\Theta = -\frac{1}{3}$ (i.e. $\Theta = 109.5°$), the factor is $2^{\frac{1}{2}}$. Therefore:

$$R_{\text{rms}} = 2^{\frac{1}{2}}N^{\frac{1}{2}}l, \qquad R_{\text{g}} = (\tfrac{1}{3})^{\frac{1}{2}}N^{\frac{1}{2}}l. \tag{25.2.4}$$

For example, in the case of a polyethylene chain with $N_{\text{r}} = 56\,000$, corresponding to $N = 4000$, since $l = 154$ pm for a C—C bond, we find $R_{\text{rms}} = 4.4$ nm and $R_{\text{g}} = 1.8$ nm, Fig. 25.12. The figure for the radius of gyration means that, on average, the coils rotate like hollow spheres of radius 1.8 nm and mass equal to the molecular mass.

The model of a randomly coiled molecule is still an approximation even after the bond angles have been restricted. This is because it does not take into account the impossibility of two or more atoms occupying the same place. Such self-avoidance tends to swell the coil, and so it is better to regard R_{rms} and R_{g} as lower bounds to the actual values. Furthermore, the model totally ignores the role of the solvent: a poor solvent will tend to cause the coil to tighten so that solute–solvent contacts are minimized; a good solvent does the opposite.

The random coil is the least structured conformation of a polymer chain, and so it corresponds to the state of maximum *conformational entropy*. Any stretching of the coil introduces order, and so corresponds to a reduction of entropy. Conversely, the formation of a random coil from a more extended

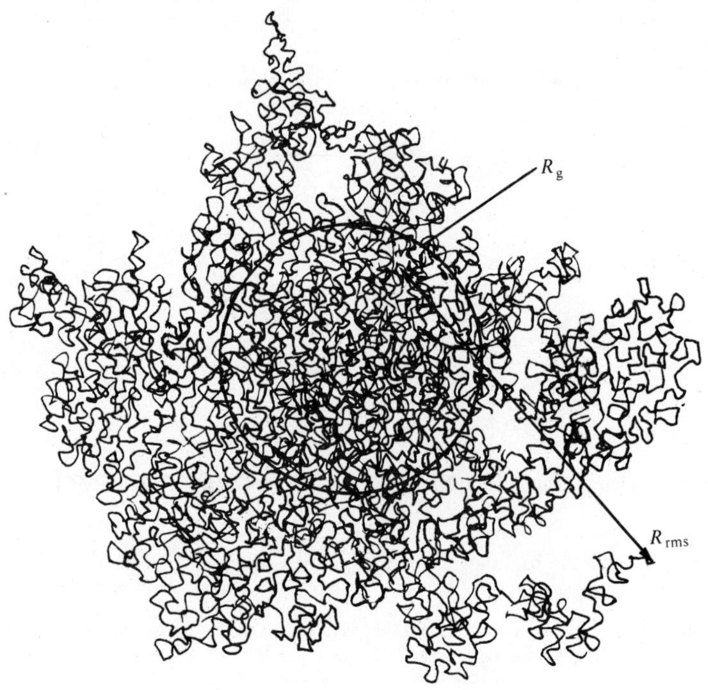

Fig. 25.12. A random coil in three dimensions. This one contains about 4000 units. The root mean square distance between the ends (R_{rms}) and the radius of gyration (R_{g}) are indicated.

form is a spontaneous process (so long as enthalpy contributions do not interfere). The elasticity of a *perfect rubber* (one in which the internal energy is independent of the extension, just as the internal energy of a perfect gas is independent of its volume) may be discussed in these terms, Appendix 25.1. The random coil model is also a helpful starting point for the discussion of the orders of magnitude of the hydrodynamic properties (e.g. sedimentation rates) of polymers and denatured proteins in solution.

25.2 (b) Helices and sheets

Natural macromolecules need a precisely maintained conformation in order to function. This is the major remaining problem in protein synthesis, for although primary structures can be built, the product is inactive because the secondary structure cannot yet be produced.

The fundamental features of the secondary structures of proteins are summarized by the rules formulated by Linus Pauling and Robert Corey, the essential feature being the stabilization of structures by hydrogen bonds involving the —CO—NH— peptide link. The latter can act both as a donor of the H atom (the NH part of the link) and as an acceptor (the CO part). The rules are as follows:

(1) The atoms of the peptide link lie in a plane, Fig. 25.13.

(2) The N, H, O atoms of a hydrogen bond lie in a straight line (with displacements of H tolerated up to not more than 30° from the N—O vector).

(3) All NH and CO groups are engaged in bonding.

These rules lead to two structures. One, in which H-bonding occurs between peptide links of the same chain, is the *α-helix*. The other, in which H-bonding links different chains, is the *β-pleated sheet*; this is the secondary structure of the protein fibroin, the constituent of silk.

The α-helix is illustrated in Fig. 25.14. Each turn of the helix contains 3.6 amino acid residues, and so the period of the helix corresponds to 5 turns (18 residues). The pitch of a single turn is 544 pm. The N—H · · · O bonds lie parallel to the axis and link every fourth group (so that residue i is linked

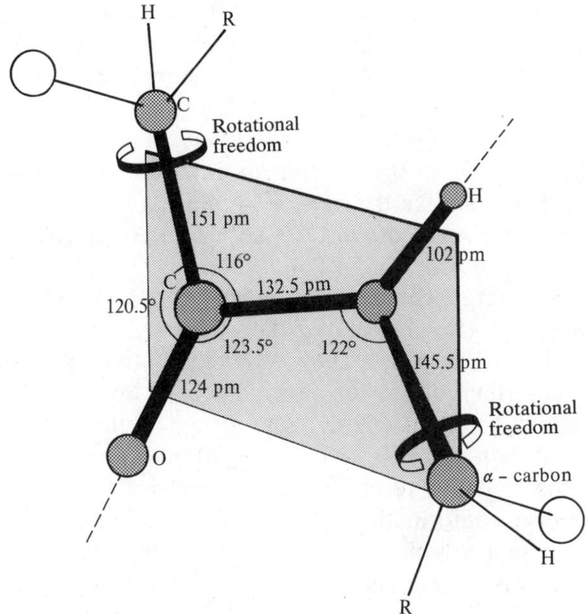

Fig. 25.13. The dimensions that characterize the peptide link. The C—CO—NH—C atoms define a plane (the C—N bond has partial double-bond character), but there is rotational freedom around the C—CO and N—C bonds.

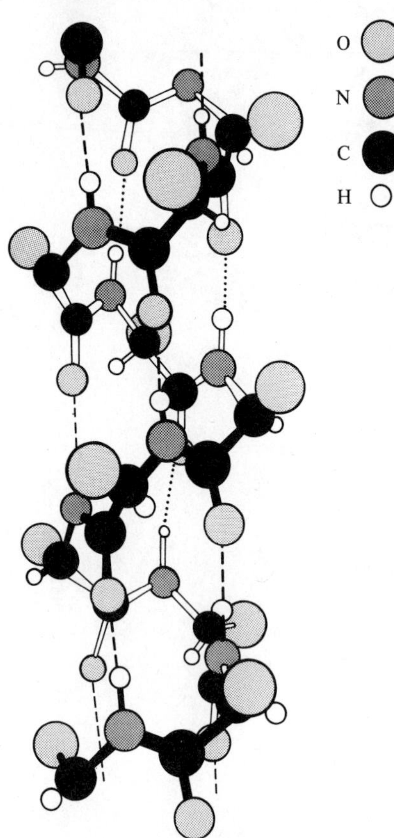

Fig. 25.14. The polypeptide α-helix. There are 3.6 residues per turn, and a translation along the helix of 150 pm per residue, giving a pitch of 540 pm. The diameter (ignoring side chains) is about 600 pm.

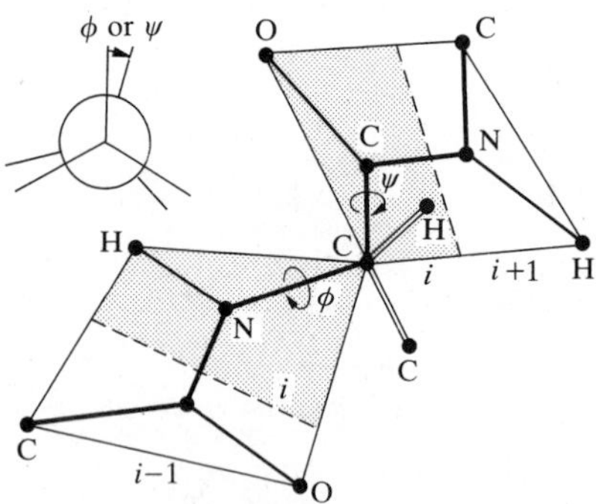

Fig. 25.15. The definition of the torsional angles ψ and ϕ between two peptide units. In this case (an α-L-polypeptide) the chain has been drawn in its all-*trans* form, with $\psi = \phi = 180°$.

to residues $i-4$ and $i+4$). There is freedom for the helix to be arranged as either a right- or a left-handed screw, but the overwhelming majority of natural polypeptides are right-handed on account of the preponderance of the L-configuration of the naturally occurring amino acids, as we explain below. The reason for the preponderance of L-amino acids is uncertain, but it is currently being speculated that it is related to the symmetries of fundamental particles and the non-conservation of parity (the fact that this universe behaves differently from its hypothetical mirror-image).

25.2 (c) Conformational energy

The reason why a right-handed α-helix is normally found is that it is marginally more stable than the left-handed version. The stabilities of different polypeptide geometries can be investigated by calculating the total potential energy of all the interactions between the atoms, and looking for a minimum. The energies for different conformations can be plotted as a contour map constructed as follows.

The geometry of the chain can be specified by two angles, ϕ (the torsional angle for the N—C bond) and ψ (the torsional angle for the C—C bond). The α-L-polypeptide shown in Fig. 25.15 defines these angles, and shows the all-*trans* form of the chain, in which all ϕ and ψ are 180°. (The sign convention is that a positive angle means that the front atom must be rotated clockwise to bring it into an eclipsed position relative to the rear atom.) A helix is obtained when all the ϕs are equal, and when all the ψs are equal. For a right-handed α-helix $\phi = -57°$, $\psi = -47°$. For a left-handed α-helix both angles are positive. Since only two angles are needed to specify the conformation of a helix, and they range from $-180°$ to $+180°$, the potential energy of the entire molecule can be represented by a point on a square plane, one axis representing ϕ, and the other ψ.

The potential energy for a given conformation (ϕ, ψ) can be calculated using the expressions developed in Section 24.2. For example, the interaction energy of two atoms separated by a distance R (which can be calculated from the geometry of the molecule once ϕ and ψ are specified) can be given the Lennard–Jones (12, 6) form, eqn (24.2.9). If the partial charges on the atoms (arising from ionic character in the bonds) are known, then a Coulombic contribution of the form $1/R$ can be included. This is sometimes done by ascribing charges $-0.28e$ and $+0.28e$ to N and H respectively, and $-0.39e$ and $+0.39e$ to O and C respectively. There is also a torsional contribution arising from the barrier to internal rotation of one bond relative to another (just like the barrier to internal rotation in ethane), and which is normally expressed as $A(1+\cos 3\phi)+B(1+\cos 3\psi)$, A and B being constants of the order of 1 kJ mol^{-1}.

The potential energy contours for helical forms of polypeptide chains formed from glycyl (R = H) and L-alanyl ($R = CH_3$) are shown in Fig. 25.16(a) and (b) respectively. They were computed by summing all the contributions described above for each choice of angles, and then plotting contours of equal potential energy. The glycyl result is symmetrical, with minima of equal depth at $\phi = -80°$, $\psi = +90°$ and at $\phi = +80°$, $\psi = -90°$. In contrast, the L-alanyl result is not symmetrical, and there are three distinct low-energy conformations (I, II, III). The minima of regions I and II lie close to the angles typical of right- and left-handed α-helices, but the former has a lower minimum. This is consistent with the formation of right-handed helices from the naturally occurring L-amino acids.

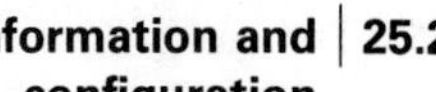

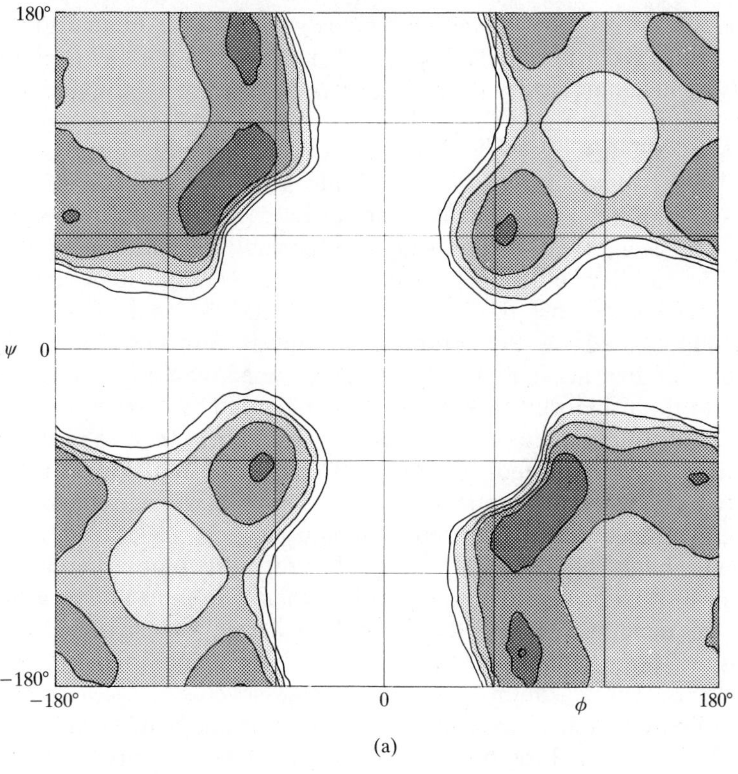

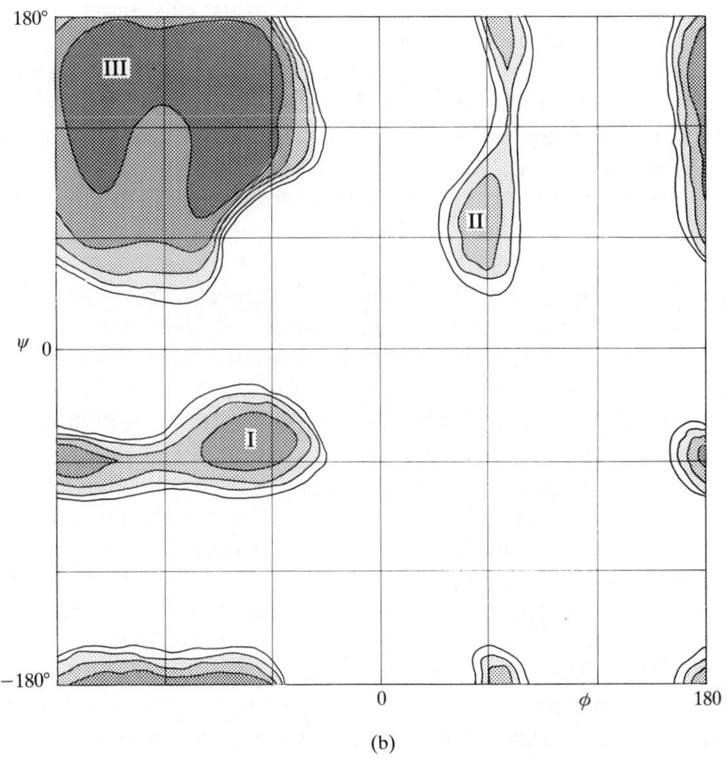

Fig. 25.16. Energy contour diagrams for (a) a glycyl residue of a polypeptide chain and (b) an L-alanyl residue. The denser the shading the lower the potential energy. The glycyl diagram is symmetrical, but regions I and II in the L-alanyl diagram, which correspond to right- and left-handed helices, are unsymmetrical, and the minimum in region I lies lower than that in region II. (After D. A. Brant and P. J. Flory, *J. mol. Biol.* **23,** 47, 1967.)

This type of *conformational energy analysis* can be extended to explorations of the structural rigidity of proteins, and is used to study the biological activity of macromolecules.

25.2 (d) Higher-order structure

Helical polypeptide chains are folded into a tertiary structure if there are other bonding influences between the residues of the chain that are strong enough to overcome the interactions responsible for the secondary structure. The folding influences include —S—S— links, ionic interactions (which depend on the pH), and strong H-bonds (such as O—H···O$^-$). This is illustrated by the structure of myoglobin, Fig. 25.17, the full structure (of 2600 atoms) having been determined by X-ray diffraction. The folding of the α-helix caused by —S—S— links can be seen in the structure: about 77% of the structure is α-helix, the rest being involved in the folds.

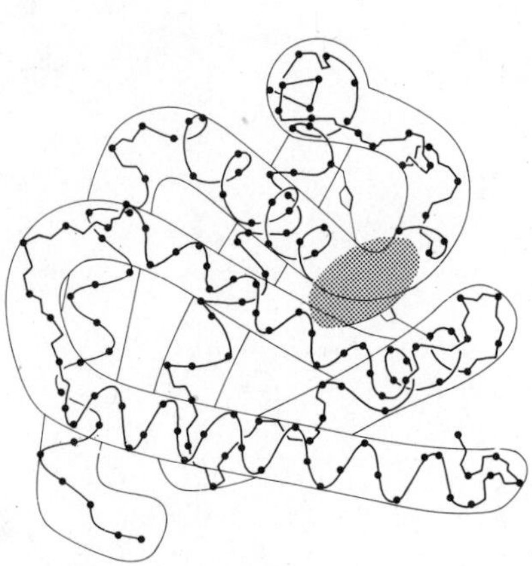

Fig. 25.17. The structure of myoglobin. Only the α-carbon atom positions are shown. The *haem* group, the oxygen binding group, is shown as a shaded region. (Based on M. F. Perutz, copyright *Scientific American*, 1964; with permission.)

Proteins with M_r exceeding about 50 000 are often found to be aggregates of two or more polypeptide chains. The possibility of such quaternary structure often confuses the determination of their RMM, since different techniques might give values differing by factors of two or more. Haemoglobin, which consists of four myoglobin-like chains, is an example.

Protein denaturation can be caused by several means, and different aspects of structure may be affected. The permanent waving of hair, for example, is reorganization at the quaternary level. Hair is a form of the protein keratin, and its quaternary structure is thought to be a multiple helix, with the α-helices bound together by —S—S— links and H-bonds (although there is some dispute about its precise structure). The process of permanent waving consists of disrupting these links, unravelling the keratin quaternary structure, and then reforming it into a more fashionable disposition. The 'permanence' is only temporary, however, because the structure of the newly formed hair is genetically controlled. Incidentally, normal hair grows at a rate that requires at least ten twists of the keratin helix to be produced each second, and so very close inspection of the human scalp would show it to be literally writhing with activity.

Denaturation at the secondary level is brought about by agents that destroy H-bonds. Thermal motion may be sufficient, in which case denaturation is a kind of intramolecular melting. When eggs are cooked the albumin is denatured irreversibly, and the protein collapses into a structure resembling a random coil. The *helix–coil transition* is sharp, like ordinary melting. This is because it is a cooperative process: when one H-bond has been broken it is easier to break its neighbours, and then even easier to break theirs, and so on. The disruption cascades down the helix, and the transition occurs sharply. Denaturation may also be brought about chemically. For instance, a solvent that forms stronger H-bonds than those within the helix will compete successfully for the NH and CO groups. Acids and bases can cause denaturation by protonation or deprotonation of various groups. All these processes, including minor conformational changes stopping short of full denaturation, can be investigated using the techniques described earlier in the chapter.

25.3 Colloids

Colloids are dispersions of small particles of one material in another. 'Small' means something around or less than about 500 nm in diameter (about the wavelength of light). In general they are aggregates of numerous atoms or molecules, but are too small to be seen with an ordinary optical microscope. They pass through most filter papers, but can be detected by light-scattering, sedimentation, and osmosis.

25.3 (a) Classification

The name given to the colloid depends on the two phases involved. *Sols* are dispersions of solids in liquids (such as clusters of gold atoms in water) or of solids in solids (such as ruby glass, which is a gold-in-glass sol, and achieves its colour by scattering). *Aerosols* are dispersions of liquids in gases (like fog and many sprays) and of solids in gases (such as smoke): the particles are often large enough to be seen with a microscope. *Emulsions* are dispersions of liquids in liquids (such as milk). Sometimes *foams*, dispersals of gases in liquids (such as beer) or of gases in solids (such as pumice), are included.

A further classification of colloids is as *lyophilic* (solvent attracting) and *lyophobic* (solvent repelling), or, in the case of water as solvent, *hydrophilic* and *hydrophobic*. Lyophobic colloids include the metal sols. Lyophilic colloids generally have some chemical similarity to the solvent, such as hydroxyl groups able to form H-bonds. A *gel* is a semi-rigid mass of a lyophilic sol in which all the dispersion medium has been absorbed by the sol particles.

25.3 (b) Preparation and purification

Preparation can be as simple as sneezing (which produces an aerosol). Laboratory and commercial methods make use of several techniques. Material (e.g. quartz) may be ground in the presence of the dispersion medium. Passing a heavy electric current through a cell may lead to the crumbling of an electrode into colloidal particles; arcing between electrodes immersed in the support medium also produces a colloid. Chemical precipitation sometimes results in a colloid. A precipitate (e.g. silver iodide) already formed may be dispersed by the addition of *peptizing agent* (e.g. potassium iodide). Clays may be peptized by alkalis, the OH^- ion being the active agent.

Emulsions are normally prepared by shaking the two components together, although some kind of *emulsifying agent* has to be used in order to stabilize the product. This emulsifier may be a soap (a long chain fatty acid), a detergent, or a lyophilic sol that forms a protective film around the dispersed phase. In the case of milk, which is an emulsion of fats in water, the emulsifying agent is casein, a protein containing phosphate groups. That casein is not completely successful in stabilizing milk is apparent from the formation of cream on the surface: the dispersed fats coalesce into oily droplets which float to the surface. This may be prevented by ensuring that the emulsion is dispersed very finely initially: violent agitation with ultrasonics brings this about, the product being *homogenized milk*.

Aerosols are formed when a spray of liquid is torn apart by a jet of gas. The dispersal is aided if a charge is applied to the liquid, for then the electrostatic repulsions blast the jet apart into droplets. This procedure may also be used to produce emulsions, for the charged liquid phase may be squirted into another liquid.

Colloids are often purified by dialysis. The aim is to remove much (but not all, for reasons explained later) of the ionic material that may have accompanied their formation. As in the discussion of the Donnan effect, a membrane (e.g. cellulose) is selected which is permeable to solvent and ions, but not to the colloid particles. Dialysis is very slow, and is normally accelerated by applying an electric field and making use of the charge carried by many colloids, the technique then being *electrodialysis*.

25.3 (c) Surface, structure, and stability

The principal feature of colloids is the very great surface area of the dispersed phase in comparison with the same amount of ordinary material. For example, a 1 cm cube of material has a surface area of 6 cm^2, but when it is dispersed as 10^{18} little 10 nm cubes the total surface area is 6×10^6 cm^2 (about the size of a tennis court). This dramatic increase in area means that surface effects are of dominating importance in colloid chemistry.

As a result of their great surface area, colloids are thermodynamically unstable with respect to the bulk: since $dG = \gamma\, d\sigma$ where γ is the surface tension, dG is negative for a decrease in area. The apparent stability must therefore be a consequence of the kinetics of collapse: colloids are kinetically, not thermodynamically, stable.

At first sight even the kinetic argument seems to fail: colloidal particles attract each other over large distances, and so there is a long-range force tending to collapse them down into a single blob. The reasoning behind this remark is as follows. The energy of attraction between two individual atoms, one in each colloidal particle, varies as their separation as $1/R_{ij}^6$ (Section 24.2). The sum of all these pairwise interactions, however, decreases as only $1/R^2$, where R is the separation of the centres of the particles (the precise dependence is given in Problem 25.36), and this is of much greater range than the $1/R^6$ dependence characteristic of individual particles and small molecules.

There are various factors working against the long-range dispersion attraction. There may be a protective film at the surface of the colloid particles that stabilizes the interface and cannot be penetrated when two particles touch. For example, the surface atoms of a platinum sol in water react chemically and are turned into —$Pt(OH)_3H_3$, and this layer encases the particle like a shell. A fat can be emulsified by a soap because the long hydrocarbon tails penetrate the oil droplet but the —CO_2^- head groups (or other hydrophilic groups in detergents) surround the surface, form hydrogen bonds with water, and give rise to a shell of negative charge which repels a possible approach from another similarly charged particle.

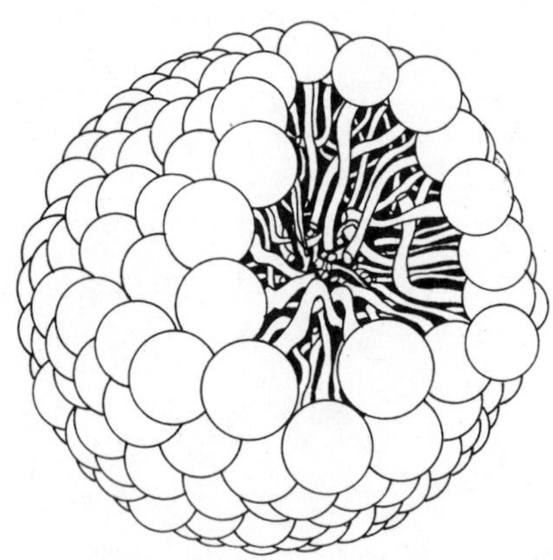

Fig. 25.18. A representation of a spherical micelle. The hydrophilic groups are represented by spheres, and the hydrophobic hydrocarbon chains are represented by the stalks: the latter are mobile.

Soap molecules can group together as *micelles* even in the absence of grease droplets, for their hydrophobic tails tend to congregate, and their hydrophilic heads provide protection, Fig. 25.18. Micelles form only above the *critical micelle concentration* (CMC) and above the *Krafft temperature*. Non-ionic detergent molecules may cluster together in swarms of 1000 or more, but ionic species tend to be disrupted by the electrostatic repulsions between head groups and are normally limited to groups of between 10 and 100 molecules. The micelle population is often polydisperse, and the shapes of the individual micelles vary with concentration. For instance, while spherical micelles do occur, they are more commonly flattened spheres close to the CMC, and rod-like at higher concentrations. The interior of a micelle is like a droplet of oil, and magnetic resonance shows that the hydrocarbon tails are mobile, but slightly more restricted than in the bulk.

Micelles are important in industry and biology on account of their solubilizing function: matter can be transported by water after it has been dissolved in their hydrocarbon interiors. For this reason, micellar systems are used as detergents and drug carriers, and for organic synthesis, froth flotation, and petroleum recovery.

The thermodynamics of micelle formation shows that the enthalpy of formation in aqueous systems is probably positive (that is, that they are endothermic) with ΔH_m around 1–2 kJ per mole of detergent species. That they do form above the CMC indicates that their entropy of formation must then be positive, and measurements suggest a value of about $+140\ \mathrm{J\ K^{-1}\ mol^{-1}}$ at room temperature. That the entropy change is positive even though the molecules are clustering together indicates a contribution to the entropy from the solvent: its molecules are less constrained once the detergent molecules have herded into small clusters and individual molecules no longer have to be held in solvent cages, Fig. 25.19. This is the origin of the *hydrophobic interaction*, where the increasing entropy of the solvent results in hydrophobic sections of molecules grouping together; a similar process affects the structures of proteins, where the hydrophobic groups tend to congregate for the same reason: it is an example of an *entropy driven* interaction.

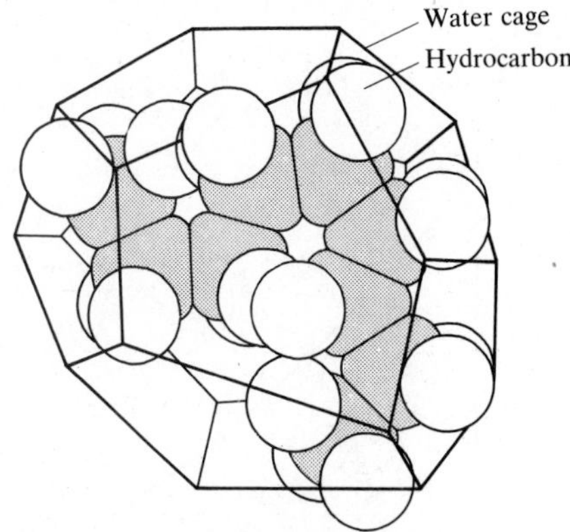

Fig. 25.19. When a hydrocarbon molecule is surrounded by water, the water molecules form a clathrate cage. As a result of this acquisition of structure, the entropy of the water decreases, and so the dispersal of the hydrocarbon into water is entropy-opposed; its coalescence is entropy-favoured.

25.3 (d) The electric double-layer

Apart from the physical stabilization of colloid particles, a major source of kinetic stability is the existence of an electric charge on their surfaces. On account of this charge, ions of opposite charge tend to cluster nearby, and an ionic atmosphere is formed, just as was described for ions in Section 11.2(a).

Two regions of charge must be distinguished. First, there is a fairly immobile layer of ions that stick tightly to the surface of the colloidal particle, which may include water molecules (if that is the support medium). The radius of the sphere that captures this rigid layer is called the *radius of shear*, and is the major factor determining the mobility of the particles. The electric potential at the radius of shear relative to its value in the distant, bulk medium is called the *ζ-potential* or the *electrokinetic potential*. The charged unit attracts an oppositely charged ionic atmosphere. The inner shell of charge and the outer atmosphere is called the *electric double layer*. The structure of the atmosphere can be described in the same way as in the Debye–Hückel theory of ionic solutions, and its thickness is represented by a Debye length, as in Section 11.2(a).

The Debye length decreases as the ionic strength of the medium increases. At high ionic strengths the atmosphere is dense and the potential falls to its bulk value within a short distance. In this case there is little electrostatic repulsion to hinder the close approach of two colloid particles. As a result, coagulation (or *flocculation*) readily occurs as a consequence of the van der Waals forces. Since the ionic strength is increased by the addition of ions, particularly those of high charge type, such ions act as flocculating agents. This is the basis of the empirical *Schultze–Hardy rule*, that hydrophobic colloids are flocculated most efficiently by ions of opposite charge type and high charge number. The Al^{3+} ions in alum are very effective, and are used in styptic pencils for inducing the congealing of blood. When river water containing colloidal clay flows into the sea, the brine induces coagulation: this is a major cause of silting in estuaries.

It is found that metal oxide sols tend to be positively charged while sulphur and the noble metals tend to be negatively charged. Naturally occuring macromolecules also acquire a charge when dispersed in water, and an important feature of proteins and other natural macromolecules is that their overall charge depends on the pH of the medium. For instance, in acid environments protons attach to basic groups, and the net charge of the macromolecule is positive; in basic media the net charge is negative as a result of proton loss. At the *isoelectric point* the pH is such that there is no net charge on the macromolecule.

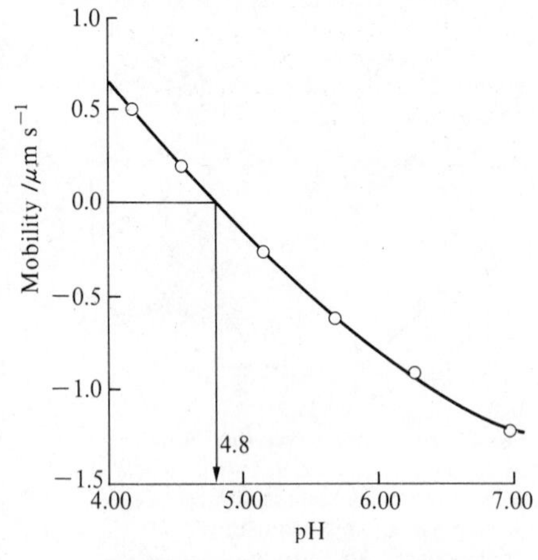

Fig. 25.20. The plot of mobility against pH allows the isolectric point to be detected as the pH at which the mobility is zero.

Example 25.8

The mobility of BSA in aqueous solution was monitored at several values of pH, and the data are listed below. What is the isoelectric point of the protein?

pH	4.20	4.56	5.20	5.65	6.30	7.00
mobility/μm s^{-1}	0.50	0.18	−0.25	−0.65	−0.90	−1.25

• *Method*. The protein has zero electrophoretic mobility when it is uncharged. Therefore, the isoelectric point is the pH at which it does not migrate in an electric field. Plot mobility against pH and find by interpolation the pH of zero mobility.

• *Answer*. The data are plotted in Fig. 25.20. There is zero mobility at pH = 4.8; hence this is the isoelectric point.

• *Comment*. Sometimes the isoelectric point must be obtained by extrapolation because the macromolecule might not be stable over the whole pH range.

• *Exercise*. The following data were obtained for another protein:

pH	4.5	5.0	5.5	6.0
mobility/μm s^{-1}	−0.10	−0.20	−0.30	−0.35

Estimate the isoelectric point. [4.3]

The primary role of the electric double layer is to confer kinetic stability. Colliding colloidal particles break through the double layer and coalesce only if the collision is sufficiently energetic to disrupt the layers of ions and solvating molecules, or if thermal motion has stirred away the surface accumulation of charge. This may happen at high temperatures, which is why sols precipitate when they are heated. The protective role of the double layer is the reason why it is important not to remove all the ions when a colloid is being purified by dialysis, and why proteins coagulate most readily at their isoelectric point.

The presence of charge on colloidal particles and natural macromolecules also permits us to control their motion, such as in dialysis and electrophoresis. Apart from its application to the determination of RMM (Section 25.1(g)) electrophoresis has several analytical and technological applications. One analytical application is to the separation of different macromolecules, and a typical apparatus is illustrated in Fig. 25.21. Technical applications include silent ink-jet printers, the painting of objects by airborne charged paint droplets, and electrophoretic rubber forming by deposition of charged rubber molecules on anodes formed into the shape of the desired product (e.g. surgical gloves).

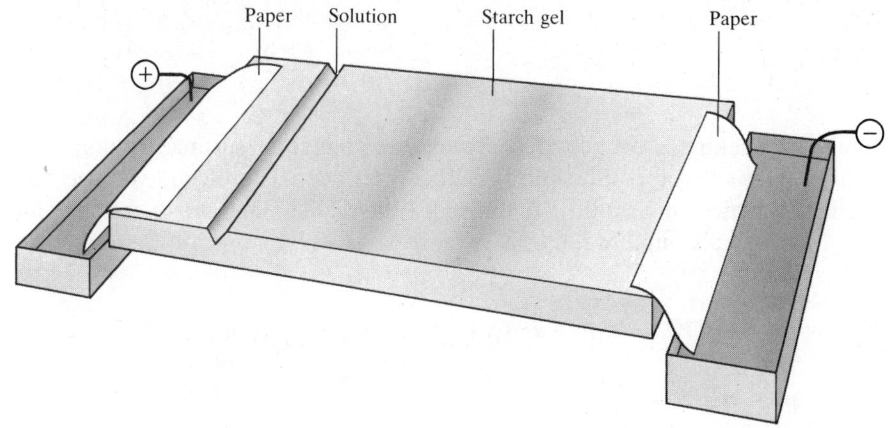

Fig. 25.21. The layout of a simple electrophoresis apparatus. The sample is introduced into the trough in the gel, and the different components form separated bands under the influence of the potential difference.

25.4 Surface tension and detergents

A *surfactant* (or *surface-active agent*) is a species that is active at the interface between two phases, such as a detergent, which is active at the interface between hydrophilic and hydrophobic phases. A detergent accumulates at the interface, and modifies the surface tension. The effect can be discussed quantitatively as follows.

Consider two phases α and β in contact; the system consists of several components J, each one present in an overall amount n_{J}. If the components were distributed uniformly through the phases right up to the interface, which is taken to be a plane of surface area σ, the total Gibbs function (G) would be the sum of the Gibbs functions of each phase, $G^{(\alpha)}+G^{(\beta)}$. But the components are not uniform, and the sum of the two Gibbs functions differs from G by an amount called the *surface Gibbs function* (symbol: $G^{(\sigma)}$):

$$G^{(\sigma)}=G-\{G^{(\alpha)}-G^{(\beta)}\}. \qquad (25.4.1)$$

Similarly, if the bulk contains an amount $n_{\mathrm{J}}^{(\alpha)}$ of J in phase α and an amount $n_{\mathrm{J}}^{(\beta)}$ in phase β, the total amount of J differs from their sum by the amount

$$n_{\mathrm{J}}^{(\sigma)}=n_{\mathrm{J}}-\{n_{\mathrm{J}}^{(\alpha)}+n_{\mathrm{J}}^{(\beta)}\}. \qquad (25.4.2)$$

This excess amount is expressed as an amount per unit area of the surface by introducing the *surface excess* (symbol: Γ_{J}):

$$\Gamma_{\mathrm{J}}=n_{\mathrm{J}}^{(\sigma)}/\sigma. \qquad (25.4.3)$$

Note that both $n_{\mathrm{J}}^{(\sigma)}$ and Γ_{J} may be either positive (accumulation of J at the interface) or negative.

A general change in G is brought about by changes in T, p, σ, and the n_{J}:

$$\mathrm{d}G=-S\,\mathrm{d}T+V\,\mathrm{d}p+\gamma\,\mathrm{d}\sigma+\sum_{\mathrm{J}}\mu_{\mathrm{J}}\,\mathrm{d}n_{\mathrm{J}}.$$

When this is applied to G, $G^{(\alpha)}$, and $G^{(\beta)}$, and eqn (25.4.1) is used, we find

$$\mathrm{d}G^{(\sigma)}=-S^{(\sigma)}\,\mathrm{d}T+\gamma\,\mathrm{d}\sigma+\sum_{\mathrm{J}}\mu_{\mathrm{J}}\,\mathrm{d}n_{\mathrm{J}}^{(\sigma)} \qquad (25.4.4)$$

because at equilibrium the chemical potential of each component is the same in every phase ($\mu_{\mathrm{J}}^{(\alpha)}=\mu_{\mathrm{J}}^{(\beta)}=\mu_{\mathrm{J}}^{(\sigma)}$). Just as in the discussion of partial molar quantities (Section 8.1), this expression integrates at constant temperature to

$$G^{(\sigma)} = \gamma\sigma + \sum_{J} \mu_J n_J^{(\sigma)}. \tag{25.4.5}$$

We are seeking a connection between the change of surface tension ($d\gamma$) and the change of composition at the interface. Therefore, we use the argument which in Section 8.1(b) led to the Gibbs–Duhem equation, eqn (8.1.5), but this time we compare eqn (25.4.4) at constant temperature ($dT = 0$) with

$$dG^{(\sigma)} = \gamma\, d\sigma + \sigma\, d\gamma + \sum_{J} n_J^{(\sigma)}\, d\mu_J + \sum_{J} \mu_J\, dn_J^{(\sigma)}.$$

This implies that

$$\sigma\, d\gamma + \sum_{J} n_J^{(\sigma)}\, d\mu_J = 0, \quad \text{at constant } T.$$

Division by σ then gives the *Gibbs surface tension equation:*

$$d\gamma = -\sum_{J} \Gamma_J\, d\mu_J. \tag{25.4.6}$$

The Gibbs equation can be simplified in the case of a detergent D distributed between the two phases of the system by making the approximation that the 'oil' and 'water' phases are separated by a geometrically flat surface. This means that only the detergent accumulates at the surface, and so both Γ_{oil} and Γ_{water} are zero. Then the Gibbs equation becomes $d\gamma = -\Gamma_D\, d\mu_D$. For dilute solutions $d\mu_D = RT\, d\ln c_D$, where c_D is the concentration of the detergent. It follows that $d\gamma = -(RT/c_D)\Gamma_D\, dc_D$ at constant temperature, or

$$(\partial\gamma/\partial c_D)_T = -RT\Gamma_D/c_D. \tag{25.4.7)°$$

The implication of this equation is as follows. If the detergent accumulates at the interface its surface excess is positive, and so $(\partial\gamma/\partial c_D)_T$ is negative. That is, the surface tension *decreases* when a solute accumulates at a surface. Conversely, if the concentration dependence of γ is known, the surface excess may be predicted. The predictions have been tested by the simple (but technically elegant) procedure of slicing thin layers off the surfaces of solutions and analysing their compositions. Note that the result summarized by eqn (25.4.7) depends on the assumption that the solutions are ideal; there may be marked deviations at the concentrations of detergents used in practice.

This section (and this part) can be rounded off by linking the properties of an ideal surface phase, as expressed by eqn (25.4.7), with the properties of perfect gases that began Part 1. At small detergent concentrations the surface tension can be expected to vary linearly, and we can write $\gamma = \gamma^* - Kc_D$, with K some constant. Then, eqn (25.4.7) leads to

$$\Gamma_D = Kc_D/RT = (\gamma^* - \gamma)/RT.$$

If the *surface pressure* is defined as $\pi = \gamma^* - \gamma$, then with eqn (25.4.3) this equation becomes

$$\pi\sigma = n_D^{(\sigma)}RT, \tag{25.4.8}$$

which is the equation for a two-dimensional perfect gas. The excess solute at the interface of dilute, ideal solutions can therefore be pictured as a perfect gas confined to a two-dimensional surface.

Appendix 25.1 The elasticity of rubber

Consider a one-dimensional freely-jointed polymer. The conformation can be expressed in terms of the number of bonds pointing to the right (N_R) and the number pointing to the left (N_L). The distance between the ends of the chain is $(N_R - N_L)l$, where l is the length of an individual bond. We write $n = N_R - N_L$. The total number of bonds is $N = N_R + N_L$.

The number of ways of forming a chain with a given end-to-end distance nl is the number of ways of having N_R right-pointing and N_L left-pointing bonds. This is given by the binomial coefficient

$$W = N!/N_R!\,N_L! = N!/\{\tfrac{1}{2}(N+n)\}!\,\{\tfrac{1}{2}(N-n)\}!.$$

The *conformational entropy* of the chain, $S = k \ln W$, is therefore

$$S/k = \ln N! - \ln N_R! - \ln N_L!.$$

Since the factorials are large (except for large extensions), the more accurate form of Stirling's approximation (footnote, p. 509), $\ln x! \approx \ln (2\pi)^{\frac{1}{2}} + (x+\tfrac{1}{2}) \ln x - x$, can be used to obtain

$$S/k = -\ln (2\pi)^{\frac{1}{2}} + (N+1) \ln 2 + (N+\tfrac{1}{2}) \ln N \\ -\tfrac{1}{2} \ln \{(N+n)^{N+n+1}(N-n)^{N-n+1}\}.$$

The most probable conformation of the chain is the one with the ends close together ($n = 0$), as may be confirmed by differentiation. Therefore, the maximum entropy is

$$S_{max}/k = -\ln (2\pi)^{\frac{1}{2}} + (N+1) \ln 2 - \tfrac{1}{2} \ln N.$$

The change of entropy when the chain is stretched from its most probable conformation so that the distance between its ends is nl is therefore

$$\Delta S = S - S_{max} = \tfrac{1}{2}k\{\ln N^{N+1}N^{N+1} - \ln (N+n)^{N+n+1}(N-n)^{N-n+1}\} \\ = -\tfrac{1}{2}kN \ln \{(1+\nu)^{1+\nu}(1-\nu)^{1-\nu}\},$$

with $\nu = n/N$. The molar entropy change is therefore

$$\Delta S_m = -\tfrac{1}{2}NR \ln \{(1+\nu)^{1+\nu}(1-\nu)^{1-\nu}\}.$$

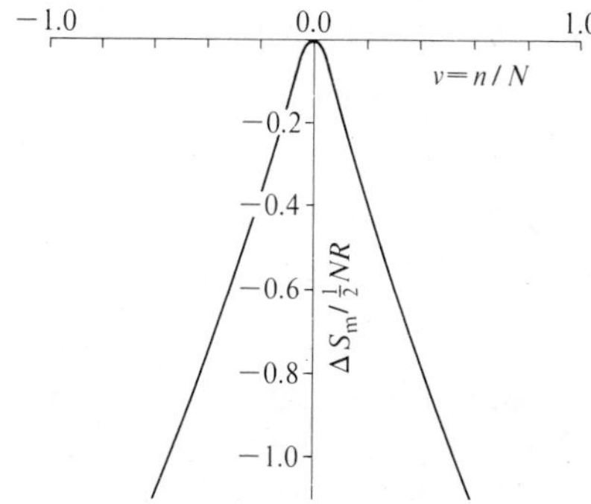

Fig. 25.22. The change in molar entropy of a perfect rubber as its extension changes. $\nu = 1$ corresponds to complete extension; $\nu = 0$, the conformation of highest entropy, corresponds to the random coil.

This is plotted in Fig. 25.22. Note that it is negative for all extensions, and so we conclude that adiabatic contraction of the chain to its fully coiled state is spontaneous.

The calculation can be taken further. The work done on a piece of rubber when it is extended through a distance dx is $f\,dx$, where f is the restoring force. The First Law is therefore $dU = T\,dS - p\,dV + f\,dx$. It follows that

$$(\partial U/\partial x)_{T,V} = T(\partial S/\partial x)_{T,V} + f.$$

In a perfect rubber, as in a perfect gas, the internal energy is independent of the dimensions (at constant temperature), and so $(\partial U/\partial x)_{T,V} = 0$. The restoring force is therefore

$$f = -T(\partial S/\partial x)_{T,V}.$$

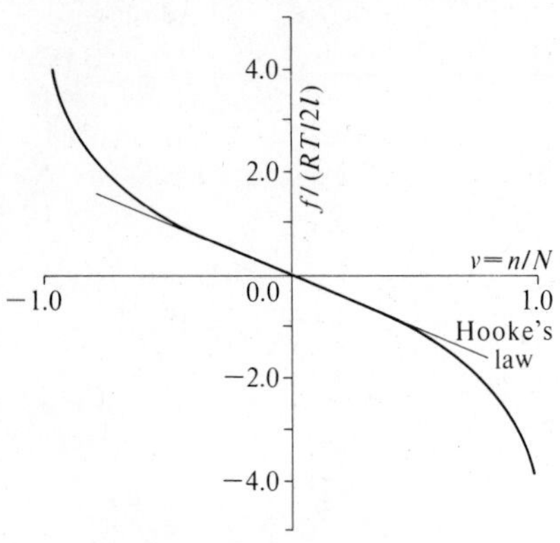

Fig. 25.23. The restoring force, f, of a one-dimensional perfect rubber. For small deflections, f is linearly proportional to extension, corresponding to Hooke's Law.

If the statistical expression for the entropy is introduced into this equation (and we evade problems arising from the constant volume constraint by assuming that the sample contracts laterally when it is stretched), we obtain

$$f = -(T/l)(\partial S/\partial n)_{T,V} = -(T/Nl)(\partial S/\partial \nu)_{T,V}$$
$$= (RT/2l)\ln\{(1+\nu)/(1-\nu)\},$$

which is plotted in Fig. 25.23. At low extensions ($\nu \ll 1$), $f \approx RTn/Nl$, and so the sample obeys Hooke's Law (restoring force proportional to the displacement), but it departs from it at higher extensions.

Further reading

Biological macromolecules:

Physical biochemistry. K. E. van Holde; Prentice-Hall, Englewood Cliffs, 1971.
Physical biochemistry (2nd edn). D. Freifelder; W. H. Freeman & Co., San Francisco, 1982.
Biophysical chemistry. A. G. Marshall; Wiley-Interscience, New York, 1978.
Biophysical chemistry (Part 1). C. R. Cantor and P. R. Schimmel; W. H. Freeman & Co., San Francisco, 1980.
The structure and action of proteins. R. E. Dickerson and I. Geiss; Benjamin, Menlo Park, 1969.

Polymers:

Physical chemistry of macromolecules. C. Tanford; Wiley, New York, 1961.
Principles of polymer chemistry. P. Flory; Cornell University Press, 1953.

Colloids:

Colloids and interparticle forces. D. Eagland; *Contemporary physics* **14,** 119 (1973): J. N. Israelichvili: *ibid*. **15,** 159 (1974).
Physical chemistry of surfaces (3rd edn). A. W. Adamson; Wiley, New York, 1976.
Surfactants and interfacial phenomena. M. J. Rosen; Wiley-Interscience, New York, 1978.

Introductory problems

A25.1. The concentration dependence of the osmotic pressure of solutions of a macromolecule at 20 °C was found to be as follows:

$c/\text{g dm}^{-3}$	1.21	2.72	5.08	6.60
Π/Pa	134	321	655	898

Find the molar mass of the macromolecule and the osmotic virial coefficient.

A25.2. At the start of a membrane equilibrium experiment, the first compartment contains 1.00 dm^3 of solution with a concentration of 0.100 mol dm^{-3} of NaX, where X^- cannot pass through the membrane. The second compartment has 2.00 dm^3 of 0.030 M NaCl solution. Find the concentration of Cl^- ion in the first compartment after equilibrium is established.

A25.3. At the start of a Donnan membrane experiment, the first compartment contains 2.00 dm^3 of solution which is 0.015 M in the polyelectrolyte Na_2P and 0.010 M in NaCl. The second compartment has 2.00 dm^3 of solution which is 0.005 M in NaCl. What is the electrode potential associated with the sodium ion concentration difference across the membrane at 300 K?

A25.4. The data from a sedimentation equilibrium experiment performed at 300 K on a macromolecular solute in aqueous solution show that a graph of ln c vs. r^2 is a straight line with a slope of 729 cm^{-2}. The rotational speed of the centrifuge was 50 000 r.p.m. The specific volume of the solute is 0.61 cm^3 g^{-1}. Find the molar mass of the solute.

A25.5. A polymer chain consists of 700 segments, each 0.90 nm long. If the chain were ideally flexible and were randomly positioned, what would be the r.m.s. separation of the ends of the chain?

A25.6. The radius of gyration of a long chain molecule is found to be 7.3 nm. The chain consists of C—C links. Assume the chain is randomly coiled and estimate the number of links in the chain.

A25.7. The concentration dependence of the viscosity of a polymer solution is found to be as follows:

$c/\text{g dm}^{-3}$	1.32	2.89	5.73	9.17
$\eta/\text{g m}^{-1}\text{s}^{-1}$	1.08	1.20	1.42	1.73

The viscosity of the solvent is $0.985\ \text{g m}^{-1}\text{s}^{-1}$. Find the intrinsic viscosity of the polymer.

A25.8. In a sedimentation experiment the position of the boundary as a function of time was found to be as follows:

t/minutes	15.5	29.1	36.4	58.2
r/cm	5.05	5.09	5.12	5.19

The rotational speed of the centrifuge was 45 000 r.p.m. Calculate the sedimentation constant of the solute.

A25.9. At 20 °C the diffusion coefficient of a macromolecule is found to be $8.3\times10^{-7}\ \text{cm}^2\,\text{s}^{-1}$. Its sedimentation constant is 3.2×10^{-13} s in a solution of density $1.06\ \text{g cm}^{-3}$. The specific volume of the macromolecule is $0.656\ \text{cm}^3\,\text{g}^{-1}$. Calculate the RMM of the macromolecule.

A25.10. A solution consists of solvent, a dimer with $M_r = 30\,000$ and its monomer. Thirty per cent of the total mass of solute is dimer. What average RMM would be obtained from measurement of: (a) osmotic pressure, (b) light scattering?

Problems

25.1. The osmotic pressures of solutions of polystyrene in toluene were measured at 25 °C, with the following results:

$c_P/\text{mg cm}^{-3}$	3.2	4.8	5.7	6.9	7.8
h/cm	3.11	6.22	8.40	11.73	14.90

h is the height of the solution of density $0.867\ \text{g cm}^{-3}$ corresponding to the osmotic pressure. Obtain the RMM of the polymer by plotting h/c_P against c_P.

25.2. The osmotic pressure of a fraction of polyvinylchloride in a ketone solvent was measured at 25 °C. The density of the solvent (which is virtually equal to the density of the solution) was $0.798\ \text{g cm}^{-3}$. Calculate the molar mass and the virial coefficient B of the fraction from the following data:

$c_P/(\text{g}/100\ \text{cm}^3)$	0.200	0.400	0.600	0.800	1.000
h/cm	0.48	1.12	1.86	2.76	3.88

25.3. Calculate the number average RMM and the mass average RMM of a mixture of equal amounts (n) of two polymers, one having $M_r = 62\,000$ and the other $M_r = 78\,000$.

25.4. A polymerization process produced a Gaussian distribution of polymers in the sense that the proportion of molecules having an RMM in the range M_r to $M_r + dM_r$ was proportional to $\exp\{-(M_r - \bar{M}_r)^2/2\Gamma\}\,dM_r$. What is the number average RMM when the distribution is narrow?

25.5. Calculate the excluded volume in terms of the molecular volume on the basis that the molecules are spheres of radius a. (The calculation is much more difficult in the case of rigid rods: see Tanford in *Further reading*.) Evaluate the osmotic virial coefficient in the case of bushy stunt virus, $a \approx 14.0$ nm, and haemoglobin, $a \approx 3.2$ nm.

25.6. Evaluate the percentage deviation of the osmotic pressures of $1.00\ \text{g}/100\ \text{cm}^3$ solutions of bushy stunt virus ($M_r \approx 1.07\times10^7$) and haemoglobin ($M_r \approx 66\,500$) from the ideal solution values.

25.7. The effective radius of a random coil a_{eff} is related to its radius of gyration R_g by $a_{\text{eff}} \approx \gamma R_g$, with $\gamma \approx 0.85$. Deduce an expression for the osmotic virial coefficient B in terms of the number of chain units for (a) a freely jointed chain, (b) a chain with tetrahedral interbond angles. Evaluate B for $l = 154$ pm and $N = 4000$.

25.8. Estimate the osmotic virial coefficient B for a randomly coiled polyethylene chain of arbitrary M_r and evaluate it for $M_r = 56\,000$.

25.9. Confirm the eqn (25.1.11) for the differences in ion concentrations on either side of a membrane follows from the preceding equations.

25.10. Consider the effect of adding a salt $(M^+)_2X^{2-}$ to a solution of a polyelectrolyte $(M^+)_\nu P^{\nu-}$. Find an expression for the differences of ion concentrations on either side of a membrane permeable to everything except the polyanion.

25.11. Show that the ratio $[Na^+]_L/[Na^+]_R$ in the Donnan equilibrium is equal to $x + (1+x^2)^{\frac{1}{2}}$, where $x = \nu[P]/2[Na^+]_R$ and sketch the ratio as a function of the polyelectrolyte concentration.

25.12. A polyelectrolyte $Na_{20}P$ with $M_r = 100\,000$ at a concentration $1\ \text{g}/100\ \text{cm}^3$ was equilibrated in the presence of 0.0010 M aqueous NaCl (i.e., $[Na^+]_R = 0.0010$ M). What is the value of $[Na^+]_L$ at equilibrium?

25.13. Discuss the role of the deviations from ideality ($\gamma_\pm \neq 1$) in determining the ratio of ion concentrations in the Donnan equilibrium.

25.14. Investigation of the composition of the solutions used to study the osmotic pressure due to a polyelectrolyte with $\nu = 20$ showed that at equilibrium the concentrations corresponded to $[Cl^-] \approx 0.02\ \text{mol dm}^{-3}$. Calculate the value of the osmotic virial coefficient for $\nu = 20$. Does it dominate the effect of excluded volume?

25.15. Calculate the radial acceleration (as so many g) in a cell placed at 6.0 cm from the centre of rotation in an ultracentrifuge operating at 80 000 r.p.m.

25.16. Calculate the speed of operation (in r.p.m.) of an ultracentrifuge needed in order to obtain a readily measurable concentration gradient in a sedimentation equilibrium experiment. Take that to be a concentration at the bottom of the cell about 5 times greater that at the top. Use $r_{top}=5.0$ cm, $r_{bottom}=7.0$ cm, $M_r\approx 10^5$, $\rho v_s\approx 0.75$, $T=298$ K.

25.17. In an ultracentrifuge experiment at 20 °C on bovine serum albumin the following data were obtained: $\rho=1.001$ g cm^{-3}, $v_s=1.112$ cm^3 g^{-1}, $\omega/2\pi=322$ Hz,

r/cm	5.0	5.1	5.2	5.3	5.4
c/mg cm^{-3}	0.536	0.284	0.148	0.077	0.039.

Evaluate M_r.

25.18. Sedimentation studies on the haemoglobin in water gave a sedimentation constant $S=4.5\times 10^{-13}$ s at 20 °C. The diffusion coefficient is 6.3×10^{-7} cm^2 s^{-1} at the same temperature. Calculate the RMM of haemoglobin using $v_s=0.75$ cm^3 g^{-1} for its partial specific volume and $\rho=0.998$ g cm^{-3} for the density of the solution.

25.19. Estimate the effective radius of the haemoglobin molecule by combining the data in the last Problem with the information that the viscosity of the solution is 1.00×10^{-3} kg m^{-1} s^{-1}.

25.20. The diffusion coefficient for bovine serum albumin, a prolate ellipsoid, is 6.97×10^{-7} cm^2 s^{-1} at 20 °C, its partial specific volume is 0.734 cm^3 g^{-1}, and its sedimentation constant is 5.01×10^{-13} s in a solution of density 1.0023 g cm^{-3} and viscosity 1.00×10^{-3} kg m^{-1} s^{-1}. Estimate its dimensions.

25.21. The rate of sedimentation of a recently isolated protein was monitored at 20 °C and with a rotor speed of 50 000 r.p.m. The boundary receded as follows:

t/s	0	300	600	900
r/cm	6.127	6.153	6.179	6.206
t/s	1200	1500	1800	
r/cm	6.232	6.258	6.284	

Calculate the sedimentation constant S and the molar mass of the protein on the basis that its partial specific volume (which was measured in a pyknometer) is 0.728 cm^3 g^{-1} and its diffusion coefficient is 7.62×10^{-7} cm^2 s^{-1} at 20 °C, the density of the solution then being 0.9981 g cm^{-3}.

25.22. Suggest a shape for the protein considered in the last Problem, given that the viscosity of the solution is 1.00×10^{-3} kg m^{-1} s^{-1} at 20 °C.

25.23. The viscosities of solutions of polyisobutylene in benzene were measured at 24 °C (the θ-temperature for the system) with the following results:

c/(g/100 cm^3)	0	0.2	0.4
$\eta/10^{-3}$ kg m^{-1} s^{-1}	0.647	0.690	0.733
c/(g/100 cm^3)	0.6	0.8	1.0
$\eta/10^{-3}$ kg m^{-1} s^{-1}	0.777	0.821	0.865

Use the information in Table 25.2 to deduce the RMM of the polymer.

25.24. The radius of gyration is defined in eqn (25.1.28). Show that an equivalent definition is that R_g is the average root mean square distance of the atoms or groups (all assumed to be of the same mass); that is, that $R_g^2=(1/N)\sum_j R_j^2$, where R_j is the distance of atom j from the centre of mass.

25.25. Use eqn (25.2.1) to deduce expressions for (a) the root mean square separation of the ends of the chain, (b) the mean separation of the ends, and (c) their most probable separation. (Integrals over Gaussians are given in Box 26.1). Evaluate these three quantities for an $N=4000$, $l=154$ pm fully flexible chain.

25.26. Construct a two-dimensional random walk either using tables of random numbers (e.g. Abramowitz and Stegun, *Handbook of mathematical functions*) or using a random number generating program of a computer. Construct a walk of 50 and 100 steps. If there are many people working on the problem, investigate the mean and most probable separations in the plots by direct measurement. Do they vary as $N^{\frac{1}{2}}$?

25.27. Evaluate the radius of gyration of (a) a solid sphere of radius a, (b) a long straight rod of radius a and length l. (Use the expression deduced in Problem 25.24.) Show that in the case of a solid sphere of specific volume v_s, $R_g/\text{nm}\approx 0.056\,902\times\{(v_s/\text{cm}^3\,\text{g}^{-1})M_r\}^{\frac{1}{3}}$. Evaluate R_g for species with $M_r=100\,000$, $v_s=0.750$ cm^3 g^{-1}, and, in the case of the rod, a radius 0.5 nm.

25.28. Use the information below and the expression for R_g of a solid sphere derived in the last Problem, to classify the species below as globular or rod-like:

	M_r	v_s/cm^3 g^{-1}	Measured R_g/nm
Serum albumin	66×10^3	0.752	2.98
Bushy stunt virus	10.6×10^6	0.741	12.0
DNA	4×10^6	0.556	117.0

25.29. In formamide as solvent, poly(γ-benzyl-L-glutamate) is found by light scattering experiments to have a radius of gyration proportional to M_r; in contrast, polystyrene in butanone has R_g proportional to $M_r^{\frac{1}{2}}$. Present arguments to show that the first polymer is a rigid rod, while the second is a random coil.

25.30. Evaluate eqn (25.1.26) for $P(\theta)$ in the case of a long rigid rod of N identical scattering units with separation l, and assume that the units are so closely spaced that sums may be replaced by integrals.

25.31. Plot $P(\theta)$ as calculated in the last Problem as a function of θ on polar graph paper in the case $L\approx\lambda$. The integral $\int_0^z(\sin x/x)\,dx$ is the 'sine-integral' Si(z), and is tabulated in Abramowitz and Stegun, *Handbook of mathematical functions*.

25.32. Evaluate the rotational correlation time for serum albumin in water at 25 °C on the basis that it is a sphere of radius 3.0 nm. What is the value for a CCl_4 molecule in carbon tetrachloride at 25 °C? (Viscosity data in Table 24.5; take $a(CCl_4)\approx 250$ pm.)

25.33. We now turn attention to the thermodynamic description of stretching rubber. The observables are the tension t and length l (like p and V for gases). Since $dw = t\,dl$, the basic equation is $dU = T\,dS + t\,dl$. ($p\,dV$ terms are supposed negligible throughout.) If $G = U - TS - tl$, find expressions for dG and dA and deduce the Maxwell relations $(\partial S/\partial l)_T = -(\partial t/\partial T)_l$ and $(\partial S/\partial t)_T = (\partial l/\partial T)_t$.

25.34. Continue the thermodynamic analysis by deducing the equation of state for rubber, $(\partial U/\partial l)_T = t - T(\partial t/\partial T)_l$.

25.35. On the assumption that the tension required to keep a sample at a constant length is proportional to the temperature ($t = aT$, the analogue of $p \propto T$), show that the tension can be ascribed to the dependence of the entropy on the length of the sample. Account for this result in terms of the molecular nature of the sample.

25.36. The energy of interaction between two spheres of radius a composed of atoms at a number density $\mathcal{N}$ that individually interact as $-C_6/R^6$ is $V(R) = -\frac{1}{6}\pi^2 C_6 \mathcal{N}^2 f(s)$, with $f(s) = 2/(s^2-4) + 2/s^2 + \ln(1-4/s^2)$ and $s = R/a$; see A. W. Adamson, *Physical chemistry of surfaces*, p. 318. Plot $f(s)$ from $s = 4$ up to $s = 10$ in order to appreciate the range of the interaction energy.

25.37. The surface tensions of a series of aqueous solutions of a surfactant were measured at 20 °C, with the following results:

[A]/M	0	0.10	0.20	0.30	0.40	0.50
γ/mN m^{-1}	72.8	70.2	67.7	65.1	62.8	59.8

Calculate the surface excess concentration.

25.38. Evaluate the surface pressure π exerted by the surfactant in the last Problem, and investigate whether eqn (25.4.8) is satisfied.

25.39. The surface tensions of aqueous salt solutions are normally greater than that of water itself. Does the salt accumulate at the surface?

25.40. The surface tensions of solutions of salts in water at concentration c can be expressed in the form $\gamma = \gamma^* + (c/\text{M})\Delta\gamma$. The values of $\Delta\gamma$ at 20 °C and near $c = 1$ M are as follows:

$$\Delta\gamma/\text{mN m}^{-1} = 1.4(\text{KCl}), \quad 1.64(\text{NaCl}), \quad 2.7(\text{Na}_2\text{CO}_3).$$

Calculate the surface excess concentrations when the bulk concentrations are 1.0 M.

Part 3 · Change

26

Molecules in motion: the kinetic theory of gases

Learning objectives

After careful study of this chapter you should be able to:

(1) Specify the *kinetic theory* model of a perfect gas, Introduction.

(2) Use the kinetic theory to calculate the *pressure* exerted by a perfect gas, eqn (26.1.2).

(3) Define the *mean value* of discrete, eqn (26.1.6), and continuous, eqn (26.1.7), distributions.

(4) Derive and use the *Maxwell–Boltzmann distribution of velocities*, eqn (26.1.8), and the *Maxwell distribution of speeds*, eqn (26.1.9) and Example 26.1.

(5) Calculate the *mean speed*, eqn (26.1.10), the *r.m.s. speed*, eqn (26.1.11), and the *most probable speed*, eqn (26.1.12), of particles in a gas.

(6) Define *collision cross-section*, Section 26.2(a), and calculate the *collision frequency*, eqns (26.2.2) and (26.2.5) and Example 26.2, and the *mean free path*, eqn (26.2.6) and Example 26.3, of particles in a gas.

(7) Calculate the frequency of collisions with a surface, eqn (26.2.9).

(8) Explain the term *transport property* and define *flux*, Section 26.3(a).

(9) State *Fick's First Law* of diffusion, eqn (26.3.1).

(10) Calculate the *rate of effusion* of a gas through a hole, eqn (26.3.4), state and justify *Graham's Law*, and use *Knudsen's method* to measure vapour pressure, Example 26.4.

(11) Derive Fick's Law and calculate the *diffusion coefficient* of a perfect gas, eqn (26.3.6).

(12) Calculate the *thermal conductivity* of a gas, eqn (26.3.11) and Example 26.5, and account for its properties.

(13) Calculate the *viscosity* of a gas, eqn (26.3.13), and account for its properties.

(14) Describe how gas viscosities are measured, Section 26.3(e) and Example 26.6.

Introduction

We saw in the Introduction (Section 0.2(e)) that the *kinetic theory*, in which the particles are assumed to move freely, allows us to calculate a number of properties of a perfect gas. That brief outline is the starting point of this chapter, and should be reviewed. We sharpen the argument in this chapter, and go on to find expressions for some of the more interesting properties of gases.

The kinetic theory is based on three assumptions:

(1) The gas consists of a swarm of particles of mass m and diameter d in continual random motion.

(2) The size of the particles is negligible (in the sense that their diameters are much smaller than the average distance travelled between collisions).

(3) The particles do not interact, except that they make perfectly elastic collisions when the separation of their centres is equal to d.

An 'elastic collision' means that the total translational kinetic energy of a pair is the same before and after a collision: no energy is transferred to their rotational or vibrational modes.

The collisions ensure that the particles constantly change their speed and direction. The *collision frequency* (symbol: z) is the average number of collisions per unit time made by a single particle. The *mean free path* (symbol: λ) is the average distance a particle travels between collisions. This means that Assumption 2 can be expressed as $d \ll \lambda$. It should be noted that since any mechanical property can be expressed as a combination of quantities with dimensions of mass, length, and time, and that since m, λ, and $1/z$ are quantities with these dimensions, once they have been calculated we can find expressions for any mechanical property of a gas by forming suitable combinations.

26.1 The basic calculations

There are two basic calculations in kinetic theory: one leads to an expression for the pressure, the other to an expression for the distribution of velocities of the particles.

26.1 (a) The pressure of a gas

Pressure is force per unit area. The kinetic theory accounts for the steady pressure exerted by a gas in terms of collisions of the particles with the walls of the container (but this is an inessential simplification: the pressure can be defined and calculated in the absence of walls). These collisions are so numerous that the walls experience a virtually constant force.

Consider the system in Fig. 26.1. When a particle of mass m collides with the shaded wall its component of momentum parallel to the x-axis changes from mv_x to $-mv_x$, its other components remaining unchanged. The momentum therefore changes by $2|mv_x|$ on each collision. The number of collisions in an interval Δt is equal to the number of particles able to reach the wall in that time. Since a particle with velocity component v_x can travel a distance $|v_x|\Delta t$ in a time Δt, all the particles within a distance $|v_x|\Delta t$ of the wall will strike it if they are travelling towards it. If the wall has area A, all the particles in a volume $A|v_x|\Delta t$ will reach the wall (if they are

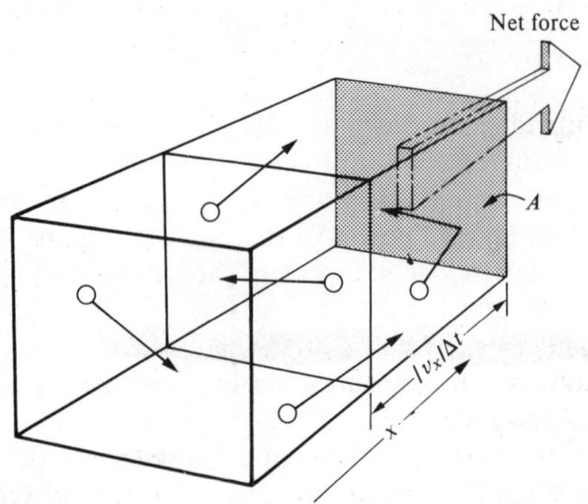

Fig. 26.1. The calculation of the pressure of a gas considers the total force exerted on the shaded wall by the colliding particles. In an interval Δt, all particles within the volume $A|v_x|\Delta t$ and moving towards the wall strike it and undergo a reversal of their x-component of momentum.

travelling towards it). If the number of particles per unit volume (the *number density*) is $\mathcal{N}$, the number in the volume $A|v_x|t$ is $\mathcal{N}A|v_x|\Delta t$. On average, half these particles are moving to the right, and half are moving to the left, and so the average number of collisions with the wall during the interval Δt is $\frac{1}{2}A|v_x|\mathcal{N}\Delta t$. The total momentum change in that interval is the product of this number and the change $2m|v_x|$:

$$\textit{Momentum change} = \{\tfrac{1}{2}\mathcal{N}A|v_x|\Delta t\}\{2m|v_x|\} = m\mathcal{N}Av_x^2\Delta t.$$

The *rate of momentum change* is therefore $m\mathcal{N}Av_x^2$. The rate of change of momenum is equal to the *force* (by Newton's Law), and so the force exerted by the gas on the wall is $m\mathcal{N}Av_x^2$. It follows that the *pressure*, the force per unit area, is $m\mathcal{N}v_x^2$.

Not all the particles travel with the same velocity, and so the detected pressure is the average (denoted $\langle\ldots\rangle$) of the quantity just calculated:

$$p = m\mathcal{N}\langle v_x^2\rangle.$$

Since the particles are moving randomly, the average of v_x^2 is the same as the average of the corresponding quantities in the y and z directions. Furthermore, since $v^2 = v_x^2 + v_y^2 + v_z^2$, the *mean square speed* (symbol: c^2) of the particles is

$$c^2 = \langle v^2\rangle = \langle v_x^2\rangle + \langle v_y^2\rangle + \langle v_z^2\rangle = 3\langle v_x^2\rangle. \qquad (26.1.1)$$

The square root of this quantity is the *root mean square speed* (symbol: c; $c = \langle v^2\rangle^{\frac{1}{2}}$). Therefore,

$$p = \tfrac{1}{3}\mathcal{N}mc^2. \qquad (26.1.2)^\circ$$

This is one of the key equations of kinetic theory. (The convention in Parts 1 and 2 is that a superscript ° signifies a perfect gas; by extension it now also means a result stemming from kinetic theory.)

The number density is equal to N/V, where N is the total number of particles present in the volume V. Then, since $N = nN_A$, where N_A is Avogadro's constant,

$$pV = \tfrac{1}{3}nN_Amc^2. \qquad (26.1.3)^\circ$$

Since we know that a perfect gas satisfies the equation of state $pV = nRT$, we can conclude at once that

$$c = \{3kT/m\}^{\frac{1}{2}}, \qquad (26.1.4)^\circ$$

and so obtain an explicit expression for the root mean square speed of the particles at any temperature.

26.1 (b) Mean values and distributions

The expression for c gives only a mean value; in the gas the particles move with a variety of different speeds. In this section we derive the *Maxwell distribution*, which gives the probability that particles have any given speed.

Suppose we want the *mean value* (symbol: $\langle X\rangle$) of a property X which may take any of the values $X_1, X_2, \ldots, X_z$ (these are the possible *outcomes* of the observation), and in a series of N measurements we find that X_1 occurs N_1 times, X_2 occurs N_2 times, and so on. Then the mean value is given by

$$\langle X\rangle = (1/N)\{N_1X_1 + N_2X_2 + \ldots + N_zX_z\} = \sum_i (N_i/N)X_i. \quad (26.1.5)$$

This can be expressed in terms of the *probability* (symbol: P_i) that an outcome X_i is obtained by writing $P_i = N_i/N$, for then

$$\langle X\rangle = \sum_i P_iX_i, \qquad P_i = N_i/N, \quad (26.1.6)$$

the sum ranging over all possible outcomes.

Now consider the case where the outcomes may take any of a *continuous* range of values, as in the height of a population or the speed of a particle in gas. We divide up the continuous range of possible values into segments, Fig. 26.2, and count 1 each time a measurement has an outcome which falls anywhere in a given segment. For example, consider the segment of length ΔX at X. If in a series of 300 observations an outcome lying in this range is obtained in six of them, we write $N(X) = 6$. If the total number of observations is N, the probability that the outcome of any single observation lies in the range X to $X + \Delta X$ is $P(X) = N(X)/N$, which in this case is $\frac{1}{50}$. The value of $N(X)$, and therefore of $P(X)$, is proportional to the length of the segment ΔX (so long as X is small), and so we write $P(X) = f(X)\Delta X$.

The bunching procedure has turned the continuous problem into one resembling the discrete problem described earlier, and we can continue as we did there. The *approximate* average value of X is found by taking the value of X for each segment, multiplying it by the probability that the observation will give an outcome in that segment, and then summing over all the segments:

$$\langle X\rangle \approx \sum_{\text{segments}} XP(X) = \sum_{\text{segments}} Xf(X)\Delta X.$$

This is onl: approximate, because $f(X)$ might vary appreciably over the width of the segment. However, the relation becomes exact if we allow each

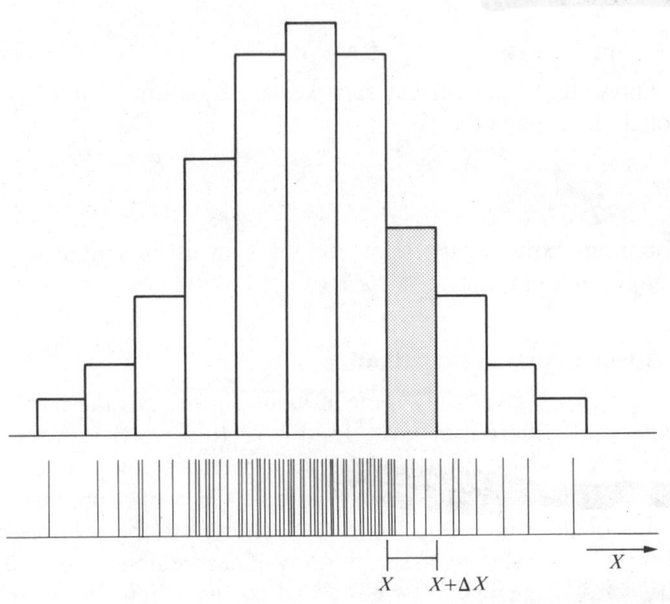

Fig. 26.2. A sequence of events: each line represents one observation of the corresponding value of X. An approximation to the average value is obtained by dividing the range of outcomes into packets of range ΔX (one is shown), and lumping together all the observations that have an outcome in that range. In this case six observations yield the outcome X.

segment to become infinitesimal, for then $f(X)$ will be constant over its range. We therefore replace the finite range ΔX by the infinitesimal range dX, and simultaneously convert the discrete sum into an integral over all possible outcomes:

$$\langle X \rangle = \int X f(X)\, dX. \qquad (26.1.7)$$

This is a very important expression: it lies at the heart of the development in this chapter.

The function $f(X)$ is called the *distribution* of the property X. From its original definition $P(X) = f(X)\,\Delta X$, which becomes $dP(X) = f(X)\,dX$ now that we are dealing with infinitesimal ranges, we see that it gives the probability that a property lies in the range X to $X + dX$. For example, $f(v_x)$ is the distribution of the x-component of velocity, and $f(v_x)\,dv_x$ tells us the probability that the velocity lies in the range v_x to $v_x + dv_x$.

We now turn to the problem of dealing with several properties simultaneously. *If two properties X and Y are independent of each other*, the probability of an observation having both X_i and Y_j as outcomes is the *product* of the individual probabilities: $P(X_i, Y_j) = P(X_i)P(Y_j)$. For example, if the probability of a person being a man in a given population is 0.495, and the probability of a person (man or woman) being left-handed is 0.110, then the probability of selecting a left-handed man by random choice from a crowd is $(0.110) \times (0.495) = 0.054$, or 1 in 18.5. However, if left-handedness were a male characteristic, this calculation would be false.

The same technique can be used for continuous properties. If the probability of X lying in the range X to $X + dX$ is $f(X)\,dX$, and the probability that an *independent* property Y lies in the range Y to $Y + dY$ is $f(Y)\,dY$, then the probability of X and Y lying simultaneously in these ranges is the product of the individual probabilities: $f(X, Y)\,dX\,dY = f(X)f(Y)\,dX\,dY$.

26.1 (c) The distribution of molecular velocities

We now have enough background to find the distribution of the components of velocities of particles in a perfect gas; this will prove to be useful when we are doing calculations relating to the rates of reactions and to the properties of molecular beams. The three components v_x, v_y, and v_z are independent of each other, and so the probability $F(v_x, v_y, v_z)\,dv_x\,dv_y\,dv_z$ that a particle has a velocity with components in the range v_x to $v_x + dv_x$, v_y to $v_y + dv_y$, and v_z to $v_z + dv_z$ is the product of the individual probabilities:

$$F(v_x, v_y, v_z)\,dv_x\,dv_y\,dv_z = f(v_x)f(v_y)f(v_z)\,dv_x\,dv_y\,dv_z.$$

Next, we assume that the probability of a particle having a particular range of velocity components is independent of its direction of flight. That is, we assume that $F(v_x, v_y, v_z)$ depends on the *speed* but not the individual components. So, for example, the probability of having a velocity with components $(1\ \mathrm{km\,s^{-1}}, 2\ \mathrm{km\,s^{-1}}, 3\ \mathrm{km\,s^{-1}})$ and speed $\sqrt{14}\ \mathrm{km\,s^{-1}}$ is the same as the probability of having one with components $(2\ \mathrm{km\,s^{-1}}, 1\ \mathrm{km\,s^{-1}}, 3\ \mathrm{km\,s^{-1}})$, or any other set corresponding to a speed of $\sqrt{14}\ \mathrm{km\,s^{-1}}$. It follows that F depends only on $v^2 = v_x^2 + v_y^2 + v_z^2$, and so we denote it $F(v_x^2 + v_y^2 + v_z^2)$. Then the last equation becomes

$$F(v_x^2 + v_y^2 + v_z^2) = f(v_x)f(v_y)f(v_z).$$

Only an exponential function satisfies this equation (because $e^{a+b+c} = e^a e^b e^c$), and so we can write

$$f(v_x) = K e^{\pm\zeta v_x^2},$$

with K and ζ constants. This is a *unique* solution, and

$$f(v_x)f(v_y)f(v_z) = K^3 e^{\pm\zeta(v_x^2+v_y^2+v_z^2)} = F(v_x^2 + v_y^2 + v_z^2),$$

as required. The $\pm$ ambiguity can be resolved on physical grounds: the probability of extremely high velocities must be very small; therefore, the negative sign must be taken.

The values of K and ζ are obtained as follows. Since the particle must have *some* velocity in the range $-\infty < v_x < \infty$, the total probability of the velocity being in that range is unity:

$$\int_{-\infty}^{\infty} f(v_x)\,\mathrm{d}v_x = 1.$$

Substitution of the expression given above leads to

$$\int_{-\infty}^{\infty} f(v_x)\,\mathrm{d}v_x = K\int_{-\infty}^{\infty} e^{-\zeta v_x^2}\,\mathrm{d}v_x = K(\pi/\zeta)^{\frac{1}{2}}.$$

Therefore, $K = (\zeta/\pi)^{\frac{1}{2}}$, and so

$$f(v_x) = (\zeta/\pi)^{\frac{1}{2}} e^{-\zeta v_x^2}.$$

The mean value of v_x^2 is

$$\langle v_x^2 \rangle = \int_{-\infty}^{\infty} v_x^2 f(v_x)\,\mathrm{d}v_x = (\zeta/\pi)^{\frac{1}{2}}\int_{-\infty}^{\infty} v_x^2 e^{-\zeta v_x^2}\,\mathrm{d}v_x.$$

The integral on the right is standard (Box 26.1), and is equal to $\frac{1}{2}(\pi/\zeta^3)^{\frac{1}{2}}$. It follows that

$$\langle v_x^2 \rangle = \tfrac{1}{2}(\zeta/\pi)^{\frac{1}{2}}(\pi/\zeta^3)^{\frac{1}{2}} = 1/2\zeta.$$

Therefore, the mean square speed is $c^2 = 3/2\zeta$. However, we have already established a value for c, eqn (26.1.4), and so we conclude that $\zeta = m/2kT$.

Box 26.1 Integrals over Gaussian functions

Let $I_n = \int_0^\infty x^n e^{-ax^2}\,\mathrm{d}x$, then:

n	0	1	2	3	4	5	6
I_n	$\frac{1}{2}(\pi/a)^{\frac{1}{2}}$	$1/2a$	$\frac{1}{4}(\pi/a^3)^{\frac{1}{2}}$	$1/2a^2$	$\frac{3}{8}(\pi/a^5)^{\frac{1}{2}}$	$1/a^3$	$\frac{15}{16}(\pi/a^7)^{\frac{1}{2}}$

The *error function* is the integral

$$\operatorname{erf} z = (2/\pi^{\frac{1}{2}})\int_0^z e^{-x^2}\,\mathrm{d}x = 1 - (2/\pi^{\frac{1}{2}})\int_z^\infty e^{-x^2}\,\mathrm{d}x.$$

It is listed in tables (such as M. Abramowitz and I. A. Stegun, *Handbook of mathematical functions*, Dover, 1965), and the following is a brief abstract (see also Table 14.1):

z	0	0.20	0.40	0.60	0.80	0.90	1.00	1.20	1.40	1.60	1.80	1.90
erf z	0	0.223	0.428	0.604	0.742	0.797	0.843	0.910	0.952	0.976	0.989	0.993

Therefore, the complete form of the velocity distribution is

$$f(v_x) = (m/2\pi kT)^{\frac{1}{2}} \exp(-mv_x^2/2kT). \qquad (26.1.8)^\circ$$

The right-hand side has the form of a Boltzmann distribution (Section 0.1(d)) for the proportion of particles with a kinetic energy $\frac{1}{2}mv_x^2$, and an alternative route to this equation is through the Boltzmann expression. For this reason the result is known as the *Maxwell–Boltzmann distribution* of molecular velocities, representing both Maxwell's contribution (he derived it originally) and Boltzmann's (he proved it rigorously).

The distribution of the *speeds* of the particles irrespective of their direction of motion is derived as follows. The probability that a particle has velocity components in the range v_x to $v_x + dv_x$, v_y to $v_y + dv_y$, and v_z to $v_z + dv_z$ is

$$F(v_x, v_y, v_z)\,dv_x\,dv_y\,dv_z = f(v_x)f(v_y)f(v_z)\,dv_x\,dv_y\,dv_z.$$
$$= (m/2\pi kT)^{\frac{3}{2}} e^{-mv^2/2kT}\,dv_x\,dv_y\,dv_z.$$

The probability $F(v)\,dv$ that the particle has a speed in the range v to $v + dv$ is the sum of the probabilities that it lies in any of the volume elements $dv_x\,dv_y\,dv_z$ in a spherical shell of radius v, Fig. 26.3. The sum of the volume elements on the right-hand side of the last expression is the volume of this shell, which is $4\pi v^2\,dv$. Therefore,

$$F(v) = 4\pi(m/2\pi kT)^{\frac{3}{2}} v^2 e^{-mv^2/2kT}, \qquad (26.1.9)^\circ$$

which is the *Maxwell distribution of speeds*.

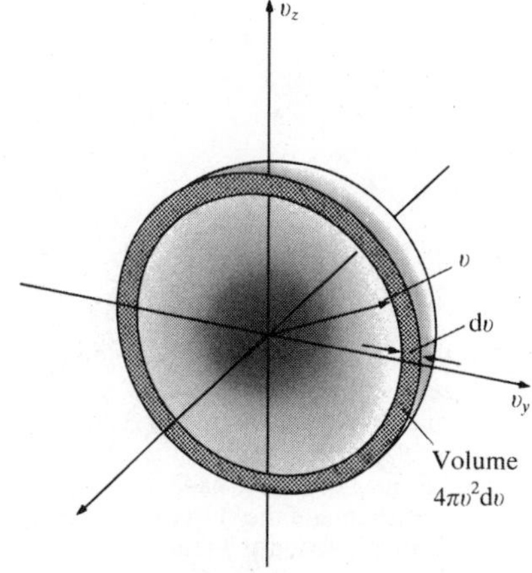

Fig. 26.3. The density of shading represents the probabilities of having each velocity. For the distribution of molecular speeds, the total probability of having a value in the shell of thickness dv and radius v is calculated.

Example 26.1

What is the mean speed of Cs atoms in an oven heated to 500 °C? What is the mean velocity of particles travelling in the positive x-direction?

- *Method*. The mean speed is obtained by evaluating the integral $\bar{c} = \int_0^\infty vF\,dv$ with F the distribution in eqn (26.1.9). Then substitute $M_r = 132.9$ and $T = 773$ K. The mean velocity towards positive x is obtained by evaluating the integral $\langle v_x \rangle = \int_0^\infty v_x f'\,dv_x$, where $f' = 2f$, with f taken from eqn (26.1.8) but multiplied by two because we are concerned with only the distribution relating to the range $0 \leqslant v_x < \infty$, which is half the range used to normalize f in the equations leading to eqn (26.1.8). Integrals are listed in Box 26.1.

- *Answer*. $\bar{c} = \int_0^\infty vF\,dv = 4\pi(m/2\pi kT)^{\frac{3}{2}} \int_0^\infty v^3 e^{-mv^2/2kT}\,dv$

$$= 4\pi(m/2\pi kT)^{\frac{3}{2}}\{\tfrac{1}{2}(2kT/m)^2\} = (8kT/\pi m)^{\frac{1}{2}}.$$

Then for $m = 132.9m_u$, $\bar{c} = 351$ m s^{-1}.
For the mean velocity towards $x = +\infty$,

$$\langle v_x \rangle = \int_0^\infty v_x f'\,dv_x = 2(m/2\pi kT)^{\frac{1}{2}} \int_0^\infty v_x e^{-mv_x^2/2kT}\,dv_x$$
$$= (2kT/\pi m)^{\frac{1}{2}}.$$

For Cs at 500 K this expression evaluates to 175 m s^{-1}.

- *Comment*. These are mean values, and the spread of individual values may be very wide.

- *Exercise*. Evaluate the root mean square speed of the atoms. $[(3kT/m)^{\frac{1}{2}}$, 380 m s$^{-1}]$

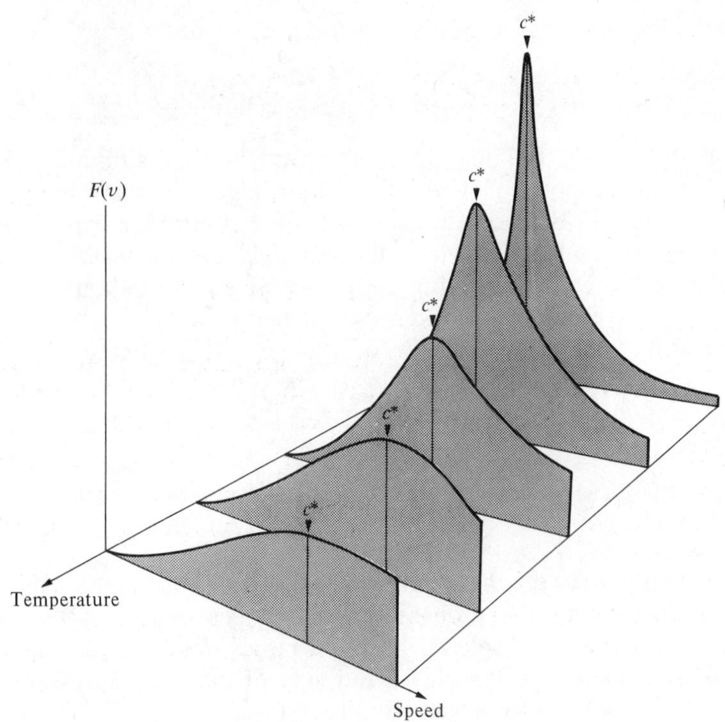

Fig. 26.4. The Maxwell distribution of speeds and its dependence on temperature calculated using eqn (26.1.9). Higher temperatures lie at the front: note the broadening of the distribution and the shift of the most probable speed at the temperature is increased.

Some properties of the Maxwell distribution were mentioned in the Introduction (Section 0.1(e)). The main features are illustrated in Fig. 26.4, which shows that the distribution of speeds broadens as the temperature increases, and that the *most probable speed* (symbol: c^*), where the distribution is a maximum, shifts to higher values. The *mean speed* (symbol: $\bar{c}$) is calculated from the Maxwell distribution as explained in *Example* 26.1:

$$\bar{c} = \langle v \rangle = \int_0^\infty vF(v)\,dv = (8kT/\pi m)^{\frac{1}{2}}. \qquad (26.1.10)°$$

It differs slightly from the root mean square speed

$$c = (3kT/m)^{\frac{1}{2}} \qquad (26.1.11)°$$

and the most probable speed

$$c^* = (2kT/m)^{\frac{1}{2}}. \qquad (26.1.12)°$$

(Numerically, $c = 1.225c^*$ and $\bar{c} = 1.128c^*$.) Light particles have higher mean speeds than heavy particles. Some values are given in Table 26.1.

Table 26.1. Average speeds at 25 °C, $\bar{c}$/m s^{-1}

He	1256
N_2	475
CO_2	379
C_6H_6	284

The Maxwell distribution has been verified experimentally. For example, molecular speeds can be measured directly with a velocity-selector of the sort shown in Fig. 26.5 (the same kind of device as used in molecular beam work, Section 24.3). The spinning discs have slots which permit the passage of only those particles moving through them at the appropriate speed, and the number of particles can be determined by collecting them at a detector. Indirect methods make use of the Doppler effect on the wavelength of emitted light, as discussed in Section 18.1(c).

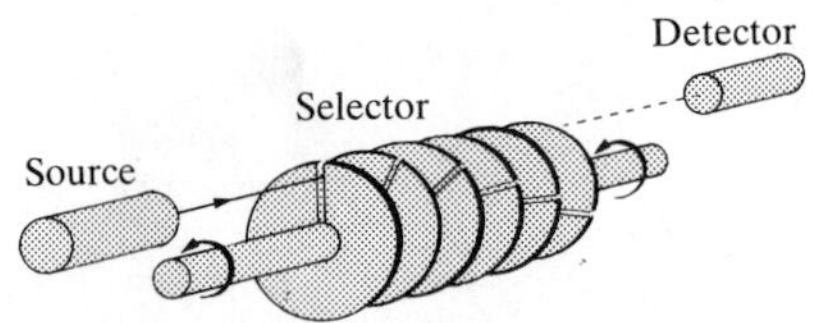

Fig. 26.5. The experimental determination of molecular speeds makes use of a series of rotating discs: the response of the detector is proportional to the number of particles having a speed that enables them to pass through the slots in the selector when it is rotating at a known rate.

26.2 Collisions

The time-scale of events in a gas, such as the rates of chemical reactions and the speed with which physical disturbances spread through a sample, is determined by the collision frequency, the number of collisions a particle makes per unit time either with other particles or with a wall. The following material is the background to, among other applications, the calculation of rates of reactions.

26.2 (a) Intermolecular collisions

We count a 'hit' whenever the centres of two particles come within some distance d of each other, where d is of the order of their diameters (for hard spheres d is the diameter). The simplest approach is to freeze the positions of all the particles except one, and then to note what happens as that mobile particle travels through the gas with a mean speed $\bar{c}$ for a time Δt. In doing so it sweeps out a 'collision tube' of cross-sectional area $\sigma = \pi d^2$ and length $\bar{c}\Delta t$, and therefore of volume $\sigma\bar{c}\Delta t$, Fig. 26.6. The area σ is called the *collision cross-section*. The number of stationary particles with centres inside the colision tube is given by its volume times the number density, $\sigma\bar{c}\Delta t\mathcal{N}$. The number of hits scored in the time Δt is equal to this number, and so the *collision frequency*, the number of collisions per unit time, is $\sigma\bar{c}\mathcal{N}$. However, the particles are not stationary, and so we should use the *relative* speed of the colliding particles. In the case of collisions between different types of molecule, the mean relative speed is

$$\bar{c}_{\text{rel}} = (8kT/\pi\mu)^{\frac{1}{2}}, \qquad \mu = m_A m_B/(m_A + m_B), \qquad (26.2.1)^\circ$$

where μ is the *reduced mass* (we saw in Appendix 15.1 that the reduced mass occurs when the relative motion of particles is being considered). For identical molecules $\mu = m/2$, and so $\bar{c}_{\text{rel}} = 2^{\frac{1}{2}}\bar{c}$. Therefore, the collision frequency is

$$z = 2^{\frac{1}{2}}\sigma\bar{c}N/V = 2^{\frac{1}{2}}\sigma\bar{c}p/kT \qquad (26.2.2)^\circ$$

(since $N/V = p/kT$).

The collision frequency z gives the number of collisions made by a single particle. The total collision frequency (the rate of collisions between all the

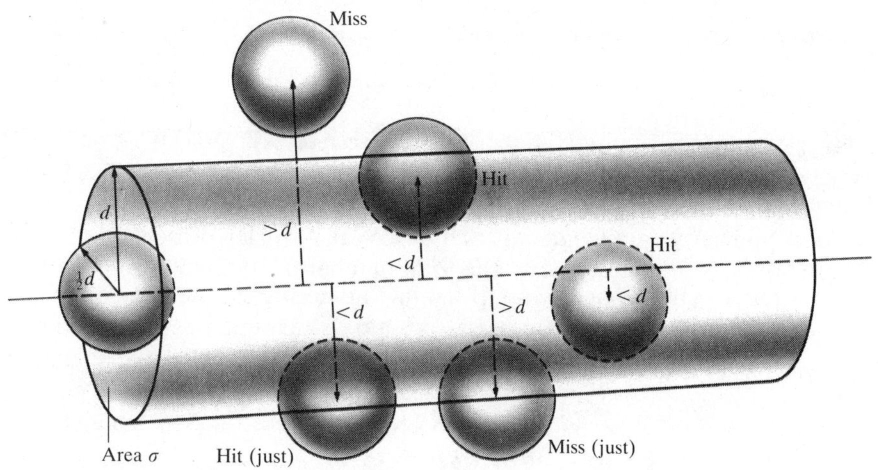

Fig. 26.6. The collision cross-section (σ) and the collision tube. A 'hit' is scored when two particles come within a distance equal to their diameter.

particles in the gas) is obtained by multiplying z by $\frac{1}{2}N$ (the factor $\frac{1}{2}$ ensures that the A . . . A′ and A′ . . . A collisions are counted as one). Therefore, the *number of collisions per unit time per unit volume is*

$$Z_{AA} = \tfrac{1}{2}zN/V = (\tfrac{1}{2})^{\frac{1}{2}}\sigma\bar{c}(N/V)^2. \qquad (26.2.3)^\circ$$

The mean speed is given by eqn (26.1.10), and so

$$Z_{AA} = \sigma(4kT/\pi m)^{\frac{1}{2}}(N/V)^2 = \sigma(4kT/\pi m)^{\frac{1}{2}}N_A^2[\mathrm{A}]^2, \qquad (26.2.4)^\circ$$

with $\sigma = \pi d^2$ and $[\mathrm{A}] = n_A/V$. Some typical collision cross-sections are given in Table 26.2 (they are measured by relating the collision frequency to various properties, as we shall see later). Collision frequencies may be very large. For example, in nitrogen at room temperature and pressure, with $d = 280$ pm, $Z = 5 \times 10^{34}\ \mathrm{s^{-1}\,m^{-3}}$.

Table 26.2. Collision cross-sections, σ/nm^2

He	0.21
N_2	0.43
CO_2	0.52
C_6H_6	0.88

Example 26.2

We continue to explore the properties of Cs atoms in an oven at 500 K. Calculate the number of collisions a single Cs atom makes per second. If the volume of the oven is 50.0 cm^3, what is the total number of collisions, inside the oven per second? The VP of Cs(s) at 500 K is 80 Torr; the collision diameter of a Cs atom is 540 pm.

• *Method*. Use eqn (26.2.2) for the collision frequency of a single atom, and eqn (26.2.3) for the total collision frequency. Use $\bar{c}$ from *Example* 26.1 (350 m s^{-1}).

• *Answer*. The collision cross-section is

$$\sigma = \pi d^2 = 9.1_6 \times 10^{-19}\ \mathrm{m^2},$$

and the pressure (80 Torr) is $1.0_7 \times 10^4$ Pa. Therefore,

$$z = 2^{\frac{1}{2}}\bar{\sigma}cp/kT = 4.5 \times 10^8\ \mathrm{s^{-1}}.$$

From eqn (26.2.3) written in the form

$$Z_{\mathrm{Cs,Cs}} = zp/2kT$$

(using $N/V = p/kT$) we find $Z_{\mathrm{Cs,Cs}} = 2.3 \times 10^{32}\ \mathrm{s^{-1}\,m^{-3}}$. Hence, in a 50.0 cm^3 container the total collision frequency is $1.1 \times 10^{28}\ \mathrm{s^{-1}}$.

• *Comment*. In order to study individual, isolated atoms and molecules it is essential to reduce the collision frequency so that they spend appreciable periods in free flight. This involves working at very low pressures.

• *Exercise*. What is the interval between collisions for a single CO_2 molecule at 298 K and 760 Torr? [1.5×10^{-10} s]

In the case of collisions between different types of molecule, the mean relative speed is given by eqn (26.2.1) and the collision cross-section is still given by $\sigma = \pi d^2$, but now $d = \frac{1}{2}(d_A + d_B)$. The number of collisions an A particle makes per unit time with all the N' B particles present is $\sigma\bar{c}N'/V$. There are N A particles, and so the total number of A–B collisions per unit time is $(\sigma cN'/V)N$, and the total number of A–B collisions per unit time per unit volume is

$$Z_{AB} = \sigma(8kT/\pi\mu)^{\frac{1}{2}}(NN'/V^2) = \sigma(8kT/\pi\mu)^{\frac{1}{2}}N_A^2[\mathrm{A}][\mathrm{B}]. \qquad (26.2.5)^\circ$$

Once we have the collision frequency, we can calculate the mean free path. If a particle moving with a mean speed $\bar{c}$ collides with a frequency z, it spends a time $1/z$ in free flight between collisions, and therefore travels a distance $(1/z)\bar{c}$. Therefore, the mean free path is simply

$$\lambda = \bar{c}/z = (1/2^{\frac{1}{2}}\sigma)kT/p, \qquad (26.2.6)^{\circ}$$

and is inversely proportional to the pressure.

Example 26.3

Calculate the mean free path of Cs atoms under the conditions specified in *Example* 26.2.

- *Method*. Use eqn (26.2.6) with $p = 80$ Torr and $\sigma = 0.92\ \text{nm}^2$.
- *Answer*. By substitution of $p = 1.1 \times 10^4$ Pa and $\sigma = 9.2 \times 10^{-19}\ \text{m}^2$, $\lambda = 770$ nm.
- *Comment*. The mean free path corresponds to about 1400 diameters, and so the gas is under conditions where the kinetic theory is likely to be reliable.
- *Exercise*. Calculate the mean free path of CO_2 molecules at 298 K and 1 atm. [55 nm]

26.2 (b) Collisions with walls and surfaces

Consider a wall of area A perpendicular to the x-axis. If a particle has a velocity between 0 and $+\infty$ it will strike the wall within a time Δt if it lies within a distance $v_x\Delta t$ of it (if v_x lies between 0 and $-\infty$ it is moving in the wrong direction). Therefore, all particles in the volume $Av_x\Delta t$, and with positive velocities, will strike the wall in the interval Δt. The total average number of collisions in this interval is therefore the average of this quantity multiplied by the number density of particles

$$\textit{Number of collisions} = \mathcal{N}A\Delta t\int_0^{\infty} v_x f(v_x)\,\mathrm{d}v_x.$$

The integral can be evaluated easily using the velocity distribution:

$$\int_0^{\infty} v_x f(v_x)\,\mathrm{d}v_x = (m/2\pi kT)^{\frac{1}{2}}\int_0^{\infty} v_x \mathrm{e}^{-mv_x^2/2kT}\,\mathrm{d}v_x = (kT/2\pi m)^{\frac{1}{2}}.$$

Therefore, the number of collisions per unit time per unit area is

$$Z_{\mathrm{W}} = (N/V)(kT/2\pi m)^{\frac{1}{2}} = \tfrac{1}{4}\bar{c}N/V. \qquad (26.2.8)^{\circ}$$

Then, since $N/V = p/kT$, we can express this as

$$Z_{\mathrm{W}} = \tfrac{1}{4}p\bar{c}/kT = p/(2\pi mkT)^{\frac{1}{2}}. \qquad (26.2.9)^{\circ}$$

When $p = 1$ atm and $T = 300$ K, a surface in air receives about 3×10^{23} collisions per second per cm^2. A knowledge of how Z_{W} depends on the pressure and the mass of the particle will turn out to be very important when we consider processes occurring at solid surfaces (in Chapter 31).

26.3 Transport properties

Transport properties are properties in which something (such as matter or energy) is transferred from one place to another. For example, a gas *effuses* through a hole into a region of lower pressure, and particles (in gases, liquids, and solids) *diffuse* down a concentration gradient until the composition is uniform. *Thermal conduction* is the transport of energy by the spreading of thermal motion down a temperature gradient, and *electric conduction* is the transport of charge (by ions or electrons) in a potential gradient. *Viscosity*, we shall see, depends on the transport of linear momentum.

We shall look at some of the general aspects of transport initially, and then calculate the quantities we need using the kinetic theory of gases. We return to the subject in the next chapter, where we deal with transport in liquids in a more general way.

26.3 (a) Flux

The rate of migration of a property is measured by its *flux* (symbol: J), which is the quantity passing through unit area per unit time. If mass is flowing, we speak of a flux of so many $\mathrm{kg\,m^{-2}\,s^{-1}}$; if the property is energy, then the energy flux is expressed in $\mathrm{J\,m^{-2}\,s^{-1}}$, and so on.

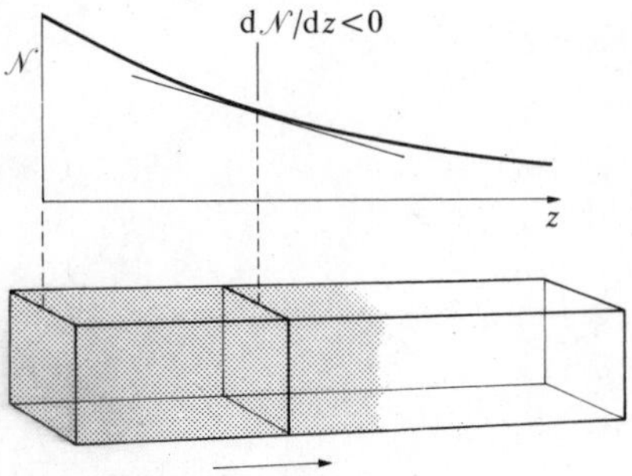

Experimental observations on transport processes show that the flux of a property is usually proportional to the gradient of a related property of a system. For example, the rate of diffusion (the flux of matter) parallel to some axis z (which we denote J_z) is found to be proportional to the concentration gradient along that axis, $J_z\,(\text{matter}) \propto \mathrm{d}\mathcal{N}/\mathrm{d}z$, and the rate of thermal diffusion (the flux of energy) is found to be proportional to the temperature gradient, $J_z\,(\text{energy}) \propto \mathrm{d}T/\mathrm{d}z$.

J_z is the component of a vector, and a positive sign indicates flow towards increasing z while a negative sign indicates flow to the left. Matter flow occurs *down* a concentration gradient, and so if $\mathrm{d}\mathcal{N}/\mathrm{d}z$ is negative (concentration decreasing to the right, Fig. 26.7), J_z is found to be positive (flow to the right). Therefore, the coefficient of proportionality in the matter flux expression must be negative, and we denote it $-D$, where D is called the *diffusion coefficient*. This leads to *Fick's First Law of diffusion:*

Fig. 26.7. The flux of particles down a concentration gradient. Fick's Law states that the flux of matter (the number of particles per unit area per unit time) is proportional to the density gradient at that point.

$$J_z(\text{matter}) = -D(\mathrm{d}\mathcal{N}/\mathrm{d}z). \qquad (26.3.1)$$

Energy flows down a temperature gradient, and so the same reasoning leads to

$$J_z(\text{energy}) = -\kappa(\mathrm{d}T/\mathrm{d}z), \qquad (26.3.2)$$

where κ is the *coefficient of thermal conductivity*.

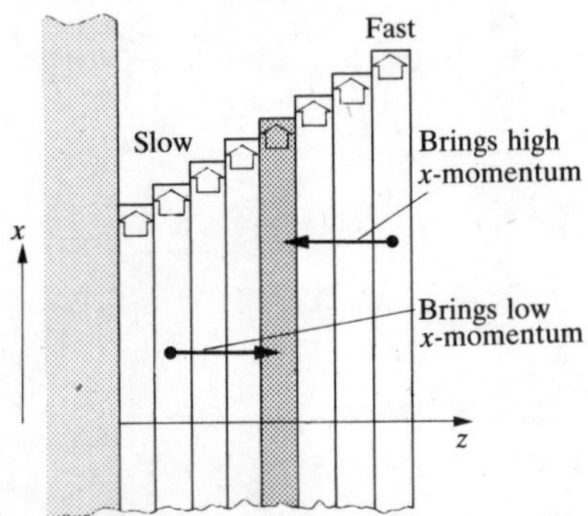

Fig. 26.8. The viscosity of a gas arises from the transport of linear momentum. In this illustration the fluid is undergoing *laminar flow*, and particles bring their initial momentum when they enter a new layer. If they arrive with high x-component of momentum they accelerate the layer; if with low x-component of momentum they retard it.

In order to see the connection between the flux of momentum and the viscosity, consider a fluid in a state of *Newtonian flow;* that is, a fluid in which the velocity of the fluid varies linearly and can be imagined as flowing as a series of layers or *laminas*, Fig. 26.8, with a stationary layer next to the wall of the vessel. Particles are continuously moving between the laminas, and bringing with them the x-component of momentum they possessed as a result of travelling with the velocity of their initial lamina. A lamina is retarded by particles arriving from the left, because they have a low momentum in the x-direction, and is accelerated by particles arriving from the right. As a result, the laminas tend towards a uniform velocity, and we interpret the retarding effect of the slow layers on the fast as the fluid's viscosity. Note, however, that the effect depends on the transfer of x-momentum into the lamina of interest, and so the viscosity depends on the flux of x-momentum in the z-direction. Furthermore, this flux depends on the gradient of the x-component of velocity of the fluid (there is no net flux when all the laminas move at the same velocity), and so we write

$$J_z(\text{momentum along } x) \propto (\mathrm{d}v_x/\mathrm{d}z), \quad \text{or} \quad J_x = -\eta(\mathrm{d}v_x/\mathrm{d}z), \qquad (26.3.3)$$

where η is the *coefficient of viscosity* (or simply the *viscosity*).

26.3 (b) The rate of effusion

When a gas at a pressure p and temperature T is separated from a vacuum by a very small hole, the rate of escape of its particles is equal to the rate at which they strike the area of the hole, eqn (26.2.9). Therefore, if the area of the hole is A_0, the number of particles that escape per unit time is

$$Z_W A_0 = pA_0/(2\pi mkT)^{\frac{1}{2}}. \qquad (26.3.4)^\circ$$

The rate of effusion is therefore inversely proportional to the square-root of the molecular mass: this is *Graham's law of effusion*. This expression is the basis of the *Knudsen method* for the determination of RMM, or, if that is known, the vapour pressure (e.g. of a solid). Thus, if the VP of a solid is p, and is enclosed in a cavity with a small hole, the rate of loss of mass from the container is proportional to p (see Problems 26.23 and 26.24). The method is reliable if the mean free path of the atoms is long compared with the diameter of the hole.

Example 26.4

When an 0.50 mm hole was opened in the oven used in previous *Examples* a mass loss of 385 mg was measured in the course of 100 s. Calculate the VP of caesium at 500 K.

- *Method*. The mass loss Δm in an interval Δt is related to the collision frequency by $\Delta m = Z_W A_0 m \Delta t$, A_0 being the area of the hole and m the molecular mass. The VP is constant inside the oven because liquid caesium is present in equilibrium with its vapour. Since Z_W is related to the pressure by eqn (26.3.4), we can write

$$p = (2\pi kT/m)^{\frac{1}{2}}(\Delta m/A_0\Delta t).$$

- *Answer*. Since $m = 132.9m_u$, substitution of the data gives $p = 1.1 \times 10^4$ Pa (using $1\ \mathrm{Pa} = 1\ \mathrm{N\,m^{-2}} = 1\ \mathrm{J\,m^{-1}}$). This corresponds to 83 Torr.

- *Comment*. When only a gas is present, the pressure decreases as it escapes, and the mass loss must be found by integration of the expression $\mathrm{d}m = Z_W(p)mA_0\,\mathrm{d}t$; this results in an exponential time dependence (Problem 26.22).

- *Exercise*. How long would it take 1.0 g of Cs atoms to effuse out of the oven? [255 s]

26.3 (c) The rate of diffusion

We now investigate the theoretical basis for Fick's Law, and obtain an expression for the diffusion coefficient of a perfect gas. We have to show that the flux of particles is proportional to their concentration gradient.

Consider the arrangement depicted in Fig. 26.9. On average, the particles passing through the area A at $z = 0$ have travelled about one mean free path, and so the number density where they originated is $\mathcal{N}(z)$ evaluated at $z = -\lambda$: this is approximately $\mathcal{N}(0) - \lambda(\mathrm{d}\mathcal{N}/\mathrm{d}z)_0$, where the subscript 0

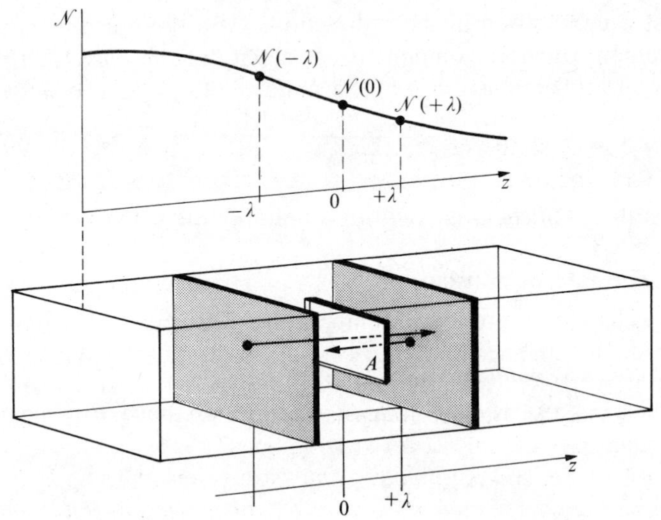

Fig. 26.9. The calculation of the rate of diffusion of a gas calculates the net flux through a plane of area A as a result of arrivals from on average a distance $\pm\lambda$ away, where λ is the mean free path.

indicates that the gradient should be evaluated at $z=0$. Since the average number of impacts on the imaginary window from the left during an interval Δt is $\frac{1}{4}\mathcal{N}(-\lambda)\bar{c}A\Delta t$, eqn (26.2.8), the flux from left to right arising from the supply of particles on the left is

$$J(L\rightarrow R)=\tfrac{1}{4}\bar{c}\mathcal{N}(-\lambda).$$

There is also a flux of particles from right to left. On average, the particles making the journey have originated from $z=\lambda$ where the number density is $\mathcal{N}(\lambda)$, and so

$$J(L\leftarrow R)=\tfrac{1}{4}\bar{c}\mathcal{N}(\lambda).$$

The average number density at $z=\lambda$ is approximately $\mathcal{N}(0)+\lambda(\mathrm{d}\mathcal{N}/\mathrm{d}z)_0$. The flow from the more concentrated region on the left dominates the backwash, and so the net flux is positive (from left to right) and of magnitude

$$\begin{aligned}J_z&=J(L\rightarrow R)-J(L\leftarrow R)\\&=\tfrac{1}{4}\{[\mathcal{N}(0)-\lambda(\mathrm{d}\mathcal{N}/\mathrm{d}z)_0]-[\mathcal{N}(0)+\lambda(\mathrm{d}\mathcal{N}/\mathrm{d}z)_0]\}\bar{c}\\&=-\tfrac{1}{2}\lambda\bar{c}(\mathrm{d}\mathcal{N}/\mathrm{d}z)_0.\end{aligned}$$

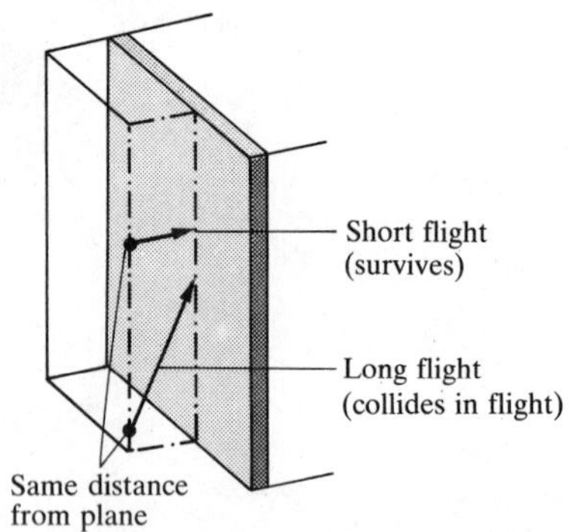

Fig. 26.10. One approximation ignored in the simple treatment is that some particles might make a long flight to the plane even though they are only a short perpendicular distance away, and therefore they have a higher chance of colliding during their journey.

This gives the flux as proportional to the gradient of concentration, in agreement with Fick's Law. Moreover, we can pick out a value of the diffusion coefficient: $D=\frac{1}{2}\lambda\bar{c}$. It must be remembered, however, that the calculation is quite crude, and is little more than an assessment of the order of magnitude of D. One aspect that has not been taken into account is illustrated in Fig. 26.10: this shows that although a particle may have begun its journey very close to the window, it could have a long flight before it gets there. Since the path is long, the particle is likely to collide before reaching the window, and so it ought to be added to the graveyard of other particles that have collided. Taking this effect into account involves a lot of work, but the end result is the appearance of a factor of $\frac{2}{3}$, representing the lower flux. This modification results in

$$J_z=-\tfrac{1}{3}\lambda\bar{c}(\mathrm{d}\mathcal{N}/\mathrm{d}z)_0. \qquad (26.3.5)^\circ$$

When this is compared with $J_z = -D(\mathrm{d}\mathcal{N}/\mathrm{d}z)_0$ we find

$$D = \tfrac{1}{3}\lambda\bar{c}. \qquad (26.3.6)°$$

Since λ decreases as the pressure is increased, so does D. The mean speed increases with the temperature, and so, therefore, does D. Specifically:

$$\begin{aligned} D &= \tfrac{1}{3}\{(1/2^{\frac{1}{2}}\sigma)(kT/p)\}\{(8kT/\pi m)^{\frac{1}{2}}\} \\ &= (2k^{\frac{3}{2}}/3\pi^{\frac{1}{2}})(1/\sigma m^{\frac{1}{2}})T^{\frac{3}{2}}/p, \end{aligned} \qquad (26.3.7)°$$

where σ is the collision cross-section πd^2.

26.3 (d) Thermal conductivity

The flux of interest is now that of energy: we have to account for its proportionality to a temperature gradient, and find a value of the thermal conductivity κ; some experimental values are given in Table 26.3. Thermal conductivities are made use of in the *Pirani gauge*, in which pressure is measured by monitoring the temperature of a heated wire, and in *katharometer detectors* in chromatography (GLC), where changes of composition are detected similarly.

Table 26.3. Transport properties of gases at 1 atm

	$\kappa/\mathrm{mJ\,cm^{-2}\,s^{-1}\,(K\,cm^{-1})^{-1}}$	$\eta/\mu\mathrm{P}$	
	273 K	273 K	293 K
He	1.442	187	196
Ar	0.163	210	223
N_2	0.240	166	176
CO_2	0.145	173	182

$1\,\mu\mathrm{P} = 10^{-7}\,\mathrm{kg\,m^{-1}\,s^{-1}}$.

Suppose that each monatomic particle carries an average energy $\varepsilon = \tfrac{3}{2}kT$ (from the equipartition theorem, Sections 0.1(f) and 22.3(a)) as a result of its thermal motion. When one passes through the imaginary window it transports that amount of energy. Although the number density is uniform (that is an approximation), the temperature is not, and so the energy fluxes are

$$\begin{aligned} J(L \rightarrow R) &= \{\tfrac{1}{4}\mathcal{N}\bar{c}\}\varepsilon(-\lambda), \qquad & \varepsilon(-\lambda) &= \tfrac{3}{2}k\{T - \lambda(\mathrm{d}T/\mathrm{d}z)_0\} \\ J(L \leftarrow R) &= \{\tfrac{1}{4}\mathcal{N}\bar{c}\}\varepsilon(\lambda), & \varepsilon(\lambda) &= \tfrac{3}{2}k\{T + \lambda(\mathrm{d}T/\mathrm{d}z)_0\} \end{aligned}$$

because, on average, particles arrive from the left after travelling a mean free path from a hotter region ($\mathrm{d}T/\mathrm{d}z$ is negative), and therefore with a higher energy, and from the right after travelling a mean free path from a cooler region. The net energy flux is therefore

$$J_z = J(L \rightarrow R) - J(L \leftarrow R) = -\tfrac{3}{4}\mathcal{N}\bar{c}k\lambda(\mathrm{d}T/\mathrm{d}z)_0.$$

As before, we multiply by $\tfrac{2}{3}$ in order to take long flight paths into account, and so arrive at

$$J_z = -\tfrac{1}{2}\mathcal{N}\bar{c}k\lambda(\mathrm{d}T/\mathrm{d}z)_0. \qquad (26.3.8)°$$

This flux is proportional to the temperature gradient, as we wanted to show. On comparing the equation with $J_z = -\kappa(\mathrm{d}T/\mathrm{d}z)_0$, we deduce that

$$\kappa = \tfrac{1}{2}\lambda\bar{c}k\mathcal{N}. \qquad (26.3.9)°$$

In the case of a gas of polyatomic molecules, we replace $\tfrac{3}{2}k$ by $C_{V,\mathrm{m}}/N_{\mathrm{A}}$, and use

$$\kappa = \tfrac{1}{3}\lambda\bar{c}C_{V,\mathrm{m}}(n/V). \qquad (26.3.10)°$$

In either case, notice that because the mean free path is inversely proportional to n/V, κ is independent of the pressure of the gas. Specifically:

$$\kappa = (1/3\surd 2)\bar{c}C_{V,\mathrm{m}}/\sigma N_{\mathrm{A}}. \qquad (26.3.11)°$$

The physical reason is that the thermal conductivity is large when many particles are available to transport the energy (hence the proportionality of κ to n in eqn (26.3.10)), but the presence of many particles limits their mean free path, and they cannot carry the energy over a great distance: these two effects balance. The thermal conductivity is found experimentally to be independent of the pressure, except when the pressure is very low, in which case the dependence is linear. This is because at very low pressures λ is greater than the dimensions of the apparatus, and so the distance over which the energy is transported is determined by the size of the vessel and not by the other particles present. The flux is still proportional to the number of carriers, but the length of the journey no longer depends on λ, and so $\kappa \propto n \propto p$.

Example 26.5

Estimate the thermal conductivity of air at room temperature.

• *Method*. Use eqn (26.3.11). The molar heat capacity of a gas of diatomic molecules is $\frac{5}{2}R$, eqn (22.3.14). The RMM of air is about 30. Take $\sigma \approx 0.42\ \text{nm}^2$ for O_2 and N_2 from Table 26.2. Estimate $\bar{c}$ from eqn (26.1.10).

• *Answer*. From eqn (26.1.10), $\bar{c} = 640\ \text{m s}^{-1}$ for particles of mass $m = 30m_u$. Then, from eqn (26.3.11), substituting $\sigma = 4.2 \times 10^{-19}\ \text{m}^2$, $\kappa = 9.9 \times 10^{-3}\ \text{J K}^{-1}\ \text{m}^{-1}\ \text{s}^{-1}$.

• *Comment*. A more cumbersome but more useful form of the answer is $\kappa = 9.9\ \text{mJ cm}^{-2}\ \text{s}^{-1}\ (\text{K cm}^{-1})^{-1}$. Then, in a temperature gradient of $1\ \text{K cm}^{-1}$ the heat flux would be about $10\ \text{mJ cm}^{-2}\ \text{s}^{-1}$.

• *Exercise*. Would carbon dioxide be a better insulator under the same conditions?
[$\kappa = 6\ \text{mJ cm}^{-2}\ \text{s}^{-1}\ (\text{K cm}^{-1})^{-1}$; yes]

26.3 (e) Viscosity

We have seen that viscosity is related to the flux of momentum. Particles travelling from the right in Fig. 26.11 (from a fast layer to a slower) transport a momentum $mv_x(\lambda)$ to their new layer at $z = 0$, and those travelling from the left transport $mv_x(-\lambda)$ to it. If the density is uniform (an approximation), the number of impacts per unit area per unit time on the imaginary window is $\frac{1}{4}\mathcal{N}\bar{c}$. Those from the right on average carry a momentum $mv_x(0) + \lambda m(dv_x/dz)_0$, and those from the left bring a momentum $mv_x(0) - \lambda m(dv_x/dz)_0$. The net flux of x-momentum in the z-direction is therefore

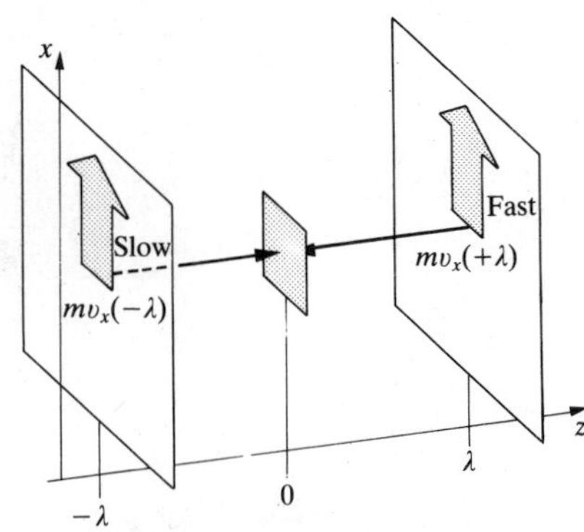

Fig. 26.11. The calculation of the viscosity of a gas examines the net x-component of momentum brought to a plane from faster and slower layers on average a mean free path away.

$$J_z = \tfrac{1}{4}\mathcal{N}\bar{c}\{[mv_x(0) - \lambda m(dv_x/dz)_0] - [mv_x(0) + \lambda m(dv_x/dz)_0]\}$$
$$= -\tfrac{1}{2}m\mathcal{N}\lambda\bar{c}(dv_x/dz)_0.$$

Hence, the flux is proportional to the velocity gradient, as we wished to show. Comparison of this expression with eqn (26.3.3), and the replacement of the factor $\frac{1}{2}$ by $\frac{1}{3}$, leads to

$$\eta = \tfrac{1}{3}m\lambda\mathcal{N}\bar{c} = \tfrac{1}{3}m\lambda\bar{c}N_A(n/V). \qquad (26.3.12)°$$

As in the case of thermal conductivity, the viscosity is *independent of the pressure*. This can be seen explicitly by introducing the expression for the mean free path:

$$\eta = \tfrac{1}{3}m\bar{c}N_A(n/V)/2^{\frac{1}{2}}\sigma N_A(n/V) = (\tfrac{1}{18})^{\frac{1}{2}}m\bar{c}/\sigma. \qquad (26.3.13)°$$

The physical reason is the same: more particles are available to transport the momentum, but carry it less far on account of the shorter mean free path. It should also be noticed that the viscosity of a gas *increases with temperature* (as $T^{\frac{1}{2}}$ because $\bar{c} \propto T^{\frac{1}{2}}$). This is because the particles travel more quickly, and the flux of momentum is greater. (In a liquid, on the other hand, the viscosity *decreases* with temperature. This is because the intermolecular interactions have to be overcome, as explained in Section 24.4(e).)

There are two principal techniques for the measurement of viscosities. One depends on the rate of damping of the torsional oscillations of a disc hanging in the gas, the time constant for the decay of the harmonic motion depending on the viscosity and the design of the apparatus. The other is based on *Poiseuille's formula* for the rate of flow of a fluid through a tube of radius r:

$$dV/dt = (p_1^2 - p_2^2)\pi r^4/16l\eta p_0, \qquad (26.3.14)$$

where V is the volume flowing, p_1 and p_2 being the pressures at each end of the tube of length l, and p_0 the pressure at which the volume is measured. Such experiments confirm that the viscosities of gases are independent of pressure over a wide range: the results for argon from 10^{-3} atm to 10^2 atm are shown in Fig. 26.12(a), and η is constant from about 0.01 atm to 50 atm. The $\sqrt{T}$ temperature dependence is roughly confirmed, Fig. 26.12(b), the dotted line being the calculated values using $\sigma = 22 \times 10^{-20}\,m^2$ (implying a collision diameter of 260 pm, in contrast to the van der Waals diameter of 335 pm obtained from the density of the solid). The agreement is not too bad, considering the simplicity of the model and the neglect of intermolecular forces.

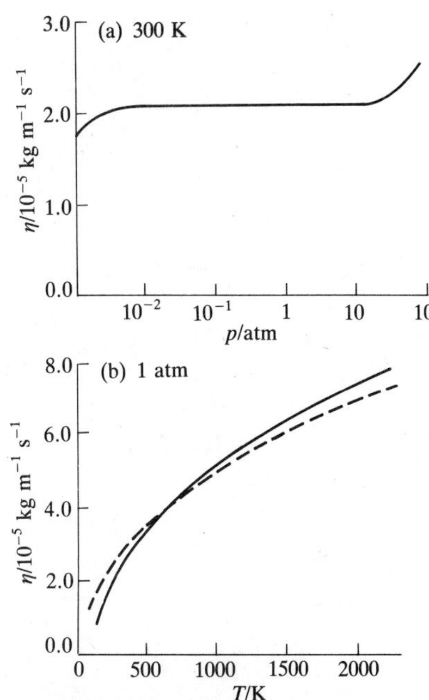

Fig. 26.12. The experimental results for (a) the pressure dependence of the viscosity of argon, and (b) its temperature dependence. The broken line in the latter is the calculated value. Fitting observed and calculated curves is one way of determining the collision cross-section.

Example 26.6

In a Poiseuille's flow experiment to measure the viscosity of air at 298 K, the sample was allowed to flow through a 100 cm tube of bore 1.0 mm. The high-pressure end was at 765 Torr and the low-pressure end was at 760 Torr. The volume was measured at the latter pressure. In 100 s a volume of 90.2 cm^3 passed through the tube. What is the viscosity of air at this temperature?

- *Method*. Use eqn (26.3.14). Convert pressures to Pa using 1 Torr = 133.3 Pa.

- *Answer*. Under the conditions stated,

$$dV/dt = (90.2\,cm^3)/(100\,s) = 9.02 \times 10^{-7}\,m^3\,s^{-1}.$$
$$p_1^2 - p_2^2 = 1.35_5 \times 10^8\,Pa^2,$$
$$\pi r^4/16lp_0 = 1.21 \times 10^{-19}\,N^{-1}\,m^5.$$

Hence, from eqn (26.3.14), $\eta = 1.8 \times 10^{-5}\,kg\,m^{-1}\,s^{-1}$.

- *Comment*. The kinetic theory expression, eqn (26.3.13), gives $1.4 \times 10^{-5}\,kg\,m^{-1}\,s^{-1}$, and so the agreement is quite good. Viscosities are often expressed in centipoise (cP) or (for gases) micropoise, the conversion being 1 cP = $10^{-3}\,kg\,m^{-1}\,s^{-1}$: the viscosity of air is 180 μP.

- *Exercise*. What volume would be collected if the pressure gradient were doubled, other conditions remaining constant? [180 cm^3]

Some experimental results are given in Table 26.3, and the theoretical formulas are collected (together with some exact expressions from more detailed calculations) in Box 26.2.

Box 26.2. Transport properties of perfect gases

Property	Transported quantity	Simple kinetic theory	Refined theory	Units
Diffusion	Matter	$D=\frac{1}{3}\lambda\bar{c}$	$\frac{3}{16}\lambda\bar{c}$	$m^2 s^{-1}$
Thermal conductivity	Energy	$\kappa=\frac{1}{3}\lambda\bar{c}C_{V,m}(n/V)$ $=(1/3\sqrt{2})\bar{c}C_{V,m}/\sigma N_A$	$(25\pi/64)\lambda\bar{c}C_{V,m}(n/V)$	$J K^{-1} m^{-1} s^{-1}$
Viscosity	Momentum	$\eta=\frac{1}{3}\lambda\bar{c}m\mathcal{N}$ $=(\frac{1}{18})^{\frac{1}{2}}m\bar{c}/\sigma$	$0.499\lambda\bar{c}m\mathcal{N}$	$kg\, m^{-1} s^{-1}$

Mean free path: $\lambda=(1/2^{1/2}\sigma)kT/p$

Further reading

Kinetic theory of gases. R. D. Present; McGraw-Hill, New York, 1958.

Gases, liquids, and solids (2nd edn). D. Tabor; Cambridge University Press, 1979.

Three phases of matter (2nd edn). A. J. Walton; Clarendon Press, Oxford, 1983.

Transport phenomena. R. B. Bird, W. E. Stewart, and E. N. Lightfoot; Wiley, New York, 1960.

The molecular theory of gases and liquids. J. O. Hirschfelder, C. F. Curtiss, and R. B. Bird; Wiley, New York, 1954.

Introductory problems

A26.1. Calculate the mean speed of Xe atoms at 650 K.

A26.2. Use the Maxwell distribution of speeds to estimate the percentage of N_2 molecules at 500 K which have speeds in the range 290 to 300 m s^{-1}.

A26.3. A solid surface with dimensions 2.5 mm × 3.0 mm is exposed to Ar gas at a pressure of 90 Pa and a temperature of 500 K. How many collisions do Ar molecules make with this surface in 15 seconds?

A26.4. A 5.00 dm^3 container holds 1.50 millimole of N_2 and 3.00 millimoles of H_2 at a total pressure of 2.00 kPa. Calculate the number of N_2–H_2 collisions which occur in this gas mixture in 1 ms.

A26.5. At what temperature is the mean free path of CH_4 molecules 1200 nm if the gas pressure is 3.65 kPa?

A26.6. An effusion cell has a circular hole which is 2.50 mm in diameter. If the solid in the cell has an RMM of 260 and a vapour pressure of 0.835 Pa at 400 K, how much will the mass of the solid decrease in a period of 2 hours?

A26.7. Find the flux of heat associated with a temperature gradient of 2.5 K m^{-1} in an Ar sample for which the mean temperature is 273 K.

A26.8. Use the experimental value of the thermal conductivity of Ne at 273 K to estimate a collision cross section for Ne atoms. Compare the result with the value in Table 26.2.

A26.9. Use the experimental value of the coefficient of viscosity for Ne to estimate a collision cross section for Ne atoms at 273 K. Compare it with the result of the previous problem and with the value in Table 26.2.

A26.10. Find the inlet pressure required to maintain a flow rate of 9.5×10^5 dm^3 hr^{-1} for N_2 at 293 K flowing through a pipe of length 8.50 m and diameter 1 cm. The pressure of gas as it exits the tube is 1.00 bar. The volume of the gas is measured at that pressure.

Problems

26.1. Calculate the *mean free path* in air ($\sigma\approx0.43$ nm^2) at 25 °C and (a) 10 atm, (b) 1 atm, (c) 10^{-6} atm.

26.2. At what pressure does the mean free path of argon at 25 °C become comparable to the size of the 1 dm^3 bulb that contains it? Take $\sigma\approx0.36$ nm^2.

26.3. At what pressure does the mean free path of argon at 25 °C become comparable to the size of the atoms themselves?

26.4. At an altitude of 20 km the temperature is 217 K and the pressure 0.05 atm. What is the mean free path of nitrogen molecules? ($\sigma \approx 0.43\ \mathrm{nm}^2$.)

26.5. Now we turn to the time scale of events in gases. How many collisions does a single argon atom make in 1 s when the temperature is 25 °C and the pressure is (a) 10 atm, (b) 1 atm, (c) 10^{-6} atm?

26.6. What is the total number of molecular collisions per second in a 1 dm^3 sample of argon under the same conditions as in Problem 26.5?

26.7. The frequency of molecular collisions is an important quantity in atmospheric chemistry. How many collisions per second does an excited nitrogen molecule make at an altitude of 20 km? (See Problem 26.4.)

26.8. Calculate the number of collisions per cm^3 in 1 s in air at 25 °C and 1 atm (a) between oxygen molecules, (b) between oxygen and nitrogen molecules. Take $R(O_2) \approx$ 178 pm and $R(N_2) \approx 185$ pm.

26.9. How many collisions per cm^3 are made in 1 s in the interior of the envelope of a Dewar flask where the remaining air is at a pressure of 1.2 Torr and 25 °C?

26.10. What is the mean speed of (a) helium atoms, (b) methane molecules at (i) 77 K, (ii) 298 K, (iii) 1000 K?

26.11. What is the mean translational kinetic energy (in kJ mol^{-1}) of (a) hydrogen molecules, (b) iodine molecules in a gas at 300 K and 1 atm pressure?

26.12. In an experiment to measure the speed of molecules by a rotating slotted-disc experiment, the apparatus consisted of five coaxial 5 cm diameter discs separated by 1 cm, the slots in their rims being displaced by 2° between neighbours. The relative density I of the detected beam of krypton atoms for two different temperatures and at a series of rotation rates were as follows:

ν/Hz	20	40	80	100	120
I(40 K)	0.846	0.513	0.069	0.015	0.002
I(100 K)	0.592	0.485	0.217	0.119	0.057

Find the distributions of molecular velocities $f(v_x)$ at these temperatures, and check that they conform to the theoretical prediction for a one-dimensional system.

26.13. The Maxwell–Boltzmann distribution was derived from arguments about probability. But it was stated that it could also be derived from the Boltzmann distribution itself. Do so.

26.14. A specially constructed velocity-selector accepts a beam of molecules from an oven at a temperature T but blocks the passage of molecules with a speed greater than the mean. What is the mean speed of the emerging beam, relative to the initial value, treated as a one-dimensional problem?

26.15. Cars are timed by police radar as they pass in both directions below a bridge. Their velocities (m.p.h., numbers of cars in brackets) to the east and west are as follows: 50 E (40), 55 E (62), 60 E (53), 65 E (12), 70 E (2); 50 W (38), 55 W (59), 60 W (50), 65 W (10), 70 W (2). What are (a) the mean velocity, (b) the mean speed, (c) the root mean square speed?

26.16. A population consists of people of the following heights (in feet and inches, numbers of individuals in brackets): 5′5″ (1), 5′6″ (2), 5′7″ (4), 5′8″ (7), 5′9″ (10), 5′10″ (15), 5′11″ (9), 6′0″ (4), 6′1″ (0), 6′2″ (1). What is (a) the mean height, (b) the root mean square height of the population.

26.17. What is the proportion of gas molecules having (a) more than, (b) less than the root mean square speed? What are the proportions having speeds greater and smaller than the mean speed ($\bar{c}$)?

26.18. What are the proportions of molecules in a gas that have a speed in a range at the speed nc^* relative to those in the same range at c^* itself? This calculation can be used to estimate the proportion of very energetic molecules (which is important for reactions). Evaluate the ratio for $n = 3$ and $n = 4$.

26.19. What is the *escape velocity* from the surface of a planet of radius R? What is the value for (a) the Earth, $R = 6.37 \times 10^6$ m, $g = 9.81\ \mathrm{m\ s^{-2}}$, (b) Mars, $R = 3.38 \times 10^6$ m, $m_{\mathrm{Mars}}/m_{\mathrm{Earth}} = 0.108$. At what temperatures do hydrogen, helium, and oxygen have *mean* speeds equal to their escape speeds? What proportion of the molecules have enough speed to escape when the temperature is (a) 240 K, (b) 1500 K? Calculations of this kind are very important in considering the composition of planetary atmospheres.

26.20. A manometer was connected to a bulb containing carbon dioxide under slight pressure. The gas was allowed to escape through a small pinhole, and the time for the manometer reading to drop from 75 cm to 50 cm was 52 s. When the experiment was repeated using nitrogen (M_r = 28) the same fall took place in 42 s. What is the RMM of carbon dioxide?

26.21. A particular make of electric light bulb contains argon at a pressure of 50 Torr and has a tungsten filament of radius 0.1 mm and length 5 cm. When operating, the gas close to the filament has a temperature of about 1000 °C. How many collisions are made with the filament in each second?

26.22. A space vehicle of internal volume 3.0 m^3 is struck by a meteor and a hole 0.1 mm radius is formed. If the oxygen pressure within the vehicle is initially 0.8 atm and its temperature 298 K, how long will it take to fall to 0.7 atm?

26.23. The *Knudsen cell* makes use of the expression for the frequency of collisions with an area in order to measure the vapour pressure of slightly volatile solids. A weighed amount of the sample is heated inside a container, in the wall of which there is a small hole. The mass loss over a period of time can be related to the vapour pressure at the temperature of the experiment. If Δm is the mass lost in time t through a circular hole of radius R, find an expression relating the vapour pressure p to Δm and t.

26.24. A Knudsen cell was used to determine the vapour pressure of germanium at 1000 °C. During an interval of 7200 s the mass loss through a hole of radius 0.50 mm amounted to 4.3×10^{-2} mg. What is the VP of germanium at 1000 °C? Assume the gas to be monatomic.

26.25. In a study of the catalytic properties of a titanium surface it was necessary to maintain the surface free from contamination. Calculate the collision frequency per cm^2 of surface made by oxygen at a pressure of (a) 1 atm, (b) 10^{-6} atm, (c) 10^{-10} atm and a temperature of 300 K. Estimate the number of collisions made with a single surface atom in each second. This underlines the importance of working at very low pressures in order to study the properties of uncontaminated surfaces. Take the nearest neighbour distance as 291 pm.

26.26. The nuclide $^{244}_{97}$Bk (berkelium) decays by producing α-particles, which capture electrons and form helium atoms. Its half-life is 4.4 h. A 1 mg sample was placed in a 1 cm^3 container that was impermeable to α-radiation, but there was also a hole of radius 0.002 mm in the wall (that is not quite a paradox). What is the pressure of helium, at 298 K, inside the container after (a) 1 hr, (b) 10 hr?

26.27. A spherical glass bulb of diameter 10 cm has a tube of radius 3 mm attached to it. While the bulb is maintained at 300 K the tube is at 77 K (it is immersed in liquid nitrogen). Initially the gas in the bulb is damp, the partial pressure of the water vapour being 1.0 Torr. Estimate the time required for condensation in the tube to lower the partial pressure to 10 μTorr if effusion were rate-determining.

26.28. An atomic beam is designed to function with (a) cadmium, (b) mercury. The source is an oven maintained at 380 K, there being a small slit of dimensions 1 cm $\times$ 10^{-3} cm. The vapour pressure of cadmium is 0.13 Pa and that of mercury is 152 kPa at this temperature. What is the atomic current (the number of atoms per unit time) in the beams?

26.29. The reaction $H_2 + I_2 \rightarrow 2HI$ depends on collisions between a variety of species in the reaction mixture, and we examine it in more detail in Chapter 28. Calculate the collision frequencies for the encounters (a) $H_2 + H_2$, (b) $I_2 + I_2$, (c) $H_2 + I_2$ for a gas with partial pressures of 0.5 atm for both species, the temperature being 400 K. $\sigma(H_2) \approx 0.27$ nm^2, $\sigma(I_2) \approx 1.2$ nm^2.

26.30. What is the viscosity of air at (a) 273 K, (b) 298 K, (c) 1000 K? Take $\sigma \approx 0.40$ nm^2. (The experimental value is 173 μP at 273 K, 182 μP at 20 °C, and 394 μP at 600 °C.)

26.31. Calculate the thermal conductivities of (a) argon, (b) helium at 300 K and 1.0 mbar. Each gas is confined in a cubic vessel of 10 cm side, one wall being at 310 K and the one opposite at 295 K. What is the rate of flow of heat from one wall to the other in each case?

26.32. The viscosity of carbon dioxide was measured by comparing its rate of flow through a long narrow tube (using Poiseuille's formula) with that of argon. For the same pressure differential, the same volume of carbon dioxide passed through the tube in 55 s as argon in 83 s. The viscosity of argon at 25 °C is 208 μP; what is the viscosity of carbon dioxide? Estimate the molecular diameter of carbon dioxide.

26.33. Calculate the thermal conductivity of argon ($C_{V,m} = 12.5$ J K^{-1} mol^{-1}, $\sigma \approx 0.36$ nm^2) and air ($C_{V,m} = 21.0$ J K^{-1} mol^{-1}, $\sigma \approx 0.40$ nm^2) at room temperature.

26.34. What is the ratio of the thermal conductivities of gaseous hydrogen at 300 K to gaseous hydrogen at 10 K? Be circumspect.

26.35. In a double-glazed window the panes of glass are separated by 5 cm. What is the rate of transfer of heat by conduction from the warm room (25 °C) to the cold exterior (−10 °C) through a window of area 1 m^2? What power of heater is required to make good the loss of heat?

26.36. Calculate the diffusion constant D for argon at 25 °C and (a) 10^{-6} atm, (b) 1 atm, (c) 100 atm. If a pressure gradient of 0.1 atm cm^{-1} is established in a pipe, what is the flow of gas due to diffusion?

26.37. The rate of growth of droplets of lead from a vapour has been studied in the laboratories of an oil company (J. B. Homer and A. Prothero, *J. chem. Soc. Faraday Trans.* I **69**, 673 (1973)). Virtually all the lead condensed within 0.5 ms of the initiation of the run, and the concentration in the gas phase was no more than about 3×10^{15} atoms/cm^3. Find an expression for the rate of growth of the radius of the spherical particles. Take $T = 935$ K and assume that every atom sticks to the growing surface when it collides with it.

27

Molecules in motion: ion transport and molecular diffusion

Learning objectives

After careful study of this chapter you should be able to:

(1) Define the *conductivity*, eqn (27.1.1), and the *molar conductivity*, eqn (27.1.2), of electrolyte solutions, and explain how they are measured, Example 27.1.

(2) Distinguish between *weak electrolytes* and *strong electrolytes*, Section 27.1(a).

(3) State *Kohlrausch's Law*, eqn (27.1.3), and the *law of the independent migration of ions*, eqn (27.1.4) and Example 27.2.

(4) State, derive, and use *Ostwald's Dilution Law*, eqn (27.1.7) and Example 27.3.

(5) Define and calculate the *drift speed* of ions, eqn (27.1.9), and the *mobility* of an ion, eqn (27.1.10).

(6) Relate an ion's mobility to its molar conductivity, eqn (27.1.12).

(7) Define the *transport number* of an ion, eqn (27.1.15), and describe how it may be measured.

(8) Describe the *relaxation effect* and the *electrophoretic effect* on the conductivities of ions, and outline and use the *Debye–Hückel–Onsager theory*, Section 27.1(d) and Example 27.4.

(9) Describe the *Wien effect* and the *Debye–Falkenhagen effect*, Section 27.1(d).

(10) Explain what is meant by a *thermodynamic force*, eqn (27.2.1).

(11) Derive *Fick's First Law of diffusion* from the chemical potential, Section 27.2(a).

(12) Deduce and use the *Einstein relation*, eqn (27.2.5), the *Nernst–Einstein relation*, eqn (27.2.7), and the *Stokes–Einstein relation*, eqn (27.2.9) and Example 27.5.

(13) State, justify, and criticize *Walden's rule*, Section 27.2(a).

(14) Derive the *diffusion equation*, eqn (27.3.1), and indicate the general features of its solutions, Section 27.3(a and b).

(15) Explain how *diffusion coefficients* are measured, Section 27.3(b).

(16) Explain the meaning of *convection* and write the *generalized diffusion equation*, eqn (27.3.10).

(17) Express the diffusion of particles in terms of a one-dimensional *random walk*, Section 27.3(d) and Appendix 27.1.

(18) Deduce and use the *Einstein–Smoluchowski relation*, eqn (27.3.12) and Example 27.6.

Introduction

When an electric potential difference is applied across two electrodes immersed in an ionic solution the net motion of the ions is towards one or other of the electrodes, and an electric current is conducted through the solution. So long as we do not enquire too closely into the details of the motion, this drift can be discussed very simply. We have to take into account the roles of the solvent and the other ions, but this can also be done quite simply if the analysis is not pressed too far. The study of the electrical *conductances* of ionic solutions (their ability to conduct electricity, which we define more precisely shortly) turns out to have several important applications. For example, since conductance is related to ionic concentrations, its measurement can be used to determine equilibrium constants and to study the rates of reactions involving ions.

In this chapter we investigate a type of motion that at first sight appears to be entirely different from the kind treated in Chapter 26. However, there is a very deep connection between the motion of ions and the motion of neutral particles in liquids and gases. This relation can be expressed thermodynamically, as we see in Section 27.2, and it leads to very general techniques for discussing the motion of all kinds of particles in fluids.

27.1 Ion transport

In solution, cations move towards a negatively charged electrode and anions move towards a positively charged electrode. This migration is a transport process of the kind considered in Chapter 26, the principal differences being that the ions drift under the influence of an electric field and are supported by a solvent. The presence of the field is a simplification, because the *average* motion of the ions is easy to describe. The presence of the solvent (and the ion–ion interactions) is a complication which is still the subject of current research.

27.1 (a) Ion motion: the empirical facts

The conductivity of a solution is determined by measuring its electrical resistance, and the standard method is to incorporate a *conductivity cell* of the type shown in Fig. 27.1 into one arm of a Wheatstone bridge and to search for the balance point, as explained in standard texts on electricity. The main complication is that alternating current must be used because a direct current would lead to electrolysis and to *polarization*, the charging of the layers of solution in contact with the electrodes. The use of alternating current (with a frequency of about 1 kHz) avoids this polarization because the charging that occurs on one half of the cycle is undone during the second half.

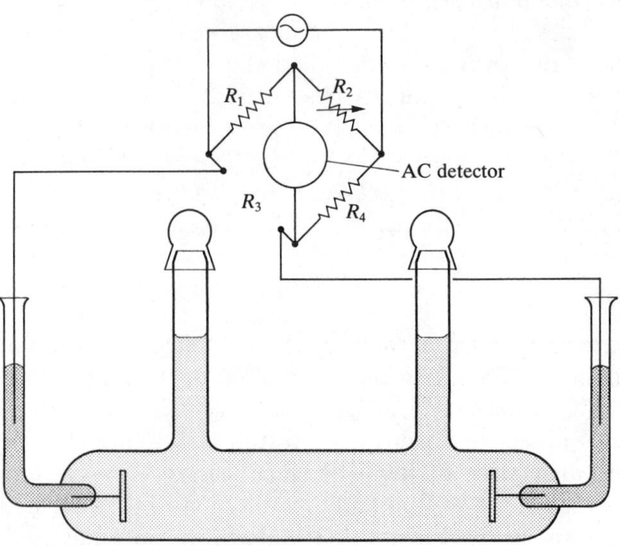

Fig. 27.1. The conductivity of an electrolyte solution is measured by making the cell one arm of a Wheatstone bridge and measuring its resistance. The balance point is obtained when the resistances satisfy $R_1/R_2 = R_3/R_4$ (or, since the current is alternating, a similar relation for the impedances).

The *resistance* (symbol: R) of a sample increases with its length l and decreases with its cross-sectional area A: $R \propto l/A$. The proportionality coefficient is called the *resistivity* (symbol: ρ), and we write $R = \rho l/A$. The *conductivity* (symbol: κ) is the inverse of the resistivity, and so

$$R = (1/\kappa)l/A, \quad \text{or} \quad \kappa = l/RA. \tag{27.1.1}$$

Resistance is expressed in ohm (Ω). The reciprocal ohm, Ω^{-1}, used to be called the mho, but its official designation is now the *siemens* (symbol: S, 1 S = 1 Ω^{-1}). It follows that the units of conductivity are S m^{-1} (or S cm^{-1}).

Calculating the conductivity from the resistance of the sample and the cell dimensions is unreliable because the current distribution is complicated. In practice, the cell is calibrated using a solution of known conductivity κ^* (typically an aqueous solution of potassium chloride) and a *cell constant C* is determined from $\kappa^* = C/R^*$, where R^* is the observed resistance. Then, if the sample has a resistance R in the same cell, its conductivity is $\kappa = C/R$.

The conductivity of a solution, which is due to contributions from both cations and anions, depends on the number of ions present, and it is normal to introduce the *molar conductivity* (symbol: Λ_m) defined as

$$\Lambda_m = \kappa/c, \tag{27.1.2}$$

c being the (molar) concentration of the added electrolyte. The molar conductivity is normally expressed in S cm^2 mol^{-1}‡. The *equivalent conductivity* is also sometimes used. This takes into account the fact that some ions are multiply charged, and so might transport more charge even though they may be moving more slowly: the equivalent conductivity treats all ions as being singly charged. For example, in NaCl(aq), each ion carries unit charge in each direction, and the equivalent conductivity is the same as the molar conductivity. In $CuSO_4$(aq), however, each ion transports two units of charge, and the equivalent conductivity is *half* the molar conductivity.

Example 27.1

The molar conductivity of 0.100 M KCl(aq) at 298 K is 129 S cm^2 mol^{-1}. The measured resistance in a conductivity cell was 28.44 Ω. When the same cell contained 0.050 M NaOH(aq) the resistance was 31.6 Ω. Calculate the molar conductivity of NaOH(aq) at that temperature and concentration.

• *Method*. Establish the cell constant using $C = R^*\kappa^*$; then use C and R to find the conductivity of the test solution. Convert to molar conductivity using eqn (27.1.2).

• *Answer*. The conductivity of an 0.100 M KCl(aq) solution is

$$\kappa = c\Lambda_m = (0.100 \text{ mol dm}^{-3}) \times (129 \text{ S cm}^2 \text{ mol}^{-1}) = 1.29 \times 10^{-2} \text{ S cm}^{-1},$$

and so $C = (1.29 \times 10^{-2} \text{ S cm}^{-1}) \times (28.44\ \Omega) = 0.367 \text{ cm}^{-1}$. Therefore, for NaOH(aq),

$$\kappa = C/R = (0.367 \text{ cm}^{-1})/(31.6\ \Omega) = 0.0116 \text{ S cm}^{-1}.$$

The molar conductivity is therefore

$$\Lambda_m = (0.0116 \text{ S cm}^{-1})/(0.050 \text{ mol dm}^{-3}) = 232 \text{ S cm}^2 \text{ mol}^{-1}.$$

‡ The conductivity is normally available in S cm^{-1} and the concentration in mol dm^{-3}, and so the practical relation is

$$\Lambda_m = 1000\{(\kappa/\text{S cm}^{-1})/(c/\text{mol dm}^{-3})\} \text{ S cm}^2 \text{ mol}^{-1}.$$

• *Comment*. It is good practice to calibrate cells with solutions having conductances close to the conductance of the test solution.

• *Exercise*. The same cell was used to measure the conductivity of 0.100 M $NH_4Cl(aq)$, and a resistance of 28.5 was measured. Calculate the molar conductivity of $NH_4Cl(aq)$ at this concentration. [$129\ \mathrm{S\ cm^2\ mol^{-1}}$]

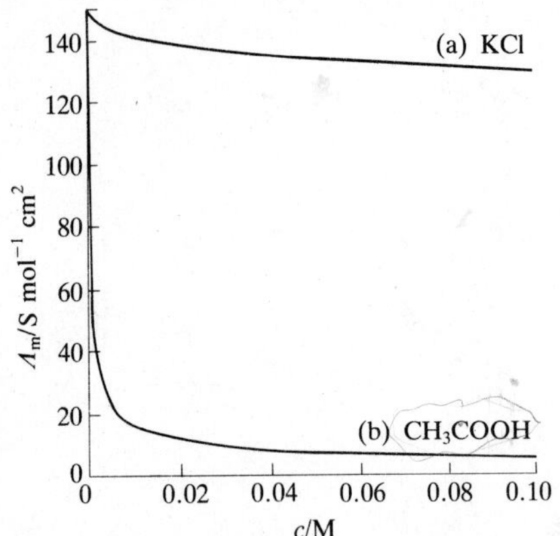

Fig. 27.2. The concentration dependence of the molar conductivities of (a) a typical strong electrolyte and (b) a typical weak electrolyte.

Molar conductivities depend on the concentration of the electrolyte. (They would be independent of concentration only if the conductivity were exactly proportional to the concentration, and for such strongly interacting particles as ions this is unlikely to be the case.) Measurement of the concentration dependence of molar conductivities shows that there are two classes of electrolyte. In one, the *strong electrolytes*, the molar conductivity decreases slightly as the concentration is increased, Fig. 27.2, but the effect is not large. In the other, the *weak electrolytes*, the molar conductivity is normal at concentrations close to zero, but drops sharply to low values when the concentration departs from zero, Fig. 27.2. The classification depends on the solvent employed as well as the solute: lithium chloride, for example, is a strong electrolyte in water, but a weak electrolyte in propanone.

Strong electrolytes are substances that are fully ionized in solution. As a result, the concentration of ions in solution is proportional to the concentration of electrolyte added. They include ionic solids and strong acids. In an extensive series of measurements, Friedrich Kohlrausch showed that at low concentrations the molar conductivities of strong electrolytes are given by

$$\Lambda_m = \Lambda_m^0 - \mathscr{K}\sqrt{c}. \tag{27.1.3}$$

This is now called *Kohlrausch's Law*. Λ_m^0 is called the *limiting molar conductivity* and $\mathscr{K}$ is a coefficient that is found to depend more on the nature of the electrolyte (i.e. whether it is of the form MA, or M_2A, etc.) than on its specific identity.

Kohlrausch was also able to confirm that the value of Λ_m^0 for any electrolyte can be expressed as the sum of contributions from its individual ions. If the molar conductivity of the cations is denoted λ_+, and that of the anions λ_-, then his law of the *independent migration of ions* is

$$\Lambda_m^0 = \nu_+\lambda_+ + \nu_-\lambda_-, \tag{27.1.4}$$

where ν_+ and ν_- are the numbers of cations and anions per formula unit of electrolyte (e.g. $\nu_+ = \nu_- = 1$ for HCl, NaCl, and $CuSO_4$, but $\nu_+ = 1$, $\nu_- = 2$ for $MgCl_2$). This simple result, which can be understood on the grounds that the ions behave independently when the solution is infinitely dilute, lets us predict the molar conductivity of any strong electrolyte (at infinite dilution) using the data in Table 27.1.

Table 27.1. Limiting ionic conductivities in water at 298 K, $\lambda/\mathrm{S\ cm^2\ mol^{-1}}$

H^+	349.6	OH^-	199.1
Na^+	50.1	Cl^-	76.4
K^+	73.5	Br^-	78.1
Zn^{2+}	105.6	SO_4^{2-}	160.0

Example 27.2

Predict the limiting molar conductivities of aqueous LiBr and $BaCl_2$ at 298 K.

• *Method*. Use eqn (27.1.4) with data from Table 27.1.

• *Answer*. For LiBr,

$$\Lambda_m^0 = (38.7 + 78.1)\ \mathrm{S\ cm^2\ mol^{-1}} = 116.8\ \mathrm{S\ cm^2\ mol^{-1}}.$$

For $BaCl_2$,

$$\Lambda_m^0 = (127.2 + 2 \times 76.3)\ \mathrm{S\,cm^2\,mol^{-1}} = 279.8\ \mathrm{S\,cm^2\,mol^{-1}}.$$

• *Comment*. The limiting *equivalent* conductances are 116.8 S cm^2 $equiv^{-1}$ and 139.9 S cm^2 $equiv^{-1}$, because twice as much charge per mole of $BaCl_2$ is transported than is transported per mole of LiBr.

• *Exercise*. Predict the limiting molar and equivalent conductances of $ZnSO_4(aq)$.
[266 S cm^2 mol^{-1}, 133 S cm^2 $equiv^{-1}$]

Weak electrolytes are substances that are not fully ionized in solution, and include weak acids and bases. The marked concentration dependence of their conductivities arises from the displacement of the $MA(aq) \rightleftharpoons M^+(aq) + A^-(aq)$ equilibrium towards the right at high dilutions. This can be demonstrated explicitly as follows. The conductivity depends on the number of ions in the solution, and therefore on the *degree of ionization*, α, of the electrolyte. If the ionization equilibrium constant of some weak 1:1 electrolyte MA is denoted K, then when it is present at a concentration c,

$$K = [M^+(aq)][A^-(aq)]/[MA(aq)] = (\alpha c)(\alpha c)/(1-\alpha)c = \{\alpha^2/(1-\alpha)\}c.$$

(For simplicity, we have written c in place of the dimensionless c/M, and we have ignored activity coefficients.) The degree of ionization when the concentration is c is therefore

$$\alpha = \tfrac{1}{2}(K/c)\{(1+4c/K)^{\frac{1}{2}} - 1\}. \qquad (27.1.5)$$

If the molar conductivity of the hypothetical *fully* ionized electrolyte is Λ'_m, then since only a fraction α is ionized in the actual solution, the measured molar conductivity (Λ_m) is given by

$$\Lambda_m = \alpha\Lambda'_m, \qquad (27.1.6)$$

with α given by eqn (27.1.5). The value of Λ'_m may be approximated by Λ_m^0, and so, if K is known, eqn (27.1.5) can be used to predict the concentration dependence of the molar conductivity. The dependence agrees quite well with the experimental curve shown in Fig. 27.2.

Alternatively, the concentration dependence of Λ_m can be used in measurements of the limiting molar conductivity. It is quite easy to manipulate the equilibrium constant expression into the form

$$1/\alpha = 1 + \alpha c/K,$$

and so, with $\alpha = \Lambda_m/\Lambda_m^0$,

$$1/\Lambda_m = 1/\Lambda_m^0 + \Lambda_m c/K(\Lambda_m^0)^2. \qquad (27.1.7)$$

This is called *Ostwald's Dilution Law*, and it shows that if $1/\Lambda_m$ is plotted against $c\Lambda_m$, the intercept is $1/\Lambda_m^0$. Once Λ_m^0 is known, the equilibrium constant may be found, as illustrated by the following *Example*.

Example 27.3

The resistance of an 0.010 M $CH_3COOH(aq)$ solution was measured at 298 K in the same cell as used in *Example* 27.1, and found to be 2220 Ω. Find the ionization constant K_a, the value of pK_a, and the degree of ionization of the acid.

• *Method*. Calculate Λ_m using the cell constant determined in *Example* 27.1 ($C = 0.367$ cm^{-1}). Find Λ_m^0 from the data in Table 27.1. Use eqn (27.1.6) for α with $\Lambda_m' = \Lambda_m^0$, and then find K_a from the expression above. Form $pK_a = -\lg K_a$.

• *Answer*. From Table 27.1,

$$\Lambda_m^0 = (349.6 + 40.9)\ \mathrm{S\ cm^2\ mol^{-1}} = 390.5\ \mathrm{S\ cm^2\ mol^{-1}}.$$

From $\Lambda_m = \kappa/c = C/Rc$, $\Lambda_m = 16.5\ \mathrm{S\ cm^2\ mol^{-1}}$. Therefore, $\alpha = 16.5/390.5 = 0.0423$. The ionization constant is

$$K_a = \{\alpha^2/(1-\alpha)\}c = (0.0423^2/0.9577) \times 0.010 = 1.9 \times 10^{-5}.$$

Therefore, $pK_a = 4.72$.

• *Comment*. The thermodynamic value of K_a would be obtained by repeating the determination with different concentrations and extrapolating to zero concentration.

• *Exercise*. The resistance of an 0.025 M HCOOH(aq) solution was measured as 444 Ω; evaluate pK_a of the acid. [3.74]

27.1 (b) The mobilities of ions

The experimental observations summarized above can be used to obtain insight into the way ions move in solution. We need to know why they move at different rates, why they have different molar conductivities, and why the molar conductivities of strong electrolytes depend on the square-root of the concentration. The fundamental idea is that the greater the mobility of an ion in solution, the greater its contribution to the conductivity. We must therefore examine the abilities of ions to migrate through solutions.

Ions in solution are subject to a variety of forces. They experience the random bombardment of the solvent particles, but since these occur in random directions they do not result in net motion if the solution is uniform. In an electric field they also experience an electric force. When two electrodes a distance l apart are at a potential difference $\Delta\phi$, there is a uniform electric field of magnitude $E = \Delta\phi/l$ between them. An ion of charge ze experiences a force of magnitude

$$\mathscr{F} = zeE = ze\Delta\phi/l. \tag{27.1.8}$$

A cation responds by accelerating in the direction of the negative electrode and an anion accelerates towards the positive electrode.

As the ion rubs through the solvent a frictional force, $\mathscr{F}'$, retards it with a strength that increases with its speed. If we assume that the Stokes formula (Section 25.1, $\mathscr{F}' = fs$, $f = 6\pi\eta a$) for a sphere of radius a and speed s applies even on a microscopic scale (and independent evidence from magnetic resonance suggests that it often gives at least the right order of magnitude), then a terminal speed, the *drift speed*, is reached when the accelerating force ($\mathscr{F} = zeE$) is balanced by the viscous drag ($\mathscr{F}' = fs$). This occurs when

$$s = zeE/f. \tag{27.1.9}$$

Since the drift velocity governs the rate at which charge is transported, it follows that we might expect the conductivity to decrease with (a) increasing solution viscosity and (b) increasing ion size.

Experimental results confirm these predictions for bulky ions (such as

R_4N^+ and $RCOO^-$), but not for small ions. For example, the molar conductivities of the alkali metal ions *increase* from Li^+ to Cs^+ (Table 27.1), even though the ionic radii increase (Table 23.1). The discrepancy is resolved when we realize that the radius in Stokes' formula is the *hydrodynamic radius* of the ion, its *effective* radius in the solution taking into account all the solvent molecules it carries. Small ions give rise to stronger electric fields than large ones (it is a result from electrostatics that the electric field at the surface of a sphere of radius R is proportional to ze/R^2, and so the smaller the radius the stronger the field), and so small ions are more extensively solvated than big ions. This means that an ion of small ionic radius has a large hydrodynamic radius because it drags many solvent molecules through the solution as it migrates; therefore, it also has a lower molar conductivity than a bigger ion of the same charge.

The proton, however, although it is very small, has a very high molar conductivity (Table 27.1)! This is because it conducts by a mechanism (the *Grotthus mechanism*) that does not involve its actual motion through the solution. Instead of a single, highly solvated proton moving through the solution, it is believed that there is an *effective* motion of a proton which involves the rearrangement of bonds through a long chain of water molecules, Fig. 27.3, and the conductivity is governed by the rates at which the water molecules can rotate into orientations in which they can accept or donate protons.

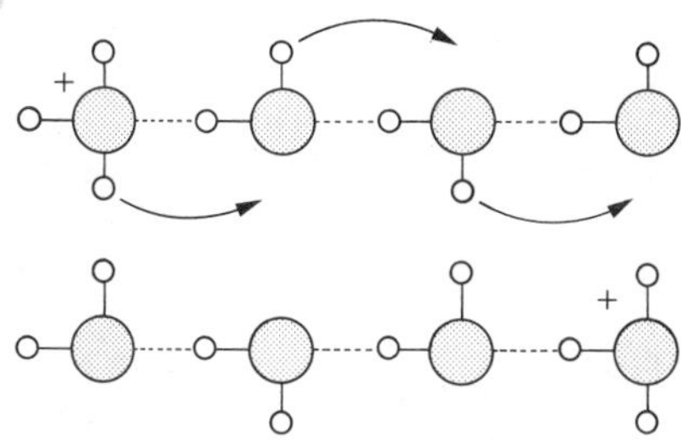

Fig. 27.3. The mechanism of conduction in water. Because protons can be transferred between molecules, the current is transported very rapidly down a chain by an effective, not an actual, migration of proton as each H_2O molecule passes one on to the next.

The drift speed is proportional to the strength of the applied field. The coefficient of proportionality is called the *mobility* of the ion (symbol: u), and we write

$$s = uE. \tag{27.1.10}$$

Some values of ionic mobilities are given in Table 27.2 (we show below how they are related to conductivities and hence how they may be measured), and are typically around $6\times10^{-4}\,\mathrm{cm^2\,s^{-1}\,V^{-1}}$. This means that at a stage in an AC cycle when there is a potential difference of 1 V across a solution of length 1 cm (so that $E = 1\,\mathrm{V\,cm^{-1}}$), the drift speed is typically about $6\times10^{-4}\,\mathrm{cm\,s^{-1}}$. This might seem slow, but not when expressed on a molecular scale, for it corresponds to an ion passing about 10 000 solvent molecules per second.

Table 27.2. Ionic mobilities in water at 298 K, $u/10^{-4}\,\mathrm{cm^2\,s^{-1}\,V^{-1}}$

H^+	36.23	OH^-	20.64
Na^+	5.19	Cl^-	7.91
K^+	7.62	Br^-	8.09
Zn^{2+}	5.47	SO_4^{2-}	8.29

The usefulness of ionic mobilities is that they provide a link between measurable and theoretical quantities. As a first step we establish the relation between mobility and conductivity. In order to keep the calculation simple, we ignore signs in the following, and concentrate on the magnitudes of quantities: the direction of flux can always be decided by common sense.

Consider a solution of a strong electrolyte at a concentration c. Let each formula unit give rise to ν_+ cations of charge z_+e and ν_- anions of charge z_-e. The concentration of each type of ion is therefore νc (with $\nu = \nu_+$ or ν_-), and the number density of each type is $\nu c N_A$. The number of cations that pass through an imaginary window of area A, Fig. 27.4, during an interval Δt is equal to the number within the distance $s_+\Delta t$, and therefore to the number in the volume $s_+\Delta t A$. (The same sort of argument was used in Section 26.1(a).) The number in this volume is equal to $(s_+\Delta t A)(\nu_+ c N_A)$. The number passing through the window per unit area per unit time, the cation flux, is therefore

$$J(\text{cations}) = (s_+\,\Delta t A)(\nu_+ c N_A)/\Delta t A = s_+\nu_+ c N_A.$$

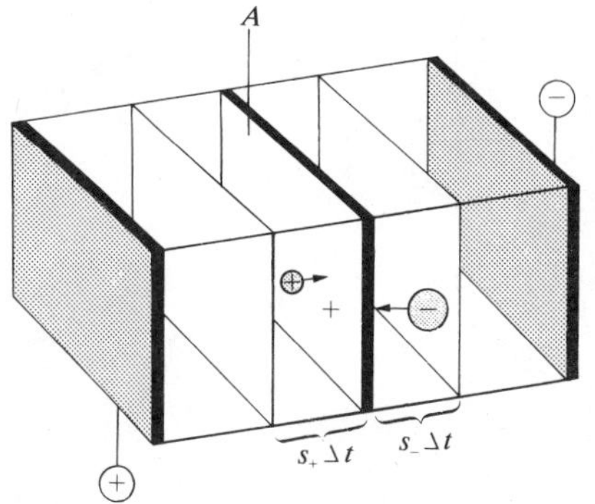

Fig. 27.4. In the calculation of the current, all the cations within a distance $s_+\Delta t$ (i.e. those in the volume $s_+A\Delta t$) will pass through the area A. Likewise, the anions in the corresponding volume the other side of the window will also contribute to the current.

Each cation carries a charge z_+e, and so the flux of charge is

$$J(\text{charge}) = z_+s_+\nu_+ceN_A = z_+s_+\nu_+cF,$$

where $F = eN_A$ is Faraday's constant. Since $s = uE$ this becomes

$$J(\text{charge}) = z_+u_+\nu_+cFE.$$

The current through the window due to the cations (I_+) is the charge flux times the area, $I_+ = JA$. The electric field is the potential gradient $\Delta\phi/l$. Therefore,

$$I_+ = z_+u_+\nu_+cFA\Delta\phi/l. \tag{27.1.11}$$

Now we are almost there. Current and potential difference are related by Ohm's Law, $I = \Delta\phi/R$, and the resistance R is related to the conductivity by eqn (27.1.1). Therefore, for the cations, $I_+ = \kappa_+(\Delta\phi A/l)$. Comparison of this with the last expression gives

$$\kappa_+ = z_+u_+\nu_+cF,$$

and so the molar ionic conductivity is

$$\lambda_+ = z_+u_+F. \tag{27.1.12a}$$

(The *equivalent* conductivity, as defined above, is this quantity divided by the charge number, or u_+F.) Exactly the same argument leads to the corresponding expression for the anion molar conductivity:

$$\lambda_- = z_-u_-F. \tag{27.1.12b}$$

It follows that the limiting molar conductivity of a solution is related to the individual ion mobilities by

$$\Lambda_m^0 = \{z_+u_+\nu_+ + z_-u_-\nu_-\}F. \tag{27.1.13}$$

In the case of a symmetrical $z:z$ electrolyte (e.g. $CuSO_4$), this simplifies to

$$\Lambda_m^0 = z(u_+ + u_-)F. \tag{27.1.14}$$

Ions that are very mobile carry a large fraction of the total current. The *transference number* or *transport number* (symbol: t) is defined as the fraction of total current carried by the ions of a specified type. In the case of a solution of two kinds of ion, the cation transport number is defined as

$$t_+ = I_+/I, \tag{27.1.15}$$

with a similar expression for the anion transport number t_-. Since the total current I is the sum of the cation and anion currents, $t_+ + t_- = 1$. The cation current is related to the cation mobility by eqn (27.1.11), with a similar expression for the anions, and so the relation between the limiting value t^0 and u is

$$t_+^0 = z_+\nu_+u_+/(z_+\nu_+u_+ + z_-\nu_-u_-). \tag{27.1.16}$$

In the case of a symmetrical electrolyte this simplifies to

$$t_+^0 = u_+/(u_+ + u_-), \qquad t_-^0 = u_-/(u_+ + u_-). \tag{27.1.17}$$

Moreover, since the ionic conductivities are related to the mobilities, eqn (27.1.12), it follows that

$$t_+^0 = \nu_+\lambda_+/(\nu_+\lambda_+ + \nu_-\lambda_-), \qquad t_-^0 = \nu_-\lambda_-/(\nu_+\lambda_+ + \nu_-\lambda_-), \quad (27.1.18)$$

and therefore that

$$\nu_+\lambda_+ = t_+^0\Lambda_m^0, \qquad \nu_-\lambda_- = t_-^0\Lambda_m^0. \quad (27.1.19)$$

Consequently, since there are independent ways of measuring transport numbers of ions, we can determine the individual ionic conductivities and (through eqn (27.1.12)) the ionic mobilities.

27.1 (c) The measurement of transport numbers

One of the most accurate methods for measuring transport numbers is the *moving boundary method* in which the motion of a boundary between two ionic solutions having a common ion is observed as a current flows.

Let MX be the salt of interest and NX (the *indicator*) a salt giving a denser solution. The indicator solution occupies the lower part of a vertical tube, Fig. 27.5, and the MX solution, the *leading solution*, occupies the upper part with a sharp boundary between them. The indicator solution must be denser than the leading solution, and the mobility of the N ions must be greater than the mobility of the M ions. When a current I is passed for a time t the boundary moves from AB to CD, and so all the M ions in the volume between AB and CD must have passed through CD. That number is cVN_A, and so the charge they transfer through the plane is $z_+cVeN_A = z_+cVF$. However, the total quantity of charge transferred when a current I flows for a time t is It. Therefore, the fraction due to the motion of the M ions, which is their transport number, is

$$t_+ = z_+cVF/It. \quad (27.1.20)$$

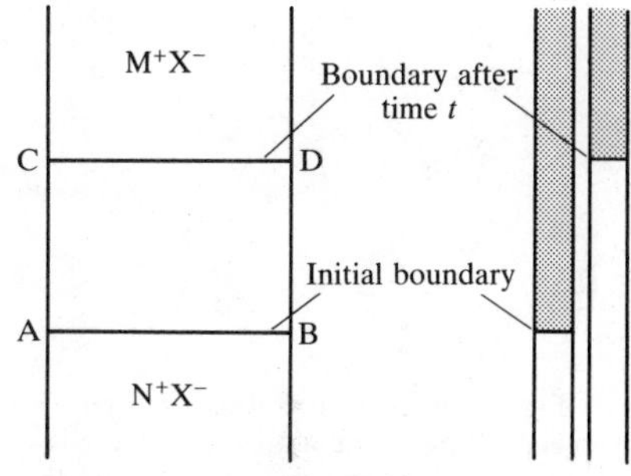

Fig. 27.5. In the *moving boundary* method for the measurement of transference numbers the distance moved by the boundary is observed as a current is passed. All the M ions in the volume between AB and CD must have passed through CD if the boundary moves from AB to CD.

Hence, knowing the diameter of the tube and the distance moved by the boundary in the time t, the transport number, and hence the ion conductivity and mobility, may be determined.

In the *Hittorf method* an electrolytic cell is divided into three compartments and an amount It of electricity is passed. An amount It/z_+F of cations are discharged at the cathode but an amount $t_+(It/z_+F)$ of cations migrate into the cathode compartment. The net change in the amount of cations in that compartment is therefore $(t_+ - 1)(It/z_+F)$, or $-t_-(It/z_+F)$. Therefore, by measuring the change of composition in the cathode compartment, t_-, the *anion* transport number, can be deduced. Likewise, the change in composition of the anions in the anode compartment is $-t_+(It/z_-F)$.

Finally, e.m.f. measurements may also be used. The e.m.f. of a cell with transference (E_t) having an electrode reversible with respect to anions, for example

$$\text{Ag}\,|\,\text{AgCl}\,|\,\text{HCl}(m_1)\,|\,\text{HCl}(m_2)\,|\,\text{AgCl}\,|\,\text{Ag},$$

is related to the e.m.f. of a cell with the same overall reaction but without transference (E), such as

$$\text{Ag}\,|\,\text{AgCl}\,|\,\text{HCl}(m_1)\,|\,\text{H}_2(\text{g})\,|\,\text{Pt}\,|\,\text{H}_2(\text{g})\,|\,\text{HCl}(m_2)\,|\,\text{AgCl}\,|\,\text{Ag}$$

by $E_t = t_+E$. (The argument is similar to that for the Hittorf method; see the solution to Problem 27.18.) Therefore, comparison of the two e.m.f.s gives the transport number, in this case of Cl^-.

27.1 (d) Conductivities and ion–ion interactions

Having accounted for the form of the law of independent migration of ions, the remaining problem is to account for the $c^{\frac{1}{2}}$ dependence of the Kohlrausch Law, eqn (27.1.3). Note that we have to account for a *reduction* of the conductivity below its value at zero ionic strength. In Chapter 11 we saw something similar: the activity coefficients of ions also depend on $c^{\frac{1}{2}}$, and depend on their charge type rather than their specific identities. That was explained in terms of the properties of the ionic atmosphere around each ion, and we should suspect that the same explanation applies here.

When an electric field is applied to a solution of ions and all the ions drift in a definite direction, we have to modify the picture of the ionic atmosphere as being a spherically symmetrical haze of charge around each ion. In the first place, the ions do not adjust to the moving ion infinitely quickly, and the atmosphere is incompletely formed in front of the moving ion and incompletely decayed in its wake, Fig. 27.6. The overall effect is the displacement of the centre of charge of the atmosphere a short distance behind the moving ion. Since the two charges are opposite, the result is a retardation of the moving ion. This reduction of the ions' mobility is called the *relaxation effect*, because the rearrangement of the ionic atmosphere is a relaxation of the ions into an equilibrium distribution.

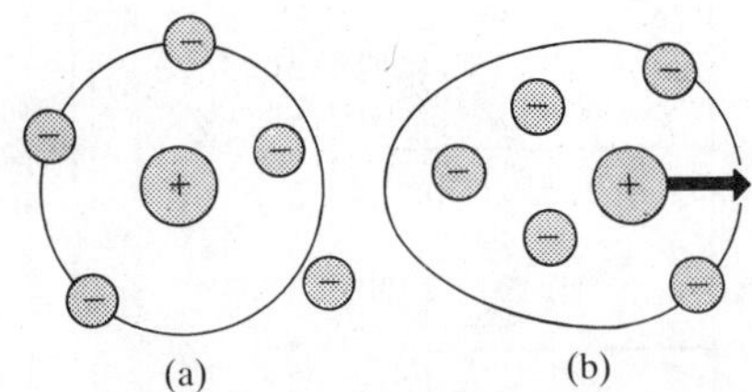

Fig. 27.6. (a) In the absence of an applied field, the ionic atmosphere is spherically symmetric, but (b) when a field is present it is distorted and the centres of negative and positive charge no longer coincide. This retards the motion of the central ion.

The *electrophoretic effect* is a second consequence of the ionic atmosphere. As we have seen, the central ion experiences a viscous drag as it moves through the solution. When the ionic atmosphere is present this effect is enhanced because the ionic atmosphere moves in an opposite direction to the central ion. The enhanced viscous drag reduces the mobility of the ions, and hence also reduces their conductivities.

The quantitative formulation of these effects is far from simple, but the *Debye–Hückel–Onsager theory* is an attempt to obtain quantitative expressions at about the same level of sophistication as the Debye–Hückel theory itself. The theory leads to a Kohlrausch-like expression, eqn (27.1.3), with

$$\mathcal{K} = A + B\Lambda_m^0, \tag{27.1.21a}$$

$$A = (z^2eF^2/3\pi\eta)(2/\varepsilon RT)^{\frac{1}{2}}, \qquad B = q(z^3eF^2/24\pi\varepsilon RT)(2/\pi\varepsilon RT)^{\frac{1}{2}}, \tag{27.1.21b}$$

where ε is the electric permittivity of the solvent and $q = 0.50$ for a 1:1 electrolyte. The numerical values for 1:1 electrolytes in several solvents are listed in Table 27.3. The slopes of the conductivity curves are predicted to depend on the charge type of the electrolyte, as the Kohlrausch measurements require, and some comparisons between theory and experiment are shown in Fig. 27.7. The agreement is quite good in the limit of extremely low concentration.

Table 27.3. Debye–Hückel–Onsager coefficients for (1,1)-electrolytes at 298 K

Solvent	A	B
Water	60.20	0.229
Methanol	156.1	0.923
Propanone	32.8	1.63

Example 27.4

The molar conductivity of aqueous potassium chloride at 298 K depends on concentration as reported below. Use the Kohlrausch Law to find the limiting molar conductivity of the salt.

c/M	0.001	0.005	0.010	0.020
Λ_m/S cm^2 mol^{-1}	146.9	143.5	141.2	138.2

• *Method*. The Kohlrausch expression, eqn (27.1.3), is valid only at very low concentrations, and so we take it to be the start of a more complicated expression, with the next term

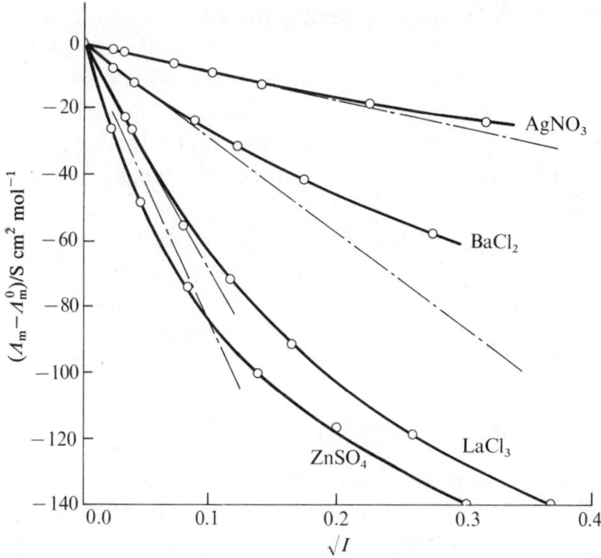

Fig. 27.7. The dependence of molar conductivities on the square root of the ionic strength, and comparison (broken lines) with the dependence predicted by the Debye–Hückel–Onsager theory, eqn (27.1.21).

proportional to c:

$$\Lambda_m = \Lambda_m^0 - (A + B\Lambda_m^0)c^{\frac{1}{2}} + xc.$$

On rearranging to

$$\Lambda_m + (A + B\Lambda_m^0)c^{\frac{1}{2}} = \Lambda_m^0 + xc$$

we see that a plot of the expression on the left against c should give a straight line with intercept Λ_m^0 at $c = 0$. Take the values of A and B from Table 27.3, and make a preliminary estimate of Λ_m^0 from the information in Table 27.1.

• *Answer*. From Table 27.1, $\Lambda_m^0 = 149.8$ S cm^2 mol^{-1}. Then, from Table 27.3,

$$\begin{aligned}\mathscr{K} &= 60.20 \text{ S cm}^2 \text{ mol}^{-1} \text{ M}^{-\frac{1}{2}} + (0.229 \text{ M}^{-\frac{1}{2}}) \times (149.8 \text{ S cm}^2 \text{ mol}^{-1}) \\ &= 94.50 \text{ S cm}^2 \text{ mol}^{-1} \text{ M}^{-\frac{1}{2}}.\end{aligned}$$

Now draw up the following table, with $X = \{\Lambda_m + (A + B\Lambda_m^0)c^{\frac{1}{2}}\}/\text{S cm}^2 \text{ mol}^{-1}$:

c/M	0.001	0.005	0.010	0.020
X	149.9	150.2	150.7	151.6

These points are plotted in Fig. 27.8. The intercept is at $X = 149.8$, and so $\Lambda_m^0 = 149.8$ S cm^2 mol^{-1}, in perfect agreement with the preliminary value.

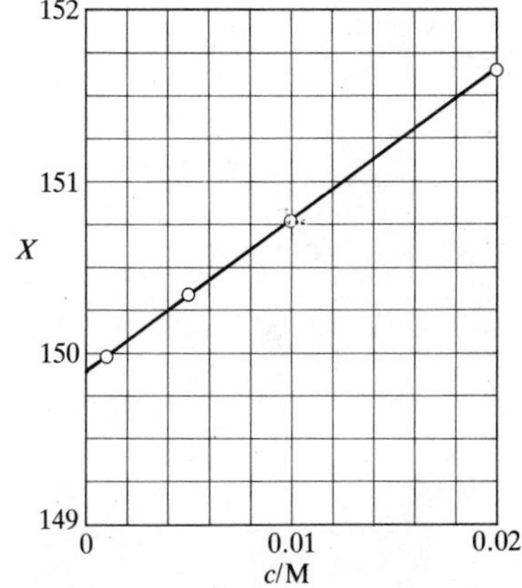

Fig. 27.8. The determination of the limiting molar conductance using an extrapolation based on the Debye–Hückel–Onsager theory; see *Example* 27.4.

• *Comment*. Had the agreement been poor, the new estimate of the limiting conductivity could have been used in repeats of the procedure until self-consistency was reached.

• *Exercise*. The following results were obtained for NaI(aq) at 298 K:

c/M	0.001	0.005	0.010	0.020
Λ_m/S cm^2 mol^{-1}	124.2	121.2	119.2	116.6

Find the limiting molar conductivity of the salt. [126.9 S cm^2 mol^{-1}]

The success of the Debye–Hückel–Onsager theory suggests that the model is substantially correct. A further test is to examine what happens when the effects of the ionic atmosphere are eliminated in some way. One way of doing this is to measure the conductivity at very high frequency, for when the central ion moves backwards and forwards very rapidly the retarding effect of the asymmetric atmosphere ought to average to zero. The

Debye–Falkenhagen effect is the observation of this increase in conductivity. Another way of eliminating the atmosphere is to cause the ions to move so quickly that they outrun it. The *first Wien effect* is the observation of higher conductivities at high electric fields, as this predicts. (The *second Wien effect* is the observation of an increased conductivity as a result of the enhancement of the degree of ionization of a weak electrolyte when a strong field is applied.)

The simple picture of ions surrounded by a diffuse ionic atmosphere fails at higher concentrations because ions stick together as pairs, triples, and so on. This can be seen from X-ray diffraction by ionic solutions, where peaks in the diffraction pattern can be interpreted in terms of definite ion–ion distances.

27.2 Diffusion and transport

In this section we draw together the discussions in this chapter and the last. This we do using thermodynamics and statistics, and discover that we can find several useful relations between transport properties.

27.2 (a) Diffusion: the thermodynamic view

In Part 1 it is shown that the thermodynamic quantity that governs the direction of spontaneous change under conditions of constant temperature and pressure is the chemical potential (Section 6.3). When unit amount of solute particles moves from a region where their chemical potential is $\mu(1)$ to one where it is $\mu(2)$, the work required is $w = \mu(2) - \mu(1)$. Suppose the chemical potential depends on the position x in the system, then the work required to move material from x to $x + \mathrm{d}x$ is

$$\mathrm{d}w = \mu(x + \mathrm{d}x) - \mu(x) = \{\mu(x) + (\mathrm{d}\mu/\mathrm{d}x)\,\mathrm{d}x\} - \mu(x) = (\mathrm{d}\mu/\mathrm{d}x)\,\mathrm{d}x.$$

(The derivatives ought to be partial, under the constraints of constant pressure and temperature, but we shall remember that rather than writing it explicitly.) In classical mechanics the work required to move an object through a distance $\mathrm{d}x$ against a force $\mathcal{F}$ is $\mathrm{d}w = -\mathcal{F}\,\mathrm{d}x$ (recall Box 2.2). Therefore, by comparing these expressions, we see that *the slope of the chemical potential is an effective force*. We shall write this *thermodynamic force* as

$$\mathcal{F} = -(\partial\mu/\partial x)_{p,T}. \qquad (27.2.1)$$

There is no real force pushing the particles down the slope of the chemical potential, for that is their natural drift as a consequence of the Second Law and the hunt for maximum entropy; but thinking in terms of these *effective* forces can be very helpful, as we shall see. Another point is that the quantity is a *molar force*: the effective force per unit amount of substance (its units are $\mathrm{N\,mol^{-1}}$ because the units of μ are $\mathrm{J\,mol^{-1}}$ and $1\,\mathrm{J} = 1\,\mathrm{N\,m}$).

In a solution where the activity of the particles is a, the chemical potential is $\mu = \mu^{\ominus} + RT\ln a$ (Box 8.1). If the solution is not uniform the activity depends on the position and we can write

$$\mathcal{F} = -(\mathrm{d}/\mathrm{d}x)\{\mu^{0} + RT\ln a\} = RT(\partial\ln a/\partial x)_{p,T}.$$

If the solution is ideal a may be replaced by the concentration c, and so

$$\mathcal{F} = -(RT/c)(\partial c/\partial x)_{p,T} \qquad (27.2.2)^{\circ}$$

because $(\mathrm{d}/\mathrm{d}x)\ln c = (1/c)\,\mathrm{d}c/\mathrm{d}x$. This shows that the thermodynamic force pushes in the direction of decreasing concentration ($\mathcal{F}$ is positive when $\mathrm{d}c/\mathrm{d}x$ is negative).

The last equation lets us derive Fick's First Law of diffusion (that flux, the number per unit area per unit time, is proportional to the concentration gradient, eqn (26.3.1)) from a thermodynamic viewpoint. We suppose that the flux of particles is a response to a thermodynamic force due to the concentration gradient. As a result of the balancing of the force $\mathcal{F}$ and the viscous drag on the particles, they settle down to a drift speed s. This drift speed is proportional to the thermodynamic force, and so we can write $s \propto \mathcal{F}$. If the number density of particles is $\mathcal{N}$, the flux of particles through an imaginary window in the fluid is equal to $s\mathcal{N}$. (This comes from the argument that we have used several times before: all the particles within a distance $s\Delta t$, and therefore in a volume $s\Delta tA$, pass through the window of area A in the time Δt.) That is, $J = s\mathcal{N} = scN_A$. But since s is proportional to $\mathcal{F}$, which in turn is proportional to $(1/c)(\mathrm{d}c/\mathrm{d}x)$, it follows that J is proportional to $\mathrm{d}c/\mathrm{d}x$, as we set out to show.

Instead of simply reproducing Fick's Law we can go further. Since $J = scN_A$, eqn (26.1.3) can be written

$$scN_A = -D(\mathrm{d}\mathcal{N}/\mathrm{d}x) = -DN_A(\mathrm{d}c/\mathrm{d}x).$$

Expressing $\mathrm{d}c/\mathrm{d}x$ in terms of $\mathcal{F}$ using eqn (27.2.2) gives

$$s = -(D/c)(\mathrm{d}c/\mathrm{d}x) = (D/RT)\mathcal{F}. \qquad (27.2.3)^\circ$$

Therefore, once we know the effective force and D, we can calculate the drift speed of the particles (and vice versa), and this will be true whatever the origin of the force.

There is one case where we already know the drift speed and the effective force acting on a particle: an ion in solution has a drift speed $s = uE$ when it experiences a force ezE, so that the force per mole is zFE. Therefore, substituting these known values into eqn (27.2.3) gives

$$uE = (D/RT)(zFE), \quad \text{or} \quad u = ezD/kT. \qquad (27.2.4)^\circ$$

This rearranges into the very important result known as the *Einstein relation* between the diffusion coefficient and the ionic mobility:

$$\textit{Einstein relation:}\ D = ukT/ez. \qquad (27.2.5)^\circ$$

The Einstein relation can be developed in two ways. In the first place, it can be used to find the *Nernst–Einstein relation* between the molar conductivity of an electrolyte and the diffusion coefficients of its ions. First we write

$$\lambda_+ = u_+z_+F = ez_+^2D_+F/kT = z_+^2D_+F^2/RT, \qquad (27.2.6)^\circ$$

and a similar expression for the anion. Then the sum of the two weighted by the numbers of ions from each formula unit, eqn (27.1.4), is the limiting molar conductivity:

$$\textit{Nernst–Einstein relation:}\ \Lambda_m^0 = (F^2/RT)\{\nu_+z_+^2D_+ + \nu_-z_-^2D_-\}. \qquad (27.2.7)^\circ$$

One application of this expression is to the determination of ionic diffusion coefficients from conductivity measurements; another is to the prediction of conductivities using models of ionic diffusion (see below).

The other development of the Einstein relation leads to the *Stokes–Einstein relation* between the diffusion coefficient and the viscosity of the solution. By combining the expressions $s = uE$ and $s = ezE/f$ (which comes from eqn (27.1.9)), we obtain

$$u = ez/f \qquad (27.2.8)$$

as a relation between an ion's mobility and its hydrodynamic radius. However, we have already deduced that $u = ezD/kT$, and so we obtain the

$$\textit{Stokes–Einstein relation: } D = kT/f. \qquad (27.2.9)^\circ$$

Therefore, since $f = 6\pi\eta a$, if we know the hydrodynamic radius of the ion we can calculate its diffusion coefficient from the viscosity of the solution. An important feature of this result, however, is that *it is independent of the charge of the diffusing particle*. Therefore, it also applies in the limit of vanishingly small charge; that is, *it also applies to neutral particles*. This means that we may use viscosity measurements to estimate the diffusion coefficients for molecules in solution (and this relation was in fact used in Section 25.1(e) in the discussion of macromolecules). Some diffusion coefficients are listed in Table 27.4 and the relations developed above are collected in Box 27.1.

Table 27.4. Diffusion coefficients at 298 K, $D/10^{-5}\,\mathrm{cm^2\,s^{-1}}$

I_2 in hexane	4.05
Sucrose in water	0.521
H^+ in water	9.31
Na^+ in water	1.33

Box 27.1 Transport properties in solution

Einstein–Smoluchowski relation between jump length d and jump time τ:

$$D = d^2/2\tau.$$

Stokes–Einstein relation between the diffusion coefficient D and the solution viscosity η:

$$D = kT/f, \qquad f = 6\pi\eta a.$$

a is the effective hydrodynamic radius of the spherical particle; expressions for other shapes are given in Box 25.1.

Einstein relation between the diffusion coefficient and ion mobility u:

$$D = ukT/ez = uRT/zF.$$

z is the charge number of the ion.

Nernst–Einstein relation between the diffusion coefficient and the ion conductivity λ:

$$\lambda = (z^2F^2/RT)D.$$

Example 27.5

Evaluate the diffusion coefficient, limiting molar conductivity, and effective hydrodynamic radius of SO_4^{2-} in water at 298 K.

• *Method*. Take the mobility from Table 27.2 and obtain D using the eqn (27.2.5). Use eqn

(27.1.12b) for the ionic conductivity. Obtain a from eqn (27.2.9) and $f = 6\pi\eta a$. The viscosity of water is 1.00 cP (Table 24.5).

• *Answer*. Using $u = 8.29 \times 10^{-4}\,\text{cm}^2\,\text{s}^{-1}\,\text{V}^{-1}$,

$$D = ukT/ez = 1.1 \times 10^{-5}\,\text{cm}^2\,\text{s}^{-1}.$$

$$\lambda_- = z_- u_- F = 160\,\text{S cm}^2\,\text{mol}^{-1}.$$

From eqn (27.2.9), with $\eta = 1.00 \times 10^{-3}\,\text{kg m}^{-1}\,\text{s}^{-1}$,

$$a = kT/6\pi\eta D = 2.1 \times 10^{-10}\,\text{m} = 210\,\text{pm}.$$

• *Comment*. The bond length in SO_4^{2-} is 144 pm, and so the radius calculated here is plausible and compatible with a small degree of solvation.

• *Exercise*. Repeat the calculation for the NH_4^+ ion.

[$1.96 \times 10^{-5}\,\text{cm}^2\,\text{s}^{-1}$, $73.6\,\text{S cm}^2\,\text{mol}^{-1}$, 110 pm]

Some experimental support for these results comes from conductivity measurements. *Walden's rule* is the empirical observation that the product $\Lambda_m^0\eta$ is very approximately constant for the same ions in different solvents. Since $\Lambda_m^0 \propto D$, and we have just seen that $D \propto 1/\eta$, we do indeed predict that $\Lambda_m^0 \propto 1/\eta$ as Walden's rule requires. The usefulness of the rule, however, is muddied by the role of solvation: different solvents solvate the same ions to different extents, and so the hydrodynamic radius, as well as the viscosity, also changes with the solvent. Some experimental values are given in Table 27.5: they show that great care must be used in applying the rule and that it is, at best, only of very approximate validity.

Table 27.5. Viscosity and conductivity

		Solvent	
		Water	Ethanol
$\eta/\text{kg m}^{-1}\,\text{s}^{-1}$		8.9×10^{-4}	1.08×10^{-3}
$\lambda^0\eta/\text{S kg m}^{-1}\,\text{s}^{-1}\,\text{mol}^{-1}$	(a)	6.5×10^{-6}	2.5×10^{-6}
	(b)	1.7×10^{-6}	2.1×10^{-6}

(a) K^+, (b) ${}^t\text{Bu}_4\text{N}^+$

27.3 Diffusion

We now turn to the discussion of time-dependent diffusion processes, where we are interested in the spreading of inhomogeneities. One example is the temperature of a metal bar that has been suddenly heated at one end: if the source of heat is removed, then the bar settles down into a state of uniform temperature; if the source is maintained and the bar can radiate, then it settles down into a *steady state* of non-uniform temperature. Another example is the concentration distribution in a solvent to which a solute is added.

27.3 (a) The diffusion equation

We concentrate on the diffusion of particles, but similar arguments apply to the diffusion of properties. Our aim is to obtain an equation for the rate of change of the concentration of particles in some inhomogeneous region.

Consider a thin slab of cross-sectional area A and extending from x to $x + \Delta x$, Fig. 27.9. Let the number density at x be $\mathcal{N}$ at the time t. JA is the number of particles that enter the slab per unit time, and so the increase in concentration inside the slab (which has volume $A\Delta x$) on account of the flux from the left is

$$\partial\mathcal{N}/\partial t = J(x)A/A\Delta x = J(x)/\Delta x.$$

There is also an outflow through the right-hand window. The flux there is $J(x + \Delta x)$, and so the change of concentration due to this efflux is

$$\partial\mathcal{N}/\partial t = -J(x + \Delta x)A/A\Delta x = -J(x + \Delta x)/\Delta x.$$

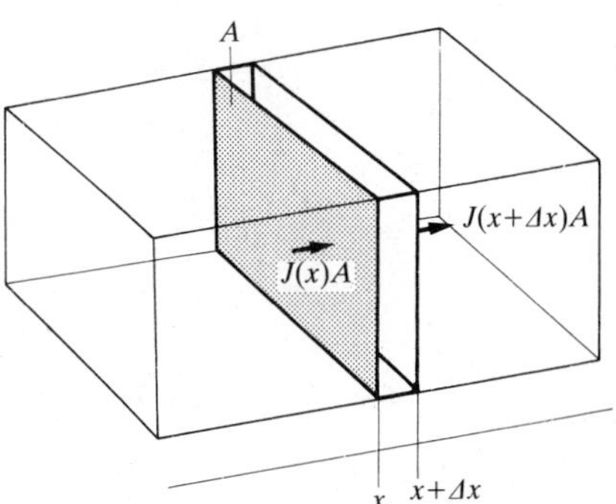

Fig. 27.9. The net flux in a region is the difference between the flux entering from the region of high concentration (on the left) and the flux leaving to the region of low concentration (on the right).

The net rate of change of concentration is therefore

$$\partial \mathcal{N}/\partial t = \{J(x) - J(x+\Delta x)\}/\Delta x.$$

The fluxes are proportional to the concentration gradients at the two windows. Using Fick's Law we write

$$\begin{aligned} J(x) - J(x+\Delta x) &= -D\left(\frac{\partial \mathcal{N}(x)}{\partial x}\right) + D\left(\frac{\partial \mathcal{N}(x+\Delta x)}{\partial x}\right) \\ &= -D\left(\frac{\partial \mathcal{N}}{\partial x}\right) + D\frac{\partial}{\partial x}\left\{\mathcal{N} + \left(\frac{\partial \mathcal{N}}{\partial x}\right)\Delta x\right\} = D(\partial^2 \mathcal{N}/\partial x^2)\,\Delta x. \end{aligned}$$

(Unless otherwise specified, $\mathcal{N}$ means $\mathcal{N}(x,t)$.) Substitution of this relation into the expression for the rate of change of concentration in the slab leads to the

$$\textit{Diffusion equation: } \partial \mathcal{N}/\partial t = D(\partial^2 \mathcal{N}/\partial x^2). \qquad (27.3.1)$$

This is also sometimes called *Fick's Second Law of Diffusion*.

First, a word about the general form of the diffusion equation. It shows that the rate of change of concentration is proportional to the *curvature* (more precisely, the second derivative) of the concentration with respect to distance. If the concentration changes sharply from point to point (if the distribution is highly wrinkled) then the concentration changes rapidly with time. If the curvature is zero, then the concentration is constant in time. For example, if the concentration decreases linearly with distance, then the concentration at any point is constant because the inflow of particles is exactly balanced by the outflow. The diffusion equation can be regarded as a mathematical formulation of the intuitive notion that there is a natural tendency for the wrinkles in a distribution to disappear ('Nature abhors a wrinkle').

The diffusion equation is a second-order differential equation with respect to space and a first-order differential equation with respect to time. Therefore, in order to arrive at a solution, we must specify two boundary conditions for the spatial dependence and a single initial condition for the time-dependence. This can be illustrated by the specific example of a solvent in which the solute is initially coated on one surface (e.g. a layer of sugar on the bottom of a deep beaker of water). The initial condition is that at $t=0$ all N_0 particles are concentrated on the yz-plane (of area A) at $x=0$. The boundary conditions are (1) that the concentration must everywhere be finite and (2) that the total number of particles present is N_0 at all times. The solution of the diffusion equation under these conditions is

$$\mathcal{N} = \{N_0/A(\pi D t)^{\frac{1}{2}}\}e^{-x^2/4Dt}, \qquad (27.3.2)$$

as may be verified by direct substitution. The shape of the concentration distribution at various times is shown in Fig. 27.10, and it is clear that the concentration spreads and tends to uniformity. Another solution is for the case of a localized concentration of solute in a three-dimensional solvent (a sugar lump suspended in a large flask of water): the concentration distribution is spherically symmetrical, and at a radius r is

$$\mathcal{N} = \{N_0/8(\pi D t)^{\frac{3}{2}}\}e^{-r^2/4Dt}. \qquad (27.3.3)$$

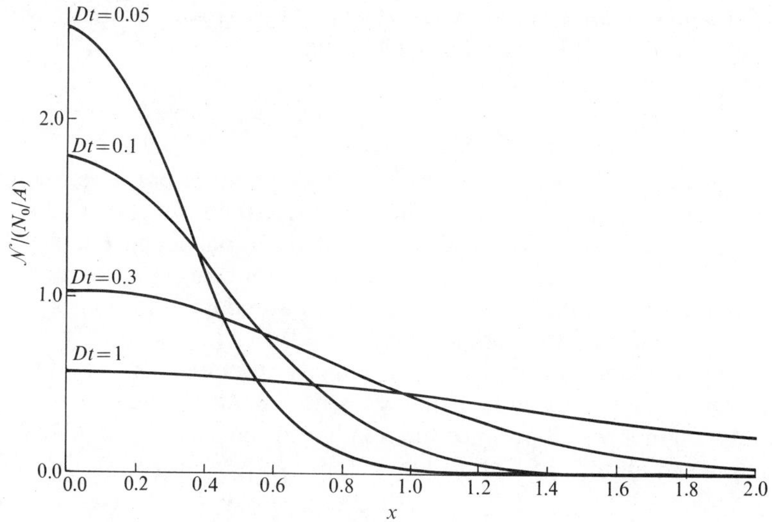

Fig. 27.10. The concentration profiles above a plane from which a solute is diffusing. These are plots of eqn (27.3.2). The units of Dt and x are arbitrary, but related so that Dt/x^2 is dimensionless. For example, if x is in cm, Dt would be in cm^2, and so for $D = 10^{-5}$ cm^2 s^{-1}, '$Dt = 0.1$' corresponds to $t = 10^4$ s.

Other chemically (and physically) interesting arrangements can be treated, but the solutions are more cumbersome to write down. The solutions are useful for experimental determinations of diffusion coefficients. In the *capillary technique* a capillary tube, open at one end and containing a solution, is immersed in a well stirred larger quantity of solvent, and the change of concentration in the tube is measured at a series of times. The solute diffuses from the open end of the capillary at a rate that can be calculated by solving the diffusion equation with the appropriate boundary conditions, and so D may be determined. In the *diaphragm technique* the diffusion occurs through the capillary pores of a sintered glass diaphragm separating the well-stirred solution and solvent. The concentrations are monitored and then related to the solutions of the diffusion equation corresponding to this arrangement.

27.3 (b) Properties of the solutions

The solutions of the diffusion equation, such as those above, can be used to predict the concentration of particles (or the value of some other physical quantity, such as the temperature in a non-uniform system) at any location. The solutions may also be used to calculate the *mean distances* through which the particle diffuse in a given time. The calculation runs as follows.

The number of particles in a slab of thickness dx at x is $\mathcal{N}A$dx, and so the probability that any of the N_0 particles is there is $\mathcal{N}A\,\mathrm{d}x/N_0$. If the particle *is* there, then it has travelled a distance x from the origin. Therefore, the mean distance travelled by all the particles is the sum of each x weighted by the probability of its occurrence:

$$\langle x\rangle = \int_0^\infty x\mathcal{N}A\,\mathrm{d}x/N_0 = (1/\pi Dt)^{\frac{1}{2}}\int_0^\infty x\mathrm{e}^{-x^2/4Dt}\,\mathrm{d}x = 2(Dt/\pi)^{\frac{1}{2}}. \quad (27.3.4)$$

The average distance of diffusion varies as the *square root* of the lapsed time. This is a general result to which we shall return. If we use the Stokes–Einstein relation for the diffusion coefficient, the mean distance travelled by particles of radius a in a solvent of viscosity η is

$$\langle x\rangle = (2kT/3\pi^2\eta a)^{\frac{1}{2}}\sqrt{t}. \quad (27.3.5)$$

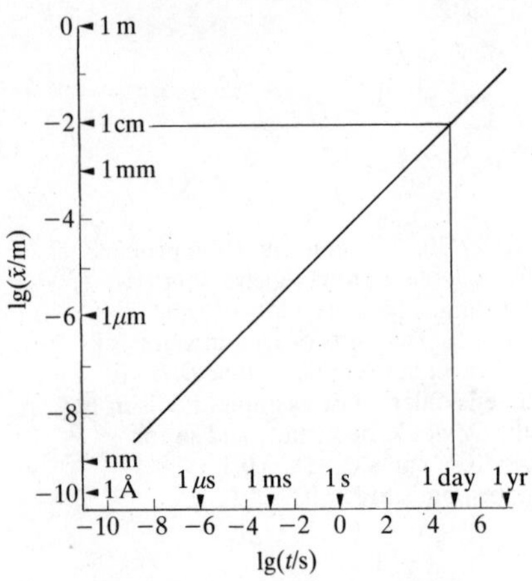

Fig. 27.11. The root mean square distance covered by particles with $D = 5 \times 10^{-6}$ cm^2 s^{-1}. Note the great slowness of diffusion.

The *root mean square distance* travelled by diffusing particles is $\bar{x} = \langle x^2 \rangle^{\frac{1}{2}}$, and its value may be calculated similarly:

$$\bar{x} = \left\{ \int_0^\infty x^2 \mathcal{N} A \, dx / N_0 \right\}^{\frac{1}{2}} = (2Dt)^{\frac{1}{2}}. \tag{27.3.6}$$

This quantity is a valuable measure of the spread of particles when they can diffuse in both directions from the origin (for then $\langle x \rangle = 0$ at all times). The value of $\bar{x}$ for particles with a typical diffusion coefficient ($D = 5 \times 10^{-6}$ cm^2 s^{-1}) is shown in Fig. 27.11: this shows how long it takes for diffusion to increase x to about 1 cm in an unstirred solution: diffusion is a very slow process (which is why solutions are stirred).

The *proportion* of particles that remain within a distance $\bar{x}$ of the origin is a useful number because it indicates how many particles remain close to their original position. Since the number in the slab at x is $\mathcal{N} A dx$, the total number in all the slabs up to the one at $\bar{x}$ is the sum (integral)‡

$$N(x \leqslant x) = \int_0^{\bar{x}} \mathcal{N} A \, dx = 0.68 N_0. \tag{27.3.7}$$

It follows that the proportion of particles inside the range $0 \leqslant x \leqslant \bar{x}$ is 0.68. Therefore, over two-thirds of the particles are still close to the origin and only 32% have escaped beyond $\bar{x}$ (but do not forget that $\bar{x}$ increases with time).

27.3 (c) Diffusion with convection

Because diffusion is so slow, reaction vessels are generally stirred and the slow diffusion of particles is dominated by brute-force mixing. The transport of particles arising from the motion of a streaming fluid is called *convection*. If we ignore diffusion, the flux of particles through an area A in a time Δt when the fluid is streaming at a velocity v can be calculated in the way we have used several times before (counting the particles within a distance $v \, \Delta t$), and is

$$J = (\mathcal{N} A v \Delta t) / A \Delta t = \mathcal{N} v. \tag{27.3.8}$$

J is called the *convective flux*. This flux contributes to the concentration of particles in a region, and a calculation similar to the one already done gives the rate of change of concentration in a slab of thickness Δx and area A as

$$\begin{aligned} \partial \mathcal{N} / \partial t &= \{AJ(x) - AJ(x + \Delta x)\} / A \Delta x \\ &= v\{\mathcal{N} - [\mathcal{N} + (\partial \mathcal{N} / \partial x) \Delta x]\} / \Delta x = -v(\partial \mathcal{N} / \partial x). \end{aligned} \tag{27.3.9}$$

(We have assumed that the velocity is constant.)

When both diffusion and convection are of similar importance the total change of concentration in a region is the sum of the two effects, and so the

‡ The integral is simplified by substituting $y = x/2(Dt)^{\frac{1}{2}}$, when it becomes

$$N = \{N_0 / (\pi D t)^{\frac{1}{2}}\} \int_0^{\bar{x}} e^{-x^2/4Dt} \, dx = N_0 (2/\pi^{\frac{1}{2}}) \int_0^{1/\sqrt{2}} e^{-y} \, dy.$$

The integral is standard, and proportional to the error function (Box 26.1) erf z evaluated at $z = 1/2^{\frac{1}{2}}$:

$$N = N_0 \operatorname{erf}(1/2^{\frac{1}{2}}),$$

with $\operatorname{erf}(1/2^{\frac{1}{2}}) = \operatorname{erf} 0.71 = 0.68$.

generalized diffusion equation is

$$\partial\mathcal{N}/\partial t = D(\partial^2\mathcal{N}/\partial x^2) - v(\partial\mathcal{N}/\partial x). \qquad (27.3.10)$$

A further important refinement, which is important in chemistry, is the possibility that the concentrations of particles may change as a result of reaction. When this is incorporated into the last equation (Section 30.2(c)) we have a powerful differential equation for discussing the properties of reacting, diffusing, convecting systems: it is the basis of reactor design in chemical industry and of the utilization of resources in living cells.

27.3 (d) The statistical view of diffusion

An intuitive picture of diffusion is that the particles move in a series of small steps and gradually migrate from their original positions. We shall explore this idea by building a model by assuming that the particles can jump through a distance d in a time τ, so that the distance travelled by a particle in a time t is $(t/\tau)d$. This does not mean that the particle will be found at that distance from the origin. The direction of each step may be different, and the net distance travelled must take this into account. If we simplify the discussion by allowing the particles to travel only along a straight line (the x-axis), and for each step (to the left or the right) to be through the same distance d each time, we obtain the *one-dimensional random walk*, a process first met in Appendix 25.1.

When the calculation is carried through (Appendix 27.1), it turns out that the probability of being at a distance x from the origin after an interval t is

$$P = (2\tau/\pi t)^{\frac{1}{2}}e^{-x^2\tau/2td^2}. \qquad (27.3.11)$$

This has precisely the same form as eqn (27.3.2), the difference of detail arising from the fact that in this calculation the particles can migrate in either direction from the origin and may reside only at discrete points separated by d instead of being anywhere on a continuous line. This suggests that diffusion can indeed be interpreted as the outcome of a large number of steps in random directions. Furthermore, the details of the calculation also show that the diffusion equation is not valid when the particles have had time to take only very few steps.

Comparison of the two exponents in eqns (27.3.2) and eqn (27.3.11) leads to the *Einstein–Smoluchowski relation* between the diffusion coefficient and the step length and rate. From the identification $2d^2/\tau = 4D$ we obtain:

$$\textit{Einstein–Smoluchowski relation: } D = d^2/2\tau. \qquad (27.3.12)$$

Example 27.6

Suppose that the SO_4^{2-} ion jumps through its own diameter each time it makes a move in an aqueous solution. How often does it change position at 25 °C?

• *Method*. The diffusion coefficient was calculated in *Example* 27.5; the effective radius was also estimated there. Combine them using eqn (27.3.12) to find τ.

• *Answer*. From *Example* 27.5, $D = 1.1 \times 10^{-5}\ \mathrm{cm^2\,s^{-1}}$ and $a = 210$ pm. Set $d = 2a$ and find

$$\tau = d^2/2D = 8 \times 10^{-11}\ \mathrm{s}.$$

• *Comment*. If the jump length were taken to be the diameter of a water molecule (about 150 pm), the jump time would be about 1×10^{-11} s.

• *Exercise*. Repeat the calculation for the NH_4^+ ion. [13 ps]

The Einstein–Smoluchowski relation is the central connection between the microscopic details of particle motion and the macroscopic parameters relating to diffusion (e.g. the diffusion coefficient and, through the Stokes–Einstein relation, the viscosity). It also brings us back full circle to the properties of gases. For if d/τ is interpreted as a mean speed of particles ($\bar{c}$), and d is interpreted as a mean free path (λ), then the Einstein–Smoluchowski relation turns into $D = \frac{1}{2}\lambda\bar{c}$, which is the same expression as we obtained from the kinetic theory of gases in Section 26.3(c). This shows that the diffusion of a perfect gas can be interpreted as a random walk with an average step size equal to the mean free path.

Appendix 27.1 The random walk

Our task is to calculate the probability that a particle will be found at a distance x from the origin after a time t. During that time it will have taken n steps, with $n = t/\tau$. If n_R of these are steps to the right, and n_L are steps to the left (with $n_L + n_R = n$), then the net distance travelled is $x = (n_R - n_L)d$. That is, to arrive at x, we must ensure that

$$n_R = \tfrac{1}{2}(n+s) \quad \text{and} \quad n_L = \tfrac{1}{2}(n-s) \qquad (27.\text{A}1.1)$$

with $s = x/d$. The probability of being at x after n steps of length d is therefore the probability that, when n random steps are taken, the numbers to the right and the left are as given above.

The *total number of different journeys* for a walk of n steps is equal to 2^n, because each step can be in either of two directions. The *number of journeys in which exactly n_R steps are taken to the right* is equal to the number of ways of choosing n_R objects from n possibilities irrespective of the order: this is $n!/n_R!\,(n-n_R)!$. As a check, consider all journeys of four steps: there are 2^4 possible journeys (see **1**). There are six ways of taking two steps to the right and two to the left, which tallies with the expression $4!/2!2! = 6$. The probability that the particle is back at the origin after four steps is therefore 6/16. The probability that it is out at $x = +4d$ is only 1/16 because, in order to be there, all four steps must be to the right, and since $4!/4!0! = 1$ the chance of that occurring is 1/16.

LLLL	LLLR	LLRR	LRRR	RRRR
	LLRL	LRLR	RLRR	
	LRLL	LRRL	RRLR	
	RLLL	RLLR	RRRL	
		RLRL		
		RRLL		

1

Returning now to the general case, the probability of being at x after n steps is

$$P = \frac{\textit{Number of journeys with } n_R \textit{ steps to the right}}{\textit{Total number of journeys}}$$
$$= \{n!/n_R!\,(n-n_R)!\}/2^n$$
$$= \{n!/[\tfrac{1}{2}(n+s)]!\,[\tfrac{1}{2}(n-s)]!\}/2^n. \qquad (27.\text{A}1.2)$$

At this stage the expression for the probability does not seem to resemble the Gaussian, eqn (27.3.2), obtained by solving the diffusion equation. However, consider the case when so much time has elapsed that the particles have taken many steps. Then the factorials may be approximated using Stirling's approximation (in the more precise form given in the footnote on p. 509):

$$\ln N! = (N + \tfrac{1}{2}) \ln N - N + \ln (2\pi)^{\frac{1}{2}}. \qquad (27.\text{A}1.3)$$

Taking logarithms of eqn (27.A1.2) then leads (after quite a lot of algebra) to

$$\begin{aligned}\ln P &= \ln n! - \ln [\tfrac{1}{2}(n+s)]! - \ln [\tfrac{1}{2}(n-s)]! - n \ln 2 \\ &= \ln (2/\pi n)^{\frac{1}{2}} - \tfrac{1}{2}(n+s+1) \ln (1+s/n) - \tfrac{1}{2}(n-s+1) \ln (1-s/n).\end{aligned}$$

So long as $s/n \ll 1$ (which is equivalent to x not being at great distances from the origin) we can use the approximation $\ln (1+z) \approx z$, and obtain

$$\ln P = \ln (2/\pi n)^{\frac{1}{2}} - s^2/2n, \quad \text{or} \quad P = (2/\pi n)^{\frac{1}{2}} e^{-s^2/2n}. \qquad (27.\text{A}1.4)$$

Finally, replace s by x/d and n by t/τ, and obtain

$$P = (2\tau/\pi t)^{\frac{1}{2}} e^{-x^2\tau/2td^2}. \qquad (27.\text{A}1.5)$$

This is the equation used in the text (eqn (27.3.11)).

Further reading

Ion transport:

Conductivity. T. Shedlovsky; *Techniques of chemistry* (A. Weissberger and B. W. Rossiter, eds.) IIA, 163, Wiley-Interscience, New York, 1971.

Determination of transference numbers. M. Spiro; *Techniques of chemistry* (A. Weissberger and B. W. Rossiter, eds.) IIA, 205, Wiley-Interscience, New York, 1971.

Electrolyte solutions (2nd edn). R. A. Robinson and R. H. Stokes; Academic Press, New York and Butterworth, London, 1959.

Ionic solution theory. H. L. Friedman; Wiley-Interscience, New York, 1962.

Electrolytic conductance. R. M. Fuoss and F. Accascina; Wiley-Interscience, New York, 1959.

Diffusion

Experimental methods for studying diffusion in liquids, gases, and solids. P. J. Dunlop, B. J. Steel, and J. E. Lane; *Techniques of chemistry* (A. Weissberger and B. W. Rossiter, eds.) IV, 205, Wiley-Interscience, New York, 1972.

Diffusion in solids, liquids, and gases. W. Jost; Academic Press, New York, 1960.

The mathematics of diffusion (2nd edn). J. Crank; Clarendon Press, Oxford, 1975.

Introductory problems

A27.1. The molar conductivity of an electrolyte solution is known to be 135.5 S cm^2 mol^{-1} at 25 °C and a concentration of 5.35×10^{-2} M. Calculate the resistivity of the solution.

A27.2. A conductivity cell has plane parallel electrodes, each 2.2 cm × 2.2 cm in area, spaced 2.75 cm apart. When the cell is filled with an electrolytic solution, the measured resistance is found to be 351 Ω. What is the conductivity of the solution?

A27.3. The molar conductivity of a strong electrolyte in water at 25 °C was found to be 109.9 S cm^2 mol^{-1} for a concentration of 6.2×10^{-3} M and 106.1 S cm^2 mol^{-1} for a concentration of 1.50×10^{-2} M. What is the limiting molar conductivity of the electrolyte?

A27.4. The mobility of the negative ion in a 1:1 electrolyte in aqueous solution at 25 °C is determined experimentally to be 6.85×10^{-8} m^2 s^{-1} V^{-1}. Calculate the molar conductivity of the ion.

A27.5. The mobility of the Rb^+ ion in aqueous solution is 7.92×10^{-8} m^2 s^{-1} V^{-1} at 25 °C. The potential difference between two electrodes placed in the solution is 35.0 V. If the electrodes are 8.00 mm apart, what is the drift speed of the Rb^+ ion?

A27.6. What fraction of the total current is carried by Li^+ when current flows through an aqueous solution of LiBr at 25 °C?

A27.7. The diffusion constant of a certain globular protein

in water at 20 °C is 7.1×10^{-11} m^2 s^{-1}. The partial specific volume of the protein is 0.75 cm^3 g^{-1}. The viscosity of water at 20 °C is 1.00×10^{-3} kg m^{-1} s^{-1}. Estimate the RMM of the protein.

A27.8. The mobility of the nitrate ion in aqueous solution at 25 °C is 7.40×10^{-8} m^2 s^{-1} V^{-1}. Find its diffusion constant in water at 25 °C.

A27.9. A macromolecule has a diffusion coefficient of 4.0×10^{-11} m^2 s^{-1} in water at 20 °C. The viscosity of water at that temperature is 1.00×10^{-3} kg m^{-1} s^{-1}. Assume that the macromolecule is spherical and determine its effective radius.

A27.10. The diffusion coefficient of CCl_4 in n-heptane at 25 °C is 3.17×10^{-9} m^2 s^{-1}. Estimate the time required for a CCl_4 molecule to have a root mean square displacement of 5 mm.

Problems

27.1. Conductivities are often measured by comparing the resistance of a cell filled with the sample to its resistance when filled with some standard solution, such as aqueous potassium chloride. The conductivity of water is 7.6×10^{-4} S cm^{-1} at $\mathcal{T}$ and the conductivity of 0.100 M aqueous KCl is 1.1639×10^{-2} S cm^{-1}. A cell had a resistance of 33.21 Ω when filled with 0.100 M KCl solution and 300 Ω when filled with 0.100 M acetic acid. What is the molar conductivity of acetic acid at that concentration and temperature?

27.2. A 0.0200 M aqueous KCl solution has a molar conductivity of 138.3 S cm^2 mol^{-1} at $\mathcal{T}$, and in the cell its resistance was found to be 74.58 Ω. Find the cell constant $C = \kappa R$. This is 'the cell' of subsequent Problems.

27.3. The resistances of a series of aqueous NaCl solutions, formed by successive dilution of a sample, were measured in the cell. The following resistances were found:

c/M	0.0005	0.001	0.005
R/Ω	3314	1669	342.1
c/M	0.010	0.020	0.050
R/Ω	174.1	89.08	37.14

Check that the molar conductivity follows the Kohlrausch equation, eqn (27.1.3) and find the limiting molar conductivity. Determine the coefficient $\mathcal{K}$.

27.4. Use the value of $\mathcal{K}$ (which should depend only on the nature, not the identity of the ions) from the last Problem and the information that $\lambda^0_+(Na^+) = 50.1$ S cm^2 mol^{-1} and $\lambda^0_-(I^-) = 76.8$ S cm^2 mol^{-1}, to predict (a) the molar conductivity, (b) the conductivity, (c) the resistance it would show in the cell, of an 0.010 M aqueous solution of NaI at $\mathcal{T}$.

27.5. The limiting molar conductivities of KCl, KNO_3, and $AgNO_3$ are 149.9 S cm^2 mol^{-1}, 145.0 S cm^2 mol^{-1}, and 133.4 S cm^2 mol^{-1}, respectively (all at $\mathcal{T}$). What is the limiting molar conductivity of AgCl at this temperature?

27.6. After correction for the water conductivity, the conductivity of a saturated solution of AgCl in water at 25 °C was found to be 1.887×10^{-6} S cm^{-1}. Use the results of the last Problem to find the solubility and the solubility product at this temperature.

27.7. The limiting molar conductivities of aqueous sodium acetate, hydrochloric acid, and sodium chloride are 91.0 S cm^2 mol^{-1}, 425.0 S cm^2 mol^{-1}, and 128.1 S cm^2 mol^{-1} respectively at $\mathcal{T}$. What is the limiting molar conductivity of acetic acid?

27.8. The resistance of an 0.020 M solution of acetic acid in the cell of Problem 27.2 was found to be 888 Ω. What is the degree of ionization of the acid at this concentration?

27.9. What is the pH of the acid in the last Problem? Calculate this in two ways. First ignore activity coefficients. Then estimate them from the Debye–Hückel limiting law (Chapter 11).

27.10. The resistances of aqueous acetic acid solutions were measured at $\mathcal{T}$ in the cell with the following results:

c/M	0.000 49	0.000 99	0.001 98
R/Ω	6146	4210	2927
c/M	0.015 81	0.063 23	0.2529
R/Ω	1004	497	253

Draw the appropriate graph to obtain values for the acid ionization constant K_a, and pK_a.

27.11. What is the molar conductivity, the conductivity, and the resistance (in the cell) of an 0.040 M solution of acetic acid at $\mathcal{T}$? Use p$K_a = 4.72$, $\gamma_\pm = 1$.

27.12. At 25 °C the molar ionic conductivities of Li^+, Na^+, and K^+ are 38.7 S cm^2 mol^{-1}, 50.1 S cm^2 mol^{-1}, and 73.5 S cm^2 mol^{-1} respectively. What are their mobilities?

27.13. What are the drift speeds of the ions in the last Problem when a potential difference of 10 V is applied across a 1 cm conductivity cell? How long would it take an ion to move from one electrode to the other? In conductivity measurements it is normal to use alternating current: what are the displacements of the ions in (a) cm, (b) solvent diameters, about 300 pm, during a half cycle of 1 kHz applied potential?

27.14. Show how the ratio of two transport numbers t', t'' for two cations in a mixture depends on their concentrations c', c'', and their mobilities u', u''.

27.15. The mobilities of H^+ and Cl^- at 25 °C in water are 3.623×10^{-3} cm^2 s^{-1} V^{-1} and 7.91×10^{-4} cm^2 s^{-1} V^{-1}, respectively. What proportion of the current is carried by the protons in 1.0 mM hydrochloric acid? What fraction do they carry when the NaCl is added to the acid so that it is 1.0 M in the salt? Note how concentration as well as mobility governs the transport of current.

27.16. In a moving boundary experiment on KCl the apparatus consisted of a tube of bore 4.146 mm, and it contained aqueous KCl at a concentration of 0.021 M. A steady current of 18.2 mA was passed, and the boundary advanced as follows:

t/s	200	400	600	800	1000
x/mm	64	128	192	254	318

Find the transport numbers of K^+.

27.17. From the information in Problem 27.5, find the mobility of K^+ in aqueous solution and then its ionic conductivity.

27.18. A third technique for measuring transport numbers (and therefore ion mobilities) is by an e.m.f. method. Show that the e.m.f. of the cell Ag|AgCl|HCl(m_1)|HCl(m_2)| AgCl|Ag with transference E_t is related to its e.m.f. without transference E by $E_t = t_+E$. By the same argument show that for electrodes reversible with respect to the cation, the e.m.f. is given by $E_t = t_-E$.

27.19. The proton possesses abnormal mobility in water, but does it behave normally in liquid ammonia? In order to investigate this question, a moving-boundary technique was used to determine the transport number of NH_4^+ in liquid ammonia (the analogue of H_3O^+ in liquid water) at -40 °C (J. Baldwin, J. Evans, and J. B. Gill, *J. chem. Soc.* A, 3389 (1971)). A steady current of 5.0 mA was passed for 2500 s, during which time the boundary formed between mercury(II) iodide and ammonium iodide ammoniacal solutions moved 286.9 mm in a 0.013 65 mol kg^{-1} solution and 92.03 mm in a 0.042 55 mol kg^{-1} solution. Calculate the transport number of NH_4^+ at these concentrations, and comment on the mobility of the proton in liquid ammonia.

27.20. The conductivity of the purest water prepared is 5.5×10^{-8} S cm^{-1}. What would be the resistance measured for a sample of this water in the conductivity cell of Problem 27.2? What is the ionization constant of water? What are the values of pK_w and pH for pure water? (Take the mobilities of H^+ and OH^- from Table 27.2.)

27.21. In Problem 27.3 we found an experimental value for $\mathscr{K}$ in the Kohlrausch expression for the conductivity of NaCl. Calculate the values of A and B that appear in the Debye–Hückel–Onsager equation, eqn (27.1.21), and estimate a theoretical value of $\mathscr{K}$.

27.22. A dilute solution of potassium permanganate in water at 25 °C was prepared. The solution was in a horizontal 10 cm tube, and at first there was a linear gradation of intensity of the purple solution from the left (where the concentration was 0.10 M) to the right (where the concentration was 0.05 M). What is the magnitude and sign of the thermodynamic force acting on the solute (a) close to the left face of the container, (b) in the middle, (c) close to the right face. Give the force per mole and force per molecule in every case.

27.23. The diffusion coefficient for sucrose in water at 25 °C is 5.2×10^{-6} cm^2 s^{-1}. Suppose that the permanganate in the last Problem were replaced by sucrose and we had some way (what, for instance?) of monitoring its concentration. What is the drift velocity of the sugar at the three points (left, middle, right)? What effect on the concentration distribution do these different drift velocities bring about? Sketch the concentration distribution at several later times. What is the drift velocity after an infinite time?

27.24. Estimate the effective radius of a sugar (sucrose) molecule in water at 25 °C given that its diffusion coefficient is 5.2×10^{-6} cm^2 s^{-1} and that the viscosity of water is 1.00 cP.

27.25. The diffusion coefficient for molecular iodine in benzene is 2.13×10^{-5} cm^2 s^{-1}. How long does a molecule take to jump through about one molecular diameter (approximately the fundamental jump length for translational motion)?

27.26. What is the root mean square distance travelled by (a) an iodine molecule in benzene, (b) a sucrose molecule in water at 25 °C in 1 s?

27.27. About how long, on average, does it take for the molecules in the last Problem to drift to a point (a) 1 mm, (b) 1 cm from their starting points?

27.28. Estimate the diffusion coefficients and the effective hydrodynamic radii of the alkali metal cations in water from their mobilities at 25 °C.

27.29. Estimate the approximate number of water molecules that are dragged along by the cations in the last Problem. Ionic radii are given in Table 23.1.

27.30. Nuclear magnetic resonance can be used to determine the mobility of molecules in liquids. A set of measurements on methane in carbon tetrachloride showed that its diffusion coefficient is 2.05×10^{-5} cm^2 s^{-1} at 0 °C and 2.89×10^{-5} cm^2 s^{-1} at 25 °C. Deduce what information you can about the mobility of methane in carbon tetrachloride.

27.31. Confirm that eqn (27.3.2) is a solution of the diffusion equation with the correct initial value.

27.32. A concentrated sucrose solution is poured into a 5 cm diameter cylinder. Take it as 10 g of sugar in 5 cm^3 of water. A further 1 dm^3 of water is then poured very carefully on top of the layer, without disturbing it. Ignore gravitational effects, and pay attention only to diffusional processes. Find the concentration at 5 cm above the lower layer after a lapse of (a) 10 s, (b) 1 year.

27.33. The diffusion equation is valid when many elementary steps are taken in the time interval of interest; but the random walk calculation lets us discuss distributions for short times as well as for long. Use eqn (27.A1.2) to calculate the probability of being six paces from the origin (that is, at $x = 6d$) after (a) four, (b) six, (c) twelve steps.

27.34. Write a program for calculating $P(x)$ in a one-dimensional random walk, and evaluate the probability of being at $x = 6d$ for $n = 6, 10, 14, \ldots, 60$. Compare the numerical value with the analytical value in the limit of a large number of steps. At what value of n is the discrepancy no more than 0.1%?

27.35. In a series of observations on the displacement of rubber latex spheres of radius 2.12×10^{-5} cm the mean square displacements after selected time intervals were on average as follows:

t/s	30	60	90	120
$10^8\langle x^2 \rangle/\text{cm}^2$	88.2	113.5	128	144

These results were originally used to find the value of Avogadro's constant (a remarkably eye-straining procedure for such a large quantity!) but there are now better ways of determining it, and so the data can be used to find another quantity. Find the effective viscosity of water at the temperature of this experiment (25 °C).

28

The rates of chemical reactions

Learning objectives

After careful study of this chapter you should be able to:

(1) Describe how changing concentrations are monitored, Section 28.1.

(2) Define the *reaction rate*, eqn (28.2.2), and relate it to the rate of change of concentrations, eqns (28.2.4) and (28.2.5) and Example 28.1.

(3) Explain what is meant by the *rate constant* of a reaction, eqn (28.2.6), and by *rate law*, Section 28.2(b).

(4) Define *reaction order*, Section 28.2(b).

(5) Justify and use the *method of initial slopes* to determine the order of a reaction, Example 28.2.

(6) Integrate *first-order*, eqn (28.2.14), and *second-order*, eqn (28.2.17), rate laws for concentration as a function of time.

(7) Determine reaction order from the time-dependence of concentration. Example 28.3.

(8) Explain the *isolation method* and state what is meant by a *pseudo-first-order reaction*, Section 28.2(e).

(9) Define the *half-life* of a reaction, relate it to the rate constant, eqns (28.2.21) and (28.2.22), and use it to determine reaction order, Example 28.4.

(10) Explain the meaning of *molecularity*, Section 28.3.

(11) Write the rate laws for *simple unimolecular*, eqn (28.3.2), and *simple bimolecular*, eqn (28.3.1), reactions.

(12) Write the *Arrhenius equation*, eqn (28.3.4), and use it to measure the *activation energy* and *pre-exponential factor* of a reaction, Example 28.5.

(13) Define the general *activation energy*, eqn (28.3.6), and use it to account for *negative activation energies*, eqn (28.3.28).

(14) Write and solve the rate equations for reactions approaching equilibrium, eqn (28.3.8), and express the *equilibrium constant* of a composite reaction in terms of the individual rate constants of its stages, eqn (28.3.12).

(15) Write and solve the rate equations for *consecutive reactions*, eqn (28.3.14) and Example 28.6.

(16) Justify and use the *steady state approximation*, eqn (28.3.18).

(17) Explain what is meant by a *pre-equilibrium*, eqn (28.3.21), and use it to relate a rate law to a mechanism.

(18) Describe the *Michaelis–Menten mechanism* of enzyme action, and deduce the rate laws, eqn (28.3.33).

(19) Define the term *unimolecular reaction*, describe the *Lindemann–Hinshelwood mechanism*, eqn (28.3.35), and deduce the rate law, eqn (28.3.38).

Introduction

There are two main reasons for studying the rates of reactions. The first is the practical importance of being able to predict how quickly a reaction mixture approaches equilibrium. This rate might depend on variables under our control, such as the pressure, the temperature, and the presence of a catalyst, and we may be able to optimize it by the appropriate choice of conditions. The second reason is that the study of reaction rates leads to an understanding of the *mechanisms* of reactions, the analysis of a chemical reaction into a sequence of elementary steps. An elementary reaction step may be a reactive collision in a gas phase reaction, or a reactive encounter in solution, and is not necessarily related to the stoichiometry of the overall reaction. For example, we might discover that the reaction of hydrogen and bromine proceeds by a series of steps involving the dissociation of Br_2, the attack of a Br atom on H_2, and so on.

28.1 Empirical chemical kinetics

The basic data of chemical kinetics are the concentrations of the reactants and products at different times. The method used to monitor the concentrations depends on the substances involved and the rapidity with which they change. Many reactions go to completion (i.e. reach thermodynamic equilibrium) over periods of minutes or hours, and one of the following techniques may be used (the special techniques for studying faster reactions are described in Section 29.5).

(1) *Pressure changes*. A reaction in which at least one component is gaseous might result in a change of pressure, and so its progress may be followed by recording the pressure as a function of time. An instance of this is the decomposition of nitrogen(V) oxide, $2N_2O_5(g) \rightarrow 4NO_2(g) + O_2(g)$. For each mole of reactant molecules destroyed, $\frac{5}{2}$ mol gas molecules are formed, and so the total pressure increases as the reaction proceeds (if the volume is constant). A disadvantage of this method is that it is not specific: all gas phase particles contribute to the pressure.

(2) *Spectroscopy*. This is widely applicable, and is especially useful when a substance has a strong characteristic absorption in a conveniently accessible region of the spectrum. For example, the reaction $H_2(g) + Br_2(g) \rightarrow 2HBr(g)$ can be followed by measuring the visible light absorption by the bromine.

(3) *Electrochemical methods*. If the reaction changes the number or type of ions present in a solution, then it may be followed by monitoring the conductivity. The pH of the solution may also be monitored. One important class of reactions consists of those taking place at electrodes (Chapter 32).

(4) *Miscellaneous methods*. Other important methods of determining composition include titration, mass spectrometry, gas chromatography, and magnetic resonance. Polarimetry, the observation of the optical activity of a reaction mixture, is occasionally applicable.

There are three ways of using these analytical techniques. In *real time analysis* the composition of the system is analysed while the reaction is in progress (by withdrawing a small sample or monitoring the bulk). In the *quenching* method the reaction is stopped after it has been allowed to proceed for a certain time, and the composition is analysed at leisure. The quenching (of the entire mixture or of a sample drawn from it) can be achieved either by cooling suddenly or by adding the mixture to a large amount of solvent. The method is suitable only for reactions that are slow enough for there to be little reaction during the time it takes to quench the mixture. In the *flow method* the reactants are mixed as they flow together into a chamber, Fig. 28.1. The reaction continues as the thoroughly mixed solutions flow through the outlet tube, and observation of the composition at different positions along the tube (e.g. spectroscopically) is equivalent to

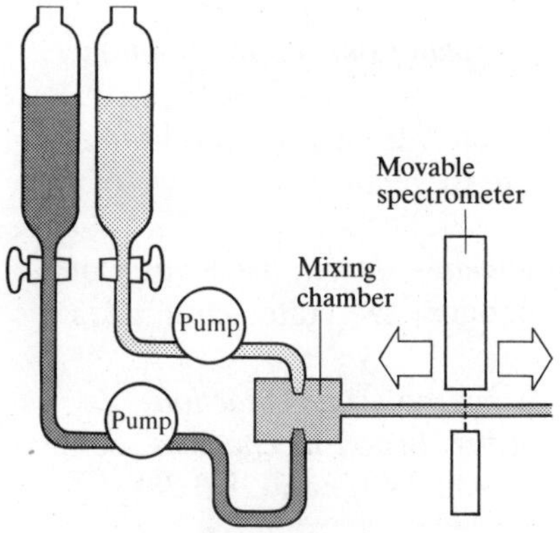

Fig. 28.1. The arrangement used in the *flow technique* for studying reaction rates. The reactants are pumped into the mixing chamber at a steady rate by the peristaltic pumps (i.e., pumps that squeeze the fluid through flexible tubes). The location of the spectrometer corresponds to different times after initiation.

observing the reaction mixture at different times after mixing. Reactions complete within a few milliseconds can be studied in this way, but large volumes of solutions are needed (an improvement, the *stopped-flow method*, is described in Section 29.5(b)).

Since the rates of chemical reactions are generally sensitive to the temperature (for reasons we explore later), the temperature of the reaction mixture must be held constant throughout the course of the reaction, for otherwise the observed rate would be a meaningless average of rates at different temperatures. This requirement puts severe demands on the design of an experiment. Gas phase reactions, for instance, are often carried out in a vessel held in contact with a substantial block of metal, and liquid phase reactions, including flow reactions, must be carried out in an efficient thermostat.

The general outcome of these experiments is the observation that the rates of reactions depend on the composition and the temperature of the reaction mixture. The next few sections look at these observations in more detail.

28.2 The rates of reactions

Consider a reaction of the form $A + B \rightarrow C$, the concentrations of the participants being [A], [B], and [C] at some instant. The *rate of change of concentration of a specified species* (symbol: v_J), $v_J = d[J]/dt$, is one measure of the instantaneous rate of a reaction. In the present case the reaction stoichiometry implies that

$$d[C]/dt = -d[B]/dt = -d[A]/dt$$

because whenever a C molecule is formed, one A and one B molecule are destroyed. In the case of a reaction with a more complicated stoichiometry, such as $A + 2B \rightarrow 3C + D$, the reaction rate may still be expressed as $d[J]/dt$, where J is one of the participants, but the relation between the various rates is more complicated. In this case

$$d[C]/dt = -3\,d[A]/dt, \qquad d[D]/dt = \tfrac{1}{3}\,d[C]/dt,$$

and so on. Since the rate can be expressed in so many ways, it is important to specify to which component a numerical value refers.

28.2 (a) Reaction rate

A more elegant way of expressing the rate, which does away with the ambiguity and has further advantages, is to express the chemical reaction in the general form introduced in Section 4.1(b):

$$0 = \sum_J \nu_J J, \qquad (28.2.1)$$

where the stoichiometric coefficients are positive for products and negative for reactants. The changes in the *amounts* (n_J) of each component that occur during the reaction can then be expressed in terms of the *advancement* (ξ) of the reaction, as explained in Section 10.1. For example, when the reaction advances by an infinitesimal amount $d\xi$, the amount of component J changes by $\nu_J\,d\xi$. (In the case of the reaction $A + 2B \rightarrow 3C + D$, $\nu_A = -1$, $\nu_B = -2$, $\nu_C = 3$, and $\nu_D = 1$. For an advancement $d\xi$, the amount of A changes by $dn_A = -d\xi$; similarly $dn_B = -2\,d\xi$, $dn_C = 3\,d\xi$, and

$dn_D = d\xi$.) The *reaction rate* (symbol: $\dot{\xi}$) is now defined as

$$\textit{Reaction rate: } \dot{\xi} = d\xi/dt, \tag{28.2.2}$$

and is the *rate of change of advancement of the reaction*. In order to measure the rate, we measure the rate of change of the amount of any participant in the reaction. For example, since $dn_J = \nu_J\, d\xi$, it follows that

$$\dot{\xi} = (1/\nu_J)\, dn_J/dt. \tag{28.2.3}$$

In this way, the same numerical value is obtained whichever component we select, and there is a *single* rate for a reaction (for a given composition and temperature of the reaction mixture).

It is important to note that the reaction rate as defined in eqn (28.2.2) refers to the *amounts* of the substances present, not their concentrations. This is another advantage of the definition, because in systems where the volume changes during the reaction the concentration might change even though the numbers of molecules of each kind remain the same. In many applications of kinetics (e.g. to reactions taking place in closed vessels), however, the volume of the system is constant (at V), and so the reaction rate can be related directly to the rate of change of concentration. The explicit relation is obtained by dividing

$$dn_J/dt = \nu_J(d\xi/dt) = \nu_J\dot{\xi}$$

by the volume, when the left-hand side becomes

$$(dn_J/dt)/V = d(n_J/V)/dt = d[J]/dt = v_J,$$

and so, for constant-volume systems,

$$v_J = (\nu_J/V)\dot{\xi}. \tag{28.2.4}$$

This equation is also useful for converting a rate of change of concentration to a reaction rate, because it rearranges to

$$\dot{\xi} = (V/\nu_J)v_J, \tag{28.2.5}$$

and v_J is obtained experimentally by following the concentration of J. In the following sections we shall always assume that the volume of the system is constant, and deal with rates of change of concentration.

Example 28.1

The rate of change of concentration of NO(g) in the reaction $2NOBr(g) \rightarrow 2NO(g) + Br_2(g)$ was reported as $1.6 \times 10^{-4}\ M\ s^{-1}$. The reaction vessel was a $1.00\ dm^3$ closed flask. What is the reaction rate and the rate of change of concentration of NOBr?

• *Method*. The reaction rate ($\dot{\xi}$) is related to the rate of change of concentration by eqn (28.2.5). Identify the stoichiometric coefficient by writing the reaction in the form of eqn (28.2.1). The rate of change of concentration of NOBr is obtained from the true rate using eqn (28.2.4).

• *Answer*. The reaction is formally

$$0 = -2NOBr(g) + 2NO(g) + Br_2(g),$$

and so $\nu_{NO} = +2$. Therefore, the reaction rate is obtained from eqn (28.2.5) with $v_{NO} =$

$1.6\times10^{-4}\,\mathrm{M\,s^{-1}}$:

$$\dot{\xi}=(1.00\,\mathrm{dm^3}/2)\times(1.6\times10^{-4}\,\mathrm{M\,s^{-1}})=8.0\times10^{-5}\,\mathrm{mol\,s^{-1}}.$$

Then, since $\nu_{\mathrm{NOBr}}=-2$, from eqn (28.2.4):

$$\mathrm{d[NOBr(g)]}/\mathrm{d}t=(-2/1.00\,\mathrm{dm^3})\times(8.0\times10^{-5}\,\mathrm{mol\,s^{-1}})=-1.6\times10^{-4}\,\mathrm{M\,s^{-1}}.$$

• *Comment*. When $\dot{\xi}$ is used, a single quantity expresses the rate of the reaction, which is why is called *the* reaction rate. In this example the rate of change of concentration of the NOBr is trivially related to that of NO, but we see how to proceed systematically.

• *Exercise*. The rate of change of concentration of CH_3 radicals in the reaction $2CH_3(g)\rightarrow CH_3CH_3$ was reported as $-1.2\,\mathrm{M\,s^{-1}}$ under particular conditions in a $5.0\,\mathrm{dm^3}$ vessel. What is (a) the reaction rate and (b) the rate of change of concentration of CH_3CH_3?

[(a) $3.0\,\mathrm{mol\,s^{-1}}$, (b) $+0.60\,\mathrm{M\,s^{-1}}$]

28.2 (b) Rate laws and rate constants

The measured rate at some stage of a reaction is often found to be proportional to the concentrations of the reactants raised to some power. For example, it may be found that the rate of change of the concentration of A as a result of the forward reaction $A+2B\rightarrow 3C+D$ depends on its own concentration and that of B as follows:

$$\mathrm{d[A]}/\mathrm{d}t=-k[\mathrm{A}][\mathrm{B}], \qquad (28.2.6)$$

the *rate constant* (or *rate coefficient*) k being independent of the concentrations (but dependent on the temperature). This empirically determined equation is called the *rate law* of the reaction.

The determination and statement of the rate law serves three purposes. First, it enables us to predict the reaction rate, given the composition of the mixture and the experimental value of the rate constant. Second, the rate law is a guide to the mechanism for the reaction, and any proposed mechanism must lead to the observed rate law. Third, the rate law lets us classify reactions according to their *order*. The *order with respect to some component* is the power to which the concentration of that component is raised in the rate law. For example, a reaction with the rate law in eqn (28.2.6) is first-order in A and also first-order in B. The *overall order* of a reaction is the sum of the orders of all the components. The reaction in eqn (28.2.6) is therefore second-order overall.

The overall orders of simple reactions (i.e. those corresponding to a small number of steps, and typically occurring in solution, less commonly in gases) are generally in the range 0–4, and the orders with respect to each participant are generally in the range 0–2. Moreover, in these simple cases only the reactant concentrations normally occur in the rate law. However, a reaction need not have an integral order, and many gas-phase reactions do not. For example, if a reaction is found to have the rate law

$$\mathrm{d[A]}/\mathrm{d}t=-k[\mathrm{A}]^{\frac{1}{2}}[\mathrm{B}], \qquad (28.2.7)$$

it is half-order in A, first order in B, and three-halves order overall. When a rate law is not of the form $[\mathrm{A}]^x[\mathrm{B}]^y[\mathrm{C}]^c\ldots$, the reaction does not have an order. For example, the experimentally determined rate-law for the gas-phase reaction $H_2+Br_2\rightarrow 2HBr$ is

$$\mathrm{d[HBr(g)]}/\mathrm{d}t=\frac{k[\mathrm{H_2(g)}][\mathrm{Br_2(g)}]^{\frac{3}{2}}}{[\mathrm{Br_2(g)}]+k'[\mathrm{HBr(g)}]}, \qquad (28.2.8)$$

and although it is first-order in H_2, it has an indefinite order with respect to both Br_2 and HBr and overall.

In some cases a rate law reflects the stoichiometry of the reaction, but it is important to recognize that this is not generally the case: *the rate law is arrived at experimentally, and cannot be inferred from the reaction equation*. The reaction of hydrogen and bromine, for example, has a very simple stoichiometry, but its rate law, eqn (28.2.8), is very complicated. Similarly, for the thermal decomposition of nitrogen(V) oxide

$$2N_2O_5(g) \rightarrow 4NO_2(g) + O_2(g), \qquad d[O_2]/dt = k[N_2O_5(g)], \quad (28.2.9)$$

and so it is a first-order reaction. In some cases, however, the rate law does happen to reflect the reaction stoichiometry. This is the case with the oxidation of nitrogen(II) oxide:

$$2NO(g) + O_2(g) \rightarrow 2NO_2(g), \qquad d[NO_2]/dt = k[NO(g)]^2[O_2(g)]. \quad (28.2.10)$$

These remarks point to three problems. First, we must see how to arrive at a rate law, and the value of the rate constant, from the experimental data. Second, we must see how to construct reaction mechanisms that are compatible with the rate law. Third, we must account for the values of the rate constants and explain their temperature dependence.

28.2 (c) The determination of the rate law

The raw kinetic data are the concentrations of substances at various times. There are various ways of extracting the rate law.

In the method of *initial rates* the rate is measured *at the beginning of the reaction* for several different initial concentrations of reactants. Then suppose that the rate law for a reaction involving A and B is guessed to be

$$v_A = k[A]^a[B]^b, \qquad v_A = d[A]/dt,$$

then its initial rate ($v_{A,0}$) is given by the initial values of the concentrations, $[A]_0$ and $[B]_0$ as

$$v_{A,0} = k[A]_0^a[B]_0^b.$$

Taking logarithms:

$$\log v_{A,0} = \log k + a \log [A]_0 + b \log [B]_0. \quad (28.2.11)$$

Therefore, a plot of the logarithm of the initial rate of change of concentration against the logarithm of the initial concentration of A (with $[B]_0$ held constant) should be a straight line with slope a. Similarly, the order with respect to B can be determined by plotting $\log v_{A,0}$ against $\log [B]_0$ with $[A]_0$ held constant.

Example 28.2

The recombination of iodine atoms in the gas phase in the presence of argon was investigated and the order of the reaction was determined by the method of initial slopes. The initial rates of formation of I_2 by the reaction $2I(g) + Ar(g) \rightarrow I_2(g) + Ar(g)$ were as follows:

$[I(g)]_0/10^{-5}$ M	1.0	2.0	4.0	6.0
$v_{I_2,0}/M\,s^{-1}$ (a)	8.7×10^{-4}	3.48×10^{-3}	1.39×10^{-2}	3.13×10^{-2}
(b)	4.35×10^{-3}	1.74×10^{-2}	6.96×10^{-2}	1.57×10^{-1}
(c)	8.69×10^{-3}	3.47×10^{-2}	1.38×10^{-1}	3.13×10^{-1}

The argon concentrations are (a) 1.0×10^{-3} M, (b) 5.0×10^{-3} M, and (c) 1.0×10^{-2} M. Find the orders of reaction with respect to the iodine and argon atom concentrations and the rate constant.

• *Method.* Plot $\lg(v_{I_2,0}/\mathrm{M\,s^{-1}})$ against $\lg([\mathrm{I}]_0/\mathrm{M})$ for a given $[\mathrm{Ar}]_0$ and against $\lg([\mathrm{Ar}]_0/\mathrm{M})$ for a given $[\mathrm{I}]_0$. Intercepts give $\lg k$ and slopes the orders.

• *Answer.* The slopes are 2 and 1 respectively, and so the (initial) rate law is

$$v_{I_2,0} = k[\mathrm{I}]_0^2[\mathrm{Ar}]_0,$$

signifying that the reaction is second-order in [I], first-order in [Ar], and third-order overall. The intercept corresponds to $k = 9.9\,\mathrm{M^{-2}\,s^{-1}}$.

• *Comment.* The third-order reaction suggests a three-body collision mechanism for this reaction: the third body (Ar) mops up the energy of the forming I_2 molecule, and prevents its immediate dissociation. Note the units of k: they come automatically from the calculation, and are always such as to convert the product of concentrations to concentration per unit time (e.g. $\mathrm{M\,s^{-1}}$).

• *Exercise.* The initial rate of formation of a substance J depended on concentration as follows:

$[\mathrm{J}]_0/10^{-3}\,\mathrm{M}$	5.0	8.2	17	30
$v_{J,0}/10^{-7}\,\mathrm{M\,s^{-1}}$	3.6	9.6	41	130

Find the order of the reaction and the rate constant. [2, $1.4 \times 10^{-2}\,\mathrm{M^{-1}\,s^{-1}}$]

The method of initial slopes might not reveal the full rate law, for in a complex reaction the products themselves might affect the rate. This is the case with the HBr synthesis: eqn (28.2.8) shows that the full rate law depends on the concentration of HBr, none of which is present initially. In order to avoid this difficulty the rate law should be fitted to the data throughout the reaction. This may be done, in simple cases at least, by using a proposed rate law to predict the concentration of any component at any time, and comparing it with the data. Since the rate laws are differential equations, this means that we must solve them (i.e. integrate them) for the concentrations as functions of time. Now that computers are so widely available, even the most complex rate laws may be solved numerically, and the concentrations fitted to the data by adjusting the values of the rate constants. However, in a number of simple cases analytical solutions are easily obtained, and prove to be very useful. We shall examine a few of these simple cases here, and illustrate the computational approach in Section 29.1(a).

28.2 (d) First-order reactions

The first-order rate law for the disappearance of some reactant A is

$$-\mathrm{d}[\mathrm{A}]/\mathrm{d}t = k[\mathrm{A}]. \qquad (28.2.12)$$

(A common example of this type of process is the radioactive decay of a nuclide.) This differential equation rearranges to

$$(1/[\mathrm{A}])\,\mathrm{d}[\mathrm{A}] = -k\,\mathrm{d}t,$$

which can be integrated directly. Since initially (at $t = 0$) the concentration of A is $[A]_0$, and at a later time t it is $[A]_t$, we write

$$\int_{[A]_0}^{[A]_t} \frac{d[A]}{[A]} = -\int_0^t k\, dt, \quad \text{or} \quad \ln [A]_t - \ln [A]_0 = kt.$$

The solution can be expressed in two useful forms:

$$\ln [A]_t/[A]_0 = -kt, \qquad (28.2.13)$$
$$[A]_t = [A]_0 e^{-kt}. \qquad (28.2.14)$$

The latter shows that *in a first-order reaction the reactant concentration decreases exponentially with time*, with a rate determined by k. The former shows that if $\ln [A]_t/[A]_0$ is plotted against t, then a first-order reaction will give a straight line. If the plot *is* straight, then the reaction is first-order, and the value of k may be obtained from the slope (the slope is $-k$). The values of some first-order rate constants are given in Table 28.1.

Table 28.1. Kinetic data for first-order reactions

Reaction	Phase	$\theta/°C$	k/s^{-1}	$t_{\frac{1}{2}}$
$2N_2O_5 \rightarrow 4NO_2 + O_2$	g	25	3.14×10^{-5}	6.1 h
$C_2H_6 \rightarrow 2CH_3$	g	700	5.46×10^{-4}	21.2 m
$2N_2O_5 \rightarrow 4NO_2 + O_2$	$Br_2(l)$	55	2.08×10^{-3}	333 s

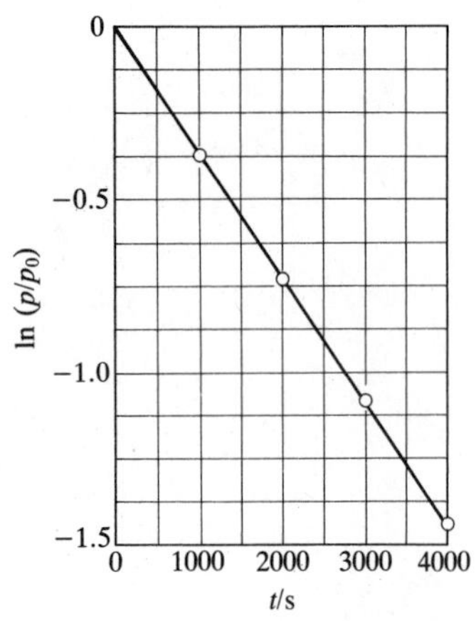

Fig. 28.2. The determination of the rate constant of a first-order reaction: a straight line is obtained when ln [A] (or ln p) is plotted against t; the slope gives k. See *Example* 28.3.

Example 28.3

The partial pressure of azomethane, $CH_3N_2CH_3$ (denoted A in the following), was observed as a function of time at 600 K, with the results given below. Confirm that the decomposition $A(g) \rightarrow CH_3CH_3(g) + N_2(g)$ is first-order in A, and find the rate constant at this temperature.

t/s	0	1000	2000	3000	4000
$p/10^{-2}$ Torr	8.20	5.72	3.99	2.78	1.94

• *Method*. Plot in (p/p_0) against t. If the reaction is first-order, the graph will be straight; its slope is $-k/s^{-1}$.

• *Answer*. The data are plotted in Fig. 28.2. The plot is straight, and its slope is -3.6×10^{-4}. Therefore, $k = 3.6 \times 10^{-4}\ s^{-1}$.

• *Comment*. First-order gas phase reactions have some peculiar features, in particular, how they can occur at all. We consider their mechanism later (Section 28.3(g)).

• *Exercise*. The concentration of N_2O_5 in bromine varied with time as follows:

t/s	0	200	400	600	1000
$[N_2O_5]/M$	0.110	0.073	0.048	0.032	0.014

Determine the order and rate constant. [1, $2.1 \times 10^{-3}\ s^{-1}$]

28.2 (e) Second-order reactions

In the case of a reaction that is *second-order* in A, the rate law is

$$d[A]/dt = -k[A]^2, \qquad (28.2.15)$$

and it integrates easily:

$$-\int_{[A]_0}^{[A]_t} \frac{d[A]}{[A]^2} = \int_0^t k\,dt, \quad \text{or} \quad \frac{1}{[A]_t} - \frac{1}{[A]_0} = kt. \tag{28.2.16}$$

This rearranges to

$$[A]_t = \frac{[A]_0}{1 + kt[A]_0}. \tag{28.2.17}$$

The first of these expressions shows that in order to test for a second-order reaction we should plot $1/[A]_t$ against t and expect a straight line. If it is straight, then the reaction is second-order in A and the slope of the line is equal to the rate constant. The second expression, eqn (28.2.17), lets us predict the concentration of A at any time after the start of the reaction.

When a reaction is *second-order overall*, but first-order in each of two reactants A and B, the rate law is

$$d[A]/dt = -k[A][B], \tag{28.2.18}$$

and we cannot integrate it until we know how the concentration of B is related to that of A. This depends on the stoichiometry of the reaction, and for simplicity we consider $A + B \rightarrow$ *Products*. If the initial concentrations are $[A]_0$ and $[B]_0$, then when the concentration of A has fallen to $[A]_0 - x$, the concentration of B will have fallen to $[B]_0 - x$ because each A that disappears entails the disappearance of one B. It follows that

$$-d[A]/dt = k[A][B] = k\{[A]_0 - x\}\{[B]_0 - x\}.$$

Then, since $d[A]\,dt = d\{[A]_0 - x\}/dt = -dx/dt$, because $[A]_0$ is a constant, the rate law is

$$dx/dt = k\{[A]_0 - x\}\{[B]_0 - x\}.$$

Since $x = 0$ when $t = 0$,

$$kt = \int_0^x \frac{dx}{\{[A]_0 - x\}\{[B]_0 - x\}}$$

$$= \frac{-1}{[A]_0 - [B]_0} \int_0^x \left\{\frac{1}{[A]_0 - x} - \frac{1}{[B]_0 - x}\right\} dx$$

$$= \frac{-1}{[A]_0 - [B]_0} \left\{\ln\left(\frac{[A]_0}{[A]_0 - x}\right) - \ln\left(\frac{[B]_0}{[B]_0 - x}\right)\right\}.$$

This can be simplified by combining the two logarithms and noting that $[A]_t = [A]_0 - x_t$ and $[B]_t = [B]_0 - x_t$, for then

$$kt = \frac{1}{[A]_0 - [B]_0} \ln\left\{\frac{[A]_t[B]_0}{[A]_0[B]_t}\right\}. \tag{28.2.19}$$

The integrated forms of the rate laws for several types of reaction are listed in Box 28.1.

At this stage we have expressions which, when rearranged, give the concentrations of A and B at any stage after the beginning of the reaction.

Box 28.1 Integrated rate laws

Order	Reaction	Rate law ($x = [P]$)	$t_{\frac{1}{2}}$
0	$A \to P$	$dx/dt = k$ $kt = x$ for $0 \leqslant x \leqslant [A]_0$	$[A]_0/2k$
1	$A \to P$	$dx/dt = k[A]$ $kt = \ln\{[A]_0/([A]_0 - x)\}$	$(\ln 2)/k$
2	$A \to P$	$dx/dt = k[A]^2$ $kt = x/[A]_0([A]_0 - x)$	$1/k[A]_0$
	$A + B \to P$	$dx/dt = k[A][AB]$ $kt = \{1/([B]_0 - [A]_0)\}$ $\times \ln\{[A]_0([B]_0 - x)/([A]_0 - x)[B]_0\}$	
	$A + 2B \to P$	$dx/dt = k[A][B]$ $kt = \{1/([B]_0 - 2[A]_0)\}$ $\times \ln\{[A]_0([B]_0 - 2x)/([A]_0 - x)[B]_0\}$	
	$A \to P$ with auto-catalysis	$dx/dt = k[A][P]$ $kt = \{1/([A]_0 + [P]_0)\}$ $\times \ln\{[A]_0([P]_0 + x)/([A]_0 - x)[P]_0\}$	
3	$A + 2B \to P$	$dx/dt = k[A][B]^2$ $kt = \{1/(2[A]_0 + [B]_0)\}\{2x/([B]_0 - x)[B]_0\}$ $+ \{1/(2[A]_0 - [B]_0)\}^2$ $\times \ln\{[A]_0([B]_0 - 2x)/([A]_0 - x)[B]_0\}$	
$n \geqslant 2$	$A \to P$	$dx/dt = k[A]^n$ $kt = \{1/(n-1)\}\{1/([A]_0 - x)^{n-1} - 1/[A]_0^{n-1}\}$	$2^{n-1}/(n-1)k[A]_0^{n-1}$

However, they are too complicated to use as a direct test of second-order kinetics or to find the rate constant. A better approach is to use the *isolation method*, in which all concentrations except one are in large excess, and therefore change little during the course of the reaction. If B is in large excess, for example, it is a good approximation to write $[B] = [B]_0$ throughout the reaction, and to write $k' = k[B]_0$; then the rate law in eqn (28.2.18) becomes

$$d[A]/dt = -k'[A], \qquad (28.2.20)$$

which has the form of a first-order rate law. Since it has been forced into first-order form by assuming a constant B concentration, it is called a *pseudo-first-order rate law*. Its integrated form has already been derived, and the value of k' can be found in the way already described. The dependence of the rate on the other substances may be found by isolating them in turn (by having all the other substances present in large excess), and so a complete picture of the overall rate law can be constructed. The values of some second-order rate constants are given in Table 28.2.

Table 28.2. Kinetic data for second-order reactions

Reaction	Phase	$\theta/°C$	$k/M^{-1}\,s^{-1}$
$2NOBr \to 2NO + Br_2$	g	10	0.80
$2I \to I_2$	g	23	7×10^9
$CH_3Cl + CH_3O^-$	$CH_3OH(l)$	20	2.29×10^{-6}

28.2 (f) Half-lives

A useful indication of the rate of a chemical reaction is the time it takes for the concentration of a substance to fall to half its initial value. This is the *half-life* of the substance (symbol: $t_{\frac{1}{2}}$). The half-life depends on the initial concentration of the substance in a characteristic way that depends on the order of the reaction, and so its measurement is a guide to the reaction order.

In the case of a first-order reaction, the time for [A] to decrease from $[A]_0$ to $\frac{1}{2}[A]_0$ is given by eqn (28.2.13) as

$$kt_{\frac{1}{2}} = -\ln\{\tfrac{1}{2}[A]_0/[A]_0\} = -\ln\tfrac{1}{2} = \ln 2,$$

so that

$$t_{\frac{1}{2}} = (\ln 2)/k = 0.693/k. \qquad (28.2.21)$$

The main point to note about this result is that *for a first-order reaction, the half-life of a reactant is independent of its initial concentration*. This means that if the concentration of A at some *arbitrary* stage of the reaction is [A], then it will have fallen to $\frac{1}{2}[A]$ after a further interval of $0.693/k$. Some half-lives are given in Table 28.1.

In the case of a second-order reaction with eqn (28.2.15) as its rate-law, we can use eqn (28.2.16) to find the half-life by substituting $t = t_{\frac{1}{2}}$ and $[A]_t = \frac{1}{2}[A]_0$. This leads to

$$t_{\frac{1}{2}} = 1/k[A]_0. \qquad (28.2.22)$$

In this case, the half-life depends on the initial concentration, and the greater the initial concentration the less time it takes to fall to half its value. This gives another way of checking for second-order kinetics: $t_{\frac{1}{2}}$ is measured for a series of different initial concentrations and plotted against $1/[A]_0$. A straight line confirms second-order kinetics, and the slope gives k.

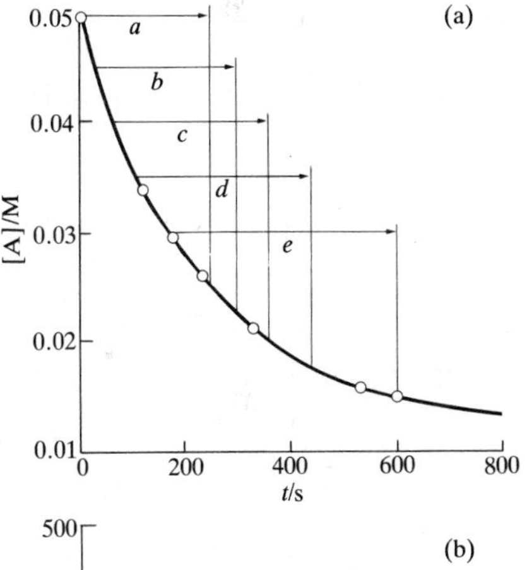

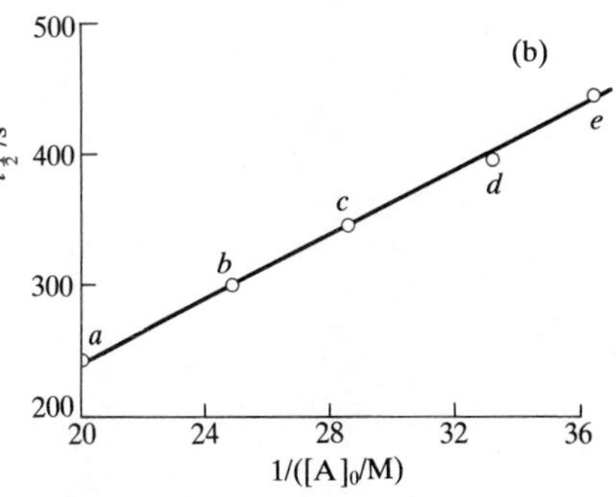

Fig. 28.3. The determination of the rate constant of a second-order reaction by the half-life method. (a) The original data; a, b, . . . , f are the half-lives for selected initial concentrations. (b) A straight line is obtained when the half-lives are plotted against $1/[A]_0$; its slope is $1/k$. See *Example* 28.4.

Example 28.4

The results of the alkaline hydrolysis of ethyl nitrobenzoate (A) are reported below. Determine the order of reaction by the half-life method, and find the rate constant.

t/s	0	100	200	300	400	500	600	700	800
$[A]/10^{-2}$ M	5.00	3.55	2.75	2.25	1.85	1.60	1.48	1.40	1.38

• *Method*. Plot [A]/M against t/s. Choose a sequence of times to regards as 'initial' times, and note the corresponding 'initial' concentrations. Find the time for each concentration to fall to half its initial value. If $t_{\frac{1}{2}}$ is independent of $[A]_0$ the reaction is first-order; if not, try plotting $t_{\frac{1}{2}}$ against $1/[A]_0$, eqn (28.2.22).

• *Answer*. The data are plotted in Fig. 28.3(a). Initial times are chosen at a, b, . . . , f. We find

	a	b	c	d	e	f
$[A]_0/10^{-2}$ M	5.00	4.50	4.00	3.50	3.00	2.75
$t_{\frac{1}{2}}$/s	240	270	300	345	400	450

Clearly, $t_{\frac{1}{2}}$ is not independent of $[A]_0$. Plot $t_{\frac{1}{2}}$/s against $1/([A]_0/M)$, as in Fig. 28.3(b). This gives a straight line, and so the reaction is second-order. The slope is 12.5; Therefore $k = (1/12.5)\ M^{-1}\,s^{-1} = 8.0 \times 10^{-2}\ M^{-1}\,s^{-1}$.

• *Comment*. If the second plot had not given a straight line, then $1/[A]_0^2$ could have been tried.

• *Exercise*. Use the same technique to determine the order and rate constant of the reaction given the following data:

t/s	0	100	200	300	400	500	600	700	800
$[A]/10^{-3}$ M	4.40	3.55	2.98	2.57	2.26	2.01	1.81	1.65	1.52

[2, 0.54 $M^{-1} s^{-1}$]

In the case of a general nth-order reaction, where $d[A]/dt = -k[A]^n$, the same procedures lead to the result that the half-life is proportional to $1/[A]_0^{n-1}$. The explicit expression is given in Box 28.1.

28.2 (g) A summary of the general approach

The investigation of a reaction aims to determine the rate law and the value of the rate constant, often at several temperatures. Ideally, the first step is to identify all the products, and to investigate whether transient intermediates and side reactions are involved. The isolation method may then be used to examine the role of each component in turn, and to determine the order with respect to each one. The order with respect to each substance can be gauged from the method of initial slopes or the dependence of the half-life on the concentration, and then the order confirmed, and k determined, by a plot of the appropriate function of the concentration against time using one of the integrated law expressions (such as eqn (28.2.13) or eqn (28.2.16)). However, since all the laws considered so far disregarded the possibility that the reverse reaction is important, none of them is reliable when the reaction is close to equilibrium. Therefore, all plots can be expected to acquire some curvature for times so long that the reactions involving the products become important. In the case of more intricate rate laws (such as those we encounter later) the concentrations of reactants, intermediates, and products are computed numerically and the rate constants are varied until the experimental data are reproduced.

28.3 Accounting for the rate laws

We now move on to the second stage of the analysis of kinetic data, accounting for them in terms of a postulated reaction mechanism. Most reactions can be broken down into a sequence of steps each of which involves only one or two molecules. The *molecularity* of an individual step is the number of molecules taking part in it. In a *unimolecular* step a single molecule shakes itself apart or its atoms into a new arrangement. In a *bimolecular* step a pair of molecules collide and exchange energy, atoms, or groups of atoms, or undergo some other kind of change. It is most important to distinguish molecularity from order: the order is an empirical quantity, and obtained from the experimental rate law; the molecularity refers to an individual step in some postulated mechanism.

28.3 (a) Simple reactions

Although a second-order reaction need not take place in a single bimolecular step, when an individual step *is* bimolecular the rate law *for that step* is second-order. This is because the rate of a bimolecular step depends on the rate at which the reactants meet, and that is proportional to their concentrations:

$$A+B \rightarrow \textit{Products}, \qquad d[A]/dt = -k[A][B]. \qquad (28.3.1)$$

Therefore, if we believe (or simply postulate) that a reaction is a single step, bimolecular process, then we can write down the rate law (and then go on to test it). This is the case with many homogeneous reactions, such as the dimerizations of alkenes and dienes, and reactions such as

$$CH_3I + CH_3CH_3ONa \rightarrow CH_3OCH_2CH_3 + NaI$$

in alcoholic solution.

In the case of a unimolecular reaction step, we can also be confident that the rate law is first-order. This is because the number of A molecules that decay in a short interval is proportional to the number available to decay (ten times as many decay in the same interval when there are initially 1000 A molecules than when there are only 100 present), and so the rate of change of the amount of A is proportional to its concentration. Therefore, for a simple unimolecular reaction step, we can write

$$A \rightarrow \textit{Products}, \qquad d[A]/dt = -k[A]. \qquad (28.3.2)$$

The interpretation of a rate law is full of pitfalls, partly because a second-order rate expression, for instance, can also result from a reaction scheme more complex than a simple bimolecular encounter. We shall see below how to string simple steps together into a complex reaction scheme and to arrive at the corresponding rate law, but for the present we emphasize that *if* the reaction is a simple, bimolecular, single step process, then it has second-order kinetics, *but* if the kinetics are second-order, then the reaction *might* be complex. The postulated mechanism can be explored only by detailed detective work on the system, and by investigating whether side products or intermediates appear during the course of the reaction. This was one of the ways, for example, in which the reaction $H_2 + I_2 \rightarrow 2HI$ was shown to proceed by a complex reaction after many years during which people had accepted on good, but insufficiently meticulous evidence, that it was a fine example of a simple bimolecular reaction in which atoms exchanged partners during a collision.

28.3 (b) The temperature dependence of reaction rates

The rates of most reactions increase as the temperature is raised. Many reactions fall somewhere in the range spanned by the hydrolysis of methyl ethanoate (where the rate coefficient at 35 °C is 1.82 times that at 25 °C) and the hydrolysis of sucrose (where the factor is 4.13). An empirical observation is that many reactions have rate constants that follow the *Arrhenius equation*,

$$\ln k = \ln A - E_a/RT. \qquad (28.3.3)$$

That is, for many reactions it is found that a plot of $\ln k$ against $1/T$ often gives a straight line. The Arrhenius equation is often written as

$$k = Ae^{-E_a/RT}. \qquad (28.3.4)$$

Table 28.3. Arrhenius parameters

(1) First-order	A/s^{-1}	E_a/kJ Mol^{-1}
$CH_3NC \rightarrow CH_3CN$	3.98×10^{13}	160
$2N_2O_5 \rightarrow 4NO_2 + O_2$	6.31×10^{14}	88
(2) Second-order	$A/M^{-1}\,s^{-1}$	E_a/kJ mol^{-1}
$OH + H_2 \rightarrow 2H_2 + O$	8×10^{10}	42
$C_2H_5ONa + CH_3I$ in ethanol	2.42×10^{11}	81.6

A is called the *pre-exponential factor* and E_a the *activation energy*; some experimental values are given in Table 28.3.

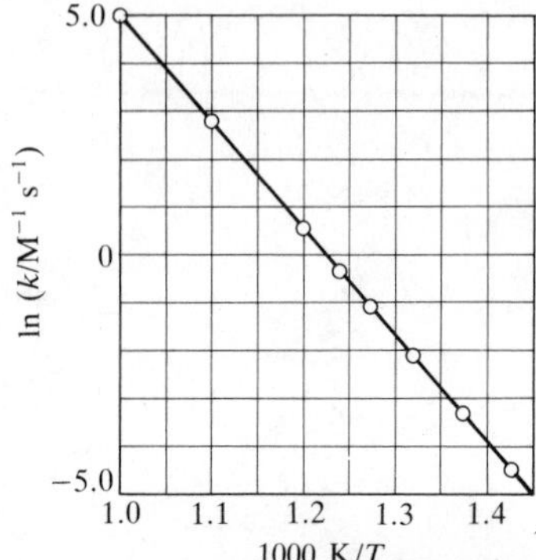

Fig. 28.4. The *Arrhenius plot* of ln k against $1/T$ for the decomposition of CH_3CHO, and the best straight line. The slope gives $-E_a/R$ and the intercept at $1/T = 0$ gives ln A. See *Example* 28.5.

Example 28.5

The rate of the second-order decomposition of acetaldehyde (ethanal, CH_3CHO) was measured over the temperature range 700–1000 K, and the rate constants are reported below. Find the activation energy and the pre-exponential factor.

T/K	700	730	760	790	810	840	910	1000
$k/M^{-1}\,s^{-1}$	0.011	0.035	0.105	0.343	0.789	2.17	20.0	145

• *Method*. Plot $\ln(k/M^{-1}\,s^{-1})$ against $1/(T/K)$; the slope is $-E_a/R$ and the intercept at $1/T = 0$ is $\ln A$. Evaluate the slope and extrapolate to the intercept analytically by selecting a point $(1/T, \ln k)$ on the best straight line and evaluating A from eqn (28.3.3) in the form

$$\ln A = \ln k + E_a/RT.$$

(Ideally, do a least-squares best fit of the data.)

• *Answer*. The points are plotted in Fig. 28.4. The least-squares best fit of the line is with slope -2.21×10^4 and intercept 27.0. Therefore,

$$E_a = (2.21 \times 10^4\,K) \times (8.314\,J\,K^{-1}\,mol^{-1}) = 184\,kJ\,mol^{-1},$$
$$A = e^{27.0}\,M^{-1}\,s^{-1} = 5.3 \times 10^{11}\,M^{-1}\,s^{-1}.$$

• *Comment*. Note that A has the same units as k. For this reaction, the reaction rate doubles on increasing the temperature from 700 K to 716 K or from 900 K to 926 K, etc.

• *Exercise*. Find A and E_a from the following data:

T/K	300	350	400	450	500
$k/M^{-1}\,s^{-1}$	7.9×10^6	3.0×10^7	7.9×10^7	1.7×10^8	3.2×10^8

[$8 \times 10^{10}\,M^{-1}\,s^{-1}$, 23 kJ mol^{-1}]

In some cases the temperature dependence is not Arrhenius-like, but it is still possible to express the strength of the dependence in terms of an activation energy by *defining* one as

$$E_a = RT^2(\partial \ln k/\partial T)_V. \tag{28.3.5}$$

This definition reduces to the earlier one (as the slope of an Arrhenius plot) in the case of a temperature-independent activation energy because it can be rearranged using $d(1/T) = -dT/T^2$ into

$$E_a = -R\,d\ln k/d(1/T), \tag{28.3.6}$$

which integrates to eqn (28.3.4) if E_a is independent of temperature.

However, it is more general, because it allows E_a to be obtained from the slope (at the temperature of interest) of a plot of $\ln k$ against $1/T$ even if the Arrhenius plot is not a straight line. In any case, eqn (28.3.5) shows that *the higher the activation energy, the steeper the temperature dependence of the rate constant*.

The Arrhenius equation can be justified in a general way in terms of the *collision theory* of chemical reactions and by more sophisticated theories of reaction rates, as we shall see in Chapter 30. In the collision theory of gas-phase reactions it is assumed that the reaction rate is proportional (a) to the frequency with which the reactants collide with each other (this is the collision frequency, Z, introduced in Section 26.2) and also (b) to the probability that the collision is sufficiently energetic for a reaction to occur. We suppose that for a collision to be effective it must occur with a kinetic energy along the direction of approach of at least E_a (hence the name 'activation energy' for this quantity). According to the Boltzmann distribution (Section 0.1(d)), when the temperature is T the proportion of such collisions is $e^{-E_a/RT}$. Consequently the reaction rate is of the form

$$rate = Ze^{-E_a/RT}.$$

(We supply the details in Section 30.1(a) and take care there of the temperature dependence of Z itself; here we are concerned only with the general form of the expression.) This has the form of the Arrhenius equation, and becomes identical with it if we write $Z = A[\mathrm{A}][\mathrm{B}]$ on the right and $rate = k[\mathrm{A}][\mathrm{B}]$ on the left. Therefore, we can conclude that the Arrhenius temperature dependence of rate coefficients reflects the number of encounters that are sufficiently energetic to lead to reaction: the higher the temperature the greater the proportion of vigorous encounters, Fig. 28.5.

The temperature dependence of reaction rates has important consequences for industrial reaction vessels, or *chemical reactors*. Chemical reactors include single batch vessels, *continuous stirred tank reactors* (CSTR) into which there is a continuous flow of reactants and from which products are withdrawn, and various types of *tubular reactors*, in which there is a steady flow of reacting material flowing through a tubular vessel. In the case of a single batch vessel, the temperature may rise in the course of an exothermic reaction if heat is liberated faster than it can be conducted

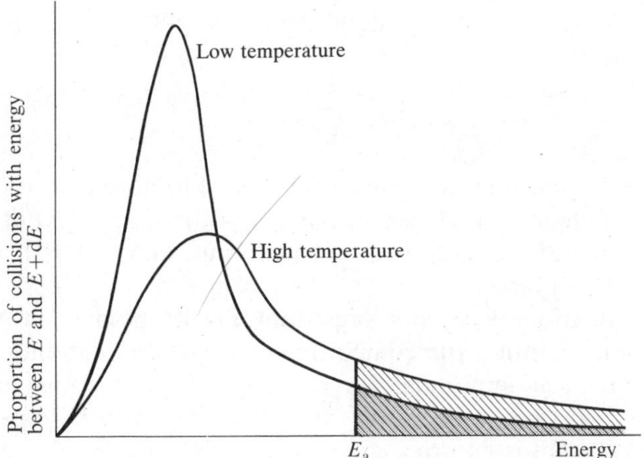

Fig. 28.5. The proportion of molecules having a kinetic energy along their line of approach of at least E_a (the shaded regions) increases strongly with temperature. The area is proportional to $e^{-E_a/RT}$.

away, and so the rates of the reactions taking place in it will change. If there are several competing reactions, or a multi-stage reaction mechanism, then there will be important consequences for the yield of product, which the process designer must take into account. This is normally done by supposing that each reaction has an Arrhenius temperature dependence, measuring the activation energies, and then modelling the reactions by taking account of their exothermicities. In a tubular reactor there is an additional complication: the temperature might not be uniform, and the reactions proceed at different rates in different locations. In some cases (e.g. in hydrocarbon cracking) the temperature may differ by over a hundred degrees along a reactor and across even quite narrow tubes, and a major feature of reactor design is the assessment and control of the effects of these differences.

28.3 (c) Reactions approaching equilibrium

In this and the following sections we consider increasingly complex reaction schemes, and see how to construct and solve the corresponding rate laws. As a first example, we consider $A \rightleftharpoons B$, in which both forward and reverse reactions are important, both reactions being first-order. The rate of change of [A] has two contributions: it is depleted by the forward reaction at a rate $k[A]$ but is replenished by the reverse reaction at a rate $k'[B]$. The net rate of change is therefore

$$d[A]/dt = -k[A] + k'[B]. \qquad (28.3.7)$$

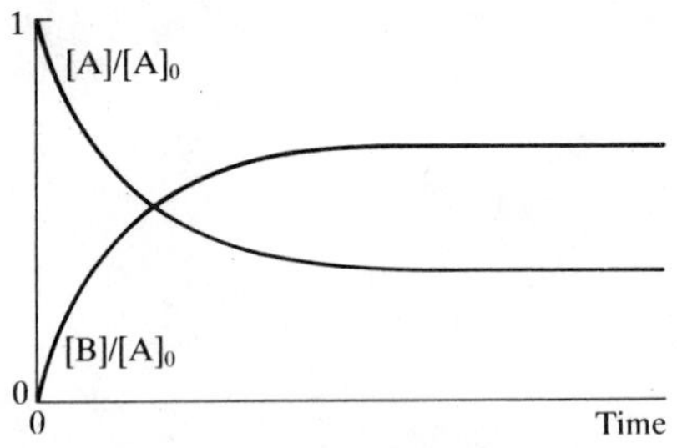

Fig. 28.6. The approach of concentrations to their equilibrium values as predicted by eqn (28.3.8) for reaction $A \rightleftharpoons B$ that is first-order in each direction. and for which $k' = 2k$.

If the initial concentration of A is $[A]_0$, and there is no B present initially, at all times $[A] + [B] = [A]_0$, and so

$$d[A]/dt = -k[A] + k'\{[A]_0 - [A]\} = -(k + k')[A] + k'[A]_0.$$

The solution of this first-order differential equation, with the initial condition $[A] = [A]_0$, is

$$[A]_t = [A]_0\left\{\frac{k' + ke^{-(k+k')t}}{k + k'}\right\} \qquad (28.3.8)$$

The time dependence predicted by this equation is drawn in Fig. 28.6. When t approaches infinity the concentrations reach their *equilibrium* values:

$$[A]_\infty = k'[A]_0/(k + k'), \qquad [B]_\infty = [A]_0 - [A]_\infty = k[A]_0/(k + k'), \qquad (28.3.9)$$

and so the ratio of these equilibrium concentrations, which is the *equilibrium constant* of the $A \rightleftharpoons B$ reaction, is

$$K_c = [B]_\infty/[A]_\infty = k/k'. \qquad (28.3.10)$$

Another way of reaching the same conclusion is to note that at equilibrium the net rate of change of [A] is zero, and so we may set $d[A]/dt = 0$ in eqn (28.3.7), and hence conclude that at equilibrium $k[A] = k'[B]$, which then gives eqn (28.3.10) directly.

The last equation is a very important result, because it relates the thermodynamic quantity, the equilibrium constant, to quantities relating to rates. The practical importance of the result is that if one of the rate constants (e.g. k) can be measured, then the other may be obtained if the equilibrium constant is known.

The same type of calculation may be made for other types of equilibria. In the case of a reaction $A+B \rightleftharpoons C+D$ that is simple bimolecular and therefore second-order in both directions, the rates of change of the concentration of A as a result of the forward and reverse reactions are

$$A+B \rightarrow C+D, \quad v_A = -k[A][B]; \quad C+D \rightarrow A+B, \quad v_A = k'[C][D].$$

At equilibrium the net rate of change is zero. Hence, at equilibrium, $-k[A][B]+k'[C][D]=0$, and so

$$K_c = \{[C][D]/[A][B]\}_{eq} = k/k', \tag{28.3.11}$$

as before. In the case of a reaction which proceeds by a sequence of simple reactions, such as

$$\begin{aligned} A+B &\rightleftharpoons C+D \quad v_{A,\text{forward}} = -k_a[A][B], \qquad v_{A,\text{reverse}} = k_a'[C][D] \\ C &\rightleftharpoons E+F \quad v_{C,\text{forward}} = -k_b[C], \qquad v_{C,\text{reverse}} = k_b'[E][F], \end{aligned}$$

at equilibrium all the reactions are individually at equilibrium, so that

$$\{[C][D]/[A][B]\}_{eq} = k_a/k_a', \qquad \{[E][F]/[C]\}_{eq} = k_b/k_b'.$$

The overall reaction equilibrium is

$$A+B \rightleftharpoons D+E+F,$$

$$\begin{aligned} K &= \{[D][E][F]/[A][B]\}_{eq} = \{[C][D][E][F]/[A][B][C]\}_{eq} \\ &= \{[C][D]/[A][B]\}_{eq}\{[E][F]/[C]\}_{eq} = k_ak_b/k_a'k_b', \end{aligned}$$

and so *when the overall reaction is the sum of a sequence of steps*,

$$K = k_ak_b\ldots/k_a'k_b'\ldots \tag{28.3.12}$$

where the ks are the rate constants for the individual steps and k' refers to the corresponding reverse reaction step.

28.3 (d) Consecutive reactions

Some reactions proceed through the formation of an intermediate, as in the *consecutive first-order reactions* $A \rightarrow B \rightarrow C$. An example is the decay of a radioactive family, such as

$$^{239}\text{U} \xrightarrow{23.5\text{ min}} {}^{239}\text{Np} \xrightarrow{235\text{ days}} {}^{239}\text{Pu}.$$

(The times are half-lives.) The characteristics of this type of reaction can be studied by setting up the rate laws for the net rate of change of the concentration of each substance. A decays at a rate equal to

$$d[A]/dt = -k_a[A], \tag{28.3.13a}$$

and is not replenished. The intermediate B is formed from A (at a rate $k_a[A]$) but decays to C (at a rate $k_b[B]$):

$$d[B]/dt = k_a[A] - k_b[B]. \tag{28.3.13b}$$

C is formed by the unimolecular, first-order decay of B:

$$d[C]/dt = k_b[B]. \tag{28.3.13c}$$

We suppose that initially only A is present, and its concentration is $[A]_0$.

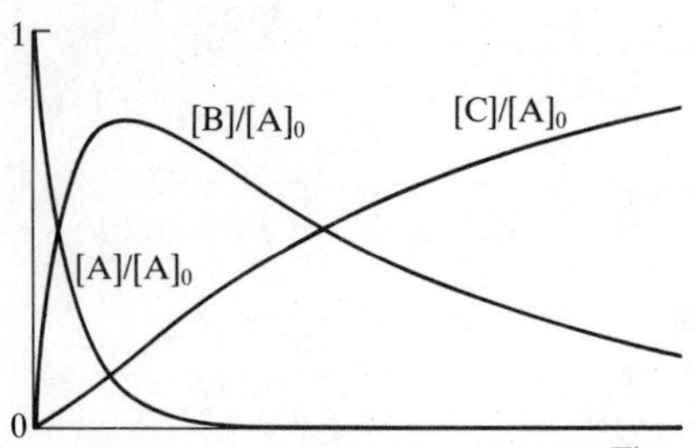

Fig. 28.7. The concentrations of A, B, and C in the *consecutive reaction* scheme $A \rightarrow B \rightarrow C$. The curves are plots of eqns (28.3.14a-c) with $k_a = 10k_b$. If B is the desired product, it is important to be able to predict when its concentration is greatest; see *Example* 28.6.

The first of the rate laws is an ordinary first-order decay, and so

$$[A]_t = [A]_0 e^{-k_a t}. \tag{28.3.14a}$$

When this is substituted in eqn (28.3.13b), and we impose $[B]_0 = 0$, the solution is

$$[B]_t = k_a[A]_0\left\{\frac{e^{-k_a t} - e^{-k_b t}}{k_b - k_a}\right\}. \tag{28.3.14b}$$

Since at any time $[A] + [B] + [C] = [A]_0$, it follows that

$$[C]_t = [A]_0\left\{1 + \left(\frac{k_a e^{-k_b t} - k_b e^{-k_a t}}{k_b - k_a}\right)\right\}. \tag{28.3.14c}$$

These concentrations are plotted in Fig. 28.7. We see that the intermediate's concentration rises to a maximum, and then falls to zero. The concentration of final product C rises from zero and reaches $[A]_0$.

Example 28.6

Suppose that in an industrial batch process a substance A produces the desired product B which goes on to decay to a worthless product C, each stage of the reaction being first-order. At what times will product B be present in greatest concentration?

• *Method*. The time-dependence of the concentration of B is given by eqn (28.3.14b). Find, by differentiation, the time at which it passes through a maximum.

• *Answer*. Differentiating B with respect to t gives

$$d[B]_t/dt = -k_a[A]_0\left\{\frac{k_a e^{-k_a t} - k_b e^{-k_b t}}{k_b - k_a}\right\}.$$

This is equal to zero when

$$k_a e^{-k_a t} = k_b e^{-k_b t}.$$

Therefore, the time at which B is at maximum concentration is

$$t = \{\ln (k_a/k_b)\}/(k_a - k_b).$$

• *Comment*. For a given value of k_a, as k_b increases the time of maximum [B] increases. The *yield* of B also increases.

• *Exercise*. Calculate the maximum concentration of B and justify the last remark.
$[[B]_{max}/[A]_0 = (k_a/k_b)^c,\ c = k_b/(k_a - k_b)]$

Suppose now that $k_b \gg k_a$, then whenever a B molecule is formed it decays rapidly into C. Since $e^{-k_b t} \ll e^{-k_a t}$, and $k_b - k_a$ is close to k_b, eqn (28.3.14c) reduces to

$$[C]_t = [A]_0\{1 - e^{-k_a t}\}, \tag{28.3.15}$$

which shows that the formation of C depends on only the *smaller* of the two rate constants. That is, the rate of formation of C depends on the rate at which B is formed, not on the rate at which B converts into C. For this reason, the step $A \rightarrow B$ is called the *rate-determining step* of the reaction. Its existence has been likened to building a six-lane highway up to a single-lane bridge: the traffic flow is governed by the rate of crossing the bridge. Similar remarks apply to more complicated reactions mechanisms, and in general *the rate-determining step is the one with the smallest rate coefficient*.

28.3 (e) The steady-state approximation

One feature of the calculation so far has probably not gone unnoticed: there is a considerable increase in mathematical complexity as soon as the reaction mechanism acquires even the slightest complexity. We can anticipate that a reaction scheme involving many steps will be unsolvable analytically, and therefore that we should look for alternative methods of solution. One approach is to solve the rate laws numerically (Section 29.1(a)). An alternative, which continues to be widely used because it leads to convenient expressions, is to make the *steady-state approximation*.

The steady state approximation assumes that *during the major part of the reaction, the concentration of all reactive intermediates are constant and small*. This can be justified by considering the rate law for B in the consecutive first-order scheme. On substituting the explicit solutions for [A] and [B] into eqn (28.3.13b) we obtain

$$\mathrm{d}[\mathrm{B}]/\mathrm{d}t = k_{\mathrm{a}}[\mathrm{A}]_0\left\{\frac{k_{\mathrm{b}}\mathrm{e}^{(k_{\mathrm{a}}-k_{\mathrm{b}})t} - k_{\mathrm{a}}}{k_{\mathrm{b}} - k_{\mathrm{a}}}\right\}\mathrm{e}^{-k_{\mathrm{a}}t}. \qquad (28.3.16)$$

This is exact. Now assume that (a) $k_{\mathrm{b}} \gg k_{\mathrm{a}}$ and (b) it is so long after the start of the reaction that $t \gg 1/k_{\mathrm{a}}$. Then

$$\mathrm{d}[\mathrm{B}]/\mathrm{d}t = (k_{\mathrm{a}}/k_{\mathrm{b}})[\mathrm{A}]_0\{k_{\mathrm{b}}\mathrm{e}^{-k_{\mathrm{b}}t} - k_{\mathrm{a}}\}\mathrm{e}^{-k_{\mathrm{a}}t} \approx 0 \qquad (28.3.17)$$

(because $\mathrm{e}^{-k_{\mathrm{a}}t} \approx 0$ when $k_{\mathrm{a}}t \gg 1$). This means that once B is formed (during the *induction period* when t is not large compared with $1/k_{\mathrm{a}}$), the rate of change of its concentration can be neglected in comparison with the rates of change of the reactants and products.

The steady-state approximation greatly simplifies the discussion of reaction schemes. For example, when we apply it to the consecutive first-order mechanism, we set $\mathrm{d}[\mathrm{B}]/\mathrm{d}t = 0$ in eqn (28.3.13b), which becomes $k_{\mathrm{a}}[\mathrm{A}] - k_{\mathrm{b}}[\mathrm{B}] = 0$. Then

$$[\mathrm{B}] = (k_{\mathrm{a}}/k_{\mathrm{b}})[\mathrm{A}], \qquad (28.3.18)$$

which shows that $[\mathrm{B}] \ll [\mathrm{A}]$ at all stages of the reaction. On substituting this value of [B] into eqn (28.3.13c) that equation becomes

$$\mathrm{d}[\mathrm{C}]/\mathrm{d}t = k_{\mathrm{b}}[\mathrm{B}] = k_{\mathrm{a}}[\mathrm{A}], \qquad (28.3.19)$$

and we see that C is formed by a first-order decay of A, with a rate constant k_{a} (i.e. the rate constant of the slower, rate-determining, step). The solution of this equation can be written down at once by substituting the solution for [A], eqn (28.3.14a) and integrating:

$$[\mathrm{C}]_t = k_{\mathrm{a}}[\mathrm{A}]_0\int_0^t \mathrm{e}^{-k_{\mathrm{a}}t}\,\mathrm{d}t = [\mathrm{A}]_0(1 - \mathrm{e}^{-k_{\mathrm{a}}t}), \qquad (28.3.20)$$

the same (approximate) result as before, but much more quickly obtained.

28.3 (f) Pre-equilibria

A similar technique can be used to simplify the discussion of a consecutive reaction in which an intermediate reaches an equilibrium with the reactants:

$$\mathrm{A} + \mathrm{B} \underset{k_{\mathrm{a}}'}{\overset{k_{\mathrm{a}}}{\rightleftharpoons}} (\mathrm{AB}) \xrightarrow{k_{\mathrm{b}}} \mathrm{C}, \qquad (28.3.21)$$

where (AB) denotes the intermediate. This is called a *pre-equilibrium*, and arises when the rates of formation of the intermediate and its decay back into reactants are much faster than its rate of formation of products. Since we assume that A, B, and (AB) are in equilibrium, we can write

$$K = \{[(AB)]/[A][B]\}_{eq}, \quad \text{with} \quad K = k_a/k'_a. \tag{28.3.22}$$

In writing these equations, we are ignoring the fact that (AB) is slowly leaking away as it forms C. The rate of formation of C may now be written:

$$d[C]/dt = k_b[(AB)] = k_b K[A][B] = k[A][B], \tag{28.3.23}$$

with $k = k_a k_b / k'_a$. The reaction therefore has second-order kinetics overall, and a rate coefficient which is a composite of the coefficients for each step.

Two examples of consecutive reactions with pre-equilibria will help to show how kinetic studies elucidate mechanisms. The oxidation of nitrogen (II) oxide is found to be third-order overall:

$$2NO(g) + O_2(g) \rightarrow 2NO_2(g), \quad d[NO_2(g)]/dt = k[NO(g)]^2[O_2(g)]. \tag{28.3.24}$$

One explanation might be that the reaction is a single termolecular simple step, but that requires the simultaneous collision of three particles, which occurs very infrequently. Furthermore, it is observed that the reaction rate *decreases* as the temperature is raised. This points to a complex reaction mechanism because simple reactions almost always go faster at higher temperatures. A mechanism that accounts for the rate law and the temperature dependence is a pre-equilibrium

$$\text{(a)} \quad 2NO(g) \rightleftharpoons N_2O_2(g), \qquad K = \{[N_2O_2(g)]/[NO(g)]^2\}_{eq} \tag{28.3.25a}$$

followed by a simple bimolecular reaction

$$\text{(b)} \quad N_2O_2(g) + O_2(g) \rightarrow 2NO_2(g),$$
$$d[NO_2(g)]/dt = k_b[N_2O_2(g)][O_2(g)]. \tag{28.3.25b}$$

The reaction rate is obtained by combining the two equations into

$$d[NO_2(g)]/dt = k_b K[NO_2(g)]^2[O_2(g)], \tag{28.3.26}$$

which has the observed overall third-order form. Furthermore, it also accounts for the anomalous temperature dependence. This is because K decreases with temperature (the dimerization of NO is exothermic, and thermodynamics shows that K therefore decreases as the temperature is raised, Section 10.2(c)), and so, provided k_b does not increase more sharply, the rate coefficient $k = k_b K$ decreases as the temperature is increased and the effective activation energy of the reaction is *negative*. This can be shown formally as follows.

From the definition of activation energy, eqn (28.3.5) we write

$$E_a = RT^2(\partial \ln k/\partial T)_V = RT^2(\partial \ln k_b K/\partial T)_V$$
$$= RT^2(\partial \ln k_b/\partial T)_V + RT^2(\partial \ln K/\partial T)_V.$$

The first term is the activation energy of the second-order step, which we denote E'_a. The second term can be expressed in terms of the standard reaction internal energy ($\Delta_r U^{\ominus}$) for the dimerization using the constant-volume version of eqn (10.2.4):

$$d \ln K/dT = \Delta_r U^{\ominus}/RT^2.$$

Then we obtain

$$E_a = E'_a + \Delta_r U^\ominus. \quad (28.3.27)$$

If all the substances are treated as perfect gases, the dimerization reaction enthalpy and internal energy are related by $\Delta_r H = \Delta_r U - RT$, and so

$$E_a = E'_a + RT + \Delta_r H^\ominus. \quad (28.3.28)$$

Therefore, if $\Delta_r H^\ominus$ is strongly negative (an exothermic reaction), the activation energy of the reaction may be negative, as we set out to show.

Another example of a pre-equilibrium reaction is the *Michaelis–Menten mechanism* of enzyme action. The rate of an enzyme-catalysed reaction in which a *substrate* S is converted into products P is found to depend on the concentration of the enzyme E even though the enzyme undergoes no net change. The proposed mechanism is

$$\mathrm{E} + \mathrm{S} \rightleftharpoons (\mathrm{ES}) \rightarrow \mathrm{P} + \mathrm{E}, \qquad \mathrm{d}[\mathrm{P}]/\mathrm{d}t = k_b[(\mathrm{ES})]. \quad (28.3.29)$$

(ES) denotes a bound state of the enzyme and its substrate. In order to relate [(ES)] to the enzyme concentration we write its rate law and then impose the steady-state approximation:

$$\mathrm{d}[(\mathrm{ES})]/\mathrm{d}t = \underbrace{k_a[\mathrm{E}][\mathrm{S}]}_{\text{Formation from E, S}} - \underbrace{k'_a[(\mathrm{ES})]}_{\text{Decay to E, S}} - \underbrace{k_b[(\mathrm{ES})]}_{\text{Decay to P}} = 0. \quad (28.3.30)$$

This solves to

$$[(\mathrm{ES})] = \{k_a/(k_b + k'_a)\}[\mathrm{E}][\mathrm{S}]. \quad (28.3.31)$$

[E] and [S] are the concentrations of the *free* enzyme and *free* substrate, and if $[\mathrm{E}]_0$ is the total concentration of enzyme, $[\mathrm{E}] + [(\mathrm{ES})] = [\mathrm{E}]_0$, a constant. Since only a little enzyme is added, the free substrate concentration is almost the same as the total substrate concentration, and we can ignore the fact that [S] differs slightly from $[\mathrm{S}]_{\mathrm{total}}$. Therefore,

$$[(\mathrm{ES})] = \{k_a/(k_b + k'_a)\}\{[\mathrm{E}]_0 - [(\mathrm{ES})]\}[\mathrm{S}],$$

which rearranges to

$$[(\mathrm{ES})] = \frac{k_a[\mathrm{E}]_0[\mathrm{S}]}{k_b + k'_a + k_a[\mathrm{S}]}. \quad (28.3.32)$$

It follows that the rate of formation of product is

$$\mathrm{d}[\mathrm{P}]/\mathrm{d}t = \frac{k_b k_a[\mathrm{E}]_0[\mathrm{S}]}{k_b + k'_a + k_a[\mathrm{S}]} = \frac{k_b[\mathrm{E}]_0[\mathrm{S}]}{K_M + [\mathrm{S}]} \quad (28.3.33)$$

where the *Michaelis constant*, K_M, is

$$K_M = (k_b + k'_a)/k_a. \quad (28.3.34)$$

According to eqn (28.3.33), the rate of enzymolysis depends *linearly* on the enzyme concentration, but in a more complicated way on the concentration of substrate. Thus, when $[\mathrm{S}] \gg K_M$ the rate-law in eqn (28.3.33) reduces to $\mathrm{d}[\mathrm{P}]/\mathrm{d}t = k_b[\mathrm{E}]_0$, and it is *zeroth-order* in S (this means that the rate is constant because there is so much S present that it remains at effectively the same concentration even though products are being formed). However, when so little S is present that $[\mathrm{S}] \ll K_{\mathrm{M}}$, then the rate of formation of products is proportional to [S] as well as to $[\mathrm{E}]_0$.

28.3 (g) Unimolecular reactions

A number of gas phase reactions follow first-order kinetics and are believed to proceed through a unimolecular rate-determining stage. These are called *unimolecular reactions*. Presumably a molecule acquires enough energy to react as a result of its collisions with other molecules; but collisions are simple *bimolecular* events, and so how can they result in first-order kinetics?

The first successful explanation of unimolecular reactions was provided by Frederick Lindemann and Cyril Hinshelwood. In the *Lindemann–Hinshelwood mechanism* it is supposed that a reactant molecule A collides with another M, a diluent gas molecule, and becomes energetically excited at the expense of M's translational kinetic energy:

$$\mathrm{A} + \mathrm{M} \rightarrow \mathrm{A}^* + \mathrm{M}, \qquad \mathrm{d}[\mathrm{A}^*]/\mathrm{d}t = k_{\mathrm{a}}[\mathrm{A}][\mathrm{M}]. \qquad (28.3.35\mathrm{a})$$

The energized molecule might lose its excess energy by collision with another:

$$\mathrm{A}^* + \mathrm{M} \rightarrow \mathrm{A} + \mathrm{M}, \qquad \mathrm{d}[\mathrm{A}^*]/\mathrm{d}t = k_{\mathrm{a}}'[\mathrm{A}^*][\mathrm{M}]. \qquad (28.3.35\mathrm{b})$$

(We are taking M to be in excess, and ignoring A^*,A collisions.) Alternatively, the excited molecule might shake itself apart and form products (P). That is, it might undergo the unimolecular decay

$$\mathrm{A}^* \rightarrow \mathrm{P}, \qquad \mathrm{d}[\mathrm{P}]/\mathrm{d}t = k_{\mathrm{b}}[\mathrm{A}^*], \qquad \mathrm{d}[\mathrm{A}^*]/\mathrm{d}t = -k_{\mathrm{b}}[\mathrm{A}^*]. \qquad (28.3.35\mathrm{c})$$

If the unimolecular step is slow enough to be the rate-determining step, we can expect it to dominate the overall process. In this case the overall reaction $\mathrm{A} \rightarrow \mathrm{P}$ will have first-order kinetics, as we require. We can demonstrate this explicitly by applying the steady-state approximation to the net rate of formation of A^*:

$$\mathrm{d}[\mathrm{A}^*]/\mathrm{d}t = k_{\mathrm{a}}[\mathrm{A}][\mathrm{M}] - k_{\mathrm{a}}'[\mathrm{A}^*][\mathrm{M}] - k_{\mathrm{b}}[\mathrm{A}^*] = 0.$$

This solves to

$$[\mathrm{A}^*] = \frac{k_{\mathrm{a}}[\mathrm{A}][\mathrm{M}]}{k_{\mathrm{b}} + k_{\mathrm{a}}'[\mathrm{M}]},$$

and so the rate law for the formation of P is

$$\mathrm{d}[\mathrm{P}]/\mathrm{d}t = k_{\mathrm{b}}[\mathrm{A}^*] = \frac{k_{\mathrm{a}}k_{\mathrm{b}}[\mathrm{A}][\mathrm{M}]}{k_{\mathrm{b}} + k_{\mathrm{a}}'[\mathrm{M}]}. \qquad (28.3.36)$$

At this stage the rate law is not first-order. However, if the rate of deactivation by A^*,M collisions is much greater than the rate of unimolecular decay, so that

$$k_{\mathrm{a}}'[\mathrm{A}^*][\mathrm{M}] \gg k_{\mathrm{b}}[\mathrm{A}^*], \quad \text{or} \quad k_{\mathrm{a}}'[\mathrm{M}] \gg k_{\mathrm{b}},$$

then we can neglect k_{b} in the denominator and obtain

$$\mathrm{d}[\mathrm{P}]/\mathrm{d}t = k_{\mathrm{a}}k_{\mathrm{b}}[\mathrm{A}][\mathrm{M}]/k_{\mathrm{a}}'[\mathrm{M}] = \{k_{\mathrm{a}}k_{\mathrm{b}}/k_{\mathrm{a}}'\}[\mathrm{A}], \qquad (28.3.37)$$

a first-order rate law, as we set out to show.

The Lindemann–Hinshelwood mechanism can be tested because it predicts that as the concentration of M is reduced (in practice, that means as the pressure of M is reduced) the reaction should switch to overall second-order kinetics. This is because when $k_a'[M] \ll k_b$, the rate law in eqn (28.3.36) is approximately

$$d[P]/dt = k_a k_b[A][M]/k_b = k_a[A][M]. \tag{28.3.38}$$

The physical reason for the change of order is that at low pressures the rate-determining step is the bimolecular formation of A*. If we write the full rate law in eqn (28.3.36) as

$$d[P]/dt = k_{eff}[A], \qquad k_{eff} = \frac{k_a k_b[M]}{k_b + k_a'[M]} \tag{28.3.39}$$

then the expression for the effective rate constant can be rearranged to

$$1/k_{eff} = 1/k_a[M] + k_a'/k_a k_b. \tag{28.3.40}$$

This shows that a test of the theory is to plot $1/k_{eff}$ against $1/[M]$, and to expect a straight line. A typical result is shown in Fig. 28.8 (the decomposition of N_2O_5). The graph has a pronounced curvature, corresponding to a *larger* value of k_{eff} (smaller value of $1/k_{eff}$) at high pressures (low $1/[M]$) than would be expected by extrapolation of the reasonably linear low pressure (high $1/[M]$) data.

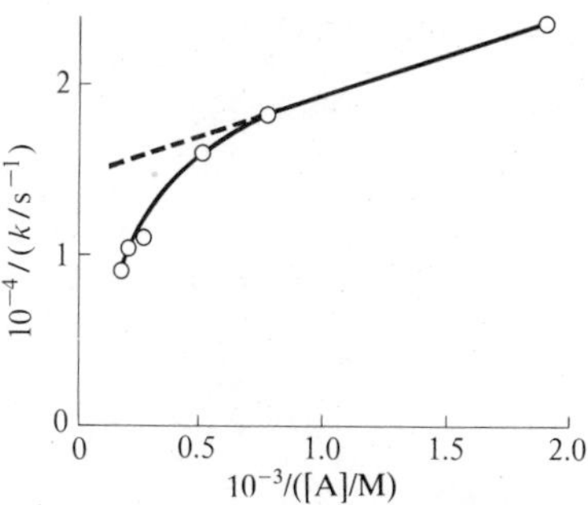

Fig. 28.8. The pressure-dependence of the unimolecular isomerization of *trans*-CHD=CHD showing a pronounced departure from the straight line predicted by eqn (28.3.40) based on the Lindemann–Hinshelwood mechanism. (M. J. Pilling, *Reaction kinetics*, Clarendon Press, Oxford, 1975.)

One of the reasons for the discrepancy is that the Lindemann–Hinshelwood mechanism fails to recognize that a *specific* excitation of the molecule may be required before reaction occurs. For example, in the isomerization of cyclobutene (**1**→**2**), which is known to be first-order, the crucial step is the breaking of one C—C bond, and occurs when that bond is highly vibrationally excited. In the collision leading to excitation, however, the excess energy is shared among all the bonds and the rotation of the molecule, and so the isomerization occurs only after the energy accumulates in the critical bond. This suggests that we should distinguish between the *energized molecule* A*, where the excess energy is dispersed over many modes, and the *activated state* $A^{\ddagger}$, where the excitation is specific and the molecule is poised for reaction. The unimolecular part of the mechanism therefore ought to be modified to

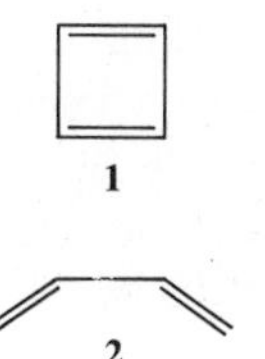

$$A^* \rightarrow A^{\ddagger} \rightarrow P,$$

with a rate coefficient for each step. The values of the various rate coefficients are related to the numbers and frequencies of the vibrational modes available by the *Rice–Ramsperger–Kassel theory* (RRK theory) of unimolecular reactions, and by the more sophisticated *RRKM theory* (where M is Marcus), which also takes into account the effects of molecular rotation. Descriptions of these theories will be found in the books mentioned in *Further reading*.

Further reading

Chemical kinetics (2nd edn). K. J. Laidler; McGraw-Hill, New York, 1965.
Chemical kinetics. J. Nicholas; Harper & Row, New York, 1976.
Rates and mechanisms of chemical reactions. W. C. Gardner; Benjamin, New York, 1969.
Foundations of chemical kinetics. S. W. Benson; McGraw-Hill, New York, 1960.

Homogeneous gas phase reactions. A. Maccoll; *Techniques of chemistry* (E. S. Lewis, ed) VIA, 47, Wiley-Interscience, New York, 1974.

Kinetics in solution. J. F. Bunnett; *Techniques of chemistry* (E. S. Lewis, ed) VIA, 129, Wiley-Interscience, New York, 1974.

Comprehensive chemical kinetics (*Vols. 1–20*). C. H. Bamford and C. F. Tipper (eds); Elsevier, Amsterdam, 1969–1980.

Data:

Kinetic data on gas phase unimolecular reactions. S. W. Benson and H. E. O'Neal; NSRDS-NBS-21, US Department of Commerce, Washington, 1970.

Tables of bimolecular gas phase reactions. A. F. Trotman–Dickenson and G. S. Milne; NSRDS-NBS-9, US Department of Commerce, Washington, 1967.

Introductory problems

A28.1. For a reaction in liquid solution, the stoichiometry is $A + 2B \rightarrow P + Q$. The rate is given by $d[P]/dt = k_1[A][B]$, where $k_1 = 3.67 \times 10^{-3}\ M^{-1}\,s^{-1}$. At the start of the reaction, 0.255 mole of A is mixed with 0.605 mole of B in 1.7 dm^3 of solvent. Calculate the initial value of each of the following: $d[A]/dt$, $d[B]/dt$, $d[P]/dt$, dn_B/dt, and the reaction rate.

A28.2. A substance decomposes according to the mechanism $2A \rightarrow$ products with a rate constant $2.62 \times 10^{-3}\ M^{-1}\,s^{-1}$. In a rate experiment, the initial concentration of A is 1.70 M. What is the half-life of the reaction?

A28.3. A reaction with the mechanism $2A \rightarrow$ products has a rate constant $3.50 \times 10^{-4}\ M^{-1}\,s^{-1}$. Find the time required for the concentration of A to change from 0.260 M to 0.011 M.

A28.4. The rate constant for the decomposition of a certain substance is $2.80 \times 10^{-3}\ M\,s^{-1}$ at 30 °C and is $1.38 \times 10^{-2}\ M^{-1}\,s^{-1}$ at 50 °C. Find the Arrhenius energy of activation and the pre-exponential factor.

A28.5. The reaction $2H_2O_2(aq) \rightarrow 2H_2O + O_2(g)$ is catalysed by certain ions, M^-. If the mechanism is:

$$H_2O_2(aq) + M^-(aq) \rightarrow H_2O(l) + MO^-(aq) \qquad \text{(slow)}$$

$$MO^-(aq) + H_2O_2(aq) \rightarrow H_2O(l) + O_2(g) + M^-(aq) \qquad \text{(fast)}$$

give the order of the reaction with respect to the various participants.

A28.6. The reaction mechanism:

$$A_2 \rightleftharpoons 2A \qquad \text{(fast)}$$

$$A + B \rightarrow \text{products} \qquad \text{(slow)}$$

involves an intermediate A. Determine the dependence of the rate of reaction on the concentration of A_2 and of B.

A28.7. Consider the following mechanism for renaturation of a double helix from its strands A and B:

$$A + B \rightleftharpoons \text{unstable helix with 2 base pairs} \qquad \text{(fast)}$$

$$\text{two-base-pair helix} \rightarrow \text{stable double helix} \qquad \text{(slow)}.$$

Find the rate equation for the formation of the double helix. Show the relationship of the rate constant of the renaturation reaction to the rate constants for the individual steps.

A28.8. The enzymatic conversion of a substrate at 25 °C has a Michaelis constant of 0.035 M. The rate of the reaction is $1.15 \times 10^{-3}\ M\,s^{-1}$ when the substrate concentration is 0.110 M. What is the maximum rate of this enzymolysis?

A28.9. For an enzymatic reaction which follows Michaelis–Menten kinetics, find the condition for which the reaction rate is half its maximum value.

A28.10. The effective rate constant for a gaseous reaction which has a Lindemann–Hinshelwood mechanism has the following values at 150 °C: $2.50 \times 10^{-4}\ s^{-1}$ for $p = 1.30$ kPa and $2.10 \times 10^{-5}\ s^{-1}$ for $p = 12$ Pa. Find the rate coefficient for the activation step in the mechanism.

Problems

28.1. The reaction rate of $A + 2B \rightarrow 3C + D$ was reported as 1.0 mol s^{-1}. State the rates of change of concentration of each of the participants in a 1.0 dm^3 vessel.

28.2. The rate of change of concentration of C in the reaction $2A + B \rightarrow 2C + 3D$ was reported as 1.0 M s^{-1}. State the reaction rate, and the rates of change of concentrations of A, B, and D.

28.3. The rate law for the reaction in Problem 28.1 was expressed as $\dot{\xi} = k[A][B]$. What are the units of k? Express the rate law in terms of the rate of change of concentration of (a) A, (b) C.

28.4. The rate law for the reaction in Problem 28.2 was reported as $d[C]/dt = k[A][B][C]$. Express it in terms of the reaction rate; what are the units for k in each case?

28.5. The rate constant for the first-order decomposition of N_2O_5 has the value $4.8 \times 10^{-4}\ s^{-1}$. What is the half-life of the reaction? What will be the pressure, initially 500 Torr, after (a) 10 s, (b) 10 min after initiation of the reaction?

28.6. If the rate laws are expressed in (a) concentrations in M, (b) pressures in atmospheres, what are the units of the second-order and third-order rate constants?

28.7. The half-life for radioactive decay of ^{14}C is 5730 yr (it emits β-rays with an energy of 0.16 MeV). An archaeological sample contained wood that had only 72% of the ^{14}C found in living trees. What is its age?

28.8. One of the hazards of nuclear explosions is the generation of ^{90}Sr and its subsequent incorporation in place of calcium in bones. This isotope emits β-rays of energy 0.55 MeV, and has a half-life of 28.1 yr. Suppose 1.00 μg was absorbed by a newly born child. How much will remain after (a) 18 yr, (b) 70 yr if none is lost metabolically?

28.9. The second-order rate constant for the reaction AcOEt(aq) + NaOH(aq) $\rightarrow$ AcONa(aq) + EtOH(aq) where AcO is the acetate group is 0.11 $M^{-1} s^{-1}$. What is the concentration of ester after (a) 10 s, (b) 10 min when ethyl acetate is added to sodium hydroxide so that the initial concentrations are [NaOH] = 0.050 M and [AcOEt] = 0.100 M?

28.10. Derive an integrated expression for a second-order rate law $d[P]/dt = k[A][B]$ for a reaction of stoichiometry $2A + 3B \rightarrow P$.

28.11. Derive the integrated form of a third-order rate law $-d[A]/dt = k[A]^2[B]$ in which the stoichiometry is $2A + B \rightarrow P$ and the reactants are initially present in their stoichiometric proportions.

28.12. Repeat the last Problem, but with B present initially in twice the amount.

28.13. Derive an expression for the half-life of the reaction in Problem 28.11 defined (a) as the time for A to halve its initial concentration, (b) defined as the time for B to halve its concentration, (c) defined as the time for the advancement of the reaction to reach $\xi = \frac{1}{2}$ mol.

28.14. Show that $t_{\frac{1}{2}} \propto 1/[A]_0^{n-1}$ for a reaction nth-order in A.

28.15. Show that the ratio $t_{\frac{1}{2}}/t_{\frac{3}{4}}$, where $t_{\frac{1}{2}}$ is the half-life and $t_{\frac{3}{4}}$ is the threequarter-life (the time for the concentration of A to drop to $\frac{3}{4}$ of its initial value, implying $t_{\frac{3}{4}} < t_{\frac{1}{2}}$) can be written as a function of n alone, and so it can be used as a rapid assessment of the order of a reaction.

28.16. The composition of a liquid phase reaction $2A \rightarrow B$ was followed as a function of time by a spectroscopic method with the following results:

t/min	0	10	20	30	40	∞
[B]/M	0	0.089	0.153	0.200	0.230	0.312

What is the order of the reaction? What is the value of the rate constant?

28.17. In an experiment to investigate the stability of substituted allyl radicals (A. B. Trenwith, *J. chem. Soc. Faraday Trans.* I, 1737 (1973)) the rate of formation of water in the reaction $CH_3CH(OH)CH{=}CH_2 \rightarrow H_2O + CH_2{=}CHCH{=}CH_2$ was monitored. At 810 K the results were as follows:

t/min	0.5	1.0	1.5	2.0	2.5	∞
V/cm^3	1.0	1.4	1.6	1.7	1.8	2.0

V is the volume of water produced. Find the order and rate coefficient. The C_4H_6 did not appear to form as rapidly as the water: suggest a reason.

28.18. The experiment described in the last Problem was repeated at several temperatures and the following values of the rate constant were obtained:

T/K	773.5	786	797.5	810	810	824	834
k/units	1.63	2.95	4.19	8.13	8.19	14.9	22.2

What is the activation energy and pre-exponential factor A for the reaction? (Decide on the units and magnitude of k by referring to the previous Problem.)

28.19. Cyclopropane isomerizes into propene when heated to 500 °C in the gas phase. The extent of conversion for various initial pressures has been followed by gas chromatography by allowing the reaction to proceed for a time with various initial pressures:

p_0/Torr	200	200	400	400	600	600
t/s	100	200	100	200	100	200
p/Torr	186	173	373	347	559	520

where p_0 is the initial pressure and p is the final pressure of cyclopropane. What is the order and rate constant for the reaction under these conditions?

28.20. A second-order gas phase reaction is of the form $2A \rightarrow B$, both A and B being gases. Find an expression for the time-dependence of the total pressure of the reacting system. Let the initial pressure, when no B is present, be p_0; then plot p/p_0 as a function of $x = p_0kt$. What time is needed for the pressure to drop half way towards its final value? What is the advancement of the reaction at that time?

28.21. The composition of the gas phase reaction $2A \rightarrow B$ was monitored by measuring the total pressure as a function of time. The following are the results:

t/s	0	100	200	300	400
p/Torr	400	322	288	268	256

What is the order of the reaction, and what is the value of the rate constant?

28.22. The radioactive decay of a family of nuclei was specified in Section 28.3(d). Calculate the abundance of the nuclides as a function of time, and plot the results as a graph.

28.23. The addition of hydrogen halides to olefins has played a basic role in the investigation of organic reaction mechanisms. In one modern study (M. J. Haugh and D. R. Dalton, *J. Am. chem. Soc.* **97**, 5674 (1975)) high pressures of hydrogen chloride (up to 25 atm) and propene (up to 5 atm) were examined over a range of temperatures and the amount of 2-chloropropane formed was determined by NMR. Show that if the reaction $A+B \rightarrow P$ proceeds for a short time Δt, the concentration of product follows $[P]/[A] = k[A]^{m-1}[B]^n \Delta t$ if the reaction is mth-order in A and nth-order in B. In a series of runs the ratio [chloropropane]/[propene] was independent of [propene] but the ratio [chloropropane]/[HCl] for constant amounts of propene depended on [HCl], and for $\Delta t \approx 100$ h (which is short on the time scale of the reaction) the ratio rose from zero to 0.05, 0.03, 0.01 for $p(HCl) = 10$ atm, 7.5 atm, 5.0 atm. What are the orders of the reaction with respect to each reactant?

28.24. Show that the following mechanism can account for the rate law of the reaction in the last Problem:

$$2HCl \rightleftharpoons (HCl)_2 \qquad K_1$$
$$HCl + CH_3CH{=}CH_2 \rightleftharpoons \text{complex} \qquad K_2$$
$$(HCl)_2 + \text{complex} \rightarrow CH_3CHClCH_3 + 2HCl \qquad k, \text{ slow}$$

What further tests could you apply to check this mechanism?

28.25. In the experiments described in the last two Problems, an inverse temperature dependence of the reaction rate was observed, the overall rate of reaction at 70 °C being roughly one-third that at 19 °C. Estimate the apparent activation energy and the activation energy of the rate determining step given that the enthalpies of the two equilibria are both of the order of $-14\ \text{kJ mol}^{-1}$.

28.26. The second-order rate constants for the reaction of oxygen atoms with aromatic hydrocarbons have been measured (R. Atkinson and J. N. Pitts, *J. phys. Chem.* **79**, 295 (1975)). In the reaction with benzene the rate constants are 1.44×10^7 at 300.3 K, 3.03×10^7 at 341.2 K, and 6.9×10^7 at 392.2 K, all in $M^{-1}\,s^{-1}$. Find the pre-exponential factor and activation energy of the reaction.

28.27. In a study of the autoxidation of hydroxylamine (M. N. Hughes, H. G. Nicklin, and K. Shrimanker, *J. chem. Soc.* A 3845 (1971)), the rate constant k_{obs} in the rate equation $-d[NH_2OH]/dt = k_{obs}[NH_2OH][O_2]$ was found to have the following temperature-dependence:

$\theta/°C$	0	10	15	25	34.5
$10^4 k_{obs}/M^{-1}\,s^{-1}$	0.237	0.680	1.02	2.64	5.90

What is the activation energy of the reaction? This analysis is taken further in Problem 28.30.

28.28. The equilibrium $A \rightleftharpoons B$ is first-order in both directions. Derive an expression for the concentration of A as a function of time when the initial amounts of A and B are $[A]_0$ and $[B]_0$. What is the final composition of the system?

28.29. In Problem 28.19 the isomerization of cyclopropane over a limited pressure range was examined. If the Lindemann mechanism of first-order reactions is to be tested we also need data at low pressures. These have been obtained (H. O. Pritchard, R. G. Sowden, and A. F. Trotman-Dickenson, *Proc. R. Soc.* **A217**, 563 (1953)):

p/Torr	84.1	11.0	2.89	0.569	0.120	0.067
$10^4 k_{eff}/s^{-1}$	2.98	2.23	1.54	0.857	0.392	0.303

Test the Lindemann theory with these data (see eqn (28.3.40)).

28.30. The kinetics of autoxidation of hydroxylamine in the presence of EDTA (ethylenediamine tetra-acetic acid) have been studied in the hydroxide ion concentration range 0.5–3.2 M (reference in Problem 28.27). The reaction proceeds by the mechanism

$$NH_2OH + OH^- \rightarrow NH_2O^- + H_2O \quad \text{(fast)}$$
$$NH_2O^- + O_2 \rightarrow \text{products} \quad \text{(slow)},$$

with the rate laws

$$-d[(NH_2OH)]/dt = k_{obs}[(NH_2OH)][O_2]$$
$$-d[(NH_2OH)]/dt = k[NH_2O^-][O_2],$$

where $[(NH_2OH)]$ denotes the total hydroxylamine, $[NH_2OH]+[NH_2O^-]$, and $k = k_{obs}/f$, where f is the fraction of hydroxylamine present as NH_2O^-. Show that a plot of $1/k_{obs}$ against $[H^+]$ should give a straight line, and that the acid dissociation constant of NH_2OH can be determined from the slope.

28.31. The data below relate to the reaction system described in the last Problem. Find pK_a for the hydroxylamine, and the amount of NH_2O^- present at each OH^- concentration.

$[OH^-]/M$	0.50	1.00	1.6	2.4
$10^4 k_{obs}/s^{-1}$	2.15	2.83	3.32	3.54

28.32. The initial rate of oxygen production by the action of an enzyme on a substrate was measured for a range of substrate concentrations; the data are below. What is the value of the Michaelis constant for the reaction?

[S]/M	0.050	0.017	0.010	0.005	0.002
rate/$mm^3\,min^{-1}$	16.6	12.4	10.1	6.6	3.3

29

The kinetics of complex reactions

Learning objectives

After careful study of this chapter you should be able to:

(1) Describe the meaning of *initiation*, *propagation*, *inhibition*, and *termination* steps in a chain reaction, Section 29.1(a).

(2) Write the rate laws for individual steps in a chain reaction, and use the steady-state approximation to deduce the overall rate law, Section 29.1(a).

(3) Distinguish between *chain polymerization* and *stepwise polymerization*, and give examples of each, Section 29.1(b).

(4) Define the *kinetic chain length*, eqn (29.1.9), relate it to the kinetic scheme, and show how the RMM of a polymer depends on the rates of initiation and propagation, eqn (29.1.15).

(5) Deduce an expression for the degree of stepwise polymerization, eqn (29.1.12) and use it to calculate the average properties of a macromolecule, Example 29.1.

(6) Explain the meaning of a *chain-branching step* and its role in explosions, Section 29.1(c).

(7) Describe and explain the occurrence of the *explosion limits* of a reaction, Fig. 29.3.

(8) State the *Einstein–Stark law* and define the *quantum efficiency* of a photochemical reaction, eqn (29.2.1) and Example 29.2.

(9) State the processes that may occur after photochemical initiation of a reaction, Box 29.1.

(10) Write rate laws for photochemical reactions, eqn (29.2.2).

(11) Describe *laser isotope separation*, Section 29.2(b).

(12) Explain the significance and mechanism of *photosensitization* of reactions, Section 29.2(c).

(13) Explain how *catalysis* increases the rates of reactions and classify the different kinds, Section 29.3.

(14) State the meaning of *autocatalysis* and explain how it affects the rates of reactions, Example 29.3.

(15) Describe the *Lotka–Volterra mechanism* of chemical oscillation, eqn (29.4.1), and account for the periodic variation of concentrations to which it leads, Figs. 29.7 and 29.8.

(16) Explain what is meant by a *limit cycle*, Fig. 29.9.

(17) State the mechanism of the *brusselator*, eqn (29.4.2), and the *oregonator*, eqn (29.4.3).

(18) State the conditions necessary for chemical oscillation, and explain the meaning and significance of *bistability*, Fig. 29.10.

(19) Describe how *flash photolysis* and *flow techniques* may be used to study fast reactions, Sections 29.5(a and b).

(20) Describe how *relaxation methods* are used to study fast reactions, and derive expressions for *relaxation time* in terms of rate constants, eqn (25.5.7) and Example 29.4.

Introduction

Many reactions take place by mechanisms that involve many steps, and some take place at a useful rate only if a catalyst is present. In some cases, when the reaction system is far from equilibrium, the concentrations vary in peculiar ways. In this chapter we see how the ideas introduced in Chapter 28 can be developed to deal with these special kinds of reactions.

29.1 Chain reactions

Many gas phase and polymerization reactions are *chain reactions*, in which a reactive intermediate (which is often a *free radical*, an atom, or a fragment of a stable molecule with an unpaired electron) produced in one step of the mechanism takes part in subsequent steps.

29.1 (a) The structure of chain reactions

The steps in a chain reaction can be classified into five types. In the *initiation step* the radicals are formed from ordinary (spin-paired) molecules. For example, Cl atoms are formed by the dissociation of Cl_2 molecules either as a result of collisions (in a *thermal reaction*, one driven by heat) or as a result of absorption of light (in a *photochemical reaction*). The radicals produced in the initiation step attack other spin-paired molecules in the *propagation steps*, and each attack gives rise to a new radical, as in the attack of a CH_3 radical on ethane:

$$\cdot CH_3 + CH_3CH_3 \rightarrow CH_4 + \cdot CH_2CH_3.$$

(The dot signifies the unpaired electron and marks the radical.) In some cases the attack results in the production of more than one radical, and we deal with these *branching reactions* in Section 29.1(c). In other cases the radical might attack a product molecule formed earlier in the reaction. This reduces the net rate of formation of product, and is called an *inhibition step*. For example, in a photochemical reaction in which HBr is formed from H_2 and Br_2, an H atom might attack an HBr molecule, leading to $H_2 + Br$. This does not break the chain, because one radical ($\cdot H$) gives rise to another ($\cdot Br$). Steps in which radicals are removed are called *termination steps* because they end the chain of reactions. For example, when two $\cdot CH_2CH_3$ radicals dimerize, the chain terminates. Some radicals react with the walls of the vessel and terminate the chain. Others may be *scavenged* by reaction with other radicals introduced into the reaction vessel. The nitrogen(II) oxide molecule, NO, has an unpaired electron and is a very efficient radical scavenger. The observation that a reaction is quenched when NO is introduced is a good indication that a chain mechanism is in operation.

A chain reaction often leads to a complicated rate law. As a first example, consider the thermal reaction between hydrogen and bromine. The overall reaction and the observed rate law are as follows:

$$H_2(g) + Br_2(g) \rightarrow 2HBr(g),$$

$$d[HBr]/dt = \frac{k[H_2(g)][Br_2(g)]^{\frac{3}{2}}}{[Br_2(g)] + k'[HBr(g)]}. \qquad (29.1.1)$$

The complexity of the rate law suggests that a complicated mechanism is involved. The following chain mechanism has been proposed:

(a) *Initiation:* $Br_2 + M \rightarrow 2Br + M$,

$$d[Br]/dt = 2k_a'[Br_2][M] = 2k_a[Br_2]. \qquad (29.1.2a)$$

(M is an inert molecule; we absorb its concentration into the constant k_a' and write $k_a = k_a'[M]$.)

(b) *Propagation:*

$$\cdot\mathrm{Br} + \mathrm{H_2} \rightarrow \mathrm{HBr} + \cdot\mathrm{H}, \qquad \mathrm{d[HBr]/d}t = k_\mathrm{b}[\mathrm{Br}][\mathrm{H_2}], \qquad (29.1.2\mathrm{b})$$

$$\cdot\mathrm{H} + \mathrm{Br_2} \rightarrow \mathrm{HBr} + \cdot\mathrm{Br}, \qquad \mathrm{d[HBr]/d}t = k'_\mathrm{b}[\mathrm{H}][\mathrm{Br_2}]. \qquad (29.1.2\mathrm{b}')$$

(c) *Inhibition:* $\cdot\mathrm{H} + \mathrm{HBr} \rightarrow \mathrm{H_2} + \cdot\mathrm{Br}$,

$$\mathrm{d[HBr]/d}t = -k_\mathrm{c}[\mathrm{H}][\mathrm{HBr}]. \qquad (29.1.2\mathrm{c})$$

(d) *Termination:* $\cdot\mathrm{Br} + \cdot\mathrm{Br} + \mathrm{M} \rightarrow \mathrm{Br_2} + \mathrm{M}$,

$$\mathrm{d[Br]/d}t = -2k'_\mathrm{d}[\mathrm{Br}]^2[\mathrm{M}] = -2k_\mathrm{d}[\mathrm{Br}]^2. \qquad (29.1.2\mathrm{d})$$

The net rate of formation of the product HBr is

$$\mathrm{d[HBr]/d}t = k_\mathrm{b}[\mathrm{Br}][\mathrm{H_2}] + k'_\mathrm{b}[\mathrm{H}][\mathrm{Br_2}] - k_\mathrm{c}[\mathrm{H}][\mathrm{HBr}]. \qquad (29.1.3)$$

There are now two ways of proceeding. One is to solve the set of differential equations numerically, without making further approximations.‡ In this way the concentrations of all the participants in the reaction may be calculated and the rate constants determined by varying them until a best-fit is obtained. An alternative procedure (the only one available until recently) is to look for approximate solutions and see if they agree with the empirical rate law. This involves making the steady state approximation to find the concentration of the intermediates ·H and ·Br, and therefore to set their net rates of change of concentration equal to zero:

$$\mathrm{d[H]/d}t = k_\mathrm{b}[\mathrm{Br}][\mathrm{H_2}] - k'_\mathrm{b}[\mathrm{H}][\mathrm{Br_2}] - k_\mathrm{c}[\mathrm{H}][\mathrm{HBr}] = 0,$$

$$\mathrm{d[Br]/d}t = 2k_\mathrm{a}[\mathrm{Br_2}] - k_\mathrm{b}[\mathrm{Br}][\mathrm{H_2}] + k'_\mathrm{b}[\mathrm{H}][\mathrm{Br_2}] + k_\mathrm{c}[\mathrm{H}][\mathrm{HBr}] - 2k_\mathrm{d}[\mathrm{Br}]^2 = 0.$$

Solving these two simultaneous equations for [H] and [Br], and substituting the solutions into eqn (29.1.3), gives

$$\mathrm{d[HBr]/d}t = \frac{2k_\mathrm{b}(k_\mathrm{a}/k_\mathrm{d})^{\frac{1}{2}}[\mathrm{H_2}][\mathrm{Br_2}]^{\frac{3}{2}}}{[\mathrm{Br_2}] + (k_\mathrm{c}/k'_\mathrm{b})[\mathrm{HBr}]}. \qquad (29.1.4)$$

‡ The general method of solving differential equations numerically is to approximate differentials by small but finite differences. Suppose we have the differential equation $\mathrm{d}f/\mathrm{d}x = g(x)$, then instead of $\mathrm{d}f = g(x)\,\mathrm{d}x$ we write $f(x + \delta x) - f(x) = g(x)\,\delta x$, with δx a small increment of x. Then

$$f(x + \delta x) = f(x) + g(x)\,\delta x.$$

This is exact when δx is infinitesimal. Now we just move along the x coordinate in steps of magnitude δx, using the known initial value of f for the first step, and evaluating $g\,\delta x$ at that initial value of x, which gives the value of f at $x + \delta x$; then we use this new value of f as the initial value for the next step together with the value of $g\,\delta x$, at the new value of x, to calculate f two steps away from the origin, and so on, until we have reached the value of x of interest. This is easy to implement on a microcomputer, and may be extended to the case where the values to use in one equation are determined by another (i.e. when the differential equations are *coupled*).

In the case of eqn (29.1.4), for example, we write $[\mathrm{H_2}] = a - x$, $[\mathrm{Br_2}] = a - x$ (assuming equal concentrations initially), and $[\mathrm{HBr}] = 2x$ (assuming that the intermediates are at such low concentrations that they can be neglected). Then, with $y = x/a$, the rate law becomes

$$\mathrm{d}y/\mathrm{d}t = \frac{c_1(1-y)^{\frac{5}{2}}}{1 + c_2 y}$$

where $c_1 = \frac{1}{2}ka^{\frac{1}{2}}$ and $c_2 = 2k' - 1$. This may be integrated by stepping along the t axis and calculating successive values of y.

This is the same as the empirical law, and the two empirical rate constants can be identified as

$$k = 2k_b(k_a/k_d)^{\frac{1}{2}}, \qquad k' = k_c/k_b'. \tag{29.1.5}$$

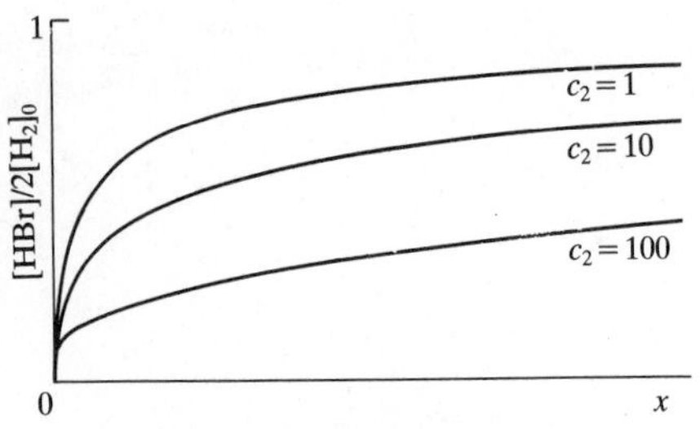

Fig. 29.1. The numerical integration of the HBr rate law, eqn (29.1.4), can be used to explore how the concentration of HBr changes with time. These runs began with stoichiometric proportions of H_2 and Br_2; $x = c_1 t$; $c_1 = \frac{1}{2}k[H_2]_0$, $c_2 = 2k' - 1$.

Whereas classically this was the end of the calculation, we can now make use of computers at this stage even though we did not use them to find the rate law, and go on to integrate this *approximate* rate law numerically, as explained in the footnote. This lets us predict the time dependence of the HBr concentration. Some typical results are shown in Fig. 29.1: they show how the rate of formation of products is slow when the inhibition step is important.

Some chain reactions turn out to give rise to simple rate laws, which underlines yet again the care that has to be taken when analysing kinetic data. For example, the dehydrogenation of ethane to ethene (E) follows first-order kinetics:

$$CH_3CH_3(g) \rightarrow CH_2{=}CH_2(g) + H_2(g), \qquad d[E]/dt = k[CH_3CH_3(g)], \tag{29.1.6}$$

and the 'obvious' interpretation is that the reaction occurs by a Lindemann–Hinshelwood mechanism. Further experiments, however, show that methyl and ethyl radicals occur during the reaction, and so a chain mechanism is involved. The *Rice–Herzfeld mechanism* has been proposed:

(a) *Initiation:*

$$CH_3CH_3 \rightarrow 2CH_3\cdot, \qquad d[CH_3]/dt = 2k_a[CH_3CH_3].$$

(b) *Propagation:*

$$\cdot CH_3 + CH_3CH_3 \rightarrow CH_4 + CH_3CH_2\cdot, \qquad d[CH_3]/dt = -k_b[CH_3][CH_3CH_3],$$

$$CH_3CH_2\cdot \rightarrow CH_2{=}CH_2 + H\cdot, \qquad d[E]/dt = k_b'[CH_3CH_2],$$

$$\cdot H + CH_3CH_3 \rightarrow H_2 + CH_3CH_2\cdot, \qquad d[CH_3CH_2]/dt = k_b''[H\cdot][CH_3CH_3].$$

(c) *Termination:*

$$\cdot H + CH_3CH_2\cdot \rightarrow CH_3CH_3, \qquad d[CH_3CH_2]/dt = -k_c[H][CH_3CH_2].$$

When the steady state approximation is used and certain conditions about the sizes of the rate-coefficients are fulfilled (Problem 29.2), the rate law becomes

$$d[E]/dt = (k_a k_b' k_b''/k_c)^{\frac{1}{2}}[CH_3CH_3],$$

in agreement with the observed first-order kinetics.

29.1 (b) Polymerization

There are two general types of polymerization reaction, chain polymerization and stepwise polymerization.

In *chain polymerization* an activated monomer (M) attacks another monomer, links to it, then that unit attacks another monomer, and so on, so that the process results in the rapid growth of an individual polymer chain for each activated monomer. Examples include the *addition polymerizations* of ethene, methyl methacrylate, and styrene. There are three basic reaction steps:

(a) *Initiation:* $M \rightarrow M\cdot$, $\quad d[M\cdot]/dt = k_i[M][I]$, where I is the initiator; we have shown a reaction in which a radical is produced, but in some polymerizations the initiation step leads to the formation of a cation or an anion chain carrier.

(b) *Propagation:* $M\cdot + M \rightarrow M_2\cdot$

$$M_2\cdot + M \rightarrow M_3\cdot$$

$$\cdots$$

$$M_{n-1}\cdot + M \rightarrow M_n\cdot.$$

(c) *Termination:* $M_n\cdot + M_m\cdot \rightarrow M_{n+m}.$

If we suppose that the rate of termination is independent of the length of the chain, its rate law can be written

$$d[R]/dt = -2k_t[R]^2,$$

where [R] is the total radical concentration. In practice, other termination steps may intervene, and side reactions occur, such as *chain transfer*, in which the reaction $R\cdot + M \rightarrow R + M\cdot$ initiates a new chain at the expense of the one currently growing.

Throughout the main part of the polymerization the total radical concentration is approximately constant (but the chain lengths are growing), and so we can use the steady state assumption to write

$$d[R]/dt = k_i[M][I] - 2k_t[R]^2 = 0.$$

The steady state concentration of radical chains is therefore

$$[R] = \{(k_i/2k_t)[M][I]\}^{\frac{1}{2}}. \tag{29.1.7}$$

The rate of propagation of the chains (the rate at which the monomer is consumed) is

$$d[M]/dt = -k_P[R][M].$$

Therefore,

$$d[M]/dt = -k_P(k_i/2k_t)^{\frac{1}{2}}[I]^{\frac{1}{2}}[M]^{\frac{3}{2}}. \tag{29.1.8}$$

The *kinetic chain length* (symbol: χ) is defined as the ratio of the rate of propagation to the rate of initiation:

$$\chi = \textit{Propagation rate/initiation rate}. \tag{29.1.9}$$

In the steady-state approximation the initiation rate is equal to the termination rate, and we can write

$$\begin{aligned}\chi &= k_P(k_i/2k_t)^{\frac{1}{2}}[I]^{\frac{1}{2}}[M]^{\frac{3}{2}}/k_i[M][I] \\ &= \{k_P/(2k_tk_i)^{\frac{1}{2}}\}[M]^{\frac{1}{2}}/[I]^{\frac{1}{2}}.\end{aligned}$$

Furthermore, the average polymer length is long when the termination steps occur more slowly than the propagation steps; if they are equal in rate, then the polymers do not grow and the average length is about unity. Therefore, the polymer length can be identified with the kinetic chain length, and so we see from the last equation that the lower the rate of initiation (as controlled by the rate constant k_i and the initiator concentration [I]), the longer the average chain length, and therefore the higher the RMM of the polymer.

Some of the consequences of RMM were explored in Chapter 25: now we see how we can exercise kinetic control over them.

The second type of polymerization is *stepwise polymerization* in which the polymer is built up by a series of stepwise reactions, often with each step involving the elimination of a small molecule (typically H_2O). This is the mechanism for the production of polyamides (nylon):

$$\begin{aligned}&H_2N\text{—}(CH_2)_6\text{—}NH_2 + HOOC\text{—}(CH_2)_4\text{—}COOH\\ &\quad\rightarrow H_2N\text{—}(CH_2)_6\text{—}NH\text{—}CO\text{—}(CH_2)_4\text{—}COOH + H_2O\ldots\\ &\quad\rightarrow H\text{+}NH\text{—}(CH_2)_6\text{—}NH\text{—}CO\text{—}(CH_2)_4\text{—}CO\text{+}_nOH.\end{aligned}$$

Polyesters and polyurethanes are formed similarly (the latter without elimination). A polyester, for example, can be regarded as the result of the stepwise condensation of a hydroxyacid HO . . . COOH. We shall consider the formation of a polyester from the monomer A—B, where A is the acid group and B the base group, and show how to use a reaction scheme to predict RMM distributions.

We assume that the rate constant for the condensation is independent of the chain length. The condensation can be expected to be second-order in the concentration of one of the functional groups (e.g. A):

$$d[A]/dt = -k[A]^2.$$

This will be the case if the polymerization is acid-catalysed by a constant concentration of added acid: if it is acid-catalysed by the monomer itself, then the rate law will be third-order in [A]. The solution of this rate law, eqn (28.2.17), is

$$[A] = [A]_0/(1 + kt[A]_0). \qquad (29.1.10)$$

The fraction of A groups that have condensed at time t is

$$p = ([A]_0 - [A])/[A]_0, \qquad (29.1.11)$$

and so from eqn (29.1.10)

$$1/(1-p) = 1 + kt[A]_0. \qquad (29.1.12)$$

The next step is to relate p to the details of the polymer chain length. p is equal to the probability that a group (e.g. A) has formed a link to another molecule (at the start of the reaction $p=0$; at the end $p=1$). The probability that a polymer consists of n monomers (i.e. that it is an n-mer) is therefore equal to the probability that it has $n-1$ reacted A groups *and* one unreacted A group. The former probability is p^{n-1}; the latter probability is equal to $1-p$; therefore, the total probability of finding an n-mer is

$$P_n = p^{n-1}(1-p). \qquad (29.1.13)$$

This is the key equation for predicting RMM values.

First, check that the probability is correctly normalized. That is, the total probability of selecting any n-mer, for n from 1 to ∞, must be unity:

$$\sum_{n=1}^{\infty} P_n = (1-p)\sum_{n=1}^{\infty} p^{n-1} = \{(1-p)/p\}\sum_{n=1}^{\infty} p_n$$

$$= \{(1-p)/p\}\left\{\left(\sum_{n=0}^{\infty} p^n\right) - 1\right\} = 1,$$

where we have used

$$\sum_{n=0}^{\infty} p^n = 1 + p + p^2 + \ldots = 1/(1-p).$$

Next, calculate the average chain length (χ). This is the average value of n, which is therefore the sum of ns, each one weighted by the probability that it occurs, eqn (26.1.6):

$$\chi = \langle n \rangle = \sum_{n=1}^{\infty} nP_n = (1-p)\sum_{n=1}^{\infty} np^{n-1}$$

$$= (1-p)(\mathrm{d}/\mathrm{d}p)\sum_{n=0}^{\infty} p^n = (1-p)(\mathrm{d}/\mathrm{d}p)\{p/(1-p)\}$$

$$= 1/(1-p). \tag{29.1.14}$$

The shape of this function is drawn in Fig. 29.2. It is clear that we need p very close to unity in order to have high molecular weight polymers. By combining this expression with the solution to the rate law, eqn (29.1.12), we arrive at the simple relation

$$\chi = 1 + kt[\mathrm{A}]_0 \tag{29.1.15}$$

for the growth of the average length of the polymer chain.

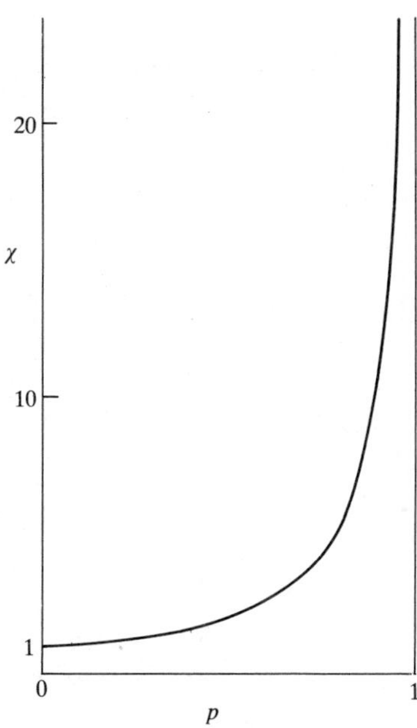

Fig. 29.2. The average chain length of a polymer as a function of the fraction of reacted monomers (p). Note that p must be very close to unity for the chains to be long.

Example 29.1

Deduce an expression for the mass-average RMM of the condensation polymer.

• *Method*. The mass-average RMM is defined in eqn (25.1.23). The mass of an n-mer is nM_{mon}, where M_{mon} is the mass of the monomer (we neglect small end-of-chain effects). The mass fraction in the sample is $nM_{\mathrm{mon}}P_n/M$, where M is the total mass of the sample (the sum over $nM_{\mathrm{mon}}P_n$). Evaluate the sums using the expression for P_n in eqn (29.1.13). Use the differentiation trick for relating sums of np^n to sums of p^n.

• *Answer*. The total mass is

$$M = M_{\mathrm{mon}}\sum_n nP_n = M_{\mathrm{mon}}\chi = M_{\mathrm{mon}}/(1-p).$$

The mass-weighted RMM is therefore

$$\langle M_{\mathrm{r}}\rangle_M = (1/M)\sum_n M_n M_{\mathrm{r},n} = \{(1-p)/M_{\mathrm{mon}}\}\sum_n \{nM_{\mathrm{mon}}P_n\}\{nM_{\mathrm{mon}}\}$$

$$= M_{\mathrm{mon}}(1-p)\sum_n n^2P_n = M_{\mathrm{mon}}(1-p)^2\sum_n n^2p^{n-1}.$$

$$= M_{\mathrm{mon}}(1-p)^2\{(1+p)/(1-p)^3\} = M_{\mathrm{mon}}(1+p)/(1-p).$$

• *Comment*. The mass-weighted mean RMM is determined by one of the methods outlined in Chapter 25 (e.g. light scattering).

• *Exercise*. Evaluate (a) the number-average RMM and (b) the ratio of mass-weighted and number-weighted RMMs. [(a) $M_{\mathrm{mon}}/(1-p)$, (b) $1+p$]

29.1 (c) Explosions

Some reactions proceed explosively, either by design or by accident. There are two types of explosion. A *thermal explosion* is due to the rapid increase of reaction rate with temperature. If the energy of an exothermic explosion cannot escape, the temperature of the reaction system rises, and the reaction goes faster. This results in a faster rise of temperature, and so the reaction goes even faster . . . catastrophically fast. A *chain-branching explosion* may occur when there are chain branching steps in a reaction, for then the number of radicals propagating the chain increases and the rate of reaction may cascade into an explosion.

An example of both types of explosion is provided by the hydrogen/oxygen reaction, $2H_2(g) + O_2(g) \rightarrow 2H_2O(g)$. Although the net reaction is very simple, the mechanism is very complex and has not yet been fully elucidated. It is known that a chain reaction is involved, and that the *chain carriers* (the radicals involved in the propagation steps) include H, O, HO, and HO_2. Some steps are:

Initiation: $H_2 + O_2 \rightarrow HO_2\cdot + H\cdot$

Propagation: $H_2 + HO_2\cdot \rightarrow HO\cdot + H_2O$

$H_2 + HO\cdot \rightarrow H\cdot + H_2O$

$H\cdot + O_2 \rightarrow HO\cdot + \cdot O\cdot$ (*branching*)

$\cdot O\cdot + H_2 \rightarrow HO\cdot + H\cdot$ (*branching*).

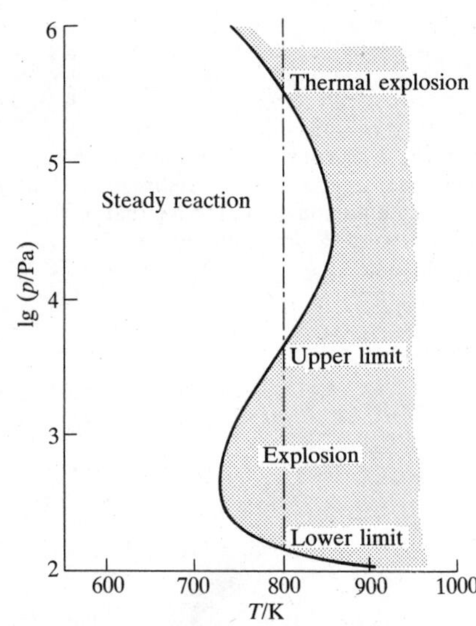

Fig. 29.3. The explosion limits for the $H_2 + O_2$ reaction. In the explosive regions the reaction proceeds explosively when heated homogeneously.

The last two branching steps are potentially capable of leading to a chain-branching explosion. The occurrence of an explosion, however, depends on the temperature and pressure of the system, and the explosion regions for the reaction are shown in Fig. 29.3. At very low pressures the system is outside the explosion region and the mixture reacts smoothly. At these very low pressures the chain carriers produced in the branching steps can reach the walls of the container where they combine (with an efficiency that depends on the composition of the walls). Raising the pressure (along the broken line in the illustration) takes the system through the *first explosion limit* (but note that the temperature must be greater than about 730 K). The mixture then explodes because the chain carriers react before reaching the walls and the branching reactions are explosively efficient. When the pressure is higher and above the *second explosion limit* the reaction is smooth. Now the pressure is so great that the radicals produced in the branching reaction combine in the body of the gas, and reactions such as $H + O_2 \rightarrow HO_2$ can occur, the excess energy being removed by another molecule colliding simultaneously in a three-particle collision:

$$H + O_2 + M \rightarrow HO_2 + M^*.$$

At low pressures three-particle collisions are unimportant and reactions such as this cannot occur; at higher pressures, when they become important, the explosive propagation of the chain by the radicals produced in the branching step is inhibited because HO_2 is not a direct participant in the branching. If the pressure is increased still further (above the *third explosion limit*) the reaction rate increases so much that a thermal explosion occurs.

29.2 Photochemical reactions

Many reactions can be initiated by the absorption of light. The most important of all are the photochemical processes that capture the sun's

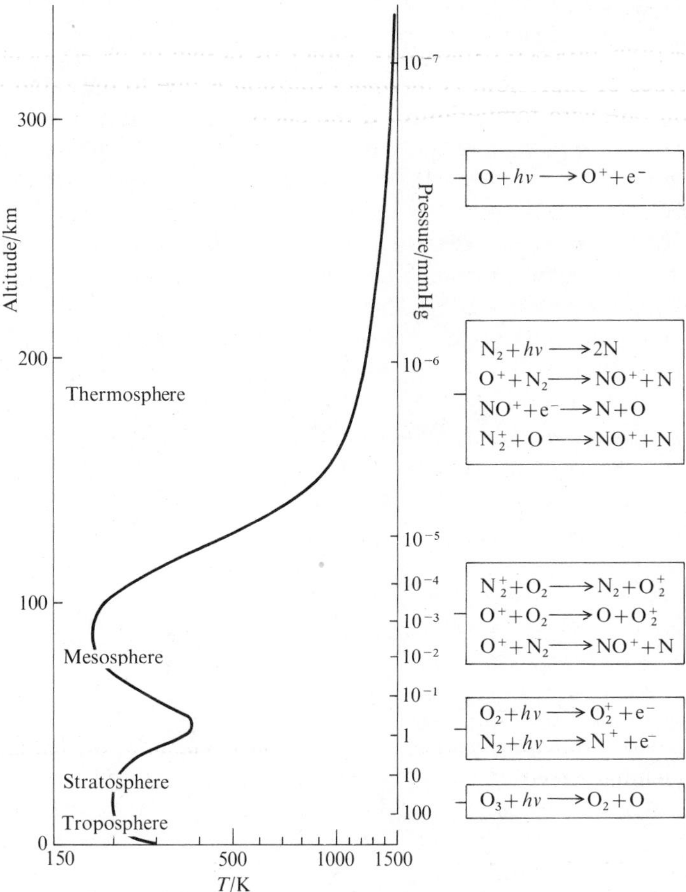

Fig. 29.4. The temperature profile through the atmosphere and some of the reactions that occur. The temperature peak at about 50 km is due to the absorption of solar radiation by the O_2 and N_2 ionization reactions.

radiant energy. These range from the ones that lead to the heating of the atmosphere during the daytime by absorption in the ultraviolet region as a result of reactions (such as those depicted in Fig. 29.4) to the absorption of red and blue light by chlorophyll and its subsequent use of the energy to bring about the synthesis of carbohydrates. Photochemical processes are also the basis of undesirable effects, such as the degradation of polymers and the formation of smog, and of leisure activities, such as photography. Without photochemical processes the world would be simply a warm, sterile, rock. A summary of the processes that can occur following photochemical excitation is given in Box 29.1 (*p*. 722).

29.2 (a) Quantum efficiency

A molecule acquires enough energy to react by absorbing photons. The *Einstein–Stark Law* states that *one photon is absorbed by each molecule responsible for the primary photochemical process*. This law is not valid under all conditions, because in a laser beam the density of photons is so high that a single molecule may absorb several. Indeed, in the laser isotope separation techniques we describe below, a single molecule may absorb dozens of infrared photons from an intense CO_2 laser beam. Under normal conditions the Einstein–Stark Law is valid, but it does not imply that only one reactant molecule is consumed for each photon absorbed. This is

Box 29.1 Photochemical processes

In the following, S denotes a singlet state and T a triplet state; A, B, and M are arbitrary.

Primary absorption: $S \rightarrow S^*$ followed by vibrational and rotational relaxation

Physical processes:

Fluorescence: $S^* \rightarrow S + h\nu$
Collision-induced emission: $S^* + M \rightarrow S + M + h\nu$
Stimulated emission: $S^* + h\nu \rightarrow S + 2h\nu$
Intersystem crossing (ISC): $S^* \rightarrow T^*$
Phosphorescence: $T^* \rightarrow S + h\nu$
Internal conversion (IC): $S^* \rightarrow S'$
Singlet electronic energy transfer: $S^* + S \rightarrow S + S^*$
Energy pooling: $S^* + S^* \rightarrow S^{**} + S$
Triplet electronic energy transfer: $T^* + S \rightarrow S + T^*$
Triplet–triplet absorption: $T^* + h\nu \rightarrow T^{**}$

Ionization:

Penning ionization: $A^* + B \rightarrow A + B^+ + e^-$
Dissociative ionization: $A^* + B\text{–}C \rightarrow A\text{–}B^+ + C + e^-$
Collisional ionization: $A^* + B \rightarrow A^+ + B + e^-$ (or B^-)
Associative ionization: $A^* + B \rightarrow AB^+ + e^-$

Chemical processes:

Dissociation: $A\text{–}B^* \rightarrow A + B$
Addition or insertion: $A^* + B \rightarrow AB$
Abstraction or fragmentation: $A^* + B \rightarrow C + D$
Isomerization: $A \rightarrow A'$
Dissociative excitation: $A^* + C\text{–}D \rightarrow A + C^* + D$

because, although one photon dissociates (or excites) one reactant molecule, a subsequent chain reaction can lead to the removal of several reactant molecules (and the formation of several product molecules, possibly of different kinds). The ratio of the number of molecules reacting to the number of photons absorbed is called the *quantum efficiency* (symbol: Φ) of the reaction:

$$\Phi = \frac{\textit{Number of reactant molecules destroyed}}{\textit{Number of photons absorbed}}. \qquad (29.2.1)$$

In the photolysis of HI, for example, the processes are $HI + h\nu \rightarrow H + I$, $H + HI \rightarrow H_2 + I$, $2I \rightarrow I_2$, and so the quantum efficiency is two because the absorption of one photon leads to the destruction of two HI molecules. In a chain reaction the quantum efficiency may be very large, and values of about 10^4 are common. In these cases the chain reaction acts as a chemical amplifier of the initial absorption step.

Example 29.2

The quantum efficiency for the formation of ethene from heptan-4-one with 313 nm light is 0.21. How many molecules of the ketone per second, and how many moles per second, are destroyed when the sample is irradiated with a 50 W 313 nm source under conditions of total absorption?

- *Method.* Calculate the number of photons emitted by the lamp per second; all are absorbed (by assertion); the number of molecules destroyed per second is the number of photons absorbed multiplied by the quantum efficiency.

• *Answer*. The energy of a 313 nm photon is

$$hc/\lambda = 6.35 \times 10^{-19}\ \text{J}.$$

A 50 W source therefore generates photons at a rate of

$$(50\ \text{J s}^{-1})/(6.35 \times 10^{-19}\ \text{J}) = 7.9 \times 10^{19}\ \text{s}^{-1}.$$

The number of heptan-4-one molecules destroyed per second is therefore 0.21 times this quantity, or $1.7 \times 10^{19}\ \text{s}^{-1}$. This corresponds to $2.8 \times 10^{-5}\ \text{mol s}^{-1}$.

• *Comment*. Note that the quantum efficiency depends on the wavelength of light used. 1 mol of photons is called an *einstein*.

• *Exercise*. The quantum efficiency for another molecule at 290 nm is 0.30. For what length of time must irradiation with a 100 W source continue in order to destroy 1 mol of molecules?

[3.8 h]

As an example of how to incorporate the photochemical activation step into a reaction scheme, consider the photochemical activation of the $H_2 + Br_2 \rightarrow 2HBr$ reaction. In place of eqn (29.1.2a) of the thermal reaction we have

$$Br_2 \xrightarrow{h\nu} 2Br, \qquad d[Br]/dt = 2I_{\text{absorbed}}, \tag{29.2.2}$$

where I_{absorbed} is the number of photons of the appropriate frequency absorbed per unit time. I_{absorbed} therefore takes the place of $k_a[Br_2]$ in the thermal reaction scheme, and so from eqn (29.1.4) we can write

$$d[HBr]/dt = \frac{2k_b(1/k_d)^{\frac{1}{2}}[H_2][Br_2]I^{\frac{1}{2}}_{\text{absorbed}}}{[Br_2] + (k_c/k_b')[HBr]}. \tag{29.2.3}$$

This predicts that the reaction rate should depend on the square root of the absorbed light intensity, which is confirmed experimentally.

29.2 (b) Isotope separation

Laser radiation is so monochromatic that it may be absorbed by one molecule but not by an isotopically substituted version of the same molecule because its frequency lies between the latter's absorption lines. This is the basis of *laser isotope separation*, in which molecules containing different isotopes are irradiated with an intense beam of infrared photons from a CO_2 laser. The absorption of the first few infrared photons depends critically on the match between their frequency and the vibration–rotation energy separations of the molecule (which depend on the masses of the atoms, and hence differ between different isotopic species). However, after the initial excitation has occurred, several dozen more photons may be absorbed, leading to the dissociation and reaction of the molecule and its fragments. Since the initial absorption is the doorway to the later dissociation, the isotopes can be separated. An alternative approach is to excite the vibrations of a selected isotopic version of a molecule (e.g. UF_6) with an infrared laser beam, and then to use an ultraviolet beam to photodissociate that vibrationally excited species.

29.2 (c) Photosensitization

When a molecule does not absorb directly, its reactions can be *photosensitized* if another absorbing molecule is present because the latter may be able to transfer its energy during a collision. An example is the reaction often used to generate atomic hydrogen by irradiation of hydrogen containing a trace of mercury using 254 nm light from a mercury discharge lamp. The mercury atoms are excited by resonant absorption of the radiation, and then collide with hydrogen molecules. Two reactions then take place:

$$Hg^*(g) + H_2(g) \rightarrow Hg(g) + 2H(g); \qquad Hg^*(g) + H_2(g) \rightarrow HgH(g) + H(g),$$

the latter, which is monitored by detecting the HgH spectroscopically, accounting for 67% of the process (but 76% when deuterium is used in place of hydrogen). This reaction is the initiation step for other mercury-photosensitized reactions, such as the synthesis of formaldehyde (HCHO) from carbon monoxide and hydrogen:

$$H + CO \rightarrow HCO, \qquad HCO + H_2 \rightarrow HCHO + H, \qquad 2HCO \rightarrow HCHO + CO.$$

Photosensitization also plays an important role in solution kinetics, and molecules containing the carbonyl chromophore, such as benzophenone ($C_6H_5{\cdot}CO{\cdot}C_6H_5$) are often used to trap the incident light and to transfer it to some potentially reactive species.

29.3 Catalysis

If the activation energy of a reaction is high, at normal temperatures only a small proportion of molecular encounters are sufficiently energetic to result in reaction. A *catalyst* lowers the activation energy of the reaction (often by providing an alternative path that avoids the slow, rate-determining step of the uncatalysed reaction), and results in a higher reaction rate at the same temperature. (We saw in Section 9.2 that a catalyst does not change the position of equilibrium.) Catalysts can be very effective; for instance, the activation energy for the decomposition of hydrogen peroxide in solution is 76 kJ mol^{-1}, and the reaction is slow at room temperature. When a little iodide is added the activation energy falls to 57 kJ mol^{-1}, and the rate increases by a factor of 2000. Enzymes, biological catalysts, are very specific and can have a dramatic effect on the reaction they control. The activation energy for the acid hydrolysis of sucrose is 107 kJ mol^{-1}, but the enzyme saccharase reduces it to 36 kJ mol^{-1}, corresponding to an acceleration of the reaction by a factor of 10^{12} at blood temperature (310 K).

There are two classes of catalyst. A *homogeneous catalyst* is in the same phase as the reaction mixture (e.g. an acid added to an aqueous solution). A *heterogeneous catalyst* is in a different phase (e.g. a solid catalyst for a gas-phase reaction). We examine heterogeneous catalysis in Chapter 31, and consider only homogeneous catalysis here.

29.3 (a) Homogeneous catalysis

Some idea of the mode of action of homogeneous catalysts can be obtained by examining the kinetics of the bromine-catalysed decomposition of hydrogen peroxide, the net reaction being $2H_2O_2(aq) \rightarrow 2H_2O(aq) + O_2(g)$.

The reaction is believed to proceed through the following steps:

(a) $$Br_2 + 2H_2O \rightarrow HOBr + H_3O^+ + Br^-,$$

(b) $$H_3O^+ + HOOH \rightleftharpoons HOOH_2^+ + H_2O,$$

$$K = \{[HOOH_2^+][H_2O]/[HOOH][H_3^+O]\}_{eq},$$

(c) $$HOOH_2^+ + Br^- \rightarrow HOBr + H_2O,$$

$$d[HOBr]/dt = k[HOOH_2^+][Br^-],$$

(d) $$HOBr + HOOH \rightarrow H_3O^+ + O_2 + Br^-.$$

We take (c) to be the rate determining step, and (b) to be a pre-equilibrium. By combining (b) and (c) the rate law of the overall reaction is predicted to be

$$d[O_2]/dt = kK[HOOH][H_3^+O][Br^-],$$

in agreement with the observed dependence of the rate on the Br^- concentration and the pH of the solution. The observed activation energy is the activation energy corresponding to the effective rate coefficient kK; in the absence of Br_2 the reaction cannot proceed through the path set out above, and a different and much larger activation energy is observed.

Two important types of homogeneous catalysis are *acid catalysis* and *base catalysis*, and many organic reactions involve one or the other (and sometimes both). Acid catalysis is the transfer of a hydrogen ion to the substrate:

$$X + HA \rightarrow HX^+ + A^-, \qquad HX^+ \text{ then reacts.}$$

This is the primary process in the solvolysis of esters, keto–enol tautomerism, and the inversion of sucrose. Base catalysis is the transfer of a hydrogen atom from the substrate to a base:

$$XH + B \rightarrow X^- + BH^+, \qquad X^- \text{ then reacts.}$$

This is the primary step in the isomerization and halogenation of organic compounds, and of the Claisen and aldol reactions. The relation between the activities of these catalysts and their strengths (as measured by pK_a or pK_b, Section 12.4(c)) is a major link between physical and organic chemistry.

29.4 Autocatalysis and oscillating reactions

Odd things can happen when the products of a reaction (or some intermediate) can take part in its earlier stages.

29.4 (a) Autocatalysis

The phenomenon of *autocatalysis* is the increase of the rate of a reaction by the *products*. For example, in a reaction $A \rightarrow P$ it may be found that the rate law is $d[P]/dt = k[A][P]$, and so the rate increases as products are formed. (The reaction gets started because there are usually other reaction routes which lead to the formation of some P initially, which then takes part in the autocatalytic reaction proper.) An example is provided by two steps in the *Belousov–Zhabotinskii reaction* (BZ reaction) which will figure in discussions later in the section:

$$BrO_3^- + HBrO_2 + H_3O^+ \rightarrow 2BrO_2 + 2H_2O,$$
$$2BrO_2 + 2Ce(III) + 2H_3O^+ \rightarrow 2HBrO_2 + Ce(IV) + 2H_2O,$$

where the product $HBrO_2$ is a reactant in the first step.

Example 29.3

Set up and solve the rate equation for the A → P autocatalytic reaction referred to above.

• *Method*. Write $[A] = [A]_0 - x$, $[P] = [P]_0 + x$, and integrate the rate equation (by partial fractions) with $x = 0$ at $t = 0$.

• *Answer*. The rate law is

$$dx/dt = k([A]_0 - x)([P]_0 + x).$$

Integration by partial fractions, using

$$1/([A]_0 - x)([P]_0 + x) = \{1/([A]_0 + [P]_0)\}\{1/([A]_0 - x) + 1/([P]_0 + x)\}$$

gives

$$\{1/([A]_0 + [P]_0)\} \ln \{([P]_0 + x)[A]_0/([A]_0 - x)[P]_0\} = kt.$$

Hence

$$x/[P]_0 = (e^{at} - 1)/(1 + be^{at}),$$

with $a = ([A]_0 + [P]_0)k$ and $b = [P]_0/[A]_0$.

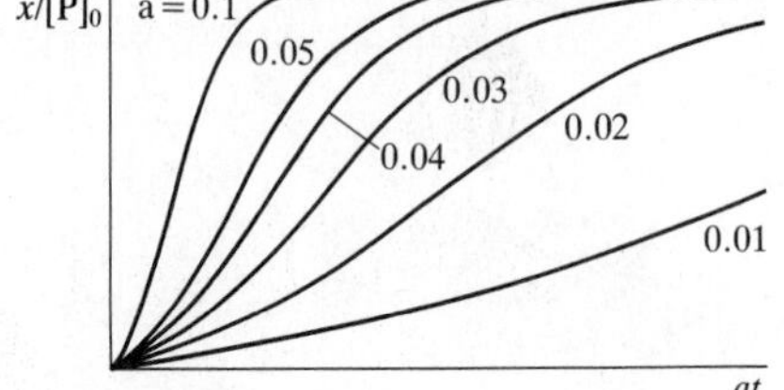

Fig. 29.5. The concentration of product during the autocatalysed A → P reaction discussed in *Example* 29.3 (using $b = 0.1$).

• *Comment*. The solution is plotted in Fig. 29.5. The rate of reaction is slow initially (little P present), then fast (when P and A are both present), and finally slow again (when A has disappeared).

• *Exercise*. At what time is the reaction rate a maximum? $[t_{max} = (1/ka) \ln (1/b)]$

The industrial importance of autocatalysis (which occurs in a number of reactions, such as oxidations) is that the rate of the reaction can be maximized by ensuring that the optimum concentrations of reactant and product are always present. This can be achieved in a stirred reactor (a CSTR, Section 28.3(b)) because the flow rates can be adjusted and the stirring ensures that product is continuously mixed with the incoming reactant.

29.4 (b) Oscillations

Another consequence of autocatalysis is the possibility that the reaction will *oscillate* (this is like the effect of positive feedback in electrical circuits), and that the concentrations of reactants and products vary periodically either in space or in time, Fig. 29.6. An autocatalytic reaction of the *Lotka–Volterra form* illustrates how these oscillations may occur (although the actual chemical examples that have been discovered so far have a different mechanism, as we shall see). The Lotka–Volterra mechanism is as follows:

(a)	$A + X \rightarrow 2X$,	$d[A]/dt = k_a[A][X]$,	(29.4.1a)
(b)	$X + Y \rightarrow 2Y$,	$d[X]/dt = -k_b[X][Y]$,	(29.4.1b)
(c)	$Y \rightarrow B$,	$d[B]/dt = k_c[Y]$.	(29.4.1c)

Steps (a) and (b) are autocatalytic. The concentration of A is held constant by supplying it to the reaction vessel as needed. (B plays no part in the reaction once it has been produced, and so it is unnecessary to remove it; in practice it would normally be removed.) This leaves [X] and [Y], the concentrations of the intermediates, as variables. Note that we are considering a *steady state*, which is maintained by the flow of A into the reactor. This steady state must not be confused with the steady state approximation made earlier: in the present case we shall solve the rate equations *exactly* for the variable concentrations of X and Y, but hold [A] at an arbitrary but constant value. Since the steady state is artificially maintained, we are dealing with a *non-equilibrium condition* of the reaction, and we should be prepared for strange things to happen.

The Lotka–Volterra equations can be solved numerically in the way described previously, and the results can be depicted in two ways. One way is to plot [X] and [Y] against time, Fig. 29.7. The same information can be displayed more succinctly by plotting one concentration against the other: this gives the series of closed curves shown in Fig. 29.8.

The periodic variation of the concentrations of the intermediates can be explained as follows. At some stage there may be only little X, but reaction (a) provides more, and it autocatalyses the production of even more X. There is therefore a surge of X. However, as X is formed, reaction (b) can begin. It occurs slowly initially because [Y] is small, but autocatalysis leads to a surge of Y. This surge, though, removes X, and so reaction (a) slows, and less X is produced. Since less X is now available reaction (b) slows. As less Y becomes available to remove X, X has a chance to surge forward again, and so on.

Another very interesting set of equations is called the *brusselator* (because it is an oscillator that originated with Ilya Prigogine's group in Brussels):

(a)	$A \rightarrow X,$	$d[X]/dt = k_a[A]$	(29.4.2a)
(b)	$2X + Y \rightarrow 3X,$	$d[Y]/dt = -k_b[X]^2[Y]$	(29.4.2b)
(c)	$B + X \rightarrow Y + C,$	$d[Y]/dt = k_c[B][X]$	(29.4.2c)
(d)	$X \rightarrow D,$	$d[X]/dt = -k_d[X].$	(29.4.2d)

The reactants (A and B) are maintained at constant concentration, and so the two variables are the concentrations of X and Y. These may be calculated by solving the rate equations numerically, and the results are plotted in Fig. 29.9. The interesting feature is that *whatever the initial concentrations of X and Y, the system settles down into the same periodic variation of concentrations*. The common trajectory is called a *limit cycle*, and its period depends on the values of the rate coefficients.

One of the first oscillating reactions to be reported and studied systematically was the BZ reaction mentioned earlier, which takes place in a mixture of potassium bromate, malonic acid, and a cerium(IV) salt in an acid solution. The mechanism has been elucidated by Richard Noyes, and involves 18 elementary steps and 21 different chemical species. The main features of this awesomely complex mechanism can be reproduced by the following *oregonator* (Noyes and his group worked in Oregon), where X stands for $HBrO_2$, Y for Br^-, and Z for $2Ce^{4+}$:

$$A + Y \rightarrow X, \quad X + Y \rightarrow C, \quad B + X \rightarrow 2X + Z, \quad 2X \rightarrow D, \quad Z \rightarrow Y, \qquad (29.4.3)$$

A, B, C, D being held constant (in the model). The oscillations arise in a

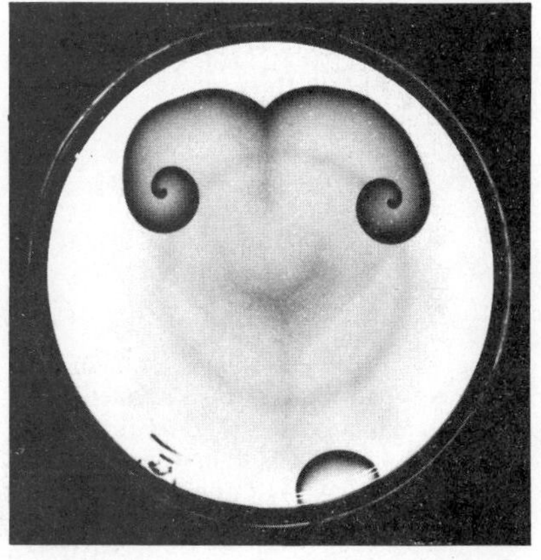

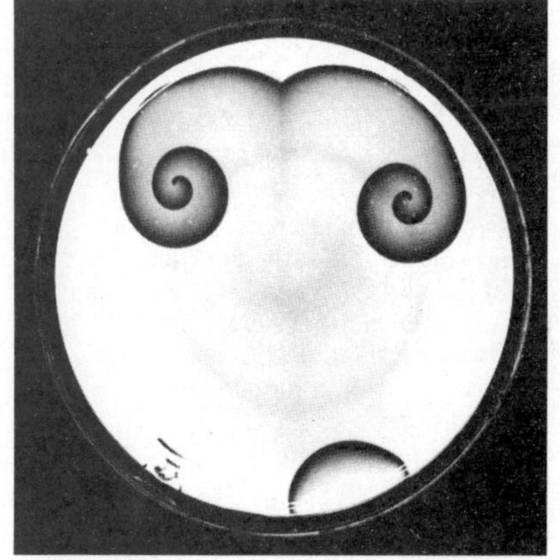

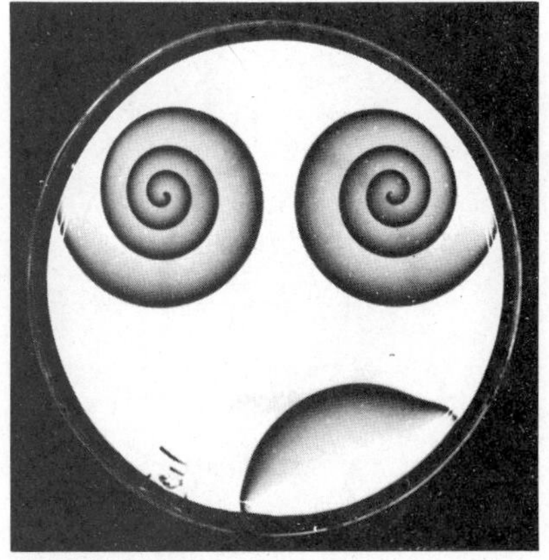

Fig. 29.6. Some reactions show oscillations in time; some show spatially periodic variations. This sequence of photographs shows the emergence of a spatial pattern.

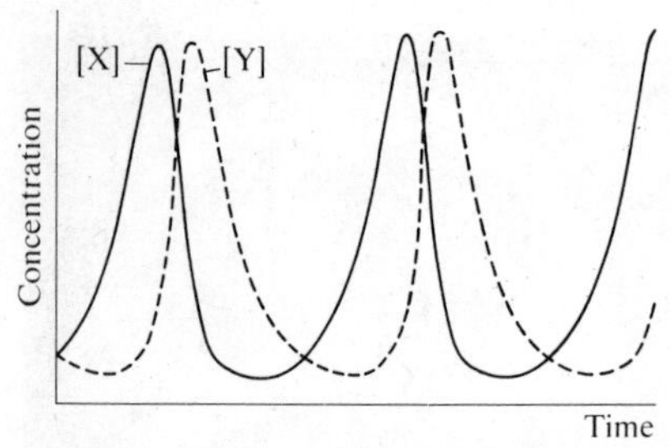

Fig. 29.7. The periodic variation of the concentrations of the intermediates X and Y in a Lotka–Volterra reaction, eqn (29.4.1). The system is in a steady state, but not at equilibrium.

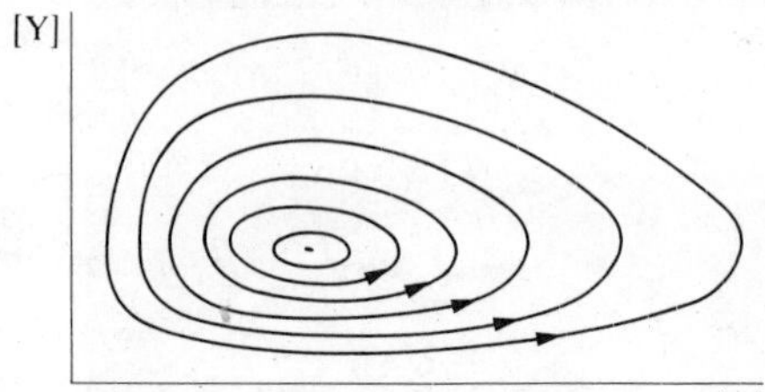

Fig. 29.8. An alternative representation of the periodic variation of [X] and [Y] is to plot one against the other. Then the system describes closed orbits, different ones being obtained with different starting conditions.

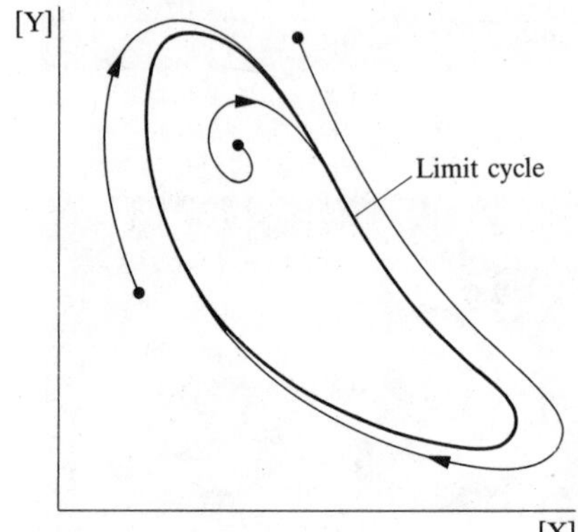

Fig. 29.9. Some oscillating reactions approach a closed trajectory whatever their starting conditions. The closed trajectory is called a *limit cycle*.

similar way to the brusselator, and can be traced to the autocatalysis in step 3 and the linkage between the reactions provided by the other steps.

Oscillating reactions are much more than a laboratory curiosity. While they are known to occur in only a few cases in industrial processes, there are many examples in biochemical systems, where a cell plays the role of a chemical reactor. Oscillating reactions, for example, maintain the rhythm of the heartbeat. They are also known to occur in the glycolytic cycle, in which one molecule of glucose is used to produce (through enzyme catalysed reactions involving ATP) two molecules of ATP. All the metabolites in the chain oscillate under some conditions, and do so with the same period but with different phases.

Such is the present interest in oscillating reactions that attempts have been made to discover the underlying causes more deeply than simply recognizing the role of autocatalysis. It appears that three conditions must be fulfilled in order to obtain oscillations:

(1) The reactions must be far from equilibrium.
(2) The reactions must have autocatalytic steps.
(3) The system must be able to exist in two steady states.

The last criterion is called *bistability*, and is a property that takes us well beyond what is familiar from the equilibrium properties of systems.

29.4 (c) Bistability

The property of bistability is a chemical version of supercooling, in which the temperature of a liquid may be lowered beneath its freezing point without it solidifying. Consider a reaction in which there are two intermediates X and Y. If the concentration of Y is at some high value in a reactor, and X is added, then the concentration of Y might decrease as shown by the upper line in Fig. 29.10. If X is at some high value, then as Y is added the reaction might result in the slow increase of Y as shown by the lower line. However, in each case a concentration may be reached at which the concentration will jump from one curve to the other (just as a supercooled liquid might suddenly solidify). The two curves represent the two stable states of the bistable system. Note that neither is an equilibrium state in the thermodynamic sense: they occur in steady states that are well removed from equilibrium, and the concentrations of X and Y represent the consequences of reactants continuously flowing into and of products flowing out of the reactor.

Now consider what happens when a third type of intermediate, Z, is present. Suppose Z reacts with both X and Y. In the absence of Z the flows of material might correspond to the stable state c on the upper curve of Fig. 29.11. However, as Z reacts with Y to produce X the state of the system moves along the curve (to the right, as Y decreases and X increases) until the sudden transition occurs to the lower curve. Then Z reacts with X and produces Y, which means that the composition moves to the left along the lower curve. There comes a point, however, when the concentration of X has been reduced so much, and that of Y has risen so much, that there is a sudden transition to the upper curve, when the process begins again. It is the leaping from one stable state to another that we see as the sudden surge or depletion of the concentration of a species, Fig. 29.12.

By studying the regions of concentrations and the rate constants for the individual steps of a reaction it is now becoming possible to predict the occurrence of bistable chemical systems and to anticipate the occurrence of oscillations. We are still far, however, from being able to use these ideas to account for gene expression, the patterns on tigers and butterflies, and the oscillation of cool flames, in all of which it is thought these processes play a part.

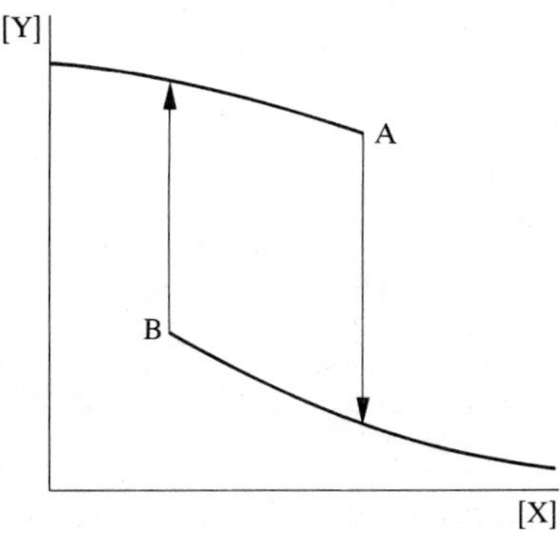

Fig. 29.10. A system showing *bistability*. As the concentration of X is increased (by adding it to a reactor) the concentration of Y decreases along the upper curve but at A drops sharply to a low value. If X is then decreased, the concentration increases along the lower curve, but rises sharply to a high value at B.

29.5 Fast reactions

All reactions are fast. At least, the individual steps of a reaction, when molecules rearrange in a unimolecular step or exchange atoms during a bimolecular encounter, occur on an atomic time scale and are complete in less than about 10^{-9} s. The slowness of the net reaction is due to the slowness with which molecules are activated or encounter, but even the net rate may become very fast when the activation barrier can be overcome readily and when the molecules encounter each other frequently.

In recent years notable advances have been made in the study of *fast reactions*, which we shall take to be reactions complete in less than about 1 s (and often very much less), and the present thrust of chemical kinetics is to ever shorter time-scales. With special laser techniques it is now possible to observe processes occurring in a few tens of femtoseconds ($1\text{ fs} = 10^{-15}$ s). Some of the techniques now available are designed to monitor concentrations and to measure rate coefficients. These include flash photolysis flow techniques, and relaxation methods. Others, including flash photolysis, shock tubes, and molecular beams, are used to explore the dependence of the reaction rate on the energy states of individual reactant molecules. In this section we consider only the first group, and see what kinetic information they provide. The second group are dealt with in the next chapter.

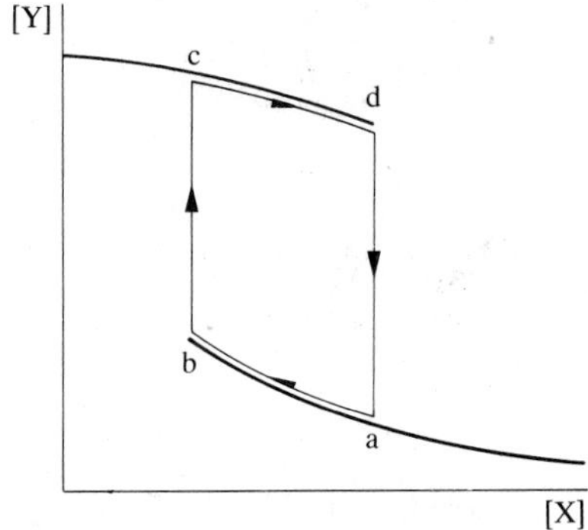

Fig. 29.11. Chemical oscillation in a bistable system occurs as a result of the effect of a third substance Z which can react with X to produce Y and with Y to produce X. As a result, the system switches periodically between the upper and lower curves.

29.5 (a) Flash photolysis

In flash photolysis the sample is exposed to a brief photolytic flash of light, and then the contents of the reaction chamber are monitored spectroscopically. Although discharge lamps can be used for flashes of about 10^{-5} s, most work is now done with lasers with flashes of about 10^{-9} s duration, but mode-locking (Section 19.4) has made picosecond pulses available ($1\text{ ps} = 10^{-12}$ s), and many studies are now carried out on that time-scale. The world record for pulse shortness at the time of writing stands at about 16 fs. Both emission and absorption spectroscopy may be used to monitor the reaction, and the spectra are observed electronically or photographically at a series of times following the flash.

Nanosecond flash photolysis, like its slower forerunner, has been applied to many problems. As an example we consider the photolysis of halogen molecules and their subsequent recombination. The photolytic flash causes $X_2 \rightarrow 2X$. The only fate for the atoms is recombination, but the energy released in $2X \rightarrow X_2$ must be removed for otherwise the newly formed X_2 would dissociate immediately. One way of removing the excess energy is for there to be a *three-body collision*: $X + X + M \rightarrow X_2 + M^*$. The third body M might be the wall of the container or some added gas. Nitrogen(II) oxide, NO, is an efficient third body, and we have already seen, eqn (28.3.24), that it plays a role in a pre-equilibrium. The same type of process, in which NO forms NOX which then collides with another X, can be expected here. In the latter collision the second X displaces the NO from its partner, and X_2 is formed. In other words, NO acts as a sticky third body. When the spectrum

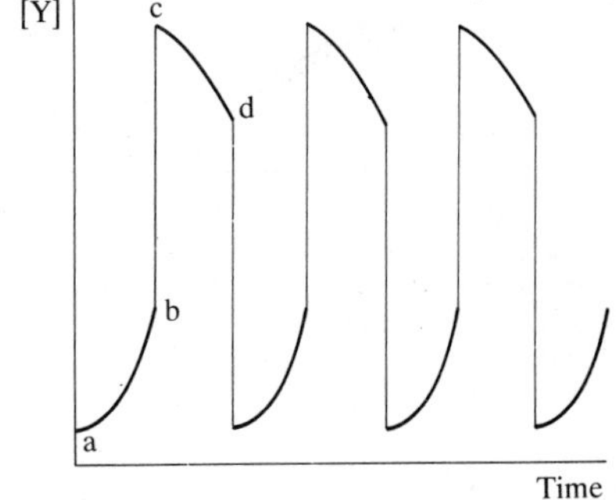

Fig. 29.12. The leaping from one branch to another of a bistable system appears to the observer as a periodic surge and depletion of concentration.

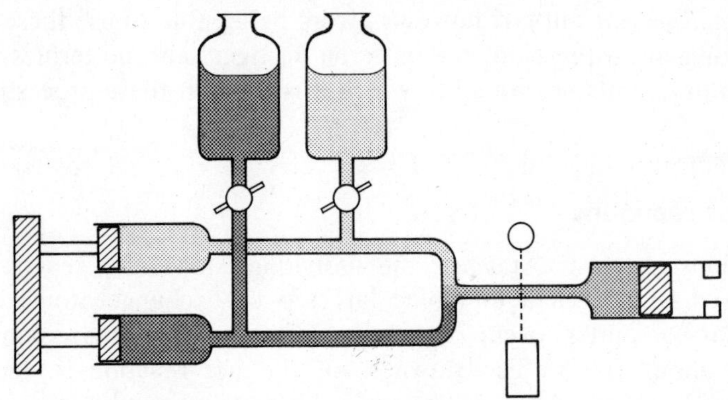

Fig. 29.13. In a *stopped-flow* experiment, the reactant solutions are driven quickly into the mixing chamber, and then the time dependence of the concentrations is observed (e.g. spectroscopically or electrochemically).

of the photolysed system is observed immediately after the photolytic flash, the presence of NOX can be detected, and the kinetic scheme constructed.

29.5 (b) Flow techniques

The idea behind conventional flow techniques was described in Section 28.1. The disadvantage mentioned there was that a large volume of reactant solution is necessary. This is particularly important for fast reactants, because in order to spread the reaction over a length of tube the flow must be rapid. The *stopped-flow technique* avoids this disadvantage. The apparatus is illustrated in Fig. 29.13. The two solutions are mixed very rapidly by injecting them into the reaction chamber. This chamber is fitted with a plunger that moves back as the liquids flood in, but comes up against a stop. The reaction then continues in the thoroughly mixed solution. Observations, which are usually spectroscopic, are made at a series of times after the mixing is complete.

The ability of the stopped-flow technique to study small samples means that it is very suitable for biochemical reactions, and it has been widely used to study the kinetics of enzyme action. As we saw in Section 28.3(f), the Michaelis–Menten mechanism of enzyme action leads to a rate law, eqn (28.3.33), which can be written

$$\mathrm{d}[\mathrm{P}]/\mathrm{d}t = k_{\mathrm{eff}}[\mathrm{E}]_0, \qquad k_{\mathrm{eff}} = k_{\mathrm{b}}[\mathrm{S}]/(K_{\mathrm{M}} + [\mathrm{S}]), \tag{29.5.1}$$

where $K_{\mathrm{M}} = (k_{\mathrm{b}} + k_{\mathrm{a}}')/k_{\mathrm{a}}$. The expression for the effective rate coefficient k_{eff} can be rearranged to

$$1/k_{\mathrm{eff}} = 1/k_{\mathrm{b}} + K_{\mathrm{M}}/k_{\mathrm{b}}[\mathrm{S}]. \tag{29.5.2}$$

It follows that a plot of $1/k_{\mathrm{eff}}$ against $1/[\mathrm{S}]$ will give the value of k_{b} and K_{M}, but not the values of the individual rate coefficients k_{b} and k_{a}' that appear in K_{M}. The stopped-flow technique can give the additional data needed, because the rate of formation of the enzyme–substrate complex can be found by monitoring its concentration after mixing enzyme and substrate. This gives k_{a}, and k_{a}' can then be found from the value of K_{M}.

29.5 (c) Relaxation methods

The term *relaxation* denotes the return of a system to equilibrium. It is used in chemical kinetics to indicate that some externally applied influence has shifted the equilibrium position of a reaction, normally very suddenly, and that the reaction is adjusting to the new equilibrium composition.

We take as an example a simple $\mathrm{A} \rightleftharpoons \mathrm{B}$ equilibrium, which is first-order in each direction. Under a certain set of conditions (e.g. temperature) the rate of change of [A] is

$$d[A]/dt = -k'_a[A] + k'_b[B]. \tag{29.5.3}$$

At equilibrium under these conditions, $d[A]/dt = 0$, and the concentrations are $[A]'_{eq}$ and $[B]'_{eq}$. Therefore,

$$k'_a[A]'_{eq} = k'_b[B]'_{eq}. \tag{29.5.4}$$

Now let the conditions be changed (e.g. the temperature is suddenly increased) so that the rate constants change to k_a and k_b. The concentrations of A and B are still, momentarily, at their old equilibrium values, but the system is no longer at equilibrium and readjusts to the new equilibrium concentrations, which are now given by

$$k_a[A]_{eq} = k_b[B]_{eq}, \tag{29.5.5}$$

and it does so at a rate that depends on the new rate constants.

We write the deviation of [A] from its new equilibrium value as x, so that $[A] = x + [A]_{eq}$. At $t = 0$ (when the jump in conditions took place), $[A]_0 = [A]'_{eq}$, so that $x_0 = [A]'_{eq} - [A]_{eq}$. The concentrations then change as follows:

$$d[A]/dt = -k_a\{x + [A]_{eq}\} + k_b\{-x + [B]_{eq}\} = -\{k_a + k_b\}x \tag{29.5.6}$$

because the two terms involving the equilibrium concentrations cancel (by eqn (29.5.5)). Since $d[A]/dt = dx/dt$ this is a simple first-order differential equation with the solution

$$x_t = x_0 e^{-t/\tau}, \qquad 1/\tau = k_b + k_a. \tag{29.5.7}$$

This result shows that the concentration of A (and of B) relaxes into the new equilibrium at a rate determined by the *sum* of the two new rate constants. Since the equilibrium constant under the new conditions is $K = k_a/k_b$ its value may be combined with the relaxation time measurement to find the individual k_a and k_b.

One of the most important relaxation techniques uses a *temperature jump*. The equilibrium is changed by causing a sudden change of temperature. One way of doing this is to discharge an electric current through a sample made conducting by the addition of ions, but laser discharges can also be used. Temperature jumps of between 5 and 10 K can be achieved in about 10^{-7} s.

An important application of the temperature-jump technique has been to the determination of the rate of $H_3O^+(aq) + OH^-(aq) \rightarrow 2H_2O(l)$ by monitoring the conductivity of the sample. A relaxation time of 40 μs was measured at room temperature, corresponding to $k = 1.4 \times 10^{11}\ M^{-1}\ s^{-1}$, making it one of the fastest liquid-phase reactions known (but the reaction is faster in ice, where $k = 8.6 \times 10^{12}\ M^{-1}\ s^{-1}$).

Example 29.4

The $H_2O \rightleftharpoons H^+ + OH^-$ equilibrium relaxes in 37 μs at 298 K and $pK_w = 14.01$. Calculate the rate constants for the forward and backward reactions.

• *Method*. Derive an expression for τ in terms of k_1 (forward, first-order reaction) and k_2 (reverse, second-order reaction). Proceed as above, but make the assumption that the deviation from equilibrium (x) is so small that terms in x^2 can be neglected. Relate k_1 and k_2 through the equilibrium constant, using $K = K_w/([H_2O]/M)$.

• *Answer*. The forward rate is $k_1[H_2O]$ and the reverse rate is $k_2[H^+][OH^-]$. The same procedure as before leads to

$$1/\tau = k_1 + k_2\{[H^+] + [OH^-]\}_e.$$

The equilibrium condition is

$$k_1[H_2O]_e = k_2[H^+]_e[OH^-]_e,$$

so that $K = k_1/(k_2\,M)$ with $K = 0.98 \times 10^{-14}/55.5 = 1.8 \times 10^{-16}$. Therefore,

$$\begin{aligned} 1/\tau &= k_2\{(K\,M) + [H^+]_e + [OH^-]_e\} = k_2\{K + K_w^{\frac{1}{2}} + K_w^{\frac{1}{2}}\}\,M \\ &= 2.0 \times 10^{-7} k_2\,M. \end{aligned}$$

Hence, $k_2 = 1/(37 \times 10^{-6}\,s) \times (2.0 \times 10^{-7}\,M) = 1.4 \times 10^{11}\,M^{-1}\,s^{-1}$. It follows that $k_1 = k_2 K\,M = 2.4 \times 10^{-5}\,s^{-1}$.

• *Comment*. Notice how we keep track of units: K and K_w are dimensionless; k_2 is expressed in $M^{-1}\,s^{-1}$ and k_1 in s^{-1}.

• *Exercise*. Derive an expression for the relaxation time of a concentration when the reaction $A + B \rightleftharpoons C + D$ is second-order in both directions. $[1/\tau = k\{[A] + [B]\}_e + k'\{[C] + [D]\}_e]$

The equilibrium constant of a reaction is temperature dependent so long as the reaction enthalpy is non-zero (Section 10.2), and so the temperature-jump method is widely applicable. The equilibrium composition also depends on pressure in some cases, and so *pressure-jump techniques* may sometimes be used. Other relaxation techniques include magnetic resonance (as described in Section 20.1) and the propagation of ultrasonic radiation through a medium.

Further reading

Chain reactions:

Chain reactions. V. N. Kondratiev; *Comprehensive chemical kinetics*, *Vol. 2* (C. H. Bamford and C. F. Tipper, eds.), Elsevier, Amsterdam, 1970.

Kinetics of chemical chain reactions. F. G. R. Gimblett; McGraw-Hill, New York, 1970.

Polymer chemistry. J. C. Bevington; *Photochemistry and reaction kinetics* (P. G. Ashmore, F. S. Dainton, and T. M. Sugden, eds.), Cambridge University Press, 1967.

Principles of polymer chemistry. P. Flory; Cornell University Press, 1953.

Photochemistry:

Photochemistry. R. P. Wayne; Butterworth, London, 1970.

Introduction to molecular photochemistry. C. H. J. Wells; Chapman and Hall, London, 1972.

Photochemistry. J. G. Calvert and J. N. Pitts; Wiley, New York, 1966.

Chemistry of the atmosphere. M. J. McEwan and L. F. Phillips; Edward Arnold, London, 1975

Chemistry of atmospheres. R. P. Wayne; Clarendon Press, Oxford, 1985.

Oscillating reactions:

Oscillating chemical reactions. I. R. Epstein, K. Kustin, P. De Kepper, and M. Orban; *Scientific American* **248** (3), 96 (1983).

Chemical reactor theory (3rd edn). K. G. Denbigh and J. C. R. Turner; Cambridge University Press, 1984.
The physical chemistry of biological organization. A. R. Peacocke; Clarendon Press, Oxford, 1983.

Fast reactions:

Fast reactions. J. N. Bradley; Clarendon Press, Oxford, 1974.
Flash photolysis. G. Porter and M. A. West; *Techniques of chemistry* (G. G. Hammes, ed.) VIB, 367, Wiley-Interscience, New York, 1974.
Rapid flow methods. B. B. Chance; *Techniques of chemistry* (G. G. Hammes, ed.) VIB, 5, Wiley-Interscience, New York, 1974.
Temperature-jump methods. G. G. Hammes; *Techniques of chemistry* (G. G. Hammes, ed.) VIB, 147, Wiley-Interscience, New York, 1974.
Pressure-jump methods. W. Knoche; *Techniques of chemistry* (G. G. Hammes, ed.) VIB, 187, Wiley-Interscience, New York, 1974.
Photostationary methods. R. M. Noyes; *Techniques of chemistry* (G. G. Hammes, ed.) VIB, 343, Wiley-Interscience, New York, 1974.

Introductory problems

A29.1. A saturated hydrocarbon undergoes pyrolysis at 800 K according to the mechanism:

$$A \underset{(2)}{\overset{(1)}{\rightleftharpoons}} CH_3 + R$$

$$R \underset{(4)}{\overset{(3)}{\rightleftharpoons}} B + H$$

$$H + A \xrightarrow{(5)} H_2 + G$$

$$G \underset{(7)}{\overset{(6)}{\rightleftharpoons}} H + M$$

$$G + G \xrightarrow{(8)} G_2.$$

The stable products of the reaction are B, H_2, M, and G_2. Identify the chain initiation step, the chain propagation steps, and the chain termination steps.

A29.2. Photolysis of $Cr(CO)_6$ in the presence of certain molecules M, can give rise to the following reaction sequence:

$$Cr(CO)_6 + h\nu \rightarrow Cr(CO)_5 + CO \quad (1)$$

$$Cr(CO)_5 + CO \rightarrow Cr(CO)_6 \quad (2)$$

$$Cr(CO)_5 + M \rightarrow Cr(CO)_5M \quad (3)$$

$$Cr(CO)_5M \rightarrow Cr(CO)_5 + M. \quad (4)$$

Suppose that the absorbed light intensity is so weak that $I \ll k_4[Cr(CO)_5M]$. Find the factor f in the equation: $d[Cr(CO)_5M]/dt = -f[Cr(CO)_5M]$. Show that a graph of f^{-1} vs. [M] should be a straight line.

A29.3. Consider the following mechanism for the thermal decomposition of R_2:

$$R_2 \rightarrow 2R \quad (1)$$

$$R + R_2 \rightarrow P_B + R' \quad (2)$$

$$R' \rightarrow P_A + R \quad (3)$$

$$2R \rightarrow P_A + P_B \quad (4)$$

where R_2, P_A, P_B are stable hydrocarbons. R and R′ are free radicals. Find the dependence of the rate of decomposition of R_2 on the concentration of R_2.

A29.4. Refer to Fig. 29.3 and determine the pressure range for branching chain explosion in the hydrogen–oxygen reaction at 700 K, 800 K, and 900 K.

A29.5. In a photochemical reaction, the stoichiometry for which is $A \rightarrow 2B + C$, the quantum efficiency with 500 nm light is 2.1×10^2 mol einstein^{-1}. After exposure of 3.00 mmol of A to the light, 2.28 mmol of B are formed. How many photons were absorbed by A?

A29.6. In an experiment to measure the quantum efficiency for a photochemical reaction, the absorbing substance was exposed to 490 nm light from a 100 W source for a period of 45 minutes. The intensity of the transmitted light was 40% of the intensity of the incident light. As a result of irradiation, 0.344 mole of the absorbing substance decomposed. Find the quantum efficiency.

A29.7. The condensation reaction of acetone, $(CH_3)_2CO$, in aqueous solution is catalysed by bases, B, which react reversibly with acetone to form the carbanion $C_3H_5CO^-$. The carbanion then reacts with a molecule of acetone to give the product. A simplified version of the mechanism can be written:

$$AH + B \rightarrow BH^+ + A^- \quad (1)$$

$$A^- + BH^+ \rightarrow AH + B \quad (2)$$

$$A^- + A \rightarrow \text{product} \quad (3)$$

where AH stands for acetone and A^- its carbanion. Use the steady state method to find the concentration of the carbanion and derive the rate equation for the formation of the product.

A29.8. The reaction system described by the following

mechanism:

$$HA + H^+ \underset{(2)}{\overset{(1)}{\rightleftharpoons}} HAH^+ \qquad \text{(fast)}$$

$$HAH^+ + B \xrightarrow{(3)} BH^+ + AH \qquad \text{(slow)}$$

shows general acid catalysis. Find the rate equation and show that it can be made independent of the specific term $[H^+]$.

A29.9. The pK_a of NH_3 is 4.74 at 25 °C. The rate constant at 25 °C for the reaction of NH_4^+ and OH^- to form aqueous NH_3 is $4.0 \times 10^{10}\, M^{-1}\, s^{-1}$. Calculate the rate constant for the ionization of aqueous NH_3. What relaxation time would be observed if a temperature jump were applied to a solution of 0.15 M NH_3(aq) at 25 °C?

A29.10. The equilibrium system $A \underset{(2)}{\overset{(1)}{\rightleftharpoons}} B + C$ at 25 °C is subjected to a temperature jump which slightly increases the concentrations of B and C. The measured relaxation time is 3.0 microseconds. The equilibrium constant for the system is 2.0×10^{-16} at 25 °C, and the equilibrium concentrations of B and C at 25 °C are both 2.0×10^{-4} M. Calculate the numerical values of the rate constants for steps (1) and (2).

Problems

29.1. Consider the following chain mechanism:
(a) $AH \rightarrow A\cdot + H\cdot$
(b) $A\cdot \rightarrow B\cdot + C$
(c) $AH + B\cdot \rightarrow A\cdot + D$
(d) $A\cdot + B\cdot \rightarrow P$.
Identify the initiation, propagation, and termination steps, and use the steady-state approximation to deduce that the decomposition of AH is first order in AH.

29.2. The *Rice–Herzfeld mechanism* for the dehydrogenation of ethane was specified on p. 716, and it was noted there that it led to first-order kinetics. Confirm this remark, and find the approximations that lead to the rate law quoted there. How may the conditions be changed so that the reaction shows different orders?

29.3. The following mechanism has been proposed for the thermal decomposition of acetaldehyde (ethanal):

$$\begin{aligned} CH_3CHO &\rightarrow \cdot CH_3 + CHO, & k_a \\ \cdot CH_3 + CH_3CHO &\rightarrow CH_4 + \cdot CH_2CHO, & k_b \\ \cdot CH_2CHO &\rightarrow CO + \cdot CH_3, & k_c \\ \cdot CH_3 + \cdot CH_3 &\rightarrow CH_3CH_3, & k_d. \end{aligned}$$

Find an expression for the rate of formation of methane and the rate of disappearance of acetaldehyde.

29.4. The reaction $Cl_2 + CO \rightarrow COCl_2$ is believed to proceed by the following mechanism:

$Cl_2 \rightarrow 2Cl,$	k_a forward,	k_a' reverse
$Cl + CO \rightarrow COCl$	k_b forward,	k_b' reverse
$COCl + Cl_2 \rightarrow COCl_2 + Cl$	k_c forward.	

Deduce a rate law for the formation of $COCl_2$ on the assumption that the first two steps are fast. Integrate the rate law for equal initial concentrations of Cl_2 and CO.

29.5. An improved rate law is obtained by assuming not that the first two reactions in Problem 29.3 are fast, but by making the steady state approximation for Cl and COCl. Deduce a rate law for the formation of $COCl_2$ in this way.

29.6. Finally, obtain an 'exact' version of the rate law for the formation of $COCl_2$ by numerical integration of the rate laws. Write a computer program for the mechanism, and use it to explore the time dependences of all the components in the reaction.

29.7. Consider a radical chain polymerization reaction in which the rate of initiation is $k_i f[I]$, where f is the fraction of initiating radicals that succeed in generating a radical $M\cdot$. Use the steady-state approximation to deduce the rate law for the consumption of monomer, and show that the kinetic chain length is proportional to [M] and inversely proportional to the square root of the rate of initiation.

29.8. The derivation of eqn (29.1.8) assumed that the rates of termination and propagation were independent of the polymer chain length. Suppose that both reactions are slower as the chain length increases, and that k_t depends on the monomer concentration as $k_t = k_t^0(1 + a[M])$ and k_P depends as $k_P = k_P^0(1 + b[M])$. Derive an expression for the rate of change of concentration of M.

29.9. The derivation of the rate law leading to eqn (29.1.12) assumes that the polymerization is catalysed by added acid. Repeat the derivation assuming that the reaction is self-catalysed by A, and that the condensation is third order in A.

29.10. Use eqn (29.1.12) and the expression for the mass average RMM derived in *Example* 29.1 to show how the mass average RMM changes with time.

29.11. Repeat the calculation for the time-dependence of the number average and of the ratio of the number and mass average RMMs.

29.12. Express the r.m.s. deviation of the RMM of the condensation polymer in terms of p, where $\delta M_r = \{\langle M_r^2 \rangle - \langle M_r \rangle^2\}^{\frac{1}{2}}$, and deduce its time-dependence.

29.13. Calculate the ratio of the mean cube RMM to the mean square RMM in terms of (a) p, (b) χ.

29.14. Use the steady-state approximation with the following reaction scheme to calculate the concentration of hydrogen atoms in the hydrogen/oxygen reaction, and show that under some circumstances the concentration can become (explosively) high.

(a) $H_2 + O_2 \rightarrow 2OH$, rate r_a
(b) $H_2 + OH\cdot \rightarrow H_2O + H\cdot$
(c) $O_2 + H\cdot \rightarrow OH\cdot + \cdot O\cdot$
(d) $H_2 + \cdot O\cdot \rightarrow OH\cdot + H\cdot$
(e) $H\cdot \rightarrow P$

29.15. The number of photons falling on a sample can be determined by a variety of methods, of which the classical one is *chemical actinometry*. The decomposition of oxalic acid $(COOH)_2$, in the presence of uranyl sulphate, $(UO_2)SO_4$, proceeds according to the sequence

$$UO_2^{2+} + h\nu \rightarrow (UO_2^{2+})^*,$$
$$(UO_2^{2+})^* + (COOH)_2 \rightarrow UO_2^{2+} + H_2O + CO_2 + CO$$

with a quantum efficiency of 0.53 at the wavelength used. The amount of oxalic acid remaining after exposure can be determined by titration (with $KMnO_4$) and the extent of decomposition used to find the number of incident photons. In a particular experiment, the actinometry solution consisted of 5.232 g anhydrous oxalic acid in 25.0 cm^3 water (together with the uranyl salt). After exposure for 300 s the remaining solution was titrated with 0.212 M $KMnO_4$, and 17.0 cm^3 were required for complete oxidation of the remaining oxalic acid. What is the rate of incidence of photons at the wavelength of the experiment? Express the answer in photons/second and einstein/second, an *einstein* being 1 mol of photons.

29.16. The intensity of fluorescence or phosphorescence from an excited molecule M will depend on the efficiency of any competitive chemical quenching process. Consider the gas phase fluorescence from some excited molecule M* generated by $M + h\nu_i \rightarrow M^*$, the light intensity absorbed being I_a. This absorption is followed by $M^* + Q \rightarrow M + Q$, the second-order quenching reaction with rate coefficient k_q, in competition with the fluorescence intensity I_f arising from $M^* \rightarrow M + h\nu_f$, with rate coefficient k_f. Show that this scheme leads to the *Stern–Volmer relation*,

$$1/I_f = (1/I_a)\{1 + (k_q/k_f)[Q]\}$$

between the intensity and the quencher concentration. Thus the ratio k_q/k_f can be determined from a plot of $1/I_f$ against quencher concentration. Show, furthermore, that if the rate of disappearance of fluorescence is measured, k_q itself may be found.

29.17. When benzophenone is illuminated with ultraviolet light it is excited into a singlet state. This singlet changes rapidly into a triplet, which phosphoresces. Triethylamine acts as a quencher for the triplet. In an experiment in methanol as solvent, the phosphorescence intensity varied with amine concentration as shown below. A flash photolysis experiment had also shown that the half-life of the fluorescence in the absence of quencher is 2.9×10^{-7} s. What is the value of k_q?

[Q]/M	0.001	0.005	0.010
I_f/arbitrary units	0.41	0.25	0.16

29.18. Find an expression for the rate of disappearance of a species A in a photochemical reaction where the mechanism is (a) initiation with light of intensity I, $A \rightarrow 2R\cdot$, (b) propagation, $A + R\cdot \rightarrow R\cdot + B$, rate coefficient k_p, (c) termination, $R\cdot + R\cdot \rightarrow R_2$, rate, coefficient k_t. Hence show that rate measurements will give only a combination of k_p and k_t if a steady state is reached, but that both may be obtained if a steady state is not reached. The latter is the basis of the *rotating sector method* of studying photochemical reactions (see Further Reading).

29.19. The photochemical chlorination of chloroform in the gas has been found to follow the rate law $d[CCl_4]/dt = k_{\frac{1}{2}}[Cl_2]^{\frac{1}{2}}I_a^{\frac{1}{2}}$. Devise a mechanism that leads to this rate law when the chlorine pressure is quite high.

29.20. Conventional equilibrium considerations do not apply when a reaction is being driven by light absorption. Thus the steady-state concentration of products and reactions might differ significantly from equilibrium values: this is the *photostationary state*. For instance, suppose the reaction $A \rightarrow B$ is driven by light absorption, and that its rate is I_a, but that the reverse reaction $B \rightarrow A$ is bimolecular and second-order with a rate $k[B]^2$. What is the photostationary state concentration of B? Why does the photostationary state differ from the equilibrium state?

29.21. When anthracene is dissolved in benzene and exposed to ultraviolet light it dimerizes. In concentrated solutions the dimerization occurs with high quantum efficiency (and little fluorescence), but in dilute solutions fluorescence deactivates the excited molecules and the dimerization proceeds with low quantum efficiency. Calculate the quantum efficiency as a function of concentration, basing your derivation on the sequence $A + h\nu \rightarrow A^*$, $A^* + A \rightarrow A_2$, $A^* \rightarrow A + h\nu_f$, and account for the observations.

29.22. An autocatalytic reaction $A \rightarrow P$ is observed to have the rate law $d[P]/dt = k[A]^2[P]$. Solve the rate law for initial concentrations $[A]_0$ and $[P]_0$.

29.23. Another reaction with the same stoichiometry has the rate law $d[P]/dt = k[A][P]^2$; integrate the rate law for initial concentrations $[A]_0$ and $[P]_0$.

29.24. Calculate the times at which the autocatalysed rates in Problems 29.22 and 29.23 reach a maximum.

29.25. Write a program for the integration of the Lotka–Volterra equations, eqn (29.4.1), and arrange for it to plot the concentration of Y against that of X. Explore the consequences of varying the starting concentrations for the integration.

29.26. Identify the conditions (the concentrations of X and Y) corresponding to the steady state of the Lotka–Volterra equations (the point at the centre of the orbits drawn in Problem 29.25).

29.27. Integrate the brusselator equations, eqn (29.4.2), numerically, with an adaptation of the computer program devised for Problem 29.25, and plot [Y] against [X] for a selection of initial concentrations of X and Y: the solutions will show the existence of a limit cycle, Fig. 29.9.

29.28. Set up the overall rate equations for the concentrations of X and Y in terms of the oregonator, eqn (29.4.3), and explore the periodic properties of the solutions.

29.29. Studies of combustion reactions depend on knowing the concentrations of hydrogen atoms and hydroxyl radicals. Measurements on a flow system using ESR for the detection of radicals gave information on the reactions

$$H + NO_2 \rightarrow OH + NO \qquad k = 2.9 \times 10^{10},\ M^{-1}\,s^{-1}$$

$$OH + OH \rightarrow H_2O + O \qquad k' = 1.55 \times 10^{9}\ M^{-1}\,s^{-1}$$

$$O + OH \rightarrow O_2 + H \qquad k'' = 1.1 \times 10^{10}\ M^{-1}\,s^{-1}$$

(J. N. Bradley, W. Hack, K. Hoyermann, and H. G. Wagner, *J. chem. Soc. Faraday Trans.* I, 1889 (1973)). Using initial hydrogen atom and NO_2 concentrations of 4.5×10^{-10} mol cm^{-3} and 5.6×10^{-10} mol cm^{-3} respectively, compute and plot curves showing the O, O_2, and OH concentrations as a function of time in the range 0–10 ns.

29.30. In a flow study of the reaction between oxygen atoms and chlorine (J. N. Bradley, D. A. Whytock, and T. A. Zaleski, *J. chem. Soc. Faraday Trans.* I, 1251 (1973)) at high chlorine pressures, plots of $\ln [O]_0/[O]$ against distances along the flow tube, where $[O]_0$ is the oxygen concentration at zero chlorine pressure, gave straight lines. Given the flow velocity as 6.66 m s^{-1} and the data below, find the rate coefficient for the reaction $O + Cl_2 \rightarrow ClO + Cl$.

distance/cm	0	2	4	6	8
$\ln [O]_0/[O]$	0.27	0.31	0.34	0.38	0.45

distance/cm	10	12	14	16	18
$\ln [O]_0/[O]$	0.46	0.50	0.55	0.56	0.60

$[O]_0 = 3.3 \times 10^{-8}$ M, $[Cl_2] = 2.54 \times 10^{-7}$ M, $p = 1.70$ Torr.

29.31. Show that the reaction $A \rightleftharpoons B + C$, first-order forwards, second-order backwards, relaxes exponentially for small displacements from equilibrium. Find an expression for the relaxation time in terms of k_1 and k_2.

30

Molecular reaction dynamics

Learning objectives

After careful study of this chapter you should be able to:

(1) Specify the *collision theory* of bimolecular gas phase reactions, Section 30.1, and derive and use an expression for the rate constant, eqn (30.1.5).

(2) Define, explain, and deduce the *P-factor* and the *reactive cross-section*, eqn (30.1.7) and Example 30.1.

(3) Calculate the steric factor for the *harpoon mechanism*, eqn (30.1.8).

(4) Distinguish between *diffusion-controlled* and *activation-controlled* reactions in solution, Section 30.2(a).

(5) Relate the rate constant of a diffusion-controlled reaction to the diffusion coefficient, eqn (30.2.8), and the viscosity, eqn (30.2.9) and Example 30.2.

(6) State the *material balance equation*, eqn (30.2.12), and indicate the form of its solutions, Fig. 30.2.

(7) Explain the meaning of *reaction coordinate*, *activated complex*, and *transition state*, Section 30.3(a).

(8) Describe the *activated complex theory* of reaction rates, Section 30.3(b).

(9) Derive the *Eyring equation* for the rate constant, eqn (30.3.11).

(10) Use the Eyring equation to calculate a rate constant for the reaction between two structureless particles, eqn (30.3.14), and to estimate the magnitude of the steric factor, Example 30.3.

(11) Use the Eyring equation to calculate the effect of deuteration on the magnitude of a rate constant, eqn (30.3.16).

(12) Define the *Gibbs function*, *entropy*, and *enthalpy* of activation, eqns (30.3.17) and (30.3.19).

(13) Relate the enthalpy of activation to the activation energy, eqn (30.3.21), and the entropy of activation to the pre-exponential factor, eqn (30.3.23) and Example 30.4.

(14) Describe, and calculate the magnitude of, the *kinetic salt effect*, eqn (30.3.28) and Example 30.5.

(15) Indicate how molecular beams are used to study reactive collisions, Section 30.4(a).

(16) Sketch and explain the *potential energy surface* of a simple reaction, Section 30.4(b) and Fig. 30.9.

(17) Explain how potential energy surfaces are used to account for the energy demands of reactive encounters, Section 30.4(c).

(18) Distinguish between *attractive* and *repulsive* surfaces, and explain how they determine the energy requirements of a reaction, Section 30.4(d) and Fig. 30.14.

Introduction

Now we are at the heart of chemistry. Here we examine the details of what happens to molecules at the climax of reactions. Massive changes of form are taking place and energies the size of dissociation energies are being redistributed among bonds: old bonds are being ripped apart and new bonds are being formed.

As may be imagined, the calculation of rate coefficients from first principles is a very difficult undertaking. Nevertheless, like so many intricate problems, the broad features can be established quite simply, and only when we enquire more deeply do the complications emerge. In this chapter we look at three levels of approach to the calculation of a rate constant for a bimolecular simple reaction: *collision theory* (which has been sketched in outline already, Section 28.3(b)), *activated complex theory*, which uses statistical thermodynamics to estimate the rate, and *molecular reaction dynamics*, which uses information from molecular beam studies and detailed calculations to study individual collisions in great detail. Although a great deal of information can be obtained from gas phase reactions, most reactions of interest take place in solution, and we shall also see to what extent their rates can be predicted.

30.1 Collision theory

The basic concepts of collision theory were explained in Section 28.3(b): particles collide with a frequency calculated from the kinetic theory of gases, and reaction occurs if the collision is sufficiently energetic. We now use this picture to build a quantitative expression for the rate coefficient for the simple reaction A + B → *Products* (P).

30.1 (a) The basic calculation

The rate of the reaction above is

$$v_A = -d[A]/dt. \tag{30.1.1}$$

The number of A, B collisions per unit time per unit volume is Z_{AB} (Section 26.2(a)). The rate of change of the number of A particles per unit volume is equal to Z_{AB} multiplied by the proportion of collisions that occur with a kinetic energy along their line of approach in excess of some threshold value E_a'. The rate of change of (molar) concentration of A is this rate divided by Avogadro's constant N_A. Therefore,

$$d[A]/dt = -(Z_{AB}/N_A)e^{-E_a'/RT}. \tag{30.1.2}$$

The collision frequency is given by eqn (26.2.5):

$$Z_{AB} = \sigma(8kT/\pi\mu)^{\frac{1}{2}}N_A^2[A][B], \tag{30.1.3}$$

where σ is the collision cross-section which, for hard spheres of radii R_A and R_B is $\pi(R_A + R_B)^2$, and μ is the reduced mass, $1/\mu = 1/m_A + 1/m_B$, of the colliding particles. Therefore,

$$d[A]/dt = -\sigma(8kT/\pi\mu)^{\frac{1}{2}}N_A[A][B]e^{-E_a'/RT}. \tag{30.1.4}$$

Since the rate coefficient of this simple binary reaction is defined through $-d[A]/dt = k_a[A][B]$, it follows that

$$k_a = \sigma(8kT/\pi\mu)^{\frac{1}{2}}N_A e^{-E_a'/RT}. \tag{30.1.5}$$

The expression just obtained has the form of the Arrhenius Law $k = Ae^{-E_a/RT}$ so long as the exponential temperature dependence dominates the weak square-root temperature dependence of the pre-exponential factor. From the general definition of activation energy, eqn (28.3.5):

$$E_a = RT^2(\partial \ln k_a/\partial T)_V = E_a' + \tfrac{1}{2}RT, \tag{30.1.6}$$

Table 30.1. Arrhenius parameters for gas-phase reactions

	$A/\text{M}^{-1}\,\text{s}^{-1}$		$E_a/\text{kJ mol}^{-1}$	P
	Experiment	Theory		
$2NOCl \rightarrow 2NO + Cl$	9.4×10^9	5.9×10^{10}	102.0	0.16
$2ClO \rightarrow Cl_2 + O_2$	6.3×10^7	2.5×10^{10}	0.0	2.5×10^{-3}
$H_2 + C_2H_4 \rightarrow C_2H_6$	1.24×10^6	7.3×10^{11}	180	1.7×10^{-6}
$K + Br_2 \rightarrow KBr + Br$	1.0×10^{12}	2.1×10^{11}	0.0	4.8

and so the activation energy is weakly temperature dependent. Generally E_a' is much larger than $\frac{1}{2}RT$, and the latter can be neglected.

When calculating rate coefficients on the basis of eqn (30.1.5) the simplest procedure is to use for σ the values obtained for non-reactive collisions (e.g. typically from viscosity measurements) or from tables of molecular radii. Table 30.1 compares some calculated values of the pre-exponential factor A with values obtained from Arrhenius plots of $\ln k$ against $1/T$ (Section 28.3). One of the reactions shows fair agreement between theory and experiment, but for others there are major discrepancies. In some cases the experimental values are orders of magnitude smaller than the calculated ones, which suggests that the collision energy is not the only criterion for reaction and that some other feature, such as the relative orientation of the colliding species, is important. Moreover, one reaction has a pre-exponential factor larger than theory, which seems to indicate that the reaction occurs more quickly than the particles collide! In this case it may be that the distance of approach required for reaction is quite different from the distance that leads simply to a deflection of the direction of motion, and therefore that it is wrong to use the non-reactive collision cross-section.

30.1 (b) The steric requirement

The disagreement between experiment and theory can be expressed by replacing σ in eqn (30.1.5) by the *reactive cross-section* (symbol: σ^*). Sometimes it is convenient to express σ^* as some multiple of σ by introducing a *steric factor* (symbol: P) and writing $\sigma^* = P\sigma$. Then the rate-coefficient becomes

$$k_a = \sigma^*(8kT/\pi\mu)^{\frac{1}{2}}N_A e^{-E_a'/RT} = P\{\sigma(8kT/\pi\mu)^{\frac{1}{2}}N_A\}e^{-E_a'/RT}. \quad (30.1.7)$$

P is normally found to be several orders of magnitude less than unity.

Example 30.1

Estimate the P-factor for the reaction $H_2 + C_2H_4 \rightarrow C_2H_6$ at 628 K given that the pre-exponential factor is $1.24 \times 10^6\,\text{M}^{-1}\,\text{s}^{-1}$.

• *Method*. Use eqn (30.1.7), identifying the experimental pre-exponential factor with $\sigma^*(8kT/\pi\mu)^{\frac{1}{2}}N_A$. Use $M_r(H_2) = 2.016$ and $M_r(C_2H_4) = 28.05$, and estimate σ from the mean of the values given in Table 26.2: $\sigma(H_2) = 0.27\,\text{nm}^2$, $\sigma(C_2H_4) = 0.64\,\text{nm}^2$.

• *Answer*. The reduced mass is

$$\mu = mm'/(m+m') = M_r M_r' m_u/(M_r + M_r) = 3.12 \times 10^{-27}\,\text{kg}.$$

Hence $(8kT/\pi\mu)^{\frac{1}{2}} = 2.66 \times 10^3\,\text{m s}^{-1}$. The mean collision cross-section is $\sigma = 0.46\,\text{nm}^2$. Therefore,

$$A = \sigma^*(8kT/\pi\mu)^{\frac{1}{2}}N_A = (7.37 \times 10^{11} \times 10^8\,\text{m}^3\,\text{mol}^{-1}\,\text{s}^{-1})P$$
$$= (7.37 \times 10^{11}\,\text{M}^{-1}\,\text{s}^{-1})P.$$

Then since the experimental value of A is $1.24 \times 10^6\ M^{-1}\ s^{-1}$, it follows that $P = 1.7 \times 10^{-6}$.

• *Comment*. As a rough guide, the more complex the molecules, the smaller the value of P.

• *Exercise*. For the reaction $NO + Cl_2 \rightarrow NOCl + Cl$ it is found that $A = 4.0 \times 10^9\ M^{-1}\ s^{-1}$ at 298 K. Use $\sigma(NO) = 0.42\ nm^2$ and $\sigma(Cl_2) = 0.93\ nm^2$ to estimate the P factor. [0.02]

Equation (30.1.7) displays very clearly the three contributions to the rate coefficient. The final factor is the *energy criterion*; the second factor is the *transport property* governing how often the particles come together; the first factor (P) takes care of the *local properties* of the reaction, such as the orientations necessary for reaction and the details of how close they must come. The theory would be all very well if we could calculate the steric factor.

In some cases we can. Take, for example, the reaction $K + Br_2 \rightarrow KBr + Br$, for which $P = 4.8$, indicating an 'anomalously' large reaction cross-section, $\sigma^* = 4.8\sigma$. It has been proposed that the reaction proceeds by a *harpoon mechanism*. This brilliant name is based on a model of the reaction that pictures the K atom as approaching the Br_2 molecules, and when the two are close enough an electron (the harpoon) flips across to the Br_2 molecule. In place of two neutral particles there are now two ions, and so there is a Coulombic attraction between them: this is the line on the harpoon. Under its influence the ions move together (the line is wound in), the reaction takes place, and $KBr + Br$ emerge. The harpoon extends the cross-section for the reactive encounter, and we greatly underestimate the reaction rate by taking for the collision cross-section the value for simple mechanical contact between $K + Br_2$. The value of P can be estimated by calculating the distance at which it is energetically favourable for the electron to leap from K to Br_2. There are three contributions to the energy of the process $K + Br_2 \rightarrow K^+ + Br_2^-$. The first is the ionization energy of K, I. The second is the electron affinity of Br_2 (E_A; energy is released in this step). The third is the Coulombic interaction energy between the ions when they have been formed: when their separation is R this energy is $-e^2/4\pi\varepsilon_0 R$. Therefore, the net change of energy when the transfer occurs at separation R is

$$E = I - E_A - e^2/4\pi\varepsilon_0 R.$$

I is larger than E_A, and so E becomes negative only when R has decreased to less than some critical value R^* given by

$$e^2/4\pi\varepsilon_0 R^* = I - E_A.$$

When the particles are at this separation the harpoon shoots across, and so the reactive cross-section can be identified as $\sigma^* = \pi R^{*2}$. This implies that the steric factor is

$$P = \sigma^*/\sigma = R^{*2}/d^2 = \{e^2/4\pi\varepsilon_0 d[I - E_A]\}^2, \qquad (30.1.8)$$

where $d = R(K) + R(Br_2)$. Using $I = 420\ kJ\ mol^{-1}$, $E_A < 300\ kJ\ mol^{-1}$, and $d = 310\ pm$, we find $P < 12$, which is consistent with the experimental value.

The example shows two things. First, the P-factor is not wholly useless because in some cases it can be estimated. Second (and more pessimistically) many reactions are much more complex than $K + Br_2$, and we cannot expect to obtain P so easily. What we need is a more powerful theory which lets us calculate, and not merely guess, its value. We go some way to setting up that theory in Section 30.3.

30.2 Diffusion-controlled reactions

Encounters between reactants in solutions occur in a very different way from those in gases. Particles have to jostle their way through the solvent, and so their encounter frequency is drastically less than in a gas. However, since a particle also migrates only slowly *away* from a location, two particles encountering each other stay near each other for much longer than in a gas. This means that the chance of them reacting is increased. Furthermore, the activation energy of a reaction is a much more complicated quantity in solution than in a gas, because the encounter pair is surrounded by solvent. We look at these aspects in this section.

30.2 (a) Classes of reaction

We begin to divide the complicated overall picture into simpler parts by setting up a simple kinetic scheme. We suppose that the rate of formation of an encounter pair AB, is second-order in the reactants A and B:

$$\mathrm{A}+\mathrm{B}\rightarrow\mathrm{AB},\qquad \mathrm{d}[\mathrm{AB}]/\mathrm{d}t=k_{\mathrm{d}}[\mathrm{A}][\mathrm{B}]. \tag{30.2.1a}$$

As we shall see, k_d is determined by the diffusion characteristics of A and B. The encounter pair can break up without reaction or it can go on to form products (P). If we suppose that these are pseudo-first-order reactions (with the solvent perhaps playing a role), we can write

$$\mathrm{AB}\rightarrow\mathrm{A}+\mathrm{B},\quad \mathrm{d}[\mathrm{AB}]/\mathrm{d}t=-k_{\mathrm{d}}'[\mathrm{AB}];\qquad \mathrm{AB}\rightarrow\mathrm{P},\quad \mathrm{d}[\mathrm{AB}]/\mathrm{d}t=-k[\mathrm{AB}]. \tag{30.2.1b}$$

The stationary-state concentration of AB can now be found from the rate law:

$$\mathrm{d}[\mathrm{AB}]/\mathrm{d}t=k_{\mathrm{d}}[\mathrm{A}][\mathrm{B}]-k_{\mathrm{d}}'[\mathrm{AB}]-k[\mathrm{AB}]=0;$$
$$[\mathrm{AB}]=\{k_{\mathrm{d}}/(k_{\mathrm{d}}'+k)\}[\mathrm{A}][\mathrm{B}].$$

The overall rate law for the formation of products is therefore

$$\mathrm{d}[\mathrm{P}]/\mathrm{d}t=k[\mathrm{AB}]=k_{\mathrm{eff}}[\mathrm{A}][\mathrm{B}],\qquad k_{\mathrm{eff}}=\frac{kk_{\mathrm{d}}}{k_{\mathrm{d}}'+k}. \tag{30.2.2}$$

Two limits can now be distinguished. If the rate of break-up of the encounter pair is much slower than the rate at which it forms products, then $k_d' \ll k$ and the effective rate coefficient is

$$k_{\mathrm{eff}}=kk_{\mathrm{d}}/k=k_{\mathrm{d}}. \tag{30.2.3}$$

In this limit the rate of reaction is governed by the rate at which the reactant particles diffuse through the medium: it is called the *diffusion-controlled limit*, and the reaction is *diffusion-controlled*. (In due course we shall see that a signal that a reaction is diffusion-controlled is that its rate constant is of the order of $10^9\ \mathrm{M^{-1}\,s^{-1}}$ or greater.) Because the combination of radicals with unpaired spins involves very little activation energy, radical and atom recombination reactions are often diffusion-controlled.

The second limit of an *activation-controlled reaction* arises when a

Table 30.2. Arrhenius parameters for reactions in solution

	Solvent	$A/\mathrm{M^{-1}\,s^{-1}}$	$E_a/\mathrm{kJ\,mol^{-1}}$
$(CH_3)_3CCl$ solvolysis	Water	7.1×10^{16}	100
	Ethanol	3.0×10^{13}	112
	Chloroform	1.4×10^{4}	45
$CH_3CH_2Br + OH^-$	Ethanol	4.3×10^{11}	89.5

substantial activation energy is involved in the reaction of AB. Then $k \ll k'_d$ and

$$k_{\mathrm{eff}} = k(k_d/k'_d) = kK. \tag{30.2.4}$$

K is the equilibrium constant for $A + B \rightleftharpoons AB$. In this limit the reaction rate depends on the accumulation of energy in the encounter pair as a result of its interaction with the surrounding solvent. Some experimental data are given in Table 30.2.

30.2 (b) Diffusion and reaction

The rate of a diffusion-controlled reaction can be calculated by considering the rate at which the reactants diffuse together. We begin by treating the case of a stationary A particle in a solvent which also contains B particles.

Consider an imaginary sphere of radius r surrounding A, Fig. 30.1. Since the flux (symbol: J) of B through the surface is the number per unit area per unit time, the total *flow* (symbol: $\mathcal{J}$), the number per unit time passing through the entire surface of area $4\pi r^2$ is $\mathcal{J} = 4\pi r^2 J$. From Fick's First Law of diffusion (eqn (26.3.1)), the flux is proportional to the concentration gradient $d\mathcal{N}/dr = N_A\, d[B]/dr$, where $\mathcal{N}$ is the number density of B particles ($\mathcal{N} = N_A[B]$, N_A is Avogadro's constant). Therefore,

$$\mathcal{J} = 4\pi r^2 J = 4\pi r^2 D_B N_A(d[B]/dr), \tag{30.2.5}$$

D_B being the diffusion coefficient for B. The concentration of B at any distance from A can be calculated by integrating this equation. We need two pieces of information. First, when r is infinite, the concentration of B is the same as in the bulk, [B]. Second, the flow through a shell is the same whatever the radius because no reaction occurs until A and B touch. Fig. 30.1. Therefore, under steady-state conditions, $\mathcal{J}$ is a constant independent of r. It follows that

$$\int_{[B]_r}^{[B]} d[B] = (\mathcal{J}/4\pi D_B N_A)\int_r^{\infty}(1/r^2),$$

and so

$$[B]_r = [B] - (\mathcal{J}/4\pi D_B N_A)(1/r), \tag{30.2.6}$$

and the average concentration of B varies inversely with distance. Now suppose that at some critical distance R^* the reactants touch, reaction occurs, and B is destroyed. This means that when $r = R^*$, $[B]_r = 0$. When this is substituted into eqn (30.2.6) we obtain an expression for the flow of B towards A:

$$\mathcal{J} = 4\pi R^* D_B N_A[B]. \tag{30.2.7}$$

This flow is the average number of B particles per unit time passing through any spherical surface centred on any A.

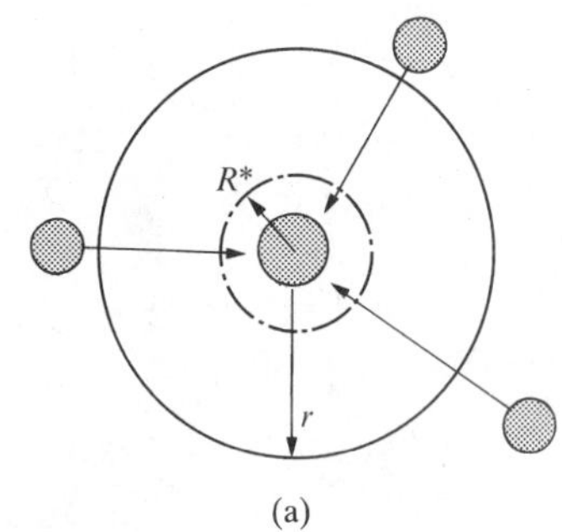

(a)

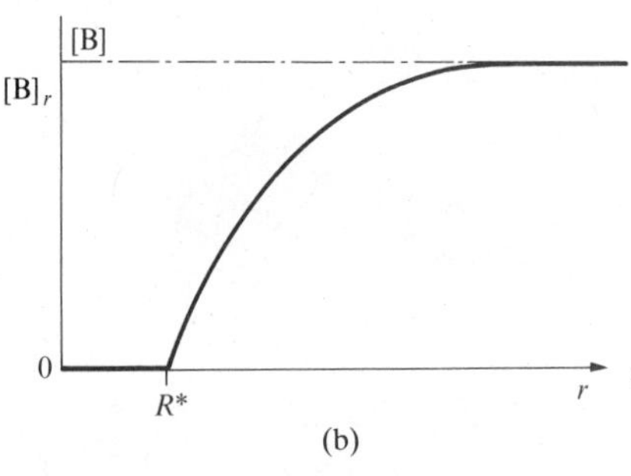

(b)

Fig. 30.1. (a) The flux through a spherical surface of B particles on their approach to a stationary A particle. The total flow through a surface is independent of r; reaction occurs when the particles are at a separation R^*. The corresponding concentration profile is shown in (b).

The rate of the diffusion-controlled reaction is equal to the average flow of B particles to all the A particles in the sample. If the bulk concentration of A is [A], the number in the sample of volume V is $N_A[A]V$, and so the global flow of all B to all A is equal to $\mathscr{J}N_A[A]V = 4\pi R^* D_B N_A^2[A][B]V$. This is the flow in terms of the *numbers* of B: to obtain the flow in terms of the concentration of B we divide by $N_A V$. Of course, it is unrealistic to suppose that all A are stationary; this feature is removed by replacing D_B by $D = D_A + D_B$. Then the rate of change of concentration of AB is

$$\mathrm{d}[AB]/\mathrm{d}t = 4\pi R^* D N_A[A][B],$$

and by comparing this expression with eqn (30.2.1a) we can identify the diffusion-controlled rate constant as

$$k_d = 4\pi R^* D N_A. \quad (30.2.8)$$

The last expression can be taken further by incorporating the Stokes–Einstein relation, Box 27.1, for each particle:

$$D_A = kT/6\pi\eta R_A, \qquad D_B = kT/6\pi\eta R_B,$$

where R_A and R_B are their effective hydrodynamic radii. Since this relation is quite crude, little extra error is introduced if we write $R_A = R_B = \frac{1}{2}R^*$, which leads to

$$k_d = 8N_A kT/3\eta = 8RT/3\eta. \quad (30.2.9)$$

Note that the radii have cancelled, and so in this approximation the rate constant is independent of the identities of the reactants, and depends only on the temperature and the viscosity of the solvent.

Example 30.2

Estimate the second-order rate constant for the recombination of iodine atoms in hexane at 298 K.

• *Method*. Use eqn (30.2.9). The viscosity of the solvent is 0.326 cP at this temperature ($1\ \mathrm{cP} = 10^{-3}\ \mathrm{kg\ m^{-1}\ s^{-1}}$).

• *Answer*. From eqn (30.2.9),

$$k_d = 8RT/3\eta = (6.61 \times 10^9\ \mathrm{M^{-1}\ s^{-1}})/(\eta/\mathrm{cP})$$
$$= 2.0 \times 10^{10}\ \mathrm{M^{-1}\ s^{-1}} \quad \text{when} \quad \eta = 0.326\ \mathrm{cP}.$$

• *Comment*. The experimental value is $1.3 \times 10^{10}\ \mathrm{M^{-1}\ s^{-1}}$, and so the agreement is very good considering the approximations involved.

• *Exercise*. Estimate the rate in benzene at the same temperature. [$1.1 \times 10^{10}\ \mathrm{M^{-1}\ s^{-1}}$]

30.2 (c) The details of diffusion

The diffusion of reactants plays an important role in many chemical processes, such as the diffusion of oxygen into red blood corpuscles and of a gas towards a catalyst. We can have a glimpse of the kinds of calculations involved by considering the generalized diffusion equation, eqn (27.3.10), but generalized even more to take into account the possibility that the diffusing, convecting particles are also reacting.

Consider a small volume element in a chemical reactor (or a biological cell), then during some small time interval the total change of the number of particles of some substance J must satisfy the following *material balance equation:*

> *Net change of number of* J *in the volume element*
> = (*Number entering*) − (*Number leaving*)
> + (*Number formed by reaction*) − (*Number destroyed by reaction*).

The rate of change of the number density, $\partial\mathcal{N}_J/\partial t$, is this expression divided by the volume and the time interval. The net rate at which J particles enter the region by diffusion and convection is given by eqn (27.3.10):

$$\partial\mathcal{N}_J/\partial t = D(\partial^2\mathcal{N}_J/\partial x^2) - v(\partial\mathcal{N}_J/\partial x), \qquad (30.2.10a)$$

and the net rate in terms of concentration is obtained by dividing by N_A:

$$\partial[J]/\partial t = D(\partial^2[J]/\partial x^2) - v(\partial[J]/\partial x). \qquad (30.2.10b)$$

The net rate of change due to chemical reaction is

$$\partial[J]/\partial t = -k[J] \qquad (30.2.11)$$

if we suppose that J is disappearing by a pseudo-first-order reaction (other rate laws could be used here, including those depending on other substances that might be present). Therefore, the mathematical form of the material balance equation is

$$\partial[J]/\partial t = D(\partial^2[J]/\partial x^2) - v(\partial[J]/\partial x) - k[J]. \qquad (30.2.12)$$

A similar equation, with different reaction terms, applies to each species present. This important equation is the basis of the theory of chemical reactor design, because it allows us to explore how inhomogeneities of concentration and the convective flow of matter affects the rate at which substances are produced.

The material balance equation is a second-order partial differential equation, and it is far from easy to solve in general. Some ideas of the difficulty of dealing with it can be seen by considering the special case in which there is no convective motion:

$$\partial[J]/\partial t = D(\partial^2[J]/\partial x^2) - k[J]. \qquad (30.2.13)$$

If the solution in the absence of reaction (i.e. $k=0$ in this equation) is $[J]_t$, then it is easy to show that the solution in the presence of reaction ($k\neq 0$) is

$$[J]_t^* = k\int_0^t [J]_t e^{-kt}\,dt + [J]_t e^{-kt}. \qquad (30.2.14)$$

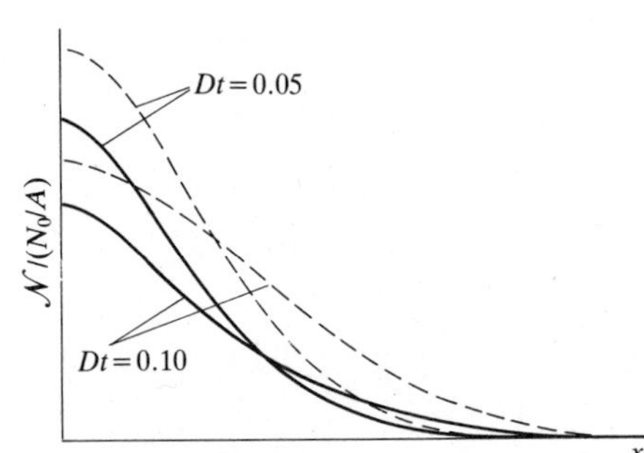

Fig. 30.2. The concentration profiles for a diffusing, reacting system (e.g. a column of solution) in which one reactant is initially in a layer at $x=0$. In the absence of reaction (dotted lines) the concentration profiles are the same as in Fig. 27.10. (Arbitrary values for D and k have been taken; detailed calculations should be made by substitution of eqn (30.2.15) into eqn (30.2.14), followed by numerical integration.)

We have already met one solution of the diffusion equation (Section 27.3(a)), and eqn (27.3.2) is the solution for a system in which initially a layer of $N_0 = n_0 N_A$ particles are spread over a plane of area A:

$$[J]_t = \{n_0/A(\pi Dt)^{\frac{1}{2}}\}e^{-x^2/4Dt}. \qquad (30.2.15)$$

Therefore, when we substitute this expression into the one above, and perform the integration, we obtain the concentration of J as it diffuses away from its initial surface layer and undergoes reaction in the solution above, Fig. 30.2.

Even this example has led to a very difficult equation to compute, and only in some special cases can the full material balance equation be solved analytically. Most modern work on reactor design and cell kinetics now uses numerical methods to solve the equation, and detailed solutions for realistic environments can be obtained reasonably easily. Some of the interesting applications include the exploration of the spatial periodicity of autocatalytic reactions, Fig. 29.3, as mentioned in Section 29.4(b).

30.3 Activated complex theory

We now turn to the calculation of rate constants in terms of a statistical thermodynamic description of a simple reaction. This approach has the advantage that a quantity corresponding to the steric factor appears automatically, and P does not need to be grafted on in an *ad hoc* manner. That is not to say that the theory is complete, or even very reliable: it still remains an attempt to identify the principal features governing the magnitude of the rate constant, and it takes little account of the time-dependent, dynamic features of the actual process.

30.3 (a) The reaction coordinate and the transition state

The general features of how the energy of the reactants A and B change in the course of a simple bimolecular reaction are shown in Fig. 30.3. Initially only A and B are present, and their total potential energy has some value. As the reaction event proceeds, A and B come into contact, distort, and begin to exchange or discard atoms. The potential energy rises to a maximum, falls as the product particles separate, and then reaches a value characteristic of the products. The horizontal axis of the diagram represents the course of the individual reaction event and is called the *reaction coordinate*. The climax of the reaction is at the peak of the potential energy. Here two reactant particles have come to such a degree of closeness and distortion that a small further distortion will send them in the direction of products. This crucial configuration is called the *transition state* of the reaction. Although some particles entering the transition state might revert to reactants, if they pass *through* this configuration it is inevitable that products will emerge from the encounter.

As an example, consider the approach of a hydrogen atom to a fluorine molecule, and for simplicity imagine the approach as occurring along the F—F bond direction. At great distances the potential energy is the sum of the potential energies of H and F_2. When H and F_2 are so close that their orbitals start to overlap, the F—F bond begins to stretch and rise in energy, and a bond begins to form between H and the nearer F. The H atom comes closer, the F—F bond lengthens, and the H—F bond shortens and strengthens. There is a point where the composite molecule, which is called the *activated complex*, has maximum potential energy. It is then poised at the transition state. An infinitesimal compression of the H—F bond and a stretch of F—F takes the activated complex through the transition state. Distances further to the right along the reaction coordinate represent stages at which the H—F bond forms more fully and the F—F bond breaks. Motion along the reaction coordinate from left to right therefore represents the evolution of H and F_2 through these configurations. Whether or not a colliding pair actually crosses the potential barrier depends on the *kinetic*

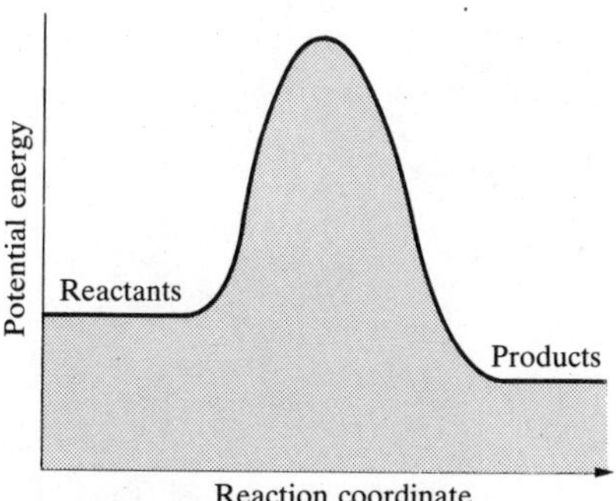

Fig. 30.3. A reaction profile. The horizontal axis is the *reaction coordinate*, and the vertical axis is potential energy. The activated complex is the region near the potential maximum, and the transition state corresponds to the maximum itself.

energy the particles have initially, because they must be able to climb the barrier and reach the transition state.

In an actual reaction, H atoms approach F_2 molecules from all angles, and the exact specification of the reaction coordinate is a difficult problem. We therefore regard it simply as an indication of the distortions in the reactant particles as the activated complex is formed, the critical transition state is reached, and the product particles emerge. At the transition state the reaction coordinate corresponds to some complicated collective vibration-like motion of all the atoms.

30.3 (b) The formation and decay of the activated complex

Activated complex theory (ACT) pictures a reaction between A and B as proceeding through the formation of an activated complex $C^{\ddagger}$ which falls apart into products (P) with a rate constant $k^{\ddagger}$:

$$C^{\ddagger} \rightarrow P, \qquad d[P]/dt = k^{\ddagger}[C^{\ddagger}]. \tag{30.3.1}$$

There are two problems: to find $k^{\ddagger}$ and the concentration $[C^{\ddagger}]$. The latter is likely to be proportional to the concentration of the reactants, and later we show explicitly that

$$[C^{\ddagger}] = K^{\ddagger}[A][B], \tag{30.3.2}$$

where $K^{\ddagger}$ is some proportionality constant (with dimensions 1/concentration). It follows that

$$d[P]/dt = k_{\text{eff}}[A][B], \qquad k_{\text{eff}} = k^{\ddagger}K^{\ddagger}. \tag{30.3.3}$$

The coefficient $k^{\ddagger}$ can be related to a property of the activated complex as follows. The criterion for the complex turning into products is that it must pass through the transition state. If this vibration-like motion occurs with a frequency ν, then the frequency with which the complex swings towards the transition state is also ν. It is possible that not every oscillation along the reaction coordinate takes the complex through the transition state (e.g. because the centrifugal effect of rotations might also be an important contribution to the breaking up of the complex, and in some cases the complex might be rotating slowly or about the wrong axis). Therefore, we suppose that the rate of passage of the complex through the transition state is *proportional* to the vibrational frequency along the reaction coordinate, and write

$$k^{\ddagger} = \kappa\nu. \tag{30.3.4}$$

κ is called the *transmission coefficient:* in many cases it is about unity.

Now we estimate the concentration of the activated complex. The simplest procedure‡ is to assume that there is a pre-equilibrium between the reactants and their activated complex, and to write

$$A + B \rightleftharpoons C^{\ddagger}, \qquad K_p = \{(p_C/p^{\ominus})/(p_A/p^{\ominus})(p_B/p^{\ominus})\}_{\text{eq}} = \{p_C p^{\ominus}/p_A p_B\}_{\text{eq}}. \tag{30.3.5}$$

‡ In earlier editions a different line of argument was used: we recognized that almost nothing is known about the populations of the levels of the activated complex. Since nothing is known, the most honest approach is to assume that the populations depend on the energy and not on the identity of the level (i.e. whether it belongs to A, B, or $C^{\ddagger}$). This 'least-prejudiced' approach leads to the same results as here, but has the advantage that it avoids the presumption of equilibrium between the reactants and their activated complex. Its disadvantage is that it is slightly less direct. The *Second edition* of this text should be consulted for details.

The partial pressures can be expressed in terms of the concentrations using $p_J = RT[J]$, and so

$$[C^{\ddagger}] = (RT/p^{\ominus})K_p[A][B]. \tag{30.3.6}$$

On comparison with eqn (30.3.2) we see that

$$K^{\ddagger} = (RT/p^{\ominus})K_p, \tag{30.3.7}$$

and so the remaining task is the calculation of the equilibrium constant K_p.

The calculation of equilibrium constants was discussed in Section 22.3(d). We can use eqn (22.3.21) directly, which in this case gives

$$K_p = \{N_A q^{\ominus}_{C,m}/q^{\ominus}_{A,m}q^{\ominus}_{B,m}\}e^{-\Delta E_0^{\ddagger}/RT}. \tag{30.3.8}$$

$\Delta E_0^{\ddagger}$ is the energy separation between the zero-point levels of $C^{\ddagger}$ and $A + B$, eqn (22.3.10), and the $q^{\ominus}_{J,m}$ are the molar partition functions as defined in Section 22.2.

In the final step of the calculation we focus attention on the partition function of the activated complex. We have already assumed that a vibration of $C^{\ddagger}$ tips it through the transition state. The partition function for this vibration is $1/(1 - e^{-h\nu/kT})$, where ν is its frequency (the same frequency that determines $k^{\ddagger}$). This frequency is much lower than for an ordinary molecular vibration because in this case the motion corresponds to the complex falling apart. Therefore, since $h\nu/kT \ll 1$, the exponential may be expanded and the partition function reduces to $kT/h\nu$. We can therefore write

$$q_{C,m} = (kT/h\nu)\bar{q}_{C,m}, \tag{30.3.9}$$

where $\bar{q}$ denotes the partition function for all the *other* modes in the complex. The coefficient $K^{\ddagger}$ is therefore

$$K^{\ddagger} = (kT/h\nu)K, \tag{30.3.10a}$$

$$K = (RT/p^{\ominus})\bar{K}_p, \tag{30.3.10b}$$

$$\bar{K}_p = \{N_A\bar{q}^{\ominus}_{C,m}/q^{\ominus}_{A,m}q^{\ominus}_{B,m}\}e^{-\Delta E_0^{\ddagger}/RT}, \tag{30.3.10c}$$

with $\bar{K}_p$ a kind of equilibrium constant, but with one vibrational mode of $C^{\ddagger}$ discarded.

All the parts of the calculation can now be combined into

$$k_{\text{eff}} = k^{\ddagger}K^{\ddagger} = (\kappa\nu)(kT/h\nu)K,$$

which, because the unknown frequencies ν cancel, reduces to the *Eyring equation:*

$$k_{\text{eff}} = \kappa(kT/h)K. \tag{30.3.11}$$

Since K is given by eqn (30.3.10b) in terms of the partition functions of A, B, and $C^{\ddagger}$, in principle we have an explicit expression for calculating the second-order rate constant for a simple bimolecular reaction in terms of the molecular parameters for the reactants and the activated complex.

30.3 (c) How to use the Eyring equation

The partition functions for the reactants can normally be calculated quite readily, using either spectroscopic information about their energy levels or the approximate expressions set out in Box 22.1. The difficulty with the

Eyring equation lies in calculating the partition function for the activated complex, since $C^{\ddagger}$ cannot normally be investigated spectroscopically. We are normally forced to make assumptions about its size, shape, and structure.

As a first example, consider the case of two structureless particles A and B colliding to give an activated complex in the form of a diatomic molecule, which then breaks up to give products. Since the reactants are structureless 'atoms' the only contributions to their partition functions are the translational terms:

$$q^{\ominus}_{A,m} = V^{\ominus}_m/\Lambda^3_A, \qquad q^{\ominus}_{B,m} = V^{\ominus}_m/\Lambda^3_B, \tag{30.3.12}$$

with $\Lambda_J = h(\beta/2\pi m_J)^{\frac{1}{2}}$ and $V^{\ominus}_m = RT/p^{\ominus}$. The activated complex is a diatomic molecule of mass $m_C = m_A + m_B$ and moment of inertia I. It has one vibrational mode, but that corresponds to motion along the reaction coordinate, and so it does not appear in $\bar{q}_C$. It follows that the molar partition function of the activated complex is

$$\bar{q}^{\ominus}_{C,m} = (2IkT/\hbar^2)V^{\ominus}_m/\Lambda^3_C. \tag{30.3.13}$$

Since the moment of inertia of a diatomic molecule of bond length R_{AB} is equal to μR^2_{AB}, where μ is the reduced mass, the expression for the rate constant is

$$\begin{aligned} k_{\text{eff}} &= (RT/p^{\ominus})\left\{\frac{(\kappa kT/h)N_A\Lambda^3_A\Lambda^3_B V^{\ominus}_m(2IkT/\hbar^2)}{V^{\ominus 2}_m p^{\ominus}\Lambda^3_C}\right\}e^{-\Delta E^{\ddagger}_0/RT} \\ &= \kappa(RT/h)(\Lambda_A\Lambda_B/\Lambda_C)^3(2IkT/\hbar^2)e^{-\Delta E^{\ddagger}_0/RT} \\ &= N_A(8kT/\pi\mu)^{\frac{1}{2}}\pi\kappa R^2_{AB}e^{-\Delta E^{\ddagger}_0/RT}, \end{aligned} \tag{30.3.14}$$

and by identifying $\kappa\pi R^2_{AB}$ as σ^*, the reactive cross-section, we arrive at precisely the same expression as obtained from simple collision theory, eqn (30.1.5).

Example 30.3

Obtain an order-of-magnitude estimate of the P-factor for the reaction between two non-linear molecules.

- *Method*. Use the Eyring equation twice, first for two structureless particles, and then for two molecules with internal structure (so that they can rotate and vibrate). The ratio of the two is P. For the order-of-magnitude estimate, assume that all the translational partition functions are identical, likewise all rotational and all vibrational partition functions. An N-atomic nonlinear molecule has three translational, three rotational, and $3N-6$ vibrational modes.

- *Answer*. For no internal modes of the colliding particles,

$$q^{\ominus}_{A,m} = V^{\ominus}_m/\Lambda^3, \qquad q^{\ominus}_{B,m} = V^{\ominus}_m/\Lambda^3, \qquad \bar{q}^{\ominus}_{C,m} = q^2_r V^{\ominus}_m/\Lambda^3.$$

Therefore, from eqn (30.3.10),

$$\begin{aligned} k_{\text{eff}} &= \kappa(kT/h)(RT/p^{\ominus})\{N_A q^2_r\Lambda^6 V^{\ominus}_m/\Lambda^3 V^{\ominus}_m\}e^{-\Delta E^{\ddagger}_0/RT} \\ &= \kappa(RT/h)q^2_r\Lambda^3 e^{-\Delta E^{\ddagger}_0/RT}. \end{aligned}$$

When all the modes are active,

$$q^{\ominus}_{A,m} = q^3_r q^{3N-6}_v V^{\ominus}_m/\Lambda^3, \qquad q^{\ominus}_{B,m} = q^3_r q^{3N'-6}_v V^{\ominus}_m/\Lambda^3,$$
$$\bar{q}^{\ominus}_{C,m} = q^3_r q^{3(N+N')-7}_v V^{\ominus}_m/\Lambda^3.$$

Note that one vibrational mode has been subtracted from the modes of the activated complex.

It follows that

$$k_{\text{eff}} = \kappa(kT/h)(RT/p^{\ominus})\left\{\frac{N_A q_r^3 q_v^{3(N+N')-7} \Lambda^6 V_m^{\ominus}}{\Lambda^3 q_r^6 q_v^{3(N+N')-12} V_m^{\ominus}}\right\} e^{-\Delta E_0^{\ddagger}/RT}$$

$$= \kappa(RT/h)\{q_v^5 \Lambda^3 / q_r^3\} e^{-\Delta E_0^{\ddagger}/RT}.$$

Comparison of the two results gives

$$P = q_v^5 q_r^5.$$

Since $q_v/q_r \approx \frac{1}{50}$, $P \approx 3 \times 10^{-9}$.

• *Comment*. The calculation suggests that, for reactions with similar activation energies, the rate constants for reactions between complex molecules in the gas phase should be much slower than reactions between simple molecules.

• *Exercise*. Estimate the P-factor for two linear molecules that form a non-linear activated complex. $[P = q_v/q_r \approx \frac{1}{50}]$

As a second example, consider the effect of *deuteration* on a reaction in which the rate determining step is the breaking of a C—H bond. The reaction coordinate corresponds to the stretching of the C—H bond, and the potential energy profile is shown in Fig. 30.4. On deuteration, the dominant change is the reduction of the zero-point energy of the bond (because the deuterium atom is heavier). The whole reaction profile is not lowered, however, because the relevant vibration in the activated complex has a very low force constant, and so there is little zero-point energy in either the proton or the deuterium forms of the complex.

We assume that the deuteration affects only the reaction coordinate, and hence that the partition functions for all the other internal modes remain unchanged. The translational partition functions are changed by deuteration, but the mass of the rest of the molecule is normally so great that the change is insignificant. The value of $\Delta E_0^{\ddagger}$ changes on account of the change of zero-point energy, and

$$\Delta E_0^{\ddagger}(\text{C–D}) - \Delta E_0^{\ddagger}(\text{C–H}) = \tfrac{1}{2}\hbar\omega(\text{C–H}) - \tfrac{1}{2}\hbar\omega(\text{C–D})$$
$$= \tfrac{1}{2}\hbar k_f\{(1/\mu_{\text{CD}})^{\frac{1}{2}} - (1/\mu_{\text{CH}})^{\frac{1}{2}}\}, \quad (30.3.15)$$

where k_f is the force constant of the bond and μ the reduced mass. Since all the partition functions are the same (by assumption), the rate coefficients for the two species should be in the ratio

$$k_{\text{C-D}}/k_{\text{C-H}} = \exp\{(\hbar k_f^{\frac{1}{2}}/2kT)[(1/\mu_{\text{CD}})^{\frac{1}{2}} - (1/\mu_{\text{CH}})^{\frac{1}{2}}]\}. \quad (30.3.16)$$

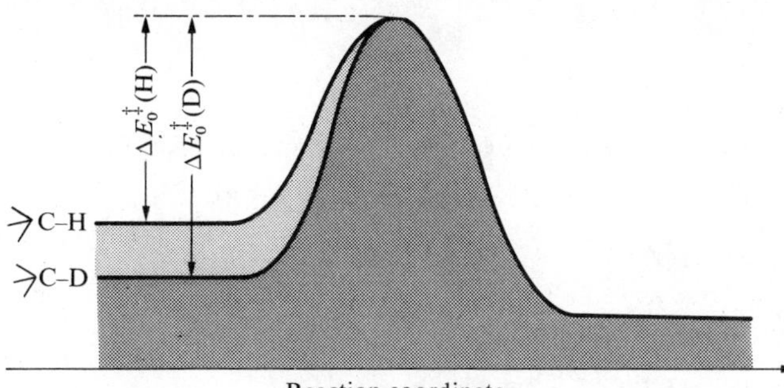

Fig. 30.4. Changes in the reaction profile when a bond undergoing cleavage is deuterated. The only significant change is to the zero-point energy of the reactants, which is lower for C—D than for C—H. As a result, the activation energy is greater for C—D than for C—H.

This predicts that at room temperature C—H cleavage should be about seven times faster than C—D cleavage, other conditions being equal, the principal reason being the smaller activation energy in the case of C—H on account of its greater zero-point energy.

30.3 (d) Thermodynamic aspects

If we accept that $\bar{K}_p$ is an equilibrium constant (even though one mode of $C^{\ddagger}$ has been discarded), we can express it in terms of a molar *Gibbs function of activation* through

$$\Delta G^{\ddagger} = -RT \ln \bar{K}_p. \tag{30.3.17}$$

Then the rate constant becomes

$$k_{\text{eff}} = \kappa(kT/h)(RT/p^{\ominus})e^{-\Delta G^{\ddagger}/RT}. \tag{30.3.18}$$

Since $G = H - TS$, the Gibbs function of activation can be divided into a molar *entropy of activation* ($\Delta S^{\ddagger}$) and a molar *enthalpy of activation* ($\Delta H^{\ddagger}$) through

$$\Delta G^{\ddagger} = \Delta H^{\ddagger} - T\Delta S^{\ddagger}. \tag{30.3.19}$$

Gibbs functions, enthalpies, and entropies of activation are widely used to report experimental reaction rates, especially for organic reactions in solution, and they are encountered when relationships between equilibrium constants and rates of reaction are explored using *correlation analysis*. In correlation analysis $\ln K(=-\Delta_r G^{\ominus}/RT)$ is plotted against $\ln k(\propto -\Delta G^{\ddagger}/RT)$, and in many cases the relation is linear: this is the origin of the name *linear free energy relation* (LFER; see *Further reading*).

When eqn (30.3.19) is used in eqn (30.3.18), and κ is absorbed into the entropy term, we obtain

$$k_{\text{eff}} = (kT/h)(RT/p^{\ominus})e^{\Delta S^{\ddagger}/R}e^{-\Delta H^{\ddagger}/RT}. \tag{30.3.20}$$

This equation can be expressed in Arrhenius form by identifying the activation energy E_a using the definition in eqn (28.3.5): after some calculation‡ we find

$$\Delta H^{\ddagger} = E_a - 2RT. \tag{30.3.21}$$

(This is the case for a bimolecular gas phase reaction; for a reaction in solution $\Delta H^{\ddagger} = E_a - RT$; see footnote.) Consequently

$$k_{\text{eff}} = Ae^{-E_a/RT}, \qquad A = (kTe^2/h)(RT/p^{\ominus})e^{\Delta S^{\ddagger}/R}. \tag{30.3.22}$$

‡ Write $RT/p^{\ominus} = V_m^{\ominus}$ and $k_{\text{eff}} = (kT/h)V_m^{\ominus}\bar{K}_p$ and obtain

$$\begin{aligned} E_a &= RT^2(\partial \ln k_{\text{eff}}/\partial T)_V = (RT^2/k_{\text{eff}})(\partial k_{\text{eff}}/\partial T)_V \\ &= (RT/\bar{K}_p)(\partial T\bar{K}_p/\partial T)_V = (RT/\bar{K}_p)\{\bar{K}_p + T(\partial \bar{K}_p/\partial T)_V\} \\ &= RT + RT^2(\partial \ln \bar{K}_p/\partial T)_V = RT + \Delta U^{\ddagger}, \end{aligned}$$

where $\Delta U^{\ddagger}$ is the molar *activation internal energy*. Since $\Delta H^{\ddagger} = \Delta U^{\ddagger} - RT$ for a reaction $A + B \rightarrow C^{\ddagger}$ involving perfect gases, it follows that $E_a = \Delta H^{\ddagger} + 2RT$. For a reaction in solution $\Delta H^{\ddagger} \approx \Delta U^{\ddagger}$, and then $E_a = \Delta H^{\ddagger} + RT$.

Hence,

$$\Delta S^{\ddagger} = R \ln \{hAp^{\ominus}/N_{A}e^{2}k^{2}T^{2}\}. \qquad (30.3.23)$$

A practical form of this expression is given in the following *Example*.

Example 30.4

Calculate the activation Gibbs function, enthalpy, and entropy of the second-order hydrogenation of ethane at 355 °C using information from Table 30.1.

• *Method*. From the Table, $A = 1.24 \times 10^{6}\ \mathrm{M^{-1}\,s^{-1}}$ and $E_a = 180\ \mathrm{kJ\,mol^{-1}}$. For a simple bimolecular gas-phase reaction $\Delta H^{\ddagger}$ is related to E_A by eqn (30.3.21). The entropy of activation is obtained using eqn (30.3.23), which (with $p^{\ominus} = 1$ bar) can be expressed as followed:

$$hAp^{\ominus}/N_{A}e^{2}k^{2}T^{2} = 7.8119 \times 10^{-11}(A/\mathrm{M^{-1}\,s^{-1}})/(T/\mathrm{K})^{2},$$

so that

$$\Delta S^{\ddagger} = R \ln \{7.8119 \times 10^{-11}(A/\mathrm{M^{-1}\,s^{-1}})/(T/\mathrm{K})^{2}\}.$$

Finally, form $\Delta G^{\ddagger}$ using eqn (30.3.19).

• *Answer*. From eqn (30.3.21):

$$\Delta H^{\ddagger} = E_a - 2RT = (180 - 2 \times 5.2)\ \mathrm{kJ\,mol^{-1}} = +170\ \mathrm{kJ\,mol^{-1}}.$$

From the expression above,

$$\Delta S^{\ddagger} = R \ln \{(7.8119 \times 10^{-11}) \times (1.24 \times 10^{6})/628^{2}\} = -184\ \mathrm{J\,K^{-1}\,mol^{-1}}.$$

From eqn (30.3.19),

$$\Delta G^{\ddagger} = \Delta H^{\ddagger} - T\Delta S^{\ddagger} = +286\ \mathrm{kJ\,mol^{-1}}.$$

• *Comment*. Notice the large negative entropy of activation. The simple collision theory magnitude of A calculated in *Example* 30.1 corresponds to a much less negative value ($-73\ \mathrm{J\,K^{-1}\,mol^{-1}}$).

• *Exercise*. Repeat the calculation for the $\mathrm{K + Br_2 \rightarrow KBr + Br}$ reaction at 298 K.
[$-2.5\ \mathrm{kJ\,mol^{-1}}$, $-58\ \mathrm{J\,K^{-1}\,mol^{-1}}$, $+15\ \mathrm{kJ\,mol^{-1}}$]

We have now made contact with the collision theory expression for the rate coefficient. In that theory A is determined by the frequency of collisions in the gas. But collisions correspond to a decrease in entropy (because they correspond to particles coming together and therefore to a reduction of randomness in the gas). Hence the negative value of $\Delta S^{\ddagger}$ reflects the occurrence of collisions. Furthermore, collisions with well-defined relative orientations correspond to an even greater reduction of entropy than is brought about by collisions in general, and so the entropy of activation should then be even more negative. This reduces the value of A, a feature taken into account in collision theory by the steric factor P.

30.3 (e) Reactions between ions

The application of ACT to reactions in solution is very complicated because the solvent plays a role in the activated complex. It is much simpler to use the thermodynamic version of the theory, and to combine

$$\mathrm{d[P]}/\mathrm{d}t = k^{\ddagger}[\mathrm{C}^{\ddagger}] \qquad (30.3.24)$$

with

$$K=\{a_C/a_Aa_B\}_{eq}=K_\gamma\{[C^\ddagger]/[A][B]\}_{eq}, \qquad (30.3.25)$$

K_γ being the ratio of activity coefficients, $\gamma_C/\gamma_A\gamma_B$. Then

$$d[P]/dt=k_{eff}[A][B], \qquad k_{eff}=k^\ddagger(K/K_\gamma). \qquad (30.3.26)$$

If k^0_{eff} is the rate coefficient when the activity coefficients are unity,

$$k_{eff}=\{\gamma_A\gamma_B/\gamma_C\}k^0_{eff}. \qquad (30.3.27)$$

This discussion can be taken further by expressing the activity coefficients in terms of the ionic strength of the solution using the Debye–Hückel theory (Section 11.2, particularly eqn (11.2.11)) in the form:

$$\lg\gamma_J=-Az_J^2I^{\frac{1}{2}},$$

for then

$$\lg k_{eff}=\lg k^0_{eff}-A\{z_A^2z_B^2-(z_A+z_B)^2\}I^{\frac{1}{2}},$$
$$\lg k_{eff}=\lg k^0_{eff}+2Az_Az_BI^{\frac{1}{2}}. \qquad (30.3.28)$$

(The charges of A and B are z_A and z_B, and so the charge of the activated complex is z_A+z_B; the z_J are positive for cations and negative for anions.)

The last equation shows that the rate constant for a reaction between ions depends on the ionic strength of the solution: this is the *kinetic salt effect*, Fig. 30.5. One consequence is that it is therefore essential to state the ionic strength when reporting a rate constant for an ionic reaction. Furthermore, if the ions have the same sign (e.g. a reaction between cations), then increasing the ionic strength (e.g. by the addition of inert ions) *increases* the rate coefficient. This is because the formation of a single, highly charged ionic complex from two less highly charged ions is favoured by a high ionic strength because the new ion has a denser ionic atmosphere. Conversely, ions of opposite charge react more slowly in solutions of high ionic strength. This is because the charges cancel and the complex has a less favourable interaction with its atmosphere than the separated ions.

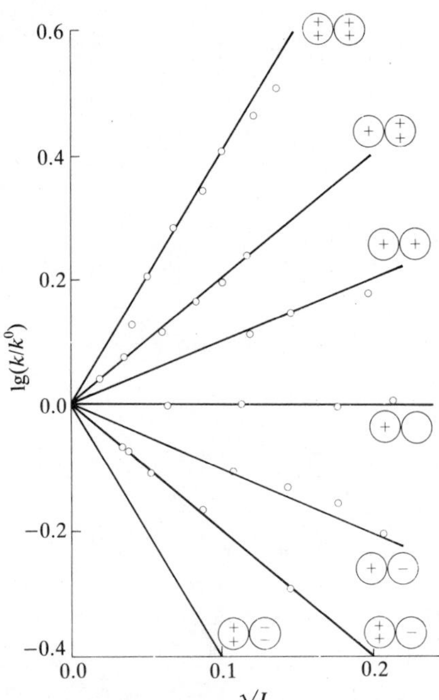

Fig. 30.5. Experimental tests of the kinetic salt effect for reactions in water at 298 K. The ion types are shown as spheres, and the slopes of the lines are those given by the Debye–Hückel limiting law via eqn (30.3.28).

Example 30.5

The rate constant for the alkaline hydrolysis of $[Co(NH_3)_5Br]^{2+}$ depends on ionic strength as tabulated below. What can be said about the charge of the activated complex in the rate-determining stage.

I	0.005	0.010	0.015	0.020	0.025	0.030
k/k^0	0.718	0.631	0.562	0.515	0.475	0.447

• *Method*. Plot $\lg(k/k^0)$ against $I^{\frac{1}{2}}$; the slope will be $1.02z_Az_B$. For the OH^- ion, $z_A=-1$; therefore infer the charge number of the other ion participating in the activated complex.

• *Answer*. Form the following table:

I	0.005	0.010	0.015	0.020	0.025	0.030
$I^{\frac{1}{2}}$	0.071	0.100	0.122	0.141	0.158	0.173
$\lg(k/k^0)$	−0.14	−0.20	−0.25	−0.29	−0.32	−0.35

These points are plotted in Fig. 30.6. The slope of the (least squares) line is −2.1, indicating that $z_B=+2$. Hence, the complex ion appears to be involved in the activated complex with the OH^-.

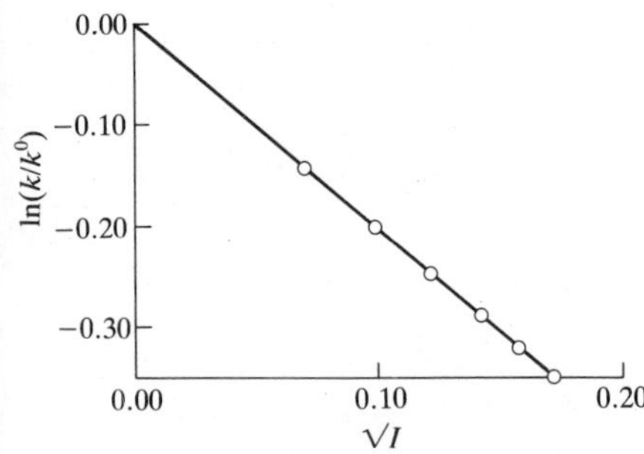

Fig. 30.6. The experimental ionic strength dependence of the rate constant of a hydrolysis reaction: the slope gives information about the charge types involved in the activated complex of the rate determining step. See *Example* 30.5.

• *Comment*. The rate constant is also influenced by the relative permittivity of the medium (see Problem 30.29).

• *Exercise*. One ion of charge number +1 is known to be involved in the activated complex of a reaction. Deduce the charge number of the other ion from the following data:

I	0.005	0.010	0.015	0.020	0.025	0.030
k/k^0	0.98	0.97	0.97	0.96	0.96	0.95

[−1]

30.4 The dynamics of molecular collisions

We now come to the third level of our examination of the factors that govern the rates of reactions: the description of the individual events. *Molecular beam* techniques allow us to study collisions between molecules in preselected states, and can be used to determine the states of the products of a reactive collision. Information of this kind is essential if a full picture of the reaction is to be built, because the rate constant is an *average* over events in which reactants in different initial states evolve into products in different final states. However, we are still far from having a complete theory because most of the reactions we want to know about take place in solution, and our ability to study individual processes in liquids is still rudimentary.

30.4 (a) Reactive encounters

Detailed experimental information comes from molecular beams, especially crossed molecular beams (Section 24.3). The detector for the products of the collision of two beams can be moved to different angles, and so the *angular distribution* of the products can be determined. Furthermore, the detector can distinguish between different energy states of the products. Therefore, since the molecules in the incoming beams can be prepared with different energies (e.g. different translational energies using rotating sectors and supersonic nozzles, and different vibrational energies using selective excitation with lasers) and different orientations (using electric fields), it is possible to study the dependence of the success of collisions on these variables and to study how they affect the properties of the outcoming product particles. A helpful method for examining the energy distribution in the products is *infrared chemiluminescence*, in which vibrationally excited molecules emit infrared radiation as they return to their ground states. By studying the line intensities in the infrared emission spectrum, the populations of the vibrational states may be determined, Fig. 30.7.

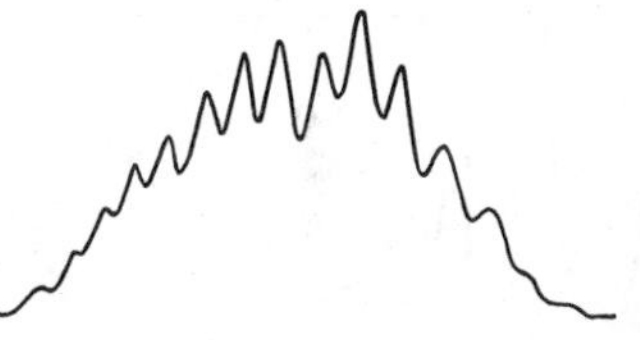

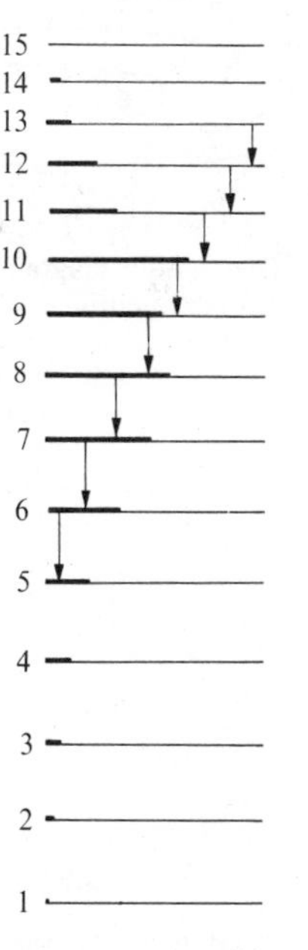

Fig. 30.7. Infrared chemiluminescence from CO produced in the reaction $O + CS \rightarrow CO + S$ arises from the non-equilibrium populations of the vibrational states of CO and the radiative relaxation to equilibrium.

30.4 (b) Potential energy surfaces

One of the most important concepts for discussing beam results is the *potential energy surface* of a reaction. In the case of a collision between an H atom and an H_2 molecule, for instance, the potential energy surface is the plot of the potential energy for all relative locations of the three hydrogen nuclei. However, at the outset we are confronted with a major difficulty: three coordinates are needed to specify the separations of three nuclei, and so the depiction of the energy requires a diagram in four dimensions. That is absurdly complicated, and in order to draw a surface we need to make a major simplification (but a computer, of course, does not need the simplification: it can perform calculations in any number of dimensions, and in modern calculations the collisions are treated with all their degrees of freedom intact). Detailed calculations show that the approach of an atom along the H—H axis requires less energy for reaction than any other approach, and so initially we confine our attention to a collinear collision. Now only *two* parameters are required to define the nuclear separations: one is the H_a—H_b separation R_{ab}, and the other is the H_b—H_c separation R_{bc}.

At the start of the collision R_{ab} is infinite and R_{bc} is the H_2 equilibrium bond length. At the end of a successful reactive collision R_{ab} is equal to the bond length and R_{bc} is infinite. The total energy of the three-atom system depends on their relative separations, and can be found by doing a normal molecular structure calculation using the Born–Oppenheimer approximation (Section 16.1(a)), where all atom locations are regarded as frozen. The plot of the total energy of the system against each frozen value of R_{ab} and R_{bc} gives the *potential energy surface* of the reaction, Fig. 30.8. This is normally depicted as a contour diagram, Fig. 30.9.

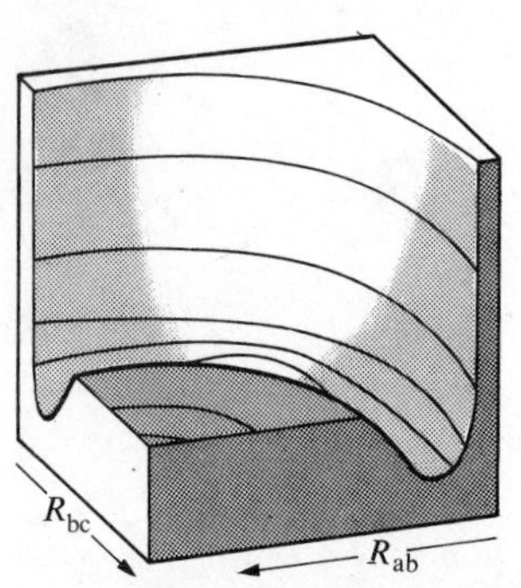

Fig. 30.8. The potential energy surface for the $H + H_2$ reaction when the atoms are constrained to be collinear.

When R_{ab} is very large, the variation in energy represented by the surface as R_{bc} changes are those of an isolated H_2 molecule as its bond length is altered. The slice through the surface at $R_{ab} = \infty$, for example, is the same as the H_2 potential energy curve drawn in Fig. 16.2. On the other hand, at the edge of the diagram where R_{bc} is very large, the slice through the surface is the molecular potential energy curve of an isolated H_a—H_b molecule.

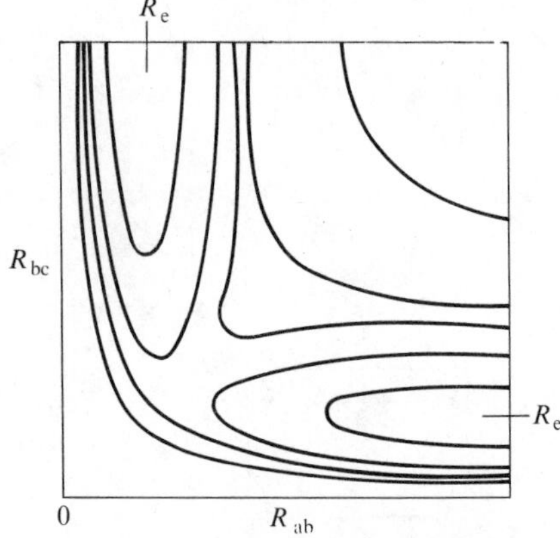

Fig. 30.9. The contour diagram (with contours of equal potential energy) corresponding to the surface in Fig. 30.8. R_e marks the equilibrium bond length of an H_2 molecule (strictly, it relates to the arrangement when the third atom is at infinity).

The actual path of the atoms depends on their *total* energy, which depends on their kinetic energies as well as their potential energies. However, a general idea of the paths available to the system can be obtained by considering the potential energy surface alone, and looking for paths that correspond to least potential energy. For example, consider the changes in potential energy as H_a approaches an H_b—H_c molecule. If the H_b—H_c bond length is treated as constant during the initial approach of H_a, the potential energy of the H_3 system would rise along the path marked A in Fig. 30.10. This shows that the potential energy rises to a high value as H_a is pushed into the molecule, and then decreases sharply as H_c breaks off and is taken to a great distance. An alternative reaction path can be imagined (B) in which the H_b—H_c bond length increases while H_a is still far away. It is

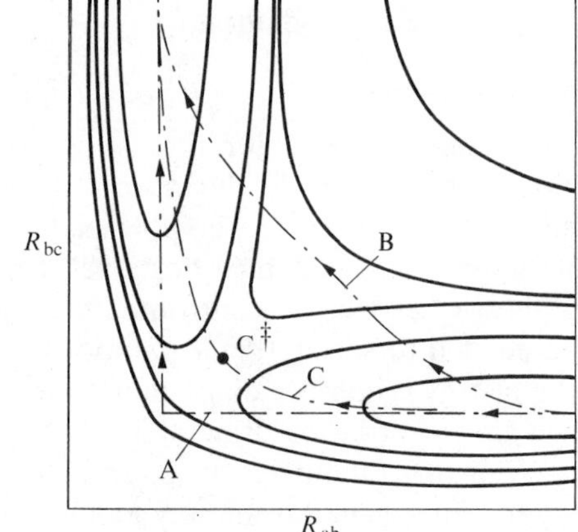

Fig. 30.10. Various trajectories through the potential energy surface shown in Fig. 30.9. A corresponds to a path in which R_{bc} is held constant as H_a approaches; B corresponds to a path in which R_{bc} lengthens at an early stage during the approach of H_a. C is the path along the floor of the potential valley.

clear that both paths, although feasible if the particles have sufficient initial kinetic energy, take the atoms to regions of high potential energy.

The path corresponding to least potential energy is the one marked C. It corresponds to R_{bc} lengthening as H_a approaches and begins to form a bond with H_b. The H_b—H_c bond relaxes at the demand of the incoming atom, and although the potential energy rises, it climbs only as far as the *saddle point* marked $C^{\ddagger}$. The reaction path involving least potential energy is the route C up the floor of the valley, through the saddle point, and down the floor of the other valley as H_c recedes and the new H_a—H_b bond sinks to its equilibrium length. This path is the *reaction coordinate* we met in Section 30.3(a).

30.4 (c) Motion through the surface

The kinds of questions explored by molecular beam studies can be introduced by considering how the particles move through their potential energy surface. In order to trace the path from reactants to products by making the journey along C, the incoming particles have to possess sufficient kinetic energy to be able to climb to the saddle point. By changing the relative speed of approach, and determining the kinetic energy at which reaction occurs, the shape of the surface can be explored experimentally. If the collision occurs with a large amount of kinetic energy, the saddle point might be overshot and a path may be taken that leads to vibrational excitation of the product molecule, Fig. 30.11. By observing the degree of excitation of the products, details of the shape of the potential energy surface can be investigated.

The kind of practical question that can be answered is whether it is better

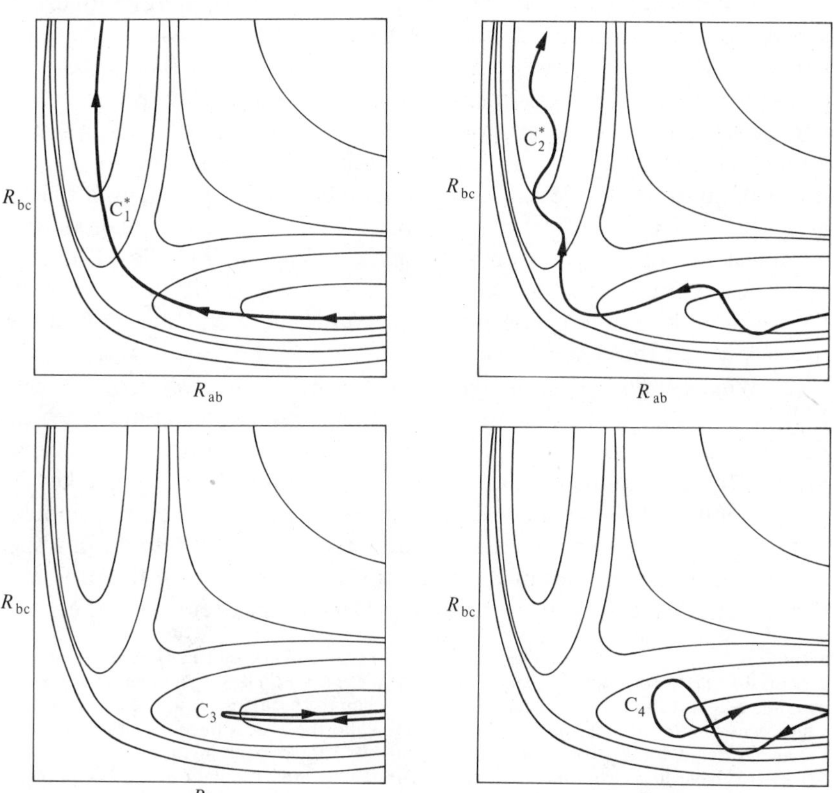

Fig. 30.11. Some successful (*) and unsuccessful encounters. C_1^* corresponds to the path along the foot of the valley; C_2^* corresponds to an approach of A to a vibrating BC molecule, and the formation of a vibrating AB molecule as C departs. C_3 corresponds to A approaching a non-vibrating BC molecule, but with insufficient translational kinetic energy; C_4 corresponds to A approaching a vibrating BC molecule, but still the energy, and the phase of the vibration, is insufficient for reaction.

to smash the reactants together with a lot of translational kinetic energy or to ensure that they approach in highly excited vibrational states instead. For example, is trajectory C_2^* in Fig. 30.11, where the H_b—H_c molecule is initially vibrationally excited, more efficient at leading to reaction than the trajectory C_1^*, in which the total energy is the same but has a large amount of translational kinetic energy? We see below something of the answers to questions like this.

A related question is how the information given by molecular beam studies is related to the value of the rate constant of the reaction. This is the most complex part of the analysis, because in an actual reaction the particles collide with many different energies and in many different rotational and vibrational states. Each individual collision can be imagined as a trajectory on the potential energy surface. Some of these trajectories will be successful (C_1^* and C_2^* in Fig. 30.11) and some unsuccessful (C_3 and C_4) either because they lack sufficient energy or because the energy is distributed inappropriately. The rate of reaction is an average over all these trajectories, and so the calculation of the rate coefficient means that many trajectories must be calculated, and then averaged in some way. One of the techniques for doing the average in a manner that is consistent with the Boltzmann distribution of populations over the states of the system is the Monte Carlo method, which was described in Section 24.4(c).

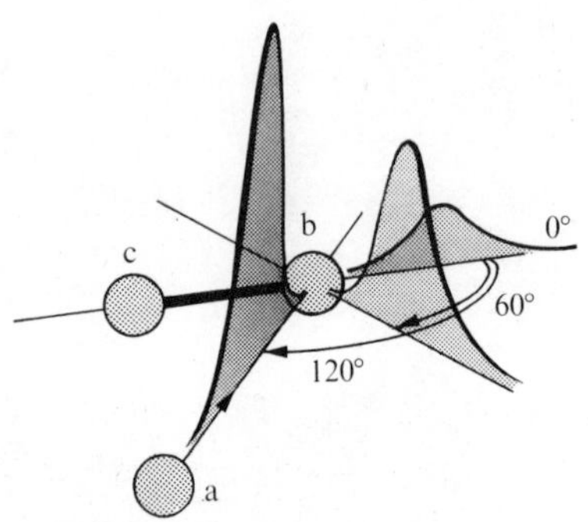

Fig. 30.12. An indication of the anisotropy of the potential energy changes as H approaches H_2 with different angles of attack. The collinear attack has the lowest potential barrier to reaction. The curves show the potential energy profiles along the reaction coordinate for each configuration.

30.4 (d) Some results from experiments and calculations

In this section we see how some of the questions just raised are answered, and how the study of collisions and the calculation of potential energy surfaces gives some insight into the course of reactions.

(1) *Is collinear approach the path of least potential energy?* Figure 30.12 shows the results of a calculation of the potential energy as a hydrogen atom approaches a hydrogen molecule from different angles, the H_2 bond being allowed to relax to the optimum length in each case. The potential energy barrier is least for collinear attack, as we assumed earlier. (But we must be aware that other lines of attack are feasible and contribute to the overall rate.) In contrast, Fig. 30.13 shows the potential energy changes that occur as a Cl atom approaches an HI molecule. The lowest barrier occurs for approaches within a cone of half-angle 30° surrounding the H atom. The relevance of this result to the calculation of the steric factor of collision theory should be noted: not every collision is successful, because not every one lies within the reactive cone.

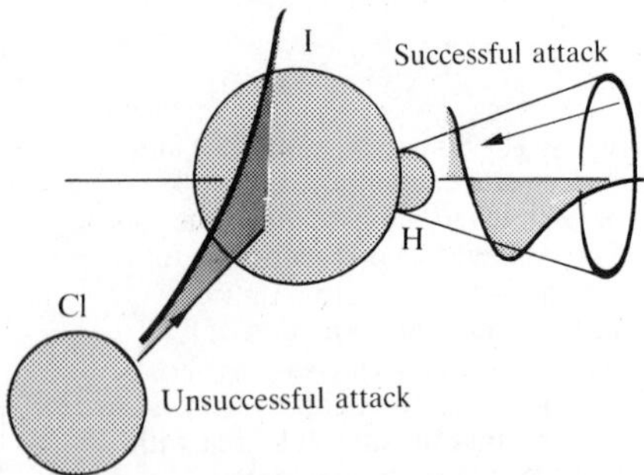

Fig. 30.13. Partial potential energy barriers for approach of Cl to HI. In this case successful encounters occur only when Cl approaches within a cone surrounding the H atom.

(2) *What determines the angular distribution of products?* If the collision is sticky, so that when the reactants collide they rotate around each other, the products can be expected to emerge in random directions because all memory of the approach direction has been lost. A rotation takes about 10^{-12} s, and so if the collision is over in less than that time the complex will not have had time to rotate and the products will be thrown off in a specific direction. In the collision of K and I_2, for example, most of the products are thrown off in the forward direction.‡ This is consistent with the harpoon

‡ There is a subtlety here. In molecular beam work the remarks normally refer to directions in a *centre-of-mass coordinate system*. The origin of the coordinates is the centre of mass of the colliding reactants, and the collision takes place when the particles are at the origin. The way in which centre-of-mass coordinates are constructed and the events in them interpreted involves too much detail for our present purposes, but we should bear in mind that 'forward' and 'backward' have unconventional meanings. The details are explained in the books in *Further reading*.

mechanism (Section 30.1(b)) because the transition takes place at long range. In contrast, the collision of K with CH_3I leads to reaction only if the particles approach each other very closely. This is like K bumping into a brick wall, and the KI product bouncing out in the backward direction. The detection of this anisotropy in the angular distribution of products gives an indication of the distance and orientation of approach needed for reaction, as well as showing that the event is complete in less than 10^{-12} s.

(3) *Is it better to have the collisional energy in translation or vibration?* Some reactions are very sensitive to whether the energy has been predigested into a vibrational mode or left as the relative translational kinetic energy of the colliding particles. For example, if two HI molecules are hurled together with more than twice the activation energy of the reaction, no reaction occurs if all the energy is translational. In the case of $F + HCl \rightarrow Cl + HF$ it has been found that the reaction is about five times as efficient when the HCl is in its first vibrational excited state than when, although it has the same total energy, it is in its vibrational ground state.

The reasons for these requirements can be found by examining the potential energy surfaces of the reactions. Two important cases are shown in Fig. 30.14. In Fig. 30.14(a) the saddle point occurs early in the reaction coordinate: this is called an *attractive surface*. In Fig. 30.14(b) the saddle point occurs later: this is a *repulsive surface*. Note that a reaction that is attractive in one direction is repulsive in the reverse direction.

Consider first the attractive surface, Fig. 30.14(a). If the original molecule is vibrationally excited, a collision with an incoming particle takes the system along C. This path is bottled up in the region of the reactants, and does not take the system to the saddle point. If, however, the same amount of energy is present solely as translational kinetic energy, the system moves along C* and travels smoothly over the saddle point into products. We can therefore conclude that *reactions with attractive potential energy surfaces proceed more efficiently if the energy is in relative translational motion*. Moreover, the potential surface shows that once past the saddle point the trajectory runs up the steep wall of the product valley, and then rolls from side to side as it falls to the foot of the valley as the products separate. In other words, the products emerge in a vibrationally excited state.

Now consider the repulsive surface, Fig. 30.14(b). On trajectory C the collisional energy is largely in translation. As the reactants approach, the potential energy rises. Their path takes them up the opposing face of the valley, and they are reflected back into the reactant region. This corresponds to an unsuccessful encounter, even though the energy is sufficient for reaction. On C* some of the energy is in the vibration of the reactant molecule. This vibration causes the trajectory to weave from side to side up the valley as it approaches the saddle point. This motion may be sufficient to tip the system round the corner to the saddle point and then on to products. In this case, the product molecule is expected to be in an unexcited vibrational state. It follows that *reactions with repulsive potential surfaces can be expected to proceed more efficiently if the excess energy is present as vibrations*. This is the case with the $H + Cl_2 \rightarrow HCl + Cl$ reaction, for instance.

(4) *Can we calculate trajectories showing the reaction in progress?* A clear picture of the reaction event can be obtained using classical mechanics to calculate the trajectories of the atoms taking place in a reaction. Figure 30.15 shows the result of such a calculation of the positions of the three

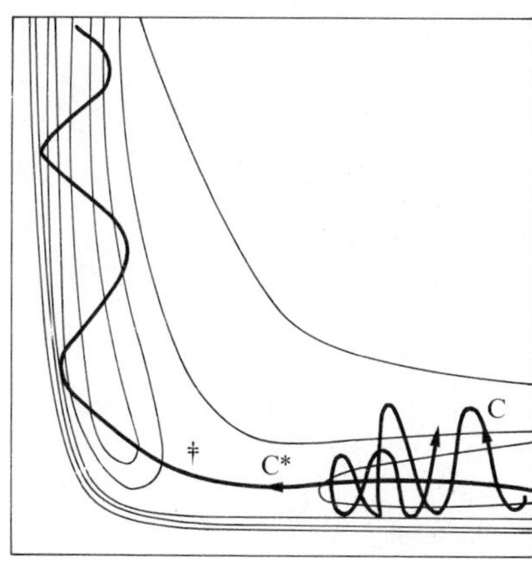

(a) Attractive surface

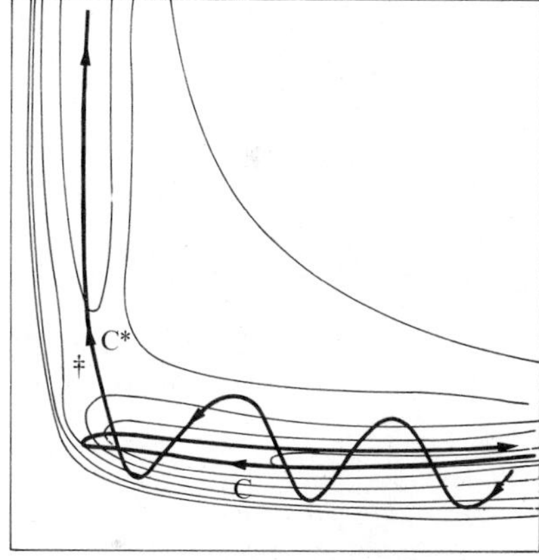

(b) Repulsive surface

Fig. 30.14. (a) An *attractive* potential energy surface. A successful encounter (C*) involves high translational kinetic energy and results in a vibrationally excited product. (b) A *repulsive* potential energy surface. A successful encounter (C*) involves initial vibrational excitation and the products have high translational kinetic energy. A reaction that is attractive in one direction is repulsive in the reverse direction.

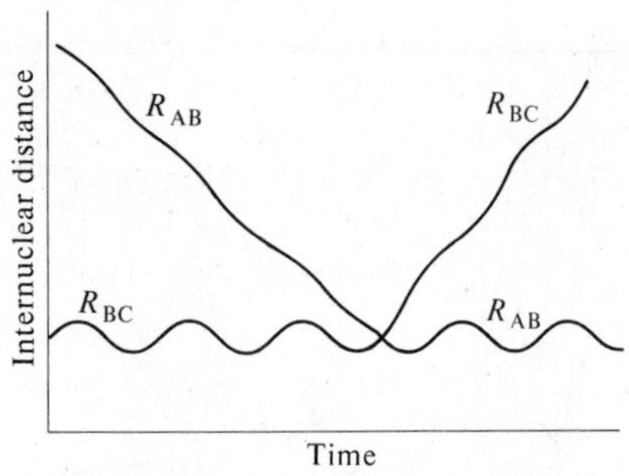

Fig. 30.15. The calculated trajectories for a reactive encounter between A and a vibrating BC molecule leading to the formation of a vibrating AB molecule.

atoms in the reaction $A + BC \rightarrow AB + C$, the horizontal coordinate now being time and the vertical coordinate the separations. This shows clearly the vibration of the original molecule and the approach of the attacking atom. The reaction itself, the switch of partners, takes place very rapidly. Then the new molecule shakes, but quickly settles down to steady, harmonic vibration as the expelled atom departs.

Although this kind of calculation gives a good sense of what happens during a reaction, its limitations must be kept in mind. In the first place, a real gas-phase reaction occurs with a wide variety of different speeds and angles of attack. In the second place, the motion of the atoms, electrons, and nuclei, is governed by quantum mechanics. The concept of trajectory then fades and is replaced by the unfolding of a wavefunction that represents initially reactants and finally products. Nevertheless, recognizing these limitations should not be allowed to obscure the fact that recent advances in molecular reaction dynamics have given us a first glimpse of the processes going on at the core of reactions.

Further reading

General texts:

Reaction kinetics. M. J. Pilling; Clarendon Press, Oxford, 1974.

Theory of chemical reaction rates. K. J. Laidler; McGraw-Hill, New York, 1969.

Gas phase reaction rate theory. H. S. Johnstone; Ronald, New York, 1966.

Kinetics and dynamics of elementary gas reactions. I. W. M. Smith; Butterworth, London, 1980.

Energetic principles of chemical reaction. J. Simons; Jones and Bartlett, Portola Valley, 1983.

Correlation analysis in organic chemistry. J. Shorter; Clarendon Press, Oxford, 1973.

Activated complex theory:

Activated complex theory: current status, extensions, and applications. R. A. Marcus; *Techniques of chemistry* (E. S. Lewis, ed) VIA, 13, Wiley-Interscience, New York, 1974.

Diffusion with reaction:

Chemical reactor theory (3rd edn). K. G. Denbigh and J. C. R. Turner; Cambridge University Press, 1984.

The mathematics of diffusion (2nd edn). J. Crank; Clarendon Press, Oxford, 1975.

Molecular beams:

Molecular beams in chemistry. M. A. D. Fluendy and K. P. Lawley; Chapman and Hall, London, 1974.

Molecular beams in chemistry. J. E. Jordan, E. A. Mason, and I. Amdur; *Techniques of chemistry* (A. Weissberger and B. W. Rossiter, eds) IIID, 365, Wiley-Interscience, New York, 1972.

Reactive scattering. R. Grice; *Adv. chem. Phys*. **30,** 249 (1975).

Molecular reaction dynamics. R. D. Levine and R. B. Bernstein; Clarendon Press, Oxford, 1974.

Chemical dynamics via molecular beam and laser techniques. R. B. Bernstein; Clarendon Press, Oxford, 1982.

Introductory problems

A30.1. The temperature dependence of the rate constant for a certain bimolecular reaction in the gas phase is given by the expression: $3.72 \times 10^{12} \exp(-8600\,\mathrm{K}/T)$ $\mathrm{dm^3\,mol^{-1}\,min^{-1}}$ in the temperature range 10–90 °C. The RMM values for the two reacting species are 16 and 100. Use collision theory to find the value of the reactive cross-section for the reaction at 25 °C.

A30.2. For the gaseous reaction $A + B \rightarrow$ products, the reactive cross-section obtained from the experimental value of the pre-exponential factor has the value $9.2 \times 10^{-22}\ m^2$. The collision cross-sections of A and B estimated from the transport properties are 0.95 and 0.65 nm^2 respectively. Find the *P*-factor for the reaction.

A30.3. Substances A and B, which are neutral species with diameters 588 pm and 1650 pm respectively, undergo the diffusion controlled reaction $A + B \rightarrow P$ in a liquid solvent with viscosity $2.37 \times 10^{-3}\ kg\ m^{-1}\ s^{-1}$ at 40 °C. Calculate the initial rate $d[P]/dt$ if the initial concentrations of A and B are 0.150 and 0.330 $mol\ dm^{-3}$ respectively.

A30.4. Find the rate coefficient at 30 °C for the diffusion controlled reaction of two anions: $A^- + B^- \rightarrow$ products. The viscosity of the solvent at 30 °C is $3.70 \times 10^{-3}\ kg\ m^{-1}\ s^{-1}$ and $P = 0.19$.

A30.5. The reaction of n-propylxanthate ion in acetic acid buffer solutions has the mechanism: $A^- + H^+ \rightarrow$ products. Near 30 °C the rate constant is given by the empirical expression $2.05 \times 10^{13} \exp(-8681\ K/T)\ M^{-1}\ s^{-1}$. Find the Arrhenius energy of activation and the entropy of activation at 30 °C.

A30.6. When the reaction in the previous problem occurs in a dioxane–water mixture which is 30 weight % dioxane, the rate coefficient is given by the expression $7.78 \times 10^{14} \exp(-9134\ K/T)$ for temperature near 30 °C. The entropy of activation at 30 °C is $31.7\ J\ K^{-1}\ mol^{-1}$. Calculate $\Delta H^{\ddagger}$ and $\Delta G^{\ddagger}$ for the reaction at 30 °C.

A30.7. The gas phase association reaction between F_2 and IF_5 is first order in each of the reactants. The Arrhenius energy of activation for the reaction is $58.6\ kJ\ mol^{-1}$. At 65 °C the rate coefficient is $7.84 \times 10^{-3}\ kPa^{-1}\ s^{-1}$. Calculate the entropy of activation at 65 °C.

A30.8. The rate coefficient for the reaction $H_2O_2 + I^- + H^+ \rightarrow H_2O + HIO$ is sensitive to the ionic strength of the aqueous solution in which the reaction occurs. The rate coefficient measured at 25 °C and at an ionic strength of 0.0525 is $12.2\ M^{-2}\ min^{-1}$. Use the Debye Limiting Law to estimate the value of the rate coefficient at zero ionic strength.

A30.9. The rate coefficient for the reaction $I^- + H_2O_2 \rightarrow H_2O + IO^-$ in aqueous solution varies slowly with ionic strength, even though the Debye Limiting Law predicts no effect. Use the following data for 25 °C to find the dependence of $\lg k_r$ on the ionic strength:

I	0.0207	0.0525	0.0925	0.1575
$k_r/M^{-1}\ min^{-1}$	0.663	0.670	0.679	0.694

What is the value of k_r at zero ionic strength? What does the result suggest for the functional form of the dependence of $\lg \gamma$ on ionic strength for a neutral molecule in an electrolytic solution?

A30.10. Use the Debye Limiting Law to show that changes in ionic strength can affect the rate of a reaction catalysed by H^+ from the ionization of a weak acid. Consider the mechanism: $H^+ + B \rightarrow$ products, where H^+ comes from the ionization of the weak acid, HA. The weak acid has a fixed concentration. First show that $\lg [H^+]$, derived from the ionization of HA, depends on the activity coefficients of ions and thus depends on the ionic strength. Then find the relationship between lg (rate) and $\lg [H^+]$ to show that the rate also depends on the ionic strength.

Problems

30.1. Calculate the collision frequencies z and Z in gases of (a) ammonia, $R \approx 190$ pm, (b) carbon monoxide, $R \approx 180$ pm at 25 °C and 1 atm pressure. What is the percentage increase when the temperature is raised by 10 K at constant volume?

30.2. The Boltzmann distribution gives the number of molecules in the energy range dE at the energy E as proportional to $\exp(-E/kT)\,dE$. What is the constant of proportionality on the assumption of a uniform density of states? What is the proportion of molecules that have energies of at least E_a?

30.3. Collision theory depends on knowing the proportion of molecular collisions having at least the threshold energy E_a. What is this proportion when (a) $E_a = 10\ kJ\ mol^{-1}$, (b) $E_a = 100\ kJ\ mol^{-1}$ at 200 K, 300 K, 500 K, 1000 K?

30.4. Calculate the percentage increase in the proportions in the last Problem when the temperature is raised by 10 K.

30.5. In the dimerization of methyl radicals at 25 °C the experimental pre-exponential factor is $2.4 \times 10^{10}\ M^{-1}\ s^{-1}$. What is (a) the reactive cross-section, (b) the *P*-factor for the reaction on the basis of a C–H bond length of 154 pm.

30.6. In Problem 28.26 the pre-exponential factor was determined. Estimate the *P*-factor and the reactive cross-section σ^* for the reaction of oxygen atoms with benzene at 340 K. Take $R(O) \approx R(Ne) \approx 78$ pm and $R(C_6H_6) \approx 265$ pm.

30.7. Nitrogen dioxide reacts bimolecularly in the gas phase to give $2NO + O_2$. The temperature dependence of the second-order rate constant for the rate law $d[\text{Product}]/dt = k\,[NO_2]^2$ is given below. What is the *P*-factor and the reactive cross-section for the reaction?

T/K	600	700	800	1000
$k/cm^3\ mol^{-1}\ s^{-1}$	4.6×10^2	9.7×10^3	1.3×10^5	3.1×10^6

Take $\sigma \approx 0.60\ nm^2$.

30.8. The diameter of the methyl radical is about 308 pm. What is the maximum rate constant in the expression $d[C_2H_6]/dt = k[CH_3]^2$ for second-order recombination of radicals at room temperature?

30.9. 10 per cent of a 1.0 dm^3 sample of ethane at 298 K and 1 atm pressure is dissociated into methyl radicals. What is the minimum time for 90 per cent recombination?

30.10. Calculate the magnitude of the diffusion-controlled rate coefficient at 298 K for a species in (a) water, (b) *n*-pentane, (c) *n*-decylbenzene. The viscosities are $1.00\ g\ m^{-1}\ s^{-1}$, $0.22\ g\ m^{-1}\ s^{-1}$, $3.36\ g\ m^{-1}\ s^{-1}$ (1.00 cP, 0.22 cP, 3.36 cP), respectively.

30.11. Confirm that eqn (30.2.14) is a solution of eqn (30.2.13) where $[J]_t$ is a solution of the same equation but with $k = 0$ and for the same initial conditions.

30.12. Evaluate $[J]_t^*$ numerically by writing a program for the numerical integration of eqn (30.2.14) with $[J]_t$ given by eqn (30.2.15), and explore the effect of increasing reaction rate constant on the spatial distribution of J.

30.13. Estimate the orders of magnitude of the partition functions involved in the rate expression. State the order of magnitude of $q_m^{\ominus T}/N_A$, q^R, q^V, q^E for typical molecules. Check that in the collision of two structureless molecules the order of magnitude of the pre-exponential factor is of the same order as that predicted by collision theory.

30.14. Using the order-of-magnitude partition functions derived in the last Problem, estimate the *P*-factor for a reaction in which $A + B \rightarrow P$, and A and B are non-linear triatomic molecules.

30.15. The base-catalysed bromination of nitromethane-d_3 in water at room temperature (298 K) proceeds 4.3 times more slowly than the bromination of the undeuterated material. Account for this difference. Use $k_f(C\text{–}H) \approx 450\ N\ m^{-1}$.

30.16. Predict the order of magnitude of the isotope effect on the relative rates of displacement of (a) ^1H and ^3H, (b) ^{16}O and ^{18}O. Will raising the temperature enhance the difference? Take $k_f(CH) \approx 450\ N\ m^{-1}$, $k_f(CO) \approx 1750\ N\ m^{-1}$.

30.17. The major difficulty in applying ACT (and, it must be admitted, in devising straightforward problems to illustrate it) is to decide on the structure of the activated complex and to ascribe appropriate bond strengths and lengths to it. The following exercise gives some familiarity with the difficulties involved, yet leads to a numerical result for a reaction of some interest. Consider the attack of H on D_2, which is one step in the $H_2 + D_2$ reaction. Suppose that the H approaches D_2 from the side and forms a complex in the form of an isosceles triangle. Take the H–D distance as 30 per cent greater than in H_2 (74 pm) and the D–D distance as 20 per cent greater than in H_2. Let the critical coordinate be the antisymmetric stretching vibration in which one H–D bond stretches as the other shortens. Let all the vibrations be at about $1000\ cm^{-1}$. Estimate *k* for this reaction at 400 K using the experimental activation energy of about $35\ kJ\ mol^{-1}$.

30.18. Now change the model of the activated complex in the last Problem so that it is linear. Use the same estimated molecular bond lengths and vibrational frequencies to calculate k_{eff} for this choice of model.

30.19. Clearly, there is much scope for modifying the parameters of the models of the activated complex in the last pair of Problems. Write and run a program that allows you to vary the structure of the complex and the parameters in a plausible way, and look for a model (or more than one model) that gives a value of *k* close to the experimental value, $4 \times 10^5\ M^{-1}\ s^{-1}$.

30.20. Calculate the molar entropy of activation for a collision between two essentially structureless particles at 300 K, taking $M_r \approx 50$, $\sigma \approx 0.4\ nm^2$.

30.21. Evaluate the entropies of activation for the $H + D_2$ reaction using the models described in Problems 30.17 and 30.18.

30.22. The Eyring equation can also be applied to physical processes. As an example, consider the rate of diffusion of an atom stuck to the surface of a solid. Suppose that in order to move from one site to another it has to reach the top of the barrier where it can vibrate classically in the vertical direction and in one horizontal direction, but vibration along the other horizontal direction takes it into the neighbouring site. Find an expression for the rate of diffusion, and evaluate it for tungsten atoms on a tungsten surface ($E_a = 60\ kJ\ mol^{-1}$). Suppose that the vibration frequencies at the transition state are (a) the same as, (b) one half the value for the adsorbed atom. What is the value of the diffusion coefficient *D* at 500 K? (Take the site separation as 316 pm and $\nu \approx 10^{11}$ Hz.)

30.23. Suppose now that the adsorbed, migrating species is a spherical molecule, and that it can rotate classically as well as vibrate at the top of the barrier, but that at the adsorption site itself it can only vibrate. What effect does this have on the diffusion constant? Take the molecule to be methane, for which $B = 5.24\ cm^{-1}$.

30.24. ESR results showed that a nitroxide radical trapped in a solid was rotating at about 1.0×10^8 Hz at 115 K (see Problem 20.36). Use the Eyring equation to estimate the activation Gibbs function for the rotation.

30.25. The pre-exponential factor for the gas-phase decomposition of ozone at low pressures is $4.6 \times 10^{12}\ M^{-1}\ s^{-1}$ and its activation energy is $10.0\ kJ\ mol^{-1}$. What is (a) the entropy of activation, (b) the enthalpy of activation, (c) the Gibbs function of activation at 298 K?

30.26. The rates of thermolysis of a variety of *cis*- and *trans*-azoalkanes have been measured over a range of temperatures because of a controversy concerning the mechanism of the reaction. In ethanol an unstable *cis*-azoalkane decomposed at a rate that was followed by observing the nitrogen evolution, and this led to the rate constants listed below (P. S. Engel and D. J. Bishop, *J. Am. chem. Soc.* **97,** 6754 (1975)). Calculate the enthalpy, entropy, energy, and Gibbs function of activation at −20 °C.

θ/°C	−24.82	−20.73	−17.02	−13.00	−8.95
$10^4 \times k/s^{-1}$	1.22	2.31	4.39	8.50	14.3

30.27. If the activated complex is formed from ions of charges $z'e$ and $z''e$, and there is some characteristic distance $R^{\ddagger}$ between them in the activated complex, then the Gibbs function of activation will contain a term proportional to $z'z''/R^{\ddagger}\varepsilon_r$, where ε_r is the relative permittivity of the solvent. Deduce the expression $\ln \bar{k}_{eff} = \ln k_{eff} - z'z''B/\varepsilon_r$, $B = e^2/4\pi\varepsilon_0 R^{\ddagger}kT$, for the dependence of the rate coefficient on ε_r. (Note that this Problem does not involve questions of ionic strength.)

30.28. The model just constructed can be tested using the following data. Bromophenol blue fades when OH^- is added, the reaction rate being controlled by the step that can be symbolized as $B^{2-} + OH^- \rightarrow BOH^{3-}$. The reaction between an azodicarbonate ion (A^{2-}) and H^+ has a rate-determining step that may be symbolized as $A^{2-} + H^+ \rightarrow AH^-$. Both reactions were carried out in solvents of different relative permittivities, and the results are below. Do they support the model and calculation in the last Problem?

Bromophenol blue reaction:

ε_r	60	65	70	75	79
$\lg k_{eff}$	−0.987	0.201	0.751	1.172	1.401

Azodicarbonate ion reaction:

ε_r	27	35	45	55	65	79
$\lg k_{eff}$	12.95	12.22	11.58	11.14	10.73	10.34

30.29. In an experimental study of a bimolecular reaction in aqueous solution, the second-order rate constant was measured at 25 °C and at a variety of ionic strengths and the results are tabulated below. It is known that a singly charged ion is involved in the rate-determining step. What is the charge on the other ion involved?

I	0.0025	0.0037	0.0045	0.0065	0.0085
$k/M^{-1}\,s^{-1}$	1.05	1.12	1.16	1.18	1.26

30.30. Show that the intensities of a molecular beam before and after passing through a chamber of length l containing inert scattering atoms are related by $I(l) = I(0)\exp(-\mathcal{N}_s \sigma l)$, where σ is the collision cross-section and $\mathcal{N}_s$ the number density of scattering atoms.

30.31. In a molecular beam experiment to measure collision cross-sections it was found that the intensity of a CsCl beam was reduced to 60% of its intensity on passage through CH_2F_2 at 10 μTorr, but that when the target was argon at the same pressure the intensity was reduced only by 10 per cent. What are the relative cross-sections of the two types of collision? Why is one much larger than the other?

30.32. The total cross-sections for reactions between alkali metal atoms and halogen molecules are given in the table below (R. D. Levine and R. B. Bernstein, *Molecular reaction dynamics*, Clarendon Press, Oxford, p. 72 (1974)). Assess the data in terms of the harpoon mechanism.

σ^*/nm^2	Cl_2	Br_2	I_2
Na	1.24	1.16	0.97
K	1.54	1.51	1.27
Rb	1.90	1.97	1.67
Cs	1.96	2.04	1.95

Electron affinities are approximately 1.3 eV (Cl_2), 1.2 eV (Br_2), and 1.7 eV (I_2), and ionization energies are 5.1 eV (Na), 4.3 eV (K), 4.2 eV (Rb), and 3.9 eV (Cs).

31

Processes at solid surfaces

Learning objectives

After careful study of this chapter you should be able to:

(1) Describe how crystal surfaces grow and explain the importance of *surface defects*, Section 31.1(a).

(2) Explain why *ultra-high vacuum techniques* must be used to in order to study clean surfaces, Section 31.1(b).

(3) Describe the application of *photoelectron spectroscopy*, *Auger spectroscopy*, and *photoelectron spectromicroscopy* to the study of surfaces, Section 31.1(b).

(4) Explain the application of *LEED* techniques to surface studies, Section 31.1(c).

(5) Describe *field emission microscopy* and *field ionization microscopy*, Section 31.1(c).

(6) Define *fractional coverage*, eqn (31.2.1), and list the techniques for its measurement.

(7) Describe the information available from *photoemission spectroscopy* and *EELS*, Section 31.2(a).

(8) Distinguish between *physisorption* and *chemisorption*, Section 31.2(b).

(9) Account for the *activation energy* of chemisorption and desorption, Fig. 31.18.

(10) Define and use the *sticking probability*, eqn (31.2.3) and Example 31.1.

(11) Describe *flash desorption spectroscopy* and show how it may be used to measure desorption activation energies, eqn (31.2.5).

(12) Discuss the *mobilities* of adsorbed species, Section 31.2(f).

(13) Derive and use the *Langmuir isotherm* for non-dissociative adsorption, eqn (31.3.3) and Example 31.2, and dissociative adsorption, eqn (31.3.4).

(14) Define and measure the *isosteric adsorption enthalpy*, eqn (31.3.5) and Example 31.3.

(15) Derive and use the *BET isotherm*, eqn (31.3.14) and Example 31.4.

(16) State and use the *Freundlich* and *Temkin isotherms*, eqns (31.3.16–17) and Example 31.5.

(17) Describe the *Eley–Rideal* and *Langmuir–Hinshelwood mechanisms* of heterogeneously catalysed reactions, and write the corresponding rate laws, eqns (31.4.1–5) and Example 31.6.

(18) Describe the information provided by molecular beams about heterogeneously catalysed reactions, Section 31.4(a).

(19) Explain the significance of a *volcano curve*, Fig. 31.30.

(20) Describe the processes involved in catalytic *hydrogenation*, *oxidation*, *cracking*, and *reforming*, Sections 31.4(b, c, and d).

Introduction

Processes at surfaces govern most aspects of daily life, including life itself. Even if we limit our attention to solid surfaces the importance of the processes is hardly reduced. Processes at solid surfaces govern the viability of industry either constructively, as in catalysis, or destructively, as in corrosion.

A fundamental dynamical aspect of a surface is its role as a region where a solid grows, evaporates, and dissolves. We therefore begin with a brief look at the growth of solids and the structures of simple surfaces. Then we see how surfaces are contaminated by the deposition of foreign material, and how that contamination is responsible for heterogeneous catalysis. The special case of electrode surfaces is treated in the next chapter.

31.1 The growth and structure of surfaces

In this section we see how surfaces are extended and crystals grow, and begin to picture the structures of surfaces responsible for catalysis.

31.1 (a) How crystals grow

A simple idea of the form of a perfect crystal surface is as a cobbled street. A gas particle that collides with the surface can be imagined as a ball bouncing erratically over the cobbles. The atom loses energy as it bounces, but it is likely to escape from the surface before it has lost enough kinetic energy to be trapped. The same is true of an ionic crystal in contact with a solution: in this case there is little energy advantage for an ion in solution to shed some of its solvating molecules and stick at an exposed position on the surface.

The picture changes when the surface has *defects*, for then there are ridges of incomplete layers of atoms or ions. A typical type of surface defect is a *step* between two otherwise flat *terraces*, Fig. 31.1. This step defect might itself be defective: it might have *kinks*. When an atom settles on a terrace it bounces under the influence of the intermolecular potential, and might come to a step or a corner formed by a kink. Instead of interacting with a single terrace atom it now interacts with several, and the interaction may be strong enough to stop and trap it. Likewise, in the case of the deposition of ions from solution, the loss of the solvation interaction is compensated by a strong Coulombic interaction between the arriving ions and *several* ions at the surface defect.

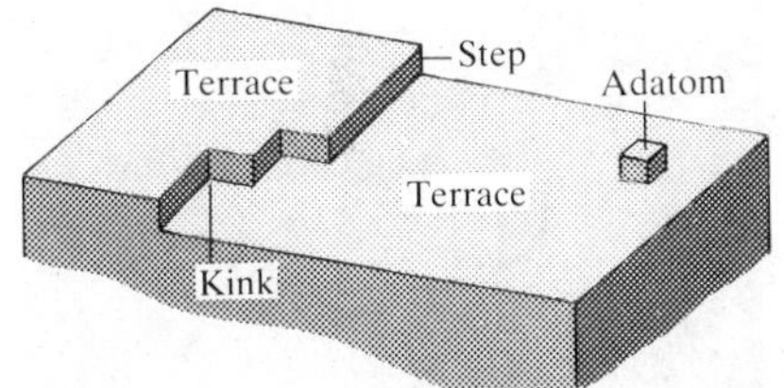

Fig. 31.1. Some of the kinds of defects that may occur on otherwise perfect terraces. Defects play an important role in surface growth and catalysis.

This picture shows that surface defects are necessary for deposition and growth, but not all kinds of defect are suitable. As the process of settling into ledges and kinks continues, there comes a stage when an entire lower terrace has been covered. At this stage the surface defects have been eliminated, and so growth will cease. For continuing growth, a surface defect is needed that propagates itself as the crystal grows. We can see what form of defect this must be by considering the types of dislocations that exist in the bulk of a crystal.

Dislocations take several forms and contribute to the mechanical properties of the solid, such as its ductility and brittleness. They arise when particles lay down in a way that disrupts the regularity of the packing of the lattice. One reason may be that the crystal grows so quickly that its particles do not have time to settle into states of lowest potential energy before being trapped in position by the deposition of the next layer. Another reason is that an impurity atom may have distorted the lattice nearby. A special kind of dislocation is the *screw dislocation* depicted in Fig. 31.2. In order to picture it, imagine a cut in the crystal, with the particles to the left of the cut pushed up through the distance of one unit cell. The unit cells now form a continuous spiral around the end of the cut, the *screw axis*. A path circulating about the screw axis spirals up to the top of the crystal, and

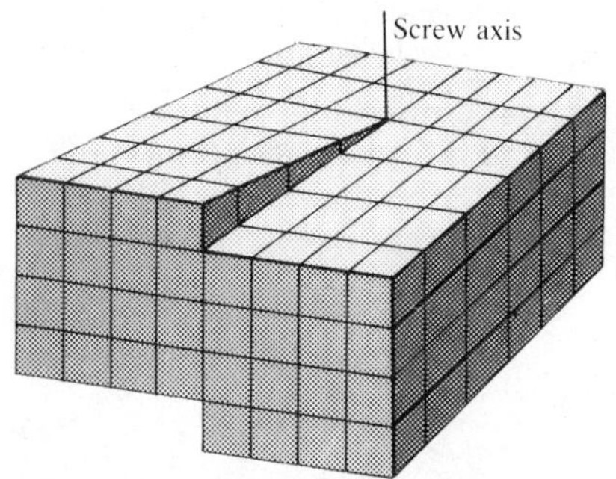

Fig. 31.2. A *screw dislocation* occurs where one region of the crystal is pushed up through one or more unit cells relative to another region. The cut extends to the *screw axis*. As atoms lie along the step the dislocation rotates round the screw axis, and is not annihilated.

where the dislocation breaks through to the surface it takes the form of a spiral ramp, Fig. 31.2.

The surface defect formed by a screw dislocation is a step, possibly with kinks, where growth can occur. The incoming particles lie in ranks on the ramp, and successive ranks reform the step at an angle to its initial position. As deposition continues the step rotates around the screw axis, and is not eliminated. Growth may therefore continue indefinitely. However, the whole length of the step might not grow at the same rate, and instead of covering the entire surface with new atoms a spiral might be formed. Several layers of deposition may occur, and so the edges of the spirals might be cliffs several atoms high, Fig. 31.3.

Fig. 31.3. The spiral growth arising from the propagation of a screw axis is clearly visible in this photograph of an *n*-alkane crystal. (B. R. Jennings and V. J. Morris. *Atoms in contact*, Clarendon Press, Oxford, 1974; courtesy of Dr A. J. Forty.)

Propagating spiral ledges can also give rise to flat terraces, Fig. 31.4. This happens if growth occurs simultaneously at left- and right-handed screw dislocations, Fig. 31.5. The growth shown there, which occurs as counter-rotating defects collide on successive circuits, need not take place uniformly over the whole length of the edges, and so successive tables of atoms may decrease in size. The terraces formed in this way may then fill up by further deposition at their edges, and more or less flat crystal planes can result.

The rapidity of growth depends on the crystal plane concerned, and the *slowest* growing faces dominate the appearance of the crystal. This is explained in Fig. 31.6, where we see that although the horizontal face grows forward most rapidly, it grows itself out of existence, and the slower-growing faces survive.

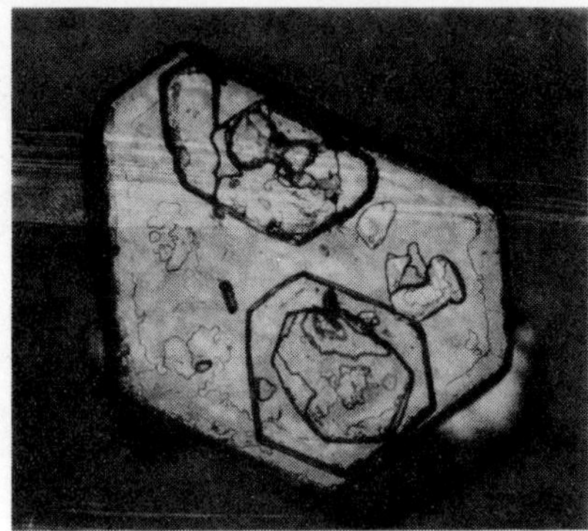

Fig. 31.4. The spiral growth pattern is sometimes concealed because the terraces are subsequently completed by further deposition. This accounts for the appearance of this cadmium iodide crystal. (H. M. Rosenberg, *The solid state* (2nd edn), Clarendon Press, Oxford, 1978.)

31.1 (b) Surface composition

We consider solid surfaces in contact with gases. The first step in seeing how surfaces catalyse reactions is to characterize the clean surface. In this context 'clean' means much more than scrubbing the sample and handling it with care. Under normal conditions a surface is constantly bombarded with gas particles and a freshly prepared surface is covered very quickly. How quickly can be estimated using the kinetic theory of gases (Section 26.2), which gives the following expression for the number of collisions per unit time per unit area when the pressure is p:

$$Z_W = p/(2\pi mkT)^{\frac{1}{2}}. \qquad (31.1.1)$$

(This is eqn (26.2.9).) A practical form of this equation is

$$\begin{aligned} Z_W/\text{cm}^{-2}\,\text{s}^{-1} &= 3.51\times10^{22}(p/\text{Torr})/\{(T/\text{K})M_r\}^{\frac{1}{2}} \\ &= 2.03\times10^{21}(p/\text{Torr})/\surd M_r \text{ at } 298\,\text{K}. \qquad (31.1.2)\end{aligned}$$

M_r is the RMM (molecular weight) of the gas. For air ($M_r = 29$) at atmospheric pressure ($p = 760$ Torr), the collision frequency is $3\times10^{23}\,\text{cm}^{-2}\,\text{s}^{-1}$. Since $1\,\text{cm}^2$ of metal surface consists of about 10^{15} atoms, each atom is struck about 10^8 times each second. Even if only a few collisions leave a molecule stuck to the surface, the time for which a freshly prepared surface remains clean is very short.

The obvious solution is to reduce the pressure. When it is reduced to 10^{-6} Torr (as in a simple vacuum system) the collision frequency falls to about $10^{14}\,\text{cm}^{-2}\,\text{s}^{-1}$, corresponding to one hit per surface atom in each 0.1 s. Even that is too brief in most experiments, and in *ultra-high vacuum* (UHV) techniques pressures as low as 10^{-9} Torr ($Z_W = 10^{11}\,\text{cm}^{-2}\,\text{s}^{-1}$) are reached on a routine basis and 10^{-11} Torr ($Z_W = 10^9\,\text{cm}^{-2}\,\text{s}^{-1}$) with special care.

These collision frequencies correspond to each surface atom being hit once every 10^5–10^6 s, or about once a day.

The layout of a typical UHV apparatus is such that the whole of the evacuated part can be heated to 200–300 °C for several hours in order to drive gas molecules from the walls. All the taps and seals are of metal in order to avoid contamination from greases. The sample is usually in the form of a thin foil, a filament, or a sharp point. Where there is interest in the role of specific crystal planes the sample is in the form of a single crystal with a freshly cleaved face. Initial cleaning of the surface is achieved either by heating it electrically or by bombarding it with accelerated gaseous ions. The latter demands care because ion bombardment can shatter the surface structure and leave it an amorphous jumble of atoms. High temperature annealing is then required to return the surface to an ordered state.

The *surface composition* (and the detection of any remaining contamination after cleaning, or the detection of layers of material adsorbed later in the experiment) can be determined by a variety of ionization techniques. Their common feature is that the *escape depth* of the electrons is in the range 0.1–1.0 nm, which ensures that only surface species contribute.

One technique is *photoelectron spectroscopy* (Section 19.5). X-PES seems to be better than UV-PES for the analysis of composition because it is able to fingerprint the materials present. Figure 31.7 shows the result of an X-PES study of a sample of gold foil on which some mercury has been deposited: the presence of the latter is clearly visible. UV-PES is more suited to establishing the bonding characteristics and the details of valence shell electronic structures of substances on the surface. A recent advance has been to focus the ejected electron and to build an image of the surface from which they have been ejected. This is the technique of *photoelectron spectromicroscopy* (PESM), and a PESM image of a silicon diode is shown in Fig. 31.8.

A very important technique, which is now widely used in the microelectronics industry, is *Auger spectroscopy*. The *Auger effect* is the emission of a second electron after high energy radiation has expelled another. The first electron to depart leaves a hole in a low-lying orbital, and an upper electron falls into it. The energy this releases may result either in the generation of radiation, which is called *X-ray fluorescence*, Fig. 31.9(a), or in the ejection of another electron, Fig. 31.9(b), the *secondary electron* of the Auger effect. The energies of the secondary electrons are characteristic of the material present, and so the Auger effect takes a fingerprint of the sample. Figure 31.10 shows an Auger spectrum of the same sample as in Fig. 31.7: the presence of the mercury is easily visible. In practice the Auger spectrum is normally obtained by irradiating with an electron beam rather than electromagnetic radiation.

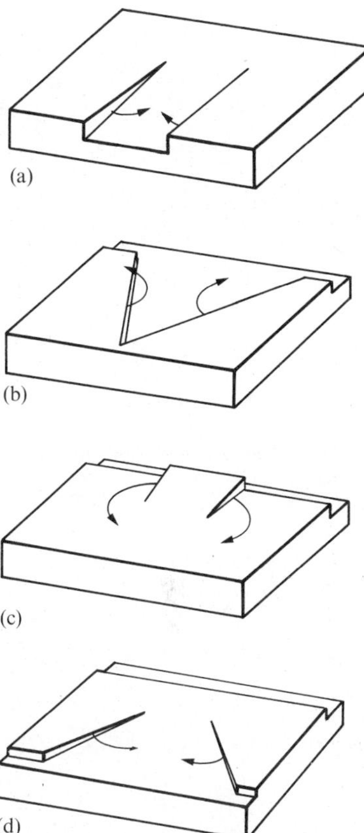

Fig. 31.5. Counter-rotating screw dislocations on the same surface lead to the formation of terraces. Four stages of one cycle of growth are shown here. Subsequent deposition can complete each terrace.

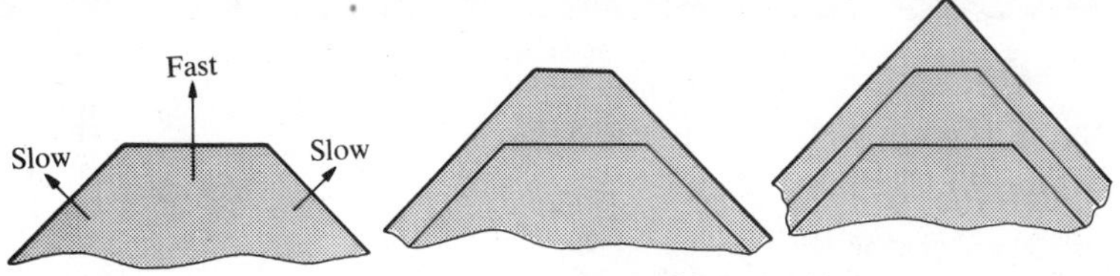

Fig. 31.6. The slower-growing faces of a crystal dominate its final external appearance. Three successive stages of the growth are shown.

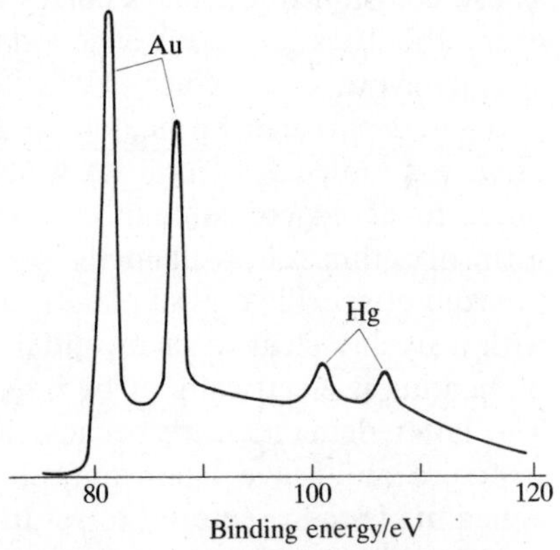

Fig. 31.7. The X-ray photoelectron emission spectrum of a sample of gold contaminated with a surface layer of mercury. (M. W. Roberts and C. S. McKee, *Chemistry of the metal–gas interface*, Oxford, 1978.)

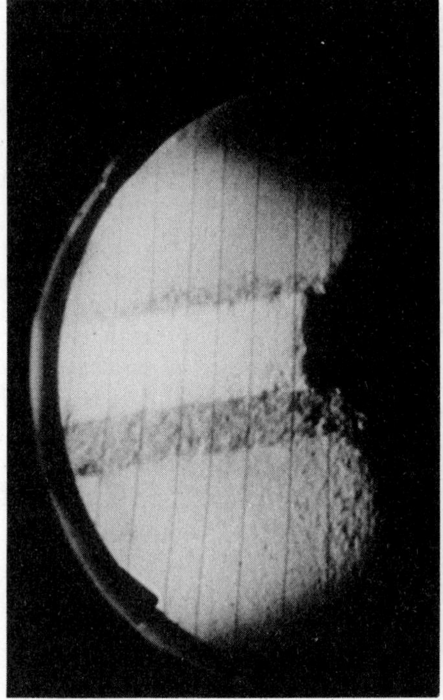

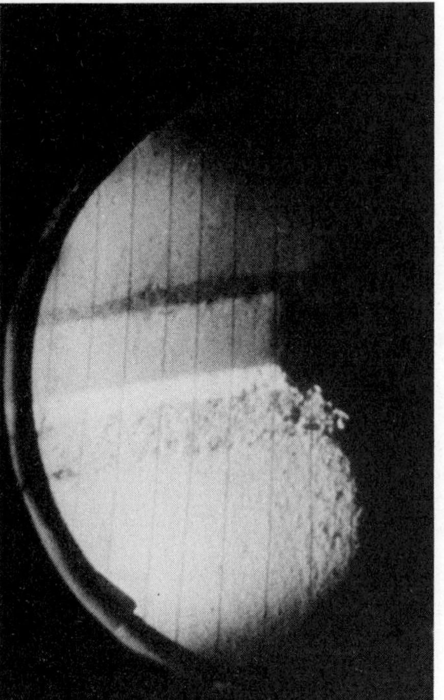

Fig. 31.8. He(I) photoelectron spectromicroscopy images of a sectional silicon diode (left) unbiased and (right) with −30 V reverse bias. (Photograph provided by Dr D. W. Turner.)

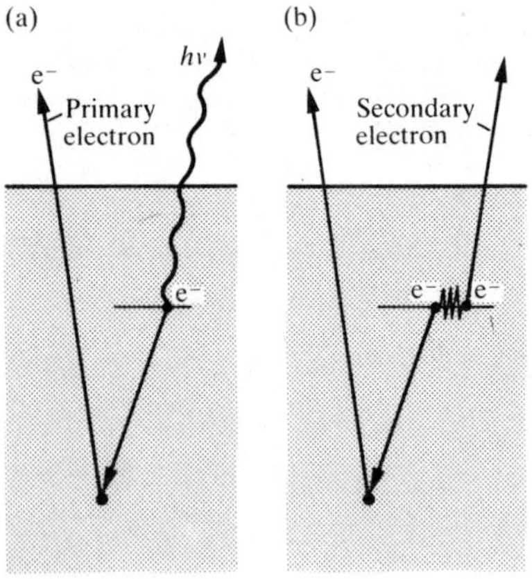

Fig. 31.9. When an electron is expelled from a solid (a) an electron of higher energy may fall into the vacated orbital and emit an X-ray photon: this is *X-ray fluorescence*. Alternatively (b) the electron falling into the orbital may give up its energy to another electron, which is ejected; this is the *Auger effect*.

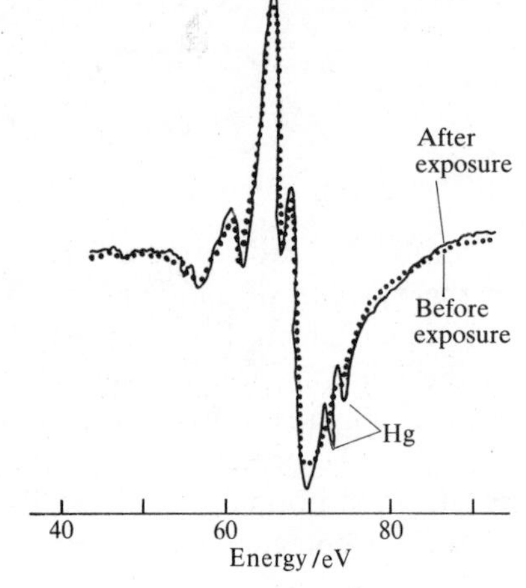

Fig. 31.10. An Auger spectrum of the same sample used for Fig. 31.7 taken before and after deposition of mercury. (M. W. Roberts and C. S. McKee, *Chemistry of the metal–gas interface*, Oxford, 1978.)

31.1 (c) Surface structure

Once the surface composition has been determined, or once it has been established that the surface is clean, attention can turn to the arrangement of the atoms. One of the most informative techniques is *low energy electron diffraction* (LEED). This is basically the electron diffraction technique described in Section 23.5, but the sample is now the surface of a solid. The use of low energy electrons ensures that the diffraction is caused by atoms on the surface, not atoms in the bulk. The experimental arrangement is shown in Fig. 31.11, and typical LEED patterns, obtained by photographing the fluorescent screen through the viewing port, are shown in Fig. 31.12.

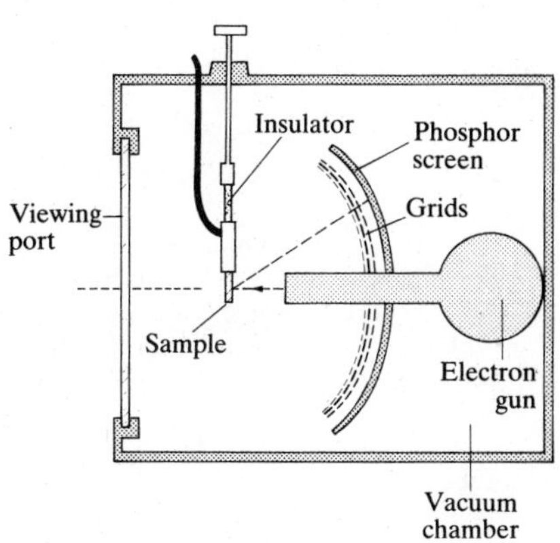

Fig. 31.11. A schematic diagram of the apparatus used for a LEED experiment. The electrons diffracted by the surface layers are detected by the fluorescence they cause on the phosphor screen.

The LEED pattern portrays the two-dimensional structure of the surface, but by studying how the diffraction intensities depend on the energy of the electron beam it is also possible to infer some details about the vertical location of the atoms and to measure the thickness of the surface layer. The pattern is sharp if the surface is well-ordered for distances long compared with the wavelength of the incident electrons, which in practice means that sharp patterns are obtained for surfaces ordered for about 20 nm and more. Diffuse spots indicate either a poorly ordered surface or the presence of impurities. If the spots do not correspond to what can be predicted by extrapolating the bulk surface to the surface, then either a reconstruction of the surface has occurred (such as in LiF, where the Li^+ and F^- ions apparently lie on slightly different planes) or there is order in the arrangement of an adsorbed layer. One outcome of LEED experiments is that they show how important it is not to assume that the surface of a crystal has exactly the same form as a slice through the bulk. As a general rule, it is found that metal surfaces *are* simply truncations of the bulk lattice, but the distance between the top layer of atoms and the one below is contracted by around 5%. On the other hand, semiconductors generally have surfaces reconstructed to a depth of several layers.

The presence of terraces, steps, and kinks in a surface shows up in LEED patterns and their densities can be estimated. The importance of this will emerge later. Three examples of how steps and kinks affect the pattern are shown in Fig. 31.13. The samples used were obtained by cleaving a crystal at different angles to a plane of atoms: only terraces are produced when the cut is parallel to the plane, and the density of steps increases as the angle of the cut increases. The observation of additional structure in the LEED patterns, not merely blurring, shows that the steps are arrayed regularly.

By far the most spectacular results are obtained from two closely related techniques. In *field emission microscopy* (FEM) the sample is in the form of a filament etched to a sharp tip. This is enclosed in a chamber fitted with a fluorescent screen. When a large potential difference is applied between the sample and the screen, electrons are stripped out towards the screen and give a flash of light where they strike it. The ease with which the electrons can escape from the metal depends on its surface structure (and, in particular, the variation of the *work function*, Section 13.2, with the composition and structure of the surface), and so the screen shows a corresponding variation of intensity. A typical result is shown in Fig. 31.14. The change in the FEM pattern when material has been deposited can be used to detect the places where atoms are most likely to stick.

Field ionization microscopy (FIM) is a development of FEM. The apparatus is virtually the same, but the potential difference is reversed, the

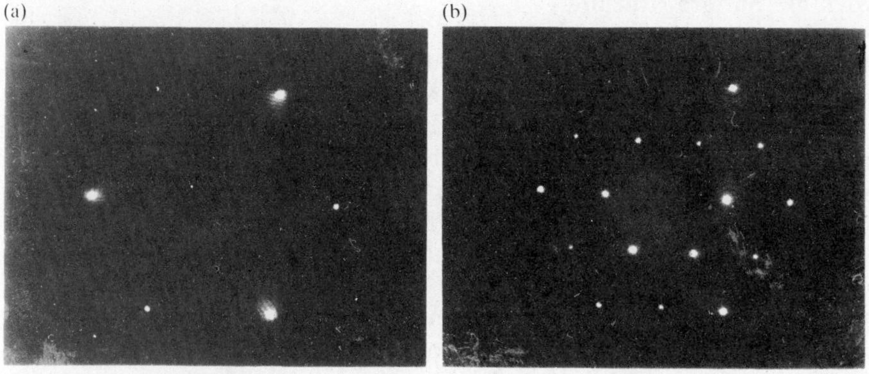

Fig. 31.12. LEED photographs of (a) a clean platinum surface and (b) after its exposure to propyne, $CH_3C{\equiv}CH$. (Photographs provided by Professor G. A. Samorjai.)

fluorescent screen being made negative relative to the tip. In the experiment a little gas (e.g. helium), is admitted. An atom strikes the tip and bounces over its terraced, cobbled surface until it strikes a protruding atom, such as one on the rim of a ledge, Fig. 31.15. Protruding atoms are able to ionize the gas atom, and immediately the positive ion (e.g. He^+) is formed, the potential difference drags it off towards the screen, where its collision generates fluorescence.

The resolution of FIM depends on the transverse motion of the gas ions. This motion can be reduced by cooling the tip to about 20 K, when the resolution is of the order of atomic dimensions and the positions of individual atoms can be discerned, Fig. 31.16. This remarkable picture, in which the small bright spots are caused by individual terrace atoms, shows

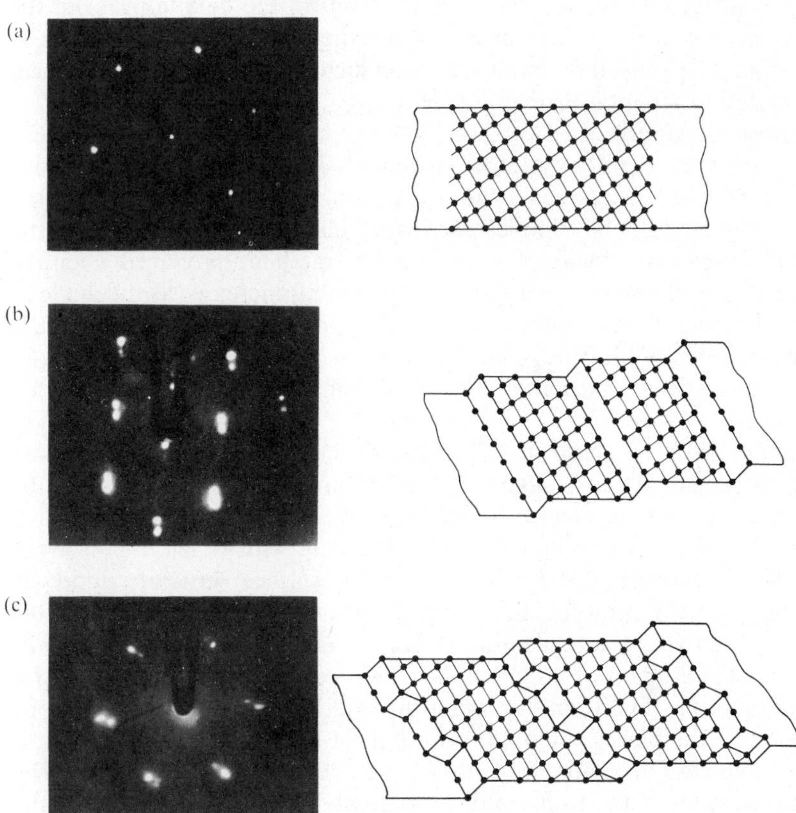

Fig. 31.13. LEED patterns may be used to assess the defect density of a surface. The photographs correspond to a platinum surface with (a) low defect density, (b) regular steps separated by about six atoms, and (c) regular steps with kinks. (Photographs provided by Professor G. A. Samorjai.)

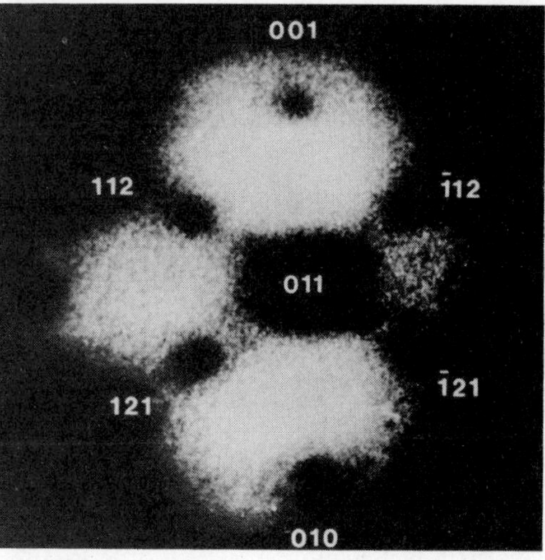

Fig. 31.14. A field emission photograph of a tungsten tip of radius 210 nm and the assignment to the exposed crystal faces. (M. Prutton, *Surface physics*, Clarendon Press, Oxford, 1975.)

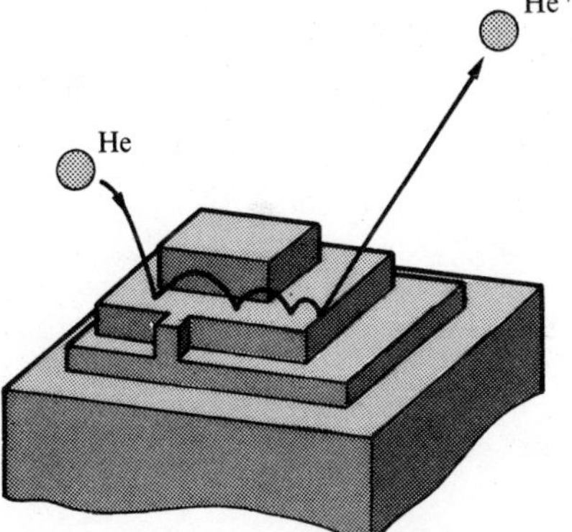

Fig. 31.15. The events leading to an FIM image of a surface. The He atom migrates across the surface until it is ionized at an exposed atom, when it is pulled off by the externally applied potential. (The bouncing motion is due to the intermolecular potential, not gravity!)

Fig. 31.16. An FIM image of an iridium tip showing the details of the exposed crystal faces. (Photograph provided by Professor E. W. Müller.)

the power of the technique, but we should not forget its limitations. First, ionization occurs unequally at different atoms, and many of the atoms on the surface and even on the edges are insufficiently exposed to cause ionization, and so remain invisible. Second, the sample must be in the form of a tip, and be made of material strong enough to withstand the very high electric fields required. Despite these limitations, FIM is remarkable for its ability to portray the positions of individual atoms.

An elegant refinement of FIM identifies individual atoms on an otherwise clean tip. *Atom-probe FIM* is the ultimate in surgery. The FIM image of an adsorbed atom is brought into coincidence with a hole in the fluorescent screen. The imaging gas is removed, and a pulse of potential difference plucks off the atom (as an ion). It moves in the same direction as did the gas ions, and passes through the hole in the screen. Behind the screen is a mass spectrometer, and so the atom can be identified. Apart from being the ultimate analytical technique (because as well as knowing what the atom is we also know exactly where it was in the sample) events can be observed which on an atomic scale are really dramatic. For example, the analysing pulse lasts for about 2 ns, and during that time the evaporation of about 10 atomic layers is sometimes observed. This corresponds to a rate of evaporation equivalent to the surface receding at about $1\ \mathrm{m\,s^{-1}}$.

31.2 Adsorption at surfaces

The accumulation of particles at a surface is called *adsorption*; the substance that adsorbs is the *adsorbate*; the underlying material is the *adsorbent* or *substrate*. The reverse of adsorption is *desorption*.

31.2 (a) The experimental analysis of the surface layer

The extent of surface coverage is normally expressed as the *fractional coverage* (symbol: θ) defined as follows:

$$\theta = \frac{\text{Number of adsorption sites occupied}}{\text{Number of adsorption sites available}}. \tag{31.2.1}$$

The *rate of adsorption* is the rate of change of surface coverage ($\mathrm{d}\theta/\mathrm{d}t$), and can be determined by observing the change of fractional coverage with time.

Among the principal techniques for measuring θ are the following:

(1) *Flow methods*. The sample acts as a pump because adsorption removes particles from the gas. One commonly used technique is therefore to monitor the rates of flow of gas into and out of the system: the difference is the rate of gas uptake by the sample. Integration of this rate then gives the extent of coverage at any stage.

(2) *Flash desorption*. The sample is suddenly heated (electrically) and the rise of pressure this leads to is interpreted in terms of the amount originally on the sample. The interpretation may be confused by the desorption of a compound (e.g. WO_3 from oxygen on tungsten).

(3) *Gravimetry*. The sample is weighed on a microbalance during the experiment.

(4) *Radioactive tracers*. The radioactivity of the sample is measured after exposure to an isotopically labelled gas.

While many important studies have been carried out simply by exposing a surface to a gas, modern work is increasingly making use of beam techniques. One advantage is that the activity of specific crystal planes can be investigated by directing the beam on to an orientated surface with known step and kink densities (as measured by LEED). Furthermore, if the adsorbate reacts at the surface the products (and their angular distributions) can be analysed as they are ejected from the surface and pass into a mass spectrometer. Another advantage is that the time of flight of a particle may be measured and interpreted in terms of its *residence time* on the surface. In this way a very detailed picture can be constructed of the events taking place during reactions at surfaces.

Apart from extent, rate, and crystallographic specificity, the other experimental information needed is the state of the adsorbate and the nature of its bonds to the substrate. One technique is photoelectron spectroscopy (especially UV-PES, Sections 19.5(b and c)), which in surface studies is normally called *photoemission spectroscopy*. Its usefulness is its ability to reveal which orbitals of the adsorbate are involved in the bond to the substrate. For instance, the principal difference between the PES results on free benzene and benzene adsorbed on palladium is the energies of the π-electrons. This is interpreted as meaning that the molecules lie parallel to the surface and are attached to it through their π-orbitals. In contrast, pyridine is known to stand more or less perpendicular to the surface, and is attached to it by means of a σ-bond formed by the nitrogen lone pair.

Several kinds of vibrational spectroscopy have been developed to study adsorbates and to show whether dissociation has occurred. Infrared and Raman spectroscopy have been improved by the development of laser and Fourier transform techniques, but still suffer from difficulties arising from low intensities on account of the low surface coverages normally encountered under laboratory conditions. However, under conditions typical of industrial catalytic operations (high pressures and high coverages) these are the only viable techniques. A hybrid of PES and vibrational spectroscopy is *electron energy loss spectroscopy* (EELS) in which the energy loss suffered by a beam of electrons is monitored when they are reflected from a surface. As in optical Raman spectroscopy, the spectrum of energy loss can be interpreted in terms of the vibrational spectrum of the adsorbate. High resolution and sensitivity are attainable, and the technique is sensitive to light elements (to which X-ray techniques are insensitive). Very tiny amounts of adsorbate can be detected, and one report estimates that about 48 atoms of phosphorus were detected in one sample. As an example, Fig. 31.17 shows the EELS result for CO on the (111) face of a Pt crystal as the extent of surface coverage increases. The main peak arises from CO attached perpendicular to the surface by a single Pt atom. As the coverage increases the smaller peak next to it increases in intensity. This peak is due to CO at a *bridge site*, where it is attached to two Pt atoms, as in **1**.

O
||
C
M M

1

31.2 (b) Physisorption and chemisorption

Molecules and atoms can stick to surfaces in two ways. In *physisorption* (an abbreviation of *physical adsorption*), there is a van der Waals interaction (e.g. a dispersion or a dipolar interaction) between the adsorbate and the substrate. This is a long-range but weak interaction, and the energy released when a particle is physisorbed is of the same order of magnitude as the enthalpy of condensation. This energy can be absorbed as vibrations of the lattice and dissipated as heat, and a particle bouncing across the cobble-like

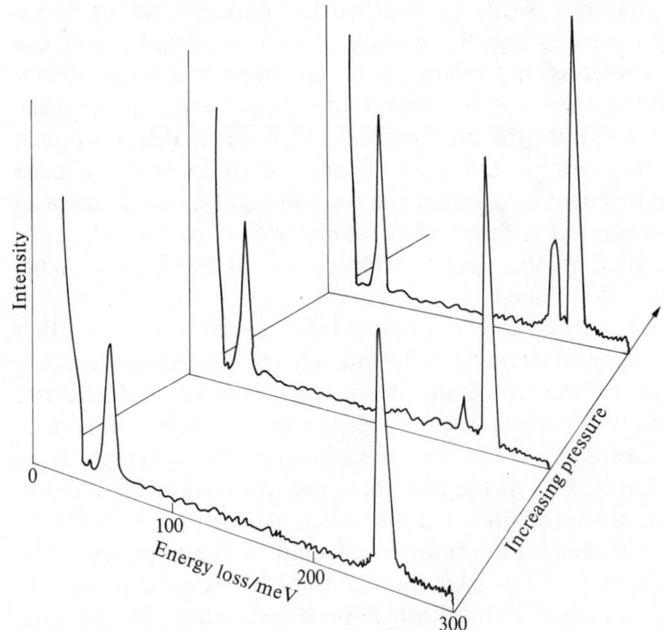

Fig. 31.17. The *electron energy loss spectrum* of CO adsorbed on Pt(111). The results for three different pressures are shown, and the growth of the additional peak at about 200 meV ($1600\ \mathrm{cm^{-1}}$) should be noted. (Spectra provided by Professor H. Ibach.)

surface will gradually lose its energy and finally stick to it (this is the process of *accommodation*). The enthalpy of physisorption can be measured by noting the rise in temperature of a sample of known heat capacity, and typical values are in the region of $20\ \mathrm{kJ\ mol^{-1}}$, Table 31.1. This is insufficient to lead to bond breaking, and so a physisorbed molecule retains its identity, although it might be distorted on account of the presence of the surface.

Table 31.1. Maximum observed enthalpies of physisorption, $-\Delta_{ad}H^{\ominus}/\mathrm{kJ\ mol^{-1}}$

H_2	84
N_2	21
H_2O	57
CH_4	21

A physisorbed particle vibrates in its shallow potential well, and might shake itself off the surface after a short time. The temperature-dependence of the first-order rate of departure can be expected to be Arrhenius-like, with an activation energy comparable to the enthalpy of physisorption. Since $k_{\mathrm{desorption}} = A\mathrm{e}^{-E_a/RT}$, the half-life for remaining on the surface has a temperature dependence

$$t_{\frac{1}{2}} = (\ln 2)/k_{\mathrm{desorption}} = \tau_0 \mathrm{e}^{E_0/RT}, \qquad (31.2.2)$$

with $\tau_0 = 0.693/A$. Taking $1/\tau_0$ equal to the vibrational frequency of the weak particle–surface bond (about 10^{12} Hz) and $E_a \approx 25\ \mathrm{kJ\ mol^{-1}}$, residence half-lives of around 10^{-8} s are predicted at room temperature. Lifetimes close to 1 s are obtained only by lowering the temperature to about 100 K.

In *chemisorption* (an abbreviation of *chemical adsorption*) the particles stick to the surface by forming a chemical (usually covalent) bond, and tend to find sites that maximize their coordination number with the substrate. The enthalpy of chemisorption is very much greater than for physisorption, and typical values are in the region of $200\ \mathrm{kJ\ mol^{-1}}$, Table 31.2. A chemisorbed molecule may be torn apart at the demand of the unsatisfied valencies of the surface atoms, and the existence of molecular fragments on the surface as a result of chemisorption is one reason why surfaces catalyse reactions.

Table 31.2. Enthalpies of chemisorption, $-\Delta_{ad}H^{\ominus}/\mathrm{kJ\ mol^{-1}}$

Adsorbate	Adsorbent (substrate)		
	Cr	Fe	Ni
H_2	188	134	
CO	192		
NH_3	188	155	
C_2H_4	427	285	209

31.2 (c) Chemisorbed species

Except in special cases, chemisorption must be *exothermic*. The argument runs as follows. A spontaneous process requires a negative ΔG. Since the translational freedom of the adsorbate is reduced when it is adsorbed, ΔS is negative. Therefore, in order for $\Delta G = \Delta H - T\Delta S$ to be negative, ΔH must be negative (exothermic). Exceptions may occur if the adsorbate dissociates and has high translational mobility on the surface. For example, hydrogen adsorbs *endothermically* on glass because there is a large increase of translational entropy accompanying the dissociation of the molecules into atoms that move quite freely over the surface, the net change of entropy for the process $H_2(g) \rightarrow 2H(glass)$ being sufficiently positive to overcome the small positive enthalpy change.

The principal test for distinguishing chemisorption from physisorption used to be the size of the enthalpy of adsorption, values less negative than $-25\ kJ\ mol^{-1}$ being taken to signify physisorption, and values more negative than around $-40\ kJ\ mol^{-1}$ signifying chemisorption. However, this is by no means foolproof and spectroscopic techniques that identify the adsorbed species are now available.

The enthalpy of adsorption depends on the extent of surface coverage, mainly because the adsorbate particles interact. If the particles repel each other (as for CO on Pd) the enthalpy of adsorption becomes less exothermic (less negative) as coverage increases. Moreover, LEED studies show that such species settle on the surface in a disordered way until packing requirements demand order. If the adsorbate particles attract each other (as for O_2 on W) they tend to cluster together in islands, and growth occurs at the borders. These adsorbates also show order–disorder transitions when they are heated enough for thermal motion to overcome the particle–particle interactions, but not so much that they are desorbed.

The dependence of the potential energy of a particle on its distance from the substrate surface is depicted in Fig. 31.18. As the particle approaches the surface its energy falls as it becomes physisorbed into the *precursor state* for chemisorption. Dissociation into fragments often takes place as a molecule moves into its chemisorbed state, and after an initial increase of energy as the bonds stretch there is a sharp decrease as the adsorbate–substrate bonds reach their full strength. Even if the molecule does not fragment there is likely to be an initial increase of potential energy as the substrate atom adjusts its bonds in response to the incoming particle. In all cases, therefore, we can expect there to be a potential energy barrier separating the precursor and chemisorbed states. This barrier, though, might be low, and might not rise above the energy of a distant, stationary particle, as in Fig. 31.18(a). In this case, chemisorption is not an activated process and can be expected to be rapid. Many gas adsorptions on to clean metals appear to be non-activated. In some cases the barrier rises above the zero axis, as in Fig. 31.18(b), and such chemisorptions are activated and slower than the non-activated kind. An example is H_2 on Cu, which has an activation energy somewhere in the region of 20–40 $kJ\ mol^{-1}$.

One point that emerges from this discussion is that rates are not good criteria for distinguishing between physisorption and chemisorption. Chemisorption can be fast if the activation energy is small or zero; but it may be slow if the activation energy is large. Physisorption is usually fast, but it can appear to be slow if adsorption is taking place on a porous medium.

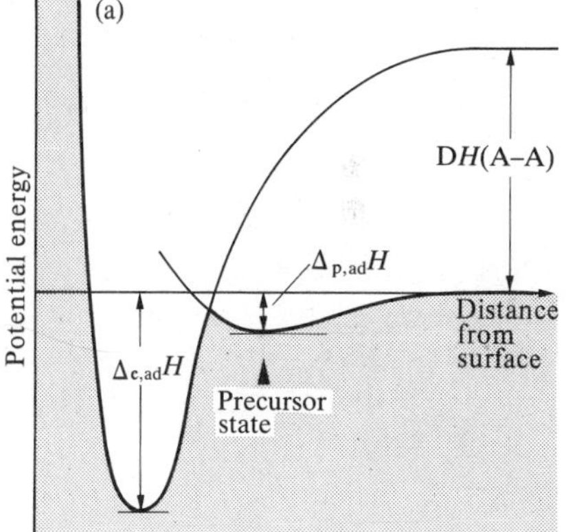

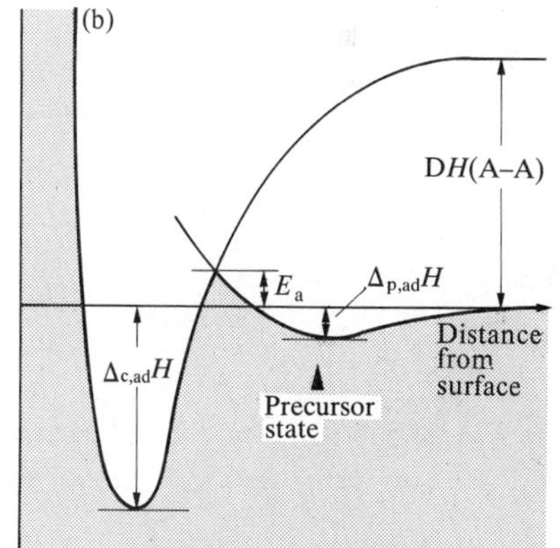

Fig. 31.18. The potential energy profiles for the dissociative chemisorption of an A—A molecule. In each case $\Delta_{p,ad}H$ is the enthalpy of (non-dissociative) physisorption and $\Delta_{c,ad}H$ that for chemisorption (at $T = 0$). The relative locations of the curves determines whether the chemisorption is (a) not activated or (b) activated.

31.2 (d) The sticking probability

The rate at which a surface is covered by adsorbate depends on the substrate's ability to dissipate the energy of the incoming particle as thermal motion as it crashes on to the surface. If the energy is not dissipated quickly, the particle migrates over the surface until a vibration kicks it into the overlying gas or it reaches an edge. The proportion of collisions with the surface that successfully lead to adsorption is called the *sticking probability* (symbol: s):

$$s = \frac{\textit{Rate of adsorption of particles by the surface}}{\textit{Rate of collision of particles with the surface}}. \tag{31.2.3}$$

The denominator can be calculated from kinetic theory, and the numerator can be measured by observing the rate of change of pressure.

Values of s vary widely. For example, at room temperature CO has s in the range 0.1–1.0 for several transition metal surfaces, but for N_2 on rhenium $s < 10^{-2}$, indicating that more than a hundred collisions are needed before one molecule sticks successfully. Sticking probabilities are generally low on non-metal surfaces, but they can be low on metals too (O_2 on silver has $s < 10^{-4}$). Beam studies on specific crystal planes show a pronounced specificity: for N_2 on tungsten s ranges from 0.74 on the (320) faces down to less than 0.01 on the (110) faces at room temperature.

Example 31.1

Calculate the time for 10% of the sites on a (100) surface to be covered with nitrogen at 298 K when the pressure is 2.0×10^{-9} Torr and the sticking probability is 0.55.

• *Method*. Begin by calculating the number of atoms per unit area of the (100) surface. The lattice constant of the b.c.c. unit cell is 316 pm; calculate Z_W from eqn (31.1.2).

• *Answer*. The distance between lattice sites on the (100) face is 316 pm; therefore, each atom accounts for $(316\,\text{pm})^2$ of the surface. The number of atoms in $1\,\text{m}^2$ is therefore $(1\,\text{m}^2)/(316\,\text{pm})^2 = 1.00 \times 10^{19}$ so that the surface density of sites is $1.00 \times 10^{15}\,\text{cm}^{-2}$. From eqn (31.1.2), $Z_W = 7.7 \times 10^{11}\,\text{cm}^{-2}\,\text{s}^{-1}$. Therefore, the time required for 10% coverage is $t = (1/10) \times (1.00 \times 10^{15}\,\text{cm}^{-2})/0.55Z_W = 240\,\text{s}$.

• *Comment*. Covering the entire surface will take longer than ten times 240 s because the sticking probability decreases with increasing coverage.

• *Exercise*. Suppose that the sticking probability depends on surface coverage as $s = (1-\theta)s_0$, with s_0 given above. Calculate the surface coverage after 2400 s. [0.64]

The sticking probability decreases as the surface coverage increases, Fig. 31.19. A simple model would be obtained by assuming that s is proportional to $1-\theta$, the fraction uncovered, and it is common to write $s = s_0(1-\theta)$, where s_0 is the sticking probability on a perfectly clean surface. The results in the illustration do not fit this expression because they show that s remains close to s_0 until the coverage has risen to about 6×10^{13} molecules/cm^2, and then falls steeply. The explanation is probably that the colliding molecule does not enter the chemisorbed state at once, but moves over the surface until it encounters an empty site.

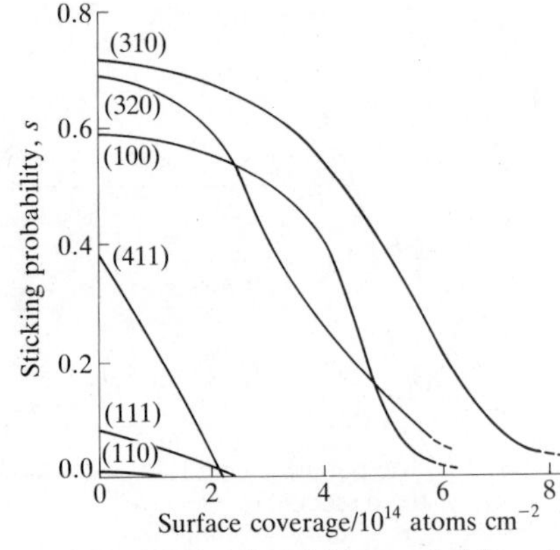

Fig. 31.19. The *sticking probability* of N_2 on various faces of a tungsten crystal and its dependence on surface coverage. Note the very low sticking probability for the (110) and (111) faces. (Data provided by Professor D. A. King.)

31.2 (e) Desorption

The desorption of an adsorbate is always activated because the particles have to be raised from the foot of a potential well. On the basis of eqn (31.2.2), using $E_a = 100\,\text{kJ mol}^{-1}$ and guessing that $\tau_0 = 10^{-14}\,\text{s}$ (because the adsorbate–substrate bond is quite stiff), we expect a residence half-life of about 3×10^3 s (about an hour) at room temperature, decreasing to 1 s at about 350 K.

The desorption activation energy can be measured in several ways. However, we have to be guarded in its interpretation because it often depends on the extent of surface coverage (and so may change as desorption proceeds). Moreover, the transfer of concepts such as 'reaction order' and 'rate constant' from bulk studies to surfaces is hazardous, and there are few examples of strictly first-order or second-order desorption kinetics.

If we disregard these complications, then one way of measuring the desorption activation energy is to monitor the rate of increase in pressure when the sample is maintained at a series of temperatures, and to attempt an Arrhenius plot of the rate coefficient. A more sophisticated technique is to determine the *flash desorption spectrum* of the sample. In this, the basic observation is that in a pumped vessel there is a pressure surge when the temperature is raised to the value at which the activation energy is overcome and desorption occurs rapidly; but once the desorption has occurred there is no more to escape from the surface, and so the pressure falls again as the temperature continues to rise. The flash desorption spectrum, the plot of pressure against temperature, therefore consists of a peak, the location of which depends on the desorption activation energy. In the example shown in Fig. 31.20 three peaks are shown, indicating the presence of three sites with different activation energies.

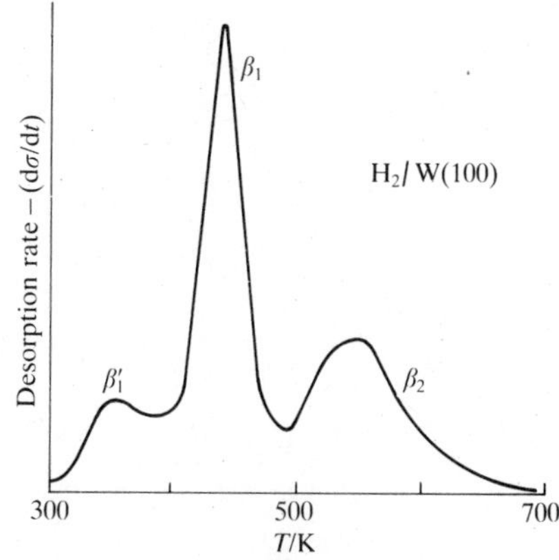

Fig. 31.20. The flash desorption spectrum of H_2 on the (100) face of tungsten. The three peaks indicate the presence of three sites with different adsorption enthalpies and therefore different desorption activation energies. (P. W. Tamm and L. D. L. D. Schmidt, *J. chem. Phys.* **51,** 5352, 1969.)

The effect can be analysed quantitatively as follows. Suppose for simplicity that the desorption is first-order in the surface concentration (σ) of the adsorbate:

$$-\mathrm{d}\sigma/\mathrm{d}t = k\sigma = (A\mathrm{e}^{-E_a/RT})\sigma. \tag{31.2.4}$$

Let the temperature be increased linearly with time from some initial value T_i, so that $T = T_i + \kappa t$. Then the temperature T^* at which the desorption rate is a maximum is the solution of

$$(\mathrm{d}/\mathrm{d}T)(-\mathrm{d}\sigma/\mathrm{d}t) = 0.$$

Since $\mathrm{d}T = \kappa\,\mathrm{d}t$, this condition can be developed as follows:

$$\begin{aligned}(\mathrm{d}/\mathrm{d}T)(-\mathrm{d}\sigma/\mathrm{d}t) &= \mathrm{d}(k\sigma)/\mathrm{d}T = k(\mathrm{d}\sigma/\mathrm{d}t)(\mathrm{d}t/\mathrm{d}T) + \sigma(\mathrm{d}k/\mathrm{d}T)\\ &= -(A^2/\kappa)\sigma\mathrm{e}^{-2E_a/RT} + (E_aA\sigma/RT^2)\mathrm{e}^{-E_a/RT} = 0.\end{aligned}$$

Therefore, the temperature at which the pulse occurs is the solution of

$$(A/\kappa)\mathrm{e}^{-E_a/RT} = E_a/RT^{*2}.$$

This can be turned into a practical form by rearranging it to

$$\ln(T^{*2}/\kappa) = \ln(E_a/RA) + E_a/RT^*. \tag{31.2.5}$$

Therefore, in order to determine E_a, a series of different heating rates (different values of κ) are used and the temperature of the desorption surge (T^*) is determined. A plot of $\ln(T^{*2}/\kappa)$ against $1/T^*$ should give a straight line. If it does, then its slope is E_a/R and A can be obtained from its intercept at $1/T^* = 0$.

In many cases only a single activation energy (and a single peak in the flash desorption spectrum) is observed. In the cases when several peaks are observed they might correspond to adsorption on different crystal planes or to multilayer adsorption. For instance, cadmium on tungsten shows two activation energies, one of 18 kJ mol^{-1} and the other of 90 kJ mol^{-1}. The explanation is that the more tightly bound Cd atoms are attached directly to the substrate, and the less strongly bound are in a layer (or layers) above the primary overlayer. Chemisorption cannot normally exceed monolayer coverage because a hydrocarbon gas, for instance, cannot chemisorb on to a surface already covered with its fragments; a metal, on the other hand, as in this case, can provide another surface capable of further adsorption.

Another example of a system showing two desorption activation energies is CO on tungsten, the values being 120 kJ mol^{-1} and 300 kJ mol^{-1}. The explanation is believed to be the existence of two types of metal–adsorbate binding site, one involving a simple M—CO bond, the other adsorption with dissociation (into individually adsorbed C and O atoms). In some cases two desorption peaks may occur even though there is only one type of site. This complication arises when there are significant interactions between the adsorbate particles so that at low surface coverages the enthalpy of adsorption is significantly different from its value at high coverages.

31.2 (f) Mobility on surfaces

A further aspect of the strength of the interactions between adsorbate and substrate is the former's mobility. This is often a vital featurè of a catalyst's activity, because it might be impotent if the reactant molecules adsorb so strongly that they cannot migrate. The activation energy for diffusion over a surface need not be the same for desorption because the particles may be able to move through valleys between potential peaks without leaving the surface completely. In general, it turns out that the activation energy for migration is about 10–20% of the energy of the surface–adsorbate bond, but it depends on the extent of coverage. The defect structure of the sample (which depends on the temperature) may also play a dominant role because the adsorbed particles might find it easier to skip across a terrace than to roll along the foot of a step, and they might become trapped in vacancies in an otherwise flat terrace. Diffusion may also be easier across one crystal face than another, and so the surface mobility depends on which lattice planes are exposed.

There are two very elegant FIM-based methods for determining the diffusion characteristics of an adsorbate. One involves depositing an adsorbate on a surface plane at low temperature, taking an FIM image before and after the temperature is raised, and monitoring the migration of the boundary as the adsorbate floods across the crystal faces at different rates. In a modification of the technique an *individual* atom is imaged, the temperature is raised, and then lowered after a definite interval. A new image is then taken, and the new position of the atom measured, Fig. 31.21. A sequence of pictures shows that the atom makes a random walk across the surface, and the diffusion coefficient D can be inferred from the mean

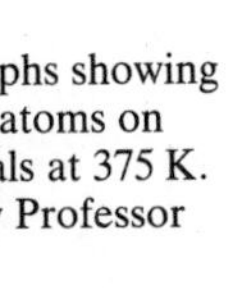

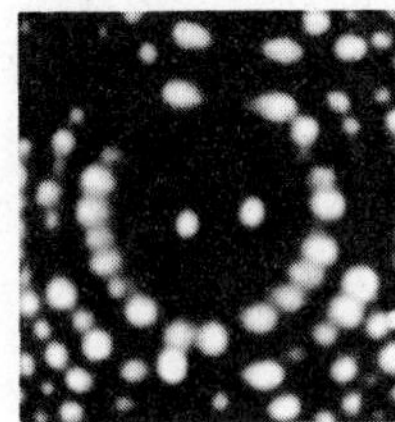

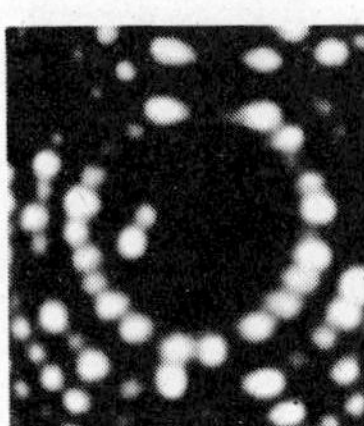

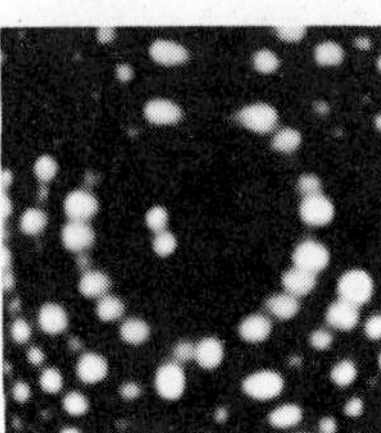

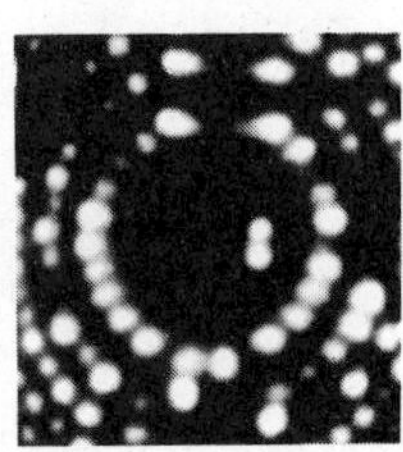

Fig. 31.21. FIM micrographs showing the migration of rhenium atoms on rhenium during 3 s intervals at 375 K. (Photographs provided by Professor G. Ehrlich.)

distance d travelled in an interval τ using the two-dimensional random walk expression $d = (D\tau)^{\frac{1}{2}}$. The value of D for different crystal planes at different temperatures can be determined directly in this way, and the activation energy for migration over each plane obtained from the Arrhenius-like expression

$$D = D_0 e^{-E_a/RT}. \qquad (31.2.6)$$

Typical values for W atoms on W have E_a in the range 57–87 kJ mol^{-1} and $D_0 \approx 3.8 \times 10^{-7}$ cm^2 s^{-1}. For CO on W, the activation energy falls from 144 kJ mol^{-1} at low surface coverage to 88 kJ mol^{-1} when the coverage is high.

31.3 Adsorption isotherms

The free and the adsorbed gas are in dynamic equilibrium, and the question that now arises is how the fractional coverage of the surface depends on the pressure of the overlying gas. The dependence of θ on the pressure at a set temperature is called the *adsorption isotherm*.

31.3 (a) The Langmuir isotherm

The simplest calculation of an isotherm is based on the assumption that every site is equivalent, and the ability of a particle to bind there is independent of whether or not nearby sites are occupied.

The dynamic equilibrium is A(g) + M(surface) $\rightleftharpoons$ AM, with rate constants k_a for adsorption and k_d for desorption. The rate of change of surface coverage due to adsorption is proportional to the pressure of A (p) and the number of vacant sites, $N(1-\theta)$, N being the total number of sites:

$$d\theta/dt = k_a p N(1-\theta). \qquad (31.3.1)$$

The rate of change of θ due to desorption is proportional to the number of adsorbed species, $N\theta$:

$$d\theta/dt = k_d N\theta. \qquad (31.3.2)$$

At equilibrium the two rates are equal, and solving for θ gives the *Langmuir isotherm*:

$$\theta = Kp/(1+Kp), \qquad K = k_a/k_d. \qquad (31.3.3)$$

Example 31.2

The data below are for the adsorption of CO on charcoal at 273 K. Confirm that they fit the Langmuir isotherm, and find (a) the constant K and (b) the volume corresponding to complete coverage. In each case V has been corrected to 1 atm.

p/Torr	100	200	300	400	500	600	700
V/cm^3	10.2	18.6	25.5	31.4	36.9	41.6	46.1

• *Method*. From eqn (31.3.3), $Kp\theta + \theta = Kp$. Write $\theta = V/V_\infty$, where V_∞ is the volume corresponding to complete coverage. Then

$$p/V = p/V_\infty + 1/KV_\infty.$$

Hence, a plot of p/V against p should give a straight line of slope $1/V_\infty$ and intercept $1/KV_\infty$.

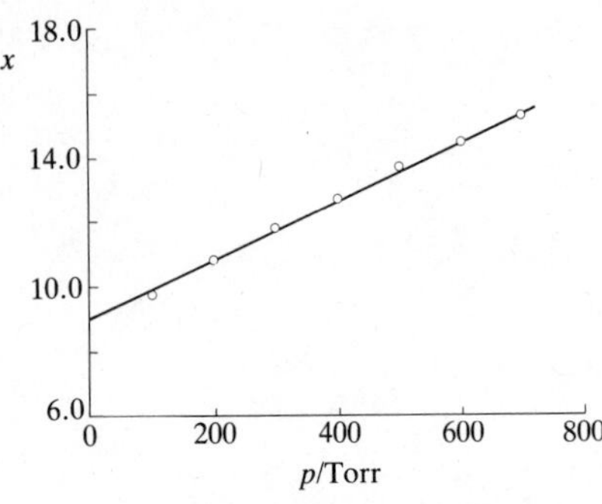

Fig. 31.22. The Langmuir isotherm predicts that a straight line should be obtained when p/V is plotted against p. This is a test of the isotherm using the information in *Example* 31.2.

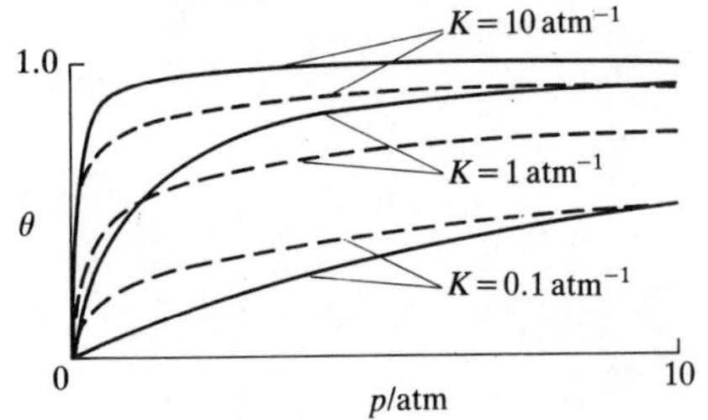

Fig. 31.23. The Langmuir isotherm for non-dissociative (full lines) and dissociative (dotted lines) adsorption for different values of K. The curves are plots of eqns (31.3.3) and (31.3.4) respectively.

• *Answer*. Draw up the following table with $x = (p/\text{Torr})/(V/\text{cm}^3)$:

p/Torr	100	200	300	400	500	600	700
x	9.80	10.8	11.8	12.7	13.6	14.4	15.2

The points are plotted in Fig. 31.22. The slope is 0.0090. and so $V_\infty = 110\ \text{cm}^3$. The intercept at $p = 0$ is 9.0, and so

$$K = 1/(110\ \text{cm}^3) \times (9.0\ \text{Torr cm}^{-3}) = 1.0 \times 10^{-3}\ \text{Torr}^{-1}.$$

• *Comment*. Note the deviation from a straight line at high surface coverages and that the dimensions of K are 1/pressure.

• *Exercise*. Repeat the calculation for the following data:

p/Torr	100	200	300	400	500	600	700
V/cm^3	10.3	19.3	27.3	34.1	40.0	45.5	48.0

[$150\ \text{cm}^3$, $7.4 \times 10^{-3}\ \text{Torr}^{-1}$]

In the case of adsorption with dissociation the rate of adsorption is proportional to the pressure and to the probability that both atoms will find sites, and therefore proportional to $p\{N(1-\theta)\}^2$. The rate of desorption is proportional to the frequency of encounters of atoms on the surface, and therefore proportional to $\{N\theta\}^2$. The condition for these two rates to be equal leads to the isotherm

$$\theta = (Kp)^{\frac{1}{2}}/\{1 + (Kp)^{\frac{1}{2}}\}. \qquad (31.3.4)$$

The surface coverage now depends more weakly on pressure.

The form of the Langmuir isotherm with and without dissociation is shown in Fig. 31.23. The surface coverage increases with increasing pressure, and approaches unity only at very high pressure, when the gas is effectively squashed on to the surface. Different curves (and therefore values of K) are obtained at different temperatures, and the temperature dependence of K can be used to determine the *isosteric enthalpy of adsorption* ($\Delta_{\text{ad}}H^\ominus$, the enthalpy of adsorption at a fixed surface coverage). To do so we recognize that K is an equilibrium constant, and then use eqn (10.2.4) to write

$$(\partial \ln K/\partial T)_\theta = \Delta_{\text{ad}}H^\ominus/RT^2. \qquad (31.3.5)$$

Example 31.3

The data below show the pressures of CO needed for the volume of adsorption (corrected to 1 atm and 273 K) to be $10.0\ \text{cm}^3$ using the same sample as in *Example* 31.2. Calculate the adsorption enthalpy at this surface coverage.

T/K	200	210	220	230	240	250
p/Torr	30.0	37.1	45.2	54.0	63.5	73.9

• *Method*. The Langmuir isotherm rearranges to

$$Kp = \theta/(1-\theta).$$

Therefore, when θ is constant, $\ln K + \ln p = \text{constant}$, and so from eqn (31.3.5)

$$(\partial \ln p/\partial T)_\theta = -(\partial \ln K/\partial T)_\theta = -\Delta_{\text{ad}}H^\ominus/RT^2.$$

Using $d(1/T)/dT = -1/T^2$ this rearranges to

$$(\partial \ln p/\partial[1/T])_\theta = \Delta_{ad}H^\ominus/R,$$

and so a plot of $\ln p$ against $1/T$ should be a straight line of slope $\Delta_{ad}H^\ominus/R$.

• *Answer*. Draw up the following table:

T/K	200	210	220	230	240	250
1000/(T/K)	5.00	4.76	4.55	4.35	4.17	4.00
ln (p/Torr)	3.40	3.61	3.81	3.99	4.15	4.30

The points are plotted in Fig. 31.24. The slope (of the least squares fitted line) is -0.90, and so $\Delta_{ad}H^\ominus = -(0.90 \times 10^3\ \text{K})R = -7.5\ \text{kJ mol}^{-1}$.

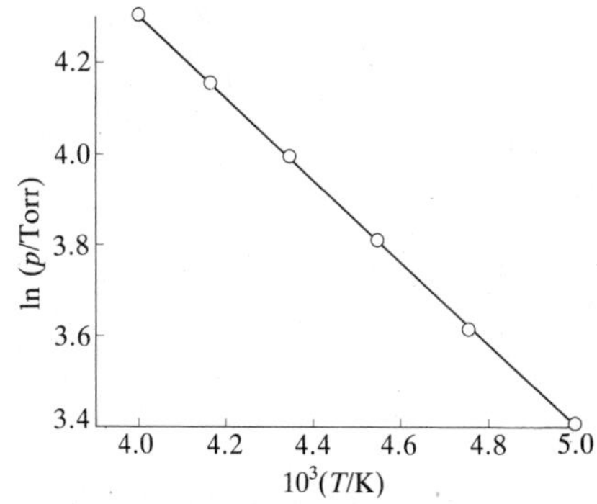

Fig. 31.24. The isosteric enthalpy of adsorption can be obtained from the slope of the plot of ln p against $1/T$, where p is the pressure needed to achieve the specified surface coverage. The data used are from *Example* 31.3.

• *Comment*. The value of K can be used to obtain a value of $\Delta_{ad}G^\ominus$, and then that value combined with $\Delta_{ad}H^\ominus$ to obtain an entropy of adsorption; however, that raises some tricky features of interpretation.

• *Exercise*. Repeat the calculation using the following data:

T/K	200	210	220	230	240	250	
p/Torr	32.4	41.9	53.0	66.0	80.0	96.0	[9.0 kJ mol^{-1}]

31.3 (b) The BET isotherm

The Langmuir isotherm ignores the possibility that the initial overlayer may act as a substrate for further (e.g. physical) adsorption. In this case, instead of the isotherm levelling off to some saturated value at high pressures, it can be expected to rise indefinitely. The most widely used isotherm dealing with multilayer adsorption is due to Stephen Brunauer, Paul Emmett, and Edward Teller, and is called the *BET isotherm*. It can be derived as follows.

Consider Fig. 31.25, which shows a region of the surface covered in monolayers, bilayers, etc. The rate constants for adsorption and desorption of the primary layer are k_a and k_d, and those of the overlayers are all k_a' and k_d'. The numbers of sites corresponding to zero, monolayer, bilayer, . . . coverage at any stage are N_0, N_1, N_2, etc. and N_i in general. The condition for equilibrium of the initial layer is the equality of the rates of its formation and desorption: $k_aN_0 = k_dN_1$. The condition for equilibrium of the next layer is $k_a'N_1 = k_d'N_2$, and in general

$$k_a'N_{i-1} = k_d'N_i, \qquad i = 2, 3, \ldots. \tag{31.3.6}$$

This condition may be expressed in terms of N_0 as follows:

$$\begin{aligned} N_i &= (k_a'/k_d')pN_{i-1} = (k_a'/k_d')^2p^2N_{i-2} = \ldots \\ &= (k_a'/k_d')^{i-1}(k_a/k_d)p^iN_0. \end{aligned}$$

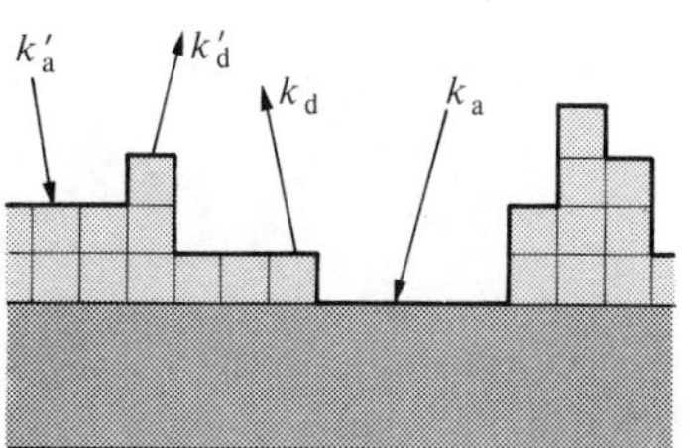

Fig. 31.25. In the derivation of the BET isotherm, different adsorption and desorption rate constants are used depending on whether the site is already occupied or not. The total volume adsorbed is proportional to the number of particles adsorbed, which is obtained by summing the column heights.

Now write $k_a'/k_d' = x$ and $k_a/k_d = cx$; then with this simpler notation

$$N_i = c(xp)^iN_0. \tag{31.3.7}$$

We now calculate the total volume, V, of adsorbed material. V is proportional to the total number of particles adsorbed, and so

$$V \propto N_1 + 2N_2 + 3N_3 + \ldots = \sum_{i=1}^{\infty} iN_i \tag{31.3.8}$$

because a monolayer site contributes one particle, a bilayer site two, and so on. If there were complete monolayer coverage the volume adsorbed would be V_{mon}, with

$$V_{\text{mon}} \propto N_0 + N_1 + N_2 + \ldots = \sum_{i=1}^{\infty} N_i \tag{31.3.9}$$

because each site in the actual sample then contributes one particle to the total. It follows that

$$V/V_{\text{mon}} = \left\{\sum_{i=1}^{\infty} iN_i\right\} \Big/ \left\{\sum_{i=1}^{\infty} N_i\right\} = cN_0\left\{\sum_{i=1}^{\infty} i(xp)^i\right\} \Big/ \left\{N_0 + cN_0 \sum_{i=1}^{\infty} (xp)^i\right\}.$$

Both sums can be evaluated using

$$\sum_{i=1}^{\infty} y^i = y/(1-y), \qquad \sum_{i=1}^{\infty} iy^i = y/(1-y)^2 \tag{31.3.10}$$

with $y = xp$. After a little rearrangement this leads to

$$V/V_{\text{mon}} = pxc/(1-xp)(1-xp+cxp). \tag{31.3.11}$$

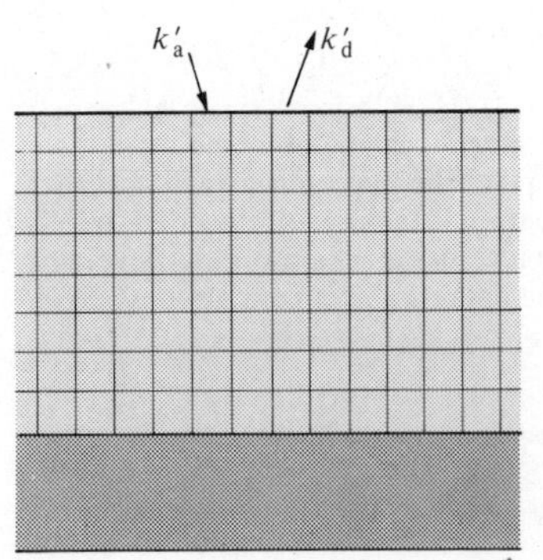

Fig. 31.26. When the surface is deeply covered, the equilibrium pressure is the vapour pressure of the bulk liquid because the substrate surface plays no role.

The final step is to equate $1/x$ to the vapour pressure p^* of the bulk liquid adsorbate. Consider the equilibrium between a gas and a sample in which all N surface sites are deeply and uniformly buried, Fig. 31.26. The condition for equilibrium is now

$$k_a' Np = k_d' N, \quad \text{or} \quad k_a' p = k_d'. \tag{31.3.12}$$

Such an equilibrium applies at the liquid surface irrespective of whether a surface is deeply buried under it, and so p, the equilibrium pressure, can be identified with p^*, the bulk VP. It follows from the last equation that $x = k_a'/k_d' = 1/p^*$, as we set out to show. When this result is introduced into eqn (31.3.11) we obtain the *BET isotherm*:

$$V/V_{\text{mon}} = cz/(1-z)\{1-(1-c)z\}, \qquad z = p/p^*. \tag{31.3.13}$$

This is usually reorganized into

$$z/(1-z)V = 1/cV_{\text{mon}} + (c-1)z/cV_{\text{mon}}. \tag{31.3.14}$$

$(c-1)/cV_{\text{mon}}$ can therefore be obtained from the slope of a plot of the expression on the left against z, and cV_{mon} can be found from the intercept at $z = 0$, the results then being combined to give c and V_{mon}.

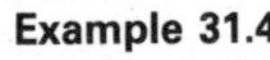

Example 31.4

The data below relate to the adsorption of nitrogen on rutile at 75 K. Confirm that they fit a BET isotherm in the range of pressures reported, and find V_{mon} and c.

p/Torr	1.20	14.0	45.8	87.5	127.7	164.4	204.7
V/cm^3	601	720	822	935	1046	1146	1254

At 75 K $p^* = 570$ Torr. The volumes have been corrected to 1 atm and 273 K and refer to 1 g of substrate.

• *Method*. Use eqn (31.3.14), and plot $z/(1-z)V$ against z. The intercept at $z = 0$ is equal to $1/cV_{\text{mon}}$ and the slope is $(c-1)/cV_{\text{mon}}$.

• *Answer*. Draw up the following table with $x = 10^4 z/(1-z)(V/\text{cm}^3)$

p/Torr	1.20	14.0	45.8	87.5	127.7	164.4	204.7
$1000z$	2.11	24.6	80.4	154	224	288	359
x	0.035	0.350	1.06	1.95	2.76	3.53	4.47

These points are plotted in Fig. 31.27. The least squares best line has an intercept at 0.034, and so $1/cV_{\text{mon}} = 0.034 \times 10^{-4}\,\text{cm}^{-3} = 3.4 \times 10^{-6}\,\text{cm}^3$. The slope of the line is 1.23×10^{-2}, and so $(c-1)/cV_{\text{mon}} = 1.23 \times 10^{-2} \times 10^3 \times 10^{-4}\,\text{cm}^{-3} = 1.23 \times 10^{-3}\,\text{cm}^{-3}$. Solving these equations gives $c - 1 = 362$, or $c = 363$ and $V_{\text{mon}} = 810\,\text{cm}^3$.

• *Comment*. At 1 atm and 273 K, $810\,\text{cm}^3$ corresponds to 0.036 mol, or 2.2×10^{22} atoms. Since each molecule occupies an area of about $0.16\,\text{nm}^2$, the surface area of the sample is about $3500\,\text{m}^2$ per gram.

• *Exercise*. Repeat the calculation for the following data:

p/Torr	1.20	14.0	45.8	87.5	127.7	164.4	204.7
V/cm^3	235	559	649	719	790	860	950

[290, 620 cm^3]

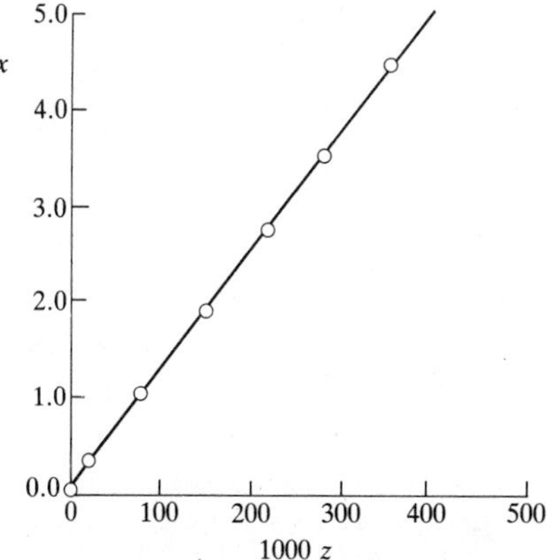

Fig. 31.27. The BET isotherm can be tested, and the parameters determined, by plotting the left-hand side of eqn (31.3.14) against $z = p/p^*$. The data are from *Example* 31.4.

The shapes of BET isotherms are drawn in Fig. 31.28. They rise indefinitely as the pressure is increased, as anticipated. When the coefficient c is large, the isotherm takes the simpler form

$$V/V_{\text{mon}} = 1/(1-z). \qquad (31.3.15)$$

This is applicable to unreactive gases on polar surfaces, for then $c \approx 10^2$. The BET isotherm fits experimental observations moderately well over restricted pressure ranges, but it errs by underestimating the extent of adsorption at low pressures, and overestimates it at high pressures.

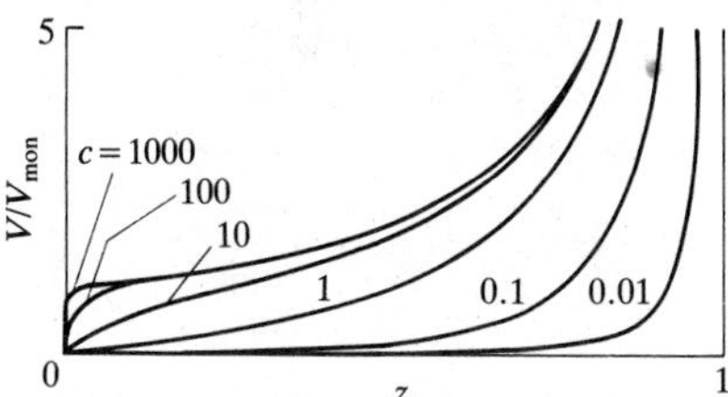

Fig. 31.28. Plots of the BET isotherm for different values of c. The value of V/V_{mon} rises indefinitely because the adsorbate may condense on the covered substrate surface.

31.3 (c) Other isotherms

Another assumption of the Langmuir isotherm is the independence and equivalence of the adsorption sites. Deviations from the Langmuir expression can often be traced to the failure of these assumptions. For example, the enthalpy of adsorption often becomes less negative as θ increases, which suggests that the energetically most favourable sites are occupied first. Various attempts have been made to take these variations into account. The *Temkin isotherm*,

$$\theta = c_1 \ln(c_2 p), \qquad (31.3.16)$$

where c_1 and c_2 are constants, corresponds to supposing that the adsorption enthalpy changes linearly with pressure. The *Freundlich isotherm*,

$$\theta = c_1 p^{1/c_2}, \qquad (31.3.17)$$

corresponds to a logarithmic change. Different isotherms agree with experiment more or less well over restricted ranges of pressure, but they remain largely empirical. Empirical, however, does not mean useless, for if the parameters of a reasonably reliable isotherm are known, reasonably reliable results can be obtained for the extent of surface coverage under various conditions. This kind of information is essential for any discussion of heterogeneous catalysis.

Example 31.5

Examine whether the Freundlich isotherm is a better representation than the Langmuir for the data in *Example* 31.2.

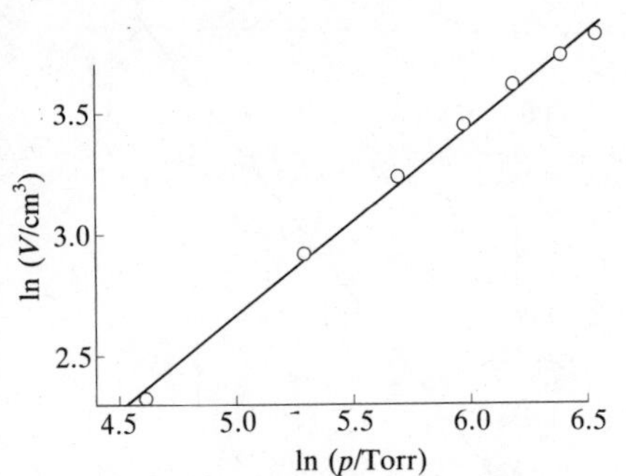

Fig. 31.29. The Freundlich isotherm may be tested by plotting ln V against ln p because it predicts that a straight line should then be obtained. The data are from *Example* 31.5.

• *Method*. Write $\theta = V/V_{\text{mon}}$, and rearrange eqn (31.3.17) into

$$\ln V = \ln c_1 V_{\text{mon}} + (1/c_2)\ln p.$$

Plot ln (V/cm^3) against ln (p/Torr). Compare the least squares fit of the points with the plot in *Example* 31.2. The best way of doing this is to quote the coefficient of determination of the least-squares lines. See the Appendix of the *Solutions Manual*.

• *Answer*. Draw up the following table.

p/Torr	100	200	300	400	500	600	700
ln (p/Torr)	4.61	5.30	5.70	5.99	6.21	6.40	6.55
ln (V/cm^3)	2.32	2.92	3.24	3.45	3.61	3.73	3.83

The points are plotted in Fig. 31.29. The coefficient of determination of the least-squares best line is 0.9980. The value for the Langmuir plot is 0.9978. Hence the former is slightly better.

• *Comment*. The Freundlich isotherm is often used in discussions of adsorption from liquid solutions, when it is written $w = c_1 c^{1/c_2}$, where w is the weight fraction adsorbed (the mass of solute adsorbed per unit mass of adsorbent) and c is the solution's concentration.

• *Exercise*. Examine the suitability of the Temkin isotherm to fit the data.

[$V \propto \ln p$, 0.9731]

31.4 Catalytic activity at surfaces

Table 31.3. Activation energies of catalysed reactions

Reaction	Catalyst	E_a/kJ mol^{-1}
$2HI \rightarrow H_2 + I_2$	None	184
	Au	105
	Pt	59
$2NH_3 \rightarrow N_2 + 3H_2$	None	350
	W	162

A heterogeneous catalyst acts by providing an alternative reaction path with a lower activation energy, Table 31.3. It does not disturb the thermodynamically determined equilibrium composition of the system, only the rate at which that equilibrium is attained.

31.4 (a) Adsorption and catalysis

Heterogeneous catalysis normally depends on at least one reactant being adsorbed (usually chemisorbed) and modified to a form in which it readily undergoes reaction. Often this modification takes the form of a fragmentation of the reactant molecules.

When a reaction takes place by a gas phase particle colliding with a molecule adsorbed on the surface, the rate of formation of product is expected to be proportional to the partial pressure of the non-adsorbed gas B and the extent of surface coverage by the adsorbed gas A. This is the *Eley–Rideal mechanism*, the corresponding rate law being

$$A + B \rightarrow P, \qquad d[P]/dt = kp_B\theta_A. \tag{31.4.1}$$

The rate constant k might be much larger than for the uncatalysed gas-phase reaction because the reaction on the surface has a low activation energy (and the adsorption itself is often non-activated). If the adsorption isotherm for A is known, the rate law can be expressed in terms of its partial pressure, p_A. For example, if the adsorption of A follows a Langmuir isotherm in the pressure range of interest, then the rate law is

$$d[P]/dt = kKp_Ap_B/(1 + Kp_A). \tag{31.4.2}$$

If A is a diatomic molecule that adsorbs as atoms, then the Langmuir isotherm is given by eqn (31.3.4) and the rate law is obtained by substituting this expression into eqn (31.4.1).

In the case represented by eqn (31.4.2), when the partial pressure of A is high there is almost complete surface coverage, and the rate is equal to kp_B because the rate-determining step is the rate of collision of B with the adsorbed fragments. When the pressure of A is less, perhaps because of its reaction, Kp_A is small, and the rate becomes equal to kKp_Ap_B; now the extent of surface coverage is important.

Example 31.6

The decomposition of phosphine on tungsten is first-order at low pressures and zeroth-order at high pressures. Account for these observations.

• *Method*. The reaction is $PH_3 \rightarrow$ *Products*. The rate is proportional to the extent of surface coverage by PH_3. Substitute the Langmuir isotherm for θ, and take the low and high pressure limits.

• *Answer*. The rate equation is

$$dp/dt = -k\theta = -kKp/(1+Kp),$$

where p is the pressure of phosphine.

$$\text{When } Kp \ll 1\text{:} \quad dp/dt = -kKp,$$

and the decomposition is first-order.

$$\text{When } Kp \gg 1\text{:} \quad dp/dt = k,$$

and the decomposition is zeroth-order.

• *Comment*. Many heterogeneous reactions are first-order, which indicates that the rate-determining stage is the adsorption process.

• *Exercise*. Suggest the form of the rate law for the deuteration of NH_3 in which D_2 adsorbs dissociatively and extensively (i.e. $Kp \gg 1$, with p the partial pressure of D_2), and NH_3 (with partial pressure p') adsorbs at different sites. $[dp_{NH_2D}/dt = k(Kp)^{\frac{1}{2}}K'p'/(1+K'p')]$

Many catalysed reactions, including alkene hydrogenation, involve collisions between particles (e.g. molecular fragments and atoms) adsorbed on the surface, and so can be expected to be second-order in the extent of surface coverage. The rates of these *Langmuir–Hinshelwood reactions* follow expressions of the form

$$A + B \rightarrow P, \qquad d[P]/dt = k\theta_A\theta_B. \tag{31.4.3}$$

Insertion of the appropriate isotherms for A and B then gives the reaction rate in terms of the partial pressures of the reactants. For example, if A and B follow Langmuir isotherms, and adsorb without dissociation, then

$$\theta_A = K_Ap_A/(1+K_Ap_A+K_Bp_B), \qquad \theta_B = K_Bp_B/(1+K_Ap_A+K_Bp_B) \tag{31.4.4}$$

and the rate law is

$$d[P]/dt = kK_AK_Bp_Ap_B/(1+K_Ap_A+K_Bp_B)^2. \tag{31.4.5}$$

(This rate law can be integrated to give the time-dependence of the partial pressures: see Problem 31.21.) Since the parameters in the isotherms and the rate constant k are all temperature dependent, the overall temperature dependence of the rate may be strongly non-Arrhenius.

Molecular beam studies are able to give detailed information about catalysed reactions. It has become possible to investigate how the catalytic activity of a surface depends on its structure as well as its composition. For instance, the cleavage of C–H and H–H bonds appears to depend on the presence of steps and kinks, and a terrace often has only minimal catalytic activity. The reaction $H_2 + D_2 \rightarrow 2HD$ has been studied in detail, and it is found that terrace sites are inactive, but one molecule in ten reacts when it strikes a step. While the step itself might be the important feature, it may be that the presence of the step merely exposes a more reactive crystal face (the step face itself). Likewise, the dehydrogenation of hexane to hexene depends strongly on the kink density, and it appears that kinks are needed to cleave C–C bonds. These observations suggest a reason why even small amounts of impurities may poison a catalyst: they are likely to attach to step and kink sites, and so impair the activity of the catalyst entirely. A constructive outcome is that the extent of hydrogenolysis may be controlled relative to other types of reactions by seeking impurities that adsorb at kinks and act as specific poisons.

31.4 (b) Examples of catalysis

Almost the whole of modern chemical industry depends on the development, selection, and application of catalysts, Box 31.1. All we can hope to do in this section is to give a brief indication of some of the problems involved. Other than the ones we consider, these include the danger of the catalyst being poisoned by by-products or impurities and economic considerations relating to cost and lifetime.

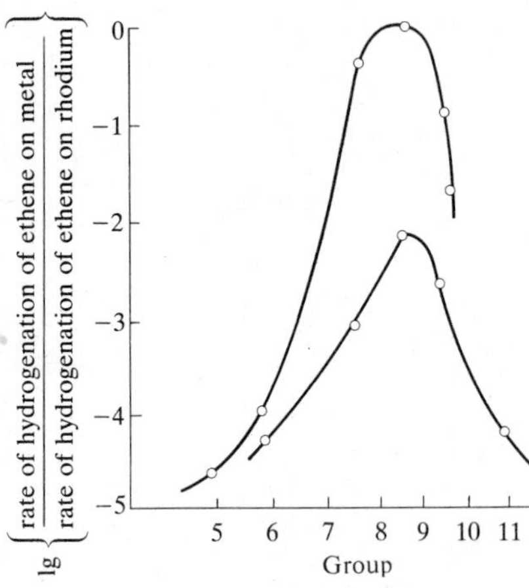

Fig. 31.30. A *volcano curve* of catalytic activity arises because although the reactants must adsorb reasonably strongly, they must not adsorb so strongly that they are immobilized. The lower curve refers to the first series of d-block metals, the upper curve to the second and third series metals. The group numbers relate to the Periodic Table inside the back cover.

The activity of a catalyst depends on the strength of chemisorption as indicated by the 'volcano' curve in Fig. 31.30. In order to be active, the catalyst should be extensively covered by adsorbate, which is the case if chemisorption is strong. On the other hand, if the strength of the substrate–adsorbate bond becomes too great the activity declines either because the other reactant molecules cannot react with the adsorbate or because the adsorbate molecules are immobilized on the surface. This suggests that the activity of a catalyst initially should increase with strength of adsorption (as measured, for instance, by the enthalpy of adsorption) and then decline, and that the most active catalysts should be those lying near the peak of the volcano.

Many metals are suitable for adsorbing gases, and the general order of adsorption strengths decreases along the series O_2, C_2H_2, C_2H_4, CO, H_2, CO_2, N_2. Some of these molecules adsorb dissociatively (e.g. H_2). Transition metals such as Fe, V, and Cr show a strong activity towards all these gases, but Mn and Cu are unable to adsorb N_2 and CO_2, and adsorb H_2 only very weakly. Metals towards the left of the Periodic Table (e.g. Mg and Li) can adsorb only the most active gas (O_2). These trends are also summarized in Box 31.1.

31.4 (c) Hydrogenation

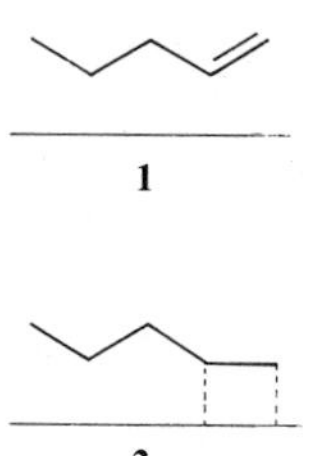

An example of catalytic action is the hydrogenation of alkenes. The alkene (**1**) adsorbs by forming two bonds with the surface (**2**), and on the same surface there may be adsorbed H atoms. When an encounter occurs one of the alkene-surface bonds is broken (**2** → **3** or **4**) and later an encounter with a second H atom releases the fully hydrogenated hydrocarbon, the thermodynamically more stable species. The evidence for a two-stage reaction is the appearance of different isomeric alkenes in the mixture. This

Box 31.1 Properties of catalysts

Catalyst	Function	Examples
Metals	Hydrogenation Dehydrogenation	Fe, Ni, Pt, Ag
Semiconducting oxides and sulphides	Oxidation Desulphurization	NiO, ZnO, MgO, Bi_2O_3/MoO_3
Insulating oxides	Dehydration	Al_2O_3, SiO_2, MgO
Acids	Polymerization Isomerization Cracking Alkylation	H_3PO_4, H_2SO_4, SiO_2/Al_2O_3

Chemisorption abilities:

	O_2	C_2H_2	C_2H_4	CO	H_2	CO_2	N_2
Ti, Cr, Mo, Fe	+	+	+	+	+	+	+
Ni, Co	+	+	+	+	+	+	−
Pd, Pt	+	+	+	+	+	−	−
Mn, Cu	+	+	+	+	±	−	−
Al, Au	+	+	+	+	−	−	−
Li, Na, K	+	+	−	−	−	−	−
Mg, Ag, Zn, Pb	+	−	−	−	−	−	−

+ Strong chemisorption, ± chemisorption, − no chemisorption. See G. C. Bond, *Heterogeneous catalysis*, Oxford, 1986, for further information.

comes about because while the hydrocarbon chain is waving about over the surface of the metal, it might chemisorb again (**4** → **5**) which might desorb to **6**, an isomer of the original **1**. The new alkene would not be formed if the two hydrogen atoms attached simultaneously.

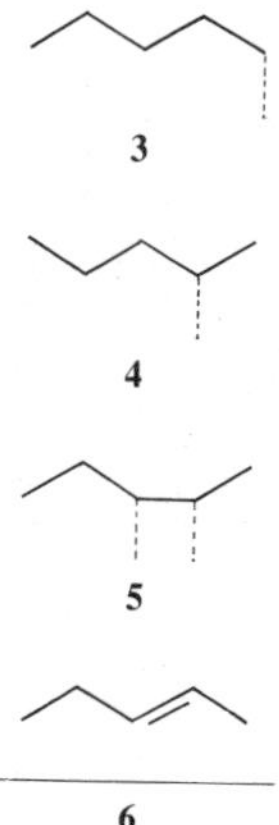

A major industrial application of catalytic hydrogenation is to the formation of edible fats from vegetable and animal oils. Raw oils obtained from sources such as the soya bean have the structure $CH_2(O_2CR)CH(O_2CR')CH_2(O_2CR'')$, where R, R′, and R″ are long-chain hydrocarbons with several double bonds. One disadvantage of the presence of many double bonds is that the oils are susceptible to atmospheric oxidation, and therefore are liable to become rancid. The geometrical configuration of the chains is responsible for the liquid nature of the oil, and in many applications (such as on sandwiches) a solid fat is at least much better and often necessary. Controlled partial hydrogenation of an oil with a catalyst carefully selected so that hydrogenation is incomplete and so that the chains do not isomerize, is used on a wide scale to produce edible fats. The process, and the industry, is not made any easier by the seasonal variation of the number of double bonds in the oils.

31.4 (d) Oxidation

Catalytic oxidation is also widely used in industry and in pollution control. Although in some cases it is desirable to achieve complete oxidation (as in the production of nitric acid from ammonia), in others partial oxidation is the aim. For example, the complete oxidation of propene to carbon dioxide and water is wasteful, but its partial oxidation to acrolein, $CH_2{=}CHCHO$, is the start of important industrial processes. Likewise, the controlled oxidations of ethene to ethanol, ethanal (acetaldehyde), and (in the presence of acetic acid or chlorine) to vinyl acetate or vinyl chloride, are the initial stages of very important chemical industries.

Not all these reactions proceed on a simple metal catalyst, for some depend on oxides of various kinds. The physical chemistry of these surfaces is very obscure, as can be appreciated by considering what happens during the oxidation of propene, $CH_2{=}CHCH_3$, to acrolein, $CH_2{=}CHCHO$, on bismuth molybdate. The first stage is the adsorbation of the propene molecule with loss of a hydrogen to form the allyl radical, $CH_2{=}CHCH_2\cdot$. The oxygen in the surface can transfer to the allyl, leading to the formation of acrolein and its desorption from the surface. The H atom also escapes with a surface O, and goes on to form H_2O, which leaves the surface. The surface is left with the charges the oxide ions discarded as they formed O atoms, and these are attacked by O_2 molecules in the overlying gas which then chemisorb as oxide ions, so reforming the catalyst. This sequence of events involves great upheavals of the surface, and some materials break up under the stress.

31.4 (e) Cracking and reforming

Many of the small organic molecules used in the preparation of all kinds of chemical products come from oil. These small building blocks of polymers, perfumes, and petrochemicals in general, are often cut from the long-chain hydrocarbons squeezed out of the earth. The catalytically induced fragmentation of the long-chain hydrocarbons is called *cracking*, and is often brought about on silica–alumina catalysts. These act by forming unstable carbonium ions, which fall apart and rearrange to more highly branched isomers. These branched isomers burn more smoothly and efficiently in internal combustion engines, and are used to produce higher 'octane' fuels.

Catalytic cracking has been largely superseded by *catalytic reforming* using a *dual-function catalyst*, such as a mixture of platinum and alumina. The platinum provides the *metal function*, and brings about dehydrogenation and hydrogenation. The alumina provides the *acidic function*, being able to form carbonium ions.

The sequence of events in catalytic reforming shows up very clearly the complications that must be unravelled if a reaction as important as this is to be understood and improved. The first step is the attachment of the long-chain hydrocarbon by chemisorption to the platinum. In this process first one and then a second H atom is lost, and an alkene is formed. The alkene migrates to an acid site, where it accepts a proton and attaches to the surface as a carbonium ion. This carbonium ion can undergo several different reactions. It can break into two, isomerize into a more highly branched form, or undergo varieties of ring-closure. Then it loses a proton, escapes from the surface, and migrates (possibly through the gas) as an alkene to a metal part of the catalyst where it is hydrogenated. We end up with a rich selection of smaller molecules which can be withdrawn, fractionated, and then used as raw materials for other products.

Further reading

Adsorption:

On physical adsorption. S. Ross and J. P. Oliver; Interscience, New York, 1964.
Physical chemistry of surfaces (3rd edn). A. W. Adamson; Wiley-Interscience, New York, 1975.
Principles of surface chemistry. G. Somorjai; Prentice-Hall, Englewood Cliffs, 1972.
Chemisorption. D. O. Haywood and B. M. W. Trapnell; Butterworth, London, 1964.
The dynamical character of adsorption. J. de Boer; Clarendon Press, Oxford, 1953.
An introduction to chemisorption and catalysis. R. P. H. Gasser; Clarendon Press, Oxford, 1985.
Chemistry of the metal–gas interface. M. W. Roberts and C. S. McKee; Clarendon Press, Oxford, 1978.

Catalysis:

Chemistry in two dimensions: surfaces. G. Somorjai; Cornell University Press, 1981.
Heterogeneous catalysis: principles and applications (2nd edn). G. C. Bond; Clarendon Press, Oxford, 1986.
Introduction to the principles of heterogeneous catalysis. J. M. Thomas and W. J. Thomas; Academic Press, New York, 1967.
Kinetics of heterogeneous catalysis reactions. M. Boudart and G. Djéga-Mariadassou; Princeton University Press, 1984.
Heterogeneous catalysis in practice. C. Satterfield; McGraw-Hill, New York, 1980.

Introductory problems

A31.1. What pressure of Ar gas is required to produce a collision rate of 4.5×10^{20} atoms s^{-1} at 425 K on a circular surface with diameter 1.5 mm?

A31.2. Find the average rate at which He atoms strike a Cu atom in a surface formed by exposing a (100) plane in metallic Cu to He gas at 80 K and a pressure of 35 Pa. Crystals of copper are face centred cubic with a cell edge of 361 pm.

A31.3. The enthalpy of adsorption of CO on a surface is found to be -120 kJ mol^{-1}. Is this physisorption or chemisorption? Estimate the mean lifetime of a CO molecule on the surface at 400 K.

A31.4. The adsorption of a gas is described by the Langmuir isotherm with $K = 0.85$ kPa^{-1} at 25 °C. Find the pressure at which the surface coverage is (a) 15%, (b) 95%.

A31.5. A certain solid sample adsorbs 0.44 mg of CO when the pressure of the gas is 26.0 kPa and the temperature is 300 K. The amount adsorbed when the pressure is 3.0 kPa and the temperature is 300 K is 0.19 mg. The Langmuir isotherm is known to describe the adsorption. Find the per cent coverage of the surface at the two pressures given above.

A31.6. A solid in contact with a gas at a pressure of 12 kPa and a temperature of 25 °C adsorbs 2.5 mg of the gas. The system follows the Langmuir isotherm. The amount of heat absorbed when 1.00 mmol of the adsorbed gas is desorbed is 10.2 J. What is the equilibrium pressure for the adsorption of 2.5 mg of gas at a temperature of 40 °C?

A31.7. The gas HI is very strongly adsorbed on gold. Under the same temperature and pressure conditions, HI is only slightly adsorbed on platinum. Assume the adsorption follows the Langmuir isotherm and predict the order of the HI decomposition reaction on each of the two metal surfaces.

Problems

31.1. Calculate the frequency of molecular collisions per cm^2 of surface in a vessel containing (a) hydrogen, (b) propane at 25 °C when the pressure is (i) 100 Pa, (ii) 10^{-7} Torr?

31.2. Knowing the frequency of collisions per unit area is not particularly informative unless we can translate it into collisions per atom on the surface. That can be done using the material of Chapter 23. Nickel is face-centred cubic with a unit cell of side 352 pm. What is the number of atoms per cm^2 exposed on a surface formed by (a) (100), (b) (110), (c) (111) planes?

31.3. Tungsten crystallizes into a body-centred cubic form with a unit cell of side 316 pm. Calculate the number of atoms exposed per cm^2 on the same three surfaces as in the last Problem. In filaments of tungsten (110) and (100) planes predominate: estimate the average number of atoms exposed per unit area.

31.4. Translate the frequencies in Problem 31.1 into frequencies per atom for the surfaces encountered in the last two Problems.

31.5. For how long on average would a hydrogen atom remain on a surface at 298 K if its desorption activation energy is (a) $15\ \text{kJ mol}^{-1}$, (b) $150\ \text{kJ mol}^{-1}$? Take $\tau_0 = 10^{-13}$ s.

31.6. For how long on average would the same atoms remain at 1000 K?

31.7. What would be the corresponding times in the last two Problems if hydrogen is replaced by deuterium?

31.8. The average time for which an oxygen atom remains adsorbed to a tungsten surface is 0.36 s at 2548 K and 3.49 s at 2362 K. Find the activation energy for desorption. What is the pre-exponential factor for these tightly chemisorbed atoms?

31.9. The chemisorption of hydrogen on manganese is activated, but only weakly so. Careful measurements have shown that it proceeds 35 per cent faster at 1000 K than at 600 K. What is the activation energy for chemisorption?

31.10. The deposition of atoms and ions on a surface depends on their ability to stick, and therefore on the energy changes that occur. As an illustration, consider a two-dimensional square lattice of univalent positive and negative ions separated by 200 pm, and consider a cation approaching the upper terrace of this array from the top of the page. Calculate, by direct summation, its Coulombic interaction when it is in an empty lattice point directly above an anion. Now consider a high step in the same lattice, and let the approaching ion go into the corner formed by the step and the terrace. Calculate the Coulombic energy for this position, and decide on the likely settling point for a deposited cation.

31.11. In the case of a Langmuir isotherm, show that a plot of p/V_a, where V_a is the volume of gas adsorbed (corrected here, and throughout the following, to a standard temperature and pressure) against p should give a straight line from which both V_a^0, the volume corresponding to complete coverage, and the constant K may be determined. Also show that for small surface coverages another test is to plot $\ln(\theta/p)$ against θ, when a line of gradient -1 should be obtained. What should the slope be when $\ln V_a/p$ is plotted against V_a at small coverages?

31.12. The data below are for the chemisorption of hydrogen on copper powder at 25 °C. Confirm that they fit the Langmuir isotherm. Then find the value of K for the adsorption equilibrium and the adsorption volume corresponding to complete coverage.

p/Torr	0.19	0.97	1.90	4.05	7.50	11.95
V_a/cm^3	0.042	0.163	0.221	0.321	0.411	0.471

31.13. Suppose it is known that ozone adsorbs on a particular surface in accord with a Langmuir isotherm. How could you use the pressure dependence of the fractional coverage to distinguish between adsorption (a) without dissociation, (b) with dissociation into $O + O_2$, (c) with dissociation into $O + O + O$?

31.14. The data for the adsorption of ammonia on barium fluoride are reported below. Confirm that they fit a BET isotherm and find values of c and V_{mon}.

$T = 0\ °\text{C}$, $p^* = 3222$ Torr;

p/Torr	105	282	492	594	620	755	798
V/cm^3	11.1	13.5	14.9	16.0	15.5	17.3	16.5

$T = 18.6\ °\text{C}$, $p^* = 6148$ Torr;

p/Torr	39.5	62.7	108	219	466	555	601	765
V/cm	9.2	9.8	10.3	11.3	12.9	13.1	13.4	14.1

31.15. The *Freundlich isotherm* can be written $V_a = c_1 p^{1/c_2}$, where c_1 and c_2 are constants. The following data were obtained for the adsorption of methane on 10 g of carbon black at 0 °C. Which isotherm, the Langmuir or the Freundlich fits the data better?

p/Torr	100	200	300	400
V_a/cm^3	97.5	144	182	214

31.16. Carbon monoxide adsorbs on mica, and the data for 90 K are given below. Decide whether the Langmuir or the Freundlich isotherm is a better representation of the system. What is the value of K? Given that the total sample area is $6.2 \times 10^3\ \text{cm}^2$, calculate the area occupied by each adsorbed molecule.

p/Torr	100	200	300	400	500	600
V_a/cm^3	0.130	0.150	0.162	0.166	0.175	0.180

31.17. What volume of carbon monoxide would be adsorbed by the mica at 90 K when the pressure is 1 atm?

31.18. The designers of a new industrial plant wanted to use a catalyst code-named CR-1 in a step-involving the fluorination of butadiene. As a first step in the investigation they determined the form of the adsorption isotherm. The volume of butadiene adsorbed per gram of CR-1 at 15 °C depended on the pressure as given below. Is the Langmuir isotherm suitable at this pressure?

p/Torr	100	200	300	400	500	600
V_a/cm^3	17.9	33.0	47.0	60.8	75.3	91.3

31.19. Investigate whether the BET isotherm gives a better description of the adsorption of butadiene on CR-1. At 15 °C $p^*(\text{butadiene}) = 200$ kPa. Find V_{mon} and c.

31.20. Confirm eqn (31.4.4) for the extent of surface coverage in the presence of two adsorbing gases.

31.21. Integrate the rate law in eqn (31.4.5) in the case where $p_A(0) = p_B(0)$. Sketch the time-dependence of the pressure of the product in the case of $K_A \approx K_B \approx 1\ \text{atm}^{-1}$ and $p_A(0) \approx p_B(0) \approx 1$ atm.

31.22. Fluorine adsorbs on CR-2 in accord with the Langmuir isotherm, but butadiene adsorbs in accord with the Freundlich isotherm with $c_2 = 2$. Deduce an expression for the reaction rate on the basis that fluorination depends on encounters of F and butadiene on the catalyst surface. Assume that the adsorption of the two species occurs at different types of site.

31.23. Nitrogen gas adsorbed on charcoal to the extent of $0.921\ \text{cm}^3\ \text{g}^{-1}$ at a pressure of 4.8 atm and a temperature of 190 K, but at 250 K the same amount of adsorption was achieved only when the pressure was increased to 32 atm. What is the molar enthalpy of adsorption of nitrogen on charcoal?

31.24. In an experiment on the adsorption of oxygen on tungsten it was found that the same volume of oxygen was desorbed in 27 min at 1856 K, 2 min at 1978 K, and 0.3 min at 2070 K. What is the activation energy of desorption? How long would it take for the same amount to desorb at (a) 298 K, (b) 3000 K?

31.25. Ammonia was introduced into a bulb at a pressure 27 kPa. At 856 °C it was found that a tungsten catalyst brought about a pressure change of 8 kPa in 500 s, and 15 kPa in 1000 s. What is the order of the catalysed decomposition? Account for the result.

31.26. In some catalytic reactions the products may adsorb more strongly than the reacting gas. This is the case, for instance, in the catalytic decomposition of ammonia on platinum at 1000 °C. As a first step in examining the kinetics of this type of process, show that the rate of ammonia decomposition should follow $-\text{d}p(NH_3)/\text{d}t = k_c p(NH_3)/p(H_2)$ in the limit of very strong adsorption of hydrogen. Start by showing that when a gas J adsorbs very strongly, and its pressure is $p(\text{J})$, that the fraction of uncovered sites is approximately $1/Kp(\text{J})$.

31.27. Solve the rate equation for the catalytic decomposition of ammonia on platinum and show that a plot of $F(t) = (1/t)\ln(p/p_0)$ against $G(t) = (p - p_0)/t$ where p is the pressure of ammonia, should give a straight line from which k_c can be determined. Check the rate law on the basis of the data below, and find k_c for the reaction.

t/s	0	30	60	100	160	200	250
p/Torr	100	88	84	80	77	74	72

31.28. Now suppose that the retarding gas is less weakly adsorbed and that the full form of the Langmuir isotherm has to be used. The reactant gas is still supposed not to adsorb. Deduce the rate law $-\text{d}p/\text{d}t = k_c p/(1 + Kp')$, where p is the pressure of the reactant gas and p' that of the product. Integrate the rate law for the simple decomposition $\text{A} \rightarrow \text{B} + \text{C}$, where B adsorbs and inhibits but C does not.

31.29. An example of this type of reaction is the decomposition of dinitrogen oxide, N_2O, on platinum at 750 °C. In this case the oxygen produced by the reaction adsorbs strongly and inhibits the catalyst. Confirm that the data below fit the integrated rate law and determine K and k_c for the reaction.

t/s	0	315	750	1400	2250	3450	5150
p/Torr	95	85	75	65	55	45	35

31.30. Although the attractive van der Waals interaction between individual molecules varies as R^{-6}, the interaction of a molecule with a nearby solid (a collection of molecules) varies as R^{-3}, where R is its vertical distance above the surface. Confirm this assertion.

31.31. Calculate the interaction energy between an argon atom and the surface of solid argon on the basis of a Lennard–Jones (6, 12)-potential. Estimate the equilibrium distance for an atom above the surface.

31.32. The adsorption of solutes on solids from liquids often follows a Freundlich isotherm. Check the applicability of this isotherm to the following data for the adsorption of acetic acid on charcoal at 25 °C and find the values of the parameters c_1 and c_2.

[acid]/M	0.05	0.10	0.50	1.0	1.5
w_a/g	0.04	0.06	0.12	0.16	0.19

w_a is the mass adsorbed per unit mass of charcoal. Investigate whether the adsorption of acetic acid on charcoal would be better represented by a Langmuir isotherm.

31.33. We now turn to the application of the Gibbs isotherm, Section 25.4, to the adsorption of gases. Show by a directly analogous argument to that used in Chapter 25 that the volume adsorbed per unit area of solid, V_a/σ, is related to the pressure of the gas by $V_a = -(\sigma/RT)(\text{d}\mu/\text{d}\ln p)$, where μ is the chemical potential of the adsorbed gas.

31.34. If the dependence of the chemical potential of the gas on the extent of surface coverage is known, the Gibbs isotherm can be integrated to give a relation between V_a and p, as in a normal adsorption isotherm. For instance, suppose that the change in the chemical potential of a gas when it adsorbs is of the form $\text{d}\mu = -c_2(RT/\sigma)\,\text{d}V_a$, where c_2 is a constant of proportionality: show that the Gibbs isotherm leads to the Freundlich isotherm in this case.

31.35. Finally we come full circle and return to the Langmuir isotherm. Find the form of $\text{d}\mu$ that, when inserted in the Gibbs isotherm, leads to the Langmuir isotherm.

32

Dynamic electrochemistry

Learning objectives

After careful study of this chapter you should be able to:

(1) Describe the structure of the *electric double-layer* and define the *inner* and *outer potentials*, Fig. 32.3.

(2) Describe the *Helmholtz*, *Gouy–Chapman*, and *Stern* models of the electric double-layer, Section 32.1(a).

(3) Explain the meaning of *cathodic* and *anodic current density*, and indicate their significance, Fig. 32.4.

(4) Define *overpotential*, eqn (32.1.5) and *exchange current density*.

(5) Derive and use the *Butler–Volmer equation*, eqn (32.1.7).

(6) Derive, explain, and use the low-overpotential form of the Butler–Volmer equation, eqn (32.1.9).

(7) Derive and use the high-overpotential form of the Butler–Volmer equation, eqn (32.1.10).

(8) Describe and follow the procedure for making a *Tafel plot*, Example 32.2.

(9) Indicate when large or small exchange current densities can be expected. Section 32.1(c).

(10) Describe *concentration polarization* and the *Nernst diffusion layer*, Fig. 32.11, and derive and use an expression for the *limiting current density*, eqn (32.2.8).

(11) Explain the basis of *polarography* and the use of the *rotating-disc electrode*, Section 32.2(b).

(12) Describe how the overpotential and the exchange current density affect the relative rates of gas evolution and metal deposition in cells, Section 32.3(a) and Example 32.4.

(13) Describe the contributions to the reduction of EMF of a working cell, Section 32.3(b), and indicate how they affect its performance, Fig. 32.15.

(14) Describe the operation of *fuel cells*, Section 32.4(a) and Example 32.5.

(15) Indicate the factors involved in the storage of electrical power, Section 32.4(b).

(16) Describe the processes involved in *corrosion*, Section 32.4(c), and derive an approximate expression for its rate, eqn (32.4.3).

(17) Describe how corrosion may be inhibited, Section 32.4(d).

Introduction

Most of electrochemistry depends on processes that occur at a metal in contact with an ionic solution, and the search for efficient forms of power generation and storage relies on a thorough knowledge of what is involved. The economic consequences of electrochemistry are almost incalculable. Most of the modern methods of generating electricity are inefficient, and the development of fuel cells could revolutionize our production and deployment of energy. Today we produce energy inefficiently to produce goods that then decay by corrosion. Each step of this wasteful sequence could be improved by discovering more about the processes that go on at electrodes. Similarly, the emerging techniques of organic and inorganic electrosynthesis, where an electrode is an active component of an industrial process, depend on a detailed knowledge of the factors that affect their rates. One example is in nylon production, where the adiponitrile is synthesized commercially by the electrolytic reductive coupling (hydrodimerization) of acrylonitrile.

The problem we focus on is the rate at which ions can be discharged at electrodes. A measure of this rate is the *current density* (symbol: j), the current per unit area, and most of the discussion will aim to find the properties that affect its magnitude. In an electrolytic cell (i.e. an electrochemical cell in which a chemical reaction is driven by an external supply of electricity), deposition and gas evolution occur significantly only when the applied PD exceeds the equilibrium e.m.f. by an amount called the *overpotential* (symbol: η). We shall establish a relation between the overvoltage and the current, and see what determines their values. In the case of a galvanic cell (i.e. an electrochemical cell that produces electricity) the e.m.f. it develops when producing current is smaller than the equilibrium e.m.f. by an amount equal to the overpotential, and we shall develop relations that enable us to calculate the working e.m.f.

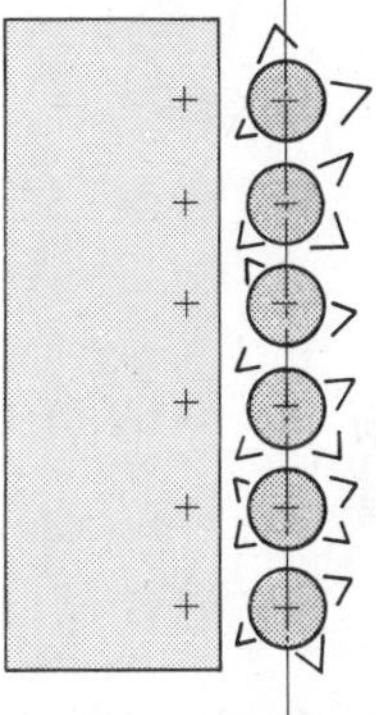

Fig. 32.1. A simple model of the electric double-layer treats it as two rigid planes of charge, one plane, the *outer Helmholtz plane* being due to the ions in solution, the other, the *inner Helmholtz plane*, being due to the charge in the electrode.

32.1 Processes at electrodes

When only equilibrium properties are of interest there is no need to know the details of the charge separation responsible for the PD at an interface (just as we do not need to propose a reaction mechanism when discussing equilibria). However, when we are interested in the rate of charge transfer, a description of the interface becomes essential.

32.1 (a) The double-layer at the interface

The PD across the interface arises from a separation of charge. If electrons leave the electrode and decrease the local cation concentration, the electrode becomes positively charged relative to the solution. The most primitive model of the interface is therefore that it is an *electric double-layer* consisting of a sheet of positive charge at the surface of the electrode and a sheet of negative charge next to it in the solution (or vice versa).

A more detailed picture of the double-layer can be constructed by speculating about the arrangement of ions in the solution. The simplest view, the *Helmholtz model*, is to suppose that the solvated ions range themselves along the surface of the electrode, and are held away from it only by the presence of their hydration spheres, Fig. 32.1. The position of the sheet of ionic charge, which is called the *outer Helmholtz plane*, can be identified as the plane running through the solvated ions. The Helmholtz model ignores the disrupting effect of thermal motion which tends to break up and disperse the rigid wall of charge and make it diffuse. In the *Gouy–Chapman model* of the *diffuse double-layer* the effect of thermal motion is taken into account by a calculation similar to the Debye–Hückel model of the ionic atmosphere of an ion, Section 11.2(a), but their single central ion is replaced by an infinite, plane electrode.

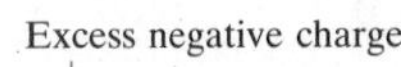

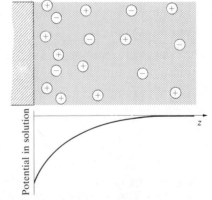

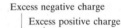

Fig. 32.2. The *Gouy–Chapman model* of the electric double-layer treats it as an atmosphere of counter charge, similar to the Debye–Hückel theory of ion atmospheres.

We shall not describe the Gouy–Chapman calculation, but simply illustrate the results. Figure 32.2 shows how the local concentrations of cations and anions differ from the bulk values: the clustering close to the electrode of ions of opposite charge, and the repulsion of like-charged ions, can be seen. It also shows how misleading it might be to use activity coefficients characteristic of the bulk to discuss the thermodynamic properties of ions lying close to electrodes. This is one of the reasons why dynamic electrochemical experiments are almost always done using a large excess of supporting electrolyte (e.g. a 1 M solution of some salt, acid, or base), for then the activity coefficients are almost constant because the inert ions swamp the effects of local changes due to any reactions taking place.

Neither the Helmholtz nor the Gouy–Chapman model is a very good representation of the structure of the double-layer. The former overemphasizes the rigidity of the local solution; the latter underemphasizes its structure. The two are combined in the *Stern model*, in which the ions closest to the electrode are constrained into a rigid Helmholtz plane while outside that plane the ions are dispersed as in the Gouy–Chapman model.

The most important property of the double-layer is the effect it has on the electric potential near the electrode. Imagine separating the electrode from the solution, but with the charges of the metal and the solution frozen in position. Now consider the electrode and the solution separately.

At great distances from the electrode a positive test charge experiences a Coulomb potential which varies inversely with distance, Fig. 32.3. As the test charge approaches the electrode it enters a region where the potential changes more slowly on account of the surface charge being not point-like but spread over an area. At about 10^{-5} cm from the surface the potential is only weakly distance-dependent and in this region it is called the *Volta potential* or the *outer potential* (symbol: ψ). As the test charge is taken through the skin of electrons on the surface of the electrode the potential it measures changes until it reaches the inner, bulk metal environment. This extra potential is called the *surface potential* (symbol: χ). The total potential inside the electrode is called the *Galvani potential* (symbol: ϕ_M).

A similar sequence of changes of potential is observed as a positive test charge is brought up to and through the solution surface. The potential changes to its Volta value as the charge approaches the charged medium, then to its Galvani value (ϕ_S) as it is taken through into the bulk.

When the electrode and solution are reassembled (without any change of charge distribution) the PD between points in the bulk metal and the bulk solution is the *Galvani potential difference* (symbol: $\Delta\phi$). This is the quantity used in the discussion of electrode potentials in Chapters 11 and 12.

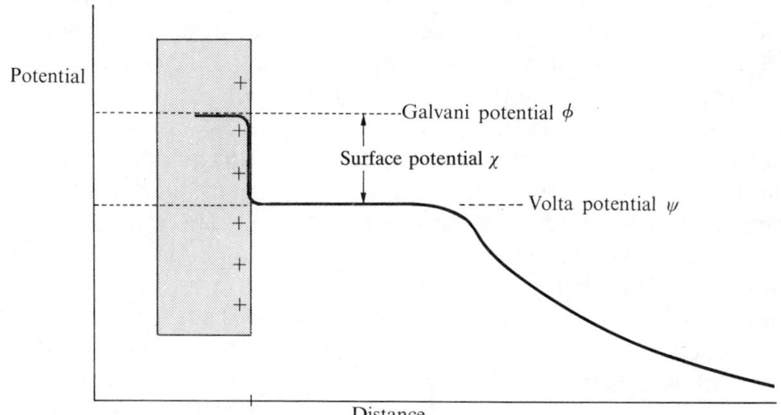

Fig. 32.3. The variation of potential with distance from an electrode that has been separated from the electrolyte solution without there being an adjustment of charge. A similar diagram applies to the separated solution.

32.1 (b) The rate of charge transfer

In order for an ion (or neutral molecule) to participate in charge transfer at an electrode it has to discard any solvating molecules and migrate through the electric double-layer and adjust its hydration sphere as it receives or discards electrons. Likewise, an ion or molecule already at the inner plane must be detached and migrate into the bulk. Both processes are activated and their rate constants can be written

$$k = B\mathrm{e}^{-\Delta G^{\ddagger}/RT}, \tag{32.1.1}$$

where $\Delta G^{\ddagger}$ is the activation Gibbs function (Section 30.3(d)) and B is some constant. The magnitude of $\Delta G^{\ddagger}$, and therefore the flow of current, depends on the identity of the migrating particle and the PD between its initial location and the transition state. The flows of current towards and away from the electrode are equal when the PD has its equilibrium value. The *net* flow of current therefore depends on the extent to which the PD at the interface differs from its equilibrium value. That is, the net current depends on the overpotential. We now set about finding this dependence.

The first step is to be more precise about the rate constant. We are dealing with a heterogeneous process, and so it is natural to express the rate as the amount of material produced per unit time per unit area of surface. A first-order heterogeneous rate law therefore has the form

Amount produced per unit area per unit time $= k \times$ *(Concentration)*.

If the concentration is in moles per unit volume, then since the expression on the left is expressed in moles per unit area per unit time, the rate constant has dimensions of length per unit time (e.g. $\mathrm{cm\,s^{-1}}$). The constant B therefore also has these dimensions.

Now consider a reaction at the electrode in which a particle is reduced by the transfer of a single electron in the rate-determining step. (The last phrase is important: in the deposition of cadmium, for instance, only one electron is transferred in the rate-determining step even though overall the deposition involves the transfer of two.) Let the molar concentration of the oxidized and reduced materials outside the double layer be [Ox] and [Red] respectively. Then the net current at the electrode is the difference of the currents arising from the reduction of Ox and the oxidation of Red, the rates of these processes being k_c[Ox] and k_a[Red] respectively (the notation k_c and k_a is explained below).

The redox processes involve the transfer of one electron per reaction event, and so the magnitude of the charge transferred per mole of events in either direction is $eN_A = F$, where F is Faraday's constant. The current densities arising from the redox processes are the rates (expressed as above) multiplied by the charge associated with the material generated. Therefore, there is a *cathodic current density* of magnitude $j_c = Fk_c[\mathrm{Ox}]$ arising from the reduction, and an opposing *anodic current density* of magnitude $j_a = Fk_a[\mathrm{Red}]$ arising from the oxidation. The magnitude of the *net current density* at the electrode, Fig. 32.4, is therefore the difference

$$\begin{aligned} j = j_a - j_c &= Fk_a[\mathrm{Red}] - Fk_c[\mathrm{Ox}] \\ &= FB_a[\mathrm{Red}]\mathrm{e}^{-\Delta G_a^{\ddagger}/RT} - FB_c[\mathrm{Ox}]\mathrm{e}^{-\Delta G_c^{\ddagger}/RT}. \end{aligned} \tag{32.1.2}$$

We have introduced eqn (32.1.1) and have allowed for the activation Gibbs functions to be different for the cathodic and anodic processes. That they *are* different is the central feature of the remaining discussion. Note that when $j_a > j_c$, so that $j > 0$, the current is *anodic*, and when $j_c > j_a$, so that $j < 0$, it is *cathodic*, Fig. 32.4.

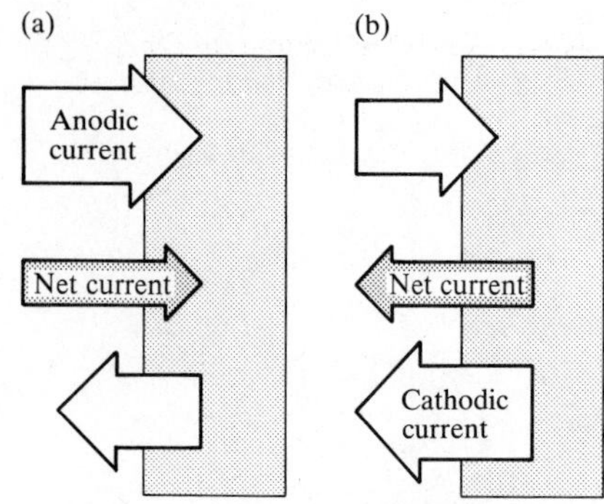

Fig. 32.4. The net current is defined as the difference $j_a - j_c$. (a) When $j_a > j_c$ the net current is *anodic*, and there is a net oxidation of the species in solution. (b) When $j_c > j_a$ the net current is cathodic, and the net process is reduction.

32.1 (c) The current density

Now we relate j to the Galvani potential difference, which varies across the double-layer as shown in Fig. 32.5.

Consider the reduction reaction. An electron is transferred from the electrode where the potential is ϕ_M to the solution where it is ϕ_S. There is therefore an electrical contribution to the work of magnitude $e\Delta\phi$. If the transition state corresponds to Ox being very close to the electrode, Fig. 32.6, the activation Gibbs function is changed from $\Delta G^{\ddagger}$ to $\Delta G^{\ddagger} + F\Delta\phi$. Therefore, if the electrode is *more positive* than the solution ($\Delta\phi > 0$), more work has to be done to bring Ox to its transition state; in this case the activation Gibbs function is increased. If the transition state corresponds to Ox being close to the outer plane of the double-layer, Fig. 32.7, then the value of $\Delta G^{\ddagger}$ is independent of $\Delta\phi$. In a real system the transition state lies at some intermediate location between these extremes. Fig. 32.8, and so we write the activation Gibbs function for reduction as $\Delta G^{\ddagger} + \alpha F\Delta\phi$. The parameter α is called the *transfer coefficient* or *symmetry factor*. It lies in the range 0 to 1, and experimentally is often found to be about $\frac{1}{2}$.

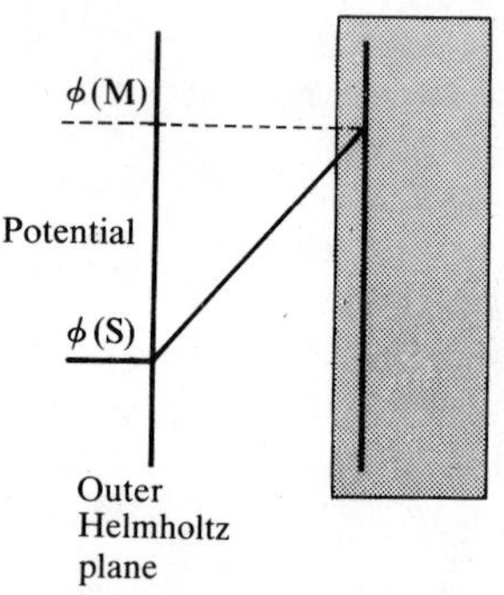

Fig. 32.5. The electric potential varies linearly between two plane parallel sheets of charge, and its effect on the Gibbs function of the transition state depends on the latter's location between the inner and outer planes.

Now consider the oxidation of Red. In this case Red discards an electron to the electrode and so the extra work of reaching the transition state is zero if that state lies close to the electrode, Fig. 32.6, the full $-F\Delta\phi$ if it lies at the outer plane, Fig. 32.7, and $\Delta G^{\ddagger} - (1-\alpha)F\Delta\phi$ in general, Fig. 32.8.

Now we take the main step in the argument. The two activation Gibbs functions can be inserted in place of the values used in eqn (32.1.2) with the result that

$$j = \{FB_a[\text{Red}]e^{-\Delta G_a^{\ddagger}/RT}\}e^{(1-\alpha)F\Delta\phi/RT} - \{FB_c[\text{Ox}]e^{-\Delta G_c^{\ddagger}/RT}\}e^{-\alpha F\Delta\phi/RT}, \tag{32.1.3}$$

which is an explicit, if complicated, expression for the net current density in terms of the potential difference.

The appearance of eqn (32.1.3) can be simplified as follows. First, we

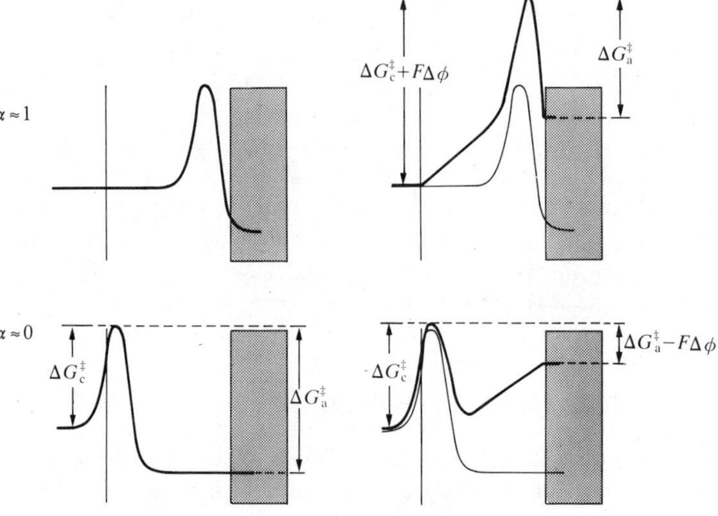

Fig. 32.6. When the transition state is very close to the inner plane the activation Gibbs function for the anodic current is almost unchanged, but the full effect applies to the cathodic current.

Fig. 32.7. When the transition state is very close to the outer plane the cathodic current activation Gibbs function is almost unchanged but the anodic current activation Gibbs function is strongly affected.

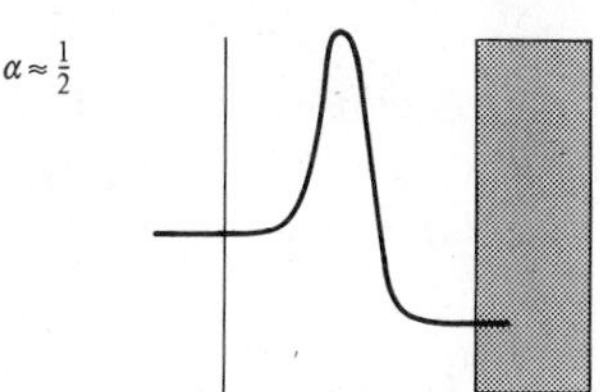

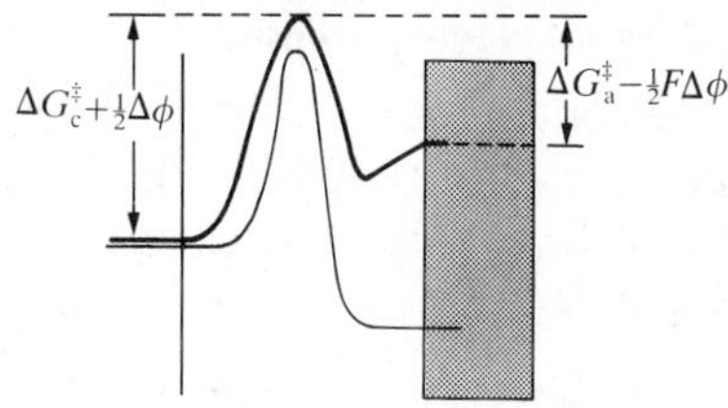

Fig. 32.8. When the transition state is located at an intermediate position as measured by α (with $0 < \alpha < 1$), both activation Gibbs functions are affected.

identify the individual cathodic and anodic current densities:

$$\left.\begin{aligned} j_a &= \{FB_a[\text{Red}]e^{-\Delta G_a^\ddagger/RT}\}e^{(1-\alpha)F\Delta\phi/RT} \\ j_c &= \{FB_c[\text{Ox}]e^{-\Delta G_c^\ddagger/RT}\}e^{-\alpha F\Delta\phi/RT} \end{aligned}\right\} j = j_a - j_c. \qquad \begin{aligned}(32.1.4a)\\(32.2.4b)\end{aligned}$$

When $\Delta\phi$ has its equilibrium value $\Delta\phi_{eq}$ the net current is zero, and so the equilibrium current densities $j_{a,e}$ and $j_{c,e}$ are equal (and have a value obtained by substituting $\Delta\phi_{eq}$ for $\Delta\phi$ in either of these expressions). It follows that when the PD differs from its equilibrium value by the overpotential η, so that

$$\eta = \Delta\phi - \Delta\phi_{eq}, \qquad (32.1.5)$$

the two current densities are

$$j_a = \{FB_a[\text{Red}]e^{-\Delta G_a^\ddagger/RT}\}e^{(1-\alpha)F\Delta\phi_{eq}/RT}e^{(1-\alpha)\eta F/RT} = j_{a,e}e^{(1-\alpha)\eta F/RT}, \qquad (32.1.6a)$$

$$j_c = \{FB_c[\text{Ox}]e^{-\Delta G_c^\ddagger/RT}\}e^{-\alpha F\Delta\phi_{eq}/RT}e^{\alpha\eta F/RT} = j_{c,e}e^{-\alpha\eta F/RT}. \qquad (32.1.6b)$$

Since the two equilibrium current densities are equal, we can drop the a, c subscripts and call each one j_0. Then from eqn (32.1.3) we obtain the *Butler–Volmer equation:*

$$j = j_0\{e^{(1-\alpha)\eta F/RT} - e^{\alpha\eta F/RT}\}. \qquad (32.1.7)$$

j_0 is called the *exchange current density*.

A small note may be added in passing. The expression for j_c (or j_a) can be used to calculate by how much the current density changes when the PD at an electrode changes from 1 V to 2 V. Taking $\alpha = \frac{1}{2}$ in eqn (32.1.6b) gives

$$\frac{j_c(2\text{ V})}{j_c(1\text{ V})} = \exp\left\{\frac{-\frac{1}{2}\times(1\text{ V})\times(9.65\times10^4\text{ C mol}^{-1})}{(8.314\text{ J K}^{-1}\text{ mol}^{-1})\times(298\text{ K})}\right\} = 3\times10^8.$$

This huge change of current density occurs for a very mild and easily applied change of conditions. An understanding of why it is so great is obtained by realizing that a change of PD by 1 V changes the activation Gibbs function by (1 V)F/RT, or about 50 kJ mol^{-1}, which has an enormous effect on the rates. An explanation in terms of atomic events is obtained by expressing the change in PD in terms of its effect on the electric field acting on an ion stationed at the outer plane of the double-layer. A change of 1 V across a 1 nm thick layer changes the field by 10^9 V m^{-1}. It is hardly surprisingly that huge changes in current arise from electric fields of such magnitudes. Innocuous as electrochemistry may appear from the outside, scrutiny at the molecular level reveals the great electrical stresses involved as electrodes tear electrons from ions.

32.1 (d) Overpotential

The content of the Butler–Volmer equation can be clarified by examining its limiting forms. When the overpotential is very small (formally when $\eta F/RT \ll 1$; in practice less than about 0.01 V) the exponentials can be expanded using $e^x = 1 + x + \ldots$ to give

$$j = j_0\{[1 + (1-\alpha)\eta F/RT + \ldots] - [1 - (\alpha\eta F/RT) + \ldots]\} = j_0\eta F/RT. \tag{32.1.8}$$

This shows that *the current density is proportional to the overpotential*, and so at low overpotentials the interface behaves like an ohmic conductor (current $\propto$PD). When there is a small positive overpotential the current is anodic ($j > 0$ when $\eta > 0$), and when the overpotential is small and negative the current is cathodic ($j < 0$ when $\eta < 0$). Furthermore, as in an ohmic conductor, we can reverse the relation and calculate the PD that must exist if a current density j has been established by some external circuit:

$$\eta = (RT/F)j/j_0. \tag{32.1.9}$$

The importance of this interpretation will become apparent below.

Example 32.1

The exchange current density of a Pt $|H_2(g)|$ $H^+(aq)$ electrode at 298 K is 0.79 mA cm^{-2}. What current flows through a standard electrode of total area 5.0 cm^2 when the PD across the interface is +5.0 mV?

- *Method*. The equilibrium PD is zero, and so the overpotential is +5.0 mV. Use eqn (32.1.8) and convert to current by multiplying the current density by the electrode area. Note that $R\mathcal{T}/F = 25.68$ mV.

- *Answer*. From eqn (32.1.8)

$$j = j_0\eta F/R\mathcal{T} = (0.79 \text{ mA cm}^{-2}) \times (+5.0 \text{ mV}/25.58 \text{ mV})$$
$$= +0.15 \text{ mA cm}^{-2}.$$

The current is therefore 0.75 mA.

- *Comment*. The current density is anodic, which means that the oxidation reaction dominates when the overpotential is positive (remember Box 11.2). Since $\eta F/R\mathcal{T} = 0.1$, the linear approximation is valid.

- *Exercise*. What would be the current at pH = 2.0, the other conditions being the same?

[−17 mA]

When the overpotential is large (greater than about 0.1 V) the Butler–Volmer equation takes a different limiting form. When the overpotential is large and positive, corresponding to the electrode being the anode in electrolysis (Box 11.2), the second exponential in eqn (32.1.7) is much smaller than the first and may be neglected. Then

$$j = j_0 e^{(1-\alpha)\eta F/RT}, \quad \text{and so} \quad \ln j = \ln j_0 + (1-\alpha)\eta F/RT. \tag{32.1.10a}$$

When the overpotential is large but negative (corresponding to the cathode in electrolysis), the first exponential in eqn (32.1.7) may be neglected. Then

$$j = -j_0 e^{-\alpha\eta F/RT}, \quad \text{and so} \quad \ln(-j) = \ln j_0 - \alpha\eta F/RT. \quad (32.1.10b)$$

The plot of the logarithm of the current density against the overpotential is called a *Tafel plot:* the slope gives the symmetry parameter α and the intercept at $\eta = 0$ gives the exchange current density.

The experimental arrangement used for a Tafel plot is shown in Fig. 32.9. The electrode of interest is called the *working electrode*, and the current flowing through it is controlled externally. If its area is A and the current is I, then the current density across its surface is I/A. The PD across the interface cannot be measured directly, but the potential of the working electrode relative to a third electrode, the *reference electrode* can be measured with a high impedance digital voltmeter, and no current flows in that half of the circuit. The reference electrode is in contact with the solution close to the working electrode through a *Luggin capillary*, which helps to eliminate any ohmic PD that might arise accidentally. Changing the current flowing through the working circuit causes a change of potential of the working electrode, which is measured with the voltmeter. The overpotential is then obtained by taking the difference between the potentials measured with and without a flow of current through the working circuit.

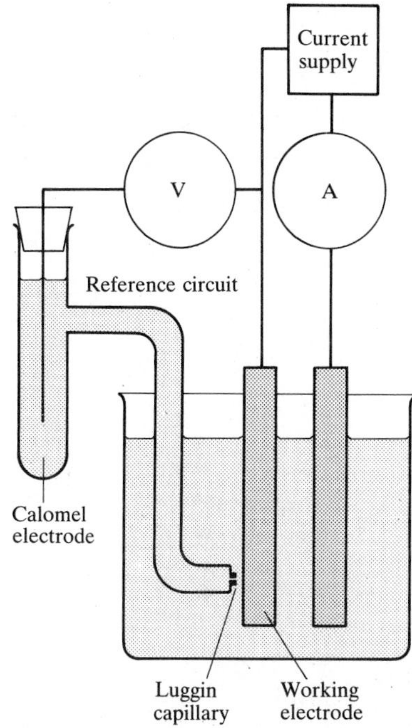

Fig. 32.9. The general arrangement for electrochemical rate measurements. The external source establishes a current between the working electrodes, and its effect on the potential difference of either of them relative to the reference electrode is observed. No current flows in the reference circuit.

Example 32.2

The data below refer to the anodic current through a 2.0 cm^2 platinum electrode in contact with an Fe^{2+}, Fe^{3+} aqueous solution at 298 K. Calculate the exchange current density and the transfer coefficient for the electrode process.

η/mV	50	100	150	200	250
I/mA	8.8	25.0	58.0	131	298

• *Method*. The anodic process is the oxidation $Fe^{2+} \rightarrow Fe^{3+} + e^-$. Make a Tafel plot ($\ln j$ against η) using the anodic form, eqn (32.1.10a). The intercept at $\eta = 0$ is $\ln j_0$ and the slope is $(1-\alpha)F/R\mathcal{T}$.

• *Answer*. Draw up the following table:

η/mV	50	100	150	200	250
j/mA cm^{-2}	4.4	12.5	29.0	56.6	149
$\ln(j/\text{mA cm}^{-2})$	1.50	2.53	3.37	4.18	5.00

The points are plotted in Fig. 32.10. The high overpotential region gives a straight line of intercept 0.92 and slope 0.0163. From the former it follows that $\ln(j_0/\text{mA cm}^{-2}) = 0.92$, so that $j_0 = 2.5$ mA cm^{-2}. From the latter,

$$(1-\alpha)F/R\mathcal{T} = 0.0163\,\text{mV}, \quad \text{so that} \quad \alpha = 0.58.$$

• *Comment*. Note that the Tafel plot is non-linear for $\eta < 150$ mV. In this region $\eta F/R\mathcal{T} = 3$, which is not much larger than unity.

• *Exercise*. Repeat the analysis using the following cathodic current data:

η/mV	−50	−100	−150	−200	−250	−300
I/mA	−0.3	−1.5	−6.4	−27.6	−118.6	−510

[$\alpha = 0.75$, $j_0 = 0.040$ mA cm^{-2}]

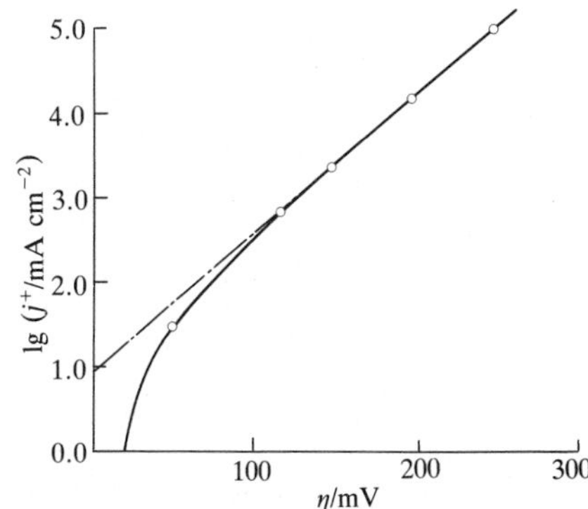

Fig. 32.10. A *Tafel plot* is used to measure the exchange current density (given by the extrapolated intercept at $\eta = 0$) and the transfer coefficient (from the slope). The data are from *Example* 32.2.

Some experimental values for the Butler–Volmer parameters are given in Table 32.1. Exchange current densities vary over a very wide range. For

Table 32.1. Exchange current densities and transfer coefficients at 298 K

Reaction	Electrode	$j_0/\mathrm{A\,cm^{-2}}$	α
$2H^+ + 2e^- \rightarrow H_2$	Pt	7.9×10^{-4}	
	Ni	6.3×10^{-6}	0.58
	Pb	5.0×10^{-12}	
$Fe^{3+} + e^- \rightarrow Fe^{2+}$	Pt	2.5×10^{-3}	0.58

example, the N_3^-,N_2 couple on Pt has $j_0 = 10^{-76}\ \mathrm{A\,cm^{-2}}$ whereas the H_2,H^+ couple on Pt has $j_0 = 8 \times 10^{-4}\ \mathrm{A\,cm^{-2}}$, a range of 73 orders of magnitude. Exchange currents are generally large when the redox process involves no bond breaking (as in the $[Fe(CN)_6]^{3-}, Fe(CN)_6]^{4-}$ couple) or if only weak bonds are broken (as in Cl_2,Cl^-). They are generally small when more than one electron needs to be transferred, or when multiple or strong bonds are broken, as in the N_3^-,N_2 couple and in redox reactions of organic compounds.

Electrodes with potentials that change only slightly when a current passes through them are called *non-polarizable*. Those with strongly current-dependent potentials are called *polarizable*. From the linearized equation, eqn (32.1.9), it is clear that the criterion for low polarizability is high exchange current density (so that may be small even though j is large). The calomel and H_2/Pt electrodes are both highly non-polarizable, which is one reason why they are so extensively used in equilibrium electrochemistry measurements.

32.2 More aspects of polarization

One of the assumptions in the derivation of the Butler–Volmer equation is the uniformity of concentration near the electrode. This assumption fails at high current densities because the diffusion towards the electrode from the bulk is slow and may become rate-determining. A larger overpotential is then needed to produce a given current, the effect being called *concentration polarization*. The extra overpotential due to concentration polarization is called the *polarization overpotential* (symbol: η^c).

32.2 (a) The polarization overpotential

The value of the concentration overpotential can be estimated as follows. For simplicity we consider (a) a case where the concentration polarization dominates all the rate processes and (b) an electrode couple of the type M^{z+},M. At equilibrium, when the net current density is zero, the PD across the double layer is related to the activity a of the ions in the solution by eqn (11.3.6):

$$\Delta\phi_{eq} = \Delta\phi^{\ominus} + (RT/zF)\ln a. \qquad (32.2.1)$$

As has been remarked, electrode kinetics are normally studied using a large excess of support electrolyte, so that the concentrations of ions, and therefore mean activity coefficients, remain approximately constant. Therefore the constant activity coefficient in $a = \gamma c$ may be absorbed into $\Delta\phi^{\ominus}$, and $\Delta\phi^{\circ} = \phi^{\ominus} + (RT/zF)\ln\gamma$ is called the *formal potential difference*. When the electrode is not at equilibrium, the active ion concentration just outside the double-layer changes to c' and the PD changes to

$$\Delta\phi = \Delta\phi^\circ + (RT/zF)\ln c'. \qquad (32.2.2)$$

The concentration overpotential is therefore

$$\eta^c = \Delta\phi - \Delta\phi_{eq} = (RT/zF)\ln(c'/c). \qquad (32.2.3)$$

The next problem is to estimate the concentration just outside the double-layer. The difference of c' from c depends on the diffusion coefficient, D, of the ions, for if they are very mobile the concentration at the electrode will be close to that of the bulk even when a current is flowing. The value of c' can be calculated using Fick's Law, eqn (26.3.1), relating the particle flux to the concentration gradient:

$$J = -D(\partial\mathcal{N}/\partial x). \qquad (32.2.4)$$

$\mathcal{N}$ is the number density, $\mathcal{N} = cN_A$. We now suppose that the solution has its bulk concentration up to a distance δ from the outer Helmholtz plane, and then falls linearly to c' at the plane itself. This *Nernst diffusion layer* is illustrated in Fig. 32.11; note that the thickness of the layer (typically 0.1 mm, and strongly dependent on the condition of hydrodynamic flow) is quite different from that of the electric double-layer (typically less than 1 nm, and unaffected by stirring). The concentration gradient through the Nernst layer is

$$\partial\mathcal{N}/\partial x = (\mathcal{N}' - \mathcal{N})/\delta = (c' - c)N_A/\delta, \qquad (32.2.5)$$

which implies that the particle flux towards the electrode is

$$J = -(N_A D/\delta)(c' - c).$$

The current density towards the electrode is the product of the particle flux and the charge ze each particle carries: $j = ezJ$. Therefore, by combining this relation with the last we can express c' in terms of the current density at the double layer:

$$c' = c - (\delta/zFD)j. \qquad (32.2.6)$$

When this is substituted into eqn (32.2.3) we obtain the following expressions for the overpotential in terms of the current density:

$$\eta^c = (RT/zF)\ln\{1 - (j\delta/zcFD)\}, \qquad (32.2.7a)$$
$$j = (zcFD/\delta)\{1 - e^{zF\eta^c/RT}\}. \qquad (32.2.7b)$$

The maximum rate of diffusion across the Nernst layer occurs when the gradient is steepest, when $c' = 0$. This concentration occurs when an electron from an ion that diffuses across the layer is snapped over the activation barrier, through the double-layer, and on to the electrode. No flow of current can exceed the *limiting current density* (symbol: j_L) that then occurs, and which is given by

$$j_L = ezJ_L = (zFD/\delta)c. \qquad (32.2.8)$$

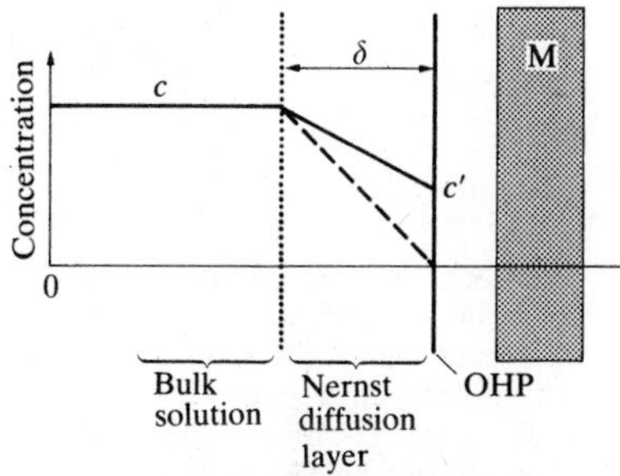

Fig. 32.11. In a simple model of the *Nernst diffusion layer* there is a linear variation in concentration between the bulk and the outer Helmholtz plane; the thickness of the layer depends strongly on the state of flow of the fluid.

Example 32.3

Estimate the limiting current density at 298 K for an electrode in a 0.10 M Cu^{2+}(aq) unstirred solution.

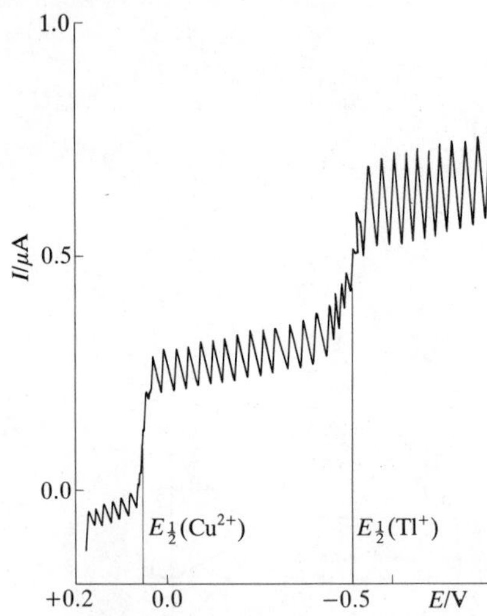

Fig. 32.12. A *polarograph* is obtained when the current flowing through a cell is plotted against the applied potential. This polarograph shows the presence of Cu^{2+} and Tl^{+} ions by their characteristic half-wave potentials, $E_{\frac{1}{2}}$.

Table 32.2. Half-wave potentials at 298 K, $E_{\frac{1}{2}}/V$

Zn^{2+}	−1.00
Cd^{2+}	−0.60
Fe^{3+}	0.0
Cu^{2+}	+0.04

Values refer to ions in 0.1 M KCl(aq), relative to the calomel electrode.

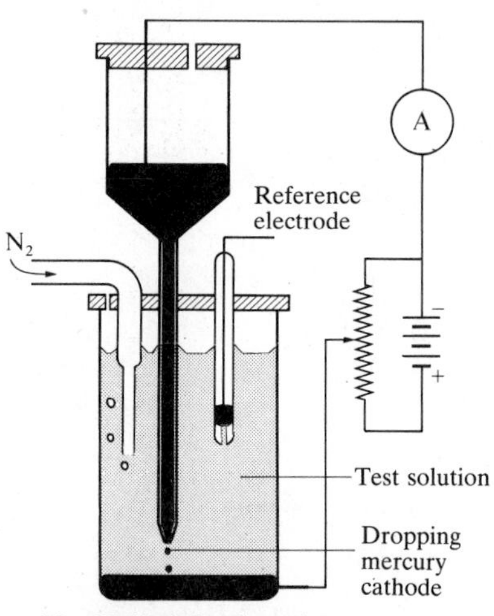

Fig. 32.13. The *dropping mercury electrode* ensures that a renewed, clean surface is provided continuously. The growth of the droplets accounts for the periodic variation of the current in Fig. 32.12.

• *Method*. Use eqn (32.2.8). Take $\delta = 0.3$ mm as a typical value for an unstirred solution. Estimate D from the ionic conductivity $\lambda = 107\ \mathrm{S\ cm^2\ mol^{-1}}$ (Table 27.1) and the Nernst–Einstein relation, Box 27.1.

• *Answer*. Substitution of the Nernst–Einstein relation, $\lambda = (z^2F^2/RT)D$ into eqn (32.2.8) gives

$$j_L = \lambda cRT/zF\delta.$$

Therefore, with $\delta = 0.3$ mm, $c = 0.10$ M, $z = 2$, and $T = 298$ K, $j_L = 0.5\ \mathrm{mA\ cm^{-2}}$.

• *Comment*. This result implies that the current towards a 1 cm^2 electrode cannot exceed 0.5 mA in this solution. If the solution is stirred, or if the electrode surface is moving (as in polarography) the Nernst layer is much thinner.

• *Exercise*. Evaluate the limiting current density for an $Ag^+|Ag$ electrode in a stirred 0.01 M Ag^+(aq) solution at 298 K. Take δ as about 0.03 mm. [5 mA cm^{-2}]

32.2 (b) Polarography

The magnitude of the limiting current density depends on the charge, the mobility, and the concentration of the ions, and its measurement is the basis of the analytical technique of *polarography*.

In a polarography experiment the current is measured as the PD between two electrodes is changed. The current climbs with the PD until the limiting value is reached, the value being characteristic of the ion present. If several ions are present the current rises to a series of characteristic values, and they are identified by measuring their *half-wave potentials*, Fig. 32.12 and Table 32.2. The magnitude of the limiting current density is used to find the concentrations of the ions.

One of the problems with the procedure is the contamination of the electrode surface. This can be avoided by using a *dropping mercury electrode*, Fig. 32.13, in which the electrode is constantly renewed as drops form at the end of the capillary and then fall off. The growth of the drop, and therefore of the surface area of the electrode, accounts for the oscillations in the polarogram.

The problems of dealing with the complicated hydrodynamics of flow towards a growing electrode are reduced by using the *rotating-disc electrode*, Fig. 32.14. The electrode is a small flat disc set in a vertical rotating axle. The rotation of its surface sets up a steady hydrodynamic flow that circulates the solution over its face. The flow pattern can be calculated, and the limiting current related to the speed of rotation.

Various modifications of the rotating-disc electrode have been developed. One involves pulsing the PD and then observing the growth of current towards its limiting value. Another is the *ring-disc electrode*, where the central spinning disc is surrounded by an electrode in the form of a narrow ring. As the disc rotates the flow lines bring the solution towards it and then over the ring. Analysis of the currents at both electrodes and knowledge of the time it takes for the products formed at the disc to flow to the ring gives very detailed information about electrode kinetics, such as the identities of the reactants and the rates of electron transfer.

32.3 Electrochemical processes

In order to induce current to flow through an electrolytic cell the applied PD must exceed the equilibrium cell e.m.f. by at least the *cell overpotential*, the sum of the overpotentials at the two electrodes and the ohmic drop (IR) due to the current through the electrolyte, and this extra potential may need to be large when the exchange current density at the electrodes is small. For similar reasons a galvanic cell producing current generates a smaller e.m.f. than at equilibrium. In this section we see how to cope with both aspects of the overpotential.

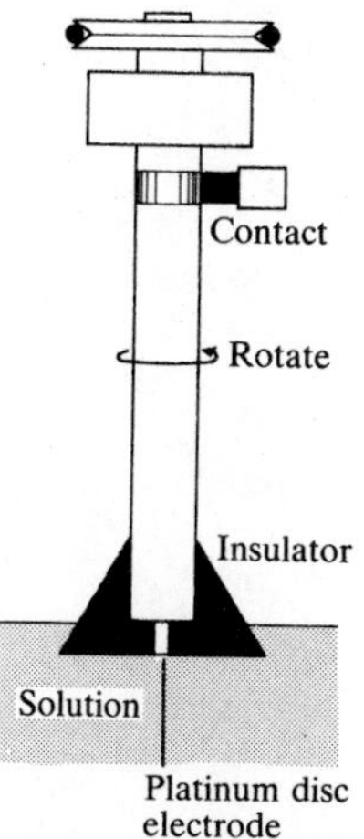

Fig. 32.14. The dropping mercury electrode has been largely replaced by the *rotating-disc electrode*, because the hydrodynamic flow over its face can be calculated more reliably.

32.3 (a) Evolution and deposition at electrodes

The rate of gas evolution or metal deposition can be estimated from the Butler–Volmer equation and tables of exchange current densities. The exchange current density is strongly dependent on the nature of the electrode surface, and changes during the electrodeposition of one metal on another.

Example 32.4

Derive an expression for the relative rates of electrodeposition and hydrogen evolution in a solution in which both may occur.

• *Method*. Calculate the ratio of the cathodic currents using eqn (32.1.10b). Write j for the hydrogen cathodic current and j' for the metal. For simplicity, assume equal transfer coefficients.

• *Answer*. From eqn (32.1.10b)

$$j'/j = (j_0'/j_0)e^{\alpha(\eta-\eta')F/RT}.$$

• *Comment*. This shows that metal deposition is favoured by a large exchange current density and relatively high hydrogen evolution overvoltage (so that $\eta - \eta'$ is positive and large).

• *Exercise*. Deduce an expression for the ratio when the hydrogen evolution is limited by transport across a diffusion layer. $[j'/j = (\delta c j_0'/FD)e^{-\alpha\eta'/RT}]$

A glance at Table 32.1 shows the wide range of exchange current densities for a metal/hydrogen electrode. The most sluggish exchange currents occur for lead and mercury, and the value of 10^{-12} A cm^{-2} corresponds to a monolayer of atoms (about 10^{15} cm^{-2}) being replaced in about 5 years. For such systems, a large overpotential is needed to induce significant hydrogen evolution. In contrast, the value for platinum (10^{-3} A cm^{-2}) corresponds to a monolayer being replaced in 0.1 s, and so gas evolution occurs for a much lower overpotential.

The exchange current density also depends on the crystal face exposed. For the deposition of copper on copper, the (100) face has $j_0 = 1$ mA cm^{-2}, and so for the same overpotential grows at 2.5 times the rate of the (111) face for which $j_0 = 0.4$ mA cm^{-2}.

32.3 (b) Current and cell e.m.f.

We expect the e.m.f. of a cell to decrease as current is generated because it is then no longer working reversibly and can therefore do less than maximum work.

Consider the cell $M|M^+(aq)||M'^+(aq)|M'$. We ignore all the complications arising from liquid junctions. The working e.m.f. of the cell is $E = \Delta\phi' - \Delta\phi$. Since the working potential differences differ from their equilibrium values by overpotentials, we can write

$$\Delta\phi = E_{\mathrm{M,M^+}} + \eta, \qquad \Delta\phi' = E_{\mathrm{M',M'^+}} + \eta', \tag{32.3.1}$$

$E_{\mathrm{M',M'^+}}$ and $E_{\mathrm{M,M^+}}$ being the electrode potentials. Then

$$E = E_e + \eta' - \eta, \tag{32.3.2}$$

E_e being the equilibrium cell e.m.f. Subtracted from this expression should be the ohmic potential difference IR_s, where R_s is the cell's internal resistance. (The ohmic drop is a contribution to the cell's irreversibility—it is a thermal dissipation term—and so the sign of IR_s is always such as to reduce the e.m.f. in the direction of zero.)

The overpotentials in eqn (32.3.2) can be calculated from the Butler–Volmer equation for a given current I being drawn. Supposing that the areas (A) of the electrodes are the same, that only one electron is transferred in the rate-determining steps at the electrodes, that the transfer coefficients are both $\frac{1}{2}$, and that the high-overpotential limit of the Butler–Volmer equation may be used, the relation between E and I is

$$E = E_e - (4RT/F)\ln\{I/A\bar{j}_0\} - IR_s, \tag{32.3.3}$$

where $\bar{j}_0 = \{j_0/j_0'\}^{\frac{1}{2}}$.

The concentration overpotential also reduces the cell e.m.f. If we use the Nernst diffusion layer model for each electrode the total change of e.m.f. due to concentration polarization is given by eqn (32.2.7a) as

$$E = E_e + (RT/zF)\ln\{(1 - I/Aj_L)(1 - I/Aj_L')\}. \tag{32.3.4}$$

This contribution can be added to the one in eqn (32.3.3) to obtain a full (but still very approximate) expression for the e.m.f. of a cell when a current I is being drawn:

$$E = E_e - IR_s + (2RT/zF)\ln f, \tag{32.3.5a}$$

$$f = \frac{\{I/A\bar{j}_0\}^{2z}}{\{(1 - I/Aj_L)(1 - I/Aj_L')\}^{\frac{1}{2}}}. \tag{32.3.5b}$$

This expression depends on a lot of parameters, but an example of its general form is given in Fig. 32.15. Notice the very steep decline of working

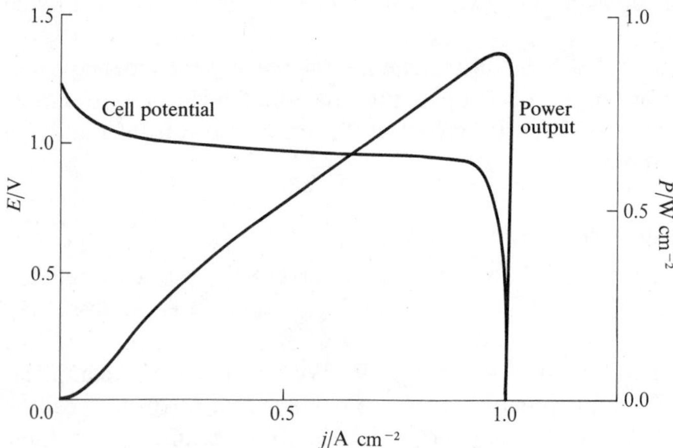

Fig. 32.15. The dependence of the e.m.f. of a working cell on the current being drawn (full line) and the corresponding power output (broken line) calculated using eqns (32.3.5) and (32.3.6) respectively. Notice the sharp decline in power just after the maximum.

e.m.f. when the current is high and close to the limiting value for one of the electrodes.

The *power* (P) supplied by a cell is IE, and so

$$P = IE_e - I^2R_s + (2IRT/zF)\ln f. \qquad (32.3.6)$$

The first term on the right is the power that would be produced if the cell retained its equilibrium e.m.f. when delivering current. The second term is the power generated uselessly as heat as a result of the resistance of the electrolyte. The third term is the lowering of PD at the electrodes as a result of drawing current. The general dependence of power output (the main purpose of a cell) is illustrated in Fig. 32.15 for the e.m.f./current curve drawn earlier. Notice how maximum power is achieved just before the concentration polarization quenches the e.m.f. Information of this kind is essential if the optimum conditions for operating electrochemical devices are to be found and their performance improved.

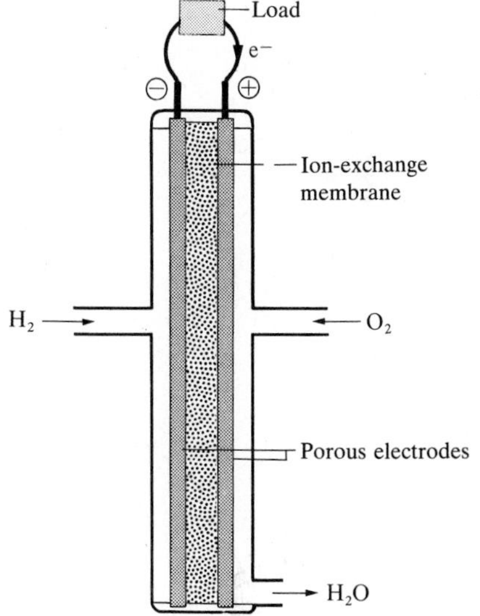

Fig. 32.16. A single cell of the *Bacon fuel cell*. In practice a battery of many cells are combined.

32.4 Applications

The previous sections have outlined the principles involved in dynamic electrochemical processes. We now see what light they shed on some aspects of power generation and storage and on corrosion.

32.4 (a) Fuel cells

A *fuel cell* operates just like a conventional galvanic cell with the exception that the reactants are supplied from outside rather than forming an integral part of its construction. A fundamental and important example of a fuel cell is the *hydrogen/oxygen cell*, Fig. 32.16, which is also called the *Bacon cell* after its inventor. This is the cell used to power the Apollo moon missions; as well as providing electricity its by-product was the water the astronauts drank. The electrolyte is concentrated aqueous potassium hydroxide maintained at 200 °C and 20–40 atm; the electrodes are porous nickel in the form of sheets of compressed powder. The cathode reaction is the reduction $O_2 + 2H_2O + 4e^- \rightarrow 4H_2O$, with $E^\ominus(\mathcal{T}) = +0.40$ V and the anode is the oxidation $H_2 + 2OH^- \rightarrow 2H_2O + 2e^-$, with $E^\ominus(\mathcal{T}) = -0.83$ V. The overall reaction $2H_2 + O_2 \rightarrow 2H_2O$, $E^\ominus(\mathcal{T}) = 1.2$ V, is exothermic, and so it is less favourable at 200 °C than at 25 °C, and so the e.m.f. is lower. However, the increased pressure compensates for the increased temperature, and at 200 °C and 40 atm the e.m.f. is approximately 1.2 V.

One advantage of the hydrogen/oxygen system is the large exchange current density of the hydrogen reaction. Unfortunately, the oxygen reaction has an exchange current density of only about 10^{-10} A cm^{-2}, and so this limits the current available from the cell. One way round the difficulty is to use a catalytic surface (to increase j_0) with a large surface area. *Hydrocarbon/air fuel cells* require platinum electrodes, which makes them prohibitively expensive. Therefore, in practice hydrocarbons are cracked with steam, the hydrogen is separated using a silver/palladium alloy membrane, and then the separated hydrogen is used in a Bacon cell.

Example 32.5

Calculate the e.m.f. of a reversible propane/oxygen fuel cell operating under standard conditions at 298 K.

• *Method*. Write the cell reaction (a combustion) and identify the number of electrons transferred (ν). The standard Gibbs function of reaction is obtained from the standard Gibbs functions of formation, Table 5.3. Then use eqn (12.3.2) to convert $\Delta_r G^\ominus$ to $E^\ominus$.

• *Answer*. The cell reaction is

$$C_3H_8(g) + 5O_2(g) \rightarrow 3CO_2(g) + 4H_2O(l),$$

which is the combination of the following two half-reactions:

$$C_3H_8(g) + 6H_2O(l) \rightarrow 3CO_2(g) + 20H^+ + 20e^-, \qquad 5O_2(g) + 20H^+ + 20e^- \rightarrow 10H_2O(l).$$

Therefore, $\nu = 20$. The standard Gibbs function of the combustion is

$$\Delta_r G^\ominus = \{3(-394.4) + 4(-237.2) - (-23.5)\}\ \text{kJ mol}^{-1} = -2108\ \text{kJ mol}^{-1},$$

and so from eqn (12.3.2),

$$E^\ominus = -(-2108\ \text{kJ mol}^{-1})/20F = +1.09\ \text{V}.$$

• *Comment*. The standard *enthalpy* of combustion at 298 K is -2220 kJ mol^{-1}, and so in a cell working with perfect efficiency, 112 kJ mol^{-1} of energy is released as heat.

• *Exercise*. Repeat the calculation for a methanol-powered fuel cell under the same conditions. [1.21 V]

32.4 (b) Power storage

Electric *storage cells* operate as galvanic cells while they are producing electricity but as electrolytic cells while they are being charged by an external supply. We shall look at two of the most common, the lead storage battery and the nickel/cadmium cell.

The *lead storage battery* is an old device, but one quite well suited to the job of starting cars. During charging the cathode reaction is the reduction of Pb^{2+} and its deposition as lead on the lead electrode: this occurs instead of the reduction of the acid to hydrogen because the latter has a low exchange current density on lead. The anode reaction during charging is the oxidation of Pb(II) to Pb(IV), which is deposited as the oxide PbO_2. On discharge the two reactions run in reverse. Because they have such high exchange current densities the discharge can occur rapidly. That is why the lead battery can produce large currents on demand.

The *nickel/cadmium cell* is used when small, steady currents are required. One electrode is coated with $Cd(OH)_2$, the other with $Ni(OH)_2$, the electrolyte being KOH. On charging the cathode reaction is the reduction $Cd(OH)_2 + 2e^- \rightarrow Cd + 2OH^-$ and the anode reaction is the oxidation $2Ni(OH)_2 + 2OH^- \rightarrow 2NiOOH + 2H_2O + 2e^-$, the discharge corresponding to these reactions running in reverse.

32.4 (c) Corrosion

The thermodynamic warning of the likelihood of corrosion is given by comparing the standard electrode potentials of the metal reduction, such as

$$Fe^{2+}(aq) + 2e^- \rightarrow Fe(s), \qquad E^\ominus = -0.44\ \text{V}$$

with the values for one of the following reactions:

In acid solution: (a) $2H^+(aq) + 2e^- \rightarrow H_2(g), \quad E^\ominus = 0,$

(b) $4H^+(aq) + O_2(g) + 4e^- \rightarrow 2H_2O(l), \quad E^\ominus = 1.23\ \text{V}.$

In basic solution: (c) $2H_2O(l) + O_2(g) + 4e^- \rightarrow 4OH^-(aq), \quad E^\ominus = 0.40\ \text{V}.$

Since all three reactions have standard potentials more positive than

$E_{Fe^{2+},Fe}$, all three can drive its oxidation. Of course, these are only *standard* values, and they change with the pH of the medium. In the case of the first two:

$$E_a = E_a^{\ominus} + (RT/F)\ln a_{H^+} = (-0.059\text{ V})\,\text{pH},$$
$$E_b = E_b^{\ominus} + (RT/F)\ln a_{H^+} = 1.23\text{ V} - (0.059\text{ V})\,\text{pH}.$$

These values, following the discussion in Chapter 12, let us judge at what pH the iron will have a tendency to oxidize. A thermodynamic discussion of corrosion, however, only indicates whether a *tendency* to corrode exists. Given that there is a thermodynamic tendency, we have to examine the kinetics of the processes involved in order to see whether the process is significantly fast.

A model of a corrosion system is shown in Fig. 32.17(a). It can be taken to be a drop of slightly acidic (or basic) water containing some dissolved oxygen in contact with the metal. The oxygen near the edges of the droplet is reduced by electrons donated by the iron over an area A. Those electrons are replaced by others released elsewhere as $Fe \rightarrow Fe^{2+} + 2e^-$. This oxidative release occurs over an area A' under the oxygen-deficient inner region of the droplet. The droplet is acting as a short-circuited electrochemical cell, Fig. 32.17(b).

(a)

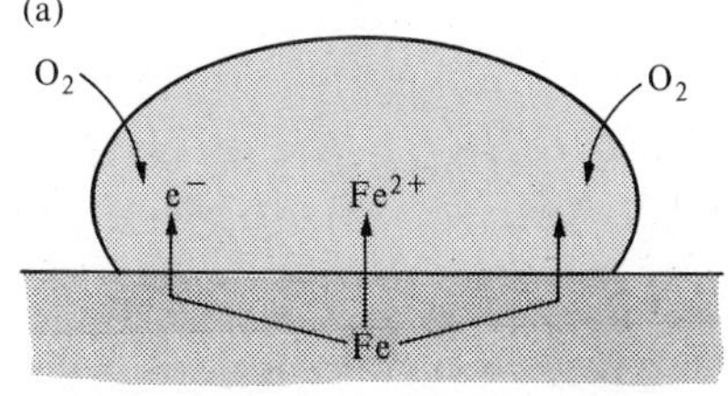

(b)

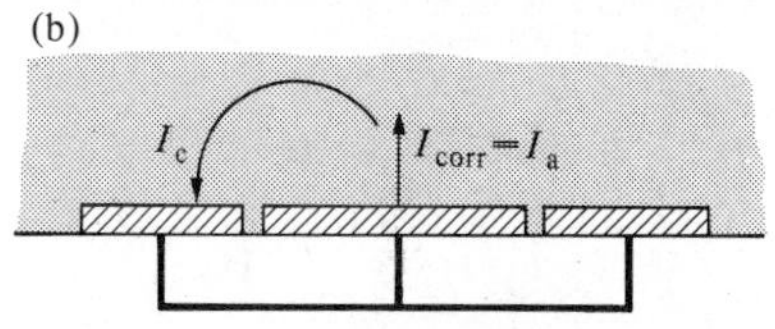

Fig. 32.17. (a) A simple version of the corrosion process is that of a droplet of water, which is oxygen rich near its boundary with air. The oxidation of the iron takes place in the region away from the oxygen because the electrons are transported through the metal. (b) The process may be modelled as a short-circuited electrochemical cell.

The rate of corrosion is measured by the current of metal ions leaving the metal surface in the anodic region. This gives rise to the *corrosion current*, I_{corr}, which can be identified with the anodic current, I_a. Since any current emerging from the anodic region must find its way to the cathodic region, the cathodic current, I_c, must also be equal to the corrosion current. In terms of the current densities at the oxidation (j) and reduction (j') sites, we can write

$$I_{corr} = jA = j'A' = (jj'AA')^{\frac{1}{2}} = \bar{j}\bar{A}, \qquad (32.4.1)$$

where $\bar{A} = (AA')^{\frac{1}{2}}$ and $\bar{j} = (jj')^{\frac{1}{2}}$. The current densities can be expressed in terms of the overpotentials using the Butler–Volmer equation. For simplicity we assume that the overpotentials are large enough for the high-overpotential limits to apply, that polarization overpotential can be neglected, that the rate-determining step is the transfer of a single electron, and that the transfer coefficients are $\frac{1}{2}$. We also assume that since the droplet is so small, there is negligible PD between the cathode and anode regions of the solution. Moreover, since it is short-circuited by the metal, the potential of the metal is the same in both regions, and so the PD between the metal and the solution is the same in both regions too. This common PD is called the *corrosion potential difference* or *mixture potential*, $\Delta\phi_{corr}$. The overpotentials in the two regions are therefore $\eta = \Delta\phi_{corr} - \Delta\phi_{eq}$ and $\eta' = \Delta\phi_{corr} - \Delta\phi'_{eq}$, and so the current densities are

$$j = j_0 e^{\eta F/2RT} = j_0 e^{\Delta\phi_{corr}F/2RT} e^{-\Delta\phi_{eq}F/2RT}, \qquad (32.4.2a)$$
$$j' = j'_0 e^{-\eta' F/2RT} = j'_0 e^{-\Delta\phi_{corr}F/2RT} e^{\Delta\phi'_{eq}F/2RT}. \qquad (32.4.2b)$$

This is very nearly the end of the calculation because these expressions can be substituted into eqn (32.4.1) and $\Delta\phi' - \Delta\phi$ replaced by the difference of electrode potentials, E, to give

$$I_{corr} = \bar{A}\bar{j}_0 e^{FE/4RT}, \qquad (32.4.3)$$

where $\bar{j}_0 = (j_0 j'_0)^{\frac{1}{2}}$.

Several conclusions can be drawn from this approximate expression. First, the rate of corrosion depends on the areas exposed: if either A or A' is zero, then the corrosion current is zero. This points to a trivial, yet often effective, method of slowing corrosion: cover the surface with a coating, such as paint. (Paint also increases the effective solution resistance between the cathode and anode patches on the surface.) Second, other things being equal (i.e. for corrosion reactions with similar exchange current densities), the rate of corrosion is high when E is large. That is, rapid corrosion can be expected when the oxidizing and reducing couples have widely differing electrode potentials.

The effect of the exchange current density on the corrosion rate can be seen by considering the specific case of iron in contact with acidified water. Thermodynamically either oxygen reduction reaction (a) or (b) on p. 804 is effective. However, the exchange current density of reaction (b) on iron is only about $10^{-14}\,\mathrm{A\,cm^{-2}}$, while for (a) the value is $10^{-6}\,\mathrm{A\,cm^{-2}}$. The latter therefore dominates kinetically, and iron corrodes by hydrogen evolution in acid solution.

32.4 (d) The inhibition of corrosion

Several techniques for inhibiting corrosion are available. Coating the surface with some impermeable layer, such as paint, may prevent the access of damp air. Unfortunately, this protection fails disastrously if the paint has defects. The oxygen then has access to the exposed metal and corrosion continues beneath the paintwork. Another form of surface coating is provided by *galvanizing*, the coating of an iron object with zinc. Since the latter's electrode potential is $-0.76\,\mathrm{V}$, which is more negative than the iron couple, the corrosion of zinc is thermodynamically favoured and the iron survives (the zinc survives because it is protected by a hydrated oxide layer). In contrast, tin plating leads to a very rapid corrosion of the iron once its surface is scratched and the iron exposed because the tin couple ($E^\ominus = -0.14\,\mathrm{V}$) oxidizes the iron couple ($E^\ominus = -0.44\,\mathrm{V}$). Some oxides are stable kinetically in the sense that they adhere to the metal surface and form an impermeable layer over a fairly wide pH range. This is the reason why aluminium is stable in air even though its reduction potential is strongly negative ($-1.66\,\mathrm{V}$). This type of *passivation* can be seen as a way of decreasing the exchange currents by sealing the surface.

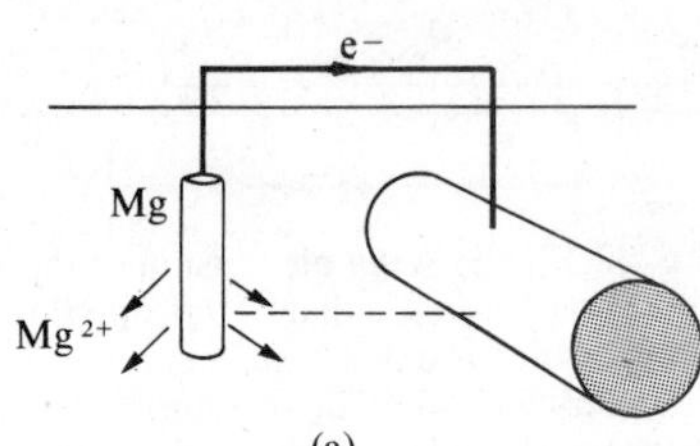

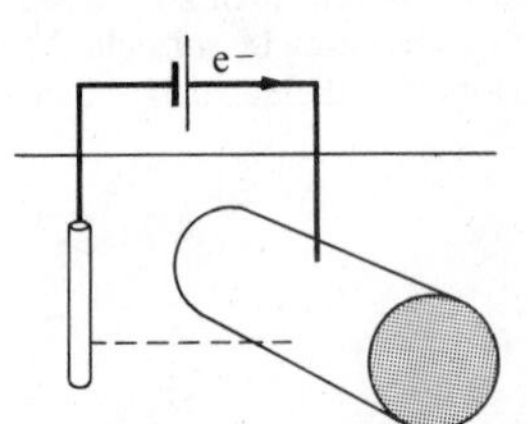

Fig. 32.18. (a) In *cathodic protection* an anode of a more strongly reducing metal is sacrificed to maintain the integrity of the protected object (e.g. a pipeline, bridge, or boat). (b) In *impressed current cathodic protection* electrons are supplied from an external cell so that the object itself is not oxidized.

Another method of protection is to change the potential of the subject by pumping in electrons, which can be used to satisfy the demands of the oxygen reduction without involving the oxidation of the metal. In *cathodic protection* the object is connected to a metal with a more negative electrode potential (such as magnesium, $-2.36\,\mathrm{V}$). The magnesium acts as a *sacrificial anode*, supplying its own electrons to the iron and decaying to Mg^{2+} in the process, Fig. 32.18(a). A block of magnesium, replaced occasionally, is much cheaper than the ship, building, or pipeline for which it is being sacrificed. Another technique is *impressed current cathodic protection*, Fig. 32.18(b), in which an external cell supplies the electrons and eliminates the need for iron to transfer its own.

Further reading

Electrode processes:

Electrode processes. G. J. Hills; *Essays in chemistry* (J. N. Bradley, R. D. Gillard, and R. F. Hudson, eds) **2,** 19 (1971).

Experimental approach to electrochemistry. N. J. Selley; Edward Arnold, London, 1977).

Modern electrochemistry. J. O'M. Bockris and A. K. N. Reddy; Plenum, New York, 1970.

Electrode kinetics. W. J. Albery; Clarendon Press, Oxford, 1974.

Reactions of molecules at electrodes. N. S. Hush; Wiley-Interscience, New York, 1971.

Charge transfer processes in condensed media. J. Ulstrup; Springer, Berlin, 1979.

Electroanalytical techniques:

Potentiometry: oxidation–reduction potentials. S. Wawzonek; *Techniques of chemistry* (A. Weissberger and B. W. Rossiter, eds) IIA, 1, Wiley-Interscience, New York, 1971.

Electrochemical techniques. A. J. Bard and L. R. Faulkner; Wiley-Interscience, New York, 1979.

Fuel cells:

Fuel cells: their electrochemistry. J. O'M. Bockris and S. N. Srinivasan; McGraw-Hill, New York, 1969.

Fuel cells. A. McDougall; Macmillan, London, 1976.

Corrosion:

Advances in corrosion science and technology. M. G. Fontana and R. W. Staehle (eds); Plenum, New York, 1980.

Introductory problems

A32.1. For an electrode in contact with two cations, M^{3+} and M^{4+}, in aqueous solution at 25 °C, the transfer coefficient is 0.39. The current density is found to be 55.0 mA cm^{-2} when the overvoltage is 125 mV. What is the overvoltage required for a current density of 75 mA cm^{-2}?

A32.2. From the information given in the previous problem, determine the exchange current density.

A32.3. Estimate the limiting current density at an electrode in which the concentration of Ag^+ ions is 2.5 mmol dm^{-3} at 25 °C. The thickness of the Nernst diffusion layer is 0.40 mm. The ionic conductivity of Ag^+ at infinite dilution and 25 °C is 61.9 S cm^2 mol^{-1}.

Problems

32.1. The key to the entire chapter is the Butler–Volmer equation, and so in the first few Problems we gain some familiarity with the way it can be used and manipulated. First, take $\alpha = \frac{1}{2}$ and plot the current density, j/j_e, as a function of the overpotential η at 298 K.

32.2. A typical exchange current density, that for H^+ discharge at platinum, is 0.79 mA cm^{-2} at 25 °C. What is the current density at an electrode when its overpotential is (a) 10 mV, (b) 100 mV, (c) −5.0 V? Take $\alpha = 0.5$.

32.3. The exchange current density for a Pt|Fe^{3+},Fe^{2+} electrode is 2.5 mA cm^{-2}. The standard electrode potential is +0.77 V. Calculate the current flowing through an electrode of surface area 1 cm^2 as a function of the potential of the electrode. Take unit activity for both ions.

32.4. Suppose that the electrode potential is set at 1.00 V. Calculate the current that flows for the ratio of activities $a(Fe^{2+})/a(Fe^{3+})$ in the range 0.1 to 10.0 and a temperature $\mathcal{T}$.

32.5. What overpotential is needed to sustain a 20 mA current at the Fe^{3+}, Fe^{2+}|Pt electrode in which both ions are at a mean activity $a = 0.1$?

32.6. How many electrons or protons are transported through the double layer in each second when the Pt|H_2|H^+, Pt|Fe^{3+},Fe^{2+}, and Pb|H_2|H^+ electrodes are at equilibrium at $\mathcal{T}$? Take the area as 1 cm^2 in each case.

32.7. In order to appreciate the magnitude of these quantities on an atomic scale, estimate the number of times each second a single atom on the surface takes part in an electron transfer event, assuming an electrode atom occupies about (280 pm)^2 of the surface.

32.8. What is the effective resistance at $\mathcal{T}$ of an electrode interface when the overpotential is small? Evaluate it for 1 cm^2 (a) Pt|H_2|H^+, (b) Hg|H_2|H^+ electrodes. (Data in Table 32.1.)

32.9. If $\alpha = \frac{1}{2}$, an electrode interface is unable to rectify alternating current because the current density curve is symmetrical about $\alpha = 0$ (Problem 32.1). When $\alpha \neq \frac{1}{2}$ the magnitude of the current density depends on the sign of the overpotential, and so some degree of rectification may be obtained: this is called *faradaic rectification*. Suppose that the overpotential varies as $\eta = \eta_0 \cos \omega t$. Find an expression for the mean flow of current (averaged over a cycle) for general α, and confirm that the mean current is zero when $\alpha = \frac{1}{2}$. In each case work in the limit of small η_0, but you will have to work to second order in $\eta_0 F/RT$.

32.10. Calculate the mean direct current at $\mathscr{T}$ for a 1 cm² hydrogen–platinum electrode with $\alpha = 0.38$ when the overpotential varies between ±10 mV at 50 Hz.

32.11. Now suppose that the overpotential is in the high overpotential region at all times even though it is oscillating. What waveform will the current across the interface show if η varies linearly and periodically (as a sawtooth waveform) between η_- and η_+ around η_0? Take $\alpha = \frac{1}{2}$.

32.12. The Tafel plot gives a means of determining both the exchange current density and the transfer coefficient α for electrodes, and so it lies at the root of the information we need when discussing electrode processes. In an experiment on the $Pt|H_2|H^+$ electrode in dilute H_2SO_4 the following current densities were observed at 25 °C. Find α and j_e for the electrode.

η/mV	50	100	150	200	250
j^+/mA cm^{-2}	2.66	8.91	29.9	100	335

32.13. How would the current density at this electrode depend on the overpotential of the same set of magnitudes but of opposite sign?

32.14. State what happens when a platinum electrode in an aqueous solution containing both Cu^{2+} and Zn^{2+} ions at unit activity is made the cathode of an electrolysis cell.

32.15. What are the minimum (thermodynamically determined) potentials at which (a) zinc, (b) copper can be deposited from aqueous molar solutions? What are the corresponding potentials when the concentrations are 0.01 M?

32.16. What are the conditions that allow a metal to be deposited from aqueous acidic solution before hydrogen evolution occurs significantly? Why may silver be deposited from aqueous silver nitrate? Why may cadmium be deposited from aqueous cadmium sulphate? (The overpotential for hydrogen evolution on cadmium is about 1 V at current densities of 1 mA cm^{-1}.)

32.17. The exchange current density for H^+ discharge at zinc is about 5×10^{-11} A cm^{-2}. Can zinc be deposited from a unit activity aqueous solution of a zinc salt?

32.18. The standard potential of the $Zn^{2+}|Zn$ electrode is −0.76 V at 25 °C. The exchange current density for H^+ discharge at platinum is 0.79 mA cm^{-2}. Can zinc be plated on to platinum at that temperature? (Take unit activities.)

32.19. Can magnesium be deposited on a zinc electrode from a unit activity acid solution at 25 °C?

32.20. Find an expression for the current density at an electrode where the rate process is diffusion controlled and η^c is known. Sketch the form of j/j_L as a function of η^c. What changes occur if anion currents are involved?

32.21. The limiting current density for the reaction $I_3^- + 2e^- \rightarrow 3I^-$ at a platinum electrode is 28.9 μA cm^{-2} when the concentration of KI_3 is 6.6×10^{-4} M and the temperature 25 °C. The diffusion coefficient of I_3^- is 1.14×10^{-5} cm s^{-1}. What is the thickness of the diffusion layer?

32.22. The limiting current density is given in terms of the ionic diffusion constant D_+ which can be related to the ionic conductivity λ_+. Derive an expression for the limiting current in terms of λ_+, the concentration c_+, and the thickness of the diffusion layer δ.

32.23. The ionic conductivity of Fe^{2+} is 40 S cm² mol^{-1}. The limiting current at a platinum electrode of area 40 cm² dipping in to a solution of iron(II) chloride at 25 °C was measured at various concentrations, the results are given below. What is the thickness of the diffusion layer at each concentration? (The answer you should get is a quite typical thickness.) Can you think of an independent non-electrochemical way of measuring λ?

$[FeCl_2]$/M	0.250	0.125	0.063	0.031
I/mA	215	107	49	23

32.24. The standard electrode potentials of lead and tin are −126 mV and −136 mV respectively at 25 °C, and the overvoltages for their deposition are close to zero. What should their relative activities be in order to ensure simultaneous deposition from a mixture?

32.25. The standard electrode potentials of silver, copper, and zinc are 799 mV, 337 mV, and −763 mV, respectively. When they are present in a mixture with cyanide ions they may be deposited simultaneously. Account for this observation.

32.26. We now turn to power generation. As a first step, find the maximum (reversible) potential difference of a nickel–cadmium cell, and the maximum possible power output when 100 mA is drawn, the temperature being 25 °C.

32.27. Calculate the thermodynamic limit to the e.m.f. of fuel cells operating on (a) hydrogen and oxygen, (b) methane and air. Use the Gibbs function information in Table 4.1, and take the species to be in their standard states at 25 °C.

32.28. Estimating the power output and e.m.f. of a cell under operating conditions is very difficult, but eqn (32.3.5) summarizes, in an approximate way, some of the parameters involved. As a first step in manipulating this expression, identify all the quantities that depend on the ionic concentrations. Express E in terms of the concentration and conductivities of the ions present in the cell.

32.29. Estimate the parameters in the last Problem for $Zn|ZnSO_4(aq)||CuSO_4(aq)|Cu$. Take electrodes of area

5 cm^2 separated by 5 cm. Ignore both potential differences and resistance of the liquid junction. Take the concentration as 1 M, the temperature 25 °C, and neglect activity coefficients. Plot E as a function of the current drawn.

32.30. On the same graph, plot the power output of the cell. What current corresponds to maximum power?

32.31. Consider a cell in which the current is activation controlled. Show that the current for maximum power can be estimated by plotting $\lg(I/I_0)$ and $c_1 - c_2 I$ against I (where $I_0^2 = A^2 j_0 j_0'$ and c_1 and c_2 are constants), and looking for the point of intersection of the curves. Carry through this analysis for the cell in Problem 32.29 ignoring all concentration overpotentials.

32.32. If copper and mild steel are in contact (as in a badly designed domestic water system) which will tend to corrode?

32.33. Find the pH dependence of the electrode potential of the alkaline corrosion reaction, reaction (c) in Section 32.4(c). The $H^+ + e^- \rightleftharpoons \frac{1}{2}H_2$ reaction has a more negative potential at any pH: why, then, does it every play a role in corrosion?

32.34. Which of the following metals has a thermodynamic tendency to corrode in moist air at pH 7: Fe, Cu, Pb, Al, Ag, Cr, Co? Take as a criterion of corrosion a metal ion concentration of at least 10^{-6} M.

32.35. Estimate the magnitude of the corrosion current for a patch of zinc of area 0.25 cm^2 in contact with a similar area of iron in an aqueous environment at 25 °C. Take the exchange current densities as 10^{-6} A cm^{-2} and the local ion concentrations as 10^{-6} M.

Tables index

Tables index

Tables of data

The following tables reproduce and expand the data given in the short tables in the text, and follow their numbering. The notation xPy means $x \times 10^{+y}$ and xNy means $x \times 10^{-y}$. $\mathcal{T}$ denotes the *conventional temperature*, 298.15 K. Standard states refer to a pressure of 1 bar (as far as possible). The general references are as follows:

AIP: D. E. Gray, ed. (1972). *American Institute of Physics handbook*. McGraw-Hill, New York.

AS: M. Abramowitz and I. A. Stegun, eds. (1963). *Handbook of mathematical functions*. Dover, New York.

HCP: R. C. Weast, ed. (1984). *Handbook of chemistry and physics*. CRC Press, Boca Raton.

KL: G. W. C. Kaye and T. H. Laby, edn. (1973). *Tables of physical and chemical constants*. Longman, London.

LR: G. N. Lewis and M. Randall (1961). *Thermodynamics*. (Revised by K. S. Pitzer and L. Brewer.) McGraw-Hill, New York.

NBS: *NBS Tables of chemical thermodynamic properties*. Published as Suppl. 2 to Vol. 11 of *J. Phys. Chem. Ref. Data*, 1982.

RS: R. A. Robinson and R. H. Stokes (1959). *Electrolyte solutions*. Butterworths, London.

Table 0.1. Physical properties of selected materials

	ρ/g cm^{-3}	θ_m/°C	θ_b/°C
Elements			
Aluminium(s)	2.698	660.4	2350
Argon(g)		−189.4	−185.9
Boron(s)	2.466	2030	370
Bromine(l)	3.120	−7.1	58.9
Carbon(s, graphite)	2.266	3700 s	
Carbon(s, diamond)	3.52		
Chlorine(g)		−101	−34.0
Copper(s)	8.933	1084.5	2580
Fluorine(g)		−219.6	−188.1
Gold(s)	19.281	1064.4	2850
Helium(g)		3.5 K	4.22 K
Hydrogen(g)		−259.2	−252.87
Iodine(s)	4.953	113.6	184
Iron(s)	7.873	1540	2760
Krypton(g)		−157.3	−153.4
Lead(s)	11.343	327.502	1760
Lithium(s)	0.533	180	1360
Magnesium(s)	1.738	650	1100
Mercury(l)	13.546	−38.862	356.862
Neon(g)		−248.6	−246.05
Nitrogen(g)		−210	−195.80
Oxygen(g)		−218.8	−182.96
Phosphorus(s, white)	1.820	44.2	280.4
Potassium(s)	0.862	63.2	777
Silver(s)	10.500	961.93	2160
Sodium(s)	0.966	97.8	900
Sulphur(s)	2.086	115	444.674
Uranium(s)	19.050	1135	4000
Xenon(g)		−111.9	−108.1
Zinc(s)	6.507	1850	4400
Inorganic compounds			
$CaCO_3$(s, calcite)	2.71		
$CuSO_4.5H_2O$(s)	2.284		
HBr(g)		−86.82	−66.73
HCl(g)		−114.2	−82.05
HI(g)		−50.80	−35.36
H_2O(l)	0.997	0	100
D_2O(l)	1.104	3.8	101.42
NH_3(g)		−77.7	−34.35
KBr(s)	2.750	730	1435
KCl(s)	1.984	776	1500 (s)
NaCl(s)	2.165	801	1413
H_2SO_4(l)	1.841	10.35	338

Table 0.1 (Contd)

	ρ/g cm^{-3}	θ_m/°C	θ_b/°C
Organic compounds			
Acetaldehyde, CH_3CHO(l, g)	0.788 (18 °C)	−121	20
Acetic acid, CH_3COOH(l)	1.049	16.7	118
Acetone, $(CH_3)_2CO$(l)	0.787	−95	56
Aniline, $C_6H_5NH_2$(l)	1.026	−6	184
Anthracene, $C_{14}H_{10}$(s)	1.243	217	342
Benzene, C_6H_6(l)	0.879	5.5	80.1
Carbon tetrachloride, CCl_4(l)	1.63	−23	76.7
Chloroform, $CHCl_3$(l)	1.499 (15 °C)	−63.5	61
Ethanol, C_2H_5OH(l)	0.789	−117	78.3
Formaldehyde, HCHO(g)		−92	−19.1
Glucose, $C_6H_{12}O_6$(s)	1.544	142	
Methane, CH_4(g)		−182.5	−161.5
Methanol, CH_3OH(l)	0.791	−93.9	64.5
Naphthalene, $C_{10}H_8$(s)	1.145	80.2	218
Octane, C_8H_{18}(l)	0.703	−56.8	125.7
Phenol, C_6H_5OH(s)	1.073	41.0	181.9
Sucrose, $C_{12}H_{22}O_{11}$(s)	1.588	184 (d)	

g: gas, l: liquid, s: solid (and denotes the normal state at room temperature). ρ is the density at $\mathcal{T}$, θ_m the normal melting point and θ_b the normal boiling point.
Data: AIP, HCP, KL.

Table 1.1. Second virial coefficients, B/cm^3 mol^{-1}

	100 K	273 K	373 K	600 K
He	11.4	12.0	11.3	10.4
Ne	−4.8	10.4	12.3	13.8
Ar	−187.0	−21.7	−4.2	11.9
Kr		−62.9	−28.7	2.0
Xe		−153.7	−81.7	−19.6
H_2	−2.5	13.7	15.6	
N_2	−160.0	−10.5	6.2	21.7
O_2	−197.5	−22.0	−3.7	12.9
CO_2		−149.7	−72.2	−12.4
CH_4		−53.6	−21.2	8.1
Air	−167.3	−13.5	3.4	19.0

Data: AIP. The values relate to the expansion in eqn (1.3.2); convert to eqn (1.3.1) using $B' = B/RT$.

Table 1.2. Critical constants

	p_c/atm	$V_{c,m}$/cm^3 mol^{-1}	T_c/K	Z_c	T_B/K
He	2.26	57.76	5.21	0.305	22.64
Ne	26.86	41.74	44.44	0.307	122.1
Ar	48.00	75.25	150.72	0.292	411.5
Kr	54.27	92.24	209.39	0.291	575.0
Xe	58.0	118.8	289.75	0.290	768.0
H_2	12.8	64.99	33.23	0.305	110.0
N_2	33.54	90.10	126.3	0.292	327.2
O_2	50.14	78.0	154.8	0.308	405.9
F_2	55		144		
Cl_2	76.1	124	417.2	0.276	
Br_2	102	135	584	0.287	
HCl	81.5	81.0	324.7	0.248	
HBr	84.0		363.0		
HI	80.8		423.2		
CO_2	72.85	94.0	304.2	0.274	714.8
H_2O	218.3	55.3	647.4	0.227	
NH_3	111.3	72.5	405.5	0.242	
CH_4	45.6	98.7	190.6	0.288	510.0
C_2H_4	50.50	124	283.1	0.270	
C_2H_6	48.20	148	305.4	0.285	
C_6H_6	48.6	260	562.7	0.274	

Data: AIP, KL.

Table 1.3. Van der Waals constants

	a/dm^6 atm mol^{-1}	b/dm^3 mol^{-1}
He	3.412N2	2.37N2
Ne	0.2107	1.709N2
Ar	1.345	3.219N2
Kr	2.318	3.978N2
Xe	4.194	5.105N2
H_2	0.2444	2.661N2
N_2	1.390	3.913N2
O_2	1.360	3.183N2
Cl_2	6.493	5.622N2
CO	1.485	3.985N2
CO_2	3.592	4.267N2
SO_2	6.714	5.636N2
H_2O	5.464	3.049N2
H_2S	4.431	4.287N2
NH_3	4.170	3.707N2
CH_4	2.253	4.278N2
C_2H_4	4.471	5.714N2
C_2H_6	5.489	6.380N2
C_6H_6	18.00	11.54N2

Data: HCP.

Table 2.1. Heat capacities at $\mathcal{T}$, 1 atm (see also Table 4.1)

	$C_{V,m}$/J K^{-1} mol^{-1}	$C_{p,m}$/J K^{-1} mol^{-1}	γ
Gases			
He, Ne, Ar, Kr, Xe	12.48	20.79	1.667
H_2	20.44	28.82	1.410
N_2	20.74	29.12	1.404
O_2	20.95	29.36	1.401
CO_2	28.46	37.11	1.304
Liquids			
H_2O		75.29	
CH_3OH		81.6	
C_2H_5OH		111.5	
C_6H_6		136.1	

Table 2.1 (Contd)

	$C_{V,m}$/J K^{-1} mol^{-1}	$C_{p,m}$/J K^{-1} mol^{-1}	γ
Solids			
C (diamond)		6.11	
Cu		244.4	
Fe		25.1	
SiO_2		44.4	

Specific heat capacity of various substances, C/J K^{-1} g^{-1}

Brass	0.38	Steel	0.51	Asbestos	0.84
Glass (Pyrex)	0.78	Granite	0.82	Rubber	1.1–1.2
Sea-water	3.93	Glycerol	2.43	Ice	2.0–2.1

Data: NBS; KL.

Table 3.1. Isobaric volume expansivities and isothermal compressibilities

	α/K^{-1}	κ/atm^{-1}
Liquids		
Water	2.1N4	4.96N5
Carbon tetrachloride	1.24N3	9.05N5
Benzene	1.24N3	9.21N5
Ethanol	1.12N3	7.68N5
Mercury	1.82N4	3.87N5

Table 3.1 (Contd)

	α/K^{-1}	κ/atm^{-1}
Solids		
Lead	8.61N5	2.21N6
Copper	5.01N5	7.35N7
Diamond	3.0N6	1.87N7
Iron	3.54N5	5.97N7

The values refer to 20 °C.
Data: AIP(α), KL(κ).

Table 3.2. Inversion temperatures and Joule–Thomson coefficients

	T_I/K	μ_{JT}/K atm^{-1}
Helium	40	−0.060
Neon	231	
Argon	723	
Krypton	1090	
Hydrogen	202	
Nitrogen	621	0.25
Oxygen	764	0.31
Carbon dioxide	1500	1.11 at 300 K
Methane	968	
Air	603	0.189 at 50 °C

Maximum values for μ_{JT} are quoted.
Data: AIP and M. W. Zemansky (1957). *Heat and thermodynamics*. McGraw-Hill, New York.

Table 4.1. Thermodynamic data for inorganic compounds (all values relate to $\mathcal{T}$)

	M_r	$\Delta_f H^\ominus$/kJ mol^{-1}	$\Delta_f G^\ominus$/kJ mol^{-1}	$S^\ominus$/J K^{-1} mol^{-1}	C_p/J K^{-1} mol^{-1}
Aluminium					
Al(s)	26.98	0	0	28.33	24.35
Al(g)	26.98	326.4	285.7	164.54	21.38
Al^{3+}(aq)	26.98	−524.7	−481.2		
Al_2O_3(s, α)	101.96	−1675.7	−1582.3	50.92	79.04
$AlCl_3$(s)	133.24	−704.2	−628.8	110.67	91.84
Argon					
Ar(g)	39.95	0	0	154.84	20.786
Antimony					
Sb(s)	121.75	0	0	45.69	25.23
SbH_3(g)	153.24	145.11	147.75	232.78	41.05
Arsenic					
As(s, α)	74.92	0	0	35.1	24.64
As_4(g)	299.69	143.9	92.4	314	
AsH_3(g)	77.95	66.44	68.93	222.78	38.07
Barium					
Ba(s)	137.34	0	0	62.8	28.07
Ba^{2+}(aq)	137.34	−537.64	−560.77	9.6	
BaO(s)	153.34	−553.5	−525.1	70.43	47.78
Beryllium					
Be(s)	9.01	0	0	9.50	16.44

Table 4.1. (Contd)

	M_r	$\Delta_f H^\ominus$/kJ mol^{-1}	$\Delta_f G^\ominus$/kJ mol^{-1}	$S^\ominus$/J K^{-1} mol^{-1}	C_p/J K^{-1} mol^{-1}
Bromine					
Br_2(l)	159.82	0	0	152.23	75.689
Br_2(g)	159.82	30.907	3.110	245.46	36.02
Br(g)	79.91	111.88	82.396	175.02	20.786
Br^-(aq)	79.91	−121.55	−103.96	82.4	−141.8
HBr(g)	90.92	−36.40	−53.45	198.70	29.142
Cadmium					
Cd(s, γ)	112.40	0	0	51.76	25.98
Cd^{2+}(aq)	112.40	−75.90	−77.612	−73.2	
CdO(s)	128.40	−258.2	−228.4	54.8	43.43
Caesium					
Cs(s)	132.91	0	0	85.23	32.17
Cs^+(aq)	132.91	−258.28	−292.02	133.05	−10.5
Calcium					
Ca(s)	40.08	0	0	41.42	25.31
Ca(g)	40.08	178.2	144.3	154.88	20.786
Ca^{2+}(aq)	40.08	−542.83	−553.58	−53.1	
CaO(s)	56.08	−635.09	−604.03	39.75	42.80
$CaCO_3$(s, calcite)	100.09	−1206.9	−1128.8	92.9	81.88
$CaCO_3$(s, aragonite)	100.09	−1207.1	−1127.8	88.7	81.25
$CaCl_2$(s)	110.99	−795.8	−748.1	104.6	72.59
$CaBr_2$(s)	199.90	−682.8	−663.6	130	
Carbon (for 'organic' compounds of carbon, see Table 4.1(a))					
C(s, graphite)	12.011	0	0	5.740	8.527
C(s, diamond)	12.011	1.895	2.900	2.377	6.113
C(g)	12.011	716.68	671.26	158.10	20.838
C_2(g)	24.022	831.90	775.89	199.42	43.21
CO(g)	28.011	−110.53	−137.17	197.67	29.14
CO_2(g)	44.010	−395.51	−394.36	213.74	37.11
CO_2(aq)	44.010	−413.80	−385.98	117.6	
H_2CO_3(aq)	62.03	−699.65	−623.08	187.4	
HCO_3^-(aq)	61.02	−691.99	−586.77	91.2	
CO_3^{2-}(aq)	60.01	−677.14	−527.81	−56.9	
CCl_4(l)	153.82	−135.44	−65.21	216.40	131.75
CS_2(l)	76.14	89.70	65.27	151.34	75.7
HCN(g)	27.03	135.1	124.7	201.78	35.86
HCN(l)	27.03	108.87	124.97	112.84	70.63
CN^-(aq)	26.02	150.6	172.4	94.1	
Chlorine					
Cl_2(g)	70.91	0	0	223.07	33.91
Cl(g)	35.45	121.68	105.68	165.20	21.840
Cl^-(g)	35.45	−167.16	−131.23	56.5	−136.4
HCl(g)	36.46	−92.31	−95.30	186.91	29.12
HCl(aq)	36.46	−167.16	−131.23	56.5	−136.4
Copper					
Cu(s)	63.54	0	0	33.150	24.44
Cu^+(aq)	63.54	71.67	49.98	40.6	
Cu^{2+}(aq)	63.54	64.77	65.49	−99.6	
Cu_2O(s)	143.08	−168.6	−146.0	93.14	63.64
CuO(s)	79.54	−157.3	−129.7	42.63	42.30
$CuSO_4$(s)	159.60	−771.36	−661.8	109	100.0
$CuSO_4.H_2O$(s)	177.62	−1085.8	−918.11	146.0	134
$CuSO_4.5H_2O$(s)	249.68	−2279.7	−1879.7	300.4	280

Table 4.1. (Contd)

	M_r	$\Delta_f H^\ominus$/kJ mol^{-1}	$\Delta_f G^\ominus$/kJ mol^{-1}	$S^\ominus$/J K^{-1} mol^{-1}	C_p/J K^{-1} mol^{-1}
Deuterium					
$D_2(g)$	4.028	0	0	144.96	29.20
$D_2O(g)$	20.028	−249.20	−234.54	198.34	34.27
$D_2O(l)$	20.028	−294.60	−243.44	75.94	84.35
HDO(g)	19.022	−245.30	−233.11	199.51	33.81
HDO(l)	19.022	−289.89	−241.86	79.29	
HD(g)		0.318	−1.464	143.80	29.196
Fluorine					
$F_2(g)$	38.00	0	0	202.78	31.30
F(g)	19.00	78.99	61.91	158.75	22.74
$F^-(aq)$	19.00	−332.63	−278.79	−13.8	−106.7
HF(g)	20.01	−271.1	−273.2	173.78	29.13
Gold					
Au(s)	196.97	0	0	47.40	25.42
Helium					
He(g)	4.003	0	0	126.15	20.786
Hydrogen (see also *Deuterium*)					
$H_2(g)$	2.016	0	0	130.684	28.824
H(g)	1.008	217.97	203.25	114.71	20.784
$H^+(aq)$	1.008	0	0	0	0
$H_2O(l)$	18.015	−285.83	−237.13	69.91	75.291
$H_2O(g)$	18.015	−241.82	−228.57	188.83	33.58
$H_2O_2(l)$	34.015	−187.78	−120.35	109.6	89.1
Iodine					
$I_2(s)$	253.81	0	0	116.135	54.44
$I_2(g)$	253.81	62.44	19.33	260.69	36.90
I(g)	126.90	106.84	70.25	180.79	20.786
$I^-(aq)$	126.90	−55.19	−51.57	111.3	−142.3
HI(g)	127.91	26.48	1.70	206.59	29.158
Iron					
Fe(s)	55.85	0	0	27.28	25.10
$Fe^{2+}(aq)$	55.85	−89.1	−78.90	−137.7	
$Fe^{3+}(aq)$	55.85	−48.5	−4.7	−315.9	
Fe_3O_4(s, magnetite)	231.54	−1118.4	−1015.4	146.4	143.43
Fe_2O_3(s, haemetite)	159.69	−824.2	−742.2	87.40	103.85
FeS(s, α)	87.91	−100.0	−100.4	60.29	50.54
$FeS_2(s)$	119.98	−178.2	−166.9	52.93	62.17
Krypton					
Kr(g)	83.80	0	0	164.08	20.786
Lead					
Pb(s)	207.19	0	0	64.81	26.44
$Pb^{2+}(aq)$	207.19	−1.7	−24.43	10.5	
PbO(s, yellow)	223.19	−217.32	−187.89	68.70	45.77
PbO(s, red)	223.19	−218.99	−188.93	66.5	45.81
$PbO_2(s)$	239.19	−277.4	−217.33	68.6	64.64
Lithium					
Li(s)	6.94	0	0	29.12	24.77
$Li^+(aq)$	6.94	−278.49	−293.31	13.4	68.6
Magnesium					
Mg(s)	24.31	0	0	32.68	24.89
Mg(g)	24.31	147.70	113.10	148.65	20.786
$Mg^{2+}(aq)$	24.31	−466.85	−454.8	−138.1	
MgO(s)	40.31	−601.70	−569.43	26.94	37.15

Table 4.1. (Contd)

	M_r	$\Delta_f H^\ominus$/kJ mol^{-1}	$\Delta_f G^\ominus$/kJ mol^{-1}	$S^\ominus$/J K^{-1} mol^{-1}	C_p/J K^{-1} mol^{-1}
Mercury					
Hg(l)	200.59	0	0	76.02	27.983
Hg(g)	200.59	61.32	31.82	174.96	20.786
Hg^{2+}(aq)	200.59	171.1	164.40	−32.2	
Hg_2^{2+}(aq)	401.18	172.4	153.52	84.5	
HgO(s)	216.59	−90.83	−58.54	70.29	44.46
Hg_2Cl_2(s)	472.09	−265.22	−210.75	192.5	102
$HgCl_2$(s)	271.50	−224.3	−178.6	146.0	
Neon					
Ne(g)	20.18	0	0	146.33	20.786
Nitrogen					
N_2(g)	28.013	0	0	191.61	29.125
N(g)	14.007	472.70	455.56	153.30	20.786
NO(g)	30.01	90.25	86.55	210.76	29.844
N_2O(g)	44.01	82.05	104.20	291.85	38.45
NO_2(g)	46.01	33.18	51.31	240.06	37.20
N_2O_4(g)	92.01	9.16	97.89	304.29	77.28
HNO_3(l)	63.01	−174.10	−80.71	155.60	109.87
HNO_3(aq)	63.01	−207.36	−111.25	146.4	−86.6
NO_3^-(aq)	62.01	−205.0	−108.74	146.4	−86.6
NH_3(g)	17.03	−46.11	−16.45	192.45	35.06
NH_3(aq)	17.03	−80.29	−26.50	111.3	
NH_4^+(aq)	18.04	−132.51	−79.31	113.4	79.9
NH_2OH(s)	33.03	−114.2			
HN_3(g)	43.03	294.1	328.1	238.97	43.68
HN_3(l)	43.03	264.0	327.3	140.6	
NH_4Cl(s)	53.49	−314.43	−202.87	94.6	84.1
N_2H_4(l)	32.05	50.63	149.43	121.21	98.87
Oxygen					
O_2(g)	31.999	0	0	205.138	29.355
O(g)	15.999	249.17	231.73	161.06	21.912
O_3(g)	47.998	142.7	163.2	238.93	39.20
OH^-(aq)	17.007	−229.99	−157.24	−10.75	−148.5
Phosphorus					
P(s, white)	30.97	0	0	41.09	23.840
P(g)	30.97	314.64	278.25	163.19	20.786
P_2(g)	61.95	144.3	103.7	218.13	32.05
P_4(g)	123.90	58.91	24.44	279.98	67.15
PH_3(g)	34.00	5.4	13.4	210.23	37.11
PCl_3(g)	137.33	−287.0	−267.8	311.78	71.84
PCl_5(g)	208.24	−374.9	−305.0	364.6	112.8
PCl_5(s)	208.24	−443.5			
H_3PO_4(s)	82.00	−964.4			
Potassium					
K(s)	39.10	0	0	64.18	29.58
K^+(aq)	39.10	−252.38	−283.27	102.5	21.8
KOH(s)	56.11	−424.76	−379.08	78.9	64.9
KF(s)	58.10	−567.27	−537.75	66.57	49.04
KCl(s)	74.56	−436.75	−409.14	82.59	51.30
KBr(s)	119.01	−393.80	−380.66	95.90	52.30
KI(s)	166.01	−327.90	−324.89	106.32	52.93
Silicon					
Si(s)	28.09	0	0	18.83	20.00
SiO_2(s, α)	60.09	−910.94	−856.64	41.84	44.43

Table 4.1. (Contd)

	M_r	$\Delta_f H^\ominus$/kJ mol^{-1}	$\Delta_f G^\ominus$/kJ mol^{-1}	$S^\ominus$/J K^{-1} mol^{-1}	C_p/J K^{-1} mol^{-1}
Silver					
Ag(s)	107.87	0	0	42.55	25.351
Ag^+(aq)	107.87	105.58	77.11	72.68	21.8
AgBr(s)	187.78	−100.37	−96.90	107.1	52.38
AgCl(s)	143.32	−127.07	−109.79	96.2	50.79
Ag_2O(s)	231.74	−31.05	−11.20	121.3	65.86
$AgNO_3$(s)	169.88	−124.39	−33.41	140.92	93.05
Sodium					
Na(s)	22.99	0	0	51.21	28.24
Na^+(aq)	22.99	−240.12	−261.91	59.0	46.4
NaOH(s)	40.00	−425.61	−379.49	64.46	59.54
NaCl(s)	58.44	−411.15	−384.14	72.13	50.50
NaBr(s)	102.90	−361.06	−348.98	86.82	51.38
NaI(s)	149.89	−287.78	−286.06	98.53	52.09
Sulphur					
S(s, rhombic)	32.06	0	0	31.80	22.64
S(s, monoclinic)	32.06	0.33	0.1	32.6	23.6
S(g)	32.06	278.81	238.25	167.82	23.673
S_2(g)	64.13	128.37	79.30	228.18	32.47
S^{2-}(aq)	32.06	33.1	85.8	−14.6	
SO_2(g)	64.06	−296.83	−300.19	248.22	39.87
SO_3(g)	80.06	−395.72	−371.06	256.76	50.67
H_2SO_4(l)	98.08	−813.99	−690.00	156.90	138.9
H_2SO_4(aq)	98.08	−909.27	−744.53	20.1	−293
SO_4^{2-}(aq)	96.06	−909.27	−744.53	20.1	−293
HSO_4^-(aq)	97.07	−887.34	−755.91	131.8	−84
H_2S(g)	34.08	−20.63	−33.56	205.79	34.23
H_2S(aq)	34.08	−39.7	−27.83	121	
HS^-(aq)	33.072	−17.6	12.08	62.08	
SF_6(g)	146.05	−1209	−1105.3	291.82	97.28
Tin					
Sn(s, white)	118.69	0	0	51.55	26.99
Sn^{2+}(aq)	118.69	−8.8	−27.2	−17	
SnO(s)	134.69	−285.8	−256.9	56.5	44.31
SnO_2(s)	150.69	−580.7	−519.6	52.3	52.59
Xenon					
Xe(g)	131.30	0	0	169.68	20.786
Zinc					
Zn(s)	65.37	0	0	41.63	25.40
Zn^{2+}(aq)	65.37	−153.89	−147.06	−112.1	46
ZnO(s)	81.37	−348.28	−318.30	43.64	40.25

Table 4.1(a). Thermodynamic data for organic compounds (all values relate to $\mathcal{T}$)

	M_r	$\Delta_f H^\ominus$/kJ mol^{-1}	$\Delta_f G^\ominus$/kJ mol^{-1}	$S^\ominus$/J K^{-1} mol^{-1}	C_p/J K^{-1} mol^{-1}
C(s, graphite)	12.011	0	0	5.740	8.527
C(s, diamond)	12.011	1.895	2.900	2.377	6.113
CO_2(g)	44.01	−393.51	−394.36	213.74	37.11
Hydrocarbons					
CH_4(g), methane	16.04	−74.81	−50.72	186.26	35.31
CH_3(g), methyl	15.04	145.69	147.92	194.2	38.70
C_2H_2(g), ethyne	26.04	226.73	209.20	200.94	43.93
C_2H_4(g), ethene	28.05	52.26	68.15	219.56	43.56

Table 4.1(a). (Contd)

	M_r	$\Delta_r H^\ominus$/kJ mol^{-1}	$\Delta_f G^\ominus$/kJ mol^{-1}	$S^\ominus$/J K^{-1} mol^{-1}	C_p/J K^{-1} mol^{-1}
Hydrocarbons					
C_2H_6(g), ethane	30.07	−84.68	−32.82	229.60	52.63
C_3H_6(g), propene	42.08	20.42	62.78	267.04	63.89
C_3H_6(g), cyclopropane	42.08	53.30	104.45	237.55	55.94
C_3H_8(g), propane	44.10	−103.85	−23.49	269.91	73.5
C_4H_8(g), 1-butene	56.11	−0.13	71.39	305.71	85.65
C_4H_8(g), cis-2-butene	56.11	−6.99	65.95	300.94	78.91
C_4H_8(g), trans-2-butene	56.11	−11.17	63.06	296.59	87.82
C_4H_{10}(g), butane	58.13	−126.15	−17.03	310.23	97.45
C_5H_{12}(g), pentane	72.15	−146.44	−8.20	348.40	120.2
C_5H_{12}(l)	72.15	−173.1			
C_6H_6(l), benzene	78.12	49.0	124.3	173.3	136.1
C_6H_6(g)	78.12	82.93	129.72	269.31	81.67
C_6H_{12}(l), cyclohexane	84.16	−37.3	6.4	204.3	156.5
C_6H_{14}(l), hexane	86.18	−198.7			
$C_6H_5CH_3$(g), methylbenzene	92.14	50.0	122.0	320.7	103.6
C_7H_{16}(l), heptane	100.21	−224.4	1.0	328.6	224.3
C_8H_{18}(l), octane	114.23	−249.9	6.4	361.1	
C_8H_{18}(l), iso-octane	114.23	−255.1			
$C_{10}H_8$(s), naphthalene	128.18	78.53			
Alcohols and phenols					
CH_3OH(l), methanol	32.04	−238.66	−166.27	126.8	81.6
CH_3OH(g)	32.04	−200.66	−161.96	239.81	43.89
C_2H_5OH(l), ethanol	46.07	−277.69	−174.78	160.7	111.46
C_2H_3OH(g)	46.07	−235.10	−168.49	282.70	65.44
C_6H_5OH(s), phenol	94.12	−165.0	−50.9	146.0	
Carboxylic acids, hydroxy acids, and esters					
HCOOH(l), formic	46.03	−424.72	−361.35	128.95	99.04
CH_3COOH(l), acetic	60.05	−484.5	−389.9	159.8	124.3
CH_3COOH(aq)	60.05	−485.76	−396.46	178.7	
CH_3COO^-(aq)	59.05	−486.01	−369.31	86.6	−6.3
$(COOH)_2$(s), oxalic	90.04	−827.2			117
C_6H_5COOH(s), benzoic	122.13	−385.1	−245.3	167.6	146.8
$CH_3CH(OH)COOH$(s), lactic	90.08	−694.0			
$CH_3COOC_2H_5$(l) ethyl acetate	88.11	−479.0	−332.7	259.4	170.1
Alkanals and alkanones					
HCHO(g), methanal	30.03	−108.57	−102.53	218.77	35.40
CH_3CHO(l), ethanal	44.05	−192.30	−128.12	160.2	
CH_3CHO(g)	44.05	−166.19	−128.86	250.3	57.3
CH_3COCH_3(l), propanone	58.08	−248.1	−155.4	200.4	124.7
Sugars					
$C_6H_{12}O_6$(s), α-D-glucose	180.16	−1274			
$C_6H_{12}O_6$(s), β-D-glucose	180.16	−1268			
$C_6H_{12}O_6$(s), β-D-fructose	180.16	−1266			
$C_{12}H_{22}O_{11}$(s), sucrose	342.30	−2222		360.2	
Nitrogen compounds					
$CO(NH_2)_2$(s), urea	60.06	−333.51	−197.33	104.60	93.14
CH_3NH_2(g), methylamine	31.06	−22.97	32.16	243.41	53.1
$C_6H_5NH_2$(l), aniline	93.13	31.1			
$CH_2(NH_2)COOH$(s), glycine	75.07	−532.9	−373.4	103.5	99.2

Table 4.2. Standard enthalpies of combustion, $-\Delta H_m^\ominus(\mathcal{T})$/kJ mol^{-1}; for enthalpies of formation, see Table 4.1(a)

	M_r	$-\Delta H_m^\ominus$/kJ mol^{-1}
CH_4(g), methane	16.04	890
C_2H_2(g), ethyne	26.04	1300
C_2H_4(g), ethene	28.05	1411
C_2H_6(g), ethane	30.07	1560
C_3H_6(g), cyclopropane	42.08	2091
C_3H_6(g), propene	42.08	2058
C_3H_8(g), propane	44.10	2220
C_4H_{10}(g), butane	58.12	2877
C_5H_{12}(g), pentane	72.15	3536
C_6H_{12}(l), cyclohexane	84.16	3920
C_6H_{14}(l), hexane	86.18	4163
C_6H_6(l), benzene	78.12	3268
C_7H_{16}(l), heptane	100.21	4854
C_8H_{18}(l), octane	114.23	5471
C_8H_{18}(l), iso-octane	114.23	5461
$C_{10}H_8$(s), naphthalene	128.18	5157
CH_3OH(l), methanol	32.04	726
CH_3CHO(g), ethanal	44.05	1193
CH_3CH_2OH(l), ethanol	46.07	1368
CH_3COOH(l), ethanoic acid	60.05	874
$CH_3COOC_2H_5$(l), ethyl ethanoate	88.11	2231
C_6H_5OH(s), phenol	94.11	3054
$C_6H_5NH_2$(l), aniline	93.13	3393
C_6H_5COOH(s), benzoic acid	122.12	3227
$(NH_2)_2CO$(s), urea	93.13	632
NH_2CH_2COOH(s), glycine	75.07	964
$CH_3CH(OH)COOH$(s), lactic acid	90.08	1344
$C_6H_{12}O_6$(s), α-D-glucose	180.16	2802
$C_6H_{12}O_6$(s), β-D-glucose	180.16	2808
$C_{12}H_{22}O_{11}$(s), sucrose	342.30	5645

Data: AIP.

Table 4.3. Molar heat capacities, $C_{p,m} = a + bT + c/T^2$

	a/J K^{-1} mol^{-1}	b/J K^{-2} mol^{-1}	c/J K mol^{-1}
Monatomic gases	20.78	0	0
Other gases			
H_2	27.28	3.26N3	0.50P5
O_2	29.96	4.18N3	−1.67P5
N_2	28.58	3.77N3	−0.50P5
F_2	34.56	2.51N3	−3.51P5
Cl_2	37.03	0.67N3	−2.85P5
Br_2	37.32	0.50N3	−1.26P5
I_2	37.40	0.59N3	−0.71P5
CO_2	44.22	8.79N3	−8.62P5
NH_3	29.75	2.51N2	−1.55P5
Liquids (from melting to boiling)			
I_2	80.33	0	0
H_2O	75.48	0	0
$C_{10}H_8$ (naphthalene)	79.5	4.075N4	0
Solids			
C (graphite)	16.86	4.77N3	−8.54P5
Cu	22.64	6.28N3	0
Al	20.67	12.38N3	0
Pb	22.13	11.72N3	0.96P5
I_2	40.12	49.79N3	0
NaCl	45.94	16.32N3	0
$C_{10}H_8$ (naphthalene)	−115.9	3.920	0

Source: LR.

Table 4.4. Bond dissociation enthalpies, $\Delta H_m^\ominus$(A—B)/kJ mol^{-1} at $\mathcal{T}$

Diatomic molecules

H—H 436	F—F 155	Cl—Cl 242	Br—Br 193	I—I 151
O=O 497	C≡O 1074	N≡N 945		
H—O 428	H—F 565	H—Cl 431	H—Br 366	H—I 299

Polyatomic molecules

H—CH_3	435	H—NH_2	431	H—OH	492	H—C_6H_5	469
H_3C—CH_3	368	H_2C=CH_2	699	HC≡CH	962		
HO—CH_3	377	Cl—CH_3	452	Br—CH_3	293	I—CH_3	234
O=CO	531	HO—OH	213	O_2N—NO_2	57		

Data: HCP, KL.

Table 4.5. Bond enthalpies, E(A—B)/kJ mol^{-1}

	H	C	N	O	F	Cl	Br	I	S	P	Si
H	436										
C	412	348(i) 612(ii) 518(a)									
N	388	305(i) 613(ii) 890(iii)	163(i) 409(ii) 945(iii)								
O	463	360(i) 743(ii)	(157)	146(i) 497(ii)							
F	565	484	270	185	155						
Cl	431	338	200	203	254	242					
Br	366	276				219	193				
I	299	238				210	178	151			
S	338	259				250	212		264		
P	322									172	
Si	318			374							176

(i) Single bond, (ii) double bond, (iii) triple bond, (a) aromatic.
Data: HCP and L. Pauling (1960). *The nature of the chemical bond*. Cornell University Press.

Table 4.6. Standard enthalpies of atomization, $\Delta_a H^\ominus(\mathcal{T})$/kJ mol^{-1}

Main group elements

H 217.97						
Li 161	Be 321	B 590	C 716.682	N 472.70	O 249.17	F 78.99
Na 108.4	Mg 150	Al 314	Si 455.6	P 314.6	S 278.81	Cl 121.68
K 89.8	Ca 193	Ga 289	Ge 377	As 290	Se 202	Br 111.88
Rb 85.8	Sr 164	In 244	Sn 301	Sb 254	Te 199	I 106.84
Cs 78.7	Ba 176	Tl 186	Pb 195.8	Bi 208	Po 144	

Selected d-block elements

Ti 469	Cr 398	Mn 279	Fe 404.5	Co 427	Ni 431
Cu 339.3	Zn 130.5	Ag 286.2	Hg 60.84		

Source: AIP, LR, NBS.

Table 4.7. Enthalpies of fusion and vaporization at the transition temperature, $\Delta H^\ominus_m$/kJ mol^{-1}

	T_f/K	Fusion	T_b/K	Vaporization
Elements				
He	3.5	0.021	4.22	0.084
Ar	83.81	1.188	87.29	6.51
Xe	161	2.30	165	12.6
H_2	13.96	0.117	20.38	0.916
N_2	63.15	0.719	77.35	5.586
O_2	54.36	0.444	90.18	6.820
F_2	53.6	0.26	85.0	3.16
Cl_2	172.1	6.41	239.1	20.41
Br_2	265.9	10.57	332.4	29.45
I_2	386.8	15.52	458.4	41.80
Hg_2	234.3	2.292	629.7	59.30
Na	371.0	2.601	1156	98.01
Ag	1234	11.30	2436	250.6
Inorganic compounds				
H_2O	273.15	6.008	373.15	40.656 44.016 at $\mathcal{T}$

	T_f/K	Fusion	T_b/K	Vaporization
Inorganic Compounds				
H_2S	187.6	2.377	212.8	18.67
NH_3	195.4	5.652	239.7	23.35
CO_2	217.0	8.33	194.6	25.23 (s)
CCl_4	250.3	2.47	349.9	30.00
CS_2	161.2	4.39	319.4	26.74
H_2SO_4	283.5	2.56		
Organic compounds				
CH_4	90.68	0.941	111.7	8.18
C_2H_6	89.85	2.86	184.6	14.7
C_6H_6	278.6	10.59	353.2	30.8
CH_3OH	175.2	3.16	337.2	35.27 37.99 at $\mathcal{T}$
C_2H_5OH	156	4.60	352	43.5

Data: AIP.

Table 4.8. Standard molar enthalpies of formation of ions in solution at infinite dilution, $\Delta_f H^\ominus(\mathcal{T})/\text{kJ mol}^{-1}$

Cations

H^+	0	Mg^{2+}	−466.85	Al^{3+}	−524.7
Li^+	−278.49	Cu^{2+}	+64.77	Fe^{3+}	−48.5
Na^+	−240.1	Fe^{+2}	−89.1		
K^+	−252.38	Zn^+	−153.89		
Rb^+	−251.17	Cd^{2+}	−75.90		
Cs^+	−258.28	Hg^{2+}	+171.1		
NH^{4+}	−132.51	Pb^{2+}	−1.7		
Cu^+	+71.67	Sn^{2+}	−8.8		
Ag^+	+105.58	Ca^{2+}	−542.83		

Anions

OH^-	−229.99	CO_3^{2-}	−677.14	PO_4^{3-}	−1277.4
F^-	−332.63	S^{2-}	+33.1		
Cl^-	−167.16	SO_4^{2-}	−909.27		
Br^-	−121.55	HPO_4^{2-}	−1292		
I^-	−55.19				
NO_3^-	−205.0				
ClO_4^-	−129.33				
HS^-	−17.6				
HCO_3^-	−691.99				
CH_3COO^-	−486.01				

Data: NBS, KL.

Table 4.9. First and second ionization energies, $I/\text{kJ mol}^{-1}$

H 1312							He 2372 5251
Li 520.3 7298	Be 899.5 1757	B 800.6 2427	C 1086 2353	N 1402 2856	O 1341 3388	F 1681 3375	Ne 2081 3952
Na 495.8 4562	Mg 737.7 1451	Al 577.6 1817	Si 786.6 1577	P 1012 1903	S 1000 2258	Cl 1251 2296	Ar 1521 2665
K 418.9 3069	Ca 589.8 1145					Br 1140 2084	Kr 1351 2370
Rb 402 2650	Sr 548 1060					I 1010 1840	Xe 1170 2050
Cs 376 2420	Ba 502 966						

Data: AIP.

Table 4.10. Electron affinities, $E_A/\text{kJ mol}^{-1}$

H 74						He 18
Li 56	B 29	C 121	N 26	O 141 O^- −844	F 333	
Na 75	Al 47	Si 134	P 75	S 200 S^- −532	Cl 349	Ar −29
K 79					Br 328	
					I 314	

Data: Principally C. A. McDowell (1969). *Physical chemistry* (ed. H. Eyring, D. Henderson, and W. Jost) Vol. 3. Academic Press, London.

Table 4.10(a). Electronegativities (Pauling)

Main group elements

H 2.20						
Li 0.98	Be 1.57	B 2.04	C 2.55	N 3.04	O 3.44	F 3.98
Na 0.93	Mg 1.31	Al 1.61	Si 1.90	P 2.19	S 2.58	Cl 3.16
K 0.82	Ca 1.10	Ga 2.01	Ge 2.01	As 2.18	Se 2.55	Br 2.96
Rb 0.82	Sr 0.95	In 1.78	Sn 1.96	Sb 2.05	Te 2.1	I 2.66
Cs 0.7	Ba 0.9	Tl 1.8	Pb 1.8	Bi 1.9	Po 2.0	At 2.2

d-block elements

Sc 1.3	Ti 1.5	V 1.6	Cr 1.6	Mn 1.5	Fe 1.8	Co 1.8	Ni 1.8	Cu 1.9	Zn 1.6
Y 1.2	Zr 1.4	Nb 1.6	Mo 1.8	Tc 1.9	Ru 2.2	Rh 2.2	Pd 2.2	Ag 1.9	Cd 1.7
La 1.1	Hf 1.3	Ta 1.5	W 1.7	Re 1.9	Os 2.2	Ir 2.2	Pt 2.2	Au 2.4	Hg 1.9

f-block elements
All 1.1–1.3

Dipole moment: $\mu/\text{D} = |\chi_A - \chi_B|$
Ionicity/% $= 16\,|\chi_A - \chi_B| + 3.5(\chi_A - \chi_B)^2$
Covalent–ionic resonance energy/eV $= (\chi_A - \chi_B)^2$.
Data: Principally A. L. Allred (1961). *J. inorgan. nucl. Chem.* **17,** 215.

Table 4.11. Standard molar enthalpies of hydration at infinite dilution, $-\Delta H^\ominus_{\text{solv,m}}(\mathcal{T})/\text{kJ mol}^{-1}$

	Li^+	Na^+	K^+	Rb^+	Cs^+
F^-	1026	911	828	806	782
Cl^-	884	784	685	664	640
Br^-	856	742	658	637	613
I^-	815	701	617	586	572

Data: Principally J. O'M. Bockris and A. K. N. Reddy (1970). *Modern electrochemistry*, Vol. 1. Plenum Press, New York.

Table 4.12. Standard molar ion hydration enthalpies, $\Delta H^{\ominus}_{\mathrm{solv,m}}(\mathcal{T})/\mathrm{kJ\ mol^{-1}}$

Cations					
H^+	(1090)	Ag^+	464	Mg^{2+}	1920
Li^+	520	NH_4^+	301	Ca^{2+}	1650
Na^+	405			Sr^{2+}	1480
K^+	321			Ba^{2+}	1360
Rb^+	300			Fe^{2+}	1950
Cs^+	277			Cu^{2+}	2100
				Zn^{2+}	2050
				Al^{3+}	4690
				Fe^{3+}	4430
Anions					
OH^-	460				
F^-	506	Cl^-	364	Br^-	337
I^-	296				

Data: Principally J. O'M. Bockris and A. K. N. Reddy (1970). *Modern electrochemistry*, Vol. 1. Plenum Press, New York.

Table 5.1. Standard entropies (and temperatures) of phase transitions at 1 atm pressure, $\Delta_t S^{\ominus}_m(T_t)/\mathrm{J\ K^{-1}\ mol^{-1}}$

	Melting	Boiling
He	6.0 (at 3.5 K)	19.9 (at 4.22 K)
Ar	14.17 (at 83.8 K)	74.53 (at 87.3 K)
H_2	8.38 (at 14.0 K)	44.96 (at 20.38 K)
N_2	11.39 (at 63.2 K)	75.22 (at 77.4 K)
O_2	8.17 (at 54.4 K)	75.63 (at 90.2 K)
Cl_2	37.22 (at 172.1 K)	85.38 (at 239.0 K)
Br_2	39.76 (at 265.9 K)	88.61 (at 332.4 K)
H_2O	22.00 (at 273.2 K)	109.0 (at 373.2 K)
H_2S	12.67 (at 187.6 K)	87.75 (at 212.0 K)
NH_3	28.93 (at 195.4 K)	97.41 (at 239.73 K)
CH_3OH	18.03 (at 175.2 K)	104.6 (at 337.2 K)
CH_3COOH	40.4 (at 289.8 K)	61.9 (at 391.4 K)
C_6H_6	38.00 (at 278.6 K)	87.19 (at 353.2 K)

Data: AIP.

Table 5.2. Standard Third Law entropies at $\mathcal{T}$: see Tables 4.1 and 4.1(a).

Table 5.3. Standard Gibbs functions of formation at $\mathcal{T}$: see Tables 4.1 and 4.1(a).

Table 6.1. The fugacity of nitrogen at 273 K

p/atm	f/p	p/atm	f/p
1	0.99955	300	1.0055
10	0.9956	400	1.062
50	0.9812	600	1.239
100	0.9703	800	1.495
150	0.9672	1000	1.839
200	0.9721		

Data: LR.

Table 7.1. Surface tensions of liquids at 293 K

	$\gamma/\mathrm{N\ m^{-1}}$		$\gamma/\mathrm{N\ m^{-1}}$
Water	7.275N2	Methanol	2.26N2
	7.20N2 at 25 °C	Ethanol	2.28N2
	5.80N2 at 100 °C	Benzene	2.888N2
Mercury	47.2N2	Hexane	1.84N2
		Carbon tetrachloride	2.70N2

Data: KL.

Table 8.1. Henry's Law constants for gases at $\mathcal{T}$, K/Torr

	Water	Benzene
H_2	5.34P7	2.75P6
N_2	6.51P7	1.79P6
O_2	3.30P7	
CO_2	1.25P6	8.57P4
CH_4	3.14P5	4.27P5

Data: F. Daniels and R. A. Alberty (1980). *Physical chemistry*. Wiley, New York.

Table 8.2. Cryoscopic and ebullioscopic constants

	$K_f/\mathrm{K\ kg\ mol^{-1}}$	$K_b/\mathrm{K\ kg\ mol^{-1}}$
Acetic acid	3.90	3.07
Benzene	5.12	2.53
Camphor	40	
Carbon disulphide	3.8	2.37
Carbon tetrachloride	30	4.95
Naphthalene	6.94	5.8
Phenol	7.27	3.04
Water	1.86	0.51

Data: KL.

Table 10.1. Giauque functions, $-\Phi_0/\mathrm{J\ K^{-1}\ mol^{-1}}$

	$\mathcal{T}$	500 K	1000 K	1500 K	2000 K	$\{H^\ominus_m(\mathcal{T})-H^\ominus_m(0)\}$ /kJ mol^{-1}
$H_2(g)$	102.2	116.9	137.0	148.9	157.6	8.468
$N_2(g)$	162.4	177.5	197.9	210.4	219.6	8.669
$O_2(g)$	176.0	191.1	212.1	225.1	234.7	8.682
$Cl_2(g)$	192.2	208.6	231.9	246.2	256.6	9.180
$Br_2(g)$	212.8	230.1	254.4	269.1	279.6	9.728
C(s)	2.2	4.85	11.6	17.5	22.5	1.050
CO(g)	168.4	183.5	204.1	216.6	225.9	8.673
$CO_2(g)$	182.3	199.5	226.4	244.7	258.8	9.364
$CH_4(g)$	152.5	170.5	199.4	221.1	239	10.029
$NH_3(g)$	159.0	176.9	203.5	221.9	236.6	9.92
$H_2O(g)$	155.5	172.8	196.7	211.7	223.1	9.908
HCl(g)	157.8	172.8	193.1	205.4	214.3	8.640

Data: LR.

Table 11.1. Relative permittivities (dielectric constants) at $\mathcal{T}$

Non-polar molecules		Polar molecules	
Methane	1.70 (at −173 °C)	Water	78.54; 80.37 at 20 °C
		Ammonia	16.9; 22.4 at −33 °C
Carbon tetrachloride	2.228	Hydrogen sulphide	9.26 at −85 °C
Cyclohexane	2.015	Methanol	32.63
Benzene	2.274	Ethanol	24.30
		Nitrobenzene	34.82

Data: HCP.

Table 11.2. Mean activity coefficients in water at $\mathcal{T}$

$m/m^\ominus$	HCl	KCl	$CaCl_2$	H_2SO_4	$LaCl_3$	$In_2(SO_4)_3$
0.001	0.966	0.966	0.888	0.830	0.790	
0.005	0.929	0.927	0.789	0.639	0.636	0.16
0.01	0.905	0.902	0.732	0.544	0.560	0.11
0.05	0.830	0.816	0.584	0.340	0.388	0.035
0.10	0.798	0.770	0.524	0.266	0.356	0.025
0.50	0.769	0.652	0.510	0.155	0.303	0.014
1.00	0.811	0.607	0.725	0.131	0.387	
2.00	1.011	0.577	1.554	0.125	0.954	

Data: RS, HCP, and S. Glasstone (1942). *Introduction to electrochemistry*. Van Nostrand Reinhardt, New York.

Table 12.1. Standard electrode potentials at $\mathcal{T}$

Couple	$E^\ominus$/V	Couple	$E^\ominus$/V
$Sm^{2+}+2e^-\rightarrow Sm$	−3.12	$Pb^{2+}+2e^-\rightarrow Pb$	−0.13
$Li^++e^-\rightarrow Li$	−3.05	$O_2+H_2O+2e^-\rightarrow HO_2^-+OH^-$	−0.08
$K^++e^-\rightarrow K$	−2.93	$Fe^{3+}+3e^-\rightarrow Fe$	−0.04
$Rb^++e^-\rightarrow Rb$	−2.93	$Ti^{4+}+e^-\rightarrow Ti^{3+}$	0.00
$Cs^++e^-\rightarrow Cs$	−2.92	$2H^++2e^-\rightarrow H_2$	0
$Ra^{2+}+2e^-\rightarrow Ra$	−2.92	$AgBr+e^-\rightarrow Ag+Br^-$	0.07
$Ba^{2+}+2e^-\rightarrow Ba$	−2.91	$Sn^{4+}+2e^-\rightarrow Sn^{2+}$	0.15
$Sr^{2+}+2e^-\rightarrow Sr$	−2.89	$Cu^{2+}+e^-\rightarrow Cu^+$	0.16
$Ca^{2+}+2e^-\rightarrow Ca$	−2.87	$Bi^{3+}+3e^-\rightarrow Bi$	0.20
$Na^++e^-\rightarrow Na$	−2.71	$AgCl+e^-\rightarrow Ag+Cl^-$	0.2223
$La^{3+}+3e^-\rightarrow La$	−2.52	$Hg_2Cl_2+2e^-\rightarrow 2Hg+2Cl^-$	0.27
$Ce^{3+}+3e^-\rightarrow Ce$	−2.48	$Cu^{2+}+2e^-\rightarrow Cu$	0.34
$Mg^{2+}+2e^-\rightarrow Mg$	−2.36	$O_2+2H_2O+4e^-\rightarrow 4OH^-$	0.40
$Be^{2+}+2e^-\rightarrow Be$	−1.85	$NiOOH+H_2O+e^-\rightarrow Ni(OH)_2+OH^-$	0.49
$U^{3+}+3e^-\rightarrow U$	−1.79	$Cu^++e^-\rightarrow Cu$	0.52
$Al^{3+}+3e^-\rightarrow Al$	−1.66	$I_3^-+2e^-\rightarrow 3I^-$	0.53
$Ti^{2+}+2e^-\rightarrow Ti$	−1.63	$I_2+2e^-\rightarrow 2I^-$	0.54
$V^{2+}+2e^-\rightarrow V$	−1.19	$Hg_2SO_4+2e^-\rightarrow 2Hg+SO_4^{2-}$	0.62
$Mn^{2+}+2e^-\rightarrow Mn$	−1.18	$Fe^{3+}+e^-\rightarrow Fe^{2+}$	0.77
$Cr^{2+}+2e^-\rightarrow Cr$	−0.91	$AgF+e^-\rightarrow Ag+F^-$	0.78
$Fe(OH)_2+2e^-\rightarrow Fe+2OH^-$	−0.88	$Hg_2^{2+}+2e^-\rightarrow 2Hg$	0.79
$2H_2O+2e^-\rightarrow H_2+2OH^-$	−0.83	$Ag^++e^-\rightarrow Ag$	0.80
$Cd(OH)_2+2e^-\rightarrow Cd+2OH^-$	−0.81	$2Hg^{2+}+2e^-\rightarrow Hg_2^{2+}$	0.92
$Zn^{2+}+2e^-\rightarrow Zn$	−0.76	$Pu^{4+}+e^-\rightarrow Pu^{3+}$	0.97
$Cr^{3+}+3e^-\rightarrow Cr$	−0.74	$Br_2+2e^-\rightarrow 2Br^-$	1.09
$U^{4+}+e^-\rightarrow U^{3+}$	−0.61	$Pt^{2+}+2e^-\rightarrow Pt$	1.20
$O_2+e^-\rightarrow O_2^-$	−0.56	$MnO_2+4H^++2e^-\rightarrow Mn^{2+}+2H_2O$	1.23
$In^{3+}+e^-\rightarrow In^{2+}$	−0.49	$O_2+4H^++4e^-\rightarrow 2H_2O$	1.23
$S+2e^-\rightarrow S^{2-}$	−0.48	$Cr_2O_7^{2-}+14H^++6e^-\rightarrow 2Cr^{3+}+7H_2O$	1.33
$In^{3+}+2e^-\rightarrow In^+$	−0.44	$Cl_2+2e^-\rightarrow 2Cl^-$	1.36
$Fe^{2+}+2e^-\rightarrow Fe$	−0.44	$Au^{3+}+3e^-\rightarrow Au$	1.40
$Cr^{3+}+e^-\rightarrow Cr^{2+}$	−0.41	$Mn^{3+}+e^-\rightarrow Mn^{2+}$	1.51
$Cd^{2+}+2e^-\rightarrow Cd$	−0.40	$MnO_4^-+8H^++5e^-\rightarrow Mn^{2+}+4H_2O$	1.51
$In^{2+}+e^-\rightarrow In^+$	−0.40	$Ce^{4+}+e^-\rightarrow Ce^{3+}$	1.61
$Ti^{3+}+e^-\rightarrow Ti^{2+}$	−0.37	$Pb^{4+}+2e^-\rightarrow Pb^{2+}$	1.67
$PbSO_4+2e^-\rightarrow Pb+SO_4^{2-}$	−0.36	$Au^++e^-\rightarrow Au$	1.69
$In^{3+}+3e^-\rightarrow In$	−0.34	$Co^{3+}+e^-\rightarrow Co^{2+}$	1.81
$Co^{2+}+2e^-\rightarrow Co$	−0.28	$Ag^{2+}+e^-\rightarrow Ag^+$	1.98
$V^{3+}+e^-\rightarrow V^{2+}$	−0.26	$S_2O_8^{2-}+2e^-\rightarrow 2SO_4^{2-}$	2.05
$Ni^{2+}+2e^-\rightarrow Ni$	−0.23	$F_2+2e^-\rightarrow 2F^-$	2.87
$AgI+e^-\rightarrow Ag+I^-$	−0.15		
$Sn^{2+}+2e^-\rightarrow Sn$	−0.14		
$In^++e^-\rightarrow In$	−0.14		

Data: M. S. Antelman (1982). *The encyclopedia of chemical electrode potentials*. Plenum, New York.

Table 12.2. Standard thermodynamic functions of ions in aqueous solution at $\mathcal{T}$ (see also Table 4.1)

	$\Delta_f H^\ominus$/ kJ mol^{-1}	$\Delta_f G^\ominus$/ kJ mol^{-1}	$S^\ominus$/ J K^{-1} mol^{-1}	$C_{p,m}$/ J K^{-1} mol^{-1}
Cations				
H^+	0	0	0	0
Li^+	−278.5	−293.3	13.4	68.6
Na^+	−240.1	−261.9	59.0	46.4
K^+	−252.4	−283.3	102.5	21.8
Rb^+	−251.2	−284.0	121.5	
Cs^+	−258.3	−292.0	133.0	−10.5
NH_4^+	−132.5	−79.3	113.4	79.9
Ag^+	105.6	77.1	72.7	21.8
Cu^+	71.7	50.0	40.6	
Mg^{2+}	−466.8	−454.8	−138.1	
Ca^{2+}	−542.8	−553.6	−53.1	
Cu^{2+}	64.8	65.5	−99.6	
Zn^{2+}	−153.9	−147.1	−112.1	
Sn^{2+}	−8.8	−27.2	−17	
Pb^{2+}	−1.7	−24.4	10.5	
Hg^{2+}	171.1	164.4	−32.2	
Fe^{2+}	−89.1	−78.9	−137.7	
Fe^{3+}	−48.5	−4.7	−315.9	
Al^{3+}	−531.4	−485.3	−321.7	
Anions				
OH^-	−230.0	−157.2	−10.8	−148.5
F^-	−332.6	−278.8	−13.8	−106.7
Cl^-	−167.2	−131.2	56.5	−136.4
Br^-	−121.6	−104.0	82.4	−141.8
I^-	−55.2	−51.6	111.3	54.4
CN^-	151.0	165.7	−49.3	
HS^-	−17.6	12.1	62.8	
S^{2-}	33.1	85.8	−14.6	
SO_4^{2-}	−909.3	−744.5	2.01	−293
SO_3^{2-}	−486.6	−635.5	−29.3	
PO_4^{3-}	−1018.8	−1277.4	−221.8	
HPO_4^{2-}	−1089.3	−1292.1	−33.5	
$H_2PO_4^-$	−1130.4	−1296.3	90.4	
CO_3^{2-}	−677.1	−527.8	−56.9	
HCO_3^-	−692.0	−586.8	91.2	
ClO_4^-	−129.3	−8.5	182.0	
ClO_3^-	−3.3	−99.3	162.3	
CH_3COO^-	−486.0	−369.3	86.6	−6.3

Data: NBS, KL.

Table 12.3. Ionization constants in water at $\mathcal{T}$

	pK_{a1}	pK_{a2}	pK_{a3}
Inorganic acids			
$B(OH)_3$	9.14	12.74	13.80
H_3CO_3	6.37	10.25	
HCN	9.31		
HF	3.25		
H_2S	7.05	12.92	
HOCl	7.43		
HOBr	8.70		
HOI	10.64		
H_3PO_4	2.12	7.21	12.36
H_2SO_4		1.92	
H_2SO_3	1.92	7.21	
Organic acids			
HCOOH, formic acid	3.75		
CH_3COOH, acetic acid	4.75		
CH_3CH_2COOH, propionic acid	4.87		
$CH_3CH_2CH_2COOH$, butyric acid	4.82		
$CH_3CH(OH)COOH$, lactic acid	3.86		
NH_2CH_2COOH, glycine	9.78		
$(COOH)_2$, oxalic acid	1.23	4.19	
$HOOCCH_2CH_2COOH$, succinic acid	4.16	5.61	
Inorganic bases			
$NH_3(NH_4^+)$	9.25		
$NH_2NH_2(NH_2NH_3^+)$	8.23		
Organic bases			
CH_3NH_2, methylamine	10.66		
$(CH_3)_2NH$, dimethylamine	10.73		
$(CH_3)_3N$, trimethylamine	9.81		
$C_2H_5NH_2$, ethylamine	10.81		
$(C_2H_5)_2NH$, diethylamine	10.99		
$(C_2H_5)_3N$, triethylamine	10.76		
$C_6H_5NH_2$, aniline	4.63		

Data: HCP.

Table 14.1. The error function

z	erf(z)	z	erf(z)
0	0	0.45	0.475 48
0.01	0.011 28	0.50	0.520 50
0.02	0.022 56	0.55	0.563 32
0.03	0.033 84	0.60	0.603 86
0.04	0.045 11	0.65	0.642 03
0.05	0.056 37	0.70	0.677 80
0.06	0.067 62	0.75	0.711 16
0.07	0.078 86	0.80	0.742 10
0.08	0.090 08	0.85	0.770 67
0.09	0.101 28	0.90	0.796 91
0.10	0.112 46	0.95	0.820 89
0.15	0.168 00	1.00	0.842 70
0.20	0.222 70	1.20	0.910 31
0.25	0.276 32	1.40	0.952 28
0.30	0.328 63	1.60	0.976 35
0.35	0.379 38	1.80	0.989 09
0.40	0.428 39	2.00	0.995 32

Data: AS.

Table 15.1. Effective atomic numbers

	H							He
1s:	1							1.70
	Li	Be	B	C	N	O	F	Ne
1s:	2.70	3.70	4.70	5.70	6.70	7.70	8.70	9.70
2s:	1.30	1.95	2.60	3.25	3.90	4.55	5.20	5.85
2p:			2.60	3.25	3.90	4.55	5.20	5.85
	Na	Mg	Al	Si	P	S	Cl	Ar
1s:	10.70	11.70	12.70	13.70	14.70	15.70	16.70	17.70
2s:	6.85	7.85	8.85	9.85	10.85	11.85	12.85	13.85
2p:	6.85	7.85	8.85	9.85	10.85	11.85	12.85	13.85
3s:	2.20	2.85	3.50	4.15	4.80	5.45	6.10	6.75
3d:			3.50	4.15	4.80	5.45	6.10	6.75

Data: P. W. Atkins (1983). *Molecular quantum mechanics*. Oxford University Press, Oxford.

Table 18.1. Properties of diatomic molecules

	$\tilde{\nu}_0$/cm^{-1}	B/cm^{-1}	R/pm	k/N m^{-1}	D/kJ mol^{-1}
$^1H_2^+$	2321.8	29.8	106	160.0	255.8
$^1H_2$	4400.39	60.864	74.138	574.9	432.1
$^2H_2$	3118.46	30.442	74.154	577.0	439.6
$^1H^{19}F$	4138.32	20.956	91.680	965.7	564.4
$^1H^{35}Cl$	2990.95	10.593	127.45	516.3	427.7
$^1H^{81}Br$	2648.98	8.465	141.44	411.5	362.7
$^1H^{127}I$	2308.09	6.511	160.92	313.8	294.9
$^{14}N_2$	2358.07	1.9987	109.76	2293.8	941.7
$^{16}O_2$	1580.36	1.4457	120.75	1176.8	493.5
$^{19}F_2$	891.8	0.8828	141.78	445.1	154.4
$^{35}Cl_2$	559.71	0.2441	198.75	322.7	239.3
$^{127}I_2$	214.57	0.0374	266.67	172.1	148.8

Data: AIP.

Table 18.2. Typical vibration wavenumbers, $\tilde{\nu}$/cm^{-1}

C—H stretch	2850–2960	C—F stretch	1000–1400
C—H bend	1340–1465	C—Cl stretch	600–800
C—C stretch, bend	700–1250	C—Br stretch	500–600
C=C stretch	1620–1680	C—I stretch	500
C≡C stretch	2100–2260	CO_3^{2-}	1410–1450
O—H stretch	3590–3650	NO_3^-	1350–1420
H—bonds	3200–3570	NO_2^-	1230–1250
C=O stretch	1640–1780	SO_4^{2-}	1080–1130
C≡N stretch	2215–2275	Silicates	900–1100
N—H stretch	3200–3500		

Data: L. J. Bellamy (1958, 1968). *The infrared spectra of complex molecules* and *Advances in infrared group frequencies*. Chapman and Hall, London.

Table 19.1. Colour, frequency, and energy of light

Colour	λ/nm	ν/Hz	$\tilde{\nu}$/cm^{-1}	E/eV	E/kJ mol^{-1}	E/kcal mol^{-1}
Infrared	1000	3.00P14	1.00P4	1.24	120	28.6
Red	700	4.28P14	1.43P4	1.77	171	40.8
Orange	620	4.84P14	1.61P4	2.00	193	46.1
Yellow	580	5.17P14	1.72P4	2.14	206	49.3
Green	530	5.66P14	1.89P4	2.34	226	53.9
Blue	470	6.38P14	2.13P4	2.64	254	60.8
Violet	420	7.14P14	2.38P4	2.95	285	68.1
Near ultraviolet	300	1.00P15	3.33P4	4.15	400	95.7
Far ultraviolet	200	1.50P15	5.00P4	6.20	598	143

Data: J. G. Calvert and J. N. Pitts (1966). *Photochemistry*. Wiley, New York.

Table 19.2. Absorption characteristics of some groups and molecules

Group	$\tilde{\nu}_{max}/cm^{-1}$	λ_{max}/nm	$\varepsilon_{max}/M^{-1}\,cm^{-1}$
C=C	5.5P4	180	2.5P2
	5.7P4	175	1.7P4
	5.9P4	170	1.7P4
	6.2P4	160	1.0P4
C=O	3.4P4	295	10
	5.4P4	185	strong
—NO_2	3.6P4	280	10
	4.8P4	210	1.0P4
—N=N—	2.9P4	350	15
	>3.9P4	<260	strong
Benzene ring	3.9P4	255	2.0P2
	5.0P4	200	6.3P3
	5.5P4	180	1.0P5
Cu^{2+}(aq)	1.2P4	810	10
$Cu(NH_3)_4^{2+}$	1.7P4	600	50

Data: principally J. G. Calvert and J. N. Pitts (1966). *Photochemistry*. Wiley, New York.

Table 20.1. Nuclear spin properties

Nuclide	Natural abundance %	Spin I	Magnetic moment μ/μ_N	g-value	NMR frequency at 1 T, ν/MHz
^{1}n*		$\frac{1}{2}$	−1.913 10	−3.826 0	29.167
^{1}H	99.9844	$\frac{1}{2}$	2.792 85	5.585 7	42.576
^{2}H	0.0156	1	0.857 45	0.857 45	6.536
^{3}H*		$\frac{1}{2}$	−2.127 65	−4.255 3	32.434
^{13}C	1.108	$\frac{1}{2}$	0.702 3	1.404 6	10.705
^{14}N	99.635	1	0.403 56	0.403 56	3.076
^{17}O	0.037	$\frac{5}{2}$	−1.893	−0.757 2	5.772
^{19}F	100	$\frac{1}{2}$	2.628 35	5.256 7	40.054
^{31}P	100	$\frac{1}{2}$	1.131 7	2.263 4	17.238
^{33}S	0.74	$\frac{3}{2}$	0.643 4	0.428 9	3.266
^{35}Cl	75.4	$\frac{3}{2}$	0.821 8	0.547 9	4.171
^{37}Cl	24.6	$\frac{3}{2}$	0.684 1	0.456 1	3.472

* Radioactive.
μ is the magnetic moment of the spin state with the largest value of m_I: $\mu = g_I\mu_N I$; μ_N is the
nuclear magneton (see inside front cover).
Data: KL.

Table 20.2. Hyperfine coupling constants for atoms, a/mT

Nuclide	Spin	Isotropic coupling	Anisotropic coupling
^{1}H	$\frac{1}{2}$	50.8(1s)	
^{2}H	1	7.8(1s)	
^{13}C	$\frac{1}{2}$	113.0(2s)	6.6(2p)
^{14}N	1	55.2(2s)	4.8(2p)
^{19}F	$\frac{1}{2}$	1720(2s)	108.4(2p)
^{31}P	$\frac{1}{2}$	364(3s)	20.6(3p)
^{35}Cl	$\frac{3}{2}$	168(3s)	10.0(3p)
^{37}Cl	$\frac{3}{2}$	140(3s)	8.4(3p)

Data: P. W. Atkins and M. C. R. Symons (1967). *The structure of inorganic radicals*. Elsevier, Amsterdam.

Table 23.1. Ionic radii, R/pm

Main group elements

Li^+	Be^{2+}	B^{3+}		O^{2-}	F^-
59(4)	27(4)	12(4)		135(2)	128(2)
74(6)				138(4)	131(4)
				140(6)	133(6)
Na^+	Mg^{2+}	Al^{3+}		S^{2-}	Cl^-
102(6)	58(4)	39(4)		184	181
116(8)	72(6)	53(6)			
K^+	Ca^{2+}	Ga^{3+}			Br^-
138(6)	100(6)	47(4)			196
151(8)	112(8)	62(6)			
160(12)	135(12)				
Rb^+	Sr^{2+}	In^{3+}	Sn^{4+}		I^-
149(6)	116(6)	79(6)	69(6)		220
160(8)	125(8)	92(8)			
173(12)	144(12)				
Cs^+	Ba^{2+}	Tl^+	Pb^{4+}		
170(6)	136(6)	88(6)	78(6)		
182(8)	142(8)		94(8)		
188(12)	160(12)				

d-block elements (high-spin ions)

Sc^{3+}	Ti^{4+}	Cr^{3+}	Mn^{3+}	Fe^{2+}	Co^{3+}	Cu^{2+}	Zn^{2+}
73(6)	60(6)	61(6)	65(6)	63(4)	61(6)	73(6)	75(6)

Data: KL.

Table 23.2. Van der Waals radii, R/pm

H 120				He 120
	N 150	O 140	F 135	Ne 234a, 320b
	P 190	S 185	Cl 180	Ar 286a, 383b
	As 200	Se 200	Br 195	
	Sb 220	Te 220	I 215	

a: gas viscosity. b: close packing.
Data: principally HCP.

Table 23.3. Covalent radii, R/pm

H	37						
C	77(i) 67(ii) 60(iii) 69.5(a)	N	74(i) 65(ii)	O	74(i) 57(ii)	F	72
		P	110(i)	S	104(i) 95(ii)	Cl	99
		As	121(i)	Se	117(i)	Br	114
		Sb	141(i)	Te	137(i)	I	133

(i) Single bond, (ii) double bond, (iii) triple bond, (a) aromatic.
Data: principally KL.

Table 24.1. Dipole moments, polarizabilities, and polarizability volumes

	μ/C m	μ/D	α'/cm^3	α/J^{-1} C^2 m^2
He	0	0	2.0N25	2.2N41
Ar	0	0	1.66N24	1.85N40
H_2	0	0	8.19N25	9.11N41
N_2	0	0	1.77N24	1.97N40
CO	3.90N31	0.117	1.98N24	2.20N40
CO_2	0	0	2.63N24	2.93N40
HF	6.37N30	1.91	5.1N25	5.7N41
HCl	3.60N30	1.08	2.63N24	2.93N40
HBr	2.67N30	0.80	3.61N24	4.01N40
HI	1.40N30	0.42	5.45N24	6.06N40
H_2O	6.17N30	1.85	1.48N24	1.65N40
NH_3	4.90N30	1.47	2.22N24	2.47N40
CCl_4	0	0	1.05N23	1.17N39
$CHCl_3$	3.37N30	1.01	8.50N24	9.46N40
CH_2Cl_2	5.24N30	1.57	6.80N24	7.57N40
CH_3Cl	6.24N30	1.87	4.53N24	5.04N40
CH_4	0	0	2.60N24	2.89N40
CH_3OH	5.70N30	1.71	3.23N24	3.59N40
C_2H_5OH	5.64N30	1.69		
C_6H_6	0	0	1.04N23	1.16N39
$C_6H_5CH_3$	1.20N30	0.36		
o-$C_6H_4(CH_3)_2$	2.07N30	0.62		

Data: HCP and C. J. F. Böttcher and P. Bordewijk (1978). *Theory of electric polarization*. Elsevier, Amsterdam.

Table 24.2. Refractive indices relative to air at 293 K

	434 nm	589 nm	656 nm
Water	1.3404	1.3330	1.3312
Methanol	1.3362	1.3290	1.3277
Ethanol	1.3700	1.3618	1.3605
Benzene	1.5236	1.5012	1.4965
Methylbenzene	1.5170	1.4955	1.4911
Carbon tetrachloride	1.4729	1.4676	1.4579
Carbon disulphide	1.6748	1.6276	1.6182
KCl(s)	1.5050	1.4904	1.4973
KI(s)	1.7035	1.6664	1.6581

Data: AIP.

Table 24.3. Molar refractivities at 589 nm, R_m/cm^3 mol^{-1}

C—H 1.65	C—C 1.20 C=C 2.79 C≡C 4.79	C—O 1.41 C=O 3.34	C≡N 4.69	O—H 1.85

						He 0.5
Li^+ 0.07	Be^{2+} 0.20			O^{2-} 7	F^- 2.65	Ne 1.00
Na^+ 0.46	Mg^{2+} 0.24	Al^{3+} 0.17			Cl^- 9.30	Ar 4.14
K^+ 2.12	Ca^{2+} 1.19				Br^- 12.12	Kr 6.26
Rb^+ 3.57					I^- 18.07	Xe 10.16

Data: E. A. Moelwyn-Hughes (1961). *Physical chemistry*. Pergamon, Oxford.

Table 24.4. Lennard–Jones (12,6)-potential parameters

	(ε/k)/K	σ/pm
He	10.22	258
Ne	35.7	279
Ar	124	342
Xe	229	406
H_2	33.3	297
N_2	91.5	368
O_2	113	343
Cl_2	357	412
Br_2	520	427
CO_2	190	400
CH_4	137	382
CCl_4	327	588
C_2H_4	205	423
C_6H_6	440	527

Data: J. O. Hirschfelder, C. F. Curtiss, and R. B. Bird (1954). *Molecular theory of gases and liquids*. Wiley, New York.

Table 24.5. Viscosities of liquids at 293 K, $\eta/\text{kg m}^{-1}\text{s}^{-1}$

Water*	1.0019N3
Methanol	5.94N4
Ethanol	1.197N3
Pentane	2.34N4
Benzene	6.47N4
Mercury	1.552N3
Carbon tetrachloride	9.72N4
Sulphuric acid	2.7N2

* The viscosity of water over its entire liquid range is represented with less than 1% error by the expression

$$\lg(\eta_{20}/\eta) = A/B,$$
$$A = 1.370\,23(t-20) + 8.36\times10^{-4}(t-20)^2,$$
$$B = 109 + t, \quad t = \theta/°\text{C}.$$

Convert $\text{kg m}^{-1}\text{s}^{-1}$ to centipoise (cP) by multiplying by 10^3 (so that $\eta \approx 1$ cP for water).
Data: AIP, KL.

Table 24.6. Madelung constants

Lattice	$\mathcal{M}$
Rock salt	1.747 56
Caesium chloride	1.772 67
Zinc blende	1.638 05
Wurtzite	1.641 32
Fluorite	2.519 39
Rutile	2.408
Cuprite	2.221 24
Corundum	4.171 9

Data: HCP.

Table 24.7. Lattice enthalpies, $-\Delta H_m^{\ominus}(\mathcal{T})/\text{kJ mol}^{-1}$

	F	Cl	Br	I
Li	1022	846	800	744
Na	902	771	733	684
K	801	701	670	629
Rb	767	675	647	609
Cs	716	645	619	585

Data: J. G. Stark and H. G. Wallace (1982). *Chemistry data book*. Murray, London.

Table 24.8. Magnetic susceptibilities at $\mathcal{T}$

	κ	χ
Water	−9.0N5	−9.0N9
Benzene	−7.2N6	−8.2N9
Cyclohexane	−7.9N6	−1.02N8
Carbon tetrachloride	−8.9N6	−5.4N9
NaCl (s)	−1.39N5	−6.4N9
Cu (s)	−9.6N5	−1.07N9
S (s)	−1.29N5	−6.2N9
Hg (l)	−2.85N5	−2.1N9
$CuSO_4.5H_2O$ (s)	1.76N4	7.7N8
$MnSO_4.4H_2O$ (s)	2.64N3	8.12N7
$NiSO_4.7H_2O$ (s)	4.16N4	2.01N7
$FeSO_4(NH_4)_2SO_4.6H_2O$ (s)	7.55N4	4.06N7
Al (s)	2.2N5	8.2N9
Pt (s)	2.62N4	1.22N8
Na (s)	7.3N6	7.5N9
K (s)	5.6N6	6.5N9

The CGS values of the susceptibility (per gram) are obtained by forming $1000\chi/4\pi$, and the value per gram mole from $1000\chi M_r/4\pi$.
Data: KL.

Table 25.1. Diffusion coefficients of macromolecules in water at 293 K

	M_r	$D/\text{cm}^2\text{s}^{-1}$
Sucrose	342	4.586N6
Ribonuclease	13 683	1.19N6
Lysozyme	14 100	1.04N6
Serum albumin	65 000	5.94N7
Haemoglobin	68 000	6.9N7
Urease	480 000	3.46N7
Collagen	345 000	6.9N8
Myosin	493 000	1.16N7

Data: C. Tanford (1961). *Physical chemistry of macromolecules*. Wiley, New York.

Table 25.2. Intrinsic viscosity

Macromolecule	Solvent	θ/°C	$K/\text{cm}^3\text{g}^{-1}$	a
Polystyrene	Benzene	25	9.5N3	0.74
	Cyclohexane	34*	8.1N2	0.50
Polyisobutylene	Benzene	23*	8.3N2	0.50
	Cyclohexane	30	2.6N2	0.70
Amylose	0.33 M KCl(aq)	25*	1.13N1	0.50
Various proteins†	Guanidine hydrochloride + β-mercaptoethanol		7.16N3	0.66

* θ-temperature.
† Use $[\eta] = KN^a$, N the number of amino acid residues.
Data: K. E. Van Holde (1971). *Physical biochemistry*. Prentice Hall, Englewood Cliffs, NJ.

Table 25.3. Radius of gyration of some macromolecules

	M_r	R_g/nm
Serum albumin	66 000	2.98
Myosin	493 000	46.8
Polystyrene	3.2P6	49.4 (in poor solvent)
DNA	4P6	117.0
Tobacco mosaic virus	3.9P7	92.4

Data: C. Tanford (1961). *Physical chemistry of macromolecules*. Wiley, New York.

Table 26.1. Average speeds at $\mathcal{T}$, $c/\mathrm{m\,s^{-1}}$

He	1256	H_2	1770	H_2O	592	C_6H_6	284
Ar	398	N_2	475	NH_3	609	Hg	177
		O_2	444	CO_2	379	Air	466
		Cl_2	298				

Data: Calculated using eqn (26.1.10) in the form $c/\mathrm{m\,s^{-1}} = 145.51(T/\mathrm{K})^{\frac{1}{2}}/M_r^{\frac{1}{2}} = 2512.48 M_r^{\frac{1}{2}}$ at $\mathcal{T}$.

Table 26.2. Collision cross-section, σ/nm^2

He	0.21	H_2	0.27	CO_2	0.52	CH_4	0.46
Ne	0.24	N_2	0.43	SO_2	0.58	C_2H_4	0.64
Ar	0.36	O_2	0.40			C_6H_6	0.88
		Cl_2	0.93				

Data: KL.

Table 26.3. Transport properties of gases at 1 atm

	$\kappa/\mathrm{mJ\,cm^{-2}\,s^{-1}\,(K\,cm^{-1})^{-1}}$	$\eta/\mu\mathrm{P}$	
	273 K	273 K	293 K
He	1.442	187	196
Ne	0.465	298	313
Ar	0.163	210	223
Kr	0.087	234	250
Xe	0.052	212	228
H_2	1.682	84	88
N_2	0.240	166	176
O_2	0.245	195	204
Cl_2	0.79	123	132
CO_2	0.145	136	147
CH_4	0.302	103	110
C_2H_4	0.164	97	103
Air	0.241	173	182

Data: KL.

Table 27.1. Limiting ionic conductivities in water at $\mathcal{T}$, $\lambda/\mathrm{S\,cm^2\,mol^{-1}}$

H^+	349.6	OH^-	199.1
Li^+	38.7	F^-	55.4
Na^+	50.10	Cl^-	76.35
K^+	73.50	Br^-	78.1
Rb^+	77.8	I^-	76.8
Cs^+	77.2	NO_3^-	71.46
NH_4^+	73.5	ClO_4^-	67.3
Mg^{2+}	106.0	SO_4^{2-}	160.0
Ca^{2+}	119.0	CH_3COO^-	40.9
Sr^{2+}	118.9	$(COO)_2^{2-}$	148.2
Ba^{2+}	127.2	$[Fe(CN)_6]^{3-}$	302.7
Cu^{2+}	107.2	$[Fe(CN)_6]^{4-}$	442.0
Zn^{2+}	105.6	CO_3^{2-}	138.6
$[N(CH_3)_4]^+$	44.9		
$[N(C_2H_5)_4]^+$	32.6		

Data: KL, RS.

Table 27.2. Ionic mobilities in water at $\mathcal{T}$, $u/\mathrm{cm^2\,s^{-1}\,V^{-1}}$

H^+	3.623N3	OH^-	2.064N3
Li^+	4.01N4	F^-	5.70N4
Na^+	5.19N4	Cl^-	7.91N4
K^+	7.62N4	Br^-	8.09N4
Rb^+	7.92N4	I^-	7.96N4
Ag^+	6.42N4	NO_3^-	7.40N4
NH_4^+	7.63N4	CO_3^{2-}	7.46N4
$[N(CH_3)_4]^+$	4.65N4	SO_4^{2-}	8.29N4
Ca^{2+}	6.17N4	CH_3COO^-	4.24N4
Cu^{2+}	5.56N4	$[Fe(CN)_6]^{3-}$	1.05N3
Zn^{2+}	5.47N4	$[Fe(CN)_6]^{4-}$	1.14N3

Data: principally Table 27.1 via eqn (27.1.12a), $u = \lambda/zF$.

Table 27.3. Debye–Hückel–Onsager coefficients for (1,1)-electrolytes at $\mathcal{T}$

solvent	a	b
Water	60.20	0.229
Methanol	156.1	0.923
Ethanol	89.7	1.83
Acetonitrile	22.9	0.716
Acetone	32.8	1.63
Nitromethane	125.1	0.708
Nitrobenzene	44.2	0.776

Data: J. O'M. Bockris and A. K. N. Reddy (1970). *Modern electrochemistry*. Plenum, New York.

Table 27.4. Diffusion coefficients at $\mathcal{T}$, $D/\mathrm{cm^2\,s^{-1}}$

I_2 in hexane	4.05N5	H_2 in CCl_4(l)	9.75N5
in benzene	2.13N5	N_2 in CCl_4	3.42N5
CCl_4 in heptane	3.17N5	O_2 in CCl_4(l)	3.82N5
Glycine in water	1.055N5	Ar in CCl_4(l)	3.63N5
Dextrose in water	6.73N6	CH_4 in CCl_4(l)	2.89N5
Sucrose in water	5.21N6	H_2O in water	2.26N5
		CH_3OH in water	1.58N5
		C_2H_5OH in water	1.24N5

Ions in water

H^+ 9.31N5 Li^+ 1.03N5 Na^+ 1.33N5 K^+ 1.96N5
OH^- 5.30N5 F^- 1.46N5 Cl^- 2.03N5 Br^- 2.08N5 I^- 2.05N5

Data: AIP and (for the ions) eqn (27.2.6) and Table 27.2.

Table 27.5. Viscosity and conductivity

	Solvent				
	Methanol	Water	Ethanol	Formamide	Glycol
$\eta/\mathrm{kg\,m^{-1}\,s^{-1}}$	0.54N3	0.89N3	1.08N3	3.30N3	16.8N3
$\lambda^0\eta/\mathrm{S\,kg\,m^{-1}\,s^{-1}\,mol^{-1}}$ (a)	2.9N6	6.5N6	2.5N6	4.2N6	7.8N6
(b)	2.1N6	1.7N6	2.1N6	2.3N6	2.6N6

(a) K^+; (b) $^tBu_4N^+$.
Data: Dr M. Spiro.

Table 28.1. Kinetic data for first-order reactions

	Phase	$\theta/°C$	k/s^{-1}	$t_{\frac{1}{2}}$
$2N_2O_5 \rightarrow 4NO_2 + O_2$	g	25	3.14N5	6.1 h
		55	1.42N3	8.2 min
	HNO_3(l)	55	9.27N5	125 min
	Br_2(l)	55	2.08N3	333 s
$C_2H_6 \rightarrow 2CH_3$	g	700	5.46N4	21.2 min
Cyclopropane→ propene	g	500	6.71N4	17.2 min

g: High pressure gas-phase limit.
Data: principally K. J. Laidler (1965). *Chemical kinetics*, McGraw-Hill, New York. M. J. Pilling (1974). *Reaction kinetics*. Clarendon Press, Oxford. J. Nicholas (1976). *Chemical kinetics*. Harper and Row, New York.

Table 28.2. Kinetic data for second-order reactions

	Phase	$\theta/°C$	$k/\mathrm{M^{-1}\,s^{-1}}$
$2NOBr \rightarrow 2NO + Br_2$	g	10	0.80
$2NO_2 \rightarrow 2NO + O_2$	g	300	0.54
$H_2 + I_2 \rightarrow 2HI$	g	400	2.42N2
$D_2 + HCl \rightarrow DH + DCl$	g	600	0.141
$2I \rightarrow I_2$	g	23	7P9
	hexane	50	1.8P10
$CH_3Cl + CH_3O^-$	methanol	20	2.29N6
$CH_3Br + CH_3O^-$	methanol	20	9.23N5
$H^+ + OH^- \rightarrow H_2O$	water	25	1.5P11

Data: principally K. J. Laidler (1965). *Chemical kinetics*. McGraw-Hill, New York. M. J. Pilling (1974). *Reaction kinetics*. Clarendon Press, Oxford. J. Nicholas (1976). *Chemical kinetics*. Harper and Row, New York.

Table 28.3. Arrhenius parameters

First-order reactions	A/s^{-1}	E_a/kJ mol^{-1}
Cyclopropane → propene	1.58P15	272
$CH_3NC \rightarrow CH_3CN$	3.98P13	160
cis-CHD=CHD → *trans*-CHD=CHD	3.16P12	256
Cyclobutane → $2C_2H_4$	3.98P15	261
$C_2H_5I \rightarrow C_2H_4 + HI$	2.51P13	209
$C_2H_6 \rightarrow 2CH_3$	2.51P17	384
$2N_2O_5 \rightarrow 4NO_2 + O_2$	6.31P14	88
$N_2O \rightarrow N_2 + O$	7.94P11	250
$C_2H_5 \rightarrow C_2H_4 + H$	1.0P13	167
Second-order, gas phase	$A/M^{-1} s^{-1}$	E_a/kJ mol^{-1}
$O + N_2 \rightarrow NO + N$	1P11	315
$OH + H_2 \rightarrow H_2O + H$	8P10	42
$Cl + H_2 \rightarrow HCl + H$	8P10	23
$2CH_3 \rightarrow C_2H_6$	2P10	*ca.* 0
$NO + Cl_2 \rightarrow NOCl + Cl$	4P9	85
$SO + O_2 \rightarrow SO_2 + O$	3P8	27
$CH_3 + C_2H_6 \rightarrow CH_4 + C_2H_5$	2P8	44
$C_6H_5 + H_2 \rightarrow C_6H_6 + H$	1P8	*ca.* 25
Second-order, solution	$A/M^{-1} s^{-1}$	E_a/kJ mol^{-1}
$C_2H_5ONa + CH_3I$ in ethanol	2.42P11	81.6
$C_2H_5Br + OH^-$ in water	4.30P11	89.5
$C_2H_5I + C_2H_5O^-$ in ethanol	1.49P11	86.6
$CH_3I + C_2H_5O^-$ in ethanol	2.42P11	81.6
$C_2H_5Br + OH^-$ in ethanol	4.30P11	89.5
$CO_2 + OH^-$ in water	1.5P10	38
$CH_3I + S_2O_3^{2-}$ in water	2.19P12	78.7
Sucrose + H_2O in acid water	1.50P15	107.9
$(CH_3)_3CCl$ solvolysis		
in water	7.1P16	100
in methanol	3.2P13	107
in ethanol	3.0P13	112
in acetic acid	4.3P13	111
in chloroform	1.4P4	45
$C_6H_5NH_2 + C_6H_5COCH_2Br$		
in benzene	91	34

Data: principally J. Nicholas (1976). *Chemical kinetics*, Harper and Row, New York. A. A. Frost and R. G. Pearson (1961). *Kinetics and mechanism*. Wiley, New York.

Table 30.1. Arrhenius parameters for gas-phase reactions

	$A/M^{-1} s^{-1}$			
	Experiment	Theory	E_a/ kJ mol^{-1}	P
$2NOCl \rightarrow 2NO + Cl_2$	9.4P9	5.9P10	102.0	0.16
$2NO_2 \rightarrow 2NO + O_2$	2.0P9	4.0P10	111.0	5.0N2
$2ClO \rightarrow Cl_2 + O_2$	6.3P7	2.5P10	0.0	2.5N3
$H_2 + C_2H_4 \rightarrow C_2H_6$	1.24P6	7.3P11	180	1.7N6
$K + Br_2 \rightarrow KBr + Br$	1.0P12	2.1P11	0.0	4.8

Data: principally M. J. Pilling (1974). *Reaction kinetics*. Clarendon Press, Oxford.

Table 30.2. Arrhenius parameters for reactions in solution. See Table 28.3.

Table 31.1. Maximum observed enthalpies of physisorption, $-\Delta_{ad}H^{\ominus}$/kJ mol^{-1}

H_2 84	N_2 21	O_2 21	Cl_2 36	CO 25
CO_2 25	H_2O 29	H_2O 57	NH_3 38	
CH_4 21	C_2H_2 38	C_2H_4 34		

Data: D. O. Haywood and B. M. W. Trapnell (1964). *Chemisorption*. Butterworths, London.

Table 31.2. Enthalpies of chemisorption, $-\Delta_{ad}H^{\ominus}$/kJ mol^{-1}

Adsorbate	Adsorbent (substrate)											
	Ti	Ta	Nb	W	Cr	Mo	Mn	Fe	Co	Ni	Rh	Pt
H_2		188			188	167	71	134			117	
N_2		586						293				
O_2						720					494	293
CO	640							192	176			
CO_2	682	703	552	456	339	372	222	225	146	184		
NH_3				301				188		155		
C_2H_4		577		427	427			285		243	209	

Data: D. O. Haywood and B. M. W. Trapnell, *Chemisorption*, Butterworths, London.

Table 31.3. Activation energies of catalysed reactions

	Catalyst	E_a/kJ mol^{-1}
$2HI \rightarrow H_2 + I_2$	None	184
	Au (s)	105
	Pt (s)	59
$2NH_3 \rightarrow N_2 + 3H_2$	None	350
	W (s)	162
$2N_2O \rightarrow 2N_2 + O_2$	None	245
	Au (s)	121
	Pt (s)	134
$(C_2H_5)_2O$ pyrolysis	None	224
	I_2 (g)	144

Data: G. C. Bond (1986). *Heterogeneous catalysis*. Clarendon Press, Oxford.

Table 32.1. Exchange current densities and transfer coefficients at $\mathcal{T}$

Reaction	Electrode	j_0/A cm^{-2}	α
$2H^+ + 2e^- \rightarrow H_2$	Pt, HCl	7.9N4	
	Cu, HCl	1N6	
	Ni, HCl	6.3N6	0.58
	Hg, HCl	7.9N13	0.50
	Pb, H_2SO_4	5.0N12	
$Fe^{3+} + e^- \rightarrow Fe^{2+}$	Pt	2.5N3	0.58
$Ce^{4+} + e^- \rightarrow Ce^{3+}$	Pt	4.0N5	0.75

Data: principally J. O'M. Bockris and A. K. N. Reddy (1970). *Modern electrochemistry*, Plenum, New York.

Table 32.2. Half-wave potentials at $\mathcal{T}$, $E_{\frac{1}{2}}$/V

Cd^{2+}	−0.60	Co^{2+}	−1.4	Cr^{3+}	−0.91	Cu^{2+}	+0.04
Fe^{3+}	0.0	Fe^{2+}	−1.3	Tl^+	−0.46	Zn^{2+}	−1.0

Values are relative to the calomel electrode, and depend on the supporting electrolyte. Data: KL.

Character tables

C_{2v}, 2*mm*	E	C_2	σ_v	σ'_v	$h = 4$	
A_1	1	1	1	1	z, z^2, x^2, y^2	
A_2	1	1	−1	−1	xy	R_z
B_1	1	−1	1	−1	x, xz	R_y
B_2	1	−1	−1	1	y, yz	R_x

C_{4v}, 4*mm*	E	C_2	$2C_4$	$2\sigma_v$	$2\sigma_d$	$h = 8$	
A_1	1	1	1	1	1	$z, z^2, x^2 + y^2$	
A_2	1	1	1	−1	−1		R_z
B_1	1	1	−1	1	−1	$x^2 - y^2$	
B_2	1	1	−1	−1	1	xy	
E	2	−2	0	0	0	$(x, y), (xz, yz)$	(R_x, R_y)

C_{3v}, 3*m*	E	$2C_3$	$3\sigma_v$	$h = 6$	
A_1	1	1	1	$z, z^2, x^2 + y^2$	
A_2	1	1	−1		R_z
E	2	−1	0	$(x, y), (xy, x^2 - y^2)(xz, yz)$	(R_x, R_y)

C_{5v}	E	$2C_5$	$2C_5^2$	$5\sigma_v$	$h = 10$, $\alpha = 72°$	
A_1	1	1	1	1	$z, z^2, x^2 + y^2$	
A_2	1	1	1	−1		R_z
E_1	2	$2\cos\alpha$	$2\cos 2\alpha$	0	$(x, y), (xz, yz)$	(R_x, R_y)
E_2	2	$2\cos 2\alpha$	$2\cos\alpha$	0	$(xy, x^2 - y^2)$	

C_{6v}, 6*mm*	E	C_2	$2C_3$	$2C_6$	$3\sigma_d$	$3\sigma_v$	$h = 12$	
A_1	1	1	1	1	1	1	$z, z^2, x^2 + y^2$	
A_2	1	1	1	1	−1	−1		R_z
B_1	1	−1	1	−1	−1	1		
B_2	1	−1	1	−1	1	−1		
E_1	2	−2	−1	1	0	0	$(x, y), (xz, yz)$	(R_x, R_y)
E_2	2	2	−1	−1	0	0	$(xy, x^2 - y^2)$	

$C_{\infty v}$	E	C_2	$2C_\phi$	σ_v	$h = \infty$	
$A_1(\Sigma^+)$	1	1	1	1	$z, z^2, x^2 + y^2$	
$A_2(\Sigma^-)$	1	1	1	−1		R_z
$E_1(\Pi)$	2	−2	$2\cos\phi$	0	$(x, y), (xz, yz)$	(R_x, R_y)
$E_2(\Delta)$	2	2	$2\cos 2\phi$	0	$(xy, x^2 - y^2)$	
...	.	.	.	.	...	

$D_2, 222$	E	C_2^z	C_2^y	C_2^x	$h=4$	
A	1	1	1	1	x^2, y^2, z^2	
B_1	1	1	−1	−1	z, xy	R_z
B_2	1	−1	1	−1	y, xz	R_y
B_3	1	−1	−1	1	x, yz	R_x

$D_3, 32$	E	$2C_3$	$2C_2'$	$h=6$	
A_1	1	1	1	z^2, x^2+y^2	
A_2	1	1	−1	z	R_z
E	2	−1	0	$(x, y), (xz, yz)(xy, x^2-y^2)$	(R_x, R_y)

$D_4, 422$	E	C_2	$2C_4$	$2C_2'$	$2C_2''$	$h=8$	
A_1	1	1	1	1	1	z^2, x^2+y^2	
A_2	1	1	1	−1	−1	z	R_z
B_1	1	1	−1	1	−1	x^2-y^2	
B_2	1	1	−1	−1	1	xy	
E	2	−2	0	0	0	$(x, y), (xz, yz)$	(R_x, R_y)

$D_{3h}, \bar{6}m2$	E	σ_h	$2C_3$	$2S_3$	$3C_2'$	$3\sigma_v$	$h=12$	
A_1'	1	1	1	1	1	1	z^2, x^2+y^2	
A_2''	1	1	1	1	−1	−1		R_z
A_1''	1	−1	1	−1	1	−1		
A_2'	1	−1	1	−1	−1	1	z	
E′	2	2	−1	−1	0	0	$(x, y), (xy, x^2-y^2)$	
E″	2	−2	−1	1	0	0	(xz, yz)	(R_x, r_y)

$D_{\infty h}$	E	$2C_\phi$	C_2'	i	$2iC_\phi$	iC_2'	$h=\infty$	
$A_{1g}(\Sigma_g^+)$	1	1	1	1	1	1	z^2, x^2+y^2	
$A_{1u}(\Sigma_u^+)$	1	1	1	−1	−1	−1		
$A_{2g}(\Sigma_g^-)$	1	1	−1	1	1	−1		R_z
$A_{2u}(\Sigma_u^-)$	1	1	−1	−1	1	1	z	
$E_{1g}(\Pi_g)$	2	$2\cos\phi$	0	0	$-2\cos\phi$	0	(xz, yz)	(R_x, R_y)
$E_{1u}(\Pi_u)$	2	$2\cos\phi$	0	0	$2\cos\phi$	0	(x, y)	
$E_{2g}(\Delta_g)$	2	$2\cos 2\phi$	0	0	$2\cos 2\phi$	0	(xy, x^2-y^2)	
$E_{2u}(\Delta_u)$	2	$2\cos 2\phi$	0	0	$-2\cos 2\phi$	0		

$T_d, \bar{4}3m$	E	$8C_3$	$3C_2$	$6\sigma_d$	$6S_4$	$h=24$	
A_1	1	1	1	1	1	$x^2+y^2+z^2$	
A_2	1	1	1	−1	−1		
E	2	−1	2	0	0	$(3z^2-r^2, x^2-y^2)$	
T_1	3	0	−1	−1	1		
T_2	3	0	−1	1	−1	(x, y, z) (xy, xz, yz)	(R_x, R_y, R_z)

$O, 432$	E	$8C_3$	$3C_2$	$6C_2'$	$6C_4$	$h=24$	
A_1	1	1	1	1	1	$x^2+y^2+z^2$	
A_2	1	1	1	−1	−1		
E	2	−1	2	0	0	$(x^2-y^2, 3z^2-r^2)$	
T_1	3	0	−1	−1	1	(x, y, z)	(R_x, R_y, R_z)
T_2	3	0	−1	1	−1	(xy, yz, zx)	

$O_h, m3m$	E	$8C_3$	$6C_2$	$6C_4$	$3C_2$	i	$6S_4$	$8S_6$	$3\sigma_h$	$6\sigma_d$	$h = 48$	
A_{1g}	1	1	1	1	1	1	1	1	1	1	$x^2 + y^2 + z^2$	
A_{2g}	1	1	−1	−1	1	1	−1	1	1	−1		
E_g	2	−1	0	0	2	2	0	−1	2	0	$(2z^2 - x^2 - y^2, x^2 - y^2)$	
T_{1g}	3	0	−1	1	−1	3	1	0	−1	−1		(R_x, R_y, R_z)
T_{2g}	3	0	1	−1	−1	3	−1	0	−1	1	(xz, yz, xy)	
A_{1u}	1	1	1	1	1	−1	−1	−1	−1	−1		
A_{2u}	1	1	−1	−1	1	−1	1	−1	−1	1		
E_u	2	−1	0	0	2	−2	0	1	−2	0		
T_{1u}	3	0	−1	1	−1	−3	−1	0	1	1	(x, y, z)	
T_{2u}	3	0	1	−1	−1	−3	1	0	1	−1		

Answers to introductory problems

1 **(1.1) (a)** 2.57×10^3 Torr; **(b)** 3.38 atm. **(1.2)** 401 K. **(1.3)** 4.27 kPa. **(1.4) (a)** 3.14 dm^3; **(b)** 3.47 kPa; **(c)** 28.3 kPa. **(1.5)** 169. **(1.6) (a)** 0.88; **(b)** 1.2 dm^3; attractive. **(1.7) (a)** 9.0 cm^3, **(b)** -0.15 $dm^3\ mol^{-1}$. **(1.8)** 206 K, 277 pm. **(1.9) (a)** 12.5 $dm^3\ mol^{-1}$; **(b)** 12.4 $dm^3\ mol^{-1}$. **(1.10) (a)** 1.407×10^3 K, 0.281 nm.

2 **(2.1)** $w' = -w = -123$ J, $q = -123$ J. **(2.2)** $w = -87.8$ J, $q = 87.8$ J, $\Delta U = 0$; -167 J, 167 J, 0. **(2.3)** 28.6 J. **(2.4)** $q = 2.20$ kJ, $\Delta H = 2.20$ kJ, $\Delta U = 1.58$ kJ. **(2.5)** 403 K. **(2.6)** $C_{p,\mathrm{m}} = 29.9$ J $K^{-1}\ mol^{-1}$, $C_{V,\mathrm{m}} = 21.6$ J $K^{-1}\ mol^{-1}$. **(2.7)** $w = -37.0$ J, $q = 120$ J, $\Delta H = 353$ J. **(2.8)** $w = 0$, $\Delta U = 2.35$ J, $\Delta H = 3.04$ kJ. **(2.9)** $q = -1200$ J, $\Delta H = -1200$ J, $C_p \approx 80$ J K^{-1}. **(2.10)** $q = 13.0$ kJ, $w = -1.04$ kJ, $\Delta H = 13.0$ kJ, $\Delta U = 12.0$ kJ.

3 **(3.3)** $(\partial H/\partial U)_p = 1 + p(\partial V/\partial U)_p$. **(3.4)** $\mathrm{d}\ln V = -\kappa \mathrm{d}p + \alpha \mathrm{d}T$. **(3.5)** 1.3×10^{-3} K^{-1}. **(3.6)** 1.09×10^3 atm **(3.7)** -7.2 J $atm^{-1}\ mol^{-1}$, 8.1 kJ. **(3.8)** $q = 0$, $w = -3.19$ kJ, $\Delta T = -37.8$ K, $\Delta U = -3.19$ kJ, $\Delta H = -4.45$ kJ. **(3.9)** $q = 0$, $w = 4.12$ kJ, $\Delta U = 4.12$ kJ, $\Delta H = 5.37$ kJ, $V_\mathrm{f} = 11.8$ dm^3, $p = 5.28 \times 10^5$ Pa. **(3.10)** $V_\mathrm{f} = 9.47$ dm^3, $T_\mathrm{f} = 288$ K, $w = -4.58 \times 10^2$ J.

4 **(4.1)** -392.1 kJ mol^{-1}, -946.6 kJ mol^{-1}, $+52.4$ kJ mol^{-1}. **(4.2)** -4.564×10^3 kJ mol^{-1}. **(4.3)** -126 kJ mol^{-1}. **(4.4)** -432 kJ mol^{-1}. **(4.5)** -87.29 kJ mol^{-1}. **(4.6)** 1.58 kJ K^{-1}, 2.05 K. **(4.7)** 8.48×10^4 kJ. **(4.8)** 3.66×10^{-2} m^3. **(4.9)** -146 kJ. **(4.10)** 65.4 kJ mol^{-1}.

5 **(5.1)** 54.8 kJ, -194 J K^{-1}. **(5.2)** 26.0 J K^{-1}. **(5.3)** 6.57 dm^3. **(5.4)** 87.8 J $K^{-1}\ mol^{-1}$ **(5.5)** 9.83 J K^{-1}, -9.43 J K^{-1}. **(5.6)** (a) -385.8 J K^{-1}, (b) 92.6 J K^{-1}. **(5.7)** -1007 kJ. **(5.8)** -160.2 kJ mol^{-1}. **(5.9)** (b). **(5.10)** -49.7 kJ mol^{-1}.

6 **(6.1)** $(\partial S/\partial p)_T = -\alpha V$, $(\partial S/\partial V)_T = \alpha/\kappa$. **(6.2)** $(\partial p/\partial S)_V = \alpha T/\kappa C_V$, $(\partial V/\partial S)_p = \alpha TV/C_p$. **(6.3)** -36.5 J. **(6.4)** -3.8 J. **(6.5)** 8.8×10^2 kg m^{-3}. **(6.6)** -501.1 kJ. **(6.7)** $p_\mathrm{f} = 15.8$ atm, 8.25 kJ. **(6.8)** 5.9 kJ mol^{-1}. **(6.9)** 7.27 kJ mol^{-1}. **(6.10)** -2.64×10^{-8} Pa^{-1}, 0.875.

7 **(7.1)** 45.23 J $K^{-1}\ mol^{-1}$, 15.86 kJ mol^{-1}. **(7.2)** 20.80 kJ mol^{-1}, $2501.8(T/\mathrm{K})^2$. **(7.3)** 303.0 K. **(7.4)** 5.40 kPa K^{-1}, 2.88%. **(7.5)** 201 K. **(7.6)** 3.15 kJ. **(7.7)** 1.69×10^{-2} J. **(7.8)** 2.05×10^{-2} N m^{-1}. **(7.9)** 208 nm. **(7.10)** 99.3 kPa.

8 **(8.1)** 6.41×10^3 kPa. **(8.2)** 8.397 kJ mol^{-1}, solution $\rightarrow$ gas. **(8.3)** $K_\mathrm{A} = 15.58$ kPa, $K_\mathrm{B} = 47.03$ kPa. **(8.5)** 272.41 K. **(8.6)** 381 g mol^{-1}. **(8.7)** 273.06 K. **(8.8)** 440 Torr, $x_\mathrm{A} = 0.268$, $x_\mathrm{B} = 0.732$. **(8.9)** $\mu_1 = \mu_1^\ominus + RT\ln x_1 + RTg_0x_2^2$. **(8.10)** $-1.841\,RT$.

9 **(9.10) (b)** Between 29 and 83 mol% $C_2H_5NO_2$.

10 **(10.1)** -242 kJ mol^{-1}. **(10.2)** 3.02. **(10.3)** 1.28×10^3 K. **(10.4)** 2.77 kJ mol^{-1}, -16.5 J $K^{-1}\ mol^{-1}$. **(10.5)** 12 kJ mol^{-1}. **(10.6)** -41.0 kJ mol^{-1}. **(10.7)** $\Delta_\mathrm{f}H^\ominus = -\frac{3}{2}R\{1.464 \times 10^4 - 5.65(T/\mathrm{K})\}$, $8.475R$. **(10.8)** $\Delta_\mathrm{r}G^\ominus = \{7.814 \times 10^4 - 161.5(T/\mathrm{K})\}$ J mol^{-1}. **(10.9) (a)** -50%; **(b)** 0. **(10.10)** 0.904.

11 **(11.1)** 0.32. **(11.2)** 2.07. **(11.3) (a)** 2.7g; **(b)** 2.9 g. **(11.4)** 6.216. **(11.5)** 0.650. **(11.6)** 34.6. **(11.7)** 1.34×10^4. **(11.8) (a)** 1.75; **(b)** 1.67; **(c)** 1.57. **(11.9)** -1103 kJ mol^{-1}. **(11.10)** 34.2 mV.

12 **(12.1)** Mn | $MnCl_2$(aq) | Cl_2(g) | Pt, -1.18 V. **(12.2) (a)** Zn(s) + 2Ag^+(aq) $\rightarrow$ Zn^{2+}(aq) + 2Ag(s); **(b)** Cd(s) + 2H^+(aq) $\rightarrow$ Cd^{2+}(aq) + H_2(g); **(c)** 3$Fe(CN)_6^{4-}$(aq) + Cr^{3+}(aq) $\rightarrow$ 3$Fe(CN)_6^{3-}$ + Cr(s); **(d)** Ag_2CrO_4(s) + 2Cl^-(aq) $\rightarrow$ 2Ag(s) + Cl_2(g) + CrO_4^{2-}(aq). **(12.3)** -1.18 V. **(12.4) (a)** 0.324 V; **(b)** 0.45 V. **(12.5)** -0.32 V. **(12.6)** -1.24 V, 233.1 kJ mol^{-1}, 300.3 kJ mol^{-1}; -1.20 V, 230.8 kJ mol^{-1}. **(12.7) (a)** 9.4×10^{-8} [mol kg^{-1}], 0.13 V. **(12.8) (a)** 6.33×10^9; **(b)** 1.47×10^{12}; **(c)** 2.8×10^{-16}. **(12.9) (a)** 4.90×10^{-10}; **(b)** 4.69; **(c)** 4.96; **(d)** 9.09. **(12.10)** 3.98×10^{-6}, 3.01.

13 **(13.1)** 1.72 MW. **(13.2)** 5.2×10^{16} Hz. **(13.3)** 1.6×10^6 m s^{-1}. **(13.4) (a)** 8.84×10^{-28} kg m s^{-1}; **(b)** 9.5×10^{-24} kg m s^{-1}; **(c)** 3.4×10^{-35} kg m s^{-1}. **(13.5)** 50.6 nm. **(13.6)** 72.38 pm. **(13.7)** 702 pm. **(13.8)** 6.10 nm. **(13.9)** $8.05 \times 10^{-2}L$ or $0.920L$. **(13.10)** 5.9×10^{-24} J.

14 **(14.1)** 7.7×10^{-2}. **(14.2)** 3. **(14.3)** 24%. **(14.4)** 4.31×10^{-21} J. **(14.5)** 277 N m^{-1}. **(14.6)** 2.63 μm. **(14.7)** 1.09 μm. **(14.8)** $\pm\infty$, $y = 1$. **(14.9)** 146 pm. **(14.10)** 1.49×10^{-34} J s; 0, $\pm\hbar$.

15 **(15.1)** 486.276. **(15.2)** 6, $J = 2$. **(15.3)** 6842 cm^{-1}. **(15.4)** $4a_0$. **(15.5)** 376 pm. **(15.6)** 1.49×10^{-34} J s. **(15.7)** $L = 2$, $S = 0$, $J = 2$. **(15.8)** 18 pm. **(15.9) (a)** 110 pm; **(b)** 86.3 pm and 29 pm. **(15.10)** 2.

16 **(16.1)** 1, 0, 2. **(16.2)** C_2. **(16.3)** $S = \frac{1}{2}$, $\Lambda = 0, \ldots, (2p\sigma_g)^1$. **(16.4)** 3, u. **(16.5)** NO. **(16.6)** $(1s\sigma_g)^1(2p\pi_u)^1$. **(16.8)** $\psi \propto (\rho/\sqrt{2}) \sin\theta(-\cos\phi + 3^{\frac{1}{2}}\sin\phi)$. **(16.9)** 3.7%, 92.6%.

17 **(17.1)** 4. **(17.2)** C_3, $3\sigma_v$. **(17.3)** $C_2H_5NO_2$, CH_3Cl. **(17.4)** 0. **(17.7)** Yes. **(17.9)** $D_{2h}(i)$, $C_{3h}(C_3, \sigma_h = S_3)$, $T_h(i)$, $T_d(S_4)$.

18 **(18.1)** **(a)** 1.6×10^{-27} kg; **(b)** 3.1×10^{-27} kg; **(c)** 4.6×10^{-26} kg. **(18.2)** 3.45×10^{-45} kg m^2. **(18.3)** 2.45×10^{-45} kg m^2. **(18.4)** 20 475 cm^{-1}. **(18.5)** 2598.19 cm^{-1}. **(18.6)** 1.1%. **(18.7)** 330 N m^{-1}. **(18.8)** $1 \rightarrow 0$; 381.3 cm^{-1}, 378 cm^{-1}. **(18.9)** 2.177 eV. **(18.10)** $4A_1$ (ir, raman) + A_2 (raman) + $2B_1$ (ir, raman) + $2B_2$ (ir, raman).

19 **(19.1)** 79.8%. **(19.2)** 1.79×10^6 cm^2 mol^{-1}. **(19.3)** 3.36 mmol dm^{-3}. **(19.4)** 4.8×10^{-4} M. **(19.5)** 0.102. **(19.6)** 0.0104. **(19.7)** strong: 0.91, 0.75; weak: 2.9×10^{-2}; forbidden: 6.2×10^{-6}, 3.2×10^{-9}. **(19.10)** H_2^+.

20 **(20.1)** $\pm 8.124 \times 10^{-27}$ J. **(20.2)** 6.116×10^{-26} J. **(20.3)** 3.523 T. **(20.4)** Merge. **(20.5)** 0.231 T. **(20.6)** $\frac{3}{2}$. **(20.8)** 215 K. **(20.9)** 1.41 mm s^{-1}. **(20.10)** -22.1 MHz, Sn(IV).

21 **(21.1)** **(a)** 2.58×10^{27}; **(b)** 3.97×10^{27}. **(21.2)** 2.83. **(21.3)** 3.156. **(21.4)** 2.46 kJ mol^{-1}. **(21.5)** 354 K. **(21.7)** 146 J K^{-1}. **(21.8)** 1.14 J K^{-1} mol^{-1}. **(21.9)** 151 J K^{-1} mol^{-1}.

22 **(22.1)** **(a)** 1; **(b)** 2; **(c)** 2; **(d)** 12. **(22.2)** 72.114. **(22.3)** 1.116×10^4. **(22.4)** $\tilde{\nu} \geqslant 2400$ cm^{-1}. **(22.5)** 267.2, 1.088; 14.15 kJ mol^{-1}. **(22.6)** 1.96 J K^{-1} mol^{-1}, 1.60 J K^{-1} mol^{-1}. **(22.7)** 3, -3.65 kJ mol^{-1}. **(22.8)** 11.5 J K^{-1} mol^{-1}. **(22.9)** 83.1 J K^{-1} mol^{-1}. **(22.10)** 0.2500.

23 **(23.1)** 70.7 pm. **(23.2)** 0.21 cm. **(23.3)** 3.96×10^{-28} m^3. **(23.4)** 4, 4.01 g cm^{-3}. **(23.5)** 190 nm. **(23.6)** 240 pm, 606 pm, 395 pm. **(23.7)** (111), (200), (220), (311); 402 pm. **(23.9)** 2.68×10^{-20} J. **(23.10)** 4.64 kV.

24 **(24.1)** 220 pF. **(24.2)** 1.66 D, 1.01×10^{-39} J^{-1} C^2 m^2. **(24.3)** 4.75. **(24.4)** Structure 2. **(24.5)** $2\frac{1}{2}\mu$. **(24.7)** 1.34. **(24.8)** 1.42×10^{-39} J^{-1} C^2 m^2. **(24.9)** 3. **(24.10)** 0, 4.9 μ_B.

25 **(25.1)** 2.32×10^4 g mol^{-1}, 1.03×10^3 M^{-1}. **(25.2)** 6.7×10^{-3} M. **(25.3)** 29 mV. **(25.4)** 3.39×10^6 g mol^{-1}. **(25.5)** 23.8 nm. **(25.6)** 6.74×10^3. **(25.7)** 0.0715 dm^3 g^{-1}. **(25.8)** 4.81×10^{-13} s. **(25.9)** 31 000 g mol^{-1}. **(25.10)** **(a)** 1.76×10^4; **(b)** 1.95×10^4.

26 **(26.1)** 324 m s^{-1}. **(26.2)** 0.91%. **(26.3)** 1.9×10^{20}. **(26.4)** 1.95×10^{26}. **(26.5)** 206 K. **(26.6)** 104 mg. **(26.7)** -41 mJ m^{-2} s^{-1}. **(26.8)** 0.0563 nm^2. **(26.9)** 0.142 nm^2. **(26.10)** 205 kPa.

27 **(27.1)** 1.38 Ω m. **(27.2)** 1.62 mS cm^{-1}. **(27.3)** 116.7 S cm^2 mol^{-1}. **(27.4)** 66.1 S cm^2 s^{-1}. **(27.5)** 0.347 mm s^{-1}. **(27.6)** 0.331. **(27.7)** 9.9×10^4. **(27.8)** 1.90×10^{-9} m^2 s^{-1}. **(27.9)** 5.5 nm. **(27.10)** 3.94×10^3 s.

28 **(28.1)** -1.96×10^{-4} M s^{-1}, -3.92×10^{-4} M s^{-1}, 1.96×10^{-4} M s^{-1}, -6.70×10^{-4} mol s^{-1}, 3.35 mol s^{-1}. **(28.2)** 225 s. **(28.3)** 2.49×10^5 s. **(28.4)** 64.9 kJ mol^{-1}, 4.3×10^8 M^{-1} s^{-1}. **(28.5)** $[H_2O]^1[M^-]^1$. **(28.6)** rate = $k_2(k_1/2k_{-1})^{\frac{1}{2}}[A_2]^{\frac{1}{2}}[B]$. **(28.7)** rate = $k_2(k_1/k_{-1})[A][B]$. **(28.8)** 1.52×10^{-3} M s^{-1}. **(28.9)** $[S] = K_M$. **(28.10)** 1.89×10^{-6} Pa^{-1} s^{-1}.

29 **(29.1)** Initiation: 1; Propagation: 3, 4, 5, 6, 7; Termination: 2, 8. **(29.2)** $1/f = 1/k_4 + \{k_3/k_2k_4[CO]\}$ [M]. **(29.3)** $d[R_2]/dt = -k_1[R_2] - k_2(k_1/k_4)^{\frac{1}{2}}[R_2]^{\frac{3}{2}}$. **(29.4)** 282–2240 Pa, 141–8910 Pa, $\geqslant$112 Pa. **(29.5)** 3.27×10^{18}. **(29.6)** 0.521 mol einstein^{-1}. **(29.7)** $d[P]/dt = k_1k_3[A][AH][B]/(k_2[BH^+] + k_3[A])$. **(29.8)** $d[AH]/dt = k_3(k_1/k_2) \times K_i[HA][HB^+]$. **(29.9)** 7.2×10^5 s^{-1}, 7.7 ns. **(29.10)** 1.66×10^{-7} s^{-1}, 8.3×10^8 M^{-1} s^{-1}.

30 **(30.1)** 0.152 nm^2. **(30.2)** 1.2×10^{-3}. **(30.3)** 1.87×10^8 M s^{-1}. **(30.4)** 3.50×10^8 M^{-1} s^{-1}. **(30.5)** 72.17 kJ mol^{-1}, 1.5 J K^{-1} mol^{-1}. **(30.6)** 73.42 kJ mol^{-1}, 63.81 kJ mol^{-1}. **(30.7)** -90 J K^{-1} mol^{-1}. **(30.8)** 20.9 M^{-2} min^{-1}. **(30.9)** 0.66 M^{-1} min^{-1}. **(30.10)** lg (rate) = lg $(k_r[B])$ + lg $[H^+]$.

31 **(31.1)** 13 kPa. **(31.2)** 3.4×10^5 s^{-1}. **(31.3)** Chemisorption, 48 s. **(31.4)** **(a)** 0.21 kPa; **(b)** 22 kPa. **(31.5)** 82.9%, 35.8%. **(31.6)** 14 kPa. **(31.7)** 0, 1.

32 **(32.1)** 138 mV. **(32.2)** 2.83 mA cm^{-2}. **(32.3)** 0.99 A m^{-2}.

Answers to main problems

1 **(1.1)** 10 atm. **(1.2) (a)** 15 m^2; **(b)** 150 m^3. **(1.3)** 0.5 m^3. **(1.4)** 1.5%. **(1.5)** 30 K. **(1.6)** 29 lb in^{-2}. **(1.7)** 0.03 atm. **(1.8)** $p = \rho RT/M_m$, plot p/ρ vs p, 45.9. **(1.9)** 4622 mol, 129 kg, 120 kg, 124 kg (H_2), 115 kg (He). **(1.10) (a)** 2.7×10^{-26} kg; **(b)** 16.4. **(1.11)** 102, CH_2FCF_3. **(1.12)** 0.184 Torr, 68.6 Torr, 0.184 Torr. **(1.13)** H_2: 0.67, 2.0 atm; N_2: 0.33, 1.0 atm; $p = 31.0$ atm. **(1.14)** H_2: 0; N_2: 0.33 atm; NH_3: 1.33 atm. **(1.15)** 24 atm. **(1.16)** 22 atm. **(1.17) a(i)** 1.00 atm; **a(ii)** 821 atm; **b(i)** 1.003 atm; **b(ii)** 1468 atm. **(1.18) (a)** 8.69 atm, 3640 K; **(b)** 4.53 atm, 2600 K; **(c)** 0.18 atm, 46.7 K. **(1.19)** 54.5 atm, 67.8 $cm^3\ mol^{-1}$ 120 K. **(1.20)** 0.055 nm^3, 0.24 nm. **(1.21)** 226 pm (He), 202 pm (Ne), 246 pm (Ar), 287 pm (Xe). **(1.22)** 5.65 dm^6 atm mol^{-2}, 59.4 $cm^3\ mol^{-1}$, 20.6 atm. **(1.23)** $B = b - a/RT$, $C = b^2$. **(1.24)** $B = b - a/RT$, $C = b^2 - ab/RT + a^2/2R^2T^2$. **(1.25)** 25.8 atm, 87.8 $cm^3\ mol^{-1}$, 102 K. **(1.26)** $p_c = B^3/27C^2$, $V_{m,c} = 3C/B$, $T_c = B^2/3RC$, $Z_c = \frac{1}{3}$. **(1.27)** $B' = B/RT$, $C' = (C - B^2)/R^2T^2$. **(1.28)** a/Rb, 1000 K. **(1.29) (a)** 27/8; **(b)** 4. **(1.30)** $B' = 8.3 \times 10^{-7}\ Pa^{-1}$, $B = 2.07\ dm^3\ mol^{-1}$. **(1.32) (a)** 0.999 98p_0; **(b)** 0.95 atm. **(1.34)** $1.50 \times 10^{-23}\ J\ K^{-1}$, $5.5 \times 10^{23}\ mol^{-1}$.

2 **(2.1) (a)** 9.8 J; **(b)** 1.6 J. **(2.2)** 2.0 kJ. **(2.3)** 10 J. **(2.4)** 200 kJ, 200 W, 68°C. **(2.5) (a)** $2Fa/\pi$; **(b)** 0. **(2.6)** 100 J. **(2.7)** 4.9 J. **(2.8)** 4.9 J. **(2.9) (a)** −0.27 kJ; **(b)** −0.94 kJ. **(2.10)** −8.1 kJ. **(2.11)** −8.1 kJ. **(2.12)** −1.53 kJ. **(2.13)** $w = -nRT\{\ln (V_f/V_i) - nB[(1/V_f) - (1/V_i)] + \ldots\}$. **(2.14) (a)** −1.53 kJ; **(b)** −51 J; **(c)** −1.57 kJ. **(2.15)** $w = -nRT \ln\{(V_f - nb)/(V_i - nb)\} + n^2a(V_f - V_i)/V_iV_f$. **(2.17)** $w = -(8nT_r/9) \ln\{(3V_{r,f} - 1)/(3V_{r,i} - 1)\} + n(V_{r,f} - V_{r,i})/V_{r,i}V_{r,f}$. **(2.18) (a)** p/T, $-p/V$; **(b)** $(1 + na/RTV)p/T$, $\{(na/RTV) - V/(V - nb)\}p/V$. **(2.19) (a)** 1%; **(b)** 2%; **(c)** 3%. **(2.20) (a)** $\frac{3}{2}nR$; **(b)** $3nR$. **(2.21)** 18.02 cal $K^{-1}\ mol^{-1}$, 75.4 J $K^{-1}\ mol^{-1}$. **(2.22) (a)** 0.12 kJ; **(b)** 0.21 kJ, 0, −80 J. **(2.23)** 670 kJ, 670 s. **(2.24)** 320 kJ, 320 s. **(2.25) (a)** −170 kJ; **(b)** 2.3 MJ; **(c)** 2.1 MJ; **(a)** 2.3 MJ. **(2.26) (a)** −150 kJ; **(b)** 2.5 MJ; **(c)** 2.3 MJ; **(d)** 2.5 MJ. **(2.27) (a)** 3.8 kW, 24 km; **(b)** 4.2 kW, 26 km. **(2.28)** 0.32 MJ. **(2.29)** 22.2 kJ, 1.7 kJ, 20.5 kJ, 22.2 kJ, 40 kJ mol^{-1}. **(2.30)** 98.9 kJ mol^{-1}, 96.2 kJ mol^{-1}.

3 **(3.1)** Mass, enthalpy, heat capacity. **(3.2)** 0, 0. **(3.3)** $2.3 \times 10^{-11}\ N^{-1}\ m^2$; **(a)** −7.0 mm^3; **(b)** −2.1 cm^3. **(3.4)** −2.6 cm^3; **(a)** −2.6 cm^3; **(b)** −47 cm^3. **(3.6)** $p = p_i e^{-t/\tau}$, $1/\tau = 1/\tau_T - 1/\tau_V$, $p = p_i$ when $\tau_T = \tau_V$. **(3.7)** $d \ln p = -d \ln (V - nb) + d \ln T + (n^2a/pV^2)\{(1 - 2nb/V)\, d \ln (V - nb) + d \ln T\}_S$, $V \to nb$. **(3.9)** $\kappa = V^2(V - nb)^2/\{nRTV^3 - 2n^2a(V - nb)^2\}$, $\alpha = RV^2(V - nb)/\{RTV^3 - 2an(V - nb)^2\}$. **(3.10)** $\mu_{JT}C_p = (nb\zeta/V - 1)V/(\zeta - 1)$, $\zeta = RTV^3/2an\,(V - nb)^2$, 1.4 K atm^{-1}. **(3.11)** $T_I = (2a/bR)(1 - nb/V)^2$, $T_{I,r} = (27/4)(1 - 1/3V_r)^2$; **(a)** 224 K; **(b)** 2053 K. **(3.12)** −4.2 atm. **(3.13)** 1.3 K atm^{-1}, yes. **(3.15)** 0.29 K atm^{-1}. **(3.16) (a)** 0.9 mm^3; **(b)** 0.02 mm^3. **(3.17)** 9.2 J $K^{-1}\ mol^{-1}$. **(3.18) (a)** 0.731 J $K^{-1}\ mol^{-1}$; **(b)** 39.8 J $K^{-1}\ mol^{-1}$; **(a)** 0.29 kJ; **(b)** 12.7 kJ. **(3.19) (a)** $2.06 \times 10^9\ J\ m^{-3}$; **(b)** $4.07 \times 10^8\ J\ m^{-3}$. **(3.20) (a)** 0.75 kJ; **(b)** 0.75 kJ. **(3.21)** $w = -20$ J, $q = 0$, $\Delta U = -20$ J, $\Delta H = -26$ J, $\Delta T = -0.35$ K. **(3.22) (a)** 226 K; **(b)** 238 K. **(3.23)** 31.6 J $K^{-1}\ mol^{-1}$. **(3.24)** 1.31, 41.3 J $K^{-1}\ mol^{-1}$. **(3.25)** −1.57 kJ mol^{-1}, −1.98 kJ mol^{-1}. **(3.26)** $\Delta H = C_p\Delta T$. **(3.27)** 5/3, 4/3. **(3.28)** $1 + 2\lambda/3$, $1 + \lambda/3$. **(3.29) (a)** 1.671; **(b)** 1.336. **(3.30) (a)** 1.02 km s^{-1}; **(b)** 346 m s^{-1}. **(3.31)** 1.24, 34.6 J $K^{-1}\ mol^{-1}$. **(3.32)** 357 Hz.

4 **(4.1) (a)** Exothermic; **(b, c)** endothermic. **(4.2) (a)** $\nu(CO_2) = 1$, $\nu(H_2O) = 2$, $\nu(CH_4) = -1$, $\nu(O_2) = -2$; **(b)** $\nu(C_2H_2) = 1$, $\nu(C) = -2$, $\nu(H_2) = -1$; **(c)** ν(NaCl, aq) = 1, ν(NaCl, s) = −1. **(4.3) (a)** −57.20 kJ mol^{-1}; **(b)** −176.01 kJ mol^{-1}; **(c)** −32.88 kJ mol^{-1}; **(d)** −55.84 kJ mol^{-1}. **(4.4)** 641 J K^{-1}. **(4.5) (a)** −2800 kJ mol^{-1}; **(b)** −2800 kJ mol^{-1}; **(c)** −1270 kJ mol^{-1}. **(4.6)** 17.7 kJ mol^{-1}, 72.5 kJ mol^{-1}. **(4.7)** −2130 kJ mol^{-1}, −2130 kJ mol^{-1}, −1270 kJ mol^{-1}. **(4.8)** 78.6 kJ mol^{-1}. **(4.9)** −383 kJ mol^{-1}. **(4.10) (a)** −6.20 kJ mol^{-1}; **(b)** −7.73 kJ mol^{-1}; **(c)** +7.78 kJ mol^{-1}. **(4.11)** −95.2 kJ mol^{-1}. **(4.12) (a)** 2877, 3536, 5471 kJ mol^{-1}; **(b)** 49.50, 49.01, 47.91 kJ g^{-1}. **(4.13) (a)** 2871, 3529, 5501 kJ mol^{-1}; **(b)** 49.40, 48.91, 48.17 kJ g^{-1}. **(4.14) (a)** 0.47 g; **(b)** 0.47 kg. **(4.15)** 11.3 kJ mol^{-1}. **(4.16)** 1.90 kJ mol^{-1}. **(4.17)** −56.98 kJ mol^{-1}. **(4.18)** $\Delta H(T_2) = \Delta H(T_1) + \Delta a(T_2 - T_1) + \frac{1}{2}\Delta b(T_2^2 - T_1^2) + \Delta c(T_2 - T_1)/T_1T_2$. **(4.19) (a)** −292.6 kJ mol^{-1}; **(b)** −242.7 kJ mol^{-1}. **(4.20)** −283.5, −283.2 kJ mol^{-1}. **(4.21)** $\Delta U(T_2) = \Delta U(T_1) + \int_{T_1}^{T_2} \Delta C_V(T)\, dT$. **(4.22) (a)** −1060 kJ mol^{-1}; **(b)** −2340 kJ mol^{-1}. **(4.23)** −5376 kJ mol^{-1} vs. −268 kJ mol^{-1}. **(4.24)** −25 kJ, 10 m. **(4.25)** 2.5 MJ, 160 g, 81°F. **(4.26)** −2205, −2200 kJ mol^{-1}. **(4.27)** −2203, −2198 kJ mol^{-1}. **(4.28)** 150 kJ mol^{-1}. **(4.29) (a)** KF: ΔH_m/kJ $mol^{-1} = -35.4 + 6.2m$, $\Delta H_m^\infty = -35.4$ kJ mol^{-1}; **(b)** KF.AcOH: ΔH_m/kJ $mol^{-1} = 3.1 + 0.9m$, $\Delta H_m^\infty = 3.1$ kJ mol^{-1}. **(4.30)** 1173 kJ mol^{-1}. **(4.31)** 13 kJ mol^{-1}. **(4.32)** −1890.3 kJ mol^{-1}. **(4.33)** −46 kJ mol^{-1}.

5 **(5.1) (a)** 92 J K^{-1}; **(b)** 67 J K^{-1}. **(5.2)** 45.4 J K^{-1}. **(5.3) (a)** 0; **(b)** 51.2 J K^{-1}. **(5.4)** 152.68 J $K^{-1}\ mol^{-1}$. **(5.5) (a)** 200.6 J $K^{-1}\ mol^{-1}$; **(b)** 231.9 J $K^{-1}\ mol^{-1}$. **(5.6)** +2.83 J K^{-1}. **(5.7)** +17.0 J K^{-1}. **(5.8)** +36.0 J K^{-1}. **(5.9) (a)** 2.88, −2.88, 0 J K^{-1}; **(b)** 2.88, 0, 2.88 J K^{-1}; **(c)** 0, 0, 0.

(5.10) **(a)** $\Delta S^{\text{sys}} = -21.3$ J K^{-1} mol^{-1}, $\Delta S^{\text{univ}} = +0.4$ J K^{-1} mol^{-1}; **(b)** $\Delta S^{\text{sys}} = +109.7$ J K^{-1} mol^{-1}, $\Delta S^{\text{univ}} = -1.5$ J K^{-1} mol^{-1}. **(5.11)** **(a)** -0.11 kJ mol^{-1}; **(b)** $+0.55$ kJ mol^{-1}. **(5.12)** **(a)** 109, -109, 0 J K^{-1} mol^{-1}; **(b)** 87.3, -87.3, 0 J K^{-1} mol^{-1}. **(5.13)** 11.8, 84.8 J K^{-1} mol^{-1}. **(5.14)** **(a)** 0.11 kJ mol^{-1}; **(b)** 0.11 kJ mol^{-1}. **(5.15)** 798, 803 kJ mol^{-1}. **(5.16)** **(a)** 63.88 J K^{-1} mol^{-1}; **(b)** 66.08 J K^{-1} mol^{-1}. **(5.17)** 152.0 J K^{-1} mol^{-1}. **(5.20)** 0.11, 0.38. **(5.21)** 0.10, 0.03 kg. **(5.22)** 6.9 km. **(5.24)** $\Delta S = C_p \ln (T_f^2/T_h T_c)$ with $T_f = \frac{1}{2}(T_h + T_c)$; 22.6 J K^{-1}. **(5.25)** 0.95 J K^{-1} mol^{-1}. **(5.26)** 12.1, -12.1, 0 J K^{-1} mol^{-1}. **(5.27)** **(a)** -153.1 J K^{-1} mol^{-1}; **(b)** -21.0 J K^{-1} mol^{-1}; **(c)** $+512.0$ J K^{-1} mol^{-1}. **(5.28)** **(a)** -178.7 kJ mol^{-1}; **(b)** -212.4 kJ mol^{-1}; **(c)** -5798 kJ mol^{-1}. **(5.29)** **(a)** -163.35 J K^{-1} mol^{-1}, -285.83, -237.13 kJ mol^{-1}; **(b)** $+360.8$ J K^{-1} mol^{-1}, -205.2, -117.9 kJ mol^{-1}; **(c)** -193.1 J K^{-1} mol^{-1}, -318.31, -261.0 kJ mol^{-1}. **(5.30)** 96.864, 76.9 J K^{-1} mol^{-1}. **(5.33)** 243 J K^{-1} mol^{-1}, 34.4. kJ mol^{-1}, -128 J K^{-1} mol^{-1}.

6 **(6.5)** **(a)** $(\partial H/\partial p)_T = 0$; **(b)** $(\partial H/\partial p)_T = \{nb - (2na/RT)\lambda^2\}/\{1 - (2na/RTV)\lambda^2\}$, $\lambda = 1 - nb/V$; -8.2 J atm^{-1}, -8.2 J. **(6.6)** **(a)** 300 J m^{-3}; **(b)** 30 kJ m^{-3}. **(6.8)** $(\partial C_{V,m}/\partial V)_T = (RT/V_m^2)\{(\partial^2/\partial T^2)BT\}_V$, 6×10^{-2} J K^{-1} mol^{-1}. **(6.10)** $\mu_J C_V = p - \alpha T/\kappa$. **(6.11)** $a/V_m^2 \to 0$; **(a)** 0.23 kJ m^{-3}; **(b)** 23 kJ m^{-3}. **(6.12)** anp/RTV, $p_r/T_r V_r$. **(6.14)** $nRT \ln \{(V_f - nb)/(V_i - nb)\}$. **(6.15)** $T\,dS = C_p\,dT - \alpha TV\,dp$, $q = -\alpha TV\Delta p$, -0.50 kJ. **(6.16)** **(a)** -17.0 kJ; **(b)** 8.87 kJ; **(c)** -8.1 kJ. **(6.17)** 1.0 kJ. **(6.18)** $G(p_f) - G(p_i) = \{V(p_i)/\kappa\}\{1 - e^{-\kappa(p_f - p_i)}\} \approx V\,\Delta p - \frac{1}{2}\kappa V(\Delta p)^2$. **(6.19)** **(a)** 72 J mol^{-1}; **(b)** 7.2 kJ mol^{-1}; 0.01%. **(6.20)** **(a)** 183 J mol^{-1}; **(b)** 13.8 kJ mol^{-1} vs. 18.3 kJ mol^{-1}. **(6.21)** $\Delta G(T_f) \approx \tau\,\Delta G(T_i) + (1 - \tau)\{\Delta H(T_i) - T_i\,\Delta C_p\} - T_f\,\Delta C_p \ln \tau$, $\tau = T_f/T_i$. **(6.22)** -231.9 kJ mol^{-1}, -53 J mol^{-1}. **(6.23)** **(a)** 10.7 kJ mol^{-1}; **(b)** 114 kJ mol^{-1}. **(6.24)** 5 kJ mol^{-1}. **(6.25)** 11 kJ. **(6.26)** $G(\bar{p}) = G(p_i) + p^* V_0\{1 - e^{-\bar{p}/p^*}\}$, expansion. **(6.27)** 73.1 atm. **(6.28)** $f = \gamma p$, $\gamma = Bp/RT + \{(C - B^2)/2R^2T^2\}p^2$. **(6.29)** **(a)** 0.9991 atm, **(b)** 0.999 99 atm. **(6.30)** 1.000 95 atm. **(6.31)** **(a)** 0.989 atm; **(b)** 34.4 atm. **(6.32)** $f = \gamma p$, $\gamma = 2e^{a-1}/(a + 1)$, $a^2 = 1 + 4pB/R$.

7 **(7.1)** d $\to$ g, g $\to$ d. **(7.2)** **(a, b)** 2.8668 kJ mol^{-1}. **(7.3)** **(a)** -22.00 J K^{-1} mol^{-1}; **(b)** -109.0 J K^{-1} mol^{-1}. **(7.4)** **(a)** -1.65×10^{-4} kJ mol^{-1} atm^{-1}; **(b)** $+3.04$ kJ mol^{-1} atm^{-1}. **(7.5)** 110 J mol^{-1}. **(7.6)** 0.55, 0.61 kJ mol^{-1}. **(7.7)** -7.5°C. **(7.8)** 8.7°C, **(7.9)** $+0.36$ K. **(7.10)** $\int d(\Delta H/T) = \int \Delta C_p\,d \ln T$, $\Delta H(T)/T = \Delta H(T^*)/T^* + \Delta C_p \ln (T/T^*)$. **(7.11)** 234.4 K. **(7.13)** 9.8 Torr. **(7.14)** 60.0 kJ mol^{-1}, 498 K. **(7.15)** 22 °C. **(7.16)** 86 kg/hr. **(7.17)** **(a)** 1.7 kg; **(b)** 30.8 kg; **(c)** 1.4 g. **(7.18)** **(a)** 1.7 Torr; **(b)** 1.2 kg. **(7.19)** **(a)** 357 K; **(b)** 38.2 kJ mol^{-1}. **(7.20)** **(a)** 55 kJ mol^{-1}; **(b)** 227.5 °C. **(7.21)** $1/T_h = 1/T_b + M_m gh/T_{\text{atmos}}\,\Delta H_{\text{vap,m}}$; 363 K. **(7.22)** $\Delta H_{\text{vap,m}} = a$ J mol^{-1}; 0.039 mmHg, 63.1 kJ mol^{-1}. **(7.23)** Yes, 3 mmHg. **(7.28)** **(a)** 13.7; **(b)** 8.8. **(7.29)** 6.10 kJ, 61s. **(7.30)** $w_{\min} = C_p\{(T_f - T_i) - T_h \ln (T_f/T_i)\}$. **(7.31)** 6.85 kJ, 69s, 6.85 kJ. **(7.32)** 3.7×10^{-7} J. **(7.33)** 6.7×10^{-6} J. **(7.34)** 8 s. **(7.35)** **(a)** 0; **(b)** 9.6 kJ. **(7.36)** 17 W. **(7.37)** **(a)** 1:21; **(b)** 1:21 000; **(c)** 1:2.1 $\times 10^6$. **(7.38)** 9.5 J, 9.5 J. **(7.39)** **(a)** 1.0002; **(b)** 1.02. **(7.40)** 4.8 g, 4.9 g. **(7.41)** **(a)** 1.5, 1.2 cm; **(b)** 15, 12 cm. **(7.42)** 5.8 cm, 440 Pa. **(7.43)** 15 cm. **(7.44)** $h = (2\gamma/\rho gr) \cos \theta$. **(7.46)** **(a)** 4 m s^{-1}; **(b)** 2 m s^{-1}.

8 **(8.1)** NaCl: 17.5 cm^3 mol^{-1}; H_2O: 18.1 cm^3 mol^{-1}. **(8.2)** Salt: -1.4 cm^3 mol^{-1}; water: 18.0 cm^3 mol^{-1}. **(8.5)** 886.4, 887.7 cm^3. **(8.6)** 45.7 g of each, 0.97 cm^3. **(8.10)** $V_{B,m}(x_A, x_B) = V_{B,m}(0, 1) - \int_{V_{A,m}(0,1)}^{V_{A,m}(x_A,x_B)} \{x_A/(1 - x_A)\}\,dV_{A,m}$. **(8.11)** 1.2 J K^{-1}, -350 J. **(8.12)** 2.4 J K^{-1}, -700 J; -130 J when same. **(8.13)** 4.69 J K^{-1} mol^{-1}. **(8.14)** -18.5 kJ, 62.0 J K^{-1}, 0. **(8.15)** **(a)** $x_A = x_B = \frac{1}{2}$; **(b)** $m_{\text{hex}} = 0.86 m_{\text{hept}}$. **(8.16)** **(a)** 3.4 mmol kg^{-1}; **(b)** 34 mmol kg^{-1}. **(8.17)** N_2: 0.5 mmol kg^{-1}; O_2: 0.3 mmol kg^{-1}. **(8.18)** 0.3 M. **(8.19)** 3.9×10^5 mmHg. **(8.20)** $x_T/x_X = 11.5$; $y_T/y_X = 30.3$. **(8.21)** $K_f = 5.22$ K/mol kg^{-1}, $K_b = 32.0$ K/mol kg^{-1}. **(8.22)** -1.3 °C. **(8.23)** 82. **(8.24)** ± 0.05 K. **(8.26)** $\Delta T = (a/2b)\{[1 - 4(b/a^2) \ln x_A]^{1/2} - 1\}$, $a = \Delta H_{\text{melt,m}}/RT^{*2}$, $b = \{\Delta H_{\text{melt,m}} - 2T^* \times \Delta C_{p,m}\}/RT^{*3}$. **(8.27)** -0.17 K. **(8.28)** $CaCl_2$. **(8.30)** 4. **(8.31)** $x = 0.92$. **(8.32)** 24g/kg. **(8.33)** 87 000. **(8.34)** 14 000. **(8.35)** **(a)** $y_T = 0.36$; **(b)** $y_T = 0.82$. **(8.37)** 119 °C, $y_T = 0.410$, 0.3. **(8.38)** 4. **(8.39)** 9.

9 **(9.1)** $F = 0$. **(9.2)** **(a)** 2; **(b)** 2; **(c)** 3. **(9.4)** 2, 2. **(9.5)** 1, 2; 2, 2. **(9.6)** 3, 2, 1. **(9.7)** 2, 2, 2. **(9.9)** **(a)** 2150 °C; **(b)** $y(\text{MgO}) = 0.18$, $x(\text{MgO}) = 0.35$, $l'/l = 0.42$; **(c)** 2650 °C. **(9.11)** **(a)** 5; **(b)** no liquid. **(9.13)** $T_{uc} = 122$ °C, $T_{lc} = 8$ °C. **(9.18)** **(a)** 80% by mass; **(b)** Ag_3Sn decomposes. **(9.24)** $P = 2$, 81 g water. **(9.25)** $P = 2$; (0.06, 0.82, 0.12) and (0.62, 0.16, 0.22) in ratio 3.7:1. **(9.27)** **(a)** $P = 2$, $(NH_4)_2SO_4$(s) + liq; **(b)** $P = 3$, NH_4Cl(s) + NH_4SO_4(s) + liq; **(c)** $P = 1$, liq; **(d)** invariant, saturated solution. **(9.28)** **(a)** 19.5 mol kg^{-1}; **(b)** 23.8 mol kg^{-1}.

10 **(10.1)** **(a)** -91.12 kJ mol^{-1}, $\rightarrow$; **(b)** $+594.7$ kJ mol^{-1}, $\leftarrow$; **(c)** -66.8 kJ mol^{-1}, $\rightarrow$; **(d)** $+99.8$ kJ mol^{-1}, $\leftarrow$; **(e)** -476.00 kJ mol^{-1}, $\rightarrow$. **(10.2)** b, d. **(10.3)** **(a)** 53 kJ mol^{-1}; **(b)** -53 kJ mol^{-1}. **(10.4)** % change $= 9.5/\ln K$ **(10.5)** **(a)** 775 (K); **(b)** 6.0×10^5 (K^2); **(c)** 1.3×10^{-3} ($1/K$). **(10.6)** -14 kJ mol^{-1}. **(10.7)** 1.68×10^{-5}. **(10.8)** **(a)** 9.01; **(b)** -12.8 kJ mol^{-1}; **(c)** 160 kJ mol^{-1}; **(d)** 250 J K^{-1} mol^{-1}. **(10.9)** $K_1 = 0.731$, $x_1(A_2) = 0.44$; $K_2 = 5.7$, $x_2(A_2) = 0.13$; 100 kJ/mol dimer. **(10.10)** 1.77×10^{-10}, 1.3×10^{-5} mol kg^{-1}. **(10.11)** 57, ± 2 kJ mol^{-1}. **(10.12)** 14.7, 18.8 kJ mol^{-1} **(10.13)** **(a)** 320 K; **(b)** 399 K. **(10.14)** 0.01 mol H_2, 0.11 mol I_2, 0.78 mol HI. **(10.16)** $K_x = 0.33$ **(10.17)** $K_p = \{16(2 - \xi)^2\xi^2/27(1 - \xi)^4\}(p^{\ominus}/p)^2$. **(10.18)** $\xi_e = 1 - 1/\{1 + \frac{3}{4}(3K_p p/p^{\ominus})^{\frac{1}{2}}\}$. **(10.19)** N_2: 0.0033; H_2: 0.01; NH_3: 0.99. **(10.20)** -24.5, -66.1 kJ mol^{-1}, -142 J K^{-1} mol^{-1}. **(10.21)** 1.8×10^{-3}, 1.1×10^{-2}, 4.8×10^{-2}; 157.6 kJ mol^{-1}. **(10.22)** 8.3; -5.2, -55.6 kJ mol^{-1}; -169 J K^{-1} mol^{-1}. **(10.23)** 105. **(10.24)** -137 kJ mol^{-1}. **(10.25)** -98.2 kJ mol^{-1}, -301 J K^{-1} mol^{-1}, 1.04×10^{-6}. **(10.26)** $K_p(550\text{ atm})/K_p(500\text{ atm}) \approx 1.2$. **(10.27)** 213, 219 kJ mol^{-1}. **(10.28)** 3.3, 0.94 GJ. **(10.29)** $\Delta G(T_2) = \Delta G(T_1) + (T_1 - T_2)\,\Delta S(T_1) + \alpha\,\Delta a + \beta\,\Delta b + \gamma\,\Delta c$, $\alpha = T_2 - T_1 - T_2 \ln (T_2/T_1)$,

$\beta = \frac{1}{2}(T_2^2 - T_1^2) - T_2(T_2 - T_1)$, $\gamma = 1/T_1 + 1/T_2 + \frac{1}{2}T_2\{(1/T_2^2) - (1/T_1^2)\}$. **(10.30)** -225.4 kJ mol^{-1}. **(10.31)** $S(T) = \{H(T) - H(0)\}/T - \Phi_0$. **(10.32)** **(a)** 125.9 kJ mol^{-1}; **(b)** -20.3, $+25.2$ kJ mol^{-1}. **(10.33)** **(a)** 2.7×10^{-7}; **(b)** 130, 0.22. **(10.34)** $\Delta\Phi(T) = \Delta\Phi(\mathcal{T}) + \alpha'\Delta a + \beta'\Delta b + \gamma'\Delta c$, $\alpha' = (T - \mathcal{T})/T - \ln(T/\mathcal{T})$, $\beta' = \frac{1}{2}(T^2 - \mathcal{T}^2)/T - (T - \mathcal{T})$, $\gamma' = 1/T\mathcal{T} - 1/T^2 + \frac{1}{2}\{(1/T^2) - (1/\mathcal{T}^2)\}$.

11 **(11.1)** $I = \{1/2(\rho/\mathrm{g\ cm^{-3}})\}\sum_j (c_j/\mathrm{M})z_j^2$. **(11.3)** 0.9. **(11.4)** 2.52. **(11.6)** 0.25 mol kg^{-1}. **(11.7)** $\gamma_\pm^2(m/m^\ominus)^2$, $4\gamma_\pm^3(m/m^\ominus)^4$, $27\gamma_\pm^{44}(m/m^\ominus)^4$, $\gamma_\pm^2(m/m^\ominus)^2$, $108\gamma_\pm^5(m/m^\ominus)^5$. **(11.8)** **(a)** -1.92 kJ; **(b)** -53 kJ. **(11.9)** **(a, b, c)** 3×10^{-23} N. **(11.10)** **(a)** -2×10^{-23} N; **(b)** -1×10^{-26} N; **(c)** 0. **(11.11)** **(a)** 5.5 nm; **(b)** 5.2 nm. **(11.12)** 3.2 nm. **(11.13)** 4.0. **(11.14)** 0.964, 0.949, 0.920, 0.889, 0.847. **(11.19)** 0.77. **(11.20)** $\lg K_c = \lg K + 2A(\alpha c)^{\frac{1}{2}}$. **(11.21)** 4.719. **(11.22)** 1.80×10^{-10}, 9.04×10^{-7}. **(11.23)** 7.4×10^{-7} mol kg^{-1}. **(11.26)** 6.3×10^{-11} M. **(11.27)** $c \approx K_{sp}^{\frac{1}{2}}10^{Ac^{\frac{1}{2}}}$. **(11.28)** **(a)** 3.8×10^{-9} M; **(b)** 2.3×10^{-8} M; **(c)** 1.5×10^{-5} M. **(11.29)** $K_{sp} = 1.58 \times 10^{-10}$, $\gamma_\pm = 0.800$. **(11.30)** 390 kJ mol^{-1}. **(11.31)** -0.83 V. **(11.32)** $\Delta\phi = \Delta\phi^\ominus + (RT/zF)\ln(f^{\frac{1}{2}}/a)$. **(11.33)** $\Delta\phi = \Delta\phi^\ominus - (RT/F)\ln a_{OH^-}$; -38 mV. **(11.34)** $\Delta\phi = \Delta\phi^\ominus + (RT/6F)\ln\{a_{Cr_2O_7^{2-}}a_{H^+}^{14}/a_{Cr^{3+}}^2\}$.

12 **(12.1)** **(a)** $H_2 + 2AgCl \rightarrow 2Ag + 2HCl$; L: $2H^+ + 2e^- \rightarrow H_2$; R: $2AgCl + 2e^- \rightarrow 2Ag + 2Cl^-$; 0.22 V; **(b)** $2Fe^{2+} + Sn^{4+} \rightarrow 2Fe^{3+} + Sn^{2+}$; L: $2Fe^{3+} + 2e^- \rightarrow 2Fe^{2+}$; R: $Sn^{4+} + 2e^- \rightarrow Sn^{2+}$; -0.62 V; **(c)** $Cu + MnO_2 + 4H^+ \rightarrow Mn^{2+} + 2H_2O + Cu^{2+}$; L: $Cu^{2+} + 2e^- \rightarrow Cu$; R: $MnO_2 + 4H^+ + 2e^- \rightarrow Mn^{2+} + 2H_2O$; 0.89 V; **(d)** $AgBr + HCl \rightarrow AgCl + HBr$; L: $AgCl + e^- \rightarrow Ag + Cl^-$; R: $AgBr + e^- \rightarrow Ag + Br^-$; -0.15 V. **(12.2)** **(a)** Zn |ZnSO$_4$| |CuSO$_4$| Cu, 1.10 V; **(b)** Pt |H$_2$| HCl |AgCl| Ag, 0.22 V; **(c)** Pt |H$_2$| HCl |O$_2$| Pt, 1.23 V; **(d)** Na |NaOH| H$_2$ |Pt, or better: Na| NaI(ethylamine) |Na, Hg| NaOH |H$_2$| Pt, 1.88 V; **(e)** Pt |H$_2$| HI |I$_2$| Pt, 0.54 V. **(12.3)** See entries above. **(12.4)** **(a)** -363 kJ mol^{-1}; **(b)** -203 kJ mol^{-1}, **(c)** -291 kJ mol^{-1}, **(d)** $+122$ kJ mol^{-1}. **(12.5)** **(a)** 1.2×10^{16}; **(b)** 2.4×10^{83}; **(c)** 8.3×10^{-7}. **(12.6)** **(a)** 1.10 V; **(b)** -212 kJ mol^{-1}; **(c)** 1.5×10^{37}; **(d)** K. **(12.7)** $2Al + 3Sn^{4+} \rightarrow 2Al^{3+} + 3Sn^{2+}$; **(a)** 1.83 V, 1.81 V; **(b)** -1050 kJ mol^{-1}; **(c)** 10^{183}; Sn^{4+}, Sn^{2+} is positive. **(12.8)** 1.85 V. **(12.9)** 0.95 V. **(12.10)** 18 mV, -12 mV. **(12.11)** 0.44 V. **(12.12)** 1.03, 26.1, 33.5 atm; 1.03, 0.52, 0.34. **(12.13)** $E(t) = E^\ominus + (2RT/F)\ln(1 + \kappa t)$, $\partial E/\partial t = 2\kappa(RT/F)/(1 + \kappa t)$. **(12.14)** 1.23 V, 237 kJ mol^{-1}. **(12.15)** 2877, 2738, 2746 kJ mol^{-1}. **(12.17)** **(a, b)** 0.2223 V. **(12.18)** **(a)** 0.0796; **(b)** 0.7957; **(c)** 1.10. **(12.19)** Cell reaction: -21.46, -40.03 kJ mol^{-1}, -62.30 J K^{-1} mol^{-1}; Cl^- ion: -131.25, -167.10 kJ mol^{-1}, 56.78 J K^{-1} mol^{-1}. **(12.20)** 0.927. **(12.21)** 0. **(12.22)** **(a)** 9.19×10^{-9} mol kg^{-1}; **(b)** 8.45×10^{-17}. **(12.23)** 0.572, 0.533, 0.492, 0.469, 0.444, 0.486. **(12.24)** 0.26835 V (0.99895). **(12.25)** $pK_w = 14.23$, 14.01, 13.78; 74.9 kJ mol^{-1}; -17.1 J K^{-1} mol^{-1}. **(12.26)** 0.2916 V. **(12.27)** $A_{calc} = 0.419$, $A_{exp} = 0.317$. **(12.28)** **(a)** 5.13; **(b)** 8.88; **(c)** 9.38; **(d)** 2.88. **(12.29)** 8.43, 8.43. **(12.30)** 9.14. **(12.31)** **(a)** Na_2HPO_4/H_3PO_4; **(b)** NaH_2PO_4/Na_2HPO_4. **(12.34)** No. **(12.35)** -1.1 V.

13 **(13.1)** **(a)** 3.31×10^{-19} J, 199 kJ mol^{-1}; **(b)** 3.61×10^{-19} J, 218 kJ mol^{-1}; **(c)** 4.97×10^{-19} J, 299 kJ mol^{-1}; **(d)** 9.93×10^{-19} J, 598 kJ mol^{-1}; **(e)** 1.32×10^{-15} J, 79.8×10^4 kJ mol^{-1}; **(f)** 1.99×10^{-23} J; 0.12 kJ mol^{-1}. **(13.2)** p/kg m s^{-1} = **(a)** 1.10N27; **(b)** 1.20N27; **(c)** 1.66N27; **(d)** 3.31N27; **(e)** 4.42N24; **(f)** 6.63N32; v/m s^{-1} = **(a)** 0.66N5; **(b)** 0.72N5; **(c)** 0.99N5; **(d)** 1.98N5; **(e)** 2640; **(f)** 3.96N5. **(13.3)** 21 m s^{-1}. **(13.4)** **(a)** 2.8×10^{18} s^{-1}; **(b)** 2.8×10^{20} s^{-1}. **(13.5)** **(a)** 1.6×10^{-33} J m^{-3}; **(b)** 2.5×10^{-4} J m^{-3}. **(13.6)** $\lambda_{max}T = hc/5k$. **(13.7)** 6.51×10^{-34} J s. **(13.8)** 6000 K. **(13.9)** **(a)** 7.5×10^{-29} J m^{-3}; **(b)** 4.6×10^{-14} J m^{-3}; **(c)** 1.6×10^{-8} J m^{-3}. **(13.10)** **(a)** 19.7 J K^{-1} mol^{-1}; **(b)** 22.4 J K^{-1} mol^{-1}; **(c)** 24.4 J K^{-1} mol^{-1}; 24.9 J K^{-1} mol^{-1}. **(13.11)** $\theta_E = h\nu/k$, 341 K. **(13.12)** **(a)** No ejection; **(b)** 837 km s^{-1}. **(13.13)** 6.93 keV. **(13.14)** **(a)** 2.426 pm; **(b)** 1.321 fm. **(13.16)** **(a)** 400 kJ mol^{-1}; **(b)** 40 kJ mol^{-1}; **(c)** 4×10^{-13} kJ mol^{-1}. **(13.17)** **(a)** 7×10^{-29} m; **(b)** 7×10^{-36} m; **(c)** 73 pm; **(d)** 123, 40, 4 pm. **(13.18)** 1×10^{-28} m s^{-1}, 1×10^{-27} m. **(13.19)** 5×10^{-25} kg m s^{-1}, 5×10^5 m s^{-1}. **(13.20)** N = **(a)** $(1/L)^{\frac{1}{2}}$; **(b)** $1/c(2L)^{\frac{1}{2}}$; **(c)** $1/(\pi a_0^3)^{\frac{1}{2}}$; **(d)** $1/(32\pi a_0^5)^{\frac{1}{2}}$. **(13.21)** **(a)** 2%; **(b)** 0.7%; **(c)** 6.6×10^{-6}; **(d)** 0.5; **(e)** 61%. **(13.22)** **(a)** 9.0×10^{-6}; **(b)** 1.2×10^{-6}. **(13.23)** N = **(a)** $(1/32\pi a_0^3)^{\frac{1}{2}}$; **(b)** $(1/32\pi a_0^3)^{\frac{1}{2}}$. **(13.24)** a, c. **(13.25)** a, b, c; b, d. **(13.26)** **(a)** $\cos^2\chi$; **(b)** $\sin^2\chi$; $0.95e^{ikx} \pm 0.32e^{-ikx}$. **(13.27)** $k^2\hbar^2/2m$. **(13.28)** **(a)** $k\hbar$; **(b, c)** 0. **(13.29)** **(a)** $6a_0$, $42a_0^2$; **(b)** $5a_0$, $30a_0^2$; $-e^2/4\pi\varepsilon_0 a_0$. **(13.30)** 0, 0, 0, $i\hbar$, 0, no, yes, no. **(13.31)** No, yes, yes. **(13.32)** No.

14 **(14.1)** **(a)** 1.8×10^{-19} J, 110 kJ mol^{-1}, 1.1 eV, 9100 cm^{-1}; **(b)** 6.6×10^{-19} J, 400 kJ mol^{-1}, 4.1 eV, 33000 cm^{-1}. **(14.2)** 7.5×10^{-19} kJ mol^{-1}, $n \approx 2 \times 10^9$, 9.6×10^{-10} kJ mol^{-1}. **(14.3)** $E_{n_1n_2n_3} = (h^2/8m)\{(n_1/L_x)^2 + (n_2/L_y)^2 + (n_3/L_z)^2\}$, $E_{n_1n_2n_3} = h^2n^2/8mL^2$, $n^2 = n_1^2 + n_2^2 + n_3^2$. **(14.4)** $P = |A|^2/\kappa$, $\langle x\rangle = P/\kappa$. **(14.5)** $P = 1/\{\frac{1}{4}(1 + \lambda)^2 + G\}$, $G = \{[1 + (\lambda^2 - 1)\varepsilon]/16\varepsilon(1 - \varepsilon)\}\{e^{\kappa L} - e^{-\kappa L}\}^2$. **(14.6)** $P = 1/(1 + G)$; $G = \{(V/E)^2\sin^2 k'L\}/4(1 - V/E)$. **(14.7)** $P = G/(1 + G)$; $G = \sin^2\{[(2mV/\hbar^2)(1 + \lambda)]^{\frac{1}{2}}L\}/4\lambda(1 + \lambda)$, $\lambda = E/V$. **(14.8)** **(a)** 0.146; **(b)** 0.150. **(14.9)** $g = \frac{1}{2}(mk/\hbar^2)^{\frac{1}{2}}$, $E = \frac{1}{2}\hbar\omega$, $\omega = (k/m)^{\frac{1}{2}}$, $\Delta E = \hbar\omega$. **(14.10)** **(a)** 3×10^{-34} J; **(b)** 3×10^{-33} J; **(c)** 2.2×10^{-29} J; **(d)** 3.1×10^{-20} J. **(14.11)** CO > NO > HCl > HBr > HI. **(14.12)** $\frac{1}{2}\hbar\omega(v + \frac{1}{2})$. **(14.13)** $\langle x^3\rangle = 0$, $\langle x^4\rangle = \frac{3}{4}(2v^2 + 2v + 1)\alpha^4$. **(14.14)** $v + 1 \leftarrow v$: $(\alpha/2^{\frac{1}{2}})(v + 1)^{\frac{1}{2}}$; $v - 1 \leftarrow v$: $(\alpha/2^{\frac{1}{2}})v^{\frac{1}{2}}$. **(14.15)** $\langle T\rangle = -\frac{1}{2}\langle V\rangle$. **(14.16)** 78.2×10^{-3} kJ mol^{-1}, 6.54 cm^{-1}, 1.055×10^{-34} J s **(14.17)** **(a, b)** uniform. **(14.18)** **(a)** $\hbar$, $\hbar^2/2I$; **(b)** $-2\hbar$, $2\hbar^2/I$; **(c)** 0, $\hbar^2/2I$; **(d)** $\hbar\cos 2\chi$, $\hbar^2/2I$. **(14.19)** $\langle l_z\rangle = \hbar(a^2 + 2b^2 + 3c^2)/(a^2 + b^2 + c^2)$; $\langle T\rangle = (\hbar^2/2I)(a^2 + 4b^2 + 9c^2)/(a^2 + b^2 + c^2)$; $\delta l_z = \hbar(4a^2 + b^2)^{\frac{1}{2}}c/(a^2 + b^2 + c^2)$. **(14.20)** 0(1), $2^{\frac{1}{2}}(3)$, $6^{\frac{1}{2}}(5)$, $12^{\frac{1}{2}}(7)$. **(14.21)** 0, 0.158, 0.473, 0.945 kJ mol^{-1}; 0, 13.2, 39.5, 79.0 cm^{-1}. **(14.25)** $\theta = \arccos m_s/\{s(s + 1)\}^{\frac{1}{2}}$, 54°44′.

15 **(15.1)** $n_2 \rightarrow n_1 = 6$; 12370, 7503, 5908, 5129, . . . , 3908, . . . , 3282 nm. **(15.2)** 397.13 nm, 3.40 eV. **(15.3)** 987 663 cm^{-1}. **(15.4)** 131 175, 185 187 cm^{-1}. **(15.5)** 3.3594×10^{-27} kg. **(15.6)** 7621, 10288, 11522 cm^{-1}, 6.80 eV. **(15.7)** $E = -hcR_\infty/n^2$ **(15.9)** $r^* = a_0/Z$; **(a)** 26

pm; **(b)** 5.9 pm. **(15.10)** 0.42 pm. **(15.11)** **(a)** 0, 0, 0; **(b)** 0, 2, 0; **(c)** $6\frac{1}{3}\hbar$, 0, 2; **(d)** $2\frac{1}{2}\hbar$, 0, 1; **(e)** $2\frac{1}{2}\hbar$, 1, 1; N(radial) = $n-l-1$; N(angular) = l. **(15.12)** $\langle r\rangle_{2p} = 5a_0/Z$; $\langle r\rangle_{2s} = 6a_0/Z$; 700 pm. **(15.13)** $-Z^2R_H$, -1102 eV. **(15.14)** $z = \pm 106$ pm. **(15.15)** **(a)** 1; **(b)** 9; **(c)** 25 **(15.16)** **(b, c, e).** **(15.17)** **(a)** 2; **(b)** 6; **(c)** 10; **(d)** 18. **(15.19)** 122.5 eV. **(15.20)** 5.39 eV. **(15.21)** **(a)** 14.0 eV; **(b)** 4.1 eV. **(15.22)** $I_D = 1.000\,272\, I_H$ **(15.23)** $\frac{7}{2}, \frac{5}{2}$; $\frac{3}{2}\sqrt{7}\hbar$, $\frac{1}{2}\sqrt{35}\hbar$. **(15.24)** 41/9, 39/2. **(15.25)** **(a, b)** $J = 8, 7, 6, 5, 4, 3, 2$; use $\{J(J+1)\}^{\frac{1}{2}}\hbar$. **(15.26)** ${}^2P_{\frac{3}{2}} \rightarrow {}^2S_{\frac{1}{2}}$, ${}^2P_{\frac{1}{2}} \rightarrow {}^2S_{\frac{1}{2}}$; 38.50 cm^{-1}. **(15.27)** **(a)** $S = 1, 0$; **(b)** $S = \frac{3}{2}, \frac{1}{2}, \frac{1}{2}$; **(c)** $S = 2, 1, 1, 1, 0, 0$; use $2S+1$ for $S \leqslant L$. **(15.28)** 1(1); $\frac{3}{4}$(4), $\frac{1}{2}$(2); 2(5), 1(3), 0(1); 3(7), 2(5), 1(3); $\frac{5}{2}$(6), $\frac{3}{2}$(4); 2(5); $\frac{7}{2}$(8), $\frac{5}{2}$(6), $\frac{3}{2}$(4), $\frac{1}{2}$(2). **(15.29)** ${}^2P_{\frac{3}{2}} \leftrightarrow {}^2D_{\frac{5}{2}}$, ${}^2D_{\frac{3}{2}}$; ${}^2P_{\frac{1}{2}} \leftrightarrow {}^2D_{\frac{3}{2}}$; ${}^2P_2 \leftrightarrow {}^3D_3$, ${}^3D_2$, ${}^3D_1$; ${}^3P_1 \leftrightarrow {}^3D_2$, ${}^3D_1$; ${}^3P_0 \leftrightarrow {}^3D_1$.

16 **(16.5)** $N = 1/\{1 + 2\lambda S + \lambda^2\}^{\frac{1}{2}}$. **(16.6)** $\psi = -\mu 1s_A + 1s_B$, $\mu = -(\lambda + S)/(1 + \lambda S)$. **(16.9)** **(a)** 8.6×10^{-7}; **(b)** 8.6×10^{-7}; **(c)** 3.7×10^{-7}; **(d)** 4.9×10^{-7}. **(16.10)** **(a)** 2.0×10^{-6}; **(b)** 2.0×10^{-6}; **(c)** 0; **(d)** 5.5×10^{-7}. **(16.11)** **(a)** 1.9 eV; **(b)** 130 pm. **(16.13)** $N = 1/\{2(1+S)\}^{\frac{1}{2}}$. **(16.15)** $2.1a_0$. **(16.17)** ***(a)*** C_2, CN; **(b)** NO, O_2, F_2. **(16.18)** $\sigma, \pi, \pi, \delta, \delta$; **(a)** $d\sigma^2$; **(b)** $d\sigma^2\, d\pi^4$; **(c)** $d\sigma^2\, d\pi^4\, d\delta^2$. **(16.19)** Yes. **(16.20)** **(a)** g; **(b)** n.a.; **(c)** g; **(d)** u; **(e)** ugug. **(16.21)** 472 kJ mol^{-1}. **(16.23)** 1. **(16.24)** $P = 3\cos\Theta/(\cos\Theta - 1) = 0.67$ for NH_3. **(16.25)** CO_2, NO_2^+, H_2O^{2+}. **(16.26)** NH_3^{2+}, (CH_3), NO_3^-, CO_3^{2-}. **(16.29)** $5(h^2/8m_eL^2)$. **(16.30)** 2.7 eV, 460 nm, orange. **(16.31)** **(a)** $-hcR_H$; **(b)** $-(8/3\pi)hcR_H$. **(16.33)** **(a)** $a_{2u}^2 e_{1g}^4 e_{2u}^1$, $E_\pi = 7\alpha + 7\beta$; **(b)** $a_{2u}^2 a_{1g}^3$, $E_\pi = 5\alpha + 7\beta$. **(16.34)** ***(a)*** H(1)F(2) + H(2)F(1); **(b)** F(1)F(2); **(c)** 0.89[H(1)F(2) + H(2)F(1)] + 0.45F(1)F(2).

17 **(17.1)** R_3, C_{2v}, D_{3h}, $D_{\infty h}$, $C_{\infty v}$, D_3, D_{6h}, C_{4v}, C_s. **(17.2)** C_{2v}, C_{3v}, C_{3v}, D_{2h}, C_{2v}, C_{2h}, D_{2h}, D_{2h}, C_{2v}. **(17.3)** D_{3d}, D_{3d}, D_{2h}, $D_{\infty h}$, D_3, D_{4d}. **(17.4)** trans-CHCl=CHCl. **(17.5)** $C_2\sigma_h = i$. **(17.6)** NO_2, CH_3Cl, CCl_3H, cis-CHCl=CHCl, C_6H_5Cl. **(17.7)** $Co(en)_3^{3+}$. **(17.10)** $D(\sigma_v) = D(\sigma_d) = \pm 1$, $D(\sigma_v) = -D(\sigma_d) = \pm 1$. **(17.11)** No p-orbitals; d_{xy} spans A_2. **(17.12)** $B_1(x)$, $B_2(y)$, $A_1(z)$. **(17.13)** A_2. **(17.14)** **(a)** $E_{1u}(x, y)$, $A_{2u}(z)$; **(b)** $B_1(x)$, $B_2(y)$, $A_1(z)$. **(17.15)** $A_1 + T_2$; $s(A_1)$, $p(T_2)$, $d(E + T_2)$. **(17.16)** **(a)** All d-orbitals; **(b)** all except $d_{xy}(A_2)$. **(17.17)** $\psi(A_1) = c_H(H_1 + H_2) + c_{O1}\psi(O2s) + c_{O2}\psi(O2p_z)$, $\psi(B_1) = \psi(O2p_x)$, $\psi(B_2) = c'_H(H_1 - H_2) + c'_{O1}\psi(O2p_y)$. **(17.19)** **(a)** $2A_1 + A_2 + 2B_1 + B_2$; **(b)** $A_1 + 3E$; **(c)** $A_1 + T_1 + T_2$; **(d)** $A_{2u} + T_{1u} + T_{2u}$. **(17.20)** **(a)** $T_1 + T_2 + A_1$; **(b)** $T_{1u} + T_{2u} + A_{2u}$. **(17.21)** **(a)** i, ii; **(b)** ii. **(17.23)** **(a)** $B_1 + B_2 + A_1$, $B_2 + B_1 + A_2$; **(b)** $E + A_1$, $E + A_2$; **(c)** $E_1 + A_1$, $E_1 + A_2$; **(d)** T_{1u}, T_{1g}. **(17.24)** **(a)** p, f; **(b)** p, d, f. **(17.25)** **(a)** $B_2(x)$, $B_1(y)$, $A_2(z)$; **(b)** A_{2g}, E_g, T_{1g}, T_{2g}; **(c)** A_2. **(17.27)** **(a)** Yes; **(b)** no; **(c)** yes. **(17.28)** $3A_1 + 2A_2 + 2B_1 + 3B_2$. **(17.29)** $4A_1 + 2B_1 + 3B_2 + A_2$.

18 **(18.1)** HCl, CH_3Cl, CH_2Cl_2, H_2O, H_2O_2, NH_3, NH_4Cl. **(18.2)** HCl, CO_2, H_2O, CH_3CH_3, CH_4, CH_3Cl, N_3^-. **(18.3)** H_2, HCl, CH_3Cl, CH_2Cl_2, CH_3CH_3, H_2O. **(18.4)** All. **(18.5)** $0.999\,999\,925 \times 660$ nm, 1.4×10^8 m.p.h. **(18.6)** 2.4×10^4 km s^{-1}, 8.4×10^5 K. **(18.7)** $\Delta\lambda/\lambda = 5.04 \times 10^{-19}/(m/\text{kg})^{\frac{1}{2}}$; **(a)** 1.3 MHz, 0.006 cm^{-1}; **(b)** 3.3 kHz, 3.7×10^{-4} cm^{-1}. **(18.8)** **(a)** 50 ps; **(b)** 5 ps; **(c)** 2 ns. **(18.9)** **(a)** 50 cm^{-1}; **(b)** 0.5 cm^{-1}. **(18.10)** $\tau \approx (kT/p)(\pi m/8kT)^{\frac{1}{2}}/2^{\frac{1}{2}}\sigma$; 0.2 ns, 700 MHz, 0.8 Torr. **(18.4)** **(a)** 0.052; **(b)** 0.067; **(c)** 0.200. **(18.12)** 30. **(18.13)** 6. **(18.14)** **(a)** 1.6266×10^{-27} kg, 2.6422×10^{-47} kg m^2; **(b)** 3.1622×10^{-27} kg, 5.1368×10^{-47} kg m^2; **(c)** 1.6291×10^{-27} kg, 2.6462×10^{-47} kg m^2. **(18.15)** 596 GHz, 19.9 cm^{-1}, 0.25 mm and less. **(18.16)** $2B(J+1)$ with $B = 1.9318$ cm^{-1}; 0.1 cm^{-1}. **(18.17)** 2.728×10^{-47} kg m^2, 129.5 pm. **(18.18)** $2B(J+1)$ with $B = 5.278$ cm^{-1}. **(18.19)** R(HCl) = 128.5 pm, R(DCl) = 128.2 pm. **(18.20)** **(a)** 4.599×10^{-48} kg m^2; **(b)** 7.495×10^{-45} kg m^2. **(18.21)** 160.5 pm. **(18.22)** 218 pm. **(18.23)** $1.379\,98 \times 10^{-45}$ kg m^2. **(18.24)** R(CO) = 116.28 pm, R'(CS) = 155.97 pm. **(18.26)** 206 pm. **(18.27)** $HgCl_2$: 0.0459 cm^{-1}; $HgBr_2$: 0.0184 cm^{-1}; HgI_2: 0.0103 cm^{-1}. **(18.28)** 967, 516, 412, 314 N m^{-1}. **(18.29)** 3002, 2144, 1886, 1640 cm^{-1}. **(18.32)** 5.1 eV. **(18.33)** 93.8 N m^{-1}, 142.8 cm^{-1}, 3.4 eV. **(18.34)** **(a)** 5.15 eV; **(b)** 5.20 eV. **(18.35)** P: $\bar{\nu}_0 - (B_v + B_{v+1})J + (B_{v+1} - B_v)J^2$; Q: $\bar{\nu}_0 + (B_{v+1} - B_v)J(J+1)$; R: $\bar{\nu}_0 + 2B_{v+1} + (3B_{v+1} - B_v)J + (B_{v+1} - B_v)J^2$. **(18.36)** 480.7 N m^{-1}, 128 pm, 130 pm. **(18.37)** 198.9 pm. **(18.38)** **(a)** All; **(b)** ν_1: Raman; ν_2, ν_3 infrared. **(18.39)** **(a)** Yes; **(b)** no.

19 **(19.1)** 450 M^{-1} cm^{-1}. **(19.2)** 160 M^{-1} cm^{-1}. **(19.3)** **(a)** 0.09, 0.45, 0.90, 4.5; **(b)** 0.32. **(19.4)** **(a)** 69%, 1.0×10^{-14}%; **(b)** 90%, 3.2×10^{-3}%. **(19.5)** **(a)** 88 cm; **(b)** 290 cm. **(19.6)** **(a)** 1.5×10^{18} cm^2 s^{-1} mmol^{-1}; **(b)** 7.5×10^{16} cm^2 s^{-1} mmol^{-1}. **(19.7)** **(a)** 0.22; **(b)** 0.011. **(19.9)** **(a)** 1.4×10^{15} cm^2 s^{-1} mmol^{-1}; $f = 2.0 \times 10^{-4}$; **(b)** same. **(19.10)** **(a)** f(280 nm) = 2.6×10^{-4}, f(430 nm) = 3×10^{-4}; **(b)** f(280 nm) = 2.5×10^{-4}, f(430 nm) = 2.9×10^{-4}; **(c)** $\mathscr{A} = c\varepsilon_{max}(2\pi\gamma)^{\frac{1}{2}}/\lambda_0^2$, $\gamma = \Delta\lambda_{\frac{1}{2}}^2/8\ln 2$; f(280 nm) = 3.9×10^{-4}, f(430 nm) = 2.9×10^{-4}. **(19.11)** $f = 4.8 \times 10^{-3}$ forbidden; A_1, B_1, B_2. **(19.12)** $f_{n+1,n} = (64/3\pi^2)\{n^2(n+1)^2/(2n+1)^3\}$, $f_{n+2,n} = 0$. **(19.13)** 7400 cm^{-1}, 1.4×10^5 M^{-1} cm^{-1}, 2×10^{-6} cm. **(19.16)** $2.1a_0$. **(19.17)** **(a)**. **(19.18)** **(a)** $\bar{\nu} = 1765$ cm^{-1}; **(b)** nothing. **(19.21)** pH = pK_{In} − lg $\{(1-\alpha)/\alpha\}$. **(19.23)** 6.9. **(19.24)** 0.848 eV. **(9.25)** 5.1 eV. **(19.26)** **(a)** 3 + 1, 3 + 3; **(b)** 4 + 4, 2 + 2. **(19.29)** Non-bonding. **(19.30)** Bonding.

20 **(20.1)** **(a)** 1.41, 9.18, 5.61, 19.5, 1.50, 3.48; **(b)** 7.05, 45.9, 28.0, 97.5, 7.5, 17.4. **(20.2)** **(a)** 7.7×10^{-5}, 3.8×10^{-4}, 1.8×10^{-3}; **(b)** 1.0×10^{-6}, 5.1×10^{-6}, 2.4×10^{-5}. **(20.3)** 210 T, 2.1 mT. **(20.4)** 2.06 T, -5×10^{-6}, β. **(20.5)** **(a)** -11 μT; **(b)** -53 μT. **(20.6)** **(a)** 460 Hz, **(b)** 2.3 kHz. **(20.8)** 158 pm. **(20.9)** 78 μT. **(20.10)** 500 s^{-1}. **(20.11)** 56 kJ mol^{-1}. **(20.12)** $J(\omega) \propto \tau/\{1 + (\omega - \omega_0)^2\tau^2\}$ for $\omega \approx \omega_0$. **(20.15)** 1.3 T. **(20.16)** 2×10^{-11} M, $\delta N/N = 6.8 \times 10^{-4}$. **(20.17)** **(a)** 0.05; **(b)** 0.0007. **(20.18)** 2.002 24. **(20.19)** **(a)** 331.9 mT; **(b)** 1.201 T. **(20.20)** 1.3 mT, 4.8 mT. **(20.21)** 1.9920, 2.0022. **(20.22)** 50.7 mT. **(20.23)** 2.3 mT. **(20.24)** 2.0025, 64 MHz. **(20.25)** 330.2, 332.2, 332.8, 334.8 mT; 1:1:1:1. **(20.26)** 1.1×10^8 s^{-1}. **(20.27)** **(a)** 1:3:3:1; **(b)** 1:3:6:7:6:3:1. **(20.28)** 0.69, 0.21 mT. **(20.30)** 1:2:3:2:1 quintet of 1:4:6:4:1

quintets. **(20.31)** 10%. **(20.32)** 38%. **(20.33)** $P(\mathrm{N}) = 0.48$, $P(\mathrm{O}) = 0.52$, $\lambda = 3.8$ ($\mathrm{sp}^{3.8}$), yes, 131°. **(20.36)** 1.6 ns. **(20.37)** **(a)** 80 m s^{-1}, 65 GHz; **(b)** 7.7×10^{-20} m s^{-1}, 6.2×10^{-11} Hz. **(20.39)** Increasing ionic character.

21

(21.1) 1. **(21.2)** {2, 2, 0, 1, 0, 0} and {2, 1, 2, 0, 0, 0} jointly. **(21.4)** {4, 2, 2, 1, 0, 0, 0, 0, 0, 0}, $N = 3780$. **(21.5)** 160 K. **(21.6)** $T = \varepsilon/k \ln 2$; 104 K; $q = 1 + a$. **(21.8)** 2.00, 1.95, 2.06. **(21.9)** **(a, b, d, f, [g])**. **(21.10)** **(a)** 4.76×10^{25}; **(b)** 2.45×10^{26}; **(c)** 4.76×10^{28}; **(d)** 1. **(21.11)** 3.5×10^{-15} K, 18.6. **(21.12)** **(a)** 5.00; **(b)** 6.25. **(21.13)** **(a)** 1.00, 0.80; **(b)** 6.5×10^{-11}, 0.12. **(21.14)** $q = 2 + 2 \exp\{-174.4/(T/\mathrm{K})\}$; 0.64, 0.36. **(21.15)** 0.518 kJ mol^{-1}. **(21.16)** **(a)** 1.049; **(b)** 1.56. **(21.17)** **(a)** 0.954, 0.0443, 0.002; **(b)** 0.643, 0.230, 0.083. **(21.18)** **(a)** 125 J mol^{-1}; **(b)** 1.40 kJ mol^{-1}. **(21.19)** $q = 1 + \mathrm{e}^{-x}$, $\langle\varepsilon\rangle = 2\mu_{\mathrm{B}}B\mathrm{e}^{-x}/(1 + \mathrm{e}^{-x})$, $x = 2\mu_{\mathrm{B}}B\beta$; 0.904, 0.999. **(21.20)** $q = 1 + \mathrm{e}^{-x} + \mathrm{e}^{-2x}$, $x = g_I\mu_{\mathrm{N}}B\beta$; $\langle\varepsilon\rangle = -g_I\mu_{\mathrm{N}}Bf$, $f = (1 - \mathrm{e}^{-2x})/(1 + \mathrm{e}^{-x} + \mathrm{e}^{-2x})$. **(21.20)** **(a)** -298 K; **(b)** -10 K; **(c)** '-0 K'. **(21.24)** **(a)** 13.38 J K^{-1} mol^{-1} (R ln 5); **(b)** 18.07 J K^{-1} mol^{-1}. **(21.25)** **(a)** 11.2 J K^{-1} mol^{-1}; **(b)** 11.4 J K^{-1} mol^{-1}. **(21.26)** **(a)** 1.65 J K^{-1} mol^{-1}; **(b)** 8.37 J K^{-1} mol^{-1}. **(21.27)** $S_{\mathrm{m}}/R = x\mathrm{e}^{-x}/(1 + \mathrm{e}^{-x}) + \ln(1 + \mathrm{e}^{-x})$; **(a)** 0; **(b)** R ln 2. **(21.28)** 155 J K^{-1} mol^{-1}. **(21.29)** **(a)** 7.7 J K^{-1} mol^{-1}; **(b)** 43 J K^{-1} mol^{-1}.

22

(22.1) **(a)** 6.56; **(b)** 19.6; **(c)** 32.8. **(22.2)** **(a)** 6.9051; **(b)** 19.889. **(22.3)** 43.10, $T \gg 40$ K. **(22.4)** **(a)** 36.3; **(b)** 78.9. **(22.5)** **(a)** 36.6055, **(b)** 79.1716. **(22.6)** $\Delta S_{\mathrm{m}} = R \ln f$, $f = (\sigma/V)(h^2\beta/2\pi m\mathrm{e})^{\frac{1}{2}}$ **(22.7)** 89, -62 J K^{-1} mol^{-1}. **(22.9)** **(a)** 2; **(b)** 1; **(c)** 12; **(d)** 12; **(e)** 3. **(22.10)** 250 J K^{-1} mol^{-1}. **(22.11)** 0.444 at 500 K, 5.774 at 5000 K. **(22.12)** **(a)** 3.917; **(b)** 2.433. **(22.13)** 3.2×10^{-3}. **(22.14)** $K_p^{\mathrm{true}} = 1.004\,K_p$. **(22.15)** $K_p(B) = \{1 - (20/9)(\mu_{\mathrm{B}}B\beta)^2\}K_p(0)$; 100 T. **(22.16)** $U - U(0) = H - H(0) = N\hbar\omega\mathrm{e}^{-x}/(1 - \mathrm{e}^{-x})$, $S/Nk = x\mathrm{e}^{-x}/(1 - \mathrm{e}^{-x}) - \ln(1 - \mathrm{e}^{-x})$, $A - A(0) = G - G(0) = NkT \ln(1 - \mathrm{e}^{-x})$. **(22.17)** $R \ln(1 - \mathrm{e}^{-x})$. **(22.18)** **(a)** -4.31 J K^{-1} mol^{-1}; **(b)** -6.52 J K^{-1} mol^{-1}. **(22.19)** **(a)** -136 J K^{-1} mol^{-1}; **(b)** -232 J K^{-1} mol^{-1}; **(c)** -203 J K^{-1} mol^{-1}; **(d)** -198 J K^{-1} mol^{-1}; **(e)** -217 J K^{-1} mol^{-1}. **(22.20)** 4.1×10^{-7}. **(22.21)** $U - U(0) = nRT\dot{q}/q$, $S/nR = \dot{q}/q + \ln(\mathrm{e}q/N)$, $C_V/nR = \ddot{q}/q - (\dot{q}/q)^2$. **(22.22)** $U_{\mathrm{int}} - U_{\mathrm{int}}(0) = nRT\,\dot{q}_{\mathrm{int}}/q_{\mathrm{int}}$, $S_{\mathrm{int}}/nR = \dot{q}_{\mathrm{int}}/q_{\mathrm{int}} + \ln q_{\mathrm{int}}$, $C_{V,\mathrm{int}}/nR = (\ddot{q}/q)_{\mathrm{int}} - (\dot{q}/q)^2_{\mathrm{int}}\}$. **(22.23)** **(a)** 4.322 kJ mol^{-1}; **(b)** -0.138 J K^{-1} mol^{-1}; **(c)** 5.405 J K^{-1} mol^{-1}. **(22.24)** -5.771 J K^{-1} mol^{-1}, $0.060\,R$. **(22.26)** $K_p = 8.65$, $x(\mathrm{Na_2}) = 0.095$, $x(\mathrm{Na}) = 0.905$. **(22.27)** **(a, b, f)** 21 J K^{-1} mol^{-1}; **(c, d, e)** 25 J K^{-1} mol^{-1}. **(22.28)** $C_{V,\mathrm{m}}/R = x^2\mathrm{e}^{-x}/(1 - \mathrm{e}^{-x})^2$; **(a)** 27.2, 27.3 J K^{-1} mol^{-1}; **(b)** 33.2, 38.1 J K^{-1} mol^{-1}. **(22.29)** **(a)** 35.72 J K^{-1} mol^{-1}; **(b)** 46.44 J K^{-1} mol^{-1}. **(22.30)** $C_{V,\mathrm{m}}/R = x^2\mathrm{e}^{-x}/(1 + \mathrm{e}^{-x})^2$, $x = \Delta/kT$. **(22.31)** **(a)** 2.94 J K^{-1} mol^{-1}; **(b)** 0.654 J K^{-1} mol^{-1}; **(c)** 0.244 J K^{-1} mol^{-1}. **(22.32)** **(a)** $+3.72 \times 10^{-2}$ J K^{-1} mol^{-1}; **(b)** $+1.06 \times 10^{-3}$ J K^{-1} mol^{-1}. **(22.33)** 4.16, 15.1 J K^{-1} mol^{-1}. **(22.35)** **(a)** $c_s = \{1.40RT/M_{\mathrm{m}}\}^{\frac{1}{2}}$; **(b)** $c_s = \{5(1 + a)RT/(3 + 5a)M_{\mathrm{m}}\}^{\frac{1}{2}}$, $a = 2(6B/kT)^2\mathrm{e}^{-6\beta B}/(1 + 5\mathrm{e}^{-6B\beta})^2$. **(22.36)** 346 m s^{-1}. **(22.37)** **(a)** 9.13; **(b)** 13.4; **(c)** 14.9 J K^{-1} mol^{-1}. **(22.39)** 191.4 J K^{-1} mol^{-1}.

23

(23.1) $O_{\mathrm{h}}(m3m)$. **(23.2)** Cubic, $T_{\mathrm{d}}(\bar{4}3m)$. **(23.3)** Triclinic, $C_{\mathrm{i}}(\bar{1})$. **(23.4)** Monoclinic. **(23.5)** Cubic, O_{h} $(m3m)$. **(23.6)** 54° 44′. **(23.9)** **(a)** $d = ab/(a^2 + b^2)^{\frac{1}{2}}$; $d = (ab/3^{\frac{1}{2}})/2(a^2 + b^2 - ab)^{\frac{1}{2}}$. **(23.10)** (326), (111), (122), $(3\bar{2}\bar{2})$. **(23.12)** 249, 176, 432 pm. **(23.13)** 58.9 pm. **(23.14)** Yes. **(23.15)** 8° 10′, 4° 49′, 11° 46′. **(23.16)** F.c.c. **(23.17)** B.c.c. **(23.18)** 564, 326, 252 pm. **(23.21)** 834, 606, 870 pm. **(23.22)** 6.05×10^{23} mol^{-1}. **(23.23)** **(a)** 1; **(b)** 2; **(c)** 4; **(d)** 8. **(23.24)** **(a)** $\pi/6 = 0.5236$; **(b)** $\pi/8\sqrt{3} = 0.6802$; **(c)** $\pi/3\sqrt{2} = 0.7405$, f.c.c. or h.c.p. **(23.25)** $\pi/2\sqrt{3}$, 0.9069. **(23.26)** 9.32 g cm^{-3}, 13.2 g cm^{-3}. **(23.27)** B.c.c. **(23.28)** $\theta_{hkl} = 21°\,41'$ (111), 25° 15′ (200), . . . ; 8.97 g cm^{-3}. **(23.29)** 4.8×10^{-5} K^{-1} (volume), 1.6×10^{-5} K^{-1} (linear). **(23.30)** 312 pm, cubic P. **(23.31)** F.c.c., 412 pm. **(23.32)** **(a)** B.c.c., $a = 316$ pm, $R = 137$ pm; **(b)** F.c.c., $a = 361$ pm, $R = 128$ pm. **(23.33)** 597, 1270, 434 pm; 4. **(23.35)** $F_{hkl} = f$. **(23.36)** **(a)** $F_{hkl} = f$; **(b)** $F_{hkl} = \frac{1}{2}f_{\mathrm{A}}$(if $h + k + l$ odd), $\frac{3}{2}f_{\mathrm{A}}$(if $h + k + l$ even); **(c)** $F_{hkl} = 0$ (if $h + k + l$ odd), $2f$ (if $h + k + l$ even). **(23.39)** **(a)** 39 pm; **(b)** 12 pm; **(c)** 6.1 pm. **(23.40)** 7.9 km s^{-1}, 145 pm. **(23.42)** 177 pm.

24

(24.1) 0, 0.7, 0.4 D; p-xylene. **(24.2)** $\mu = (3.02\ \mathrm{D}) \cos\frac{1}{2}\phi$; 90.1°. **(24.3)** **(a)** 1.1×10^8 V m^{-1}; **(b)** 4.1×10^9 V m^{-1}, **(c)** 4.1×10^3 V m^{-1}. **(24.4)** 2.27 nm. **(24.5)** 1.03 (HCl), 0.80 (HBr), 0.36 D (HI); 3.5×10^{-24}(HCl), 3.7×10^{-24}(HBr), 5.6×10^{-24}(HI) cm^3. **(24.6)** 1.34. **(24.7)** 1.19×10^{-23} cm^3, 0.9 D. **(24.8)** 1.38×10^{-23} cm^3, 0.35 D. **(24.9)** $\alpha' = (3kT/4\pi p)\{(n_{\mathrm{r}}^2 - 1)/(n_{\mathrm{r}}^2 + 2)\}$. **(24.11)** 13.1. **(24.13)** Steam: 1.000 19; $\mathrm{CaCl_2}$: 1.69; NaCl: 1.65; Ar: 1.23. **(24.14)** 1.58 D, 2.24×10^{-24} cm^3. **(24.15)** 5.66 cm^3 mol^{-1}, 1.578 D. **(24.16)** $a = (2\pi/3)N_{\mathrm{A}}^2C_6/d$. **(24.17)** $B \approx (2\pi/3)N_{\mathrm{A}}d^3\{1 - C_6/kTd^6\}$. **(24.18)** $B = b - a/RT$, $a = (2\pi/3)N_{\mathrm{A}}^2\varepsilon(\sigma_2^3 - \sigma_1^3)$, $b = (2\pi/3)N_{\mathrm{A}}\sigma_1^3$. **(24.20)** $F = nC_n/R^{n+1} - 6C_6/R^7$; $R = \{(n/6)C_n/C_6\}^{1/(n-6)}$. **(24.21)** 230 cm^3 mol^{-1}. **(24.22)** $V_{\mathrm{g}}/V_{\mathrm{e}} \approx 1.8 \times 10^{-30}$. **(24.24)** 1.74 kJ mol^{-1}. **(24.25)** $\alpha = (2\pi/3)N_{\mathrm{A}}^2(C_6/d^3)$; $V_{\mathrm{m,c}} = (\pi/2)N_{\mathrm{A}}d^3$; $p_{\mathrm{c}} = (8/9\pi) \times (C_6/d^9)$; $T_{\mathrm{c}} = (32/27)C_6/kd^6$. **(24.26)** $\theta = \pi - 2 \arcsin\{b/(R_1 + R_2)\}$. **(24.28)** 2 ln 2 = 1.386. **(24.29)** $1/\kappa = V(\partial^2U/\partial V^2)_T$, $R^* = \frac{2}{5}R_0 + (18\pi/5)\varepsilon_0R_0^2V_{\mathrm{m}}/\kappa N_{\mathrm{A}}\mathrm{e}^2\mathcal{M}$. **(24.30)** -709, -150 kJ mol^{-1}. **(24.31)** -910 kJ mol^{-1}. **(24.32)** -605 kJ mol^{-1}. **(24.33)** $\xi = -3.946 \times 10^{-29}$ C^2 m^2 kg^{-1}, $\chi = -2.962 \times 10^{-8}$. **(24.34)** 1.572×10^{-5}, 1.569×10^{-5}. **(24.35)** 1.0×10^{-6}.

25

(25.1) 88 000. **(25.2)** 147 000, 11 m^3 mol^{-1}. **(25.3)** 70 000, 71 000. **(25.4)** $M_{\mathrm{r}} + (2\Gamma/\pi)^{\frac{1}{2}}$. **(25.5)** 28 m^3 mol^{-1}, 0.33 m^3 mol^{-1}. **(25.6)** 2.6%, 50%. **(25.7)** **(a)** 4 22 × 10^{23} mol^{-1} $(l\sqrt{N})^3$; **(b)** 1.19×10^{24} mol^{-1} $(l\sqrt{N})^3$; **(a)** 0.39 m^3 mol^{-1}; **(b)** 1.10 m^3 mol^{-1}. **(25.8)** 0.11 m^3 mol^{-1}. **(25.10)** $\mathrm{M}_{\mathrm{R}}^+ - \mathrm{M}_{\mathrm{L}}^+ = -\nu[\mathrm{M}^+]_{\mathrm{L}}^2[\mathrm{P}]/f^+$, $\mathrm{X}_{\mathrm{R}}^{2-} - \mathrm{X}_{\mathrm{L}}^{2-} = \nu[\mathrm{X}^{2-}]_{\mathrm{L}}[\mathrm{P}] \times \{[\mathrm{X}^{2-}]_{\mathrm{L}} + \frac{1}{4}\nu[\mathrm{P}]\}/f^-$; $f^+ = [\mathrm{M}^+]_{\mathrm{R}}^2 + [\mathrm{M}^+]_{\mathrm{R}}[\mathrm{M}^+]_{\mathrm{L}} + [\mathrm{M}_{\mathrm{L}}^+]^2$, $f^- = [\mathrm{X}^{2-}]_{\mathrm{R}}^2 + [\mathrm{X}^{2-}]_{\mathrm{R}}[\mathrm{X}^{2-}]_{\mathrm{L}} + [\mathrm{X}^{2-}]_{\mathrm{L}}^2$.

(25.12) 2.4 mmol dm^{-3}. **(25.13)** $[Na^+]_L/[Na^+]_R \approx x + (1.041 + x^2)^{\frac{1}{2}}$. **(25.14)** 5 m^3 mol^{-1}, No. **(25.15)** 4.3×10^5 g. **(25.16)** 3500 rpm. **(25.17)** 66 400 **(25.18)** 69 100. **(25.19)** 3.4 nm. **(25.20)** $a \approx 6.2$ nm, $b \approx 1.8$ nm. **(25.21)** 5.16×10^{-13} s, 60 500. **(25.22)** $a/b \approx 2.8$. **(25.23)** 158 000. **(25.25)** **(a)** $lN^{\frac{1}{2}}$, 9.74 nm; **(b)** $l(8N/3\pi)^{\frac{1}{2}}$, 8.97 nm; **(c)** $l(2N/3)^{\frac{1}{2}}$, 7.95 nm. **(25.27)** **(a)** $a(3/5)^{\frac{1}{2}}$, 2.40 nm; **(b)** $l/12^{\frac{1}{2}}$, 46 nm. **(25.28)** SA and BSV spheres, DNA rod-like. **(25.30)** $P(\theta) \approx (2s\mathscr{L})\mathrm{Si}(s\mathscr{L}) - \{\sin(\frac{1}{2}s\mathscr{L})/\frac{1}{2}s\mathscr{L}\}^2$, $\mathscr{L} = Nl$. **(25.32)** 25 ns, 14 ps. **(25.33)** $dG = -S\,dT - l\,dt$, $dA = -S\,dT + t\,dl$. **(25.35)** $t = -T \times (\partial S/\partial l)_T$. **(25.39)** No. **(25.40)** $\Gamma = -(4.145 \times 10^{-11}$ mol $cm^{-2}) \times (c/M) \times (\Delta\gamma/mN\ m^{-1})$.

26 **(26.1)** **(a)** 6.7 nm; **(b)** 67 nm; **(c)** 67 mm. **(26.2)** 8×10^{-7} atm. **(26.3)** 130 atm. **(26.4)** 970 nm. **(26.5)** **(a)** 5×10^{10} s^{-1}; **(b)** 5×10^9 s^{-1}; **(c)** 5×10^3 s^{-1}. **(26.6)** **(a)** 6×10^{33} s^{-1}; **(b)** 6×10^{31} s^{-1}; **(c)** 6×10^{19} s^{-1}. **(26.7)** 4×10^8 s^{-1}. **(26.8)** **(a)** 3×10^{27}; **(b)** 3×10^{28}. **(26.9)** 2.0×10^{23} s^{-1}. **(26.10)** **(a)** 640, 1260, 2300 m s^{-1}; **(b)** 320, 630, 1150 m s^{-1}; **(26.11)** **(a, b)** 3.7 kJ mol^{-1}. **(26.14)** $\langle v_x \rangle_f / \langle v_x \rangle_i = (1 - e^{-1/\pi})/\mathrm{erf}\,(1/\pi^{\frac{1}{2}}) = 0.47$. **(26.15)** **(a)** 1.8 mph; **(b, c)** 56 mph. **(26.16)** **(a, b)** 5′9½″. **(26.17)** **(a)** 0.39; **(b)** 0.61; 0.47, 0.53. **(26.18)** $F(nc^*)/F(c^*) = n^2 e^{(1-n^2)}$; 3.02×10^{-3}, 4.9×10^{-6}. **(26.19)** $v = (2Gm/R)^{\frac{1}{2}} \approx (2gR)^{\frac{1}{2}}$; **(a)** 11.2 km s^{-1}; **(b)** 5.0 km^{-1}. **(26.20)** 43. **(26.21)** 2.5×10^{21}. **(26.22)** 30 hr. **(26.23)** $p = (\Delta m/St) \times (2\pi kT/m)^{\frac{1}{2}}$. **(26.24)** 7.3 mPa. **(26.25)** **(a)** 2.7×10^{23}, 2.3×10^8 s^{-1}; **(b)** 2.7×10^{17}, 230 s^{-1}; **(c)** 2.7×10^{13} s^{-1}, 0.02 s^{-1}. **(26.26)** **(a)** 100 Pa; **(b)** 24 Pa. **(26.27)** 1.4 s. **(26.28)** **(a)** 1.7×10^{14} s^{-1}; **(b)** 1.4×10^{20} s^{-1}. **(26.29)** **(a)** 3.3×10^{34} m^{-3} s^{-1}; **(b)** 1.3×10^{34} m^{-3} s^{-1}; **(c)** 1.1×10^{35} m^{-3} s^{-1}. **(26.30)** **(a)** 1.3×10^{-5} kg m^{-1} s^{-1}, **(b)** 1.3×10^{-5} kg m^{-1} s^{-1}; **(c)** 2.4×10^{-5} kg m^{-1} s^{-1}. **(26.31)** **(a)** 5.4 mJ K^{-1} s^{-1}, 8 mJ s^{-1}; **(b)** 30 mJ K^{-1} m^{-1} s^{-1}, 40 mJ s^{-1}. **(26.32)** 140 μP, 390 pm. **(26.33)** **(a)** 5 mJ K^{-1} m^{-1} s^{-1}; **(b)** 9.6 mJ K^{-1} m^{-1} s^{-1}. **(26.34)** 9.1. **(26.35)** 7 J s^{-1}, 7 W. **(26.36)** **(a)** 10 m^2 s^{-1}, 0.3 mol cm^{-2} s^{-1}; **(b)** 1.1×10^{-5} m^2 s^{-1}, 3×10^{-7} mol cm^{-2} s^{-1}; **(c)** 1.1×10^{-7} m^2 s^{-1}, 3×10^{-9} mol cm^{-2} s^{-1}. **(26.37)** $dr/dt = (s\mathcal{N}/N_A\rho)(M_m RT/\pi)^{\frac{1}{2}} \approx 7 \times 10^{-4}$ cm s^{-1}.

27 **(27.1)** 5.3 S cm^2 mol^{-1}. **(27.2)** 20.63 m^{-1}. **(27.3)** 126 S cm^2 mol^{-1}, $\mathscr{K} = 76.5$ S cm^2 $mol^{-1}/M^{\frac{1}{2}}$. **(27.4)** **(a)** 119.2 S cm^2 mol^{-1}; **(b)** 1.192 mS cm^{-1}; **(c)** 173.1 Ω. **(27.5)** 138.3 S cm^2 mol^{-1}. **(27.6)** 1.36×10^{-5} M, 1.86×10^{-10}. **(27.7)** 387.9 S cm^2 mol^{-1}. **(27.8)** 0.030. **(27.9)** 3.22, 3.23. **(27.10)** 1.86×10^{-5}, 4.73. **(27.11)** 8.59 S cm^2 mol^{-1}, 3.44×10^{-4} S cm^{-1}, 600 Ω. **(27.12)** 4.01×10^{-4}, 5.19×10^{-4}, 7.62×10^{-4} cm^2 s^{-1} V^{-1}. **(27.13)** 4.0×10^{-3}, 5.2×10^{-3}, 7.6×10^{-3} cm s^{-1}; 250, 190, 130 s; **(a)** 1.3×10^{-6}, 1.7×10^{-6}, 2.4×10^{-6} cm; **(b)** 43, 55, 81. **(27.14)** $t'/t'' = c'u'/c''u''$. **(27.15)** 0.82, 0.0028. **(27.16)** 0.48. **(27.17)** 7.5×10^{-4} cm^2 $s^{-1}$$V^{-1}$, 72 S cm^2 mol^{-1}. **(27.19)** $t_+ = 0.407$. **(27.20)** 3.75 MΩ, 1.0×10^{-14}, 14.0, 7.0. **(27.21)** $a = 60.4$ S cm^2 $mol^{-1}/M^{\frac{1}{2}}$, $b = 0.13/M^{\frac{1}{2}}$, $\mathscr{K} = 76.8$ S cm^2 $mol^{-1}/M^{\frac{1}{2}}$. **(27.22)** **(a)** 1.2×10^4 N mol^{-1}, 2.1×10^{-20} N/molecule; **(b)** 1.7×10^4 N mol^{-1}, 2.8×10^{-20} N/molecule; **(c)** 2.5×10^4 N mol^{-1}, 4.1×10^{-20} N/molecule. **(27.23)** 2.7, 3.5, 5.2 nm s^{-1}. **(27.24)** 420 pm. **(27.25)** 21 ps. **(27.26)** **(a)** 65 μm; **(b)** 32 μm. **(27.27)** I_2: **(a)** 240s; **(b)** 2.4×10^4 s; sucrose: **(a)** 960 s; **(b)** 9.6×10^4 s. **(27.28)** 212, 164, 112, 107 pm. **(27.29)** *ca.* 3 for Li^+ **(27.30)** 78 pm, 9.3 kJ mol^{-1}. **(27.32)** **(a)** 0; **(b)** 0.06 M. **(27.33)** **(a)** 0; **(b)** 1/64 = 0.016; **(c)** $55/2^{10} = 0.054$. **(27.34)** $n > 60$. **(27.35)** 1.2 g m^{-1} s^{-1}.

28 **(28.1)** −1.0, −2.0, +3.0, +1.0 M s^{-1}. **(28.2)** 0.50 mol s^{-1}; −1.0, −0.50, +1.5 M s^{-1}. **(28.3)** M^{-1} s^{-1}; $v_A = -k'[A][B]$, $k' = k/V$; $v_C = k''[A][B]$, $k'' = 3k/V$. **(28.4)** $\dot{\xi} = k'[A][B][C]$, $k' = kV/2$; $[k] = M^{-2}\ s^{-1}$, $[k'] = mol^{-2}\ dm^9\ s^{-1}$. **(28.5)** 1.4×10^3 s, $p_{total} =$ **(a)** 503 Torr; **(b)** 688 Torr. **(28.6)** **(a)** M^{-1} s^{-1}, M^{-2} s^{-2}; **(b)** atm^{-1} s^{-1}, atm^{-2} s^{-1}. **(28.7)** 2720 yr. **(28.8)** **(c)** 0.64 μg; **(b)** 0.18 μg. **(28.9)** **(a)** [NaOH] = 0.045 M, [AcOEt] = 0.095 M; **(b)** [NaOH] = 0.001 M, [AcOEt] = 0.051 M. **(28.10)** $kt = -\{1/6(2B_0 - A_0)\}\ \ln\{(2x - A_0)B_0/(3x - B_0)A_0\}$. **(28.11)** $kt = 4x(A_0 - x)/A_0^2(A_0 - 2x)^2$. **(28.12)** $kt = \{4x/A_0^2(A_0 - 2x)\} + (1/A_0^2)\ \ln\{(A_0 - 2x)/(A_0 - x)\}$. **(28.13)** **(a, b, c)** $t_{\frac{1}{2}} = 3/kA_0^2$. **(28.15)** $t_{\frac{1}{2}}/t_{\frac{3}{4}} = (2^{n-1} - 1)/\{(4/3)^{n-1} - 1\}$. **(28.16)** 1st; 5.6×10^{-4} s^{-1}. **(28.17)** 1st; 1.3×10^{-2} s^{-1}; C_4H_6 reactive. **(28.18)** 240 kJ mol^{-1}, 1.1×10^{13} s^{-1}. **(28.19)** 1st; 7.2×10^{-4} s^{-1}. **(28.20)** $p/p_0 = (1 + x/2)/(1 + x)$; $x = p_0 kT$; $t = 1/p_0 k$; $\xi = \frac{1}{2}$. **(28.21)** 2nd; 1.6×10^{-5} $Torr^{-1}$ s^{-1}. **(28.23)** $[HCl]^3$[propene]. **(28.24)** $k_{eff} = kK_1K_2$. **(28.25)** −18, +5 kJ mol^{-1}. **(28.26)** 1.14×10^{10} M^{-1} s^{-1}, 16.7 kJ mol^{-1}. **(28.27)** 66 kJ mol^{-1}, 8.7×10^7 M^{-1} s^{-1}. **(28.28)** $[A] = a + be^{-(k+k')t}$, $a = k'(A_0 + B_0)/(k + k')$, $b = (kA_0 - k'B_0)/(k + k')$; $([B]/[A])_\infty = k/k'$. **(28.29)** Plot $1/k_{eff} = 1/k_a p + k_a'/k_a k_b$. **(28.30)** $1/k_{obs} = 1/k + [H^+]/kK_a$. **(28.31)** 13.7. **(28.32)** 0.010 M.

29 **(29.1)** **(a)** I; **(b, c)** P; **(d)** T; $d[AH]/dt = -k_{eff}[AH]$, $k_{eff} = k_a(1 + k_c/2k_d)$. **(29.2)** $k_a/4k_c^2 \ll k_d/k_c k_e$. **(29.3)** $d[CH_4]/dt = k_b(k_a/2k_d)^{\frac{1}{2}}[CH_3CHO]^{\frac{3}{2}} \approx -d[CH_3CHO]/dt$. **(29.4)** $d[COCl_2]/dt = k_{eff}[CO][Cl_2]^{\frac{3}{2}}$, $k_{eff} = k_c K' K^{\frac{1}{2}}$; $3k_{eff}t/2 = 1/(a - x)^{\frac{3}{2}} - 1/a^{\frac{3}{2}}$. **(29.5)** $k_{eff} = k_c K' K^{\frac{1}{2}}/\{1 + (k_c/k_b')[Cl_2]\}$. **(29.7)** $d[M]/dt = -k_p(2k_i f/2k_t)^{\frac{1}{2}}[M][I]^{\frac{1}{2}}$; $\chi \propto 1/[I]^{\frac{1}{2}}$. **(29.8)** $d[M]/dt = [29.1.8] \times (1 + b[M])/(1 + a[M])^{\frac{1}{2}}$. **(29.9)** $P = 1 - 1/x$, $x^2 = 1 - 2[A]_0^2 kt$. **(29.10)** $\langle M \rangle_M = M_{mon}(1 + 2kt[A]_0)$. **(29.11)** $\langle M \rangle_N = M_{mon}(1 + kt[A]_0)$; $\langle M \rangle_N/\langle M \rangle_M = (1 + kt[A]_0)/(1 + 2kt[A]_0)$. **(29.12)** $\delta M_r/M_{mon} = p^{\frac{1}{2}}/(1 - p) = \{kt[A]_0 + k^2t^2[A]_0^2\}^{\frac{1}{2}}$. **(29.13)** **(a)** $M_{mon}(1 + 4p + p^2)/(1 - p^2)$; **(b)** $(6\chi^2 - 6\chi + 1)/(2\chi - 1)$. **(29.14)** $[H\cdot] = r_a/(k_e - 2k_c[O_2])$. **(29.15)** 1.9×10^{20} s^{-1}, 3.1×10^{-4} einstein s^{-1}. **(29.17)** 5.1×10^8 M^{-1} s^{-1}. **(29.18)** $d[A]/dt = -I_a - k_p(I_a/k_t)^{\frac{1}{2}}[A]$. **(29.19)** $Cl_2 + h\nu \rightarrow 2Cl$, $Cl + CHCl_3 \rightarrow CCl_3 + HCl$, $CCl_3 + Cl_2 \rightarrow CCl_4 + Cl$, $2CCl_3 + Cl_2 \rightarrow 2CCl_4$. **(29.20)** $[B]_{eq} = (I_a/k)^{\frac{1}{2}}$. **(29.21)** $\Phi \approx 2k[A]/(k_f + k[A])$. **(29.22)** $[A]_0([A]_0 + [P]_0)kt = y/(1 - y) + \{1/(1 + p)\}\ \ln\{(p + y)/(1 - y)p\}$, $y = x/[A]_0$, $p = [P]_0/[A]_0$, $x = [P] - [P]_0$. **(29.23)** $[A]_0([A]_0 + [P]_0)kt = y/p(p + y) + \{1/(1 + p)\}\ \ln\{(p + y)/(1 - y)p\}$. **(29.24)** **(a)** $([A]_0 + [P]_0)^2 kt_{max} = \frac{1}{2} - p + \ln(1/2p)$; **(b)** $([A]_0 + [P]_0)^2 kt_{max} = 1/p - 2 + \ln(2/p)$. **(29.25)** See Fig. 29.8. **(29.26)** $[X] = k_c/k_b$, $[Y] = k_a[A]/k_b$. **(29.27)** See Fig. 29.9. **(29.30)** 5.0×10^7 M^{-1} s^{-1}. **(29.31)** $1/\tau = k_1 + k_2[B]_e + k_2[C]_e$.

30 **(30.1)** **(a)** $9.6 \times 10^9\ s^{-1}$, $1.2 \times 10^{29}\ s^{-1}\ cm^{-3}$; **(b)** $6.7 \times 10^9\ s^{-1}$, $8.3 \times 10^{28}\ s^{-1}\ cm^{-3}$; 1.7%. **(30.2)** $K = N/kT$, $P = e^{-\beta E_a}$ **(30.4)** $100\ E_a\ \delta T/RT^2$. **(30.5)** **(a)** $4.3 \times 10^{-20}\ m^2$; **(b)** 0.14. **(30.6)** $0.026\ nm^2$, $P \approx 0.07$. **(30.7)** 0.007, $4.0 \times 10^{-21}\ m^2$. **(30.8)** $2.5 \times 10^{11}\ M^{-1}\ s^{-1}$. **(30.9)** 22 ns. **(30.10)** **(a)** $6.6 \times 10^9\ M^{-1}\ s^{-1}$; **(b)** $3.0 \times 10^{10}\ M^{-1}\ s^{-1}$; **(c)** $2.0 \times 10^9\ M^{-1}\ s^{-1}$. **(30.13)** 1.4×10^7, 200 (linear), 900 (non-linear), 1, 1; $6.5 \times 10^9\ M^{-1}\ s^{-1}$. **(30.14)** 5×10^{-6}. **(30.15)** C-H, C-D cleavage rate-determining. **(30.16)** **(a)** 1/15; **(b)** 1/1.1; NO; reduced. **(30.17)** $1.8 \times 10^6\ M^{-1}\ s^{-1}$. **(30.18)** $1.5 \times 10^6\ M^{-1}\ s^{-1}$. **(3.20)** $-76\ J\ K^{-1}\ mol^{-1}$. **(30.21)** -86, $-87\ J\ K^{-1}\ mol^{-1}$. **(30.22)** **(a)** $2.7 \times 10^{-11}\ cm^2\ s^{-1}$; **(b)** $1.1 \times 10^{-10}\ cm^2\ s^{-1}$. **(30.23)** **(a)** $2 \times 10^{-9}\ cm^2\ s^{-1}$; **(b)** $8 \times 10^{-9}\ cm^2\ s^{-1}$. **(30.24)** $9.6\ kJ\ mol^{-1}$. **(30.25)** **(a)** $-46\ J\ K^{-1}\ mol^{-1}$; **(b)** $5\ kJ\ mol^{-1}$; **(c)** $19\ kJ\ mol^{-1}$. **(30.26)** $88.2\ kJ\ mol^{-1}$, $44\ J\ K^{-1}\ mol^{-1}$, $90.7\ kJ\ mol^{-1}$, $77\ kJ\ mol^{-1}$. **(30.29)** 2. **(30.31)** $\sigma(CH_2F_2)/\sigma(Ar) = 5$; CH_2F_2 polar.

31 **(31.1)** **(a)** 1.4×10^{21}, $1.4 \times 10^{14}\ cm^{-2}\ s^{-1}$; **(b)** 3.1×10^{20}, $3.1 \times 10^{13}\ cm^{-2}\ s^{-1}$. **(31.2)** **(a)** 1.61×10^{15}; **(b)** 1.14×10^{15}; **(c)** 1.86×10^{15}. **(31.3)** **(a)** 1.00×10^{15}; **(b)** 1.41×10^{15}; **(c)** 1.16×10^{15}; 1.20×10^{15}. **(31.5)** **(a)** 42 ps; **(b)** 600 000 yr. **(31.6)** **(a)** 0.61 ps; **(b)** 7.3 μs. **(31.7)** $\tau_D/\tau_H = 1.8$, 1.5. **(31.8)** $610\ kJ\ mol^{-1}$, 0.11 ps. **(31.9)** $-3.7\ kJ\ mol^{-1}$. **(31.11)** $p/V_a = p/V_a^0 + 1/KV_a^0$; $\ln(\theta/p) \approx \ln K - \theta$; $-1/V_a^0$. **(31.12)** $0.52\ Torr^{-1}$, $0.48\ cm^3$. **(31.14)** $c =$ **(a)** 164.3; **(b)** 263.5; $V_{mon} =$ **(a)** $13.1\ cm^3$; **(b)** $12.5\ cm^3$. **(31.15)** Freundlich. **(31.16)** Langmuir, $K = 0.017\ Torr^{-1}$, $0.13\ nm^2$. **(31.17)** $0.19\ cm^3$. **(31.18)** No. **(31.19)** $75.4\ cm^3$, 3.98. **(31.21)** $At = x/p(p-x) + K^2x + 2K \ln\{p/(p-x)\}$, $A = kK_AK_B$, $K = K_A + K_B$. **(31.22)** rate $= kp_F^{\frac{1}{2}}p_B^{\frac{1}{2}}/(1 + K^{\frac{1}{2}}p_F^{\frac{1}{2}})$. **(31.23)** $-13\ kJ\ mol^{-1}$. **(31.24)** **(a)** 2×10^{104} min; **(b)** 50 μs. **(31.25)** 0. **(31.27)** $G(t) = k_c + F(t)$, $k_c = 0.02\ Torr\ s^{-1}$. **(31.28)** $G(t) = k_c/K + (1 + Kp_0)F(t)/Kp_0$, $G(t) = (p - p_0)/t$, $F(t) = (p_0/t) \ln (p/p_0)$. **(31.29)** $K = 0.033\ Torr^{-1}$, $k_c = 4.3 \times 10^{-4}\ s^{-1}$. **(31.30)** $U = -\pi\mathcal{N}C_6/6R^3$. **(31.31)** $U = 2\pi\mathcal{N}C_{12}/90R^9 - \pi\mathcal{N}C_6/6R^3$; 0.858σ. **(31.32)** $w_s = c_1[A]^{1/c_2}$, $c_1 = 0.16$, $c_2 = 2.4$. **(31.33)** $d\mu = (RTV_a^0/\sigma)\, d\ln(1-\theta)$.

32 **(32.2)** **(a)** $0.31\ mA\ cm^{-2}$; **(b)** $5.4\ mA\ cm^{-2}$; **(c)** $-1.4 \times 10^{42}\ mA\ cm^{-2}$. **(32.5)** 110 mV. **(32.6)** 4.9×10^{15}, 1.6×10^{16}, 3.1×10^7. **(32.7)** 3.8, 12, 2.4×10^{-8}. **(32.8)** $R_\Omega = RT/SFj_0$; **(a)** 33 Ω; **(b)** $3.3 \times 10^{10}\ \Omega$. **(32.9)** $\langle j \rangle = (1 - 2\alpha)(F/2RT)^2 j_0 \eta_0^2$. **(32.10)** 7.2 μA. **(32.12)** 0.38, $0.78\ mA\ cm^{-2}$. **(32.13)** $j = -(0.78\ mA\ cm^{-2})e^{-0.015(\eta/mV)}$. **(32.15)** **(a)** 0.34 V, 0.28 V; **(b)** -0.76 V, -0.82 V. **(32.17)** Yes. **(32.18)** No. **(32.19)** No. **(32.20)** $j = j_L(1 - e^{F\eta_{corr}/RT})$. **(32.21)** 0.25 mm. **(32.22)** $j_L = (RT/zF)(\lambda_+/\delta)c$. **(32.23)** 0.23 mm. **(32.24)** $a(Sn^{2+})/a(Pb^{2+}) \approx 2.2$ **(32.26)** -1.30 V, 0.13 W. **(32.27)** **(a)** 1.23 V; **(b)** 1.06 V. **(32.30)** 120 mA, 60 mW, 0.6 V. **(32.31)** 116.5 mA, 57.7 mW. **(32.32)** Steel. **(32.33)** $E = E^\ominus + (2.303\ RT/F)\{pK_w - pH\}$. **(32.34)** Fe, Al, Co, Cr if O_2 absent; all if O_2 present. **(32.35)** 6 μA.

Copyright acknowledgements

The permissions of the respective copyright holders to reproduce tables of data and extracts from them is gratefully acknowledged as follows: Academic Press (4.10 [© 1969]), Addison Wesley (Boxes 17.1, 23.1 [© 1962]), Butterworths Ltd. (31.1, 31.2 [© 1964]), Professor J. G. Calvert (19.1, 19.2 [© 1966]), Chapman and Hall Ltd. (18.2 [© 1975]), CRC Press Inc. (1.1, 4.4, 11.1, 12.1, 12.3, 23.2, 24.1, 24.6 [© 1979]), Cornell University Press (4.5, [© 1960]), Dover Publications Inc. (Box 14.1, 14.1 [© 1965]), Elsevier Scientific Publishing Company (24.1 [© 1978]), Professor A. A. Frost (28.3), Longman (1.2, 2.1, 3.1, 4.4, 4.8, 7.1, 8.2, 12.2, 12.3, 23.1, 23.3, 24.8, 26.2, 26.3, 27.1, 27.2, 32.2 [© 1973]), McGraw-Hill Book Company (1.2, 1.1, 3.1, 3.2, 4.2, 4.6, 4.7, 4.9, 5.1, 18.1, 24.2, 24.5, 27.4 [© 1972]; 4.3, 4.6, 6.1, 10.1 [© 1961]; 28.1, 28.2 [© 1965]; Box 14.2, Box 15.1 [© 1935]; 3.2 [© 1968]; Box 18.1 [© 1955]), National Bureau of Standards (4.1 [© 1982]), Dr. J. Nicholas (28.1, 28.2, 28.3 [© 1976]), Oxford University Press (28.1, 28.2, 28.3 [© 1975]; 31.3, Box 31.1 [© 1974] Pergamon Press Ltd. (24.3 [© 1961]), Prentice Hall Inc. (Box 25.1, 25.2 [© 1971]), John Wiley and Sons inc. (8.1 [© 1975]; Box 16.1 [© 1944]; 24.4 [© 1954]; 25.1, 25.3 [© 1961]). The sources of the material are quoted at the foot of each table.

The permission of the following individuals and journals for reproduction of illustrations is gratefully acknowledged: *Acta crystallogr*. (0.2), Professor B. J. Alder (0.13), Dr R. F. Barrow (13.7, 18.13), Professor G. C. Bond (31.30), Professor A. V. Crewe (25.1), Professor G. Ehrlich (31.21), Dr A. J. Forty (31.3), Professor D. Freifelder (24.6), Professor H. Ibach (31.17), Professor D. A. King (31.19), Professor D. A. Long (18.22), Dr G. Morris (20.10), Professor E. W. Müller (0.1, 31.16), Nicolet XRD Corpn (23.21), Dr A. H. Narten (24.24), Open University Press (19.24 [© 1976]), Oxford University Press (31.3, 31.4), Dr M. Prutton (31.14), Professor M. W. Roberts (31.10), Dr H. M. Rosenberg (31.4), Professor G. A. Samorjai (31.12, 31.13), *Scientific American* (9.1, 25.17), Dr D. W. Turner (31.8).

Subject index

(T) after a page number refers to a Table in the text (in a Box); T before a number refers in a Table in the data section.

Subject index

Subject index

Subject index

Subject index